SCIENCE

SECOND EDITION

Sybil P. Parker
Editor in Chief

Some of the material in this volume has been published previously in the McGRAW-HILL ENCYCLOPEDIA OF SCIENCE AND TECHNOLOGY, Fourth Edition.

McGRAW-HILL ENCYCLOPEDIA OF ENVIRONMENTAL SCIENCE, Second Edition.

Library of Congress Cataloging in Publication Data

McGraw-Hill encyclopedia of environmental science.

"Some of the material in this volume has been published previously in the McGraw-Hill encyclopedia of science and technology, fourth edition."
Includes bibliographies and index.
1. Ecology—Dictionaries. 2. Man—Influence on nature—Dictionaries. 3. Environmental protection—Dictionaries. I. Parker, Sybil P. II. McGraw-Hill Book Company. III. McGraw-Hill encyclopedia of science and technology. IV. Title: Encyclopedia of environmental science.
QH540.4.M3 1980 304.2'03 79-28098
ISBN 0-07-045264-4

Contents

Editorial Staff

Consulting Editors

Preface

Organisms exist under the influence of external conditions which in total constitute the environment. The physical or abiotic component of the environment includes all the nonliving aspects; the biotic component consists of the organisms which interact with each other and with their abiotic environment. Any factor that disturbs the delicate balance between the two components causes a chain reaction that may end in drastic permanent changes.

The consequences of disturbing the balance of an ecosystem are well documented throughout history in the extinction of species . . . the devastation of floods and earthquakes . . . the depletion of natural resources . . . the deaths and crop losses from disease epidemics . . . and the widespread effects of pollution — the product of modern technology.

While evolution documents the successes and failures of species to adapt to changing environments, civilization attests to the human ability to implement environmental change. In either case, it is clear that not all changes are beneficial. Environmental science is concerned with evaluating these changes, considering both natural and human activities as distinct but inseparable.

The study of environmental science encompasses the fields of ecology, geophysics, geochemistry, forestry, public health, meteorology, agriculture, oceanography, soil science, and mining, civil, petroleum, and power engineering. Among the problems confronting environmentalists are land reclamation, eutrophication, desertification, climate modification, energy sources, urban planning, crop production, and pollution. Solutions to these problems will provide the foundation for the ultimate protection and preservation of our environment.

The *Encyclopedia of Environmental Science* treats all of these topics and gives an insight into the present state of knowledge, directions that must be taken, and laws and conservation practices required to deal with the problems. Coverage is complete and up to date in accessible, authoritative articles written for the general reader as well as the professional.

This edition of the Encyclopedia has been thoroughly revised and expanded. The more than 250 articles are organized in two sections — a section containing five feature articles on topics of broad, general interest, and a section of alphabetically arranged articles dealing with the basic scientific and technical concepts. Each article was prepared by a specialist. Many were written especially for this volume, and some were taken from the widely acclaimed *McGraw-Hill Encyclopedia of Science and Technology* (4th ed., 1977). The articles were selected and reviewed by the consulting editors and the editorial staff. There are 650 photographs, diagrams, charts, graphs, and line drawings to supplement the text; bibliographies for further reading; cross-references to guide the reader to related articles; and an analytical index for easy access to the information.

Sybil P. Parker
EDITOR IN CHIEF

Surveying the Environment

Environmental Protection

Emil T. Chanlett

Emil T. Chanlett is professor of sanitary engineering in the Department of Environmental Sciences and Engineering at the University of North Carolina, Chapel Hill, where he has taught since 1946. Author of numerous articles, he has served as consultant to many organizations, including the World Health Organization, the Office of the Surgeon General, Department of the Army, the U.S. Public Health Service, and the North Carolina State Board of Health.

Environmental protection is the system of procedures which limit the impairment of the quality of water humans use, of the air they breathe, and of the land that sustains them. It includes the means to control the physical energies of ionizing radiations, nonionizing radiations, sound, air pressure changes, and heat and cold. Human activities produce wastes that are vapors or gases, solids, liquids, or energy states. Humans seek to disperse these to the open environment of water, air, or land. The receptors are all forms of life on Earth, with people the primary concern.

Three objectives. Environmental protection has three objectives. The first is to protect people from physiological damage from pathogenic organisms, from toxic chemicals, and from excesses of physical energies. The second is to spare humans annoyance, irritation, and discomfort from offensive conditions in water, in air, and on the land. The physical energies have a role in this second objective when there are excesses of noise, heat, cold, and even electromagnetic transmission interference disturbing radio and television reception. Uncontrolled insect and rodent populations may be more a source of discomfort, disgust, and fear than a real risk of disease transmission. The evident corollary is the provision of an environment which adds to comfort, pleasure, and productivity. Air cooling for summer comfort and cleanliness of recreational areas are examples of positive actions to meet the second objective. The third objective is to safeguard the balances in the Earth's ecosystems and to conserve natural resources. Many people strongly advocate that this should be the primary goal of environmental protection. Fortunately, the three objectives are not incompatible, although conflicts arise. The drainage of a swamp which is a breeding place for anopheline mosquito vectors of malaria obviously changes the ecosystem that has existed there. Thus there are differences of opinions on which environmental actions should be given priority when the three objectives are not compatible.

Assimilative capacity of the environment. When humans' waste loads on the water, air,

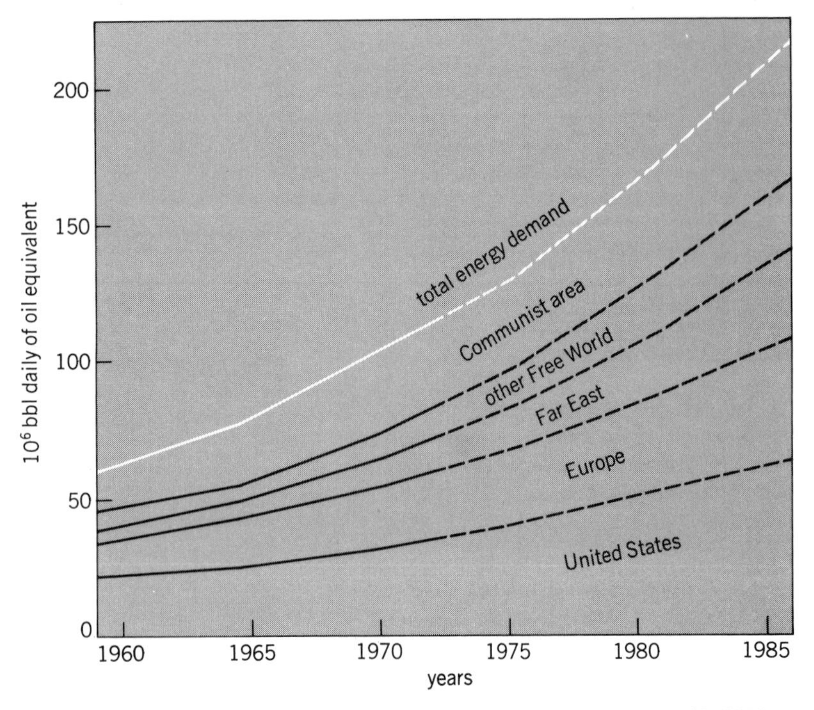

Fig. 1. Exponential rise of the world's energy demand from 1960 projected to 1985.

and land overwhelm the natural processes of assimilation of such wastes, pollution occurs. In the case of the physical energies, it is usually human or animal tolerance that is overwhelmed. The condition of pollution may jeopardize one or more of the three objectives of environmental protection. For example, the management of stream pollution is directed at all three objectives. Overwhelming the assimilative capacity of a local environment by human wastes is not only a recent phenomenon. English literature notes the foul air of London in the 17th century due to the burning of sea-coal. Samuel Coleridge viewing Cologne, Germany, at the end of the 18th century wrote: "The River Rhine, it is well known,/ Doth wash your city of Cologne,/ But tell me, nymphs, what power divine,/ Shall henceforth wash the River Rhine?"

Today's large urban-industrial areas with the massive use of individual internal combustion engines for transportation have both chronic and acute conditions of air pollution. The airborne waste loads exceed the capacity of horizontal and vertical air movement to disperse the materials. Additionally, conditions of inversion, stagnation, and ultraviolet radiation produce reactants from the primary pollutants. On land and in water, not all wastes are usable as a food for the natural biota; these are labeled nonbiodegradable. The chlorinated hydrocarbon pesticides are high on the list of persistent contaminants which change slowly in the open environment. It is this property that makes DDT an effective anopheline killer even 4 to 6 months after spraying on home interiors.

Increased energy use, production, and population. The human ability to produce prodigious amounts of wastes depends on energy use and on numbers of people. James Watt's improvement in 1769 on Newcomen's atmospheric engine broke the energy limits of human and animal muscle, wind, and water power. The capacity to do work had been in those bounds since humans' start on Earth. The steam engine was soon moving boats and then trains, and driving factory machinery. About 100 years later, petroleum became a fuel and made the internal combustion engine possible. Almost concurrently, electricity became another magical energy. And finally there developed nuclear energy, with which society has not yet come to terms. Human energy uses started their exponential climb with Watt's invention. The curve rises slowly through the 19th century and the first half of the 20th. The exponential character of the curve becomes drastically marked about 1950, not only for the United States but for the world.

Obviously energy use is for a purpose, for producing goods, for transportation, and for people's comfort and convenience. Expressed in some suitable units, these uses of energy also trace exponentially increasing curves (Fig. 1). Human population itself exponentially increased through the 19th century and continues to do so (Fig. 2). All of this accelerated human activity produces more wastes. The capacity of the open environment to assimilate the wastes remains essentially static. The result is an erosion and change of the original ecosystem when the waste loading continues an unabated rise. The wastewater discharges of New York City, Boston, and Baltimore increased during the 20th century to surpass the dilution and assimilative capacity of the receiving waters. Rivers have been overwhelmed by the cumulative wastes as the water moved downstream (Figs. 3 and 4). The Ohio, the Rhine, and the Danube cannot cope with the multiple sources of wastes along their courses. Segments of the Mediterranean Sea are threatened. The limiting meteorological characteristics of the Los Angeles area resulted in photochemical smog when gasoline-fueled automobile emissions skyrocketed after World War II (Fig. 5). However, Los Angeles is no longer unique in smog pollution.

Options for control of pollutants. To keep the balance, the alternatives are: eliminate the source; eliminate the waste; treat the waste to reduce the deleterious load on the open environment; or augment the environmental capacity to assimilate the waste. All of the alternatives are applied in one way or another to manage liquid wastes, solid wastes, airborne wastes, and the excesses of physical energies. Abatement action is taken to prevent injury to humans and animal and plant life, to protect property and resource values, and to limit conditions that are offensive to people apart from

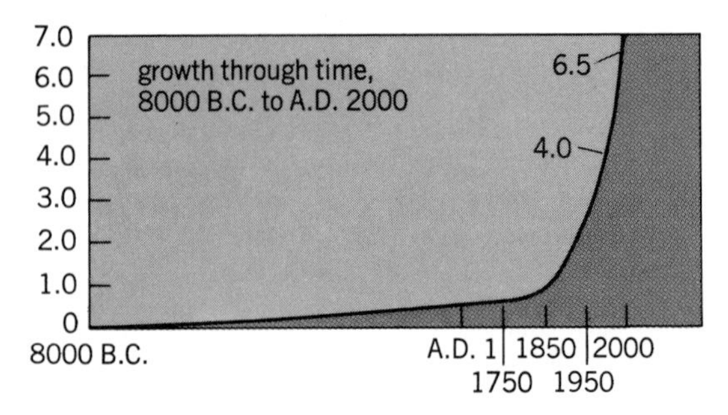

Fig. 2. Exponential growth of the world's population from 8000 B.C. to A.D. 2000. The vertical axis refers to billions of people.

Fig. 3. Potomac River below Key Bridge, Washington, DC.

physiological damage. Abatement of wastewater effects began in England and Germany toward the end of the 19th century. The purposes were to maintain the quality of streams for downstream water users, to protect fish life, and to add in some measure to the prevention of waterborne infectious diseases. Water source selection and water purification, along with the sanitary management of human excreta, are the primary environmental controls of waterborne diseases. By the turn of the century, municipal sewage was being treated in some places in the United States. The devices used were coarse screens, septic tanks, and later Imhoff tanks in combination with intermittent sand filters. Primary settling then came into use.

The work at the Lawrence Experiment Station of the Massachusetts State Health Department in the 1890s and the U.S. Public Health Service at Cincinnati in the 1920s provided the scientific bases for sewage treatment and stream pollution control. Consulting sanitary engineers such as Allen Hazen, Leonard Metcalf, and Harrison P. Eddy made important contributions during the first decades of this century. University research began in the 1920s by George Whipple and Gordon Fair at Harvard, Charles Gilman Hyde at Berkeley, Harold Babbit at the University of Illinois, and William Sedgwick and Murray Horwood at the Massachusetts Institute of Technology. Airborne wastes received less attention, although F. G. Cottrell of the University of California developed and applied an electrostatic precipitator to control sulfuric acid mists and vapors at a Du Pont sulfuric acid plant near Berkeley, CA, in 1907. For the most part, air pollution efforts up through World War II were directed at smoke and dust control.

The mainstay of wastewater treatment is biooxidation (Fig. 6). Trickling filters, activated sludge, and stabilization ponds depend on the same natural process as a well-aerated stream. That process is the biological feeding, primarily by bacteria, on the carbohydrates, proteins, and fats in the wastewater in the presence of ample oxygen. The settled solids, or sludge, are separated and made the food of anerobic organisms in sludge digestors. Sewage sludge is an example of removing a waste from a mainstream and then having to deal with it further. The same holds for removing particulates from an airstream or from flue gas. The captured material requires further handling. There are always possibilities of recovery of the isolated material for economically useful purposes.

Governmental action. In the United States, governmental action in environmental protection began in the 19th century with municipal services of water supply, sewers, street cleaning, and solid waste collection. Slowly, as the need arose, these extended to sewage treatment and solid waste incineration or ocean dumping. Beginning in Massachusetts, state health departments developed sanitary engineering divisions. Their functions were both advisory and regulatory. State health department laboratory services for the quality of drinking water began before World War I. After that war,

Fig. 4. Scum and foam from a paper plant on the Androscoggin River, fall 1969.

the laboratories became an instrument for defining stream pollution conditions and sources. State health departments established design and operational standards for water and sewage treatment plants.

The U.S. Public Health Service issued its first Drinking Water Standards in 1914. Although these had the force of law through the Federal power over interstate and foreign commerce, hence on the drinking water provided for passengers on common carriers, states and cities varied widely in how they followed the standards.

Without specific Federal legislation, the Public Health Service was the source of national action in environmental protection for the first 60 years of the 20th century. Its activities extended to all phases of environmental protection, including industrial hygiene, vector control, and milk and food sanitation. It carried on basic research, extensive field studies, demonstration projects, model code and standard development, technical assistance, and help in staffing and training. A notable Federal participation in environmental protection came in the depression years from 1934 to 1940. The Public Works Administration funded the construction of water and sewer systems, water and sewage treatment plants, and solid waste incinerators. The Works Progress Administration funded labor-intensive contruction of small-town sewers, antimalaria drainage, and privy-building. The first sign of Federal legislation on environmental protection was a 1948 water pollution control act to provide loans to municipalities to build sewage treatment facilities. The act was never funded, but it was a sign of things to come.

After World War II, the exponential growth of the national economy, of energy use, and of waste loads gained astonishing momentum. The population shifts from farm to urban and from urban to suburban were accelerated. Through the 1950s and 1960s, matters of water and air pollution and land abuse, including solid waste dumping, became direct experiences for American families. Septic tanks failed, recreational waters were dirty, normal seasonal stagnant air became polluted air, passing vehicle exhaust became a personal assault, pesticides became suspect—indeed there were doubts cast on the blessings of technology. With a few exceptions, the states were not politically capable of moving effectively even against water pollution. In a state, the vested interests of the municipalities and industry were too strong. Legislative initiative moved to the U.S. Congress despite accomplishments of control of water and air problems in such states as California, New York, and Oregon. Air quality had improved in St. Louis, Pittsburgh, and New York City with the change from coal to oil for heating, a significant source reduction.

Congress was in tune with the growing popular concern for environmental quality as it passed new legislation and amended old laws. The principal ones dealing with water, air, solid wastes, and general objectives from 1960 to the present are:

Air:
- Clean Air Act of 1963 (PL 88–206)
- Motor Vehicle Air Pollution Control Act of 1965
- Air Quality Act of 1967 (PL 90–148)
- Clean Air Act of 1970 (PL 90–604)
- Energy Supply and Environmental Coordination Act of 1974 (PL 93–319)
- Clean Air Act Amendments of 1977 (PL 95–96)

Water:
- Safe Drinking Water Act of 1974 (PL 93–523)

Water pollution:
- Water Pollution Control Act of 1965 (PL 89–234)

Fig. 5. Low-lying belt of photochemical smog in Los Angeles, CA.

Fig. 6. High-rate trickling filter at Burlington, NC, for use in biooxidation process for organics in wastewater.

Clean Water Restoration Act of 1966 (PL 89–753)
Water Quality Improvement Act of 1970 (PL 91–224)
Federal Water Pollution Control Act Amendments of 1972 (PL 92–500)

Solid wastes:
Solid Waste Disposal Act of 1965, Title 1 of PL 89–272
Amended Resources Recovery Act of 1970 (PL 91–512)
Marine Protection, Research and Sanctuaries Act of 1972 (PL 92–532)
Resources Recovery and Conservation Act of 1976 (PL 94–580)

General objectives:
National Environment Protection Act of 1969 (PL 91–190)
Occupational Safety and Health Act of 1970 (PL 91–596)
Toxic Substances Control Act of 1976 (PL 94–469)
Council of Environmental Quality created under NEPA, 1969

Environmental Protection Agency. The initiatives of environmental protection in the United States have passed to the Federal government. The National Environmental Protection Act (NEPA) consolidated all Federal activities on air and water pollution, solid wastes, pesticides, noise, and environmental radiation in a new organization, the Environmental Protection Agency (EPA). Many states followed suit by organizing all or most of the environmental protection work in an independent or autonomous unit. States follow the EPA leads as surrogates to keep out direct Federal intervention and to qualify for various forms of monetary subsidies. When the subsidies go to local projects, the state agency is the control gate. An extensive and sometimes intensive bureaucratic process has evolved to move the authorities and mandates of Congress to EPA, sometimes directly and sometimes through the states to the point where the problem can be defined and solved.

Congressional mandates of specific reductions of pollutants by specific dates have deprived the EPA and its surrogates of flexibility. The Clean Air Act of 1970 made it law that by 1975 automobile exhaust emissions on new cars were to be reduced by 90% compared with 1970 levels. That was not accomplished. There was an extension to 1979. The 1977 Amendments to the Clean Air Act postponed compliance stepwise to 1979, 1980, and 1981. The Federal Water Pollution Control Act Amendments of 1972 require that there be zero discharge of pollutants to the waters of the United States by 1983. Unless there is some remarkable definition of "zero," the law will not be obeyed. The EPA has a very strong legal approach in its procedures. Hearings, petitions, and appeals abound. Regulations are numerous and often lengthy. Administration channels flowing from local government or industry through the states to regional offices to Washington are choked with paper. Relations with direct clients in industry are not different.

EPA regulations, standards, and requirements are nationwide. In some instances, national standards are meritorious and necessary, as for drinking water quality or automobile exhaust. Requirements for specific means of implementation can be wasteful such as wastewater treatment without consideration of the natural assimilative capacity of the receiving waters. The intent to require granulated activated carbon filters at water purification plants to reduce trihalomethanes whether they exist or not, or whether they can be reduced by other means, is a wasteful procedure. The requirement of high-efficiency flue gas cleaners on all coal-burning electricity-generating plants without considering coal quality is another example. Bans are a seemingly complete administrative solution to a problem substance. The EPA has used the ban on a few pesticides and one herbicide. The implementation of the Toxic Substances Control Act of 1976 may lead to wider use of the ban. One unfortunate result of these procedures is that professional judgment of scientists and engineers is removed from the analysis and solution of problems at the scene.

Nevertheless, much has been accomplished. Environmental protection in governmental agencies is firmly institutionalized. It has greater strength and support than ever. Pollutant sources are being controlled. Many streams and air sheds are showing improvement. Ocean dumping has been greatly reduced. Solid wastes are being managed much better than in the past. Air loading of particulates, hydrocarbons, and carbon monoxide on a national scale has been reduced. State agencies for environmental protection have been strengthened.

The environmental movement. The years on Federal legislation listed above indicate that congressional action did not lag behind the environmental movement. The long-established professional societies such as the American Water Works Association, the Water Pollution Control Federation, the American Public Health Association, the reoriented Air Pollution Control Association, and the American Public Works Association worked steadily through the years to intensify environmental protection. In 1962 Rachel Carson's book *Silent Spring* produced a wide public awakening on the nature and consequences of environmental pollution with persistent DDT residuals and their effect on ecosystems as a model. Existing conservation organizations such as the Audubon Society, the Wilderness Society, the National Wildlife Federation, and the Sierra Club became active participants in environmental concerns. New groups have been organized such as Environmental Action, Natural Resources Defense Council, the Friends of Earth, and the Environmental Defense Fund. Two significant events of the environmental movement were Earth Day in April 1970 in the United States, and the United Nations conference of official governmental delegations and numerous unofficial groups in Stockholm, Sweden, in the summer of 1972.

This United Nations Conference on the Human Environment resulted in the creation in 1972 of the United Nations Environmental Program (UNEP), with headquarters in Nairobi, Kenya. Its mission is education and dissemination of information. Its view of the environment is broader than that stated in this article. There is great concern for the advancing deserts, for forest disappearance as the wood is cut for fuel and as land is cleared for crops, and for preventing the overuse of ground and surface water sources. Global ecological and climatological changes and endangered animal species concern UNEP. Health matters of environment remain with the environmental health staff of the World Health Organization, which has had a productive program since 1950 and an enviable reputation among international agencies and its client nations.

From biological to chemical hazards. Environmental protection began about 100 years ago as specific communicable diseases were identified with water, milk, food, and insect and rodent vectors. The activities were identified as sanitation and sanitary engineering. These problems are by no means laid to rest. They are now overshadowed by the threats, real and alleged, of toxic chemicals in the open environment. Epidemiological evidence for human injury is substantial for some 600 toxic substances known to produce specific pathology in the workplace, namely, occupational diseases. Effects on the general population from infancy to old age have been identified in acute episodes of environmental contamination. Examples of air pollution episodes are: Donara, PA, in 1948; London in 1952; and Seveso, Italy, in 1976. Fish from mercury-polluted waters caused poisoning in Japan at Minamata, 1953–1960, and Nigata, 1964–1965. Infants suffered severe brain and nervous system damage from methyl mercury passed from the fish to the mother through the placenta to the fetus. Japan also provided the evidence of the toxicity of polychlorinated biphenyls (PCB). A leaking heating system contaminated rice oil. From 1968 to 1973, 1200 cases of "yusho," the oil disease, were attributed to the mishap. Again placental transfer from mothers with yusho damaged their children. The infants were small at birth and below normal in subsequent growth and development.

Low-level intakes through a lifetime. The matter of low-level intakes of known or suspected toxicants through a life of 65 to 75 years cannot be resolved by present toxicological and epidemiological information or methods. Decisions on the use and control of such substances cannot be made solely on existing scientific evidence. These issues are politically and socially sensitive and highly emotional when laboratory animal tests indicate the possibility of the substances being carcinogenic, mutagenic, or teratogenic. The continued use of saccharin, which has been found to be a weak carcinogen in tumor-susceptible strains of rats, is a dilemma, as no satisfactory alternate is presently available; yet many people benefit from the substitution of saccharin for sugar. Two things are evolving from these situations. One is the increased effort to know more fully the biochemistry of cancers so that susceptible people can be identified. Another is the recognition of risks and benefits and the developments of methods to make such judgments. Many are made implicitly. The American people accept 50,000 motor vehicle deaths per year as a fair exchange for their automobile use. Energy source choices may be forced upon them that will impair environmental quality or possibly heighten environmental risks. A well-informed people is required to understand the stakes in such risk-benefit judgments.

[EMIL T. CHANLETT]

Bibliography: E. Chanlett, *Environmental Protection*, 2d ed., 1979; M. Eisenbud, *Environment, Technology and Health*, 1978; R. H. Wagner, *Environment and Man*, 3d ed., 1978.

Precedents for Weather Extremes

David E. Parker

David E. Parker is a Principal Scientific Officer at the United Kingdom Meteorological Office, Bracknell, and is involved in research and climatic change.

The weather varies on time scales ranging from seconds to geological ages, and on space scales ranging from the microscopic to the area of the whole Earth. The climate at any place is determined both by the average of the weather and by its variability.

Extremes of weather must be viewed in the context of where, at what time of year, and in what climatic period they occur; how long they last; and what area they cover. A temperature of 30°C is more unusual in Scotland than in England; and a foot of snow is far less likely in South Carolina than in New Hampshire. Not only is snow less likely in summer than in winter, but if the climate at some location became drier and warmer, snow would become a rarer, more extreme event even in winter. Thus the likelihood of a particular weather extreme can vary in time as well as in space. Also, persistent extremes are rarer than transient ones: a New York heatwave is unlikely to last for 2 months without a break. And many extremes are more likely to cover small areas than large ones: the whole of India is never flooded simultaneously, though droughts are very extensive in some parts of the world, such as the Sahel zone of western Africa.

The incidence of extremes, defined in accordance with recent experience, will change not only if the average changes but also if the variability or standard deviation changes. For example, for a gaussian distribution of growing-season temperature, if the mean temperature is 18°C and the standard deviation is 1.02°C, 2.5% of seasons will be warmer than 20°C and 2.5% will be colder than 16°C. If the mean decreases from 18 to 17°C, the corresponding 2.5% criteria will be 19 and 15°C and the likelihood of extreme cold, as defined by values less than 16°C, will increase to 16.3%. Similarly the likelihood of extreme warmth (over 20°C) will decrease to less than 0.2%. However, if the mean remains 18°C but the standard deviation increases from 0.98 to 1.5°C, the probabilities of extreme warmth or cold will both rise to 9.2%. Finally, if the mean falls to 17°C and the standard deviation rises to 1.5°C, the probabilities of extreme warmth and extreme cold will become 2.3 and

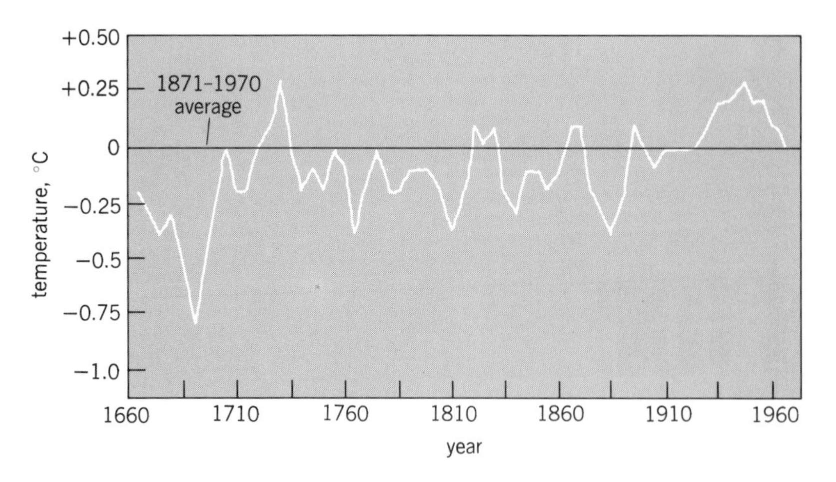

Fig. 1. Five-year means of Northern Hemisphere temperature anomalies estimated from central England temperature.

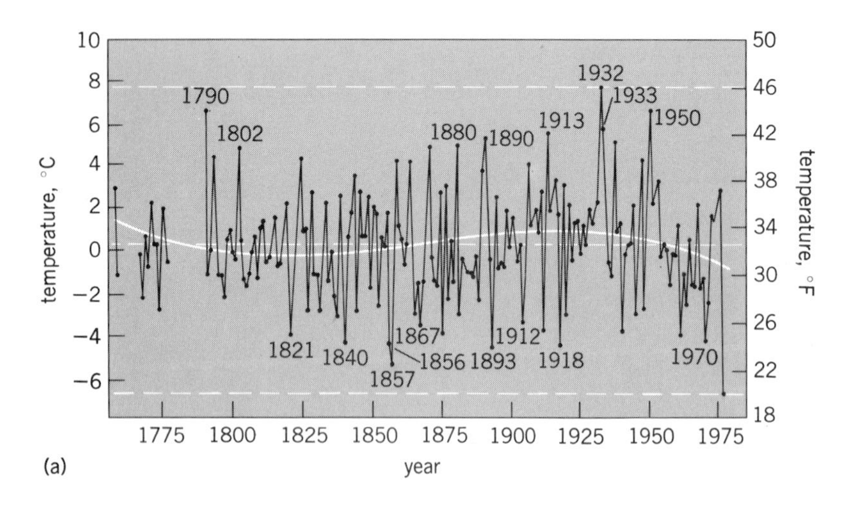

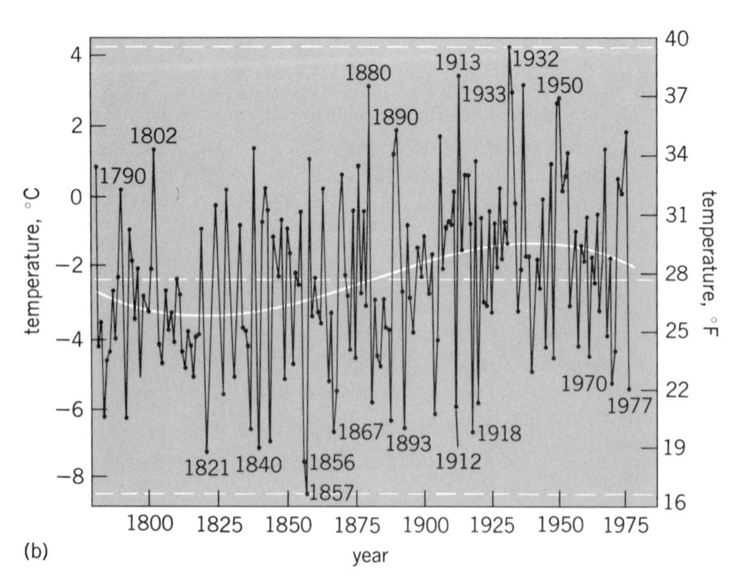

Fig. 2. Long-term average January temperature for: *(a)* Philadelphia, PA (1758–1977); the smooth trend line is a third-order polynomial least-squares fit; *(b)* New Haven, CT (1781–1977); the trend line is a third-order polynomial least-squares fit. *(From H. F. Diaz and R. G. Quayle, The 1976–77 winter in the contiguous United States in comparison with past records, Mon. Weath. Rev., 106(10):1393–1421, 1978)*

25.2% respectively. For rainfall the distribution is usually skew, and not gaussian, and a realistic example would be more complex. But the general principle stands: the likelihood of extremes, as judged by present standards, depends on whether the average or the variability, or both, is subject to change. If such changes can be predicted, the forecast can be used to estimate the magnitude of future extremes. Otherwise, the best approach is actuarial: one assumes that extremes are as probable in the future as they have been in the past, and plans accordingly.

CLIMATIC CHANGE

The study of climatic change includes the movement of the average and also of the variability about that average. It is usual to try to separate changes due to natural causes from changes caused by human activities. A necessary basis for these studies is an understanding of the mechanisms whereby the various components of the climatic system interact. These components include the atmosphere, the oceans, the polar ice and snow, and the vegetation on the surface of the Earth.

Natural climatic change. Possible causes of past and future natural climatic change are the following.

Variations in solar radiation. The most important variation would be a change in total energy. Prior to about 1970, the solar constant, which is about 1370 W/m², was known only to an accuracy of 1%, and claims that it oscillated by about 0.1% of its magnitude following the 11-year sunspot cycle may have been premature. The supporting data may have reflected changes in the transparency of the atmosphere related to variations in the ultraviolet portion of the solar beam affecting the stratosphere by changing the quantity of ozone there. Recent measurements from Mariner spacecraft did not detect a variation of as much as 0.03% in the solar constant as a result of sunspots or the surrounding bright faculae. It would appear therefore that, based on the most up-to-date data available, there is no reason to expect any significant change in the solar constant in the foreseeable future.

Variations in Earth's orbit. In the 1930s M. Milankovitch documented periodicities of 96,000 years in the elliptical shape of the Earth's orbit (that is, the orbit becomes alternately more circular or more elliptical), 40,000 years in the tilt of the Earth (that is, there is a cycle in the highest latitude at which the sun passes overhead in summer, at present about 23½°N or S), and 21,000 years in the relation of the seasons to the distance of the Earth from the Sun (that is, the Earth is at present nearest to the Sun in the Northern Hemisphere winter, but in 10,000 years' time it will be farthest at that season). These orbital changes will affect the amount of solar radiation available at a given latitude at a given season, and may therefore affect the climate. Evidence on past ice ages, particularly from cores drilled in seabed sediments, has supported oscillations in climate on about these time scales. However, the changes are too slow to be capable of influencing the Earth's climate within the time scale of human generations.

Variations in transparency of atmosphere. Volcanic dust, when injected into the atmosphere in large quantities, will reduce the amount of solar radiation reaching the Earth's surface, so that after periods of intense volcanic activity the Earth may be expected to cool. However, it is observed that the dust usually clears from the stratosphere within a few years and from the troposphere in a few weeks, so unless volcanic activity is sustained, the changes induced will not be long-lasting. Moreover, the dust, by absorbing outgoing long-wave radiation, may counteract the loss of solar input and thus prevent cooling of the Earth. Observations following the eruption of Mount Agung in Bali in 1963 suggest that the troposphere cooled (because of reduced insolation) and the stratosphere warmed (because of increased absorption by the dust). M. K. Miles and P. B. Gildersleeves found volcanic dust to be a possible partial explanation of temperature changes over the Northern Hemisphere in recent years.

Since volcanic activity cannot yet be forecast, neither can its consequences for the climate.

Changes in albedo. It has been postulated that if the area covered by snow and ice increases, more solar radiation is reflected and the Earth cools; as a result, more precipitation falls as snow, the area covered increases further, and the cycle is repeated in an amplifying cascade until glaciation become extensive. Conversely, according to this concept, a decrease of snow and ice cover could lead to a warmer Earth. The fact that this has not happened in historical times is a result of the strong domination of climate by the annual radiation cycle: the largest regional winter snow or ice anomalies fail to survive the following summer; on the other hand, the reduced insolation in winter always allows new ice and snow fields to form.

The physical effect of increased snow cover, in terms of its influence on the Earth's total radiation balance, is probably small in comparison with the effect of color changes (due, for example, to destruction of forests by fires, or to spread of deserts) in low latitudes, where amounts of radiation available to be absorbed or reflected are much larger.

Interactions between atmosphere and ocean. The dynamics of ocean-atmosphere interaction are only partially understood, and a complete description of such interactions would require an account of the links between the upper and lower layers of the ocean, which are even less understood. However, ocean-atmosphere interaction appears to be important in regard to variations in climate on all time scales, and there are clear indications of its relevance to changes on time scales of a few years. Examples of the latter are the Southern Oscillation and the associated El Niño, and the severe winters in North America and Europe.

Human-induced changes of climate. The wastes of human activity can have an effect on climate over a period of time.

Carbon dioxide. It is estimated that industrial emission of carbon dioxide, combined with reduction of the Earth's vegetation cover by humans, has already resulted in an increase in carbon dioxide in the atmosphere from 260 to over 330 parts per million, and it seems likely that it will reach 400 ppm by the end of the century. Since carbon dioxide absorbs outgoing long-wave radiation, the temperature of the atmosphere may be expected to rise. The average global warming for a doubling of carbon dioxide in the atmosphere is generally estimated at about 2°C. This value includes the effect of a resulting increase in water vapor content which enhances the warming by further absorption of outgoing long-wave radiation.

Any global climatic change is likely to affect also the wind-flow patterns causing regional differences in degrees of warming or cooling, and changes in rainfall. These interactions within the atmosphere will be substantially modified by the oceans. The extent to which the oceans may absorb excess carbon dioxide is as yet unknown, and could delay significant climatic changes until well into the 21st century.

Dust and other pollutants. Human activities induce pollutants from fires and chimneys, and also a change in soil erosion and dust-raising by the wind (such as in the dust bowl of the United States in the 1930s), but estimates of the extent of the pollution vary widely. Dust and other pollutants in the lower atmosphere reduce both insolation and outgoing radiation. The changes may or may not balance, and it is uncertain whether the amount of dust is significant in relation to natural quantities.

Alteration of albedo. Removal of forests and denudation of vegetation by overgrazing increase the proportion of solar radiation reflected and lost to space (the albedo), and local changes in climate may be considerable. If the areas affected are large enough, the total radiation budget, and therefore the climate of the Earth, may be influenced.

Heat input or thermal pollution. Heat input from human activity in some cities is already as great as that from solar radiation, but the areas affected are globally insignificant. The total world heat input from human activity is only 0.01% of that from the Sun, and is unlikely to reach 0.5% within the next 100–150 years. However, concentrated local heating (of hundreds of millions of megawatts) may result in circulation changes far downwind.

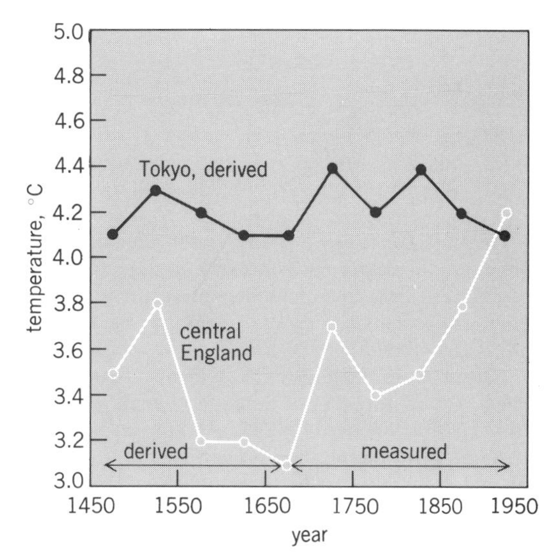

Fig. 3. The 50-year averages for Tokyo derived winter temperatures and central England winter temperatures. *(From B. M. Gray, Japanese and European winter temperatures, Weather, 30(11):359–368, 1975)*

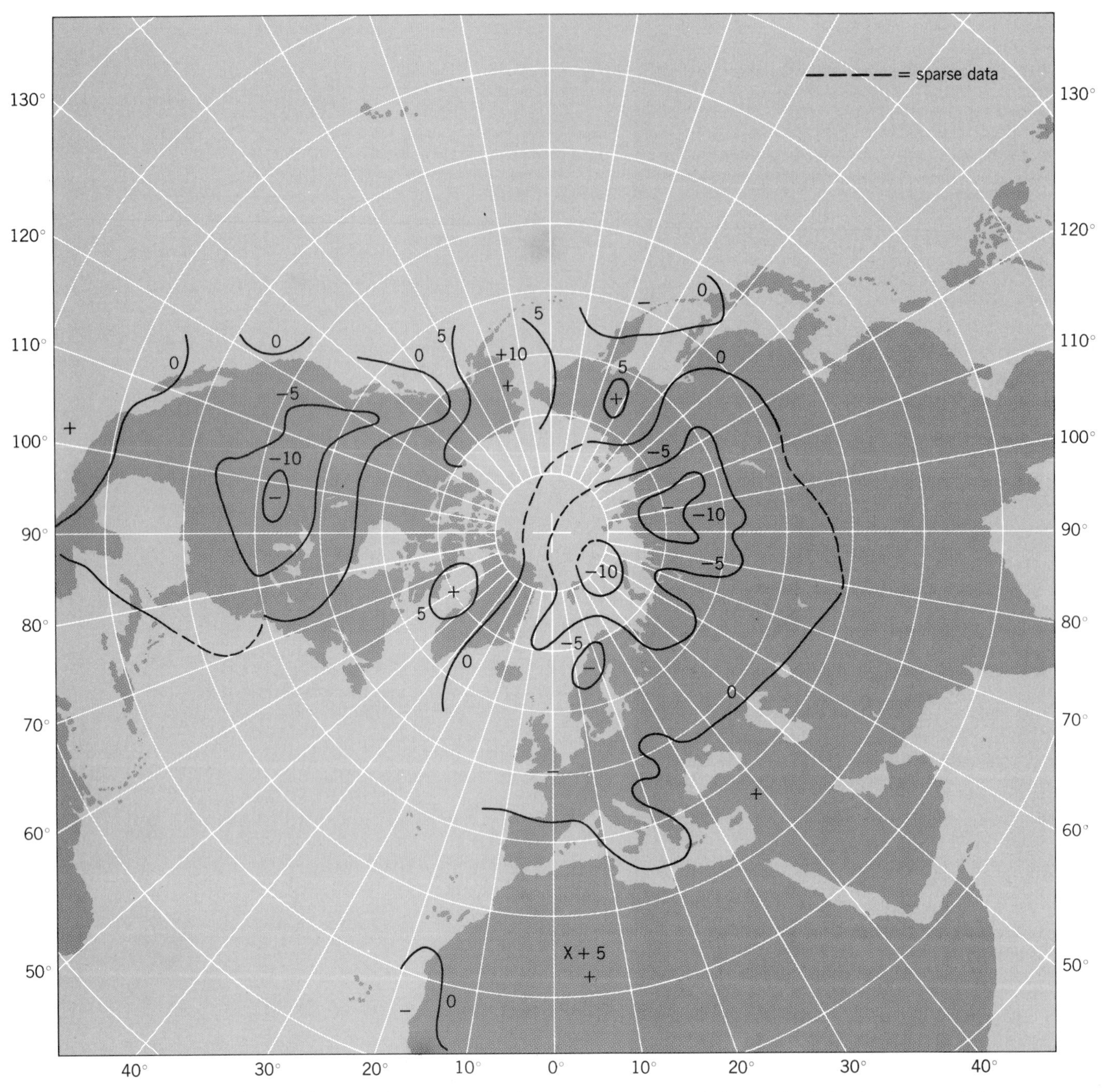

Fig. 4. Difference of temperature (labeled contours) from normal, Jan. 1979 (°C). X + 5 indicates a +5°C maximum.

Central city areas already experience weather which, in the historical context, must be classified as extreme in terms of high temperatures. There may also be local increases of rainfall because of increased convection.

Future expectations. There is as yet no firm basis for any forecast of climatic change resulting from natural causes. Extrapolation of apparently regular cycles cannot be justified scientifically until their causes have been established, and until their existence has been recorded for a sufficiently long time for their stability to be confirmed. So far the first condition is fulfilled only for annual and shorter-term oscillations, and the second in respect of 2-year and shorter time scales, and for the long-term changes of the Earth's orbit.

The effects of human activity are already evident in local climates, especially in cities. There are good reasons to expect significant changes on a global scale due to increased carbon dioxide, but even these may not be detectable before the end of the century, and the global patterns of changes in temperature and rainfall cannot be forecast with any certainty.

For the time being, therefore, the best estimate that can be made regarding the future expectation of weather extremes must be based on past experience.

CLIMATE AND WEATHER IN THE RECENT PAST

If the future expectation of climatic extremes is to be based only on evidence from the past, it is necessary to examine the record to establish the period of time which can give a statistically reliable base for these expectations. It is immediately apparent that the standard period of human memory, or one generation (30 years), is too short. Over the whole period of instrumental records, neither the average nor the incidence of extreme events has remained stationary. Therefore it is necessary to allow for the possibility of extremes which have not been observed recently but which did occur further back in time. Up-to-date trends of temperature or rainfall could be used as a guide in this respect without a specific climatic forecast being made.

Temperature patterns. Figure 1 shows a Northern Hemisphere mean temperature series since 1660 derived from temperatures in England and from correlations between these and variations elsewhere. Notable features are the coldness before about 1700 and the peak warmth about 1940; the recent cooling is probably still continuing, and this suggests that future extreme weather of the severity experienced in the 17th and 18th centuries cannot be ruled out (as exemplified by the winters 1976/1977, 1977/1978, and 1978/1979 in the United States). However T. P. Barnett has emphasized the spatial variability of temperature fluctuations, and the hemispheric temperatures in the data-sparse earlier years in Fig. 1 are not firmly established. The information for populated areas in the 19th century and earlier does sometimes, but not always, support the pattern indicated in Fig. 1. Mean January temperatures in the eastern United States did follow similar trends (Fig. 2), but those in Japan did not (Fig. 3). The local urban heating effects have been removed from Fig. 3 but not from Fig. 2, so the real recent cooling in the eastern United States may have been greater. An overall change of hemispheric temperature will involve change of wind-flow patterns, bringing temperature changes in opposite senses in different places. This was shown clearly by H. van Loon and J. Williams. Figure 4 illustrates the pattern of temperature differences from normal in the month of January 1979, which had unusual wind circulation features.

Wind-flow patterns. Variability and trend in wind-flow pattern are illustrated in Fig. 5, which shows surface pressure differences, Azores minus Iceland, for summers and for winters since the 1860s. If the difference is large, the Azores anticyclone (high-pressure area) and Icelandic depression (low-pressure area) are strong and in their normal position, and there are strong westerly winds between them, which often affect Europe; but if the difference is small, the high and low, and the corresponding westerlies, are weak or displaced. The summer graph shows little trend, though sometimes there are successions of years of high- or low-pressure differences. The winter graph shows a decreasing trend since 1920, again with successions of years of low or high differences; but recent years with low values are not unprecedented in the light of the earlier years shown (for example, 1881, a very severe winter in northern Europe). In neither graph is there a clear change of variability.

Another interesting measure of atmospheric circulation is the difference in the height of the 500-mb (50-kilopascal) pressure level over the northern Rockies and over the southern Hudson Bay. When this difference is large, especially in winter, with

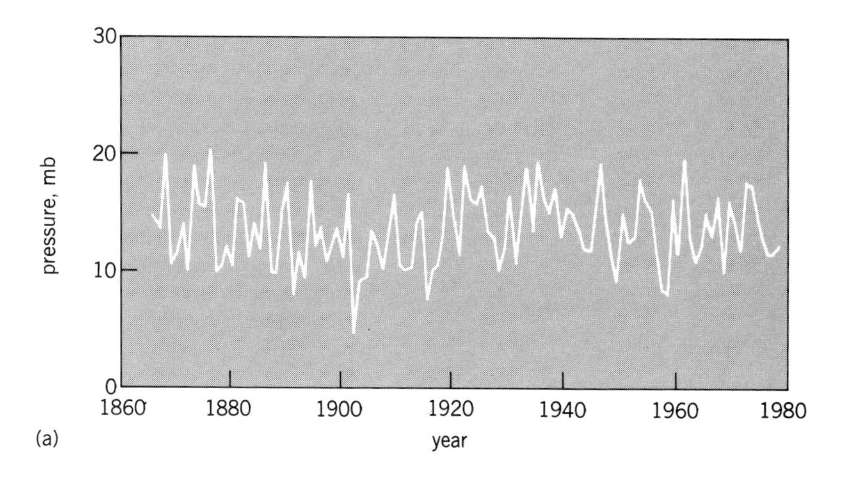

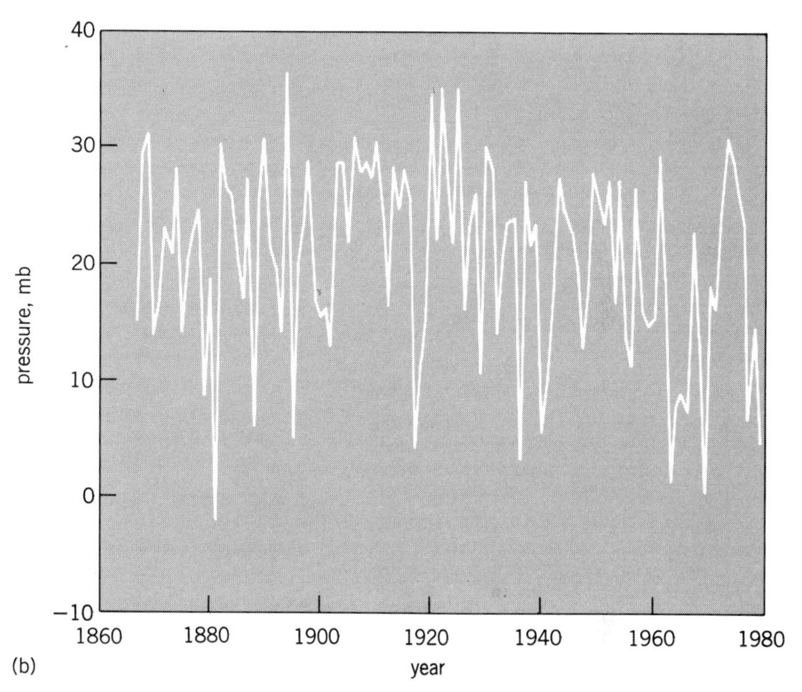

Fig. 5. Pressure difference, Ponta Delgada (Azores) minus Stykkisholman (Iceland), for *(a)* summers, 1867–1978, and *(b)* winters, 1867–1979; 10 mb = 1.0 kilopascal.

the Rockies value higher, the west of North America experiences drought and central parts experience cold. Eastern parts may be snowy or wet; the circulation much further afield may meander more than usual (likely low values in Fig. 5*b*), with consequential changes over the hemisphere. Figure 6 shows a sequence of the winter height differences. The cause of the high values is not certain, but ocean temperatures in the Pacific may be important. In the light of the earlier parts of Fig. 5*b*, the record in Fig. 6 is too short to indicate that the 1977 value has never been exceeded in historical times.

Changes of variability. Changes in mean temperature and circulation are thus well documented; but the position is less clear where changes of variability are concerned. J. K. Angell and J. Korshover have found evidence for a recent increase of variability in many parts of the world; but R. A. S. Ratcliffe and coworkers found no trend toward increased variability over the past 100 years when they considered surface pressure over the Northern Hemisphere and pressure, temperature, and

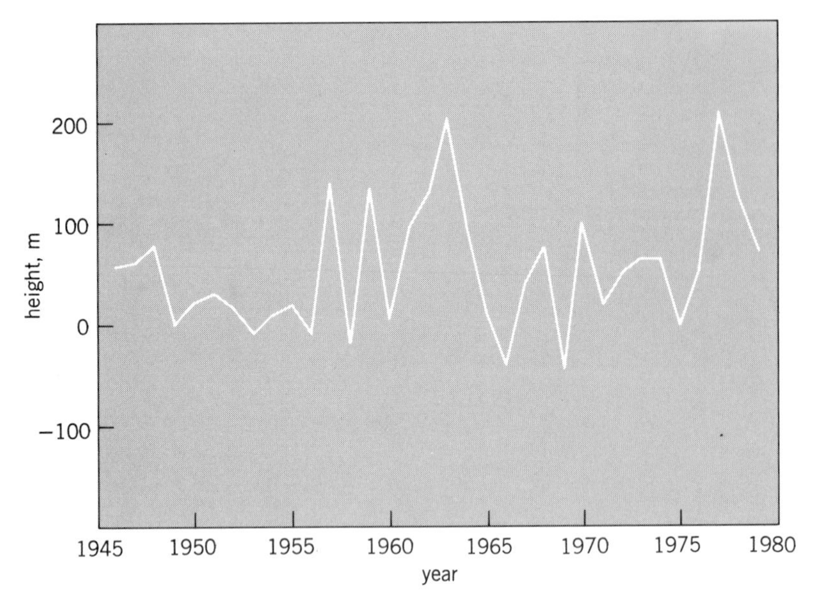

Fig. 6. Height of the 500-mb pressure level, Whitehorse (Ontario; 61°N 135°W) minus Moosonee (Yukon; 51°N 91°W) in winter.

rainfall over Britain. However, Ratcliffe and coworkers did note the unusual nature of the 1975–1976 drought in southern England and of temperature fluctuations in Britain in these 2 years. Figure 7*a* also illustrates the unusual nature of the 1976 summer in England by showing its extreme position on a temperature-rainfall scatter diagram for 1871–1978. But 1826 summer was 0.1°C hotter than 1976, and 1800 was slightly drier, so there was a precedent for summer 1976, even though such a combination of heat and drought had not been recorded before (temperature values are established since 1659 and rainfall since 1727).

The very cold and wet winter of 1978–1979 in Britain was not unprecedented either for temperature or for total rainfall alone; but the combination on first sight appears extreme (Fig. 7*b*) because in Britain, as in many other areas, cold winters are usually dry, and wet winters are usually mild. In fact, it was the only winter since rainfall records began in 1727 that occurs in the upper left portion of Fig. 7*b* bounded by the dashed line. One cannot deduce from this alone that the weather is changing its character: the wettest month, December, was only slightly colder than average, while the very cold January and February had near-average rainfall (though more than most very cold winter months). It is scientifically unacceptable to extrapolate any trend which may be observed in these records.

It is possible to reconcile the results of Angell and Korshover with those of Ratcliffe's group by noting that Angell and Korshover have used upper-air data since 1958, and that while Ratcliffe's group finds no overall increase in variability, their graph for 30–85°N surface pressure (Fig. 8) does show a slight minimum of variability in the 1950s, and a recent slight rise to levels nearer but not above those of 1900–1920. It may be noted that in Fig. 7*b* few of the 1970s or the 1910s appear near the center of the distribution (though more of 1900–1909 do).

Conclusions. Evidently then, there is no immunity from extremes of weather which have occurred within the period of the historical record. At the same time, and considering the whole of the period for which instrumental records are available—about 300 years in Britain—rather than recent living memory, there is no indication that humankind is about to experience weather the like of which has never happened before. These are the only conclusions which are scientifically justified

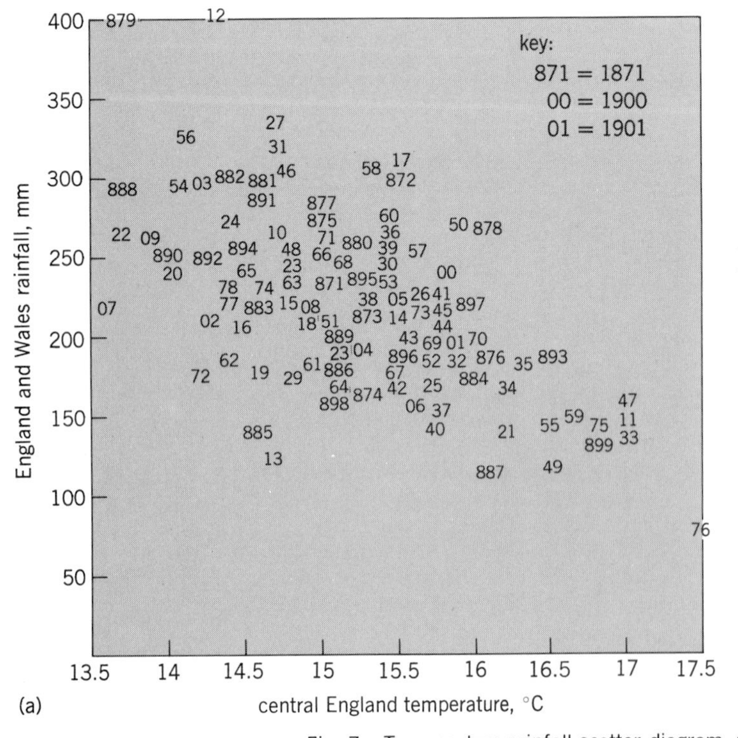

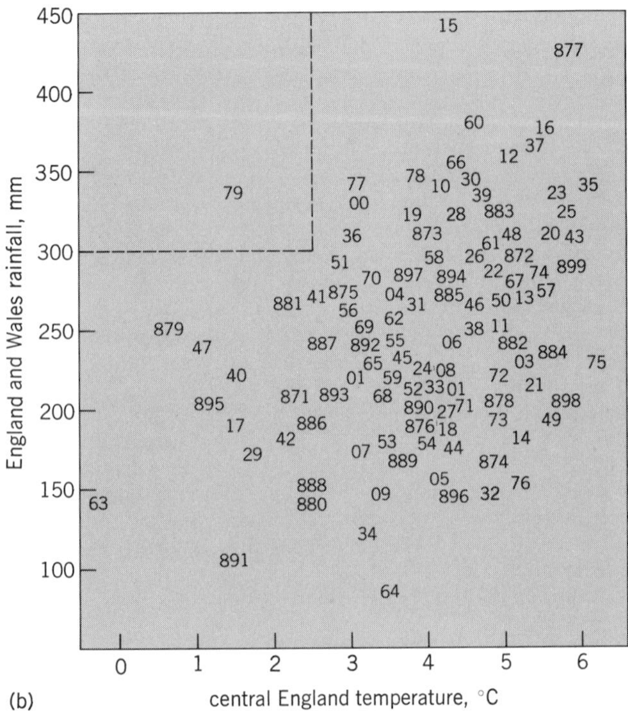

Fig. 7. Temperature-rainfall scatter diagram, showing (*a*) summers, 1871–1978, and (*b*) winters, 1871–1979. The numbers correspond to years, for example, 871 = 1871, 00 = 1900, 01 = 1901.

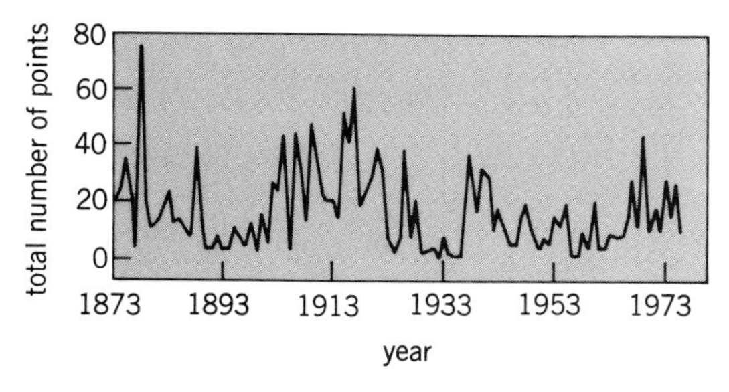

Fig. 8. Total number of points on an annual surface pressure grid of 344 points (30–85°N) with anomaly greater than ±2 standard deviations, for period 1873–1976. *(After R. A. S. Ratcliffe, J. Weller, and P. Collison, Variability in the frequency of unusual weather over approximately the last century, Quart. J. Roy. Meteorol. Soc., 104:243–255, 1978)*

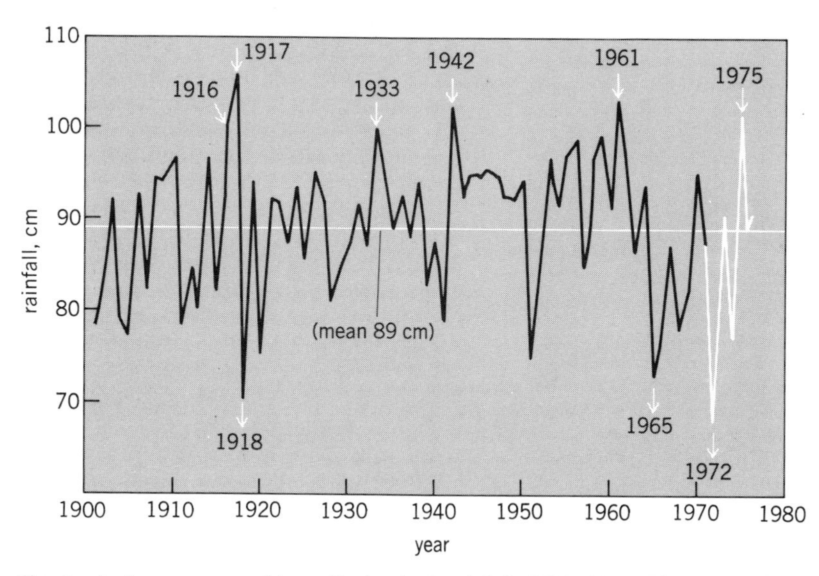

Fig. 9. Indian monsoon (June–September) rainfall, 1901–1977. The values for 1971–1977 are approximate. *(From B. Parthasarathy and D. A. Mooley, Some features of a long homogeneous series of Indian summer monsoon rainfall, Mon. Weath. Rev., 106(6): 771–781, 1978)*

on the basis of acceptable statistical analysis, and they apply to all time and space scales of weather extremes: short-period local events such as flash floods, thunderstorms, and tornadoes; somewhat longer-period and larger-scale phenomena such as hurricanes; and long-period large-area occurrences such as droughts, severe winters, and sequences of unusually warm or cold years. It is the general circulation of the atmosphere which controls all these effects, and as the present circulation is not expected to behave in an unprecedented manner in the near future, it follows that its local and large-scale effects will not be unprecedented either.

Exceptions to the general conclusion that unprecedented weather extremes cannot be expected in the near future are the inner-city areas where artificially high temperatures are expected to continue and even to increase because of heat production and retention in the buildings.

The impact of extreme weather on developed nations may be inconvenient and occasionally serious, particularly where new life-styles have developed in periods when extremes of weather have been lacking; but developing nations are much more vulnerable. The next section will therefore consider extremes of weather in certain key areas of the tropics.

MONSOONS AND THE WALKER CIRCULATION

In 1972 a series of natural disasters drew considerable attention to the problems caused by weather extremes and variability of climate. In the Sahel zone south of the Sahara, several years of drought culminated in an almost total failure of the rains, with consequent deaths of people by starvation and enormous losses of livestock. Drought affected eastern Africa also. The Indian monsoon rains failed over large areas. The Soviet grain crop was poor, with far-reaching economic consequences. And a sudden warming of the Peruvian coastal waters caused the collapse of the anchovy fishing industry when the fish died en masse, again with worldwide economic repercussions.

Walker circulation. There has been discovered a worldwide pattern of wind flow which may relate the behavior of the rains over Africa, India, South America, and the western tropical Pacific. This so-called Walker circulation is also connected with

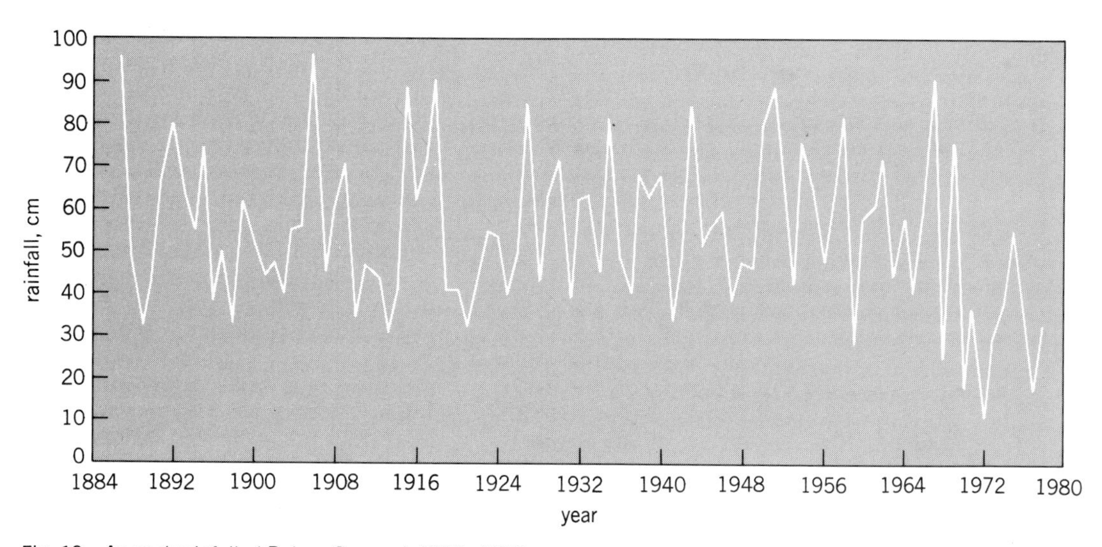

Fig. 10. Annual rainfall at Dakar, Senegal, 1887–1978.

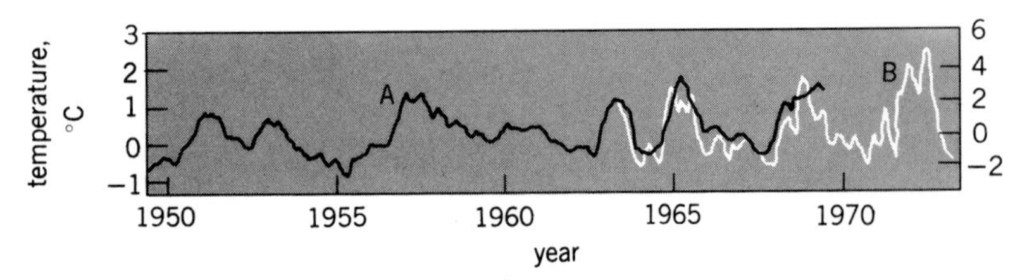

Fig. 11. Sea-surface temperature anomalies: A, 5°N–5°S, 80°W–180°W; B, Puerto Chicama, Peru. *(From P. R. Julian and R. M. Chervin, A study of the Southern Oscillation and Walker circulation phenomena, Mon. Weath. Rev., 106(10):1433–1451, 1978)*

variations in atmospheric pressure over Indonesia and the eastern South Pacific, known as the Southern Oscillation. These atmospheric variations interact mutually with the ocean currents in the Pacific in such a way that the coastal waters off Peru, which are normally cold (compared with most equatorial water), suddenly become much warmer, a phenomenon known as El Niño. There are less certain connections with weather in temperate latitudes, as those noted by P. R. Julian and R. M. Chervin. In addition, weather conditions in many parts of the world tend to oscillate between alternate years or in a 3-year cycle. For example, B. Parthasarathy and D. A. Mooley found a quasi-biennial oscillation in Indian summer monsoon rainfall; this is visible in parts of the time series shown in Fig. 9.

It is thus important, if scientists are to predict weather extremes in the tropics, that the Walker circulation and the quasi-biennial oscillation be thoroughly understood. El Niño can now be predicted to some extent up to 6 months in advance because it is most closely related with the wind flow over the Pacific in the Walker circulation, but relationships in regard to the Indian and African rains are more complex, and as yet inadequately understood. Progress is being made as data coverage improves, but in the meantime the characteristic sequences of monsoon rainfalls must be used as an indication of the extremes, such as sequences of drought years, to be expected in future.

Rainfall. Figure 10 shows annual rainfall since 1887 at Dakar, Senegal, at the western end of the Sahel zone of western Africa. Very nearly all the rain falls in July to September. The dry years since 1968 form an unprecedented sequence in the context of the last 100 years, but there were also dry periods in 1896–1903, 1910–1914, and 1919–1922. Harmonic analysis shows a tendency to a 3-year cycle, but this is not strong enough to be used for forecasting. Statistical tests on the whole data series to 1978 do not provide clear support even now for a theory that the climate is becoming irreversibly drier; and conversely, even if the next few years are wet, frequent sequences of 3 or 4 years around the 40-cm mark must be expected, and severer sequences like 1968–1978 must not be ruled out.

Parthasarathy and Mooley found no cause for fear of any disastrous decreasing trend in monsoon rainfall setting in over India. They analyzed 1841–1970, but the period since then has included not only the very dry 1972 but also the very wet 1978 with its serious floods. Their values for 1901–1977 are given in Fig. 9.

Figure 11 illustrates the variations in tropical Pacific Ocean temperature which affect the anchovy fisheries. The warmth in 1972–1973 exceeded anything since 1950, but sudden warmings occurred in 1951, 1957, 1963, 1965, 1968, and 1976, and can be expected to recur in the future. It would be wise to plan the economy accordingly, and to use the limited forecasting capability for El Niño, referred to above.

CONCLUSIONS AND OUTLOOK

Although it has not been proved that climate is definitely changing or weather becoming more variable, it is beyond doubt that recent extreme weather events have often exceeded the lesser extremes of some quieter interludes of history, but they have not gone beyond some of the more remarkable events. What may be more important is that at least some parts of the world have become more sensitive to extreme weather as a result of increased population and other cultural and economic factors. In future planning, it is important to take into account knowledge about weather extremes known to have occurred in the past, since similar events cannot be ruled out in the future.

[DAVID E. PARKER]

Bibliography: J. K. Angell and J. Korshover, Global temperature variation, surface–100 millibars: An update into 1977, *Mon. Weath. Rev.*, 106(6):755–770, 1978; T. P. Barnett, Estimating variability of surface air temperature in the Northern Hemisphere, *Mon. Weath. Rev.*, 106(9):1353–1367, 1978; H. F. Diaz and R. G. Quayle, The 1976-77 winter in the contiguous United States in comparison with past records, *Mon. Weath. Rev.*, 106(10):1393–1421, 1978; B. M. Gray, Japanese and European winter temperatures, *Weather*, 30(11):359–368, 1975; P. R. Julian and R. M. Chervin, A study of the Southern Oscillation and Walker Circulation phenomena, *Mon. Weath. Rev.*, 106(10):1433–1451, 1978; H. E. Landsberg, *An Analysis of the Annual Rainfall at Dakar (Senegal), 1887–1972*, University of Maryland, Institute of Fluid Dynamics and Applied Mathematics, 1973; M. Milankovitch, Canon of Isolation and the Ice Age Problem, K. Serb. Akad., Spec. Publ. 132, Sect. Math. Nat. Sci., 1941; M. K. Miles and P. B. Gildersleeves, A statistical study of the likely causative factors in the climatic fluctuations of the last 100 years, *Meteorol. Mag. (HMSO)*, 106:314–322, 1977; B. Parthasarathy and D. A. Mooley, Some features of a long homogeneous series of Indian summer monsoon rainfall, *Mon. Weath. Rev.*, 106(6):771–781, 1978; R. A. S. Ratcliffe, J. Weller, and P. Collison, Variability in the frequency of unusual weather over approximately the last century, *Q. J. Roy. Meteorol. Soc.*, 104:243–255, 1978; H. van Loon and J. Williams, The connection between trends of mean temperature and circulation at the surface, pt. 1: Winter, *Mon. Weath. Rev.*, 104(4):365–380, 1976.

Environmental Satellites

E. Paul McClain

E. Paul McClain is the director of the Environmental Sciences Group in the National Environmental Satellite Service of the National Oceanic and Atmospheric Administration. Holder of a Ph.D. degree in meteorology from Florida State University, he has conducted and directed research on meteorological, oceanographic, and hydrologic applications of the data collected by environmental satellites since 1962.

The first TIROS (television and infrared observation satellite) vehicles were placed in Earth orbit in 1960. From these relatively crude meteorological satellites have steadily evolved the increasingly sophisticated operational environmental satellite systems such as GOES (geostationary operational environmental satellite) and ITOS (improved TIROS operational satellite). Several series of specialized research spacecraft such as Nimbus and Landsat, and the more recent Seasat, have been developed as well. This article gives brief descriptions of these satellites and discusses some of their meteorological and oceanographic applications.

OPERATIONAL ENVIRONMENTAL SATELLITES

The National Oceanic and Atmospheric Administration (NOAA) maintains both the GOES and ITOS.

ITOS. The latest generation of ITOS is designated TIROS-N (see Fig. 1 and Table 1), the first of which was launched in 1978. Although the previous ITOS vehicles were in circular, Sun-synchronous, near-polar orbits around the Earth at an altitude of about 1450 km, the TIROS-N series will orbit at 835-870 km. With each circuit of its orbit (pass), data are collected along a swath about 2700 km wide, and global coverage is achieved twice daily as the spacecraft completes 14 or so passes each 24 hr of Earth rotation.

All geographic areas are viewed by the line-scanning advanced very-high-resolution radiometer (AVHRR) at about the same local time (near 3 A.M. and 3 P.M. for the first TIROS-N). The AVHRR takes over the functions of the two-channel (visual and thermal-infrared [IR]) scanning radiometer (SR) and very-high-resolution radiometer (VHRR) on the earlier NOAA polar orbiters. The AVHRR features on-board digitizing of the imaging data before they are transmitted from the satellite. The on-board data processor will also "degrade" the stored 1-km-resolution data to 4-km resolution to provide global coverage. The AVHRR has four channels of 1.1-km-resolution data: visible, reflected-infrared, and two thermal (that is, emitted) infrared. The data from the two thermal-IR channels are used

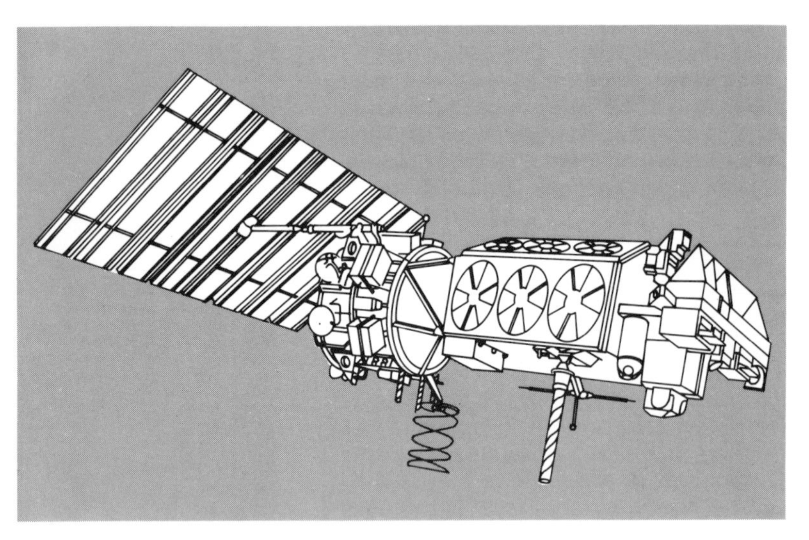

Fig. 1. Schematic diagram of TIROS-N spacecraft. *(From A. Schwalb, The TIROS-N/NOAA A-G Satellite Series, NOAA Tech. Mem. NESS 95, 1978)*

for improved surface-temperature computations at night. A later version of the AVHRR will have a third thermal-IR channel to enable accurate surface temperature determination in the daytime as well. The radiometers view on-board reference-level targets and cold outer space for purposes of infrared calibration.

One of the other primary systems on TIROS-N is the TIROS operational vertical sounder (TOVS). The vertical profiles of atmospheric temperature and water vapor derived from the TOVS are used to supplement the soundings of the atmosphere made by means of conventional meteorological radiosondes sent up from the Earth's surface on small balloons. The stations of the global radiosonde network, however, are irregularly distributed over the globe, with few or no stations over large, relatively uninhabited land areas and over the major oceans.

GOES. Besides the relatively low-altitude polar-orbiting satellites, NOAA also maintains several very-high-altitude geostationary satellites, the GOES (see Table 1). These spacecraft orbit in the Earth's equatorial plane, and at an altitude of 35,787 km their west-to-east motion is matched to that of the rotation of the Earth beneath, that is, they can be made stationary at the desired longitude. Two such United States geosynchronous spacecraft are now on station, one at 75°W and the other at 135°W. From such a great altitude the Earth's disk viewed from the satellite extends toward each pole about 60° in latitude and 60° in longitude to the west and to the east. Being fixed in location relative to the Earth also permits a high frequency of observations each day, and enables the spacecraft to have a data collection and relay function.

The several United States geostationary satellites are part of an international constellation of such operational spacecraft. Japan has stationed one over the western Pacific Ocean, and the European Space Agency (ESA) has one over the extreme eastern Atlantic Ocean; the Soviet Union and India are both planning for one in the Indian Ocean area within the next few years. When all five GOES-type satellites have become operational, all but the polar regions will be under essentially continuous (about half-hourly) environmental watch.

The visible/infrared spin scan radiometer (VISSR) on board the GOES is designed primarily to provide frequent visual and IR images of cloud patterns to the meteorologist, but VISSR data are also proving to be of increasing value for other environmental applications (such as snow mapping and surface-temperature mapping). The normal mode of operation results in full-disk visual and infrared images every 30 min. Special, very-high-frequency (3–10 min intervals) coverage can be provided, but the north-to-south extent of the images is reduced accordingly.

Table 1. Operational environmental satellites and their instrumentation

Spacecraft and equipment	Spectral band	Resolution at nadir, km	Swath width	Repeat coverage, hr	Remarks
Improved TIROS operational satellite (ITOS/TIROS-N)					
Advanced very-high-resolution radiometer (AVHRR)	0.55–0.90 μm 0.72–1.10 3.55–3.93 10.50–11.50	1.1 and 4.0	2700 km	12	Spectral band reduced to 0.55–0.68 on later satellites
High-resolution infrared sounder (HIRS-2)	15.0 μm (CO_2) 11.0 (window) 9.7 (O_3) 6.7 (H_2O) 4.3 (CO_2) 3.7 (window) 0.7 (visible)	20	2240 km	12	These three systems comprise TIROS operational vertical sounder (TOVS), designed to obtain temperature profiles from Earth's surface to 10 mb (1.0 kilopascal); water vapor content at three levels; and total ozone content
Stratospheric sounder unit (SSU)	14.97 μm	147	1485 km	12	
Microwave sounder unit (MSU)	50.30 GHz 53.74 54.96 57.05	110	2150 km	12	
Geostationary operational environmental satellite (GOES)					
Visible-infrared spin-scan radiometer (VISSR)	0.54–0.70 μm 10.50–12.60	1–8 8	120° lat./long. disk	½	Data relay system

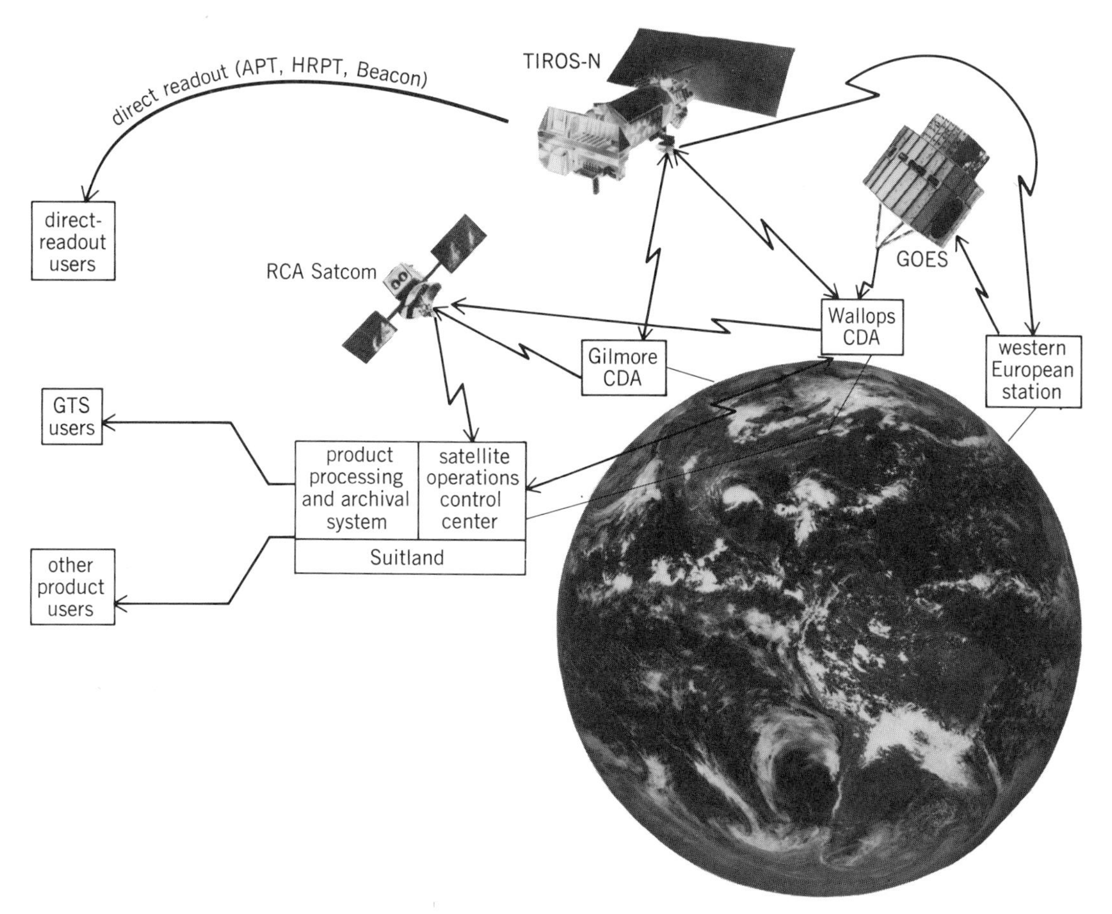

Fig. 2. Elements of TIROS-N operational system. *(From W. J. Hussey and E. L. Heacock, The Economic Benefits of Environmental Satellites, U.S. Department of Commerce, 1977)*

Solar activity monitors. In addition to their visual and infrared scanners, TIROS-N and GOES are equipped with solar activity monitors and data collection systems. The space environment monitor (SEM) on TIROS-N consists of three separate and independent components: (1) the total-energy detector (TED) to determine the intensity of particles in the energy bands from 0.3 to 20 keV; (2) the medium-energy proton and electron detector (MEPED), which senses protons, electrons, and ions in the range from 30 to 60 keV; and (3) the high-energy proton and alpha detector (HEPAD) to detect flux densities from about 370 keV to about 850 MeV. These data supplement those from the solar environment monitor (also SEM) on the GOES, which likewise includes three separate sensor systems: (1) the energetic particle monitor to measure flux densities in seven energy ranges for protons, six ranges for alphas, and one for electrons; (2) the magnetometer to measure the magnitude and direction of the ambient magnetic field; and (3) the solar x-ray sensor, which responds to x-ray fluxes ranging from 10^{-6} to 10^{-1} erg$\cdot$ cm$^{-2}\cdot$s^{-1} (10^{-9} to 10^{-4} W$\cdot$m$^{-2}\cdot$5^{-1}) in the 0.5–3.0-angstrom (0.05–0.3-nm) band and from 10^{-5} to 1 erg$\cdot$cm$^{-2}\cdot$s^{-1} (10^{-8} to 10^{-3} W$\cdot$m$^{-2}\cdot$s^{-1}) in the 1.0–8.0-angstrom (0.1–0.8-nm) band. All these data are relayed to NOAA's Space Environment Laboratory in Boulder, CO, for monitoring the changes in solar activity that affect terrestrial communications, electrical power distribution, and high-altitude supersonic flight.

Data collection system. The data collection system (DCS) on the GOES is designed to collect environmental information from more than 10,000 data collection platforms (DCPs) every 6 hr and relay it to NOAA's command and data acquisition (CDA) stations. Among the possible types of DCPs are ships remote weather stations, hydrological or agricultural sensors, fire weather stations, ocean buoys, and seismic sea sensors. There are two types of DCP radio receiver-transmitters: those that are interrogated through the spacecraft and commanded to send their stored data at a given time; and those that are self-timed and equipped to transmit data at predetermined times. The data collection and platform location system (DCPLS) on TIROS-N provides a means to locate or collect data from fixed and free-floating buoy or balloon platforms. The DCPLS can interrogate or locate up to 459 platforms within view of the satellite at any given time, or up to 4000 globally, and can handle 4 to 32 eight-bit sensors for environmental data. The platform position and velocity determination accuracies are 3–5 km and 0.5-1.5 m$\cdot$s^{-1} rms (root mean square), respectively.

TIROS-N Operational System. Figure 2 depicts the major elements of the TIROS-N Operational System. CDA stations are located near Gilmore Creek, AK, and Wallops Island, VA. Programming and commanding the spacecraft originates at the Satellite Operational Control Center (SOCC) in Suitland, MD, with commands and spacecraft telemetry data being relayed through a commercial

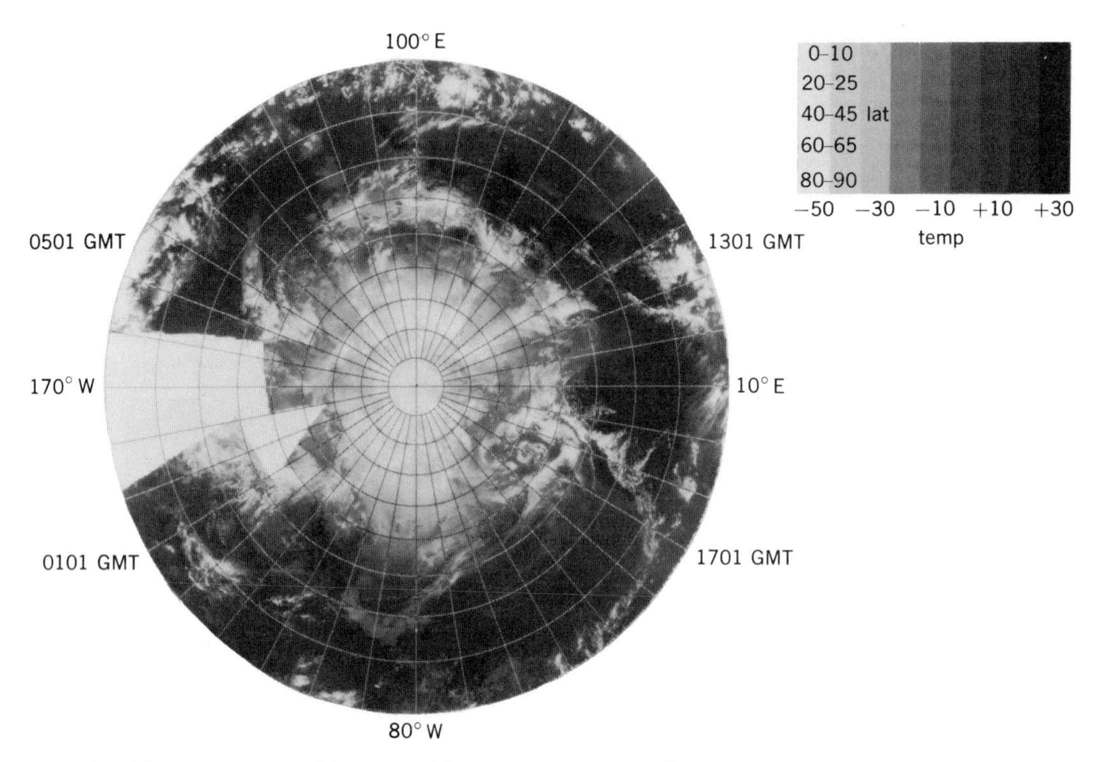

Fig. 3. AVHRR-IR polar mosaic (GAC) from TIROS-N on Feb. 9, 1979. *(NOAA)*

communications satellite network. Data cannot be acquired from the spacecraft by either CDA station during 3, or occasionally 4, out of the 14 sequential passes each day. These data must be stored on board the spacecraft for delayed transmission. In order to eliminate this delay in the receipt of high-priority sounding data, a western European station has been established in cooperation with France, and this station relays the data to the United States by using the DCS of the eastern GOES.

The stored global data from the AVHRR and other instruments on TIROS-N are recorded at the CDA stations and then played back at a high data rate via commercial satellite links to Suitland. There the raw satellite data are ingested, preprocessed, and stored by computer along with auxiliary information such as Earth location and quality control parameters. Subsequently these data are

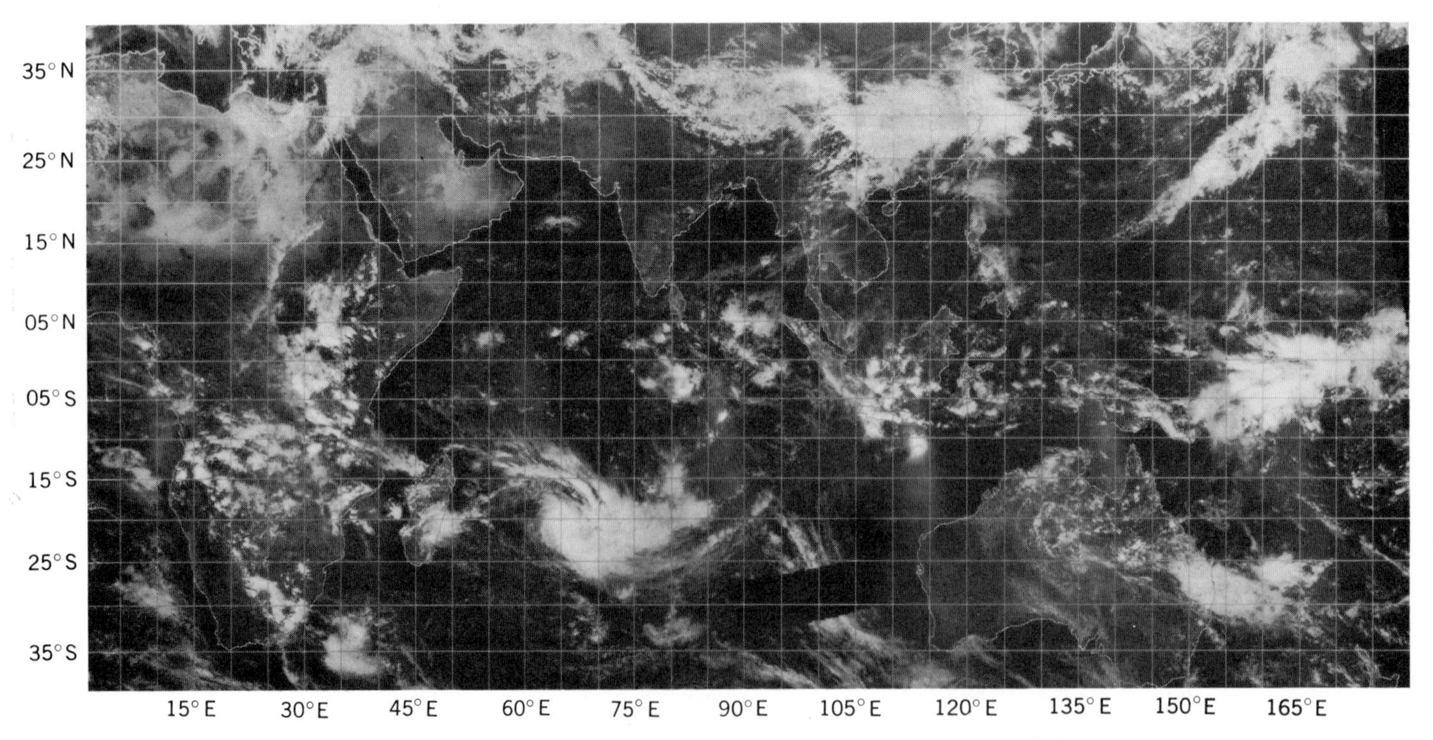

Fig. 4. AVHRR-VIS tropical mosaic (GAC) from TIROS-N on Feb. 9, 1979. *(NOAA)*

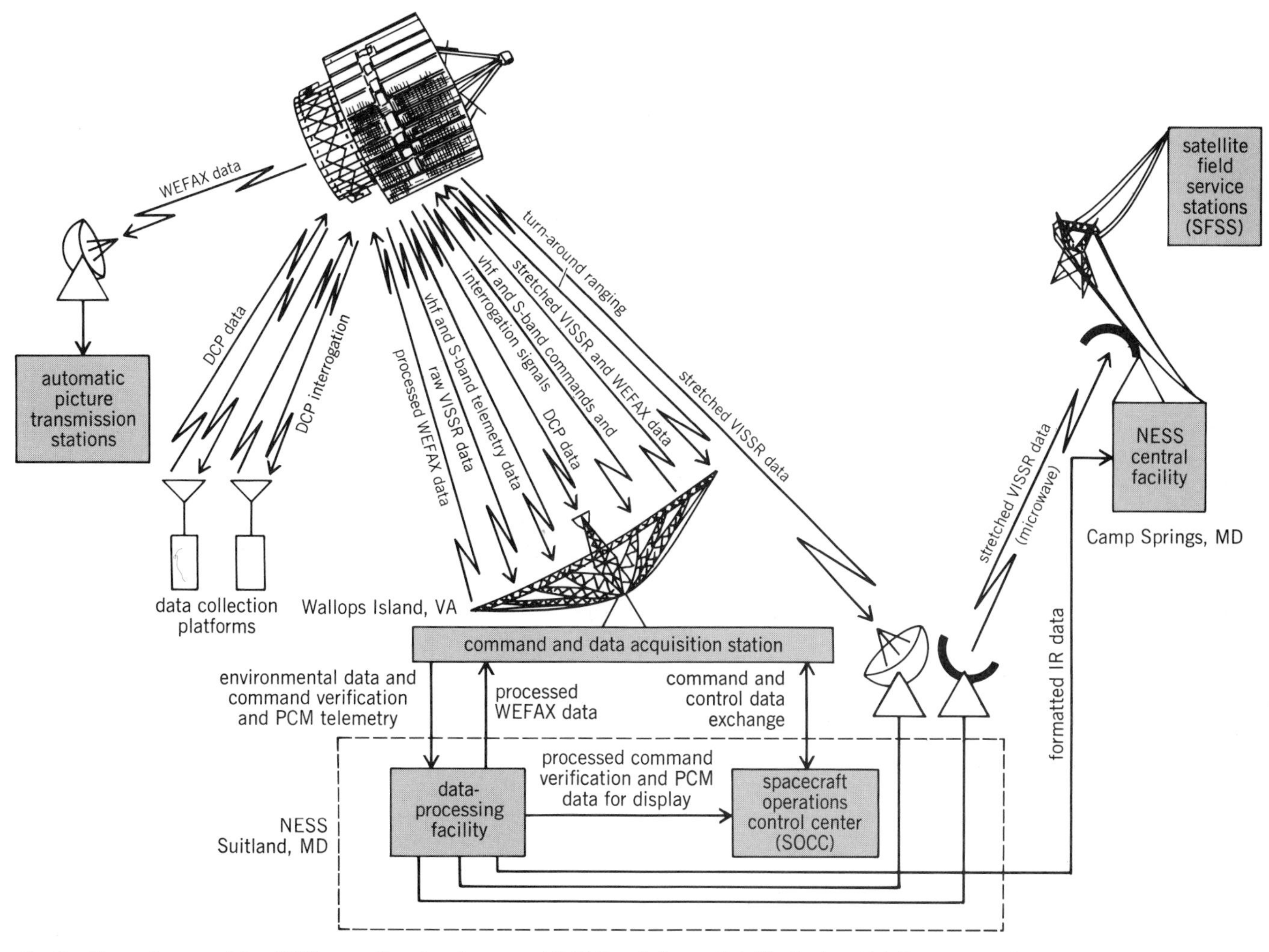

Fig. 5. Elements comprising GOES operational system. *(From G. J. Ensor, Users Guide to the Operation of the NOAA Geostationary Satellite System, U.S. Department of Commerce, 1978)*

archived on digital tape and also passed on to large computers for the generation of quantitative products, many of which are distributed to the user community via the Global Telecommunication System (GTS). Large high-speed computers are required for mapping each day's satellite observations and displaying them with readily usable map projections at reduced spatial resolution. Examples of such daily mosaics of satellite-observed cloudiness for the polar-temperate zone and for the subtropical-tropical zone are given in Figs. 3 and 4, respectively.

In addition to the stored global area coverage (GAC) data collected from TIROS-N at the CDA stations, limited-area coverage (LAC) data from the AVHRR and data from the TOVS instruments are available from two separate real-time data transmission links, one being digital and the other analog in mode. The direct broadcast of 1.1-km-resolution AVHRR digital image data is designated high-resolution picture transmission (HRPT) service, whereas analog transmission of reduced-resolution (namely, 4-km) image data, the reception of which requires less complex and less costly equipment, is a continuation of the automatic picture transmission (APT) service long familiar to many hundreds of satellite data users the world over. There are perhaps only a dozen or so stations in the world presently equipped to receive digital AVHRR and TOVS data, however.

GEOS operational system. The elements of the GOES system are shown schematically in Fig. 5. The primary ground station is the Wallops CDA. The SOCC in Suitland, near Washington, DC, provides overall command and control of the spacecraft through the CDA and also monitors the spacecraft housekeeping telemetry. Unprocessed VISSR image data are transmitted from both the eastern and western GOES to the Wallops CDA at a high data rate. Here a so-called line stretcher reduces the data transmission rate required by about 8 to 1 and transmits the 1-km-resolution stretched visual data back to the spacecraft for relay to the National Environmental Satellite Service (NESS) data processing facility, also in Suitland. The 8-km VISSR-IR data are transmitted by telephone lines from the Wallops CDA directly to NESS's satellite field service stations (SFSSs) in various parts of the United States, as well as digitally by land line to the NESS facility at Suitland. The Suitland facility further relays the data to NESS's central data distribution center in nearby Camp Springs, MD, via a dedicated microwave link.

Fig. 6. Full-disk VISSR-VIS image from the eastern GOES at 1700 GMT on Apr. 27, 1978. *(NOAA)*

At Camp Springs the full-disk VISSR data are processed digitally into a variety of subareas, called sectors. These sectors are selectable with respect to a variety of geographical areas, spatial resolutions, and IR enhancements. Enhancement refers to a special photographic display of the image in which selected parts of the total temperature range are emphasized at the expense of others for particular meteorological, hydrological, or oceanographic purposes. The sectorized image data are routed to the SFSSs via telephone lines for their own use and for further dissemination to forecast offices of the National Weather Service.

Figure 6 is an example of a full-disk VISSR-VIS (VIS indicates visual) image from the GOES stationed at 75°W longitude. The half-hourly GOES images, that is, full-disk images and sectors, are also assembled into time-lapse movie loop sequences for time periods covering several hours to days. This animation of the movement of clouds and weather patterns is used in forecasting, research, and training, and to derive wind information (Fig. 7).

A service known as Weather Facsimile (WEFAX) is available to the APT receiving community having S-band receivers. Mapped AVHRR/GAC images and unmapped VISSR images are broadcast via the GOES. Another extension of the GOES image distribution service is the GOES-TAP system, which allows Federal, state, and local agencies, television stations, universities, and industry to receive a limited inventory of images directly from the nearest SFSS.

METEOROLOGICAL APPLICATIONS

A great number and variety of image-type and other products are derived from observations made by NOAA's operational satellites, both polar-orbiting and geostationary. These products, which are supplied to governmental, military, and industrial users, and which aid in preparation and display of weather analyses and forecasts for the public, are summarized in Tables 2 and 3.

Disaster warnings. Operational environmental satellites have been very effective in improving disaster warnings in connection with tornadoes, hurricanes, tropical storms, flooding, severe local thunderstorms and massive winter storms, and other, less destructive weather phenomena. Until several years ago, costly aircraft reconnaissance was the primary method of finding and tracking dangerous tropical storms in ocean areas. Day and night observations from operational satellites have virtually replaced aircraft reconnaissance except when hurricanes or typhoons approach landfall. Furthermore, satellite tracking has enabled more efficient and cost-effective aircraft mission planning. As severe tropical storms near a coastal area, weather analyses based on satellite data play a major role in keeping the hurricane warning areas to a minimum, which significantly reduces personal and governmental costs of protective measures. Satellite tropical disturbance summaries are issued twice daily, and satellite weather bulletins are sent directly to affected foreign meteorological agencies whenever tropical storms detected by United States satellites threaten the populace anywhere in the world.

High-resolution visual and enhanced IR imagery from GOES at short intervals (as often as 15 min or less during critical periods) has enabled forecasters to better pinpoint areas where tornadoes are highly probable and to reduce the size of the severe weather watch areas that are disseminated to the general public. Figure 8 is a portion of a VISSR-IR image showing the clouds associated with an outbreak of severe weather in the southeastern United States in March 1975.

Frost forecasting. When frost is expected in Florida citrus groves, specially enhanced GOES IR images and interpretations are provided to the National Weather Service office responsible for statewide fruit frost forecasting. Heating citrus groves is an expensive operation, and it has been demonstrated that 1–2 hr of heating protection per "cold" night are saved as a result of the half-hourly satellite tracking of the "freeze line," and this program became operational several years ago.

Hawaii. The method of harvesting of sugarcane in Hawaii is highly dependent upon local weather conditions, and these are often difficult to forecast with sufficient accuracy in areas surrounded by thousands of kilometers of open ocean. Satellite imagery provided by the Honolulu SFSS has proved to be economically very valuable to the sugar companies in the islands.

Search and rescue. Tested in 1975 and used extensively since 1976, operational satellite support of official search and rescue missions of the Californian Civil Air Patrol has resulted in an average decrease of 62% in the number of hours flown per mission. Similar savings have been realized by the Aerospace Rescue and Recovery Service of the U.S. Air Force.

Snow cover information. Among the kinds of operations requiring snow cover information are flood warnings; municipal and regional water supply management, including irrigation systems management; and hydroelectric power scheduling.

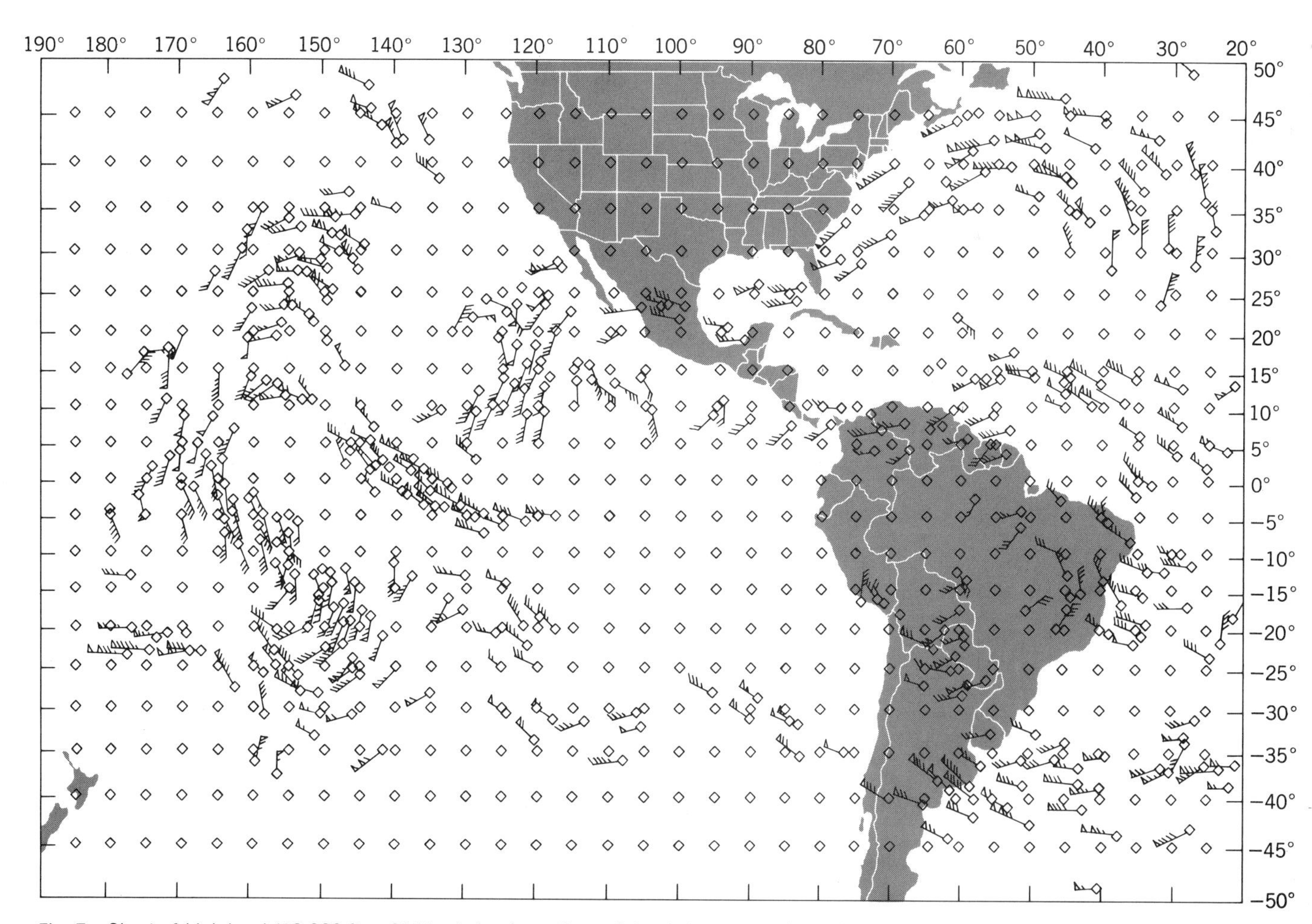

Fig. 7. Chart of high-level (30,000-ft or 9144-m) cloud motion vectors derived from animated sequences of half-hourly GOES images by manually fixing the positions of cloud elements at the beginning and end of each sequence. (*NOAA*)

Table 2. Mapped-gridded image products from the AVHRR on TIROS-N

Product description	Accuracy goals	Coverage and spatial resolution	Processing	
			Format	Schedule
Hemispheric-mapped polar mosaics, IR and VIS mosaics	Nominal ± 10 km for polar and mercator mapping location	Northern and Southern hemispheres 1024 × 1024 km; Equator 12.8 km; Poles 15.8 km	Mapped imagery; CCT (computer compatible tape)	Daily
Mercator-mapped mosaics IR/VIS	Same as above	360° long. 40°N 40°S; Equator 8.1 km, increasing poleward	Mapped imagery	Daily
Polar-mapped composites IR/VIS (minimum brightness/maximum temperature)	Same as above	North and South Pole regions 1024 × 1024 km; Equator 12.8 km; Poles 25.8 km	Mapped imagery	Conforms to compositing period (7-day)
Pass-by-pass gridded imagery VIS/IR (one satellite)	Nominal ± 10 km grid placement	Global 8 km	Gridded imagery	Orbit-by-orbit
Imagery from limited-area coverage (LAC) data: both recorded and direct readout (ungridded)	Not available	Recorded data: selectable; two 11.5-min segments per orbit. Direct readout: continental U.S. resolution = 1.1 km.	Imagery	Recorded variable; two 11.5-min segments per orbit. Direct readout: all continental U.S.

In the mountainous regions of the United States, millions of dollars are spent each year to measure snowpack conditions at fixed locations, and snow cover by aerial mapping, for water supply and discharge forecasts. It has been shown that areal mapping of snow cover can be done 200 times more cheaply by satellite than by aerial surveys, and the fractional snow cover of a large number of western mountain basins is now routinely surveyed from space. A VHRR visual image of mountain snow is given in Fig. 9.

OCEANOGRAPHIC APPLICATIONS

A variety of oceanographic-type applications, both in operations and research, have been developed since the late 1960s. Table 4 summarizes the oceanographic and hydrologic products being generated with TIROS-N data.

Sea-surface temperatures. Only satellites have the capability to measure Earth-surface temperatures over regional, hemispheric, or global areas on a repetitive and timely basis. The most serious constraint to mapping surface temperatures from space is cloud cover, and the newer microwave sensors may remove that limitation. Studies have shown that differences between satellite-derived sea-surface temperatures and those measured by commercial ships range from about 1 to 2°C rms. A number of operational products are presently derived from satellite observations of surface-water temperature or temperature contrast. The Gulf Stream Analysis is a depiction of the water masses and their associated thermal fronts obtained by photo-interpretation of AVHRR-IR images. The *Gulf Stream North Wall Bulletin* provides mariners with outline depictions of the Gulf Stream

Table 3. Sounding products from the TOVS on TIROS-N*

Product description	Accuracy goals	Coverage and spatial resolution
Layer mean temperatures (K) for the layers below:	(Accuracy indicators to be appended)	Global coverage
SFC – 850 mb, 300 – 200 mb, 30 – 20 mb	SFC – 850 mb ± 2.5 K	Nominal 250 km near the subsatellite track. Spacing and resolution increase with scan angle. 250-km spacing may be increased to 500 km if data volume or computing time is excessive. Resolution will be fixed at 250 km nominal.
850 – 700 mb, 200 – 100 mb, 10 – 05 mb	850 – tropopause ± 2.25 K	
700 – 500 mb, 100 – 70 mb, 05 – 01 mb		
500 – 400 mb, 70 – 50 mb, 02 – 01 mb	Tropopause –2.0 mb ± 3 K	
400 – 300 mb, 50 – 30 mb, 01 – 0.4 mb	2.0 – 0.4 mb ± 3.5 K	
Layer precipitable water (mm) for these layers: Surface – 700 mb, 700 – 500 mb, above 500 mb	±30%	
Tropopause pressure and temperature	P: ±50 mb T: ±2.5 K	
Total ozone (Dobson units)	±15% tropical ±50% polar	
Equivalent blackbody temperatures (K) for 20 HIRS-2 stratospheric channels, 4 MSU channels, 3 SSU channels	±2 K	
Cloud cover	±20%	

*1 mb = 100 pascals.

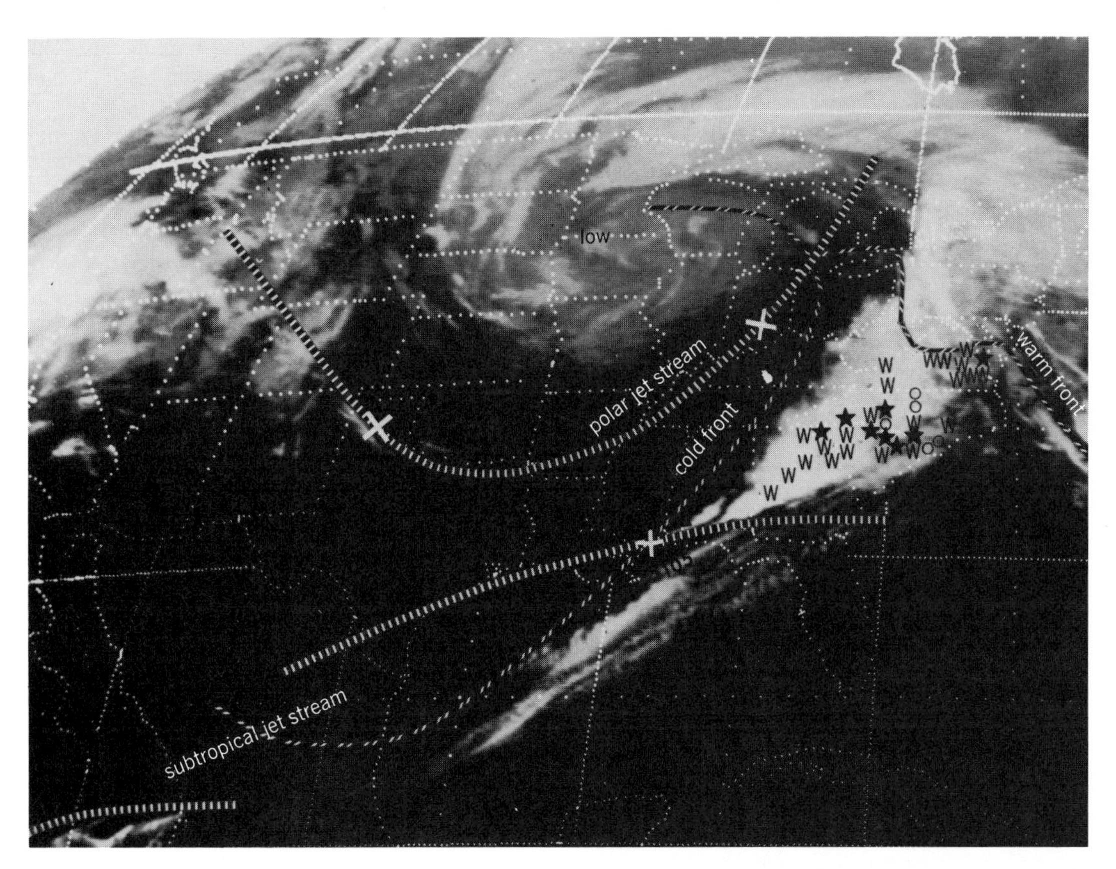

Fig. 8. Sectorized VISSR-IR image at 1200 CST on Mar. 24, 1975. Severe weather events near the time of this image were tornadoes (indicated by stars), funnels (circles), and high winds (W). *(From L. F. Whitney, Jr., Relationship of the subtropical jet stream to severe local storms, Mon. Weath. Rev., 105(4):398–412, 1977)*

through analysis of sea-surface temperatures from satellites, ships, and bathythermographs. An experiment conducted jointly by a major oil company and NESS demonstrated that significant fuel savings can be realized by using timely satellite analyses of the Gulf Stream position to aid oil tankers along the eastern seaboard, and this has now become a standard navigating procedure for this company.

Upwelling processes. The West Coast Frontal Chart is generated from VHRR-IR image interpretation and depicts the thermal patterns associated chiefly with upwelling processes. It has been documented by marine biologists and commercial fishers that certain species of fish are temperature-sensitive and tend to congregate in the nutrient-rich waters associated with ocean thermal fronts and upwelling. Enhanced satellite IR imagery has been used since 1975 to produce ocean frontal charts to aid fishing crews in locating productive areas, and the fisheries industry has reported significant savings in fuel from decreased search times.

Great Lakes and global. The Great Lakes Surface Temperature Chart is produced from unmapped AVHRR-IR data. These analyses are used as input for the freeze-up forecasts of the Great Lakes and by scientists and others interested in upwelling phenomena. Global Sea Surface Temperature Computation (GOSSTCOMP) is a method of routinely producing daily ocean-surface temperature charts for the entire world at a resolution of several hundred kilometers. Typically 5000–7000 observations are generated daily.

Ocean current systems. Oceanographic research with the aid of satellite IR data has led to a variety of important findings. Long-term VHRR and AVHRR sequences of infrared images have enabled the thermal monitoring of such major current systems as the Gulf Stream, which exhibits complex meander and eddy structure (Fig. 10). Thermal-IR images from geostationary satellite data collected hourly have been used to construct time-lapse movie loops of the Gulf Stream and Gulf of Mexico Loop Current for periods of a week or more. Investigations using these have led to the measurement of the wavelength, frequency, and speed of northward-propagating cyclonic eddies along the inshore Gulf Stream boundary. Satellite imagery enabled the detection and study of a series of large (300-km-diameter), previously unknown, anticyclonic gyres in the ocean that are associated with extensive and well-defined upwelling episodes induced by norther-type winds blowing out of the Gulf of Mexico and across the Gulfs of Tehuantepec and Papagayo. Another discovery credited to satellite observations are the long-wave-type perturbations on the sea-surface thermal front between the South Equatorial Current and the North Equatorial Countercurrent. These have a wavelength of the order of 1000 km and propagate westward at speeds of about 50 km per day.

Ice mapping. The usefulness of visual pictures taken by meteorological satellites for relatively crude ice mapping was recognized quickly after launch of the first TIROS. Visual imagery from the vidicons and scanners on Nimbus and NOAA

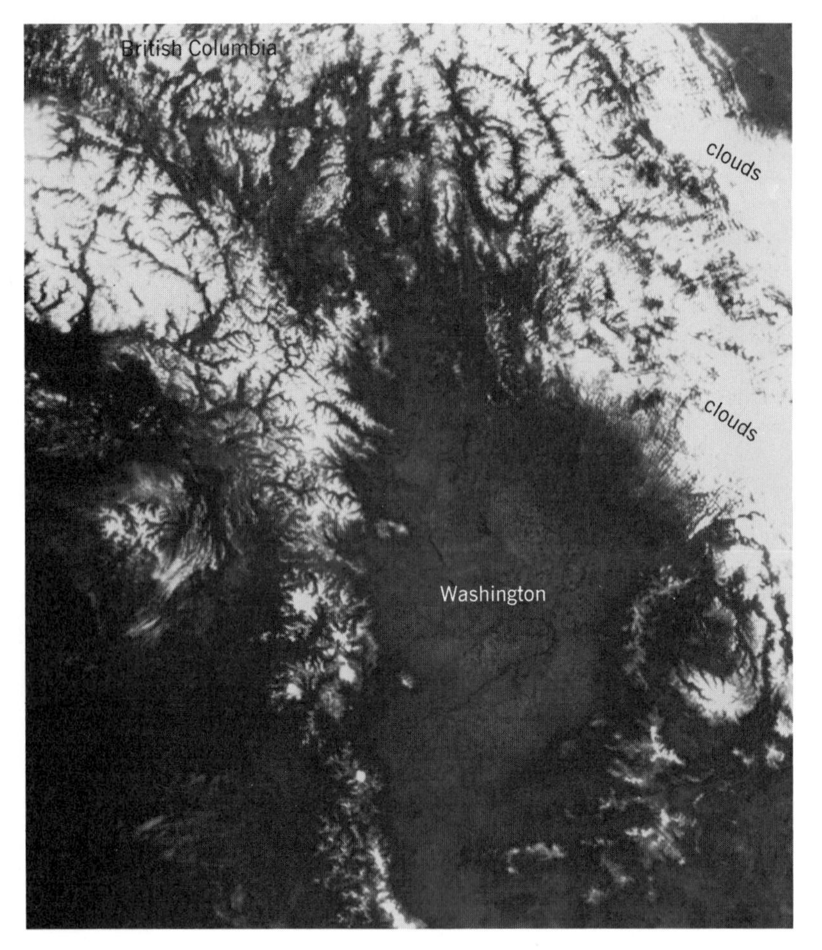

Fig. 9. VHRR-VIS image of mountain snow in the northwestern United States and southwestern Canada on Apr. 22, 1974. *(NOAA)*

spacecraft have long been used in operational and research applications for sea ice surveillance. The improved 1-km-resolution visual and thermal-IR data introduced with the VHRR on ITOS in 1972 represented a fourfold improvement in resolution over previous visual observations and a tenfold one over previous IR measurements, and this enabled a substantial gain in pack ice mapping capability from space (Fig. 11). This capability continues with the AVHRR on TIROS-N and is augmented by the new 0.72–1.10-μm data, which permit detection of thawing conditions.

Special satellite data coverage is regularly obtained for areas of Antarctica to assist U.S. Navy ice forecasters in their support of resupply missions. Some of these data have also found use in research, including a study of ice movement and the calving of tabular icebergs. Satellite images have been used to document the movement of several huge tabular icebergs in the Weddell Sea area over enormous distances and over periods as long as 11 years—a data-gathering feat generally impossible or impracticable by other means. Digitized satellite IR temperatures have been used to investigate the thickness information content of ice temperature variations. The scanning microwave radiometers on *Nimbus 5* and *6* represented a signigicant advance in the capability to monitor pack ice in virtually all weather, although spatial resolutions were only 25–30 km, for unlike radiation at visual and infrared wavelengths, that in the microwave band at 0.81 and 1.55 cm readily penetrates most clouds.

Several operational sea ice products produced from photo-interpretation of AVHRR images are distributed by NESS via facsimile or mail. One is a composite weekly ice type and ice concentration chart for the Bering, Chukchi, and Beaufort seas bordering Alaska; another is a twice-weekly composite satellite analysis of ice conditions in the Great Lakes. Arctic and Great Lakes ice monitoring by satellite has proved so valuable to Canada's ice reconnaissance service that it has installed its own receiving stations for HRPT service. It uses the images as an aid to preflight planning and in locating areas of improved visibility, resulting in cost reductions of the order of $5,000,000 annually from decreases in costly flight hours. Satellite support has been useful to and cost-effective for Canadian geophysical survey ships operating off the north coast of Canada, as well as to the U.S. Military Sea Lift Command in the course of routing its ships through Arctic waters. Satellite imagery of ice conditions has aided in extending the commercial shipping season in the Great Lakes, each additional day having an estimated cost benefit of $1,000,000.

Table 4. Oceanographic products from TIROS-N

Product description	Accuracy goals	Spatial resolution and geographical coverage	Format and schedule
Sea-surface temperature observations	±1.5°C abs., ±1.5°C rel.	50 km (nominal); global	CCT weekly
Sea-surface temperature regional-scale analysis	Same as above	50 km grid; 3 regions of 10,000 grid points	Image weekly; CCT monthly
Sea-surface temperature global-scale analysis	Same as above	100 km; global lat./long.; grid	Contour chart weekly; CCT semimonthly; image daily
Sea-surface temperature climatic-scale analysis	Same as above	500 km; global lat./long.	Contour chart weekly; CCT monthly; image daily
Sea-surface temperature monthly observation mean	Same as above	250 km; global lat./long.; grid	CCT monthly; contour chart monthly

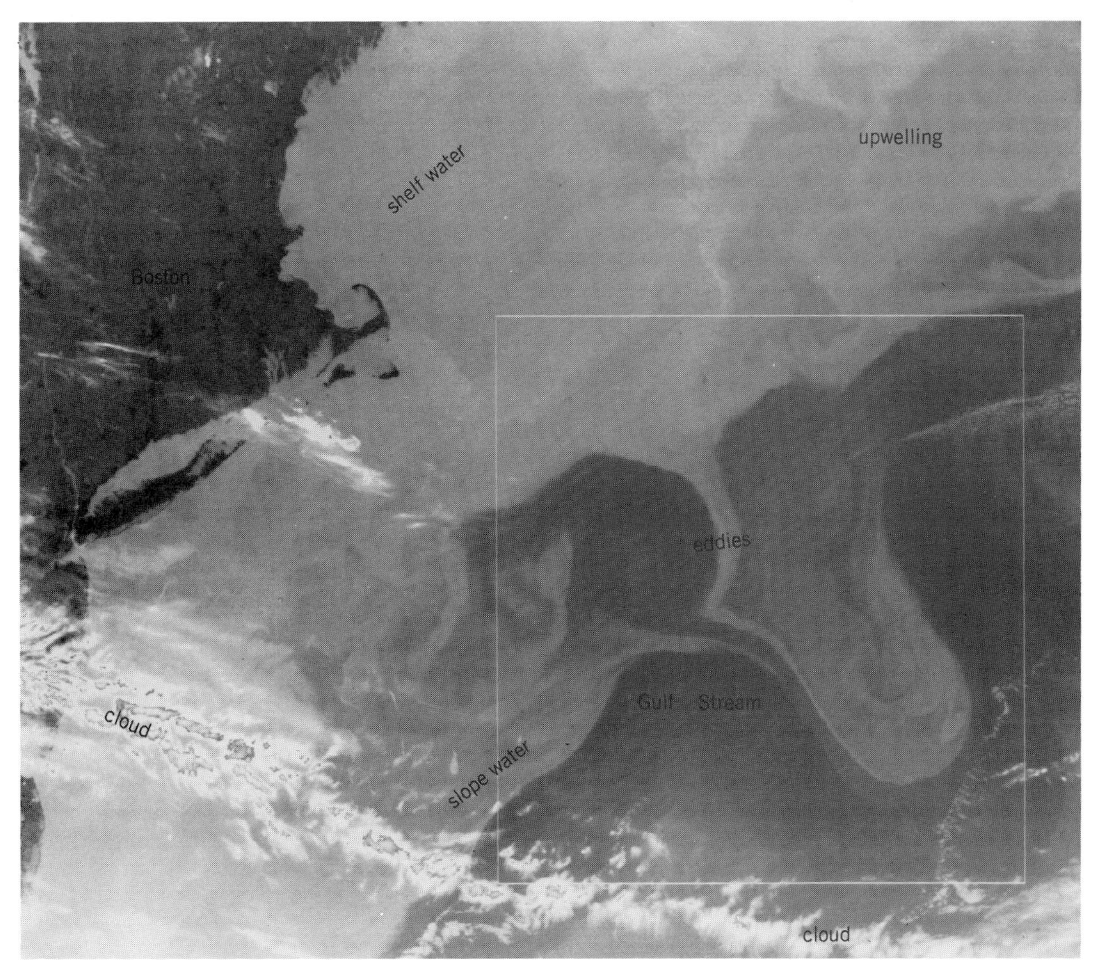

Fig. 10. VHRR-IR image of Atlantic Ocean just east of Long Island and Cape Cod on May 23, 1978. A complex interaction between the warm Gulf Stream (dark tone) and cooler slope and shelf waters (progressively lighter gray tones), as well as upwelling, is evident. The urban heat island effect is easily seen in the case of Boston and New York City areas (dark splotches). *(NOAA)*

RESEARCH ENVIRONMENTAL SATELLITES

A variety of research-type spacecraft for meteorological and other environmental studies have been launched by the National Aeronautics and Space Administration (NASA), and these are described in Table 5.

Landsat 3. One of these, the third in a series of Earth resources technology satellites, is *Landsat 3*. Put in a 920-km-high, near-polar, Sun-synchronous orbit in March 1978, this spacecraft is designed for extremely-high-resolution (40 and 80 m), synoptic, and repetitive coverage of the Earth's surface at about 10 A.M. local time. Although the repeat cycle (18 days) of a single Landsat is too long for most meteorological and oceanographic uses, the data have found valuable use in agriculture, forestry, geology, geography, cartography, energy and mineral resources surveys, and ecology. *Landsat 3* is also equipped with a data collection system for relay of environmental data from ground-based or airborne instrument platforms, fixed or moving, to central processing stations on the ground.

Nimbus 7. The seventh in NASA's series of Nimbus satellites is equipped with a large variety of imagers, sounders, and other equipment for meteorological and oceanological experiments (Fig. 12). *Nimbus 7* was put in a 955-km-high, Sun-synchronous, near-polar orbit in October 1978, which provides for essentially global coverage for many of the sensors every 72 hr.

Coastal zone color scanner (CZCS). The CZCS on *Nimbus 7* is the first spacecraft instrument dedicated to quantitative measurement of ocean color variations (Table 5). Although radiometers on other satellites, such as the multispectral scanner (MSS) on Landsat, have sensed ocean color, their spectral bands, spectral resolution, and dynamic range are far from optimum for this purpose. The CZCS channel that measures emitted thermal energy is registered with the channels measuring reflected solar energy and has the same spatial resolution. The first four spectral bands have four separate gain levels that are changed on command to allow for the range of Sun angles encountered seasonally throughout an orbital pass. Furthermore, the CZCS scan mirror can be tilted up to 20° from nadir in order to avoid or minimize sun glitter contamination of the measurements.

The scientific objective of the CZCS experiment is to attempt to discriminate between organic and inorganic materials suspended or dissolved in the water, to determine within useful limits the quantity of these materials, and in some cases to identify

Fig. 11. VHRR-VIS image of Labrador ice stream on Mar. 3, 1973. Snow covers both the forested (dark) and unforested (light) areas of Canada. *(From E. P. McClain, Some new satellite measurements and their application to sea ice analysis in the Arctic and Antarctic, Advanced Concepts & Techniques in the Study of Snow & Ice Resources, National Academy of Sciences, pp. 457–466, 1974)*

these materials. If it is established that ocean color measurements can be used to derive such products as chlorophyll and sediment concentrations, and that measurements on this scale are useful to oceanographers and fisheries biologists, they can serve to guide future development in this area, including possibly the carrying of an advanced ocean color scanner on a future operational satellite. The archived CZCS magnetic tapes contain both calibrated radiances from the solar channels and equivalent blackbody temperatures from the thermal-IR channel, plus certain derived products related to the content of the water. An example of images generated from several of the CZCS channels is given in Fig. 13.

Scanning multichannel microwave radiometer (SMMR). This experiment on *Nimbus 7* is identical to the SMMR carried by *Seasat 1* (Table 5), and it is an outgrowth of the electrically scanning microwave radiometers operated on *Nimbus 5* and *6*. The SMMR is a 10-channel instrument delivering orthogonally polarized (horizontal and vertical) antenna temperatures at five microwave wavelengths. By combining the variously polarized brightness temperatures from different combinations of the five wavelengths, the following types of meteorological, hydrological, and oceanographic information can be extracted: near-surface wind speeds; sea ice type and concentration; sea-surface temperature; a mesoscale soil-moisture index; snow accumulation rates over continental ice sheets; and (over the open ocean only) total water vapor and total nonprecipitating liquid water in the atmospheric column, and medium to heavy rain rates. Although the minimum spatial resolution of the measurements is rather large and varies considerably from one wavelength to another, microwave observations have an advantage over those in the visual and IR spectral regions in that they are not constrained by cloudiness or even by rainfall unless it is relatively heavy.

Temperature-humidity-infrared radiometer (THIR). This subsystem on *Nimbus 7* is identical to those carried by *Nimbus 4*, *5*, and *6* with one significant exception: the measurements are now digitized on board the spacecraft. The THIR is a two-channel scanning radiometer designed to image cloud cover and provide temperatures of cloud tops, land, and oceans (11.5-μm channel), and to provide information on the water vapor and cirrus cloud content of the upper troposphere and lower stratosphere (6.7-μm channel). In-flight calibration of the thermal-IR channel is accomplished by having the detector view a blackbody target (part of the radiometer housing) of known temperature as well as cold outer space.

Earth radiation budget (ERB). This experiment on *Nimbus 7*, which is a continuation of an experi-

Fig. 12. Schematic diagram of *Nimbus 7* spacecraft. *(From C. R. Madrid, ed., The Nimbus 7 Users' Guide, Goddard Space Flight Center, 1978)*

Table 5. Research environmental satellites and their instrumentation

Spacecraft and equipment	Spectral band	Resolution at nadir	Swath width, km	Repeat coverage	Remarks
Landsat 3					
Multispectral scanner system (MSS)	0.5–0.6 μm	80 m	185	18 days	Coverage limited to 82° lat. north and south, data collection system
	0.6–0.7				
	0.7–0.8				
	0.8–1.1				
	10.5–12.5	250 m			
Return beam vidicon (RBV)	0.55–0.72 μm	40 m	185	18 days	
Nimbus 7					
Coastal zone color scanner (CZCS)	0.433–0.453 μm	825 m	1565	4 days	Limited-area coverage
	0.510–0.530				
	0.540–0.560				
	0.660–0.680				
	0.700–0.800				
	10.5–12.5				
Scanning multichannel microwave radiometer (SMMR)	4.55 cm	118 × 103 km	780	36 hr	Global coverage
	2.81	73 × 68			
	1.67	47 × 41			
	1.36	37 × 32			
	0.81	22 × 19			
Temperature-humidity infrared radiometer (THIR)	6.5–7.0 μm	20.0 km	2800	12 hr	Global coverage
	10.5–12.5	6.7			
Earth radiation budget experiment (ERB)	10 solar channels		3376	12 hr	Global coverage
	4 Earth-viewing	Earth disk			Fixed bands
	4 Earth-viewing	84 × 42 km			Scanning (reflected) bands
	4 Earth-viewing	84 × 42			Scanning (emitted) bands
Stratospheric and mesospheric sounder (SAMS)	25–100 μm (H_2O)		2800	24 hr	Global coverage; not all six parameters at all 16 pressure levels every day
	14.4–15.7 (CO_2)				
	7.6– 7.8 (N_2O, CH_4)				
	4.1– 5.4 (CO, NO, CO_2)				
	2.5– 2.6 (H_2O)				
Solar backscatter ultraviolet and total ozone mapping spectrometer (SBUV/TOMS)	SBUV: 12 narrow bands from 250 to 340 nm; 160–400 nm; 343 nm;	200 km	2800	24 hr	Global coverage
	TOMS: 6 bands from 312.5 to 380 nm	50 km			
Limb infrared monitor of the stratosphere (LIMS)	6.2–6.3 μm (NO_2)	28 km horiz. 3.6 km vert.		24 hr	Coverage limited to 64–84° lat. north and south
	6.5–7.2 (H_2O)	28 3.6			
	9.1–10.6 (O_3)	18 1.8			
	11.1–11.6 (HNO_3)	18 1.8			
	13.5–16.8 (CO_2)	18 1.8			
	14.9–15.5 (CO_2)	18 1.8			
Stratospheric aerosol measurement experiment (SAM II)	0.98–1.02 μm	1.0 vert.		7–30 days	Coverage limited to 64–80° lat. north and south
Seasat 1					
Compressed pulse altimeter (CPA)	13.9 GHz	1.6–12.0 km	10	6 months	Polar coverage of all Seasat sensors limited to about 72° lat. north and south
Seasat-A satellite scatterometer (SASS)	14.595 GHz	50 km	1900	36 hr	
Synthetic aperture radar (SAR)	21.5 cm	25 m	100	36 hr	Direct readout only; limited swath lengths
Scanning multichannel microwave radiometer (SMMR)	Essentially the same as the SMMR on *Nimbus 7*		650	36 hr	
Visible-infrared radiometer (VIRR)	0.5–0.9 μm	9 km	2280	36 hr	
	10.5–12.5				
Heat capacity mapping mission *(HCMM 1)*					
Heat capacity mapping radiometer (HCMR)	0.55–1.10 μm	0.5 km	700	12 hr	
	10.5–12.5				

ment from *Nimbus 6*, has two objectives: (1) to determine the radiation budget of the Earth on synoptic and planetary scales over a year; and (2) to develop angular models of the reflection and emission of radiation from clouds and various Earth surfaces. The first objective is being met by making simultaneous measurements of incoming solar radiation and both outgoing Earth-reflected (short-wave) and Earth-emitted (long-wave) radiation. The ERB experiment measures the solar radiation in 10 spectral channels, whereas Earth-reflected and emitted energy is measured with four fixed wide-angle (121°, or horizon-to-horizon) sensors. The second objective is being met by use of eight narrow-angle (0.25 × 5.12°) channels designed to obtain a large number of angularly independent views of the same geographical area as the spacecraft passes overhead. Characteristic angular dis-

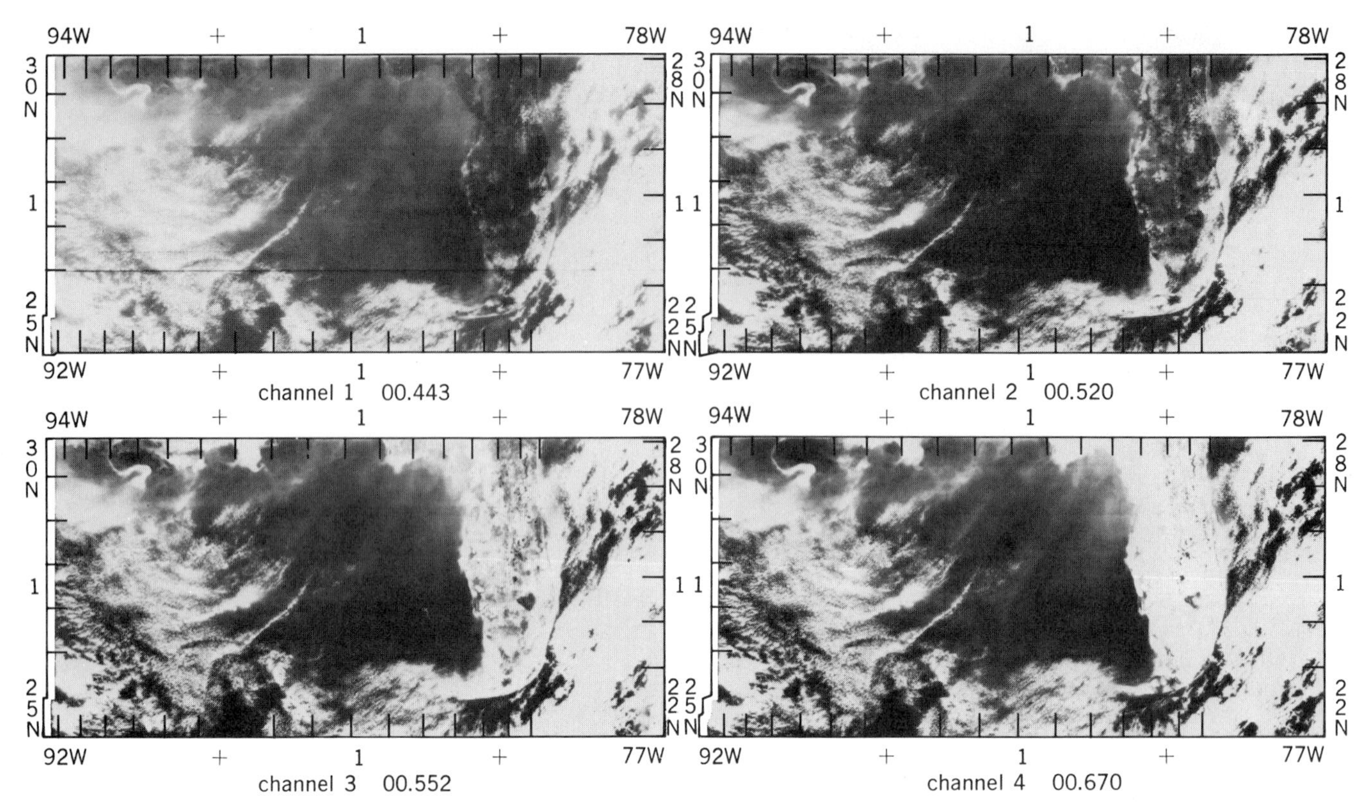

Fig. 13. Simultaneous images from four CZCS channels (centered at 0.433, 0.520, 0.552, and 0.670 μm) in the vicinity of Florida on Nov. 2, 1978. Only ocean area free of clouds is the Gulf of Mexico just west of Florida, where a radiance front is visible in channel 1. *(NASA)*

tribution models will be derived for a variety of reflecting surface conditions (different types of clouds, ice, snow, and land surfaces) from a composite of the scanning channel observations of each area. These models are used together with the scanning channel observations to generate radiation budgets with a resolution of about 500 km.

Stratospheric and mesospheric sounder (SAMS). This experiment on *Nimbus 7* is the fourth instrument in a series of multichannel IR radiometers designed to measure emission from the upper atmosphere. Preceding the SAMS were selective chopper radiometers on *Nimbus 4* and *5* and a pressure modulator radiometer on *Nimbus 6*. The SAMS is a 12-channel instrument observing thermal emission and solar resonance fluorescence from the atmospheric limb. In addition to conventional radiometry, SAMS also makes use of gas correlation spectroscopy. An absorption cell of the gas in question is placed in the optical path of the radiometer. This gas has an absorption spectrum that matches, line for line, the emission spectrum of that gas in the atmosphere. The amount of absorption and the shapes of the absorption lines in the cell are determined by the pressure and temperature in the cell, which can be varied. SAMS global measurements, when interpreted with results from the LIMS and SBUV/TOMS instruments, provide extensive data for chemical and dynamic models of the stratosphere and mesosphere. Specific objectives of the SAMS experiment are to derive the following: (1) temperature from emission in the 15-m CO_2 band from 15 to 80 km altitude; (2) vibrational temperature of CO_2 bands from 50 to 140 km; (3) distribution of CO, NO, CH_4, N_2O, and H_2O from 15 to 60 km; and (4) distributions of CO_2 (4.3 μm) and CO (4.7 μm) from 100 to 140 km and H_2O from 60 to 100 km to study dissociation in the lower thermosphere.

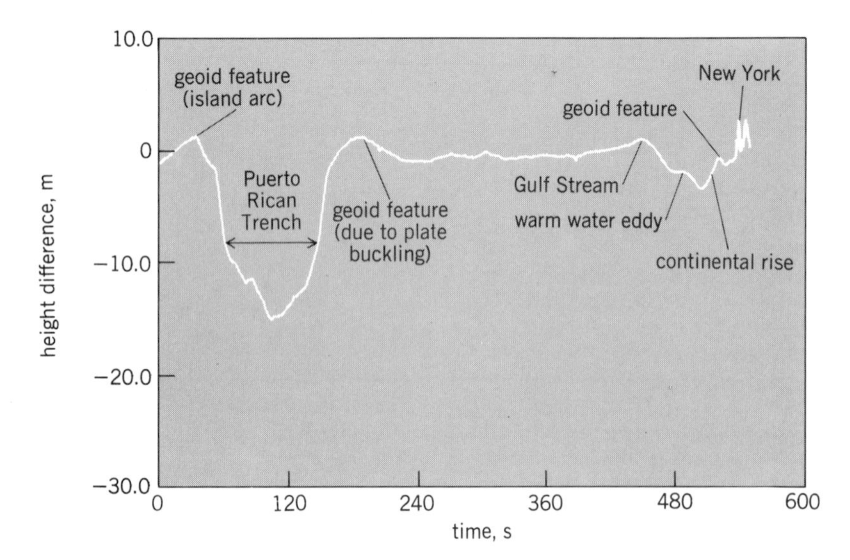

Fig. 14. Difference of *Seasat 1* altimeter measurement of sea surface height and the GEM 10B geoid for a North Atlantic pass on July 10, 1978. Clearly evident is the dip in the sea surface over the Puerto Rican Trench and the slopes associated with ocean current features. *(NASA)*

Solar backscatter ultraviolet (SBUV) and total ozone mapping spectrometer (TOMS). This experiment on *Nimbus 7* is an expanded and improved version of the backscatter ultraviolet experiment on *Nimbus 4*. The SBUV subsystem consists of a

Fig. 15. Near-surface winds obtained from SASS (*Seasat 1*) measurements near Hurricane FICO. Vector lengths are proportional to wind speed, which ranged from 4 to 50 knots (2 to 26 m/s). The SASS wind vectors are superimposed on a VHRR-VIS image from the *NOAA 5* operational satellite. (*NASA*)

double Ebert-Fastie spectrometer and a filter photometer. Both channels simultaneously view identical fields of solar radiation scattered by the terrestrial atmosphere in the nadir of the solar flux scattered from the instrument diffuser plate that is deployed on command. The spectrometer serially monitors 12 narrow wavelength bands from 250 to 340 nm, or scans the waveband continuously from 160 to 400 nm, while the photometer measures the energy in a fixed band centered at 343 nm. The TOMS subsystem, an essentially independent instrument, employs a cross-scanning single monochromator to sequentially sample backscattered radiation at six wavelengths from 312.5 to 380 nm. The objectives of the SBUV experiment are: (1) to determine the total amount of atmospheric ozone in the vertical column; (2) to determine the vertical profile of ozone above the ozone maximum; and (3) to measure the ultraviolet solar spectral irradiance (from 160 to 400 nm with a spectral resolution of 1 nm) and monitor its temporal variability. The objective of the TOMS experiment is to obtain contiguous global mapping of total ozone.

Limb infrared monitor of the stratosphere (LIMS). This experiment on *Nimbus 7* is a follow-on to the limb radiance inversion radiometer experiment on *Nimbus 6* to measure O_3, H_2O, and temperature. LIMS is designed to determine profiles of O_3, NO_2, HNO_3, H_2O, and temperature with high vertical resolution from the lower stratosphere to the lower mesosphere. These quantities are determined by inverting measured limb radiance profiles obtained by a scanning IR radiometer with channels in six spectral regions (see Table 5). A programmed scannning mirror in the radiometer causes the field-of-view of the six detectors to make coincident vertical scans across the Earth's horizon. The measured limb radiance profiles of the CO_2 channels are inverted to determine the vertical profile of temperature. This, together with the radiance profiles in the other channels, is used to infer the vertical distribution of the trace gases. The ability to monitor time variations of the global distributions of temperature and trace gases in the 10–70-km region is important to understanding of the dynamics there because the chemistry largely determines the global stratospheric temperature distribution. This temperature distribution, in turn, is almost totally responsible for the global wind fields in the stratosphere; and the winds, in turn, transport the chemical species; and so on.

Stratospheric aerosol measurement (SAM II). This experiment on *Nimbus 7* comprises a one-spectral-channel Sun photometer that views a small portion of the Sun through the Earth's atmosphere during spacecraft sunrise and sunset. This time-dependent radiance is combined with the spacecraft ephemeris data and local atmospheric density profile, then inverted to obtain a vertical profile of aerosol extinction above the Earth tangent point. SAM II provides aerosol data from polar regions in the 64–80° latitude band, where little such data exist. Aerosols have the potential to significantly modify climate, and their microphysical and chemical interactions enter into important environmental processes. The measurement goal of the SAM II experiment is to determine a vertical profile of aerosol extinction from cloud tops (about 10 km) to heights of 40 km with a vertical resolution of 1 km or better, and with an accuracy of about 10% over the height range where aerosol 1.0-μm extinction exceeds about 50% of molecular 1.0-μm extinction (about 10–20 km).

Seasat 1. Another NASA research satellite, this is the first dedicated to observation of the oceans (see Table 5). It was placed at a nominal orbital altitude of 800 km in June 1978, and the orbit is of high inclination but is not Sun-synchronous. Unfortunately, *Seasat 1* suffered total failure from a massive electrical short circuit after collecting high-quality data for 99 days. Preliminary geophysical evaluation of the measurements from the various sensors indicates that, in large part, the intended proof of concept will be established.

Compressed pulse altimeter (CPA). This is a radar-type instrument that views only at nadir. The CPA was constructed to obtain measurements of the absolute height of the ocean surface precise to 10 cm rms over seas ranging from calm to significant wave heights of about 20 m (Fig. 14). Precise spacecraft tracking, which was accomplished with the aid of a laser retroreflector and tracking beacons, is needed for geodesy, currents, and storm surge information. Processing of altimeter pulse data should yield significant wave height estimates to within 0.5 m or 10%.

Seasat-A satellite scatterometer (SASS). This is also an active microwave instrument. It illuminated the ocean surface with four fan-shaped beams, and the amount of energy returned to the spacecraft enables estimating the magnitude and direction of near-surface winds (Fig. 15). The goal is to determine these winds, in a band 230–730 km to either side of the spacecraft, to within 2 $m \cdot s^{-1}$ or 10% in speed and to within 20° in direction over a range of 3–25 $m \cdot s^{-1}$.

Synthetic aperture radar (SAR). This system views a 100-km-wide area centered 20° off nadir, and the maximum length of the data swath is about 4000 km because data can be received by a ground station only while the satellite is above the horizon. The types of oceanic information expected from the SAR are: waves and wave spectra for ocean waves 50 m or more in length; images of sea ice features, with great detail; possible iceberg detection; and extremely-high-resolution images of wave patterns as modified by currents, shoaling, internal waves, and oil spills.

Scanning multichannel microwave radiometer (SMMR). This was identical to the one carried by *Nimbus 7*. The differing orbital altitudes of the two spacecraft, however, result in a different swath width on the Earth and slightly different footprints (spatial resolutions) for the various channels.

Visible/infrared radiometer (VIRR). This is a slightly modified scanning radiometer from the earlier-generation NOAA/ITOS. Its primary function was to supply cloud and other feature identification for use in interpretation of the microwave observations. The VIRR data were digitized on board the spacecraft, enabling good-quality estimates of cloud-top, land, and ocean-surface temperatures. Unfortunately, the VIRR ceased to operate after 61 days in orbit.

Heat capacity mapping radiometer (HCMR). This is the only sensor carried on NASA's heat capacity mapping mission *(HCMM 1)*. The HCMR is a two-channel, very-high-resolution radiometer intended for obtaining diurnal temperature variations for identification of various types of Earth surfaces. The *HCMM 1* is in a relatively low Sun-synchronous orbit; further details are given in Table 5. [E. PAUL McCLAIN]

Bibliography: C. L. Bristor, (ed.), *Central Processing and Analysis of Geostationary Satellite Data*, NOAA Tech. Mem. NESS 64, 1975; G. J. Ensor, *Users Guide to the Operation of the NOAA Geostationary Satellite System*, U.S. Department of Commerce, 1978; Environmental Data and Information Service, NOAA, *Satellite Data Users Bull.*, vol. 1, no. 1, 1979; W. J. Hussey, *The TIROS-N Polar Orbiting Environmental Satellite System*, U.S. Department of Commerce, 1977; W. J. Hussey & E. L. Heacock, *The Economic Benefits of Environmental Satellites*; U.S. Department of Commerce, 1978; R. Legeckis, A survey of worldwide sea surface temperature fronts detected by environmental satellites, *J. Geophys. Res.*, 83 (C9): 4501–4522, 1978; E. P. McClain, Monitoring Earth surface characteristics important to weather and climate with Earth satellites, *Proceedings of the Symposium on Meteorological Observations from Space*, Committee on Space Research of the International Council of Scientific Unions, Philadelphia, June 8–10, 1976; C. R. Madrid (ed.), *The Nimbus 7 Users' Guide*, Goddard Space Flight Center, 1978; A. Schwalb, *The TIROS-N/NOAA A-G Satellite Series*, NOAA Tech. Mem. NESS 95, 1978; J. W. Sherman, III, Seasat-A and Nimbus-G, *Mariners Weath. Log*, 21 (6):355–361, 1977; L. F. Whitney, Jr., Relationship of the subtropical jet stream to severe local storms, *Mon. Weath. Rev.*, 105 (4):398–412, 1977; H. J. Zwally and P. Gloersen, Passive microwave images of the polar region and research applications, *Polar Rec.* 18:431–450, 1977.

Urban Planning

Sigurd Grava

Sigurd Grava, educated in New York in civil engineering and urban transportation planning, has taught urban planning at Columbia University for close to 20 years. He has also maintained an active consulting practice, and currently is the technical director for planning with Parsons Brinckerhoff, the oldest continuing firm of its type in the United States. His professional work has been largely concentrated in the area of public works and transportation service planning and development. Besides projects in the United States, Grava has undertaken assignments in about 15 other countries.

While the first cities can be traced back 10,000 years, urban planning as an organized activity is only a few generations old. There is still a debate as to what it should encompass and how far it should go in guiding the structure and operations of cities.

In the United States in particular, there has been a long process to reconcile the need by the society as a whole or by a given community to rationally control its development with the presumed freedom of choice and action by individuals. Constraints can be accepted by the majority of the public because of a gradually evolving and maturing consensus that positive and systematic guidelines for city development support the common good. These controls have not been imposed from the outside; there have been extended discussion, review, and the opportunity to change and adjust. Practically any urban issue will stimulate some vocal national or local group to question the popular wisdom and seek protection for those adversely affected, or to try to expand the field of concern.

CONTROL OF LAND DEVELOPMENT

There have always been problems associated with city life, and different responses and emphases have characterized different historical periods. The one issue that has remained fundamental and still dominates urban planning is the organization of the physical environment: the determination of a structure for the community by defining a circulation network, by placing the important activity centers, and by identifying locations where residences and workplaces are to be built.

An ancient Greek polis was different from a medieval burg, an English factory town, and an American suburb because of differing contemporary capabilities of technology, demands of life-styles, constraints of economic and political systems, and characteristics of the surrounding natural environment.

Yet, beyond these largely functional determinants of city form, there have also been powerful expressions of otherwise abstract societal ideas. An ancient Chinese capital was a replica of heavenly order, the monumental axes of a baroque town glorified the absolute ruler, the monotonous grid of an American town expedited real estate transactions, the structured neighborhoods of post-World War II new towns embodied an idealized social concept. It is not easy to identify such a driving force in urban planning today, except perhaps personal comfort and overall efficiency.

Thus, physical urban planning—or, looking at it narrowly, the transformation of raw land into a built environment and the reconstruction of these districts—cannot ever be examined outside its social-economic-political context. It is, however, possible to construct cities without a grand, long-range development plan guided by a specific master planner. Most cities, indeed, have grown in this "natural" and anonymous fashion and continue to do so today because the forces outlined above achieve a form of contemporary order—by trial and error.

EXPANSION OF AMERICAN CITIES

The opening of the New World provided an opportunity to develop communities with a clear-cut pattern. In the areas of the Spanish conquest, a basic rectangular plan was used which focused on a central square. In most of North America, however, many influences were at play, and only a few comprehensive plans expressing a unified concept were implemented. Prominent examples are Philadelphia (founded by William Penn, 1682) and Savannah, GA (James Oglethorpe, 1733), with their orderly progression of squares and streets, while in many other designs the New England village green was a recurring element (Fig. 1).

Washington, DC, is a separate case with its unique system of monumental diagonals and circles (designed by Pierre L'Enfant, 1791), but expressing also the then popular ideas of European town organization. There was experimentation with many other original or borrowed city structural concepts, and there was a separate stream of utopian building often based on extreme religious or social precepts.

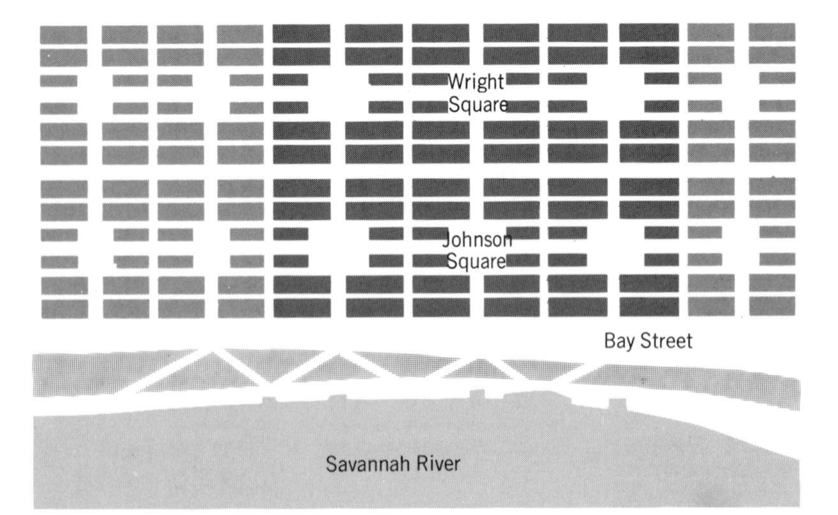

Fig. 1. Savannah, GA, plan of 1733. James Oglethorpe laid out the new town with a generous system of green squares among the grid streets which are differentiated by function. A high degree of livability and urban order was achieved. *(From S. Maholy-Nagy, Matrix of Man, Praeger, 1968)*

Nevertheless, the gridiron street pattern was the dominant characteristic of American cities. The 1811 plan of New York gave the form legitimacy, but it was also the most convenient means of organizing a community as settlements marched westward and each town was assumed to have an unlimited future. The gridiron was an expandable framework into which any activity could be plugged with minimal effort and within which parcels for development (or further sale) could be easily delineated. Such schemes were found suitable for new state capitals, industrial centers, railroad towns, and purely speculative urbanization ventures. The gridiron map, with some embellishments such as a few parks and diagonal avenues, by and large represented the sum total of city planning in American cities in the 19th century.

This, sadly, was not enough because here as well as in Europe the Industrial Revolution had a critical impact on urban life. Surplus labor from the farms was drawn to the industrial concentrations, trade and commerce flourished, and cities expanded in every direction, disrupting all functional services and social balances. The second half of the 19th century was perhaps the worst period for city dwellers who were not reasonably wealthy. The modern city or metropolis can trace its problems to this period of change in the functions and character of urban areas.

Corrective responses had to be made, but for a long time they were made within distinct sectors (housing, transportation, education, utilities, and so on) by single-orientation specialists or concerned citizens. Many decades passed before the need for coordination and integration was accepted—the earmarks of good urban planning.

The nearly intolerable conditions in the cities of industrialized countries brought forth reformers concerned not only with sanitation and housing but also with prevailing development practices, blight, and ugliness. Among the latter, the preeminent name is Daniel Burnham, who took advantage of the Chicago Columbian Exposition of 1893 to inspire American civic leaders to be concerned with urban appearance. The result was the "city beautiful" movement that dealt primarily with public spaces, boulevards, and other facade improvements, but also established the practice of master planning as an attempt to address a multitude of issues and their relationships.

In 1898 an English accountant and social reformer, Ebenezer Howard, published his ideas about "garden cities"—communities that were to be largely self-contained, with adequate and properly located space for each activity, of finite size, and at tolerable densities. His far-reaching impact has colored most planning concepts up to the present.

CONTEMPORARY GUIDANCE

These early efforts were scattered and voluntary, and were implemented so long as a consensus existed or a local leader or group had the power to carry things through. General urban development was still largely uncontrolled and often resulted in unpleasant or destructive conditions: houses in the shadows of large factories, good retail facilities submerged by opportunists, excessive densities

Fig. 2. Effect of zoning on the New York skyline. Above a fixed height of street canyon wall, buildings are allowed to climb by stepped setbacks, or towers of limited area can reach up even further—thereby containing total density and presumably allowing sun and air to reach the ground level.

creating sanitary and health problems, disappearance of any open space, overloaded services and utilities. It was gradually realized that regulatory controls would have to be enacted to prevent at least the worst abuses that unchecked development can cause. In 1916 the New York City zoning ordinance came into effect, specifying district by district what land uses were to be permitted and at what densities (Fig. 2). Gradually this type of code was adopted by other cities, and it became the single most powerful tool of city planning and management.

Yet this is an imperfect tool: state laws allow cities to enact zoning ordinances, they do not mandate them (Houston is the largest American city still without one); these codes are often overly restrictive and uncompromising; sometimes they are poorly enforced and administered; and, above all, since they are purely local ordinances with jurisdictions limited by municipal boundaries, the result is a patchwork of regulations in metropolitan areas that consist of many adjoining cities, towns, and villages. One of the tasks facing urbanists in the near future might be to rethink the entire concept of zoning ordinances, to make development controls more responsive to contemporary conditions, and to achieve equitable and comprehensive direction of urban forces.

In the United States today, planning at the municipal level is still a matter of choice, even though the Federal and some state governments try to encourage this activity as much as possible—often tying eligibility for funds to the existence of plans.

Local governments and city planners can shape their communities by positive actions, such as the construction and operation of public services and utilities to the extent that resources are available, and by restraining actions such as zoning ordinances, building and health codes, and subdivision regulations (governing the construction of new residential sections), and by taxation which can be adjusted to favor or constrain some private activities. Even together with outright powers of condemnation (the taking of land or buildings for a public purpose after compensation), all these municipal tools and devices do not appear to be adequate to cope with contemporary urban conditions. But that is as far as society deems it appropriate to go.

BUILDING OF RESIDENCES

One of the major sectors of planning concerns is housing, which not only constitutes the bulk of the built environment but also represents the immediate interest of any family in the city. Unlike the situation in Europe, the construction of residences is still perceived in the United States as primarily a private responsibility. But there is a growing awareness that this vital activity has to be regulated and supported by public efforts since the product has never satisfied the entire spectrum of income groups.

The working assumption has always been that each family in the New World would build its own home and be fully responsible for its management, not only the homesteaders in the wilderness but also city residents. This precept has generally been valid for the well-to-do and usually also for the middle class, as exemplified by the suburban boom of single-family residences as late as the 1970s.

Traditional practices. Most of the acres that constitute American metropolitan areas today have been changed from raw land to urban blocks by the process of subdivision: an individual business or corporation acquiring a piece of land, laying out the street and providing at least the basic service lines, and then retailing individual parcels. This has gone on for several centuries, and the differences in the practice between the early and recent periods have been basically the extent to which the process has been controlled by the public (very little until World War II), the degree of improvement provided by the developer (usually dependent upon market demand), and whether the houses are built by the families themselves or by the subdivider (the common practice today). All this happened in the older cities of the eastern seaboard as they became large conurbations; it took place across the continent as entrepreneurs established, promoted, and expanded towns and communities.

Before actual subdivision and building take place, speculation in land on a large scale or with individual parcels has been a time-honored practice which has brought wealth to a few but also left serious disruptions and inefficiencies in city development.

Before this century, besides the ostentatious and Europe-derived homes of the rich, certain indigenous housing types evolved in the older centers, such as the triple-deckers of New England, the brownstones of New York, the white-stooped row houses of Baltimore, the gallery houses of Charleston. But the problem was providing shelter for the working class and the poor immigrants.

The immigrants did what they could with makeshift arrangements—shanty towns or converted, dilapidated structures—but they also generated a new industry: tenement house construction. Tenements were the cheapest and most rudimentary high-density accommodations possible. This housing—characterizing the entire social milieu—sparked a gradually growing reform movement that is as responsible as anything at a fundamental level for improving American cities.

Contemporary efforts. Out of this came tangible upgrading by regulation or example of the housing stock as new construction took place. But the efforts went beyond the structure itself, and serious attempts were made to build better communities, at least on the peripheries of cities, in such places as Riverside near Chicago, Roland Park in Baltimore, Forest Hill Gardens in New York City, Shaker Heights outside Cleveland, Palos Verdes Estate near Los Angeles. Their design consolidated the garden town ideas from Europe, the previous experience with some exemplary factory towns, and the American concept of the private home for every family. Albeit these developments did not touch the housing problem of the poor, they established the planning standards to be emulated in the succeeding decades.

Public housing. In the meantime, "public housing" programs were established to provide decent shelter for those who could not keep pace with the costs of apartments built by the private sector. These were units subsidized by the national and local governments, and in the 1940s, 50s, and 60s numerous projects were built, often on a massive scale, in the old sections of central cities.

From today's perspective, it is believed that these public housing efforts were too heavy-handed because they did not address the fundamental causes of poverty, did not accommodate the lifestyles of the residents, and quite frequently stigmatized the neighborhoods where they were located.

The urban renewal efforts of the same period, which involved wholesale razing of slum districts, have gone through a similar reevaluation process. Some bad buildings were removed, but it is taking many years to fill the empty spaces that were left near the city cores. This change of attitude has resulted in a more careful approach that tries to rebuild and upgrade residences where possible, to recycle buildings for new uses, and generally to preserve the established urban fabric in a social and functional sense.

Now it is believed that public housing requires a more sensitive treatment and better integration with the rest of the city. Indeed, the emphasis has definitely shifted to a policy that attempts to strengthen the ability of families to pay for decent housing rather than to build separate projects for them (except for the elderly who need special facilities). There are many programs supporting this concept, but clearly they have not succeeded in erasing the pervasive and apparently chronic housing problem. The general recognition that standard quality housing is not reachable even by moderate-income families without substantial public assistance and participation is a major breakthrough. Industrialized countries in Europe have worked under this assumption for some time.

Suburbia. Another major component of the housing stock is the suburban sprawl that has ringed all American cities since the end of World War II. These classical subdivisions beyond the municipal boundaries of the core cities represent an escape for many million families from the problems of the old urban communities. It is an interesting speculation whether the private automobile is a cause or effect for this type of development, but a relationship is certainly there. There can be little question that this living environment approaches the ideal for many individual families, at least in their early years when they have small children.

But there are problems, and they might very well be the crucial concerns for American cities in the next decade or two. The residents of the inner-city neighborhoods feel with some justification that they have been deliberately excluded; the municipal operations within the limits of small political units are very constraining and inefficient; the low densities result in extraordinarily high costs of providing public services; the sameness of the social-economic-demographic composition demands cyclic adjustments, the complete dependence on the automobile is hazardous. At the same time that many of the urban ills have started to emigrate too, the central cities have been drained of much of their strength, and a more "normal" composition of urban activities is being established in the suburbs as they mature.

It will probably be necessary in the near future for urbanists and politicians to reexamine the present patterns and operations of suburbia and its relationships with the core.

NEW TOWNS

It is frequently assumed that urban planners mostly build new towns. As a matter of fact, such construction of largely self-contained communities is a rare occurrence. In recent decades there have been important programs of new town building in Great Britain, Sweden, Israel, the Soviet Union, and elsewhere to relieve population pressures on old centers or to establish industrial or economic activity poles. Also, the tradition of establishing new capitals on formerly open sites (Brazil, Pakistan, Punjab, Nigeria, Alaska) has continued.

In the continental United States, after a promising Depression-era effort which established several greenbelt towns (in Maryland, Ohio, and Wisconsin), there have been mostly theoretical discussion and imaginative but unimplemented regional sketch planning.

There are, however, dozens of large-scale residential developments on the periphery of metropolitan areas that choose to call themselves planned new communities, often because this label is now recognized as a good marketing tool, but this usage creates a definitional problem. Some could be regarded as new towns because of their size but, then again, they might not qualify because they include few workplaces and there is not physical or administrative separation from the adjoining communities. Yet, among the many endeavors there are a few privately sponsored developments that truly deserve the new-town designation because of the planning that has gone into the design, the superiority of their character and appearance, and the fact of large-scale effectuation. These are Reston, VA (Fig. 3); Columbia, MD; Irvine, CA; and perhaps several more.

There also has been a Federal effort since 1970 (Title VII program) which has provided mortgage guarantees to about 14 selected settlements. It is generally agreed that this encouragement was too little and came too late, and most of these projects are now bankrupt. They frequently made poor locational choices, they were undercapitalized, they were not well enough managed, or they were caught by the general slowing down of the real estate market.

OPERATION OF FUNCTIONAL SERVICES

The high concentrations of people who live or work in cities can exist only if they are supported and supplied by technological systems providing practically everything from drinking water to mechanical means of mobility. The quality of life is a direct function of the adequacy of these services, and thus they fall properly under the purview of city management and urban planning. These public responsibilities became critical after cities reached a size where individual families could no longer tap natural supplies, when travel distances became too long for walking, and as the environment could no longer absorb the wastes generated. For most large communities in Europe and North America this turning point occurred during the 19th century.

In the United States, when the needs first became apparent, entrepreneurs and corporations started to offer services through water companies, omnibus enterprises, rubbish collectors, coal and ice distributors, and many other specialized business concerns. In most of the industrialized countries of Europe the government assumed direct responsibility for almost all of these vital activities; in America, on the other hand, the evolution of public services has been exceedingly complex.

For example, railroads are still run by private companies, but not passenger service; mail service is a Federal responsibility, but not telephone; electricity is supplied by public utility companies, but

Fig. 3. Reston, VA, a satellite town of Washington, DC, offers an attractive residential environment. *(a)* The higher-density, urban-character village center provides a focus to *(b)* the housing groups.

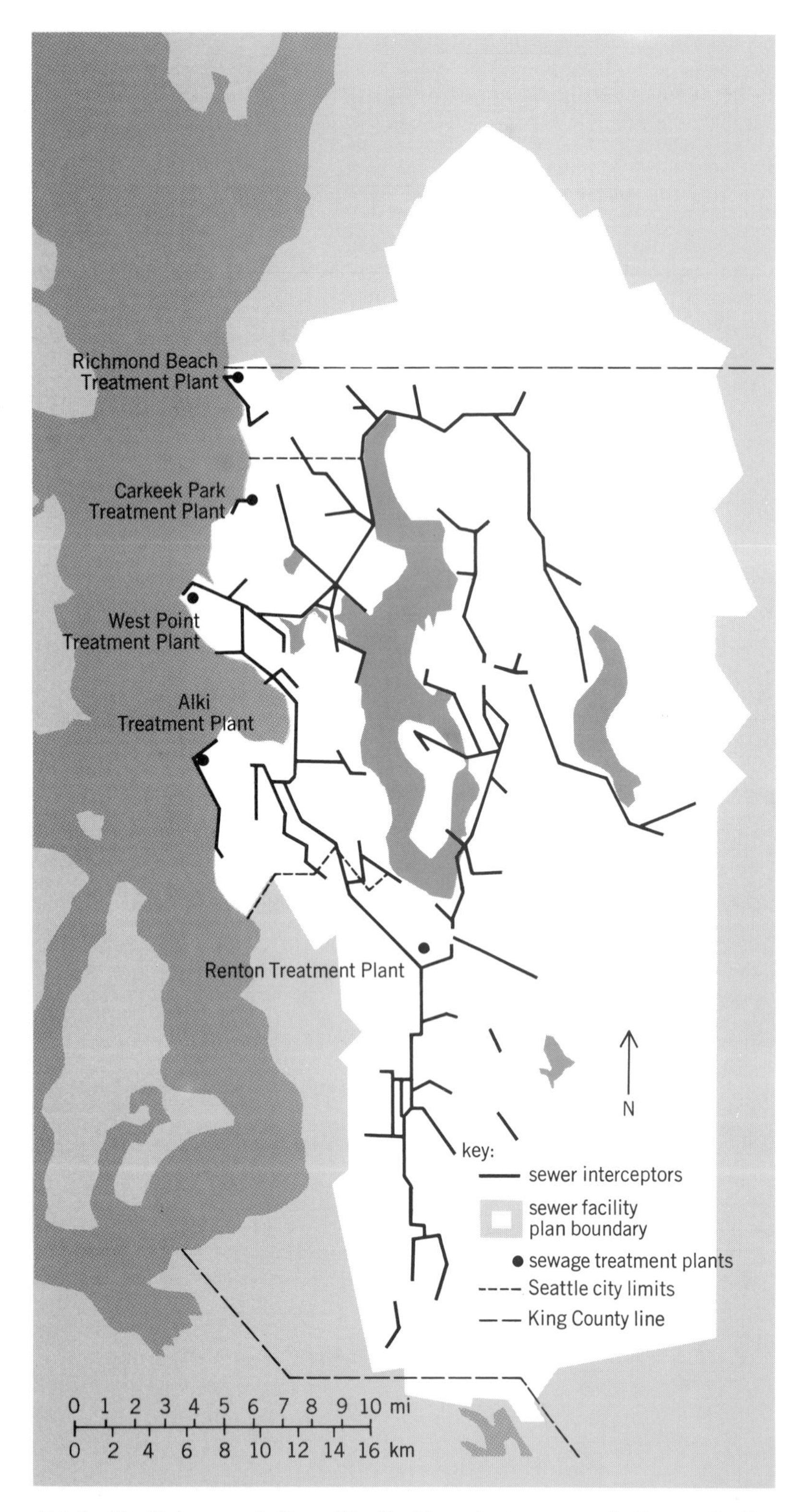

Fig. 4. The Metro organization of the Seattle region has succeeded in coordinating liquid waste control efforts in very tangible benefits to the surrounding bodies of water. The work is integrated functionally and administratively across municipal boundaries.

is sometimes generated by state authorities. It is therefore dangerous to generalize, but a few basic conditions can be identified. First, there are services that are regarded to be outside direct government responsibility—intercity transportation, communications, power and energy supply, food distribution—but they are critical to the public welfare, and therefore, are closely regulated, with varying degrees of success, to assure a reliable and affordable supply. Second, there are services that have gradually been taken over by public bodies: water supply, sewerage and solid waste disposal, intracity transportation. But even in these service areas, there are many variations in management from place to place. The general trend, however, has been for municipalities to assume control, then to regionalize the systems as the demands and operations have grown beyond the original political jurisdictions, and lately to seek and rely on Federal assistance.

It should be noted that the second group of services are the truly essential ones and also those where the collection of full user charges is either impractical or contrary to public objectives, that is, their unhampered availability to everybody.

There is no question that until very recently all the public services were implemented in a strictly sectoral fashion, worrying only about the direct foreseeable demands, construction feasibility, and functional efficiency, and often regarding the entire operation from the viewpoint of a single agency or department. This approach has been changed completely, and the provision, expansion, or upgrading of any of these services is regarded from a more comprehensive perspective. This includes an evaluation of the benefits, the secondary and nonquantifiable costs, the impacts on communities and the environment, the relationships to other systems and operations. It is expected, and indeed required, that all alternatives to solve a problem will be examined and that full public participation will take place. Also, it is an article of faith among planners that the rational control of the expansion of these systems, particularly if they are planned in combination, is one of the most powerful tools available to guide the extent and location of urban development. Planning thus almost becomes a self-fulfilling prophecy: only land that is accessible and equipped with utilities can be readily built upon. It is of some interest to take a closer look at several of the primarily service systems, even though there are dangers of oversimplification in a brief analysis.

Utilities. Water supply was one of the first activities that not only became a direct public responsibility but also moved into the regional dimension. The drinking water supply had to be reliable, safe, and accessible to everybody. But for any sizable settlement, it is rarely possible to find an adequate source within the municipal boundaries. Thus, water systems for larger cities had to extend far into the hinterland.

From the urban planning viewpoint, this action established the principles that many services can be organized only at a metropolitan level, that central cities have the (limited) right to extend geographically their regulatory powers (to protect watersheds), and that businesslike cooperation among municipalities is necessary.

European cities have always been a step ahead of comparable American cities in liquid waste disposal and control—primarily because their densities are higher and therefore the need for workable sewerage systems is much more critical. In North America the sprawling urban development and

Fig. 5. The Central Artery of Boston was one of the early Interstate Highway entries in an old city. It was one of the largest and heaviest such intrusions and became a target for extensive criticism; it is now slated for replacement. *(Parsons, Brinckerhoff, Quade & Douglas, Inc.)*

the relatively open spaces between settlements gave the impression that the natural environment (rivers, lakes, or the subsoil) could absorb the wastes. Also, it is almost impossible to generate any income from sewage, and no community will ever voluntarily deal with the material coming from its neighbors. Thus, systems to control these wastes were always implemented at the last moment, at the least cost, and within the constraining boundaries of each political jurisdiction. This resulted in the pollution of the water bodies in and around almost all cities.

These deplorable conditions, however, were the driving force of the worldwide ecological and environmental control movement. It has been recognized for only about a decade that extensive (and expensive) physical control systems and treatment facilities must be built rapidly, and that coordinated planning is necessary to achieve efficient and acceptable results.

Today, in this and the solid waste sectors, the traditional local piecemeal improvements are being replaced by regional efforts, strict controls from higher levels of government, and evaluations of all consequences (Fig. 4). As a result, many rivers and lakes have been measurably upgraded and in some places fish have returned. There are, of course, still problems: the backlog of work is very large, the cooperative arrangements do not always work smoothly, the facilities are very costly, and local communities appear to have become overly dependent on national programs and aid to the point of losing their own initiative.

Transportation. Attaining adequate urban mobility is a large and complicated task. Cities have always assumed the responsibility for streets as public rights-of-way, and as mentioned previously, the structuring of this network has been one of the principal traditional tools for shaping each community. The provision of transport services, however, is another issue.

It can be concluded that the explosive growth of cities in the 19th century was made possible by the development of both horizontal and vertical mechanical movement systems (rail and elevators). The ability to overcome distances rapidly and relatively cheaply permitted great concentrations of people and activities and also encouraged separation of uses and segregation of districts along economic, social, and even ethnic lines. A new texture of the urban pattern emerged because people did not have to live next to their working places, and service establishments did not have to be distributed on a strictly local basis. The economic consequences were positive because great productive efficiency was achieved; the social consequences were frequently negative. Just to give a few examples: people of an "undesirable" race or class could be pushed into large ghettos; great concentrations of mutually supporting managerial, professional, and communications offices emerged; even the neighborhood breweries with their attached beer gardens that had served as social centers were replaced by a large-scale production and distribution system.

None of this was planned; it happened in response to unguided and powerful forces. Neither was the automobile anticipated, and the private car soon overwhelmed the capacity of the old districts and dominated completely the structure of the zones built after its ascendancy.

A response came after World War II, and its most profound manifestation was the U.S. Interstate Highway network that permeated metropolitan areas with radial and circumferential major limited-access channels (Fig. 5). The system was purposefully structured, but people no longer accept the procedures under which most of the urban links were implemented. It is sobering to note that a complete turnover in the practices came in the 1960s after community groups in many cities independently rebelled against these mammoth facilities that sliced apart usually the more vulnerable neighborhoods. The public pressure was powerful enough to stop many projects in their tracks and, even more importantly, to generate an entire inventory of approaches that advanced the art and science of urban planning at a fundamental level.

Fig. 6. Pedestrian mall in Minneapolis. The Nicollet project has shown how the creation of an exclusive area for buses and pedestrians can help to rejuvenate a city center. The environment is attractive for people and businesses, and the example has inspired many other similar efforts.

In the meantime, drastic changes have taken place in the public transportation field: buses, subways, streetcars, and even jitneys. Europeans, who have long had strong mass transit systems, feel threatened by the unchecked growth of their automobile fleets; Americans are recognizing that workable cities cannot exist without public service and are now at least trying to rejuvenate systems that have deteriorated if not disappeared. A major current planning activity which would have been unthinkable only 15 years ago is the designation of automobile-free blocks or zones (Fig. 6).

Historically, the initial efforts to provide communal transport were private responses to an existing demand. Some of the heavier modes, such as the New York City subway, received municipal assitance in construction, but the operations were intended to be profit-making.

The peak utilization of public transit in North American cities occured during the 1920s and again during World War II, but even then revenue incomes started to fall short of what they should be for a self-supporting business enterprise. To maintain service, municipalities, through their own departments or special authorities, had to take over the operations and either cut back service or seek subsidies. Expansion of lines or replacing rolling stock became almost impossible until the 1950s when the Federal government began assuming a greater role in the allocation of capital funds.

It was not until the late 1960s that it became politically acceptable for the national and state governments to provide operational assistance as well. These subsidy programs continue to grow, recognizing the vital public purpose associated with mobility for all city residents.

There has also been a change in attitude toward various public modes since they became popular again in planning analyses. In the 1950s, subways (heavy rail systems) were again seen as the proper solutions to urban transportation problems. After a hiatus of several decades, all the cities with existing services resumed or contemplated additional construction, and completely new systems were built in the San Francisco–Bay Area (Fig. 7), Washington, DC, and Atlanta. In many other cities, however, exhaustive studies showed that the high costs and the low densities precluded effective utilization of such high-capacity service.

In the aftermath of successes in outer space, attention turned to automated-guideway, computer-controlled people movers (horizontal elevators). Though work is going on in a few locations, it is apparent that such devices are applicable only to special, high-density districts.

Then the bus in its various incarnations became the most actively studied choice. This renewed popularity is highly deserved because of the versatility, ruggedness, and simplicity of the vehicle and its mode of operation. Effective programs ranging from minibus loops to massive exclusive busways can be found across the country. But there are also communities without any public transit service whatsoever.

Other alternatives include special small-scale

services—now called paratransit—such as dial-a-ride, carpooling, and shared taxis. Lastly, transportation planners have rediscovered the streetcar with its relatively high capacity and low costs, energy and labor efficiency, absence of pollution, and general compatibility with the urban environment.

For many years, planning for public transport service was a purely functional and financial effort, and it was a separate sectorial endeavor until recently when transportation planning assumed its integrative character. Presently, public transit service—as contrasted with automobiles and highways—is a full partner in any regional or local planning effort, at least in terms of attention if not in the amount of funds available for implementation. The other major development—parallel to other urban services—has been the growing establishment of metropolitan-level operations with corresponding administrative and managerial structures.

PROVISION OF SOCIAL SERVICES

Cities are made endurable, workable, and attractive because of the multitude of choices and opportunities that they offer. This includes not only residences, workplaces, and the functional systems, but also such social services as fire and police protection, medical and educational facilities, libraries and cultural establishments, parks and recreational places. Again, these are sectors each of which has had its own pattern of evolution in American cities.

The common thread has been the similarities in the change of responsibility for these services in the United States. Each service began at different periods to satisfy public demand and was usually first provided by volunteers or philanthropic organizations under a self-generated control and administrative body. As the service became a regular operation with a steady clientele, municipalities assumed direct responsibility but generally kept a supervising board or commission made up by private, and frequently leading, citizens with a personal interest in the service. Many of these bodies still exist and presumably provide the objective and impartial control that these vital activities demand. They are theoretically removed from day-to-day politics—which, however, can be a mixed blessing if administrative coordination is to be achieved.

These uniquely American arrangements have preserved a great amount of independence for these services, and the organizations have done their own planning without too much integration among them or with the overall city planning effort, such as it has existed. It is still an open question whether practical and political constraints make efforts at further coordination counterproductive, an issue which comes to the forefront at least once a year in every municipality at budget preparation time. But any resolution or accommodation then is definitely short-range planning.

Another heritage of these typical administrative and policy formulation arrangements, created when city planning became recognized and formalized in American cities earlier in this century, is that a citizens' board was adopted as the control mechanism for planning. It is still the pattern today, even though in most places a specialized civil service staff is permanently engaged and professional leadership is provided by a director who is responsible both to the board and to the chief elected executive. There is a legitimate question whether planning should be more of an operational part of city government or more of an independent force keeping tabs on other departments.

Fig. 7. Rapid transit in San Francisco. The regional rail system of the Bay Area is efficient and provides a new level of communal service. But it does not extend to all communities and has not changed the established life-styles and development activity to a great extent.

OTHER CHALLENGES FOR THE FUTURE

For the very large metropolitan areas, the cleavage between the old core city and the post-automobile era suburbs may carry the greatest danger of future disruption. The center and the doughnut around it are not only different in their appearance and composition, but are also split by political boundaries that do not allow resources to be equalized between them. The core cities are old and largely inhabited by lower-income families, but they also contain the vital facilities that hold the entire urban complex together. Nobody at the present time is seriously advocating regional government, but there is a need for cooperation on a sectorial basis or in further-reaching arrangements. The few promising examples of such cooperation are the Twin Cities in Minnesota, Nashville, Jacksonville, and Miami.

At the same time that a trend toward regionalization and greater public responsibility is taking place, there is also an apparently contrary effort to strengthen the capability and authority of local communities and neighborhoods to deal with their own space and services. It is not really a paradox, because the metropolitan area is a functional entity with many activities being meaningful only at that level (the location of workplaces and water supply, for example), while many of the true needs of the neighborhood can be handled only on the spot (location of playgrounds and street cleaning, for example). The challenge is to restructure planning and managerial responsibilities so that each entity does what it is best equipped to do.

Most cities of the United States have always grown until very recently, and throughout this history, expansion has been equated with progress. It took a number of years for many of the older cities to accept the fact that they are actually losing population, and even more time will elapse before this new condition can be translated into appropriate planning ideas: contraction, rebuilding, repair, reorganization, upgrading.

There are now entire metropolitan regions which are losing population, when previously any loss in the center was more than offset by growth on the periphery. It will take some time for this trend to penetrate the public's consciousness and for adjustments in attitudes and programs to be made. With the entire country approaching zero population growth and the median age steadily climbing, certain actions are indicated for those places which will not expand due to migration. These actions include a purposeful reevaluation of needs and capabilities, the modification of the available facilities, the correction under an easing pressure of prior mistakes, the abandonment of those elements or parts that no longer serve or that are inefficient.

But thinking even on a metropolitan scale may no longer be sufficient for urban planners. For some time, the term "exurban" has been used to denote those remote areas that are beyond convenient daily commuting distance but are still linked to a central core. While the science fiction concept of instantaneous and cheap movement of material or bodies is not likely to be achieved, instantaneous and cheap transfer of information is a fact. In recent years the effects of those capabilities, long talked about, are becoming visible: the selective dispersal of establishments that have always been regarded as characteristic of high-density centers (offices and entertainment, not to mention retail). Also, despite the energy crisis, recreation opportunities hundreds of miles distant are being considered accessible by more and more city residents.

All this does not spell the end of the city, but it brings into practical use such geographical concepts as the "metropolitan field" extending considerably beyond the old metropolitan area, and even such terms as the "global village" which is intended to highlight further the interdependency of contemporary life among centers far removed from each other.

Some painful adjustments and some dislocations may be facing people in the near future, but the overall prospects for urban areas are still bright. An opportunity to rebuild and enhance the old city environment is present; the task of dealing with the expanded scale of linked activities opens up new challenges. Perhaps in the process it will even be possible to redress the social and economic inequities that have come down from the past. Besides some money, it will take imagination and good planning. [SIGURD GRAVA]

Environmental Analysis

David F. S. Natusch

Professor David F. S. Natusch is chairperson of the Department of Chemistry at Colorado State University. After obtaining a doctoral degree in physical chemistry as a Rhodes Scholar at Oxford University, he worked in his native New Zealand, then went to the United States in 1970 as a Fulbright Fellow. His current interests include characterization of material emitted from fossil-fueled energy sources and elucidation of the environmental behavior of trace elements, sulfur species, and organic compounds present in the atmosphere.

Environmental analysis involves the performance of chemical, physical, and biological measurements in an environmental system. This system may involve either the natural or the polluted environment, although the term "environmental analysis" is increasingly used to refer only to situations in which measurements are made of pollutants.

Normally, four types of measurements are made: qualitative analysis to identify the species present; quantitative analysis to determine how much of the species is present; speciation or characterization to establish details of chemical form and the manner in which the pollutant is actually present (such as being adsorbed onto the surface of a particle); and impact analysis in which measurements are made for the specific purpose of determining the extent to which an environmental impact is produced by the pollutants in question.

The overall objective of environmental analysis is to obtain information about both natural and pollutant species present in the environment so as to make a realistic assessment of their probable behavior. In the case of pollutants, this involves assessment of their actual or potential environmental impact, which may be manifest in several ways. Thus, a pollutant species may present a toxicological hazard to plants or animals. It may also cause contamination of resources (such as air, water, and soil) so that they cannot be utilized for other purposes. The effects of pollutants on materials, especially building materials, may be another area of concern and one which is often very visible and displeasing (for example, the defacing of ancient statues by sulfur dioxide in the atmosphere). A further area of environmental impact involves esthetic depreciation such as reduced visibility, dirty skies, and unpleasant odors. Finally, it is important to recognize that an environmental impact may not always be discernible by normal human perception so that detection may require sophisticated chemical or physical analyses (for example, depletion of the stratospheric ozone layer with consequent increase in the intensity of short-wavelength ultraviolet solar radiation reaching the Earth's surface).

Several different approaches to environmental analyses are encountered. Thus, it may be necessary to determine the natural background concentrations of a number of chemical species present in a pristine or remote area. Such information is required to enable determination of the extent of any increases in pollution which may result from the operation of some proposed activity (say, installation of a power plant). A second rationale for environmental analysis involves situations where one or more potentially polluting species are known to be exhausted into the environment so that it is necessary to determine the rate and extent to which introduction is occurring. Third, there are situations where it is apparent that pollution of the environment is taking place, but it is necessary to identify those chemical and physical species which are responsible. Finally, there is the situation where it is apparent that a significant adverse environmental impact is occurring. In such instances it is normal to obtain information not only about the identity, amounts, and rates of introduction of pollutant species but also about the nature and extent of their impact.

In order to provide a meaningful description of the general field of environmental analysis, this field may be considered from three points of view; the basic concepts underlying the reasons for and choice of the analyses which are normally performed; the available techniques and methodology commonly used; and the current status of capabilities in environmental analysis. The following sections are devoted to these topics.

BASIC CONCEPTS

Some of the philosophical concepts which form the bases for environmental analyses are as follows.

Purpose. Collection and analysis of an environmental sample may be undertaken for the purpose of research or monitoring, or as a spot check. A spot check analysis is used to obtain rapid information about the approximate extent or nature of an environmental problem. If, however, an environmental problem has already been established, analyses are performed in a monitoring mode to observe continuously the level of one or more of the pollutants present. Analyses conducted for research purposes are designed to provide a full understanding of the relevant physical and chemical processes, and may frequently involve unique methodology which cannot be employed in the field.

Authenticity. Whatever the information being sought, it is vitally important to obtain an authentic sample which represents the particular system being investigated. In fact, the ability to obtain an authentic sample is probably one of the most difficult aspects of any environmental analysis due to the considerable complexity and heterogeneity of most environmental systems. For example, collection of a small localized sample from a sanitary landfill will clearly result in an extremely untypical example of the chemical and physical species present. It is important, therefore, to collect a sufficiently large sample so that deviations in composition within the system can be reasonably accounted for.

Detection limits. A statement of the detection limits which can be attained by the analytical method being employed must always be included in providing the results of any environmental analysis. This is because considerable confusion exists about the meaning of a "zero" level of concentration. In fact, it is probably never possible to state that none of the atoms or molecules of the species in question are present so that no true zero exists.

Precision and accuracy. In reporting an environmental analysis it is also necessary to specify the precision and accuracy associated with the measurements. Thus, many environmental measurements involve comparison of results obtained in different systems or under different conditions (of temperature, time, pollutant concentration, and so on) in the same system so that it is necessary to establish whether two numbers which are different are in fact indicative of different conditions. Experience in environmental analysis has established that precisions are rarely better than ±5% and are frequently only good to ±30%. Accuracy is dependent on the ability to calibrate any analytical technique against some standard. In making such a comparison it is necessary to choose a standard whose matrix composition and general physical and chemical characteristics are closely related to those of the sample. This is because matrix effects can introduce both systematic and random errors into analytical measurement in environmental systems.

State of matter. In making an environmental analysis it is necessary to designate the physical form of the species being analyzed. Most simply, this involves the actual state of matter in which the species exists (whether it is solid, liquid, or gas) since many species (both inorganic and organic) may exist concurrently in different states. For example, certain organic gases can exist either as gases or adsorbed onto the surface of solid particles, and the analytical procedures employed for determination of each form are quite different.

Element/compound distinction. One of the most strongly emphasized aspects of environmental analysis involves the distinction between a chemical element and the chemical compound in which that element exists. For instance, arsenic is often encountered in polluted waters in the form of both arsenate and arsenite; however, the potential environmental impact of these two species is significantly different insofar as arsenite is both more soluble and considerably more toxic than arsenate. It is appropriate, therefore, to establish the chemical form in which a given element exists in an environmental sample rather than simply to specify the fact that the element is present at a given concentration. While such a concept is philosophically acceptable, analytical methodology has not reached the stage where specification of inorganic compounds present at trace levels can readily be achieved.

Particle surfaces. Where pollutant species are present in or associated with a condensed phase, it is sometimes necessary to establish whether the pollutant is part of the bulk system or present on its surface. Such a consideration is particularly meaningful since material present on the surface of an airborne particle, for example, comes into immediate contact with the external environment, whereas that which is distributed uniformly throughout the particle is effectively present at a much lower concentration and can exert a much lower chemical intensity at the particle surface. Since airborne particles can be inhaled, surface predominance can result in high localized concentrations of chemical species at the points of particle deposition in the lung.

Availability. While not one of the analyses normally performed by analytical chemists, determination of the availability of a chemical species is often necessary. To exert a meaningful environmental effect, a pollutant species must almost always enter solution. For instance, toxic compounds associated with airborne particles must be dissolved by lung fluids before either a local or systemic effect can be produced. Thus, laboratory-scale simulation of the availability characteristics of a chemical species can provide a necessary link between its presence and its eventual impact.

Environmental effect. The final link in the analytical/environmental effects chain involves determination of the actual environmental effect produced. In most cases, this involves some form of biological measurement or bioassay which determines the toxicological effect upon a biological organism. Due to the expense, complexity, and time-consuming nature of bioassays, it is usual to substitute a chemical analysis for the purposes of monitoring toxicological effects, and in setting standards for compliance. In doing so, however, it is necessary to establish a so-called dose-response relationship between the level of one or more pollutant species and the degree of toxic impact.

ANALYTICAL TECHNIQUES AND METHODOLOGY

Any chemical analysis consists of four clearly definable steps: (1) collection and preservation of the sample; (2) preparation or pretreatment of the sample for analysis; (3) analytical evaluation; (4) presentation of the resulting data in a meaningful form. The present status of environmental analysis is discussed in terms of these steps in the following.

Sample collection. The type of sample collected depends entirely on the type of information required. Thus, a monitoring operation may require collection and analysis of a sample in real time so that little opportunity exists for preconcentration of the chemical species being determined. In such cases, considerable sensitivity constraints are placed on the analytical method employed. Alternatively, it may be satisfactory if the sample is collected and accumulated over a period of several hours or even days prior to removal to the laboratory for analytical evaluation. In such situations, the requirements placed upon the subsequent analytical technique are less stringent insofar as sufficient material may be collected for even a quite insensitive technique to function effectively.

Relatively few environmental analyses are actually conducted in real time due to the fact that the levels of most pollutants in the environment are too low for presently available instrumentation to detect. Probably the most effective real-time analyzer presently available is a nuclei counter which determines the number of aerosol nuclei or particles present in the atmosphere. Such an instrument can be mounted on an aircraft and used to locate the presence of a plume emitted from some air-polluting source.

A much more common approach to sample collection involves accumulation of the chemical species to a level where a subsequent analytical technique is capable of performing a precise and accurate analytical evaluation. Such cumulative collection procedures have the advantage that they can, at least in theory, collect as much material as may be required for some subsequent operation. On the other hand, they have the responsibility of maintaining the sample in a form such that the subsequent analysis measures all of the material actually present in the original environmental sample. This second consideration is not trivial since sample stabilization is probably one of the more important (and indeed most neglected) aspects of an environmental analysis. As a general rule, preservation of the authenticity of a solid sample is comparatively straightforward; however, changes in gaseous and liquid samples following collection are extremely common so that considerable emphasis must be placed upon stabilization and preservation of these types of samples.

Solids. Bulk collection of solids is comparatively straightforward since a selected portion of material such as soil, tree roots, leaves, or biological organisms may readily be collected by using some predetermined criterion for the selection. It is important, however, to specify the size or amount of sample which should be collected. Such decisions are determined almost entirely by the heterogeneity of the sample in question, and the only general statement that can be made is that the sampling statistics should be designed to ensure that this heterogeneity is averaged out by the sampling method (such as filtration) and criteria (such as particle size distribution) employed.

1. Filtration. A significant fraction of the solid sampling undertaken involves the collection of particulate matter from the atmosphere or from an aquatic system. The technique most commonly employed involves some form of filtration whereby the air or water is passed through a filter which retains solid material. This retention is normally of two types. The first occurs when the solid particle is physically larger than the hole in the filter so that only the fluid medium passes through the holes. The second occurs in situations where the momentum or inertia of the particles is sufficiently great for them to impact upon the structure of the filter medium and be held there even though the holes in the filter are sufficiently large for them to pass through.

Filtration can be undertaken by using high-volume or low-volume conditions. Thus, high-volume conditions (which are employed only in collection of particles from air) involve the passage of large quantities of air (20 to 60 ft^3 per minute or 10 to 30 liters per second) through a filter to collect large amounts of material which can be weighed. Low-volume collection utilizes a much lower flow rate (1 to 5 liters per minute). It is generally employed for collecting particles on some specialized substrate for subsequent chemical analysis which has a sensitivity significantly greater than that of mass determination.

2. Particle size distribution. In some instances, it is important to determine the distribution of a solid material as a function of size. This is particularly important with respect to airborne particles whose size distribution determines the extent to which the particles may penetrate into the innermost regions of the lung when inhaled. A variety

of techniques have been developed for the determination of airborne particle size distributions. The two most important involve direct determination, which frequently utilizes the principle of light scattering, and some form of sequential impaction whereby particles are collected in different regions of the impactor depending upon their aerodynamic (or hydrodynamic) size. Thus, a differentiation between particles of different effective size can be achieved.

At this point it is appropriate to establish the concept of aerodynamic and hydrodynamic size. Such a definition is important because the motional behavior of the particle in any type of fluid medium depends upon both its density and its physical size as determined by the relationship in the equation below. Here D_{aer}, D_r; ρ_{aer}, ρ_p; and

$$D^2_{aer} = D^2_r \frac{C(D_r)\,\rho_r}{C(D_{aer})\,\rho_{aer}}$$

$C(D_{aer})$, $C(D_r)$ are the particle diameters, particle densities, and Cunningham slip correction factors for aerodynamic equivalent and real particles, respectively. For this purpose, the aerodynamic equivalent diameter of a particle is defined as being equal to the diameter of a spherical particle of unit density whose aerodynamic behavior in an airstream is the same as that of the particle in question.

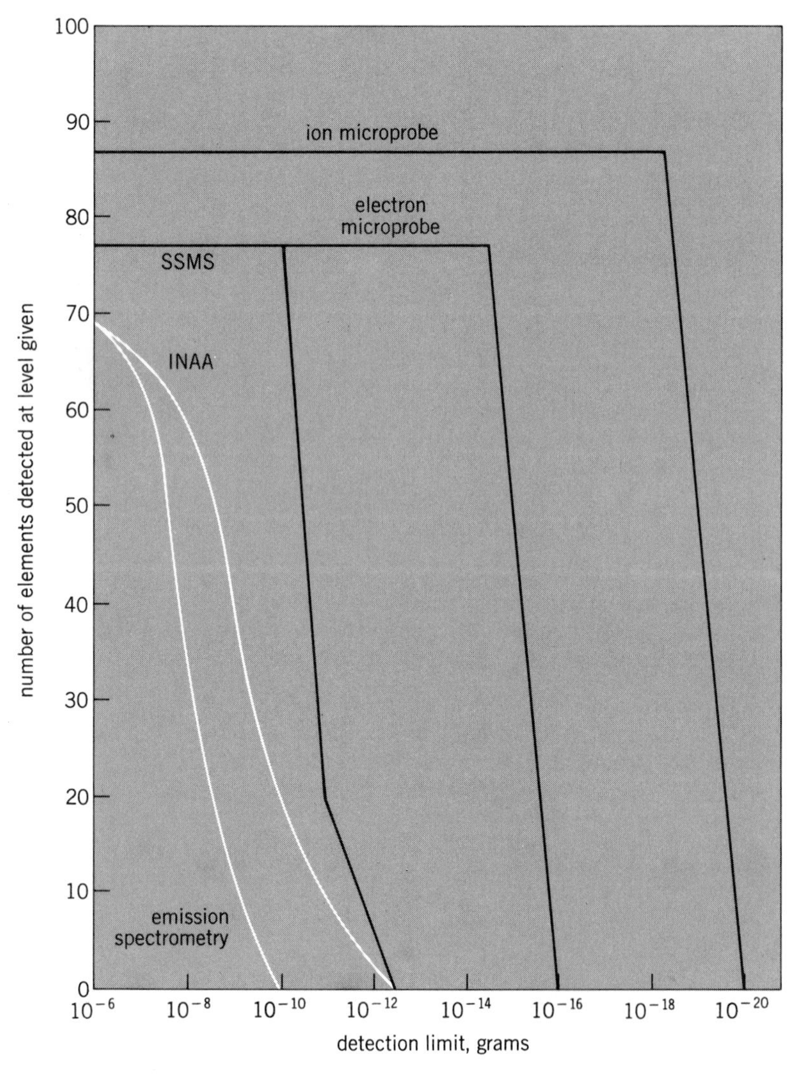

Fig. 1. Comparison of the number of elements that can be detected at given levels by ion microprobe mass spectrometry, electron microprobe x-ray spectrometry, instrumental thermal neutron activation analysis (INAA), spark source mass spectrometry (SSMS), and dc arc emission spectrometry. *(From D. F. S. Natusch, C. F. Bauer, and A. Loh, Collection and analysis of trace elements in the atmosphere, in W. Strauss, ed., Air Pollution Control, Wiley Interscience, 1978)*

Liquids. Collection of liquid samples is comparatively straightforward insofar as such samples are usually collected in bulk. The main considerations involved are the extent to which the sample is representative of the system in question and the nature of the container employed for collecting the liquid. In this latter regard, it is noted that many dissolved inorganic ions (such as the mercuric ion) show a strong tendency to adsorb on glass container walls; organic species, on the other hand, show comparatively little tendency to adsorb on glass container walls but adsorb quite readily on many types of plastic.

In some instances, the techniques of filtration or adsorption are employed in sampling liquid systems. Filtration, such as that described above for solid sampling, is commonly employed to separate particulate material from the liquid; however, a more sophisticated approach involves ultrafiltration whereby entities having large molecular dimensions are retained by a filter. At the present time such collection of environmental samples is comparatively rare.

Adsorption of selected molecular and ionic species from solution is rapidly increasing in popularity as a means of sampling. Specifically, filters or columns packed with ion-exchange resins are employed to collect either anionic or cationic species present in solution. Such filters may be highly selective so as to collect one or more specified ions, or they may have a broad spectrum of adsorptive capability so as to collect all ions carrying a positive or negative charge. Selective adsorption of organic material present in the solution may be achieved by using adsorbents (such as resins). Such a method of isolating organic species is employed more for the purposes of research than for routine analysis or for monitoring, and all adsorption methods of this type are considered to be still in the developmental stage.

Gases. Collection of gaseous samples can be achieved in a number of ways. The most obvious is simply to collect a gaseous sample (usually polluted air or some atmospheric emission stream) in a container which may consist of an evacuated glass or steel bulb or possibly a small balloon made of a substance which exhibits little tendency to adsorb species from the sample.

A more common approach to gas sampling involves collection and preconcentration of selected species by utilizing the principles of adsorption and absorption. Adsorption involves collection of selected species (organic gases such as butane and propane, for example) onto a solid substrate. One of the most universally useful adsorbents of this type is a commercially available resin called Tenax, which is an effective broad-spectrum adsorbent for a wide variety of nonpolar or moderately polar gases.

The technique of absorption involves collection of a gaseous species in solution. In practice, such collecting solutions generally contain some chemical species which will promote the retention and stabilization of the gas. For example, one of the most common methods for the determination of sulfur dioxide in the atmosphere involves absorption of this gas into a solution of tetrachloromercurate wherein the sulfur dioxide is complexed, stabilized, and accumulated.

Fulfilling basic concepts. Overall, therefore, the methods employed for collection of environmental samples vary depending upon the physical state of the material in question. In all cases, however, the basic philosophical concept is to collect, accumulate, and stabilize material over a period of time for subsequent analysis, or, in the case of real-time monitoring, to present the sample essentially as it occurs in the environment to some form of selective analytical sensing or evaluation device.

Probably the most neglected yet most important aspect of an environmental analysis involves the selection, collection, and analysis of a sample which is representative of the system being investigated. Thus, a small sample may not represent a heterogeneous system. Similarly, a small sample may result in poor precision and accuracy of the eventual analytical evaluation technique because of the presence of only a small number of atoms or molecules of the species being determined. It is of vital importance, therefore, to ensure that the size of the sample collected, together with its method and point of collection, be given serious consideration in any environmental analysis.

Considerable research has been conducted into the question of sampling statistics, and sound guidelines have been established for the procedures which must be employed in obtaining an authentic sample.

Sample pretreatment. It is often necessary to modify a collected sample in some way so as to prepare it for subsequent analysis. Usually, this involves separation of matrix material or of individual molecular species.

Solids. In the case of solids, stabilization of the chemical species present is generally not a problem. For the purposes of analysis, however, it may be necessary to present the sample in the form of very small particles or fused with some added matrix material. In cases where the analytical method employed requires introduction of a solution, the original solid sample must be digested (normally in a highly acid medium) or it may be reduced to an inorganic ash by oxidation at elevated temperature. The ash is then normally soluble in acidic aqueous solution. Organic species associated with solids are usually separated therefrom by one or more solvent extraction stages.

Liquids. Pretreatment of liquid samples for subsequent analysis usually requires chemical stabilization of the species of interest. This can be achieved by addition of a chemical which interacts with the species so as to render it effectively stable. Alternatively, it is sometimes possible to stabilize a potential analyte by freezing the liquid sample. The most common form of chemical adjustment involves modification of the pH or acidity of the sample or addition of a chelating agent which is capable of binding metallic ions and preventing their loss by such processes as adsorption onto the walls of glass containers.

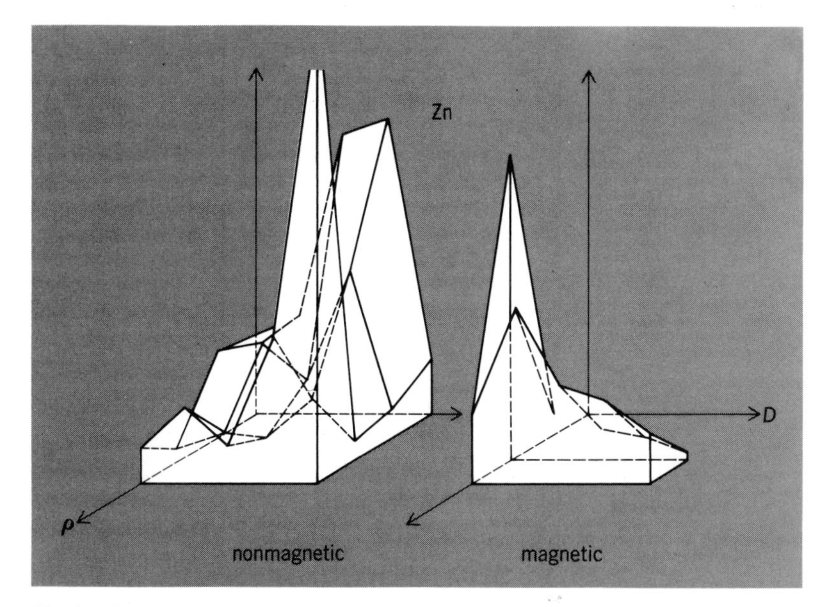

Fig. 2. Three-dimensional graphical representation of distribution of zinc as a function of particle diameter D and particle density ρ in magnetic and nonmagnetic fractions of urban roadway dust.

In addition to stabilization of the analyte, it is often necessary to adjust the sample matrix. Such adjustment is required for most electrochemical analysis and is frequently employed in atomic spectroscopic analysis.

Another common method of liquid sample pretreatment involves extraction of the dissolved species into an alternative solvent. For example, low levels of organic species present in aqueous solution can be significantly concentrated into a much smaller volume of organic solution either by direct solution-solution partitioning or by liquid-liquid Sohxlet cumulative extraction. In both cases, the net effect is to remove the species of interest from one solution where it occurs in low concentration to another solvent in which it can be attained at high concentration suitable for subsequent analytical evaluation.

Gases. The sample pretreatment procedures employed for gaseous samples are similar, in principle, to those employed for liquids. This is because species present in both fluid media have considerable diffusional motion and so can be adsorbed onto container walls and react with other species (such as oxygen) which are present. Consequently, emphasis, in both cases, is based on chemical stabilization in the solid or solution phase, or adsorption onto a solid—possibly in conjunction with low temperature. For example, as mentioned above, sulfur dioxide is stabilized by formation of a complex with tetrachloromercurate in solution, and organic gases are stabilized and accumulated by adsorption onto resins such as Tenax.

In order to analyze for adsorbed species, it is necessary normally to remove them from the adsorbent substrate. This is achieved either thermally by rapidly heating the sample to drive off in vapor form the adsorbed species for subsequent collection (at a much higher concentration than was originally present in the atmospheric sample collected) or to extract the adsorption system so as to remove the species of interest into the extract-

Table 1. Atomic spectroscopic methods employed in environmental analysis

Method	Detection limits, μg	Precision, %	Applications
Atomic absorption spectroscopy			Most metallic elements
Flame	0.04–50.0	1–5	Solution samples
Flameless (carbon rod or furnace)	0.001–0.2	5–15	Solution or solid samples
Cold vapor	0.001	1–5	Mercury Solution or gaseous samples
Atomic emission spectroscopy			Most metallic elements
DC arc	0.01–1.0	10–20	Solid samples
Microwave plasma	0.005–0.5	2–5	Solution samples
Inductively coupled plasma	0.001–0.5	1–5	Solution samples
Fluorescence	0.01–10.0	1–5	Solution samples
X-ray emission spectroscopy			
X-ray tube excitation	0.05–5	1–5	Solid samples
^{141}Am isotope excitation	1–50	1–5	Solid samples
Proton excitation	0.005–0.010	1–5	Solid samples

ing solvent. In both cases a considerable preconcentration factor is achieved.

Analytical evaluation. The actual analytical evaluation step of the overall analytical scheme is generally the most widely recognized and highly emphasized aspect of an environmental analysis. Nevertheless, the analytical evaluation can only be successful in providing the numbers necessary to describe an environmental system provided that the other steps in the analytical sequence are suitably designed and performed.

The analytical evaluation process is usually required to recognize the analyte species of interest and to determine the amount of that species which is present. Two general approaches are normally employed. The first involves analysis of a composite sample utilizing techniques (primarily spectroscopic) which are capable of recognizing specific characteristics of elements and compounds present in a complex mixture. The second approach involves physical separation of molecular species into separate compartments in time or space followed by selective or specific identification of the components thus separated.

Analysis of composite sample. Identification of individual chemical species or groups of species may be based upon one of several characteristics. Probably the most common involves the atomic or molecular spectroscopic characteristics of a compound whereby that compound is capable of absorbing or emitting specific wavelengths of electromagnetic radiation. The intensity of the absorption or emission is directly related to the number of atoms or molecules which are present. Such techniques are given together with their detection limits, associated precision, and areas of application, in Tables 1 and 2.

Identification and quantitation of chemical species can also be achieved in terms of their electrochemical characteristics whereby their oxidation or reduction is determined in terms of the electrical current or voltage which results. The techniques of this type which are employed in environmental analyses are given in Table 3.

The classic but still widespread approach to environmental analysis for chemical species involves so-called chemical identification of those species. This approach is based upon the fact that certain chemical reactions will only proceed provided that a given reactant species (in this case, the analyte of interest) is present. The extent or rate to which the reaction proceeds is directly dependent on the amount of the chemical species present. A number of reactions of this type which are still routinely used in environmental analyses are given in Table 4.

Although not used very extensively in environmental analyses, determination of the thermal or thermodynamic characteristics of chemical species of interest can provide a means for their detection and quantitation. For example, the temperature at which heat is evolved or adsorbed by a sample can provide information about the nature and amount of a species which may be present.

Some of the most precise and sensitive methods for environmental analysis involve detection of radioactive emissions from the species of interest. The energy of the emissions provides qualitative information about the identity of the emitting species, and the number of emissions which occur in a given period of time establish the amount of that species which is present. Several methods based

Table 2. Molecular spectroscopic methods employed in environmental analysis

Method	Detection limits	Precision, %	Applications
Ultraviolet/visible spectroscopy	Highly variable		Organic and inorganic species having ultraviolet or visible absorbance
Absorption	10^{-5} to 10^{-3} M	3–5	Solution samples
Fluorescence	10^{-7} to 10^{-3} M	3–5	Organic fluorescent species Solution samples
Infrared spectroscopy	0.1 to 1.0 mg	3–5	Most molecules Solid, solution, and gaseous samples
Raman spectroscopy	0.1 to 1.0 mg	3–5	Most molecules Solid, solution, and gaseous samples
Nuclear magnetic resonance spectroscopy	10^{-3} to 1 M	3–5	Most molecules Solution samples

Table 3. Electrochemical methods employed in environmental analysis

Method	Detection limits, moles/liter	Precision, %	Applications
Ion selective electrodes	10^{-6} to 10^{-6}	1–5	Any electroactive species, such as metal ions, NH_4^+ Solution samples
Polarography	10^{-7} to 10^{-4}	5–10	Any electroactive species, such as metal ions, organics Solution samples
Anodic stripping voltammetry	10^{-9} to 10^{-6}	5–10	Electroactive metals and metalloids Solution samples

upon the principle of radioactive decay are given in Table 5.

Although used comparatively rarely for quantitative determination of very small amounts of a species of environmental concern, several techniques, such as x-ray diffraction and nuclear magnetic resonance spectrometry (Table 6), are capable of identifying the chemical compounds present as opposed to the individual elements which compose those compounds. Consideration of Table 6 shows clearly that the levels at which inorganic compound identification can be achieved are much higher than those at which identification and determination of individual elements are possible. Thus, one may in effect recognize the paradox between the well-developed ability of the environmental analytical chemist to determine the presence and amounts of individual chemical elements as opposed to the essentially poor ability to determine the compounds in which those elements exist.

Separation of species. The techniques listed in Tables 1 through 6 are for the most part applicable to situations where the environmental sample is a composite mixture of a number of compounds. As indicated earlier, however, a more definitive approach to analysis of complex mixtures involves the separation of individual components present in the mixture prior to the actual analytical evaluation. A number of techniques are available whereby such separations can be achieved. These include static solvent partitioning, solution precipitation, vaporization, several types of chromatography, and mass spectrometry (which is not truly a spectroscopic technique but rather a separation technique).

These methods all enable physical differentiation of individual compounds in both time and space, thereby giving rise to the possibility of actually taking a complex mixture and separating it into its individual components each of which may be stored in a separate container. Not unexpectedly, the several types of chromatography which are applicable to environmental analyses come close to accounting for approximately 50% of the analyses currently performed on environmental samples.

Separation techniques are provided in Table 7, which includes a brief description of the separation characteristics of the several chromatographic techniques currently used in environmental analyses. A very large number of detectors have been developed for use in gas and liquid chromatography, however. These fall into two classes: universal detectors are capable of determining essentially any chemical compound which is present in the gas stream emerging from the separation column; selective detectors are designed to be selective (and in a few cases, even specific) for compounds of a certain type.

Microscopic analysis. Environmental analysis, as defined herein, most commonly involves determination of species present in bulk samples of air, water, or soil. As discussed earlier, however, there are many situations in which it is important to determine the chemical elements or compounds which are present in certain regions of a condensed sample. The techniques available for such analyses are generally based upon the principle of bombarding the sample with a microscopic-size beam of high-energy irradiation. Such analyses are proving to be extremely important since they provide very specific information about very localized regions. Some of the techniques commonly used in the area of microscopic, regional or surface analysis are presented in Table 8. As shown in Fig. 1, such techniques have extreme intrinsic sensitivity and detection capabilities for a large number of elements. Due to the very small analytical region involved, however, actual concentrations detected are quite high.

Calibration. Calibration of analytical methods is both a vital and difficult procedure primarily due to the fact that the matrix associated with most environmental samples is extremely complex, poorly defined, and not precisely reproducible. Since the actual measurement may be influenced significantly by the nature of the matrix in

Table 4. Examples of specific chemical reactions employed in environmental analysis

Analyte	Method
Sulfur dioxide	West-Gaeke: Complexation with $HgCl_4^{--}$ and reaction with *p*-rosaniline
Hydrogen sulfide	Methylene blue: Collection as CdS and reaction with *p*-amino dimethylaniline
Chemical oxygen demand	COD: Oxidation with potassium dichromate
Cyanide	Liebig: Treatment with potassium iodide and ammonium sulfate and titration with silver nitrate
Phenols	Total phenols: Bromination and back titration with potassium iodide
Detergents	Methylene blue: Complexation with basic methylene blue and extraction into chloroform
Sulfate	Barium chloride: Precipitation as barium sulfate

Table 5. Radiochemical methods employed in environmental analysis

Method	Detection limit, ng	Precision, %	Applications
Instrumental activation analysis			
Thermal neutrons	0.001–10	2–3	Most elements of environmental interest except Be, Cd, Pb, P, S, and Tl Solid samples
Fast neutrons	1–100	2–3	O, Cl, Si Solid samples
Photons	0.005–10	2–3	As, Br, Ca, Ce, Cl, Cr, I, Na, Ni, Sb, Ti, Zn, Zr Solid samples

which the analyte is presented, there is a need to obtain calibrating standards which closely represent the real environmental matrix. Undoubtedly, the best procedure for achieving such matrix identity is to use the technique of standard addition whereby a known amount of the analyte in question is added to the actual environmental sample under study. Even here, however, difficulties are encountered since the chemical compounds in which the analyte is present (if it is an element) may not be known.

Data presentation. It is usually necessary to modify raw analytical results so as to present them in a form which can provide meaningful information to the user. Commonly, either tabular or graphical representation is used. In either case, however, it is essential to include both averages and ranges of multiple measurements and to estimate the precision associated with single measurements.

Sometimes useful insights into data sets can be obtained by modifying units or by utilizing three, or more, dimensional graphs. An example of three-dimensional graphical presentation is shown in Fig. 2, in which the concentration of zinc in an urban roadway dust is presented as a function of particle size and density.

A more sophisticated approach to data handling involves statistical manipulation of data. Such techniques as curve fitting and determination of statistical means, averages, and standard deviations are essentially routine. However, a more useful and productive approach, which is extremely appropriate to environmental systems, involves the utilization of multivariate statistics. This approach considers the interactive effects of a large number of variables which together determine the current status of a given system.

Multivariate statistics require that a large number of samples be selected at random in the area being considered and that this random selection should represent variations in all of the variables being considered. It is also generally important that the number of samples collected should exceed the number of variables within those samples which are being taken into account. While such multivariate statistical treatments cannot provide categorically definitive information, they do provide significant insights into the probable behavior and the controlling variables associated with the species of interest.

PRESENT STATUS

By way of a summary, a number of statements can be made about the present status of environmental analysis:

1. There is no single analytical method which can provide information about all of the chemical species of interest in the environment.

2. Environmental analysis is a developing field wherein the most well-developed capabilities are associated with specific determination of individual inorganic (and in a few cases, organic) species; with determination of the mass and elemental content of airborne particles; with determination of trace elements in water; and with the physical measurement of meteorological conditions.

3. The areas which are least well developed at this time include the specific identification of individual compounds in which inorganic elements are

Table 6. Instrumental methods which identify chemical compounds

Method	Detection limits, g	Applications
Mass spectrometry	$(1-100) \times 10^{-9}$	Identification of organic molecules on the basis of their characteristic fragmentation patterns and charge-to-mass ratios of the ions produced Solid, solution, and gaseous samples
Nuclear magnetic resonance spectroscopy	$(0.01-10) \times 10^{-3}$	Identification of organic, and some inorganic, molecules on the basis of their characteristic nuclear magnetic resonance spectra Solution samples
Infrared spectroscopy	$(0.01-1) \times 10^{-3}$	Identification of organic and inorganic molecules on the basis of their characteristic infrared spectra Solid, solution, and gaseous samples
X-ray powder diffraction	$(0.01-1) \times 10^{-3}$	Identification of crystalline compounds on the basis of their characteristic x-ray diffraction behavior Crystalline solid samples

Table 7. Atomic and molecular separation methods employed in environmental analysis

Method	Detection limits, ng	Applications
Mass spectrometry		
Organic mass spectrometry	1 – 100	Separation and identification of ionic molecular fragments of organic molecules for determination and identification of the parent compounds Solid, solution, and gaseous samples
Spark source mass spectrometry	0.01 – 0.1	Separation and identification of atomic ions derived from inorganic compounds for elemental analysis Solid and solution samples
Chromatography		
Gas-liquid and gas-solid	(Table 8)	Separation of volatile organic and inorganic compounds based on their vapor pressure differences between mobile and stationary phases Solution and gaseous samples
Liquid-liquid	0.001 – 10 (injected)	Separation of soluble organic compounds based on their solubility differences between two liquids of different polarity Solution samples
Liquid-solid	0.01 – 10 (injected)	Separation of soluble organic compounds based on the interactions of their substituent groups with the stationary phase Solution samples
Ion	0.001 – 10 (injected)	Separation of inorganic and organic anions (notably Cl^-, F^-, NO_3^-, SO_4^{--}), the alkali metal cations, and NH_4^+ based on interactions between the analyte and an ionized stationary phase Solution samples

Table 8. Methods employed for surface and microscopic analysis in environmental systems

Method	Applications
Microscopy	
Optical	Observation of particles and suspended material larger than about 1 μm
Scanning electron	Observation of particles and suspended material larger than about 10 nm
Transmission electron	Observation of particles and thin biological sections larger than about 10 nm
Electron microprobe	Elemental analysis of localized sample regions (>1 μm) to depths of 5 – 100 μm
Ion microprobe	Surface elemental analysis of localized sample regions (>1 μm); depth resolution ~1 molecular layer
Auger electron spectrometry	Surface elemental analysis of localized sample regions (>1 μm); depth resolution 1 – 2 nm; sensitivity similar to the electron microprobe
Electron spectroscopy for chemical analysis	Surface elemental and compound analysis of nonlocalized regions (>1 mm); depth resolution 1 – 2 nm; sensitivity similar to the electron microprobe
Ion scattering spectroscopy	Surface elemental analysis of localized regions (>1 μm); depth resolution 1 – 2 nm; sensitivity about 10 times better than the electron microprobe

present, and definition and identification of chemical species which are associated with particular regions of an environmentally important condensed phase. For example, determination of the chemical species present in the surface microlayer of ocean or inland waters is an area of important environmental concern which requires considerable development.

4. The future development of environmental analyses should proceed in the direction of obtaining a series of simple, portable devices capable of sensing and determining the amounts of a large number of environmental pollutants. Such a device is currently unrealistic. Nevertheless, there is a real need for low-cost specific and precise analytical instrumentation to determine the presence and amounts of environmental pollutants at the levels encountered. More realistically, areas which are currently receiving major emphasis and which should in the near future continue to do so include the development of analytical techniques capable of identifying the actual chemical compounds in which potentially toxic trace elements exist; improvement of techniques and methodology which can be used for surface analysis; and further development of capabilities of determining individual organic compounds and classes of compounds present at trace levels in complex systems.

[DAVID F. S. NATUSCH]

Bibliography: American Public Health Association, *Standard Methods for the Examination of Water and Waste Water*, 14th ed., 1975; T. R. Keyser et al., Characterizing the surfaces of environmental particles, *Environ. Sci. Technol.*, 12(7): 768 – 773, 1978; W. Liethe, *The Analysis of Organic Pollutants in Water and Waste Water*, 1973; S. E. Manahan, *Environmental Chemistry*, 3d ed., 1979; National Academy of Sciences, Committee on Biological Effects of Atmospheric Pollutants, *Particulate Polycyclic Organic Matter*, 1972; D. F. S. Natusch, C. F. Bauer, and A. Loh, Collection and analysis of trace elements in the atmosphere, in W. Strauss (ed.), *Air Pollution Control*, 1978; M. Pinta, *Detection and Determination of Trace Elements*, 1972; D. A. Skoog and D. M. West, *Principles of Instrumental Analysis*, 1971.

A–Z Environmental Science and Engineering

A–Z

Aeolian landforms

Topographic features generated by the wind. The most commonly seen aeolian landforms are sand dunes created by transportation and accumulation of windblown sand. Blankets of wind-deposited loess, consisting of fine-grained silt, are less obvious than dunes, but cover extensive areas in some parts of the world.

Movement of particles by wind. Although wind is not capable of moving large particles, it can transport large amounts of sand, silt, and clay, especially in desert or coastal areas where loose sand and finer particles are exposed at the surface as a result of the scarcity of vegetation or the continuous supply of loose material.

The size of particles which can be moved by wind depends primarily on wind velocity. Swirls and eddies associated with wind turbulence have upward components of movement which pick up loose material. The velocity of upward gusts is usually not very constant, but generally averages about 20% of mean wind velocity. Where turbulence is strong enough to overcome the force of gravity, the particles remain suspended in the air and are carried downwind. Because the maximum size of a particle suspended in the air varies with the square of its radius, wind is generally limited to the movement of material of sand size or smaller, and suspended particles are quite sensitive to changes in wind velocity. Thus, the wind is an effective winnowing agent, separating finer from coarser material, and grain size of aeolian deposits is typically quite uniform.

Windblown sand seldom rises more than a few feet above the ground. However, fine silt and dust may rise to altitudes of hundreds or thousands of feet during desert windstorms. Individual sand grains rise and fall as they are blown downwind and travel in a bouncing fashion known as saltation. Sand grains bouncing along the ground within a few feet of the surface abrade materials with which they collide. Such natural sand blasting produces polished, pitted, grooved, and faceted rocks known as ventifacts, and mutual abrasion of the sand results in highly rounded, spherical grains.

Silt- and clay-sized particles can be held in suspension much longer than sand grains, and thus may travel long distances before settling out.

Sand dunes. Where abundant loose sand is available for the wind to carry, sand dunes develop. As soon as enough sand accumulates in one place, it interferes with the movement of air and a wind shadow is produced which contributes to the shaping of the pile of sand. Sand grains bounce up the windward side of the sand pile until they reach the crest, then tumble down the lee side in the wind shadow behind the crest. Sand trapped in the wind shadow accumulates until the slope reaches

Fig. 1. Barchan dunes at Moses Lake in Washington.

the angle of repose for loose sand, where any additional increase in slope causes sliding of the sand and development of a slip face. Dunes advance downwind by erosion of sand on the windward side and redeposition on the slip face. Dunes may have a variety of shapes, depending on wind conditions, vegetation, and sand supply.

Barchan dunes. These are crescent-shaped forms in which the ends of the crescent point downwind and the steep slip face of the dune is concave downwind (Fig. 1). The crescent shape is maintained as the dune advances downwind because the rate of movement of sand is somewhat slower in the central part of the crescent where dune height is greatest. Barchans are commonly found on barren desert floors, where they may occur singly or in clusters.

Fig. 2. Star-shaped dune at Death Valley in California.

Parabolic dunes. Parabolic dunes also have crescent-shaped forms, except that the crescent faces the opposite direction and the steep slip face is on the convex rather than the concave side of the dune. They typically occur where vegetation impedes the advance of the points of the crescent, allowing the higher vegetation-free central part of the dune to move at a faster rate, leaving the ends of the dune trailing behind.

Transverse dunes. These are elongate forms whose long axes are at right angles to the prevailing wind direction. They frequently develop as a result of coalescence of other dune types.

Longitudinal dunes. These consist of long ridges parallel to the prevailing wind direction, often found where sand is blown through a gap in a high ridge. However, very large elongate dunes up to 700 ft (210 m) high and 50 mi (80 km) long also occur in regular rows in the desert regions of Africa and the Middle East.

Star-shaped dunes. Dunes of this type are found in some areas where the direction of prevailing winds shifts from season to season, piling up sand in forms with long radial arms extending from a central high point (Fig. 2).

Fine-grained deposits. The fine silt and clay winnowed out from coarser sand is often blown longer distances before coming to rest as a blanket of loess mantling the preexisting topography. Thick deposits of loess are most often found in regions downwind from glacial outwash plains or alluvial valleys such as the Mississippi Valley, southeastern Washington, and portions of Europe, China, and the Soviet Union.

[DON J. EASTERBROOK]

Bibliography: R. A. Bagnold, *The Physics of Blown Sand and Desert Dunes*, 1941; W. S. Cooper, *Coastal Sand Dunes of Oregon and Washington*, Geol. Soc. Amer. Mem. 72, 1958; D. J. Easterbrook, *Principles of Geomorphology*, pp. 288–303, 1969; J. A. Mabutt, *Desert Landforms*, 1977.

Aerobiology

The word aerobiology came into use in the 1930s as a collective term for studies of air spora—airborne fungus spores, pollen grains, and microorganisms. During the International Biological Program (IBP) from 1964 to 1974, the aerobiology component extended the term to include investigations of airborne materials of biological significance. The more common problems of aerobiology today are concerned with airborne plant and animal disease organisms, pollen, spores, algae, protozoa, and minute arthropods such as aphids and the smaller spiders; but in some instances pollution gases and particles that exert specific biological effects are included. The discipline of aerobiology is held together by common principles of atmospheric dispersion, some common biological qualities such as viability and source strength and qualities that permit these atmospheric transport events to be treated as similar ecological systems.

The 10-year term of the IBP afforded a wholly new set of opportunities for multinational and intercontinental studies that had been hampered previously by lack of direct communication between scientists and by absence of cooperation between governments. In 1974 an International Association for Aerobiology was founded at The Hague, and it is linked with a Commission on Aerobiology of the International Union of Biological Sciences (IUBS). Aerobiology has emerged as a distinct interdisciplinary field, with new capabilities, such as calculating the probabilities for spread of plant or animal diseases in various directions, or determining loads of airborne allergens in indoor environments, or tracing aerial pathways of minute organisms for various scientific and practical purposes.

Long-distance aerial transport. Long-range dispersal of many kinds of minute organisms and spores underscores the necessity for intercontinental coordination of certain aerobiology studies. The northward spread of black stem rust of wheat every year from Mexico and Texas to the Northern states and Canada was discovered by E. C. Stakman and colleagues in the 1920s. Similar great seasonal wind-driven infections of grain crops occur also in the plains of eastern Europe, the Soviet Union, and India. Evidence of such long-range dispersals is coming to light with increasing frequency. U. Hafsten in 1960 reported finding pollen grains in sediments on Tristan da Cunha and Gough Island—grains that could have been transported only from continental areas to those isolated tiny islands in the South Atlantic Ocean. In 1964 L. Maher showed that pollen of the desert shrub *Ephedra*, whose growth in North America is restricted to the Great Basin and the southwestern deserts, is blown as far east as the central Great Lakes region. Still more recently, N. Moar identified pollen grains of the Australian beefwood tree *(Casuarina)* in snow samples at 2100 m altitude on the Tasman Glacier in the Mount Cook region of New Zealand's South Island and from two other localities which also could receive wind drift only from the west. At one of these localities the pollen was associated with fine brick-red dust typical of Australian dust storms.

The improved quality and availability of meteorological data are making it possible to reconstruct the atmospheric trajectories that bring exotic biological matter to an area. During November and December 1967, R. C. Close found the grain aphid *(Macrosiphum miscanthi)* to be common in wheat crops throughout the Canterbury Plain of the South Island of New Zealand, but more particularly in coastal areas. The source was not local, as this aphid had been found only once during a trapping period from 1957 to 1965 and never in wheat crops of this region, although the crops are regularly surveyed for pests. Close, a phytopathologist, worked with A. I. Tomlinson of the New Zealand Meteorological Service to reconstruct the weather sequences of October and November 1967. The weather from Melbourne to Canterbury was found to provide excellent launching conditions and the proper trajectories for dispersal to the South Island from October 29 until November 3. Although circumstantial, the evidence is believed to validate dispersal of these small and delicate insects in living condition across the Tasman Sea.

Rothamsted Experimental Station in England has been the site of many fundamental investigations of fungus spore production and dispersal in the atmosphere, especially through the work of P. H. Gregory and J. M. Hirst. In recent years scientists there have examined the diurnal surges of spore release by different species of saprophytic and parasitic fungi associated with crops and grassland. Some of the daily surges of identifiable spores were traced downwind, eastward, onto the Continent by sampling from airplanes. Although they gradually disintegrate, these spore clouds could still be detected at 1000 m altitude after crossing the North Sea, provided they had not been washed out by rain.

It is possible, although not yet proved, that the outbreak of coffee leaf rust in Bahia, northeastern Brazil, in January 1970 was the result of long-range wind transport of the spores of the fungus from western Africa more than 2000 km distant. The wind-disseminated rust fungus *Hemileia vastatrix* is an endemic disease of the coffee trees of Ethiopia. Since the 15th century, coffee trees were introduced in other tropical montane regions, and it took considerable time for the disease to catch up to the new areas of coffee plantations. Ceylon (now Sri Lanka) was a major producer of coffee, most of it going to the British, until 1869 when the leaf rust appeared. In 5 years the plantations of Ceylon were ruined, and the disease had spread to India and Malaya, causing the collapse of many plantation-based companies and some banking institutions. Many of the plantations in Ceylon and elsewhere in southeastern Asia were replanted to tea, which is resistant to the rust fungus; this is said to have made Britons mostly tea drinkers. The local spread of the coffee rust is dependent upon raindrop splash to dislodge and launch the spores; but once airborne, the spores can be transported long distances in viable condition. How far they travel is the result of all the factors governing the transport of the moving air mass, and how long the spores remain viable is determined by temperature, humidity, and solar irradiation. Such information is supplied by cooperating agricultural meteorologists and biometeorologists.

Many organisms get rides on dust particles. For example, H. E. Schlichting, Jr., showed that the number of algae and protozoa in the air over north-central Texas usually ranges from 0 to 283 organisms per cubic meter, whereas during a dust storm the air contained 3000 organisms per cubic meter. Dusts from dry grasslands and desert areas are capable of carrying large numbers of minute organisms by this "airborne raft" process. Saharan dust has been identified by meteorological tracking and by mineral identification in the Caribbean region in recent years during the Sahel drought, but no assessment of the biological component has been attempted.

Predicting disease outbreaks. Aerial transport of disease spores and microorganisms and of insects is becoming predictable within certain limits. Studies of airborne fungus diseases of crop plants led the way. In the northern Plains states, spore trapping has been used for 25 years for forecasting the advent of wheat stem rust. In South Africa, spore trapping is an aid in predicting outbreaks of citrus black spot. The blister blight of tea, caused by the fungus *Exobasidium vexans*, has been known in Assam since 1868, but it suddenly ap-

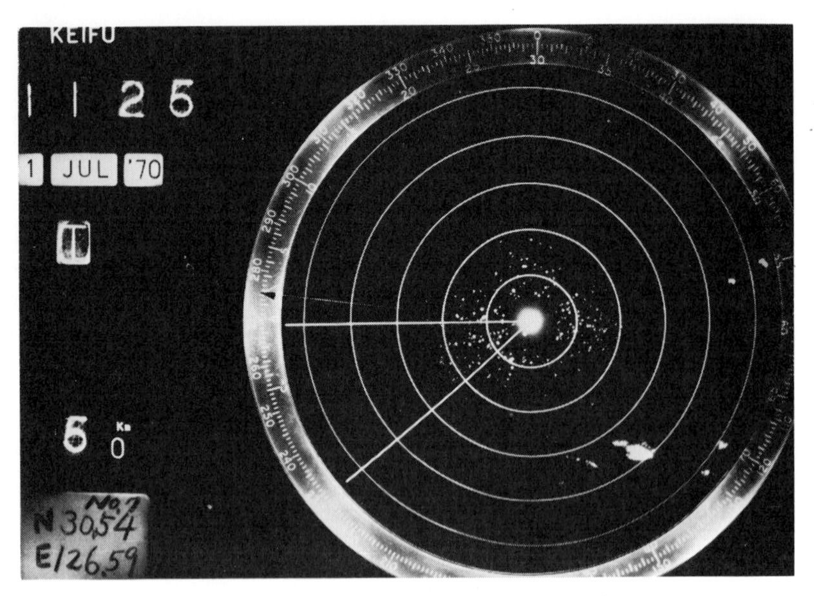

Fig. 1. Radar screen indicates swarms of plant hoppers as bright "angels" above the sea south of Kyushu, Japan, July 1, 1970. *(Courtesy of Eiichi Inoue, National Institute of Agricultural Sciences, Tokyo)*

peared in southern India and Sri Lanka in 1949. The blister blight was brought under control by use of copper fungicides, and now spore trapping is used to warn against attacks and to signal spraying of the fungicides.

Japan has an elaborate and effective organization of local agricultural observers reporting crop disease conditions to regional and national centers where remedial actions are planned and initiated. Japanese entomologists, phytopathologists, and meteorologists are also engaged in vigorous research on insect pests of crops, especially the sucking insects which transmit diseases from one plant to another and from one field to another. These scientists have amassed much evidence in support of the view that pheromones, volatile substances given off by organisms to the air, are used to guide these insects to the crop plants. Although these insects are weak fliers and tiny, not more than about 2 mm long, and must depend upon wind for travel over any appreciable distance, they can in some manner control where they fall out of the wind stream. Plant hoppers blown from mainland China across the East China Sea in early summer bring rice crop diseases that then spread northward from Kyushu to the other islands of the Japanese Archipelago. Rainstorms over the sea do not entirely stop these migrations -- plant hoppers can become airborne from the sea surface. The economic importance of this insect vector of rice diseases led Eiichi Inoue and a team of other meteorologists to conduct Operation Plant Hopper in 1973 as part of a synoptic study of the changing characteristics of continental air masses from the Asiatic mainland as they come over the East China Sea. The meteorologists used shipborne and aircraft radar to locate and track swarms of plant hoppers (Fig. 1). They trapped the insects from shipboard with kite-suspended nets, and from aircraft with towed nets (Fig. 2).

Fig. 2. Japanese aircraft tows a net to catch airborne plant hoppers over the sea south of Kyushu. *(Courtesy of Eiichi Inoue, National Institute of Agricultural Sciences, Tokyo)*

The Anti-Locust Control Centre in London is an example of an effective union between basic research and immediate application of findings. For several decades this center has directed a program of research and forecasting by fieldworkers in locust-plagued areas across Africa between the Sahara and the equatorial forest and through the Middle East into southwestern Asia. Every workday in the London laboratories, the staff gathers to review situation reports and to draft forecasts, much as an army general staff would gather at headquarters for intelligence briefings and battle directives. This organization has saved untold tonnages of crops, and thereby many human lives, through knowledge of the way locusts behave in response to different weather and crop conditions. Now the scientists are employing radar and other electronic remote-sensing devices to track swarms and to monitor positions of weather fronts, which strongly influence locust swarm movements. The center's methods will undoubtedly be emulated in approaches to problems of other diseases and pests.

Aeroallergens. Pollinosis (hay fever) affects an estimated 10% of the population of the United States, and is probably responsible for more discomfort and disabling illness than the direct effects of all forms of air pollution. In temperate North America, ragweed *(Ambrosia)* is a leading cause of allergic responses along with sage *(Artemisia)*, grasses, and certain wind-pollinated trees such as elm *(Ulmus)*. The biology and ecology of ragweed are relatively well known, so that it is now possible to eliminate the plant locally by land management practices. Spores of fungi, especially of common molds on dead vegetable matter and soils (such as *Cladosporium* and *Alternaria*), are receiving increased study as aeroallergens; but advances are slowed by difficulties in collecting measured samples of these small spores (mostly 4–15 μm in diameter) and in identifying the spores with the producing plant bodies. Allergic responses associated with walking through dry grass or digging in dry soil are in large part due to exposure to fungus spores. Assessments of weather and land-use conditions that affect atmospheric concentrations of aeroallergens are continuing. Such studies, exemplified by those of E. C. Ogden, G. Raynor, and J. V. Hayes in New York State and A. M. Solomon in New Jersey and Arizona, permit allergists to assist their pollinosis patients in planning for avoidance or protection.

In view of the ubiquity of microscopic algae, and of the aerial dispersal phase of many species, it is appropriate that these minute plants be investigated as allergens. Causal relations between algae and allergy were first demonstrated for direct exposure to algae-infested waters, but subsequent recognition of aerial transport has opened the possibility of microscopic algae as aeroallergens. Bursting bubbles from breaking waves on a pond surface or beach, or in a sewage treatment lagoon, are likely sites for launching minute algae into the atmosphere. A green alga, *Chlorella*, has been found to be common in trickling filter sewage disposal systems in Ohio, and H. E. Schlichting in 1969 reported collecting varying concentrations of microorganisms including algae in the atmosphere near such installations.

Although much has been learned about the air spora of the free atmosphere, especially near the ground, little has been done until very recently in studying the air of indoor environments other than in health care facilities and industrial and research "clean room" spaces. Penetrating microbiological studies of the air in hospitals and dentists' offices have provided new standards for building and ventilating designs and management policies, as indicated in publications of M. A. Chatigny, R. L. Dimmick, A. B. Akers, and others at the Naval Biological Laboratory in California. House dusts have been found to contain plant and animal debris, pollen, spores, and algae. I. L. Bernstein and R. S. Safferman found viable algae, notably *Chlorella* and *Chlorococcum*, in many samples of house dust from Los Angeles and Milwaukee, and from Caracas; both of these green algae can cause skin sensitivity and respiratory allergy for some persons. Air-conditioning and humidifying systems of buildings may harbor sources of dust and mold spores which may be causes of allergic reactions in some people. As indoor environments reflect many of the air-pollution characteristics of the ambient outdoor air, there are possibilities of synergistic, or reinforcing, effects of air pollutants with indoor allergens. Although investigations of the great variety of indoor environments have barely begun, these studies promise to bring considerable benefits in human health and comfort.

Current approaches. During the IBP, aerobiologists were exposed to concepts and methods of other disciplines, and innovative techniques of collateral sciences. Improved devices for sampling particles in the atmosphere, more precise identifications of the particles, and more coordinated reporting from stations are making possible earlier and more reliable predictions of disease and allergy outbreaks due to airborne agents. Japan, for example, has an elaborate system of local and regional stations for observing and reporting crop diseases, as noted above. Most human and animal disease microbes collected in air samples cannot be identified by simple microscopic examination, but must be cultured in special media and often under strict conditions of temperature. However, organizations such as the U.S. Center for Disease Control in Atlanta, and task groups of the World Health Organization, are able to join current information about the geographic distribution of a disease outbreak with meteorological conditions and knowledge of the etiology of the disease to produce progressively improved predictions of the outbreak's course. Within the confines of buildings, airborne diseases are being traced with considerable confidence once the etiology is known, as in recent instances of Legionnaires' disease.

Both analysis and prediction of the transmission and impact aspects of airborne diseases are greatly aided by the ecological systems approach, which was given considerable impetus during the IBP. This approach views a process, and in aerobiology this is some aspect of aerial transport, as a machine in which various parts function together, with certain feedback controls that regulate the speed of internal reactions and the kind and magnitude of responses to the external environment. The complexities and great numbers of iterative steps in tests of the mathematical models of these systems would scarcely be possible without the use of electronic computers. For example, in 1969 Paul E. Waggoner and J. G. Horsfall of the Connecticut Agricultural Experiment Station brought out a simulation model of a generalized plant disease of the kind that reaches epidemic proportions on row crops. They called this computer model EPIDEM and demonstrated how it could be used to predict the progress of a disease if appropriate values are entered for source strength of inoculum, meteorological variables such as wind speed and direction, and age and condition of the exposed crop. In response to two summers when the southern corn leaf blight threatened the maize crop of the Central states, Waggoner and his colleagues in 1972 brought out a modified version of the model called EPIMAY which was applied, but after the blight was subsiding due to natural causes. The excellent promise of simulation modeling, especially for prediction, has convinced many biologists that it will be a valuable management tool in the future. In view of the fact that cereal crops are so vulnerable to stem rust diseases that new resistant varieties must be bred and introduced every 10 to 12 years, the early detection and prediction of rust disease spread may play a vital role in future grain production.

The systems approach has also permitted improved methods for estimating crop losses. A group led by L. Calpouzos in the U.S. Aerobiology Program extended the reliability of estimating losses by taking into account different environments and geographical areas. The United Nations Food and Agriculture Organization is concerned with world reports on crop losses, but is in need of improved information from some parts of the world. Remote-sensing imagery by aircraft and satellite (such as Eros and Landsat) permits better measurements of crop areas and detection of some diseases and other crop losses. As human consumption of food approaches the limits of crop yields, the accurate prediction of crop losses from disease will become increasingly important.

Global and regional monitoring. Since the 1950s, citizens of industrialized countries have become increasingly aware of the burden of undesirable and toxic wastes in the atmosphere, waters, and soils of populated areas. Conifer forests have been decimated by air pollution in coastal California, eye-stinging smog often hangs over urban areas, and friezes on stone of centuries-old buildings dissolve in a few years under acidified rains. Measures to control amounts of sulfur oxides and nitrogen oxides released into the atmosphere by industry and motor vehicles have resulted in decreased pollution since about 1970. However, there are many air pollutants not yet under control, and many effects of those pollutants on human health and on other organisms are still unknown. Some subtle, but nonetheless significant, effects of gaseous pollutants are being discovered, such as the killing of some airborne fungus spores by sulfur dioxide at normal pollution levels in humid air. Such effects raise other and more serious questions. What does pollution do to the airborne pollen necessary for seed set in flowering plants? Where do the pollutants come to rest? How much can the world biogeochemical system stand before some irreversible change occurs? Monitoring of the atmosphere's pollutants was initiated primarily by governments to protect against direct and immediate damage to human health. Since about 1970, longer-term changes in the atmosphere have been recognized as resulting from human activity, some of these are seen to have potentially grave consequences for human life, and many have significant relationships with airborne biological particles.

Monitoring stations. With the International Council of Scientific Unions (ICSU) taking the scientific lead, international organizations are moving toward regional and global networks of stations to monitor such items as the ozone and carbon dioxide content of the atmosphere, atmospheric turbidity (an index to the concentration of microscopic particles and droplets in the air), the electric field gradient in the atmosphere, and fluorocarbon and organohalide residues from various industrial sources. Other kinds of monitoring stations make regular observations of fish, bird, and mammal populations, green plant productivity, dissolved salts in lakes and streams, and so on.

One of the meteorological stations, on Mauna Loa, Hawaii, has been making measurements of ozone, carbon dioxide, turbidity, and other factors longer than any other station (Fig. 3). Tests have been made of a suction-type spore trap and an insect trap at the Mauna Loa station, and a program of regular sampling is being considered. H. Nichols of the University of Colorado recently completed a year-long sampling of the pollen and spore fallout or "rain" with Tauber sedimentation traps in a zone across the Canadian Subarctic and Low Arctic.

Fig. 3. General view of the Mauna Loa Observatory of the National Oceanic and Atmospheric Administration (NOAA), a unique location for making observations of the chemistry and particulate matter of the atmosphere. (*Environmental Research Laboratories of NOAA*)

A leading palynological laboratory in Montpellier, southern France, has organized an International Group for Studies of Applied Aeropalynology with participation by scientists in Algeria, Spain, France, Switzerland, the Netherlands, and Sweden. Started in 1977, the stations employ a newly designed, sticky gauze filter 25 cm square which is faced into the wind by a vane. In fact, two filters are exposed simultaneously, side by side, so that one may be archived for future reference. Analyses of the air spora are performed in Montpellier and Stockholm, and the data are processed by computer and stored in Montpellier. The data are being used continuously by allergy clinics of medical institutes and hospitals in various European countries.

Plans are being laid by a group of Southern Hemisphere scientists working within the Scientific Committee on Antarctic Research (SCAR) to establish a series of stations for regular exposure of sedimentation traps for airborne biological materials on the margins of the Antarctic continent and on Subantarctic islands.

Testing for viability of spores and microbes may be feasible within a few years, but routine culturing by microbiological techniques of particles trapped from the atmosphere is probably many years away.

Particles as nuclei. Another reason for monitoring organic particles in the atmosphere is that these particles participate as nuclei for ice crystal and droplet formation. The particles range in size from a fraction of a micrometer, as with the condensed terpenes evaporated from plant foliage, to nearly 100 μm, as with fragments of ash from grass fires and forest fires. Gabor Vali and Russell Schnell at the University of Wyoming in the early 1970s discovered that ice nucleation requires less heat extraction around a nucleus of the bacterium *Pseudomonas syringae* than around many other living and dead nuclei tested. That bacterium is especially common on dead vegetation in the autumn, but the significance of the lessened heat extraction requirement is not understood. Dust storms, such as those of the High Plains of interior North America, the mistral down the Rhone Valley to the western Mediterranean, and the strong east winds across the Sahara-Sahel region of northern Africa, carry fragments of humus (rafts) upon which smaller organic particles and microorganisms ride. Sahel or Saharan dust has been identified mineralogically in fallout over Florida. Two instances of "red snow" in southeastern Sweden and southern Finland in 1969 were traced by pollen to the dry steppes north of the Black Sea, and meteorological reconstructions confirmed the conclusion. Thus, the long-distance dispersal of small particles, living and dead, in the atmosphere and the potential of those particles as nuclei for ice and water droplet formation have been established. Systematic observations in the pattern of monitoring schemes are needed to improve understanding of the processes and the consequences to weather, climate, biogeography, and human health.

Particle electrostatics. Recently attention has been drawn to the fact that airborne dusts, like wind-driven snow crystals, commonly carry a positive electrostatic charge, increasing the vertical electric field gradient in the atmosphere from the usual 50 volts per meter (or thereabouts) to several times that value (or even 100 times that value in the cold dry polar regions). The charged dust particles may induce a like charge on insulated or strongly dielectric stationary bodies, which would then repel the dust; or the charged dust particles will tend to be attracted to grounded (negative) bodies such as steel fences set in moist soil or wet soil surfaces. These effects are being studied especially in polar regions, where the electrostatic charges are consistently higher than in low latitudes. It would seem that, at least under dry or cold weather conditions (hot or cold deserts), the suspension of small particles in the lower atmosphere and the tendency to fall out onto the solid earth are influenced by their charges and the electrostatic environment. Electrostatic (Cottrell) precipitators work on this principle to remove fly ash and similar material from industrial chimney gases, or on a much smaller scale to remove pollen and house dust from the air in dwellings. Better understanding of the influence of electrostatics on aerobiological processes will give improved methods of controlling airborne particles. *See* AIR POLLUTION CONTROL.

Knowledge of dispersal of minute organisms, spores, and pollen grains in the atmosphere, over both short- and long-range trajectories, will serve many useful purposes. Benefits to human health and protection of food crops against disease are obvious. On a more theoretical level, biogeographers, for example, will be able to improve the probability calculations for gene flow in populations, or to assess quantitatively the migration ability of those species in which aerial transport is part of the life cycle. With such knowledge, aerobiological monitoring can become an important device in the protection and management of selected species and biotic communities. *See* AIR POLLUTION; PLANT DISEASE.

[WILLIAM S. BENNINGHOFF]

Bibliography: W. S. Benninghoff and A. S. Benninghoff, Airborne particles and electric fields near the ground in Antarctica, *Antarct. J. U.S.*, 13(4): 163–164, 1978; W. S. Benninghoff and R. L. Edmonds (eds.), *Proceedings of a Conference on Aerobiology Objectives in Atmospheric Monitoring*, US/IBP Aerobiol. Prog. Handb. no. 1, 1971; R. L. Dimmick and A. B. Akers (eds.), *An Introduction to Experimental Aerobiology*, 1969; R. L. Edmonds (ed.), *Aerobiology: The Ecological Systems Approach*, US/IBP Synthesis Ser. 10, 1979; R. B. Felch and G. L. Barger, *U.S. Weath. Bur. Week. Weath. Crop Bull.*, 58(43):13–17, 1971; J. F. P. Hers and K. C. Winkler (eds.), *Airborne Transmission and Airborne Infection*, 1973; E. C. Ogden et al., *Manual for Sampling Airborne Pollen*, 1974; R. W. Romig and L. Calpouzos, *Phytopathology*, 60:1801–1805, 1970; P. E. Waggoner and J. G. Horsfall, *Conn. Agr. Exp. Sta. Bull.*, no. 698, 1969; P. E. Waggoner, J. G. Horsfall, and R. J. Lukens, EPIMAY, *Conn. Agr. Exp. Sta. Bull.*, no. 729, 1972.

Agricultural chemistry

The science of chemical compositions and changes involved in the production, protection, and use of crops and livestock. As a basic science, it embraces in addition to test-tube chemistry all the life processes through which man obtains food and fiber for himself and feed for his animals. As an applied science or technology, it is directed toward control of those processes to increase yields, improve quality, and reduce costs. One important branch of it, chemurgy, is concerned chiefly with utilization of agricultural products as chemical raw materials.

Scope of field. The goals of agricultural chemistry are to expand man's understanding of the causes and effects of biochemical reactions related to plant and animal growth, to reveal opportunities for controlling those reactions, and to develop chemical products that will provide the desired assistance or control. So rapid has progress been that chemicalization of agriculture has come to be regarded as a 20th-century revolution. Augmenting the benefits of mechanization (a revolution begun in the mid-19th century and still under way), the chemical revolution has advanced farming much further in its transition from art to science.

Every scientific discipline that contributes to agricultural progress depends in some way on chemistry. Hence agricultural chemistry is not a distinct discipline, but a common thread that ties together genetics, physiology, microbiology, entomology, and numerous other sciences that impinge on agriculture. Chemical techniques help the geneticist to evolve hardier and more productive plant and animal strains; they enable the plant physiologist and animal nutritionist to determine the kinds and amounts of nutrients needed for optimum growth; they permit the soil scientist to determine a soil's ability to provide essential nutrients for the support of crops or livestock, and to prescribe chemical amendments where deficiencies exist. *See* FERTILIZER; SOIL.

Chemical materials developed to assist in the production of food, feed, and fiber include scores of herbicides, insecticides, fungicides, and other pesticides, plant growth regulators, fertilizers, and animal feed supplements. Chief among these groups from the commercial point of view are manufactured fertilizers, synthetic pesticides (including herbicides), and supplements for feeds. The latter include both nutritional supplements (for example, minerals) and medicinal compounds for the prevention or control of disease. *See* HERBICIDE; PESTICIDE.

Important chemicals. Chemical supplements for animal feeds may be added in amounts as small as a few grams or less per ton of feed, but the tremendous tonnage of processed feeds sold, coupled with the high unit value of some of the chemical supplements, makes this a large market.

Of increasing importance since their commercial introduction have been chemical regulators of plant growth. Besides herbicides (some of which kill plants through overstimulation rather than direct chemical necrosis), the plant growth regulators include chemicals used to thin fruit blossoms, to assist in fruit set, to defoliate plants as an aid to mechanical harvesting, to speed root development on plant cuttings, and to prevent unwanted growth, such as sprouting of potatoes in storage. *See* DEFOLIANT AND DESICCANT.

Striking effects on growth have been observed in plants treated with gibberellins. These compounds, virtually ignored for two decades after they were first isolated from diseased rice in Japan, attracted widespread research attention in the United States in 1956; first significant commercial use began in 1958. The gibberellins are produced commercially by fermentation in a process similar to that used to manufacture penicillin.

In the perennial battle with insect pests, chemicals that attract or repel insects are increasingly important weapons. Attractants (usually associated with the insect's sexual drive) may be used along with insecticides, attracting pests to poisoned bait to improve pesticidal effectiveness. Often highly specific, they are also useful in insect surveys; they attract specimens to strategically located traps, permitting reliable estimates of the extent and intensity of insect infestations.

Repellents have proved valuable, especially in the dairy industry. Milk production is increased when cows are protected from the annoyance of biting flies. Repellents also show promise as aids to weight gain in meat animals and as deterrents to the spread of insect-borne disease. If sufficiently selective, they may protect desirable insect species (bees, for instance) by repelling them from insecticide-treated orchards or fields.

Agricultural chemistry as a whole is constantly changing. It becomes more effective as the total store of knowledge is expanded. Synthetic chemicals alone, however, are not likely to solve all the problems man faces in satisfying his food and fiber needs. Indeed, many experts are coming to the view that the greatest hope for achieving maximum production and protection of crops and livestock lies in combining the best features of chemical, biological, and cultural approaches to farming. *See* AGRICULTURAL SCIENCE (ANIMAL); AGRICULTURAL SCIENCE (PLANT).

[RODNEY N. HADER]

Agricultural geography

Farming systems are an amalgam of climatic, geographical, historical, and cultural features. The evolution of these systems has led to the characterization of distinct types of agricultural activity, each with its own developmental trends that sometimes diminish or even erase differences between systems.

Agricultural systems involve one-third of the Earth's land area—one-third of which is tilled land (1.47×10^9 ha) and two-thirds of which is pasture (2.9×10^9 ha). Rarely, however, are more than $8.5-9 \times 10^8$ ha actually harvested in any year, due to floods, droughts, pests, and diseases. In addition, it is anticipated that 3×10^8 ha of tilled land will become unsuitable prior to the year 2000 due to ongoing soil erosion, salination, spread of waterborne diseases, and the mounting force of urbanization. Another loss involves rangeland and pasture, more than half of which are currently overgrazed and are evolving into steppe or desert.

Every society, however, is dependent directly or indirectly on agriculture for survival. On a global scale more people (half the working population) are engaged in agriculture than in any other economic activity. Without agriculture the type of culture with which most of the world's inhabitants are familiar could not exist.

ORIGIN OF AGRICULTURAL SYSTEMS

Most present-day agricultural systems can be traced back 10,000 years to the beginnings of agriculture: the first domestication of plants and animals. While today most cultivated crops and domesticated animals are distributed worldwide, they were internationalized fairly late in human history, mostly within the last 4 or 5 centuries. The Indians of the Western Hemisphere contributed several major crops to the world: corn, potato, manioc, beans, groundnut, and tomato. Wheat, barley, and rye originated in the Near East, rice and bananas in southern Asia, sugarcane in India, and sorghum in Africa.

A distinction should be made at this point between "seed agriculture" and "vegeculture." Seed agriculture refers to plant reproduction from seed. In seed agriculture, cereals dominate as staple food crops in all but a few parts of the world. Such crops are predominantly annuals and are harvested in one short period, necessitating storage between crops. Seed agriculture formed the basis of the major agricultural civilizations of history. There were three major centers of cereal domestication in the Old World: southwestern Asia, northern China, and southeastern Asia; and possibly two in the New World: southern Mexico and Peru.

Vegeculture refers to plant reproduction by vegetative propagation. The principal crops of vegeculture are tropical roots such as taro, manioc, yams, sweet potatoes, and arrowroot. In vegeculture, rhizomes are cut from the growing plant and individually planted. There are two distinct advantages of this system: First, there is less need to completely clear the natural vegetation, because if a mixture of roots is grown, crops may be harvested when needed; and second, storage is less imperative.

Vegeculture developed in the tropics, on the boundary between forest and grassland, in the Americas, Africa, and southeastern Asia. Over the years, it has contracted as seed agriculture expanded from the heartlands of these regions. The best-known tropical vegeculture is that of southeastern Asia and includes not only the mainland, but also the Malaysian archipelago, Assam (India), and southern China. Root and tree crops indigenous to this region include taro, the greater yam, breadfruit, sago palm, bamboo, coconuts, and bananas.

The Middle East was the origin of all domesticated food-producing animals except the water buffalo and the llama. In the 1530s European pioneers brought cattle to the Western Hemisphere, initially close to present-day Buenos Aires. This early trial failed, but a renewed effort was made in 1569 when horses were also introduced. Cattle reached the North American continent with the early European settlers in 1610, and sheep in 1633. Australia, today's leading sheep-producing nation, received its first sheep as late as 1797 (possibly 1788).

CLIMATIC FEATURES

Climate, as defined by temperature and precipitation patterns, is the major factor in the selection of specific crops to satisfy human needs. Agricultural geography tries to define these physical limits for crop cultivation. The limits of cultivation have persistently been extended by developing crop strains which are resistant to cold, drought, or excessive moisture. Where economic forces or incentives are sufficiently strong, adverse climatic conditions may be overcome by irrigation, pest control, and other means. Production is basically determined by the vegetation period, mostly the frost-free period. In warmer latitudes this period is extended, allowing for multicropping.

The environment as shaped through climate is the basis for each kind of agriculture as it has evolved in various regions of the world. Soils have been shaped by topography and the composition of parent materials in an intricate interplay with climate. Soils have also been affected by the vegetation cover. This pattern has been the operational basis for agricultural pursuits. With time, agricultural systems evolved which were profoundly modified by plants and soils to increase production, originally with the double aim of feeding humans and livestock; more recently these systems have become increasingly specialized for the production of crops or the raising of livestock. Specialization itself has evolved into monocultures, as well as into mass raising of a single category of livestock. *See* AGRICULTURAL METEOROLOGY.

SYSTEM CLASSIFICATION

The most widely accepted classification of agricultural systems dates back to the 1930s when D. Whittlesey recognized two main factors as basic: environmental conditions (such as climate, soil, and slope) and human and social factors (such as population density, technological advances, and traditional practices). These factors interact to create circumstances that provide the functional basis for every type of agriculture.

There are several areas of the world in which the land is totally unsuitable for agricultural use. Such environments, termed restrictive or negative, include high mountains, with rugged surfaces, shallow stony soils, restricted flat land, and inhospitable climates; deserts, where water for irrigation is nonexistent; icecaps; and tundras, with permafrost and poor drainage.

Thirteen distinct types of agricultural activity were originally identified. Later developments somewhat modified this early classification, and also reduced the validity of distinctive features. This review identifies 11 major types: (1) subsistence tillage – shifting cultivation, transition forms, and intensive variants; (2) subsistence crop and livestock farming; (3) wet rice cultivation; (4) pastoral nomadism; (5) Mediterranean agriculture; (6) mixed farming (crop and livestock farming); (7) dairying; (8) plantation agriculture; (9) specialized horticulture and truck farming; (10) large-scale grain production; and (11) livestock ranching.

Subsistence tillage: shifting cultivation. Shifting cultivation is mostly practiced in isolated areas in the tropics. Three major subtypes may be identified.

The least advanced type usually consists of hunting-gathering tribes that supplement their major sources of livelihood with primitive agricultural pursuits. Few if any tools are employed, and the ground is not plowed. Wherever there is a natural clearing in the forest, seeds are placed in holes dug by hand or stick. The growers do not till or care for their crops and return only for harvesting. Due to pests and diseases, crop yields are low.

In a slightly more advanced type of shifting cultivation, the tribes occupy semipermanent dwellings for about 3 to 4 years, and agricultural activities occur in the immediate vicinity of the settlement.

In the third and most common type of shifting agriculture, the tribe stays in the same area for periods of up to 20 to 30 years. The forest is cleared in patches with one clearing used at a time. Within a year or two, however, weeds and pests become serious or fertility declines, at which time the land is abandoned and a new clearing is made. After several clearings are made and their soil resources depleted, the dwellings and fields are abandoned and a new area is sought. This category of subsistence growers numbers approximately 54×10^6, with the approximate number (in millions of tillers) per region as follows: Africa, 30; Indonesia, 10; Southeast Asia, 5.5; India, 5; South America, 1.1; Madagascar, 1; New Guinea, 0.5; Central America, 0.25; and the Philippines, 0.2.

As a rule only one or two vegetable crops are planted per family in the wet tropics, although a wide variety of tree crops (coconuts, breadfruit, oil palm, mangos, avocados, bananas, plantains, cacao, and guavas) and vegetable crops (yams, sweet potatoes, pumpkins, beans, and peanuts) are encountered. Some rice and corn are grown in a few places when introduced from outside. Here and there sugarcane and millet are raised. Bread is made from cassava or manioc, which frequently is a staple in the diet. Bananas, yams, and sweet potatoes provide abundant starch; coconuts and oil palms supply needed fats.

Transitional forms. Throughout areas inhabited by hunter-gatherers practicing shifting cultivation, some changes have gradually taken place, evolving into rudimentary and sedentary tillage systems. The main driving forces for change are a lowered standard of living resulting in a switch to other forms of resources; migration to other areas; and changes in socioeconomic activities.

Production of foodstuffs and other plants solely for local use takes many forms in terms of products (rice, tubers, fruits), techniques (hand tools, irrigation, use of work animals), intensity (settled agriculture, minor gardening), organization (ownership of land, financing, work methods), and other variables. Native communities often supplement their subsistence cropping to some extent with marketable products. For example, growing rubber or pepper or making copra from coconuts provides communities with money to help fill minimum needs for imported goods.

Much of the sedentary subsistence farming in Latin America is still characterized by extensive employment of human labor, little use of machinery, forest destruction, and soil depletion or erosion. Three or four crops, such as corn, beans, rice, manioc, and even cotton, are planted and harvested with a hoe or stick. After the soil is depleted, grasses are planted to be grazed until scubby trees choke them out, or the land is simply abandoned for a newly cleared plot.

In African sedentary subsistence farming, the cultivated crops of the wet-dry savanna lands are mainly shrubs such as the cotton bush, small plants such as tobacco and peanut, or domesticated grasses such as maize and millet. Other primary crops include corn, wheat, barley, rice, sorghum, sweet potatoes, and various types of beans which are diet staples. Secondary crops commonly grown in the savanna lands are manioc, sesame, artichokes, peppers, tomatoes, and pumpkins.

Intensive variants. India and China are classic examples of intensively cultivated, densely populated lands with the bulk of their labor force engaged in subsistence agriculture. Although cultivated intensively, much of interior India, dry inland portions of southeastern Asia, and regions of China north of the Yangtze River differ in the nature and methods of crop production. Rice is dominant where water is supplied either through irrigation systems or by natural rainfall and flooding. Other grains become the staples of diet where precipitation is inadequate or undependable and irrigation is too costly.

In India, most cultivators remain involved in subsistence agriculture. The land tenure system provides little incentive to the cultivator. However, recent land reform, agricultural extension services, better seed, and the rapidly expanding use of fertilizer have resulted in an increase in food production which is slightly ahead of the increase in population. However, there are still extensive problems and the population pressure on the land is getting worse. There are an estimated 45×10^6 landless agricultural laborers in India; crop failures in various sections have been a recurrent feature; and approximately four-fifths of the cultivatable land area is dependent on the rainy seasons or monsoons.

The major subsistence grains cultivated in the drier parts of India are sorghum, millet, and wheat. Sorghum is widely distributed over the Deccan Plateau but is most heavily concentrated on the black soils in the central part of the Indian peninsula. Millet is the major grain in the extreme northwest and Rajasthan portions of India bordering Pakistan. Wheat is grown mainly in northwestern and central India. Secondary crops include gram or chickpea, grown from the Punjab plains eastward to the delta region; peanuts, which are highly concentrated in the southern half of the peninsula; and cotton, produced mostly in the northwestern and central parts of the Deccan Plateau.

A clear-cut separation exists in China between the areas north and south of the Yangtze Valley. Rice predominates to the south in a climate which has a more dependable rainfall and is moist and warm throughout most of the year. To the north is a region of fertile but often barren plains and bare mountains, except for a short-lived green revival where the Hwang River approaches the sea. Like the southern region, there is much rich soil, but rainfall is infrequent. Prolonged droughts may parch the soil and burn crops out before harvest. Sudden outbursts of rain produce devastating floods.

Despite this climate, northern China is a highly productive region. The area near the coast, the famous North China Plain, is one of the most densely populated and most intensively cultivated parts of the world. This is the winter wheat–kaoliang region and it comprises most of the provinces of Hopei, Shantung, and Honan. Due to a high percentage of usable land (60% or more under cultivation), this region produces a wide variety of crops without irrigation. Fortunately, the maximum rainfall of 24 in. (61 cm) occurs during the hot summer. Winter wheat is the principal crop and occupies almost half the cultivated land, while barley and peas are grown in a few districts. Kaoliang, a type of millet, is the chief summer crop, with soybeans, cotton, corn, and sweet potatoes also covering a considerable acreage. Persimmons and hard pears are key fruits; peanuts are grown in several districts, especially in Shantung and Honan; and tobacco in a few localities.

To the west of this region are the Loess Highlands, an area of windblown silt deposits 300 or more feet (90 meters) thick in places, with fertile soils, steep slopes, and marginal rainfall. More than one-third of the cropland is terraced to check erosion of the steep hillsides. The best agricultural districts are situated along the Hwang River in southern Shansi and northern Honan, extending west into Shensi. Winter wheat, the major crop, is confined to the plains and valleys; other winter crops are rapeseed, peas, and barley. Millet is the most widely grown summer crop, followed by corn, soybeans, sesame, and buckwheat; cotton is important in some of the more productive agricultural districts.

The spring wheat region forms a fringe along the Mongolian frontier, lying on either side of the Great Wall. The percentage of cultivable land is small, owing to the cold climate and limited rainfall. Spring wheat is the principal crop, interspersed with millet, oats, peas, flax, and buckwheat.

Subsistence crop and livestock farming. Subsistence crop and livestock farming occupies large tracts of semidesert, hill, or mountainous lands in the Anatolian and Iranian plateaus in southwestern Asia, portions of Soviet Siberia bordering the Mongolian People's Republic, and small segments in Soviet Europe north of the 55th parallel. The Carpathian and Rhodope mountain areas of east-central and southeastern Europe also have small tracts of land that are still classified as subsistence crop and livestock farming areas. The only area of corresponding size outside the Eurasian realm is the Mexican plateau.

Few people are engaged in this type of agriculture. In the past, most of them followed a nomadic way of life, but inroads are constantly being made on the occupied lands through government intervention, improvement in agricultural techniques, and resettlement programs. The Kurds of eastern Turkey and Armenia, the Buryats of southeastern Siberia, and the Yakuts north and west of Lake Baikal in the Soviet Union are former nomadic herders who have become sedentary yet still dependent on livestock farming.

For the most part, agriculture and animal husbandry tend to split into separate occupations. The mixed type of farming characteristic of many parts of the world is practiced only on a minor scale, and many villages maintain one or more shepherds to look after a common herd, while the great majority of owners devote themselves entirely to crop cultivation. All tillable land is devoted to growing cereals, and vegetables and fruits are cultivated wherever possible.

Herding tends to be restricted to less favorable regions. People who still engage in this endeavor have become seminomadic villagers who move their flocks in summer to hillsides or mountain slopes where the animals graze until cold weather threatens.

Wet rice cultivation. Rice is the predominant crop in the dry and cool monsoon regions. The staple food of well over half the world's people, rice is so prevalent that an agricultural system has been developed around it.

The task of clearing and maintaining rice fields, which are small (often averaging less than 2 acres or 0.8 ha per farm), is relatively easy. The young trees and dry grass are burned near the close of the dry season and the ground is hoed. Prior to monsoon rains, the fields are tilled and flooded with water. The rice is then planted and the fields remain flooded until a few days before harvest.

The soils are generally alluvial in nature and mainly hold key drainage basins, such as those of the Irrawaddy, Salween, Menam, Mekong, Red, Yangtze, and Si river basins. These soils are naturally fertile, but owing to intensive tillage they require large amounts of fertilizing from animal manure and night soil (human excrement). The latter is by far the most widely used fertilizer in southeastern Asia, where the population is so dense that animals are relatively few.

Rice cultivation of this kind covers much of Sri Lanka, southern India, Bangladesh, the delta of mainland southeastern Asia, Java, Sumatra, the Philippines, and southern China and Japan. The total area under wet rice cultivation exceeds 10^8 ha.

As rice exists in so many varieties, it can be grown under widely varying conditions, from brackish or even saline soils to deeply flooded deltas like those of Thailand and southern Vietnam. In many places, given enough water, rice can supply two crops from the same land each year. It can give worthwhile yields on the same land year after year for generations without fertilizer, although yields may be greatly increased by the judicious use of manure. Dry crops under similar conditions in the tropics tend to give declining yields that often stabilize at an uneconomic level. *See* RICE.

Pastoral nomadism. The principal regions of pastoral nomadism are in the arid lands of the Old World, extending from northern Africa through Saudi Arabia to inner Asia. This type of highly specialized livelihood involves frequent movement of livestock in response to the need for grazing lands and water. Settlement is generally characterized by small clusters of homesteads temporarily established near a common watering point and pasture. Subsistence is maintained almost entirely by domestic stock, such as cattle, camels, goats, and sheep.

In many areas, nomadism is a product of cultural associations that may be as dominant a factor in the activity as the physical landscape. Many Moslem Arabs prefer nomadism to a sedentary way of life because of long tradition. In other areas, without the benefits of modern technology and capital investment, nomadism is the only economic system that can survive.

There has been a conscious effort on the part of some governments to make nomads sedentary. Egypt has attempted this with moderate success. Iran and Iraq have taken measures to permanently settle more than 3,000,000 nomads. Somali, Israel, Tanzania, and the Soviet Union also have experimented with settling nomads.

Mediterranean agriculture. This refers to a diverse crop-livestock economic system that is practiced most extensively around the Mediterranean Sea and in four other non-Mediterranean areas with similar climatic patterns: southern California, central Chile, southwest of Capetown in South Africa, and southern parts of South Australia and the area around Spencer Gulf. From a geographic standpoint, this system is the most clearly defined. The climate is characterized by mild, wet winters and hot, dry summers. Unless crops are irrigated, they must either be sown in the autumn and harvested by early summer or be drought-resistant.

Mediterranean agriculture has many distinctive features. Winter grains, such as wheat and barley, and all-year crops, such as olives, grapes, and carob, are grown without irrigation. All-year or summer crops that are generally irrigated include oranges, lemons, deciduous fruits, corn, rice, and vegetables. Livestock, mainly sheep, goats, and some cattle, are brought to the highlands for grazing in summer and kept on the lowland plains in winter. This transhumance is less significant today but still is practiced in parts of Greece, Italy, Yugoslavia, Spain, and North Africa.

Both subsistence and cash crops are important in the economy of every region of Mediterranean agriculture, although not on every farm. The relative emphasis on the several products varies with the amount of rainfall. Thus, northern Africa produces more barley and goatskins, and southern Europe more wheat and sheepskins. Citrus and vine crops have long been important in world trade (Table 1). Southern France, Italy, California, and

Table 1. Global production of grapes and citrus fruits in top producers, in 10^6 metric tons*

Producer	1961–1965	1973–1974	Percent change
Grapes			
France	9.6	13.1	+ 36.5
Italy	9.8	11.3	+ 15.2
Spain	4.2	6.2	+ 47.5
Soviet Union	2.8	4.7	+ 67.5
Turkey	3.1	3.2	+ 3.2
United States	3.3	3.8	+ 15.1
Argentina	2.4	3.1	+ 29.5
World total	50.6	63.2	+ 25.0
Oranges			
United States	4.5	8.9	+ 19.8
Brazil	2.0	3.2	+ 60.0
Spain	1.6	2.0	+ 25.0
Mexico	1.2	2.0	+ 66.5
Italy	0.9	1.5	+ 66.8
Israel	0.6	1.2	+100.0
World total	17.5	28.5	+ 63.2
Lemons and limes			
Italy	0.55	0.84	+ 52.5
United States	0.55	0.72	+ 31.0
India	0.44	0.45	+ 2.3
Argentina	0.08	0.28	+250.0
Mexico	0.17	0.23	+ 35.3
Spain	0.11	0.24	+118.0
World total	2.8	4.1	+ 46.5

*From *FAO Production Yearbook*, 1974.

Chile are important wine producers; California and Spain stress oranges.

Despite the importance of agriculture, only about one-fifth of the land in the Mediterranean climatic regions is intensively cultivated. Due to the restricted rainfall, no less than 40% of the grainlands are kept in fallow each year. Much land is too steep, dry, or rocky. Due to these constraints, a definite pattern of land use has evolved over the centuries: small fruits and vegetables occupy the lowlands; wheat and barley are grown in the more arid portions of the lower slopes; and tree and vine crops dominate the upper slopes. Frequently, interculture is practiced to make the most of the usable land.

Livestock grazing is based primarily on natural vegetation. Generally, cattle tend to predominate in the wetter areas; sheep, which far outnumber other domestic animals, dominate vast areas of the arid zone; and goats predominate in the driest, more rocky areas. During the wet winter season, the lush green lowlands with mild temperatures are ideal for grazing livestock. In summer, when the lowland pastures are dry and dormant, the cool grassy upper slopes are utilized.

Livestock farming in Mediterranean agriculture is based on irrigated meadows planted in alfalfa, hay, clover, sown grasses, or other forage crops. The animals are stall-fed. Meat, milk, and cheese are produced both for local consumption and export in northern Algeria, northern Italy, southern France, and southern and eastern Spain. Dairying is especially important near the larger cities. In other areas of the Mediterranean Basin, such as southern Italy, central Spain, northern Greece, Yugoslavia, Turkey, and Syria, subsistence livestock farming prevails. Large-scale commercial livestock farming is important in California. Irrigated pastures and imported grain support huge dairies and meat-packing plants near Los Angeles and San Francisco. The San Joaquin Valley is a major milk-producing center.

Crop and livestock farming. This mode of agriculture, also called mixed farming, is based on a combination of crop cultivation and animal husbandry. It is extensive and occupies the humid middle latitudes of all continents. The monetary return is generally less than that for commercial dairy farming or specialized agriculture. Periodic rather than daily shipment to market is the rule. Farm population density is high, and large urban centers are dispersed within the mixed farming belts. Smaller cities, towns, and villages dot the landscape, providing market outlets and services.

China and India are top-ranking in total number of livestock, but the Soviet Union and the United States lead in commercial production. The two largest commercial livestock and crop farming regions are in the United States and Eurasia. In the United States, this activity is prevalent in the Midwest (Ohio, Indiana, Illinois, Iowa, and Nebraska), the South (Virginia, Tennessee, and Georgia), and Southwest (Oklahoma and much of Texas). In Eurasia, this system is encountered in much of northern Portugal and Spain, the Po River plain of Italy, a large portion of the Danube basin countries of Hungary, Yugoslavia, and western Romania, and the Amur Valley in the Soviet Union. Minor regions are found in Argentina, southeastern Brazil, south-central Chile, and South Africa.

The commercial crop and livestock farming system concentrates on breeding and plant selection and a well-established crop rotation in which legumes and hay play an important part in proper soil management. Some farmers produce grains such as wheat, corn, or soybeans for sale as a cash crop; others produce only corn as feed for livestock. Farm animals use roughage, which may be produced in any mixed farming system. On a typical midwestern farm with a rotation of corn, oats, and clover, approximately 1.75 tons of roughage are produced for every ton of grain (1.95 metric tons for every metric ton). In Eurasia, root crops planted in rotation with grain crops in order to maintain soil fertility are fed to farm animals. Hay is a substitute for corn in most of Eurasia; corn is an important feed crop only in the Danube Valley. Elsewhere potatoes, turnips, sugarbeets, and oats are the major feed crops.

Farms in Eurasia are smaller and less mechanized than their United States couterparts, with higher production per acre but lower output per worker. Average livestock-grain operations in the United States range from 120 to 200 acres (50 to 80 ha); in South America and South Africa, they are about the same size; and in Eurasia, about half as large.

Livestock and crop production are complementary activities. They make possible a more even seasonal distribution of labor. Cattle raising plays an important role in the intensive agriculture of northwestern Europe and the American Midwest.

Dairying. Commercial dairy farming is profitable only where the products can be sold to an urban market, and is generally located near densely populated industrial areas. Fresh milk cannot be shipped distances farther than 12 hr away and requires refrigerated tank cars; some milk is presently flown into major cities, such as New York. Refrigerated butter can be shipped great distances. Cheese will keep for up to 3 or 4 years, depending on the type.

Dairying is an intensive form of agriculture. It is elaborately mechanized, and the capital investment in buildings and equipment on high-grade dairy farms exceeds that in any other type of agriculture. The labor requirement for dairying is exceeded only in few other agricultural activities.

Table 2. Global production of cow milk in top producers, in 10^6 metric tons*

Producer	1961–1965		1973–1974		Percent change
Soviet Union	63.8		89.8		+40.1
United States	57.0		52.3		− 8.2
France	25.1		29.1		+11.6
India (includes buffalo)	19.6		24.1		+12.3
West Germany	20.6		21.4		+ 3.5
Poland	12.9		16.6		+28.8
United Kingdom	12.0		14.3		+19.1
Canada	8.4		7.6		− 9.5
Brazil	5.9		7.3		+23.7
World total	324.4		382.2		+11.8
Europe	135.7		158.9		+11.7
North America	65.4		60.0		− 8.3
		%		%	
Satisfied world	282.6	87	329.8	86	
Hungry world	41.8	13	52.4	14	

*From *FAO Production Yearbook*, 1974.

Every step in feeding, milking, and processing is critical since dairy cattle must be milked twice a day to produce their maximum capacity.

The Soviet Union has become the world's leader in producing milk, followed by the United States (Table 2). The major European producers of dairy products are France, West Germany, Poland, the United Kingdom, Italy, the Netherlands, East Germany, Czechoslovakia, and Denmark. The chief dairy regions of the Soviet Union are in the Baltic countries. In North America, the north-central United States and adjacent parts of Canada, in particular the state of Quebec, contribute most of the milk. Production in Australia and New Zealand is about equally divided.

Plantation agriculture. Plantation agriculture is usually a corporate enterprise that organizes the production of valuable commercial crops, such as cotton, rubber, rice, sugarcane, sugarbeets, pineapple, bananas, coconut, tea, cacao, and citrus fruits. This system is not characterized by a predominant crop, but by the manner in which crops are produced and the land is managed. Plantation agriculture is not strictly a phenomenon of the tropics or subtropics; the system is employed in a wide range of extratropical regions. The type of plantation ranges from those producing bananas or rubber in the wet tropics to state and collective enterprises in the Soviet Union, China, and Cuba.

The tropical plantation system initially evolved to meet the world demand for certain staple crops. As part of the colonial era, it rested on European acquisition of land areas that would make economically attractive units. Manual labor was held cheap, so that the product could be sold at a price that would ensure large and growing volumes. These conditions no longer prevail. Some plantation crops are grown and even harvested by small individually owned farms, but the output is channeled into major marketing enterprises; examples are cacao and peanuts in Nigeria and Senegal, rubber in Amazonas and Malaysia, bananas in Central America, and tea in Ceylon. Monoculture no longer prevails. Crop diversification has taken place in order to minimize crop failures, hurricane destruction, or attacks by pests and diseases. For example, rubber and bananas are produced in tandem on plantations in British Honduras; cacao and oil palms in other parts of Central America; and sugarcane, citrus fruits, and cattle in Florida.

The three most important commerical tropical plantation crops are bananas, sugarcane, and rubber. Brazil leads the world in production of bananas, but Ecuador is the major exporter. While India is the largest producer of sugarcane, the bulk of it is consumed domestically. Cuba, once the major producer of sugarcane, has reduced its acreage because of economic pressures and loss of the United States market. Malaysia continues to dominate the production of natural rubber.

Specialized crop production on a much smaller scale is encountered in some dry regions where irrigation is employed. Cotton is grown in this manner in parts of Central Asia, southern Kazakhstan and Transcaucasus (Soviet Union), the coastal oases of Peru, sections of the lower Colorado Basin, and the Argentine Chaco. In the same manner sugarbeets are produced in the Platte River and Salt Lake oases of the western United States, and sugarcane in coastal Peru and northern Argentina.

Specialized horticulture and truck farming. Early types of commercial fruit and vegetable production hinged on either closeness to city markets or a rapid transportation system in order to meet the demand for perishables. Until about 1900, such gardening was mainly restricted to an area lying within 10–15 mi (16–24 km) of a city, but since then production has been extended to distances of several hundred miles. The increased use of trucks—mostly refrigerated—and the development of improved roads have been important factors in this widened range. The development of such specialized horticulture is most clearly evident in parts of the Po Valley (northern Italy), the Rhone Valley of southern France, Sicily, and the Campania—often a result of railroads and highways linking production areas and large commerce centers.

Market gardening, a very intensive form of cultivation, is carried out on land adjacent to cities. Such production close to the market benefits from lower transportation costs and less packing, compared with production at greater distances. The market gardener supplies principally in-season vegetables; out-of-season vegetables may be raised in hothouses. Most major cities have outlying market garden areas. For example, Long Island and northern New Jersey serve the New York City region. London receives a large portion of its vegetables from the nearby counties of Middlesex and Bedfordshire and from the Isle of Ely. All of England receives early potatoes from the Channel Islands. East and West Ridings of Yorkshire are important sources of vegetables for the northern sections of Great Britain. In the Netherlands vegetables are grown over such wide areas that it would be difficult to define those supplying each city.

Truck farming is characterized by high specialization, mostly a single crop, applies less intensive methods, and as a rule is located further from the markets. Each producer concentrates on a crop particularly adapted to the soil conditions and seasonal climate of the region. In each trucking region, crops are grown at a time of year when there is little competition with crops grown in other sections. For instance, vegetables from Mexico and the West Indies reach the eastern cities of the United States first, and then crops are successively received from Florida, Georgia, South and North Carolina, and as far north as Canada.

The most important trucking sections of the United States are the Atlantic Coastal Plain from southern New Jersey to Florida; the Gulf Coastal Plain from Alabama to Texas; California; and certain north-central states. Important trucking areas are also found in Algeria, Tunisia, Egypt, and the French Riviera. Other such areas are located near the major cities of South Africa and southwest of Australia.

Large-scale grain production. Commercial grain farms lie on the border between humid and semiarid climates, where summers are short and winters cold. Most of these regions are deep in the interior of continental land masses. Wheat is the dominant grain crop, being well adapted to the conditions of scarce labor and extensive cultivation on vast tracts of land, distant from markets. These factors explain the importance of wheat in

Table 3. Average 1973–1974 production of main grain producers, in 10^6 metric tons

Producer	Amount
Cereals	
World total	1,345.9
China	227.1
United States	220.9
Soviet Union	200.3
India	113.3
France	41.4
Asia	515.3
North America	255.2
Europe	229.7
Latin America	75.4
Africa	61.5
Australia and New Zealand	17.8
Wheat	
World total	368.8
Soviet Union	96.8
United States	47.6
China	36.5
India	23.4
France	18.4
Canada	15.3
Asia	89.5
Europe	86.4
North America	63.0
Rice	
World total	323.9
China	113.6
India	63.6
Indonesia	22.3
Bangladesh	18.3
Japan	15.8
Thailand	14.0
United States	4.7
Asia	296.0
Latin America	11.8
Corn	
World total	301.7
United States	130.8
China	30.7
Brazil	15.1
Soviet Union	12.7
Argentina	9.8
France	9.8
North America	133.5
Asia	49.9
Europe	44.9
Latin America	38.2
Africa	22.3
Barley	
World total	170.1
Soviet Union	54.6
China	20.3
France	10.4
Canada	9.4
Great Britain	9.0
West Germany	6.8
Denmark	5.7
Europe	58.7
Asia	30.8
North America	17.4

the North American Great Plains region (United States and Canada), Argentina, Australia, the Ukraine, and Kazakhstan.

Three technological developments which made possible the cultivation of the vast grasslands of the world's chief wheat-growing regions are transportation facilities (such as railroads, highway trucks, and barges) for handling the wheat; agricultural machinery for cultivating large tracts of land; and well-drilling machinery to make water available in semiarid regions.

The major regions of commercial grain farming are concentrated in the middle latitudes of the Northern and Southern hemispheres, between 30 and 55°. Eurasia's commercial grain area stretches from west to east for some 2000 mi (3200 km) and north to south for 700 mi (1120 km). It includes all of the Ukraine except its extreme southeastern corner, the Crimea; much of the northern Caucasus Mountains and the irrigated regions of Central Asia and Transcaucasia; the middle and lower Volga Basin; the southern Ural Mountains; western Siberia; and Kazakhstan.

The Soviet Union is by far the world's leading wheat producer—surpassing the United States in most years by 60–100%. Canada's wheat production declined drastically from 1963 to 1972, while India showed the greatest percentage of gain (Table 3).

Livestock ranching. In general, livestock ranching involves extensive grazing by cattle or sheep. Ranching is practiced in the humid margins of steppelands in the middle latitudes and the savanna areas of the tropics, tending to occupy regions where land values are low and population is sparse. This type of economic activity is characterized by the use of relatively large land areas, as opposed to smaller areas for field agriculture and livestock farming. Economic factors play a more immediate role in the magnitude of operation than in other systems.

Major ranching areas are the semiarid parts of the Great Plains, stretching from Texas through the prairie provinces of Canada, and throughout the intermontane basins and plateaus between the Rocky Mountains and the Sierra Nevada–Cascades from Canada to central Mexico: the llanos of Venezuela and Colombia, the sertão of Brazil, the pampa of Uruguay, the southeastern Argentine pampa, the Chaco, and Patagonia; the Karoo of South Africa; the big arid interior of Australia; and the high parts of the South Island of New Zealand. Livestock ranching is largely absent from the Eurasian realm, and has largely been the mode of occupying expanding frontiers.

Carrying capacity varies considerably. In deserts, 40 ha or more is required for forage to support one steer. Steppes and mountain meadows vary from 10 to 30 ha, depending on availability of moisture. In the subhumid part of the Great Plains, each steer averages 4 to 6 ha. When carrying capacity is low, large acreages are required. In western Texas many ranches have 8000 ha, and in southern Texas even more land is required. In Arizona and New Mexico many ranches have from 12,000 to 16,000 ha. The world's largest ranches are found in Australia.

The United States is the chief producer of beef in the world, followed by the Soviet Union (Table 4). The considerable percentage of increase in beef

Table 4. Total production of beef and veal in top producers, in 10^6 metric tons*

Producer	1961–1965	1973–1974	Percent change
United States	8.1	10.2	+25.9
Soviet Union	3.5	6.1	+73.5
Argentina	2.2	2.2	–
Brazil	1.4	2.2	+37.5
France	1.4	1.6	+ 7.3
China	1.4	1.5	+ 7.3
Australia	0.88	1.37	+56.0
World	31.0	40.8	+31.5

*From *FAO Production Yearbook*, 1974.

production in the Soviet Union, Australia, and Brazil from 1961–1965 to 1973–1974 should be noted.

WORLD HUNGER AND AGRICULTURE

No issue confronting modern society is more complex than that of ensuring an adequate global food supply in the decades which lie immediately ahead. More than 500,000,000 people live under near-famine conditions in various parts of the world. Double that number of people live on a critically low standard which becomes still more precarious in poor crop years or between growing seasons.

The risk of expanded starvation will become greater toward the end of the century with another 3,000,000,000 expected to be added to a current world population of 4,000,000,000. It is generally agreed that the food crisis is global in nature and requires a comprehensive well-coordinated program of urgent, cooperative worldwide action. Food has become a central element of the international economy. A world characterized by energy shortages, rampant inflation, and a weakening trade and monetary system will be plagued by food scarcities as well.

During the 1950s and 1960s global food production consistently increased. Per capita output expanded even in the food-deficit nations, and the world's total output increased by more than half. But at the precise moment when growing populations and rising expectations made a continuation of this trend essential, a dramatic change occurred. Since 1974 world cereal production has fallen, and reserves have dropped to the point where major crop failures will be disastrous. In 1976 some recovery was noticeable.

The long-term picture remains grim. Since increases in food production are not evenly distributed, the absolute number of malnourished people is in fact probably greater today than ever before.

The world faces a challenge unprecedented in severity, pervasiveness, and global dimension. The minimum objective of the next quarter century must be to more than double world food production and to improve its quality; yet even this would not remove the "hunger gap" unless drastic measures are taken to change current distribution patterns and to channel less grain into the excessive animal production of the affluent world, which now utilizes three times more tilled land to feed each person than the developing world.

[GEORG BORGSTROM]

Bibliography: J. R. Borchert, The Dust Bowl in the 1970's, *Ann. Ass. Amer. Geogr.*, 61:1–22, 1971; J. C. Dickinson, Alternatives to monoculture in the humid tropics of Latin America, *Prof. Geogr.*, 24: 217–222, 1973; A. N. Duckham and G. B. Masefield, *Farming Systems of the World*, 1971; D. B. Grigg, *The Agricultural Systems of the World: An Evolutionary Approach*, 1974; F. Hart, The Middle West, *Ann. Ass. Amer. Geogr.*, 62:258–282, 1972; M. U. Igbozurike, Ecological balance in tropical agriculture, *Geogr. Rev.*, 61:519–529, 1971; E. Mather, The American Great Plains, *Ann. Ass. Amer. Geogr.*, 62:237–257, 1972; W. B. Morgan, *Agriculture in the Third World*, 1978; J. E. Spencer and N. R. Stewart, The nature of agricultural systems, *Ann. Ass. Amer. Geogr.*, 63:529–544, 1973; H. L. Trueman, The hungry seventies, *Can. Geogr. J.*, 83:114–129, 1971; D. Whittlesey, Major agricultural regions of the Earth, *Ann. Ass. Amer. Geogr.*, 26:199–240, 1936; G. C. Wilken, Microclimate management by traditional farmers, *Geogr. Rev.*, 62:544–560, 1972.

Agricultural meteorology

The study and application of meteorology and climatology (a branch of meteorology) to the specific problems of agriculture. Agriculture is the production of food and fiber in all its forms—crops for human and animal consumption, pasture and range for animal grazing, crops (including trees) as raw materials for manufactured products. Hence, agricultural meteorology deals with farming, ranching, and forestry as well as with the transportation of substances required for production—water for irrigation, fertilizer, and agricultural chemicals—to the producer and transportation of the products to markets.

Problems. Agricultural meteorologists deal with a variety of problems in cooperation with many other types of specialists. Examples of these follow.

Predicting weather effects. Ultimate crop yields are influenced by weather factors. Statistical regression models treat yield as a function of such factors as preplanting soil moisture supply, date of plant seeding or emergence, and temperature and rainfall occurrences as the season progresses. Other models of a more deterministic nature predict dry matter accumulation as the difference between photosynthesis and respiration. The latter processes require knowledge of the solar radiation, carbon dioxide concentration, tissue temperature, plant size, physiological condition, nutritional status of the crop, and many other complex factors of the environment.

Weather effects influence the demand for water in crop fields, both irrigated and unirrigated, and in pastures, rangeland, and forests. Predictions of these effects require knowledge of the processes of turbulent transport of heat and water vapor in the lowest layers of the atmosphere. *See* CROP MICROMETEOROLOGY; EVAPOTRANSPIRATION.

Water use efficiency. Cultural practices (such as row spacings, population densities, plant architectures, and land shaping) must be defined and tested to best increase water use efficiency—defined as photosynthetic production of dry matter/water consumption or photosynthesis/evapotranspiration.

Weather-related diseases and pests. To predict the outbreak of weather-related crop and animal diseases or insect pests, the ways in which organism life cycles are affected by weather must be learned. When these relations are understood, the climatologist can determine the probabilities of an outbreak during any part of the growing season. Agricultural meteorologists can prepare special advisories on the basis of the actual weather conditions occurring and the forecast of coming weather. This information guides the grower in preparing control measures for the periods of especial danger and provides for additional alerts when specific pest control actions will be needed. Climatologists, weather forecasters, and agricultural meteorologists work closely with pathologists and entomologists to refine these techniques.

Growing season. It is important to predict the probable length of the growing season for regions in which historical weather records are available. This prediction aids in planning planting and harvest operations and helps in the choice of new crops for introduction to an area. *See* CLIMATOLOGY.

Frost. Agrometeorologists are involved in predicting the timing and severity of frosts and their probable impacts on crops, especially high-value fruits, vegetables, and ornamentals. Climatologists use historical weather records to specify the degree of frost risk for a general area. Micrometeorologists refine these specifications by study of the topography of the specific orchard, the patterns of cold air drainage into low spots, and the likely soil temperature differences due to slope and aspect since the "lay of the land" determines exposure to the Sun's rays during daytime. Forecasters can predict the severity of frost on a given night from knowledge of general weather conditions, particularly sky cover (clouds trap much of the Earth's thermal radiation to space) and windspeed (in calm conditions thermal inversions develop and cold air piles up near the surface). Engineers and agricultural meteorologists work on techniques for frost protection.

Agricultural meteorologists are involved in developing means of frost protection. Coverings are used to protect low-growing crops against frost. Space heating and thermal radiation devices are used for warming orchard trees during a frost night. Sprinkling devices are used to keep the frost-threatened vegetation ice-covered as insulation against temperatures lower than freezing and to liberate heat of fusion to warm the air slightly. Agricultural meteorologists have also developed foams to cover small plants and insulate them against freezing temperatures. Foliar sprays of aluminum powder have been used to lower the rate of thermal emission from citrus trees so that they cool more slowly during a frost.

Wind. The problem lies in predicting the usual or predominant speed and direction of winds for locating houses, barns, or other farm buildings. Odor problems may be minimized by properly locating the farmstead in the lee of natural obstructions to the wind or planted windbreaks. Energy costs for heating and cooling can also be reduced by proper placement of buildings.

The impact of windiness on animal comfort and on the growth of crops is another area of study. Windbreaks of trees (shelterbelts) or of construction materials can be designed to provide effective shelter for animals against extreme chilling winds in winter or hot, desiccating winds in summer. Agricultural meteorologists assist in the design of windbreaks for this purpose.

Crops generally grow better in the lee of windbreaks, especially in the arid, semiarid, and subhumid regions of the world. The more uniform spread of snow in winter results in a favorable water supply when crops are grown in spring. During the growing season a more favorable microclimate – generally warmer and more humid – results in less frequent and less severe moisture stress in the sheltered plants. Agricultural meteorologists conduct studies to determine optimum design of tree windbreaks and of ways to use annual or perennial tall crops (such as corn, elephant grass, wheat grass, and sunflowers) to shelter shorter crops.

Severe weather conditions. Agrometeorologists work with animal husbandry specialists, animal physiologists, and engineers to develop protective techniques against, and prepare advisory alerts on, severe weather conditions which can endanger animals on the open range or in exposed feedlots. Growers need warnings, as far in advance as possible, of blizzards which are likely to cut cattle off from supplies or feed or which may cause animals to wander into low-lying areas where they may become buried by snow. Forecasting periods of very hot weather is also important since animals on feedlots suffer greatly when very high temperatures occur in conjunction with high humidity. Under such conditions animals are unable to dissipate their body heat effectively. Emergency rations can be fed during these periods to reduce body respiration rates, and emergency cooling can be effected with sprinklers. Shade can also be provided to minimize radiation load on the animals.

Turbulence and remote sensing. Some other areas of research and applications in agrometeorology are in turbulence and remote sensing. The Earth's surfaces (terrestrial and marine) exchange radiation and mass with the atmosphere and absorb momentum of the wind. Understanding the mechanisms of these exchanges is the function of micrometeorology. The exchanges of mass (CO_2, water vapor, pollen, dust, pollutants, and so on) is effected through the processes of turbulent transport. Agrometeorologists are necessarily interested in the development and testing of theoretical explanations of the mechanisms of turbulent transport and in its applications, which include predicting the rates at which heat and water vapor will be removed from or delivered to the surface, and the rates at which CO_2 is delivered to the plant while it is photosynthesizing, and is removed at night when only respiration is occurring. Effectiveness of pollination and of the applications of dusts and sprays are also determined by the intensity of turbulence. *See* MICROMETEOROLOGY.

The space age has been a boon to agricultural meteorologists in providing techniques for remotely sensing the condition of the land and the vegetation growing on it. Multispectral scanners on a number of National Oceanic and Atmospheric Administration (NOAA) and National Aeronautics and Space Administration (NASA) satellites and on aircraft provide data on the extent, density, and vigor of vegetation. The flux density of radiation in the visible-wave-band range can indicate whether tissues are healthy or necrotic. The near-infrared waveband provides information on whether or not leaves are hydrated and turgid. The thermal wavebands provide information on temperature of the emitting surfaces. Plants short of water may be considerably warmer than others not so stressed. The microwave and radar bands can be used to determine wetness of the surface soil. Agricultural meteorologists have been especially active in conducting "ground truth" studies, which aid in the interpretation of remotely sensed information.

Major agencies. The State Agricultural Experiment Station network and the Science and Education Administration of the U.S. Department of Agriculture (USDA) are active in agricultural meteorology research. Applications of the research and the provision of advice to farmers, ranchers, agricultural industries, and the general public are functions of the Cooperative Extension Services in each state. The NOAA provides special forecasts of interest to agriculturalists through the National Weather Service networks and also special weather advisories and other advisory services in limited areas of the country served by Environmental Science Service Centers. The World Food Outlook and Situation Board, operated by USDA, and the Center for Climatic and Environmental Assessment of NOAA attempt to monitor the impacts of weather on food production worldwide. A world weather and crop update is published in the *Weekly Weather and Crop Bulletin*, a joint publication of NOAA and USDA. In some projects—for example, the recently completed Large Area Crop Inventory Experiment (LACIE) to predict wheat production in countries around the world—several government agencies cooperate. In the LACIE project and continuing efforts to evaluate worldwide production of other crops, NASA, NOAA, and USDA are involved.

The Commission for Agricultural Meteorology (CAgM) of the World Meteorological Organization (WMO) carries out an international exchange of information on all facets of agricultural meteorology. Standardization of instrumentation and procedures and compilation of information on specific problems (such as drought) are carried out by CAgM. The organization also engages in training agrometeorologists for the developing countries and assists in the creation of agrometeorological services which function within ministries of agriculture, development, water resources, and such. A recent program with this objective is AGRHYMET, aimed at providing agrometeorological and operational hydrology services to the countries of Sahelian Africa. [NORMAN J. ROSENBERG]

Bibliography: G. Campbell, *An Introduction to Environmental Physics*, 1977; R. Geiger, *The Climate near the Ground*, 4th ed., 1965; R. Lee, *Forest Microclimatology*, 1978; J. L. Monteith, *Principles of Environmental Physics*, 1973; J. L. Monteith, *Vegetation and the Atmosphere*, vol. 2; *Case Studies*, 1976; R. E. Munn, *Descriptive Micrometeorology*, 1966; N. J. Rosenberg, *Microclimate: The Biological Environment*, 1974; O. G. Sutton, *Micrometeorology*, 1953.

Agricultural science (animal)

The science which deals with the selection, breeding, nutrition, and management of domestic animals for economical production of meat, milk, eggs, wool, hides, and other animal products. Horses for draft and pleasure, dogs and cats and rabbits for meat production, and bees for honey production may also be included in this group.

When primitive man first domesticated animals, they were kept as means of meeting his immediate needs for food, transportation, and clothing. Sheep probably were the first and most useful animals to be domesticated, furnishing milk and meat for food, and hides and wool for clothing.

As chemistry, physiology, anatomy, genetics, nutrition, parasitology, pathology, and other sciences developed, their principles were applied to the field of animal science. Since the beginning of the 20th century, great strides have been made in livestock production. Today, farm animals fill a highly important place in the life of man. They convert raw materials, such as pasture herbage, which are of little use to man as food, into animal products having nutritional values not directly available in plant products.

Ruminant animals (those with four compartments or stomachs in the fore portion of their digestive tract, such as cattle and sheep) have the ability to consume large quantities of roughages because of their particular type of digestive system. They also consume large tonnages of grains, as well as mill feeds, oil seed meals, industrial and agricultural by-products, and other materials not suitable for human food.

Products of the animal industry furnish raw materials for many important processing industries, such as meat packing, dairy manufacturing, poultry processing, textile production, and tanning. Many services are based on the needs of the animal industry, including livestock marketing, milk deliveries, poultry and egg marketing, poultry hatcheries, artificial insemination services, feed manufacturing, pharmeutical industry, and veterinary services. Thus, animal science involves the application of scientific principles to all phases of animal production, furnishing animal products efficiently and abundantly to consumers. Products from animals are often used for consumer products other than food, for example, hides for leather, and organ meats for preparation of drugs and hormones.

Livestock breeding. The breeding of animals began thousands of years ago. During the last half of the 19th century, livestock breeders made increasing progress in producing animals better suited to the needs of man by simply mating the best to the best. However, in the 20th century animal breeders began to apply the scientific principles of genetics and reproductive physiology. Some progress made in the improvement of farm animals resulted from selected matings based on knowledge of body type or conformation. This method of selection became confusing due to the use of subjective standards which were not always related to economic traits. This error was corrected, however, and many breeders of dairy cattle, poultry, beef cattle, sheep, and swine in the mid-1970s made use of production records or records of performance. Some of their breeding plans were based on milk fat production or egg production, as well as on body type or conformation. The keeping of poultry and dairy cow production records began in a very limited way late in the 19th century. The first Cow-Testing Association in the United States was organized in Michigan in 1906. Now over 1,500,000 cows are tested regularly in the United States. *See* BREEDING (ANIMAL).

Many states now have production testing for beef cattle, sheep, and swine, in which records of rate of gain, efficiency of feed utilization, incidence of twinning, yield of economically important carcass cuts, and other characteristics of production are maintained on part or all of the herd or flock. These records serve as valuable information in the selection of animals for breeding or sale.

Breeding terminology. A breed is a group of animals that has a common origin and possesses characteristics that are not common to other individuals of the same species.

A purebred breed is a group that possesses certain fixed characteristics, such as color or markings, which are transmitted to the offspring. A record, or pedigree, is kept which describes their ancestry for five generations. Associations have been formed by breeders primarily to keep records, or registry books, of individual animals of the various breeds. Purebred associations have taken a more active role in promoting and improving the breeds.

A purebred is one that has a pedigree recorded in a breed association or is eligible for registry by such an association. A crossbred is an individual produced by utilizing two or more purebred lines in a breeding program. A grade is an individual having one parent, usually the sire, a purebred and the other parent a grade or scrub. A scrub is an inferior animal of nondescript breeding. A hybrid is one produced by crossing parents that are genetically pure for different specific characteristics. The mule is an example of a hybrid animal produced by crossing two different species, the American jack, *Equus asinus*, with a mare, *E. caballus*.

Systems of breeding. The modern animal breeder has genetic tools which he may apply, such as selection and breeding, and inbreeding and outbreeding. Selection involves directly the retaining or rejecting of a particular animal for breeding purposes, being based largely on qualitative characteristics. Inbreeding is a system of breeding related animals. Outbreeding is a system of breeding unrelated animals. When these unrelated animals are of different breeds, the term crossbreeding is usually applied. Crossbreeding is in common use by commercial swine producers. About 80–90% of the hogs produced in the Corn Belt states are now crossbred. Crossbreeding is also used extensively by commercial beef and sheep producers.

Grading-up is the process of breeding purebred sires of a given breed to grade females and their female offspring for generation after generation. Grading-up offers the possibility of transforming a nondescript population into one resembling the purebred sires used in the process. It is an expedient and economical way of improving large numbers of animals.

Formation of new breeds. New breeds of farm animals have been developed from crossbred foundation animals. Montadale, Columbia, and Targhee are examples of sheep breeds so developed. The Santa Gertrudis breed of beef cattle was produced by crossing Brahman and Shorthorn breeds on the King Ranch in Texas. In poultry, advantage has been taken of superior genetic ability through the development of hybrid lines.

Artificial insemination. In this process spermatozoa are collected from the male and deposited in the female genitalia by instruments rather than by natural service. In the United States this practice was first used for breeding horses. Artificial insemination in dairy cattle was first begun on a large scale in New Jersey in 1938. In 1958 over 6,000,000 cows were bred artificially in the United States. Freezing techniques for preserving and storing spermatozoa have been applied with great success to bull semen, and it is now possible for outstanding bulls to sire calves years after the bulls have died. The use of artificial insemination for beef cattle and poultry (turkeys) has become more common since 1965. Drugs are being developed which stimulate beef cow herds to come into heat (ovulate) at approximately the same time. This will permit the insemination of large cow herds without the individual handling and inspection which is used with dairy cattle. Although some horses are bred by artificial insemination, many horse breed associations allow only natural breeding.

Livestock feeding. Scientific livestock feeding involves the systematic application of the principles of animal nutrition to the feeding of farm animals. The science of animal nutrition has advanced rapidly since 1930, and the discoveries are being utilized by most of those concerned with the feeding of livestock. The nutritional needs and responses of the different farm animals vary according to the functions they perform and to differences in the anatomy and physiology of their digestive systems. Likewise, feedstuffs vary in usefulness depending upon the time and method of harvesting the crop, the methods employed in drying, preserving, or processing them, and the forms in which they are offered to the animals.

Chemical composition of feedstuffs. The various chemical compounds that are contained in animal feeds have been divided into groups called nutrients. These include proteins, fats, carbohydrates, vitamins, and mineral matter. Proteins are made up of amino acids. Twelve amino acids are essential for all nonruminant animals and must be supplied in their diets. Fats and carbohydrates provide mainly energy. In most cases they are interchangeable as energy sources for farm animals. Fats furnish 2.25 times as much energy per pound as do carbohydrates because of their higher proportion of carbon and hydrogen to oxygen. Thus the energy concentration in poultry and swine diets can be increased by inclusion of considerable portions of fat. Ruminants cannot tolerate large quantities of fat in their diets, however.

Vitamins essential for health and growth include fat-soluble A, D, E, and K, and water-soluble vitamins thiamine, riboflavin, niacin, pyrodoxine, pantothenic acid, and cobalamin. *See* VITAMIN.

Mineral salts that supply calcium, phosphorus, sodium, chlorine, and iron are often needed as supplements, and those containing iodine and cobalt may be required in certain deficient areas. Zinc may also be needed in some swine rations. Many conditions of mineral deficiency have been noted in recent years by using rations that were not necessarily deficient in a particular mineral but in which the mineral was unavailable to the animal because of other factors in the ration or imbalances with other minerals. For example, copper deficiency can be caused by excess molybdenum in the diet.

By a system known as the "proximate analysis," developed prior to 1895 in Germany, feeds have long been divided into six fractions including moisture, ether extract, crude fiber, crude protein, ash, and nitrogen-free extract. The first five fractions are determined in the laboratory. The nitrogen-free extract is what remains after the percentage sum of these five has been subtracted from 100%. Although proximate analysis serves as a guide in the classification, evaluation, and use of feeds, it gives very little specific information about particular chemical compounds in the feed.

The ether extract fraction includes true fats and certain plant pigments, many of which are of little nutritional value.

The crude fiber fraction is made up of celluloses and lignin. This fraction, together with the nitrogen-free extract, makes up the total carbohydrate content of a feed.

The crude protein is estimated by multiplying the total Kjeldahl nitrogen content of the feed by the factor 6.25. This nitrogen includes many forms of nonprotein as well as protein nitrogen.

The ash, or mineral matter fraction, is determined by burning a sample and weighing the residue. In addition to calcium and other essential mineral elements, it includes silicon and other nonessential elements.

The nitrogen-free extract (NFE) includes the more soluble and the more digestible carbohydrates, such as sugars, starches, and hemicelluloses. Unfortunately, most of the lignin, which is not digestible, is included in this fraction.

A much better system of analysis has been developed for the crude fiber fraction of feedstuffs. This system separates more specifically the highly digestible cell-soluble portion and the less digestible fibrous portion of plant cell walls. This system of nonnutritive residue analysis was developed at the U.S. Department of Agriculture laboratories in Maryland.

Digestibility of feeds. In addition to their chemical composition or nutrient content, the nutritionist and livestock feeder should know the availability or digestibility of the different nutrients in feeds. The digestibility of a feed is measured by determining the quantities of nutrients eaten by an animal over a peroid of time and those recoverable in the fecal matter. By assigning appropriate energy values to the nutrients, total digestible nutrients (TDN) may be calculated. These values have been determined and recorded for a large number of feeds.

Formulation of animal feeds. The nutritionist and livestock feeder finds TDN values of great use in the formulation of animal feeds. The TDN requirements for various classes of livestock have been calculated for maintenance and for various productive capacities. However, systems have been developed for expressing energy requirements of animals or energy values of feeds in units which are more closely related to the body process being supported (such as maintenance, growth, and milk or egg production). New tables of feeding standards are being published using the units of metabolizable energy and net energy, which are measurements of energy available for essential body processes. Recommended allowances of nutrients for all species of livestock and some small animals (rabbit, dogs, and mink) are published by the National Academy of Sciences–National Research Council. These are assembled by experts in the field of animal science and are available for distribution through the Superintendent of Documents, Washington, D.C.

Nutritional needs of different animals. The nutritional requirements of different classes of animals are partially dependent on the anatomy and physiology of their digestive systems. Ruminants can digest large amounts of roughages, whereas hogs and poultry, with simple stomachs, can digest only limited amounts and require more concentrated feeds, such as cereal grains. In simple-stomached animals the complex carbohydrate starch is broken down to simple sugars which are absorbed into the blood and utilized by the body for energy.

Microorganisms found in the rumen of ruminant animals break down not only starch but the fibrous carbohydrates of roughages, namely, cellulose and hemicellulose, to organic acids which are absorbed into the blood and utilized as energy. Animals with simple stomachs require high-quality proteins in their diets to meet their requirements for essential amino acids. On the other hand, the microorganisms in ruminants can utilize considerable amounts of simple forms of nitrogen to synthesize high-quality microbial protein which is, in turn, utilized to meet the ruminant's requirement for amino acids. Thus, many ruminant feeds now contain varying portions of urea, an economical simple form of nitrogen, which is synthesized commercially from nonfeed sources. Simple-stomached animals require most of the vitamins in the diet. The microorganisms in the rumen synthesize adequate quantities of the water-soluble vitamins to supply the requirement for the ruminant animal. The fat-soluble vitamins A, D, and E must be supplied as needed to all farm animals. Horses and mules have simple stomachs but they also have an enlargement of the cecum (part of the large intestine), in which bacterial action takes place similar to that in the rumen of ruminants. The requirements for most nutrients do not remain the same throughout the life of an animal but relate to the productive function being performed. Therefore, the requirements are much higher for growth and lactation than they are for pregnancy or maintenance.

Livestock judging. The evaluation, or judging, of livestock is important to both the purebred and the commercial producer.

Show-ring judging. The purebred producer usually is much more interested in show-ring judging, or placings, than is the commercial producer. Because of the short time they are in the show-ring, the animals must be placed on the basis of type or appearance by the judge who evaluates them. The show-ring has been an important influence in the improvement of livestock by keeping the breeders aware of what judges consider to be desirable types. The shows have also brought breeders together for exchange of ideas and breeding stock and have helped to advertise breeds of livestock and the livestock industry. The demand for better meat-animal carcasses has brought about more shows in which beef cattle and swine are judged, both on foot and in the carcass. This trend helps to promote development of meat animals of greater carcass value and has a desirable influence upon show-ring standards for meat animals. The standards used in show rings have shifted toward traits in live animals which are highly related to both desirable carcass traits and high production efficiency.

Selection of animals for breeding. The evaluation or selection of animals for breeding purposes is of importance to the commercial as well as to the purebred breeder. In selecting animals for breeding, desirable conformation or body type is given careful attention. The animals are also examined carefully for visible physical defects, such as blindness, crooked legs, jaw distortions, and abnormal udders. Animals known to be carriers of genes for heritable defects, such as dwarfism in cattle, should be discriminated against.

When they are available, records of performance or production should be considered in the selection of breeding animals. Some purebred livestock record associations now record production performance of individual animals on their pedigrees.

Grading of market animals. The grading on foot of hogs or cattle for market purposes requires special skill. In many modern livestock markets, hogs are graded as no. 1, 2, 3, or 4 according to the estimated values of the carcasses. Those hogs grading no. 3 are used to establish the base price, and sellers are paid a premium for better animals. In some cases the grade is used also to place a value of the finished market product. For example, the primal cuts from beef cattle are labeled prime, choice, or good on the basis of the grade which the rail carcass received. A trend is underway toward pricing market animals on the basis of grade and yield, which also takes into account factors associated with yield of lean cuts.

Livestock pest and disease control. The innumerable diseases of farm livestock require expert diagnosis and treatment by qualified veterinarians. The emphasis on intensive animal production has increased stresses on animals and generally increased the need for close surveillance of herds or flocks for disease outbreaks. Both external and internal parasites are common afflictions of livestock but can be controlled by proper management of the animals. Sanitation is of utmost importance in the control of these pests, but under most circumstances sanitation must be supplemented with effective insecticides, ascaricides, and fungicides. *See* FUNGISTAT AND FUNGICIDE; INSECTICIDE; PESTICIDE.

Internal parasites. Internal parasites, such as stomach and intestinal worms in sheep, cannot be controlled by sanitation alone under most farm conditions. They are a more critical problem under intensive management systems and in warm, humid climates. For many years the classic treatment for sheep was drenching with phenothiazine and continuous free choice feeding of one part phenothiazine mixed with nine parts of salt. New drugs have been developed which more effectively break the life cycle of the worm and have a broader spectrum against different classes of parasites.

Control of gastrointestinal parasites in cattle can be accomplished in many areas by sanitation and the rotational use of pastures. In areas of intensive grazing, animals, especially the young ones, may become infected. Regular and timely administration of antiparasite drugs is the best means of controlling the pests. Otherwise their effects will seriously decrease the economic productivity of animals.

The use of drugs to control gastrointestinal parasites and also certain enteric bacteria in hogs is commonplace. Control is also dependent on good sanitation and rotational use of nonsurfaced lots and pastures. Similarly, because of intensive housing systems, the opportunities for infection and spread of both parasitism and disease in poultry flocks are enhanced by poor management conditions. The producer has a large number of materials to choose from in preventing or treating these conditions, including sodium fluoride, piperazine salts, nitrofurans, and arsenicals and antibiotics such as penicillin, tetracyclines, hygromycin, and tylosin.

External parasites. Control of horn flies, horseflies, stable flies, lice, mange mites, ticks, and fleas on farm animals has been in the process of rapid change with the introduction of many new insecticides. Such compounds as DDT, methoxychlor, toxaphene, lindane, and malathion were very effective materials for the control of external parasites. The use of these materials was wisely restricted to certain conditions and classes of animals by the provisions of Public Law 518, which is the Miller Amendment to the Federal Food, Drug, and Cosmetic Act. For example, use of DDT was not permitted on dairy animals. Reliable information should be obtained before using these materials for the control of external parasites. Pressure from various nonscientific lobby groups has prompted legislative prohibition of an increasing number of these chemicals.

Control of cattle grubs, or the larvae of the heel fly, may be accomplished by dusting the backs of the animals with powders or by spraying them under high pressure. Systemic insecticides for grub control have been given approval if used according to the manufacturer's recommendation.

Fungus infections. Actinomycosis is a fungus disease commonly affecting cattle, swine, and horses. In cattle this infection is commonly known as lumpy jaw. The lumpy jaw lesion may be treated with tincture of iodine or by local injection of streptomycin in persistent cases. Most fungus infections, or mycoses, develop slowly and follow a prolonged course. A veterinarian should be consulted for diagnosis and treatment.

General animal health care. Numerous other disease organisms pose a constant threat to livestock. Although many of these organisms can be treated therapeutically, it is much more advisable economically to establish good preventive medicine and health care programs under the guidance of a veterinarian.

Management. Economic changes have continued to narrow the profit margins for economic livestock producers. This has increased the necessity for attention to good management practices in all aspects of production.

[RONALD R. JOHNSON]

Bibliography: J. R. Campbell and J. F. Lasley,

The Science of Animals That Serve Mankind, 1975; J. R. Campbell and R. T. Marshall, *The Science of Providing Milk for Man*, 1975; M. E. Ensminger, *Animal Science*, 1977; M. E. Ensminger, *Beef Cattle Science*, 1976; M. E. Ensminger, *Horses and Horsemanship*, 1977; M. E. Ensminger, *Sheep and Wool Science*, 1970; J. L. Krider and W. E. Carroll, *Swine Production*, 4th ed., 1971; Merck and Co., Inc., *The Merck Veterinary Manual*, 4th ed., 1973; M. O. North, *Commercial Chicken Production Manual*, 1978; R. R. Snapp and A. L. Newmann, *Beef Cattle*, 1977.

Agricultural science (plant)

The pure and applied science that is concerned with botany and management of crop and ornamental plants for utilization by humankind. Crop plants include those grown and used directly for food, feed, or fiber, such as cereal grains, soybeans, citrus, and cotton; those converted biologically to products of utility, such as forage plants, hops, and mulberry; and those used for medicinal or special products, such as digitalis, opium poppy, coffee, and cinnamon. In addition, many plant products such as crambe oil and rubber are used in industry where synthetic products have not been satisfactory. Ornamental plants are cultured for their esthetic value.

The ultimate objective of plant agriculture is to recognize the genetic potential of groups of plants and then to manipulate and utilize the environment to maximize that genetic expression for return of a desirable product. Great advancements in crop culture have occurred by applying knowledge of biochemistry, physiology, ecology, morphology, anatomy, taxonomy, pathology, and genetics of plants. Contributions of improved plant types by breeding, and the understanding and application of principles of atmospheric science, soil science, and animal and human nutrition, have increased the efficiency and decreased the risks of crop production.

Domestication of crop plants. All crops are believed to have been derived from wild species. However, cultivated plants as they are known today have undergone extensive modification from their wild prototypes as a result of the continual efforts to improve them. These wild types were apparently recognized as helpful to humans long before recorded history. Desirable plants were continually selected and replanted in order to improve their growth habit, fruiting characteristics, and growing season. Selection has progressed so far in cases such as cabbage and corn (maize) that wild ancestors have become obscure.

Centers of origins of most crop plants have been determined to be in Eurasia, but many exceptions exist. This area of early civilization apparently abounded with several diverse plant types that led to domestication of the crops known today as wheat, barley, oats, millet, sugarbeets, and most of the cultivated forage grasses and legumes. Soybeans, lettuce, onions, and peas originated in China and were domesticated as Chinese civilization developed. Similarly, many citrus fruits, banana, rice, and sugarcane originated in southern Asia. Sorghum and cowpeas are believed to have originated in Africa. Crops which were indigenous to Central and South America, but which migrated to North America with Indian civilization, include corn, potato, sweet potato, pumpkin, sunflower, tobacco, and peanut. Thus, prior to 1492 there was little, if any, mixing of crop plants and cultural practices between the Old and the New Worlds. Most of the major agricultural crops of today in the United States awaited introduction by early settlers, and later by plant explorers.

During domestication of crop plants the ancient cultivators in their geographically separated civilizations must have had goals similar to present-day plant breeders. Once valuable attributes of a plant were recognized, efforts were made to select the best types for that purpose. Desirable characters most likely included improved yield, increased quality, extended range of adaptation, insect and disease resistance, and easier cultural and harvesting operations. About 350,000 species of plants exist in the world, yet only about 10,000 species can be classified as crops using the broadest of definitions. Of these, about 150 are of major importance in world trade, and only 15 make up the majority of the world's food crops. On a world basis, wheat is grown on the most acreage, followed by rice, but rice has a higher yield per area than wheat so their total production is about equal. Other major crops are corn, sorghum, millet, barley; sugarcane and sugarbeet; potato, sweet potato, and cassava; bean, soybean, and peanut; and coconut and banana.

Redistribution of crop plants. Crop distribution is largely dictated by growth characters of the crop, climate of the region, soil resources, and social habits of the people. As plants were domesticated by civilizations in separate parts of the world, they were discovered by early travelers. This age of exploration led to the entrance of new food crops into European agriculture. The potato, introduced into Spain from the New World before 1570, was to become one of the most important crops of Europe. When introduced into Ireland, it became virtually the backbone of the food source for that entire population. Corn was introduced to southern Europe and has become an important crop. Rice has been cultivated in Italy since the 16th century. Tobacco was also introduced to European culture, but continued to provide a major source of export of the early colonial settlements in America.

European agriculture reciprocated by introducing wheat, barley, oats, and several other food and feed crops into the New World. In the new environment, plants were further adapted by selection to meet the local requirements. Cultural technology regarding seeding, harvesting, and storing was transferred along with the crops. This exchange of information oftentimes helped allow successful culture of these crops outside of their center of origin and domestication. Today the center of production of a crop such as wheat in the United States and potato in Europe is often markedly different from its center of origin.

The United States recognized early the need for plant exploration to find desirable types that could be introduced. Thomas Jefferson wrote in 1790, "The greatest service which can be rendered to any country is to add a useful plant to its culture." Even today the U.S. Department of Agriculture conducts many plant explorations and maintains plant introduction centers to evaluate newly found plants. Explorations are also serving a critical need for preservation of germplasm for plant breeders, as many of the centers of origin are becoming agriculturally intensive, and wild types necessary to increase genetic diversity will soon be extinct.

Adaptation of crop plants. Crop plants of some sort exist under almost all environments, but the major crops on a world basis tend to have rather specific environmental requirements. Furthermore, as crop plants are moved to new locations the new environment must be understood, and cultural or management changes often must be made to allow best performance. More recently, varieties and strains of crops have been specifically developed that can better cope with cold weather, low rainfall, diseases, and insects to extend even further the natural zone of adaptation.

Temperature. Temperature has a dominant influence on crop adaptation. The reason is that enzyme activity is very temperature-dependent and almost all physiological processes associated with growth are enzymatically controlled. Crops differ widely in their adapted temperature range, but most crops grow best at temperatures of 15 to 32°C. Optimum day temperature for wheat, however, is 20 to 25°C, while for corn it is about 30°C and for cotton about 35°C.

The frost-free period also influences crop adaptation by giving an indication of the duration of the growing season. Thus a growing season for corn or soybeans might be described as one with mean daily temperatures between 18 and 25°C, and with an average minimum temperature exceeding 10°C for at least 3 months. Small grains such as wheat, barley, and oats tolerate a cooler climate with a period of only 2 months when minimum temperatures exceed 10°C. Because of optimum growth temperatures and frost-free periods, it is easily recognized why the spring wheat belt includes North Dakota and Montana, the corn belt Iowa and Illinois, and the cotton belt Mississippi and Alabama. Farmers use planting dates and a range in maturity of varieties to match the crop to the growing season.

In many winter annual (the plants are sown in fall, overwinter, and mature in early summer), biennial, and perennial crops, cold temperatures also influence distribution. The inherent ability of the crop to survive winter limits distribution of crops. As a generalized example, winter oats are less able to survive cold winters than winter barley, followed by winter wheat and then winter rye. Thus oats in the southern United States are mostly winter annual type, while from central Arkansas northward they are spring type. The dividing line for barley is about central Missouri and for wheat about central South Dakota. Winter rye can survive well into Canada.

A cold period may be required for flowering of winter annual and perennial crops. This cold requirement, termed vernalization, occurs naturally in cold environments where the dormant bud temperature is near 0°C for 4 to 6 weeks during winter. Without this physiological response to change the hormonal composition of the terminal bud, flowering of winter wheat would not occur the following spring. Bud dormancy is also low-temperature-induced and keeps buds from beginning spring growth until a critical cold period has passed. The flower buds of many trees such as peach, cherry, and apple require less chilling than do vegetative buds, and therefore flower before the leaves emerge. The intensity and duration of cold treatment necessary to break dormancy differs with species and even within a species. For example, some peaches selected for southern areas require only 350 hr below 8°C to break dormancy, while some selected for northern areas may require as much as 1200 hr. This cold requirement prevents production of temperate fruit crops in subtropical regions, but in temperate climates its survival value is clear. A physiological mechanism that prevents spring growth from occurring too early helps decrease the possibility of cold temperature damage to the new succulent growth.

Temperature of the crop environment can be altered by date of planting of summer annuals, by proper site selection, and by artificial means. In the Northern Hemisphere the south- and west-facing slopes are usually warmer than east- or north-facing slopes. Horticulturists have long used mulches for controlling soil temperature, and also mists and smoke for short-term low-temperature protection.

Water. Water is essential for crop production, and natural rainfall often is supplemented by irrigation. Wheat is grown in the Central Plains states of the United States because it matures early enough to avoid the water shortage of summer. In contrast, corn matures too late to avoid that drought condition and must be irrigated to make it productive. In the eastern United States where rainfall is higher, irrigation is usually not needed to produce good yields. *See* IRRIGATION OF CROPS.

Crops transpire large amounts of water through their stomata. For example, corn transpires about 350 kg of water for each kilogram of dry weight produced. Wheat, oats, and barley transpire about 600 kg, and alfalfa about 1000 kg for each kilogram of dry weight produced. Fallowing (allowing land to be idle) every other season to store water in the soil for the succeeding crop has been used to overcome water limitations and extend crops further into dryland areas. *See* EVAPOTRANSPIRATION.

Light. Light intensity and duration also play dominant roles. Light is essential for photosynthesis. The yield of a crop plant is related to its efficiency in intercepting solar radiation by its leaf tissue, the efficiency of leaves in converting light energy into chemical energy, and the transfer and utilization of that chemical energy (usually sugars) for growth of an economic product. Crops differ markedly in the efficiency of their leaves. Photosynthetic rate of corn or sorghum leaves may be as high as 60 mg CO_2/dm^2 of leaf area/hr with full sunlight. Photosynthesis of wheat, soybeans, or rice is about 25–35 mg/dm^2/hr, while that of pineapple or tree nut crops is about 8–15 mg/dm^2/hr.

Crops also differ markedly in leaf area and leaf arrangement. Humans greatly influence the photosynthetic area by altering number of plants per area and by cutting and pruning. Corn producers have gradually increased plant population per area about 50% since about 1950 as improved varieties and cultural methods were developed. This has increased the leaf area of the crop canopy and the amount of solar energy captured. Defoliation and pruning practices also influence solar energy capture and growth rate. Continued defoliation of pastures by grazing may reduce the photosynthetic area to the point that yield is diminished. In these perennial plants, carbohydrates are stored in underground organs such as roots, rhizomes, and stem bases to furnish food for new shoots in spring and following cutting. Availability of water and nitrogen fertilizer also influences the amount of leaf area developed.

Photoperiodism, the response of plants to day length, also has a dramatic effect on plant distribution. Such important adaptive mechanisms as development of winter hardiness, initiation and maintenance of bud dormancy, and floral initiation are influenced by photoperiod. Plants have been classified as long-day, those that flower when day lengths exceed a critical level; short-day, which is opposite to long-day; and day-neutral, those that are not affected by day length. Photoperiod also has a controlling influence on formation of potato tubers, onion bulbs, strawberry runners, and tillers of many cereal grains and grasses. Farmers select varieties bred for specific areas of the country to ensure that they flower properly for their growing season. Horticulturists, in a more intensive effort, provide artificial lighting in greenhouses to lengthen the photoperiod, or shorten it by shading, to induce flowering and fruit production at will; for example, to ready poinsettias for the Christmas holiday trade or asters for Easter. Natural photoperiods are important in plant breeding when varities of day-length-sensitive crops must be developed for specific localities. Soybeans are day-length-sensitive and have been classified into several maturity groups from north to south in latitude.

Pathogens. Pathogens of plants that cause diseases include fungi, bacteria, viruses, and nematodes. These organisms are transmitted from plant to plant by wind, water, and insects and infect the plant tissue. Organisms infect the plant and interfere with the physiological functions to decrease yield. Further, they infect the economic product and decrease its quality. Pathogens are most economically controlled by breeding resistant varieties or by using selective pesticides. Insects decrease plant productivity and quality largely by mechanical damage to tissue and secretion of toxins. They are also usually controlled by resistant varieties or specific insecticides. *See* PLANT DISEASE; PLANT DISEASE CONTROL.

Soil. The soil constitutes an important facet of the plant environment. Soil physical properties such as particle size and pore space determine the water-holding capacity and influence the exchange of atmospheric gases with the root system. Soil chemical properties such as pH and the ability to supply nutrients have a direct influence on crop productivity. Farmers alter the chemical environment by addition of lime or sulfur to correct acidic or basic conditions, or by addition of manures and chemical fertilizers to alter nutrient supply status. Soil also is composed of an important microbiological component that assists in the cycling of organic matter and mineral nutrients. *See* SOIL.

Management. During and following domestication of crop plants, humankind has learned many cultural or management practices that enhance production or quality of the crop. This dynamic process is in operation today as the quest continues to improve the plant environment to take advantage of the genetic potential of the crop. These practices have made plant agriculture in the United States one of the most efficient in the world. Some major changes in technology are discussed in the following sections.

Mechanization. In 1860 an average United States farm worker produced enough food and fiber for fewer than 5 other persons. In 1950 it was enough for 25, and today exceeds 50 other persons. Mechanization, which allowed each farm worker to increase the area managed, is largely responsible for this dramatic change.

A model of a grain reaper in 1852 demonstrated that nine men with the reaper could do the work of 14 with cradles. By 1930 one man with a large combine had a daily capacity of 20–25 acres (1 acre = 4047 m^2), and not only harvested but threshed the grain to give about a 75-fold gain over the cradle-and-flail methods of a century earlier. In 1975 one man with a faster-moving combine could harvest 50–75 acres per day. The mechanical cotton picker harvests a 500-lb (1 lb = 0.45 kg) bale in 75 min, 40–50 times the rate of a hand picker. The peanut harvester turns out about 300 lb of shelled peanuts per hour, a 300-hr job if done by hand labor. Hand setting 7500 celery plants is a day's hard labor for one person. However, a modern transplanting machine with two people readily sets 40,000, reducing labor costs by 67%. Today crops such as cherries and tomatoes are machine-harvested. When machines have been difficult to adapt to harvesting of crops such as tomatoes, special plant varieties that flower more uniformly and have tougher skins on the fruit have been developed.

Fertilizers and plant nutrition. No one knows when or where the practice originated of burying a fish beneath the spot where a few seeds of corn were to be planted, but it was common among North American Indians when Columbus discovered America and is evidence that the value of fertilizers was known to primitive peoples. Farm manures were in common use by the Romans and have been utilized almost from the time animals were first domesticated and crops grown. It was not until centuries later, however, that animal fertilizers were supplemented by mineral forms of lime, phosphate, potassium, and nitrogen. Rational use of these substances began about 1850 as an outgrowth of soil and plant analyses.

Justus von Liebig published *Chemistry and Its Application to Agriculture and Physiology* in 1840, which led to concepts on modifying the soil by fertilizers and other amendments. These substances soon became the center of crop research. In 1842 John Bennet Lawes, who founded the famous Rothamsted Experiment Station in England, obtained a patent for manufacture of superphosphates and introduced chemical fertilizers to agriculture. As research progressed, crop responses to levels of phosphate, lime, and potassium were worked out, but nitrogen nutrition remained puzzling. The nitrogen problem was clarified in 1886, when H. Hellriegel and H. Wilfarth, two German chemists, reported that root nodules and their associated bacteria were responsible for the peculiar ability of legume plants to use atmospheric nitrogen. These findings collectively allowed rational decision-making regarding nutrition of crop plants and were understandably very significant in increasing crop productivity.

By the early 20th century 10 elements had been identified as essential for proper nutrition. These were carbon, hydrogen, and oxygen, which are supplied by the atmosphere; and nitrogen, potassium, phosphorus, sulfur, magnesium, calcium, and iron, supplied by the soil. The first 40 years of the 20th century witnessed the addition of manganese, boron, copper, zinc, molybdenum, and chlorine to the list of essential mineral nutrients. These 6 are required only in very small amounts as compared with the first 10 and have been classified as micronutrients. From a quantitative view they are truly minor, but in reality they are just as critical as the others for plant survival and productivity. Many early and puzzling plant disorders are now known to be due to insufficient supplies of micronutrients in the soil, or to their presence in forms unavailable to plants.

An important result of discoveries relating to absorption and utilization of micronutrients is that they have served to emphasize the complexity of soil fertility and fertilizer problems. In sharp contrast to early thoughts that fertilizer practices should be largely a replacement in the soil of what was removed by the plant, it is now recognized that interaction and balance of mineral elements within both the soil and plant must be considered for efficient crop growth. Usage of chemical fertilizers has become more widespread with the passage of time. *See* FERTILIZER; PLANT, MINERALS ESSENTIAL TO.

Pesticides. Total destruction of crops by swarms of locusts and subsequent starvation of many people and livestock have occurred throughout the world. Pioneers in the Plains states suffered disastrous crop losses from hordes of grasshoppers and marching army worms. An epidemic of potato blight brought hunger to much of western Europe and famine to Ireland in 1846–1847. The use of new insecticides and fungicides has done much to prevent such calamities. Various mixtures, really nothing more than nostrums (unscientific concoctions), were in use centuries ago, but the first really trustworthy insect control measure appeared in the United States in the mid-1860s, when paris green was used to halt the eastern spread of the Colorado potato beetle. This was followed by other arsenical compounds, culminating in lead arsenate in the 1890s.

A major development occurred during World War II, when the value of DDT (dichlorodiphenyltrichloroethane) for control of many insects was discovered. Although this compound was known to the chemist decades earlier, it was not until 1942 that its value as an insecticide was definitely established and a new chapter written in the continual contest between humankind and insects. Three or four applications of DDT gave better season control of many pests at lower cost than was afforded by a dozen materials used earlier. Furthermore, DDT afforded control of some kinds of pests that were practically immune to former materials. In the meantime other chemicals have been developed that are even more effective for specific pests, and safer to use from a human health and ecological viewpoint.

Dusts and solutions containing sulfur have long been used to control mildew on foliage, but the first highly effective fungicide was discovered accidentally in the early 1880s. To discourage theft, a combination of copper sulfate and lime was used near Bordeaux, France, to give grape vines a poisoned appearance. Bordeaux mixture remained the standard remedy for fungus diseases until relatively recently when other materials with fewer harmful side effects were released.

Today new pesticides are being rapidly developed by private industry and are carefully monitored by several agencies of the Federal government. Besides evaluation for its ability to repel or destroy a certain pest, influence of the pesticide on physiological processes in plants, and especially long-range implications for the environment and for human health, are carefully documented before Federal approval for use is granted. *See* PESTICIDE.

Herbicides. Because they are readily visible, weeds were recognized as crop competitors long before microscopic bacteria, fungi, and viruses. The time-honored methods of controlling weeds have been to use competitive crops or mulches to smother them, or to pull, dig, hoe, or cultivate them out. These methods are still effective and most practical in many instances. However, under many conditions weeds can be controlled chemically much more economically. A century ago regrowth of thistles and other large weeds was prevented by pouring salt on their cut stubs. Salt and ashes were placed along areas such as courtyards, roadsides, and fence rows where all vegetation needed to be controlled. However, until the 1940s selective weed control, where the crop is left unharmed, was used on a very limited scale. The first of these new selective herbicides was 2,4-D (2,4-dichlorophenoxyacetic acid), followed shortly by 2,4,5-T (2,4,5-trichlorophenoxyacetic acid) and other related compounds. This class of herbicides is usually sprayed directly on the crop and weeds and kills susceptible plants by upsetting normal physiological processes, causing abnormal increases in size, and distortions that eventually lead to death of the plant. As a group, these herbicides are much less toxic to grasses than to broad-leaved species, and each one has its own character of specificity. For instance, 2,4-D is more efficient for use against herbaceous weeds, whereas 2,4,5-T is best for woody and brushy species.

Today's herbicides have been developed by commercial companies to control a vast array of weeds in most crop management systems. Some herbicides such as atrazine are preemergence types that are sprayed on the soil at time of corn planting. This herbicide kills germinating weed seeds by interfering with their photosynthetic system so they starve to death. Meanwhile metabolism of the resistant corn seedlings alters the chemical slightly to cause it to lose its herbicidal activity. Such is the complexity that allows specificity of herbicides. The place that herbicides have come to occupy in agriculture is indicated by the fact that farmlands treated in the United States increased from a few thousand acres in 1940 to over 150,000,000 acres in 1974, not including large areas of swamp and overflow lands treated for aquatic plant control and thousands of miles of treated highways, railroad tracks, and drainage and irrigation ditches. *See* HERBICIDE.

Growth regulators. Many of the planters' practices in the past hundred years may be classified as methods of regulating growth. Use of specific substances to influence particular plant functions, however, has been a more recent development, though these modern uses, and even some of the substances applied, had their antecedents in century-old practices employed in certain parts of the world.

Since about 1930 many uses have been discovered for a considerable number of organic compounds having growth-regulating influences. For instance, several of them applied as sprays a few days to several weeks before normal harvest will prevent or markedly delay dropping of such fruits as apples and oranges. Runner development in strawberries and sucker (tiller) development in tobacco can be inhibited by sprays of maleic hydrazide. Another compound called CCC (2-chloroethyl trimethyl ammonium chloride) is used to shorten wheat plants in Europe to allow higher levels of fertilizer. Many greenhouse-grown flowers are kept short in stature by CCC and other growth regulators. Striking effects from gibberellins and fumaric acid have also been reported. The first greatly increase vegetative growth and the latter causes dwarfing. In practice, growth-inhibiting or growth-retarding agents are finding wider use than growth-stimulating ones. In higher concentration many growth-retarding agents become inhibiting agents.

There are marked differences between plant species and even varieties in their response to most plant growth regulators. Many, if not most, growth regulators are highly selective; a concentration of even 100 times that effective for one species or variety is necessary to produce the same response in another. Furthermore, the specific formulation of the substances, for example, the kind or amount of wetting agent used with them, is important in determining their effectiveness. In brief, growth regulators are essentially new products, though there are century-old instances of the empirical use of a few of them. Some influence growth rate, others direction of growth, others plant structure, anatomy, or morphology. With the discovery of new ones, indications are that it is only a matter of time before many features of plant growth and development may be directly or indirectly controlled by them to a marked degree. Applications of these substances in intensive agriculture are unfolding rapidly, and their use is one of the many factors making farming more of a science and less of an art.

Plant improvement. From earliest times humans have tried to improve plants by selection, but it was the discovery of hybridization (cross-mating of two genetically different plants) that eventually led to dramatic increases in genetic potential of the plants. Hybridization was recognized in the early 1800s, well before Mendel's classic genetic discoveries, and allowed the combination of desirable plants in a complementary manner to produce an improved progeny. Plant breeding had a dramatic flourish in the early 20th century following the rediscovery of Mendel's research and its implications, and has had much to do with the increased productivity per area of present-day agriculture.

Corn provides the most vivid example of how improvement through genetic manipulation can occur. Following the commercial development of corn hybrids about 1930, only a few acres were planted, but by 1945 over 90% of the acreage was planted to hybrids, and today nearly 100% is planted. It has been conservatively estimated that hybrids of corn have 25% more yield potential than old-style varieties. Subsequently, plant breeders have utilized hybridization for development of modern varieties of most major crop species.

While very significant changes through crop breeding have occurred in pest resistance and product quality, perhaps the character of most consequence was an increase in the lodging (falling over) resistance of major grain crops. Corn, wheat, and rice have all been bred to be shorter in stature and to have increased stem resistance to breaking. These changes have in turn allowed heavier fertilization of crops to increase photosynthetic area and yield. The impact of this was recognized when Norman Borlaug was awarded the Nobel Peace Prize in 1970 for his breeding contribution to the "green revolution." His higher-yielding wheats were shorter and stiff-strawed so they could utilize increased amounts of nitrogen fertilizer. They were also day-length-insensitive and thus had a wide adaptation. New varieties of rice such as IR-8 developed at the International Rice Research Institute in the Philippines are shorter, are more responsive to nitrogen fertilizer, and have much higher potential yield than conventional varieties. With those new wheat and rice varieties many countries gained time in their battle between population and food supply. *See* BREEDING (PLANT).

Often in tree crops and high-value crops it is not feasible to make improvements genetically, and other means are utilized. Grafting, or physically combining two or more separate plants, is used as a method of obtaining growth control in many fruit trees, and also has been used to provide disease resistance and better fruiting characteristics. This technique is largely limited to woody species of relatively high value. Usually tops are grafted to different rootstocks to obtain restricted vegetative growth, as in dwarf apple and pear trees which still bear normal-sized fruit. Alternatively, junipers and grapes are grafted to new rootstocks to provide a better root system. In both cases the desirability of the esthetic or economic portion warrants the cost and effort of making the plant better adapted to environmental or management conditions.

[CURTIS J. NELSON]

Bibliography: R. J. Delorit, L. J. Greub, and H. L. Ahlgren, *Crop Production*, 1974; J. R. Harlan, *Crops and Man*, 1975; J. Janick et al., *Plant Science: An Introduction to World Crops*, 1974; J. H. Martin et al., *Principles of Field Crop Production*, 3d ed., 1976.

Agricultural wastes

The food and fiber industry, including production and processing, produced over 2×10^9 tons (1.81×10^9 metric tons) of waste in 1970, or about 58% of all the solid wastes produced in the country. These wastes include the obvious manure and organic residues from farms and forests, as well as various solid materials and water discharged from processing or manufacturing plants that use any form of an agricultural product as a raw material. Some of these wastes are utilized or recycled, but most of them require disposal. Many of the present disposal techniques simply move the waste to another place rather than solve the problem. Therefore the only true solution to the waste problems of agriculture is to consider the material as a national resource to be recovered by recycling and utilization.

Solid wastes. The following sections discuss the recycling of various types of solid waste.

Animal wastes. Disposal of animal wastes poses a major problem to the agricultural industry. Tremendous amounts of manure are produced (Table 1), but it is the concentration of this manure in small areas that constitutes the real problem. Agricultural science and technology met the challenge to increase animal production by developing procedures to raise large numbers of animals in confinement, as in cattle feedlots. Unfortunately the technology for managing animal waste has not developed as rapidly. Where usual disposal techniques are being curtailed by social pressure or by possible danger to the environment, the animal producer is forced to operate under a narrowing profit margin and higher overhead costs. Recycling and utilization may help to alleviate these problems.

Many systems for injecting manure into the subsoil have been investigated. Such systems reduce the insect and odor problems, but the operations are not economical and probably will be used only in areas where the social pressures are extreme. At present, land spreading must be considered as a least expensive rather than economical or profit-making method of waste handling.

Many methods have been investigated for the potential utilization of animal wastes, for example, selling as fertilizer, composting, animal feeding, and energy (methane) production. The primary purpose of composting is to eliminate putrescible organic matter while conserving much of the original plant nutrients. Manures may be composted alone but frequently are combined with high-carbon, low-nitrogen wastes such as sawdust, corn cobs, paper, and municipal refuse. The resulting material is suitable for use as a soil conditioner and organic fertilizer. Full-scale operations have been technically successful with poultry, beef, and dairy manure in a unique regional situation, but in general the market for this product is very small. Without a market, essentially all the dry matter remains for further disposal. Manure, either as a fertilizer or soil conditioner, is difficult to sell profitably, regardless of preparation procedures, because of the low cost of commercial fertilizers.

Dried and dehydrated manure is sometimes sold as a soil conditioner, organic fertilizer, or animal feed supplement, but the cost of drying and dehydration is generally greater than the return realized.

Use of animal wastes in the feed of animals has recently received much publicity as a potential method for utilizing agricultural wastes. When nutritional principles are followed, the technique has produced good results, especially if the waste of a single-stomached animal is added to the rations of ruminants or if the waste of the ruminant is chemically treated before use. However, a Food and Drug Administration regulation (1970) prohibits the use of animal wastes as feed supplements because of the possible transmission of drugs, feed additives, and pesticides to another animal and to some agricultural products, such as milk and eggs. On-going research should clarify the situation.

Much of the nutritional value of animal wastes might be captured by growing housefly larvae and insects as a source of protein for animal feed supplement. The method has been tested and shows considerable promise. Algae can be commercially grown on a manure substrate and can also be used as a feed supplement. The remaining growth media could be used as a soil conditioner or fertilizer.

Manure can be used directly as a fuel or as the substrate to produce methane anaerobically. Manure must be collected and dried to be used as a fuel, a costly operation. Therefore it is doubtful that any appreciable quantity will ever be utilized in this manner.

Regardless of the disposal procedure or recovery process, the soil will be the ultimate disposal site of most of the bulk of animal wastes. Therefore the challenge is to develop techniques of recycling to incorporate animal wastes in land management programs without damaging the environment or causing a nuisance to the human population. This will most likely be accomplished by convincing the farmer that manure improves the physical condition of a soil as well as supplies plant nutrients.

Crop and orchard residues. The tonnage of plant residues left on the farms exceeds by far the tonnage of the crops taken to market. These wastes

Table 1. Numbers of livestock and their total waste production in the United States in 1970

Livestock	Total population, millions	Solid wastes, millions of metric tons/year	Liquid wastes, millions of metric tons/year	Total wastes, millions of metric tons/year
Cattle	107	1015.0	391.0	1406.0
Hogs	57	62.5	36.5	99.0
Sheep	21	9.6	5.8	14.4
Horses	3	17.5	4.4	21.9
Chickens	2950	54.0	–	54.0
Turkeys	106	21.7	–	21.7
Ducks	11	1.6	–	1.6
Total	–	1181.9	–	1618.6

Table 2. Wastes produced by selected agricultural industries in the United States during 1970

Type of processing industry	Total solid waste discharged, millions of metric tons/year	Solid waste salvaged, millions of metric tons/year	Solid waste requiring disposal, millions metric tons/year	Comment
Meat	17.5	17.5	0.003	Values include all livestock and broilers
Vegetables and fruit	12.7	8.7	4.0	Organic and inorganic wastes in disposed fraction
Dairy	0.2	*	*	As fat-free solids
Forestry	36.0	*	*	Logging residue from lumber mill
	55.0	*	*	As sawdust and edging
	41.5	*	*	From pulp and paper mill

*No estimates are made because the amount of salvaged and disposed by-product depends upon the location and size of the plant or operation.

consist of straw, stubble, leaves, hulls, vines, tree limbs, and similar trash. Most of the residue is burned to eliminate troublesome plant diseases, pests, and weeds. Unfortunately this pollutes the air with smoke and volatile organic compounds. A small part of these materials is being used for mulch and as a soil builder when the mulch is plowed under, ensilage, bedding for animals, and as bulk material in the manufacture of corrugated cartons, insulated board, and specialty paper. Although the handling of crop residues is not a major problem in terms of amount, these wastes can be a focal point of major infestation by harboring insects and plant diseases. Obviously, new and better methods are needed to handle these wastes without polluting the environment.

Food processing wastes. The meat processing industry produces few wastes, other than water, that require disposal (Table 2). This industry has illustrated that the utilization of wastes can be profitable. In recent years over 35×10^9 lb (15.89×10^9 kg) of meat and similar amounts of other animal parts, or wastes, have been processed annually. These wastes, including those collected during water disposal procedures, are the raw material for the manufacture of soap, leather goods, glue, gelatin, and animal feeds. Glands and organs are processed to produce hormones, vitamins, enzymes, liver products, bile acids, and sterols. Other animal parts are the source of certain fatty acids, oils, grease, and glycerine. Bones can be processed to produce proteins and fats, as well as bone meal. Finally fuels and solvents can be made from the waste products of the above industries. The waste waters leaving the processing plants still contain small amounts of solids, and disposal procedures are necessary. The utilization of these suspended materials must be preceded by a more efficient use of water which will result in concentrating the wastes and improving the economics of recovery.

Although the fruit and vegetable processing industry does not produce large amounts of solid wastes relative to other segments of agriculture, these wastes are difficult to handle because of their varied nature. Peels, skins, pulp, seeds, and fibers are suspended in billions of gallons of water which may be saline, alkaline, or acidic and may contain a wide variety of soluble organic compounds. Of the 12.7×10^6 tons (11.52×10^6 metric tons) of solid waste produced, only about 4×10^6 tons require disposal—but at a cost of $\$25 \times 10^9$. Most of the remaining 8.7×10^6 tons are utilized as animal feed and consist principally of the by-products of the processing of citrus fruits, potatoes, and corn. These by-products are usually given away, resulting in a cost-free disposal procedure which is very valuable to the industry. Increasing the efficiency of the in-plant use of water is a necessity before any new salvage operations will be economical.

Approximately 43% of all milk produced goes into fluid milk and cream; the remainder is used for butter, cheese, and other products. The amount of solid wastes coming from this industry is small compared with the amounts of other agricultural wastes, but they require costly handling. Milk solids have very high pollution potentials. Therefore these solids must be removed from millions of gallons of water daily before this water can be released to streams or used for irrigation. The collected milk solids do have value as animal feed, feed supplements, raw material for the production of some chemicals such as alcohol, and as a basic ingredient in growth medium for microorganisms in the production of pharmaceutical chemicals. Lowering the cost of production and increased utilization are, as before, related to a more efficient use of water.

Forest waste products. A large part of a tree harvested for pulp paper or lumber becomes waste. It has been estimated that 19% of the tree is left in the forest. Approximately 16% of the log is wasted as sawdust and 34% as slabs and edgings during milling. Finally pulp and paper mills discharge effluents containing about 50% of the log. Wastes from the wood industry have great pollution potential and are usually destroyed by burning because other disposal procedures are too costly. Burning of forest debris is deemed to be necessary in the control of forest diseases and insects, but it results in air pollution.

There have been many processes developed by wood science and technology laboratories to recover a large variety of by-products, but these processes have been only sparingly applied, primarily because of economic limitations. Wood is composed primarily of lignin and cellulose. Lignin can be processed to obtain many valuable chemicals, such as artificial vanilla, or used directly as a binder, a dispersant, an emulsion stabilizer, and a sequestrant. Lignin can also be utilized as the raw material for plastic production. Cellulose is readily processed to simple sugars, alcohol, fodder, yeasts, and chemicals such as furfural. Utilization methods that may be economically feasible in the future include the use of wastes for the production of specific chemicals, through fermentation, and in new building products such as board, paper, and blocks.

Recycling of paper does not seem to have a promising future. Now less than 20% is recycled as compared to about 30% 20 years ago. If the cost of removing the adulterations of the paper, such as ink, plastic, and metal clips, is decreased, waste paper recycling probably will be increased, thereby decreasing the overall wastes produced by the wood industry.

Liquid wastes. The agricultural industry uses water in almost every operation of production, processing, and manufacturing. Irrigation and food processing use the largest amounts. These waters must be considered agricultural wastes in the same sense as are solid wastes and must either be recovered or disposed of. In the past the axiom "control pollution by dilution" was readily accepted. Consequently, large volumes of water were used to dilute the agricultural wastes before they were discharged into lagoons or other surface waters, sprinkled on a field, or discharged into a municipal sewage system. The great increase of soluble and solid wastes to be transported, the lack of usable water in some areas, and social pressures are causing a reevaluation of water use for this purpose. Processing plant procedures are being changed to increase recycling of water in the plant by segregating highly contaminated water and using the clean discharged water to irrigate crops. The highly contaminated water is handled by conventional disposal procedures before it is utilized or recycled.

Irrigation of agricultural crops uses vast quantities of water. In the past this was not carefully controlled and streams were contaminated by runoff and drainage waters containing sediment, soluble salts, pesticides, and so forth. However, important changes are taking place, such as better controls in the ditches and fields and reuse of runoff irrigation waters. Irrigation water is reused by collecting the normal runoff water in shallow ponds and then pumping it to other fields. Less than 10% of the liquid water is lost by this method. Also, most of the soluble and solid materials usually lost in runoff are kept on the farm, thereby decreasing the possible spread of a polluting agent, although salt buildup may be a problem in some areas.

Agricultural waters can also be used to recharge groundwater. Uncontaminated water can be discharged directly into the groundwater. Contaminated water can be processed, or in some cases the soil can be used as a filter. *See* WATER POLLUTION.

[WALTER R. HEALD]

Bibliography: P. N. Hobson and A. M. Robertson, *Waste Treatment in Agriculture*, 1978; N. H. Wooding, Jr., *Spec. Circ. No. 113*, Pennsylvania State University Extension Service, 1970.

Air mass

A term applied in meteorology to an extensive body of the atmosphere which approximates horizontal homogeneity in its weather characteristics. An air mass may be followed on the weather map as an entity in its day-to-day movement in the general circulation of the atmosphere. The expressions air mass analysis and frontal analysis are applied to the analysis of weather maps in terms of the prevailing air masses and of the zones of transition and interaction (fronts) which separate them. The relative horizontal homogeneity of an air mass stands in contrast to the sharp horizontal changes in a frontal zone. The horizontal extent of important air masses is reckoned in millions of square miles. In the vertical dimension an air mass extends at most to the top of the troposphere, and frequently is restricted to the lower half or less of the troposphere. The frontal zones between air masses usually slope in such a manner that the colder air mass underlies the warmer as a wedge. In the vertical direction the properties of an air mass, specifically its content of heat and moisture, may vary between a high degree of stratification and one of homogeneity produced by vertical mixing. *See* FRONT; METEOROLOGY.

Development of concept. Practical application of the concept to the air mass and frontal analysis of daily weather maps for prognostic purposes was a product of World War I. A contribution of the Norwegian school of meteorology headed by V. Bjerknes, this development originated in the substitution of close scrutiny of weather map data from a dense local network of observing stations for the usual far-flung internation network. The advantage of air-mass analysis for practical forecasting became so evident that during the three decades following World War I the technique was applied in more or less modified form by nearly every progressive weather service in the world. However, the rapid increase of observational weather data from higher levels of the atmosphere during and since World War II has resulted in a progressive tendency to drop the careful application of air-mass analysis techniques in favor of those involving the kinematical or dynamic analysis of upper-level air-flow patterns.

Origin. The occurrence of air masses as they appear on the daily weather maps depends upon two facts, the existence of air-mass source regions, and the large-scale character of the branches or elements of exchange of the general circulation. Air-mass source regions consist of extensive areas of the Earth's surface which are sufficiently uniform so that the overlying atmosphere acquires similar characteristics throughout the region; that is, it approximates horizontal homogeneity. The designation of an area of the Earth's surface as a source region assumes that the overlying atmosphere in that area normally remains there long enough to approximate thermodynamic equilibrium with respect to the underlying surface, or in other words, to acquire the weather characteristics that typify that particular source region.

The large-scale character of the elements by which the general circulation is accomplished is observed on the daily weather maps in the major atmospheric currents of polar or tropical origin, whose southward or northward progress can be traced from day to day. These major currents, together with the associated polar and tropical air masses, are the means by which surplus tropical heat is effectively transported to polar latitudes. *See* ATMOSPHERIC GENERAL CIRCULATION.

Weather significance. The thermodynamic properties of air mass determine not only the general character of the weather in the extensive area covered by the air mass but also, to some extent, the severity of the weather activity in the frontal zone of interaction between air masses. Those properties which determine the primary weather characteristics of an air mass are defined by the vertical distribution of the two elements, water vapor and heat (temperature). On the vertical distribution of water vapor depend the presence or absence of condensation forms and, if present, the elevation and thickness of fog or cloud layers. On the vertical distribution of temperature depend the relative warmth or coldness of the air mass and, more importantly, the vertical gradient of temperature, known as the lapse rate. The lapse rate determines the stability or instability of the air mass for thermal convection and, consequently, the stratiform or convective cellular structure of the cloud forms and precipitation. The most unstable moist air mass, in which the vertical lapse rate may approach 1°C/100 m, is characterized by severe turbulence and heavy showers or thundershowers. In the most stable air mass there is observed an actual increase (inversion) of temperature with increase of height at low elevations. With this condition there is little turbulence, and if the air is moist there is fog or low stratus cloudiness and possible drizzle, but if the air is dry there will be low dust or industrial smoke haze. *See* TEMPERATURE INVERSION.

Classification. A wide variety of systems of classification and designation of air masses was developed by different weather services around the world. The usefulness of a system with type designators to be applied in the analysis of weather maps is directly proportional to its effectiveness in accurately expressing the thermodynamic properties which determine the weather characteristics of the air mass. These properties are imparted to the air masses primarily by the particular source region of its origin, and secondarily by the modifying influences to which it is subjected after leaving the source region. Consequently, most systems of air-mass classification are based on a designation of the character of the source region and the subsequent modifying influences to which the air mass is exposed. Probably the most effective and widely applied system of classification is a modification of the original Norwegian system that is based on the following four designations.

Polar versus tropical origin. All primary air-mass source regions lie in polar (*P* in Figs. 1 and 2) or in tropical (*T*) latitudes. In middle latitudes there occur the modification and interaction of air masses initially of polar or tropical origin. This difference of origin establishes the air mass as cold or warm in character.

Maritime versus continental origin. To be homogeneous, an air-mass source region must be exclusively maritime or exclusively continental in character. On this difference depends the presence or absence of the moisture necessary for extensive condensation forms. However, a long trajectory over open sea transforms a continental to a maritime air mass, just as a long land trajectory, particularly across major mountain barriers, transforms a maritime to a continental air mass. On Figs. 1 and 2, *m* and *c* are used with *P* and *T* (*mP*, *cP*, *mT*, and *cT*) to indicate maritime and continental character, respectively.

Heating versus cooling by ground. This influence

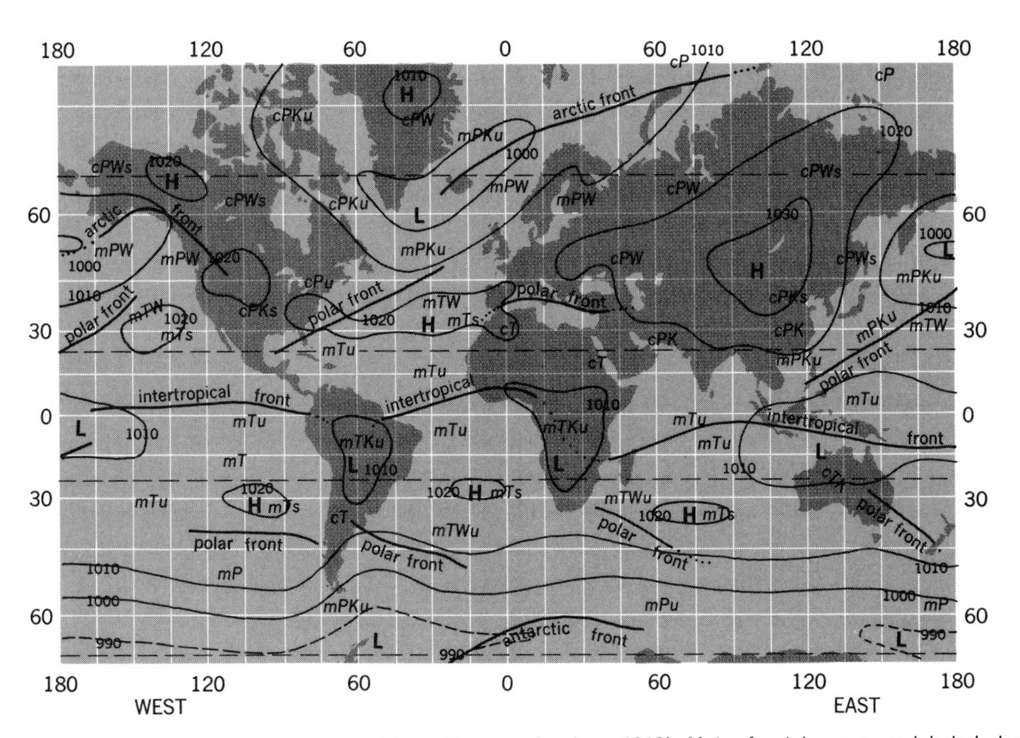

Fig. 1. Air-mass source regions, January. High- and low-atmospheric-pressure centers are designated *H* and *L* within average pressure lines numbered in millibars (such as 1010). Major frontal zones are labeled along heavy lines. (*From H. C. Willett and F. Sanders, Descriptive Meteorology, 2d ed., Academic Press, 1959*)

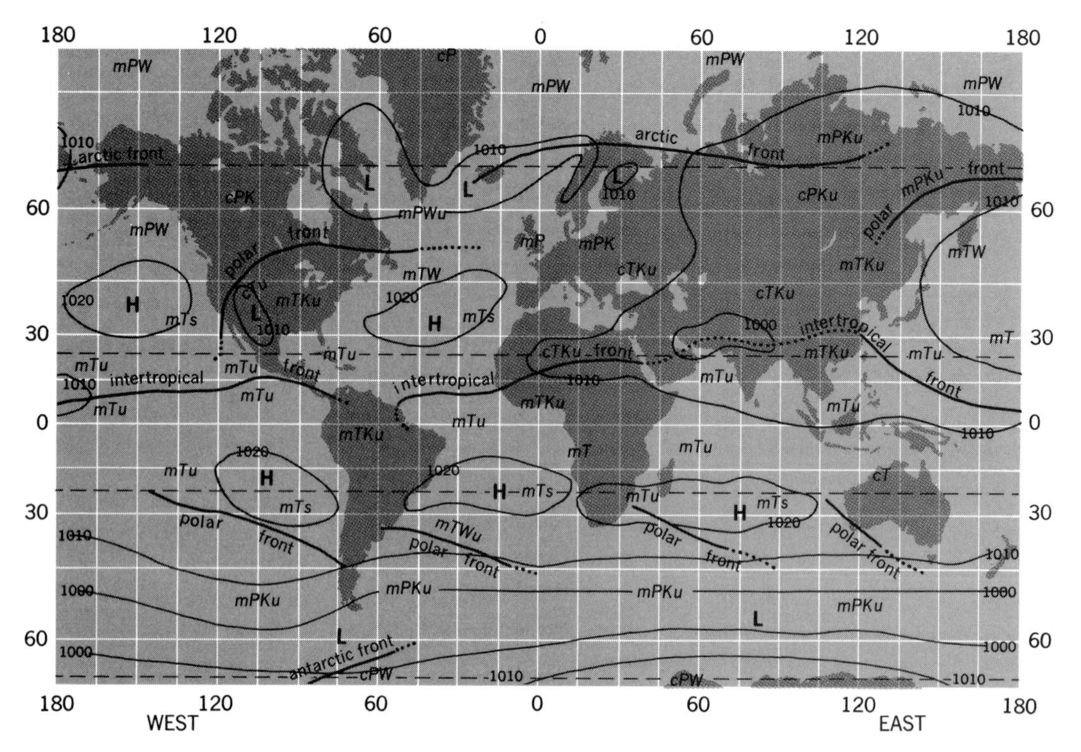

Fig. 2. Air-mass source regions, July. The symbols which are used in this figure are the same as those for Fig. 1. *(From H. C. Willett and F. Sanders, Descriptive Meteorology, 2d ed., Academic Press, 1959)*

determines whether the air mass is vertically unstable or stable in its lower strata. In a moist air mass it makes the difference between convective cumulus clouds with good visibility on the one hand and fog or low stratus clouds on the other. Symbols *W* (warm) and *K* (cold) are used on maps—thus, *mPK* or *mPW*.

Convergence versus divergence. Horizontal convergence at low levels is associated with lifting and horizontal divergence at low levels with sinking. Which condition prevails is dependent in a complex manner upon the large-scale flow pattern of the air mass. Horizontal covergence produces vertical instability of the air mass in its upper strata (*u* on maps), and horizontal divergence produces vertical stability (*s* on maps). On this difference depends the possibility or impossibility of occurrence of heavy air-mass showers or thundershowers or of heavy frontal precipitation. Examples of the designation of these tendencies and the intermediate conditions for maritime polar air masses are *mPWs, mPW, mPWu, mPs, mPu, mPKs, mPK,* and *MPKu.*

[HURD C. WILLETT]

Bibliography: W. L. Donn, *Meteorology*, 4th ed., 1975; H. C. Willett and F. Sanders, *Descriptive Meteorology*, 2d ed., 1959.

Air pollution

Alteration of the atmosphere by the introduction of natural and artificial particulate contaminants. Most artificial impurities are injected into the atmosphere at or near the Earth's surface. The atmosphere cleanses itself of these quickly, for the most part. This occurs because in the troposphere, that part of the atmosphere nearest to the Earth, temperature decreases rapidly with increasing altitude (Fig. 1), resulting in rapid vertical mixing: the rainfall sometimes associated with these conditions also assists in removing the impurities. Exceptions, such as the occasional temperature inversion layer over the Los Angeles Basin, may have notably unpleasant results. *See* ATMOSPHERE.

In the stratosphere, that is, above the altitude of the temperature minimum (tropopause), either temperature is constant or it increases with altitude, a condition that characterizes the entire stratosphere as a permanent inversion layer. As a result, vertical mixing in the stratosphere (and hence, self-cleansing) occurs much more slowly than that in the troposphere. Contaminants introduced at a particular altitude remain near that altitude for periods as long as several years. Herein lies the source of concern: the turbulent troposphere cleanses itself quickly, but the relatively stagnant stratosphere does not. Nonetheless, the pollutants injected into the troposphere and stratosphere have impact on humans and the habitable environment.

All airborne particulate matter, liquid and solid, and contaminant gases exist in the atmosphere in variable amounts. Typical natural contaminants are salt particles from the oceans or dust and gases from active volcanoes; typical artificial contaminants are waste smokes and gases formed by industrial, municipal, household, and automotive processes, and aircraft and rocket combustion processes. Another postulated important source of artificial contaminants is certain fluorocarbon compounds (gases) used widely as refrigerants, as propellants for aerosol products, and for other applications. Pollens, spores, rusts, and smuts are natural aerosols augmented artificially by humans' land-use practices. *See* ATMOSPHERIC CHEMISTRY; SMOG.

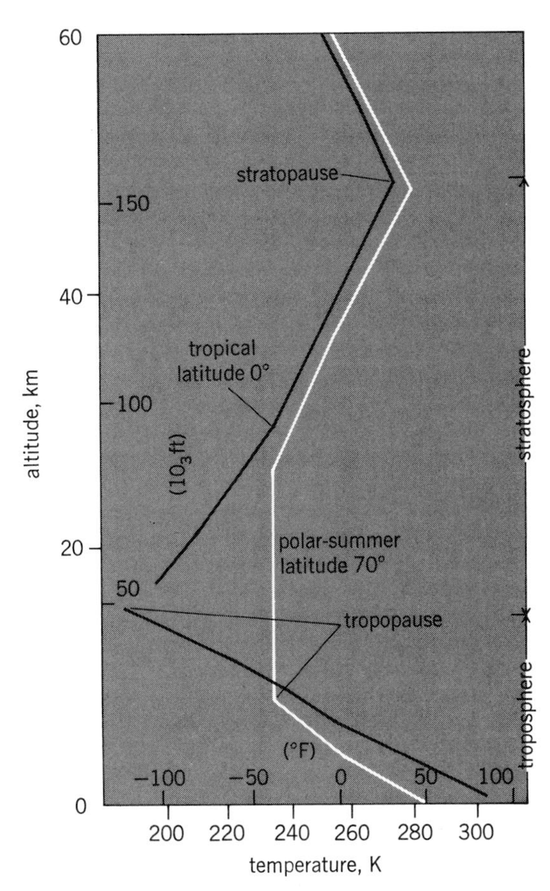

Fig. 1. The atmosphere's temperature-altitude profile. *(Adapted from R. E. Newell, Radioactive Contamination of the Upper Atmosphere, in Progress in Nuclear Energy, ser. 12: Health Physics, vol. 2, p. 538, Pergamon Press, 1969)*

Sources and types. Sources may be characterized in a number of ways. A frequent classification is in terms of stationary and moving sources. Examples of stationary sources are power plants, incinerators, industrial operations, and space heating. Examples of moving sources are motor vehicles, ships, aircraft, and rockets. Another classification describes sources as point (a single stack), line (a line of stacks), or area (a city).

Different types of pollution are conveniently specified in various ways: gaseous, such as carbon monoxide, or particulate, such as smoke, pesticides, and aerosol sprays; inorganic, such as hydrogen fluoride, or organic, such as mercaptans; oxidizing substances, such as ozone, or reducing substances, such as oxides of sulfur and oxides of nitrogen; radioactive substances, such as iodine-131, or inert substances, such as pollen or fly ash; or thermal pollution, such as the heat produced by nuclear power plants.

Air contaminants are produced in many ways and come from many sources. It is difficult to identify all the various producers. For example, it is estimated that in the United States 60% of the air pollution comes from motor vehicles and 14% from plants generating electricity. Industry produces about 17% and space heating and incineration the remaining 9%. Other sources, such as pesticides and earth-moving and agricultural practices, lead to vastly increased atmospheric burdens of fine soil particles, and of pollens, pores, rusts, and smuts; the latter are referred to as aeroallergens because many of them induce allergic responses in sensitive persons.

The annual emission over the United States of many contaminants is very great (Fig. 2). As mentioned, motor vehicles contribute about 60% of total pollution: nearly all the carbon monoxide, two-thirds of the hydrocarbons, one-half of the nitrogen oxides, and much smaller fractions in other categories.

Pollution in the stratosphere. Sources of contaminants in the stratosphere are effluents from high-altitude aircraft such as the supersonic transport (SST), powerful nuclear explosions, and volcanic eruptions. There are also natural and artificial sources of gases which diffuse from the troposphere into the stratosphere.

Table 1 lists the natural burden of gases and particles injected into the stratosphere by high-flying aircraft, assuming the consumption of 2×10^{11} kg of fuel during a period of one year. It is to be noted that the percentage increase over the natural burden is substantial for NO, NO_2, HNO_3, and SO_2. Since the concentrations are substantial, they may adversely affect humans' living environment.

Other manufactured pollutants which diffuse from the troposphere into the stratosphere are the halogenated hydrocarbons, specifically dichloromethane (CF_2Cl_2) and trichlorofluoromethane ($CFCl_3$) gases which are used as propellants in many of the so-called aerosol spray cans for deodorants, pesticides, and such, and as refrigerants. These gases have an average residence time (residence time is the time required for a substance to reduce its concentration by $1/e$, approximately 1/3) of 1000 years or more. $CFCl_3$ is one of a family of halogenated hydrocarbons (also known as fluorocarbons) which are widely used. Of the total amount of these compounds produced worldwide, almost all are ultimately released to the atmosphere.

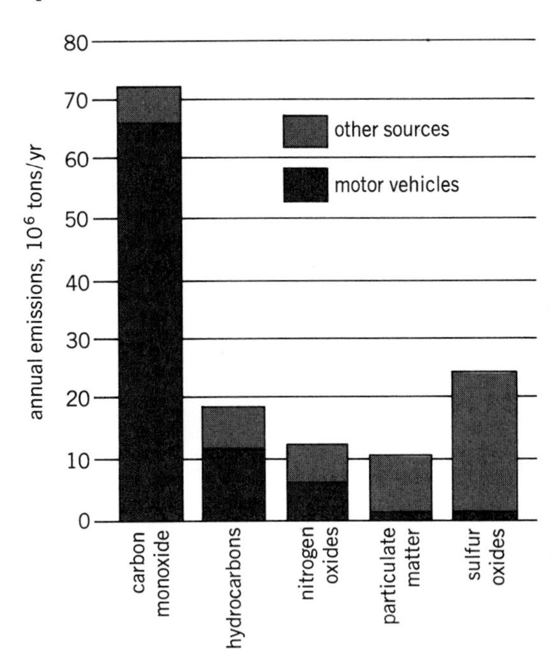

Fig. 2. Motor vehicles' contribution to five atmospheric contaminants in the United States.

Table 1. The natural stratospheric background of several atmospheric gases from 13 to 24 km compared to engine emissions

Gas	Mass mixing ratio	Natural burden, kg		Increase in mass due to aircraft emission, %‡
		IDA*	Penndorf†	
CO_2	480 E-6	500 E-12§	480 E-12	0.1
H_2	2.7 E-6	2 E-12	2.7 E-12	9
CH_4	0.55 E-6	1 E-12	0.55 E-12	0.02
CO	0.05 – 0.1 E-6	30 E-9	50 – 100 E-9	0.6 – 1.2
NO	0.5 E-9	1 E-9	0.52 E-9	100
NO_2	1.6 E-9	3 E-9	1.8 E-9	100
HNO_3	4 E-9	<10 E-9	3.6 E-9	85
SO_2	1 E-9	4 E-9 (?)	1.4 E-9	10 – 40
Aerosol ($\alpha > 0.1\ \mu m$)	2 E-9	0.3 E-9	2 E-9	10

*Estimation by R. Oliver, Institute for Defense Analyses, 1974.
†Estimation by R. Penndorf, *CIAP Atmospheric Monitoring and Experiments, The Program and Results*, DOT-TST-75-106, pp. 4 – 7, 1975.
‡One-year fuel consumption by stratospheric aircraft of 2×10^{11} kg.
§Read 500 E-12 as 5×10^{12}.

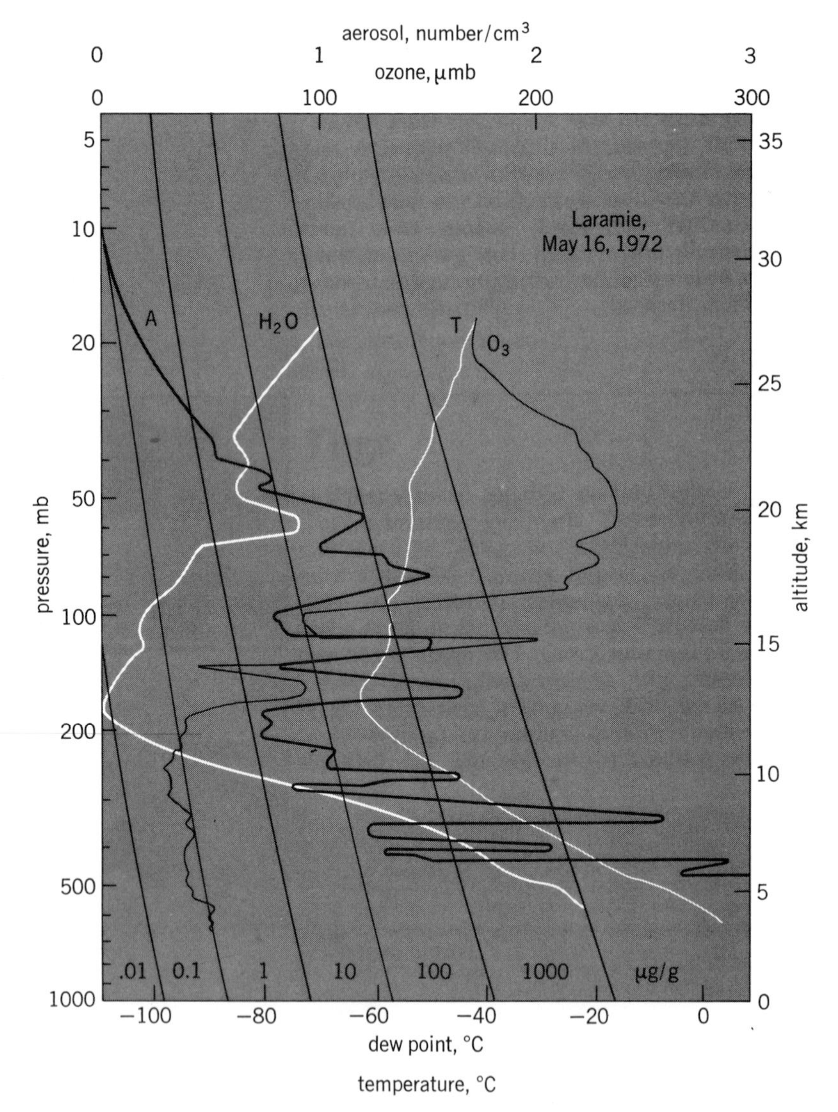

Fig. 3. A simultaneous measurement of the vertical distribution of ozone (O_3), water vapor (H_2O), and temperature (T). The aerosol (A) sounding was made about 5 hr before the ozone–water vapor soundings. The smooth curves are lines of constant water vapor mixing ratio. *(From T. J. Pepin, J. M. Rosen, and D. H. Hoffman, The University of Wyoming Global Monitoring Program, in Proceedings of the AIAA/AMS International Conference on the Environmental Impact of Aerospace Operations in the High Atmosphere, AIAA Pap. no. 73-521, 1973)*

The nitrous oxide (N_2O) and the halogenated hydrocarbons reach the upper regions of the stratosphere, where they are photodissociated by the Sun's radiation to produce nitric oxide (NO) and chlorine (Cl). The NO and Cl react with the ozone, as in Eqs. (1)–(6). The ozone is destroyed

$$N_2O + O \rightarrow 2NO \quad (1)$$

$$NO + O_3 \rightarrow NO_2 + O_2 \quad (2)$$

$$NO_2 + O \rightarrow NO + O_2 \quad (3)$$

$$CF_2Cl_3 + h\nu \text{ (solar energy)} \rightarrow CFCl_2 + Cl \quad (4)$$

$$Cl + O_3 \rightarrow ClO + O_2 \quad (5)$$

$$ClO + O \rightarrow Cl + O_2 \quad (6)$$

by NO and Cl respectively, whereas NO and Cl are conserved. Stratospheric ozone is valuable as a filter for solar ultraviolet (uv) radiation. A decrease of its concentration results in an increase in the amount of uv impinging on the surface of the Earth. As an illustration based on theoretical considerations, an injection 2×10^9 kg/yr of NO_x($NO_x = NO + NO_2$) at 17 km (see Fig. 1) will result in a reduction of the total amount of ozone by about 3%. This represents an increase of vertically incident uv on the Earth of 6%. An increase of uv could adversely affect both humans and plants.

SO_2-aerosol-climate relations. Aircraft engine effluents contain SO_2, as shown in Table 1, and could add a considerable amount of aerosol particles at the end of this century for the predicted aircraft fleet sizes. Table 1 indicates aircraft effluents could increase the natural background by 10–40%—seemingly large, yet small compared to volcanic injections, for which estimates range up to 10,000%.

Why are particles so important? They scatter and absorb (in specific wavelength regions) solar radiation, and thereby influence the radiative budget of the Earth-atmospheric system, and finally perhaps the climate on the ground. The particles formed from aircraft effluents may increase the optical thickness of the layer, the upwelling and downwelling infrared radiation, and the albedo of the Earth. The average global albedo of the Earth-atmospheric system has been measured as 28% with probably some short-term variation of unknown but small magnitude. It has been calculated that for an additional mass of 0.1 $\mu g/m^3$ particles over a 10-km layer from 15 to 25 km (equivalent to 5.1×10^8 kg for the whole Earth, or about 20% of the "natural" background concentration), the albedo increases by about 0.05% (from 28 to 28.05%) at low latitudes all year and at high latitudes in summer, but by about 0.1–0.15% from September to February in high latitudes. If the added mass is larger than 0.1 $\mu g/m^3$, the albedo increases proportionally to the cited numbers. For the optical thickness of the stratosphere, a value of 0.02 is generally assumed. While the present subsonic flights increase this value by a very small amount (10^{-4}), a large fleet of high-flying aircraft (Table 1) could increase it by about 10%.

The chemistry of the natural stratospheric aerosols is dominated by sulfate, presumably of volcanic origin. These naturally occurring aerosols are concentrated in thin layers at altitudes between 15 and 25 km. R. Cadle has described the chemical composition of stratospheric particles at an altitude of 20 km during the period 1969–1973 as consisting of 48%, by mass, of sulfate; 24% of stony elements (such as silicon, aluminum, calcium, and magnesium); and 20% of chlorine; other constituents make up the remainder. The amount of sulfate introduced into the stratosphere as a result of a large fleet of SSTs operating at about 18–20 km may, in a worst-case estimate, equal the total amount occurring naturally. The influence of dynamic motions of the stratosphere on the distribution of aerosols is indicated by Fig. 2, which illustrates high correlation of the aerosol and ozone-rich layers in the 15-km region. Moreover, the water vapor mixing ratio also increases in layers at about 15–20 km. Since there is no known chemical link between the production of aerosols and that of water vapor and ozone, the observation illustrated by Fig. 3 may be a dynamic rather than a chemical effect, the implication being that the dynamic effects are far more important in determining the relative profiles of these constituents than any chemical effect at this altitude.

The perturbation of the lower stratosphere by the engine effluents of a large fleet of vehicles may strongly increase its optical thickness to visual-band solar radiation, which has a natural value of about 0.02. An increase in optical thickness results in a reduction of solar radiation, the principal source of atmospheric heating, by about the same fraction. This effect can be likened to a reduction in solar constant by about the same amount at the subsolar point, and about double that value when the solar zenith angle is 60° or greater, as it may be at high latitudes. The sensitivity of the troposphere to changes in the solar constant has been studied by M. MacCracken. In the modeling for which the computed precipitation is illustrated by Fig. 4, a 10% reduction of solar constant leads to a reduction of average temperatures, from 3°C at the Equator to 10°C average from latitude 40° to the pole. The winds are substantially weakened, and total precipitation reduced, mainly in the summer. Snowfall increases, and the winter snow line moves lower in latitude by about 5°.

An increase in the solar constant of 13% results in an annual mean temperature increase of 2–5°C at all latitudes. Although the overall precipitation increases, as illustrated in Fig. 4, the relative humidity decreases slightly, and thus cloudiness decreases. The total snowfall decreases, and the snow line moves higher in latitude by 10°. Extensive melting of the polar ice caps also begins. The effects of doubling of the stratospheric optical thickness, for example, are an order of magnitude smaller than the changes of precipitation and snowfall depicted in Fig. 4.

Sinks. A sink is defined as a process by which gases or particles are removed from a given volume of atmosphere. It could be chemical, homogeneous or heterogenous (gas-solid reactions), or dynamical, such as dispersion (transport), diffusion, gravity, or precipitation.

In the stratosphere, three important contaminants are NO, Cl, and SO_2. The sink for NO_2 is a chemical reaction by which NO_2 is combined by a complex method with (OH)—a derivative of water (H_2O) present in the stratosphere in parts per million (ppm)—to form nitric acid (HNO_3).

The SO_2 reacts chemically with oxygen (O), water, and its derivatives (OH, HO_2) to form H_2SO_4. Then, water molecules are absorbed by H_2SO_4 to form $nH_2O{\cdot}(H_2SO_4)$ cluster or aerosol (where n, number, is equal to about 2). Aggregates of these polymolecules, growing larger with each successive collision, eventually become aerosols; when the diameter is greater than 0.1 μm, they act as scatterers of the visual band of sunlight.

The chlorine (Cl) chemically reacts with oxygen and water to form Cl, ClO, and HCl. In the stratosphere, these reactions involving small concentrations have not been measured adequately.

In the troposphere, the predominant sinks are dispersion, transport, and precipitation. The chemical reactions in the troposphere are less active in general than in the stratosphere.

Dispersion. Dispersion of pollution is dependent on atmospheric conditions. Winds transport and diffuse contaminants; rain may wash them to the

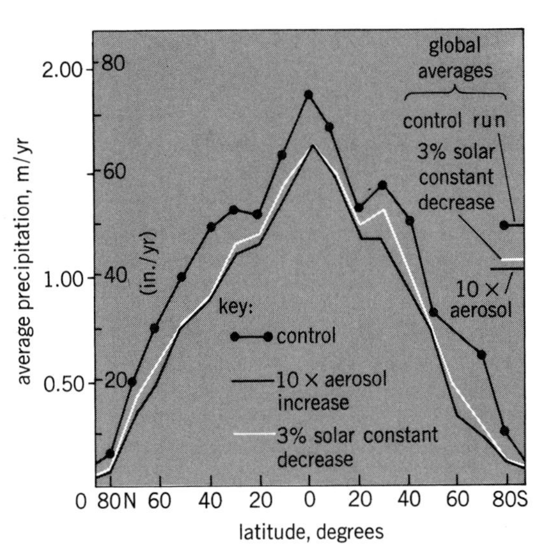

Fig. 4. Latitudinal distribution of total precipitation. *(From M. C. MacCracken, in Report of Findings: Effects of Stratospheric Pollution by Aircraft, DOT-TST-75-50, 1974)*

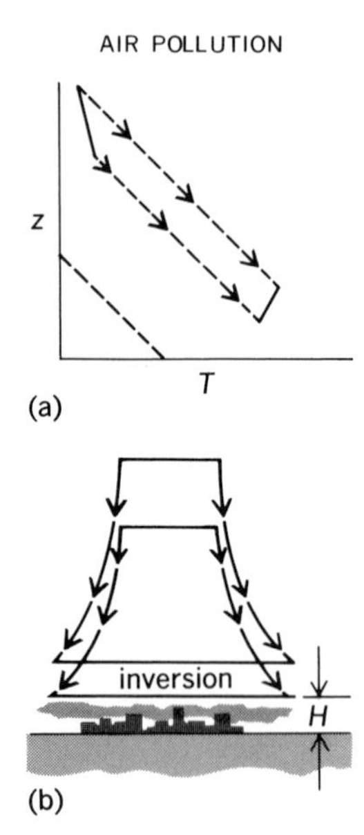

Fig. 5. Subsidence inversion. (*a*) Solid lines show temperature *T* and height *z* before and after dry adiabatic descent of air; dashed lines represent dry adiabatic rate of heating. (*b*) Inversion limits mixing height *H* over city.

surface; and under cloudless skies, solar radiation may induce important photochemical reactions.

Wind direction, speed, and turbulence influence atmospheric pollution. Wind direction determines the area into which the pollution is carried. Dilution of contaminants from a source is directly proportional, other factors being constant, to wind speed, which also determines the intensity of mechanical turbulence produced as the wind flows over and around surface objects, such as trees and buildings.

Eddy diffusion by wind turbulence is the primary mixing agency in the troposphere; molecular diffusion is negligible in comparison. In addition to mechanical turbulence, there is thermal turbulence which occurs in an unstable layer of air. Thermal turbulence and associated intense mixing develop in an unsaturated layer in which the temperature decreases with height at a rate greater than 1°C/100 m, the dry adiabatic rate of cooling. When the temperature decreases at a lower rate, the air is stable, and turbulence and mixing, now primarily mechanical, are less intense. If the temperature increases with height—its normal behavior in the stratosphere, creating a condition known as an inversion—the air is stable, and horizontal turbulence and mixing are still appreciable, but vertical turbulence and mixing are almost completely suppressed. *See* TEMPERATURE INVERSION.

Inversion. Precipitation, fog, and solar radiation exert secondary meteorological influences. Falling raindrops may collect particles with radii greater than 1 μm or may entrain gases and smaller particles and carry them to the ground. Gas reactions with aerosols also occur; neutralizing cations in fog droplets or traces of ammonia (NH_3) in the air act as catalysts to accelerate reaction rates leading to rapid oxidation of sulfur dioxide (SO_2) in fog droplets. For highly polluted city air, it is estimated that in the presence of NH_3, the oxidation of the SO_2 to ammonium sulfate, $(NH_4)_2SO_4$, is completed in 1 hr for fog droplets 10 μm in radius. Photochemical oxidation of hydrocarbons in sunlight is frequent. Most hydrocarbons do not have appropriate absorption bands for a direct photochemical reaction; nitrogen dioxide (NO_2), when present, acts as an oxidation catalyst by absorbing solar radiation strongly and subsequently transferring the light energy to the hydrocarbon and thereby oxidizing it.

Natural ventilation in the atmosphere is best when the winds are strong and turbulent so that mixing is good, and when the volume in which mixing occurs is large so that dilution of pollution is rapid. As cities have grown in size, air pollution has become more widespread. It has become necessary to think of whole urban complexes as large area sources of pollution. The rate of natural ventilation of an urban area is dependent on two quantities: the wind speed and the mixing volume over the city. Active mixing upward is often limited by a stable layer, perhaps even a very stable inversion layer, aloft. The upward extent of this region of active mixing, known as the mixing height, determines the magnitude of the mixing volume of the city.

The number of air changes per unit time in this mixing volume specifies the rate of natural ventilation of the urban area. The problems of air pollution become highly complex, however, because the mixing height is rarely constant for long. Some of the factors causing it to vary are described below.

At night when the sky is clear and the wind light, Earth's surface loses heat by long-wave radiation to space. As a result, the ground cools and a surface radiation inversion is formed. The inversion inhibits mixing, so that pollution accumulates. Solar heating of the ground causes a reversal of the lapse rate, which may exceed the dry adiabatic rate of cooling and enhances active mixing in the unstable layer.

The mixing may bring pollution from aloft, causing a temporary peak in the surface concentrations, a process known as an inversion breakup fumigation. By midafternoon, the height of mixing is a maximum for the day, and surface concentrations tend to be low as the natural ventilation improves. In the evening, the lapse rate becomes stable, and accumulation of contaminants may begin again.

Subsidence inversion. The accumulation of pollution for longer periods of time is especially likely to occur if a persistent inversion aloft exists. Such an inversion aloft is the subsidence inversion formed by the sinking and vertical convergence of air in an anticyclone, illustrated in Fig. 5. A layer of air at high levels descends, diverging horizontally and hence converging in the vertical, and warms at the dry adiabatic rate of heating of 1°C/100 m. Figure 5*a* shows how a low-level inversion may result from this process, while Fig. 5*b* depicts how the mixing height *H* is limited in vertical extent by the subsidence inversion aloft, so that pollution accumulates within and just above the city. It is the presence of such a subsidence inversion aloft associated with the Pacific subtropical anticyclone which is the primary cause of Los Angeles and other California smogs; these are made even worse by local mountain and valley sides which prevent horizontal dispersion.

Fog. The worst pollution occurs when, in addi-

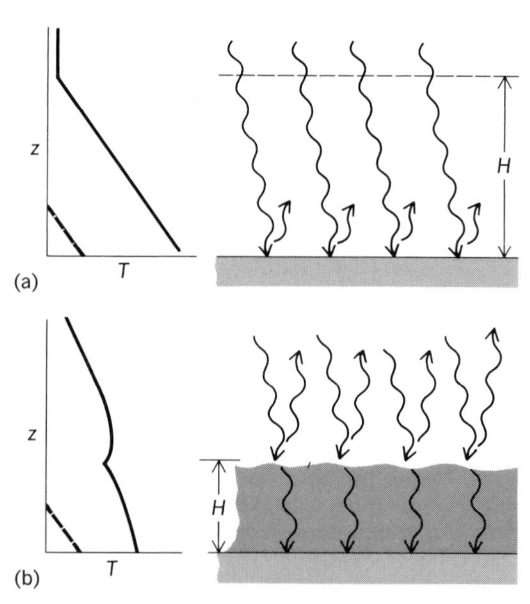

Fig. 6. The influence of fog in reducing mixing height *H*. (*a*) Without fog. (*b*) With fog. The dashed lines represent the dry adiabatic rate of cooling. Arrows represent solar and reflected radiation.

tion to subsidence inversions accompanying slowly moving or stationary anticyclones, fog also develops. All the major air pollution disasters, such as those listed in a later section, took place when fog persisted during protracted stagnant anticyclonic conditions. The reasons for the adverse influence of fog are shown in Fig. 6. When there is no fog (Fig. 6*a*), solar radiation heats the ground, which in turn causes a lapse rate equal to, or greater than, the dry adiababic rate of cooling, with good mixing and hence a substantial mixing height H. On the other hand, with a fog layer (Fig. 6*b*), up to 70% of the solar radiation incident at the top of the fog is reflected to space, with relatively little left to heat the fog and ground below. With the cloudless skies characteristic of anticyclonic weather, there is a continuous loss of heat to outer space from the upper surface of the fog bank, which acts radiatively as an elevated ground surface. More heat is lost to space than is gained from the Sun, and an inversion develops above the fog and persists night and day until the anticyclone dissipates or moves away. If the air is polluted, the fog particles may become acids and salts in solution; the saturation vapor pressure over such particles may decrease to 90 or 95% of the pure water value, so the smog becomes even more persistent than if it remained as a pure water fog. Disastrous concentrations of contaminants may accumulate during prolonged foggy conditions of this kind.

Warm fronts. Another significant inversion aloft is associated with a slowly moving warm frontal surface. Consider two cities, one lying to the southwest and the other lying to the northeast of a warm front extending from the southeast to the northwest (and moving in a northeast direction), as illustrated in Fig. 7. City B lies in the cool air, with the warm frontal above it. In the cool air ahead of the warm front, the pollution from City B is trapped below the warm frontal inversion and may travel for many miles with large surface concentrations. On the other hand, the prevailing southwest winds in the warm sector will carry pollution from City A up and above the warm frontal inversion, which effectively prevents its diffusion downward to the surface. This situation brings out an important point: an inversion layer may be advantageous, not disadvantageous, if it inhibits diffusion down to the ground.

Stack dispersion. Dispersion from an elevated point source, such as a stack, is conveniently expressed by Eq. (7), where χ is ground level concentration of contaminant in mass per unit volume;

$$\chi = \frac{Q}{\pi \sigma_y \sigma_z \bar{u}} \exp\left[-\frac{1}{2}\left(\frac{y^2}{\sigma_y{}^2} + \frac{h^2}{\sigma_z{}^2}\right)\right] \quad (7)$$

Q is the source strength in mass per unit time; $\bar{u}$ is mean wind speed; y is the horizontal direction perpendicular to the mean wind; σ_y and σ_z are diffusion coefficients expressed in length units in the y and z directions, respectively, the z direction being vertical; and h is the height of the source above ground.

This diffusion equation should be used only under the simplest conditions, for example, in flat uniform terrain and well away from hills, slopes, valleys, and shorelines. Table 2 lists various meteorological categories, and Table 3 gives values of the diffusion coefficients appropriate for each

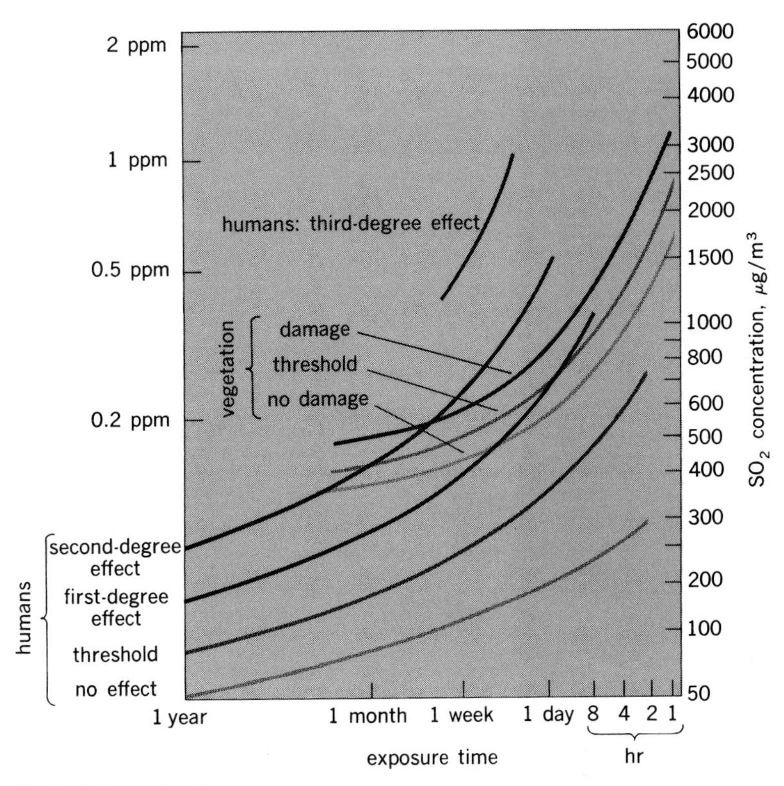

Fig. 7. Effects of sulfur dioxide on humans and vegetation. (*L. J. Brasser et al.*)

category. It should be noted that the values to be used depend on distance from the source at which concentrations are to be calculated. A variety of other forms is available for more complex conditions of terrain and meteorology.

Natural cleansing processes. Pollution is removed from the atmosphere in such ways as washout, rain-out, gravitational settling, and turbulent impaction. Washout is the process by which contaminants are washed out of the atmosphere by raindrops as they fall through the contaminants; in rain-out the contaminants unite with cloud droplets, which may later grow into precipitation.

Gravitational settling is significant mainly for large particles, those having a diameter greater than 20 μm. Agglomeration of finer particles may result in larger ones which settle out by gravitation. Fine particles may also impact on surfaces by centrifugal action in very small turbulent eddies. Gases may be converted to particulates, as by photochemical action of sunlight in Los Angeles, Denver, and Mexico City. These particulates may then be removed by settling or impaction.

The rate of natural cleansing may be slower than the rate of injection of pollutants into the atmosphere, in which case pollution may increase on a global scale. There is evidence that the concentration of atmospheric carbon dioxide has been increasing slowly since the beginning of the century because of combustion of fossil fuels. The tropospheric burden of very small particles and of Freon gas may also be increasing.

Effects of stratospheric pollution. Pollution of the stratosphere with nitrogen oxide causes reduction of stratospheric ozone. Ozone reduction in the stratosphere has been linked to biological effects such as skin cancer in two steps: (1) reduced ozone in the stratosphere causes an in-

Table 2. Meteorological categories*

Surface wind speed, m/s	Daytime insolation			Thin overcast or ≥ 4/8 cloudiness†	≤ 3/8 cloudiness
	Strong	Moderate	Slight		
<2	A	A-B	B		
2	A-B	B	C	E	F
4	B	B-C	C	D	E
6	C	C-D	D	D	D
<6	C	D	D	D	D

*A = Extremely unstable conditions. D = Neutral conditions (applicable to heavy overcast, day or night).
B = Moderately unstable conditions. E = Slightly stable conditions.
C = Slightly unstable conditions. F = Moderately stable conditions.
†The degree of cloudiness is defined as that fraction of the sky above the local apparent horizon which is covered by clouds.

crease in uv radiation reaching the Earth's surface, and (2) increased uv radiation enhances the normal biological effects of natural uv radiation.

Biological damage. In step 1, the relation between the reduction of stratospheric ozone and the increase in solar uv flux effective in causing sunburn and, presumably, also skin cancer is readily calculable. The factors of interest here are the uv wave band of 290–320 nanometers (nm), the middle latitudes (where the summer sun is nearly overhead and where the worst cases of skin cancer occur), and small decreases in ozone. For these factors of interest, the percent increase in solar ultraviolet flux is about twice the percent decrease in ozone.

Step 2, from uv radiation to the enhancement of skin cancer incidence, involves the assumption, supported by some scientific evidence but not proven by experiments on humans, that skin cancer in humans is induced by exposure to uv radiation of the same wavelength (290–320 nm) that causes erythema (sunburn), and that the relative effectiveness of the various wavelengths for carcinogenesis is the same as that for sunburn. Estimates of biological damage to humans from exposure to uv radiation are based on inferences from the statistics of epidemiological surveys of humans and of laboratory experiments with animals. Nonmelanomic skin cancer, which is almost never fatal if given proper care, occurs primarily on sun-exposed areas of the skin, especially the face and hands. It is relatively common—about 250 cases per 100,000 persons occur in fair-skinned Caucasians in the United States. The incidence of nonmelanomic skin cancer is correlated with latitude—and, therefore, with sunlight, including uv flux, since average sunlight varies with latitude.

Climatic changes. Pollution of the stratosphere may involve also a climate chain of cause-and-effect relations by which aircraft engine effluents, notably sulfur dioxide (SO_2), and to a lesser degree water vapor (H_2O) and nitrogen oxides (NO_x), affect climatic variables such as temperature, wind, and rainfall.

If enough particles larger than 0.1 μm in diameter were added to the stratosphere, they could alter the radiative heat transfer of the Earth-Sun system, and thereby influence climate. Particles of this size are produced by several constituents of engine emissions, in particular those of SO_2. When considering the large numbers of aircraft postulated for future operations in the stratosphere, the amount of particles developed from the SO_2 engine emissions are potentially serious, unless the fuels used have a sulfur content smaller than that of today's fuels.

The sequence in the climate cause-and-effect chain is proposed as follows. Stratospheric SO_2, after first being oxidized, interacts with the abundant water vapor exhaust from jet engines to produce solid sulfuric acid particles that build up to sizes greater than 1 μm. These particles disperse within the stratosphere, principally within the hemisphere in which they are injected, where they remain for periods as long as 3 years, depending on their altitude.

The effects of SO_2 emissions are summarized in Table 4, where the stratosphere's opacity to sunlight is represented by optical thickness. The natural optical thickness of the stratosphere is about 0.02, which means that sunlight and heat are reduced by about 2% while passing through the stratosphere.

The present subsonic fleet of about 1700 air-

Table 3. Values of diffusion coefficients

Distance from source, m	Diffusion coefficient, m², in various meteorological categories					
	A	B	C	D	E	F
10^2, σ_y	22	16	12	8	6	4
10^3	210	150	105	75	52	36
10^4	1700	1300	900	600	420	360
10^5	11,000	8500	6300	4100	2800	2000
10^2, σ_z	14	11	7.6	4.8	3.6	2.2
10^3	500	120	70	32	24	14
10^4	—	—	420	140	90	46
10^5	—	—	2100	440	170	92

Table 4. Estimated increase in stratospheric optical thickness per 100 aircraft

Subsonic aircraft type*	Fuel burned, kg/yr†	Altitude, km (10^3 ft)	Maximum SO_2 EI‡ without controls, g/kg fuel	Percent change in stratospheric optical thickness in Northern Hemisphere: Without controls	With future EI controls achieving only 1/20 of emission of present-day aircraft
707/DC-8	1×10^9	11 (36)	1	0.023	0.0012
DC-10/L-1011	1.5×10^9	11 (36)	1	0.032	0.0016
747	2×10^9	11 (36)	1	0.044	0.0022
747-SP	2×10^9	13.5 (44)	1	0.10	0.0050

*The present subsonic fleet consists of 1217 707/DC-8s, 232 DC-10/L-1011s, and 232 747s flying at a mean altitude of 11 km (36,000 ft), and is estimated to cause an increase in stratospheric optical thickness of 0.5%.

†Subsonics are assumed to operate at high altitude, 5.4 hr per day, 365 days per year.

‡EI is emission index, which is defined as grams of pollutants per kilogram of fuel used.

craft operates at times in the low stratosphere, burning a total of about 2×10^{10} kg (2×10^7 tons) of fuel per year and producing an estimated increase in the stratosphere's optical thickness of about 0.0001 or 0.5% of the natural value.

As a matter of history, the variability of temperature due to natural causes is a substantial fraction of 1°C, even over a few decades. The year-to-year variation is several tenths of a degree. In the 1890–1940 period, a warming of ½°C occurred. During the 1940–1960 period, a general cooling of ⅕°C took place.

Effects of tropospheric pollution. A number of the many effects of tropospheric pollution are described briefly below.

Humans. The effects of many pollutants on human health under most ordinary circumstances of living, rural or urban, are difficult to specify with confidence. In the United States, a number of animal experiments under controlled conditions and epidemiological studies have been made, but the results are difficult to interpret in terms of human health. A study on the lower East Side of New York City indicated that, in children less than 8 years old, the occurrence of respiratory symptoms was associated with the levels of particulate matter and of carbon monoxide in the atmosphere. With heavy smokers, however, eye irritation and headache were directly related to increasing concentrations of carbon monoxide.

Air pollution is suspected as a causative agent in the occurrence of chronic bronchitis, emphysema, and lung cancer, but the evidence is not clear cut. On the other hand, from mid-August to late September the potent aeroallergen, ragweed pollen, is a substantial cause of allergic rhinitis and bronchial asthma for over 10,000,000 persons living east of the Rockies. In Los Angeles County the effects of smog on health are becoming better understood. For example, people living in less smoggy areas of the county survive heart attacks more readily than others: In 1958 in high-pollution areas the mortality rate per 100 hospital admissions for heart attacks averaged 27.3 in comparison with 19.1 for low-smog areas. Other studies show a small but significant relation between motor vehicle accidents and oxidant levels. Medical authorities in Los Angeles are becoming concerned about the long-term influences of various pollutants, including photochemical smog, despite the lack of comprehensive knowledge of the nature of such effects.

In Great Britain there is a similar lack of precise knowledge. There are indications that emphysema, bronchitis, and other respiratory diseases are not caused primarily by increased atmospheric pollution, but because more people are living longer. Despite a substantial reduction in the concentrations of atmospheric particulates since the Clean Air Act was passed in 1956, the respiratory disease rate continues to rise. These facts do not prove that air pollution is not a factor, but that its influence may be synergistic and therefore difficult to identify precisely. For example, it is known that the combined effect of sulfur oxides and particulates is substantially greater than the sum of the two separate effects, and many other such synergistic effects doubtless occur.

In the Netherlands special efforts have been made to relate SO_2, both concentration values and exposure times, to effects on humans and on vegetation. The results of these studies, based on investigations in England, the United States, West Germany, Italy, and the Netherlands, are illustrated in Fig. 7. Influences on humans are shown in the lower family of curves: A first-degree effect is a small increase in functional disturbances, symptoms, illnesses, diseases, and deaths; a second-degree effect is a more prevalent or more pronounced effect of the same kind; and a third-degree effect is a substantial increase in the number of deaths. It should be emphasized that the exposures were, in general, to SO_2 in dusty and sooty atmospheres.

Under extreme circumstances when stagnant atmospheric conditions with persistent low wind and fog exist, major disasters involving many deaths occurred, as in and around London in 1873, 1880, 1891, 1948, 1952, 1956, and 1962. Similar diasters occurred in the Meuse Valley of Belgium in 1930 and at Donora, Pa., in 1948.

Atmospheric pollution has a substantial influence on the social aspects of human life and activity. For example, the distribution of urban populations is being increasingly affected by such pollution, and recreational patterns are similarly influenced. The atmospheric burden of pollution is thus becoming more and more important as a determinant in social decision making.

Animals. Studies of the response of laboratory animals to specified concentrations of pollutants have been conducted for many years, but the interpretation of the results in terms of corresponding human response is most difficult. Assessment of the effects of certain contaminants on livestock is relatively straightforward, however. Thus contamination of forage by airborne fluorides and arsenicals from certain industrial operations has led to the loss of large numbers of cattle in the areas adjacent to such chemical industries.

Plants. Damage to vegetation by air pollution is of many kinds. Sulfur dioxide may damage such field crops as alfalfa, and trees such as pines, especially during the growing season; some general relations are presented in Fig. 7. Both hydrogen fluoride and nitrogen dioxide in high concentrations have been shown to be harmful to citrus trees and ornamental plants which are of economic importance in central Florida. Ozone and ethylene are other contaminants which cause damage to certain kinds of vegetation.

Materials. Corrosion of materials by atmospheric pollution is a major problem. Damage occurs to ferrous metals; to nonferrous metals, such as aluminum, copper, silver, nickel, and zinc; to building materials; and to paint, leather, paper, textiles, dyes, rubber, and ceramics.

Weather. Tropospheric pollution may affect weather in a number of ways. Heavy precipitation at Laporte, Ind., is attributed to a substantial source of air pollution there, and similar but less pronounced effects have been observed elsewhere. Industrial smoke reduces visibility and also ultraviolet radiation from the Sun, and polluted fogs are more dense and more persistent than natural fogs occurring under similar conditions. Possible major effects of air pollution on Earth's climate have been mentioned earlier. *See* CLIMATE MODIFICATION.

Controls. Four main methods of air-pollution control are indicated below.

Prevention. This method was originally applied mainly to reduce pollution from combustion processes. Improved equipment design and smokeless fuels have reduced pollution both from industrial and motor vehicle sources.

Collection. Collection of contaminants at the source has been one of the important methods of control. Many types of collectors have been employed successfully, such as settling chambers, cyclone units employing centrifugal action, bag filters, liquid scrubbers, gas-solid adsorbers, ultrasonic agglomerators, and electrostatic precipitators. The optimum choice for a given industrial process depends on many factors. A major problem is disposal of the collected materials. Sometimes they can be used in by-product manufacture on a profitable or a break-even basis.

Containment. This method is useful for pollutants whose noxious characteristics may decrease with time, such as radioactive contaminants from nuclear power plants. For contaminants with a short half-life, containment may allow the radioactivity to decay to a level which permits their release to the atmosphere. Containment, with destruction or conversion of the offending substances, often malodorous or toxic, is used in certain chemical, oil refining, and metallurgical processes and in liquid scrubbing.

Dispersion. Atmospheric dispersion as a control method has a number of advantages, especially for industrial processes which can be varied to take advantage of the periods when dispersion conditions are so good that contaminants may be distributed very widely in such small concentrations that they inconvenience no one. Some coal-burning electrical power stations are building high stacks, up to 1000 ft (300 m), to lift the SO_2-bearing stack gases well above the ground. Some plants store low-sulfur anthracite coal for use when atmospheric dispersion is poor. *See* AIR-POLLUTION CONTROL.

Laws. Many laws designed to limit air pollution have been enacted. Major steps forward were taken by the Netherlands in 1952, by Great Britian in 1956, by Germany in 1959 and 1962, by France in 1961, by Norway in 1962, by the United States in 1963 and 1967, and by Belgium in 1964.

Efforts to control air pollution by legal means commenced many years ago in Great Britain. In 1906 the Alkali Act consolidated and extended previous similar acts, the first of which was passed in 1863. This calls for the annual registration of scheduled industrial processes, and requires that the escape of contaminants to the atmosphere from scheduled processes must be prevented by the "best practicable means." The Alkali Act functions by interpretation and not by statutory requirement, the Alkali Inspector being the sole judge of the "best practicable means." The Clean Air Act of 1956 provided more effective ways of limiting air pollution by domestic smoke, industrial particulates, gases and fumes from the processes registrable under the Alkali Act, and smoke from diesel engines. This legislative program has had considerable success in alleviating air-pollution problems in Great Britain.

In the United States, air-pollution control had been considered to be a matter of local concern only. By 1963 only one-third of the states had air-pollution control programs, most of which were relatively ineffective. Only in California were local programs, at the city and county level, supported adequately. The Clean Air Act of 1963 brought the Federal government into a regulatory position of increased scope by granting the Secretary of Health, Education, and Welfare specific abatement powers under certain circumstances. It also established a Federal program of financial assistance to local control agencies and recommended more vigorous action to combat pollution by motor vehicle exhausts and by smoke from incinerators.

The Air Quality Act of 1967 brought the Federal government into a more substantial regulatory role. One of its important effects has been to change the emphasis in legislation from standards based on emissions from sources, such as stacks, to standards based on concentrations of contaminants in the ambient air which result from such emissions. The Air Quality Act of 1967 consists of three main portions, listed below.

Title I: Air-Pollution Prevention and Control. The first section of the Air Quality Act amends the Clean Air Act to encourage cooperative activities by states and local governments for the prevention and control of air pollution and the enactment of uniform state and local laws; to establish new and more effective programs of research, investigation, training, and related activities; to give special emphasis to research related to fuel and vehicles; to make grants to agencies to support their programs; to provide strong financial support for interstate air-quality agencies and commissions; to define atmospheric areas and to assist in establishing air quality control regions, criteria, and control techniques; to provide for abatement of pollution of the air in any state or states which endangers the health or welfare of any persons and to establish the necessary procedures; to establish the President's Air Quality Advisory Board and Advisory Committees; and to provide for control of pollution from Federal facilities.

Title II: National Emission Standards Act. This section is concerned mainly with pollution from motor vehicles which accounts for some 60% of the total for the United States. The act covers such matters related to motor vehicle emissions as the following: establishment of effective emission standards and of procedures to ensure compliance by means of prohibitions, injunction procedures, penalties, and programs of certification of new motor vehicles or motor vehicle engines and registration of fuel additives. The act also calls for a comprehensive report on the need for, and the effect of, national emission standards for stationary sources.

Title III. The final section is general and covers matters such as comprehensive economic cost studies, definitions, reports, and appropriations.

There is no doubt that this far-reaching legislative program, stimulating new approaches at the local, state, and Federal levels, will play a major role in controlling air pollution within the United States. The other industrial nations of the world are preparing to meet their growing air-pollution problems by initiatives appropriate to their own particular cirumstances.

Other acts. The Clean Air Act of 1970 set specific deadlines for the reduction of certain hazardous automobile emissions. That year the Environmental Protection Agency (EPA) assumed control over air-pollution programs formerly administered by the Department of Health, Education, and Welfare.

In April 1973 EPA granted a 1-year extension of the strict auto emission standards. In 1975 the Energy Supply and Environmental Coordination Act relaxed emission requirements for another year.

The Clean Air Act Amendments of 1977 set new deadlines for compliance with emission limits for both industrial and automobile emissions.

[A. J. GROBECKER; S. C. CORONITI; E. WENDELL HEWSON]

Bibliography: L. J. Brasser et al., *Sulphur Dioxide: To What Level Is It Acceptable?*, Research Institute of Public Health Engineering, Delft, Netherlands, Rep. no. G300, 1967; J. H. Chang and H. Johnston, *Proceedings of the 3d CIAP Conference*, DOT-TSC-OST-74-15, pp. 323–329, 1974; R. E. Dickinson, in *Proceedings of the AIAA/AMS International Conference on the Environmental Impact of Aerospace Operations in the High Atmosphere*, AIAA Pap. no. 73-527, 1973; Federal Task Force on Inadvertent Modification of the Stratosphere, *IMOS Report*, prepared for the Federal Council for Science and Technology, 1975; J. Friend, R. Liefer, and M. Tichon, *Atmos. Sci.*, 30:465–479, 1973; D. Garvin and R. F. Hampson, *Proceedings of the AIAA/AMS International Conference on the Environmental Impact of Aerospace Operations in the High Atmosphere*, AIAA Pap. no. 73-500, 1973; A. J. Grobecker, S. C. Coroniti, and R. H. Cannon, *Report of Findings: The Effects of Stratospheric Pollution by Aircraft*, U.S. Department of Transportation, DOT-TST-75-50, 1974; P. A. Leighton, *Photochemistry of Air Pollution*, 1961; M. C. MacCracken, *Tests of Ice Age Theories Using a Zonal Atmospheric Model*, UCRL-72803, Lawrence Livermore Laboratory, 1970; A. R. Meetham, *Atmospheric Pollution*, 1964; M. J. Molina and F. S. Rowland, *Geophys. Rev.*, pp. 810–812, 1974; National Academy of Sciences, *Environmental Impact of Stratospheric Flight*, pp. 128–129, 1975; P. A. O'Connor (ed.), *Congress and the Nation*, vols. 3 and 4, 1973, 1977; R. Scorer, *Air Pollution*, 1968; A. R. Smith, *Air Pollution*, 1966; A. C. Stern (ed.), *Air Pollution*, 1968; R. S. Stolarski and R. J. Cicerone, *Can. J. Chem.*, 52:1610–1615, 1974; S. C. Wofsy and M. B. McElroy, *Can. J. Chem.*, 52:1582–1591, 1974.

Air-pollution control

Air pollution, according to the definition developed by the Engineers Joint Council, means the presence in the outdoor atmosphere of one or more contaminants, such as dust, fumes, gas, mist, odor, smoke, or vapor, in quantities, of characteristics, and of duration such as to be injurious to human, plant, or animal life or to property, or to interfere unreasonably with the comfortable enjoyment of life and property. The sources of airborne wastes are many. They may be roughly divided into natural, industrial, transportation, agricultural activity, commercial and domestic heat and power, municipal activities, and fallout.

Sources of pollution. Natural sources include the pollen from weeds, water droplet or spray evaporation residues, wind storm dusts, meteoritic dusts, and surface detritus. Industrial sources include ventilation products from local exhaust systems, process waste discharges, and heat, power, and waste disposal by combustion processes. Transportation sources include motor vehicles, rail-mounted vehicles, airplanes, and vessels. Agricultural activity sources include insecticidal and pesticidal dusting and spraying, and burning of vegetation. Commercial heat and domestic heat and power sources include gas-, oil-, and coal-fired furnaces used to produce heat or power for individual dwellings, multiple dwellings, commercial establishments, utilities, and industry. Municipal activity sources include refuse disposal, liquid waste disposal, road and street plant operations, and fuel-fired combustion operations. Fallout is a term applied to radioactive pollutants in mass atmosphere resulting from thermonuclear explosion.

The sources are so varied that pollution of the atmosphere is a matter of degree. Pollution from natural sources is in effect a base line of pollution. The major problems of pollution are associated with community activity as opposed to rural activity, because community air is generally more grossly polluted and may contain harmful and dangerous substances affecting property, plant life, and, on occasion, health. Environment is made less desirable by the polluting influence, and there is ample reason to conserve the air resource in many ways parallel to the need for conservation of the water resource. In actuality, the engineer is concerned with engineering management of the air

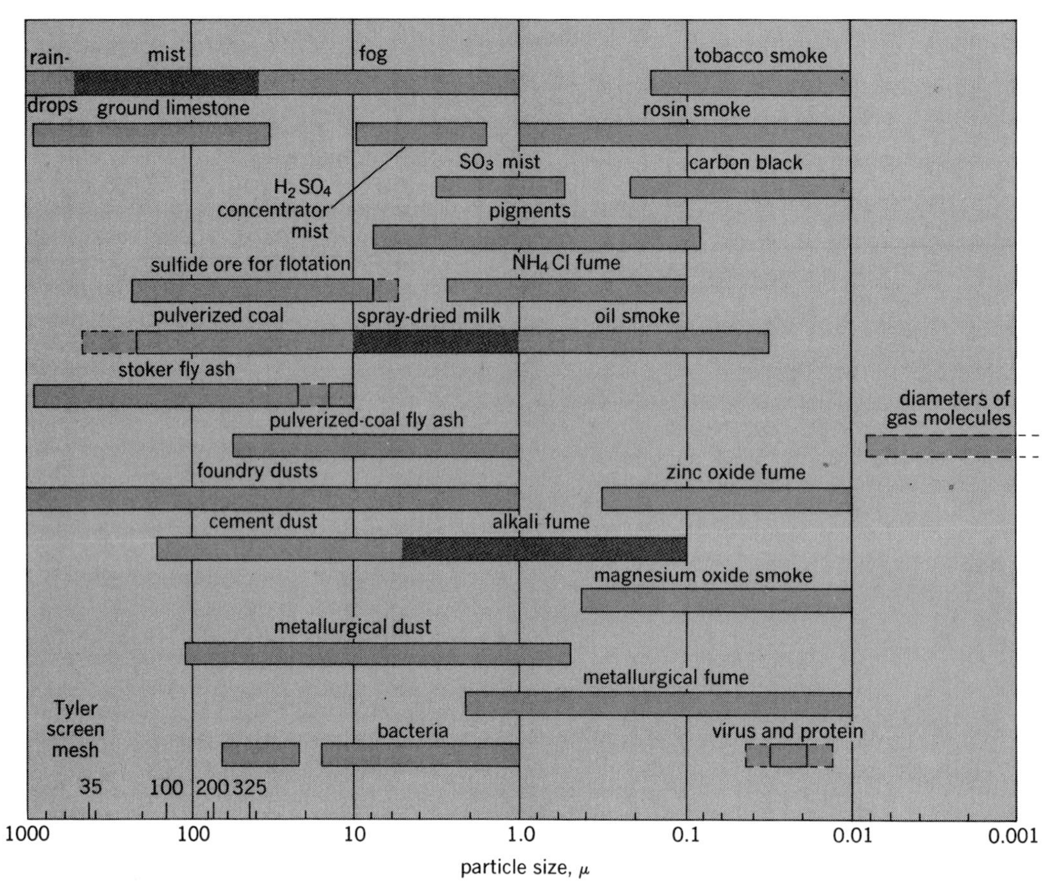

Fig. 1. Particle size ranges applicable to aerosols, dusts, and fumes. (*From W. L. Faith, Air Pollution Control, copyright © 1959 by John Wiley & Sons, Inc.; reprinted with permission*)

resource, a broader concept than the control of air pollution.

Control. Air-pollution control suggests in its simplest form a background of knowledge concerning ideal atmospheres and criteria of clean air, the existence of specific standards setting limits on the allowable degree of pollution, means of precise measurement of pollutants, and practical means of treating polluting sources to maintain the desired degree of air cleanliness. There are many areas in the above listing that are under research at the present time. University research foundations, Federal, state, and municipal air-pollution control agencies, and all of the professional engineering societies are actively engaged in the development of criteria, standards, design factors, and equipment for the control of air pollution.

Reduced visibility has been a focal point of air pollution for over 700 years. The burning of soft coal in England combined with the fog of the atmosphere forms a particularly opaque mixture which may at times reduce visibility to zero. The word smog has been coined for this mixture.

Microscopic water droplets condense about nucleating substances in the air to form aerosols. An aerosol is a liquid or solid submicron particle dispersed in a gaseous medium. An atmosphere having an aerosol concentration of about 1 mg/m^3 has been estimated to limit visibility to 1600 ft (488 m). The mass would contain perhaps 16,000 particles/cm^3. Restriction in visibility is the result of light scattering by these particles. Chemical condensation of reaction products in the air may also nucleate and grow to size that will bring about light scattering. Sulfur dioxide is also a nucleating substance as it oxidizes and hydrolyzes to form sulfuric acid mist.

Elimination of sources of pollution has been one

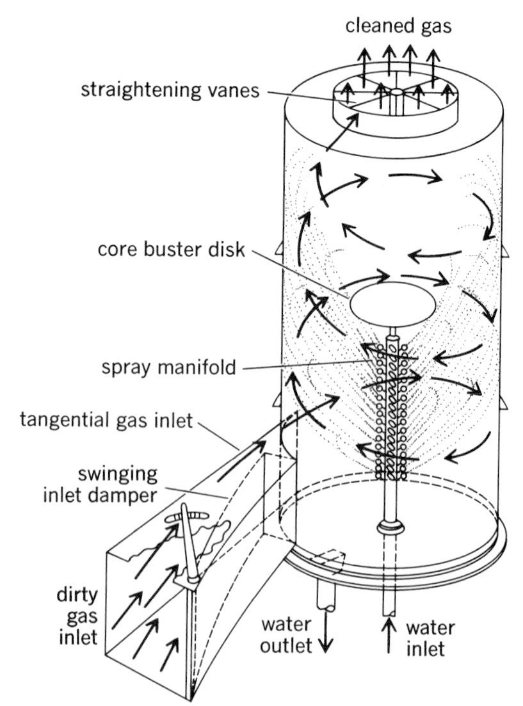

Fig. 2. Typical cyclonic spray scrubber. (*From W. L. Faith, Air Pollution Control, copyright © 1959 by John Wiley & Sons, Inc.; reprinted with permission*)

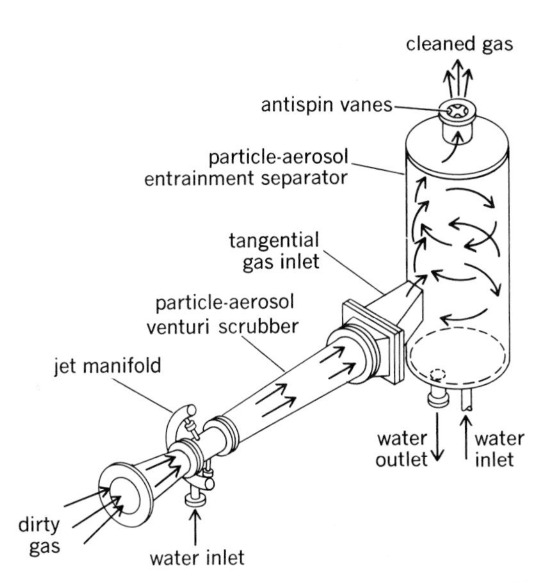

Fig. 3. Typical venturi scrubber. (*From W. L. Faith, Air Pollution Control, copyright © 1959 by John Wiley & Sons, Inc.; reprinted with permission*)

of the favored means of controlling pollution. There are many means of accomplishing the reduction of pollution, but complete elimination is not always practicable. Sulfur dioxide release can be reduced by choosing low sulfur-bearing fuel. An industrial process with a gaseous effluent can be changed to eliminate the gaseous waste. Gases and particulates can be removed from a gas stream by air-cleaning equipment.

Air-cleaning devices. Air-cleaning devices to remove particulates are selected to remove particles and aerosols on the basis of their size (Fig. 1). Screens will remove coarse solids. Settling chambers are containers which by expanded cross section reduce velocity below 10 feet per second (fps; 3m/s) and thereby allow particles to settle. Particles down to 10 μm in size may be recovered with such chambers. Cyclone separators operate by injecting a gas stream tangentially at the top of a cylindrical chamber. A high-velocity spiral motion is created. Particles are centrifuged out of the gas stream, hit the side wall, and fall to a conical bottom out of the airflow, which turns up through the core or vortex beginning at the bottom and flows to the top through a pipe inserted into the core and extending into the body of the cyclone. Particles from 10 to 200 μm are removed with 50–90% efficiency. Filters are made of cloth, fiber, or glass. Air velocities are low and efficiency is about 50% for dry fiber filters. Efficiency is increased by using a low volatile oil viscous coating. Cloth filters are usually tubular and a number of bags are enclosed in a large chamber. Particles are trapped as air passes through the cloth from inside to outside. Dust is knocked down by shaking and falls to a hopper. Bag filters remove 99% of particles above 10-μm size. Wet collectors, or scrubbers, operate by passing and contacting the gas with a liquid. Water is sprayed, atomized, or distributed over a geometric shape. Deflectors may be added to provide an impinging surface. Scrubbers are efficient on 1- to 5-μm size particles (Figs. 2 and 3). Electrostatic precipitators operate by charging or ionizing particles as the gas flow passes through the unit (Fig. 4). Opposite-pole high-voltage plates, or electrodes, are provided to trap particles. Precipitators operate at 80–99% efficiency of ionizable aerosols down to 0.1-μm size.

Scrubbers may also remove water-soluble gases. Chemicals may be added to the liquid to provide improved absorption. Filters packed with activated charcoal are used to adsorb gases.

Packed towers, plate towers, and spray towers are also used to absorb gaseous pollutants from a gas stream. These devices provide for mixing a gas stream under treatment with water or a chemical solution, so that gases are taken into solution and possibly converted chemically as well.

Atmospheric dilution. This provides another means of reducing air pollution. Meteorology of a region, local topography, and building configuration are critical factors in determining suitability of atmosphere as a dispersal, diffusion, and dilution medium. Basic meteorological conditions of atmosphere that must be considered include wind speed and direction, gustiness of wind, and vertical temperature distribution. Humidity is also important under certain circumstances.

In general, diffusion theories predict that the ground concentration of a gas or a fine particle effluent with very low subsidence velocity is inversely proportional to the mean wind speed. Vertical temperature distribution is an important factor, determining the distance from stack of known

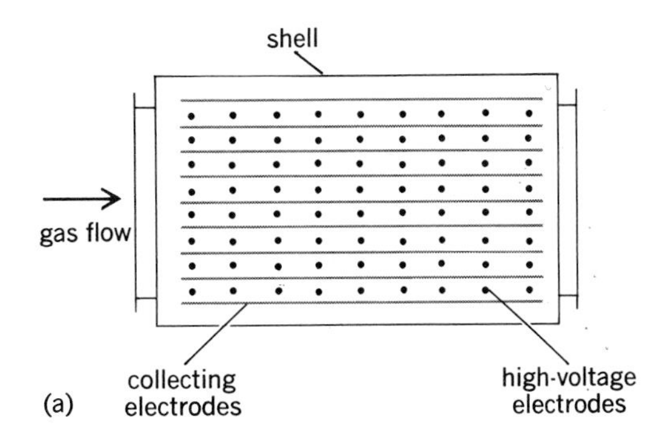

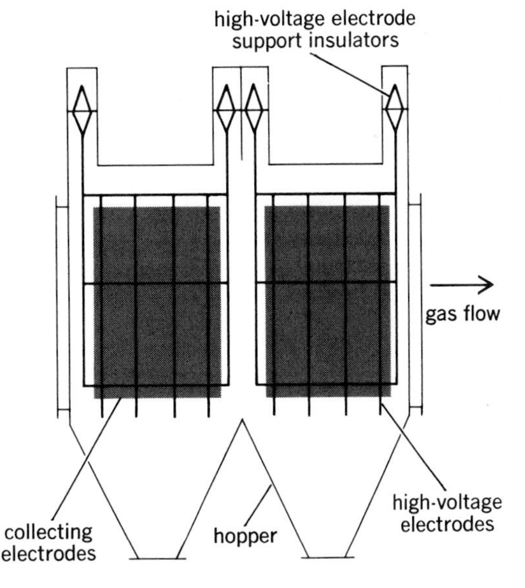

Fig. 4. Diagram of horizontal-flow electrostatic precipitator. (*a*) Plan. (*b*) Elevation. (*From W. L. Faith, Air Pollution Control, copyright © 1959 by John Wiley & Sons, Inc.; reprinted with permission*)

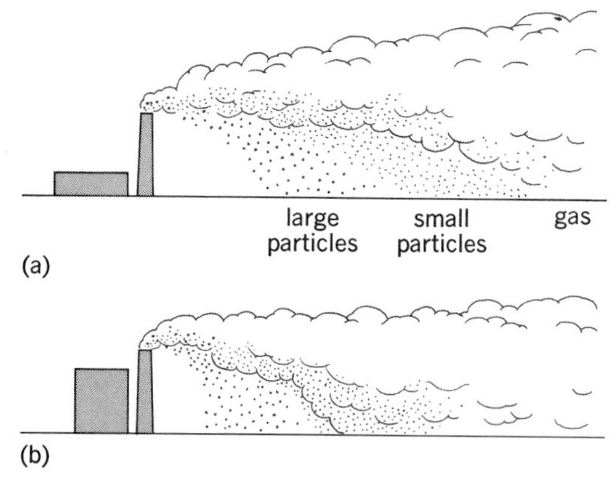

Fig. 5. Effect of building configuration on dispersal of gas plume. (*a*) Favorable configuration. (*b*) Unfavorable configuration. (*Research Division, New York University School of Engineering and Science*)

height at which maximum ground concentration occurs. Temperature of the stack gas has the effect of increasing stack height, as does stack gas velocity. Gas does not normally come to the ground under inversion conditions, but may accumulate aloft under calm or near calm conditions and be brought down to the surface as the Sun heats the ground in the early morning. Effect of building configuration is shown in Fig. 5. The turbulence introduced by buildings and topography is so complex that it is difficult to make theoretical calculations of effect. Model studies in wind tunnels have been used successfully to make predictions based on measurements of gas concentration and visible pattern of smoke (Fig. 6).

Nonventilating conditions may be present over an area for several days as a result of certain meteorological phenomena. During such periods the pollution emitted from various sources, such as fuel-fired combustion and automobile exhaust, continues to increase in concentration until ventilation sufficient to dilute the accumulated gases and particulates takes place. Figure 7 illustrates the record of sulfur dioxide–concentration measurement in the atmosphere over New York City during one such period of poor ventilation lasting for several days.

Incineration. The need for municipalities to find a means of disposing of refuse when land values are high and little land is available for sanitary landfill has resulted in increased use of incineration for refuse disposal. Incineration introduces problems of air pollution that are quite different from those of fuel-fired combustion. The material is not homogeneous, and has a wide variation in fuel value ranging from 600 to 6500 Btu/lb (1.4 to 15.1 MJ/kg) of refuse as fired. Volatiles are driven off by destructive distillation and ignite from heat of the combustion chamber. Gases pass through a series of oxidation changes in which time-temperature relationship is important. The gases must be heated above 1200°F (650°C) to destroy odors. End products of refuse combustion pass out of the stack at 800°F (430°C) or less after passing through expansion chambers, fly ash collectors, wet scrubbers, and in some instances electrostatic precipitators. The end products include carbon dioxide; carbon monoxide; water; oxides of nitrogen; aldehydes; unoxidized or unburned hydrocarbons; particulate matter comprising unburned carbon, mineral oxides, and unburned refuse; and unused or excess air. Particulates are reduced in quantity. Normally only micrometer-size and submicrometer particles should escape with the flue gases. Care in operation is required to hold down

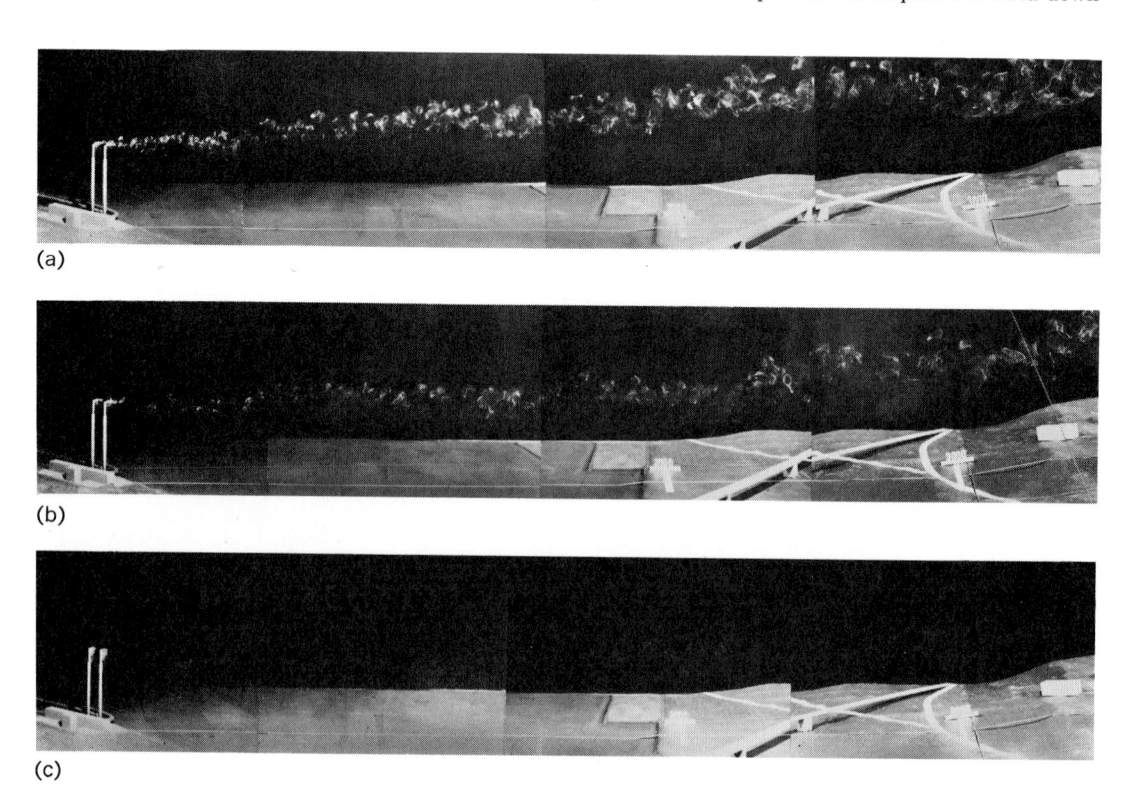

Fig. 6. Photographs of wind tunnel demonstration of dispersal patterns of smoke at three specified wind speeds: (*a*) 20 mph or 8.9 m/s. (*b*) 25 mph or 11.2 m/s. (*c*) 30 mph or 13.4 m/s.

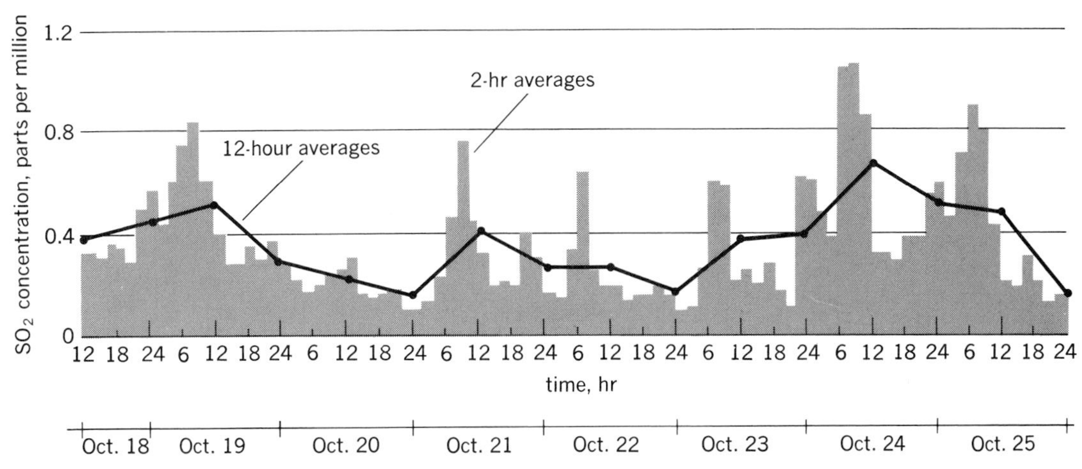

Fig. 7. Air-pollution episode in New York City, Oct. 18–25, 1963. The SO_2 values were at Christodora Station, 189 ft (57.6 m) above ground. (*Research Division, New York University School of Engineering and Science*)

particulate loading. Dust emissions in stacks may be in the range of 2–3 lb/ton/hr (1–15 kg/metric ton/hr) of refuse charged at a well-operated unit equipped with air-cleaning devices.

Incinerator design. There are several types of incinerator design promoted by manufacturers of incinerator equipment. Kiln shape may be round, rectangular, or rotary. The hearth may be horizontal fixed with grates, traveling with grates, multiple, step movement, or barrel-type rotary (Fig. 8). Drying hearths are provided on some types. Feed into the incinerator may be continuous, stoker, gravity, or batch.

It is necessary to know or estimate water content, percentage combustible material and inert material, Btu content, and weight of refuse to complete a rational design of incinerators. Heat balance can be calculated from several estimates based on averages. Available heat from the refuse must be balanced against the heat losses due to radiation, as well as from moisture, excess air, flue gas, and ash. Each type design has recommended sizings suggested by the manufacturer. There is fair agreement on the need for over 100% excess air. An allowance of 20,000 Btu/ft^3 (745 MJ/m^3) has been suggested for approximating chamber volume, and an allowance of 300,000 Btu/ft^2 (3.4 GJ/m^2) for grate area. Incinerator loading rates of 40–70 lb/(ft^2 grate area) (hr) [195–341 kg/(m^2 grate area) (hr)] have been used. Small incinerators for apartment houses and institutions are loaded at much lower rates. The Incinerator Institute of America in its standards has suggested loading rates for household or domestic-type refuse from 20 lb/(ft^2) (hr) [98 kg/(m^2) (hr)] in 100 lb/hr [45 kg/hr] burning units up to 30 lb/(ft^2) (hr) [146 kg/(m^2) (hr)] in 1000 lb/hr [454 kg/hr] units.

The Building Research Advisory Board (National Academy of Sciences, National Research Council) suggests that apartment house single-chamber incinerators should be sized on the basis of 0.375 ft^3 (10.62 liters) capacity per person, 0.075 ft^2 (69.7 cm^2) grate area per person, and heat release rate of not more than 18,000 Btu/ft^3 (670 MJ/m^3) of capacity, where the burning period is 10 hr or less.

Air-monitoring instruments. Air-sampling methods may be classified generally as those for sampling particulates or gases or both concurrently. The samples may be analyzed for specific pollutants or for general pollution levels. Sampling devices have been constructed with many variants. Generally, however, they follow reasonably well-defined principles which include gravity and suction-type collection, with passage through thermal and electrostatic precipitators; impingers and impactors; cyclones; absorption and adsorption trains; scrubbing apparatus; filters of various materials, such as paper, glass, plastic, membrane, and wool; glass plates; and impregnated papers. Combination instruments that measure wind direction and velocity and direct air samples into multiple sample units, each of which represents a wind sector, are used for general sampling and locating of emission sources. Samples may be taken as single samples, or as a composite over a predetermined time period, or as a continuing monitoring operation. Some instruments are designed to extract a sample from the air, analyze it automatically, and record the result on a chart. Others take a sample which must be examined in a laboratory.

Many instruments have been developed during the 1950s and 1960s that are mechanized, automatic, and recording, so that they can be used with a minimum of attendance and manipulation. Such instruments require careful initial calibration with standard test substances that are to be measured, and a continuing field check with recalibration at frequent intervals to maintain accuracy.

Particulate samples may be analyzed for weight, size range, effect on visibility, chemical character,

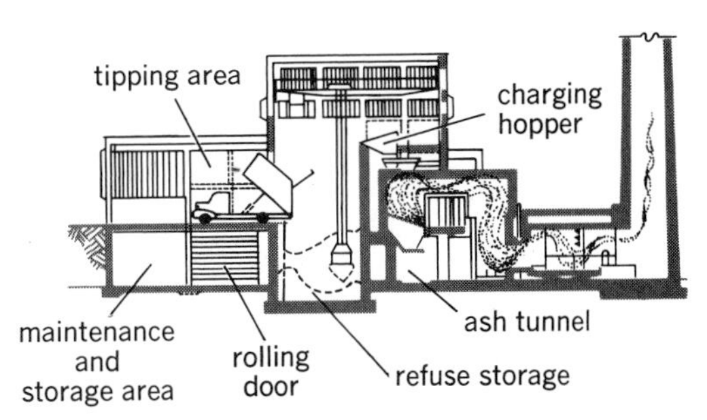

Fig. 8. Diagram of incinerator with rectangular grate. (*American Society of Civil Engineers*)

shape, and other specific information required. Gas samples may be analyzed to determine the presence of a specific gas or of a group of gases of the same chemical family; that is, nitrogen oxide may be determined or the concentration of all oxides of nitrogen may be established by analysis, or total hydrocarbons may be measured, or by more specific analysis the fractions of several specific hydrocarbons making up the total may be found.

Several types of units with air pumps drawing air through paper tapes mounted on a spool have been developed. The tape is moved automatically so that successive samples on fresh paper are taken at timed intervals.

High-volume samplers are used at many sampling network stations in the United States. The electron microscope has been employed for the examination of aerosols and fine particles. Spectrographic instruments are used for analyzing hydrocarbons and oxides of nitrogen and carbon. Automatically operated units take a sample and then pump chemicals into it at the appropriate time to produce a succession of chemical reactions; they are used to obtain a continuous record of concentration fluctuation of gaseous pollutants, such as sulfur dioxide and others where a wet-chemical analytical method is appropriate. Other instruments use the principle of conductance for measurement of a gas dissolved in a liquid medium. The sample is passed into the liquid medium and a change in electrical energy is measured and recorded. Such instruments measure the effect of any substance that is ionizable in the medium. Orsat analyses are made on flue gases. Photoelectric cells are used to control alarm systems connected to stacks.

Analytical instruments that use several principles of measurement have made it possible to take more data at less cost and manpower. Systems are being developed whereby data from a number of monitoring stations can be transmitted to a central point, transferred to computer program operations, and become statistical information concerning air-pollution concentration. Other systems under development provide for measurement of certain index pollutants while in motion by utilizing automatic instruments mounted in vehicles.

Methods of sampling and methods of analysis have yet to achieve widespread agreement or standardization. The American Society for Testing and Materials (ASTM) has published some 17 standard methods of tests applicable to atmospheric analysis. ASTM has also published definitions of terms relating to air sampling and analysis. Numerous industrial associations and professional organizations are in the process of bringing together in published form the multitude of sampling and testing methods in use.

Air-quality control. This is predicated on standards or guides that take form in official regulations and laws according to three control approaches: restriction of all sources of pollution so that pollution levels in community air are not in excess of certain levels chosen as a safe standard of air quality; limitation on the amount of a specific pollutant that may be present in the exhaust gas from a duct or a stack; or limitation on the amount of impurity in raw materials whose residues reach the community air. These approaches are frequently combined in an attempt to achieve maximum control. The guides or standards may consist of one or more of the following: ambient air-quality standards, emission standards, and material-quality standards.

Official control agencies at local, state, and Federal levels are now in the process of establishing ambient air-quality levels for such pollutants as sulfur dioxide and particulates. Many municipal control agencies have specified the limits of pollutants such as sulfur dioxide, particulates, and solvents that may be emitted from a single source. Many control agencies at municipal and state levels have adopted standards limiting the amount of sulfur in fuel. The U.S. Department of Health, Education, and Welfare publishes from time to time a digest of state air-pollution control laws. All major cities of the United States have adopted laws and regulations based on one or more of the quality control approaches mentioned.

Standards of ambient air quality may vary. In the state of New York, for example, there is a recognized difference in the kind and quantity of pollutants that may be emitted in rural areas, as opposed to highly urbanized areas. Within any region, land use may vary as to its industrial, commercial, residential, or rural components. Subregions based on the predominant land use and on air-quality objectives to be obtained may be established, each having its own air-quality guides. *See* AIR POLLUTION.

[WILLIAM T. INGRAM]

Bibliography: Air Quality Committee of the Manufacturing Chemists Association, *Source Materials for Air Pollution Control Laws*, 1968; American Industrial Hygiene Association, Air Pollution Committee, *Air Pollution Manual*, pt. 1: Evaluation, 1960, pt. 2: Control equipment, 1968; American Society for Testing and Materials Standards, *ASTM Standards on Methods of Atmospheric Sampling and Analysis*, pt. 23, 1967; M. R. Bethea, *Air Pollution Control*, 1978; R. J. Bibbero and I. G. Young, *Systems Approach to Air Pollution Control*, 1974; W. T. Ingram et al., Adaption of Technicon Auto-Analyzer for continuous measurement while in motion, *Technicon Symposia, 1967*, vol. 1: *Automation in Analytical Chemistry*, 1968; W. T. Ingram, C. Simon, and J. McCarroll, *Air Research Monitoring Station System*, J. Sanit. Eng. Div., Proc. Amer. Soc. Chem. Eng. 93, no. SA2, 1967; Interbranch Chemical Advisory Committee, U.S. Department of Health, Education, and Welfare, *Selected Methods for the Measurement of Air Pollutants*, Environmental Health Series, 1965; W. C. McCrone et al., *The Particle Atlas*, 6 vols., 1973–1978; A. C. Stern (ed.), *Air Pollution*, vols. 2 and 5; 3d ed., 1977; C. D. Yaffe et al. (eds.), *Air Sampling Instruments for Evaluation of Atmospheric Contaminants*, 2d ed., 1962.

Air pressure

The force per unit area that the air exerts on any surface in contact with it, arising from the collisions of the air molecules with the surface. It is equal and opposite to the pressure of the surface against the air, which for atmospheric air in normal motion approximately balances the weight of the atmosphere above. It is the same in all directions and is the force that balances the weight of the column of mercury in the Torricellian barometer, commonly used for its measurement.

Units. In 1975 the American Meteorological Society adopted the International System of Units (SI). In this system the basic unit of pressure is the pascal (Pa), which is defined as equal to 1 newton per square meter. Since this is a very small unit, air pressures are normally measured in kilopascals; 1 kilopascal (kPa) equals 1000 pascals. Before the adoption of SI, the units of pressure most frequently used in meteorology were those based on the bar, which is defined as equal to 10^5 pascals or 10^6 dynes/cm^2; 1 bar equals 1000 millibars (mb) equals 100 centibars (cb). Such units are still permitted, but tolerance for their use is likely to be transitory.

Also widely used in practice are units based on the height of the mercury barometer under standard conditions, expressed commonly in millimeters or in inches. The standard atmosphere (760 mmHg) is also used as a unit, mainly in engineering where large pressures are encountered. The following equivalents show the conversions between the commonly used units of pressure, where $(\text{mmHg})_n$ and $(\text{in. Hg})_n$ denote the millimeter and inch of mercury, respectively, under standard (normal) conditions, and where $(\text{kg})_n$ and $(\text{lb})_n$ denote the weight of a standard kilogram and pound mass, respectively, under standard gravity.

$$\begin{aligned}
1\ \text{kPa} &= 1000\ \text{Pa} = 10^4\ \text{dynes/cm}^2 = 10\ \text{mb} \\
&= 7.50062\ (\text{mmHg})_n = 0.295300\ (\text{in. Hg})_n \\
1\ \text{mb} &= 100\ \text{Pa} = 0.1\ \text{kPa} = 1000\ \text{dynes/cm}^2 \\
&= 0.750062\ (\text{mmHg})_n = 0.0295300\ (\text{in. Hg})_n \\
1\ \text{atm} &= 101.3250\ \text{kPa} = 1013.250\ \text{mb} \\
&= 760\ (\text{mmHg})_n = 29.9213\ (\text{in. Hg})_n \\
&= 14.6959\ (\text{lb})_n/\text{in.}^2 = 1.03323\ (\text{kg})_n/\text{cm}^2 \\
1\ (\text{mmHg})_n &= 1\ \text{torr} = 0.1333224\ \text{kPa} \\
&= 1.333224\ \text{mb} = 0.03937008\ (\text{in. Hg})_n \\
1\ (\text{in. Hg})_n &= 3.38639\ \text{kPa} = 33.8639\ \text{mb} \\
&= 25.4\ (\text{mmHg})_n
\end{aligned}$$

Variation with height. Because of the almost exact balancing of the weight of the overlying atmosphere by the air pressure, the latter must decrease with height, according to the hydrostatic equation, Eq. (1), where P is air pressure, ρ is air

$$dP = -g\rho\, dZ \tag{1}$$

density, g is acceleration of gravity, Z is altitude above mean sea level, dZ is infinitesimal vertical thickness of horizontal air layers, dP is pressure change which corresponds to altitude change dZ, P_1 is pressure at altitude Z_1, and P_2 is pressure at altitude Z_2. The expressions on the right-hand side of Eq. (2) represent the weight of the column of air between the two levels Z_1 and Z_2.

$$P_1 - P_2 = \int_{Z_1}^{Z_2} \rho g\, dZ \tag{2}$$

In the special case in which Z_2 refers to a level above the atmosphere where the air pressure is nil, one has $P_2 = 0$, and Eq. (2) yields an expression for air pressure P_1 at a given altitude Z_1 for an atmosphere in hydrostatic equilibrium.

By substituting in Eq. (1) the expression for air density based on the well-known perfect gas law and by integrating, one obtains the hypsometric equation for dry air under the assumption of hydrostatic equilibrium, Eq. (3), valid below about 90

$$\log_e\left(\frac{P_1}{P_2}\right) = \frac{10^{-3}M}{R}\int_{Z_1}^{Z_2}\frac{g}{T}\,dZ \tag{3a}$$

$$Z_2 - Z_1 = \frac{10^3 R}{M}\int_{P_2}^{P_1}\frac{T}{g}\frac{dP}{P} \tag{3b}$$

km, where g is the gravitational acceleration in m/s^2; M is the gram-molecular weight, 28.97 for dry air; R is the gas constant for 1 mole of ideal gas, or 8.31470 joules/(mole)(K); T is the air temperature in K; and the altitude Z is expressed in meters.

Equation (3) may be used for the real moist atmosphere if the effect of the small amount of water vapor on the density of the air is allowed for by replacing T by T_v, the virtual temperature given by Eq. (4), in which e is partial pressure of water va-

$$T_v = T\left[1 - \left(1 - \frac{M_w}{M}\right)\frac{e}{P}\right]^{-1} \tag{4}$$

por in the air, M_w is gram-molecular weight of water vapor (18.0160 g/mole), and $(1 - M_w/M) = 0.37803$.

Equation (3a) is used in practice to calculate the vertical distribution of pressure with height above sea level. The temperature distribution in a standard atmosphere, based on mean values in middle latitudes, has been defined by international agreement. The use of the standard atmosphere permits the evaluation of the integrals of Eqs. (3a) and (3b) to give a definite relation between pressure and height. This relation is used in all altimeters which are basically barometers of the aneroid type. The difference between the height estimated from the pressure and the actual height is often considerable; but since the same standard relationship is used in all altimeters, the difference is the same for all altimeters at the same location, and so causes no difficulty in determining the relative position of aircraft. Mountains, however, have a fixed height, and accidents have been caused by the difference between the actual and standard atmospheres.

Horizontal and time variations. In addition to the large variation with height discussed in the previous paragraph, atmospheric pressure varies in the horizontal and with time. The variations of air pressure at sea level, estimated in the case of observations over land by correcting for the height of the ground surface, are routinely plotted on a map and analyzed, resulting in the familiar "weather map" representation with its isobars showing highs and lows. The movement of the main features of the sea-level pressure distribution, typically from west to east, produces characteristic fluctuations of the pressure at a fixed point, varying by a few percent within a few days. Smaller-scale variations of sea-level pressure, too small to appear on the ordinary weather map, are also present. These are associated with various forms of atmospheric motion, such as small-scale wave motion and turbulence. Relatively large variations are found in and near thunderstorms, the most intense being the low-pressure region in a tornado. The pressure drop within a tornado can be a large fraction of an atmosphere, and is the principal cause of the explosion of buildings over which a tornado passes.

It is a general rule that in middle latitudes at localities below 1000 m (3280 ft) in height above sea level, the air pressure on the continents tends to be slightly higher in winter than in spring, summer, and autumn; whereas at considerably greater heights on the continents and on the ocean surface, the reverse is true.

Various maps of climatic averages indicate certain regions where systems of high and low pressure predominate. Over the oceans there tend to be areas or bands of relatively high pressure, most marked during the summer, in zones centered near latitude 30°N and 30°S. The Asiatic landmass is dominated by a great high-pressure system in winter and a low-pressure system in summer. Deep low-pressure areas prevail during the winter over the Aleutian, the Icelandic-Greenland, and Antarctic regions. These and other centers of action produce offshoots which may travel for great distances before dissipating.

Thus during the winter, spring, and autumn in middle latitudes over the land areas, it is fairly common to experience the passage of a cycle of low- and high-pressure systems in alternating fashion over a period of about 6–9 days in the average, but sometimes in as little as 3–4 days, covering a pressure amplitude which ranges on the average from roughly 15–25 mb (1.5–2.5 kPa) less than normal in the low-pressure center to roughly 15–20 mb (1.5–2.0 kPa) more than normal in the high-pressure center. During the summer in middle latitudes the period of the pressure changes is generally greater, and the amplitudes are less than in the cooler seasons (see table).

Within the tropics where there are comparatively few passages of major high- and low-pressure systems during a season, the most notable feature revealed by the recording barometer (barograph) is the characteristic diurnal pressure variation. In this daily cycle of pressure at the ground there are, as a rule though with some exceptions, two maxima, at approximately 10 A.M. and 10 P.M., and two minima, at approximately 4 A.M. and 4 P.M., local time.

The total range of the diurnal pressure variation is a function of latitude as indicated by the following approximate averages (latitude N and range in millibars): 0°, 0.3 kPa; 30°, 0.25 kPa; 35°, 0.17 kPa; 45°, 0.12 kPa; 50°, 0.09 kPa; 60°, 0.04 kPa. These results are based on the statistical analysis of thousands of barograph records for many land stations. Local peculiarities appear in the diurnal variation because of the influences of physiographic features and climatic factors. Mountains, valleys, oceans, elevations, ground cover, temperature variation, and season exert local influences; while current atmospheric conditions also affect it, such as amount of cloudiness, precipitation, and sunshine. Mountainous regions in western United States may have only a single maximum at about 8–10 A.M. and a single minimum at about 5–7 P.M., local time, but with a larger range than elsewhere at the same latitudes, especially during the warmer months (for instance, about 4 mb difference between the daily maximum and minimum).

At higher levels in the atmosphere the variations of pressure are closely related to the variations of temperature, according to Eq. (3*a*). Because of the lower temperatures in higher latitudes in the lower 10 km, the pressures at higher levels tend to decrease toward the poles. The figure shows a typical pattern at approximately 10 km above sea level. As is customary in representing pressure patterns at upper levels, the variation of the height of a surface of constant pressure, in this case 300 mb (30 kPa), is shown, rather than the variation of pressure over a horizontal surface.

Besides the latitudinal variation, the figure also shows the characteristic wave pattern in the pressure field, and the midlatitude maximum in the wind field known as the jet stream, with its "waves in the westerlies." In the stratosphere the temperature variations are such as to reduce the pressure variations at higher levels, up to about 80 km, except that in winter at high latitudes there are relatively large variations above 10 km. At altitudes above 80 km the relative variability of the pressure increases again. Although the pressure and density at these very high levels are small, they are important for rocket and satellite flights, so that their variability at high altitudes is likewise important.

Mean atmospheric pressure and temperature in middle latitudes, for specified heights above sea level*

Altitude above sea level			
Standard geopotential meters, m′	m at latitude 45°32′40″	Air pressure, kPa	Assumed temperature, K
0	0	1.01325×10^{3}	288.15
11,000	11,019	2.2632×10^{2}	216.65
20,000	20,063	5.4747×10^{1}	216.65
32,000	32,162	8.6798×10^{0}	228.65
47,000	47,350	1.1090×10^{0}	270.65
52,000	52,429	5.8997×10^{-1}	270.65
61,000	61,591	1.8209×10^{-1}	252.65
79,000	79,994	1.0376×10^{-2}	180.65
88,743	90,000	1.6437×10^{-3}	180.65†

*Approximate annual mean values based on radiosonde observations at Northern Hemisphere stations between latitudes 40 and 49°N for heights below 32,000 m and on observations made from rockets and instruments released from rockets. Some density data derived from searchlight observations were considered. Values shown above 32,000 m were calculated largely on the basis of observed distribution of air density with altitude. In correlating columns 1 and 2, G is 9.80665 m^2/s^2 per standard geopotential meter (m′). Data on first three lines are used in calibration of aircraft altimeters.

†Above 90,000 m there occurs an increase of temperature with altitude and a variation of composition of the air with height, resulting in a gradual decrease in molecular weight of air with altitude.

Relations to wind and weather. The practical importance of air pressure lies in its relation to the wind and weather. It is because of these relationships that pressure is a basic parameter in weather forecasting, as is evident from its appearance on the ordinary weather map.

Horizontal variations of pressure imply a pressure force on the air, just as the vertical pressure variation implies a vertical force that supports the weight of the air, according to Eq. (1). This force, if unopposed, accelerates the air, causing the wind to blow from high to low pressure. The sea breeze is an example of such a wind. However, if the pressure variations are on a large scale and are changing relatively slowly with time, the rotation of the

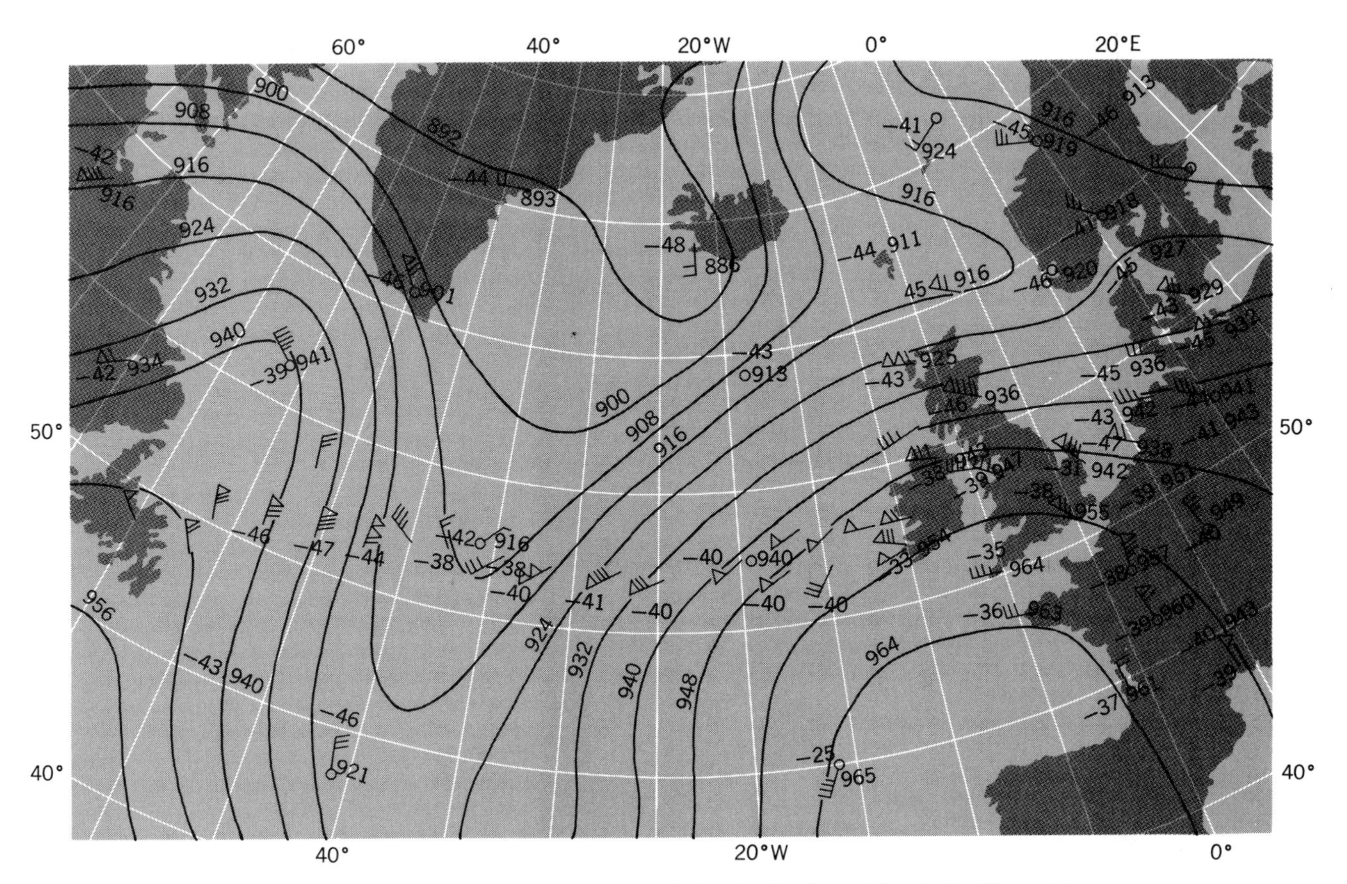

Contours of 300-mb (30-kPa) surface, in tens of meters, with temperature in °C, and measured winds at the same level, on June 16, 1960. Winds are plotted with arrow pointing in direction of the wind, with each bar of the tail representing 10 m/s. Triangle represents 50 m/s.

Earth gives rise to geostrophic or gradient balance such that the wind blows along the isobars. This situation occurs when the pressure variations are due to the slow-moving lows and highs that appear on the ordinary weather map, and to the upper air waves shown in the figure, in which the relationship is well illustrated.

The wind near the ground, in the lowest few hundred meters of the atmosphere, is retarded by friction with the surface to a degree that depends on the smoothness or roughness of the surface. This upsets the balance mentioned in the previous paragraph, so that the wind blows somewhat across the isobars from high to low pressure.

The large-scale variations of pressure at sea level shown on a weather map are associated with characteristic patterns of vertical motion of the air, which in turn affect the weather. Descent of air in a high heats the air and dries it by adiabatic compression, giving clear skies, while the ascent of air in a low cools it and causes it to condense and produce cloudy and rainy weather. These processes at low levels, accompanied by others at higher levels, usually combine to justify the clear-cloudy-rainy marking on the household barometer.

[RAYMOND J. DELAND]

Bibliography: H. R. Byers, *General Meteorology*, 1974; R. G. Fleagle and J. A. Businger, *An Introduction to Atmospheric Physics*, 1963; A. Miller, *Meteorology*, 1976; O. G. Sutton, *Challenge of the Atmosphere*, 1961; J. Van Mieghem, *Atmospheric Energetics*, 1973.

Animal community

Any assemblage of animals occurring in a given place at a given time. Animal communities are held together by interaction with the physical environment and with the associated vegetation, and by the internal relationships of predation, parasitism, mutual dependence, and competition. Because animals are in general more mobile and smaller than the plants with which they live, animal communities are usually less visible and less clearly defined than are plant communities, and the species composition of animal communities is typically more diverse and subject to change in response to alteration of the environment. *See* PLANT COMMUNITY.

The basic dependence of animals upon plants for food, and the importance of animals to plants as agents of pollination and dispersal of propagules, suggest that an animal community and its associated plant community can best be understood as components of a single unit, the biotic community or biocenose. The biotic community together with its physical environment functions as an ecological system, or ecosystem, involving the movement and exchange of matter and energy through biogeochemical cycles, with the green plants serving as primary producers of organic materials and the animal community as the consumers. For recycling of matter to be effective in such a system, the dead products of both plants and animals must be reduced to relatively simple physical-chemical states; this is accomplished largely by the action of decomposers, including small to microscopic species of fungi, bacteria, and a variety of detritivorous animals.

The modern study of animal communities focuses on their structure and functioning. Features of animal communities that seem of greatest importance include species composition and diversity; spatial distributions of the component populations; temporal patterns of activity; functional organization; replacement of species over time (succession); and productivity in terms of numbers of individuals, biomass, or energy equivalent. *See* ECOLOGICAL INTERACTIONS; ECOSYSTEM.

Species composition and diversity. Any animal community has a certain species composition, but most communities are too large or too diverse to make a complete list of species feasible. For the better-known taxonomic groups of animals, however, the number of species present (species richness) can be counted and compared among different communities. Latitudinal gradients in species richness have been demonstrated for breeding land birds, fresh-water fishes, ants, and various marine invertebrates; in such groups, tropical communities often show greater richness than their temperate or boreal counterparts. The observation that species richness often increases on islands of successively larger size provides the basis for a major theory of biogeography developed by R. H. MacArthur and E. O. Wilson.

When animal communities are sampled, it is usually observed that a few species are relatively common, some are very rare, and the rest are intermediate in abundance. In some cases, the distribution of species abundances approaches that of a lognormal series, in others that of a geometric series, and in still others that of the "broken stick" model proposed by MacArthur. Such diversity can be assessed by an index that considers both the species richness and the differing abundance of species, as in the equation below, where S is the

$$H = -\sum_{i=1}^{S} (p_i)\log(p_i)$$

total number of species and p_i is the proportion of the total number of individuals that belongs to the ith species. *See* SPECIES DIVERSITY.

Spatial distribution patterns. The structure of animal communities is affected by the spatial dispersions of their species populations. Few animal populations are randomly dispersed, and although competitive or territorial behavior may produce a regular distribution, some form of aggregation or grouped dispersion is probably the most common pattern among animals. *See* POPULATION DISPERSION.

Species differences in tolerance to local conditions of microclimate or vegetation may result in a continuous distribution of overlapping populations along an environmental gradient or in one or more points at which several to many species have their optima in common. Distinct zones, each with its characteristic species, can sometimes be recognized in animal communities. Transitions (ecotones) between adjacent communities are often marked by "edge" species that prefer habitats of intermediate type. Methods of gradient analysis and ordination, developed for vegetation by R. H. Whittaker and others, have as yet been little applied to animal communities.

Animal communities also show vertical distribution patterns, with different groups of species in each vegetation stratum or soil horizon. Within a single stratum, competition is reduced by spatial and behavioral separation; for example, five species of warbler (*Dendroica*) in the spruce-fir forests of New England feed in distinct parts of the canopy, move in different directions through the trees, and obtain their food in different ways.

Functions in the animal community may differ from stratum to stratum. In a deciduous forest, many of the herbivores will be concentrated in the canopy, where most new leaf tissue is being produced, whereas accumulated litter on the floor is the focus of activity by decomposers.

Temporal activity patterns. Animals which are active by day, such as hawks and meadow voles, are often replaced by nocturnal forms, such as owls and deer mice. The summer residents of temperate-forest bird communities are similarly replaced by species which are active through the winter. Some invertebrates move to the forest canopy by day and return to the floor at night. Such daily and seasonal shifts in species composition and in kinds of activities impart an annual rhythm to the functioning of the animal community and its utilization of matter and energy, creating a highly dynamic system.

Functional organization. A considerable body of theory about the structure of animal communities is based on the concept of the niche, defined for each species by its particular environmental requirements and its unique set of relationships to other organisms. When niches overlap, competition for space or other resources may result, and unsuccessful species may be eliminated. Ecologists are now investigating the ways in which species respond to niche overlap and the factors that may limit the number of species that can coexist in a single community.

Food relationships are generally held to be the major organizing force in animal communities and are considered in terms of food chains, trophic levels, and ecological pyramids. Food resources are often exploited in different ways by distinct groups (guilds) of animals; for example, herbaceous plants may be attacked by some insects which chew holes in the leaves, by other insects which scrape tissue from the leaf surfaces, and by still others which suck juices from the stems. *See* ECOLOGICAL SYSTEMS, ENERGY IN; FOOD CHAIN.

Community structure is sometimes strongly influenced by the activity of single "keystone" species; for example, a starfish of the rocky intertidal zone along the coast of western North America has been found to keep mussel populations in check sufficiently to allow the coexistence of other invertebrates and algae which the mussel would otherwise displace.

Some animal species exert significant influence over the community simply by virtue of their great numbers or biomass, in which case they are called dominants. In many temperate lakes the zooplankton community is dominated by large species when fish predators are absent but by small species when fish are present.

Succession. The direct dependence of many animals on specific plants for food or shelter often results in a gradual replacement of animal species contingent on that of the plant community. Thus, in southern Michigan the meadowlark and vesper sparrow are replaced by the indigo bunting and towhee as grassland changes to shrubs and saplings, and these birds are in turn replaced by the red-eyed vireo and downy woodpecker as the vegetation becomes a mature forest. However, animals frequently influence the course of plant succession through modification of the physical environment, grazing of the vegetation, or the spread of destructive disease. Widespread vegetational changes took place in England in the 1950s after an epizootic of myxomatosis virtually eliminated rabbits from the countryside: plants which had been held back by rabbit browsing spread rapidly and increased in vigor over wide areas. *See* ECOLOGICAL SUCCESSION.

Animal succession also occurs in the communities of such temporary microhabitats as rotting logs, the fecal droppings of large herbivores, and the carcasses of animals, where a sequence of decomposer insects and other invertebrates, fungi, and many microorganisms is associated with the successive stages of decay. These microsuccessions differ from plant successions in that they are of relatively short duration, do not progress to a relatively stable climax state, and appear to be controlled more by exhaustion of particular food materials and accumulation of toxic wastes, and less by factors such as light and climate.

Productivity. The quantity of organic matter synthesized by animals over a given time is referred to as secondary productivity. The productivity of an animal community depends upon the quantity of plant food that is available and upon the efficiency with which each animal species consumes the available food and converts it into new protoplasm. To ascertain such productivity, estimates must be made of the population density, food intake and assimilation, and growth and reproduction of each major species, and it is usually necessary to supplement field measurements with laboratory studies of metabolism. *See* BIOLOGICAL PRODUCTIVITY.

It has been estimated that, on average, animals consume 2–3% of net plant production in tundras and deserts, 4–7% in forests, 10–15% in grasslands, and up to 30–40% in marine communities. There is also great variation in the extent to which animals convert into new protoplasm. When consumption values are multiplied by estimates of conversion efficiency, not much more than 1% of net plant production on land is turned into animal tissue, as compared with 5–6% in the oceans. Animal production, then, is in general marked by low efficiencies, by a much higher rate in the sea than on land, and by the diversity and complexity of its forms and of the pathways it provides for the dispersion and transfer of matter and energy. *See* COMMUNITY. [FRANCIS C. EVANS]

Bibliography: M. L. Cody and J. M. Diamond, (eds.), *Ecology and the Evolution of Communities*, 1975; L. R. Dice, *Natural Communities*, 1952; C. S. Elton, *The Pattern of Animal Communities*, 1966; R. H. Whittaker, *Communities and Ecosystems*, 1975.

Applied ecology

Applied ecology involves the application of ecological principles to the solution of human problems and the maintenance of a quality life. It is assumed that humans are an integral part of ecological systems and that they depend upon healthy, well-operating, and productive systems for their continued well-being. For these reasons, applied ecology is based on a knowledge of ecosystems, and the principles and techniques of ecosystem ecology are used to interpret and solve specific environmental problems and to plan new management systems in the biosphere. Although a variety of management fields, such as forestry, agriculture, wildlife management, environmental engineering, and environmental design, are concerned with specific parts of the environment, applied ecology is unique in taking a view of the whole system, and attempting to account for all inputs to and outputs from the system—and all impacts. In the past, applied ecology has been considered as being synonymous with the above applied sciences.

Ecosystem ecology. Ecological systems, or ecosystems, are complexes of plants, animals, microorganisms, and humans, together with their environment. Environment includes all those factors, physical, chemical, biological, sociocultural, which affect the ecosystem. The complex of life and environment exists as a balanced system and is unique for each part of the Earth. The unique geological features, soils, climate, and availability of plants, animals, and microorganisms create a variety of different types of ecosystems, such as forests, fields, lakes, rivers, and oceans. Each ecological system may be composed of hundreds to thousands of biological species which interact with each other through the transfer of energy, chemical materials, and information. The interconnecting networks which characterize ecosystems are often called food webs (Fig. 1). It is obvious from this structural feature of interaction that a disturbance to one population within an ecosystem could potentially affect many other populations. From another point of view, ecosystems are composd of chemical elements, arranged in a variety of organic complexes. There is a continual process of loss and uptake of chemicals to and from the environment as populations are born, grow and die, and are decomposed. Ecosystems operate on energy derived from photosynthesis (called primary production) and from other energy exchanges. The functional attributes of ecosystems, such as productivity, energy flow, and cycling of chemical elements, depend upon the biological species in the ecosystem and the limiting conditions of the environment. *See* BIOLOGICAL PRODUCTIVITY; ECOLOGICAL SYSTEMS, ENERGY IN.

Ecological systems develop in accord with the regional environment, and maintain a stability of structure and function. Although these systems have evolved to resist the normal expected perturbations encountered in the environment, unusual disturbances and catastrophic events can upset stability and even destroy the system. In this case, recovery can occur after the disturbance stops. Recovery is termed ecological succession since it comprises a sequence of communities which succeed each other until the steady state is reesta-

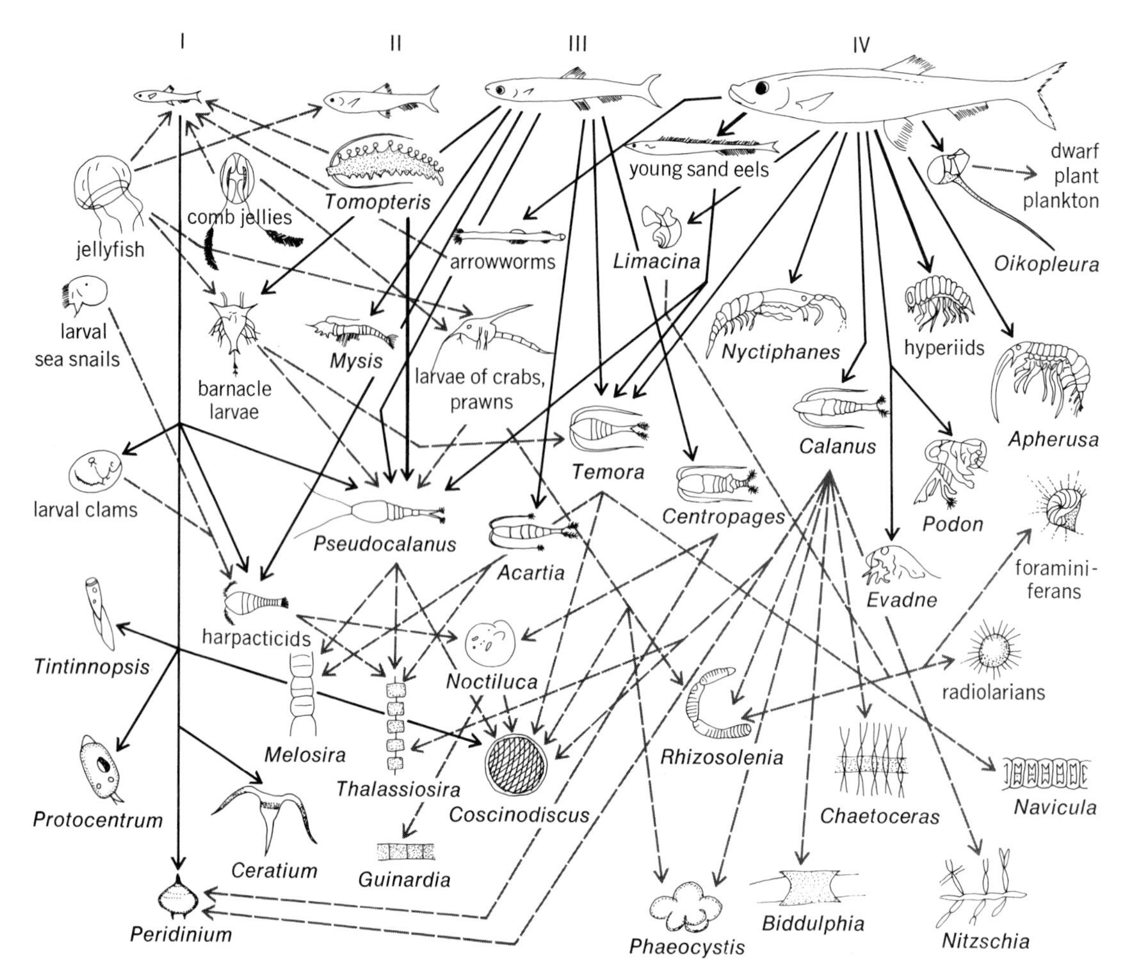

Fig. 1. The food relations of the herring at different stages in its life. Sizes of herring are (I) 0.6 to 1.3 cm, (II) 1.3 to 4.5 cm, (III), 4.5 to 12.5 cm, and (IV) over 1.25 cm. Solid lines indicate food eaten directly by herring. (*From Wells, Huxley, and Wells, 1939; Copyright A. P. Watt and Son, and Messrs. Cassel and Co., Ltd.*)

blished. *See* ECOLOGICAL SUCCESSION.

Populations in these ecosystems evolve so that they fill a unique structural and functional role within the system. Often, groups of populations coevolve, so that they form a more or less isolated subunit. For example, the pollinators of a plant species, and their predators and parasites, form such a guild. Populations continually adapt and develop through natural selection, expanding to the limit of their resources. Population growth is, therefore, due to an increase in resources or a relaxation of limiting factors. *See* ECOSYSTEM; POPULATION DYNAMICS.

Ecosystem management theory. The objective of applied ecology management is to maintain the stability of the system while altering its inputs or outputs. Often, ecology management is designed to maximize a particular output or the quantity of a specific component. Since outputs and inputs are related (Fig. 2), maximization of an output may not be desirable; rather, the management objective should be the optimum level. Optimization of systems can be accomplished through the use of systems ecology methods which consider all parts of the system rather than a specific set of components. In this way, a series of strategies or scenarios can be evaluated, and the strategy producing the largest gain for the least cost can be chosen for implementation.

The applied ecology management approach has been partially implemented through the U.S. National Environmental Policy Act, which requires an environmental-impact analysis for major Federal actions significantly affecting the quality of the human environment. According to the act, the environmental-impact analysis must be carried out by an interdisciplinary team of specialists representing the subjects required for an ecosystem analysis. This team is required to evaluate a project on the basis of its environmental, social, and cultural features. One alternative that must be considered is that of no alteration of the system.

A variety of general environmental problems within the scope of applied ecology relate to the major components of the Earth: the atmosphere, water, and land, and the biota. The ecological principles used in applied ecology are discussed elsewhere; a sequence of environmental problems of special importance to applied ecology is discussed below.

Atmospheric problems. The atmosphere is one of the most important components of the environment to consider from the viewpoint of applied ecology since it connects all portions of the Earth into one ecosystem. The atmosphere is composed of a variety of gases, of which oxygen and nitrogen make up the largest percentage. It is not uniform in its depth or its composition, but is divided into several layers or zones which differ in density and composition. Although most interaction with humans occurs in the zone nearest Earth, the most distant parts of the atmosphere are also important since they affect the heat balance of the Earth and the quality of radiant energy striking the surface. Disturbances to these portions of the atmosphere could affect the entire biosphere. *See* ATMOSPHERE.

The composition of the atmosphere varies according to location. The qualities of minor gases such as carbon dioxide, the amounts of various metallic elements, and the quantity of water vapor and dust all may differ, depending on the relative distance from land or sea. But, in addition, the atmospheric composition may change in time. For example, over the history of the planet, the percentage composition of oxygen has changed from a very oxygen-poor environment to the present atmosphere, with 20.95% oxygen by volume. *See* ATMOSPHERIC CHEMISTRY.

Human activities may introduce a variety of pollutants into the atmosphere. The principal pollutants are carbon dioxide, sulfur compounds, hydrocarbons, nitrogen oxides, solid particles (particulates), and heat. The amounts of pollutants that are produced may be quite large, especially in local areas, and have increased in amount as industrialization has become more widespread. For example, carbon gas output has been estimated to have increased from about 6,400,000 tons per year in 1950 to 13,400,000 tons per year in 1967 (1 short ton = 0.9 metric ton). Industrial and domestic activities have also been estimated to put 218×10^6 tons of sulfur into the atmosphere per year. In most cases, these pollutants have increased during the recent past, and in many areas have become a serious problem. *See* AIR POLLUTION.

Atmospheric problems can have a variety of impacts on humans. Numerous observers have attributed climatic change to atmospheric alteration, since any change in the gaseous envelope of the Earth could alter the heat balance and the climate. The Earth's climate is not constant, and it is difficult to establish an exact correlation between pollution and variation in temperature or solar radiation at the Earth's surface. The pollutants most likely to have an effect on climate are carbon dioxide and solid particles. Besides this possible effect

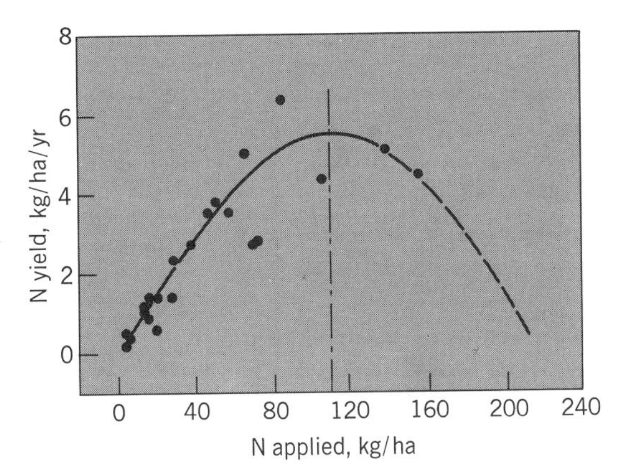

Fig. 2., The relationship between N yield and N consumption (1964–1969). (*Data from FAO, Fertilizer Report, 1969*)

on overall climate, pollution may affect the chemical balance of regions of the Earth. These effects may be extremely complicated. For example, it has been reported that sulfur oxides produced in the industrial districts of northern Europe have moved north in the atmosphere over Scandinavia. The sulfur oxides react with water to form sulfuric acid, which rains out in precipitation. The acid rain changes the acidity of the soil in this region and depresses the activity of blue-green algae, which fix nitrogen from the atmosphere. The reduction in nitrogen fixation causes serious reduction in the growth of trees. Other pollutants may act in a similar complicated fashion through the network of interactions in the Earth ecosystem.

Finally, atmospheric pollution has direct effects on plants and animals and human activities. Pollutants, like other materials, can act as limiting influences on the growth, reproduction, and survival of plants and animals. A variety of plants, such as lichens and mosses, that are extremely sensitive to pollution can be used to indicate the degree of atmospheric deterioration. In some severe cases, all vegatation and animal life may be destroyed in the vicinity of the polluting industry. Gases and solids are taken into the lungs of humans, and cause disease or discomfort. In cities, such pollutants as asbestos and lead are exceedingly dangerous to the population. And finally, the impact of pollutants on buildings, clothes, artwork, and machines is costly.

Control of atmospheric pollution requires interception of pollutants at the point of discharge. Industrial control can be achieved by the use of special filters, precipitators, and other devices. Control of pollution for automobiles also may involve special equipment, as well as redesign of engines and fuel. Reduction in dust and similar general sources of air pollution may demand a change in the operation producing the problem. *See* AIR-POLLUTION CONTROL.

Water problems. The aquatic environment is of equal importance to applied ecology since most of the Earth's surface is covered by the oceans, and the land is connected to the oceans by streams and rivers. Thus, like the atmosphere, the waters are a connection between distant parts of the biosphere and can carry a disturbance from its origin to another region.

The composition of waters varies widely, and it is essential in evaluating aquatic health to establish the base-line conditions which are stable under the normal or undisturbed conditions. Water pollution arises from a variety of sources—industry, domestic sewage, agricultural fertilizers and feedlots, construction activities, and forest practices. Principal pollutants are sediments, organic pollutants containing nitrites, and phosphates, lead, mercury, hydrocarbons, pesticides, and various synthetic chemicals. The impact of the pollutant depends both on its chemical nature and on the quantity released. All water bodies receive quantities of chemicals and solid materials; a variety of organisms which break down and utilize these inputs have evolved. Serious problems arise when the inputs to the water body become unusually large or contain synthetic materials which cannot be decomposed by the extant organisms. Further problems may develop through concentrations of pollutants in the food web of the aquatic system. For example, if a chemical is not metabolized by organisms, but is concentrated in their tissues, as are some pesticides, then as each organism is eaten by another, the body burden of chemicals can increase. In this way, predators may obtain very large and dangerous amounts of pollutants. The decline in population of certain fish-eating birds has been attributed to this process of transfer and concentration.

Aquatic pollution has many consequences of significance to humans. An excess of chemical materials which enrich plant and animal growth can cause rapid increase in life. This process, termed eutrophication, may entail dramatic increases in the algal, planktonic, or rooted aquatic plant populations, with the result that the water body appears green in color or becomes clogged with vegetation. Toxic chemicals released to water bodies may directly kill aquatic life or, if present in sublethal amounts, may change the species of plants and animals present. Often aquatic pollution is not a dramatic either-or proposition, with all fish or other aquatic life killed; rather, more commonly a trend toward an increase in the more resistant species is seen, with the elimination of those forms which are especially susceptible to the pollutant. Aquatic pollution also involves heat, especially that derived from industrial activities, including nuclear power and fossil fuel plants. In these instances, water is used to cool the machines or reactors and is exhausted to the environment at elevated temperatures. Since all metabolic and chemical processes are influenced by heat, thermal pollution should have a significant effect on aquatic systems, but thus far it has been difficult to prove that such an impact occurs. *See* WATER POLLUTION.

Other aquatic problems of interest to applied ecology concern alteration of water channels by impoundments or channelization and irrigation. In each instance the natural pattern of water movement is altered, and deterioration of the environment may result. Impoundments limit the natural movement of sediment and chemical elements; production patterns in the water and lands below the impoundment may be altered, other changes may occur. But, on the plus side, impoundments often provide fisheries, electrical energy, recreation, and other advantages. Irrigation problems may involve the movement of salts from depths in the soil, with deposition near the surface. Disturbance of the chemical equilibrium of the soil, in turn, interferes with plant growth.

Terrestrial and soil problems. Terrestrial environments constantly undergo a degrading and decomposing process owing to the action of water, frost, wind, and other environmental processes on the surface which involve the linkages joining land, water, and atmosphere. Human activities may accelerate these natural processes. In addition, the use of chemical materials on the land may have effects similar to those resulting from their addition to water. Most terrestrial environmental problems are caused by agricultural, grazing, or forest practices. Probably the most serious effect concerns practices which increase the rate of surface erosion. Only a small percentage of the Earth's surface is suitable for agriculture, and the loss of soil from these areas is extremely serious. As a consequence, certain regions have been denuded and are no longer productive. Overgrazing may also remove the cover of vegetation and allow water and wind to erode the soil. Deserts have increased in extent almost everywhere because of overgrazing; and in India, the increase in the Rajasthan Desert can be measured in feet per year. Dust from this desert blows as far east as Thailand. Overcutting trees and lack of reforestation programs also may increase soil erosion and nutrient losses in forest regions. These impacts are not solely the mark of modern civilization. Misuse of the land has been noted in many past civilizations and can even be a problem for present-day primitive societies. *See* EROSION; FOREST AND FORESTRY.

However, modern agriculture has added new problems to those of the primitive farmer. Various chemicals such as fertilizers, pesticides, and herbicides are used to increase agricultural production. These chemicals may be needed because of past misuse of the land or because of economic demands in a society that does not recognize the need to maintain and protect terrestrial resources. Organic or ecological agricultural practices seek to reestablish a pattern of land use without causing deterioration of the soil and biotic resources. Although reestablishment of the pattern may result in somewhat lower productivity, proponents of "ecoagriculture" argue that high productivity can be maintained without loss of soil through erosion and without a reduction in fertility.

Probably the most serious short-term impact arises from the use of chemicals on the land. In the most extreme cases, the health of the agriculturalist may be affected by the materials. But more commonly, the pesticide or chemical in the soil is taken up by the crop and then enters the human food chain. Many modern governments maintain agencies to advise farmers on the proper amounts of chemicals to apply so that buildup does not occur; other agencies periodically sample and analyze foodstuffs for residues. In this way, the consumer can be protected from misuse of chemicals or from excessive concentrations. Unfortunately, these agencies seldom consider the impacts of agricultural chemicals on other animal food chains. For example, the soil fauna and the natural nitrogen-fixing organisms present in the soil, as well as the terrestrial faunas living near the agricultural or forest plantations, can be significantly affected; and populations of animals, even beneficial species, may be reduced through the misuse of chemical materials. However, the extinction of plants and animals, which is also a serious applied ecology problem, usually is due to the destruction of their habitat. Pollution, disturbance of the land, and overhunting may provide the final cause leading to the destruction of a particular living species.

Nuclear energy. Industrialized societies require large quantities of energy. Energy production from nuclear reactors has been enthusiastically developed in many regions of the biosphere. However, nuclear energy also has environmental consequences that are of concern to applied ecology and that must be considered when these facilities are designed and operated, so that negative impacts on the environment do not occur.

Nuclear energy has three primary environmental consequences: the storage of radioactive products, the release of radioactive material to the environment, and, as mentioned above, the release of heat.

There are various kinds of radioactivity associated with the particular elements used in the reactor. In the process of the generation of energy, these fuel elements are changed into a suite of radioactive materials. Although these materials, in turn, form a new source of chemicals, the process of separation and concentraton is very costly and dangerous. In either case, however, the processes result in radioactive waste that must be stored for periods of hundreds, even thousands of years. The fact that the potential danger of these wastes will require technical attention for periods of time longer than the histories of many modern societies is a prime argument against the widespread use of nuclear energy. However, proponents of the use of nuclear energy state that certain geological structures such as salt mines can be safely used for storage indefinitely.

A second environmental problem concerns the loss of relatively small quantities of radioactive materials to the environment during chemical processing in reactors or chemical plants. If these materials enter the body, they can cause disease and death. Like pesticides, radioactive chemical materials may be concentrated in food chains and can appear in relatively large concentrations. In certain fragile environments such as the arctic tundra, the food chains are very short. Thus, radioactive chemicals derived from testing atomic weapons pass through lichens or reindeer or caribou to humans. Concentrations in certain localities may be high enough to cause concern to public health authorities.

Population problems. Applied ecology also is concerned with the size of the human population, since many of the impacts of human activities on the environment are a function of the number and concentration of people. The human population on Earth has increased exponentially, and in many countries this increase poses almost insurmountable problems. Control of environmental degradation, even a concern for environment, is nonexistent when the population is undernourished, starving, ill-housed, and underemployed. Social disorder, alienation, psychological disturbances, physical illness, and other sorts of problems have been correlated with overpopulation. Population also places demands on the resources of the Earth; as the standard of living rises, these demands increase.

The human population problem is exceedingly complicated because the growth of population is controlled largely by the decisions of individual families. The family may visualize several different strategies—the number of children that is best for the family, best for their social group or tribe, or best for the human race—or have no strategy at all. Families may decide that a large family is best even under serious conditions of overpopulation. Considerable evidence indicates that the size of the family declines as the population becomes less rural and more industrial. Thus, some specialists urge economic development, regardless of environmental impact, as a means of solving the population problem. Others urge that the family be more directly influenced to reduce the number of children. Direct action might entail birth-control advice, medical abortion and sterilization, and taxation. Yet others argue that no measures such as these can be significant and that the human population will be controlled by famine, war, or disease. Each of these positions leads toward a set of social policies, all of which have an environmental impact which, in turn, affects the human society.

Environmental planning and design. The foregoing discussion suggests that there is an optimum environment for the human race which is influenced by a variety of population densities and activities. Thus, although the population of the United States is relatively sparse, it has a large environmental impact. This, in turn, suggests that the human environment and society could be designed in such a way to minimze the negative impacts and provide a satisfactory productive life for the population. Environmental design considers economic and social policy, as well as the impact of designed rural, urban, transport, industrial, and other systems. It also considers the design of the individual environment of house, furniture, clothes, and so on. Considering the often violent impact modern society has had on the environment, a design revolution is required to reorganize the environment created by society so that these impacts can be reduced.

Environmental planning and design obviously have a deep political component, since the methods used to redesign society depend upon the control of individual demands. At one extreme, individual demand is allowed to express itself without limit, and education is used to create in the individual a realization that environmental constraints must be recognized. At the other extreme, the government or party controls demand through regulation. Most societies operate somewhere between these extremes.

Throughout human history, utopian designs have been developed for human societies and environments. Today these designs pay more attention to environmental features of society and are often labeled as ecological. Applied ecology, thus, considers not only the alteration of specific features of the modern industrial society to correct some environmental defect but also the fundamental reorientation of society to achieve a balance between humans and the natural world on which they depend. *See the feature articles* ENVIRONMENTAL PROTECTION; URBAN PLANNING.

[FRANK B. GOLLEY]

Aquifer

A subsurface zone that yields economically important amounts of water to wells. The term is synonymous with water-bearing formation. An aquifer may be porous rock, unconsolidated gravel, fractured rock, or cavernous limestone. Economically important amounts of water may vary from less than a gallon per minute (6.3×10^{-5} m^3/s) for cattle water in the desert to thousands of gallons per minute for industrial, irrigation, or municipal use.

Among the most productive are the sand and gravel formations of the Atlantic and Gulf Coastal plains of the southeastern United States. These layers extend for hundreds of miles and may be several hundred feet thick. Also highly productive are deposits of sand and gravel washed out from the continental glaciers in the northern United

States; the outwash gravel deposits from the western mountain ranges; certain cavernous limestones such as the Edwards limestone of Texas and the Ocala limestone of Florida, Georgia, and South Carolina; and some of the volcanic rocks of the Snake River Plain in Idaho and the Columbia Plateau.

Aquifers are important reservoirs storing large amounts of water relatively free from evaporation loss or pollution. If the annual withdrawal from an aquifer regularly exceeds the replenishment from rainfall or seepage from streams, the water stored in the aquifer will be depleted. This "mining" of groundwater results in increased pumping costs and sometimes pollution from sea water or adjacent saline aquifers. Lowering the piezometric pressure in an unconsolidated artesian aquifer by overpumping may cause the aquifer and confining layers of silt or clay to be compressed under the weight of the overlying material. The resulting subsidence of the ground surface may cause structural damage to buildings, altered drainage paths, increased flooding, damage to wells, and other problems. Subsidence of 10 to 15 ft (3.0 to 4.6 m) has occurred in Mexico City and parts of the San Joaquin Valley of California. Careful management of aquifers is important to maintain their utility as a water source. *See* ARTESIAN SYSTEMS; GROUNDWATER. [RAY K. LINSLEY]

Arctic biology

The Arctic, with its small land areas surrounding the frozen arctic seas, has been exploited from the south for its yield of oil and furs of animals. Until the recent prospect of arctic petroleum and minerals, only traders and explorers from temperate lands came as transients among the scant population of indigenous arctic people. Terence Armstrong described how the recent development of arctic natural resources has multiplied invasion of the Asian Arctic by strangers from the south. Many new residents are now moving northward to exploit American arctic petroleum and minerals. This new and still unsettled movement of people into the Arctic encounters life that is strange to them in the long cold and darkness of arctic winters. Interesting biological, social, and economic conditions, as well as problems of adjustment to the Arctic, face the newcomers.

Life on land. Only about 20 species of mammals of the more than 3000 species in the world live on the treeless arctic tundra, where production of plants is so sparse and specialized that few kinds of animals can find a living. The only large arctic herbivores are musk-ox and caribou or reindeer, but many small herbivorous mice and lemmings live in obscure ways under winter snow. These small herbivores sustain carnivorous bears, wolves, foxes, weasels, and a surprising variety of small shrews.

The sparse vegetation on arctic lands must accomplish the annual production on which all arctic life depends during a short cool summer in which periodic freezing requires that all plants must retain resistance to frost and adjust reproduction opportunistically to brief spells of sufficient warmth. Spiders, insects, and many cold-blooded animals meet requirements of the Arctic by devices that could not be projected by imagination based on southern experience. Their success demonstrates the surprising adaptability of some species. Their few kinds show the rigorous selection by the arctic seasonal regimes.

No vertebrate animal seems to endure freezing, as do so many plants and invertebrates. Arctic flowers protruding through snow in early spring freeze to brittle hardness in repeated episodes of cold with only brief interruptions of the progress of forming seeds. John Baust finds that a small boreal beetle that winters in stumps above snow contains over 20% of glycerol in its fluids. The attractive view that the substance serves as an antifreeze is spoiled by the observed freezing of the beetle in arctic temperatures. In warm summer the beetle contains no glycerol, and is killed at the temperature at which water freezes. It is possible that glycerol protects against the destructive consequences of freezing as glycerol and some other cryoprotective agents allow long frozen storage of bull's sperm for artificial insemination and tissues like cornea for surgical repair of damaged eyes. Natural protection from injury by freezing occurs by diverse ways in the Arctic. Results of research into natural adaptation to arctic temperatures may lead to cryoprotective ways for preservation of cells and tissues.

Life on seas. Peter Freuchen and Finn Salomonsen narrated how the arctic seas support the wandering polar bear and the small arctic fox that often accompanies the bear far out on winter ice-covered seas. These ice-going mammals nevertheless return to shore to breed. The main food of the bear is the small arctic ringed seal, but it also takes eagerly to carrion remains of sea mammals. On sea ice the arctic foxes take remnants of the prey of bears or any flesh. On land in summer their prey is mice and birds.

The large bowhead whale drew whalers into the margins of Atlantic and Pacific ice with such success that the Atlantic bowhead is very nearly extinct. Fur seals and sea lions do not enter arctic life, but large numbers of hair seals breed at the margins of arctic ice. In fact, about 90% of the world's hair seals bear their pups on antarctic or arctic ice, attesting the productivity of polar seas and the capability of warm-blooded seals for living in icy waters.

Preservation of warmth. The preservation of warmth in arctic winter is a matter of real concern. Early explorers recorded that some arctic birds and mammals were as warm as those of warmer regions. L. Irving and J. Krog found that arctic and tropical birds and mammals are similarly warm in their interiors. Exposed experimentally to a temperature of −40° an arctic fox is comfortable in its thick fur. It generates no more heat in cold and so conserves heat by its insulation. This model of adaptation to arctic cold through insulation indicates an economy of heat that sustains a large arctic mammal at an expenditure of metabolism no greater than that required for mammals in a milder climate. The large arctic mammal is thus not handicapped by cold.

Arctic people. Arctic man, with meager natural insulation, wore fur clothing skillfully made from arctic animals, built shelters, and utilized fire. From Greenland across North America to the edge of Siberia, Eskimos lived with a similar language and a culture distinct from those of northern American Indians.

Across arctic Eurasia, people of different cultures and languages reflected the fact that they were derived from peoples of steppes and forests who had migrated northward along great rivers and coastal routes. Indigenous Eurasian people are now far exceeded in numbers by fresh migrants who perform technical tasks for exploitation of minerals. Migration of strangers into arctic America seems to have just begun.

Migration of life into Arctic. Throughout the last Pleistocene glaciation, arctic plants and animals were confined to limited refuge areas free from ice. David Hopkins showed that 10,000 years ago melting of the ice sheets allowed spread of plants and animals from arctic refuges and invasion from the south by plants and animals that had retained versatility to adapt for arctic life. The distribution and many forms of life are modern characteristics of the recently changing Arctic.

Spectacular in the arctic spring are migrations of birds coming to nest for a few summer months on arctic lands and coasts after wintering on temperate lands and seas and even over South America and Africa. These great migrations visibly demonstrate the inclinations of birds to occupy opportunistically suitable parts of the world and their capability for organized flights and long navigation. In the seas fishes, whales, and seals annually migrate northward to breed and harvest the production of the arctic year. On land a residue of American caribou still migrate in summer into the Arctic. As arctic migrants return south for winter, they export from the Arctic the annual increment in their populations, an exploitation of arctic production.

It is an interesting speculation that these visible annual migrations in some ways recapitulate the postglacial resettlement of the Arctic. Plant, animal, and human life seem to have been ever pressing their capability for life in the Arctic and developing adaptability for the changing conditions of the arctic seasons. *See* BIOCLIMATOLOGY; BIOGEOGRAPHY; VEGETATION ZONES, WORLD.

[LAURENCE IRVING]

Bibliography: T. E. Armstrong, *The Russians in the Arctic*, 1972; P. Freuchen and F. Salomonsen, *The Arctic Year*, 1958; D. M. Hopkins (ed.), *The Bering Land Bridge*, 1967; L. Irving, Adaptations to cold, *Sci. Amer.*, 214:94-101, 1966; L. Irving, *Arctic Life of Birds and Mammals Including Man*, 1972; M. Martna (ed.), *Arctic Bibliography*, vol. 16, 1975.

Artesian systems

Groundwater conditions formed by water-bearing rocks (aquifers) in which the water is confined above and below by impermeable beds. These systems are named after the province of Artois in France, where artesian wells were first observed. *See* GROUNDWATER.

Because the water table in the intake area of an artesian system is higher than the top of the aquifer in its artesian portion, the water is under sufficient head to cause it to rise in a well above the top of the aquifer. Many of the systems have sufficient head to cause the water to overflow at the surface, at least where the land surface is relatively low. Flowing artesian wells were extremely important during the early days of the development of groundwater from drilled wells, because there was no need for pumping. Their importance has diminished with the decline of head that has occurred in many artesian systems and with the development of efficient pumps and cheap power with which to operate the pumps. When they were first tapped, many artesian aquifers contained water that was under sufficient pressure to rise 100 ft or more above the land surface. Besides furnishing water supplies, many of the wells were used to generate electric power. With the increasing development of the artesian aquifers through the drilling of additional wells, the head in most of them has decreased and it is now from a few feet to several hundred feet below the land surface in many areas of former artesian flow. A majority of artesian wells are now equipped with pumps.

Perhaps the best-known artesian aquifer in the United States is the Dakota sandstone, of Cretaceous age, which underlies most of North Dakota, South Dakota, and Nebraska, much of Kansas, and parts of Minnesota and Iowa at depths ranging from 0 to 2000 ft. The water is highly mineralized, as a general rule, but during the latter part of the 19th century, when these areas were being settled, the Dakota sandstone provided a valuable source of water supply under high pressure. Few wells in this aquifer flow more than a trickle of water today. The St. Peter sandstone and deeper-lying sandstones of early lower Paleozoic age, which underlie parts of Minnesota, Wisconsin, Iowa, Illinois, and Indiana, form another well-known artesian system. Formerly, wells on low ground flowed abundantly, but now wells have to be pumped throughout most of the area. Some of the water is highly mineralized, but in many places it is of good quality. In New Mexico, in the Roswell artesian basin, cavernous limestone of Permian age provides water to irrigate thousands of acres of cotton and other farm crops. Although the head has been steadily declining, many wells still have large flows and others yield copious supplies by pumping. Among the most productive artesian systems are the Cretaceous and Tertiary aquifers of the Atlantic and Gulf Coastal plains. These provide large quantities of water for irrigation and industrial use and supply large cities, such as Savannah, Ga., Memphis, Tenn., and Houston and San Antonio, Tex. Numerous artesian basins are found in intermontane valleys of the West. Some of the best known are in the Central Valley, Calif., where confined aquifers provide water to irrigate millions of acres of farmland, and the San Luis Valley, Colo. Numerous other lesser artesian systems are found in all parts of the United States.

[ALBERT N. SAYRE/RAY K. LINSLEY]

Atmosphere

The gaseous envelope surrounding a celestial body. The terrestrial atmosphere, by its composition, control of temperature, and shielding effect from harmful wavelengths of solar radiation, makes possible life as known on Earth. The atmosphere, which is retained on the Earth by gravitational attraction and in a large measure rotates with it, is a system whose chemical and physical properties and fields of motion constitute the sub-

Table 1. Composition of the atmosphere*

Molecule	Fraction by volume near surface	Vertical distribution
Major constituents		
N_2	7.8084×10^{-1}	Mixed in homosphere; photochemical dissociation high in thermosphere
O_2	2.0946×10^{-1}	Mixed in homosphere; photochemically dissociated in thermosphere, with some dissociation in mesosphere and stratosphere
A	9.34×10^{-3}	Mixed in homosphere with diffusive separation increasing above
Important radiative constituents		
CO_2	3.1×10^{-4}	Mixed in homosphere; photochemical dissociation in thermosphere
H_2O	highly variable	Forms clouds in troposphere; little in stratosphere; photochemical dissociation above mesosphere
O_3	variable	Small amounts, 10^{-8}, in troposphere; important layer, $10^{-6}-10^{-5}$, in stratosphere; dissociated above
Other constituents		
Ne	1.82×10^{-5}	Mixed in homosphere with diffusive separation increasing above (Ne, He, Kr)
He	5.24×10^{-6}	
Kr	1.14×10^{-6}	
CH_4	1.5×10^{-6}	Mixed in troposphere; dissociated in upper stratosphere and above
H_2	5×10^{-7}	Mixed in homosphere; product of H_2O photochemical reactions in lower thermosphere, and dissociated above
NO	$\sim 10^{-8}$	Photochemically produced on stratosphere and mesosphere

*Other gases, for example, CO, N_2O, and NO_2, and many by-products of atmospheric pollution also exist in small amounts.

ject matter of meteorology. The changing atmospheric conditions which affect man's environment, particularly temperature, wind, humidity, cloudiness, and precipitation, constitute weather, and the synthesis of these conditions over a period determines the climate at any place. *See* CLIMATOLOGY; METEOROLOGY.

The average atmospheric pressure at the Earth's surface is about 1013 mb and the density 1.2 kg m^{-3}, and these vary by only a few percent over the globe. They both decrease rapidly and roughly exponentially with height, and at several earth radii the density can be said to have fallen to that of interplanetary space.

Composition. The atmosphere is thought to have developed as the result of chemical and photochemical processes combined with differential escape rates from the Earth's gravitational field. The chemical abundances in the atmosphere, therefore, are not directly related to the cosmic abundances; in particular, the atmosphere is highly oxidized and contains very little hydrogen.

The atmosphere, apart from its highly variable water-vapor content in the troposphere, its liquid droplets and solid matter in suspension, and its variable ozone content in the stratosphere, is well mixed and constant in composition up to about 100 km. This region is termed the homosphere. At higher levels where there is little mixing, diffusive separation tends to take place, with the lighter elements becoming progressively more dominant with height. Moreover, in this region oxygen and the minor constituents, such as carbon dioxide and water vapor, are dissociated by solar ultraviolet radiation. At about 300 km atomic oxygen probably becomes the most important constituent, until about 800 km where helium and hydrogen in turn predominate. This region of highly variable composition is termed the heteorosphere. Ionization of the various constituents also due to absorption of ultraviolet solar radiation becomes a major factor above about 60 km, the base of the ionosphere, which is of major importance to radio communications. Above 600–800 km collisions between atmospheric particles become so infrequent that some traveling outward may escape from the atmosphere. This region is termed the exosphere. *See* AIR POLLUTION; ATMOSPHERIC CHEMISTRY.

Table 1 gives a summary of the composition of the atmosphere. The water vapor is a very variable constituent having a mass mixing ratio of 10^{-3} to 10^{-2} gram per gram of dry air (gg^{-1}) in the lower troposphere and about 2.10^{-6} gg^{-1} in the stratosphere. The ozone layer in the stratosphere has its maximum concentration of about 10^{-5} gram per gram of air at 30–35 km, and is formed by molecular-atomic collisions following dissociation of molecular oxygen by solar radiation wavelengths below about 2400 A. The minor constituents water vapor, carbon dioxide, and ozone play a vital role in the atmosphere's primary biological, heating, and shielding functions.

Thermal structure and circulation. A convenient division of the atmosphere is by spherical shells, "spheres," each characterized by the way its temperature varies in the vertical, and with tops denoted by "pauses," as in the figure. Table 2 lists characteristic values for temperature, pressure, density, and mean molecular weight at a selection of levels within these shells.

Table 2. Atmospheric structure, a selection of mean midlatitude values*

Height, km	Pressure, mb	Temperature, K	Density, kg m^{-3}	Mean molecular weight	Layer
0	1.01×10^{3}	288	1.23×10^{0}	28.96	Troposphere
5	5.40×10^{2}	256	7.36×10^{-1}	28.96	
10	2.65×10^{2}	223	4.14×10^{-1}	28.96	
20	5.53×10^{1}	217	8.89×10^{-2}	28.96	Stratosphere
40	2.87×10^{0}	250	4.00×10^{-3}	28.96	
60	2.20×10^{-1}	247	3.10×10^{-4}	28.96	Mesosphere
80	1.05×10^{-2}	199	1.85×10^{-5}	28.96	
100	3.20×10^{-4}	195	5.60×10^{-7}	28.40	Thermosphere
150	4.54×10^{-6}	634	2.08×10^{-9}	24.10	
200	8.47×10^{-7}	855	2.54×10^{-10}	21.30	
300	8.77×10^{-8}	976	1.92×10^{-11}	17.73	
400	1.45×10^{-8}	996	2.80×10^{-12}	15.98	
500	3.02×10^{-9}	999	5.22×10^{-13}	14.33	
600	8.21×10^{-10}	1000	1.14×10^{-13}	11.51	

*Based on United States Standard Atmosphere, 1976.

Troposphere. The average temperature in the troposphere decreases with height from the surface (~300 K in low latitudes and 260 K in high latitudes) to its upper level, the tropopause, which is around 16 km (temperature 180 K) in the tropics and 10 km (230 K) near the poles. The troposphere includes the layer in which man lives and is the seat of all the important weather phenomena affecting the environment. Its thermal structure is primarily due to the heating of the Earth's surface by solar radiation, followed by upward heat transfer by turbulent mixing and convection. Heat is also transferred toward the poles by atmospheric motions from the more strongly heated equatorial regions.

The processes involved include evaporation of water from the surface and condensation with release of latent heat, leading to clouds and precipitation. The general circulation of the troposphere includes wind systems on all scales—prevailing winds, monsoons, long waves, anticyclones and depressions, fronts, hurricanes, thunderstorms, and shower clouds—each system being associated with characteristic weather patterns. The application of physical and dynamical principles to prognosis of atmospheric motions and weather is one of the main tasks of modern meteorology. *See* CLOUD PHYSICS; WEATHER FORECASTING AND PREDICTION.

Stratosphere. The stratosphere extends from about 10–16 km to about 50 km. Its thermal structure is mainly determined by its radiation balance, in contrast to the troposphere, where convective turbulent exchange is of major importance. It is generally very stable with low humidity and no weather in the popular sense. The only clouds in this region are the "mother-of-pearl" clouds seen infrequently at 20–30 km in high latitudes in winter. The ozone layer strongly absorbs solar radiation of 2000–3000 A in the mesosphere and upper stratosphere, which results in the high temperature (250–290 K) region around the stratopause. The main energy source driving the circulation at these levels is thought to be the excess of absorbed energy over infrared energy emitted by the atmosphere (mainly by carbon dioxide and ozone) in the summer hemisphere, compared with the deficit or sink in the winter hemisphere. This produces an atmospheric circulation system in the upper stratosphere and mesosphere which is separate from that of the lower stratosphere, whose circulation is broadly driven by that of the troposphere below. However, in winter large-scale wave motions also propagate upward into the upper stratosphere and mesophere and have important effects on the circulation there.

Mesosphere. There is a decrease of temperature with height throughout the mesosphere from 55 km to about 80 km, the mesopause, where there is a temperature minimum. In summer its value is about 150 K at high latitudes where occasionally noctilucent clouds are formed, while in winter it has a higher value, around 220 K. This distribution is probably mainly due to dynamic effects, as there is comparatively little direct absorption of solar ultraviolet radiation at this level. The lowest ionized region, the D layer, with $10-10^3$ electrons cm^{-3}, is from about 60 to 90 km.

Thermosphere. The thermosphere is a very high temperature region and extends from 80 km to the outer edge of the atmosphere. It receives its energy by the direct absorption of solar radiation below about 2000 A. The response to change in solar radiation, for example, from night to day, with solar activity and with sunspot cycle, is very marked, and this region is under direct solar control. The atmospheric motions seem to be mainly thermally induced solar tides, but the dynamics of this region are not well understood. The physical phenomena also are complicated and include excitation, dissociation, and ionization of the constituents and effects, such as the aurora following corpuscular radiation from the Sun.

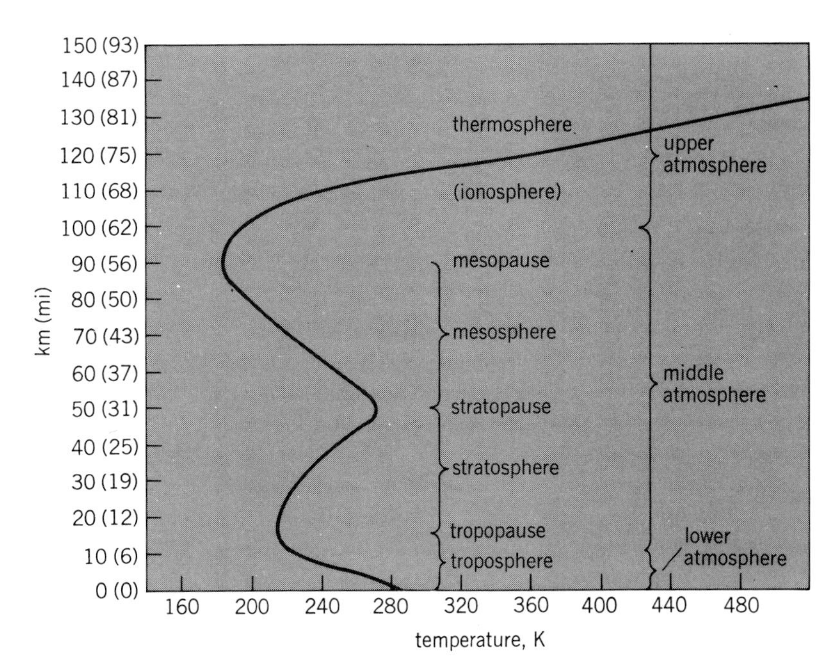

Thermal structure of the atmosphere, showing major divisions.

The increasing importance of ionization with height also means that electrical and magnetic forces have important effects on the atmospheric motions and these, in turn, produce measurable geomagnetic effects at the Earth's surface. The principal ionized layers or wedges are the E layer at 90–160 km, the F1 layer at 160–200 km, and the F2 layer above 200 km with 10^4-10^6 electrons cm^{-3}. The effects of viscosity increasing with height also become manifest, and above 100–110 km, the turbopause, atmospheric flow no longer appears to be turbulent, and diffusive separation of the atmospheric constituents becomes increasingly important with height.

[R. J. MURGATROYD]

Bibliography: R. A. Craig, *The Upper Atmosphere: Meteorology and Physics*, 1965; R. M. Goody and J. C. G. Walker, *Atmospheres*, 1972; F. K. Hare, *The Restless Atmosphere*, 1966; E. N. Lorenz, *The Nature and Theory of the General Circulation of the Atmosphere*, World Meteorol. Organ. Tech. Publ. no. 115, 1967; E. Palmen and C. W. Newton, *Atmospheric Circulation Systems*, 1969; J. A. Ratcliffe (ed.), *Physics of the Upper Atmosphere*, 1960; K. Rawer (ed.), *Winds and Turbulence in Stratosphere, Mesosphere and Ionosphere*, 1968; A. N. Strahler, *The Earth Sciences*, 1963; J. M. Wallace and P. V. Hobbs, *Atmospheric Science: An Introductory Survey*, 1978.

Atmospheric chemistry

A subdivision of atmospheric science concerned with the chemistry and physics of atmospheric constituents, including studies of their sources, circulation, and sinks and their perturbations caused by anthropogenic activity. *See* AIR POLLUTION.

Known gaseous constituents have mixing ratios with air by volume (or equivalently by number), f, ranging from 0.78 for N_2 to 6×10^{-20} for Rn. Known particulate constituents (solid or liquid) have mixing ratios with air by mass, χ, ranging from about 10^{-3} for liquid water in raining clouds to about 10^{-16} for large hydrated ions in otherwise clear air. These constituents are involved in cyclic processes of varying complexity which, in addition to the atmosphere, may involve the hydrosphere, biosphere, lithosphere, and even the deep interior of the Earth. The subject of atmospheric chemistry has assumed considerable importance in recent years because a number of the natural chemical cycles in the atmosphere may be particularly sensitive to perturbation by the industrial and related activities of humans.

Atmospheric composition. A summary of the important gaseous constituents of tropospheric air is given in the table. The predominance of N_2 and O_2 and the presence of the inert gases ^{40}Ar, Ne, ^{4}He, Kr, and Xe are considered to be the result of a very-long-term evolutionary sequence in the atmosphere. These seven gases have extremely long atmospheric lifetimes, the shortest being 10^6 years for ^{4}He, which escapes from the top of the atmosphere. In contrast, all the other gases listed in the table participate in relatively rapid chemical cycles and have atmospheric residence times of a few decades or less.

Particles in the atmosphere range in size from about 10^{-3} to more than 10^2 μm in radius. The term "aerosol" is usually reserved for particulate material other than water or ice. A summary of tropospheric aerosol size ranges and compositions is given in Fig. 1. Concentrations of the very smallest aerosol particles in the atmosphere are limited by coagulation to form larger particles, and the concentrations of the larger aerosols are restricted by sedimentation, the rate of which increases as the square of the aerosol radius. In addition, the size distributions of water droplets and ice crystals are affected by evaporation, condensation, and coalescence processes. Some dry aerosols are water-soluble and can also grow by condensation. Aerosols are important in the atmosphere as nuclei for the condensation of water droplets and ice crystals, as absorbers and scatterers of radiation, and as participants in various chemical cycles.

Composition of tropospheric air

Gas	Volume mixing ratio
Nitrogen, N_2	0.781 (in dry air)
Oxygen, O_2	0.209 (in dry air)
Argon, ^{40}Ar	9.34×10^{-3} (in dry air)
Water vapor, H_2O	Up to 4×10^{-2}
Carbon dioxide, CO_2	2 to 4×10^{-4}
Neon, Ne	1.82×10^{-5}
Helium, ^{4}He	5.24×10^{-6}
Methane, CH_4	1 to 2×10^{-6}
Krypton, Kr	1.14×10^{-6}
Hydrogen, H_2	4 to 10×10^{-7}
Nitrous oxide, N_2O	2 to 6×10^{-7}
Carbon monoxide, CO	1 to 20×10^{-8}
Xenon, Xe	8.7×10^{-8}
Ozone, O_3	Up to 5×10^{-8}
Nitrogen dioxide, NO_2	Up to 3×10^{-9}
Nitric oxide, NO	Up to 3×10^{-9}
Sulfur dioxide, SO_2	Up to 2×10^{-8}
Hydrogen sulfide, H_2S	2 to 20×10^{-9}
Ammonia, NH_3	Up to 2×10^{-8}
Formaldehyde, CH_2O	Up to 1×10^{-8}
Nitric acid, HNO_3	Up to 1×10^{-9}
Methyl chloride, CH_3Cl	Up to 3×10^{-9}
Hydrochloric acid, HCl	Up to 1.5×10^{-9}
Freon-11, $CFCl_3$	About 8×10^{-11}
Freon-12, CF_2Cl_2	About 10^{-10}
Carbon tetrachloride, CCl_4	About 10^{-10}

Above the tropopause, the composition of the atmosphere begins to change, primarily because of the decomposition of molecules by ultraviolet radiation and the subsequent chemistry. For example, decomposition of O_2 produces an O_3 layer in the stratosphere. Although the peak value of f for O_3 in this layer is only about 10^{-5}, this amount of ozone is sufficient to shield the Earth's surface from biologically lethal ultraviolet radiation. About 100 km above the surface, ultraviolet dissociation of O_2 is so intense that the predominant atmospheric constituents become N_2 and O. There is also a layer of aerosols in the lower stratosphere composed primarily of sulfuric acid and dust particles. The sulfuric acid is probably produced by oxidation and hydration of sulfur gases. This same process, but strongly amplified, is the probable source of the much thicker clouds of sulfuric acid recently identified on the planet Venus.

A number of radioactive nuclides are formed naturally in the atmosphere by decay of Rn and by cosmic radiation. Radon, which is produced by decay of U and Th in the crust, enters the atmosphere, where it in turn decays to produce a number of radioactive heavy metals. These metal atoms become attached to aerosol particles and sediment out. Cosmic rays striking N_2, O_2, and Ar principally in the stratosphere give rise to a number of radioactive isotopes, including ^{14}C, ^{7}Be, ^{10}Be, and ^{3}H. The incorporation of ^{14}C into organic matter, where it decays with a half-life of about 5600 years, forms the basis of the radiocarbon dating method. In addition to naturally occurring radioactivity, large quantities of radioactive material have been injected into the atmosphere as a result of nuclear bomb tests. The most dangerous isotope is ^{90}Sr, which can be incorporated into human bones, where it radioactively decays with a half-life of about 28 years. Both natural and anthropogenic radioisotopes have been used as tracers for the study of tropospheric and stratospheric motions.

Atmospheric chemical models. In order both to adequately understand the present chemistry of the atmosphere and to predict the effects of anthropogenic perturbations on this chemistry, it has been necessary to construct quantitative chemical models of the atmosphere. In general, gas and particle mixing ratios show considerable variation with space and time. This variability can be quantitatively analyzed using the continuity equation for the particular species. In terms of χ, time t, wind velocity v, average particle sedimentation

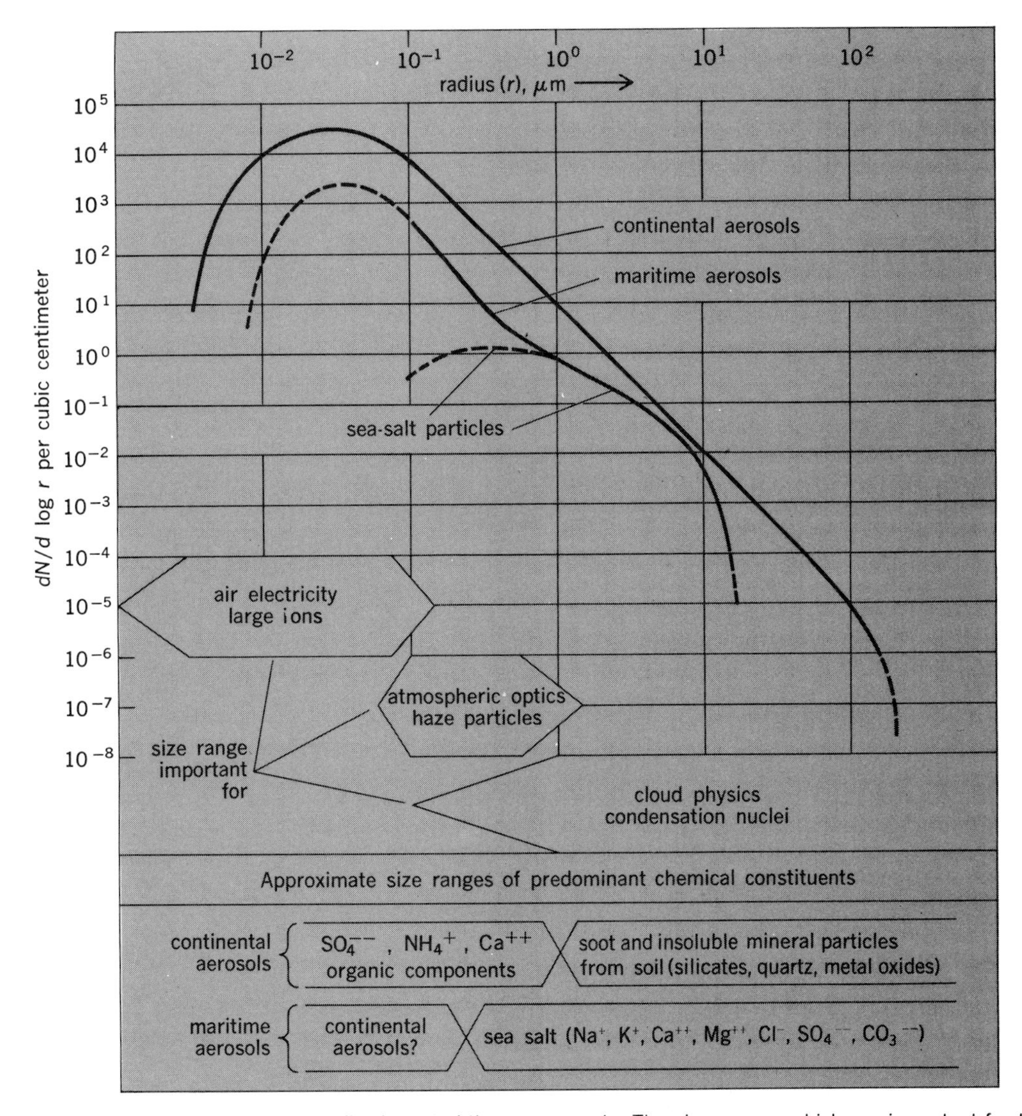

Fig. 1. Chart of the average size distributions and the predominant chemical constituents of some natural aerosols. The size ranges which are important for the various fields of meteorology are shown.

velocity W_p, and air density ρ, this equation is conveniently written as shown below, where an over-

$$\frac{\partial \bar{\chi}}{\partial t} \simeq -\bar{\mathbf{v}} \cdot \nabla \bar{\chi} - W_p \frac{\partial \bar{\chi}}{\partial z} - \frac{1}{\bar{\rho}} \nabla \cdot (\bar{\rho} \overline{\chi' \mathbf{v}'}) + \frac{d\bar{\chi}}{dt}$$

bar denotes an average over a time scale that is long compared to that associated with turbulence, and a prime denotes an instantaneous fluctuation from this average value. The first term on the right-hand side of the equation describes the changes in $\bar{\chi}$ due to the mean atmospheric circulation; the second term gives the $\bar{\chi}$ alternation due to sedimentation which is, of course, zero for gases; the third term denotes fluctuations in $\bar{\chi}$ due to turbulence or eddies, and in this term the eddy flux $\bar{\rho}\overline{\chi' \mathrm{v}'}$ is often roughly approximated by $-K\bar{\rho}\nabla\bar{\chi}$, where K is a three-dimensional matrix of eddy diffusion coefficients; and the last term describes changes in $\bar{\chi}$ caused by chemical production or destruction which may involve simple condensation or evaporation or more complex chemical reactions.

From the equation above, it is seen that the variability of gas or aerosol concentrations is generally due to a combination of transport and true production or destruction. If a constituent has a destruction time T_0, then $d\bar{\chi}/dt = -\bar{\chi}/T_0$, and the equation then implies that the larger the T_0 value the less variability one expects to see in the atmosphere. A constituent for which $\bar{\chi}$ is completely independent of space and time is said to be well mixed. For example, the very-long-lived gases N_2, O_2, ^{40}Ar, Ne, ^{4}He, Kr, and Xe are essentially well mixed in the lower atmosphere.

Atmospheric chemical models involve simultaneous solution of the above equation for each atmospheric constituent involved in a particular chemical cycle. Often, simplified versions of this equation can be utilized, for example, when chemical lifetimes are much shorter than typical atmospheric transport times, or vice versa. Models for the ozone layer have now progressed from historical one-dimensional models which neglected transport to sophisticated three-dimensional models which, in addition to solving the above equation, including transport, solve the equations of motion to obtain v as a function of position and time.

Chemical cycles. In studying the chemical cycles of atmospheric gases, it is important to con-

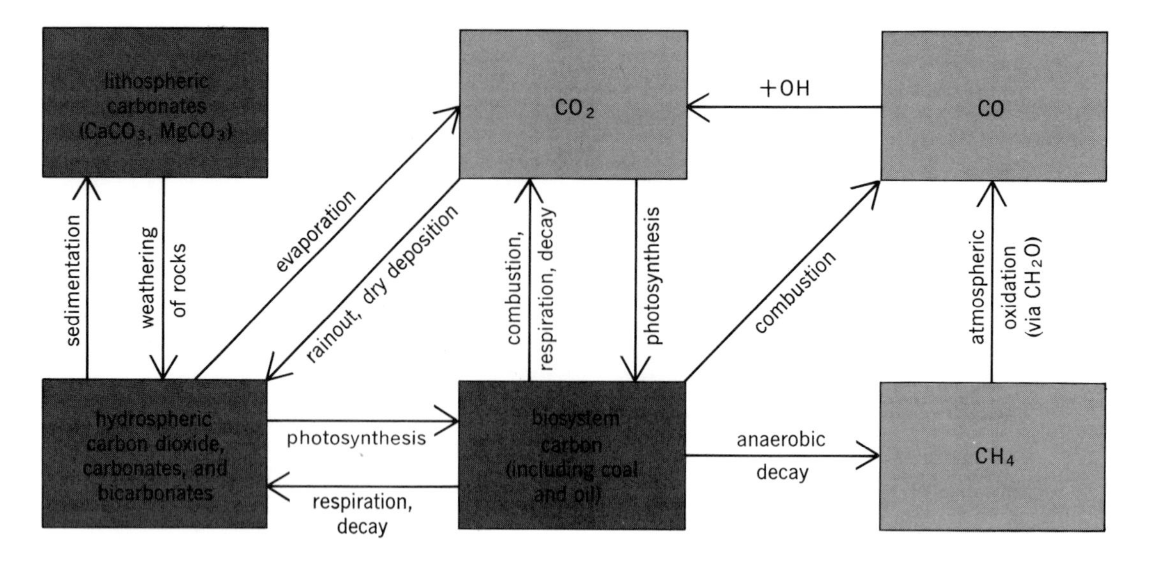

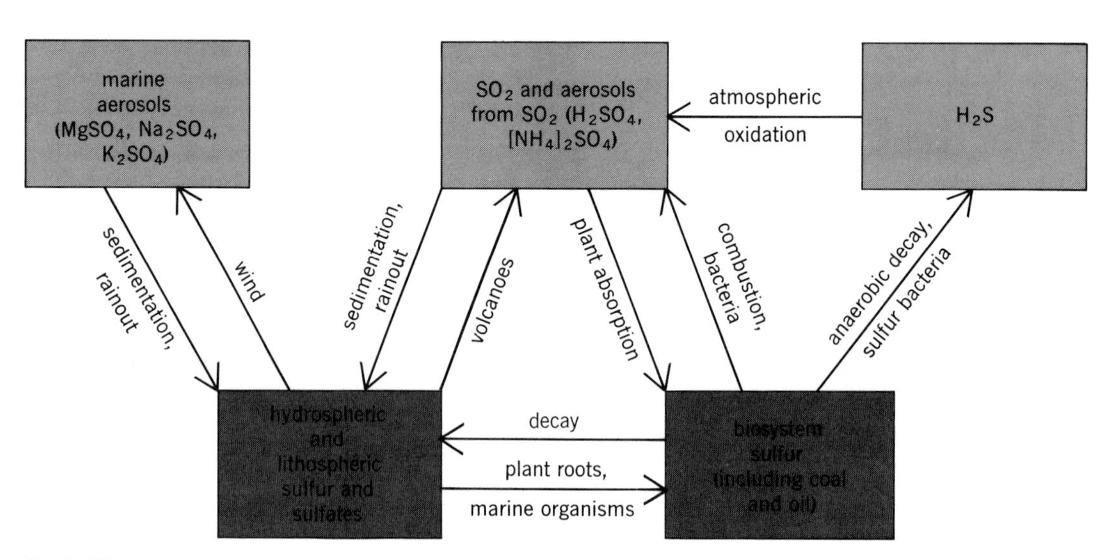

Fig. 2. The carbon and sulfur cycles.

sider both the overall budgets on a global scale and the kinetics of the elementary chemical reactions on a local scale. The study of chemical cycles is in its infancy. Some of the minor details in the cycles outlined below may be subject to change, but this present lack of definition does not detract from their importance.

Carbon cycle. The atmospheric cycle which is of primary significance to life on Earth is that of carbon, which is illustrated in Fig. 2. The CO_2 content of the oceans is about 60 times that of the atmosphere and is controlled by the temperature and acidity of sea water. Release of CO_2 into the atmosphere over tropical oceans and uptake by polar oceans result in a CO_2 residence time of about 5 years. On the other hand, the cycle of CO_2 through the biosphere has a turnover time of a few decades. The amount of CO_2 buried as carbonate in limestone, marble, chalk, dolomite, and related deposits is about 600 times that in the ocean-atmosphere system. If all this CO_2 were released to the atmosphere, a massive CO_2 atmosphere similar to that on the planet Venus would result. Because CO is a poisonous gas, its production from automobile engines in urban areas must be closely monitored. On a global scale, the principal sources of CO are combustion of oil and coal and oxidation of the CH_4 produced naturally during anaerobic decay.

Sulfur cycle. The important aspects of the sulfur cycle are also illustrated in Fig. 2. Sulfur dioxide is produced in the atmosphere by combustion of high-sulfur fuels, by plants and bacteria, and by oxidation of H_2S introduced by anaerobic decay. On a global scale, the SO_2 budget is perturbed significantly by the anthropogenic source, and this perturbation is even more pronounced in urban localities. Oxidation of SO_2 produces sulfuric acid, which is a particularly noxious pollutant; thus regulation of high-sulfur fuel combustion, at least on the local scale, is now required.

Nitrogen cycle. The nitrogen cycle is shown in Fig. 3. The nitrogen oxides (NO, NO_2) are important because they are presently the major compounds governing ozone concentrations in the stratosphere. The main natural source of strato-

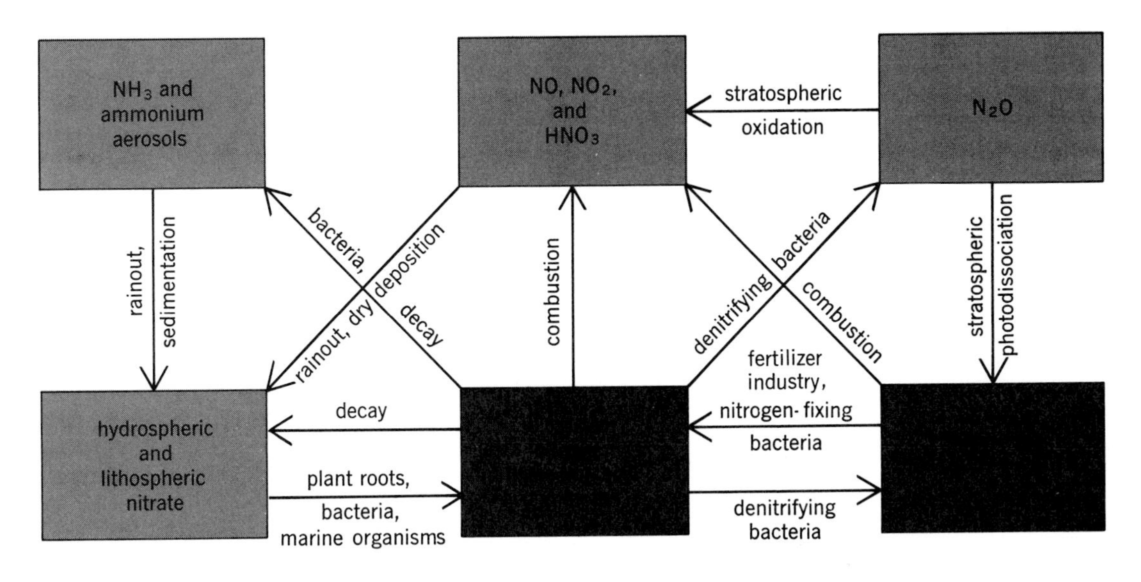

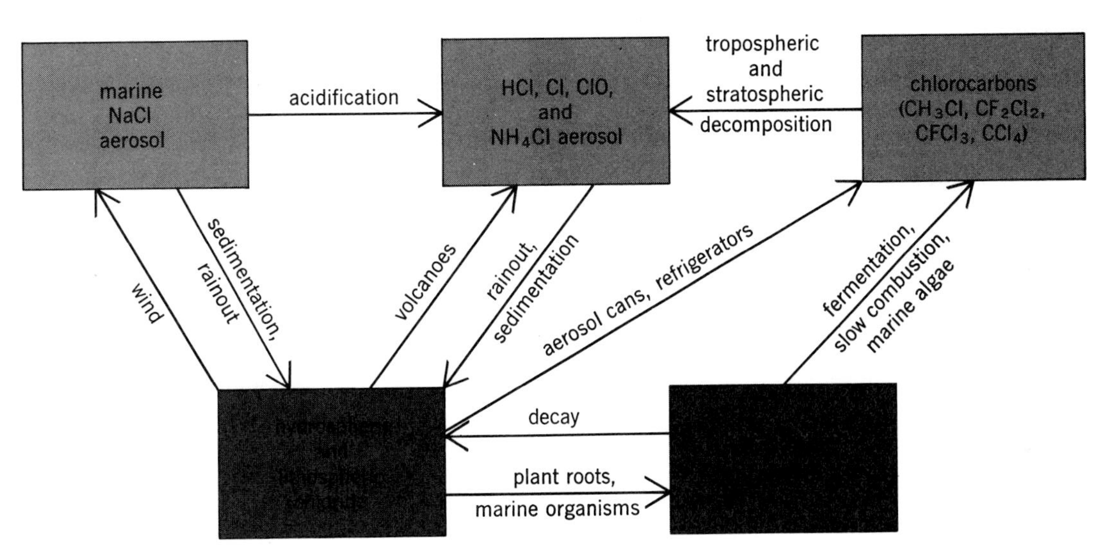

Fig. 3. The nitrogen and chlorine cycles.

spheric NO and NO_2 is decomposition of the N_2O produced from soil nitrate ions (NO_3^-) by denitrifying bacteria. There has been considerable concern about injection of nitrogen oxides directly into the stratosphere by supersonic aircraft. Projected fleet levels for the year 2000 suggest a future anthropogenic source for these oxides comparable to their natural source. This anthropogenic perturbation would cause a decrease of about 10% in total atmospheric ozone and would significantly increase the ultraviolet radiation dosage at the Earth's surface.

Chlorine cycle. The main natural source for atmospheric chlorine is acidification of chloride aerosols producing HCl near the surface. The chlorine cycle (Fig. 3) has received considerable attention because any significant concentrations of Cl and ClO in the stratosphere will lead to depletion of ozone in a manner similar to that caused by NO and NO_2. Because the HCl produced at the ground is severely depleted by rain-out, it does not give rise to significant stratospheric chlorine concentrations. However, the chlorocarbons CCl_4, $CFCl_3$, CF_2Cl_2, and CH_3Cl are relatively insoluble and are not rained out. They appear to decompose in the stratosphere, releasing chlorine; and although the influence of this chlorine on the present stratospheric ozone budget is small, an unchecked buildup of these compounds could lead to significant ozone depletion. The compounds CF_2Cl_2 and $CFCl_3$ are manufactured for use as propellants in aerosol cans, for use in refrigerators and air conditioners, and for manufacture of plastic foams. Methyl chloride is produced naturally by microbial fermentation and by combustion of vegetation, and CCl_4 is probably derived from both natural and industrial sources.

[RONALD G. PRINN]

Bibliography: S. S. Butcher and R. J. Charlson, *An Introduction to Air Chemistry*, 1972; J. Heicklen, *Atmospheric Chemistry*, 1976; C. E. Junge, *Air Chemistry and Radioactivity*, 1963; M. J. McEwan and F. F. Phillips, *The Chemistry of the Atmosphere*, 1975; L. T. Pryde, *Chemistry of the Air Environment*, 1973; E. Robinson and R. E. Robbins, in W. Strauss (ed.), *Air Pollution Control*, pt. 2, 1972.

Atmospheric electricity

The electrical processes constantly taking place in the lower atmosphere. This activity is of two kinds, the intense local electrification accompanying storms, and the much weaker fair-weather electrical activity over the entire globe, which is produced by the many electrified storms continuously in progress over the Earth.

The mechanisms by which storms generate electric charge are presently unknown, and the role of atmospheric electricity in meteorology has not been determined. Some scientists believe that electrical processes may be of importance in precipitation formation and in severe tornadoes.

Disturbed-weather phenomena. Almost all precipitation-producing storms throughout the year are accompanied by energetic electrical activity. The most intense of these are the thunderstorms, in which the electrification attains values sufficient to produce lightning. Measurements show that most other storms, even though they do not give lightning, are also quite strongly electrified.

Thunderstorms begin as little fair-weather clouds that usually form and disappear without producing rain or electrical effects. When the air is sufficiently thermally unstable, a few of these clouds undergo a rapid and dramatic change. Suddenly, for no obvious reason, one will begin a very rapid growth, increasing in height as fast as 15 m/s. When this happens, other significant changes occur. Often in only a few minutes, strong electric fields and rain appear; shortly after, if the cloud is sufficiently vigorous, lightning appears.

The usual height of a thunderstorm is about 10 km; however, they can be as low as 4 km or as high as 20 km. A common feature of these storms is their strong updrafts and downdrafts, which often have speeds in excess of 30 m/s.

The electric fields are most intense within the cloud, reaching values as high as 5000 V/cm. Above the top of the cloud, values in excess of 1000 V/cm have been measured. Over land surfaces beneath the storm, the fields seldom exceed 100 V/cm, but over water surfaces, which lack points to produce point discharge, they can be as high as 500 V/cm. Although the distribution of the electric fields in and about the thunderstorm is complex and variable, most storms approximate a tipped dipole with positive charge above and negative charge below. A few storms appear to have just the reverse polarity.

The origin and nature of the charged regions responsible for the electric fields of thunderstorms are not understood. A variety of explanations has been proposed, most of which are based on the idea that electrification is caused by the falling of precipitation particles electrified within the cloud by such processes as ion capture, contact electrification, freezing, and drop breakup. Alternatively it is suggested that the transport of charged cloud particles by updrafts and downdrafts produces the regions of charge in the cloud. In the absence of reliable data on charge-carrying particles and their motions within the cloud, there is no general agreement on the relative contributions of the various mechanisms that have been postulated.

The electric fields of thunderstorms cause three currents to flow, each of a few amperes: lightning, point discharge from the ground beneath, and conduction in the surrounding air. Because the external field and conductivity are greatest over the top of the cloud, most of the conduction current flows to the ionosphere, the upper, highly conductive layer of the atmosphere.

Fair-weather field. Fair-weather measurements, irrespective of place and time, show the invariable presence of a weak negative electric field caused by the estimated several thousand electrified storms continually in progress. Together these storms cause a 2000-A current from the earth to the ionosphere that raises the ionosphere to a positive potential of about 300,000 V with respect to the earth. This potential difference is sufficient to cause a return flow of positive charge to the earth by conduction through the intervening lower atmosphere equal and opposite to the thunderstorm supply current. The fair-weather field is simply the voltage drop produced by the flow of this current through the atmosphere. Because the electrical resistance of the atmosphere decreases with altitude, the field is greatest near the Earth's surface and gradually decreases with altitude until it vanishes at the ionosphere.

Because the ionosphere is a good electrical conductor, it is often assumed that at any time the ionosphere is everywhere at the same potential. Observations show, however, that geomagnetic activity produces horizontal electric fields that can, over large distances, give rise to potential differences of tens of kilovolts. Influences of solar activity on the ionospheric potential and on the electrical conductivity of the upper atmosphere have been suggested as possible links between solar activity and the Earth's weather.

Human activities affect atmospheric electricity. Measurements indicate that in the Northern Hemisphere the electrical conductivity of the lower atmosphere has been decreasing, probably as the result of manufactured aerosols. Local temporary increases in conductivity have been noted as the result of nuclear tests. Calculations indicate that, if continued, the release into the atmosphere of krypton-85 gas, a radioactive by-product of nuclear power plants, will in 50 years have decreased the resistance between the Earth and the ionosphere by 15%.

The fair-weather field at the Earth's surface is observed to fluctuate somewhat with space and time, largely as a result of local variations in atmospheric conductivity. However, in undisturbed locations far at sea or over the polar regions, the field is observed to have a diurnal cycle independent of position or local time and quite similar to the diurnal variation of thunderstorm activity over the globe. No importance is presently attached to fair-weather atmospheric electricity except that according to some theories it is responsible for the initiation of the thunderstorm electrification process. *See* CLOUD PHYSICS; STORM DETECTION; THUNDERSTORM; TORNADO.

[BERNARD VONNEGUT]

Bibliography: J. A. Chalmers, *Atmospheric Electricity*, 2d ed., 1967; H. Israel, *Atmospheric Electricity*, vol. 1: *Fundamentals, Conductivity, Ions*, 1971, vol. 2: *Fields, Charges, Currents*, 1973; V. P. Kolokolov et al. (eds.), *Studies in Atmospheric Electricity*, 1974.

Atmospheric general circulation

The statistical description of atmospheric motions over the Earth, their role in transporting energy, and the transformations among different forms of energy. Through their influence on the pressure distributions that drive the winds, spatial variations of heating and cooling generate air circulations, but these are continually dissipated by friction. While large day-to-day and seasonal changes occur, the mean circulation during a given season tends to be much the same from year to year. Thus, in the long run and for the global atmosphere as a whole, the generation of motions nearly balances the dissipation. The same is true of the long-term balance between solar radiation absorbed and infrared radiation emitted by the Earth-atmosphere system, as evidenced by its relatively constant temperature. Both air and ocean currents, which are mainly driven by the winds, transport heat. Hence the atmospheric and oceanic general circulations form cooperative systems. *See* OCEAN-ATMOSPHERE RELATIONS.

Owing to the more direct incidence of solar radiation in low latitudes and to reflection from clouds, snow, and ice, which are more extensive at high latitudes, the solar radiation absorbed by the Earth-atmosphere system is about three times as great in the equatorial belt as at the poles, on the annual average. Infrared emission is, however, only about 20% greater at low than at high latitudes. Thus in low latitudes (between about 35°N and 35°S) the Earth-atmosphere system is, on the average, heated and in higher latitudes cooled by

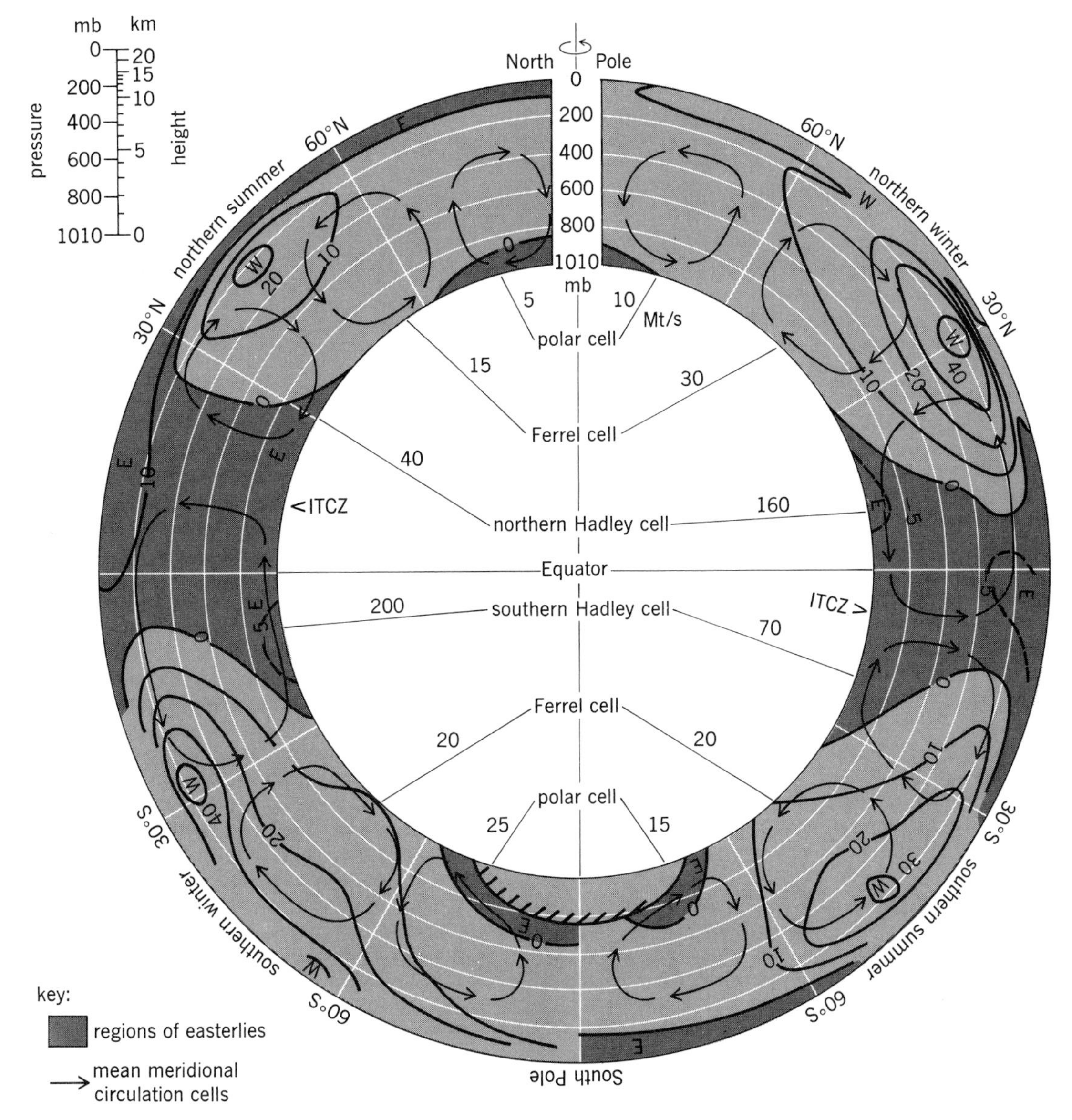

Fig. 1. Pattern of mean zonal (west-east) wind speed averaged over all longitudes, as a function of latitude, height, and season. Height, greatly exaggerated relative to Earth radius, is shown on a linear pressure scale with geometrical equivalent given at upper left. Mean zonal wind (westerly positive, easterly negative, shaded regions) has the same value along any one line and is shown in meters per second (1 m/s ≈ 2 knots). Values are for the geostrophic wind; in the lowest kilometer or so, the actual zonal wind is somewhat weaker. Mean meridional circulation cells in each season are named and intensities are given in terms of mass flow in megatons per second (Mt/s). ITCZ denotes mean latitude of intertropical convergence zone in each season.

radiation. The Earth's surface receives more radiative heat than it emits, whereas the reverse is true for the atmosphere. Therefore, heat must be transferred generally poleward and upward through processes other than radiation. At the Earth-atmosphere interface, this transfer occurs in the form of turbulent flux of sensible heat and through evapotranspiration (flux of latent heat). In the atmosphere the latent heat is released in connection with condensation of water vapor. *See* CLIMATOLOGY.

Considering the atmosphere alone, the heat gain by condensation and the heat transfer from the Earth's surface exceed the net radiative heat loss in low latitudes. The reverse is true in higher latitudes. The meridional transfer of energy, necessary to balance these heat gains and losses, is accomplished by air currents. These take the form of organized circulations, whose dominant features are notably different in the tropical belt (roughly the half of the Earth between latitudes 30°N and 30°S) and in extratropical latitudes. It is convenient to discuss these circulations in terms of the zonal mean circulation (air motions averaged at all longitudes around the Earth) and the eddies (waves, cyclones, and so forth, representing deviations from the mean circulation). *See* METEOROLOGY; STORM.

Mean circulation. In a mean meridional cross section the zonal (west-east) wind component is almost everywhere dominant and in quasigeostrophic balance with the mean meridional pressure gradient. The pressure gradient changes with height in accordance with the distribution of air density, which at a given pressure is inverse to temperature. Hence the distribution of zonal wind is related to that of temperature, as expressed by Eq. (1), where u denotes zonal wind component; z,

$$\frac{\partial \bar{u}}{\partial z} \approx -\frac{g}{2(\Omega \sin \phi + \bar{u}a^{-1} \tan \phi)\bar{T}} \frac{\overline{\partial T}}{\partial y} \quad (1)$$

height above sea level; g, acceleration of gravity; Ω, angular velocity of the Earth; a, Earth radius; ϕ, latitude; T, Kelvin temperature; and y, distance northward. Overbars denote values averaged over longitude and time.

Only in the lowest kilometer or so, where surface friction disturbs the geostrophic balance, and in the vicinity of the Equator is the mean meridional (south-north) component comparable to the zonal wind. Because of the nature of the atmosphere as a shallow layer, the mean vertical wind component is weak. Whereas the magnitude of the mean zonal wind varies between 0 and 45 m/s, and the mean meridional wind varies up to 3 m/s, the mean vertical wind nowhere exceeds 1 cm/s. The vertical component cannot be observed directly, but can be calculated from the distribution of horizontal motions.

In the troposphere and lower stratosphere, the zonal circulation is similar in winter and summer, with easterlies in low latitudes and westerlies in higher latitudes, except in small regions of low-level easterlies around the poles (Fig. 1). The strongest west winds, about 40 m/s, are in the winter hemispheres observed near latitudes 30° at about 12 km. In summer, the west-wind maxima are weaker and located farther poleward.

In the troposphere, the zonal wind increases upward according to Eq. (1) and with a general poleward decrease in temperature. In the lower stratosphere over most of the globe, this temperature gradient is reversed, and the wind decreases with height. Above about 20 km, separate wind systems exist, with prevailing easterly winds in summer and westerlies in winter that attain speeds up to 60–80 m/s near 60 km height. *See* JET STREAM.

The much weaker meridional circulation consists of six separate cells. Their general locations and nomenclatures are shown in Fig. 1, along with the approximate circulations in terms of mass flux. For each cell, only central streamlines are shown, but these represent flows that are several kilometers deep in the horizontal branches, while each vertical branch represents gentle ascending or descending motions over latitude belts some thousands of kilometers wide. The tropical Hadley cells, best developed in the winter hemisphere, are mainly responsible for maintaining the westerly winds that are strongest near the poleward bounds of their upper branches, in subtropical latitudes.

Balance requirements. Equation (1) and certain principles related to the generation of air motions prescribe that the wind distribution over the globe must be consistent with the thermal structure of the atmosphere. Sources and sinks (physical processes such as friction and radiation that act to increase or decrease the momentum or heat content of the air) continually tend to change the existing distributions. The sources and sinks are largely a function of latitude and elevation, such that meridional and vertical fluxes of heat energy and momentum are required. Two methods by which these fluxes are calculated from different kinds of observations give results that offer a check upon one another.

In the first method estimates are made of the rate of change of any property X per unit area of the Earth's surface due to sources and sinks. To maintain an unchanged condition, their integrated value, over the area north of a latitude ϕ, must equal the northward flux F_ϕ of the property across the latitude circle. This requirement is expressed by Eq. (2), in which t denotes time.

$$F_\phi = -2\pi a^2 \int_\phi^{90°\mathrm{N}} \frac{d\overline{X}}{dt} \cos \phi d\phi \quad (2)$$

The second method is to compute the fluxes directly from aerological observations (made by balloon-borne radiosondes). If x denotes a given property per unit mass of air, the flux is given by Eq. (3), where v is the meridional wind component

$$F_\phi = \frac{2\pi a \cos \phi}{g} \int_0^{p_0} (\bar{x}\bar{v} + \overline{x'v'})dp \quad (3)$$

(positive northward). The integration, with pressure p as a vertical coordinate (related to height z as in Fig. 1), is extended from the bottom ($p = p_0$) to the top of the atmosphere ($p = 0$). Here $\bar{x}$ and $\bar{v}$ denote values averaged over time and longitude, and x' and v' are deviations from these mean values at a given pressure surface. A corresponding expression can be written for the vertical fluxes.

The first term of the integrand may be called the circulation flux, and the second term, the eddy flux. The circulation flux arises from a correlation,

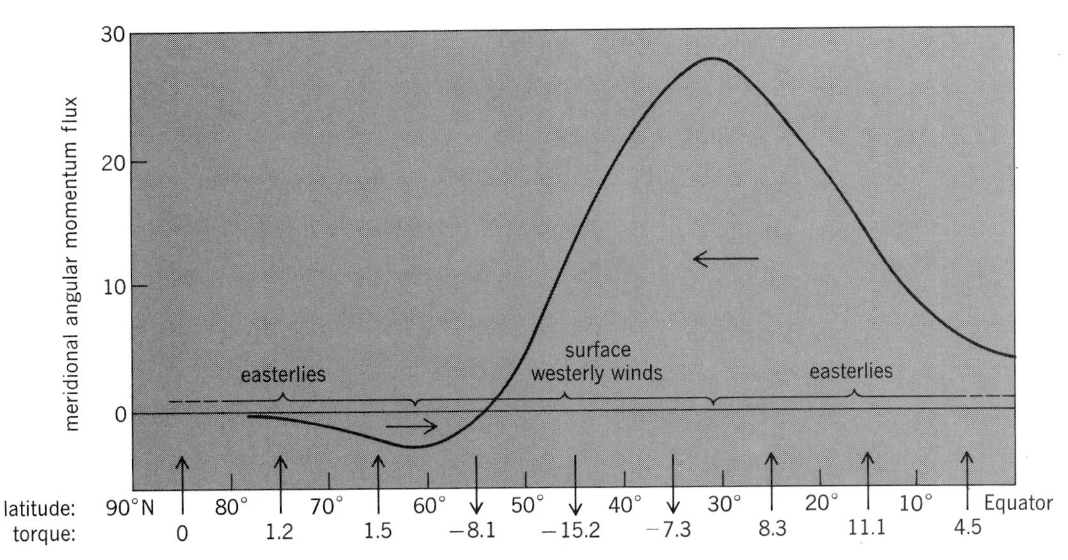

Fig. 2. Average annual meridional flux (positive values northward) of angular momentum in the Northern Hemisphere, with corresponding surface torques in belts of latitude. Units are 10^{18} kg m^{2}s^{-2}. Maximum flux is 50% larger in winter and 75% smaller in summer than the value shown; seasonal variations are much smaller in the Southern Hemisphere.

in the vertical, between the meridional wind component $\bar{v}$ and the atmospheric variable $\bar{x}$. Eddy flux results from correlations between the fluctuations x' and v' at given pressure levels. The eddy fluxes generally dominate in the meridional transfer of properties, except in the central parts of the tropical Hadley cells.

Eddy fluxes can be divided into two groups: large-scale disturbances (such as cyclones) and small-scale eddies associated either with mechanical turbulence or with thermal convection (especially cumulus clouds). Only the large-scale disturbances play a significant role in the meridional eddy fluxes of properties. Concerning the vertical flux, however, the entire spectrum of eddies has to be considered. Generally, mechanical turbulence tends to transport heat and zonal momentum downward but moisture upward, whereas convective turbulence transports heat and moisture upward and momentum downward. Especially in the tropics, convective turbulence is very active in this respect. In the equatorial belt (ITCZ in Fig. 1), practically all of the ascending motion in the Hadley cells actually takes place in the warm updrafts of convective clouds, wherein intense rising currents occupy a small part of the total region. In extratropical regions, where convective phenomena are less prominent especially in winter, the large-scale eddies, associated with polar-front cyclones, dominate in transporting heat and moisture upward as well as poleward. *See* FRONT; THUNDERSTORM.

Table 1. Sources of atmospheric properties

Different sources	$\frac{dX}{dt}$
Atmospheric heat	$\bar{R}_a + \bar{Q}_s + L\bar{P}$
Latent heat	$L(\bar{E} - \bar{P})$
Heat of Earth's surface	$\bar{R}_e - \bar{Q}_s - L\bar{E}$
Heat of atmosphere and Earth	$\bar{R}_a + \bar{R}_e$

Angular momentum balance. In an absolute framework, the motion of the atmosphere plus the eastward speed of a point on the Earth's surface represent the total motion relative to the Earth's axis. The angular momentum of a unit mass of air is given by Eq. (4). Considering the mean value in a

$$M = (u + \Omega a \cos \phi) a \cos \phi \tag{4}$$

zonal ring around a latitude circle, this quantity is conserved unless the ring is subjected to a torque.

The surface easterlies in low latitudes and westerlies in middle latitudes (Fig. 1) exert the principal torques upon the atmosphere, due to frictional drags that are opposite to the direction of the surface winds. The frictional stress τ_x can be estimated from the surface wind velocity by an empirical formula. The torque per unit area, $\tau_x a \cos \phi$, can then be inserted in Eq. (2) to give estimates of the meridional flux of angular momentum.

Since the torques would tend to diminish the westerlies and easterlies, and this is not observed to occur over the long run, it follows that angular momentum has to be transferred from the belts of surface easterlies to the zones of westerlies. Calculations using Eq. (3), with x replaced by M from Eq. (4), show that this meridional flux occurs mainly in the upper troposphere. Hence angular momentum has to be brought upward from the surface layer in low latitudes, transferred poleward, primarily in higher levels, and ultimately brought down to the Earth in the belt of westerlies. Figure 2 shows the annual average meridional flux of angular momentum, and the surface torque in each belt.

Heat energy balance. The energy balance may be calculated by substituting the quantities summarized in Table 1 into Eq. (2). The listed heat sources comprise R_a, net (absorbed minus emitted) radiation of the atmosphere; R_e, net radiation at the Earth's surface; Q_s, flux of sensible heat from the surface to the atmosphere; LE, flux of latent heat from the surface (E denoting rate of evapotranspiration and L, heat of vaporization);

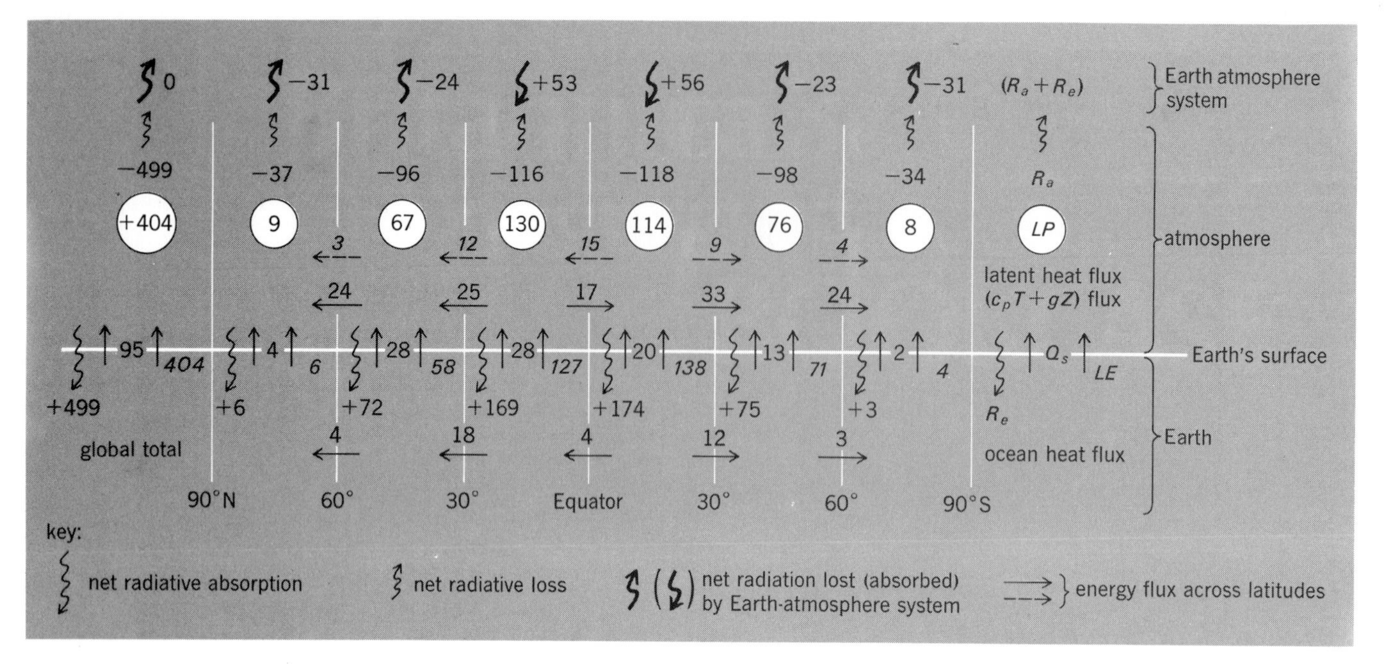

Fig. 3. Annual heat balance for the Earth as a whole and for each 30° latitude belt. Units are 10^{14} W. All values are keyed to the column at right. Italic numbers apply to water vapor (latent heat) flux.

and *LP*, release of latent heat in the atmosphere as estimated from the observed rate of precipitation *P*.

Alternatively, direct calculations of the fluxes of quantities in Table 2 can be made from aerological observations, by use of Eq. (3). The atmospheric energy comprises the sum (c_pT+gz), the two quantities being interchangeable during vertical air movement; decompression of rising air diminishes its sensible heat content by an amount equal to the increase of potential energy associated with change of elevation. Almost everywhere (c_pT+gz) increases upward, and Lq generally but not always decreases upward.

Various components of the energy balance are summarized, by 30° belts of latitude, in Fig. 3. Of the net radiation absorbed at the Earth's surface over the whole globe, 81% is expended in evaporation. Correspondingly, 81% of the net radiative loss by the atmosphere is compensated by release of latent heat when water vapor condenses and falls out as rain, snow, or hail, and 19% by transfer of sensible heat from the Earth. In the tropical belt 30°N–30°S, the Earth-atmosphere system gains heat by radiation; the excess is exported to higher latitudes as atmospheric heat and latent heat, and by ocean currents. Considering the tropical- and temperate-latitude belts of the two hemispheres, significant differences in the apportionments of evaporation and sensible heat transfer from Earth to atmosphere, and of the meridional transports of energy in the various forms, arise from the greater dominance of continents in the Northern Hemisphere. On the annual average, water-vapor-laden trade winds (lower branches of Hadley cells in Fig. 1) converge at about 5°N, where the greatest precipitation is observed. Minima of precipitation occur in the belts near 30°N, where mean descending motions occur. Secondary maxima of precipitation, in latitudes 40–50°, are associated with frequent extratropical cyclones. *See* HYDROMETEOROLOGY.

Table 2. Atmospheric properties used for flux computations

Property	x (per unit mass)*
Atmospheric energy	
Sensible heat	c_pT
Potential energy	gz
Total atmospheric energy	c_pT+gz
Latent heat	Lq

*Here c_p is specific heat of air; q is specific humidity, or mass of water vapor per unit mass of air.

The frequency and intensity of cyclones, and the contrasts between their cold and warm air masses, are much greater in winter than in summer in the Northern Hemisphere; these variations are much less pronounced in the oceanic Southern Hemisphere. Thus there is a fourfold greater poleward transport of sensible heat in middle latitudes of the Northern Hemisphere in winter than in summer, contrasted with only about a 30% seasonal variation in the Southern Hemisphere. In the tropics, large seasonal changes in the intensities of the Hadley circulation, together with a migration of the rain belt of the intertropical convergence zone (Fig. 1), are most pronounced in the monsoon regions of Asia-Australia and Africa.

Between late spring and early autumn a given hemisphere receives solar radiation far in excess of the amount it loses by infrared radiation. The excess heat is stored mainly in the oceans during the warm seasons and given up to the atmosphere as sensible and latent heat during the cooler seasons. Thus the oceans serve as an energy reservoir that tempers the seasonal changes of atmospheric temperature over them and over neighboring land areas invaded by marine air masses.

[C. W. NEWTON]

Bibliography: J. R. Holton, *An Introduction to Dynamic Meteorology*, 1972; S. Petterssen, *Introduction to Meteorology*, 3d ed., 1969; W. D. Sellers, *Physical Climatology*, 1965.

Atmospheric water vapor

Water in the form of vapor diffused in the atmosphere. Water vapor is a relatively small and variable constituent of the atmosphere, but because water may exist in the liquid, solid, and vapor states within the usual range of temperatures which occur in the atmosphere, water vapor plays an important role in the world environment. Water vapor is usually less than 4% of the atmospheric gas, but this small amount helps to determine the temperature distribution over the Earth and is the source of the world's fresh water.

The primary source of atmospheric water vapor is the oceans, which cover the major part of the Earth. Evaporation from lakes, rivers, ice, snow, and soil, and transpiration by plants also contribute large amounts of water vapor to the atmosphere, although these sources are small in comparison with evaporation from the oceans, estimated to be about 492,000 km^3 of water annually.

Heat of vaporization. The heat of vaporization of water at 21°C is 2445 joules/gram (584 cal/g or 1047 Btu/lb). Because of this heat requirement, water vapor in air masses transports large quantities of heat from warmer to cooler regions of the world. When an air mass containing vapor is cooled, the vapor condenses to form clouds of small water droplets. Under some conditions the released heat may be important in increasing the buoyancy of the air, which causes the air to rise and cool, leading to more condensation.

Evaporation cools a water surface because the escaping water carries with it the heat of vaporization. As the water cools, its vapor pressure decreases and the evaporation rate drops. Unless additional heat is provided, evaporation will cease. In a deep water body the cooled surface water may sink and be replaced by warmer water from below.

The transformation from ice to vapor at 0°C (called sublimation) requires a heat exchange of 2834 J/g (677 cal/g), which is the sum of the heat of vaporization at 0°C of 2500 J/g (597 cal/g) and the heat of fusion of 335 J/g (80 cal/g). Conversely, the condensation of a gram of vapor on a snow surface releases 2500 J (597 cal), which is sufficient heat to melt 7.5 g of snow.

Evapotranspiration. Evaporation from the continents returns the greater portion of the precipitation directly to the atmosphere. Transpiration by plants plays an important role in this process. Evaporation from bare soil becomes very small after the surface layer has been dried, because of the slow movement of water upward in the soil. Plants whose roots may extend several meters into the soil are important in removing moisture deep in the soil. Water taken up by the roots moves through the plant system and, except for the small amount incorporated in the plant, is released to the atmosphere through the leaves. *See* EVAPOTRANSPIRATION.

Measurement of water vapor. The vapor content of the atmosphere (humidity) is expressed quantitatively in several ways. The actual vapor content in g/kg of moist air is called specific humidity. The temperature at which moisture begins to condense is known as the dew-point temperature and is extensively used in meteorology as a measure of humidity. The partial pressure exerted by the water vapor in the atmosphere independently of the other gases is called the vapor pressure. The amount of water in a vertical column of the atmosphere from the ground to the outer limits of the atmosphere is the precipitable water, a unit often used in studies of rainfall. The relative humidity is probably the most widely used measure of humidity. It is the ratio of the actual moisture content of the air at some location to that which would be present if the space were saturated with vapor at prevailing temperature and pressure. Since the saturation value is a function of temperature, relative humidity varies with temperature even though the specific humidity remains constant. Hence, relative humidity is not a very useful measure of humidity for scientific purposes. It is, however, an index to human comfort.

Water vapor content of the atmosphere can be measured with various devices. A psychrometer consists of two thermometers, one with its bulb covered with wet muslin. When the device is ventilated, the temperature of the wet bulb is depressed below that of the dry bulb by evaporation of water from the muslin. The humidity is a function of the two temperatures. The dew-point temperature can be measured by measuring the temperature of a polished metal surface at the moment when condensation begins to form on it. Devices which indicate relative humidity by measuring the change in length of human hair or chemically treated paper are also available. Such devices are cheap but not very reliable. By measuring the attenuation of light at a wavelength where there is high absorption by water vapor, it is possible to calculate the mass of water vapor in the light path. With the Sun as a light source, the precipitable water can be determined in this fashion.

Estimating evaporation. Evaporation can result in serious losses of water from reservoirs. Hence, methods of estimating evaporation from existing or proposed reservoirs have been developed. J. Dalton in 1802 first pointed out that evaporation is proportional to the difference in vapor pressure between the water surface and the air above it. Hence, evaporation is possible only when the dew-point temperature of the air is less than the temperature of the water surface. The greater this temperature difference, the greater is the evaporation rate. In absolutely still air a vapor blanket forms at the water surface, and evaporation decreases rapidly to rates limited by the diffusion of vapor from the blanket. With wind the vapor is carried away and replaced by dryer air, and evaporation continues.

Another basis for estimating evaporation is the energy balance method. If the total radiation input to the water body less the heat loss by long-wave radiation and convection can be determined, the excess of heat input over heat outgo divided by the latent heat of vaporization indicates the volume of evaporation.

Evaporimeters are pans of water from which evaporation can be measured. Because of heat transfers through the sides and bottoms of such pans, evaporation from pans is always higher than from lakes. The National Weather Service class A pan is 1.2 m (4 ft) in diameter and 25 cm (10 in.) deep. It is supported a few inches above the ground on a timber grid. Evaporation from the class A pan must be reduced about 30% to estimate lake evaporation.

Salt, ice, and snow factors. Addition of salts to water reduces the vapor pressure at the water surface. Thus evaporation per unit area from the oceans and salt lakes is slightly less than from a fresh-water body. Roughly, evaporation is reduced about 1% for each percent of dissolved salts, so that evaporation from the oceans is about 3.5% less than from an equivalent fresh-water surface. Evaporation from ice and snow is also less than from water at the same temperature, because the vapor pressure over ice is somewhat less than over water. Evaporation from snow and ice is also low, because vapor pressure (and vapor-pressure differences) are very low at temperatures below freezing.

Based largely on evaporimeter data, it can be said that evaporation varies from 5 to 7 m (15 to 20 ft) of water per year in tropical deserts to near zero in polar regions.

[RAY K. LINSLEY]

Bibliography: T. A. Blair and R. C. Fite, *Weather Elements*, 5th ed., 1965; R. K. Linsley, M. A. Kohler, and J. L. H. Paulhus, *Hydrology for Engineers*, 2d ed., 1975; R. C. Ward, *Principles of Hydrology*, 2d ed., 1975.

Avalanche

In general, a large mass of snow, ice, rock, earth, or mud in rapid motion down a slope or over a precipice. In the English language, the term avalanche is reserved almost exclusively for snow avalanche. Minimal requirements for the occurrence of an avalanche are snow and an inclined surface, usually a mountainside. Most avalanches occur on slopes between 30 and 45°.

Types. Two basic types of avalanches are recognized according to snow cover conditions at the point of origin. A loose-snow avalanche originates at a point and propagates downhill by successively dislodging increasing numbers of poorly cohering snow grains, typically gaining width as movement continues downslope. This type of avalanche commonly involves only those snow layers near the surface. The mechanism is analogous to dry sand. The second type, the slab avalanche, occurs when a distinct cohesive snow layer breaks away as a unit and slides because it is poorly anchored to the snow or ground below. A clearly defined gliding surface as well as a lubricating layer may be identifiable at the base of the slab, but the meteorological conditions which create these layers are complex. The thickness and areal extent of the slab may vary greatly, and those slab avalanches with larger dimensions pose the greatest threat to life and property. Both loose and slab types may occur in dry or wet snow. Dry avalanches often entrain large amounts of air within the moving mass of snow and are thus referred to as powder avalanches. Velocities for dry-snow avalanches may exceed 150 mph (67 m/s). Wet-snow avalanches occur when liquid water is present in the snow cover at the point of origin. While the wet avalanches move at lower velocities, they often involve greater masses of snow and therefore significant destructive forces. Theoretical calculations and empirical evidence indicate the general range of maximum impact forces to be between 5 and 50 tons/m^2 with extreme values reaching 100 tons/m^2. It is frequently the wet-snow avalanche which damages the soil and vegetation cover.

Release mechanism. In the case of the loose avalanche, release mechanisms are primarily controlled by the angle of repose, while slab releases involve complex strength-stress problems. A release may occur simply as a result of the overloading of a slope during a single snowstorm and involve only snow which accumulated during that specific storm, or it may result from a sequence of meteorological events and involve snow layers comprising numerous precipitation episodes. In the latter case, large avalanches may not necessarily be restricted to storms with large amounts of precipitation, but can result from lesser amounts of precipitation falling on older snow layers underlain by an extremely weak structure.

Defense methods. Where snow avalanches constitute a hazard, that is, where they directly threaten human activities, various defense methods have evolved. Attempts are made to prevent the avalanche from occurring by artificial supporting structures or reforestation in the zone of origin. The direct impact of an avalanche can be avoided by construction of diversion structures, dams, sheds, or tunnels. Hazardous zones may be temporarily evacuated while avalanches are released artificially, most commonly by explosives. Finally, attempts are made to predict the occurrence of avalanches by studying relationships between meteorological and snow cover factors. In locations where development has yet to occur, zones of known or expected avalanche activity can be mapped, allowing planners to avoid such areas entirely. Avalanche hazard is small when compared with certain other natural hazards such as floods and tornadoes, but it continues to rise as the popularity of wintertime mountain recreation increases.

[RICHARD L. ARMSTRONG]

Bibliography: C. Fraser, *Avalanches and Snow Safety*, 1978; E. R. La Chapelle, Snow avalanche: A review of current research and applications, *J. Glaciol.*, 19(81):313–324, 1977; M. Mellor, *Avalanches*, Cold Regions Science and Engineering, Pt. III, Sec. A3: Snow Tech., U.S. Army, CRREL, Hanover, NH, 1968; R. I. Perla, Failure of Snow on Slopes, in B. Voight (ed.), *Rockslides and Avalanches*, vol. 1 of *Natural Phenomena*, 1977; R. I. Perla and M. Martinelli, Jr., *Avalanche Handbook*, USDA Forest Service, Agriculture Handb. 489, 1976.

Bioclimatology

A study of the effects of the natural environment on living organisms. These effects may be direct, such as the influence of ambient temperature on body heat, or indirect, such as the influences on composition of food. Only direct effects are discussed in this article. Natural and artificial elements cannot be sharply distinguished. For example, smoke from a lightning-induced forest fire is natural, but smoke from a chimney is artificial.

Bioclimatology encompasses biometeorology, climatophysiology, climatopathology, air pollution, and other fields. The interplay between disciplines, such as meteorology and physiology, is emphasized. Concerning the time scale, the study of events enduring for hours to days is called biometeorology; for years to centuries, bioclimatology; for millennia and more, paleobioclimatology. For intrinsic reasons bioclimatology should serve as an overall term. Bioclimatology divides naturally into the two broad areas of plants and of animals and humans.

In plant bioclimatology solution of any problem involves study of the natural climatic values, such as air temperature, precipitation, and wind speed and its variations; the transfer mechanism, such as the eddy diffusivity; and the effect of these agents on the plant. The most important mechanisms are (1) photochemical effects, such as photosynthesis and (blue) phototaxis; photosynthesis needs blue and red light, water from the roots (and possibly from the air), and CO_2 from the air; all these components vary with weather and climate; (2) evapotranspiration through integument and stomata of plants, a process depending highly on availability of water, transfer of liquids from root to leaf, temperature of the leaf, water vapor of the air, and ventilation; (3) picking up compounds of N, K, P, Ca, and others from the ground; and (4) avoidance of destructive conditions, such as freezing, drying out of leaves (if water supply is smaller than evaporative demand), and overheating. *See* AGRICULTURAL METEOROLOGY; MICROMETEOROLOGY.

Although considerable work has been done on bioclimatology of domestic animals, the following discussion focuses on humans.

Photochemical bioclimatology. Essentially this is the investigation of the effects of light from the Sun and sky. Sunburn of the skin and cornea of the eye is initiated by denaturization of nucleic acid and skin proteins, causing a local histaminelike action. In nature the effect is restricted and is induced by ultraviolet radiation with wavelengths around 0.3–0.31 μ. Sunburn is delayed or prevented by three screeening agents in the skin: pigment, horny layer, and urocanic acid. The pigment of permanently brown or black races, as well as the radiation-induced pigment in variably colored races (the so-called white race), acts as a protective filter. The horny layer of the skin reportedly grows in thickness by ultraviolet exposure. It is still debated how much protection is offered this way. The third screen is urocanic acid, a substance derived by enzymatic action from the animo acid L-histidine in the sweat. A 1-mm sweat layer absorbs 50% of the natural ultraviolet at 300 mm and nearly 100% of the mercury lamp ultraviolet at 250 mm. Artificial ultraviolet sources, such as the cold mercury lamp, act very differently from these sources. Solar erythema can be seen within minutes of exposure. A simple rule to avoid solar overexposure is this: As soon as the dividing line between exposed and shielded (clothed) areas can be discerned, stop sun bathing.

There are two kinds of solar pigmentation, late and direct. The late seems more prevalent in fair races; it occurs in conjunction with sunburns days later. The direct type is found more in southern Europeans and Japanese. It occurs, without sunburn, during a 30-min exposure, and is caused by long-ultraviolet (0.32–0.4 μ) and possibly visible light.

Frequent exposure over years leads to skin elastosis (sailor's skin) and finally skin cancer. Skin carcinomas occur more frequently on facial skin exposed to sun. Proof that exposure to sun is carcinogenic comes from statistical evidence and from the results of animal experiments, in which rodents were exposed to very strong artificial ultraviolet radiation.

Although bacteria are easily killed with artificial ultraviolet radiation of short wavelength, natural sunlight probably has very little bactericidal action because it lacks these short wavelengths.

Vitamin D is produced from natural sterols in plant and animal foodstuffs and in human skin by short-wavelength solar ultraviolet radiation (0.3 μ).

It has been claimed that solar ultraviolet radiation favorably affects blood circulation and general health, but sunbather faddism has not been conductive to serious research. It is difficult to separate the effects of ultraviolet radiation, thermohygric exposure, and mental stimuli on a sunbather.

Most sunlight and sky light received by the eye is that reflected by clouds and surfaces. The intensity of the incoming light, its angle of incidence, and the amount and kind (specular or diffuse) of albedo control the amount of light received by the eye. The specular reflections of water, ice, metals, snow, clouds, and white sand are bright enough to irritate the eye. Most sunglasses dampen the whole visible spectrum uniformly and eliminate ultraviolet and infrared rays.

All the important ultraviolet effects mentioned above occur at about 0.3 μ, near the end of the solar spectrum. This ultraviolet radiation is controlled by absorption in stratospheric ozone and scattering by air molecules, clouds, and smog. Dependence of ultraviolet radiation on solar and geometric altitude is pronounced. Usually, the scattered sky ultraviolet radiation exceeds that of the Sun. Snow reflects both ultraviolet rays and visible light to cause snow blindness and sunburn below the chin. No other natural substance reflects more than a few percent of natural ultraviolet; however, metals reflect highly.

Natural penetrating ionizing radiation is composed of β- and γ-rays from radioactive minerals, including their emanations and secondary rays from cosmic radiation, mainly mesons and neutrons. Ionization by both effects at altitudes below 18 km is less than 40 milliroentgens per day, probably an insignificant figure.

Air bioclimatology. Gaseous air constitutents, such as oxygen and water vapor, influence body chemistry and heat balance. The oxygen partial pressure (pO_2) of inhaled air is vital for blood oxygenation and depends mainly on altitude. At or above an altitude of about 3000 m, the pO_2 is low enough to cause air sickness, or mountain sickness. Residence at an altitude of 1–2 km is supposed to benefit circulation, but may be harmful for some heart patients.

Low water vapor pressure may cause drying of the skin and of the mucous membranes of upper respiratory organs. Water vapor pressure, even indoors, is especially low when outdoor air temperatures are low. Many skin and respiratory complaints in winter are caused more by dryness than by cold. Water vapor influence on heat balance is discussed below.

Carbon dioxide, CO_2, water, H_2O, and organic vapors emitted by man contribute to the unpleasantness of crowded and ill-ventilated rooms, which may also be overheated.

Man's industries, volcanoes, fires, dust storms, and other more or less violent events spew large amounts of matter into the air known as fallout, smog, air pollution, and so forth. No bioclimatic problem in this field can be reported as solved unless source, transfer mode, change in transfer, concentration near the sink, the mode of intake by the sink, and the biological and chemical action of the sink are clear.

Ozone, O_3, may serve as an example. It is produced by solar ultraviolet radiation in the higher stratosphere, by lightning, and by sunlight falling on smog (Los Angeles). Stratospheric ozone is brought down by turbulence, and might reach toxic levels for crews of airplanes at about 15-km altitude. Ozone is blamed for some smog-induced injuries, for example, ocular pain in Los Angeles. A very important sink action is the inhalation into the respiratory organs, where especially the lung's surface is changed by oxidation. This causes coughing at first and tiredness later. Amounts of O_3 less than 0.5 part per million are toxic.

Carbon monoxide, unburned hydrocarbons, NO, NO_2, SO_2, and other by-products are prevalent where there is incomplete combustion. Many by-products are also highly hygroscopic, and serve as nuclei of condensation to cause the low-humidity type of smog, such as that occurring in Los Angeles. This smog condition is further complicated by the occurrence of light-induced reactions between hydrocarbons and NO_2 and other pollutants. The high-moisture type of smog seems typical for London. *See* SMOG.

Very few gases, in fact very little of any gas, pass the skin, with the exception of more easily permeated areas such as chapped lips, scrotum, and labia.

Aerosol bioclimatology. Solid or liquid suspensions in the air affect breathing organs or skin. Widely discussed elements of the atmospheric aerosols are the ions, which are particles containing one or more electrical charges. For a long time beneficial or dangerous effects of an abundance of particles of one charge type have been claimed. These electrical space charges reportedly have influenced results of physiological tests, growing of living cells, and severity of hay fever attacks. As a rule, negative space charge of the order of 1000–10,000 ions/cm³ is reportedly beneficial. These results are expected to explain some observations of statistical bioclimatology. *See* AIR POLLUTION; ATMOSPHERIC CHEMISTRY.

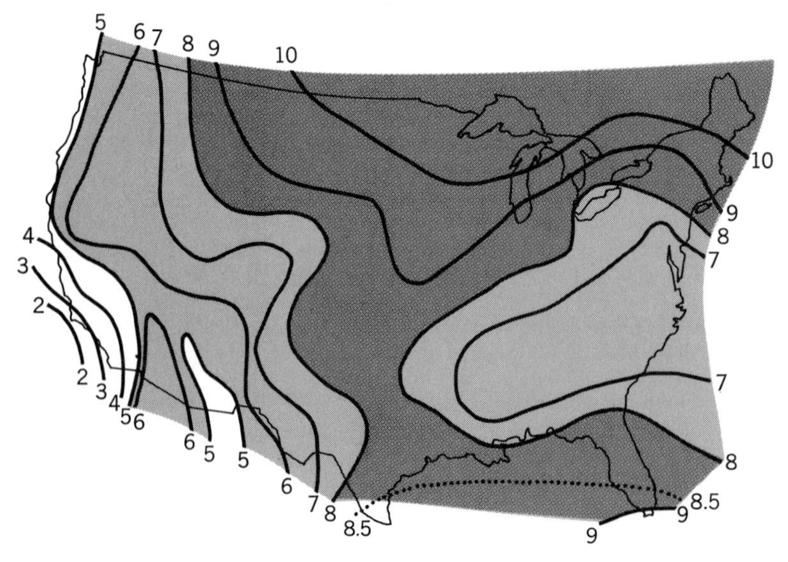

Fig. 1. Sum of heating degree days plus two times cooling degree days. The larger the sum, the larger is the integrated year-round outdoors discomfort, and also total fuel and power bill for heating and air conditioning. Units 1000°F times days. (*From K. J. K. Buettner, Human aspects of bioclimatological classification, in S. W. Tromp, ed., Biometeorology, Pergamon Press, 1962*)

Thermal bioclimatology. This subdiscipline concerns the heat balance of man as controlled by his environment. The basic relation is shown by Eq. (1), where M is metabolic heat production; C is

$$M - Cd\theta_b/dt + H_h = H_r + H_a + H_e + R \qquad (1)$$

body heat capacity; H is heat exchanged through skin; R is respiratory heat exchange; θ is temperature (°C); b refers to total body; h refers to solar radiation; r refers to infrared radiation; s refers to skin; a refers to air (convection); e refers to evaporation; and t is time. The H terms are defined in Eq. (2), where S is visible and near-infrared radiant

$$\begin{array}{ll} H_h = S \cdot \epsilon_h \cdot A_h & H_r = h_r \cdot A_r(\theta_s - \theta_r) \\ H_a = h_a A_a(\theta_s - \theta_a) & H_e = h_e \cdot A_e(e_s - e_a) \end{array} \qquad (2)$$

heat flow from the Sun, sky, and environment; and ϵ_h is absorptivity of skin (0.6 for white and 0.9 for black skin). The A factors, that is, the respective surface areas, depend on body posture and on the process involved, such as radiation or convection; these areas are always smaller than the geometric surface. The terms e_s and e_a are the vapor pressures of skin and air, respectively.

The h factors are heat conductances at absolute temperature T, as shown by Eq. (3), where ϵ_r is

$$h_r = \epsilon_r \sigma (T_s^4 - T_r^4)/(\theta_s - \theta_r) \qquad (3)$$

0.98 (infrared absorptivity of skin), and σ is Stefan's constant, 4.9 kcal-m²/(hr)(deg⁴). From experiments with men, and using kilocalorie units (kcal), $h_a = 6.3\sqrt{v}$ kcal/(m²)(hr)(deg) for wind velocity v m/sec; further, $h_a = 3.3$ in calm. Equation (4)

$$h_e = 1.63 h_a \text{ kcal/(m}^2\text{)(hr)(mb)(deg)} \qquad (4)$$

follows from h_a. These data are valid for supine adults at sea level, and refer to the heat- or vapor-exchanging area.

Equation (5) applies for the clothed body, if h_c is

$$\theta_s - \theta_a = M/A_a h_c + (M + H_h)/A_a(h_a + h_r) \qquad (5)$$

conductance of clothes, and if $\theta_r = \theta_a$, $A_r = A_a$, $H_c = 0$, and $d\theta_b/dt = 0$. The absorptivity of the clothing is inserted for ϵ_h. If $h_a \gg h_e$ (strong wind, thick clothes), solar heating H_h becomes unimportant.

H_e can be measured directly with a good balance. The relative skin humidity $r_s = e_s/e^*_s$, where e_s^* is saturation vapor pressure at skin temperature, can be measured with a hair hygrometer or can be derived from H_e, h_e, e_a, and e_s. During sweating r_s is 100%; otherwise, it is usually below 60%. If skin relative humidity is below 10%, the skin may crack.

Skin water transfer is accomplished by sweating and diffusion. The latter corresponds usually to 1 kcal/(m²)(hr); it may reverse, so that water or vapor is transferred into the skin. Sweat is a powerful emergency measure capable of producing $H_e = 600$ kcal/(m²)(hr) or more. Bioclimatic sweat control works two ways, via body temperatures and via water coverage. The prime control for the amount secreted seems to be the temperature of a section of the hypothalamus, skin temperature acting as a moderator. These temperatures are of course bioclimatologically controlled. Skin totally covered with sweat or bath water lowers its sweat water loss strongly, about 4:1. The effect is absent in saline or with sweat highly concentrated by evaporation.

The respiratory heat and vapor loss is small except during hyperventilation at reduced pressure. Ordinarily, this loss varies little because the exhaled temperature drops as θ_a and e_a fall.

Of the climatic elements, S, h_a (wind), and precipitation are easiest to control. High values of θ_a, θ_r, and e_a are much harder to influence than low values. To ensure a constant θ_b, the body alters the peripheral blood flow and thus alters θ_s. In low temperatures M is raised by shivering or work; in a hot environment H_e is raised by sweat. These emergency regulations are effective only for certain periods. Short-time limits are set by variations of θ_b and finally θ_s (skin burn or freezing). All body controls vary with age, sex, health, exercise, and adaptation. Reportedly, frequent limited exposures to adverse thermal conditions, particularly cold, invigorate many body functions, especially if exposure is combined with exercise.

The most important climatic element is local air temperature; in the cold, wind increases the rate of body cooling. For a well-insulated man or house, Eq. (5), air temperaute alone is the deciding factor. The house heating bill is proportional to the number of degree days, that is, time multiplied by $(\alpha - \theta_a)$, where α is the preferred room temperature, about 22°C. No usable formula for the cooling effect of open air on average individuals can be derived because clothing styles change.

At temperatures greater than 25°C sweating starts. Its evaporation is restricted by high values of e_a. Hence, the combination of θ_a and e_a describes livability in the heat. Tests on sensation experienced by young men exposed to different pairs of θ_a and e_a led to the effective temperature (ET), which approximately equals (θ_a + dew-point temperature)/2. The area of high ET or extreme summer discomfort in the United States is the Gulf Coast, especially the lower Rio Grande Valley (average ET, 26–27°C). In the Indus Valley ET is 28.5°C; and the world maximum is in Zeila, Somali Coast, where ET is 29.7°C. Figure 1 shows a United States map of the sum of heating and cooling degree days. Since cooling also involves drying, the cooling degree days are multiplied by two. Large numbers of the sum mean climatic hardship for the outdoors man and high bills for heating fuel and cooling wattage. *See* SENSIBLE TEMPERATURE.

The microclimate can enhance discomfort. The values of θ_a and θ_r are much below normal on calm clear nights, especially when the terrain is concave. The Sun can raise the surface temperatures of such features as sand, walls, and vehicles to as much as 40° above the air temperature.

Extreme climates and microclimates. Certain climatic conditions can be tolerated only for a limited time, since they either overrun the physiological defense mechanism or injure and destroy body parts directly. The first mechanism usually leads to a breakdown of temperature regulation and of the body circulation, and the latter usually involves skin injuries.

The safe survival time depends primarily on air and wall temperature, if there is no artificial ventilation and no radiation other than the walls. Other conditions may suitably be converted into equivalent temperature data. Figure 2 contains such data, listing safe times between seconds (in a fire) to life-span. Some case studies are listed below.

1. The hot desert: Temperatures are in the fifties, vapor pressure is 10–15 mb, and there is sunshine and wind. Over a period of a few hours no equilibrium of body data is reached, and especially heart rate and core temperatures rise. The skin is dry and salt covered. The skin water loss is much below the level needed to compensate by evaporation for the large heat input by solar and infrared radiation and convection.

2. The jungle: Normal jungle is the habitat of a large segment of people. However, extreme jungle conditions are around 35°C and 40 mb vapor pressure. The skin is totally wet, but evaporation becomes insufficient due to external moisture. The effective temperature is a good measure for this danger signal. Worst in this respect are locally wet areas surrounded by a hot desert.

3. Fire: A forest or house fire has flame and glowing fuel temperatures of 600–900°C. Injury comes from skin burns via contact, convection, or radiation. The latter can be warded off by aluminum-covered suits. Near a large fire radiant heat of 40,000kcal m^{-2} hr^{-1} may cause skin pain in

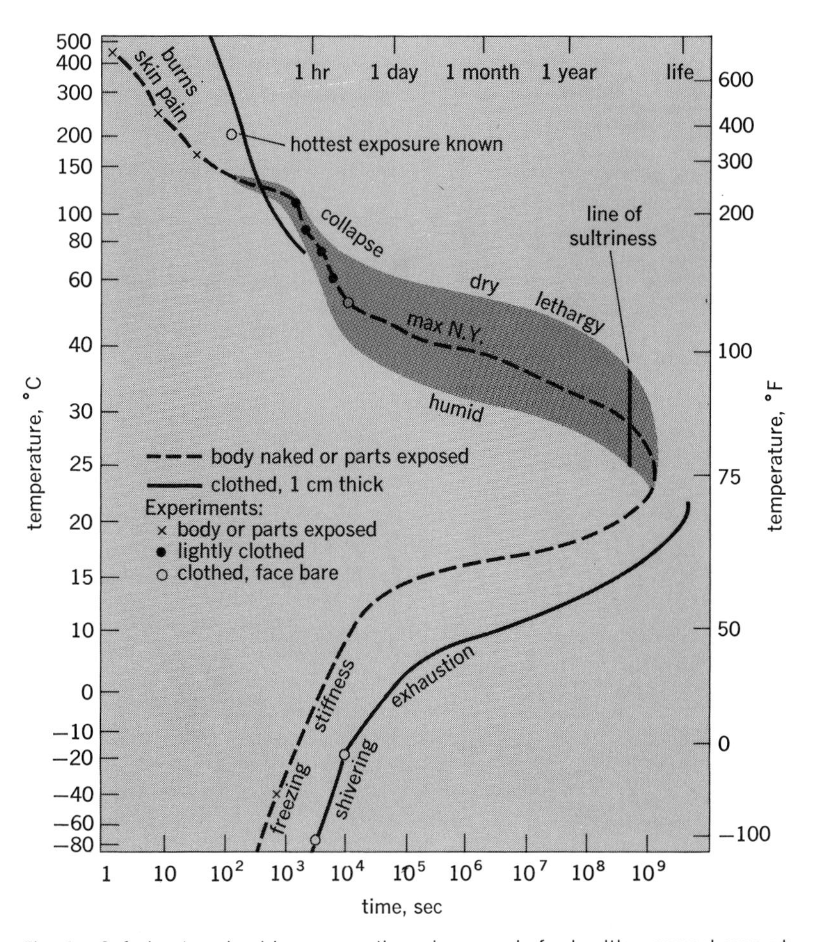

Fig. 2. Safe heat and cold exposure times in seconds for healthy, normal men at rest, with body wholly or partly exposed. Room is free from artificial ventilation and radiation; air and walls have identical temperature. Humidity influence is shown by shading. Max N.Y.= highest data of temperature and humidity in New York City. *(From K. J. K. Buettner, Space medicine of the next decade as viewed by an environmental physicist, U.S. Armed Forces Med. J., 10:416, 1959)*

1 sec and burns in a few seconds. This heat transfer is superior to flame contact. Inhaling of hot air and lack of O_2 are frequently minor problems compared to poisoning by CO.

4. Frostbite and freezing fast: Strong convective cooling by wind and low temperatures may overrun the local defenses, causing defective circulation in the skin and breakdown of small vessels and blood particles. Developing of ice crystals may puncture cells. Skin touching metals below −30°C freezes on contact, causing mechanical loss of skin layers upon removal.

5. Weather accident and disaster: Yearly average of fatalities by weather catastrophes for the United States during 1901–1960 are as follows: hurricanes, 100; tornadoes, 150; lightning, 175; floods, 80; and snowstorms, blizzards, and ice storms, 200. There are probably deaths of similar numbers from heat waves and smog periods. Weather-caused farming disasters have produced many mass-killing famines. These figures fortunately have declined, as have death figures from hurricanes and tornadoes and, probably, lightnings. All others mentioned here have risen.

Statistical bioclimatology. This aspect of bioclimatology investigates correlation between weather and climate phenomena and their effects on man. The most important work concerns clinical data correlated with daily weather. Two weather types seem to be involved: Fronts, central low, and ascending air coincide with frequent attacks of angina pectoris and embolism; incidence of foehn and cyclogenesis to the west are found to correlate with increasing circulatory disorders, mental troubles, increase of accidents, and kidney colic. It has been claimed that there is a correlation between increased solar activity and all deaths in big cities, especially deaths from mental diseases. *See* CLIMATOLOGY.

Climatotherapy. This health treatment is more venerated than exact. Going to a resort is intimately connected with nonclimatic changes of housing, mode of living, food, exercise or rest, and phychological stimuli. The vacation effect itself may be quite helpful.

A health climate location should avoid extremes of temperature, especially high effective temperature felt as sultriness. Air should be free of smoke, smog, allergens, ozone, car exhausts, and industrial and volcanic effluents.

Solar ultraviolet and natural radioactive substances and rays are generally harmful. Beneficial outdoor living usually involves exposure of parts of the skin to solar ultraviolet, and the ensuing pigment is then taken as proof of health, a rather doubtful conclusion. A certain moderate thermal stimulation by wind, waves, and temperature changes may be helpful to most people. Adaptation to the new time zone may severely interfere with adaptation of intercontinental East-West travelers.

If a physician prescribes a very balmy climate for a weak elderly patient, the patient should be aware of the difference between the dry summer heat of for example, El Paso, Texas, and the moist heat of the Lower Rio Grande.

Paleobioclimatology. This deals with possible environmental influences on the development of species, especially of man. The local microclimate has to be considered. A cave, for example, has a very constant temperature and humidity and no sunlight. Fire, as an adjunct, was introduced in China and Hungary more than 500,000 years ago. In a tropical jungle one is not exposed to solar rays. The reduction of fur in man seems correlated with the perfection of eccrine sweating as part of a well-developed cranial temperature control. It might also permit man to utilize his skin temperature sensors to function as infrared "eyes" in the dark of a cave.

In most animals skin and fur color seem to promote camouflage or its opposite. Should this be true for man, one would find dark skin in the tropical jungle; while in a temperate maritime clime white in winter and brown in summer seems adequate. Thermally, a black skin is less suited for sunny hot climate than is a lighter one. Fair-skinned but unprotected man at a latitude of 40–50° may seasonally adapt himself to the varying solar ultraviolet. Sunburn is unlikely if daily exposure is routine. During the ultraviolet-rich summer season a sufficient deposit of vitamin D can be produced in the body to last for the darker seasons. There might have been sufficient vitamin D in the food of Stone Age man. So neither sunburn nor vitamin D seems to have had an effect on evolution.

Ideal bioclimate. There is no ideal climate for all men. One's physiological and psychological preference seems to be set in childhood. History starts, long before effective room heating, in such subtropical climates as Egypt, Mesopotamia, and southern China, or in tropical lowlands, for example, Yucatan, or in tropical highlands, as Peru. It moves in the Old World to higher latitudes, where Romans developed heating. Later, civilized power centers around the 45° latitude, where the house microclimate was adequate the year around, even before air cooling.

Statements that culture best develops in a particular climate are of little value.

In the United States people entirely free to settle where they like the climate seem to prefer the Mediterranean climate of Southern California and the desert of Arizona. A large immigration proves this preference.

[KONRAD J. K. BUETTNER/H. E. LANDSBERG]

Bibliography: F. Becker et al., *A Survey of Human Biometeorology*, World Meteorol. Organ. Tech. Note no. 65, 1964; K. Buettner, *Physikalische Bioklimatologie*, 1938; K. Buettner et al., Biometeorology today and tomorrow, *Bull. Amer. Meteorol. Soc.*, 48:378–393, 1967; G. E. Folk, *Introduction to Environmental Physiology*, 2d ed., 1974; H. E. Landsberg, *Weather and Health*, 1969; S. Licht (ed.), *Medical Climatology*, Physical Medicine Library, vol. 8, 1964; J. H. Prince, *Weather and the Animal World*, 1975; S. W. Tromp and W. H. Weihe (eds.), Biometeorology Two, in *Proceedings of the 3d International Bioclimatological Congress*, pts. 1 and 2, 1967.

Biogeochemical balance

Chemical elements that are essential components of living organisms, and others which are caught up in the biochemistry of organisms, may be influenced in their abundance and distribution by basic biological processes, such as photosynthesis, assimilation, respiration, and excretion. At the same time, the abundance and distribution of species and communities of organisms may be influenced by the availability of the elements essential to them. Those elements of particular significance to natural ecosystems include nitrogen, phosphorus, sulfur, oxygen, and some essential trace elements. Because of human interactions with the biosphere, there is also concern with the cycles of carbon, lead, mercury, cadmium, and both natural and synthetic organic compounds. Measurable changes in the biogeochemical balance of several of the essential elements have resulted from combustion of fossil fuels, manufacturing processes, and other aspects of high technology. However, biogeochemists are learning to appreciate the amount and nature of some of the naturally occurring cycles of the essential elements.

Carbon dioxide. The carbon cycle has substantial biological components. Most of the world's carbon is in the form of limestone, dissolved marine humus (see illustration), and fossil fuels, all of which are of biogenic origin. A small amount of carbon is in the atmosphere and in living biomass at any instant. Atmospheric carbon dioxide is in equilibrium with bicarbonate dissolved in the ocean, and in the long run the ocean controls the concentration of CO_2 in the atmosphere. However, the mixing time of the ocean is on the order of 1000 years. Therefore, the CO_2 content of the atmosphere is increasing as the result of both deforestation and burning of fossil fuels. Over the last century those processes have had about equal impact. Once the tropical forests have been largely harvested, the rate of deforestation will decline; but the burning of fossil fuels continues to increase, and a doubling of the CO_2 in the atmosphere is expected early in the next century. This estimate is based on the rate of increase observed, not the rate of CO_2 production, which is substantially higher. It is known that some CO_2 is being taken up by the ocean, but other sinks which must exist have not been identified. There is a small net production of CO_2 by the biosphere at present. Because CO_2 absorbs infrared radiation, a doubling of atmospheric CO_2 would cause a mean increase in the Earth's surface temperature of 2–3°C, causing dramatic climatic changes. This could be offset by increased smoke and dust resulting from burning and deforestation, so the outcome cannot be predicted, but there is cause for concern. *See* AIR POLLUTION.

Oxygen. L. Van Valen proposed that the oxygen content of the atmosphere is not controlled primarily by photosynthesis and respiration, which very nearly balance. Oxygen is being added to the upper atmosphere as the result of photolysis of water, with subsequent loss of hydrogen to outer space. Presumably this process has operated since water vapor entered the atmosphere in the early Precambrian. Evolutionary trees constructed on the basis of changes in amino acid sequences in biologically essential compounds, such as ferredoxin, ribonucleic acid, and cytochrome C, suggest that primitive organisms developed the ability to respire before oxygen-producing photosynthesis evolved. Moreover, the oxygen demand associated with chemical oxidations and carbonate formation will deplete the atmosphere in 1,000,000 years. Therefore, the atmosphere has been replenished at least 1000 times. The most quantitatively significant biological process in the oxygen cycle appears to be carbonate formation rather than respiration or photosynthesis. Most of the key processes in the cycle appear to be purely physicochemical. It is not known how constant these processes are, or if there are any homeostatic mechanisms which tend to maintain the concentration of oxygen in the atmosphere. However, if photosynthesis stopped completely and all organic matter of the biosphere were oxidized, the process would use up about 1% of the atmospheric oxygen. Burning fossil fuels will have used up another 0.2% by the year 2000. Therefore, there is no concern about oxygen supply for the next million years. *See* ATMOSPHERIC CHEMISTRY.

Sulfur. Much of the cycle of sulfur is biologically controlled. Native sulfur deposits are of biological origin, having been laid down in anaerobic basins of epicontinental seas. There is a substantial amount of sulfate in solution in the sea, and it is synthesized into organic sulfur compounds by marine plants. Dimethyl sulfide released by the plants moves from the sea to the atmosphere, where it is oxidized to sulfuric acid, causing rain to be acid. Atmospheric sulfuric acid is also produced from the release of sulfates in smoke plumes from burning coal and oil. There is evidence that the pH of rain is decreasing in much of the Northern Hemisphere as a result of burning fossil fuels. At the same time, acid rain has been reported in remote sites, such as the central Amazon Basin, where the sources must be natural ones. Acid rain lowers the pH of lakes and streams, sometimes resulting in the elimination of sensitive fishes and invertebrates. It also increases the weathering rate of rocks. This may prove to be a situation in which naturally occurring rates of cycling of sulfur are

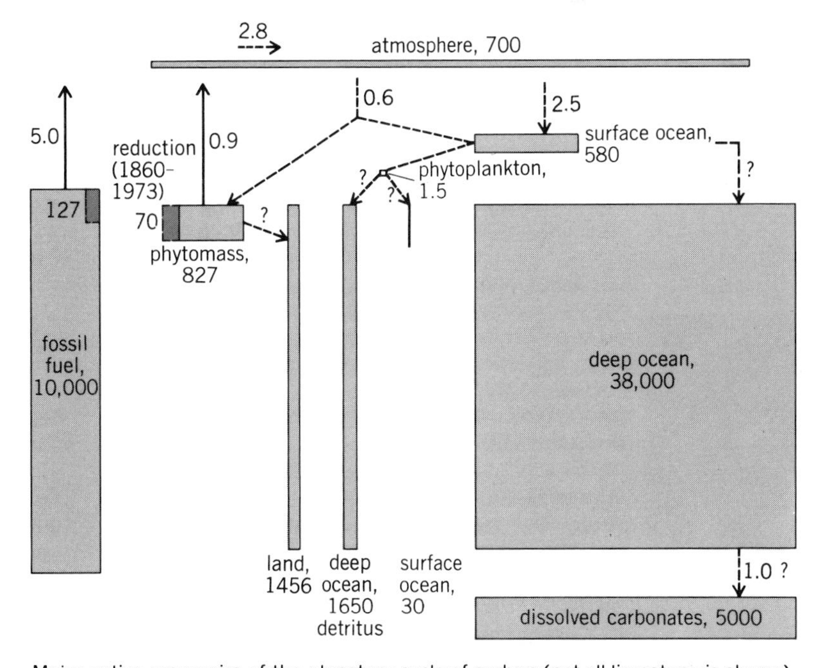

Major active reservoirs of the planetary cycle of carbon (not all limestone is shown). Standing stocks are in units of 10^{15} g C, and fluxes in units of 10^{15} g C/year. (*From C. S. Wong, Carbon dioxide: A global environment problem into the future, Mar. Pollut. Bull., 9:257–264, 1978*)

near the limit of tolerance for some species, and the additional sulfur from fuels upsets a delicate balance. *See* HYDROSPHERIC GEOCHEMISTRY.

Nitrogen and phosphorus. Much attention has focused on N and P as elements which limit the rate of growth of plants, particularly in aquatic and marine systems. Both N and P are in limiting supply in much of the surface waters of the ocean. N tends to be relatively more limiting in coastal waters and estuaries, while P tends to be relatively more limiting in lakes. Where there is an abundance of N and P, as in those parts of the ocean where subsurface water is upwelled or where there is terrestrial enrichment of lakes and streams, plant and animal growth is luxuriant, and is described as eutrophic. Naturally eutrophic waters are sources of major fisheries and are esthetically pleasing. However, when naturally N- and P-depleted waters are enriched by inputs of sewage, phosphate detergents, or industrial wastes, the result is cultural eutrophication. Dramatic changes in the aquatic plants are usually followed by loss of, or changes in, fish populations. The result is usually both esthetically and economically unsatisfactory. *See* NITROGEN CYCLE.

Cultural eutrophication of water bodies is in principle a reversible process. In the case of Lake Washington, in Seattle, cultural eutrophication was eliminated by the diversion of sewage, and the lake returned to its former condition in a few years. Where the influx of wastes cannot be diverted, they must be stripped of N and P, which requires expensive, three-stage waste treatment plants. This has proved necessary, particularly in sensitive, highly oligotrophic lakes, such as Lake Tahoe (California-Nevada). Cultural eutrophication is a local or regional problem. The ocean is sufficiently large so that only a few localities near sewage outfalls of large cities show the effects of added nutrients. *See* EUTROPHICATION.

Heavy metals. Although the heavy metals, such as mercury and lead, are highly toxic to all organisms, their natural abundance and rates of cycling are such that they rarely pose a threat to natural populations. Some large, old, predatory fishes have been found to contain substantial quantities of mercury, acquired naturally over their lifetimes. The cycles of mercury, lead, and other metals have been enhanced by industrial and agricultural processes. Fossil-fuel combustion releases an amount equal to about 2% of the natural flux, and other processes release the equivalent of another 5%. However, the release tends to be highly localized in some cases. Bacteria and some higher organisms convert mercury to methyl mercury, which is accumulated by some fishes. These can be toxic to humans.

Most of the anthropogenic input of lead is from antiknock additives to gasoline. Although concentrations tend to be high near roadways, lead from automobile exhausts becomes widely distributed in soils and waters. However, it precipitates readily and presents relatively local problems. These should be alleviated as the use of the alkyl lead in gasoline is phased out. *See* HEAVY METALS, PATHOLOGY OF.

Organic toxins. Certain synthetic organic compounds enter biogeochemical cycles, sometimes with deleterious results. DDT was an excellent and widely used insecticide which proved to degrade very slowly. It accumulated in soils and sediments and was absorbed in the fatty tissues of organisms. It also inhibited eggshell production in birds, with striking effects on the populations of some hawks and seabirds. Because it is distributed primarily in the atmosphere, DDT is now present on essentially all of the planet, but probably it is of concern only where the concentrations are high enough to affect birds or aquatic invertebrates.

Polychlorinated biphenyls (PCBs) are compounds used as insulators and lubricants. Although they are not intentionally released into the environment, much of the PCBs eventually find their way into soils and waters, where they are accumulated in fatty tissues of organisms and are toxic. PCBs are not biodegraded and so, like DDT, they are an environmental problem. Their uses have been restricted somewhat, but there does not seem to be a way to prevent them from eventually entering the environment. They are still widely used as electrical insulating materials. *See* INDUSTRIAL WASTES; PESTICIDE; POLYCHLORINATED BIPHENYLS.

Analysis of cycles. Biogeochemical cycles are difficult to describe and understand, because they involve processes operating over a range of scales of space and time, from seconds to millions of years, and from bacteria to the cosmos. Moreover, it is often necessary to consider interactions between two or more elements in order to predict the ultimate effects on the biosphere. Modeling with large digital computers is now a standard tool, as are a number of highly sensitive chemical methods. Biogeochemistry is a rapidly progressing field and one of great significance, since it is the only means of predicting deleterious or possibly catastrophic consequences of human technology. *See* BIOSPHERE.

[LAWRENCE R. POMEROY]

Bibliography: E. D. Goldberg, in J. P. Riley and G. Skirrow (eds.), *Chemical Oceanography*, 2d ed., vol. 3, pp. 39–89, 1975; W. W. Kellogg et al., *Science*, 175:587–596, 1972; J. E. Lovelock et al., *Nature*, 237:452–453, 1972; L. R. Pomeroy, in F. G. Howell, J. B. Jentry, and M. H. Smith (eds.), *Mineral Cycling in Southeastern Ecosystems*, ERDA CONF-740513, 1975; R. M. Schwartz and M. O. Dayoff, *Science*, 199:395–403, 1978; U. Siegenthaler and H. Oeschger, *Science*, 199:388–395, 1978; M. Stuiver, *Science*, 199:253–258, 1978; L. Van Valen, *Science*, 171:439–443, 1971; G. M. Woodwell et al., *Science*, 199:141–146, 1978.

Biogeography

The science concerned with distribution of life on the Earth. Plants and animals are irregularly distributed, both on the continents and in the oceans. Some areas have a great abundance and variety of life forms, whereas others are relatively sterile. Biogeographic studies are concerned with learning the manner in which living organisms are arranged on the Earth and the causative factors of this arrangement. All animals are dependent in the final analysis on plants to supply food. Therefore, the distribution of plant life is the basic component of biogeography. Animal life is the secondary or dependent component.

Plant geography. Many factors regulate the growth of plants, but of particular importance are climatic conditions and the supply of nutrient substances on which plants feed. Driving across the western United States, during the course of a few miles, one may pass from a barren desert through wooded foothills, then up the slopes of a heavily timbered mountain to the crest which is above timberline and sparsely covered with low vegetation. This stratification of vegetation is controlled largely by conditions of temperature and moisture, which vary from the flat lowlands to the mountaintop. The nature of the underlying rock, from which plant nutrients are derived by weathering, may strongly influence the growth of plants, but the effect is less striking than that of climate. *See* TERRESTRIAL ECOSYSTEM.

Animal life in turn is stratified in conformance with the zonation of plants. There are typical animals of the desert, the foothills, the forest zone, and the alpine crest. Each of these major communities of plants and animals may be called a life zone, a vegetation zone, a biome, or a climax formation according to various systems of classification. All of these communities are part of the temperate region of the world. To the north is the boreal region, which extends generally from the North Pole to the latitude of southern Canada, and south of the temperate region is the tropical region, extending from the Equator roughly to central Mexico. The Southern Hemisphere may be similarly divided into biotic regions, and these in turn into vegetation zones or communities. *See* COMMUNITY; PLANT GEOGRAPHY.

As an example of the subdivision of a landscape into component types, Fig. 1 shows the distribution of vegetation zones in Mexico. This country spans the border of the temperate and tropical regions; the vegetation zones are arranged correspondingly in two series. *See* PLANT FORMATIONS, CLIMAX.

Tropical region. The greatest variety of plant life exists in the tropics, especially in wet forested areas. In southeastern Mexico, for example, there are tall, dense rainforests made up of many species of trees stratified in layers. The trees are heavily laden with climbing vines and with epiphytic plants, such as orchids and bromeliads. The ground beneath has a complex flora of shrubs, herbs, and lesser plants. On a few acres there may be hundreds of different kinds of plants. In general, as one proceeds north or south from the tropics, the variety of plant life decreases and fewer kinds dominate the vegetation. In polar latitudes, as on islands of the Arctic or Antarctic seas, floras are very simple.

Temperate regions. Within temperate zones, although the variety of plants is less than in the trop-

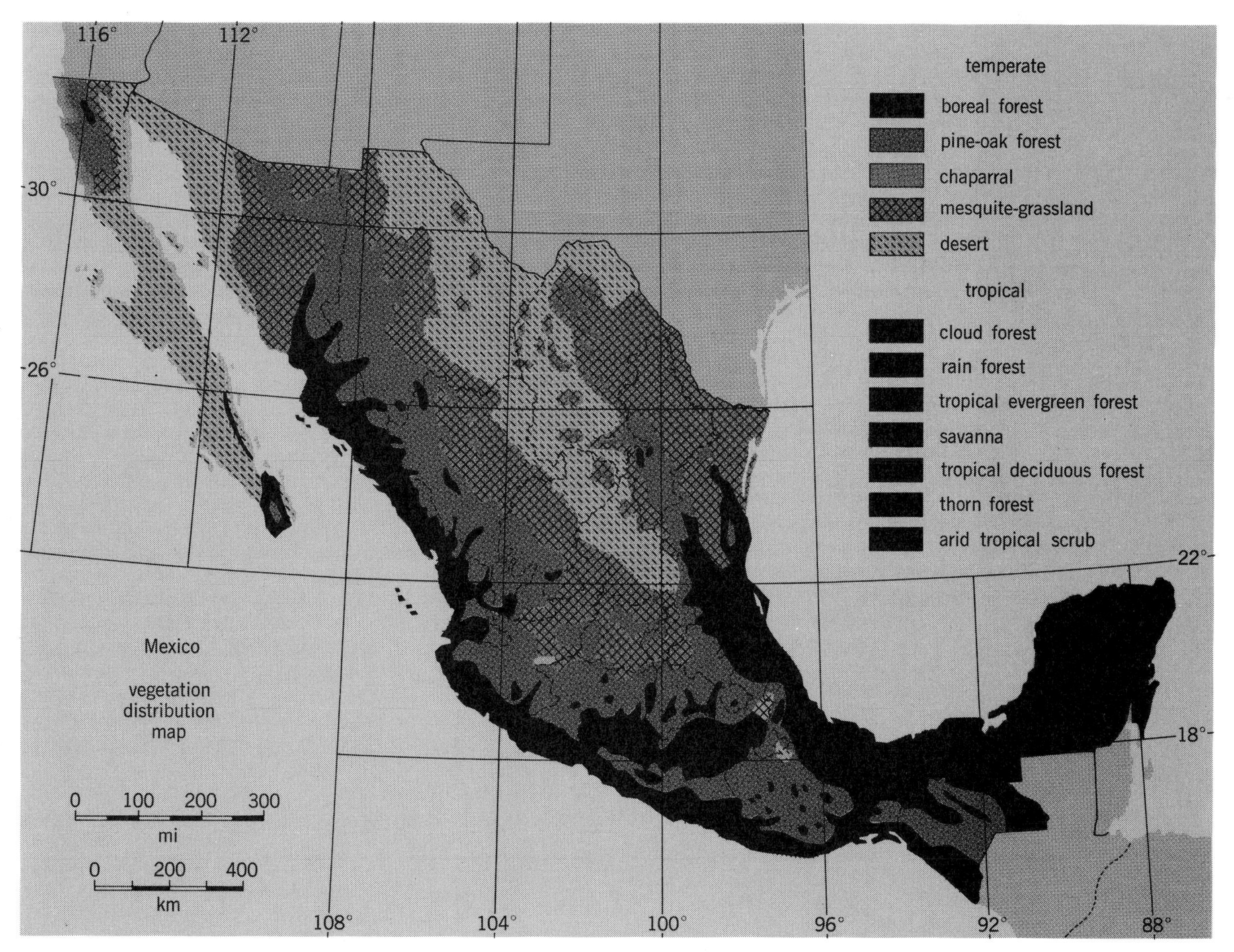

Fig. 1. Vegetation zones of Mexico. (*Museum of Vertebrate Zoology, University of California*)

ics, the total mass of vegetation may be as great, at least in wet climates. The coniferous forests of coastal Washington or the original hardwood forests of Ohio or central Europe were as massive as the great tropical rainforests of Mexico or the Philippines.

Arid zones. Arid zones in any region of the world support relatively little vegetation, for the obvious reason that lack of water limits plant growth. The sparse development of plants in polar regions may be due as much to frozen ground as to the cold itself.

Changes in distribution. The distribution of plants on the surface of the Earth is never static, but is constantly changing with shifts in climate, with emergence, submergence, and movement of land masses, and with such surface disturbances as volcanic activity or geologic erosion. As environmental conditions change, floras tend to migrate and in so doing some species become extinct and new species evolve. Certain floristic communities have moved long distances in geologic time. During the Cretaceous Period, redwood trees dominated a rich forest in the American Arctic. Redwood fossils are abundant along the Colville River in Alaska. Cold and drought of subsequent periods pushed the redwood flora southward, both in the Americas and in Asia, so that now redwoods are found in California and in China but not in their original northern homeland. *See* POPULATION DISPERSAL.

Thus the study of plant geography must consider the historic background of plant communities as well as problems of ecologic adaptation to existing conditions.

Animal geography. As stated in the discussion of plant geography, terrestrial animal populations are distributed largely in accordance with vegetation patterns. Every kind of animal is to some degree specific in the type of habitat it requires. The distribution of individual species of animals often conforms rather closely with the distribution of particular vegetation zones. As an example, Fig. 2 shows the range of the Montezuma quail (*Cyrtonyx montezumae*) in Mexico. Reference to the vegetation map (Fig. 1) will show that this quail occupies chiefly the pine-oak zone. This association is so close that the Montezuma quail may be taken as an indicator of pine-oak vegetation.

Not all animals are as specific as the Montezuma quail in their habitat requirements. Most species are adaptable enough to occupy two or more vegetation zones, although they may achieve highest density in one favored plant association. But in all the world there is no known animal that is completely nonspecific in its environmental needs. Requirements for particular types of food, shelter or cover, and amounts of water are the main components of habitat that limit the range, and locally the density, of every kind of animal.

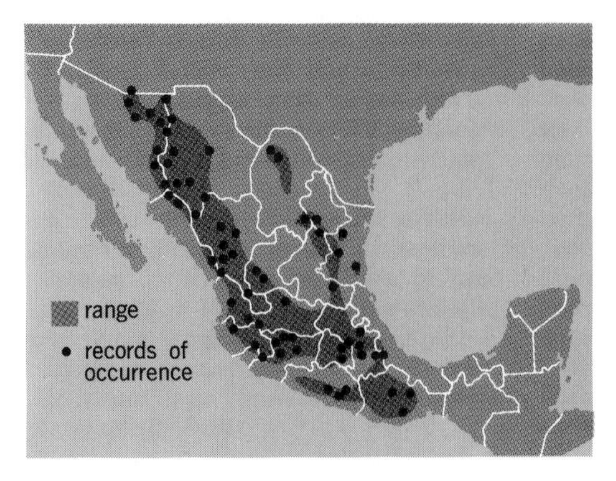

Fig. 2. Range of the Montezuma quail in Mexico.

Some animals make seasonal use of different habitats by migrating. Thus many species of migratory birds nest in northern boreal zones in summer but return southward to temperate or even tropical habitats in winter. On a more local scale, some species of both birds and mammals migrate vertically in mountainous terrain, that is, upward in summer and downward in winter. The device of seasonal migration permits utilization of environments habitable only part of the year.

Distribution in time. In geologic time, whole faunas have moved up or down the continental land masses in response to major climatic changes or to shifting of the continents. As is true of the plants, many cases of discontinuous distribution of families or orders of animals are best explained on the basis of past movements of land masses or of the animals themselves. Thus the marsupial mammals are found in Australia and in the Americas, but nowhere else. Fossil evidence places the origin of marsupials in the Far North. The animals moved southward as the climate turned colder and as the continents drifted apart. Now they are found in two southerly areas, separated by the Pacific Ocean. The ranges of other animal groups have similarly altered over geologic time.

Oceanic distribution. Animal life in the oceans is distributed as irregularly as on land. The temperature of the water and the concentration of mineral nutrients varies with air temperature, ocean currents, and especially with points of upwelling of enriched waters from the ocean depths. In favorable waters, floating microscopic plants and animals, collectively called plankton, are abundant and serve as rich pastures for larger organisms which in turn feed the fishes and even large mammals such as baleen whales. Fish-eating birds, seals, and porpoises thrive in turn. Areas rich in oceanic life are situated as a rule in cold currents and along continental shelves. Thus the great oceanic fisheries of the world and the centers of whaling are mostly in northern or southern waters, or along continental shelves such as the Peruvian coast where cold, rich waters of the ocean floor upwell to the surface.

Human biogeography. The distribution of human populations on the Earth bears a close relationship to the distribution of natural plant and animal life. Throughout history, man has thrived best in areas that are highly productive of foodstuffs. By and large the most favorable habitats for man are situated in the temperate regions of the world, where climates are moderate and soils are rich. The highest development of agricultural and pastoral industries today is in the regions of temperate climate and moderate rainfall. Various human societies, however, have become adapted to many specialized environments. Eskimos and Lapps, for example, live successfully in the icy northern wastelands, Incas and Sherpas in towering mountains, and Mayans and Pygmies in wet tropical lowlands.

Unlike most other animals, man has to a considerable degree altered his environment in supplying his needs. Thus, forests have been cleared for crops, deserts irrigated with water brought from distant mountains, marshes filled and cities built thereon. Foodstuffs, ores, and fuels are transported all over the world. The system which man calls civilization has become so complex that many people lose track of the fact that it still depends upon the function of natural resources, just as it always did. Urban dwellers in particular come to think that industry supports us, forgetting what supports industry.

Through blind destruction of agricultural soils and exploitation of other resources, many past civilizations have become impoverished or have even destroyed themselves. Since Biblical times, the rich lands of the Mediterranean region have largely been reduced to semideserts. The glory that was the Mayan Empire persists only as ruins overgrown with jungle. The great experiments of civilization have not been unqualified successes. In the last analysis, man is still an animal dependent upon many natural features of the environment for his welfare. He can modify and use these features but he cannot with impunity destroy them. *See* APPLIED ECOLOGY; HUMAN ECOLOGY.

Perhaps the outstanding aspect of human biogeography today is the explosive increase of population since 1900. Modern medicine has reduced the normal death rate among people and technological improvements have intensified the processes of resource extraction, permitting, for the moment at least, more people than ever before to live upon the Earth. How long the resources of the Earth will support such spiraling demands is conjectural. *See* ZOOGEOGRAPHIC REGION; ZOOGEOGRAPHY. [A. STARKER LEOPOLD]

Bibliography: H. G. Andrewartha and L. C. Birch, *The Distribution and Abundance of Animals*, 1954; *Cold Spring Harbor Symposia on Quantitative Biology*, vol. 22, 1957; P. M. Dansereau, *Biogeography*, 1957; P. J. Darlington, *Zoogeography: The Geographic Distribution of Animals*, 1957; W. George, *Animal Geography*, 1962; R. Hesse, W. C. Allee, and K. P. Schmidt, *Ecological Animal Geography*, 2d ed., 1951; C. L. Hubbs (ed.), *Zoogeography*, AAAS Symposium, vol. 51, 1959; N. V. Polunin, *Introduction to Plant Geography and Some Related Sciences*, 1960.

Biological productivity

The quantity of organic matter (in the form of living matter, stored food, waste products, and material taken by consumers) or its equivalent in dry matter, carbon, or energy content which is accumulated during a given time period. This quantity may apply to a single organism, a single species population, or a group of organisms. Following discussions within the International Biological Program, the term biological productivity is now applied in more general and nonquantitative contexts, while biological production has been defined as "increase in biomass (weight of living matter in a population); this includes any reproductive products released during the period concerned." In practice, studies of production are always made in the context of a particular time period and are thus expressed as production rate. They are also related to a defined population or a defined area (or volume) of the environment. *See* BIOMASS.

Productivity. This may be a useful parameter of a single population, a group of organisms belonging to the same trophic level, or to a whole ecosystem. In all cases it refers to rate of gain of material, as distinct from the amount actually present (measured as standing crop of biomass) or the turnover rate (the proportion of biomass incorporated in unit time). *See* ECOSYSTEM; FOOD CHAIN.

The general term may be qualified by giving the further definitions: (1) Production is used by fisheries biologists to exclude released reproductive products. (2) Gross production (confined to plants) is the total product of photosynthesis, much of which will be oxidized by plant respiration. (3) Primary production is the production by plants (autotrophic organisms). (4) Secondary production is production by animals (or other heterotrophic organisms). (5) Gross primary production is the total primary production by photosynthesis. (6) Net primary production is gross production less losses within the plant due to respiration. (7) Potential production is the maximum possible production within certain clearly defined environmental circumstances.

Related concepts, all of which relate to amounts of matter (or equivalent energy or carbon content) and which have sometimes been confused with production include (1) yield, which is that proportion of production that is exploited by one particular predator of the organism in question (often by man); (2) consumption, represented by C, the amount which is eaten by a heterotrophic organism (this may be much less than that killed because many animals are wasteful feeders and many also cause additional damage to their food organisms); and (3) assimilation, or A, the amount actually digested or absorbed. This equals consumption minus production of feces, or F (egesta), and other rejected material (rejecta), such as feeding pellets. It follows that assimilation is the sum of production P, (of food P_f, gametes P_g, sloughed skins P_s, and other products) and of respiration, represented by R, (loss of assimilate due to the liberation of energy from organic material and of inorganic materials including carbon dioxide and nitrogenous compounds). These relationships can be expressed as $C=A-F$, $A=P+R$, and $P=P_f+P_g+P_s$.

Matter and energy in ecosystem. All foodstuffs are organic compounds which on combustion yield large amounts of energy (around 5000 cal/g). The energy and matter remain linked until separated by respiration. In the context of the whole ecosystem, energy and matter are incorporated together in photosynthesis and separated whenever respiration occurs. In a system which is isolated and in balance, the losses of energy, whenever oxidation occurs, equate with the gain at photosynthesis. The magnitude of the energy flow (in calories per unit time and unit area) is a precise measure of the activity of the system and can be used for quantitative comparisons between whole systems and between their parts.

Matter, whether as carbon or inorganic elements, is associated with the energy flow but recy-

Table 1. Some representative figures of biological productivity

Type of productivity	Rate, kcal/(m²)(yr)
Marine	
Solar energy reaching surface of North Sea (55°N lat.)	884,000
Gross productivity, organic matter	
North Sea	360
Sargasso Sea	800
Long Island Sound	4,700
Coral reefs	26,600
Fish yield	
North Sea fisheries	2.3
Tropical seas, maximum	192
Fresh water	
Gross productivity	
Temperate lakes, oligotrophic	1,020
Temperate lakes, eutrophic	3,060
Florida stream, Silver Springs	25,600
Fish yield	
German rivers	1.9
German carp ponds, maximum	3.8
Chinese fish ponds, maximum	27
Land	
Deserts, gross productivity, maximum	700
Temperate grassland, gross productivity, maximum	14,000
Yield of temperate crops (edible part only)	
Wheat, Europe	890
Potatoes, Europe	2,280
Sugar beet, Europe	3,240
Timber, Europe	1,000
Alfalfa, United States	372
Carrot (very productive), Europe	2,250
Asparagus (low productivity), Europe	140
Yield of tropical crops	
Sugarcane	16,100
Banana	24,000
Cassava	34,800
Meat	
Beef, United States*	29.8
Beef, Germany†	10.8
Pig, Germany†	16.8
Human metabolism, Pennsylvania, United States (0.3 men/acre)	67.5

*E. P. Odum, *Fundamentals of Ecology*, 2d ed., Saunders, 1959.

†K. Kalle, *Deut. Hydrograph. Z.*, vol. 1, no. 1, 1948.

cles to an indeterminate extent. Thus the cycles of carbon, sulfur, phosphorus, or nitrogen cannot be considered a universal measure of activity in the same way that energy can.

Measurement of production. Practical methods for the measurement of primary production depend greatly on the type of organism involved. In the case of plankton, for instance, production is estimated by comparing the effect of the algae on dissolved oxygen and carbon dioxide concentrations in closed containers. One container is held under conditions of respiration only (darkened) and another under conditions of respiration plus photosynthesis (exposed to sunlight). Labeled C^{14} can be used also to measure carbon uptake. Such experiments take but a few hours and must be repeated throughout the year.

Production in forests, on the other hand, is measured by adding together gain in biomass, loss due to stem, leaf, and litter fall, and loss due to predators. This may involve repeated seasonal observations, but use can also be made of girth increment, revealed by growth rings, and of regression analysis, which is used to link different measurements of height, weight, and leaf area. In nearly all terrestrial plants reliable data are lacking for root production.

Secondary production can be estimated from measurements of growth rate and reproductive rate. It can also be derived from measurements of mortality because these equate with growth in a balanced situation. Finally, estimates can be derived from the difference between respiration and assimilation. A classical study is that of K. R. Allen, who was the first to show that annual production of a fish population may greatly exceed the biomass of food organisms available to it at any one time. This is because the invertebrate food animals grow quickly and constantly renew their numbers.

In many young, and most mature, animals the greater part of assimilation contributes not to growth but to respiration. It follows that, from the point of view of whole ecosystem studies (and since assimilation and respiration are nearly equal and metabolism is usually easier to measure than production or assimilation), respiration rates provide a more practicable index for determining the main paths of energy flow. In all cases, however, it may be very difficult to extrapolate from measurements of confined animals in the laboratory to populations in the fields. Many of the quantities (assimilation, consumption, and production) and their ratios vary greatly between species, age groups, and sexes. They also vary seasonally with food quality, with climate, with the type of food eaten, and with many other factors.

Efficiency. The ratios between quantities or calorific contents of foodstuffs at different stages of treatment by the organism are termed efficiencies. These are best designated in terms of the two quantities being compared. Important examples are: (1) Production-consumption efficiency; this is of major interest to the farmer and other predators. In nature this is usually of the order of 10% or less, but special breeding of domestic animals has raised it to 30% or more. (2) Food chain efficiency; this is the ratio of food available to two successive trophic levels.

Table 2. Calorie content of foodstuffs and organic substances

Type	Calorie content, kcal/g
Pure substances	
Fats	8.4–9.4
Carbohydrates	4.5–5.7
Proteins	3.9–4.2
Plant products	
Wood	4.8
Leaf litter	4.8
Cereals	3.6–4.3
Vegetables, root	3.4–3.8
Vegetables, leguminous	3.2–3.8
Vegetables, green	2.6–3.1
Fruits	3.3–3.8
Animal products	
Meats	6.4–7.4
Poultry	5.9–7.0
Fish	4.0–4.4
Milk, cow's	5.2
Cheese	6.3

(3) Growth-assimilation efficiency, or growth efficiency; in this case assimilation includes contributions from the mother to the embryo via the placenta and to the young animal in milk. *See* BREEDING (ANIMAL).

The photosynthetic process itself is not very efficient; in ideal laboratory conditions the yield of carbohydrate from light of the correct wavelength may reach 30%, but the efficiency of plants under field conditions is seldom above 1%.

Units of measurement. Productivity figures are usually quoted in terms of calories per square meter per annum where calorific contents are known or can reasonably be derived (Tables 1 and 2). In some cases dry weights are substituted, and it may be more appropriate to use an area of a hectare. Anglo-Saxon measurements of pounds and acres are obsolescent and the SNU (standard nutritional unit) of 10^9 calories or 10^6 kcal has been suggested, but not widely adopted. This has considerable merit, being about one human being's annual dietary requirement and of the order of the yield per acre of many plant crops.

[AMYAN MACFADYEN]

Bibliography: K. R. Allen, *The Horokiwi Stream*, N.Z. Mar. Dep. Fish. Bull. no. 10, 1951; A. Macfadyen, *Animal Ecology: Aims and Methods*, 2d ed., 1963; P. J. Newbould, *Methods for Estimating the Primary Production of Forests*, I.B.P. Handb. no. 2, International Biological Program, London, 1967; E. P. Odum, C. E. Connell, and L. B. Davenport, Population energy flow of three primary consumer components of old-field ecosystems, *Ecology*, 43:88–96, 1962; E. P. Odum, *Fundamentals of Ecology*, 1971; K. Petrusewicz, *Secondary Productivity of Terrestrial Ecosystems: Principles and Methods*, International Biological Program and Institute of Ecology, Warsaw, 1967; J. Phillipson (ed.), *Methods of Study in Quantitative Soil Ecology*, 1972; W. E. Ricker (ed.), *Methods for Assessment of Fish Production in Fresh Waters*, I.B.P. Handb. no. 3, International Biological Program, London, 1968.

Biomass

The dry weight of living matter, including stored food, present in a species population and expressed in terms of a given area or volume of the habitat. Biomass is an expression used chiefly in relation to food chains. For example, a comparison of herbivores feeding on grass should include cows as well as leaf hoppers, and weight of flesh is a fairer basis for comparison than numbers of individuals.

Measurement. The conversion of census figures to biomass strictly requires data on age distribution of weight and mortality, but these vary with biotic and physical factors and with season and have been computed only for some fish populations. The synthesis of biomass figures for whole communities is usually based on adult animals only.

Metabolic rate. The ultimate criterion of biological productivity is metabolic rate, and this is not simply related to biomass. Smaller species have higher metabolic rates per unit mass according to a 2/3 power law in general, but there are many exceptions to this law. Also, some forms, such as the mollusks, contain nonliving minerals (an objection which can be overcome by measuring total body nitrogen and converting results to weight of protoplasm). The biomass of a species at a given time (which is a measure of standing crop) is a poor index of productivity. As an example, the biomass of marine animals in lower in tropical than artic plankton of equal productivity, but the rate of synthesis and breakdown of organic matter is higher. Again, metabolism is faster at some seasons than at others, and restriction of nutrients may result in accumulation of biomass but reduction in productivity. *See* BIOLOGICAL PRODUCTIVITY.

Biomass pyramids. Biomass pyramids can be constructed on the same lines as pyramids of numbers to summarize the biomass structure of communities. Excellent examples are given by E. Odum, who shows that the shape of the pyramid depends upon whether the community is self-contained (as regards primary production of food substances), or imports or exports matter to or from the different trophic levels of the food chain. *See* FOOD CHAIN. [AMYAN MACFADYEN]

Bibliography: E. P. Odum, *Ecology*, 1975; E. P. Odum, *Fundamentals of Ecology*, 1971.

Biome

A complex biotic community covering a large geographic area and characterized by the distinctive life forms of important climax species of plants and animals. A life form is the common morphological features that characterize a group of organisms. On the land, a biome is identified by the life form of the dominant climax plants, as well as by the distinctive types of vegetation and landscape in which they grow. In the ocean, the life forms of the predominant animals serve as the criterion. The biome incorporates all seral (successional) as well as climax stages of the community, and its distribution is controlled by climate or, in the ocean, by other physical factors of the environment. Principal terrestrial biomes are the temperate deciduous forests, coniferous forest, woodland, chaparral, tundra, grassland, desert, tropical savanna, and tropical broad-leaved forest. Principal marine biomes are the pelagic, the barnacle–gastropod–brown algae, the sea urchin–large snail, the bivalve-annelid, the coral reef, and the abyssal-benthic. *See* GRASSLAND ECOSYSTEM; SAVANNA; TUNDRA.

Terrestrial biomes are divisible into plant associations, which are distinguished by distinctive combinations of climax dominant species. For instance, the temperate deciduous forest biome includes the beech-maple, oak-hickory, and other associations. Both terrestrial and marine biomes are divisible into animal biociations. A biociation is distinguished by the distinctiveness of the predominant animal species. Some of the biociations within the temperate deciduous forest biome are the North American deciduous forest, with gray squirrel and eastern chipmunk among predominant animals; European deciduous forest, with the red deer; and Eurasian deciduous forest, with the Manchurian tiger, musk deer, and true wolf. *See* COMMUNITY; ECOLOGY. [S. CHARLES KENDEIGH]

Bibliography: P. Dansereau, *Biogeography*, 1957; L. R. Dice, *Natural Communities*, 1952; S. C. Kendeigh, *Ecology with Special Reference to Animals and Man*, 1974.

Biosphere

The thin film of living organisms and their environments at the surface of the Earth. Included in the biosphere are all environments capable of sustaining life above, on, and beneath the Earth's surface as well as in the oceans. Consequently, the biosphere includes virtually the entire hydrosphere and portions of the atmosphere and outer lithosphere.

Neither the upper nor lower limits of the biosphere are sharp. A variety of organisms inhabit the ocean depths. Spores of microorganisms can be carried to considerable heights in the atmosphere, but these are resting stages that are not actively metabolizing. Evidence exists for the presence of bacteria in oil reservoirs at depths of about 2000 m within the Earth. The bacteria are apparently metabolically active, utilizing the normal paraffinic hydrocarbons of the oils as an energy source. These are extreme limits to the biosphere; most of the mass of living matter is within the upper 100 m of the lithosphere and hydrosphere, although there are places even within this zone that are too dry or too cold to support much life. Most of the biosphere is within the zone which is reached by sunlight and where liquid water exists.

ORIGIN OF THE BIOSPHERE

For over 50 years after Louis Pasteur disproved the theory of spontaneous generation, scientists believed that life was universal and was transplanted to other planets and solar systems by the spores of microorganisms. This is the theory of panspermia. In the 1920s the modern theory of chemical evolution of life on Earth was proposed independently by A. I. Oparin, a Soviet biochemist, and J. B. S. Haldane, a British biochemist. The basic tenet of the theory is that life arose through a series of chemical steps involving increasingly complex organic substances that had been chemically synthesized and had accumulated on the prebiotic Earth. The first organisms originated under reducing conditions in an atmosphere devoid of free molecular oxygen. Because they were incapable of photosynthesis, a complex process which evolved undoubtedly after life originated, the oxygen in the present atmosphere arose as a secondary addition.

Evidence for chemical evolution. Two kinds of evidence support the theory of chemical evolution: laboratory experimentation and analyses of meteorites.

Laboratory experiments. Experimental evidence indicates that a variety of simple organic compounds can be formed in a closed system by the input of energy to a mixture of inorganic gases thought to compose the primitive atmosphere. The first experiment was conducted by Stanley Miller in 1953 at the University of Chicago. Utilizing a closed glass system containing the gases CH_4, NH_3, H_2, and water vapor, subjected to a continuous electric discharge, Miller was able to synthesize a number of amino acids and organic acids. Subsequently, other scientists have conducted similar experiments substituting different gaseous constituents, such as CO_2, CO, and N_2, and different energy sources, including ultraviolet radiation and heat, with much the same results. The conclusion that can be drawn from these experiments is that even though the composition of the primitive atmosphere or the energy sources likely to have been present are not known with certainty, simple biochemical monomers could have been formed from various possible compositions and energy sources.

Meteorites. Direct evidence for abiogenic chemical synthesis of simple biochemical compounds is obtained from meteorites. The Murchison Meteorite, a type II carbonaceous chondrite that fell in Australia in 1969, contains a variety of amino acids as well as other organic compounds. Analyses of the Murchison and other recent falls have shown that at least 35 amino acids are present. Protein amino acids are present in greatest abundance, but nonprotein amino acids make up the greatest number of species. Unlike the amino acids in proteins of organisms, which are entirely in the L configuration, the amino acids recovered from the meteorite are present as a racemic mixture, with equal amounts of L and D isomers present. The characteristics of this suite of compounds—the large number of nonprotein amino acids and their racemic nature—strongly suggest an origin by abiogenic processes. Neither extraterrestrial organisms nor terrestrial contamination is a likely source of these substances.

Cell development. The major intermediary steps from simple monomers to a living cell must have included polymerization of monomers to larger molecules, the aggregation of these macromolecules and their eventual separation from other aggregates by some sort of membrane (a precursor to a cell wall), and the development of metabolic activity and the capability to reproduce. Scientists have experimentally demonstrated plausible schemes for forming polymers, some with membranes, under possible primitive Earth conditions. Because of the complexity of even a primitive organism, however, the remaining possible pathways leading to the first cell are speculative and, indeed, it may never be known which of these was actually followed.

Anoxic environment. Various lines of indirect evidence (astronomic, meteoritic, geologic) indicate that surface environments were anoxic on the Earth during the time that life emerged, except for trace amounts of oxygen formed by photodissociation of water vapor; consequently, the first organisms were probably anaerobic heterotrophs comparable to modern anaerobic bacteria. These organisms obtained their food from the coexisting biochemicals formed by abiogenic processes.

First organisms. The time of appearance of the first organisms is not firmly established, but it could be prior to 3.3 billion years (b.y.). Evidence for possible life-forms is found in rocks of this age from the Swaziland Sequence in South Africa. If life did exist by this time, then the preceding period of chemical evolution leading up to the first cells lasted for about 0.5 to 1.0 b.y., or about the time span between 3.3 b.y. and the cooling of the Earth's surface sufficiently to contain liquid water, an event that may have occurred within a few hundred million years of the formation of the Earth about 4.6 b.y. ago. Although it is not certain that the fossillike objects from the Swaziland Sequence are indeed remains of life-forms, other evidence also suggests the interval between 3.0 and 3.3 b.y. as the time of emergence of life.

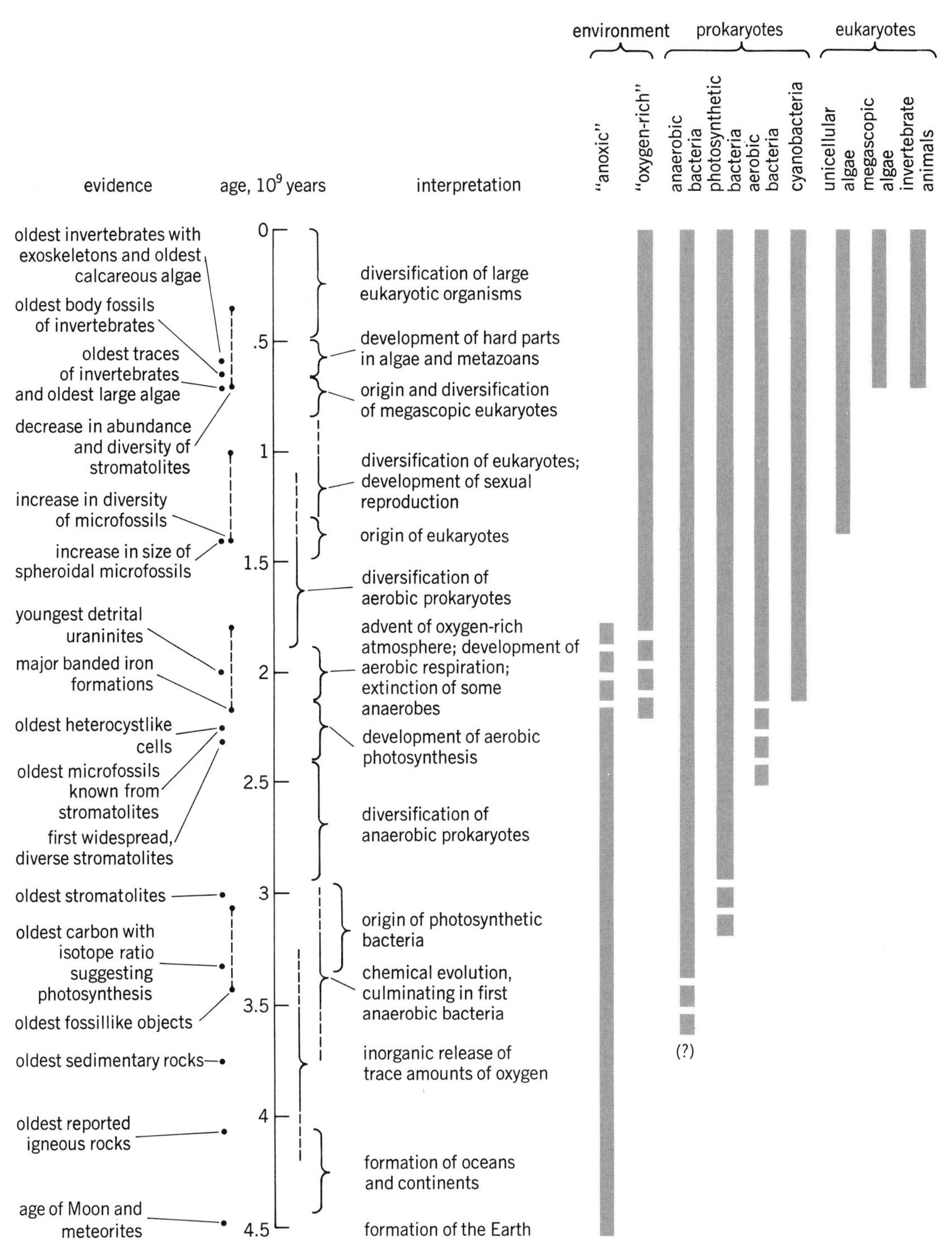

Fig. 1. Chart of major events in Precambrian evolution. (*From J. W. Schopf, The evolution of the earliest cells, Sci. Amer., 239:110–138, copyright © 1978 by Scientific American, Inc.; all rights reserved*)

One line of evidence is that furnished by carbon isotopes. The $^{13}C/^{12}C$ ratio and the $\delta^{13}C$ of the insoluble organic carbon in the rocks of the Swaziland Sequence are within the range of reduced carbon found in all younger rocks, an indication that organisms, which are responsible for the isotopic fractionation of carbon, had already appeared. Another line of evidence pinpointing the time of emergence of the first cells is the presence of stromatolites that range in age from 2.5 to 3.0 b.y. in a number of localities. Stromatolites are distinctive, thinly laminated deposits formed by mats of blue-green algae in association with calcium carbonate. The existence of these features no later than 2.7 b.y. ago places a lower limit on the time of origin of the biosphere.

EVOLUTION OF THE BIOSPHERE

The biosphere, once in being, experienced Darwinian evolution; the changes wrought during the evolution of the biosphere had profound effects on the atmosphere, hydrosphere, and outer lithosphere.

Photosynthesis. Probably the first significant event in the evolution of the biosphere was the development of photosynthesis, a process that led to autotrophic organisms capable of synthesizing organic matter from inorganic constituents. This evolutionary advance freed the organisms from a dependence on the coexisting abiogenic organic matter, which was being depleted by heterotrophic organisms. These autotrophs were anaerobic organisms similar to modern photosynthetic bacteria, and they did not release oxygen as a by-product. Some time later, oxygen-releasing photosynthesis, another major milestone in the evolution of the biosphere, evolved.

Eukaryotes. A profound evolutionary event was the emergence of eukaryotic organisms. All early life-forms were prokaryotic, a classification that currently includes only blue-green algae and bacteria. All other organisms, including other algae, higher plants, and all animals, are eukaryotic, characterized by the presence of a nucleus and capable of reproducing by meiosis and mitosis. The origin of eukaryotic organisms was an important evolutionary event, as it represented the emergence of sexual reproduction and set the stage for the later evolution of multicellular organisms. The time of first appearance of undisputed eukaryotic organisms is not firmly established, but most scientists accept the interval between 1.0 and 1.5 b.y. ago as the time for this event.

Geochemical cycles. Organisms have had an effect on the geochemistry of their environment for about 3 b.y. Although details of the conditions existing and types of reactions occurring on the Earth for the first 0.5 to 1.0 b.y. after life appeared are not known in detail, a plausible setting can be presented. Prior to the development of oxygen-releasing photosynthesis, the Earth's surface was anoxic, all organisms were anaerobic, and chemical elements existed in their reduced forms. The gradual release of free oxygen permanently changed the Earth's surface from a reducing environment to an oxidizing one, and this change led to the evolution of eukaryotic organisms and of multicellular life-forms.

Oxidation of iron and sulfur. Elements that had previously existed in reduced form were converted to their oxidized state. As a consequence of the conversion, the geochemical cycles of many elements capable of existing in different oxidation states (such as iron, sulfur, manganese, carbon, and uranium) were significantly modified. Under reducing conditions iron would be present in the more soluble state as Fe^{2+}, and the oceans and surface waters would contain much more dissolved iron than is currently in solution. Sulfur, which forms sparingly soluble heavy-metal sulfides, such as pyrite, FeS_2, would be present in solution in low concentration because of the large excess of Fe^{2+}. The atmosphere probably contained some H_2S, HCN, and other gases toxic to aerobic organisms. With the advent of oxygen production by photosynthetic organisms, oxygen would initially react with the large quantity of reduced chemical species before invading the atmosphere to a significant extent. For several hundred million years, as the Fe^{2+} and S^{2-} were gradually oxidized to Fe^{3+} and SO_4^{2-}, these elements served as oxygen sinks to prevent an oxygen buildup in the atmosphere.

Evidence for the gradual oxidation of the sinks is found in the massive banded iron formations (BIF), widespread deposits of iron oxide that represent the major world iron ore reserves. These deposits are essentially limited to rocks older than about 1.8 b.y., with most of them being formed between 2.2 and 1.8 b.y. ago. Consequently, organisms capable of oxygen-releasing photosynthesis had emerged prior to 2.2 b.y. ago. BIF usually consist of alternating bands or layers of iron-rich and silica-rich minerals, with each pair of layers possibly formed seasonally or annually. Subsequent to BIF deposition, redbeds, composed of detrital iron oxide–coated grains, first appeared. Redbeds are nonmarine to marginal marine sediments formed under oxidizing conditions, and their presence indicates that free oxygen was beginning to accumulate in the atmosphere and shallow portions of the hydrosphere.

The state of sulfur changed along with that of iron. Sulfur concentration in the reducing ocean would be maintained at low levels by the formation of insoluble metal sulfides and the greater concentration of iron. Because iron is much more abundant in the weathering zone than sulfur, more iron than sulfur would be delivered to the oceans, where the excess iron would cause the removal of most of the sulfur as pyrite. With the transition to oxidizing conditions and the removal of the iron into BIF, the reduced sulfur would be oxidized to the more soluble sulfate salts. As the concentration of sulfate ion increased, evaporite deposits, consisting of gypsum and anhydrite, began to form in restricted evaporating basins. Metal sulfides were, henceforth, limited in their occurrence to restricted anoxic environments in sediments or to hot metalliferous fluids originating within the crust.

Carbon. Concurrently with the oxidation of iron and sulfur, a greater amount of the reduced carbon was converted to CO_2, increasing the HCO_3^- and CO_3^{2-} concentrations in the oceans and resulting in the precipitation of $CaCO_3$ to form the massive limestones and dolomites of the later Precambrian. With a further increase in the oxygen level of the atmosphere, metazoans appeared during the latest Precambrian and within a short span of geologic time had developed the ability to calcify. The advent of calcareous organisms transferred major control of the carbonate cycle in the oceans to the biosphere. Virtually all $CaCO_3$ deposits today, and probably all since the beginning of the Phanerozoic, are the products of calcareous organisms. Major events in Precambrian evolution are given in Fig. 1. *See* BIOGEOCHEMICAL CYCLES.

ORGANISMS OF THE BIOSPHERE

Although the biosphere is the smallest in mass, it is one of the most reactive spheres. In the course of its reactions, the biosphere has important influences on the outer lithosphere, hydrosphere, and atmosphere.

Ecosystem. The biosphere is characterized by the interrelationship of living things and their environments. Communities are interacting systems of organisms tied to their environments by the transfer of energy and matter. Such a coupling of living organisms and the nonliving matter with which they interact defines an ecosystem. An ecosystem may range in size anywhere from a small pond to a tropical forest to the entire biosphere. *See* ECOSYSTEM.

Metabolic processes. The major metabolic processes occurring within the biosphere are photosynthesis and respiration. Green plants, through the process of photosynthesis, form organic compounds composed essentially of C, H, O, and N from CO_2, H_2O, and nutrients, with O_2 being released as a by-product. The O_2 and organic compounds are partially reconverted into CO_2 and H_2O through respiration by plants and animals. Driving this cycle is energy from the Sun; respiration releases the Sun's energy which has been stored by photosynthesis.

Types of organisms. The fundamental types of organisms engaged in these activities are: producers, or green plants that manufacture their own food through the process of photosynthesis and respire part of it; consumers, or animals that feed on plants directly or on other animals by ingestion of organic matter; and decomposers, or bacteria that break down organic substances to inorganic products.

Producers. Productivity, the rate of formation of living tissue per unit area in a given time, is one of the fundamental attributes of an ecosystem. All organisms in the community are dependent on the energy obtained through gross primary productivity. The productivity of an ecosystem is determined by a number of environmental variables, particularly temperature and the availability of water and nutrients. In marine environments, availability of nutrients is the limiting factor, while the productivity in terrestrial environments is limited by the availability of water. The productivity of different ecosystems is given in Table 1. The biomass, or the total mass of living organisms at one time in an ecosystem, may or may not correlate with productivity. The productivities of tropical rain forests and reefs are nearly identical, but their biomasses differ by a factor of 600. *See* BIOLOGICAL PRODUCTIVITY.

Food chains. The portion of gross primary productivity that is not respired by green plants is the net primary productivity that is available to consumer and reducer organisms. Organic matter and energy are passed from one organism to another along food chains. The steps, or groupings of organisms, in a chain are trophic levels. Green plants, the producers, occupy the first trophic level, followed by herbivores (primary consumers), first carnivores, secondary carnivores, and tertiary carnivores. The number of links in a food chain is variable, but three to five levels are common. Only about 10% or less of the matter and energy is passed from one trophic level to the next because of the utilization of energy in respiration at each level. *See* FOOD CHAIN.

Decomposers. Living matter that is not consumed by higher trophic levels is decomposed as dead tissue through bacterial action. Decomposers play a major role in the flow of matter and energy in ecosystems because they are the final agents to release the photosynthetic energy from the organic compounds which they utilize. Decomposers recycle the chemical components, but not energy, back into the ecosystem. *See* ECOLOGICAL INTERACTIONS.

CHEMICAL COMPOSITION

Over 60 elements are reported to occur in organisms. Of these, over 30 play a vital role in cellular processes. On an atomic basis hydrogen, the lightest element, constitutes almost 50% of the content

Table 1. Primary production and biomass estimates for the biosphere*

(1)	(2)	(3)	(4)	(5)	(6)	(7)	(8)
Ecosystem type	Area, 10^6 km²	Mean net primary productivity, g C/m²/year	Total net primary production, 10^9 metric tons C/year	Combustion value, kcal/g C	Net energy fixed, 10^{15} kcal/year	Mean plant biomass, kg C/m²	Total plant mass, 10^9 metric tons C
Tropical rain forest	17.0	900	15.3	9.1	139	20	340
Tropical seasonal forest	7.5	675	5.1	9.2	47	16	120
Temperate evergreen forest	5.0	585	2.9	10.6	31	16	80
Temperate deciduous forest	7.0	540	3.8	10.2	39	13.5	95
Boreal forest	12.0	360	4.3	10.6	46	9.0	108
Woodland and shrubland	8.0	270	2.2	10.4	23	2.7	22
Savanna	15.0	315	4.7	8.8	42	1.8	27
Temperate grassland	9.0	225	2.0	8.8	18	0.7	6.3
Tundra and alpine meadow	8.0	65	0.5	10.0	5	0.3	2.4
Desert scrub	18.0	32	0.6	10.0	6	0.3	5.4
Rock, ice, and sand	24.0	1.5	0.04	10.0	0.3	0.01	0.2
Cultivated land	14.0	290	4.1	9.0	37	0.5	7.0
Swamp and marsh	2.0	1125	2.2	9.2	20	6.8	13.6
Lake and stream	2.5	225	0.6	10.0	6	0.01	0.02
Total continental	149	324	48.3	9.5	459	5.55	827
Open ocean	332.0	57	18.9	10.8	204	0.0014	0.46
Upwelling zones	0.4	225	0.1	10.8	1	0.01	0.004
Continental shelf	26.6	162	4.3	10.0	43	0.005	0.13
Algal bed and reef	0.6	900	0.5	10.0	5	0.9	0.54
Estuaries	1.4	810	1.1	9.7	11	0.45	0.63
Total marine	361	69	24.9	10.6	264	0.0049	1.76
Full total	510	144	73.2	9.9	723	1.63	829

*All values in columns 3 to 8 expressed as carbon on the assumption that carbon content approximates dry matter × 0.45.

SOURCE: R. H. Whittaker and G. E. Likens, in G. M. Woodwell and E. V. Pecan (eds.), *Carbon in the Biota*, in *Carbon and the Biosphere*, U.S. Atomic Energy Commission, Technical Information Center, 1973.

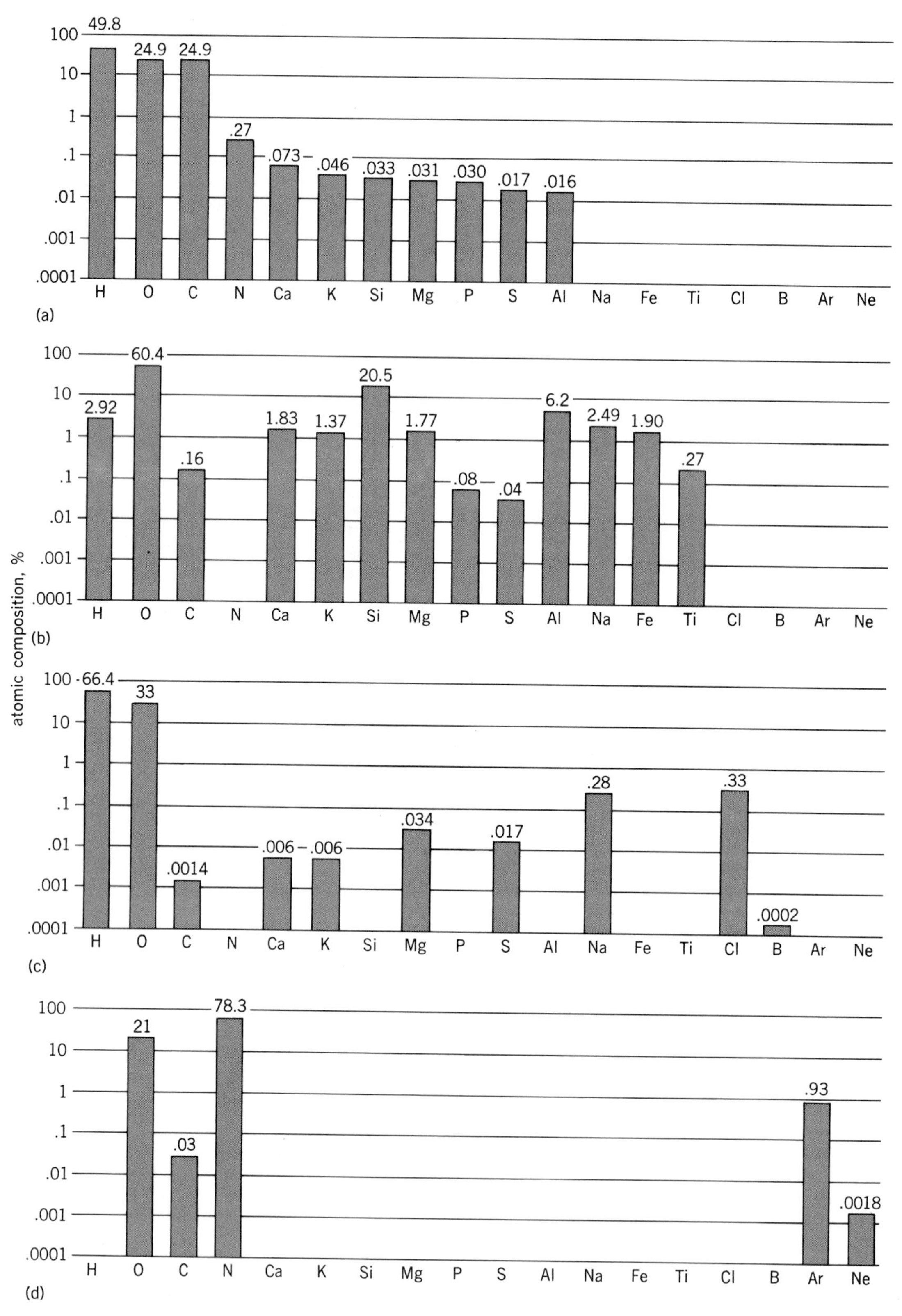

Fig. 2. Minerals required by living organisms: (*a*) biosphere, (*b*) lithosphere, (*c*) hydrosphere, and (*d*) atmosphere. (*From E. S. Deevey, Jr., Mineral cycles, Sci. Amer., 223:149–158, 1970*)

of organisms in the biosphere, with carbon and oxygen combined making up the other rough half. Taken together, these three elements make up over 99% of all the atoms in the biosphere. Other important constituents such as nitrogen, phosphorus, and sulfur total only about 3 atoms per 1000. This composition reflects the dominance of the woody plants, which are rich in the carbohydrate cellulose, a compound of C, H, and O. Nitrogen and sulfur are important constituents of proteins,

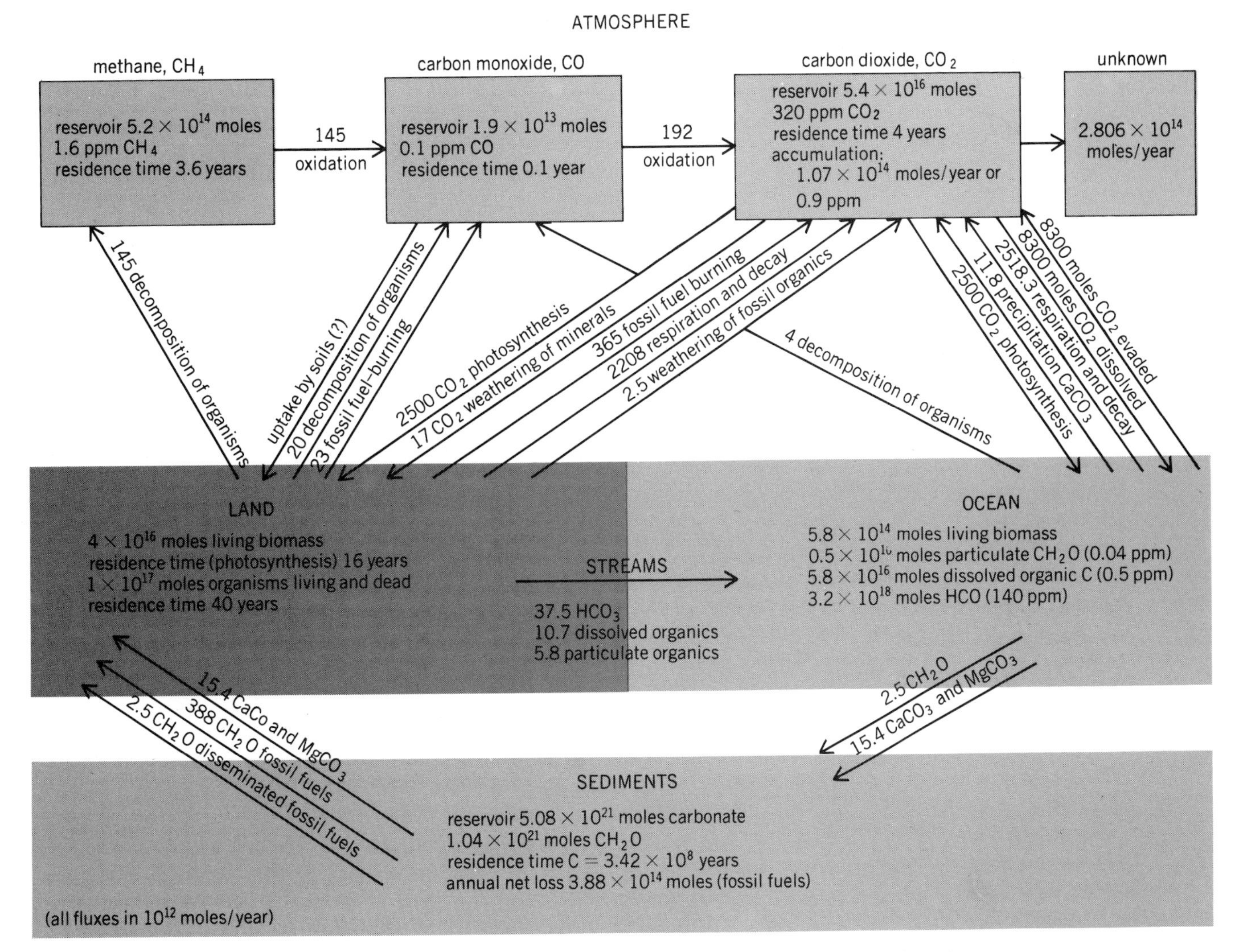

Fig. 3. Diagram illustrating the steps in the carbon cycle. (*From R. M. Garrels, F. T. Mackenzie, and C. Hunt, Chemical Cycles and the Global Environment, W. Kaufmann, 1975*)

while phosphorus is responsible for energy transfer within the cell through formation of adenosinetriphosphate (ATP). The six elements H, C, O, N, S, and P, although present in widely varying amounts, are necessary ingredients of living organisms (Fig. 2).

Some essential constituents are present only in trace quantities. Important examples are the magnesium in the chlorophyll molecule and the iron in hemoglobin. Most of the required elements are needed by both plants and animals, but a few are essential only for one or the other. The necessary constituents are generally those of lower atomic number through zinc, element 30, although some with higher atomic numbers such as iodine and molybdenum also perform vital functions. A number of trace elements, for example, cobalt, copper, selenium, and zinc, have a narrow range of concentrations tolerable for metabolic requirements; greater quantities are toxic or lead to physiological impairments. In some parts of the world, animals have been poisoned by grazing on metal-enriched plants, although the high metal contents did not apparently affect the plant growth. *See* PLANTS, MINERALS ESSENTIAL TO.

CARBON CYCLE

The $CO_2 - HCO_3^- - CaCO_3$ system is one of the most important systems affected by organisms because of its effect on the pH balance in the ocean, the content of CO_2 in the atmosphere, and the production of $CaCO_3$ sediment. Despite the dominance of the biosphere in the production of $CaCO_3$, the atmosphere and hydrosphere contain important components of the carbonate system. The inorganic carbon cycle is also an integral part of the larger carbon cycle of the Earth and, in fact, most of the carbon in the cycle is stored in the limestones and dolomites initially precipitated for the most part by organisms (Fig. 3).

Hydrosphere. Surface ocean waters are typically supersaturated with respect to the $CaCO_3$ minerals calcite and aragonite, while deep waters (constituting about 80% of the oceans) are demonstrably undersaturated with respect to $CaCO_3$. In warm shallow waters such as coral reef and shelf environments, where carbonate productivity is high, $CaCO_3$ forms extensive deposits. In the open ocean, the $CaCO_3$ precipitated primarily by coccoliths and foraminifera sinks to the deep ocean floor. However, calcareous organisms form $CaCO_3$ in much greater quantity than is preserved in the sediments. Because dissolved CO_2 increases toward the bottom of the oceans due to respiration, deeper waters are more acidic than shallow waters, and a major portion of the $CaCO_3$ is dissolved as it descends through these undersaturated waters or after it reaches the ocean floor. Consequently, about 80–85% of the $CaCO_3$ precipitated by organisms in surface waters is dissolved before burial in the sediment. The 15–20% that is preserved falls on the shallower portions of the ocean floor projecting into the waters that are saturated, or only slightly undersaturated, with respect to $CaCO_3$.

The depth zone in the ocean separating sediments rich in $CaCO_3$ from those low in $CaCO_3$ is called the calcite compensation depth (CCD). This level varies slightly from ocean to ocean because of differences in the CO_2 content of deep waters; the CCD is about 4500 m in the Atlantic and about 3500 m in the Pacific. In addition, the compensation level for aragonite is much shallower than the level for calcite because of the greater solubility of aragonite.

The location of the CCD is established by the carbon budget of the oceans. The input of carbon from rivers must equal the carbon removed so that the carbon balance in the ocean can be maintained. Presently about 90% of the incoming carbon is removed from the ocean in the form of $CaCO_3$ and 10% as organic matter. Changes in the amount of carbon brought in by rivers, in the amount of $CaCO_3$ formed by organisms, in the area of continental shelves occupied by carbonate-secreting organisms, or in a number of other factors could cause the CCD to rise or fall so that only the amount of $CaCO_3$ required to maintain the carbon balance is removed. *See* SEA WATER.

Carbon reservoirs. The carbonate cycle is the oxidized portion of the larger carbon cycle of the Earth with the deposits of $CaCO_3$ now in the form of limestones and dolomites forming the major reservoir of carbon. The photosynthetic fixation of CO_2 by plants, the primary producers, leads to the reduced portion of the carbon cycle. Through a number of steps in the food chain, the plants are consumed by herbivores, who are in turn the prey of carnivores. In this manner, organic matter and energy are passed through a number of levels of food chains. At each level it is estimated that about 90% of the matter is required for energy and converted back to CO_2, with the remainder forming part of the body mass. Organic matter that is not immediately consumed by higher organisms may be attacked by bacteria, converting an additional portion of it to CO_2. Under some conditions, when the supply of organic matter exceeds the ability of microorganisms to metabolize it, or the organic matter is buried in sediments faster than microoorganisms can oxidize it, a portion of the organic matter may escape oxidation and be stored in sediments. This sedimentary organic matter eventually makes up the dispersed organic matter called kerogen, which is the source material of hydrocarbons in petroleum and oil shales. Less than 1% of the organic matter escapes annually from the biosphere (about 2.5×10^{12} moles), with the release of an equal amount of free oxygen to the atmosphere. To maintain a steady state in the carbon and oxygen cycles, the loss of organic carbon is balanced by the oxidation to CO_2 of a comparable amount of organic matter in sedimentary rocks exposed to weathering.

The various reservoirs contain unequal amounts of carbon, with widely different transfer rates from one reservoir to another. The total mass and the mass of carbon in the outer spheres of the Earth are given in Table 2, together with the approximate residence time (time for one complete turnover) of carbon in each sphere. The atmosphere (6.5×10^{17} g carbon) is dominated by CO_2, with minor amounts of CH_4 and CO. The ocean, with HCO_3^- in greatest abundance and small amounts of dissolved and particulate carbon, contains 3.9×10^{19} g of carbon. The largest reservoir of carbon is sedimentary rocks (7.35×10^{22} g), with carbonate carbon about five times as abundant as organic carbon. Only about 0.002% of the carbon is in living biomass, with the terrestrial biomass about 80–90 times greater than the marine biomass. The total living biomass is about 4.8×10^{17} g of carbon.

Residence time. Carbon is cycled fastest through the atmosphere and biosphere. The residence time of CO_2 in the atmosphere is only about 4 years. Through photosynthesis alone, the entire CO_2 content of the atmosphere would cycle through plants in 11 years. Relative rates of carbon fixation or net productivity by the terrestrial and marine ecosystems are uncertain. Estimates vary from approximately equal net productivity to ratios of 2:1 to 3:1 for terrestrial to marine productivity. Although the rate of fixation of carbon may be comparable for both ecosystems, carbon is turned over much faster in the oceans. The residence time of carbon in the biosphere is controlled by the slower turnover of carbon in terrestrial plants. The oceanic portion of the biosphere contains only about 1% of the total biomass, and the residence time of carbon is about 3 months.

Carbon dioxide balance. Of special importance in the carbon cycle is the balance of CO_2 between the atmosphere and the oceanic and terrestrial ecosystems. Since about 1850 the burning of fossil

Table 2. Total mass, mass of carbon, and residence time of carbon in the outer spheres of the Earth's surface*

Sphere	Total mass, $\times 10^{20}$ g	Carbon mass, $\times 10^{15}$ g	Residence time of carbon years
Atmosphere	52	650	4
Hydrosphere	17,000	39,600	385
Sedimentary rocks	25,000	7,350,000	342×10^6
Biosphere (living)	0.1855	487	–
Terrestrial	0.1852	480	16
Marine	0.0003	7	0.23

*From R. M. Garrels, F. T. Mackenzie, and C. Hunt, *Chemical Cycles and the Global Environment: Assessing Human Influences*, W. Kaufmann, 1975.

fuels has released a significant amount of CO_2 into the atmosphere. This input by human activities continues to be a source of concern because it is thought that the added CO_2 will lead to an overall warming of the Earth, a phenomenon called the greenhouse effect. The atmospheric CO_2 concentration has been monitored since 1958; the CO_2 content increased from 315 parts per million in 1958 to about 330 ppm in 1975, an increase at a rate of about 0.8 ppm per year. The increase in atmospheric CO_2, however, accounts for only about one-half of the total world CO_2 production during this interval. Part of the unaccounted half has entered the oceans, which is a major sink for CO_2 on a long time scale.

The problem with balancing the CO_2 budget of the last 100 years is that there is disagreement about whether the oceans, on the short time scale of 100 years, could remove the required amount of CO_2. While the surface ocean is well mixed and can quickly remove some CO_2, the water volume is insufficient to remove the required CO_2. The deep ocean has much greater volume, but the circulation time is on the order of 1000 years, not tens of years. Consequently, very little of the missing CO_2 could have found its way to the deep sea.

Plants might respond to an increase in CO_2 by increasing their photosynthetic fixation of carbon. Such a mechanism could remove some of the added CO_2 from the atmosphere. Estimates of the changes in volume of the terrestrial or marine biomasses are subject to large uncertainties, however, and it is difficult to determine if the biomass has increased or decreased in response to the burning of fossil fuels. Due to deforestation and wood burning, the terrestrial biomass may even have decreased, with a net transfer of CO_2 from the biomass to the atmosphere. If this occurred, the biomass would be a source rather than a sink for CO_2. Consequently, the amount of CO_2 injected into the atmosphere would be greater than estimated from combustion of fossil fuels alone, and the fraction of CO_2 removed from the atmosphere, and still largely not accounted for, would also be greater than estimated from fossil-fuel sources. An accurate assessment of the current fluxes and reservoir masses is necessary to understand the effect of human activity on the carbon cycle. *See* ATMOSPHERIC CHEMISTRY; BIOMASS.

NITROGEN CYCLE

The nitrogen cycle is considerably complex, and portions of it are only incompletely known. The number of different nitrogen species that exist in the atmosphere, oceans, and terrestrial environments, the highly reactive nature of some of these species, and the role of different microorganisms in the oxidation and reduction of nitrogen compounds account for this complexity (Fig. 4). While almost all the carbon is contained in the Earth's crust, a significant portion of the nitrogen is stored in the atmosphere, largely as N_2 but also as reactive nitrogen oxides and ammonia. *See* NITROGEN CYCLE.

Biological nitrogen fixation. Nitrogen-fixing microorganisms, for example, *Azotobacter*, some species of blue-green algae, and the legume symbiont *Rhizobium*, convert atmospheric N_2 to a form that can be readily assimilated by the plants for the synthesis of protein. However, only a fraction of the total nitrogen used by plants comes directly from atmospheric N_2 or from nitrogen-fixing organisms. Although the figures are, at best, approximate, about 4.7×10^{15} g/year of nitrogen are taken up by marine plants, while the amount of nitrogen addition to the oceans from all sources is only about 8.5×10^{13} g/year. The large difference between the amount used and the amount added represents the quantity of nitrogen that is apparently efficiently recycled in the marine biosphere. The annual nitrogen input balances losses due to burial of organic matter in the sediments and to evasion of NH_3 and nitrogen oxides to the atmosphere.

The rate of incorporation of nitrogen into terrestrial plants is only about 10–15% of the rate of assimilation by marine plants. The C/N ratio in terrestrial plants is about 80, while in marine plants it is about 7. The difference reflects the greater protein content of marine plants. About one-sixth of the nitrogen annually incorporated into the terrestrial biosphere is obtained by nitrogen fixation, while approximately two-thirds is recycled from decaying organisms.

The nitrogen of proteins is released as NH_3 or NH_4^+. Some of this escapes to the atmosphere directly, and some is converted to nitrite and nitrate by nitrification reactions employed by bacteria such as *Nitrosomonas* and *Nitrobacter*. A major portion of the nitrate and ammonia is recycled through plants. Bacterial denitrification reactions convert part of the nitrate to molecular nitrogen, which returns to the atmosphere. A significant fraction of nitrogen is returned to the atmosphere as nitrous oxide, N_2O, also by microbial denitrification. Automobile exhaust, industrial activity, and natural sources all make contributions to the N_2O. In the atmosphere, this gas is converted to NO and N_2 by photochemical reactions. NO is oxidized to NO_2 by ozone, a reaction that leads to the destruction of the ozone layer.

Industrial nitrogen fixation. Since 1950, industrial fixation of nitrogen for chemical fertilizers has increased tenfold, and all other industrial sources, including automobile exhaust, have increased about threefold. At present, the amount of nitrogen annually fixed for the production of chemical fertilizer equals the amount naturally fixed in soils by bacterial processes. Estimates for industrial nitrogen fixation for fertilizer use in the year 2000 are 30 times the 1950 level, or three times the natural rate of terrestrial bacterial fixation. All of the nitrogen fixed by microorganisms or applied as fertilizer is returned eventually to the atmosphere as either NH_3, N_2, or N_2O. Thus, chemical fertilizers and industrial activity have increased the N_2O flux to the atmosphere, and this heavy interference by humans in the nitrogen cycle has severe implications with regard to the destruction of the ozone layer.

INFLUENCE OF THE PRESENT BIOSPHERE

Organisms, through their metabolic activities, continually influence the processes and reactions occurring in the environments of the Earth's surface. Microorganisms, because of their large numbers, biochemical versatility, large surface-to-volume ratio, and rapid growth rate, are the most active agents in these influences. Consequently, numerous reactions affecting the cycling of ele-

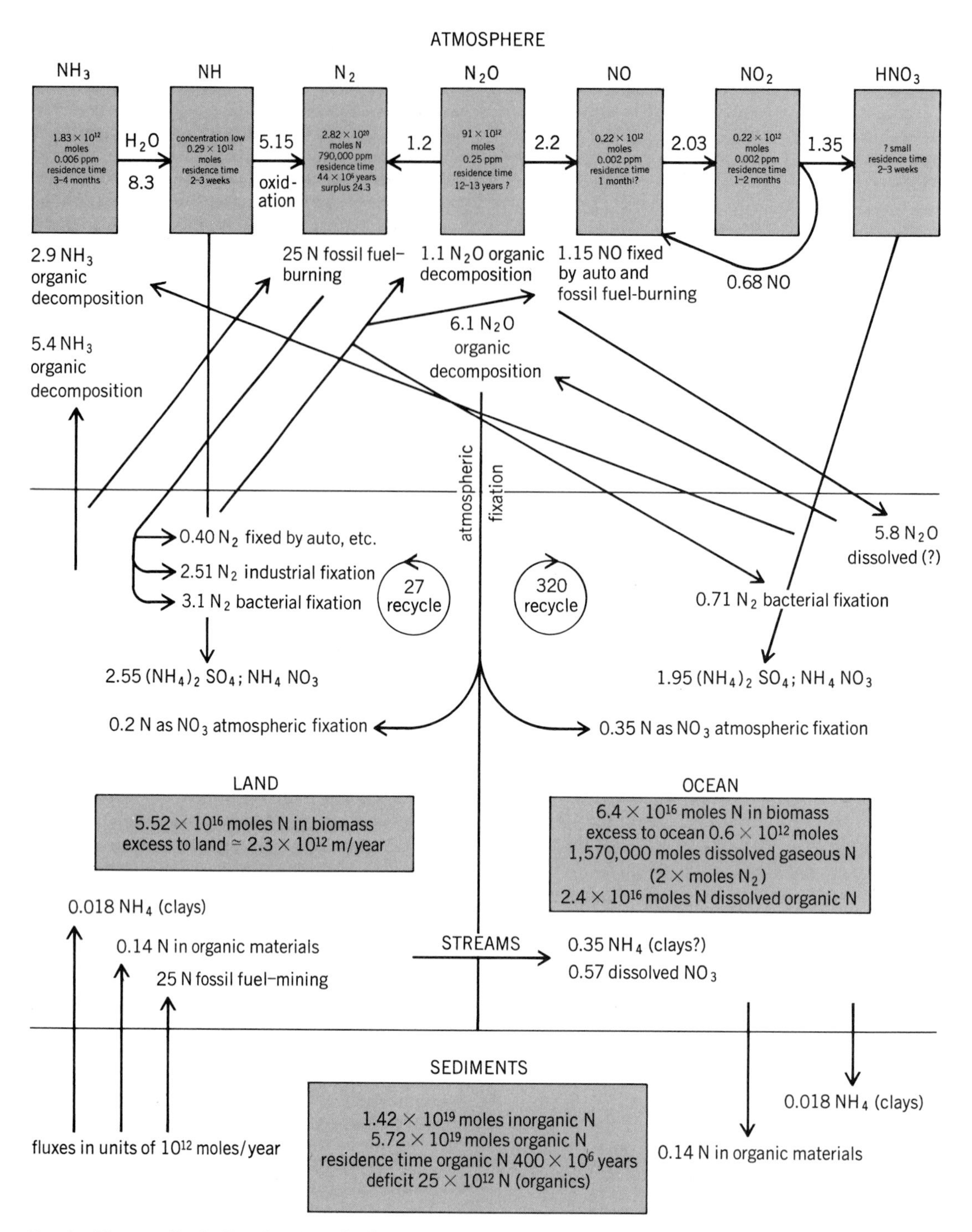

Fig. 4. Diagram illustrating the steps in the complex nitrogen cycle. (*From R. M. Garrels, F. T. Mackenzie, and C. Hunt, Chemical Cycles and the Global Environment, W. Kaufmann, 1975*)

ments through the biosphere are frequently controlled by microbial processes. Organisms also indirectly influence the chemical behavior of an element or species by altering the concentration of other ions or elements such as H^+ or O_2, which in turn affects equilibrium concentrations of the substance of interest. Mineral surfaces may acquire organic coatings from the dissolved organic matter, a process which may affect the rate and extent of equilibration of the mineral with the water.

Chemical weathering. Microorganisms are active in the soil zone of humid regions, where they metabolize the soil humus, converting it to CO_2. In arid environments, productivity is much less, soil is poorly developed, and the amount of humus is insufficient to support an extensive microbial population. The land biota plays an important role in the chemical weathering of carbonate and silicate minerals. The rate of dissolution and destruction of these minerals is related to the dissolved CO_2 content of water in contact with them. Through the metabolic activities of microorganisms and tree roots, the partial pressure of CO_2 (pCO_2) in soils is from 10 to 100 times higher than that of the atmosphere (0.03% CO_2). While soil CO_2 contents as high as 10% have been measured, most values range from 0.15 to 0.65% CO_2. The greater dissolved CO_2 content in groundwater enhances the rate of chemical weathering in humid regions. Because the rate of plant decay is related to the rate of photosynthesis, the rate of chemical weathering is, in turn, roughly dependent on plant productivity.

Metal poisoning. Terrestrial vegetation accumulates certain metals in various tissues and organs. In some cases this metal uptake is a normal requirement of the plant metabolism, and the particular metals involved and their concentrations are directly related to the nutritional needs of the plant. Other plants are accumulators of metals because their metal content is not related to growth requirements but to the metal content of the soil or bedrock substrate on which the plants are growing. Many plants are particularly susceptible to metal poisoning in regions of metallic ore bodies where the concentrations of certain metals in soils above ore bodies is high. Vegetation in such areas may be sparse and limited to metal-tolerant or accumulator plants.

The flora surrounding known ore bodies has anomalously higher concentrations of the ore metals than does vegetation further removed. Discerning such halos of heavy metal accumulation in plants can assist in locating unknown or hidden ore deposits. The presence of a plant thought to be indicative of copper mineralization was used as an exploration aid in Scandinavia as early as the 17th century. Modern biogeochemical exploration techniques for locating ore bodies were developed in the 1930s.

Stress from metal poisoning produces a recognizable pattern in aerial infrared color photographs of vegetation. Successful attempts have been made to locate metal-poisoned regions by use of Landsat data. Significant differences in multichannel reflectance ratios can be detected between poisoned and normal vegetation. In the future, satellite reconnaissance will permit mineral exploration in remote or inaccessible regions and even in less remote regions that were overlooked previously. *See* REMOTE TERRAIN SENSING.

Silica cycle. The carbonate minerals are not the only minerals precipitated in large quantity by organisms. Diatoms are major producers of opaline silica, SiO_2, in oceans and lakes, and the marine and lacustrine silica cycles are dominated by the productivity of these organisms. In highly productive waters, diatom blooms deplete the surface waters of silica until it becomes a limiting nutrient. Most of the silica is transferred to deeper waters by dissolution of the opaline tests as they settle to the bottom. This process creates a gradient of increasing dissolved silica concentration toward the ocean bottom. Upwelling conditions return this silica to the surface, where it is again available for uptake by phytoplankton. The amount of silica preserved in the sediments is probably equal, on a long-term basis, to the river input of silica from the continents so that the overall silica content of the oceans is at steady state.

The silica cycle is influenced by the carbon cycle. The amount of silica released from continental rocks by chemical weathering processes is related to the CO_2 content of the groundwater in contact with the rocks. Consequently, the productivity of the continental biota, which affects the CO_2 content of the weathering solutions, influences the amount of silica delivered to the oceans. Any changes in the productivity of terrestrial plants may affect the oceanic silica budget and the productivity of marine diatoms.

Organic matter. In marine and lacustrine sediments, microorganisms exert an important influence on the chemistry of the pore waters. Aerobic organisms metabolize the organic matter until oxygen is depleted. When the supply of organic matter exceeds the amount of oxygen available, anoxic conditions are established and the activities of anaerobic microorganisms prevail. Bacteria of the genus *Desulfovibrio* reduce sulfate to sulfide in anoxic marine sediments. If iron is available in sufficient quantities, pyrite forms. Rapid sulfate depletion and pyrite formation are generally associated with a high organic carbon content of the sediments. In nearshore sedimentary environments, measurements show that most sulfate is reduced to sulfide within 1 m of the sediment-water interface. In the sediments of the deep ocean basin, where the organic carbon content and the sedimentation rate are lower, the activity of sulfate reducers is much less and sulfate is not completely depleted. In addition to the production of sulfide, the HCO_3^- content increases due to the increase in dissolved CO_2 from oxidation of the organic matter. Beneath the zone of sulfate reduction and after complete reduction of the dissolved sulfate, some of the CO_2 is reduced to methane by methanogenic bacteria. In marine environments, the sulfate reduction and methane production zones do not overlap, and much of the methane that diffuses into the overlying sulfate-reducing zone is apparently reoxidized to CO_2 by the bacterial population. In fresh-water marshes, bogs, and swamps, where sulfate concentration is normally very low, methane, referred to as swamp gas, escapes directly to the atmosphere, where it is eventually oxidized to CO_2.

The diminished microbial activity in deeper portions of the ocean basins was demonstrated when the research submarine *Alvin* accidentally sank in 1500 m of water off Massachusetts. When the submersible was recovered almost a year later, the crew's lunch was still virtually intact. After refrigeration in the laboratory at sea-bottom temperatures, most of the food subsequently spoiled in a few weeks.

Most of the marine and terrestrial biomass is quickly recycled to CO_2 by bacterial processes. Of the one part in a thousand of living matter produced annually that is not oxidized, most is preserved in sedimentary rocks of marine origin. The organic carbon content in marine sediments is variable and generally reflects the productivity of overlying waters. Shales contain about 0.8% organic carbon, while sandstones and carbonates have about 0.3% organic carbon.

The bulk of the organic matter buried in sediments and sedimentary rocks is in disseminated form and consists of complex polymeric materials termed humic substances and kerogen. Under appropriate conditions some of the kerogen will be converted to liquid and gaseous hydrocarbons. Migration and trapping of these hydrocarbons will result in an oil or gas deposit. The generation and accumulation of petroleum hydrocarbons is a very inefficient process. Only 0.01% of sedimentary organic matter composes discrete oil and gas reservoirs. The hydrocarbons of petroleum are generated primarily from the marine lipid fraction of the kerogen, while natural gas hydrocarbons are generally derived from the lignin-rich terrestrial organic matter.

ISOTOPES IN ORGANISMS

Many of the light elements involved in biological processes are isotopically fractionated by organisms as well as by physical processes in the environment. Important examples of well-studied isotope systems are the pairs $^{12}C-{}^{13}C$,$^{16}O-{}^{18}O$, and $^{32}S-{}^{34}S$. Generally, organisms preferentially utilize the lighter isotopes in their metabolic processes. The isotopic signatures created by biochemical or physical reactions are frequently useful in establishing the source of, and the reactions leading to, a particular product. Enrichments and depletions of any isotopes are always determined by comparison with an established standard.

Carbon and sulfur. Two major isotopic reservoirs of carbon have been formed because of biological fractionation: organic carbon relatively enriched in carbon-12; and inorganic carbon relatively enriched in carbon-13. The tissues of marine and terrestrial plants are enriched in carbon-12 relative to inorganic carbon by about 15–20 atoms per thousand carbon atoms. Both of these reservoirs have retained their isotopic values since at least the late Precambrian, an indication that the biosphere had become well established by that time. Bacteria preferentially reduce sulfur-32 in forming sulfides, thereby enriching the remaining sulfate in sulfur-34. The two major reservoirs are: reduced sulfur species occurring in iron sulfide minerals and dissolved in anoxic waters; and sulfate dissolved in seawater.

Oxygen. Important physical processes involved in fractionation of oxygen isotopes are temperature and evaporation. Water evaporated from the oceans and falling as rain on the continents is relatively enriched in oxygen-16 compared with sea water. Minerals precipitating from either fresh or marine waters can be differentiated on the basis of their different oxygen isotopic signatures. The degree of isotopic fractionation is also a function of temperature; there is less fractionation at higher temperatures because the vibrational energies of the two isotopes becomes nearly identical. Consequently, the temperature of mineral precipitation can be calculated for minerals forming in isotopic equilibrium with the water. This technique has been applied to the determination of paleotemperatures in ancient oceans and to the chronology of the ice advances during the Pleistocene Epoch.

HUMAN IMPACT ON BIOSPHERE

Human beings, of course, are part of the biosphere. Some of their activities have an adverse impact on many ecosystems and on themselves by the addition of toxic or harmful substances to the outer lithosphere, hydrosphere, and atmosphere. Many of these materials are eventually incorporated into or otherwise affect the biosphere. The major types of environmental pollutants are sewage, trace metals, petroleum hydrocarbons, synthetic organic compounds, and gaseous emissions.

Sewage. Bulk sewage is generally disposed of satisfactorily and is normally not a problem unless it contains one or more of the other types of substances. Under certain conditions, either when the load of organic matter is at a high level or nutrients lead to eutrophication, dissolved oxygen is depleted, creating undesirable stagnation of the water.

Trace metals. Some metals are toxic at low concentrations, while other metals that are necessary for metabolic processes at low levels are toxic when present in high concentrations. Examples of metals known to be toxic at low concentrations are lead, mercury, cadmium, selenium, and arsenic. Industrial uses and automobile exhaust are the chief sources of these metals. Additions of these metals through human activity are, in some cases, about equal to the natural input. The content of mercury in the Greenland ice sheet has doubled since 1952 because of increasing use of the metal. Even greater increases of metals are observed near major sources. Plants and soils near highways or lead-processing facilities contain as much as 100 times the unpolluted value of lead.

Petroleum hydrocarbons. Petroleum hydrocarbons from runoff, industrial and ship discharges, and oil spills pollute coastal waters and may contaminate marine organisms. Oil slicks smother marine life by preventing the diffusion of oxygen into the water or by coating the organisms with a tarry residue. Aromatic constituents of petroleum are toxic to all organisms. Large doses can be lethal, while sublethal quantities can have a variety of physiological effects, especially carcinogenicity. The hydrocarbons are lipophilic and, because they are not metabolized or excreted, tend to accumulate in fatty tissues. Even when marine life has not been subjected to massive contamination by an oil spill, the organisms may be gradually contaminated by accumulating the hydrocarbons to which they have been continuously exposed at low concentrations. Contamination of shellfish, in particular, is of concern because they are consumed in great quantities by humans.

Synthetic organic compounds. Manufacture and use of synthetic organic compounds, such as insecticides and plasticizers, have become commonplace worldwide. Many of these chemicals are chlorinated hydrocarbons, a category of substances that is known to be generally toxic to organisms. Of even greater concern is that chlorinated hydrocarbons are carcinogenic and long-term exposure to sublethal doses increases the risk of cancer. The widespread use of chlorinated hydrocarbons in industrial society leads to an increasing background of these compounds in natural waters, including drinking water. DDT residues in marine phytoplankton in the waters off California tripled between 1955 and 1969.

Synthetic organic compounds and petroleum hydrocarbons are generally not metabolized or excreted by organisms. Both groups tend to be retained in the fatty tissues and passed on to other organisms in the next level of the food chain. Consequently, predators high in the food chain are exposed to much greater concentrations of toxic or carcinogenic substances through their ingested food than is present at background levels in the environment. Prior to banning its widespread use in the United States, DDT was found in concentrations as high as tens of parts per million in fish-eating birds such as gulls and grebes, even though the concentration of DDT in the lower organisms of their food chain and in the bottom sediment was as low as .01 ppm. The high concentrations led to thinning of egg shells and decline in reproduction rate in many predator bird populations.

Gaseous emissions. Gaseous emissions to the atmosphere from industrial sources and automobiles represent both short-term and long-term hazards. Short-term problems include the addition of sulfurous gases, leading to more acidic rainfall, the emission of nitrogen oxides which form smog, and carbon monoxide from automobile exhaust which leads to impairment of breathing in cities. Of more long-term concern is the increase in the CO_2 content of the atmosphere, which will lead to a gradual warming of the Earth—the so-called greenhouse effect—with its attendant problems of melting of the icecaps, shifting of climatic zones, and possibly other unanticipated effects; and the decrease in the ozone layer, which shields organisms from lethal ultraviolet radiation. Combustion of fossil fuels is the source of the CO_2, while N_2O, one of the chief agents leading to the destruction of ozone, is derived by denitrification of organic nitrogen of soil humus and animal wastes as well as from automobile exhaust. Increasing reliance on chemical fertilizers to increase crop productivity is generating a larger N_2O flux to the atmosphere.

Humans must continue to be concerned about the potential hazards, not only to themselves but to other organisms in the biosphere, of the products of an industrialized society. Of course, the human is not the first creature to affect the environment; organisms have been doing so since life began in the Precambrian. *See* AIR POLLUTION; ATMOSPHERE; HUMAN ECOLOGY; HYDROSPHERE; WATER POLLUTION.

[RICHARD M. MITTERER]

Bibliography: N. R. Andersen and A. Malahoff (eds.), *The Fate of Fossil Fuel CO_2 in the Oceans*, 1977; B. Bolin, The carbon cycle, *Sci. Amer.*, 223: 124–132, 1970; P. Cloud, Beginnings of biospheric evolution and their biogeochemical consequences, *Paleobiology*, 2:351–387, 1976; J. L. Cox, DDT residues in marine phytoplankton: Increase from 1955 to 1969, *Science*, 170:71–73, 1970; E. S. Deevey, Jr., Mineral cycles, *Sci. Ameri.*, 223: 149–158, 1970; C. C. Delwiche, The nitrogen cycle, *Sci. Amer.*, 223:136–146, 1970; R. E. Dickerson, Chemical evolution and the origin of life, *Sci. Amer.*, 239:70–86, 1978; R. M. Garrels, F. T. Mackenzie, and C. Hunt, *Chemical Cycles and the Global Environment*, 1975; H. D. Holland, *The Chemistry of the Atmosphere and Oceans*, 1978; G. E. Hutchinson, The biosphere, *Sci. Amer.*, 223:45–53, 1970; H. W. Jannasch et al., Microbial degradation of organic matter in the deep sea, *Science*, 171:672–675, 1971; I. R. Kaplan (ed.), *Natural Gases in Marine Sediments*, 1974; D. H. Kenyon and G. Steinman, *Biochemical Predestination*, 1969; K. A. Kvenvolden, Natural evidence for chemical and early biological evolution, *Origins of Life*, 5:71–86, 1974; M. B. McElroy, S. C. Wofsy, and Y. L. Yung, The nitrogen cycle: Perturbations due to man and their impact on atmospheric N_2O and O_3, *Roy. Soc. London Phil. Trans. Ser. B.*, 277:159–181, 1977; J. W. Schopf, The evolution of the earliest cells, *Sci. Amer.*, 239:110–138, September 1978; R. H. Whittaker, *Communities and Ecosystems*, 1970; G. M. Woodwell and E. V. Pecan (eds.), *Carbon and the Biosphere*, U.S. Atomic Energy Commission, Technical Information Center, 1973; G. M. Woodwell et al., The biota and the world carbon budget, *Science*, 199:141–146, 1978.

Biotic isolation

The occurrence of organisms in isolation from others of their species. Many organisms live in aggregations or social groups. Among many others, however, there appears an opposite tendency toward relative isolation, with each individual or pair living by itself, separated by at least a small distance from others of its kind. Thus in a desert, shrubs may grow widely scattered, each with its roots occupying an area around it larger than that of its foliage. A shrub seedling which begins to grow close to an established larger shrub encounters more intense root competition for water and nutrients, and in some cases stronger unfavorable chemical effects of the larger shrub, than does a seedling which grows at a point more distant from other shrubs. Seedlings most likely to survive are those near the center of spaces between other shrubs. The shrubs consequently grow not in clumps but in relative isolation from each other, with a spacing between individuals which may be more even than in a random distribution. They may show a degree of negative contagion in their distribution. In many animals a breeding pair may isolate themselves, occupying a definite territory from which others of the species are excluded. Such biotic isolation provides each individual or pair with a living space from which it can obtain food, or water and nutrients, more or less free from competition with other members of its species. *See* ECOLOGICAL INTERACTIONS; POPULATION DISPERSION.

[ROBERT H. WHITTAKER]

Bibliography: E. P. Odum, *Ecology*, 1975; E. P. Odum, *Fundamentals of Ecology*, 1971.

Breeding (animal)

The theory and application of quantitative and Mendelian genetics to the genetic and economic improvement of farm animals for such traits as growth rate and meat quality in beef cattle, sheep, and swine; egg yield and quality in chickens; and milk yield and composition in dairy cattle. Although farm animals have been domesticated and improved for thousands of years, the theory for genetic improvement was not developed until after Gregor Mendel's laws were rediscovered in 1901. Before then, however, Robert Bakewell (1726–1795), in England, is credited with development of several improved breeds of livestock by application of his still true axioms, "Like begets like" and "Breed the best to the best." The ideas of Charles Darwin on natural selection in the middle 1800s also anticipated the effects of selection controlled by humans on the genetic improvement of farm animals.

History. The first applications of Mendel's genetic laws involved the qualitative effects of only a pair of genes, but most economic traits are controlled by many genes, each with relatively small effects. This form of inheritance, while following Mendel's principles, is called polygenic or quantitative inheritance. Although Bakewell is sometimes called the father of animal breeding, the honor in the modern era more fittingly belongs to Jay L. Lush, who in the 1920s and 1930s developed most of the theory of modern animal breeding. The mathematical basis of these principles can be credited to Sewell Wright, certainly the grandfather of modern animal breeding. The theory for more complicated statistical situations has been developed by C. R. Henderson and Alan Robertson, both of whom studied with Lush.

Principles of selection. Selection of superior parents can be very effective in improving a single trait or combination of traits weighted by the economic values of the traits. The principles are the same in either case, although the application is more complicated for a combination of traits. Animals must show differences (variation) or selection is impossible. A fraction of the variation (heritability) must also be due to genetic differences. Gain from selection per year can be predicted as the product of accuracy of selection, a factor related to the fraction of animals selected (the selection intensity factor), and the standard deviation of the trait (a standard measure of variation) which is divided by the interval measured in years between when an animal is born and when its replacement is born. The accuracy of evaluation, which may range between 0 and 100%, increases as more records on the animal and close relatives are used for evaluation, and is higher for traits with high heritability.

Selection. Generally, mass selection (selection on the animal's own record) is efficient for traits with heritability above 25%. Progeny testing is more effective for traits with low heritability and those that are limited to only one sex such as milk production.

For most traits, male selection will contribute 90% or more to total genetic gain because fewer and thus more highly selected males are needed as compared with females. The most striking gains from selection have been made in dairy cattle because of the primary importance of a single trait, the accuracy of selection made possible by an excellent records system for progeny testing, and the

The result of a cross between a Red Sindhi bull (a Zebu breed, *Bos indicus*) and a Jersey cow (a European breed, *B. taurus*). *(From G. H. Schmidt and L. D. Van Vleck, Principles of Dairy Science, W. H. Freeman and Company; copyright © 1974)*

use of artificial insemination which makes intense selection of males practical since up to 50,000 matings per year are possible to each accurately evaluated bull.

Rapid genetic change from intense selection is also possible in swine breeding because of the rapid rate of reproduction—two litters of 8 to 10 pigs per year. Progress has been limited, however, by the necessity of selecting simultaneously for several traits and because of changing goals of selection.

Progress in beef cattle has been slow because different parts of the industry consider different traits to have primary importance. In addition, reproductive rate is low, and the potential of artificial insemination has not been exploited due to management difficulties.

Most economic traits of farm animals fall into three groups: production traits such as milk yield or growth rate; quality of product traits such as carcass quality or milk composition; and reproductive traits such as calving interval or services per conception. The quality traits generally have high heritability and exhibit little heterosis; production traits usually have moderate heritability and heterosis; and reproductive traits have low heritability and relatively high heterosis. Heterosis is defined as the difference of crossbred offspring from the average of purebred offspring of the same breeds of parents.

Crossbreeding. Crosses between breeds can be used to improve traits which exhibit positive heterosis. Crossbreeding is also used when the optimum for a trait is intermediate between the average for the two breeds (see illustration). Problems with crossbreeding involve the maintaining of pure breeds, the lack of selection possible in the pure breeds, and the difficulty of finding enough breeds with desirable characteristics to use as males on crossbred females. Thus, after three to four crosses or less of different breeds of males on succeeding generations of crossbred females, the breeds of sires are usually repeated in rotation. Most commercial swine and sheep are produced by crossbreeding because of heterosis for reproductive rate and also for growth rate. Beef cattle production has recently increased the use of crossbreeding, especially for terminal crosses where a large breed of sire is used to produce offspring, all of which are marketed, and also because of heterosis for reproductive rate and for improved mothering ability of crossbred cows.

Crossbreeding is also used to upgrade native breeds. An imported or exotic breed with more desirable production characteristics is mated to native breeds of low productivity but high adaptability to the local conditions. These crossbred offspring can then be mated to sires of the same imported breed—a process that can be continued until the genes of the population are nearly all from the imported breed. Since the imported breed is often not adapted to the local conditions, management often must be improved to allow for increased productivity of the upgraded animals. If the management is increased only slightly, the first cross, in many cases, will be more productive than crosses with three-fourths or more of imported genes; in such situations, a rotational system maintaining three-eighths to five-eighths imported genes may be optimum. Matings and selection among the crossbred animals also can be used to form a synthetic breed with the desirable characteristics of the parent breeds—productivity of the imported breed and adaptability of the native breed. *See* AGRICULTURAL SCIENCE (ANIMAL).

[DALE VAN VLECK]

Bibliography: J. F. Lasley, *Genetics of Livestock Improvement*, 1978; G. H. Schmidt and L. D. Van Vleck, *Principles of Dairy Science*, 1974; A. M. Sorenson, Jr., *Animal Reproduction: Principles and Practice*, 1979.

Breeding (plant)

The application of genetic principles to the improvement of cultivated plants, with heavy dependence upon the related sciences of statistics, pathology, physiology, and biochemistry. The aim of plant breeding is to produce new and improved types of farm crops or decorative plants, to better serve the needs of the farmer, the processor, and the ultimate consumer. New varieties of cultivated plants can result only from genetic reorganization that gives rise to improvements over the existing varieties in particular characteristics or in combinations of characteristics. In consequence, plant breeding can be regarded as a branch of applied genetics, but it also makes use of the knowledge and techniques of many aspects of plant science, especially physiology and pathology. Related disciplines, like biochemistry and entomology, are also important, and the application of mathematical statistics in the design and analysis of experiments is essential.

Plant breeding has made major contributions to increasing the yields of crops and to diminishing their susceptibility to hazards that limit their productivity. It has been estimated that the annual production of corn in the United States has been increased by 750,000,000 bu by plant breeding methods, especially by the exploitation of hybrid corn. In western Europe the yields of wheat and barley have been increased by approximately 1% per annum since the late 1940s. Perhaps the most dramatic impact of plant breeding occurred during the 1960s in Mexico where, as the result of the stimulus from a Rockefeller Foundation program, a wheat-importing country was changed into a wheat-exporting country because of a surplus of wheat.

Scientific method. The cornerstone of all plant breeding is selection. By selection the plant breeder means the picking out of plants with the best combinations of agricultural and quality characteristics from populations of plants with a variety of genetic constitutions. Seeds from the selected plants are used to produce the next generation, from which a further cycle of selection may be carried out if there are still differences. Much of the early development of the oldest crop plants from their wild relatives resulted from unconscious selection by the first farmers. Subsequent conscious acts of selection slowly molded crops into the forms of today. Finally, since the early years of the 20th century, plant breeders have been able to rationalize their activities in the light of a rapidly expanding understanding of genetics and of the detailed biology of the species studied.

Plant breeding can be divided into three main categories on the basis of ways in which the species are propagated. Species that reproduce sexually and that are normally propagated by seeds occupy two of these categories. First come the species that set seeds by self-pollination, that is, fertilization usually follows the germination of pollen on the stigmas of the same plant on which it was produced. The second category of species sets seeds by cross-pollination, that is, fertilization usually follows the germination of pollen on the stigmas of different plants from those on which it was produced. The third category comprises the species that are asexually propagated, that is, the commercial crop results from planting vegetative parts or by grafting. Consequently, vast areas can be occupied by genetically identical plants of a single clone that have, so to speak, been budded off from one superior individual. The procedures used in breeding differ according to the pattern of propagation of the species.

Self-pollinating species. The essential attribute of self-pollinating crop species, such as wheat, barley, oats, and many edible legumes, is that, once they are genetically pure, varieties can be maintained without change for many generations. When improvement of an existing variety is desired, it is necessary to produce genetic variation among which selection can be practiced. This is achieved by artificially hybridizing between parental varieties that may contrast with each other in possessing different desirable attributes. All members of the first hybrid (F_1) generation will be genetically identical, but plants in the second (F_2) generation and in subsequent generations will differ from each other because of the rearrangement and reassortment of the different genetic attributes of the parents. During this segregation period the breeder can exercise selection, favoring for further propagation those plants that most nearly match the ideal he has set himself and discarding the remainder. In this way the genetic structure is remolded so that some generations later, given skill and good fortune, when genetic segregation ceases and the products of the cross are again true-breeding, a new and superior variety of the crop will have been produced.

This system is known as pedigree breeding, and while it is the method most commonly employed, it can be varied in several ways. For example, instead of selecting from the F_2 generation onward, a bulk population of derivatives of the F_2 may be maintained for several generations. Subsequently, when all the derivatives are essentially true-breeding, the population will consist of a mixture of forms. Selection can then be practiced, and it is assumed, given a large scale of operation, that no useful segregant will have been overlooked. By whatever method they are selected, the new potential varieties must be subjected to replicated field trials at a number of locations and over several years before they can be accepted as suitable for commercial use.

Another form of breeding that is often employed with self-pollinating species involves a procedure known as backcrossing. This is used when an existing variety is broadly satisfactory but lacks one useful and simply inherited trait that is to be found in some other variety. Hybrids are made between the two varieties, and the F_1 is crossed, or back-

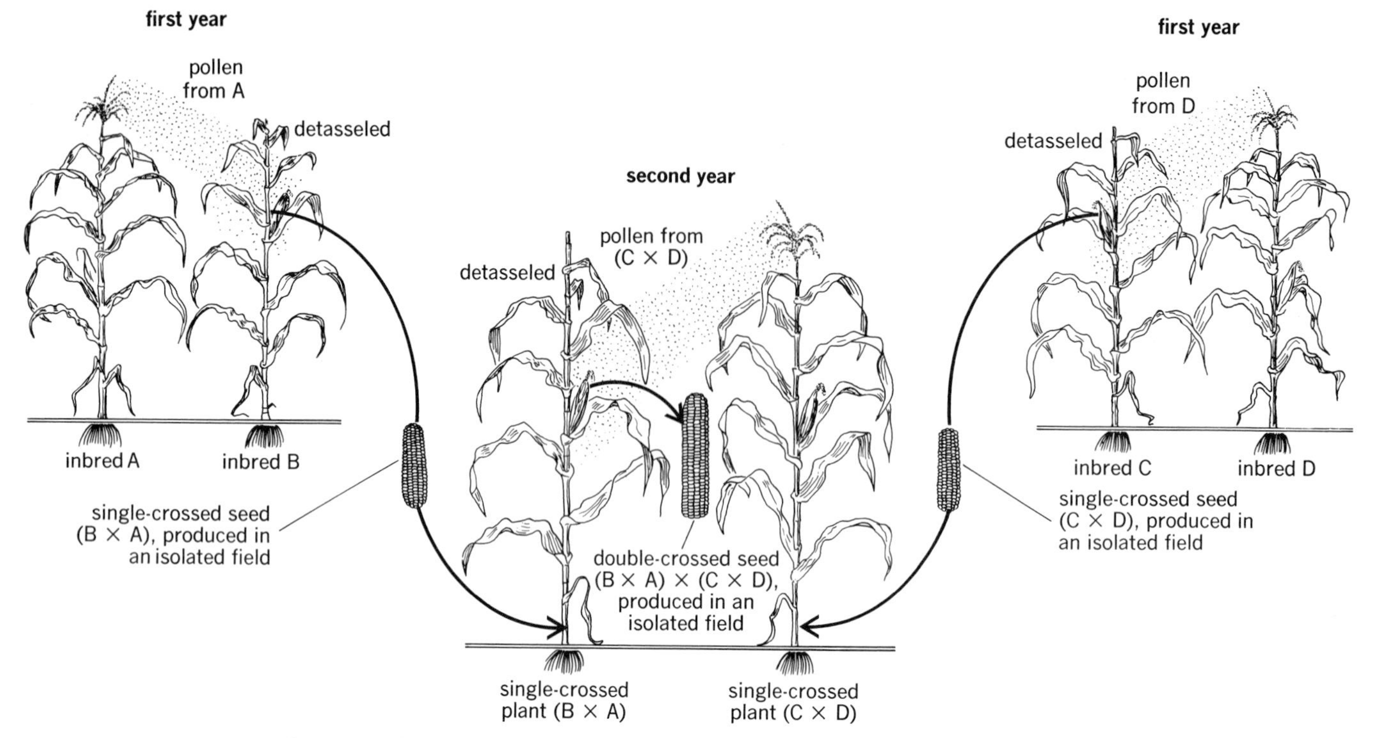

Sequence of steps in crossing inbred plants and using the resulting single-crossed seed to produce double-crossed hybrid seed. (*Crops Research Division, Agricultural Research Service, USDA*).

crossed, with the broadly satisfactory variety which is known as the recurrent parent. Among the members of the resulting first backcross (B_1) generation, selection is practiced in favor of those showing the useful trait of the nonrecurrent parent and these are again crossed with the recurrent parent. A series of six or more backcrosses will be necessary to restore the structure of the recurrent parent, which ideally should be modified only by the incorporation of the single useful attribute sought from the nonrecurrent parent. Backcrossing has been exceedingly useful in practice and has been extensively employed in adding resistance to diseases, such as rust, smut, or mildew, to established and acceptable varieties of oats, wheat, and barley. *See* PLANT DISEASE.

Cross-pollinating species. Natural populations of cross-pollinating species are characterized by extreme genetic diversity. No seed parent is true-breeding, first because it was itself derived from a fertilization in which genetically different parents participated, and second because of the genetic diversity of the pollen it will have received. In dealing with cultivated plants with this breeding structure, the essential concern in seed production is to employ systems in which hybrid vigor is exploited, the range of variation in the crop is diminished, and only parents likely to give rise to superior offspring are retained.

Inbred lines. Here plant breeders have made use either of inbreeding followed by hybridization (see illustration) or of some form of recurrent selection. During inbreeding programs normally cross-pollinated species, such as corn, are compelled to self-pollinate by artificial means. Inbreeding is continued for a number of generations until genetically pure, true-breeding, and uniform inbred lines are produced. During the production of the inbred lines rigorous selection is practiced for general vigor and yield and disease resistance, as well as for other important characteristics. In this way desirable attributes can be maintained in the inbred lines, which are nevertheless usually of poor vigor and somewhat infertile. Their usefulness lies in the vigor, high yield, uniformity, and agronomic merit of the hybrids produced by crossing different inbreds. Unfortunately, it is not possible from a mere inspection of inbred lines to predict the usefulness of the hybrids to which they can give rise. To estimate their value as the parents of hybrids, it is necessary to make tests of their combining ability. The test that is used depends upon the crop and on the ease with which controlled cross-pollination can be effected.

Tests may involve top crosses (inbred × variety), single crosses (inbred × inbred), or three-way crosses (inbred × single cross). Seeds produced from crosses of this kind must then be grown in carefully controlled field experiments designed to permit the statistical evaluation of the yields of a range of combinations in a range of agronomic environments like those normally encountered by the crop in agricultural use. From these tests it is possible to recognize which inbred lines are likely to be successful as parents in the development of seed stocks for commercial growing. The principal advantages from the exploitation of hybrids in crop production derive from the high yields produced by hybrid vigor, or heterosis, in certain species when particular parents are combined.

Economic considerations. The way in which inbred lines are used in seed production is dictated by the costs involved. Where the cost of producing F_1 hybrid seeds is high, as with many forage crops or with sugar beet, superior inbreds are combined into a synthetic strain which is propagated under conditions of open pollination. The commercial crop then contains a high frequency of superior hybrids in a population that has a similar level of variability to that of an open-pollinated variety. However, because of the selection practiced in the isolation and testing of the inbreds, the level of yield is higher because of the elimination of the less productive variants.

When the cost of seed is not of major significance relative to the value of the crop produced, and where uniformity is important, F_1 hybrids from a single cross between two inbred lines are grown. Cucumbers and sweet corn are handled in this way. By contrast, when the cost of the seeds is of greater significance relative to the value of the crop, the use of single-cross hybrids is too expensive and then double-cross hybrids (single cross A × single cross B) are used, as in field corn. As an alternative to this, triple-cross hybrids can be grown, as in marrow stem kale, in the production of which six different inbred lines are used. The commercial crop is grown from seeds resulting from hybridization between two different three-way crosses.

Recurrent selection. Breeding procedures designated as recurrent selection are coming into limited use with open-pollinated species. In theory, this method visualizes a controlled approach to homozygosity, with selection and evaluation in each cycle to permit the desired stepwise changes in gene frequency. Experimental evaluation of the procedure indicates that it has real possibilities. Four types of recurrent selection have been suggested: on the basis of phenotype, for general combining ability, for specific combining ability, and reciprocal selection. The methods are similar in the procedures involved, but vary in the type of tester parent chosen, and therefore in the efficiency with which different types of gene action (additive and nonadditive) are measured. A brief description is given for the reciprocal recurrent selection.

Two open-pollinated varieties or synthetics are chosen as source material, for example, A and B. Individual selected plants in source A are self-pollinated and at the same time outcrossed to a sample of B plants. The same procedure is repeated in source B, using A as the tester parent. The two series of testcrosses are evaluated in yield trials. The following year inbred seed of those A plants demonstrated to be superior on the basis of testcross performance are intercrossed to form a new composite, which might be designated A_1. Then B_1 population would be formed in a similar manner. The intercrossing of selected strains to produce A_1 approximately restores the original level of variability or heterozygosity, but permits the fixation of certain desirable gene combinations. The process, in theory, may be continued as long as genetic variability exists. In practice, the hybrid $A_n \times B_n$ may be used commercially at any stage of the process if it is equal to, or superior to, existing commercial hybrids.

Asexually propagated crops. A very few asexually propagated crop species are sexually sterile, like the banana, but the majority have some sexual fertility. The cultivated forms of such species are usually of widely mixed parentage, and when propagated by seed, following sexual reproduction, the offspring are very variable and rarely retain the beneficial combination of characters that contributes to the success of their parents. This applies to such species as the potato, to fruit trees like apples and pears that are propagated by grafting, and to raspberries, grapes, and pineapples.

Varieties of asexually propagated crops consist of large assemblages of genetically identical plants, and there are only two ways of introducing new and improved varieties. The first is by sexual reproduction and the second is by the isolation of sports or somatic mutations. The latter method has often been used successfully with decorative plants, such as chrysanthemum, and new forms of potato have occasionally arisen in this way. When sexual reproduction is used, hybrids are produced on a large scale between existing varieties with different desirable attributes in the hope of obtaining a derivative possessing the valuable characters of both parents. In some potato-breeding programs many thousands of hybrid seedlings are examined each year. The small number that have useful arrays of characters are propagated vegetatively until sufficient numbers can be planted to allow the agronomic evaluation of the potential new variety.

Special techniques. So far attention has been concentrated on the general principles of plant breeding and on methods of general applicability. However, there are a few special procedures that should be mentioned.

Cytoplasmic male sterility. In several crop species variants have been found that do not produce fertile pollen. However, pollen fertility can be restored in hybrids that result when the male sterile lines are pollinated by other, so-called restorer, lines. Using crosses between male sterile lines and restorer lines, F_1 hybrid seed supplies can easily be obtained on a large scale. Systems of this kind are used in the commercial production of hybrids in corn, sorghum, sugar beet, and onions. Unfortunately, hybrid corn varieties based on a single cytoplasm, causing male sterility, led to widespread susceptibility to the southern corn blight disease in the United States in 1971. Cytoplasmic male sterility has since been abandoned in the production of hybrid corn. However, the potentialities of using male sterility to produce hybrid wheat are still being explored.

Polyploids. Many crop species are naturally polyploid, and polyploidy can be induced artificially by colchicine treatment and in other ways. Polyploids are often characterized by a more sturdy growth habit and by larger roots, leaves, and flowers than the related diploids. Artificial polyploids are grown commercially in clover, watercress, sugar beet, and forage grasses. In watermelons crosses between normal diploids and artificial polyploids are used to produce hybrids that are sterile and so produce fruit without seeds—obviously a desirable attribute.

Multilinear varieties. Varieties of self-pollinating species like wheat generally consist of genetically identical plants. Multilinear varieties are also made up of genetically similar plants but contain several lines that differ in having genetically different forms of disease resistance. Each component line is produced by backcrossing different forms of resistance into a common recurrent parental variety. The advantage of the multilinear constitution is that, if the resistance of one of the constituent lines breaks down, production will be maintained by the remaining resistant lines.

Haploids. Normally, in the life cycle of higher plants, nuclei of the pollen grains and the egg cells have half the number of chromosomes (the haploid number) of the number in the nuclei of the plant on which they are produced (the diploid number). The diploid number is restored on fertilization by the fusion of egg and pollen haploid nuclei. Means have been discovered of producing essentially normal plants (sporophytes) of several crop species with the haploid chromosome number. The controlled production of such haploids, followed by doubling of the chromosome number using colchicine, enables the immediate fixation of the first products of segregation and recombination. Thus there is a very rapid return to complete homozygosity and homogeneity from hybrids of self-pollinated species. Alternatively, the procedure can give rise to the equivalents of inbred lines in outbreeding species. Haploidy may be induced by the culture on a nutrient medium of anthers or pollen grains (tobacco) by the elimination of the set of chromosomes of one parent in interspecific hybrids (barley and wheat) or by the parthenogenetic functioning of an egg which has not been fertilized by a pollen parent carrying a distinct and dominant genetic marker (corn and potato).

Cell biology. The understanding of methods by which plant cells can be cultured is now considerable. Moreover, unlike animal cells, plant cells in culture can be induced to regenerate into entire organisms. This has made possible the use of cell biological methodology in plant breeding. Two procedures are being explored. First, since very large numbers of cells can be cultured in a relatively small space, and inexpensively, many more potentially distinct genetic variants can be examined than would be the case if entire plants were used. Cells in culture can be exposed to an environment (say, a drug or an amino acid analog), and only those which are resistant will survive. The survivors can be multiplied and regenerated into entire plants after more rigorous selection than would otherwise have been possible.

The ability to regenerate entire plants from cells has also made it possible to contemplate the fusion of somatic plant cells. Such fusion can lead to the production of hybrids between species that are incapable of sexual hybridization. The first step in this process is the enzymatic degradation of the cellulose, hemicellulose, and pectin walls that surround all plant cells. This releases a naked protoplast; fusion of protoplasts is induced by chemical treatments, such as by polyethylene glycol. Following protoplast fusion, nuclear fusion will occur in some instances. Removal of the wall-degrading enzymes then permits regeneration of the cell wall and the establishment of a culture of hybrid cells which are subsequently induced to regenerate into complete plants. The value of these techniques is not yet fully visualized. The possibility of transfer of genetic material between species and genera is one possibility. It may provide the plant breeder with new reservoirs of genetic variation. Another possibility is in the improvement of asexually propagated crop plants.

[RALPH RILEY]

Bibliography: F. N. Briggs and P. F. Knowles, *Introduction to Plant Breeding*, 1967; K. J. Kasha (ed.), *Haploids in Higher Plants*, 1974; F. G. Maxwell and P. Jennings, *Breeding Plants Resistant to Insects*, 1979; A. Müntzing, *Genetics: Basic and Applied*, 1967; R. R. Nelson (ed.), *Breeding Plants for Disease Resistance*, 1973.

Climate modification

Twenty thousand years ago the places where many of the world's great cities now stand were deeply covered by ice, as were regions where the food for millions of people is now grown. The reasons why the ice receded are not known, but it did so with extraordinary speed, in a length of time comparable with that for which there exists written record of the doings of civilized people. Some of the ice remains. In Greenland and in the Antarctic there are land-borne masses of ice several thousand feet thick. If the Greenland ice alone were to melt with a speed comparable with that reached during the last retreat of the northern continental ice, the sea level would be raised by about 25 ft and many major seaports and the spaces where millions of people now live would be inundated. There is much more ice in the Antarctic, carrying the potential for sea-level changes of several hundred feet.

There is no reason to suggest that the human race would not survive either the return of the Quaternary ice caps or the melting of the ice caps existing today, but current social and political organizations would not survive. A change of climate could cause either catastrophe, and in recent years each has been put forward as a possible long-term consequence of man's modification of the atmosphere and the Earth's surface. There are, as shall be discussed later, other consequences of climatic change which, although less universal and less spectacular, are nonetheless undesirable.

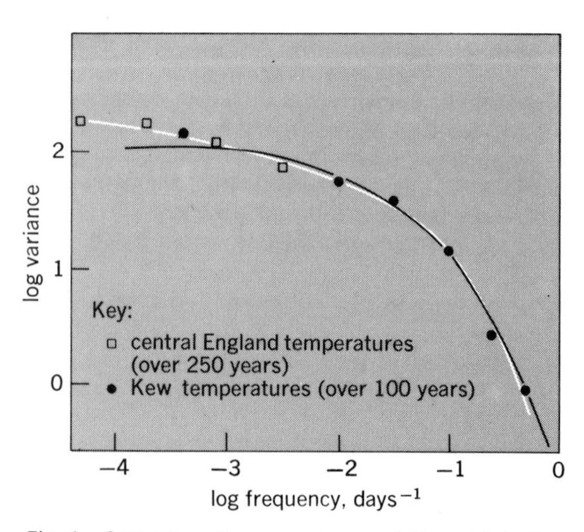

Fig. 1. Spectrum of temperature variations. Light line shows the relative variance associated with periodicities of 2 days to 40 years in the temperatures of southern and central England; due to J. M. Craddock. Dark line indicates the variance in a stationary series having the same statistics as the temperatures in the range of 2 days to 2 years. Diurnal and annual periodicities are removed.

"Climate" is not readily defined, except loosely as "weather" averaged over a period of time. Statistics of temperature and wind, rain, snow, cloud, sunshine, and humidity are the usual measures of the climate of an area. On the scale of the whole planet, the concept of climate is closely tied to the idea of the "general circulation of the atmosphere," a concept equally difficult to define but embodying the regularities and variations of the motion and temperature of the whole atmosphere. *See* CLIMATOLOGY.

CHANGE OF CLIMATE

One undoubted attribute of climate, local or global, is that it changes. This is evident from the records of measuring instruments, from historical texts, from prehistoric artifacts, and from geological evidence.

The longest instrumental record which has been appropriately analyzed is that of temperature in southern and central England. Change is clearly evident in this record, after it has been processed mathematically to remove diurnal and annual periodic variations. Figure 1 shows variance calculations for two series of monthly and daily mean temperature measurements (one series is for central England over 250 years, and the other is for Kew over 100 years). The two series are not stationary (a stationary curve is included for comparison), and the variance increases with observation time. Figure 2 is more direct evidence of change, from temperatures reported for 1870–1967 over extensive areas of the globe; data are due to J. M. Mitchell and R. Scherhag.

An example of historical evidence of climatic change (and one that clearly shows the effect of such change on human society) is the Norse settlement of southern Greenland about 1000 years ago. A pastoral community appears to have been well established, but it vanished with little trace as the land ice readvanced over the pastures and the sea ice cut communications with Iceland and Europe. It is estimated that when the settlement was established, temperatures in the area were 2–4°C higher than now.

Modern techniques of isotope analysis have produced a remarkable confirmation and extension of fragmentary records from various other sources. (When water evaporates or condenses there is temperature-dependent fractionation of the isotopic constituents. Thus the isotopic makeup of an ice sample is a clue to the temperature that prevailed when it was formed.) W. Dansgaard and others have examined the annual layers in the Greenland ice cap. Figure 3 shows their findings concerning the O^{16}/O^{18} ratio, which in turn relates to surface temperatures over the seas of the Northern Hemisphere. The long sequence of warm years ending about A.D. 1000 and the "little ice age" that followed show clearly. There is some indication of comparatively large recent fluctuations, which will be discussed when the possibility of human intervention is considered here.

Only one aspect of the fossil evidence of climatic change need be mentioned here. Interpretation of the fossil record is complicated by geographical changes accompanying the drift of the land masses, but it is accepted that for the great bulk of its lifetime the Earth has been free of land ice. The present anomalous situation appears to have been initiated about 5,000,000 years ago. This conclusion is the more remarkable since the Sun is regarded as a star in the main sequence whose output of radiation should have been appreciably lower than it is at present at epochs when geologists assert the planet to have been free of ice.

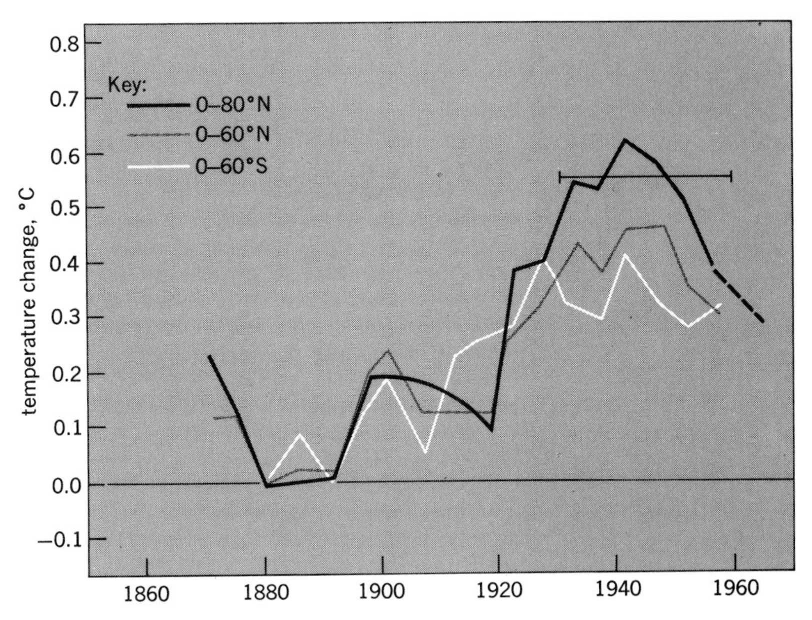

Fig. 2. Mean annual temperature for various latitudes for 1870–1967. Horizontal bar shows mean value for 0 to 80°N for 1931–1960. (*From Inadvertent Climate Modification, M.I.T. Press, 1971*)

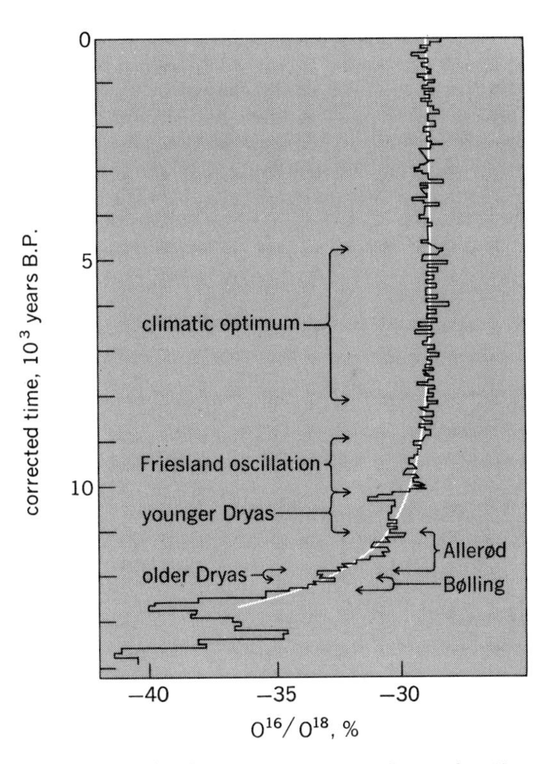

Fig. 3. The ratio of oxygen-16 to oxygen-18 as a function of depth in an ice core sample from the Greenland ice sheet. An increase in this ratio accompanies an increase in temperature at middle and high latitudes in the Northern Hemisphere. (*From Inadvertent Climate Modification, M.I.T. Press, 1971*)

THEORIES AND MODELS OF CLIMATE

The atmosphere is often compared to a heat engine. Solar energy falling on the planet is partially reflected to space; the remainder is absorbed, to some extent in the atmosphere but mainly at the irradiated surface. Air in contact with the ground or sea is heated and water is evaporated, and the heated moist buoyant air rises. Water condenses in clouds which, together with the Earth's surface and certain trace gases, notably water vapor and carbon dioxide, radiate energy to space. There is a net absorption of solar energy at low levels, low latitudes, and high temperatures and a net radiation to space at high levels, high latitudes, and low temperatures. Heat is transported between source and sink, the "boiler" and "condenser" of the atmospheric engine, by the moving air. Pressure gradients and winds are established. The Earth's rotation provides deflecting forces. Instabilities at various scales develop in the global patterns of motion—the traveling storms of the weather map. The work generated in the thermodynamic cycle is dissipated by friction in the working fluid of the engine.

Mathematical expression. Individually and in principle, the major processes of weather and climate are well understood and in this sense there exists a "theory" of climate. This theory is expressed mathematically in terms of the continuity equations for momentum, mass, and energy, the integral equation of radiative transfer, the equations governing the phase transitions of water substance, and the equation of state of dry air. There are as many equations as variables, so that the system appears soluble in principle.

Much effort was devoted to such calculations in the 1960s and 1970s, the incentive being prediction of weather more than understanding of climate. Systems ("models") have been developed for the numerical solution of equations for the time change of weather elements (wind, temperature, pressure, and other factors) at points covering a hemisphere and as little as 100 mi apart. Such models are now the major tool of the weather forecaster. However, they treat empirically—parameterize, in the jargon of the trade—processes which are important on small space scales and on long time scales. The long-time-scale processes are important in the investigation of climate, and in practice the integration of these processes into the models has proved so complex that a satisfactory theory or even numerical model of global climate is still far from being formalized. One cannot yet assess analytically or computationally the effect of any changes which humans might impose on the atmosphere or the Earth's surface. It may be possible to start the chain of reasoning, and identify the initial reactions of the atmospheric system, but it is not possible as yet to work analytically through the feedback loops to the final result.

Model types. Currently feasible quantitative investigation of climatic change is of two types. One type employs what have been termed "global average models," which are of varying degrees of complexity but all of which neglect the horizontal components of motion of the atmosphere. It is generally conceded that the global average models can do little more than indicate the direction of the initial reaction of certain variables (for example, surface temperature) to imposed change in others (such as carbon dioxide content of the atmosphere). The second type of investigation employs empirical models of the whole atmosphere, adding to the global average models certain empirically adjusted terms in an attempt to simulate the effects of atmospheric and oceanic motions. For example, some empirical relation must be assumed between the temperature difference between two latitudes and the heat transport between them, and between the temperature of a locality and the solar radiation absorbed there (the local albedo).

The pronouncements now current concerning man's impact on the global climate are based in the main on these empirical and global average models. But although they are based on empiricism and intuitive judgment to a degree unusual in the physical sciences, they are not just idle speculation. The problem they imply deserves and is receiving serious consideration; it is as difficult as any problem which has ever required the attention of applied scientists.

It is clear that it is not going to be a simple matter to detect human-produced changes in a system which changes significantly and, so far as we now understand it, capriciously. Climate has changed and will change, with no assistance from people, and it may be that they will influence its course and never know that they have done so.

CLIMATE AND ATMOSPHERIC POLLUTION

One must consider possible mechanisms by which human activities might influence the climate. People are changing the composition of the atmosphere and the nature of the Earth's surface. These changes affect the magnitude and distribution of the sources and sinks of heat and moisture in the thermodynamic system whose workings produce the climate.

Carbon dioxide pollution. The carbon dioxide (CO_2) content of the atmosphere has increased during the 20th century, and continues to increase at a rate correlated with the consumption of fossil fuels. Figure 4 shows seasonally smoothed measurements of CO_2 concentration at Mauna Loa Observatory, Hawaii, and the South Pole. These measurements are not entirely independent, being referred to a common standard gas mixture assumed not to change, but there is much other evidence to give them credence.

CO_2 reservoirs. The annual increase in total atmospheric content of CO_2 varies between about one-third and two-thirds of the yearly output from the burning of fossil fuels. The fate of the remainder is not understood in detail. There are two major potential reservoirs—the oceans and the biomass, that is, living and decaying organic matter. Plants normally respond to an increase of ambient CO_2 by an increased growth rate, but in spite of this, it is likely that at the present time the total biomass is being reduced by human activity, particularly by clearing of the tropical forests. Current estimates are that about 100,000 km^2 of forest is newly cleared each year; the resulting burning and decay contribute about 10% of the total CO_2 emitted by human activity. The current "permanent"

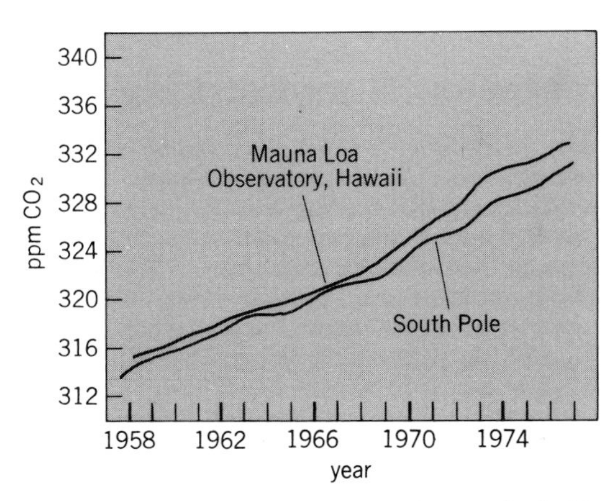

Fig. 4. Trends in the concentration of atmospheric CO_2 at Mauna Loa Observatory, Hawaii, and the South Pole; adjusted for seasonal changes. *(From C. D. Keeling and R. B. Bacastow, Impact of industrial gases on climate, in Energy and Climate: Studies in Geophysics, National Research Council, 1978)*

removal by fossilization is clearly negligible compared with destruction of the fossil store. Solution of CO_2 in the oceans—actually a complex of chemical reactions with dissolved salts—is not readily estimated, but its rate depends on near-surface temperatures among other variables. The long-term fixation of CO_2 in the ocean as carbonate rocks by biological processes and by reaction with silicate minerals is very slow, taking as long a time as major glacial and climatic changes.

Future changes of CO_2 content. When an attempt is made to predict the future atmospheric CO_2 content, the scientific uncertainties connected with the storage terms are at least matched by uncertainties in the future ecomonic and political situations which will govern the rate of fossil fuel consumption. Extrapolation of current trends seems reasonably reliable until about the year 2000, when the CO_2 concentration is expected to be between 375 and 400 parts per million by volume (ppmv). Further prediction is much more speculative, but assuming eventual maximum exploitation of fossil reserves, it is difficult to avoid the conclusion that a very high peak CO_2 concentration will be reached, estimates lying between eight times the present value about the year 2100 and six times the present value about 2300. Concentration of CO_2 is then expected to decrease slowly with a time constant of thousands of years. Increases of this magnitude could have very serious climatic consequences.

Temperature change. Figure 5 shows diagrammatically the radiative properties of atmospheric gases. It shows that CO_2 is only a weak absorber of solar radiation, so that an increase of CO_2 amount has little effect on the solar radiation reaching the ground. The gas does, however, have a major absorption band near the peak of the terrestrial spectrum, so that an increase in the atmospheric content would reduce the energy escaping from the surface of the planet to space, the net effect being an increase in temperature near the surface—the "greenhouse effect." *See* GREENHOUSE EFFECT.

Some decades ago G. S. Callendar made the first reasonably accurate quantitative estimates of the surface temperature increases to be expected from an increase of CO_2 content in an otherwise unchanged atmosphere. At that time, the late 1930s, there was also an obvious upward trend in surface temperatures, at least over much of the Northern Hemisphere and particularly the polar sea, with a marked recession of the sea ice. It was natural to associate the trends as cause and effect and to conclude that humans were polluting the air and changing the climate. It was natural that the proviso that no other change should accompany the increase of CO_2 should go relatively unnoticed by many interpreters of Callendar's work.

Water vapor content. There is no question that if nothing else changed, an increase in CO_2 concentration of the magnitude predicted for the end of the century would cause a general, climatically significant increase in near-surface temperatures, estimated at approximately 0.5°C averaged over the globe. But there is also no question that something else would change. At this time the nature and magnitude of this change cannot be decided,

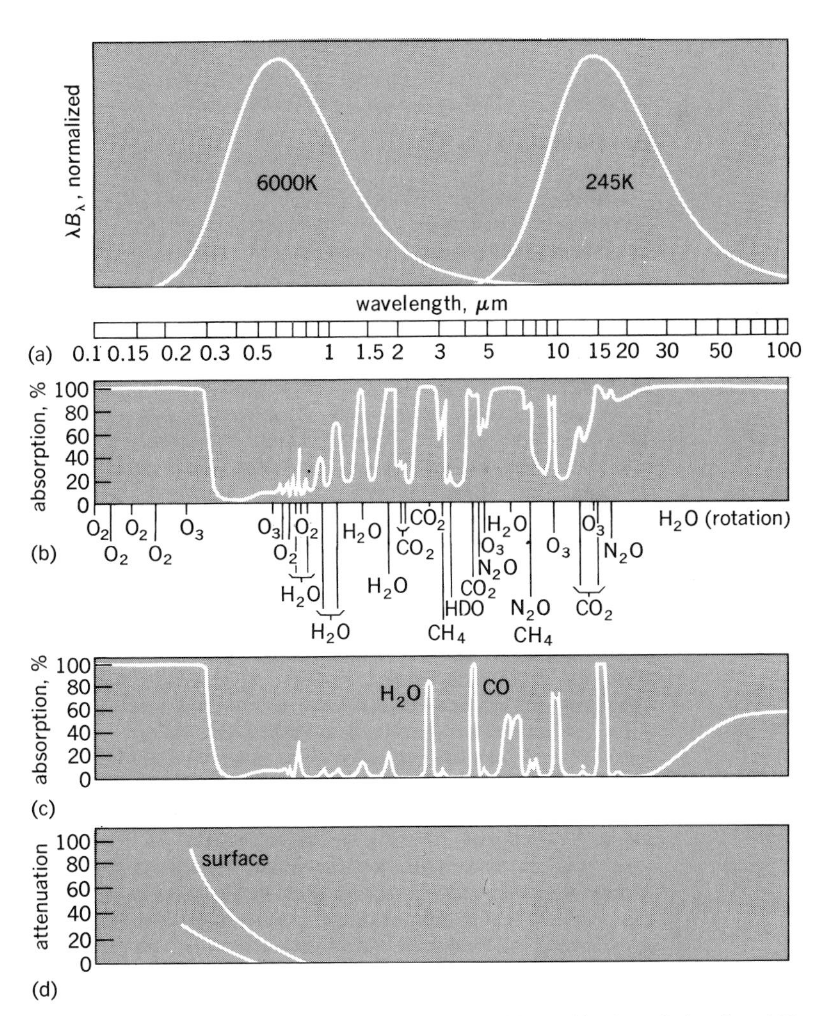

Fig. 5. Radiative properties of atmospheric gases. (*a*) Blackbody emission for 6000 and 245K, being approximate emission spectra of the Sun and Earth, respectively (since inward and outward radiation must balance, the curves have been drawn with equal areas—though in fact 30% of solar radiation is reflected unchanged). (*b*) Atmospheric absorption spectrum for a solar beam reaching the ground. (*c*) Atmospheric absorption spectrum for a beam reaching the tropopause in temperate latitudes. (*d*) Attenuation of the solar beam by Rayleigh scattering, at the ground and at the temperate tropopause. *(From R. M. Goody and G. D. Robinson, Quart. J. Roy. Meteorol. Soc., vol. 77, 1951)*

but there are certain obvious possibilities. The water content of the atmosphere would increase because of increased evaporation from the oceans. If this were a change in water vapor content alone, two things would happen. First, more solar radiation would be absorbed before it reached the ground, tending to reduce the buildup in near-surface temperatures. Second, the greenhouse effect would be enhanced, the increased H_2O content behaving in the same way as the increased CO_2 content. This would tend to augment the increase in near-surface temperatures. Examination of the net effect of these opposing trends has proved a tractable problem: There can be no doubt that the net effect of an increase of water vapor content (alone) would be to reinforce the tendency for an increase in surface temperature. This is a destabilizing condition, a positive feedback. But it is unlikely that the water vapor content of the atmosphere would increase without a corresponding increase in cloud, either in amount or depth, since an increase in evaporation implies a corresponding increase in precipitation. An increase of cloud amount would mean increased reflection of solar radiation to space and a decrease in the average temperature of the planet. But cloud is also opaque to the radiation from the Earth's surface. Increased cloudiness would enhance the greenhouse effect and tend to increase surface temperature while decreasing the temperature of the upper atmosphere. The planet Venus is completely covered by highly reflecting cloud, but has a surface temperature high enough to melt sulfur. The actual climatic outcome of the predicted CO_2 increase would be the result of a balance between opposing tendencies, and the computational resources presently available are not adequate to resolve the problem. It can be stated confidently, however, that the initial change would be in the direction of increased surface temperatures, and it is difficult to imagine how the very large predicted changes in CO_2 content could be accommodated without very significant climate change.

Estimates of particles smaller than 20-μm radius emitted into, or formed in, the atmosphere

Source	Amount, 10^6 metric tons/year
Natural	
Soil and rock debris*	100–500
Forest fires and slash-burning debris*	3–150
Sea salt	300
Volcanic debris	25–150
Gaseous emissions	
Sulfate from H_2S	130–200
Ammonium salts from NH_3	80–270
Nitrate from NO_2	60–430
Hydrocarbons from plant exudations	75–200
Subtotal	773–2200
Man-made	
Direct emissions	10–90
Gaseous emissions	
Sulfate from SO_2	130–200
Nitrate from NO_2	30–35
Hydrocarbons	15–90
Subtotal	185–415
Total	958–2615

*Includes unknown amounts of indirect contributions from human activities.

SOURCE: *Inadvertent Climate Modification*, M.I.T. Press, 1971.

Atmospheric particles. As Fig. 2 shows, the steady rise in temperature which was a feature of world weather early this century appears to have ended by about 1940. The arrangements for collecting and processing of world weather records at that time were slow and were disrupted by World War II, so that it was not until the mid-1950s that the reversal of the temperature trend was confirmed.

Those who had attributed the earlier temperature increase to increasing atmospheric CO_2 content sought a reason, and found one possibility in a postulated increase in the particle content of the atmosphere associated with the rapid increase in human populations. The argument was that atmospheric particles scatter solar radiation to space: More particles would scatter more radiation, less solar energy would be absorbed by the planet (the planetary albedo would increase), and the mean planetary temperature would fall. A probable, but not inevitable, consequence would be a drop in surface temperature. These speculations called for three lines of investigation. Is the particle load of the atmosphere increasing? Are the optical properties of the particulate products of combustion such that an increase in their number would produce an increase in the Earth's albedo? Even if the albedo were increased, would the particle layer interfere with radiation of energy from the surface sufficiently to produce a greenhouse effect. *See* AIR POLLUTION.

Particle load. The first question has proved surprisingly difficult to answer. There is no doubt that, at least until a few years ago particle concentration in the air near major cities was increasing as population and the energy conversion per head of population increased. Air pollution control technology and legislation appear to have abated the rate of increase, but reversals of the trend of particle production can be expected only in areas previously subject to intensive industrial pollution. Changes in visibility statistics provide evidence of recent increases in the particle content of air in rural districts of eastern North America, and in surface air over the northern Atlantic there is some evidence of an increase in number of very small particles of a kind produced by combustion. But similar measurements show no such increase over the central Pacific, and there is no clear evidence of an increase on the global scale of the number of particles in the atmosphere.

Furthermore, researchers have gradually realized that the proportion of the atmospheric particle load attributable to human activities is small. The extreme limits of published estimates of this proportion are 5 and 45%; most estimates are less than 10%. It is also clear that industrially produced particles are only a fraction of the total of human-produced particles. Agricultural malpractice appears to be responsible for many more particles than industry. In fact, much of the uncertainty in estimating the human contribution to the particle load stems from the near-impossibility of distinguishing between natural and human-induced forest fires and natural and human-induced wind erosion. The table summarizes what we know about the source of the particle load of the atmosphere.

Optical properties. Whether or not the present level of human activity has an appreciable impact on the particle load, it is conceivable that future developments may have an impact. The optical properties of particles, therefore, should be investigated. For example, the argument that since particles scatter sunlight, they must increase the global albedo is not necessarily true. The surface below the particles plays a key role. White particles over a black surface clearly increase the albedo, but black particles over a white surface decrease it. The effects of brown over brown, or gray over gray, or brown over green—the real-life cases—are not so easy to predict. Similarly, the exact solution for real particles of mixed composition and and irregular shape over real inhomogeneous surfaces with topographical irregularities is not possible.

Various approximations are possible, however. Figure 6 shows some of the results of an investigation using a relatively simple approach. In this figure, a particle is characterized by its absorption and backscatter coefficients (a numerical measure of the proportion of incident radiation scattered upward to space). If the characteristic point for a particle is above and to the right of the surface albedo line, the system albedo is increased. The surface albedo lines in Fig. 6 correspond to total solar radiation and new snow or thick cloud (0.75) and thickly settled areas with mixed agriculture and industry (0.15). The diagram illustrates both the paucity of data and the probability that the atmospheric particle load has little direct effect on the Earth's albedo.

It is just possible that hygroscopic particles originating in combustion may have an indirect effect on albedo by modifying the nature of cloud. A cloud forming in air polluted by combustion nuclei tends to have a relatively large number of relatively small particles, whereas a cloud forming in cleaner air but in otherwise similar circumstances concentrates the same amount of water on fewer nuclei. The "polluted" cloud is the more efficient light scatterer with the higher albedo. Considering all possibilities and the few available measurements, it seems likely that an increase in atmospheric particle load would lead to an increased albedo, but that the effect might not be large on a global scale. Confirmation by measurement is certainly required.

Ozone layer. Reference to Fig. 5 shows that solar radiation in the far ultraviolet, at wavelengths shorter than about 0.3 μm, does not reach the Earth's surface in significant amount in energy terms, that the cause is absorption by the gas ozone (O_3), and that the absorption occurs predominantly in the stratosphere, that is, above a height of about 10 km in temperate latitudes and about 18 km in the tropics. The O_3 is known to be formed at greater heights by a complex process initiated by dissociation of the oxygen molecule by solar radiation of even shorter wavelength. It became clear about 1970 that certain oxides of nitrogen (NO, NO_2), of which there is a natural source in the high atmosphere, play a key role in limiting the amount of ozone produced. These gases are also produced by a variety of human activities, mainly those which involve high-temperature combustion, but there are removal mechanisms which appear to be quite efficient when the emission is near the Earth's surface. If, however, the emission is very high in the atmosphere, as from the engines of supersonic and the most modern subsonic aircraft, it is possible that this artificial source might augment the natural source of nitrogen oxides and deplete the natural ozone content of the stratosphere. There has been much research, but not yet any firm conclusion. The current opinion is that operation of any commercial aircraft now flying or under construction would not detectably change the natural ozone layer, but that large fleets of technically feasible supersonic aircraft might do so. Plans to construct this type of aircraft were postponed some years ago, and have not yet been revived.

In the course of research on the aircraft emission problem, it became clear that oxides of chlorine, of which there is no demonstrated natural source, would be much more efficient than oxides of nitrogen in depleting the natural ozone layer, and that they could be produced by dissociation of the chlorofluorocarbon compounds (freons) used as refrigerants and as propellants in aerosol sprays. Chlorine oxides have since been detected in the high stratosphere at about the expected concentration, and the release of freons has been limited by regulation.

The major concern about depletion of the ozone layer was not one traditionally associated with the concept climate. Ultraviolet light of wavelength about 0.3 μm is absorbed by the genetic material of living cells and can modify it: the long-term effects of even a very small systematic increase of this radiation are incalculable. The immediate

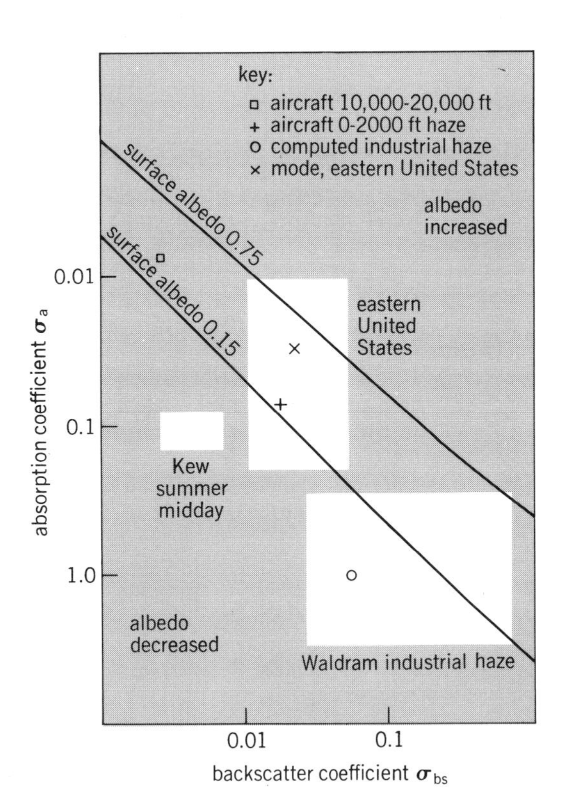

Fig. 6. Effect of atmospheric particles on the global albedo. Particles are characterized by their absorption and backscattering coefficients. They reduce or increase the global albedo according to the position of their characteristic point relative to the appropriate surface albedo line.

cause of concern was an apparent correlation between ultraviolet irradiance and the incidence of certain skin cancers. However, in addition to this biological effect, even small changes in radiative properties of the atmosphere must have some repercussion on the traditional climatic elements. Attempts to quantify these effects for the nitrogen oxide emissions suggest that they would be much smaller than those associated with CO_2 emissions. It was, however, found that the freons strongly absorb radiation of wavelengths about 10 μm, and as greenhouse effect materials are very much more efficient than CO_2. If their emission had continued at the rate of the early 1970s, they would have significantly augmented the effect of CO_2.

CLIMATE AND SURFACE CHANGES

There is evidence that, within the present century, industrial activities have begun detectably to affect significant properties of the atmosphere. The effect of people's search for, and production of, food is much more evident, and has been clear throughout his history. The significant properties are the albedo, the ability to retain and release water, and the associated ability to store heat. Together these properties dictate the amount of energy returned to the atmosphere and the partition of energy between latent heat, which can be transported and released elsewhere, and sensible heat. Major differences in albedo exist between vegetated, bare, and ice- or snow-covered surfaces. The type of vegetation, except when associated with snow cover, has a relatively small effect on albedo, but a greater effect on water retention, particularly in the case of cultivated and harvested crops. People have changed the type of vegetation over considerable areas of the globe, clearly without globally catastrophic consequences (though humans have been responsible for local catastrophes). But there is now in sight an increase of population in the tropics of such magnitude that climate changes following surface changes connected with tropical agriculture seem at least as likely to occur as any associated with industrial development.

Agriculture. People are continually changing the nature of the Earth's surface, and there is little doubt that they have already changed the climate of certain regions—for the worse from their own point of view. These known deteriorations have been the consequence of destruction of forests, mainly to create pastures and cultivate food crops but often to exploit the timber crop. The surface that replaces the forest has a higher albedo and less capacity for retaining water and returning it locally to the atmosphere by evaporation.

Overuse of cleared land makes it unsuitable for further agricultural use and accelerates the spread of cleared areas. Particularly in low-rainfall areas, "dust-bowl" conditions can develop. The effect of the dust is to stabilize the lower atmosphere and reduce the incidence of convective rainfall. This process is believed to have increased the aridity of much of the area from the southern Mediterranean shores to Afghanistan. It appears to be operative today in semiarid areas of India and Pakistan. There, however, it has been demonstrated that, at least locally, the desert-producing process is not yet irreversible and that considerably increased vegetative cover quickly follows the exclusion of grazing animals.

The situation is somewhat different in areas where the tropical rainforest is under attack by people. Here the principal destructive process is that of "slash and burn" agriculture—relatively small areas of jungle are cut, the vegetation burnt, and the clearing exploited for rather primitive agriculture. When in a few years the soil becomes uncultivable and infertile, the exploiters move on. The forest does not recover, since the soil has become as unsuitable for forest growth as for agriculture. The surface albedo has changed and the water-retention properties have changed, so that the nature and amount of cloud cover might be expected to change. Substantial change in the albedo and evaporation from the tropical land areas must be expected to affect the world's climate. As a further complication, the fires of the "slash and burn" cycle may be a major source of the human-produced particle load of the atmosphere, comparable to the direct emissions from industrial sources.

Sea ice. Arctic sea ice has assumed a prominent role in much recent speculation on climatic change because most of the simplified empirical models of climate suggest that the extent of sea ice is extremely sensitive to changes in global temperature. The underlying cause of this sensitivity is the high albedo of ice. In the formulation of the models, an increase in sea-ice cover increases the global albedo, with a consequent further reduction in global temperature and increase in ice. Decrease in sea ice decreases the albedo, increases the solar energy absorbed, and accelerates the decrease in ice cover. There is a "positive feedback" instability with no constraint in the model formulation to counteract it. However, the historical record suggests that some constraints which are not modeled do exist—there have been quite considerable trends in the area of the ice, all ultimately reversed—but the balance of instability and constraint may be delicate. People may disturb this balance by initiating changes of unprecedented magnitude or speed.

Apart from the possible global temperature changes resulting from unintentional modification of the atmosphere, there have been some projects proposed which could directly or indirectly change the area of the ice in a calculated effort to modify the climate and navigability of arctic waters. The potential dangers of such schemes are now realized.

It is generally conceded that implementation of any such projects must await a better understanding of their less immediate consequences. However, there is still occasional discussion of projects whose less immediate consequences could be a major disturbance of the arctic sea ice. One such is diversion of the waters of rivers draining into the Arctic Basin—the Ob, Yenisei, and Mackenzie. Such diversion would change the salinity of arctic waters, a change which would in turn affect the thickness and extent of the sea ice and set in train a sequence of climatic reactions perhaps trivial, perhaps catastrophic, but certainly—with present resources and knowledge—incalculable. Whatever the value to the human race of availability of the Mackenzie basin waters in Southern California, and it is debatable, the theory of climatic change in its present state suggests that the interests of more than the inhabitants of the two watersheds might be affected.

Heat release. Direct release of heat by human activities is very small indeed compared with that supplied by the Sun. The United Nations estimate of world power consumption in 1967 was about 5×10^9 kW, against an average solar supply (absorbed power) of more than 10^{14} kW. There is little prospect of humans' changing the global climate by directly heating the environment until their energy consumption attains more than 100 times its present level. There are those who believe that this level will be reached in the future, but at least we have a little time to consider the consequences.

Nevertheless, this direct heating—often accomplished by release of hot water or water vapor—is clearly changing the local climate of some cities. It is conceivable that the future growth and aggregation of cities in certain areas might produce a heat source strategically placed to trigger major climatic change. One such location is the northeastern coast of the United States, where in winter there is a strong land-to-sea temperature gradient which breeds and steers traveling storms and which might be modified by intense energy conversion in the coastal strip.

SUMMARY

People have changed the climate of cities. Cities are warmer than their surroundings, often by up to 10°C at night and in calm winter weather, and 1 or 2°C at other times. The amount of solar radiation reaching the ground in cities is lower than that in surrounding areas, sometimes by as much as 10%. There is a little, not yet conclusive, evidence of an increase, or at least a redistribution, of summer rainfall near a few large cities. Subarctic cities have their own near-permanent blanket of ice fog in winter.

While no evidence of human-induced change in global climate has been found, scientists are unable to be sure that it has not occurred, and have reason to believe that it could occur in the future. Humans' energy conversion is at present negligible in comparison with the supply of solar energy, and any influence people may have on global climate must be by way of the triggering of instabilities. Two such "positive-feedback" mechanisms have been identified—one the greenhouse effect, a surface temperature increase, initiated by added carbon dioxide and accelerated by water vapor added as the surface temperature increases; the second, a change in ice or snow cover tending to change the amount of solar radiation absorbed by the Earth in the direction calculated to augment the rate of change of ice cover. During the first 30–40 years of this century, the CO_2 content of the atmosphere increased. The temperature also increased and the ice cover diminished. These tendencies have been reversed, in spite of the continuing increase in CO_2.

Scientists, therefore, seek a constraining or negative-feedback mechanism. The possible effect of atmospheric particles added by people has not yet been confirmed—neither the fact of a significant increase nor the direction of the temperature changes which would follow it has been established. There is a possible "natural" constraint in the increase of condensed water expected in the atmosphere if surface temperatures rise and evaporation increases. This extra cloud would increase the Earth's albedo, with a corresponding tendency to reduced temperatures, but it would also augment the greenhouse effect of CO_2 and water vapor, and current climate models cannot confidently decide whether the cloud change would be a constraint or another instability.

There is no doubt that in the natural state of the Earth-biosphere-atmosphere system, constraints exist as well as instabilities. The Earth has never been completely covered by ice; instead ice has formed and spread within limits on a previously ice-free globe. The composition of the atmosphere has varied considerably and enormous amounts of CO_2 have passed through it, to be fixed temporarily in fossil fuels and permanently in carbonate rocks; yet surface conditions have always permitted life, once it appeared.

But although scientists may conclude that there are constraints on climatic instabilities, little is known about the time scale on which they may operate. It is possible that a man-induced perturbation may operate so rapidly that the unknown constraint is unable to counteract it. Humans should not be content with the argument that since life appears to have been cared for, or to have cared for itself, on Earth for 1,000,000,000 years, it is likely to continue to do so. The life of the future may not be the kind of life existing now. The situation should be watched carefully, and until there is a better understanding of the workings of the environment, excessive interference with it by people should be avoided. At the same time it should be realized that the environment cannot be preserved entirely without change. No living thing has ever done that.

[G. D. ROBINSON]

Climatic change

The long-term fluctuations in precipitation, temperature, wind, and all other aspects of the Earth's climate. The Earth's climate, like the Earth itself, has a history extending over several billion years. Climatic fluctuations have occurred at time scales ranging from the longest observable (10^8–10^9 years) to interdecadal variability (10^1 years) and interannual variability (10^0 years). Processes in the atmosphere, oceans, cryosphere (snow cover, sea ice, continental ice sheets), biosphere, and lithosphere, and certain extraterrestrial factors (such as the Sun), are part of the climate system.

The present climate can be described as an ice age climate, since large land surfaces are covered with ice sheets (Antarctica, Greenland). The origins of the present ice age may be traced, at least in part, to movement of the continental plates. With the gradual movement of Antarctica toward its present isolated polar position, ice sheets began to develop about 30,000,000 years ago. Within the past several million years, the Antarctic ice sheet reached approximately its present size, and ice sheets appeared on the lands bordering the northern North Atlantic Ocean. During the past million years of the current ice age, about 10 glacial-interglacial fluctuations have been documented. The most recent glacial period came to an end between about 15,000 and 6000 years ago with the rapid melting of the North American and European ice sheets and the associated rise in sea level. The present climate is described as interglacial. The scope of this article is limited to a discussion of climatic fluctuations within the present interglacial period and, in particular, the climatic fluctuations of the past 100 years—the period of instrumental records. *See the feature article* PRECEDENTS FOR WEATHER EXTREMES.

Evidence. Instrumental records of climatic variables such as temperature and precipitation exist for the past 100 years in many locations and for as long as 200 years in a few locations. These records provide evidence of year-to-year and decade-to-decade variability but are completely inadequate for the study of century-to-century and longer-term variability. Even for the study of short-term climatic fluctuations, instrumental records are of limited usefulness, since most observations were made from the continents (only 29% of the Earth's surface area) and limited to the Earth's surface. Aerological observations which permit the study of atmospheric mass, momentum and energy budgets, and the statistical structure of the large-scale circulation are available for only about 20 years. Again, there is a bias toward observations over the continents. It is only with the advent of satellites that global monitoring of the components of the Earth's radiation budget (planetary albedo, from which the net incoming solar radiation can be estimated; and the outgoing terrestrial radiation) has begun. *See* TERRESTRIAL ATMOSPHERIC HEAT BALANCE.

There remain important gaps in the ability to describe the present state of the climate. For example, precipitation estimates, especially over the oceans, are very poor. Oceanic circulation, heat transport, and heat storage are only crudely estimated. Also, the solar irradiance is not being monitored to sufficient accuracy to permit estimation of any variability and evaluation of the possible effect of fluctuations in solar output upon the Earth's climate. Thus, although climatic fluctuations appear in instrumental records, defining the scope of these fluctuations and diagnosing potential causes are at best difficult and at worst impossible.

In spite of the inadequacy of the instrumental records for assessing global climate, there is considerable evidence of regional climatic variations. For example, there is evidence of climatic warming in the polar regions of the Northern Hemisphere during the first 4 to 5 decades of the 20th century. During the 1960s, on the other hand, there is evidence of cooling in the polar and mid-latitude regions of the Northern Hemisphere; and, especially in the early 1970s, there were drier conditions along the northern margin of the monsoon lands of Africa and Asia.

Under the auspices of the World Meteorological Organization and the International Council of Scientific Unions, the Global Atmospheric Research Program (GARP) is developing plans for detailed observation and study of the global climate system—especially the atmosphere, the oceans, the sea ice, and the changeable features of the land surface.

Causes. Many extraterrestrial processes and terrestrial processes have been hypothesized to be possible causes of climatic fluctuations. A number of these processes are listed and described below.

Solar irradiance. It is possible that variations in total solar irradiance could occur over a wide range of time scales (10^0-10^9 years). If these variations did take place, they would almost certainly have an influence on climate. Radiance variability in limited portions of the solar spectrum has been observed, but not linked clearly to climate variability. *See* SOLAR ENERGY.

Orbital parameters. Variations of the Earth's orbital parameters (eccentricity of orbit about the Sun, precession, and inclination of the rotational axis to the orbital plane) lead to small but possibly significant variations in incoming solar radiation with regard to seasonal partitioning and latitudinal distribution. These variations occur at time scales of 10^4-10^5 years.

Lithosphere motions. Sea-floor spreading and continental drift, continental uplift, and mountain building operate over long time scales (10^5-10^9 years) and are almost certainly important in long-in long-term climate variation.

Volcanic activity. Volcanic activity produces gaseous and particulate emissions which lead to the formation of persistent stratospheric aerosol layers. It may be a factor in climatic variations at all time scales.

Internal variability of climate system. Components of the climate system (atmosphere, ocean, cryosphere, biosphere, land surface) are interrelated through a variety of feedback processes operating over time scales from, say, 10^0 to 10^9 years. These processes could, in principle, produce fluctuations of sufficient magnitude and variability to explain any observed climate change. For example, atmosphere-ocean interactions may operate over time scales ranging from 10^0 to 10^3 years, and atmosphere-ocean-cryosphere interactions may operate over time scales ranging from 10^0 to 10^5 years. Several hypotheses have been proposed to explain glacial-interglacial fluctuations as complex internal feedbacks among atmosphere, ocean, and cryosphere. (Periodic buildup and surges of the Antarctic ice sheet and periodic fluctuations in sea ice extent and deep ocean circulation provide examples.) Atmosphere-ocean interaction is being studied intensively as a possible cause of short-term climatic variations. It has been observed that anomalous ocean temperature patterns (both equatorial and mid-latitude) are often associated with anomalous atmospheric circulation patterns. Although atmospheric circulation plays a dominant role in establishing a particular ocean temperature pattern (by means of changes in wind-driven currents, upwelling, radiation exchange, evaporation, and so on), the anomalous ocean temperature distribution may then persist for months, seasons, or longer intervals of time because of the large heat capacity of the oceans. These anomalous oceanic heat sources and sinks may, in turn, produce anomalous atmospheric motions. *See* OCEAN-ATMOSPHERE RELATIONS.

Human activities. Forest clearing and other large-scale changes in land use, changes in aerosol loading, and the changing CO_2 concentration of the atmosphere are often cited as examples of possible mechanisms through which human activities may influence the large-scale climate. Because of the large observational uncertainties in defining the state of the climate, it has not been possible to establish the relative importance of human activities (as compared to natural processes) in recent climatic fluctuations. There is, however, concern that future human activities may lead to large climatic variations (for example, continued increase in atmospheric CO_2 concentration due to burning of fossil fuels) within the next several decades. *See* AIR POLLUTION; CLIMATE MODIFICATION.

It is likely that at least several of the above-mentioned processes have played a role in past climatic fluctuations (that is, it is unlikely that all climatic fluctuations are due to one factor). In addition, certain processes may act simultaneously, or in various sequences. Also, the climatic response to some causal process may depend on the particular initial climatic state, which, in turn, depends upon previous climatic states because of the long time constants of oceans and cryosphere. True equilibrium climates may not exist, and the climate system may be in a continual state of transience.

Modeling. Because of the complexity of the real climate system, simplified numerical models of climate are being used to study particular processes and interactions. Some models treat only the global-average conditions, whereas others, particularly the atmospheric models, simulate detailed patterns of climate. These models are still in early stages of development but will undoubtedly be of great importance in attempts to understand climatic processes and to assess the possible effects of human activities on climate. *See* ATMOSPHERIC GENERAL CIRCULATION; CLIMATOLOGY.

[JOHN E. KUTZBACH]

Ocean-atmosphere interaction. The atmosphere and the oceans have always jointly participated in climatic change, past and contemporary. Some of the contemporary changes can be investigated in the modern records of climatic anomalies in the atmosphere and the oceans.

The most important source of climatic change surpassing 1-year duration seems to be located along the equatorial zone of the Pacific Ocean. The prevailing winds there are easterly and maintain a westward wind drift of the surface water which diverges, under influence of the Earth's rotation, to the right of the wind direction north of the Equator and to the left south of the Equator. The resulting equatorial upwelling of cold water, and subsequent lateral mixing, ordinarily maintains a belt of cold surface water several hundred kilometers wide straddling the Equator from the coast of South America about to the dateline, about one earth quadrant farther to the west.

Analogous processes are found in the equatorial belt of the Atlantic, but the upwelling water there covers a much smaller area and is also less cold than in the Pacific. The Indian Ocean has no steady easterlies and thus no equatorial upwelling.

The equatorial upwelling process varies in intensity with the equatorial easterlies. Since that wind system is mostly fed by way of the southerlies along the west coasts of South Africa and South America, it is likely that anomalies in the Southern Hemisphere atmospheric circulation frequently are transmitted to the tropical belt. Once an impulse, for example, a strengthening of the Pacific equatorial easterlies, has occurred, the new anomaly has a built-in tendency of self-amplification, because it makes the upwelling strengthen too and thus increases the temperature deficit of the Pacific compared to the persistently warm Indonesian and Indian Ocean tropical waters. This in turn feeds back into further strengthening of the easterlies which started the anomaly in the first place.

The observational proof of this feedback system can be seen in the statistically well-substantiated "southern oscillation," which exhibits opposite contemporaneous anomalies of atmospheric pressure over the tropical parts of the Pacific Ocean on the one side and the Indonesian and Indian Ocean tropical waters on the other (nodal line on the average at 165°E). The periodicity is rather irregular, so the term oscillation should not be taken too literally. The average length of the cycles is 2–3 years.

The cycles of tropical precipitation of more than a year's length by and large agree with those of pressure wherever special local conditions do not interfere. Satellite photos confirm that in the phase of the southern oscillation with positive pressure anomaly over the Pacific, along with strong equatorial easterlies and strong upwelling, most of the Pacific equatorial belt is arid; while in the opposite phase the western and central part of that belt experiences heavy rainfall. In extreme "El Niño" years this rainfall can also extend to the normally arid coast of northern Peru.

When there is more than normal rainfall at the Equator, the general circulation of the atmosphere is supplied with more-than-normal total heat convertible into kinetic energy. The remote effect of this phenomenon, particularly in the winter hemisphere, is the occurrence of stronger-than-normal tradewinds and midlatitude westerlies. At the opposite extreme, the tradewinds are weak and the westerlies meandering. This produces cold winters in the longitude sectors with winds out of high latitudes and mild winters interspersed at longitudes where wind components from low latitudes prevail. Again, it is in the Pacific longitude sector that these teleconnections are most clearly seen, because the interannual variability of sea temperature up to a range of 3°C over a large equatorial area is found only in the Pacific.

[JACOB BJERKNES]

Bibliography: H. P. Berlage, *The Southern Oscillation and World Weather*, Kon. Ned. Meterol. Inst. Meded. Verh. no. 88, 1966; J. Bjerknes, A possible response of the Hadley circulation to variations of the heat supply from the equatorial Pacific, *Tellus*, 18:820–829, 1966; R. A. Bryson and T. J. Murray, *Climates of Hunger: Mankind and the World's Changing Weather*, 1977; M. I. Budyko, *Climate Changes*, 1977; R. W. Fairbridge (ed.), *Solar Variations, Climatic Change, and Related Geophysical Problems*, Ann. N.Y. Acad. Sci. no. 95, 1961; H. H. Lamb, *Climate: Present, Past and Future*, vol. 1: *Fundamentals and Climate Now*, 1972; *Long-Term Climatic Fluctuations: Proceedings of the WMO-IAMAP Symposium*, 1975; S. H. Schneider and R. E. Dickinson, Climate modelling, *Rev. Geophys. Space Phys.*, 12:447–493, 1974; H. Shapley (ed.), *Climatic Change*, 1953; Study of man's impact on climate, in *Inadvertent Climate Modification*, 1971; C. Tickell, *Climatic Change and World Affairs*, 1977; U.S. National Academy of Sciences, *Understanding Climatic Change: A Program for Action*, 1975; G. Walker, *World Weather VI*, Mem. Roy. Meteorol. Soc., vol. 4, no. 39, 1937.

Climatology

That branch of meteorology concerned with climate, that is, with the mean physical state of the atmosphere together with its statistical variations in both space and time, as reflected in the totality of weather behavior over a period of many years. Climatology encompasses not only the description of climate but also the physical origins and the wide-ranging practical consequences of climate and of climatic change. Thus it impinges on a wide range of other sciences, including solar system astronomy, oceanography, geography, geology and geophysics, biology and medicine, agriculture, engineering, economics, social and political science, and mathematical statistics. *See* CLIMATE MODIFICATION; CLIMATIC CHANGE.

Like meteorology, climatology is conveniently resolved into subdisciplines in a way that recognizes a hierarchy of geographical scales of climatic phenomena and their governing physics. Macroclimatology refers to the largest (planetary) scale of regimes and phenomena; regional climatology to the scale of continents and subcontinental areas; mesoclimatology to the scale of individual physiographic features such as a mountain, lake, or urban area; and microclimatology to the smallest scale, for example, a house lot or the habitat of an insect. Climatology is also resolved in another way that distinguishes between the theoretical, descriptive, and applied aspects of the science: physical and dynamic climatology, concerned with the governing physical laws; descriptive and synoptic climatology, concerned with comparisons of climatic norms and anomalies, respectively, as concurrently observed in different places; and applied climatology, concerned with the practical utilization of climatological data in engineering design, operations strategy, and activity planning. Other important subdisciplines are climatography, concerned with the comprehensive documentation of climate by means of data summaries, maps, and atlases; and statistical climatology, concerned mostly with the estimation of climatological expectancies, risks of extreme events, and probability distributions of climatic variables (including joint distributions of combinations of variables) as needed to solve problems in applied climatology.

Origins and pattern of global climate. The climate of the Earth as a whole, including its geographical and seasonal variations, is fundamentally prescribed by the disposition of solar radiant energy that is intercepted by the atmosphere. This disposition depends in part on astronomical factors and in part on terrestrial factors, such as atmospheric composition, the distribution of land and sea, and the physiographic nature of the land. At the mean Earth-Sun distance, solar energy arrives in the vicinity of the Earth at the rate of approximately 1.95 g-cal/(cm²)(min), as measured outside the atmosphere perpendicular to the incident rays. This value, known as the solar constant,

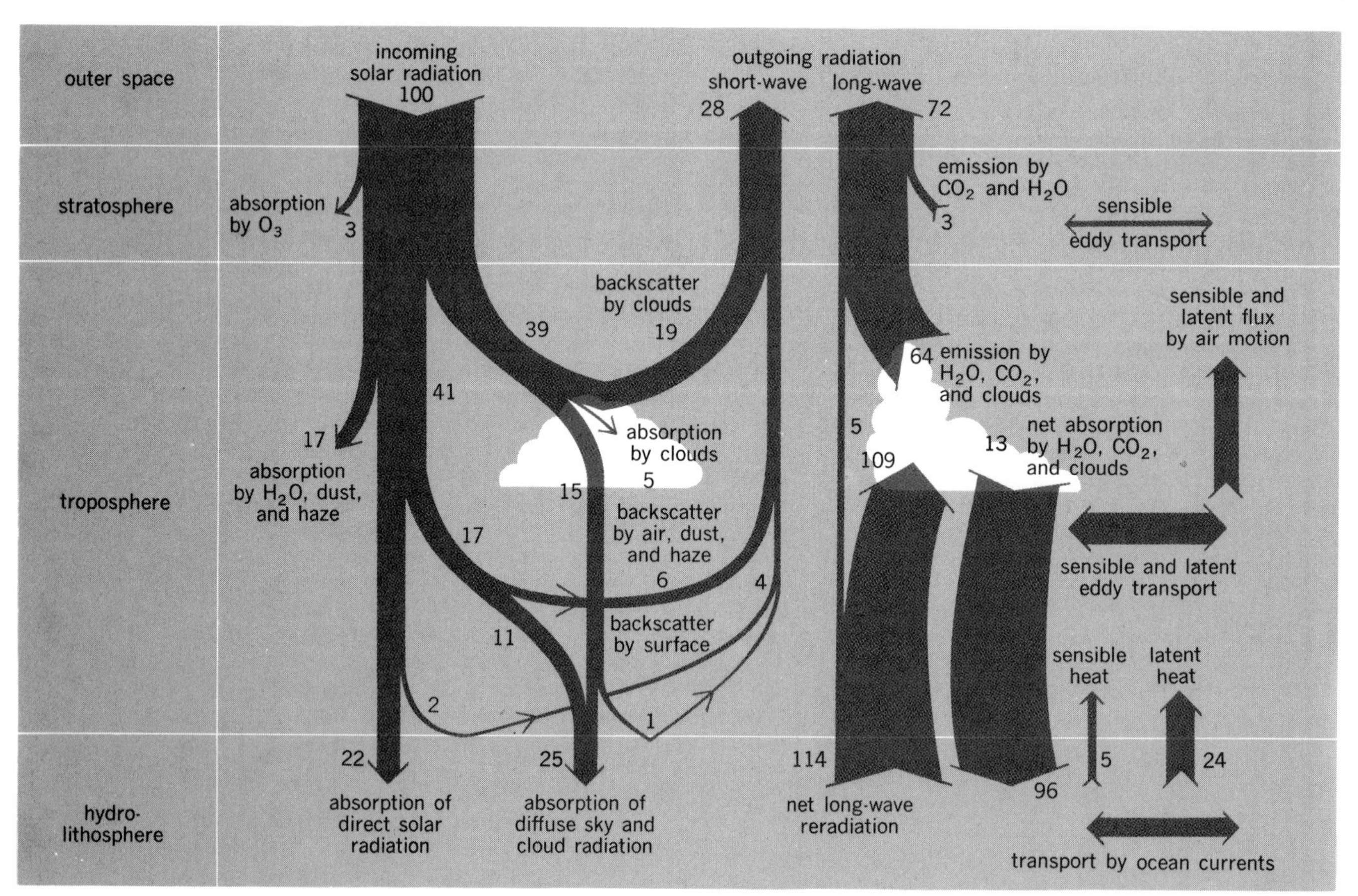

Fig. 1. Mean annual heat balance of whole Earth, showing magnitudes of principal items in percent of total solar radiation arriving at top of atmosphere. *(Based mostly on estimates for Northern Hemisphere by M. I. Budyko and K. Y. Kondratiev, from R. M. Rotty and J. M. Mitchell, Jr., in Symposium Series Volume on Air Pollution Control and Clean Energy, American Institute of Chemical Engineers, 1975).*

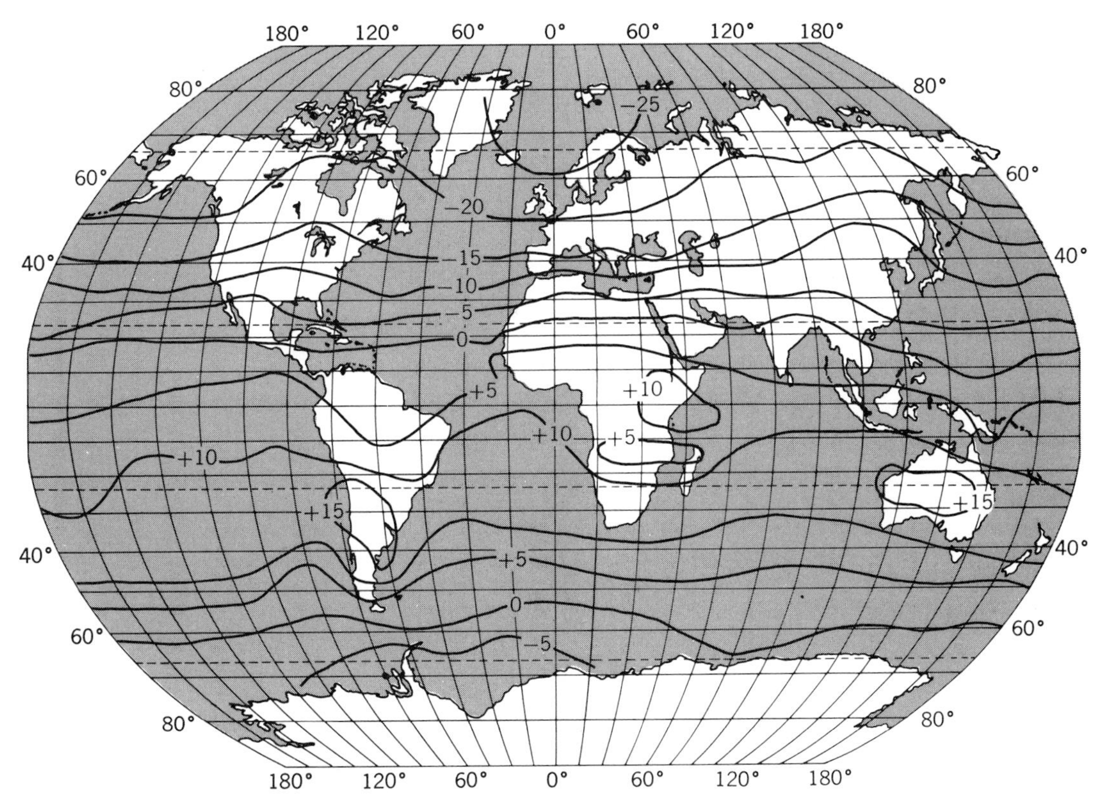

Fig. 2. Net radiation over Earth, January. Values in kg-cal/(cm²) (month). Regions of net surplus and deficit of radiation energy are shown. *(Modified after G. C. Simpson; base map copyright Denoyer-Geppert Co., Chicago)*

has been measured to an accuracy of about ±1% and is believed to vary by not more than a fraction of 1% from year to year. When reckoned per unit horizontal area, however, incoming solar radiation (or insolation) is intercepted at different rates over different parts of the Earth. The rotation of the Earth causes this rate to change locally with the time of day. In addition, the shape of the Earth, together with the seasonal changes in orbital distance and axial tilt of the Earth relative to the Sun, causes solar radiation (per unit horizontal area and per unit time) to vary with latitude and with the time of year.

Terrestrial heat balance. The disposition of solar energy after it penetrates the atmosphere is a complex process that locally depends on the solar elevation angle and on various properties of both the atmosphere and its underlying surface. Averaged over the whole Earth and over all seasons of the year, the disposition is as shown in Fig. 1. The figure represents the planetary average heat balance. This term is appropriate because in the observed absence of large changes of atmospheric temperature from year to year the income and outgo of heat energy through the Earth's surface, within the atmosphere, and through the top of the atmosphere must balance closely. In the annual mean for the whole planet, only about half of the solar energy incident on the top of the atmosphere penetrates as far as the Earth's surface. Of this amount, more than 90% is absorbed in the surface, which is used in about equal shares to heat the surface material and to evaporate water, primarily from the oceans. Approximately 28% of the incident solar energy is returned unused to space, mostly by backscatter from clouds; this corresponds to the planetary albedo. The remainder of the incident solar energy (25%) is absorbed on the way down through the atmosphere by water vapor, ozone, dust, haze, and clouds. This provides a direct source of heat to the atmosphere.

More than twice as much heat is added indirectly to the atmosphere by way of the surface. Nearly half of this indirect heating is supplied by a net upward transfer of thermal (infrared) radiation energy that is constantly being exchanged between the surface and atmospheric water vapor, carbon dioxide, and clouds. The remainder consists mainly of the latent heat content of water evaporated from the oceans, which is made available to the atmosphere at the point of condensation as clouds and precipitation. Finally, balance is achieved by a flow of thermal (infrared) radiation energy back to space (72% of the incident solar energy), nearly all of it by emission from water vapor, carbon dioxide, and the tops of clouds in the atmosphere. Less than one-tenth of this thermal radiation loss to space originates at the Earth's surface because the atmosphere is almost completely opaque to infrared radiation except in relatively narrow-wavelength bands (notably between 8 and 12 μm).

The fact that the atmosphere intercepts most of the thermal radiation from the Earth's surface, which in a transparent atmosphere would flow unimpeded into space, is very important to climate because it leads to the maintenance of much higher surface temperatures than those otherwise possible. This warming (the greenhouse effect) is largely attributable to the presence of water vapor and carbon dioxide in air. The magnitude of the

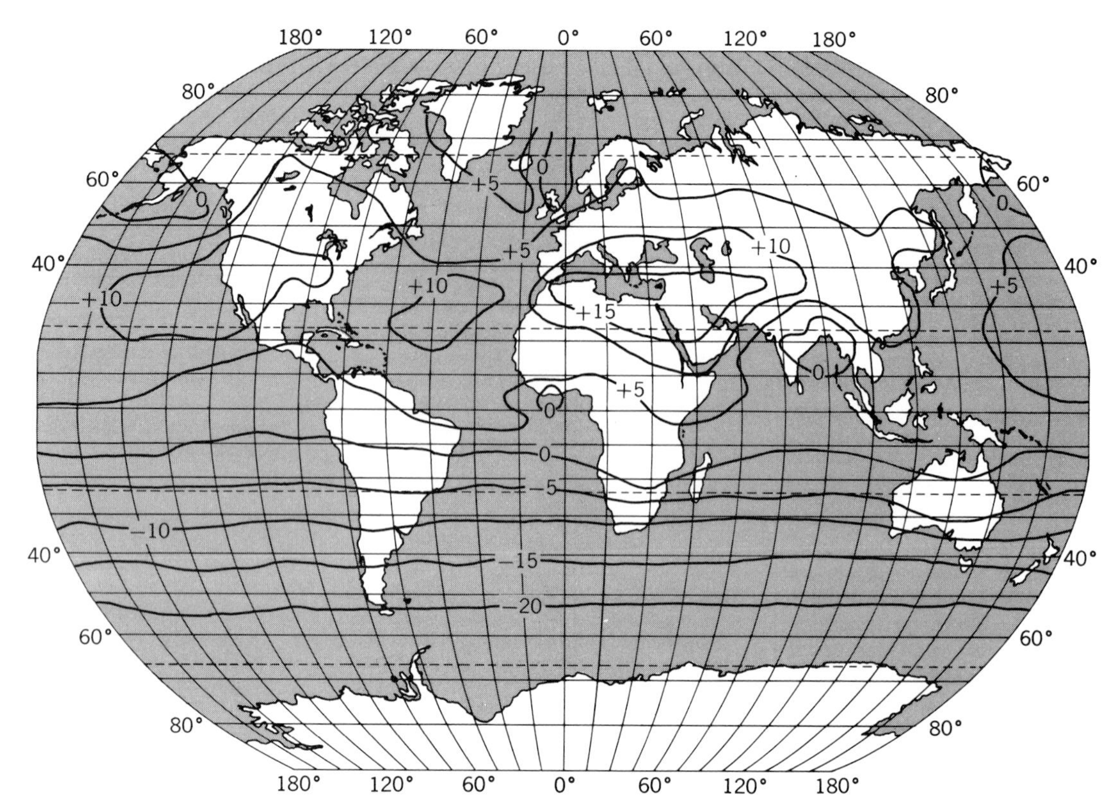

Fig. 3. Net radiation over the Earth, July. Values in kg-cal/(cm²)(month). Regions of net surplus and net deficit of radiation energy are shown. (*Modified after G. C. Simpson; base map copyright Denoyer-Geppert Co., Chicago*)

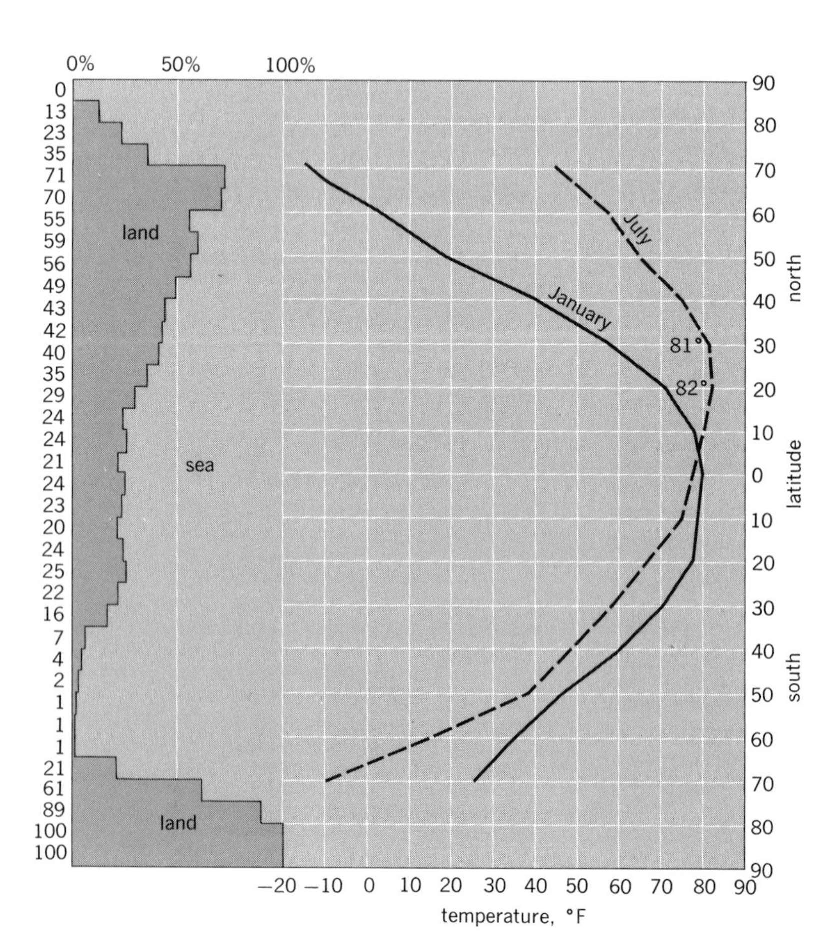

Fig. 4. The distribution of land and sea by 5° latitude zones and the mean temperature by latitude circles between 70°N and 70°S for January and July. (*Land-sea percentages after E. Kossinna; temperatures after J. Hann*)

warming effect importantly depends on the variable concentration of water vapor, and thus it changes with latitude, time of year, and atmospheric conditions generally. *See* GREENHOUSE EFFECT.

Role of atmospheric circulation. In regions of intense solar radiation, primarily in the tropics, there is a net surplus of radiation energy. Similarly in regions of relatively little solar radiation, primarily in the polar regions, there is a net deficit of radiation energy (Figs. 2 and 3). These radiative imbalances lead to differential heating and cooling of the atmosphere, and thereby generate a large-scale circulation of air that transports heat and moisture from areas of surplus to areas of deficiency, generally into higher latitudes. In this way global air currents (along with ocean currents) importantly modify the zonation of world climate otherwise dictated by the astronomical factors. The characteristic unsteadiness of these air currents causes the familiar changeability of weather whose statistics are another important aspect of climate. The pattern of atmospheric circulation, as reflected at the Earth's surface by differences of atmospheric pressure and their associated winds, is discussed in fuller detail below. *See* ATMOSPHERIC GENERAL CIRCULATION.

Land- and sea-surface influences. The climatic significance of the distribution of land and sea lies in the contrast between these two major kinds of surfaces regarding heating and cooling. Except in regions of snow cover or inland water bodies, the lands are heated to higher temperatures than are the seas, both diurnally during the daytime and seasonally during the summer. Obversely, during the nighttime and during the winter the lands are

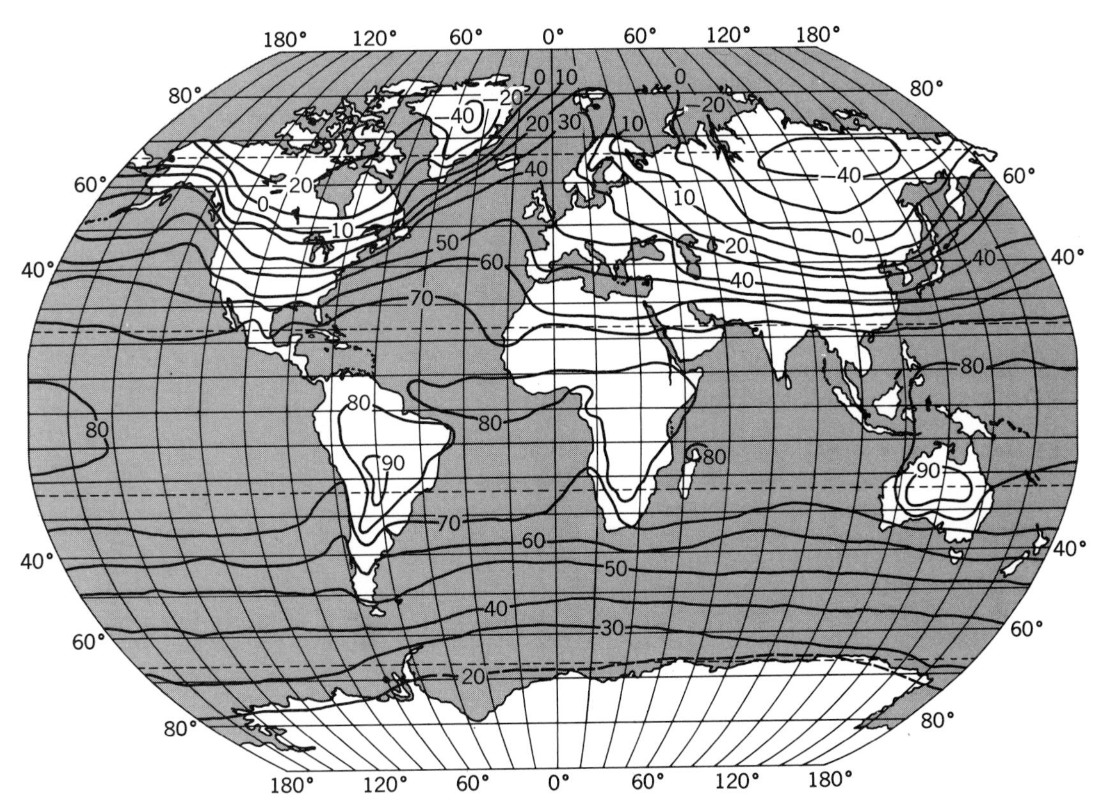

Fig. 5. Mean sea-level temperatures, °F, January. Note location of cold poles in interior of great landmasses. (*Modified from Sir Napier Shaw et al.; base map copyright Denoyer-Geppert Co., Chicago*)

cooled to lower temperatures than are the seas. As a direct result, both diurnal and annual temperature ranges are greater over the land than over the sea. *See* OCEAN-ATMOSPHERE RELATIONS.

The distribution of land and sea by 5° latitude zones is shown in Fig. 4. The importance of this factor is evident from the companion curves, showing mean temperatures by latitude circles between 70°N and 70°S for January and for July. Note, for example, that the January temperature is 19°F at 50°N, where land occupies more than half the latitude circle; and the temperature is 38° in the corresponding month of July at 50°S, where there is no land.

Land-sea distribution and the distribution of radiation are two of the three major factors that determine the broad features of the distribution of temperature. The third factor is the circulation of the atmosphere and of the oceans. In the mean, both circulations transfer energy from lower to higher latitudes. Of the transport, 65% is by the atmospheric circulation, which carries energy both in the form of sensible heat and latent heat (as water vapor that will give up heat upon condensing). The high-latitude regions of great net radiation loss (Figs. 2 and 3) are in large part fed by energy derived from latent heat. Of the transport, another 35% is accomplished through the circulation of ocean waters, with warm water moving poleward along the western sides of the oceans and cool water moving equatorward along the eastern sides. This circulation is reflected in the mean temperature maps for January and July (Figs. 5 and 6), as in the northward bulge of the isotherms in the Iceland-Spitsbergen-Norway area. Other notable features of these maps are the location of the cold poles in the interior of the great landmasses and the displacement of the warmest zone northward from the Equator, a displacement associated with the great landmass of Africa in latitudes 10–30°N. *See* WIND.

Relations with air pressure and wind. The global distribution of temperature is related to the distribution of surface air pressure and of wind. These relationships are neither simple nor direct, but there tends to be an inverse relationship between mean temperature and mean surface pressure over the continents, with the hot desert areas sustaining thermal lows and the cold polar areas sustaining thermal highs. The mean low- and high-pressure areas over the oceans are quite different. In the first instance they represent the summation of moving, dynamic lows (extratropical cyclones), which are particularly common synoptic features in the Aleutian and Icelandic areas, as well as in the waters north of Antarctica. In the second instance they represent the summation of subtropical high cells, which are persistent synoptic circulation features over the subtropical oceans, even though they vary continuously in size, intensity, and location. *See* AIR PRESSURE; METEOROLOGY.

The mean pressure patterns are approximately matched by the patterns of dominant windflow. The most constant winds are those of the trades, which lie on the equatorward side of the subtropical oceanic highs, and those of the monsoon circulation in the Asiatic-Australian area. In this great monsoon circulation, winds blow outward from Asia and inward onto Australia during the Northern Hemisphere winter, with a reversal of the circulation during the summer. Other major wind

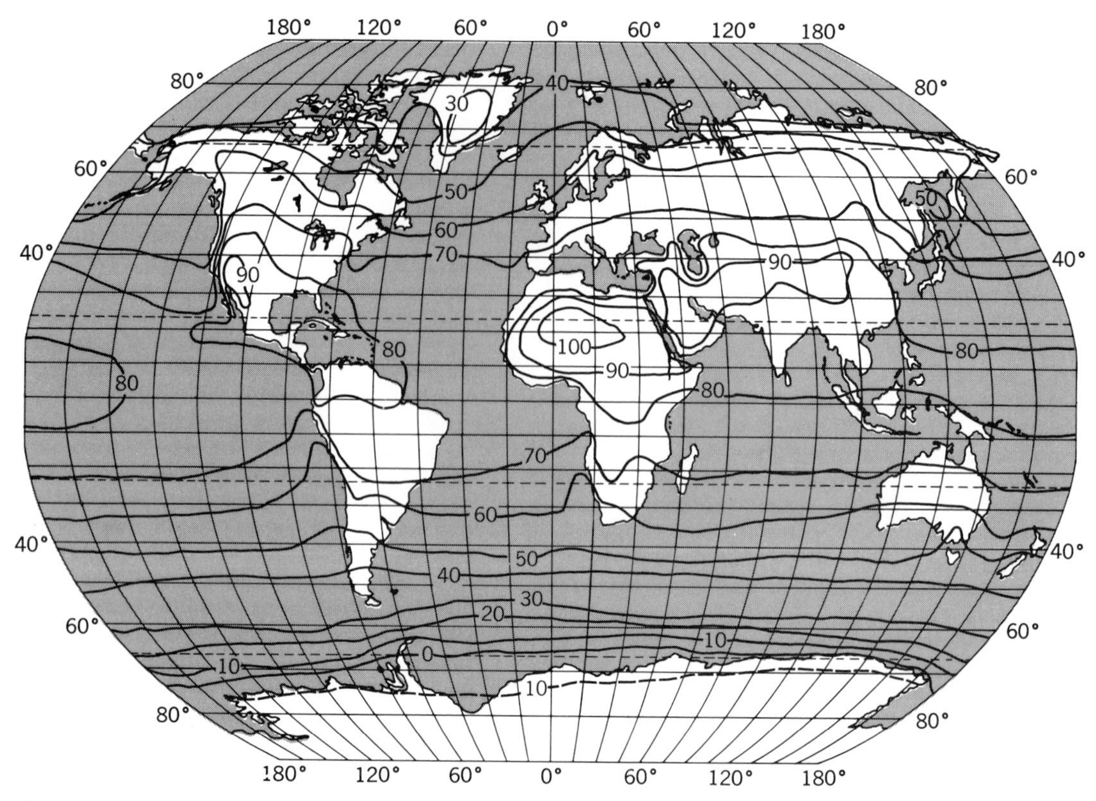

Fig. 6. Mean sea-level temperatures, °F, July. Note displacement of warmest zone northward from Equator. *(Modified from Sir Napier Shaw et al.; base map copyright Denoyer-Geppert Co., Chicago)*

flow regions are those of the westerlies, to poleward of the oceanic highs, and of the polar easterlies, off Antarctica and Greenland. Neither the westerlies nor the polar easterlies are as constant as the trades or the major monsoon winds because both occur in areas that are frequently the scene of moving cyclones, which may bring winds from any direction.

Marine atmospheric influences. Air that moves long distances across the oceans acquires great quantities of moisture through evaporation from the ocean surface. Because evaporation is greatest where cold dry air moves across much warmer ocean water, the most rapid flux of moisture from sea to air occurs on the western sides of the northern oceans in middle latitudes during the winter season, and over the waters off Antarctica during winter. For example, at 35°N in the western Atlantic an average of about 600 g-cal/(cm^2) (day) of thermal energy is expended for evaporation of sea water. In contrast, at latitudes 0–20°N in the eastern Atlantic, the value does not exceed 200 g-cal during any season.

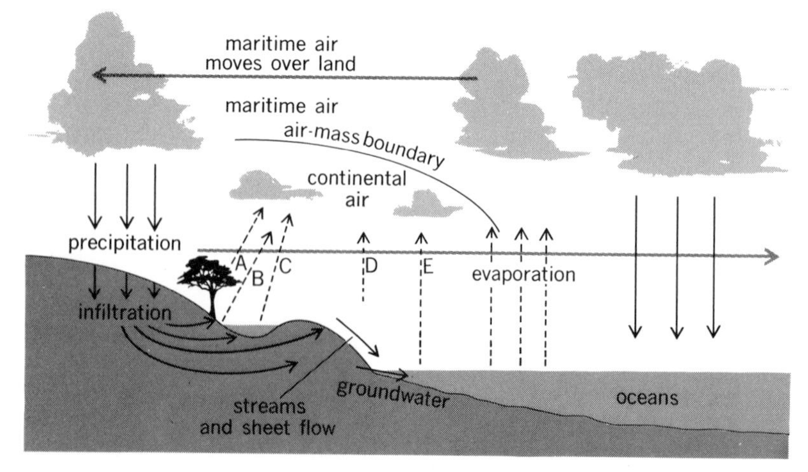

Fig. 7. Diagram of hydrologic cycle. Water returns chiefly to dry continental air through transpiration (A), evaporation from soil (B), lakes and ponds (C), and streams (D). Continental air moves over ocean to become more moist (E) with conversion to maritime air with precipitation over the oceans. *(Adapted from B. Holzman)*

Whatever the variations in detail, the oceans everywhere make a large net contribution of water to the air (except when there is ice). Subsequently, this moisture is carried onto the land by maritime air masses and there it is partly precipitated as rain, snow, hail, dew, or frost. Thereafter, the water returns to the oceans in streams, as sheet wash along the margins of the lands, as underground water, or in moving air (chiefly dry continental air) that acquires water from the land through evaporation and transpiration. This worldwide water cycle is diagrammed in Fig. 7. *See* HYDROLOGY.

Continental precipitation patterns. Figure 8 shows the mean annual precipitation over the continents. Appreciable precipitation occurs only when moist air is forced to rise and this takes place largely through convection, through orographic lifting (the forced lifting of air upslope, as up the flanks of a mountain), or through convergence and the forced ascent of air within an eddy, particularly within an extratropical or tropical cyclone. Hence the precipitation patterns can be viewed in terms of the relative frequency with which moist maritime air is present and the frequency with which the air is forced to rise to appreciable heights. The vertical structure of the air with reference to temperature and moisture is also important, because the structure may be stable and so resist vertical movement, or it may be conditionally unstable, so that when forced lifting has produced condensation, the release of latent heat causes the air to rise to still greater heights.

Annual precipitation is very high on the west coasts of the continents in high middle latitudes, in the major monsoon areas, and in equatorial areas. On the west coasts of the continents, the high totals are associated with frequent and prolonged cyclonic precipitation and with the orographic lifting of maritime air. In these areas most of the precipitation is in the winter half-year, when cyclonic storms are most common. In contrast, the monsoon areas have their maximum rains during summer, with an influx of very moist, conditionally unstable air. Here, as in the wet equatorial areas, the rainfall mechanisms are convection, orographic lifting, and convergence in minor eddy systems. However, in the equatorial areas the rainfall is fairly well distributed throughout the year.

The extremely dry areas are characterized by infrequent invasions of moist unstable air and relatively little cyclonic activity. These dry areas are the west coast deserts, such as the Sahara; the desert basins that are shielded from fresh maritime air by high mountains, such as the Tarim Basin; and the polar lands.

In absolute terms, the widest swings in annual precipitation from year to year occur in the wettest areas of the monsoon and the equatorial regions. In these regions it is not uncommon during a 20-year period for the annual rainfall to range from a minimum of 60–80 in. to a maximum of 200–300 in. In percentage terms, the greatest variability is in the driest areas—the deserts and dry polar regions. Here the minimum is less than 2 in. in many localities with maximum annual amounts reaching 15 in. or more in so-called wet years.

The water circumstance of a locality is not only a function of the annual precipitation and its variability but even more of the distribution of precipitation throughout the year. Figure 8 shows the seasonal precipitation regimes at eight selected localities. Also shown are the extreme monthly

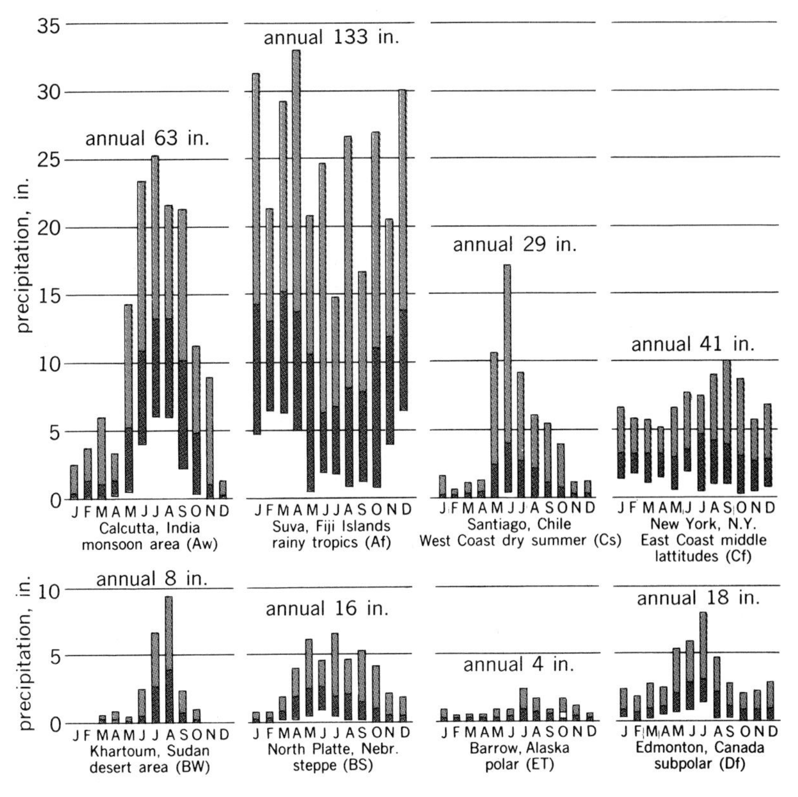

Fig. 8. Mean annual precipitation, with mean and extreme monthly precipitation bar graphs, at selected stations. Top of each bar shows extreme high value; bottom of each bar shows extreme low value; mean indicated by horizontal line within each bar. Letter symbols according to Koeppen type of climate. *(Based on 20-year data for 1921–1940, World Weather Records, Smithsonian Miscellaneous Collections)*

Principal quantitative definitions of major climatic regions*

Symbol	Name of region	Mean temperature, °F	Precipitation: Main season	Precipitation: Amount, formula for inches†
Af	Tropical rain forest	>64.4‡	All	Driest month >2.4; Annual >40
Aw	Tropical savanna	>64.4‡	Summer	Driest month <2.4; Annual >40
BW	Desert		Winter	Annual $<0.22t-7$
			Summer	Annual $<0.22t-1.5$
			Even	Annual $<0.22t-4.25$
BS	Steppe		Winter	Annual $<0.44t-14$
			Summer	Annual $<0.44t-3$
			Even	Annual $<0.44t-8.5$
Cf	Humid mesothermal	<64.4 > 26.6‡	Even	Annual $>0.44t-8.5$
Cw	Humid mesothermal, winter dry	<64.4 > 26.6‡	Summer	Annual $>0.44t-3$
Cs	Humid mesothermal, summer dry	<64.4 > 26.6‡	Winter	Annual $>0.44t-14$
Df	Humid microthermal	<26.6‡ > 50§	Even	Annual $>0.44t-8.5$
Dw	Humid microthermal, winter dry	<26.6‡ > 50§	Summer	Annual $>0.44t-3$
ET	Tundra	<50 > 32§		
EF	Perpetual frost	<32§		

*After W. Koeppen, as outlined in G. T. Trewartha, *An Introduction to Climate*, McGraw-Hill, 3d ed., 1954. Arrangement modified from one by H. E. Landsberg.

†All temperature values are in °F (including t, which is mean value). Precipitation values are in inches.

‡Temperature of coldest month. §Temperature of warmest month.

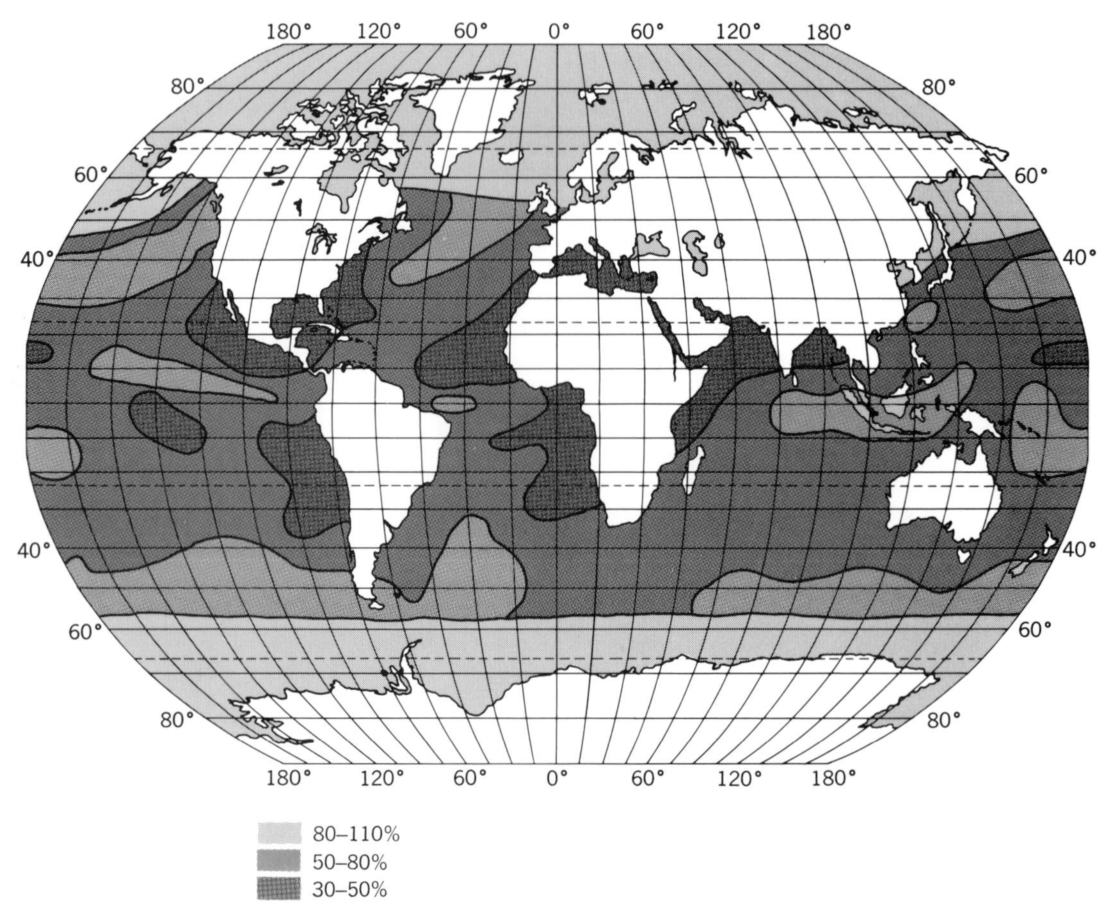

Fig. 9. Climatic regions for fresh-water supply on the oceans, December-February, according to scheme for problem solving. How well climatic conditions meet requirements of rainfall and sunshine, as related to use of solar stills, is indicated in percentages. *(After W. C. Jacobs; base map copyright Denoyer-Geppert Co., Chicago)*

precipitation amounts during the 20-year period which is covered by the graphs. The general form of these graphs is in each instance broadly representative of the precipitation regime throughout the regions which are specified in the diagrams. The variability as shown by the extreme values is typical.

Climatic regions. A wide variety of schemes is available for the definition of climatic regions. These schemes fall into two classes, those intended to bring out one or another aspect of relationships and processes within the atmosphere, and those intended to bring out relationships between areal variations in climatic conditions and corresponding variations in phenomena related to climate.

The first class of schemes is the major radiation regions of the Earth. The distinction is among regions in which the monthly radiation balance is always positive (in the mean), always negative, never strongly positive, never strongly negative, or highly variable and ranging from strongly positive to strongly negative.

The second class of schemes is represented (1) by the major climatic regions of the Earth according to W. Koeppen's classification, and (2) by Fig. 9. The Koeppen classification was intended to bring out the general coincidence between the distribution of natural vegetative formations and climatic conditions. It also serves, however, as a useful classification for purposes of general description, in terms of quantitative definitions of precipitation and temperature conditions. The major aspects of Koeppen's classification scheme are summarized in the table, whose arrangement is taken from H. E. Landsberg. Koeppen and his followers revised his classification twice and carried it beyond the broad scheme given here. For example, cold and hot deserts were distinguished, isothermal regions (mean annual temperature range less than 9°F) were identified, and subregions where fog was common were designated. Clearly a climatic classification scheme such as this could be extended in any desired way to show areal variations in greater and greater detail.

Quite different from Koeppen's scheme is a class of schemes illustrated by Fig. 9. This is a classification developed to solve a particular problem in applied climatology, that of designing life rafts for use at sea. The regions were defined primarily through consideration of the rainfall probabilities and the effective sunshine, the latter being related to the efficiency of portable solar stills. In applied climatology, special classifications of these kinds abound, each designed to serve a practical purpose.

It is feasible and instructive to distinguish minor climatic regions, not on a global scale, but within much smaller areas. Where such regions are defined upon the lands, the topographic factor becomes very important. This is especially true where the area is of the order of a few square miles or less. *See* MICROMETEOROLOGY.

[J. MURRAY MITCHELL, JR.]

Bibliography: I. P. Danilina (ed.), *Meteorology and Climatology*, vol. 2, 1977; H. Dickson, *Climate and Weather*, 1976; E. S. Gates, *Meteorology and Climatology*, 1972; J. F. Griffiths, *Applied Climatology*, 1976; H. H. Lamb, *Climate: Present, Past, and Future*, vol. 1, 1972; H. E. Landsberg, *World Survey of Climatology: General Climatology*, vol. 1, 1977; World Meteorological Organization, *Physical and Dynamic Climatology*, 1975; W. D. Sellers, *Physical Climatology*, 1965.

Climax community

A complex or organized community of plant, animal, and microbial populations which has a high probability of perpetuating itself on a specified set of land or water areas. The word "climax" carries the implication that previous communities on these areas were not so self-perpetuating, but instead prepared the biological or physical conditions which allowed the invasion or prolonged regeneration by the climax species.

For several reasons, the concept of climax has been an important and controversial topic of theoretical and applied ecology. Paradoxically, the idea gained quick attention from studies that emphasized the instability instead of the stability of the organic-inorganic complex which later was called the ecosystem.

In 1899 Henry Cowles declared, "The ecologist employs the methods of physiography, regarding the flora of a pond or swamp or hillside not as a changeless landscape feature, but rather as a panorama, never twice alike. [He] must study the order of succession of the plant societies in the development of a region, and he must endeavor to discover the laws which govern the panoramic changes. Ecology, therefore, is a study in dynamics." Cowles deduced carefully that the obviously unstable pioneer stages described in his quoted thesis on Lake Michigan sand dunes, and in many other landscapes in the Midwest and elsewhere, would tend to create conditions less favorable for their own species' survival, but more suitable for invasion by a new stage. This in turn seemed to prepare conditions favoring establishment and survival of populations of still later communities. Finally, some of these communities maintained, instead of undermining, the prerequisites for their own regeneration unless interrupted by externally imposed changes. *See* APPLIED ECOLOGY; DUNE VEGETATION; ECOLOGICAL INTERACTIONS; ECOLOGICAL SUCCESSION; ECOSYSTEM.

Stabilization. Let the vector **v** stand for as many variables as are considered necessary or feasible for describing the state of the community, or of the whole ecosystem of which it is the living part. The individualistic concept of the community can be restated by calling **v** a random variable which has very little interdependence between the component species, and only the most general constraints on selection of future combinations. The ecosystem concept implies flows (through food chains and nutrient cycles) and other dependences which combine with environmental constraints to narrow the range of probable future values for **v**, given the site and its larger environment. *See* ECOLOGY.

F. Egler's "theory of initial composition" implies that the complex prior history of areas undergoing secondary succession, plus fortuitous accidents of weather and perhaps seed supply in the year of last cultivation, mowing, or pasture, have profound influence on the set of available species and other biotic factors which will long influence later survivors. Pioneer herbs, shrubs, and short-lived trees are (by definition) genetically programmed to germinate and flourish (and often disappear) quickly. Egler's experience suggests that some of the long-lived species (like sugar maple and the more ubiquitous red maple) that epitomize certain northeastern states' self-regenerating climax communities start to seed simultaneously with the pioneers, and merely outlast them. *See* VEGETATION MANAGEMENT.

What Egler calls the "theory of floristic relays" carries more of a hint of cybernetic control—perhaps in mechanism, not just imagery. On nutrient-poor substrates, subject to extreme microclimates either most of the available germules fail to get "turned on" by the right sequence of signals, or the plant seedlings, animal larvae, or sensing tissues get "turned off" (too inactive to match the competition) or otherwise simply eliminated. It may be a long time before perceptive natural history and rigorous physiological ecology prove how these chains of conditioning and selection explain why the transient populations of a succession rise and then fall. For the idea of the climax, it presently suffices to note that members of the species mixture jointly have got together all that it takes—not only to complete their own life cycles, but to be ready for continued participation in whatever "game plan" is feasible over many generations. Practical knowledge of what can be predicted reliably or not in the transient and stable communities is important, not only for vegetation management

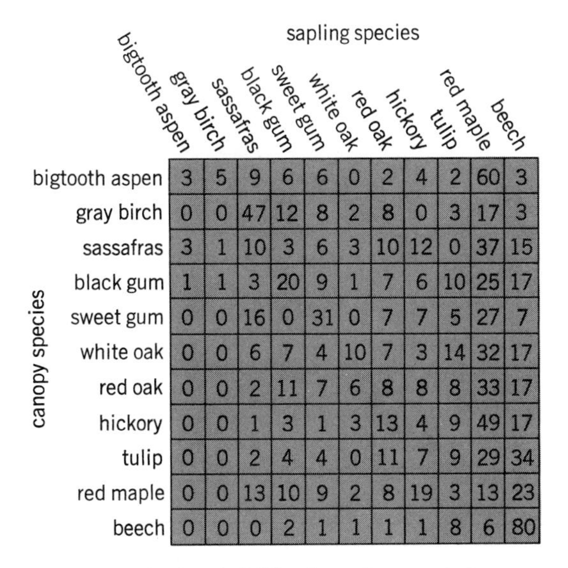

canopy species \ sapling species	bigtooth aspen	gray birch	sassafras	black gum	sweet gum	white oak	red oak	hickory	tulip	red maple	beech
bigtooth aspen	3	5	9	6	6	0	2	4	2	60	3
gray birch	0	0	47	12	8	2	8	0	3	17	3
sassafras	3	1	10	3	6	3	10	12	0	37	15
black gum	1	1	3	20	9	1	7	6	10	25	17
sweet gum	0	0	16	0	31	0	7	7	5	27	7
white oak	0	0	6	7	4	10	7	3	14	32	17
red oak	0	0	2	11	7	6	8	8	8	33	17
hickory	0	0	1	3	1	3	13	4	9	49	17
tulip	0	0	2	4	4	0	11	7	9	29	34
red maple	0	0	13	10	9	2	8	19	3	13	23
beech	0	0	0	2	1	1	1	1	8	6	80

Fig. 1. Matrix of probabilities for replacement of canopy trees in a forest by trees of the same or other species, assuming every sapling has an equal chance of replacing the tree. (*From H. S. Horn, Forest succession, Sci. Amer., 232(5):90–100; copyright 1975 by Scientific American Inc., all rights reserved*)

but for many other aspects of human ecology. *See* FOREST ECOLOGY; PHYSIOLOGICAL ECOLOGY (PLANT); WILDLIFE CONSERVATION.

If the proportions of trees in a forest represent the components of vector **v**, then Fig. 1 is simply a matrix expressing how the proportions of sapling species are conditioned by the overtopping canopy species. The last column suggests that in this case beech can arrive early (as Egler claimed), but its probability of doing so is small, that is replacing 3% of the aspen and of gray birch (both pioeers). Beech replaces 7–17% of the areas occupied by many other species: .15 × sassafras + .17 × black gum + .07 × sweet gum + .17 × white oak + .17 × red oak + .17 × hickory. In addition, .34 of the tulip tree (yellow poplar) and .23 of the red maple are ready to give way to beech, but this very shade-tolerant species replaces 80% of its own kind!

A variety of initial species distributions of **v** can then be operated upon by the matrix in Fig. 1, and can converge upon a predicted stationary climax distribution of species very much like the one actually observed by Henry Horn on the oldest forest available (about 250 years) on grounds of the Princeton Institute for Advanced Study, after longevities are taken into account (see Fig. 2). Because the actual beech stand is on a site moister, and presumably with better nutrient reserves, than the average condition of all stands contributing data to the matrix of Fig. 1, it is not surprising that the nutrient-demanding species sugar maple and white ash are also present in the actual stand. The proportions of these last species might be still higher in other New Jersey climax stands which were more loamy than sandy, with higher availability of basic elements such as calcium.

Fluctuation. One may generalize that the rate of change of species composition $d\mathbf{v}/dt$ will decrease as a climax stage is approached, along with the range of environmental extremes which must be tolerated. But exceptions to these trends are important for the concept as well as for specific interpretations of the "climax" condition.

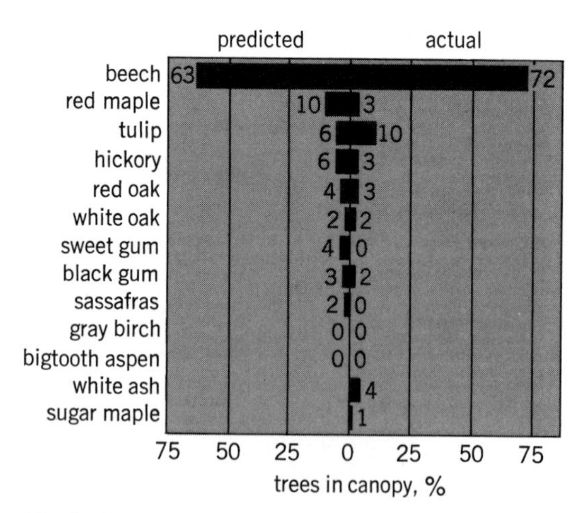

Fig. 2. Predicted and actual distribution of canopy trees in a forest. The prediction is the stationary distribution calculated from the matrix of probabilities in Fig. 1 and the weighted longevities of each species. The predicted composition and the actual one are based on completely different sets of data. (*From H. S. Horn, Forest succession, Sci. Amer., 232(5):90–100; copyright 1975 by Scientific American Inc.; all rights reserved*)

Forest canopy moderates the heat load reaching the ground or lower vegetation; but when holes in that overstory do occur (by chance or from tree death or blowdown as well as selective logging), these gaps bear the brunt of full sunlight without as much ground temperature moderation by wind as would occur in open stands. Deciduous forests often give a multilayered shade which is diffuse; but seasonally it changes abruptly, and dense canopy favors a ground layer with winter annuals as well as very shade-tolerant mosses and shrubs. Hemlocks are examples of climax evergreens so shade-tolerant that they can mix with and even replace the less tolerant predecessors in much of New England, the Appalachians, and upper Great Lakes (*Tsuga canadensis*) and wetter areas of the Pacific Northwest (*Tsuga heterophylla* and *mertensiana*), Japan, and southern China. *See* VEGETATION AND ECOSYSTEM MAPPING.

In pure stands the shade of tolerant conifers such as hemlock and spruce often become too dense for regeneration, even of their own kind. Insufficient restocking is then ready when openings are created suddenly. Also, decay-resistant litter becomes thick, making such a good heat insulator that direct solar heat is not dissipated well enough to prevent lethal temperatures at the surface of the seedbed, thereby turning off climax regeneration temporarily. Where mineral soil was brought to the surface (as by blowdown) or in the shade of fallen stems, hemlock might regenerate directly, sometimes with white pine. But pioneer species (such as gray, black, white, or yellow birch in different parts of the Northeast) also commonly get a head start. They may grow so rapidly that they dominate the overstory for a whole generation until the end of their natural lifetime, or cutting or other disturbance. Meanwhile deciduous trees provide a nurse crop for conifers (including spruce in the upper part of the hemlock–northern hardwoods zone and especially in the Boreal needleleaf forest formation dominating much of Canada).

Are such fluctuations too drastic to be accommodated within the concept of a single climax community? The openings may or may not be quite small. Observers and data focused only within the openings may suggest sudden and drastic changes in dominant species, and here in life form as well. Reversion to an early successional stage seems to be occurring followed by repetition of the climax (promptly, unless seed source or seedlings are eliminated by fire). But a larger view of the landscape mosaic, with many openings undergoing "gap-phase replacement" described by some matrix like the table, suggests merely that a climax of mixed life form be defined broadly enough to encompass such "alternation of generations" as well as less abruptly contrasting mixtures of compatible species. Parts will continue to oscillate around some mean condition which can be expressed as an average over space or time or both.

In areas where hurricanes or fire (by lightning or human causes) are frequent and of great extent, the idea of the climax is sometimes considered a hypothetical condition. Where cropping has eliminated most wild vegetation and grazing reaches remote corners of the landscape, ecologists depend on ingenious detective work and evidence from exceptional refugia to reconstruct what might conceivably be approached if the prevailing pattern of disturbance were eliminated or changed. It might even be impossible to judge the "natural" future of the landscape, and many of the plants and animals formerly available to participate are either vanished or too remote geographically to be dispersed into the area of concern. But it may be quite feasible to infer and even test ideas on the most probable state for the community (as part of the whole ecosystem) under the set of conditioning factors which prevails at present or which is projected for the future. The word "subclimax" has often been used for the systems developing under protracted disturbance, implying that it is "below" the stage of development attainable if the limiting constraint were removed.

Cowles expressed the idea of climax as a limiting condition: "in other words[,] the condition of equilibrium is never reached, and when we say that there is an approach to a mesophytic forest, we speak only roughly and approximately. As a matter of fact, we have a variable approaching a variable rather than a constant." Furthermore, the limit which is approached by the community's state variable v not only varies with time but with space as well.

Polyclimax or monoclimax. Controversy about the climax concept throughout the first half of the 20th century centered not only on semantics, but also on the admitted utility of the hypothesis in linking all (or most) parts of the landscape panorama into a coherent scheme. The "polyclimax" idea now seems to apply even in Cowles's dunes. His "basewood–[sugar] maple series" seems inherently limited to pockets and lee slopes with protected microclimate, extra inputs of moisture (from higher dunes) and nutrients (from leaves blowing in), and examples become less common rather than more so on old dunes. His "pine formation" is commonly succeeded by oak (black and white oak around southern Lake Michigan; red oak mostly in the pockets there but more widespread northward in Michigan and Wisconsin). Apparently many landscapes of the oak-hickory region of the Midwest and the oak-pine region of the Southeast would be expected now to show communities on hilltops and protected hollows rather different from those on intermediate sites. *See* PLANT FORMATIONS, CLIMAX.

However, even the "monoclimax" theory did not deny that the limiting community would differ in significant ways with topography, geologic substrate, and other factors influencing moisture. It substituted the term "preclimax" for the communities of drier sites, emphasizing their similarities in vegetation (and presumably animals) to the climax community which occupied intermediate or widespread sites typical of the next drier climatic zone (that is, westward in the Midwest where F. E. Clements and Victor Shelford popularized many of these ideas soon after Cowles raised them in a more cautious way). "Postclimax" was the term which once was used by this school for communities of the sites much moister and richer than average – with the plausible assumption that a climatic shift toward improved moisture would help these localized communities to become widespread on intermediate sites. *The* climax (often called "climatic climax") typical of a region was that supposedly approached by various successions on a wide variety of sites intermediate between the dry (xeric) or wet (hydric) conditions. *See* VEGETATION ZONES, ALTITUDINAL; VEGETATION ZONES, WORLD.

Successional convergence or divergence. Mesic (or mesophytic) conditions are, by definition, intermediate between the xeric and hydric conditions of a region, but do not necessarily provide a good preview of the limit to be approached by diverse sites when there is enough time for their typical shade-tolerant species to invade other communities. The working hypothesis of convergence of diverse successional lines toward a single climax (such as beech-maple) became a ruling theory for a while, and is still repeated by habit in many textbooks almost as a dogma. A polyclimax view has more leeway for accommodating the many parallel lines of succession which retain plant and animal indicators of habitat differences which will not be erased merely by allowing more time. Closer attention to soil conditions and nutrient budgets will probably clarify more cases (as in Cowles's classic dunes) where there may be a divergence of succession: lower probability for old black oak – blueberry dune on acidic leached sands to become a mesophytic forest than for a dune young enough to do so, and retain part of its initial calcium carbonate, as a result of recycling for the nutrients required for a basswood – sugar maple forest.

[JERRY S. OLSON]

Bibliography: D. B. Botkin, J. F. Janak, and J. R. Wallis, Some ecological consequences of a computer model of forest growth, *J. Ecol.*, 60: 849–872, 1972; H. C. Cowles, The ecological relations of the vegetation of the sand dunes of Lake Michigan, *Bot. Gaz.*, vol. 27, 1899; H. S. Horn, Forest succession, *Sci. Amer.*, 232(5): 90–100, 1975; R. H. Whittaker, *Communities and Ecosystems*, 3d ed., 1975.

Cloud

Suspensions of minute droplets or ice crystals produced by the condensation of water vapor (the ordinary atmospheric cloud). Other clouds, less commonly seen, are composed of smokes or dusts. *See* AIR POLLUTION; DUST STORM.

This article presents an introductory outline of cloud formation upon which to base an understanding of cloud classifications. For a more technical consideration of the physical character of

Fig. 1. Cirrus, with trails of slowly falling ice crystals at a high level. (*F. Ellerman, U.S. Weather Bureau*)

Fig. 2. Small cumulus. (*U.S. Weather Bureau*)

Fig. 3. An overcast of stratus, with some fragments below the hilltops. (*U.S. Weather Bureau*)

Fig. 4. View from Mount Wilson, Calif. High above is a veil of cirrostratus, and below is the top of a low-level layer cloud. (*F. Ellerman, U.S. Weather Bureau*)

Fig. 5. Cirrocumulus, high clouds with a delicate pattern. (*A. A. Lothman, U.S. Weather Bureau*)

atmospheric clouds, including the condensation and precipitation of water vapor *see* CLOUD PHYSICS.

Rudiments of cloud formation. A grasp of a few physical and meteorological relationships aids in an understanding of clouds and skies. First, if water vapor is cooled sufficiently, it becomes saturated, that is, in equilibrium with a plane surface of liquid water (or ice) at the same temperature. Further cooling in the presence of such a surface causes condensation upon it; in the absence of any surfaces no condensation occurs until a substantial further cooling provokes condensation upon random large aggregates of water molecules. In the atmosphere, even in the apparent absence of any surfaces, they are in fact always provided by invisible motes upon which the condensation proceeds at barely appreciable cooling beyond the state of saturation. Consequently, when atmospheric water vapor is chilled sufficiently, such motes, or condensation nuclei, swell into minute water droplets and form a visible cloud. The total concentration of liquid in the cloud is controlled by its temperature and the degree of chilling beyond the state in which saturation occurred, and in most clouds approximates to 1 g in 1 m^3 of air. The concentration of droplets is controlled by the concentrations and properties of the motes and the speed of the chilling at the beginning of the condensation. In the atmosphere these are such that there are usually about 100,000,000 droplets/m^3. Because the cloud water is at first fairly evenly shared among them, these droplets are necessarily of microscopic size, and an important part of the study of clouds is concerned with the ways in which they often become aggregated into drops large enough to fall as rain.

The chilling which produces clouds is almost always associated with the upward movements of air which carry heat from the Earth's surface and restore to the atmosphere that heat lost by radiation into space. These movements are most pronounced in storms, which are accompanied by thick, dense clouds, but also take place on a smaller scale in fair weather, producing scattered clouds or dappled skies. *See* STORM.

Rising air cools by several degrees Celsius for each kilometer of ascent, so that even over equatorial regions temperatures below 0°C are encountered a few kilometers above the ground, and clouds of frozen particles prevail at higher levels. Of the abundant motes which facilitate droplet condensation, very few cause direct condensation into ice crystals or stimulate the freezing of droplets, and especially at temperatures near 0°C their numbers may be vanishingly small. Consequently, at these temperatures, clouds of unfrozen droplets are not infrequently encountered (supercooled clouds). In general, however, ice crystals occur in very much smaller concentrations than the droplets of liquid clouds, and may by condensation alone become large enough to fall from their parent cloud. Even small high clouds may produce or become trails of snow crystals, whereas droplet clouds are characteristically compact in appearance with well-defined edges, and produce rain only when dense and well-developed vertically (2 km or more thick).

Fig. 6. Altocumulus, which occurs at intermediate levels. (*G. A. Lott, U.S. Weather Bureau*)

Fig. 7. Altostratus, a middle-level layer cloud. Thick layers of such cloud, with bases extending down to low levels, produce prolonged rain or snow, and are then called nimbostratus. (*C. F. Brooks, U.S. Weather Bureau*)

Classification of clouds. The contrast in cloud forms mentioned above was recognized in the first widely accepted classification, as well as in several succeeding classifications. The first was that of L. Howard (London, 1803), recognizing three fundamental types: the stratiform (layer), cumuliform (heap), and cirriform (fibrous). The first two are indeed fundamental, representing clouds formed respectively in stable and in convectively unstable atmospheres, whereas the clouds of the third type are the ice clouds which are in general higher and more tenuous and less clearly reveal the kind of air motion which led to their formation. Succeeding classifications continued to be based upon the visual appearance or form of the clouds, differentiating relatively minor features, but later in the 19th century increasing importance was attached to cloud height, because direct measurements of winds above the ground were then very difficult, and it was hoped to obtain wind data on a great scale by combining observations of apparent cloud motion with reasonably accurate estimates of cloud height, based solely on their form.

WMO cloud classification. The World Meteorological Organization (WMO) uses a classification which, with minor modifications, dates from 1894 and represents a choice made at that time from a number of competing classifications. It divides clouds into low-level (base below about 2 km), middle-level (about 2 to 7 km), and high-level

Fig. 8. Cumulonimbus clouds photographed over the upland adjoining the upper Colorado River valley. Note the rain showers which appear under some of the clouds. (*Lt. B. H. Wyatt, U.S.N., U.S. Weather Bureau*)

Cloud classification based on air motion and associated physical characteristics

Kind of motion	Typical vertical speeds, cm/sec	Kind of cloud	Name	Characteristic dimensions, km		Characteristic precipitation
				Horizontal	Vertical	
Widespread slow ascent, associated with cyclones (stable atmosphere)	10	Thick layers	Cirrus, later becoming: cirrostratus, altostratus, altocumulus	10^3	1–2	Snow trails
			nimbostratus	10^3	10	Prolonged moderate rain or snow
Convection, due to passage over warm surface (unstable atmosphere)	10^2	Small heap cloud	Cumulus	1	1	None
	10^3	Shower- and thunder-cloud	Cumulo-nimbus	10	10	Intense showers of rain or hail
Irregular stirring causing chilling during passage over cold surface (stable atmosphere)	10	Shallow low layer clouds, fogs	Stratus Stratocumulus	10^2 $<10^3$	<1	None, or slight drizzle or snow

(between roughly 7 and 14 km) forms within the middle latitudes. The names of the three basic forms of clouds are used in combination to define 10 main characteristic forms, or "genera."

1. Cirrus are high white clouds with a silken or fibrous appearance (Fig. 1).

2. Cumulus are detached dense clouds which rise in domes or towers from a level low base (Fig. 2).

3. Stratus are extensive layers or flat patches of low clouds without detail (Fig. 3).

4. Cirrostratus is cirrus so abundant as to fuse into a layer (Fig. 4).

5. Cirrocumulus is formed of high clouds broken into a delicate wavy or dappled pattern (Fig. 5).

6. Stratocumulus is a low-level layer cloud having a dappled, lumpy, or wavy structure. See the foreground of Fig. 4.

7. Altocumulus is similar to stratocumulus but lies at intermediate levels (Fig. 6).

8. Altostratus is a thick, extensive, layer cloud at intermediate levels (Fig. 7).

9. Nimbostratus is a dark, widespread cloud with a low base from which prolonged rain or snow falls.

10. Cumulonimbus is a large cumulus which produces a rain or snow shower (Fig. 8).

Classification by air motion. Modern detailed studies of clouds were stimulated by the discovery of cheap methods of seeding supercooled clouds with artificial motes, promoting ice-crystal formation, aimed at stimulating or increasing snowfall. These studies show that the external form of clouds gives only indirect and incomplete clues to the physical properties which determine their evolution. Throughout this evolution the most important properties appear to be the air motion and the size-distribution spectrum of all the cloud particles, including the condensation nuclei. These properties vary significantly with time and position within the cloud, so that cloud studies demand the intensive examination of individual clouds with expensive facilities such as aircraft and radar. The interplay of physical processes in atmospheric clouds is very complicated, and until it is better understood, no satisfactory physical classification will be possible. From a general meteorological point of view a classification can be based upon the kind of air motion associated with the cloud, as shown in the table. [FRANK H. LUDLAM]

Bibliography: H. R. Byers, *General Meteorology*, 4th ed., 1974; C. E. Koeppe and G. C. DeLong, *Weather and Climate*, 1958; R. S. Scorer and H. Wexler, *A Colour Guide to Clouds*, 1964; World Meteorological Association, Manual on the Observation of Clouds and Other Meteors, rev. ed., WMO no. 47, 1970.

Cloud physics

The study of the physical and dynamical processes governing the structure and development of clouds and the release from them of snow, rain, and hail (collectively known as precipitation).

The factors of prime importance are the motion of the air, its water-vapor content, and the numbers and properties of the particles in the air which act as centers of condensation and freezing. Because of the complexity of atmospheric motions and the enormous variability in vapor and particle content of the air, it seems impossible to construct a detailed, general theory of the manner in which clouds and precipitation develop. However, calculations based on the present conception of laws governing the growth and aggregation of cloud particles and on simple models of air motion provide reasonable explanations for the observed formation of precipitation in different kinds of clouds.

Cloud formation. Clouds are formed by the lifting of damp air which cools by expansion under continuously falling pressure. The relative humidity increases until the air approaches saturation.

Then condensation occurs (Fig. 1) on some of the wide variety of aerosol particles present; these exist in concentrations ranging from less than 100 particles per cubic centimeter (cm^3) in clean, maritime air to perhaps 10^6/cm^3 in the highly polluted air of an industrial city. A portion of these particles are hygroscopic and promote condensation at relative humidities below 100%; but for continued condensation leading to the formation of cloud droplets, the air must be slightly supersaturated. Among the highly efficient condensation nuclei are the salt particles produced by the evaporation of sea spray, but it now appears that particles produced by man-made fires and by natural combustion (for example, forest fires) also make a major contribution. Condensation onto the nuclei continues as rapidly as the water vapor is made available by cooling of the air and gives rise to droplets of the order of 0.01 millimeter (mm) in diameter. These droplets, usually present in concentrations of a few hundreds per cubic centimeter, constitute a nonprecipitating water cloud.

Mechanisms of precipitation release. Growing clouds are sustained by upward air currents, which may vary in strength from a few centimeters per second (cm/sec) to several meters (m) per second. Considerable growth of the cloud droplets (with falling speeds of only about 1 cm/sec) is therefore necessary if they are to fall through the cloud, survive evaporation in the unsaturated air beneath, and reach the ground as drizzle or rain. Drizzle drops have radii exceeding 0.1 mm, while the largest raindrops are about 6 mm across and fall at nearly 10 m/sec. Production of a relatively few large particles from a large population of much smaller ones may be achieved in one of two ways.

Coalescence process. Cloud droplets are seldom of uniform size for several reasons. Droplets arise on nuclei of various sizes and grow under slightly different conditions of temperature and supersaturation in different parts of the cloud. Some small drops may remain inside the cloud for longer than others before being carried into the drier air outside.

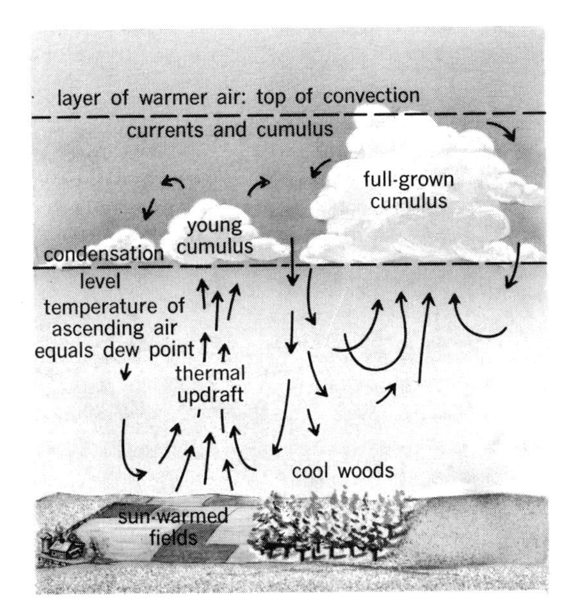

Fig. 1. Conditions leading to birth of a cumulus cloud. *(Based on photodisplay in Willetts Memorial Weather Exhibit, Hayden Planetarium, New York City)*

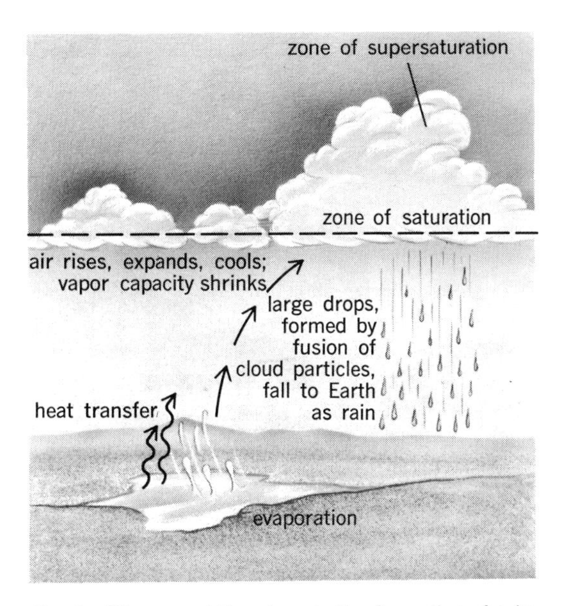

Fig. 2. Diagram of the steps in the formation of rain. *(Based on photodisplay in Willetts Memorial Weather Exhibit, Hayden Planetarium, New York City)*

A droplet appreciably larger than average will fall faster than the smaller ones, and so will collide and fuse (coalesce) with some of those which it overtakes (Fig. 2). Calculations show that, in a deep cloud containing strong upward air currents and high concentrations of liquid water, such a droplet will have a sufficiently long journey among its smaller neighbors to grow to raindrop size. This coalescence mechanism is responsible for the showers that fall in tropical and subtropical regions from clouds whose tops do not reach the 0°C level and therefore cannot contain ice crystals which are responsible for most precipitation. Radar evidence also suggests that showers in temperate latitudes may sometimes be initiated by the coalescence of waterdrops, although the clouds may later reach to heights at which ice crystals may form in their upper parts.

Initiation of the coalescence mechanism requires the presence of some droplets exceeding 20 μm in diameter. Over the oceans and in adjacent land areas they may well be supplied as droplets of sea spray, but in the interiors of continents, where so-called giant salt particles of marine origin are probably scarce, it may be harder for the coalescence mechanism to begin.

Ice crystal process. The second method of releasing precipitation can operate only if the cloud top reaches elevations where temperatures are below 0°C and the droplets in the upper cloud regions become supercooled. At temperatures below −40°C the droplets freeze automatically or spontaneously; at higher temperatures they can freeze only if they are infected with special, minute particles called ice nuclei. As the temperature falls below 0°C, more and more ice nuclei become active, and ice crystals appear in increasing numbers among the supercooled droplets. But such a mixture of supercooled droplets and ice crystals is unstable. The cloudy air, being usually only slightly supersaturated with water vapor as far as the droplets are concerned, is strongly oversaturated for the ice crystals, which therefore grow more rapidly than the droplets. After several minutes the growing crystals will acquire definite falling speeds, and several of them may become joined together to form a snowflake. In falling into the warmer regions of the cloud, however, the snowflake may melt and reach the ground as a raindrop.

Precipitation from layer-cloud systems. The deep, extensive, multilayer-cloud systems, from which precipitation of a usually widespread, persistent character falls, are generally formed in cyclonic depressions (lows) and near fronts. Such cloud systems are associated with feeble upcurrents of only a few centimeters per second, which last for at least several hours. Although the structure of these great raincloud systems, which are being explored by aircraft and radar, is not yet well understood, it appears that they rarely produce rain, as distinct from drizzle, unless their tops are colder than about −12°C. This suggests that ice crystals may be responsible. Such a view is supported by the fact that the radar signals from these clouds usually take a characteristic form which has been clearly identified with the melting of snowflakes.

Production of showers. Precipitation from shower clouds and thunderstorms, whether in the form of raindrops, pellets of soft hail, or true hailstones, is generally of greater intensity and shorter duration than that from layer clouds and is usually composed of larger particles. The clouds themselves are characterized by their large vertical depth, strong vertical air currents, and high concentrations of liquid water, all these factors favoring the rapid growth of precipitation elements by accretion.

In a cloud composed wholly of liquid water, raindrops may grow by coalescence with small droplets. For example, a droplet being carried up from the cloud base would grow as it ascends by sweeping up smaller droplets. When it becomes too heavy to be supported by the vertical upcurrents, the droplet will then fall, continuing to grow by the same process on its downward journey. Finally, if the cloud is sufficiently deep, the droplet will emerge from its base as a raindrop.

In a dense, vigorous cloud several kilometers deep, the drop may attain its limiting stable diameter (about 5 mm) before reaching the cloud base and thus will break up into several large fragments. Each of these may continue to grow and attain breakup size. The number of raindrops may increase so rapidly in this manner that after a few minutes the accumulated mass of water can no longer be supported by the upcurrents and falls out as a heavy shower. The conditions which favor this rapid multiplication of raindrops occur more readily in tropical regions.

In temperate regions, where the 0°C level is much lower in elevation, conditions are more favorable for the ice-crystal mechanism. However, many showers may be initiated by coalescence of waterdrops.

The ice crystals grow initially by sublimation of vapor in much the same way as in layer clouds, but when their diameters exceed about 0.1 mm, growth by collision with supercooled droplets will usually predominate. At low temperatures the impacting droplets tend to freeze individually and quickly to produce pellets of soft hail. The air spaces between the frozen droplets give the ice a relatively low density; the frozen droplets contain large numbers of tiny air bubbles, which give the pellets an opaque, white appearance. However, when the growing pellet traverses a region of relatively high air temperature or high concentration of liquid water or both, the transfer of latent heat of fusion from the hailstone to the air cannot occur sufficiently rapidly to allow all of the deposited water to freeze immediately. There then forms a wet coating of slushy ice, which may later freeze to form a layer of compact, relatively transparent ice. Alternate layers of opaque and clear ice are characteristic of large hailstones, but their formation and detailed structure are determined by many factors such as the number concentration, size and impact velocity of the supercooled cloud droplets, the temperature of the air and hailstone surface, and the size, shape, and aerodynamic behavior of the hailstone. Giant hailstones, up to 10 cm in diameter, which cause enormous damage to crops, buildings, and livestock, most frequently fall not from the large tropical thunderstorms, but from storms in the continental interiors of temperate latitudes. An example is the Nebraska-Wyoming area of the United States, where the organization of larger-scale wind patterns is particularly favorable for the growth of severe storms.

The development of precipitation in convective clouds is accompanied by electrical effects culminating in lightning. The mechanism by which the electric charge dissipated in lightning flashes is generated and separated within the thunderstorm has been debated for more than 200 years, but there is still no universally accepted theory. However, the majority opinion holds that lightning is closely associated with the appearance of the ice phase, and the most promising theory suggests that the charge is produced by the rebound of ice crystals or a small fraction of the cloud droplets that collide with the falling hail pellets.

Basic aspects of cloud physics. The various stages of the precipitation mechanisms raise a number of interesting and fundamental problems in classical physics. Worthy of mention are the supercooling and freezing of water; the nature, origin, and mode of action of the ice nuclei; and the mechanism of ice-crystal growth which produces the various snow crystal forms.

It has now been established how the maximum degree to which a sample of water may be supercooled depends on its purity, volume, and rate of cooling. The freezing temperatures of waterdrops containing foreign particles vary linearly as the logarithm of the droplet volumes for a constant rate of cooling. This relationship, which has been established for drops varying between 10 μm and 1 cm in diameter, characterizes the heterogeneous nucleation of waterdrops and is probably a consequence of the fact that the ice-nucleating ability of atmospheric aerosol increases logarithmically with decreasing temperature.

When extreme precautions are taken to purify the water and to exclude all solid particles, small droplets, about 1 μm in diameter, may be supercooled to −40°C and drops of 1 mm diameter to −35°C. Under these conditions freezing occurs spontaneously without the aid of foreign nuclei.

The nature and origin of the ice nuclei, which are necessary to induce freezing of cloud droplets at temperatures above −40°C, are still not clear. Measurements made with large cloud chambers on aircraft indicate that the most efficient nuclei, active at temperatures above −10°C, are present in concentrations of only about 10 in a cubic meter of air, but as the temperature is lowered, the numbers of ice crystals increase logarithmically to reach concentrations of about 1 per liter at −20°C and 100 per liter at −30°C. Since these measured concentrations of nuclei are less than one-hundredth of the numbers that apparently are consumed in the production of snow, it seems that there must exist processes by which the original number of ice crystals are rapidly multiplied. Laboratory experiments suggest the fragmentation of the delicate snow crystals and the ejection of ice splinters from freezing droplets as probable mechanisms.

The most likely source of atmospheric ice nuclei is provided by the soil and mineral-dust particles carried aloft by the wind. Laboratory tests have shown that, although most common minerals are relatively inactive, a number of silicate minerals of the clay family produce ice crystals in a supercooled cloud at temperatures above −18°C. A major constituent of some clays, kaolinite, which is active below −9°C, is probably the main source of highly efficient nuclei.

The fact that there may often be a deficiency of efficient ice nuclei in the atmosphere has led to a search for artificial nuclei which might be introduced into supercooled clouds in large numbers. Silver iodide is a most effective substance, being active at −4°C, while lead iodide and cupric sulfide have threshold temperatures of −6°C for freezing nuclei.

In general, the most effective ice-nucleating substances, both natural and artificial, are hexagonal crystals in which spacings between adjacent rows of atoms differ from those of ice by less than 16%. The detailed surface structure of the nucleus, which is determined only in part by the crystal geometry, is of even greater importance. This is strongly indicated by the discovery that several complex organic substances, notably steroid compounds, which have apparently little structural resemblance to ice, may act as nucleators for ice at temperatures as high as −1°C.

The collection of snow crystals from clouds at different temperatures has revealed their great variety of shape and form. By growing the ice crystals on a fine fiber in a cloud chamber, it has been possible to reproduce all the naturally occurring forms and to show how these are correlated with the temperature and supersaturation of the environment. With the air temperature along the length of a fiber ranging from 0 to −25°C, the following clear-cut changes of crystal habit are observed (Fig. 3):

Hexagonal plates — needles — hollow prisms — plates — stellar dendrites — plates — prisms

This multiple change of habit over such a small temperature range is remarkable and is thought to be associated with the fact that water molecules apparently migrate between neighboring faces on an ice crystal in a manner which is very sensitive to the temperature. Certainly the temperature rather than the supersaturation of the environment is primarily responsible for determining the basic shape of the crystal, though the supersaturation governs the growth rates of the crystals, the ratio of their linear dimensions, and the development of dendritic forms.

Artificial stimulation of rain. The presence of either ice crystals or some comparatively large waterdroplets (to initiate the coalescence mechanism) appears essential to the natural release of precipitation. Rainmaking experiments are conducted on the assumption that some clouds precipitate inefficiently, or not at all, because they are deficient in natural nuclei; and that this deficiency can be remedied by "seeding" the clouds artificially with dry ice or silver iodide to produce ice crystals, or by introducing waterdroplets or large hygroscopic nuclei. In the dry-ice method, pellets of about 1-cm diameter are dropped from an aircraft into the top of a supercooled cloud. Each pellet chills a thin sheath of air near its surface to well below −40°C and produces perhaps 10^{12} minute ice crystals, which subsequently spread through the cloud, grow, and aggregate into snowflakes. Only a few pounds of dry ice are required to seed a large cumulus cloud. Some hundreds of experiments, carried out mainly in Australia, Canada, South Africa, and the United States, have shown that cumulus clouds in a suitable state of development may be induced to rain by seeding them with dry ice on occasions when neighboring clouds, untreated, do not precipitate. However, the amounts of rain produced have usually been rather small.

For large-scale trials designed to modify the rainfall from widespread cloud systems over large areas, the cost of aircraft is usually prohibitive. The technique in this case is to release a silver iodide smoke from the ground and rely on the air currents to carry it up into the supercooled regions of the cloud. In this method, with no control over the subsequent transport of the smoke, it is not possible to make a reliable estimate of the concentrations of ice nuclei reaching cloud level, nor is it known for how long silver iodide retains its nucleating ability in the atmosphere. It is usually these unknown factors which, together with the impossibility of estimating accurately what would have been the natural rainfall in the absence of seeding activities, make the design and evaluation of a large-scale operation so difficult.

In published data, little convincing evidence can be found that large increases in rainfall have been produced consistently over large areas. Indeed, in temperate latitudes most rain falls from deep layer-cloud systems whose tops usually reach to levels at which there are abundant natural ice nuclei and in which the natural precipitation processes have plenty of time to operate. It is therefore not obvious that seeding of these clouds would produce a significant increase in rainfall, although it is possible that by forestalling natural processes some redistribution might be effected.

Perhaps more promising as additional sources of rain or snow are the persistent supercooled clouds

CLOUD PHYSICS

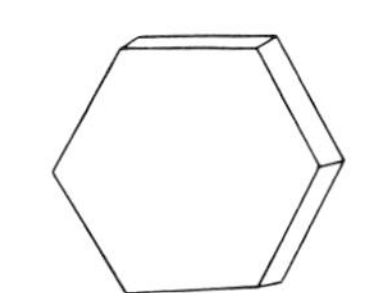

(a) thin hexagonal plate

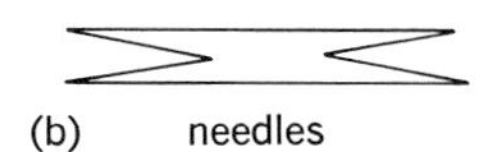

(b) needles

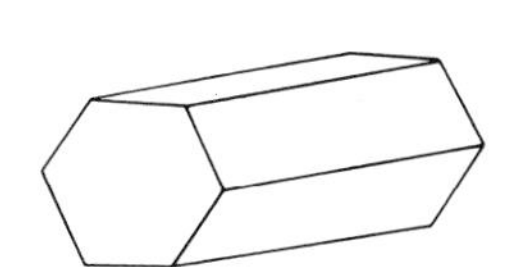

(c) hexagonal prismatic column

(d) dendritic star-shaped crystal

Fig. 3. Ice crystal types formed in various temperature ranges (°C); (*a*) 0 to −3°, −8 to −12°, −16 to −25°; (*b*) −3 to −5°; (*c*) −5 to −8°, below −25°; (*d*) −12 to −16°.

produced by the ascent of damp air over large mountain barriers. The continuous generation of an appropriate concentration of ice crystals near the windward edge might well produce a persistent light snowfall to the leeward, since water vapor is continually being made available for crystal growth by lifting of the air. The condensed water, once converted into snow crystals, has a much greater opportunity of reaching the mountain surface without evaporating, and might accumulate in appreciable amount if seeding were maintained for many hours.

The results of trials carried out in favorable locations in the United States and Australia suggest that in some cases seeding has been followed by seasonal precipitation increases of about 10%, but rarely have the effects been reproduced from one season to the next, and overall the evidence for consistent and statistically significant increases of rainfall is not impressive. Experiments of improved design will probably have to be continued for several more years before the effects of large-scale cloud seeding can be realistically assessed.

Attempts to stimulate the coalescence process by spraying waterdroplets or dispersing salt crystals into the bases of incipient shower clouds have been made in places as far apart as Australia, the Caribbean, eastern Africa, and Pakistan. The results of these experiments, though encouraging, are not yet sufficient to allow definite conclusions.

On a more optimistic note, it is relatively easy to clear quite large areas of the supercooled cloud and fog by seeding and converting it into ice crystals, which grow at the expense of the waterdrops and fall out to leave large holes.

The science and technology of cloud modification is, however, still in its infancy. It may well be that further knowledge of cloud behavior and of natural precipitation mechanisms will suggest new possibilities and improved techniques which will lead to developments far beyond those which now seem likely.

[BASIL J. MASON]

Bibliography: N. H. Fletcher, *The Physics of Rainclouds*, 1962; B. J. Mason, *Clouds, Rain and Rainmaking*, 2d ed., 1975; B. J. Mason, *The Physics of Clouds*, 2d ed., 1971.

Coal gasification

The conversion of coal, coke, or char to gaseous products by reaction with air, hydrogen, oxygen, steam, carbon dioxide, or a mixture of these. Products consist of carbon dioxide, carbon monoxide, hydrogen, methane, and some other chemicals in a ratio dependent upon the particular reactants employed and the temperatures and pressures within the reactors, as well as upon the type of treatment which the gases from the gasifier undergo subsequent to their leaving the gasifier. Strictly speaking, reaction of coal, coke, or char with air or oxygen to produce heat plus carbon dioxide might be called gasification. However, that process is more properly classified as combustion, and thus is not included in this coal gasification summary.

Industrial uses. In most coal gasification processes which are available for use or under development, the reactions are endothermic. Air or oxygen is typically supplied to the gasifier to provide the necessary heat input. From the industrial user's viewpoint, the net result is to produce either low-Btu or intermediate-Btu gas. Low-Btu gas typically has a heating value on the order of 4.5 kilojoules/m^3 (150 Btu/ft^3). Because of the absence of the nitrogen diluent, intermediate-Btu gas has twice the heating value, or about 9 kJ/m^3. Also, because the inert nitrogen is not present, intermediate-Btu gas can be upgraded to produce substitute natural gas (SNG) with a heating value of about 30 kJ/m^3 (1000 Btu/ft^3).

Each of these gas types has attracted industrial interest. The electric industry is principally investigating the production of clean, low-Btu gas from coal with the object of burning it in a combined-cycle power generation system. Such a combined-cycle system would use a gas turbine plus a waste heat boiler and steam turbine to drive separate electric generators, thereby providing overall station efficiencies approaching 45%. The natural gas industry views intermediate-Btu gas production with strong interest since this same gasification product can also be upgraded to synthetic natural gas by using relatively conventional and proved gas-processing steps. Heavy industry is studying the feasibility of using both low- and intermediate-Btu gas for many applications, thereby freeing critical volumes of natural gas and also reducing imported oil consumption. Parts of the chemical industry are examining intermediate-Btu gas as a source of hydrogen and carbon monoxide to offset potentially reduced supplies of these synthetic chemical building blocks, which are currently obtained primarily from steam reforming of natural gas, petroleum, natural gas liquids, or petroleum derivitives. Various projects are also being studied which would involve coproduction of SNG and intermediate-Btu gas for industrial or chemical use, electric power, coal-derived liquids, and various other products.

Gasification processes. Some of the basic technology for many of the coal gasification processes under active consideration today is quite old and well known. Much of it derived directly from the large body of manufactured gas technology. In 1965 Bituminous Coal Research, Inc., completed a study for the Office of Coal Research (OCR) of the U.S. Department of the Interior and published a report containing the results of a literature search on 65 different coal gasification processes. In spite of the large number of processes described and identified at that time, there continues to be a variety of new processes under development. In nearly all of the processes the chemistry of the high-temperature gasification is the same (Fig. 1). The basic reactions are:

Coal reactions

$$\text{Coal} \xrightarrow{\text{Heat}} \text{gases } (CO, CO_2, CH_4, H_2) + \text{liquids} + \text{char} \quad (1)$$

$$\text{Coal} + H_2 \xrightarrow{\text{Catalyst}} \text{liquids} + (\text{char}) \quad (2)$$

$$\text{Coal} + H_2 \text{ (from a hydrogen donor)} \rightarrow \text{liquids} + (\text{char}) \quad (3)$$

$$\text{Coal} + H_2 \xrightarrow[\text{destruction}]{\text{Noncatalytic}} CH_4 + \text{char} \quad (4)$$

Char reactions

$$C\,(\text{char}) + 2H_2 \rightarrow CH_4 \text{ exothermic} \quad (5)$$

$$C\,(\text{char}) + H_2O \rightarrow CO + H_2 \text{ endothermic} \quad (6)$$

$$C\,(\text{char}) + CO_2 \rightarrow 2CO \text{ endothermic} \quad (7)$$

$$C\,(\text{char}) + O_2 \rightarrow CO_2 \text{ exothermic} \quad (8)$$

Gaseous reactions

$$CO + H_2O \xrightarrow{\text{Catalyst}} H_2 + CO_2 \text{ exothermic} \quad (9)$$

$$CO + 3H_2 \xrightarrow{\text{Catalyst}} CH_4 + H_2O \text{ exothermic} \quad (10)$$

$$CO_2 + 4H_2 \xrightarrow{\text{Catalyst}} CH_4 + 2H_2O \text{ exothermic} \quad (11)$$

$$xCO + yH_2 \xrightarrow{\text{Fe}} \text{hydrocarbon gases and/or liquids} + zCO_2 \text{ exothermic} \quad (12)$$

Thermodynamics. From a thermodynamic standpoint in coal gasification, at least one simplifying assumption is customarily made; namely, coal can be considered as carbon in the form of coke. This assumption is made because coal is a chemically ill-defined material that does not fit into the regime of rigorous thermodynamic deduction. Errors associated with this assumption are not likely to be very large. Thus the individual high-Btu gas-producing step, reaction (5) is considered to be of great interest. Not only is this reaction exothermic, but it also has a large negative standard free energy change, indicating that it is a spontaneous reaction. Unfortunately, the rate of this reaction at ordinary temperatures and in the absence of a catalyst is nearly zero. In order to force the reaction to proceed at a fast rate, the temperature must be raised considerably. Some processes are designed to operate at as high as 1100°C. Another important consideration in regard to reaction (5) is that very large quantities of energy must be used to obtain the hydrogen for this reaction. The most widely available source of hydrogen is water, large quantities of which are necessary for a gasification plant. In fact, a plant producing 7,000,000 m^3 (250,000,000 ft^3) of pipeline-quality gas per day requires about 11,300 m^3 (3,000,000 gal) of water per day to supply hydrogen. Substantially greater volumes of water are required for associated process operations such as cooling tower makeup for low-level heat rejection.

Reaction (13) is highly endothermic and also

$$H_2O \rightarrow H_2 + \tfrac{1}{2}O_2 \quad (13)$$

requires large quantities of energy. As a result, nearly all of the gasification processes under investigation today do not decompose water directly according to reaction (13) and then follow with a reaction of hydrogen with coal or char. Rather, the water is decomposed according to reaction (6). Products of this reaction are then treated as shown in reactions (9), (10), and (11). It is currently necessary to use a multistep process as outlined in Fig. 1, although some processes are under development which rearrange these blocks in order to enhance methane formation in the gasifier.

Gasification step. The various gasification processes which have already been developed or which are in the development stage vary with respect to the gasification step. In fact, to a large degree, the mechanical and engineering variations of this step, particularly the features designed for supplying the heat for the endothermic reaction $C + H_2O$ and for handling solids in the gasifier, characterize the processes. Heat can be furnished by one of several methods: by partial combustion of coal or char with air or oxygen, by an inert (pelletized ash, ceramic spheres, and so on) heat carrier, by heat released from the reaction of a metal oxide with CO_2, by high-temperature waste heat from a nuclear reactor, or by heat produced when an electric current passes through the coal or char bed between electrodes immersed in the bed. Processes have been investigated for gasification at atmospheric and elevated pressures by using molten-media, moving-bed, fluidized-bed, or entrained suspended, dilute-phase operations in the gasifier.

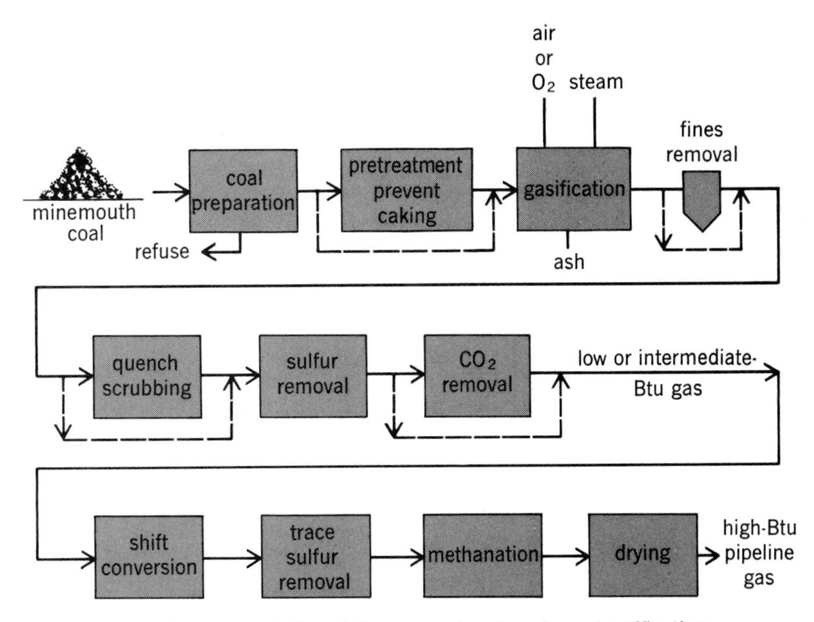

Fig. 1. Schematic representation of the processing steps in coal gasification.

Research and development. The cost of gas made by coal gasification is sufficiently high to justify continuing research and development to try to decrease capital and operating costs of gasification plants. In the United States, calculated costs for a plant capable of producing 7,000,000 m^3 (250,000,000 ft^3) per day of high-Btu gas are in the range of $1,500,000,000. Most development work on processes, or on parts of processes, is directed toward lowering the plant costs or increasing the efficiency of certain process steps. One of the major programs for developing improved coal gasification is funded jointly by the Gas Research Institute (GRI) and DOE.

In addition to the GRI-DOE program, other research and development is underway. Slagging gasifier studies have been carried out on a Lurgi unit at Westfield, Scotland, and on a COGAS unit being developed by the Coal Utilisation Research Laboratory (formerly BCURA) at Leatherhead, England. Texaco and Shell-Koppers have pilot plants under commissioning in Europe. This area is being emphasized because it is believed that slagging gasifiers will produce gas at a faster rate than nonslagging gasifiers can.

A discussion of some of the coal gasification processes, as well as some of their pertinent features, follows.

COGAS process. In the COGAS process, low-pressure coal pyrolysis is combined with char gasification to produce liquid fuel and synthetic pipeline gas. Synthesis gas is produced from the char, with air used instead of oxygen to provide the heat for the steam-carbon reaction through a separately fired combustor which heats a circulating stream of gasifier char. The operation may be carried out at the relatively low pressure of 450 kPa absolute (50 psig).

Commercially available process technology is reportedly used for hydrogen generation, gas purification and dehydration, sulfur and oil recovery, and oil hydrotreating. The COGAS technology is being considered by DOE for the construction of an 1800-metric-ton-per-day SNG demonstration plant to be built in Illinois.

Consolidated synthetic gas (CSG) process. A basic feature of this process, also known as the CO_2 acceptor process, is the use of lime or calcined dolomite to react exothermally with the CO_2 liberated in gasification to supply a portion of the heat needed for the carbon-steam reaction. The removal of CO_2 by this method enhances the water-gas shift reaction and methane formation, both of which are exothermic. The combination of these effects allows the gasification system to operate without an oxygen supply. The spent "acceptor" (lime or dolomite) is withdrawn from the gasification vessel and calcined separately, with residual char and air supplying the necessary heat. The process also embodies the feature of contacting the incoming raw coal with the hydrogen-rich synthesis gas to form a portion of the methane directly.

The operating conditions of the CO_2 acceptor process are limited to about 870°C (1600°F) and 2000 kPa (300 psia) because of melt-formation problems in the dolomite system. Since the rate of the uncatalyzed steam-carbon reaction for gasification of bituminous coals is prohibitively slow—870°C or lower—the process is not competitive for the gasification of these types of coals.

Exxon catalytic process. The Exxon catalytic process differs significantly from other SNG gasification methods in that a potassium catalyst solution is sprayed on the feed coal in order to prevent agglomeration and catalyze reactions (6) and (10) in a fluid-bed gasifier with dry ash removal. The catalyst enables the gasifier to operate at the attractively mild process conditions of 700°C (1300°F) and 3550 kilopascals (500 psig). The resulting gasification reactions are almost thermally neutral with the minor amount of required energy being supplied by preheating the fluidizing gas, which is steam plus recycled H_2 and CO. This eliminates the need for oxygen injection or other indirect heat transfer to supply heat to the gasifier. The gasifier off-gas is treated by conventional acid gas removal processing and the methane product cryogenically separated, with H_2 and CO recycled to the gasifier. The catalyst-char-ash material which is removed from the gasifier is digested at 150°C (300°F) and washed countercurrently with water to recover about 90% of the potassium catalyst for reuse.

The process has been studied in a fluid-bed test gasifier, and Exxon is constructing a fully integrated 1-metric-ton-per-day process development unit close to its refinery in Baytown, TX. Construction of a 90-metric-ton-per-day large pilot plant is under consideration by DOE.

HYGAS. Pretreated coal is slurried with a light oil (produced as a by-product of the process), and the slurry is pumped to 3550–10,450 kPa (500–1500 psig) and injected into a fluidized bed at the top of the vertical multistage unit, where the oil is driven off and recovered for reuse.

The oil-free coal then passes downward through two hydrogasification stages which provide countercurrent treatment (coal passes down, gases produced pass upward and are drawn off).

In the first hydrogasification stage, dried coal is flash-heated to the reaction temperature 620–705°C (1200–1300°F) by dilute-phase contact with hot reaction gas and recycled hot char. Volatile matter in the coal and active carbon are converted to methane in a few seconds. Active carbon, which is present during the first moments of gasification, gasifies at a rate in excess of 10 times that of the carbon in the less reactive char.

The solids then pass down to the second HYGAS stage and enter a dense-phase, fluidized-bed reactor at temperatures of 925–980°C (1700–1800°F), where formation of methane from partially depleted coal char continues simultaneously with a steam-carbon reaction that produces hydrogen and carbon monoxide.

Hot gases produced in the lower hydrogasification stage rise, passing through the first-stage hydrogasification reactor and into the fluidized drying bed, where much of their heat is used to dry the feed coal. After leaving the hydrogasifier, the raw gas is shifted to the proper CO/H_2 ratio in preparation for methanation; the oils, carbon dioxide, unreacted steam, sulfur compounds, and other impurities are removed; and the purified gas is catalytically methanated. Sulfur is recovered in elemental form.

The partially depleted coal char is used to produce a hydrogen-rich stream required in the process by steam-oxygen gasification of the char. Ash is withdrawn from the system as high-mineral-matter char.

Lurgi. The dry-bottom Lurgi gasifier employs a bed of crushed coal traveling downward through the gasifier, and operates at pressures up to 3200 kPa (450 psig). Steam and oxygen are admitted through a revolving steam-cooled grate which also removes the ash produced at the bottom of the gasifier. The gases are made to pass upward through the coal bed, carbonizing and drying the coal. Steam is used to prevent the ash from clinkering and the grate from overheating, and a hydrogen-rich gas is produced. Because of the pressure, some of the coal is hydrogenated into methane (in addition to that distilled from the coal), releasing heat which in turn is given up to the coal, minimizing the oxygen requirements.

Most gas produced by coal gasification comes from Lurgi reactors of the general type shown in Fig. 2. The internal design and operating conditions of these reactors may be altered for coals of different caking characteristics.

The dry-bottom Lurgi is considered particularly attractive for western coals and lignite. A number of gasification projects utilizing dry-bottom Lurgi technology have been proposed or are in various stages of consideration.

A number of moving-bed gasifiers which are similar to the dry-bottom Lurgi—but of atmospheric pressure—such as Wellman-Galusha and Woodall-Duckham, have been used worldwide and have been utilized or considered for on-site industrial gasification in the United States. Dry-bottom Lurgis are used at the world's largest gasification facility, a multibillion-dollar SASOL coal-to-oil plant near Johannesburg, South Africa.

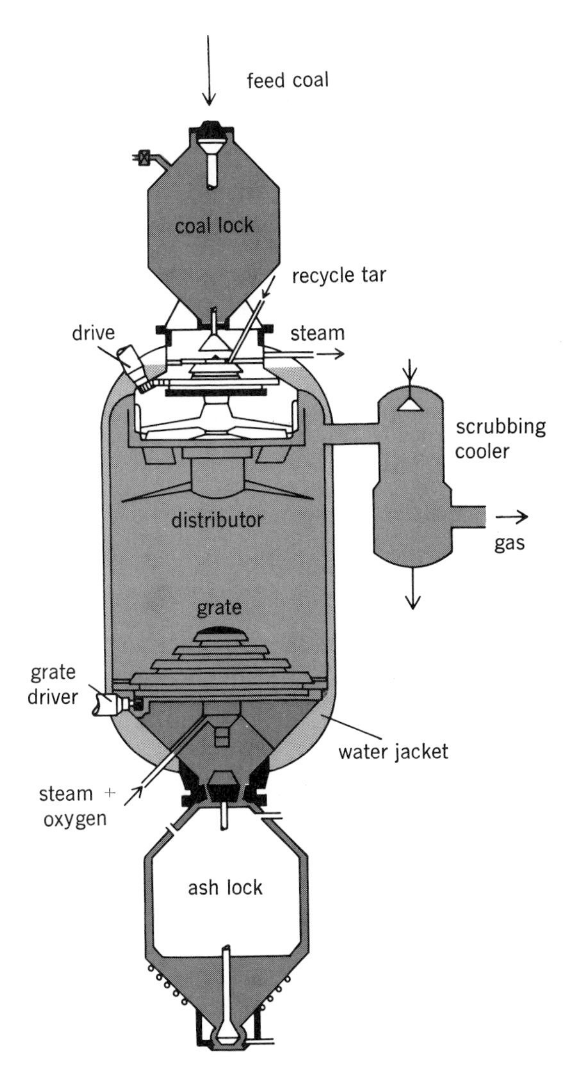

Fig. 2. Diagram of a Lurgi pressure gasifier.

An extension of this technology is the BGC-Lurgi slagging gasifier, which is under development at Westfield, Scotland. In this process the grate assembly and dry ash lock shown in Fig. 2 are replaced by a series of tuyeres firing into the lower side of the gasifier, a slag tap hole at the bottom of the gasifier, and a water-filled quench tank and slag lockhopper system. Steam and oxygen are injected through the tuyeres into a 1260°C (2300°F) 2800 kPa (400-psig) reaction zone at the bottom of the moving bed. Ash in the form of molten slag is removed through the tap hole and granulated by water in the quench vessel. This process is considered to be far more attractive than the dry-bottom Lurgi for eastern coals, which generally have lower ash fusion temperatures and reactivities. The BGC-Lurgi slagging gasification technology is being considered by DOE for an SNG demonstration plant to be built in Noble County, OH.

Rocketdyne hydrogasification process. The Rocketdyne process uses rocket engine injector technology to rapidly mix unpretreated coal in dense-phase transport with 1100°C (2000°F) hydrogen at up to 10,400 kPa (1500 psig) pressure for use in a short-residence (about 1500-ms) entrained flow reactor tube. Conventional heat exchange technology is used to heat the inlet hydrogen to 815°C (1500°F), followed by partial combustion with injected oxygen to reach required reactor inlet temperatures. Depending on reactor temperatures, pressures, and residence times, the reactor product consists of various mixtures of unconsumed excess hydrogen, light hydrocarbon gases which have included up to 90% methane, vapors of light and heavy hydrocarbon liquids, and a dry char.

The process is being developed under DOE sponsorship at 0.22- and 0.9-metric-ton-per-hour experimental reactors which are operated at Rockwell International's laboratory in Canoga Park, CA. Designs are also under way for a 90-metric-ton-per-day experimental reactor system. In commercial concepts, hydrogen to operate the Rocketdyne process is projected to be supplied from a separate oxygen-based char-coal gasification system. Commercial concepts have generally been based on the coproduction of SNG and benzene-like liquids.

Texaco process. In the Texaco process, a coal-water slurry is reacted with oxygen in an entrained gasifier with ash removal as a granulated slag. The pilot test gasifier at Montebello, CA, is a downward-fired, two-zone gasifier. The top zone is a reaction chamber, and the bottom zone a water-filled quench vessel. The molten slag and any ungasified carbon soot by-product are removed by the water. Saturated product gas exits the quench chamber at about 925°C (1700°F). The granulated slag is removed periodically by a lockhopper and the soot by an aqueous slurry which can be concentrated and recycled to the gasifier. The clarified quench water stream reportedly contains no tars or phenols.

The Texaco coal gasification process is the result of a development program based on the Texaco synthesis gas generation process, which has been licensed to numerous worldwide plants for feedstocks ranging from natural gas to asphalts. Several different-sized generators for these feedstocks are available which can be operated at over 8400 kPa (1200 psig) and produce more than 3,000,000 m^3 (110,000,000 ft^3) per day of hydrogen and carbon monoxide in a single generator. The Electric Power Research Institute (EPRI) has used Texaco's Montebello test unit to demonstrate gasification of coal liquefaction residues. A Texaco coal gasification pilot plant of 136-metric-tons-per-day capacity is being demonstrated in West Germany.

Westinghouse process. The Westinghouse gasification process consists of a fluid-bed devolatilizer operating at 870°C (1600°F) and 1650 kPa (225 psig). The devolatilizer is fluidized by hot reducing gases from a gasifier in which ash is removed as an agglomerate. Unpretreated coal enters the devolatilizer, where it is rapidly mixed with char from the fluidized bed in a draft tube in order to prevent agglomeration as the coal is heated through its plastic stage. Char is continuously withdrawn from the devolatilizer to feed the gasifier-agglomerator. In the bottom portion of this reactor, a portion of the char is combusted with air at a temperature of 1065°C (1950°F), at which ash particles agglomerate and defluidize. The spherical ash agglomerates are continuously removed from the gasifier after being cooled with steam, which is used to gasify the remainder of the char in the gasifier-agglomerator and to moderate the combustion zone temperature.

The Westinghouse process is being developed under the auspices of DOE at a 13.6-metric-ton-per-day pilot unit at Waltz Mill, PA. The process has been initially developed for air gasification to supply fuel for electric production from combustion turbines or combined cycle systems. However, tests have been run using oxygen in both the two-stage integrated configuration of coupled devolatilizer and gasifier-agglomerator vessels and in an advanced single-vessel configuration.

Other processes. A number of other gasification processes are either in use worldwide or under development. (These gasification systems are characterized parenthetically in this section by their reaction method/pressure/type of ash removal.) Examples of systems in worldwide use include Koppers-Totzek (entrained/atmospheric/combination fritted slag and char-ash) and Winkler (fluid-bed/atmospheric/char-ash). Examples of developmental processes which are of interest either because of their industrial support or because of their technological features include: Atomics International (molten salt bath/pressurized/aqueous slurry ash), Combustion Engineering (entrained-flow air-fired/pressurized/fritted slag), Otto Szarberg (molten slag bath/pressurized/fritted slag), and Shell-Koppers (entrained/pressurized/fritted slag).

Underground gasification. At the request of the Bureau of Mines, a detailed survey of underground gasification technology was prepared in 1971. This survey of worldwide activities revealed no large-scale active program of either basic or applied research. However, there has been a renewal of interest in the United States on the part of industrial concerns, and there are indications of one or two exploratory experimental programs on a modest scale. Underground coal gasification has been practiced on commercial scale in the Soviet Union's steeply dipping coal beds.

The Bureau of Mines has begun work in this field again, with particular emphasis on the utilization of new technology developed since the mid-1950s. This new technology includes an understanding of the nature and direction of subsurface fracture systems and the means to calculate underground fluid movement. Directional drilling techniques have been advanced; and chemical explosive fracturing, a new method of preparing underground formations, has been introduced by the Bureau of Mines. DOE has conducted field tests using linked vertical wells near Hanna, WY. Underground coal gasification is generally considered most applicable to low-rank nonswelling coals found in the western United States. Additionally, field experiments have established the technical possibility of in-place recovery of crude oil and shale oil, and modern methods of surface-processing of coal to high-Btu gases have been demonstrated.

At the same time, an active program on fracture characterization and gas flow in coal beds is under way at the Morgantown Energy Research Center.

[RICHARD H. MC CLELLAND]

Bibliography: American Gas Association–GRI-DOE-IGU, *Proceedings of the 7th, 8th, 9th and 10th Synthetic Gas–Pipeline Gas Symposiums*, 1975–1978; Coal Technology *Conference Papers*, Vols. 1 and 2, 1978; DOE–Morgantown Energy Research Center, *An Engineering Assessment of Entrainment Gasification*, Rep. MERC/RI-78/2, 1978; Dravo-DOE, *Handbook of Gasifiers and Gas Treatment Systems*, 1976, Rep. FE-1772-11, 1976.

Coastal engineering

A branch of civil engineering concerned with the planning, design, construction, and maintenance of works in the coastal zone. The purposes of these works include control of shoreline erosion; development of navigation channels and harbors; defense against flooding caused by storms, tides, and seismically generated waves (tsunamis); development of coastal recreation; and control of pollution in nearshore waters. Coastal engineering usually involves the construction of structures or the transport and possible stabilization of sand and other costal sediments.

The successful coastal engineer must have a working knowledge of oceanography and meteorology, hydrodynamics, geomorphology and soil mechanics, statistics, and structural mechanics. Tools that support coastal engineering design include analytical theories of wave motion, wave-structure interaction, diffusion in a turbulent flow field, and so on; numerical and physical hydraulic models; basic experiments in wave and current flumes; and field measurements of basic processes such as beach profile response to wave attack, and the construction of works. Postconstruction monitoring efforts at coastal projects have also contributed greatly to improved design practice.

Environmental forces. The most dominant agent controlling coastal processes and the design of coastal works is usually the waves generated by the wind. Wind waves produce large forces on coastal structures, they generate nearshore currents and the alongshore transport of sediment, and they mold beach profiles. Thus, a primary concern of coastal engineers is to determine the wave climate (statistical distribution of heights, periods, and directions) to be expected at a particular site. This includes the annual average distribution as well as long-term extreme characteristics. In addition, the nearshore effects of wave refraction, diffraction, reflection, breaking, and run-up on structures and beaches must be predicted for adequate design.

Other classes of waves that are of practical importance include the astronomical tide, tsunamis, and waves generated by moving ships. The tide raises and lowers the nearshore water level and thus establishes the range of shoreline over which coastal processes act. It also generates reversing currents in inlets, harbor entrances, and other locations where water motion is constricted. Tidal currents which often achieve a velocity of 1–2 m/s can strongly affect navigation, assist with the maintenance of channels by scouring sediments, and dilute polluted waters.

Tsunamis are quite localized in time and space but can produce devastating effects. Often the only solution is to evacuate tsunami-prone areas or suffer the consequences of a surge that can reach elevations in excess of 10 m above sea level. Some attempts have been made to design structures to withstand tsunami surge or to plant trees and construct offshore works to reduce surge velocities and runup elevations. *See* TSUNAMI.

The waves generated by ships can be of greater importance at some locations than are wind-generated waves. Ship waves can cause extensive bank erosion in navigation channels and undesirable disturbance of moored vessels in unprotected marinas.

On coasts having a relatively broad shallow offshore region (such as the Atlantic and Gulf coasts of the United States), the wind and lower pressures in a storm will cause the water level to rise at the shoreline. Hurricanes have been known to cause storm surge elevations of as much as 5 m for periods of several hours to a day or more. Damage is caused primarily by flooding, wave attack at the raised water levels, and high wind speeds. Defense against storm surge usually involves raising the crest elevation of natural dune systems or the construction of a barrier-dike system. *See* STORM SURGE.

Other environmental forces that impact on coastal works include earthquake disturbances of the sea floor and static and dynamic ice forces. Direct shaking of the ground will cause major structural excitations over a region that can be tens of kilometers wide surrounding the epicenter of a major earthquake. Net dislocation of the ground will modify the effect of active coastal processes and environmental forces on structures. *See* EARTHQUAKE.

Ice that is moved by flowing water and wind or raised and lowered by the tide can cause large and often controlling forces on coastal structures. However, shore ice can prevent coastal erosion by keeping wave action from reaching the shore.

Coastal processes. Wind-generated waves are the dominant factor that causes the movement of sand parallel and normal to the shoreline as well as the resulting changes in beach morphology. Thus, structures that modify coastal zone wave activity can strongly influence beach processes and geometry.

Active beach profile zone. Figure 1 shows typical beach profiles found at a sandy shoreline (which may be backed by a cliff or a dune field). The backshore often has one or two berms with a crest elevation equal to the height of wave runup during high tide. When low swell, common during calm conditions, acts on the beach profile, the beach face is built up by the onshore transport of sand. This accretion of sand adds to the seaward berm. On the other hand, storm waves will attack the beach face, cut back the berm, and carry sand offshore. This active zone of shifting beach profiles occurs primarily landward of the −10-m depth contour. If the storm tide and waves are sufficiently high, the berms may be eroded away to expose the dunes or cliff to erosion. The beach profile changes shown in Fig. 1 are superimposed on any longer-term advance or retreat of the shoreline caused by a net gain or loss of sand at that location.

Any structure constructed along the shore in the active beach profile zone may retain the sand behind it and thus reduce or prevent erosion. However, wave attack on the seaward face of the structure causes increased turbulence at the base of the structure, and usually increased scour which must be allowed for or prevented, if possible, by the placement of stone or some other protective material.

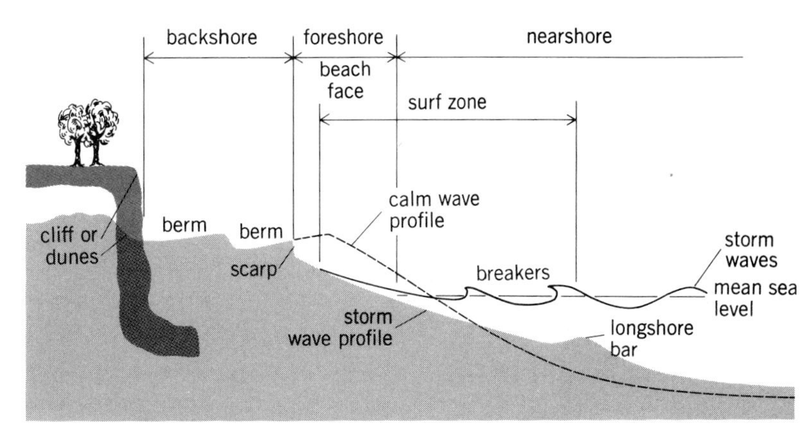

Fig. 1. Typical beach profiles (vertical scale is exaggerated).

It is desirable to keep all construction of dwellings, recreational facilities, and such landward of the active beach profile zone, which usually means landward of the frontal dunes or a good distance back from retreating cliffs. It is also desirable to maintain and encourage the growth of the frontal dune system by planting grass or installing sand fencing.

In addition to constructing protective structures and stabilizing the dune system, it is common practice to nourish a beach by placing sand on the beach face and nearshore area. This involves an initial placement of sand to develop the desired profile and periodic replenishment to make up for losses to the profile. A common source for sand, which should be clean and at least as coarse as the native sand, is the offshore area near the nourishment site.

Alongshore current and transport. Waves arriving with their crest oriented at an angle to the shoreline will generate a shore-parallel alongshore current in the nearshore zone. The current flows in the direction of the alongshore component of wave advance and has the highest velocity just inside the breaker line. It may be assisted or hindered by the wind and by tidal action, particularly along sections of the shore adjacent to tidal inlets and harbor entrances. There is a continuous accumulation of flow in the downcoast direction which may be relieved by seaward-flowing jets of water known as rip currents.

The alongshore current transports sand in suspension and as bed load, and is assisted by the breaking waves, which place additional sand in suspension. Also, wave runup and the return flow transports sand particles in a zigzag fashion along the beach face. Coastal structures that obstruct these alongshore transport processes can cause the deposition of sand. They do this by blocking wave action from a section of the shore and thus removing the wave energy required to maintain the transport system; by interfering with the transport process itself; or by directly shutting off a source of sand that feeds the transport system (such as a structure that protects an eroding shoreline).

The design of most coastal works requires a determination of the volumetric rate of alongshore sand transport at the site—both the gross rate (upcoast plus downcoast transport) and the net rate (upcoast minus downcoast transport). The most reliable method of estimating transport rates is by measuring the rate of erosion or deposition at an artificial or geomorphic structure that interrupts the transport. Also, field studies have developed an approximate relationship between the alongshore transport rate and the alongshore component of incident wave energy per unit time. With this, net and gross transport rates can be estimated if sufficient information is available on the annual wave climate. Typical gross transport rates on exposed ocean shorelines often exceed 500,000 yd^3 (382,000 m^3) per year.

Primary sources of beach sediment include rivers discharging directly to the coast, beach and cliff erosion, and artificial beach nourishment. Sediment transported alongshore from its sources will eventually be deposited at some semipermanent location or sink. Common sinks include harbors and tidal inlets; dune fields; offshore deposition; spits, tombolos, and other geomorphic formations; artificial structures that trap sand; and areas where beach sand is mined.

By evaluating the volumetric transports into and out of a segment of the coast, one can develop a sediment budget for the coastal segment. If the supply exceeds the loss, shoreline accretion will occur, and vice versa. When a coastal project modifies the supply or loss to the segment, geomorphic changes can be expected. For example, when a structure that traps sediment is constructed upcoast of a point of interest, the shoreline at the point of interest can be expected to erode as it resupplies the longshore transport capacity of the waves.

Harbor entrance and tidal inlet control structures built to improve navigation conditions, stabilize navigation channel geometry, and assist with the relief of flood waters will often trap a large portion of the alongshore transport. This can result in undesirable deposition at the harbor or inlet entrance and subsequent downcoast erosion. The solution usually involves designing the entrance structures to trap the sediment at a fixed and acceptable location and to provide protection from wave attack at this location so a dredge can pump the sand to the downcoast beach.

Coastal structures. Coastal structures can be classified by the function they serve and by their structural features. Primary functional classes include seawalls, revetments, and bulkheads; groins; jetties; breakwaters; and a group of miscellaneous structures including piers, submerged pipelines, and various harbor and marina structures.

Seawalls, revetments, and bulkheads. These structures are constructed parallel or nearly parallel to the shoreline at the land-sea interface for the purpose of maintaining the shoreline in an advanced position and preventing further shoreline recession. Seawalls are usually massive and rigid, while a revetment is an armoring of the beach face with stone rip-rap or artificial units. A bulkhead acts primarily as a land-retaining structure and is found in a more protected environment such as a navigation channel or marina.

A key factor in the design of these structures is that erosion can continue on adjacent shores and flank the structure if it is not tied in at the ends. Erosion on adjacent shores also increases the exposure of the main structure to wave attack. Structures of this class are prone to damage and possible failure caused by wave-induced scour at the toe. In order to prevent this, the toe must be stabilized by driving vertical sheet piling into the beach, laying stone on the beach seaward of the toe, or maintaining a protective beach by artificial nourishment.

Revetments that are sufficiently porous will allow leaching of sand from below the structure. This can lead to structure slumping and failure. A proper stone or cloth filter system must be developed to prevent damage to the revetment.

Groins. A groin is a structure built perpendicular to the shore and usually extending out through the surf zone under normal wave and surge-level conditions. It functions by trapping sand from the alongshore transport system to widen and protect a beach or by retaining artificially placed sand. The resulting shoreline positions before and after construction of a series of groins and after nourishment is placed between the groins are shown schematically in Fig. 2. Typical groin alongshore spacing-to-length ratios vary from 1.5:1 up to 4:1.

There will be erosion downcoast of the groin field, the volume of erosion being approximately equal to the volume of sand removed by the groins from the alongshore transport system. Groins must be sufficiently tied into the beach so that downcoast erosion superimposed on seasonal beach profile fluctuations does not flank the landward end of a groin. Even the best-designed groin system will not prevent the loss of sand offshore in time of storms.

Jetties. Jetties are structures built at the entrance to a river or tidal inlet to stabilize the entrance as well as to protect vessels navigating the entrance channel. Stabilization is achieved by eliminating or reducing the deposition of sediment coming from adjacent shores and by confining the river or tidal flow to develop a more uniform and hydraulically efficient channel. Jetties improve navigation conditions by eliminating bothersome crosscurrents and by reducing wave action in the entrance.

At many entrances there are two parallel (or nearly parallel) jetties that extend approximately to the seaward end of the dredged portion of the channel. However, at some locations a single updrift or downdrift jetty has been used, as have

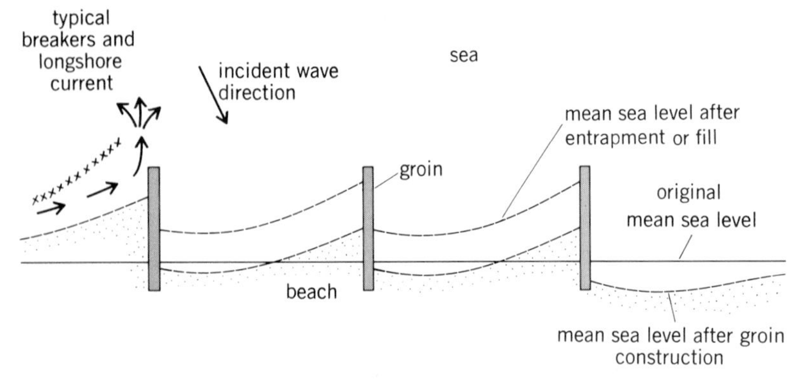

Fig. 2. Groin system and beach response.

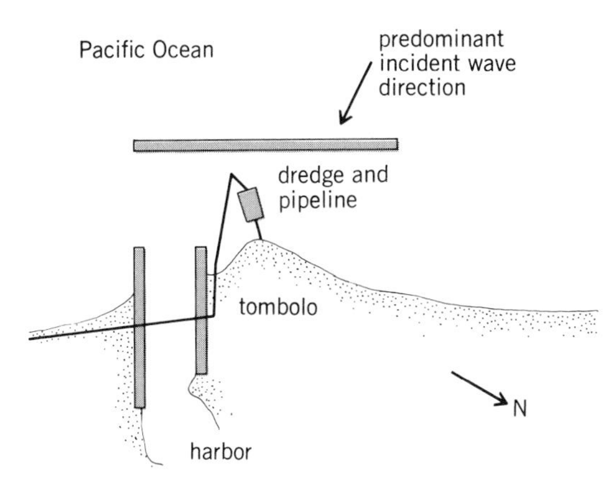

Fig. 3. Channel Islands Harbor, in California.

other arrangements such as arrowhead jetties (a pair of straight or curved jetties that converge in the seaward direction). Jetty layouts may also be modified to assist sediment-bypassing operations. A unique arrangement is the weir-jetty system in which the updrift jetty has a low section or weir (crest elevation about mean sea level) across the surf zone. This allows sand to move over the weir section and into a deposition basin for subsequent transport to the downcoast shore by dredge and pipeline.

Breakwaters. The primary purpose of a breakwater is to protect a shoreline or harbor anchorage area from wave attack. Breakwaters may be located completely offshore and oriented approximately parallel to shore, or they may be oblique and connected to the shore where they often take on some of the functions of a jetty. At locations where a natural inland site is not available, harbors have been developed by the construction of shore-connected breakwaters that cover two or three sides of the harbor.

Figure 3 shows the breakwater and jetty system at the entrance to Channel Islands Harbor in southern California. The offshore breakwater intercepts incident waves, thus trapping the predominantly southeastern longshore sand transport; it provides a protected area where a dredge can operate to bypass sediment; and it provides protection to the harbor entrance. A series of shore-parallel offshore breakwaters (with or without artificial nourishment) has been used for shore protection at a number of locations. If there is sufficient fill or trapped material, the tombolo formed in the lee of each breakwater may grow until it reaches the breakwater.

Breakwaters are designed to intercept waves, and often extend into relatively deep water, so they tend to be more massive structures than are jetties

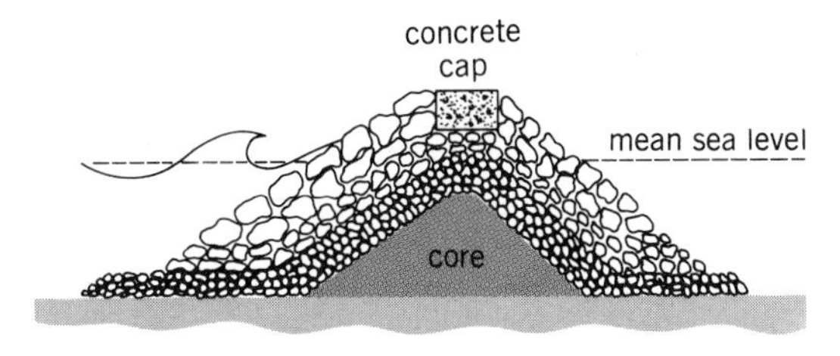

Fig. 4. Rubble mound breakwater cross section.

or groins. Breakwaters constructed to provide a calm anchorage area for ships usually have a high crown elevation to prevent overtopping by incident waves and subsequent regeneration of waves in the lee of the breakwater.

Groins, jetties, and breakwaters are most commonly constructed as rubble mound structures. Figure 4 shows the cross section of a typical rubble mound breakwater placed on a sand foundation. The breakwater has an outer armor layer consisting of the largest stones or, if sufficiently large stones are not available, the armor units may be molded of concrete with a special shape. Stone sizes decrease toward the core and base in order to develop a filter system so that the fine core stone and base sand are not removed by wave and current action. The core made of fine stone sizes is provided to diminish wave transmission through the structure. Jetties and groins have a simpler cross section, consisting typically of only armor and core layers. Breakwaters, groins, and jetties have also been constructed of steel or concrete caissons with sand and gravel fill; wood, steel, and concrete sheet piles; and sand-filled bags.

A different type of breakwater that can be effective where incident wave periods are short and large water-level fluctuations occur (as in a reservoir marina) is the moored floating breakwater. This type has been constructed of hollow concrete prisms, scrap tires, logs, and a variety of other materials.

In an attempt to develop low-cost shore protection, a number of novel materials have been used for shoreline revetments, including cinder blocks, tires, sand-filled rubber tubes, woven-fiber mattresses, and soil-cement paving.

[ROBERT M. SORENSEN]

Bibliography: American Society of Civil Engineers, *Journal of the Waterway, Port, Coastal and Ocean Engineering Division* and *Proceedings of the International Conferences on Coastal Engineering*; P. D. Komar, *Beach Processes and Sedimentation*, 1976; A. D. Quinn, *Design and Construction of Ports and Marine Structures*, 1972; R. Silvester, *Coastal Engineering*, 2 pts., 1974; R. M. Sorensen, *Basic Coastal Engineering*, 1978; U.S. Army Coastal Engineering Research Center, *Shore Protection Manual*, 3 vols., 1977.

Comfort heating

The maintenance of the temperature in a closed volume, such as a home, office, or factory, at a comfortable level during periods of low outside temperature. Two principal factors determine the amount of heat required to maintain a comfortable inside temperature: the difference between inside and outside temperatures and the ease with which heat can flow out through the enclosure.

Heating load. The first step in planning a heating system is to estimate the heating requirements. This involves calculating heat loss from the space, which in turn depends upon the difference between outside and inside space temperatures and upon the heat transfer coefficients of the surrounding structural members.

Outside and inside design temperatures are first selected. Ideally, a heating system should maintain the desired inside temperature under the most severe weather conditions. Economically, however, the lowest outside temperature on record for a locality is seldom used. The design temperature selected depends upon the heat capacity of the structure, amount of insulation, wind exposure, proportion of heat loss due to infiltration or ventilation, nature and time of occupancy or use of the space, difference between daily maximum and minimum temperatures, and other factors. Usually the outside design temperature used is the median of extreme temperatures.

The selected inside design temperature depends upon the use and occupancy of the space. Generally it is between 66 and 75°F (19 and 24°C).

The total heat loss from a space consists of losses through windows and doors, walls or partitions, ceiling or roof, and floor, plus air leakage or ventilation. All items but the last are calculated from $H_l = UA\,(t_i - t_o)$, where heat loss H_l is in British thermal units per hour (or in watts), U is overall coefficient of heat transmission from inside to outside air in Btu/(hr)(ft²)(°F) (or J/s · m² · °C), A is inside surface area in square feet (or square meters), t_i is inside design temperature, and t_o is outside design temperature in °F (or°C).

Values for U can be calculated from heat transfer coefficients of air films and heat conductivities for building materials or obtained directly for various materials and types of construction from heating guides and handbooks.

The heating engineer should work with the architect and building engineer on the economics of the completed structure. Consideration should be given to the use of double glass or storm sash in areas where outside design temperature is 10°F (−12°C) or lower. Heat loss through windows and doors can be more than halved and comfort considerably improved with double glazing. Insulation in exposed walls, ceilings, and around the edges of the ground slab can usually reduce local heat loss by 50–75%. Table 1 compares two typical dwellings. The 43% reduction in heat loss of the insulated house produces a worthwhile decrease in the cost of the heating plant and its operation. Building the house tight reduces the normally large heat loss due to infiltration of outside air. High heating-energy costs may now warrant 4 in. (10 cm) of insulation in the walls and 8 in. (20 cm) or more in the ceiling.

Humidification. In localities where outdoor temperatures are often below 36°F (2°C), it is advisable to provide means for adding moisture in heated spaces to improve comfort. The colder the outside air is, the less moisture it can hold. When it is heated to room temperature, the relative humidity in the space becomes low enough to dry out nasal membranes, furniture, and other hygroscopic materials. This results in discomfort as well as deterioration of physical products.

Various types of humidifiers are available. The most satisfactory type provides for the evaporation of the water to take place on a mold-resistant treated material which can be easily washed to get rid of the resultant deposits. When a higher relative humidity is maintained in a room, a lower dry-bulb temperature or thermostat setting will provide an equal sensation of warmth. This does not mean, however, that there is a saving in heating fuel, because heat from some source is required to evaporate the moisture.

Some humidifiers operate whenever the furnace fan runs, and usually are fed water through a float-controlled valve. With radiation heating, a unitary humidifier located in the room and controlled by a humidistat can be used.

Insulation and vapor barrier. Good insulating material has air cells or several reflective surfaces. A good vapor barrier should be used with or in addition to insulation, or serious trouble may result. Outdoor air or any air at subfreezing temperatures is comparatively dry, and the colder it is the drier it can be. Air inside a space in which moisture has been added from cooking, washing, drying, or humidifying has a much higher vapor pressure than cold outdoor air. Therefore, moisture in vapor form passes from the high vapor pressure space to the lower pressure space and will readily pass through most building materials. When this moisture reaches a subfreezing temperature in the structure, it may condense and freeze. When the structure is later warmed, this moisture will thaw and soak the building material, which may be harmful. For example, in a house that has 4 in. (10 cm) or more of mineral wool insulation in the attic floor, moisture can penetrate up through the second floor ceiling and freeze in the attic when the temperature there is below freezing. When a warm day

Table 1. Effectiveness of double glass and insulation*

Heat-loss members	Area, ft²†	Heat loss, Btu/hr‡ — With single-glass weather-stripped windows and doors	Heat loss, Btu/hr‡ — With double-glass windows, storm doors, and 2-in. (5.1-cm) wall insulation
Windows and doors	439	39,600	15,800
Walls	1,952	32,800	14,100
Ceiling	900	5,800	5,800
Infiltration		20,800	20,800
Total heat loss		99,000	56,500
Duct loss in basement and walls (20% of total loss)		19,800	11,300
Total required furnace output		118,800	67,800

*Data are for two-story house with basement in St. Louis, Mo. Walls are frame with brick veneer and 25/32-in. (2.0-cm) insulation plus gypsum lath and plaster. Attic floor has 3-in. (7.6-cm) fibrous insulation or its equivalent. Infiltration of outside air is taken as a 1-hr air change in the 14,400 ft³ (408 m³) of heating space. Outside design temperature is −5°F (−21°C); inside temperature is selected as 75°F (24°C). †1 ft² = 0.0929 m². ‡1 Btu/hr = 0.293 W.

comes, the ice will melt and can ruin the second floor ceiling. Ventilating the attic helps because the dry outdoor air readily absorbs the moisture before it condenses on the surfaces. Installing a vapor barrier in insulated outside walls is recommended, preferably on the room side of the insulation. Good vapor barriers include asphalt-impregnated paper, metal foil, and some plastic-coated papers. The joints should be sealed to be most effective.

Thermography. Remote heat-sensing techniques evolved from space technology developments related to weather satellites can be used to detect comparative heat energy losses from roofs, walls, windows, and so on. A method called thermography is defined as the conversion of a temperature pattern detected on a surface by contrast into an image called a thermogram (see illustration). Thermovision is defined as the technique of utilizing the infrared radiation from a surface, which varies with the surface temperatures, to produce a thermal picture or thermogram. A camera can scan the area in question and focus the radiation on a sensitive detector which in turn converts it to an electronic signal. The signal can be amplified and displayed on a cathode-ray tube as a thermogram.

Normally the relative temperature gradients will vary from white through gray to black. Temperatures from −22° to 3540°F (−30° to 2000°C) can be measured. Color cathode-ray tubes may be used to display color-coded thermograms showing as many as 10 different isotherms. Permanent records are possible by using a photographic camera or magnetic tape.

Infrared thermography is used to point out where energy can be saved, and comparative insulation installations and practices can be evaluated. Thermograms of roofs are also used to indicate areas of wet insulation caused by leaks in the roof.

Infiltration. In Table 2, the loss due to infiltration is large. It is the most difficult item to estimate accurately and depends upon how well the house is built. If a masonry or brick-veneer house is not well caulked or if the windows are not tightly fitted and weather-stripped, this loss can be quite large. Sometimes, infiltration is estimated more accurately by measuring the length of crack around windows and doors. Illustrative quantities of air leakage for various types of window construction are shown in Table 2. The figures given are in cubic feet of air per foot of crack per hour.

Design. Before a heating system can be designed, it is necessary to estimate the heating load for each room so that the proper amount of radiation or the proper size of supply air outlets can be selected and the connecting pipe or duct work designed.

Heat is released into the space by electric lights and equipment, by machines, and by people. Credit to these in reducing the size of the heating system can be given only to the extent that the equipment is in use continuously or if forced ventilation, which may be a big heat load factor, is not used when these items are not giving off heat, as in a factory. When these internal heat gain items are large, it may be advisable to estimate the heat requirements at different times during a design day under different load conditions to maintain inside temperatures at the desired level.

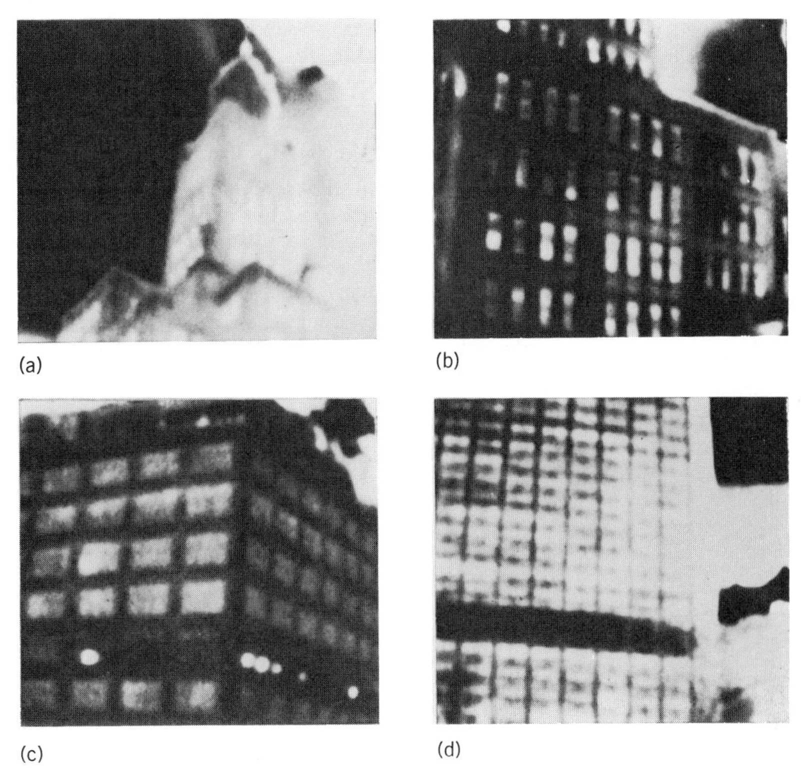

Thermograms of building structures: *(a–c)* masonry buildings; *(d)* glass-faced building. Black indicates negligible heat loss; gray, partial loss; and white, excessive loss. *(Courtesy of A. P. Pontello)*

Cost of operation. Design and selection of a heating system should include operating costs. The quantity of fuel required for an average heating season may be calculated from

$$F = \frac{Q \times 24 \times \mathrm{DD}}{(t_i - t_o) \times \mathrm{Eff} \times H}$$

where F = annual fuel quantity, same units as H
Q = total heat loss, Btu/hr (or J/s)
t_i = inside design temperature, °F (or °C)
t_o = outside design temperature, °F (or °C)
Eff = efficiency of total heating system (not just the furnace) as a decimal
H = heating value of fuel
DD = degree-days for the locality for 65°F (19.3°C) base, which is the sum of 65 (19.3) minus each day's mean temperature in °F (or °C) for all the days of the year.

Table 2. Infiltration loss with 15-mph outside wind

Building item	Infiltration, ft³/(ft)(hr)
Double-hung unlocked wood sash windows of average tightness, non-weather-stripped including wood frame leakage	39
Same window, weather-stripped	24
Same window poorly fitted, non-weather-stripped	111
Same window poorly fitted, weather-stripped	34
Double-hung metal windows unlocked, non-weather-stripped	74
Same window, weather-stripped	32
Residential metal casement, 1/64-in. crack	33
Residential metal casement, 1/32-in. crack	52

If a gas furnace is used for the insulated house of Table 1, the annual fuel consumption would be

$$F = \frac{56{,}500 \times 24 \times 4699}{[75-(-5)] \times 0.80 \times 1050} = 94{,}800 \text{ ft}^3 \text{ (2684 m}^3\text{)}$$

For a 5°F (3°C), 6- to 8-hr night setback, this consumption would be reduced by about 5%.

[GAYLE B. PRIESTER]

Bibliography: American Society of Heating, Refrigerating, and Air Conditioning Engineers, *Handbook of Fundamentals*, 1977; A. P. Pontello, Thermography: Bringing energy waste to light, *Heat./Piping/Air Condit.*, 50(3):55–61, 1978.

Community

An aggregation of organisms characterized by a distinctive combination of species, such as deciduous forest, grassland, pond, or mud flat. An ecological community is sometimes called a biocenose. Each community occupies a particular habitat and together with its habitat constitutes an ecosystem. A community may be composed of any distinctive combination of plant or animal species. When both are included, the term biotic community is commonly used.

Major communities together with their habitats are self-sustaining units. Minor communities, sometimes called societies, are secondary combinations of species within major communities. *See* ECOSYSTEM.

Dominants. Within most but not all major communities, certain organisms, called dominants, exert a commanding role in determining the character and composition of the community. On land, plants are commonly dominants because they receive the full impact of the climate and environment. By reactions on the habitat plants establish the conditions of light, moisture, air movement, space, and other factors under which all other organisms must live. The plant dominants are ordinarily the most prominent species and serve as the major source of food, substrate, and shelter for the animals that are present. Thus trees are the dominant species in a forest community, while grasses dominate in a prairie. Animals sometimes exert dominance, especially in aquatic habitats. For instance, carp and suckers in ponds may root out all submerged vegetation and make the water so turbid by stirring up the bottom mud that some other species of fish cannot exist. Dominance may thus be affected through coactions (interaction of one organism on another) as well as by reactions, although this is less common. *See* ECOLOGICAL INTERACTIONS.

Structure. The structure of a community is evident in various ways in addition to the relation between the dominants and other constituent species. Terrestrial vegetation is commonly stratified into subterranean, ground, herb, shrub, and one or more tree layer societies, although tree and shrub strata are absent from some communities, for instance, the grassland. The microclimate of each stratum is somewhat different and this, together with the variation in substrate, divides the animal distribution into two principal layer societies: subterranean-ground and herb-shrub-tree.

Community changes. Except for tropical rainforests, seasonal changes of climate during the year bring important changes in the kinds of plants that are active and in the activities of animals. In humid north temperate climates, different aspects may be recognized: hiemal, from mid-November to mid-March; vernal, to late May; estival, to mid-August; and autumnal. In addition, the community changes between day and night, especially in the predominant animals that are active. Species are commonly divided into diurnal, crepuscular (evening, dawn), nocturnal, and arhythmic, or those exhibiting no periodicity.

Species composition. Species may be classified in respect to their fidelity to a particular community into exclusive, if confined to one community; characteristic, if significantly more abundant in one community than in any other; or ubiquitous in several communities. In respect to abundance, they may be predominant and member constituents. Two aggregations of species in different localities or habitats are not considered as belonging to distinct communities unless they are more different than they are alike; that is, more than 50% of the predominant species in each one are characteristic or exclusive to it alone. Communities are commonly named after two or three dominant or predominant species or the type of vegetation or habitat. *See* BIOME; PLANT FORMATIONS, CLIMAX; PLANT GEOGRAPHY; VEGETATION ZONES, ALTITUDINAL; VEGETATION ZONES, WORLD.

Community balance. All organisms fit into food chains, which may have only three links, or trophic levels—such as grass, antelope, coyote—or as many as five, for instance, bacteria, protozoan, rotifer, small fish, large fish. In each community the food chains anastomose in various ways to form a characteristic food web. *See* ECOLOGY; FOOD CHAIN.

The number, biomass or number times weight, and rate of reproduction of organisms decrease progressively at higher trophic levels. There is an inverse relation between number and size of organisms, the pyramid of numbers. Since each trophic level depends on the surplus production of food by lower trophic levels, a balance of nature becomes established within the community. Thus, the abundance of each species tends to become stabilized or to fluctuate mildly around some definite level. A temporary excess or diminution in numbers of any species may reverberate throughout the community until a new balance becomes established. *See* BIOMASS.

Succession. Because of fixation of nitrogen by certain bacteria or influx of nutrients from outside, the habitat may increase in fertility, and new species may invade it. These species may modify the physical characteristics of the habitat by decreasing light intensity or monopolizing space or by other factors, such as eliminating species already present. This may be sufficiently extensive in the course of time, decades or centuries, to bring a biotic succession of different communities replacing each other until a climax is reached which becomes stabilized under the particular climate. Communities do not remain permanently constant because each is dependent upon climate, physiography, and organic evolution, factors which are themselves continuously changing. *See* CLIMAX COMMUNITY; ECOLOGICAL SUCCESSION.

Concepts of structure. The organismic concept of the community is that it is a complex organic entity representing the highest stage in the organization of living matter, that is, cell, tissue, organ, system, organism, community. It behaves as a unit in its relation to other communities, response to climate, local and geographic distribution, and in its evolution. It possesses such emergent characteristics as density, population dynamics, trophic balance, dominance, and succession beyond those of the individual organisms of which the community is composed. On the other hand, the individualistic concept considers the community as a chance combination of species that occur together because of similarity in their environmental relations. There is no intrinsic interrelation between the species in respect to distribution or evolution. The population density of each species is distributed in a form of a more or less normal curve when plotted along the gradient of a given environmental factor, and the curves of different species overlap in a heterogeneous or random manner without exact agreement between any two species. The community is simply an arbitrary portion of a continuum of constantly changing composition of plant and animal life along environmental gradients that extend from one extreme to another. Adherents of both concepts agree, however, that there exist functional relations between species that occur in the same place, and, regardless of one's point of view, the community is a useful unit for detailed ecological investigation. *See* ANIMAL COMMUNITY; PLANT COMMUNITY; POPULATION DYNAMICS.

[S. CHARLES KENDEIGH]

Conservation of resources

Conservation is concerned with the utilization of resources—the rate, purpose, and efficiency of use. This article emphasizes integrated conservation trends and policies. For treatment of individual resources *see* FISHERY CONSERVATION; FOREST MANAGEMENT AND ORGANIZATION; LAND-USE PLANNING; MINERAL RESOURCES CONSERVATION; RANGELAND CONSERVATION; SOIL CONSERVATION; WATER CONSERVATION; WILDLIFE CONSERVATION.

Nature of resources. Universal natural resources are the land and soil, water, forests, grassland and other vegetation, fish and wildlife, rocks and minerals, and solar and other forms of energy. Some natural resources, such as metallic ores, coal, petroleum, and stone, are called fund or stock resources. They usually are referred to as nonrenewable natural resources because extraction from the stock depletes the usable quantity remaining and, even if some is being formed, the rate of formation is too slow for practical meaning. Other natural resources, such as living organisms and their products and solar and atomic radiation, are called flow resources. They usually are referred to as renewable natural resources because they involve organic growth and reproduction or because they are relatively quickly recycled or renewed in nature, as in the case of water in the hydrologic cycle and certain atmospheric phenomena.

Some natural resources are difficult to fit into such a simple system. Soil, for example, is commonly thought of as renewable, as erosion and nutrient depletion can in some cases be rather quickly corrected, but if the upper layers of the soil are removed or bedrock is exposed, renewal may take thousands of years. Water also is commonly renewable, but rapid extraction of water by wells from deep aquifers may be equivalent to mining minerals. *See* HYDROLOGY.

Human resources are of two types: the people themselves (their numbers, qualities, knowledge, and skills) and their culture (the tools and institutions of society). The three broad classes of resources correspond to the economic factors of production: land (natural resources), labor (personal human resources), and capital (cultural tools and institutions). The natural resources have meaning only as there is human ability to make use of them.

Nature of conservation. Conservation has received many definitions because it has many aspects, concerns issues arising between individuals and groups, and involves private and public enterprise. Conservation receives impetus from the social conscience aware of an obligation to future generations and is viewed differently according to one's social and economic philosophy. To some extent, the meaning of conservation changes with the time and place. It is understood differently when approached from the natural sciences and technologies than when it is approached from the social sciences. Conservation for the petroleum engineer is largely the avoidance of waste from incomplete extraction; for the forester it may be sustained yield of products; and for the economist it is a change in the intertemporal distribution of use toward the future. In all cases, conservation deals with the judicious development and manner of use of natural resources of all kinds.

No definition of conservation exists that is satisfactory to all elements of the public and applicable to all resources. In its absence, an operational or functional definition can be arrived at by considering a series of conservation measures.

1. Preservation is the protection of nature from commercial exploitation to prolong its use for recreation, watershed protection, and scientific study. It is familiar in the establishment and protection of parks and reserves of many kinds.

2. Restoration, another widely familiar conservation measure, is essentially the correction of past willful and inadvertent abuses that have impaired the productivity of the resources base. This measure is familiar in modern soil and water conservation practices applied to agricultural land.

3. Beneficiation is the upgrading of the usefulness or quality of something, for instance, the utilization of ores that were formerly of uneconomic grade. Modern technology has provided many examples of this type of conservation.

4. Maximization includes all measures to avoid waste and increase the quantity and quality of production from resources.

5. Reutilization, in industry commonly called recycling, is the reuse of waste materials, as in the use of scrap iron in steel manufacture or of industrial water after it has been cleaned and cooled.

6. Substitution, an important conservation measure, has two aspects: the use of a common resource instead of a rare one when it serves the same end and the use of renewable rather than nonrenewable resources when conditions permit.

7. Allocation concerns the strategy of use—the best use of a resource. For many resources and products from them, the market price, as deter-

mined by supply and demand, establishes to what use a resource is put, but under certain circumstances the general welfare may dictate usage and resources may be controlled by government through the use of quotas, rationing, or outright ownership.

8. Integration in resources management is a conservation measure because it maximizes over a period of time the sum of goods and services that can be had from a resource or a resource complex, such as a river valley; this is preferable to maximizing certain benefits from a single resource at the expense of other benefits or other resources. This is one of the meanings of multiple use, and integration is a central objective of planning.

A generalized definition that fits many but not all meanings of conservation is "the maximization over time of the net social benefits in goods and services from resources." Although it is technologically based, conservation cannot escape socially determined values. There is an ethic involved in all aspects of conservation. Certain values are accepted in conservation, but they are the creation of society, not of conservation.

Conservation trends. There has been an important trend in conservation from an almost exclusive interest in production from individual natural resources to a balanced interest in that need and in the human resources and social goals for which resources are managed. The conservation movement originated with the realization that the economic doctrine of laissez-faire and quick profits —whether from forest, farm, or oil field—was resulting in tremendous waste that was socially harmful, even if it seemed to be good business. The beginning of the conservation movement stemmed from revulsion against destructive and wasteful lumbering. In time, the movement spread to farm and grazing lands, water, wildlife, and oil and gas. It was gradually learned that conservation management was good business in the long run.

The second trend in conservation was toward integrated management of resources. Students and administrators of resources—in colleges, business, and government— were discovering that the way one resource was handled affected the usability of others. Forestry broadened its interest to include forest influences on the watershed and the relations of the forest to wildlife and human uses for recreation. For many industries working directly with natural resources or their products, as in paper and chemical manufacturing and in coal mining, it was discovered that waste products produced costly and dangerous pollution of streams and air. Engineering on great river systems moved on from problems of flood hazard abatement and hydropower development to the design of structures with regard to fisheries and recreational values, and it was slowly realized that the way the land of a valley was managed affected erosion, siltation, water retention, and flooding.

Conservation in its third phase extends the ecological or integrated approach to resources management to include a more complete acceptance of the force of societal factors (such as economics, government, and social conditions) in determining resource management. There also is a closer examination of social costs and benefits and of human goals for which resources are employed.

Conservation of human resources. Many students of conservation prefer for practical reasons to limit conservation to the management of natural resources and to leave problems of human resources to other disciplines and fields of action. However, because of the inevitable interplay of natural resources and the resourcefulness of man in utilizing them, others emphasize the role of man himself as a resource. This leads to the application of conservation measures to man and his institutions. The many measures that tend to preserve, rehabilitate, renew, maximize, allocate, and integrate human abilities are coming to be referred to as conservation measures. These measures tend to be organized and institutionalized so that the institutions themselves become means to an end within the society and thus are considered to be resources as truly as water, petroleum, or the labor force. *See* HUMAN ECOLOGY.

Conservation of recreation resources. More than 30,000,000 acres of rural public land in the United States are managed for recreational use or to preserve scenic, historical, or scientific and natural history values, and additional space is allocated for more intensive recreational activities by units of local government. There has been in recent decades a tremendous growth in the use of such lands for all forms of recreation, and the forecast is that outdoor recreation will increase more rapidly than population growth because of the increase in urban living and personal incomes, more and better roads, the shorter workweek, and paid vacations. Rapidly growing new trends in recreation require special uses of space, such as bow and arrow shooting, skiing, and motorboating. With increasing numbers of persons camping, picnicking, and swimming, the provision of adequate facilities and physical maintenance of crowded sites have become problems.

Except for the more intensive recreational uses, rural and wildland areas serve multiple purposes. An abundance of game can be raised in agricultural regions. Watershed protection and the maintenance of scenic values go naturally together. Stream impoundments for multiple purposes create new recreational facilities. Yet some land uses are incompatible: municipal, manufacturing, and mining pollution destroys water recreation values; commercial developments destroy wilderness values.

Many trends indicate that there are not enough acres dedicated to recreation, especially smaller areas near strongly urbanized regions. There is a growing need for research on trends in wildland usage by people and for better use of space in large parks and forests so as to avoid wearing out the sites by the persistent concentration of people, as at campsites. It is clear that certain human facilities must be provided where many people use wildlands and that inappropriate ones, such as amusement facilities, must be kept separate from activities requiring special terrain. Some growing recreational demands are being met by private enterprise, such as in ski resorts and privately owned public hunting grounds. Although game is owned by the public, it lives and breeds largely on private land. In time more farm and ranch owners will be paid for hunting privileges. Further development in recreation is coming from increased knowledge of the biology of fish and wildlife and the consequent improvement of management arising from it and from a better understanding by the sportsman of the management problems.

Conservation policies. Individuals, corporations, and governmental entities at all levels have policies pertaining to resources. There could be a single national policy concerning a natural resource, such as water, only if the central government had complete authority over all governmental agencies and private enterprises. In the United States, however, many Federal and state departments, bureaus, agencies, and commissions have some authority over natural resources. There usually are several policies concerning the use and management of soil, water, forests, rangelands, fish, wildlife, minerals, and space, and they are not necessarily uniform as to objectives or program. One exception is the Atomic Energy Commission, which has centralized authority in the creation and execution of policy regarding radioactive resources. This authority is granted by Congress and can be modified by congressional action.

Because each resource is capable of being utilized for a variety of goods and services, and because individuals, enterprises, and regions tend to value certain uses more than others, conflicts arise in the allocation of a resource among competing uses when the supply is inadequate for all desirable uses. The demand for water, for example, may be for rural domestic needs, urban and industrial needs, irrigation, power development, and recreation in its many aspects.

Because situations such as this exist or are potential with respect to every resource, two outstanding needs arise in the conservation of natural resources. Detailed information is needed concerning the location, quantity, and quality of each resource and of the interrelations among the resources. As a result, each agency of government needs to strengthen its own fact-finding, analysis, and programming machinery. The second need is for coordinating machinery that will improve the efficiency of allocation of resources so as to maximize the net private and public benefits from them. This is the goal of conservation policy, and it must, in a democratic society, be approached as far as possible through the voluntary cooperation of government and private enterprise. However, as pressures on society increase, whether because of actual depletion of resources or because of an increase in critical demand, as in war, authority to allocate resources tends to be delegated by the citizenry to the central government.

Federal conservation legislation. National resources policy in the United States is framed by acts of Congress and in the states by acts of legislatures. The language of specific acts, however, usually permits some freedom for administrative decisions and also permits different interpretations that must be settled in the courts of law. As the country has developed and conditions have changed, a sequence of laws has been passed to deal with the exigencies of natural resource conditions. Also, there has occurred some evolution of political philosophy to meet the changing conditions. In addition to legislative and court actions, some natural resources are subject to treaties and other international agreements.

Water. In 1824 Congress assigned to the Army Corps of Engineers responsibility for improving rivers and harbors for navigation, and this assignment has become its principal civilian activity. The Inland Waterways Commission in 1907 and the National Waterways Commission in 1909 were created by Congress with broad responsibilities for planning the development and conservation of water and related resources. The Flood Control Act of 1927 provided for Federal surveys, and the act of 1936 provided for Federal construction of projects. Subsequent amendments led to the 1954 Watershed Protection and Flood Prevention Act, which clearly recognized the relation of upland management practices to erosion, siltation, streamflow, and other water-related resource problems. Interest in irrigation has resulted in a series of acts from 1866 onward (the Desert Land Act of 1877 being significant), while water facilities, drainage, hydroelectric power, and water pollution also have received repeated attention. The Water Power Act of 1920 set up the Federal Power Commission and aimed to safeguard the rights of government and hydroelectric companies. Water pollution control is largely state responsibility, but the Federal Water Pollution Control Act of 1948, amended in 1956, enables financial and other assistance to the states.

Soil. Although soil surveys had been carried out for a century, it was the Soil Conservation Act of 1935 that created the Soil Conservation Service and the autonomous Soil Conservation Districts. Soil conservation payments were started under the Agricultural Adjustment Act of 1933. Research, education, cooperation, and financial inducements have been stressed more than regulatory measures for soil and water conservation on the land.

Minerals. In 1866 Congress provided for the sale of mineral lands, but a general minerals policy did not exist until later. The Minerals Leasing Act of 1920 provided that nonmetalliferous minerals on public lands could be utilized only under lease. In 1953 Congress confirmed jurisdiction of the states over minerals under navigable inland waters and on the continental shelf to state boundaries.

Forests. A long series of acts has been concerned with the disposal of the public lands and the exploitation of timber and forage. An act in 1891 empowered the President to set aside forest reserves. In 1901 a Forestry Division was created in the General Land Office in the Department of Interior, and in 1905 forestry activities were transferred to the Department of Agriculture. An act in 1907 changed the reserves to the national forests. The Forest Service soon became recognized as one of the most efficient bureaus in Washington as it dealt, on a decentralized basis, with the conservational use of all resources, not just timber. The Weeks Act in 1911 inaugurated cooperation with the states and is best known because it started the policy of land acquisition. Today management of the nation's forest lands has become a balanced cooperation between government and private enterprise on a mosaic of ownership.

Wildlife. The first Federal recognition of responsibility for fisheries and wildlife was in regard to research, with the establishment in 1905 of the Bureau of Biological Survey. An international convention for protection of fur seals was signed in

1911, with protection of other marine animals coming later. The Migratory Bird Act was passed in 1913. Federal refuges were started in 1903, and sizable funds for refuges became available with the Migratory Bird Conservation Act of 1929 and at an accelerated rate from 1933 to 1953. State activities in the fields of wildlife and fishery management have been greatly facilitated by Federal aid under the Pittman-Robertson Act of 1937 and the Dingell-Johnson Act of 1950. [STANLEY A. CAIN]

Bibliography: S. W. Allen and J. W. Leonard, *Conserving Natural Resources: Principles and Practice in a Democracy*, 2d ed., 1966; G. Borgstrom, *The Hungry Planet: The Modern World at the Edge of Famine*, 1965; C. H. Callison (ed.), *America's Natural Resources*, 1957; S. V. Ciriacy-Wantrup, *Resource Conservation: Economics and Policies*, 1952; D. C. Coyle, *Conservation: An American Story of Conflict and Accomplishment*, 1957; S. T. Dana, *Forest and Range Policy*, 1956; R. F. Dasman, *Environmental Conservation*, 1959; E. S. Helfman, *Rivers and Watersheds in America's Future*, 1965; G.-H. Smith, *Conservation of Natural Resources*, 3d ed., 1965; E. W. Zimmermann, *World Resources and Industries*, rev. ed., 1951.

Continental shelf and slope

The continental shelf is the zone around the continent, extending from the low-water line to the depth at which there is a marked increase in slope to greater depth. The continental slope is the declivity from the edge of the shelf extending down to great ocean depths. The shelf and slope comprise the continental terrace, which is the submerged fringe of the continent, connecting the shoreline with the 2½-mi-deep (4-km) abyssal ocean floor (see illustration).

Continental shelf. This comparatively featureless plain, with an average width of 45 mi (72 km), slopes gently seaward at about 10 ft/mi (1.9m/km). At a depth of about 70 fathoms (128 m) there generally is an abrupt increase in declivity called the shelf break, or the shelf edge. This break marks the limit of the shelf, the top of the continental slope, and the brink of the deep sea. However, some shelves are as deep as 200–300 fathoms (180–550 m), especially in past or presently glaciated regions. For some purposes, especially legal, the 100-fathom line or 200-m line is conventionally taken as the limit of the shelf. Characteristically, the shelves are thinly veneered with clastic sands, silts, and silty muds, which are patchily distributed. Geologically, the shelf is an extension of, and in unity with, the adjacent coastal plain. The position of the shoreline is geologically ephemeral, being subject to constant prograding and retrograding, so that its precise position at any particular time is not important. Genetically, the origin of the shelf seems to be primarily related to shallow wave cutting (waves cut effectively as breakers and surf only down to about 5 fathoms (9 m), the depth of vigorous abrasion), shoreline deposition, and oscillations of sea level, which have been especially strong during the Pleistocene and Recent. Although worldwide in distribution and comprising 5% of the area of the Earth, shelves differ considerably in width. Off the east coast of the United States the shelf is about 75 mi wide (120 km), while off the west coast it is about 20 mi wide (32 km). Especially broad shelves fringe northern Australia, Argentina, and the Arctic Ocean. As along the eastern United States, continental shelves commonly acquire a prism of sediments as the continental margin downflexes. Such capping prisms appear to be nascent miogeoclines.

Continental slope. The drowned edges of the low-density "granitic" or sialic continental masses are the continental slopes. The continental plateaus float like icebergs in the Earth's mantle with the

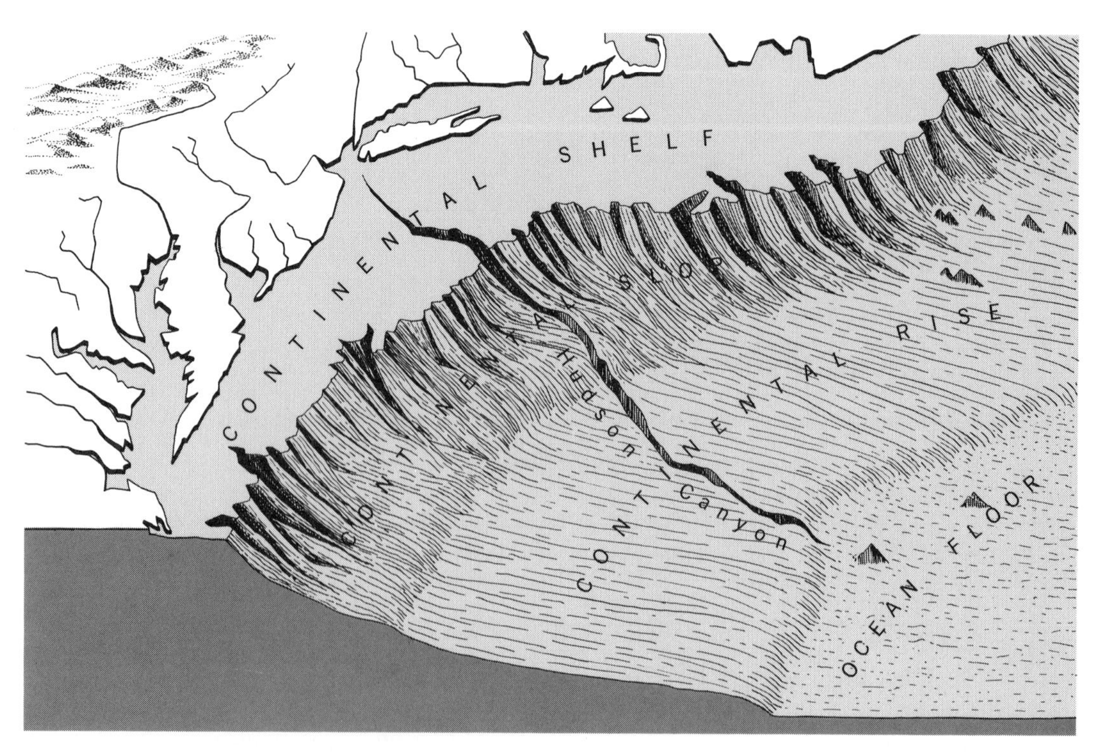

Continental margin off Northeastern United States.

slopes marking the transition between the low-density continents and the heavier oceanic segments of the Earth's crust. Averaging 2½ mi high (4 km) and in some places attaining 6 mi (10 km), the continental slopes are the most imposing escarpments on the Earth. The slope is comparatively steep with an average declivity of 4.25° for the upper 1000 fathoms (1.8 km). Most slopes resemble a straight mountain front but are highly irregular in detail; in places they are deeply incised by submarine canyons, some of which cut deeply into the shelf. Usually the slope does not connect directly with the sea floor; instead there is a transitional area, the continental rise, or apron, built by the shedding of sediments from the continental block.

[ROBERT S. DIETZ]

Bibliography: J. L. Culliney, *The Forests of the Sea: Life and Death on the Continental Shelf*, 1977; R. McQuillin, *Exploring the Geology of Shelf Seas*, 1975; F. P. Shephard, *Geological Oceanography*, 1977; J. Waters, *The Continental Shelves*, 1975.

Crop micrometeorology

Crop micrometeorology deals with the interaction of crops and their immediate physical environment. Especially, it seeks to measure and explain net photosynthesis (photosynthesis minus respiration) and water use (transpiration plus evaporation from the soil) of crops as a function of meteorological, crop, and soil moisture conditions. These studies are complex because the intricate array of leaves, stems, and fruits modifies the local environment and because the processes of energy transfers and conversions are interrelated. As a basic science, crop micrometeorology is related to plant anatomy, plant physiology, meteorology, and hydrology. Expertise in radiation exchange theory, boundary-layer and diffusion processes, and turbulence theory is needed in basic crop micrometeorological studies. A practical goal is to provide improved plant designs and cropping patterns for light interception, for reducing infestations of diseases, pests, and weeds, and for increasing crop water-use efficiency. Shelter belts are modifications that have been used in arid or windy areas to protect crops and seedlings from a harsh environment. *See* AGRICULTURAL METEOROLOGY; MICROMETEOROLOGY.

Unifying concepts. Conservation laws for energy and matter are central to crop micrometeorology. Energy fluxes involved are solar wavelength radiation, consisting of photosynthetically active radiation (0.4–0.7 μm) and near-infrared radiation (0.7–3 μm); far-infrared radiation (3–100 μm); convection in the air; molecular heat conduction in and near the plant parts and in the soil; and the latent heat carried by water vapor. The main material substances transported to and from crop and soil surfaces are water vapor and carbon dioxide. However, fluxes of ammonia, sulfur dioxide, pesticides, and other gases or pollutants to or from crop or soil surfaces have been measured. These entities move by molecular diffusion near the leaves and soil, but by convection (usually turbulent) in the airflow. During the daytime generally, and sometimes at night, airflow among and above crops is strongly turbulent. However, often at night a stable air layer forms because of surface cooling caused by emission of far-infrared radiation back to space, and the air flow becomes nonturbulent. Fog or radiation frosts may result. The aerodynamic drag and thermal (heat-absorbing) effects of plants contribute to the pattern of air movement and influence the efficiency of turbulent transfer.

Both field studies and mathematical simulation models have dealt mostly with tall, close-growing crops, such as maize and wheat, which can be treated statistically as composed of infinite horizontal layers. Downward-moving direct-beam solar and diffuse sky radiation are partly absorbed, partly reflected, and partly transmitted by each layer. Less photosynthetically active radiation than near-infrared radiation is transmitted to ground level and reflected from the crop canopy because photosynthetically active radiation is strongly absorbed by the photosynthetic pigments (chlorophyll, carotenoids, and so on) and near-infrared radiation is only weakly absorbed. The plants act as good emitters and absorbers of far-infrared radiation. Transfers of momentum, heat, water vapor, and carbon dioxide can be considered as composed of two parts; a leaf-to-air transfer and a turbulent vertical transfer. As a bare minimum, two mean or representative temperatures are needed for each layer: an average air temperature and a representative plant surface temperature. Because some leaves are in direct sunlight and some are shaded, a representative temperature is difficult to obtain. Under clear conditions, traversing solar radiation sensors show a bimodal frequency distribution of irradiances in most crop communities; that is, most points in space and time are exposed to either high irradiances of direct-beam radiation or low irradiances characteristic of shaded conditions. Models of radiation interception have been developed which predict irradiance on both shaded leaves and on exposed leaves, depending on the leaf inclination angle with respect to the rays. The central concept of both experimental studies and simulation models is that radiant energy fluxes, sensible heat fluxes, and latent heat fluxes are coupled physically and can be expressed mathematically. This interdependence applies to a complex crop system as well as to a single leaf.

Photosynthesis. Studies of photosynthesis of crops using micrometeorological techniques do not consider the submicroscopic physics and chemistry of photosynthesis and respiration, but consider processes on a microscopic and macroscopic scale. The most important factors are the transport and diffusion of carbon dioxide in air to the leaves and through small ports called stomata to the internal air spaces. Thence it diffuses in the liquid phase of cells to chloroplasts, where carboxylation enzymes speed the first step in the conversion of carbon dioxide into organic plant materials. Solar radiation provides the photosynthetically active radiant energy to drive this biochemical conversion of carbon dioxide. Progress has been made in understanding the transport processes in the bulk atmosphere, across the leaf boundary layer, through the stomata, through the cells, and eventually to the sites of carboxylation. Transport resistances have been identified for this catenary process: bulk aerodynamic resistance, boundary-layer resistance, stomatal diffusion resistance, mesophyll resistance, and carboxylation resistance. All these resistances are plant factors which control the rate of carbon dioxide uptake by leaves of a crop; however, boundary-layer resistance and especially bulk aerodynamic resistance are determined also by the external wind flow.

Carbon dioxide concentration and photosynthetically active radiation are two other factors which control the rate of crop photosynthesis. Carbon dioxide concentration does not vary widely from about 315 microliters per liter. Experiments have revealed that it is not practical to enrich the air with carbon dioxide on a field scale because it is rapidly dispersed by turbulence. Therefore carbon dioxide concentration can be dismissed as a practical variable. Solar radiation varies widely in quantity and source distribution (direct-beam or diffuse sky or cloud sources). Many species of crop plants have leaves which can utilize solar radiation having flux densities greater than full sunlight (tropical grasses such as maize, sugar cane, and Burmuda grass, which fix carbon dioxide through the enzyme phosphoenolpyruvate carboxylase). Other species have leaves which may give maximum photosynthesis rates by individual leaves at less than full sunlight (such as soybean, sugarbeet, and wheat, which fix carbon dioxide through the enzyme ribulose 1,5-diphosphate carboxylase). However, in general, most crops show increasing photosynthesis rates with increasing irradiances for two reasons. First, more solar energy would become available to the shaded and partly shaded leaves deep in the crop canopy. Second, many of the well-exposed leaves at the top of a crop canopy are exposed to solar rays at wide angles of inclination so that they do not receive the full solar flux density. These leaves will respond to increasing irradiance also. Furthermore, increased diffuse to direct-beam ratios of irradiance (which could be caused by haze or thin clouds) may increase the irradiance on shaded leaves and hence increase overall crop photosynthesis.

If crop plants lack available soil water, the stomata may close and restrict the rate of carbon dioxide uptake by crops. Stomatal closure will protect plants against excessive dehydration, but will at the same time decrease photosynthesis by restricting the diffusion of carbon dioxide into the leaves.

Transpiration and heat exchange. Transpiration involves the transport of water vapor from inside leaves to the bulk atmosphere. The path of flow of water vapor is from the surfaces of cells inside the leaf through the stomata, through the leaf aerodynamic boundary layer, and from the boundary layer to the bulk atmosphere. Sensible heat is exchanged by convection directly from plant surfaces; therefore there is no stomatal diffusion resistance associated with this exchange. Stomatal diffusion resistance does affect heat exchange from leaves, however, because when stomata are open wide (low resistance) much of the heat exchanged from leaves is in the form of latent heat of evaporation of water involved in transpiration.

Small leaves, such as needles, convect heat much more rapidly than large leaves, such as banana leaves. Engineering boundary-layer theory suggests that boundary-layer resistance should be proportional to the square root of a characteristic dimension of a leaf and inversely proportional to the square root of the airflow rate past a leaf. Experiments support these relationships.

Under high-irradiance conditions, low air humidity, high air temperature, and low stomatal diffusion resistance will favor high transpiration, whereas high air humidity, low air temperature, and high stomatal diffusion resistance will favor sensible heat exchange from leaves. The function of wind is chiefly to enhance the transport rather than determine which form of convected energy will be most prominent. In arid environments, the latent energy of transpiration from crops may exceed the net radiant energy available, because heat from the dry air may actually be conducted to crops which will cause transpiration to increase. In those areas, crop temperature is lower than air temperature.

Flux methods. At least three general methods have been employed to measure flux density of carbon dioxide, water vapor, and heat to and from crop surfaces on a field scale. These methods are restricted to use in the crop boundary layer immediately above the crop surface, and they require a sufficient upwind fetch of a uniform crop surface free of obstructions. Flux densities obtained by these methods will reflect the more detailed interactions of crop and environment, but will not explain them.

The principle of the energy balance methods is to partition the net incoming radiant energy into energy associated with latent heat of transpiration and evaporation, sensible heat, photochemical energy involved in photosynthesis, heat flux into the soil, and heat stored in the crop. Measurement of net input of radiation to drive these processes is obtained from net radiometers, which measure the total incoming minus the total outgoing radiation. The most important components—latent heat, sensible heat, and photochemical energy—are determined by average vertical gradients of water vapor concentration (or vapor pressure), air temperature, and carbon dioxide concentration.

The principle of the bulk aerodynamic methods is to relate the vertical concentration gradients of those transported entities to the vertical gradient of horizontal wind speed. The transports are assumed to be related to the aerodynamic drag (or transport of momentum) of the crop surface. Corrections are required for thermal instability or stability of the air near the crop surface.

The eddy correlation methods are direct methods which correlate the instantaneous vertical components of wind (updrafts or downdrafts) to the instantaneous values of carbon dioxide concentration, water vapor concentration, or air temperature. Under daytime conditions, turbulent eddies, or whorls, transport air from the crop in updrafts, which are slightly depleted in carbon dioxide, and conversely, turbulence transports air to the crop in downdrafts which are representative of the atmospheric content of these entities. More basic and applied research is being done on eddy correlation methods because they measure transports through direct transport processes.

Plant parameters. The stomata are the most important single factor in interactions of plant and environment because they are the gateways for gaseous exchange. Soil-to-air transfers are also very important while crops are in the seedling stage until a large degree of ground cover is attained. Coefficients of absorption, transmission, and reflection by leaves of photosynthetically active, near-infrared, and long-wavelength infrared radiation are not very different among crop species, but the geometric arrangement and stage of growth of plants in a crop may affect radiation exchange greatly. The crop geometry also interacts with radiation-source geometry (diffuse to direct-beam irradiance, solar elevation angle). Crop micrometeorology attempts to show how the plant parameters interact with the environmental factors in crop production and water requirements of crops under field conditions. *See* EVAPOTRANSPIRATION. [L. H. ALLEN, JR.]

Bibliography: J. Goudriaan, *Crop Micrometeorology*, 1977; E. Lemon, D. W. Stewart, and R. W. Shawcroft, The sun's work in a cornfield, *Science*, 174:371–378, 1971; J. L. Monteith (ed.), *Vegetation and the Atmosphere*, vol. 1, 1975, vol. 2, 1976; R. E. Munn, *Biometeorological Methods*, 1971; N. J. Rosenberg, *Microclimate. The Biological Environment*, 1974; W. D. Sellars, *Physical Climatology*, 1965; L. P. Smith (ed.), *The Application of Micrometeorology to Agricultural Problems*, 1972; O. G. Sutton, *Micrometeorology*, 1977.

Dam

A structure that bars or detains the flow of water in an open channel or watercourse. Dams are constructed for several principal purposes. Diversion dams divert water from a stream; navigation dams raise the level of a stream to increase the depth for navigation purposes (Fig. 1); power dams raise the level of a stream to create or concentrate hydrostatic head for power purposes; and storage dams store water for municipal and industrial use, irrigation, flood control, river regulation, recreation, or power production. A dam serving two or more purposes is called a multiple-purpose dam. Dams are commonly classified by the material from which they are constructed, such as masonry, concrete, earth, rock, timber, and steel. Most dams now are built either of concrete or of earth and rock.

Concrete dams. Concrete dams may be typed as gravity, arch, or buttress type. Gravity dams depend on weight for stability against overturning and for resistance to sliding on their foundations (Figs. 2 and 3). An arch dam may have a near-vertical face or, more usually, one that curves concave downstream (Figs. 4 and 5). The dam acts as an arch to transmit most of the horizontal thrust from the water pressure against the upstream face of the dam to the abutments of the dam. The buttress type of concrete dam includes the slab-and-buttress, or Ambursen, type; round- or diamond-head buttress type; multiple-arch type; and multiple-dome type. Buttress dams depend on the weight of the structure and of the water on the dam to resist overturning and sliding.

Forces acting on concrete dams. Principal forces acting on a concrete dam are (1) vertical forces from weight of the structure and vertical component of water pressure against the upstream and downstream faces of the dam, (2) uplift pressures under the base of the structure, (3) horizontal forces from the horizontal component of the water pressure against the upstream and downstream faces of the dam, (4) forces from earthquake accelerations in regions subject to earthquakes, (5) temperature stresses, (6) pressures from silt deposits and earth fills against the structure, and (7) ice pressures.

The uplift pressure under the base of a dam var-

Fig. 1. John Day Lock and Dam, looking upstream across the Columbia River at the Washington shore. In the foreground the navigation lock may be seen, then the spillway dam beyond it, and then the powerhouse. The John Day multiple-purpose project boasts the highest single-lift navigation lock in the United States. (*U.S. Army Corps of Engineers*)

ies with the effectiveness of the foundation drainage system and the perviousness of the foundation.

Earthquake loads are usually selected after consideration of the accelerations which may be expected at the site as indicated by the geology, proximity to major faults, and the earthquake history of the region. Conventionally, earthquake forces have been treated as static forces representing the effects of the acceleration of the dam itself and the hydrodynamic force produced against the dam by water in the reservoir. Such horizontal forces often are assumed to equal 0.05–0.10 the force of gravity, with a somewhat smaller vertical force. Dynamic analysis procedures have been developed which determine the structure's response to combined effects of the contemplated ground motion and the structure's dynamic properties.

Stresses resulting from temperature changes must be considered in analyzing arch dams. These stresses are usually disregarded in the design of concrete gravity dams, but must be controlled to acceptable limits by concreting and curing methods, discussed below.

Pressure from silt deposited in the reservoir against the dam is considered only after sedimentation studies indicate that it may be a significant factor. Backfill pressures are important where a concrete gravity dam ties into an embankment.

Ice pressure, applied at the maximum elevation at which the ice will occur in project operations, is considered when conditions indicate that it would be significant. The pressure, commonly assumed to be 10,000–20,000 lb per linear foot, results from the thermal expansion of the ice sheet and varies with the rate and magnitude of temperature rise and thickness of the ice.

Fig. 2. Green Peter Dam, a concrete gravity type on the Middle Santiam River, Willamette River Basin, OR. Gate-controlled overflow-type spillway is constructed through crest of dam; powerhouse is at downstream toe of dam. (*U.S. Army Corps of Engineers*)

Stability and allowable stresses. Stability of a concrete gravity dam is evaluated by analyzing the available resistance to overturning and sliding. To satisfy the former, the resultant of forces is required to fall within the middle third of the base under normal load conditions. Sliding stability is assured by requiring available shear and friction resistance to be greater by a designated safety factor than the forces tending to produce sliding. The strengths used in computing resistance to sliding are based on investigation and tests of the foundation. Bearing strength of the foundation for a gravity dam is a controlling factor only for weak foundations or for high dams. Because an arch dam depends on the competency of the abutments, the rock bearing strength must be sufficient to provide an adequate safety factor for the compressive stresses, and the resistance to sliding along any weak surface must be great enough to provide an adequate safety factor.

Concrete stresses control the design of arch dams, but ordinarily not gravity dams. Stresses adopted for concrete arch dams are conservative. A safety factor of 4 on concrete compressive strength is commonly used for normal load conditions.

Concrete temperature control. Volume changes accompanying temperature changes in a concrete dam tend to cause the development of tensile stresses. A major factor in development of temperature changes within a concrete mass is the heat developed by chemical changes in the concrete after placement. Uncontrolled temperature changes can cause cracking which may endanger the stability of a dam, cause leakage, and reduce durability. Temperatures are controlled by using cementing materials having low heat of hydration, and by artificial cooling by precooling the concrete mix or circulating cold water through pipes embedded in the concrete or both.

Concrete dams are constructed in blocks, with the joints between the blocks serving as contraction joints (Fig. 6). In arch dams the contraction joints are filled with cement grout after maximum shrinkage has occurred to assure continuous bearing surfaces normal to the compressional forces set up in the arch when the water load is applied to the dam.

Quality control. During construction, continuing testing and inspection are performed to ensure that the concrete will be of required quality. Tests are also made on materials used in manufacture of the concrete, and concrete batching, mixing, transporting, placing, curing, and protection are continuously inspected.

Earth dams. Earth dams have been used for water storage since early civilizations. Improvements in earth-materials techniques, particularly the development of modern earth-handling equipment, have brought about a wider use of this type of dam, and today as in primitive times the earth embankment is the most common dam (Figs. 7 and 8). Earth dams may be built of rock, gravel, sand, silt, or clay in various combinations.

Most earth dams are constructed with an inner impervious core, with upstream and downstream zones of more pervious materials, sometimes including rock zones. Earth dams limit the flow of

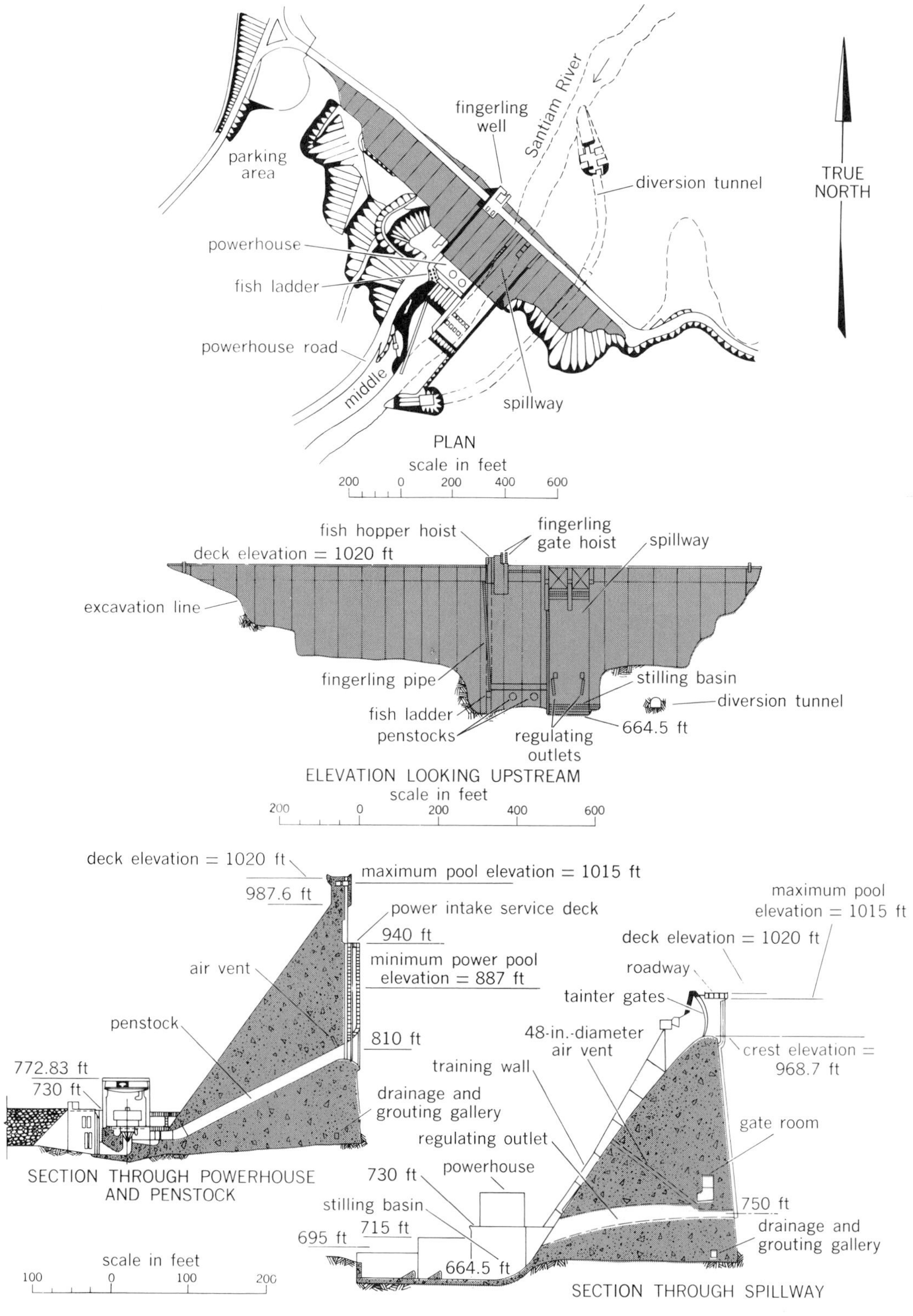

Fig. 3. Plan and sections of Green Peter Dam. 1 in.= 2.54 cm; 1 ft= 0.3 m. *(U.S. Army Corps of Engineers)*

water through the dam by use of fine-grained soils. Where possible, these soils are formed into a relatively impervious core. When there is a sand or gravel foundation, the core may be connected to bedrock by a cutoff trench backfilled with compacted soil. If such cutoffs are not economically feasible because of the great depth of pervious foundation soils, then the central impervious core is connected to a long horizontal upstream impervious blanket that increases the length of the seepage path. The impervious core is often encased in pervious zones of sand, gravel, or rock fill for stability. When there is a large difference in the particle sizes of the core and pervious zones, transition zones are required to prevent the core material from being transported into the pervious zones by seeping water. In some cases where pervious soils are scarce, the entire dam may be a homogeneous fill of relatively impervious soil. Downstream pervious drainage blankets are provided to collect seepage passing through, under, and around the abutments of the dam.

Materials can be obtained from required excavations for the dam and appurtenances or from borrow areas. Rock fill is generally used when large quantities of rock are available from required excavation or when soil borrow is scarce.

Earth-fill embankment is placed in layers and compacted by sheepsfoot rollers or heavy pneumatic-tire rollers. Moisture content of silt and clay soils is carefully controlled to facilitate optimum compaction. Sand and gravel fills are compacted in slightly thicker layers by pneumatic-tire rollers, vibrating steel drum rollers, or placement equipment. The placement moisture content of pervious fills is less critical than for silts and clays. Rock fill usually is placed in layers 1–3 ft (0.3–0.9 m) deep and is compacted by placement equipment and vibrating steel drum rollers.

Spillways. A spillway releases water in excess of storage capacity so that the dam and its foundation are protected against erosion and possible failure. All dams must have a spillway, except small ones where the runoff can be safely stored in the reservoir without danger of overtopping the dam. Ample spillway capacity is of particular importance for large earth dams, which would be destroyed or severely damaged by being overtopped. Failure of a large dam could result in severe hazards to life and property downstream.

Types. Spillways are of two general types: the overflow type, constructed as an integral part of the dam; or the channel type, located as an independent structure discharging through an open chute or tunnel. Either type may be equipped with gates to control the discharge. Various control structures have been used for channel spillways, including the simple overflow weir, side-channel overflow weir, and drop or morning-glory inlet where the water flows over a circular weir crest and drops directly into a tunnel.

Unless the discharge end of a spillway is remote from the toe of the dam or erosion-resistant bedrock exists at shallow depths, some form of energy dissipator must be provided to protect the toe of the dam and the foundation from spillway discharges. For an overflow spillway the energy dissipator may be a stilling basin, a sloping apron downstream from the dam, or a submerged bucket. When a channel spillway terminates near the dam, it usually has a stilling basin. A flip bucket is used for both overflow and channel spillways when the flow can be deflected far enough downstream, usually onto rock, to prevent erosion at the toe of the dam or end of the spillway.

Gates. Several types of gates may be used to regulate and control the discharge of spillways (Fig. 9). Tainter gates are comparatively low in cost and require only a small amount of power for operation, being hydraulically balanced and of low friction. Drum gates, which are operated by reservoir pressure, are costly but afford a wide, unobstructed opening for passage of drift and ice over the gates. Vertical-lift gates of the fixed-wheel or roller type are sometimes used for spillway regulation, but are more difficult to operate than the others. Floating ring gates control the discharge of morning-glory spillways. Like the drum gate, this type offers a minimum of interference to the passage of ice or drift over the gate and requires no external power for operation.

Reservoir outlet works. These are used to regulate the release of water from the reservoir; they consist essentially of an intake and an outlet connected by a water passage, and are usually provided with gates. Outlet works usually have trashracks at the intake end to prevent clogging by debris. Bulkheads or stop logs are commonly provided to close the intakes so that the passages may be unwatered for inspection and maintenance. A stilling basin or other type of energy dissipator is usually provided at the outlet end.

Locations. Outlets may be sluices through concrete dams with control valves located in chambers in the dam or on the downstream end of the sluices, tunnels through the abutments of the dam, or cut-and-cover conduits extending along the foundation through an earth-fill dam. In the last case, the control valves are usually located within the dam or at the upstream end of the conduit, and special precautions must be taken to prevent leakage of water along the outside of the conduit.

Outlet control gates. Various gates and valves are used for regulating the release of water from reservoirs, including high-pressure slide gates, tractor gates (roller or wheel), and radial or tainter gates (Fig. 10); also needle valves of various kinds, butterfly valves, fixed cone dispersion valves, and cylinder or sleeve valves. They must be capable of operating, without excessive vibration and cavitation, at any opening and at any head up to the maximum to which they may be subjected. They also must be capable of opening and closing under the

Fig. 4. East Canyon Dam, a thin-arch concrete structure on the East Canyon River, UT. Note uncontrolled over-flow-type spillway through crest of dam at right center of photograph. (*U.S. Bureau of Reclamation*)

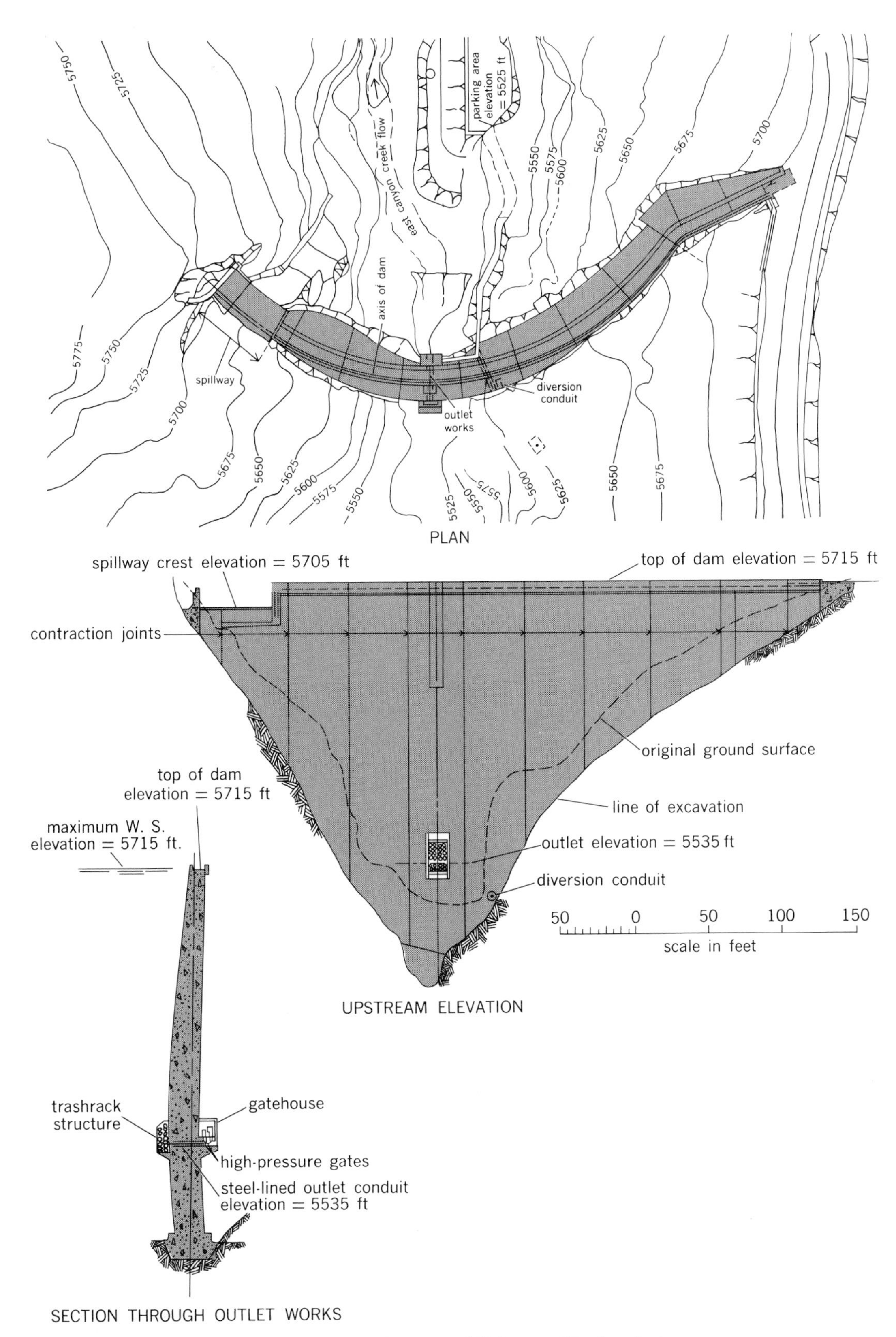

Fig. 5. Plan and sections of East Canyon Dam. 1 ft = 0.3 m. (*U.S. Bureau of Reclamation*)

maximum operating head. Emergency gates generally are used upstream of the operating gates, where stored water is valuable, so that closure can be made if the service gate should fail to function.

The slide gate, which consists of a movable leaf that slides on a stationary seat, is the most commonly used control gate. The high-pressure slide gate is of rugged design, having corrosion-resisting metal seats on both the movable rectangular leaf and the fixed frame. This gate has been used for

Fig. 6. Block method of construction on a typical concrete gravity dam. *(U. S. Army Corps of Engineers)*

regulating discharges under heads that exceed 600 ft (180 m).

Provision of low-level outlet. The usual storage reservoir has low-level outlets near the elevation of the stream bed to enable release of all the stored water. Some power and multiple-purpose dams have relatively high-level dead storage pools and do not require low-level outlets for ordinary operation. In such a dam, provision of a capability for emptying the reservoir in case of an emergency must be weighed against the additional cost.

Penstocks. A penstock is a pipe that conveys water from a forebay, reservoir, or other source to a turbine in a hydroelectric plant. It is usually made of steel, but reinforced concrete and wood-stave pipe have also been used. Pressure rise and speed regulation must be considered in the design of a penstock.

Pressure rise, or water hammer, is the pressure change that occurs when the rate of flow in a pipe or conduit is changed rapidly. The intensity of this pressure change is proportional to the rate at

Fig. 7. Aerial view of North Fork Dam, a combination earth and rock embankment on the North Fork of Pound River, VA. Channel-type spillway (left center) has simple overflow weir. *(U.S. Army Corps of Engineers)*

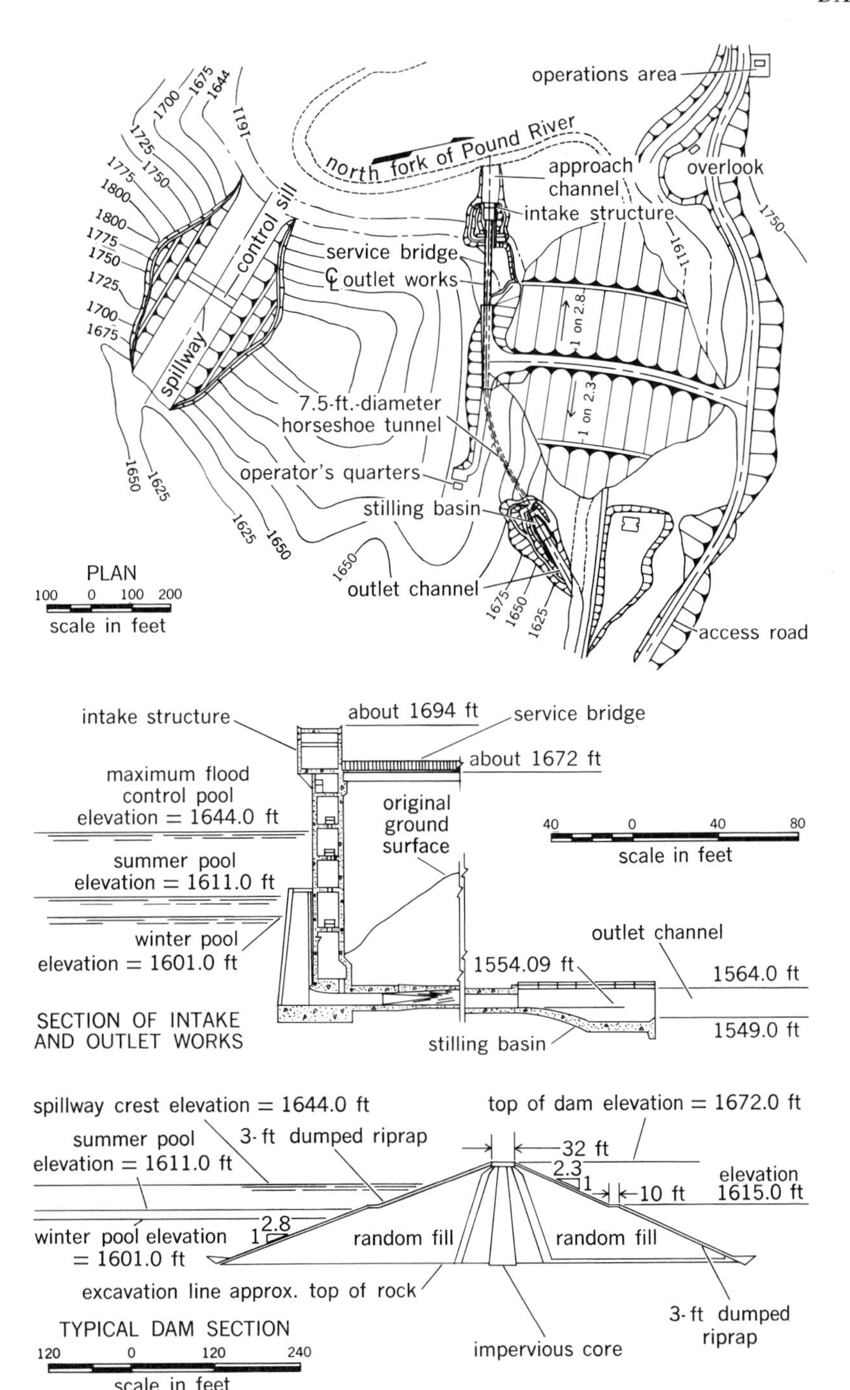

Fig. 8. Plan and sections of North Fork of Pound Dam. 1 ft= 0.3 m. (*U.S. Army Corps of Engineers*)

which the velocity of the flow is accelerated or decelerated. Accurate determination of the pressure changes that occur in a penstock involves consideration of all operating conditions. For example, one important consideration is the pressure rise that occurs in a penstock when the turbine wicket gates are closed after the loss of load.

Selection of dam site. This depends upon such factors as hydrologic, topographic, and geologic conditions; storage capacity of reservoir; accessibility; cost of lands and necessary relocations of prior occupants or uses; and proximity of sources of suitable construction materials. For a storage dam the objective is to select the site where the desired amount of storage can be most economically developed. Power dams must be located to develop the desired head and storage. For a diversion dam the site must be considered in conjunction with the location and elevation of the outlet canal or conduit. Site selection for navigation dams involves special factors such as desired navigable depth and channel width, slope of river channel, natural river flow, amount of bank protection, amount of channel dredging, approach and exit conditions for tows, and locations of other dams in the system.

Unless topographic and geologic conditions for a proposed storage, power, or diversion dam site are satisfactory, hydrological features may need to be subordinated. Important topographic characteristics include width of the floodplain, shape and height of valley walls, existence of nearby saddles for spillways, and adequacy of reservoir rim to retain impounded water. Controlling geologic conditions include the depth, classification, and engineering properties of soils and bedrock at the dam site, and the occurrence of sinks, faults, and major landslides at the site or in the reservoir area. The elevation of the groundwater table is also significant because it will influence the construction operations and suitability of borrow materials. The beneficial effect of reservoir water on groundwater recharge may become an important consideration, as well as the adverse effects on existing or potential mineral resources and developments that would be destroyed or require relocation at the site or within the reservoir.

Selection of type of dam. This is made on the basis of the estimated costs of various types. The most important factors are topography, foundation conditions, and the accessibility of construction materials. In general, a hard-rock foundation is suitable for any type of dam, provided the rock has no unfavorable jointing, there is no danger of movement in existing faults, and foundation underseepage can be controlled at reasonable cost. Rock foundations of high quality are essential for arch dams because the abutments receive the full thrust of the water pressure against the face of the dam. Rock foundations are necessary for all medium and high concrete dams. An earth dam may be built on almost any kind of foundation if properly designed and constructed.

The chance of an embankment dam being most economical is improved if large spillway and outlet capacities are required and topography and foundation are favorable. In a wide valley a combination of an earth embankment dam and a concrete dam section containing the spillway and outlets often is economical. Availability of suitable construction materials frequently determines the most economical type of dam. A concrete dam requires adequate quantities of suitable concrete aggregate and reasonable availability of cement, while an earth dam requires sufficient quantities of both pervious and impervious earth materials. If quantities of earth materials are limited and enough rock is available, a rock-fill dam with an impervious earth core may be the most economical.

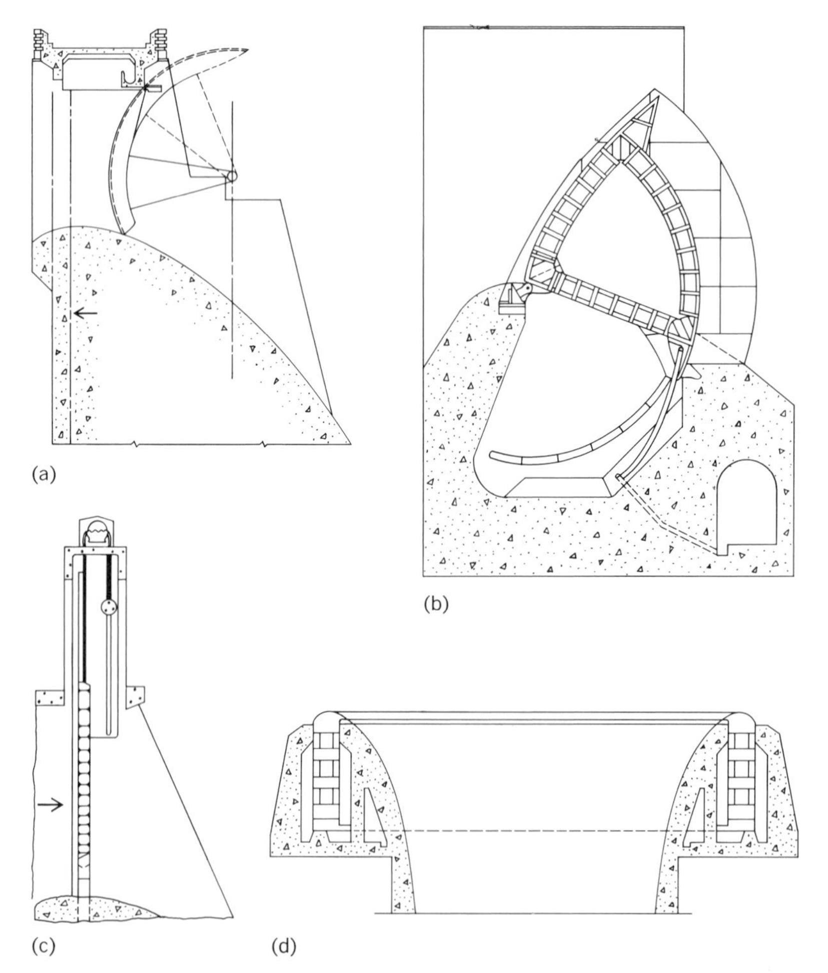

Fig. 9. Spillway gates. (*a*) Tainter gate. (*b*) Drum gate. (c) Vertical lift gate. (*d*) Ring gate. (*U.S. Army Corps of Engineers and U.S. Bureau of Reclamation*)

Determination of dam height. The dam must be high enough to (1) store water to the normal full-pool elevation required to meet intended functions of the project, (2) provide for the temporary storage needed to route the spillway design flood through the dam, and (3) provide sufficient freeboard height above the maximum surcharge elevation to assure an acceptable degree of safety against possible overtopping from waves and runup.

Physical characteristics of the dam and reservoir site or existing developments within the reservoir area may impose upper limits in selecting the normal full-pool level. In other circumstances economic considerations govern.

With the normal full-pool elevation established, flood flows of unusual magnitude may be passed by providing spillways and outlets large enough to discharge the probable maximum flood or other spillway design flood without raising the reservoir above the normal full elevation or, if it is more economical, by raising the height of the dam and obtaining additional lands to permit the reservoir to temporarily attain surcharge elevations above the normal pool level during extreme floods. Use of temporary surcharge storage capacity also serves to reduce the peak rates of spillway discharge.

Freeboard height is the distance between the maximum reservoir level and the top of the dam. Usually 3 ft or more of freeboard is provided to avoid overtopping the dam by wind-generated waves. Additional freeboard may be provided in order to accommodate possible effects of surges induced by earthquakes, landslides, or other unpredictable events.

Diversion of stream. During construction the dam site must be unwatered so that the foundation may be prepared properly and materials in the structure may be moved easily into position. The stream may be diverted around the site through tunnels, passed through or around the construction area by flumes, passed through openings in the dam, or passed over low sections of a partially completed concrete dam. Diversion may be conducted in one or more stages, with a different method used for each stage. Initial diversion is conducted during a period of low flow to avoid the necessity for passing large flows.

Foundation treatment. The foundation of a dam must support the structure under all operating conditions. For concrete dams, following removal of unsatisfactory materials to a sound foundation surface, imperfections such as adversely oriented rock joints, open bedding planes, localized soft seams, and faults lying on or beneath the foundation surface receive special treatment. Necessary foundation treatment prior to dam construction may include "dental excavation" of surface weaknesses, or shafting and mining to remove deeper localized weaknesses, followed by backfilling with concrete or grout. Such work is sometimes supplemented by pattern grouting of foundation zones after construction of the dam. Foundation features such as rock joints, bedding planes, or faults that do not require preconstruction treatment are made relatively water-tight by curtain grouting from a line of deep grout holes located near the upstream heel and extending the full length of the dam. Although a grout curtain controls seepage at depth, the effectiveness of the grouting or its permanency cannot be relied upon alone to reduce hydrostatic pressures acting on the base of the dam. As a result, drain holes are drilled into the foundation just downstream of the grout curtain to intercept seepage passing through it and to reduce hydrostatic pressure. Occasionally chemical solutions such as acrylamide, sodium silicate, chromelignin, and polyester and epoxy resins are used for consolidating soils or rocks with fine openings.

The foundation of an earth dam must safely support the weight of the dam, limit seepage of stored water, and prevent transportation of dam or foundation material away or into open joints or seams in the rock by seepage. Earth-dam foundation treatment may include removal of excessively weak surface soils to prevent both potential sliding and excessive settlement of the dam, excavation of a cutoff trench to rock, and grouting of joints and seams in the bottom and downstream side of the cutoff trench. The cutoff trenches and grouting extend up the abutments, which are first stripped of weak surface materials.

When weak soils in the foundation of an earth dam cannot be removed economically, the slopes of the embankment must be flattened to reduce shear stress in the foundation to a value less than the soil strength. Relief wells are installed in pervious foundations to control seepage uplift pressures and to reduce the danger of piping when the depth of the pervious material is such as to preclude an economical cutoff.

Instrumentation. Instruments are installed at dams to observe structural behavior and physical conditions during construction and after filling, to check safety, and to provide information for design improvement.

In concrete dams instruments are used to measure stresses either directly or to measure strains from which stresses may be computed. Plumb lines are used to measure bending, and clinometers to measure tilting. Contraction joint openings are measured by joint meters spanning between two adjacent blocks of a dam. Temperatures are measured either by embedded electrical resistance thermometers or by adapting strain, stress, and joint measuring instruments. Water pressure on the base of a concrete dam at the contact with the foundation rock is measured by uplift pressure cells. Interior pressures in a concrete dam are measured by embedded pressure cells. Measurements are also made to determine horizontal and vertical movements; strong-motion accelerometers are being installed on and near dams in earthquake regions to record seismic data.

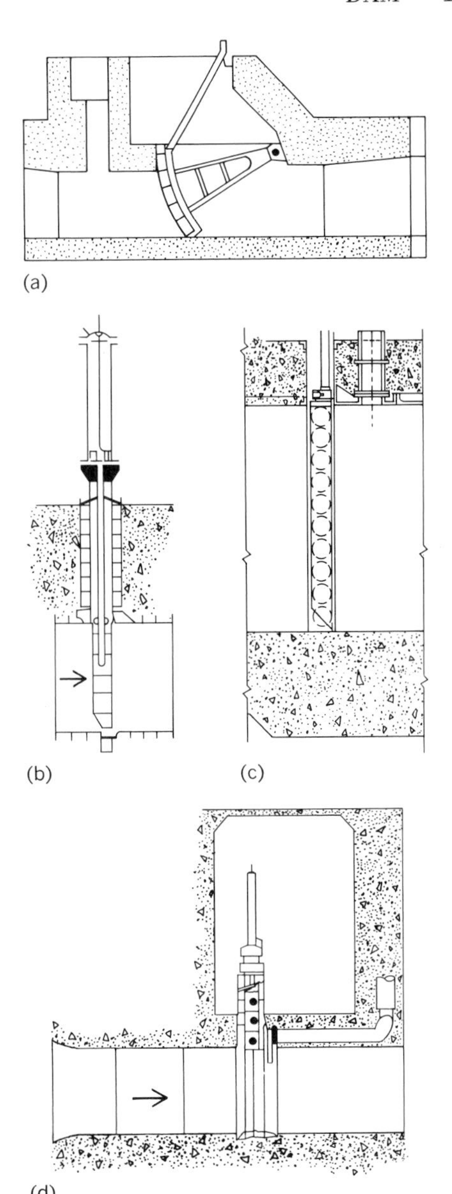

Fig. 10. Outlet gates. *(a)* Tainter gate. *(b)* High-pressure slide gate. *(c)* Tractor gate. *(d)* Jet flow gate. *(U.S. Army Corps of Engineers and U.S. Bureau of Reclamation)*

Instruments installed in earth-dam embankments and foundations are piezometers to determine pore water pressure in the soil or bedrock during construction and seepage after reservoir impoundments; settlement gages to determine settlements of the foundation of the dam under dead load; vertical and horizontal markers to determine movements, especially during construction; and inclinometers to determine horizontal movements along a vertical line.

Inspection of dams. Because failure of a dam may result in loss of life or property damage in the downstream area, it is essential that dams be inspected systematically both during construction and after completion. The design of dams should be reviewed to assure competency of the structure and its site, and inspections should be made during construction to ensure that the requirements of the design and specifications are incorporated in the structure.

After completion and filling, inspections may vary from cursory surveillance during day-to-day operation of the project to regularly scheduled comprehensive inspections. The objective of such inspections is to detect symptoms of possible distress in the dam at the earliest time. These symptoms include significant sloughs or slides in embankments; evidence of piping or boils near embankments; abnormal changes in flow from drains; unusual increases in seepage quantities; unexpected changes in pore water pressures or uplift pressures; unusual movement or cracking of embankments or abutments; significant cracking of concrete structures; appearance of sinkholes or localized subsidence near foundations; excessive deflection, displacement, erosion, or vibration of concrete structures; erratic movement or excessive deflection or vibration of outlet or spillway gates or valves; or any other unusual conditions in the structure or surrounding terrain.

Detection of any such symptoms of distress should be followed by an investigation of the causes, probable effects, and remedial measures required. Inspection of a dam and reservoir is particularly important following significant seismic events in the locality. Systematic monitoring of the instrumentation installed in dams is essential to the inspection program.

[JACK R. THOMPSON]

Bibliography: American Concrete Institute, *Symposium on Mass Concrete*, Spec. Publ. SP-6, 1963; C. V. Davis, *Handbook of Applied Hydraulics*, 1969; J. Hinds and W. P. Creager, *Engineering for Dams*, vols. 1 and 2, 1964; J. L. Sherard et al., *Earth and Earth-Rock Dams*, 1963; G. B. Sowers and G. F. Sowers, *Introductory Soil Mechanics and Foundations*, 1970; U.S. Bureau of Reclamation, *Design of Small Dams*, 2d ed., 1973; U.S. Bureau of Reclamation, *Trial Load Method of Analyzing Arch Dams*, Boulder Canyon Proj. Final Rep., pt. 5, Bull. no. 1, 1938; H. M. Westergaard, Water pressures on dams during earthquakes, *Trans. ASCE*, 98:418–472, 1933.

Defoliant and desiccant

Defoliants are chemicals that cause leaves to drop from plants; defoliation facilitates harvesting. Desiccants are chemicals that kill leaves of plants; the leaves may either drop off or remain attached; in the harvesting process the leaves are usually shattered and blown away from the harvested material. Defoliants are desirable for use on cotton plants because dry leaves are difficult to remove from the cotton fibers. Desiccants are used on many seed crops to hasten harvest; the leaves are cleaned from the seed in harvesting.

True defoliation results from the formation of an abscission layer at the base of the petiole of the leaf. Most of the chemicals bring about this type of defoliation, and the leaves abscise and drop from the plant. Certain chemicals kill plant leaves at low application rates, resulting in desiccation but not abscission.

The most common agency of defoliation in nature is frost; frosted leaves of cotton dry up and fall off after a few days. With the rapid increase in mechanical harvesting of crops, new chemicals and new processes have been introduced; these are termed harvest aids. Most harvest-aid chemicals are growth regulators that function at low concentrations or at low application rates; in some way they stimulate or inhibit growth. Some serve to hold fruits such as apples and pears upon the trees, thus preventing preharvest drop. Others cause loosening of fruits, facilitating harvest by shaking; examples are sour cherries and grapes.

Because of the advantages of harvest aids, much effort has gone into the search for new chemicals.

A list of active ingredients in currently used harvest aids includes (trade names are given in parentheses) ametryn, (Evik), amino triazole, ammonia gas, ammonium nitrate, ammonium thiocyanate, arsenic acid, cacodylic acid, calcium cyanamide, diethyl dithio-bis-thionoformate, diquat, endothal (Accelerate, Des-I-Cate), ethephon (Ethrel), 4,6-dinitro-*o*-*sec*-butylphenol, hexachloroacetone, magnesium chlorate, nitrophen, oxyfluorfen, paraquat, pentachlorophenol, petroleum solvents, sodium borate, sodium cacodylate (Bollseye), sodium chlorate (Tumbleleaf), sodium-*cis*-3-chloroacrylate, sodium naphtalene acetate (Fruitone N), tributylphosphorothioate (Folex), tributylphosphorotrithioate (Def-6), 2,4,5-TP (Silvex), triethanolamine salt of Silvex (Fruitone T).

Crops upon which harvest aids are being used include alfalfa, apples, blackberries, blueberries, clovers (Alsike, Ladino), cantaloupes, castor beans, cherries, cranberries, flax, figs, filbert nuts, grapes, guar, hops, lemons, onions, oranges, peas, potatoes, rice, safflower, sudan grass, sorghum, soybeans, sunflowers, tangerines, tomatoes, walnuts, and wheat. Tomato plants are sometimes cut below the soil surface and allowed to dry out; this brings the fruits to a uniform ripeness for mechanical harvesting.

Defoliants and desiccants have also been used during war to destroy crops.

Modern agricultural technology, which has greatly increased production in the United States, is dependent upon harvest-aid chemicals. These aids bring fruits to uniform ripeness for harvest by machinery or by well-managed crews. They reduce or eliminate much labor, save time, and make for reliable management of harvest. They, along with fertilizers and improved crop varieties, are responsible for the great abundance of foods evidenced in modern American supermarkets.

It was in an effort to find such chemical agents that 2,4-D was discovered in 1942 by E. J. Kraus. For more than 30 years 2,4-D and 2,4,5-T have been important compounds used for weed control.

[ALDEN S. CRAFTS]

Desert ecosystem

As in the analysis of other ecosystems, desert research focuses on such functional characteristics as energy flow and chemical cycling and such structural patterns as community composition. Energy parameters of particular interest are the rates at which the vegetation fixes solar energy, the rates and efficiencies with which that energy is exchanged between vegetation and other organisms, and the environmental constraints limiting these rates. Chemical-cycling questions of interest are the magnitudes of elemental pools in the soil and biota, the rates at which these pools are incremented and decremented by input and output processes and, again, the environmental constraints affecting these rates. A knowledge of the physical and biotic influences impinging on individual species is essential to predicting the effects of changes in these influences on community structure.

Patterns of primary production. The low net aboveground primary production of desert ecosystems is well known. Most values fall between 100 and 1000 kg/ha/yr of air-dried material. The median of the range is roughly one-fifth or one-sixth of the production levels of grasslands and temperate forests. Although rates of root production have not been studied as extensively as aboveground growth, one may approximate root growth on the assumption that ratios of root/shoot biomass reflect similar production ratios. Since root/shoot ratios in deserts commonly range from 1:1 to 6:1, net primary production may range from two to six times the range given above.

Less well known is the extreme variability of desert primary production. Over a period of years, production in the most favorable years may exceed that in the least favorable by a factor of 6 to 8. In a temperate forest system, high values may exceed low values by no more than 20 or 30%. The source of this variation is twofold: (1) Because desert production is so profoundly constrained by moisture, it varies markedly between years with annual variations in precipitation (Fig. 1). Where moisture is no longer a significant constraint, as in temperate forests, annual variations in precipitation no longer produce the marked variations in production. (2) Precipitation in arid areas is relatively more variable than in mesic areas. At one weather station near Milford in southwestern Utah, the standard deviation of the mean annual precipitation was 30%. At a station near Oak Ridge, TN, the standard deviation was 17% of the mean. Thus, because precipitation is more variable in deserts and because primary production is so closely correlated with precipitation, desert production is more variable than in grasslands and forest.

Role of annuals. In the low-latitude (<40°) "hot" deserts, annual plants occupy a uniquely important position in undisturbed, climax vegetation. These plants are diverse, commonly constituting two-thirds or more of the plant species in a limited area. Their production is an important fraction of the total primary production for a site, commonly making up one-half or more in a year of high precipitation. Since their production may fail completely in a very dry year, it is more variable than that of perennials, a factor which further contributes to variability of desert primary production.

The biotic interrelationships between annuals and perennials in hot deserts is not entirely clear, with some workers postulating that annuals exist as commensals with perennials. If this hypothesis is correct, the implication follows that any disturbance producing a reduction in perennial abundance and diversity would also reduce annual abundance and diversity. A second school of thought holds that hot-desert annuals operate virtually as an autonomous flora, largely independent of perennials.

The position and function of annuals in high-latitude "cold" deserts has not been studied as widely as in hot deserts. However, in North America, native species are less diverse than in hot deserts, and in many areas a small number of exotic species are the only conspicuous annuals. Furthermore, these exotic species enter the flora as disturbance or seral forms with undisturbed, fully populated perennial communities largely devoid of annuals.

One hypothesis suggests that the difference in position and function of annuals between the two desert types lies in the difference in perennial root structure. Root/shoot ratios in North American hot deserts are commonly of the order of 1:1, while those in cold deserts may reach 6:1. On one Sonoran Desert site in Arizona, standing-crop biomass of roots was measured at 10,929 kg/ha; on one Great Basin Desert site in northern Utah, it was 27,499 kg/ha. Furthermore, nearly one-half of the greater root mass in the Utah area was in the upper 20 cm of soil, while only one-fourth of the lesser root mass in the Arizona area was in this same layer. Root structure also differs. In southern deserts, woody roots may extend laterally for considerable distances before they subdivide into small, actively growing forms which penetrate the soil to absorb moisture. In the north, the root system is a dense mass of fine, fibrous roots which permeate the soil. In short, the surface 20 cm of soil in the southern deserts appears to offer a niche into which an abundant flora has evolved. In the northern deserts, this zone is permeated by fine, actively growing perennial roots which may have blocked the evolution of an extensive ephemeral flora.

Role of consumers. As appears true of many other terrestrial ecosystems, herbivorous animals in the desert consume less than 10–20% of the primary production while it is living phytomass. However, seeds constitute one subset of primary production which is a notable exception to this generalization. The profuse annuals along with many perennials produce a seed resource which

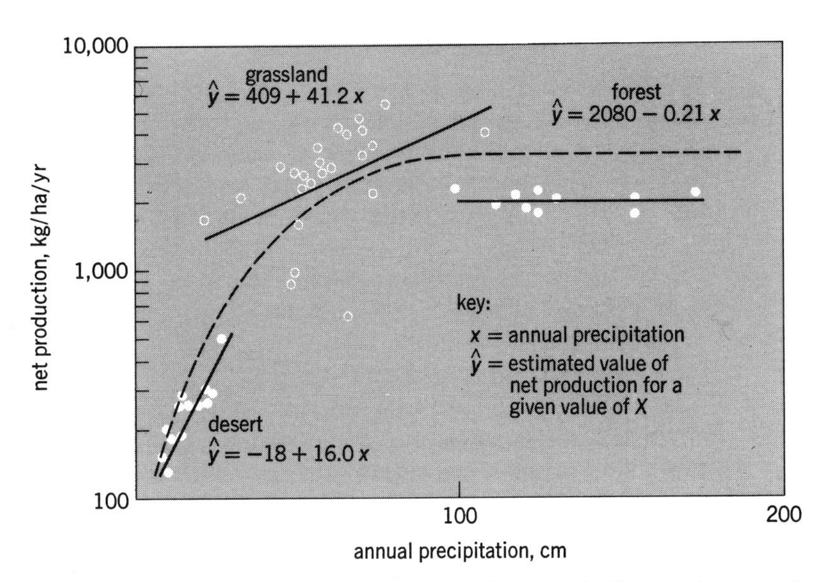

Fig. 1. Regressions of net annual aboveground primary production on mean annual precipitation for desert, grassland, and deciduous forest sites. (*From F. H. Wagner, Integrating and control mechanisms in arid and semiarid ecosystems: Considerations for impact assessment, Proceedings of the Symposium on the Biological Evaluation of Environmental Impact, 27th Annual AIBS Meeting, New Orleans, 1977*)

Wood consumption by termites in the desert

Area	Deadwood produced, kg/ha/yr	Wood consumed, kg/ha/yr
Southern Arizona	450	414
Southern New Mexico playa	1030	528
Southern New Mexico bajada	410	336

may aggregate to as much as 20% of aboveground primary production. A diverse fauna of rodents, ants, and wintering birds has evolved to use this resource, and appears to exploit it so fully that it may limit population densities. Evidence from the Sonoran Desert indicates definite interspecific competition between rodents and ants.

Since most of the primary production is not utilized by herbivores and since most undisturbed desert ecosystems are presumed to be in approximate bioenergetic equilibrium, it follows that most of the plant material produced must die and become detritus. In fact, a high percentage of dead phytomass characterizes many desert systems. In one Utah desert, the ratio of live aboveground phytomass to standing dead material was 2.6:1, while in three southern deserts the same ratio varied from 3:1 to 1:1. This dead material constitutes a resource for which a diverse termite fauna has evolved in the southern deserts. Studies in New Mexico and Arizona indicate that annual deadwood consumption by termites is roughly equivalent to the annual rate of deadwood production (see table).

Nitrogen cycling. The nitrogen content of desert soils is lower than that in grassland and forest systems, and nitrogen appears to be second to water as a limiting factor on primary production in deserts. While the source of this nitrogen shortage and the characteristics of the nitrogen cycle

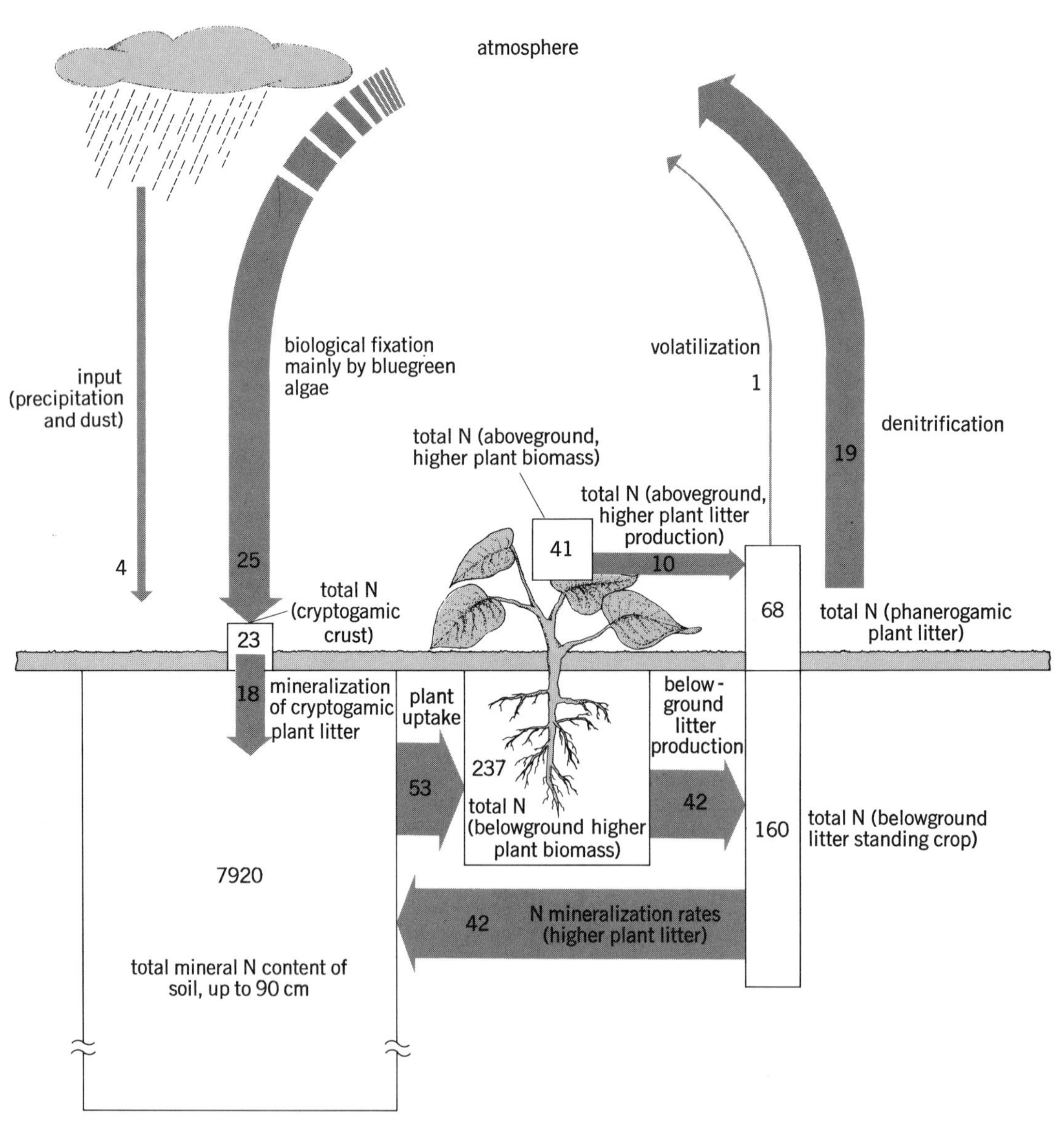

Fig. 2. Schematic representation of the annual nitrogen balance for an *Atriplex confertifolia*- dominated ecosystem in Curlew Valley, UT. Arrows are annual fluxes (kg N/ha/yr), their widths approximating the magnitudes of the numbers inside arrows. Boxes represent the highest annual values of the components (kg N/ha). (*From N. E. West and J. Skujins, The nitrogen cycle in North American cold-winter semi-desert ecosystems, Oecol. Plant, 12(1): 45–53, Gauthier-Villars, 1977*)

have not been studied intensively in all deserts, recent concerted effort in North American cold deserts has elucidated the patterns in this type. During the rainy season, nitrogen is fixed at relatively high rates (25 kg/ha/yr) by blue-green algae in soil cryptogamic crusts (Fig. 2). An additional 4 kg/ha/yr enters in dust and precipitation. However, over 70% of this is lost through volatilization and denitrification, apparently because the low carbon content of the soil serves as a limiting factor on the development of an abundant soil microflora which could mineralize the nitrogen. Because primary production and standing-crop phytomass are low, return of dead organic matter to the soil is meager.

Consequently, nitrogen pools in desert soils do not build up to substantial levels, both because mineralization rates are low and because fixed nitrogen is not held. In contrast, little nitrogen is fixed in deciduous forest soils, but denitrification loss is also minimal. Primary production and phytomass are high, as are soil organic matter and mineralization rates. The forest nitrogen cycle is thus a closed cyclic entity with major flow through the biota. The desert cycle is open with much of the nitrogen which flows through the system bypassing the macroflora. The practical implication of the desert findings is that in order to reduce the nitrogen constraint on the desert ecosystem by fertilizing, one must add a carbon source rather than a nitrogen fertilizer, since without added carbon the fertilizer would simply be denitrified.

For background information *see* BIOLOGICAL PRODUCTIVITY; BIOMASS; ECOSYSTEM; NITROGEN CYCLE; PLANTS, LIFE FORMS OF; TERRESTRIAL ECOSYSTEM.

[FREDERIC H. WAGNER]

Bibliography: R. A. Perry and D. W. Goodall (eds.), *Structure, Function and Management of Arid Land Ecosystems*, 1977; F. H. Wagner, Integrating and control mechanisms in arid and semiarid ecosystems: Considerations for impact assessment, *Proceedings of the Symposium on the Biological Evaluation of Environmental Impact*, 27th Annual AIBS Meeting, New Orleans, 1977; N. E. West and J. Skujins, *Oecol. Plant.* 12(1): 45–53, 1977.

Desertification

Desertification is the spread of desertlike conditions in arid and semiarid areas, due to human influence or climatic change. Natural vegetation is being lost over increasingly larger areas because of overgrazing, especially in Africa and southwestern Asia. Where the original state of vegetation has been retained, for example, in military compounds, the contrast is dramatic. In such a case, near Nefta in southern Tunisia, the coverage of vegetation inside an area fenced 60 years ago is 85%, in contrast to 5% outside the area; here the original dry steppe has changed into a semidesert, without any appreciable variation of precipitation (about 80 mm per year). Another dramatic example of this type is revealed in a NASA Earth Resources and Technology Satellite (ERTS) photograph of the Sinai-Negev desert region. The political boundary established in the 1948 armistice between these two regions is clearly visible, with the lighter Sinai region to the west of the boundary and the darker Negev region to the east. The sharp demarcation is due to the fact that Bedouin Arabs' goats in the Sinai have defoliated enough land on the Egyptian side to make the boundary visible from space. Similar conditions have been observed in India.

In many areas the deserts seem to be spreading, with an apparent speed of about one or more kilometers per year (the Sahara Desert appeared to be advancing into the Sahel—consisting of Mauritania, Senegal, Mali, Upper Volta, Niger, and Chad—during the 1968–1973 drought at the rate of 50 km per year), depending on the density of the population, as a consequence of grazing animals (especially goats). However, recent desertification is the result of the interaction of naturally recurring drought with unwise land-use practices. In order to understand to what extent each of these factors plays a part in the total desertification process, and to consider whether the process can be slowed down or reversed, it is helpful to review natural causes and human causes of deserts, as follows.

Natural causes of deserts. The world distribution of dry climates depends mainly on the subsidence associated with the subtropical high-pressure belts, which migrate poleward in summer and equatorward in winter. These migrations are connected with the atmospheric general circulation. There results a threefold structure in the arid zone: a Mediterranean fringe, with rains occurring only in winter; a desert core (in about 20–30° latitude), with little or no rain; and a tropical fringe, with rains mainly in the high sun season. Throughout the arid zone, rainfall variability is high. Aridity arises from persistent widespread subsidence or from more localized subsidence in the lee of mountains. In some regions, such as the northern Negev, the combination of widespread and local subsidence is clearly visible as clouds on the windward side of mountains form and immediately dissipate on the leeward side. Aridity may also be caused by the absence of humid airstreams and of rain-inducing disturbances. Clear skies and low humidities in most regions give the dry climates very high solar radiation (averaging over 200 W/m^2), which leads to high soil temperatures. The light color and high reflectivity (albedo) of many dry surfaces cause large reflectional losses, and long-wave cooling is also severe. Hence net radiation incomes are relatively low—of the order of 80–90 W/m^2.

Climatic variation takes place on many time scales. The processes that may cause climatic variation, plotted against their characteristic time scale, are shown in the illustration, in which autovariation means internal behavior. The world's deserts and semideserts are very old, although they have shifted in latitude and varied in extent during geological history. The modern phase of climate began with a major change about 10,000 years before present, when a rapid warming trend removed most of the continental ice sheets. The Sahara and Indus valleys were at first moist, but since about 4000 years before present, when severe natural desiccation took place, aridity has been profound.

Recent climatic variations, such as the Sahelian drought, are natural in origin, and are not without precedent. Statistical analysis of rainfall shows a distinct tendency for abnormal wetness or drought

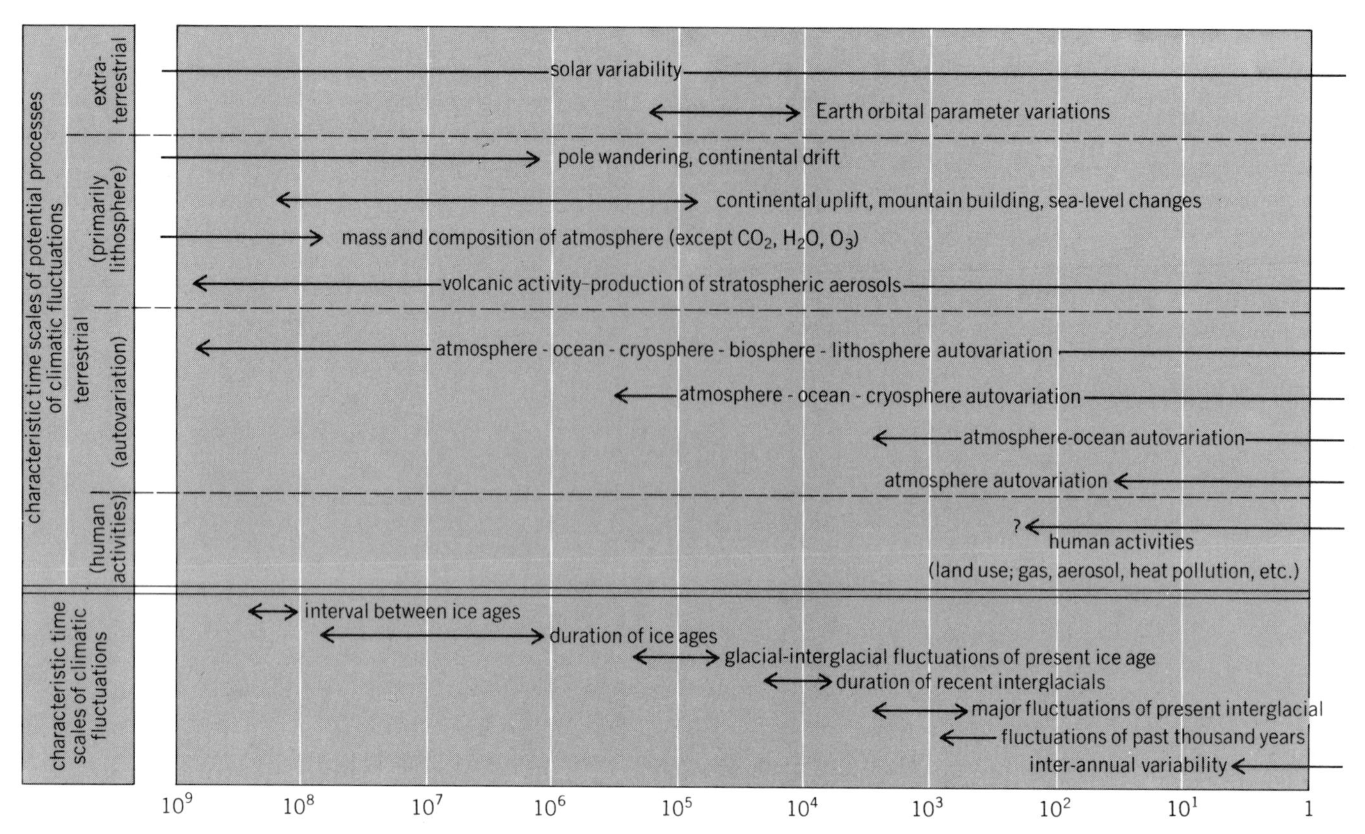

Processes that may cause climatic variation, plotted against their characteristic time scale. *(From United States Committee for the Global Atmospheric Research Program, Understanding Climatic Change: A Program for Action, National Academy of Sciences, Washington, DC, 1975)*

to persist from year to year, especially in the Sahel. Prolonged desiccation, lasting a decade and more, is common and often ends abruptly with excessive rainfall. This persistence suggests that feedback mechanisms may be operating, whereby drought feeds drought and rain feeds rain.

Human causes of deserts. Recent desertification has resulted from the spatially uneven pressure exerted by humans on soil and vegetation, especially at times of drought or excessive rainfall. The Dust Bowl in the Great Plains of the United States and the degeneration in the Sahel, the Ethiopian plateaus, and the Mendoza Province of Argentina are all manifestations of human misuse of the land at times of climatic stress.

A common mechanism of desertification includes the following steps: (1) expansion and intensification of land use in marginal dry lands during wet years, including increased grazing, plowing, and cultivation of new lands, and wood collection around new camps or settlements; and (2) wind erosion during the next dry year, or water erosion during the next maximum rainstorm.

This desertification process has the following climatological implications: (1) Increased grazing during wet years tends to compact the soil near water holes, and increased livestock numbers cause pressure on perennial plants during dry seasons. The result is to expose surface soil to erosion by wind. Runoff may be increased, and so may albedo (that is, reflectivity of the ground with respect to solar radiation). (2) Increased cultivation during wet years greatly improves the chances of wind erosion of fine soil materials during the dry season, and possibly increases evapotranspiration (in the case of water-demanding crops). (3) Removal of wood increases direct solar heating and may considerably decrease evapotranspiration. (4) In the ensuing dry years, the acceleration of wind erosion by the above processes further reduces water storage capacity by removal of topsoil. The loss of some or all perennial plant cover lowers infiltration rates and hence the potential percolation. During subsequent rains, surface runoff is increased, with an attendant loss of water for subsequent use by shrubs, herbage, or crops.

In order to test hypotheses of both natural and human-caused effects on the climate, one must have effective mathematical models of the dry climate.

Modeling attempts. In the mathematical model, the particular physical phenomenon to be studied is described in mathematical terms, and the equations are then solved by means of high-speed computers. In this way, predictions may be made of the effects of changes in external forcing, including the inadvertent or intentional changes caused by humans. Experiments thus far comprise:

1. General circulation models (GCMs) applied to the specific problems of the causes of climatic fluctuations in the dry climates. These models simulate in three dimensions the general circulation of the hemisphere or globe, and can be subjected to chosen perturbations, such as sea surface temperature anomalies, to determine the effects.

2. Investigations of specific feedback processes, such as changed albedo or variations of soil moisture storage. The models seek to predict the

effect of feedbacks on circulation and precipitation over the arid zone. One recent experiment of this type was carried out in the United Kingdom. A model investigation tested the speculation that rainfall variations over north and central Africa were related to sea surface temperature anomalies over the tropical Atlantic. The experiment showed that introduction of an extensive sea surface temperature anomaly over the tropical Atlantic was related to higher precipitation amounts over North Africa. While the experiment did not actually prove the causal relationship, it changed the speculation into a credible hypothesis.

Most recent modeling attempts have been related to specific feedback processes that may augment or retard naturally induced climatic variations along the desert margin, especially the effect of changed albedo and other consequences of the degradation of vegetation cover.

The albedo feedback hypothesis argues that the destruction of vegetation and exposure of soil increase albedo and hence lower surface temperatures and suppress convective shower formation. This hypothesis has been considered to be a mechanism for desertification. Since certain regions of the central and northern Sahara, eastern Saudi Arabia, and southern Iraq have a negative radiation balance at the top of the atmosphere on hot summer days, in spite of the intense input of solar radiation through the cloudless atmosphere, the following argument has been advanced: Since the ground stores little heat, it is the air that loses heat radiatively. In order to maintain thermal equilibrium, the air must descend and compress adiabatically. Since the relative humidity then decreases, the desert increases its own dryness. A biogeophysical feedback mechanism of this type could lead to instabilities or metastabilities in borders themselves, which might conceivably be set off or maintained by anthropogenic influences.

This hypothesis has been tested by means of a dynamical model, in which the surface albedo was increased over large desert regions. Sharp reductions of cloud and rain followed the increase in albedo. Possible mechanisms by which albedo is changed are removal of vegetation by drought, overstocking, cultivation, or all of these, or by desiccation of the soil itself, soil albedo being related to soil water content. In practice, the three mechanisms are likely to occur simultaneously, so that soil moisture content may itself serve as a positive feedback for drought, a wet soil favoring renewed rainfall which is derived from local evapotranspiration.

All of the above hypotheses work by influencing the overall dynamics of the desert margin climates, essentially via their effect on rates of subsidence and hence stability. There is also the possibility that cloud microphysics may be affected by surface conditions. It has been suggested that cumulus and cumulonimbus clouds of the Sahelian and Sudanian belts of western Africa are "seeded" by organic ice nuclei raised from the vegetated surface below. Removal of the vegetation destroys the local source of organic nuclei. Hence, well-formed clouds remain unseeded and rainless, thereby accelerating the decay of surface vegetation.

In summary, modeling experiments suggest that there may well be a positive feedback process along the desert margin, operating via the increase of surface albedo attendant upon the destruction of the surface vegetation layer, and possibly also upon the decrease of soil moisture and surface organic litter. It thus appears possible that widespread destruction of the vegetation cover of the dry world may tend to further reduce rainfall over these areas.

Possible solutions. Suggestions for solution of the desertification problem include using techniques for weather and climate modification, control of surface cover, and maximum application of modern technology.

Weather and climate modification. Cloud seeding, establishing green belts, and flooding of desert basins are a few of the weather and climate modification techniques under consideration.

Special conditions are required for rainfall augmentation by cloud seeding. If the proper type of seedable cloud exists, it may be possible to increase the rainfall locally by this method.

Establishment of green belts along the northern and southern margins of the Sahara is considered to be of dubious value climatologically, as desertification does not spread outward from the desert. Thus the green belt would not serve as a "shelter belt." However, this hypothesis should be tested by means of a model.

Precipitation depends largely on water vapor which has traveled great distances, along with upward motion of the air. The experience with artificial oases to date has shown little or no change of climate in their vicinity. Modeling experiments carried out by simulating the "flooding" of Lake Sahara predicted no significant change of rainfall around the lake, although rainfall was increased over an isolated mountain region 900 km from the shore. However, these results do indicate the possibility of creating artificial bodies of water judiciously with respect to the areas where rainfall augmentation is desired. This hypothesis may also be tested by model experiments.

Control of surface cover. The key to the control of the desertification process is the control of surface cover. If the relatively secure wind-stable surfaces of some desert areas, or a reasonably complete vegetation cover (even if dead) can be maintained, soil drifting and deflation are minimized. Overstocking, unwise cultivation, and the use of overland vehicles weaken and ultimately destroy these protective covers. The proposed green belt is of value because of the added protection it affords the soil.

The usefulness of land depends on the surface microclimate. The ability to conserve this microclimate rests on transformation of desert technology, rather than on transformation of climate. The ability of the surface to respond quickly and generously to renewed abundance of rainfall depends on the soil's capacity to retain nutrients, organic substances, and fine materials; on high infiltration capacity; and, of course, on viable seeds as well as a surviving root system. A surface litter of organic debris may also be important for precipitation mechanisms, and has some effect on surface radiation balance.

Use of modern technology. Satellite data could be used in tracking the major rainstorms of the rainy season to study their habits and, if possible, to predict their displacement. Other important satellite data, for example, radiation, could be collected and used in connection with feedback studies.

Future research. More research is needed into the relationship between climate and the desertification process. There should be a concerted international research effort, with major attention focused on deserts as well as oceans, and on surface physical and bioclimatic processes as well as atmospheric dynamics.

Further experimentation is needed that specifically examines the problems of the general circulation of the Earth's atmosphere, and even of the smaller-scale circulation, over the dry land areas of the subtropics. There is also a need for more detailed study of the physical climatology and bioclimatology of dry land surfaces, and particularly for a closer synthesis of climatology with geomorphology, soil science, hydrology, and ecology.

For background information *see* CLIMATE MODIFICATION; CLIMATIC CHANGE; DESERT ECOSYSTEM; DROUGHT.

[LOUIS BERKOFSKY]

Bibliography: M. H. Glantz (ed.), *The Politics of Natural Disaster*, 1976; F. K. Hare, T. Gal-Chen, and K. Hendrie, *Climate and Desertification*, Institute for Environmental Studies, University of Toronto, 1976; Massachusetts Institute of Technology, *Inadvertent Climate Modification: Report of the Study of Man's Impact on Climate (SMIC)*, 1971; S. H. Schneider and L. E. Mesirow, *The Genesis Strategy*, 1976.

Detergent

A substance used to enhance the cleansing action of water. A detergent is an emulsifier, which penetrates and breaks up the oily film that binds dirt particles, and a wetting agent, which helps them to float off. Emulsifier molecules have an oillike nonpolar portion which is drawn into the oil, and a polar group that is water-soluble; by bridging the oil-water interface, they break the oil into dispersible droplets (emulsion). As a surfactant, a detergent decreases the surface tension of water and helps it penetrate soil.

Soap, the sodium salt of long-chain acids, was the principal detergent until superseded in 1954 by synthetic detergents (syndets) which, unlike soap, do not form insoluble products with the calcium in hard water. Most syndets are of the anionic type, that is, sodium salts of alkyl sulfates or sulfonates which contain a chain of 7–18 carbon atoms; an example is sodium lauryl sulfate, $CH_3(CH_2)_{11}$-OSO_3^-,Na^+. Alkyl benzene sulfonates (ABS) with branched carbon chains were found to persist in wastewater and have been replaced by linear alkyl benzene sulfonates (LAS), which are biodegradable by bacterial action. Anionic detergents are best for water-absorbing fibers such as cotton, wool, and silk. Nonionic detergents are polyethers made by combining ethylene oxide with a 12-carbon lauryl alcohol obtained from tall oil, a waste product from the paper industry. They are used for water-repelling "permanent press" fabrics, and their low-foaming property is desirable for automatic washers. Cationic syndets are quarternary base compounds of type $(NR_4)^+$,X^-. They are more expensive, but some are germicidal; some are used as fabric softeners, and they are good metal cleaners.

Detergents must contain alkaline "builders" to bind dissolved metal ions and support emulsification. Sodium pyrophosphate or polyphosphate is preferred because of low cost and high cleansing effectiveness. However, when discharged with laundry wastewater, these compounds supply nutrient to phosphate-deficient lakes and streams and thus lead to eutrophication, and their use is now banned by law. For a brief time an effective builder designated NTA was promoted as a phosphate substitute, until it was found that it solubilizes metal and may lead to metal poisoning. Less harmful, but less effective, builders such as sodium carbonate are now widely used in detergents. *See* EUTROPHICATION.

Many additives are used in detergents to provide scent, brightening (usually through fluorescent action), or bleaching action. For a time, enzymes which cleave protein molecules (proteolytic) or fats (lipolytic) were widely used as additives, but they do not add significantly to cleansing action. Biodegradability is essential for detergents; it ensures that components of detergents will be broken down by bacterial action before undesirable aftereffects can occur. Nonbiodegradable detergents can prevent effective bacterial action in septic tanks and sewage treatment plants, and can cause undesirable persistent foaming in rivers.

[ALLEN L. HANSON]

Bibliography: J. O'M. Bockris, *Environmental Chemistry*, 1977; R. E. Burk and O. Grummitt, *Frontiers in Colloid Chemistry*, 1950; J. P. Sisley and P. J. Wood, Encyclopedia of Surface-Active Agents, 2 vols., 1952, 1965.

Dew point

The temperature at which air becomes saturated when cooled without addition of moisture or change of pressure. Any further cooling causes condensation; fog and dew are formed in this way.

Frost point is the corresponding temperature of saturation with respect to ice. At temperatures below freezing, both frost point and dew point may be defined because water is often liquid (especially in clouds) at temperatures well below freezing; at freezing (more exactly, at the triple point, +.01°C) they are the same, but below freezing the frost point is higher. For example, if the dew point is −9°C, the frost point is −8°C. Both dew point and frost point are single-valued functions of vapor pressure.

Determination of dew point (or frost point) can be made directly by cooling a flat polished metal surface until it becomes clouded with a film of water or ice; the dew point is the temperature at which the film appears. In practice, the dew point is usually computed from simultaneous readings of wet- and dry-bulb thermometers. *See* HUMIDITY.

[J. R. FULKS]

Bibliography: R. J. List (ed.), *Smithsonian Meteorological Tables*, 6th ed. rev., 1951; M. Thaller, Instrument Development Inquiry, 2d ed., 1977.

Disease

An alteration of the dynamic interaction between an individual and his environment sufficient to be deleterious to the well-being of the individual. The cause of a disease may be environmental, or an altered reactivity of the individual to his environment, or a combination of both. The environmental change may be a physical, chemical, or biological agent.

A disease state is life under altered conditions; thus it represents a deviation from a norm and is a summation of the characteristics of the deviation from normal structure and function. The norm, called homeostasis, is the constellation of dynamic bodily activities that serve to maintain the functional integrity of an individual. These activities range from the biochemical subcellular level of organization to the cellular, tissue, organ, and organismal levels. The dynamic equilibrium of homeostasis for one person is not necessarily the same for another, varying with age, sex, race, and environment. The individual, as a biological system, is an open system, which means that homeostasis is a steady state in a dynamic system. Open systems have two principal characteristics: They are inseparable from, and constantly interact with, their external environment and they do not maintain unique stationary levels. Therefore there is an infinite variety of normal homeostatic systems, all of which have a common feature—the constancy of change. *See* HOMEOSTASIS.

Recognition. The terms health or disease are relative, and the condition of perfect health probably does not exist. Consequently, the recognition of a disease is a subjective concept, the limits of which cannot be precisely defined. The term disease is used in many ways and is often employed synonymously with illness, sickness, or a specific condition. It should be recognized that there is no sharp division between extremes of physiological response to stimuli and the ill-defined beginnings of a disease process. Consideration must be given to the individual—sex, age, habits, health pattern, genetic background, and many other pertinent factors.

The existence of disease in a person is arbitrarily defined, based on certain recognizable objective (signs) and subjective (symptoms) changes. Examples of the signs of disease are abnormal changes in temperature, pulse rate, and respiratory rate and laboratory measurements of blood and other body fluids. Symptoms such as pain, lassitude, restlessness, or emotional upset cannot be recorded or quantitated, though they may be no less real to the ill individual. Methods for observing and recording objective signs are based on the application of the principles of biochemistry, biophysics, and morphology. To compare the results of a test with a norm, a physician must note the environmental agent, as well as the mechanisms by which the individual adapts to its influence. In some instances the response of the individual may be so characteristic for a particular agent that by observing the response the cause of the disease may be reliably inferred. The goal of the medical scientist is to recognize all fundamental factors related to the initiating agent of disease (cause or etiology) and the mechanisms of adaptation to the agent (pathogenesis).

Death occurs when the rate or degree of interaction is so altered that the individual can no longer maintain homeostasis and irreversible changes ensue. The individual enters a steady state of equilibrium with his environment. Without dynamic change, life ceases.

In common usage the term disease indicates a constellation of specific signs and symptoms which can be attributed to altered reactions in the individual (form and function) produced by certain agents that affect the body or its parts. Alterations of body structure or function are often called lesions or pathologic lesions, which represent the reactions of cells, tissues, organs, or the organism to injury. The physician observes signs and symptoms in an individual, evaluates their relative significance, and compares them with known patterns of signs and symptoms. Therapeutic measures either protect the individual from the etiologic agent or alter the body's response to it.

Etiology. The causative agents of disease may be classified as exogenous (environmental) and endogenous (altered capacity of the individual to adapt to environmental changes). Whether environmental agents can produce disease in an individual depends on the amount, rate, and duration of exposure to the agent, as well as the individual's capacity to react (resistance) in an appropriate manner. Exogenous agents include virtually every feature of the environment. A partial list of exogenous and endogenous causes of disease is given in the table.

Frequently the interaction of multiple endogenous and exogenous factors is involved in the genesis of a disease. For example, excessive exposure to ultraviolet rays (physical agent) of 2800–3100-A wavelengths produces a disease called erythema solare (sunburn). The intensity of the radiation per minute varies with the latitude north or south of the Equator, season of the year, altitude, and atmospheric conditions (for example, dust and moisture in the air may act as a filter). Individuals vary greatly in their reaction and sensitivity to the ultraviolet rays (hereditary factors). Blondes and the rufous Celtic types are more susceptible than darker or black-skinned individuals or those previously exposed to small doses. Men react more than women, and infants more than adults (age factor). Women have increased sensitivity during menses. Coexistence of diseases such as hyperthyroidism or tuberculosis increases sensitivity, as do some medicines. Therefore, a given exposure to sunshine may produce the disease in one individual and not in another.

Environmental factors. Though some diseases may have an endogenous basis, the expression of the defect may depend on the environment of the individual. For example, nutritional diseases are a general class of exogenous diseases. If the individual fails to ingest sufficient quantities of ascorbic acid (vitamin C), a chemical substance found in abundance in citrus fruits and tomatoes and many other raw vegetables, scurvy develops. Ascorbic acid is essential to the metabolism of certain amino acids (components of proteins), especially in connective tissue fibroblasts.

Common exogenous and endogenous causes of disease

Causative agent	Disease
Exogenous factor	
Physical	
Mechanical injury	Abrasion, laceration, fracture
Nonionizing energy	Thermal burns, electric shock, frostbite, sunburn
Ionizing radiation	Radiation syndrome
Chemical	
Metallic poisons	Intoxication from methanol, ethanol, glycol
Nonmetallic inorganic poisons	Intoxication from phosphorus, borate, nitrogen dioxide
Alcohols	Intoxication from methanol, ethanol, glycol
Asphyxiants	Intoxication from carbon monoxide, cyanide
Corrosives	Burns from acids, alkalies, phenols
Pesticides	Poisoning
Medicinals	Barbiturism, salicylism
Warfare agents	Burns from phosgene, mustard gas
Hydrocarbons (some)	Cancer
Nutritional deficiency	
Metals (iron, copper, zinc)	Some anemias
Nonmetals (iodine, fluorine)	Goiter, dental caries
Protein	Kwashiorkor
Vitamins:	
A	Epithelial metaplasia
D	Rickets, osteomalacia
K	Hemorrhage
Thiamine	Beriberi
Niacin	Pellagra
Folic acid	Macrocytic anemia
B_{12}	Pernicious anemia
Ascorbic acid	Scurvy
Biological	
Plants (mushroom, fava beans, marijuana, poison ivy, tobacco, opium)	Contact dermatitis, systemic toxins, cancer, hemorrhage
Biological (cont.)	
Bacteria	Abscess, scarlet fever, pneumonia, meningitis, typhoid, gonorrhea, food poisoning, cholera, whooping cough, undulant fever, plague, tuberculosis, leprosy, diphtheria, gas gangrene, botulism, anthrax
Spirochetes	Syphilis, yaws, relapsing fever, rat bite fever
Virus	Warts, measles, German measles, smallpox, chickenpox, herpes, roseola, influenza, psittacosis, mumps, viral hepatitis, poliomyelitis, rabies, encephalitis, trachoma
Rickettsia	Rocky Mountain spotted fever, typhus
Fungus	Ringworm, thrush, actinomycosis, histoplasmosis, coccidiomycosis
Parasites (animal)	
Protozoa	Amebic dysentery, malaria, toxoplasmosis, trichomonas vaginitis
Helminths (worms)	Hookworm, trichinosis, tapeworm, filariasis, ascariasis
Endogenous factor	
Hereditary	Phenylketonuria, alcaptonuria, glycogen storage disease, Down's syndrome (trisomy 21), Turner's syndrome, Klinefelter's syndrome, diabetes, familial polyposis
Developmental	Many congenital anomalies
Hypersensitivity	Asthma, serum sickness, eczema, drug idiosyncrasy

For some animals, such as birds, exogenous vitamin C is not a requirement. The body tissue of the animal can synthesize its own ascorbic acid and thus the animal is not subject to scurvy. Other animals, such as guinea pigs and humans, are unable to synthesize ascorbic acid (the endogenous defect) and must ingest it from the environment. In those environments where vitamin C is plentiful, scurvy does not exist. One may hypothesize that during the evolution of man the capacity of the cells to synthesize this substance was lost. However, the loss was not detrimental to man's survival because of the plentifulness of vitamin C in many environments.

Whether a particular disease is harmful or not may also depend on the environment. Sickle-cell trait is an inherited feature of hemoglobin in some individuals, usually those of African Negro descent. Under conditions of decreased blood oxygen the red blood cells assume a sickle shape rather than the usual round shape; these cells clump and may occlude (thrombose) blood vessels, producing death of the adjacent tissues; red blood cell destruction results in severe anemia. If this trait occurs in an individual in North America, it is obviously deleterious to his health and well-being. However, the same trait is protective in the tropical zones, for individuals with the sickle-cell trait have increased resistance to malaria. Thus the trait may be regarded as a disease in one environment and an improved adaptation to maintain homeostasis in another.

Cytological response. Regardless of the etiologic agent and the functional changes that ensue, cells are the basic units of body structure that are altered by disease. Cells can react to injury in a limited number of ways. They can undergo degeneration and death (necrosis) or increase in size (hypertrophy) or number (hyperplasia). Some cells can change their functional specialization (metaplasia), or acquire new growth properties (neoplasia). Furthermore, cells may decrease in size (atrophy) or number (hypoplasia).

[N. KARLE MOTTET]

Bibliography: D. T. Purtilo, *A Survey of Human Diseases*, 1978.

Drought

A general term implying a deficiency of precipitation of sufficient magnitude to interfere with some phase of the economy. Agricultural drought, occurring when crops are threatened by lack of rain, is the most common. Hydrologic drought, when reservoirs are depleted, is another common form. The Palmer index has become popular among agriculturalists to express the intensity of drought as a function of rainfall and hydrologic variables.

The meteorological causes of drought are usually associated with slow, prevailing, subsiding motions of air masses from continental source regions. These descending air motions, of the order of 200 or 300 m/day, result in compressional warming of the air and therefore reduction in the relative humidity. Since the air usually starts out dry, and the relative humidity declines as the air descends, cloud formation is inhibited—or if clouds are formed, they are soon dissipated. The area over which such subsidence prevails may involve several states, as in the 1977 drought over much of the Far West, the 1962–1966 Northeast drought, or the dust bowl of the 1930s over the Central Plains.

The atmospheric circulations which lead to this subsidence are the so-called centers of action, like the Bermuda High, which are linked to the planetary waves of the upper-level westerlies. If these centers are displaced from their normal positions or are abnormally developed, they frequently introduce anomalously moist or dry air masses into regions of the temperate latitudes. More important, these long waves interact with the cyclones along the Polar Front in such a way as to form and steer their course into or away from certain areas. In the areas relatively invulnerable to the cyclones, the air descends, and if this process repeats time after time, a deficiency of rainfall and drought may occur. In other areas where moist air is frequently forced to ascend, heavy rains occur. Therefore, drought in one area, say the northeastern United States, is usually associated with abundant precipitation elsewhere, like over the Central Plains.

After drought has been established in an area, there seems to be a tendency for it to persist and expand into adjacent areas. Although little is known about the physical mechanisms involved in this expansion and persistence, some circumstantial evidence suggests that numerous "feedback" processes are set in motion which aggravate the situation. Among these are large-scale interactions between ocean and atmosphere in which variations in ocean-surface temperature are produced by abnormal wind systems, and these in turn encourage further development of the same type of abnormal circulation. Atmospheric interactions called teleconnections can then establish drought-producing upper-level high-pressure areas over North America. Then again, if an area, such as the Central Plains, is subject to dryness and heat in spring, the parched soil appears to influence subsequent air circulations and rainfall in a drought-extending sense. A theory supported by numerical modeling studies suggests that over very dry areas, particularly deserts, the loss of heat by radiation relative to the surroundings creates a slow descending motion. This results in compressional warming, lowered relative humidity in the descending air, and inhibition of rain. These processes are probably partly responsible for the tropical droughts observed in various parts of the Sahel (sub-Saharan) region during the 1970s. Of course, interactions with other parts of the atmosphere's prevailing circulation as indicated earlier also play a role in tropical droughts. *See* CLIMATIC CHANGE.

Finally, it should be pointed out that some of the most extensive droughts, like those of the 1930s dust bowl era, require compatibly placed centers of action over both the Atlantic and the Pacific.

In view of the immense scale and complexity of drought-producing systems, it will be difficult for man to devise methods of eliminating or ameliorating them.

[JEROME NAMIAS]

Bibliography: J. G. Charney, Dynamics of deserts and drought in the Sahel, *Quart. J. Roy. Meteorol. Soc.*, 101(428):193–202, 1975; J. Namias, *Factors in the Initiation, Perpetuation and Termination of Drought*, Int. Union Geod. Geophys. Ass. Sci. Hydrol. Publ. no. 51, 1960; J. Namias, Multiple causes of the North American abnormal winter 1976–77, *Mon. Weath. Rev.*, 106 (3): 279–295, March 1978; W. C. Palmer, *Meteorological Drought*, U. S. Weather Bur. Res. Pap. no. 45, 1965.

Dune vegetation

The plants which occupy dunes and the swales and flats between dunes. Along many shores these plants form distinct zones of vegetation. Plants also grow over desert dunes during short periods of precipitation but seldom form any distinct zones of vegetation except around oasis basins.

Succession and zonation. The vegetation on coastal dunes usually becomes denser and the plants become larger and more persistent as the dunes become stabilized—a process known as plant succession. Generally there are three main zones of this succession: (1) a pioneer zone composed mostly of herbaceous plants on the foredunes near the beaches; the plants in this zone are very instrumental in forming and stabilizing the succeeding dunes further inland; (2) a scrub zone of woody shrubs and vines and dwarfed trees; this zone continues stabilization of the dunes and adds humus to the sands; and (3) a forest zone, dominated by trees, which is known as the mesophytic climax of the dune succession. The plants of the pioneer and scrub zones are usually adapted to dry conditions and are xerophytes. Those plants near salt waters, the halophytes, are also resistant to saline conditions. Some types of vegetation in a dune succession are illustrated in Figs. 1, 2, and 3. *See* ECOLOGICAL SUCCESSION.

The vegetation of dunes along salt waters and along fresh waters is essentially similar. Some plants of desert dunes are similar to those of coastal dunes. Nearly all withstand periods of dryness, and many withstand shifting sands. Those plants growing near salt water withstand salt spray which, together with winds, often deforms them, especially the trees and shrubs. This complex of plants, their forms and adaptations, the successional stages, and their zonation have been described for many coastal strand areas. The follow-

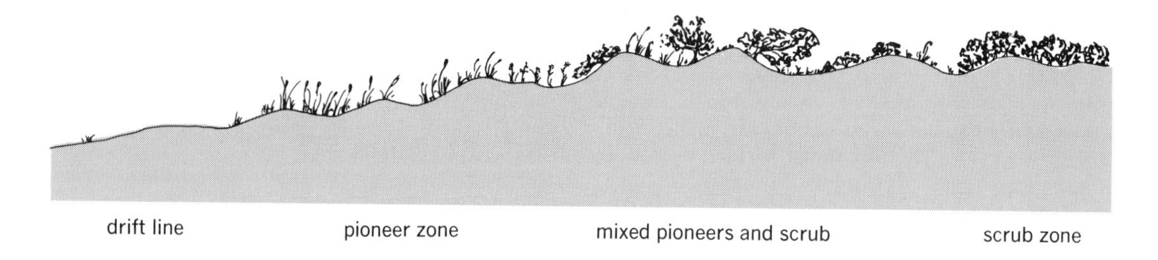

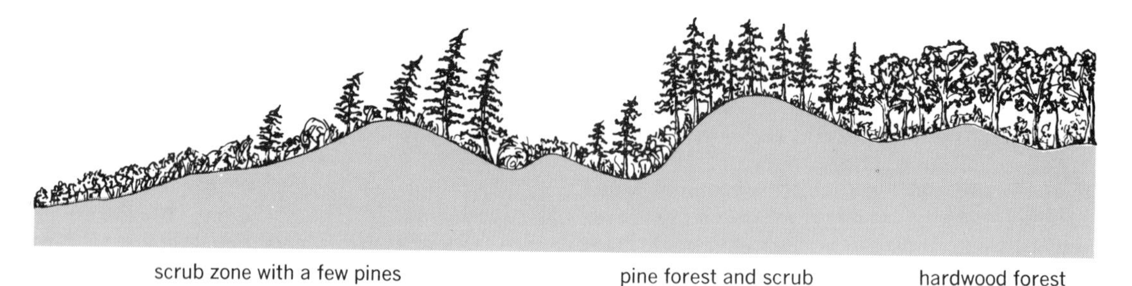

Fig. 1. Diagrammatic cross section showing types of vegetation over broad coast. (*Adapted from J. H. Davis, Dune Formation and Stabilization by Vegetation and Plantings, Beach Erosion Board, U.S. Army Corps of Engineers, Tech. Memo. no. 101, 1957*)

ing is a brief summary of North American dune vegetation.

Atlantic and Gulf coasts. Along the Atlantic Ocean and Gulf of Mexico there is a gradual change in dune plants and features of the vegetation from the areas of temperate climate to those of tropical climate. In the north, the herbs of the pioneer zone are mostly annuals, and severe winters present the development of dense scrub and forest zones. Southward the variety and number of species increase from an average of less than 60 to more than 150 (on subtropical dunes in Florida). This abundance and variety of plants on subtropical dunes have the effect of creating more dune ridges of greater stability. Many evergreen shrubs and trees, fibrous and succulent plants, and palms increase toward the tropics. A fringe of palm forest, especially coconut trees, is common in the tropics on the upper beach and foredunes. The following plants are common along coasts of middle latitudes, from Maryland to Palm Beach, Fla.

Pioneer zone. Plants commonly found in this zone include beach grass (*Ammophila*), sea rocket (*Cakile*), sea oats (*Uniola paniculata*), panic grass (*Panicum amarum*), salt wort (*Salsola kali*), orach (*Atriplex arenaria*), and in subtropical areas the railroad vine (*Ipomoea pes-caprae*) and sea purslane (*Sesuvium portulacastrum*).

Scrub zone. Plants that are typical of this zone include march elder (*Iva imbricata*), myrtles (*Myrica*), scrub forms of oaks (*Quercus*), cedars (*Juniperus*), holly (*Ilex opaca*), groundsel bush (*Baccharis*), buckthorns (*Bumelia*), and alders (*Alnus*). Northward, sand myrtle (*Leiophyllum buxifolium*), birches (*Betula*), false heather (*Hudsonia tomentosa*), and some true heaths (*Erica*) are locally abundant. Southward, the saw palmetto (*Serenoa repens*) is very abundant and is

Fig. 2. Pioneer zone composed of sea oats. These plants assist in formation of the foredunes.

important as a dune stabilizer. In Florida the sand pine (*Pinus clausa*) is typical of the scrub vegetation. *Agave* and species of cactus (*Opuntia*) are common along dry coasts, and the sea grape (*Coccoloba uvifera*) becomes very frequent in subtropical areas. Many of these plants, especially the oaks, are densely branched shrubs and dwarfed, wind-formed trees that often form thickets (Fig. 4). Between the thickets are bare migrating sands or partly stabilized sands populated by the pioneer plants.

Forest zone. Many different trees form the dune forests. Along northern Atlantic coasts, poplars and aspens (*Populus*), pines (*Pinus*), birches (*Betula*), cherries (*Prunus*), and oaks (*Quercus*) are common. Southward, the pines and oaks, particularly the evergreen live oak (*Quercus virginiana*), are common. Hickories (*Carya*) are frequent, and the palms begin to increase toward the tropics. The cabbage palm (*Sabal palmetto*) is very frequent from North Carolina to the tip of Florida and along much of the Gulf of Mexico coast. The coconut palm (*Cocos nucifera*) is now naturalized and abundant along American tropical coasts.

Pacific Coast. The dune vegetation along the Pacific Coast varies from a dry semidesert type in Mexico and southern California to a humid-temperature-climate type along coastal areas north of central California. Introduced European and American beach or marram grass (*Ammophila arenaria* and *A. breviligulata*) have been used to stabilize foredune ridges and parts of the very large blowout areas of bare sand, and are spreading from areas planted in the 1930s more vigorously than native pioneer species. Trees resistant to salt spray include lodgepole pine (*Pinus contorta*), which persists or regenerates on dry, leached soils after burning, and sitka spruce (*Picea sitchensis*).

Southern dry type. Along the dry parts of the coast are *Yucca gloriosa, Opuntia serpentina*, sand ambrosia (*Franseria chamissonis*), sagebrush (*Artemesia pychocephala*), mariposa (*Calochortus venustus*), and *Abronia latifolia*, with a few seasonal grasses and other herbs over the young forming dunes. The distinct pioneer and scrub zones are usually present with *Haplopappus erocoides* and lupine (*Lupinus chamissonis*). The shore shrubs and dwarfed trees include Monterey pine (*Pinus radiata*), a live oak (*Quercus agrifolia*), leather oak (*Q. durata*), coffeeberry (*Rhamnus californica*), *Yucca, Opuntia*, and *Ceanothus dentatus*. These plants seldom form a dense, distinct forest zone.

Northern moist type. In contrast to the southern dry-climate coast, the northern moist-climate coastal areas have a forest zone on stabilized dunes, and scrub and pioneer zones where there are active dunes. A few plants are the same as, or similar to, those of the northern Atlantic Coast. Redwoods (*Sequoia*), pines, spruces (*Picea*), aspens, and manzanitas (*Arctostaphylus*) are common.

Great Lakes coasts. Dunes of the Great Lakes shores became famous as examples of successional change shown by progressively older dunes. The dune's external environment did not have a gradient of salt spray confounded with gradients of age since surface stabilization. Both beaches and blowout sands have annuals: mostly sea rocket (*Cakile*), bugseed (*Corispermum*), winged pigweed

Fig. 3. Scrub zone with bare sand of a blowout area in foreground. Dwarfed evergreen magnolias are a part of this zone, near Panama City, Fla.

(*Cycloloma*), and beach pea (*Lathyrus*). Dune-building grasses include marram (*Ammophila breviligulata*) and sand reed (*Calamovilfa longifolia*), in order of tolerance to rapid sand burial on new dune ridges along the beach and on lee slopes and crests of the U- or V-shaped blowout ridges. The latter open toward the lake or the dominant westerly winds. Little bluestem bunchgrass typically invades within a few years after deposition stops or becomes reduced to a few centimeters per year. Sand cherries (*Prunus pumila*), several willows (*Salix*), and poplars (*Populus deltoides* or *P. balsamifera* and hybrids) germinate in hollows of young dunes but tolerate burial well, as do grape vines (*Vitis*) and basswood (*Tilia americana*) on lee slopes. The latter is joined by red oak (*Quercus rubra*), maple (*Acer saccharum* and *A. rubrum*), and in Michigan by beech (*Fagus grandifolia*) and hemlock (*Tsuga canadensis*) within a few hundred or thousand years after stabilization on protected slopes and hollows that are relatively favorable for rapid succession. However, the majority of old Indiana, Wisconsin, and even Michigan and Ontario dunes follow alternative successions through pines (*Pinus banksiana* and *P. strobus*, in addition to *P. resinosa* northward) to oak (*Quercus velutina, Q. alba*, and northward *Q. rubra*). Evidence of replacement of the oak-blueberry vegetation, on acid soils, by the mesophytic forests dominated by the

Fig. 4. Typical deformed trees, live oak, caused by winds and salt spray, on edge of dune forest.

trees mentioned earlier is rare or conjectural, if not absent, even on old dunes (stabilized 8000–12,000 years ago). *See* CLIMAX COMMUNITY.

[JOHN H. DAVIS]

Bibliography: S. G. Boyce, The salty spray community, *Ecol. Monogr.*, 24:29-67, 1954; H. C. Cowles, The ecological relations of the vegetation on sand dunes of Lake Michigan, *Bot. Gaz.*, 27: 95–117, 167–202, 281–308, 361–391, 1899; J. H. Davis, *Dune Formation and Stabilization by Vegetation and Plantings*, Beach Erosion Board, U.S. Corps of Engineers, Tech. Memo. no. 101, 1957; J. S. Olson, Rates of succession and soil changes on southern Lake Michigan sand dunes, *Bot. Gaz.*, 119:125–170, 1968; H. J. Oosting, Ecological processes and vegetation of the maritime strand in the southeastern United States, *Bot. Rev.*, 20(4): 226–262, 1954; D. S. Ranwell, *Ecology of Salt Marshes and Sand Dunes*, 1973; E. J. Salisbury, *Downs and Dunes: Their Plant Life and Environment*, 1952; L. A. Starker and *Life* (eds.), *The Desert*, 1961.

Dust and mist collection

The physical separation and removal of particles, either solid or liquid, from a gas in which they are suspended. Such separation is required for one or more of the following purposes: (1) to collect a product which has been processed or handled in gas suspension, as in spray-drying or pneumatic conveying; (2) to recover a valuable product inadvertently mixed with processing gases, as in kiln or smelter exhausts; (3) to eliminate a nuisance, as a fly-ash removal; (4) to reduce equipment maintenance, as in engine intake air filters; (5) to eliminate a health, fire, explosion, or safety hazard, as in bagging operations or nuclear separations plant ventilation air; and (6) to improve product quality, as in cleaning of air used in processing pharmaceutical or photographic products. Achievement of these objectives involves primarily gas-handling equipment, but the design must be concerned with the properties and relative amounts of the suspended particles as well as with those of the gas being handled.

All particle collection systems depend upon subjecting the suspended particles to some force which will drive them mechanically to a collecting surface. The known mechanisms by which such deposition can occur may be classed as gravitational, inertial, physical or barrier, electrostatic, molecular or diffusional, and thermal or radiant. There are also mechanisms which can be used to modify the properties of the particles or the gas to increase the effectiveness of the deposition mechanisms. For example, the effective size of particles may be increased by condensing water vapor upon them or by flocculating particles through the action of a sonic vibration. Usually, larger particles simplify the control problem. To function successfully, any collection device must have an adequate means for continuously or periodically removing collected material from the equipment.

Devices for control of particulate material may be considered, by structural or application similarities, in eight categories as follow.

Gravity settling chamber. In this, the simplest type of device but not necessarily the least expensive, the velocity of the gas is reduced to permit particles to settle out under the action of gravity. Normally, settling chambers are useful for removing particles larger than 50 μm in diameter, although with special configurations they may be used to remove particles as small as 10 μm.

Inertial device. The basis of this type of device is that the particles have greater inertia than the gas. The cyclone separator, typical of this type of equipment, is one of the most widely used and least expensive types of dust collector. In a cyclone, the gas usually enters a conical or cylindrical chamber tangentially and leaves axially. Because of the change of direction, the particles are flung to the outer wall, from which they slide into a receiving bin or into a conveyor, while the gases whirl around to the central exit port (Fig. 1). A large variety of configurations is available. For large air-handling capacities, an arrangement of multiple small-diameter units in parallel is often used to attain high collection efficiencies and to permit lower headroom requirements than a single unit would.

Mechanical inertial units are similar to cyclones except that the rotational motion of the gas is induced by the action of a rotating member. Some such units are designed to act as fans in addition to their dust-collecting function. There are also a wide variety of other units; many are called im-

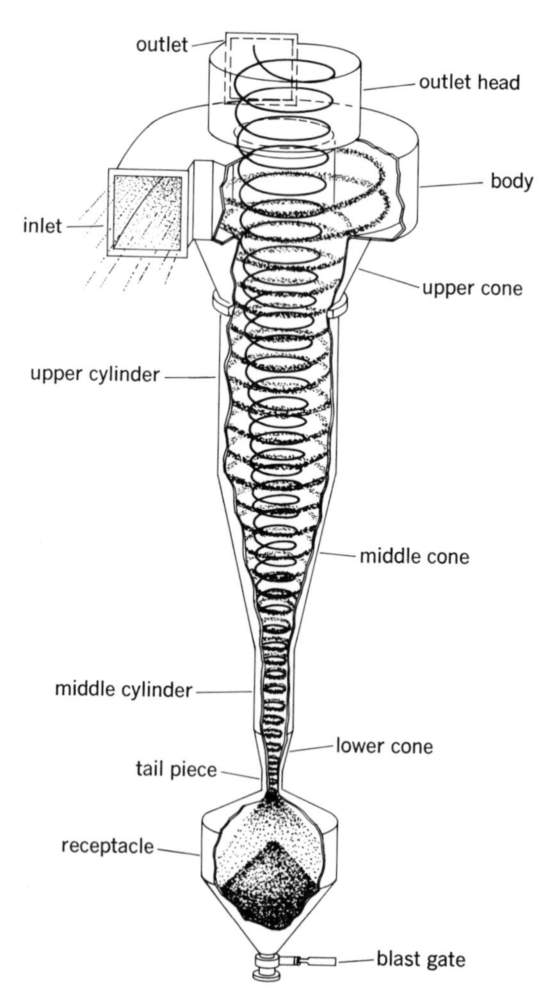

Fig. 1. Cyclone dust separator, an inertial device. (*American Standard Industrial Division*)

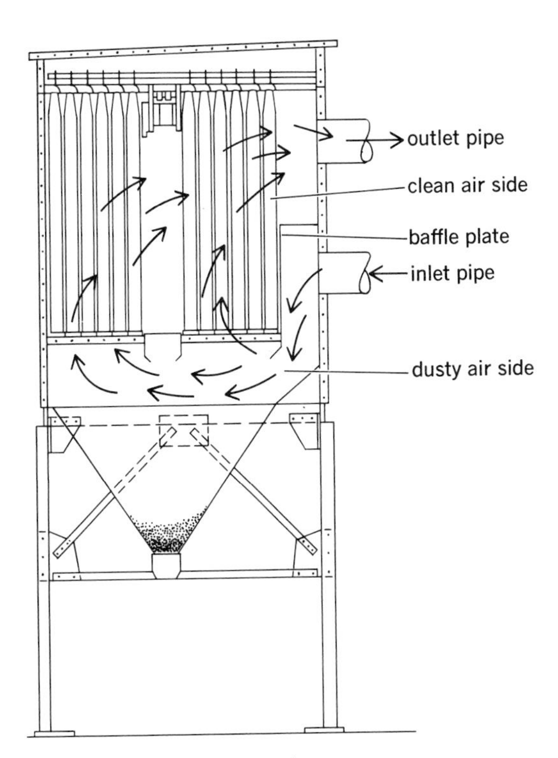

Fig. 2. Cloth collector. (*Wheelabrator-Frye Inc.*)

pingement separators. Most separators used to remove entrained liquids from steam or compressed air fall into this category.

Packed bed. A particle-laden gas stream may be cleaned by passing it through a bed or layer of packing composed of granular materials such as sand, coke, gravel, and ceramic rings, or fibrous materials such as glass wool, steel wool, and textile staples. Depending on the application, the bed depth may range from a fraction of an inch to several feet. Coarse packings, which are used at relatively high throughput rates (1–15 ft/sec or 0.3–5 m/s superficial velocity) to remove large particles, rely primarily on the inertial mechanism for their separating action. Fine packings, operated at lower throughput rates (1–50 ft/min or 5–250 mm/s superficial velocity) to remove relatively small suspended particles, usually depend on a variety of deposition mechanisms for their separating effect. Packed beds, because of a gradual plugging caused by particle accumulation, are usually limited in use to collecting particles present in the gas at low concentration, unless some provision is made for removing the dust—for example, by periodic or continuous withdrawal of part of the packing for cleaning. Depending on the application and design, the collection efficiencies of packed beds range widely (50–99.999%).

Cloth collector. In such a collector, also known as a bag filter, the dust-laden gas is passed through a woven or felted fabric upon which the gradual deposition of dust forms a precoat, which then serves as a filter for the subsequent dust. These units are analogous to those used in liquid filtration and represent a special type of packed bed. Because the dust accumulates continuously, the resistance to gas flow gradually increases. The cloth must, therefore, be vibrated or flexed periodically to dislodge accumulated dust (Fig. 2). A wide variety of filter media is available. Cotton or wool sateen or felts are usually used for temperatures below 212°F (100°C). Some of the synthetic fibers may be used at temperatures up to 300°F (149°C). Glass and asbestos or combinations thereof have been used for temperatures up to 650°F (343°C). For special high-temperature applications, metallic screens and porous ceramics or stainless steel have been employed. Collection efficiencies of over 98% are attained readily with cloth collectors, even with very fine dusts.

Scrubber. A scrubber uses a liquid, usually water, to assist in the particulate collection process. An extremely wide variety of equipment is available, ranging from simple modifications of corresponding dry units to permit liquid addition, to devices specifically designed for wet operation only (Fig. 3). When properly designed for a given application, scrubbers can give very high collection efficiency, although the mere addition of water to a gas stream is not necessarily effective. For a

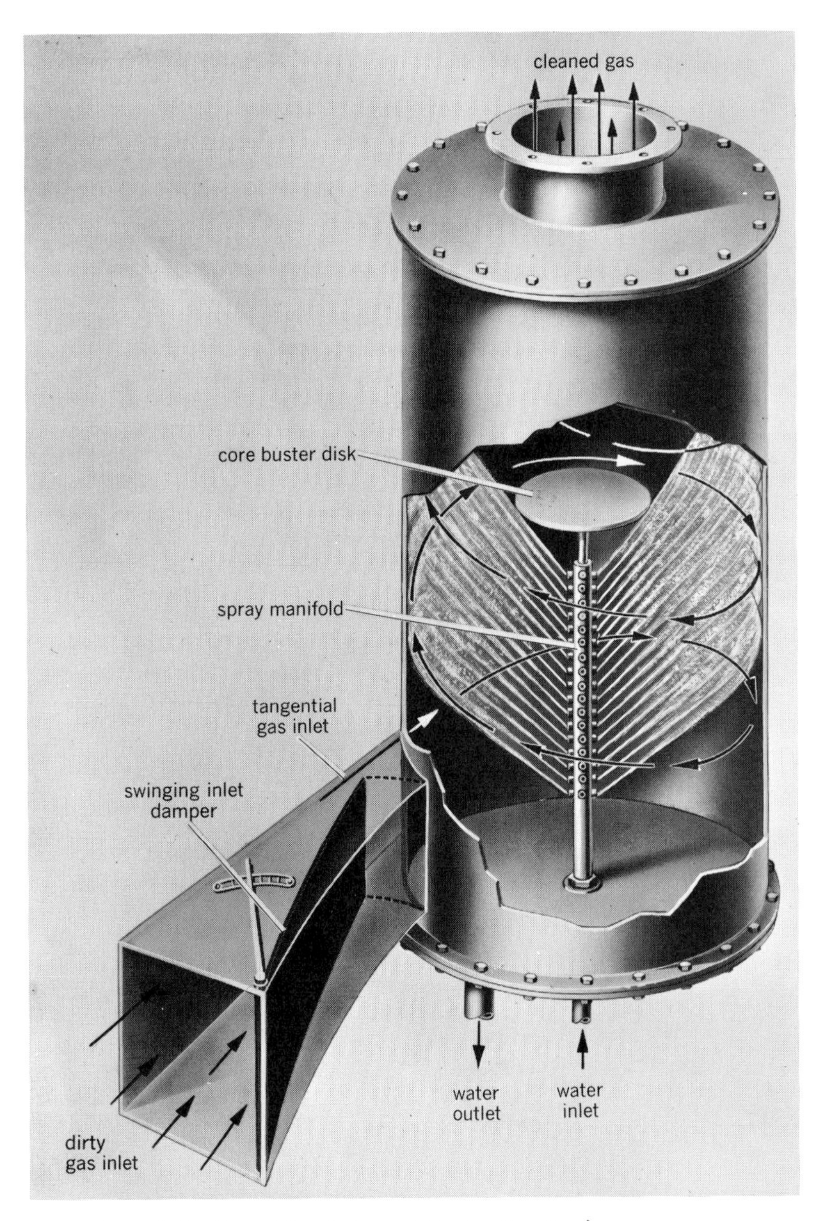

Fig. 3. Cyclonic liquid scrubber. (*Chemical Construction Corp.*)

given application, collection efficiency is primarily a function of the amount of power supplied to the gas stream, in the form of gas pressure drop, water pressure, or mechanical energy. With scrubbers it is important that proper attention be given to liquid entrainment separation in order to avoid a spray nuisance. Consideration must also be given to the liquid-sludge disposal problem.

Electrostatic precipitator. Particles may be charged electrically by a corona discharge and caused to migrate to a collecting surface. The single-stage unit, which is commonly known as a Cottrell precipitator, and in which the charging and collecting proceed simultaneously, is the type generally used for industrial or process applications. These units normally employ direct current at voltages ranging from 30,000 to 100,000 volts. The two-stage unit, in which charging and collection are carried out successively, is commonly used for air conditioning applications. These units also employ direct current, ranging from 5,000 to 13,000 volts, and involve close internal clearances (0.25–0.5 in.). Electrical precipitators are capable of high collection efficiency of fine particles. The reentrainment of collected material as flocs in the exhaust gas, a phenomenon known as snowing, must be avoided to prevent a possible accentuated nuisance or vegetation damage problem.

Air filter. This is a unit used to eliminate very small quantities of dust from large quantities of air, as in air conditioning applications. Although units in this class actually fall into one of the previous classes, they are given a special category because of wide usage and common special features. In this category are viscous-coated fiber-mat filters and dry filters. These are actually a form of packed bed and are frequently known as unit filters; they are available as standard packaged units from a large number of manufacturers. The domestic furnace filter is an example of a viscous-coated unit filter. Automatic filters provided with continuous and automatic cleaning arrangements are available in both the viscous-coated and dry forms, as well as with electrostatic provisions.

Miscellaneous equipment. Acoustic or sonic vibrations imparted to a gas stream cause particulates to collide and flocculate, forming larger particles that are more readily collected in conventional apparatus. This principle has been employed, but has had extremely limited application because of economic and other practical considerations. In thermal precipitation, suspended particles are caused to migrate toward a cold surface or away from a heated surface by the action of a temperature gradient in the gas stream. This principle has found extensive use in atmospheric sampling work. *See* AIR POLLUTION; AIR-POLLUTION CONTROL.

[CHARLES E. LAPPLE]

Bibliography: R. E. Kirk and D. F. Othmer (eds.), *Encyclopedia of Chemical Technology*, vol. 7, 1951; C. E. Lapple, Elements of dust and mist collection, *Chem. Eng. Progr.*, 50(6):283–287, 1954; K. E. Lunde and C. E. Lapple, Dust and mist collection, *Chem. Eng. Progr.*, 53(8):385–391, 1957; G. Nonhebel, *Gas Purification Processes*, 1975; R. H. Perry (ed.), *Chemical Engineers' Handbook*, 5th ed., 1973; M. Sittig, *Particulates and Fine Dust Removal*, 1977; W. Strauss, *Industrial Gas Cleaning*, 1976.

Earth resource patterns

The physical character and distribution of natural resources at the face of the Earth. No section of the Earth is exactly like any other in its resource endowment. Nevertheless, the fundamental differences between land and ocean, latitudinal diferences in insolation, spatial variations in receipts of precipitation, and patterns of geological composition and deformation of the Earth's crust together provide the basis for distinguishing definite geographical patterns of resource availability over the world.

Delineation of the Earth's resource patterns begins with differentiation between continental and marine resources. Although the resources of the oceans and seas have been used by people since earliest times, the more than 4,000,000,000 on the Earth today are primarily dependent upon the resources of the land for their existence.

Five principal resources associated with the land are soils, forests, grasslands, fresh-water resources, and minerals. Although other resources such as native animal life may be of local importance, and although the very concept of "resources" in recent years has been extended to include such complexes as recreation resources, these five land resources remain of fundamental importance for the material support of human life.

Two sets of underlying causes in particular engender the spatial patterns of resources over the Earth. One set of causes consists of the basic climatic controls, including latitude, distribution of land and water, the wind and pressure system of the rotating Earth, the major landforms of the continents, and the elevation of the land surface above sea level. A second, independent set of causes consists of the tectonic and rock-forming processes which have operated over the Earth. The climatic controls account for a system of regional climates over the continents, and these climates in turn provide keys for understanding the spatial distribution of forests, grasslands, and fresh-water resources, as well as some of the fundamental attributes of soils and the agriculture they support. The second set of causes may be regarded as even more fundamental since the movements of the Earth's plates and their associated continents in conjunction with the Earth's plate tectonics account not only for the position on the Earth and hence the latitudinal location of each continent, but also for the global distribution of land and water and continental landforms, all with consequences for regional climatic patterns. Moreover, rock composition and surface configuration also produce overlays of difference on forests, grasslands, water resources, and soils which alter the patterns within the climatic regions, and they are fundamental to an understanding of the global patterns of minerals on and beneath the Earth's surface.

Eleven regional climatic types (numbered consecutively on the accompanying map) in four groups are recognized in describing the Earth's resource patterns of forests, grasslands, soils, and fresh water. This number of types is fewer than that normally employed to describe regional climates, but is considered adequate to outline the basic global patterns of each of these four major renewable resources. Distinction is made between

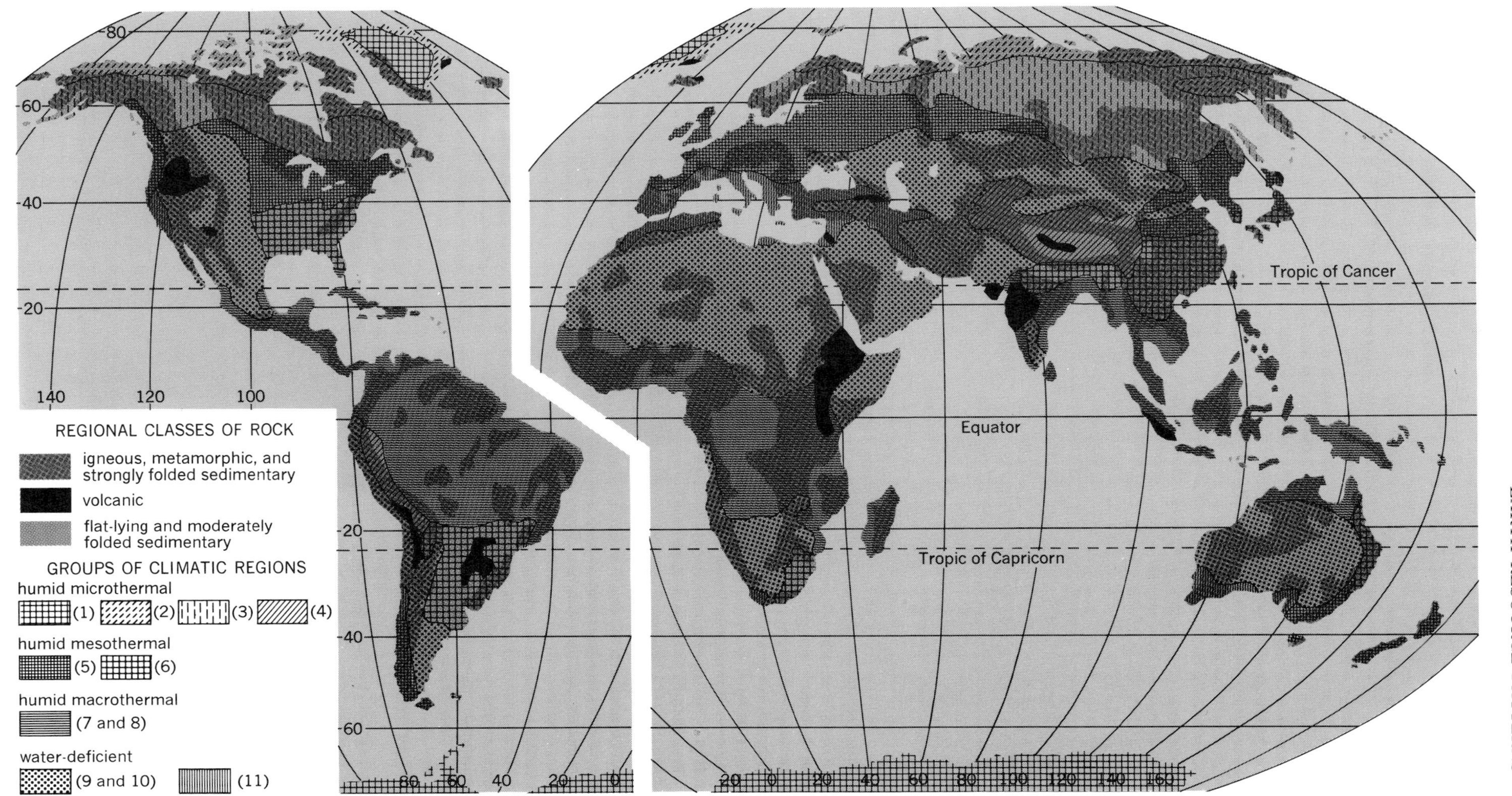

Predominant Earth resource patterns. Climate types identified as: (1) polar and ice cap, (2) tundra, (3) taiga, (4) puna, (5) upper midlatitude, (6) humid subtropical, (7 and 8) wet-and-dry and rainforest, (9 and 10) desert and semiarid, (11) Mediterranean. The map is a flat polar quartic equal-area projection.

the so-called humid climates and the water-deficient climates, with subtypes as follows:

Humid microthermal
1. Polar and ice cap
2. Tundra
3. Taiga
4. Puna

Humid mesothermal
5. Upper midlatitude
6. Humid subtropical

Humid macrothermal
7. Wet-and-dry
8. Rainforest

Water-deficient
9. Desert
10. Semiarid
11. Mediterranean

Humid microthermal regions. These areas of predominantly low temperature are so unfavorable to soil formation and use in agriculture that under present techniques their population-carrying capacity is low even in those regions with the warmest summers.

1. In polar and ice cap areas where soils and vegetative growth are essentially absent, available resources necessarily are dominantly marine and land animal life, on which the sparse native human settlement is almost wholly dependent. Despite the enormous size of the Antarctic, settlements are exclusively in the Arctic, except for special government-supported Antarctic stations.

2. Tundra, except for minor alpine locations, is entirely within the Northern Hemisphere. The principal renewable resources are lichens and the native animal life, such as reindeer and caribou, which can use these as food. Parts of the tundra may be considered a grazing land, as managed herds of reindeer are pastured nomadically. The natural resource significance of tundra lands for the larger world may be greatly enhanced locally where minerals are being extracted, such as the oil field adjacent to Prudhoe Bay on Alaska's North Slope. *See* TUNDRA.

3. Millions of acres of boreal coniferous forest, or taiga, located in a broad curving zone from Scandinavia across the northern, European part of the Soviet Union, and Siberia, and, east of the Bering Sea, across much of Alaska and northern Canada, has given its name to this climatic belt. Varieties of spruce, fir, and larch, which are of particular significance to the pulping industries, constitute the most valuable known renewable resource of the humid microthermal regions. The zone of taiga climate is of scant importance for agriculture; soils are predominantly thin, stony, and infertile, and development for agriculture is further discouraged by the possible incidence of frost in every month of the year and by the short growing season of 80 days or less. As in the case of the areas of tundra climate, the resource significance of particular localities within the taiga zone is locally enhanced by deposits of minerals, including gold, iron, and uranium. Fresh-water resources are large, and some hydroelectric power is developed. *See* TAIGA.

4. The puna type of climate is found at much lower latitudes but at elevations generally 10,000 ft (3048 m) or more above sea level, and is characteristically cold. Certain plateaus, particularly the high plateau of Tibet and the intermontane Andean plateau (Altiplano) of southeastern Peru and western Bolivia, belong to this group. Low temperatures preclude tree growth, and the principal renewable resource is low-productivity grazing land. Thin, stony soils are limited in production to hardy small grains and root crops. The resource significance of the South American puna is greatly enhanced by metal deposits, particularly copper, tin, lead, and zinc.

Humid mesothermal regions. Considered in the light of present-day technology, the heart of the world's renewable natural resource base is in the humid mesothermal regions with their generally adequate precipitation and generally intermediate temperatures. They contain a large share of the world's most productive soils which support both crop and pasture lands. Some productive coniferous and broad-leaf forests occur in these climatic regions. Concentrated surface and subsurface fresh-water supplies are relatively abundant, but owing to the high population densities and major urban and industrial complexes in these regions, the water resources are often intensively developed to the point where both quantitative and qualitative problems have developed.

5. Upper midlatitude climate contains most of the lands adapted to the raising of wheat, barley, rye, and oats. In addition, certain areas within this climatic type, and particularly the North American corn belt and the Middle Danube Basin and the North Italian Plain, also produce maize. Further crops, including soybeans, are also important, and extensive forage cropping supports dairying and meat animal raising in east-central North America, western and central Europe, and the nonarid part of the southern European Soviet Union. The lands which support the agricultural types of this climatic region possess the most extensive areas of superior soils on Earth. They are relatively deep and are often moderately to highly endowed with plant nutrients and humus. Some of the best soils have developed from glacial drift; others have benefited from the deposition of fine-grained windblown loess deposits during the glacial period. Extensive preagricultural grasslands also contributed high humus content to some of the soils. Finally, the extensive plains under which these soils are deployed, and the generally favorable growing season conditions of temperature and rainfall, also underlie the present high productivity of the agriculture supported by the soils in this climatic type. It is within the upper midlatitude climate that the region of greatest surplus food production (the North American interior) at present exists.

6. Humid subtropical areas have an ample water supply and relatively long growing season (200 days or longer), making the best of these lands potentially very productive for crops or for forest products. Soils other than in floodplains tend to be less fertile than those of upper midlatitudes, however, owing to the effects of relatively high rainfall and temperatures on the removal of plant nutrients, and generally require fertilization for sustained cultivation over long periods. Where cropping is on alluvial soils of floodplains large and small, as in central and southern China and in southern Japan, and in the Ganges Plain of India, soils are more easily maintained at high levels of

productivity. High rural population desities generally preclude the generation of large exportable food surpluses from regions with the best soils. Although the favorable combination of temperature and moisture can result in a high timber growth, as on the best-managed tracts of forest in the American South, the potential is not realized in other areas of the Earth having this type of climate, owing to previous large-scale deforestation. Water resources are generally abundant; in part of southern and eastern Asia, fresh-water resources underlie the most extensive humid land irrigation in the world—namely, for paddy rice.

Humid macrothermal regions. These predominantly winterless regions of warm to hot temperatures are divided according to the regime of rainfall: (7) wet-and-dry, with a pronounced dry season, and (8) rainforest, with year-round precipitation. Macrothermal regions have year-round growing seasons, but over extensive areas their soils are lateritic with a high iron content, and harden irreparably under use for cropping. Problems induced by fungal growth, bacterial disease, and insect abundance also handicap the use of the soil resources. Within the great alluvial valleys, flooding may also be disruptive. Owing to these adverse factors, shifting cultivation is common, and much land is not in production at any particular time. Within the rainforest regions, extensive and rapid-growing forests occur, including the largest such forest area on the Earth in the Amazon Basin, but are mostly unexploited commercially in the 20th-century economy. Extensive clearing of the Amazon forest for agricultural development appears likely during the last 2 decades of the 20th century. Sites of enormous potential hydroelectric generation are largely undeveloped. *See* RAINFOREST.

Water-deficient regions. Receipts of moisture are scant or lacking during much of the year. Except where exotic water supplies are available for irrigation, soils inherently are less productive than those of any humid region with comparable land surface. There are three major subdivisions.

9. Deserts differ strikingly in their form and in temperature conditions, but everywhere present meager resources for agriculture. Where water is available, desert oases blossom, but vast areas contain only scrub growth, ephemerals, or virtually no vegetation. Livestock-carrying capacity of desert scrub is meager. Sparsity of vegetation has made mineral prospecting and exploration somewhat easier than in vegetation-covered humid areas, and in this century desert occupancy often started with mineral discoveries. *See* DESERT ECOSYSTEM.

10. The semiarid regions are basically grasslands, which have grazing as their characteristic resource use. Because of cyclic rainfall variability, people have converted the inherently fertile soils, as in China and the United States, to cereal growing during periods of higher rainfall. Rainfall fluctuations, however, make the soils unstable under cultivation. For this reason, these regions have suffered consistently from the accelerated and destructive erosion resulting from unsuited cultivation and overgrazing. Semiarid lands are responsive to and most productive under irrigation, but neither surface-water nor groundwater resources are adequate to irrigate more than a fraction of the area. *See* GRASSLAND ECOSYSTEM.

11. Regions of Mediterranean climate, because of their winter rainfall, generally are classed as humid lands. However, the greater part of the growing season is water-deficient, and the most productive agricultural lands depend on irrigation. Soils are the major resource since water deficiency is pronounced enough to discourage forest productivity. Major mineral deposits may complement agricultural lands in a few areas.

Rock composition and surface configuration. Imposed on the basic resource pattern induced by climatic differences are variations in rock composition and surface configuration which cause intraregional differences within the pattern. Although not exactly the same in their effects, the variations caused by these two geographical elements are often concomitant and may be treated together as follows:

1. Rock composition and structure
   a. Flat-lying and moderately folded sedimentary rocks
   b. Igneous, metamorphic, strongly folded sedimentary rocks
   c. Volcanic rocks
2. Surface configuration
   a. Floodplains and other flat or gently sloped surfaces
   b. Mountains and maturely dissected hill lands, plateau faces, or faces of cuestas

These elements of crustal variation produce the six following geographical differences in resource endowment:

First, all major agricultural lands are on flat-lying or moderately folded sediments and have gentle slopes, well exemplified in such alluvial valleys as the Mississippi, Nile, Huang, and Ganges-Brahmaputra, and such other outstanding agricultural areas as the North American corn belt or the Paris Basin.

Second, productive secondary agricultural lands, particularly in the tropics, are located on volcanic areas where soils have been formed through weathering or wind action. Examples include the Deccan Plateau in west-central peninsular India, and volcanic soils fringing much of the central mountain backbone of Java.

Third, agricultural lands are extremely limited on igneous rock areas, no matter what the surface configuration, in regions north of 40°N latitude, as illustrated in the Laurentian Shield area of Canada and the Fenno-Scandian Shield in northern Europe. In the humid tropical and subtropical climates, however, where weathering has proceeded long enough to produce a substantial soil mantle as on the Piedmont of the southeastern United States, underlying igneous rocks do not have the same negative effect on agricultural development.

Fourth, forest lands are not limited in their extent (although limited in productivity) by either crustal rock composition or surface configuration.

Fifth, mountains are important catchment areas and sources of fresh water and the services which may be derived from water. Most hydroelectric generation, or generation potential, is associated with mountains. In arid and semiarid regions, mountains are sources of water for irrigation, domestic and industrial supply, wood products, and warm-season grazing lands.

Sixth, mineral resources have definite patterns which are associated with rock structure and composition. Major deposits of coal and lignite, petroleum, and natural gas are, with few exceptions, found in flat-lying or gently folded and faulted sedimentary rocks, as in Texas oil fields and the coal fields of the Allegheny Plateau. Sedimentary nonmetallics (the phosphates, potash, sulfur, nitrates, and limestone) as well as bauxite and uranium (carnotite) also are associated with sedimentary rocks.

Associated with the igneous and metamorphic rock areas are most metals, for example, iron (usually), lead, copper, tin, the ferroalloys, gold, and silver. Most gems and some nonmetallic minerals (mica and asbestos) are found in the same associations. Uranium (pitchblende) occurs in these rocks.

Whereas these associations are well recognized, the mineral deposits themselves have a highly erratic geographical occurrence owing to the great variety of processes involving both mineral enrichment and dispersal which have occurred. The broad global pattern of rock classes with which minerals are related is delineated on the accompanying map.

Employed and potential resources. Resources have meaning insofar as they are placed in use or are available for future exploitation. Distinction must be made between the employed and the potential resources. In practice this distinction is complex, but here only the simple geographical distinction will be noted. Employed resources are those which are significant to the present support of humans, at least locally. In general, the denser the population and the more advanced the technical arts of an area, the greater the need for production from resources, and employed resources become more nearly synonymous with all known resources. Thus, the recognized resources of the European peninsula and of the northeastern United States are mainly employed resources. On the other hand, the natural resources of the Amazon Basin or of much of Siberia are still largely potential resources.

Marine resources. Although the physical and biotic geography of the oceans is much less fully explored than that of the continents, enough is known to indicate that both living and mineral resources extend far beyond those presently exploited. The employed resources are rather sharply localized. The principal exploitation of marine animals and vegetation occurs: over the continental shelves; in the vicinity of the mixing of warm and cold currents; near large upwellings which occur particularly off the west coast of continents in lower middle latitudes; and adjacent to densely populated countries. Thus, the North Atlantic near Europe and from New England to Newfoundland contains heavily exploited fishing grounds, as do the seas near Japan, Korea, and southern California, and also waters of the Gulf of Mexico.

Minerals are derived from three separate types of marine sources: from sedimentary deposits underlying the continental shelves; from inshore deposits on the surface of the continental shelves; and from sea water. By far the most valuable of the mineral resources exploited from marine environments is petroleum; there has been a rapid expansion of exploration, drilling, and pumping from beneath continental shelf waters, as off the Gulf coast of Louisiana and in the middle of the North Sea between Scotland and Norway. Offshore placer deposits on the surface of the continental shelves yield gold, platinum, and tin. Common salt, magnesium, and bromine are derived directly from sea water.

Potential marine resources include: the population of life-forms now exploited in some parts of the world, but not in others; animal and vegetative species now unused; fresh water from desalted sea water; and minerals so far unexploited which are in solution, are precipitated to the ocean bottom, or lie within rock below the surface of the continental shelf. One of the interesting speculative resources appears in the large quantities of so-called manganese nodules that cover some sections of sea bottoms at intermediate depths. *See* MARINE RESOURCES.

Resources of the continents. The resource pattern of the Earth may be summarized in a brief description of that for each continent and its neighboring waters.

Eurasian continent. As the largest landmass in the world, the Eurasian continent has the largest area of agricultural land in use, a very extensive total forest land area in which the softwood coniferous forest belt from Scandinavia to eastern Siberia predominates, and a wide variety of mineral resources. Great differences mark the major sections of the continent. The most productive agricultural areas are generally near the edge of the landmass, in western and central Europe and extending eastward into much of the central and southern sections of the Soviet Union in Europe; in the Indian subcontinent; and in mainland east Asia from the Red River valley in Vietnam northward to the Great Wall of China. The aggregate mineral endowment is outstanding and includes the largest known aggregate iron ore reserves in the world within the Soviet Union, located particularly in districts adjacent to the Urals; very substantial coal deposits, including the Ruhr field and the extensive coal beds in northern China and Manchuria; and what increasingly appears to be one of the two great concentrations of petroleum fields on Earth, namely, the Persian Gulf fields shared by a number of separate states in the Middle East. The southeastern and eastern borders of the Eurasian heartland, moreover, have some of the great, but still undeveloped, hydroelectric generation sites of the world. Off the coasts of western Europe and Japan are the two most productively employed fisheries of the world.

Africa and Australia. Much of the entire area of Africa north of approximately 12°N must be classed as desert or semiarid, with few exotic water sources other than the Nile. Much of the remainder of the continent has wet-and-dry or rainforest climates, with the former predominating greatly in area. Seasonal drought, soil infertility, and widespread problems of laterization handicap agricultural exploitation. The east African highlands from Ethiopia southward, the high veld in South Africa, and the loftier sections of Zimbabwe Rhodesia, and the Nile Valley and Delta are noteworthy exceptions. Except along the Nile there are still potential agricultural land resources, but they are comparatively minor. Associated with the extensive areas of igneous and metamorphic rocks which underlie much of the continent and particularly its southern half are outstanding deposits of metalliferous minerals, such as the copper ores astride the Zaire-Zambia boundary, the chrome-bearing ores of Zimbabwe Rhodesia, and the gold deposits of the Witwatersrand in South Africa. Other mineral resources include diamonds, uranium, and, in Nigeria and the far north of the continent (Libya, Algeria), petroleum and natural gas. The water resources of mid-Africa include the largest potential hydroelectric power on Earth.

Similar general remarks may be made about Australia, whose much smaller area is covered mostly by desert, semiarid, and tropical wet-and-dry environments. Most agricultural productivity is peripheral, especially in the southeast. Metallic minerals at currently exploitable levels of size and richness support a substantial number of mining operations on the continent.

South America. The land resource is dominated by the unbroken extent of rainforest and wet-and-dry climates stretching east of the Andes from Colombia to northern Argentina and by substantial areas of water-deficient territory along the west coast and in the south and northeast of the South American continent. Some highly fertile soils in flat humid subtropical lands both west and east of the Paraná – La Plata river system are of minor extent by comparison with the whole. The Amazon Basin contains the largest stand of tropical hardwood forest on Earth. Metallic mineral resources are abundant in three general regions: the Andes, the largely crystalline rock highlands of eastern and southeastern Brazil, and the low plateau south of the lower Orinoco River. The Caribbean coast of Venezuela has the most productive petroleum reserves on the continent.

North America. Large sections of North American lands benefit from the advantages which characterize midlatitude humid-land resources under present technology. Disadvantages of desert and semiarid environments in much of the western half of the North American continent are tempered somewhat by the interspersal of mountain ranges throughout these drier regions. Taiga and other northern climatic environments are in considerable part coincident with the igneous and metamorphic rocks of the Laurentian Shield; and tropical environments are of small extent. In sum, this continent may be considered to have one of the best-balanced sets of resources, considering its substantial endowment in minerals of many different kinds, extensive forest lands, large annually renewed fresh-water supplies, great and varied agricultural lands, and the productive fisheries off both Atlantic and Pacific coasts. In addition, evidence has accumulated that in the southern, Latin American section of the continent there may exist a major concentration of petroleum and natural gas resources extending south-eastward from eastern Mexico and the contiguous continental shelf beneath the waters of the Gulf of Mexico to the fringes of the western Caribbean, with eventual productivity on the order of magnitude of the cluster of fields in the Middle East. Finally, North America has the highest ratio of employed resources to land area of all continents. In addition, it still contains considerable potential resources.

Summary comment. The Earth's resource pattern has certain general characteristics. (1) Minerals usable under present technology are found in every environment, although mineral types differ according to location in sedimentary or igneous and metamorphic rock areas. Mineral exploration will continue indefinitely in all land areas, but the mineral resource possibilities of North America and the European part of the Eurasian continent have been examined in greater detail than those of any other large area. Ocean basins are the least-known part of the world as to mineral possibilities. (2) Agricultural lands and forest lands usable under present technology are dominated by those lying in midlatitudes. Sections of the taiga are becoming more important as forest resources, (3) The great potential agricultural and forest resources, if some technological improvement is assumed, lie within the humid tropical environments. *See* CLIMATOLOGY; LAKE; RIVER; SOIL ZONALITY; VEGETATION ZONES, WORLD.

[DONALD J. PATTON]

Bibliography: P. R. Cresson and K. D. Frederick, *The World Food Situation: Resource and Environmental Issues in the Developing Countries and the United States*, 1977; J. F. McDivitt and G. Manners, *Minerals and Men: An Exploration of the World of Minerals and Metals, Including Some of the Major Problems That Are Posed*, 1974; I. G. Simmons, *The Ecology of Natural Resources*, 1980; W.L. Thomas, *Man's Role in Changing the Face of the Earth*, 1956.

Earthquake

A phenomenon during the occurrence of which the Earth's crust is set shaking for a period of time. The shaking is caused by the passage through the Earth of seismic waves—low-frequency sound waves that are emanated from a point in the Earth's interior where a sudden, rapid motion has taken place. It is more proper today to use the term earthquake to refer to the source of seismic waves, rather than the shaking phenomenon, which is an effect of the earthquake. Earthquakes vary immensely in size, from tiny events that can be detected only with the most sensitive seismographs, to great earthquakes that can cause extensive damage over widespread areas.

Although thousands of earthquakes occur every day, and have for billions of years, a truly great earthquake occurs somewhere in the world only once every 2 or 3 years. When a great earthquake occurs near a highly populated region, tremendous destruction can occur within a few seconds. In 1556, in the Shensi Province of China, 800,000 people were killed in a single earthquake. The city of Lisbon, one of the principal capitals of that day, was utterly destroyed, with high loss of life, in 1755. In the 20th century such cities as Tokyo and San Francisco have been leveled by earthquakes. In these more modern cases, much of the damage was not due to the shaking of the earthquake itself, but was caused by fires originating in the gas and electrical lines which interweave modern cities, and by damage to fire-fighting capability which rendered the cities helpless to fight the conflagrations.

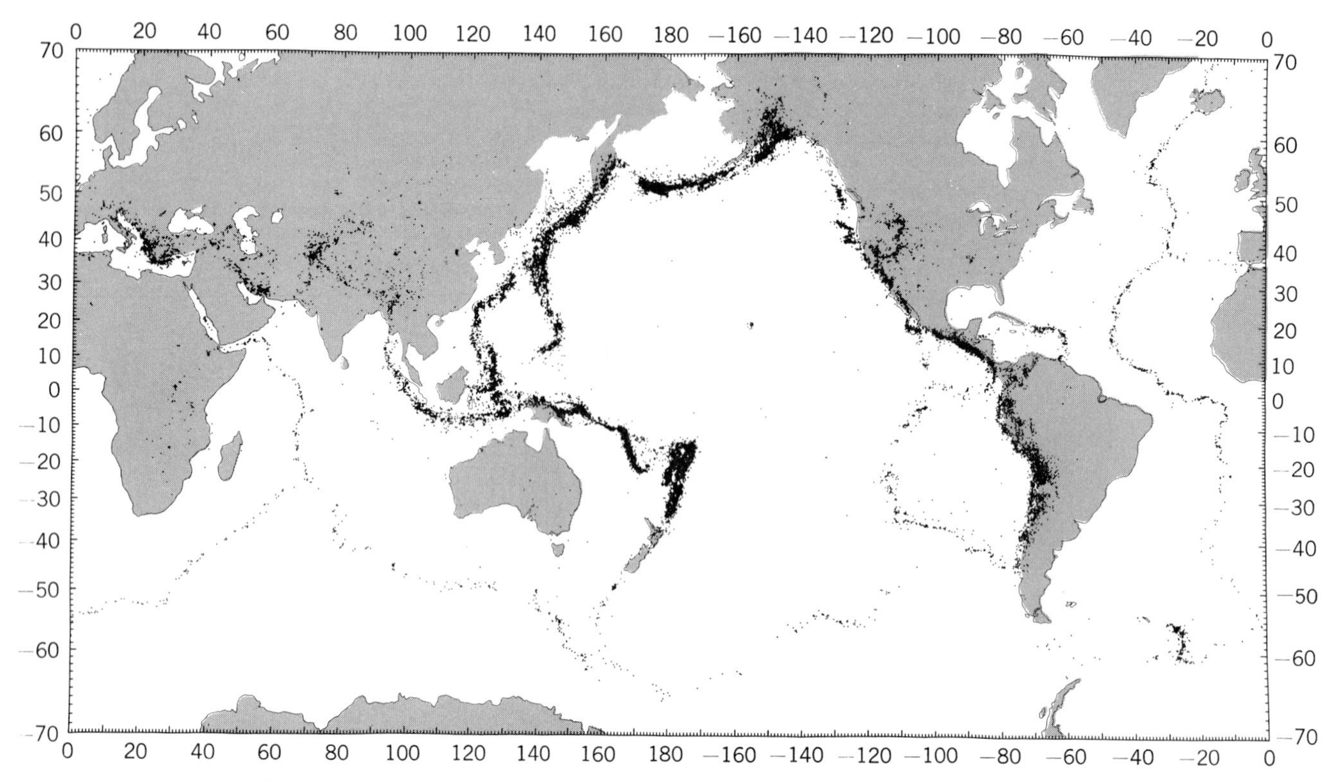

Fig. 1. Seismicity of the Earth from 1961 to 1967; depths to 700 km. The earthquake belts mark the plate boundaries. The number scales indicate latitude and longitude. *(From M. Barazangi and J. Dorman, World seismicity maps compiled from ESSA, Coast and Geodetic Survey, Epicenter Data, 1961–1967, Bull. Seis. Soc. Amer., 59:369–380, 1969)*

Cause. The locations of earthquakes which occurred between 1961 and 1967 are shown on the map in Fig. 1. The map shows that earthquakes are not distributed randomly over the globe but tend to occur in narrow, continuous belts of activity. These earthquake belts link up so that they encircle large seismically quiet regions, which are known as plates. The plates are in continuous motion with respect to one another at rates on the order of centimeters per year; this plate motion is responsible for most geological activity, including earthquakes.

Plate motion occurs because the outer cold, hard skin of the Earth, the lithosphere, overlies a hotter, soft layer known as the asthenosphere. Heat from decay of radioactive minerals in the Earth's interior sets the asthenosphere into thermal convection. This convection has broken the lithosphere into plates which move about in response to the convective motion in a manner shown schematically in Fig. 2. The plates move apart at oceanic ridges. Magma wells up in the void created by this motion and solidifies to form new sea floor. This process, in which new sea floor is continually created at oceanic ridges, is called sea-floor spreading. Since new lithosphere is continually being created at the oceanic ridges by sea-floor spreading, a like amount of lithosphere must be destroyed somewhere. This occurs at the oceanic trenches, where plates converge and the oceanic lithosphere is thrust back down into the asthenosphere and remelted. The melting of the lithosphere in this way supplies the magma for the volcanic arcs which occur behind the trenches. Where two continents collide, however, the greater bouyancy of the less dense continental material prevents the lithosphere from being underthrust, and the lithosphere buckles under the force of the collision, forming great mountain ranges such as the Alps and Himalayas. Where the relative motion of the plates is parallel to their common boundary, slip occurs along great faults which form that boundary, such as the San Andreas fault in California.

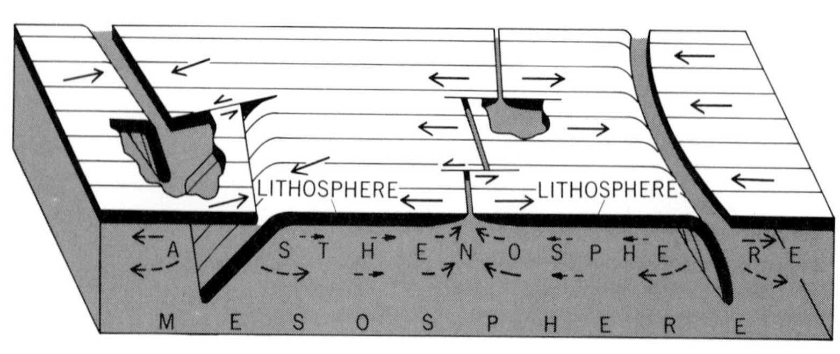

Fig. 2. Movement of the lithosphere over the more fluid asthenosphere. In the center, the lithosphere spreads away from the oceanic ridges. At the edges of the diagram, it descends again into the asthenosphere at the trenches. *(From B. Isacks, Oliver, and L. R. Sykes, Seismology and the new global tectonics, J. Geophys. Res., 73 5855–5899; copyright 1968 by American Geophysical Union)*

According to the theory of plate tectonics, the motion of the plates is very similar to the movement of ice floes in arctic waters. Where floes diverge, leads form and water wells up, freezing to the floes and producing new floe ice. The formation of pressure ridges where floes converge is analogous to the development of mountain ranges where plates converge.

Stick-slip friction and elastic rebound. As the plates move past each other, the motion at their boundaries does not occur by continuous slippage but in a series of rapid jerks. Each jerk is an earthquake. This happens because, under the pressure and temperature conditions of the shallow part of the Earth's lithosphere, the frictional sliding of rock exhibits a property known as stick-slip, in which frictional sliding occurs in a series of jerky movements, interspersed with periods of no motion—or sticking. In the geologic time frame, then, the lithospheric plates chatter at their boundaries, and at any one place the time between chatters may be hundreds of years.

The peroids between major earthquakes is thus one during which strain slowly builds up near the plate boundary in response to the continuous movement of the plates. The strain is ultimately released by an earthquake when the frictional strength of the plate boundary is exceeded. This pattern of strain buildup and release was discovered by H. F. Reid in his study of the 1906 San Francisco earthquake. During that earthquake, a 250-mi-long (400 km) portion of the San Andreas fault, from Cape Mendicino to the town of Gilroy, south of San Francisco, slipped an average of 12 ft (3.6 m). Subsequently, the triangulation network in the San Francisco Bay area was resurveyed; it was found that the west side of the fault had moved northward with respect to the east side, but that these motions died out at distances of 20 mi (32 km) or more from the fault. Reid had noticed, however, that measurements made about 40 years prior to the 1906 earthquake had shown that points far to the west of the fault were moving northward at a slow rate. From these clues, he deduced his theory of elastic rebound, illustrated schematically in Fig. 3. The figure is a map view, the vertical line representing the fault separating two moving plates. The unstrained rocks in Fig. 3*a* are distorted by the slow movement of the plates in Fig. 3*b*. Slippage in an earthquake, returning the rocks to an unstrained state, occurs as in Fig. 3*c*.

Classification. Most great earthquakes occur on the boundaries between lithospheric plates and arise directly from the motions between the plates. Although these may be called plate boundary earthquakes, there are many earthquakes, sometimes of substantial size, that can not be related so simply to the movements of the plates.

Near many plate boundaries, earthquakes are not restricted to the plate boundary itself, but occur over a broad zone—often several hundred miles wide—adjacent to the plate boundary. These earthquakes, which may be called plate boundary-related earthquakes, do not reflect the plate motions directly, but are secondarily caused by the stresses set up at the plate boundary. In Japan, for example, the plate boundaries are in the deep ocean trenches offshore of the Japanese islands, and that is where the great plate boundary earthquakes occur. Many smaller events occur scattered throughout the Japanese islands, caused by the overall compression of the whole region. Although these small events are energetically minor when compared to the great offshore earthquakes, they are often more destructive, owing to their greater proximity to population centers.

Although most earthquakes occur on or near plate boundaries, some also occur, although infrequently, within plates. These earthquakes, which are not related to plate boundaries, are called intraplate earthquakes, and can sometimes be quite destructive. Although intraplate earthquakes are probably caused by the same convective forces which drive the plates, their immediate cause is not understood. Some of them can be quite large. One of the largest earthquakes known to have occurred in the United States was one of a series of intraplate earthquakes which took place in the Mississippi Valley, near New Madrid, MO, in 1811 and 1812. Another intraplate earthquake, in 1886, caused moderate damage to Charleston, SC.

In addition to the tectonic types of earthquakes described above, some earthquakes are directly associated with volcanic activity. These volcanic earthquakes result from the motion of undergound magma that leads to volcanic eruptions.

Sequences. Earthquakes often occur in well-defined sequences in time. Tectonic earthquakes are often preceded, by a few days to weeks, by several smaller shocks (foreshocks), and are nearly always followed by large numbers of aftershocks. Foreshocks and aftershocks are usually much smaller than the main shock. Volcanic earthquakes often occur in flurries of activity, with no discernible main shock. This type of sequence is called a swarm.

Size. Earthquakes range enormously in size, from tremors in which slippage of a few tenths of an inch occurs on a few feet of fault, to the greatest events, which may involve a rupture many hundreds of miles long, with tens of feet of slip. Accelerations as high as 1 *g* (acceleration due to gravity) can occur during an earthquake motion. The velocity at which the two sides of the fault move during an earthquake is only 1–10 mph, but the rupture front spreads along the fault at a velocity of nearly 5000 mph. The earthquake's primary damage is due to the generated seismic waves, or sound waves which travel through the Earth, excited by the rapid movement of the earthquake. The energy radiated as seismic waves during a large earthquake can be as great as 10^{12} cal, ($10^{12} \times 4.19$ J) and the power emitted during the few hundred seconds of movement as great as a billion megawatts.

The size of an earthquake is in terms of a scale of magnitude based on the amount of seismic waves generated. Magnitude 2.0 is about the smallest tremor that can be felt. Most destructive earthquakes are greater than magnitude 6; the largest shock known measured 8.9. The scale is logarithmic, so that a magnitude 7 shock is about 30 times more energetic than one of magnitude 6, and 30×30, or 900 times, more energetic than one of magnitude 5. Because of this great increase in size with magnitude, only the largest events (greater than magnitude 8) significantly contribute to plate movements. The smaller events occur much more often but are almost incidental to the process.

The intensity of an earthquake is a measure of the severity of shaking and its attendant damage at a point on the surface of the Earth. The same earthquake may therefore have different intensities at different places. The intensity usually decreases away from the epicenter (the point on the surface directly above the onset of the earthquake), but its value depends on many factors in addition to earthquake magnitude. Intensity is usually higher in areas with thick alluvial cover or landfill than in areas of shallow soil or bare rock. Poor

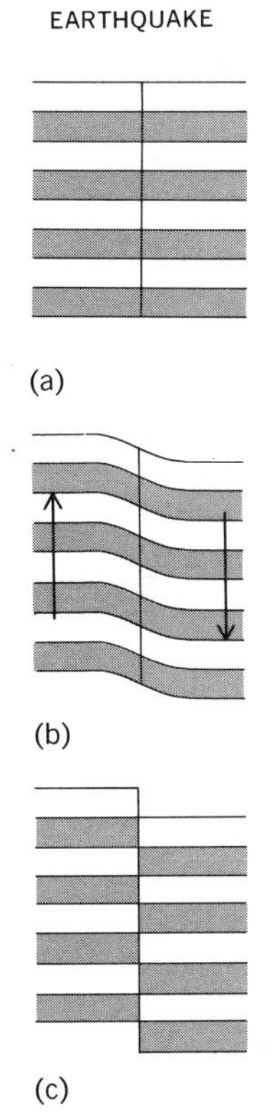

Fig. 3. Schematic of elastic rebound theory. (*a*) Unstrained rocks (*b*) are distorted by relative movement between the two plates, causing strains within the fault zone that finally become so great that (*c*) the rocks break and rebound to a new unstrained position. (*From C. R. Allen, The San Andreas Fault, Eng. Sci. Mag., Calif. Inst. of Technol., pp. 1–5, May 1957*)

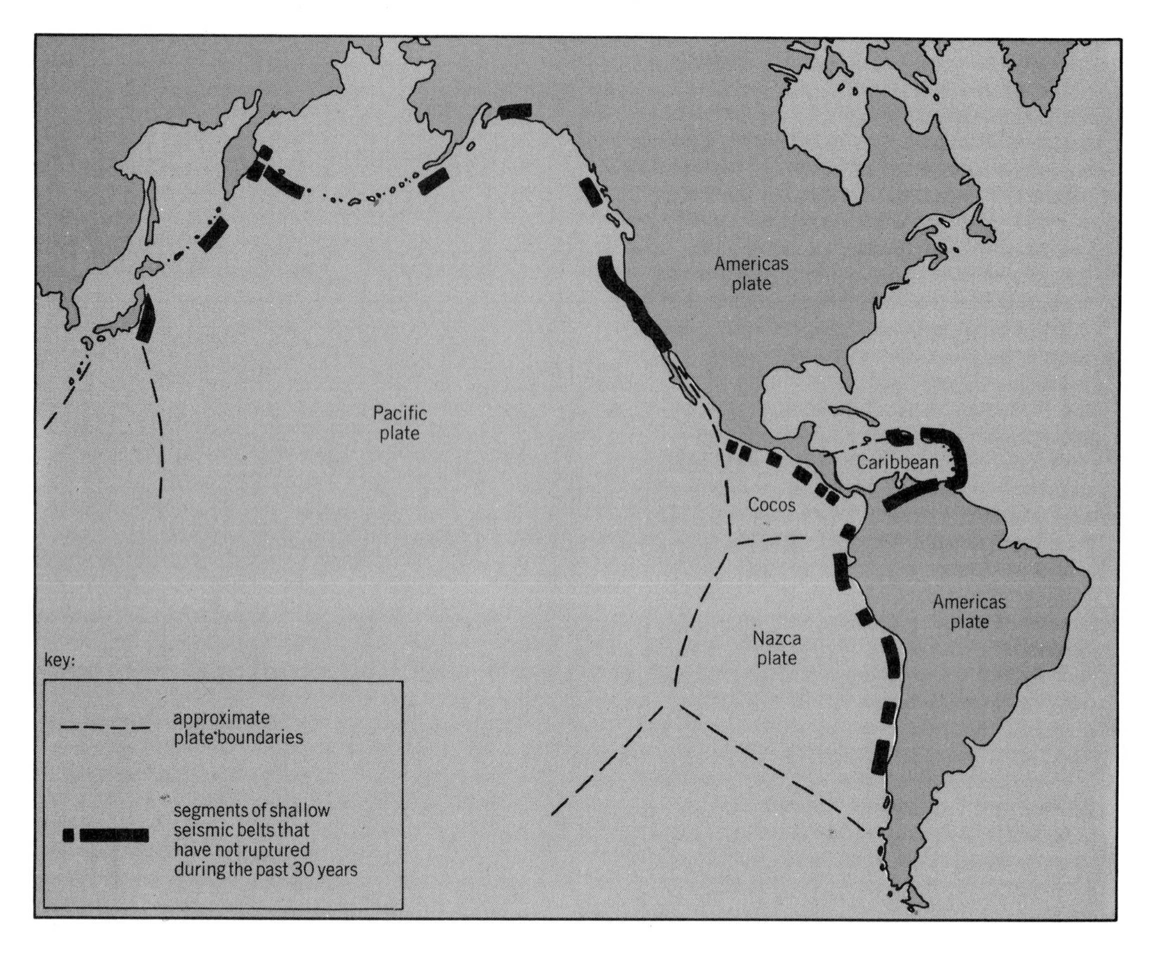

Fig. 4. Major seismic gaps which occur in the western Pacific Ocean. *(From J. Kelleher et al., J. Geophys. Res., 78:2547–2585; copyright 1973 by American Geophysical Union)*

building construction leads to high intensity ratings because the damage to structures is high. Intensity is therefore more a measure of the earthquake's effect on humans than an innate property of the earthquake.

Effects. Many different effects are produced by earthquake shaking. Although the fault motion that produced the earthquake is sometimes observed at the surface, often other earth movements, such as landslides, are triggered by earthquakes. On rare occasions the ground has been observed to undulate in a wavelike manner, and cracks and fissures often form in soil. The flow of springs and rivers may be altered, and the compression of aquifers sometimes causes water to spout from the ground in fountains. Undersea earthquakes often generate very-long-wavelength water waves, which are sometimes called tidal waves but are more properly called seismic sea waves, or tsunami. These waves, almost imperceptible in the open ocean, increase in height as they approach a coast and often inflict great damage to coastal cities and ports.

Prediction. The largest earthquakes occur along the boundaries between moving lithospheric plates, and are caused by a sudden slip of portions of these boundaries when the pent-up stresses exceed the frictional resistance to motion. Each segment of a plate boundary must therefore periodically experience great earthquakes, the frequency of which will depend on the relative rate of plate motion between the two plates in contact, and on the nature of the boundary itself. It is thus possible to state in general that, on a given plate boundary, the most likely sites of the next great earthquakes are those segments, called seismic gaps, in which no large earthquakes have occurred in the longest time. Known seismic gaps in the western Pacific are shown in Fig. 4.

Recognizing seismic gaps is a crude means of predicting earthquakes, although it really provides only a rough estimate of earthquake risk, since it does not permit a precise estimation of the time, place, and magnitude of future shocks. Furthermore, there are many smaller earthquakes, many of which are destructive, which occur on or adjacent to plate boundaries; they do not play such a direct role in plate motion, and hence cannot be anticipated in terms of seismic gaps. These earthquakes occur almost randomly in time, and the hazard that they present can be estimated only statistically. Thus it is known that areas near plate boundaries, such as the San Andreas Fault, are more prone to earthquakes than are other areas, say, Colorado, which are far from currently active plate margins; thus it is possible to statistically rate the difference in hazard between the two areas.

Precursory phenomena. Neither the seismic gap approach nor the statistical approach provides a means of actual prediction of individual earthquakes. Research on earthquake prediction has concentrated instead in discovering phenomena which are precursory to earthquakes, and which

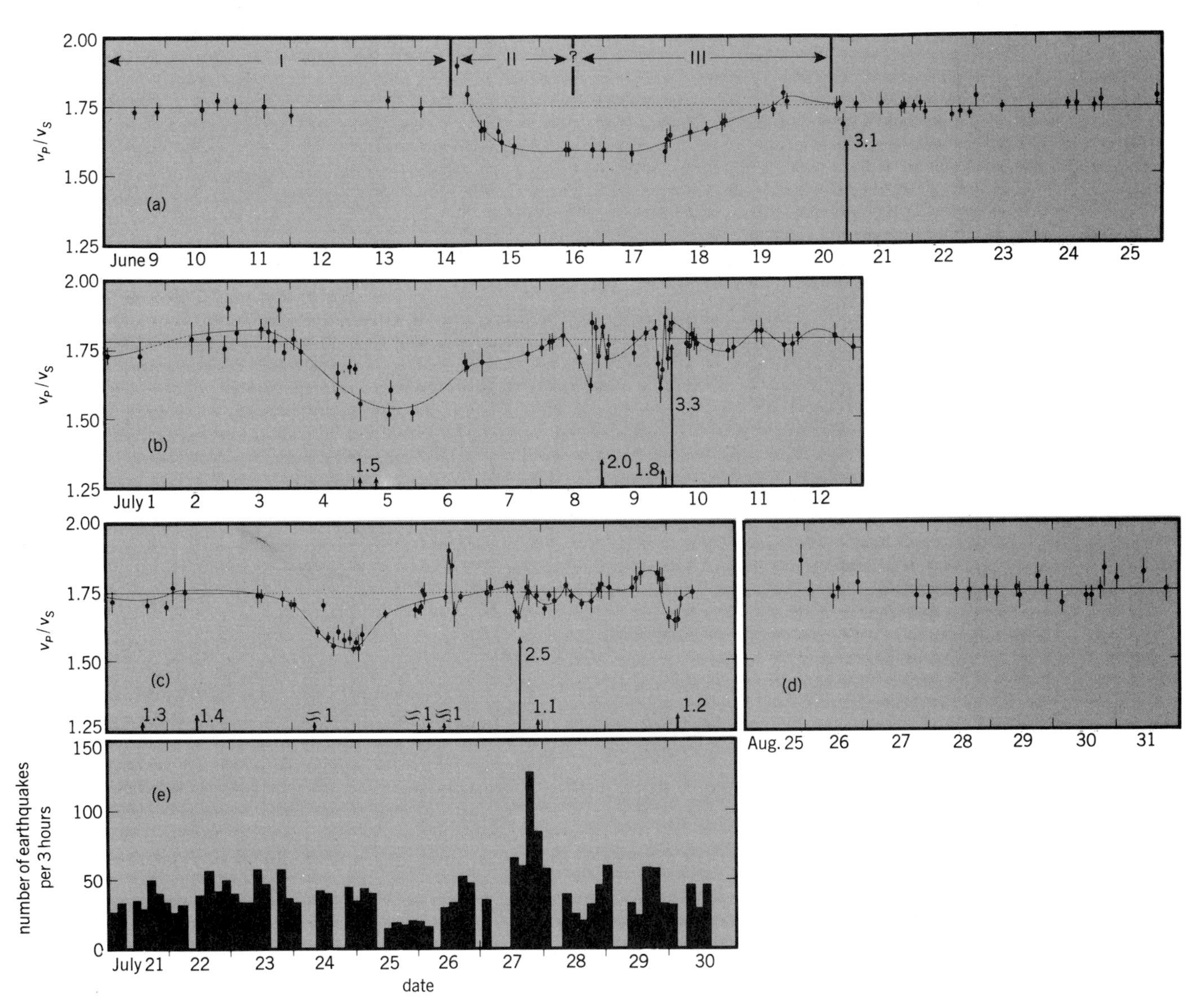

Fig. 5. (*a–c*) Diagrams of changes in the ratio v_p/v_s preceding three of the largest events in the Blue Mountain Lake, NY, earthquake swarm of 1971. All quakes of magnitude greater than 1 are shown as arrows. (*d*) A similar plot during a time when no large events took place. (e) A plot of the frequency of occurrence of small earthquakes in the swarm plotted on the same time scale as c. (*From C. H. Scholz et al., Earthquake prediction: A physical basis, Science, 181:803–810; copyright 1973 by American Association for the Advancement of Science*)

can therefore be used as warnings. A number of such phenomena have been discovered to precede earthquakes by days or months or even years. These precursors are usually changes in some property of the Earth in the vicinity of the coming earthquake. They include such phenomena as anomalous uplift of the ground, changes in electrical conductivity of rock, changes in the isotopic composition of deep well water, and changes in the nature of small earthquake activity.

The most often observed precursory event is the relative change in the velocity at which the two types of seismic waves, the p (or pressure) and the s (or shear) waves (also written as P waves and S waves) travel through the source region of an impending earthquake. The velocity of the p wave v_p is always greater than that of the s wave v_s. Usually it is about 1.75 times that of the s wave. However, when the p and s waves travel through a region in which an earthquake is about to occur, it has been observed that the p wave travels more slowly than it normally does, only about 1.5 times faster than the s wave. This anomalous behavior persists for some time, and then the velocities return to normal; shortly thereafter, the earthquake occurs. This phenomenon was found to occur repeatedly before small earthquakes took place in the Adirondack Mountains of New York (Fig. 5); one earthquake was successfully predicted in advance on the basis of this precursor.

Dilatancy. These precursors indicate that some fundamental change takes place in the Earth's crust just before the occurrence there of an earthquake. Many scientists now believe that these precursory effects are caused by a property of rock called dilatancy, which is an increase in rock volume caused by an increase in distortional strain on the rock. When stress builds up prior to fracture, numerous cracks open up within a rock, causing it to swell.

According to the dilatancy theory, stress slowly builds up over many years in an earthquake-prone

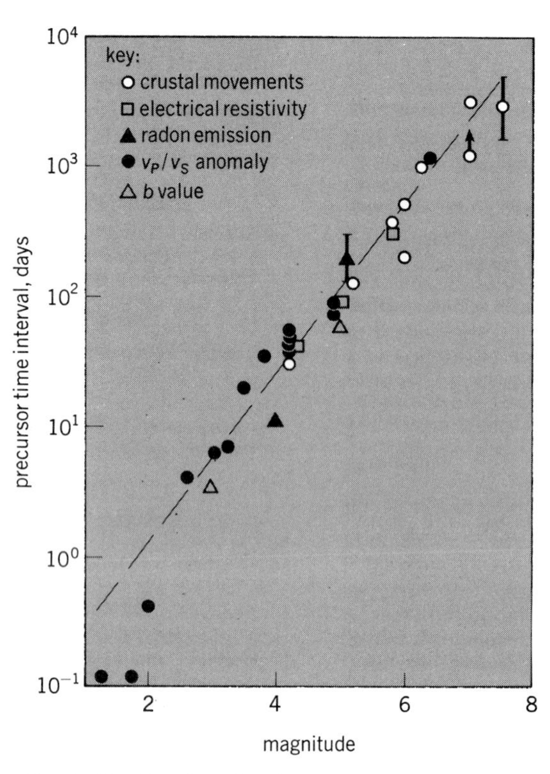

Fig. 6. Duration time of various precursory phenomena as a function of the magnitude of the following earthquake. Here *b* is a parameter that describes the ratio of small to large shocks. (*From C. H. Scholz et al., Earthquake prediction: A physical basis, Science, 181:803–810; copyright 1973 by American Association for the Advancement of Science*)

area. Eventually, in some part of that area, the stress becomes high enough to cause cracks to open; hence dilatancy occurs. Since the cracks and pores of the rock have been filled with water and other fluids at some pressure, known as the pore pressure, the increase in total crack volume reduces this pore pressure, thereby strengthening the rock. At the same time, the various premonitory phenomena begin to occur, because properties such as electrical conductivity and seismic wave velocity are very sensitive to the amount of open cracks in the rock. At a later time, these effects begin to disappear because water flows in from adjacent areas, filling the cracks and bringing the pore pressure back to normal, which weakens the rock again and triggers the earthquake.

An important consequence of this process is that the larger the upcoming earthquake, the larger the dilatant volume, and hence the longer the time required for the influx of water. Thus precursory phenomena last longer for larger earthquakes, as evidenced in Fig. 6. This means that the magnitude, as well as the time and place, of the earthquake can be predicted.

These precursory phenomena have been found to precede about 50 earthquakes. It is not known if they occur before all earthquakes, or if the phenomena are always the same. Many aspects of the dilatancy theory remain to be proved. This line of research, however, holds out the greatest promise for providing a practical means of predicting earthquakes.

[CHRISTOPHER H. SCHOLZ]

Ecological entomology

Ecological entomology is the study of insects' relation to their environment. By number of species alone, insects exceed all other species of animals and plants combined. In total living weight, insects probably outweigh any other animal group, including humans, and play an important role in keeping the life system functioning. Insects also exert a significant impact on the quality of humans' food, homes, and health. In the United States insects cause a greater loss of crops than any other pest group. On a worldwide basis, they are vectors of some of the most serious diseases infecting humans, such as malaria.

Role in cycle of matter. Insects help concentrate and convert plant and animal material into food for other animals. Insects are the prime source of food for many species of fish, including trout and salmon. Toads, frogs, and other amphibians consume insects. The basic food items of many birds, such as the bluebird and house wren, consist primarily of insects. Moles, shrews, skunks, and other small mammals feed almost exclusively on insects. *See* FOOD CHAIN.

Several groups of insects feed on dead and decaying animals and plants and thereby help to degrade these wastes. Because of this role and the tremendous living mass of insects, they play a dominant role in recycling the vital elements of life for reuse in the life system.

Pollination role. Insects not only keep the life system functioning by eating and being eaten, but also serve a most important role in plant reproduction—pollination. Despite many technological advances, substitutes for general insect pollination of plants, cultivated and wild, have not been found. The production of fruits, vegetables, and forages depends upon the activity of bees and other insects which pollinate or fertilize the blossoms.

The role of honeybees alone in pollination is impressive. For example, a single honeybee may visit and pollinate 1000 blossoms in a single day (10 trips with 100 blossoms visited per trip). In New York State, with about 3 million beehives and each with 10,000 worker bees, honeybees could visit 30 trillion blossoms in a day. Wild bees pollinate a number of blossoms equal to or more than that pollinated by honeybees. In fact, on a bright, sunny day, more than 60 trillion blossoms may be pollinated.

Honey and wax production. Insects make several products of value to humans, of which honey and wax are the most valuable. About 200 million pounds (90 million kilograms) of honey and 4 million pounds (1.8 million kilograms) of wax are produced in the United States annually, with a value of $100 million and $4 million respectively. To produce 1 lb (0.45 kg) of honey, 1500 bees work 10 hr, and to make 1 lb of beeswax (an ingredient of most lipsticks), the honeybee must consume about 8 lb of its honey.

Biological control. In addition to pollinating plants, some insects, because of their tremendous numbers, are useful in controlling and preventing outbreaks of pest populations of insects and plants. The impact of insect feeding pressure for weed control is illustrated throughout nature. For example, the plant pest known as Klamath weed was introduced into California about 1900, and by

1944 the weed had become a dominant plant in about 2 million acres (8 km^2) of pasture and range land. In 1945 a small beetle *(Chrysolina quadrigemina)* which fed on the weed was introduced to the area. This beetle population has increased, dispersed, and eliminated 99% of the weed. The previous forage vegetation has regrown, with the pasture and range land once again suitable for cattle production.

In United States agriculture about 95% of control of pest insects depends upon various natural controls, including insect predators and parasites. Surprisingly, only 9% of total agricultural acres in the United States is treated with insecticides. Therefore, the prime means of pest insect control is biological control. *See* INSECT CONTROL, BIOLOGICAL.

Insecticide use. What about the pest insects and insecticides used to control them? Nearly a half billion pounds (227 million kilograms) of insecticides is used annually in the United States for insect control. Three-fourths of this insecticide is used directly in agriculture, with the remainder used by government agencies, industries, and homeowners.

Insecticides used in agriculture are not evenly distributed over all crops. For example, 50% of all insecticide applied in agriculture is used on the nonfood crops of cotton and tobacco. Of the food crops, corn, fruit, and vegetables receive the largest amounts of insecticide (see table). Citrus, apples, and potatoes have more than 85% of their acreages treated with insecticides. Many acres of small grains and pastures receive little or no insecticide treatment. This is the reason that only 6% of the crop acres receives any insecticide treatment at all.

Crop loss due to insects. Despite a 20-fold increase in the use of insecticides during the past 30 years, crop losses due to insects have nearly doubled; they have increased from about a 7% loss in the 1940s to about a 13% loss currently. Although crop losses due to insects have generally increased despite the significant use of insecticides, important advances have been made in reducing crop losses from certain pests. For example, losses from potato insects have declined from about 22% previous to 1935 to about 14% today. In contrast, corn losses due to insects have been increasing. In the 1940s corn losses due to insects averaged about 3.5%; however, this loss has increased significantly and now averages about 12%. Factors contributing to increased corn losses due to insects include the continuous culture of corn on the same land and the planting of insect-susceptible types instead of resistant types.

If the 9% of crop acres now treated with insecticides were no longer treated, it is estimated that crop losses would increase from the current total of 13% to about 18%. This would amount to a loss of about $5 billion annually. If no insecticides were used, the increased losses would probably increase retail food costs about 6%. Based on these estimates, the nation would not starve if insecticides were not used; supplies of food for the nation would be ample, but quantities of certain fruits and vegetables such as apples, peaches, plums, onions, potatoes, and cabbages would be significantly reduced. Because of this, substitutes for some fruits and vegetables that people like to eat would have to be found.

Actually, the loss in some fruits and vegetables could be reduced if "cosmetic standards" were modified. Although safe and nutritionally sound, some fruits and vegetables with skin blemishes are not sold because wholesalers and retailers generally rate them unacceptable. Small blemishes do not adversely affect these products' overall palatability and nutrition.

Home and health hazards. Insects not only attack crops, but damage homes and more importantly are health hazards. As nuisances, they bite and sting humans, but of greater significance on a worldwide basis is the fact that insects are vectors of some of the most serious diseases of humankind.

In much of the world the ancient afflictions—malaria, filariasis, yellow fever, and dengue—continue to limit progress. In terms of numbers affected, mosquito-transmitted malaria continues to be the number one disease of humans. About 400 million people live in areas where malaria is highly

Some examples of percentages of crop acreas treated, of pesticide amounts used on crops, and of acres planted to this crop

Crops	Insecticides		Herbicides		Fungicides		% of total crop acres
	% acres	% amount	% acres	% amount	% acres	% amount	
Nonfood	NA*	50	NA	NA	<0.5	NA	NA
Cotton	61	47	82	9	4	1	11.1
Tobacco	77	3	7	NA	7	NA	0.11
Field crops	NA	33	NA	NA	NA	15	NA
Corn	35	17	79	45	1	NA	7.43
Peanuts	87	4	92	2	85	11	0.16
Rice	35	1	95	3	0	NA	0.22
Wheat	7	1	41	5	0.2	NA	6.11
Soybeans	8	4	68	16	2	NA	4.19
Pasture hay + range	0.5	2	1	4	0	NA	68.40
Vegetables	NA	7	NA	2	NA	24	NA
Potatoes	77	2	51	NA	49	10	0.16
Fruit	69	9	NA	1	NA	60	NA
Apples	91	3	35	NA	61	18	0.07
Citrus	72	2	22	NA	47	24	0.08
All crops	9	35	22	51	1	14	NA

*Not available

endemic; it is estimated that at least 100 million cases occur annually and result in 1 million deaths. *See* DISEASE.

At present, mosquito control is in a state of crisis because people have become heavily dependent upon insecticides. The very properties of the insecticides that made them so useful (persistent residues and high toxicity) have caused serious environmental problems.

Environmental impact. Despite the fact that only a small percentage (9%) of the crop acres in the United States is treated, serious insecticide pollution problems have occurred. Insecticides have destroyed natural enemies of other pests, resulting in other pest outbreaks and thereby requiring additional pesticide sprays. This has occurred in several crops, but especially in fruit and cotton pest control.

About half of the food that is purchased in the supermarket is contaminated with low levels of pesticide residues, according to the Food and Drug Administration. Most of these residues are below the acceptable tolerance levels and therefore are considered to present no direct hazard to human health. However, the quantifiable effects of long-term, low-level exposure of humans to pesticides are unknown at this time.

Pesticides are drifting in the atmosphere and contaminating the environment. Large quantities enter the atmosphere during application. For instance, 65% of all pesticides are sprayed and dusted by aircraft. Only 40 to 80% of the pesticide lands in the target area; the remainder drifts off into the environment, where it may be hazardous to humans and other creatures.

Systems approach to pest management. Pest population numbers are limited by many factors. The new objective in pest management is simply to make the environment unfavorable enough that the numbers of pests cause little economic damage. This ecosystem or systems approach includes three parts: source reduction of pests by environmental manipulations; integrated pest control employing combinations of pesticides and natural enemies where necessary; and cost-benefit analyses of the pest management options available to humans and the environment.

Source reduction requires a thorough knowledge of the ecology of the pest to understand which cultural technologies employed in raising the crop are causing pest problems. Once this is known, the agroecosystem can be carefully manipulated to make the crop or livestock environments less favorable to the pest. The environmental manipulations might include: rotating crops with nonhost crops; increasing crop diversity; increasing genetic diversity of crops; determining climate effects on crop and pest; altering fertilizer combinations and quantities; changing plant spacings; altering planting times; selecting appropriate crop plant associations; and determining whether pesticides are either harming natural enemies or altering the physiology of the crop, making it more susceptible to pest attack.

Integrated pest control requires that pest and natural enemy populations be monitored to make sure that pesticide applications are used when essential. This requires having sufficient knowledge so that the pesticide has mimimal impact upon the natural enemy population.

A cost-benefit analysis of the pest management strategy must be conducted to measure the total direct and indirect costs to humans and the environment, as well as the benefits as measured primarily by increased crop yield. By such analyses, maximal benefits with minimal costs can be determined for all pest control strategies.

[DAVID PIMENTEL]

Bibliography: D. J. Borror and D. M. DeLong, *An Introduction to the Study of Insects*, 1964; D. Pimentel, *Ecological Effects of Pesticides on Nontarget Species*, 1971; D. Pimentel et al., Benefits and costs of pesticide use in U.S. food production, *BioScience*, 28:772, 778–784, 1978.

Ecological interactions

Relationships between members of different species belonging to a particular ecological community. Ecologists generally classify interactions according to the effect that members of one species have on the population growth rate of a second species. If an increase in one species' population increases the growth rate of a second species' population, the effect is positive or beneficial (denoted by +). If an increase in one species' population decreases the growth rate of a second species' population, the effect is negative or harmful (denoted by −). If neither occurs, the effect is neutral (denoted by 0). Effects may be indirect, in which case they involve one or more additional species in the community. For example, earthworms increase plant growth and thereby provide more food for caterpillars. Alternatively, effects may be direct; for example, injuries or death may result from territorial fighting between members of different species.

An ecological interaction between two species A and B can be denoted by a pair of signs, each member of which may be +, −, or 0. One sign denotes the effect of A on B, and the other denotes the effect of B on A. In this designation, direct effects are usually used; but in the case of competition for food, indirect effects involving the food species are also frequently considered. Effects between species that are more indirect than the latter are not discussed here.

Pairwise combinations of signs give six possible types of interactions. The interaction (+,+), in which each species benefits the other, is mutualism. Some authorities still use the term symbiosis to describe the (+,+) interaction, but this term is now generally defined as any kind of nonneutral interaction among species, positive or negative. The interaction (−,−), in which each species harms the other, is called competition. The interaction (+,−), in which one species benefits and the other is harmed, is called predation. [Parasitism, also (+,−), in which the benefitted species is much smaller than the harmed species and usually derives food or shelter from it, is sometimes distinguished from other types of predation.] The interaction (+,0), in which one species benefits and the other is unaffected, is called commensalism. The interaction (−,0), in which one species is harmed and the other is unaffected, is called amensalism. Finally, the interaction (0,0), in which neither species is affected, is called neutralism. One may argue that species in the same ecological community always have some effect, however small, on each other, so that these last three types of interactions cannot really occur. One may counterargue that if the effect is small enough, it can be ignored, so that it is still useful to think in terms of interactions with zero effect. Nonetheless, examples of amensalism are virtually unreported, and examples of commensalism are relatively rare. Of the first three interactions, ecologists have concentrated much more on competition and predation than on mutualism, possibly because of mutualism's lesser importance. However, as illustrated below, numerous examples of mutualism do exist.

Competition. Interspecific competition occurs when two species negatively affect one another. Two principal kinds of competition are recognized. Exploitative competition between species occurs when individuals of one species, by consuming some resource, usually food, deprive individuals of another species of that resource and thereby lower the second species' population growth rate. Interference competition is more direct; it occurs when individuals of one species directly harm individuals of a second species. The degree to which such an effect is harmful can be quite varied. Organisms may kill other organisms by producing toxins or by fighting; fighting may also produce injuries. More subtle types of interference competition occur when individuals interact with one another in such a way as to use energy that might have been used for the production of offspring, or take up time that might have been used to gather additional energy for the production of offspring.

Frequency: interference versus exploitative. The relative frequencies of these two types of competition vary among different kinds of organisms. Interference seems most common among organisms that compete for space. For example, a myriad of fouling organisms settle on bare space in marine habitats such as rock, coral surfaces, and the bottoms of boats; examples are barnacles, sponges, tunicates, and various algae. Such organisms compete by secretion of toxins and by physical overgrowth and overcrowding. Among vertebrates, competition for space between individuals of different species is rarer, though it is common in coral reef fishes and is known in birds and lizards. In these vertebrates, the mechanism of competition is fighting or more subtle modes of aggression. In all cases, individuals of the competing species require much the same resource: the fouling organisms require space on which to settle and gather food or light, and the vertebrates eat much the same types of food in the same habitats. Such common requirements explain the adaptive significance of interference mechanisms; were it not for the consequent acquisition of resources necessary for the production of offspring, the time and energy spent in interference would not be worthwhile.

Interspecific exploitative competition is probably at least as widespread in nature as interference competition, owing to the reason just stated: where interference behavior occurs, the adaptive rationale is to secure resources that would otherwise be depleted by members of another species. However, because the mechanisms are more indirect, it is more difficult to observe in its entirety. In contrast to interference behavior such as fighting, which is directly observable, to show exploitative competition one must observe resource depletion by members of one species and then demonstrate that those resources would probably have been used by members of the other species. For both types of competition, one must also demonstrate that the interference or resource depletion adversely affects population growth rate. The difficulty of doing so without elaborate experiments implies that most evidence for competition is inferential. For example, it is frequently assumed that if two species live in the same area and use much the same resources, they are in competition; indeed, the intensity of such competition is frequently measured by the degree of overlap in resource types, such as foods. If this assumption is usually true, the implication is that exploitative competition is very common, since species frequently overlap in resource use. However, such competition may not be very intense, since the overlap is frequently small. Indeed, a low overlap in resource use is frequently taken as evidence of past competition, and may be the result of evolutionary changes developing from selection to avoid competition.

Adaptations of competitors. Species similar in many ways frequently differ in certain structures or behaviors that allow them to exploit most efficiently different types of foods or other resources. For example, three species of ground-inhabiting finches occur in the same community on the Galapagos Islands. As was first noted by Darwin, these species differ greatly in the sizes of their beaks, and several ecologists have recently demonstrated that the species take different sizes of seeds. The Galapagos finches represent a group that has evolved from a common ancestral species in virtual isolation, and it is thought that the adaptive significance of different-sized beaks is to avoid interspecific exploitative competition.

Use of resources. How species differ in their use of resources is called resource partitioning, and numerous studies have demonstrated a variety of such differences. Most frequently, animals differ in the habitats in which they feed. For example, certain lizard species differ in the places in vegetation in which they forage; some inhabit leaves, others twigs, others trunks, and others foliage near the ground. Differences in food type (such as size or hardness) are almost as common, as in the Galapagos finches. Least common are differences in the time of resource use, though insect or lizard species frequently differ in daily activity time.

Experimental demonstrations. In part because of their greater difficulty, experimental demonstrations of ongoing competition are much rarer than inferences from observations of the sort just discussed. Such demonstrations are perhaps most common for plants and have led to the conclusion that plant height is usually the most important characteristic correlated with competitive success. This is because plants compete most frequently for light, and taller individuals can shade other individuals, making it difficult for the latter to photosynthesize and ultimately leading to their deaths. Among animals, competition has been most frequently demonstrated in species that compete for space. A classical study showed that the depth zonation of two species of barnacles in an intertidal habitat results from a combination of competition and physiological limitation. Larvae of the higher species could settle and grow at the depth occupied by the lower species, but under natural conditions were prevented from doing so:

the lower species grew against the other and slowly "bulldozed" it off the rock. The lower species, which is the superior competitor, does not take over the entire habitat, however, since it cannot survive the desiccation to which the higher species is exposed. Another well-studied example involves two starfish species co-occurring over much of the northwestern American coast. These species are carnivorous and are of greatly different sizes. Apparently, they coexist largely by taking different-sized foods, though interference also occurs: one pinches the other with numerous structures called pedicillariae. The resource overlap is significant enough, however, so that experiments have shown substantial ongoing competition: when the larger species population is artificially depressed, the other increases; when it is increased, the other decreases. Although competition experiments have usually been performed with closely related species, an interesting exception involves rodents and ants from southwestern American deserts. Using huge enclosures, ecologists have recently demonstrated that members of these two groups compete for seeds.

Population dynamics. By definition, when two species of competitors meet, they negatively affect one another's population growth rate. The ultimate outcome of this process, in terms of the survival or extinction of the species in the area of competition, is varied. The outcome may, but need not, depend on the population sizes of the species when they first come together. Three cases are distinguished: (1) One or the other species always wins. (2) The species coexist, but their equilibrium population sizes are below those found

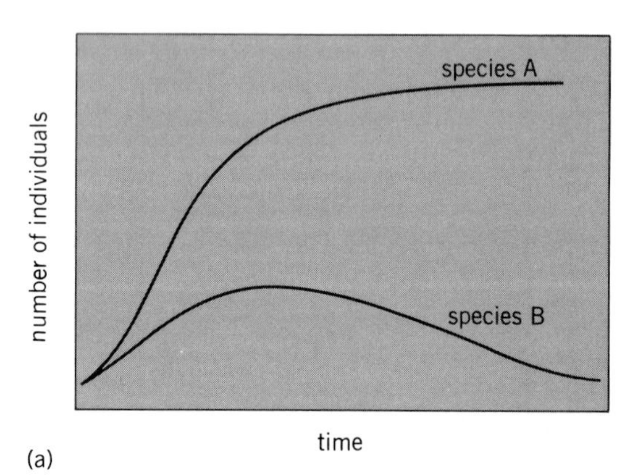

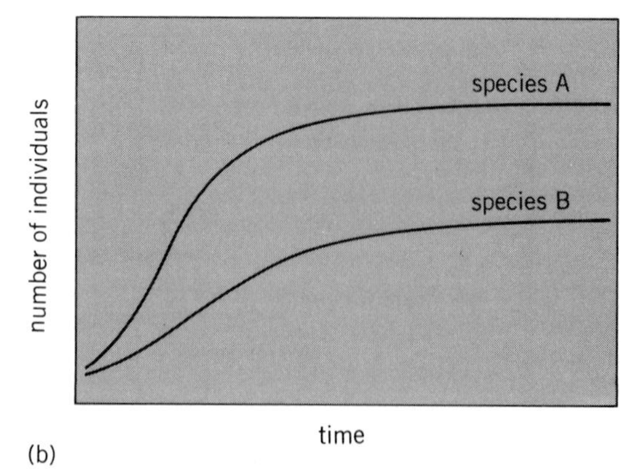

Fig. 1. Time course of competition. (*a*) A outcompetes B. (*b*) A and B coexist.

in the absence of competition. (3) One or the other species wins, depending on the initial population sizes—roughly, the more abundant a species is initially, the more likely it is to win. All these cases have been documented in the laboratory, using easily reared organisms such as certain beetles, flies, yeast, or protozoa. If one species is very much superior to another, case 1 occurs. If the species are fairly equal in competitive ability, but individuals harm members of their own species more than members of the competing species, case 2 occurs. Such would be true, for example, were the principal resources used by each species different (that is, were there a high degree of resource partitioning), or were intraspecific aggression higher than interspecific, as it usually is. If the species are fairly equal in competitive ability, but individuals harm members of the competing species more than members of their own species, case 3 occurs. Such would be true, for example, were each species to secrete a toxin that it was insensitive to but that harmed the other species. In general, case 3 is much more likely to occur when competition is mainly interference rather than exploitative.

The time course (change in population size with time) of competition between two species usually looks as illustrated in Fig. 1: oscillations or cycles usually do not occur. Three or more species systems may oscillate, however.

Predation. Predation occurs when individuals of one species, the predator, consume part or all of individuals of another species, the prey. In this interaction, the predator benefits and the prey is harmed. Broadly defined, the predator and prey may be animal or plant, and the predator may be a parasite. Examples of predation include (1) a hawk eating a sparrow, (2) a lizard eating an insect, (3) a sparrow eating a seed, (4) a sheep eating grass, (5) a Venus' flytrap eating an insect, (6) a tick drawing blood from a dog, and (7) a pathogenic bacterium afflicting a human. Some ecologists exclude plants from the definition, in which case examples 3–5 do not apply, or exclude parasitism, in which case examples 6 and 7 do not apply.

Observations and experiments. The existence of predation is often less disputed than that of competition, because the process can usually be directly observed, at least when the prey are lower organisms. Moreover, exploitative competition implies predation, by definition. For this reason, it is perhaps fruitless to argue whether predation or competition is most important in general. However, one may ask whether the size of a particular species' population is more controlled by competitors or predators, and whether that species' evolutionary adaptations are more a result of past competitive or past predatory interactions. The answer to the first question has been worked out experimentally for certain marine communities, and is thought to be true in general. Animals at the top of the food chain or web (such as predatory starfish, lions, or owls) are more likely affected by competition, whereas those lower down are usually more likely affected by predation. *See* FOOD CHAIN.

Predator adaptations. Adaptations that enhance the efficiency of predators are numerous and sometimes spectacular. Prey pursuit is facilitated by great bursts of speed, such as the downward flight (or stoop) of peregrine falcons which attains 275 mph (123 m/s). The capture and consumption of prey are facilitated by structures such as beaks, claws, and teeth, as well as by the production of toxins as in many snakes and spiders. Sometimes predators use elaborate morphological lures to capture prey. The anglerfish has an extraordinary fishing-rod–like structure (a modified fin) protruding from its dorsal surface; this is apparently used to attract prey. Social grouping may facilitate either the location of prey, as in certain grain-eating birds, or the capture of prey, as in wolves or wild dogs. Camouflage can prevent a predator from being detected by prey, as in the tropical praying mantids that resemble flowers.

Prey adaptations. Adaptations that enhance the ability of prey to escape are no less spectacular. These include great speed as well as more elaborate feigning and maneuvering; a gazelle, for example, will jump into the air and change direction (stotting behavior) to confuse a predator. Morphological aggressive structures of prey are sometimes formidable in their own right; hoofs and horns of large herbivorous mammals can be fatal to a predator. Social grouping also sometimes facilitates prey escape: prey detection is increased in certain birds, and group defense occurs in certain baboons. Prey, both animal and plant, sometimes produce toxins rendering them distasteful or even fatal to predators. An example is the monarch butterfly, which has incorporated into its own tissues chemicals distasteful to birds from the milkweed plants on which it feeds. Elaborate protective resemblance occurs among prey; many insects, such as walkingsticks or leaf katydids, resemble parts of plants on which they feed. Sometimes, distasteful animals or plants possess a striking coloration or pattern; this can cause a predator to remember a particular prey as distasteful, even after one encounter. Potential prey sometimes mimic the pattern or coloration of a distasteful species.

Population dynamics. Mathematical models of predation suggest that its time course will frequently be more oscillatory than that of competition. Figure 2 gives a hypothetical example. Intuitively, the oscillation can be seen to occur as follows. Suppose a predator is introduced into an area having a prey population. At first, because many more prey than predators exist, the predator population will increase markedly. The prey population declines, and the per-predator rate of prey capture simultaneously declines. However, because the predator population at any moment depends on the number of prey caught in the past, the total rate of prey capture will still be rather high. This causes a marked decline in the prey population at the time the predators are still increasing (Fig. 2). Eventually, the prey are so rare that the birth rate of the predators is lowered severely enough to cause their population to decline. This decline eventually allows the prey to increase their population size, and the cycle returns to or near its starting point.

The outcomes of the predation process are: (1) extinction of the prey and then the predator; (2) extinction of the predator and continued existence of the prey; or (3) coexistence of both the predator and prey. Case 1 is likely when the predator is very efficient. Case 2 is likely when the prey is very good at escape. In case 3, the predator and prey can coexist at fixed population sizes, or they can coexist in a stable oscillation, sometimes called a stable limit cycle. In this case, predator and prey populations change in numbers through time, but the numbers repeat themselves periodically; this is the sort of oscillation illustrated in Fig. 2. This stable oscillation is most likely when individual predators are relatively limited in their capacity to find and consume prey; the less these limits are, the more likely a stable equilibrium will exist. Cases 1 and 2 have been demonstrated in the laboratory, as have what appear to be stable limit cycles, though the latter are hard to distinguish from a system slowly oscillating to extinction. In nature, it is much harder to identify the various cases. Systems in which extinction of one or both populations has occurred (cases 1 and 2) leave little evidence of the interaction. Certain systems, such as the lynx-hare interaction in North America, do resemble predator-prey limit cycles (case 3) in some of their properties. In that system, the cycle seems to have a period of about 10 years. But because hares sometimes cycle in the places where lynxes are largely or entirely absent, alternative explanations are needed, at least for those places. One of the most recent and attractive is that hares act as predators and the foliage upon which they browse acts as prey; lynxes, where they occur, then have their numbers affected by the hare-foliage cycle. *See* POPULATION DYNAMICS.

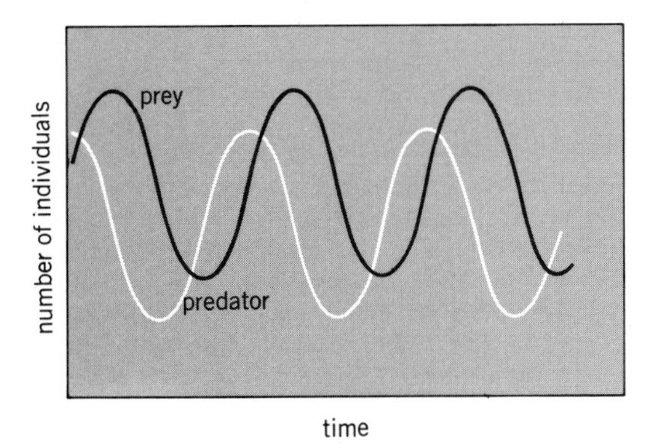

Fig. 2. Predator-prey limit cycle.

Mutualism. Mutualism occurs when two species positively affect one another. The degree to which the interaction is necessary for the survival of one or both species is sometimes emphasized; when the interaction is not necessary for either species, it is called protocooperation.

Several well-known examples of mutualistic interactions that are necessary for the survival of both species exist. Various animals, such as termites and ruminants, that consume woody plant parts are dependent on the protozoa in their guts to digest cellulose. Various bacteria live in nodules on the roots of leguminous plants such as peas and clover. These bacteria are able to extract nitrogen directly from air spaces in the soil and make it available to the plant, which otherwise would have to acquire the nitrogen from nitrates. A third example is the lichen, which is composed of two kinds of plants, algae and fungi, living in a mutualistic association. The algae can photosynthesize and so provide various organic compounds. The fungi are believed to be able to dissolve nutrients from the rock on which the lichen lives; they also provide a structural matrix. This relationship is usually but not always obligatory. *See* NITROGEN CYCLE.

A well-documented case of mutualism is the ant-acacia system of tropical America. The ants which inhabit acacia plants protect them from being eaten by herbivorous insects, and they also decrease the density of surrounding plants. In turn, the acacias have various elaborate structures that provide food (such as nectaries) or shelter (large thorns) for the ants. Depending on the species, this system exemplifies protocooperation or more obligatory mutualism.

An extremely important type of mutualism is the plant-pollinator system. While pollination is obligatory for the reproduction of many plants, frequently a variety of species can act as pollinators. One conspicuous exception occurs in a certain genus of orchids: each member, by producing a distinct fragrance, attracts a unique species of bee. Even more complicated is the interaction of yucca plants and a particular moth: the moth is the unique pollinator for the yucca, and the yucca provides the unique habitat, its ovary with developing seeds, for the development of the larval moths.

Commensalism. Commensalism occurs when one species positively affects another while being little affected itself. Several examples meet this criterion to various degrees. Plants such as bromeliads and orchids grow on other plants such as trees; the former are benefitted, while the latter are not usually seriously affected. Cattle egrets, which are insectivorous heronlike birds, have become adapted to following cattle and consuming the insects that the cattle disturb. The egrets are benefitted, and their actions seem to bother the cattle little. Certain fish *(Amphiprion percula)* are specialized to live among the poisonous tentacles of sea anemones. The anemones feed on other kinds of fish, and the *Amphiprion* feed on the leftovers from the anemones' meals. *See* ECOLOGY: ECOSYSTEM. [THOMAS W. SCHOENER]

Bibliography: S. J. McNaughton and L. L. Wolf, *General Ecology*, 1979; E. R. Pianka, *Evolutionary Ecology*, 1978; R. E. Ricklefs, *Ecology*, 1979; T. W. Schoener, Competition and the Niche, in C. Gans and D. W. Tinkle (eds.), *Biology of the Reptilia*, vol. 7, 1977.

Ecological succession

A gradual process brought about by the change in the number of individuals of each species of a community and by the establishment of new species populations which may gradually replace the original inhabitants. Succession depends upon several major factors: (1) the floristic and faunistic resources of the general region, that is, what organisms are in a position to invade a site; (2) the rate of change of the habitat and its receptivity to these potential invaders; and (3) chance factors which may influence these interactions. *See* PLANT COMMUNITY; POPULATION DISPERSAL.

Terminology. The term sere is used to describe all of the temporary communities which occur during a successional sequence on a given site. Eventually succession ends, changes in species composition of the community cease or fluctuate within some bounds, and the result is a climax community.

Successional sequences may be classified on the basis of starting habitat. Thus, hydroseres are communities in which pioneer plants invade open water, eventually forming some kind of soil such as peat or muck; xeroseres are communities on dry, sterile ground, such as rock, sand, or clay.

If an open habitat has never been occupied before, the way is opened for primary succession. If the habitat has been disturbed, leaving vestiges of soil, seeds, and other organic debris from previous occupancy, a secondary succession will begin.

Stages of succession. As an area is invaded and taken over by successive populations of plants and animals, the physiognomy of the community changes. Isolated pioneer plants eventually give way to a consolidation of the plant cover. With continued exploitation of the habitat, new species take over which have life forms able to utilize more fully the resources of the habitat than previous life forms until finally the climax community is established. Such a community can perpetuate itself indefinitely, at least in theory, because each species reproduces and maintains a stable number of individuals. *See* PLANTS, LIFE FORMS OF.

It was traditional to divide succession into pioneer, consolidation, subclimax, and climax stages on the basis of physiognomy. However, such discrete stepwise changes are not always shown by close study of the populations, because there is a continuously changing array of species. Therefore, the subdivision of a successional continuum must depend on local conditions and purposes of study. *See* APPLIED ECOLOGY; VEGETATION MANAGEMENT.

The following characteristics of ecological succession have been listed by R. Whittaher: Its nature is continuous; comparable habitats are not always settled at the same rate or by the same group of species; and certain stages may be prolonged, telescoped into one another, or omitted entirely in different habitats or in different parts of the same habitat.

Methods of study. Any unexploited or undisturbed habitat to which organisms have access is suitable for succession studies. These habitats include microhabitats, such as dead pine cones with

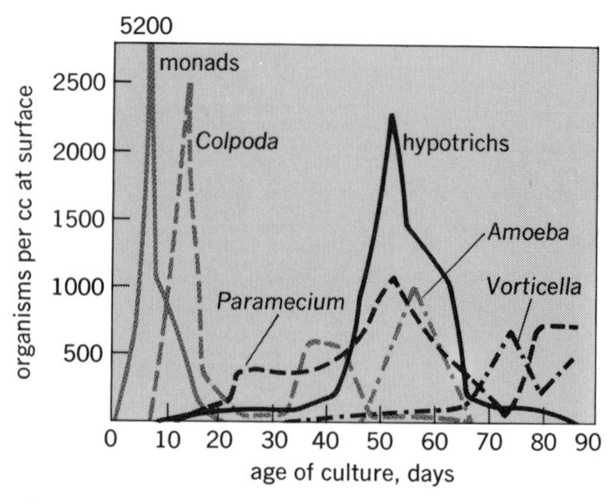

Fig. 1. Succession of protozoan populations in hay infusion. (*After Woodruff, 1912, from E. P. Odum, Fundamentals of Ecology, 2d ed., Saunders, 1959*)

changing fungal populations, dead animal carcasses with changing insect and microorganism populations, and hay infusions (Fig. 1) and other types of laboratory microcosms, and the more extensive habitats such as ponds, sections of rivers, mud flats, sand dunes, abandoned fields, lava flows, or other terrain types. *See* ECOSYSTEM.

To study the population changes occurring during natural succession, permanent sample plots may be used which have been established in the area being studied and which are observed year after year with the changes recorded. However, succession in such plots tends to slow down after a few years as more and more perennial plants become established. It is then necessary to deduce the course of succession from a comparison with similar habitats, which were invaded at earlier times and hence are further advanced in succession. This procedure introduces a number of possible errors. Environmental conditions of the compared plots may be slightly different; also, the availability of disseminules is never exactly the same in different places and at different times. Because of these difficulties, knowledge of succession is partly extrapolated from observed short-term changes and partly synthesized from isolated examples.

Studies of succession have been carried out in the laboratory in carefully controlled, small habitats known as aquarium ecosystems. These offer advantages in terms of size, types and numbers of flora and fauna, environmental control, and the time element, which is greatly telescoped. Such systems, however, are not similar to natural ones in that they are closed; that is, there is no means for constant addition of new species and there is no natural means for removing accumulated organic substances. Despite this, on the basis of such studies many useful hypotheses are being framed, and these can be tested against natural systems.

Examples of succession. The general patterns which occur during succession are illustrated by several examples.

Primary succession is exemplified by the development of forests following deglaciation in southeastern Alaska. After the ice has retreated, mosses, fireweed, and herbs pioneer on the raw, calcareous glacial till. Willows then begin to form a prostrate mat which eventually becomes an erect, dense scrub. Alder dominates the next stage, eventually forming pure thickets; these are succeeded by trees, with sitka spruce appearing first and western hemlock later. On well-drained sites the hemlocks are the climax, but wet sites may revert to more or less open muskeg.

In the southeastern United States, secondary succession has been widespread where cultivated fields were abandoned (Fig. 2). During the first summer after cultivation horseweed and crabgrass dominate the fields. In the second year, dominance abruptly changes to aster. Beginning with the third year there is a shift toward a grass called broomsedge (*Andropogon virginicus*). By the fifth year this plant may be replaced by aster or it may persist as brambles and shrubs invade. Young pines may invade fields at any stage, but they typically become large enough to shade and litter the ground only after the field reaches the broomsedge stage. The pines mature rapidly, shading out the

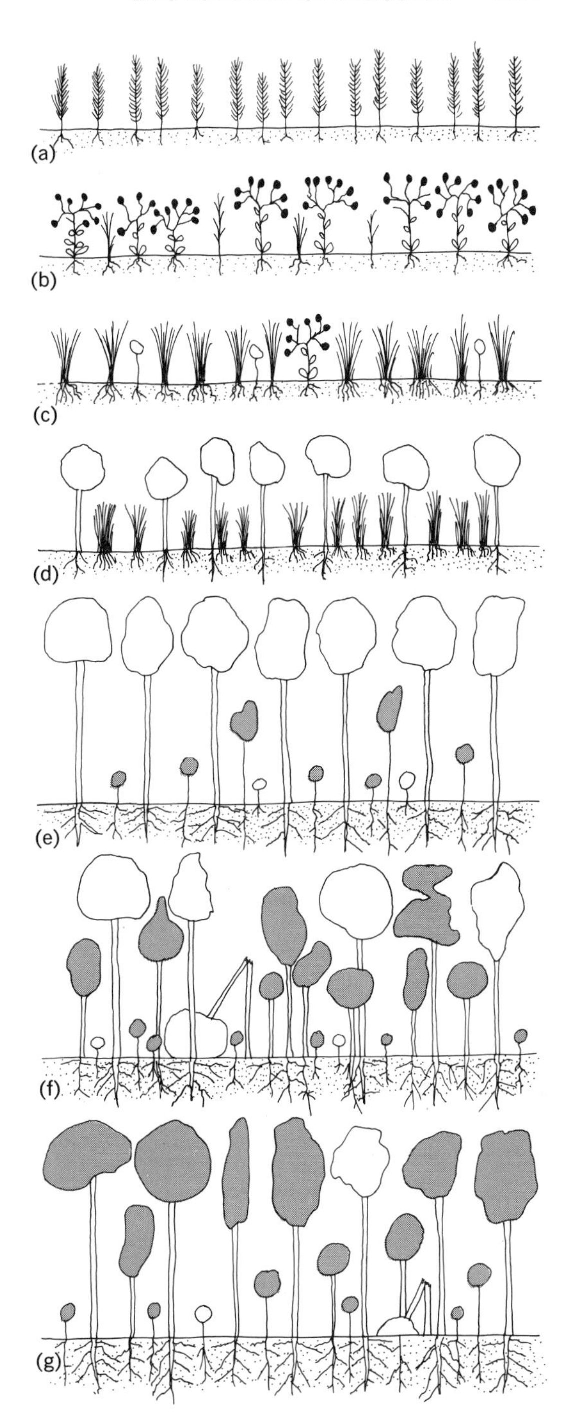

Fig. 2. Plant succession on abandoned field in North Carolina Piedmont. (*a*) First year dominated by horseweed. (*b*) Second year dominated by aster with dead horseweed. (*c*) Third to fifth year dominated by broomsedge and young pine. (*d*) Young pines overtopping broomsedge. (*e*) Pure stand of pine with hardwood seedlings after 25–30 years. (*f*) Old-growth pine being replaced by vigorous young hardwoods. (*g*) Climax hardwoods of oak and hickory.

broomsedge so that by the tenth to twelfth year what was once a field has become a young pine forest. Pines typically do not reproduce in their own shade and litter unless suitable fires and site conditions occur; but young hardwoods (oaks, hickories, and sweet gum) do become established under the pine canopy. As the pines mature and die, the young hardwoods replace them. Thus, within a period of 150–200 years (or less if pines are cut without regenerating well), an oak-hickory forest can be reestablished on the abandoned field. *See* PLANT FORMATIONS, CLIMAX.

Causes and nature of succession. Although there are variations in successions among different geographic areas and different habitats, certain general characteristics can be observed.

Many successional changes can be explained by changes in the physical environment occurring in conjunction with the changes in species composition. Some of these changes may be induced by the activities of the organisms themselves and are referred to as autogenic; that is, as a community exploits the resources of its habitat, physical and chemical changes are brought about. For example, in the sere following deglaciation, the alder stage is accompanied by a rapid increase in the amount of available soil nitrogen, due to the activity of nitrogen-fixing organisms living in nodules on the alder roots. Trees enter the community when there has been a significant increase in nitrogen in the surface soil and woody debris from pioneer trees. The resources needed by one population are altered, and conditions become favorable for the invasion of others; communities pave the way for one another in ideal autogenic succession, but its time phasing may depend on rates of aging, death, and regeneration in each "wave."

In some cases a factor not directly controlled by the organisms in the community acts to modify the environment. Such forces are called allogenic and include continued deposition of sand along riverbanks or silt on lake bottoms. Autogenic and allogenic forces operate together in most successions. *See* BIOSPHERE; FOOD CHAIN.

In some cases successional changes seem to occur without habitat change and appear to be explained by characteristics of the life cycles of the organisms involved. In the old fields of the southeastern United States, horseweed dominates first-year fields because its seed require no cold treatment and because they ripen early enough to germinate in the autumn of the year in which the field is abandoned. Aster seed, although not requiring cold treatment either, ripen too late to germinate in the autumn. Important trends may depend on chemical inhibition or on allelopathic relations.

Mosaic patterns. Succession proceeds at different rates in different parts of the habitat. New invaders, at first randomly occupying small spots, soon become centers for the establishment of clusters of some new species, or for inhibition of others. Thus a mosaic of vegetation pattern comes about which may long persist into a climax. Other patterns may develop as individual trees die and as gap phase replacement occurs in sunny and shaded sides or openings.

Fig. 3. Zonation of vegetation in the succession on open sand flat with jack pine as climax.

Zonation. Succession frequently progresses from the edge to the center of an open habitat, creating belts of different species. The belts in the center of such areas are still in the early stages of succession, while the edge has progressed to a later stage. Thus, communities which will succeed one another in time become laid out in spatial arrangements, making possible the study of successional stages across the different zones (Fig. 3). *See* VEGETATION.

Zonation, however, is not always evidence of succession, nor do the communities found in successive zones necessarily reflect the true course of succession. On a riverbank with a zonation of willows, cottonwoods, sycamores, ashes, and maples, each species occupies a slightly different site, with different degrees of flooding, drainage, and soil deposition. Hence they are not always successive stages in a sere but merely adjacent stands of trees in different habitats.

Succession-retrogression cycles. The progress of succession is often interrupted by natural or man-made disturbances which open up closed communities, clear much of the habitat, or kill off certain key species of the community. If this disturbance does not recur, succession is simply set back (retrogression) and starts again from an earlier stage. Several examples of regular alternation of succession and retrogression in a cyclic pattern have been discovered. Such is the case in raised bogs, dunes, heaths, and certain arctic areas where the permafrost level fluctuates. However, if the disturbance continues or recurs at regular or irregular intervals, the community will adjust permanently and will stabilize in a kind of disturbance-controlled climax or mosaic pattern of alternating communities.

Convergence toward the climax. A key observation about the process of succession is that, within a given climatic region, the initial stages of various seres may differ greatly in environment and species composition; but as succession proceeds, later stages bear a progressively greater resemblance to one another, first in their physiognomy and later in species composition. This convergence is largely dependent upon the development of the soil, and research on succession must be firmly based upon a study of soil in different stages of genesis.

Succession and soil development. In most primary successions the initial substrate is not differentiated into soil horizons. In the course of succession, plants take up certain minerals and deposit organic matter on the surface, thus starting the process of soil development. Products of decay may infiltrate the soil to various depths, where they are deposited and can be reutilized. If the initial substrate is rocky, plant roots and rhizoids may break it up while their stems and leaves trap dust and sand particles. Gradually a layered soil is formed. *See* SOIL.

The climax. Ecological succession eventually leads to a community, the climax, with greater stability and permanence than any of the successional stages because such a community results from the ability of the species which compose it to reproduce and exclude potential dominants under the conditions which prevail in the community.

Climate and related factors. The basic characteristics of the climax are determined by climate. In humid climates some type of forest will develop with its basic properties determined by interactions of temperature and moisture. In more arid climates savanna, grassland, or even desert may be climax. Within a given climatic region the termination of succession may be related to various specific combinations of environmental conditions, such as repeated fires, drainage patterns, and topographic factors. Thus a mosaic of several communities of similar appearance, but with different species composition, make up a regional complex of climaxes. *See* CLIMAX COMMUNITY.

It is known from the analysis of pollen grains found in lake deposits that for many thousands of years the vegetation of the Northern Hemisphere has been in continuous change as a result of increasing mildness of climate. Similar, but less notable, changes of climate have occurred during the past 100 years. These have led some ecologists to conclude that the necessary climatic stability for attainment of the climax does not exist. Such a position would lead to viewing the climax as a community in which the rate of change has slowed down to the point where it is essentially nondetectable with current methods.

Monoclimax concept. Early in the 20th century F. Clements proposed that, within a given climatic area, all seres would end in a single kind of climax community. Thus the regional climax would be identical for all seres within the area. This concept, known as the monoclimax concept, was based primarily on the observation that successions in different habitats tended to converge on a similar sort of climax community. Clements felt that over a long period of time the earlier differences in site conditions, such as soil texture and chemical composition of the parent rock, would be overcome and what were originally very different sites would then be sufficiently similar to support an identical regional climax community. Close examination of the theory showed that, although there were trends of the sort Clements proposed, certain environments never could become sufficiently similar to support the same climax community. For example, there are certain unalterable properties peculiar to each soil. A soil deficient in calcium will not improve no matter how many plants grow on it; in fact, calcium reserves as carbonate may disappear by solution. Although the texture of a soil may be slightly modified by additional organic matter, some basic properties are not changed and others may become less favorable. For these reasons convergence toward a uniform habitat with a uniform climax is never complete. Thus the monoclimax concept fell into disfavor.

Polyclimax concept. An alternative theory, the polyclimax concept, recognized that a variety of different sites might exist in an area and that each of these could potentially support a climax community. The polyclimax theory took note of four factors: (1) the effect of local relief, creating a mosaic rather than a homogeneous community cover; (2) topographic differences responsible for different topographic climaxes on various exposures in mountains; (3) soil inhibitions creating edaphic climaxes; and (4) climatic changes leaving relic communities as climaxes of the past, persisting in isolated refuges. Thus only a portion of the landscape may actually support the regional climax as viewed by Clements. Despite this, within a climatic area there generally is a complex of species which tend to be present frequently enough on similar sites so that the concept of a regional climax has a degree of validity. However, it should not be expected to develop on all sites within the area. Such regional climaxes are the units often portrayed on maps of community types of large sections of continents (Fig. 4).

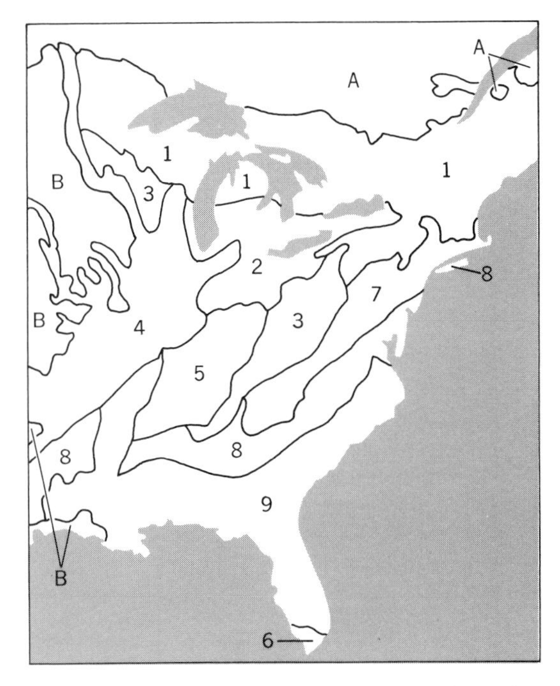

Fig. 4. Climax regions within the deciduous forest formation: 1, hemlock, northern hardwoods region; 2, beech-maple region; 3, maple-basswood region; 4, oak-hickory region; 5, western mesophytic region; 6, mixed mesophytic region; 7, oak-chestnut region; 8, oak-pine region; 9, southeastern evergreen forest region; A, boreal needleleaf forest formation; B, the grassland formation. *(Redrawn after E. L. Braun, 1950)*

Trends in succession. E. P. Odum has outlined a series of trends which summarizes important functional and compositional changes during succession. Species composition changes rapidly during the early stages of a sere and more slowly as the climax is approached. The total weight of living and nonliving organic matter increases through early succession but it may decrease later. Feeding relationships, generally simple at first, become more complex in some climax and many mixed stages. Early in some stages of succession a greater amount of organic matter tends to be produced by organisms than is used by them as food; consequently, organic matter accumulates. As succession proceeds, however, the number or activity (or both) of heterotrophic forms increases, causing an increase in the total amount of community respiration. Thus community production exceeds community respiration in some pioneer or transitional stages, and the two processes are theoretically balanced (averaging over space and tine) in a climax community *See* BIOMASS; PHYSIOLOGICAL ECOLOGY (PLANT).

Importance of succession. An understanding of succession is important for effective management of communities by man.

Man's activities generally have the effect of stopping succession at an early stage. Fire was used by most primitive people to keep forests suitable for hunting by clearing away the underbrush or to clear land for agriculture. Since almost all major crop plants are annuals of the pioneer stage, regular plowing is needed to hold down succession. Other methods of checking succession include mowing and grazing, which favor sod-forming grasses and kill most tree seedlings, and selective weed control with chemicals. *See* LAND-USE PLANNING.

Succession can be either retarded or accelerated by manipulation of drainage. Such manipulation may be used to establish a cranberry bog or a stand of productive timber. Many timber trees occur naturally as successional species rather than in climaxes. Thus the forest must be managed so as to permit these species to grow free from the competition of the climax species which will eventually replace them. Fire, herbicides, and cutting practices are all useful in regulating the growth and composition of forest stands so that early successional stages are maintained. *See* CONSERVATION OF RESOURCES; ECOLOGY; FOREST MANAGEMENT AND ORGANIZATION; VEGETATION MANAGEMENT; VEGETATION ZONES, WORLD.

[ARTHUR W. COOPER]

Bibliography: B. W. Allred and E. S. Clements (eds.), *Dynamics of Vegetation*, 1949; E. L. Braun, *Deciduous Forests of Eastern North America*, 1950; F. E. Clements, *Plant Succession and Indicators*, (reprint) 1973; R. Daubenmire, *Plant Communities*, 1968; K. A. Kershaw, *Quantitative and Dynamic Plant Ecology*, 2d ed., 1974; E. P. Odum, *Fundamentals of Ecology*, 3d ed., 1971; R. L. Smith, *Ecology and Field Biology*, 2d ed., 1974.

Ecological systems, energy in

Energy is defined as equivalent to work, having the dimensions L^2M/T^2 and being measured in ergs. It occurs as potential and as kinetic energy in static and dynamic systems, respectively. Its liberation results in its appearance as heat, light, mechanical work, chemical change, or electric current. All forms of energy are potentially interconvertible and subject to the laws of thermodynamics. These laws state that energy is neither created nor destroyed and that such conversions occur spontaneously only when the result is to produce energy in a more dispersed form than before. This is equivalent to saying that the overall result of all energy changes is to increase the entropy of the system as a whole. Entropy is a measure of uniformity, increased probability, or lack of organization.

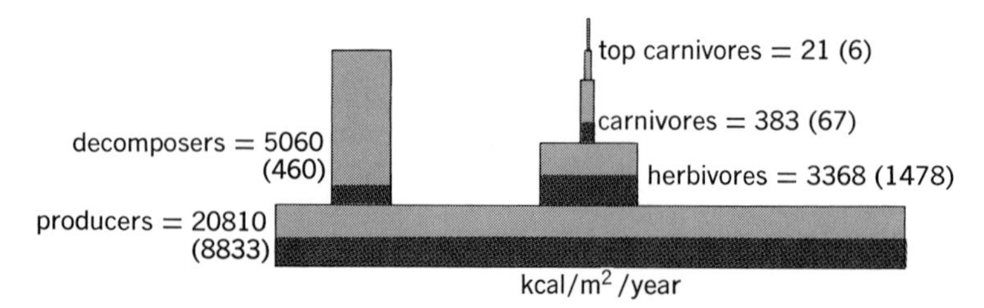

Energy pyramid for Silver Springs, Fla., on annual basis. Portion of total energy flow actually fixed as organic biomass and potentially available as food for other populations in next trophic level is shown by figures in parentheses and by the darker portion of each tier. About one-half the total assimilation ends up in bodies of organisms (the rest is lost as heat or export), except for decomposers level, where percentage loss in respiration is much greater. (*After H. T. Odum, in E. P. Odum, Fundamentals of Ecology, 2d ed., Saunders, 1959*)

Ecological systems are highly improbable ones in which entropy production of the universe as a whole is used to create local pockets of concentrated energy—the organisms and their stored foods. The accumulation of energy in this way is another aspect of the accumulation of information; the techniques of information theory can be applied to the study of energy flow in ecosystems (B. Patten).

Measurement of energy flow. Because energy liberation and oxidation of food are quantitatively equivalent, the rates of energy transfer through an organism, heat liberation, oxygen consumption, and carbon dioxide production are related and can all be used to measure energy flow, provided that the type of food (carbohydrate, fat, or protein) is known. When it is not, the respiratory quotient (RQ) must be determined from any two of these quantities.

Metabolism and size of organism. Metabolism is approximately proportional to the 2/3 power of weight, so that equal weights of small organisms respire faster than larger ones. However, increases of specific metabolic weight have occurred several times during evolution so that, for instance, worms, insects, and mammals of equal sizes respire at very different rates. One result of this is that biomass is a poor indicator of the energy flow capacity of organisms. *See* BIOMASS.

Path of energy flow. Because energy enters ecological systems only through photosynthesis, and because the energy is lost to the system after it has been liberated by oxidation of food, energy measurement has a unique importance in the quantitative study of food chains. The partition of energy flow can be illustrated by energy flow diagrams and energy pyramids (see illustration). Energy flow diagrams show the sequential fate of energy contained in food whereas energy pyramids, which were introduced by E. P. Odum, demonstrate the quantitative division of metabolism between the trophic levels and the relation between biomass and energy. *See* BIOLOGICAL PRODUCTIVITY; FOOD CHAIN.

[AMYAN MACFADYEN]

Bibliography: S. Brody, *Bioenergetics and Growth*, (reprint) 1964; M. Gabel, *Earth, Energy, and Everyone*, 1979; K. A. Kershaw, *Quantitative and Dynamic Ecology*, 2d ed., 1974; E. P. Odum, *Ecology*, 2d ed., 1975; E. P. Odum, *Fundamentals of Ecology*, 3d ed., 1971.

Ecology

A study of the relation of organisms to their environment, or in more simple terms, environmental biology. Ecology is concerned especially with the biology of groups of organisms and with functional processes on the lands, in the oceans, and in fresh waters. Ecology is the study of the structure and function of nature (mankind being considered part of nature).

Ecology is one of the basic divisions of biology which are concerned with principles or fundmentals common to all life. Since biology may also be divided into taxonomic divisions which deal with specific kinds of organisms, ecology is not only a basic division of biology but also an integral part of any and all of the taxonomic divisions. Both approaches are profitable. If it often productive to restrict certain studies to taxonomic groups, insects, for example, because different kinds of organisms require different methods of study. (Some groups of organisms are of greater interest to man than others.) It is also important to seek and to test unifying principles which may be applicable to nature as a whole. *See* ANIMAL COMMUNITY; PLANT COMMUNITY.

Approaches to ecology. From the standpoint of general ecology, an instructive subdivision is in terms of the concept of level of organization. For convenience, a biological spectrum can be visualized as follows: protoplasm, cells, tissues, organs, organ systems, organisms, populations, communities, and ecosystems. Ecology is concerned largely with the levels beyond that of the individual organism. In ecology the term of population, originally used to denote a group of people, is broadened to include groups of individuals of any one species of organism. Community in the ecological sense, sometimes designated as biotic community, includes all of the species populations of a given situation. The community or individual and the nonliving environment function together as an ecological system or ecosystem. The Earth as an ecosystem is conveniently designated as the biosphere. The term autecology is often used to refer to the study of environmental relations of individuals or species, whereas the term synecology refers to the study of groups of organisms such as communities. Mathematical concepts have permeated the basic foundations of ecology and have provided new means of exploring qualitative and quantitative relations at all levels. This new brand of ecology is known as systems ecology. *See* BIOSPHERE; COMMUNITY; POPULATION DYNAMICS.

For convenience, subdivisions of ecology may also be based on the kind of environment or habitat to be considered or studied. Marine ecology, fresh-water ecology, and terrestrial ecology are the three broad divisions from this point of view; estuarine ecology, stream ecology, or grassland ecology represent more restricted interests. *See* FRESH-WATER ECOSYSTEM; MARINE ECOSYSTEM; TERRESTRIAL ECOSYSTEM.

Some attributes, obviously, become more complex and variable during the procession from cells to ecosystems; however, it is often overlooked that other attributes may become less complex and less variable from the small to the large unit. The reason is that homeostatic mechanisms, which may be defined as checks and balances or forces and califorces, operate all along the line to produce a certain amount of integration, and smaller units function within larger units. For example, the rate of photosynthesis of a whole forest or a whole cornfield may be less variable than that of the individual trees or corn plants within the communities, because when one individual or species slows down, another may speed up in a compensatory manner. *See* HOMEOSTASIS.

Furthermore, it is important that findings at any one level aid in the study of another level, but never completely explain the phenomena occurring at that level. The old saying that the forest is more than a collection of trees will illustrate very well what is meant here. For a full understanding of the forest it is necessary to study both the trees as separate units and also the forest as a whole. During recent years ecology has made the most progress, not from intensive study at one level, but by coordinated, simultaneous attack at all levels.

Coral reef study. The importance of studying both the part and the whole may be illustrated by the following example. A coral reef represents one of the most beautiful and best-adapted ecosystems. Corals are small animals that have tentacles adapted to seize small animals called zooplankton which are in the water. Embedded in the tissues of the coral animal and in the calcareous skeleton are numerous small plants or algae.

Some years ago C. M. Yonge carried out a carefully planned series of experiments with isolated coral colonies in tanks in an effort to clarify the relationship between corals and their contained algae. He found that when the corals were supplied with abundant zooplankton they thrived and grew normally, if all of the algae were killed by keeping the colonies in the dark. On the basis of these experiments Yonge concluded that the algae do not contribute to the well-being of the coral, and therefore are of no importance in the reef-building activities of corals. Some years later H. T. Odum and E. P. Odum were able to measure the metabolism of an intact coral reef and thus determine the amounts of food which corals require. It was soon evident that there were not enough zooplankton in the surrounding infertile oceans to account for the large population and rapid growth of the coral reef. It was suggested, therefore, that the corals must indeed obtain some of their food from the algae. Other investigators became interested in the problem, and L. Muscatine and C. Hand (1958) showed by the use of carbon-14 isotopic tracer that food does indeed diffuse from the algae, which manufacture food by photosynthesis, to the coelenterate. Still other investigators have shown that activities of the symbiotic algae are important in skeleton formation. Thus it was proved that what was true of an isolated colony in a tank was not entirely true for the coral living in its natural ecosystem.

Functional ecology. Returning to the definition of ecology as the study of structure and function of nature, it should be noted that until fairly recently ecologists often described how nature looked. Now equal emphasis is being placed on what nature does, because the changing face of nature can never be understood unless her metabolism is also studied. Consideration of function brings the small organisms, which may be inconspicuous but very active, into true perspective with the large organisms, which may be conspicuous but relatively inactive. As long as a purely descriptive viewpoint was maintained, there seemed little in common between such structurally diverse organisms as trees, birds, and bacteria. In real life, however, all these are intimately linked functionally in ecological systems according to well-defined laws. Likewise from a descriptive standpoint, a forest and a pond seem to have little in common, yet both function according to exactly the same principles. Thus general ecology is essentially an ecology of function.

THE ECOSYSTEM

Any area or volume of nature that includes living organisms and nonliving (abiotic) substances interacting so that a flow of energy leads to characteristic trophic structure and material cycles. *See* ECOSYSTEM.

Biotic components. From a functional standpoint, an ecosystem has two biotic components: an autotrophic component able to fix light energy and manufacture food from simple inorganic substances, and a heterotrophic component which utilizes, rearranges, and decomposes the complex materials synthesized by the autotrophs. From a structural standpoint it is convenient to recognize four constituents composing the ecosystem, as shown in Fig. 1: (I) abiotic substances, basic elements and compounds of the environment; (II) producers, the autotrophic organisms, largely the green plants such as phytoplankton algae or terrestrial vegetation; (III) the large consumers or macroconsumers (also called biophages), heterotrophic organisms, chiefly animals, which ingest other organisms or particulate organic matter; and (IV) the decomposers or microconsumers (also called saprophages or saprophytes), heterotrophic organisms, including the bacteria and fungi, which break down the complex compounds of dead protoplasm, absorb some of the decomposition products, and release simple substances usable by the producers. Microorganisms also release growth-promoting and growth-inhibiting substances that are important in regulating the metabolism of the ecosystem as a whole. As shown in Fig. 1, a pond is a good example of an ecosystem which exhibits a recognizable unity both in regard to function and structure.

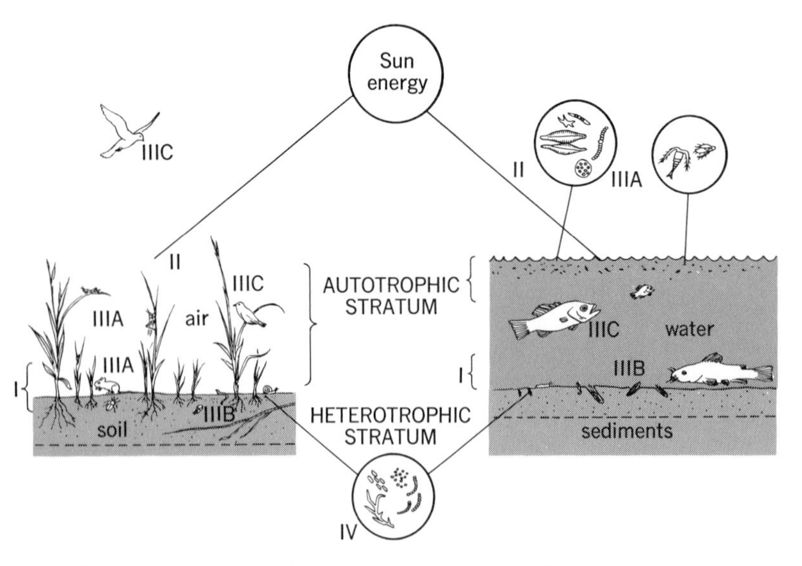

Fig. 1. Comparison of trophic structure of a simple terrestrial ecosystem (grassland) with an open-water aquatic ecosystem (either fresh water or marine). Basic units of the ecosystem are I, abiotic substances; II, producers; III, macroconsumers (A= direct herbivores, B= detritus eaters or saprovores, C= carnivores); and IV, decomposers (bacteria and fungi of decay).

The stratification of autotrophic and heterotrophic functions and basic trophic levels is the same for land and water systems, but the kinds and relative sizes of the organisms are different. On land, the individual producer organisms tend to be relatively large in relation to consumers, whereas in water the producers are often microscopic in size. Although the biomass of the land vegetation per unit of area may be much greater than that of the biomass of aquatic phytoplankton, photosynthesis of the latter can be as great, if light and nutrients are equivalent. Marshes, ponds, and shallow margins of lakes and the oceans have a trophic structure intermediate between these two systems, because both large and small producers may be present.

The concept of the ecosystem is and should be a broad one, its main function in ecological thought being to emphasize obligatory relationships, interdependences, and casual relationships. Ecosystems may be conceived and studied in various sizes. A small pond, a large lake, a tract of forest, or even a small aquarium may provide a convenient unit of study. As long as the major components are present and operate together the entity may be considered an ecosystem. Throughout the entire biosphere, ecosystems have a very similar functional makeup, though they may differ markedly in structural features and degree of stability.

Very frequently the basic functions and the organisms responsible for the basic processes in an ecosystem are partially separated in space and also in time. The autotrophic and heterotrophic components are often stratified one above the other, and there is often a considerable delay in the heterotrophic utilization of the products of the autotrophic organisms. For example, photosynthesis predominates in the aboveground portion of a forest ecosystem. Only a part, often very small, of the food manufactured in the upper layers of the forest is immediately and directly used by the plants and by herbivores and parasites which feed on foliage and new wood. Consumption of much of the synthesized food material in the form of leaves, wood, and stored food in seeds and roots is delayed until it eventually falls to the ground and becomes part of the litter and soil, which constitute a well-defined heterotrophic stratum. A similar functional stratification and time lag may be observed in the pond ecosystem (Fig. 1); the phytoplankton comprise the well-defined autotrophic layer, whereas the sediments constitute the heterotrophic system.

The ecological categories which have been discussed above are ones of function rather than of species. There are no hard and fast lines between such categories as producers, consumers, and decomposers, because some species of organisms occupy intermediate positions in the series and others are able to shift their mode of nutrition according to environmental circumstances. For example, certain species of algae and bacteria are able to function either as autotrophs under certain conditions, or at least partly as heterotrophs under other conditions. The separation of heterotrophs into large consumers and small decomposers is arbitrary. Many decomposers (bacteria and fungi) are relatively immobile (usually embedded in the medium being decomposed) and are very small with high rates of metabolism and turnover. They obtain their energy by heterotrophic absorption of decomposition products. Their specialization is more evident biochemically than morphologically; consequently, their role in the ecosystem cannot be determined by such direct methods as looking at them or counting their numbers. Rather, their actual functions must be measured. Macroconsumers obtain their energy by heterotrophic ingestion of particulate organic matter. These are largely the animals in the broad sense. In contrast to the decomposers the macroconsumers are larger, have slower rates of metabolism, and are more readily studied by direct means. They tend to be morphologically adapted for active food seeking or food gathering, with the development in higher forms of complex sensory and neuromotor as well as digestive, respiratory, and circulatory systems. Much may be inferred about the functioning of the macroconsumer groups by observing them and counting their numbers. Even in this case it is necessary to devise means of assaying rate of functions to fully understand their role in the ecosystem.

A sizable ecosystem may contain only producers and microorganism decomposers (pioneer aquatic communities, for example). Almost everywhere on Earth, however, the macroconsumers or animals invade sooner or later and play a prominent role in the functioning of the ecosystem. It is convenient to consider soil, fallen logs, or deep-sea basins as ecosystems (because they show consistent structural and functional characteristics), provided it is recognized that these systems consist only of the heterotrophic components and are therefore ecologically incomplete, as subsystems of a larger system.

Biotic influence on environment. The physical environment controls the organisms, and organisms also influence and control the abiotic environment in many ways. Organisms return new compounds and isotopes to the nonliving environment; for example, the chemical content of sea water and of air is largely determined by the action of organisms. Plants growing on a sand dune build up a soil radically different from the original substrate. Thus the biosphere is important not only as a place in which living organisms can exist, but also as a region in which the incoming radiation energy of the Sun brings about fundamental chemical and physical changes in the inert material of the Earth, chiefly through the functioning of various ecosystems.

Homeostatic mechanisms. Some equilibriums between organisms and environment may be maintained by a balance of nature which tends to resist change in the ecosystem as a whole. Much has been written about this, but the fundamentals involved are not yet clearly understood. These homeostatic mechanisms include those which regulate the storage and release of nutrients, as well as those which regulate the growth of organisms and the production and decomposition of organic substances. Many organisms, particularly decomposer and producer groups, release organic substances into the environment during their growth processes. These substances often have a profound influence on other organisms and on the regulation of function of the whole ecosystem. Some substances are antibiotic in that they inhibit the growth of other organisms, whereas other substances may be stimulatory, as, for example, various vitamins and other growth-promoting substances. Such external metabolites may be considered to be ectocrines or environmental hormones, in that there is growing evidence that many such substances influence and control the functioning of the ecosystem in the same general manner that endocrines control metabolic rates within individual organisms. Much needs to be learned about the specific action of these substances.

As a result of the evolution of the central nervous system and brain, man has gradually become a most powerful organism, as far as the ability to modify the ecosystem is concerned. Man's power to change and control, unfortunately, seems to be increasing faster than his realization and understanding of the profound changes of which he is capable. Although nature has remarkable resilience, the limits of homeostatic mechanisms can easily be exceeded by the actions of man. When treated sewage is introduced into a stream at a moderate rate, for example, the system is able to purify itself and return to the previous condition within a comparatively few miles downstream. However, if the pollution is great or if toxic substances for which no natural homeostatic mechanisms have been evolved are included, the stream may be permanently altered or even destroyed as far as usefulness to man is concerned. Consequently, the concept of the ecosystem and the realization that mankind is, and must always be, part of an ecosystem and that he has increasing power to modify these systems are concepts basic to modern ecology. These also are points of view of extreme importance to human affairs generally. Conservation of natural resources, one of the most important practical applications of ecology, must be built around these points of view. Thus if understanding of ecological systems and moral responsibility among mankind can keep pace with man's power to effect changes, the old practice of unlimited exploitation of resources will give way to unlimited ingenuity in perpetuating a cyclic abundance of resources. *See* APPLIED ECOLOGY; CONSERVATION OF RESOURCES; WATER POLLUTION.

ENERGY FLOW

Everyone is familiar with the fact that the kinds of organisms to be found in any particular part of the world depend on the local conditions of existence and on the geography, because each major region of the Earth, especially if isolated from other regions, tends to have its own special flora and fauna. Often, however, ecologically similar or ecologically equivalent species have evolved in different parts of the globe where the basic environment is similar. The species of grasses in the temperate, semiarid part of Australia are largely different from those of a similar climatic region in North America, but they perform the same basic function as producers in the ecosystem. Likewise, the grazing kangaroos of the Australian grasslands are ecological equivalents of the grazing bison, or the cattle which have replaced them, on North American grasslands. The kangaroo and bison, although not closely related taxonomically, occupy the same ecological niche in the sense that they have a similar functional position in the ecosystem in a similar type of habitat. It is also true that the same species may function differently; that is, they may occupy different niches, in different habitats.

The point to emphasize is that a list of species to be found in an area is not sufficient information in itself to determine how the biotic community works. For a full understanding of nature, rate functions must also be investigated. *See* BIOGEOGRAPHY; BIOTIC ISOLATION; GRASSLAND ECOSYSTEM; PLANT FORMATIONS, CLIMAX; POPULATION DISPERSAL; VEGETATION ZONES, WORLD.

Energy and materials. In any ecosystem the number of organisms and the rate at which they live depend on the rate at which energy flows through the system, and the rate at which materials circulate within the system and are exchanged with adajcent systems, or both. Materials circulate, but energy does not. Nitrogen, carbon, water, and other materials of which living organisms are composed may circulate many times between living and nonliving entities; that is to say, the atoms may be used over and over again. On the other hand, free energy is used once by a given organism or population and then is converted into heat and lost from the ecosystem. Life is kept going by the continuous inflow of solar energy from the outside. The interaction of energy and materials in the ecosystem is of primary concern to ecologists. The one-way flow of energy and the circulation of materials are the two great principles or laws of general ecology, because these principles are equally applicable to any type of environment and any type of organism, including man. *See* HUMAN ECOLOGY.

Energy flow diagrams. A simplified energy-flow diagram which might, in principle, be applied to any ecosystem is shown in Fig. 2. The "boxes" represent the population mass or biomass; the "pipes" depict the flow of energy between the living units. Only about one-half the average sunlight impinging upon green plants (producers) is absorbed by the photosynthetic machinery, and only a small portion of absorbed energy, about 1–5% in productive vegetation, is converted into food energy. The assimilation rate of producers in an ecosystem is designated as primary production or primary productivity. Gross primary production (P_G in Fig. 2) is the total amount of organic matter fixed, including that used up by plant respiration during the measurement period; net primary production is organic matter stored in plant tissues in excess of plant respiration during the period of measurement. Net production represents food potentially available to heterotrophs. In Fig. 2, net primary production is represented by the flow P which leaves the producer component. When plants are growing rapidly under favorable light and temperature conditions, plant respiration may account for as little as 10% of gross production. However, under most conditions in nature net production is much less than 90% of gross. *See* BIOLOGICAL PRODUCTIVITY; BIOMASS.

Energy transfer. The transfer of food energy from the source in plants through a series of organisms with repeated eating and being eaten is known as the food chain or web. In complex natural communities, organisms whose food is obtained from plants by the same number of steps are said to belong to the same trophic level. Thus green plants occupy the first trophic level, that is, the producer level; ideally, plant-eaters (herbivores), the second level; carnivores which eat the herbivores, the third level; and secondary carnivores, a fourth level. This trophic classification is one of function and not of species as such; a given species population may occupy one trophic level or more, according to the source of energy actually assimilated, and this may change with time.

Energy degradation. At each transfer of energy from one organism to another, or from one trophic level to another, a large part of the energy is degraded into heat as required by the second law of thermodynamics. The shorter the food chain or the nearer the organism to the beginning of the food chain, the greater will be the available food energy. As shown in Fig. 2, the energy flows are greatly reduced with each successive trophic level, whether considering the total flow or the production P or respiration R components. The reduction with each link in the food chain is about one order of magnitude. Thus, for every 100 cal of net plant production in a stable community about 10 kcal would probably be reconstituted into primary con-

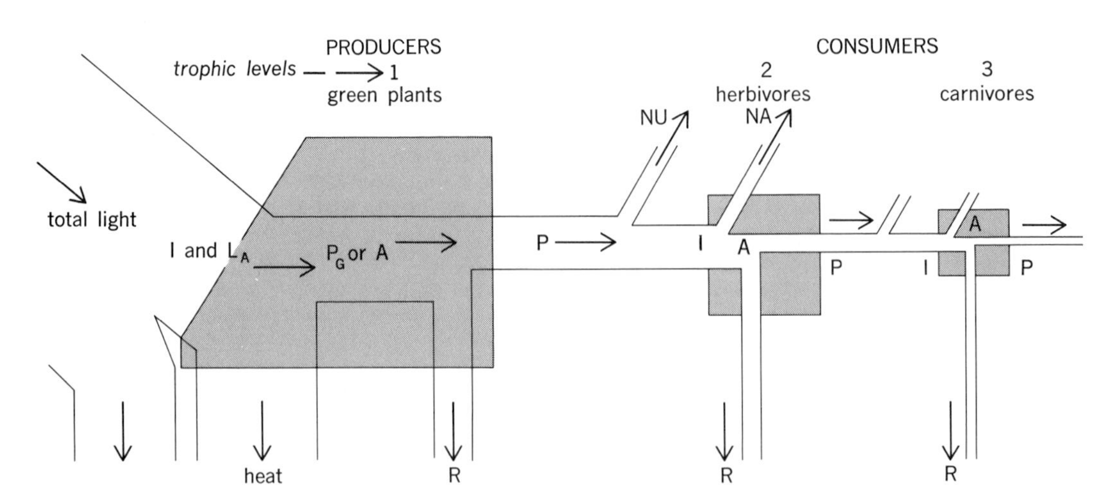

Fig. 2. A simplified energy-flow diagram of an ecosystem. L_A, absorbed light; P_G, total photosynthesis (gross production); *P*, net primary production; *I*, energy intake; *R*, respiration; *A*, assimilation; *NA*, ingested but not assimilated; and *NU*, not used by trophic level shown in diagram. (*After E. P. Odum, Fundamentals of Ecology, 3d ed., Saunders, 1971*)

sumers, the P/P_{T-1} efficiency ratio shown in Fig. 2, which then becomes available to carnivores, which in turn might convert 1 kcal of the 10 into new organic materials. Although a small amount of the primary food energy might be involved in a number of transfers, it is evident that the amount of food remaining after two or three successive transfers is so small that few organisms could be supported if they had to depend entirely on food available at the end of a long food chain. For all practical purposes, then, the food webs are limited to three or four links along most pathways.

Ecological efficiencies. Ecological efficiencies may be expressed by formulas which indicate the ratios between and within trophic levels:

$$E_I = \frac{I_T}{I_{T-1}} \quad \text{or} \quad \frac{I_2}{I_1} \qquad E_E = \frac{A_T}{A_{T-1}}$$

$$E_P = \frac{P_T}{P_{T-1}} \qquad E_U = \frac{I_T}{P_{T-1}} \quad \text{and} \quad \frac{A_T}{P_{T-1}}$$

Intake, growth, production, and utilization efficiencies between trophic levels are expressed by the above equations. The Lindemann efficiency or the efficiency of trophic level intake E_I is expressed as the ratio between two successive trophic levels; the subscripts indicate the trophic levels being compared. Growth efficiency E_E is the ratio between the assimilation at one trophic level with that of the preceding trophic level. Trophic level production E_P is the ratio of production rates at different trophic levels. The efficiency of utilization E_U is the ratio of intake at one level to the production rate at the preceding level, or it may be expressed as the assimilation rate A_T to the production rate P_{T-1}. Ecological efficiencies within trophic levels are measured as tissue-growth and assimilation efficiencies. Tissue-growth efficiency E_G is the ratio between the production and assimilation rates at the same trophic level or between production and intake rates. Assimilation efficiency E_A is expressed as a ratio of the rate of assimilation to the energy intake or consumption:

$$E_G = \frac{P_T}{A_T} \quad \text{or} \quad \frac{P_T}{I_T} \qquad E_A = \frac{A_T}{I_T}$$

Energy-flow models. In the simplified diagram of Fig. 2, bacteria and fungi which decompose plant tissues and stored plant food would be placed in the primary consumer "box" along with herbivorous animals; likewise, microorganisms decomposing animal remains would go along with the carnivores. However, because there is usually a considerable time lag between direct consumption of living plants and animals and the ultimate utilization of dead organic matter, not to mention the metabolic differences between animals and microorganisms, a more realistic energy-flow model is obtained if the decomposers are placed in a separate box connected by appropriate energy flows to the other components. For example, the *NU* (not utilized) and *NA* (not assimilated) flow components, if not exported from the system, would ultimately be utilized by decay organisms, which in turn might supply at least part of the food used by such macroconsumers as detritus-feeding or filter-feeding animals. Organic excretions, sugars, amino acids, and other organic materials which leak out of organisms into the environment may be considered to be a part of production, because these materials are ultimately consumed by microorganisms aggregated in bubbles or particles and possibly returned through consumer food chains.

Standing crop and energy flow. The boxes in Fig. 2 represent the energy in the dry biomass of organisms functioning at the trophic level indicated. The number of organisms per unit area at any one time, or the average quantity over a period of time, may be informally designated as the standing crop. The relationship between the "boxes" and the "pipes," that is, between standing-crop energy and the energy flows *P*, *A*, or *I*, is of great interest and importance. The energy flow must always decrease with each successive trophic level. Likewise, in many situations the energy in biomass of standing crop also decreases. However, standing-crop biomass is much influenced by the size of the individual organisms making up the trophic group in question. In general, the smaller the organism, the greater will be the rate of metabolism per gram of weight (inverse size – metabolism law). Thus, 1 g of bacteria may have an energy flow equal to many grams of cow, or 1 g of small algae may be equal in metabolism to many grams of tree leaves. Consequently, if the producers of an ecosystem are composed largely of very small organisms and the consumers are large, then the standing-crop biomass of consumers may be greater than that of the producers; of course, the energy flow through the latter must average greater, assuming that food used by consumers is not being imported from another ecosystem. Such a situation often exists in marine environments of moderate depths because bottom invertebrate consumers (clams, crustacea, and echinoderms) and fish often outweigh the phytoplankton (microscopic floating algae) on which they depend. By harvesting at frequent intervals, man, as well as the clam, may obtain as much food (net production) from mass cultures of small algae as he obtains from a grain crop which is harvested after a long interval of time. However, the standing crop of algae at any one time would be much less than that of a mature grain crop. To summarize, standing-crop biomass is usually expressed in terms of grams of organic matter, grams of carbon, or kilocalories per unit of area or volume. Productivity is expressed as grams or calories per unit of time. As indicated by the above examples, these two quantities should not be confused; the relationship between the two depends on the kind of organisms involved.

Gross production and respiration. The relationship between gross production P_G and total community respiration (the sum of all the *R*s in Fig. 2) is important in the understanding of the total function of the ecosystem and in predicting future events. One kind of ecological climax or steady state exists if the annual production of organic matter equals total consumption, that is, $P/R = 1$, and if exports and imports of organic matter are either nonexistent (as in a self-sufficient climax) or equal. In a mature tropical rainforest the balance may be almost a day-by-day affair, whereas in mature temperate forests an autotrophic regime in spring and summer is balanced by a heterotrophic regime in autumn and winter. Another type of steady state exists if gross production plus imports equals total respiration, as in some types of stream ecosystems, or if gross production equals respiration plus exports, as in stable agriculture. *See* ECOLOGICAL SYSTEMS, ENERGY IN.

Species, individuals, and energy flow. An important and little-known area of ecology deals with the relationship between the number of species, the number of individuals, and the energy flow. Most natural communities contain a few species in each trophic group which are abundant and account for most of the energy flow. Usually, however, the few common species are associated with a large number of rare species. The number of species relative to the number of individuals (species diversity) often increases with the maturity of stability of the ecosystem. Increasing diversity is not necessarily accompanied by increasing total productivity (in fact, the reverse may be the case). One advantage of diversity, for the survival value to the community, lies in increased protection against perturbations in physical environment (drought stress, for example). The more species present, the greater are the possibilities for adaptation of the given type of ecosystem to changing conditions, whether these be short-term or long-term changes in climate or other factors.

Primary production in the world. The world distribution of primary production is shown schematically in Fig. 3. Values represent the average gross production rate, in grams of dry organic matter per square meter per day, to be expected over an annual cycle. For an estimate of total annual production, multiply by 365. To visualize these values in terms of approximate kilocalories of potential food, multiply by 5. Much less than 90% of gross production may be available to heterotrophs. Man or any other single species cannot assimilate all energy fixed as dry matter by plants. For example, corn stalks and wheat stubble and roots would be included in the total production of these crops, but only the grain is currently consumed by man. Figure 3 shows about three orders of magnitude in potential biological fertility of the world: (1) some parts of the open oceans and land deserts, which range around 0.1 g/(m^2) (day) or less; (2) semiarid grasslands, coastal seas, shallow lakes, and ordinary agriculture, which range between 1 and 10; (3) certain shallow-water systems such as estuaries, coral reefs, and mineral springs, together with moist forests, intensive agriculture (such as year-round culture of sugarcane or cropping on irrigated deserts), and natural communities on alluvial plains, which may range from 10 to 20 g/(m^2) (day). Production rates higher than 20 have been reported for experimental crops, polluted waters, and limited natural communities, but these values are based on short-term measurements; values higher than 25 have not been obtained for extensive areas over long periods of time.

Productivity controls. Two tentative generalizations may be made from the data at hand. First, basic primary productivity is not necessarily a function of the kind of producer organism or of the kind of medium (whether air, fresh water, or salt water), but is controlled by local supply of raw materials and solar energy, and the ability of local communities as a whole (including man) to utilize and regenerate materials for continuous reuse. Terrestrial systems are not inherently different from aquatic situations if light, water, and nutrient conditions are similar. However, large bodies of water are at a disadvantage because a large portion of light energy may be absorbed by the water before it reaches the site of photosynthesis. Second, a very large portion of the Earth's surface is open ocean or arid and semiarid land and thus is in the low-production category, because of lack of nutrients in the former and lack of water in the latter. Many deserts can be irrigated successfully and it is theoretically possible, and perhaps feasible in the future, to bring up nutrients from the bottom of the sea and thus greatly increase production. Such an upwelling occurs naturally in some coastal areas which have a productivity many times that of the average ocean.

Efficiency limits. It now seems clear that there is a rather definite upper limit to the efficiency with which light may be converted into organic matter on any large scale; this maximum has apparently been achieved by some natural communities (coral reefs, for example) as well as by the most efficient agriculture. In the former, of course, production is consumed by a large variety of organisms, whereas in the latter a large portion of the net is temporarily stored and then harvested by man. Average agriculture is far below the maximum; world average grain production, for example, is 2 g/(m^2) (day). The best immediate possibilities for increasing

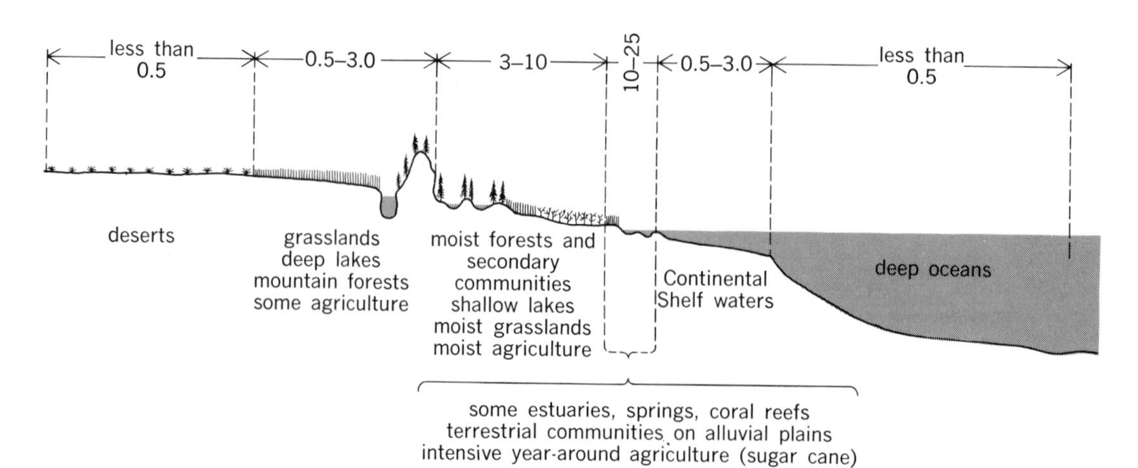

Fig. 3. World distribution of primary production showing gross production rate (grams of dry matter per square meter per day) expected. (*After E. P. Odum, Fundamentals of Ecology, 3d ed., Saunders, 1971*)

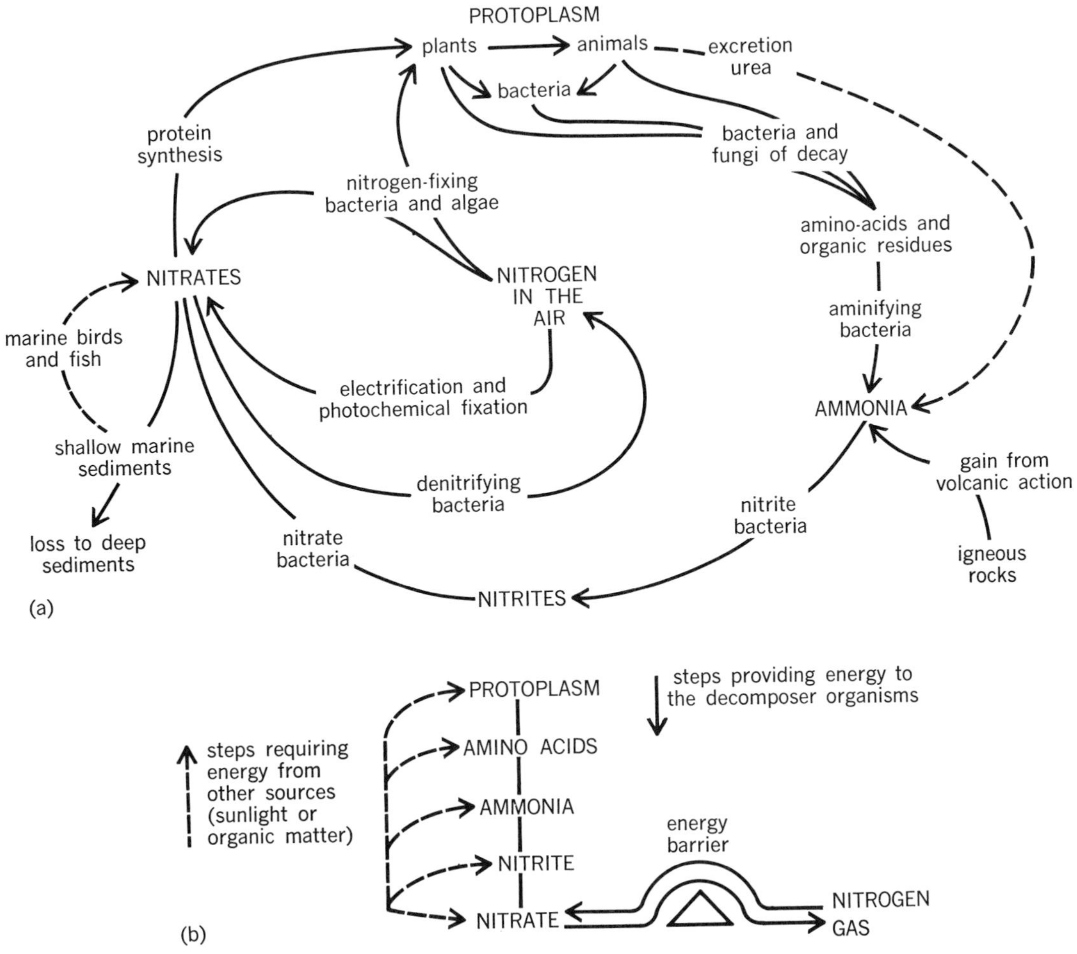

Fig. 4. Nitrogen biogeochemical cycle. (*a*) Circulation of nitrogen between organisms and environment, with microorganisms responsible for key steps. (*b*) Same basic steps, with the high-energy forms on top to distinguish steps which require energy from those which release energy. (*After E. P. Odum. Fundamentals of Ecology, 2d ed., Saunders, 1959*).

food production for man lie in measures which reduce physical limiting factors and increase the season of growth so that sunlight is utilized for as large a part of the annual cycle as possible.

BIOGEOCHEMICAL CYCLES

The more or less circular paths of chemical elements passing back and forth between organisms and environment are known as biogeochemical cycles. The rate at which vital elements become available to biological components of the ecosystem is more important in determining primary and secondary productivity than flow of solar energy. If an essential element or compound is in short supply in terms of potential growth, the substance may be said to be a limiting factor. The productivity of an entire ecosystem is sometimes limited by one material available in least amount in terms of need, according to the principle of Liebig's law of the minimum. Thus water limits the desert ecosystem, and nitrogen or phosphorus often limits ocean ecosystems. However, nature has considerable powers of adaptation and compensation. In many environments, species and varieties have evolved which have low requirements for scarce materials. Also, the amount of one substance which, in itself, may not be limiting often greatly affects the requirement for another substance which is approaching a critical minimum. Consequently, it is usually necessary to consider the interaction of the essential materials if the limiting factors operating in a given situation are to be determined.

Nutrients. Dissolved salts essential to life may be conveniently termed biogenic salts or nutrients. They may be divided into two groups, the macronutrients and the micronutrients. The macronutrients include elements and their compounds needed in relatively large quantities, for example, carbon, hydrogen, oxygen, nitrogen, potassium, calcium, phosphorus, and magnesium. The micronutrients include those elements and their compounds necessary for the operation of living systems but which are required only in minute amounts. At least 10 micronutrients are known to be necessary for primary production: iron, manganese, copper, zinc, boron, sodium, molybdenum, chlorine, vanadium, and cobalt. Several others such as iodine are essential for certain heterotrophs. Because minute requirements are often associated with an equal or even greater minuteness in environmental occurrence, the micronutrients deserve equal consideration along with the macronutrients as possible limiting factors.

Types of cycle. From the standpoint of the biosphere as a whole, biogeochemical cycles fall into two groups: the gaseous-type cycles, as illustrated by the nitrogen cycle in Fig. 4; and the sedimentary-type cycles illustrated in Fig. 5. Cycles of oxygen, carbon, and water resemble the cycle of nitrogen, in that a large gaseous pool is important in the continuous flow between inorganic and organic states. As shown in Fig. 4, the nitrogen of protoplasm is broken down from organic to inorganic form by a series of decomposer bacteria, each specialized for a particular part of the job. The nitrogen ends up as nitrate or other form usable by green plants in the synthesis of new organic matter. The air is the great reservoir and safety valve of the system. Nitrogen is continually entering the air by action of denitrifying bacteria and continually returning to the cycle through the action of nitrogen-fixing organisms and also through nonbiological fixation. Only certain bacteria and algae (which, however, are abundant in water and on land) can fix nitrogen. No so-called higher plant or animal has this ability; legumes fix nitrogen only because of the symbiotic bacteria which live in their roots. The steps from protoplasm down to nitrates provide energy for the decomposer organisms, whereas the return steps require energy from other sources, such as from organic matter or sunlight. Likewise, nitrogen fixers must use up some of their carbohydrate or other energy stores in order to transform atmospheric nitrogen into nitrates. *See* NITROGEN CYCLE.

Feedback mechanisms. The self-regulating feedback mechanisms, as shown in a simplified manner in Fig. 4, make the nitrogen and the other gaseous-type cycles relatively perfect when large areas of the biosphere are considered. Thus increased movement of materials along one path is quickly compensated for by adjustments along other paths. However, nitrogen often becomes a limiting factor locally, either because regeneration is too slow, or because loss from the local system is too rapid.

Most biogenic substances are more earthbound than nitrogen, and their cycles follow the pattern of erosion, sedimentation, mountain building, and volcanic activity, as shown in Fig. 5. Biological activity on land and in the upper layers of water results in local cycles from which there is usually a continual loss downhill and replacement from uphill runoff and from solid matter moving in the air as dust, that is, natural fallout. Man often disrupts the sedimentary cycles because he tends to increase the downhill movement. The phosphorus cycle is a good example. Phosphorus is relatively rare in the surface materials of the Earth in terms of biological demand. At present more phosphorus is apparently escaping to the deep ocean sediments (where it is unavailable to producers) than is being replaced by natural processes. For the time being, man is able to mine the considerable reserves of underground phosphate rock and make up some of this loss, but eventually a means may have to be found to recover phosphorus from the sea.

Nonessential and radioactive elements. Biogeochemical cycles involve elements essential to life. The nonessential elements pass back and forth between organisms and environment and many of them are involved in the general sedimentary cycle, as pictured in Fig. 5. Although they have no known value to the organism, many of these elements become concentrated in tissues, apparently because of similarity to specific vital elements. The ecologist would have little interest in most of the nonessential elements were it not for the fact that atomic bombs and nuclear power operations produce radioactive isotopes of some of these elements, which then find their way into the environ-

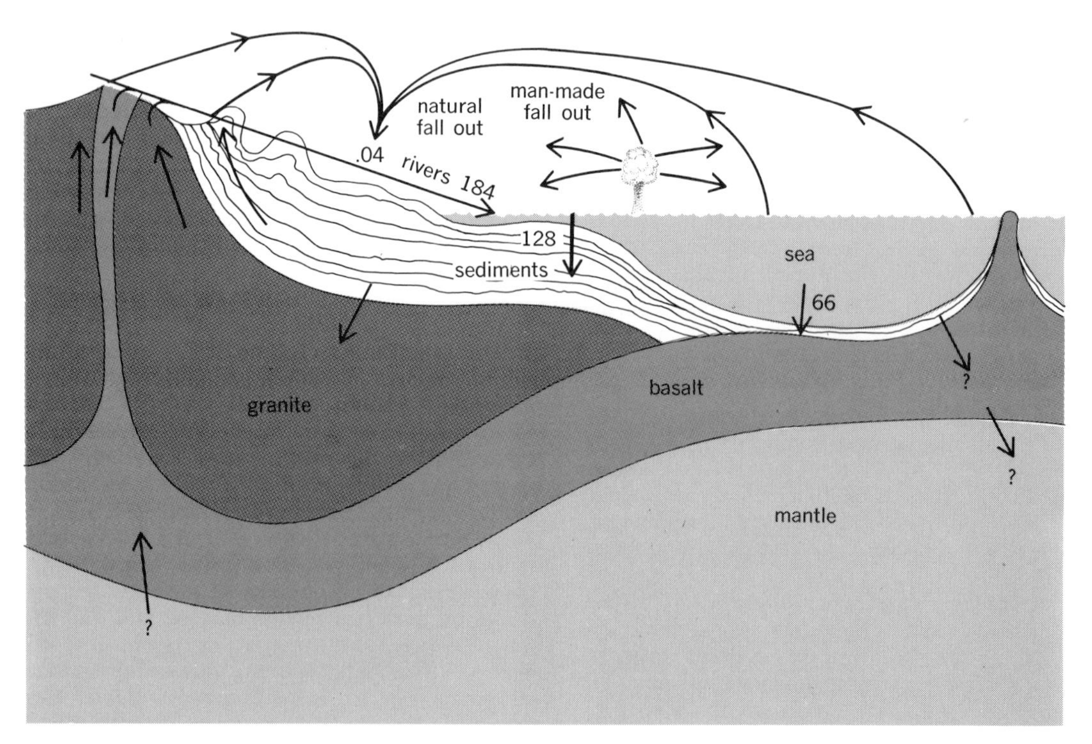

Fig. 5. Diagram of the sedimentary cycle involving movement of the more earthbound elements. Where estimates are possible, the amounts of material are estimated in geograms/10^6 years (1 geogram= 10^{20} g).

ment and into food chains. Even a rare element, in the form of a radioactive isotope, can be of biological concern because a very small amount of material from a geochemical standpoint can have marked biological effects. Thus the cycling of such things as strontium, cesium, cerium, ruthenium, and many others may be of great concern in coming years. Strontium is receiving special attention; it behaves like calcium and follows it into biological systems. None of the strontium which is naturally involved in the calcium cycle is radioactive, but radioactive strontium is becoming widespread in the biosphere as a result of fallout from atomic weapons tests. Small amounts of radiostrontium have now followed calcium from soil and water into vegetation, animals, milk, and other human food, and into human bones in almost all parts of the world. In 1958 several hundred micromicrocuries (1 micromicrocurie $= 10^{-12}$ curie) of radiostrontium were present for every gram of calcium in some soils. The bones of children in North America and Europe averaged 1.2 micromicrocuries/g of calcium. There is considerable controversy as to whether these small amounts are detrimental, but most scientists agree that it would be desirable to keep as much radioactivity out of the food chain as possible.

Tracers. There is a bright side to the atomic-age production of radioactive isotopes. Tiny amounts of such isotopes provide convenient tracers whereby the movement of materials in ecosystems can be accurately measured. Much has already been learned about the cycling and turnover rates of phosphorus through the use of the isotope P^{32}. The radioactive isotope of carbon, C^{14}, has proved invaluable in measuring the rate of primary production in the ocean. It is certain that intelligent use of radioisotopes as tools can help solve problems of the atomic age. Radioisotopes not only offer unprecedented means of studying flow in ecological networks, but are used as sources of high-energy radiation for experimental manipulation. *See* BIOSPHERE; RADIOECOLOGY.

ECOSYSTEM DEVELOPMENT

Ecosystems undergo orderly development as do individual organisms because the biotic components are capable of modifying and controlling the physical environment to varying degrees. This process of change is commonly called ecological succession or ecosystem development. The table summarizes the major changes that occur in the development of an ecosystem, irrespective of the physiographic site or the climatic region. Trends are emphasized by contrasting the situation in early development with that at maturity. The table, in essence, contrasts "young nature" with "mature nature" and lays the foundation for an ecological solution to the growing conflicts between man and nature.

As shown, youthful types of ecosystems, such as agricultural crops or the early stages of forest development, have characteristics that contrast to those of adult systems, such as mature forests or climax prairies, for example:

1. High net productivity that is available for harvest or storage.
2. Relatively small standing crop at any one time with a consequent low ratio between biomass and productivity.
3. Unbalanced metabolism with an excess of production over utilization.
4. Linear grazing food chains (that is, grass-cow-man) in contrast to complex weblike energy flows involving animal-microbial detritus utilization that characterize older systems.
5. Low biotic diversity (for example, few species but large numbers of individuals).
6. Rapid one-way flux of inorganic nutrients in contrast to recycling and retention within the organic structure.
7. Selection for species with high birth rates, rapid growth rates, and simple life histories.

These attributes combine to produce a general lack of stability in that young systems are more easily disrupted by drought, storms, disease, or other perturbations.

These contrasts underline the basic conflict between the strategy of man and nature. For example, the goal of agriculture or intensive forestry, as now generally practiced, is to achieve high rates of production of readily harvestable products with little standing crop left to accumulate on the landscape or, in other words, a high production-biomass efficiency. Nature's strategy, on the other hand, as seen in the outcome of the ecosystem

Trends in the ecological development of landscapes.

Ecosystem attribute	Developmental stages	Mature stages
Gross primary productivity (total photosynthesis)	Increasing	Stabilized at moderate level
Net primary community productivity (yield)	High	Low
Standing crop (biomass)	Low	High
Ratio growth to maintenance (production-respiration)	Unbalanced	Balanced
Ratio biomass to energy flow (growth + maintenance)	Low	High
Utilization of primary production by heterotrophy (animals and man)	Predominantly via linear grazing food chains	Predominantly weblike detritus food chains
Diversity	Low	High
Nutrients	Inorganic (extrabiotic)	Organic (intrabiotic)
Mineral cycles	Open	Closed
Selection pressure	For rapidly growing species adapted to low density	For slow growing species adapted to equilibral density
Stability (resistance to outside perturbations)	Low	High

development process, is directed toward the reverse efficiency, namely, a high biomass-production ratio. The natural strategy of maximum protection (that is, optimizing for the support of a complex living structure that buffers the physical environment) conflicts with man's goal of maximum production (that is, maximizing for high efficiency in production). Since the environment is man's living space as well as a supply depot for food and fiber, its gas exchange, waste purification, climate control, and recreational and esthetic capabilities are as vitally important in the long run as its capacity to yield consumable products. Many "protective" or "living space" functions are best provided by the more stable or mature-type ecosystems. It is clear, then, that there must be both "productive" and "protective" landscapes in reasonable balance to safeguard the gaseous, water, and mineral cycles on which life depends.

As the human population of the world increases and more and more changes in the face of the Earth are contemplated, increasing attention must be given to the total, and not just the immediate, effect of the changes. It is necessary to maintain a moderate amount of diversity in nature, and to preserve a maximum amount of "open space" if the quality of man's environment is to be protected. Attention must be directed toward conservation of the ecosystem, and not just conservation of individual organisms which are in demand at the moment.

[EUGENE P. ODUM]

Bibliography: S. J. McNaughton and L. LeWolf, *General Ecology*, 1979; E. P. Odum, *Ecology*, 2d ed., 1975; E. P. Odum, *Fundamentals of Ecology*, 3d ed., 1971; H. J. Oosting, *The Study of Plant Communities*, 2d ed., 1956; R. E. Ricklefs, *Ecology*, 1979; V. E. Shelford, *The Ecology of North America*, 1978; R. L. Smith, *Ecology and Field Biology*, 2d ed., 1974.

Ecosystem

A functional system which includes the organisms of a natural community together with their environment. The term, a contraction of ecological system, was coined by the British ecologist A. G. Tansley in 1935. The concept has been of increasing importance in the development of ecology and related fields. All natural communities, or assemblages of organisms which live together and interact with one another, are closely related to their environments. It is consequently appropriate to conceive of community and environment as forming a single, complex whole—the ecosystem. *See* ANIMAL COMMUNITY; APPLIED ECOLOGY; ECOLOGY; ENVIRONMENT; PLANT COMMUNITY.

The term ecosystem can be applied to any size or level of environment chosen for study. Communities of small and distinctive environments, such as the organisms inhabiting an animal's intestine, a decaying log, or the foliage of a particular plant, are sometimes referred to as microcommunities, and their environments, as microcosms or microenvironments. These are generally parts of larger ecosystems, such as a forest and its environment or a lake and its organisms, for which the terms macrocommunity and macroenvironment are appropriate. At the opposite extreme from microcommunities, all the organisms of the world together may be regarded as a world community which, with its environment, forms a single world ecosystem. The living part of this world ecosystem is often termed the biosphere. The transfer and concentration of materials in the world ecosystem by both organisms and physical processes are the subject of biogeochemistry. Since man himself is part of the world ecosystem, man's place and effects on it are part of the study of human ecology. *See* BIOSPHERE; HUMAN ECOLOGY.

Any organism lives in a natural context which includes an environment and a community of other organisms. The natural context of the life of an organism is thus an ecosystem. Two aspects of an organism's relation to the ecosystem in which it lives are often distinguished, though they cannot be sharply separated. Its location, as described in terms of physical and chemical factors, and often, kind of community, constitute its functional habitat. Its place within the community, including relations to other organisms, is its niche.

Within an ecosystem certain major fractions, or divisions, may be recognized. A terrestrial ecosystem is often thought of as having four or five such subdivisions: (1) the physical environment, such as climate; (2) the soil, if distinguished from physical environment; (3) the vegetation or plant community; (4) the animal community; and (5) the saprobe community, which consists of bacteria and fungi. These same divisions may be recognized in a marine or fresh-water ecosystem, with the substitution of substrate or bottom material for soil. Aquatic ecosystems, however, are often divided in a different way. These divisions are the communities and environments of the open water, those of the shores, and those of the deeper bottom. *See* FRESH-WATER ECOSYSTEM; MARINE ECOSYSTEM; TERRESTRIAL ECOSYSTEM.

Some major features of ecosystems important in ecological understanding are discussed here.

Complexity and interrelations. The scientist studying an ecosystem is confronted by a bewildering variety of interrelations. The environmental factors which can be specified and measured are variously interrelated in their effects on one another, and in their significance to the physiology of the organisms. Thus, the effect of humidity cannot simply be measured by relative humidity, for the effect on any given organism is conditioned also by the existing temperature and wind velocity, by the position in the community and characteristics of the organism concerned, and by the results of water loss on the physiology of that organism. The environment affecting the organisms in a community is often conceived as the environmental complex, that is, the sum of all distinguishable environmental factors together with the interrelations among these. The factors may, in fact, be regarded as isolates from the whole environmental complex, as a subset of those particular features abstracted from the whole and chosen for measurement and study. Different organisms in the community are variously affected by different factors and combinations of factors. Organisms are also most variously affected by major kinds of interrelations, such as food relations, shelter, competition for space and other resources of environment, shading, chemical influences, and others. These interrelations among species in the community, which so link together the species that almost any one may directly or indirectly affect almost any other, are referred to as the web of life. The species' position in the web of life, in relation to other species and the community as a whole, is one way of conceiving of the species' niche. Such is the complexity of an ecosystem that it is generally not possible to study all its features at once or to seek complete understanding of all its interrelations. *See* ECOLOGICAL INTERACTIONS.

Environmental dependence and change. Environment and community are always intimately interrelated. The community is necessarily dependent on environment and cannot be understood apart from it. The relation between environment and community is not one of simple cause and effect, however, for environmental factors are, in various ways and to varying degrees, modified by the community. Some factors, such as salinity of sea water, temperature of water of a pond, and others, may be scarcely affected by the presence of organisms; others, such as humidity and wind velocity inside a forest and concentration of phosphates in the water, are strongly modified by organisms, or are determined by the changing activities of organisms. Still others, for example, organic matter in water and soil, exist only because of the activities of organisms. Some communities, such as a highly developed, stable forest, modify the local environments in which their component organisms live more than do other communities, such as those of deserts and the first stages of successions. In general, however, there is a pervasive reciprocity between environment and community in determining the characteristics of both environment and community.

Community adaptation. Characteristics of environment are reflected in characteristics of the community. The community may be said to express the environment surrounding the local ecosystem. Certain broad correlations between kinds of environments and kinds of communities around the world can be observed.

Among terrestrial communities climate is expressed in the overall structure and composition in terms of major kinds of plants, the physiognomy of vegetation. Thus, wherever tropical climates with high rainfall throughout the year occur, these climates support the kind of community known as tropical rainforest. Wherever temperate continental climates with sufficient summer rainfall occur, these support temperate deciduous forests. One may thus recognize the general biological phenomenon of adaptation to environment on two levels: first, the organism, which must have structure and function suited to survival in its niche and the environmental complex of its ecosystem; and second, the community, which must have structure and function appropriate to utilization of the resources of, and persistence in, its environment. *See* BIOME; PLANT FORMATIONS, CLIMAX; RAINFOREST.

Spatial patterning. Many environmental factors form gradients in space, as they are followed from one point to another on the Earth's surface. The many factors of the environment change in different directions through space, some in correlation with one another, others in partial or complete independence. At each point in this environmental pattern a natural community develops in dependence upon the environment of that point. To the pattern of environments in space there consequently corresponds a pattern of communities, and the populations and other characteristics of communities form complex patterns of gradients in space, as do the factors of environment. Often this patterning in space has a self-repeating character, as in an area of low mountains in New York, where one may find a pattern of hemlock forests in ravines and on north-facing slopes, oak forests on most other slopes, and grassy oak openings on dry south slopes, with this pattern repeating itself across a series of valleys and ridges. Environments and communities, in general, grade continuously into one another along spatial gradients of environment. Unlike organisms, ecosystems and communities consequently are not distinct, clearly bounded, and separate from each other. There are exceptions to this. A lake may be regarded as an ecosystem with a fairly clear boundary. Relative discontinuities between communities occur because of disturbance, and where environment is relatively discontinuous, as at the shores of lakes and seas. When such relative discontinuities, or ecotones, are studied, they are often found to represent not merely the meeting place of two communities but distinctive communities in their own rights.

Rhythms. Environments vary also in time, in part irregularly, but more generally and more significantly in regular, periodic rhythms. Only the abyssal communities of the ocean depths live in a constant environment. Other communities are subject to annual and diurnal cycles, and communities of the seashore, to more complex rhythms of tides. Organisms of the community respond in various ways in their cycles of activity to these rhythms of environment. Consequently the environmental rhythms impose self-repeating patterns of change in time on the community. Similar activities may be pursued by different species at different times in these cycles. Thus, flycatchers feeding in the daytime on diurnal insects may be replaced at dusk by nighthawks feeding on nocturnal insects; spring flowers, and the insects visiting them, may be replaced in fall by quite different groups of flowers and insect visitors. Daily and yearly rhythms thereby permit specialization by time of function among the species of a community, and differentiation in time of the community as a whole.

Community differentiation. A natural community comprises a large number of species, but seldom are two species exact competitors. The species of the community are specialized to fill different, though often overlapping, niches. Species with similar niche requirements may be active at different times or in different places in the community. The specialization of species and differentiation of the community are illustrated in the stratification of terrestrial communities. A forest may include five or more strata: the upper and lower tree layers, and shrub, herb, and moss layers, with each stratum having its characteristic plant and animal species. Parallel activities may be pursued by different species in different strata. For example, one bird species may nest and feed on insects in the tree stratum, another on the ground. Such vertical differentiation in space is a

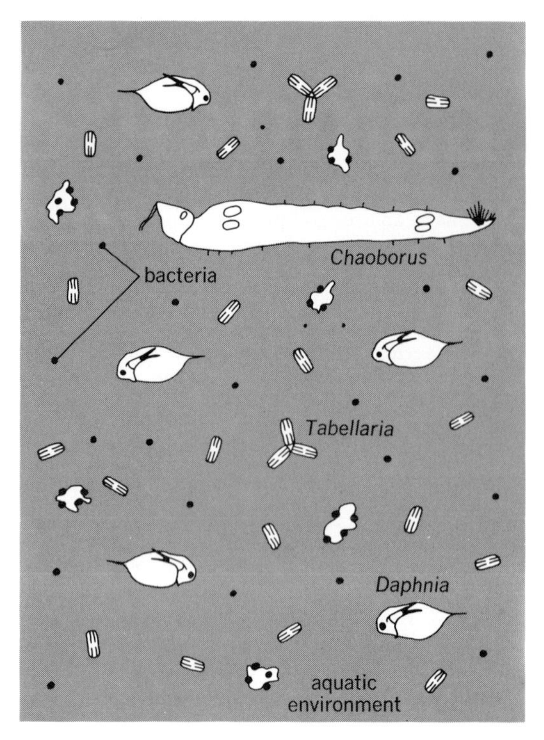

Fig. 1. Simplified ecosystem consisting of water and plankton organisms. Components of this ecosystem are environment, or water and its dissolved material; green plants, or producers (*Tabellaria*, diatom); herbivorous animals, or primary consumers (*Daphnia*, water flea); carnivorous animals, or secondary consumers (*Chaoborus*, fly larva); and bacteria, or decomposers. The various organisms are not drawn in correct size relations.

basis of one of the major characteristics of communities—their species diversity, or relative richness in numbers of species. One may generalize that species diversity, differentiation into strata, and productivity of terrestrial communities are maximal in most favorable environments, such as that of the tropical rainforest, and decrease toward environments which are unfavorable, unstable, or extreme. There are, however, various limitations to this statement, and species diversity and productivity do not simply parallel one another.

Productivity relations. Productivity is, even more than species diversity, a fundamental characteristic of communities and ecosystems. It is perhaps the most important single feature of the community. Productivity may be defined as the rate at which energy is bound or matter combined into organic compounds by organisms, per unit time per unit of the Earth's surface. Energy is dissipated back to environment and organic matter reduced to inorganic forms by the respiration of organisms in the community. Productivity is thus counterbalanced by various loss rates. Although particular environmental factors may effectively limit productivity, it is in general a complex resultant of all factors of the environment, an expression in biological activity of the environmental resources and of other factors affecting utilization of these resources by the community. Major factors affecting productivity include those of temperature, nutrient availability, and water availability. Certain broad correlations of productivity with environments may be observed, such as the decline along the temperature gradient from tropical rainforest to the treeless arctic tundra, and along the moisture gradient from tropical rainforest to desert. Productivity thus underlies such other characteristics of communities as their structure or physiognomy and stratification. *See* BIOGEOGRAPHY; BIOLOGICAL PRODUCTIVITY; COMMUNITY.

Food chains and trophic levels. Productivity of a community is based wholly on the activities of green plants, except for certain autotrophic bacteria, and the communities of such lightless environments as the depths of oceans and lakes, which are dependent on other communities for their food. Green plants use the energy of sunlight to produce organic substances from carbon dioxide and water. These plants are then eaten by animals, and these animals by other animals. In the community illustrated in Fig. 1, organic matter and food energy might be passed along from the diatoms, to the water fleas, to the *Chaoborus* larva, and from this to a small fish, and to a heron. Such a sequence of organisms is referred to as a food chain. Since many species feed on a number of other species, food chains, as they are variously interrelated, form an important part of the web of life in a community. Certain major steps (trophic levels) in food webs are the green plants, or producers; the herbivorous animals; the carnivorous animals; and carnivorous animals feeding on carnivorous animals. At each level much of the energy available to organisms is expended in the life activities of those organisms. Only a fraction of this energy can be harvested and used by the next trophic level. There is, consequently, a stepwise decrease of productivity through the sequence of consumer trophic levels. This relation is known as the pyramid of life. *See* FOOD CHAIN.

Cycling and functional kingdoms. Food chains and pyramids are part of a broader phenomenon, the cycling of materials between organisms and environment in the ecosystem. A given substance, such as phosphate in the soil, may be taken up by the roots of green plants and used in organic syntheses and then passed along food chains through animals until, with the death and decomposition of these organisms, the phosphate is released and returned to the soil, where it is again available to green plants. Decomposition of organic material results largely from the activities of certain organisms, such as bacteria and fungi. One may thus recognize three major nutritional groupings and evolutionary directions as the functional kingdoms in communities. These are the true or green plants, or producers, characterized by photosynthesis as their mode of nutrition; the animals, or consumers, characterized by consumption or ingestion of organic food; and the bacteria and fungi, the saprobes, decomposers, or reducers, characterized by absorption of organic food and by their contribution to the breakdown of dead organic matter. The generalized relation among these in cycling of materials through the ecosystem is illustrated in Fig. 2. When the movement of particular substances is studied in detail, a great complexity of routes of movement in the ecosystem may be revealed. This is best known for the nitrogen cycle. Concentrations of many substances in the environment are determined by the cycling of the material through the community. An experiment was con-

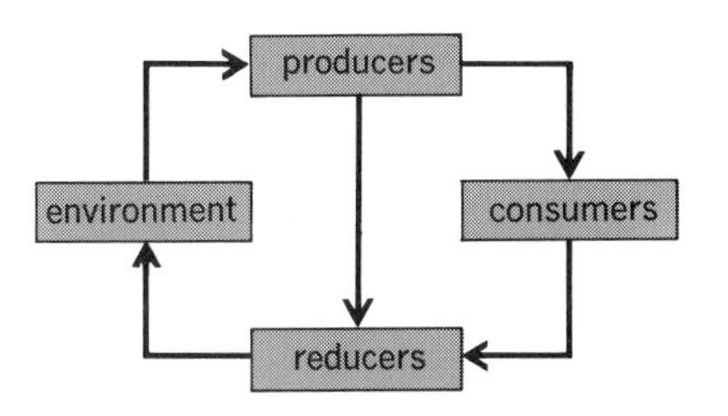

Fig. 2. Generalized cycle of materials in an ecosystem.

ducted by R. H. Whittaker, who used a radioactive phosphorous (^{32}P) tracer to determine the pattern of phosphate movement in a pond (Fig. 3). The amount of ^{32}P in water at a given time depends on the various rate values, and the temporary steady state of ^{32}P distribution between water and organisms. Productivity relations also are determined, not simply by the amounts of nutrients, and in terrestrial communities by the amount of water in the environment, but by the rate and manner of cycling of these materials through the ecosystem.

Energy flow and steady state. Energy also flows in a cycle of uptake from the environment, movement through the community, and dissipation back to the environment. The community and the ecosystem are thus, like organisms, open energy systems. In a stable ecosystem, the community possesses a stable pool of available free energy of organic compounds, and the flow of energy into this pool by photosynthesis is balanced by the outward flow by respiration, which dissipates energy into environment. With regard to both energy and matter, the community is in steady state, a dynamic equilibrium of apparent constancy underlying which there is a continuing flow of energy and matter. As an open energy system in steady state, the community, like the organism, resists running down to maximum entropy, according to the second law of thermodynamics, by its continuing intake of energy and passing of this energy through the trophic levels as the negative entropy of food. Rates of uptake of nutrients and other substances from environment and rates of return of these substances to environment are in balance. The biomass, or mass of organic materials making up the community, and the chemical composition of this biomass are consequently in steady state. Not only the community as a whole, but also the species populations of a stable community may be in steady state. The steady state for the species involves a balance of individual births and deaths, while the population itself remains essentially stable. *See* POPULATION DYNAMICS.

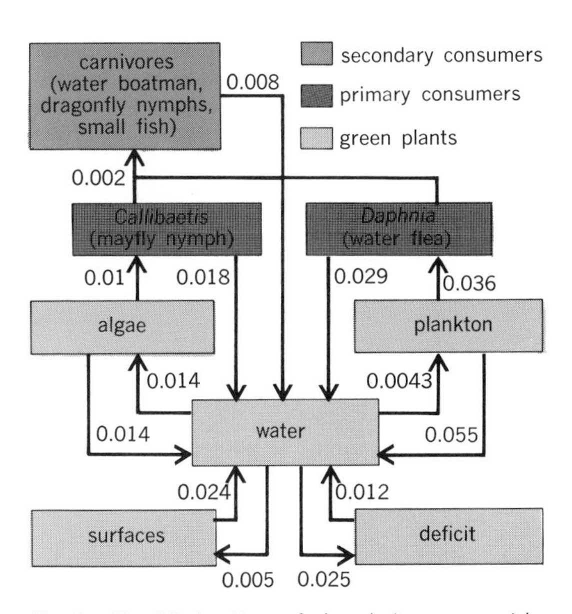

Fig. 3. Simplified pattern of phosphate movement in pond, based on an experiment by R. H. Whittaker with radioactive phosphorus (^{32}P) tracer. Numbers are transfer rates (that fraction of ^{32}P) in box at tail of arrow moving in direction indicated per hour). Surface is film of bacteria and other microorganisms on rock surfaces; deficit refers to ^{32}P, mainly in depths of rocky substrate, not otherwise accounted for.

Succession. Not all communities are fully stabilized in this sense, and many become so only approximately, after an extended process of community development, or succession. If a bare area of the Earth's surface is exposed, as by a landslide, this bare area may be occupied first by such simple plants as lichens and mosses. These plants contribute, along with physical processes, to the breakdown of the rock and formation of soil until higher plants, such as grasses, may occupy the environment. The grasses may in turn make it possible for shrubs to grow, and the shrubs make growth possible for trees, until finally a stable forest may result. Through the course of succession there is usually increasing productivity, species diversity, and stability of the community, with increasing development of the soil, increasing stocks of material in circulation, and increasing modification of the environment by the community. There are, however, exceptions to these trends in particular successions. *See* ECOLOGICAL SUCCESSION; VEGETATION MANAGEMENT.

Climax. The stable ecosystem which ends the succession is termed climax. Climax communities are characterized by self-maintenance and a considerable degree of permanence when free from external disturbance. That is, the populations in the climax stage reproduce and maintain themselves, as they do not during the successional stages. The climax thus represents a steady state of energy flow, materials circulation, and population reproduction. It may represent a maximum of sustained productivity and biomass for the ecosystem. (Or it can be a system of degraded productivity, after irreversible losses of nutrients, for example.) The climax has a self-stabilizing character, and tends to return to normal after disturbance of its equilibria. For instance, a heavy phosphate fertilization of a lake produces only a temporary increase in productivity, and phosphate in the water rapidly declines back to the original level. Thus a temporary overpopulation of a given species is counteracted by increased mortality from predation, disease, or other factors, until the population returns to its normal level. It appears that these stability mechanisms, in general, depend not so much on any single factor or interaction as on the reestablishment of a steady-state balance in relation to various and complexly interrelated factors. A pattern of climax communities may correspond to the pattern of environments in any large area. Within a given area, however, many environments may be similarly affected by the general climate of the area. Many of the successions of a given area consequently converge toward similar climax communities which occupy the largest part of the landscape surface. In the low mountains referred to previously, the oak forests would occupy the greatest part of the mountain slopes, give the vegetation its predominant character, and express the climate. Such a community may be termed a climatic climax or prevailing climax. *See* CLIMAX COMMUNITY. [ROBERT H. WHITTAKER]

Bibliography: G. L. Clarke, *Elements of Ecology*, rev. ed., 1965; E. P. Odum, *Ecology*, 2d ed., 1975; E. P. Odum, *Fundamentals of Ecology*, 3d ed., 1971; R. H. Whittaker, *Communities and Ecosystems*, 2d ed., 1975.

Ecosystem models

To the environmental scientist a system is a collection of interacting components, some of which are living individuals, populations, or communities, operating within any arbitrarily defined space during some specified time. Thus nature can be viewed as a series of ecosystems between which there are strong or weak mutual dependencies, but within which the interactions are even more strongly expressed. The major goal of environmental science is formulation and testing of theories about how these systems operate and interact.

A traditional way to formulate theories in science is by an explanatory model. Initially the model must be verbal. Eventually, however, the verbalization must be translated into a set of mathematical equations. These, upon simultaneous solution through time, form a dynamical simulation model, capable of predictions that are in turn subject to experimental or observational verification.

This article outlines the steps necessary to construct realistic theoretical simulation models and describes the uses for which they may be employed. Whatever use is to be made of the simulation model, either development of theory or applied prediction of response to a perturbation, nature must always be used as a strict guide in the construction of the model. To do otherwise is valueless, and at worst positively dangerous, for today more and more decisions affecting the health and economics of human societies are shaped by the outcomes of ecosystem simulations.

Model construction. All simulation models of ecosystems are constructed according to the same general sequence of steps. In the first step, the ecosystem to be modeled is defined, that is, boundaries are set and the biological and abiotic components of interest are enumerated. The model may, of course, specify variables such as single life-history stages, species populations, taxonomic groups, or trophic or feeding groups. These entities, called the state variables of the model, must be defined first, since the remaining numerical or quantitative parts of the model, called parameters, will be defined on the basis of one or more state variables.

Because the simulation model will be predicting changes in the dynamics of the biotic and abiotic state variables, these must be expressed in some common unit. Numbers and biomass of organisms have been used in such models, but they are not conserved. One oak tree or 1 g of acorns is not equivalent to one adult squirrel or 1 g of newborn squirrels. In models employing numbers or biomass as primary units, conversion factors must be used in every equation concerning transfer of material.

A better solution is to employ units that are conserved, such as elemental mass and energy content, and then to recover such units as numbers and biomass at any particular time, thus necessitating very few conversion multiplications. Conserved units imply a balanced budget. Thus a model using energy transfer to express the amount of each component which changes through time can account for all energy entering and leaving each component, according to the conservation-of-energy principle in the first law of thermodynamics. The same is true for elemental transfers.

Food web. Once the boundaries, state variables, and units have been chosen, the second step in the construction of a simulation model is to determine the trophic or food web. In more general terms—because models deal with elemental or energy units and with abiotic as well as biotic state variables—this web should simply be called a transfer diagram. It is a qualitative statement of the movement of matter-energy in the simulation model, indicating both the presence and the direction of a transfer (Fig. 1).

Control factors. The factors which control the magnitude of all these flows are at the heart of the model construction. The simulation model must realistically compute the time course of these matter-energy flows, constantly adding to and subtracting from the state variables in order to portray change.

First the description of the controlling factors must be qualitative, showing which variables affect which flows. In Fig. 2, for example, the ingestion by consumer species is a function of (1) the amount of food (producer 1) available; (2) the number of individuals of consumer 1 potentially available to interfere with each other; and (3) any number of outside influences, both biotic (other species which may interfere with ingestion or reproduction by consumer 1) and abiotic (such as temperature,

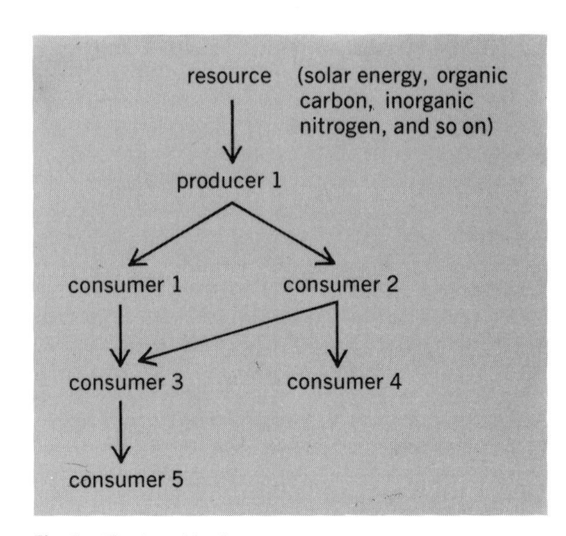

Fig. 1. The trophic diagram, a general statement of the paths and direction of material transport due to the ecological interactions between the component groups of an ecosystem.

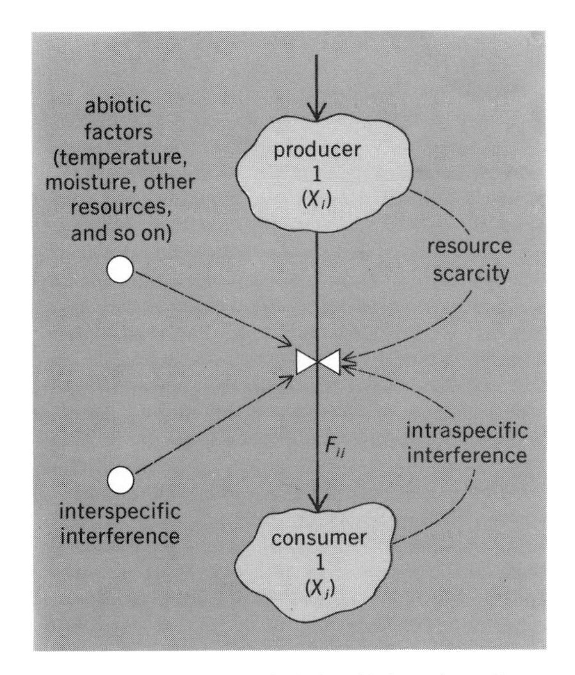

Fig. 2. An example of a single trophic transfer pathway showing several possible factors (indicated by dashed lines) that could be expected, under certain conditions, to affect the flow of matter-energy from producer 1 to consumer 1.

available water, or other nutrients). Although, in theory, category 3 is virtually endless, in practice relatively few outside influences are important control factors. The ecologist, acting alone or in consultation with the model builder, must decide, on the basis of the data available, which influences are important and how they operate.

Figure 2 indicates a fundamental distinction between control factors. The material resource factor operates through scarcity of that resource. The space resource term operates by direct interference with ingestion or reproduction because of crowding, that is, limited space.

The interactions of these factors is illustrated by the relationship between predator and prey. The predator is endowed with a genetically determined (physiologically expressed) capacity to grow and reproduce at some maximum rate under optimum conditions. For example, the food resource or prey must be abundant enough so that the work of capture does not detract from optimum rates of growth and reproduction. As the food density drops below this threshold level, the predator must compensate for decreased ingestion by slower growth or less reproduction. Continued lowering of the food resource density will eventually result in a situation where the predator population obtains only enough prey to just meet its maintenance needs. The population growth rate will be zero, a steady state. Prey density may, of course, continue to decline and may occasionally reach a second, lower threshold level below which the predator cannot capture any prey no matter how much effort is expended or, more commonly, this lower threshold may simply represent the prey density at the point where potential rewards to the predator are so low that it simply switches to another food source. In any case, the important result is that predator ingestion of the prey is reduced to zero when prey population is at or below this lower level.

Conversely, if prey are superabundant, the predator population grows at the maximum rate. Eventually, crowding leads to some direct interference between predators, which results in lowered ingestion and individual growth or reproduction. At a particular predator density, there will be enough interference—for example, time spent fighting instead of hunting—to reduce ingestion to a maintenance level.

Equations of interaction. The ecological description must now be translated into mathematical terms, a step which has many logical and philosophical pitfalls.

Functional forms. Translating the verbal ecological description into a mathematical expression requires choice of the functional form. The donor-controlled linear relationship is represented quite well by Eq. (1), where X_i is the donor component of

$$F_{ij} = \lambda_{ij} X_i \tag{1}$$

flux F_{ij}, and λ_{ij} is the coefficient of transfer, a specific rate with units of time^{-1}. However, with nonlinear and especially discontinuous interactions involving feedback controls, the proper form is not immediately obvious.

Some models employ "off-the-shelf" functional forms such as the logistic and the Michaelis-Menten enzyme kinetics equations. Unfortunately, the logistic equations make no distinction between material resource and space limitations, are limited to one functional form of feedback relationship, and make no provision for discontinuities. The Michaelis-Menten equations were developed to handle the dynamics of enzyme kinetics, and wholesale application to populations of multicellular organisms is hardly justified except where individuals are small and of uniform size. Thus the most common application of these equations is to phytoplankton and zooplankton populations in models of open-water aquatic systems.

Although choosing the proper mathematical function to express an ecological process is not simple, by far the best procedure is to begin with a realistic general basis for the equations and tailor this to suit each individual situation. This is illustrated best by example.

Realistic base equation. The general base equation must be able to incorporate the temporal and spatial heterogeneity, thresholds, and feeding preferences peculiar to the specific ecosystem to be modeled.

Consider the transfer diagrammed in Fig. 2. Assume a specific rate τ_{ij} as the maximum physiological rate of ingestion that can be achieved by component X_j under an optimun specified environment. If the environment does not vary and if evolution of physiological characteristics of X_j is negligible, then τ_{ij} is constant. Otherwise it is a variable.

The flux of matter-energy into a population at any instant is given by Eq. (2). When the product

$$F_{ij} = \tau_{ij} X_j \cdot f(X_i) \cdot f(X_j) \tag{2}$$
$$0 \le f(X_i) \le 1;\ 0 \le f(X_j) \le 1$$

$f(X_i) \cdot f(X_j)$ is unity, the population is ingesting at the maximum rate τ_{ij}; when $f(X_i) \cdot f(X_j)$ is zero, ingestion is also zero.

The control function $f(X_i)$ [Eq. (3)] requires the

$$f(X_i) = \left(1 - \frac{(\alpha_{ij} - X_i)_+}{(\alpha_{ij} - \gamma_{ij})}\right)_+^{\phi} \qquad (3)$$

following: (1) the level of X_i where limitation due to scarcity of X_i first begins (saturation density, α_{ij}); (2) the lowest density of X_i at which X_i is still available (refuge density, γ_j); and (3) the functional relationship between a unit change in the density of X_i and the change in the value of $f(X_i)$.

Thus, whenever $X_i \leq \gamma_{ij}$, then $f(X_i) = 0$, and whenever $X_i \geq \alpha_{ij}$, then $f(X_i) = 1$. The change in $f(X_i)$ for changing X_i can be linear (simplest case), $\phi = 1$, or nonlinear, $\phi \neq 1$. The subscript + in Eq. (3) places the additional constraints through the convention that $(\cdot) = 0$ if $(\cdot) \leq 0$, and $(\cdot) = (\cdot)$ if $(\cdot) > 0$. When $\phi = 1$, a unit change in $f(X_i)$ is observed for a unit change in X_i for any X_i within the range $\gamma_{ij} \rightarrow \alpha_{ij}$. When $\phi > 1$, $f(X_i)$ changes more rapidly per unit change in X_i when X_i is large, that is, a predator whose effort and catch drop dramatically as soon as prey show the slightest scarcity (switching behavior). When $\phi < 1$, $f(X_i)$ changes less rapidly per unit change in X_i when X_i is large, that is, a predator which searches harder as prey become scarce, thus maintaining close to the maximum capture rate.

The space-limiting control function $f(X_j)$ [Eq. (4)]

$$f(X_j) = \left(1 - \left\{1 - \frac{\lambda_j}{\tau_{ij}(1-\epsilon_{ij})}\right\} \frac{(X_j - \alpha_{jj})_+}{(\gamma_{jj} - \alpha_{jj})}\right)_+^{\phi} \qquad (4)$$

requires the following: (1) a lower density threshold of X_j where the effects of intraspecific interference on ingestion are first evident (a threshold response density, α_{jj}); (2) an upper threshold of X_j where the effects of crowding become severe enough to reduce ingestion to the maintenance requirements and thus reduce growth to zero (an equilibrium threshold or carrying capacity γ_{jj}); and (3) the functional relationship between a unit change in X_j and the change in $f(x_j)$.

The additional term with λ_j (specific rate of maintenance cost) and ϵ_{ij} (proportion of F_{ij} egested) is necessary because the upper asymptotic density γ_{jj} is defined as the equilibrium point, that is, growth rate = 0, not as the point where $F_{ij} = 0$. Then when $X_j \leq \alpha_{jj}$ $f(X_j) = 1$, and when $X_j = \gamma_{jj}$, then Eq. (5) applies.

$$f(X_j) = \left(\frac{\lambda_j}{\tau_{ij}(1-\epsilon_{ij})}\right) \qquad (5)$$

Note in Eq. (4) that the interspecific interference effect of a competitor species can be easily included by writing the numerator $(X_j - \beta_{jk}X_k - \alpha_{jj})_+$, where β_{jk} is the competition coefficient and X_k is the competitor population. Also, if when $\phi = 1$, the response threshold $\alpha_{jj} = 0$ (an assumption of the logistic equation), Eq. (4) reduces to the logistic. As with Eq. (3), the exponent ϕ is used to specify whether the change in $f(X_j)$ with change in X_j is constant ($\phi = 1$), decreasing with increasing X_j ($\phi > 1$) or increasing with decreasing X_j ($\phi < 1$). These nonlinear responses could be related directly to changes in the behavior of competitors with change in density.

When both scarcity of resources and direct interference as a result of crowding are acting as limiters, the simplest assumption is that the effect is multiplicative, and the overall flux is given in Eq. (2).

Under the simplest assumption of multiplicitivity, for example, a predator whose prey was so scarce that catch per unit time was half the maximum [$f(X_i) = 0.5$] and that intragroup strife consumed half the time normally devoted to searching for prey [$f(X_j) - 0.5$] would show an ingestion rate 0.25 of the maximum. Additivity of control functions instead of a multiplicative effect could be used if indicated, or more complex relationships could be incorporated into the basic Eq. (2). In particular, when more than one resource is utilized, some allocation preference factor must be used in each flux equation.

The use of such as equational structure produces models that combine aspects of generality, reality, and precision. Additional modifications in the mathematical functions can simulate the effects of time delays, spatial and temporal heterogeneity, variable parameters, and stochastic terms. Such equations are nonlinear and discontinuous, yet models constructed in this manner are stable in the sense that compartments cannot increase to infinity or decrease to extinction without a biological reason. Most important, every state variable and parameter employed has a biological analog and a procedure outlined for its observational or experimental measurement.

Overview. As the pressure of human-induced detrimental changes in ecosystems intensifies via pollution, exploitation, or outright destruction, the need increases for fast, accurate allocation of research effort to assess the effects of such perturbations. Simulation models, properly constructed and used, can help to meet this need.

[RICHARD G. WIEGERT]

Electric power generation

The production of bulk electric power for industrial, residential, and rural use. Although limited amounts of electricity can be generated by many means, including chemical reaction (as in batteries) and engine-driven generators (as in automobiles and airplanes), electric power generation generally implies large-scale production of electric power in stationary plants designed for that purpose. The generating units in these plants convert energy from falling water, coal, natural gas, oil, and nuclear fuels to electric energy. Most electric generators are driven either by hydraulic turbines, for conversion of falling water energy; or by steam or gas turbines, for conversion of fuel energy. Limited use is being made of geothermal energy, and developmental work is progressing in the use of solar energy in its various forms. Electric power generating plants are normally interconnected by a transmission and distribution system to serve the electric loads in a given area or region.

An electric load is the power requirement of any device or equipment that converts electric energy into light, heat, or mechanical energy, or otherwise consumes electric energy as in aluminum reduction, or the power requirements of electronic and control devices. The total load on any power system is seldom constant; rather, it varies widely with hourly, weekly, monthly, or annual changes in the requirements of the area served. The minimum system load for a given period is termed the base load or the unity load-factor component. Maximum loads, resulting usually from temporary conditions, are called peak loads. Electric energy cannot feasibly be stored in large quantities; therefore the operation of the generating plants must be closely coordinated with fluctuations in the load.

Actual variations in the load with time are recorded, and from these data load graphs are made to forecast the probable variations of load in the future. A study of hourly load graphs (Figs. 1 and 2) indicates the generation that may be required at a given hour of the day, week, or month, or under unusual weather conditions. A study of annual load graphs and forecasts indicates the rate at which new generating stations must be built. Load graphs and forecasts are an inseparable part of utility operation and are the basis for decisions that profoundly affect the financial requirements and overall development of a utility.

Generating plants. Often termed generating stations, these plants contain apparatus that converts some form of energy to electric energy in bulk. Three significant types of generating plants are hydroelectric, fossil-fuel-electric, and nuclear-electric.

Hydroelectric plant. This type of generating plant utilizes the potential energy released by the weight of water falling through a vertical distance called head. Ignoring losses, the power, in horsepower (hp) and kilowatts (kW), obtainable from falling water is shown in the equations below (metric quantities in brackets).

$$\text{hp} = \frac{\begin{pmatrix}\text{quantity of water}\\ \text{in ft}^3/\text{s } [\text{m}^3/\text{s}]\end{pmatrix}\begin{pmatrix}\text{vertical head}\\ \text{in ft } [\text{m}]\end{pmatrix}}{8.8\ [0.077]}$$

$$\text{kW} = 0.746 \text{ hp}$$

A plant consists basically of a dam to store the water in a forebay and create part or all of the head, a penstock to deliver the falling water to the turbine, a hydraulic turbine to convert the hydraulic energy released to mechanical energy, an alternating-current generator (alternator) to convert the mechanical energy to electric energy, and all accessory equipment necessary to control the power flow, voltage, and frequency, and to afford the protection required (Fig. 3).

Pumped storage hydroelectric plants are being used increasingly. Under suitable geographical and geological conditions, electric energy can, in effect, be stored by pumping water from a low to a higher elevation and subsequently releasing this water to the lower elevation through hydraulic turbines. These turbines and their associated generators are reversible. The generators, operating in reverse direction as motors, drive their turbines as pumps to elevate the water. When this water is released through the turbines, electric power is produced by the generators. A relatively high overall cycle efficiency can be attained, usually of the order of 65–75%.

Since system peak loads are usually of relatively short duration (Figs. 1 and 2), the high output available for a short time from pumped storage can be used to supply this peak. During off-peak hours, that is, 10 P.M. to 7 A.M., the surplus generating capacity of the most economical system energy resources can be used to return the water by pumping to the elevated storage space for use on the next peak. This type of operation assists in maintaining a high capacity factor on prime generation with resulting best economy.

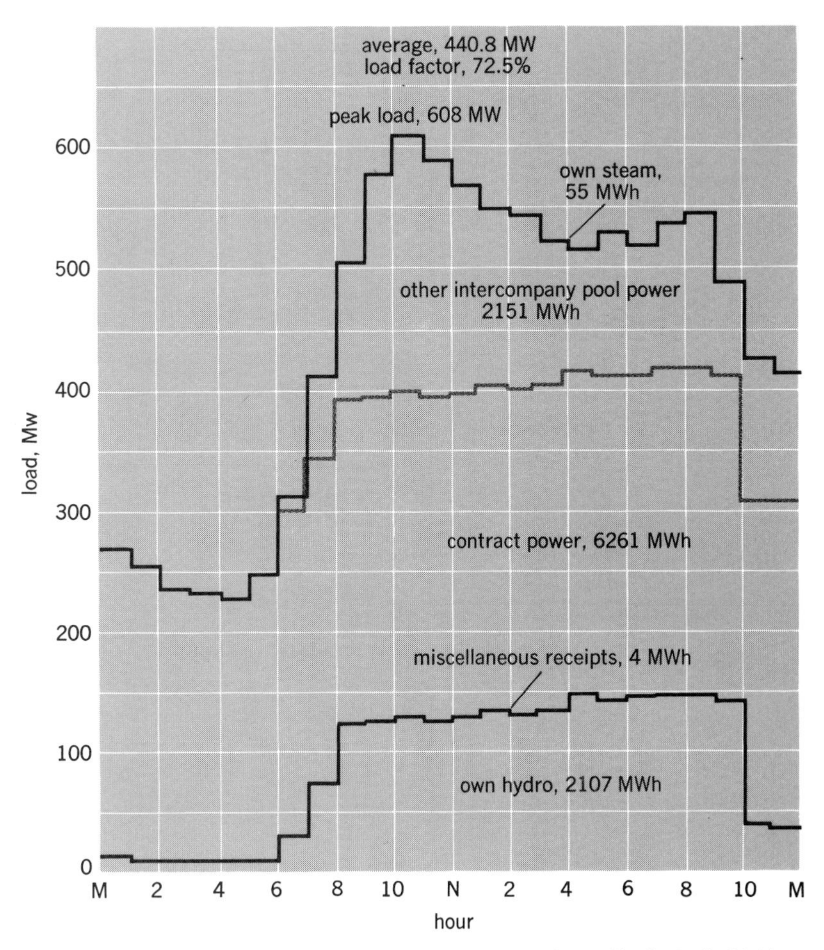

Fig. 1. Load graph indicates net system load of a metropolitan utility for typical 24-hr period (midnight to midnight), totaling 10,578 MWh. Such graphs are made to forecast probable variations in power required.

Pumped storage plants can be brought up to load much faster than large steam plants and, hence, contribute to system reliability by providing an immediately available reserve against the unscheduled loss of other generation.

Fossil-fuel-electric plant. This type utilizes the energy of combustion from coal, oil, or natural gas. A typical large plant (Fig. 4) consists of fuel processing and handling facilities, a combustion furnace and boiler to produce and superheat the steam, a steam turbine, an alternator, and the accessory equipment required for plant protection and for control of voltage, frequency, and power flow. A steam plant can frequently be built near a convenient load center, provided an adequate supply of cooling water and fuel is available, and is usually readily adaptable to either base loading or intermediate or peak loading. Environmental constraints require careful control of stack emissions with respect to sulfur oxides and particulates. Cooling towers or ponds are often required for

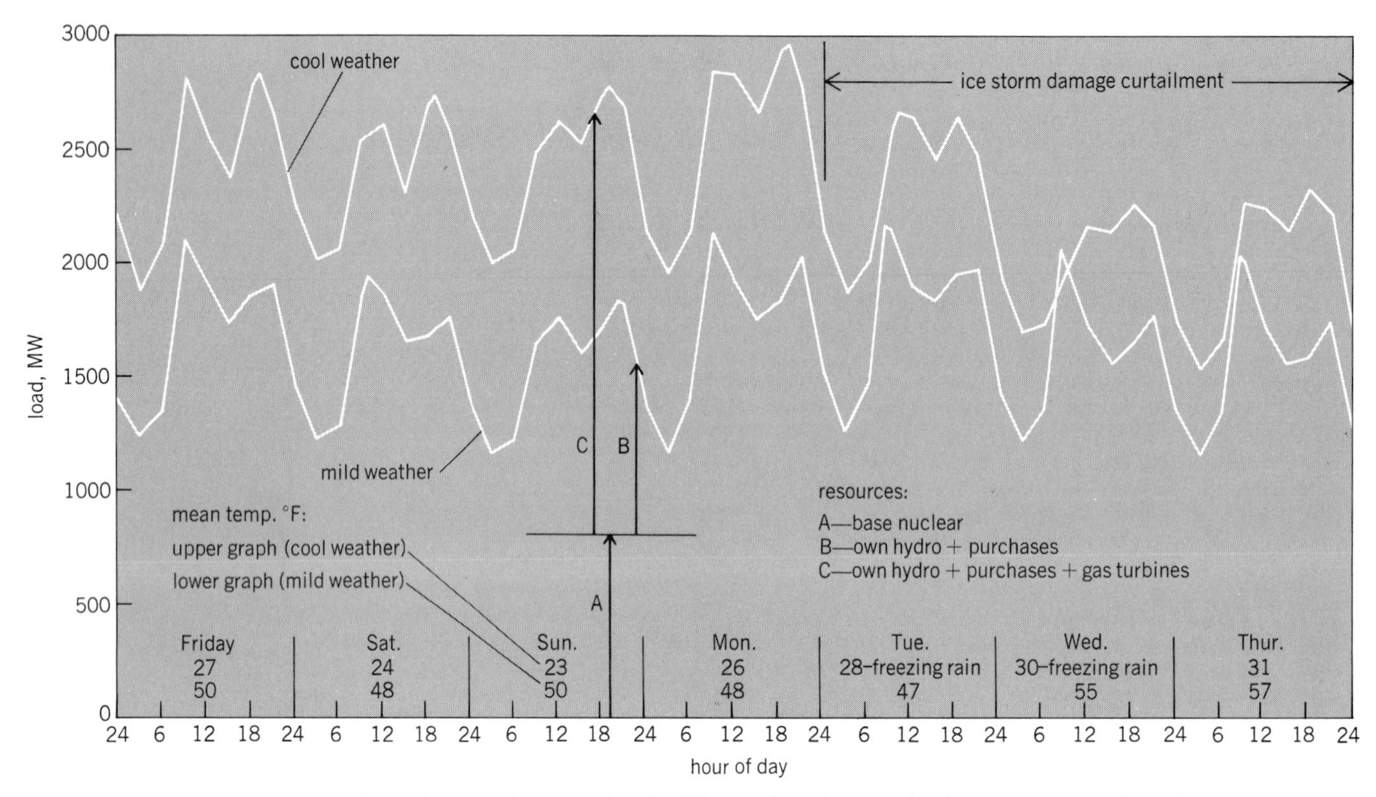

Fig. 2. Examples of northwestern electric utility weekly load curves showing same-year weather influence.

waste heat dissipation. Gas turbine plants do not require condenser cooling water (unless combined with a steam cycle), have a relatively low unit capital cost and relatively high unit fuel cost, and are widely used for peaking service. Progress is being made in the development of magnetohydrodynamic (MHD) "topping" generators to be used in conjuction with normal steam turbines to improve the overall thermal conversion efficiency.

Nuclear electric plant. In this type of plant one or more of the nuclear fuels are utilized in a suitable type of nuclear reactor, which takes the place of the combustion furnace in the typical steam electric plant. The heat exchangers and boilers (if not combined in the reactor), the turbines, and alternating-current generators, complete with controls, accessories, and auxiliaries, make up the atomic electric plant. Large-scale fission reaction plants have been developed to the point where they are economically competitive in much of the United States, and many millions of kilowatts of capacity are under construction and more on order. The current and projected future growth of the nuclear power industry in the United States is shown graphically in Figs. 5 and 6. Although in 1978 only approximately 9% of the total generating capacity was nuclear, by 1989 it is predicted to become over 21%. However, in 1978 about 14% of the net electric energy production was nuclear and is forecast to become more than 22% by 1988. The coal-fired share in 1978 was approximately 45% and is forecast to become about 50% in 1988. Conventional hydro generation as a source of prime electric energy had reached near-saturation by 1978, and growth in fossil-fuel generation shows a declining trend. However, nuclear generation exhibits a doubling time of about 51/2 years during a considerable period after 1978. Recent operating events may slow this growth somewhat.

Three types of nuclear plants are in general use and a fourth is being developed actively. The pressurized and boiling light-water types, shown schematically in Figs. 7 and 8, have been highly developed in the United States. The pressurized heavy-water type has been well developed in Canada. The gas-cooled type, one version of which is shown schematically in Fig. 9, has been highly developed and extensively used in Great Britain. The liquid-metal (sodium) cooled type is under intensive development and will be featured prominently in the fast or breeder reactor field. It is shown schematically in Fig. 10. As in the high-temperature gas-cooled reactor system, steam temperatures and pressures in the liquid-metal systems can be essentially the same as those recorded in the plants powered by fossil fuel and can give similar thermal efficiencies.

Several other types of nuclear fission reactor systems are receiving attention, and some may be expected to become commercially competitive. Fusion reaction nuclear plants are in the early research and development stage with possible commercialization early in the next century. Direct conversion from nuclear reaction energy to electric energy on a commercial scale for power utility service is a future possibility but is not economically feasible at present.

Generating unit sizes. The size or capacity of electric utility generating units varies widely, depending upon type of unit; duty required, that is base-, intermediate-, or peak-load service; and system size and degree of interconnection with neighboring systems. Base-load nuclear or coal-fired units may be as large as 1200 MW each, or more. Intermediate-duty generators, usually coal-, oil-, or gas-fueled steam units, are typically of 200 to 600 MW capacity each. Peaking units, combustion turbines or hydro, range from several tens of megawatts for the former to hundreds of megawatts for the latter. Hydro units, in both base-load and intermediate service, range in size up to 700 MW.

The total installed generating capacity of a system is typically 20 to 30% greater than the annual predicted peak load in order to provide reserves for maintenance and contingencies.

Power-plant circuits. Both main and accessory circuits in power plants can be classified as follows:

1. Main power circuits to carry the power from the generators to the step-up transformers and on to the station high-voltage terminals.
2. Auxiliary power circuits to provide power to the motors used to drive the necessary auxiliaries.
3. Control circuits for the circuit breakers and other equipment operated from the control room of the plant.
4. Lighting circuits for the illumination of the plant and to provide power for portable equipment required in the upkeep and maintenance of the plant. Sometimes special circuits are installed to supply the portable power equipment.
5. Excitation circuits, which are so installed that they will receive good physical and electrical protection because reliable excitation is necessary for the operation of the plant.
6. Instrument and relay circuits to provide values of voltage, current, kilowatts, reactive kilovolt-amperes, temperatures, and pressures, and to serve the protective relays.
7. Communication circuits for both plant and system communications. Telephone, radio, transmission-line carrier, and microwave radio may be involved.

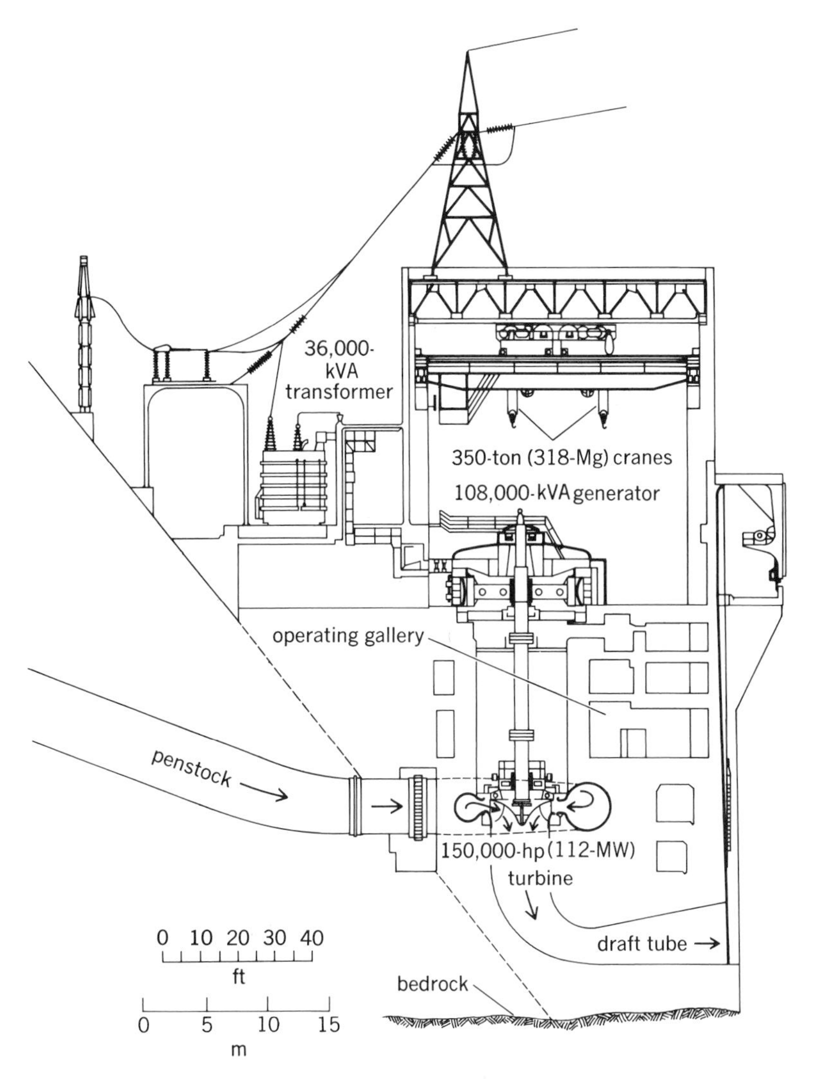

Fig. 3. Typical layout and apparatus arrangement in a hydroelectric generating plant. Cross section is made through the powerhouse and dam at a main generator position in the original Grand Coulee plant.

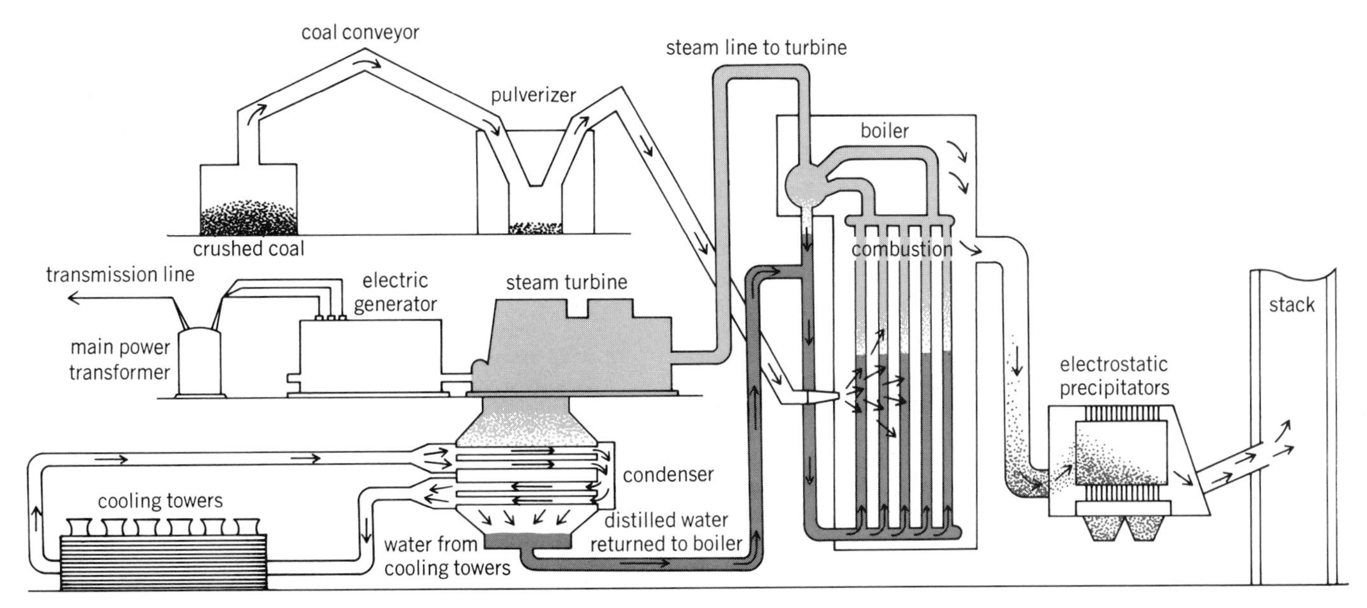

Fig. 4. Schematic of typical coal-fired steam electric power plant. *(Pacific Power & Light Co.)*

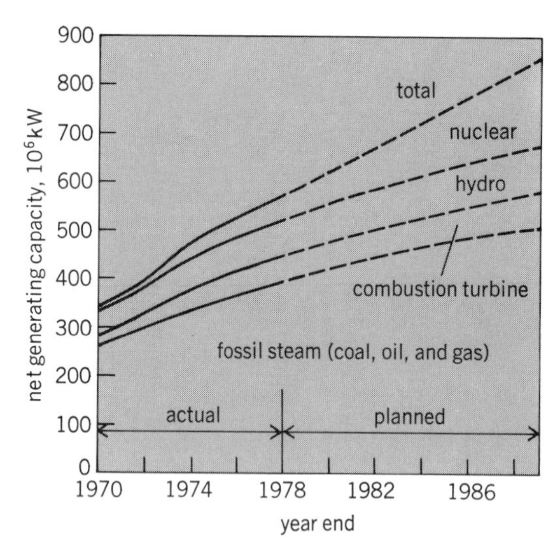

Fig. 5. Net generating capacity of the United States electric power industry. Hydro includes pumped storage. Combustion turbine includes internal combustion.

It is important that reliable power service be provided for the plant itself, and for this reason station service is usually supplied from two or more sources. To ensure adequate reliability, auxiliary power supplies are frequently provided for start-up, shut-down, and communication services.

Generator protection. Necessary devices are installed to prevent or minimize other damage in cases of equipment failure. Differential-current and ground relays detect failure of insulation, which may be due to deterioration or accidental overvoltage. Overcurrent relays detect overload currents that may lead to excessive heating; overvoltage relays prevent insulation damage. Loss-of-excitation relays may be used to warn operators of low excitation or to prevent pulling out of synchronism. Bearing and winding overheating may be detected by relays actuated by resistance devices or thermocouples. Overspeed and lubrication failure may also be detected.

Not all of these devices are used on small units or in every plant. The generator is immediately deenergized for electrical failure and shut down for any over-limit condition, all usually automatically.

Voltage regulation. This term is defined as change in voltage for specific change in load (usually from full load to no load) expressed as percentage of normal rated voltage. The voltage of an electric generator varies with the load and power factor; consequently, some form of regulating equipment is required to maintain a reasonably constant and predetermined potential at the distribution stations or load centers. Since the inherent regulation of most alternating-current generators is rather poor (that is, high percentagewise), it is necessary to provide automatic voltage control. The rotating or magnetic amplifiers and voltage-sensitive circuits of the automatic regulators, together with the exciters, are all specially designed to respond quickly to changes in the alternator voltage and to make the necessary changes in the main exciter output, thus providing the required adjustments in voltage. A properly designed automatic regulator acts rapidly, so that it is possible to maintain desired voltage with a rapidly fluctuating load without causing more than a momentary change in voltage even when heavy loads are thrown on or off.

Electronic voltage control has been adapted to some generator and synchronous condenser installations. Its main advantages are its speed of operation and its sensitivity to small voltage variations. As the reliability and ruggedness of electronic components are improved, this form of voltage regulator will become more common.

Generation control. Computer-assisted (or on-line controlled) load and frequency control and economic dispatch systems of generation supervision are being widely adopted, particularly for the larger new plants. Strong system interconnections greatly improve bulk power supply reliability but require special automatic controls to ensure adequate generation and transmission stability. Among the refinements found necessary in large, long-distance interconnections are special feedback controls applied to generator high-speed excitation and voltage regulator systems.

Synchronization of generators. Synchronization of a generator to a power system is the act of matching, over an appreciable period of time, the instantaneous voltage of an alternating-current generator (incoming source) to the instantaneous voltage of a power system of one or more other generators (running source), then connecting them together. In order to accomplish this ideally the following conditions must be met:

1. The effective voltage of the incoming generator must be substantially the same as that of the system.

2. In relation to each other the generator voltage and the system voltage should be essentially 180° out of phase; however, in relation to the bus to which they are connected, their voltages should be in phase.

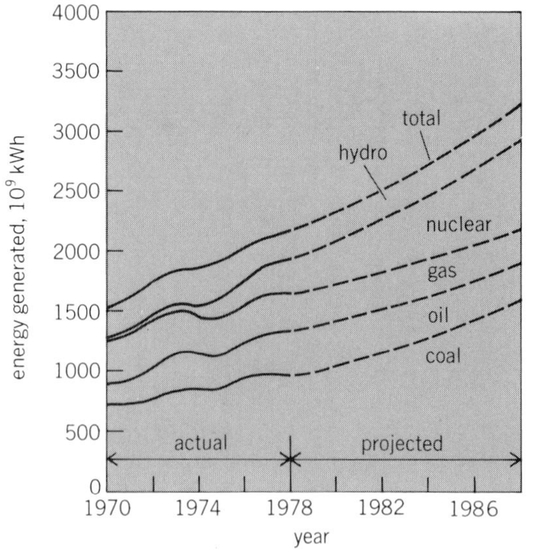

Fig. 6. Net energy production of the United States electric power industry.

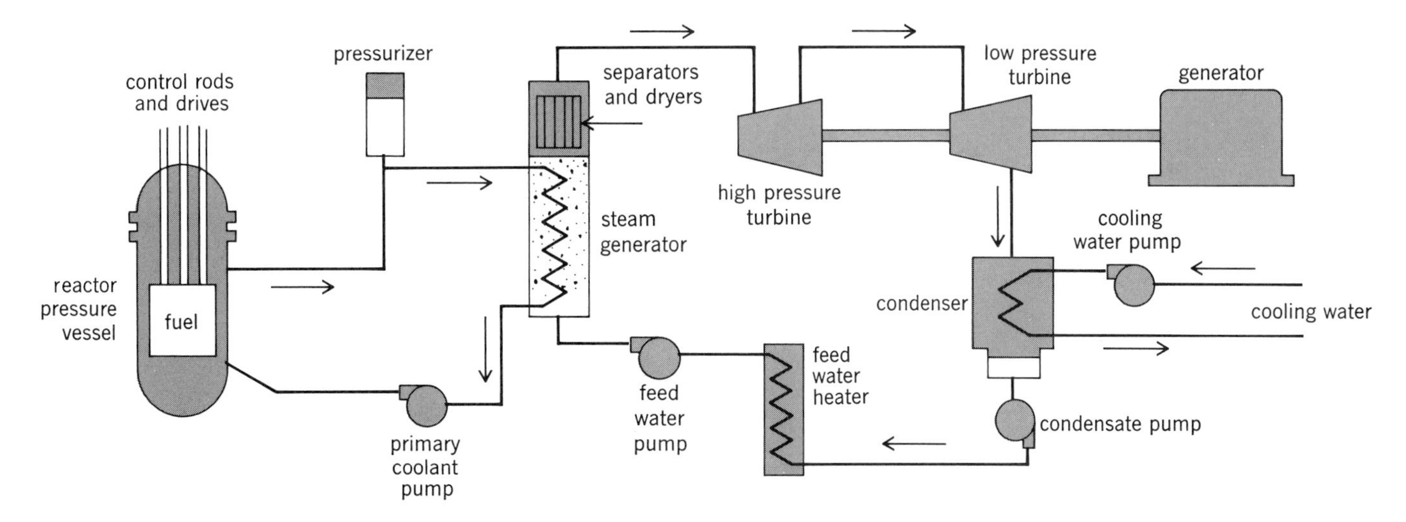

fuel: slightly enriched uranium oxide with zirconium alloy
moderator: water coolant: water pressure of primary system: 2250 psi (15.5 MPa)
reactor outlet temperature: 605°F (318°C)

Fig. 7. Schematic of a pressurized-water reactor plant. *(From the Nuclear Industry, USAEC, WASH 1174–73, 1973)*

3. The frequency of the incoming machine must be near that of the running system.
4. The voltage wave shapes should be similar.
5. The phase sequence of the incoming polyphase machine must be the same as that of the system.

Synchronizing of ac generators can be done manually or automatically. In manual synchronizing an operator controls the incoming generator while observing synchronizing lamps or meters and a synchroscope, or both. Voltage (potential) transformers may be used to provide voltages at lamp and instrument ratings. Lamps properly connected between the two sources are continuously dark when voltage, phase, and frequency are properly matched. Wave shape and phase sequence are determined by machine design and rotation or terminal sequence. Large units generally are provided with voltmeters and frequency meters for matching these quantities, and a synchroscope connected to both sources to indicate phase relationship. Lamps may also be included. The standard synchroscope needle revolves counterclockwise when the incoming machine is slow and clockwise when fast. The needle points straight up when the two sources are in phase. The operator closes the connecting switch or circuit breaker as the synchroscope needle slowly approaches the in-phase position.

Automatic synchronizing provides for automatically closing the breaker to connect the incoming machine to the system, after the operator has properly adjusted voltage (field current), frequency (speed), and phasing (by lamps or synchroscope). A fully automatic synchronizer will initiate speed changes as required and may also balance voltages as required, then close the breaker at the proper time, all without attention of the operator. Automatic synchronizers can be used in unattended

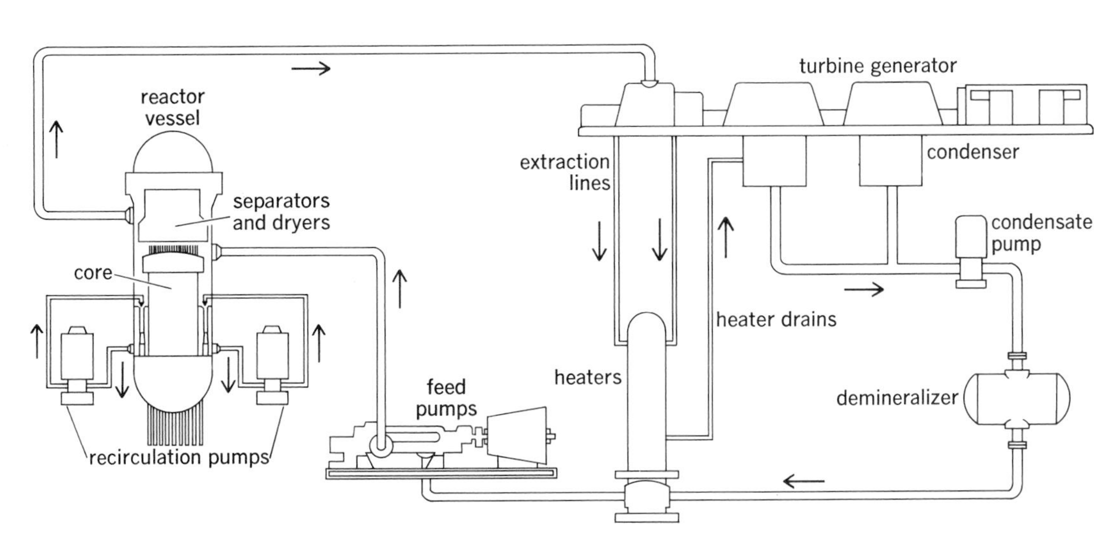

Fig. 8. Single-cycle boiling-water reactor system flow diagram. *(General Electric Co.)*

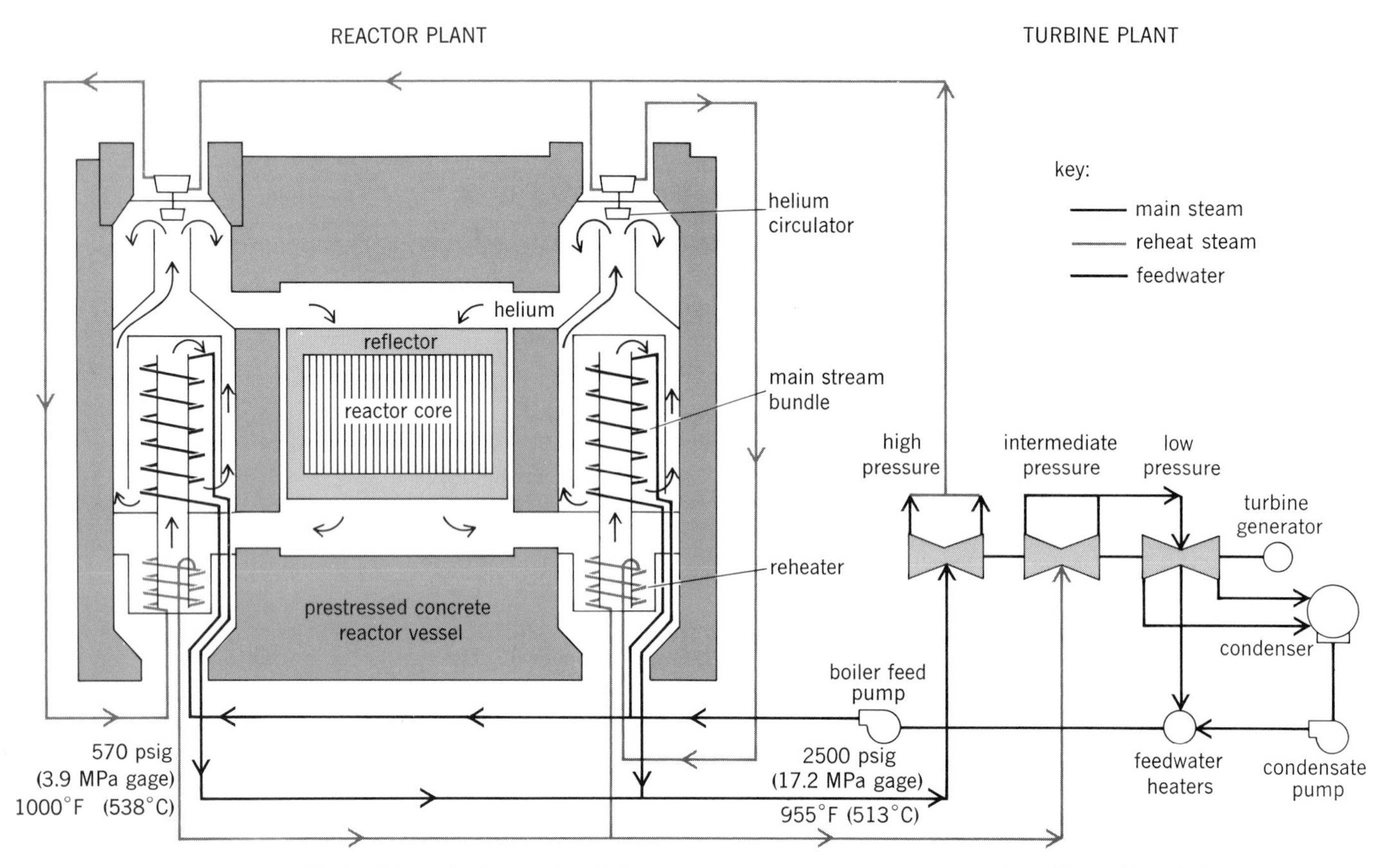

Fig. 9. Schematic diagram for a high-temperature gas-cooled reactor power plant. (*General Atomic Corp.*)

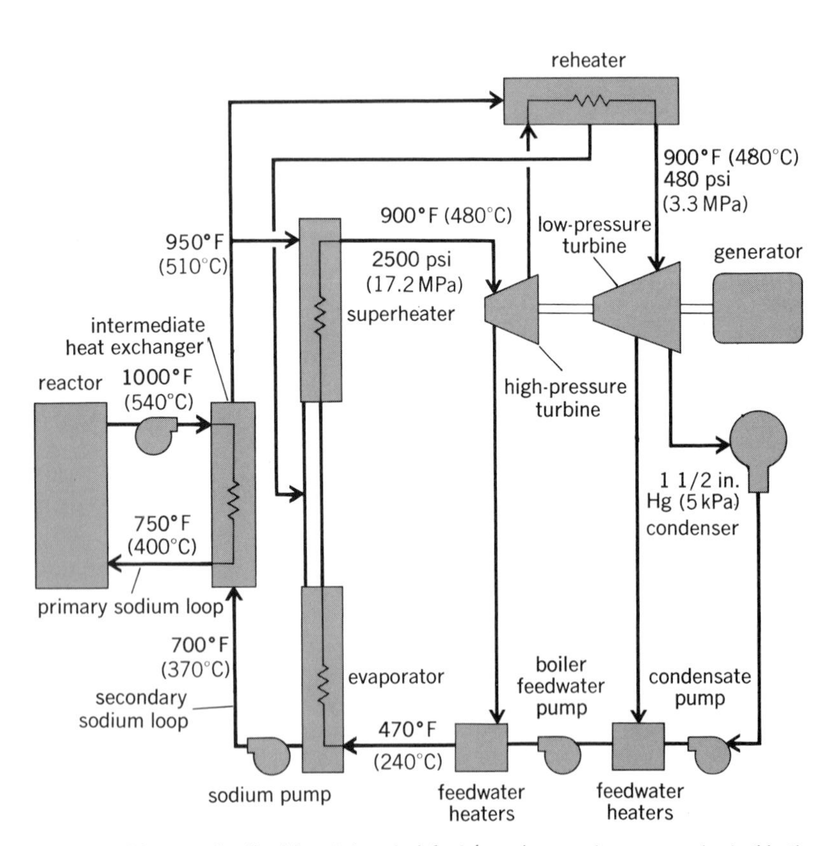

Fig. 10. Diagram for liquid-metal-cooled fast breeder reactor power plant. (*North American Rockwell Corp.*)

stations or in automatic control systems where units may be started, synchronized, and loaded on a single operator command. *See* ELECTRIC POWER SYSTEMS.

[EUGENE C. STARR]

Bibliography: H. C. Barnes et al., Alternator-rectifier exciter for Cardinal Plant 724-MVA generation, *IEEE Trans. Power App. Syst.*, PAS-87 (4): 1189, 1968; J. G. Brown, *Hydro-Electric Engineering Practice*, 1958; P. H. Cootner, *Water Demand for Steam Electric Generation*, 1966; K. Fenton, *Thermal Efficiency and Power Production*, 1966; S. Glasstone and A. Sesonske, *Nuclear Reactor Engineering*, 1967; R. F. Grundy (ed.), *Magnetohydrodynamic Energy for Electric Power Generation*, Noyes Data Corporation, 1978; G. E. Kholodovskii, *Principles of Power Generation*, 1965; A. H. Lovell, *Generating Stations: Economic Elements of Electrical Design*, 4th ed., 1951; F. T. Morse, *Power Plant Engineering: The Theory and Practice of Stationary Electric Generating Plants*, 3d ed., 1953; F. R. Schleif et al., Excitation control to improve powerline stability, *IEEE Trans. Power App. Sys.*, PAS-87(6):1426–1432, 1968; B. G. A. Skrotzki, *Electric Generation: Hydro, Diesel, and Gas-Turbine Stations*, 1956; P. Sporn, *Energy: Its Production, Conversion and Use in the Service of Man*, 1963; *Standard Handbook for Electrical Engineers*, 10th ed., sect. 6, 8, 9, and 10, 1968; C. D. Swift, *Steam Power Plants: Starting, Testing and Operation*, 1959; H. G. Tak, *Economic Choice Between Hydroelectric and Thermal Power Developments*, 1966.

Electric power systems

A complex assemblage of equipment and circuits for generating, transmitting, transforming, and distributing electrical energy. In the United States, electrical energy is generated to serve more than 87,625,000 customers. The investment represented by these facilities has grown rapidly over the years until in 1978 it was close to $270,000,000,000, with about 75% spent on the power systems of investor-owned utilities and the remainder spread among systems built by governmental agencies, municipal electric departments, and rural electric cooperatives. Principal elements of a typical power system are shown in Fig. 1.

Generation. Electricity in the large quantities required to supply electric power systems is produced in generating stations, commonly called power plants. Such generating stations, however, should be considered as conversion facilities in which the heat energy of fuel (coal, oil, gas, or uranium) or the hydraulic energy of falling water is converted to electricity. See ELECTRIC POWER GENERATION.

Steam stations. About 89% of the electric power used in the United States is obtained from generators driven by steam turbines. The largest such unit in service in 1978 was rated 1300 MW, equivalent to about 1,730,000 hp. But 650-, 800- and 950-MW units are commonplace for new fossil-fuel–fired stations, and 1150–1300-MW units are the most commonly installed units in nuclear stations.

Coal was the fuel for nearly 51% of the steam turbine generation in 1978, and its share should increase somewhat because of the projected long-term shortage of natural gas and both the sharp rise in the cost of fuel oil and governmental policy of restricting firing of oil. Natural gas, used extensively in the southern part of the United States, fueled about 16% of the steam turbine generation, and heavy fuel oil, 19%, largely in power plants able to take delivery from ocean-going tankers or river barges. The remaining 14% was generated from the radioactive energy of slightly enriched

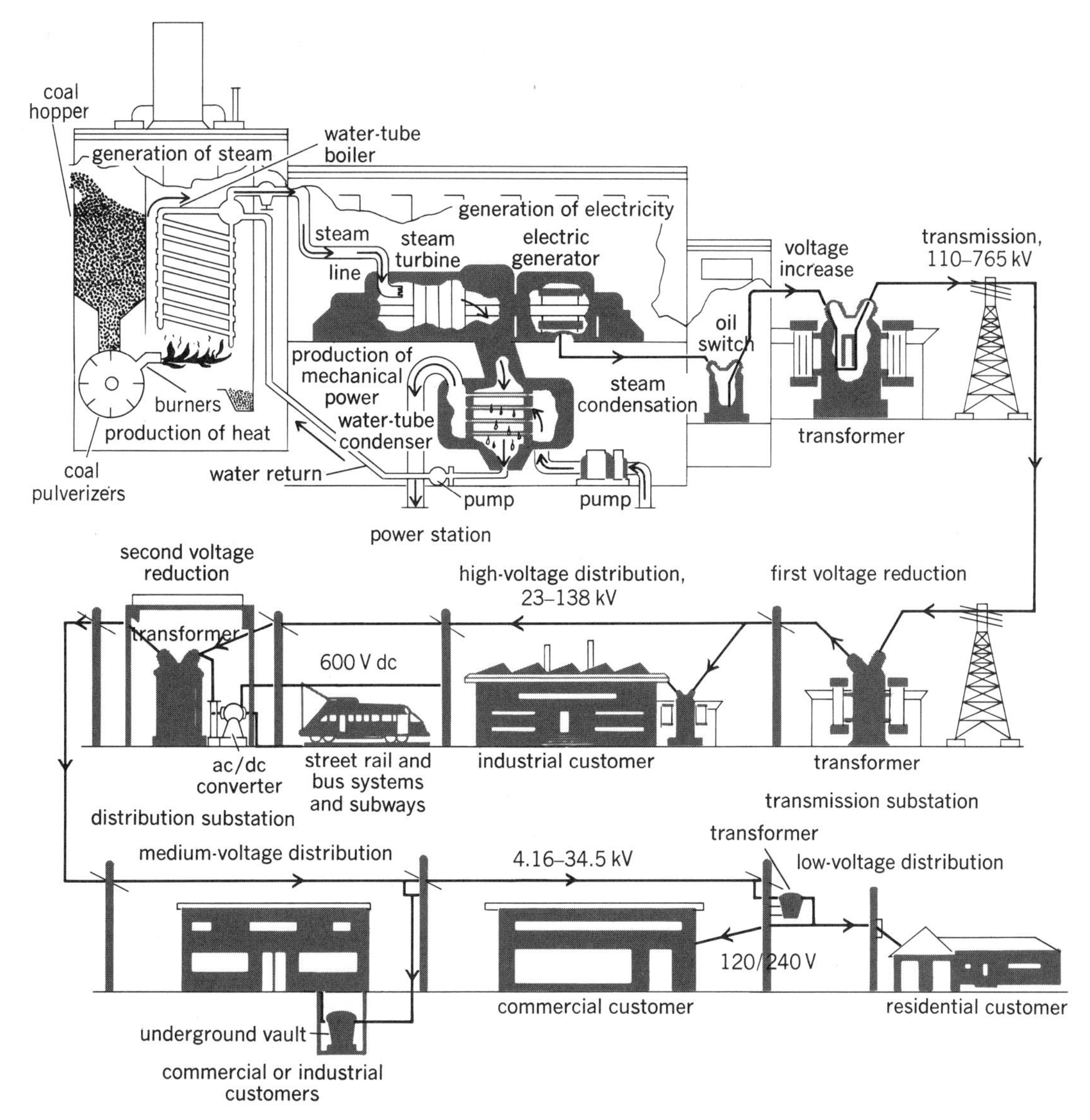

Fig. 1. Major steps in the generation, transmission, and distribution of electricity.

uranium, which, for many power systems, produces electricity at a lower total cost than either coal or fuel oil at current delivered prices. As a consequence, some 46% of the generating capability additions planned as of Jan. 1, 1979, according to an *Electrical World* survey, will be nuclear and largely in the 900- to 1300-MW-per-unit range. As these nuclear units go into commercial operation, the contribution of uranium to the electrical energy supply will rise, probably to more than 25% of the total fuel generated output by the mid-1980s.

Nuclear steam systems used by United States utilities are mostly of the water-cooled-and-moderated type, in which the heat of a controlled nuclear reaction is used to convert water into steam to drive a conventional turbine generator; such units are presently limited to about 1300 MW by the thermal limit placed on nuclear reactors by the Nuclear Regulatory Commission of the U.S. Department of Energy.

Hydroelectric plants. Waterpower during 1978 supplied about 10.3% of the electric power consumed in the United States. But this share can only decline in the years ahead because very few sites remain undeveloped where sufficient water drops far enough in a reasonable distance to drive reasonably sized hydraulic turbines. Consequently, the generating capability of hydro plants, 10.4% of the utility industry's total as of Dec. 31, 1978, is slated to fall off to 8.9% by the mid-1980s because of its very small share of the planned additions. Much of this additional hydro capability will be used at existing plants to increase their effectiveness in supplying peak power demands, and as a quickly available source of emergency power.

Some hydro plants totaling 10,640 MW as of Dec. 31, 1978, actually draw power from other generating facilities during light system-load periods to pump water from a river or lake into an artificial reservoir at a higher elevation from which it can be drawn through a hydraulic station when the power system needs additional generation. These pumped-storage installations consume about 50% more energy than they return to the power system and, accordingly, cannot be considered energy sources. Their use is justified, however, by their ability to convert surplus power that is available during low-demand periods into prime power to serve system needs during peak-demand intervals—a need that otherwise would require building more generating stations for operation during the relatively few hours of high system demand. Installations now planned should double the existing capacity by the late 1980s to 22,684 MW.

Combustion turbine plants. Gas-turbine-driven generators, now commonly called combustion turbines because of the growing use of light oil as fuel, have gained wide acceptance as an economical source of additional power for heavy-load periods. In addition, they offer the fastest erection time and the lowest investment cost per kilowatt of installed capability. Offsetting these advantages, however, is their relatively less efficient consumption of more costly fuel. Typical unit ratings in the United States have climbed rapidly in recent years until some units operating in 1979 are rated at 80 MW. Some turbine installations involve a group of smaller units totaling, in one case, 260 MW. Combustion turbine units, even in the larger ratings, offer extremely flexible operation and can be started and run up to full load in as little as 10 min. Thus they are extremely useful as emergency power sources, as well as for operating during the few hours of daily load peaks. Combustion turbines totaled 8.5% of the total installed capability of United States utility systems at the close of 1978 and supplied less than 3% of the total energy generated.

In the years ahead, however, combustion turbines are slated for an additional role. Several installations in the 1970s have used their exhaust gases to heat boilers that generate steam to drive steam turbine generators. Such combined-cycle units offer fuel economy comparable to that of modern steam plants and at considerably less cost per kilowatt. In addition, because only part of the plant uses steam, the requirement for cooling water is considerably reduced. A number of additional combined-cycle installations have been planned, but wide acceptance is inhibited by the doubtful availability of light fuel oil for them. This barrier should be resolved, in time, by the successful development of systems for fueling them with gas derived from coal. *See* COAL GASIFICATION.

Internal combustion plants. Internal combustion engines of the diesel type drive generators in many small power plants. In addition, they offer the ability to start quickly for operation during peak loads or emergencies. However, their small size, commonly about 2 MW per unit although a few approach 10 MW, has limited their use. Such installations account for about 1% of the total power-system generating capability in the United States, and make an even smaller contribution to total electric energy consumed.

Three-phase output. Because of their simplicity and efficient use of conductors, three-phase 60-Hz alternating-current systems are used almost exclusively in the United States. Consequently, power-system generators are wound for three-phase output at a voltage usually limited by design features to a range from about 11 kV for small units to 30 kV for large ones. The output of modern generating stations is usually stepped up by transformers to the voltage level of transmission circuits used to deliver power to distant load areas.

Transmission. The transmission system carries electric power efficiently and in large amounts from generating stations to consumption areas. Such transmission is also used to interconnect adjacent power systems for mutual assistance in case of emergency and to gain for the interconnected power systems the economies possible in regional operation. Interconnections have expanded to the point where most of the generation east of the Rocky Mountains, except for a large part of Texas, regularly operates in parallel, and over 90% of all generation in the United States, exclusive of Alaska and Hawaii, and in Canada can be linked.

Transmission circuits are designed to operate up to 765 kV, depending on the amount of power to be carried and the distance to be traveled. The permissible power loading of a circuit depends on many factors, such as the thermal limit of the conductors and their clearances to ground, the voltage drop between the sending and receiving end and the degree to which system service reliability depends on it, and how much the circuit is needed to hold various generating stations in synchronism. A widely accepted approximation to the voltage appropriate for a transmission circuit is that the permissible load-carrying ability varies as the square of the voltage. Typical ratings are listed in Table 1.

Transmission as a distinct function began about 1886 with a 17-mi (27-km) 2-kV line in Italy. Transmission began at about the same time in the United States, and by 1891 a 10-kV line was operating (Fig. 2). In 1896 an 11-kV three-phase line brought electrical energy generated at Niagara Falls to Buffalo, 20 mi (32 km) away. Subsequent lines were built at successively higher levels until 1936, when the Los Angeles Department of Water and Power energized two lines at 287 kV to transmit 240 MW the 266 mi (428 km) from Hoover Dam on the Colorado River to Los Angeles. A third line was completed in 1940.

For nearly 2 decades these three 287-kV lines were the only extra-high-voltage (EHV). lines in North America, if not in the entire world. But in 1946 the American Electric Power (AEP) System inaugurated, with participating manufacturers, a test program up to 500 kV. From this came the basic design for a 345-kV system, the first link of which went into commercial operation in 1953 as part of a system overlay that finally extended from Roanoke, VA, to the outskirts of Chicago. By the late 1960s the 345-kV level had been adopted by many utilities interconnected with the AEP System, as well as others in Illinois, Wisconsin, Minnesota, Kansas, Oklahoma, Texas, New Mexico, Arizona and across New York State into New England.

The development of 500-kV circuits began in 1964, even as the 345-kV level was gaining wide acceptance. One reason for this was that many utilities that had already adopted 230 kV could gain only about 140% in capability by switching to 345 kV, but the jump to 500 kV gave them nearly 400% more capability per circuit. The first line energized at this new level was by Virginia Electric & Power Company to carry the output of a new mine mouth station in West Virginia to its service area. A second line completed the same year provided transmission for a 1500-MW seasonal interchange between the Tennessee Valley Authority and a group of utilities in the Arkansas-Louisiana area. Lines at this voltage level now extend from New Jersey to Texas and from New Mexico via California to British Columbia.

The next and latest step-up occurred in 1969 when the AEP System, after another cooperative test program, completed the first line of an extensive 765-kV system to overlay its earlier 345-kV system. The first installation in this voltage class, however, was by the Quebec Hydro-Electric Commission to carry the output at 735 kV (the expected international standard) from a vast hydro project to its load center at Montreal, some 375 mi (604 km) away.

Table 1. Power capability of typical three-phase open-wire transmission lines

Line-to-line voltage, kV	Capability, MVA
115 ac	60
138 ac	90
230 ac	250
345 ac	600
500 ac	1200
765 ac	2500
800 dc*	1500

*Bipolar line with grounded neutral.

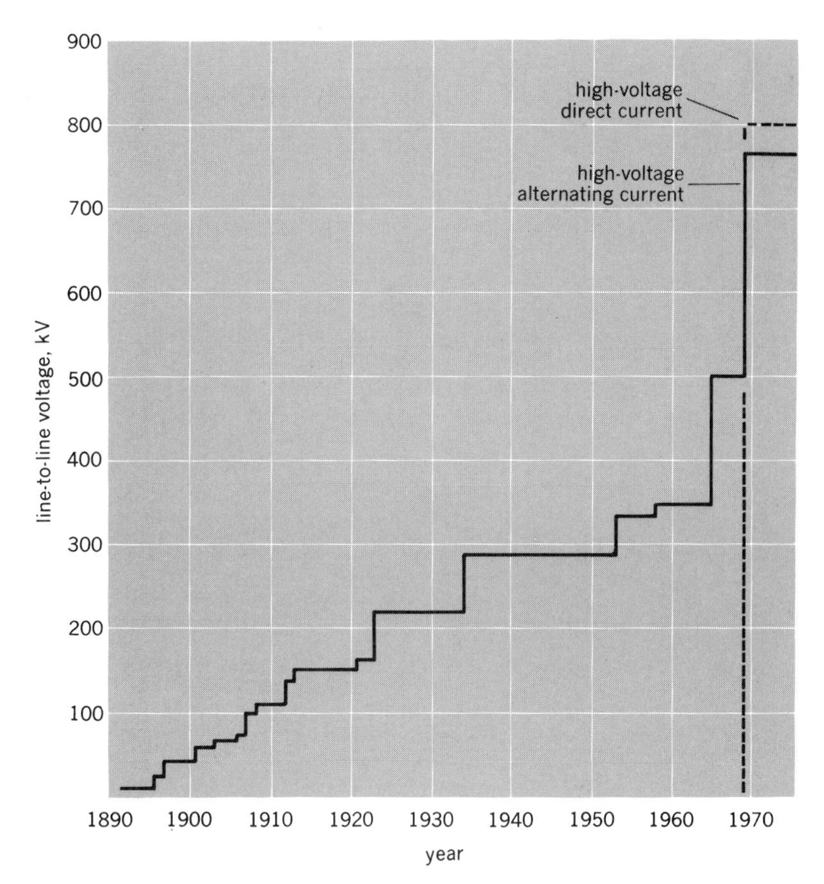

Fig. 2. Growth of ac transmission voltages from 1890.

Transmission engineers are anticipating even higher voltages of 1100 to 1500 kV, but they are fully aware that this objective may prove too costly in space requirements and funds to gain wide acceptance. Experience already gained at 500 kV and 765 kV verifies that the prime requirement no longer is insulating the lines to withstand lightning discharges, but insulating them to tolerate voltage surges caused by the operation of circuit breakers. Audible noise levels, especially in rain or humid conditions, are high, requiring wide buffer zones. Environmental challenges have been brought on the basis of possible negative biological effects of the electrostatic field produced under EHV lines, though research to date has not shown any such effects.

Experience has indicated that, within about 10 years after the introduction of a new voltage level for overhead lines, it becomes necessary to begin connecting underground cable. This has already occurred for 345 kV; the first overhead line was completed in 1953, and by 1967 about 100 mi (160 km) of pipe-type cable had been installed to take power received at this voltage level into metropolitan areas. The first 500-kV cable in the United States was placed in service in 1976 to take power generated at the enormous Grand Coulee hydro plant to a major switchyard several thousand feet (1 ft = 0.3 m) away. And 765-kV cable, after extensive testing above the 500-kV level at the Waltz Mill Cable Test Center operated by Westinghouse Electric Corporation for the Electric Power Research Institute (EPRI), will be available when required.

In anticipation of the need for transmission circuits of higher load capability, an extensive research program is in progress, spread among several large and elaborately equipped research centers. Among these are: Project UHV for ultra-high-voltage overhead lines operated by the General Electric Company near Pittsfield, MA, for EPRI; the Frank B. Black Research Center built and operated by the Ohio Brass Company near Mansfield; the above-mentioned Waltz Mill Cable Test Center; and an 1100 kV, 1-mi (1.6 km) test line built by the Bonneville Power Authority. All include equipment for testing full-scale or cable components at well over 1000 kV. In addition, many utilities and specialty manufacturers have test facilities related to their fields of operation.

A relatively new approach to high-voltage long-distance transmission is high-voltage direct current (HVDC), which offers the advantages of less costly lines, lower transmission losses, and insensitivity to many system problems that restrict alternating-current systems. Its greatest disadvantage is the need for costly equipment for converting the sending-end power to direct current, and for converting the receiving-end direct-current power to alternating current for distribution to consumers. Starting in the late 1950s with a 65-mi (105-km) 100-kV system in Sweden, HVDC has been applied successfully in a series of special cases around the world, each one for a higher voltage and greater power capability. The first such installation in the United States was put into service in 1970. It operates at 800 kV line to line, and is designed to carry a power interchange of 1440 MW over a 1354-km overhead tie line between the winter-peaking Northwest Pacific coastal region and the summer-peaking southern California area. These HVDC lines perform functions other than just power transfer, however. The Pacific Intertie is used to stabilize the parallel alternating-current transmission and lines, permitting an increase in their capability; and back-to-back converters with no tie line between them are used to tie together two systems in Nebraska that otherwise could not be synchronized. The first urban installation of this technology was energized in 1979 in New York City.

In addition to these high-capability circuits, every large utility has many miles of lower-voltage transmission, usually operating at 110 to 345 kV, to carry bulk power to numerous cities, towns, and large industrial plants. These circuits often include extensive lengths of underground cable where they pass through densely populated areas. Their design, construction, and operation are based upon research done some years ago, augmented by extensive experience.

Interconnections. As systems grow and the number and size of generating units increase, and as transmission networks expand, higher levels of bulk-power-system reliability are attained through properly coordinated interconnections among separate systems. This practice began more than 50 years ago with such voluntary pools as the Connecticut Valley Power Exchange and the Pennsylvania-New Jersey-Maryland Interconnection. Most of the electric utilities in the contiguous United States and a large part of Canada now operate as members of power pools, and these pools (except one in Texas) in turn are interconnected into one gigantic power grid known as the North American Power Systems Interconnection. The operation of this interconnection, in turn, is coordinated by the North American Power Systems Interconnection Committee (NAPSIC). Each individual utility in such pools operates independently, but has contractual arrangements with other members in respect to generation additions and scheduling of operation. Their participation in a power pool affords a higher level of service reliability and important economic advantages.

Regional and national coordination. The Northeast blackout of Nov. 9, 1965, stemmed from the unexpected trip-out of a key transmission circuit carrying emergency power into Canada and cascaded throughout the northeastern states to cut off electric service to some 30,000,000 people. It spurred the utilities into a chain reaction affecting the planning, construction, operation, and control procedures for their interconnected systems. They soon organized regional coordination councils, eventually nine in number, to cover the entire contiguous United States and four Canadian provinces (Fig. 3). Their stated objective was "to further augment reliability of the parties' planning and operation of their generation and transmission facilities."

Then, in 1968, the National Electric Reliability Council (NERC) was established to serve as an effective body for the collection, unification, and dissemination of various reliability criteria for use by individual utilities in meeting their planning and operating responsibilities. NAPSIC was reorganized shortly afterward to function as an advisory group to NERC.

Increased interconnection capability among power systems reduces the required generation reserve of each of the individual systems. In most utilities the loss-of-load probability (LOLP) is used to measure the reliability of electric service, and it is based on the application of probability theory to unit-outage statistics and load forecasts. A common LOLP criterion is 1 day in 10 years when load may exceed generating capability. The LOLP decreases (that is, reliability increases) with increased interconnection between two areas until a saturation level is reached which depends upon the amount of reserve, unit sizes, and annual load shape in each area. Any increase in interconnection capability beyond that saturation level will not cause a corresponding improvement in the level of system reliability.

Traditionally, systems were planned to withstand all reasonably probable contingencies, and operators seldom had to worry about the possible effect of unscheduled outages. Operators' normal security functions were to maintain adequate generation on-line and to ensure that such system variables as line flows and station voltages remained within the limits specified by planners. However, stronger interconnections, larger generating units, and rapid system growth spread the transient effects of sudden disturbances and increased the

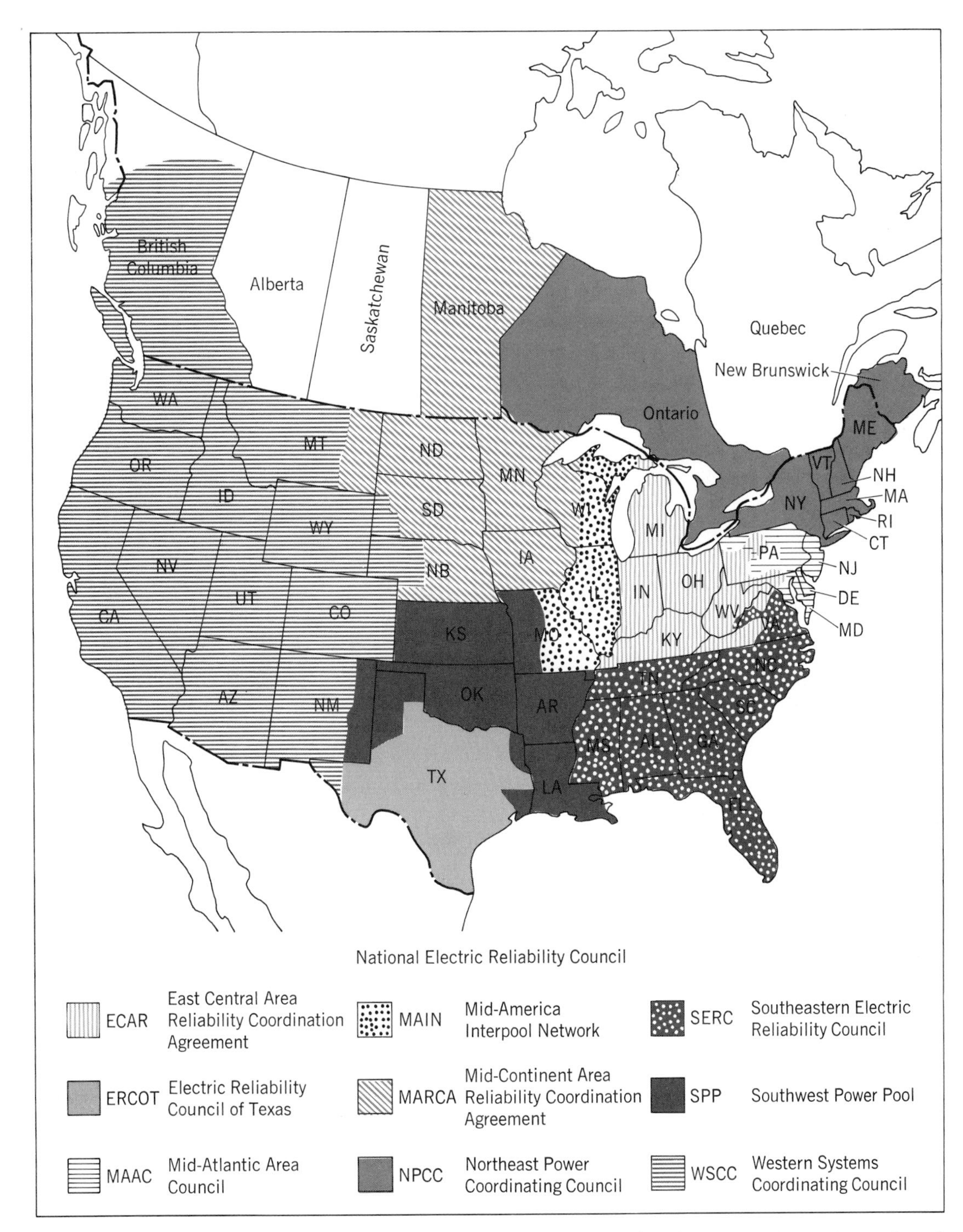

Fig. 3. Areas served by the nine regional reliability councils, coordinated by the National Electric Reliability Council, that guide the coordination and operation of generation and transmission facilities.

responsibilities of operators for system security.

System security is concerned with service continuity at standard frequency and voltage levels. The system is said to be insecure if a contingency would result in overloading some system components, in abnormal voltage levels at some stations, in change of system frequency, or in system instability, even if there is adequate capability as indicated by some reliability index.

Energy control centers. The majority of large electric systems have computerized energy centers whose functions are to control operation of the system so as to optimize economy and security. In some large interconnected networks of individual utility companies, such as those in the Pennsylvania–New Jersey–Maryland power pool, regional control centers have been established to perform the same function for a group of companies. New generation centers now use digital computers, or hybrid analog-digital units, rather than the analog machines previously used, and redundant systems are required to ensure security.

Control can be be broken into two phases: operations and security. In the operational area, a typical center provides a completely updated analysis of power flows in the system at 10-min intervals. In

Fig. 4. Matrix mimic board and CRT displays. *(Courtesy of Ferranti-Packard Inc. and Cleveland Electric Illuminating Co.)*

real time, this means calculation of the cost penalty factors for line losses, machine efficiencies, and fuel costs, thereby minimizing cost of power. Modern systems also include predictive programs that produce probabilistic forecasts of hourly system load for several days into the future, using historical data on weather-load correlations. This permits scheduling startup and shutdown of available generating units to optimize operating economy and to establish necessary maintenance schedules.

System security and analysis programs permit the system operators to simulate problems on the system as it actually exists currently, thereby preparing them for appropriate actions should such emergencies occur. The most advanced centers today use techniques such as state estimation, which processes system data to calculate the probabilities of emergencies in the near-term future, thus permitting the operators to take preventive action before the event occurs.

Interface between the operators and computers are through data loggers, cathode-ray tubes (CRTs), plotters, or active mimic boards (Fig. 4). Dynamic mimic boards, in which the displays are driven by the computer output, have generally displaced the older static representation systems which used supervisory lights to display conditions. The CRTs display the system in up to seven colors, and the operator can interact with them by using a cursor or light pen either to analog the effect of changes to the system or actually to operate the equipment.

Substations. Power delivered by transmission circuits must be stepped down in facilities called substations to voltages more suitable for use in industrial and residential areas. On transmission systems, these facilities are often called bulk-power substations; at or near factories or mines, they are termed industrial substations; and where they supply residential and commercial areas, distribution substations.

Basic equipment in a substation includes circuit breakers, switches, transformers, lightning ers and other protective devices, instrumen control devices, and other apparatus related to specific functions in the power system.

Distribution. That part of the electric power system that takes power from a bulk-power substation to customers' switches, commonly about 35% of the total plant investment, is called distribution. This category includes distribution substations, subtransmission circuits that feed them, primary circuits that extend from distribution substations to every street and alley, distribution transformers, secondary lines, house service drops or loops, metering equipment, street and highway lighting, and a wide variety of associated devices.

Primary distribution circuits usually operate at 4160 to 34,500 V line to line, and supply large commercial institutional and some industrial customers directly. The lines may be overhead open wire on poles, spacer or aerial cable, or underground cable. Legislation in more than a dozen states now requires that all new services to devel-

Table 2. Electric utility industry statistics, 1971–1978*

Category	1971	1972	1973	1974	1975	1976	1977	1978
Generating capacity installed at year's end, 10^3 kW†	367,396	399,606	438,492	474,574	504,393	531,162	557,174	574,366
Electric energy output, 10^6 kWh	1,617,500	1,754,900	1,878,500	1,884,000	1,919,912	2,037,674	2,124,026	2,232,774
Energy sales, 10^6 kWh								
Total	1,466,440	1,577,714	1,703,203	1,700,769	1,733,024	1,849,625	1,950,791	2,020,610
Residential	179,080	511,423	554,171	554,960	591,108	613,072	652,345	680,874
Small light and power	333,752	361,859	396,903	392,716	421,088	440,625	469,227	481,561
Large light and power	592,699	639,467	687,235	689,435	656,440	725,169	757,168	783,682
Other	60,909	64,965	64,894	63,558	67,270	70,758	72,051	74,492
Customers at year's end, $\times 10^3$	74,265	76,150	78,461	80,102	81,845	83,613	85,590	87,629
Revenue, $\$10^6$	24,725	27,921	31,663	39,127	46,853	53,463	62,610	69,865
Average residential use, kWh/year	7,380	7,691	8,079	7,907	8,176	8,360	8,693	8,873
Average residential rate, cents/kWh	2.19	2.29	2.38	2.83	3.21	3.45	3.78	4.02
Coal (and equivalent) burned, lb/kWh‡	0.918	0.911	0.918	0.946	0.952	0.949	0.968	0.984
Capital expenditures, $\$10^6$	15,130	16,651	18,723	20,556	20,155	25,189	27,711	30,250

*From *Elec. World*, p. 61, Sept. 15, 1978, and p. 51, Mar. 15, 1979; and Edison Electric Institute.
†Fuel and hydro.
‡1 lb = 454 g.

opments of five or more residences be put underground. The bulk of existing lines are overhead, however, and will remain so for the indefinite future.

At conveniently located distribution transformers in residential and commercial areas, the line voltage is stepped down to supply low-voltage secondary lines, from which service drops extend to supply the customers' premises. Most such service is at 120/240 V, but other common voltages are 120/208 V, 125/216, and, for larger commercial and industrial buildings, 240/480, 265/460, or 277/480 V. These are classified as utilization voltages.

Electric utility industry. In the United States, which has the third highest per-capita use of electricity in the world and more electric power capability than any other nation, the electric capability systems as measured by some criteria are the largest industry (Table 2). Total plant investment, as of Dec. 31, 1978, was about $270,000,000,000. The electric utility industry spends approximately $30,000,000,000 a year for new plants to supply the growing load, and collects nearly $70,000,000,000 per year from more than 87,000,000 customers. This industry comprised about 3115 public and investor-owned systems producing electricity in about 3100 generating plants with a combined operating capability of 574,365 MW at the close of 1978.

The 3115 systems included 301 investor-owned companies operating 78.3% of the generation and serving 77.5% of the ultimate customers. The remaining 22.5% were served by 1769 municipal systems, about 928 rural electric cooperatives, 58 public power districts, 7 irrigation districts, 40 United States government systems, 9 state-owned authorities, 1 county authority, and 2 mutual systems.

The industry's annual output reached 2,232,-774,000,000 kWh in 1978, and its sales to ultimate customers totaled 2,020,610,000,000 kWh at an average of 3.46 cents/kWh. Of this, about 39% was consumed by industrial and other large power customers, the remainder going mostly to residential (34%) and commercial customers. Residential usage has climbed steadily for many years, reaching a record 8873 kWh/average customer in 1978.

[WILLIAM C. HAYES]

Bibliography: A. S. Brookes et al., *Proceedings of International Conference on Large High Voltage Electric Systems*, Paris, Aug. 21–29, 1974; J. F. Dopazo, State estimator screens incoming data, *Elec. World*, 185(4):56–57, Feb. 15, 1976; *EHV Transmission Line Reference Book*, Edison Electric Institute, 1968; The electric century, 1874–1974, *Elec. World*, 181(11):43–431, June 1, 1974; Electric Research Council, *Electric Transmission Structures*, Edison Electric Institute, 1968; *Electrical World Directory of Electric Utilities, 1977–1978*, 1977; *Electrostatic and Electromagnetic Effects of Ultra-High-Voltage Transmission Lines*, Electric Power Research Institute, 1978; L. W. Eury, Look one step ahead to avoid crises, *Elec. World*, 187(10):50–51, May 15, 1977; G. F. Friedlander, 15th Steam Station Design Survey, *Elec. World*, 190(10):73–88, Nov. 15, 1978; G. F. Friedlander, 20th Steam Station Cost Survey, *Elec. World*, 188(10):43–58, Nov. 15, 1977; 1979 Annual Statistical Report, *Elec. World*, 191(6):51–82, Mar. 15, 1979; W. P. Rades, Convert to digital control system, *Elec. World*, 186(1):50–51, July 1, 1976; N. D. Reppon et al., *Proceedings of the Department of Energy's System Engineering for Power: States and Prospects Conference*, Henniker, NH, Aug. 17–22, 1975; W. D. Stevenson, Jr., *Elements of Power System Analysis*, 1975; R. L. Sullivan, *Power System Planning*, 1977; *Transmission Line Reference Book, HVDC to ±600 kV*, Electric Power Research Institute & Bonneville Power Administration, 1977; 29th Annual Electric Utility Industry Forecast, *Elec. World*, 190(6):62–76, Sept. 15, 1978.

Engineering

Most simply, the art of directing the great sources of power in nature for the use and the convenience of humans. In its modern form engineering involves people, money, materials, machines, and energy. It is differentiated from science because it is primarily concerned with how to direct to useful and economical ends the natural phenomena which scientists discover and formulate into acceptable theories. Engineering therefore requires above all the creative imagination to innovate useful applications of natural phenomena. It is always dissatisfied with present methods and equipment. It seeks newer, cheaper, better means of using natural sources of energy and materials to improve people's standard of living and to diminish toil.

Types of engineering. Traditionally there were two divisions or disciplines, military engineering and civil engineering. As knowledge of natural phenomena grew and the potential civil applications became more complex, the civil engineering discipline tended to become more and more specialized. Practicing engineers began to restrict their operations to narrower channels. For instance, civil engineering came to be concerned primarily with static structures, such as dams, bridges, and buildings, whereas mechanical engineering split off to concentrate on dynamic structures, such as machinery and engines. Similarly, mining engineering became concerned with the discovery of, and removal from, geological structures of metalliferous ore bodies, whereas metallurgical engineering involved extraction and refinement of the metals from the ores. From the practical applications of electricity and chemistry, electrical and chemical engineering arose.

This splintering process continued as narrower specialization became more prevalent. Civil engineers had more specialized training as structural engineers, dam engineers, water-power engineers, bridge engineers; mechanical engineers as machine-design engineers, industrial engineers, motive-power engineers; electrical engineers as power and communication engineers (and the latter divided eventually into telegraph, telephone, radio, television, and radar engineers, whereas the power engineers divided into fossil-fuel and nuclear engineers); mining engineers as metallic-ore mining engineers and fossil-fuel mining engineers (the latter divided into coal and petroleum engineers).

As a result of this ever-increasing utilization of technology, humans and their environment have been affected in various ways—some good, some bad. Sanitary engineering has been expanded from treating the waste products of humans to also treating the effluents from technological processes. The increasing complexity of specialized machines and their integrated utilization in automated processes has resulted in physical and mental problems for the operating personnel. This has led to the development of bioengineering, concerned with the physical effects upon people, and management engineering, concerned with the mental effects.

Integrating influences. While the specialization was taking place, there were also integrating influences in the engineering field. The growing complexity of modern technology called for many specialists to cooperate in the design of industrial processes and even in the design of individual machines. Interdisciplinary activity then developed to coordinate the specialists. For instance, the design of a modern structure involves not only the static structural members but a vast complex including moving parts (elevators, for example); electrical machinery and power distribution; communication systems; heating, ventilating, and air conditioning; and fire protection. Even the structural members must be designed not only for static loading but for dynamic loadings, such as for wind pressures and earthquakes. Because people and money are as much involved in engineering as materials, machines, and energy sources, the management engineer arose as another integrating factor.

Typical modern engineers go through several phases of activity during their career. Their formal education must be broad and deep in the sciences and humanities underlying their field. Then comes an increasing degree of specialization in the intricacies of the particular discipline, also involving continued postscholastic education. Normal promotion thus brings interdisciplinary activity as they supervise the specialists under their charge. Finally, they enter into the management function as they interweave people, money, materials, machines, and energy sources into completed processes for the use of humans.

[JOSEPH W. BARKER]

Engineering, social implications of

The rapid development of human ability to bring about drastic alterations of the environment has added a new element to the responsibilities of the engineer. Traditionally, the ingredients for sound engineering have been sound science and sound economics. Today, sound sociology must be added if engineering is to meet the challenge of continued improvement in the standard of living without degradation of the quality of the environment.

Despite the fact that present and evolving engineering practices must meet the criteria of scientific and economic validity, these same practices generally cause societal problems of new dimensions. Consider, for example, exhaust gases emitted from tens of millions of internal combustion engines, both stationary and moving; stack gases from fossil-fuel-burning plants generating steam or electric power; gaseous and liquid effluents and solid waste from incinerators and waste-treatment systems; strip mining of coal and mineral ores; noise issuing from automotive vehicles, aircraft, and factory and field operations; toxic, nondegradable or long-lived chemical and particulate residues from ore reduction, chemical processing, and a broad spectrum of factory and mill operations; dust storms, soil erosion, and disruption of groundwater quality and quantity accompanying intensified mechanized farming in conjunction with massive irrigation and fresh-water diversion. *See* APPLIED ECOLOGY.

Progress often results in the substitution of one set of problems for another. For example, in nuclear electric power generating plants, replacement of fossil fuels by nuclear fuels relieves the burden of atmospheric pollution from stack gas emissions. Lower thermal efficiency of a nuclear plant, however, results in higher heat rejection rates and increased thermal pollution of sources of cooling water or air. The attendant consequences on atmospheric conditions or on the viability of aquatic life in the affected water are of great concern in the short and long terms. Ultimately, the cost and benefit considerations of nuclear power must be all-inclusive; in addition to usual considerations of economic length of plant life and so forth, one must account for all the economic and societal costs of the entire fuel cycle, from mining and refinement through use and ultimate recycling or safe disposal. The long-term effects of very low levels of radiation exposure (as such studies become available) will be an additional factor to consider.

The modern engineer must be increasingly conscious of the societal consequences of technological innovation. *See* ENGINEERING.

[EUGENE A. AVALLONE]

Environment

Ecologically, the environment is the sum of all external conditions and influences affecting the life and development of organisms. Various ecological principles and concepts have been developed in regard to the environment. Two main aspects of the environment are usually considered, the abiotic and the biotic. These divisions are artificial in the sense that neither can be separated when organisms are studied. All environmental aspects and their influences on living organisms must be considered together.

Nature of the environment. Holocoenosis, the nature of the action of the environment, pertains to those factors which exist as a vast complex and therefore do not act separately and independently, but as a whole. This principle, which lies at the core of all ecological thinking, is shown in the illustration.

The particular environment occupied by an or-

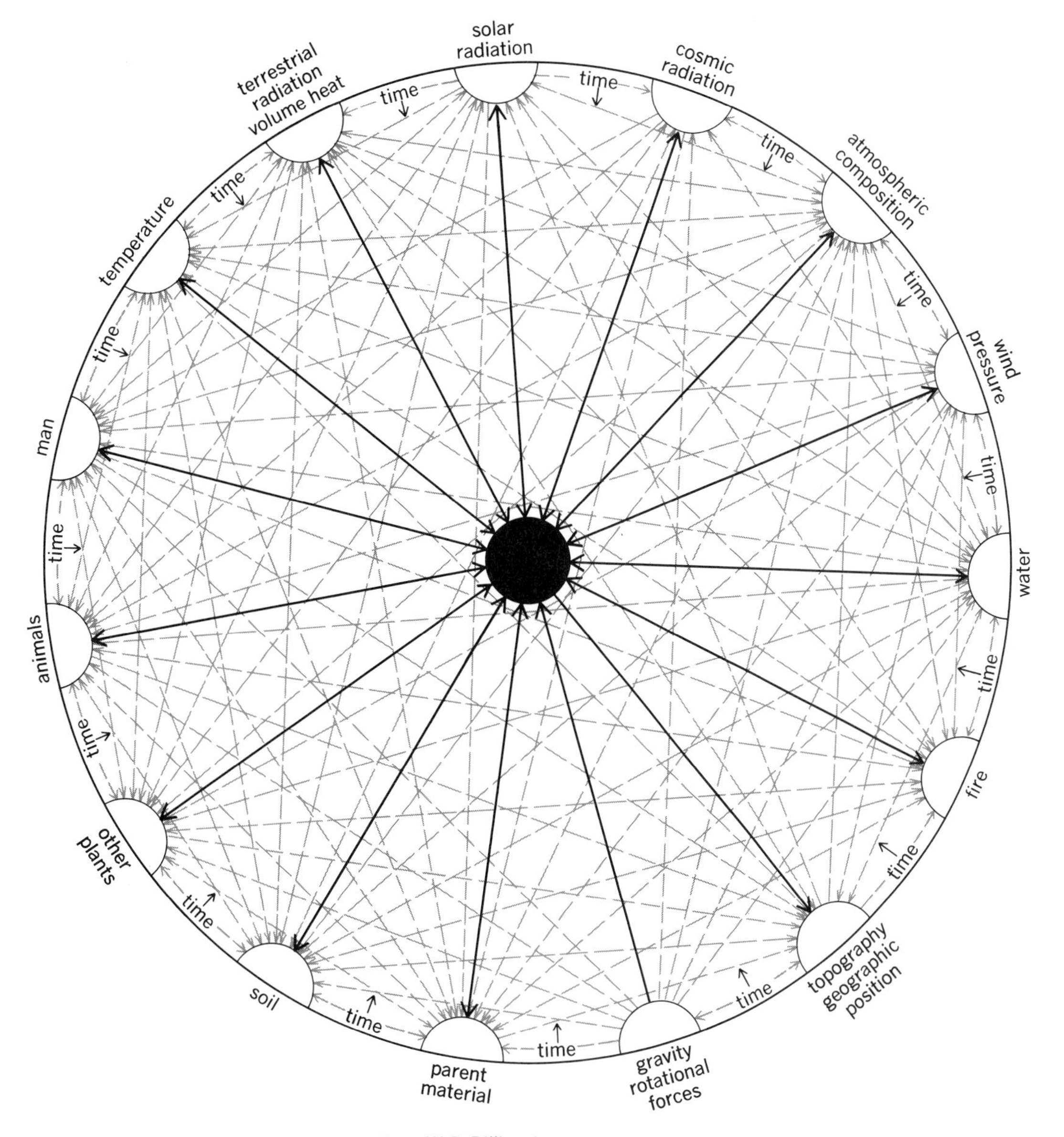

A holocoenotic environmental complex. (*From W. D. Billings*)

ganism or by groups and communities of organisms is the habitat. Environmental conditions generally exist in an area independently of the occurrence of any species of plant or animal that may depend on them, and area is used to express this concept, for example, in contrasting humid versus dry climates. The whole system of interaction between a particular organism and its physical and biotic environment is the niche of that organism. The term indicates not only the habitat, or microenvironment, but also the activity of the organism. The environment of small areas in contrast to that of larger ones, or of particular organisms in contrast to generalized environments of communities, is recognized as the microenvironment. Other terms commonly applied to this concept, but much more restricted in scope, are microclimate and bioclimate. Being relative, the term microenvironment may signify the environment of a pine stand or, equally well, a lichen which is located within that stand.

Dynamic concept. Variations in space and time are responsible for a dynamic concept of the environment. In space, there is continual change from the Equator to the poles, from sea level to outer space or to ocean depths, and from the forest floor to the treetops. Variations in time may be geological, as climates, carbon dioxide content of the air, and salt content of the ocean, or be concerned with present-day cycles and periodicities. The more familiar include day and night, ocean tides, Moon phases, and seasonal cycles.

The change or rate of change in an environmental variable is the gradient. This is pertinent, through all units of space and time, to an understanding of the continuous adjustment life must make to its environment.

Action systems provide a convenient way of referring to the interaction between organisms and their complex environment. Ecological action is the effect of the physical environment on organisms; reaction is the reciprocal effect of organisms on the physical environment; and coaction is the interaction of organisms on each other.

Limiting factors. The concept of limiting factors recognizes that any factor of the environmental complex which approaches or exceeds the limits of tolerance of any organism or group of organisms acts as a control on them. This concept is derived from two laws. The law of the minimum, originally given by Justus Liebig in 1840, has been restated by W. P. Taylor as follows: "The growth and functioning of an organism is dependent upon the amount of the essential environmental factor presented to it in minimal quantity during the most critical season of the year, or during the most critical years of a climatic cycle." In 1913, by adding the concept of an ecological maximum, V. Shelford showed, through his law of tolerance, that all organisms or groups of organisms must live in a range of conditions between the maximum and minimum, which then represent the limits of tolerance. A series of terms has come into general use attaching the prefixes "steno-" meaning narrow, and "eury-" meaning wide, to indicate the relative degree of tolerance, for example, stenothermal, which indicates a narrow tolerance to temperature.

Other factors. Those factors which upset the delicate balance of an ecosystem are termed trigger factors. They cause a chain reaction which may not end until drastic changes have occurred. For example, the permanent addition of water to a desert environment may ultimately change the whole ecosystem.

The substitution of elevation for latitude is an example of a compensating factor which allows plants of northern distribution to grow far southward on high mountain ranges. Some factors such as temperature may substitute for one another without apparent effect on the organism, but factors such as day length are different.

Environmental substances such as nutrients are necessary for the growth and development, or nutrition, of organisms. Thus, inorganic substances provide for the nutrition of green plants, which in turn provide the organic nutrients of nongreen plants and animals. The characteristic pathways in which these substances circulate from the environment to organisms and back again are known as biogeochemical cycles. Oxygen, carbon, and nitrogen are the principal gases involved, while more than 40 minerals are obtained from the crust of the Earth, among them phosphorus and potassium. The synthesis of protoplasm by living organisms and their decomposition after death from the energy basis for these cycles and provide for a classification of organisms into producers, consumers, and reducers. *See* BIOSPHERE.

ABIOTIC ENVIRONMENT

The physical or abiotic environment includes all those physical and nonliving chemical aspects which exert an influence on living organisms. Among these factors are soils, water, and the atmosphere, as well as the influence of energy from various sources.

Various forms of energy exert an influence on, or modify, the environment in which organisms are found. With minor exceptions, the ecologist is concerned with radiant energy received directly from the Sun, infrared radiation or heat, visible radiation, and ionizing radiation. These energies, coming from the same source, are subject to the same modifying factors, the primary one being daily and seasonal cycles which in turn are modified by latitude, altitude, slope, exposure, atmospheric conditions, and the color, texture, and cover of the ground surface.

Temperature. The intensity aspect of heat energy is temperature, and, with moisture and light, it is one of the most familiar of environmental conditions. In nature, temperature ranges from a low of −70°C (−94°F) recorded in Siberia, to highs of 60°C (140°F) in deserts and almost 100°C (212°F) in hot springs. With few exceptions, however, almost all life exists within the relatively narrow range of −17.8°C (0°F) to 45°C (113°F). The specific responses developed by organisms to temperature include metabolic activity, behavior, abundance, and distribution. With the exception of birds and mammals, the body temperature of plant and animal life is determined primarily by the external environment. *See* HOMEOSTASIS.

Light. The source of light may be the Sun, Moon, stars, or luminescence, but usually only the first two are of sufficient intensity to influence life. Light varies according to wavelength, intensity, and duration. Sunlight may vary at noon in the open from a few hundred footcandles (ftc) on overcast days to more than 14,000 ftc on clear days in southern latitudes. In woods, it may vary from a few to 1000–2000 ftc under the same conditions. Maximum moonlight is about 0.05 ftc. Light penetration in relatively clear water is about 30% at 5 m and 10% at 10 m. In addition to the well-known uses of light for photosynthesis in plants and vision in more advanced animals, it also affects growth, development, and survival, and through many types of photoperiodicities, affects the behavior, life cycle, and distribution of many plants and animals.

Ionizing radiation. Ultraviolet, nuclear, and cosmic radiations have sufficient energy to damage protoplasm on relatively short exposures. The potential significance of ionizing radiation as an environmental factor was not recognized until the mid-1950s, when worldwide attention was focused on ionizing radiation received from fallout. The atmosphere is opaque to all ultraviolet light shorter than 2900 A, so that little or no injury results at moderate altitudes. In the stratosphere and above, the lethal action of ultraviolet light is important, the bacterial effects beginning at 3650 A. Nuclear radiation, composed of alpha and beta particles and gamma rays, comes from radioactive substances naturally occurring in the crust of the Earth, and from living and nonliving materials which have these incorporated in them. The intensity of this background radiation varies on an average from 0.1 to 0.5 roentgen unit (r) per year, although many areas give much higher readings. Because of fallout, this natural radiation is being added to each year. Cosmic rays come from space and vary appreciably with altitude. Some geneticists estimate that 10% of the mutations in geological time are attributable to these ionizing radiations. Very little is yet known about their direct effects at natural levels on organisms.

Water. Water and air are the fundamental media in which life exists. As such, they provide the basis for the division of the world into two major environments, aquatic and terrestrial. Water not only covers 70% of the Earth's surface, but also provides for the existence of life throughout its depths. Water has universal solubility, low chemical activity, high surface tension, high ionization, high freezing point, latent heat of fusion, latent heat of evaporation, and high specific heat. As the chief components of protoplasm, water and air have many common physicochemical characteristics.

The distribution of water through the atmosphere and over the terrestrial environment is controlled by the hydrological cycle. Principal marine environments are tidal zones (neritic), and oceanic zones (euphotic, pelagic, and benthic). Fresh-water environments are standing water (lakes, ponds, swamps, and bogs) and running water (springs, streams, and rivers). On land, the term xeric refers to relatively dry, hydric to relatively wet, and mesic to moist environments. *See* HYDROLOGY.

Atmospheric moisture may exist in three states, solid (hail, sleet, or snow), liquid (rain), or gas (water vapor). Water vapor content of the atmosphere varies with temperature and is referred to as humidity. The ratio of precipitation to evaporation is the most important factor in the distribution of vegetation zones, such as desert, grassland, woodland, and forest.

Soil moisture varies with the structure and texture of the soil, as well as with precipitation-evaporation factors. Soil is saturated when all pore spaces are filled. The pore spaces make up 40–60% of the volume of soil. Water which subsequently drains away is gravitational water, while that retained by capillary forces is capillary water. In very dry soil, the extremely thin film which cannot move as a liquid is hygroscopic water, and that chemically bound with soil materials is combined water. All gravitational and most capillary water is available to plants, while bound water and hygroscopic water cannot be utilized. Plants which can no longer obtain moisture from the soil reach a physiological state at which wilting occurs. This varies among different plant species and is known as the wilting coefficient of the soil.

Atmosphere. The atmosphere is composed of 78% nitrogen, 21% oxygen, 0.03% carbon dioxide, several other gases in very small quantities, and varying amounts of water vapor. All exist in a simple physical mixture. In addition to the physical characteristics of the atmosphere as a medium, its oxygen and carbon dioxide chemically affect all forms of life through photosynthesis and respiration, that is, through their reciprocal relationship of synthesis and decomposition. Concentration of these gases varies with organic and industrial activity, with altitude in the atmosphere, with organic activity, and with their solubility in fresh and salt water. Carbon dioxide also influences the hydrogen ion concentration of soil and water.

Wind is a significant atmospheric phenomenon acting through its physical force. It exerts an influence on the rate of transpiration in plants and evaporation of moisture. It modifies weather and climate, and is a disseminating agent for pollen, spores, and microscopic organisms.

Substratum. This is any solid on whose surface an organism rests or moves, or within which it lives, thus providing anchorage, shelter, and food. Aquatic substrata are rock, sand, and mud, while terrestrial substrata are rock, sand, and soil, soil being the most complex.

Soil is the link between the mineral core of the Earth and life on its surface. It consists of decomposed mantle rock and organic matter, with spaces for water and air, and is organized into distinct layers or horizons. The A horizon, or surface soil, is the zone of most abundant life and also of leaching. The B horizon, as subsoil, is a zone of deposition from the layer above, while the C horizon is composed of the parent materials. The major factors in soil formation are climate, topography, living things, and time. The differential nature of these factors is responsible for the great soil groups of the world. *See* SOIL.

Other abiotic factors. These include such conditions as gravity, pressure, sound, and fire. Gravity affects the environment through its action in isostasy and stratification in air and water. Its effects on organisms are through polarity, orientation, structure, distribution, and behavior. Pressure acts mechanically, and environmental pressures range from a half atmosphere pressure at 5800 m elevation to 1000 atm at a depth of 10,000 m in the ocean. Since pressure changes much more rapidly in water than in air, it becomes of paramount importance in the ocean through its effects on the structure, physiology, and distribution of organisms.

Sounds are produced and conveyed by mechanical vibrations, substratal vibrations, and mechanical shock, which are interconnected phenomena. While aerial, and usually audible, vibrations are important to man and many animals, particularly birds and mammals, the substratal vibrations such as earthquakes, especially when sudden and intense, are the ones of prime importance in ecology.

Fire has always been an important factor in the terrestrial environment. Classified as crown, surface, and ground types, fires exert direct effects through injury and destruction, and indirect effects through often drastic alterations of environmental conditions and changes in community relationships. The beneficial values of controlled burning in forestry and game management are well established, and fire is an important tool in shifting agriculture in the humid tropics.

BIOTIC ENVIRONMENT

The biotic environment consists of living organisms, which both interact with each other and are inseparably interrelated with their abiotic environment.

Interactions between organisms. Within a population, interactions include such aspects as density, birth rate, death rate, age distribution, dispersion, and growth forms. *See* ECOLOGICAL INTERACTIONS.

Interactions between organisms at the interspecies population level include competition, which arises from utilization of the same thing when it is in short supply. Predation and parasitism are negative interactions which result in harm or destruction of a prey or host population by the predator or parasite population. Positive interactions also occur, such as commensalism, in which one population is benefited; cooperation, in which both populations are benefited; and mutualism, in which both populations are benefited and are completely dependent on each other. At the community levels, interactions include dominance, where one or several species control the habitat. In succession, there is an orderly process of community change. Rhythmic change in community activity is termed periodicity.

Interactions with abiotic environment. These interactions include (1) the effects of organisms on the microenvironment through temperature, water, wind, and light, (2) their modification of the substrate, as through soil building, and (3) their modification of the medium, as in aquatic habitats. *See* ECOSYSTEM. [ROBERT B. PLATT]

Bibliography: A. H. Benton and W. E. Wagner, Jr., *Field Biology and Ecology*, 3d ed., 1974; P. Farb, *Ecology*, 1970; R. Geiger, *The Climate near the Ground*, 4th ed., 1965; S. J. McNaughton and L. LeWolf, *General Ecology*, 1979; E. P. Odum, *Fundamentals of Ecology*, 3d ed., 1971; R. B. Platt and J. Griffiths, *Environmental Measurement and Interpretation*, reprint 1972; R. L. Smith, *Ecology and Field Biology*, 2d ed., 1974.

Environmental engineering

The discipline which evaluates the effects of humans on the environment and develops controls to minimize environmental degradation.

In the 1960s the United States became acutely aware of the deterioration of its air, water, and land. The roots of the problem lie in the rapid growth of the national population and the industrial development of natural resources which has given Americans the highest standard of living in the world. Since 1964 there has been enactment of national, state, and local legislation directed toward the preservation of these resources.

The technology which provided society with all the necessities and luxuries of life is now expected to continue providing these services without degradation to the environment. How well this goal is attained will depend in great measure on the environmental engineers, who must cope with the enormous challenges presented by society.

It is the feeling of industry and governmental agencies that the ultimate goals should be the design of processes and systems which need minimal treatment for pollution control and the ultimate recycling of all wastes for reuse. This philosophy is both logical and necessary in a society that is rapidly depleting its nonrenewable natural resources.

Governmental policy. The Federal government enacts environmental air, water, and land-use laws which in most cases require the individual states to develop programs and enforcement policies to ensure that the goals are fulfilled. Much environmental legislation was enacted in the 1970s by both the state and Federal governments.

The significant Federal legislation has included the Wilderness Act of 1964; the Air Quality Act of 1967, followed by the Clean Air Act of 1970 and the very significant Clean Air Act Amendments of 1977; the Water Quality Act of 1965, replaced by the Water Pollution Control Act of 1972 with amendments to the act in 1977; the National Environmental Policy Act of 1969; Executive Order No. 11574 of December 1970, restating the Refuse Act of the Rivers and Harbors Act of 1899; the Noise Control Act and the Coastal Zone Management Act, both of 1972; the Resource Conservation and Recovery Act of 1976; the Surface Mining Control and Reclamation Act and the Toxic Substances Control Act, both of 1977.

Environmental legislation enacted from 1970 to 1977 has had an unmistakable impact upon the community. Standards and limitations governing the release of pollutants to the environment have placed severe restrictions on industrial operations. The sweeping Federal legislation program for environmental protection has also led to the establishment of complementary legislation, regulations, and requirements at the state level. All states now have extensive programs for environmental control and protection. The confrontation between government and industry as a result of the legislation must be eliminated and replaced by a cooperative effort toward developing the technology necessary for environmental improvement in the best interests of all concerned.

The National Environmental Policy Act is an example of the all-inclusiveness of government regulation. This act makes it mandatory that an in-depth study of environmental impacts be made in connection with any new industrial or government activity that may involve the Federal government, directly or indirectly. Similarly, state environmental policy acts, patterned after this Federal act, often require more detailed studies which must withstand more vigorous scrutiny.

These requirements extend to a wide range of activities such as:

Construction of electric power plants and transmission lines, gas pipelines, railroads, highways, bridges, nuclear facilities, airports, and mine facilities.

Any industrial facility releasing effluents to navigable waters and their tributaries.

Rights-of-way, drilling permits, mining leases, and other uses of Federal lands.

Applications to the Interstate Commerce Commission for the approval of transportation rate schedules.

In December 1970, by presidential order, the Environmental Protection Agency (EPA) was formed. Under Public Law 91-604, which extended the Clean Air and Air Quality acts, this agency now embodies under one administrator the responsibility for setting standards and a compliance timetable for air and water qualities improvement. This agency also administers the Noise Control Act, the Resource Conservation and Recovery Act, and the Toxic Substances Control Act. In addition, the agency administers grants to state and local governments for construction of wastewater treatment facilities and for air and water pollution control.

Industrial policy. Through its interdisciplinary environmental teams, industry is directing large amounts of capital and technological resources both to define and resolve environmental challenges. The solution of the myriad complex environmental problems requires the skills and experience of persons knowledgeable in health, sanitation, physics, chemistry, biology, meteorology, engineering, and many other fields.

Each air and water problem has its own unique approach and solution. Restrictive standards necessitate high retention efficiencies for all control equipment. Off-the-shelf items, which were applicable in the past, no longer suffice. Controls must now be specificially tailored to each installation. Liquid wastes can generally be treated by chemical or physical means, or by a combination of the two, for removal of contaminants with the expectation that the majority of the liquid can be recycled. Air or gaseous contaminants can be removed by scrubbing, filtration, absorption, or adsorption and the clean gas discharged into the atmosphere. The removed contaminants, either dry or in solution, must be handled wisely, or a new water- or air-pollution problem may result.

Industries that extract natural resources from the earth, and in so doing disturb the surface, are being called on to reclaim and restore the land to a condition and contour that is equal to or better than the original state. The Surface Mining Control and Reclamation Act, which now covers only the mining of coal, requires the states to have an EPA-approved program for controlling surface mining operations and the reclaiming of abandoned mined lands. Most states already have reclamation laws which cover all types of mining activities; most require an approved restoration plan and bonding to assure that restoration is accomplished.

Air quality management. The air contaminants which pervade the environment are many and emanate from multiple sources. A sizable portion of these contaminants are produced by nature, as witnessed by dust that is carried by high winds across desert areas, pollens and hydrocarbons from vegetation, and gases such as sulfur dioxide and hydrogen sulfide from volcanic activity and the biological destruction of vegetation and animal matter.

The greatest burden of atmospheric pollutants resulting from human activity comprises carbon monoxides, hydrocarbons, particulates, sulfur oxides, and nitrogen oxides, in that order. Public and private transportation using the internal combustion engine is a major source of these contaminants. It has been conservatively estimated that 50% of the major pollutants in the United States come from the use of the internal combustion engine; restrictive regulations by EPA to control these emissions have been imposed on the automobile industry. Expensive emission control measures such as catalytic converters and the mandatory use of lead-free gasoline are the two most significant requirements of the Clean Air Act of 1970 directed toward control of automobile emissions.

Industrial and fuel combustion sources (primarily utility power plants) together contribute approximately 30% of the major pollutants. Interestingly, it has been reported that sulfur dioxide and suspended particulate levels in New York City have been reduced in the last 15 years by nearly 90%; however, no demonstrable improvement has been seen in the occurrence of sickness and death from respiratory diseases. This would suggest, possibly, that the annual expenditure of $10,000,000,000 to $15,000,000,000 by the public for achieving the goals of the Clean Air Act may benefit esthetics by improving visibility and reducing damage to materials and crops more than providing the planned health benefits.

All states have ambient air and emission standards directed primarily toward the control of industrial and utility power plant pollution sources. The general trend in gaseous and particulate control is to limit the emissions from a process stack to a specified weight per hour based on the total material weight processed to assure compliance with ambient air regulation. Process weights become extremely large in steel and cement plants and in large nonferrous smelters. The degree of control necessary in such plants can approach 100% of all particulate matter in the stack. Retention equipment can become massive both in physical size and in cost. The equipment may include high-energy venturi scrubbers, fabric arresters, and electrostatic precipitators. Each application must be evaluated so that the selected equipment will provide the retention efficiency desired.

Sulfur oxide retention and control present the greatest challenges to industrial environmental engineers. Ambient air standards are extremely low, and the emission standards calculated to meet these ambient standards place an enormous challenge on the affected industries. Many copper smelters and all coal-fired utility power plants have large-volume, weak-sulfur-dioxide effluent gas streams. Scrubbing these weak-sulfur-oxide gas streams with limestone slurries or caustic solutions is extremely expensive, requires prohibitively large equipment, and creates water and solid waste disposal problems of enormous magnitude. Installations employing dry scrubbing have been used on very-low-sulfur-oxide gas streams. A number of utilities had elected to burn low-sulfur coal to meet the existing sulfur dioxide regulations. This may not be an attractive alternative for utility plants constructed or modified after June 11, 1979. EPA promulgated on that date a minimum 70% reduction in potential sulfur oxide emissions for utilities burning low-sulfur coal.

Copper smelters are required to remove 85–95% of the sulfur contained in the feed concentrate. Smelters using old-type reverbatory furnaces produce large volumes of gas containing low concentrations of sulfur dioxide which is not amenable to removal by acid making. However, gas streams from newer-type flash and roaster-electric furnace operations can produce low-volume gas streams containing more than 4% sulfur dioxide which can be treated more economically to obtain elemental sulfur, liquid sulfur dioxide, or sulfuric acid. Smelters generally have not considered the scrubbing of weak-sulfur-dioxide gas streams as a viable means of attaining emission limitations because of the tremendous quantities of solid wastes that would be generated. *See* ATMOSPHERIC POLLUTION.

The task of upgrading weak smelter gas streams to produce products which have no existing market has led to extensive research into other methods of producing copper. A number of mining companies piloted, and some have constructed, hydrometallurgical plants to produce electrolytic-grade copper from ores by chemical means, thus eliminating the smelting step. These plants have generally experienced higher unit costs than smelters, and a number have been plagued with operational problems. It does not appear likely that hydrometallurgical plants will replace conventional smelting in the foreseeable future. Liquid ion exchange, followed by electrowinning, is also being used more extensively for the heap leaching of low-grade copper. This method produces a very pure grade of copper without the emission of sulfur dioxide to the atmosphere.

Water quality management. The Federal effort in water-pollution control has been assumed by the EPA and its office of water quality under the National Pollutant Discharge Elimination System (NPDES). The essential function of NPDES is compliance with effluent standards adopted for municipal and industrial waste dischargers. Many states already have a permit program in effect and have been granted authority to administer these programs; however, permit conditions and compliance schedules are still subject to Federal approval. Originally, EPA guidelines called for universal standards, but some flexibility has been permitted under sections of the law addressing receiving-water-quality criteria, and at least in some cases permit conditions are being altered to treat more site-specific situations when permits expire and are renewed. In the case of new source permitting, however, flexibility in permit conditions continues to be lacking even in the face of good background survey information. Some mining industry environmental engineers can be expected to take issue with the philosophy of wasting a portion of the assimilative capacity of receiving waters by not utilizing part of this natural phenomenon and thus producing finished products at a lower cost to the consumer.

The state of the art of treatment of wastewaters containing metals has barely advanced beyond neutralization and chemical precipitation. As water quality standards continue to become more stringent, especially concerning dissolved solids, it is apparent that chemical treatment of metals wastes will be unacceptable. Chemical treatment in many instances substitutes a different molecular species for another with no reduction in total dissolved solids.

Some segments of the mining industry are considering physical treatment methods of desalination techniques such as evaporation, reverse osmosis, and electrodialysis. These techniques require vast quantities of energy and are prohibitive in both capital and operational costs. Continued experimentation will undoubtedly improve the benefit-cost ratio.

For the most part the mining industry tends toward a policy of complete recycle of water. Some metallurgical processes are amenable to reuse of process waters with only minimal treatment, such as removal of suspended and settleable solids. Other processes require higher-quality water with lower dissolved-solids content. Even this generally requires removal of only a portion of the solids from a waste stream in order to maintain an acceptable process water quality. Many of these solids can also be reclaimed, thus recovering values now lost. This approach is not only logical but economical. If receiving-water quality criteria are more stringent than process-water treatment requirements, it is impractical to comply with those standards and then waste the water to the nearest surface drainage. In the more arid parts of the world, recycle of water has become a practice by necessity; in water-rich areas it will become practice by governmental decree.

Originally, the Federal Clean Water Act required best practicable treatment technology by 1977, best available technology (BAT) by 1983, and (stated as a goal) no pollutant discharge by 1985. The deadline for achieving BPT was extended under limited conditions to Apr. 1, 1979. The deadline applicable to BAT was extended to July 1, 1984. In addition to better definition of treatment of conventional pollutants, recent changes in regulations have focused EPA attention on toxic and hazardous materials. Most mining and milling wastewaters contain many of these "priority pollutants," and more stringent permit conditions are certain to be a result. Hopefully, in the case of both BAT and ND, definition and application of this requirement and goal can be tempered by demonstrated need and economic practicality. As stated before, a significant portion of the nation's energy could be needlessly wasted, especially in efforts to attain ND. *See* WATER POLLUTION.

Land reclamation. It has been stated that strategic mineral development is the most productive use of land because of the great values that are received by the nation from such small land areas. Open-pit and strip mining have recently been criticized for their impact on the surrounding ecosystems. Some preservationists and conservationists feel that the Earth's surface should not be disturbed by either exploration or mining activities. Mining companies must reverse the spoiler image that has been created in the past, especially in the large-scale stripping of coal in the eastern and mid-central states. Most states now have regulations that require that all stripped land be reclaimed. Strip mining of coal has often produced an attendant acid mine drainage problem. *See* OPEN-PIT MINING; STRIP MINING.

Ecosystem studies are now being conducted by mining companies during exploration and prior to the commencement of mining operations. These studies lead to the effective planning for the most desirable method of mining that will least disturb the environment and yet lend itself to later reclamation.

Much controversy erupted with the passage of the Federal Surface Mining Control and Reclamation Act of 1977, pitting states, which had already enacted reclamation legislation, against the Office of Surface Mining, which was charged with the administration of the act. Many states are attempting to retain control of reclamation and are busily engaged in amending legislation and rewriting regulations to ensure compliance with the Federal act.

Even prior to state or Federal legislation, many mining companies had taken the initiative in planning operations so as to limit the adverse impact on the environment. Notable examples of this forward-looking and concerned approach are programs carried on by American Cyanamid Company, American Metal Climax, Inc., Anamax Company (formerly Anaconda), Bethlehem Steel Company, and Peabody Coal Company.

Typical of these industrial programs is that of Anamax which undertook the development of the Twin Buttes Copper Mine in the desert area near Tucson, AZ. In excess of 240,000,000 tons of alluvium were removed from this open-pit site before production could begin. This material and future waste will be used to form large dikes that will impound tailings. The dikes are terraced and planted with vegetation indigenous to the area and, when completed, will be 1000 ft wide (1 ft = 0.3 m) at the base and 250 ft wide at the apex, with a maximum height of 230 ft. These dikes take on the appearance of mesas and blend into the desert landscape.

Moreover, American Cyanamid has found that restoration of Florida's phosphate-mined lands can best be accomplished by reclaiming the major portions of those areas simultaneously during mining operations.

Land reclamation is made an integral part of mine planning by advanced consideration and decisions regarding what the area should look like upon completion of mining activity. By systematically forming eventual lakes and distributing stripped wastes into previously mined cuts, grading the area and restoring it to desirable land become relatively easy. Many of these reclaimed areas have become useful as parks, recreational areas, wildlife sanctuaries, and agricultural and residential development sites.

Again, if a reasonable and empirical approach to mined-land reclamation is allowed, following the dictates of physical, chemical, and biological laws, the very small area of the Earth's surface disturbed by mining can be restored to beneficial use at minimal cost to the consumer and optimum conservation of energy. *See* LAND RECLAMATION.

[LEWIS N. BLAIR; JOHN C. SPINDLER, WALTER H. UNGER]

Bibliography: M. Eisenbud, *Environment, Technology, and Health: Human Ecology in Historical Perspective*, 1978; J. E. McKee and H. W. Wolf (eds.), *1963 State of California Water Quality Criteria*, California Water Quality Board and U.S. Public Health Service, 2d ed., 1963; *Modern Pollution Control Technology*, vols. 1 and 2, Research and Education Association, 1978; F. W. Schaller and P. Sutton (eds.), *Reclamation of Drastically Disturbed Lands*, American Society of Agronomy, Crop Science Society of America, and Soil Science Society of America, 1978.

Environmental geology

The study of the relationship between humans and their geologic habitat. It is concerned with the adjustment of humans to the Earth and the reaction of the Earth to human use of it. The Earth is dynamic and constantly being subjected to modification by seismic, volcanic, tectonic, chemical, radiological, thermal, climatic, erosive, and sedimentological processes. The study of the processes themselves are in such disciplines as seismology, volcanology, plate tectonics, geomorphology, and sedimentology. The other major disciplines of geology that are closely related to environmental geolo-

Fig. 1. Eratini area, central Greece, showing landslide with typical scar. Slide debris has gone through olive orchard in foreground. The hills are more slide-prone since removal of the forests during classic times.

Fig. 2. Southern California is naturally slide-prone, but disturbance of the land by construction and loss of brush cover by burning increase the problem. *(Courtesy of James L. Ruhle)*

gy are petroleum geology, economic (or mining) geology, hydrology, geochemistry, and engineering geology. These disciplines are concerned primarily with people's use of the Earth, but only incidentally with the reaction of the Earth to that use. The practice of environmental geology is a process whereby scientists bring to bear their special insights to help in making decisions about how to use the Earth and its resources for maximum benefit.

Geologic processes affecting humans. Virtually all geologic processes affect humans. Although the effects on people may be sudden and catastrophic, by far the greatest influence on human civilizations has been the relatively subtle geologic processes, especially if either ignored or enhanced by the activities of people.

Catastrophic processes. The human race has always lived with various forms of geologic violence, with the toll in human lives so great at times that whole towns, cities, or, perhaps, even civilizations have been destroyed. For instance, in about 1470 B.C. the violent explosion of the volcano Thera (or Santorini) in the Aegean Sea along with the accompanying tidal wave may have been responsible for the sudden disappearance of the highly advanced Minoan civilization in the eastern Mediterranean. A single earthquake may cause hundreds of thousands of human deaths, as in the case of the earthquake of July 27, 1976, in Hopeh Province of China which took 650,000 lives.

Less violent processes. Human activity is influenced by many less spectacular and broadly predictable Earth processes which may have a persistent, widespread influence or which may have locally violent forms. For instance, river floodplains exist because of the processes of erosion and sedimentation and, unless these processes are altered by structures such as dams, flooding in such areas is inevitable. Similarly, in many areas with bowl-shape erosional scars (Fig. 1), bent trees, tilted fence lines or power poles, and slump topography are ample evidence to the trained eye of

Fig. 3. Nilo Pecanha power plant, located west of Rio de Janeiro, was made inoperative by a single rain which caused debris flows, largely from deforested slopes. *(Courtesy of Fred O. Jones)*

land instability commonly leading to landslides and debris flows. The problem is so common in parts of the Coast Ranges of California that geologic reports analyzing slide potential are required in advance of any construction.

More subtle in terms of the impact on humans are the effects of natural dust. Dust from volcanic eruptions, such as that of Krakatoa in 1883 or of Tambora, Indonesia, in 1815, may remain in the atmosphere for years and cause worldwide temperatures to drop by several degrees. As atmospheric dust increases because of the expansion of the arid regions of the Earth, the long-term impact of silica or other mineral particles on the human system could increase. For instance, in central Turkey, the death rate from lung cancer is extremely high. The cancer appears to be environmentally caused and related to a large quantity of the elongated mineral erionite, a zeolite that occurs in the local volcanic material and is carried in the dust of this arid area. Natural dust deposits (loess) formed after the last glaciation cover vast areas of the world and may be up to hundreds of feet thick.

Natural radiation is somewhat similar to dust in its prevalence. The average human exposure from cosmic radiation is 41–45 millirem per year, but dose rates may be up to five times greater at 8000 ft (2400 m) than at sea level. Natural radiation from rocks appears to average from 30 to 95 millirem per year, but the residents of many areas get significantly higher natural exposure. The Denver area, for instance, receives from 300 to 400 millirem per year, with large sections of France and Brazil receiving even higher natural terrestrial radiation. The highest natural radiation appears to be in the Karala area of India, where people receive up to 5000 millirem annually. Other sources of radiation are internal dosages from air, food, and water, which appear to average about 18 to 25 millirem per year. Fallout from atomic testing is about 2 millirem per year, and the average dose per person annually from medical x-rays in the United States is about 103 millirem. Epidemiological studies in areas of high terrestrial radiation do not appear to indicate higher cancer or death rates than normal. *See* RADIATION BIOLOGY.

Human adjustment to geologic processes. Through history it has rarely been in human nature to solve or anticipate geologic problems. For instance, even though the nature and causes of floodplains have long been known, dwellings and other structures continue to be built on these lowlands, resulting in major annual loss of lives and property. Similarly, the geologic landforms which indicate landslide conditions have often not been observed, and structures have been built on or below potentially active earth masses (Fig. 2). In general, the civilizations of the world have rarely anticipated or planned for violent geological events such as earthquakes and tidal waves sufficiently well to minimize their effects. A classic example of ignoring inevitable catastrophe is the placing of housing developments or other structures within or across major active faults.

Earth resources. The resources of the Earth can be grouped into two large categories: those which are limited and considered nonrenewable, and those which are regenerated, that is, are renewable.

Nonrenewable resources. Those components of mineral resources which are of current value and have been precisely defined as to quality, quantity, and location, and whose extraction is planned or is actually occurring, are called reserves. By definition, mineral reserves are limited. They are also nonrenewable. In a sense, however, most mineral resources cannot really be depleted by human activities because, once used, they are still on the Earth, merely in a different form. Fossil fuels are in a chemically reduced form and, when used, are altered to oxides (carbon dioxide and water) from which energy can no longer be extracted. Some other materials, such as aluminum, iron, silicon, potassium, sodium, and magnesium, occur abundantly in nature in the form of oxides, and cannot really be exhausted because they are basic constituents of the Earth itself. Because mineral resources are critical not only to the industrial segment of the society but to environmental improvement, the long-range management of mineral resources is within the purview of environmental geology. *See* MINERAL RESOURCES CONSERVATION.

Renewable resources. Those resources which are regenerated, such as plants and animals, are called renewable resources. The classic renewable geologic resource is water because the supply is replenished in the form of precipitation. Soils may also be considered renewable in some cases where reforestation or other cultural processes may be

Fig. 4. Steep slope tilled for corn in Petionville area, Haiti. Once forested and stable, the topsoil is now being washed away, largely as debris flows.

increasing the quantity of soil. Somewhat paradoxically, the renewable resources appear to be depleting in many areas, while the available nonrenewable resources appear to be increasing. *See* CONSERVATION OF RESOURCES.

Deleterious human influence. As long as humans have inhabited the Earth, their presence has caused modifications in the resources and processes of the Earth, often in an undesirable way. The human race itself is a major geologic agent, and virtually all civilizations have influenced their geologic habitats and variously depleted their earth resources. The earliest, and in some ways the best, work on the subject of human influence on the geologic surroundings was by G. P. Marsh. His classic work was largely ignored by geologists, who were more concerned with utilization of the Earth's surface along with its mineral and water resources. Thus, the long-term impacts of human activities on land forms, erosion and sedimentation rates, the quantity of surface water and subsurface water and mineral resources, soil and groundwater quality, climate, and the size of deserts and prairies are only now being systematically analyzed.

Desertification. Humans have long had enormous impact on the rates of natural geologic processes, although in the span of existence of individuals the human impacts have often not been noticed. Marsh showed that deforestation (Fig. 3), overtilling (Fig. 4), overgrazing, improper irrigation practices, and burning of grasslands are responsible for much of the world's deserts, grasslands, and parklands. M. Kassas estimates that of the 43% of the Earth that is vegetative desert area, 6.7% (an area about the size of Brazil) may have been created by human activity. This is probably a minimum figure. *See* DESERTIFICATION; FOREST MANAGEMENT AND ORGANIZATION; VEGETATION MANAGEMENT.

Soil depletion. The human process of exhausting the land has a long history. For instance, the salination of soils in lower Mesopotamia has been thoroughly documented by archeologists, where wheat yields were prolific in about 3500 B.C. but became negligible by 1700 B.C. because of salination. Mute testimony of the pervasive nature of such problems exists in the form of abandoned cities such as Ankor Wat in Thailand and those on the Yucatan Peninsula in Mexico. In the Mideast, the ancient city mounds and tells are found in areas where agriculture is no longer possible. Depletion of soil nutrients, erosion, and salination caused by poor agricultural practices made it more difficult for the ancient civilizations to produce sufficient food nearby. This, coupled with expanding populations and extended primitive transportation lines, must have created enormous logistic problems and must have contributed significantly to the demise of these cultures. *See* AGRICULTURAL GEOGRAPHY.

In recent years, the processes of deleterious modification of the Earth by human activities have probably accelerated because of expanded human population coupled with the sudden increases in the cost of energy. High energy prices have placed multiple pressures on the land because the poorer people of the world have been forced to cut more timber for fuel while simultaneously not being able to afford as much fertilizer, the cost of which is largely a function of energy prices. In addition, llama and cattle dung which formerly were returned to the land are now used to an ever greater extent for cooking. Thus, more land is being tilled to produce the same amount of food. The multiple pressures are more rapidly expanding the uninhabitable areas of the Earth than at any time in history. Virtually all of southern Asia, South and Central America, and Africa are affected.

Climate modification. A subtle but possibly significant change that could alter the rate of change of surficial processes is the increase of CO_2 in the Earth's atmosphere because of the burning of such carbon compounds as coal, oil, and wood for energy. Carbon dioxide in the atmosphere has increased 10% in the 20 years that systematic records have been kept, and could double by the year 2050. The feared result is a buildup of heat—the so-called hothouse or greenhouse effect, which could radically alter climatic processes by increasing worldwide temperatures by an average of 6°C. One major possible impact is partial melting of ice sheets, a phenomenon which could easily raise ocean water levels by 15 to 25 ft (4.6 to 7.6 m) and threaten vast coastal areas of the world. The atmospheric heating could begin within the next few decades. *See* ATMOSPHERIC POLLUTION; GREENHOUSE EFFECT.

Reversing the trend does not appear easy. Reduction of burning of fossil fuels is a partial answer. But the burning of wood for fuel may be a far greater cause because burning wood not only contributes directly to the CO_2, but also depletes growing vegetation which removes CO_2 from the air as a normal part of plant growth (photosynthesis).

Waste disposal. The way in which civilizations handle their wastes also influences the nature of earth resources. The industrial nations, largely because of their wealth, have been able to concentrate, and often treat, their waste materials (in contrast to less developed nations, where wastes tend to be pervasive and untreated). However, the results of concentrations may cause local problems. Such things as industrial waste and ordinary garbage need to be disposed of. In addition, environmental treatment wastes such as sludges from sewage and water treatment plants, sulfate sludges which result from removing SO_2 gas from burning sulfur-rich fuels, and fly ash from burning coal are also disposal problems. The quality of surface and groundwaters may be adversely influenced locally if the environmental controls are inadequate at disposal sites. *See* INDUSTRIAL WASTES.

Drinking water contamination. A larger-scale problem is the introduction of blood anoxia-causing nitrate into the groundwater of large areas where cattle concentrate or where high nitrate fertilizers are improperly used. The natural geologic environment does not readily take up excess nitrate, particularly where vegetation may be sparse as at cattle feedlot areas. Thousands of wells in such areas as the American Midwest are known to be contaminated. Nitrate contamination of drinking water may currently be a worldwide problem, and may well have caused the sapping of

Fig. 5. The wooded areas of the Pocono Mountains in northeastern Pennsylvania were once used for agriculture. The land is now relatively stabilized, with agriculture restricted to relatively flat areas.

the energy of the populations of some ancient city-states. However, systematically gathered information is not available for most of the world. *See* AGRICULTURAL WASTES; EUTROPHICATION; WATER POLLUTION.

Deforestation. As pointed out by E. P. Eckholm, the forests in virtually all of the poorer countries of the world are in jeopardy and could all but disappear in the next few decades in most of the Southern Hemisphere. In the absence of a radical reduction of fossil-fuel prices, a reversal of the deforestation trend does not appear likely because the alternatives of solar, wind, tidal, or geothermal energy are not generally practical. Even water power is of limited value, largely because devegetation increases erosion rates and causes reservoirs used for hydroelectric power generation to fill rapidly with debris. For instance, the Lower Anchicaya Reservoir in Columbia filled with erosion debris from stripped steep slopes in just 7 years.

Human beneficial influences. In contrast to the less developed nations, the industrial nations in the past half century have successfully reversed some of the major trends toward deleterious alteration of their geological environments. The efficient production of food by the industrialized nations because of the expanded use of fertilizer, agricultural lime, pesticides, and herbicides has allowed reforestation of steep slopes (Fig. 5) which were formerly needed for agriculture. The reduced erosion caused by revegetation and good farming practice has greatly stabilized the physical environment (Fig. 6). The possibility of sustained yield from the Earth's resources appears very likely for these nations, especially in that their populations are also stabilizing. The continued high cost of energy, however, will have an uncertain long-term effect.

Some wastes from human activities may be reused or may be put to other human uses so that the wastes themselves may become resources. Recycling of metals or glass is being done to a greater degree to conserve mineral resources. Somewhat similarly, sewage treatment sludges may be used as soil supplements or sulfate sludges and fly ash may be used as a component of construction materials.

Recovering minerals from the earth invariably affects the earth itself and, improperly done, the influences may be deleterious. Contamination of waters by heavy metals, iron, acidic sulfate, and mineral particles may result, and the landscape may be left in a form which has little potential value. However, modern reclamation practices involving the concepts of multiple land use may often—perhaps nearly always—leave land in a form that is equal to or better than its state before mining (in terms of the ability to use the land for hu-

Fig. 6. In north-central Maryland, contour plowing, leaving steeper, more erodable slopes in woods, and the farm pond (in the center of the picture) are stabilizing influences. Two very heavy rains—Hurricane Agnes, 1972, and Hurricane Eloise, 1975—had little effect on the land.

man purposes). Thus, flatland suitable for agriculture or for construction may be created in hilly country; recreational or scenic lakes, water storage, water spreading, or flood control sites may be created; valuable underground space and needed space for waste disposal may be other by-products of mineral extraction. Thus mining, properly done, may not only supply important minerals but often creates environmental or economic values. Such values are possible in well over three-quarters of the area which is disturbed by mining in the United States. *See* LAND RECLAMATION.

Geology and planning. Environmental geology is needed by decision makers who are involved with integrating human needs with the realities imposed by geologic dynamics and resource distribution. Populations in general, have, in a sense, long been mining the environment in that they have been depleting soil, forest, and groundwater resources. It is axiomatic that if the depletion of resources continues and the world population continues to expand, much of the Earth will be unable to sustain humans. Analysis by environmental geologists of the long-term affects of the human population on the geologic environment and adjusting human activities to minimize problems and maximize benefits may potentially lead to the sustained yield of the Earth for the benefit of people.

Working with planners in the United States, environmental geologists help to determine areas which are most amenable to such land uses as construction, water resource recharge and retrieval, farming, mineral resource development, and waste disposal. Analysis by environmental geologists of land use and general resource utilization by the poorer nations could contribute to reversing the trend toward the destruction of their environments. *See the feature article* ENVIRONMENTAL PROTECTION; *see also* HUMAN ECOLOGY; LAND-USE PLANNING. [JAMES R. DUNN]

Bibliography: E. P. Eckholm, *Losing Ground: Environmental Stress and World Food Prospects*, 1976; K. L. Feldman et al., *Radiological Quality of the Environment in the United States*, EPA 520/1-77-009, 1977; M. Kassas, A brief history of land use in Mareotic Region, Egypt, *Minerva Biol.*, 1(4):167, September–October 1972.

Epidemiology

The study of the mass aspects of disease. The word epidemic literally means "upon the people," and was coined to describe the way an infectious disease, in spreading through a group of people, gives the impression of an affliction that has been placed upon the community as a whole. An observer reflecting on such an epidemic would ask questions about the circumstances which allowed the disease to develop and flourish, what permitted some members to escape, and what brought the epidemic to an end. These are questions about the disease as it affects groups of people, rather than individuals in the group, and they are the typical problems of epidemiology.

In recent years diseases other than infectious diseases have been included in the subject matter of epidemiology. It is quite natural, therefore, that these massive noninfectious diseases should be studied as group phenomena, and that one should speak of the epidemiology of accidents, cancer, heart disease, or mental disease. For both infectious and noninfectious diseases there are similar general problems such as identifying cases in the population at risk, and finding which circumstances lead to an increase and which to a decrease in the disease. In spite of the differences in detail arising in a study of Asian influenza and a study of suicide in a particular population, both are fundamentally investigations of disease rates and a comparison of the way these rates vary from one subgroup to another; for example, is there a difference in the rates for the two sexes, the various age groups, or different socioeconomic classes? There are traditionally a number of rates that have proved useful in the study of epidemiology. Some of the most important are the mortality rate, morbidity rate, prevalence ratio and incidence rate, and case fatality ratio.

Mortality rate. The mortality rate for a given year is the number of deaths occurring in the year per 1000 total midyear population. In the United States in 1946 there were 1,395,600 deaths and the population as of July 1 was 139,893,000. Hence the mortality rate was

$$\frac{1{,}395{,}600 \times 1000}{139{,}893{,}000} = 10.0$$

This rate is sometimes referred to as the crude death rate because no account is taken of such factors as the age and sex of the members of the population. Since France has a higher percentage of old people than the United States and Israel a lower percentage, it is not surprising to find that the crude death rate in France is considerably higher than that of the United States and in Israel considerably lower. These objections to the crude death rate can be met by computing separate rates for particular age groups in each sex, for example, the rate for males aged 15–24, another rate for males 25–34, and so on. The process can be carried further by computing separate death rates for individual diseases such as tuberculosis, or groups of diseases such as those associated with child bearing. The rates for particular age groups, particular causes of death, or specified sex are known as specific rates. They are often essential for the thorough study of an epidemic. However, it is true in general that the more specific a rate, the more difficult it is to collect accurately the information necessary for the computation. Thus to compute the death rate from accidents in 1959 among male children aged 5–14 years, it is necessary to know both the number of deaths from accidents among the group in 1959 and also what is called the population at risk, that is, the number of boys in the given age group at the time and place under consideration. Specific rates are more difficult to obtain than crude rates, and it is for this reason that crude rates are still used and relied upon in the methodology of epidemiology. An example of the effect of a major epidemic on this rate is shown by the death rate for the United States during the influenza epidemic of 1918. The figure for that year was 18.1 deaths per 1000 population; for the previous year it was 14.0, and for the succeeding year 12.9.

Morbidity rates. There are certain diseases which cannot be studied by a means of mortality rates of any kind. Mental diseases, arthritis, and the common cold are examples of conditions the epidemiology of which is reflected inadequately by death rates. A more useful approach is to consider morbidity rates which give the number of cases of disease per 100,000 population, either as cases existing at a particular point in time or as cases occurring in a particular period of time. Knowledge of these rates for the population in general is inadequate. There are many countries where the reporting of death is virtually 100%; but there is no comparable record by which one can ascertain how many people suffer from disease. It is true that there is compulsory reporting in many countries of all discovered cases of certain infectious diseases, such as cholera; that schools, hospitals, factories, and health insurance groups have records giving useful information about the diseases found in special segments of the population; and that morbidity surveys such as the United States National Health Survey, based on household inquiry by skilled interviewers, have proved useful. But, in general, whenever the information about morbidity covers the whole population or a representative sample of it, the data are limited in detail. Thus the course of the epidemic of Asian influenza in the United States in 1957–1958 can be traced accurately, as to the number of cases involved, from the returns of the National Health Survey, but there is no information on the finer details about the disease, such as can be studied when samples of blood are available for special examinations.

Incidence and prevalence. In the investigation of a particular epidemic, ascertainment of disease rates is of great importance. It is necessary to distinguish two main kinds of rates, the one based on the incidence of the disease and the other on its prevalence.

Incidence rate. The incidence rate is concerned with the cases of disease that develop or, more strictly, that come under diagnosis in a given period of time. The incidence rate per 100,000 population for a given year is as shown in the equation here. This rate measures the risk of developing the disease in the given time period; for convenience, it is expressed in terms of so many cases per 100,000 population at risk per year.

$$\text{Incidence rate} = \frac{\text{number of new cases occurring during year} \times 100{,}000}{\text{midyear population}}$$

Prevalence. The prevalence ratio is concerned with the cases of disease that exist at a particular point in time. The concept is clearest when the point of time is thought of as being instantaneous; for example, one might ask how many cases of tuberculosis per 100,000 population existed at a specified moment, and this would then be a prevalence ratio for the disease. Prevalence ratios are also computed, especially for chronic diseases, over periods as long as a year. The essential characteristic is still maintained, however, that the count is based on cases existing at some part of the given period, and not necessarily developing within the given period. The prevalence ratio takes in cases that developed prior to the period under consideration, provided the disease persisted until the prevalence count was made.

In the case of diseases which last for several years, such as tuberculosis and diabetes, the prevalence may be much greater than the incidence. In fact, there is some evidence that the modern effective treatment of diabetes, which holds the disease in check without curing it in a fundamental sense, has actually increased the prevalence of the disease. A similar point arose during World War II in connection with the epidemiology of tuberculosis. Since it is well known that in Tennessee the death rate from tuberculosis is higher among the Negroes than the whites, it was expected that in the Selective Service examinations a higher prevalence of the disease would be found in Negroes than in whites. This was not so, presumably because for various reasons the disease ran a longer but less fatal course in the whites.

The distinction between incidence and prevalence is crucial for much epidemiological work. In particular, if one wishes to decide whether a factor is associated with a particular disease, for example, baldness with disease of the coronary arteries, one should make the correlation on cases that arise in a given period of time. The cases of disease that exist at some point in time represent a biased sample of the total, since they include only those who have survived for a period of time, and if survival were better for bald men than for the rest, one would be in danger of concluding that baldness predisposed to coronary artery disease even if, in fact, this were not true.

During the period an incidence rate is being measured, a person may have more than one attack or may suffer from more than one disease. Consequently it is necessary to state whether the unit of measurement is a person, an episode of illness, or a particular disease; the count will vary according to the definition. There is the further obstinate problem of defining when a person is sick. For some purposes this may be done by the subject himself, in terms of the number of days of absence from work. In other instances, especially when long-lasting diseases such as diabetes are involved, periodic examinations may be made by a physician. The problem of the careful definition of cases is important in epidemiological research and this has led to a close cooperation in modern work among physicians, who identify cases by clinical means; laboratory workers, who identify cases by microbiological, biochemical, or other laboratory examinations; and statisticians, who make the comparisons of the disease rates in various subgroups of the population at risk.

Case fatality ratio. In addition to studying the presence or absence of a disease, the epidemiologist often uses some measure of the severity of the disease. One index of severity is the case fatality ratio, which is defined as the number of deaths expressed as a percentage of the number of cases.

Experimental epidemiology. The observational methods so far discussed do not exhaust the possibilities of studying the factors influencing epidemics. The experimental approach has also been used and experimental epidemiology has great potential for improving knowledge of the subject. An epidemic is started by introducing into an animal colony a certain number of infected animals. The course of the subsequent epidemic can then be followed and a study made of the influence of factors such as diet, the percentage of susceptible animals, and the intimacy of the contact between the animals. It is easy to show by such methods, for example, that the introduction of fresh susceptibles into the population from time to time has a marked effect in sustaining an epidemic. *See* DISEASE. [COLIN WHITE]

Erosion

The loosening and transportation of rock debris at the Earth's surface. Erosion is one of the most important geologic processes at the surface and operates to move earth materials to lower levels. Agents of erosion include moving surface, ground, and ocean waters; ice; wind; organisms; temperature changes; and gravity. They work by moving earth materials in two ways: physically, that is, without change in chemical composition of the earth materials; and chemically, in which the rock materials have been dissolved or decomposed. The energies driving the forces of erosion are dominantly solar (which includes temperature change, raising moisture for precipitation, wind, and waves); gravitational, aided by rotation of the Earth; and chemical. Counteracting the effects of erosion are geologic processes which raise the Earth's surface, such as volcanism (volcanoes and intrusions) and diastrophism (folding, faulting, plate tectonics).

Erosion develops and alters landscapes and scenery, and modifies agricultural practice and human habitats. It slowly lowers the surface of the Earth, on the average 1 m in the order of several thousands of years; but local movements, such as landslides and mud slumps, may occur very rapidly. As a high region is lowered by erosion, it gradually progresses through an erosion cycle in topographic stages conceptually characterized by distinctive groups of landforms. These stages were named youthful, mature, and old age by W. M. Davis. He considered the lowering to be simultaneously effective over the whole region. Alternatively, W. Penck and L. King described the erosional process as one in which laterally directed planation and parallel retreat of slopes were the most prominent mechanisms.

Within a given region, distinctive landforms, enjoyed as scenery, are produced by stream erosion, mountain glaciation, continental glaciation, wind (desert) erosion, landslides and mass wastage, and groundwater solution (karstic) features. *See* STREAM TRANSPORT AND DEPOSITION.

Not only does erosion affect human beings, but the human race in turn is an almost violent agent of erosion, with its practices of agriculture, overgrazing, lumbering, and urban development on continental land slopes and coastal topographic features. For ages—from ancient Greece, Egypt, Babylonia, and China, to modern nations with their dustbowls and intensive tillage practices—the human race has been the cause of rapid (in geologic time) and sometimes disastrous episodes of erosion. *See* SOIL CONSERVATION.

In long enough time, erosion would tend to reduce the Earth's surface to a subdued plain—except that geologic processes driven by internal earth energy presumably will reactivate the geomorphic cycle by reelevating the land. *See* GEOMORPHOLOGY. [W. D. KELLER]

Estuarine oceanography

The study of the physical, chemical, biological, and geological characteristics of estuaries. An estuary is a semienclosed coastal body of water which has a free connection with the sea and within which the sea water is measurably diluted by fresh water derived from land drainage. Many characteristic features of estuaries extend into the coastal areas beyond their mouths, and because the techniques of measurement and analysis are similar, the field of estuarine oceanography is often considered to include the study of some coastal waters which are not strictly, by the above definition, estuaries. Also, semienclosed bays and lagoons exist in which evaporation is equal to or exceeds fresh-water inflow, so that the salt content is either equal to that of the sea or exceeds it. Hypersaline lagoons have been termed negative estuaries, whereas those with precipitation and river inflow equaling evaporation have been called neutral estuaries. Positive estuaries, in which river inflow and precipitation exceed evaporation, form the majority, however.

Topographic classification. Embayments are the result of fairly recent changes in sea level. During the Pleistocene ice age, much of the sea water was locked up in continental ice sheets, and the sea surface stood about 100 m below its present level. In areas not covered with ice, the rivers incised their valleys to this base level; and during the ensuing Flandrian Transgression, when the sea level rose at about 1 m per century, these valleys became inundated. Much of the variation in form of the resulting estuaries depends on the volumes of sediment that the river or the nearby coastal erosion has contributed to fill the valleys.

Where river flow and sediment discharge were high, the valleys have become completely filled and even built out into deltas. Generally, deltas are best developed in areas where the tidal range is small and where the currents cannot easily redistribute the sediment the rivers introduce. They occur mainly in tropical and subtropical areas where river discharge is seasonally very high. The distributaries, or "passes," of the delta are generally shallow, and often the shallowest part is a sediment bar at the mouths of the distributaries. The Mississippi and the Niger are examples of this type of delta.

Where sediment discharge was less, the estuaries are unfilled, although possibly they are still being filled. These are drowned river valleys or coastal plain estuaries, and they still retain the topographic features of river valleys, having a branching, dendritic, though meandering, outline and a triangular cross section, and widening regularly toward the mouth, which is often restricted by spits. River discharge tends to be reasonably steady throughout the year, and sediment discharge is generally small. These estuaries occur in areas of high tidal range, where the currents have helped to keep the estuaries clear of sediment. They are typical of temperate regions such as the eastern coast of North America and northwestern Europe, examples being the Chesapeake Bay system, the Thames, and the Gironde.

In areas where glaciation was active, the river valleys were overdeepened by glaciers, and fiords were created. A characteristic of these estuaries is the rock bar or sill at the mouth that can be as little as a few tens of meters deep. Inside the mouth, however, they can be at least 600 m deep and can extend hundreds of kilometers inland. Fiords are typical of the Norwegian and the Canadian Pacific coasts.

There is another estuarine type, called the bar-built estuary. These are formed on low coastlines where extensive lagoons have narrow connecting passages or inlets to the sea. Within the shallow lagoons the tidal currents are small, but the deep inlets have higher currents. Again, a sediment bar is generally present across the entrance. In tropical areas the lagoons can be hypersaline during the hot season. They are typical of the southern states of the United States and of parts of Australia.

Estuaries are ephemeral features since great alterations can be wrought by small changes in sea level. If the present ice caps were to melt, the sea level would rise an estimated 30 m, and the effect on the form and distribution of estuaries would be drastic.

Physical structure and circulation. Within estuaries, the river discharge interacts with the sea water, and river water and sea water are mixed by the action of tidal motion, by wind stress on the surface, and by the river discharge forcing its way toward the sea. The difference in salinity between river water and sea water—about 35 parts per thousand—creates a difference in density of about 2%. Even though this difference is small, it is sufficient to cause horizontal pressure gradients within the water which affect the way it flows. Density differences caused by temperature variations are comparatively smaller. Salinity is consequently a good indicator of estuarine mixing and the patterns of water circulation. Obviously, there are likely to be differences in the circulation within estuaries of the same topographic type which are caused by differences in river discharge and tidal range. The action of wind on the water surface is an important mixing mechanism in shallow estuaries, particularly in lagoons; but generally its effect on estuarine circulation is only temporary, although it can produce considerable variability and thus make interpretation of field results difficult. *See* SEA WATER.

Salt-wedge estuaries. Fresh water, being less dense than sea water, tends to flow outward over the surface of sea water, which penetrates as a salt wedge along the bottom into the estuary (Fig. 1). This creates a vertical salinity stratification, with a narrow zone of sharp salinity change, called a halocline, between the two water masses which can reach 30 parts per thousand in 1/2 m. If the sea is tideless, the water in the salt wedge is almost motionless. However, if the surface layer flowing to-

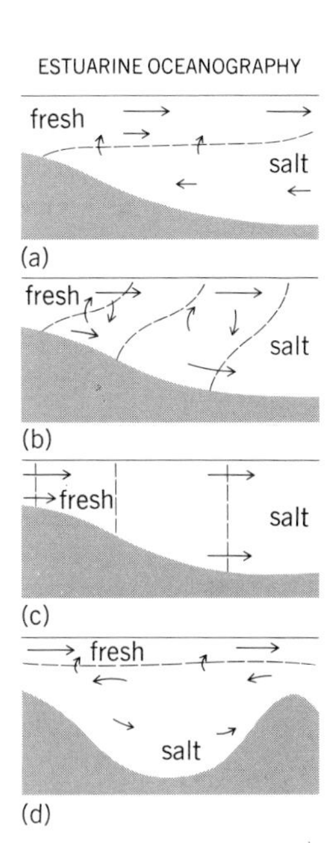

Fig. 1. Diagrammatic representation of mixing in estuaries. (*a*) Salt-wedge type (*b*) Partially mixed type. (*c*) Well-mixed type. (*d*) Fiord.

ward the sea has a sufficiently high velocity, it can create interfacial waves on the halocline. These waves break, ejecting small parcels of salt water into the fresher surface layer; this process is called entrainment, and it occurs all along the halocline. No fresh water is mixed downward; thus the salinity within the salt wedge is almost constant along the estuary. However, the salt wedge loses salt water which is mixed into the surface layer and discharged into the sea. Consequently, for this loss to be replaced, there must be a compensatory flow of salt water toward the head of the estuary within the salt wedge, but of a magnitude much less than that of the flow in the surface layer. There is a considerable velocity gradient near the halocline as a result of the friction between the two layers. Consequently, the position of the salt wedge will change according to the magnitude of the flow in the surface layer, that is, according to the river discharge. The Mississippi River is an example of a salt-wedge estuary. When the flow in the Mississippi is low, the salt wedge extends more than 100 mi (160 km) inland, but with high discharge, the salt wedge only extends a mile (1.6 km) or so above the river mouth. Some bar-built estuaries, in areas of restricted tidal range and at times of high river discharge, as well as deltas, are typical salt-wedge types.

Partially mixed estuaries. When tidal movements are appreciable, the whole mass of water in the estuary moves up and down with a tidal periodicity of about 12-1/2 hr. Considerable friction occurs between the bed of the estuary and the tidal currents, and this causes turbulence. The turbulence tends to mix the water column more thoroughly than entrainment does, although little is known of the relationship of the exchanges to the salinity and velocity gradients. However, the turbulent mixing not only mixes the salt water into the fresher surface layer but also mixes the fresher water downward. This causes the salinity to decrease toward the head of the estuary in the lower layer and also to progressively increase toward the sea in the surface layer. As a consequence, the vertical salinity gradient is considerably less than that in salt-wedge estuaries. In the surface, seaward-flowing layer, the river discharge moves toward the sea; but because the salinity of the water has been increased by mixing during its passage down the estuary, the discharge at the mouth can be several times the river discharge. To provide this volume of additional water, the compensating inflow must also be much higher than that in the salt-wedge estuary. The velocities involved in these movements are only on the order of a few centimeters per second, but the tidal velocities can be on the order of a hundred centimeters per second. Consequently, the only way to evaluate the effect of turbulent mixing on the circulation pattern is to average out the effect of the tidal oscillation, which requires considerable precision and care. The resulting residual or mean flow will be related to the river discharge, although the tidal response of the estuary can give additional contributions to the mean flow. The tidal excursion of a water particle at a point will be related to the tidal prism, the volume between high- and low-tide levels upstream of that point; and the instantaneous cross-sectional velocity at any time will be related to the rate of change of the tidal prism upstream of the section. In details, the velocities across the section can differ considerably. It has been found that in the Northern Hemisphere the seaward-flowing surface water keeps to the right bank of the estuary, looking downstream, and the landward-flowing salt intrusion is concentrated on the left-hand side (Fig. 2). This is caused by the Coriolis force, which deflects the moving water masses toward the right. Of possibly greater importance, however, is the effect of topography, because the curves in the estuary outline tend to concentrate the flow toward the outside of the bends. Thus, in addition to a vertical circulation, there is a horizontal one, and the halocline slopes across the estuary. Because the estuary has a prismatic cross section, the saline water is concentrated in the deep channel and the fresher water is discharged in the shallower areas. Examples of partially mixed estuaries are the rivers of the Chesapeake Bay system.

Well-mixed estuaries. When the tidal range is very large, there is sufficient energy available in the turbulence to break down the vertical salinity stratification completely, so that the water column becomes vertically homogeneous. In this type of estuary there can be lateral variations in salinity and in velocity, with a well-developed horizontal circulation; or if the lateral mixing is also intense, the estuary can become sectionally homogeneous (also called a one-dimensional estuary). Because there is no landward residual flow in the sectionally homogeneous estuary, the upstream movement of salt is produced during the tidal cycle by salty water being trapped in bays and creeks and bleeding back into the main flow during the ebb. This mechanism spreads out the salt water, but it is

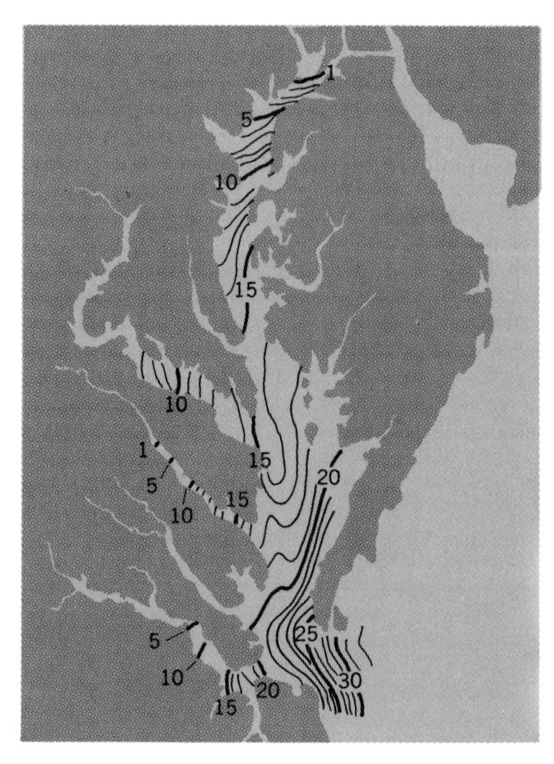

Fig. 2. Typical surface salinity distribution in Chesapeake Bay. (*From D. W. Pritchard, Estuarine Hydrography, in H. E. Landsberg, ed., Advances in Geophysics, vol. 1, Academic Press, 1952*)

probably effective only for a small number of tidal excursions inland.

Fiords. Because fiords are so deep and restricted at their mouths, tidal oscillation affects only their near-surface layer to any great extent. The amount of turbulence created by oscillation is small, and the mixing process is achieved by entrainment. Thus fiords can be considered as salt-wedge estuaries with an effectively infinitely deep lower layer. The salinity of the bottom layer will not vary significantly from mouth to head, and the surface fresh layer is only a few tens of meters deep. When the sill is deep enough not to restrict circulation, the inflow of water occurs just below the halocline, with an additional slow outflow near the bottom. When circulation is restricted, the replenishment of the deeper water occurs only occasionally, sometimes on an annual cycle; and between the inflows of coastal water, the bottom layer can become anoxic.

The descriptive classification of estuaries outlined above depends on the relative intensities of the tidal and river flows and the effect that these flows have on stratification. A quantitative comparison between estuaries can be made using the diagram of Fig. 3, which is based on a stratification and a circulation parameter.

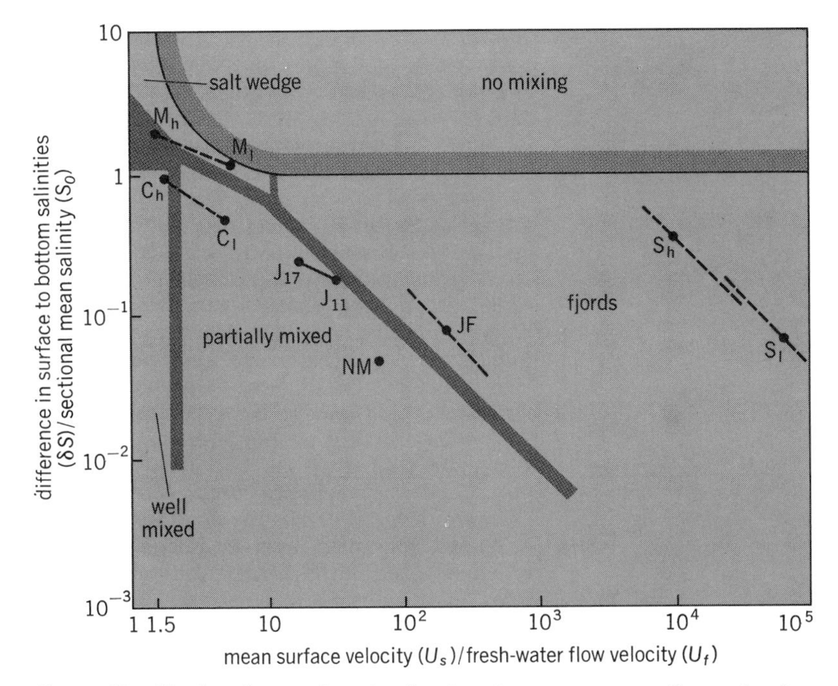

Fig. 3. Classification diagram for estuaries. An estuary appears as a line on the diagram; the upper reaches are less well mixed than the lower sections. Subscript letters refer to high (h) and low (l) river discharge; subscript numbers are distances from the mouth. J= James River; M= Mississippi; C= Columbia River; NM= Narrows of the Mersey; S= Silver Bay; JF= Strait of Juan de Fuca. *(After D. V. Hansen and M. Rattray, Jr., New dimensions in estuary classification, Limnol. Oceanogr., 11:319–326, 1966)*

Flushing and pollution-dispersal prediction. Much research into estuarine characteristics is aimed at predicting the distribution of effluents discharged into estuaries. Near the mouth of a partially mixed estuary, the salt water is only slightly diluted, and in order for a volume of fresh water equivalent to the river flow to be discharged, a much greater volume of mixed water must flow seaward. Consequently, estuaries are more effective in diluting and removing pollutants than rivers. It has been observed that increased river flow causes both a downstream movement of the saline intrusion and a more rapid exchange of water with the sea. The latter effect occurs because increased river discharge increases stratification; increased stratification diminishes vertical mixing and enhances the flow toward the sea in the surface layer. Thus, increased river discharge has the effect of increasing the volume of fresh water accumulated in the estuary, but to a lesser extent than the increase of the discharged volume. Obviously, it takes some time for the fresh water from the river to pass through the estuary. The flushing time can be determined by dividing the total volume of fresh water accumulated in the estuary by the river flow. For most estuaries the flushing time is between 5 and 10 days.

If a conservative, nondecaying pollutant is discharged at a constant rate into an estuary, the effluent concentration in the water moving past will vary with the tidal current velocity, and will spread out by means of turbulent mixing. The concentrations will be increased during the next half cycle as the water passes the discharge point again. After several tidal cycles a steady-state distribution will be achieved, with the highest concentration near the discharge point. Concentrations will decrease downstream, but not as quickly as they decrease upstream. However, the details of the distribution will depend largely on whether the discharge is of dense or light fluid and whether it is discharged into the lower or upper layer. Since its movement will be modified by the estuarine circulation, the effluent will be more concentrated in the lower layer upstream of the discharge point, and it will be more concentrated in the upper layer downstream. To obtain maximum initial dilution, a light effluent would have to be discharged near the estuary bed so that it would mix rapidly as it rose.

For nonconservative pollutants, such as coliform sewage bacteria, prediction becomes more difficult. The population of bacteria dies progressively through the action of sunlight, and concentrations diminish with time as well as by dilution. The faster the mixing, the larger the populations at any distance from the point of introduction, since less decay occurs.

Because of the poor mixing of fresh water into a salt-wedge estuary, an effluent introduced in the surface layer will be flushed from the estuary before it contaminates the lower layer, provided that it is not too dense.

Mathematical modeling. Increasingly, mathematical modeling is being used, with reasonable success in many instances, to predict effluent dispersal with the minimum amount of field data. Although the governing mathematical equations can be stated, they cannot be solved in their full form because there are too many unknowns. Consequently, to reduce the number of unknowns, various assumptions are made, including some form of spatial averaging to reduce a three-dimensional problem to two dimensions or even one dimension. The exchange ratios, about which little is known, are assumed to be constant, or they are considered as a simple variable in space, and are altered so that the model fits the available prototype data.

The first step is usually to model the flow and salinity distribution. Because the density field is important in determining the flow characteristics, density and flow are interlinked problems. Then, for pollutant studies, the pollutant is assumed to act in the same manner as fresh or salt water, or the flow parameters are used with appropriate exchange coefficients to predict the distribution. Simple models consider the mean flow to be entirely the result of river discharge, and tidal flow to be given by the tidal prism. Segmentation is based on simple mixing concepts and crude exchange ratios. Salinity and pollutant concentrations can then be calculated for cross-sectionally averaged and vertically homogeneous conditions by using the absolute minimum of field data. These models are known as tidal prism models. One-dimensional models are very similar, but use a finer grid system and need better data for validation. Two-dimensional models either assume vertical homogeneity and allow lateral variations, or vice versa. There are difficulties in including the effects of tidally drying areas and junctions; the models become increasingly costly and require extensive prototype data, but they are more realistic. The ideal situation of modeling the flow and salinity distribution accurately simply on the basis of knowledge about the topography, the river discharge, and the tidal range at a number of points is still a long way off.

Estuarine environments. Estuarine ecological environments are complex and highly variable when compared with other marine environments. They are richly productive, however. Because of the variability, fewer species can exist as permanent residents in this environment than in some other marine environments, and many of these species are shellfish that can easily tolerate short periods of extreme conditions. Motile species can escape the extremes. A number of commercially important marine forms are indigenous to the estuary, and the environment serves as a spawning or nursery ground for many other species.

River inflow provides a primary source of nutrients such as nitrates and phosphates which are more concentrated than in the sea. These nutrients are utilized by plankton through the photosynthetic action of sunlight. Because of the energetic mixing, production is maintained throughout, in spite of the high levels of suspended sediment which restrict light penetration to a relatively thin surface layer. Plankton concentrations can be extremely high, and when they are, higher levels of the food web—filter-feeding shellfish and young fish—have an ample food source. The rich concentrations provide large quantities of organic detritus in the sediments which can be utilized by bottom-feeding organisms and which can be stirred up into the main body of the water by tidal action. For a more complete treatment of the ecology of estuarine environments from the biological viewpoint *see* MARINE ECOSYSTEM.

There is a close relationship between the circulation pattern in estuaries and the faunal distributions. Several species of plankton peculiar to estuaries appear to confine their distribution to the estuary by using the water-circulation pattern; pelagic larvae of oysters are transported in a similar manner. The fingerling fish (*Micropogon undulatus*), spawned in the coastal waters off the eastern coast of the United States, are carried into the estuarine nursery areas by the landward residual bottom flow.

Estuarine sediments. The patterns of sediment distribution and movement depend on the type of estuary and on the estuarine topography. The type of sediment brought into the estuary by the rivers, by erosion of the banks, and from the sea is also important; and the relative importance of each of these sources may change along the estuary. Fine-grained material will move in suspension and will follow the residual water flow, although there may be deposition and re-erosion during times of locally low velocities. The coarser-grained material will travel along the bed and will be affected most by high velocities and, consequently, in estuarine areas, will normally tend to move in the direction of the maximum current.

Fine-grained material. Fine-grained clay material, about 2 μm in size, brought down the rivers in suspension, can undergo alterations in its properties in the sea. Base (cation) exchange with the sea water can alter the chemical composition of some clay minerals; also, because the particles have surface ionic charges, they are attracted to one another and can flocculate. Flocculation depends on the salinity of the water and on the concentration of particles. It is normally complete in salinities in excess of 4 parts per thousand, and with suspended sediment concentrations above about 300 ppm (mg l^{-1}), and has the effect of increasing the settling velocities of the particles. The flocs have diameters larger than 30 μm, but effective densities of about 1.1 g cm^{-3} because of the water closely held within. If the material is carried back into regions of low salinity, the flocculation is reversed, and the flocs can be disrupted by turbulence. In sufficiently high concentrations, the suspended sediment can suppress turbulence. The sediment then settles as layers which can reach concentrations as high as 300,000 ppm and which are visible as a distinctive layer of "fluid mud" on echo-sounder recordings. At low concentrations, aggregation of particles occurs mainly by biological action.

Turbidity maximum. A characteristic feature of partially mixed estuaries is the presence of a turbidity maximum. This is a zone in which the suspended sediment concentrations are higher than those either in the river or farther down the estuary. This zone, positioned in the upper estuary around the head of the salt intrusion and associated with mud deposition in the so-called mud reaches, is often related to wide tidal mud flats and saltings. The position of the turbidity maximum changes according to changes in river discharge, and is explained in terms of estuarine circulation. Suspended sediment is introduced into the estuary by the residual downstream flow in the river. In the upper estuary, mixing causes an exchange of suspended sediment into the upper layer, where there is a seaward residual flow causing downstream transport. In the middle estuary, the sediment settles into the lower layer in areas of less vigourous mixing to join sediment entering from the sea on the landward residual flow. It then travels in the salt intrusion back to the head of the estuary. This recirculation is a very effective mechanism for sorting the sediment, which is of exceedingly uniform mineralogy and settling velocity. Flocs with low settling velocities tend to be swept out into the coastal regions and onto the continental shelf. The heavier or larger flocs tend to be deposited.

The concentrations change with tidal range and during the tidal cycle, and fluid muds can occur within the area of the turbidity maximum if concentrations become sufficiently high. During the tidal cycle, as the current diminishes, individual flocs can settle and adhere to the bed, or fluid muds can form. The mud consolidates slightly during the slack water period, and as the current increases at the next stage of the tide, erosion may not be intense enough to remove all of the material deposited. A similar cycle of deposition and erosion occurs during the spring-neap tidal cycle. Generally, there is more sediment in suspension in the turbidity maximum than is required to complete a year's sedimentation on the estuary bed.

Mud flats and tidal marshes. The area of the turbidity maximum is generally well protected from waves, and there are often wide areas of mud flats and tidal marshes (Fig. 4). These areas also exchange considerable volumes of fine sediment with sediment in suspension in the estuary. At high water the flats are covered by shallow water, and there is often a long stand of water level which gives the sediment time to settle and reach the bottom, where it adheres or is trapped by plants or by filter-feeding animals. The ebb flow is concentrated in the winding creeks and channels. At low water there is not enough time for the sediment to settle, and it is distributed over the tidal flats during the incoming tide. Thus there is a progressive movement of fine material onto the mud flats by a process that depends largely on the time delay between sediment that is beginning to settle and sediment that is actually reaching the bed. The tidal channels migrate widely, causing continual erosion. Consequently, there is a constant exchange of material between one part of the marshes and another by means of the turbidity maximum. As the muds that are eroded are largely anaerobic, owing to their very low permeability, the turbidity maximum is an area with reduced amounts of dissolved oxygen in the water.

Fig. 4. Aerial photograph of tidal flats showing the areas of pans, marshes, and vegetation between the channels, Scolt Head Island, England. (*Photograph by J. K. St. Joseph, Crown copyright reserved*)

Coarse-grained material. Coarser materials such as quartz sand grains that do not flocculate travel along the bed. Those coming down the river will stop at the tip of the salt intrusion, where the oscillating tidal velocities are of equal magnitude at both flood and ebb. Ideally, coarser material entering from the sea on the landward bottom flow will also stop at the tip of the salt intrusion, which becomes an area of shoaling, with a consequent decrease of grain size inland. However, normally the distribution of the tidal currents is too complex for this pattern to be clear. Especially in the lower part of the estuary, lateral variations in velocity can be large. The flood and ebb currents preferentially take separate channels, forming a circulation pattern that the sediment also tends to follow. The channels shift their positions in an apparently consistent way, as do the banks between them. This sorts the sediment and restricts the penetration of bed-load material into the estuary.

Salt-wedge patterns. In salt-wedge estuaries the river discharge of sediment is much larger, though generally markedly seasonal. Both suspended and bed-load material are important. The bed-load sediment is deposited at the tip of the salt wedge, but because the position of the salt wedge is so dependent on river discharge, the sediments are spread over a wide area. At times of flood, the whole mass of accumulated sediment can be moved outward and deposited seaward of the mouth. Because of the high sedimentation rates, the offshore slopes are very low, and the sediment has a very low bearing strength. Under normal circumstances, the suspended sediment settles through the salt wedge, and there is a zonation of decreasing grain size with distance down the salt wedge, but changes in river flow seldom allow this process to occur.

Fiord sediments. Sedimentation often occurs only at the heads of fiords, where river flow introduces coarse and badly sorted sediment. The sediment builds out into deltalike fans, and slumping on the fan slopes carries the sediment into deep water. Much of the rest of the fiord floor is bare rock or only thinly covered with fine sediment.

Bar-built estuary sediments. Bar-built estuaries are a very varied sedimentary environment. The high tidal currents in the inlets produce coarse lag deposits, and sandy tidal deltas are produced at either end of the inlets, where the currents rapidly diminish. In tropical areas, the muds that accumulate in the lagoons can be very rich in chemically precipitated calcium carbonate.

[K. R. DYER]

Bibliography: R. S. K. Barnes and J. Green (eds.), *The Estuarine Environment*, 1972; Council on Education in the Geological Sciences and F. F. Wright, *Estuarine Oceanography*, 1974; K. R. Dyer, *Estuaries: A Physical Introduction*, 1973; A. T. Ippen (ed.), *Estuary and Coastline Hydrodynamics*, 1966; G. H. Lauff (ed.), *Estuaries*, Amer. Ass. Advan. Sci. Publ. no. 83, 1967; B. W. Nelson (ed.), *Environmental Framework of Coastal Plain Estuaries*, Geol. Soc. Amer. Mem. no. 133, 1973; C. B. Officer, *Physical Oceanography of Estuaries (and Associated Coastal Waters)*, 1976.

Eutrophication

The deterioration of the esthetic and life-supporting qualities of natural and man-made lakes and estuaries, caused by excessive fertilization from effluents high in phosphorus, nitrogen, and organic growth substances. Algae and aquatic plants become excessive, and, when they decompose, a sequence of objectionable features arise. Water for consumption from such lakes must be filtered and treated. Diversion of sewage, better utilization of manure, erosion control, improved sewage treatment and harvesting of the surplus aquatic crops alleviate the symptoms. Prompt public action is essential. *See* WATER CONSERVATION.

Extent of problem. In inland lakes this problem is due in large part to excessive but inadvertent introduction of domestic and industrial wastes, runoff from fertilized agricultural and urban areas, precipitation, and groundwaters. The interaction of the natural process with the artificial disturbance caused by the activities of man complicates the overall problem and leads to an accelerated rate of deterioration in lakes. Since population increase necessitates an expanded utilization of lakes and streams, cultural eutrophication has become one of the major water resource problems in the United States and throughout the world. A more thorough understanding must be obtained of the processes involved. Without this understanding and the subsequent development of methods of control, the possibility of losing many of the desirable qualities and beneficial properties of lakes and streams is great.

Cultural eutrophication is reflected in changes in species composition, population sizes, and productivity in groups of organisms throughout the aquatic ecosystem. Thus the biological changes caused by excessive fertilization are of considerable interest from both the practical and academic viewpoints.

Phytoplankton. One of the primary responses to eutrophication is apparent in the phytoplankton, or suspended algae, in lakes. The nature of this response can be examined by comparing communities in disturbed and undisturbed lakes or by following changes in the community over a period of years during which nutrient input is increased.

The former approach was utilized in studies at the University of Wisconsin, in which the overall structure of the phytoplankton communities of the eutrophic Lake Mendota, at Madison, and the oligotrophic Trout Lake, in northern Wisconsin, was analyzed. These investigations showed that in the eutrophic lake the population of species is slightly lower, although the average size of organisms is considerably larger, indicating higher levels of production than in the oligotrophic lake. When compared in terms of an index of species diversity, the community of the eutrophic lake displayed values lower than those observed in the oligotrophic lake. Seasonal changes and bathymetric differences in the index of diversity were also more apparent in the eutrophic lake.

Often the low species diversity of the phytoplankton in eutrophic lakes is a result of high populations of blue-green algae, such as *Aphanizomenon flosaquae* and *Anabaena spiroides* in Lake Mendota. Frequently, however, species of diatoms such as *Fragillaria crotonensis* and *Stephanodiscus astrae* also attain high degrees of dominance in the community. W. T. Edmondson reported that dense populations of the blue-green algae *Oscillatoria rubescens* were indicative of deteriorating conditions in Lake Washington. The same species has been observed in several European lakes that have undergone varying degrees of cultural eutrophication. The relatively high nutrient concentrations in eutrophic waters appear to be capitalized on by one or two species that outcompete other species and periodically develop extremely high population levels. Because of the formation of gas vacuoles during metabolism, senescent forms of the blue-green algae rise to the surface of the lake, causing nuisance blooms.

In addition to nuisance scums in the pelagial, or open-water, regions, the rooted aquatic plants and the attached algae of the littoral, or shoreward, region often prove to be equally troublesome in eutrophic lakes. Species of macrophytes, such as *Myriophyllum* and *Ceratophyllum*, and algal forms, such as *Cladophora*, frequently form dense mats of vegetation, making such areas unsuited for both practical and recreational uses.

Bottom fauna. Often in eutrophic lakes the bottom fauna display characteristics similar to those observed in the algal community. Changing environmental conditions appear to allow one or two species to attain high degrees of dominance in the community. Generally higher levels of production are associated with the change in structure of the community—the result being nuisance populations of organisms. In Lake Winnebago, in Wisconsin, for example, the lake fly or midge *Chironomus plumosus* develops extremely high populations, which as adults create an esthetic as well as an economic problem in nearby cities.

Great Lakes. It was originally thought that eutrophication would not be a major problem in large lakes because of the vast diluting effect of their size. However, evidence is accumulating that indicates eutrophication is occurring in the lower Great Lakes. Furthermore, the undesirable changes in the biota appear to have been initiated in relatively recent years. Charles C. Davis, utilizing long-term records from Lake Erie, has observed both qualitative and quantitative changes in the phytoplankton of that large body of water owing to cultural eutrophication. Total numbers of phytoplankton have increased more than threefold since 1920, while the dominant genera have changed from *Asterionella* and *Synedra* to *Melosira*, *Fragillaria*, and *Stephanodiscus*.

Other biological changes usually associated with the eutrophication process in small lakes have also been observed in the Great Lakes. Alfred M. Beeton has summarized the literature pertaining to the trophic status of the Great Lakes in terms of their biological and physicochemical characteristics and indicated that, of the five lakes, Lake Erie has undergone the most noticeable changes due to eutrophication. In terms of annual harvests, commercially valuable species of fish, such as the lake herring or cisco, sauger, walleye, and blue pike, have been replaced by less desirable species, such as the freshwater drum or sheepshead, carp, and smelt. Similarly, in the organisms living in the bottom sediments of Lake Erie, drastic changes in species composition have been observed. Where formerly the mayfly nymph *Hexagenia* was abundant to the extent of 500 organisms per square meter, it presently occurs at levels of 5 and less per square meter. Chironomid midges and tubificid worms now are dominant members in this community.

Oxygen demand. It is apparent that the increase in organic matter production by the algae and plants in a lake undergoing eutrophication has ramifications throughout the aquatic ecosystem. Greater demand is placed on the dissolved oxygen in the water as the organic matter decomposes at the termination of life cycles. Because of this process, the deeper waters in the lake may become entirely depleted of oxygen, thereby destroying fish habitats and leading to the elimination of desirable species. The settling of particulate organic matter from the upper, productive layers changes the character of the bottom muds, also leading to the replacement of certain species by less desirable organisms. Of great importance is the fact that nutrients inadvertently introduced to a lake are for the most part trapped there and recycled in accelerated biological processes. Consequently, the damage done to a lake in a relatively short time requires a many-fold increase in time for recovery of the lake.

Action programs and future studies. Lake eutrophication represents a complex interaction of biological, physical, and chemical processes. The problem, therefore, necessitates basic research in a wide variety of scientific disciplines. Moreover, to be profitable, such research must be coordinated into well-integrated team-research efforts requiring extensive monetary support.

Studies are needed in such areas as the identification of those nutrients that reach critical levels in lake waters and lead to the development of nuisance growths of plants and algae. Nitrogen and phosphorus are undoubtedly important; however, biologically active substances such as vitamin B_{12} and other organic growth factors may also play an important role.

In Lake Mendota sewage effluent is the major contributor of nitrogen and phosphorus to lakes, followed by runoff from manured and fertilized land. Considering nitrogen alone, rain adds more than any single source. Its nitrogen content comes from combustion engines and smokestacks.

Where sewage effluent has been diverted from lakes (as at Lakes Monona, Waubesa, and Kegonsa in Wisconsin, Lake Washington near Seattle, and two lakes in Germany), an improvement in nuisance conditions occurs. Hence this treatment is the first step in alleviation.

The conditions observed in lakes and streams reflect not only the processes operating within the body of water but also the metabolism and dynamics of the entire watershed or drainage basin. After precise identification of the critical nutrient compounds, it is necessary to determine the nutrient budget of the whole drainage basin before action can be taken to alleviate the undesirable fertilization of a lake or stream.

New methods of treating sewage plant effluent are being explored, and further support for these efforts is justified. For example, the complete evaporation of effluent to a powder would alleviate to a great extent the present problem of dealing with large volumes of these materials. Such methods will be expensive, but this may be inevitable for some pollution problems. *See* SEWAGE TREATMENT.

It is known that aquatic plants concentrate nutrients from the lake waters in their tissues. The removal or harvesting of aquatic plants in eutrophic lakes, consequently, is a good potential method for reducing nutrient levels in these lakes. Similarly, significant amounts of nitrogen and phosphorus are concentrated in fish flesh. Efficient methods of harvesting these organisms are important in impoverishing a well-fertilized lake.

Utilization or land disposal of farm manure is a major problem. Animal manures are largely unsewered, yet in the Midwest it is equivalent to the sewage of 350,000,000 people.

In addition to improvements in waste disposal, more research and development are needed on the profitable utilization of surplus algae, aquatic plants, fish, manure, and sewage. The over-fertilized lake needs to be impoverished of its nutrients as well as protected from inflowing sources. Chemicals have been used to poison the plants and algae, but this is not a good conservation practice because the plants and algae rot and provide more nutrients. Moreover, eventual harm to other species has not been assessed.

It would seem desirable to set aside certain lake areas for research purposes. More information is needed to decide upon the best plans for allowing the domestic development of these areas with the least disturbance to the water resources. Steps will be necessary to devise optimum zoning laws and multiple-use programs in light of the intense economic and recreational uses made of water resources. Legislation and law enforcement in relation to public interactions undoubtedly will be a complex problem to overcome in this respect.

Cultural eutrophication is a paradoxical condition, since it is in large part due to the economic and recreational activities of man and at the same time eventually conflicts with these same activities of society in general. *See* FRESH-WATER ECOSYSTEM; LAKE. [ARTHUR D. HASLER]

Bibliography: D. G. Frey and F. E. J. Fry (eds.), *Fundamentals of Limnology*, 1963; A. D. Hasler, Eutrophication of lakes by domestic drainage, *Ecology*, 28(4):383–395, 1947; A. D. Hasler and B. Ingersoll, Dwindling lakes, *Natural History*, Nov. 1968; A. D. Hasler and M. E. Swenson, Eutrophication, *Science*, 158(3798):278–282, 1967; G. E. Hutchinson, *A Treatise on Limnology*, vol. 1: *Geography, Physics, and Chemistry*, 1957, and vol. 2: *Introduction to Lake Biology and the Limnoplankton*, 1967; O. T. Lind, *Handbook of Common Methods in Limnology*, 1979; G. A. Rohlich and K. M. Stewart, *Eutrophication: A Review*, Water Qual. Contr. Board Calif. Publ. no. 34, 1967.

Evapotranspiration

The total process of water vapor transfer into the atmosphere from vegetated land surfaces. Evaporation is the change of liquid water into gaseous water vapor; it occurs from bodies of water such as lakes and streams or from other wet surfaces such as wet soil surfaces. Transpiration is the process whereby water is absorbed by plant roots, transported through the plant, and evaporated from plant surfaces, especially from leaves into the air above. The energy required to change liquid water into water vapor comes from sources traceable to the Sun. Energy for evapotranspiration may come from the transfer of heat from warm air to a cooler plant or soil surface, but the major source of energy is generally net radiation. Net radiation is the difference between the incoming and the outgoing shortwave and long-wave radiation streams. Incoming shortwave radiation is composed of direct and diffuse solar radiation, and outgoing shortwave radiation is that amount of solar radiation which is reflected by the surface. The incoming and outgoing long-wave radiation streams are the radiation emitted by the atmosphere and the Earth's surface, respectively, as a function of the atmospheric or the surface temperature.

Rate and amount. The evapotranspiration rate is governed by the total quantity of absorbed energy, by the amount of moisture that is available both at the soil surface and within the plant root zone, and sometimes by the plants themselves. In humid regions the maximum evapotranspiration rate will seldom exceed the net amount of radiative energy available. However, warm dry air emanating from arid regions often becomes a major additional source of energy for evapotranspiration.

Net radiation is also generally greater in arid than in humid regions because of less cloud cover. These two factors combine to cause evapotranspiration rates in arid areas to generally exceed those in humid areas, unless, of course, available water becomes limiting. Evapotranspiration rates in humid areas seldom exceed about 7 mm per day, but rates as high as 12–14 mm per day have been reported from arid, semiarid, and subhumid regions.

In general, the amount of evapotranspiration will be less than or equal to the amount of precipitation that falls in a given region. Exceptions arise when precipitation is supplemented with irrigation, or in instances when plant roots may extract water from underground water tables located within a meter or two of the soil surface. Some plants, particularly those classified as phreatophytes, have rooting systems that can extend several meters, perhaps as deep as 20 to 30 m, into the soil and can draw water from these depths. Phreatophytes are generally located adjacent to streams, canals, or riverbeds and are responsible for nonbeneficial losses of extremely large quantities of water. This is especially serious in arid regions where water is at a premium for irrigation.

It has been estimated that approximately 1.1×10^{10} m^3 (3×10^{12} gal) of water are returned daily to the atmosphere by evapotranspiration from the continental United States. Reduction in the evapotranspiration rate by only 1% for a single day would result in saving of enough water to supply the yearly needs of a city of 2,000,000 people. Much current research is aimed at finding techniques to reduce evapotranspiration losses or to improve the efficiency with which water is utilized by plants. Use of windbreaks, proper scheduling of irrigation, minimum tillage practices, and planting at optimal plant populations in proper row widths are among those techniques which have been shown to conserve moisture or increase the water-use efficiency. Water-use efficiency is the amount of dry matter produced divided by the amount of water used.

Measurement. Direct measurement of evapotranspiration is made with lysimeters. Lysimeters are large containers of soil in which plants are grown; loss of water from the container is due to evapotranspiration and is determined by weighing or by accounting for the quantities of water needed to replenish the lysimeter. Some lysimeters are very accurate and can detect evapotranspiration losses as small as 0.01 mm of water over periods as short as 15 min.

There are also many mathematical models developed to estimate evapotranspiration losses. These consider the effects of various weather factors, principally radiation, temperature, humidity, and wind. The models vary in complexity and in the accuracy with which they estimate actual evapotranspiration rates. Some provide estimates that are reliable only on a monthly basis, and others provide acceptable information for periods shorter than 1 hr. Refinements continue to be made in evapotranspiration models, especially those which can make use of information acquired through satellites and other modern systems for collecting meteorological data. *See* ATMOSPHERIC WATER VAPOR. [BLAINE L. BLAD]

Bibliography: J. L. Monteith, *Vegetation and the Atmosphere*, vol. 1, 1975, N. J. Rosenberg, *Microclimate: The Biological Environment*, 1974; B. Yaron, E. Danfors, and Y. Vaadia, *Arid Zone Irrigation*, 1973.

Fertilizer

Materials added to the soil to supply elements needed for plant nutrition. The principal elements required are nitrogen, phosphorus, and potassium. Several others—calcium, magnesium, sulfur, boron, iron, zinc, manganese, copper, molybdenum, and chlorine—are needed in lesser amounts. They are supplied in such various ways as materials produced for the purpose, as incidental components of fertilizers supplying other nutrients, and as compounds already present in the soil. *See* PLANT, MINERALS ESSENTIAL TO.

Fertilizers may be products manufactured for the purpose, by-products from other chemical manufacturing operations, or by-product natural materials. By-products, particularly natural organic materials, were important nutrient sources in the early days of the industry. The growing need for fertilizers, however, has outstripped the supply of by-products. Today manufactured materials are the major type by far and by-products have only minor significance.

The fertilizer industry once was mainly a simple materials-handling and mixing technology. However, it has become a major segment of the chemical industry, with giant plants embodying advanced developments in chemical engineering producing very large quantities of fertilizer. Several factors have contributed to its change in status: a maturing in agricultural practice in developed countries, rapidly increasing use of fertilizer in countries that had used little before, and a growing realization that massive production of fertilizer is the first line of defense against the problems of growing populations. The increase in world fertilizer consumption is indicated in the illustration.

Fertilizer types. The nutrient elements cannot be supplied to plants as such; they must be combined with other elements in the form of suitable compounds. Phosphorus, for example, is toxic to plants in the elemental form but is a good fertilizer when combined with oxygen and ammonia to form ammonium phosphate. For each of the elements there are several compounds that can be used. The choice between them, in most cases, is based on economic factors. Material is used that gives lowest cost per unit of nutrient applied to the soil.

Nitrogen, phosphorus, and potassium are called the macronutrients because of the relatively large amounts needed. The usual requirement range per acre per cropping season is 50–200 lb of nitrogen, 10–40 lb of phosphorus, and 30–150 lb of potassium.

Nitrogen is supplied in the ammoniacal (NH_4^+) or nitrate (NO_3^-) form or as urea, $(NH_2)_2CO$. Except in unusual situations, one form is as good as the other because in the soil the ammoniacal and urea forms are rapidly converted by microorganisms to nitrate. The principal fertilizer nitrogen materials are ammonium nitrate, urea, ammonia, ammonium sulfate, and ammonium phosphate.

Phosphorus is supplied mainly as ammonium phosphates and calcium phosphates. Ammonium phosphates are the leader and are increasing in importance. The ammonium phosphates have advantages of higher nutrient concentration and higher water solubility, which favor them in regard to handling and shipping costs and, in some instances, in agronomic value.

Potassium is supplied almost entirely as potassium chloride, a mineral widely available in various parts of the world and usable without altering its chemical structure.

Sulfur, calcium, and magnesium are also essential elements but are needed in lesser quantities than the macronutrients. Therefore they are called secondary nutrients. Typical quantities of the secondary nutrients needed per acre per cropping season are as follows: 5–50 lb of sulfur, 5–25 lb of calcium, 3–30 lb of magnesium. In the past, substantial quantities of the secondary nutrients were supplied incidentally in macronutrient fertilizers, such as from calcium phosphates, from soil minerals, and from power plant emissions of sulfur oxides. However, with the trend to ammonium phosphates and other high-concentration products and to control of stack gas emissions, there is an increasing requirement for specifically formulating secondary nutrients into fertilizer products.

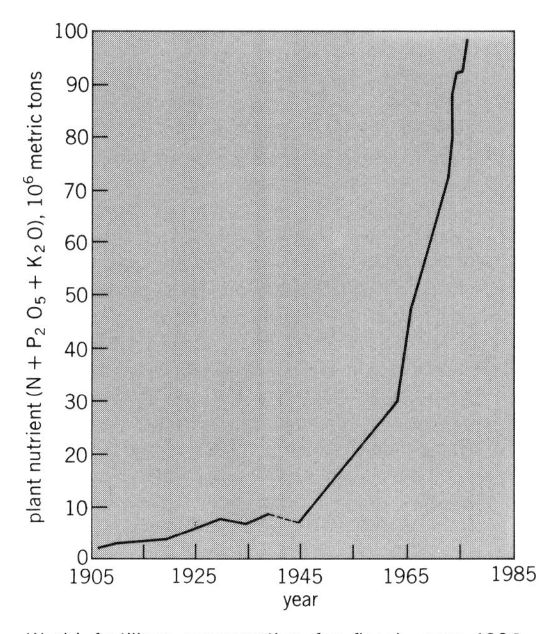

World fertilizer consumption for fiscal years 1906–1977. *(Food and Agriculture Organization)*

The remaining essential elements are called micronutrients; the amounts required are very small, from a trace to 1 lb/acre. They are usually supplied as cations or anions of salts such as borates or sulfates.

The nutrient amounts given refer to the quantity taken up from the soil by various crops. This does not mean necessarily that such an amount added to the soil will supply the plant adequately; there are many ways in which nutrient can be lost before the plant can take it up. Moreover, some nutrient is obtained from materials already present in the soil. Therefore, the amount of fertilizer actually applied depends on several factors that vary widely. Beyond the factor of plant need, a major consideration is the financial ability and technological status of the farmer; these are major factors in the very wide variation in fertilizer application per acre in various parts of the world. Highly developed and crowded countries use the most; in Belgium, for example, average macronutrient application is about 200 lb/acre. In contrast, less-developed areas, such as India, use as little as 1 lb/acre.

Nitrogen. This nutrient is required by plants in the largest quantity. Its use is expanding more rapidly than for any of the other nutrients. It has been found that, if adequate amounts of phosphorus, potash, and other nutrients are applied, very large amounts of nitrogen, coupled with high plant population per acre, give yields much higher than those attainable under previous fertilization practice. In 1956 world consumption was higher for phosphate and potash than for nitrogen (tonnage is on the oxide basis, P_2O_5 and K_2O, the common reporting method in the industry). But by 1961 nitrogen had passed both and continues to climb rapidly; in 1976–1977 the estimated world use of fertilizer nitrogen was 45,884,000 metric tons.

The fetilizer nitrogen supply is divided among several materials as shown in Table 1.

Ammonia. The basic nitrogen fertilizer material is ammonia, used both as a primary fertilizer and as the starting material for all the other leading nitrogen fertilizers. Ammonia is also an important general chemical, but its principal use (almost 85% of the total on a worldwide basis) is in the fertilizer industry. In 1977 there were about 104 complexes in the United States, with a production capacity of more than 21,768,000 tons per year.

In past decades the fertilizer industry used large quantitites of natural nitrogen materials, such as animal by-products, ammonia recovered from coal during coke production, and sodium nitrate from localized natural deposits in Chile. Today these sources are inadequate or too expensive; elemental nitrogen, present in inexhaustible quantities in the atmosphere, is the major source by far, ac-

Table 1. Fertilizer nitrogen supplied by various materials

Material	Nitrogen content, %	Supply, % of total nitrogen*	
		United States†	World‡
Urea	45–46	8.7	18.0
Ammonium nitrate	33.5	10.7	22.6
Complex fertilizers	Varies	–	8.4
Anhydrous ammonia (used as such)	82	40.1	–
Ammonium sulfate	20.5	5.0	9.3
Nitrogen solutions (used as such)	Varies	22.4	–
Calcium nitrate	15.5	–	0.7
Sodium nitrate	16	–	0.4
Ammonium sulfate nitrate	Varies	–	1.0
Calcium cyanamide	28	–	0.4
Ammonium phosphate	Varies	–	5.9
Other nitrogen fertilizers	Varies	13.1	33.3

*In 1976–1977. †USDA data. ‡FAO data.

counting for more than 95% of the fertilizer nitrogen used in the United States.

Ammonia production technology changed rapidly in the 1960s. Natural gas became firmly established as the major feedstock, particularly after discovery and development of gas bodies in Europe and Japan. The use of elevated pressure in reforming, that is, reaction of gas with steam to give synthesis gas (hydrogen plus carbon monoxide), added considerable economy to the process. Also, low-pressure shift conversion, that is, reaction of synthesis gas with steam to eliminate carbon monoxide and produce more hydrogen, was widely adopted.

After purification the hydrogen is combined with nitrogen (introduced earlier in the process as air) under pressure (2000 psig or more) and high temperature (425–650°C) in the presence of promoted iron catalyst to form ammonia.

The synthesis step in ammonia production has undergone major changes. Centrifugal compressors, used to compress the nitrogen-hydrogen mixture to synthesis pressure, have reduced plant cost. Single converters produce up to 3000 tons of ammonia per day, and converter design has been simplified to make the large size feasible.

Use of ammonia directly as a fertilizer requires use of pressure equipment since ammonia is a gas at atmospheric pressure. The material is carried in pressure tanks on applicators and injected into the soil through injector knives. Care is necessary to avoid loss by volatilization, and with deep placement, the ammonia is quickly sorbed by the clay and other constituents of the soil.

In 1958 ammonia became the leading nitrogen fertilizer in the United States. Use in other parts of the world, although growing, is relatively small.

Urea. Urea soon is expected to surpass ammonium nitrate as the world's leading form of nitrogen fertilizer, and to continue to gain in importance. This is due to the more favorable economics in producing urea, to its high nutrient content (45–46%), and to its explosive-free properties in comparison with ammonium nitrate. With increasing costs for handling and shipping, the high concentration is assuming more and more importance.

Some urea is used in chemical industry, mainly in plastics production and in animal feeds, but the major proportion by far goes into fertilizers, either for direct application or as a constituent in mixes.

Urea is made by reacting ammonia and carbon dioxide under pressure and at elevated temperature but without a catalyst. The carbon dioxide is readily available because it is a by-product, usually wasted, in manufacture of ammonia from carbonaceous materials. Thus there is little or no cost for the carbon dioxide, but the urea plant must be built close to an ammonia plant.

The urea reaction does not go to completion; the product from the reactor is a solution of urea and ammonium carbamate, an unstable compound of ammonia, water, and carbon dioxide. The solution must be heated to decompose the carbamate and drive off the resulting ammonia and carbon dioxide, which are then recycled back to the reactor. The urea solution is concentrated to a melt containing less than 1% water in an evaporator. Practice differs in the method used for converting the melt to the final solid product. Prilling, which involves solidification of melt droplets by air-cooling in a tall tower, is the most widely used method. There is an increasing trend to granulation of the melt instead of prilling in the United States, because the granules are stronger and larger than prills. Also, air pollution abatement is less difficult for granulation.

Urea processes differ mainly in the method of removing and recycling the unreacted gases and in means of energy recovery. The processes have become increasingly efficient due to innovations for recovery and reuse of process heat for preheating feeds and for generation of steam which is used in the process.

Corrosion is a major problem in urea manufacture. Reactors lined with stainless steel are widely used. Some add air or oxygen to decrease corrosiveness of the stainless steel. A few reactors have been lined with highly resistant materials such as glass or titanium.

Another problem is formation of biuret, $(NH_2CO)_2NH$, during manufacture. Formation can be kept to a minimum by reducing the time during which the urea solution is exposed to elevated temperature. Biuret is harmful to some types of plants.

The use of urea presents the added problem that surface application may result in nitrogen loss because of urea decomposition (by hydrolysis) back to the gases from which it was made. The decomposition is accelerated by high soil temperature, low pH, and an enzyme (urease) found in the soil. By proper attention to conditions of application, losses can be kept to an acceptable level.

Urea is less hygroscopic than ammonium nitrate and gives less trouble with caking. It is also nonexplosive.

Ammonium nitrate. The most important nitrogen fertilizer, from the standpoint of world consumption, is ammonium nitrate. Large tonnages are used directly as a fertilizer, and considerable amounts go into mixtures with other materials. The material plays the somewhat remarkable double role of fertilizer and explosive. The explosive tendency has been somewhat of a problem in handling and shipping for fertilizer use. In modern practice, however, the hazard has been adequately controlled. In some countries ammonium nitrate fertilizer is diluted to a concentration of less than 70% by adding limestone so that the mixture is not detonable.

Ammonia and nitric acid are the raw materials for ammonium nitrate production. Since nitric acid is made by oxidizing ammonia with air, the only basic raw material needed is ammonia. This gives ammonium nitrate some advantage because other major nitrogen fertilizers require some raw material in addition to ammonia.

Nitric acid plants differ mainly in the pressure used at various stages of the process. Ammonia is oxidized catalytically with air by passing the mixture through a platinum catalyst screen at high temperature, after which the product gases are cooled and then passed through a scrubber, where water absorbs the nitrogen oxides to form nitric acid. The absorption and sometimes the oxidation are carried out at pressures up to about 100 psig.

Ammonium nitrate is made by neutralizing nitric acid with ammonia, usually in pressure-type neutralizers so that steam can be recovered from the highly exothermic reaction. The neutralized solution is then concentrated to a melt of low moisture content in an evaporator. Practice differs in the method used for converting the melt to the final solid product. Like urea, most ammonium nitrate is prilled. However, there is an increasing tendency to granulation, primarily to minimize air pollution problems.

Finally, ammonium nitrate is usually conditioned to reduce moisture absorption and caking. The main practice is either to coat with up to about 2% of a parting agent, such as clay, or to add a lesser amount of one of several chemical additives.

Ammonium sulfate. Ammonium sulfate is declining in popularity in relation to urea and ammonium nitrate. A major reason for the decline is the low nutrient concentration of 20–21% for ammonium sulfate versus 33–34% for ammonium nitrate and 45–46% for urea.

Almost all of the ammonium sulfate produced in the United States and much of that in the remainder of the world are from by-product sources. Much of this is derived from coke manufacture (by scrubbing ammonia out of the oven gases with sulfuric acid) and from production of caprolactam. Coke ovens, for example, supply about 500,000 tons of ammonium sulfate annually in the United States. A considerable tonnage is made also by reacting ammonia with sulfuric acid, which often is a by-product from petroleum refining or organic syntheses.

Manufacture of ammonium sulfate is relatively simple. Solution, either by-product or made by neutralizing sulfuric acid, is concentrated and then crystallized in one of several crystallization processes or else compacted into flakes.

Nitrogen solutions. In the United States liquid forms of fertilizer are quite popular because they can be handled and applied with a minimum amount of labor. The use of anhydrous ammonia, already described, is an example of this. Ammonia, however, must be handled under high pressure and must be applied carefully to prevent loss. To avoid these problems a class of products called nitrogen solutions has been developed. These are of various types, but in general they can be classified as water solutions of nitrogen salts (plus ammonia in some), all characterized by low vapor pressure (above atmospheric) or, in some instances, no gage pressure at all. The nonpressure solutions can be handled in ordinary tanks without special precautions and can be sprayed on the soil surface without loss. The pressure type requires some care in handling and must be injected into the soil, but the pressure is so low, only a few pounds, that requirements are much less rigorous than for anhydrous ammonia.

Nitrogen solutions are popular in the United States but are little used in other parts of the world. In 1976–1977 consumption in the United States was more than 5,796,000 tons, or about 1,685,000 tons of nitrogen—second only to anhydrous ammonia.

The leading nitrogen solution is a combination of urea and ammonium nitrate. These two salts are soluble separately only to the extent of about 20% nitrogen, but they have a mutual solubility effect on each other that produces a 32% nitrogen content when the two are dissolved together. About 80 lb of the combined salts can be dissolved by 20 lb of water.

Pressure-type solutions that contain ammonia and urea or ammonium nitrate, or in some products all three together, are also popular. The advantage of the ammonia is that it increases the total nitrogen content of the solution. Several of the pressure solutions contain over 40% nitrogen, with vapor pressures on the order of 10–15 psig.

Aqua ammonia, a solution of ammonia in water, also has the advantage of easier handling than anhydrous ammonia, but nitrogen content must be reduced to about 20% to get a satisfactorily low vapor pressure. Nevertheless, large quantities of aqua ammonia are used (about 655,000 tons in 1976–1977).

Other compounds. Numerous other nitrogen compounds are used as fertilizers, but the amounts are much smaller than for the leading nitrogen fertilizers described and are not expected to gain in the foreseeable future.

Sodium nitrate was once a major fertilizer but in 1976–1977 supplied only about 0.4% of world fertilizer nitrogen. The relative low nitrogen content (16%) is a major disadvantage in view of rising labor and shipping costs.

Calcium nitrate is used mainly in Europe and in 1976–1977 supplied about 0.7% of world nitrogen. Almost half of the production is in Norway, where the material was made in the past as a method of using nitric acid made by the now-obsolete arc process for nitrogen fixation; production still continues from nitric acid made by ammonia oxidation.

Calcium cyanamide, made from calcium carbide and atmospheric nitrogen, in 1976–1977 made up about 0.4% of the nitrogen supply. Most of the production is in Europe and Japan.

Organic nitrogen materials once were major nitrogen fertilizers but now are estimated to supply less than 1% of the nitrogen. Organics may regain a significant place, however, because of the growing problem in disposing of urban organic waste. When the problem becomes severe enough in a situation to warrant waste processing, fertilizer usually is the most appropriate product to make.

Controlled-release nitrogen materials are not used in large quantities but may become significant in the future. By using a slowly soluble nitrogen compound or depositing an impermeable coating on a soluble one, release of nitrogen in the soil through dissolution of the fertilizer is delayed, thus reducing leaching and other losses that can be incurred with soluble nitrogen materials.

Multinutrient fertilizers supply the remainder of the fertilizer nitrogen, about 11 and 19%, respectively, of world and United States consumption. These are classed with phosphate or mixed fertilizers rather than with nitrogen products.

Phosphate. World consumption of fertilizer phosphate is not as large as that of nitrogen. In 1976–1977 the estimated world usage was 27,299,000 metric tons, with about 20% of the total used in the United States. The principal phosphatic fertilizers and their consumption and phosphate content are listed in Table 2.

Phosphoric acid. Modern high-analysis phosphatic fertilizers, such as ammonium phosphate and concentrated superphosphate, require phosphoric acid in their manufacture. The acid in turn is made from phosphate rock, an ore found mainly in the Soviet Union, North Africa, and the United States (principally in Florida).

The leading method for phosphoric acid manufacture involves treatment of phosphate rock with sulfuric acid. The insoluble calcium phosphate in the ore dissolves in the acid, and crystals of calcium sulfate form. After separation of the calcium sulfate by filtration, the acid is concentrated to the level required in making the various fertilizer phosphates. This is called the wet process for phosphoric acid.

In the United States wet-process acid plants, the calcium sulfate is precipitated as a hydrated form called gypsum. The gypsum is discarded in storage piles and becomes a problem because of the large amount produced; the tonnage is larger than that of the phosphate rock used. It is simply not economical to reuse the waste gypsum. In Japan and some European countries, the by-product gypsum is reused to produce building materials, such as wallboard and cement.

Production of phosphate fertilizer uses over half of the sulfur consumed in the United States. In 1976 about 60% of this sulfur was supplied from elemental sulfur mined by the Frasch (hot-water) process and the remainder from sulfur removed from oil and natural gas. The trend is to increasing use of recovered sulfur.

Table 2. Fertilizer phosphate supplied by various materials*

Material	Phosphate content, %	Supply, % of total phosphate† United States	World
Ordinary superphosphate	16–22	4.8	23.0
Concentrated superphosphate	44–47	22.3	17.9
Complex fertilizers	Varies	—	13.1
Basic slag	17.5	—	1.1
Ammonium phosphate	20–46	51.7	30.1
Other phosphate fertilizers	Varies	21.2	14.8

*FAO data. †As P_2O_5; in 1976–1977.

Another method of phosphoric acid production is the electric furnace process. Phosphate rock, mixed with coke and silica, is heated to a high temperature in the furnace to reduce the phosphate to elemental phosphorus. The phosphorus is then burned with air and the resulting oxide absorbed in water to give phosphoric acid. The furnace method is not competitive with the sulfuric acid process for producing fertilizers except in unusual situations involving high sulfur cost and unusually low electrical power cost. The capital investment required for the furnace process is much higher than for the wet process. The furnace process yields a higher-purity product acid, but pure acid is not required for fertilizers.

A major development is superphosphoric acid which contains less water than ordinary phosphoric acid. Superphosphoric acid is made either by concentrating wet-process acid to a much higher concentration than usual (70–72% P_2O_5 versus 54%) or by using less water in the furnace method. The superphosphoric acid has the advantages of higher concentration, lower suspended solids (in the wet-process type), and certain chemical properties which make it preferable for production of some types of fertilizers. The superphosphoric acid is especially desirable for production of liquid products because it provides higher concentration and improved quality.

Ammonium phosphate. Ammonium phosphates are the leading forms of phosphate fertilizer and are gaining in importance. Over half the fertilizer P_2O_5 in the United States is applied in this form. Major reasons for the increasing popularity of the ammonium phosphates include: (1) high nutrient concentration, (2) good storage and handling properties, (3) high water solubility, (4) relative ease of production in large plants, and (5) favorable economics in comparison with competing products. The most popular ammonium phosphate product is diammonium phosphate, which contains 18% nitrogen and 46% P_2O_5.

Several processes have been developed for making ammonium phosphates. In one of the more popular ones, the acid is treated with part of the ammonia in a tank-type vessel, the partially ammoniated slurry flows onto a rolling bed of solids (solidified ammonium phosphate) in a rotary granulating drum, the rest of the ammonia is injected under the bed of solids, the granules are dried, and the finished product of the desired particle size is screened out. Fine material passing through the screen is recycled to the drum to provide the bed of solids.

Important new products in the ammonium phosphate family include solid ammonium polyphosphate and urea-ammonium phosphate. Urea-ammonium phosphate is made by granulating urea into ammonium phosphate to give very-high-concentration products such as 28% nitrogen and 28% P_2O_5. Ammonium polyphosphate is made in either granular or liquid forms by contacting phosphoric acid and ammonia in a simple pipe reactor to produce a high-temperature, anhydrous melt. This anhydrous melt is either granulated or dissolved in water so as to give products with comparatively high nutrient concentration and good storability.

Superphosphate. The oldest of the phosphatic fertilizers, is normal superphosphate. The manufacturing process is quite simple: Phosphate rock is mixed with sulfuric acid (a smaller amount than in phosphoric acid production), the resulting slurry is held in a container for a few minutes until it solidifies, and the material is removed to a storage pile, where it cures for approximately 3 weeks in order for the reaction between rock and acid to reach completion.

The cured superphosphate is used to some extent as a fertilizer without further treatment, but most of it serves as one of the starting materials in making other fertilizers. The status of the material is declining, mainly because of its low nutrient concentration (about 20% P_2O_5). In the period 1961–1962 through 1976–1977, the percentage of world phosphate supplied by normal superphosphate declined from 49 to 23%.

Concentrated superphosphate (usually called triple superphosphate) is in a much better position because of its higher concentration (about 46% P_2O_5), over twice that of normal superphosphate. Its popularity has been growing in the United States for several years, and in 1964 it passed normal superphosphate, the leader since the beginning of fertilizer history. On a worldwide basis, triple superphosphate has not yet surpassed normal superphosphate in popularity; during 1976–1977 it supplied 18% of the phosphate, and the normal type supplied 23%. However, triple superphosphate is expected to continue to gain in importance.

Triple superphosphate is made by treating phosphate rock with phosphoric acid rather than with sulfuric acid. Therefore there is no sulfate present to dilute the product; instead, the acid supplies more nutrient phosphate than does the phosphate rock, and a very high nutrient content is realized.

Triple superphosphate is manufactured in much the same manner as the normal type. The rock and acid are mixed, the slurry held in a "den" until it solidifies, and the moist mass transferred to storage for curing. Unlike normal superphosphate, much of the triple is granulated; that is, finely divided material cut from the curing pile is agglomerated to particles of fairly large size, on the order of $\frac{1}{16}$–$\frac{1}{8}$ in. in diameter. In this form it is easier to handle in direct application and in mixing with other granular materials to make a multinutrient, nondusty fertilizer.

The present trend is to directly granulate triple superphosphate without any denning step. Phosphoric acid and phosphate rock are continuously mixed in tanks to produce a fluid slurry which is then granulated, dried, and screened to give a hard, spherical granular product.

Nitric phosphate. In making ammonium phosphates or the superphosphates, sulfuric acid is required. This can be a major problem during periods when sulfur is in short supply or for countries that have no indigenous sulfur sources. Therefore there is interest in acidulation of phosphate rock with nitric acid, rather than with sulfuric.

Substitution of nitric acid for sulfuric or phosphoric acids, however, causes some special problems. Reactions with phosphate rock gives phosphoric acid plus calcium nitrate, rather than the phosphoric acid–calcium sulfate combination formed when sulfuric acid is used in making phosphoric acid. Calcium sulfate precipitates and can be separated, but calcium nitrate remains dissolved in the acid. If allowed to remain through further processing, it has a diluting effect and, more seriously, makes the product hygroscopic.

Various methods have been developed for coping with the calcium nitrate problem. In Europe it is crystallized out (by cooling the solution) and used directly as a fertilizer. Hygroscopicity is a problem, but with moistureproof bags the poor physical properties can be tolerated. Other acids that give an insoluble calcium salt can be used along with the nitric to convert the calcium nitrate, for example, to calcium sulfate or calcium phosphate.

In a typical process, phosphate rock is treated with 20 moles of nitric acid and 4 moles of phosphoric per mole of calcium phosphate in the rock. The acidulate slurry is then neutralized with ammonia and granulated. The phosphoric acid and ammonia convert the calcium nitrate to dicalcium phosphate and ammonium nitrate, both acceptable products.

In the crystallization version, enough calcium nitrate is crystallized out so that, when the solution is separated and ammoniated, the remaining calcium precipitates as dicalcium phosphate and the nitrate is converted to ammonium nitrate.

One disadvantage to nitric phosphate is reduced nutrient content because part or all of the calcium is left in the product. A typical product contains 20% nitrogen and 20% phosphate; in comparison, complete removal of the calcium would give a product containing about 24% of each nutrient.

Nitric phosphates are popular in continental Europe, but production in other major fertilizer-producing areas, the United Kingdom, the United States, and Japan, is relatively low. These products are most popular in countries that have limited supplies of native sulfur.

Other compounds. A few other phosphates contribute to the fertilizer supply. The most important of these is basic slag, fourth in world phosphate supply but losing ground to the newer highly concentrated materials. Basic slag, which contains only about 17% P_2O_5, is popular because it is a byproduct of steel production in Europe and therefore sells at a relatively low price. Over 430,000 metric tons of P_2O_5 was supplied by basic slag in Europe in 1976–1977, but only a minor proportion was made in other parts of the world.

Phosphate rock itself is also a leading phosphate fertilizer. Although the ore is so insoluble that acidulation or other treatment normally is considered necessary to make it usable, there are special soil-crop situations in which finely ground phosphate rock gives enough crop response to make its use justifiable. Although phosphate utilization is much lower than for processed phosphates, elimination of the processing cost makes it economical.

Generally the crop response to ground phosphate rock is better when it is applied to acid soils. Even then, the effectiveness is dependent on the mineral structure of the phosphate, with some types being much more effective than others.

Phosphate rock consumption as a direct fertilizer is quite large, about 1,071,000 metric tons of P_2O_5 in 1976–1977 (world usage). The full amount is not counted when ranking phosphate rock with other phosphate fertilizers, however, because much of the P_2O_5 in the rock is inert and unused.

Thermal phosphates are used to a limited extent, mainly in Germany and Japan. Phosphate rock is heated, usually to fusion, with some solid material that reacts with the rock at high temperature. In Japan, for example, the reactant is a magnesium silicate ore. In Germany soda ash plus silica has been used. The product has low concentration (19–24% P_2O_5) and is declining in popularity.

Bones and other organic sources of phosphate, once the leading type, have little significance in phosphate supply.

Potash. Tonnage of potash is the lowest of any of the macronutrients, estimated at 25,262,000 metric tons (of K_2O) in 1976–1977 as compared with 27,299,000 for P_2O_5 and 45,884,000 for nitrogen. The United States leads in consumption, with France, Germany, and the Soviet Union also major consumers.

The potash industry is quite simple in comparison to the nitrogen and phosphate industries. Atmospheric nitrogen and phosphate rock must be subjected to expensive processing to make them usable as fertilizers, but potash ore is soluble and can be used directly as mined without any treatment other than removing impurities. Hence, potash supply is more of a mining industry than a chemical processing.

The principal potash ore, supplying more than 90% of the world total, is potassium chloride. The natural deposits are tremendous; Canadian reserves in 1975 were estimated to contain about 74,000,000,000 short tons of recoverable K_2O.

The Soviet Union leads in potash production, followed by Canada, East Germany, West Germany, the United States, and France. Most of the potash mined, over 90%, is in the form of potassium chloride. Potassium sulfate, also a soluble, readily available material, is the only other potash ore of significance.

Potassium chloride is a high-grade material with more than 60% potassium after impurities are removed. Potassium sulfate is somewhat lower in nutrient content, about 50% K_2O. Both are normally used in fertilizer mixtures without any intended reaction with other constituents, although reaction does occur incidentally in some cases, as in mixtures with ammonium nitrate, in which reaction with potassium chloride produces ammonium chloride and potassium nitrate in the mixture.

Minor quantities of potassium nitrate and potassium phosphate fertilizers are produced by treating potassium chloride with either nitric acid or phosphoric acid.

Mixed fertilizers. Most fertilizers are of the mixed type; that is, they contain more than one nutrient. This is mainly a convenience for the farmer since his soils need nutrients in certain proportions. Although he could buy single-nutrient materials and apply them in the desired ratios, it is more convenient and usually less expensive to buy them already mixed.

A mixed fertilizer may be simply a mechanical mixture, or the constituent materials may be reacted to form a "chemical" mixture. In the early days of the industry, the mechanical mixture was the prevalent type; today chemically reacted products occupy an important place in the industry.

The chemically mixed fertilizers usually are made by processes involving ammoniation of some combination of phosphoric acid, sulfuric acid, and superphosphates. Superphosphates, both the normal and concentrated types, are acidic in nature and will take up a considerable amount of ammonia. Either ammonia or ammoniating solutions may be used. The solutions have the advantage that they supply ammonium nitrate or urea as well as ammonia, and therefore give a higher content of nitrogen in the mix.

Much of the mixed fertilizer is produced in granular form; the powdery, dusty mixtures of the past have met with increasing resistance on the part of farmers. Granulation, usually carried out in a rotary drum or a paddle mixer, is accomplished by moistening the mix with water or with solution until the dry solids agglomerate. The modern practice is to carefully adjust the amount and proportion of soluble materials used (urea, ammonium nitrate, or acids), so that, at the elevated reaction temperature reached in the granulator, the proper proportion between liquid and solid phases will be reached and granulation accomplished at low water content. This method reduces the expense of drying, which is relatively high when water alone is used for granulation.

Trends in granulation are toward prereaction of acids and ammonia in separate tank or pipe reactors before feeding them to the granulation step. Heat and moisture are driven off during the prereaction, thereby decreasing the liquid phase in the granulator. This permits use of high proportions of acids, lowering cost of feed materials. The prereactor also can completely eliminate the need for drying some products.

Liquid mixed fertilizers are also an important fertilizer type. Low handling cost has popularized the liquid type to the extent that 17% of all mixed fertilizers in the United States is supplied in liquid form, and liquids are continuing to gain in popularity. In other parts of the world, however, liquids are not used as extensively.

The basic step in liquid fertilizer production is reaction of phosphoric acid to make an ammonium phosphate solution. Other materials, such as urea–ammonium nitrate solution and potassium chloride, are then added to give the nutrient ratio desired.

Although liquid fertilizers are simple to handle with pumps and through pipelines, they have the drawback that the water required to keep the constituents in solution dilutes the product. Development of polyphosphates has improved this situation; for example, a solution containing 11% nitrogen and 37% phosphate can be made from superphosphoric acid, but with the usual type of acid the contents are 8% nitrogen and 24% phosphate.

The most popular process for producing ammonium phosphate liquid mixtures involves contact of wet-process superphosphoric acid containing about 20% of the phosphate in polyphosphate form in a pipe reactor. The heat of ammoniation drives off more water so that a hot melt with 70–80% of the phosphate as polyphosphate is discharged from the pipe. The melt is caught in a circulating stream of liquid fertilizer and cooled before storing. The increase in polyphosphate content substantially improves the storage and handling properties of the liquid.

However, if the liquid mix contains a large proportion of potash, the polyphosphate is not nearly so effective in increasing concentration. For such products, suspension fertilizers, a relatively new fertilizer type, give a very high nutrient content. Suspensions are made by restricting the amount of water and carrying the resulting crystallized salts in suspension by use of 1% or so of clay as a suspending agent.

Bulk blending. The trend to granulation has brought about a revival of the mechanical-mixing practice prevalent in the early days of the industry, but with the difference that granular rather than powdered materials are used and the product is handled mainly in bulk rather than in bags. The main advantage is that a very simple mixing plant is adequate and the resulting low investment makes small, community-type plants feasible. Materials brought into such a plant do not have to be shipped very far after they are mixed, in contrast to "chemical-mixing" plants, which must be relatively large because of higher investment. Such plants must distribute the product over a larger area, and therefore raw materials may be hauled back over part of the route they traveled initially.

The principal materials used in bulk blending are ammonium nitrate, ammonium sulfate, urea, superphosphate, ammonium phosphate, and potassium chloride—all granular. They should be of uniform size to minimize segregation in handling and shipping.

The favorable economics of bulk blending have caused the practice to grow rapidly. It is estimated that more than one-third of the mixed fertilizer consumed in the United States is in the bulk blend form.

Micronutrients. Supplying micronutrients is not yet a very significant activity in the fertilizer industry. Although such nutrients are as essential as the macronutrients, natural supplies in the soil are adequate in most instances. The number of identified deficient areas is growing, however, and use of micronutrient materials is increasing.

From the standpoint of amount used, the principal micronutrient appears to be zinc, followed by boron, iron, manganese, and copper. Consumption was about 485,000 tons of total material in the United States in 1976–1977.

Practice is split between "shotgun" and prescription application. Most agronomists and technical people prefer the prescription method, in which soil analysis and crop response are used as the basis for prescribing application in a particular situation. The "shotgun" approach involves mixing small amounts of micronutrient material into standard mixed fertilizers, on the basis that the micronutrient is needed generally as insurance against the development of deficiency.

Micronutrient salts can be incorporated in liquid fertilizers if polyphosphate is a constituent. The polyphosphate sequesters the micronutrient metals (zinc, iron, copper, and manganese), holding them in solution when they would otherwise precipitate.

There are often problems in incorporating micronutrients into solid fertilizers of the bulk blend type. The micronutrient must be finely divided because such a small amount is required; if granules were used they would be too far apart when applied to the soil. Mixing the fine material with granular mixed fertilizer, however, is a problem because the different sizes segregate in handling. Methods have been developed to cause the micronutrient to adhere to the surface of the granules.

[C. H. DAVIS]

Bibliography: British Sulphur Corporation Limited, *Phosphorus and Potassium No. 68*, p. 19, November-December 1973; Food and Agriculture Organization, *Annual Fertilizer Review*, United Nations; Statistical Reporting Service, *Consumption of Commercial Fertilizers*, U.S. Department of Agriculture, annual reports; Statistical Reporting Service, *Consumption of Commercial Fertilizers by Class*, U.S. Department of Agriculture, annual reports; Tennessee Valley Authority, *North American Production Capacity Data, January 1975*, National Fertilizer Development Center, Muscle Shoals, AL, Circ. Z-57; Tennessee Valley Authority, *TVA Fertilizer Bulk Blending Conference* (held Aug. 1–2, 1973, Louisville, KY), National Fertilizer Development Center, Muscle Shoals, AL, Bull. Y-62; U.S. Bureau of Mines, *1972 Minerals Yearbook*, vol. 3, U.S. Department of Interior, 1974.

Fishery conservation

The term fishery is used here to mean the taking or propagation of fishes or other aquatic life in inland or oceanic waters. A fishery may be operated for subsistence, for pleasure (sport fishery), or for profit (commercial fishery). The "fishes" include: fin fishes; invertebrate shellfishes such as the mollusks (including clams, mussels, and oysters) and the shrimps and their relatives, the marine lobsters and the fresh-water crayfishes; aquatic plants; sponges; coral; sea cucumbers; amphibians, chiefly frogs; reptiles such as turtles, alligators, and crocodiles; and mammals, including whales, seals, and walruses. *See* MARINE RESOURCES.

Fisheries are important for the production of food and of raw materials for industry, and for recreation. They occur in all the many kinds of waters that together cover nearly three-fourths of the Earth's surface. In the international accounting system of world fish catch maintained by the UN Food and Agriculture Organization (FAO), there are three major fisheries—marine (salt-water), fresh-water, and diadromous (involving fishes that ascend fresh-water streams to spawn, like salmon, or descend from streams to reproduce in the oceans, like certain eels). The 1976 world catch totaled 73,500,000 metric tons, having ranged from 66,200,000 to 70,900,000 MT during the previous 6 years. This total catch can be increased in the future, but the limit of increase continues to be debated. Some estimates of potential exceed 300,000,000 MT, but involve harvesting small organisms such as the 2-cm-long crustacean krill of the Antarctic which are strained from the water in great numbers for food by some of the whales. To all of the foregoing must be added the huge but unrecorded take by subsistence fishers and by sport fishers. In the United States alone there were an estimated 33,000,000 sport fishers in 1975. Still to be added is the also large and growing production from aquaculture.

Marine fisheries are mostly commercial and are located predominantly in the Northern Hemisphere. The most important fishery centers and the principal products they yield are the northeastern Atlantic (flatfish including halibut and flounder, cod, haddock, coalfish, herring, and shrimp); the northern Pacific (salmon, flatfish, herring, crab, shrimp, lobster, squid, and octopus); the northern Atlantic (flatfish, cod, haddock, herring, lobster, and crab); and the Indo-Pacific (herring, bonito, mackerel, and shrimp).

The major fresh-water fishing areas and their products are Asia (ayu, salmon, milkfish, and carp); Soviet Union (salmon and whitefish); Africa (many kinds of fishes); and central and northern North America (trout, whitefish, bass, and perch and their relatives). Fresh-water fisheries in the more highly developed parts of the world are primarily for sport. However, they are important largely for food in developing regions, including parts of Africa and South America, and in densely populated lands, especially those on a rice economy, such as India, China, Japan, and the smaller nations of Southeast Asia.

Problems. The major problem in fishery conservation is how to manage both humans (and their impact on water quality) and the aquatic ecosystem to ensure the production and the harvest of aquatic crops for the present and for the future when the demand for them will be greater than now. Lack of knowledge or care by fishers may make their fishing efforts inefficient, their treatment of captured fishes wasteful, and their methods of capture destructive to stocks needed as "seed" for future harvests. Handlers, marketers, and consumers cause waste by careless preservation (refrigeration, salting, and canning) and by inefficient preparation for the table. Lack of information may also be responsible for inadequate harvest of many underexploited segments of the resource. Consumers need to be educated to depart from highly preferential buying habits that accelerate the demand and price for certain species but lead to the discard or wastage of other kinds of aquatic organisms with equally sound food values.

Destruction of habitat. The destruction of the aquatic habitat by humans accentuates fishery problems, especially in inland waters. Deforestation and destructive agriculture, resulting in soil erosion and excessive warming of waters, have changed the fish-producing characteristics of many streams, usually for the worse. Sewage and industrial wastes reduce the quality of water suited for desired aquatic life. Organic pollutants such as domestic sewage may remove life-important oxygen from the water (waters generally are undesirable for aquatic life if the dissolved oxygen in them falls below 3 ppm). Pollutants may also be directly poisonous to fishes and other water organisms (for example, many pesticides and other chemicals that are poisonous to people, such as cyanides, are also toxic to desirable aquatic life). Habitats may be destroyed by smothering them with silt (washings from mineral refineries such as coal and iron plants).

Changing water cycles. Humans have likewise created problems for aquatic life in continental waters by changing phases of the water cycles. Deforestation and agricultural land drainage have lowered the groundwater table and have lessened stream flows during dry periods. The construction of dams for water storage, power, navigation, industry, flood control, domestic water supply, and irrigation has interfered with the movements (usually for spawning) of native migratory fishes. Outstanding problems created by dams are those affecting the migrations of the Columbia River salmon in the western United States and the Atlantic salmon throughout its range in both northeastern North America and Europe. To date, there has been only modest success in developing devices (fish ladders) that enable fishes to progress over high dams or other obstacles to upstream spawning areas. Irrigation channels may lead fishes to doom by leaving them stranded when the channels are drained, and fishes may be destroyed by being jammed against intake screens of power developments or damaged by heated effluents from industrial or thermoelectric installations.

Fishing. Problems of fishery conservation also arise from fishing. The greatest problem is that of managing each fishery so as to provide sustained yields of desired species. The solutions of these problems lie in research, public education, legislation and law enforcement, and continuing reevaluation of management procedures. Fishing itself may be destructive; the gear may injure the young while capturing harvestable adults, or it may destroy animals other than those being harvested (such as porpoises in the purse-seine tuna fishery of the eastern Pacific). Fishing may exceed the capacity of a species to maintain itself through reproduction and growth. Although the species may not be exterminated by such overfishing, the fishery may become unattractive economically and for recreation. If it collapses entirely, this may bring considerable hardship to the fishers. Similarly, underfishing is wasteful. In small inland waters, underfishing may destroy the quality of a fishery by leading to overpopulation and thus to dense stands of undersized fishes, which are unattractive to fishers. Selective fishing for preferred species, often for predatory ones, may cause coarse, unwanted kinds to usurp the fish-producing capacity of a body of water. *See* POPULATION DYNAMICS.

Management measures. Judged historically as well as by the geographic extent of application, legal measures to regulate the taking of fishes lead all other forms of fishery management. This is particularly true in the commercial fisheries as evidenced in the 1970s by the action of seacoast countries to extend their jurisdiction to 200 mi (322 km) offshore. Commonly, laws control who may fish, as well as where, when, and how they may fish. These laws regulate the kind, size, and amount of fishes that may be taken during a prescribed period. Legal measures such as antipollution legislation also protect the habitat of aquatic organisms. Because of changing conditions, the efficacy of laws needs to be reviewed continually.

There are two basic conceptual frameworks and related mathematical models on which to base management decisions that regulate fishing effort: maximum sustainable yield (MSY) and maximum economic yield (MEY). Although different in concept, these are not mutually exclusive. MSY is the greatest catch which growth in a population can maintain while the population is being fished. Growth to replace losses due to harvesting depends on the number and age of remaining individuals in the population. At some point, if more fishes are removed than are required to provide the harvested growth, the level of the catch will decline and overfishing will result. Of course, the catch level can be maintained by increasing the fishing effort, but ultimately the fishing will become uneconomical and the stock may be destroyed. MEY emphasizes that a maximum difference be reached between the total revenue from the fishery and the total cost of earning that revenue. On the basis of this concept, effort to fish a population should be expanded until the marginal value of the catch equals the marginal increase in the cost of the effort required to produce the catch. This point of equality is the MEY, and guarantees that the net contribution of the fishery to the economy is maximized. Actually, the two foregoing concepts converge in that any point on a sustained yield curve relating catch and effort is both a biological and an economic equilibrium point.

Aquaculture. The artificial propagation of fishes and other aquatic organisms, reputedly originating in China several centuries B.C., is a means of increasing the quantities of preferred species. In Southeast Asia, aquaculture combined with rice culture is an important source of fishes for human food. Preferred food fishes in Europe (carp, and trout) and North America (trout and catfish) are also produced for sale at fish farms or hatcheries. In addition, bait fishes are propagated (and sold) extensively in North America. Aquatic farm production methods are also applied to shrimps, prawns, oysters (as food or for pearls), frogs, turtles, ornamental fishes (such as goldfish), and water plants (ornamental, as the lotus, or edible, as the water chestnut).

Introduction of new species. Some artificially propagated fishes are used in the management of sport and commercial fisheries. Many kinds have been established successfully in waters far removed from their native ranges or in newly created water areas. Still others have been stocked successfully to bolster numbers available to fishers. However, countless others have been placed in waters not suited to them and have soon disappeared, or they have been planted where adequate spawning stocks of the species were already present, in which cases they were wasted or they aggravated an already bad situation of overpopulation. In other situations, introduced fishes have increased so rapidly that they have destroyed or seriously menaced desirable species already present, such as the Asiatic common carp in North America.

Improving the environment. Aquatic life in continental lakes, ponds, and streams requires a stable supply of water, with chemical and physical factors differing according to species requirements. Consequently, much effort is spent in learning the best conditions for life of preferred organisms and in managing habitat to provide the prescribed conditions. Often the environment can be improved. Starting in the headwaters of a stream system, land-use practices may be adjusted with the water world in mind, for example, by retarding surface runoff and encouraging percolation of water into the ground. In the water area itself, chemical, physical, and biological changes can be made to enhance stability, productivity, and yield to fishers. Pollution and erosion may be controlled and improvement made in food, feeding, and shelter conditions. Species composition, competition, predation, and fishing pressure can be regulated to some extent.

Farm ponds. Ponds have been built for many centuries to provide fishes and fishing in areas having few natural surface-water bodies. In the mid-20th century there was a surge of artificial pond or lake construction in the United States, and hundreds of thousands of small impoundments resulted from the "farm pond" program, especially in states not bordering the Great Lakes. By the 1960s this program resulted in a greater surface area of artificial lakes than natural lakes (except the Great Lakes).

Education. Fishery management measures also include educational and extension–demonstration programs on fish conservation. They likewise encompass basic and developmental research in government and university laboratories. Moreover, these programs include the training of professional fishery scientists to develop and apply the most effective fishery management methods.

Government regulation and aids. In general, governments regulate inland, coastal, and boundary-water fisheries; extended jurisdiction of the 1970s to include marine resources to 322 km offshore is an illustration. Fishers are commonly required to buy licenses to fish. In the United States, regulation is primarily at the state level, with the state retaining ownership of the fishes in public waters. In many international waters, international agreements and treaties are used; often these are in the form of commissions such as the North Pacific Fishery Commission, Northwest Atlantic Fishery Commission, and the Great Lakes Fishery Commission.

There are numerous governmental aids to the conservation of fisheries at the local, national, and international levels. Educational and extension programs constitute one of these. Another aid is in the form of exploration for new stocks and the development of more effective gear and methods for handling, storing, processing, shipping, and marketing the products. There is also direct Federal aid to states for research and the development of fishing. This is particularly true in the United States, where certain taxes (Dingell-Johnson funds) are earmarked for the purpose. The FAO and many individual donor nations extend international aid and mutual assistance in fishery conservation to developing countries. *See* WILDLIFE CONSERVATION. [KARL F. LAGLER]

Food chain

The scheme of feeding relationships which unites the member species of a biological community. The idea was introduced to modern ecology by C. Elton to describe the linear series of species, usually involving plants (autotrophs), herbivores,

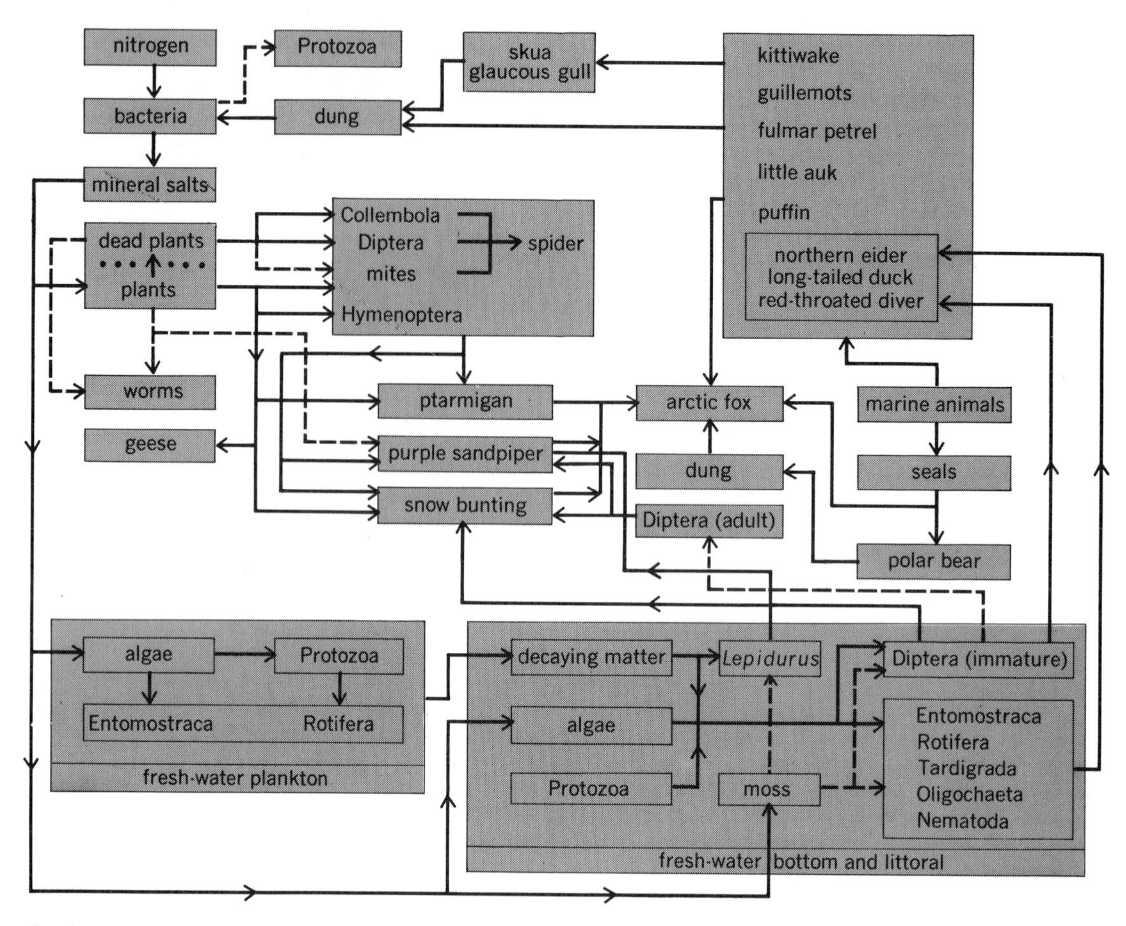

Food web among the animals on Bear Island, a barren spot in the Arctic Zone, south of Spitsbergen (the broken lines represent probable food relations that have not yet been proved). To read the diagram, start at the marine animals and follow the arrows. *(From C. S. Elton, Animal Ecology, Wiley, 1927)*

and one or two successive sets of predators (a predator chain), or alternatively, the series of parasites and hyperparasites exploiting a host (a parasite chain). Saprophytic chains, exploiting dead tissues, are now known to be most important. The successive categories in such a chain are widely known as trophic levels.

The number of stages in a given chain does not usually exceed five and the number of organisms (or better, their biomass) diminishes rapidly. Although the species involved in a given food chain vary in space and time, similar sets of relationships recur in different habitats. For instance, the Arctic fox feeds on guillemot eggs in winter and seal carrion left by polar bears in summer; the spotted hyena has similar habits in Africa involving ostrich eggs and zebra killed by lions. Such professions were termed niches by Elton. *See* BIOMASS.

Food web. A neat classification by trophic levels often fails because a species may occupy more than one level during its life cycle or in response to changes in availability of food. For instance, hover flies change from decomposers to herbivores, whereas many Diptera, when adult, suck both plant and animal juices. Therefore, the system of feeding relationships often resembles a web rather than a chain (see illustration).

Quantitative studies. In view of the complexity of food chains, attempts have been made to measure the size of the feeding links between species in order to select the most important. The only universal currency for such measurements is the caloric content of the food. A number of quantitative food-chain studies of this type permit comparison between different levels in a chain and also the chains in different communities. However, it must be stressed that certain species have an importance in the community which is out of proportion to their energy intake; for instance, the removal of pollinating insects may greatly affect the productivity of orchards and meadows; the excretion of external metabolites and antibiotics in minute amounts may change the balance among microorganisms. *See* ECOLOGY. [AMYAN MACFADYEN]

Bibliography: J. W. Krutch, *The Great Chain of Life*, 1978; E. P. Odum, *Ecology*, 2d ed., 1975, E. P. Odum, *Fundamentals of Ecology*, 3d ed., 1971.

Forest and forestry

Forestry involves the management of forest lands for wood, forage, water, wildlife, and recreation. A forest is much more than an assemblage of trees. Rather, it is a community—technically an ecosystem—consisting of plants and animals and their environment. Trees are the dominant form of vegetation, but shrubs, herbs, mosses, fungi, insects, reptiles, birds, mammals, soil, water, and air are essential parts of the community, and each reacts with all the others. Therefore, the first prerequisite in forestry is a knowledge of the characteristics of the individual members of the community, of the environment in which they live, and of the interrelations between organisms and environment. *See* FOREST ECOLOGY; FOREST SOIL.

A second prerequisite is a knowledge of the economic and social values and costs of the many goods and services produced by the forest. Sometimes these values can be expressed without much difficulty in dollars and cents, as is the case with wood and forage. Often, however, this is not the case, as with wildlife, recreational use, amelioration of climate, protection of watersheds, and preservation of scenery. This fact does not mean that intangible values and costs should be ignored, but that the difficulty of expressing them in monetary terms should be recognized, and that subjective judgment must often be taken into consideration in making policy decisions as to the purposes for which forest lands should be managed.

With respect to both ecology and economics, the fact that forests are a long-term crop is a complicating factor. Predicting the long-term environmental effects of any specific treatment of the forest requires much more knowledge than is now possessed, and estimates of future costs and returns can be little more than educated guesses.

Policy. The primary objective of all forest owners is to obtain the highest net return from using their forest properties for the purposes in which they are chiefly interested. This requires the adoption of a policy, or course of action, which will give due weight to biological potentialities and to the economic and social factors involved. As indicated in the preceding paragraphs, this is a complex matter involving not only current ecological and economic knowledge but also estimates of the future, which are difficult to make.

A basic aspect of forest policy is that it requires a decision as to the relative emphasis to be placed on the various goods and services provided by the forest. Producing wood for manufacture into innumerable materials, such as lumber, plywood, pulp, and paper, and for direct use, such as pilings, poles, posts, and fuel, has so far been the chief concern of forest management (Fig. 1).

The less tangible services of the forest are receiving increasing attention. For example, forests reduce the surface runoff of water and tend to prevent erosion and to regulate the flow of springs and streams. By decreasing wind movement and reducing evaporation from the surrounding area, forests temper the local climate and, in dry regions increase production of crops in their lee. Forests also furnish highly attractive sites for hunting, fishing, picnicking, camping, and other recreational activities. *See* CONSERVATION OF RESOURCES.

Large areas of primitive forest and related lands are being set aside as "wilderness areas," in which roads, habitations, and commercial use are prohibited, and people are only transient visitors. Legislation providing for the establishment of such areas on Federal lands was enacted in 1964.

Multiple use. Management may be centered on any one or a combination of these various goods and services. The variety of goods and services gives special emphasis to the principle of multiple use. In essence, this principle requires the use of different parts of the forest for the purpose, or combination of purposes, for which each is best suited. Often there are areas in the forest in which management for the simultaneous use of timber, forage, wildlife, water, and recreation is both feasi-

Fig. 1. Loading Douglas fir log onto logging truck, Olympic National Forest, Wash. (*U.S. Forest Service*)

ble and desirable (Fig. 2). Multiple use, however, has definite limitations. Some uses are incompatible under certain conditions; maximum production of all kinds of goods and services on the same area is impossible.

Because of the close relationship between the production and utilization of wood, the forester is concerned with the properties of wood and the processes by which it is manufactured into thousands of products. Although wood technology is not strictly a part of forestry, the two are so closely related that they are commonly treated together in education and research.

Fig. 2. Forest opening providing forage for livestock, Challis National Forest, Idaho. (*U.S. Forest Service*)

Fig. 3. Mechanized planting to speed up reforestation, DeSoto National Forest, Miss. (*U.S. Forest Service*)

Fig. 4. Timber stand improvement by light thinning in dense areas in order to forestall mortality and concentrate growth on better trees. McCleary Experimental Forest, Wash. (*U.S. Forest Service*)

Sustained yield. The principle of sustained yield involves the continuous production of forest goods and services. In timber management, not only must reforestation, either by natural or artificial means, follow cutting, but the forest property as a whole must be organized so that approximately the same amount of wood can be harvested annually or periodically (Fig. 3). Growth and use of the product must be sufficiently balanced so that expected future demands can be met.

The same principle that provides for continuous wood production without reduction in quantity or impairment in quality applies to other products, such as water, wildlife, and recreation. As population increases, and standards of living rise, pressure on the forest pushes to increasingly higher levels the sustained yield which will be adequate to meet mounting needs (Fig. 4). The higher the level of sustained yield, and the greater the number of products and services involved, the more difficult the task of management becomes. *See* FOREST PLANTING AND SEEDING.

In 1960, through passage of the Multiple-Use–Sustained-Yield Act, the U.S. Congress made these practices compulsory on all national forests.

Techniques. Policies can only define general objectives and programs. Their translation into action in the forest is the acid test of their validity. The task is monumental. For any forest of considerable size, permanent improvements (such as roads, trails, buildings, fences, and telephone lines) must be installed; all resources (tangible and intangible) must be inventoried; the best techniques for assuring multiple use and sustained yield must be determined; forests must be protected from fire, insects, and disease; logging, camping, picnicking, and other activities must be supervised; and cooperation between users and the general public must be maintained.

The successful discharge of these responsibilities requires professional competence of a high order in many fields. In general, the broadly trained professional forester is particularly well equipped to supervise the operation as a whole; but the forester is increasingly dependent on the staff assistance of engineers, biologists, landscape architects, economists, and others. As the techniques of forest management become increasingly complex, the need for interdisciplinary cooperation becomes more urgent.

Government participation. Forests play so vital a role in so many ways in the well-being of a nation that governmental action to assure their effective management has long been recognized as desirable virtually throughout the world. That action may take the form of public ownership, public regulation, or public cooperation.

In the United States, the emphasis of government activity has been chiefly in the fields of ownership and cooperation, with few attempts at regulation. As of 1973 the Federal government owned 107,000,000 acres (1 acre = 4047 m^2), or 21% of the 500,000,000 acres of commercial forest land in the United States. About 85% of this area was in national forests. The states owned 21,000,000 acres, more than half of which was in the four states of Pennsylvania, Michigan, Minnesota, and Washington; and counties and municipalities owned

Fig. 5. Fire detection tower in the Olympic National Forest, Wash. (*U.S. Forest Service*)

Fig. 6. Airplane spraying insecticide, a modern technique to control insect epidemics in the forest, Boise National Forest, Idaho. (*U.S. Forest Service*)

8,000,000 acres, making a total of 27% of the commercial forest area in public ownership.

The Federal government and the state governments cooperate in a wide variety of ways. Most prominent is the protection of forests from fire (Fig. 5), an activity handled largely by the states, with the assistance of grants-in-aid from the Federal government. Other cooperative activities include participation in the protection of forests from insects and diseases (Fig. 6), sale of planting stock at cost of production, and educational and service assistance to private owners in the production, harvesting, and marketing of forest crops.

Some states have special legislation relating to the taxation of forest lands which is intended to encourage their improved management. A few states exercise some control over the management of private lands by establishing minimum standards of forest practice.

Conclusion. In short, forestry is a highly complex activity requiring sound judgment in the formulation and interpretation of policy, and great managerial and administrative skills in its implementation. Professional competence in a wide variety of interdisciplinary fields is essential. *See* FOREST MANAGEMENT AND ORGANIZATION.

[SAMUEL T. DANA]

Bibliography: A. A. Brown and K. P. Davis, *Forest Fire: Control and Use*, 1973; Marion Clawson, *Forests for Whom and for What*, 1975; H. Clepper, *Crusade for Conservation*, 1975; H. Clepper, *Professional Forestry in the United States*, 1973; E. Eckholm, *Planting for the Future: Forestry for Human Needs*, 1979; R. D. Forbes, *Woodlands for Profit and Pleasure*, 1971; R. C. Hawley and D. M. Smith, *The Practice of Silviculture*, 1962; H. L. Shirley, *Forestry and Its Career Opportunities*, 1973; H. L. Shirley and P. F. Graves, *Forestry for Pleasure and Profit*, 1967; S. H. Spurr, *American Forest Policy in Development*, 1976.

Forest ecology

The science that deals with the relationship of forest trees to their environment, to one another, and to other plants and animals in the forest. The term "silvics" is synonymous. The science includes the fundamental biological knowledge upon which rests the manipulation of forests for humanity's own purposes.

Forest ecology is subdivided into forest autecology, the study of trees as individuals in relation to their environment, and into forest synecology, the study of the forest as a community. In addition, forest ecology includes the study of genetic variation of tree species in relation to environment (genecology) and the functional dynamics of forest ecosystems (ecosystem analysis).

Forest tree variation. Like other organisms, the forest tree is a product of its genetic constitution (genotype) and the environment in which it lives. A given genotype may be modified by different environmental conditions to assume different forms or phenotypes. This plasticity has substantial adaptive value to trees because they are rooted in their environments and have long life-spans.

In addition, trees maintain themselves and persist under changing conditions because of substantial genetic variation among individuals and populations. Through natural selection, tree populations become adapted genetically to their environments. This adaptation is typically continuous, paralleling continuous environmental gradients. Gross climatic gradients such as length of growing season, relative length of day and night (photoperiod), and temperature elicit the strongest genetic responses.

Forest environment. The forest environment or site consists of the physical environment surrounding the aerial portions of the tree (climatic factors) and that surrounding the subterranean portion (edaphic or soil factors). Site is subject to continual change. Atmospheric conditions within the forest fluctuate diurnally, seasonally, annually, and over a period of years. The forest soil owes its original nature to the parent geological material and to the weathering processes that form the soil. It, too, is affected by the climate, the vegetation and animal life it supports, and the passage of time. External influences, particularly fire, grazing and browsing animals, and humans, affect markedly the nature of forest sites and their capacity to support tree growth. Around the world, grazing by livestock has probably been more important than any other factor in reducing the productive capacity of uncultivated land.

Site evaluation. Forest site quality, the capacity of an area to produce forests, is best measured by

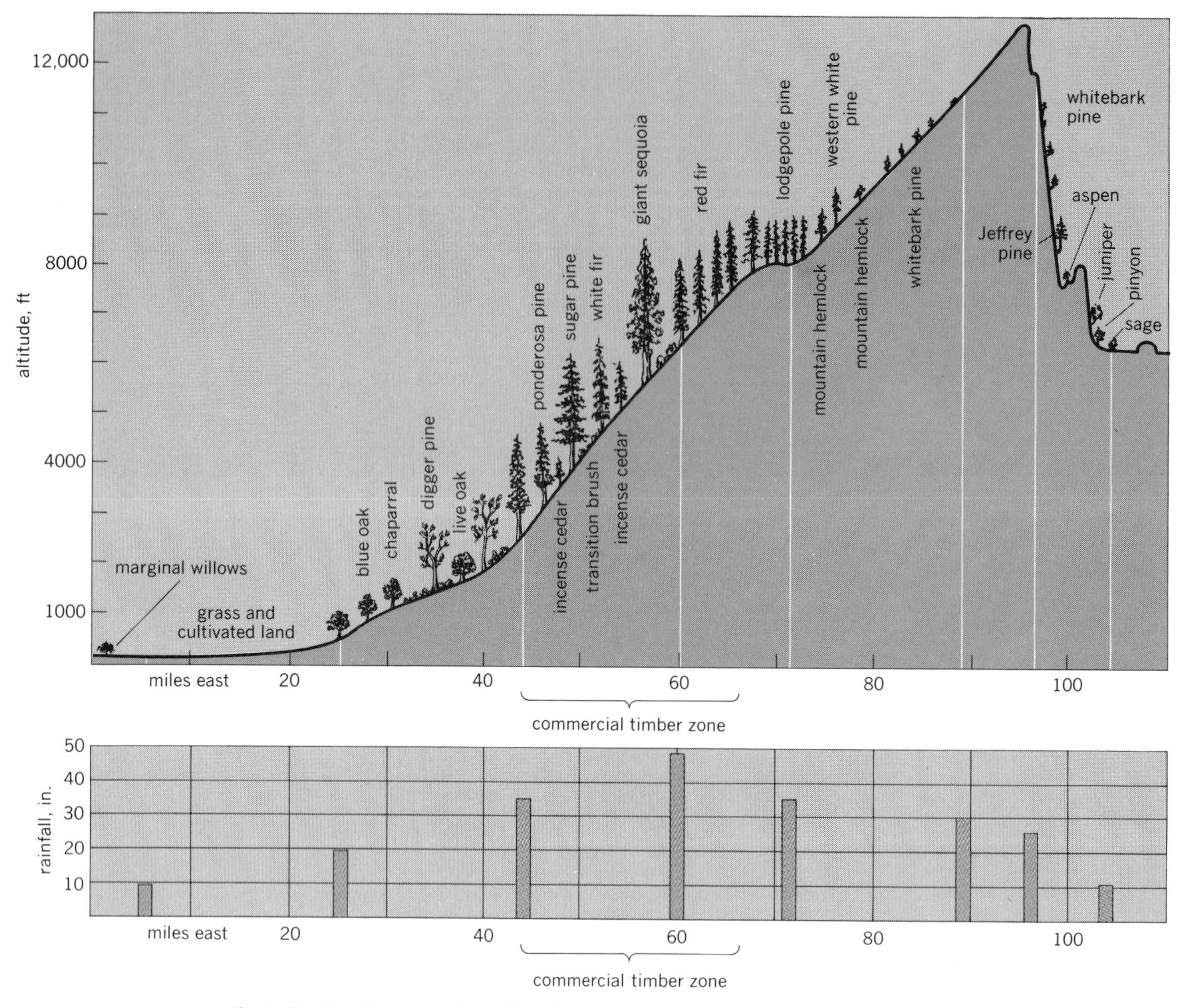

Fig. 1. Profile of the central Sierra Nevada showing altitudinal limits of the principal forest types. (*USDA*)

the recorded growth of trees under forest conditions. Permanent sample plots, measured periodically for many decades, provide the most accurate measure of productivity. Such plots, however, are relatively few. The height of trees grown free from overhead shade to a specified age (usually 50 or 100 years) provides the best general index of site. In the boreal forest the presence of individual shrubs and herbs, "plant indicators," has shown excellent correlation with site quality. Elsewhere, groups of species denoting different site conditions, "ecological species groups," have found widespread acceptance.

In situations where forest trees and associated plants cannot be used to indicate site quality, reliance must be placed on soil properties and physiography. Among the properties that have been successfully used as indices of site quality in given situations are the parent geological material, relative topographic position, depth of soil, water-holding capacity of the soil, and depth to soil water table. *See* FOREST SOIL.

Because site quality is the net result of many factors affecting the capacity of land to produce forests, the accuracy of the estimate of site productivity improves as more factors are taken into account. Various methods integrating many of these factors, including landform, topography, soils, vegetation, and forest history are increasingly employed.

Influence of site upon life history. From seed to old age, forest trees are greatly influenced by site. Although many seeds may be produced, very few—usually less than 1%—actually grow into trees. Many seeds are destroyed by insects, fungi, rodents, and adverse weather. Only those seeds that come to rest under suitable conditions of moisture and temperature will germinate. Most of the resulting seedlings will be killed by soil fungi, solar heating at the ground line, animals, and drought. The most favorable sites for successful forest regeneration usually have moist but well-drained and exposed mineral soil with sufficient amounts of dead material on the surface to provide some shade for the seedlings. *See* FOREST PLANTING AND SEEDING.

As the seedlings develop, site plays an important part in determining the growth rate of different species and the outcome of the competition. Some trees, including most of the pines and oaks, will grow well under a wide variety of site conditions. Others, such as black walnut and sycamore, can thrive only under specific site conditions. Consequently, members of the former group are found on many sites, whereas trees in the latter group are less widely distributed.

Different groupings of forest tree species (forest types or communities) thus tend to develop on different sites. In the northeastern United States, for example, a spruce-fir type is characteristic of higher elevations and of cold, wet areas. Pine types predominate on drier sites, whether sandy or rocky. Hardwoods occupy the moister lowland soils, the oaks frequently predominating in the somewhat warmer and drier situations, and the birches, beech, and maples on the somewhat cooler and moister sites. Eastern hardwoods such as willows, cottonwood, sycamore, and elms are adapted to periodic flooding and form river floodplain communities. *See* VEGETATION ZONES, WORLD.

Wherever mountains occur, forest communities form in relatively distinct zones. The tree species are adapted to different positions along gradients of moisture, temperature, and exposure to wind and sun. The Sierra Nevada range in California offers a distinctive example of this species-site interaction (Fig. 1). At low elevations scant precipitation and high temperatures limit trees and favor grasslands. The adjacent zone is a dry, open woodland of scattered oaks, occasional pines, and shrubby chaparral. Above this savanna-woodland, ponderosa pines form parklike stands on grassy slopes in dry, fire-dominated sites. Unlike zones below it, the next or montane zone (4000 to about 7000 ft, or 1200–2100 m) is characterized by magnificent mixed stands, including incense cedar, sugar pine, white fir, and others. Cool temperatures and abundant moisture favor increased species diversity and tree size. At higher elevations, temperature and exposure become limiting, and few trees are adapted to the increasingly severe site conditions. On the east side of the Sierra Nevada range, precipitation declines markedly. Moisture there is limiting, and only a few tree species (junipers and pinyon pines) can persist in the hot, dry, desertlike environment. *See* VEGETATION ZONES, ALTITUDINAL.

Influence of forest on site. As the forest becomes established and develops, the site itself is greatly changed. Forest cover moderates the extreme diurnal temperature regime of open sites resulting in more uniform conditions. Wind velocity is slowed in the vicinity of tree crowns and becomes negligible within the forest. Tree crowns intercept sunlight and alter the quantity and quality of radiation reaching the forest floor compared with that reaching open sites.

On the forest floor, accumulating layers of leaves, twigs, and other litter attract a characteristic grouping of plants and animals that live on decaying organic matter and on each other. The organic matter is gradually incorporated into the top layer of mineral soil, creating a soil horizon rich in nutrients and humus. Water, leaching downward through the soil, tends to deplete the topsoil of soluble materials, depositing them in a lower soil horizon. Some of these materials are absorbed by deeply penetrating tree roots, utilized in the development of leaves, and later recycled to the surface soil in the annual leaf fall.

Earthworms, insects, spiders, mites, and springtails play important roles in decomposition and thus influence forest sites. Mammals, too, are significant in decomposition and soil development. Tunnels and burrows of ground-inhabiting animals materially aid in loosening compact soils, in incorporating organic matter in deeper layers, and in allowing downward movement of rainwater containing nutrients.

The influences of the forest on the site result in a characteristic pattern of streamflow differing from that on unforested land; peak streamflow is lower, and the flow after storms is prolonged. Since much of the water needed for urban and agricultural uses originates from precipitation falling on forested lands, the management of forests for water production is of great importance (Fig. 2). In mountainous areas of high snowfall, the snowpack can have an important influence on water yield. Manipulating forest stands by clearcutting, for example, may increase snow-water storage and delay snow melt. *See* HYDROLOGY.

Ecosystem analysis. The term "ecosystem" is the most concise formulation of the concept of an interacting system comprising living organisms together with their site. The goal of the study of whole ecosystems is to understand their processes and functions. Such an understanding is urgent because we need to predict the long-term as well as the short-term consequences of humans' actions in ecosystems. Greatest emphasis has been placed on photosynthesis and respiration, on cycling of water and nutrients, and on the relationship of these to forest productivity.

Studies of nutrient cycling in hardwood forests of the northeastern United States show the enormous amount of nutrients available for uptake and regeneration of a new forest after cutting of the mature forest. Conventional forest cuttings pose little threat to site productivity, but short-rotation

Fig. 2. Much urban and agriculture water flows from forested lands, particularly from those at higher elevations. (*U.S. Forest Service*)

Fig. 3. Effects of fire on forest stands. *(a)* In presettlement times, fires maintained open, grassy parklike stands of ponderosa pine such as this one in western Montana. *(b)* A light surface fire burns grass and litter of this stand in central Idaho. *(U.S. Forest Service)*

cropping and whole-tree harvesting can lead to nutrient deficiencies. This kind of integrated study will increasingly provide the basis for management of terrestrial and adjacent aquatic ecosystems. *See* ECOSYSTEM; FOREST MANAGEMENT AND ORGANIZATION.

Forest succession. Since the developing forest changes the forest site, it follows that the changed site may itself be more favorable to a new group of tree species other than those currently occupying it. As a result, successive forest communities occupy the site in the absence of fire, logging, windstorm, or other disturbance. Thus, in the northeastern United States early successional or pioneer communities characterized by aspens, cherries, birches, and pines occupy open sites. These in time are replaced on fertile sites by midsuccessional communities containing oaks, ash, and other species until, in several hundred years, the late successional or climax community is attained. This climax contains species, such as hemlock, sugar maple, and beech, that are capable of reproducing under their own cover in the absence of disturbance. On dry, sandy sites, especially, periodic fires enable pines to succeed themselves over long time periods, and a fire climax prevails. Thus, species composition of late successional communities is dependent on prevailing site factors; several climax types are typically recognized in each forest region.

Throughout the world, fire greatly influences forest succession, setting it back or maintaining fire species as climax types. In the southeastern United States, pine types predominate and constitute the most important commercial forests. Pines owe their existence to periodic fires; without fire they would be replaced successionally on many sites by mixed hardwood forests. Similarly, in western North America, the extensive pure stands of Douglas-fir owe their existence to fire and periodic felling by wind. They would be replaced by less valuable, mixed coniferous types if these disturbances were eliminated. Also in western North America the esthetically pleasing and commercially important parklike stands of ponderosa pine are perpetuated and shaped by fire (Fig. 3). In present-day silviculture humans increasingly use controlled fire together with cutting, chemicals, and mechanical treatments to maintain the forest on a given site at a stage of forest succession most valuable to humans. *See* ECOLOGICAL SUCCESSION; ECOLOGY; FOREST AND FORESTRY.

[BURTON V. BARNES]

Bibliography: S. H. Spurr and B. V. Barnes, *Forest Ecology*, 1973.

Forest fire control

In the decade 1964–1973, more than a million forest fires burned over 26,000,000 acres of forest land in the United States. In 1974, government and private forestry organizations spent more than $415,000,000 in preventing, preparing for, and suppressing wildfires. In addition, several million acres of forest land are intentionally burned each year under controlled conditions to accomplish some silvicultural or other land-use objective (Fig. 1).

Combustion of forest fuels. Successful control of forest fires depends on adequate knowledge of the combustion process and the environmental factors which influence it. In view of the importance of combustion to man, it is surprising how little is known about the fundamental physicochemical processes involved. This is especially true of unconfined fires.

Combustion of forest fuels is an exothermic oxidation chain reaction requiring for its initiation the application of sufficient heat to a suitable body of fuel in the presence of oxygen. These three conditions (heat, fuel, and oxygen) can be considered to constitute three legs of a "fire triangle." All three

Fig. 1. Controlled fire conditions. (*a*) Closed jack pine (*Pinus banksiana*) cones, which require heat in order to open. (*b*) Prescribed burning after seed tree cutting so that jack pine cones can open for natural regeneration. (*Photographs by W. R. Beaufait*)

must be present for a fire to burn; removal of any one will extinguish the fire.

In addition, combustion of forest fuels also involves the process of pyrolysis, the cracking of complex wood molecules into simpler, more volatile, and more flammable substances. In general, these processes precede the oxidation reactions and may occur without the presence of oxygen. In the early stages of combustion, pyrolytic reactions are endothermic; subsequent reactions are exothermic and contribute heat to keep the combustion reactions going. The reactions are strongly influenced by the presence or absence of catalytic substances.

Determinants of forest fire behavior. The manner in which fuel ignites, flame develops, and fire spreads and exhibits other phenomena constitutes the field of fire behavior. Factors determining forest fire behavior may be considered under four headings: attributes of the fuel, the atmosphere, topography, and ignition.

Fuel. Live and dead vegetation in the natural forest complex constitutes the fuel for forest fires. The amount of moisture in relation to the heat content of the fuel determines its flammability. Moisture in living vegetation varies with species and with time of year. Moisture content of dead fuels, the more important component of the fuel complex, responds to precipitation, relative humidity, and to a lesser extent, temperature. It shows important daily fluctuations.

Rate of combustion and heat output are also determined by quantity of fuel, fuel arrangement, and the thermal and chemical properties of individual fuel particles.

Atmosphere. Major direct effects of the atmosphere on fire behavior are through variations in oxygen supply and in flame angle caused by wind, and the stability of the atmosphere. Secondary effects on combustion rate are through air and fuel temperature, and air relative humidity (Fig. 2). *See* ATMOSPHERE; HUMIDITY.

Topography. Heated air adjacent to a slope tends to flow up the slope, creating a chimney effect. Unburned fuels above a fire advancing up a slope are closer to approaching flames because of the angle of the slope. Both of these effects act similarly to wind on flat ground in increasing rate of fire spread.

Ignition. An igniting agent of sufficient intensity and duration is necessary to start a forest fire. The agent may be human-caused, such as a match or cigarette, or natural, such as lightning. During 1973, 65% of the 92,000 wildfires occurring in the United States were started accidentally by humans, 24% were of incendiary origin, and 11% were started by lightning. These percentages vary widely by region, however. In the 12 Rocky Mountain states, 44% were started by lightning and 3% were incendiary; in the 13 Southern states, 2% were started by lightning and 36% were of incendiary origin.

Once a fire has started, its spread is by successive ignition of unburned fuel adjacent to the burning area. This fuel is brought to ignition temperature by heat radiated and convected from the flame front. Generally, fuels must be very close to or even enveloped by flame or superheated gases in the convection column before igniting. This mechanism results in more or less continuous advancement of a fire. Fire spread may also be caused by spotting, that is, ignition by burning brands carried outside the fire area by wind and turbulence.

Fire behavior. A forest fire may burn primarily in the crowns of trees and shrubs, a crown fire (Fig. 3*a*); primarily in the surface litter and loose debris of the forest floor and small vegetation, a surface fire (Fig. 3*b*); or in the organic material beneath the surface litter, a ground fire. The most common type is a surface fire. Crown fires are usually accompanied and sustained by a surface fire; occasionally a fire will burn over an area twice, once as a surface fire, then later as a crown fire.

Since fire is a self-sustaining chain reaction, its behavior is determined not only by existing environmental conditions but also by the conditions produced by the fire itself. For example, the vertical convection column produced by the fire creates indrafts. These indrafts increase the rate of com-

Fig. 2. Observer checking weather readings at Canyon Village, Alaska. (*U. S. Forest Service*)

Fig. 3. Characteristic fire behavior. (*a*) Crown fire. (*b*) Surface fire. (*U.S. Forest Service*)

bustion, which leads to stronger vertical convection and consequently stronger indrafts. Because of this feedback, a slight change in environmental conditions, such as fuel becoming slightly drier, may produce a large change in fire behavior. The feedback principle holds for all levels of fire behavior.

Occasionally fires exhibit erratic phenomena not readily explained on the basis of known weather-fire relationships. These phenomena include flashovers, apparently in accumulations of unburned volatile matter, and extreme rates of spread under conditions thought to be comparatively safe. Spectacular increases in fire intensity in a short time are usually termed blowups.

Fire danger rating. In order to satisfy the needs of fire control organizations for estimates and predictions of potential protection work load, numerous systems have been developed which attempt to integrate selected determinants of fire behavior into numerical indexes of fire danger. These expressions of fire danger are related to such fire phenomena as rate of fire spread and probability of fire occurrence.

The U.S. Forest Service has developed a National Fire Danger Rating System (NFDRS) to provide fire control personnel with numerical ratings that will help them with the tasks of fire control planning and the suppression of specific fires. The system includes three basic indexes: an occurrence index, a burning index, and a fire load index. Each of these is related to a specific part of the fire control job.

Measurements of precipitation, relative humidity, temperature, and cloud amount are integrated in various ways to estimate the moisture content of various sizes of forest fuels, a basic fire behavior determinant. Further integration with wind speed, slope, fuel type, and condition of herbaceous vegetation leads to predictions of potential fire behavior. With knowledge of probable number of fire starts, estimates can be made of the probable size of the "fire control task" (fire load) for any particular area and day. These indexes are used by dispatchers in making decisions on how and where to station fire-fighting forces, manning lookout systems, and so forth.

Fire seasons. Because of annual cycles in precipitation and in growth of vegetation, most sections of the country have a well-defined fire season during which most forest and wild-land fires occur. This fire season varies widely in length and duration from two short seasons in the Northeast, spring and fall, to frequently a year-long season in southern California. Figure 4 shows the normal peak fire seasons for the United States.

Fire suppression. Because of the damage done by fire to forests and wild lands in terms of timber, watershed, range, and aesthetic values and because of the threat to human life, efforts have been made since the turn of the century to prevent and control wildfires. Although protection of wild lands from fire is a more recent development than protection of urban areas, the two fields share common fundamentals. Development and application of these fundamentals differ in the two areas; but these differences are largely in degree rather than in kind, and many of them stem from economic considerations rather than from the nature of fire.

Simply stated, the problem is to prevent potentially damaging fires from occurring and to suppress those that do occur efficiently with minimum cost. Certain elements are common to all solutions: fires may be prevented; those that start must be detected; this information must be communicated to the suppression forces; those forces must be transported to the fire; the fire perimeters must be determined and mapped; and specific suppres-

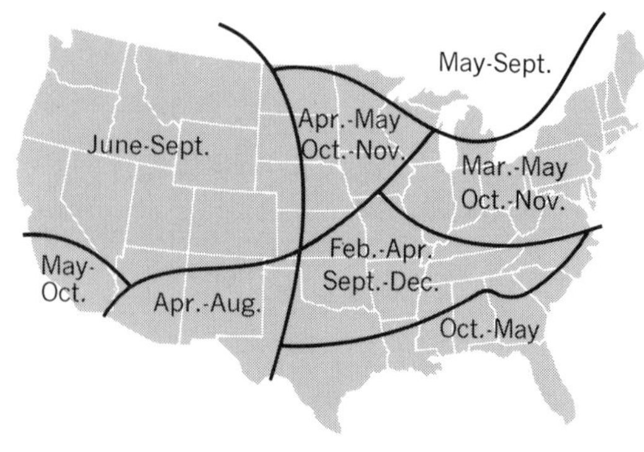

Fig. 4. Normal peak fire seasons in United States.

Fig. 5. Mobile fire-weather forecasting unit operated by the National Weather Service.

Fig. 6. Fire fighters rappelling to the scene of a fire.

sion methods must be employed in a strategic manner to effect control of the fire.

Prevention. Emphasis can be placed either on keeping igniting agents away from forest fuels (risk reduction) or reducing or eliminating the flammable materials (hazard reduction). Efforts directed toward reducing the incidence of man-caused fires encompass law enforcement and educational activities. The latter include campaigns aimed at the mass audience, for example, the "Smokey Bear" and "Keep Green" programs, and efforts by forest patrolmen directed at educating individuals.

In areas of intensive use, many forestry activities are aimed at reducing fuel accumulations in order to reduce the chance of a fire starting or to reduce the severity of one that might be started. Hazard reduction activities include roadside burning, slash disposal, controlled burning of forest areas to reduce fuel accumulations, and certain silvicultural practices.

Detection. Although many forest fires are detected and reported by local residents and transient forest users, primary reliance for their detection is placed on specialized detection systems (Fig. 5). Most fire-control agencies use a combination of lookout towers manned during the fire season by trained observers and regularly scheduled aerial patrols. Use of aircraft for routine detection patrol has increased greatly, paralleling the increase in aircraft use in other protection activities.

Infrared scanners developed originally by the military for battlefield use have been adapted for use in detecting forest fires. Aircraft-mounted detectors are routinely used to detect fires started by intense lightning storms. Such scanners have capability to detect fires the size of campfires from an altitude of 12,000 ft. Other electronic equip-

Fig. 7. Parachutes used for the delivery of men and materials to fire areas. (*a*) Two fire fighters descending to a forest fire. (*b*) Air drop of fire-fighting equipment, showing drop location marker (X). (*U.S. Forest Service*)

Fig. 8. Land fire suppression methods. (*a*) Digging a trench ahead of an approaching fire. (*b*) Constructing a fire line to stop the advancing fire. (*c*) Bulldozer making fire lane near the edge of the fire. (*d*) Using a tank truck to suppress fire. (*e*) Fighting fire with hose attached to a portable pump. (*U.S. Forest Service*)

ment, including television, radar, and sferics (lightning discharges detectable by radio methods), has been used experimentally.

Communication. Rapid and accurate communication is essential not only to the detection network but also to the forces engaged in actual suppression work. Landline telephones are being rapidly supplanted or supplemented by radios. In many organizations all motorized units are equipped with two-way radios. As the size of portable, self-contained equipment is reduced, more and more supervisory personnel on actual fires are also being equipped.

Transportation. Emphasis on fast attack of discovered fires has steadily increased over the years. Many fire-control organizations, both urban and forest, divide their protection area into zones of allowable attack times. These are based on times shown by experience as necessary for successful control to acceptable standards. Location of suppression forces and methods of transportation are planned to meet these objectives.

Because of the great distances involved, forest-fire-control organizations have placed special emphasis on increasing the speed and mobility of their forces. Most personnel and equipment arrive at the majority of fires by a combination of ground-transportation methods, including the most elementary of all, walking. Therefore, in areas of high value and high fire incidence an extensive network of roads and trails is usually maintained. The development of reliable, lightweight gasoline engines has made possible the use of trail scooters and motorized equipment carriers. These require trails intermediate in quality between horse and foot trails and the more expensive truck trails.

In remote areas with relatively low and widely scattered fire incidence, the airplane and helicopter are the primary means of delivering both men and material to the scene of a fire (Fig. 6). Even in areas with good road and trail networks, it is often cheaper and faster to supply suppression forces by air. Successful parachute jumping by fire fighters, started by the U.S. Forest Service in 1940, demonstrated the feasibility of wartime use of paratroops. Postwar development by the Forest Service of techniques and equipment for delivery of men and material by parachute has led to increased use of airborne fire fighters in the forests of the western United States (Fig. 7).

Helicopters are being used more and more for transportation of men and equipment as their capacity and service ceiling are raised. Because of their ability to land on hastily prepared heliports

Fig. 9. Aerial fire suppression. (*a*) Mixing and loading fire retardants in preparation for an air drop. (*b*) A helicopter with a detachable 40-gal tank drops borate slurry on a pinpoint target. (*U.S. Forest Service*)

and because the men so delivered require no special training such as that required by parachute jumpers, helicopters are being used increasingly in all phases of fire control work.

Mapping. Since 1966 specialized heat-sensitive infrared mapping devices have been used operationally to map fire perimeters from the air. These are somewhat similar to television systems that are sensitive to radiation in the far-infrared rather than to visible light. They have the great advantage of being able to see through the smoke of a fire and locate the regions of active combustion on the ground. The image, which can be photographed, indicates the temperature of the fire in shades of black and white and thus enables fire control personnel to map the fire edge, determine the intensity of the fire along the perimeter, and locate spot fires outside the main fire at night or when the fire itself is completely obscured by smoke.

Suppression methods. As previously indicated, a fire can be suppressed by removing one or more legs of the fire triangle. Thus a fire may be robbed of its fuel or its oxygen supply, or enough heat may be removed to stop the chain reaction, and the fire goes out. Because of the influence of catalysts and anticatalysts on the processes of pyrolysis and oxidation, combustion may also be suppressed by the addition of small amounts of certain chemicals to the combustion zone.

Fuel removal. One of the most common methods of controlling a forest fire is by creating a fuelless barrier. This may be done by digging a trench to mineral soil ahead of an approaching fire of sufficient width to prevent the fire from crossing (Fig. 8*a*). Since this may require a line of considerable width (some evidence is available that this width is approximately equal to the square of the height of the flames), it is customary for the unburned fuel between the prepared line and the approaching fire to be intentionally fired, thereby effectively widening the line.

Techniques of constructing fire lines vary with terrain, soil character, and type of vegetation. Organized crews with hand tools, such as axes, shovels, rakes, and other specially constructed tools, are used in many areas, especially those too rugged or remote to permit use of mechanized equipment (Fig. 8*b*). Specially developed motor-driven flails, similar to garden-tractor rotary tillers, are sometimes employed, although their usefulness is limited to terrain and soil types on which they can be maneuvered satisfactorily. Bulldozers are used extensively (Fig. 8*c*), and tractor-drawn plows are standard line-construction equipment in the relatively flat terrain of the southeastern United States.

Occasionally, existing roads or previously prepared firebreaks are used as control lines, usually as a line from which a counterfire is set. Natural firebreaks, such as rocky ridges or rivers, are used similarly.

Heat removal. The second major method used in suppressing fires is to remove enough heat to break the chain reaction. Water, with its high specific heat and heat of vaporization, is an ideal substance for this purpose. Long the favorite with urban fire control organizations, water is becoming more popular with forestry organizations as techniques and equipment for handling it are improved. An extraordinarily large amount of burning material can be extinguished with a small amount of water when it is used with maximum efficiency.

The effectiveness of water as a suppression agent can be increased by the addition of several types of chemicals. Water thickeners such as sodium alginate and carboxymethylcellulose are frequently added to water prior to its use on a fire to increase the quantity that will adhere to burning fuel, particularly standing trees and brush. Wetting agents are also added to water to increase its ability to penetrate smoldering litter and duff.

Hand-operated pumps with a 5-gal tank carried on the back are most satisfactory when used on fires in light fuels, for suppressing small spot fires, and for mop-up, the tedious work of extinguishing burning fuel after the fire has been contained within a fireproof boundary.

Gasoline-operated pumps are used extensively wherever water supplies are convenient or road networks permit the use of tank trucks (Fig. 8*d*). Portable pumps are available that can be carried by one man (Fig. 8*e*), making hose lays of several thousand feet or even several miles feasible. Techniques have been developed for laying hose rapidly from helicopters, cutting time required to lay a mile of hose from hours to minutes, even seconds. For use in inaccessible regions, small units including a pump, a 50- or 100-gal tank, and several hundred feet of hose have been developed that can be delivered by helicopter to the scene of a fire.

Chemicals that inhibit the combustion process are used regularly in the suppression of large forest fires. These are generally prepared in the form of water solutions or slurries and are dropped from aircraft or applied to fires from ground-based tank trucks. More than 15,000,000 gal of such fire-retardant chemicals are used annually in the United States.

Although surplus World War II aircraft were originally used in the development of airborne delivery systems (Fig. 9*a*), newer aircraft have permitted more flexible and efficient release systems. Four-engine cargo planes with pressurized tanks permit the delivery of as much as 3000 gal of retardant mixture to a fire. The retardant is released at low altitude either directly on the burning material at or near the fire edge, or some distance in front of an advancing fire. A single pass may cover an area several yards wide and several hundred yards long. Helicopters are also used, especially in initial attack and mop-up operations (Fig. 9*b*). While helicopters carry smaller amounts of retardant materials (50 to 100 gal), their great accuracy makes them excellent for strategic use of expensive materials.

Retardant drops are used primarily to cool down hot spots and help hold a fire pending arrival of ground forces and to control spot fires outside the main fire perimeter. The air tanker has limited value in attacking the head of a rapidly moving crown fire.

Little direct use is made of the third possibility of removing the oxygen supply to a fire, although water has this effect to some extent and soil is sometimes used to smother flames.

Suppression strategy. The manner in which the various suppression techniques are utilized on a fire is determined by the fire boss, who is comparable to the fire chief in urban organizations. It is his responsibility to determine the strategy appropriate to the particular fire. Depending on the size of the fire, he may have a few men or a few hundred, and occasionally more than a thousand, to assist him in carrying out his strategy. Forestry organizations maintain intricate paper organizations and large caches of fire-fighting materiel which can be mobilized rapidly in time of urgent need. *See* FOREST AND FORESTRY. [WILLIAM E. REIFSNYDER]

Bibliography: A. A. Brown, and K. P. Davis, *Forest Fire: Control and Use*, 1973; National Fire Protection Association, *Chemicals for Forest Fire Fighting*, 1977; W. E. Reifsnyder, *Proceedings of the 10th Annual Tall Timbers Fire Ecology Conference*, 1970; World Meteorological Organization, *Systems for Evaluating and Predicting*, Special Environmental Report Series, no. 11, 1978.

Forest management and organization

Since forests are renewable natural resources, the central purpose of forest management is to organize the forest for continued production of its many goods and services. This is particularly exemplified in management for wood production, from which, for many reasons, a sustained yield of forest products is desired. *See* FOREST RESOURCES.

The essential requirements of a regulated forest, that is, one that can maintain a sustained yield of harvestable products, is that tree age and size classes be maintained in balanced proportion so that an approximately equal annual or periodic yield of products of desired size and quality may be harvested. There must be a progression of tree age of size classes from small to large so that merchantable trees are regularly available for cutting in approximately equal volume.

From a management standpoint, there are only two general kinds of forests: those composed of even-aged and those composed of uneven-aged groups or stands of trees. Each is considered separately.

Even-aged stands. An even-aged forest stand is one in which the individual trees composing it originated at about the same time and occupy an area large enough to grow under essentially full-light conditions. Such a stand consequently has a beginning and end in time. A stand is established either naturally or by planting or seeding. Thinnings or other intermediate cuttings may be made during its life. The stand is finally harvested, a new one is established, and the growth process is repeated. A typical lifeline of an even-aged stand managed on a rotation or total life of 30 years is shown in Fig. 1. *See* FOREST PLANTING AND SEEDING.

Total harvest. The total harvest obtained from the stand over its span of life, shown in Fig. 1 as the cumulative bar at the right, is the sum of intermediate cuts made plus those from the final harvest or regeneration cuttings. This is the total growth of the stand that is taken in harvestable timber. The number of intermediate cuttings made and their kind and intensity are a function of biology, economics, the quality of the land, and the tree species. The objective is to keep the stand growing vigorously and to capture as much of the total growth potential of the land as possible in usable wood products.

Age-class organization. The organization of a forest area, made up of many such even-aged stands, for continuous production is indicated in Fig. 1 by the vertical broken lines. The spacing between the lines denotes age-class groups of individual stands as well as points in the age of a single stand. The total forest unit could be of any size. In a forest fully regulated for continuous production, there would be an approximately equal area of stands classed by age groups such as 0–5, 5–10, and up to 25–30. The uniform spacing of the vertical broken lines in the diagram indicates equal areas in each age group. In practice, the spacing would not be equal if it were necessary to adjust areas for differences in site quality between age classes. The basic aim is to have age groups of equal productivity, not necessarily of equal area.

Rotation and annual yield. Each year, consequently, an approximately equal area of timber at the rotation age of 30 years is available for final harvest cutting. The area, on the average, would be total area divided by rotation age. The final harvest volume cut would be this area multiplied by the average volume per unit of area at this age. Similarly, there would be an approximately equal area available for intermediate cuts through thinnings, taken at 15, 20, and 25 years in this instance. For thinning at each of these ages, there would again be total area divided by rotation units of area.

The decision regarding the length of the forest rotation is of key importance because the whole forest structure is organized around it; the average rotation of a forest unit cannot be quickly changed. The rotation length is determined by an integration of the biological growth capacity of the forest for different tree species and production costs with the value of forest products produced and the particular purposes of management by the owner. Similarly, the number, timing, and kind of intermediate cuts that may be taken, methods of stand regeneration, desired species, and the general intensity of management practiced are determined by a complex of biological and managerial objectives. Forests are managed not only for timber but for watershed, recreation, wildlife, and other uses. The particular "mix" can vary considerably but this variation does not invalidate the basic principles of even-aged forest management for continuous timber production.

Uneven-aged stands. An uneven-aged stand is one in which trees originated at different times so that the stand includes trees of varying ages. The age-class distribution is seldom perfect, nor need it be. The essential need is that there be a good distribution of trees from small to harvestable size that are capable of making normal growth. Specific age is not particularly important. The distinguishing feature of an uneven-aged stand, large enough to be of practical importance as a management unit, is that it has no beginning in point of time. Nor has it an end, because if it did the next stand would necessarily be even-aged.

Stock structure maintenance. The lifeline of an uneven-aged stand is shown by the solid line in Fig. 2. As shown, the stand has a certain volume of reserved growing stock. It grows for a certain number of years, termed a cutting cycle, a harvest cut is again made, and the process indefinitely repeated. If the reserve growing stock, measured by tree ages and diameters composing it, is considered to be the correct size, the volume cut is equal to the net growth made during the cutting cycle. If the amount of growing stock is to be decreased or increased for economic and silvicultural reasons, the cut is accordingly more or less than current growth.

The trees composing the residual growing stock should be well distributed by species, sizes, and

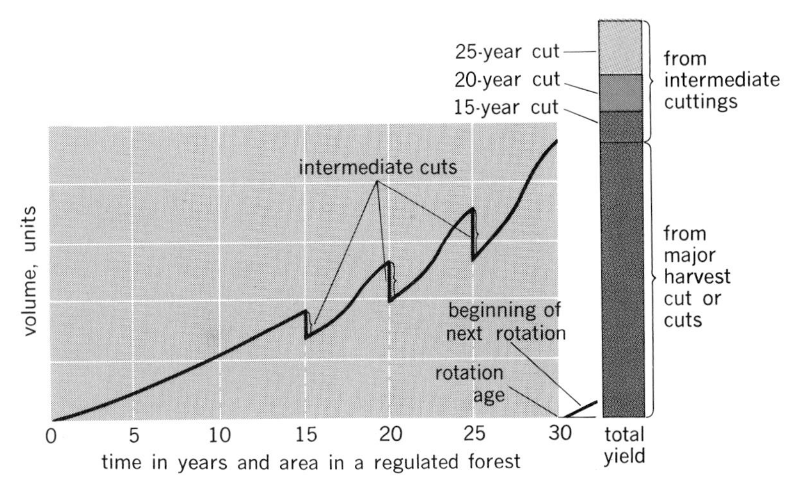

Fig. 1. Lifeline of even-aged stand and regulated forest of even-aged stands.

ages so that a continuing crop of harvestable trees keeps coming along. There should be more small trees than large to allow for mortality and to give opportunity to select the best trees for future growth. In contrast with even-aged management, cuts are uniformly made at regular cyclic intervals only and there is no major single harvest cut nor a guiding rotation. Intermediate and harvest cuts are merged in the single cyclic cut in even-aged management.

Cutting cycle. The organization of a large forest unit for continuous production is diagrammatically shown in Fig. 2. A 5-year cutting cyle is assumed, which means that the various stands constituting the forest unit are grouped into five areas of approximately equal productivity, as shown by number in the diagram.

Each year, a harvest cutting is made covering one of these five areas, equal in volume to 5 years growth, plus or minus whatever changes are considered appropriate to maintain the average amount of reserve growing stock. In Fig. 2 a small increase in the amount of reserve growing stock over the two cycles shown is indicated. The entire forest management unit is cut over every 5 years.

Determining the length of the cutting cycle is a decision of critical importance, comparable to the decision concerning rotation length in even-aged

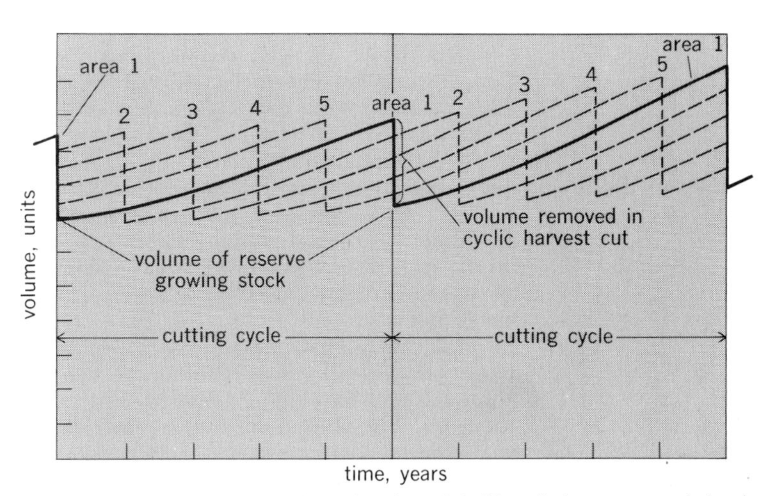

Fig. 2. Lifeline of uneven-aged stand and regulated forest of uneven-aged stands.

management, as the organization of the forest is built around the cutting cycle and quick changes are not possible. As with rotation length, the answer is derived from an integration of biological and financial factors with the objectives of management. Establishing and maintaining a desirable level of reserve growing stock and control of its composition by tree sizes and species of desired quality are other important problems.

Application in practice. In practice, forest management units are usually far from being fully regulated and especially when unmanaged areas are put under management. Continual adjustment in age- and size-class distribution, species composition, and growing stock structure is necessary in working toward a better-organized forest in consonance with changing economic conditions and management objectives. Basically, the purpose is to attain the degree of harvest continuity that is desired and considered attainable. A well-managed forest has a high degree of flexibility. Except to salvage mortality, trees do not have to be harvested at a particular time as with crop agriculture. An overall objective is to keep forest stands well stocked and growing. More can be cut at one time than at another without upsetting the balance of a well-managed forest, provided the overcuts and undercuts average out reasonably.

Even-aged management is applied, worldwide, in commercial timber production considerably more than is uneven-aged management. The reasons are that species control in regeneration, stand and area treatments, and harvest cuts can be more cheaply and effectively applied uniformly over fairly large areas, as is possible in even-aged management. In uneven-aged management, cutting is much more dispersed over the total management unit, and uniform areas treatment is less practicable. *See* FOREST AND FORESTRY.

[KENNETH P. DAVIS]

Bibliography: K. P. Davis, *Forest Management, Regulation and Valuation*, 2d ed., 1966; W. A. Duerr et al., *Forest Resource Management: Decision-Making Principles and Cases*, 2 vols., 1975.

Forest planting and seeding

The establishment of a new forest by seeding or, more generally, by planting of nursery-grown trees on forest land that fails to restock naturally with adequate numbers of the right species of trees (reforestation), or by such planting on nonforested land (afforestation).

Since artificial reforestation began about 1368 in Europe, it has become an integral part of forest practice on all continents on which trees grow. By 1974 almost 40×10^6 acres (10^6 acres = 4046 km^2) had been planted in the United States. The leading states were Florida (3.7×10^6 acres) and Georgia (3.4×10^6 acres). Other states with more than 2×10^6 acres of forest plantings were Alabama, Louisiana, Mississippi, and Oregon. Probably 30×10^6 acres still require planting or seeding in the United States.

The principal purpose of forest planting is production of wood for such products as lumber, structural timber, pulpwood, posts, and poles. About 1.5×10^6 acres of the total acreage planted to 1974 in the United States was for windbreaks and shelterbelts to protect homes, fields, and feed lots from severe winds. Lesser but increasing amounts of tree planting have been to protect watersheds, reclaim strip-mined areas, provide food and cover for wildlife, and improve the scenic values of road-sides, parks, and other recreational areas (Fig. 1). Tree planting also is being tested for noise abatement along freeways and to give early indication of air pollution in industrial areas. *See* FOREST ECOLOGY; FOREST RESOURCES; SOIL CONSERVATION; WATER CONSERVATION.

Seed-producing areas. Until the 1950s most tree seed for forestry purposes was collected from wild stands or ordinary plantations. Since 1951 there has been an increasing development of seed-production areas (Fig. 2*a*) (high-quality stands or plantations in which the best trees are left for seed production and all poorer trees removed) and seed orchards (Fig. 2*b*) (established with seedlings or grafted trees from selected parents). As of 1970, there were more than 10,000 acres of seed-production areas and almost 7000 acres of seed orchards in the United States, largely in the South. To facilitate control of these improved seeds, by 1974 some 16 states had laws requiring labeling of tree seeds, and 9 states had seed certification standards. As seed from seed orchards becomes more available, more trees from genetically improved seed are being planted.

Seed handling. Reforestation by planting or direct seeding requires an adequate supply of good seed of species and origins well adapted to the planting site.

Collection. To assure high germinability and keeping qualities, seeds should be collected when they are ripe and before they have suffered deterioration on the tree or on the ground. Often the best time to collect is when the first seeds begin to fall naturally, but large-scale operations must begin

Fig. 1. A plantation of white pine established in 1896, one of the oldest in Michigan, photographed in 1952. It has provided wood products and is also providing recreational values. (*U.S. Forest Service*)

Fig. 2. Seed-producing areas. (*a*) Stand of shortleaf pine converted to a seed-production area. Selected trees are given wide growing space so they will yield heavy crops of cones. (*b*) Seed orchard established with grafts of superior loblolly pine. After progeny testing, such an orchard will produce genetically improved tree seed. *(North Carolina State University, Industry Cooperative Tree Improvement Program)*

sooner than that to avoid substantial losses of good seed. The best time for seed gathering varies for each species from season to season and place to place. Seeds of most species are best collected in the fall, but those of the aspens, cottonwoods, most elms, red maple, silver maple, poplars, and willows are collected in the spring. Seeds of cherries, choke cherries, Douglas fir, mulberries, Siberian pea tree, and plums can be gathered in the summer, and those of some ashes, yellow birch, box elder, Osage orange, black spruce, Norway spruce, sycamore, and walnuts can also be collected in the winter.

Forest tree seeds commonly are gathered from standing trees. Sometimes, felled trees provide a cheap source. In the Pacific Northwest cones often are gathered from squirrel hoards.

Seeds of many tree species must be extracted from the fruits, such as cones, berries, or pods, and cleaned to facilitate seed storage and handling.

Storage. Commonly, forest tree seeds must be stored for a few months up to several years. Seeds of many trees can be kept reasonably viable for 5–10 years if stored at subfreezing temperatures. Seeds of most legumes can be stored dry at room temperature for several years.

Pretreatment. Of some 700 species of woody plants studied, 28% have seeds that germinate readily, but 8% have seeds with impermeable coats, 50% have seeds with internal dormancy, and 14% have more than one kind of seed dormancy. The dormant seeds require special pretreatment to induce reasonably prompt germination.

Seed-coat impermeability often can be overcome by treatment with acids or abrasives. Internal dormancy usually requires cold, moist treatment or fall sowing to promote germination. Seeds with more than one kind of dormancy often require either a combination of seed-coat softening and cold, moist treatment, or sowing soon after collection in the late summer or early fall.

Planting conditions and requirements. Except where plant competition is slight and moisture conditions good, special ground preparation usually is needed. On rough, hilly, or mountainous land with considerable grass, herbaceous, or brushy cover, a spot about 18 in. square (1 in. = 2.5 cm) is scalped off with a heavy hoe or mattock just before planting each tree. Where machinery can be operated efficiently, heavy brush plows, disks, bulldozers, root rakes, or shearing blades are used to clear off the brush and expose mineral soil in lanes or over the entire area on which to plant the trees (Fig. 3). On sod an ordinary one-bottom plow or a forestry plow may be used to open parallel

Fig. 3. A poor aspen and shrub area being prepared for planting by an Athens-type disk, which is pulled by a crawler tractor. *(U.S. Forest Service)*

Fig. 4. Tractor-drawn tree-planting machines reforesting sandy old field in Michigan. (*U.S. Forest Service*)

furrows 6 to 10 ft (1 ft = 0.3 m) apart in which trees are planted at 6- to 10-ft intervals; or narrow strips are treated with aminotriazole or similar herbicides that may permit planting without furrowing. On rolling land, furrows are placed on the contour to conserve moisture and reduce soil washing.

Chemical sprays of 2,4-D or 2,4,5-T, alone or in combination, applied by ground equipment or by aircraft, often have been used to eliminate unwanted shrub or tree growth. Similarly, spraying simazine will control grass and herbs. Controlled burns, sometimes in combination with chemical sprays applied afterward, are used to prepare brushy or recently logged areas. *See* HERBICIDE.

Tree size for planting. Trees usually are grown 1–4 years in nurseries so that they have 4- to 12-in. tops and 8- to 10-in. roots before they are field-planted. Some are transplanted once in the nursery before the final planting.

Tree planting season. Most tree planting is done at the beginning of the growing season when the trees are still dormant and can be moved with little damage and when the soil is moist. In most regions such conditions occur in spring, but in the southern United States and other areas of long growing season, planting often is best done in early winter.

Species planting regions. Tree species commonly planted are as follows: northern and northeastern United States—red, eastern white, and jack pines, and white, black, and Norway spruces; southern United States—shortleaf, loblolly, slash, and longleaf pines; western United States—Douglas fir, ponderosa, western white, and sugar pines, and Engelmann and Sitka spruces; Europe—Scotch and black pines, Norway spruce, larches, Douglas fir, and beech; Asia—native or well-adapted species of pine, spruce, fir, cypress, cedar, larch, eucalypts, oak, ash, and other hardwoods; Australia—Monterey and southern pines; South America—Monterey and southern pines, eucalypts, and hybrid poplars; and South Africa—Mexican pines and acacias.

Fig. 5. Jack pine seedlings established by direct seeding on an area that had been burned in northeastern Minnesota. (*U.S. Forest Service*)

Tree planting by machine. Level to moderately rolling land may be planted by special machines largely developed since 1930. Most of them are one-row or two-row units that are pulled by a crawler tractor (Fig. 4) or, on fairly smooth terrain, by a wheel tractor. Two men and a machine can plant 10,000 or more trees per day. A large part of the tree planting in the United States is now done by machine.

Tree planting by hand. On very small areas or where the land is steep, rocky, or stumpy, trees are planted with a mattock, planting hoe, or planting bar in exposed mineral soil. One man can plant about 200–500 trees per day.

Tree planting in containers. In drier regions where planting failures are frequent with bare-rooted stock, and elsewhere to obtain improved seedling survival and growth, trees are sometimes grown (often in greenhouses) in individual or multiple-unit containers made of plastic, paper, clay, or other materials, and are transported to the planting site in the containers. If water can reach the seedling roots and the container material is expected to deteriorate readily, the seedlings are planted with the containers intact, but more often the containers are removed before the seedlings are planted, with their roots holding together a block of soil. Planting usually is done throughout the growing season by hand with special tools, but mechanized planting methods are under development. Major advantages of container planting are: facilities for production of seedlings can be expanded more readily than in bare-root nurseries; seedlings survive and grow better; species sensitive to bare-root handling (hemlocks, true firs, and others) can be grown; the planting season can be extended; production and planting efficiencies are improved (including more efficient use and control of genetically improved seed); and seedlings can be planted with intact root systems. Major disadvantages are that successful production of seedlings in containers requires a higher level of technical knowledge and more constant attention than that required for production of bare-root nursery stock, and the cost per tree is higher and may more than offset reduced planting costs or better survival and early growth. In 1974 about 5% of the seedlings produced in public and industrial forest nurseries in the United States were grown in containers. This almost doubled the amount grown in 1973.

Seeding. Often the nut species such as oak, walnut, and beech are sown directly in the field by dropping the seeds or by drilling them in furrows, prepared terraces, scalped spots, or recently burned-over areas. Pine, spruce, and Douglas fir seeds also have been direct-seeded successfully (Fig. 5), but since the results usually are sufficiently uncertain, planting is more economical in most cases. Of the total area reforested in the United States to 1974, about 9% was direct-seeded. More than two-thirds of this area is in the South and in the Pacific Northwest, where milder climatic conditions, the use of seed treated to protect it from birds and rodents, and large-scale operations with aircraft and helicopters have given reasonably good success at moderate cost.

Plantation care. Tools or chemicals are used to remove overtopping weeds, brush, or inferior trees that frequently retard plantation development. Plantations also may require sprays to protect them against insects or diseases. Some diseases can be controlled by removal of the alternate host plants. Plantations need protection from forest fires. Injury from hares, rabbits, rodents, and larger mammals may be reduced by use of rodenticides, repellents, or fences. *See* FOREST AND FORESTRY; FOREST ECOLOGY; FOREST MANAGEMENT AND ORGANIZATION. [PAUL O. RUDOLF]

Bibliography: Great Plains Agricultural Council, *Proceedings of the North American Containerized Forest Tree Seedling Symposium*, 1974; G. W. Sharpe et al., *Introduction to Forestry*, 1976; D. M. Smith, *The Practice of Silviculture*, 1962; C. H. Stoddard, *Essentials of Forestry Practice*, 1978; U.S. Forest Service, *Seeds of Woody Plants in the United States*, 1974.

Forest resources

Forest resources consist of two separate but closely related parts: the forest land and the trees (timber) on that land. In the United States, forests cover one-third of the total land area of the 50 states, in total, about 740×10^6 acres (10^6 acres = 4046 km^2). The fact that 1 of every 3 acres of the United States is tree-covered makes this land and its condition a matter of importance to every citizen. Recognizing this importance, Congress has charged the U.S. Forest Service with the responsibility of making periodic appraisals of the national timber situation, a charge that was substantially increased in scope by the Forest and Rangeland Renewable Resources Planning Act of 1974 and the National Forest Management Act of 1976. Most of the data used here are taken from the review draft of the 1978 edition of *Forest Statistics of the U.S., 1977*. A new assessment of the timber situation was scheduled for publication in 1980.

Some 488×10^6 acres, or about two-thirds of the total forest area in the United States, is classified as commercial forest land, that is, forest land capable of producing at least 20 ft^3 (10 ft^3 = 0.28 m^3) of industrial wood per acre per year, and not reserved for uses incompatible with timber production. The 252×10^6 acres of noncommercial forest includes about 24×10^6 acres which meets the growth criteria for commercial timberland but has been set aside for parks, wilderness areas, and such. The remaining 228×10^6 acres is incapable of producing a sustained crop of industrial wood, but is valuable for watershed production, grazing, and recreation use. Table 1 gives details of the distribution of forest land in the four main regions of the United States.

Most of the noncommercial forests are in public ownership, including approximately 19.5×10^6 productive forest acres legally withdrawn for such uses as national parks, state parks, and national forest wilderness areas. Another 4.6×10^6 acres, classed as "productive deferred," is under study for possible inclusion in the wilderness system. Of the remaining noncommercial forest, about 108×10^6 acres are in Alaska (part of the Pacific Coast region). Most discussions of forest resources, however, concentrate upon the commercial areas, and these lands are the primary concern of the following discussion.

The South alone has about 38.6% of the total commercial area; 35% is in the North, and the remaining 26.4% in the West. Within this general pattern are very large differences. In Maine, for example, 81% of the land surface is covered with commercial forests, while North Dakota is at the opposite extreme, with only 1% of its area similarly utilized.

For many years, changes in United States agriculture led to abandonment of marginal farms, which rapidly reverted to forest. This "new forest" was more than enough to offset those areas lost to highways, pipelines, urban development, and such. In fact, between 1943 and 1963 the total commercial forest area increased by about 31.6×10^6 acres. Between 1963 and 1970, however, the total area of commercial forest land declined about 8.4×10^6 acres, mostly in the South and in the Rocky Mountains, and by 1977 a further reduction of 12.2×10^6 acres had occurred. Some of the early reduction in the West was the result of shifting public forest land to a reserved or deferred class to meet demands for recreation uses. A later reclassification, however, restored 3.2×10^6 acres in the West to commercial status. In all regions, substan-

Table 1. Distribution of forest land in the United States*

	Area, 10^6 acres†				
Type of forest land	Total‡	Pacific Coast§	North	South	Rocky Mountains
Commercial forest	487.7	70.8	170.8	188.4	57.8
Noncommercial forest	—	—	—	—	—
Productive reserved	19.5	4.1	5.1	1.9	8.4
Productive deferred	4.6	1.2	.2	.1	3.2
Other forest	228.3	138.4	4.8	16.7	68.4
Total Noncommercial	252.4	143.7	10.1	18.7	80.0
Total forest land	740.1	214.5	180.8	207.1	137.7

*From *U.S. Forest Service, Forest Statistics of the U.S., 1977*, USDA, review draft, 1978.
†10^6 acres = 4046 km^2.
‡Totals may not agree in last figure due to rounding.
§Pacific Coast includes both Alaska and Hawaii.

tial areas have been taken over by suburban development, highways, reserves, and other nontimber uses, while in the South (where much of the early decline occurred) timberland has been cleared for crop production and for pasture.

Forest types. There are literally hundreds of tree species used for commercial purposes in the United States. The most general distinction made is that between softwoods (the conifers, or cone-bearing trees, such as pine, fir, and spruce) and hardwoods (the broad-leafed trees, such as maple, birch, oak, hickory, and aspen). Viewed nationally, 51% of United States commercial forest land is occupied by eastern hardwood types. Softwoods of various kinds make up 42%, western hardwoods only 3%; and 4% of the area is unstocked. Oak-hickory stands cover the largest area, accounting for 23% of all commercial timberland in 1977. The oak-pine type (14% of the eastern hardwood area) is mostly in the South and is primarily the residual resulting from the cutting of merchantable pine from mixed pine-hardwood forests. During the last few decades many of these stands have been converted to pine by the killing or cutting of hardwoods and, often, by the planting of pine, although little change, percentage-wise, has occurred since 1970.

Of the eastern hardwood forests, 44% are oak-hickory types, containing a large number of species but characterized by the presence of one or more species of oak or hickory. Other important eastern types are the maple-birch-beech (found throughout the New England, Middle Atlantic, and Lake states regions), the oak-gum-cypress forests (primarily in the Mississippi Delta and other southern river bottoms), and the aspen-birch type of the Lake states (relatively short-lived species that followed logging and fires). The bottomland hardwoods (oak-gum-cypress type) were reduced about 20% between 1962 and 1970, primarily by the clearing of forests for agriculture. For many years these forests have supplied much of the quality hardwoods in the United States.

Softwood types dominate the western forests, altogether occupying 83% of the region's commercial forest area. Douglas fir and ponderosa pine, the principal types, together constitute 45% of the region's commercial timberland. The western softwood types are the principal sources of lumber and plywood in the United States. Nearly all the commercial forest area of coastal Alaska is of the hemlock–Sitka spruce type. Hardwoods, mostly in Washington and Oregon, occupy only 12% of the West's commercial forest area, but have increased substantially since 1962 as the Douglas fir forests have been cut.

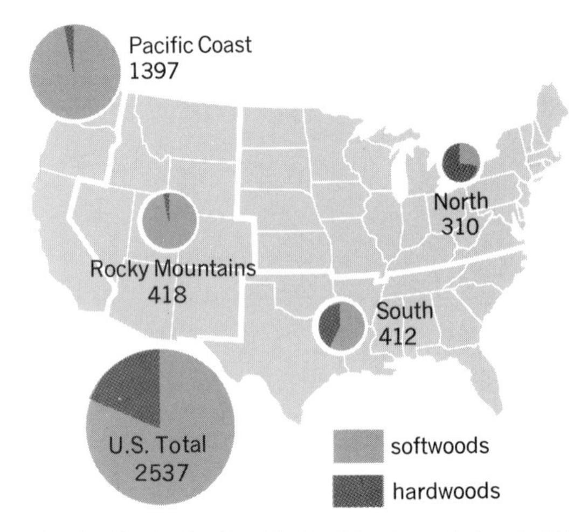

Fig. 1. Map indicating distribution of sawtimber in billions of board feet for all species as of Jan. 1, 1963. (*U.S. Department of Agriculture*)

Growth. Growth on these areas has been more than enough to match the harvest since the mid-1960s. This is commendable, but the fact that growth and drain are in approximate balance provides no assurance that all is well. Much of the growth is still on low-quality hardwoods in the East, while about two-thirds of United States demand for the raw materials of the forest is met by the softwood production of the West and South. Potentially the most productive forest land is that of the Pacific Coast states, where it is estimated that 25×10^6 acres are capable of growing more than 120 ft³/acre/year, and that another 22×10^6 acres could produce more than 85 ft³ acre/year. The next most productive area is the South, with only 12×10^6 acres of highest-quality forest (more than 120 ft³ acre/year) but some 46×10^6 acres of good-quality land (85–120 ft³/acre/year), and 101×10^6 acres producing 50–85 ft³/acre per year. The North is considerably less well off as far as forest growth rates are concerned.

Growth per acre on forests of the United States has been rising steadily—in all regions and on all ownerships. The increase has been particularly dramatic over the past quarter-century: since 1952, average per-acre growth has increased from 28 to 45 ft.³ This 61% rise is due at least in part to efforts made by American forestry on both public and private lands. But there is still far to go: net growth per acre is far below that possible under intensive management, and only three-fifths of what could be achieved from fully stocked natural stands.

Timber inventory. Only soil productivity exceeds stocking (the number of trees per acre) and the age of the trees in importance as factors determining timber production.

Stocking. The commercial forests contained a truly vast amount of sound wood—$800{,}803 \times 10^6$ ft³ ($22{,}676 \times 10^6$m³) at the beginning of 1977. Only 8.4% of this consisted of trees that were dead, diseased, or in such poor condition they were not commercially useful. Another 2.6% consisted of trees which were dead but still usable for timber. Some 63% of the total volume was in trees sufficiently large to yield at least one sawlog: such trees are called sawtimber (Fig. 1). Another 25% was in pole timber—trees ranging from 5 in. (13 cm) in diameter (breast high) up to sawtimber size. Softwoods accounted for 64% of the total growing stock (pole timber plus sawtimber) and for 73% of the portion in sawtimber sizes, with 76% of the latter in the western states. This distribution highlights an important facet of timber distribution: the western states have only 26% of the commercial forest land and 23% of the total growing stock of the nation. But these western forests have over 75% of the important softwood sawtimber, and 69% of the nation's softwood growing stock (Table 2). The East, on the other hand, has 91% of the total hardwood growing stock (about equally divided between the North and South) and 89% of the hardwood sawtimber volume.

Age. Age is the other factor that must be considered before the significance of growth (and removals) can be understood. The virgin softwood stands of the West are growing relatively slowly, but as these old stands are replaced with young, vigorous trees, growth on western forests will begin to balance the cut—although almost surely at a reduced level. In 1976, net growth of western softwood growing stock averaged 90% of softwood removals, but removals of sawtimber exceeded growth by fully 45% (nearly 10^{10} bd ft or 2.36×10^7 m^3). Harvest of western sawtimber cannot be maintained at present levels unless forest management is greatly intensified. In the East, however, net growth of softwood growing stock exceeded removals by 52%, while softwood sawtimber growth was 33% above removals. Net growth of all eastern hardwood growing stock was fully 215% above removals, and growth in sawtimber sizes was 159% of removals. Here it should be noted that hardwood harvesting is concentrated in high-valued species such as walnut, cherry, gum, yellow birch, and maple, and in trees of large diameter, while much of the growth is in trees of less preferred species, and of small size. Annual growth and removal data are summarized in Table 2.

Ownership. The future condition of forest land in the United States is dependent to a very large extent on the decisions of the people who own these areas. The key factors in understanding the forest situation, therefore, are the forest land ownership pattern and the attitude these owners have toward forest management and hence toward the future of the forest lands they hold.

About 72% of the commercial forest is held in private ownership (351.1×10^6 acres). The remaining 28% is held by various public owners, with about 18% in national forests, 2% in other Federal ownership, 5% in state holdings, 2% under county and municipal control, and slightly over 1% under native-American sovereignty. Among private owners, three major classes can be distinguished: industrial, farm, and a very heterogeneous group labeled miscellaneous. About 14% of the commercial forest land is in the hands of industry, the remainder being divided between farmers (24%) and miscellaneous private owners (34%). Table 3 gives details about forest ownership, as well as the distribution of privately owned forests, in the four major regions of the United States.

Private owners. Some of the most productive forest land of the United States is in private industrial holdings. Pulp and paper companies lead the forest-based industries, with much of their forest land concentrated in the southern states, where nearly 36×10^6 acres is industrially owned. As forest industries become integrated, it is increasingly difficult to make distinctions between pulp and paper, lumber, and plywood companies, and certainly the distinctions are much less meaningful than in former years.

The miscellaneous private category, which holds 34% of the privately owned forest land, consists of a tremendous variety of individuals and groups, ranging from housewives to mining companies. Relatively few of these owners are holding this land for commercial timber production. Some owners, such as railroad companies or oil corporations, may indeed be interested in producing timber while holding the subsurface mineral rights, but miscellaneous private owners are usually interested in other, nontimber, objectives.

Most farm forests have been cut over several times; and, because farmers are primarily interested in the production of other kinds of crops, they tend to be less concerned with the condition of their woodlands than are other owners. The condition of farm-owned woodlands has been a source of much disappointment, discussion, and considerable action on the part of public conservation agencies. Whether this situation is of crucial significance to the production of forest products in the United States is a matter of dispute. In past years

Table 2. Summary of net annual growth and removals of growing stock and sawtimber, by species group and region*

	Growing stock, 10^9 ft^3†			Sawtimber, 10^9 bd ft‡		
Section and species group	Net growth	Removals	Growth-removal ratio	Net growth	Removals	Growth-removal ratio
North						
Softwoods	1.6	0.7	2.3	4.0	2.2	1.8
Hardwoods	3.9	1.9	2.1	8.7	5.7	1.5
South						
Softwoods	6.3	4.5	1.4	24.2	19.0	1.3
Hardwoods	4.9	2.2	2.2	13.6	8.3	1.6
Rocky Mountains						
Softwoods	1.6	0.8	1.9	6.4	4.8	1.3
Hardwoods	0.1	—	18.9	0.4	0.1	3.7
Pacific Coast						
Softwoods	2.9	4.2	0.7	14.6	25.7	0.6
Hardwoods	0.5	0.1	4.3	1.7	0.4	3.9
United States						
Softwoods	12.4	10.2	1.2	49.2	51.7	0.9
Hardwoods	9.5	4.3	2.2	24.4	14.5	1.7

*From U.S. Forest Service, *An Assessment of the Forest and Range Land Situation in the United States*, USDA, review draft, 1978.

†10^9 ft^3 = 28,316,846 m^3.

‡10^9 bd ft = 2,359,737 m^3, as nominal recovered lumber.

Table 3. Area of commercial timberland in the United States by type of ownership and section, Jan. 1, 1977*

	Total United States		Region			
	Area, 10^3 acres	Proportion %	North, 10^3 acres	South, 10^3 acres	Rocky Mountains, 10^3 acres	Pacific Coast, 10^3 acres
National forests	89,007	18.3	10,121	10.955	36,436	31,496
Bureau of Land Management	5,799	1.2	15	3	1,667	4,113
Other	4,849	1.0	1,079	3,343	75	352
All Federal	99,655	20.4	11,215	14,300	38,179	35,961
Native-American	6,089	1.3	1,028	185	2,711	2,165
State	23,642	4.9	13,129	2,519	2,203	5,791
County and municipal	7,216	1.5	5,945	738	75	458
All public	136,602	28.0	31,318	17,742	43,167	44,374
Forest industry	67,976	13.9	17,777	35,754	2,096	12,349
Farm	116,785	23.9	45,384	57,217	8,311	5,872
Miscellanous private	166,364	34.1	76,290	77,720	4,191	8,163
All private	351,124	72.0	139,451	170,691	14,598	26,384
All ownerships	487,726	100.0	170,769	188,433	57,765	70,758

*From U.S. Forest Service, *Forest Statistics of the U.S., 1977*, USDA, review draft, 1978.

some professional foresters have argued that these farm woodlands should somehow be made to contribute their full share to the timber supply of the nation. Others, of a more economic persuasion, have argued that, as long as farmers have opportunities for investing time and money in ways that will yield a greater return, they should not be expected to worry about timber production. Farm owners are often unfamiliar with forest practices, usually lack the capital required for long-term investments, and in many cases are simply not interested in growing trees.

A recent study, however, indicated that when compared with other forest land of the same quality and within the same region, farm woodlot management levels were only slightly, if at all, below that given other ownership categories.

Public owners. Public agencies of several kinds hold large forest areas, the most important being the national forests. These contain 89×10^6 acres of commercial forest land which is managed and administered by the U.S. Forest Service, a bureau of the U.S. Department of Agriculture. The Bureau of Land Management oversees about 5.8×10^6 acres. Various other agencies, especially those of the armed services, administer 4.85×10^6 acres under Federal supervision. State ownerships total 23.6×10^6 acres, and counties and municipalities control another 7.22×10^6 acres. The remainder, somewhat over 6×10^6 acres, is under native-American sovereignty. Most of the public holdings are managed under multiple-use principles. As on most forest land, wood has always been the principal product of the national forests. With the passage of the 1974 Planning Act and the 1976 National Forest Management Act, the nontimber uses were given increased official attention, and the debate between those favoring timber production and those primarily concerned with range, water, wildlife, and recreation has been much intensified.

Public holdings now contain 54% of the softwood growing stock but only 17% of the hardwoods. Sixty four percent of all softwood sawtimber is publicly owned, and most is in national forests. The high concentration of sawtimber in these areas makes many wood-based industries in the United States highly dependent upon government owned raw material supplies.

World forest resources. Since much of the world's forested area has yet to be surveyed, data concerning volume, species, or even area undoubtedly contain substantial errors. Recent advances in remote sensing, particularly the information being relayed from satellites, give promise of vast improvement in knowledge of how the world's forests are faring. Currently, about 7.5×10^9 acres have a 20% or more tree crown cover—roughly one third of the world's land surface and about the same average percentage as that for the United States as a whole. Latin America and the tropical regions of Africa and Southeast Asia have most of the hardwood forests, while the softwood areas are concentrated in North America and the Soviet Union.

Latin America has about 43% of the world's hardwood growing stock, and this is nearly half again as much as the hardwood forested area of Africa—the continent next richest in hardwood. Together, Latin America and Africa account for 72% of the world's hardwoods, with Southeast Asia adding another 15%. North America has only 5% of the total. In the important softwood category, however, the Soviet Union and North America come in first and second, and together account for 83% of the world's softwoods. The Soviet Union has by far the largest share—about 2½ times as much as North America. Because of low productivity, great distance from markets, and rugged terrain, much of the softwood in the Soviet Union will be difficult to bring into commercial use.

Forests reflect differences in soil, climate, situation, and past land use and merely listing the forest types of the world would take several pages. Yet there is value in distinguishing very generalized forest types by broad locational patterns (Fig. 2).

Coniferous and temperate mixed forests. In Europe, the Soviet Union, North America, and Japan, the forests are predominantly coniferous, a fact that has been of great significance in shaping the pattern and nature of wood use in the industrialized part of the world. Closely associated with the industrialized nations are the temperate mixed

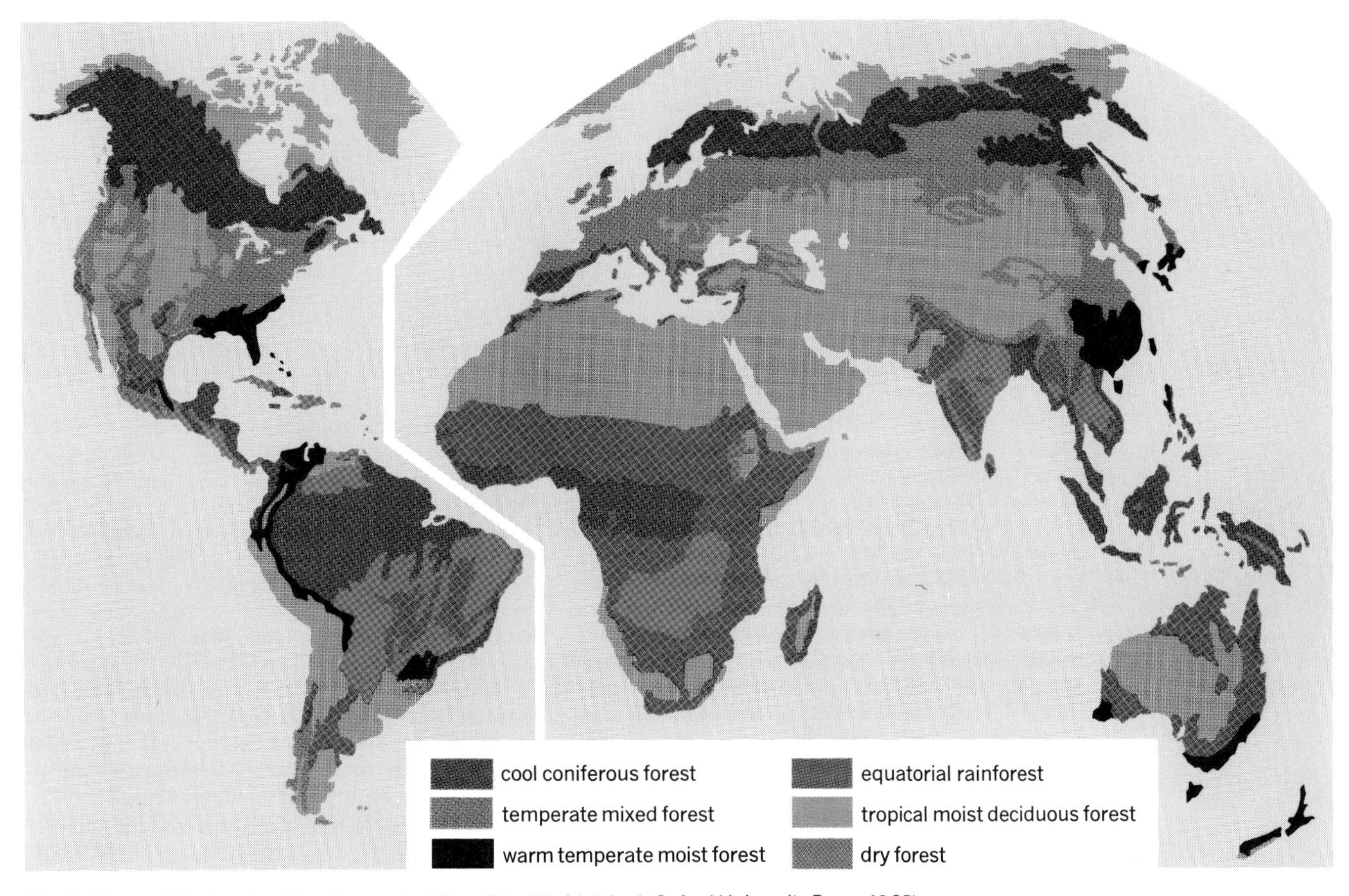

Fig. 2. The world's forests. *(From Economic Atlas of the World, 3d ed., Oxford University Press, 1965)*

forests, which normally contain a high proportion of conifers along with a few broad-leafed species.

Most of the temperate mixed forests are in use, for these heavily populated areas have well-developed transportation systems. Growth rates in both the coniferous and temperate mixed forests of the North Temperate Zone are similar to those already given for the United States, which is an excellent example of the general region.

Tropical rain forests. These forests are made up exclusively of broad-leafed species and include the bulk of the volume of the world's broad-leafed woods. They are concentrated in and around the Amazon Basin in South America, in western and west-central Africa, and in Southeast Asia. Generally characterized by sparse population and only slight industrial development, these forests have been little used. The fact that they typically contain many species within a small area (as many as a hundred species per acre) has also served to limit their use, even though tropical rain forests include some of the most valuable of all woods, such as mahogany, cedar, and greenheart (South America); okoume, obeche, lima, and African mahogany (Africa); and rosewood and teak (Asia).

The rate of growth in the tropics can be very high, and someday it may be possible to obtain a major part of the world's wood fiber needs from the more than 2×10^9 acres of these forests. Today, however, so little is known about many of the species, and so few areas are under any form of systematic management, that little reliance can be placed on these vast areas to satisfy the future wood needs of humanity.

Savanna. Most of the other forests in the tropics and subtropics are dry, open woodlands, or savannas. These forests contain low volumes per acre, mostly in small sizes, and only a few of the species are of commercial value. Much of Africa (excluding western and west-central Africa) is of this low-yielding type.

Management. In recent years intensive afforestation has received considerable attention. The use of measures such as careful soil preparation prior to planting, application of fertilizer, and even irrigation to adapt the environment to high-yielding species can result in an enormous increase in returns from forests. For example, yields of 400–500 ft^3/acre are common with eucalyptus in both South America and Africa, or with poplars in southern Europe. Only slightly smaller yields are obtained from the fast-growing pines under similarly intensive management. The high-yield potential of cultivated forests makes them much more important than their area might indicate. According to reports, most nations intend to have these forests make an even greater contribution in the future.

While fast-growing plantations could add a new component to the world's industrial wood supply, a major threat to virtually all forests in developing nations stems from the pressing need for fuelwood. Expanding populations in Africa, Latin America, and Southeast Asia are destroying forests at a truly alarming rate as increasing numbers seek the means to cook their food and warm their families. Rising petroleum prices have increased this pressure, and there is a very real danger that in some areas the deforestation could lead to ecological changes that would be difficult or impossible to reverse. Forest removal from hill regions and drought-prone areas has far-reaching consequences, since soil and wind erosion and drastic changes in stream flow are almost inevitable. These, in turn, threaten food supplies for major segments of the world's population. *See* FOREST AND FORESTRY; FOREST MANAGEMENT AND ORGANIZATION. [G. ROBINSON GREGORY]

Bibliography: S. L. Pringle, Tropical moist forests in world demand, supply and trade, *Unasylva*, vol. 28, no. 112–113, 1976, *An Assessment of the Forest and Range Land Situation in the United States*, USDA, review draft, 1979. *U.S. Forest Service*, *Forest Statistics of the U.S., 1977*, USDA, review draft, 1978.

Forest soil

The natural medium for growth of roots of trees and associated forest vegetation. This relationship with forest vegetation gives rise to characteristics that distinguish forest soils from soils formed under other vegetation systems. The most obvious feature is the humus layer, or horizon, which is peculiar to the microenvironment imposed by the forest (Fig. 1). The humus horizon affects germination of seeds, influences soil moisture distribution, serves as a reservoir of nutrient elements, influences the susceptibility of forest stands to fire, and represents the energy source for most soil-inhabiting organisms.

The process of humification starts when residues of forest plants and animals fall to the soil surface and are gradually decomposed. The course and extent of decomposition, which determines humus characteristics, depends chiefly on the microclimate, tree species, associated flora and fauna, and the nature of the underlying mineral material. Organic matter may accumulate on the soil surface with little or no mixing with mineral material. This form is called mor, raw humus, or ectohumus. In the mull form, most of the organic matter is incorporated into the mineral soil, and only coarse debris, such as twigs and petioles, remains on the surface (Fig. 2). Transitional types combining characteristics of both mor and mull also occur. Mor humus is usually associated with a cool, moist microclimate and predominantly coniferous or ericaceous vegetation growing on acidic parent material. In these situations, the animal life generally responsible for mixing organic matter with mineral soil is lacking and the population of microflora tends to be dominated by fungi.

Leaf litter associated with mull soils is usually richer in calcium and other nutrient elements and hence supports a much larger variety of small animals and microorganisms than is found in mor layers. Metabolic activity of these organisms is stimulated by somewhat higher surface soil temperatures so that litter decomposition proceeds more rapidly. Earthworms, millipedes, and similar animals mechanically break down litter and incorporate it into mineral soil. Simultaneously, the litter is inoculated with a great host of fungi and bacteria which gradually converts the organic matter into relatively stable complexes of lignin and protein. These processes darken the upper portion of the mineral soil.

Humus relations. Factors that govern humus formation are extremely complex, and their relative importance is poorly understood. A more definite relationship exists between humus characteristics and the horizons that occur deeper in the soil profile. Many of the biochemical reactions responsible for profile development originate in the humus layer, and by-products of these reactions are transmitted downward by percolating water. Mor humus is often associated with pronounced weathering of primary minerals and leaching of the breakdown products, particularly iron, to positions lower in the profile, whereas in mull soils the

Fig. 1. Soil under an uncut forest is friable and porous, and it is deeply penetrated by roots and infiltrated with organic matter. (*USDA*)

Fig. 2. Forest cover returns large quantities of organic matter to the soil each year. The depression in the center shows the depth of litter. (*USDA*)

profile transitions may be less distinct. Because of the delicate balances that function in organic matter decomposition, the humus layer responds rather quickly to changes in the microenvironment that may be caused by modification of the forest stand by wind damage, fire, or silvicultural operations. By contrast, deeper soil horizons are relatively stable and usually will show little change unless the soils are mechanically disturbed, as when trees are uprooted.

Root relations. The humus layer is the focal point of biological activity in forest soils. In addition to litter-decomposing animals and microbes, it usually contains roots of forest vegetation. Portions of the root systems of many trees and other woody plants exhibit modifications called mycorrhizae (Fig. 3), which result from invasion of the roots by certain fungi. The fungus is apparently seeking soluble carbohydrates transported from the leaves to the younger portions of the roots, and the root thus altered is evidently more efficient in absorbing water and nutrients to be used by the tree.

Because the supply of moisture and nutrients is generally favorable, most of the absorbing roots of trees and other forest vegetation are located in and immediately below the humus layer. This arrangement allows maximum utilization of nutrient elements released by decomposition of organic material, and although the root network and its associated microorganisms are extremely efficient in nutrient recovery, some leaching loss does occur. Such loss is balanced by rhizospheric weathering of rocks and minerals by roots occupying deeper soil layers. Forest soil depth is determined by root penetration which may vary from a few inches to more than 50 ft. Trees usually utilize a much larger volume of soil to maintain growth than do agricultural crops.

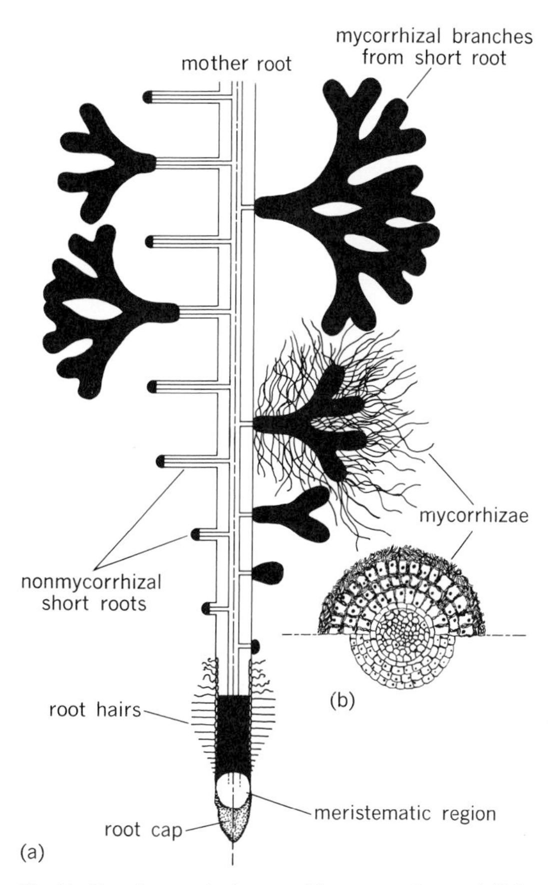

Fig. 3. Development of mycorrhizae on a pine root. Solid areas indicate absorbing surfaces. (*a*) The main axis, a mother root; (*b*) cross section, representing a mycorrhizal root. (*From P. J. Kramer, Plant and Soil Water Relationships, McGraw-Hill, 1949*)

Classification. Forest soils may be classified in various ways. The classification of the Cooperative Survey of the Soil Conservation Service is sometimes used, but since this system is based on soil features important to agriculture, it usually must be modified for forestry purposes. Forest composition, minor vegetation, parent material, topography, depth to groundwater, total soil depth, and growth and yield of forest stands all have been used, individually and in various combinations in classifying forest soils. Classifications based on forest productivity have been found to be fairly successful. In such systems, tree growth measured by height, diameter, or volume is related to soil properties, texture, structure, aeration, water infiltration, moisture-retention capacity, depth, type and distribution of organic-matter nutrient supply, and other characteristics known to influence growth of the root systems. These relationships are then used to predict potential productivity of understocked or deforested lands with similar soil characteristics. Because of the large areas involved, aerial photographs are frequently used as a mapping base on which forest cover types and soil boundaries are delineated.

Hydrologic cycle relations. In addition to their role in producing cellulose, timber, and other forest benefits, forest soils perform an important regulatory function in the hydrologic cycle. Undisturbed or well-managed forested watersheds are often characterized by soils with a high water-infiltration capacity. Such soils are usually well protected from the destructive energy of falling raindrops that cause surface sealing and promote overland flow. Excessive overland flow results in soil erosion and reduction in water quality. Where falling rainwater or meltwater from snow has maximum opportunity to penetrate into the soil, normal drainage channels such as streams and rivers are recharged gradually through underground flow, and a steady supply of high-quality water is assured. Under certain types of vegetation, burning the litter may cause vaporized organic compounds to move downward in the soil profile where they are condensed to form a water-repellent zone. If this condition is severe, as is frequently the case in the chaparral region of California, downward movement of water is retarded. The resulting increase in overland flow usually causes severe soil erosion. See HYDROLOGY.

Nutrient cycling. Much research has been done on the circulation of nutrient elements in forest ecosystems. In this concept movement of chemical elements occurs from minerals in the soil or from atmospheric sources through the root system into various tissues of forest vegetation. These tissues are utilized by consumer organisms or are ultimately deposited on the forest floor where decomposition occurs, releasing the elements for reabsorption by the roots. It has been demonstrated that in undisturbed or in well-managed forests there is little leakage from the nutrient cycles. When the forest undergoes drastic disturbance, however, as from a fire or in large-scale clearcutting, loss of nutrients from the forest ecosystem may be quite marked, and in extreme cases may cause drastic changes in downstream ecosystems. Another aspect which has arisen from the study of nutrient cycling has been the concept of limiting elements in the growth of forest vegetation. The growth rate of natural stands generally adapts itself to the availability of nutrients in the soil that supports it. In commercial forestry, however, the use of plantations is frequently prescribed for more intensive utilization of the productive capacity of the forest site. In these situations a particular element may be limiting tree growth, and it may become necessary to add chemical fertilizer to the soils supporting these trees. In intensively managed forests, encouraging results from use of fertilizers, particularly nitrogen and phosphorus, have been observed. *See* SOIL; SOIL MICROBIOLOGY.

[GARTH K. VOIGT]

Fresh-water ecosystem

An ecostystem is the functional unit of ecology, consisting of living organisms and the nonliving environment interacting upon each other to produce an exchange of energy and materials between the living and nonliving components. A lake or pond in its entirety is such an ecosystem. A stream is also an ecosystem, although less well differentiated. Limnology is the comprehensive study of all the components of inland aquatic ecosystems and their interrelationships. *See* ECOLOGY; LIMNOLOGY.

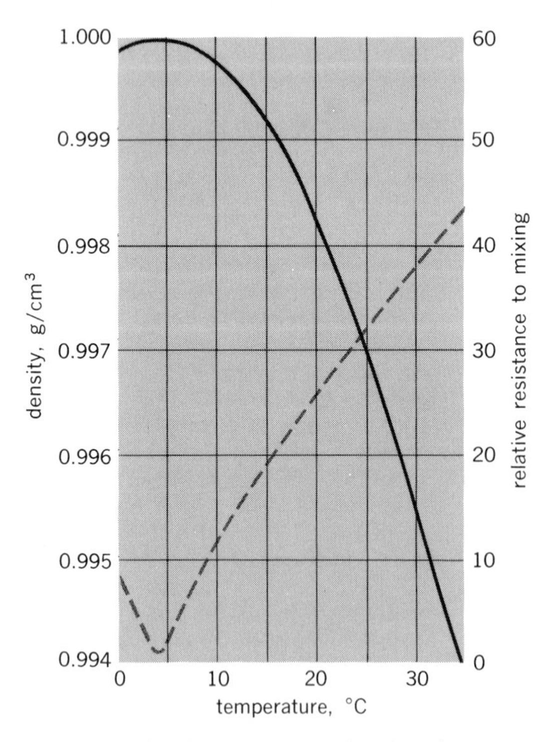

Fig. 1. Density of pure water as a function of temperature (solid line). Maximum density is at 3.94°C. The dashed line shows the relative resistance to mixing per degree difference in temperature in a water column 1 m long. Resistance to mixing between 4 and 5° is taken as unity. (*From R. C. Weast, ed., Handbook of Chemistry and Physics, 48th ed., Chemical Rubber Publications, 1967*)

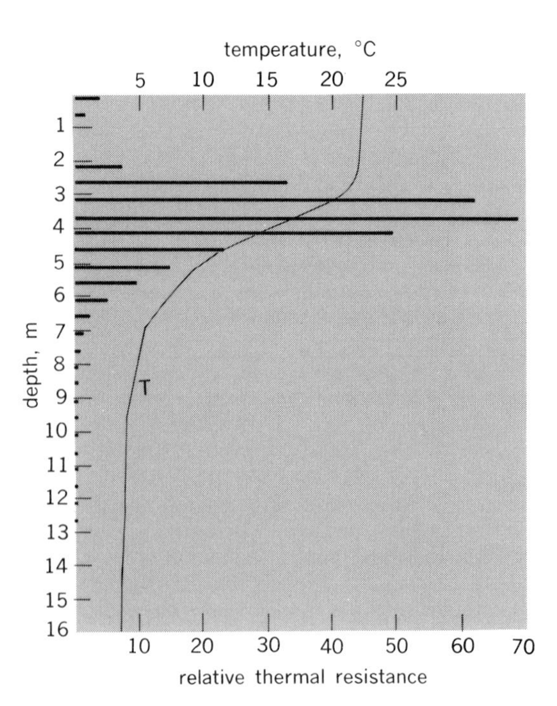

Fig. 2. A summer temperature profile (continuous curve) and relative thermal resistance to mixing (bars) for Little Round Lake, Ontario. The great thermal resistance to mixing in the metalimnion indicates a great stability. (*From J. R. Vallentyne, Principles of modern limnology, Amer. Sci., 45(3):218–244, 1957*)

The energy of fresh-water ecosystems is derived mainly from photosynthesis accomplished by the algae suspended in the water and by the higher plants and algae growing on or in the bottom. A variable proportion of the total energy available is derived from allochthonous organic matter, such as leaves and pollen, produced by terrestrial communities. The materials cycled within an ecosystem are derived ultimately from the weathering of rocks and the leaching of soil in the watershed, and to a lesser extent from the air. *See* ECOLOGICAL SYSTEMS, ENERGY IN.

Fresh-water habitats are conveniently divided into a lenitic or basin series, such as lakes, reservoirs, ponds, and bogs, and a lotic or channel series, such as rivers, streams, brooks, springs, and groundwater. The lotic series is distinguished by a continual flow of water in one direction. The limnology of the lenitic series of habitats is better understood because of the greater amount of research done on it.

LENITIC ECOSYSTEMS

The biota and biological processes of lakes and ponds operate within a framework of physical factors imposed by the external climatic environment, of which the most important are temperature stratification, light stratification, and waves and currents.

Temperature stratification. Heating of natural waters occurs mainly by the direct and rapid absorption of infrared radiation of wavelength > 0.76 μm. Because 1 m of distilled water absorbs 90% of this radiation, the direct heating of water is

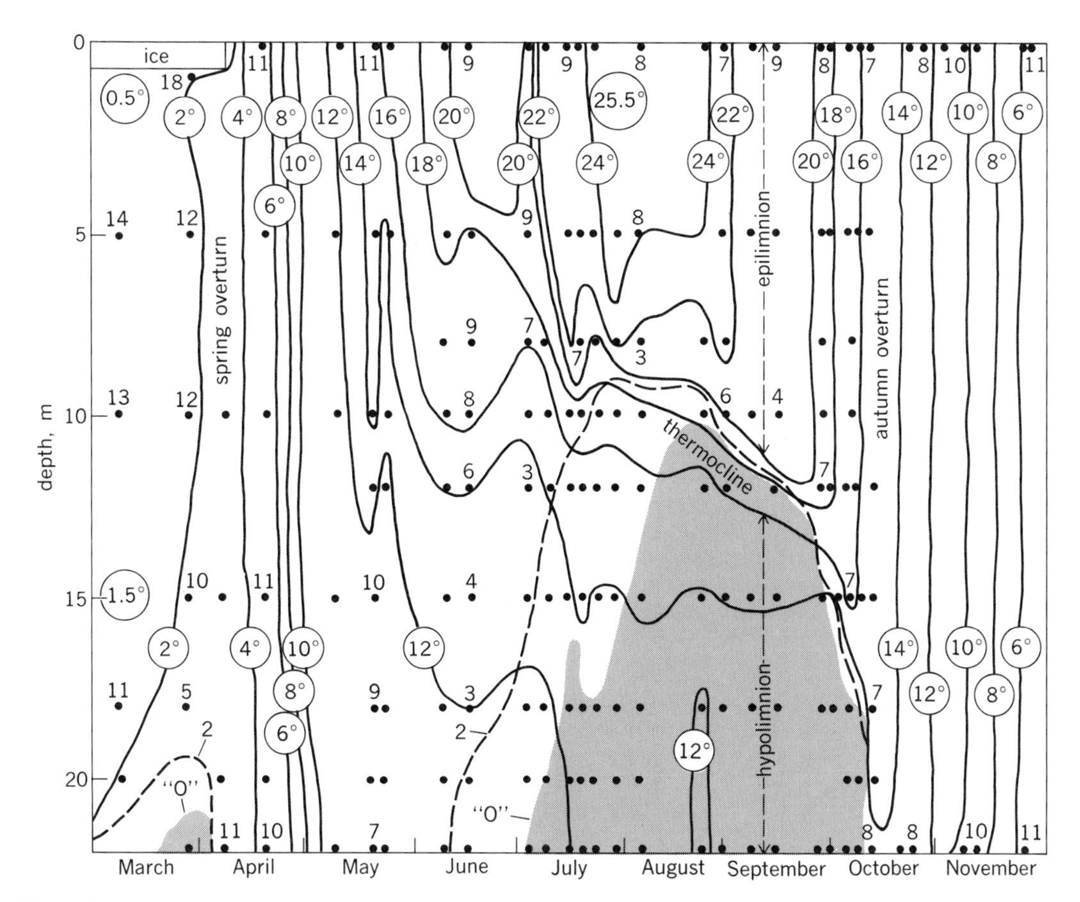

Fig. 3. Seasonal changes in water temperature and dissolved oxygen during 1906 in Lake Mendota, Wis. The figures that have no circles around them are parts per million (ppm) of oxygen. The dashed lines represent 2 ppm. The tinted area represents less than 0.2 ppm. *(After C. H. Mortimer, from G. C. Sellery, E. A. Birge, A Memoir, University of Wisconsin Press, 1956)*

important only in a relatively thin surface layer. Heat at greater depths has been transported there largely by wind-generated currents and turbulence. All heat in a lake in excess of 4°C is commonly referred to as wind-distributed heat.

Two water masses of different temperatures offer a resistance to mixing by the wind which is proportional to the difference in density between them. Because of the nature of the temperature-density relationship of water, the resistance to mixing is greater at high than at low temperatures (Fig. 1). As a consequence, in most lakes of sufficient depth, 10 m or greater, the wind is unable to circulate the entire lake in summer. Therefore, the lake becomes stratified (Fig. 2) into an upper, warm, freely circulating zone, the epilimnion; a lower, cold, relatively noncirculating zone, the hypolimnion; and an intermediate zone of rapid temperature change, the metalimnion, or thermocline in North American usage.

The metalimnion effectively isolates the hypolimnion from surface processes, and as a result biological and biochemical processes occurring in deep water are cumulative during the period of stratification. This is one of the most important interactions of the lenitic ecosystem.

Depth of the metalimnion varies with the size and shape of the lake, the degree of protection from wind action, and the wind regime during the warming phase of the annual cycle.

In temperate lakes there is typically a spring overturn, during which the lake is actively circulating from top to bottom; a period of summer stratification as described above; a period of fall overturn after the lake, by cooling, has again become homothermal; and, especially under ice cover, a period of winter stratification (Fig. 3). Winter stratification is often spoken of as an inverse stratification, because the warmest water is at the bottom rather than the top. A lake not covered with ice may continue to circulate throughout the winter.

Lake classifications. Based on these temperature relationships, an early system of lake classification was devised by F. A. Forel and later modified by G. C. Whipple. All lakes were divided into three classes, polar, temperate, and tropical. This classification depended on the annual range of surface temperature, and each class was further subdivided into three orders based on the annual range in bottom temperatures. The first, second, and third orders are, respectively, the deep, moderate, and shallow lakes. One of the chief objections to the terminology employed in this system is that some of the best examples of tropical lakes are in Scotland and British Columbia. In these lakes, the surface temperature is always above 4°C.

A more realistic system proposed by G. E. Hutchinson and H. Löffler (Fig. 4) bases the classification on the number of periods of circulation a year, and whether the circulation occurs af-

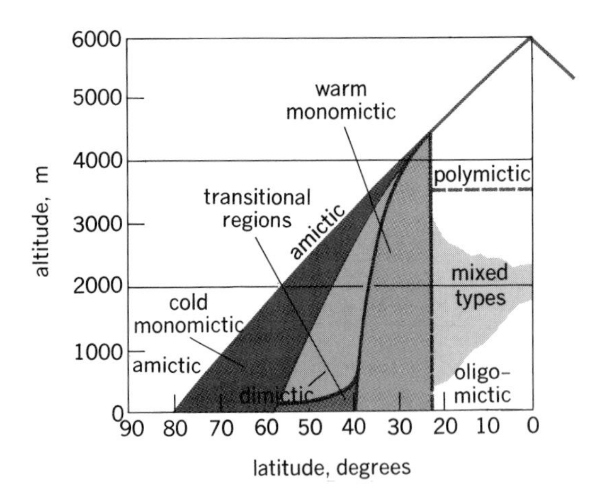

Fig. 4. The Hutchinson-Löffler system of thermal lake types. The two equatorial types occupy the unshaded areas which are labeled oligomictic and polymictic; they are separated by a region of mixed types, mainly variants of the warm monomictic type. (*From G. E. Hutchinson and H. Löffler, The thermal classification of lakes, Proc. Nat. Acad. Sci., 42(2):84–86, 1956*)

ter warming or cooling of the lake. Dimictic lakes circulate twice a year, warm monomictic circulate once a year after a period of cooling, and cold monomictic lakes circulate once a year after a period of warming. These correspond, respectively, to the temperate, tropical, and polar lakes of the Forel-Whipple system. In addition there are the permanently frozen amictic lakes, low-altitude tropical oligomictic lakes with irregular circulation, and high-altitude tropical polymictic lakes with continuous circulation.

During turnover, most lakes circulate completely from top to bottom. Such lakes are called holomictic. In a relatively small number of lakes the water nearest the bottom has a sufficiently greater density from dissolved or possibly suspended substances so that it does not become involved in the circulation. Such lakes are called meromictic. In the noncirculating layer, or monimolimnion, of meromictic lakes the same processes occur as in the hypolimnion of a holomictic lake, but they are generally more intense and extreme because of their permanence. Sometimes the monimolimnion has a higher temperature in summer than does the overlying holomictic zone, mixolimnion, an anomalous temperature condition known as dichothermy.

Light stratification. Light is important as the source of metabolic energy in the ecosystem. The energy incorporated into organic substances by green plants through the process of photosynthesis supports the activities of all the heterotrophic organisms in the ecosystem.

Light intensity is maximal at the surface, and declines exponentially as the depth increases. Since the amount of photosynthesis accomplished is related to the intensity of light, there is a depth at which the production of organic matter by photosynthesis is equal to its utilization by respiration. This is the compensation level. The zone above is the trophogenic zone, characterized by an excess of production over consumption of organic matter. The region below is the tropholytic zone, characterized by a dominance of energy-consuming processes. In most lakes the compensation level lies in the epilimnion, which because of its continuous circulation during summer is generally regarded as the trophogenic zone.

Light passing through water also undergoes a change in its spectral composition. In any water that is sufficiently deep, the light tends to become monochromatic with increasing depth.

The factors responsible for the exponential absorption of light and its changing spectral composition are the water itself, dissolved yellow-brown color, and suspended particles or seston (Figs. 5 and 6). The effect of any dissolved color or turbidity is to reduce the thickness of the trophogenic zone and to shift maximum transmission from the blue end of the spectrum in very pure water toward the red end in turbid or colored water. Dissolved color is generally more important than turbidity in producing these changes.

Waves and currents. Tides are negligible in even the largest inland lakes, being at most a few centimeters in the Great Lakes and in Lake Baikal. Wind, rather than gravity, is the important force generating waves and currents in inland waters. Such water movements erode and transport material alongshore, provide the turbulence necessary to keep microorganisms suspended, circulate the epilimnion or the entire lake, and provide for some turbulent transport across the metalimnial barrier.

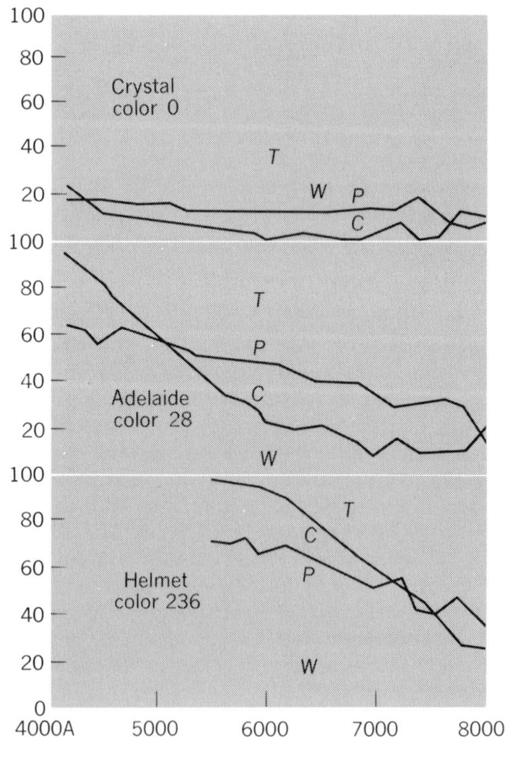

Fig. 5. Graph indicating the percentile absorption of light by water from three lakes in northeastern Wisconsin, ranging from visually uncolored (Crystal) to very dark brown (Helmet). Colors are listed as ppm of platinum units. *T*, total absorption; *P*, absorption due to suspended particulate matter; *C*, absorption due to dissolved color; *W*, absorption due to pure water. (*After H. James and E. Birge, from G. E. Hutchinson, A Treatise on Limnology, vol. 1, copyright © 1957 by John Wiley & Sons, Inc.; reprinted with permission*)

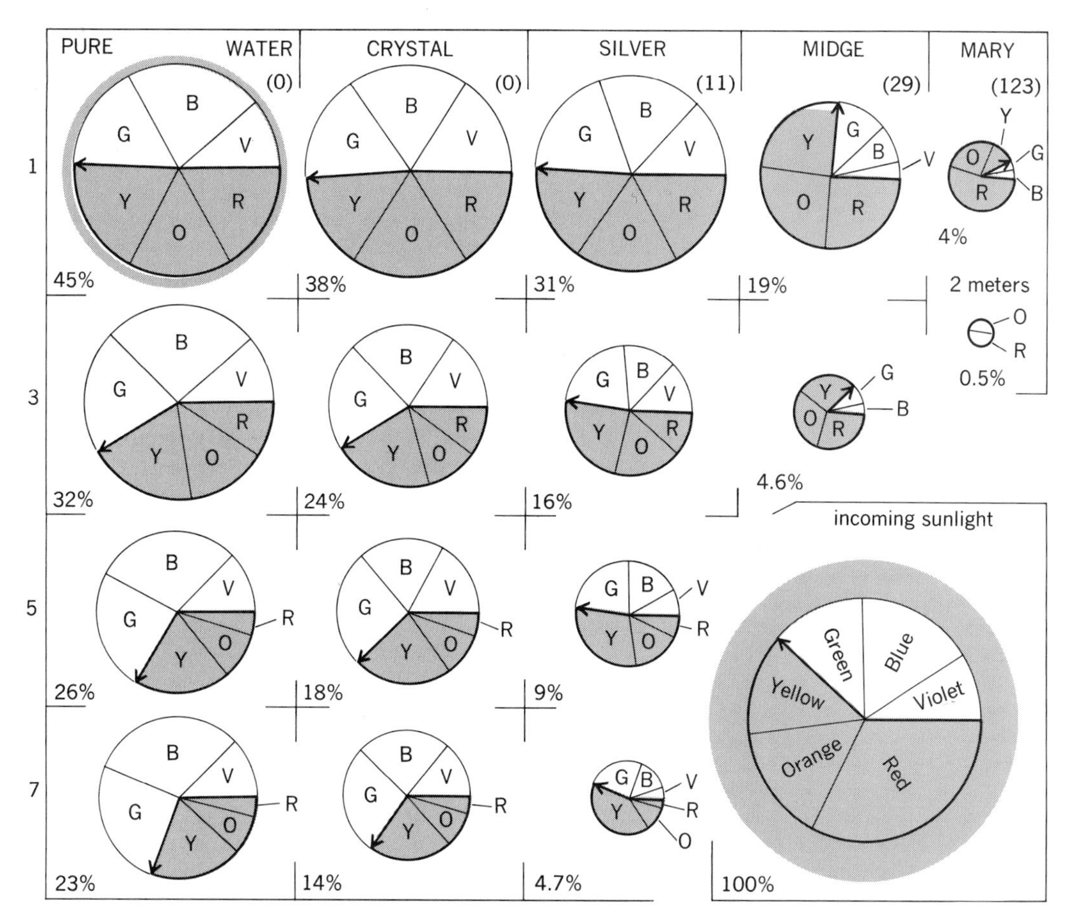

Fig. 6. Changes in color composition of light and intensity relative to the surface in a hypothetical lake of pure water and in four lakes of different dissolved color (in parentheses) in northeastern Wisconsin. Total intensities, relative to the incoming radiation, are represented by the areas of the circles and by the percentage values at their lower left. The tinted areas of the "incoming sunlight" and of "pure water" represent invisible radiation, chiefly infrared. Changing position of the clock hand in each horizontal series demonstrates selective absorption of shorter wavelengths by dissolved color in water. The numerals at extreme left show depth in meters. *(After C. H. Mortimer, from G. C. Sellery, E. A. Birge, A Memoir, University of Wisconsin Press, 1956)*

The simplest waves in deep water are waves of oscillation, in which there is vertical rotation of particles without any net forward transport. The orbit of rotation decreases by one-half for each increase in depth below the surface amounting to one-ninth of the wavelength. In most inland lakes the water is essentially still at a depth of one-half the wavelength.

When a wave of oscillation enters water that is shallower than the orbit of rotation, frictional resistance with the bottom produces an unstable configuration, and the crest of the wave now plunges shoreward as a wave of translation. It is these onshore waves that erode the headlands and transport sand and other material alongshore to form spits and bars, and build subaqueous terraces and benches.

With a steady wind from one direction the gravity-stable horizontal stratification of the lakes becomes displaced. The lake surface tilts upward slightly in a downwind direction, and the epilimnion thickens. At the same time the metalimnion tilts upward in an upwind direction, and the hypolimnion thickens here (Fig. 7). When the generating force ceases, the now unstable density surfaces begin to rock or oscillate as standing waves. The rocking motion of the lake surface, called a seiche, dampens out quickly. The rocking movement of the thermocline, known as an internal seiche, is of greater amplitude and duration and continues long after the surface may have ceased rocking (Fig. 8). Both surface and internal seiches can be uninodal, binodal, multinodal, or even rotational in some of the larger lakes. Such internal waves are important in affecting the vertical and horizontal distribution of organisms and in generating currents and turbulence in the otherwise relatively quiet hypolimnion. In extreme instances the hypolimnion can even reach the surface of a lake on the upwind side.

BIOTA

Within this physical framework the biota and its chemical environment react upon each other. Although all organisms in an aquatic ecosystem are mutually interdependent regardless of where they occur, they are conveniently grouped according to life habit. Microorganisms living free in the water and generally independent of the bottom are known collectively as plankton. Animals living on the bottom comprise the benthos, and the larger plants rooted in the bottom the phytobenthos. Fish and a few other large animals that swim actively in the water constitute the nekton. Sessile algae and

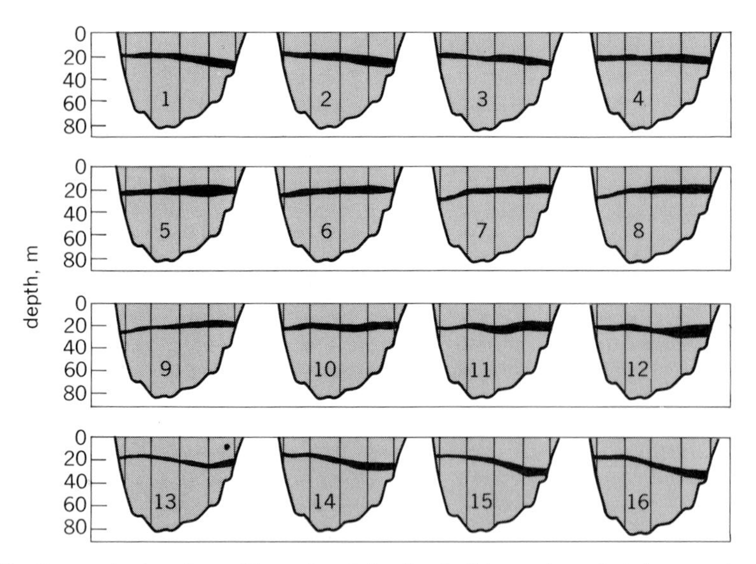

Fig. 7. Successive hourly positions of metalimnion (solid area, bounded above and below by 11 and 9° isotherms, respectively), Aug. 9, 1911, in Loch Earn, Scotland. Standing wave is an internal seiche. *(After E. Wedderburn, from C. H. Mortimer, The resonant response of stratified lakes to wind, Schweiz. Z. Hydrol., 15:94, 1953)*

fungi and their associated groups of microscopic animals constitute the periphyton or Aufwuchs. Organisms directly dependent on the surface film comprise the neuston. The algae and microanimals living in the interstices of sand are known as the psammon. *See* BIOLOGICAL PRODUCTIVITY.

Plankton. This group constitutes the most characteristic assemblage in basin ecosystems and in the oceans and is the most important in terms of overall production. Typically the plankton consists of algae, protozoa, rotifers, copepods, cladocera, and the phantom midge larva. Unlike those in the oceans, almost none of the larger invertebrates or fishes have temporary planktonic stages. Because the organisms in this assemblage have at best only limited powers of locomotion, they are kept suspended mainly by the turbulence of the water. Adaptations aiding this passive flotation are either a reduction in weight relative to the water, or an increase in relative surface area. Small size is an adaptation, and in many situations the bulk of the plankton is too small to be retained by the finest silk bolting cloth. This minute plankton is referred to as nannoplankton, in contrast to the larger net plankton.

Rate of sinking varies inversely with viscosity, which in turn varies inversely with temperature. Hence, the same organisms will sink faster in warm water than in cold. Among the planktonic cladocerans there is a summer increase in relative surface area among genetically identical generations, a phenomenon known as cyclomorphosis. The capacity for "helmet" formation is triggered by a critical temperature during development, and the helmet is maintained largely in response to turbulence in the water. Some rotifers and dinoflagellates also show increased surface area in summer.

Although a list of plankton organisms in a particular ecosystem often numbers several hundred species or more, chiefly algae, the number of species present at any one time is generally small. Thus, at any instant the majority of small to medium-sized lakes have only 1–3 species of copepods, 2–4 cladocerans, and 3–7 rotifers. In each of these groups of animals the dominant species comprises about three-fourths of the total population (Fig. 9).

Variability. A chief characteristic of plankton is its variability both from season to season and from year to year (Fig. 10). Species succeed one another often in bewildering array. Succession seems to be governed by the ever-changing combination of temperature and light, nutrient depletion, such as silica for diatoms, and by metabolites, long-chain fatty acids, carbohydrates, antibiotics, vitamins, and other substances given off to the environment that can be self-limiting to the producing species, and antagonistic or synergistic to another species. Parasitic fungi also are claimed to have a significant effect.

A peak development of a single species is known as a pulse, and when a pulse is great enough to be apparent to the unaided eye it is called a bloom. All seasons of the year exhibit pulses of at least some species, although these are often too small to affect materially the total population. Considering plankton as a whole there is generally a peak population or bloom in spring, mainly diatoms, a reduction during the summer, a second although lesser peak in fall roughly coinciding with the autumnal overturn, followed by a decline during the winter to the lowest population of the year (Fig. 11).

Vertical distribution. The vertical distribution of plankton within a lake is the result of many physical, chemical, and biotic factors. Because of the dominating influence of light, however, most spe-

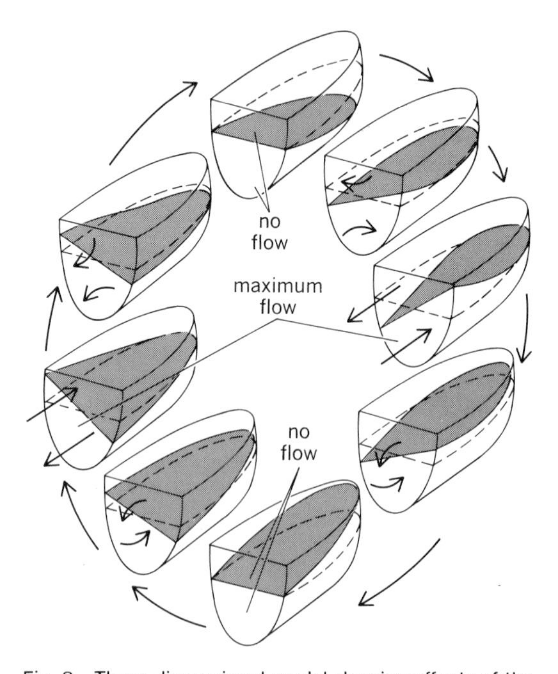

Fig. 8. Three-dimensional model showing effects of the Earth's rotation on a unimodal internal seiche in a large lake, resulting in a rotational seiche. Model fits approximately the condition observed in Loch Ness, Scotland. The metalimnion is tinted. The arrows show the direction and the relative magnitude of currents in the epilimnion and hypolimnion during various phases of the seiche. *(After C. H. Mortimer, from G. E. Hutchinson, A Treatise on Limnology, vol. 1, copyright © 1957 by John Wiley & Sons, Inc.; reprinted with permission)*

cies tend to have their maximum population density within the uppermost 10 m. Some of the larger zooplankters migrate toward the surface at night, and back into deeper water toward morning. Such diel vertical migration is particularly well developed in the deep-scattering layer of the oceans.

Plankton rain. In spite of turbulence there is a continual settling of senescent and dead plankton to the bottom of the lake. This plankton rain supplies energy to the deep-water benthos. It also serves as the chief cause of the extensive chemical differences between surface and bottom water that develop during summer or temporary stratification and under the ice in winter. The surface water tends to become depleted of nutrients, and the bottom water becomes enriched in them. Moreover, the utilization of plankton rain by bacteria and animals results in a reduction of the oxygen supply in deep water. The extent of oxygen reduction is also influenced by the amount of allochthonous organic matter that gets into deep water.

Plankton reproduction. Cladocerans and rotifers reproduce by parthenogenesis during most of the summer. This permits the rapid utilization of developing plankton pulses. During the deterioration of environmental conditions, such as the approach of winter, most of the cladocerans and many of the rotifers produce resistant bisexual or resting eggs that enable the species to endure during the unfavorable conditions. Resting eggs also facilitate the distribution of species by wind and birds. As a result, many fresh-water plankters, including species of algae, are virtually cosmopolitan in distribution. Copepods, on the other hand, reproduce almost exclusively by fertilized eggs. Some species form resistant eggs, but more commonly they survive unfavorable conditions by the encystment of immature stages. One species in Douglas Lake, Mich., produced more than 1,000,000 cysts per square meter of bottom in the hypolimnion. One species of *Daphnia* in the Arctic produces resting eggs by parthenogenesis.

Zooplankton. The majority of the zooplankton are herbivorous, and feed on bacteria, small algae, and organic detritus which they strain from the water by a variety of means. Their grazing activity can effectively reduce the standing crops of phytoplankton. A few cladocerans and rotifers, the phantom midge larva, and adult cyclopoid copepods are predacious. When abundant, the zooplankton can markedly reduce the oxygen content of restricted strata.

Phytobenthos. The rooted aquatic plants form the other major photosynthetic element in fresh water. The relative importance of the phytobenthos and phytoplankton in overall production varies according to basin morphometry. As a lake becomes shallower with increasing ecological age, the phytobenthos assumes a progressively more important role. Because of its longer life cycle it ties up nutrients for longer periods than does the phytoplankton, and it contributes more organic matter to the lake bottom and causes a faster rate of filling. *See* PHYTOPLANKTON.

The phytobenthos typically occurs as three concentric zones around a lake. At the shoreline and in shallow water is the zone of emergent plants, followed by the zone of floating leaf plants, in turn followed by the zone of submersed plants. At still greater depths sometimes occur meadows of the moss *Fontinalis*, which is able to thrive in weaker light than that which the higher plants require. In many situations, particularly in hard-water lakes, the coarse, branched algae *Chara* and *Nitella* form dense beds.

The phytobenthos and the thermal stratification are used to define the zonation of the bottom. From the shoreline to the lower limit of rooted aquatics is the littoral zone, the lower limit of which corresponds roughly to the compensation level. Between the littoral zone and the top of the metalimnion is the sublittoral zone, and at still greater depths is the profundal zone.

Periphyton. One of the chief functions of the phytobenthos in the overall economy of an ecosystem is to increase the amount of colonizable substrate in the trophogenic zone. The substrate in the littoral zone is often thickly overgrown with a mat of sessile algae and fungi and a large number of associated and dependent animals such as protozoa, *Hydra*, microcrustacea, rotifers, oligochaetes, insect larvae, and snails. This community of organisms, complete with producing and consuming elements, is almost the same as an ecosystem within an ecosystem. It helps contribute to the high productivity and the diversity of microhabitats of littoral regions.

Benthos. The great differences between the littoral and profundal benthos are controlled in part by the environmental conditions and in part by the life cycles of the individual species. In the littoral zone with its coarser sediments, wave action, a continuous supply of oxygen, warm temperatures, plus light and plants, diversity of species is the chief characteristic. Insects and mollusks are dominant, many of them confined to the littoral. All other groups of fresh-water animals also occur here. In the profundal zone, on the other hand, soft sediments, virtual absence of light, low temperatures, and potentially severe chemical conditions, especially reduction in oxygen content, severely restrict the species composition. Midge larvae, oligochaetes, and fingernail clams are the most characteristic of the larger animals. In the ooze-film assemblage there may be a great diversity of microscopic animals.

The profundal benthos is largely dependent for its livelihood on the plankton rain from above. Most of the organisms feed either directly on this detritus or on the bacteria that are decomposing it. Largely confined by steep chemical gradients to the uppermost few centimeters of the sediment, these organisms can thoroughly work through the sediments in much the same way that earthworms do through the soil. The relatively few species are sometimes present in tremendous numbers of individuals, and the total population density often exceeds that of the littoral zone.

The general absence of sessile animals in the littoral zone is one of the chief biotic differences between the oceans and inland waters. Seasonal redistribution of organisms commonly occurs. Insect larvae and nymphs move to shallow water for pupation and emergence, and mollusks move shoreward to breed. Emergence of insects in spring and summer helps to produce a minimum littoral population at this time. The maximum occurs in the winter.

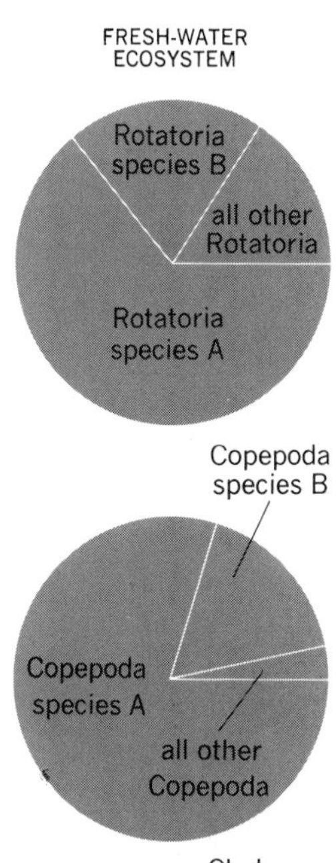

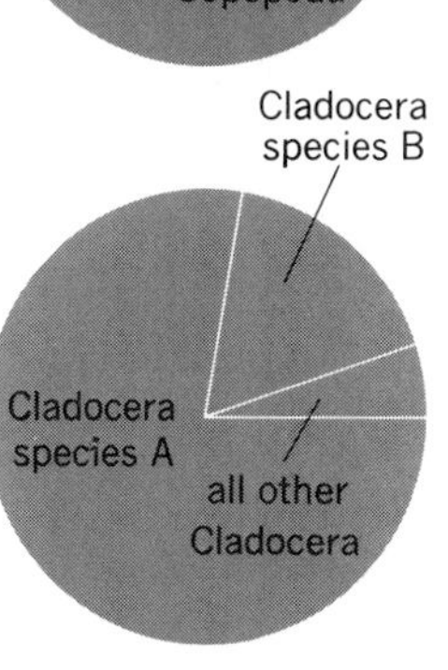

Fig. 9. Instantaneous percentage species composition of three dominant groups of zooplankters in a small- to medium-sized lake. (*From R. W. Pennak, Species composition of limnetic zooplankton communities, Limnol. Oceanogr., 2(3):222–232, 1957*)

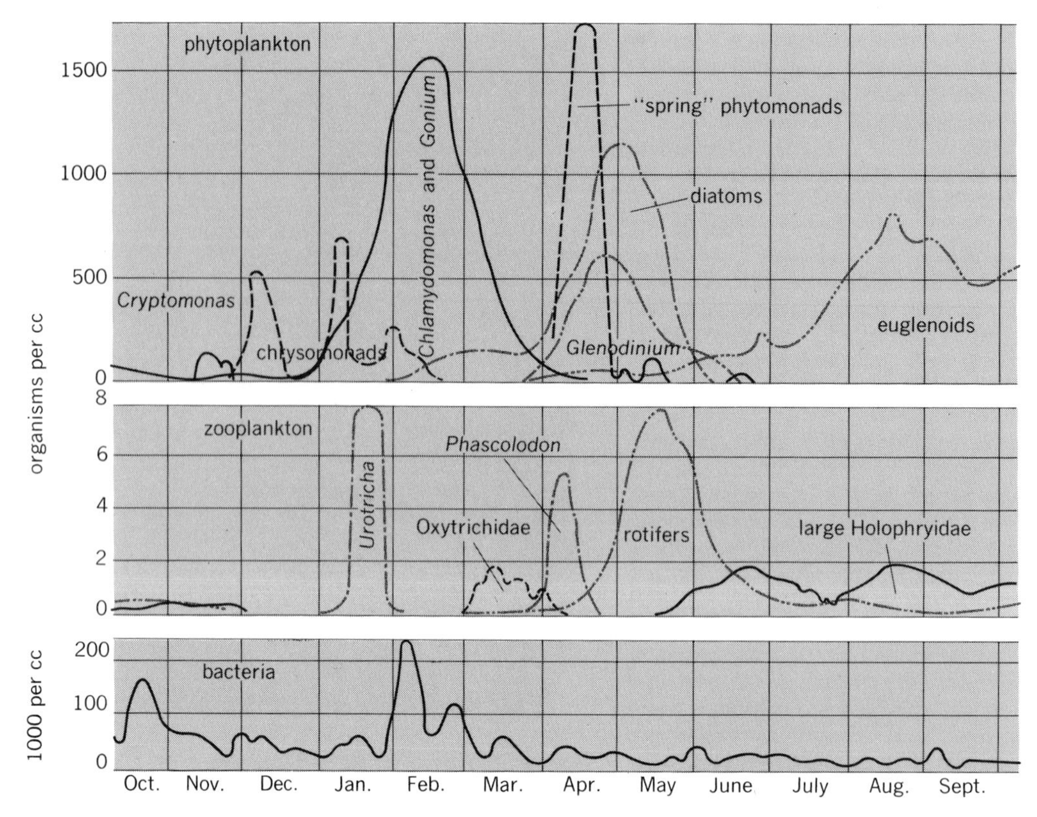

Fig. 10. Graphs illustrating seasonal succession of dominant organisms in a small permanent pond in Pennsylvania. (*From S. S. Bamforth, Ecological studies on the planktonic Protozoa of a small artificial pond, Limnol. Oceanogr., 3(4):398–412, 1958*)

Psammon. In the interstices of sand grains along lake shores is a highly specialized community consisting chiefly of pennale diatoms, bacteria, protozoa, rotifers, copepods, and primitive copepodlike crustaceans of the order Mystacocarida. Even though the sand is shifting under the influence of waves and currents, the psammon organisms are quite distinct from those in the overlying water. The shallow penetration of light into the sand enables the diatoms and other algae to make the community largely self-sustaining.

Nekton. In temperate waters fish dominate the nekton. Species requiring cold water and high levels of oxygen such as trout, coregonids, and burbot can survive in the hypolimnion of many of the larger lakes. Species tolerant of or requiring higher temperatures occur in the surface waters. Among these, certain species live offshore and others occur in the littoral region, either among weed beds or among rocks of shores exposed to strong wave action. Nearly all species, however, regardless of where the adults occur, spawn in shallow water. Here the young find an abundant food supply, and cover to help protect them from predators. In tropical lakes decapod shrimps of the families Atyidae and Palaemonidae can be important components of the nekton.

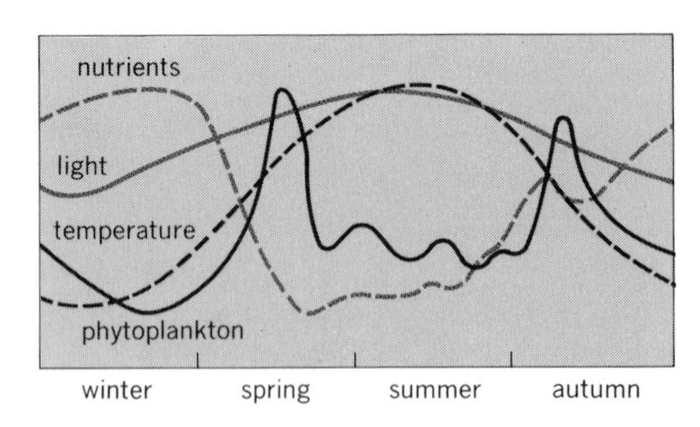

Fig. 11. Typical standing-crop curve for phytoplankton in response to temperature, light, and nutrients. Decline of the spring bloom is at least in part a response to nutrient depletion, and establishment of the autumn bloom is at least in part a response to greater nutrient availability. (*From E. P. Odum and H. T. Odum, Fundamentals of Ecology, 3d ed., Saunders, 1971*)

Higher vertebrates. Frogs, salamanders, turtles, snakes, crocodilians, various aquatic birds such as loons, herons, kingfishers, mergansers, and ospreys, and various mammals such as muskrats, beavers, racoons, minks, and otters are peripheral members of the aquatic biota. Many of them are essentially terrestrial animals that derive only a part of their sustenance from the water, and because they feed mainly on fish, their total demand on the energy resources of the ecosystem is relatively minor. For these reasons they are often omitted when aquatic ecosystems are considered.

Neuston. A minor life-habit assemblage is that associated with the surface film. Here belong the various insects that run, skate, or swim on the surface of the water, and some truly aquatic organisms such as bacteria, fungi, and certain algae that are suspended from the underside of the film. Many aquatic animals such as snails, *Hydra*, aquatic bugs, and beetles regularly or adventitiously occur on the underside of the film for brief periods, and some such as anopheline mosquito larvae feed on the microorganisms attached there.

Trophic relationships. The phytoplankton, phytobenthos, and algae of the periphyton and psammon are the producers. Their raw materials are light energy, carbon dioxide, and soluble inorganic sources of nitrogen, phosphorus, and sulfur. In the process of photosynthesis they manufacture organic compounds and give off oxygen to the environment.

The animals are the consumers. They utilize the organic compounds elaborated by the plants and modify them for their own metabolic needs. Through their activities they are continually converting chemical energy to heat energy which is irretrievably lost to the environment. Because energy can enter the system only at the producer level, there is a continuous and steady decrease in supply of bound energy with each successive level of utilization. In their metabolism animals use oxygen and give off carbon dioxide.

Bacteria are the reducers or transformers. They ultimately mineralize the metabolic wastes and dead organisms to liberate the nutrients in a form which may be used again by the producers. Thus, in an ecosystem there is a continuous recycling of materials. Aerobic bacteria utilize free oxygen and give off carbon dioxide. Anaerobic bacteria utilize chemical oxygen and give off carbon dioxide, methane, and other anaerobic gases.

CHEMISTRY

Changes in the chemistry of the water reveal much about the metabolism of the ecosystem and about general limnological relationships.

Oxygen. Dissolved oxygen, more than any other single substance, has advanced knowledge of limnology by permitting diagnosis of what is happening in an ecosystem.

The epilimnion, as a result of circulation and photosynthesis, is generally saturated with oxygen. In the hypolimnion, however, decomposition of the plankton rain and other organic matter results in a utilization of oxygen, which cannot be replenished until the next overturn. The extent of this depletion depends on the amount of organic matter being furnished to the hypolimnion, that is, the rate of production in the epilimnion, the volume of oxygen available for utilization, which is a morphometric factor roughly equivalent to the volume of the hypolimnion, and the temperature of the hypolimnion, a climate dependent factor, as influencing the rate of metabolism.

Classification based on productivity. Based on the rate of production in the epilimnion, lakes have been classified as oligotrophic or unproductive and eutrophic or productive, with a continuous spectrum between (Fig. 12). Off to one side are the brown-water, or dystrophic, lakes. A. Thienemann defined an oligotrophic lake as one that has more oxygen in the hypolimnion than in the epilimnion, and a eutrophic lake as the reverse. This, however, equates the level of production with the hypolimnetic oxygen supply and ignores the other two controlling factors. B. Aberg and W. Rodhe have resolved this situation by proposing descriptive terms for oxygen distribution without reference to level of production. An orthograde oxygen curve exhibits little or no decline in the hypolimnion. A clinograde curve shows a marked decrease, having a general shape similar to that of a temperature curve. Heterograde curves are special types, having either a maximum (plus heterograde) or a minimum (minus heterograde) oxygen content in the metalimnion. Orthograde and clinograde oxygen distributions correspond, respectively, to oligotrophy and eutrophy as the terms are still commonly used.

Change in productivity. Whatever the causes, most lakes, with time, experience a decline in the oxygen content of the hypolimnion. This is commonly the result of increasing production in the epilimnion and of the gradual accumulation of sediments, which brings about a reduction in the volume of the hypolimnion. The rate of change can be accelerated by man through pollutional and agricultural enrichment, resulting in a higher rate of production in the epilimnion.

As the oxygen content declines, various species of animals are eliminated from the profundal benthos. A lake with adequate oxygen during the summer may have as many as 200 species of profundal benthos exclusive of protozoa, whereas with reduction in oxygen content the number of species is reduced, until in the most severe cases there are no higher animals at all except perhaps the facultatively anaerobic phantom midge larva. Thienemann first demonstrated this relationship for midges in the lakes of northern Germany and pointed out that lakes with only slightly reduced oxygen content have *Tanytarsus*, those with greatly reduced oxygen *Chironomous*, and those with intermediate levels *Stictochironomus* or some other genus. *Tanytarsus* has been used as an indicator of oligotrophy and *Chironomus* of eutrophy.

With respect to fish, lakes with adequate oxygen in deep water can be called two-story lakes, meaning that there is an assemblage of cold-water fishes

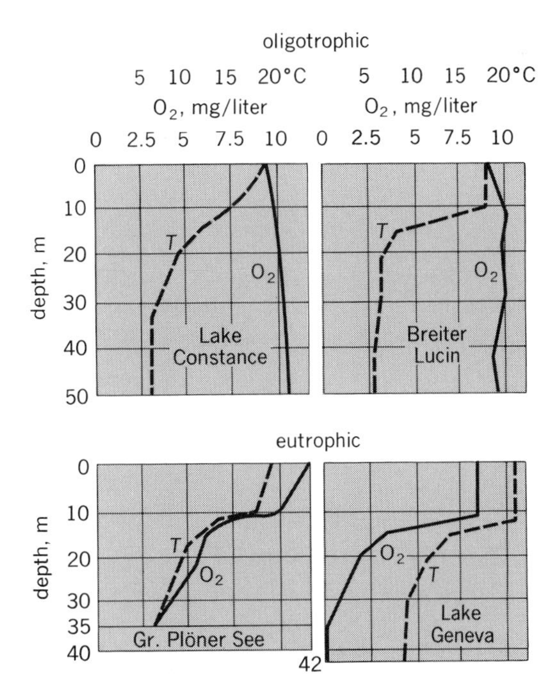

Fig. 12. Oxygen (O_2) and temperature (*T*) stratification in one Wisconsin and three European lakes. The two upper oxygen curves are orthograde, and the two lower ones clinograde. (*From F. Ruttner, Fundamentals of Limnology, 3d ed., University of Toronto Press, 1963*)

in deep water distinct from the warm-water fishes in the epilimnion. Reduction in oxygen ultimately eliminates the cold-water fishes to produce a one-story lake. In North America the cisco has been able to persist in some lakes with a plus-heterograde oxygen distribution but with no oxygen in the hypolimnion.

A critical point in lake ontogeny is the disappearance of oxygen from water in contact with the bottom. Aerobic processes are replaced by anaerobic. Methane and hydrogen sulfide are formed in quantities and get into the water. Lowering of the oxidation-reduction potential results in the conversion of previously insoluble iron and manganese to a soluble condition, allowing them to diffuse into the water. Any excess iron can create a trap for phosphate under these conditions and prevent it from returning to the epilimnion at the next overturn. Methane bubbling through the hypolimnion helps reduce the oxygen content and induces turbulence. This is known as methane circulation.

Carbon dioxide. Carbon dioxide occurs in water as the dissolved gas, as carbonic acid, and as the carbonates and bicarbonates of calcium and magnesium and sometimes iron. All of these substances are in chemical equilibrium with one another. This carbon dioxide complex is the single most important factor that controls pH level in inland waters.

Unlike the oceans, inland waters do not have a constant percentage composition. However, in most waters the world over the dominant ions occur in the same order of abundance: $Ca^{++} > Mg^{++} > Na^{+} > K^{+}$, and $HCO_3^{-} > SO_4^{--} > Cl^{-}$. As a result, the pH of most waters varies between 6.5 and 8.5. Higher pH occurs regularly in waters of arid regions and temporarily in hard-water lakes during vigorous photosynthesis. Lower pH is produced by humolimnic acids in dystrophic lakes and by inorganic acids, chiefly sulfuric acid (H_2SO_4) in special situations.

During photosynthesis carbon dioxide (CO_2) is withdrawn by the plants, causing the CO_2 equilibrium to shift with the production of the relatively insoluble calcium carbonate, which settles to the bottom. In littoral regions the precipitate tends to accumulate as marl. In the profundal region much or all of the carbonate can be reconverted to soluble bicarbonate by the excess CO_2 of the hypolimnetic metabolism.

A soft-water lake is one which has only small amounts of calcium and magnesium. Being dependent largely on the atmosphere for an additional supply of CO_2, such lakes are said to have a chronic CO_2 deficiency. Hard-water lakes have a CO_2 reserve in the calcium and magnesium bicarbonates, thus permitting a greater amount of production to be accomplished. Even they, however, can have all their reserve CO_2 utilized, resulting in an acute CO_2 deficiency. In such lakes some of the calcium carbonate can continue to be present in nonsettling colloidal form. This causes the methyl orange titration to give erroneous results as to the CO_2 reserve.

Other elements. Phosphorus and nitrogen are the two elements most commonly in short supply. Of these, phosphorus is generally more critical because of the nature of its supply. Nitrogen is fixed in the atmosphere during electrical storms, and in the aquatic ecosystem by certain blue-green algae and bacteria. Although these two elements and other nutrients a priori must exert control on production and succession in the ecosystem, the specific effects have usually been difficult to demonstrate in the field.

Production and productivity. The organisms present at any one time constitute the standing crop, which is the result of production but is not indicative of the rate of production. The quantity of plankton beneath a unit area of lake surface is approximately the same for oligotrophic and eutrophic lakes, although the quantity per unit volume of the trophogenic zone is greater in eutrophic lakes. Here the plankton is so dense that it limits its own production because it has reduced the transmission of light by its density.

Measurement of productivity. Production in planktonic diatoms has been measured by collecting and counting the cells that settle into jars suspended in the water. In several lakes greatly different in size a reproduction rate of about 10% per day was necessary to maintain the standing crop at a particular level. Thus, a low population density may be the result of a rapid rate of sinking rather than a low rate of production.

Rates of production of phytoplankton are commonly estimated by measuring chemical changes developing between clear and opaque bottles containing portions of the same phytoplankton population. In the clear bottle photosynthesis and respiration occur, in the opaque bottle only respiration. The difference between the bottles represents gross production, and the difference between the light bottle and the initial value represents net production. The latter represents the organic matter and energy available for utilization by the heterotrophic components of the ecosystem. Changes measured in the bottles are oxygen, pH,

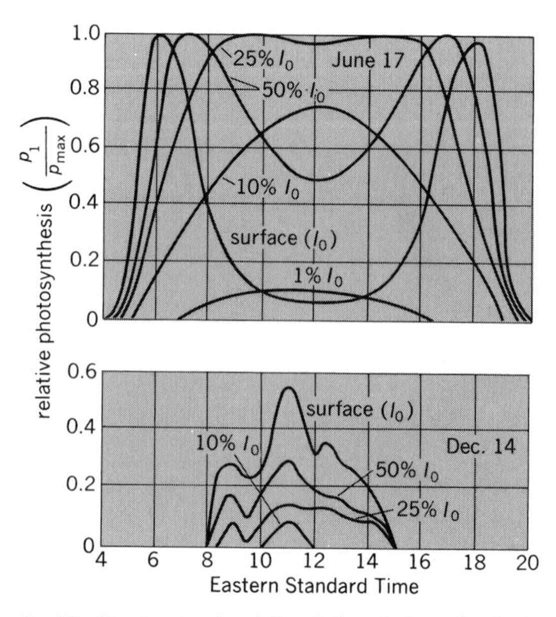

Fig. 13. Graphs showing daily relative photosynthesis at various depths in the water corresponding to stated percentages of surface light intensity (I_0) for typical days in summertime and in wintertime at Newport, R.I. (*From J. H. Ryther, The measurement of primary production, Limnol. Oceanogr., 1:72–84, 1956*)

electrical conductivity, and radiocarbon (^{14}C) uptake. The latter apparently measures net production directly, although a correction must be made for fixation of carbon in the dark.

Photosynthetic capacity. Since photosynthesis is dependent on light intensity, on bright days the trophogenic zone is thicker than on dull days, and it is thicker in summer than in winter. Because of the photoreduction of chlorophyll by bright light, maximum photosynthesis may be some distance below the surface of the water. From the photosynthetic capacity of the phytoplankton community present, it is possible to approximate the amount of photosynthesis accomplished over a 24-hr period by integrating the various subsurface light-intensity curves. In epilimnetic communities there is also a relatively constant relationship between chlorophyll content and photosynthesis. Because of this direct dependence of photosynthesis on light intensity it is not surprising that the curve for production accomplished closely follows the curve for total illumination during the year (Fig. 13). Maximum production occurs in summer, when the standing crop is low relative to spring and fall (Fig. 14). The biomass is kept at low levels by the grazing activities of herbivores in the plankton.

Consumer level. At the consumer level, production is more difficult to study because of the increased number of pathways into which the energy can be channeled. Under these conditions the concept of production is somewhat meaningless without reference to a particular product. In general, production is greater, although the standing crop is less, when the product organism is being utilized by some higher-level consumer. In two ponds studied, fish reduced the standing crop of their food benthos but increased its production (Fig. 15). Total production was about 17 times the standing crop, which compares favorably with production coefficients determined for diatoms.

Much of the organic matter fixed in the epilimnion is consumed and mineralized there to enable the nutrients to be reutilized immediately. Only a relatively small proportion of the total primary production, about one-fourth in north German lakes, gets into the hypolimnion as plankton rain. However, the chemical changes induced in the hypolimnion can be used as an index of the relative levels of production in lakes (Fig. 16). If the changes are related to unit surface area of the hypolimnion, through which the plankton sinks, the results are largely independent of the relative size of the epilimnion. The two quantities most commonly measured are the rate of generation of the hypolimnetic oxygen deficit and the rate of generation of the hypolimnetic carbon dioxide accumulation. In strongly eutrophic lakes the latter measure is better, since CO_2 continues to be produced, although at a lesser rate, under anaerobic conditions after the oxygen may have long since disappeared.

LOTIC ECOSYSTEMS

A river system consists of a treelike arrangement of small channels, joining into progressively large channels, until finally all the water flows in a single large channel, corresponding to the trunk of the tree. The distributaries in deltas and alluvial fans are morphologically analogous to a reduced root system. In limestone regions, particularly, various headwater streams can be in direct contact with an underground system of caverns and channels.

Water source. Water in a stream is derived from surface drainage during precipitation and the melting of snow and ice, and from groundwater. Surface drainage contributes the bulk of inorganic turbidity to a stream, groundwater the bulk of the dissolved solids. During low water flow the turbidity is generally low and the amount of dissolved solids high. During periods of precipitation both the volume of flow and turbidity increase, and dissolved solids decline. After peak flow is reached and the current velocity begins to subside, turbidity again declines. The response of a stream to a unit storm is known as a unit hydrograph. The as-

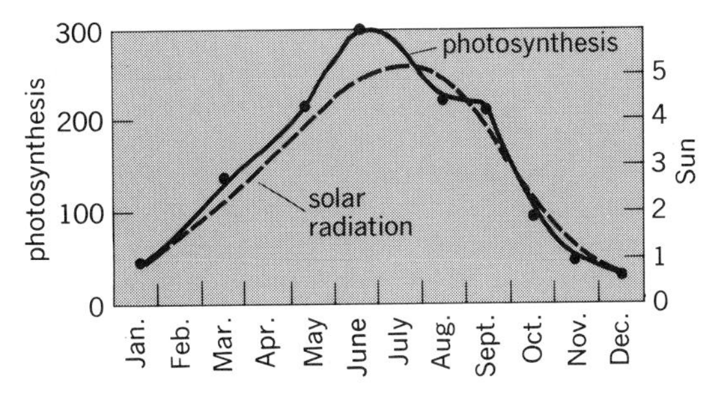

Fig. 14. Annual course of photosynthesis in relation to solar radiation under completely natural conditions in western Lake Erie. Note that maximum photosynthesis occurs in summer when the standing crop is not maximum (see Fig. 11). *(From J. Verduin, Primary production in lakes, Limnol. Oceanogr., 1:85–91, 1956)*

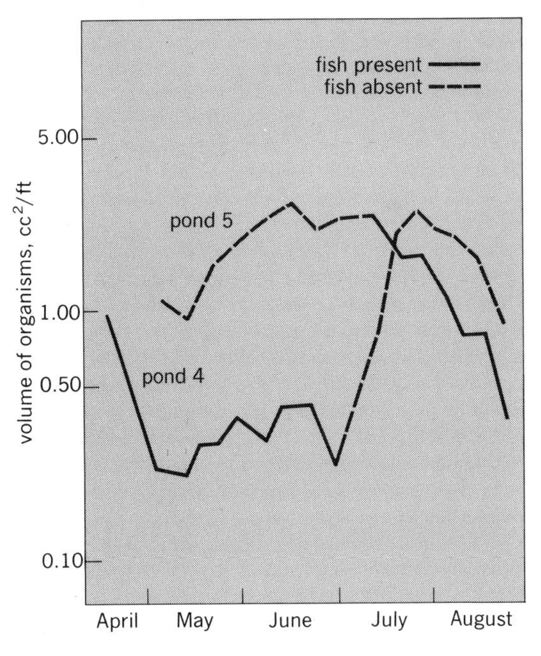

Fig. 15. Fluctuations in volume of standing crop of those benthic organisms eaten by fish in two ponds in Michigan. The action of the fish in each case was to reduce the size of the standing crop but to increase its rate of production. *(From D. W. Hayne and R. C. Ball, Benthic productivity as influenced by fish production, Limnol. Oceanogr., 1:162–175, 1956)*

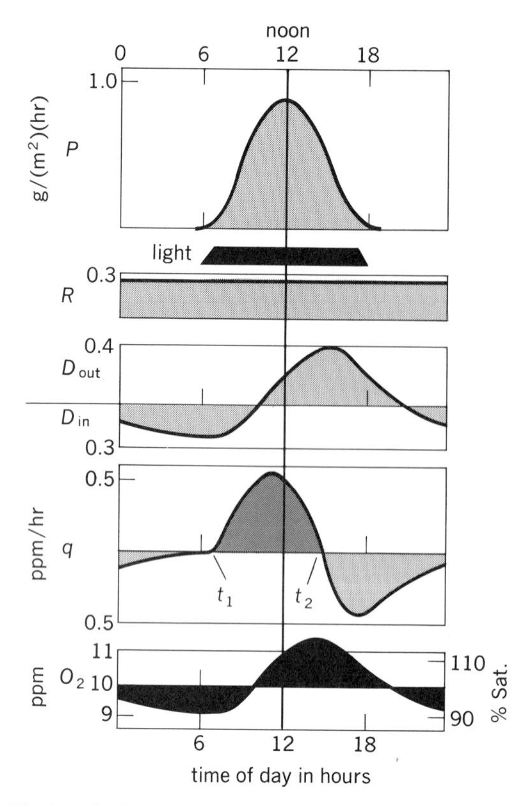

Fig. 16. Series of graphs showing component processes in the oxygen metabolism of a section of a hypothetical stream during the course of a cloudless day. The rates for production P, respiration R, and diffusion D are combined in q. The area delimited by t_1 and t_2 multiplied by the volume of flow in m^3/hr and divided by the area in square miles of the stream section studied gives gross primary production per day. The method has been successfully used in streams, coral reefs, and other marine littoral areas for measuring primary production. *(From H. T. Odum, Primary production measurements in eleven Florida springs and a marine turtle-grass community, Limnol. Oceanogr., 1(2)85–97, 1956)*

cending limb of the hydrograph is always much steeper than the descending, because of the lingering effect of increased groundwater reserves.

The headwater portion of a river is dual in nature, consisting of a stream of water and all its contained substances, and a stream of heavy materials constituting the bed load, which moves along the bottom intermittently or continuously in the nature of a giant file. The molar action of this material continually erodes the stream channel deeper, and when severe, such as during a flood, can virtually eliminate the biota from large areas. Production of organisms is directly related to stability of the bottom. By eroding, streams constantly tend to cut down their channels to base level.

Gradients. Except for light, gradients in lotic ecosystems are longitudinal rather than vertical. Most commonly the slope profile of a stream is concave upward, with the steepest slopes in the headwaters. This creates gradients of current velocity, coarseness of bottom, and other conditions. Because the headwaters are generally at considerably greater elevations and in more direct contact with the groundwater system, there tend to be gradients in temperature, such as a warming downstream in summer, and in magnitude of annual temperature variation, which is least upstream.

A stream is a continuum only in a physical sense. The various longitudinal gradients interacting bring about a longitudinal succession of species and of communities. Generally the number of species in a taxonomic group increases in a downstream direction. This is known to be true of fishes in the Rhine and Colorado rivers, and of protozoa and insects in Pennsylvania. The replacement of one species of turbellarian by another is response to a changing temperature regime was first found in western Europe, and has since been found in Japan and Brazil. Undoubtedly all groups of organisms react to these gradients.

In western Europe streams are commonly divided into four biological zones on the basis of the dominant fish species. Uppermost is a trout zone, with steep gradient, rapid flow, close alternation of pools and riffles, and low temperatures. Below this is the grayling zone, with reduced current velocity, greater annual fluctuations in temperature, and a more open valley with the stream beginning to meander in a small flood plain. The barbel zone is in some respects transitional between the two cold-water zones above and the warm-water bream zone below, dominated by cyprinids, percids, pike, and other warm-water fishes. The several zones are associated with various stages in the physiographic cycle, from the youthful V-shaped valleys and steep gradients in the headwaters to mature, relatively flat valleys and extensive flood plains downstream. By the erosional activity of a stream there is a gradual headward migration of stream zones or habitats and their associated biotas.

Biota. The various biotal components of a stream are the following:

Periphyton. In headwater streams the dominant producing element is the sessile algae on the bottom, which is constantly being broken loose by the current and drifted downstream. The microscopic material (plankton) in the overlying water is sometimes described as a "pale image" of the periphyton community.

Potamoplankton. In more sluggish sections of a stream a true plankton can develop. Most of the organisms present are the same as those occurring in lakes and ponds. Their source is backwaters, sloughs, and flood-plain pools. Plankton leaving a lake tends to be rapidly removed from the stream by entrapment in the water immediately in contact with the bottom or other substrate. The plankton of the Rhine, however, is dominated for long distances by organisms derived from the Swiss lakes via the Aare River.

Benthos. The benthos of still-water sections is similar to that of lakes. In currents the two distinctive communities are the stone fauna and the moss fauna. Organisms in the stone fauna tend to be greatly flattened for greater contact with the substrate. Many are directly attached to the substrate, or they live in webs and cases that are attached. In temperate latitudes the net-winged midges and in the tropics fishes and tadpoles have suckers for attachment. Those organisms when wrested loose and drifted by the current are known as syrton. They consitute another example of the downstream transport of organic matter, and they are important in colonizing new or depopulated areas. The tendency to be swept downstream is compensated for in part by a positive rheotaxis, which directs the animals into the current and makes them move upstream against it. Animals of the moss fauna are small and well provided with grappling structures. The moss serves primarily as a substrate, with relatively few organisms feeding on it directly.

Phytobenthos. Rooted aquatics occur abundantly in areas of reduced current. Their stems and roots further reduce the current velocity and accumulate fine sediment, which changes the nature of the stream bottom and helps delay the downstream transport of nutrients.

Nekton. Fishes that require low temperature and high oxygen tend to occur in the headwaters, and the more eurytopic fishes occur in the lowland sections. Fishes in swift currents tend to be almost round in cross section, whereas many of those in quiet water are found to be strongly compressed from side to side.

Many, perhaps most, stream fishes have a restricted home range in which they spend their entire lives. If displaced downstream, for instance during a flood, they are able to find their way back. In the long-eared sunfish this is accomplished by the sense of smell more than by sight.

Continental fish faunas are dominated by cypriniform fishes, consisting of the carplike fishes such as the characinoids, gymnotoids, and cyprinoids, and the catfishlike fishes. Approximately 80% of the fishes of Southeast Asia are of this order, 50% of the Great Lakes fishes, and more than 80% in the Amazons. Peripheral fish faunas have smaller percentages of cypriniforms.

Production relationships. R. Butcher attempted to estimate productivity by the number of algal cells settling on glass slides submerged for 28 days. Oligotrophic streams yielded fewer than 2000 cells/mm^2, whereas eutrophic streams yielded 2500–10,000 cells/mm^2. Regions with moderate organic pollution yielded higher numbers, up to 100,000/mm^2. J. Yount has found for sessile diatoms that as the population density increases, the number of species represented decreases.

In the upper sections of streams, riffles are generally more productive than pools. Current produces a eutrophication effect, and animals living in a current have a higher rate of metabolism than those in still water. In mayflies the area of gill surface decreases as the current velocity increases.

Estimates of production have been made from the 24-hr cycles of dissolved oxygen and carbon dioxide with suitable corrections for diffusion. Attempts have also been made to approximate primary production from the quantity of chlorophyll per unit area of bottom. In the outflow of a spring, production and the pathways of energy through the ecosystem have been measured by a modification of the light bottle, dark bottle principle. Because the substrate is so important in primary production in streams, the upstream shallow portions may be considered a trophogenic zone which exports its surplus production, such as plankton and syrton, to a downstream, deeper tropholytic zone. Production of detritus-feeding benthonts such as midges, mayflies, oligochaetes, and small clams in the lowermost portions of large rivers is at times tremendous.

Pollutional relationships. Quantities of organic matter entering a stream, from domestic sewage, a paper mill, or a canning factory, can markedly change the balance of biological processes. Bacteria attack the organic matter, resulting in a utilization of oxygen and a release of plant nutrients. The extent of oxygen utilization depends on the relative dilution of the organic matter by the river water. Engineers refer to the oxygen sag curve and the process of reaeration downstream. In severe cases the oxygen can be completely utilized, resulting in a septic zone or polysaprobic zone. This is a zone of bacteria, with anaerobic processes predominating. Farther downstream conditions in the stream begin to recover, forming a zone of partial recovery, the mesosaprobic zone, and still farther downstream is the zone of complete recovery or oligosaprobic zone. In this self-purification process there is a longitudinal succession of bacteria, protozoa, algae, and insects. There is also a definite and predictable longitudinal succession of chemical conditions.

Other types of pollution such as toxic wastes and inert materials may affect the biota of a stream without changing the oxygen content. A stream is considered healthy if a great variety of microhabitats in it permit establishment of a diversity of species and trophic relationships in balance with one another. Pollution of any kind eliminates various of these microhabitats and their adapted species and often results in the increased abundance of some of those remaining. The severity of pollution has been measured by comparing the number of species in several taxonomic groups with the average numbers of species present at nonpolluted sites in the same watershed. Fishes, insects, and crustaceans are the most sensitive to pollution. Blue-green algae, bdelloid rotifers, oligochaetes, leeches, and pulmonate snails are least sensitive.

Since the early work of R. Kolkwitz and M. Marsson, attempts have been made to find indicator species for different severities or kinds of pollution. Most species, however, are sufficiently adaptable in their requirements that they can occur in more than one zone. The community of organisms present is more important than the individual species in indicating community metabolism. *See* ECOSYSTEM; ENVIRONMENT; MARINE ECOSYSTEM. [DAVID G. FREY]

Bibliography: R. E. Good et al., *Freshwater Wetlands: Ecological Processes and Management Potential*, 1978; G. E. Hutchinson, *A Treatise on Limnology*, 1970; H. B. N. Hynes, *The Biology of Polluted Waters*, vols. 1–3, 1957, 1967, 1975; H. B. N. Hynes, *Stream Limnology*, 1969; A. Leadley-Brown, *Ecology of Fresh Water*, 1971; T. T. Macan, *Freshwater Ecology*, 2d ed., 1974; K. Read, *Ecology of Inland Waters and Estuaries*, 2d ed., 1976.

Front

A sloping surface of discontinuity in the troposphere, separating air masses of different density or temperature. The passage of a front at a fixed location is marked by sudden changes in temperature and wind and also by rapid variations in other weather elements, such as moisture and sky condition. *See* AIR MASS.

Although the front is ideally regarded as a discontinuity in temperature, in practice the temperature change from warm to cold air masses occurs over a zone of finite width, called a transition or frontal zone. The three-dimensional structure of the frontal zone is illustrated in Fig. 1. In typical

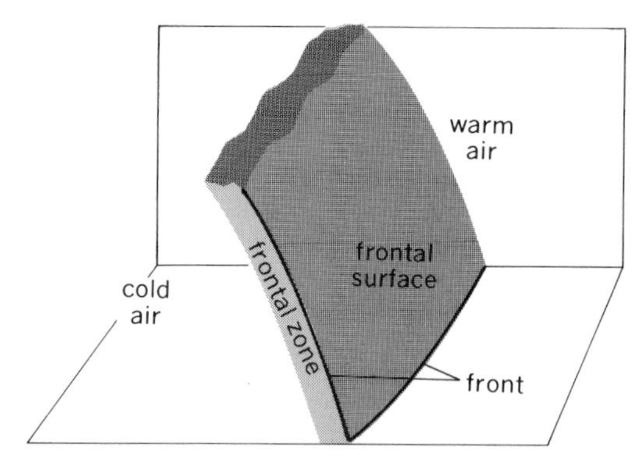

Fig. 1. Schematic diagram of the frontal zone, angle with Earth's surface much exaggerated.

cases the zone is about 3000 ft (1 km) in depth and 100 mi (100–200 km) in width, with a slope of approximately 1/100. The cold air lies beneath the warm in the form of a shallow wedge. Temperature contrasts are generally strongest at or near the Earth's surface. In the middle and upper troposphere, frontal structure tends to be diffuse, though sharp, narrow fronts of limited extent are common in the vicinity of strong jet streams. Upper-level frontogenesis is often accompanied by a folding of the tropopause and the incorporation of stratospheric air into the upper portion of the frontal zone. *See* JET STREAM.

The surface separating the frontal zone from the adjacent warm air mass is referred to as the frontal surface, and it is the line of intersection of this surface with a second surface, usually horizontal or vertical, that strictly speaking constitutes the front. According to this more precise definition, the front represents a discontinuity in temperature gradient rather than in temperature itself. The boundary on the cold air side is often ill-defined, especially near the Earth's surface, and for this reason is not represented in routine analysis of weather maps. In typical cases about one-third of the temperature difference between the Equator and the pole is contained within the narrow frontal zone, the remainder being distributed within the warm and cold air masses on either side.

The wind gradient, or shear, like the temperature gradient, is large within the frontal zone and discontinuous at the boundaries. An upper-level jet stream normally is situated above the zone, the strong winds of the jet inclining downward along or near the warm boundary.

Frontal waves. Many extratropical cyclones begin as wavelike perturbations of a preexisting frontal surface. Such cyclones are referred to as wave cyclones. The life cycle of the wave cyclone is illustrated in Fig. 2. In stage I, prior to the development, the front is gently curved and more or less stationary. In stage II the front undergoes a wavelike deformation, the cold air advancing to the left of the wave crest and the warm air to the right. Simultaneously a center of low pressure and of counterclockwise wind circulation appears at the crest. The portion of the front which marks the leading edge of the cold air is called the cold front. The term warm front is applied to the forward boundary of the warm air. During stage III the wave grows in amplitude and the warm sector narrows. In the final stage the cold front overtakes and merges with the warm front, forming an occluded front. The center of low pressure and of cyclonic rotation is found at the tip of the occluded front, well removed from the warm air source. At this stage the cyclone begins to fill and weaken. *See* STORM.

Cases have been documented in which the occluded structure depicted in panel IV (Fig. 2) forms in a different manner than described above. In such cases, sometimes referred to as pseudo-occlusions, the low-pressure center is observed to retreat into, or form within, the cold air and frontogenesis takes place along a line joining the low center and the tip of the warm sector. Cloud observations from meteorological satellites have provided visual evidence of this process. Since the classical occlusion process, in which the cold front overtakes and merges with the warm front, has never been adequately verified, it is possible that most occlusions form in this other way.

A front moves approximately with the speed of the wind component normal to it. The strength of this component varies with season, location, and individual situation but generally lies in the range of 0–50 mph; 25 mph is a typical frontal speed.

Cloud and precipitation types and patterns bear characteristic relationships to fronts, as depicted in Fig. 3. These relationships are determined mainly by the vertical air motions in the vicinity of the frontal surfaces. Since the motions are not unique but vary somewhat from case to case and, in a given case, with the stage of development, the features of the diagram are subject to considerable variation. In general, though, the motions consist of an upgliding of the warm air above the warm frontal surface, a more restricted and pronounced upthrusting of the warm air by the cold front, and an extensive subsidence of the cold air to the rear of the cold front. *See* CLOUD; CLOUD PHYSICS.

Fast-moving cold fronts are characterized by narrow cloud and precipitation systems. When

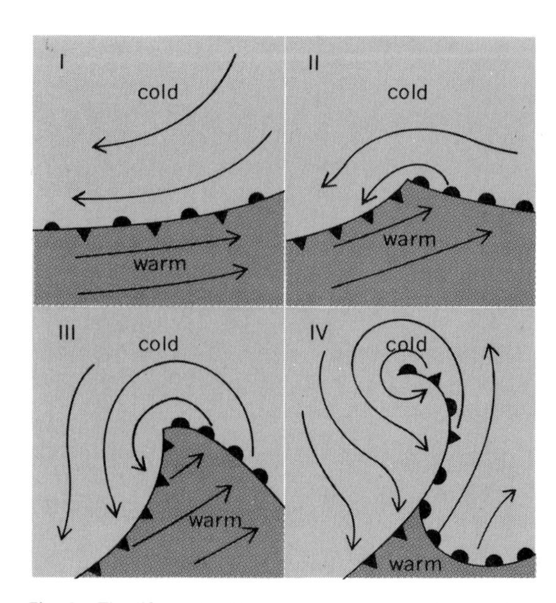

Fig. 2. The life cycle of the wave cyclone, surface projection. Arrows denote airflow. Patterns depicted are for the Northern Hemisphere, while their mirror images apply in Southern Hemisphere.

potentially unstable air is present in the warm sector, the main weather activity often breaks out ahead of the cold front in prefrontal squall lines.

Polar front. A front separating air of tropical origin from air of more northerly or polar origin is referred to as a polar front. Frequently only a fraction of the temperature contrast between tropical and polar regions is concentrated within the polar frontal zone, and a second or secondary front appears at higher latitudes. In certain locations such a front is termed an arctic front.

In winter the major or polar frontal zones of the Northern Hemisphere extend from the northern Philippines across the Pacific Ocean to the coast of Washington, from the southeastern United States across the Atlantic Ocean to southern England, and from the northern Mediterranean eastward into Asia. An arctic frontal zone is located along the mountain barriers of western Canada and Alaska. In summer the average positions of the polar frontal zones are farther north, the Pacific zone extending from Japan to Washington and the Atlantic zone from New Jersey to the British Isles. In addition to a northward-displaced polar front over Asia, an arctic front lies along the northern shore and continues eastward into Alaska.

The polar frontal zone of the Southern Hemisphere lies near 45°S in summer and slightly poleward of that latitude in winter. Two frontal bands, more pronounced in winter than in summer, spiral into the main zone from subtropical latitudes. These bands, originating east of the Andes and northeast of New Zealand, merge with the main frontal zone after making a quarter circuit of the globe.

[RICHARD J. REED]

Frontogenesis. The formation of a front or a frontal zone requires an increase in the temperature gradient and the development of a wind shift. The frontogenesis mechanism operates even when the front is in a quasi-steady state; otherwise the turbulent mixing of heat and momentum would rapidly destroy the front.

The transport of temperature by the horizontal wind field can initiate the frontogenesis process as is shown for two cases in Fig. 4. The two wind fields shown would, in the absence of other effects, transport the isotherms in such a way that they would become concentrated along the A-B lines in both cases. Since the temperature gradient is inversely proportional to the spacing of the isotherms, it is clear that frontogenesis would occur along the A-B lines. Vertical air motions will modify this frontogenesis process, but these modifications will be small near the ground where the vertical motion is small. Thus, near the ground the temperature gradients will continue to increase in the frontal zones as long as the horizontal wind field does not change. As the temperature gradient increases, a circulation will develop in the vertical plane through C-D in each case. The thermal wind relation is valid for the component of the horizontal wind which blows parallel to the frontal zone. The thermal wind, which is the change in the geostrophic wind over a specific vertical distance, is directed along the isotherms, and its magnitude is proportional to the temperature gradient. As the frontogenesis process increases the temperature

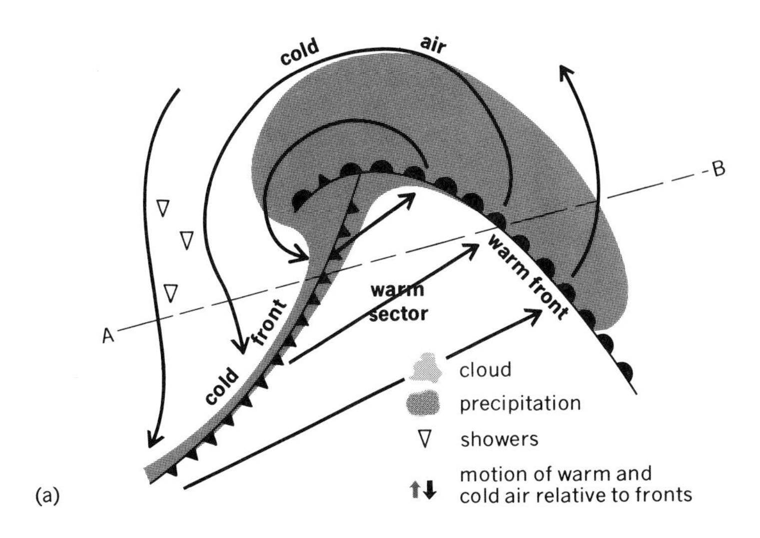

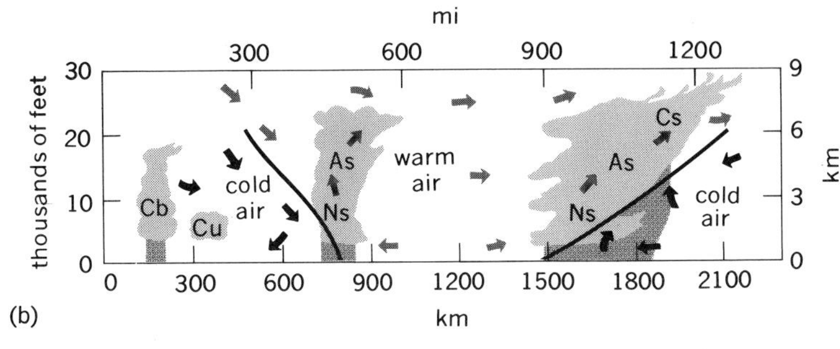

Fig. 3. Relation of cloud types and precipitation to fronts. (*a*) Surface weather map. (*b*) Vertical cross section (along A-B in diagram *a*). Cloud types: Cs, cirrostratus; As, altostratus; Ns, nimbostratus; Cu, cumulus; and Cb, cumulonimbus.

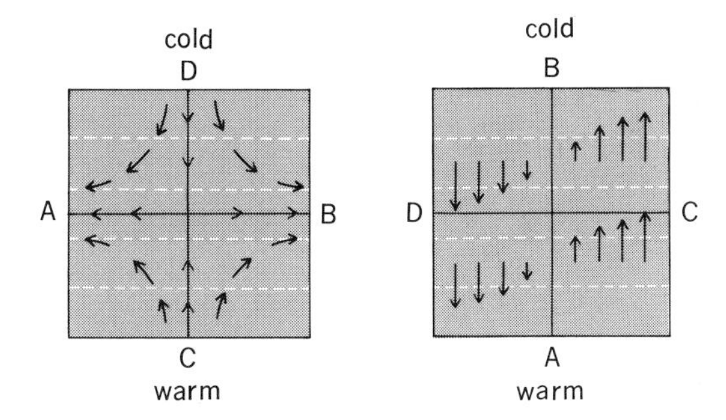

Fig. 4. Two horizontal wind fields which can cause frontogenesis. Broken lines represent isotherms and arrows show wind directions and speeds.

gradient, the thermal wind must also increase. But if the thermal wind increases, the change in the actual wind over a height interval must also increase (since the thermal wind closely approximates the actual wind change with height). This corresponding change in the actual wind component along the front is accomplished through the action of the Coriolis force. A small wind component perpendicular to the front is required if the Coriolis force is to act in this manner. This leads to the circulation in the vertical plane that is shown in Fig. 5.

This circulation plays an important role in the frontogenesis process and in determining the frontal structure. The rising motion in the warm air

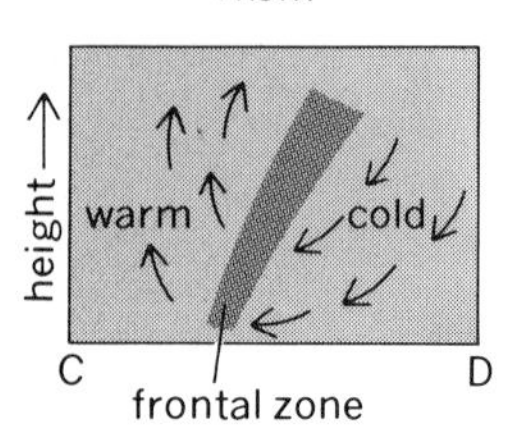

Fig. 5. The circulation in the vertical plane through C-D for both cases of Fig. 4.

and the sinking motion in the cold air are consistent with observed cloud and precipitation patterns. The circulation helps give the front its characteristic vertical tilt which leaves the relatively cooler air beneath the front. Near the ground the circulation causes a horizontal convergence of mass. This speeds up the frontogenesis process by increasing the rate at which the isotherms move together. The convergence field also carries the momentum lines together in the frontal zone in such a way that a wind shear develops across the front. This wind shear gives rise to the wind shift which is observed with a frontal passage. Eventually the front reaches a quasi-steady state in which the turbulent mixing balances the frontogenesis processes. Other frontogenesis effects are important in some cases, but the mechanism presented above appears to be the predominant one.

[ROGER T. WILLIAMS]

Bibliography: J. G. Charney, Planetary Fluid Dynamics, in P. Morel (ed.), *Dynamic Meteorology*, 1973; H. Dickson, *Climate and Weather*, 1976; J. R. Holton, *An Introduction to Dynamic Meteorology*, 1972; B. A. Hoskins, Atmospheric frontogenesis models: Some solutions, *Quart. J. Roy. Meteorol. Soc.*, 97:139–153, 1971; E. Palmén and C. W. Newton, *Atmospheric Circulation Systems*, 1969.

Frost

A covering of ice in one of several forms produced by the freezing of supercooled water droplets on objects colder than 32°F. The partial or complete killing of vegetation, by freezing or by temperatures somewhat above freezing for certain sensitive plants, also is called frost. Air temperatures below 32°F sometimes are reported as "degrees of frost"; thus, on the Fahrenheit scale, 10°F is 22 degrees of frost; on the Celsius scale, −12°C is 12 degrees of frost.

Frost forms in exactly the same manner as dew except that the individual droplets that condense in the air a fraction of an inch from a subfreezing object are themselves supercooled, that is, colder than 32°F. When the droplets touch the cold object, they freeze immediately into individual crystals. When additional droplets freeze as soon as the previous ones are frozen, and hence are still close to the melting point because all the heat of fusion has not been dissipated, amorphous frost or rime results.

At more rapid rates of condensation, the drops form a film of supercooled water before freezing, and glaze or glazed frost ("window ice" on house windows, "clear ice" on aircraft) generally follows. Glaze formation on plants, buildings and other structures, and especially on wires sometimes is called an ice storm, or a silver frost storm, or thaw.

At slower deposition rates, such that each crystal cools well below the melting point before the next joins it, true crystalline or hoar frosts form. These include fernlike assemblages on snow surfaces, called surface hoar; similar feathery plumes in cold buildings, caves, and crevasses, called depth hoar; and the common window frost or ice flowers on house windows.

So-called killing frosts occur on clear autumn nights, when radiative cooling of ground, air, and vegetation causes plant fluids to freeze. At such times, the air temperature measured in a shelter 5–7 ft above ground usually is at least 5°F below freezing, but such standard level temperatures are poor indicators of frost severity. Air temperature varies greatly with height in the first few feet above the ground, and also with topography and vegetation around the shelter.

When wind is absent, the air layer immediately above the ground, rather than the ground itself, loses the most heat by radiation. Then the lowest temperature is 2–6 in. above, and about 1°F colder than, a bare ground surface. Above this near-ground minimum, an inversion of temperature develops for several to many feet thick. Plants radiate their heat faster than the air or ground, and may be colder than this air minimum temperature.

Frost damage can be prevented or reduced by heating the lowest air layers, or by mixing the cold surface air with the warmer air in the inversion above the tops of plants or trees.

Valley bottoms are much colder on clear nights than slopes, and may have frost when slopes are frost-free. In some notable frost pockets or hollows the air temperature may be 40°F colder than at nearby stations on higher ground, because of cold air drainage and the greater radiative cooling of level areas than of slopes. *See* CLOUD PHYSICS; DEW POINT; TEMPERATURE INVERSION; WEATHER MODIFICATION.

[ARNOLD COURT]

Fungistat and fungicide

Synthetic or biosynthetic compounds used to control fungal diseases in humans, animals, and plants.

Chemotherapy in humans. Cutaneous sporotrichosis is the only mycotic disease of humans for which there is a specific chemotherapeutic drug, potassium iodide. This drug is neither fungistatic nor fungicidal. The mode of action is unknown, but it is thought that the drug alters the tissue response to the etiologic agent in such a manner that factors involved in the host's resistance to *Sporothrix schenckii* may eventually kill the fungus. Potassium iodide is unsatisfactory for the treatment of other forms of sporotrichosis.

Only two drugs are available for the treatment of progressive, life-threatening mycotic diseases. Amphotericin B, a polyene produced by *Streptomyces nodosus*, is used in the treatment of histoplasmosis, blastomycosis, coccidioidomycosis, and disseminated candidiasis, aspergillosis, and sporotrichosis. This drug binds to sterols in the fungal cell membrane and increases the permeability of the cell, resulting in loss of potassium and other essential components. Amphotericin B is administered intravenously since it is poorly absorbed from the gastrointestinal tract. Because it is nephrotoxic, it requires careful monitoring of kidney function.

Flucytosine (5-FC) is a fluorinated pyrimidine. Its mode of action is thought to be through interference with pyrimidine metabolism of the fungal cells. Flucytosine is used primarily for patients with systemic infections due to *Candida* species and *Cryptococcus neoformans*. It is used occasionally in combination with amphotericin B. The drug is administered orally. Although not as toxic as amphotericin B, kidney and liver function studies are done at regular intervals while the patient is on this medication.

Several drugs are available for treating the dermatophytoses. Miconazole is a synthetic phenylimidazole whose mode of action is not clearly understood. It is thought to affect permeability of the fungal cells in a manner similar to that of some polyenes. Miconazole is available as an ointment. It is used on skin lesions caused by *Trichophyton mentogrophytes*, *T. rubrum*, and *Epidermophyton floccosum*. Tolnaftate and haloprogin are two other fungistatic drugs used topically in the treatment of the dermatophytoses.

Griseofulvin is a fungistatic drug produced by *Penicillium griseofulvum*. Its mode of action is not completely understood, but it apparently interferes with protein synthesis of the fungal cell. It is administered orally, and subsequently is detected in the stratum corneum.

Nystatin is produced by *Streptomyces noursei*. It is used either orally or topically for treatment of candidiasis of the skin, mucous membrane, and gastrointestinal tract. It is ineffectual in the treatment of systemic candidiasis.

Pimaricin is a polyene that is considered to be an excellent antifungal agent for ocular infections, and is especially active against *Fusarium* species. It is not available commercially in the United States.

[LEANOR D. HALEY]

Agricultural fungicides. Chemical compounds are used to control plant diseases caused by fungi. Fungicides now used include both inorganic and organic compounds. Agricultural fungicides must have certain properties and conform to very strict regulations, and of thousands of compounds tested, few have reached the farmer's fields. Special equipment is needed to apply fungicides. *See* PLANT DISEASE.

Inorganic fungicides. Inorganic fungicides, such as bordeaux mixture and sulfur, are still used in the greatest amounts. Bordeaux mixture is made by mixing a solution of copper sulfate with a suspension of lime (calcium hydroxide).

Organic fungicides. Organic fungicides have become increasingly important since 1934. Some of the most useful are derivatives of dithiocarbamic acid. Examples are ferbam and ziram, the iron and zinc salts respectively of dimethyldithiocarbamic acid, and nabam, zineb, and maneb, the sodium, zinc, and manganese salts respectively of ethylenebis(dithiocarbamic acid). Another related fungicide of importance is thiram or bis-(dimethylthiocarbamoyl)sulfide. Other representative fungicides which are widely used are captan, *N*-(tri-chloromethylthio)-4-cyclohexene-1,2-dicarboximide; chlorothalonil, tetrachloro-isophthalonitrile; glyodin, 2-heptadecyl-2-imidazoline acetate; chloranil, tetrachloro-*p*-benzoquinone; dichlone, 2,3-dichloro-1,4-naphthoquinone; and dodine, *n*-dodecylguanidine acetate. An organic compound commonly used to control powdery mildews is dinocap, 4,6-dinitro-2-(1-methylheptyl)phenylcrotonate. Cycloheximide, an antibiotic, is used on cherries, ornamentals, and turf. Newer agricultural fungicides are anilazine, 4,6-dichloro-*N*-(2-chlorophenyl)-1,3,5-triazine-2-amine; fentin hydroxide, triphenyltin hydroxide; and dichloran, 2,6-dichloro-4-nitroaniline.

Requirements and regulations. The stipulations that must be met by manufacturers before they can sell fungicides and other pesticides have been drastically revised. In 1972 the Federal Environmental Protection and Control Act (FEPCA) substantially changed the Federal Insecticide, Fungicide, and Rodenticide Act (FIFRA) dating from 1947.

The FEPCA made the Environmental Protection Agency responsible for regulating the sale and use of pesticides within the country. The law states that all pesticides shall be classified according to their degree of toxicity to humans and other nontarget organisms: safer materials for "general use," and more toxic materials for "restricted use." All pesticides must still be proven effective before they may be registered for sale, and their labels must state all legal uses and the conditions of use. The law also requires that eventually each state must certify the qualifications of those who apply pesticides commercially. Fortunately, most agricultural fungicides are generally safer than the compounds used as insecticides or herbicides. *See* HERBICIDE; INSECTICIDE; PESTICIDE.

Because fungicides must now be labeled for specific uses on specific crops, the cost of development largely limits their use to such major crops as peanuts, citrus, potatoes, and apples. To fill the gap, the U.S. Department of Agriculture has established the IR-4 program to finance and facilitate label clearance for the use of pesticides on minor crops.

Formulation. The manner in which these compounds are applied, that is, as wettable powders, dusts, or emulsions, is often essential to the success of agricultural fungicides. Raw fungicides must be pulverized to uniform particles of the most effective size, mixed with wetting agents, or dissolved in solvents. These carriers or diluents must not degrade the fungicides or must not injure the plants.

Foliage fungicides. This type of fungicide is applied to aboveground parts of plants, usually to prevent disease rather than cure it. Because they are intended to form a protective coating on the plant surface that kills fungus spores before infection occurs, foliage fungicides must adhere to foliage despite weathering. Fungicides also must be sufficiently stable chemically to resist degradation by water, oxygen, carbon dioxide, and sunlight. Sometimes, as in the case of zineb, specific chemical changes by weathering are necessary to produce highly fungicidal derivatives. Protective fungicides must be insoluble in water in order to remain on foliage. Certain foliage fungicides, however, are water-soluble. These materials destroy the fungus in disease spots after infection. Fungicides of this type are called eradicant or contact fungicides. An example is dodine, *n*-dodecylguanidine acetate.

Seed and soil treatments. Seeds and seedlings are protected against fungi in the soil by treating the seeds and the soil with fungicides. Seed-treating materials must be safe for seeds and must resist degradation by soil and soil microorganisms. Some soil fungicides are safe to use on living plants. An example is pentachloronitrobenzene, which can be drenched around seedlings of cruciferous crops and lettuce to protect them against root-rotting fungi. Other soil fungicides, such as formaldehyde, chloropicrin, and methyl isothiocyanate, are injurious to seeds and living plants. These compounds are useful because they are volatile. Used before planting, they have a chance to kill soil fungi and then escape from the soil.

Systemic fungicides. Systemic fungicides are compounds that permeate plants to protect new growth or to eliminate infections that have already occurred. Since the advent of benomyl, methyl 1-(butylcarbamoyl)-benzimidazol-2-yl-carbamate, and of carboxin, 2,3-dihydro-6-methyl-5-phenylcarbamoyl-1,4-oxathiin, other effective systemic fungicides have become available. These three are representative of the newer materials: thiophanate-methyl, 1,2-bis-(3-methoxycarbonyl-2-thioureido) benzene; dodemorph, *N*-cyclododecyl-2,6-dimethylmorpholine; and ethirimol, 5-*n*-butyl-2-ethylamino-4-hydroxy-6-methyl pyrimidine.

Application methods. Dusters and sprayers are used to apply foliage fungicides. Conventional sprayers apply 300–500 gal/acre at pressures up to 600 psi. This equipment ensures the uniform, adequate coverage necessary for control. Recent developments in spray equipment are the mist blower and the low-pressure, low-volume sprayer. The mist blower uses an air blast to spray droplets onto foliage. Mist blowers have been successful for applying fungicides to trees but are less satisfactory for applying fungicides to row crops. The low-pressure, low-volume sprayers are lightweight machines that apply about 80 gal of concentrated spray liquid per acre at a pressure of about 100 psi. These have been successfully used to protect tomatoes and potatoes against diseases caused by fungi. The most recent refinement of this method is ultralow-volume (ULV) spraying. With ULV spraying, which employs special spinning cage micronizers, growers can protect certain crops by applying as little as 0.5–2 gal of spray liquid per acre. Fungicides are also applied from aircraft. [JAMES G. HORSFALL; SAUL RICH]

Bibliography: D. L. Fowler and J. N. Mahan, *The Pesticide Review*, 1975; R. W. Marsh (ed.), *Systemic Fungicides*, 2d ed., 1977; W. T. Thomson, *Agricultural Chemicals*, book IV: *Fungicides*, 1976; D. C. Torgeson (ed.), *Fungicides*, 1967; E. I. Zehr (ed.), *Methods for Evaluating Plant Fungicides, Nematicides, and Bactericides*, 1978.

Geomorphology

The study of landforms, including the description, classification, origin, development, and history of planetary surface features. Emphasis is placed on the genetic interpretation of the erosional and depositional features of the Earth's surface. However, geomorphologists also study primary relief elements formed by movements of the Earth's crust, topography on the sea floor and on other planets, and applications of geomorphic information to problems in environmental engineering.

Geomorphologists analyze the landscape, a factor of immense importance to humankind. Their purview includes the structural framework of landscape, weathering and soils, mass movement and hillslopes, fluvial features, eolian features, glacial and periglacial phenomena, coastlines, and karst landscapes. Processes and landforms are analyzed for their adjustment through time, especially the most recent portions of Earth history.

History. Geomorphology emerged as a science in the early 19th century with the writings of James Hutton, John Playfair, and Charles Lyell. These men demonstrated that prolonged fluvial erosion is responsible for most of the Earth's valleys. Impetus was given to geomorphology by the exploratory surveys of the 19th century, especially those in the western United States. By the end of the 19th century, geomorphology had achieved its most important theoretical synthesis through the work of William Morris Davis. He conceived a marvelous deductive scheme of landscape development through the action of geomorphic processes acting on the structure of the bedrock to induce a progressive evolution of landscape stages.

Perhaps the premier geomorphologist was Grove Karl Gilbert. In 1877 he published his report "Geology of the Henry Mountains." This paper introduced the concept of equilibrium to organize tectonic and erosional process studies. Fluvial erosion was magnificently described according to the concept of energy. Gilbert's monograph "Lake Bonneville" was published in 1890 and described the Pleistocene history of the predecessor to the

Fig. 1. The great bar of Pleistocene Lake Bonneville at Stockton, Ut. The eminent geomorphologist Grove Karl Gilbert is depicted in the foreground at the plane table. *(From G. K. Gilbert, U.S. Geol. Surv. Monogr. 1, 1890)*

Fig. 2. Surveying large transverse gravel bars created by flooding of the Medina River, TX, in August 1978.

Great Salt Lake (see Fig. 1). The monograph is a masterpiece of dynamic analysis. Concepts of force and resistance, equilibrium, and adjustment—these dominated in Gilbert's science. He later presented a thorough analysis of fluvial sediment transport and the environmental effects of altered fluvial systems. He even made a perceptive study of the surface morphology of the Moon.

Despite Gilbert's example, geomorphologists in the early 20th century largely worked on landscape classification and description according to Davis's theoretical framework. Toward the middle of the 20th century, alternative theoretical approaches appeared. Especially in France and Germany, climatic geomorphology arose on the premise that distinctive landforms and processes are associated with certain climatic regions. Geomorphology since 1945 has become highly diversified, with many groups specializing in relatively narrow subfields, such as karst geomorphology, coastal processes, glacial and Quaternary geology, and fluvial processes.

Process geomorphology. Modern geomorphologists emphasize basic studies of processes presently active on the landscape (Fig. 2). This work has benefited from new field, laboratory, and analytical techniques, many of which are borrowed from other disciplines. Geomorphologists consider processes from the perspectives of pedology, soil mechanics, sedimentology, geochemistry, hydrology, fluid mechanics, remote sensing, and other sciences. The complexity of geomorphic processes has required this interdisciplinary approach, but it has also led to a theoretical vacuum in the science. At present many geomorphologists are organizing their studies through a form of systems analysis. The landscape is conceived of as a series of elements linked by flows of mass and energy. Process studies measure the inputs, outputs, and transfers for these systems. Although systems analysis is not a true theory, it is compatible with the powerful new tools of computer analysis and remote sensing. Systems analysis provides an organizational framework within which geomorphologists are developing models to predict selected phenomena.

The future. Geomorphology is increasing in importance because of the increased activity of humans as a geomorphic agent. As society evolves to more complexity, it increasingly affects and is threatened by such geomorphic processes as soil erosion, flooding, landsliding, coastal erosion, and sinkhole collapse. Geomorphology plays an essential role in environmental management, providing a broader perspective of landscape dynamics than can be given by standard engineering practice.

The phenomenal achievements of 19th century geomorphology were stimulated by the new frontier of unexplored lands. The new frontier for geomorphology in the late 20th century lies in the study of other planetary surfaces (Fig. 3). Each new planetary exploration has revealed a diversity

Fig. 3. Streamlined uplands and large sinuous channels in the Chryse Planitia region of Mars. *(National Aeronautics and Space Administration)*

of processes that stimulates new hypotheses for features on Earth. Geomorphology must now solve the mysteries of meteor craters on the Moon and Mercury, great landslides and flood channels on Mars, phenomenally active volcanism on Io, and ice tectonics on Ganymede. *See* EROSION.

[VICTOR R. BAKER]

Bibliography: V. R. Baker and S. J. Pyne, G. K. Gilbert and modern geomorphology, *Amer. J. Sci.*, 278:97–123, 1978; A. L. Bloom, *Geomorphology: A Systematic Analysis of Late Cenozoic Landforms*, 1978; A. F. Pitty, *Introduction to Geomorphology*, 1971; D. F. Ritter, *Process Geomorphology*, 1978.

Geothermal power

Thermal or electrical power produced from the thermal energy contained in the Earth (geothermal energy). Use of geothermal energy is based thermodynamically on the temperature difference between a mass of subsurface rock and water and a mass of water or air at the Earth's surface. This temperature difference allows production of thermal energy that can be either used directly or converted to mechanical or electrical energy.

CHARACTERISTICS AND USE

Temperatures in the Earth in general increase with increasing depth, to 200–1000°C at the base of the Earth's crust and to perhaps 3500–4500°C at the center of the Earth. Average conductive geothermal gradients to 10 km (the depth of the deepest wells drilled to date) are shown in Fig. 1 for representative heat-flow provinces of the United States. The heat that produces these gradients comes from two sources: flow of heat from the deep crust and mantle; and thermal energy generated in the upper crust by radioactive decay of isotopes of uranium, thorium, and potassium. The gradients of Fig. 1 represent regions of different conductive heat flow from the mantle or deep crust. Some granitic rocks in the upper crust, however, have abnormally high contents of U and Th and thus produce anomalously great amounts of thermal energy and enhanced flow of heat toward the Earth's surface. Consequently, thermal gradients at shallow levels above these granitic plutons can be somewhat greater than shown on Fig. 1.

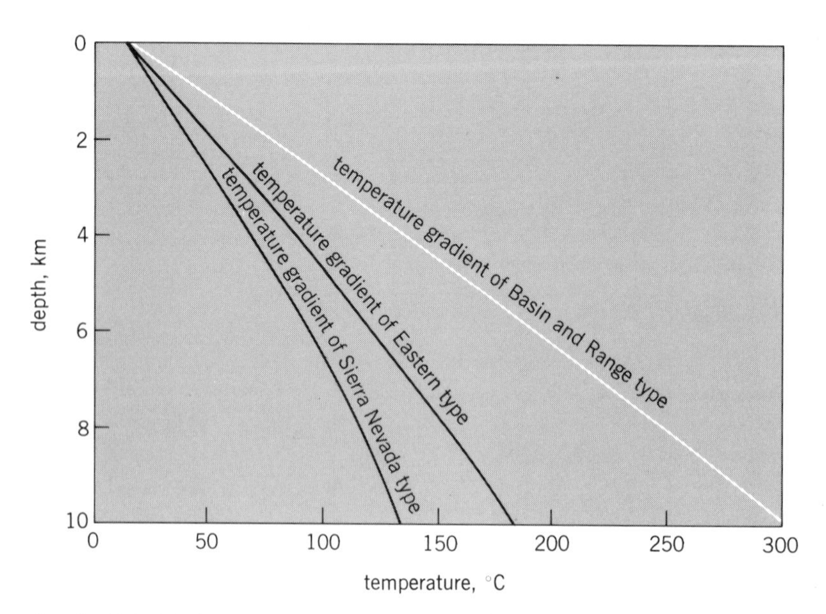

Fig. 1. Calculated average conductive temperature gradients to a depth of 10 km in representative heat-flow provinces of the United States. (*Adapted from D. E. White and D. L. Williams, eds., Assessment of Geothermal Resources of the United States—1975, USGS Circ. 726, 1975*)

The thermal gradients of Fig. 1 are calculated under the assumption that heat moves toward the Earth's surface only by thermal conduction through solid rock. However, thermal energy is also transmitted toward the Earth's surface by movement of molten rock (magma) and by circulation of water through interconnected pores and fractures. These processes are superimposed on the regional conduction-dominated gradients of Fig. 1 and give rise to very high temperatures near the Earth's surface. Areas characterized by such high temperatures are the primary targets for geothermal exploration and development.

Natural geothermal reservoirs. Commercial exploration and development of geothermal energy to date have focused on natural geothermal reservoirs—volumes of rock at high temperature (up to 350°C) and with both high porosity (pore space, usually filled with water) and high permeability (ability to transmit fluid). The thermal energy is tapped by drilling wells into the reservoirs. The thermal energy in the rock is transferred by conduction to the fluid, which subsequently flows to the well and then to the Earth's surface.

Natural geothermal reservoirs, however, make up only a small fraction of the upper 10 km of the Earth's crust. The remainder is rock of relatively low permeability whose thermal energy cannot be produced without fracturing the rock artificially by means of explosives or hydrofracturing. Experiments involving artificial fracturing of hot rock have been performed, and extraction of energy by circulation of water through a network of these artificial fractures may someday prove economically feasible.

There are several types of natural geothermal reservoirs. All the reservoirs developed to date for electrical energy are termed hydrothermal convection systems and are characterized by circulation of meteoric (surface) water to depth. The driving force of the convection systems is gravity, effective because of the density difference between cold, downward-moving, recharge water and heated, upward-moving, thermal water. A hydrothermal convection system can be driven either by an underlying young igneous intrusion or by merely deep circulation of water along faults and fractures. Depending on the physical state of the pore fluid, there are two kinds of hydrothermal convection systems: liquid-dominated, in which all the pores and fractures are filled with liquid water that exists at temperatures well above boiling at atmospheric pressure, owing to the pressure of overlying water; and vapor-dominated, in which the larger pores and fractures are filled with steam. Liquid-dominated reservoirs produce either water or a mixture of water and steam, whereas vapor-dominated reservoirs produce only steam, in most cases superheated.

Natural geothermal reservoirs also occur as regional aquifers, such as the Dogger Limestone of the Paris Basin in France and the sandstones of the Pannonian series of central Hungary. In some rapidly subsiding young sedimentary basins such as the northern Gulf of Mexico Basin, porous reservoir sandstones are compartmentalized by growth faults into individual reservoirs that can have fluid pressures exceeding that of a column of water and approaching that of the overlying rock. The pore water is prevented from escaping by the impermeable shale that surrounds the compartmented sandstone. The energy in these geopressured reservoirs consists not only of thermal energy, but also of an equal amount of energy from methane dissolved in the waters plus a small amount of mechanical energy due to the high fluid pressures. *See* AQUIFER; GROUNDWATER.

Use of geothermal energy. Although geothermal energy is present everywhere beneath the Earth's surface, its use is possible only when certain conditions are met: (1) The energy must be accessible to drilling, usually at depths of less than 3 km but possibly at depths of 6–7 km in particularly favorable environments (such as in the northern Gulf of Mexico Basin of the United States). (2) Pending demonstration of the technology and economics for fracturing and producing energy from rock of low permeability, the reservoir porosity and permeability must be sufficiently high to allow production of large quantities of thermal water. (3) Since a major cost in geothermal development is drilling and since costs per meter increase with increasing depth, the shallower the concentration of geothermal energy the better. (4) Geothermal fluids can be transported economically by pipeline on the Earth's surface only a few tens of kilometers, and thus any generating or direct-use facility must be located at or near the geothermal anomaly.

Electric power generation. The most conspicuous use of geothermal energy is the generation of electricity. Hot water from a liquid-dominated reservoir is flashed partly to steam at the Earth's surface, and this steam is used to drive a conventional turbine-generator set. In the relatively rare vapor-dominated reservoirs, superheated steam produced by wells can be piped directly to the turbine without need for separation of water. Electricity is most readily produced from reservoirs of 180°C or greater, but reservoirs of 150°C or even lower show promise for electrical generation, either by using steam directly or by transferring its heat to a working fluid of low boiling point such as isobutane or Freon. Installed geothermal electrical capacity in mid-1979 is shown in the table, and the increase in worldwide geothermal electrical capacity with time is shown in Fig. 2. The importance of geothermal electricity to a small country is illustrated by El Salvador, where in 1977 the electricity generated from the Ahuachapán geothermal field represented 32% of the total electricity generated in that country. *See* ELECTRIC POWER GENERATION.

Direct use. Equally important worldwide is the direct use of geothermal energy, often at reservoir temperatures less than 100°C. Geothermal energy is used directly in a number of ways: to heat buildings (individual houses, apartment complexes, and even whole communities); to cool buildings (using lithium bromide absorption units); to heat greenhouses and soil; and to provide hot or warm water for domestic use, for product processing (for example, the production of paper), for the culture of shellfish and fish, for swimming pools, and for therapeutic (healing) purposes.

Installed geothermal electrical generating capacity of the world

Country	Megawatts-electrical Early 1979	Projected by 1984
Italy	420.6	482
New Zealand	202.6	302
Mexico	75	235
Japan	166	216
Iceland	64	64
Soviet Union	5	28
Chile		30
China	1	?
El Salvador	60	95
Nicaragua		35
Philippines	3	548
Taiwan	0.6	5.6
Turkey	0.5	15.5
Indonesia		30
United States	608	2008
Total	1606.3	4094

Major localities where geothermal energy is directly used include Iceland (30% of net energy consumption, primarily as domestic heating), the Paris Basin of France (where 60–70°C water is used in district heating systems for the communities of Melun, Creil, and Villeneuve la Garenne), and the Pannonian Basin of Hungary.

Prospects. In any analysis of the possible contribution of geothermal energy to human energy needs, one must keep in mind that the geothermal resource (that is, the potentially usable geothermal energy) is only a fraction of the thermal energy in a

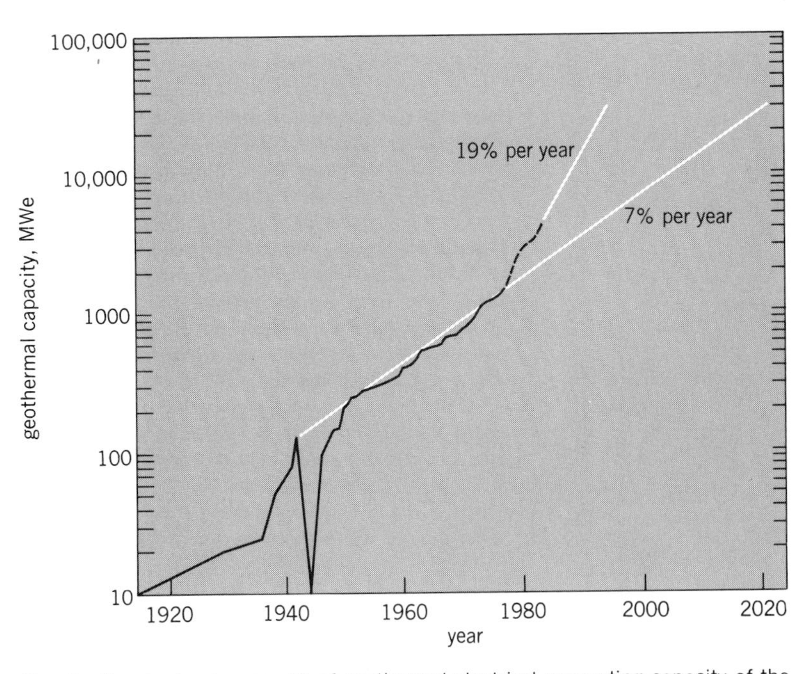

Fig. 2. Graph showing growth of geothermal electrical generating capacity of the world. *(From L. J. P. Muffler, ed., Assessment of Geothermal Resources of the United States–1978, USGS Circ. 790, 1979)*

subsurface volume of rock and water. For favorable hydrothermal convection systems, this fraction can be 25% or greater, but for systems of restricted permeability the fraction is likely to be far smaller. Only this recoverable energy can be meaningfully compared with the thermal energy equivalent of barrels of recoverable oil, cubic meters of recoverable gas, tons of minable coal, or kilograms of minable uranium.

Use of geothermal energy is likely to increase greatly in many countries as other sources of energy become less abundant and more expensive and as geothermal reservoirs become better defined through systematic exploration and resource assessment. In the United States, the U.S. Geological Survey has estimated that the geothermal resources in identified and undiscovered hydrothermal convection systems are 2400×10^{18} joules, equivalent to the energy in 430×10^9 barrels (68×10^9 m^3) of oil or 47 years of oil consumption at the projected 1980 rate. Geopressured-geothermal resources (both thermal energy and energy from dissolved methane) of the northern Gulf of Mexico Basin are estimated to be between 430 and 4400×10^{18} joules, equivalent to 74 to 780×10^9 barrels (11.8 to 124×10^9 m^3) of oil or 8 to 85 years of oil consumption at the projected 1980 rate. Any energy that might be developed in the future from rock of low permeability or from magma would be in addition to this amount. Clearly, geothermal energy can play an important role in the energy economy of the United States as well as in the economies of many other countries throughout the world.

[L. J. PATRICK MUFFLER]

PRODUCTION AND POLLUTION PROBLEMS

The chief problems in producing geothermal power involve mineral deposition, changes in hydrological conditions, and corrosion of equipment. Pollution problems arise in handling geothermal effluents, both water and steam.

Mineral deposition. In some water-dominated fields there may be mineral deposition from boiling geothermal fluid. Silica deposition in wells caused problems in the Salton Sea, California, field; more commonly, calcium carbonate scale formation in wells or in the country rock may limit field developments, for example, in Turkey and the Philippines. Fields with hot waters high in total carbonate are now regarded with suspicion for simple development. In the disposal of hot wastewaters at the surface, silica deposition in flumes and waterways can be troublesome.

Hydrological changes. Extensive production from wells changes the local hydrological conditions. Decreasing aquifer pressures may cause boiling of water in the rocks (leading to changes in well fluid characteristics), encroachment of cool water from the outskirts of the field, or changes in water chemistry through lowered temperatures and gas concentrations. After an extensive withdrawal of hot water from rocks of low strength, localized ground subsidence may occur (up to several meters) and the original natural thermal activity may diminish in intensity. Some changes occur in all fields, and a good understanding of the geology and hydrology of a system is needed so that the well withdrawal rate can be matched to the well's long-term capacity to supply fluid.

Corrosion. Geothermal waters cause an accelerated corrosion of most metal alloys, but this is not a serious utilization problem except, very rarely, in areas where wells tap high-temperature acidic waters (for example, in active volcanic zones). The usual deep geothermal water is of near-neutral pH. The principal metal corrosion effects to be avoided are sulfide and chloride stress corrosion of certain stainless and high-strength steels and the rapid corrosion of copper-based alloys. Hydrogen sulfide, or its oxidation products, also causes a more rapid degradation than normal of building materials, such as concrete, plastics, and paints.

Pollution. A high noise level can arise from unsilenced discharging wells (up to 120 decibels adjusted), and well discharges may spray saline and silica-containing fluids on vegetation and buildings. Good engineering practice can reduce these effects to acceptable levels.

Because of the lower efficiency of geothermal power stations, they emit more water vapor per unit capacity than fossil-fuel stations. Steam from wellhead silencers and power station cooling towers may cause an increasing tendency for local fog and winter ice formation. Geothermal effluent waters liberated into waterways may cause a thermal pollution problem unless diluted by at least 100:1.

Geothermal power stations may have four major effluent streams. Large volumes of hot saline effluent water are produced in liquid-dominated fields. Impure water vapor rises from the station cooling towers, which also produce a condensate stream containing varying concentrations of ammonia, sulfide, carbonate, and boron. Waste gases flow from the gas extraction pump vent.

Pollutants in geothermal steam. Geothermal steam supplies differ widely in gas content (often 0.1–5%). The gas is predominantly carbon dioxide, hydrogen sulfide, methane, and ammonia. Venting of hydrogen sulfide gas may cause local objections if it is not adequately dispersed, and a major geothermal station near communities with a low tolerance to odor may require a sulfur recovery unit (such as the Stretford process unit). Sulfide dispersal effects on trees and plants appear to be small. The low radon concentrations in steam (3–200 nanocuries/kg or 0.1–7.4 kilobecquerels/kg), when dispersed, are unlikely to be of health significance. The mercury in geothermal stream (often 1–10 μg/kg) is finally released into the atmosphere, but the concentrations created are unlikely to be hazardous. *See* AIR POLLUTION.

Geothermal waters. The compositions of geothermal waters vary widely. Those in recent volcanic areas are commonly dilute (<0.5%) saline solutions, but waters in sedimentary basins or active volcanic areas range upward to concentrated brines. In comparison with surface waters, most geothermal waters contain exceptional concentrations of boron, fluoride, ammonia, silica, hydrogen sulfide, and arsenic. In the common dilute geothermal waters, the concentrations of heavy metals such as iron, manganese, lead, zinc, cadmium, and thallium seldom exceed the levels permissible in drinking waters. However, the concentrated brines may contain appreciable levels of heavy metals (parts per million or greater).

Because of their composition, effluent geothermal waters or condensates may adversely affect potable or irrigation water supplies and aquatic life. Ammonia can increase weed growth in waterways and promote eutrophication, while the entry of boron to irrigation waters may affect sensitive plants such as citrus. Small quantities of metal sulfide precipitates from waters, containing arsenic, antimony, and mercury, can accumulate in stream sediments and cause fish to derive undesirably high (over 0.5 ppm) mercury concentrations. *See* WATER POLLUTION.

Reinjection. The problem of surface disposal may be avoided by reinjection of wastewaters or condensates back into the countryside through disposal wells. Steam condensate reinjection has few problems and is practiced in Italy and the United States. The much larger volumes of separated waste hot water (about 50 metric tons per megawatt-electric) from water-dominated fields present a more difficult reinjection situation. Silica and carbonate deposition may cause blockages in rock fissures if appropriate temperature, chemical, and hydrological regimes are not met at the disposal depth. In some cases, chemical processing of brines may be necessary before reinjection. Selective reinjection of water into the thermal system may help to retain aquifer pressures and to extract further heat from the rock. A successful water reinjection system has operated for several years at Ahuachapan, El Salvador. [A. J. ELLIS]

Bibliography: H. C. H. Armstead, *Geothermal Energy*, 1978; H. C. H. Armstead (ed.), *Geothermal Energy: Review of Research and Development*, UNESCO, 1973; R. C. Axtmann, *Science*, 187: 795–803, 1975; A. J. Ellis and W. A. J. Mahon, *Chemistry and Geothermal Systems*, 1977; L. J. P. Muffler (ed.), *Assessment of Geothermal Resources of the United States—1978*, USGS Circ. 790, 1979; *Proceedings of the First United Nations Symposium on the Development and Utilization of Geothermal Resources, Pisa, Italy, Sept. 1970*, spec. issue no. 2 of *Geothermics*, 2 vols., 1973; *Proceedings of the Second United Nations Symposium on the Development and Use of Geothermal Resources, San Francisco, May 1975*, U.S. Government Printing Office, 3 vols., 1976; E. F. Wahl, *Geothermal Energy Utilization*, 1977; D. E. White and D. L. Williams (eds.), *Assessment of Geothermal Resources of the United States—1975*, USGS Circ. 726, 1975.

Grassland ecosystem

Grassland vegetation comprises herbaceous (not woody) plant species; graminoids (grasses and grasslike plants) usually dominate. Natural grasslands once occupied at least one-third of the land surface of the Earth, occurring in regions which were too arid for forest, but where the climate was not so adverse as to prevent the development of a closed perennial herbaceous cover that is lacking in desert. In addition, grassland with scattered trees (savannas) occurs in transition zones between grassland and forest, particularly in the tropics and subtropics, and in semidesert regions shrubs are scattered throughout grassland. Humans have converted some forest areas into seminatural grasslands, and they have modified many natural grasslands by tillage and by grazing domesticated livestock. The most extensive areas of natural open grassland occur in the temperate zone (south-central North America, north-central Asia, eastern Europe, southeastern South America, and southern Africa; see Figs. 1–3). Savannas are widespread in the tropical regions of Africa, South America, and Australia; shrubby grasslands lie adjacent to deserts (Fig. 4).

Fig. 1. Natural temperate grassland in the Canadian portion of the Great Plains of North America. *(Photograph by R. T. Coupland)*

Climate and soils. The climate of natural grassland exhibits marked fluctuations in precipitation, both from season to season and from year to year. Annual droughts of several weeks to several months are typical. Severity of drought increases with distance from the forest margin. There is a tendency for years of below-average precipitation to be grouped. Mean annual precipitation in temperate grasslands ranges from about 250 to 750 mm, while in tropical and subtropical grasslands and savannas it reaches about 1500 mm. High winds cause additional water stress by increasing evaporation. The adverse effects of drought are increased by high temperature (less cloudiness, more sunshine). The length of the growing season is determined in the tropics and subtropics by the

Fig. 2. Natural grassland in the intermountain plateau of northern Mexico. *(Photograph by R. T. Coupland)*

Fig. 3. Natural grassland in the Serengeti plain of Tanzania in eastern Africa. *(Photograph by R. T. Coupland)*

length of the rainy season, and in temperate regions temperature becomes an important factor. The wide range in grassland climates results in considerable differences in the vigor of growth that is attained. Thus, short grasses dominate in areas adjacent to deserts where precipitation ranges from 100 to 300 mm per year; midgrasses and tall grasses abound in semiarid to subhumid conditions where the climate is sufficiently favorable to produce tilled crops; and high grasses (as tall as 4 m or more) occur in subtropical regions where growing conditions are very favorable during the wet season. *See* DROUGHT.

The nature of the soil-forming process in temperate grassland results in soils that are rich in nutrients (because of restricted leaching) and an abundance of organic matter. However, the soils of some tropical and subtropical grasslands are highly leached, and there are rapid decay and low levels of organic matter accumulation. Color ranges from black to brown in temperate grasslands and reddish to yellowish in tropical areas. Because of more rapid percolation (less loss through runoff and evaporation), a greater proportion of the precipitation enters sandy soils than silty or clayey types. As a result, in relatively dry regions a greater supply of soil moisture (in the total profile) is available in sandy soils, which thus support taller vegetative growth. *See* SOIL.

Fig. 4. Tropical palm savanna in the Ivory Coast in western Africa. *(Photograph by R. T. Coupland)*

Vegetation structure and composition. Layering is characteristic of grasslands, both above- and belowground. For example, where midgrasses dominate, they are typically accompanied by an understory of short grasses and forbs (nongrasslike herbs), under which the soil surface is occupied by even dwarfer species (such as mosses, clubmosses, and lichens); an overstory of forbs also occurs occasionally. Usually the grasses develop sufficiently so that their foliage covers the soil surface, although their bases may occupy only 10–25% of it.

Root layering is characteristic of a grassland community. There is a direct relationship between the height of growth and the depth of rooting. Depth of rooting also is related to the soil water, since roots penetrate only to the depth of moisture penetration. The roots of plants in sandy soil tend to grow deeper than those in clayey soil, presumably because of the deeper penetration of moisture in the former. Some forbs have roots that do not branch until they reach a depth below that of the grasses. These deep-rooted species apparently do not compete with the grasses and have remarkable abilities to endure drought.

While there are a large number of plant species in the grassland ecosystems of the world, usually very few grasses dominate in any one location: commonly 50–200 species of flowering plants occur in the same grassland, but only a few of these are considered dominants. Grassland complexity (in terms of numbers of plant species) tends to increase with increasing favorableness of habitat and to decrease with increasing aridity.

Graminoids are more resistant to grazing and trampling than are forbs, a situation that seems to be associated, among other things, with the presence of meristematic (growth) zones in each joint of the stem and in each leaf that permit grazed leaves to continue growth and bent-over stems to straighten.

The species that occupy a natural grassland can be grouped according to the season during which each develops. Thus, there are species that complete their growth cycle and form seeds before taller-growing species reach maximum activity. The changes which occur in grassland with season, as a result of the activity of different species at different times, has been referred to as aspect. *See* ECOLOGICAL SUCCESSION.

Tree-inhibiting factors. The treelessness of natural grassland is often the result of aridity. Relatively low precipitation and high evaporation restrict the supply of soil moisture. Most trees cannot compete effectively with grasses where upper soil layers are intermittently moist, but deeper layers are continuously dry. Other factors that have been cited to account for the absence of trees in various grassland areas are impermeable soil, excessive salinity, and poor drainage.

Natural grasslands have developed in areas where fires are characteristic, since herbaceous species are much better adapted than trees and shrubs to withstand the effects of fire. This is because their perennating buds are located near the soil surface where they are less exposed, and because they have no perennial stems.

Herbaceous communities above the tree line and in the Arctic are sometimes dominated by graminoids and can be considered natural grasslands. These communities exist because of the failure of trees to invade excessively cold or windy areas. *See* PLANT FORMATIONS, CLIMAX.

Animals. Natural grasslands are inhabited by a large variety of consumers. These animals exhibit various protective mechanisms against the rigorous environment. Some species migrate to escape the winter; others live most or all of their life cycle in the more stable belowground environment.

These animals may be classified functionally as herbivores, predators, and scavengers; however, in some (omnivores) the roles are mixed, and in others the role changes from season to season with the availability of different kinds of food. Plant species have become well adapted to survival under grazing, and herbivores have been inseparable components of grassland ecosystems since long before the advent of humans. Herbivores of various types exist together without competition by developing specific grazing habits in relation to choice of species eaten, part of plant consumed, or season of use.

Grazing by large herbivores, both native and domesticated, may be sufficiently intensive so as to alter the floristic composition of the vegetation. This permits plants to flourish that are preferred by smaller herbivores; thus a dynamic interaction of these two groups of herbivores is generated.

Rodents, invertebrates, and birds find suitable habitats in grassland. The passerine birds are particularly characteristic, subsisting largely on seeds or on mixtures of seeds and insects. Invertebrates (including insects) are very numerous, both above- and belowground, occupying various roles in the food webs. Some graze on leaves, while others consume roots and debris, reducing these to a form in which the microorganisms can more effectively attack them, so as to release nutrients for renewed plant growth. The total effect of many of these invertebrates on plants is not well known. Under some circumstances grasshoppers become so numerous that they compete with large herbivores for plant foliage.

Most of the reptiles and amphibians of grassland ecosystems are predators. Lizards, toads, and box turtles prey on insects, while snakes prey on rodents and other small vertebrates.

Animals in grasslands adapt to fluctuations in food supplies by various techniques. Some (such as insects) feed intensively when lush herbage is available, store up food reserves as fat, and become dormant through dry and cold periods. Others (such as most rodents) fatten during periods of vigorous plant growth and sleep during the winter in burrows or other protected niches. Still others migrate out of the grassland during periods when food is unavailable or scarce.

Grassland animals have adapted their reproductive patterns to meet expected availability of lush, green, nutritious herbage. Thus, the young of large herbivores are born in early spring to early summer. Invertebrates are sporadic in their population densities, particularly those aboveground, and are partly keyed to sequences of wet and dry years or to summer and winter extremes. The populations of rodents undergo dramatic fluctuations in population densities which are cyclic in character over several years.

Changes in the number of any group of herbivores soon induce changes in the number of predators and parasites, and changes in the vegetation. The grassland operates as an intricate feedback system, with each component affecting the others, always so as to bring population densities back to equilibrium, at least temporarily. *See* POPULATION DYNAMICS.

Microorganisms. Microbial populations occur in the soil, on the surfaces of and within the tissues of plants and animals, and in animal excreta. The most abundant groups in grassland soils are bacteria, actinomycetes, and fungi. The detritus food web is a highly complex system composed of large numbers of organisms. Several types of decomposer organisms are recognized on a functional basis. Reducers consume dead organic matter. Saprovores channel energy and nutrients from the detritus food web into the grazing food web by becoming food for carnivores. Microfloral grazers recirculate materials within the detritus food web and also transfer material back to the grazing food web. Microorganisms have other important functions in ecosystems besides their role as reducers and decomposers; in many grasslands they fix significant amounts of atmospheric nitrogen. *See* SOIL MICROORGANISMS.

Productivity. Natural grassland is more productive in terms of net biomass production, rate of energy flow, and rate of turnover of nutrients than is often inferred by the relatively low aboveground stature of the vegetation. Unlike forests and shrub lands, biomass does not accumulate aboveground and a large proportion of synthesized materials is transported below ground. *See* BIOMASS.

In natural grasslands, probably about 1% of the photosynthetically active radiation from the Sun is converted into plant materials. These most commonly amount to 700–1500 g/m^2 annually in temperate grasslands, but may exceed 4000 g/m^2 in subhumid tropical areas. In most grasslands, at least half of this material is transferred to the root system. These estimates of net production are similar to those of forest growing (under more favorable conditions) at the same latitude. However, plants protect themselves more against the adversities of the environment in grassland by transferring a larger proportion of the produced materials belowground. Grasslands can be considered as much more conservative communities than are forests.

In natural grassland, large grazing animals consume a relatively small amount (commonly 5–10%) of the total herbage produced. Some estimates have indicated that consumption by rodents, birds, and invertebrates may equal or exceed that of large herbivores. However, a very large proportion of the plant material is processed by invertebrates and microorganisms in soil and surface layers.

In grasslands the largest reservoir for cycling

plant nutrients is the soil organic matter which is undergoing various stages of decay. The mean rate of turnover is apparently of the magnitude of once every 3 or 4 years, so that the nutrients cycle rapidly, in contrast to forests where the nutrients are stored in the litter on the soil surface and in woody structures, and the average rate of turnover is measured in decades rather than years. Leaching in forest soils is sufficiently rapid so that storage of nutrients in soil organic matter is not feasible.

Exploitation and management. The two principal purposes for which natural grasslands are used are crop production and domesticated livestock grazing. Only a small fraction of the world's grasslands have been brought under tillage because of constraints imposed by climate, aridity, soil conditions (including stoniness, salinity, and erodibility), and topography. Most of the remainder are in some form of ranching enterprise.

The exploitation of natural temperate grasslands (which offer the greatest potential for dryland crop production) did not get under way on a large scale until agricultural settlement took place in the central North American grassland in the latter half of the 19th century. Only then was experience gained in cropping semiarid natural grassland regions without irrigation. Similar exploitation in the Eastern Hemisphere did not take place until the 1950s with the development of the "New Lands" of Kazakhstan. After a century or so of use, these soils have declined markedly (50% or more) in organic matter and nutrient content, and in some areas, increases in salinity have occurred. Improvements in varieties and increases in the use of commercial fertilizers have offset the effects of declining fertility, but concern is being expressed over the sustainability of production in these lands. In the subtropical and tropical grassland regions, crop production is restricted to areas of higher precipitation where leaching and rapid decay of organic matter likewise contribute to declining fertility unless management regimes are designed to offset deterioration. In order to protect the future well-being of tilled areas within the grassland zone, more constructive management techniques must be used to replace the exploitative practices that have resulted in their deterioration.

Grazing generally has an effect on native grasslands similar to that of increasing aridity. There is a tendency for the most palatable and most productive species to decrease in abundance and for the less desirable species to increase. If grazing is sufficiently heavy and prolonged, the vegetative cover thins out and is invaded by weedy species, thus reducing the carrying capacity and threatening deterioration of the soil by erosion. Range management is an applied science that is concerned with the management of natural grasslands for the sustained production of domesticated animals. This subject is closely related to that of wildlife management, particularly in respect to combined use of the same land for domesticated and game animals. The increased demand for livestock products has caused heavy grazing of many grassland areas, some of which have deteriorated to desertlike conditions. Moderate use is necessary in order to sustain their capacity as rangelands. *See* ECOSYSTEM; RANGELAND CONSERVATION.

[ROBERT T. COUPLAND; GEORGE VAN DYNE]

Bibliography: D. L. Allen, *The Life of Prairies and Plains*, 1967; A. I. Breymeyer and G. M. Van Dyne (eds.), *Grasslands, Systems Analysis and Man*, 1979; R. T. Coupland (ed.), *Grassland Ecosystems of the World: Analysis of Grasslands and Their Uses*, 1979; P. J. Darlington, *Zoogeography: The Geographical Distribution of Animals*, 1957; H. F. Heady, *Rangeland Management*, 1975; D. N. Hyder (ed.), *Proceedings of the First International Rangeland Congress*, 1978; R. M. Moore (ed.), *Australian Grasslands*, 1970; National Geographic Society, *Wild Animals of North America*, 1960; H. B. Sprague (ed.), *Grasslands*, AAAS Publ. no. 53, 1959; L. A. Stoddart, A. D. Smith, and T. W. Box, *Range Management*, 1975; J. E. Weaver, *North American Prairie*, 1954; J. E. Weaver and F. W. Albertson, *Grasslands of the Great Plains*, 1956.

Greenhouse effect

The Earth's atmosphere acts as the glass walls and roof of a greenhouse in trapping heat from the Sun. Like the greenhouse, it is largely transparent to solar radiation, but it strongly absorbs the longer-wavelength radiation from the ground. Much of this long-wave radiation is reemitted downward to the ground, with the paradoxical result that the Earth's surface receives more radiation than it would if the atmosphere were not between it and the Sun.

The absorption of long-wave (infrared) radiation is effected by small amounts of water vapor, carbon dioxide, and ozone in the air and by clouds. Clouds actually absorb about one-fifth of the solar radiation striking them, but unless they are extremely thin, they are almost completely opaque to infrared radiation. The appearance even of cirrus clouds after a period of clear sky at night is enough to cause the surface air temperature to increase rapidly by several degrees because of radiation from the cloud.

The greenhouse effect is most marked at night, and usually keeps the diurnal temperature range below 20°F (11°C). Over dry regions such as New Mexico and Arizona, however, where the water-vapor content of the air is low, the atmosphere is more transparent to infrared radiation, and cool nights may follow very hot days.

[LEWIS D. KAPLAN]

Groundwater

The water in the zone in which the rocks and soil are saturated, the top of which is the water table. The zone of saturation is the source of water for wells, which provide about one-fifth of the water supplies of the United States. It is also the source of the water that issues as springs and seeps, and maintains the dry-weather flow of perennial streams. The saturated zone is a natural reservoir which absorbs precipitation during wet periods and releases it during dry periods, thus tempering the severity of floods and droughts. The amount of groundwater stored in the rocks in the United States is estimated to be several times as great as that stored in all lakes and reservoirs, including the Great Lakes.

A knowledge of geology is essential to an understanding of the occurrence of water. For this reason, the study of groundwater is sometimes called hydrogeology or geohydrology.

Subterranean water. Water beneath the land surface occurs in the zone of aeration above the water table and in the zone of saturation below the water table. Water in the zone of aeration, also called vadose water, is divided into soil water, intermediate vadose water, and water of the capillary fringe.

Water in the capillary fringe is connected with the zone of saturation and is held above it by capillary forces. The lower part of the fringe may be saturated, but is not a part of the zone of saturation because the water is under less than atmospheric pressure and will not flow into a well. When the well reaches the zone of saturation, water will begin to enter it and will stand at the level of the water table.

Rock formations capable of yielding significant volumes of water are called aquifers. Some wells are artesian; that is, the water rises above the top of the water-bearing bed. Other wells encounter water above the saturated zone and lose their water if extended through the impermeable bed upon which the water rests. Such bodies of water are said to be perched.

The total volume of water available and the rate at which it can be released to a well are determined by the number and size of the openings in rocks and soils and the manner in which they are interconnected. Openings are practically absent in some igneous rocks. They are numerous but microscopic in clay. They are large and interconnected in many sands and gravels. There are huge caverns and tubes in many limestones and lavas. The distribution and types of openings are as diverse as the geology itself, so that general statements about them applicable to one area may be incorrect for another.

Openings in rocks. Primary openings are those which existed when the rock was formed, and secondary, those which resulted from the action of physical or chemical forces after the rock was formed. Primary openings are found in sedimentary rocks such as sand and clay and certain kinds of limestone composed of triturated shells. Openings in lava formed at the stage when the lava is partly liquid and partly solid are also considered primary. Most rocks containing primary openings are geologically young. Rocks which contain primary openings large enough to carry useful amounts of water are represented, for example, by the seaward-dipping strata of the Atlantic and Gulf coastal plains, including the coquina limestone of Florida, the intermontane valleys of the western United States, the glacial deposits of the North-Central states, and the lava rocks of the Pacific Northwest.

Secondary openings are common in older rocks. Sand and gravel that have been cemented by chemical action, limestone indurated by compression or recrystallization, schist, gneiss, slate, granite, rhyolite, basalt and other igneous rocks, and shale generally contain few primary openings; but they may contain fractures that will carry water. Limestone is subject to solution which, beginning along small cracks, may develop channels ranging from openings a fraction of an inch across to enormous caverns capable of carrying large amounts of water.

Porosity. The property of rocks for containing voids, or interstices, is termed porosity. It is expressed quantitatively as the percentage of the total volume of rock that is occupied by openings. It ranges from as high as 80% in newly deposited silt and clay down to a fraction of 1% in the most compact rocks.

Permeability. This is the characteristic capability of rock or soil to transmit water. The porosity of a rock or soil has no direct relation to the permeability or water-yielding capacity. This capacity is related to the size and degree to which the pores or openings are interconnected. If the pores are small, the rock will transmit water very slowly; if they are large and interconnected, they will transmit water readily. Permeability is expressed as the volume of flow through unit cross-sectional area under a hydraulic gradient of 100% (1 meter of head loss per meter of water travel) in unit time. It is a velocity usually expressed in feet or meters per day. Permeability varies with the viscosity of the water, hence with water temperature.

The Meinzer unit, used by the U.S. Geological Survey, is defined as the rate of flow of water at 60°F (15.6°C) in gallons per day through a cross section of 1 ft^2 (1 gallon per day through a cross section of 1 ft^2 = 4.716×10^{-7} m^3/s through a cross section of 1 m^2), under a hydraulic gradient of 100%. Under field conditions, the adjustment to standard temperature is commonly ignored.

Transmissibility. Transmissibility expresses the rate at which water moves through a saturated body of rock. It is expressed as the rate of flow of water at the prevailing temperature, in gallons per day through a vertical strip of aquifer 1 ft (0.3048 m) wide, extending the full saturated height of the aquifer under a hydraulic gradient of 100%.

Controlling forces. Water moves through permeable rocks under the influence of gravity from places of higher head to places of lower head, that is, from areas of intake or recharge to areas of discharge, such as wells or springs. Water moving through rocks is acted upon also by friction and by molecular forces. The molecular forces are the attraction of rock surfaces for the molecules of water (adhesion) and the attraction of water molecules for one another (cohesion). When wetted, each rock surface is able to retain a thin film of water despite the effect of gravity. In very-fine-grained rocks, such as clay and fine silt, the interstices may be so small that molecular attraction extends from one side of a pore to the opposite side. Molecular force then becomes dominant, and water moves through the rock very slowly under the gradients typical of natural conditions.

The amount of water that drains from a saturated rock under the influence of gravity, expressed as a percentage of the total volume of the rock, is called the specific yield. Specific yield is often called effective porosity because it represents the pore space that will surrender water to wells. The term porosity is poorly defined, and its use should be discontinued. A part of the water stored in the rocks and soil is held by molecular forces and may have only a small share in supplying springs or wells. This latter portion is of special interest to the agriculturalist because it sustains plant life. A soil that is highly permeable permits water to pass through it easily, and little is retained for the nourishment of plant life, whereas a soil that is relatively impermeable retains much of its water until it is extracted by plants or by evaporation.

Sources. The chief means of replenishment of groundwater is downward percolation of surface water, either direct infiltration of rainwater or snowmelt or infiltration from bodies of surface water which themselves are supplied by rain or snowmelt. Evidence on the replenishment of groundwater is furnished by analysis of data on the downward movement of precipitation through the soil and subsoil, the rise and fall of groundwater levels and spring discharge in response to precipitation and seepage losses from streams, and the slope of the hydraulic gradient from known areas of intake to areas of discharge. Some groundwater may originate by chemical and physical processes that take place deep within the Earth. Such water is called juvenile water to indicate that it is reaching the Earth's surface for the first time. Such water is always highly mineralized. Some water is stored in deep-lying sedimentary rocks and is a relic of the ancient seas in which these rocks were deposited. It is called connate water. The total quantity of water from juvenile and connate sources that enters the hydrologic cycle is insignificant when compared with the quantities of water derived from precipitation (meteoric water). It is balanced to some extent by withdrawal of water from the hydrologic cycle by such processes as deposition of minerals that include water in their crystalline structure. *See* HYDROLOGY.

Infiltration. Replenishment of water in the zone of saturation involves three steps: (1) infiltration of water from the surface into the rock or soil that lies directly beneath the surface, (2) downward movement through the zone of aeration of the part of the water not retained by molecular forces, and (3) entrance of this part of the water into the zone of saturation, where it becomes groundwater and moves, chiefly laterally, toward a point of discharge. Infiltration is produced by the joint action of molecular attraction and gravity. The rate of infiltration is a function of the permeability of the soil. Under conditions of unsaturated flow in the zone of aeration, it varies with moisture content as well as with pore size. It varies also with the geology. For example, in the Badlands, South Dakota, where the soil and rocks are of low permeability, the infiltration capacity of the soil is low and is reached quickly after rainfall or snowmelt begins. Hence, there is not much infiltration, and any excess of precipitation or snowmelt over infiltration runs off over the surface and enters the streams. If the excess is large, serious floods and erosion result. On the other hand, the soils of the Sand Hills, Nebraska, and the glacial outwash deposits of Long Island, New York, are so permeable that they absorb the water of the most violent storms and permit little or no direct runoff.

The permeability of the rock materials beneath the soil zone also is important. Since the soil is commonly formed by weathering of the underlying rock, the permeability of the rocks is generally comparable to that of the soil.

Water-table and artesian conditions. Water that moves downward through the soil and subsoil in excess of capillary requirements continues to move downward until it reaches a zone whose permeability is so low that the rate of further downward movement is less than the rate of replenishment from above. A zone of saturation then forms, its thickness depending on the opportunity for lateral escape of water in relation to replenishment. The top of this zone is the water table (see illustration). Under these circumstances, the water is said to be unconfined, or under water-table conditions. However, since much of the crust of the Earth has a more or less well-defined layered structure in which zones of high and low permeability alternate, situations are common in which groundwater moving laterally in a permeable rock passes between layers of relatively low permeability. Although the permeable layer contains unconfined groundwater in the area where there is no impermeable layer (confining bed) above, the part of the layer, or aquifer, that passes beneath the confining bed contains water that is pressing upward against the confining bed, and if a well is drilled in this area the water in it will rise. It tends to rise to the level of the water in the unconfined area, but fails to reach that level by the amount of pressure head lost by friction as the water moves from the unconfined area to the well. Confined water is also called artesian water, and wells tap-

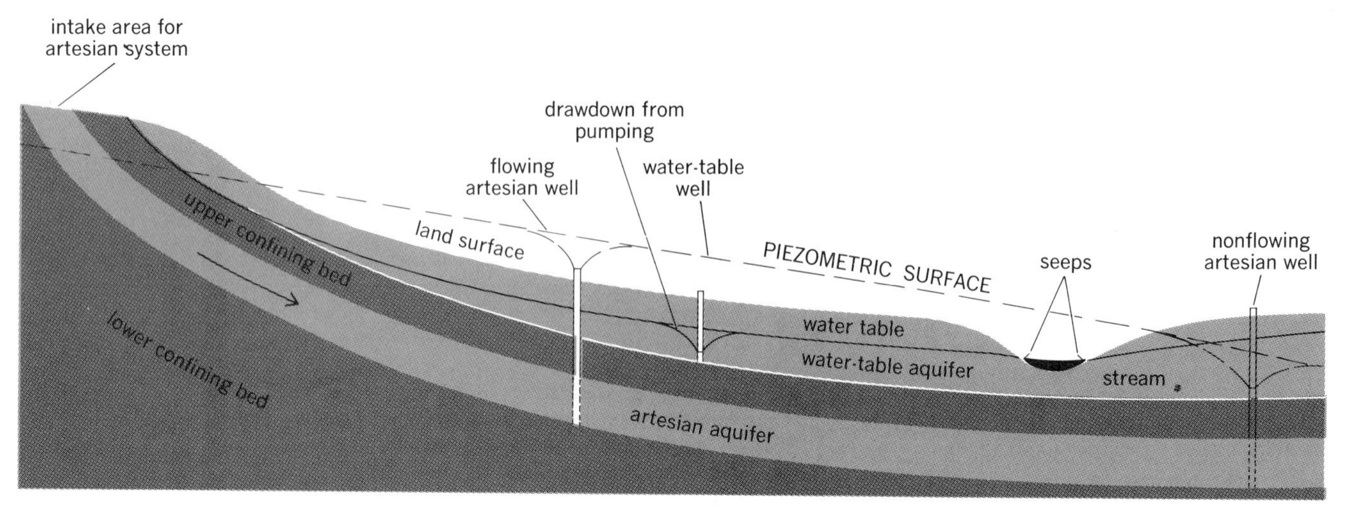

Water-table and artesian conditions.

ping it are called artesian whether or not their head is sufficient for them to flow at the land surface.

Chemical qualities. Water is said to be the universal solvent. When it condenses and falls as rain or snow, it absorbs small amounts of mineral and organic substances from the air. After falling, it continues to dissolve some of the soil and rocks through which it passes. Thus, no groundwater is chemically pure. Its most common mineral constituents are the bicarbonates, chlorides, and sulfates of calcium, magnesium, sodium, and potassium, in ionized or dissociated form. Silica also is an important constituent. Common also are small, but significant, concentrations of iron, and manganese, fluoride, and nitrate. The concentration of the dissolved minerals varies widely with the kind of soil and rocks through which the water has passed. Ordinarily, water that contains more than 1000 mg/liter of dissolved solids is considered unfit for human consumption, and water that contains more than 2000 mg/liter, unfit for stock. However, both human beings and animals may become accustomed to greater concentrations.

Pollution of groundwater. The dissolved minerals commonly found in groundwater are not harmful to humans unless present in excessive amounts. Groundwater has been generally considered safe for human consumption without treatment. However, pollution of groundwater is an increasing problem, and measures are being taken to protect major aquifers from pollution. Pollutants can reach an aquifer as a result of unwise waste disposal, overpumping the groundwater, or chemicals distributed on the land. Sanitary landfills, mine waste dumps, disposal areas for sludges from wastewater treatment or air-pollution control equipment, and industrial waste lagoons can all be sources from which pollutants, including toxic chemicals, can be leached into the underlying groundwater. Fertilizers, herbicides, and pesticides used in agriculture and forestry can be leached into the groundwater. Recharge of the groundwater with treated wastewater may transfer dissolved pollutants to the groundwater, and disposal of hazardous wastes into injection wells may, inadvertently, lead to pollution of an aquifer used as a potable water source. Overpumping of groundwater lowers the water table or piezometric surface. In coastal aquifers, sea water may be permitted to enter the aquifer. Some inland aquifers may be polluted by flow from an adjacent saline aquifer. Lowered groundwater levels may also augment seepage from any of the surface sources of pollution. Pollution, once having entered an aquifer, may be slow to disappear because of the low flow velocities encountered in some aquifers. In the hard-rock aquifers, velocity of flow in the primary openings may range from a few meters per day to a few meters per year, leaving the possibility that a pollutant, once introduced into an aquifer, might remain there for years. If pollution is induced by excessive withdrawals, inflow of pollutants will continue until water levels have risen sufficiently to reverse the gradient of flow.

[RAY K. LINSLEY]

Bibliography: H. Bouwer, *Groundwater Hydrology*, 1978; S. N. Davis and R. J. M. DeWiest, *Hydrogeology*, 1966; R. J. Kazmann, *Modern Hydrology*, 1965; W. C. Walton, *Groundwater Resource Evaluation*, 1970; R. C. Ward, *Principles of Hydrology*, 1964.

Halogen atmospheric chemistry

The halogens include the elements fluorine, chlorine, bromine, and iodine and occur in the atmosphere as both aerosols and gases. Natural sources include the dispersion of sea water through the action of breaking bubbles and the emanation of vapors from the sea surface. Pollution sources include automotive exhaust emissions and the release of synthetic organohalogen compounds. The halogens are marked by their chemical reactivity in the atmosphere, both in the troposphere and in the stratosphere. Although they occur only at trace levels, they play in important role in the total balance of chemical reactions in the atmosphere.

Sea salt particles. Sea salt is transferred to the atmosphere by bubbles of trapped air that rise to the sea's surface and break. Several jet drops of a diameter equal to one-tenth of the bubble diameter are ejected from the center of the bursting bubble and transfer material, initially accumulated within the bubble surface film, to heights of several centimeters in the atmosphere. An additional shower of smaller droplets from the breaking film cap is also transferred to the atmosphere. Once airborne, the droplets of less than 10 μm in diameter, such as are formed from small bubbles, form an aerosol and are carried far from their point of origin. They may ultimately be removed by precipitation or by fair weather deposition processes. *See* OCEAN-ATMOSPHERE RELATIONS; SEA WATER.

The composition of sea salt particles is modified from that of sea water by fractionation in the bubble-breaking process. Materials initially concentrated in the sea surface microlayer and in bubble films are preferentially transferred into the atmosphere. Measurements of chlorine, bromine, and iodine in sea spray show the weight ratio of bromine to chlorine to be approximately equal to the sea water ratio of 0.0034, although varying significantly with particle size. The ratio of iodine to chlorine, on the other hand, may be 1000 times greater in the aerosol than in sea water, especially in the smallest aerosols. The iodine enrichment is believed to be due to a combination of the concentration of surface-active organic compounds which contain iodine, in airborne droplets produced by bubble bursting and of volatile compounds of iodine emanating from the sea surface. The latter combine with aerosols by physical and chemical processes in the atmosphere.

The relative abundance of halogens at higher levels in the atmosphere is determined by their ratio in the source material and by their residence times in the atmosphere. A longer residence time, favored for elements largely in the gas phase which are removed only slowly from the atmosphere, results in a higher steady-state concentration in the atmosphere. Observations show that the ratio of bromine to chlorine is substantially greater for particles in the stratosphere than for those in the troposphere, suggesting that vapor-to-particle transfer of bromine must be important at higher altitudes.

Automotive pollution particles. The combustion of leaded gasoline generates lead chlorobromide particles and is now the major source of atmospheric particulate bromine found in populated areas. Unlike the sea salt particles, which are more abundant in larger particle-size classes than in small because of the particles' dispersive origin, automotive lead halide particles are very small, mostly less than 0.5 μm in diameter, because of their formation by condensation and coagulation of vapors in the exhaust system. Such small particles are not removed readily from the atmosphere by sedimentation, impaction, or water vapor condensation and rainout and therefore may persist in the atmosphere with longer residence times before being deposited at the Earth's surface. Similarly, small lead halide particles penetrate deeply into the human respiratory tract and may be deposited preferentially in the pulmonary instead of the upper tracheobronchial region. It has been observed that the weight ratio of bromine to lead in air is generally severalfold lower than that in leaded gasoline (0.4) and decreases with the residence time of airborne particles. Consequently, bromine from automotive pollution is a reactive element under natural conditions and is gradually released from particles as a vapor. Its subsequent chemical reactions in the atmosphere are not fully understood. *See* AIR POLLUTION.

Tropospheric halogen gases. The concentrations of gaseous compounds of chlorine, bromine, and iodine are found to be very much greater than particulate concentrations in locations remote from urban areas. By special filter techniques, which can trap inorganic and organic halogen gases, the gas-to-particle ratio has been measured in air over North America, distant from the main source. During 1974 Cl was found to be 20 to 50 times more abundant in the gas phase than in aerosols; Br 4 to 40 times; and I 6 to 15 times. Most of the gaseous halogens are believed to be organic compounds rather than inorganic vapors. Photochemical reactions may be important in their formation and transformation in the atmosphere.

Stratospheric halogen gases. Tropospheric trace gases are transferred into the stratosphere by a complex pattern of transport processes. In the stratosphere they may remain with residence times of the order of years before being transported down into the troposphere and removed at the Earth's surface. In the stratosphere, many trace gases take part in photochemical and free-radical reactions linked to the formation or destruction of ozone and therefore may influence the radiation balance of the stratosphere. Halogenated hydrocarbon compounds, derived from both natural and pollution sources in the lower atmosphere, may be dissociated in the stratosphere by the absorption of ultraviolet radiation and thereby yield reactive halogen atoms with catalytic effects on the photochemical cycle of ozone. In 1975 attention was focused on the possible significance of widespread releases of chlorofluoromethane compounds to the atmosphere. In the stratosphere, chlorine atoms formed by photochemical dissociation of chlorinated hydrocarbons may enter into reactions of the type: $Cl + O_3 \rightarrow ClO + O_2$ and $ClO + O \rightarrow Cl + O_2$.

When combined with other reactions which influence the balance of O_3 in the stratosphere, the effect of added chlorine is expected to be a decrease in the O_3 concentration. Moreover, the chlorine acts as a catalyst that is regenerated after each reaction and may transform a large number of O_3 molecules to O_2. Concern over the possibility that anthropogenic chlorofluoromethanes are the major source of chlorine atoms now in the stratosphere has stimulated research interest in the atmospheric chemistry of the halogens. *See* ATMOSPHERIC CHEMISTRY. [JOHN W. WINCHESTER]

Bibliography: G. G. Desaedeleer et al., Bromine and lead relationships with particle size and time along an urban freeway, *Trans. Amer. Nucl. Soc.*, 21:36–37, 1975; R. A. Duce, J. W. Winchester, and T. W. Van Nahl, Iodine, bromine, and chlorine in the Hawaiian marine atmosphere, *J. Geophys. Res.*, 70:1775, 1965; R. Guderian, *Air Pollution: Phytotoxicity of Acidic Gases and Its Significance in Air Pollution Control*, 1977; K. A. Rahn, R. D. Borys, and R. A. Duce, Tropospheric halogen gases: Inorganic and organic components, *Science*, 190:549–550, 1976; F. S. Rowland and M. J. Molina, Chlorofluoromethanes in the environment, *Rev. Geophys. Space Phys.*, 13:1, 1975; C. A. Willis and J. S. Handloser, *Health Physics and Operational Monitoring*, 3 vols., 1972; J. W. Winchester et al., Lead and halogens in pollution aerosols and snow from Fairbanks, Alaska, *Atmos. Environ.*, 1:105, 1967; World Meteorological Association, *Climatological Aspects of the Composition and Pollution of the Atmosphere*, 1975.

Health physics

The science that deals with problems of protection from the hazards of radiation or prevention of damage from exposure to this radiation while making it possible for humans to make full use of the various forms of energy. Initially health physics dealt only with ionizing radiations (α, β, γ, neutrons, mesons, and so forth), but it has been extended to include nonionizing radiations (ultraviolet, visible, infrared, radio-frequency, microwave, long-wave, and sonic, ultrasonic, and infrasonic radiations). Health physics is a border field of physics, biology, chemistry, mathematics, medicine, engineering, and industrial hygiene. It is concerned with radiation protection problems involving research, engineering, education, and applied activities. It involves research on the effects of ionizing radiation on matter with a goal of developing a coherent theory of radiation damage. It deals with methods of measuring and assessing radiation dose, devices for reducing or preventing radiation exposure, the effects of ionizing radiation on humans and their environment, radioactive waste disposal, the establishment of maximum permissible exposure levels, and radiation risks associated with the nuclear energy industry, medical or hospital physics, high-voltage accelerator physics, applications of radionuclides, and special problems of microwave and ultrasonic radiations.

Health physics began in 1942 along with the nuclear energy and reactor programs at the University of Chicago. It is estimated that by 1975 there were more than 4000 practicing health physicists in the United States, and 12,000 throughout the world. In 1956 the Health Physics Society was organized in the United States and Canada and now has more than 3000 members. It is one of 22 organizations affiliated with the International Radiation Protection Association, which has about 10,000 members in 65 countries.

Health physicists are employed in industry, national laboratories, state and Federal agencies, hospitals, military organizations, and space programs, and in private practice.

[KARL Z. MORGAN]

Bibliography: J. D. Abbott et al., *Protection Against Radiation*, 1961; J. S. Handloser, *Health Physics Instrumentation*, 1959; Second International Conference on Peaceful Uses of Atomic Energy, Geneva, 1958, *Progress in Nuclear Energy*, ser. 12: *Health Physics*, vol. 1, 1959.

Heavy metals, pathology of

Humans have evolved in an environment containing metals and have developed protective means of defense against natural concentrations. Indeed, some metals have become essential for life. However, industry is pouring metals, particularly heavy metals, into the environment at an unprecedented and constantly increasing rate. As a result, technological society is exposing the population to some metals in unnaturally high concentrations, in unusual physical or chemical forms, and through unusual portals of entry.

This trend is certainly likely to continue. It is vital, therefore, that the effects of metals on living cells be understood, as a first step toward controlling the problems. This article summarizes the present understanding of the effects and selectively describes the pathology of some environmentally important heavy-metal poisonings that illustrate the diversity of pathologic lesions.

NATURE OF METAL TOXICITY

Heavy metals, the principal subject of this article, are arbitrarily defined as those metals having a density at least five times greater than that of water. Although metals have many physical properties in common, their chemical reactivity is quite diverse, and their toxic effect on biological systems is even more diverse. *See* TOXICOLOGY.

A metal can be regarded as toxic if it injures the growth or metabolism of cells when it is present above a given concentration. Almost all metals are toxic at high concentrations and some are severe poisons even at very low concentrations. Copper, for example, is a micronutrient, a necessary constituent of all organisms, but if the copper intake is increased above the proper level, it becomes highly toxic. Like copper, each metal has an optimum range of concentration, in excess of which the element is toxic (Fig. 1). When the optimum range for a particular metal is narrow, the risk of toxicity increases; thus even a minor environmental increase can be serious.

Metals exert their toxicity on cells by interfering in any of several ways with cell metabolism. The most important of these affects enzyme systems. The more strongly electronegative metals (such as copper, mercury, and silver) bind with amino, imino, and sulfhydryl groups of enzymes, thus blocking enzyme activity. Another mechanism of action of some heavy metals (gold, cadmium, copper, mercury, lead) is their combination with the cell membranes, altering the permeability of the membranes. Others displace elements that are important structurally or electrochemically to cells which then can no longer perform their biologic functions.

Conditions for toxicity. The toxicity of a metal depends on its route of administration and the chemical compound with which it is bound. For example, mercury is highly toxic when injected intravenously or inhaled as a vapor, but is less toxic when taken by mouth. Cadmium, chromium, and lead are also highly toxic when they are injected intravenously, whereas a large number of metals (gold, cobalt, manganese, tin, thallium, nickel, zinc, and others) require greater intravenous doses to be toxic. When taken by mouth, gold and iron are only slightly toxic; however, copper, mercury, lead, and vanadium are much more toxic. Tantalum, on the other hand, is nontoxic no matter how it is administered.

The combining of a metal with an organic compound may either increase or decrease its toxic effects on cells. For example, the combination of mercury with a methyl organic radical makes the element more toxic, whereas the combination of the cupric ion with an organic radical, such as salicylaidoxine, makes the metal less toxic. Generally the combination of a metal with a sulfur to form a sulfide results in a less toxic compound than the corresponding hydroxide or oxide, because the sulfide is less soluble in body fluids than the oxide.

Many other heavy metals besides copper are essential to life, even though they occur only in trace amounts in the body tissues. They are taken up by the living cell in the form of cations and are admitted into an organism's internal environment in carefully regulated amounts, under normal circumstances, thus avoiding the effects of toxic levels. Toxicity results (1) when an excessive concentration is presented to an organism over a prolonged period of time, (2) when the metal is presented in an unusual biochemical form, or (3) when the metal is presented to an organism by way of an unusual route of intake.

Specificity. Metals have a remarkably specific effect; seldom can an excess of one essential metal prevent the damage caused by the deficiency of another. In fact, such an excess often increases the injurious effect of the deficiency. The heavy metals that appear to be essential for living cells

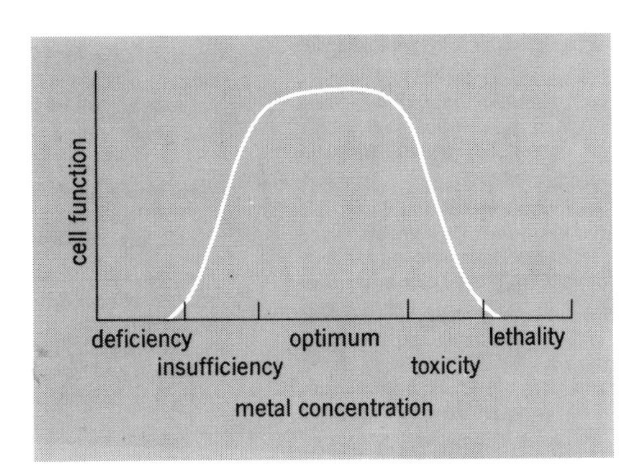

Fig. 1. Effects of metals on cells. Metals have an optimum range of concentrations within which a cell is healthy. The actual values and range vary from metal to metal.

are copper, iron, manganese, molybdenum, cobalt, and zinc. Some lighter metals, such as calcium, magnesium, sodium, and potassium, are also essential. There is some evidence that other metals, such as aluminum, cadmium, chromium, and vanadium, are essential. Many other metals found in the human body tissues have no apparent function.

Cellular response. On a molecular level of organization, metals produce diverse responses in cellular behavior. If the toxic action of a metal on a cell is interference with an essential part of cell metabolism, the cell will, of course, die. Figure 2, for example, shows two microscopic views of kidney cells. In Fig. 2*a* two glomeruli are surrounded by cross sections of normal kidney tubules, and Fig. 2*b* shows kidney tubule cell degeneration and necrosis (death) 72 hr after exposure to methyl mercury. The degenerating cells are swollen and vacuolated. The necrotic cells (indicated by arrows) are small, round, and dense and have small nuclei. Cell death provokes, in turn, an inflammatory response by the body and an attempt to repair the damage. Indeed, some metals (beryllium, for instance) provoke an excessive reparative process consisting of extensive proliferation of connective-tissue cells (histiocytes) that produce nodules of inflammatory tissue (granulomata) and extensive scarification. The microscopic views of lung tissue in Fig. 3 illustrate this effect. Figure 3*a* shows normal lung cells, with characteristically thin alveolar septa surrounding the air sacs. Figure 3*b* shows cells after exposure to beryllium. Nodular proliferations of cells and thickened septa (arrows) are evident.

Lesser degrees of injury to cells may alter their structure and function and may be associated with degenerative changes such as accumulation of fat or may limit the cells' principal biologic activity, which in turn may alter the performance of the organism as a whole. For example, degeneration of kidney tubule cells by mercury degrades the excretory function of the kidneys and leads to severe fluid and electrolytic imbalance in an individual, and possibly to death.

Carcinogens and teratogens. Less well understood but perhaps of equal significance are the carcinogenic properties of some metals. Nickel, cobalt, and cadmium have clearly been shown to produce cancers of the muscle cell type (rhabdomyosarcoma) when injected subcutaneously. However, many other metals similarly injected do not produce aberrant new cell growth (neoplasms). Nickel compounds also produce cancers of the respiratory tract. Lead has been shown to produce kidney cancers in experimental animals. The carcinogenic potential of many other metals is suspected but not proven.

Birth defects (teratogenesis) as a consequence of excessive metal intake by pregnant women do occur, but their precise nature and frequency are poorly documented. Chick embryos, when exposed to low doses of thallium or chromium, subsequently hatch with a high incidence of defective long-bone formation. Lead and cobalt have been shown to produce defective brain development in chick embryos. Many diverse metals produce growth inhibition in developing chick embryos. Manganese is a powerful mutation-producing agent for the bacterium *Escherichia coli* and the T4 bacteriophage; it is known to alter the nuclear enzyme DNA polymerase in lower forms of life, but whether it similarly affects mammals is not known. Mercury produces brain damage and mental retardation in children born to mothers exposed to excessive amounts of that metal. Lead, too, has teratogenic effects; specific congenital skeletal malformations have been induced in hamster embryos by treating the pregnant mother with various salts of lead. *See* MUTATION.

Chromosome damage. In addition to producing injury to the developing embryo, some metals produce damage to chromosomes. Both mercury and lead compounds have been shown to produce breaks in the chromosomes of somatic cells. Figure 4 shows chromosome damage from mercury. The lymphocytes have a broken chromosome and extra fragment (Fig. 4*a*), and three sister-chromatid fragments that lack centromeres (Fig. 4*b*). Steffan Skerfving, of the National Institute of Public Health, Stockholm, Sweden, cultured the lymphocytes from the blood of nine patients who were known to have excessively high tissue levels of methyl mercury from eating contaminated fish. He

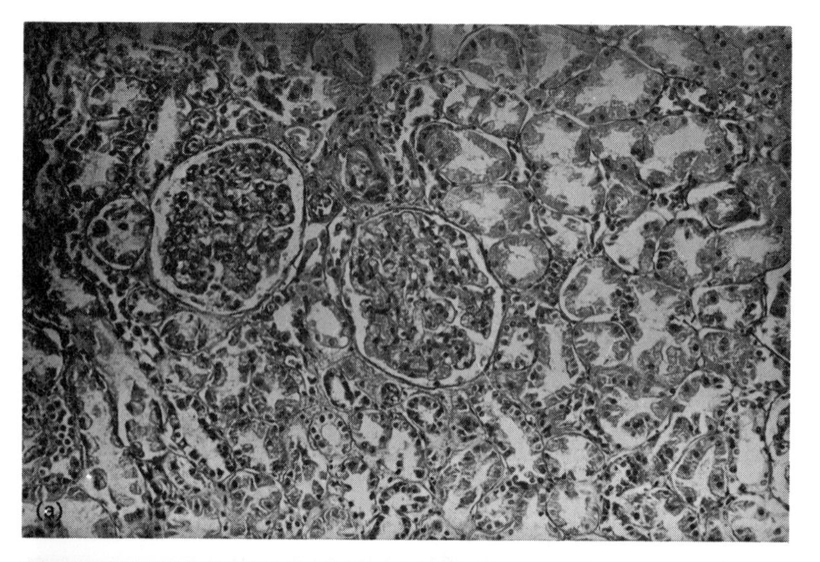

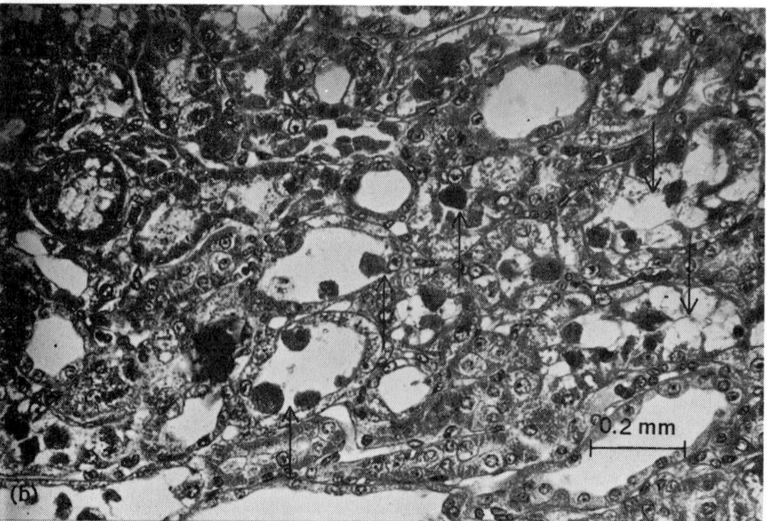

Fig. 2. Kidney cells. (*a*) Normal kidney cells. (*b*) Methyl mercury–poisoned cells. Arrows indicate dead cells.

found a statistically significant rank correlation with chromosome breaks and mercury concentration. The full significance of these findings remains to be determined.

Many questions remain to be answered regarding the effects of metals on chromosomes. Many agents produce chromosome breaks in the test tube, but do they have the same effect in living cells? Are chromosome breaks produced in germ cells as well as somatic cells? Does the genetic damage lead to cell death, carcinogenesis, or teratogenesis? Whereas the effects of the deficiency of many trace elements on the developing embryo have been extensively studied, little is known about the teratogenic effects of an excess of metals. These questions will undoubtedly be answered as research proceeds.

CADMIUM DISEASE

Cadmium occurs in trace quantities—less than 1 part per million (ppm)—throughout the Earth's crust. There are no known deposits rich enough to justify separate mining of the metal, but it is frequently found in zinc and lead ores. Industrially cadmium has important and diverse uses in electroplating and in alloys, solders, batteries, paints (heat-resistant pigments), and metal bearings. It is also used as a neutron absorber in nuclear reactors and as an insecticide for fruit trees.

Cadmium also occurs naturally in trace quantities in many plant and animal tissues (Table 1). It therefore forms part of our normal diet, although the exact amount normally ingested is not known. Cadmium is very slowly absorbed from the gastrointestinal tract and is eliminated in both urine and feces. In healthy human adults the cadmium levels are low (Table 2), except for the kidneys, which are uniquely high in cadmium. Much lower concentrations occur in the tissues of the human fetus and newborn. With increased age, the tissue cadmium level progressively increases.

Ingestion of 15 ppm of cadmium in food, or about 15 times the normal amount, can produce mild symptoms of poisoning. Food poisonings from contamination of food and drink by cadmium-plated containers have been recorded frequently. For example, several outbreaks have occurred as the result of the action of the acids in fruit juice acting on cadmium-plated ice trays. Cadmium is soluble in weak acids such as acetic acid (vinegar), citric acid, and organic acids commonly found in food. The principal human risk is from industrial exposure, namely, inhalation of cadmium fumes. Cadmium is primarily a respiratory poison and has higher lethal potential than most other metals with a mortality rate of 15% in poisoning cases.

Recent occurrence. A remarkable case occurred in 1969 when a 28-year-old Japanese girl, Takako Nakamura, who had been working as a lathe operator in a zinc fabricating factory, hurled herself before a speeding train. Her death was listed as suicide and nothing more was thought of it for a time. Several years later her body was exhumed and an autopsy performed. After correlatng the levels of cadmium within the tissues with the anatomic lesions observed, the pathologists reported that she was a victim of cadmium poisoning.

Symptoms. Mild cadmium intoxication causes smarting of the eyes, dryness and irritation of the throat, and tightness in the chest and headache. With increased exposure, the respiratory distress increases in severity with uncontrollable cough and gastrointestinal pain, nausea, retching, vomiting, and diarrhea. Takako, the suicide, recorded in her diary the early symptoms: "The doctors cannot diagnose my disease; I am afraid it is cadmium poisoning. It is running through my whole body, pain eats away at me; I feel I want to tear out my stomach; tear out all my insides and cast them away."

Tissue reaction to the inhalation of cadmium fumes results in the initial deposition of cadmium in the lungs. From there the metal is widely distributed throughout the body and accumulates in the liver, kidneys, and spleen. The high concentrations may remain in the tissues for many years after the cessation of exposure. Excretion in the urine is extremely low.

Cadmium acting on the alveoli of the lungs produces a severe proliferation of alveolar septal cells which, in some cases, may completely fill the air spaces. Chronic exposure to cadmium produces emphysema, an irreversible lesion of the lungs

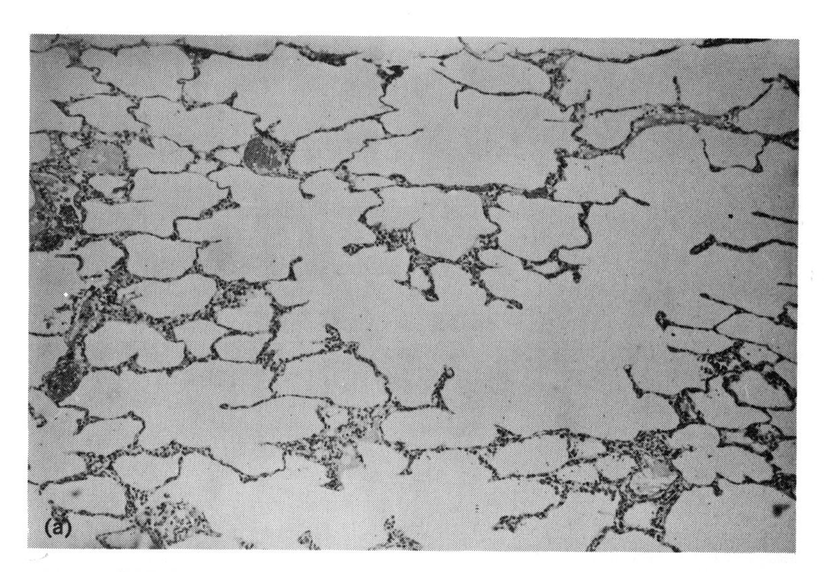

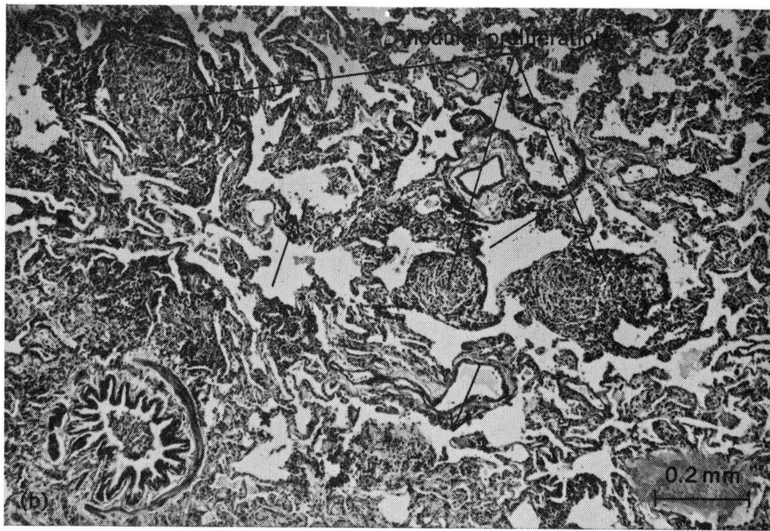

Fig. 3. Lung cells. (*a*) Normal cells. (*b*) Cells exposed to beryllium. Nodular proliferations of cells are indicated. Thickened septa are shown by arrows.

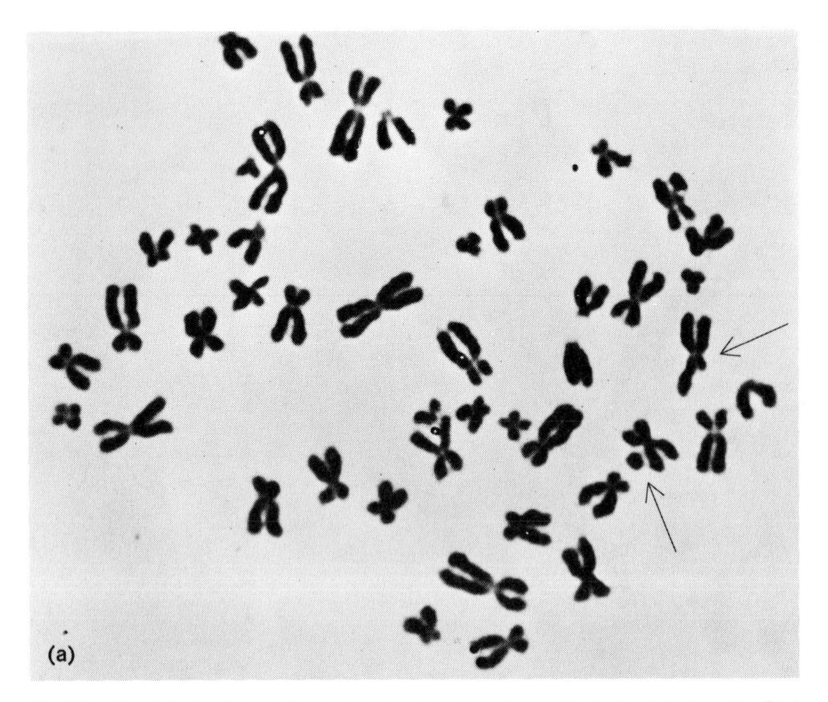

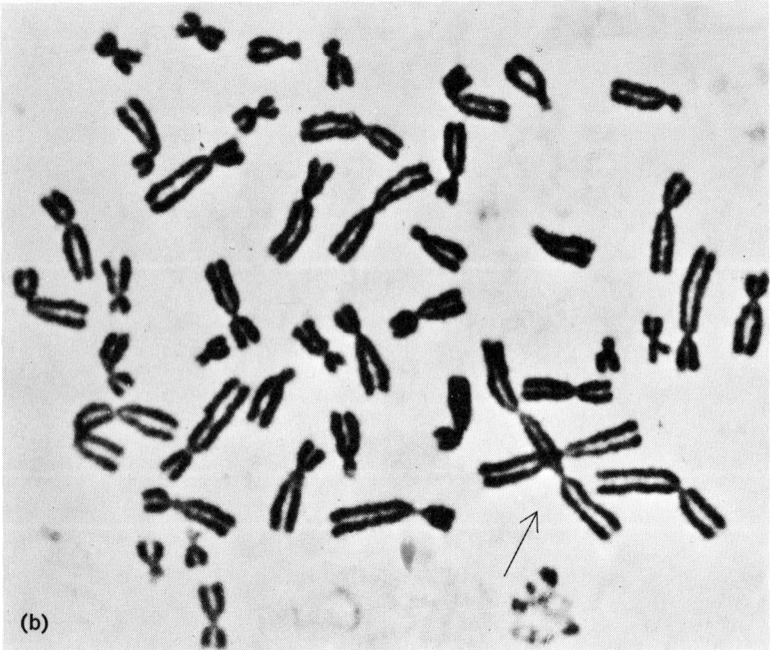

Fig. 4. Chromosomes damaged by mercury poisoning. (*a*) Arrows show a broken chromosome and an extra fragment. (*b*) Arrow shows chromatid fragments without centromeres. (*From S. Skerfving, Chromosome breakage in humans exposed to methyl mercury through fish contamination, Arch. Env. Health, 21:133–139, 1970*)

characterized by an enlargement of the air sacs and accompanied by destruction of the walls of many of the sacs. This results in the conversion of many small air sacs into fewer large ones, with the inevitable decrease in surface area for oxygen and carbon dioxide transfer.

The kidneys are also injured in chronic cadmium poisoning. Diffuse extensive scar tissue develops between the tubules and subsequently destroys many glomeruli and tubules. Serum protein is then lost in the urine.

When administered to rats in dose levels comparable to that ingested by humans, cadmium produces high blood pressure, enlargement (hypertrophy) of the muscle of the heart (left ventricle), and hard patches (sclerotic plaques) in the very small arteries (arterioles) of the internal organs. (Communities in the United States with the highest environmental cadmium content also have the highest incidence of arteriosclerosis.) Another consequence of chronic cadmium exposure is an increased brittleness of bone, to an extent that the mere act of coughing may produce fractures of the ribs. Cadmium nitrate produces gene mutation in plants, but its effects, if any, on mammalian genes remain unknown.

MERCURY POISONING (MERCURIALISM)

The biological effects of excessive intake of mercury illustrate some important similarities and differences relative to cadmium. Mercury is unique in that it is the only metal that is a liquid at ordinary temperatures. Like cadmium, mercury is a relatively rare element, ranking near the bottom of the list of elements found in abundance on Earth.

Food grown under natural conditions contains traces of mercury. Five different studies, using a variety of methods from 1934 to the present, have shown that dietary meats contain on the average less than 0.01 ppm mercury. Fish generally have twice and vegetables have half as much (the commonly accepted limit for mercury in foods is 0.5 ppm). Some algae contain more than 100 times more mercury than the sea water in which they grow. Fish eating the algae concentrate the mercury, and predators that eat the fish in turn concentrate the mercury further.

Increasing prevalence of element. Man has evolved in an environment containing mercury, and throughout his history he has ingested plants, animals, and fish containing mercury. Through his evolution he has developed a tolerance, or possibly a need, for mercury as a trace constituent of his

Table 1. Metals in our environment in parts per million

Metal	Rock	Coal	Sea water	Plants (dry wt.)	Animals (dry wt.)
Cadmium	0.2	0.25	0.0001	0.1–6.4	0.15–3.0
Chromium	100.0	60.0	0.00005	0.8–4.0	0.02–1.3
Cobalt	25.0	15.0	0.00027	0.2–5.0	0.3–4.0
Lead	12.5	5.0	0.00003	1.8–50.0	0.3–35.0
Mercury	0.08	–	0.00003	0.02–0.03	0.05–1.0
Nickel	75.0	35.0	0.0045	1.5–36.0	0.4–26.0
Silver	0.07	0.1	0.0003	0.07–0.25	0.006–5.0
Thallium	0.45	0.05–10.0	0.00001	1.0–80.0	0.2–160.0
Vanadium	135.0	40.0	0.002	0.13–5.0	0.14–2.3
Gold	0.004	0.125	0.00001	0–0.012	0.007–0.03

body. How has our urban technology affected this ecosystem?

First of all, mercury is being added to the North American environment at a rapid rate. Of greatest significance is the unmeasured increase of mercury to our atmosphere from the burning of fossil fuels. Coal and petroleum contain significant amounts of mercury; the mercury content of 36 American coals ranged from 0.10 to 33.0 ppm. In addition, organic mercurials are widely used as a fungicide and pesticide in agriculture. Other mercury compounds have been successfully used as a dressing for seed grains to prevent mildew, as a conditioner for lumber to prevent fungal discoloration, and in laundries to prevent garment mildew.

Not only is there more mercury in the environment, but evidence suggests that more of it is in a form that is most toxic to man. Liquid (metallic) mercury itself is not ordinarily toxic, since the body does not absorb mercury in this form from the digestive tract. However, many intoxications occur when metallic mercury is vaporized and inhaled. Figure 5 shows the effect of mercury vapor on lung tissue (it can be compared to the normal lung tissue in Fig. 3*a*). The tissue is from the lung of a 1-year-old child who died 6 days after exposure to the mercury vapor. The septa are swollen and thickened, and the air sacs contain fluid. Aside from such acute injuries to lung tissue, chronic low-level exposure to mercury vapor may produce symptoms or death because of its destruction of the nervous system.

Organic mercurials, too, which are readily assimilated, are becoming more prevalent. It has been proven that elemental, or inorganic, mercury, when released into the hydrosphere, is converted to an organic mercurial by the linkage of methyl or other carbon chains to the mercury by marine life. Repeated predation by animals in the food chain concentrates organic mercurials in their tissues, ultimately reaching toxic levels for man and animal.

Case histories. The increasing prevalence of organic mercurials has led to some serious episodes of intoxication of man. A case of direct ingestion of organic mercury occurred in Alamogordo, N. Mex., when several members of a family ate pork from a pig that had been inappropriately fed seed-wheat previously treated with a methyl mercury compound. Three children became seriously ill (but none died), whereas two other children who had not eaten pork were unaffected. The mother was pregnant at the time of onset of the symptoms in the three older children but she was symptom-free. The baby appeared normal at birth; however, soon thereafter this child developed convulsions and the characteristic symptom complex of chronic mercurialism, from which he seems to be gradually recovering.

Table 2. Distribution of some metals in mammalian tissues in parts per million dry weight

Metal	Skin	Lung	Liver	Kidney	Brain
Cadmium	1.0	0.08	6.7	130.0	3.0
Chromium	0.29	0.62	0.026	0.05	0.12
Cobalt	0.03	0.06	0.23	0.05	0.005
Lead	0.78	2.3	4.8	4.5	0.24
Mercury	–	0.03	0.022	0.25	–
Nickel	0.8	0.2	0.2	0.2	0.3
Silver	0.022	0.005	0.03	0.005	0.04
Thallium	0.2	0.3	0.4	0.4	0.5
Vanadium	0.02	0.05	0.04	0.05	0.3
Gold	0.2	0.3	0.0001	0.5	0.5

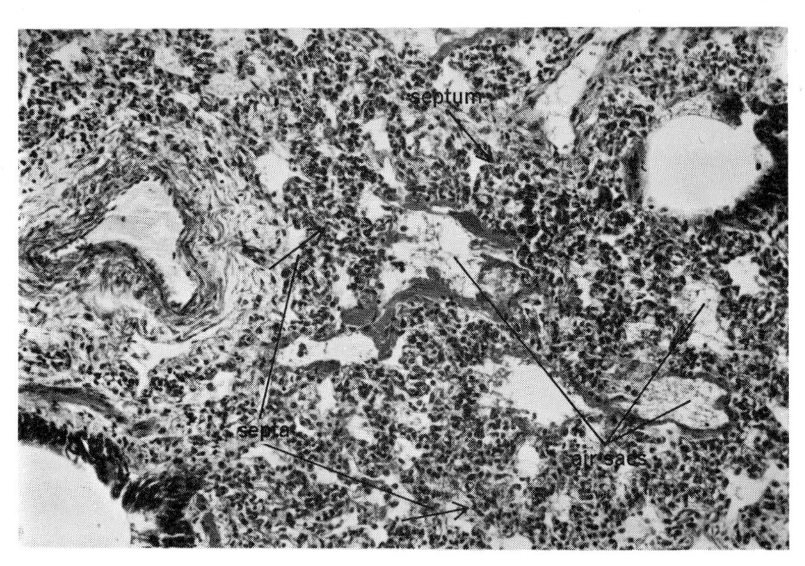

Fig. 5. Cells from lung of a child poisoned by mercury vapor. Arrows indicate swollen septa. The air sacs are filled with fluid.

Indirect chronic mercurial poisoning with a much higher morbidity and mortality rate occurred in Minamata and Niigata, Japan. The poisoning, called Minamata disease, was caused by discharge into Minamata Bay from an acetaldehyde and vinyl factory. The effluent contained large amounts of inorganic and organic mercury. There were 121 human cases of mercury poisoning recorded, with 46 deaths. About one-third of the afflicted were infants and children, some of whom had acquired mercury poisoning through the placenta prior to birth. The disease occurred mainly among fishermen and fish-eating families. Fish-eating animals such as cats, dogs, pigs, and seabirds were often affected, but herbivores such as rabbits, horses, and cows were not.

Symptoms. Chronic mercurialism presents five main symptoms: numbness, staggering gait, constriction of visual field ("tunnel vision"), garbled speech (dysarthria), and tremor. In addition, the children with Minamata disease exhibited emotional disturbances, and infants with congenital intoxication also suffered from impaired chewing or swallowing. Convulsions and mental retardation often occurred in the childhood cases. These symptoms of chronic poisoning are almost totally traceable to lesions in the central nervous system.

Autopsies of those who died soon after exposure revealed degenerative lesions in the liver, heart muscle fibers, and tubules of the kidneys. These changes were similar to those seen with inorganic mercury poisoning. Normal human liver cells (Fig. 6*a*) are of uniform size and take up stain uniformly. Mercury-poisoned liver cells (Fig. 6*b*) accumulate fat, assume various sizes, and stain unevenly. This type of fatty degeneration of cells can be seen throughout the liver and is characteristic of other metal poisonings in addition to mercury.

Acute focal inflammation without ulceration was often seen in the stomach or duodenum. The autopsies also revealed the presence of a generalized edematous swelling of the brain and a decrease in

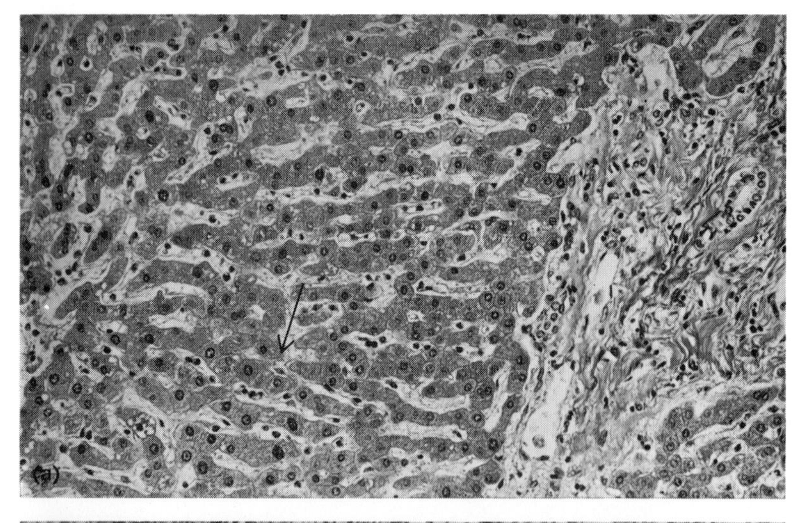

Fig. 6. Liver cells. (*a*) Normal cells. (*b*) Mercury-poisoned cells.

blood cell production by the bone marrow.

Another important early lesion involved the blood vessels, especially those of the central nervous system. Blood vessels became cuffed with white blood cells, and in some there was degeneration of the vessel wall and capillary proliferation.

The lesions of chronic mercurialism are principally in the brain. In the autopsies, destruction of the nerve cells of the gray matter of the cerebral cortex was found in several regions, especially the visual cortex. Extensive necrosis of nerve cells in the visual area and an increase in the nonneuronal supporting cells had taken place. These lesions principally account for the "tunnel vision" of methyl mercury poisoning. Similar changes were seen in the cerebellum. An astounding destruction of the granular cell layer was found, as shown in Fig. 7 (at arrows). This layer contains nerve cells that receive stimuli from the muscles, tendons, and joints throughout the body and transmit them to the Purkinje cell processes in the molecular layer, where the many stimuli are correlated and integrated. The cerebellum is the area of the brain where position sense, balance, and fine muscle coordination are controlled. In most cerebellar pathologic processes the sensitive cells are the large Purkinje cells; mercury is the only known metallic toxin from an external source that results in the preferential destruction of the granular cell layer.

LEAD POISONING (PLUMBISM)

The toxicity of lead has been known since antiquity. Lead and lead-containing products are extremely important industrially and are of great health significance. Exposure of man may result from inhalation of fumes and dust in the smelting of lead, from the manufacture of insecticides, pottery, and storage batteries, or from contact with gasoline containing lead additives. Intoxication by the inhalation of lead fumes is a most serious mode of exposure. The ingestion of soluble lead compounds accidentally, or with suicidal intent, is another portal of entry of the body. Only tetraethyllead in gasolines can be absorbed through the intact skin. Traces of lead occur in the diet; small amounts are absorbed into the body when it is present in the food as a soluble salt. Lead is absorbed mainly through the small and large intestines.

Irrespective of the route of entry, lead is absorbed very slowly into the human body. Even this slow and constant chronic absorption is sufficient to produce lead poisoning because the rate of elimination of lead is even slower and a slight excess in intake may result in its accumulation in the body. Much of the lead is taken up by red blood cells and circulated throughout the body. Organic lead compounds such as tetraethyllead become distributed throughout the soft tissues, with especially high concentrations in the liver and kidneys. Over a period of time, the lead may be redistributed, becoming deposited in bones, teeth, or brain. In bones, lead is immobilized and does not contribute to the general toxic symptoms of the patient. Organic lead compounds have an affinity for the central nervous system and produce lesions there.

Acute form. Acute plumbism is ordinarily seen as a result of the ingestion of inorganic soluble lead salts for suicidal, accidental, or abortion-inducing reasons. They produce a metallic taste, a dry burning sensation in the throat, cramps, retching, and persistent vomiting. The gastrointestinal tract is encrusted with the coagulated proteins of the necrotic mucosa, thereby hindering further absorption of the lead. Muscular spasms, numbness, and local palsy may appear.

Chronic form. Chronic lead poisoning is much more common. Two general patterns of symptoms relate to the gastrointestinal and nervous systems. One or the other may predominate in any particular patient; in general, the central nervous system changes, which predominate in children, are of greater significance. The abdominal symptoms in chronic lead poisoning are similar to those for acute cases, but are less severe: loss of appetite, a feeling of weakness and listlessness, headache, and muscular discomfort. Nausea and vomiting may result. Chronic excruciating abdominal pain, sometimes referred to as lead colic, may be the most distressing feature of plumbism.

An autopsy does not ordinarily reveal specific gross lesions, but a marked congestion and petechial hemorrhages are sometimes seen in the brain. Microscopic examination reveals inclusion bodies within cells of the kidney, brain, and liver.

Focal necrosis of the liver is found in some cases. Lesions in the central nervous system are primarily vascular and consist of scattered hemorrhages, often in the perivascular tissue; degenerative and necrotic changes in the small vessels, surrounded by a zone of edema; and sometimes a fibrinous exudate.

THALLIUM DISEASE (THALLOTOXICOSIS)

Since thallium was discovered more than 100 years ago, it has been responsible for many therapeutic, occupational, and accidental poisonings. During the first 50 years since its discovery, thallium sulfate gained general use as a treatment for syphilis, gonorrhea, gout, dysentery, and tuberculosis. It was subsequently discarded as a medicine because of its unpleasant side effects, principally the temporary loss of hair (depilation). Later, dermatologists utilized this feature as a method of removing unwanted hair.

Occurrence. By 1934 more than 700 poisonings with thallium were reported in the medical literature, with more than 90% of these the result of the use of thallium as a depilatory agent. In recent years hundreds of cases of thallotoxicosis have been related to its use as a rodenticide and insecticide. For example, a group of 31 Mexican workers in California were poisoned, six fatally, after eating tortillas made from a bag of thallium-treated barley intended for rodents.

Symptoms. The symptoms of thallotoxicosis vary extensively with the dosage, age of the patient, and the acuteness of the intoxication. Inflammation involving many nerves, loss of hair, gastrointestinal cramps, and emotional changes are the principal symptoms. For large doses, the first symptoms are gastrointestinal hemorrhage, gastroenteritis, a rapid heartbeat, and headache. Shortly thereafter abdominal pain, vomiting, and diarrhea may ensue. Neurologic symptoms appear after 2 or more days of exposure, with delirium hallucinations, convulsions, and coma occurring after severe poisoning, and death may follow in about a week due to respiratory paralysis.

When smaller doses are taken, a loss of muscle coordination (ataxia) and sensations of tingling or burning of the skin (paresthesia) may be the principal symptoms. These may be followed by weakness and atrophy of the muscles. Tremor, involuntary movements (choreic athetosis), and mental aberrations, with changes in the state of consciousness, may ensue.

Alopecia is one of the best known symptoms of chronic thallium poisoning. The hair loss begins about 10 days after the ingestion of thallium, and complete loss of hair may be reached in a month. Hair in the axillary and facial regions, including the inner one-third of the eyebrows, is usually spared. The evidence suggests thallium acts directly on the hair follicles. Cardiac symptoms—rapid heartbeat, irregular pulse, and high blood pressure, with angina-like pain—have been recorded. Heart muscle fiber degeneration accompanies these symptoms. Thallium affects the sweat and sebaceous glands, and the nails are white with transverse bands and may assume unusual shapes. A blue line may develop along the gums of an individual who has ingested thallium.

The turnover of thallium in the human body is extremely slow. Under normal circumstances the tissues contain trace amounts of the metal. The biological mechanism by which thallium produces its effects on the human body is not well understood, but as with other metals, it appears to block some enzyme systems. Autopsies on fatal human cases reveal the presence of thallium in all organs and tissues, with the highest concentrations in the kidneys, intestinal mucosa, thyroid, and testes. Fat, liver, and all types of nerve tissue are uniformly low in thallium content. The autopsies reveal bleeding at many points (punctate hemorrhages) in the gastric and upper intestinal mucosa, fatty change in the liver and kidney, and small punctate hemorrhages and focal necrosis of the surface layers of the adrenal glands. Congestion of the central nervous system blood vessels is evident, and there is some degree of cerebral edema.

Neurons show varying degrees of degenerative change, especially in the locomotor fiber pathways and associated centers. In the more chronic cases, the ganglion cells of the sensory and motor horns of the spinal cord degenerate, with chromatolysis, swelling, and fatty change. Examination of peripheral nerves shows marked degeneration in nerve cells, axons, and myelin sheath. Thallium poisoning is a serious problem not only because approximately 15% of those poisoned die, but also because there is persistent neurologic damage in over half the cases.

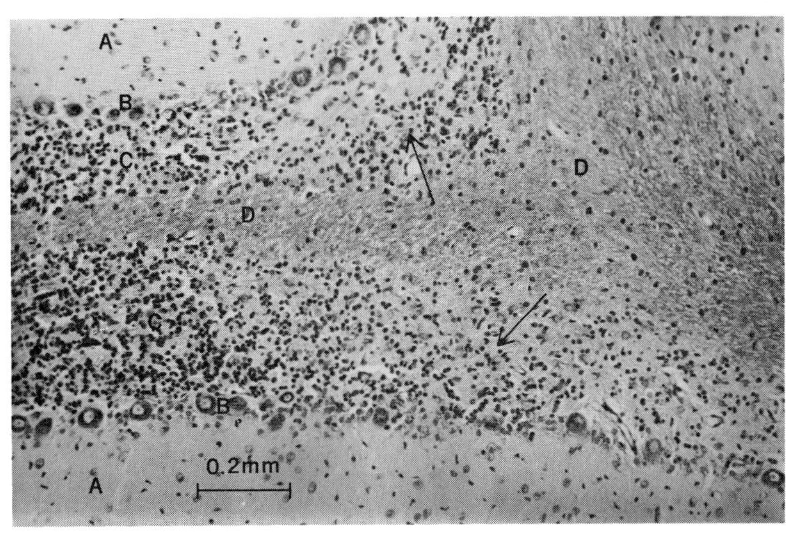

Key: A = molecular level C = granular cell level
B = Purkinje cells D = white matter

Fig. 7. Cells in cerebellum. Arrows show cells destroyed by methyl mercury.

COBALT TOXICITY

Cobalt is a metal with a very industrially important property: It is immune to attack by air or water at ordinary temperatures and imparts this property to many alloys. Under natural circumstances cobalt does not produce a toxic syndrome in man. Cobalt in most naturally occurring foodstuffs is mostly unabsorbed and is eliminated in the feces.

It takes an unusual set of circumstances to produce cobalt toxicity. Intravenously injected, large doses of cobalt have been observed to cause paralysis and enteritis, and sometimes death. When cobalt is injected into the bloodstream, the amount

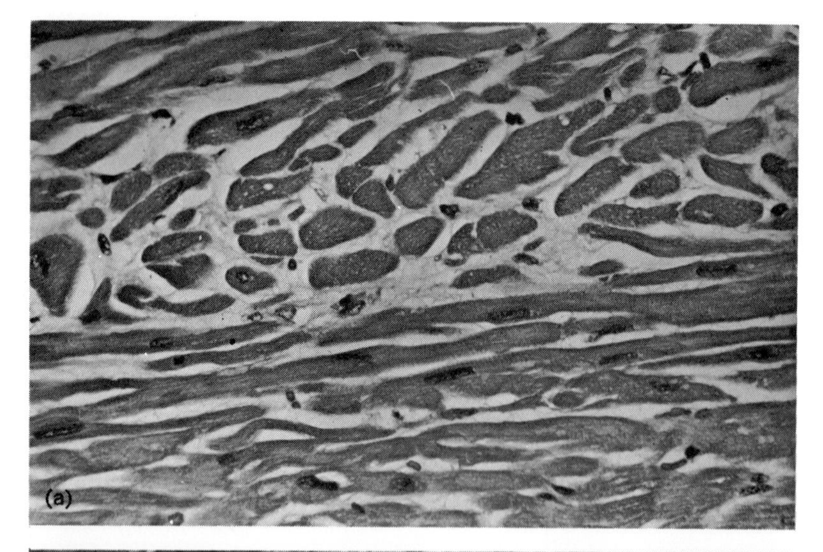

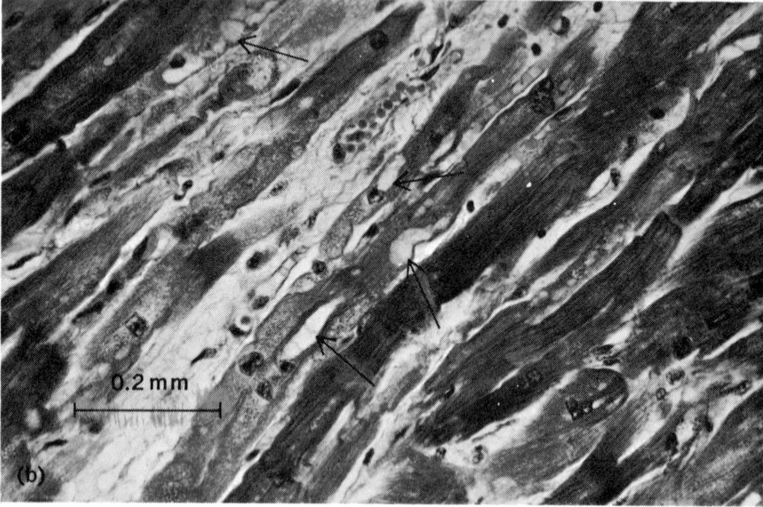

Fig. 8. Heart muscle fibers. (*a*) Normal fibers. (*b*) Fibers from a victim of beer drinker's syndrome. Arrows point to fatty vacuoles.

distributed throughout the tissues is very small; higher concentrations are present in the pancreas, liver, spleen, kidney, and bone. Elimination of cobalt is rapid.

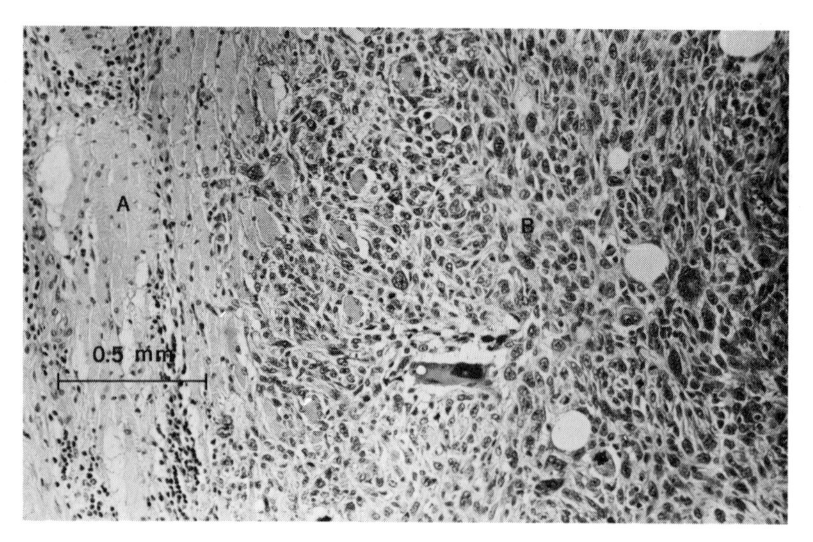

Fig. 9. Skeletal muscle fibers. Cells in region A are normal; those in region B are cancer cells.

Carcinogen. Researchers in England have shown that cobalt, like nickel and cadmium, produces a high incidence of cancer, namely rhabdomyosarcoma, when it is injected in powdered form into rats. A number of other metals, such as iron, copper, zinc, manganese, beryllium, and tungsten, are not carcinogenic under the same conditions. Cobalt has an inhibitory effect on cell oxidative metabolism and is toxic to connective-tissue cells grown in culture.

Beer drinker's syndrome. Another unusual kind of cobalt toxicity is the "beer drinker's syndrome." During the 1960s, some breweries in Nebraska and Quebec added cobalt to their beer to improve the stability of the foam. Subsequently, numerous fatalities occurred among people who consumed large quantities of beer. Following discontinuance of the cobalt additive, no new cases have been reported. Whether the cobalt acted independently or synergistically with other constituents of beer to enhance its toxicity is not known. The principal lesion was found in the heart; autopsies revealed enlarged, flabby hearts that were more than twice normal size. Microscopically the myocardial cells were vacuolated with numerous fat droplets. The microscopic view of heart muscle fibers from a normal heart is shown in Fig. 8*a*, and from a beer drinker's syndrome case in Fig. 8*b*. The fatty vacuoles (arrows) resulting from degenerative changes give a "moth-eaten" appearance to the fibers. Progressive degenerative changes in the myofibrils were also found in the autopsies, and complete necrosis and lysis of some heart muscle fibers were noted.

Cobalt chloride has been used as a therapeutic in the treatment of anemia, and dosages in excess of 100 mg per day for prolonged periods have been unattended by changes in the cardiac muscle. In contrast, the cobalt intake of the most avid beer drinker was calculated to be only 10–15 mg per day. This suggests that the production of the cardiac lesions was not due to cobalt alone. In addition to the myocardial lesions, the beer drinkers also frequently had gastrointestinal inflammation and extensive hemorrhagic necrosis of the liver. The production of similar lesions in experimental animals given cobalt in their water affirms the importance of cobalt in their genesis. However, the heart lesions are similar in appearance to those of beriberi, a thiamine vitamin deficiency disease, and other nutritional factors may have contributed to the beer drinker's syndrome.

Some experiments have linked cobalt intake to hardening of the arteries (atherosclerosis).

NICKEL POISONING

Nickel is widely distributed throughout the Earth's crust and is a relatively plentiful element. It occurs in marine organisms, is present in the oceans, and is a common constituent of plant and animal tissues. The ordinary human diet contains 0.3–0.5 mg of nickel per day; diets rich in vegetables invariably contain much more nickel than those with foods from animal origin.

Depending on the dose, the organism involved, and the type of compound involved, nickel may be beneficial or toxic. It appears to be essential for the survival of some organisms.

Industrial hazard. The use of nickel in heavy industry has increased markedly over the last few decades, principally in the production of stainless steel and other alloys and in plating. Because of the excellent corrosion resistance of nickel and high-nickel alloys, these metals are used widely in the food processing industry. In fact foods can be contaminated with nickel during handling, processing, and cooking by utensils containing large quantities of nickel. Nickel is also used frequently as a catalyst; one of its most significant catalytic applications is in the hydrogenation of fat. Nickel carbonyl, $Ni(CO)_4$, is one of the most toxic nickel compounds and is a major industrial hazard. Lesser quantities of nickel are used in ceramics, as a gasoline additive, in fungicides, in storage batteries, and in pigments.

Nickel usually is not readily absorbed from the gastrointestinal tract except as nickel carbonyl. This compound has caused most of the acute toxicity of nickel. Recently it has been shown that nickel has a carcinogenic property and may be involved in hypersensitivity reactions. Figure 9 shows normal skeletal muscle fibers (region A) and rhabdomyosarcoma cells (region B) characteristic of the type produced by nickel, cobalt, or cadmium injection. Nickel dermatitis is reported with increasing frequency in industrial workers, especially nickel platers. It produces an allergic dermatitis on almost any skin area.

Nickel workers have approximately 150 times the cancer of the nasal passages and sinuses as the general population, and approximately five times the lung cancer. Figure 10 shows a paranasal sinus carcinoma. This type of neoplasm is seen with increasing frequency in industrial employees exposed to nickel carbonyl. The numerous dividing cells (arrows) suggest rapid growth. Several reports from Great Britain have established conclusively that inhaled nickel is a carcinogenic agent, and in Great Britain it is a compensable industrial disease.

Toxic action. The mechanism of toxic action of nickel is its capacity to inhibit oxidative enzyme systems. Within the body the main storage depots of nickel are the spinal cord, brain, lungs, and heart, with lesser amounts widely distributed throughout the organ systems. Acute poisoning causes headache, dizziness, nausea and vomiting, chest pain, tightness of the chest, dry cough with shortness of breath, rapid respiration, cyanosis, and extreme weaknesses. The lesions resulting from acute exposure are mainly in the lung and brain. In the lungs hemorrhage, collapse (atelectasis), and necrosis occur. Deposits of brown-black pigments are found in the lung phagocytic cells. In the brain extensive damage to blood vessels is seen. Lung cancers due to nickel exposure are usually of the squamous carcinoma variety and are indistinguishable microscopically from lung cancer of other causes.

CONCLUSION

Some metals are extremely injurious to the body tissues, whereas others are less so or are nontoxic. Exposure to them therefore produces diverse patterns of pathologic change. Some metals may produce extensive degeneration and necrosis of the cells in a particular organ, whereas others may produce cancerous change or birth defects.

Because of the increasing incidence of such effects—and the inevitability of further additions of metals to the environment—some practical goals can be set: Man should learn more about the human biological effects of excessive metal exposure through research, and should manage the type and quantity of exposure to minimize its adverse effects.

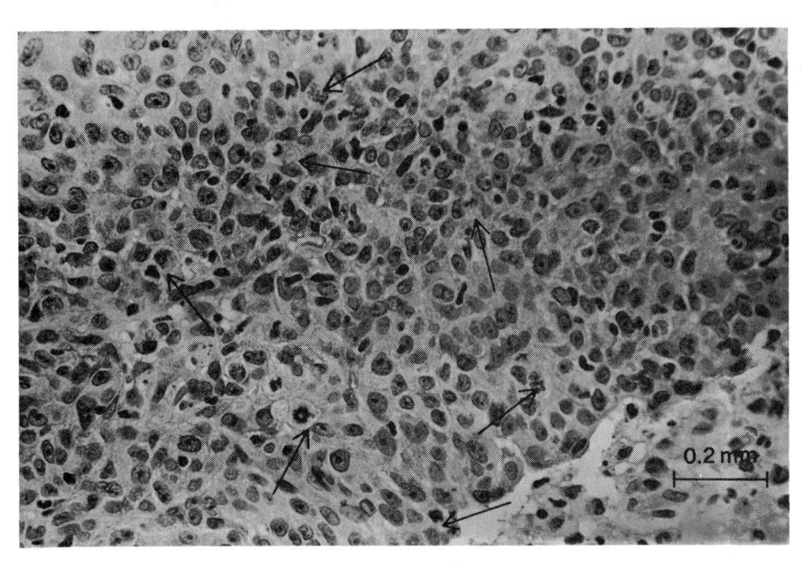

Fig. 10. Paranasal sinus carcinoma cells. Arrows indicate dividing cells.

[N. KARLE MOTTET]

Herbicide

Any chemical used to destroy or inhibit plant growth, especially of weeds or other undesirable vegetation. The concepts of modern herbicide technology began to develop about 1900 and accelerated rapidly with the discovery of dichlorophenoxyacetic acid (2,4-D) as a growth-regulator-type herbicide in 1944–1945. A few other notable events should be mentioned. During 1896–1908, metal salts and mineral acids were introduced as selective sprays for controlling broadleafed weeds in cereals; during 1915–1925 acid arsenical spray, sodium chlorate, and other chemicals were recognized as herbicides; and in 1933–1934, sodium dinitrocresylate became the first organic selective herbicide to be used in culture of cereals, flax, and peas. Since the introduction of 2,4-D, a wide variety of organic herbicides have been developed and have received wide usage in agriculture, forestry, and other industries. Today, the development of highly specific herbicides intended to control specific weed types continues. Modern usage often combines two or more herbicides to provide the desired weed control. Worldwide usage of herbicides continues to increase, making their manufacture and sale a major industry.

The control of weeds by means of herbicides has provided many benefits. Freeing agricultural crops from weed competition has resulted in higher food production, reduced harvesting costs, improved food quality, and lowered processing costs, contributing to the abundant United States supply of low-cost, high-quality food. Not only are billions of dollars saved through increased production and improved quality, but costs of labor and machinery energy necessary for weed control are reduced, livestock is saved from the effects of poisonous weeds, irrigation costs are reduced, and insect and disease control costs are decreased through the removal of host weeds for the undesirable organisms. Additional benefits due to appropriate herbicide use result as millions of people are relieved of the suffering caused by allergies to pollens and exposure to poisonous plants. Recreational areas, roadsides, forests, and parks have been freed of noxious weeds, and home lawns have been beautified. Herbicides have reduced storage and labor costs and fire hazards for industrial storage yards and warehouse areas. Modern herbicides even benefit the construction industry, where chemicals applied under asphalt prolong pavement life by preventing weed penetration of the surface.

Classification. There are well over a hundred chemicals in common usage as herbicides. Many of these are available in several formulations or under several trade names. The variety of materials are conveniently classified according to the properties of the active ingredient as either selective or nonselective. Further subclassification is by method of application, such as preemergence (soil-applied before plant emergence) or postemergence (applied to plant foliage). Additional terminology sometimes applied to describe the mobility of postemergence herbicides in the treated plant is contact (nonmobile) or translocated (mobile—that is, killing plants by systemic action). Thus, glyphosate is a nonselective, postemergence, translocated herbicide.

Selective herbicides are those that kill some members of a plant population with little or no injury to others. An example is alachlor, which can be used to kill grassy and some broad-leafed weeds in corn, soybeans, and other crops.

Nonselective herbicides are those that kill all vegetation to which they are applied. Examples are bromacil, paraquat, or glyphosate, which can be used to keep roadsides, ditch banks, and rights-of-way open and weed-free. A rapidly expanding use for such chemicals is the destruction of vegetation before seeding in the practice of reduced tillage or no tillage. Some are also used to kill annual grasses in preparation for seeding perennial grasses in pastures. Additional uses are in fire prevention, elimination of highway hazards, destruction of plants that are hosts for insects and plant diseases, and killing of poisonous or allergen-bearing plants.

Preemergence or postemergence application methods derive naturally from the properties of the herbicidal chemical. Some, such as trifluralin, are effective only when applied to the soil and absorbed into the germinating seedling, and therefore are used as preemergence herbicides. Others, such as diquat, exert their herbicidal effect only on contact with plant foliage and are strongly inactivated when placed in contact with soil. These can be applied only as a postemergence herbicide. However, the distinction between pre- and postemergence is not always clear-cut. For example, atrazine can exert its herbicidal action either following root absorption from a preemergence application or after leaf absorption from a postemergence treatment, and thus it can be used with either application method. This may be an advantage in high-rainfall areas where a postemergence treatment can be washed off the leaf onto the soil and nevertheless can provide effective weed control.

Herbicidal action. Many factors influence herbicide performance. A few are discussed below.

Soil type and organic matter content. Soils vary widely in composition and in chemical and physical characteristics. The capacity of a soil to fix or

Important herbicides

Common name*	Chemical name	Major uses†
Acrolein	2-Propenal	Aquatic weed control
Alachlor	2-Chloro-2′,6′-diethyl-*N*-(methoxymethyl) acetanilide	Sel. PE corn, soybeans, and other crops
Ametryn	2-(Ethylamino)-4-(isopropylamino)-6-(methylthio)-*s*-triazine	Sel. PE and POE pineapple, sugarcane, bananas
Amitrole	3-Amino-*s*-triazole	NS weed control in noncrop areas
AMS	Ammonium sulfate	NS POE weed and brush control
Asulam	Methyl sulfanilylcarbamate	Sel. POE sugarcane, range, forests
Atrazine	2-Chloro-4-(ethylamino)-6-(isopropylamino)-*s*-triazine	Sel. PE or POE corn, sorghum, sugarcane
Barban	4-Chloro-2-butynyl-*m*-chlorocarbanilate	Sel. POE cereals and other crops
Benefin	*N*-Butyl-*N*-ethyl-α,α, α-trifluoro-2,6-dinitro-*p*-toluidine	Sel. PE legumes, lettuce, tobacco, turf
Bensulide	*O,O*-Diisopropyl phosphonodithioate *S*-ester with *N*-(2-mercaptoethyl)benzene sulfonamide	Sel. PE turf and vegetables
Bentazon	3-Isopropyl-1*H*-2,1,3-benzothiadiazin-4-(3*H*)-1-2,2-dioxide	Sel. POE cereals and legume crops
Benzadox	(Benzamidooxy)acetic acid	Sel. POE sugar beet
Bifenox	Methyl 5-(2,4-dichlorophenoxy)-2-nitrobenzoate	Sel. PE soybean, rice, and other crops
Bromacil	5-Bromo-3-*sec*-butyl-6-methyluracil	NS PE/POE weed and brush control
Bromoxynil	3,5-Dibromo-4-hydroxybenzonitrile	Sel. POE cereals, flax, turf
Butachlor	*N*-(Butoxymethyl)2-chloro-2′,6′-diethylacetanilide	Sel. PE rice, barley
Buthidazole	3-[5-(1,1-Dimethylethyl)-1, 3, 4-thiadiazol-2-yl]-4-hydroxy-1-methyl-2-imidazolidinone	NS PE/POE weed and brush killer
Butralin	4-(1,1-Dimethyl-ethyl)-*N*-(1-methyl)-2,6-dinitrobenzeneamine	Sel. PE soybean, cotton
Butylate	*S*-Ethyl diisobutylthiocarbamate	Sel. PE corn

*Footnotes appear on page 360.

Important herbicides (cont.)

Common name*	Chemical name	Major uses†
Cacodylic acid	Hydroxydimethylarsine oxide	NS POE general weed control and cotton defoliant
Carbetamide	D-*N*-Ethyllactamide carbanilate (ester)	Sel. PE or POE alfalfa
CDAA	*N*,*N*-Diallyl-2-chloroacetamide	Sel. PE corn and vegetables
CDEC	2-Chloroallyl-diethyldithiocarbamate	Sel. PE vegetables
Chloramben	3-Amino-2,5-dichlorobenzoic acid	Sel. PE soybeans, vegetables
Chlorbromuron	3-(4-Bromo-3-chlorophenyl)-1-methoxy-l-methylurea	Sel. PE soybeans, corn, and potatoes
Chloroxuron	3-[*p*-(*p*-Chlorophenoxy) phenyl]-1,1-dimethylurea	Sel. PE/POE soybeans and other crops
Chlorpropham	Isopropyl *m*-chlorocarbanilate	Sel. PE legume and vegetable crops
Cyanazine	2-[[4-Chloro-6-(ethylamino)-*s*-triazin-2-yl]amino] 2-2 methylpropionitrile	Sel. PE corn
Cycloate	*S*-Ethyl *N*-ethylthiocyclohexanecarbamate	Sel. PE beets, spinach
Cyprazine	2-Chloro-4-cyclopropylamino)-6-(isopropylamino)-*s*-triazine	Sel. POE corn
Dalapon	2,2-Dichloropropionic acid	Sel. PE/POE sugarcane, tree crops, and other crops
Dazomet	Tetrahydro-3,5-dimethyl-2*H*-1,3,5-thiadiazine-2-thione	Sel. PE turf, tobacco seedbeds, ornamentals
DCPA	Dimethyl tetrachloroterephthalate	Sel. PE turf, ornamentals, vegetables
Desmedipham	Ethyl *m*-hydroxycarbanilate	Sel. POE sugar beet
Diallate	*S*-(2,3-Dichloroallyl) diisopropylthiocarbamate	Sel. PE sugar beet, cereals, and other crops
Dicamba	3,6-Dichloro-*o*-anisic acid	Sel. PE/POE corn, cereals, turf
Dichlorbenil	2,6-Dichlorobenzonitrile	NS PE ornamentals, tree fruits, aquatic weed control
Dichlorprop	2-(2,4-Dichlorophenoxy) propionic acid	NS brush control
Difenzoquat	1,2-Dimethyl-3,5-diphenyl-1*H*-pyrazolium	Sel. POE cereals
Dinitramine	N^4,N^4-Diethyl-α,α,α-trifluoro-3,5-dinitrotoluene-2,4-diamine	Sel. PE cotton, soybeans, and other crops
Dinoseb	2-*sec*-Butyl-4,6-dinitrophenol	Sel. PE/POE soybeans, peanuts, and other crops
Diphenamid	*N*,*N*-Dimethyl-2,2-diphenylacetamide	Sel. PE turf, peanuts, vegetables, and other crops
Dipropetryn	2,4-Bis(isopropylamino)-6-(ethylthio)-*s*-triazine	Sel. PE cotton
Diquat	6,7-Dihydrodipyrido[1,2-*a*:2′,1′-*c*]pyrazinediium ion	NS POE noncrop land and aquatic weed control
Diuron	3-(3,4-Dichlorophenyl)-1,1-dimethylurea	Sel. PE/POE cotton, sugarcane, cereals, tree crops; NS PE/POE general weed killer
DSMA	Disodium methanearsonate	Sel. POE cotton, turf; NS POE general weed killer
Endothall	7-Oxabicyclo[2.2.1]heptane 2,3-dicarboxylic acid	Sel. PE sugar beets; aquatic weed killer; cotton defoliant
EPTC	*S*-Ethyl dipropylthiocarbamate	Sel. PE corn, potatoes, and other crops
Erbon	2-(2,4,5-Trichlorophenoxy)ethyl-2,2-dichloropropionate	NS PE/POE general weed killer
Ethalfluralin	*N*-Ethyl-*N*-(2-methyl-2-propenyl)-2,6-dinitro-4-(trifluoromethyl)benzeneamine	Sel. PE cotton, beans
Fenac	(2,3,6-Trichlorophenyl)acetic acid	Sel. PE/POE sugarcane; general weed killer
Fenuron	1,1-Dimethyl-3-phenylurea	NS PE/POE general weed and brush killer
Fluchloralin	*N*-(2-Chloroethyl)-2,6-dinitro-*N*-propyl-4-(trifluoromethyl) aniline	Sel. PE cotton, soybeans
Fluometuron	1,1-Dimethyl-3-(α,α,α-trifluoro-*m*-tolyl)urea	Sel. PE/POE cotton, sugarcane
Glyphosate	*N*-(Phosphonomethyl)-glycine	NS POE general weed killer
Ioxynil	4-Hydroxy-3,5-diiodobenzonitrile	Sel. POE cereals
Isopropalin	2,6-Dinitro-*N*,*N*-dipropylcumidine	Sel. PE tobacco
Linuron	3-(3,4-Dichlorophenyl)-1-methoxy-1-methylurea	Sel. PE soybean, potatoes, and other crops
MCPA	[(4-Chloro-*o*-toly)oxy]acetic acid	Sel. POE cereals, legumes, flax
MCPB	4-[(4-Chloro-*o*-tolyl)oxy]butyric acid	Sel. POE legumes
Mecoprop	2-[(4-Chloro-*o*-tolyl)oxy]propionic acid	Sel. POE turf
Metham	Sodium methyldithiocarbamate	Sel. PE tobacco, turf
Methazole	2-(3,4-Dichlorophenyl)-4-methyl-1,2,4-oxadiazolidine-3,5-dione	Sel. PE cotton
Metolachlor	2-Chloro-*N*-(2-ethyl-6-methylphenyl)-*N*-(2-methoxy-1-methylethyl)acetamide	Sel. PE corn
Metribuzin	4-Amino-6-*tert*-butyl-3-(methylthio)-*as*-triazin-5(4H)-one	Sel. PE soybeans, potatoes, sugarcane

Important herbicides (cont.)

Common name*	Chemical name	Major uses†
Molinate	*S*-Ethyl hexahydro-1*H*-azepine 1-carbothioate	Sel. PE/POE rice
Monuron	3-(*p*-Chlorophenyl)-1, 1-dimethylurea	NS PE/POE general weed killer
MSMA	Monosodium methanearsonate	Sel. POE cotton, turf; NS POE general weed killer
Napropamide	2-(α-Naphthoxy)*N*,*N*-diethylpropionamide	Sel. PE tomatoes, tree fruits
Naptalam	*N*-1-Napthylphthalamic acid	Sel. PE soybean, peanut, and vegetables
Nitrofen	2,4-Dichlorophenyl-*p*-nitrophenyl ether	Sel. PE/POE vegetables
Norflurazon	4-Chloro-5-(methylamino)-2-(α,α,α-trifluoro-*m*-tolyl)-3(2*H*)-pyridazinone	Sel. PE cranberries
Oryzalin	3,5-Dinitro-*N*⁴,*N*⁴-dipropylsulfanilamide	Sel. PE cotton, soybeans
Paraquat	1,1′-Dimethyl-4,4′-bipyridinium ion	NS POE general weed killer
Pebulate	*S*-Propyl butylethylthiocarbamate	Sel. PE sugar beets, tobacco
Pendimethalin	*N*-(1-Ethyl)-3,4-dimethyl-2,6-dinitrobenzeneamine	Sel. PE corn, cotton, soybeans
Perfluidone	1,1,1-Trifluoro-*N*-[2-methyl-4-(phenylsulfonyl)phenyl] methanesulfonamide	Sel. PE cotton
Phenmedipham	Methyl *m*-hydroxycarbanilate *m*-methylcarbanilate	Sel. POE beets
Picloram	4-Amino-3,5,6-trichloropicolinic acid	NS PE/POE general weed and brush killer
Procyazine	2-[[4-Chloro-6-(cyclopropylamino)-1,3,5-triazine-2-yl]amino]-2-methylpropanenitrile	Sel. PE/POE corn
Profluralin	*N*-(Cyclopropylmethyl)-α,α,α-trifluoro-2,6-dinitro-*N*-propyl-p-toluidine	Sel. PE legumes, cotton, and other crops
Prometryn	2,4-Bis(isopropylamino)-6-(methylthio)-*s*-triazine	Sel. PE/POE cotton, celery
Pronamide	3,5-Dichloro(*N*-1,1-dimethyl-2-propynyl)benzamide	Sel. PE/POE legumes, turf, lettuce
Propachlor	2-Chloro-*N*-isopropylacetanilide	Sel. PE corn, milo, soybeans
Propanil	3′,4′-Dichloropropionanilide	Sel. POE rice
Propazine	2-Chloro-4,6-bis-(isopropylamino)-*s*-triazine	Sel. PE milo
Propham	Isopropyl carbanilate	Sel. PE forages, lettuce, and other crops
Pyrazon	5-Amino-4-chloro-2-phenyl-3(2*H*)-pyridazinone	Sel. PE beets
Siduron	1-(2-Methylcyclohexyl)-3-phenylurea	Sel. PE turf
Silvex	2-(2,4-5-Trichlorophenoxy)propionic acid	Sel. POE turf; NS POE general and aquatic weed control
Simazine	2-Chloro-4,6-bis-(ethylamino)-*s*-triazine	Sel. PE corn, forages, and other crops; aquatic weed control
TCA	Trichloroacetic acid	Sel. PE/POE sugar beets, sugarcane, cotton, soybeans
Tebuthiuron	*N*-[5-(1,1-Dimethyl)-1,3,4-thiadiazol-2-yl]-*N*,*N*′-dimethylurea	NS PE/POE general weed killer
Terbacil	3-*tert*-Butyl-5-chloro-6-methyluracil	Sel. PE sugarcane, tree fruits, and other crops
Terbutryn	2-(*tert*-Butylamino)-4-(ethylamino)-6-(methylthio)-*s*-triazine	Sel. PE/POE cereals, milo
Triallate	*S*-(2,3,3-Trichloroallyl)diisopropylthiocarbamate	Sel. PE cereals
Trifluralin	α,α,α-Trifluoro-2,6-dinitro-*N*,*N*-dipropyl-*p*-toluidine	Sel. PE legumes, cotton, and other crops
2,3,6-TBA	2,3,6-Trichlorobenzoic acid	NS PE/POE general weed control
2,4-D	(2,4-Dichlorophenoxy)acetic acid	Sel. POE cereals, milo, corn, and other crops; aquatic and general weed control
2,4-DB	4-(2,4-Dichlorophenoxy)butyric acid	Sel. POE legume crops
2,4,5-T	(2,4,5-Trichlorophenoxy)acetic acid	Sel. POE rice, forage grasses; general weed and brush control
Vernolate	*S*-Propyl dipropylthiocarbamate	Sel. PE peanuts, soybeans

*Major trade names can be found in *Farm Chemicals Handbook*, Meister Publishing Co., Willoughby. OH, 1977.
†Sel. = selective herbicide; NS = nonselective herbicide; PE = preemergence; POE = postemergence.

adsorb a preemergence herbicide determines how much will be available to seedling plants. For example, a sandy soil normally is not strongly adsorptive. Lower quantities of most herbicides are needed on a highly adsorptive clay soil. A similar response occurs with the organic portion of soil; higher organic matter content usually indicates that more herbicide is bound and a higher treatment rate necessary. For example, cyanazine is used for weed control in corn at the rate of 1½ lb/acre (1.7 kg/ha) on sandy loam soil of less than 1% organic matter, but 5 lb/acre (5.6 kg/ha) is required on a clay soil of 4% organic matter.

Leaching. This refers to the downward movement of herbicides into the soil. Some preemergence herbicides must be leached into the soil into the immediate vicinity of weed seeds to exert toxic action. However, excessive rainfall may leach these chemicals too deeply into the soil, thereby allowing weeds to germinate and grow close to the surface.

Volatilization. Several herbicides in use today will volatilize from the soil surface. To be effective, these herbicides must be mixed with the soil to a depth of 2 to 4 in. (5–10 cm). This process is termed soil incorporation. Once the herbicide is in contact with soil particles, volatilization loss is minimized. This procedure is commonly employed with thiocarbamate herbicides such as EPTC and dinitroaniline herbicides such as trifluralin.

Leaf properties. Leaf surfaces are highly variable. Some are much more waxy then others; many are corrugated or ridged; and some are covered with small hairs. These variations cause differences in the retention of postemergence spray droplets and thus influence herbicidal effect. Most grass leaves stand in a relatively vertical position, whereas the broad-leafed plants usually have their leaves arranged in a more horizontal position. This causes broad-leaf plants to intercept a larger quantity of a herbicide spray than grass plants.

Location of growing points. Growing points and buds of most cereal plants are located in a crown, at or below the soil surface. Furthermore, they are wrapped within the mature bases of the older leaves. Hence they may be protected from herbicides applied as sprays. Buds of many broad-leafed weeds are located at the tips of shoots and in axils of leaves, and are therefore more exposed to herbicide sprays.

Growth habits. Some perennial crops, such as alfalfa, vines, and trees, have a dormant period in winter. At that time a general-contact weed killer may be safely used to get rid of weeds that later would compete with the crop for water and plant nutrients.

Application methods. By arranging spray nozzles to spray low-growing weeds but not the leaves of a taller crop plant, it is possible to provide weed control with a herbicide normally phytotoxic to the crop. This directed spray technique is used to kill young grass in cotton with MSMA or broad-leafed weeds in soybeans with chloroxuron in an oil emulsion. A recent modification of this technique is an arrangement of nozzles spraying horizontally into a catch basin over the top of the crop in such a manner that weeds taller than the crop pass through the spray streams but the crop does not. Use of a nonselective translocated herbicide in this recirculating sprayer system will then selectively remove the weeds from the crop. Glyphosate is being used experimentally in this system to remove johnson grass from cotton.

Protoplasmic selectivity. Just as some people are immune to the effects of certain diseases while others succumb, so some weed species resist the toxic effects of herbicides whereas others are injured or killed. This results from inherent properties of the protoplasm of the respective species. One example is the use of 2,4-DB or MCPB (the butyric acid analogs of 2,4-D and MCPA) on weeds having β-oxidizing enzymes that are growing in crops (certain legumes) which lack such enzymes. The weeds are killed because the butyric acid compounds are broken down to 2,4-D or MCPA. Another example is the control of a wide variety of weeds in corn by atrazine. Corn contains a compound that removes the chlorine from the atrazine atom, rendering it nontoxic; most weeds lack this compound. A third important example of protoplasmic selectivity is shown by trifluralin, planavin, and a number of other herbicides applied through the soil which inhibit secondary root growth. Used in large seeded crops having vigorous taproots, they kill shallow-rooted weed seedlings; the roots of the crops extend below the shallow layer of topsoil containing the herbicides and the seedlings survive and grow to produce a crop.

Properties. Several factors of the commercial herbicide influence selectivity to crops.

Molecular configuration. Subtle changes in the chemical structure can cause dramatic shifts in herbicide performance. For example, trifluralin and benefin are very similar dinitroaniline herbicides differing only in the location of one methylene group. However, this small difference allows benefin to be used for grass control in lettuce, whereas trifluralin will severely injure lettuce at the rates required for weed control.

Herbicide concentration. The action of herbicides on plants is rate-responsive. That is, small quantities of a herbicide applied to a plant may cause no toxicity, or even a slight growth stimulation, whereas larger amounts may result in the death of the plants. It has long been known that 2,4-D applied at low rates causes an increase in respiration rate and cell division, resulting in an apparent growth stimulation. At high application rates, 2,4-D causes more severe changes and the eventual death of the plant.

Formulation. The active herbicidal chemical itself is seldom applied directly to the soil or plants. Because of the nature of the chemicals, it is usually necessary that the commercial product be formulated to facilitate handling and dilution to the appropriate concentration. Two common formulations are emulsifiable concentrates and wettable powders. Emulsifiable concentrates are solutions of the chemical in an organic solvent with emulsifiers added which permit mixing and spraying with water. Wettable powders are a mixture of a finely divided powder, active chemical, and emulsifiers which allow the powder to suspend in water and to be sprayed. An additional ingredient called an adjuvant is sometimes added to the formulation or to the spray tank when the spray solution is mixed. These adjuvants are normally surface-active agents (surfactants) which improve the uniformity of spray coverage on plants and the plant penetration of the herbicide. An example is the addition of surfactant to paraquat spray solutions to improve its nonselective postemergence action on weeds. Another material sometimes added to spray solutions is a nonphytotoxic oil. These oils may be used to improve postemergence action of herbicides such as diuron which normally have limited foliar absorption.

Available herbicides. In the table, some of the important herbicides used in United States are listed alphabetically by the common name approved by the Weed Science Society of America. The chemical name and some major uses of each herbicide are included. Many of the chemicals listed are used in proprietary mixtures. These normally combine two or more herbicides in a single formulation and are marketed under brand names. Such mixtures are not included in the table.

[RODNEY O. RADKE]

Bibliography: A. S. Crafts, *Modern Weed Control*, 1975; *Farm Chemicals Handbook*, 1977; J. R. Crister, Jr., *Herbicides*, 1978; *Herbicide Handbook of the Weed Science Society of America*, 3d ed., 1974; P. C. Kearney and D. Kaufman (eds.), *Herbicides: Chemistry, Degradation, and Mode of Action*, 1975–1976; G. C. Klingman and F. M. Ashton, *Weed Sciences: Principles and Practices*, 1975.

Highway engineering

Highway planning, location, design, and maintenance. Before the design and construction of a new highway or highway improvement can be undertaken, there must be general planning and consideration of financing. As part of general planning it is decided what the traffic needs of the area will be for a considerable period, generally 20 years, and what construction will meet those needs. To assess traffic needs, the highway engineer collects and analyzes information about the physical features of existing facilities, the volume, distribution, and character of present traffic, and the changes to be expected in these factors. He must determine the most suitable location, layout, and capacity of the new routes and structures. Frequently, a preliminary line, or location, and several alternate routes are studied. The detailed design is normally begun only when the preferred location has been chosen.

In selecting the best route careful consideration is given to the traffic requirements, terrain to be traversed, value of land needed for the right-of-way, and estimated cost of construction for the various plans. The photogrammetric method, which makes use of aerial photographs, is used extensively to indicate the character of the terrain on large projects, where it is most economical. On small projects ground-mapping methods are preferred.

Financing considerations determine whether the project can be carried out at one time or whether construction must be in stages, with each stage initiated as funds become available. In deciding the best method of financing the work, the engineer makes an analysis of whom it will benefit. Improved highways and streets benefit, in varying degrees, three groups: users, owners of adjacent property, and the general public.

Users of improved highways benefit from decreased cost of transportation, greater travel comfort, increased safety, and saving of time (Fig. 1).

Fig. 1. Construction of expressways in metropolitan areas requires careful planning and is costly. Often many bridges are required for an expressway in the heart of a city, as illustrated by this aerial view of St. Paul, MN. At the lower left is the State Capitol.

They also obtain recreational and educational benefits. Owners of abutting or adjacent property may benefit from better access, increased property value, more effective police and fire protection, improved street parking, greater pedestrian traffic safety, and the use of the street right-of-way for the location of public utilities, such as water lines and sewers.

Evaluation of various benefits from highway construction is often difficult but is a most important phase of highway engineering. Some benefits can be measured with accuracy, but the evaluation of others is more speculative. As a result numerous methods are used to finance construction, and much engineering work may be involved in selecting the best procedure.

Environmental evaluation. The environmental impact of constructing highways has received increased attention and importance. Many projects have been delayed and numerous others canceled because of environmental problems. The environmental study or report covers many factors, including noise generation, air pollution, disturbance of areas traversed, destruction of existing housing, and possible alternate routes.

Right-of-way acquisition. Highway engineers must also assist in the acquisition of right-of-way needed for new highway facilities. Acquisition of the land required for construction of expressways leading into the central business areas of cities has proved extremely difficult; the public is demanding that traffic engineers work closely with city planners, architects, sociologists, and all groups interested in beautification and improvement of cities to assure that expressways extending through metropolitan areas be built only after coordinated evaluation of all major questions, including the following:

(1.) Is sufficient attention being paid to beautification of the expressway itself? (2.) Would a change in location preserve major natural beauties of the city? (3.) Could a depressed design be logically substituted for those sections where an elevated expressway is proposed? (4.) Can the general design be improved to reduce the noise created by large volumes of traffic? (5.) Are some sections of the city being isolated by the proposed location?

Because of the large land areas needed for expressways, in several communities air rights above expressways are being sold to permit apartment buildings, parking garages, and similar structures to be built over the expressway, or such areas are being put to public use (Fig. 2).

At other locations, playgrounds are being built beneath elevated highways, as are parking garages. In Honolulu, Hawaii, construction of a new post office to be located under an elevated section of a freeway is being studied.

Detailed design. Detailed design of a highway project includes preparation of drawings or blueprints to be used for construction. These plans show, for example, the location, the dimensions of such elements as roadway width, the final profile for the road, the location and type of drainage facilities, and the quantities of work involved, including earthwork and surfacing.

Soil studies. In planning the grading operations the design engineer considers the type of material to be encountered in excavating or in cutting away

Fig. 2. Footbridge over vehicular roadway crossing Mississippi River in Minneapolis. (*Minnesota Dept. of Highways*)

the high points along the project and how the material removed can best be utilized for fill or for constructing embankments across low areas elsewhere on the project. For this the engineer must analyze the gradation and physical properties of the soil, determine how the embankments can best be compacted, and calculate the volume of earthwork to be done. Electronic calculating procedures are now sometimes used for the last step. Electronic equipment has also speeded up many other highway engineering calculations.

Powerful and highly mobile earth-moving machines have been developed to permit rapid and economical operations. For example, now in use in the United States is a self-propelled earthmover weighing 125 tons (113 metric tons) and capable of hauling 100 yd³ (76 m³) of earth. This unit is powered by a 600-hp (447-kW) diesel electric motor.

Surfacing. Selection of the type and thickness of roadway surfacing to be constructed is an important part of design. The type chosen depends upon the maximum loads to be accommodated, the frequency of these loads, and other factors. For some routes, traffic volume may be so low that no surfacing is economically justified and natural soil serves as the roadway. As traffic increases, a surfacing of sandy clay, crushed slag, taconic tailings, crushed stone, caliche, crushed oyster shells, or a combination of these may be applied. If gravel is used, it usually contains sufficient clay and fine material to help stabilize the surfacing. Gravel surfaces may be further stabilized by application of calcium chloride, which also aids in controlling dust. Another surfacing is composed of portland cement and water mixed into the upper few inches of the subgrade and compacted with rollers. This procedure forms a soil-cement base that can be surfaced with bituminous materials. Roadways to carry large volumes of heavy vehicles must be carefully designed and made of considerable thickness.

Drainage structures. Much of highway engineering is devoted to the planning and construction of facilities to drain the highway or street and to carry streams across the highway right-of-way.

Removal of surface water from the road or street is known as surface drainage. It is accomplished by constructing the road so that it has a crown and by sloping the shoulders and adjacent areas so as to control the flow of water either toward existing natural drainage, such as open ditches, or into a storm drainage system of catchbasins and underground pipes. If a storm drainage system is used, as it would be with city streets, the design engineer must give consideration to the total area draining onto the street, the maximum rate of runoff expected, the duration of the design storm, the amount of ponding allowable at each catchbasin, and the proposed spacing of the catchbasins along the street. From this information the desired capacity of the individual catchbasin and the size of the underground piping network are calculated.

In designing facilities to carry streams under the highway the engineer must determine the area to be drained, the maximum probable precipitation over the drainage basin, the highest expected runoff rate, and then, using this information, must calculate the required capacity of the drainage structure. Generally designs are made adequate to accommodate not only the largest flow ever recorded for that location but the greatest discharge that might be expected under the most adverse conditions for a given number of years.

Factors considered in calculating the expected flow through a culvert opening include size, length, and shape of the opening, roughness of the walls, shape of the entrance and downstream end of the conduit, maximum allowable height of water at the entrance, and water level at the outlet.

There is a trend to use designs that permit drainage structures to be assembled from standard sections manufactured at a central yard. Such procedures permit better control of the work, quicker construction, and less field work. For example, with precast concrete pipe sections it is often possible to avoid building small box culverts in the field. Circular culverts of large diameter are now also constructed in this way. However, culverts built of corrugated metal are specified for many projects for reasons of economy and to avoid placing of small volumes of concrete at numerous locations.

Numerous small bridges are being designed to permit precast beams or girders to be placed side by side across the bridge opening to form the support for the roadway. These members are frequently of prestressed concrete. When precast members are used, the need for falsework to construct the bridge deck is eliminated, an especially beneficial move if the bridge is being built over a railroad or busy street.

Planning and construction of tunnels is an important part of highway engineering. Many tunnels are being built to avoid removal of important structures aboveground and to preserve existing facilities. Long tunnels are also being built to extend highways through mountains.

The 8941-ft-long (2717-m) Eisenhower Memorial tunnel 80 mi (130 km) west of Denver, Colorado, carries Interstate Highway 70 under the Continental Divide. The Eisenhower Memorial is the world's highest highway tunnel.

The 38,280-ft-long (11,668-m) Mont Blanc tunnel in the Alps is the world's longest highway tunnel. In Japan a tunnel 28,015 ft (8539 m) long was completed in 1975 to carry a dual-lane highway under the Japan Alps. This tube is the second longest vehicular tunnel in the world.

Construction operations. Although much engineering and planning must be done preliminary to it, the actual construction is normally the costliest part of making highway and street improvements.

Staking out. With the award of a construction contract following the preparation of the detailed plans and specifications, engineers go onto the site and lay out the project. As part of this staking-out process, limits of earthwork are shown, location of drainage structures is indicated, and profiles are established.

Compaction. Heavy rollers are used to compact the soil or subgrade below the roadway in order to eliminate later settlement. Pneumatic-tired rollers and sheepsfoot rollers (steel cylinders equipped with numerous short steel teeth or feet) are often employed for this operation. Vibratory rollers have

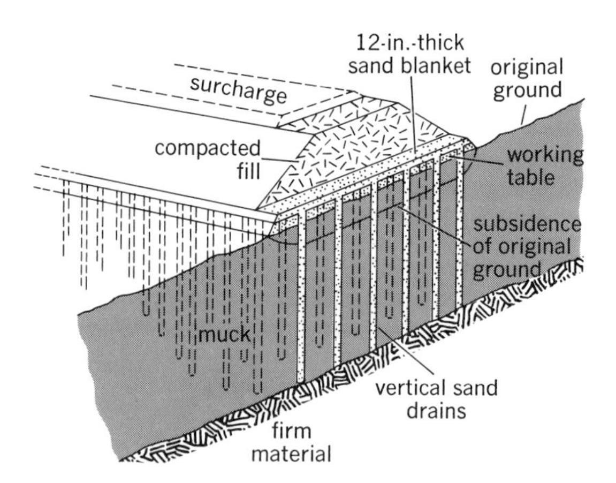

Fig. 3. Cross section of a vertical sand drain installation. Drains are 13–28 in. (33–71 cm) in diameter and are spaced 9 ft (2.7 m) apart. Broken lines represent positions of staggered sand drains. *(From L. I. Hewes and C. H. Oglesby, Highway Engineering, copyright © 1954 by John Wiley & Sons, Inc.; reprinted with permission)*

been developed and used on some projects in recent years. One type vibrates up to 3400 times/min, compacting the underlying material to an appreciable depth.

Vertical sand drains are sometimes employed to help stabilize fills or embankments constructed across wet and unstable ground (Fig. 3). Holes are drilled into the existing material and filled with sand. A horizontal layer of pervious sand or gravel is laid over the network of drains. As the fill is added over the pervious layer and the drains, the undesired water is forced upward through the columns of sand. By allowing the water to move horizontally to the edge of the embankment, the pervious layer prevents collection of water under the roadway.

To obtain quicker compaction of the unstable soil, a surcharge may be placed above the compacted fill. The surcharge material, usually intended for building the compacted fill elsewhere on the project, is removed once the unstable soil has been sufficiently compacted.

Maintenance and operation. Highway maintenance consists of the repair and upkeep of surfacing and shoulders, bridges and drainage facilities, signs, traffic control devices, guard rails, traffic striping on the pavement, retaining walls, and side slopes. Additional operations include ice control and snow removal. Because it is valuable to know why some highway designs give better performance and prove less costly to maintain than others, engineers supervising maintenance can offer valuable guidance to design engineers. Consequently, maintenance and operation are important parts of highway engineering.

[ARCHIE N. CARTER]

Bibliography: American Association of State Transportation Officials, *Highway Design and Operational Practices Related to Highway Safety*; H. W. Busching and R. Russell, American road building: Fifty years of progress, *J. Constr. Div. ASCE*, 101(CO3):565–581, September 1975; R. W. Cockfield, *A Design Method for the Preparation of a Preliminary Urban Land Use/Transport Plan*, Tech. Rep. Dep. Civil Eng., University of Waterloo, Ontario, May 1970; I. Cook and T. J. Schultz,

Highway Noise and Acoustical Buffer Zones, 1974; K. A. Godfrey, Jr., Interstate highway system, *Civil Eng.*, p. 51, March 1975; K. A. Godfrey, Jr., Urban freeways: Salvation of cities or their death, *Civil Eng.*, p. 80, May 1975; Institute of Transportation Engineers, *Introduction to Transportation Engineering: Highways and Transit*, 1978; R. S. Mayo, An introduction to fifty years of tunneling, *J. Constr. Div. ASCE*, 101(CO2):259–263, June 1975; J. N. Normann, Improved design of highway culverts, *Civil Eng.*, p. 70, March 1975; D. C. Oliver, Legal liability and highway design and maintenance, *Transp. Eng. J. ASCE*, 101(TE3): 425–435, August 1975; E. L. Priestas and T. E. Mulinazzi, Traffic volume counting recorders, *Trans. Eng. J. ASCE*, 101(TE2):211–223, May 1975; T. N. Tamburri and R. N. Smith, *The Safety Index: A Method for Measuring Safety Benefits*, State of California Department of Transportation, Analytical Studies Branch, November 1973; P. Wang, G. L. Peterson, and J. L. Schofer, Population change: An indicator of freeway impact, *Transp. Eng. J. ASCE*, 101(TE3):491–504, August 1975; P. H. Wright and R. J. Poquette, *Highway Engineering*, 1979.

Hill and mountain terrain

Land surfaces characterized by roughness and strong relief. The distinction between hills and mountains is usually one of relative size or height, but the terms are loosely and inconsistently used. Because of the prevalence of steep slopes, hill and mountain lands offer many difficulties to human occupancy. Cultivable land is scarce and patchy, and transportation routes are often difficult to construct and maintain. On the other hand, many rough lands, especially those readily accessible to centers of population, attract tourists because of their scenic quality and the opportunities they may afford for outdoor recreation.

High mountain ranges set up major disturbances in the broad pattern of atmospheric circulation, and thus affect climates over extensive areas. More locally, by inducing turbulence or forcing moist air to rise in crossing them, rough lands commonly induce condensation and precipitation that makes them moister than surrounding lowlands. Within the rough country, local differences in elevation and in exposure to sun and wind produce complex patterns of local climatic contrasts. These variations, in turn, are often accompanied by unusual variety in the vegetation and in animal life.

Development of rough terrain. Uplift of the Earth's crust is necessary to give mountain and hill lands their distinctive elevation and relief, but most of their characteristic features—peaks, ridges, valleys, and so on—have been carved out of the uplifted masses by streams and glaciers. The upraised portions of the crust may have been formed as broad, warped swells, smaller arched folds or domes, upthrust or tilted blocks, or folded and broken masses of extreme complexity. Some limited areas owe their elevation and certain of their local features to the outpouring of thick sheets of lava or the construction of volcanic cones. Hill lands, with their lesser relief, indicate only lesser uplift, not a fundamentally different course of development.

In some rough lands, for example the Appalachians or the Scottish Highlands, the most intense crustal deformation is known to have occurred hundreds of millions of years ago, while in others, such as the Alps, the Himalayas, or the California Coast Ranges, intense deformation has occurred quite recently. Since mountains can be erosionally destroyed in rather short spans of geologic time, however, the very existence of mountains and hills indicates that structural deformation, at least simple uplift, has continued in those areas until recent times. This conclusion is reinforced by the fact that the major cordilleran belts are currently foci of earthquake activity and vulcanism.

Distribution pattern of rough land. Hill and mountain terrain occupies about 36% of the Earth's land area. The greater portion of that amount is concentrated in the great cordilleran belts that surround the Pacific Ocean, the Indian Ocean, and the Mediterranean Sea. Additional rough terrain, generally low mountains and hills, occurs outside the cordilleran systems in eastern North and South America, northwestern Europe, Africa, and western Australia. Eurasia is the roughest continent, more than half of its total area and most of its eastern portion being hilly or mountainous. Africa and Australia lack true cordilleran belts; their rough lands occur in patches and interrupted bands that rarely show marked complexity of geologic structure. The broad-scale pattern of crustal disturbance, and hence of rough lands, is now known to be related to the relative movements of a worldwide system of immense crustal plates. The intense deformation of the cordilleran belts, for example, occurs where the margins of rigid continental masses are being jammed into the zones where such plates are converging. *See* TERRAIN AREAS, WORLDWIDE.

Predominant surface character. The features of hill and mountain lands are chiefly valleys and divides produced by sculpturing agents, especially running water and glacier ice. Local peculiarities in the form and pattern of these features reflect the arrangement and character of the rock materials within the upraised crustal mass that is being dissected.

Steam-eroded features. The principal features of most hill and mountain landscapes have been formed by steam erosion. The major differences between one rough land and another are in the size, shape, spacing, and pattern of the stream valleys and the divides between them. These differences reflect variations in the original form and structure of the uplifted mass and in the stage to which erosion has proceeded.

In consequence of uplift, streams have a large range of elevation through which they can cut, and as a result usually possess steep gradients early in their course of development. At this stage they are swift, have great erosive power, and as a rule are marked by many rapids and falls. Because of the rapid downcutting, the valleys are characteristically canyonlike with steep walls and narrow floors. At this same early stage of erosional development, divides are likely to be high and continuous, and broad ridge crests common. In hill lands such ridges often provide easier routes of travel than do the valleys, whereas in high mountain country the combination of gorgelike valleys and continuous high divides makes crossing unusually difficult. Thus in the Ozark Hills of Missouri, most of the

Fig. 1. Head of a glaciated mountain valley ending in a cliff-walled amphitheater, called a cirque, showing a steplike valley bottom with lakes and waterfalls. (*Photograph by Hileman, Glacier National Park*)

highways and railroads follow the broad ridge crests. In the Himalayas and the central Andes, and to a lesser degree in the Rocky Mountains of Colorado, the narrow canyons are very difficult of passage, and the divides are so continuously high as to provide no easy pass routes across the ranges.

As erosional development continues, the major streams achieve gentle gradients and their valleys continue to widen. Divides become narrower and deep notches develop at valley heads, with well-defined peaks remaining between them. At this stage, which is represented by much of the Alps and by the Cascade Mountains of Washington, the principal valleys and the relatively low passes at their heads furnish feasible routes across the mountain belts. If other conditions are favorable, the wide valleys may afford significant amounts of cultivable or grazing land, as is true in the Alps. Still further erosion continues the widening of valleys and reduction of divides until the landscape becomes an erosional plain upon which stand only small ranges and groups of mountains or hills. The mountains of New England and many of the mountains and hill groups of the Sahara and of western Africa represent late stages in the erosional sequence.

Fig. 2. The upper reaches of Susitna Glacier, Alaska, showing cirques and snowfields. The long tongues of glacier ice carry bands of debris scoured from valley walls. (*Photograph by Bradford Washburn*)

Glaciated rough terrain. Glacial features of mountain and hill lands may be produced either by the work of local glaciers in the mountain valleys or by overriding glaciers of continental size. Continental glaciation has the general effect of clearing away crags, smoothing summits and spurs, and depositing debris in the valleys. The resulting terrain is less angular than is usual for stream-eroded hills, and the characteristic glacial trademark is the numerous lakes, most of them debris-dammed, a few occupying shallow, eroded basins. Examples of rough lands overriden by ice sheets are the mountains of New England, the Adirondacks, the hills of western New York, the Laurentian Upland of eastern Canada, and the Scandinavian Peninsula.

In contrast, mountains affected by local glaciers are made rougher by the glacial action. The long tongues of ice that move slowly down the valleys are excellent transporters of debris and are able to erode actively on shattered or weathered rock material. Valleys formerly occupied by glaciers are characteristically steep-walled and relatively free of projecting spurs and crags, with numerous broad cliffs, knobs, and shoulders of scoured bedrock. At their heads they generally end in cliff-walled amphitheaters called cirques (Fig. 1). The valley bottoms are commonly steplike, with stretches of gentle gradient alternating with abrupt rock-faced risers. Especially in the lower parts of the glaciated sections are abundant deposits of rocky debris dropped by the ice. Sometimes these form well-defined ridges (moraines) that run lengthwise along the valley sides or swing in arcs across the valley floor. Lakes are strung along the streams like beads, most of them dammed by moraines but some occupying eroded basins.

Because of rapid erosion, either that effected directly by the ice or that attendant upon the exposure of the rock surface by continual removal of the products of weathering, glaciated mountains are likely to be unusually rugged and spectacular. This is true not only of such great systems as the Himalayas, the Alps, the Alaskan Range, and the high Andes, but also of such lesser ranges as those of Labrador, the English Lake District, and the Scottish Highlands (Fig. 2).

Most of the higher mountains of the world still bear valley glaciers, though these are not as large or as widespread as formerly. Dryness and long, warm summers limit glaciers in the United States to a few large groups on the higher peaks of the Pacific Northwest and numerous small ones in the northern Rockies.

Fig. 3. Hills in the western Appalachians, W.Va. *(Photograph by John L. Rich, Geographical Review)*

Effects of geologic structure. Form and extent of the elevated areas, pattern of erosional valleys and divides, and, to some extent, sculptured details of slope and crest reflect geologic structure.

Some areas, such as the Ozarks, the western Appalachians, and the coast ranges of Oregon, are simply upwarped plains of homogeneous rocks that have been carved by irregularly branching streams into extensive groups of hills or mountains (Fig. 3). Others, like the Black Hills or the ranges of the Wyoming Rockies, are domes or arched folds, deeply eroded to reveal ancient granitic rocks in their cores and upturned younger stratified rocks around their edges. The Sierra Nevada of California is a massive block of the crust that has been uplifted and tilted toward the west so that it now displays a high abrupt eastern face and a long canyon-grooved western slope. The central belt of the Appalachians displays long parallel ridges and valleys that have been hewn by erosion out of a very old structure of parallel wrinkles in the crust. The upturned edges of resistant strata form the ridges; the weaker rocks between have been etched out to form the valleys. The Alps and the Himalayas are eroded from folded and broken structures of incredible complexity involving almost all varieties of rock materials.

Most volcanic mountains, like the Cascades of the northwestern United States or the western Andes of Peru and Bolivia, are actually erosional mountains sculptured from thick accumulations of lava and ash. In these areas of volcanic activity, however, individual eruptive vents give rise to volcanic cones that range from small cinder heaps to tremendous isolated mountains. The greater cones, such as Fuji, Ararat, Mauna Loa, or Shasta, are among the most magnificent features of the Earth's surface. [EDWIN H. HAMMOND]

Homeostasis

In living substance, the maintenance of internal constancy and independence of the environment. The essence of living substance is that it differs chemically from the surrounding medium and yet maintains a dynamic equilibrium with its environment. Homeostasis occurs in all living cells and in the fluids and organ systems of multicellular organisms. Referring to homeostasis in mammals, the 19th-century French physiologist Claude Bernard coined the expression "constancy of the internal environment is the condition of life." The American physiologist W. B. Cannon, who first used the word "homeostasis," referred to the systems of checks and balances which maintain internal constancy as "the widsom of the body." Thus enzyme activities may be maintained even though internal state (temperature, sodium concentration) may vary.

Patterns of adaptive responses. The cells of multicellular organisms are bathed in fluids which are kept relatively constant in ionic composition, osmotic composition, pH, level of sugar, and organic composition. Two patterns of homeostasis of internal state in the face of environmental change are recognized: (1) An animal may alter a given property with reference to the environment; the property conforms to the medium, and homeostasis consists in cellular adjustment such that metabolism continues in the altered state. For example, in cold-blooded animals the body temperature conforms to that of the environment. In many marine animals and endoparasites the body fluids conform in osmotic concentration with the medium. (2) An animal may regulate its internal state and maintain internal constancy despite an altered environment. Such regulation is achieved by a series of automatic feedback controls as environmental stress is applied; at some environmental limits, regulation fails and the animal cannot long survive. For example, warm-blooded animals maintain a constant body temperature over a range limited by heat and cold. No aquatic organism could be as dilute as fresh water or as concentrated as some saline lakes and survive. Terrestrial organisms must be protected against desiccation; hence regulation of water content and of osmotic concentration is essential. Some aquatic animals conform over part of an environmental range and regulate in another range. An animal may conform with respect to one parameter (osmotic concentration) and regulate with respect to others (concentration of specific ions).

Two types of homeokinetic response are also distinguished: (1) Capacity adaptations are changes in cellular functions such that an intact animal can maintain normal activity over a range of internal states. (2) Resistance adaptations are changes which permit survival at extremes of some physical parameter. Resistance limits are narrow for intact integrated organisms, less restricted for isolated cells and tissues, and much wider for critical molecules such as proteins (denaturation) and lipids (melting). For example, many poikilothermic animals show little or no regulation of body temperature, but they alter energy-yielding enzyme systems to provide for nearly constant activity over a wide temperature range. Also, these animals can, by acclimation, become more or less resistant to heat or cold. Similar changes in thermal resistance occur in some homeotherms (temperature regulators).

Organ system controls. A catalog of the mechanisms of homeostasis would be a textbook of the physiology of organ systems and cells. Responses occur in a repeatable sequence, and hierarchical levels of controls and series of checks and balances are recognized. The sequence of responses of a mammal to cold illustrates the principles of sequential reactions and hierarchical controls in homeostasis. First, the stimulation of cold receptors in the skin results in constriction of peripheral blood vessels and erection of hairs. If the body is further chilled, the temperature-sensing portion of the hypothalamus is stimulated, shivering begins, and metabolism increases, partly under endocrine control. The first line of defense is insulative; the second is increase in heat production. The sequence of homeostatic responses to a severe injury-type stress varies according to the kind of animal. In mammals, a sequence of nonspecific responses may occur to any one of numerous severe stresses, the so-called adaptation syndrome of Selye. This consists, according to the severity of the stress, of an initial vascular "shock" reaction and counterreactions in which, under pituitary activation, the adrenal cortex is stimulated, lymphocytes in the blood decrease, and numerous chemical defense reactions are mobilized. In a hierarchy of controls, responses occur first in altered cells and then at several levels of endocrine and nervous system controls.

Another principle is that of checks and balances. The vagus nerves slow the heart and stimulate the intestine, while the sympathetics accelerate the heart and relax the intestine. When motor nerve centers for one set of muscles are active, those for antagonistic muscles are inhibited. Various regions of the brain counterbalance other regions. Another mechanism of homeostasis consists of morphological changes over long periods of time. Arctic birds and mammals tend to have thicker coats of feathers or fur and thicker layers of insulating fat in winter; tadpoles grow larger gills when reared in low-oxygen environments; at high altitudes human beings have increased blood hemoglobin concentration; bone structure varies according to mechanical stress. Animal behavior often compensates for environmental changes.

Cellular controls. Many animals, particularly "conforming" ones, compensate biochemically and biophysically for environmental change, with the result that a relative constancy of energy output is maintained. The metabolism of most aquatic poikilotherms acclimated to cold, for example, fish, is higher than that of individuals acclimated to warmth when both are measured at an intermediate temperature. Regulation of enzyme levels is achieved by a variety of controls on protein synthesis and degradation. In addition, enzyme activity can be modified by cofactors, membrane phospholipids, hormones, and ions. A much-used method of homeokinesis is the shift from one metabolic pathway to another, for example, the alternation between aerobic (oxidative) and anaerobic (glycolytic) metabolism according to oxygen availability. Also, numerous enzymes exist in one of several forms, as isozymes, and one form may be favored under given conditions.

An important aspect of cellular homeostasis is the control of the movement of ions and other substances into and out of cells. The first aggregates which could be called living cells must have been separated from their marine environment by a bounding layer which prevented free interchange of materials. The cells of all plants and animals differ in composition from extracellular fluids. An important mechanism of cellular individuation resides in the cell surface, which has selective permeability and which, in some instances, provides some mechanical rigidity. The cell surface consists of a relatively inert pellicle plus a plasma membrane of protein and lipid; it permits entry and exit of relatively few kinds of organic molecules and varies in its permeability to inorganic ions according to environmental conditions and level of metabolic need. The surface also has some enzymes of carriers for active transport. An example of membrane regulation is the high intracellular concentration of potassium relative to extracellular fluids which is nearly universal.

Significance. In all systems of homeostasis and homeokinesis—hierarchies of controls, checks and balances, alternate mechanisms, compensatory reactions, and negative and positive feedback—there are analogies to servo systems. Sensing elements are poised at critical levels, and, when deviation occurs, control mechanisms are brought into action to restore equilibrium either of biological state or of energy output. Homeostatic models have been provided by computer science and cybernetics.

[C. LADD PROSSER]

Human ecology

The branch of ecology that considers the relations of individual persons and of human communities with their particular environments.

Each human being, like every other kind of animal, possesses numerous regulatory mechanisms that maintain a dynamic balance between the individual's internal physiology and the constantly fluctuating conditions in the local habitat. Energy is acquired from food and is lost in radiated heat, muscular work, metabolism, and growth. People are very adaptable in their food and can exist on vegetable or animal foods alone or on various combinations of these. By acclimation, each person can to some degree adjust physiology to fluctuations in weather or season or to new climates that may be encountered by migration. The human differs from other animals, however, by being able to modify personal environment by the use of clothing, houses, and heating and cooling devices of various kinds. Humans are consequently able to thrive in many habitats that otherwise would be inhospitable to them. *See* HOMEOSTASIS.

Human community. A human ecologic community is composed of human beings and also of members of numerous associated species of animals and of plants. People are dependent upon certain of these species for food. From others they secure materials for making clothing and tools, for constructing buildings, and for numerous other uses. Wild herbivores living in the same community with humans may eat or damage some of the useful plants. These herbivores may be preyed upon by carnivores, which thereby benefit humans indirectly. Bacteria, viruses, or larger parasites may injure or destroy certain of the herbivores, carnivores, other parasites, or even humans. Scavengers and saprophytes convert dead organic matter into substances that can be reused by the plants. The numbers of the several species that together compose such a community are controlled by various ecologic regulatory mechanisms that maintain a working balance within the community.

The kinds of plants and animals that can exist in each geographic area are limited by the local climate, physiography, and soils. The plants, animals, and to some degree humans, however, may modify the soil and the microclimate and thereby may make the habitat more suited to the community. Each human community together with its local habitat thus constitutes an interacting system that may be called a human ecosystem. *See* ECOSYSTEM.

Cultural evolution. By the evolution of language and through social cooperation people have developed various types of culture. Thereby they are able to modify the immediate environment to a considerable degree and to expand greatly the resources of food and other useful materials. Each of the cultures that have evolved in the past in diverse parts of the world may be presumed to have been adapted, at least tolerably, to the climate, minerals, and other physical features of the local habitat, to the wild and domestic plants and animals of the region, and to the local diseases to which the people and their domestic crops and herds were subject. The most widespread existing cultures are based upon the domestication of plants and animals, exploitation of biologic and mineral resources, division of labor, accumulation of capital, invention of specialized technical processes, and the operation of numerous kinds of social, economic, and political institutions.

Regulatory mechanisms. Many of the natural ecosystems of the world have been greatly altered by the activities of people. Forests, for example, have been cut down or burned, swamps have been drained, pasturelands have been overgrazed, fertile soils have become eroded, fields have been planted to cultivated crops, rivers and lakes have become polluted by sewage and industrial wastes, native species of plants and animals have been reduced in numbers or locally extirpated, foreign species have been introduced, and extensive areas of productive land are covered by highways and cities. Numerous new types of ecologic communities also have developed in cultivated fields, in managed forests, about farmsteads, along roadsides, and within cities. The natural regulatory mechanisms usually are rendered ineffective in communities that are much modified by human activities. In consequence, certain native or introduced species may become destructive pests. Humans, therefore, either must establish effective new regulatory mechanisms in these disturbed communities or must themselves control the pests.

Regulatory mechanisms that maintain a working balance among the several species that compose a natural community mostly have become ineffective for controlling the density of human populations. Predators dangerous to humans largely have been extirpated from the vicinity of habitations. Communicable disease is becoming increasingly eliminated by the adoption of improved sanitary and medical practices. Human starvation in local areas usually is prevented by food storage and by the transportation of food from areas of abundance. With the natural checks to their multiplication thus greatly removed, humans have increased in numbers until they now are one of the most abundant species of mammals.

Special regulatory mechanisms have been evolved by every human society to control the activities of its members and to keep the society adjusted to the fluctuations in its resources and in the conditions in its habitat. Among these special cultural mechanisms are public opinion, punishments, rewards, supply and demand, differential taxation, and cooperative procedures of various kinds. Every social group has customs, ideals, and beliefs that govern the behavior of its members toward each other and toward strangers and regulate the utilization of the resources of the group.

War among tribes and nations in the past often has resulted from competition for land or other resources. Fortunately, other less destructive mechanisms for adjusting nations to their resources and to each other are being developed.

Scope. Human ecology is involved in many branches of natural and social science. Physiology, nutrition, hygiene, psychology, and climatology are basic to an understanding of the ecology of individuals. The ecology of human communities involves anthropology, botany, demography, economics, geography, geology, history, political science, sociology, and zoology. Also involved are the several branches of applied ecology, including agriculture, animal husbandry, conservation, forestry, game and fish management, medicine, parasitology, public health, and range management. Most of the concepts of human ecology are those of the individual disciplines concerned. The concept that each person and each human community operates as an ecologic unit that must possess effective regulatory mechanisms to maintain stability in its fluctuating habitat, however, serves to unify all branches of inquiry concerned in any way with the relations between humans and their physical, biotic, and social environment. [LEE R. DICE]

Bibliography: M. Bates, *The Forest and the Sea: A Look at the Economy of Nature and the Ecology of Man*, 1960; E. Borgstrom, *The Hungry Planet: The Modern World on the Edge of Famine*, 1965; J. B. Bresler (ed.), *Human Ecology: Collected Readings*, 1966; L. R. Dice, *Man's Nature and Nature's Man: The Ecology of Human Communities*, 1955; E. H. Graham, *Natural Principles of Land Use*, 1944; A. H. Hawley, *Human Ecology: A Theory of Community Structure*, 1950; E. S. Rogers, *Human Ecology and Health*, 1960; G. A. Theodorson, *Studies in Human Ecology*, 1961.

Humidity

Atmospheric water-vapor content, expressed in any of several measures, especially relative humidity, absolute humidity, humidity mixing ratio, and specific humidity. Quantity of water vapor is also specified indirectly by dew point (or frost point), vapor pressure, and a combination of wet-bulb and dry-bulb (actual) temperatures. *See* DEW POINT.

Relative humidity is the ratio, in percent, of the moisture actually in the air to the moisture it would hold if it were saturated at the same temperature and pressure. It is a useful index of dryness or dampness for determining evaporation, or absorption of moisture.

Human comfort is dependent on relative humidity on warm days, which are oppressive if relative humidity is high but may be tolerable if it is low. At other than high temperatures, comfort is not much affected by high relative humidity.

However, very low relative humidity, which is common indoors during cold weather, can cause drying of skin or throat and adds to the discomfort of respiratory infections. The term indoor relative humidity is sometimes used to specify the relative humidity which outside air will have when heated to a given room temperature, such as 72°F, without addition of moisture. It always has a low value in cold weather and is then a better measure of the drying effect on skin than is outdoor relative humidity. This is even true outdoors because, when air is cold, skin temperature is much higher and may approximate normal room temperature.

Absolute humidity is the weight of water vapor in a unit volume of air expressed, for example, as grams per cubic meter or grains per cubic foot.

Humidity mixing ratio is the weight of water vapor mixed with unit mass of dry air, usually expressed as grams per kilogram. Specific humidity is the weight per unit mass of moist air and has nearly the same values as mixing ratio.

Dew point is the temperature at which air becomes saturated if cooled without addition of moisture or change of pressure; frost point is similar but with respect to saturation over ice. Vapor pressure is the partial pressure of water vapor in the air. Wet-bulb temperature is the lowest temperature obtainable by whirling or ventilating a thermometer whose bulb is covered with wet cloth. From readings of a psychrometer, an instrument composed of wet- and dry-bulb thermometers and a fan or other means of ventilation, values of all other measures of humidity may be determined from tables.

[J. R. FULKS]

Hurricane

A tropical cyclone whose maximum sustained winds reach or exceed a threshold of 33 meters per second (m/s). In the western North Pacific ocean it is known as a typhoon. Many tropical cyclones do not reach this wind strength.

Maximum surface winds in hurricanes range up to about 200 mph (320 km/hr). However, much greater losses of life and property are attributable to inundation from hurricane tidal surges and riverine or flash flooding than from the direct impact of winds on structures.

Tropical cyclones of hurricane strength occur in lower latitudes of all oceans except the South Atlantic and the eastern South Pacific, where combinations of cooler sea temperatures and prevailing winds whose velocities vary sharply with height prevent the establishment of a central warm core through a deep enough layer to sustain the hurricane wind system.

In the United States, property losses resulting from hurricanes have climbed steadily in the last 2 decades because of the increasing number of seashore structures. However, the loss of life, which has been huge in many storms, has decreased markedly. This is due mainly to the fact that warnings, aided by a more complete surveillance from aircraft and satellite, and extensive programs of public education, have become more accurate and more effective. Improvements in methodology for hurricane prediction have reduced the error in pinpointing hurricane landfall and have greatly reduced the probability of large errors in prediction.

Structure and movement. At the Earth's surface the hurricane appears as a nearly circular vortex 400–800 km in diameter. Its dynamic and thermodynamic properties, however, are distributed about the vortex asymmetrically. The cyclonic circulation (counterclockwise in the Northern Hemisphere) extends through virtually the entire depth of the troposphere (15 km). In lower layers (2–4 km in depth) winds spiral inward and accelerate toward lower pressure, reaching peak velocities in a narrow annulus surrounding the pressure center at a radial distance of 20–30 km. Here there is a near balance between the pressure forces acting radially inward and the centrifugal and Coriolis forces acting outward, so that the air, no longer able to move radially, is forced upward, transporting with it the horizontal momentum it acquired in lower layers. Traveling upward through the warm core, the air parcels soon reach layers at which horizontal pressure forces begin to diminish (hydrostatically). The imbalance that results causes the air to spiral outward and join environmental circulations, carrying with it a canopy of cloud debris known as the outflow cloud shield. This circulation of mass—in, up, and out of the vortex—at the rate of some 2,000,000 metric tons per second constitutes a kind of atmospheric heat pump. It uses as fuel the latent heats of condensation and fusion released as the water vapor, brought in with the environmental air and augmented by fluxes from the ocean surface, rises, cools, and generates tall cumulus clouds in or near the annulus of maximum winds. These cumuli are the conduits for transporting mass upward to the outflow level, and in a mature hurricane they merge into a wall of nimbostratus encircling a small benign weather area known as the eye, where cloudiness is minimal, rain is absent, and winds are relatively light. The circulation in the vortex not only generates the eye-wall clouds but also a family of spiral rain bands, which move cyclonically around the vortex center (Fig. 1).

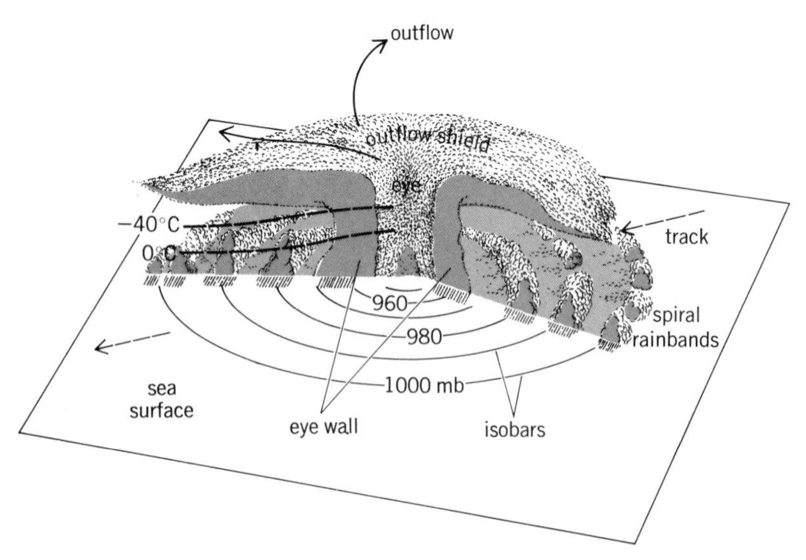

Fig. 1. Model of a hurricane circulation and cloud structure.

The hurricane system moves through its environment in response to the interactions between the circulations of the environment and the vortex. These interactions create systematic imbalances between the amount of mass flowing in and that flowing out of selective quadrants or sectors and thereby produce the pressure changes that cause the center to move. The steering of the hurricane thus involves the interaction of vortex with environment from the bottom of the inflow to the top of the outflow layers. Most tropical cyclones reach hurricane strength in the belt of sluggish but steady tradewinds and are propelled westward at 10–15 knots (5–7.7 m/s) to the extremity of the subtropical anticyclone, where they tend to recurve northward into the vigorous west winds of midlatitudes and are carried eastward, often at speeds greater than 30 knots (15.4 m/s).

Energy sources and transformations. The process by which the tall cumuli in the eye wall maintain the warm, light air in the hurricane core, which in turn generates the pressure forces that determine the strength of hurricane winds, is not a simple matter of releasing latent heat within the clouds. Early diagnostic computer models of the hurricane encountered this difficulty when simulations led to run-away development and unrealistic pressure forces. This led to the development of procedures for parameterizing cloud effects. Most procedures assumed that clouds heat the adjacent air by direct turbulent transfer of their own excess warmth.

William Gray hypothesized that the warm core of a hurricane is maintained primarily by adiabatic heating of air descending between individual clouds. If true, this would mean that the energy driving the hurricane is supplied indirectly and would call for a much different kind of parameterization. Gray's hypothesis gained support from recent studies based on observations from research aircraft flights in hurricanes, which showed that the warm core in cases studied was indeed maintained by a combination of cumulus ascent and of air descending between clouds. Nevertheless, energetically the ultimate strength developed by a hurricane still depends on the amount of latent heat liberated in the eye wall, which in turn depends on the heat content of the air rising in the eye wall and the rate of vertical transport. The rate of vertical transport depends on environmental circulations that may constrain the inflow at low levels or the efflux at the top of the vortex.

While more than three-fourths of the water vapor that fuels the hurricane is imported from outside the storm boundaries, the critical contribution to its energetics is from the flux of latent and sensible heat it derives from the warm ocean over which it travels. This is a critical source, because it is precisely this contribution that is responsible for converting a tropical cyclone from gale strength to hurricane strength. This energy flux from sea to air is sometimes constrained by an important feedback mechanism. It has been shown theoretically and experimentally that cooling of sea surface temperatures occurs in the wake of hurricanes because of the mixing and upwelling caused by hurricane wind stresses on the sea. The amount of cooling is a maximum (2–5°C) when the hurricane travels at speeds less than that of gravity waves in the ocean thermocline (4–8 knots; 2–4 m/s), which is enough to substantially reduce the sea-air heat flux and the strength of the wind system. *See* OCEAN-ATMOSPHERE RELATIONS.

The parameterization of latent heat releases in hurricanes remains one of the biggest challenges for modelers and is still unresolved. However, important progress in simulating other aspects of hurricane structure has been made, most notably by R. Anthes and by S. Rosenthal. Anthes's three-dimensional asymmetric hurricane model (Fig. 2) simulated a hurricane circulation in three dimensions, including a derived eye wall and the spiral rain bands. Such successes are necessary prerequisites to the development of an effective dynamic prediction model and lend encouragement that this task is achievable.

Hurricane formation. An average of about 70 tropical cyclones develop gale or hurricane strength somewhere on the globe each year, a figure some 15 to 20% higher than climatological records indicated before satellite weather surveillance became available in the early 1960s. About 9 occur in the North Atlantic ocean, 16 in the eastern North Pacific, and 24 in the western North Pacific. They also occur in the central and western South Pacific, the Bay of Bengal, the Arabian Sea, and the South Indian Ocean (Fig. 3).

The disturbances, sometimes referred to as

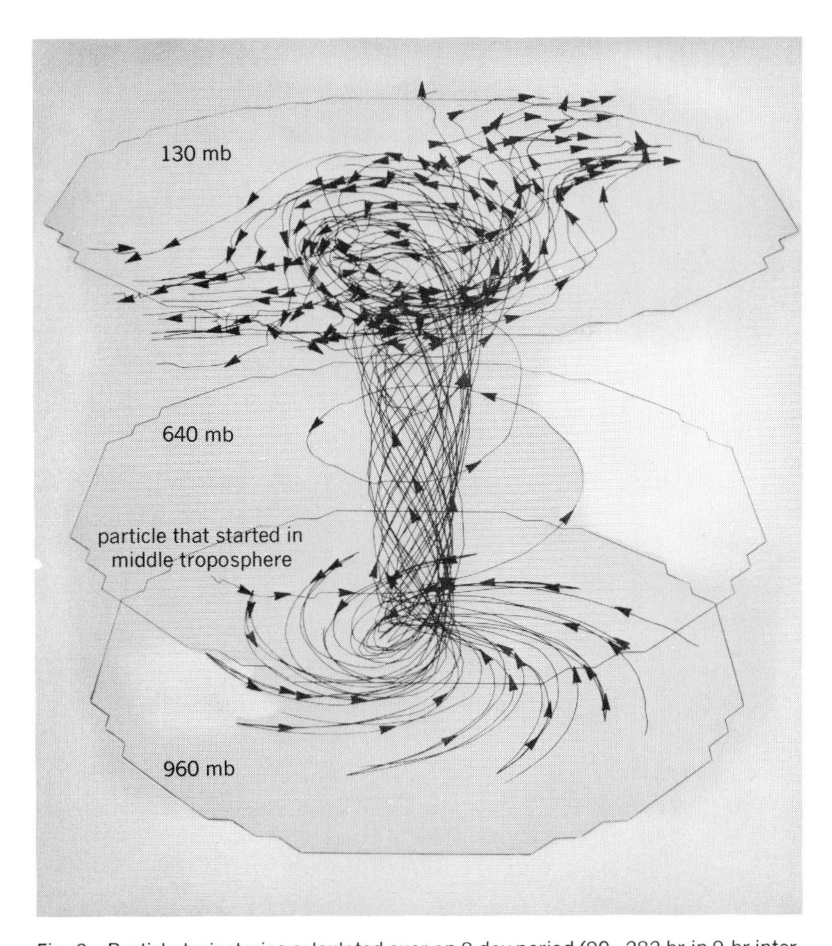

Fig. 2. Particle trajectories calculated over an 8-day period (90–282 hr in 9-hr intervals) in an experiment with a three-dimensional model hurricane. All particles start in the lower atmospheric boundary layer except one, which is started in the middle troposphere. (*From R. A. Anthes, S. L. Rosenthal, and J. W. Trout, Preliminary results from an asymmetric model of the tropical cyclone, Mon. Weather Rev., 99:744–758, 1971*)

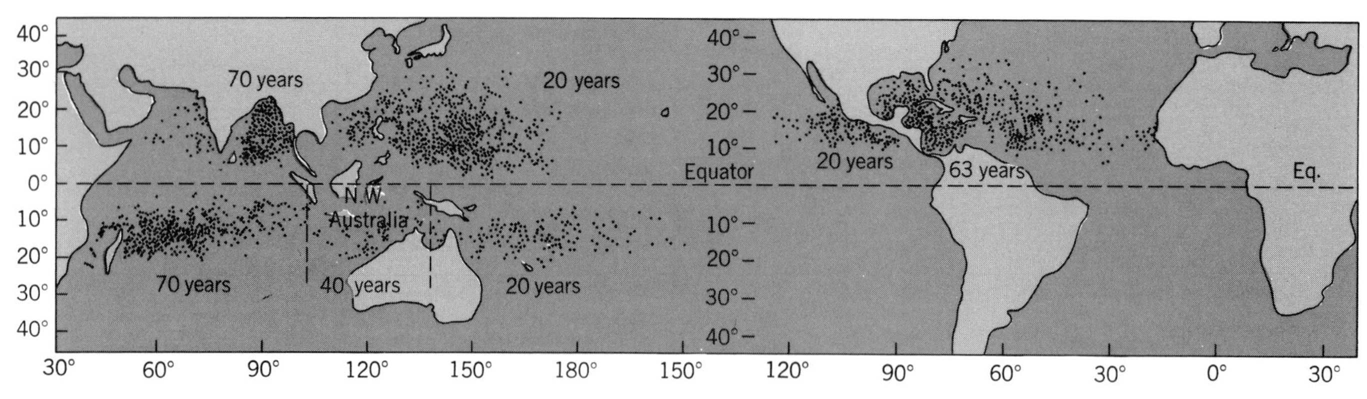

Fig. 3. Points on the globe where tropical cyclones were first detected by weather observers. (*From W. Gray, Global view of the origin of tropical disturbance, Mon. Weather Rev., 96(10):670, 1968*)

seedlings, that breed tropical cyclones of hurricane strength originate mainly in tropical latitudes where the trade winds invade the equatorial trough (sometimes referred to as the intertropical convergence zone, or ITCZ). Some have their origins over continental areas. Seedling disturbances, comprising an agglomeration of convective clouds 300–500 km in diameter, often move more than 2000 km across tropical oceans as benign rainstorms before developing closed circulations and potentially dangerous winds.

In the North Atlantic, approximately 100 seedling disturbances are tracked across the tropical belt each year, more than half of which emanate from the African continent north of the ITCZ. A few form in the subtropics from cold lows, and still others from the trailing edge of old cold fronts or shear lines in subtropical latitudes. Regardless of the sources, however, it is notable that the average of 100 seedlings per year has a standard deviation of only 8, one of the most dependable recurrences of a meterological event. It is even more notable and difficult to explain why, with 100 opportunities for tropical cyclone development each year, an average of only 9 seedlings become dangerous wind storms, and the standard deviation from year to year is 4.

Two well-known factors constrain the development of seedlings. One is the temperature of the sea environment. It was established in the early

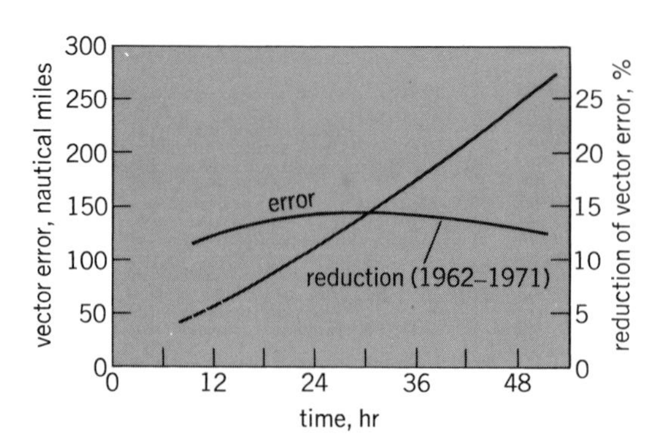

Fig. 4. Growth of the absolute magnitude of the vector error with length of forecast. The broken line is the percentage reduction in this error from 1962 through 1971. (1 nautical mile = 1.852 km). (*From R. H. Simpson, Hurricane prediction, Science, Sept. 7, 1973, copyright 1973 by the American Association for the Advancement of Science*)

1950s that to supply the heat flux from sea to air needed to generate a hurricane wind system initially, the sea surface must be warmer than about 26°C. The second factor is the variation of prevailing winds with height. If winds in the lower troposphere move at appreciably different velocities from those in the upper troposphere, the heat released by convection, or the warming by subsidence between convective clouds, cannot be stored in vertical columns of sufficient height to produce the pressure drop needed to support hurricane winds. Nevertheless, on many occasions, cyclogenesis still fails to occur when these constraints are absent.

Work by Michael Garstang based on observations from the BOMEX and GATE expeditions in 1969 and 1974 point to some additional constraints. These researches show that aggregate convection, even in the presence of a forcing influx of air from the ocean environment, can be self-defeating for several reasons. Vertical mixing by convection brings drier, sometimes cooler, air down to the layers below cloud bases, diluting the heat and moisture content. This process also increases the thermal stability of the subcloud layer. The net result is that the initial convection frequently cannot survive. This deadly cycle can be broken only if the convective mixing occurs in an environment that allows air of substantially higher momentum to be brought to the surface to increase the circulation of mass to the convection. Curiously, the latter condition seems to associate itself with a unique spacing of the convective elements of the disturbances. Further research into the environmental factors that control this spacing may supply the answer to one of the most puzzling questions in meteorology: Why are there so few hurricanes?

Prediction. Most methods for hurricane prediction are addressed exclusively to the movement and landfall of the center. In part, this is because the landfall position is the most important element in issuing warnings and in evacuating residents from areas subject to flooding. The change in strength, often a dramatic one as a hurricane approaches land, is so intimately related to cumulus parameterization and to nonlinear interactions between environment and vortex that these predictions continue to depend mainly on monitoring by reconnaissance aircraft and satellites. *See* STORM DETECTION.

Today, hurricane forecasting draws heavily on

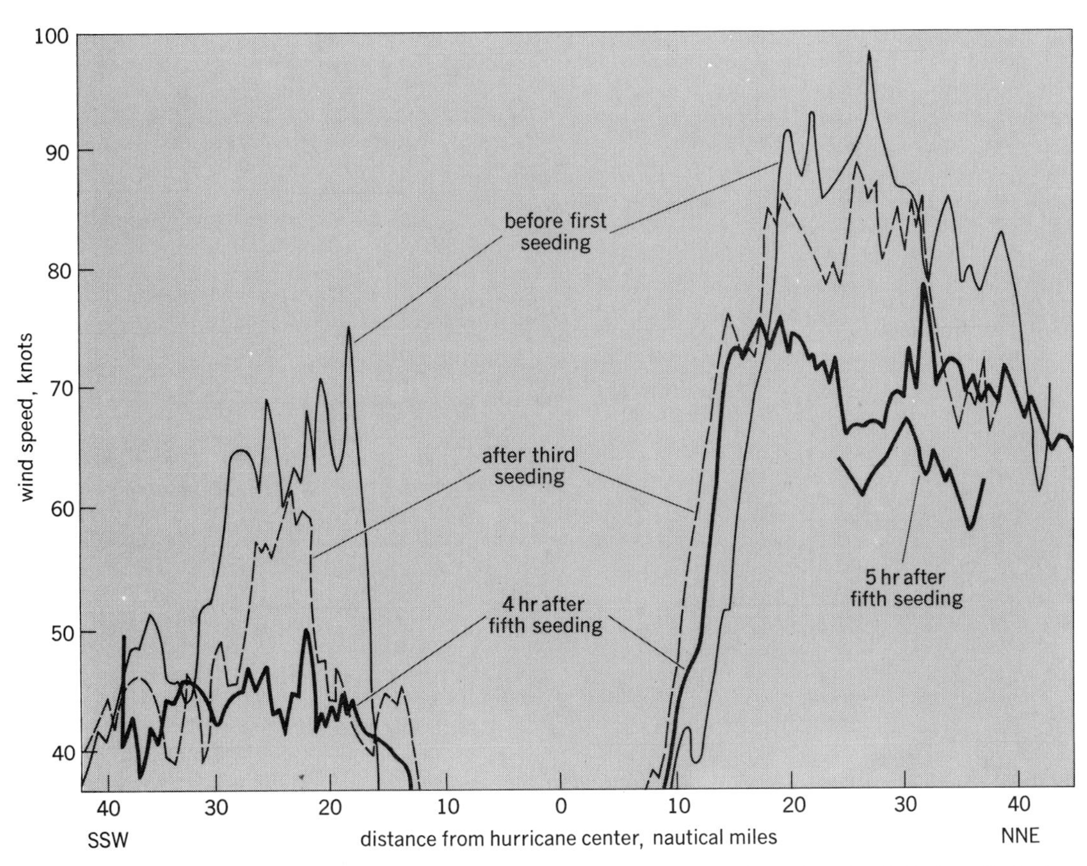

Fig. 5. Changes with time of wind speeds at 12,000 (3.7 km) in Hurricane Debbie on Aug. 18, 1969. 1 nautical mile = 1.852 km; 1 knot = 1.852 km/hr = 0.514 m/s.

machine prediction models that combine the output of dynamical short-period predictions of environmental changes with statistical and analog information on expected hurricane behavior to determine the most probable future position of the center in 12-hr increments, for up to 72 hr. The latest version of such a technique, currently in use at the National Hurricane Center in Miami, is the NHC-73 method developed by Hope and Neumann. Since the 1950s, researchers have been trying to evolve dynamical models that predict hurricane movement. The most successful of these, and the only one in operation, is known as SANBAR (Sanders barotropic model), a single-layer barotropic filtered model which uses as input the mean wind velocities for the layer 1000–100 mb. The first three-dimensional asymmetric prediction model to show useful skill was developed by B. I. Miller.

However, several problems have restricted the operational usefulness of nearly all dynamical models. Foremost is the initial value problem, which requires either that the vortex be replaced by an ersatz circulation or point vortex, or a machine analysis with high resolution using very small mesh lengths. For the latter, adequate initial data are not available. Second, if adequate physics are included, computation time for a 24-hr forecast is enormous, even on the largest computer systems.

The outlook for substantial increases in expected skill for predicting movement (Fig. 4), while uncertain, is not clearly promising. What some improved models may be able to do, whether they are dynamically or statistically founded, is to decrease the chances of large, disastrous errors. *See* WEATHER FORECASTING AND PREDICTION.

Mitigation. Systematic experimentation aimed at reducing the destructiveness of hurricanes by strategic cloud seeding began in 1961 when Atlantic hurricane Esther was seeded with silver iodide crystals on 2 days. Additional experiments were conducted in Beulah (1963), Debbie (1969), and Ginger (1971). The objectives of these experiments, sponsored by the United States government under the project name of Stormfury, were to reduce the destructive winds of the hurricane by dispersing silver iodide crystals in the eye wall. Stormfury employs a strategy that seeks a selective release of latent heat of fusion in the eye wall, designed to induce the eye wall and its annulus of maximum winds to expand, or to reform at a greater distance from the center, the results of which, from conservation of absolute angular momentum, would be a reduction in maximum wind speeds.

The initial hypothesis, advanced by Robert Simpson and Joanne Simpson, argued on hydrostatic grounds that the strategic release of latent heat of fusion in the eye wall would reduce surface pressures outward from the annulus of maximum winds, and in doing so would reduce the pressure gradient. This, it was argued, would cause a circulation imbalance and would induce the eye wall to migrate outward. After exhaustive experiments with computer simulations of this process by Stanley Rosenthal, Cecil Gentry presented a revised hypothesis that, in seeking the same basic objectives, proposed a seeding procedure designed to stimulate growth of cumulus clouds located radially outward from the annulus of maximum winds. The increases in buoyancy resulting from seeding, it was argued, should cause these clouds to penetrate and transport mass to the outflow level with the result that the eye wall should tend to reform at a greater distance from the center. The expected result, as in the earlier hypothesis, is that maximum winds would be reduced. This revision has remained the basis for Stormfury experiments in recent years, although a number of alternative and imaginative suggestions for mitigating the damage potential of hurricanes remain untested. Most notable of these is the proposal by William Gray to use carbon black seeding as a radiation sink for altering the convection strategically and thus altering the mass circulation in the vortex.

The results of each of the Stormfury seeding experiments has been a response of the kind predicted by the hypothesis, although in all but perhaps hurricane Debbie, which showed a dramatic reduction (31%) in maximum winds after a succession of seeding runs at 90-min intervals (Fig. 5), the results were inconclusive, since the changes were not of sufficient magnitude to rule out the possibility that they were caused by natural variations common in hurricanes.

The biggest handicap in conducting the Stormfury experiments has been the limited number of hurricane cases suitable for experimentation. The National Oceanic and Atmospheric Administration, in 1976, planned to resume Stormfury experiments in 1977. The biggest challenges in these follow-up experiments are (1) to determine whether there is ordinarily enough unfrozen cloud water present to generate the results predicted by the hypothesis; and (2) to establish whether the dynamical results sought would cause any significant change in storm surge levels, since inundation is the greatest source of losses of both life and property. *See* CLOUD PHYSICS; STORM.

[ROBERT SIMPSON; JOANNE SIMPSON]

Bibliography: G. E. Dunn and B. I. Miller, *Atlantic Hurricanes*, 1964; W. Hess (ed.), *Weather and Climate Modification*, chaps. 14 and 15, 1974; W. J. Kotsch and E. T. Harding, *Heavy Weather Guide*, 1965; E. Palmen and C. W. Newton, *Atmospheric Circulation Systems*, chap. 15, 1969; R. H. Simpson, Hurricane prediction, *Science*, 181:899–907, 1973.

Hydrology

The science that treats of the waters of the Earth, their occurrence, circulation, and distribution; their chemical and physical properties; and their reaction with their environment, including their relation to living things. The domain of hydrology embraces the full life history of water on the Earth.

Hydrologic cycle. The central concept of hydrology is the hydrologic cycle, a term used to describe the circulation of water from the oceans through the atmosphere to the land and back to the oceans over and under the land surface (see Fig. 1). Water vapor moves over the sea and land. It condenses into clouds and is precipitated as rain, snow, or sleet when one air mass rises to pass over another or over a mountain. Some of that which falls is evaporated while still in the air, or is intercepted by vegetation. Of the precipitation that reaches the ground surface, some evaporates quickly, some penetrates the soil, and some runs off over the land surface into streams, lakes, or ponds. Of that which penetrates the soil, some is held for a time and then returned to the atmosphere by evaporation or plant transpiration. The remainder penetrates below the soil zone to become a part of the groundwater. Study of the water in the atmosphere, although closely related to hydrology, lies more properly in the field of meteorology. The science relating to the oceans also touches closely on hydrology but is regarded as a separate science. *See* ATMOSPHERIC WATER VAPOR; EVAPOTRANSPIRATION; METEOROLOGY; OCEANOGRAPHY.

Water is important for domestic, agricultural, and industrial uses. Thus the study of water and the means by which it may be obtained and controlled for use is of utmost importance to the welfare of mankind. Water is also a destructive agent of awesome force. Great floods periodically inundate valleys, causing death and destruction. Less dramatic are the rising water tables which, especially in irrigated areas, cause deterioration of the soil and make worthless large areas that would otherwise produce crops. Erosion of soil by flowing water and ultimate deposition of the sediment in lakes, reservoirs, stream channels, and coastal harbors is also a problem for hydrologists.

Functions. Hydrology is concerned with water after it is precipitated on the continents and before it returns to the oceans. It is concerned with measuring the amount and intensity of precipitation; quantities of water stored as snow and in glaciers, and rates of advance or retreat of glaciers; discharge of streams at various points along their courses; gains and losses of water stored in lakes and ponds; rates and quantities of infiltration into the soil and movement of soil moisture; changes of water levels in wells as an index of groundwater storage; rate of movement of water in underground reservoirs; flow of springs; dissolved mineral matter carried in water and its effects on water use; quantities of water discharged by evaporation from lakes, streams, and the soil and vegetation; and sediment transported by streams. In addition to devising methods for making these diverse measurements and storing the data in usable form, hydrology is concerned with analyzing and interpreting them to solve practical water problems. Rigorous studies of all the basic data are required to determine principles and laws involved in the occurrence, movement, and work of the water in the hydrologic cycle.

Engineering hydrology studies are basic to the design of projects for irrigation, water supply, flood control, storm drainage, and navigation. The hydrologist must estimate the probability of floods and droughts, volumes of water available for use, and probable maximum floods for spillway design. Often these estimates are required at locations where streamflow has not been measured. Digital computers are playing an increasing role in these studies and electrical analog computers are widely used in groundwater studies. *See* GROUNDWATER.

[ALBERT N. SAYRE/RAY K. LINSLEY]

Snow management. Many aspects of snow management have received attention in recent years due to a continuing need for additional water sup-

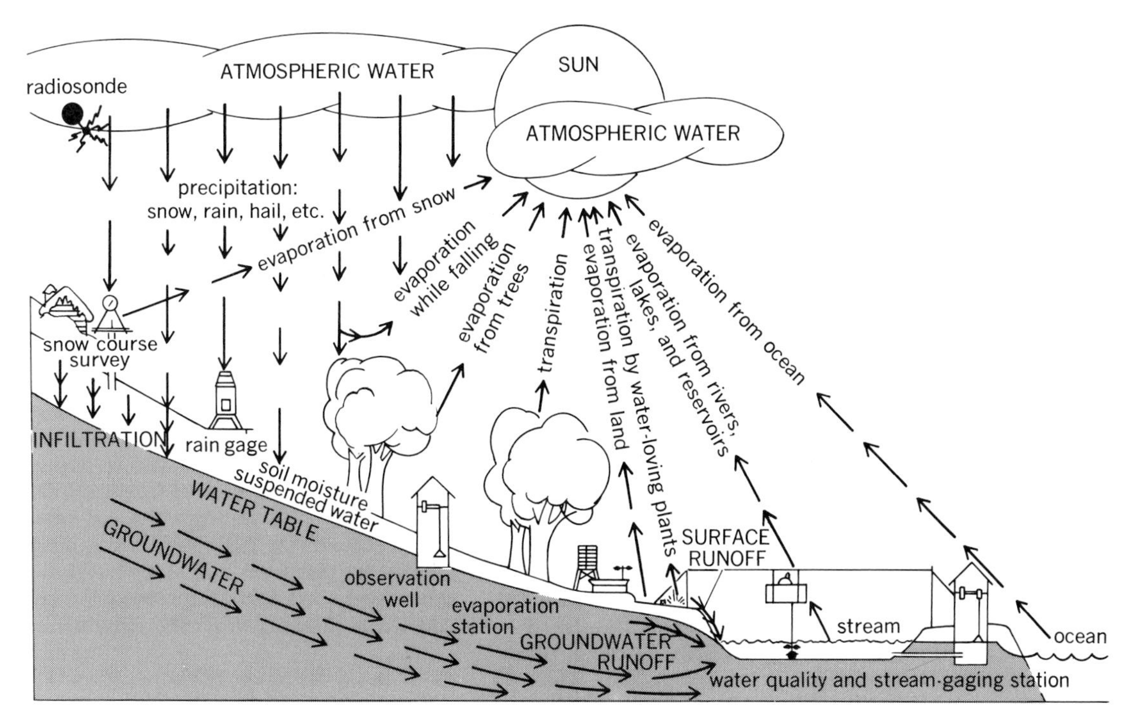

Fig. 1. Diagram of the hydrologic cycle.

plies and a need for better forecasts of existing supplies. Significant advances in snow management have occurred relative to instrumentation, snowpack modeling, and snow augmentation techniques.

Instrumentation. Improvements in the use of radioactive isotopes for monitoring the snowpack, together with a widespread use of telemetry equipment which provides direct contact with computer facilities, have made possible more detailed monitoring of the snowpack at remote locations. The use of radioactive isotopes requires the preseason installation of a gamma-radiating source in a sealed tube plus a scintillation crystal detector in another tube. A mechanism is provided to raise and lower both in their respective tubes by remote control. Each time the installation is interrogated by radio control, a profile of the snowpack water content, density, and depth is obtained. Consequently, vital forecast information can be obtained as frequently as necessary without the need for a visit to the site.

Adaptation of the lysimeter to snowpack measurements has also aided in evaluating water yields from the snow. The type of lysimeter referred to here generally consists of some type of weighing platform which supports a column of soil or snow. Provisions are made for measuring water entering or leaving the column, thus giving a measure of water gained, stored, or lost relative to the starting condition of the column. The lysimeter has been used for years to evaluate water movement through soil; however, application of the technique to snow conditions has been difficult. This is due to constant changes in the snowpack, both in dimension and internal structure, which affects the flow of water from the pack. These changes prevent an accurate computation of unit area above the lysimeter which is necessary in relating water flow from the pack with contributing area.

The snow pillow, which has been used for several years, is in effect a borderless lysimeter. That is, no vertical restraints are applied to the snow column above the surface area measured. The pillow is frequently used to obtain a continuous record of snowpack water content for forecast purposes. However, this is somewhat unreliable for determining the amount and rate of water loss as runoff from the pack.

Development of a universal surface precipitation gage has effectively taken the snow pillow one step further by including a recorder system for measuring runoff from the bottom of the snowpack. Consequently, evaluations of snowpack water content, precipitation additions, and runoff losses are continuously carried out. Problems of bridging and ice lense formation still affect unit area calculations on slopes. A simple snowmelt lysimeter has been developed for use on slopes which utilizes flexible plastic sheets to form a border around the area measured. The border is moved up or down with changes in the snowpack depth. This allows control over the unit area measured by ensuring collection of all meltwater from the area directly above the plot. Maintaining the border flush with the snow surface is necessary to prevent abnormal effects on the snow surface.

One use of the snow lysimeter has been to evaluate effects of the forest canopy on water yields from the snowpack. In the past, effects of the forest canopy on snow accumulation and distribution have been measured by monitoring the water content of the snowpack; however, such measurements have led to false interpretation of how much water is yielded from the pack. Accumulation of snow in the crowns of trees suggests a loss to the snowpack, while large openings appear to have more snow, but melt faster in the spring. Measurements of the water content of the snowpack alone suggests that at peak accumulation (approximately

April 1), the most snow exists in small openings, followed by large openings, then deciduous forests, and finally by coniferous forests. However, lysimeter tests indicate that coniferous forests produce more runoff by the time maximum accumulation occurs. The water for early runoff from the snow under a conifer stand is obtained from melting snow intercepted in the crowns of the trees and melting of the snowpack under the trees due to the heat absorbed by the dark foliage of the trees. Consequently, the snowpack under the conifer stands has less water but has already contributed more to runoff. As the season progresses, the melt rates increase in large openings due to a lack of shading and advective heat from wind movement. Consequently, the snowpack disappears in large openings first, followed by deciduous forests, then coniferous forests, and finally in small openings. The small openings appear to exhibit the best conditions for maximum accumulation of snow and delayed melt. The forest around the small opening contributes windblown snow from its crown, and shade to delay melt.

Modeling. Mathematical models of dynamic systems such as a snowpack are useful tools for predicting changes in such systems over a period of time. Because such a large number of variables are present which affect the accumulation and melt of a snowpack, use of a model was virtually impossible before computers became available to handle the computations involved. Today analog-to-digital computers make such a model very practical. The input data such as precipitation, temperature, and snowpack conditions, in the form of variable electrical resistances (analog data), are applied to the computerized model and digitized to provide discrete (digital) output in terms of runoff, stored water, or other forms of forecast information.

One useful snowpack model accounts for gains or losses of water content by a series of heat balance equations, along with precipitation input information. The snowpack can be considered a dynamic heat reservoir, which makes it possible to account for water gains or losses by balancing heat gains or losses. This is possible since water losses are primarily a result of melting or evaporation which are heat-related processes. The addition of heat and water from the atmosphere must also be accounted for once the snowpack begins to form. Once the heat balance of the pack reaches 0°C, it is considered isothermal; and further additions of heat result in melted snow, forming runoff.

Augmentation. Management of the snowpack for increased water yields has developed along two primary lines. One approach is to manipulate the forest canopy to increase the snow catch locally or alter the melt rate for a more favorable timing of runoff. The second approach is to increase winter precipitation through cloud seeding.

In the techniques of vegetation manipulation, improvements have resulted through the application of modern logging methods, such as helicopter, balloon, or skyline yarding, which provide more flexibility in the cutting patterns that are followed. This is significant on a local basis, but does not have a significant effect on basin-wide water demands.

Cloud-seeding techniques have reached the stage of providing operational systems for basin-wide increases in water yield, which have very significant ramifications on both the land in the seeded area and on downstream areas where the additional water would be used. Cloud seeding for snowpack augmentation basically consists of forming a smoke cloud of silver iodide crystals which becomes the nucleus for ice crystal formation, producing snow when the moisture-laden cloud is orographically lifted and cooled (see Fig. 2). A detailed explanation of this process has been described by M. Neiberger.

Basically, snow augmentation represents an increased water content of the snow, deeper snow, a prolonged period when snow is on the ground, increased runoff during melt, a longer period of high peak flows, and lower stream temperatures during the early summer period. The additional snow is

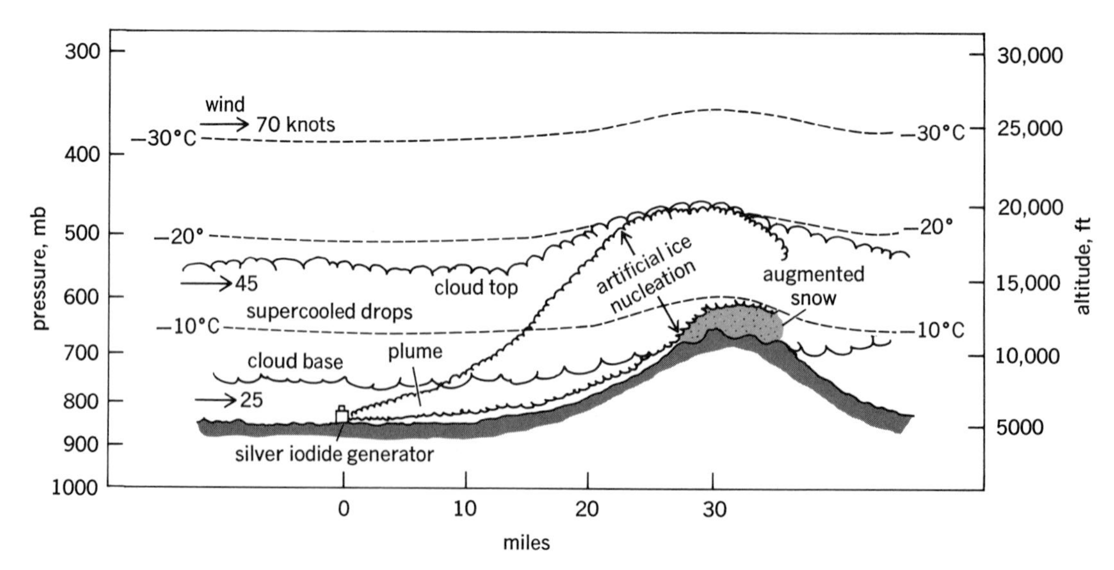

Fig. 2. Idealized model showing meterological conditions that will result in increased snowfall if clouds are seeded with silver iodide particles. 1 mb = 100 Pa; 1 knot = 1.852 km/hr = 0.514 m/s; 1 mi = 1.609 km; 1 ft = 0.3048 m. (*From L. W. Weisbecker, comp., Technology Assessment of Winter Orographic Snowpack Augmentation in the Upper Colorado River Basin, vol. 2, Stanford Research Institute, Menlo Park, CA, 1972*)

obtained by increasing snowfall during periods that normally would have produced only light snowfall.

The upper Colorado River basin is an excellent example of how a snowpack augmentation program may be applied. The target area is the zone where increased snowfall is expected, which in this case would be the high Rocky Mountains of Colorado. A heavy snowpack normally occurs in this area. Additional water in the Colorado River system would be beneficial, primarily to downstream users from Colorado to California. It is estimated that snow augmentation could produce 20–25% more water in the Colorado River system, or approximately 2,300,000 acre-feet of water. The water-rich target area would not benefit from additional water, but could benefit from the snow as a winter sports enhancement. Skiing and associated activities attract many people to this area of Colorado and adjacent states. Basin-wide snow augmentation is not without potential side effects. Changes in ecological balances, hydrologic conditions, and economic opportunities both favorable and unfavorable are possible. A detailed discussion of such potential changes has been reported by L. W. Weisbecker.

[ROBERT D. DOTY]

Water balance in North America. Demand for water has customarily been balanced against the supply on a predominantly local basis. However, in recent years water shortages have occurred in many localities, including some which have had abundant supplies in the past, and, as the population grows, more areas can be expected to experience such shortages. At present neither the North American continent as a whole nor any of its component countries has an overall water shortage. It would seem possible, therefore, to balance the supply and the demand on a larger scale so that local water shortages could be prevented or eliminated. An examination of the availability of water and the factors involved in the increasing demand for it is given below, followed by a discussion of the various means of acnieving a balance between supply and demand.

Availability of water. The availability of water in any area is determined by the balance of certain elements in the hydrologic cycle. The International Hydrological Decade has undertaken the task of studying the balance and the long-term changes among these elements throughout the world. For the purposes of this discussion, however, the long-term average runoff to the oceans is a satisfactory measure of the surplus water available for man's needs in North America.

Table 1 shows estimated water availability in terms of the long-term average runoff from the North American continent as a whole and from some selected subdivisions. The last column gives the availability on a per capita basis. The figures in Table 1 are approximate; better data should be available when the work of the Decade has been completed. The values given are long-term averages of quantities which can vary widely from day to day and from season to season, but somewhat less from year to year. The variability is greater in some areas than in others because natural and artificial storage tends to regulate discharge.

Interestingly, the average per capita availability of water is greater in Texas and New Mexico, areas which are generally associated with water shortage, than it is on the heavily populated east coast of the United States, which is usually thought of as adequately supplied. Table 1 also shows that a large per capita water supply exists when the average is taken over the entire continent or even over each individual country. This is because of the large runoff in relatively lightly populated areas such as Alaska, northern Canada, northwestern United States, and southern Mexico.

Demand for water. The demand for water is not based on man's minimal needs alone; it is a complicated function of need, availability, and custom. Demand breaks down into two categories—withdrawal uses and nonwithdrawal uses. Water withdrawals for the United States are summarized in Table 2, and these are probably not much larger, on a per capita basis, than would be the corresponding statistics for Canada and Mexico. Only a small part of the water withdrawn is lost, a little less than 10%, principally through evaporation and transpiration in connection with irrigation. The rest is returned to the lakes and streams, frequently degraded in quality. Nevertheless, much of it can be and is used again many times over, and some of the withdrawals given in Table 2 include reused water. These withdrawal data can serve as a measure of demand in the absence of a better indicator.

It is worth noting that the average water withdrawal rate per capita exceeds the availability in some areas of the United States. However, because water is reused, only a few areas have real shortages. For example, along the Eastern seaboard of the United States, where there is heavy industrial use of water, treatment of waste water generally yields enough water for reuse to satisfy all demands. On the other hand, areas with heavy demand for water for irrigation have little water for reuse, both because of losses sustained from evaporation and transpiration and because of the increased salinity of the recovered water.

Nonwithdrawal uses include hydroelectric power generation, navigation, transport of waste products, storage of floodwaters, recreation, and enhancement of aesthetic features of the landscape. In general, these uses do not degrade the quality or affect the quantity of the water in serious measure; however, some evaporation losses and quality changes may be associated with impoundments, and quality changes certainly accompany transport of waste products. There is no general way to measure the demand for nonwithdrawal uses because the demand depends so heavily on such factors as how much waste of what type will be transported, how deep the navigation channels will be, and how much water must flow in a river to make it aesthetically pleasing. These factors determine how much of the available water passes unused through a given river basin. Thus, even though water availability in Table 1 exceeds withdrawal use in Table 2, there still may be water shortages.

Although the availability of water to a given area is fairly well fixed when averaged over long time periods, the demand is not. The last column in Table 2 presents estimated increases in withdrawals. These increases are a product of the increased standard of living. Thus, whereas availability per capita is decreasing, demand per capita is increas-

Table 1. Estimated water availability for North America*

Area	Average runoff, millions of cubic feet per second†	Population, millions*	Water availability, gallons per capita per day‡
NORTH AMERICA	7.3	263	18,000
Canada	3.6	20	115,000
Conterminous United States	1.8	200	5,800
Mexico	0.4	43	6,000
Eastern seaboard from Virginia to New York	0.16	47	2,200
Texas and New Mexico	0.08	12	4,300
Southern California	0.003	10	200
Mexico, central and southern plateau	0.015	16	600
Mexico, between plateau and Gulf Coast	0.1	6.5	10,000

*In rounded values. †10^6 ft³/s = 2.83×10^4 m³/s. ‡1 gal = 3.785 liters.

ing, and sooner or later some water-rich areas will become water-poor.

When the demand factors are considered and weighed against the availability of water, the water-short areas of North America can be identified as parts of the Canadian prairie provinces, the northern plains and the southwestern part of the United States, and northern Mexico including the central and southern plateau. Other areas, such as the heavily populated part of the Eastern seaboard of the United States, may fall into this category in the foreseeable future.

Meeting the demand. Where demand threatens to outreach available water supplies, several courses of action are possible. Such courses, listed in their most probable order of application on the North American continent, include taking measures to conserve water, controlling industrial growth and population, introducing new sources of supply, and transferring water from one basin to another.

Possible conservation measures include the metering of, and adequate charge for, all water used, detection and sealing of leaks in conduits and reservoirs, control of plant growth to retard transpiration, more efficient application of water in irrigation and in industrial processes, and recycling of waste waters from industrial plants and public water supplies.

It may become desirable to plan the abandonment of certain industries in some water-short areas in favor of others that require less water. For example, it has been suggested that irrigation be abandoned in certain parts of the southwestern United States. At the same time, shortages of water may motivate certain industries which cannot accommodate themselves to conservation measures to move to areas with water surpluses, and population shifts would accompany these industrial shifts.

Possible methods of adding to the water supply include mining groundwater (as opposed to simply removing it from storage during one season or during dry years and replacing it at other times), desalinating sea water or brackish water, and mining glaciers or towing icebergs to water-short coastal areas. Controlling precipitation by seeding clouds could also be mentioned in connection with increasing the water supply; however, this process might also be characterized as a method of interbasin transfer. Indeed, the question of whether water so produced is really new water or simply water removed from another basin may present insurmountable political and legal obstacles to the practical use of precipitation control.

The data in Table 1 point toward the possibility of transferring water from areas of high per capita availability to those of low availability. Several grand schemes for such interbasin transfer of water have been proposed. Some smaller transfer projects, such as New York City's Delaware River supply and the Colorado–Big Thompson project across the Continental Divide, are already in operation. The California Water Project, which is nearing completion, and the Texas Water Plan, now stalled, are examples of larger projects. The grand schemes, however, visualize the transport of water from the northern parts of Canada, where there is a large surplus, to the central and southwestern parts of the United States and into northern areas of Mexico.

It is likely to be many years before any of these large-scale projects reaches fruition, since reasonable cost figures will need to be settled and many international political and social problems overcome. However, lesser interbasin transfer projects will probably continue to develop, and it may be through sequential interconnection of these that one of the grand schemes will eventually be realized.

[EDWARD SILBERMAN]

Bibliography: R. del Arenal C., Water resources of Mexico, *Water Resour. Bull.*, 5(1):19–38, 1969; V. T. Chow, *Handbook of Applied Hydrology*, 1964; L. M. Cox, *Proceedings of the 39th Western Snow Conference*, pp. 84–87, 1974; S. N. Davis and R. J. M. DeWiest, *Hydrogeology*, 1966; H. F. Haupt, *USDA For. Serv. Res. Pap.*, no. INT-114, 1972; R. J. Kazmann, *Modern Hydrology*, 2d ed., 1972; A. H. Laycock (ed.), *Proceedings of the Symposium on Water Balance in North America*, American Water Resources Association, 1969; W. G. McGinnies and B. J. Goldman (eds.), *Arid Lands in Perspective* (including AAAS Papers on Water Importation into Arid Lands), 1969; M. F. Meier, *J. Amer. Water Works Ass.*, 61(1):8–12, 1969; D. H. Miller, *Water at the Surface of the Earth: An Introduction to Ecosystem Hydrodynamics*, 1977; M. Neiberger, *Meteorol. Org. Tech. Note*, no. 105,

Table 2. Estimated water withdrawal in the United States

Use	Billions of gallons per day*	Gallons per capita per day*†	Annual increase, %
Public water supplies	26	130	3.0
Rural domestic supplies	7	35	
Irrigation	155	775	1.8
Steam electric cooling	127	635	2.7
Industrial and other	81	405	2.8
Total	396‡	1980	2.4

*1 gal = 3.785 liters. †Based on 200,000,000 people. ‡Includes 70,000,000,000 gal/day of groundwater.

1969; J. L. Smith, H. G. Halverson, and R. A. Jones, *U.S. Dep. Commerce Nat. Tech. Inform. Serv. Rep.*, no. TID-25987, 1972; M. Tribus, Physical view of cloud seeding, *Science*, 168:201–211, 1970; R. C. Ward, *Principles of Hydrology*, 1967; L. W. Weisbecker (comp.), *Technology Assessment of Winter Orographic Snowpack Augmentation in the Upper Colorado River Basin*, Stanford Research Institute, Menlo, Park, CA, 1972; A. Wilson and K. T. Iseri, *River Discharge to the Sea from the Shores of the Conterminous United States, Alaska, and Puerto Rico*, USGS Atlas no. HA-282, rev. ed., 1969; G. Young, Dry lands and desalted water, *Science*, 167:339–343, 1970.

Hydrometeorology

The study of the occurrence, movement, and changes in the state of water in the atmosphere. The term is also used in a more restricted sense, especially by hydrologists, to mean the study of the exchange of water between the atmosphere and continental surfaces. This includes the processes of precipitation and direct condensation, and of evaporation and transpiration from natural surfaces. Considerable emphasis is placed on the statistics of precipitation as a function of area and time for given locations or geographic regions.

Water occurs in the atmosphere primarily in vapor or gaseous form. The average amount of vapor present tends to decrease with increasing elevation and latitude and also varies strongly with season and type of surface. Precipitable water, the mass of vapor per unit area contained in a column of air extending from the surface of the Earth to the outer extremity of the atmosphere, varies from almost zero in continental arctic air to about 6 g/cm^2 in very humid, tropical air. Its average value over the Northern Hemisphere varies from around 2.0 g/cm^2 in January and February to around 3.7 g/cm^2 in July. Its average value is around 2.8 g/cm^2, an amount equivalent to a column of liquid water slightly greater than 1 in. in depth. Close to 50% of this water vapor is contained in the atmosphere's first mile, and about 80% is to be found in the lowest 2 mi.

Atmospheric water cycle. Although a trivial proportion of the water of the globe is found in the atmosphere at any one instant, the rate of exchange of water between the atmosphere and the continents and oceans is high. The average water molecule remains in the atmosphere only about 10 days, but because of the extreme mobility of the atmosphere it is usually precipitated many hundreds or even thousands of miles from the place at which it entered the atmosphere.

Evaporation from the ocean surface and evaporation and transpiration from the land are the sources of water vapor for the atmosphere. Water vapor is removed from the atmosphere by condensation and subsequent precipitation in the form of rain, snow, sleet, and so on. The amount of water vapor removed by direct condensation at the Earth's surface (dew) is relatively small.

A major feature of the atmospheric water cycle is the meridional net flux of water vapor. The average precipitation exceeds evaporation in a narrow band extending approximately from 10°S to 15°N lat. To balance this, the atmosphere carries water vapor equatorward in the tropics, primarily in the quasi-steady trade winds which have a component of motion equatorward in the moist layers near the Earth's surface. Precipitation also exceeds evaporation in the temperate and polar regions of the two hemispheres, poleward of about 40° lat. In the middle and higher latitudes, therefore, the atmosphere carries vapor poleward. Here the exchange occurs through the action of cyclones and anticyclones, large-scale eddies of air with axes of spin normal to the Earth's surface.

For the globe as a whole the average amount of evaporation must balance the precipitation. The subtropics are therefore regions for which evaporation substantially exceeds precipitation. The complete meridional cycle of water vapor is summarized in Table 1. This exchange is related to the characteristics of the general circulation of the atmosphere. It seems likely that a similar cycle would be observed even if the Earth were entirely covered by ocean, although details of the cycle, such as the flux across the Equator, would undoubtedly be different.

Complications in the global pattern arise from the existence of land surfaces. Over the continents the only source of water is from precipitation; therefore, the average evapotranspiration (the sum of evaporation and transpiration) cannot exceed precipitation. The flux of vapor from the oceans to the continents through the atmosphere, and its ultimate return to atmosphere or ocean by evaporation, transpiration, or runoff is known as the hydrologic cycle. Its atmospheric phase is closely related to the air mass cycle. In middle latitudes of the Northern Hemisphere, for example, precipitation occurs primarily from maritime air masses moving northward and eastward across the continents. Statistically, precipitation from these air masses substantially exceeds evapotranspiration into them. Conversely, cold and dry air masses tend to move southward and eastward from the interior of the continents out over the oceans. Evapotranspiration into these continental air masses strongly exceeds precipitation, especially during winter months. These facts, together with the extreme mobility of the atmosphere and its associated water vapor, make it likely that only a small percentage of the water evaporated or transpired from a continental surface is reprecipitated over the same continent. *See* ATMOSPHERIC GENERAL CIRCULATION; EVAPOTRANSPIRATION; HYDROLOGY.

Precipitation. Hydrometeorology is particularly concerned with the measurement and analysis of precipitation data. Since 1950 increasing attention

Table 1. Meridional flux of water vapor in the atmosphere

Latitude	Northward flux, 10^{10} g/sec
90°N	0
70°N	4
40°N	71
10°N	−61
Equator	45
10°S	71
40°S	−75
70°S	1
90°S	0

has been paid to the use of radar in estimating precipitation. By relating the intensity of radar echo to rate of precipitation, it has been possible to obtain a vast amount of detailed information concerning the structure and areal distribution of storms. *See* STORM DETECTION.

Deficiencies in the observational networks over the oceans and over the more sparsely inhabited land areas of the Earth are now being bridged through the use of meteorological satellite observations. Progress in the 1970s toward development of methods for estimating rainfall amounts from satellite observations of cloud type and distribution is of particular significance to hydrometeorology.

Precipitation occurs when the air is cooled to saturation. The ascent of air towards lower pressure is the most effective process for causing rapid cooling and condensation. Precipitation may therefore be classified according to the atmospheric process which leads to the required upward motion. Accordingly, there are three basic types of precipitation: (1) Orographic precipitation occurs when a topographic barrier forces air to ascend. The presence of significant relief often leads to large variations in precipitation over relatively short distances. (2) Extratropical cyclonic precipitation is associated with the traveling regions of low pressure of the middle and high latitudes. These storms, which transport sinking cold dry air southward and rising warm moist air northward, account for a major portion of the precipitation of the middle and high latitudes. (3) Air mass precipitation results from disturbances occurring in an essentially homogeneous air mass. This is a common precipitation type over the continents in mid-latitudes during summer. It is the major mechanism for precipitation in the tropics, where disturbances may range from areas of scattered showers to intense hurricanes. In most cases there is evidence for organized lifting of air associated with areas of cyclonic vorticity, that is, areas over which the circulation is counterclockwise in the Northern Hemisphere or clockwise in the Southern Hemisphere.

The availability of data from geosynchronous meteorological satellites, together with surface and upper-air data acquired as part of the Global Atmospheric Research Program (GARP), is leading to significant advances in the understanding of the character and distribution of tropical precipitation. *See* HURRICANE; STORM.

Precipitation may, of course, be in liquid or solid form. In addition to rain and snow there are other forms which often occur, such as hail, snow pellets, sleet, and drizzle. If upward motion occurs uniformly over a wide area measured in tens or hundreds of miles, the associated precipitation is usually of light or moderate intensity and may continue for a considerable period of time. Vertical velocities accompanying such stable precipitation are usually of the order of several centimeters per second. Under other types of meteorological conditions, particularly when the density of the ascending air is less than that of the environment, upward velocities may locally be very large (of the order of several meters per second) and may be accompanied by compensating downdrafts. Such convective precipitation is best illustrated by the thunderstorm. Intensity of precipitation may be extremely high, but areal extent and local duration are comparatively limited. Storms are sometimes observed in which local convective regions are embedded in a matrix of stable precipitation.

Analysis of precipitation data. Precipitation is essentially a process which occurs over an area. However, despite the experimental use of radar, most observations are taken at individual stations. Analyses of such "point" precipitation data are most often concerned with the frequency of intense storms. These data are of particular importance in evaluating local flood hazard, and may be used in such diverse fields as the design of local hydraulic structures, such as culverts or storm sewers, or the analysis of soil erosion. Intense local precipitation of short duration (up to 1 hr) is usually associated with thunderstorms. Precipitation may be extremely heavy for a short period, but tends to decrease in intensity as longer intervals are considered. Several record point accumulations of rainfall are shown in Table 2.

A typical hydrometeorological problem might involve estimating the likelihood of occurrence of a storm of given intensity and duration over a specified watershed to determine the required spillway capacity of a dam. Such estimates can only be obtained from a careful meteorological and statistical examination of large numbers of storms selected from climatological records. In the United States the U.S. Army Corps of Engineers, in cooperation with the National Weather Service, has embarked on a continuing program of analysis to make such historical depth-area-duration data available to the practicing engineer.

Evaporation and transpiration. In evaluating the water balance of the atmosphere, the hydrometeorologist must also examine the processes of evaporation and transpiration from various types of natural surfaces, such as open water, snow and ice fields, and land surfaces with and without vege-

Table 2. Record observed point rainfalls*

Duration	Depth, in.†	Station	Date
1 min	1.23	Unionville, Md.	July 4, 1956
8 min	4.96	Füssen, Bavaria	May 25, 1920
15 min	7.80	Plumb Point, Jamaica	May 12, 1916
42 min	12.00	Holt, Mo.	June 22, 1947
2 hr 45 min	22.00	Near D'Hanis, Texas	May 31, 1935
24 hr	73.62	Cilaos, La Reunion (Indian Ocean)	Mar. 15–16, 1952
1 month	366.14	Cherrapunji, India	July, 1861
12 months	1041.78	Cherrapunji, India	August, 1860 to July, 1861

*From R. K. Linsley, M. A. Kohler, and J. L. H. Paulhus, *Hydrology for Engineers*, 2d ed., 1975. †1 in. = 25.4 mm.

tation. From the point of view of the meteorologist, the problem is one of transfer in the turbulent boundary layer. It is complicated by topographic effects when the natural surface is not homogeneous. In addition the simultaneous heating or cooling of the atmosphere from below has the effect of enhancing or inhibiting the transfer process. Although the problem has been attacked from the theoretical side, empirical relationships are at present of greatest practical utility. *See* METEOROLOGY; MICROMETEOROLOGY.

[EUGENE M. RASMUSSON]

Bibliography: R. D. Fletcher, Hydrometeorology in the United States, in T. F. Malone (ed.), *Compendium of Meteorology*, 1951; R. K. Linsley, M. A. Kohler, and J. L. H. Paulus, *Hydrology for Engineers*, 2d ed., 1975; E. N. Lorenz, *The Nature and Theory of the General Circulation of the Atmosphere*, 1967; W. D. Sellers, *Physical Climatology*, 1965.

Hydrosphere

Approximately 74% of the Earth's surface is covered by water, in either the liquid or solid state. These waters, combined with minor contributions from groundwaters, constitute the hydrosphere:

World oceans	1.3×10^9 km^3
Fresh-water lakes	1.3×10^5 km^3
Saline lakes and inland seas	1.0×10^5 km^3
Rivers	1.3×10^3 km^3
Soil moisture and vadose water	6.7×10^4 km^3
Groundwater to depth of 4000 m	8.4×10^6 km^3
Icecaps and glaciers	2.9×10^7 km^3

The oceans account for about 97% of the weight of the hydrosphere, while the amount of ice reflects the Earth's climate, being higher during periods of glaciation. (Water vapor in the atmosphere amounts to 1.3×10^4 km^3.) The circulation of the waters of the hydrosphere results in the weathering of the landmasses. The annual evaporation of 3.5×10^5 km^3 from the world oceans and of 7.0×10^4 km^3 from land areas results in an annual precipitation of 3.2×10^5 km^3 on the world oceans and 1.0×10^5 km^3 on land areas. The rainwater falling on the continents, partly taken up by the ground and partly by the streams, acts as an erosive agent before returning to the seas. *See* GROUNDWATER; HYDROLOGY; RIVER; SEA WATER; TERRESTRIAL FROZEN WATER.

[EDWARD D. GOLDBERG]

Bibliography: R. L. Nace, Water resources, *Environ. Sci. Technol.*, 1:550–560, 1967.

Hydrospheric geochemistry

The oceans of the world constitute a principal reservoir for substances in the major sedimentary cycle, which involves the processes of transport of material from the Earth's crust to the sea floor. The cycle begins with the precipitation of water, acidified by the uptake of carbon dioxide in the atmosphere, onto the continents. This results in the physical and chemical breakdown of exposed surfaces. A part of the weathered material, in dissolved or solid states, is borne by the rivers to the oceans. Evaporation at the oceanic surfaces provides atmospheric water, which precipitates in part upon the continents. This latter process completes the cycle. Table 1 gives the quantitative details by contrasting the marine areas and the respective land areas draining into the world's oceans. Clearly, per unit area, the Atlantic receives the weathering products from an integrated drainage area six times as large as that of the Pacific. The interior drainage areas are responsible for such water bodies as the Great Salt Lake, the Caspian Sea, and the Dead Sea.

Table 1. Oceanic and land-drainage areas

Ocean	Area, 1000 km^2	Land area drained, 1000 km^2	Ratio, area drained/ ocean area
Atlantic	98,000	67,000	0.684
Indian	65,500	17,000	0.260
Antarctic	32,000	14,000	0.440
Pacific	165,000	18,000	0.110
Interior drainage		32,000	

Oceanic waters. The reactivities of chemical species in the oceans are reflected in the average times spent there before precipitation to the sea floor. Those elements with short residence time in the oceans engage more readily in chemical reactions that result in the formation of solid phases than those elements with long residence times.

Residence times. The calculations of residence times are based upon a simple reservoir model of the oceans, whose chemical composition is assumed to be in steady state; that is, the amount of a given element introduced by the rivers per unit time is exactly compensated by that lost through sedimentation. An elemental residence time may be defined then by $t = A/(dA/dt)$, where A is the amount of the element in the oceans and (dA/dt) is the rate of introduction or the rate of removal of the element from the marine hydrosphere. Table 2 gives values for a representative group of elements.

The element of longest residence, sodium, has a residence time within an order of magnitude of the age of the oceans, several billion years. Similarly, the more abundant alkali and alkaline-earth metals all have residence times in the range of 10^6 to 10^8 years, resulting from the relative lack of reactivity of these elements in marine waters.

Elements which pass rapidly through the marine hydrosphere in the major sedimentary cycle, such as titanium, aluminum, and iron, not only enter the oceans in part as rapidly settling solids but also are reactants in the formation of the clay and ferromanganese minerals. For a discussion of the

Table 2. Residence times of elements in the oceans

Element	Residence time, years	Element	Residence time, years
Na	2.6×10^8	Mg	4.5×10^7
Ca	8.0×10^6	Li	1.2×10^7
U	6.5×10^5	K	1.0×10^7
Cu	6.5×10^4	Rb	6.1×10^6
Si	1.0×10^4	Ba	4.0×10^4
Mn	7.0×10^3	Zn	2.0×10^4
Ti	1.6×10^2	Pb	4.0×10^2
Al	1.0×10^2	Ce	3.2×10^2
Fe	1.4×10^2	Th	1.0×10^2

inorganic regulation of the composition *see* SEA WATER.

Although the relatively low values of these residence times are significant, the absolute values are probably unrealistic inasmuch as they are in conflict with an assumption used in their derivation. In treating the oceans as a simple reservoir, the mixing times of the oceans are assumed to be much less than the residence times of the elements. Yet the oceans are believed to mix in times of the order of thousands of years. Nonetheless, it is significant that one may expect to find the concentration of such elements varying from one oceanic water mass to another.

Those elements with residence times of intermediate length, periods of the order of tens and hundreds of thousands of years, are probably of nearly uniform concentration in the oceans of the world, as are the chemical species of longer marine lives. Typical members of the group are such metals as barium, lead, zinc, and nickel, elements in extremely dilute solution but actively involved in the inorganic chemistries of the seas. Such behavior is confirmed by the observation that these elements are in states of undersaturation in oceanic waters.

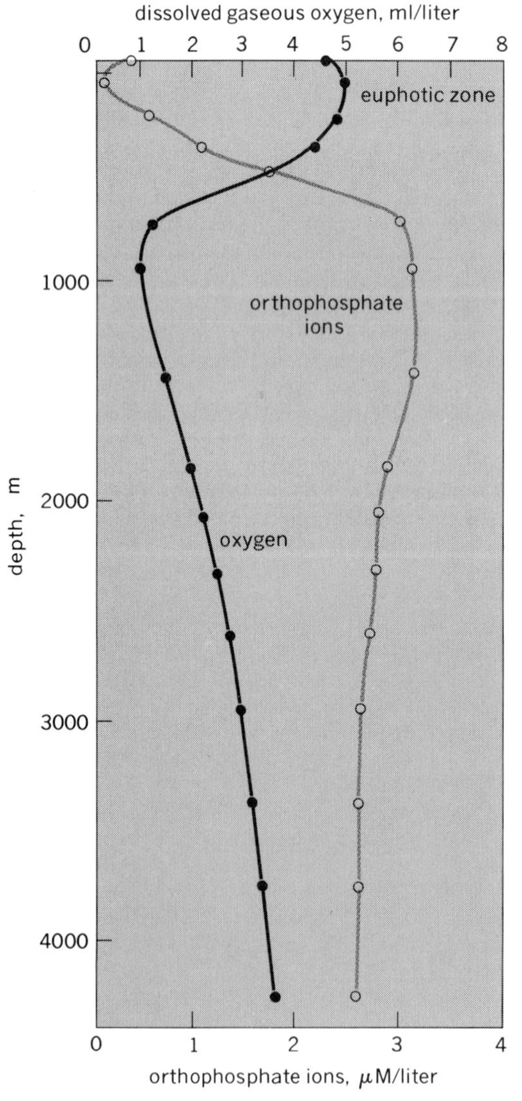

Distribution of dissolved gaseous oxygen (low values) and nutrient species (high concentrations) of orthophosphate ions at 26°22'.4N and 168°57'.5W in the Pacific Ocean. (*Data from Chinook Expedition of 1956 of the Scripps Institution of Oceanography*)

Photosynthesis. The presence of the large photosynthesizing biomass in the oceans gives rise to dramatic concentration changes. The amount of photosynthesis in the oceans, calculated to be of the order of 1×10^{17} tons of carbon dioxide consumed per year, compares with estimates for land of 2×10^{16} tons/year. The depth of the photosynthetic zone can extend downward from the surface to depths of 100 meters (m) or so, depending upon the transparency of the water, season of the year, and latitude. In waters of active plant growth, carbon dioxide and oxygen (the intake and release gases of photosynthesis) are often observed in states of depletion and supersaturation, respectively.

The photosynthesizing plants require a group of dissolved chemical species, the nutrients, which are necessary for growth and multiplication. Ions of the orthophosphoric acids, nitrate, nitrite, and ammonia, as well as monomeric silicic acid, have very low concentrations in the regions of plant productivity as compared to regions of lesser fertility. Certain other substances concerned with plant growth, such as vitamins and trace-metal ions, may have their marine concentrations governed by biological activity in surface waters, but as yet no definite relationships have been established.

The primary production of plant material furnishes the basis of nutrition for the animal domain of the oceans. The plant material which is removed from the photosynthetic zones but is not consumed by higher organisms, together with the organic debris resulting from the metabolic waste products or death of members of the marine biosphere, is oxidized principally in the oceanic waters below but adjacent to the photosynthetic zone. This results not only in low values of dissolved gaseous oxygen at such depths but also in high concentrations of the nutrient species which are released subsequent to the combustion of the organic matter (see illustration).

The dissolved organic substances, arising from life in the sea, are of the order of 0.3 mg of carbon per liter. Higher amounts of such materials are found in surface or coastal waters. *See* SEA-WATER FERTILITY.

Salt content. For many problems in the physics of the sea and in engineering, the significant parameter is the total salt content, which governs the density as a function of pressure and temperature, rather than elemental, ionic, or molecular compositions.

The total salt content is expressed by either the salinity or the chlorinity, both terms given in units of parts per thousand ‰. The salinity is defined as the weight in grams in vacuo of solids which can be obtained from a water weight of 1 kg in vacuo under the following conditions: (1) the solids have been dried to a constant weight at 480°C; (2) the carbonates have been converted to oxides; (3) the organic matter has been oxidized; and (4) the bromine and iodine have been replaced by chlorine. A weight loss from the solid phases results from such chemistries, and hence the salinity of a given sample of sea water is somewhat less than its actual salt content.

In practice, the salt content is ascertained by the precipitation of the halogens with silver nitrate or by such physical methods as electrical conductivity, sound velocity, and refractive index, with the first of these techniques having widespread use. In the chemical titration with silver nitrate the mass of halogens contained in 1 kg of sea water, assuming the bromine and iodine are replaced by chlorine, is designated as the chlorinity. The values so obtained are dependent upon the atomic weights of both silver and chlorine. Inasmuch as changes in the salt content of water are of interest when taken over long time periods, the chlorinity has been made independent of atomic weight changes by redefining it in terms of the weight of silver precipitated, as in Eq. (1). Chlorinity deter-

$$\text{Cl}\ ‰ = 0.3285234\ \text{Ag} \qquad (1)$$

minations, based either on chemical or physical methods, are related to standard sea water, a water of known salinity which is obtainable from the Hydrographic Laboratory in Copenhagen, Denmark. Chlorinity is related to the salinity, Eq. (2).

$$\text{S}\ ‰ = 0.030 + 1.8050\ \text{Cl}\ ‰ \qquad (2)$$

Salinity of open ocean waters varies regionally between 32 and 37 ‰. Areas where evaporation exceeds precipitation, such as enclosed basins, are characterized by higher values. Salinities of 38–39 ‰ are representative of the Mediterranean Sea, while the northern part of the Red Sea has values ranging from 40 to 41 ‰. Coastal bays, subject to land drainage, and those waters which mix with meltwaters from cold regions, possess salinities which have all degrees of dilution comparable to those of the open ocean. *See* OCEAN-ATMOSPHERE RELATIONS.

Although sea waters exhibit marked regional and depth differences in salt content, the ratios of the major dissolved constituents to one another, listed below, are almost invariable. (Ratios are grams of given element per kilogram of seawater divided by chlorinity in parts per thousand.)

Constituents	*Ratio*	*Constituents*	*Ratio*
Na/Cl	0.5555	S/Cl	0.0466
Mg/Cl	0.0669	Br/Cl	0.0034
Ca/Cl	0.0213	Sr/Cl	0.00040
K/Cl	0.0206	B/Cl	0.00024

Development of the hydrosphere. Many hypotheses on the origin and evolution of the Earth's hydrosphere have been advanced over the past 50 years. Most of them can be placed in one of two categories: (1) the hypothesis of an original ocean, which proposes that the present ocean has had much the same size and composition since the beginning of the geologic record; and (2) the hypothesis of continuous accumulation, which considers that the ocean has been growing continuously, but not necessarily uniformly, since its inception.

Such considerations have certain common assumptions. First, the rock-forming species in sea water, sodium, potassium, silicon, iron, magnesium, and others, have been derived from the weathering of the Earth's surface, whereas the marine quantities of water and the anionic constituents, such as chlorine and sulfur, cannot be adequately supplied by such a mechanism. These latter substances, as gases or dissolved species, have apparently evolved from the Earth's interior by the degradation of the surface rocks. This proposal has been reached through the following arguments. The abundances of the noble gases—neon, nonradiogenic argon, and krypton—are many orders of magnitudes less on the Earth than in the universe relative to other elements. It is thus assumed that they were lost by the Earth during its formative period. Therefore, substances which existed as gases and had comparable molecular weights were similarly not retained.

The hypothesis of the permanency of the oceans through geologic time complements the hypothesis of the permanence of the continents. Since old rocks are found in the basement of the central shield areas, with greater thicknesses of younger rocks at the edges, it has been assumed that the continents have grown laterally. Consequently, the reduction in area occupied by the oceans is compensated for by an increase in average depth. However, the calculated amounts of weathered materials that would form by the action of an initial ocean containing all the various anionic constituents as acids are far greater than those estimated by geologists to have been decomposed over all of geologic time. This hypothesis has been modified by some geochemists to one of an initial ocean of water with the gradual accretion of anionic substances.

The hypothesis of the slow growth of the oceans gains strength with the following observations. If only 1% of the hot-spring water is juvenile, the amounts found today, if extrapolated over geologic time, give a sufficient volume to produce the present oceans. Further, the ratios of the major anionic constituents in sea water are similar to those found in plutonic gases.

Fresh waters and rain. Fresh continental waters show enormous variations in total salt content and in the relative concentrations of their various components, and such parameters in any given water body vary seasonally as well. Chemical analyses of a number of representative waters are given in Table 3.

The factors governing the composition of freshwater bodies are many, and some are poorly under-

Table 3. The chemical compositions of fresh waters and rain (values in parts per million)

Substance	Rivers*	Irish lakes†	English rain‡
Ca^{++}	15	4.0	0.1–2.0
Na^{+}	6.3	8.6	0.2–7.5
Mg^{++}	4.1	1.0	0.0–0.8
K^{+}	2.3	0.5	0.05–0.7
CO_3^{--}	58.4	8.8	0.0–2.7
SO_4^{--}	11.2	5.2	1.1–9.6
Cl^{-}	7.8	14.7	0.2–12.6
SiO_2	13.1		
NO_3^{-}	1		
Fe	0.67		

*Daniel A. Livingston, *Data of Geochemistry*, USGS Prof. Pap. no. 440–G, 1963.

†E. Gorham, The chemical composition of some western Irish fresh waters, *Proc. Roy. Irish Acad.*, vol. 58B, 1957.

‡E. Gorham, On the acidity and salinity of rain, *Geochim. Cosmochim. Acta*, vol. 7, 1955.

stood. Significant amounts of the dissolved phases appear to come from rock weathering, organic-decomposition products, atmospheric dusts, fuel combustion, air-borne particles from the marine environment, and volcanic emanations. Such materials, with the exception of the first, carried from their sources in the atmosphere, can be taken from the air to the river and lakes below by rain. Soil and rock weathering provide the major supply of ions to rivers and lakes. This can be seen from Table 3, where rainwaters have less dissolved solids than river or lake waters by an order of magnitude.

Most fresh waters can be characterized as bicarbonate waters, with HCO_3^- exceeding all other anions, although in some instances chloride or sulfate is the dominant anion.

The sodium, chlorine, and often magnesium are dominantly of marine origin. They leave the oceans as sea spray and are air-borne to the continents. These elements show dramatic decreases in concentration in surface waters going from the coastal to the interior regions. Exceptions can be found in certain waters which derive their salts mainly by the denudation of igneous areas.

Calcium, and sometimes magnesium, can originate from drainage areas, as in the case of the Wisconsin fresh waters which drain over ancient magnesian limestones. Sulfate not only has sources in the marine environment but is also produced by the combustion of sulfur-containing fuels and from the oxidation of sulfur dioxide which results from the atmospheric burning of hydrogen sulfide, a product of the decomposition of organic matter.

The average salt content of fresh waters is of the order of 120 parts per million (ppm). Lower amounts of dissolved solids (50 ppm and less) are found in waters draining igneous rock beds, while open lakes and rivers carrying high salt contents (200 ppm and above) normally result either from the leaching of salt beds or from contamination by man.

Closed basins. The chemical compositions of closed basins, water bodies in which evaporation is the mechanism for the loss of water, are illustrated in Table 4, which contains representative examples of the three classical types, the carbonate, sulfate, and chloride waters. These classes appear in sequence during the removal of water from a system with the composition of average river or lake water. The carbonate types exist up until evaporation leads to the precipitation of calcium carbonate and to liquid phases enriched in sulfate and chloride ions. Further removal of water results in the precipitation of calcium carbonate and the subsequent precipitation of gypsum, $CaSO_4 \cdot 2H_2O$. The residual waters hence contain chloride as the dominant anion.

Table 4. The composition of waters from chloride, sulfate, and carbonate lakes (values in parts per million)*

Substance	Dead Sea	Little Manitou Lake	Pelican Lake, Ore.
Na^+	11.14	16.8	29.25
K^+	2.42	1.0	3.58
Mg^{++}	13.62	10.9	2.62
Ca^{++}	4.37	0.48	2.27
CO_3^{--}	Trace	0.47	30.87
SO_4^{--}	0.28	48.4	22.09
Cl^-	66.37	21.8	7.97
SiO_2	Trace	0.009	1.21
Al_2O_3, Fe_2O_3		0.21	0.02
Salinity	226,000	106,851	1983

*Data from G. E. Hutchinson, *A Treatise on Limonology*, vol. 1, Wiley, 1957.

[EDWARD D. GOLDBERG]

Bibliography: M. N. Hill (ed.), *The Sea*, vol. 2, 1963; G. E. Hutchinson, *A Treatise on Limnology*, vol. 1, 1957.

Imhoff tank

A sewage treatment tank named after its developer, Karl Imhoff. Imhoff tanks differ from septic tanks in that digestion takes place in a separate compartment from that in which settlement occurs. The tank was introduced in the United States in 1907 and was widely used as a primary treatment process and also in preceding trickling filters. Developments in mechanized equipment have lessened its popularity, but it is still valued as a combination unit for settling sewage and digesting sludge. *See* SEPTIC TANK; SEWAGE.

The Imhoff tank is constructed with the flowing-through chamber on top and the digestion chamber on the bottom (see illustration). The upper chamber is designed according to the principles of a sedimentation unit. Sludge drops to the bottom of the tank and through a slot along its length into the lower chamber. As digestion takes place, scum is formed by rising sludge in which gas is trapped. The scum chamber, or gas vent, is a third section of the tank located above the lower chamber and beside the upper chamber. As gases escape, sludge from the scum chamber returns to the lower chamber. The slot is so constructed that particles cannot rise through it. A triangle or sidewall deflector below the slot prevents vertical rising of gas-laden sludge. Sludge in the lower chamber settles to the bottom, which is in the form of one or more steep-sloped hoppers. At intervals the sludge can be withdrawn. The overall height of the tank is 30–40 ft (9–12 m) and sludge can be expelled under hydraulic pressure of the water in the upper tank. Large tanks are built with means for reversing flow in the upper chamber, thus making it possible to distribute the settled solids more evenly over the digestion chamber.

Design. Detention period in the upper chamber is usually about 2½ hr. The surface settling rate is usually 600 gal/(ft²)(day) [24 m³/(m²)(day)]. The weir overflow rate is not over 10,000 gal/ft (124 m³/m) of weir per day. Velocity of flow is held below 1 ft/s (0.30 m/s). Tanks are dimensioned with a length-width ratio of 5:1–3:1 and with depth to slot about equal to width. Multiple units are built rather than one large tank to carry the entire flow. Two flowing-through chambers can be placed above one digester unit. The digestion chamber is normally designed at 3–5 ft³ (0.08–0.14 m³) per capita of connected sewage load. When industrial wastes include large quantities of solids, additional allowance must be made. Ordinarily sludge withdrawals are scheduled twice per year. If these are to be less frequent, an increase in capacity is desirable. Some chambers have been provided with up to 6.5 ft³ (0.184 m³) per capita. The scum chamber should have a surface

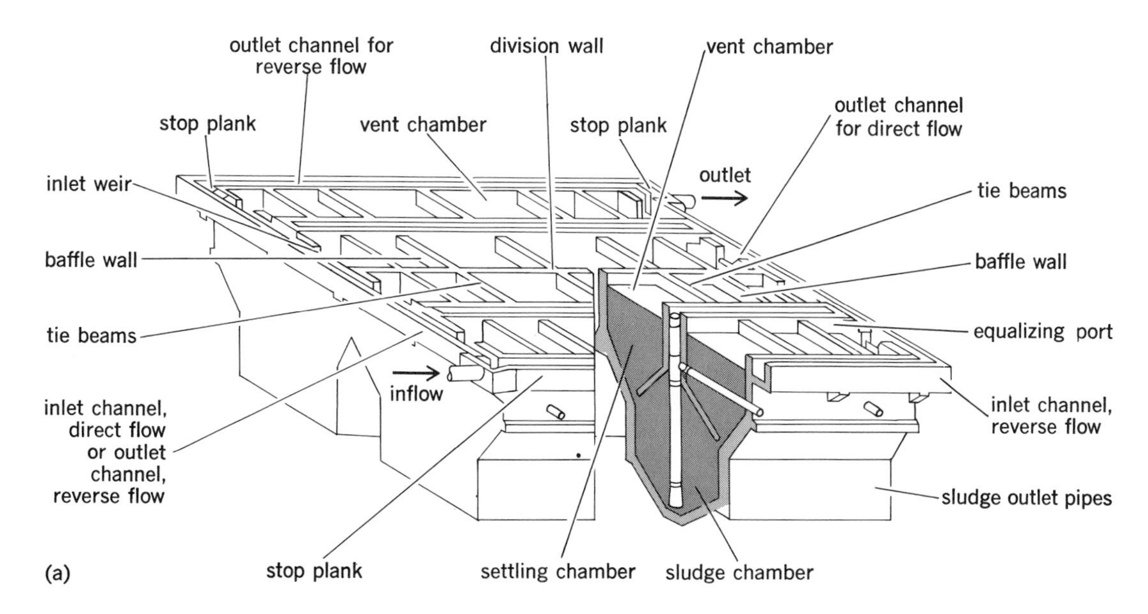

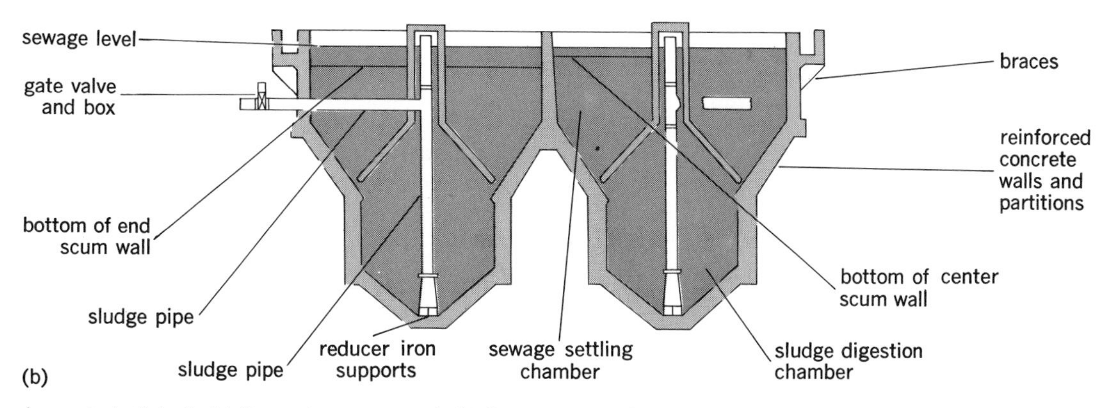

Large Imhoff tank. *(a)* General arrangement. *(b)* Cross section. *(From H. E. Babbitt and E. R. Baumann, Sewerage and Sewage Treatment, 8th ed., copyright © 1958 by John Wiley & Sons, Inc.; used with permission)*

area 25–30% of the horizontal surface of the digestion chamber. Vents should be 24 in. (61 cm) wide. Top freeboard should be at least 2 ft (61 cm) to contain rising scum. Water under pressure must be available to combat foaming and knockdown scum.

Efficiency. The efficiency of Imhoff tanks is equivalent to that of plain sedimentation tanks. Effluents are suitable for treatment on trickling filters. The sludge is dense, and when withdrawn it may have a moisture content of 90–95%. Imhoff sludge has a characteristic tarlike odor and a black granular appearance. It dries easily and when dry is comparatively odorless. It is an excellent humus but not a fertilizer. Gas vents may occasionally give off offensive odors. [WILLIAM T. INGRAM]

Industrial meteorology

The commercial application of weather information to the operational problems of business, industry, transportation, and agriculture in a manner intended to optimize the operation with respect to the weather factor. The weather information may consist of past weather records, contemporary weather data, predictions of anticipated weather conditions, or an understanding of physical processes which occur in the atmosphere. The operational problems are basically decisions in which weather exerts an influence.

The evolution of applied meteorology from primarily a governmental function to today's mixed economic approach has taken place largely since World War II. In that period there has been a gradual development in the application of meteorological information. Some of the applications are: professional meteorologists serving television or radio stations to explain to the public the changes in weather which either have occurred or are about to occur; forecasters working with airline dispatchers to utilize optimum flight paths and to avoid equipment tie-ups during lengthy adverse weather conditions; consultants who help utilities locate future power plant sites with minimal environmental impact; forecasters who advise shipping lines regarding ocean routing paths which avoid storms and decrease by a few hours the travel time for long-distance ocean freight movement; and consultant advisories which help industrial firms in the marketing of weather-sensitive products. In many cases the need for meteorological information is sufficiently important that a complete department within the operational segment of a company is established to serve the needs of that weather-affected firm.

One specialized consulting service involves snowstorm forecasting to aid in keeping city streets, highways, and especially toll roads usable. The city administrator seeks to match the use of worker-power and equipment to the size and duration of the storm. In addition to forecasting the beginning time of the snowstorm, the consultant firm typically makes additional contacts to keep the client advised of the storm's intensity, rate of movement, and total duration. The costs of such weather advisory services are nominal compared with costs of overreacting when only light snow begins, and with street or highway blockage when a little lead time could have been used to enlist supplemental equipment or worker-power.

Contract research work by professional meteorologists meets a need for intermittent special requirements of governmental agencies. Studies of this type might require the determination of atmospheric airflow frequencies related to the dispersion of air pollution throughout a broad area as it might relate to governmental regulatory policies. A distinct advantage exists when a government agency can meet a nonrecurring need with contract consulting effort.

Along with the advances of communication capacity, use of weather satellite photography, and the expanded international marketing of goods and services, consulting meteorologists within the United States have found a demand for their services among commodity-trading firms. Knowledge of recent past weather or current weather throughout the wheat-growing areas of the world is useful information to wheat users, wheat growers, and speculators in commodity futures who help establish a market value for wheat well in advance of the actual harvest.

Climatological records can be used effectively in developing statistical odds for extreme weather conditions. Climatological records also can be used for postanalyses of sales records of weather-sensitive items. For example, the sales of room air conditioners move up sharply in the months of May and June when early-season extreme hot spells occur throughout the northern areas of the United States. A series of several such hot spells will guarantee a peak sales season for all manufacturers. By contrast, a cold May and June in the same marketing areas will guarantee a carry-over of large numbers of such equipment until the next peak marketing season. Postanalyses of actual daily sequences of weather in a given season can help both the manufacturer and the marketing firms in their planning for subsequent years. An extremely cold season is not likely to be followed a year later with similar low sales.

Meteorological consulting firms range from individual consultants who operate similarly to individual consulting engineers, to large firms having nearly 100 professional meteorologists on their staff. In early 1979 the Professional Directory in the *Bulletin of the American Meteorological Society* carried a listing of 84 different firms. A high fraction of these firms serve multiple clients as compared with full-time staff members within individual companies. Approximately 10% of the 8000 professional meteorologists in the United States are engaged in some facet of industrial meteorology.

Some measure of the interest in applications of meteorology is indicated by the number of industrial corporate members within the American Meteorological Society. The nearly 100 corporate members include industrial firms, equipment suppliers, utilities, and insurance companies.

Special applications of meteorological knowledge are required in the field of weather modification. Fog dispersal at some commercial airports and field projects to increase winter snowpack on the western slopes of mountain barriers throughout the United States are examples of such use.

Forensic meteorology, the application of atmospheric information to legal cases, requires careful attention to postanalyses of factual information. Since accidents seldom happen in the immediate vicinity of weather observing stations, professional opinions are useful in determining the sequence of weather-affected events that are subject to litigation. Expert-witness testimony is the full-time emphasis of some meteorological consultants and a part-time effort by many consultants.

The National Weather Service recognizes the importance of the role of professional meteorologists serving industry. The Special Assistant for Industrial Meteorology in the National Oceanic and Atmospheric Administration is charged with coordinating the efforts of the governmental data collection service and the many users of such information, whether on a current basis or from the archives of weather records at the National Climatic Center. National Weather Service personnel are encouraged to recognize the needs of private professional practitioners. Industrial meteorologists represent an indispensable part in the government's effort to bring tailored meteorological service to those segments of commerce and industry affected by weather.

The development of weather sensing equipment fitted to the needs of industrial firms, particularly as related to air quality, has led to expanded use of professional personnel who install and maintain such measuring equipment. Weather equipment sales to private industrial firms now exceed sales to governmental agencies.

In 1968 a nonprofit organization, the National Council of Industrial Meteorologists (NCIM), was formed to further the development and expansion of industrial meteorology. In a recent 3-year period the membership of the NCIM served the following: 210 industrial firms, 52 Federal and state agencies, 296 local government agencies, and 14 university subcontract projects.

Annual conferences on industrial meteorology are sponsored by the Committee on Industrial Meteorology of the American Meteorological Society. That society has established a program for the certification of consulting meteorologists who meet rigorous standards of knowledge, experience, and adherence to ethical practice. *See* AGRICULTURAL METEOROLOGY; WEATHER FORECASTING AND PREDICTION.

[LOREN W. CROW]

Bibliography: *AMS Bull.*, vol. 59, no. 8, August 1978; W. J. Maunder, *The Value of Weather*, 1971; Papers presented at Session 4 of the 56th Annual Meeting of the American Meteorological Society, Philadelphia, Jan. 20, 1976, *AMS Bull.*, 57(11): 1318–1342, November 1976; J. A. Taylor (ed.), *Weather Forecasting for Agriculture and Industry: A Symposium*, 1972; J. C. Thompson and G. W. Brier, The economic utility of weather forecasts, *Mon. Weather Rev.*, 83(11):249–254, 1955.

Industrial wastes

The U.S. Environmental Protection Agency (EPA) estimates that 344,000,000 metric tons (wet basis) of industrial processing residues is generated annually in the United States. This represents approximately 3% of the total solid waste load in the nation, and is increasing at about 4.5% yearly. Wastes from the mining industry are estimated at approximately 2,000,000,000 metric tons, or 40% of the total solid waste. The agricultural industry produces 51% of the total waste load. *See* AGRICULTURAL WASTE.

PROBLEMS FROM THE PAST

Love Canal in Niagara Falls, NY, was used for chemical waste disposal in the 1940s by Hooker Chemicals and Plastics Corporation. In 1978 dangerous chemicals were found to be seeping into nearby homes. Higher-than-normal rates of miscarriages, birth defects, and liver ailments were found among the residents, and the canal environs were declared a disaster area, resulting in the evacuation of 235 families. More than $26,000,000 was spent within 2 years, with considerably more spending anticipated before the problem is fully resolved.

Also in 1978, residents of Stump Gap Creek, KY, complained of smelly, damaged, 55-gal (208-liter) drums of unknown chemicals floating down the Ohio River flood plain, leading to the discovery of the "Valley of the Drums" (Fig. 1).

A Federal suit was filed in 1979 seeking $1,600,000 from Kin-Buc Inc., a chemical landfill in Edison, NJ, that had received a reported 71,000,000 gal (269,000 m^3) of liquid waste over a 4-year period ending in 1976. Measurable amounts of 8 metals and 11 organic chemicals were found in tributaries of the Raritan River, and were discovered to be leaching into aquifers considered as standby water sources for the townships of Raritan and Edison.

A 5-acre (2-ha) dump along the Cedar River near Charles City, IA, is the source of groundwater seepage carrying chemicals into the Cedar Valley aquifer, which supplies drinking water to one-third of Iowa. A $5,000,000 penalty was imposed against Allied Chemical for discharge of the pesticide Kepone into the James River at Hopewell, VA. In Wood-Ridge, NJ, mercury was dumped on the 30-acre (12-ha) Meadowlands site by a company that no longer exists. Near the center of the town of Ambler, PA, officials are reluctant to disturb 40-ft-high (12-m) piles containing asbestos because of the dust that would result from the moving process. On Long Island, NY, 54 public water supply

Fig. 1. Aerial photograph of a portion of the estimated 20,000 to 100,000 barrels found at the Valley of the Drums in Shepardsville, KY. (*Environmental Protection Agency*)

Table 1. Major hazardous waste generators, among 17 industries that EPA has studied in detail (1977 estimates)*

Waste	Quantity, 10^6 metric tons (wet basis)
Organic chemicals	11.7
Primary metals	9.0
Electroplating	4.1
Inorganic chemicals	4.0
Textiles	1.9
Petroleum refining	1.8
Rubber and plastics	1.0
Miscellaneous	1.0
Total	34.5

*From *EPA J.*, vol. 5, no. 2, February 1979.

wells serving over 100,000 people have been found to be contaminated by chemicals. In Grey, ME, 750 families drank polluted groundwater for 4 years. In Hardeman County, TN, 40 families drank from wells polluted with high levels of toxic chemicals. In Grants, NM, 200 families drank radioactive water contaminated by uranium mining wastes. Illegal disposal of hazardous wastes near Byron, IL, contaminated wells serving 68 families. Several thousand drums and five tank trucks of hazardous solid and liquid wastes have posed an imminent health and environmental hazard in Chester, PA.

This is only a partial listing of the cases of mismanaged industrial wastes in EPA files. The EPA estimates that up to 51,000 sites contain some hazardous wastes—acids, toxic chemicals, corrosive metals, and other dangerous by-products generated by industry at a rate of 35,000,000 metric tons a year. The EPA is collecting detailed information on the sites posing immediate problems, and estimates that as many as 2000 of the sites could cause environmental or health problems by 1982. The cost of correcting those problems could average $26,000,000 a site.

HAZARDOUS WASTES

The EPA estimates that 10–15% of the 344,000,000 metric tons of industrial processing wastes is hazardous. The quantities of hazardous waste are expected to increase by 3% annually. A waste is proposed as hazardous by EPA if it meets any of the following characteristics: (1) ignitable (at a flash-point upper limit of 60°C); (2) corrosive (has a pH of 3 or less or of 12 or greater, or corrodes steel [SAE 1020] at a rate greater than 0.250 in. [6.35 mm] per year at 130°F [54°C]); (3) reactive (readily undergoes violent chemical change, detonation, or explosive reaction); (4) toxic (giving an extract 10 times or more concentrated than the EPA Naional Interim Primary Drinking Water Standard levels for specified contaminants); (5) radioactive (radium-226 equal to or more than 5 picocuries per gram for solids or 50 picocuries per liter for liquids); (6) infectious; (7) phytotoxic; (8) teratogenic and mutagenic.

In addition to the eight listed criteria for designating hazardous wastes, EPA specifically listed 158 industrial processes in the *Federal Register* (vol. 43, no. 243, Dec. 18, 1978) as generating hazardous wastes. The major hazardous waste generators studied in detail by EPA are listed in Table 1. Accounting for the total disposal: 80% is disposed of in nonsecure ponds, lagoons, or landfills; 10% is incinerated without proper controls; 10% is managed acceptably as compared with proposed Federal standards, that is, by controlled incineration, secure landfills, and recovery.

About 60% of the hazardous waste is in the form of liquid or sludge. The 17 industries studied by EPA—textile mill products, inorganic chemicals, organic chemicals, pesticides, explosives, petroleum refining, rubber, leather tanning and finishing, metal smelting and refining, electroplating and metal finishing, special machinery manufacturing, electronics components, batteries, paint and allied products, petroleum rerefining, plastics, and pharmaceuticals—currently spend $155,000,000 annually for hazardous waste management. EPA estimates that this will increase to an estimated $750,000,000 a year under anticipated regulations, representing about 0.28% of the annual $267,000,000,000 value of production for the 17 industry categories.

Of all hazardous wastes from these industries, 65% is generated in 10 states: Texas, Ohio, Pennsylvania, Louisiana, Michigan, Indiana, Illinois, Tennessee, West Virginia, and California (see Table 2). These states, and most of the other states, are expected to apply for authorization to operate their own waste management programs, which must be equivalent to the Federal program,

Table 2. Estimated industrial hazardous wastes generated by geographical region (1970 statistics in tons/hr)*

Region	Inorganics in aqueous		Organics in aqueous		Organics		Sludges, slurries, solids		Total		Percent of total
	Tons	Metric tons	Tons	Metric tons	Tons	Metric tons	Tons	Metric tons	Tons	Metric tons	
New England	95,000	86,000	170,000	154,000	33,000	30,000	6,000	5,450	304,000	275,450	3.1
Mid-Atlantic	1,000,000	907,200	1,100,000	1,000,000	105,000	90,600	55,000	50,000	2,260,000	2,047,800	22.9
East North-Central	1,300,000	1,180,000	850,000	770,000	145,000	132,000	90,000	81,600	2,385,000	2,163,600	24.2
West North-Central	65,000	59,000	260,000	236,000	49,500	45,000	18,500	16,800	393,000	350,800	4.0
South Atlantic	230,000	208,500	600,000	545,000	75,000	68,000	80,000	72,600	985,000	894,100	10.0
East South-Central	90,000	81,700	385,000	350,000	44,000	40,000	9,500	8,600	528,000	480,300	5.4
West South-Central	320,000	290,000	1,450,000	1,315,000	180,000	163,000	39,000	35,400	1,989,000	1,803,400	20.2
West (Pacific)	120,000	109,000	550,000	500,000	113,000	103,000	30,500	27,770	813,500	739,770	8.3
Mountain	125,000	113,500	5,000	4,540	50,000	45,400	11,500	10,400	191,500	173,840	1.9
Totals	3,345,000	3,034,900	5,370,000	4,874,540	794,500	717,000	340,000	308,620	9,849,500	8,929,060	100.0

*From N. P. Cheremisinoff et al., *Industrial Hazardous Wastes Impoundment*, p. 13, 1979.

be consistent with other state and Federal programs, and be adequately enforced. The EPA must operate a program in any state not qualifying for authorization. The keystone of the Federal program, and therefore of the state program, is control via manifests and reporting—a "cradle to grave" documentation of the history of the hazardous waste from the moment of generation to the final disposal is required of all producers of over 100 kg/month. Some 270,000 waste-generating facilities, 10,000 transporters, and 30,000 treatment, storage, and disposal sites are involved.

WASTE WATERS

More than 270,000,000,000 gal (1.022×10^9 m^3) is withdrawn daily from ground and surface water supplies in the United States; of this, industry accounts for more than 138,000,000,000 gal/day (1.38×10^{11} m^3/day), or about 51% of the total. Approximately 2% of the water used by industry is consumed, and nearly 98% has traditionally been returned to surface and underground sources. Industrial discharges affect some 72% of United States river basins; however, in many cases only one or two streams within the entire basin have noticeable problems. Actual toxic pollutants from point sources--mainly credited to industrial sources--affect 44% of the basins, with the most widespread impacts in the Northeast, North-Central, and Great Lakes regions where 62% of the river basins are affected.

With the goal of achieving water clean enough for recreation, and for the protection and propagation of wildlife, the Clean Water Act and other Federal and state legislation have been passed to eliminate, eventually, the discharge of pollutants into the national water system. The approach is to allow the discharge of various pollutants by specific industries at controlled upper concentration levels. These restrictions are enforced under the National Pollutant Discharge Elimination System (NPDES) by requiring that a permit be issued before discharges are allowed. Under NPDES requirements, all pollutants are classified as either

Table 3. EPA list of 129 priority pollutants and the relative frequency of these materials in industrial wastewaters*

Percent of samples†	Number of industrial categories‡		Percent of samples†	Number of industrial categories‡	
31 are purgeable organics					
1.2	5	Acrolein	2.1	5	1,2-Dichloropropane
2.7	10	Acrylonitrile	1.0	5	1,3-Dichloropropene
29.1	25	Benzene	34.2	25	Methylene chloride
29.3	28	Toluene	1.9	6	Methyl chloride
16.7	24	Ethylbenzene	0.1	1	Methyl bromide
7.7	14	Carbon tetrachloride	1.9	12	Bromoform
5.0	10	Chlorobenzene	4.3	17	Dichlorobromomethane
6.5	16	1,2-Dichloroethane	6.8	11	Trichlorofluoromethane
10.2	25	1,1,1-Trichloroethane	0.3	4	Dichlorodifluoromethane
1.4	8	1,1-Dichloroethane	2.5	15	Chlorodibromomethane
7.7	17	1,1-Dichloroethylene	10.2	19	Tetrachloroethylene
1.9	12	1,1,2-Trichloroethane	10.5	21	Trichloroethylene
4.2	13	1,1,2,2-Tetrachloroethane	0.2	2	Vinyl chloride
0.4	2	Chloroethane	7.7	18	1,2-*trans*-Dichloroethylene
1.5	1	2-Chloroethyl vinyl ether	0.1	2	bis(Chloromethyl) ether
40.2	28	Chloroform			
46 are base-extractable or neutral-extractable organic compounds					
6.0	9	{1,2-Dichlorobenzene 1,3-Dichlorobenzene 1,4-Dichlorobenzene	7.8	14	Pyrene
			10.6	16	{Phenanthrene Anthracene
0.5	5	Hexachloroethane	2.3	6	Benzo(a)anthracene
0.2	1	Hexachlorobutadiene	1.6	6	Benzo(b)fluoranthene
1.1	7	Hexachlorobenzene	1.8	6	Benzo(k)fluoranthene
1.0	6	1,2,4-Trichlorobenzene	3.2	8	Benzo(a)pyrene
0.4	3	bis(2-Chloroethoxy) methane	0.8	4	Indeno(1,2,3-c,d)pyrene
10.6	18	Naphthalene	0.2	4	Dibenzo(a,h)anthracene
0.9	9	2-Chloronaphthalene	0.6	7	Benzo(g,h,i)perylene
1.5	13	Isophorone	0.1	2	4-Chlorophenyl phenyl ether
1.8	9	Nitrobenzene	0	0	3,3′-Dichlorobenzidine
1.1	3	2,4-Dinitrotoluene	0.2	4	Benzidine
1.5	9	2,6-Dinitrotoluene	1.1	4	bis(2-Chloroethyl) ether
			0.8	7	1,2-Diphenylhydrazine
0.04	1	4-Bromophenyl phenyl ether	0.1	1	Hexachlorocyclopentadiene
41.9	29	bis(2-Ethylhexyl) phthalate	1.2	5	*N*-Nitrosodiphenylamine
6.4	12	Di-*n*-octyl phthalate	4.5	12	Acenaphthylene
5.8	15	Dimethyl phthalate	4.2	14	Acenaphthene
7.6	20	Diethyl phthalate	8.5	13	Butyl benzyl phthalate
18.9	23	Di-*n*-butyl phthalate	0.1	1	*N*-Nitrosodimethylamine
5.7	11	Fluorene	0.1	2	*N*-Nitrosodi-*n*-propylamine
7.2	12	Fluoranthene	1.4	6	bis(2-Chloroisopropyl) ether
5.1	9	Chrysene			

*Footnotes appear on page 390.

(continued)

Table 3. EPA list of 129 priority pollutants and the relative frequency of these materials in industrial wastewaters* (cont.)

Percent of samples†	Number of industrial categories‡		Percent of samples†	Number of industrial categories‡	
11 are acid-extractable organic compounds					
26.1	25	Phenol	1.9	8	*p*-Chloro-*m*-cresol
2.3	11	2-Nitrophenol	2.3	10	2-Chlorophenol
2.2	9	4-Nitrophenol	3.3	12	2,4-Dichlorophenol
1.6	6	2,4-Dinitrophenol	4.6	12	2,4,6-Trichlorophenol
1.1	6	4,6-Dinitro-*o*-cresol	5.2	15	2,4-Dimethylphenol
6.9	18	Pentachlorophenol			
26 are pesticides or PCB's					
0.3	3	α-Endosulfan	0.3	3	Heptachlor
0.4	4	β-Endosulfan	0.1	1	Heptachlor epoxide
0.2	2	Endosulfan sulfate	0.2	4	Chlordane
0.6	4	α-BHC	0.2	2	Toxaphene
0.8	6	β-BHC	0.6	2	Aroclor 1016
0.2	4	δ-BHC	0.5	1	Aroclor 1221
0.5	3	γ-BHC	0.9	2	Aroclor 1232
0.5	5	Aldrin	0.8	3	Aroclor 1242
0.1	3	Dieldrin	0.6	2	Aroclor 1248
0.04	1	4,4′-DDE	0.6	3	Aroclor 1254
0.1	2	4,4′-DDD	0.5	1	Aroclor 1260
0.2	2	4,4′-DDT	1	1	2,3,7,8-Tetrachlorodibenzo-*p*-dioxin (TCDD)
0.2	3	Endrin			
0.2	2	Endrin aldehyde			
13 are metals					
18.1	20	Antimony	16.5	20	Mercury
19.9	19	Arsenic	34.7	27	Nickel
14.1	18	Beryllium	18.9	21	Selenium
30.7	25	Cadmium	22.9	25	Silver
53.7	28	Chromium	19.2	19	Thallium
55.5	28	Copper	54.6	28	Zinc
43.8	27	Lead			
Miscellaneous					
33.4	19	Total cyanides	Not available		Total phenols
Not available		Asbestos (fibrous)			

*From L. H. Keith and W. A. Telliard, Priority pollutants, *Environ. Sci. Tech.*, 13:416–423, April 1979.

†The percent of samples represents the number of times that this compound was found in all samples in which it was analyzed for, divided by the total.

‡A total of 32 industrial categories and subcategories were analyzed for organics and 28 for metals.

conventional, toxic, or nonconventional. Conventional pollutants include biochemical oxygen demand (BOD), total suspended solids (TSS), and pH, with others to be proposed as data indicate the need.

Best conventional pollutant control technology (BCT) is required for conventional pollutants by July 1, 1984. Best available technology economically achievable (BAT) is required for toxic and nonconventional pollutants by the same date. Table 3 lists 129 priority pollutants that appeared on the EPA toxics list in the *Federal Register* [43(164): 4108, February 1978] and gives the relative frequency of these materials in industrial wastewaters. Nonconventional pollutants are those that are neither toxic nor conventional, such as NO_3^-, phosphorus, and oil and grease.

Effluent guidelines and standards for 42 industrial categories have been published in the *Code of Federal Regulations* (40CFR401). Each category is broken down into subcategories with guidelines given for more than 400 specific processes (Table 4). Different levels of pollutants are allowed in the effluents, depending on: (1) industrial subcategory; (2) control technology required (BCT or BAT); (3) whether the source is new or existed prior to the regulations (new sources are more severely regulated); (4) where the effluents are discharged (effluent levels discharged into publicly owned treatment works may be different from direct discharges into navigable waters). *See* WATER POLLUTION.

GASEOUS WASTES

Industry produces an estimated 17% of the air pollution resulting from gaseous waste disposal. Most gaseous wastes contain known contaminants which can be absorbed, adsorbed, oxidized, separated, or otherwise removed by existing conventional methods. However, it is not the type of contaminant so much as the total quantity of gas which must be handled that makes the gaseous

Table 4. Pollutants associated with various industries*

Point source category	Number of subcategories	Allowed discharges or parameters which may be changed
Dairy products processing	12	All subcategories—BOD_5, TSS, pH
Grain mills	10	All subcategories—BOD_5, TSS, pH
Canned and preserved fruits and vegetables	8	All subcategories—BOD_5. TSS, pH
Canned and preserved seafood	33	All subcategories—TSS, pH, oil and grease; seven subcategories—BOD_5, and three debris larger than 0.5 in. (12.7 mm)
Sugar processing	8	Two subcategories may not discharge pollutants into navigable waters; all others—BOD_5, TSS and pH; one subcategory—coliform and high-temperature
Textile industry	7	All subcategories—BOD_5, TSS, COD, fecal coliform and pH; six subcategories—phenol, color, S and Cr
Cement manufacture	3	All subcategories—TSS and pH; two subcategories—temperature
Electroplating	6	24 separate pollutants are listed; consult 40CFR413 to relate subcategory to pollutant
Organic chemicals manufacture	2	BOD_5, TSS, pH, COD
Inorganic chemicals manufacture	63	The information on this large number of subcategories is too varied to itemize in this table; consult 40CFR415
Plastics and synthetics	21	pH
Soap and detergent manufacture	19	All subcategories—BOD_5, TSS, pH, COD, oil and grease; 11 subcategories—surfactants
Fertilizer manufacture	9	The following pollutants may be discharged by industries in the various subcategories; TSS, pH, P, F, TSS, NH_3(N), org. N, NO_3(N); consult 40CFR418 for details
Petroleum refining	5	BOD_5, TSS, pH, COD, oil and grease, phenols, NH_3(N), S, Cr, Cr(VI)
Iron and steel manufacture	26	The information on these subcategories is too varied to itemize in this table; consult 40CFR420
Nonferrous metals manufacture	8	The information on these subcategories is too varied to itemize in this table; consult 40CFR421
Phosphate manufacture	6	All subcategories—TSS, pH and total P; four subcategories—F; one subcategory—As and elemental P
Steam electric power generating	4	One subcategory—TSS and pH; all others—TSS, pH, PCB, oil and grease, Cu, Fe, Cl, Zn, Cr and P
Ferroalloy manufacture	7	The most common pollutants are TSS, pH, Mn and Cr; consult 40CFR424 for details
Leather, tanning industry	6	All subcategories—BOD_5, TSS, pH, Cr, oil and grease, TKN, fecal coliform and S
Glass manufacture	13	All subcategories—TSS and pH; many may discharge oil; consult 40CFR426 for details
Asbestos manufacture	11	Two subcategories may not discharge pollutants into navigable waters; all others—TSS and pH; two—COD
Rubber manufacture	11	All subcategories—TSS, pH, oil and grease. Other pollutants which may be discharged are BOD_5, COD, Pb, Cr and Zn, in some subcategories
Lumber products processing	19	Eight subcategories may not discharge pollutants into navigable waters; BOD_5, TSS, pH, COD, oil and grease, phenols, settleable solids and debris, some subcategories
Pulp, paper and paperboard	23	All subcategories—BOD_5, TSS and pH
Builders paper and roofing	1	BOD_5, TSS, pH, settleable solids
Meat products	10	All subcategories—BOD_5, TSS, pH, oil and grease, and fecal coliform; most subcategories—NH_3
Coal mining	3	All subcategories—TSS, pH and Fe; two subcategories—Mn
Mineral mining and processing	38	Guidelines for 17 of these subcategories have not yet been promulgated; 16 may not discharge pollutants into navigable waters; five—TSS and pH; one—Fe
Pharmaceutical manufacture	5	All subcategories—BOD_5, pH and COD; three—TSS
Ore mining and dressing	7	All subcategories—TSS and pH; COD, Fe, Cu, Zn, Pb, Hg, Cd, CN, Al, As, Ra226, U, and Ni may be discharged by industries in various subcategories
Paint formulating	1	This subcategory may not discharge pollutants into navigable waters
Ink formulating	1	This subcategory may not discharge pollutants into navigable waters
Gum and wood chemicals manufacture	6	One subcategory may not discharge pollutants into navigable waters; all others—BOD_5, TSS and pH
Pesticide chemicals manufacture	5	Two subcategories may not discharge pollutants into navigable waters; all others—BOD_5, TSS, pH, COD and total pesticides; NH_3(N) and phenol—some subcategories
Explosives manufacture	2	Both subcategories—TSS and pH; BOD_5, COD, oil and grease—one subcategory
Carbon black manufacture	4	None of the subcategories may discharge pollutants into navigable waters
Photographic	1	pH, Ag, CN
Hospital	1	BOD_5, TSS, pH

*From R. D. McLaughlin et al., LBL's Pollution Instrumentation Comparability Program, *Pollution Eng*, pp. 33–38, April 1979.

waste disposal problem a difficult one. For example, a single 100,000 lb/hr (12.6 kg/s) boiler may emit 40,000 to 50,000 scfm (standard cubic feet per minute; 18.9 to 23.6 m^3s of gas at standard pressure and temperature) of flue gas. To remove the sulfur oxides resulting from fuel combustion, the physical size and cost of the treatment apparatus are dictated not by the amount of sulfur present in the flue gas, but by the total volume of flue gas that must be treated. By the end of 1977, 95% of the major polluters were in full compliance with EPA emission limits on pollutants from stacks. However, for stationary sources to be in full compliance with the control technology expected by 1985, the estimated cost to industry will be $35- to $50,000,000,000.

The criteria pollutants for which there are National Ambient Air Quality Standards (NAAQS) are: sulfur dioxide (SO_2), carbon monoxide (CO), hydrocarbons (HC), total suspended particulates (TSP), nitrogen dioxide (NO_2), and photochemical oxidants (Ox). These resulted from the Clean Air Act Amendments of 1977, and are expected to be controlled by industries that, without any air-pollution control equipment, could potentially emit 100 metric tons/year/pollutant.

These industries are the following:

1. Fossil-fuel fired steam electric plants of more than 250,000,000 Btu/hr (73 MW/hr) heat input
2. Coal cleaning plants (with thermal dryers)
3. Kraft pulp mills
4. Portland cement plants
5. Primary zinc smelters
6. Iron and steel mills
7. Primary aluminum ore reduction plants
8. Primary copper smelters
9. Municipal incinerators capable of charging more than 250 tons of refuse per day
10. Hydrofluoric acid plants
11. Sulfuric acid plants
12. Nitric acid plants
13. Petroleum refineries
14. Lime plants
15. Phosphate rock processing plants
16. Coke oven batteries
17. Sulfur recovery plants
18. Carbon black plants (furnace process)
19. Primary lead smelters
20. Fuel conversion plants
21. Sintering plants
22. Secondary metal production facilities
23. Chemical process plants
24. Fossil-fuel boilers of more than 250,000,000 Btu/hr heat input
25. Petroleum storage and transfer facilities with a capacity exceeding 300,000 bbl
26. Taconite ore processing facilities
27. Glass fiber processing plants
28. Charcoal production facilities

Contact with local regulatory agencies or trade associations is necessary for information on current NAAQS applicable for the area in which the industry is located. In addition to the NAAQS pollutants, a national ambient air quality standard for lead has been set (1978) at 1.5 μg per cubic meter of air. The cost of compliance by 1982 by nonferrous metal smelters suddenly jumped to an expected $530,000,000. The United States has six primary lead smelters, 50 secondary lead smelters, and 16 primary copper smelters. These sources will require substantial controls on their current emissions, and EPA concedes that many may not be able either technically or economically to meet the standard. *See* AIR POLLUTION.

WASTE MANAGEMENT

The following industrial waste management strategy has been suggested by EPA: (1) Minimize the quantity of waste generated by modifying the industrial process involved. (2) Concentrate the waste at the source (using evaporation, precipitation, and so on) to reduce handling and transport costs. (3) If possible, transfer the waste "as is," without reprocessing, to another facility that can use it as feedstock. (4) When a transfer "as is" is not possible, reprocess the waste for material recovery. (5) When material recovery is not possible, incinerate the waste for energy recovery and for destruction of hazardous components; or if the waste cannot be incinerated, detoxify and neutralize it through chemical treatment. (6) Use carefully controlled land disposal only for what remains.

Recycling industrial wastes. Almost all materials are recyclable, but not all materials are economically recyclable. In some, more energy will be expended in recovery than the recovered value warrants. With other wastes, geography may play a key role where optimum utilization is at a different location but transportation costs negate the opportunity. The current trend in regulatory action seeks to aid resource recovery and recycling by imposing more stringent restrictions on disposal, thereby increasing disposal costs and making alternate options more attractive.

Some industries have many years of experience in recycling because of the energy that can be conserved. Aluminum can be used again and again, each time saving about 95% of the energy required to make it from bauxite ore. Reynolds Aluminum Company reported a savings of about 1,100,000,000 kWh of electricity in 1978 because of recycling efforts. There is a 74% energy savings when iron and steel scrap is recycled instead of using virgin materials. Each 1,000,000 metric tons of ferrous scrap used instead of iron ore to make new products saves the energy equivalent of more than 3,000,000 bbl (477,000 m^3) of crude oil.

Hazardous wastes are less commonly recycled. There are five broad categories of hazardous waste being generated:

1. Unused materials of commerce, such as excess or outdated chemicals, or new materials not meeting the high quality control requirements of the purchasing industry. These are the easiest materials to recycle.

2. Process wastes that are reusable without treatment, such as cardboard for pulping, copper or other metal solutions for metal recovery, oils that can be used for fuels, and a variety of other materials that can be reused as industrial feedstocks.

3. Materials that are reusable with treatment. For example, baghouse dust from scrap steel processors containing up to 25% zinc oxide can be combined with a waste sulfuric acid to make galva-

nizer's pickle acid. The spent pickle liquor containing 8–10% zinc sulfate and some iron salts is then usable in citrus orchards or other agricultural areas deficient in zinc.

4. Waste materials that have little or no value and require some ingenuity to convert them to useful materials. An extremely dangerous mixed-acid-etch waste contains 60% nitric acid, 20% hydrofluoric acid, and 20% acetic acid. Mixed with the waste stream from the manufacture of acetylene consisting of acetylene lime plus calcium hydroxide, the acids are neutralized, making calcium nitrate, calcium acetate, and calcium fluoride. The first two products are in demand as agricultural additives, and the third is useful in cement manufacture and in hydrofluoric acid manufacture.

5. Materials that no one wants to recover. This category includes chemical carcinogens, PCBs (polychloride biphenyls), DDT, and other materials which mostly will not conveniently degrade when buried. Many of these materials should be destroyed, and thus can perhaps have some value as fuels.

Waste exchange. Industrial waste exchanges exist because material considered trash by one industry may be useful scrap feedstock for another industry. Two basic types of organizations exist to promote the transfer of waste materials from the generator to the potential user: the information clearing house and materials exchange. The former is generally a nonprofit organization, exchanging only information sufficient to make generators and users aware of each other while protecting the anonymity of the generator until contact with the potential user is desired. The latter buys, analyzes, identifies new uses, treats the waste if necessary, and actively seeks buyers to resell at a profit. In either case, money is made, raw materials and energy are saved, and less waste is fed into the environment.

The first exchanges began in Europe in 1972, and European waste "bourses" are successfully operating in Germany, Italy, Switzerland, Belgium, Great Britain, and the Scandinavian countries. Canada also has a successful nationwide Canadian Waste Materials Exchange.

The Midwest Industrial Waste Exchange, modeled after the European bourses, was initiated by the St. Louis, MO, Regional Commerce and Growth Association in 1975 as the first United States clearing house. By 1979 there were approximately 20 waste exchanges in the United States and about 15 in other parts of the world. In general, most reported transfers have occurred from larger companies using continuous processes to smaller companies using batch processes, from basic chemical manufacturers to formulators, and from industries with high purity requirements to those with lower purity requirements.

Waste disposal. As much as 80% of the hazardous wastes produced by industry in 1977 was being improperly disposed of in nonsecure ponds, land-

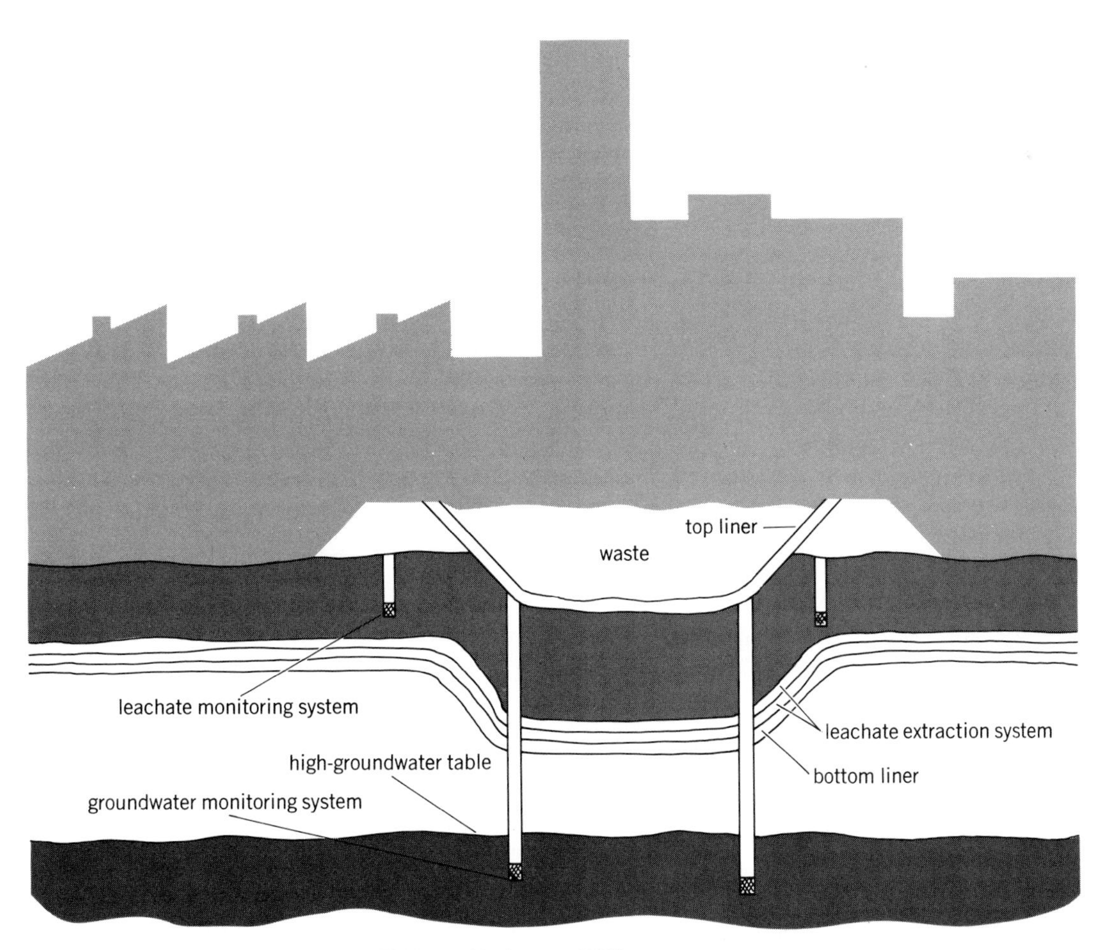

Fig. 2. Hazardous waste landfill. *(From EPRI J., p. 38, October 1978)*

fills, and lagoons, according to EPA. An additional 10% was being incinerated without proper controls. Only 10% was being managed acceptably as compared with proposed Federal standards for controlled incineration, secure landfills, and recovery. About 80% of the hazardous materials is disposed of on the manufacturer's property.

Seven methods predominate among the alternatives for ultimate waste disposal.

Incineration. This involves the oxidative conversion of combustible material to harmless gases suitable for atmospheric release. Undesired gaseous products, such as HCl, SO_2, and NO_x, must be removed prior to release.

Dispersal into contiguous water bodies. Historically this practice has been widely employed, but regulatory action has now curtailed it for most industries. Usually, the waste is temporarily stored in a lagoon where acids or bases are neutralized and suspended solids are permitted to settle prior to final discharge.

Ocean dumping. The Marine Protection, Research and Sanctuaries Act of 1972 established a permit system for control of ocean dumping, and subsequent efforts have been aimed at minimizing or ending all dumping of materials in the ocean. Usually the material is conveyed by barge out to sea and released directly or within containers.

Lagooning. Lagoons have been traditionally employed for interim storage and water treatment prior to release to contiguous water bodies. Permanent lagooning is used when the solid materials tend not to settle, as is the case with phosphate slimes; or when the material would have undesirable properties if not covered with water—for example, red mud from alumina processing creates a dust problem when dry, and the phosphorus metal in phossy water bursts into flame upon contact with air.

Sanitary landfill. This pertains to the burial of nonhazardous wastes under controls sufficient to preclude degradation of the surrounding environment. Proper site selection, exclusion of hazardous materials, and frequent soil cover are some of the required controls.

Hazardous waste landfill. This is an extension of the sanitary landfill to allow the disposal of hazardous materials (Fig. 2). The landfill should be lined with clay, plastic, or other impervious material to prevent the waste from moving through the soil and reaching water sources, and it should be at least 500 ft (152 m) from any water source. Disposed material is carefully identified, fire fighting and other safety precautions are specified, and surface water channeling and leachate treatment are required. Constant monitoring of waste movement by wells, in addition to daily visual checks, is required.

Subsurface injection. The industrial wastes are slurried and then pumped into underground cavities. Rules, mandated by the Safe Drinking Water Act, propose construction, operating, monitoring, and reporting requirements for each of five classes of underground injection wells: class I—industrial (including hazardous), municipal, and nuclear disposal wells that inject substantially below all drinking water sources in the area; class II—injection wells associated with oil and gas production; class III—special-process injection wells; class IV—wells used by generators of hazardous wastes and firms involved in hazardous waste management for injecting wastes into or above drinking water sources (EPA has proposed the complete ban of these facilities); and class V—all other injection wells.

Incompatible wastes. Many wastes, when mixed with others at a disposal facility, can produce hazardous situations through heat generation, fires, explosion, or release of toxic substances (Table 5). In 1978 a 19-year-old truck driver was

Table 5. Potentially incompatible wastes*

Group 1A	*Group 1B*
Acetylene sludge	Acid sludge
Alkaline caustic liquids	Acid and water
Alkaline cleaner	Battery acid
Alkaline corrosive liquids	Chemical cleaners
Alkaline corrosive battery fluid	Electrolyte, acid
Caustic wastewater	Etching acid liquid or solvent
Lime sludge and other corrosive alkalis	Pickling liquor and other corrosive acids
Lime wastewater	Sludge acid
Lime and water	Spent acid
Spent caustic	Spent mixed acid
	Spent sulfuric acid

Potential consequences: Heat generation, violent reaction.

Group 2A	*Group 2B*
Asbestos waste and other toxic wastes	Cleaning solvents
Beryllium wastes	Data-processing liquid
Unrinsed pesticide containers	Obsolete explosives
Waste pesticides	Petroleum waste
	Refinery waste
	Retrograde explosives
	Solvents
	Waste oil and other flammable and explosive wastes

Potential consequences: Release of toxic substances in case of fire or explosion.

Table 5. Potentially incompatible wastes* (cont.)

Group 3A	*Group 3B*
Aluminum Beryllium Calcium Lithium Magnesium Potassium Sodium Zinc powder and other reactive metals and metal hybrides	Any waste in group 1A or 1B
Potential consequences: Fire or explosion; generation of flammable hydrogen gas.	
Group 4A	*Group 4B*
Alcohols Water	Any concentrated waste in group 1A or 1B Calcium Lithium Metal hydrides Potassium Sodium SO_2Cl_2, $SOCl_2$, PCl_3, CH_3SiCl_3 and other water-reactive wastes
Potential consequences: Fire, explosion or heat generation; generation of flammable or toxic gases.	
Group 5A	*Group 5B*
Alcohols Aldehydes Halogenated hydrocarbons Nitrated hydrocarbons and other reactive organic compounds and solvents Unsaturated hydrocarbons	Concentrated group 1A or 1B wastes Group 3A wastes
Potential consequences: Fire, explosion or violent reaction.	
Group 6A	*Group 6B*
Spent cyanide and sulfide solutions	Group 1B wastes
Potential consequences: Generation of toxic hydrogen cyanide or hydrogen sulfide gas.	
Group 7A	*Group 7B*
Chlorates and other strong oxidizers Chlorine Chlorites Chromic acid Hypochlorites Nitrates Nitric acid, fuming Perchlorates Permanganates Peroxides	Acetic acid and other organic acids Concentrated mineral acids Group 2B wastes Group 3A wastes Group 5A wastes and other flammable and combustible wastes
Potential consequences: Fire, explosion or violent reaction.	

*The mixing of a group A waste with a group B waste may have the potential consequence as noted.
SOURCE: From N. P. Cheremisinoff et al., *Industrial Hazardous Wastes Impoundment*, p. 13, 1979.

asphyxiated by hydrogen sulfide fumes as he discharged hazardous wastes from his truck into an open chemical pool in Iberville Parish, LA.

The following guidelines from the California Department of Health can be used for handling and disposal of incompatible wastes:

1. Incompatible wastes should not be mixed in the same transportation or storage container.

2. A waste should not be added to an unwashed transportation or storage container that previously contained an incompatible waste.

3. Incompatible wastes should not be combined in the same pond, landfill, soil-mixing area, well, or burial container. An exception is the controlled neutralization of acids and alkalis in disposal areas. Containers that hold incompatible wastes should be well separated by soil or refuse when they are buried. Ideally, separate disposal areas should be maintained for noncompatible wastes.

4. Incompatible wastes should not be incinerated together. An exception is the controlled incineration of pesticides and other toxic substances with flammable solvents.

Industrial discharges to POTWs. There are approximately 50,000–55,000 industrial waste dischargers to publicly owned treatment works (POTW) encompassed in 21 industries. Because a POTW is required to meet permit conditions under NPDES, potential difficulties may be encountered whenever industrial pollutants enter the muncipal sewer system. These problems include the following: (1) Sludge discharges and high concentrations of certain pollutants can interfere with treatment of normal domestic and industrial wastes. (2) Metals and toxic pollutants (Table 4) can limit the sludge disposal alternatives of the POTW and thus increase the cost of sludge disposal facilities. (3) Pass-through pollutants not receiving adequate treatment from a POTW may be discharged to the environment in violation of Federal, state, and local regulations.

The majority of dischargers whose pollutants represent environmental concern are contained in the electroplating, auto, and laundry industries. EPA estimates that the cost to the approximately 6600 electroplaters alone will be an additional $461,000,000 for pretreatment under the regulations needed to safeguard POTWs.

[STEPHEN L. LAW]

Bibliography: N. P. Cheremisinoff et al., *Industrial and Hazardous Wastes Impoundment*, 1979; G. F. Hoffnagle and R. Dunlap, Industrial Expansion and the 1977 Clean Air Act Amendments, *Pollution Engineering*: *Yearbook and Product Reference Guide*, pp. 36–43, 1978; H. F. Lund, *Industrial Pollution Control Handbook*, 1971; R. B. Pojasek, *Toxic and Hazardous Waste Disposal*, 1979; P. W. Powers, *How to Dispose of Toxic Substances and Industrial Wastes*, 1976.

Insect control, biological

The term biological control was proposed in 1919 to apply to the use or role of natural enemies in insect population regulation. The enemies involved are termed parasites (parasitoids), predators, or pathogens. This remains preferred usage, although other biological methods of insect control have been proposed or developed, such as the release of mass-produced sterile males to mate with wild females in the field, thereby greatly reducing or suppressing the pests' production of progeny. Classical biological control is an ecological phenomenon which occurs everywhere in nature without aid from, or sometimes even understanding by, man. However, man has utilized the ecological principles involved to develop the field of applied biological control of insects, and the great majority of practical applications have been achieved with insect pests. Additionally, such diverse types of pest organisms as weeds, mites, and certain mammals have been successfully controlled by use of natural enemies.

History. Man observed the action of predacious insects early in his agricultural history, and a few crude attempts to utilize predators have been carried on for centuries. However, the necessary understanding of biological and ecological principles, especially those of population dynamics, did not begin to emerge until the 19th century. The first great applied success in biological control of an insect pest occurred 2 years after the importation into California from Australia in 1888–1889 of the predatory vedalia lady beetle. This insect feeds on the cottony-cushion scale, a notorious citrus pest that was destroying orange trees at that time, and the vedalia was credited with saving the citrus industry. This successful control firmly established the field or discipline of applied biological control as it is known today. *See* POPULATION DYNAMICS.

Ecological principles. Natural enemy populations have a feedback relationship to prey populations, termed density-dependence, which results in the increasing or decreasing of one group in response to changes in the density of the other group. Such reciprocal interaction prevents indefinite increase or decrease and thus results in the achievement of a typical average density or "balance." However, this balance may be either at high or low levels, depending on the inherent abilities of the natural enemies. If the natural enemy is highly density-dependent, it can regulate the prey population density at very low levels. An effective enemy achieves regulation by rapidly responding to any increase in the prey population in two different ways: killing more prey by increased feeding or parasitism, and producing more progeny for the next generation. The net effect is a more rapid increase in prey mortality as prey population tends to increase, so that first the trend is stopped and then it becomes reversed, lowering the prey population. This results in relaxing the pressure caused by the enemy so that ultimately the prey population is enabled to increase again, and the cycle is repeated over and over.

Applications of classical method. Since the vedalia beetle controlled the cottony-cushion scale, many projects in applied biological control have been undertaken and a large number of successes achieved. There are about 300 recorded cases of applied biological control of insects in 70 countries. Of these recorded cases over 85 have been so completely successful that insecticides are no longer required. In the other cases the need for chemical treatment has been more or less greatly reduced. Also, there are about 47 cases of biological control of weeds by insect natural enemies. A few of the many outstanding successes include the biological control of Florida red scale in Israel and subsequently in Mexico, Lebanon, South Africa, Brazil, Peru, Texas, and Florida; Oriental fruit fly in Hawaii; green vegetable bug in Australia; citrus blackfly and purple scale in Mexico; dictyospermum scale and purple scale in Greece; and olive scale in California.

All the cases of applied biological control mentioned above involved foreign exploration and importation and colonization of new exotic natural enemies of the insect pest in question. In the majority of cases the pest was an invader, being native to another country, and its natural enemies had been left behind. By searching the native habitat for effective enemies and sending them to the new home of the invader pest, the biological balance was reconstituted. This method is considered to be classical biological control, and its scientific application has been responsible for most of the outstanding results obtained. There are, however, two other major phases of biological control which are categorized under the headings of conservation and augmentation.

Conservation. This phase involves manipulation of the environment to favor survival and reproduction of natural enemies already established in the habitat, whether they be indigenous or exotic. In other words, adverse environmental factors are modified or eliminated, and requisites which are lacking such as food or nesting sites may be provided. Even though potentially effective natural enemies are present in a habitat, adverse factors may so affect them as to preclude attainment of satisfactory biological control. Insecticides commonly produce such adverse effects, causing so-called upsets in balance or pest population explosions.

Augmentation. This phase concerns direct manipulation of established enemies themselves. Potentially effective enemies may be periodically decimated by environmental extremes or other adverse factors which are not subject to man's control. For example, low winter temperatures may seriously decrease certain enemy populations each year. The major means of solving such problems has involved laboratory mass culture of enemies and their periodic colonization in the field, generally after the adverse period has passed. This practice is gaining rapidly in application. *Trichogramma* sp., common egg parasites of lepidopterous pests, have been utilized in this manner with reportedly good results in many countries, and various microorganisms likewise have been successfully used.

Advantages over chemical control. Although research and development costs are modest for biological control projects, they are high for new insecticides. Biological control application costs are minimal and nonrecurring, except where periodic colonization is utilized, whereas insecticide costs are high annually. There are no environmental pollution problems connected with biological control, whereas insecticides cause severe problems related to toxicity to man, wildlife, birds, and fish, as well as causing adverse effects in soil and water. Biological control causes no upsets in the natural balance of organisms, but these upsets are common with chemical control; biological control is permanent, chemical control is temporary, usually one to many annual applications being necessary. Additionally, pests more and more frequently are developing resistance or immunity to insecticides, but this is not a problem with pests and their enemies.

Both biological control and chemical control have restrictions as far as general applicability to the control of all pest insect species is concerned, although the application of biological control remains greatly underdeveloped compared to chemical control. *See* INSECTICIDE.

[PAUL DE BACH]

Bibliography: R. R. Askew, *Parasite Insects*, 1971; K. F. Baker and R. J. Cook, *Biological Control of Plant Pathogens*, 1974; C. P. Clausen, *Entomophagous Insects*, 1962; H. C. Coppel and J. W. Mertins, *Biological Insect Press Suppression*, 1977; P. DeBach (ed.), *Biological Control of Insect Pests and Weeds*, 1964; P. DeBach, *Biological Control by Natural Enemies*, 1974; C. B. Huffaker (ed.), *Biological Control*, 1974; W. W. Kilgore and R. L. Doutt, *Pest Control: Biological, Physical and Selected Chemical Methods*, 1967; L. A. Swan, *Beneficial Insects*, 1964; R. van den Bosch and P. S. Messenger, *Biological Control*, 1973.

Insecticide

A material used to kill insects and related animals by disruption of vital processes through chemical action. Insecticides may be inorganic or organic molecules. The principal source is from chemical manufacturing, although a few are derived from plants.

Insecticides are classified according to type of action as stomach poisons, contact poisons, residual poisons, systemic poisons, fumigants, repellents, attractants, insect growth regulators, or pheromones. Many act in more than one way. Stomach poisons are applied to plants so that they will be ingested as insects chew the leaves. Contact poisons are applied in a manner to contact insects directly, and are used principally to control species which obtain food by piercing leaf surfaces and withdrawing liquids. Residual insecticides are applied to surfaces so that insects touching them will pick up lethal dosages. Systemic insecticides are applied to plants or animals and are absorbed and translocated to all parts of the organisms, so that insects feeding upon them will obtain lethal doses. Fumigants are applied as gases, or in a form which will vaporize to a gas, so that they can enter the insects' respiratory systems. Repellents prevent insects from closely approaching their hosts. Attractants induce insects to come to specific locations in preference to normal food sources. Insect growth regulators are generally considered to act through disruption of biochemical systems or processes associated with growth or development, such as control of metamorphosis by the juvenile hormones, regulation of molting by the steroid molting hormones, or regulation of enzymes responsible for synthesis or deposition of chitin. Pheromones are chemicals which are emitted by one sex, usually the female, for perception by the other, and function to enhance mate location and identification; pheromones are generally highly species-specific.

Inorganic insecticides. Prior to 1945, large volumes of lead arsenate, calcium arsenate, paris green (copper acetoarsenite), sodium fluoride, and cryolite (sodium fluoaluminate) were used. The potency of arsenicals is a direct function of the percentage of metallic arsenic contained. Lead arsenate was first used in 1892 and proved effective as a stomach poison against many chewing insects. Calcium arsenate was widely used for the control of cotton pests. Paris green was one of the first stomach poisons and had its greatest utility against the Colorado potato beetle. The amount of available water-soluble arsenic governs the utility of arsenates on growing plants, because this fraction will cause foliage burn. Lead arsenate is safest in this respect, calcium arsenate is intermediate, and paris green is the most harmful. Care must be exercised in the application of these materials to food and feed crops because they are poisonous to humans and animals as well as to insects.

Sodium fluoride has been used to control chewing lice on animals and poultry, but its principal application has been for the control of household insects, especially roaches. It cannot be used on plants because of its extreme phytotoxicity. Cryolite has found some utility in the control of the

Mexican bean beetle and flea beetles on vegetable crops because of its low water solubility and lack of phytotoxicity.

The use of inorganic insecticides has declined to almost nil. Domestic production of inorganic arsenicals has apparently ceased, with current uses being supplied from existing stocks and by some limited importation. Probably the largest current uses are as constituents of formulations for wood impregnation and preservation.

Organic insecticides. These began to supplant the arsenicals when DDT [1,1,1-trichloro-2,2-bis-(*p*-chlorophenyl)ethane] became available to the public in 1945. During World War II, the insecticidal properties of γ-benzenehexachloride [γ-1,2,-3,4,5,6-hexachlorocyclohexane or lindane] were discovered independently in England and France. Historically, DDT and lindane were the two largest-volume organic insecticides. Since 1968, however, in the United States all nongovernmental and nonprescription uses of DDT have been canceled. Coincidentally, the use of lindane has declined both through development of resistant strains of insects and through official restriction.

Other chlorinated hydrocarbon insecticides which were available by or during the early 1970s included:

TDE[2,2-bis(*p*-chlorophenyl)-1,1-dichloroethane], methoxychlor [1,1′-(2,2,2-trichloroethylidene) bis-4-methoxybenzene]
Dilan [mixture of 1,1-di-(4-chlorophenyl)-2-nitrobutane and 1,1-di-(4-chlorophenyl)-2-nitropropane]
chlordane [1,2,4,5,6,7,8,8-octachloro-3*a*,4,7,7*a*-tetrahydro-4,7-methanoindane]
heptachlor [1,4,5,6,7,8,8-heptachloro-3*a*,4,7,7*a*-tetrahydro-4,7-methanoindene]
aldrin [1,2,3,4,10,10-hexachloro-1,4,4*a*,5,8,8*a*-hexahydro-1,4-*endo-exo*-5,8-dimethanonaphthalene]
dieldrin [1,2,3,4,10,10-haxachloro-6,7-epoxy-1,4,4*a*,5,6,7,8,8*a*-octahydro-1,4-*endo-exo*-5,8-dimethanonaphthalene]
endrin [1,2,3,4,10,10-hexachloro-6,7-epoxy-1,4,4*a*,5,6,7,8,8*a*-octahydro-1,4-*endo-endo*-5,8-dimethanonaphthalene]
toxaphene [camphene plus 67–69% chlorine]
endosulfan [6,7,8,9,10,10-hexachloro-1,5,5*a*,6,9,9*a*-hexahydro-6,9-methano-2,4,3-benzodioxathiepin 3-oxide]
Kepone [decachlorooctahydro-1,3,4-metheno-2*H*-cyclobuta(*cd*)pentalen-2-one]
mirex [dodecachlorooctahydro-1,3,4-metheno-1*H*-cyclobuta(*cd*)pentalene]
Perthane [1,1-dichloro-2,2-bis(*p*-ethylphenyl)-ethane]

Registrations for all uses of TDE were canceled in 1971 because it is a metabolite of DDT. Methoxychlor can be used to control insect pests associated with dairy cattle, beef cattle, and some other farm animals, crop plants, ornamental plants, and buildings. Dilan, although once having some favor for control of the Mexican bean beetle, no longer is of commercial significance. Chlordane was the first cyclopentadiene insecticide to achieve commercial status. It was once the most effective chemical available for the control of cockroaches. Related chemicals include heptachlor, aldrin, dieldrin, and endrin. The first three were particularly effective against insects inhabiting the soil. Endrin found considerable use for the control of insects attacking cotton. All were active against grasshoppers. As early as 1966, uses and registrations for aldrin and dieldrin were being curtailed. Most uses of aldrin and dieldrin were canceled by the U.S. Environmental Protection Agency (EPA) in 1975 because of their risks to human health and the environment. The principal remaining registered use for these two chemicals is for the control of subterranean termites. The production of aldrin has been stopped. Most uses of chlordane and heptachlor were similarly canceled in 1978. The principal uses still permitted for heptachlor are termite control and seed treatment, with registration of the latter to be canceled in mid-1983. A review of the registrations of uses for endrin was to be completed during 1979.

Toxaphene was the largest-volume insecticide used in 1976, with over 80% being applied to cotton. Its registrations, too, are being reviewed. Endosulfan has a broad spectrum of activity, but its major use is for the control of insects which attack cotton. Kepone was used to control cockroaches. Mirex was the pesticide of choice by the U.S. Department of Agriculture for the control of the imported fire ant in the Gulf Coast states. Unfortunately, Kepone is an exceedingly stable compound; its deleterious effects upon nontarget organisms and its high probability of being hazardous to human health were the basis for the cancellation of all uses of the chemical in 1976. Mirex is also a very stable organochlorine insecticide. Its structure is similar to that of Kepone, and mirex had special use, when incorporated in baits, as a replacement for Kepone for the control of imported fire ant. The environmental and potential health hazards of mirex, too, were judged to be untenable, and further use was prohibited after June 1978. Perthane had utility for control of a number of pests of vegetables and certain fruits and for the protection of woolens against clothes moth and carpet beetle larvae, but production of this insecticide has ceased.

Insect resistance. The resistance of insects to DDT was first observed in 1947 in the housefly. By the end of 1967, 91 species of insects had been proved to be resistant to DDT, 135 to cyclodienes, 54 to organophosphates, and 20 to other types of insecticides, including the carbamates. By 1975, 203 species were known to be resistant to DDT, 225 to cyclodienes, 147 to organophosphates, 36 to carbamates, and 35 to other insecticides. Since numerous insects are resistant to more than one type of compound, the total number of species involved is 364 worldwide. Almost every country has reported the presence of resistant strains of the housefly.

During 1957 and 1958, many growers of cotton in the Southern states changed from toxaphene and γ-BHC to organic phosphorus chemicals because of the resistance of the cotton boll weevil to chlorinated hydrocarbon insecticides. By 1967, resistant strains of the cotton bollworm and the tobacco budworm had developed and proliferated to the extent that in numerous areas the use of chlorinated hydrocarbon insecticides was of doubtful value.

During the same decade, larvae of three species of corn rootworms also developed resistance to three potent chlorinated insecticides—aldrin, dieldrin, and heptachlor—active against a wide spectrum of soil-inhabiting insects. The development of resistance to chlorinated hydrocarbon insecticides among several species of disease-transmitting mosquitoes not only continues to pose a threat to world health, but led a World Health Organization committee to report in 1976 that efforts to cope with vector-borne diseases and to eliminate malaria were being hampered most of all by problems arising from resistance to insecticides. Typhus was not a scourge during the 1970s even in the Far East, but the threat of its proliferation following development of resistant strains of lice still remains possible.

The need to replace the highly active chlorinated materials stimulated the development and use of organophosphorus and carbamate insecticides. Strains of several species developed which were resistant to widely used organophosphates. One of the notable examples was that of the cotton pest, *Heliothis virescens* (the tobacco budworm), and methyl parathion, for which laboratory tests showed a greater than 100-fold difference between susceptible and resistant strains. The production of cotton in parts of northeastern Mexico was practically abandoned and was seriously threatened in the lower Rio Grande Valley of Texas by the mid-1970s due to difficulties and costs of attempting to control this pest.

Certain of the synthetic pyrethroids have been shown under controlled laboratory conditions to induce the selection of resistant strains of mosquitoes. This knowledge has been obtained prior to their registration and release for large-scale commercial development. Perhaps with this forewarning, these products will be employed in the field in ways that will minimize selection for resistance and thus prolong their usefulness for insect control. The synthetic pyrethroids, like the natural pyrethrins, have a very high unit activity, but differ from the briefly lasting natural products in that they persist for moderate periods of time following application in the field. These two attributes—high activity and persistence—are important factors associated with the selection of resistant strains of insects, regardless of the chemical involved. According to early results from the field, the pyrethroids could be 5–10 times more active than some insecticides currently in use.

Organic phosphorus insecticides. The development of this type of insecticide paralleled that of the chlorinated hydrocarbons. Since 1947, many thousands of organophosphorus compounds have been synthesized in academic and industrial laboratories throughout the world for evaluation as potential insecticides. Parathion [*O,O*-diethyl *O*-(*p*-nitrophenyl) phosphorothioate] and methyl parathion [*O,O*-dimethyl *O*-(*p*-nitrophenyl) phosphorothioate] are estimated to have had a world production of 70,000,000 lb (32×10^6 kg) during 1966. Some 13,500,000 lb (6.1×10^6 kg) of them and 50,000,000 lb (23×10^6 kg) of other organophosphates were used just in the United States alone in 1977.

A great diversity of activity against insects is found among organophosphorus insecticides. Unfortunately, many can be quite poisonous to humans and other warm-blooded animals, as well as to insects. A few materials possess the desirable attribute of low mammalian toxicity, however. The gamut of available commercial products includes:

azinphosmethyl [*O,O*-dimethyl *S*-(4-oxo-1,2,3-benzotriazin-3(4*H*)-ylmethyl) phosphorodithioate]
bromophos [*O*(4-bromo-2,5-dichlorophenyl) *O,O*-dimethyl phosphorothioate]
carbophenothion [*S*-((*p*-chlorophenylthio) methyl) *O,O*-diethyl phosphorodithioate]
chlorpyrifos [*O,O*-diethyl *O*-(3,5,6-trichloro-2-pyridyl) phosphorothioate]
crotoxyphos [α-methylbenzyl 3-hydroxycrotonate dimethyl phosphate]
diazinon [*O,O*-diethyl *O*-(2-isopropyl-4-methyl-6-pyrimidyl) phosphorothioate]
dichlorvos [2,2-dichlorovinyl dimethyl phosphate]
EPN [*O*-ethyl *O*-(4-nitrophenyl) phenylphosphonothioate]
ethion [*S,S'*-methylene *O,O,O',O'*-tetraethyl phosphorodithioate]
ethoprop [*O*-ethyl *S,S*-dipropyl phosphorodithioate]
fensulfothion [*O,O*-diethyl *O*-(*p*-(methylsulfinyl)-phenyl) phosphorothioate]
fonofos [*O*-ethyl *S*-phenyl ethylphosphonodithioate]
malathion [diethyl mercaptosuccinate, *S*-ester with *O,O*-dimethyl phosphorodithioate]
methidathion [*S*-((2-methoxy-5-oxo-Δ^2-1,3,4-thiadiazolin-4-yl)methyl) *O,O*-dimethyl phosphorodithioate]
naled [1,2-dibromo-2,2-dichloroethyl dimethyl phosphate]
phosmet [*O,O*-dimethyl *S*-phthalimidomethyl phosphorodithioate]
stirofos [2-chloro-1-(2,4,5-trichlorophenyl)vinyl dimethyl phosphate]
sulprofos [*O*-ethyl *O*-(4-(methylthio)phenyl) *S*-propyl phosphorodithioate]
temephos [*O,O*-dimethyl phosphorothioate *O,O*-diester with 4,4'-thiodiphenol]
terbufos [*S*-[((1,1-dimethylethyl)thio)methyl] *O,O*-diethyl phosophorodithioate]
trichlorfon [dimethyl (2,2,2-trichloro-1-hydroxyethyl) phosphonate]

Schradan [di(tetramethylphosphorodiamidic) anhydride] was unique among organic insecticides in that it showed systemic properties when applied to plants. Its activity upon direct contact with insects was relatively low, but following spray application to plants, it was absorbed, transported, and altered chemically by enzymatic processes to one or more products which were toxic to pests, such as aphids, which suck juices from plants. Schradan was thus a selective insecticide because it killed pests which fed on fluids from treated plants but did not affect predators which lived on the surfaces and preyed only upon the pests; for example, aphids are the target pests, and ladybird beetles and their larvae which prey upon the aphids are not affected by the insecticide retained within the plant. Unfortunately, schradan was ultimately found to have environmental properties, the risks of which exceeded benefits to the extent that in

1976 all registrations for use in the United States were canceled.

As of early 1979, the following materials which exhibit a varied spectrum in degree and breadth of activity became available commercially for insect control through systemic action following application to plants:

acephate [*O,S*-dimethyl acetylphosphoramidothioate]
demeton [mixture of *O,O*-diethyl *S*-(and *O*)-2-((ethylthio)ethyl) phosphorothioates]
dicrotophos [3-hydroxy-*N,N*-dimethyl-*cis*-crotonamide dimethyl phosphate]
dimethoate [*O,O*-dimethyl *S*-(*N*-methylcarbamoylmethyl) phosphorodithioate]
disulfoton [*O,O*-diethyl *S*-(2-(ethylthio)ethyl) phosphorodithioate]
methamidophos [*O,S*-dimethyl phosphoramidothioate]
mevinphos [methyl 3-hydroxy-α-crotonate, dimethyl phosphate]
monocrotophos [3-hydroxy-*N*-methyl-*cis*-crotonamide dimethyl phosphate]
oxydemetonmethyl [*S*-(2-ethylsulfinyl)ethyl) *O*, *O*-dimethyl phosphorothioate]
phorate [*O,O*-diethyl *S*-((ethylthio)methyl) phosphorodithioate]
phosalone [*S*-((6-chloro-2-oxo-3-benzoxazolinyl) methyl) *O,O*-diethyl phosphorodithioate]
phosphamidon [2-chloro-*N,N*-diethyl-3-hydroxycrotonamide dimethyl phosphate]

Development of chemicals to control pests of animals through utilization of chemicals with systemic properties was quite slow during the years 1968-1978. Coumaphos [*O*-(3-chloro-4-methyl-2-oxo-2*H*-1-benzopyran-7-yl) *O,O*-diethyl phosphorothioate] was one of the first products to be discovered which possessed this property. In addition to coumaphos and dimethoate, chemicals currently useful as animal systemic insecticides include:

crufomate [4-*tert*-butyl-2-chlorophenyl methyl methylphosphoramidate]
famphur [*O*-[4-((dimethylamino)sulfonyl)phenyl] *O,O*-dimethyl phosphorothioate]
fenthion [*O,O*-dimethyl *O*-(4-(methylthio)-*m*-tolyl) phosphorothioate]
menazon [*S*-((4,6-diamino-*s*-triazin-2-yl)methyl) *O,O*-dimethyl phosphorodithioate]
ronnel [*O,O*-dimethyl *O*-2,4,5-trichlorophenyl phosphorothioate]

Activity of organic phosphate insecticides results from the inhibition of the enzyme cholinesterase, which performs a vital function in the transmission of impulses in the nervous system. Inhibition of some phenyl esterases occurs also. Inhibition results from direct coupling of phosphate with enzyme. Phosphorothionates are moderately active, but become exceedingly potent upon oxidation to phosphates.

Carbamate insecticides. The first commercial carbamate insecticide, carbaryl [1-naphthyl methylcarbamate], was introduced during the mid-1950s and continues to enjoy broad use not only in the United States but worldwide. The more important products available currently include:

aminocarb [4-dimethylamino-*m*-tolyl methylcarbamate]
bendiocarb [2,2-dimethyl-1,3-benzodioxol-4-yl methylcarbamate]
bufencarb [*m*-(1-ethylpropyl)phenyl methylcarbamate mixture (1-4) with *m*-(1-methylbutyl)-phenyl methylcarbamate]
dimetilan [1-(dimethylcarbamoyl)-5-methyl-3-pyrazolyl dimethylcarbamate]
dioxacarb [*o*-(1,3-dioxolan-2-yl)phenyl methylcarbamate]
formetanate [*N,N*-dimethyl-*N*′-[3-([(methylamino)carbonyl]oxy)phenyl] methanimidamide]
methiocarb [4-(methylthio)-3,5-xylyl methylcarbamate]
mexacarbate [4-dimethylamino-3,5-xylyl methylcarbamate]
promecarb [*m*-cym-5-yl methylcarbamate]
propoxur [*o*-isopropoxyphenyl methylcarbamate]

Considerable success has been achieved in discovering carbamates which exibit systemic action when applied to plants. The more important products with this property include:

aldicarb [2-methyl-2-(methylthio) propionaldehyde *o*-(methylcarbamoyl)oxime]
carbofuran [2,3-dihydro-2,2-dimethyl-7-benzofuranyl methylcarbamate]
methomyl [methyl *N*-[((methylamino)carbonyl)-oxy]ethanimidothioate]
oxamyl [methyl 2-(dimethylamino)-*N*-[((methylamino)carbonyl)oxy]-2-oxoethanimidothioate]
pirimicarb [2-(dimethylamino)-5,6-dimethyl-4-pyrimidinyl dimethylcarbamate]
thiofanox [3,3-dimethyl-1-(methylthio)-2-butanone *O*-(methylcarbamoyl)oxime]

Carbamates, like organophosphates, are cholinergic. Several not only interfere with cholinesterase, but may also inhibit one or more enzymes known as aliesterases. The binding between carbamate inhibitor and enzyme appears to be much more readily reversible than does that between enzyme and organophosphate inhibitor.

Other types of insecticides. During the mid-1960s, much interest developed in the possibilities of utilizing certain of their own secretions and hormones for control of insects. The secretions, named pheromones, are released by one sex, usually the female, and perceived by the other. All pheromones so far discovered are highly species-specific chemicals whose primary function is to facilitate mate finding and therefore maintenance and propagation of the species. Pheromones are volatile materials, and chemically are frequently unsaturated esters of low-molecular-weight acids in which the molecules tend to lie within a range of 8–20 carbon atoms. The number of unsaturations and the isomerization associated with each are largely responsible for both their specificity and the difficulty of chemical characterization. Attempts to use pheromones to control insects have so far met with limited success. During 1978, a combination of pheromone and a delivery system was registered for use against the pink bollworm in cotton. Its performance seems to indicate that the procedure offers considerable promise.

During the decade 1965–1974, the chemistry of

two types of hormone from insects was identified. The hormones themselves were investigated intensively and given careful consideration as potential insect control agents. These were the molting hormone and the juvenile hormone. The molting hormone is a steroidal molecule which is now generally considered too complex and costly for manufacture and use as a pesticide. The juvenile hormone, however, is a more simple aliphatic unsaturated triterpenoid methyl ester. Juvenile hormones II and III have now been discovered and identified. Structurally, they are very similar: JH−2 contains one less carbon atom than JH−1, and JH−3 contains one less than JH−2. The natural hormones have thus far not proved suitable for use in practical pest control. However, of a number of promising synthetic derivatives of the juvenile hormones, an analog known as methoprene [isopropyl (*E,E*)-11-methoxy-3,7,11-trimethyl-2,4-dodecadienoate], has been registered for the control of mosquito larvae. Its development for other uses is being explored.

Diflubenzuron [*N*-[((4-chlorophenyl)amino)carbonyl]-2,6-difluorobenzamide] is another chemical with a new and different mode of action. It interferes with the action of chitin synthetase, an enzyme necessary for the synthesis of chitin, and as a consequences of that process, disrupts the orderly deposition of that vital element of the exoskeleton. Currently, it is registered for the control of the gypsy moth in hardwood forests of the eastern United States. A conditional registration was granted for the use, beginning during the 1979 season, of diflubenzuron for the control and possibly the eradication of the boll weevil from cotton in the United States.

Pyrethrum is one of the oldest insecticides. Because it is highly active and quite safe for humans and animals, extensive effort was made during the period 1910−1945 to determine its chemical structure. The analytical problem was complicated greatly because not one but four active principles were ultimately found to be present. These were designated pyrethrins I and II and cinerins I and II. Efforts began immediately after identification of these natural products to produce chemicals with modified structures through laboratory synthesis. These efforts culminated in the discovery of some of the most active insect control agents yet devised. The first product, allethrin [2-allyl-4-hydroxy-3-methyl-2-cyclopenten-1-one ester of 2,2-dimethyl-3-(2-methylpropenyl)-cyclopropanecarboxylic acid], which was only partially synthetic, was introduced commercially during the early 1950s. Subsequent, entirely synthetic products have included:

dimethrin [2,4-dimethylbenzyl 2,2-dimethyl-3-(2-methylpropenyl) cyclopropanecarboxylate]
fenothrin [3-phenoxybenzyl benzyl ester of (±)-*cis, trans*-2,2-dimethyl-3-(2-methylpropenyl)cyclopropanecarboxylate]
resmethrin [(5-benzyl-3-furyl)methyl-2,2-dimethyl-3-(2-methylpropenyl)cyclopropanecarboxylate approx 70% trans, 30% cis isomers)]
tetramethrin [2,2-dimethyl-3-(2-methylpropenyl) cyclopropanecarboxylic acid esters with *N*-(hydroxymethyl)-1-cyclohexene-1,2-diacarboximide]

There are two analogs which are highly active with residues on plant surfaces showing significant degrees of persistence. Their structures differ considerably from those of the natural product. Both fenvalerate [±-α-cyano-*m*-phenoxybenzyl (±)-α-(*p*-chlorophenyl)isovalerate] and permethrin [3-phenoxybenzyl (±)-3-(2,2-dichlorovinyl)-2,2-dimethylcyclopropanecarboxylate] were granted conditional registrations in the United States for use in the control of pests of cotton during the season of 1979.

Insecticides obtained from plants include nicotine [(*S*)-3-(1-methyl-2-pyrrolidinyl)pyridine], rotenone, pyrethrins, sabadilla, and ryanodine, some of which are the oldest known insecticides. Nicotine was used as a crude extract of tobacco as early as 1763. The alkaloid is obtained from the leaves and stems of *Nicotiana tabacum* and *N. rustica*. It has been used as a contact insecticide, fumigant, and stomach poison and is especially effective against aphids and other soft-bodied insects.

Rotenone is the most active of six related alkaloids found in a number of plants, including *Derris elliptica*, *D. malaccensis*, *Lonchocarpus utilis*, and *L. urucu*. *Derris* is a native of eastern Asia, and *Lonchocarpus* occurs in South America. The highest concentrations of active principles are found in the roots. Rotenone is active against a number of plant-feeding pests and has found its greatest utility where toxic residues are to be avoided. Rotenone is known also as derris or cubé.

The principal sources of pyrethrum are *Chrysanthemum cinerariaefolium* and *C. coccineum*. Pyrethrins, which are purified extracts prepared from flower petals, contain four chemically different active ingredients. The pyrethrins find their greatest use in fly sprays, household insecticides, and grain protectants because they are the safest insecticidal materials available.

Synergists. These materials have little or no insecticidal activity, but increase the activity of chemicals with which they are mixed. Piperonyl butoxide [5-[(2-(2-butoxyethoxy)ethoxy)methyl]-6-propyl-1,3-benzodioxole], sulfoxide [5-(2-(octylsulfinyl)propyl)-1,3-benzodioxole], and *N*-(2-ethylhexyl)-5-norbornene-2,3-dicarboximide are commercially available. These synergists have their greatest utility in mixtures with the pyrethrins. Some have been shown to enhance the activity of insecticides as well.

Formulation and application. Formulation of insecticides is extremely important in obtaining satisfactory control. Common formulations include dusts, water suspensions, emulsions, and solutions. Accessory agents, including dust carriers, solvents, emulsifiers, wetting and dispersing agents, stickers, deodorants or masking agents, synergists, and antioxidants, may be required to obtain a satisfactory product.

Insecticidal dusts are formulated for application as powders. Toxicant concentration is usually quite low. Water suspensions are usually prepared from wettable powders, which are formulated in a manner similar to dusts except that the insecticide is incorporated at a high concentration and wetting and dispersing agents are included. Emulsifiable concentrates are usually prepared by solution of the chemical in a satisfactory solvent to which an emulsifier is added. They are diluted with water prior to application. Granular formulations are an effective means of applying insecticides to the soil to control insects which feed on the subterranean parts of plants.

Proper timing of insecticide applications is important in obtaining satisfactory control. Dusts are more easily and rapidly applied than are sprays. However, results may be more erratic, and much greater attention must be paid to weather conditions than is required for sprays. Coverage of plants and insects is generally less satisfactory with dusts than with sprays. It is best to make dust applications early in the day while the plants are covered with dew, so that greater amounts of dust will adhere. If prevailing winds are too strong, a considerable proportion of dust will be lost. Spray operations will usually require the use of heavier equipment, however. During the decade 1968–1978, the utilization of dusts as vehicles for dilution, distribution, and deposition of insecticides decreased markedly. This change was due largely to the greater likelihood of encountering drift and contamination problems from dusts as compared with sprays. In addition, technological improvements have been achieved in spray application machinery and techniques. Whatever the technique used, the application of insecticides should be correlated with the occurrence of the most susceptible or accessible stage in the life cycle of the pest involved. By and large, treatments should be made only when economic damage by a pest appears to be imminent.

During the decade 1959–1968, attention was focused sharply on the impact of the highly active synthetic insecticides upon the total environment—humans, domestic animals and fowl, soil-inhabiting microflora and microfauna, and all forms of aquatic life. Effects of these materials upon populations of beneficial insects, particularly parasites and predators of the economic species, were critically assessed. The study of insect control by biological means expanded. The concepts and practices of integrated pest control and pest management expanded rapidly.

Among problems associated with insect control which received major emphasis during the decade 1969–1978 were the development of strains of insects resistant to insecticides; the assessment of the significance of small, widely distributed insecticide residues in and upon the environment; the development of better and more reliable methods for forecasting insect outbreaks; the evolvement of control programs integrating all methods—physical, physiological, chemical, biological, and cultural—for which practicality was demonstrated; the development of equipment and procedures to detect chemicals much below the part-per-million and microgram levels; and as a consequence of the provisions of the Federal Insecticide, Fungicide, and Rodenticide Act as amended by the Federal Environmental Pesticide Control Act of 1972, a greatly amplified effort to obtain data delineating mammalian toxicology, persistence in the environment, and immediate chronic impact of pesticides upon nontarget invertebrate and vertebrate organisms occupying aquatic, terrestrial, and arboreal segments of the environment.

The registration of pesticides is a detailed, highly technical process. Pesticides must be selected and applied with care. Recommendations as to the product and method of choice for control of any pest problem—weed, insect, or varmint—are best obtained from county or state agricultural extension specialists. Recommendations for pest control and pesticide use can be obtained from each state agricultural experiment station office. In addition, it is necessary to read carefully and to follow explicitly the directions, restrictions, and cautions for use which are on the label attached to the product container. Insecticides are a boon to the production of food, feed, and fiber, and their use must not be abused in the home, garden, farm field, forest, or stream. *See* INSECT CONTROL, BIOLOGICAL; PESTICIDE.

[GEORGE F. LUDVIK]

Bibliography: J. L. Apple and R. F. Smith, *Integrated Pest Management*, 1976; G. L. Berg (ed.), *Farm Chemicals Handbook* 1978; M. C. Birch (ed.), *Pheromones*, 1974; A. B. Borkovec, *Insect Chemosterilants*, 1966; A. W. A. Brown, *Insect Control by Chemicals*, 1951; R. L. Caswell, *Pesticide Handbook—Entoma 1977–78*, 1977; Estimated losses . . . attributed to insects . . . 1976, *USDA Coop. Plant Pest Rep.*, 3(1):91–117, 1978; *Federal Insecticide, Fungicide and Rodenticide Act*, Public Law 92–516, 92d Congress, H. R. 10729, Oct. 21, 1972; *Federal Insecticide, Fungicide, and Rodenticide Act as Amended*, EPA, 1976; *Federal Insecticide, Fungicide, and Rodenticide Act as Amended*, OPA 17/9, EPA, 1978; *Federal Pesticide Act of 1978*, OPA 136/8, EPA, 1978; G. P. Georghiou and C. E. Taylor, Pesticide resistance as an evolutionary phenomenon, in *Proceedings of the 15th International Congress of Entomology*, pp. 759–784, 1976; M. Jacobson, *Insect Sex Attractants*, 1965; E. E. Kenaga and R. W. Morgan, *Commercial and Experimental Organic Insecticides (1978 revision)*, Bull. Entomol. Soc. Amer. Spec. Publ. 78–1, 1978; W. W. Kilgore and R. L. Doutt, *Pest Control: Biological, Physical and Selected Chemical Methods*, 1967; H. Martin and C. R. Worthing, *Pesticide Manual*, 4th ed., 1974; F. Matsumura, *Toxicology of Insecticides*, 1976; J. J. Menn and M. Boroza (eds.), *Insect Juvenile Hormones*, 1972; C. L. Metcalf, W. P. Flint, and R. L. Metcalf, *Destructive and Useful Insects*, 1962; R. L. Metcalf, *Organic Insecticides*, 1955; R. L. Metcalf and W. Luckman, *Introduction to Pest Management*, 1975; *Notice . . . of Products Containing Heptachlor and Chlordane*, PR Notice 78-2, EPA, 1978; R. D. O'Brien, *Insecticides: Action and Metabolism*, 1967; H. H. Shorey and J. T. McKelvey (eds.), *Chemical Control of Insect Behavior*, 1977; *Suspended and Cancelled Pesticides*, EPA, 1977; T. F. Watson, L. Moore, and G. W. Ware, *Practical Insect Pest Management*, 1975; C. F. Wilkinson (ed.), *Insecticide Biochemistry and Physiology*, 1976.

Irrigation of crops

The artificial application of water to the soil to produce plant growth. Irrigation also cools the soil and atmosphere, making the environment favorable for plant growth. The use of some form of irrigation is well documented throughout the history of humankind. Over 50,000,000 acres (202,300 km^2) are irrigated in the United States.

Use of water by plants. Growing plants use water almost continuously. Growth of crops under irrigation is stimulated by optimum moisture, but retarded by excessive or deficient amounts. Factors influencing the rate of water use by plants include the type of plant and stage of growth, temperature, wind velocity, humidity, sunlight duration and intensity, and available water supply. Plants use the least amount of water upon emergence from the soil and near the end of the growing period. Irrigation and other management practices should be coordinated with the various stages of growth. A vast amount of research has been done on the use of water by plants, and results are available for crops under varying conditions.

Consumptive use. In planning new or rehabilitating old irrigation projects, consumptive use is the most important factor in determining the amount of water required. It is also used to determine water rights.

Consumptive use, or evapotranspiration, is defined as water entering plant roots to build plant tissues, water retained by the plant, water transpired by leaves into the atmosphere, and water evaporated from plant leaves and from adjacent soil surfaces. Consumptive use of water by various crops under varying conditions has been determined by soil-moisture studies and computed by other well-established methods for many regions of the United States and other countries. Factors which have been shown to influence consumptive use are precipitation, air temperature, humidity, wind movement, the growing season, and latitude, which influences hours of daylight. Table 1 shows how consumptive use varies at one location during the growing season. When consumptive-use data are computed for an extensive area, such as an irrigation project, the results will be given in acre-feet per acre for each month of the growing season and the entire irrigation period. Peak-use months determine system capacity needs. An acre-foot is the amount of water required to cover 1 acre 1 ft deep (approx. 1214 m^2 of water).

Soil, plant, and water relationships. Soil of root-zone depth is the storage reservoir from which plants obtain moisture to sustain growth. Plants take from the soil not only water, but dissolved minerals necessary to build plant cells. How often this reservoir must be filled by irrigation is determined by the storage capacity of the soil, depth of the root zone, water use by the crop, and the amount of depletion allowed before a reduction in yield or quality occurs. Table 2 shows the approximate amounts of water held by soils of various textures.

Water enters coarse, sandy soils quite readily, but in heavy-textured soils the entry rate is slower. Compaction and surface conditions also affect the rate of entry.

Soil conditions, position of the water table, length of growing season, irrigation frequency, and other factors exert strong influence on root-zone depth. Table 3 shows typical root-zone depths in well-drained, uniform soils under irrigation. The depth of rooting of annual crops increases during the entire growing period, given a favorable, unrestricted root zone. Plants in deep, uniform soils usually consume water more slowly from the lower root-zone area than from the upper. Thus, the upper portion is the first to be exhausted of moisture. For most crops, the entire root zone should be supplied with moisture when needed.

Maximum production can usually be obtained with most irrigated crops if not more than 50% of the available water in the root zone is exhausted during the critical stages of growth. Many factors influence this safe-removal percentage, including the type of crop grown and the rate at which water is being removed. Application of irrigation water should not be delayed until plants signal a need for moisture; wilting in the hot parts of the day may reduce crop yields considerably. Determination of the amount of water in the root zone can be done by laboratory methods, which are slow and costly. However, in modern irrigation practice, soil-moisture-sensing devices are used to make rapid determinations directly with enough accuracy for practical use. These devices, placed in selected field locations, permit an operator to schedule periods of water application for best results. Evaporation pans and weather records can be used to estimate plant-water use. Computerizing these data also helps farmers schedule their irrigations. The irrigation system should be designed to supply sufficient water to care for periods of most rapid evapotranspiration. The rate of evapotranspiration may vary from 0 to 0.4 in. per day (10 mm per day) or more.

Water quality. All natural irrigation waters contain salts, but only occasionally are waters too saline for crop production when used properly. When more salt is applied through water and fertilizer than is removed by leaching, a salt buildup can occur. If the salts are mainly calcium and magnesium, the soils become saline, but if salts predominately are sodium, a sodic condition is possible. These soils are usually found in arid

Table 1. Example of consumption of water by various crops, in inches*

Crop	April	May	June	July	Aug.	Sept.	Oct.	Seasonal total
Alfalfa	3.3	6.7	5.4	7.8	4.2	5.6	4.4	37.4
Beets		1.9	3.3	5.3	6.9	5.8	1.1	24.3
Cotton	1.1	2.0	4.1	5.8	8.6	6.7	2.7	31.0
Peaches	1.0	3.4	6.7	8.4	6.4	3.1	1.1	30.0
Potatoes			0.7	3.4	5.8	4.4		14.0

*1 in. = 25.4 mm.

Table 2. Approximate amounts of water in soils available to plants

Soil texture	Water capacity, in inches for each foot of depth*
Coarse sandy soil	0.5–0.75
Sandy loam	1.25–1.75
Silt loam	1.75–2.50
Heavy clay	1.75–2.0 or more

*1 in. = 25.4 mm; 1 ft = 0.3 m.

Table 3. Approximate effective root-zone depths for various crops

Crop	Root-zone depth, ft*
Alfalfa	6
Corn	3
Cotton	4
Potatoes	2
Grasses	2

*1 ft = 0.3 m.

Fig. 1. Furrow method of irrigation. Water is supplied by pipes with individual outlets, or by ditches and siphon tubes.

areas, especially in those areas where drainage is poor. Rainfall in humid areas usually carries salts downward to the groundwater and eventually to the sea.

Saline soils may reduce yields and can be especially harmful during germination. Some salts are toxic to certain crops, especially when applied by sprinkling and allowed to accumulate on the plants. Salt levels in the soil can be controlled by drainage, by overirrigation, or by maintaining a high moisture level which keeps the salts diluted.

Sodic soils make tillage and water penetration difficult. Drainage, addition of gypsum or sulfur, and overirrigation usually increase productivity.

Ponding or sprinkling can be used to leach salts. Intermittent application is usually better and, when careful soil-moisture management is practiced, only small amounts of excess irrigation are needed to maintain healthy salt levels.

Diagnoses of both water and soil are necessary for making management decisions. Commercial laboratories and many state universities test both water and soil, and make recommendations.

Methods of application. Water is applied to crops by surface, subsurface, sprinkler, and drip irrigation. Surface irrigation includes furrow and flood methods.

Fig. 2. A side-roll sprinkler system which uses the main supply line (often more than 1000 ft, or 300 m, long) to carry the sprinkler heads and as the axle for wheels.

Furrow method. This method is used for row crops (Fig. 1). Corrugations or rills are small furrows used on close-growing crops. The flow, carried in furrows, percolates into the soil. Flow to the furrow is usually supplied by siphon tubes, spiles, gated pipe, or valves from buried pipe. Length of furrows and size of stream depend on slope, soil type, and crop; infiltration and erosion must be considered.

Flood method. Controlled flooding is done with border strips, contour or bench borders, and basins. Border strip irrigation is accomplished by advancing a sheet of water down a long, narrow area between low ridges called borders. Moisture enters the soil as the sheet advances. Strips vary from about 20 to 100 ft (6 to 30 m) in width, depending mainly on slope (both down and across), and amount of water available. The border must be well leveled and the grade uniform; best results are obtained on slopes of 0.5% or less. The flood method is sometimes used on steeper slopes, but maldistribution and erosion make it less effective.

Bench-border irrigation is sometimes used on moderately gentle, uniform slopes. The border strips, instead of running down the slope, are constructed across it. Since each strip must be level in width, considerable earth moving may be necessary.

Basin irrigation is well adapted to flatlands. It is done by flooding a diked area to a predetermined depth and allowing the water to enter the soil throughout the root zone. Basin irrigation may be utilized for all types of crops, including orchards where soil and topographic conditions permit.

Subirrigation. This type of irrigation is accomplished by raising the water table to the root zone of the crop or by carrying moisture to the root zone by perforated undergound pipe. Either method requires special soil conditions for successful operation.

Sprinkler systems. A sprinkler system consists of pipelines which carry water under pressure from a pump or elevated source to lateral lines along which sprinkler heads are spaced at appropriate intervals. Laterals are moved from one location to another by hand or tractor, or they are moved automatically. The side-roll wheel system, which utilizes the lateral as an axle (Fig. 2), is very popular as a labor-saving method. The center-pivot sprinkler system (Fig. 3) consists of a lateral carrying the sprinkler heads, and is moved by electrical or hydraulic power in a circular course irrigating an area containing up to 135–145 acres (546,200–586,700 m^2).

Extra equipment can be attached in order to irrigate the corners, or solid sets can be used. Solid-set systems are systems with sufficient laterals and sprinklers to irrigate the entire field without being moved. These systems are quite popular for irrigating vegetable crops or other crops requiring light, frequent irrigations and, in orchards, where it is difficult to move the laterals.

Sprinkler irrigation has the advantage of being adaptable to soils too porous for other systems. It can be used on land where soil or topographic conditions are unsuitable for surface methods. It can be used on steep slopes and operates efficiently with a small water supply.

Drip irrigation. This is a method of providing water to plants almost continuously through small-diameter tubes and emitters. It has the advantage of maintaining high moisture levels at relatively low capital costs. It can be used on very steep, sandy, and rocky areas and can utilize saline waters better than most other systems. Clean water, usually filtered, is necessary to prevent blockage of tubes and emitters. The system has been most popular in orchards and vineyards, but is also used for vegetables, ornamentals, and for landscape plantings.

Automated systems. Automation is being used with solid-set and continuous-move types of systems, such as the center-pivot and lateral-move. Surface-irrigated systems are automated with check dams, operated by time clocks or volume meters, which open or close to divert water to other areas. Sprinkler systems, pumps, and check dams can all be activated by radio signals or low-voltage wired systems, which, in turn, can be triggered by soil-moisture-sensing devices or water levels in evaporation pans.

Automatically operated pumpback systems, consisting of a collecting pond and pump, are being used on surface-irrigated farms to better utilize water and prevent silt-laden waters from returning to natural streams.

Multiple uses. With well-designed and -managed irrigation systems, it is possible to apply chemicals and, for short periods of time, to moderate climate. Chemicals which are being used include fertilizers, herbicides, and some fungicides. Effectiveness depends on uniformity of mixing and distribution and on application at the proper times. Chemicals must be registered to be used in this manner.

Solid-set systems are frequently used to prevent frost damage to plants and trees, since, as water freezes, it releases some heat. A continuous supply of water is needed during the protecting period. However, large volumes of water are required, and ice loads may cause limb breakage. Sequencing of sprinklers for cooling is being practiced for bloom delay in the spring and for reduction of heat damage in the summer.

Humid and arid regions. The percentage of increase in irrigated land is greater in humid areas than in arid and semiarid areas, although irrigation programs are often more satisfactory where the farmer does not depend on rainfall for crop growth. Good yields are obtained by well-timed irrigation, maintenance of high fertility, keeping the land well cultivated, and using superior crop varieties.

There is little difference in the principles of crop production under irrigation in humid and arid regions. The programming of water application is more difficult in humid areas because natural precipitation cannot be accurately predicted. Most humid areas utilize the sprinkler method.

To be successful, any irrigation system in any location must have careful planning with regard to soil conditions, topography, climate, cropping practices, water quality and supply, as well as engineering requirements.

Outlook. As mentioned, there are over 50,000,000 acres of land irrigated in the United States. Studies of land that can be developed for irrigation are becoming obsolete with the improvements in irrigation systems. Limitations of water supplies and economics will prevent future developments — not land resources. The limit of simple diversions of natural rivers and streams has been reached, with few exceptions. Future development of large acreages of irrigated land must come through extensive storage; high-lift pumping projects, characteristic of Federal programs and large corporations; transportation of supply water sources to water-poor areas; and better conservation of water supplies.

Fig. 3. Center-pivot systems are very popular in new irrigation developments.

Since the most economical irrigation projects have already been developed, future development will be more costly, depending upon many factors. Major diversions of stream flow to regions outside the watershed will involve many complicated interstate problems and compacts. As the population increases, there will be greater competition for water by industry, municipalities, power generators, recreational facilities, and wildlife reserves. Some underground water supplies are being depleted and must, at some time, be replenished by transported water if the irrigated area is to remain under cultivation.

Better water conservation could assist in expanding the irrigated acreage. It is estimated that phreatophites (water-loving plants) along streams and irrigation canals transpire 25,000,000 acre-feet of water to the atmosphere. Evaporation from reservoirs accounts for the loss of millions of additional acre-feet. Other losses include seepage from canals and water lost through the soil when more water is applied than the plants can use. *See* LAND DRAINAGE (AGRICULTURE); WATER CONSERVATION.

[MEL A. HAGOOD]

Bibliography: *Guidelines for Predicting Crop Water Requirements*, FAO Irrigation and Drainage Pap. no. 24, 1977; R. M. Hagan et al., *Successful Irrigation: Planning, Development, Management*, 1969; C. H. Pair, *Sprinkler Irrigation*, 4th ed., 1975.

Island biogeography

Islands generally have fewer species of animals and plants than do comparable continental areas, at least if the islands lie some distance from continents. Often there are rather few genera, but some genera have many local species that sometimes exhibit strong adaptive radiation. The proportion of endemic species that are present generally increases with the degree of isolation and with the complexity of the habitat. Gigantism, flightlessness, and other unusual characteristics are relatively frequent.

Island floras and faunas are of far greater interest than either the number of species or their economic importance might seem to justify. How the plants and animals came to be on the islands, why so many of them are found nowhere else, why so many have special or strange forms, and where their ancestors came from, are fascinating questions which present problems of great scientific importance to the biologist. The theory of organic evolution emerged from studies of island faunas by Charles Darwin and Alfred Russell Wallace, and islands are still among the most advantageous places to study evolutionary processes. They also offer uniquely suitable sites for the investigation of the nature and functioning of ecosystems. *See* ECOLOGY.

Nature and classification of islands. Islands are themselves very diverse, and their biotas (floras and faunas) are correspondingly varied. Islands are commonly separated into high and low islands; into volcanic, limestone, granitic, metamorphic, and mixed islands; and, perhaps most important from a biogeographic standpoint, into continental and oceanic islands. Naturally, most of these types are not sharply separated. On the contrary, as with wet and dry islands, there is a continuous series of intermediates between any pair or set of extremes. Usually significantly different biotas are associated with these categories. Low islands have small, impoverished biotas; high islands, richer ones. Oceanic islands are those considered never to have been part of, or connected with, any continental land mass. Their biotas are commonly poor in genera and unbalanced or disharmonic, that is, with the families—or larger groups—very unevenly represented, compared with those of continents or continental islands. The continental islands are believed to be the remnants of former continental land masses or at least to have had land connections with continents.

Oceanic islands or groups with outstanding or much-studied biotas are Rapa, the Hawaiian Islands, the Galapagos, Juan Fernandez, the Fiji Islands, the Marianas, the Carolines, the Marshalls, the Society Islands, the Azores, the Canary Islands, Tristan da Cunha, Kerguelen, the Seychelles, and Aldabra.

Continental islands or groups are Madagascar, Japan, Formosa, the Philippines, New Guinea, the Malay Archipelago, Ceylon, New Zealand, the Antilles, the California Islands, and the British Isles.

Endemism. On almost all islands, except those immediately contiguous to other land and most low coral islands, are species and varieties that are endemic to the particular island or group, that is, found nowhere else. Even coastal islands, such as those off southern California, have a few endemics. On high islands the percentage of the total biota that is endemic usually increases with the degree of isolation from larger land masses. Thus, the endemics form over 95% of the indigenous vascular flora in the Hawaiian Islands, which are very remote from other land. On all islands there are some widespread species. They are usually seacoast species, aquatics, or marsh dwellers; and, of course, there are many introduced weeds.

The first basic type of endemics includes species that have differentiated on the islands where they are now found, usually the products of relatively recent evolution. In certain genera there may be several to many species of a given genus adapted to particular habitats—a phenomenon called adaptive radiation. Examples of this are the famous Darwin finches (Geospizidae) of the Galapagos; *Hedyotis* (Rubiaceae) in the Hawaiian Islands; and many insect genera with species adapted to different plant hosts. Other genera have differentiated into several or many species which occupy very similar habitats, apparently an evolutionary result of geographic isolation only, for example, *Cyrtandra* (Gesneriaceae) in the Hawaiian group.

The other basic type of endemics includes relicts, isolated remnants of populations that were much more widespread in the past. A good example is *Lyonothamnus floribundus*. Though it is now confined to the California Islands, fossil evidence shows that it was widespread in California during the Tertiary Period. Islands provide refuges for these species where they may not be exposed to the competition or other unfavorable circumstances that eliminated them elsewhere. For example, islands may provide a more equable climate when severe conditions develop in continental areas. Disjunct species, those found on widely separated islands, may likewise be relicts, or they may be the result of successful long-distance transport. A case in point is possibly *Charpentiera obovata* (Amaranthaceae), a shrub found only in the Hawaiian, Cook, and Austral islands, separated by almost 3000 mi. The plants of the Austral and Cook islands have now been called a separate but very closely allied species.

Geography of floras and faunas. The geographical distribution of insular biotas shows very definite patterns, making it possible to group species into "elements" with similar distributions and, in some cases, apparently common geographic origins. Relationships of island genera and species may often be guessed at on the basis of their taxonomic relationships with groups in other areas. Thus a preponderance of the Hawaiian vascular flora and terrestrial invertebrate fauna have their affinities to the southwest in the Indo-Malaysian region. Another element has its connections, and likely its derivation, in the Australia-New Zealand area; still another, but smaller, element finds its relationships in America; another in eastern Polynesia. The numbers of species in the predominant Indo-Pacific element are very high in the western islands of the Pacific, but fall off very markedly from island to island, eastward, except where there has been strong local evolution of species, as in Hawaii. *See* BIOGEOGRAPHY.

Dispersal. One of the central problems in the study of island biotas is that of dispersal. How did the species, or their ancestors, reach the islands where they are now found? This problem is especially fascinating on oceanic islands. To solve it, those with little faith in the effectiveness of transoceanic dispersal would assume that very few islands are truly oceanic, that at different times in the past land bridges have existed in all directions, or that ancient continents foundered or drifted away, leaving the islands and their biotas as scattered remnants. A more plausible theory is that given time enough, rare events, such as accidental transport of individuals or propagules (seeds, eggs, or spores) by wind, water, or birds, will account for the ancestors of all present indigenous biotas of oceanic islands. It is also reasonable to assume that there were many former islands, now marked only by shoals and underwater mountains, that could have served as stepping stones along which some species may have traveled to their present isolated homes. *See* POPULATION DISPERSAL.

Morphological peculiarities. A notable and little-understood feature of island biotas is the frequency of unusual morphologic features—flightlessness in birds and insects, gigantism and woody habit in usually small, herbaceous groups of plants, as well as gigantism in animals, for example, the giant tortoises of the Galapagos and Aldabra and the moas of New Zealand, and the "rosette-tree" habit in many island plant genera. Some of these features are also found in the biotas of tropical mountains. Another feature of the plants of oceanic islands is that, having evolved in the absence of large herbivores, they have weak defenses against grazing and trampling, if any. Spines and thorns are infrequent on indigenous species.

Evolution on islands. The origin of these morphological peculiarities, as well as of the more ordinary diversity found in island biotas, poses evolutionary problems that are extremely interesting and that seem amenable to solution. Islands not only provided early information on which the theory of organic evolution was founded; much study has since been devoted to insular evolution to test various concepts.

In addition to generic and specific diversity, island organisms frequently show an extraordinary degree of polymorphism within species. For example, the tree *Metrosideros collina* of Polynesia, especially Hawaii, shows a complexity of forms bewildering for taxonomic arrangement.

At the other extreme are species with almost no genetic plasticity, usually existing in small populations. The extinct *Clermontia haleakalae* and *Hedyotis cookiana* of Hawaii were presumably of this nature, as are the Hawaiian hawk, crow, and goose.

The adaptive radiation and geographic speciation mentioned above are important evolutionary patterns notable on islands. Isolation, as well as the existence of many unoccupied niches and reduced competition during the early stages in evolution of insular biotas, is probably responsible for the evolutionary persistence of certain features that would be speedily eliminated in a more complex continental biotic situation.

The evolution of major ecosystems themselves may also be elucidated by careful and long-continued study of islands, as island ecosystems are simpler and better defined than those of larger land masses, and the development of their biotas can be better correlated with that of their physical features. Also, a great range of size and complexity is exhibited. *See* ECOSYSTEM.

Biotas of volcanic islands. New volcanic substrata are bare of all plant and animal life. But from the time the lava or ash surface is cool, it is subject to colonization by plants and animals. The biotas of volcanic islands are the products of a sequence: sporadic colonization; repeated partial destruction by new eruptions; isolation of local populations, with evolution of local races; breaking down of such isolation and mixing of populations, development of genetic diversity, and renewed isolation by dissection of volcanic domes which leads to further evolution of local forms, species, and even genera with time; new arrivals may occupy any open habitats available or may displace current tenants. New open habitats become available as long as volcanic activity continues and erosion makes still other habitats.

As volcanic islands grow older some subside, or get partly flooded by changing sea level. Parts or all may be eroded down to sea level by wave action. In the tropics coral reefs grow up around their shores and provide habitats for still other species, those characteristic of limestone shores and strands, mostly species of very wide distribution. As the high volcanic mountains gradually subside and wear away, the species dependent on high elevations, orographic rainfall, or rain shadows gradually disappear. New ones evolve, or colonize from elsewhere, that are adapted to the old worn-down topography and deeply weathered volcanic substrata. Even most of these finally disappear as subsidence gradually changes the island to an atoll. *See* ECOLOGICAL SUCCESSION.

Biotas of coral atolls and limestone islands. Scattered through most tropical seas are flat, often ring-shaped, islands made up of the limestone skeletons of marine plants and animals and resulting from the processes described above. Most rise just above sea level. In this relatively uniform environment, varying principally in rainfall, biotas are very impoverished and are largely made up of strand species. Numbers of species are lowest on the drier atolls. As an example of the variation, the native vascular flora of the Marshall Islands ranges from 9 species per atoll in the driest northern ones, to 75 or 100 in the very wet southern atolls. Endemics are very few. With even a slight elevation, as on Aldabra Atoll, the number of species increases very sharply. Strand species are still prominent, but species requiring moderate habitats increase in numbers. There are some endemics. High limestone islands have rich, strongly endemic biotas.

Interesting and important as insular floras and faunas are, they are disappearing with distressing rapidity. The introduction of large herbivorous animals, rats, and aggressive exotic plants capable of quickly invading disturbed situations, as well as the complete destruction of whole habitats by human activities, has resulted in the reduction or total disappearance of many species. The growth of human populations and development, disturbance, and destruction even in remote islands are accelerating.

More and more species are becoming rare and threatened with extinction as man achieves greater capacity to change his environment. Many fascinating features of island floras and faunas will not long be available for study and enjoyment unless adequate measures for protecting substantial areas of natural habitats on islands are taken at once. Conservation is an immediate necessity if the study of island biotas is to have a future. *See* PLANT GEOGRAPHY.

Theory. Up to relatively recently, investigative activity in the field of island biogeography has been almost entirely empirical—the making of observations and the accumulation of data. The observations and compiled data on occurrence and distribution of organisms provided the raw material for interpretation of events mostly so far in the past that knowledge of them is of necessity primarily speculative and, except as elucidated by new discoveries of fossils, is likely to remain so. Attempts to intercept propagules borne by air, animals, and water, though of great importance, are only more sophisticated attempts to secure new information. No real body of theory had grown up in the field.

Hence, the appearance in 1967 of *The Theory of Island Biogeography* by R. H. MacArthur and E. O. Wilson was greeted with enormous interest. It stimulated numerous investigations to determine the applicability of the theory to real situations and to test it as a guide to further thinking on the subject.

Stated simply, the proposition was that the numbers of species in the biota of an island bear a direct relation to the area of the island and its distance from the source of the biota, and that the number of species is the result of an equilibrium between the rates of immigration and extinction of the biota (or, in practice, of whatever segment of the biota is under study). These ideas were not presented simply or clearly, but were elaborated and expressed in intricate mathematical (or, better, pseudomathematical) terms. Little was gained in either clarity or understanding by this lengthy exercise.

To those with much experience with the biotas of islands, the theory proposed seems a mixture of the self-evident with the unlikely, at the very least a vast oversimplification. Clearly, there is a relation between area and size of biota. And clearly, islands of equal size, one with a desert climate, the other with 2500 mm of rainfall, or one a sea-level coral atoll, the other a substantially elevated coral platform or a mountainous volcanic island, will have very different numbers of species. The number and nature of habitats are greatly influenced by elevation, wind direction, substratum, moisture, and the accidents of early colonization. This complexity may so greatly influence the numbers of species an island can support that effects of area are usually, except perhaps in sand cays and sea-level atolls, completely masked. The appearance of new species by evolution and differentiation, and resulting increased fitness of species for island life, would also seem to have a profound role in determining the number of species inhabiting an island.

In MacArthur and Wilson's book, certain of these complicating aspects of the problem were discussed, with attempts to fit them into the theory, usually with scant success. The reason given for the lack of fit is, in almost all cases, the inadequacy or almost complete lack of data. Some of the ideas and possibilities expressed are very ingenious and perceptive. However, their varied nature, mutual discrepancies, and sheer complexity make it extremely unlikely that the equilibrium theory is applicable to cases in real life or can be used as any sort of guide or framework on which to arrange observational data and to interpret or explain their significance.

However, if the functions of a theory are to provide a framework to think and to stimulate further investigation, rather than to explain the phenomena observed, it may well be claimed that the MacArthur-Wilson theory has been much more of a success than the above remarks might suggest. Only a far more complete collection and recording of numbers of species on islands, their ecology, habitats, and geographical relationships, and the nature and direction of change in their numbers, will give a proper assessment of the validity of the MacArthur and Wilson equilibrium theory of island biogeography. [F. R. FOSBERG]

Bibliography: R. I. Bowman, *The Galapagos*, 1966; S. Carlquist, *Island Biology*, 1974; S. Carlquist, *Island Life*, 1965; F. R. Fosberg, Geography, Ecology, and Biogeography, *Ann. Ass. Amer. Geogr.*, 66:117–128, 1976; F. R. Fosberg (ed.), *Man's Place in the Island Ecosystem*, 1963; J. L. Gressitt, *Insects of Micronesia: Introduction*, 1954; J. L. Gressitt (ed.), *Pacific Basin Biogeography*, 1963; D. Lack, *Darwin's Finches*, 1947; R. H. MacArthur and E. O. Wilson, *The Theory of Island Biogeography*, 1967; R. N. Philbrick (ed.), *Biology of the California Islands*, 1967; M.-H. Sachet and F. R. Fosberg, *Island Bibliographies*, 1955, 1971; E. D. Zimmerman, *Insects of Hawaii*, 1948.

Jet stream

A relatively narrow, fast-moving wind current flanked by more slowly moving currents. Jet streams are observed principally in the zone of prevailing westerlies above the lower troposphere and in most cases reach maximum intensity, with regard both to speed and to concentration, near the tropopause. At a given time, the position and intensity of the jet stream may significantly influence aircraft operations because of the great speed of the wind at the jet core and the rapid spatial variation of wind speed in its vicinity. Lying in the zone of maximum temperature contrast between cold air masses to the north and warm air masses to the south, the position of the jet stream on a given day usually coincides in part with the regions of greatest storminess in the lower troposphere, though portions of the jet stream occur over regions which are entirely devoid of cloud.

Characteristics. The specific characteristics of the jet stream depend upon whether the reference is to a single instantaneous flow pattern or to an averaged circulation pattern, such as one averaged with respect to time, or averaged with respect both to time and to longitude.

If the winter circulation pattern on the Northern Hemisphere is averaged with respect to both time and longitude, a westerly jet stream is found at an elevation of about 13 km near latitude (lat) 25°. The speed of the averaged wind at the jet core is about 148 km/hr (80 knots). In summer this jet is displaced poleward to a position near lat 42°. It is found at an elevation of about 12 km with a maximum speed of about 56 km/hr (30 knots). In both seasons a speed equal to one-half the peak value is found approximately 15° of latitude south, 20° of latitude north, and 5–10 km above and below the location of the jet core itself.

If the winter circulation is averaged only with respect to time, it is found that both the intensity and the latitude of the westerly jet stream vary from one sector of the Northern Hemisphere to another. The most intense portion, with a maximum speed of about 185 km/hr (100 knots), lies over the extreme western portion of the North Pacific Ocean at about lat 22°. Lesser maxima of about 157 km/hr (85 knots) are found at lat 35° over the east coast of North America, and at lat 21° over the eastern Sahara and over the Arabian Sea. In summer, maxima are found at lat 46° over the Great Lakes region, at lat 40° over the western Mediterranean Sea, and at lat 35° over the central North Pacific Ocean. Peak speeds in these regions range between 74 and 83 km/hr (40–45 knots). The degree of concentration of these jet streams, as measured by the distance from the core to the position at which the speed is one-half the core speed, is only slightly greater than the degree of concentration of the jet stream averaged with respect to time and longitude. At both seasons and at all longitudes the elevation of these jet streams varies between 11 and 14 km.

Variations. On individual days there is a considerable latitudinal variability of the jet stream, particularly in the western North American and western European sectors. It is principally for this reason that the time-averaged jet stream is not well defined in these regions. There is also a great day-to-day variability in the intensity of the jet stream throughout the hemisphere. On a given winter day, speeds in the jet core may exceed 370 km/hr (200 knots) for a distance of several hundred miles along the direction of the wind. Lateral wind shears in the direction normal to the jet stream frequently attain values as high as 100 knots/300 nautical miles (185 km/hr/556 km) to the right of the direction of the jet stream current and as high as 100 knots/100 nautical miles (185 km/hr/185 km) to the left. Vertical shears below and above the jet core are often as large as 20 knots/1000 ft (37 km/305 m). Daily jet streams are predominantly westerly, but northerly, southerly, and even easterly jet streams may occur in middle or high latitudes when ridges and troughs in the normal westerly current are particularly pronounced or when unusually intense cyclones and anticyclones occur at upper levels.

Insufficiency of data on the Southern Hemisphere precludes a detailed description of the jet stream, but it appears that the major characteristics resemble quite closely those of the jet stream on the Northern Hemisphere. The day-to-day variability of the jet stream, however, appears to be less on the Southern Hemisphere.

It appears that an intense jet stream occurs at high latitudes on both hemispheres in the winter stratosphere at elevations above 20 km. The data available, however, are insufficient to permit the precise location or detailed description of this phenomenon. *See* AIR MASS; ATMOSPHERE; STORM. [FREDERICK SANDERS]

Lake

An inland body of water, small to moderately large in size, with its surface exposed to the atmosphere. Most lakes fill depressions below the zone of saturation in the surrounding soil and rock materials. Generically speaking, all bodies of water of this type are lakes, although small lakes usually are called ponds, tarns (in mountains), and less frequently pools or meres. The great majority of lakes have a surface area of less than 100 square miles (mi²) or 260 km². More than 30 well-known lakes, however, exceed 1500 mi² (3885 km²) in extent, and the largest freshwater body, Lake Superior, North America, covers 31,180 mi² (80,756 km²) (see table).

Most lakes are relatively shallow features of the Earth's surface. Even some of the largest lakes have maximum depths of less than 100 ft or 30 m (Winnipeg, Canada; Balkash, Soviet Union; Albert, Uganda), A few, however, have maximum depths which approach those of some seas. Lake Baikal in the Soviet Union is about a mile deep at its deepest point, and Lake Tanganyika, Africa, is about 0.9 mi (1.3 km).

Because of their shallowness, lakes in general may be considered evanescent features of the Earth's surface, with a relatively short life in geological time. Every lake basin forms a bed onto which the sediment carried by inflowing streams is deposited. As the sediment accumulates, the storage capacity of the basin is reduced, vegetation encroaches upon the shallow margins, and eventually the lake may disappear. Most lakes also have surface outlets. Except at elevations very near sea level, a stream which flows from such an outlet gradually cuts through the barrier forming the lake basin. As the level of the outlet is lowered, the capacity of the basin is also reduced and the disappearance of the lake assured.

Variations in water character. Lakes differ as to the salt content of the water and as to whether they are intermittent or permanent. Most lakes are composed of fresh water, but some are more salty than the oceans. Generally speaking, a number of water bodies which are called seas are actually salt lakes; examples are the Dead, Caspian, and Aral seas. All salt lakes are found under desert or semi-arid climates, where the rate of evaporation is high enough to prevent an outflow and therefore a discharge of salts into the sea. Many lesser arid-region lakes are intermittent, sometimes existing only for a short period after heavy rains and disappearing under intense evaporation. These lakes are called playas in North America, shotts in North

Dimensions of some major lakes†

Lake	Area, mi²	Volume (approx), 1000 acre-ft	Shoreline, mi	Depth	
				Av, ft	Max, ft
Caspian Sea	169,300	71,300	3,730	675	3,080
Superior	31,180	9,700	1,860	475	1,000
Victoria	26,200	2,180	2,130		
Aral Sea	26,233*	775			
Huron	23,010	3,720	1,680		
Michigan	22,400	4,660			870
Baikal	13,300*	18,700		2,300	5,000
Tanganyika	12,700	8,100			4,700
Great Bear	11,490*		1,300		
Great Slave	11,170*		1,365		
Nyasa	11,000	6,800		900	2,310
Erie	9,940	436			
Winnipeg	9,390*		1,180		
Ontario	7,540	1,390			
Balkash	7,115				
Ladoga	7,000	745			
Chad	6,500*				
Maracaibo	4,000*				
Eyre	3,700*				
Onega	3,764	264			
Rudolf	3,475*				
Nicaragua	3,089	87			
Athabaska	3,085				
Titicaca	3,200	575			
Reindeer	2,445				

*Area fluctuates.

†1 mi² = 2.59 km²; 1000 acre-ft = 1.233×10^6 m³; 1 mi = 1.609 km; 1 ft = 0.3048 m.

Africa, and other names elsewhere. In such regions the surface area and volume of permanent lakes may differ enormously from wet to dry season.

The water of the more permanent salt lakes differs greatly in the degree of salinity and the type of salts dissolved. Compared to typical ocean water (approximately 35 parts per thousand, °/oo) some salt lakes are very salty. Great Salt Lake water has a dissolved solids content about four times that of sea water (150°/oo) and the Dead Sea about seven times (246°/oo). Some of the larger salt lakes have a much lower dissolved solids content, as the Aral Sea (11°/oo) and the Caspian Sea (6°/oo). The composition of salts depends in part on the geological character of the drainage area discharging into the lake, in part on the age of the lake, and in part on the excess of evaporation over inflow. As saturation is approached, the salts common in surface waters are precipitated in such an order that magnesium chloride and calcium chloride remain in solution after other salts have precipitated.

Lakes with fresh waters also differ greatly in the composition of their waters. Because of the balance between inflow and outflow, fresh lake water composition tends to assume the composite dissolved solids characteristics of the waters of the inflowing streams – with the lake's age having very little influence. Lakes with a sluggish inflow, particularly where inflowing waters have much contact with marginal vegetation, tend to have waters with high organic content. This may be observed in small lakes or ponds in a region where drainage moves through a topography of glacial moraines. Lakes formed within drainage areas having a crystalline, metamorphic, or volcanic country rock tend to have low dissolved solids content. Thus Lake Superior, with its major drainage from the Laurentian Shield, has a dissolved solids content of 0.05°/oo. The water of Grimsel Lake in the high Alps, Switzerland, has a dissolved solids content of only 0.0085°/oo. Lakes within limestone or dolomitic drainage areas have a pronounced calcium carbonate and magnesium carbonate content. As in all surface water, dissolved gases, notably oxygen, also are present in lake waters. Under a few special situations, as crater lakes in volcanic areas, sulfur or other gases may be present in lake water, influencing color, taste, and chemical reaction of the water. *See* FRESH-WATER ECOSYSTEM; MEROMICTIC LAKE.

Basin and regional factors. Most lakes are natural, and a large proportion of them lie in depressions of glacial origin. Thus alpine locations and regions with ground moraine or glacially eroded exposures are the sites of many of the world's lakes. The lakes of Switzerland, Minnesota, and Finland are illustrations of these types.

Lakes may be formed in depressions of differing glacial origin: (1) terrain eroded by continental glaciers, with the surface differentially deepened by ice abrasion of rocks of varying hardness and resistance; (2) valleys eroded differentially by valley glaciers; (3) cirques (glacially eroded valley heads in mountains); (4) lateral moraine barriers; (5) frontal moraine barriers; (6) valley glacier barriers; (7) irregularities in the deposition of glacial drift or ground moraine.

Lakes are particularly important surface features in the peneplaned ancient rocks of the Laurentian Shield and on the Fennoscandian Shield of northern Europe. The lake region of northern North America, which centers on the Laurentian Shield, probably has one-fifth to one-quarter of its surface in lake. Many streams in these areas are interrupted over more than half their total length by lakes. Lakes on the shields as a rule are island-studded and have extremely irregular outlines. The most permanent lakes lie on the shields themselves. Many such lakes on recently ice-scoured shields have fresh hard-rock rims with high resistance to erosion at their outlets. Because of generally low stream gradients and little sediment carried, the abrading and depositing stream actions are slow. As a result, the life of all but the small lakes in these areas probably will be measured in terms of a whole geologic period.

Some lakes in glacially formed depressions, as well as in other basins, may be considered barrier lakes. These glacial lakes are formed on glacial drift behind lateral or frontal moraine barriers. In addition, depressions of sufficient depth to contain a lake may be formed by (1) sediment deposited by streams (alluvium), and also stream-borne vegetative debris, such as tree dams on the distributaries and braided river courses of the lower Mississippi Valley; (2) landslides in mountainous areas; (3) sand dunes; (4) storm beaches and current-borne sediments along the shores of large bodies of water; (5) lava flows; (6) artificial barriers.

A large percentage of lakes is found in either the glacially formed or barrier-formed depressions. However, a few other types of depressions contain lakes: (1) craters of inactive volcanoes, or calderas (Crater Lake in Oregon is a famous example); (2) depressions of tectonic or structural origin (Great Rift Valley of Africa includes Lakes Albert, Tanganyika, and Nyasa); (3) solution cavities in limestone country rock; (4) shallow depressions cause a dotting of lakes in many parts of the tundra of high latitudes.

Conservation and economic aspects. Lakes created behind manmade barriers are becoming common features and serve multiple purposes. Examples are Lake Mead behind Hoover Dam on the Colorado River, Lake Roosevelt behind Grand Coulee Dam on the Columbia, Kentucky Lake and other lakes of the Tennessee Valley, and Lake Tsimlyanskaya on the Don.

Both natural and manmade lakes are economically significant for their storage of water, regulation of stream flow, adaptability to navigation, and recreational attractiveness. A few salt lakes are significant sources of minerals. Recreational utility, long important in the alpine region of Europe and in Japan, is now a major economic attribute of many American lakes. Economic value is generally increased by location near substantial human settlement. Most of the world's lakes, however, are located in regions where they have only minor economic significance at present. *See* EUTROPHICATION. [EDWARD A. ACKERMAN]

Bibliography: B. H. Dussart, *Man-Made Lakes as Modified Ecosystems: Scope Report 2*, 1978; R. Gresswell (ed.), *Standard Encyclopedia of the World's Rivers and Lakes*, 1966; International Association for Great Lakes Research, *Proceedings of the 10th Conference on Great Lakes Research*, 1967; P. H. Kuenen, *Realms of Water*, 1955; R. H. Lowe-McConnell (ed.), *Man Made Lakes*, 1965.

Land drainage (agriculture)

The removal of water from the surface of the land and the control of the shallow groundwater table improves the soil as a medium for plant growth. The sources of excess water may be precipitation, snowmelt, irrigation water, overland flow or underground seepage from adjacent areas, artesian flow from deep aquifers, floodwater from channels, or water applied for such special purposes as leach-

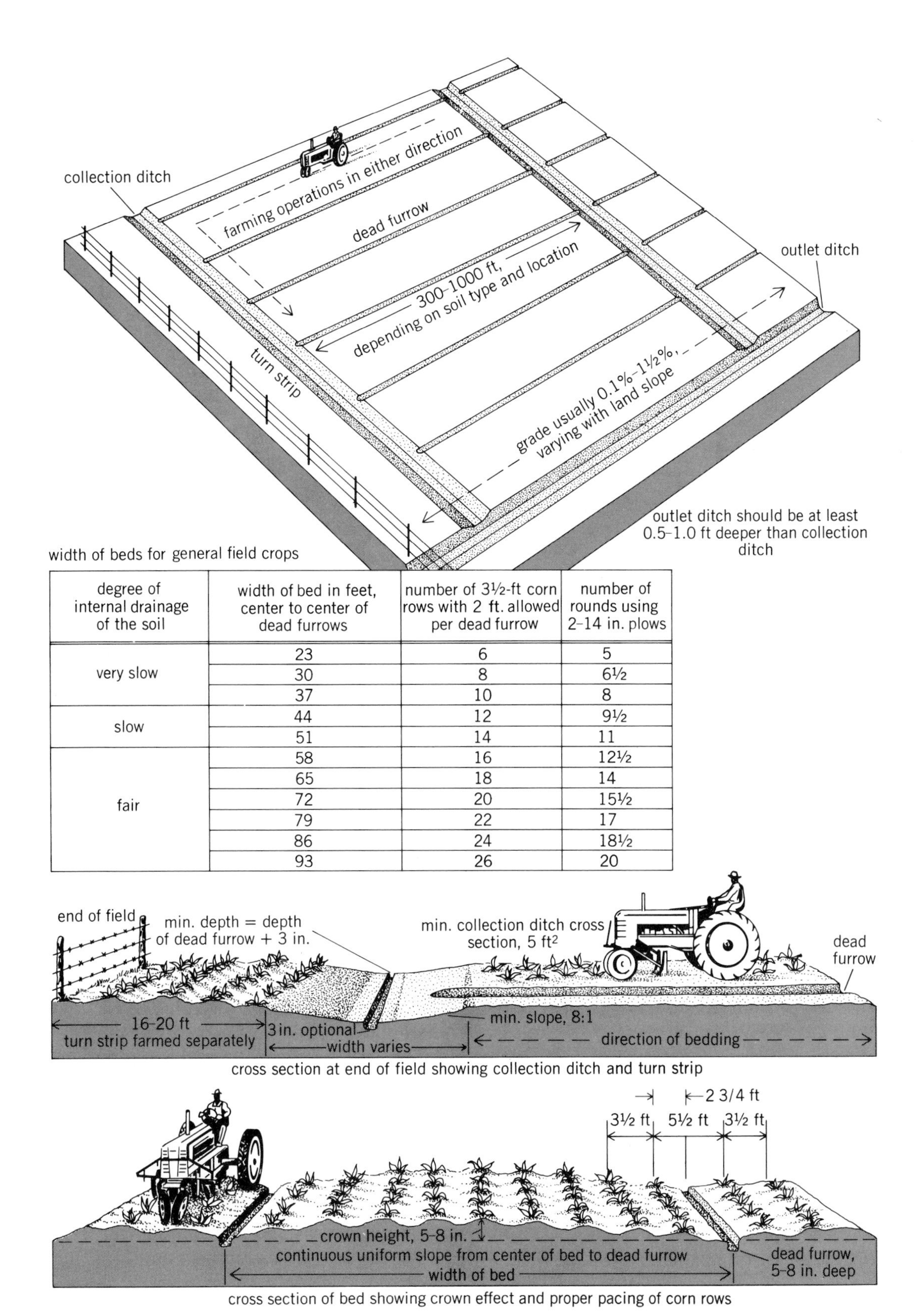

width of beds for general field crops

degree of internal drainage of the soil	width of bed in feet, center to center of dead furrows	number of 3½-ft corn rows with 2 ft. allowed per dead furrow	number of rounds using 2-14 in. plows
very slow	23	6	5
	30	8	6½
	37	10	8
slow	44	12	9½
	51	14	11
fair	58	16	12½
	65	18	14
	72	20	15½
	79	22	17
	86	24	18½
	93	26	20

Fig. 1. Surface drainage bedding system. 1 in. = 25.4 mm; 1 ft = 0.3048 m; 1 ft² = 0.0929 m². (*USDA*)

ing salts from the soil or for achieving temperature control.

The purpose of agricultural drainage can be summed up as the improvement of soil water conditions to enhance agricultural use of the land. Such enhancement may come about by direct effects on crop growth, by improving the efficiency of farming operations or, under irrigated conditions, by maintaining or establishing a favorable salt regime. Drainage systems are engineering structures that remove water according to the principles of soil physics and hydraulics. The consequences of drainage, however, may also include a change in the quality of the drainage water. Approximately 130,000,000 acres (526,000 km²), or about one-third of the cropland in the United States, is drained artificially.

Agricultural drainage is divided into two broad classes: surface and subsurface. Some installations serve both purposes.

Surface drainage. Poor surface drainage conditions exist over large areas of land in the eastern half of the United States and Canada. The condition is caused by the inability of excessive rainfall to move over the ground surface to an outlet or through the soil to a subsurface drainage system. These poor surface drainage conditions are usually associated with soils that have low hydraulic conductivity. Often the soils are very shallow over a barrier such as rock or a very dense clay pan. The impermeable subsoil prevents the water from moving downward and prevents the proper functioning of a subsurface drainage system. Often the land slope is not sufficient to permit the water to flow across the ground surface. In other cases, the areas lack adequate drainage outlets. In order to correct this problem, something must be done to eliminate the depressions and to provide sufficient slope for overland flow. In addition, it is necessary to provide channels to convey the water from the affected area.

The practice of surface drainage may be defined as the diversion or orderly removal of excess water from the land surface by means of improved natural or constructed channels. The channels may have to be supplemented by shaping and grading the land surface so that the water may flow freely into the channel.

In some instances, a subsurface drainage system of pipes is needed in conjunction with surface drains. The effectiveness of the subsurface drains is increased by the removal of the water from the soil surface. As soon as the surface water is removed, the drain pipes can act to lower the water table and to provide a satisfactory environment for the growth of plants. There are essentially five types of surface field drainage systems in common use today: bedding system (Fig. 1); random ditch system; interception system; diversion ditch system; and field ditch system. Combinations of two or more systems may be required by circumstances encountered in the field. The choice of a particular system used for surface drainage depends upon the soil type, topography, crops to be grown, and farmer preference.

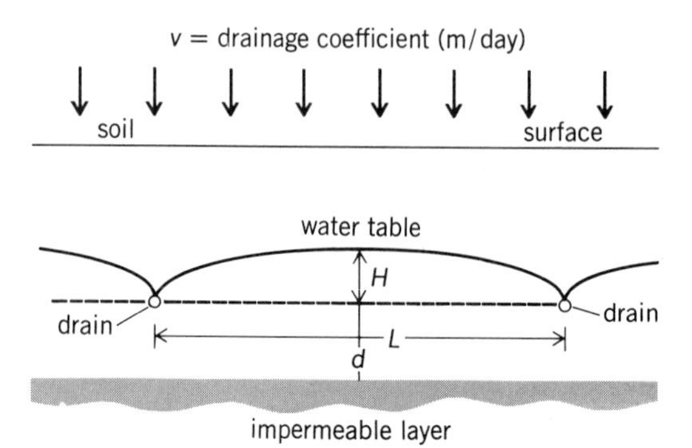

Fig. 2. Diagram of Hooghoudt's drain spacing formula. Symbols are explained in the text.

Subsurface drainage in humid regions. Subsurface drainage is required where a high water table is present. The main purpose of the drainage is to provide a root environment that is suitable for the maximum growth of plants and to sustain yields over long periods of time. One of the main reasons that poor drainage causes a decrease in crop production is the fact that the plant roots have only a limited amount of soil in which to grow. Not only do the plants lack food, but the plant roots suffer from a deficiency of oxygen which is needed for the respiratory processes, as the water that fills the soil pores displaces the air in the soil. Moreover, the water obstructs the gases which are given off by the roots, and some of these gases inhibit plant growth.

The critical need for drainage occurs in the early spring months when the plants are germinating. Lack of drainage retards the normal rise in the soil temperature, decreases plant resistance to disease, and inhibits root development. Poor drainage discourages the growth of aerobic bacteria which are needed to supply nitrogen for crops. Toxic organic and inorganic compounds develop in saturated soils.

The depth of drains in humid regions is largely determined by soil conditions. Drain depths from 60 to 120 cm are commonly used to control a shallow groundwater table. The spacing of the drains depends upon the soil hydraulic conductivity and the amount of the rainfall that must be removed. Spacings vary between 10 and 30 m.

A number of drain spacing formulas have been proposed, and one of the most successful is S. B. Hooghoudt's formula, Eq. (1),

$$L^2 = \frac{4K}{v}(H^2 + 2dH) \qquad (1)$$

where L = drain spacing (meters), K = hydraulic conductivity (meters/day), v = drainage coefficient, which is the rate of removal of water from the soil that is necessary to protect the crop from damage (meters/day), H = height of water table above the plane through drains (meters), d = distance from the plane through drains to the impermeable layer. Developed for the drainage of land in the Netherlands, the formula is based upon steady-state replenishment of the groundwater. The height of the water table halfway between the drains is calculated as a function of the soil hydraulic conductivity and the rate at which water replenished the groundwater table. The significance of the various parameters involved in the formula are presented in Fig. 2.

Drainage coefficients are of the order of 5–10 mm/day in Europe. Somewhat higher values (1–4 cm/day) are used in the United States.

Subsurface drainage in arid regions. Irrigation waters contain substantial quantities of salt, from 0.1 to 4 metric tons per 1000 m³. Irrigation water is applied at rates of 10,000 to 15,000 m³ per hectare per year, and hence between 0.1 to 60 metric tons is added to each hectare annually. Some of the applied salt precipitates in the soil; a small proportion is used by the plants, and the remainder of the salt must be removed from the soil by adding an amount of irrigation water in excess of the crop needs.

The output of salt in the drainage water must equal the input of salt in the irrigation water. If precipitation of salts in the soil and plant uptake of salts are ignored, it can be stated that salt input equals salt output.

The leaching requirement (LR) is defined as the fraction of irrigation water that must be drained in order to maintain the salt balance, as in Eq. (2),

$$\mathrm{LR} = \frac{v}{I} \tag{2}$$

where v represents the drainage coefficient and I the amount of applied irrigation water. The computation of v depends on the salt tolerance of the crop to be grown.

In arid regions, it is necessary to control the groundwater table well below the plant root zone. The plants will extract water from the soil, leaving salt behind. This results in a concentration of salts in the plant root zone; therefore, the water table must be maintained well below the plant root zone so that capillary rise of this salty groundwater into the plant root zone is reduced. The depth of the drains then is determined by capillary rise into the plant root zone (Fig. 3). Normally drains in irrigated areas are placed about 1.8 m deep. The spacing between the drains may be determined by Hooghoudt's formula. However, the drainage coefficient is determined by a consideration of the leaching requirement, precipitation of salts in soil from irrigation water, amount of leaching due to winter rainfall, and salt tolerance of crops. Typical drain spacings range from 20 to 100 m (Fig. 4).

In addition to the steady-state formula of Hooghoudt, analyses have been made of the transient

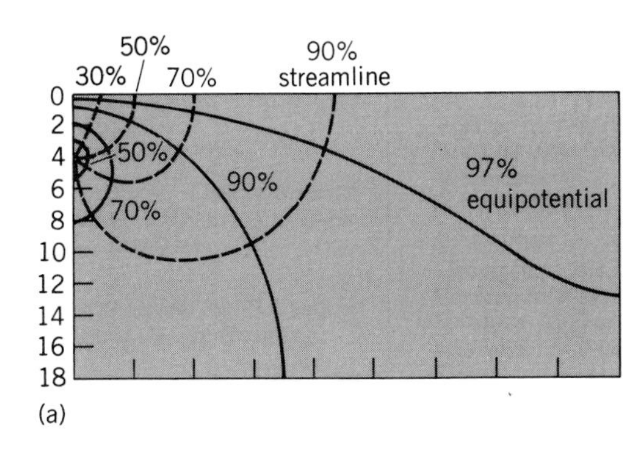

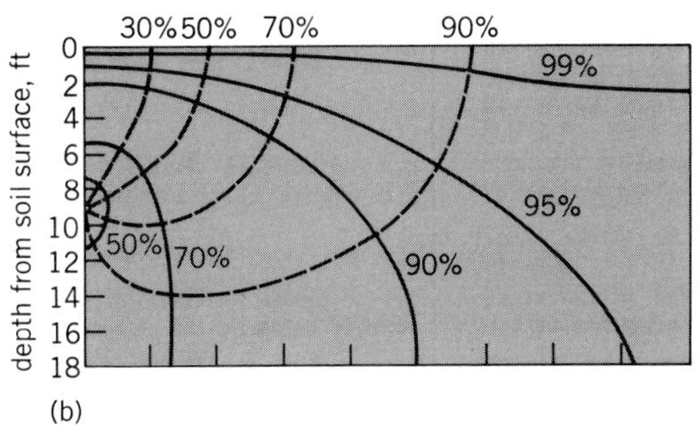

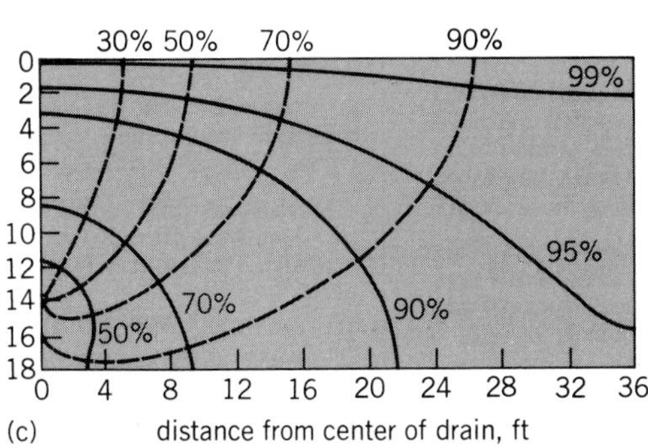

Fig. 3. Flow toward drains placed at various depths in the soil (Q is drain discharge; K is soil hydraulic conductivity). (a) $Q/K = 3.07$ ft²/ft drain (0.936 m²/m); (b) $Q/K = 5.41$ ft²/ft (1.649 m²/m); (c) $Q/K = 7.00$ ft²/ft (2.134 m²/m). *(From W. Burke and G. S. Taylor, Soil Stratification and Ponded Flow into Subsurface Drains, Ohio Agr. Exp. Sta. Res. Circ. 138, March 1965)*

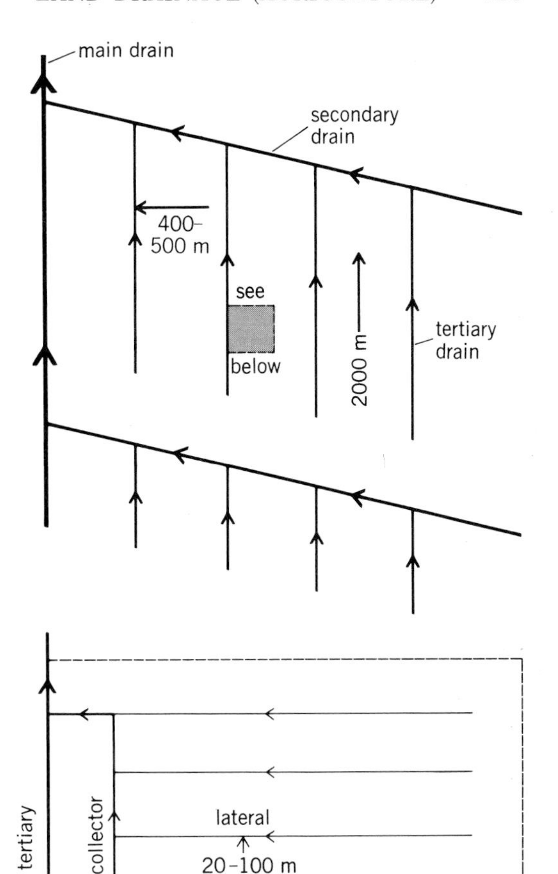

Fig. 4. Typical drainage system for an irrigation project.

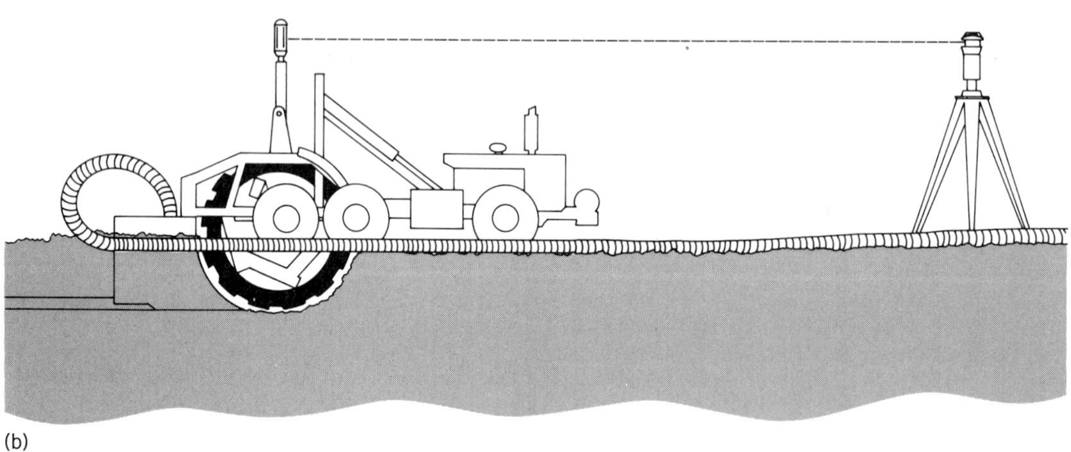

(b)

Fig. 5. Trenchless method of drain pipe installation. *(a)* Drain plow. *(b)* Grade control by laser beam.

water table situation. The U.S. Bureau of Reclamation utilizes this transient water table formula for determining the average depth of water table during the growing season. If according to the bureau's method of analysis the water table is rising during the growing season, and continues to rise, the condition indicates a serious drainage problem will occur. The bureau's drainage criteria then are based upon controlling the water table so that it either remains constant or declines over a period. The procedure is called the dynamic equilibrium method.

Construction of drains. The construction of subsurface drains was revolutionized in the years 1976–1979. In the past, a variety of materials such as rocks, clay pipe, concrete pipe, and other materials have been used for subsurface drains. With the invention of machines that make perforated corrugated plastic pipe, this pipe has largely supplanted the other materials. Corrugated plastic pipe has the advantage of light weight, and it comes in long lengths so the cost of handling is reduced.

Drain pipes either are laid in the bottom of a trench or are pulled into place by a drain plow. The development of laser beams for grade control has facilitated the use of drain plows (called the trenchless method) for installing corrugated plastic tubing (Fig. 5).

In unstable soils usually found in arid regions, it

is necessary to surround the drain pipe with gravel, sand, synthetic fibers, or organic material in order to prevent fine sands and silts from entering the pipe with the drainage water. See IRRIGATION OF CROPS. [J. N. LUTHIN]

Bibliography: J. N. Luthin, *Drainage Engineering*, rev. ed., 1978; J. N. Luthin, *Drainage of Agricultural Lands*, 1958; G. O. Schwab et al., *Soil and Water Conservation Engineering*, 1966; J. van Schilfgaarde, *Drainage for Agriculture*, in press.

Land reclamation

The process by which seriously disturbed land surfaces are stabilized against the hazards of water and wind erosion. A permanent vegetative cover usually provides the most economical means for stabilization. All seriously disturbed land areas are in need of reclamation of some kind and should be stabilized as quickly as possible after disturbance. In the United States alone, about 80,000 hectares (179,200 acres) per year is disturbed. Disturbance comes from major construction projects such as interstate highway systems, shopping centers, housing developments, and from surface mining operations for coal, stone, gravel, gold, phosphate, iron, uranium, and clay. Surface mining for coal is responsible for almost one-half of the total land area disturbed in the United States, another one-fourth is from sand and gravel, and the remainder is from mining of other materials and construction.

Surface coal mining. During the 1970s, the recovery of coal by surface mining increased dramatically because of the increased emphasis on coal as an energy source. As other forms of fossil energy become less available, the reliance on coal for energy will become even greater. Coal deposits are found in most major countries, with total world deposits estimated at 1.68×10^{13} tons. Approximately 33% of the total world coal reserves are found in the United States alone. Thirty-seven of the 50 states have known coal deposits, with 13% of the total land area in the United States being underlined by coal (see illustration). The amount of these reserves recoverable by stripping is not known since new mining technology is rapidly changing and overburden removal is basically a question of economics.

At the present time, the coal required to meet power production needs cannot be supplied from deep mining. This means that to continue meeting the power requirements demanded by the public, thousands of acres will be added annually to those already stripped. Unless precautionary measures are taken to stabilize these severely disturbed areas, serious pollution of the environment will result. Surface mining does create many environmental problems. The chemical and physical properties of the resulting spoil is drastically changed and can create a hostile environment for seed germination and subsequent plant growth. However, many of these environmental problems can be overcome or eliminated by proper planning prior to mining and by proper placement of overburden material during mining operations. Even with the best mining practices, vegetative cover must be established almost immediately or the denuded areas will be subject to both wind and water erosion that will pollute surrounding streams with sediment.

Eastern United States. Coal is surface-mined on land with topography ranging from gentle slopes to rugged mountain terrain. Strip areas in the Appalachian Region sometimes are highly variable, with

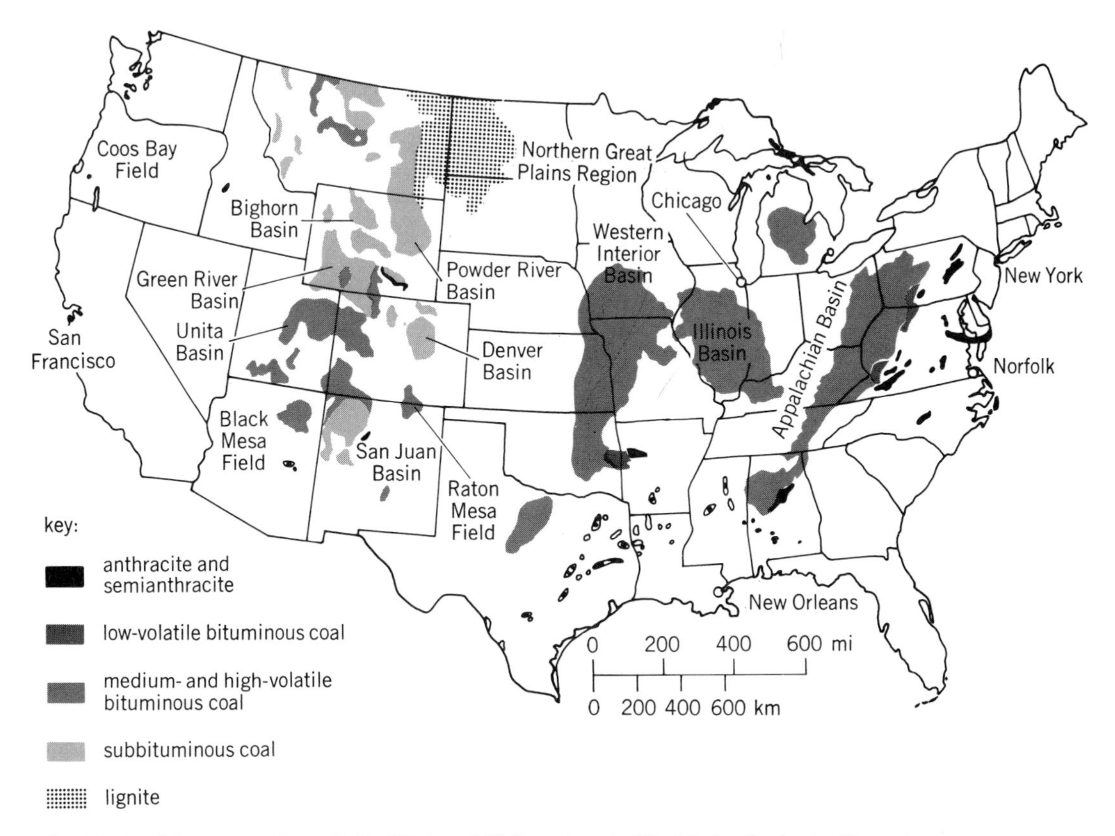

Coal fields of the conterminous United States. (*U.S. Department of the Interior, Geological Survey*)

the chemical and physical properties of spoil material being dramatically different within a few feet. Vegetative cover can be established easily on some areas, but other areas exhibit extremely hostile environments for germinations and seedling growth. Before any intelligent attempt can be made to revegetate these areas, the chemical and physical problems must be identified. In most cases, the spoil material is a conglomerate of rock fragments of various sizes mixed with a small amount of soil. The spoil materials may vary in color from very light to almost black. Because of these color differences alone, temperature and soil moisture may constitute serious problems. Temperature variations created by slope and aspect require use of specific plant species. Research has shown that mulching is usually essential for germination and seedling establishment. Some spoils are almost sterile and usually require applications of many of the nutrient elements necessary for plant growth. Phophorus, potassium, calcium, and magnesium deficiencies are major problems on some spoils in the eastern United States, while in other spoils some plant nutrients may be more abundant after mining than before due to weathering of rocks and minerals brought to the surface. In the eastern United States, chemical constraints for reclamation in humid areas are low fertility, low pH or high acidity, and toxic concentrations of elements such as aluminum, manganese, and iron.

Western United States. Under western conditions, excessive amounts of sodium and other elements may constitute serious reclamation problems. High sodic spoils may require applications of surface soil before good vegetative growth can be accomplished. Under arid and semiarid conditions, lack of available plant soil water is a factor that ultimately limits plant growth on reclaimed land, just as on unmined land. Practices that increase infiltration, reduce vaporation, and increase potential plant growth, generally improve the effectiveness of water conservation and use. Environmental problems of major concern under western conditions include salinity levels, exchangeable sodium content, nutrient deficiencies (nitrogen and phosphorus), plant toxicities (magnesium, boron, molybdemun), soil compaction, and steepness of slope.

Topsoil replacement. In 1977 the U.S. Congress passed the Federal Strip Mine Law, which requires that topsoil be removed and reapplied on the spoil surface during regrading and reclamation. This practice alone has aided materially in reclamation of strip mine spoil areas throughout the United States. Even when topsoil is reapplied, the surface may contain coarse-textured materials and rock fragments, making it difficult to establish vegetative cover. Many of the eastern mine spoils are derived from sandstone and shales and have a low water-holding capacity. These spoils tend to form crusts and thus create a water-impermeable layer. Practically all of these topsoils have low fertility and thus require extensive fertilization for reclamation and seedling establishment.

Regardless of the location, it has been shown through research that many of the chemical and physical limitations previously associated with mine spoils can be eliminated by replacing the topsoil after mining. In the western United States, appreciable quantities of topsoil are often available for spreading over spoil after final grading. As little as 2 in. (5 cm) of soil material placed over sodic spoils under western conditions produced yields of native grasses equal to 50–70% of those obtained where as much as 30 in (76 cm) or more of topsoil was replaced. They found that poor physical conditions caused by high sodium content, as well as problems related to nutrient deficiencies, toxicities, and soil-water relationships, were mostly alleviated by replacing topsoil. The situation is somewhat different under high-rainfall conditions such as found in a mountainous region of the eastern United States. These soils tend to be shallow, infertile, and stony, and have a low water-holding capacity. Even so, replacing the topsoil on the spoil material after final grading has aided materially in the reclamation process.

Placement of overburden. In the eastern United States, many of the soils underlined by coal have very low productivity. It has been shown that with proper placement of overburden during mining and use of better regrading techniques and proper surface management, many of these surface mine areas can be more productive after mining than before. Some overburden materials contain large amounts of calcium, magnesium, phosphorus, and potassium. If these materials are brought to, or near, the soil surface, weathering may provide a better soil than was there initially.

Plant species. A considerable amount of research has been conducted to determine desirable plant species for use on seriously disturbed land areas. Research has shown that almost any plant species can be grown on these sites if environmental and nutritional requirements are met. Crops tested include most of the agriculturally important grasses and legume species, horticulture species, and forest species. The law requires that herbaceous species such as grasses and legumes be established first. Tree species should be planted with, or after, stabilization with grasses and legumes.

A large number of domestic and industrial waste materials have been tested as amendments on strip mine spoils. These include digested sewage sludges, composted sewage sludge and garbage, tannery waste, bark and fiber mulches, flyash, fluidized-bed combustion waste, and scrubber sludges. Many of these materials contain considerable amounts of essential plant nutrients, but they may also contain some potentially toxic elements. Research is being conducted to determine plant uptake of potentially toxic elements and to evaluate possible risk of diseases associated with handling and disposal of these waste materials. So far, organic waste materials, such as composted sewage and garbage and sewage sludges, have been utilized effectively without any detrimental effect.

Outlook. The production capacity of many mine spoils is not yet fully known or understood. Management and wise use must be practiced in many surface mine areas before the full potential of these reclaimed lands is reached. Location and accessibility will determine agricultural, recreational, and industrial or urban development potential. However, because of the new Federal strip mine laws, future reclaimed areas will have much broader potential use than areas previously mined.

With preplanning and proper reclamation practices, more desirable plant species can be established to control erosion and sediment as well as to offer some economic potential to the landowners. The ultimate goal of research in the United States is to ensure that the nation's energy and mineral needs are met in a reasonable, selective, and orderly way without sacrificing food, fiber, quality of life in rural areas, or the quality of the total environment. At the present time, there are no assurances that all strip mine areas will be reclaimed in an acceptable manner and without environmental damage. Many problems associated with strip mining and reclamation still need answering. But, hopefully, with industry and the scientific community working together, mineral resources can be extracted while protecting the environment and the future productivity of the soil.

[ORUS L. BENNETT]

Bibliography: W. H. Armiger, J. N. Jones, Jr., and O. L. Bennett, *Revegetation of Land Disturbed by Strip Mining in Appalachia*, ARS-NE-71, 1976; O. L. Bennett, W. H. Armiger, and J. N. Jones, Jr., Revegetation and use of eastern surface mine spoils, in *Land Application of Waste Materials*, Soil Conservation Society of America, 1976; J. N. Jones, Jr., W. H. Armiger, and O. L. Bennett, Forage grasses aid the transition on spoil to soil, *Proceedings of the National Coal Association, Bituminous Coal Research Conference Exposition, II*, Louisville, KY, 1975; J. F. Power et al., Factors restricting revegetation of strip-mine spoils, W. F. Clark (ed.), *Proceedings of the Fort Union Coal Field Symposium, Montana Academy of Science*, 1975; G. E. Schuman, W. A. Berg, and J. F. Power, Management of mine wastes in the western United States, in T. M. McCalla (ed.), *Land Application of Waste Materials*, Soil Conservation Society of America, 1976; *Surface Mining Control and Reclamation Act of 1975*, H. R. 25, 94th Congress.

Land-use classes

Categories into which land areas can be grouped according to present use of suitability or potential suitability for specified use or according to limitations which restrict their use. The term is most commonly applied to land uses for productive purposes, such as agriculture or forestry, but may be applied for any use including engineering, architecture, urban development, wildlife, and recreation.

A complete description and assessment of land involves climate, land form, surface details, rock type, soil, vegetation, subsurface characteristics, hydrological features, and geographically associated factors, such as availability of water for irrigation purposes, location, and accessibility. Components of the land complex vary considerably from place to place, individually and in combination. The many attributes of land interact and are not equally important for all use purposes in all situations. This makes both the assessment of land potential and the specification of land requirements for a particular purpose difficult and often highly subjective. There is a trend toward, and a need for, more quantitative precision in methods. Approaches to land-use classification are based on present land use, land component surveys, landscape analysis, mathematical procedures, or a combination of these.

Present land use. The World Land Use Survey sponsored by the International Geographical Union aimed to encourage countries to map their own lands in terms of a uniform series of nine broad categories of use: settlements and associated nonagricultural land, horticulture, tree and other perennial crops, crop lands, improved permanent pasture, unimproved grazing land, woodland, swamps and marshes, and unproductive lands. Many countries which have reported on land-use surveys have introduced numerous subdivisions appropriate to the local situation.

More detailed classifications in terms of present land use required for statistical, administrative, or management purposes may be based on a variety of criteria, such as areas of individual crops or forest types, yields, disease and pest occurrence, climatic hazards, pasture types, animal grazing capacity, land-management systems, or purely economic factors (for example, input-output ratios and land values). If used in conjunction with potential land-use classifications, these can be used for comparative purposes or, in the area studied, to indicate where further productivity can be achieved or present land use modified.

Land-component surveys. The common basis for most methods is subdividing the land surface into unit areas which, at the scale of working, are essentially homogeneous in relation to use possibilities and follow with an assessment of use potential by comparison with known situations or responses. The characteristics used for subdivision are the inherent features of land, such as geology, land form, soil, climate, and vegetation. Although classifications on the basis of individual factors such as vegetation or land form have their use, single components of land are inadequate to determine land usefulness precisely. Most attention has been given to classification based on soil surveys, but associated features such as climate and topography are usually taken into account as well.

A widely used system for grouping soil taxonomic or mapping units into land-use classes is that of the USDA handbook *Land-Capability Classification*. The USDA system aims to assess suitability of soils for adapted or native plants and the kind of management the soils require to maintain continued productivity. Assessments are made largely on the kind and degree of hazards or limitations to productivity. This system provides for eight capability classes with a number of subclasses and units identified by these limitations. *See* SOIL, SUBORDERS OF.

The first four classes include groups of soils suitable for cultivation, but from class I to class IV the choice of suitable plants becomes more restricted or the need for more careful management or conservation practices increases. Classes V to VII soils are generally restricted to pasture, range, woodland, and wildlife use. Class VIII soils are restricted to nonagricultural purposes.

As moderately high levels of land management and of inputs and a favorable ratio of inputs to outputs are assumed, the method as a whole can be applied with precision only in areas where knowledge of land use is well advanced. In less developed areas, classification must be limited to the broader categories. In undeveloped countries with low economic ceilings, judgments need to be modified according to local standards.

The same general principles are applied, with appropriately selected criteria, for grouping soils into land classes with different degrees of suitability for a variety of purposes, such as woodland establishment, recreation, wildlife, and engineering. *See* SOIL ZONALITY.

A method adopted by the Canada Land Inventory for forestry purposes illustrates a variant of this approach. It adopts a division of the land surface into homogeneous units determined by physical characteristics, followed by a rating of these units into seven capability classes according to environmental factors which influence their inherent ability to grow commercial timber. These factors are the subsoil, soil, surface, local and regional climate, and the tree species. A feature of this sytem is that a productivity rating is set in quantitative terms for each class expressed as volume of merchantable timber per acre per year. Regional inventories include a reference to the indicator tree species present for each class.

A system of land classification for the specific purpose of establishing the extent and degree of suitability of lands for sustained profitable production under irrigation has been developed by the U.S. Bureau of Reclamation. Suitability of land is measured in terms of payment capacity and involves consideration of potential productive capacity, costs of production, and costs of land development. These are assessed from soil characteristics and topographic and drainage factors. Six basic classes of land are recognized. Classes I to III are all arable lands, but they decrease in payment capacity and become more restricted in usefulness in that order. Class IV includes lands with special uses. Class V lands are at least temporarily nonirrigable, and Class VI lands are unsuitable for irrigation use.

Landscape analysis. An alternative to the subdivision of land according to single components is the subdivision of the landscape itself into natural units, each characterized by a combination of geologic, land-form, soil, and vegetation features. This approach, referred to as the land-system approach, has been developed in a number of countries but especially in Australia. It is particularly well suited to the use of aerial photographs and has special value in the reconnaissance survey of little-known lands. The approach is based on the concept that there are discernible natural patterns of landscape covering areas with common and distinctive histories of landscape genesis. The boundaries of land systems coincide with major changes in geology or geomorphology. The pattern of a land system is formed by a number of associated and recurring land units. Each occurrence of the same land unit within a specific land system represents a similar end product of land-surface evolution from common parent material by common processes through the same series of past and present climates over the same period of time. Thus, in addition to being described similarly in terms of observable slopes and surfaces, vegetation cover, and soils, they are assumed to have a similar array of natural and potential habitats for land use. This record of the inherent features of the landscape can be interpreted in terms of land-use classes in the light of the technical knowledge available at any subsequent time. Surveys are made by concurrent and integrated studies of all the observable land features by a team of specialists in the fields of geomorphology, soils, and plant ecology. Conclusions to be drawn about immediate land use must be derived by analogy with known areas or from basic principles. The method has particular value in less developed regions. It is also being applied to special purposes such as forestry and engineering. *See* FOREST MANAGEMENT AND ORGANIZATION; GEOMORPHOLOGY.

Mathematical approaches. Land-use classification dependent upon descriptive data of land characteristics and analog processes of assessment suffer from inherent inadequacies of descriptive processes and involve a good deal of intuition and subjectivity. For this reason, effort has been made to introduce more quantitative approaches to the assessment of potentials. Foresters have developed methods of site evaluation based on the measurement of environmental factors and productivity at different sites. Key parameters are identified from multiple regressions. This information can then be applied to classification of areas in terms of potential production.

With the advent of modern computers capable of handling and storing masses of data, there is a rapidly growing trend toward quantifying land characteristics and land-use responses and using mathematical models which relate the numerous land parameters to a variety of use responses in agricultural, forestry, and engineering fields.

The automatic scanning equipment for aerial photography and the remote sensing and automatic recording devices used with Earth satellites are opening up completely new approaches to land description and subdivision and hence to land classification. *See* AGRICULTURE; LAND-USE PLANNING; LIFE ZONES. [C. S. CHRISTIAN]

Bibliography: C. S. Christian and G. A. Stewart, Methodology of integrated surveys, *Aerial Surveys and Integrated Studies: Proceedings of the Toulouse Conference 1964*, UNESCO, 1968; *Ecological Land Classification in Urban Areas*, Ecological Land Class Series, no. 3, 1978, *Field Manuals for Land Capability Classifications*, Canada Land Inventory (ARDA), Department of Forestry and Rural Development, various years; G. A. Hills, *The Ecological Basis for Land-Use Planning*, Ont. Dep. Lands Forests Res. Rep. no. 46, December, 1961; A. A. Klingebiel and P. H. Montgomery, *Land-Capability Classification*, Soil Conservation Service, USDA Handb. no. 210, 1961; D. S. Lacate, *Forest Land Classification for the University of British Columbia Research Forest*, Can. Dep. Forest. Publ. no. 1107, 1965; R. J. McCormack, *Land Capability for Forestry: Outline and Guidelines for Mapping*, Canada Land Inventory (ARDA), Department of Forestry and Rural Development, 1967; D. L. Stamp, *Land Use Statistics of the Countries of Europe*, World Land Use Surv. Occas. Pap. no. 3, 1965; D. L. Stamp, *Our Developing World*, 1960; G. A. Stewart (ed.), *Land Evaluation*, in press.

Land-use planning

Humanity has gained unprecedented physical and technical resources to regulate use and misuse of lands and resources. Planning for the distant as well as immediate consequences of man's actions is essential for long-run economy, and such investigations can warn of irreversible and irreparable deterioration in the quality of environment and life. *See* APPLIED ECOLOGY; ECOLOGY; ECOSYSTEM; VEGETATION MANAGEMENT.

The objective of land-use planning on an individual ownership may exceed maximizing the net income of the owner. The objective of land-use planning by a public agency is even more complex and is intended to maximize long-range community benefits. Three characteristics of social and economic history nurtured contemporary land-use problems.

First, strong competitive forces, promoted by a philosophy of free private enterprise, urged rapid uncoordinated exploitation of land and its resources. This exploitation was characterized by extensive use instead of intensive methods known to modern technology.

Second, early land exploitation demonstrated real but primitive understanding of natural land capability. The plow followed the ax in many areas where agriculture cannot do as well as forests; the plow broke the plains in many places where crops are not as compatible with the climate as more drought-resistant forage; and the plow has turned the sod of fields whose soil and slope cannot support cultivation without rapid erosion and soil depletion.

Third, early city development, in the absence of a land-use policy and plan, resulted in an indiscriminate mixture of residential, industrial, and commercial land uses. Residential areas often suffered deflated values, and commercial and industrial enterprises failed to achieve possible economies in transportation, power, and waste disposal. As deterioration of the central city set in, residences, stores, service establishments, and industries favored the suburbs for new location. This situation suggested needs for public decisions and controls on major aspects of land use in the central city and in the suburbs of every metropolitan center.

Rural land planning emphasizes the development of land-use patterns which reflect the physical and biological limits beyond which long-run depletion of the land resource will result. Increasingly, private land operators are learning that production according to land capability is good business. Since the great depression of the 1930s, both Federal and state governments have cooperatively developed plans for improved productivity of land and related resources of various large natural regions of the nation. The Great Lakes Cut-Over Region and the Northern Great Plains were areas early subject to such regional studies. The Tennessee Valley Authority (1933) marked a further emphasis upon regional resource planning. TVA and Appalachia development programs have had as their primary planning objective the discovery of economic opportunities and social amenities.

The importance of regional and major drainage basin resource development is indicated by rapid expansion since World War II. In addition, hundreds of local organizations such as watershed associations, various special districts, and intergovernmental planning committees have sprung up in response to land-use problems. These smaller regions may be dominated by a vigorous urban center. Here rural-type resource development problems intermingle with the problems of space allocation and design of the spreading urban area. Land-use planning during the later years of the 20th century is challenged to meet the needs of a planning area, composed of admixtures of rural and urban land-use problems, and recreation fringes.

Planning and the future. Trends in land-use planning suggest four developments in concept and practice. First, integrating the space-use considerations of urban-exurban planning and the resource-use considerations of wild-land, rural planning will demand increasing attention. More interchange and adaptation should take place between the space design orientation of the city planner, architect, and landscape designer and the resource capability orientation of the resource planner. The overlapping metropolitan areas must be planned to meet the basic regional needs for water, waste disposal, transportation, and open spaces for light, air, and recreation. More regard to the natural capability of each environment is needed to sustain an optimum level of regional economic opportunity and social values. Community requirements and the environmental capability should find reconciliation in a new type of integrated regional design.

Second, standards and criteria to guide land-use allocations must be developed by the students and practitioners of land-use planning and become incorporated in public land policies. *See* FOREST AND FORESTRY; RANGELAND CONSERVATION; SOIL CONSERVATION; WATER CONSERVATION; WILDLIFE CONSERVATION.

Third, if land planning is to be made an instrument of democracy, improvements are required in the methods by which social choices of the majority can find expression in the planning process, without ignoring individualism and diverse goals of a pluralistic society. Approving of general criteria for land-use allocation or of specific plans, should be made by people of the community through the operation of the political process. New communication channels can be more articulate in serving the political process and in responding to it. Experimentation should be applied to assure representation of the community and provide protection against misuse from self-seeking interests.

Fourth, to assure that plans will be carried out, closer relationships must be established between the planning authority and the agencies which exercise powers of implementation. A city planning commission may be established somewhat apart from the government authority available to implement the plan. Regional planning bodies, whether oriented primarily to large metropolitan regions or to large resource development regions, have a planning area and a scope of functions for which no single governmental body can serve as the implementing authority. Both of these situations make it difficult to mesh the "planning gears" with the "administrative gears." Political and administrative sciences are challenged to discover institutional forms and procedures through which the early and final planning process and the public power of implementation can interact responsibly.

By the 21st century, science and technology can wrest from the earth practically anything that is left over from prodigal use in the 20th. Critical problems in land-use planning of the future will arise from the difficulty of determining what is sought. These problems are not scientific and technological but they are human ones. They involve reconciling different human desires and organizing for cooperative decisions and programs of action. The challenge is to mold the natural sciences and technologies into an environmental framework with the applied social sciences to produce a land-use scheme to serve man best.

Zoning of land. Industrial societies often have zoned their lands (if at all) according to current economic values, forgetting long-range costs until struck by catastrophe. Residential areas and factories are built on floodplains, maybe unwittingly, because development is easier and cheaper on level lands than on hills. If costs of flood damage or of flood control are considered, major investments might well be allocated to uplands.

Farms, especially on rich soil, in large blocks suitable for mechanized treatment are frequently removed from best natural uses by urban sprawl. Many need legal protection from encroachment and fragmentation, and taxation policies which unfairly penalize open land or continued farming.

Developed parks and protected reserves, demonstrating natural plant and animal communities typical of diverse sites and regions, can include some areas too wet, dry, or shallow-soiled for other purposes. For educational and scientific as well as recreational purposes they also deserve high priority in the face of competing uses, according to each area's quality and location. National parks and monuments preserve unique scientific and scenic treasures, but even these are becoming degraded through lack of internal park zoning or control. Trampling of plants, compaction of soil, other abuse by pack animals or vehicles, wind or water erosion, noise and other disturbance, and unwise attraction of animals by feeding increasingly upset Nature's balances in the very places that attract most public use.

American legislation protecting wilderness areas in national forests and other areas as well as parks is a landmark of policy recognizing different levels and kinds of use, and sometimes calls for hard choices of clear priority. Since the first hearings under this legislation (for the Great Smoky Mountains National Park) the record of widespread public preference has commonly been for more inclusion of areas, and more exclusion of highways and resorts than were favored by local developers, and by officials who feel committed to the latter. To deflect commercialization within parks has demanded long-range plans in a larger regional context, commonly favoring wider private service to visitors while firmly limiting massive encroachment and destruction in the heartland.

At village, city, county, and state levels as well, there is a need for planning and implementing a balance of the most suitable areas for (1) direct urban and suburban uses, (2) use and, where pertinent, renewal of natural resources, and (3) acquisition or at least options allowing future control of landscapes and water areas. Nature conservancies (private in the United States, part of the Natural Resources Research Council in the United Kingdom) have been alert in finding and acting on opportunities.

Awareness and promptness. Urban blight and depressing obsolescence of huge, monotonous suburban developments are more symptoms—very expensive ones—of complex problems of a technologically oriented society. Great cities, stimulated to haphazard growth by industry in the 19th century, now may put on a front of greenery along modern roadways and developments; but this often masks failure to adapt the mix of land uses to the possibilities and real limitations of resources. *See* LAND-USE CLASSES.

Population explosions may be slower in industrial countries or regions than elsewhere; still their pressures and mobility commit space for industry, highways, airports, shopping centers, asphalt parking "deserts," and housing developments too fast for collecting and weighing relevant factors.

Democracies assert the right of each person, from the open country to the concrete city, to learn the choices that affect his life, to use expert counsel on the repercussions of the choices, and to exert his responsibility for influencing decisions. Yet even where mechanisms exist to implement these rights, irrevocable choices are often committed legally, or turned into reality, before their full consequences are either explored or identified. Even where conscientious citizens warn legislators or administrators of dangers or preferred alternatives, lobbying often speaks fastest and loudest from the sources having financial interest in a particular scheme.

Pollution of air, water, and landscapes is a problem which recently gained a public spotlight, without yet having wise enforcement against obvious abuses, or recognition of subtle ones. Disposal of residues in ways that will not upset the environment is a growing social problem. Effective management requires knowing the path and potential effects of each waste product until it becomes neutralized. So much public attention was rightly drawn to increasing radioactive waste in a nuclear-powered economy and to the analysis of "maximum credible" accidents (and of some incredible contingencies as well) that the far-reaching effects have been subjected to study as never before. *See* RADIOECOLOGY.

Modern man cannot escape costs of containment of by-products, of monitoring the inevitable releases to maintain low and tolerable levels, and of providing for emergency action in case of accidents. Such costs have been charged to governments and to increasing numbers of other producers of nuclear energy and materials. To control sulfur dioxide (from coal), exhaust effluents (from internal combustion engines), petroleum, and other chemical wastes, it is necessary to decide on passing along similar costs to consumers or taxpayers as a payment for maintaining or improving environmental quality. Land-use planning in site selection of major facilities, such as electric generators (which all produce waste heat, regardless of radioactive or combustion effluents), is essential for limiting these costs and maintaining compatibility with social values. [JERRY S. OLSON]

Bibliography: S. Chase, *Rich Land, Poor Land*, 1936; Environmental Pollution Panel, President's Science Advisory Committee, *Restoring the Quality of Our Environment*, 1965; J. G. Fabos, *Plan-*

ning the Total Landscape: A Guide to Intelligent Land Use, 1978; C. M. Harr, *Land Use Planning: A Casebook on the Use, Misuse, and Re-use of Urban Land*, 1976; R. Lord, *The Care of the Earth: A History of Husbandry*, 1962; V. Obenhaus, L. Walford, and J. Olson, Technology and man's relation to his natural environment, in C. P. Hall (compiler), *Human Values and Advancing Technology*, 1967.

Life zones

Large portions of the Earth's land area which have generally uniform climate and soil and, consequently, a biota showing a high degree of uniformity in species composition and adaptations to environment. Related terms are vegetational formation and biome.

Merriam's zones. Life zones were proposed by A. Humboldt, A. P. DeCandolle and others who emphasized plants. Around 1900 C. Hart Merriam, then chief of the U.S. Biological Survey, related life zones, as observed in the field, with broad climatic belts across the North American continent designed mainly to order the habitats of America's important animal groups. The first-order differences between the zones, as reflected by their characteristic plants and animals, were related to temperature; moisture and other variables were considered secondary.

Each life zone correlated reasonably well with major crop regions and to some extent with general vegetation types (see table). Although later studies led to the development of other, more realistic or detailed systems, Merriam's work provided an important initial stimulus to bioclimatologic work in North America.

Work on San Francisco Mountain in Arizona impressed Merriam with the importance of temperature as a cause of biotic zonation in mountains. Isotherms based on sums of effective temperatures correlated with observed distributions of certain animals and plants led to Merriam's first law, that animals and plants are restricted in northward distribution by the sum of the positive temperatures (above 43°C) during the season of growth and reproduction. The mean temperature

Characteristics of Merriam's life zones

Zone name	Example	Vegetation	Typical and important plants	Typical and important animals	Typical and important crops
Arctic-alpine zone	Northern Alaska, Baffin Island	Tundra	Dwarf willow, lichens, heathers	Arctic fox, muskox, ptarmigan	None
Hudsonian zone	Labrador, southern Alaska	Taiga, coniferous forest	Spruce, lichens	Moose, woodland caribou, mountain goat	None
Canadian zone	Northern Maine, northern Michigan	Coniferous forest	Spruce, fir, aspen, red and jack pine	Lynx, porcupine, Canada jay	Blueberries
Western division					
Humid transition zone	Northern California coast	Mixed coniferous forest	Redwood, sugar pine, maples	Blacktail deer, Townsend chipmunk, Oregon ruffed grouse	Wheat, oats, apples, pears, Irish potatoes
Arid transition zone	North Dakota	Conifer, woodland sagebrush	Douglas fir, lodgepole, yellow pine, sage	Mule deer, whitetail, jackrabbit, Columbia ground squirrel	Wheat, oats, corn
Upper Sonoran zone	Nebraska, southern Idaho	Piñon, savanna, prairie	Junipers, piñons, grama grass, bluestem	Prairie dog, blacktail jackrabbit, sage hen	Wheat, corn, alfalfa, sweet potatoes
Lower Sonoran zone	Southern Arizona	Desert	Cactus, agave, creosote bush, mesquite	Desert fox, four-toed kangaroo rats, roadrunner	Dates, figs, almonds
Eastern division					
Alleghenian zone	New England	Mixed conifer and hardwoods	Hemlock, white pine, paper birch	New England cottontail rabbit, wood thrush, bobwhite	Wheat, oats, corn, apples, Irish potatoes
Carolinian zone	Delaware, Indiana	Deciduous forest	Oaks, hickory, tulip tree, redbud	Opossum, fox, squirrel, cardinal	Corn, grapes, cherries, tobacco, sweet potatoes
Austroriparian zone	Carolina piedmont, Mississippi	Long-needle conifer forest	Loblolly, slash pine, live oak	Rice rat, woodrat, mocking bird	Tobacco, cotton, peaches, corn
Tropical zone	Southern Florida	Broadleaf evergreen forest	Palms, mangrove	Armadillo, alligator, roseate spoonbill	Citrus fruit, avocado, banana

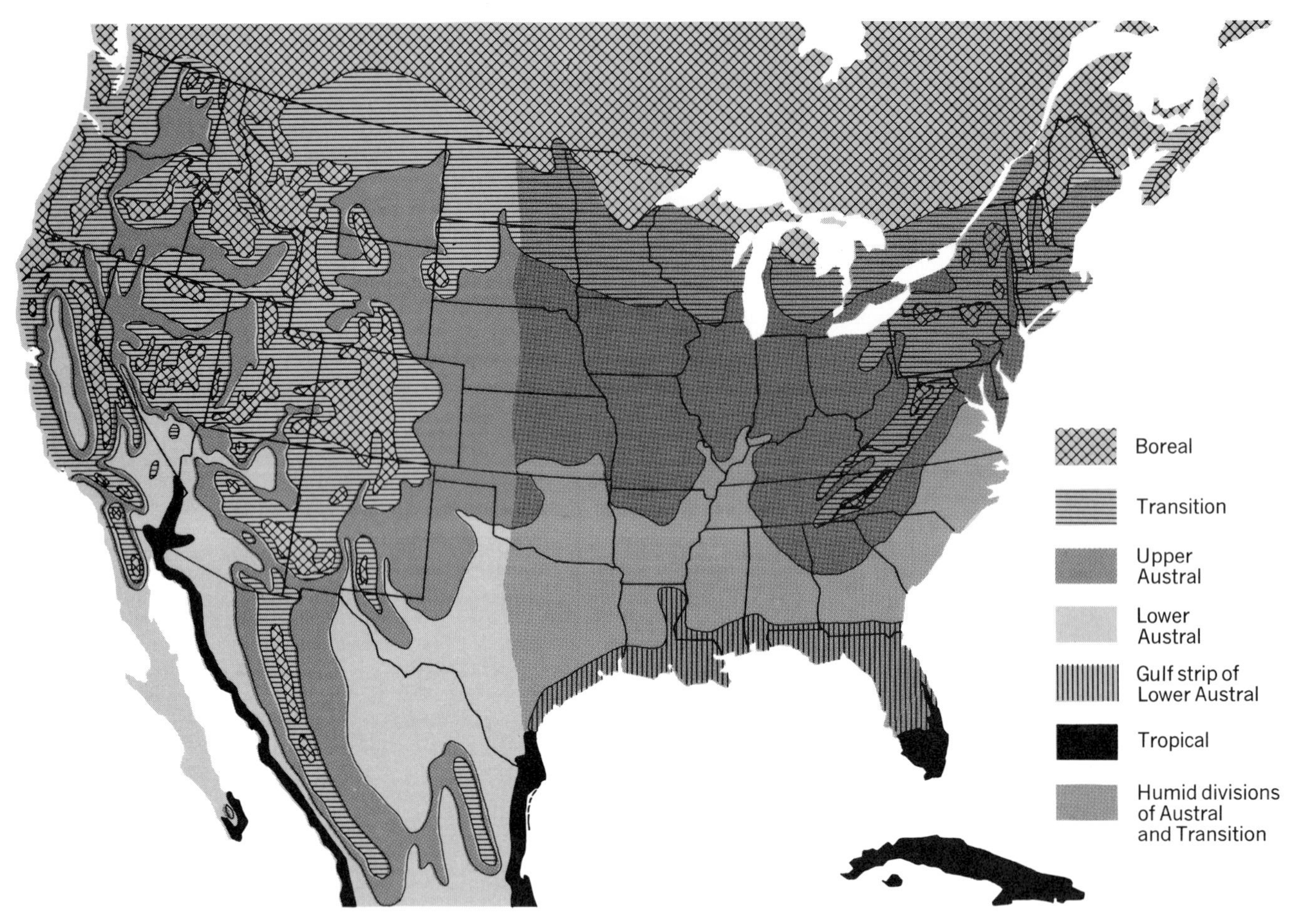

Fig. 1. Life zones of the United States. (*C. H. Merriam, 1898*)

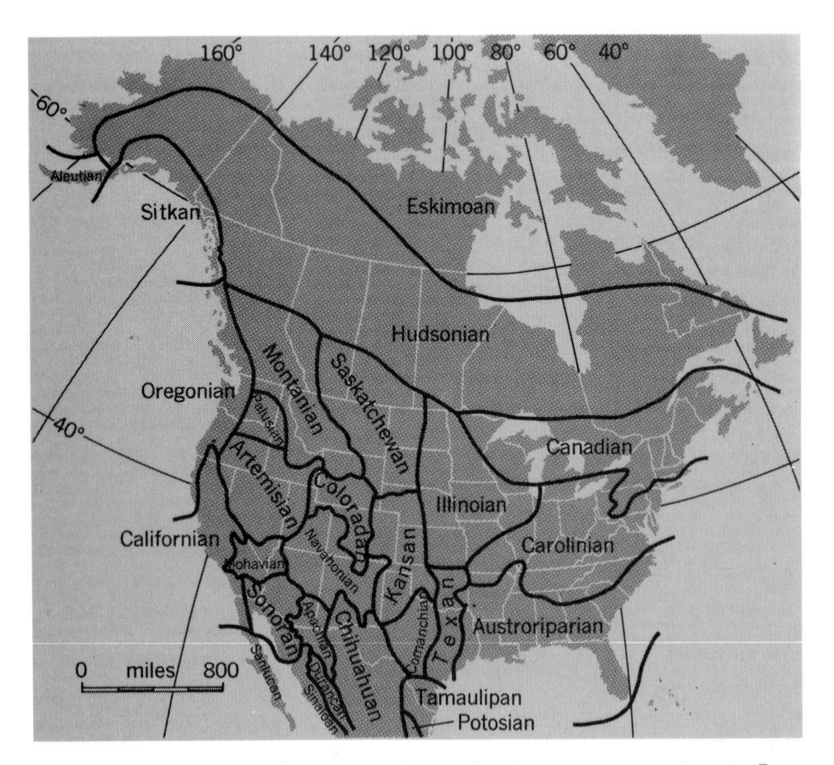

Fig. 2. Biotic provinces of part of North America (Veracruzian not shown). (*From L. R. Dice, The Biotic Provinces of North America, University of Michigan Press, 1943*)

for the six hottest weeks of the summer formed the basis for the second law, that plants and animals are restricted in southward distribution by the mean temperature of a brief period covering the hottest part of the year. Merriam's system emphasizes similarity in biota between arctic and alpine areas and between boreal and montane regions. It was already recognized that latitudinal climatic zones have parallels in altitudinal belts on mountain slopes and there is some biotic similarity between such areas of similar temperature regime.

In northern North America Merriam's life zones, the Arctic-Alpine, the Hudsonian, and the Canadian, are entirely transcontinental (Fig. 1 and table). Because of climatic and faunistic differences, the eastern and western parts of most life zones in the United States (the Transition, upper Austral, and lower Austral) had to be recognized separately. The western zones had to be further subdivided into humid coastal subzones and arid inland ones. The Tropical life zone includes the extreme southern edge of the United States, the Mexican lowlands, and Central America.

Although once widely accepted, Merriam's life zones are little used today because they include too much biotic variation and oversimplify the situation. However, much of the terminology he proposed persists, especially in North American zoogeographic literature.

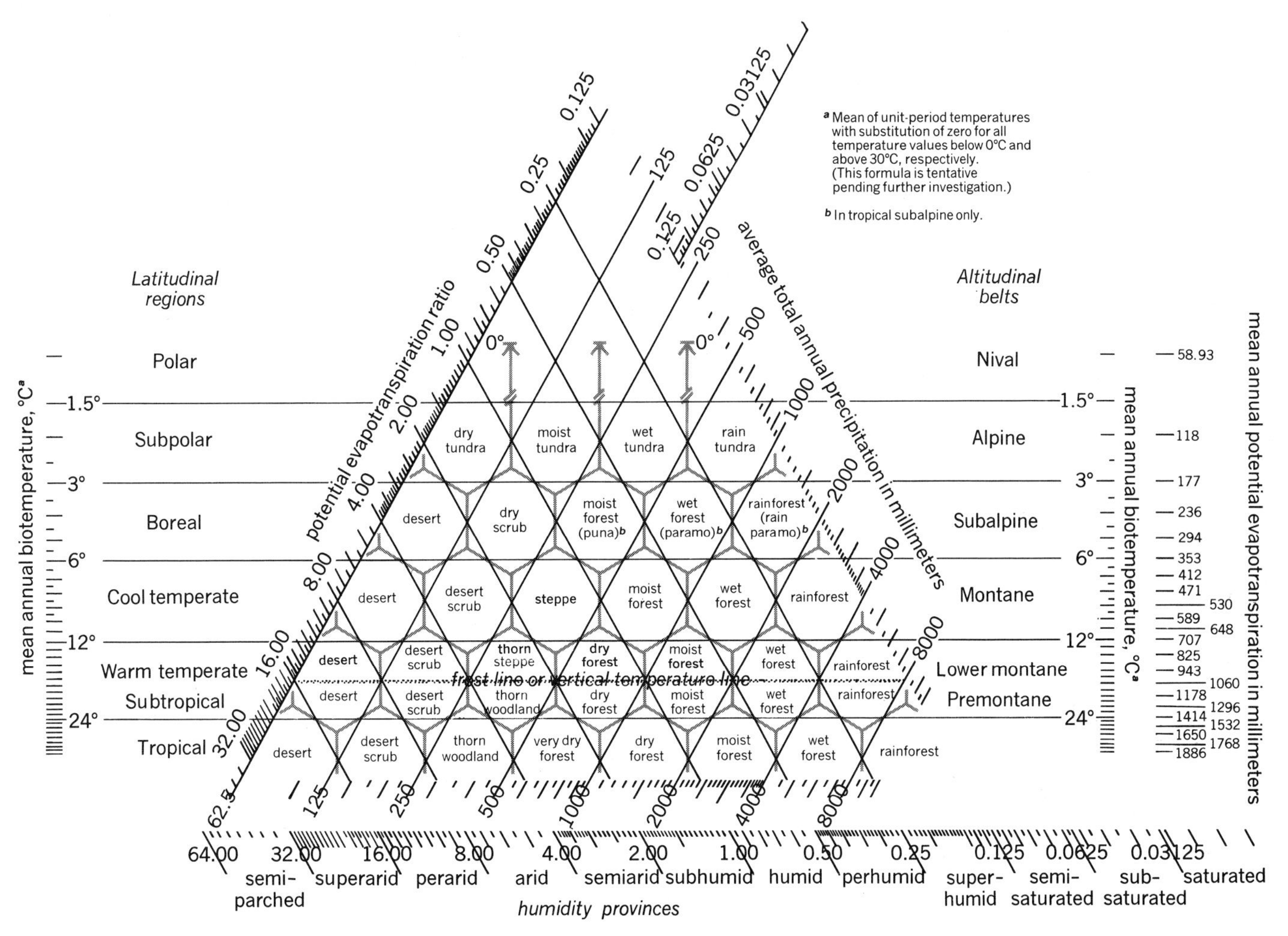

Fig. 3. World life zones. *(From L. R. Holdridge, Life Zone Ecology, rev. ed., Tropical Science Center, 1967)*

Dice's zones. Another approach to life zones in North America is the biotic province concept of L. R. Dice. Each biotic province covers a large and continuous geographic area and is characterized by the occurrence of at least one important ecological association which is distinct from those of adjacent provinces. Each biotic province is subdivided into biotic districts which are also continuous, but smaller, areas distinguished by ecological differences of lesser extent or importance than those delimiting provinces. Life belts, or vertical subdivisions, also occur within biotic provinces. These are not necessarily continuous but often recur on widely separated mountains within a province where ecological conditions are appropriate.

Boundaries between biotic provinces were largely subjective, supposedly drawn where the dominant associations of the provinces covered approximately equal areas. In practice, however, too few association data including both plants and animals were available, and vegetation generally offered the most satisfactory guide to boundaries. The Dice system recognizes 29 biotic provinces in North America (Fig. 2). The two northernmost (Eskimoan and Hudsonian) are transcontinental, reflecting broad, high-latitude climatic belts. Those in the eastern United States (Canadian, Carolinian, and Austroriparian) reflect both latitudinal temperature change and physiography, whereas province boundaries in the remainder of North America are strongly influenced by physiography.

Holdridge's zones. A life-zone system with boundaries more explicitly defined is that of L. R. Holdridge. These life zones are defined through the effects of the three weighted climatic indexes: mean annual heat, precipitation, and atmospheric moisture. Each axis of the triangle (Fig. 3) represents one climatic component, and the three sets of lines parallel to the axes define the life-zone framework. Climatic values which the lines represent progress geometrically. Mean annual biotemperature is computed by summing the monthly mean temperatures from 0° (setting negative winter months equal to 0° when they occur) to about 30°C and dividing by 12. Values increase from top to bottom in Fig. 3 and, by extension to either margin, establish equivalent latitudinal regions and altitudinal belts. Precipitation is computed as mean annual precipitation in millimeters and increases from the apex toward the lower right margin of the diagram. The humidity factor is calculated by dividing the value of the mean annual potential evapotranspiration by the mean annual precipitation in millimeters. Values of this ratio increase toward lower left side of the diagram, signifying a decrease in effective humidity.

Associations of vegetation, animals, climate, physiography, geology, and soils are interrelated in unique combinations, mostly having distinct physiognomy, in Holdridge's climatic grid. Local investigation should show how well the framework represents the most common life zones of Earth, both north and south of the Equator. The system has been applied with considerable success to the mapping of life zones, primarily in the American tropics. Names of Holdridge's life zones are based on the dominant vegetation of each climatic type; animal life of the zone is supposedly distinctive. *See* BIOME; PLANT FORMATIONS, CLIMAX; VEGETATION ZONES, ALTITUDINAL; VEGETATION ZONES, WORLD; ZOOGEOGRAPHY. [ARTHUR W. COOPER]

Bibliography: W. H. Lewis, *Ecology Field Glossary: A Naturalist's Vocabulary*, 1977; C. H. Merriam, *Life Zones and Crop Zones of the United States*, USDA Bull. no. 10, 1898; J. E. Weaver and F. E. Clements, *Plant Ecology*, 2d ed., 1938.

Limnology

The study of lakes, ponds, rivers, streams, swamps, and reservoirs that make up inland water systems. Each of these inland aquatic environments is physically and chemically connected with its surroundings by meteorologic and hydrogeologic processes (Fig. 1). For example, precipitation and runoff, combined with the gradient and watershed characteristics of a river or stream, provide the physicochemical background for the organisms of flowing water (lotic) systems. The frequency and stability of thermal stratification and mixing in standing water (lentic) systems such as lakes, ponds, and reservoirs are determined by the seasonal balance of heating and cooling, the unique temperature-density characteristics of pure water, and the energy of the wind.

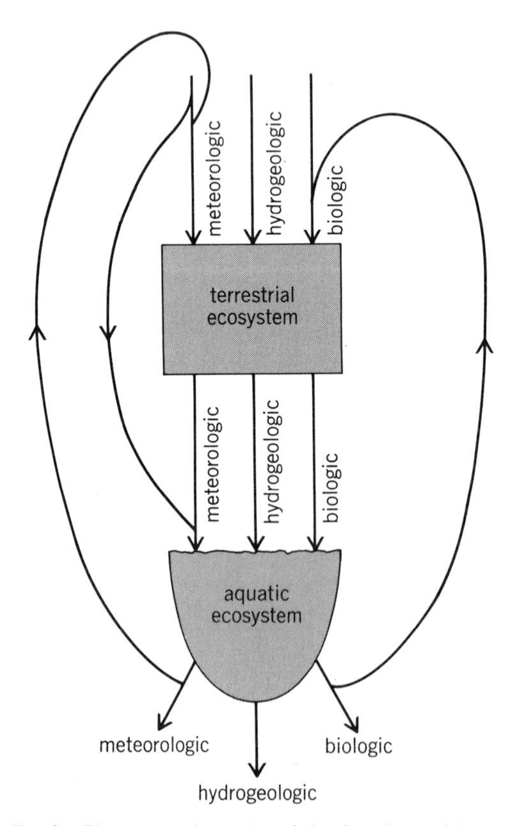

Fig. 1. Diagrammatic model of the functional linkages between terrestrial and aquatic ecosystems. Vectors may be meteorologic, hydrogeologic, or biologic components moving nutrients or energy along the pathway shown.

The interaction of solar radiation with water not only determines water color, but also provides energy for autotrophic production (such as photosynthesis). Furthermore, solar energy, nutrient substrates provided by the watershed, and mixing determine how much organic material will be produced within the aquatic environment. This organic material, along with leaves and other material produced outside of the system, supports the heterotrophic activities of aquatic communities.

Biogeochemical processes. Many interactions between plants and animals of aquatic systems have been idealized by food chains or food webs. Energy fixed by photosynthesis is traced through the various levels of heterotrophic activity within the system (Fig. 2). Dissipation of energy and production in food webs, combined with the return of minerals to inorganic nutrient pools, maintain the internal biogeochemical cycles of aquatic environments.

General habitats of aquatic environments. Aquatic organisms may be grouped into categories according to where they live within a particular system. Generally, habitats of aquatic systems are defined by unique physical conditions. Open water (limnetic zone) of lentic environments is inhabited by microscopic plants (phytoplankton) and animals (zooplankton). Shallows around the edge of the lake limited by the depth of light penetration (littoral zone) may be occupied by rooted plants and pond weeds (aquatic macrophytes). These plants provide habitats for other plants and animals. The deepest area (profundal zone) is usually bounded by the limnetic zone on the top and the littoral zone on the sides. Benthic organisms such as insects, oligochaetes, and microorganisms may occupy the profundal zone. Free-swimming animals such as fish (nekton) are usually not restricted to any particular zone of a lentic environment.

Lotic environments may also be divided into zones created by water velocity, bottom conditions, and stream gradient. In mountainous areas, rapids and turbulent water zones (riffles) are contrasted with sluggish water zones (pools); each of these habitats may have characteristic organisms. Streams and rivers in valleys and on plains may have many meanders, oxbows, and braids developed by the balance of erosion and deposition of the riverbed. Water movement and sediment characteristics, together with light penetration and oxygen conditions of the river, characterize the habitats of the lotic organisms.

Eutrophication and toxic materials. Aquatic systems with excellent physical conditions for production of organisms and high nutrient levels may show signs of eutrophication. Eutrophic lakes are generally identified by large numbers of phytoplankton and aquatic macrophytes and by low oxygen concentrations in the profundal zone. Eu-

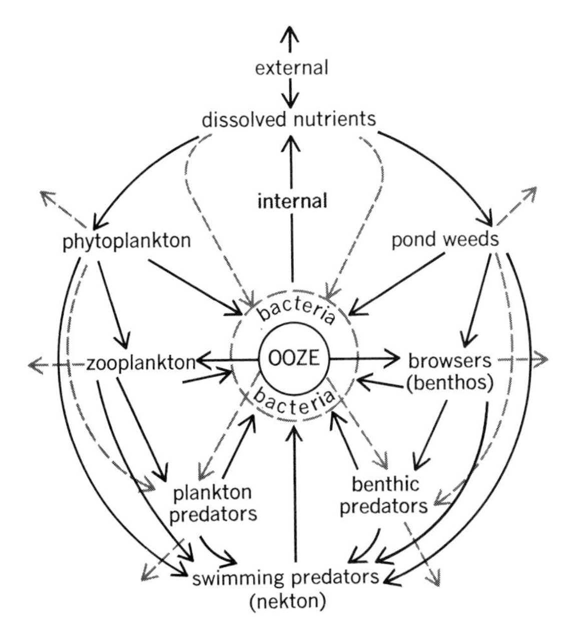

Fig. 2. Diagram of food chains in a lake, showing the interdependence of organisms. Broken lines depict the energy pathways and solid lines the food source.

trophication may be accelerated by poor watershed management, by runoff from overfertilized lands, or by poor septic facilities.

Organisms of aquatic environments may also be affected by metals such as lead and mercury, or pesticides that enter the aquatic environment from poor watershed management practices. These toxic materials may alter the food webs of aquatic systems by destroying sensitive organisms. Tolerant organisms may concentrate the toxic elements by feeding and absorption in a process called biomagnification, resulting in levels of toxicity which make them dangerous for human consumption. *See* ECOLOGY; FRESH-WATER ECOSYSTEM; HYDROLOGY; LAKE; MARINE ECOSYSTEM; OCEANOGRAPHY; RIVER; STREAM; WATER POLLUTION.

[JAMES E. SCHINDLER]

Bibliography: G. A. Cole, *Textbook of Limnology*, 1975; D. G. Frey (ed.), *Limnology in North America*, 1963; G. E. Hutchinson, *A Treatise on Limnology*, vol. 1, 1957, vol. 2, 1967; H. B. N. Hynes, *The Ecology of Running Waters*, 1970; G. E. Likens and F. H. Borman, Linkages between terrestrial and aquatic ecosystems, *Bioscience*, 24: 447–456, 1974; F. H. Rigler, Nutrient kinetics and the new typology, *Verh. Internat. Verein. Limnol.*, 19:197–210, 1975; R. G. Wetzel, *Limnology*, 1975.

Mangrove swamp

A swamp forest of low to tall trees and some shrubs, commonly associated with some salt marsh herbs. Swamp forests occur along the borders of many tropical shores where wave action is not intense and mud and peat are deposited (Fig. 1*a*). Most plants of this community are halophytes that are well adapted to salt water and fluctuations of tide level. Some have stilt or prop roots to help hold them on the shifting sediments and others have erect root structures (pneumatophores) that crop out above the surface (Fig. 1*b* and *c*). Several species have well-developed vivipary of their seeds, the hypocotyl developing while the fruit is held on the tree. These seedlings are usually so shaped and weighted that they float long distances in the sea and thus extensive migration is ensured.

The mangrove community often develops as a distinct halosere. In such areas it is zoned from open water landward in a series of different species (Fig. 2). The landward zone species develop on sediments and peats which were initially deposited in the seaward zone. These changes due to deposition in the swamp often extend the coast outward and form incipient islands in shallow, quiet waters. The swamps when dense also afford some protection against erosion resulting from violent storms. Thus, mangrove swamps have a significant geo-

Fig. 1. Mangrove swamps, Great Barrier Reef. (*a*) Mangroves at northwestern corner of King Island. (*b*) Interior of mangrove swamp, Newton Island. (c) Mangroves on north side of Howick Island. (*From J. A. Steers, Salt marshes, Endeavour, 18(70):75–82, 1959*)

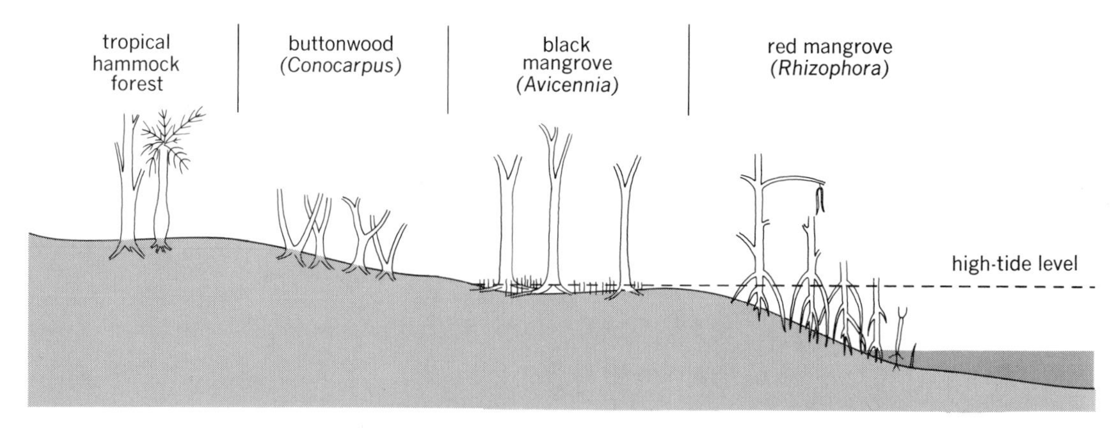

Fig. 2. Zonation of mangroves from open water landward, with their associated vegetation, in Florida. (*After J. H. Davis, The Ecology and Geologic Role of Mangroves in Florida, Carnegie Inst. Wash. Publ. No. 517, 1940*)

logic role. *See* ECOLOGICAL SUCCESSION; ECOLOGY; PLANT FORMATIONS, CLIMAX.

The composition of mangrove swamps is strikingly different in two regions. Only four (Western) species predominate in the Americas and along the west coast of Africa, but over 10 (Eastern) species are frequent in swamps of eastern Africa, Asia, and the western Pacific region. The genera belong to a number of families. They are examples of convergent adaptations to their saline habitat.

[JOHN H. DAVIS]

The Eastern mangrove swamp forms a monotonous forest between low- and high-tide marks. The vegetation is extensive, luxuriant, and tall on muddy beaches, lagoons, deltas, and estuaries; but it is narrow, dwarf, and sparse along sandy and rocky shores and old coral reefs. The exact limits of such a forest continually change through silting, colonizing, and erosion on the seaward side, and through rise in level, improved drainage, and reduced inundation in the landward direction.

The plant community, up to 40 m in height, is composed of a few viviparous and gregarious species forming an unbroken canopy with no distinct understoreys. Epiphytes, parasites, and undergrowth are either scarce or absent. Other characteristics include special root formation, such as branched stilt roots (*Rhizophora* sp.), pneumatophores (*Avicennia* and *Sonneratia* sp.), and knee-bend-like roots (*Bruguiera* and *Lumnitzera* sp.); and xerophytic features, such as thick coriaceous foliage with closely packed mesophyll, sunken stomata, and aqueous tissue. All these characteristics indicate convergent evolution in response to peculiar conditions of nutrition, assimilation, and survival in anaerobic and shifting substratum.

Fig. 3. A pioneer species (*Avicennia*) in formation of mangrove forest, colonizing mainly the exposed side, with spreading root system and asparaguslike pneumatophores. (*Courtesy of Forest Department, Malaya*)

The development and composition of a mangrove forest depend largely on soil type, salinity, duration and frequency of inundation, accretion of silt, strength of tides, exposure or shelter of the site, and unplanned exploitation. These factors interact in a complex manner; accordingly, the distribution, zonation, and succession of mangrove species are highly variable even within narrow geographical limits. Mangroves are most luxuriant and complex on the coasts of Malaysia and the surrounding islands. From here the diversity of the Eastern species decreases toward the Red Sea and southern Japan, where only a single species may be found. Of the four characteristic families, Rhizophoraceae, which forms extensive forests, is by far the most important. Other characteristic submangrove tree-forming species belong to the genera *Lumnitzera*, *Acanthus*, *Hibiscus*, and *Scyphiphora*.

The Eastern vegetation is taller and richer in diversity than its Western counterpart. All the Western genera are found in the Eastern vegetation, but their species composition, zonation, and succession are different. The pioneers are species of *Avicennia* and *Sonneratia*, the former colonizing mainly the firmer, exposed seaward side (Fig. 3) and the latter largely the soft, rich mud along sheltered river mouths. *Ceriops decandra* plays a similar role on the sheltered east coast estuaries of the Malay Peninsula. Depending on soil type and tidal height, these pioneers are normally succeeded by either pure or mixed forests of various species of *Bruguiera* and *Rhizophora*, such as *B. cylindrica* and *B. sexangula* on firmer clay beyond the reach of ordinary tides behind *Avicennia*; *B. parviflora* in pure crop or mixed with *R. apiculata* on wetter muds flooded by normal high tides; *B. gymnorhiza*, often mixed with submangrove species *Acrostichum aureum*, *Acrostichum speciosum* (ferns), and *Oncospermum filamentosa*; and *Nipah fruticans* (palms) farther inland on drier ground with less saline soil. *R. mucronata* is more tolerant of sandy, firmer bottoms than *R. apiculata* and forms tan-

Fig. 4. A mud chimney made by *Thalassina anomala*, the characteristic mangrove shrimp.

gled thickets on banks of tidal creeks and in estuaries.

Economically mangroves are a great source of timber, poles, pilings, and fuel. The bark is used in tanning and batik industries. Some species have either food or medicinal value. For better productivity, exploitation and management must be planned carefully. *See* FOREST MANAGEMENT AND ORGANIZATION.

Animals in the Eastern mangroves are abundant and are zoned both horizontally and vertically. On the mud floor brightly colored species of *Uca* and amphibious mudskipper gobies belonging to *Boliophthalmus* and *Periophthalmus* genera can be seen. There are a variety of snails on different levels of the mangrove trees, and bivalves and worms hidden away under the mud are also very common and exclusive. *Thalassina anomala*, with its characteristic mud chimneys, is found in the drier areas (Fig. 4). Other less common but interesting forms include snakes, monitor lizards, and long-tailed crab-eating monkeys. Many birds, including the Brahminy kite, green pigeons, terns, waders, sea eagles, and kingfishers, can be seen visiting the mangroves either regularly or seasonally, but only a few species are exclusively confined there, nesting commonly in tall *Sonneratia* trees. The nuisance of leeches, which are absent from mangrove forests, is more than compensated for by the myriads of small mosquitoes. *See* ECOSYSTEM.

[F. C. VOHRA]

Bibliography: V. J. Chapman, *Mangrove Vegetation*, 1976; F. C. Craighead, Land, mangroves, and hurricanes, *Fairchild Trop. Gard. Bull.*, vol. 19, no. 4, 1964; R. Dame (ed.), *Marsh-Estuarine Systems*, 1978; N. Hotchkiss, *Marsh Wealth*, 1964; D. W. Scholl, Recent sedimentary record in mangrove swamps and rise in sea level over the southwest coast of Florida, pts. 1 and 2, *Mar. Geol.*, 1:344–366, and 2:343–364, 1964.

Marine ecosystem

Marine ecology comprises the ecology of the world's oceans with their shores and estuaries. Individual organisms, groups of organisms, and the ocean environments in which they exist constitute ecological systems, termed ecosystems. Examples vary in size from a particular seaweed ecosystem, to a tidepool ecosystem, to an estuarine bay ecosystem (Figs. 1 and 2). The size limits, or boundaries, of a particular ecosystem that distinguish it from others may depend upon physical barriers to community dispersal, such as a submarine mountain; environmental factors that restrict the area of the biotic community, such as the salinity factor; the productivity cycles within the community; or the reproductive and dispersal potentials of the community. This conceptual unit of ecological science, whether large or small, possesses individuality, a degree of stability and permanence, characteristic functional cycles, and readily recognizable components, either living or nonliving or both.

The nonliving, or abiotic, materials in a marine ecosystem cycle comprise not only a variety of water-soluble inorganic nutrient salts, such as the phosphates, nitrates, and sulfates of calcium, potassium, and sodium and the dissolved gases oxygen and carbon dioxide, but also organic compounds, such as the various amino acids, vitamins, and growth substances. Most of the solid material dissolved in the sea originated from the weathering of the crust of the Earth.

MARINE ENVIRONMENT

The environment of the marine ecosystems is a subject of study by oceanographers who investigate the physical, chemical, and biological properties of ocean waters, ocean currents, and ocean basins. Although it is by far the world's largest inhabitable environment, much of its area and volume is relatively unproductive of life. Yet, it contains representatives of a few major taxonomic groups not found in terrestrial or fresh-water environments. *See* OCEANOGRAPHY.

Chemical factors. The chemical properties of ocean waters constitute an important aspect of the marine environment. Such chemical factors as the dissolved solids and gases in the ocean waters have been the subject of study. *See* SEA WATER.

Dissolved solids. The total quantity of dissolved solids in the waters of the world's oceans or in the marine ecosystem as a whole approximates

Fig. 1. A community of ribbed mussels attached to a decayed log in the intertidal zone.

Fig. 2. A marsh grass–mussel community, representing a small marine ecosystem of the intertidal zone.

5×10^{16} metric tons, enough to form a layer 153 in. thick over the land area of the Earth. Typical ocean water contains about 34.9 grams (g) per liter of dissolved materials, in which there are 19.0 g of chlorine, 10.50 g of sodium, 1.35 g of magnesium, 0.885 g of sulfur, 0.400 g of calcium, and 0.380 g of potassium. Some of the most important nutrient elements are present in very small amounts, for example, the phosphates needed for the growth of algae. Phosphorus exists in ocean waters as phosphate ions. It is an essential component of living things, and in ocean water the amount present may limit the production of plants. The inorganic form occurs in amounts varying from zero concentration to over 0.10 mg per liter. Phytoplankters, principally microscopic algae, may absorb inorganic phosphorus and reduce the amount remaining in the water to a minimum. Some is permanently lost to recycling by being bound in nonsoluble forms and deposited on the sea bottom. By their death and decay, phosphorus is returned to the aquatic environment. Certain of the 40 or more elements in sea water which, like phosphorus, exist in extremely small amounts are concentrated to important degrees. For example, macroscopic algae concentrate potassium and iodine in relatively large amounts. Of special importance is silicon, utilized by diatoms and other silica-secreting organisms.

Dissolved gases. The essential gas, carbon dioxide, is readily dissolved in sea water, but in equilibrium with the air it would contain only about 0.5 milliliters (ml) of free CO_2 per liter at 0°C. Actually, sea water contains about 47 ml of CO_2 per liter because appreciable amounts are present in the form of carbonate and bicarbonate ions. This gives great importance to the marine ecosystem as the world's reservoir of CO_2 although only about 1% is in the free form. Since, in relation to the atmosphere, its CO_2 content is 50 times greater, oceans regulate the concentration in the atmosphere. The CO_2 content of sea water originates not only from the atmosphere but also from biochemical processes in the sea—respiration and decay—and in the soil of the ocean bottom and coastal shores. The free carbon dioxide in water exists in simple solution and in the form of carbonic acid, H_2CO_3. Combined CO_2 is in the bicarbonate ions (HCO_3^-) and the carbonate ions (CO_3^{--}). These and other ions of weak acids constitute the buffer system of sea water that stabilizes its hydrogen-ion concentration, or pH, at about 8–8.4 at the surface and at 7.4–7.9 at the deeper levels. The high buffer capacity of sea water is a dominant environmental characteristic. Chemical and biochemical interactions and actions of biological agents which would otherwise produce pH changes that could seriously modify living conditions are largely negated by this buffer system. The ocean environment is thus chemically stable. Marine plants, the producers, may utilize primarily either free CO_2, or some free CO_2 and some combined CO_2. In any case, lack of CO_2 is not a limiting factor to growth of ecosystems in the sea. High concentrations, however, are a limiting factor to many marine animals such as fish.

The dissolved oxygen content in the sea-water environment varies from 0 to over 8.5 ml per liter. The oxygen in water comes from the atmosphere by diffusion and from aquatic plants by their photosynthetic action. Low temperatures and low salinities of water favor high solubilities of oxygen gas. There are low-oxygen strata in ocean waters, but in general the supply is adequate for animals. The considerable oxygen content of the deeper water layers of the ocean reached these layers when they were in the photosynthetic surface strata.

Physical factors. The physical aspects of marine ecological systems, temperature and salinity, are especially significant because of their degree of constancy away from land. The mean annual temperatures in different latitudes on the Earth remain unchanged. Seasonal variations in the ocean are small compared to those on land. Also, the entire temperature range in ocean waters is within the tolerance limits of numerous plants and animals. In polar and tropical seas, temperatures do not vary more than about 5°C during the year. Temperate seas commonly vary about 10–15°C. The extent of seasonal variations decreases in the deeper strata. In temperate and tropical regions a permanent thermal gradient, the thermocline, in which temperature decreases rapidly with depth, lies between the surface mixed layer and the deep unmixed strata. Low temperatures, around 3°C, in the deep and bottom waters of the ocean exist because the waters of greatest density, formed in high latitudes, sink to the bottom or to levels of similar density, then spread out and move toward warmer latitudes, to form a pattern of oceanic circulation. The salinity of the world's oceans varies only a few parts per thousand and averages around 35‰. Major latitudinal differences in density result from temperature differences. As expected, there is a vast system of oceanic circulation involving all depths and modified by large water masses contributed by adjoining seas. Thus, this offshore environment of the unit marine ecosystem is characterized by relative constancy of salinity over exceedingly large areas. The massive and relatively constant properties of this environment are reflected in the distribution, abundance, and morphological traits of its biota.

Biota. The organisms that comprise the biota of the marine ecosystem are characterized by a lack of diversity among the plants, in contrast to the animals. This generalization applies to both microscopic and macroscopic organisms. With few exceptions, all types of animals inhabiting land and fresh-water environments occur in the marine ecosystem. Six major animal groups, including ctenophores, starfishes, and certain worms, are restricted to the marine environment. Other major taxonomic units are predominantly marine. The kinds of plants and animals in the marine environment fall into three major groups: organisms of the plankton, nekton, and benthos.

Plankton organisms. These organisms are small, mostly microscopic, and have little or no power of locomotion, being distributed by water movements. There are two main types, the phytoplankton and zooplankton. The former includes all of the floating plants, such as the small algae, fungi, and sargassum weeds. Of these, the most important in the economy of the sea are algae—diatoms and dinoflagellates. They are the major producers in marine plankton (Fig. 3). Diatoms are microscopic, unicellular plants, some of which form chains. They possess characteristic shells composed of translucent silica, and have a great variety of form and sculpture. The shell structure consists of two nearly equal valves, one of which fits over the other and hence may be compared to a box with a telescoping lid. The valves are joined by connecting bands. The protoplasm within the shell is exposed by a slit or by small pores to permit metabolic interchanges with the environment. During reproductive division of the protoplasm of the diatom, one of the two protoplasmic daughter cells retains the larger epivalve, or lid valve, the other the hypovalve, or box valve. The daughter cells then lay down the needed complementary valves. This simple binary form of division is the most common one. It permits rapid production of vast populations in favorable environments of an ecosystem. During successive binary divisions, the size attainable by individuals is progressively reduced until a minimum limit is reached when, usually, the protoplasmic content of the shells escapes from the parted valves. It is enclosed in a flexible pectin membrane and is called an auxospore. These specialized spores grow in size and finally form the characteristic two valves. Diatoms possess one or more chromatophores, ranging in color from yellow to brown.

Diatoms occur as fossil, siliceous shell deposits, called diatomaceous earth, and as living producers in practically all habitats of the broad marine ecosystem. They are found floating in water, attached to the bottom, on larger plants, on animals, and, as spores, enclosed in Arctic ice. Free-floating diatoms possess structural adaptations that permit adjustments in depth. The bladder-type diatom is relatively larger, and in some forms such as *Planktoniella*, the shape is disklike, so that a zigzag course is followed in sinking. The hair type, such as *Rhizosolenia*, is a long and slender diatom, which sinks slowly when the long axis is horizontal to the pull of gravity, but more rapidly when oriented vertically. The ribbon-type cells, such as *Fragillaria*, are broad and flat, and are attached to form chains. The most abundant diatoms in the offshore oceanic waters are of the branched type, such as *Chaetoceros*. They possess numerous spiny projections that resist sinking (Fig. 4).

Dinoflagellates possess whiplike flagella that provide a slight degree of locomotion. Like diatoms, they possess structural modifications that indicate adaptation to environmental conditions. Some, such as *Dinophysis*, possess winglike structures that favor suspension; some have cellulose plates; others are naked cells. Many kinds are luminescent.

Fig. 3. Plankton diatoms. Their light spiny shells are suitable for floating on the surface of the water. *(Photograph by P. Conger, Smithsonian Institution)*

Phytoplankton organisms are much more abundant in nutrient-rich coastal waters than in offshore oceanic waters. They are the primary producers upon which large and small marine animals feed. The primary determinant of phytoplankton productivity is the Sun's radiant energy. Yet, less than 0.3% of the incident solar energy is converted by the diatoms and dinoflagellates to provide the basic food of the world's oceans.

Zooplankton organisms are the floating or weak-

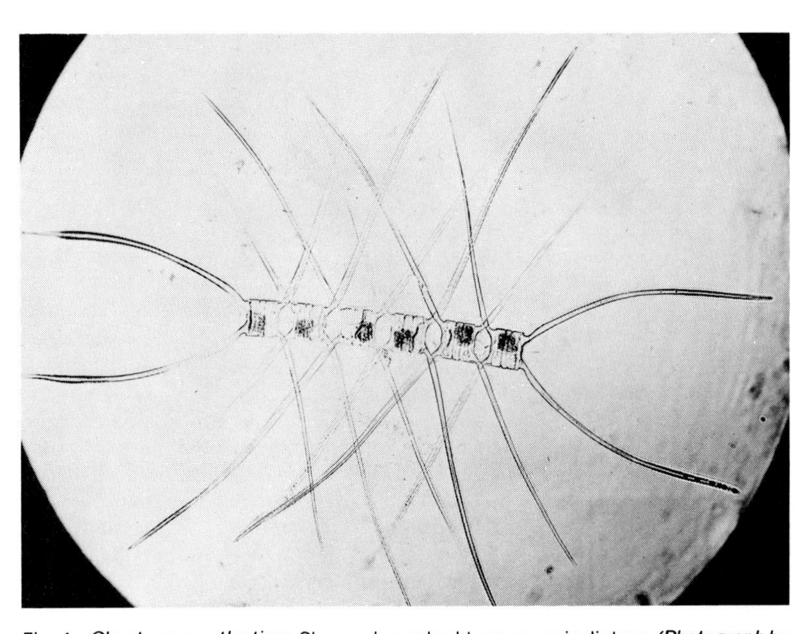

Fig. 4. *Chaetoceros atlanticus* Cleve, a branched-type oceanic diatom. *(Photograph by P. Conger, Smithsonian Institution)*

Fig. 5. A benthic community of brittle stars and isopods at a depth of 1200 m in the San Diego Trough, off California, 32°54′N and 117°36′W. (*Photograph by G. A. Shumway, U.S. Naval Electronics Laboratory*)

ly swimming animals, which include the eggs and larval stages as well as adult forms. The principal kinds include numerous Protozoa, such as Foraminifera and Radiolaria; great numbers of small Crustacea, such as ostracods and copepods, with their various larval stages various jellyfishes; numerous worms; a few mollusks; and also the eggs and early developmental stages of most of the nonplanktonic organisms in the sea. Plankton organisms are grouped on a basis of size. The smallest organisms range from about 5 to 60 μ, the next largest range up to about 1 mm in size, the next up to 1 cm, and the largest plankton group over 1 cm.

Nekton organisms. These are the actively swimming animals of the marine ecosystem. They comprise adult stages of such familiar forms as crabs, squids, fish, and whales. Some of these undergo long horizontal migrations over hundreds of miles; some migrate periodically to great depths; and a few live mostly in deep waters, for which habitat they possess marked adaptations.

Benthos organisms. These inhabit the bottom, and range from high-tide level on shore to the deep-sea bottom. There are relatively few kinds and numbers of animals on the deep-sea floor. They are mainly mud dwellers, possessing characteristic structures permitting life in a quiet, dark, muddy environment where food is scarce. Certain isopods, sponges, hydroids, brittle stars, sea urchins, and shrimp are typical animals of the deep-sea biota (Fig. 5). Inshore bottom communities at depths less than 50 m are rich in plants and animals. Although mosses and ferns are entirely absent in the sea, there are approximately 30 species of marine flowering plants. Of these, eelgrass (*Zostera*) is a characteristic shore plant. It is a perennial flowering plant, not actually a grass, that is worldwide in distribution, and is most abundant on soft bottoms in protected, coastal areas. Large brown algae, such as *Fucus* and *Ascophyllum*, are widely distributed conspicuous plants of exposed rock surfaces in the intertidal zone. *Ulva* and *Enteromorpha* are typical green algae of mud flats in quiet waters. Red algae, such as dulse (*Rhodymenia*), are commonly found below the intertidal zone. These are important food plants for the bottom-living animals of the coasts.

Conspicuous members of the benthic animal communities of the sea coast are barnacles, snails, mussels, clams, oysters, sea anemones, sea urchins, sea cucumbers, and starfish. Shore corals are largely restricted to warmer seas. The bottom animals of the marine ecosystem have structural modifications facilitating adhesion, burrowing, feeding, and protection.

MAJOR DIVISIONS

The marine ecosystem can be divided into two large areas, the pelagic and benthic divisions. Each of these consists of various zones.

Pelagic division. This division embodies all the waters of the oceans and their adjacent salt-water bodies. It is divisible into the neritic zone, extending offshore to the edge of the continental shelf to a depth of over 200 m, and the oceanic zone, embracing the remaining offshore waters. The neritic zone is rich in plant nutrients, especially phosphates and nitrates, which originate from coastal tributary waters and from bottom deposits carried upward to surface water layers by upwelling, diffusion, turbulence, or convection. The water is

Fig. 6. A littoral community with abundant sand dollars. This photograph was taken at a depth of 4 m in Mission Bay Channel, San Diego, Calif. (*Photograph by R. F. Dill, U.S. Naval Electronics Laboratory*)

more variable in density and in chemical content than oceanic waters. It is also far more productive of plankton, fish, and shellfish. Many of the inshore, coastal forms are adapted to withstanding brackish waters of coastal tributaries. The oceanic zone has a well-populated upper, lighted, 200-m stratum, and deeper, relatively dark, and sparsely populated layers, characterized by great pressures, animals modified for life in darkness and under great pressures, and very few bottom animals. Since there is less plankton and suspended organic material in the water, light pentrates further than in the neritic zone. Also the water is usually very transparent. Its salt content is uniformly high and not subject to major variations in time and space. In the upper photosynthetic zone, or zone of productivity, plant nutrients are much less concentrated, and the cycle for replenishment is much longer than in coastal waters that receive nutrients and organic detritus from the land.

Benthic division. This division of the marine environment embraces the entire ocean floor, both coastal bottom and deep-sea bottom, properly termed the littoral and deep-sea systems, respectively. The littoral system consists of the eulittoral zone and the sublittoral zone.

Eulittoral zone. This extends from high-tide level to about 60 m, the approximate depth below which attached plants do not grow abundantly. This is probably the richest zone of the marine ecosystem in respect to numbers and kinds of organisms, as well as in variety of ecological types and habitat modifications. The upper intertidal portion of the zone extends between high- and low-water marks, a vertical distance that varies on different continental shores from over 12 m to a few centimeters. Changes in such environmental factors as light, temperature, salinity, and time of exposure vary tremendously within short vertical distances across the zone. These variations are reflected in the shapes, movements, tolerances, and life histories of the characteristic animals and plants. Numerous small, partly independent ecosystems thrive in this area because the substratum includes a variety of rock exposures, gravel, sand, and mud types admixed in all degrees (Fig. 6). The lower, permanently submerged portion of the eulittoral zone is characterized by abundant sessile plants, such as conspicuous rock weeds (*Fucus*), bladder wrack (*Ascophyllum*), green sea lettuce (*Ulva*), and kelp (*Laminaria* and *Nereocystis*), many of which are common to both portions of the zone. Certain of those algae, such as the giant kelp, form productive algal forests that extend down below the typical eulittoral zone into the sublittoral zone. Here, they utilize the dimly lighted, lower, and least productive portion of the littoral system.

In tropical littoral waters, the coral reef communities abound with characteristic types and forms of both plants and animals. Visible algal growth is sparse, animals often greatly predominate over plants, and depth range is to about 60 m. Sparsely populated animal benthos extends to greater depths into the sublittoral zone, terminating at depths varying in different latitudes between about 200 to 400 m, depending upon light and temperature factors that modify the distribution of benthic animals and plants.

Deep-sea system. This portion of the benthic division is subdivided into an upper part, the archibenthic zone or, more meaningfully, the continental deep-sea zone, extending from the edge of the continental shelf (200–400 m) to depths of about 800–1100 m, and the abyssal-benthic zone that embraces the remainder of the benthic deep-sea system (Fig. 5). The zones have little or no light, relatively constant conditions of salinity and temperature, and steadily decreasing numbers and kinds of organisms. In the abyssal regions of great depth, perpetual darkness, and extremely low temperature (5 to −1°C), the extreme in monotony of environment prevails. However, bacteria and at least some animals exist at about all known depths. The remains of plants and particulate organic detritus continuously rain down to settle over the bottom as a light coating, utilized by the bottom dwellers for food. The pelagic food supply from above decreases with the increase in offshore distance for two reasons: The offshore oceanic waters contain less particulate matter than inshore waters, and the longer period required to descend the deeper strata results in greater disintegration while sinking.

Estuaries. These are coastal adjuncts of the marine ecosystem. They embrace bodies of water which, by virtue of their position, are directly subject to the combined action of river and tidal currents. Compared with offshore ocean waters, they lack constancy, possessing characteristic horizontal and vertical gradients in physical, chemical, and biological properties. These gradients are subject to pronounced changes in space and time. Estuarine waters are characterized by rapidity of response to changing external conditions. Their temperature, salinity, and turbidity conditions are distinctly not uniform. They are usually rich in plant nutrients of land origin, which they transport to coastal waters, thus fertilizing these waters. Estuarine waters contain an environmentally selected biota containing representative fresh-water and salt-water forms. *See* ESTUARINE OCEANOGRAPHY.

Marshes. These transitional land-water areas, covered part of the time at least by estuarine or coastal waters, comprise parts of the peripheral area of the marine ecosystem. Mangroves are characteristic woody marshes in tropical tidal waters of flat, muddy shores. More generally, coastal marshes are dominated by grasses and sedges. Marshes have characteristic biota for a certain latitude. In temperate regions, they are inhabited by characteristic birds such as bitterns and rails, and by crustaceans such as sand hoppers and fiddler crabs. Typical marsh animals and plants have tolerances to fresh water and salt water that differ from those of related forms living in fresh-water or salt-water habitats.

Sediments. Inanimate, particulate materials of organic and inorganic origin that have settled on the bottom in aquatic environments comprise the sediments. Vast quantities of particulates, eroded from land surfaces by natural waters, are carried to the sea by rivers and estuarine waters. They settle to the bottom and provide food for benthic animals. Sediments may be carried to the surface, or photosynthetic stratum, either by upwelling of coastal waters or by ocean currents, and thus enter a biochemical cycle of the marine ecosystem. Marine sediments are made up of microscopic fragments of weathered rock, partly decomposed plant and animal remains, skeletal remains of various organisms, inorganic precipitates from sea water, terrestrial particulates, and particulate material of volcanic origin. [CURTIS L. NEWCOMBE]

Bibliography: J. Fraser, *Nature Adrift*: *The Story of Marine Plankton*, 1962; E. D. Goldberg, The oceans as a chemical system, in *The Sea*: *Ideas and Observations on Progress in the Study of the Seas*, vol. 2, pp. 3–25, 1963; C. P. Idyll, *Abyss*: *The Deep Sea and the Creatures That Live in It*, 1964; H. B. Moore, *Marine Ecology*, 1958; C. L. Newcombe, Mussels, *Turtox News*, vol. 25, no. 1, 1947; E. P. Odum, *Fundamentals of Ecology*, 1971; H. U. Sverdrup, M. W. Johnson, and R. H. Fleming, *The Oceans*: *Their Physics, Chemistry, and General Biology*, 1942.

Marine mining

The process of recovering mineral wealth from sea water and from deposits on and under the sea floor. Unknown except to technical specialists before 1960, undersea mining is receiving increasing attention. Frequent references to marine mineral resources and marine minerals legislation by national and international policy makers, increasing activity in marine minerals exploration, and the launching of major new seagoing mining dredges for South Africa and Southeast Asia all indicate the beginnings of a viable and expanding industry.

There are sound reasons for this sudden emphasis on a previously little-known source of minerals. While the world's demand for mineral commodities is increasing at an alarming rate, most of the developed countries have been thoroughly explored for surface outcroppings of mineral deposits. The mining industry has been required to advance its capabilities for the exploration and exploitation of low-grade and unconventional sources of ore. Corresponding advances in oceanology have highlighted the importance of the ocean as a source of minerals and indicated that the technology required for their exploitation is in some cases already available.

There is a definite realization that the venture into the oceans will require large investments, and the trend toward the consortium approach is very noticeable, not only in exploration and mining activities but in research. Undersea mining has become an important diversification for many major oil and aerospace companies, and a few mining companies appear to be taking an aggressive approach. In the late 1960s over 80 separate exploration activities were reported in coastal areas worldwide. In the 1970s interest turned more to the deep-seabed deposits, subject to much negotiation at the United Nations 3d Law of the Sea Conference. Anticipating agreement, several countries, including the United States and West Germany, prepared draft interim legislation which would permit nationals of their countries and reciprocating states to mine the deep seabeds for nodules containing manganese, copper, nickel, and cobalt. Five major consortia were active in testing deep-seabed mining systems in the late 1970s including companies from the United States, Canada, United Kingdom, West Germany, France, and Japan.

While mineral resources to the value of trillions of dollars do exist in and under the oceans, their exploitation is not simple. Many environmental problems must be overcome and many technical advances must be made before the majority of these deposits can be mined in competition with existing land resources.

The marine environment may logically be divided into four significant areas: the waters, the deep sea floor, the continental shelf and slope, and the seacoast. Of these, the waters are the most significant, both for their mineral content and for their unique properties as a mineral overburden. Not only do they cover the ocean floor with a fluid medium quite different from the earth or atmosphere and requiring entirely different concepts of ground survey and exploration, but the constant and often violent movement of the surface waters combined with unusual water depths present formidable deterrents to the use of conventional seagoing techniques in marine mining operations.

The mineral resources of the marine environment are of three basic types: the dissolved minerals of the ocean waters; the unconsolidated mineral deposits of marine beaches, continental shelf, and deep-sea floor; and the consolidated deposits contained within the bedrock underlying the seas. These are described in Table 1, which shows also the subclasses of surficial and in-place deposits, characteristics which have a very great influence on the economics of exploration and mining. *See* SEA WATER.

As with land deposits, the initial stages preceding the production of a marketable commodity include discovery, characterization of the deposit to assess its value and exploitability, and mining, including beneficiation of the material to a salable product.

Exploration. Initial requirements of an exploration program on the continental shelves are a thorough study of the known geology of the shelves and adjacent coastal areas and the extrapolation of known metallogenic provinces into the offshore areas. The projection of these provinces, which are characterized by relatively abundant mineralization, generally of one predominant type, has been practiced with some success in the localization of certain mineral commodities, overlain by thick sediments. As a first step, the application of this technique to the continental shelf, overlain by water, is of considerable guidance in localizing more intensive operations. Areas thus delineated are considered to be potentially mineral-bearing and subject to prospecting by geophysical and other methods.

A study of the oceanographic environment may indicate areas favorable to the deposition of authigenic deposits in deep and shallow water. Some deposits may be discovered by chance in the process of other marine activities.

Field exploration prior to or following discovery will involve three major categories of work: ship operation, survey, and sampling.

Ship operation. Conventional seagoing vessels are used for exploration activities with equipment mounted on board to suit the particular type of operation. The use of submersibles will no doubt eventually augment existing techniques but they are not yet advanced sufficiently for normal usage.

One of the most important factors in the location of undersea minerals is accurate navigation. Ore bodies must be relocated after being found and must be accurately delineated and defined. The accuracy of survey required depends upon the phase of operation. Initially, errors of 1000 ft (300 m) or more may be tolerated.

However, once an ore body is believed to exist in a given area, maximum errors of less than 100 ft (30 m) are desirable. These maximum tolerated errors may be further reduced to a few feet in detailed ore body delineation and extraction.

There are a variety of types of electronic navigation systems available for use with accuracies from 3000 ft (900 m) down to approximately 3 ft (0.9 m). Loran, Lorac, and Decca are permanently installed in various locations throughout the world. Small portable systems are available for local use that provide high accuracy within 30–50-mi (50–80-km) ranges. For deep-ocean survey, navigational satellites have completely revolutionized the capabilities for positioning with high accuracy in any part of the world's oceans.

During sampling and mining operations, the

Table 1. Marine mineral resources

	Unconsolidated		Consolidated	
Dissolved	Surficial	In place	Surficial	In place
Metals and salts of:	Shallow beach or offshore placers	Buried beach and river placers	Exposed stratified deposits	Disseminated massive, vein, or tabular deposits
Magnesium	Heavy mineral sands	Diamonds	Coal	Coal
Sodium	Iron sands	Gold	Iron ore	Iron
Calcium	Silica sands	Platinum	Limestone	Tin
Bromine	Lime sands	Tin		Gold
Potassium	Sand and gravel		Authigenic coatings	Sulfur
Sulfur		Heavy minerals	Manganese oxide	Metallic sulfides
Strontium	Authigenic deposits	Magnetite	Associated Co, Ni, Cu	Metallic salts
Boron	Manganese nodules (Co, Ni, Cu, Mn)	Ilmenite	Phosphorite	
Uranium	Phosphorite nodules	Rutile		
And 30 other elements	Phosphorite sands	Zircon		
	Glauconite sands	Leucoxene		
Fresh water		Monazite		
	Deep ocean floor deposits	Chromite		
	Red clays	Scheelite		
	Calcareous ooze	Wolframite		
	Siliceous ooze			
	Metalliferous ooze			

vessel must be held steady over a selected spot on the ocean floor. Two procedures that have been fairly well developed for this purpose are multiple anchoring and dynamic positioning.

A three-point anchoring system is of value for a coring vessel working close to the surf. A series of cores may be obtained along the line of operations by winching in the forward anchors and releasing the stern anchor. Good positive control over the vessel can be obtained with this system, and if conditions warrant, a four-point anchoring system may be used. Increased holding power can be obtained by multiple anchoring at each point.

Dynamic positioning is useful in deeper water, where anchoring may not be practical. The ship is kept in position by use of auxiliary outboard propeller drive units or transverse thrusters. These can be placed both fore and aft to provide excellent maneuverability. Sonar transponders are held submerged at a depth of minimum disturbance, or the system may be tied to shore stations. The auxiliary power units are then controlled manually or by computer in order to keep the ranges at a constant value.

Survey. The primary aids to exploration for mineral deposits at sea are depth recorders, subbottom profilers, magnetometers, bottom sampling devices, and subbottom sampling systems. Their use is dependent upon the characteristics of the ore being sought.

For the initial topographic survey of the sea floor, and as an aid to navigation, in inshore waters, the depth recorder is indispensable. It is usually carried as standard ship equipment, but precision recorders having a high accuracy are most useful in survey work.

In the search for marine placer deposits of heavy minerals, the subbottom profiler is probably the most useful of all the exploration aids. It is one of several systems utilizing the reflective characteristics of acoustic or shock waves.

Continuous seismic profilers are a development of standard geophysical seismic systems for reflection surveys, used in the oil industry. The normal energy source is explosive, and penetration may be as much as several miles.

Subbottom profilers use a variety of energy sources including electric sparks, compressed air, gas explosions, acoustic transducers, and electromechanical (boomer) transducers. The return signals as recorded show a recognizable section of the subbottom. Shallow layers of sediment, configurations in the bedrock, faults, and other features are clearly displayed and require little interpretation. The maximum theoretical penetration is dependent on the time interval between pulses and the wave velocity in the subbottom. A pulse interval of 1/2 sec and an average velocity of 8000 ft/sec (2400 m/sec) will allow a penetration of 2000 ft (600 m), the reflected wave being recorded before the next transmitted pulse.

Penetration and resolution are widely variable features on most models of wave velocity profiling systems. In general, high frequencies give high resolution with low penetration, while low frequencies give low resolution with high penetration. The general range of frequencies is at the low end of the scale and varies from 150 to 300 Hz, and the general range of pulse energy is 100–25,000 joules for nonexplosive energy sources. The choice of system will depend very much on the requirements of the survey, but for the location of shallow placer deposits on the continental shelf the smaller low-powered models have been used with considerable success.

With the advent of the flux gate, proton precession, and the rubidium vapor magnetometer, all measuring the Earth's total magnetic field to a high degree of accuracy, this technique has become much more useful in the field of mineral exploration.

Anomalies indicative of mineralization such as magnetic bodies, concentrations of magnetic sands, and certain structural features can be detected. Although all three types are adaptable to undersea survey work, the precession magnetometer is more sensitive and more easily handled than the flux gate, and the rubidium vapor type has an extreme degree of sensitivity which enhances its usefulness when used as a gradiometer on the sea surface or submerged.

Once an ore body is indicated by geological, geophysical, or other means, the next step is to sample it in area and in depth.

Sampling. Mineral deposit sampling involves two stages. First, exploratory or qualitative sampling locate mineral values and allow preliminary judgment to be made. For marine deposits this will involve such simple devices as snappers, drop corers, drag dredges, and divers. Accuracy of positioning is not critical at this stage, but of course is dependent on the type of deposit being sampled. Second, the deposit must be characterized in sufficient detail to determine the production technology requirements and to estimate the profitability of its exploitation. This quantitative sampling requires much more sophisticated equipment than does the qualitative type, and for marine work few systems in existence can be considered reliable and accurate. However, in particular cases, systems can be put together using available hardware which will satisfy the need to the accuracy required. Specifically, qualitative sampling of any mineral deposit offshore can be carried out with existing equipment. Quantitative sampling of most alluvial deposits of heavy minerals (specific gravity, less than 8) can be carried out at shallow depths (less than 350 ft or 107 m overall) using existing equipment but cannot be carried out with reliability for the higher-specific-gravity minerals such as gold (specific gravity 19). Quantitative sampling of any consolidated mineral deposit offshore can be carried out within limits.

Any system that will give quantitative samples can be used for qualitative sampling, but in many cases heavy expenses could be avoided by using the simpler equipment.

To obviate the effects of the sea surface environment, the trend is toward the development of fully submerged systems, but it should be noted that the deficiencies in sampling of the heavy placer minerals are not due to the marine environment. Even on land the accuracy of placer deposit evaluation is not high and the controlling factors not well understood. There is a prime need for intensive research in this area.

Table 2. Production from dissolved mineral deposits offshore

Mineral	Location	Number of operations	Annual production
Sodium, NaCl	Worldwide	90+	10,000,000 tons*
Magnesium, metal, Mg, MgO, $Mg(OH)_2$	United States, United Kingdom, Germany, Soviet Union	2 25+	221,000 tons 800,000 tons
Bromine, Br	Worldwide	7	102,000 tons
Fresh water	Middle East, Atlantic region, United States	150+	†
Heavy water	Canada	1	†
Total		275+	

*1 ton = 0.907 metric ton. †Not reported.

Evaluation of surficial nodule deposits on the deep seabed is carried out by using a combination of photography, television, or acoustic imagery and sampling by dredge, corer, or clam, either towed or attached to free-fall self-surfacing devices. Analysis of nodules is done on board the vessel.

Exploitation. Despite the intense interest in undersea mining, new activities have been limited mostly to conceptual studies and exploration. The volume of production has shown little change, and publicity has tended to overemphasize some of the smaller, if more newsworthy, operations. All production comes from nearshore sources, namely, sea water, beach and nearshore placers, and nearshore consolidated deposits.

Minerals dissolved in sea water. Commercial separation techniques for the recovery of minerals dissolved in sea water are limited to chemical precipitation and filtration for magnesium and bromine salts and solar evaporation for common salts and fresh-water production on a limited scale. Other processes developed in the laboratory on pilot plant scale include electrolysis, electrodialysis, adsorption, ion exchange, chelation, oxidation, chlorination, and solvent extraction. The intensive interest in the extraction of fresh water from the sea has permitted much additional research on the recovery of minerals, but successful commercial operations will require continued development of the combination of processes involved for each specific mineral.

As shown in Table 2, three minerals or mineral suites are extracted commercially from sea water: sodium, magnesium, and bromine. Of these, salt evaporites are the most important. Japan's total production of salt products comes from the sea. Magnesium extracted from sea water accounts for 75% of domestic production of this commodity in the United States, and fresh water compares with bromine in total production value.

Unconsolidated deposits. Unconsolidated deposits include all the placer minerals, surficial and in place, as well as the authigenic deposits found at moderate to great depths.

The mining of unconsolidated deposits became widely publicized with the awareness of the potential of manganese nodules as a source of manganese, copper, nickel, and cobalt, and because of the exciting developments in the exploitation of offshore diamonds in South-West Africa in the late 1960s. Despite the fact that there are presently no operations for nodules and the offshore diamond mining operations have been suspended, unconsolidated deposits have for some years presented a major source of exploitable minerals offshore.

So far the methods of recovery which have been

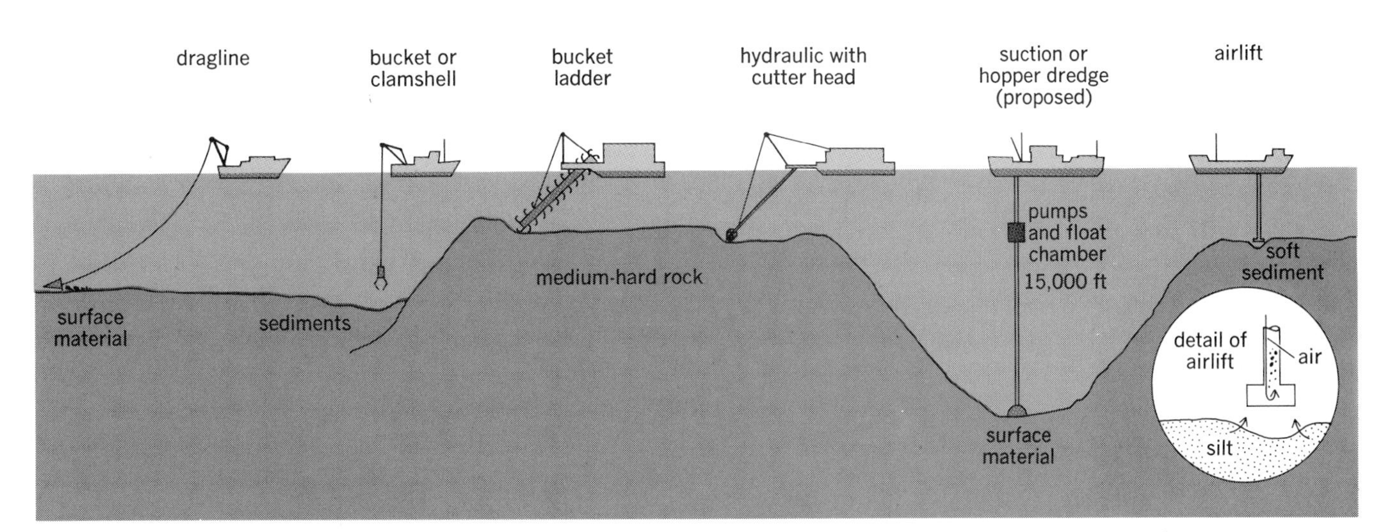

Fig. 1. Methods of dredging used in the exploitation of unconsolidated mineral deposits offshore. 15,000 ft = 4600 m.

Table 3. Production from unconsolidated mineral deposits offshore

Mineral	Location	Number of operations	Annual production
Diamonds	South-West Africa	1	221,500 yd³*
Gold	Alaska	1	–
Heavy mineral sands	North America, Europe, Southeast Asia, Australia	15	1,307,000 tons†
Iron sands	Japan	3	36,000 tons
Tin sands	Southeast Asia, United Kingdom	4	10,000 tons concentrates
Lime shells	United States, Iceland	9	20,000,000 yd³
Sand and gravel	Britain, United States	38	100,000,000 yd³
Total		71	

*1 yd³ = 0.765 m³. †1 ton = 0.907 metric ton.

used or proposed have been conventional, namely, by dredging using draglines, clamshells, bucket dredges, hydraulic dredges, or airlifts. All these methods (Fig. 1) have been used in mining to maximum depths of 200 ft (60 m), and hydraulic dredges for digging to 300 ft (90 m) are being built. Extension to depths much greater than this does not appear to present any insurmountable technical difficulties.

More than 70 dredging operations were active in the 1970s, exploiting such diverse products as diamonds, gold, heavy mineral sands, iron sands, tin sands, lime sand, and sand and gravel (Table 3). The most important of all of these commodities is the least exotic; 60% of world production from marine unconsolidated deposits, or about $100,-000,000 annually, is involved in dredging and mining operations for sand and gravel. Other major contributors to world production are the operations for heavy mineral sands (ilmenite, rutile, and zircon), mostly in Australia, and the tin operations in Thailand and Indonesia, which account for more than 10% of the world's tin.

Economics of these operations are dictated by many conditions. The costs offshore have a wide range, indicative of the effects of different environmental conditions. In general, it may be said that offshore operations are more costly than similar operations onshore.

The operations of Marine Diamond Corporation are of considerable historical interest. The first pilot dredging commenced in 1961 with a converted tug, the *Emerson K*, using an 8-in. (20-cm) airlift. The operation expanded until 1963, with the fleet consisting of 3 mining vessels, all using an air- or jet-assisted suction lift, 11 support craft, and 2 aircraft. Production totaled over $1,700,000 of stones during that year from an estimated 322,000 yd³ (246,000 m³) of gravel. At that time, the estimation of mining cost was $2.33/yd³ ($3.05/m³), showing a profit of nearly $3/yd³ ($3.92/m³). Subsequent unexpected problems, including severe storms, operating difficulties, and loss of one of the mining units, led to a reduction of profits and transfer of company control. In the year ending June 30, 1965, the company reported an operating loss of $2.02/yd³ ($2.64/m³) in the treatment of 220,000 yd³ (168,000 m³) of gravel, valued at $28/yd³ ($36.62/m³).

Production operations of Marine Diamond Corporation have fluctuated considerably. *Diamantkus*, a vessel designed to produce 7000 yd³/hr (5400 m³/hr) was withdrawn as uneconomic after only 30 months of service. Only two mining units, *Barge III* and *Colpontoon*, were in operation in 1968, both converted pipe-laying barges using combination airlifts and suction dredging equipment. A third and larger unit, the *Pomona*, a multiple-head suction dredge, was commissioned in March 1967, but was damaged by storms on the first trial run. The characteristics of the bedrock, with its many potholes and extremely irregular surface, add to the difficulty in recovering the maximum amount of diamonds. Mining operations ceased in May 1971, but large areas remained to be explored.

Liberal offshore mining laws introduced in 1962 in the state of Alaska resulted in an upsurge of exploration activity for gold, particularly in the Nome area. In 1968 a mining operation was attempted, using a 20-in. (51-cm) hydrojet dredge, in submerged gravels about 60 mi (97 km) east of Nome. No production was reported.

Over 70% of the world's heavy mineral sand production is from beach sand operations in Australia, Ceylon, and India. Only two oceangoing dredges are used. The majority being pontoon-mounted hydraulic dredges, or draglines, with separate washing plants.

The Yawata Iron and Steel Co. in Japan used a 10.5-yd³ (8.0-m³) barge-mounted grab dredge and a hydraulic cutter dredge to mine iron sand from the floor of Ariake Bay in water depths of 50 ft (15 m). These operations were suspended finally in 1966, the reason being given that the reserves had not been accurately surveyed and the cost of mining had not been competitive with Yawata's alternate sources of supply.

An interesting comparison between a clamshell dredge from Aokam Tin and a bucket-ladder dredge from Tongka Harbor, working offshore on the same deposit in Thailand was made. The clamshell was set up as an experimental unit using an oil tanker hull. It was designed for a digging depth of 215 ft (66 m), and mobility and seaworthiness were prime factors in its favor. However, in practice, it was never called upon to dig below 140 ft (43 m), its mobility was superfluous, and the ship

Table 4. Production from consolidated mineral deposits offshore

Mineral	Location	Number of operations	Annual production
Iron ore	Finland, Newfoundland*	2	1,700,000 tons†
Coal	Nova Scotia, Taiwan, United Kingdom, Japan, Turkey	57	33,500,000 tons
Sulfur	United States	1	600,000 tons‡
Total		60	

*Closed June 1966. †1 ton = 0.907 metric ton. ‡Estimated.

hull proved very unsatisfactory in terms of usable space. Although it was able to operate in sea states which prevented the operation of the neighboring bucket dredge, its mining recovery factor was low and its operating costs much higher than anticipated. It was withdrawn from service after only 9 years, in favor of the bucket dredge.

Another major operation is run by Indonesian State Mines off the islands of Bangka and Belitung. The operations are as far as 3 mi from shore in waters which are normally calm. They do have storms, however, which necessitate delays in the operation and the taking of precautions unnecessary onshore. The operations employ 12 dredges, one of which, the *Bangka I*, constructed in 1965, is the world's largest mining bucket-line dredge. With a maximum digging depth of 135 ft (41 m) it is designed to dig and treat 420,000 yd^3 (320,000 m^3) per month of 600 hr and produces over 1000 tons (907 metric tons) of tin metal per year, depending on the richness of the ground. The dredge is working up to 15 mi (24 km) offshore.

Lime shells are mined as a raw material for portland cement. Two United States operations for oyster shells in San Francisco Bay and Louisiana employed barge-mounted hydraulic cutter dredges of 16- and 18-in. (406- and 457-mm) diameter in 30–50 ft (9–15 m) of water. The Iceland Government Cement Works in Akranes uses a 150-ft (46-m) ship to dredge sea shells from 130-ft (40-m) of water, with a 24-in. (610-mm) hydraulic drag dredge.

In the United Kingdom, hopper dredges are used for mining undersea reserves of sand and gravel. Drag suction dredges up to 38 in. in diameter are most commonly used with the seagoing hopper hulls. Similar deposits have been mined in the United States, and the same type of dredge is employed for the removal of sand for harbor construction or for beach replenishment. Some sand operations use beach-mounted drag lines for removal of material from the surf zone or beyond.

For deep-seabed mining, at depths below 15,000 ft (4600 m), systems have been tested using airlift and hydraulic suction lift as well as a continuous dragline (CLB system), all towed from surface vessels. One other system incorporated a mobile gathering device on the bottom. All the tests carried out in the 1970s were 1/5 scale for 1,000,000–3,000,000 tons/year (1 ton = 0.9 metric ton) production models.

Consolidated deposits. The third and last area of offshore mineral resources has an equally long history. As Table 4 shows, the production from in-place mineral deposits under the sea is quite substantial, particularly in coal deposits. Undersea coal accounts for almost 30% of the total coal production in Japan and just less than 10% in Great Britain.

Extra costs have been due mainly to exploration, with mining and development being usually conventional. In the development of the Grand Isle sulfur mine off Louisiana, some \$8,000,000 of the \$30,000,000 expended was estimated to be due to its offshore location. There is no doubt that costs will be greater, generally, but on the other hand, in the initial years of offshore mining as a major industry, the prospects of finding accessible, high-grade deposits will be greater than they are at present on land.

Some of the mining methods are illustrated in Fig. 2. For most of the bedded deposits which extend from shore workings a shaft is sunk on land with access under the sea by tunnel. Massive and vein deposits are also worked in this manner. Normal mining methods are used, but precautions

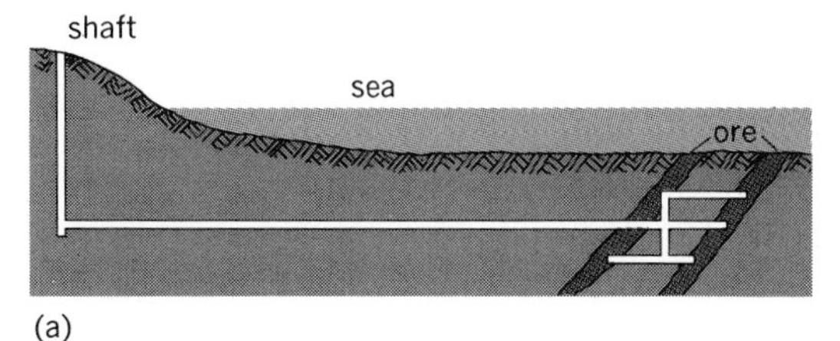

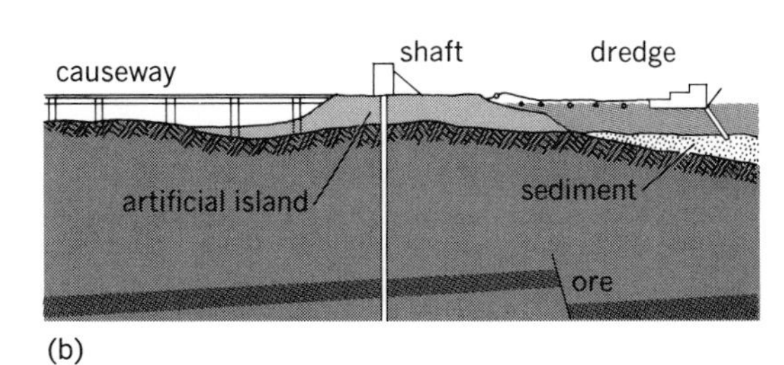

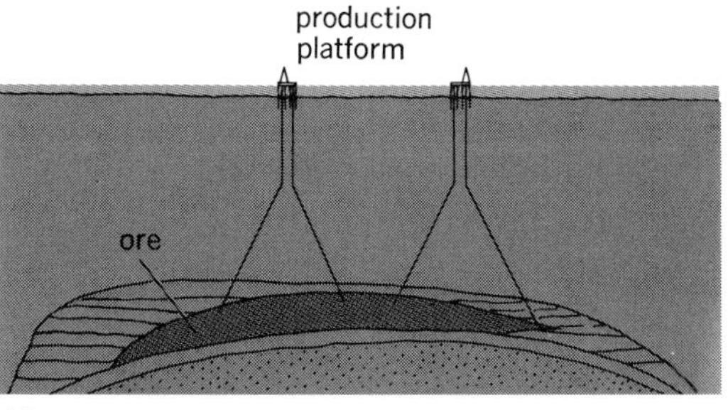

Fig. 2. Methods of mining for exploitation of consolidated mineral deposits offshore. (*a*) Shaft sunk on land, access by tunnel. (*b*) Shaft sunk at sea on artificial island. (*c*) Offshore drilling and in-place mining.

Table 5. Summary of production from mineral deposits offshore

Type	Minerals	Number of operations
Dissolved minerals	Sodium, magnesium, calcium, bromine	275+
Unconsolidated minerals	Diamonds, gold, heavy mineral sands, iron sands, tin sands, lime shells, sand and gravel	71
Consolidated minerals	Iron ore, coal, sulfur	60

must be taken with regard to overhead cover. Near land and in shallow water a shaft is sunk at sea on an artificial island. The islands are constructed by dredging from the seabed or by transporting fill over causeways. Sinking through the island is accompanied by normal precautions for loose, waterlogged ground, and development and mining are thereafter conventional. The same method is also used in oil drilling. Offshore drilling and in-place mining are used only in the mining of sulfur, but this method has considerable possibilities for the mining of other minerals for which leaching is applicable. Petroleum drilling techniques are used throughout, employing stationary platforms constructed on piles driven into the sea floor or floating drill rigs.

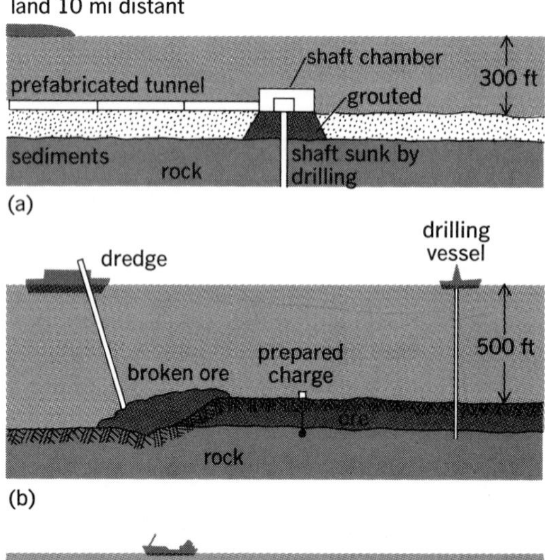

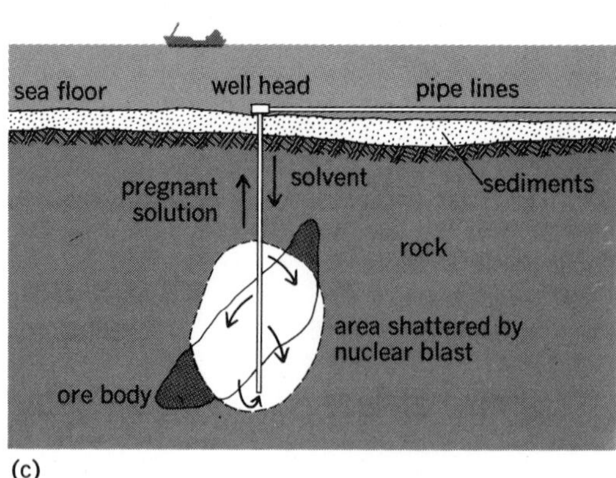

Fig. 3. Possible future methods of mining consolidated mineral deposits under the sea. (*a*) Shaft sinking by rotary drilling from tunnels laid on the sea floor. (*b*) Breaking by nuclear blasting and dredging. (*c*) Shattering by nuclear blast and solution mining. 10 mi = 16 km; 300 ft = 90 m; 500 ft = 150 m.

In summary, Table 5 shows offshore mining production, valued at $1.8 billion annually. Though this is only a fraction of world mineral production, estimated at $700 billion, the results of the extensive exploration activity taking place off the shores of all five continents may alter this considerably in the future.

The future. Despite the technical problems which still have to be overcome, the future of the undersea mining industry is without doubt as potent as it is fascinating.

Deposits of hot metalliferous brines and oozes enriched with gold, silver, lead, and copper have been located over a 38.5-mi² (100-km²) area in the middle of the Red Sea at depths of 6000–7000 ft (1800–2100 m). Similar deposits are indicated over vast areas of the East Pacific Rise.

Major problems of dissolved mineral extraction must be solved before their exploitation, and significant advances must be made in the handling of these sometimes corrosive media at such depths and distances from shore.

The mining of unconsolidated deposits will call for the development of bottom-sited equipment to perform the massive earth-moving operations that are carried out by conventional dredges today. The remarkable deposits of Co-Ni-Cu-Mn nodules covering the deep ocean floors will require new concepts in materials handling, and while some initial attempts are being made to mine them from the sea surface, it is almost certain that future operations will include some form of crewed equipment operating on the sea floor.

Consolidated deposits may call for a variety of new mining methods which will be dependent on the type, grade, and chemistry of the deposit, its distance from land, and the depth of water. Some of these methods are illustrated in Fig. 3. The possibility of direct sea floor access at remote sites through shafts drilled in the sea floor has already been given consideration under the U.S. Navy's Rocksite program and will be directly applicable to some undersea mining operations. In relatively shallow water, shafts could be sunk by rotary drilling with caissons. In deeper water the drilling equipment could be placed on the sea floor and the shaft collared on completion. The laying of large-diameter undersea pipelines has been accomplished over distances of 25 mi (40 km) and has been planned for greater distances. Subestuarine road tunnels have been built by employing prefabricated sections. The sinking of shafts in the sea floor from the extremities of such tunnels should be technically feasible under certain conditions.

Submarine ore bodies of massive dimensions and shallow cover could be broken by means of nuclear charges placed in drill holes. The resulting broken rock could then be removed by dredging. Shattering by nuclear blast and solution mining is a method applicable in any depth of water. This method calls for the contained detonation of a nuclear explosive in the ore body, followed by chemical leaching of the valuable mineral. Similar techniques are under study for land deposits.

There are many other government activities which may have a direct bearing on the advancement of undersea mining technology but possibly none as much as the International Decade of Ocean Exploration. The discovery of new deposits brings with it new incentives to overcome the multitude of problems encountered in marine mining.

[MICHAEL J. CRUICKSHANK]

Bibliography: C. F. Austin, In the rock: A logical approach for undersea mining of resources, *Eng. Mining J.*, August 1967; G. Baker, *Detrital Heavy Minerals in Natural Accumulates*, Australian Institute of Mining and Metallurgy, 1962; H. R. Cooper, *Practical Dredging*, 1958; M. J. Cruickshank, Mining and mineral recovery, *U.S.T. Handbook Directory*, pp. A/15–A/28, 1973; M. J. Cruickshank, *Technological and Environmental Considerations in the Exploration and Exploitation of Marine Mineral Deposits*, Ph.D. thesis, University of Wisconsin, 1978; M. J. Cruickshank et al., Marine mining, in *Mining Engineering Handbook*, AIME Society of Mining Engineers, sec. 20, pp. 1–200, 1973; M. J. Cruickshank, C. M. Romanowitz, and M. P. Overall, Offshore mining: Present and future, *Eng. Mining J.*, January 1968; R. H. Joynt, R. Greenshields, and R. Hodgen, Advances in sea and beach diamond mining techniques, *S.A. Mining Eng. J.*, p. 25 ff., 1977; J. L. Mero, *The Mineral Resources of the Sea*, 1965; C. G. Welling and M. J. Cruickshank. Review of available hardware needed for undersea mining, *Transactions of the Second Annual Marine Technology Society Conference*, 1966.

Marine resources

The oceans, which cover two-thirds of the Earth's surface, contain vast resources of food, energy, and minerals, and are also invaluable in many indirect ways. Capabilities to harvest these resources, and to diminish them, have become so significant that the world's nations have been forced into difficult negotiations toward an internationally accepted legal regime for the oceans, and abiding concerns have been raised about the oceans' ecological status and capacity to serve humankind.

Food. World fisheries expanded steadily from World War II to 1970, particularly the fisheries of Japan and Eastern-bloc countries. Annual harvest seems to have become stationary near 60,000,000 tons (54 teragrams); an additional 10,000,000 or perhaps 20,000,000 tons (9 or 18 teragrams) comes from fish culture in lakes, ponds, and rivers, principally in China and Southeast Asia. The ocean harvest is a very selective one, with 1% of the known oceanic species supplying nine-tenths of the total. It divides into about equal parts for food and industrial products, the latter including meal for livestock feed, fertilizers, and oils.

Seafood resources show a distinct geographic pattern, favoring the eastern edge of oceans because of geostrophic upwelling, the high latitudes through seasonal mixing and recycling of nutrients, and the proximity of large rivers. Estimates of the potential yield vary widely, from a technically optimistic high of 2,000,000,000 tons (1.8 petagrams) yearly to a pragmatically pessimistic low near the current level. The former envisages raising the productivity of wide areas by cultivation; the latter, a sharpening of institutional and environmental difficulties.

Unwanted species. Both human preferences and natural factors account for the pause in fisheries expansion. Conservative tastes and dietary preferences prevent the use of many unfamiliar species and of those whose names evoke psychological rejection. Horse mackerel, redfish, spider crabs, and dogfish are examples of undesired food that became highly accepted as Pacific mackerel, ocean perch, Alaska king crab, and grayfish. The marketing of squids, seals, jellyfishes, crabs, and other under- or unexploited species could probably benefit from similar rechristening strategies. Particularly abundant is Antarctic krill, the food of baleen whales; its harvesting and marketing are being explored by the Soviets and West Germans. Large populations of sable and lantern fishes in deep water are also attracting fishing effort. Such refocusing measures would counter the present domination of ecosystems by left-behind unwanted species.

Species diversity. The California sardine–northern anchovy imbalance seems to be a case in point, except that changes in oceanic conditions—the "climate" in terms of physical and chemical factors—have also been implicated in the ascendancy of the anchovy over the sardine. The study of this fluctuating climate and its effect on species diversity and abundance is a main thrust of oceanographic research.

Competition for resources. It also seems that the insistence on short-term profitability is incompatible with long-term yield and profitability. In whaling, the fleets have shifted from the large to the small whales, endangering the survival of most of the cetacean species as well as of the industry. Had the industry been content with a balanced and slightly reduced harvest after the lull of World War II, both the whales and fleets would still thrive. Herring, cod, tuna, and salmon fisheries also suffer from overcapacity and international competition. In both national and international arenas, means must be found to regulate the competition for clearly limited and sometimes fragile stocks and to effect a competent management of the living resources and their use.

Mariculture. To help increase the small share that seafood has in world nutrition—less than 20% in animal protein, about 1% in carbohydrates—great effort is currently invested in mariculture in the United States and Japan. There exists, of course, much skill in raising of fish and shellfish in fresh and brackish water, and techniques have been developing for thousands of years. Present work under the U.S. Sea Grant College Program emphasizes their cultivation in sea water, a sensitive process, and the hope is eventually to relieve pressure on such overexploited species as the salmon. The Japanese have been successful in the artificial propagation of the coastal sea bream and

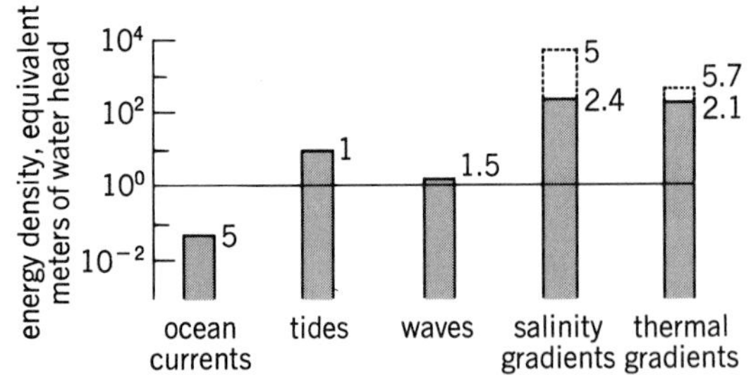

Fig. 1. Intensity or concentration of energy expressed as equivalent head of water. "Ocean currents" shows the driving head of major currents. For tides, the average head of favorable sites is given. For waves, the head represents a spatial and temporal average. The salinity-gradients head is for fresh water versus sea water, the dotted extension for fresh water versus brine (concentrated solution). The thermal-gradients head is for 12°C, and that for 20°C is dotted; both include the Carnot efficiency. *(From G. L. Wick and W. R. Schmitt, Prospects for renewable energy from the sea, Mar. Technol. Soc. J., 11(546):16–21, 1977)*

yellowtail and the eel and salmon, and also of food and pearl oysters, abalone, and scallops. Mussels are extensively cultured in some bays in Italy and France. Although natural stocks of marine organisms are not overly affected by moderate pollution and do not suffer lasting damage from most oil spills, such events would wipe out the sensitive culture work. Achievement of the goal of large-scale market penetration by the culturing industry requires consideration of its environmental and juridical needs with reference to existing uses.

Energy. The potential for energy from the sea's motion and processes has long been apparent, and designs for wave- and tidal-power devices can be traced back for hundreds of years. Wind was enlisted, of course, by the first sailor. The period from mid-1800 to mid-1900 was particularly fertile for ocean-energy technology and included discovery of the power potential from ocean-thermal gradients. World War II and the period of cheap oil and gas following it pushed many of these ideas aside. With the 1973 Oil Producing and Exporting Countries (OPEC) embargo and recognition of the limited reserves of fossil fuels, efforts into the use of alternate, renewable energy resources, including the oceanic ones, have redoubled.

Categorizing resources. The categorizing of ocean energy is somewhat arbitrary. The following are not truly marine: Offshore oil and gas are in principle not different from onshore oil and gas but require greater rigor in exploration and production; about 15% of the world's oil is produced offshore, and extraction capabilities are advancing. Coal deposits, known as extensions of land deposits, are mined under the sea floor in Japan and England. Geothermal resources are known to exist offshore; they are presently not being used, and their prospects are only dimly perceived. Biomass energy in the form of methane from giant kelp is under active investigation. Another extensive energy resource is potentially available in the fissionable and fusible elements contained in sea water; they may provide power for 10^5 and 10^9 years, respectively.

In a strict sense, ocean energy is expressed in the processes of the ocean, such as in the currents, tides, waves, thermal gradients, and the only recently recognized salinity gradients. Solar energy drives them, except the tides, which are fueled by the fossil kinetic energy of the Earth-Moon system. Estimates of the intensities of the processes are given in Fig. 1. Figure 2 shows the size of the resources.

Currents. It is evident that currents constitute the smallest and weakest resource. Low-pressure turbines of 170-m diameter have been proposed for the center of the Gulf Stream, but it is doubtful that ocean currents can be profitably harnessed. A few straits possess fast tidal currents—the Seymour Narrows in British Columbia and the Apolima Straits in Western Samoa, for instance—and offer better prospects.

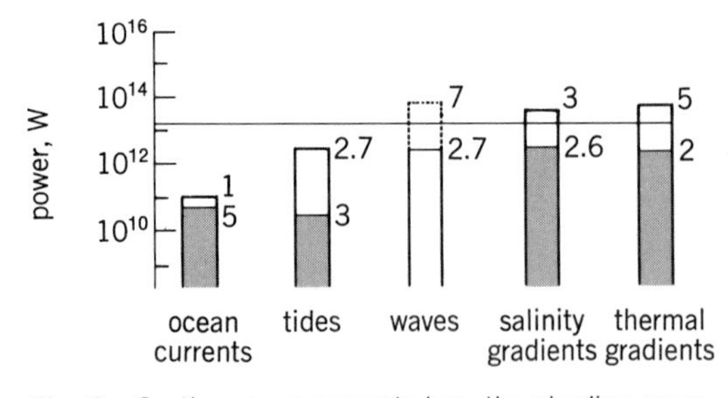

Fig. 2. On the ocean-currents bar, the shading represents the power contained in concentrated currents such as the Gulf Stream. Estimated feasible tidal power is shaded. The dotted extension on "waves" indicates that wind waves are regenerated as they are cropped. "Salinity gradients" includes all gradients in the ocean; the large ones at river mouths are shown by shading. Not shown is the undoubtedly large power if salt deposits are worked against fresh or sea water. On "thermal gradients," the shading indicates the unavoidable Carnot-cycle efficiency. The line at 1.5×10^{13} W is a projected global electricity consumption for the year 2000. *(From G. L. Wick and W. R. Schmitt, Prospects for renewable energy from the sea, Mar. Technol. Soc. J., 11(546):16–21, 1977)*

Tides. Feasible tidal power (see Fig. 2) is limited to a few sites with high tides. Tides now provide the only source of commercial ocean power in the Rance River tidal plant in Brittany, France. Since 1968 it has produced moderate amounts of power in its bank of twenty-four 10-MW turbines, and is still being fine-tuned for greater efficiency. The plant is successful technically and, increasingly, economically, and the experience gained there is being considered in renewed studies for tidal-power development in Australia, England, Canada, India, the United States, and the Soviet Union. The Soviets have an experimental 400-kW plant near Murmansk, and the Chinese have a number of very small plants. *See* TIDE.

Waves. Considerable efforts are being directed toward power from waves in England, Japan, Sweden, and the United States. Advanced lab and model testing is taking place in the British Isles, which are subjected to some of the most powerful waves of the Atlantic; the British hope to supply as much as 30% of their electricity needs from this source. As shown in Fig. 2, wave power is a unique resource in that it could expand under use, because in the open sea the waves could be rebuilt by the winds that cause them.

Salinity gradients. Salinity-gradient energy is present between aqueous solutions of different salinities. Vast amounts of this power are being dissipated in the estuaries of large rivers. Some conversion processes involve semipermeable and ion-selective membranes; they will require considerable development, however. One conversion process does not require membranes and is akin to an open-cycle thermal-gradient process; advances there will thus help. At the moment, salinity-gradient energy appears to be the most expensive of the marine options. Its high intensity and potential, especially when brines and salt deposits are included (see Figs. 1 and 2), which yield energy by the same principle, should, however, spur a strong level of research.

Thermal gradients. Thermal-gradient power, which would use the reservoirs of warm surface water and cold deep water, was first worked on by the French. Georges Claude spent much time and his own money on it in the 1920s, with generally bad luck. Now the United States government has made it the cornerstone of its ocean-energy program because, unlike tidal and wave power, it has the capacity for supplying base-load power. It also has a high intensity and potential. Floating tethered units of 100 megawatts electric are foreseen for locations in the Gulf of Mexico and the Hawaiian Islands. The design as much as possible employs conventional and proved technology, but the continuous operation of a complex system in a corrosive, befouling, and hazardous environment poses a formidable challenge.

Problems. There are some problems common to most marine alternatives. The principal one is transferring the power to shore. Indirect power harvest, such as by tanked or piped hydrogen or by energy-intensive products, from distant units or arrays is often suggested. But the detailed evaluation in the United States' thermal-gradient program points to electricity, the highest-value product, to justify the investment and to build up capacity and experience. Other problems pertain to corrosion in saline water, befouling by marine organisms, and deployment in the often violent sea.

Environmental impact. Environmental effects of ocean-energy conversion are generally negligible since the individual sources are large compared with initial demand. Only under extensive conversion would some impacts possibly be felt, such as downstream climatic changes from major ocean-current use, bay and estuarine ecosystem upsets from tidal-basin use, possible diminishing of dissolved oxygen from extensive use of waves, and short-circuiting of the ocean's internal heat transfer from thermal-gradient use. Conflicts would also arise with present uses of the sea through hazards to navigation and fisheries operations. Juridical issues will further complicate wide ocean-energy development. Some inadvertent effects may be beneficial, however. Upwelling of nutrients by thermal-gradient plants and the sheltering of organisms by ocean structures could enhance the productivity of fish and invertebrate stocks and their harvest, for instance.

Minerals. Annual production of marine minerals is worth one-tenth that of seafood or offshore oil and one-hundredth that of such nonextractive uses as maritime or military traffic. These minerals fall into two broad categories: geologic ones, or those deposited on or in the seabed; and chemical ones, or those dissolved in sea water. *See* MARINE MINING.

Geologic resources. Among the deposited resources, the manganese nodules have currently captured the spotlight in technology development and juridical deliberations, although commercial production had not commenced as of mid-1979. Far more important is the dredging of sand and gravel for construction materials, a billion-dollar-a-year industry. Produced also are aragonite in the Bahamas, monazite off Australia, and diatomaceous earth, while production of diamonds off Namibia was halted in 1971. Southeast Asia produces tin from shallow water, and phosphorite nodule deposits constitute potential fertilizer reserves off Florida and Morocco. Moreover, some hard-rock minerals are won by shaft mining, such as coal off Japan and the British Isles, and by drilling, such as barytes off Alaska.

Knowledge of the extent of mineral deposits is very scant and spotty. About 0.1% of the United States' continental shelf is surveyed in detail. The shelves of Europe, Southeast Asia, Japan, and South Africa are also explored. Often the surveys belong to private industry, and little is publicly disclosed. If and when shortages and economic prospects stimulate interest, resource surveys and extraction technology progress rapidly, as in the case of oil and gas and the manganese nodules.

The prospects are further enhanced by the work of the United States–operated international Deep Sea Drilling Program, which has drilled and cored about 500 sites thus far (mid-1979). The resultant insights have revolutionized knowledge of the geologic history and physical dynamics of the sea floor and have advanced understanding of the formation of mineral deposits and ores.

Chemical resources. Among the many elements and compounds dissolved in sea water and taken up by marine organisms, a few have become economically important. Where natural evaporation rates are high, salt and magnesium are won by distillation. Bromine yields to an oxydation-fractionation process. Currently the Japanese are planning to extract uranium, but the metals that some marine organisms strongly concentrate—strontium, copper, zinc, nickel, and vanadium—remain unused. Some marine plants are harvested for valuable compounds—a seaweed for agar, kelp for algin, Irish moss for carrageenin. Sponges, pearls, and ambergris are further products elaborated by organisms. Sea water itself is made to yield water for human use, mainly in rich arid countries and on some ships, but large-scale desalination for cities and crops seems to have become the victim of high-energy costs. Rather, the immense tabular icebergs of the Antarctic are now under study as water supplies for some arid countries such as Saudi Arabia. One modest-sized Antarctic berg could supply all of California with domestic and industrial water and absorb the waste heat of its power plants for a year.

Artifacts. In some sense, past and present artifacts that rest on the bottom of the sea may also be counted as mineral resources. These are often of very high quality in the form of processed ores, trapped oil, finished products and metals, and archeological treasures—coins, jewelry, art objects. Capabilities have grown in recent years in archeological, commercial, and military salvage, allowing increasing access to the wrecks of the past.

Passive uses. Often overlooked in resource assessments are nonextractive uses, since they are difficult to quantify. In this category belong such invaluable and indispensable items as inspiration and recreation, commercial and military traffic, and the disposal of waste materials and low-grade heat.

Besides inspiring artists and laymen, the sea has a strong recreational appeal. The crowding on coasts derives in part from the opportunities for sailing, fishing, and water sports, in part from those for trade and defense. Two-thirds of present oil production moves in tankers. Countries use the oceans for cooling water and as a receptacle for liquid and solid wastes: power plants are shore-sited wherever feasible; metropolitan areas discharge wastewater to sea and dump solid materials on a large scale. Such practices have, of course, raised concerns about environmental impact and are now under scrutiny and control; of particular concern is the planned disposal of radioactive wastes in the sub-sea floor.

Conclusions and prospects. The sea is important for intangible and practical ends. It sustains life with its materials and processes. Its organisms satisfy nearly one-fifth of the human demand for animal protein; its content of minerals helps overcome critical shortfalls in terrestrial resources; and it is beginning to yield some of its immense energy flux.

At present the overriding benefit, however, comes from its just being there. It sustains the continents by the spreading of its floor, purifies water and air by the hydrologic cycle, modifies climate zonation by its circulation, inspires humans' vision and carries their trade, and absorbs many of their wastes. It is thus crucial that the sea suffer relatively little from abuse and mismanagement often associated with development and growth, so that it can continue to provide the varied and balanced uses humankind has come to expect from it. [WALTER R. SCHMITT]

Bibliography: Committee on Merchant Marine and Fisheries, U.S. House of Representatives, *Aquaculture*, 1977; Committee on Science and Technology, U.S. House of Representatives, *Energy from the Ocean*, 1978; J. D. Isaacs, The nature of oceanic life, *Sci. Amer.*, vol. 221, no. 3, 1969; J. D. Isaacs, The sea and man, *Explorers J.*, vol. 46, no. 4, 1968; National Research Council–National Academy of Sciences, *Priorities for Research in Marine Mining Technology,* 1977; National Research Council—National Academy of Sciences, *Supporting Papers, World Food and Nutrition Study*, vol. 1, 1977; G. L. Wick and W. R. Schmitt, Prospects for renewable energy from the sea, *Mar. Technol. Soc. J.*, vol. 11, no. 546, 1977.

Meromictic lake

A lake whose water is permanently stratified and therefore does not circulate completely throughout the basin at any time during the year. In the temperate zone this permanent stratification occurs because of a vertical, chemically produced density gradient. There are no periods of overturn, or complete mixing, since seasonal fluctuations in the thermal gradient are overridden by the stability of the chemical gradient.

The upper stratum of water in a meromictic lake is mixed by the wind and is called the mixolimnion. The bottom, denser stratum, which does not mix with the water above, is referred to as the monimolimnion. The transition layer between these chemically stratified strata is called the chemocline (Fig. 1).

In general, meromictic lakes in North America are restricted to (1) sheltered basins that are proportionally very small in relation to depth and that often contain colored water, (2) basins in arid regions, and (3) isolated basins in fiords. *See* LAKE.

Prior to 1960 only 11 meromictic lakes had been reported for North America. However, during 1960–1968 that number was more than doubled. There are seven known meromictic lakes in the state of Washington, six in Wisconsin, four in New York, three in Alaska, two in Michigan, and one each in Florida, Nevada, British Columbia, Labrador, and on Baffin Island. Research activity on meromictic lakes has been focused on studies of biogeochemistry, deepwater circulation, and heat flow through the bottom sediments.

Chemical studies. Very few detailed chemical studies have been done on meromictic lakes, particularly on a seasonal basis. The most typical chemical characteristic of meromictic lakes is the absence of dissolved oxygen in the monimolimnion. Large quantities of hydrogen sulfide and ammonia may be associated with this anaerobic condition in deep water. J. Kjensmo proposed that the accumulation of ferrous bicarbonate in the deepest layers of some lakes may have initiated meromixis under certain conditions.

Meromictic lakes are exciting model systems for many important biogeochemical studies. The isolation of the monimolimnetic water makes these studies quite interesting and important. E. S. Deevey, N. Nakai, and M. Stuiver studied the biological fractionation of sulfur and carbon isotopes in the monimolimnion of Fayetteville Green Lake, N.Y. They found the fractionation factor for sulfur to be the highest ever observed. T. Takahashi, W. S. Broecher, Y. Li, and D. L. Thurber have undertaken detailed, comprehensive geochemical and hydrological studies of meromictic lakes in New York.

Sediments from meromictic lakes are among the best for studies of lake history, since there is little

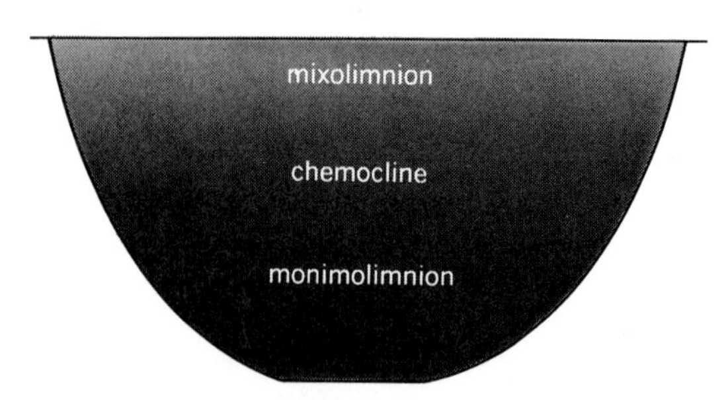

Fig. 1. Cross-sectional diagram of a meromictic lake.

decomposition of biogenic materials. G. J. Brunskill and S. D. Ludlam have obtained information on seasonal carbonate chemistry, sedimentation rates, and sediment composition and structure in Fayetteville Green Lake. Also, deuterium has been used by several workers in an attempt to unravel the history of meromictic lake water.

Physical studies. Radioactive tracers have been used by G. E. Likens and A. D. Hasler to show that the monimolimnetic water in a small meromictic lake (Stewart's Dark Lake) in Wisconsin is not stagnant but undergoes significant horizontal movement (Fig. 2). While the maximum radial spread was about 16–18 m/day, vertical movements were restricted to negligible amounts because of the strong density gradient. W. T. Edmondson reported a similar pattern of movement in the deep water of a larger meromictic lake (Soap Lake) in Washington. V. W. Driggers determined from these studies that the average horizontal eddy diffusion coefficient is 3.2 cm²/sec in the monimolimnion of Soap Lake and 17 cm²/sec in Stewart's Dark Lake.

Because vertical mixing is restricted in a meromictic lake, heat from solar radiation may be trapped in the monimolimnion and can thereby produce an anomalous temperature profile. G. C. Anderson reported monimolimnitic water temperatures as high as 50.5°C at a depth of 2 m in shallow Hot Lake, Wash.

N. M. Johnson and Likens pointed out that some meromictic lakes are very convenient for studies of geothermal heat flow because deepwater temperatures may be nearly constant. Studies of terrestrial heat flow have been made in the sediments of Stewart's Dark Lake, Wis., by Johnson and Likens and in Fayetteville Green Lake by W. H. Diment. Steady-state thermal conditions were found in the sediments of Stewart's Dark Lake, and the total heat flow was calculated to be 2.1×10^{-6} cal/cm² sec. However, about one-half of this flux was attributed to the temperature contrasts between the rim and the central portion of the lake's basin. R. McGaw found that the thermal conductivity of the surface sediments in the center of Stewart's Dark Lake was 1.10×10^{-3} cal cm/cm² sec °C, a value substantially lower than that for pure water at the same temperature but consistent with measurements on colloidal gels.

Biological studies. Relatively few kinds of organisms can survive in the rigorous, chemically reduced environment of the monimolimnion. However, anaerobic bacteria and larvae of the phantom midge (*Chaoborus* sp.) are common members of this specialized community. Using an echo sounder, K. Malueg has been able to observe vertical migrations of *Chaoborus* sp. larvae in meromictic lakes. The sound waves are reflected by small gas bladders on the dorsal surface of the larvae. The migration pattern is similar to that shown by the deep-scattering layers in the sea. The larvae come into the surface waters at night when the light intensity is low and sink into deeper waters during the daylight hours. Hasler and Likens found that biologically significant quantities of a radioisotope (iodine-131) could be transported from the deep and relatively inaccessible part of a meromictic lake to the surface and thence to the adjacent terrestrial environment by these organisms. The radiotracer appeared in flying adult *Chaoborus* sp. along the shoreline of the lake within 20 days after it had been released within the region of the chemocline. *See* FRESH-WATER ECOSYSTEM; LIMNOLOGY.

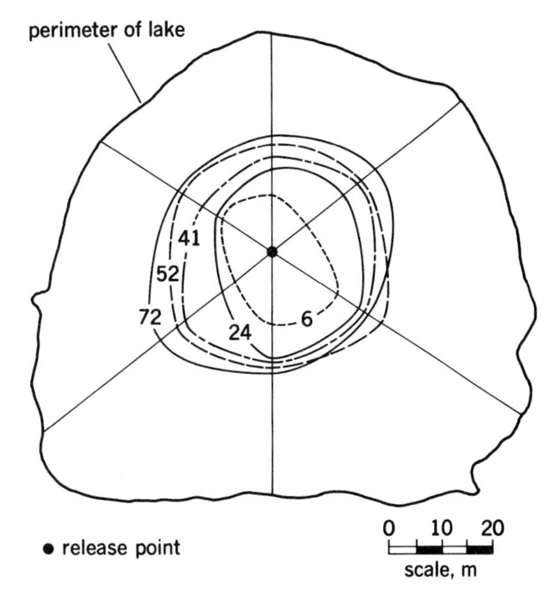

Fig. 2. The outlines show the maximum horizontal displacement of a radiotracer (sodium-24) following its release at the 8-m depth in Stewart's Dark Lake, Wis. The numbers indicate the hours elapsed after release. (*After A. D. Hasler and G. E. Likens*)

[GENE E. LIKENS]

Bibliography: E. S. Deevey, N. Nakai, and M. Stuiver, Fractionation of sulfur and carbon isotopes in a meromictic lake, *Science*, 139:407–408, 1963; W. T. Edmonson and G. C. Anderson, Some features of saline lakes in central Washington, *Limnol. Oceanogr.*, suppl. to vol. 10, pp. R87–R96, 1965; D. G. Frey (ed.), *Limnology in North America*, 1963; G. E. Hutchinson, *A Treatise on Limnology*, 2 pts., vol. 1, 1975; D. Jackson (ed.), *Some Aspects of Meromixis*, 1967; N. M. Johnson and G. E. Likens, Steady-state thermal gradient in the sediments of a meromictic lake, *J. Geophys. Res.*, 72:3049–3052, 1967; J. Kjensmo, Iron as the primary factor rendering lakes meromictic, and related problems, *Mitt. Int. Ver. Limnol.*, 14:83–93, 1968; G. E. Likens and P. L. Johnson, A chemically stratified lake in Alaska, *Science*, 153:875–877, 1966; O. T. Lind, *Handbook of Common Methods in Limnology*, 1979; F. Ruttner, *Fundamentals of Limnology*, 3d ed., 1963; K. Stewart, K. W. Malueg, and P. E. Sager, Comparative winter studies on dimictic and meromictic lakes, *Verh. Int. Ver. Limnol.*, 16:47–57, 1966.

Meteorology

The science concerned with the atmosphere and its phenomena. Meteorology is primarily observational; its data are generally "given." The meteorologist observes the atmosphere—its temperature, density, winds, clouds, precipitation, and so on—and aims to account for its observed structure and evolution (weather, in part) in terms of external influence and the basic laws of physics.

Empirical relations between observed variables, as those between the patterns of wind and

Table 1. Components of dry air

Gas	Symbol	% vol
Nitrogen	N_2	78.09
Oxygen	O_2	20.95
Argon	Ar	0.93
Carbon dioxide	CO_2	0.03
Neon	Ne	1.8×10^{-3}
Helium	He	5.2×10^{-4}
Krypton	Kr	1×10^{-4}
Hydrogen	H_2	5×10^{-5}
Xenon	Xe	8×10^{-6}
Nitrous oxide	N_2O	3.5×10^{-5}
Radon	Rn	6×10^{-16}
Methane	CH_4	1.5×10^{-4}

weather, are developed to pose more effectively the problems to be investigated and explained and to provide essential material for the application of the science. Weather forecasting serves as an example of such application because theory still remains insufficiently developed to provide more certain applications. Little controlled experiment has been made on the atmosphere, but more is probable. *See* WEATHER FORECASTING AND PREDICTION; WEATHER MODIFICATION.

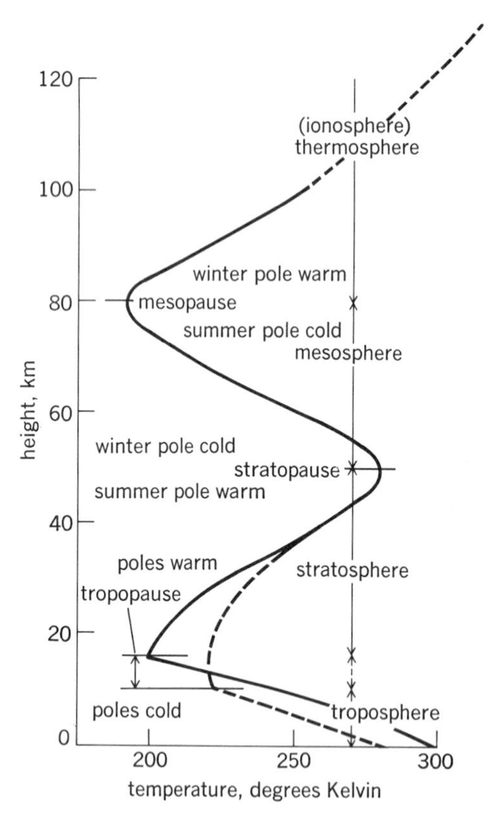

Fig. 1. Vertical structure and nomenclature of atmospheric layers in relation to temperature. Values of temperatures shown are approximate; the form of the temperature variation with height determines the nomenclature. The two curves shown for troposphere and stratosphere refer to lower (solid line) and higher (dashed line) latitude conditions. The sense of the meridional temperature gradient at various levels is shown by entries "poles cold" and "poles warm." A change of meridional gradient appears to coincide with a change of vertical gradient (pause) only at the tropopause.

This background article has a threefold organization. The first portion presents a summary of the general physics of the air; this has been the principal approach and basis for meteorological science and its applications, such as to weather phenomena (the condition of the atmosphere at any time and place) and climate (a composite generalization of weather conditions throughout the year). A second portion, synoptic meteorology, discusses the character of the atmosphere on the basis of simultaneous observations over large areas. Concurrently, and at an accelerating pace, dynamic principles (thermodynamic and hydrodynamic) are being applied to meteorological investigations. This study of naturally produced motions in the atmosphere is forming much of the scientific basis for modern weather forecasting and physical climatology. Hence, the third portion of this article deals with dynamical meteorology. *See* CLIMATOLOGY.

GENERAL PHYSICS OF THE AIR

The components of dry air, excluding ozone, and their relative volumes (mol fractions of gases) up to a height of at least 50 km are given in Table 1. Some of the rarer gases, such as CO_2, are continually entering and leaving the atmosphere through the Earth's surface, and the fractions quoted are thus mean values. The effective molecular weight of the mixture is 28.966 and its equation of state, to 1 in 10^4, is $p = R\rho T$, where p is pressure, ρ density, T absolute temperature, and R, the specific gas constant, is 2.8704×10^6 erg/(g)(°K). *See* AIR; ATMOSPHERE.

Above 50 km, O_2 becomes progressively dissociated to O, which is probably dominant above about 150 km. N_2 probably dissociates at appreciably higher levels. There is no good evidence for diffusive gravitational separation of the lighter from the heavier gases below about 100 km—nor indeed is this likely. Above 100 km, however, such separation is probable and in the levels of escape (exosphere), at several hundreds of kilometers, helium and hydrogen may be dominant with proportions varying during the solar 11-year cycle.

Water vapor and ozone are highly variable additional components of air. The fraction of the former commonly decreases rapidly with height, from about 10^{-2} at sea level to 10^{-6} or 10^{-7} at about 16 km, with dissociation at much higher levels. Ozone, the product of photochemical action in sunlight at high levels, has a maximum fraction ($\sim 10^{-8}$) at about 25 km. The importance of these two constituents is far greater than their fractions might suggest because (1) both are radiatively active, water vapor and ozone in the infrared of terrestrial emission, ozone in the near ultraviolet of solar emission; and (2) water vapor condenses to the liquid or solid, giving cloud, which reflects upward a major part of solar radiation incident upon it, and in condensing releases a large latent heat to the air. Carbon dioxide, CO_2, is the only other constituent with strong infrared activity.

Thermal structure. It is convenient to divide the atmosphere into a number of layers on the basis of its thermal structure. The first layer (Fig. 1) is the troposphere, in which temperature on average decreases with height 6°C/km (the "lapse rate") everywhere except near the winter pole. The troposphere is about 16 km deep in the tropics and about 10 km deep, with substantial variations, in higher latitudes. At any one level in the layer, the mean temperature decreases from Equator to pole, by about 40°C in winter and 25°C in summer.

The second layer is the stratosphere, in which the temperature varies at first very little with height and then increases to near the surface temperature at about 50 km. In this layer temperature increases poleward, except quite near the winter pole. The boundary between troposphere and stratosphere is termed the tropopause, which may drop abruptly in altitude at about 30° lat and again in higher latitudes.

From 50 km to 80 km the temperature again decreases with height to an absolute minimum of about −80°C, lower in summer than in winter, and in summer associated with noctilucent clouds. This layer is called the mesosphere, and its lower and upper boundaries are called the stratopause and mesopause, respectively.

Above 80 km the temperature again increases with height, at a rather uncertain rate. The layer is strongly ionized by solar ultraviolet and other radiations and is variously termed the ionosphere or thermosphere.

Radiation and thermal structure. During the passage of solar radiation downward through the atmosphere, the following, in broad outline, takes place. The shorter ultraviolet waves are absorbed by the thermosphere, more above than below, so that temperature may be expected to fall along the path of the beam. Entering the mesosphere, O_3 (ozone) is encountered, and new absorption of energy takes place in the near ultraviolet. The increase of ozone concentration along the path outweighs the depletion of the beam energy by O_3 at upper levels so that the temperature increases along this portion of the path. Beneath the stratopause, however, depletion in the wavelengths concerned has become large and the O_3 concentration itself ultimately falls so that temperature decreases along the path. There is little absorption in the troposphere—the slight near-infrared absorption by water vapor is practically uniform along the path because of the increasing concentration of vapor—but there is substantial backscatter by air molecules, haze, and clouds. The Earth's surface, except snow, absorbs the incident solar radiation strongly, and this leads to the heating of the atmosphere from below by various kinds of convection and to a lapse of temperature in the troposphere.

Because the atmosphere and underlying surface maintain their temperature over the years, the terrestrial emission of radiation to space by water vapor, carbon dioxide, and ozone in the far infrared, must nearly balance the absorption of solar energy. The balance is achieved globally but not locally, the winds providing the necessary transport of heat from the regions of net absorption (lower latitudes) to those of net emission (higher latitudes). While the broad form of the temperature profile of Fig. 1 is determined by the pattern of solar absorption, actual temperatures are also due to terrestrial emission and wind transport. *See* TERRESTRIAL ATMOSPHERIC HEAT BALANCE.

Pressure and wind. The pressure at any point in the atmosphere is the weight of a column of air of unit cross section above that point. Therefore pressure decreases with height. At any level it decreases more rapidly where the air is cold and dense than where it is warm and light. The pressure at sea level is generally a little over 1 kg/cm², varying by a few percent in space and time because of the inflow of air over some regions and its outflow over others. The horizontal pressure pattern necessarily changes with height when, characteristically, the temperature varies horizontally.

In an unheated, nonrotating atmosphere, air would flow horizontally from high pressure to low to remove the pressure difference. Other forces arise in the actual atmosphere to provide a flow which is more nearly along, than perpendicular to, the isobars (lines on a map connecting points of equal barometric pressure). Therefore pressure patterns persist for days, with gradual modification, and are translated with speeds often comparable with those of the winds blowing in them. The effect of the Earth's rotation on the wind is nearly always dominant, except very near the Equator. If the horizontal force arising from the Earth's rotation (called the Coriolis force) precisely balances the pressure force, the wind blows with a speed V along the isobars (for an observer with his back to the wind, low pressure is to the left in the Northern Hemisphere). The speed is given by the equation below, where $\partial p/\partial n$ is the horizontal

$$V = \frac{(\partial p/\partial n)}{2\omega\rho \sin\phi}$$

pressure gradient, ω the angular velocity of the Earth, ρ the air density, and ϕ the latitude. This wind V is called the geostrophic wind and is a good approximation to the actual except near the surface, where friction always intervenes. Since, as seen above, the pressure gradient generally varies with height, so do both V and the actual wind. The change of V with height is proportional to the horizontal temperature gradient, low temperature being to the left of the vector change of V in the Northern Hemisphere. *See* AIR PRESSURE; WIND.

General atmospheric circulation. The mean wind in nearly all parts of the atmosphere is predominantly along the latitude circles. The pattern of this zonal motion in latitude and height is shown in Fig. 2 and is the basis of the general circulation of the atmosphere. Surface easterlies (trade winds) are found in the tropics, surface westerlies in middle latitudes, and surface easterlies again near the poles. But everywhere in the troposphere, except near the Equator, winds increase in strength from the west with height; in the tropics and polar regions easterlies decrease before giving place aloft to increasing westerlies. Above the tropopause the westerlies decrease with height except over the winter polar cap where, at around 40 km, there occurs a winter westerly jet stream. This pattern of winds is consistent with the meridional gradient of temperature in the troposphere and its reversal (except over the winter polar cap) in the stratosphere, since the winds are quasi-geostrophic. The absolute maximum of zonal wind shown in Fig. 2 is found in the upper troposphere at about 30° latitude and is the core of the subtropical jet stream. *See* JET STREAM.

The general circulation is maintained against frictional dissipation by a "boiler-condenser" arrangement provided by the heat absorbed, directly or indirectly, from the Sun and that emitted to

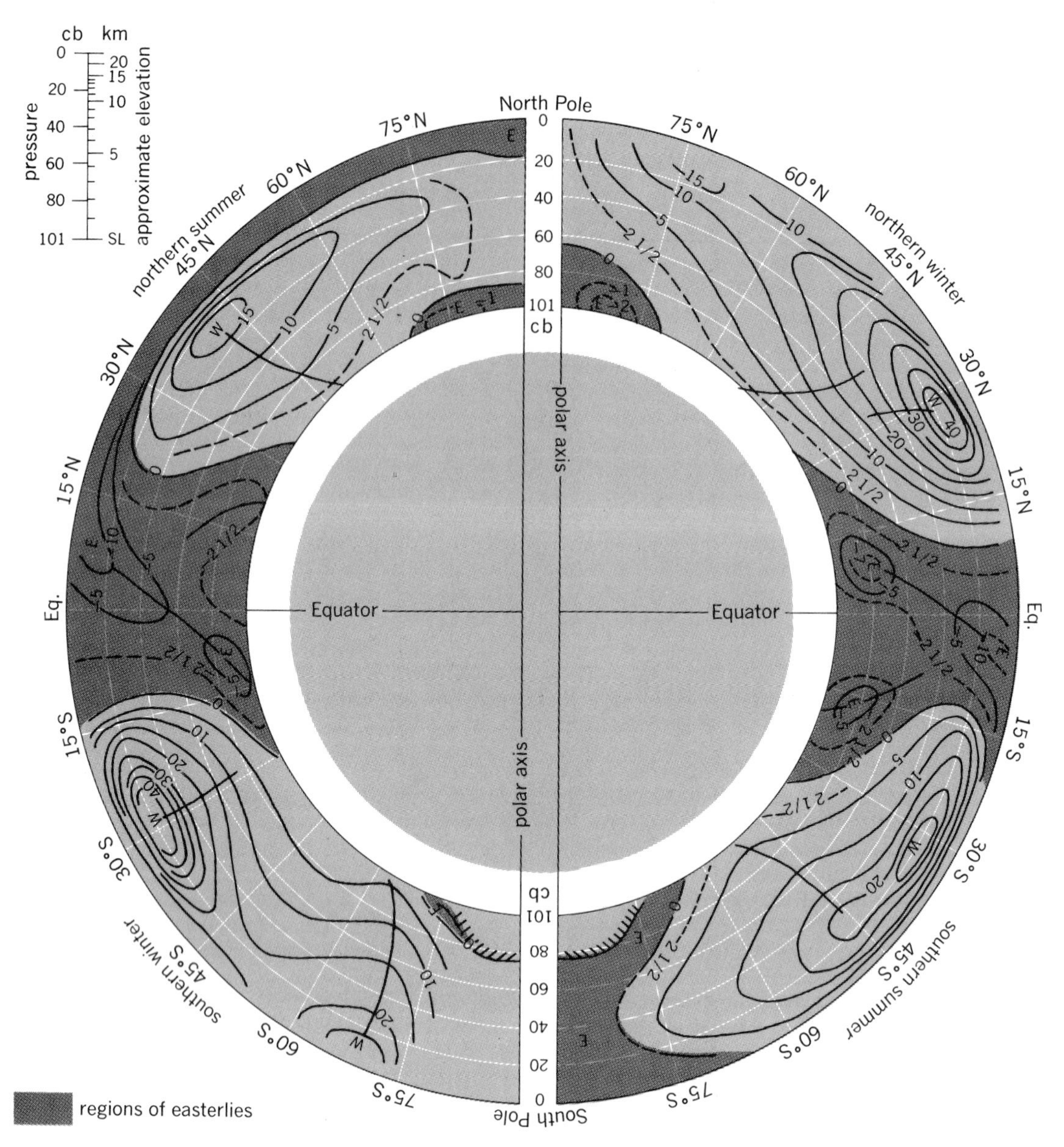

Fig. 2. The pattern of mean zonal (east-west) wind speed averaged over all longitudes as a function of latitude, height, and season (*after Y. Mintz*). Height, greatly enlarged relative to Earth radius, is shown on a linear pressure (∝ mass) scale with geometrical equivalent given at upper left. The mean zonal wind (westerly positive, easterly negative) has the same value along any one line and is shown in meters per second (1 m/sec ≅ 2 knots) on the line. Note subtropical westerly jet in high troposphere at about 30° lat (or more in northern summer). This jet should not be confused with the polar front jet of higher latitudes; the latter is migratory and so does not appear in the mean.

space by the Earth. The wind thereby generated transports heat from source to sink to maintain thermal balance, disturbances on the mean motion being important in the process. *See* ATMOSPHERIC GENERAL CIRCULATION.

Water vapor, cloud, and precipitation. The motion of the air is rarely quite horizontal; it is commonly moving upward or downward at a few centimeters per second over large areas, while locally in thunderclouds the vertical velocity may be several meters per second. Air which rises is cooled by expansion into lower pressure and that which descends is correspondingly compressed and warmed. Since air is always more or less moist, a sufficient rise will cause the temperature to drop to the saturation point and cloud will then form on nuclei present in the air. Large-scale ascent leads to extensive sheets of cloud called stratus: cirrostratus at high, ice-forming levels, altostratus at medium levels, nimbostratus at lower levels. Local, strong ascent produces heap clouds (cumulus or cumulonimbus). If the cloud base is sufficiently warm or the cloud depth sufficiently great, or both, the condensed water or ice in the cloud falls out as rain, snow, or other precipitation. Certain lenticular clouds are due to local ascent and descent brought about by hills, and these clouds do not move with the wind. Cloud formation is practically confined to the troposphere because this alone contains sufficient water vapor and sufficiently sustained vertical motion. *See* CLOUD; CLOUD PHYSICS.

Atmospheric disturbances. The mean motion, described under the general circulation of the atmosphere, is disturbed by many patterns of motion of varying scale. A chart of the flow well above the surface over the Northern Hemisphere will generally show, on any day, large meanderings of the air poleward and equatorward superposed on a general westerly flow. These are the long waves of the westerlies, with a few thousand kilometers between the turning points. They are generally associated with a jet stream in the upper troposphere. Another class of large-scale atmospheric disturbances, seasonal and most apparent in the lower troposphere, is the monsoons; these are most developed in the South Asian cyclone of summer and the Siberian anticyclone of winter.

Proceeding downward in scale are the traveling depressions (cyclones) and anticyclones, ridges and troughs of extratropical latitudes, 1000–2000 km in horizontal extent, and then the somewhat shorter waves in the easterlies (trades). Next in size is the tropical cyclone (~100 km)—the hurricane or typhoon, according to location—which is mainly confined to the tropical oceans and eastern seaboards of continents. Still smaller (~1 km) is the tornado, mainly confined to land and associated with cumulonimbus cloud, and the waterspout associated with similar cloud over the sea. The smallest "revolving storm" is the dust devil (~10 m), which occurs immediately above a hot, dry surface in a very light general wind. *See* FRONT; STORM; TORNADO.

In addition to the above well-defined patterns of flow there are randomly distributed fluctuations of flow over several orders of magnitude of scale from about 1 cm upward, particularly evident in the bottom kilometer of the atmosphere and in and around cumulus. They are related to friction, convection, and wind shear and are important agents of vertical transfer of heat, matter (water vapor, dust, ozone), and momentum.

Air masses and fronts. The horizontal variations of air temperature referred to in previous paragraphs are commonly concentrated into narrow zones, or even discontinuities, with low gradients in intervening areas. These narrow zones or discontinuities are called fronts, and the air in a region of small temperature gradient is called an air mass. If a front moves so that a warm air mass replaces a cold air mass, it is called a warm front, and conversely a cold front. Fronts may also be stationary. They slope upward at a slope ratio of about 1 in 100, with the cold air as a wedge beneath the warm air.

Most extratropical depressions first appear as dents or waves on a frontal surface, warm air rising slowly over an extended area in the forward part of the wave. This results in stratiform cloud. A stronger updraft immediately ahead of the rear part of the wave yields cumulonimbus cloud.

The tropopause is higher and its temperature lower above a warm air mass than above a neighboring cold air mass, and a break, or offset, appears at the frontal boundary with a jet stream in the warm air mass near the break in the tropopause. *See* AIR MASS.

Optical phenomena. Scattering, refraction, and reflection of light by the air or by particulate matter (dust, cloud particles, or rain) in the air give rise to a variety of optical phenomena.

Electrical properties. The atmosphere and the underlying Earth are like a leaky electrical capacitor. Positive charge is separated vertically from negative charge in thunderclouds and some other areas of disturbed weather; the net result is that the Earth's surface is left in fine-weather areas with an average negative charge σ of 2.7×10^{-4} esu/cm^2 to which corresponds a vertical field F_0 ($= 4\pi\sigma$) at the surface of 100 volts/m. The other "plate" of the capacitor is the highly conducting upper atmosphere, and the leak arises from the small conductivity of the air between the plates, produced by ionization by radioactive matter in the soil and air, and by cosmic radiation; the conductivity increases with height. The resistance R of a 1-cm^2 column of atmosphere is about 10^{21} ohms and an air-Earth current $i = V_\infty/R$ (V_∞ being the potential of the upper conducting layer) of about 2×10^{-16} amp/cm^2 flows as a discharge current. V_∞ is thus about 2×10^5 volts above Earth. The increase of conductivity with height implies a proportionate decrease of the field with height, and this in turn implies a small positive ionic space charge in the air.

The air-Earth current would discharge the capacitor in about 1/2 hr if V_∞ (or σ) were not maintained by the charge separation in thunderclouds. This separation is more than adequate, the excess providing lightning flashes within the cloud—a shorting of the generator. *See* ATMOSPHERIC ELECTRICITY. [P. A. SHEPPARD]

SYNOPTIC METEOROLOGY

This branch of meteorology comprises the knowledge of atmospheric phenomena connected with the weather, applied mainly in weather forecasting, and based on data acquired by the synoptic method. This method involves the study of weather processes through representations of atmospheric states determined by synchronous observations at a network of stations, most of which are at least 10 km apart. By international agreement, the data taken from the Earth's surface and aloft at certain international hours of observation are inserted on weather maps, upper-air maps, vertical cross sections, time sections, and sounding diagrams with international symbols according to fixed rules. These crude representations are then analyzed and critically evaluated in accordance with the knowledge of existing structure models in the (lower) atmosphere, in order to ascertain the best approximation to a three-dimensional image of the true atmospheric state at the hour of observation. From one such representation, or a series of them, future atmospheric states are then derived with the aid of empirical knowledge of their behavior and by application of the theoretical results of dynamic meteorology.

General atmospheric circulation. Figure 3 shows some of the main constituents of the average state in the bottom layer of the atmosphere as to flow and pressure patterns, main air masses and fronts. The illustration also indicates how these large-scale mechanisms form part of a general atmospheric circulation with trade winds, monsoons, high-reaching middle-latitude westerlies, and shallow polar easterlies (see the vertical cross section of Fig. 2).

The planetary high-pressure belt at 30°N and S is split into subtropical high-pressure cells mainly as a result of the joint dynamic and thermodynamic effect of the great continents and mountain ridges of the Earth, especially the Cordilleran highlands of South and North America. These cells

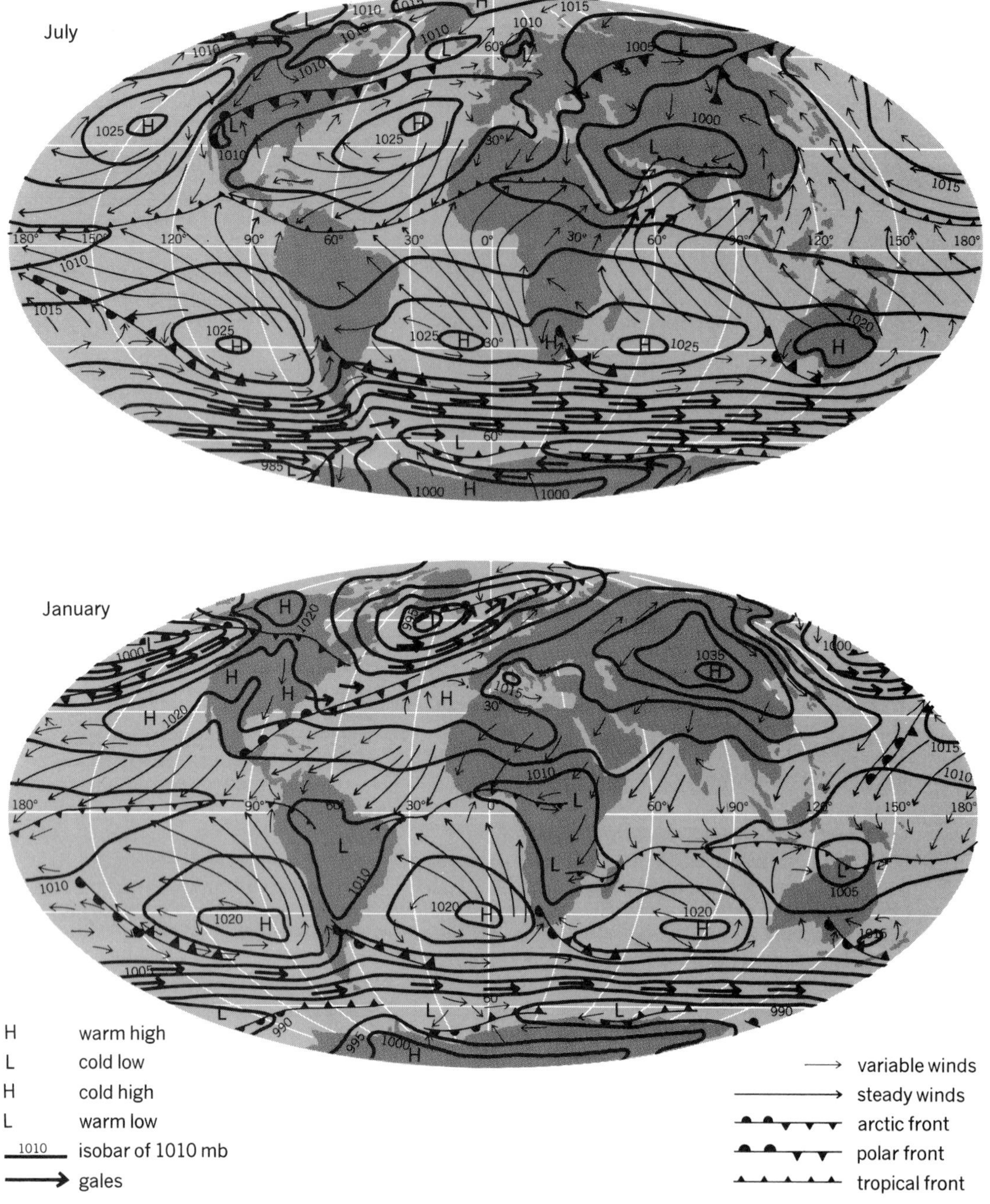

Fig. 3. Air masses and fronts as links in the general atmospheric circulation.

again determine the formation and average position of the different polar fronts and air masses at the Earth's surface.

Other general structures in the atmospheric circulation of importance to weather are the quasi-horizontal tropopause layers at different heights within different air masses and the tropical fronts, maintained within the Zones of Intertropical Convergence (ITC). The former generally mark the top of any considerable convection or upglide motion and of clouds in the atmosphere, but represent no store of potential energy. The latter form at the meeting of two opposite trade wind systems in the doldrums and have functions partly similar to those of the polar fronts, although they are much weaker and less distinct.

Air masses and front↔jet systems. Air masses may be classified in two distinctly different ways, thermodynamically and geographically.

Thermodynamic classification. This classification is based on their recent path and life history, distinguishing mainly the two opposite cases: the air being warmer or colder than the Earth's surface, resulting in warm mass and cold mass. The warm mass, usually flowing poleward, is much warmer than the seasonal normal of the region, at least aloft. With the cooling from below, it gradually acquires a stable stratification, which damps turbulence and vertical mixing. Thus, the wind is relatively steady, the visibility low, and advection fog or stratus clouds often form within it, at least at sea, sometimes even yielding drizzle.

This air mass is most typically found on the poleward side of the warm subtropical highs (Figs. 3 and 8), where the air is subsiding. These highs may get displaced poleward (Fig. 8) and will then bring periods of steady and rather dry weather—very warm in summer on land—to middle and higher latitudes.

The cold mass, mostly flowing equatorward, is by definition colder than normal, at least aloft. Due to heating from below it rapidly acquires an unstable stratification which favors turbulence and vertical mixing or convection. Thus, the wind is gusty, visibility is good, and usually where the air is moist enough convective clouds form: cumulus → cumulonimbus with showers → thunderstorms and even hail. This air mass is most typically found within (upper) cold lows (Fig. 10), where, if there is sufficient moisture, the instability, general convergence, and lifting tendency of the air may favor the formation even of nonfrontal rain areas, which are discussed later. Therefore, such a low usually brings a period of wet and stormy weather to the equatorward part of middle latitudes, and in winter to the subtropics. However, at night over land, even this air may be stabilized so efficiently that radiation fog occurs within it.

Geographic classification. The second classification is based on the geographical origin or position of the air and the values of characteristic properties (such as temperature and specific humidity).

Tropical air occupies all the space between the polar-front↔jet systems of both hemispheres; aloft it may reach far into the polar region. At the midtroposphere this air is 10–20°C warmer and much more humid than the polar air at the same level and latitude. At low levels, the tropical air emerges from the quasi-stationary subtropical highs, reaching middle latitudes from the southwest, particularly within the warm sectors of migrating cyclones (Fig. 7), as a mild or warm, moist and hazy air current (the stable warm mass). Aloft, it appears in high latitudes within the warm highs (Fig. 8). Below, the tropical air also flows equatorward and westward within the northern and southern trade wind systems. When approaching the tropical front (Fig. 3) it appears, at sea, as an unstable cold mass with intense convection and shower activity on both sides of the front: the equatorial rains. Over land, for example, in North Africa, only the southwestern monsoon (that is, the southeastern trade wind that has invaded the other hemisphere), undercutting the very hot and dry northeast trade wind, is moist enough to produce any convective clouds and rain.

Polar air is found on the polar side of the polar fronts, as shown in Fig. 3. Below, it emerges in winter from continental subpolar highs of 40–60°N, and in summer from the polar basin. The polar masses, and still more so the arctic and antarctic air masses, are mainly characterized by very low temperatures and low specific humidities aloft.

Arctic air is produced over the ice- and snow-covered parts of the arctic region during the colder seasons, when the Earth's surface is a marked cold source. In the North American sector its seat, for dynamic as well as thermal reasons, lies on the average near Baffin Island, being separated from the polar air south of it by the American arctic front (Fig. 3).

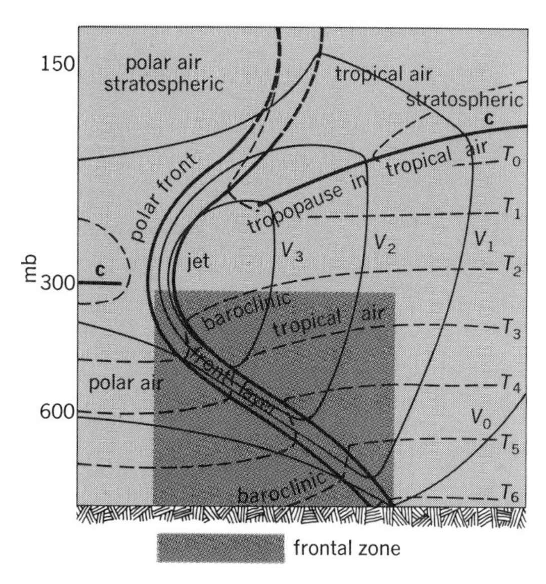

Fig. 4. Cross section of a polar-front↔jet system.

Fronts and frontogenesis. Because of the Earth's rotation a surface of density discontinuity, a front, seeks a tilted position of dynamic equilibrium. A front is defined as a dynamically important, tilting layer of transition between two air masses of markedly different origin, temperature, density, and motion. Colder air lies as a wedge below the warmer; therefore, the front will coincide with a trough or bend in the (moving) isobaric system, at whose passage the wind will veer, and as a rule with falling pressure ahead and stationary or rising pressure behind. Frictionally produced convergence, or static and inertial instability, or both, preferably within the warmer air, favors the ascent of air along the front. As a result, a vast cloud mass, a cloud system, may form, with an area of continuous precipitation near the front (Fig. 6).

The main fronts reach into the stratosphere (Fig. 4) and have a horizontal extension of several thousand kilometers. They originate as links in

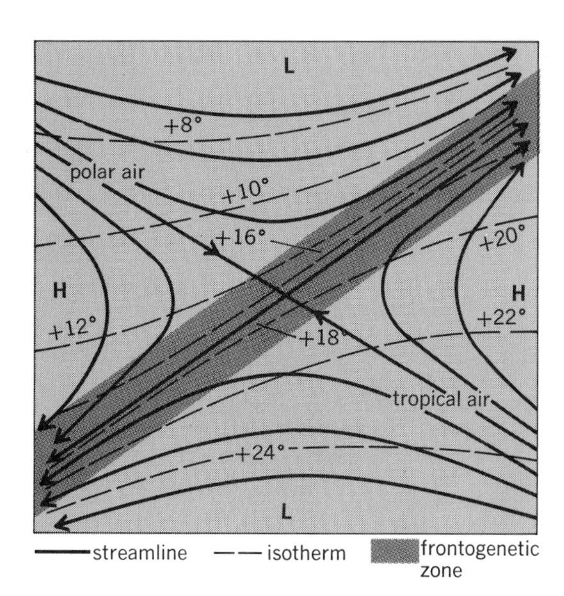

Fig. 5. Frontogenesis in a field of deformation.

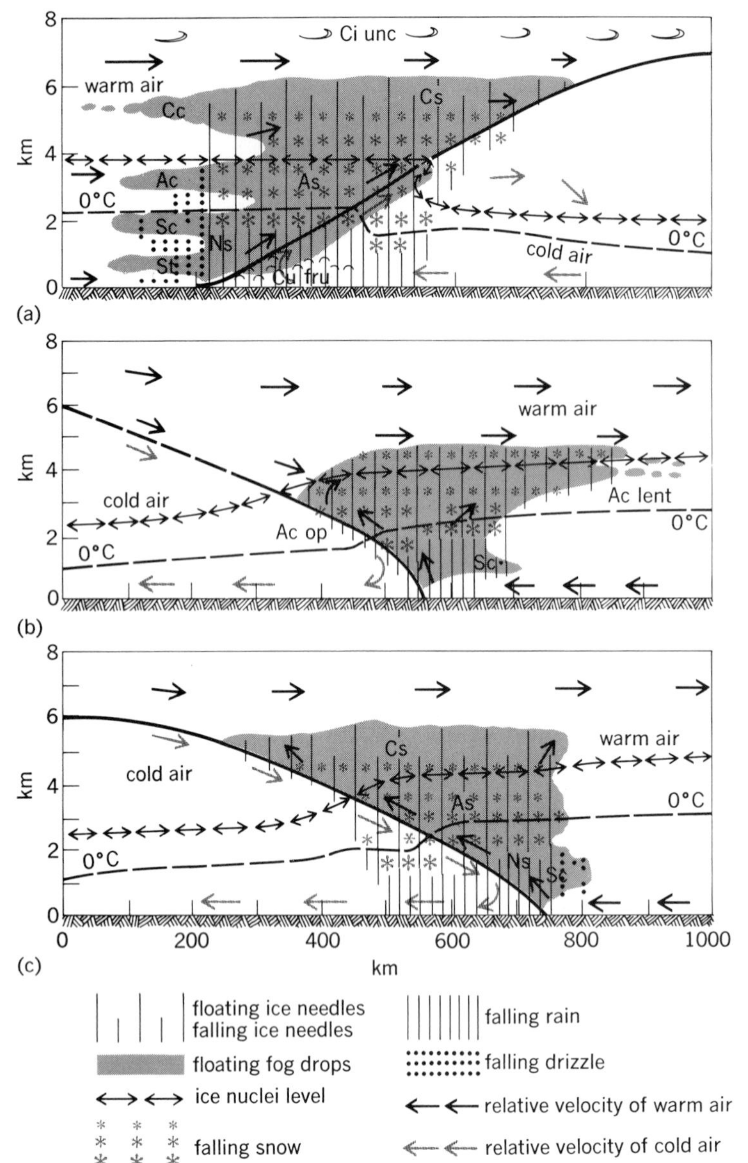

Fig. 6. Schematic cross sections of three kinds of fronts. (*a*) Warm front. (*b*) Fast-running cold front. (c) Slow-moving cold front. Ac, altocumulus; Ac lent, altocumulus lenticularis; Ac op, altocumulus opacus; As, altostratus; Ci unc, cirrus uncinus; Cc, cirrocumulus; Cs, cirrostratus; Cu fru, cumulus fractus; Ns, nimbostratus; St, stratus; and Sc, stratocumulus.

Table 2. Atmospheric waves

Name	Type*	Wavelength, km
Ultrasound	C	$<2 \times 10^{-5}$
Ordinary sound (tones)	C	2×10^{-5} to 10^{-2}
Explosion waves	C	10^{-2} to 10^{-1}
Helmholtz waves (Sc und, Ac und)	G	10^{-1} to 1
Short lee-waves (Ac lent, Cc lent)	G	1 to 2×10
Long lee-waves (nacreous clouds, precip.)	G(I)	2×10 to 10^2
Short jet-waves (frontal)	GI(V)	5×10^2 to 5×10^3
Long jet-waves	V	5×10^3 to 10^4
Tidal waves	(G)	$\leq 2 \times 10^4$

*C, compression, longitudinal; G, gravitational; I, inertial; and V, vorticity-gradient; the last three are transversal.

the atmospheric circulation, within frontogenetic zones, between two stationary anticyclones (highs) and two cyclones (lows) when the axis of stretching, or confluence, of this deformation field (Fig. 5) has some extension west to east. This pattern is in its turn determined by the large-scale orography of the Earth. Because of the north-south temperature contrast, air masses of different temperature, specific humidity, and density are then brought together in a front zone along this axis. On the warm side of the polar fronts, which together more or less encircle the hemisphere, the thermal wind, as discussed in dynamical meteorology below, corresponding to the mass distribution of the fronts, implies a narrow band of intense flow near the tropopause level, a jet stream. The thermal wind is most pronounced at middle and higher latitudes, where the front is marked, steep, and usually reaches into the stratosphere (Fig. 5). There, a main front and its jet together represent a zone of maximum potential and kinetic energy. Below, in the lower latitudes, the frontal tilt is often small, and there is an outflow of polar air into the tropics.

A front may appear as warm or cold. At a warm front, the warmer air gains ground and slides evenly upward above the retreating cold-air wedge (tilt or slope 1:200 to 1:100), producing a wedge-shaped upglide cloud system (Fig. 6*a*). At the approach of a marked warm front, therefore, a typical cloud sequence invades the sky: cirrostratus → altostratus → nimbostratus, the last yielding continuous and prolonged precipitation ahead of the front line.

At a cold front the rather steep cold-air wedge (with a slope of 1:100 to 1:50) pushes forward under the warm air and forces it upward according to one of the two flow patterns shown in Fig. 6*b* and *c*. At the approach of a cold front of the more common type (Fig. 6*b*) there is, therefore, another typical cloud sequence: altocumulus, partly lenticular, rapidly thickening into nimbostratus. The precipitation is usually more intense but of shorter duration than with the warm front. *See* CLOUD; CLOUD PHYSICS.

Waves and vortices (the weather). The actual atmospheric states affecting the weather may be regarded as composed of the general circulation and its disturbances. Together they form a multitude of weather mechanisms, some of which have already been described, in which the water-vapor cycle (evaporation → transport and lifting → condensation → precipitation) evidently has a fundamental role. The atmospheric disturbances consist of a spectrum of waves of different wavelengths λ (Table 2), corresponding circulations (vortices), or both. Only the larger of these are studied synoptically. By their size the circulations may be classified as planetary (or geographical), secondary, or tertiary:

1. The long waves forming in the front↔jet zone, that is, the region of maximum energy, may represent a steady state (see section on dynamical meteorology) when their wavelength corresponds to about four circumpolar waves at low, three at medium, and possibly two at high latitudes. The planetary waves have $\lambda \sim 10{,}000$ km. Thereby they partly determine the shape of the general circulation (Fig. 3). The shorter long waves ($3000 < \lambda < 8000$ km) propagate slowly eastward

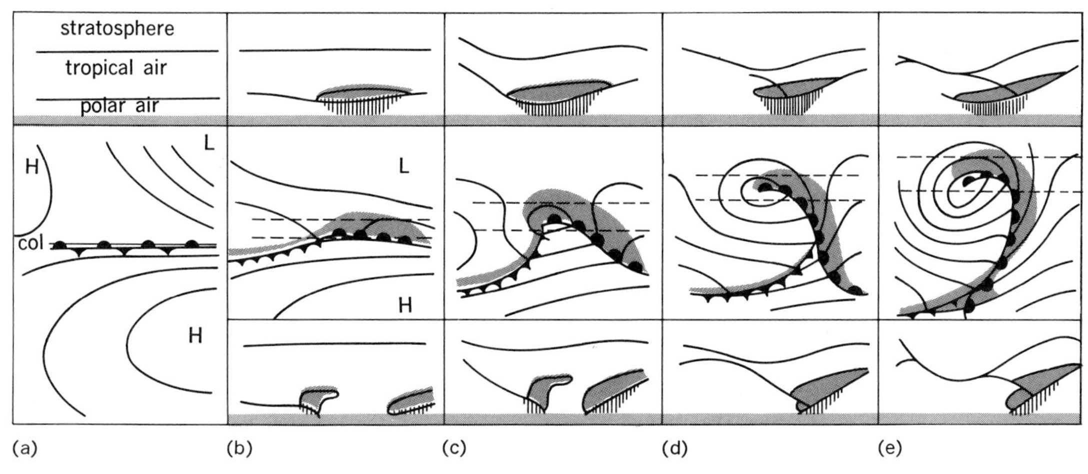

Fig. 7. Life history of cyclones from wave to vortex. The vertical sections (top and bottom) show clouds and precipitation along the two dashed lines in the map sequence. (*From C. L. Godske et al., Dynamic Meteorology and Weather Forecasting, American Meteorological Society, Waverly Press, 1957*)

and, because of inertial instability, mostly develop as shown in Figs. 8 and 10 through the stages wave → tongue → cutoff vortex. Since the air aloft flows rapidly through these quasi-stationary long-wave patterns, isotherms will approach coincidence with the streamlines and isobars. Thus, equatorward tongues and cutoffs will be cold and will tend to coincide with lows (at least aloft); the poleward ones will tend to be warm and coincide with highs. The former will contain a polar-air hourglass (at least in higher latitudes) or dome, possibly with an arctic-air dome inside it (Fig. 9). Correspondingly, the poleward tongues and cutoff vortices will contain tropical air and a much higher tropopause.

2. The secondary waves are the short waves ($1000 < \lambda < 3000$ km). The short waves in a front↔jet system, when unstable, will form secondary circulations, developing according to the scheme of Fig. 7, that is, initial front wave → young cyclone → initial occlusion → backbent occlusion. Important secondary circulations also form outside the front ↔ jet systems: the easterly wave, the tropical hurricane, and the convective system, occurring primarily in lower latitudes, the last two without an obvious preceding wave stage.

3. The tertiary circulations, barely observable by the ordinary synoptic network, require a mesoscale network ($d < 10$ km), time sections, or both, for a detailed study and forecast. Weather mechanisms of this size are land and sea breezes, mountain and valley winds, katabatic and other local winds (foehn, chinook, bora), local showers, tornadoes, orographic cloud and precipitation systems, and lee waves. Moreover, numerous different orographic factors and the daily period of radiation affect most meteorological elements and weather mechanisms. Their detailed study is therefore basic for both climatology and weather forecasting.

Life cycle of extratropical cyclones. An extratropical cyclogenesis starts as a wavelike front bulge. Ahead of the wave the front becomes a warm front, behind it a cold front. These two sections then flank a warm tongue (Fig. 7*b*) moving along the front. At first there is only a shallow low around the tip of this warm sector, but even the passage of such an initial frontal wave may cause sudden, severe, and unexpected weather changes. If the front wave has an appropriate size ($\lambda \sim 1500$ km, amplitude $\geqq 200$ km), it will usually be unstable, narrow gradually, and at last overshoot as does a breaking sea wave. The cold front overtakes the warm front at the ground, and the warm-sector air is lifted and spreads aloft. As a result, the common center of gravity of the system sinks, and potential energy is transformed into the kinetic energy of an increasing cyclonic circulation. As long as this front occluding continues, the circulation increases, and the ensuing low deepens, whereas when the warm sector has disappeared from the interior of the main cyclone, the latter will weaken, and the low fill. These features serve as good, physically comprehensible, prognostic rules. The occluded front is called an occlusion. The occluding process gives the clue to the life history of cyclones and anticyclones (Fig. 7). At its last phase the occluded front often lags behind and bends (Fig. 7*d* and *e*), a "false warm sector" forming between the direct and the backbent occlusion. Several cyclones (and lows), together constituting a cyclone series, may form on one and the same front, moving with the upper, steering current, the first one being in the most advanced stage of development (Fig. 8).

Important front↔jet systems in the Northern Hemisphere are the North Pacific polar front and the North Atlantic polar front (Fig. 3). The former extends in winter on the average from near the Philippines north of Hawaii to the northwest coast of the United States; the cyclone series forming in it then brings wet and unsteady weather to the northwestern part of North America. The latter front, extending in winter on an average from near the Bermudas to England, plays the analogous role for almost all Europe, and during certain periods for eastern United States. In summer both these systems are much weaker and lie on an average farther north. The North American arctic front and the European arctic front (Fig. 3) are fundamental for the weather in their vicinity. The former accounts for many severe weather vagaries, such as the blizzards and glaze storms of northeastern United States, and the most severe cold waves and their killing frosts farther south; it is thus of special importance to North American weather forecasting. The production of real arctic air as defined here seems to cease in summer, or at least in July. On the other hand, the corresponding air mass in the Southern Hemisphere, antarctic air, and the antarctic fronts, exist throughout the year, the antarctic ice plateau being an enormous cold source even in summer. Evidently most migrating extratropical precipitation and storm areas originate at front↔jet systems, separating air masses of radically different motion and weather type. Therefore, these systems are of utmost importance to weather forecasting, being the real atmospheric zones of danger and main sources of the salient aperiodic weather changes of synoptic size outside the tropics.

Large-scale tropical weather systems. In the tropics (and in the subtropics in summer) nonfrontal convective and other weather phenomena may grow to hundreds of kilometers in width; consequently, they can be studied and forecast individually by synoptic methods. Both on land and at sea, short waves ($\lambda \sim 500$ km) form in the frontless trade winds outside the doldrums. These easterly waves propagate slowly westward, showing a characteristic intensification of the shower activity in their eastern part.

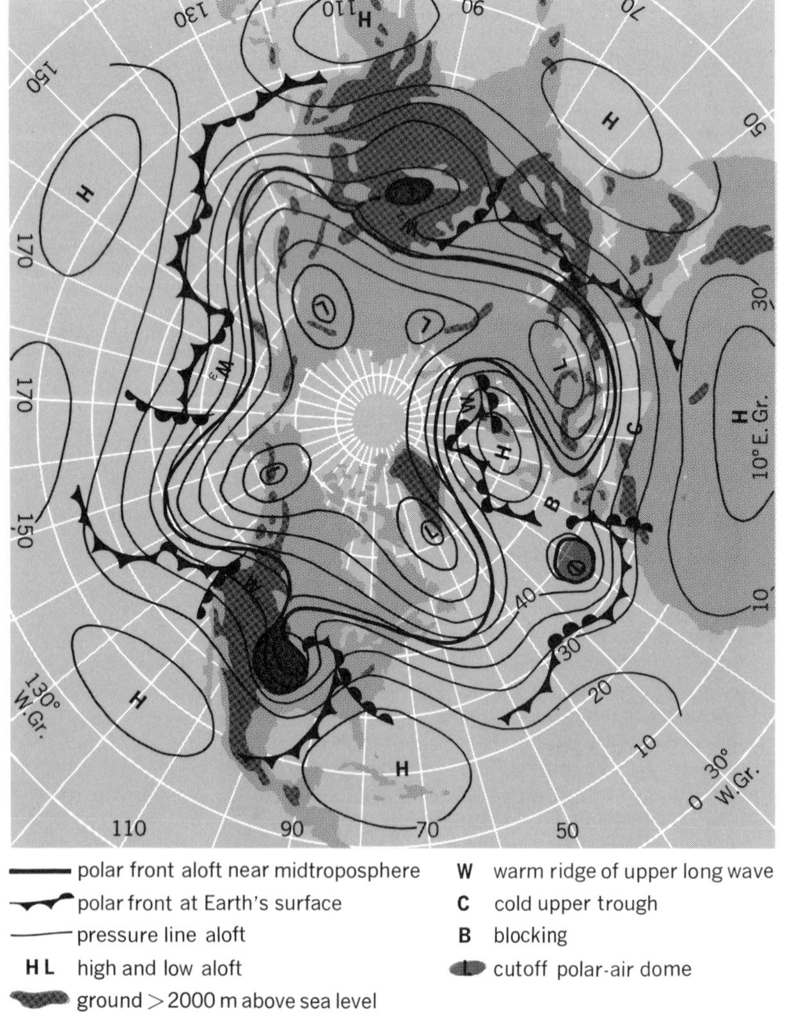

Fig. 8. Schematic circumpolar upper pressure pattern in relation to the polar fronts at the Earth's surface. (*After E. Palmén*)

Over tropical seas in late summer and early fall (Northern Hemisphere, August–October; Southern Hemisphere, February–April), conditions exist that favor the formation of tropical hurricanes. Tropical hurricanes are cyclonic vortices (smaller but much more intense than the extratropical ones) of 100–400-km width; they have wind velocities often exceeding 50 m/sec (100 knots), 100–500 mm total rainfall, and very low central pressure. They move on the whole poleward, often recurving around a subtropic high. Necessary conditions for their formation seem to be (1) air (and sea-surface) temperature above, say, 27°C, implying enough lability energy to drive such large-scale convections, (2) a preexisting cyclonic motion and frictional inflow (either at the tropical front or in an easterly wave), needed to order, and possibly trigger, the convections, (3) a divergence mechanism aloft to dispose of the air that converges and rises in their interior, possibly also triggering their formation, (4) sufficient Coriolis force (that is, the hurricanes cannot form too near the Equator), and (5) no disturbing land surface within the area of formation. Whether these conditions are sufficient is not certain. The widespread destructive power of hurricanes (shore flooding, wind pressure, downpour) makes their study a major task of tropical synoptic meteorology and weather forecasting. Tracking those already formed can be done with radar, reconnaissance flights, and satellites, but discovering their imminent formation is still an unsolved problem. *See* HURRICANE.

Over land, conditions 1 and 2 never lead to hurricane formation but may instead—within the tropics and also in warmer seasons farther poleward—favor the formation of a migrating convective system, with a forerunning pseudo cold front (in the United States also called a squall line) at its outer edge; these systems have roughly the same extent and precipitative power as a tropical hurricane. Outside the tropics they mainly form in the warmer seasons in cyclonic warm sectors, especially in the midwestern United States, where they provide the main water supply. Because of the flood hazards, soil erosion, and other aspects, and the sudden violent squalls, or even tornadoes, sometimes attending the pseudo cold front, these systems, therefore, form another major problem of weather forecasting. Within the tropics, migrating convective systems, often in the easterly waves, will cause more variation of weather from day to day than is generally recognized, whereas the small-scale showers usually have a daily period; together they constitute the equatorial rains.

Apart from the polar-front jet (meandering between 30 and 70° lat), there is a subtropical jet, and a corresponding front, at about 30° lat near the tropopause level (~15 km altitude). As a rule, it makes only small meridional excursions. Therefore, it shows up (instead of the polar-front jet) as the main wind maximum aloft (at 30° lat) in the average meridional cross section of Fig. 2. But for the same reason, and because its front is confined to the tropopause, its influence on daily weather, at least in the tropics, seems slow and diffuse.

Circumpolar aerology. Since 1940, the technical facilities of meteorology have undergone an explosive development, partly due to the great exigencies during World War II. The vast gaps for-

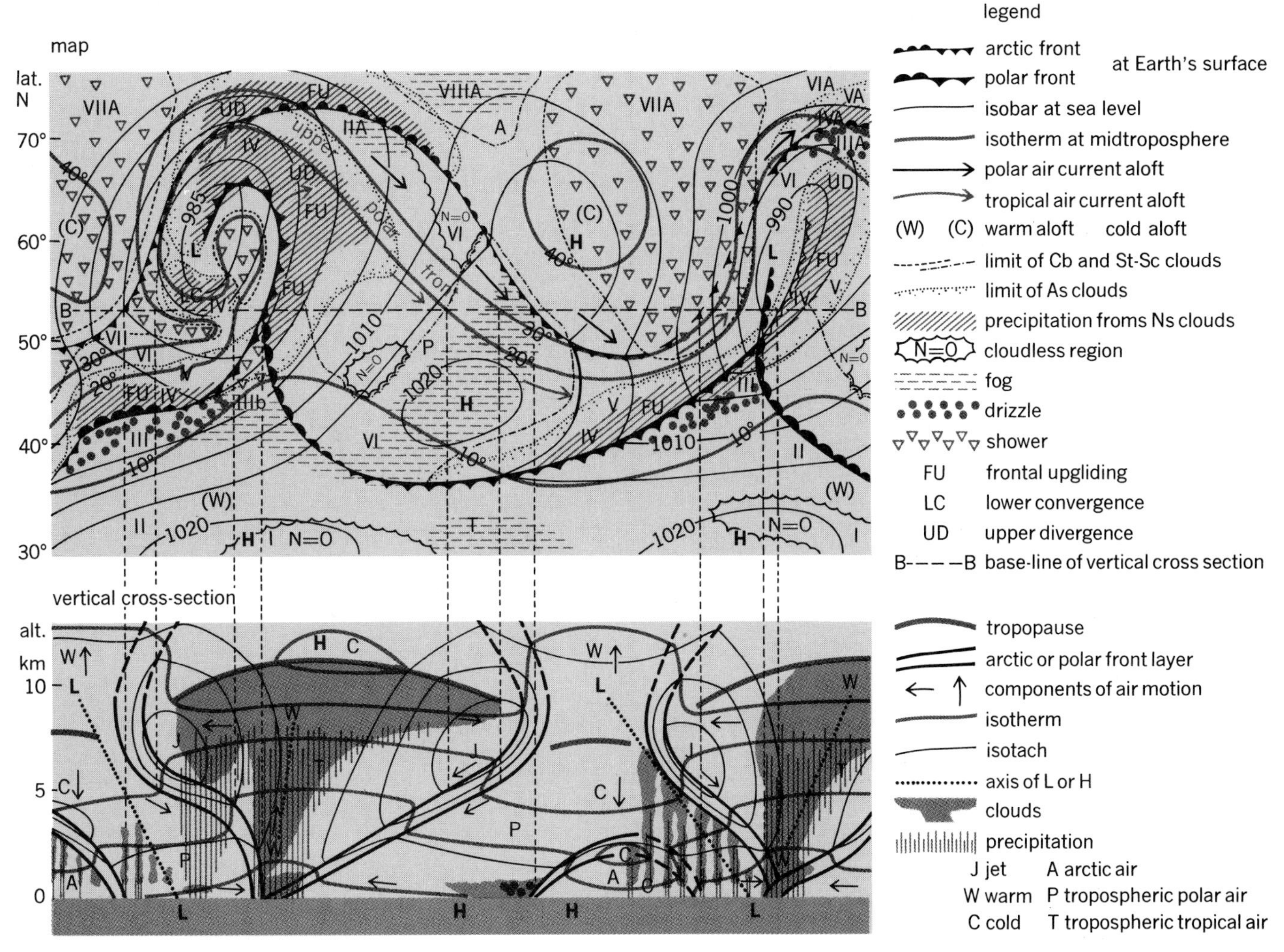

Fig. 9. Model of middle-latitude front↔jet disturbances and their weather regions (fall and spring).

merly in the meteorological network over the oceans are partly bridged by stationary weather ships, or ocean weather stations, with complete air-sounding equipment. This huge technical improvement has been followed by equally outstanding scientific achievements in the understanding of the dynamics and thermodynamics of the free atmosphere. The weather mechanisms and weather processes described above, which before 1940 had been mainly observed and studied within the lower half of the troposphere, can now be treated in their entirety and fitted into their general relationships with the rest of the atmosphere. On the whole, an intimate interaction takes place between the disturbances seen on the ordinary synoptic map (surface map) and the upper large-scale patterns shown by Fig. 8 and described above. Figure 9 shows this interdependence three dimensionally and in detail for disturbances of front↔jet systems (including the arctic front). Particularly, it shows the eight weather regions that are conditioned by such systems, these naturally being of fundamental importance to synoptic meteorology and weather forecasting.

In winter, the polar-front jet lies on an average at 25–45° lat, in summer at 40–60° lat, at 7–11 km alt; it is only a few hundred kilometers broad, with wind speed surpassing 100 m/sec (200 knots). It displays two fundamentally different patterns: the zonal or high index type, where the jet has waves of only small amplitudes; and the meridional or low index type, where its meridional excursions are huge and often are cut off. Figure 10 gives an example of the isothermal, isobaric, and flow patterns at the midtroposphere, represented by the isotherms and isohypses ("contour lines") of the 500-mb surface. It shows three long waves w_1 w_2 w_3, one deep trough T, and four cutoff lows $L_1L_2L_3L_4$. Where the isotherms diverge markedly from the parallelism with the streamlines and "contours," there will be a considerable horizontal advection of colder and denser air or of warmer air (Fig. 10), and ensuing advective changes of pressure and wind at most levels. These observations gave another powerful tool for quantitatively forecasting the thermal changes of flow patterns. Calculating the vertical motion by the divergence and vorticity equations has proved even better for this and similar purposes (as discussed below in dynamical meteorology).

Since the planetary waves represent a rather stable steady state, they offer no real forecast problem. The shorter long waves will move east, conserving most of their absolute vorticity, and may therefore be forecast numerically, using a very simplified general model of the atmosphere,

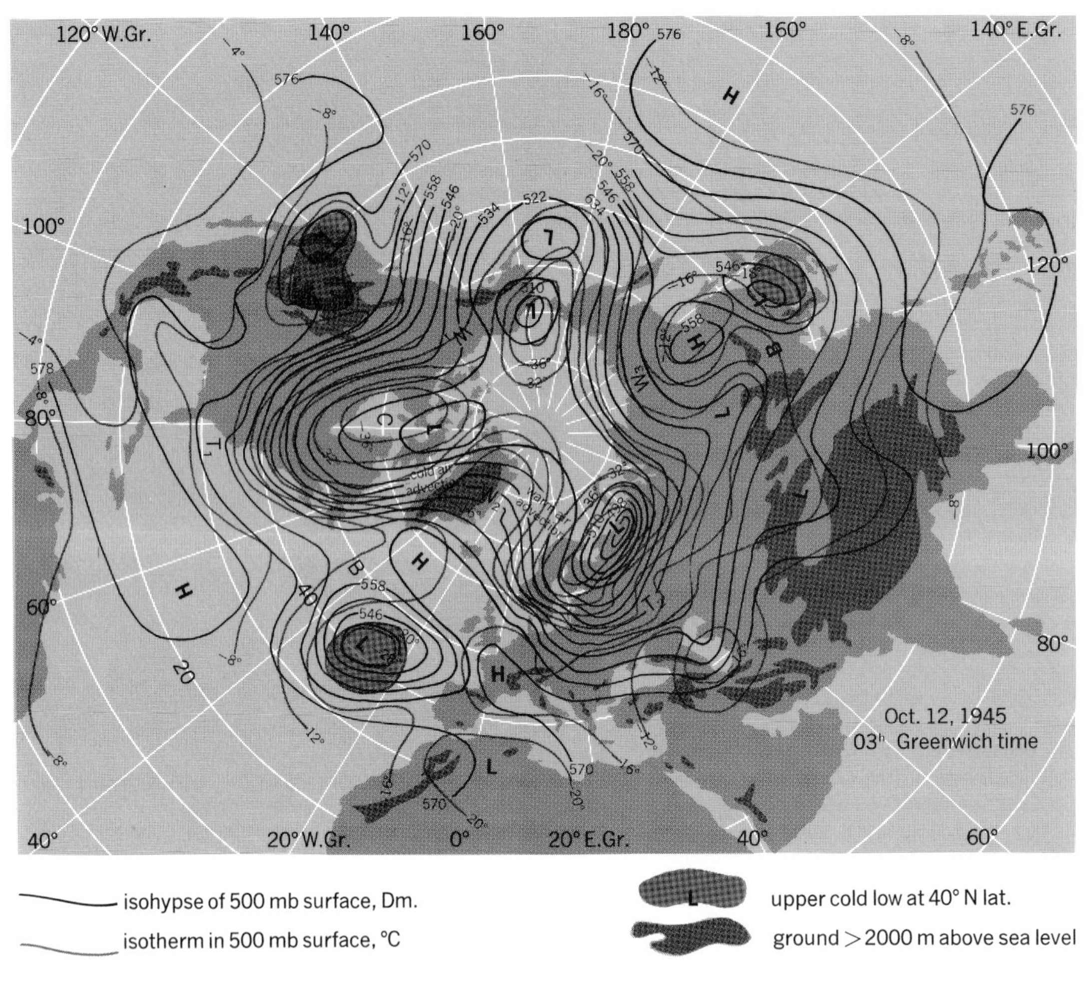

Fig. 10. Circumpolar pressure, flow and temperature pattern aloft, Oct. 12, 1945.

the barotropic model, where production and annihilation of kinetic energy is excluded. The propagation and development of the short front ↔ jet waves, that is, the weather mechanisms, will to a certain extent be steered by this long-wave pattern. However, in these mechanisms frontal potential energy is transformed into cyclonic kinetic energy, and the short jet waves will often react upon the large-scale pattern. Such processes, which may lead to intense frontal cyclogenesis and the formation of large-scale cold tongues and cutoffs, confront numerical forecasting with considerable difficulties. However, the fact that the upper steering flow patterns are so simple and move much more slowly than the lower ones has proved a great help to short-range forecasting and has made an extended forecasting (2–5 days) possible on a synoptic-physical basis. Attempts are being made, with some success, to use baroclinic models of the atmosphere as well, and to incorporate the effect of heat sources and large-scale orography into the numerical short-range and extended forecasting. A main obstacle to real long-range forecasting (7–30 days) is the fact that the causes for a definitive change from the zonal to the meridional upper flow pattern are not yet known.

Aided by radar, continuous detailed mapping and tracking of rain areas is gradually being introduced. Analogously, photographic and television cloud surveys from satellites are used in studying the large-scale development of weather regions.

[TOR BERGERON]

DYNAMICAL METEOROLOGY

This branch of meteorology is the science of naturally produced motions in the atmosphere and of the related distributions in space of pressure, density, temperature, and humidity. Based on thermodynamic and hydrodynamic theories, it forms the main scientific basis of weather forecasting and climatology. *See* CLIMATOLOGY; WEATHER FORECASTING AND PREDICTION.

Thermodynamic properties of air. Atmospheric air contains water vapor and sometimes suspended water droplets or ice particles or both. In dynamical meteorology it suffices to use a simplified cloud physics; thus it may be assumed that the water vapor pressure cannot exceed the saturation pressure over a plane water surface and that the vapor pressure equals this saturation pressure whenever the air contains water (cloud) droplets. The distinction between water and ice is ignored most of the time.

Nonsaturated air. Nonsaturated air contains no condensation products and behaves nearly as a single ideal gas. Its pressure p, density ρ, and absolute temperature T satisfy very closely gas equation (1). Its enthalpy per unit mass h is

$$\frac{p}{\rho}=RT \tag{1}$$

given by Eq. (2), and its entropy per unit mass may

$$h=c_pT+\text{constant} \tag{2}$$

be written $s=c_p \ln \theta$. The potential temperature θ is given by Eq. (3), where $p_0=1000$ mb. In these

$$\theta=T\left(\frac{p_0}{p}\right)^{R/c_p} \tag{3}$$

expressions the gas constant per unit mass R and the specific heat at constant pressure c_p both depend slightly on the specific humidity or water-vapor mass concentration m. It is often sufficiently accurate to ignore this dependence and to use the values which hold for dry air; these are $R=287\ m^2\ \sec^{-2}C^{\circ-1}$ and $c_p=1004\ m^2\ \sec^{-2}C^{\circ-1}$.

Saturated air. A sample of nonsaturated air becomes saturated if its pressure is reduced or heat is removed or both while its water-vapor concentration m is kept constant. If this process continues, cloud droplets form by condensation at a rate just sufficient to maintain the vapor pressure at the saturation value. If the process is reversed, evaporation takes place. During such phase transitions the release or binding of latent heat causes temperature changes which are important for many motion processes. This property of saturated air is expressed in the formula for its enthalpy per unit mass, Eq. (4). Here L is the latent heat, and

$$h=c_pT+Lm_s(p,T)+\text{constant} \tag{4}$$

m_s is the mass concentration of saturated water vapor, which is a known function of p and T. On the other hand, the liquid (or solid) phase contributes very little to the total mass or volume of the air, so that the gas equation (1) holds with good approximation also for saturated air. Note, however, that the approximate equations (1) and (4) are not fully consistent with the second law of thermodynamics.

Thermodynamic state. The thermodynamic state of an air sample, whether saturated or not, is defined by three state variables, for example, pressure p, density ρ, and mass concentration m of the water component (in all phases). It is often convenient to use T instead of ρ. Comparison between m and the saturation vapor concentration $m_s(p,T)$ shows whether the sample is saturated, and if so, the mass of the condensed phase.

There are many motion phenomena in which the released latent heat is unimportant. When dealing with such phenomena, the air may be considered dry and its thermodynamical state as defined by the two variables p and ρ (or T) only.

Instantaneous state of atmosphere. The instantaneous state of the atmosphere may be characterized by the distributions in space of the thermodynamic variables p, ρ, and possibly m and three components of the air velocity $\mathbf{v}$ relative to the solid Earth. These basic quantities are regarded as functions of three space coordinates and time, and as such they characterize a sequence of states or a motion process in the atmosphere.

Pressure and density are free to vary independently in space, and the density is usually variable in isobaric surfaces of constant pressure; the density field is then said to be baroclinic. Only in special cases is the density field barotropic, that is, constant in each isobaric surface.

A basic difficulty in dynamical meteorology results from the complexity of the field of motion in the real atmosphere. Superimposed upon the large-scale motion systems revealed by weather maps is a fine structure of motions of all scales down to small eddies of millimeter size. Such motion fields cannot be dealt with in all details; it is necessary to deal instead with smoothed motion fields, where details smaller than a certain scale have been left out. As a consequence of the nonlinearity of the basic equations, these details still exert a certain influence upon the larger-scale motions. This influence can be taken into account only in a statistical sense, in the form of so-called eddy terms which express the transport of momentum, enthalpy, and water substance by small-scale eddies.

Equations. The dependent variables p, ρ, m, and $\mathbf{v}$ satisfy Eqs. (5)–(8), which express, respectively,

$$\frac{\partial\rho}{\partial t}+\text{div}(\rho\mathbf{v})=0 \tag{5}$$

$$\rho\frac{D\mathbf{v}}{Dt}=-\mathbf{k}\times f\rho\mathbf{v}-\rho g\mathbf{k}-\text{grad}\,p-\frac{\partial\mathbf{F}_M}{\partial z} \tag{6}$$

$$\rho\frac{Dh}{Dt}-\frac{Dp}{Dt}=-\frac{\partial}{\partial z}(F_{\text{rad}}+F_h)+\Delta \tag{7}$$

$$\rho\frac{Dm}{Dt}=-\frac{\partial}{\partial z}(F_{\text{precip}}+F_m) \tag{8}$$

the conservation of total mass (continuity equation), Newton's second law (equation of motion), the thermodynamic energy equation (or first law), and the conservation of mass of the water component.

Here Eqs. (9) and (10) apply, and the notation

$$f=2\Omega\sin\Phi \tag{9}$$

$$\frac{D}{Dt}=\frac{\partial}{\partial t}+\mathbf{v}\cdot\text{grad} \tag{10}$$

is as follows: t denotes time; D/Dt individual time derivative or rate of change as experienced by a moving air particle; $\mathbf{k}$ a vertical unit vector; f Coriolis parameter; Φ latitude; $\Omega=0.7292\ 10^{-4}$ $\sec^{-1}$ angular velocity of the Earth's rotation; g acceleration of gravity; $\mathbf{F}_M$ vertical eddy-flux density of momentum; F_{rad} vertical radiative-heat-flux density; F_h vertical eddy-flux density of enthalpy; F_{precip} vertical flux density of water substance by precipitation; F_m vertical eddy-flux density of water substance; and Δ viscous dissipation. To these equations one must add boundary conditions, expressing the kinematic constraint at the Earth's surface, and the various fluxes into the atmosphere from below and above.

In Eqs. (6)–(8) horizontal flux densities (that is, fluxes through vertical surfaces) have been neglected. Moreover, in Eq. (6) the expression for the Coriolis force has been simplified by including only the effect of the vertical component of the Earth's rotation ($\Omega \sin \Phi$). These simplifications are common in dynamical meteorology.

In order that Eqs. (5)–(8) with boundary and initial conditions constitute a closed system of equations, it is necessary that $\mathbf{F}_M$, F_{rad}, $F_h\Delta$, F_{precip}, and F_m be expressible in terms of the dependent variables. The radiative flux F_{rad} depends upon the distribution of temperature, water vapor, cloud, and the properties of the underlying surface; it can be calculated quite accurately when these quantities are known, although the calculation is time-consuming. A major difficulty is the determination of eddy-flux densities $\mathbf{F}_M$, F_h, and F_m. It is usually assumed that a major part of these fluxes is contributed by turbulence in the atmospheric boundary layer, and this part of the fluxes can be estimated from turbulence theory. Thus the last term of Eq. (6) represents the force of turbulent friction. The calculation of F_{precip} presents another difficulty, since it depends strongly on small-scale eddy motions and microprocesses within the clouds.

Note that Eqs. (5)–(8) are nonlinear, since the operator Eq. (10) produces product terms when applied to the dependent variables. Nonlinearity may also enter in the relations between eddy fluxes and dependent variables, and in the relation between condensation heat and dependent variables.

Equations (5)–(8) can be treated either analytically or numerically, and one can distinguish between analytical and numerical dynamic meteorology.

Analytical dynamical meteorology. It is not possible to find analytical solutions of Eqs. (5)–(8) which represent motion processes of the composite kind found in the real atmosphere. Instead, one deals with various kinds of idealized motion systems, for which simplifications of the equations can be justified on the basis of scale analysis and simplified geometry.

Motion without heat sources and friction. A common simplification consists in ignoring all heat sources, including condensation heat, so that Eq. (7) may be integrated to give Eq. (11) for isen-

$$\theta = \text{constant in time for each air particle} \qquad (11)$$

tropic motion. From Eq. (3) there is then for each particle a relation between p and ρ or T; such a fluid is said to be piezotropic. Another simplification consists in ignoring the force of eddy friction ($-\partial \mathbf{F}_M/\partial z$) in Eq. (6). The bulk of the work in analytical dynamical meteorology has been done under the assumption that heat sources and friction are absent.

Equilibrium. In the absence of friction and heat sources and disregarding the slight centripetal acceleration which exists because an air current is bound to follow the Earth's curvature, Eqs. (5)–(8) permit a particularly simple solution which represents a steady, straight, horizontal current with parallel streamlines. With vanishing particle accelerations, Eq. (6) expresses balance of forces. In the vertical direction there is balance between gravity and vertical pressure force. This hydrostatic equilibrium is expressed by Eq. (12),

$$-g\rho - \frac{\partial p}{\partial z} = 0 \qquad (12)$$

where z represents height. In the horizontal direction the Coriolis force must balance the horizontal pressure force $\text{grad}_h p$. Thus this geostrophic equilibrium can be expressed by Eq. (13).

$$-\mathbf{k} \times f\rho\mathbf{v} - \text{grad}_h p = 0 \qquad (13)$$

Ignoring the Earth's curvature, one may place a cartesian coordinate system with the z axis pointing upward and the x axis along the horizontal velocity u. Then Eq. (13) gives Eq. (14). Thus the

$$u = -\frac{1}{f\rho}\frac{\partial p}{\partial y} \qquad (14)$$

geostrophic wind velocity is directed along the isobars (lines of constant pressure) in level surfaces, with pressure increasing to the right (or left) in the Northern (or Southern) Hemisphere (Fig. 11). At the Equator $f = 0$, and the formula breaks down. The dependent variables u, p, and ρ may depend upon y and z, but not on x if the current is steady. From Eqs. (12) and (14) one may derive Eq. (15), showing that the rate of change of u

$$\frac{\partial u}{\partial z} = \frac{g}{f}\frac{1}{\rho}\left(\frac{\partial \rho}{\partial y}\right)_p = -\frac{g}{f}\frac{1}{T}\left(\frac{\partial T}{\partial y}\right)_p \qquad (15)$$

with height is proportional to the baroclinicity (the subscript p indicates derivative in a surface of constant pressure). Thus in a barotropic current the velocity is constant with height; in particular, a state of persistent rest is necessarily barotropic.

Atmospheric boundary layer. Next to the Earth's surface is a turbulent boundary layer approximately 1 km thick, inside which the force of turbulent friction ($-\partial \mathbf{F}_M/\partial Z$) is strong enough to disturb significantly the geostrophic equilibrium. The vertical eddy-flux density of momentum $\mathbf{F}_M$ is given by Eq. (16), where the coefficient of eddy

$$\mathbf{F}_M = -K_M \partial \mathbf{v}_h/\partial z \qquad (16)$$

viscosity K_M depends upon the intensity of the turbulence. In the lowest 20–30 m, K_M increases rapidly with height, and the wind velocity is typically a logarithmic function of height; this wind profile is modified if the eddy flux of heat F_h is numerically large.

Overlaying this shallow layer is a much deeper layer, in which K_M changes less with height. Assuming balance between the Coriolis force, the horizontal pressure force, and turbulent friction, one obtains for constant K_M the Ekman spiral, which represents the variation of wind velocity with height (Fig. 12). As z increases, the wind

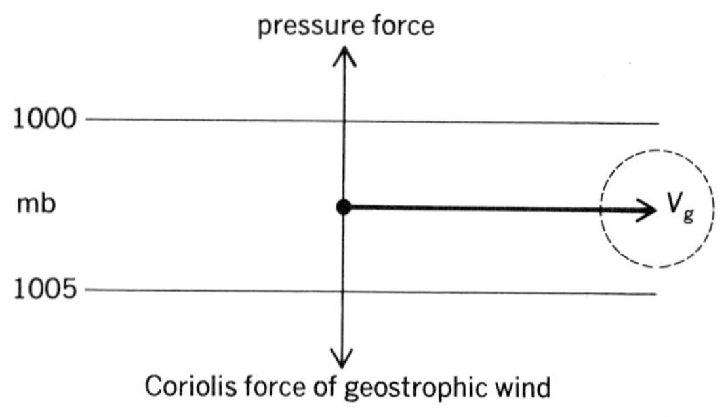

Fig. 11. The geostrophic balance. *(From S. Petterssen, Introduction to Meteorology, 3d ed., McGraw-Hill, 1969)*

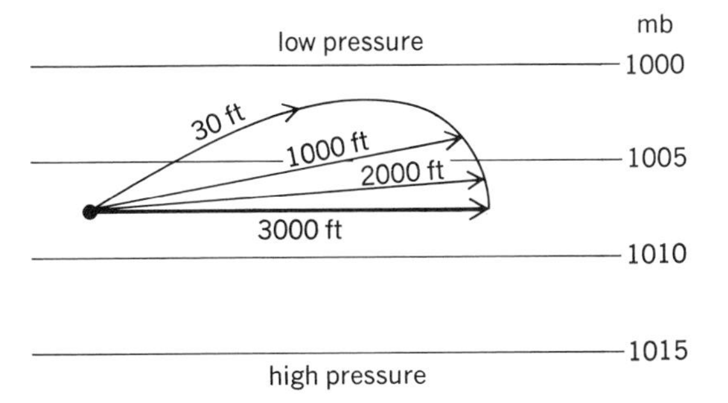

Fig. 12. Ekman spiral. Length and direction of arrows represent wind velocity for different elevations. (*From S. Petterssen, Introduction to Meteorology, 3d ed., McGraw-Hill, 1969*)

approaches the geostrophic value.

Stability. A steady state is termed stable if any small disturbance remains small indefinitely and unstable if a significant deviation from the steady state can develop from an initial disturbance, however small. In this definition it is usually understood that the motion is isentropic, as in Eq. (11).

For a state which deviates only slightly from a known steady state, the basic nonlinear equations (5)–(8) turn into a set of linear perturbation equations in the small perturbation quantities which define the deviations of the variables ($\mathbf{v}$, p, ρ) from their steady-state values. These linear equations can be treated analytically; they have solutions representing various kinds of wave motions, whose amplitudes may either remain small or grow indefinitely. Existence or nonexistence of growing wave solutions is assumed to correspond to instability or stability of the steady state, respectively. Several types of stability may be identified.

Static stability. When in an atmosphere in equilibrium a nonsaturated air particle is displaced vertically with unchanged potential temperature ($D\theta = 0$), its temperature changes at a rate $DT/Dz = -0.01\ °\text{C}\ m^{-1}$ (the dry-adiabatic lapse rate). Therefore, if the distribution of temperature with height in the atmosphere is such that at all levels $\partial\theta/\partial z > 0$ or $\partial T/\partial z > -0.01\ °\text{C}\ m^{-1}$, then a particle displaced upward (or downward) becomes colder (or warmer) than its environment, and the sum of the buoyancy and the weight acting on the particle represents a net restoring force. The state is then said to be statically stable. A rigorous proof of the stability can be given by showing that the sum of potential and internal energy of a bounded part of the atmosphere is a minimum in the state of rest if $\partial\theta/\partial z > 0$.

If $\partial\theta/\partial z < 0$ (or $\partial T/\partial z < -0.01\ °\text{C}\ m^{-1}$) at some levels, then the state is statically unstable. Saturated air is less stable than nonsaturated air with the same temperature distribution, because the temperature of a saturated particle which is displaced vertically changes at a rate which is slower than the dry-adiabatic lapse rate, the rate depending upon p and t.

Static instability is frequently caused by heating the air from the underlying surface and produces convection currents, whose horizontal scale is of the same order as the depth of the unstable layer. Above a certain condensation level, condensation takes place in the rising currents which then become visible as cumuliform clouds.

Inertial stability. In a steady current satisfying Eqs. (12) and (14), inertial stability results from excessive Coriolis forces acting on a chain of particles parallel to the current which have been displaced along the isentropic surface normal to the current. The criterion of stability, valid in both hemispheres, is $f[f-(\partial u/\partial y)_\theta] > 0$, where the subscript θ indicates that the derivative is taken in an isentropic surface ($\theta =$ constant) and f is reckoned positive in the Northern, negative in the Southern Hemisphere.

Shearing instability. If the vertical shear has a sufficiently pronounced maximum at some level, shearing instability results. The layer of maximum shear breaks up into a system of equally spaced vortices with horizontal axes normal to the flow. Condensation may occur in the upper part of the vortices and form what is known as billow clouds.

Wave motions. In a statically and inertially stable atmosphere without shearing instability there is a spectrum of possible wave motions, ranging from high-frequency acoustic waves through gravity waves at intermediate frequencies to inertia-gravity waves at frequencies comparable with the Coriolis parameter f. Typical gravity waves have orbital periods which are short compared with a pendulum day (12 hr/sin Φ) and can therefore be studied without taking the Earth's rotation into account. Their frequency cannot exceed the Väisälä-Brunt frequency $[(g/\theta)\partial\theta/\partial z]^{1/2}$, which in the troposphere mostly corresponds to a minimum orbital period of about 8 min.

Organized stationary gravity wave motions occur in the atmosphere when a stably stratified air current blows over hilly terrain. Their properties can be explained quite well from linear theory. The terrain corrugations act to draw kinetic energy from the current and transform it into wave energy, and the process may be looked upon as a radiation of wave energy upward and horizontally downstream. Depending upon the wavelength and the distributions of temperature and wind with height, the waves may either be transmitted to very high levels, absorbed, or trapped at low levels. Condensation may occur in the wave crests making them visible as stationary clouds, even on satellite photographs; their spacing is mostly 10–50 km. Such waves cause a drag (wave resistance) on the mountains and a corresponding braking force on the airstream.

Organized long gravity-inertia waves occur as forced tidal waves, the predominant forcing effect being the diurnal heating and cooling. To some extent gravity-inertia waves are also generated, as disorganized noise, by nonlinear interaction with other motion types.

Kinematics of motions in atmosphere. Apart from sound waves, which may be treated as a separate phenomenon, all motions in the atmosphere have the character of circulations along closed streamlines. The distribution of horizontal velocity $\mathbf{v}_h$ in a horizontal surface may be characterized by the fields of two scalar quantities, namely, the relative vertical vorticity $\zeta = \mathbf{k}\cdot\text{curl}\,\mathbf{v}$ and the horizontal divergence $\delta = \text{div}_h\mathbf{v}_h$. The former represents twice the angular velocity of an air particle around a vertical axis relative to the solid Earth (reckoned positive when counterclockwise). Since the solid Earth rotates around the local vertical with the angular velocity $f/2$, it follows that the absolute vorticity of the air particle relative to a nonrotating frame is $f+\zeta$. The horizontal divergence δ represents the relative rate of expansion of an infinitesimal horizontal area moving with the air.

The field of $\mathbf{v}_h$ in a horizontal surface may be broken up into one field $\mathbf{v}_\zeta$ which carries all of the vorticity but no divergence, and another field $\mathbf{v}_\delta$ which carries all of the divergence but no vorticity. This is expressed by Eqs. (17).

$$\mathbf{v}_h = \mathbf{v}_\zeta + \mathbf{v}_\delta \qquad \zeta = \mathbf{k} \cdot \text{curl}\, \mathbf{v}_\zeta \qquad \delta = \text{div}_h\, \mathbf{v}_\delta \tag{17}$$

The continuity equation may be written approximately as Eq. (18), where ρ^* represents a standard

$$\rho^*\delta + \frac{\partial(\rho^* w)}{\partial z} = 0 \tag{18}$$

density distribution, depending upon z only. It follows that the horizontal field $\mathbf{v}_\delta$ together with the vertical velocity w forms a system of circulations in the vertical, whereas $\mathbf{v}_\zeta$ represents a horizontal circulatory motion. These two parts of the motion field behave very differently; for an individual air particle, δ may change quickly as a result of divergent pressure forces, whereas ζ changes more slowly. For this reason the motions in the atmosphere may be classified into two categories: predominantly vertical circulations, in which $\mathbf{v}_\delta$ is the dominating component, and $\mathbf{v}_\zeta$ is of secondary importance; and predominantly horizontal circulations, in which $\mathbf{v}_\zeta$ is the dominating component, and $\mathbf{v}_\delta$ is secondary. All types of gravity-inertia waves and small-scale convection currents belong to the first category; in the atmosphere these motions generally have small amplitudes and carry little energy, and they are mostly too weak or of too small scale to show up on weather maps. The large-scale motion systems revealed by weather maps, which contain the bulk of the energy and are the carriers of weather systems, are all predominantly horizontal circulations and thus belong to the second category above.

Quasi-static approximation. For motion systems whose vertical scale is small compared to the horizontal scale, the vertical motion is weak and its momentum can be neglected. Disregarding also vertical eddy friction, the hydrostatic equation, Eq. (12), holds everywhere and replaces the vertical component of Eq. (6). In the atmosphere, where the vertical scale of motion systems is limited by the high static stability of the stratosphere, this quasi-static approximation applies to motion systems of horizontal scale larger than about 200 km.

Large-scale quasi-horizontal circulations. Equations (5)–(8) apply equally well to motion systems of all types. In the study of the important large-scale quasi-horizontal circulations, it is convenient to derive equations which apply specifically to such motions, whereas predominantly vertical circulations have been eliminated at the outset. This is achieved by a filtering approximation: The horizontal momentum due to the velocity component $\mathbf{v}_\delta$ in Eqs. (17) is neglected; that is, the horizontal acceleration is approximated by $D\mathbf{v}_\zeta/Dt$. It follows that the horizontal divergence of the forces on the right-hand side of Eq. (6) must balance. This gives a relation between $\mathbf{v}_\zeta$ and the pressure field, which for small Rossby numbers U/fL (U is characteristic velocity and L is horizontal scale) reduces to the geostrophic relationship, Eq. (13). In extratropical latitudes the Rossby number of the large-scale quasi-horizontal circulations is of the order 10^{-1}, and these motions are therefore also quasi-geostrophic, within 10–20% error.

The evolution of the velocity field is approximately determined by the vorticity equation, which is obtained from Eq. (6) by taking the vertical vorticity. In a somewhat simplified form, Eq. (19),

$$D(f+\zeta)/Dt = -(f+\zeta)\delta \tag{19}$$

it expresses the familiar mechanical principle that the absolute rotation of an air particle around a vertical axis speeds up when the particle contracts horizontally.

In a current moving up a mountain slope, air columns must shrink vertically and hence, by Eq. (18), diverge horizontally ($\delta > 0$); from Eq. (19) it follows that the absolute vorticity $(f+\zeta)$ must decrease numerically. The opposite process takes place when the motion is downslope. This explains the observed predominance of anticyclonic vorticity ($\zeta < 0$ in the Northern Hemisphere) over large mountain ranges which are crossed by air currents.

Over level country there is no such net vertical stretching or shrinking through the whole atmospheric column; therefore, in the mean for the column there is a tendency toward conservation of absolute vorticity. This tendency shows up at the middle level of the atmosphere about 5 km above sea level, where in a crude approximation Eq. (20)

$$f+\zeta = \text{constant in time for each particle} \tag{20}$$

holds (the barotropic model). Since f increases northward, Eq. (20) requires ζ to decrease for particles moving northward and increase for particles moving southward. As a consequence, large-scale motion systems have a tendency to move westward relative to the air.

The barotropic model has been used to study the properties of a broad zonal current, such as the westerlies in middle latitudes. It has been found that barotropic instability (a kind of shearing instability) occurs if the absolute vorticity of the undisturbed current has a pronounced maximum at some latitude; in that case large-scale vortices may grow spontaneously, feeding upon the kinetic energy of the zonal current. If the zonal current does not possess a vorticity maximum, the current is barotropically stable. In the case of a uniform current, perturbations propagate westward relative to the air as permanent Rossby waves. Such waves may be studied approximately in a cartesian coordinate system, where f increases with y (northward) at a rate $\beta = df/dy$. A wave of wavelength L, superimposed upon a uniform westerly current u, will be propagated at a speed c, satisfying the Rossby formula, Eq. (21).

$$c = u - \frac{\beta}{4\pi^2} L^2 \tag{21}$$

In spherical coordinates one obtains instead the more accurate Rossby-Haurwitz formula, Eq. (22),

$$\gamma_n = \omega - \frac{2(\omega+\Omega)}{n(n+1)} \tag{22}$$

which gives the angular velocity of propagation

around the Earth's axis (γ_n) for a horizontal motion system where the stream function is a spherical harmonic of the order n (with any longitudinal wave number), assuming that the atmosphere has a solid rotation ω relative to the solid Earth. Some of the large-scale motion systems of the atmosphere are of this kind, and there is a satisfactory agreement between their observed propagation and Eq. (22).

The barotropic model describes only a very special mode of motion; for instance, it cannot describe conversion of potential and internal energy into kinetic energy or vice versa. A more general use of the filtered equations has revealed other types of large-scale motion systems. In the case of a zonal current which is sufficiently baroclinic, the linearized perturbation equations have solutions representing growing disturbances, which are similar to extratropical cyclones with respect to size, growth rate, and structure. Such a zonal current is said to possess baroclinic instability, and it is characteristic that the growing disturbances feed on the potential and internal energy of the zonal current.

Extratropical cyclones are observed to form in the zone of middle-latitude westerlies, characterized by pronounced baroclinicity and also by a strong maximum of cyclonic wind shear. Baroclinic and barotropic instability may both operate to cause the sudden growth of cyclones that is often observed.

Numerical dynamic meteorology. High-speed electronic computers have made it possible to find approximate numerical solutions to equation systems which are untractable by analytical methods. This has opened up a new branch of dynamical meteorology. It has become possible to drop many of the simplifying assumptions which were necessary for the analytical treatment; in particular, it is no longer necessary to linearize the equations.

Weather prediction. Starting from an initial state of the atmosphere which is known from observations, later states may be calculated from Eqs. (5)–(8) by numerical integration in time. To achieve this, the fields are represented by their values in a grid of points in space, and the differential equations are approximated by finite difference equations; thus one obtains a numerical model of the atmosphere. The total amount of computation for a given integration period depends on the approximation used in Eqs. (5)–(8), on the spacing of grid points in horizontal and vertical direction, and on the area over which the computation is extended. The quasi-static approximation is always used; the filtering approximation is used in some cases, but the equations may also be applied in nonfiltered, "primitive" form.

For use in practical weather forecasting, it is necessary that the computation proceed considerably faster into the future than real weather. This strongly restricts the size of the numerical model and makes considerable simplifications necessary, depending upon the available computer capacity.

The simplest possible model is the barotropic model, which is obtained when only one grid point is used along each vertical, at the 500-mb level. The set of governing equations then reduces to Eq. (20) or a slight modification thereof, and this equation may be expressed in terms of a stream function as the single dependent variable; all other physical processes are ignored. Although simplified to the extreme, the barotropic model can give predictions of practical value up to about 3 days ahead.

By increasing the number of grid points along the vertical, one obtains baroclinic models of increasing fidelity, requiring increasing amounts of computer time. A simplification which is frequently made in such models designed for weather prediction consists in neglecting friction and heat sources (also latent heat) and omitting humidity as a variable. The error due to this simplification grows relatively slowly with the prediction period, so that predictions of value may be obtained up to about 4 days ahead. Although humidity has been omitted, such models can still be used to predict precipitation, because this is known to occur wherever there is ascending motion.

Influences in the atmosphere propagate at the speed of sound and reach one from the remotest parts of the globe in less than 20 hr. Strictly speaking, the integration should therefore be extended to the entire atmosphere, even in a prediction for only 1 day ahead. This is prohibitive in practice, partly because the amount of computation would be too large, and partly because data which define the initial state of the entire atmosphere are not yet available. Fortunately, the bulk of the influences do not propagate at sound speed, and integrations of value can be carried out for a limited part of the world.

Integrations in time carried out with baroclinic numerical models of the atmosphere have confirmed and extended many of the results of analytical dynamic meteorology, such as the growth of cyclonelike disturbances in a baroclinically unstable zonal current and the formation of the associated frontal systems.

General circulation of atmosphere. The motions in the atmosphere are caused by solar heat. The combined process of absorption and emission of radiation causes a distribution of heat and cold sources which continually disturbs the equilibrium and maintains the motion. During this process the volume and pressure of individual air particles change with time, so that expansion takes place on the average at higher pressure than contraction. As a result, heat is converted into mechanical energy, just as in a man-made heat engine. The bulk of the mechanical energy thus released is used to overcome friction within the atmosphere itself; only a very small fraction is spent to do work on the ocean surfaces, thus maintaining ocean currents and waves.

The composite motion of the entire atmosphere is termed the general circulation. It has become possible to simulate the general circulation by numerical models. Some requirements which a general-circulation model must satisfy are: (1) It must cover the entire globe (or possibly one hemisphere, assuming symmetry at the Equator); (2) have enough vertical resolution so that all energy conversions can be represented; (3) contain radiative heat sources; and (4) contain turbulent eddy fluxes of momentum, enthalpy, and water vapor.

A numerical model for study of the general circulation must necessarily be exceedingly complicated. However, it is not intended for weather prediction, and it is therefore not necessary that the numerical integration proceed faster than real time; and it is not necessary to start the integration from a real state determined by global observations. By performing long-term numerical integrations with general-circulation models, starting from a constructed initial state, it has been possible to reproduce the main statistical characteristics of the real atmosphere, including the distribution of climate. *See* ATMOSPHERIC GENERAL CIRCULATION.

Special motion systems. Another important branch of numerical dynamic meteorology is the numerical study of various mesoscale and small-scale motion systems, such as hurricanes, fronts, squall lines, cumulus clouds, and land and sea breezes. [ARNT ELIASSEN]

Bibliography: H. R. Byers, *Elements of Cloud Physics*, 1965; H. R. Byers, *General Meteorology*, 4th ed., 1974; I. P. Danilina (ed.), *Meteorology and Climatology*, vol. 2, 1977; A. Eliassen and K. Pederson, *Meteorology*, 2 vols., 1977; R. Goody and J. C. Walker, *Atmospheres*, 1972; J. Gribbin, *Forecast, Famines, and Freezes*, 1977; G. J. Haltiner and F. L. Martin, *Dynamical and Physical Meteorology*, 1957; S. L. Hess, *Introduction to Theoretical Meteorology*, 1959; G. I. Marchuk, *Numerical Methods in Weather Prediction*, 1973.

Micrometeorology

A branch of atmospheric dynamics and thermodynamics which deals primarily with the interaction between atmosphere and ground; the interchange of masses, momentum, and energy at the earth-air interface; in short, with the lower boundary conditions of atmospheric processes. Thus, there are strong interconnections with those sciences that deal with the media underlying the atmosphere. Micrometeorology, along with other branches of meteorology, is also concerned with investigating and predicting the transport and dispersion of pollution from such sources as smokestacks and automotive exhausts in the lower atmosphere. It typically deals with the transport (or flux) of air properties in the vertical direction (conduction and convection) and the vertical variation of these fluxes. *See* AIR POLLUTION; OCEAN-ATMOSPHERE RELATIONS; SEA WATER.

Scale of focus. Micrometeorology differs from general meteorology (in particular, meso- and macrometeorology) in scale and with respect to the features of the atmosphere studied. All of the various branches of meteorology deal with the temporal-spatial variations of the weather elements (such as air density, temperature, components of momentum, and mixing ratios). Meso- and macrometeorological studies are typically based on standard synoptic observations (from a nationally and internationally organized network with about 20 to 200 km distance between stations), and of primary interest are large-scale air currents which serve to transport "conservative" atmospheric properties (such as heat, admixtures, and absolute vorticity) over relatively large horizontal trajectories. *See* METEOROLOGY.

Micrometeorology is concerned with the great variety of atmospheric processes which are incompletely covered by the synoptic network. Examples of such small-scale processes are mountain and valley circulations, sea breezes, and airflow past mountains and islands. The prefix micro is justified by the fact that the detailed study of the temporal-spatial variations of air properties in the lower atmosphere requires close spacing of instruments of relatively small size and low inertia (lag time).

Instrumental needs. The development and use of nonstandard mast-supported thermometers, anemometers, and other sensing instruments for the measurement of mean vertical profiles and gradients of temperature, wind speed, and other air properties is a basic requirement of micrometeorological research. Further miniaturization of sensing instruments is necessary for the study of short-time fluctuations of meteorological variables. A significant range of such fluctuations occurs with frequencies between several cycles per second to several cycles per minute.

Another basic requirement is the development and use of instruments for the direct measurement of boundary fluxes (such as evaporation, surface stress, and energy transfer, including net radiation). Micrometeorological studies are frequently concerned with the diurnal cycle of heating and cooling, the hydrologic cycle, and the energy dissipation in large-scale air currents due to surface stress (friction).

Micrometeorology is sometimes referred to as fair-weather meteorology. This appellation can be justified by the fact that of the total solar energy intercepted in space by the Earth about 70% reaches the ground under clear skies, with only about 10% scattered back to space (the remainder being absorbed in the air), while under an average cloud deck the corresponding percentages are about 30% reaching the ground, with nearly 60% being scattered back to space. Solar energy arriving at the bottom of the atmosphere is partly reflected ("albedo" radiation, normally about 15% of incoming radiation but varying between extremes of about 5–90%, for very dark-colored ground to very bright, fresh snow cover) and partly absorbed. A primary task of micrometeorology is to analyze and predict the distribution and transformation of the absorbed solar irradiation by establishing a complete energy budget of the air-submedium interface. The respective energy balance equation takes into consideration the vertical fluxes of terrestrial (or long-wave) radiation, of sensible heat conducted into the submedium as well air, and of latent energy (most important, phase transformations of H_2O such as evaporation, sublimation, and melting, but also photosynthesis wherever significant). The partitioning of energy and the subsequent changes in surface temperature, in direct response to variation in intensity of the solar forcing function, depend strongly on the strength of the existing overall air motion in the region. Because of prevailing nonuniformity in the physical structure of natural ground and its various (vegetative or nonvegetative) covers, and the relatively strong energy flux density of solar radiation, the atmosphere within a few meters of the ground is a surprisingly complex and variable structure. Strong fluctuations and gradients of temperature, density, and moisture affect the propagation of sound and light, creating irregular scattering and refraction phenomena which account for optical mirages, shimmer, and "boiling," as well as "ducting" of radio and radar beams.

Significant applications. Micrometeorological research is important in that it supplies detailed information about the physical processes in the region of the atmosphere where life is most abundant. It closes the gaps of information from synoptic networks. It produces results useful for applications in various fields such as climatology, oceanography, soil physics, agriculture, biology, chemical warfare, and air pollution. Among the branches of atmospheric physics, micrometeorology is the one whose subjects are most amenable to fairly complete experimental description, and to testing of the theoretical models.

The characteristic scale is small enough that it is both feasible and economical to attempt control of natural processes in the lower atmosphere, for example, by artificial changes of albedo or other surface characteristics such as aerodynamic roughness; by thermal admittance of the submedium (through mulching of soil or other means); by windbreaks; by the use of sunshades and smoke screens; by heat supply and artificial stirring of air; and by irrigation. *See* AGRICULTURAL METEOROLOGY; CLIMATOLOGY; CLIMATE MODIFICATION; CROP MICROMETEOROLOGY; INDUSTRIAL METEOROLOGY; WEATHER MODIFICATION.

Turbulence research and applications. The mechanism of the vertical flux is essentially one of eddy mixing or turbulent exchange of air properties. It is convenient to distinguish two methods of approach: One deals with the mean vertical gradients caused by turbulence, while the other is based on statistical treatment of turbulent fluctuations recorded by low-inertia (fast-response) instruments. The connecting link between the two approaches is presented by the Reynolds expression of the mean flux of an air property as the average covariance of the fluctuations of vertical wind speed and the property considered.

A powerful tool for research is the spectrum analysis of fluctuations by which the distribution of the variance of a meteorological element over the various frequencies is measured. According to J. Van der Hoven the power spectrum (in the range 0.001–1000 cycles per hour) of horizontal wind speed recorded at the upper levels of the meteorological tower at Brookhaven, Long Island, N.Y., shows two major peaks of energy, one at about 1 cycle per 4 days, another at about 1 cycle per minute. A broad minimum of energy at frequencies of about 1–10 cycles per hour seems to exist under varying terrain and synoptic conditions. This spectral gap appears to separate objectively the macrometeorological from the micrometeorological scale in the atmosphere.

Owing to the relatively high Reynolds number of atmospheric flow, all air motions are turbulent, which means that individual particles do not describe straight and parallel trajectories (as in truly laminar flow) even though the originating force (pressure gradient force) may be constant and uniform and the ground smooth and flat. Turbulence is sometimes visible in the shapes assumed by smoke, and felt in the variable pressure and cooling power of the wind. The intensity of atmospheric turbulence (as measured by the ratio standard deviation of a wind component divided by mean wind speed) is normally between 0.2 and 0.4, which is significantly larger than that of wind tunnel turbulence.

Low-level turbulence derives its energy basically from large-scale air motions and depends on the surface roughness. This mechanical turbulence is intensified by surface heating and damped by nocturnal cooling. Individual gusts can be ascribed to a disturbance, or eddy, of a certain extent. There is a wide range of eddy sizes (eddy spectrum). Near the ground, eddy sizes increase in proportion to distance from the ground, then decrease with further increase in height.

Atmospheric turbulence affects the flight of airplanes and ballistic missiles (rough air). It is especially troublesome when intensified by surface heating (thermal turbulence) and additional buoyancy forces generated by latent heat release in updrafts that ascend above the condensation level (cumulus convection). There are many features of turbulence research that are of common interest to micrometeorology and aeronautics.

[HEINZ H. LETTAU]

Bibliography: R. Geiger, *The Climate near the Ground*, 4th ed., 1965; H. E. Landsberg (ed.), *Advances in Geophysics*, suppl. 1: *Descriptive Micrometeorology*, by R. E. Munn, 1966; H. Lettau and B. Davidson (eds.), *Exploring the Atmosphere's First Mile*, 2 vols., 1957; C. H. Priestley, *Turbulent Transfer in the Lower Atmosphere*, 1974; P. Schwerdtfeger, *Physical Principles of Micrometeorological Instruments*, 1976; O. G. Sutton, *The Challenge of Atmosphere*, 1961; O. G. Sutton, *Micrometeorology*, 1977; J. Van der Hoven, Power spectrum of horizontal wind speed in the frequency range from 0.0007 to 900 cycles per hour, *J. Meteorol.*, 14:160, 1957.

Mineral resources conservation

The effort to ensure to society the maximum present and future benefit from the use of mineral resources. It stresses maximum use of a commodity for the benefit of the largest numbers of people. Conservation is influenced by many economic and political factors. As costs increase, people tend to use less; they retain materials for more essential uses, thus practicing conservation. Governments influence conservation by imposing taxes, by regulating imports and exports and by controlling production and prices, particularly of gas and oil. Many governments control both the price and the sale of all raw materials. A strong sense of awareness of the needs and the values of minerals was created throughout the world by the embargo on petroleum products enforced by the Organization of Petroleum Exporting Countries (OPEC) in 1973–1974. Many underdeveloped countries from which the industrial nations obtain their needed supplies maintain close controls over their mines and oil fields, primarily to regulate production and obtain greater revenues. The worldwide result has been promotion of conservation and curtailment of waste.

Reserves and resources. Reserves represent that fraction of a commodity that can be recovered economically; resources include all of the commodity that exists in the Earth. Traces of copper and iron exist in most rocks, but their recovery is inconceivable. They are resources, but not reserves.

Reserves are classed as proved, probable, possible, and potential. When a company publishes reserve figures, it is usually referring to proved and probable reserves. Proved reserves are those known without doubt; probable reserves are those that are nearly certain but about which a slight doubt exists; possible reserves are those with an even greater degree of uncertainty but about which some favorable information is known; potential reserves are based upon geological reasoning; they represent an educated guess.

At some time in the future, certain resources may become reserves. A change of this type can be brought about by improvements in recovery techniques, either mining or metallurgical, or by improved secondary recovery methods whereby oil and gas are forced to a well and can be pumped to the surface. New methods of extraction and recovery may result in lowering costs enough to make a deposit economic. New uses may also be found for a commodity, and the increased demand may result in an increase in price; or a large deposit may become exhausted, thus forcing production from a lower-grade and higher-cost ore body.

A knowledge of reserves and resources is essential to understanding conservation problems.

Classification of raw materials. For convenience, raw materials are classified as fuels, metals (including ferrous metals, nonferrous metals, light metals, and precious metals), and industrial minerals or nonmetallics. Some metals and minerals, and their major uses, are listed in the table.

Many minerals are lost during use and cannot be recovered and recycled. Petroleum in gasoline and natural gas and oil in fuels are dissipated, as are lead and other additives in gasoline. It is not economic to recover the tiny amounts of silver on some types of photographic film. Other metals, for example, iron, steel, copper, brass, and aluminum, are recoverable and may be reused many times. The gold of the ancient Egyptian and Incan treasuries probably still reposes in the vaults of some banks; certainly, it was never destroyed.

More metals are now being recovered and reused than ever before. As the prices of primary materials increase, scrap becomes increasingly valuable, and conservation is encouraged.

Supplies of minerals. Most of the Earth's surface has been closely examined in the search for useful minerals. Future supplies will depend in large part upon techniques that permit exploration below the surface and in remote and rather inaccessible areas. Exploration is thus becoming increasingly costly and unrewarding. Petroleum is being found in the waters of the continental shelves, at depths and under conditions never before tested, and in remote parts of the Arctic. New techniques are being developed and used. Likewise, greater amounts of hard minerals are being obtained from undeveloped and politically unstable lands. The locations of mineral deposits are fixed by nature; their positions cannot be changed. Competition to obtain the materials needed to sustain industry is growing rapidly; demands are overtaking supplies, and shortages of several commodities can develop within a few years. In order to maintain their economies, the industrial nations must encourage development and conservation of all their resources; they must find substitutes for some resources and must increase recycling. As prices of commodities increase and shortages develop, the standards of living are certain to fall.

The United States imports at least part of 88 of the 100 most commonly used mineral products. For example, 100% of such essential materials as the ores of chromium (used in the manufacture of stainless steels), manganese (without which sound steel cannot be made), tin, columbium-tantalum, industrial diamonds, rutile (for the manufacture of titanium metal), platinum metals, block mica, and

Classification of selected metals and minerals and some of their major uses

Materials	Uses
	Fuels
Bituminous and anthracite coal	Direct fuel, electricity, gas, chemicals
Lignite	Electricity, gas, chemicals
Petroleum	Gasoline, heating, chemicals, plastics
Natural gas	Fuel, chemicals
Uranium	Nuclear power, explosives
	Metals
	Ferrous metals
Chromium	Alloys, stainless steels, refractories, chemicals
Cobalt	Alloys, permanent magnets, carbides
Columbium (niobium)	Alloys, stainless steels
Iron	Steels, cast iron
Manganese	Scavenger in steelmaking, batteries
Molybdenum	Alloys
Nickel	Alloys, stainless steels, coinage
Tungsten	Alloys
Vanadium	Alloys
	Nonferrous metals
Copper	Electrical conductors, coinage
Lead	Batteries, gasoline, construction
Tin	Tinplate, solder
Zinc	Galvanizing, die casting, chemicals
	Light metals
Aluminum	Transportation, rockets, building materials
Beryllium	Copper alloys, atomic energy field
Magnesium	Building materials, refractories
Titanium	Pigments, construction, acid-resistant plumbing
Zirconium	Alloys, chemicals, refractories
	Precious metals
Gold	Monetary, jewelry, dental, electronics
Platinum metals	Chemistry, catalysts, automotive
Silver	Photography, electronics, jewelry
	Industrial minerals
Asbestos	Insulation, textiles
Boron	Glass, ceramics, propellants
Clays	Ceramics, filters, absorbents
Corundum	Abrasives
Feldspar	Ceramics, fluxes
Fluorspar	Fluxes, refrigerants, acid
Phosphates	Fertilizers, chemicals
Potassium salts	Fertilizers, chemicals
Salt	Chemicals, foods, glass, metallurgy
Sulfur	Fertilizers, acid, metallurgy, paper, foods, textiles

amorphous graphite is imported. The United States imports about 90% of the ores of aluminum, cobalt, nickel, antimony, bismuth, and beryl, and more than 35% of the petroleum and 40% of the iron ores that are used. Only with a few commodities—coal, sulfur, phosphates, molybdenum, and boron—is a surplus available for export. Imports of minerals in recent years have been growing at the rate of about 1% a year.

After World War II the United States established national stockpiles of minerals in an effort to assure the country of adequate supplies during times of emergency. Over the years the needs of the country have changed, and, likewise, materials in the stockpiles have changed. Efforts to use stockpiles for purposes other than the original purpose have been made, and advocates of price regulation still attempt to buy for the stockpile when prices are low and to sell when they are high. Handling of the national stockpiles clearly influences conservation policies.

Government conservation policies. Many policies of the United States government directly influence conservation and availability of energy and other minerals. The effects of government activities are most clearly evident insofar as oil and gas are concerned. Many state governments have laws that regulate the spacing of oil wells and prevent overpumping. They establish the most effective rate of recovery and ensure maximum life of a field. On the other hand, the Federal government has established price controls that discourage oil exploration in areas where costs are high. Increases in prices charged by OPEC nations for crude oil necessitated both a search for additional oil and conversion to more abundant energy sources such as coal. Government is seeking ways to use energy more efficiently and to develop other sources such as solar power and oil shale.

Periodically the government offers blocks of public land for auction and lease. These blocks, thought to contain oil and gas, are bid on by companies or, more commonly, by groups of companies. Legislation passed by the U.S. Congress prohibits large oil companies from making joint bids except with small companies. Bids commonly total $1,000,000,000 or more, and the government also receives royalty payments from the oil and gas produced.

The depletion allowance is a tax-exemption policy based upon the rate of depletion of a natural resource and is intended to equate tax policy in the extractive industries with the depreciation allowance permitted manufacturing industries. Producers are allowed to deduct a certain percentage of their gross revenues from the part of their income subject to Federal income tax. Early in 1975 Congress canceled the depletion allowance for the large oil companies, although permitted retention of the allowance for smaller companies and the mining industry. Removal of the depletion allowance has reduced capital available for exploration and has resulted in an upward pressure on the price of petroleum products.

The political history and legal climate in a country strongly influence private investment in mines and oil fields. Many nations have expropriated raw materials properties, in some places without compensation, and in others with compensation less than the former owners thought reasonable. Such actions greatly discourage private investment, and, as a result, many governments now must finance and operate their own mines and oil fields.

Government policies also influence petroleum and mineral exploration; there is a myriad of regulations and laws concerning environment, right of access, land restoration, and other items. Open-pit or strip mining coal is cheaper, has a much higher recovery, and is safer than underground mining, which results in more than 50% of the coal being left underground. The many government bureaus that issue regulations concerning exploration and development are restrictive. Especially in the energy fields, unsightly strip mines, oil spills offshore, and worries about radioactive leakage have resulted in rigorously enforced regulations. The United States has also withdrawn from exploration extensive wilderness areas and other tracts of public lands.

Technological changes. Developments in oil well drilling, in enhanced recovery methods, and in mining, milling, and smelting methods all have profound impacts on the availability of raw materials. Oil well drilling is being carried out in the open oceans at depths of 900 ft (270 m) or more, and techniques are available that will enable wells to be drilled to considerably greater depths. Shutoff devices and other safety devices greatly reduce the dangers of oil spills and wellhead fires.

The development of large machines for handling huge tonnages of rock and the use of refined explosives have improved mining methods. Open-pit and strip mines are larger and more efficient than ever before, and lower grades of ore are being recovered. It is common for a mine to produce 50,000 tons (45,000 metric tons) of ore a day, and some yield more than 100,000 tons (90,000 metric tons). Industry is taking advantage of economies of size. Improved methods of transportation, concentration, and smelting permit less costly handling of large tonnages and improve recoveries. Newer smelting methods are cleaner, have as good or better recovery records, and may prove to be less expensive to operate.

Demand and supply problems. A serious problem facing the raw materials industry is the cyclical nature of demands for its products. When economic times are good, the markets are strong; but consumption is greatly curtailed during recessions, and excessive inventories are created. As a result, prices are cyclical and, with some commodities, for example, mercury, are highly volatile. In order to stabilize prices, many countries favor the establishment of cartels and associations. Probably the most effective association of this type is the International Tin Commission, an organization that includes both consumers and producers. A ceiling and a floor price are established: If the market price goes below the floor, the Commission buys tin; if the price goes above the ceiling, the Commission sells. In this way, the price and production of tin are maintained at reasonably stable rates.

Governments that have copper, aluminum, and iron for export have established cartels in these commodities and are making efforts to consolidate controls over rates and prices. Most such cartels aim to increase the price of the commodity and to improve the financial status of the producer at the expense of the consumer. As industrial needs increase and minerals become difficult to purchase, cartels will probably be more successful, but whether or not such associations succeed, the prices of raw materials will increase with demand. As a national policy conservation is not only desirable, it is a necessity. *See* CONSERVATION OF RESOURCES. [CHARLES F. PARK, JR.]

Bibliography: P. A. Bailly, *Conversion of Resources to Reserves*, 1975; D. A. Brobst and W. P. Pratt, *United States Mineral Resources*, U.S. Geol. Surv. Prof. Pap. no. 820, 1973; C. F. Park, Jr., *Earthbound: Minerals, Energy, and Man's Future*, 1975; A. Sutulov, *Minerals in World Affairs*, 1972; U.S. Bureau of Mines, *U.S. Minerals Yearbook*, annual.

Mining

The taking of minerals from the earth, including production from surface waters and from wells. Usually the oil and gas industries are regarded as separate from the mining industry. The term mining industry commonly includes such functions as exploration, mineral separation, hydrometallurgy, electrolytic reduction, and smelting and refining, even though these are not actually mining operations.

The use of mineral materials, and thus mining, dates back to the earliest stages of man's history, as shown by artifacts of stone and pottery and gold ornaments. The products of mining are not only basic to communal living as construction, mechanical, and raw materials, but salt is necessary to life itself, and the fertilizer minerals are required to feed a populous world.

Methods. Mining is broadly divided into three basic methods: opencast, underground, and fluid mining. Opencast, or surface mining, is done either from pits or gouged-out slopes or by strip mining, which involves extraction from a series of successive parallel trenches. Dredging is a type of strip mining, with digging done from barges. Hydraulic mining uses jets of water to excavate material.

Underground mining involves extraction from beneath the surface, from depths as great as 10,000 ft (3.05 km), by any of several methods.

Fluid mining is extraction from natural brines, lakes, oceans, or underground waters; from solutions made by dissolving underground materials and pumping to the surface; from underground oil or gas pools; by melting underground material with hot water and pumping to the surface; or by driving material from well to well by gas drive, water drive, or combustion. Most fluid mining is done by wells. A recent type of well mining, still experimental, is to wash insoluble material loose by underground jets and pump the slurry to the surface. *See* OPEN-PIT MINING; PLACER MINING; STRIP MINING; UNDERGROUND MINING.

The activities of the mining industry begin with exploration which, since mankind can no longer depend on accidental discoveries or surficially exposed deposits, has become a complicated, expensive, and highly technical task. After suitable deposits have been found and their worth proved, development, or preparation for mining, is necessary. For opencast mining this involves stripping off overburden; and for underground mining the sinking of shafts, driving of adits and various other underground openings, and providing for drainage and ventilation. For mining by wells drilling must be done. For all these cases equipment must be provided for such purposes as blasthole drilling, blasting, loading, transporting, hoisting, power transmission, pumping, ventilation, storage, or casing and connecting wells. Mines may ship their crude products directly to reduction plants, refiners, or consumers, but commonly concentrating mills are provided to separate useful from useless (gangue) minerals.

Economics. Mining is done by hand in places where labor is cheap, but in the more industrialized countries it is a highly mechanized operation. Some surface mines use the largest and most expensive machines ever developed—unless a large ship can be called a machine.

There are many small- and medium-sized mines but also a growing number of large ones. The trend to larger mines and particularly to large opencast operations is due to the great demand for mineral products, depletion of high-grade reserves, technological progress, and the need for economies of scale in mining low-grade deposits. The largest open-pit coal mine moves 300,000 tons (270,000 metric tons) of material per day and the largest metal mine, 300,000 tons. The largest underground coal mine produces 17,000 tons (15,000 metric tons) of coal per day, and the largest metal mine 50,000 tons (45,000 metric tons) of ore.

The quality of deposits which can be mined economically depends on the market value of the contained valuable minerals, on the costs of mining and treatment, and on location. Alluvial gold gravels may run as low as 1/350 oz of gold/yd^3 (0.106 g/m^3) and gold from lodes as low as 1/10 oz/ton (3.125 g/metric ton). Uranium ore containing 1.5 lb of uranium oxide per ton (0.75 kg per metric ton) is mined underground, and copper-molybdenum ore as low grade as 0.35% copper and 0.05% molybdenum is taken from open pits. On the other hand, iron ore is rarely mined below 25% iron, aluminum ore below 30% aluminum, and coal less than 90% pure. Crude petroleum is usually over 95% pure.

Use of natural resources. A unique feature of mining is the circumstance that mineral deposits undergoing extraction are "wasting assets," meaning that they are not renewable as are other natural resources. This depletability of mineral deposits not only requires that mining companies must periodically find new deposits and constantly improve their technology in order to stay in business, but calls for conservational, industrial, and political policies to serve the public interest. Depletion means that the supplies of any particular mineral, except those derived from oceanic brine, must be drawn from ever-lower-grade sources. Consciousness of depletion causes many countries to be possessive about their mineral resources and jealous of their exploitation by foreigners. Depletion also accounts for some controversial attitudes toward conservation. Some observers would reduce the scale of domestic production and increase imports in order to extend the lives of domestic deposits. Their opponents argue that encouragement of mining through tariff protection, subsidies, or import quotas is desirable, on the grounds that only a dynamic industry in being can develop the means of mining low-grade deposits or meet the needs of a national emergency. They point out that protection encourages the extraction of marginal resources that would otherwise be condemned through abandonment.

Despite its essential nature, mining today is being constrained as to where it can operate by wilderness lovers and by rapidly expanding urbanization. Concern over pollution, some of which is caused by mine water, mining wastes, and smelter effluents, has grown rapidly. More and more objection is developing to defacement of landscapes by surface mining, and many states now require restoration of the surface to a cultivable or forested condition after strip mining. People in residential areas usually resist the development of industries near their homes, particularly if such industries employ blasting, produce smoke or fumes, or cause traffic by large vehicles.

In countries with Anglo-Saxon traditions, title to minerals on private lands is vested in the private owners, but in many countries minerals are state property. Where minerals are privately owned, mining is commonly done by purchase of title or under lease and royalty contracts. State-owned minerals are mined under claims acquired through discovery and denouncement, leases, or concessions. For centuries governments have found it desirable to encourage prospecting and mining to lure the adventurous into these challenging, useful arts. [EVAN JUST]

MINING POWER

Power is applied to mining in six distinctive ways: electricity, diesel power, compressed air, hydraulic power, steam power, and hydroelectricity.

Electricity. Both alternating and direct electric current are used in modern mining. Direct current (dc) is used for locomotive haulage and for the major portion of underground coal operations because of the high torque developed in dc series motors under heavy loads and starting. Alternating current (ac) is used less extensively although some mines employ both in a dual system.

In the United States, 6000-, 4160-, and 2300-volt three-phase alternating current is generally received from service companies at the surface substation. In a well-planned mine, all the power is metered at one point and the power factor is adjusted between 90 and 95% by the use of synchronous motors and capacitors. Neoprene-covered cables, type SHD, 15,000 volts, are recommended to take the power underground through boreholes or power shafts. Each of three insulated conductors is covered with copper shielding braiding to eliminate static stresses, and a ground conductor for each power conductor is placed in the interstices of the cable.

Converting ac to dc underground may be accomplished by several methods. A 250- or 500-volt substation can be provided by converters, motor-generator sets, mercury-arc rectifiers including the glass-bulb type, and dry rectifiers made of selenium, germanium, or silicon (Fig. 1). Portable conversion units offer increased convenience and efficiency.

Trolley and feeder lines supply power throughout the main roadways. These distribution lines should not be extended more than 3500–4000 ft (1070–1220 m) from the power source to avoid low voltage at the end of the line. If local laws permit, 500-volt systems are sometimes used to minimize line voltage drop. This supplies twice the load for the same size and length of cable, or conversely, the same load can be supplied at twice the distance. Trailing cables, fastened to the trolley nipping stations or power centers, supply power to machines in sections of the mine where explosive gases may be present.

Sectionalization is a method of distributing mine power so that power cables can be isolated for reasons of faults or repairs without shutting off the main supply to several working sections. For main distribution, circuit breakers, disconnect switches, and various overcurrent protective devices are essential for properly installed sectionalization. At face areas, safety circuit centers and associated intrinsically safe circuitry make it possible to connect or disconnect short cables without danger of causing incendiary arcing which could ignite gas.

Alternating-current power for mining is increasing in popularity because its equipment is less costly than dc and its maintenance is simpler. Alternating-current motors, for example, cost one-third to one-half as much as the equivalent dc and are more compact. High voltage is transformed to 440 or 220 volts at underground substations. Portable units are also utilized. Sectionalization is applicable to ac distribution.

Alternating-current power is commonly used in strip mining. High voltage of 33,000 volts is stepped down to 7300, 6600, 4160, or 2300 volts. Permanent substations equipped with lightning arresters, circuit breakers, ground protective equipment, and other protective devices, and semiportable substations are employed for distribution. Power is distributed by pole lines or cable systems, or a combination of the two.

Diesel power. This type is rapidly gaining favor in metal and noncoal mining because of its flexibility. Some states require that underground diesel-powered equipment be approved by the U.S. Bureau of Mines. Details of these standards include explosion-proof housings, mine ventilation

Fig. 1. A 300-kW Westinghouse ignitron rectifier installed at a mine.

necessary to dilute exhaust gases, control of hot exhaust gases and surface temperatures, concentrations of toxic constituents in exhaust gases, and recommendations for the selection, handling, and storage of diesel fuel oil.

Compressed air. As a mine facility, pneumatic power is utilized in a variety of applications, mostly in the metal and noncoal fields. It is used to power drills and hoists; pneumatic tools, such as grinders, drills, riveters, chippers, pneumatic diggers, and spades; air-driven sump pumps; direct-acting and air-lift pumps; pile drivers for shaft sheathing; air pistons for unloading cars; drill-steel sharpeners; air motors; compressed-air locomotives; shank and detachable-bit grinders; mine ventilation; and in supplying air for blowing converters; starting diesel engines; and coal preparation. Compressed air is used in coal mining for blasting. This method works with 9000–10,000 pounds per square inch (psi) [62–69 MPa] and the air is released from a tube which is inserted in a hole in the face of the coal when a metal diaphragm ruptures. The force is released through slots in the tube and breaks up the coal which had been previously undercut.

Hydraulic power. For mining, hydraulic applications are rather limited in usage and may employ either water or oil as a fluid. Oil types are used in connection with small tools, lifts, and in the intricate operation of continuous mining machines and other equipment. An available waterfall may be directed to power equipment such as air compressors. A unique use of hydraulic power, called jet mining, uses air pressure to force water through 1/4-in. (6-mm) nozzles under 2000 psi (14 MPa) pressure. This jet action has been developed to cut a material called gilsonite, a solid hydrocarbon, by use of water at a rate of 300 gal/min (1.9×10^{-2} m^3/s).

Steam power. Although formerly used to drive compressors, hoists, generators, and other equipment, steam is now rarely used in mining but has definite, although limited, application in some coal mines in which there is an abundance of waste fuel or unmarketable coal. Parts of certain coal seams contain impure bands that must be rejected, or that are difficult to clean, but they can be burned under boilers with special firing equipment.

Hydroelectricity. As applied to mining, hydroelectricity is used mostly in the electrometallurgical fields, where vast amounts of power are required at a reasonable cost. Some hydroelectricity has been used for normal, electrical, mining power, but mostly it is used for processes beyond the mining operation such as beneficiation, smelting, and refining. In certain cases these are done near the mine mouth in isolated areas in order to reduce bulk before shipment. [ROBERT S. JAMES]

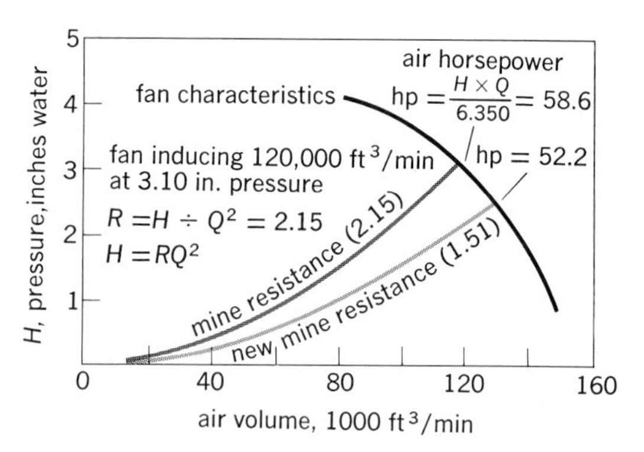

Fig. 2. Fan characteristics versus mine resistance. 1000 ft³/min = 0.472 m³/s; 1 hp = 745.7 W; pressure of 1 in. water = 249 Pa.

MINE VENTILATION

The purpose of mine ventilation is to provide comfortable, safe, and healthful atmospheric conditions at places where men work or travel.

Airflow fundamentals. The following points summarize airflow principles for mine ventilation; (1) Airflow is induced by a pressure difference between intake and exhaust; (2) the pressure created must be sufficient to overcome the system resistance and may be either negative or positive; and (3) air flows from the point of higher to lower pressure. Also, mine airflow is considered turbulent and follows the square law relationship between volume and pressure; that is, a doubled volume requires four times the pressure.

The principles of fans may be summarized as follows: (1) Air quantity varies directly as fan speed, and is independent of air density; (2) pressures induced vary directly as the square of the fan speed, and directly as the air density; (3) the fan power input varies directly as the air density and the cube of the fan speed; (4) the fan mechanical efficiency is independent of fan speed and density.

The amount of air movement induced will be dependent upon the fan characteristic and mine resistance as shown by Fig. 2. The pressure H required to pass a quantity of air Q through a mine or

Reasonable frictional coefficients

		Straight			Sinuous or curved					
					Moderate			High degree		
Type of airway	Irregularities	Clean	Slightly obstructed	Moderately obstructed	Clean	Slightly obstructed	Moderately obstructed	Clean	Slightly obstructed	Moderately obstructed
Smooth-lined	Minimum	10	15	25	25	30	40	35	40	50
	Average	15	20	30	30	35	45	40	45	55
	Maximum	20	25	35	35	40	50	45	50	60
Sedimentary rock or coal	Minimum	30	35	45	45	50	60	55	60	70
	Average	55	60	70	70	75	85	80	85	95
	Maximum	70	75	85	85	95	100	95	100	110
Timbered (5-ft centers)*	Minimum	80	85	95	95	100	110	105	110	120
	Average	95	100	110	110	115	125	120	125	135
	Maximum	105	110	120	120	125	135	130	135	145
Igneous rock	Minimum	90	95	105	105	110	120	115	120	130
	Average	145	150	160	160	165	175	170	175	195
	Maximum	195	200	210	210	215	225	220	225	235

*5 ft = 1.524 m.

segment is expressed by the formula $H = RQ^2$, where R, the mine resistance factor, may be calculated from known pressure losses, or from the common ventilation formula $R = KlO/5.2A^3$ in which K is the frictional coefficient, l is the length of the airway, O is the perimeter of the airway, and A is the cross-sectional area of the airway. For simplicity of calculation, frictional coefficients should be expressed without decimals, and the quantity Q of air in cubic feet per minute (cfm = 0.472×10^{-3} m³/s) should be divided by 100,000 before using. The table gives reasonable frictional coefficients; the table's use may be exemplified by an application to the following case: What pressure is necessary to induce 60,000 cfm through a single airway 6 ft high, 12 ft wide, and 2500 ft long in sedimentary rock? The airway is straight, has average irregularities, and is moderately obstructed. K from table is 70; l is 2500; perimeter is 36; area is 72 ft². Q is 60,000 cfm ÷ 100,000 = 0.60. The pressure H is given by Eq. (1).

$$H = RQ^2 = \frac{KlOQ^2}{5.2A^3} = \frac{70 \times 2500 \times 36 \times (0.6)^2}{5.2 \times (72)^3}$$

$$= 1.17 \text{ in. water pressure} \qquad (1)$$

Parallel air flow can be determined by the square-law relationship. For example, the pressure H required to pass 60,000 cfm through 2500 ft of single entry is 1.17 in. of water. For two identical entries the pressure is given by Eqs. (2).

$$H_1\left(\frac{1}{2}\right)^2 = H_2$$

$$H_2 = \frac{H_1}{4} = \frac{1.17}{4} = 0.292 \text{ in. water} \qquad (2)$$

When resistance factors are known or entries are not identical, the formula is Eq. (3). For example, what is the combined resistance factor of two entries 1000 ft long? One entry is substantially larger, $R_1 = 1.50$, $R_2 = 4.0$. Also, what pressure is required to pass 60,000 cfm through 2500 ft of the combined entries, assuming average conditions throughout? Computation is as follows:

$$\frac{1}{\sqrt{R}} = \frac{1}{\sqrt{R_1}} + \frac{1}{\sqrt{R_2}} + \frac{1}{\sqrt{R_3}} + \cdots + \frac{1}{\sqrt{R_n}} \qquad (3)$$

$$\frac{1}{\sqrt{R}} = \frac{1}{\sqrt{1.5}} + \frac{1}{\sqrt{4.0}} = \frac{1}{1.22} + \frac{1}{2.00}$$

$$= 0.82 + 0.50 = 1.32$$

$$\frac{1}{\sqrt{R}} = 1.32 \qquad R = 0.58 \text{ per 1000 ft entry}$$

$$R \text{ for 2500 ft} = 2.5 \times 0.58 = 1.45$$

$$\text{Pressure} = 1.45 \times (0.60)^2 = 0.52 \text{ in. water}$$

Air quantity requirements. Unless covered by state laws, the common criteria for adequate ventilation are absence of smoke and dust with moderate air temperatures in metal mines and the absence of methane, smoke, and dust in coal mines. Natural conditions of gas, rock temperatures, dust, and operating practices determine requirements. Good quality air is not deficient of oxygen and is free of harmful amounts of physiological or explosive contaminants.

Mine gases. Important contaminants of mine air are carbon dioxide, hydrogen sulfide, methane, carbon monoxide, and sulfur dioxide.

Carbon dioxide, CO_2, specific gravity 1.529, is produced by oxidation and combustion of organic compounds and is occluded in the rock strata of certain mines. It is heavy, colorless, and odorless and is usually found in low, poorly ventilated areas.

Hydrogen sulfide, H_2S, specific gravity 1.191, is the product of decomposition of sulfur compounds. It is colorless, has the odor of rotten eggs, and may be found near areas of stagnant water in poorly ventilated areas.

Methane, CH_4, specific gravity 0.554, is a natural constituent of all coals. It may be occluded in carbonaceous shales and sandstones and may infiltrate into metal mines at contacts with carbonaceous rocks. It is colorless, odorless, and may be found in high, poorly ventilated cavities.

Carbon monoxide, CO, specific gravity 0.967, is not a normal constituent of mine air, but is produced in mines by the incomplete combustion of carbonaceous matter, mine fires, or from gas or dust explosions. It is colorless and odorless.

Sulfur dioxide, SO_2, specific gravity 2.264, is not common, but may be found in sulfur mines and in mines with rich sulfide ores as the result of fires. It is a water-soluble and colorless gas with a suffocating odor.

Blackdamp is a common term applied to oxygen-deficient atmospheres; it is not a specific gas mixture but may contain any of many gases produced by oxidation and processes that use oxygen and liberate carbon dioxide.

Small quantities of air contaminants must be determined by laboratory analysis of air samples. On-the-spot safety determinations for carbon dioxide and oxygen deficiency may be made by flame safety lamp or small portable absorption instruments. Methane can be detected by flame safety lamp and commercial testers; carbon monoxide and hydrogen sulfide by special hand-held colorimetric indicators.

Dust and dust hazards. Dust is defined as the solid particulate matter thrown into suspension by mining operations. The size of particles may range upward from less than 1 micron (0.001 mm) diameter (as shown in Fig. 3, the Frank chart); particles larger than 10 μ can usually be seen by the naked eye. The dust hazard may be physiological or explosive or both physiological and explosive as is coal dust. The common physiological diseases are mostly various pneumoconioses. Preventive measures are to suppress dust at the source with water sprays, foam, fog, wetting agents, or dust collectors. Additional protection may be provided through suitable respirators. Accumulations of explosive dust should be removed and inert material, such as rock dust, applied to surface areas. The dust hazard can be determined by systematic sampling of airborne dust at critical points. The impinger, a common instrument used, draws a known volume of air through a liquid or filter to remove the dust. The dust concentrations of the sample can then be determined by count using a microscope or microprojector. *See* DUST AND MIST COLLECTION.

Air temperature and humidity. Temperature rise in mine workings is from (1) heat conducted from surrounding strata because of the thermal

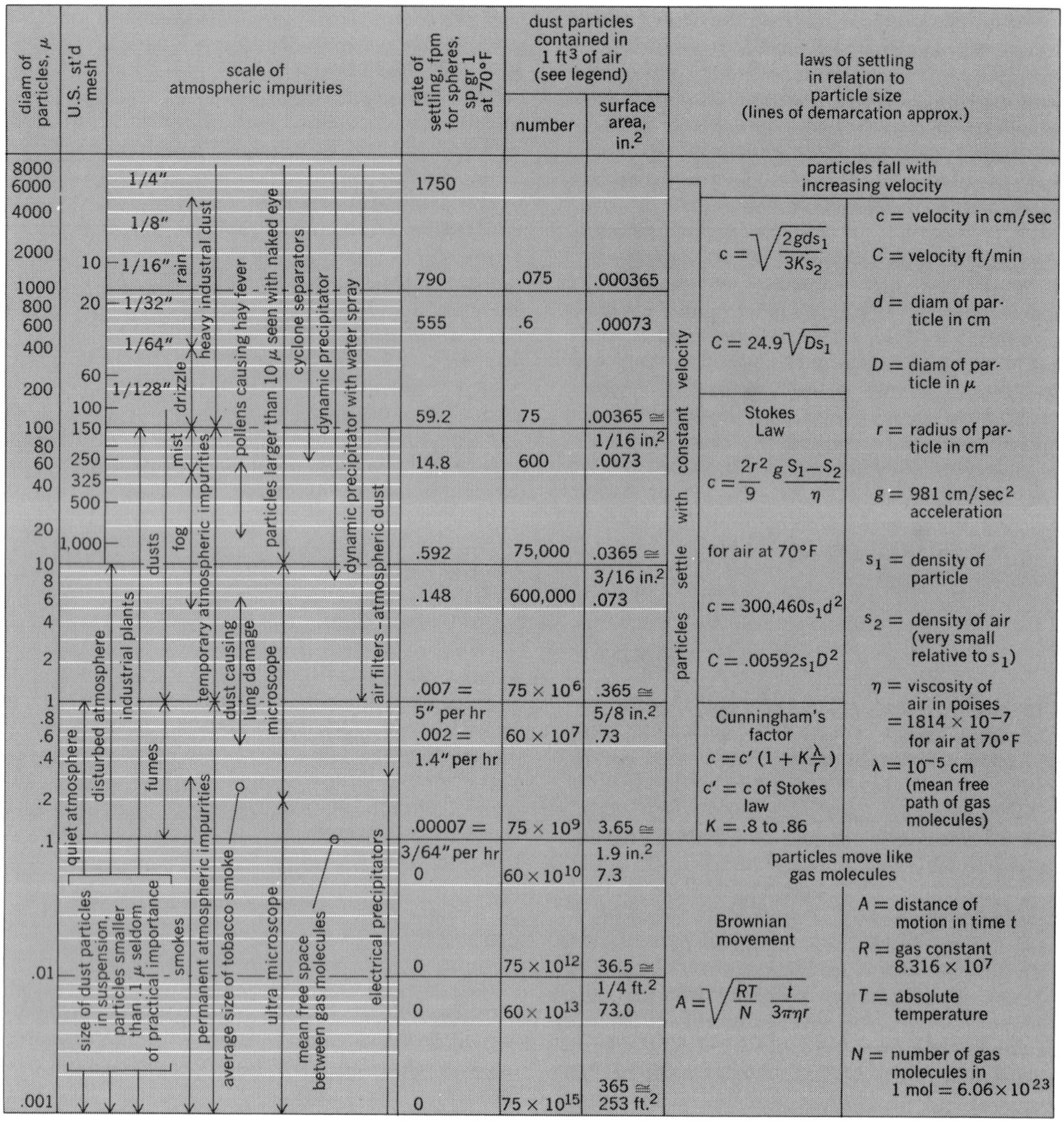

diam of particles, μ	U.S. st'd mesh	scale of atmospheric impurities	rate of settling, fpm for spheres, sp gr 1 at 70°F	dust particles contained in 1 ft³ of air (see legend): number	surface area, in.²
8000		1/4"			
6000			1750		
4000		1/8"			
2000	10	1/16"			
1000			790	.075	.000365
800	20	1/32"			
600			555	.6	.00073
400		1/64"			
200	60	1/128"			
	100		59.2	75	.00365 ≅ 1/16 in.²
100	150				
80					
60	250		14.8	600	.0073
40	325				
	500				
20	1,000				
10			.592	75,000	.0365 ≅ 3/16 in.²
8					
6			.148	600,000	.073
4					
2					
1			.007 = 5" per hr	75 × 10⁶	.365 ≅ 5/8 in.²
.8					
.6			.002 = 1.4" per hr	60 × 10⁷	.73
.4					
.2					
.1			.00007 = 3/64" per hr	75 × 10⁹	3.65 ≅ 1.9 in.²
			0	60 × 10¹⁰	7.3
.01			0	75 × 10¹²	36.5 ≅ 1/4 ft.²
			0	60 × 10¹³	73.0
.001			0	75 × 10¹⁵	365 ≅ 253 ft.²

Scale of atmospheric impurities: rain; heavy industrial dust; drizzle; mist; fog; dusts; temporary atmospheric impurities; pollens causing hay fever; particles larger than 10 μ seen with naked eye; cyclone separators; dynamic precipitator; dynamic precipitator with water spray; dynamic precipitator dust; air filters - atmospheric dust; quiet atmosphere; disturbed atmosphere; industrial plants; fumes; dust causing lung damage; microscope; smokes; permanent atmospheric impurities; average size of tobacco smoke; ultra microscope; mean free space between gas molecules; electrical precipitators; size of dust particles in suspension, particles smaller than 1 μ seldom of practical importance

Laws of settling in relation to particle size (lines of demarcation approx.)

particles fall with increasing velocity

$c = \sqrt{\dfrac{2gds_1}{3Ks_2}}$

$C = 24.9\sqrt{Ds_1}$

particles settle with constant velocity

Stokes Law

$c = \dfrac{2r^2 g}{9} \dfrac{S_1 - S_2}{\eta}$

for air at 70°F

$c = 300{,}460 s_1 d^2$

$C = .00592 s_1 D^2$

Cunningham's factor

$c = c'\left(1 + K\dfrac{\lambda}{r}\right)$

c′ = c of Stokes law

K = .8 to .86

particles move like gas molecules

Brownian movement

$A = \sqrt{\dfrac{RT}{N}\dfrac{t}{3\pi\eta r}}$

c = velocity in cm/sec

C = velocity ft/min

d = diam of particle in cm

D = diam of particle in μ

r = radius of particle in cm

g = 981 cm/sec² acceleration

s_1 = density of particle

s_2 = density of air (very small relative to s_1)

η = viscosity of air in poises = 1814 × 10⁻⁷ for air at 70°F

λ = 10⁻⁵ cm (mean free path of gas molecules)

A = distance of motion in time t

R = gas constant 8.316 × 10⁷

T = absolute temperature

N = number of gas molecules in 1 mol = 6.06 × 10²³

Fig. 3. Size and characteristics of airborne solids (compiled by W. G. Frank). It is assumed that particles are of uniform spherical shape having specific gravity 1 and that dust concentration is 0.6 gr/1000 ft³ (2.288 g/1000 m³) of air, average for metropolitan districts. 1" = 25.4 mm; 70°F = 21°C; 1 ft³ = 2.83 × 10⁻³ m³; 1 in.² = 6.45 cm²; 1 ft/min = 5.08 mm/s; 1 poise = 0.01 Pa·s.

gradient (depth per °F or °C rise in temperature); (2) adiabatic compression of descending air columns (approximately 5.5°F/1000 ft or 10°C/km); and (3) heat from oxidation of minerals.

The factors that influence humidity are: rise in dry-bulb temperature; volume changes caused by pressure and temperature changes; and moisture picked up from shafts and roadways.

The air temperature approximates the temperature of adjacent walls; seasonal temperature changes are noticeable only short distances underground. Workers' efficiency (Fig. 4) is dependent upon temperature, humidity, and motion velocity of air. A solution to excessive air temperatures is air conditioning by passing the air currents through heat exchangers filled with chilled air.

Natural ventilation. Natural ventilation is induced by differences of total weights of air columns for the same vertical distance. Natural draft may operate with or against the mechanical draft (Fig. 5) or may be the only source of pressure. Natural draft pressures at standard density can be estimated as 0.03 in. water-gage for each 10°F average temperature difference per 100 ft increment vertical elevation (or 4.2 mm water-gage for each 1°C per meter). For accurate determinations the average air densities of influencing air columns must be calculated.

The formula for determining air density is given by Eq. (4), where d is the density in lb/ft³, T is the

$$d = \frac{1.327}{460 + T}(B - 0.378\text{VP}) \qquad (4)$$

dry-bulb temperature, B is the barometric pressure, in inches of mercury, and VP is the vapor pressure of water at the dew point, in inches of mercury.

With most calculations, the vapor pressure influence can be ignored. The simplified formula then is Eq. (5).

$$d = \frac{1.327}{460 + T}(B) \qquad (5)$$

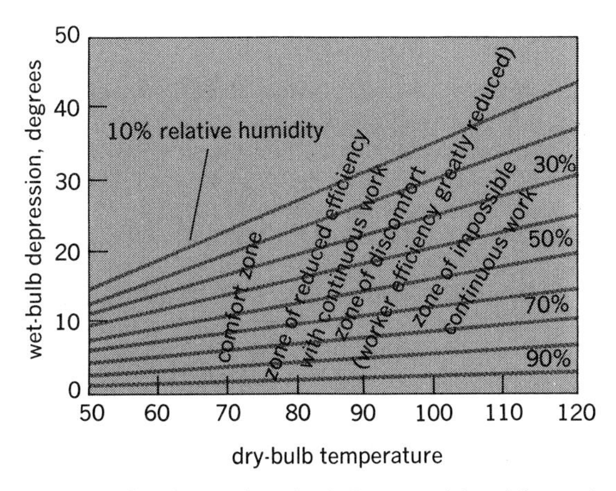

Fig. 4. Graph showing the influence of humidity and temperature on worker efficiency.

Auxiliary ventilation. This term applies to booster fans and auxiliary fans. A booster fan is placed underground and operated in series with the main fan to increase ventilating pressure of one or more splits of the ventilating current. The booster fan in effect reduces the mine resistance, thereby increasing the air quantity circulated. An auxiliary fan is a small fan installed in the air current to divert, through air tubing or ducts, a part of the ventilating current to ventilate some particular place or places. In metal mines, they are used to ventilate developing drifts, raises, crosscuts, and stopes. In coal mines, they are used to conduct air to working faces. [DONALD S. KINGERY]

Ventilation system evaluation. Accurate analysis and evaluation of a ventilation system, for the purpose of initiating improvements or projecting the system, require pressure and quantity measurements from which actual resistances can be determined. Proper utilization of such data permit efficient and economic changes in the present system and the accurate determination of requirements for the projected system. Although friction coefficients from the table enable engineers to design, with good results, ventilation systems for new mines, the continuously changing conditions following mining make it almost impossible to apply these values in evaluating older systems.

The aneroid barometer measures absolute static pressure and is the instrument usually used for mine-pressure surveys. The instrument is rugged and portable but relatively few have sufficient sensitivity and precision for pressure surveys. A variety of graduated scales are provided, the most common being "inches of mercury" and "feet of air." Instruments with the latter scale are termed altimeters. Regardless of scale on the instrument, mine pressure and pressure differentials are generally converted to inches of water for analysis of pressure data.

The surface absolute static pressure (barometer reading) is primarily a function of elevation and air density. Underground, within the ventilation system, absolute static pressure is a function of the same conditions plus the static pressure resulting from pressure generated by the fan. By compensating for elevation differences and atmospheric changes, the ventilating pressure at any point and between points resulting from fan operation can be calculated. By supplementing the pressure determinations with water-gage reading across stoppings and regulators, pressure losses for any part of the system can be readily determined.

Analysis of the pressure data by means of a pressure gradient will pinpoint high-resistance areas. The pressure gradient (Fig. 6) is constructed by plotting distance of air travel against pressure drop. Portions of the gradient with a steep slope indicate high-resistance airways.

For a complete analysis of any system, pressure data must be combined with extensive quantity determinations. Quantity flow is determined by area and velocity measurements. The more common instruments for velocity measurements include vane anemometer, pitot tube, and smoke tube. Instruments selected will depend on velocity of the airstream.

The quantity survey is useful in locating areas of excessive intake to return leakage, but more important, combined with pressure data, the actual resistance factor for each segment of the system can be calculated. These factors are necessary for accurate analysis of the effect of any major system changes or projections. Results may be determined mathematically or by analog computer.

For the economic solution of problems involving mine ventilation improvement, the U.S. Bureau of Mines employs almost exclusively an electric ana-

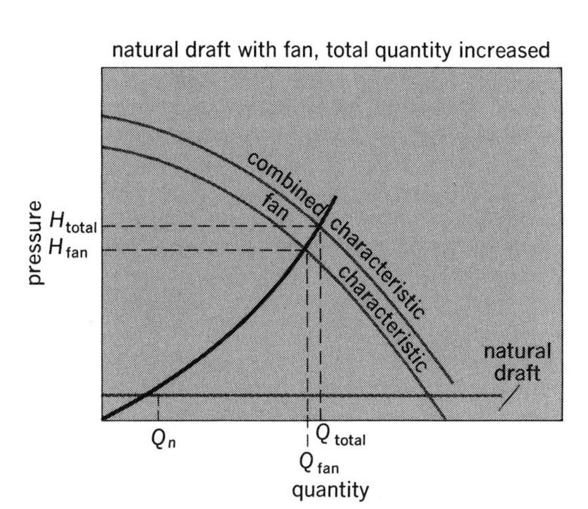

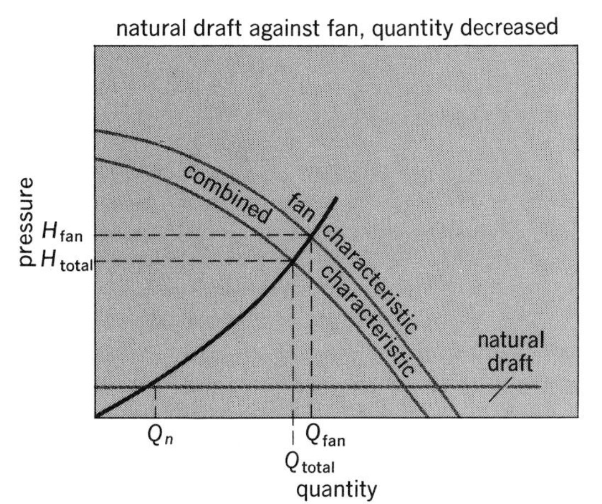

Fig. 5. Graphs of influence of natural draft to mine ventilation.

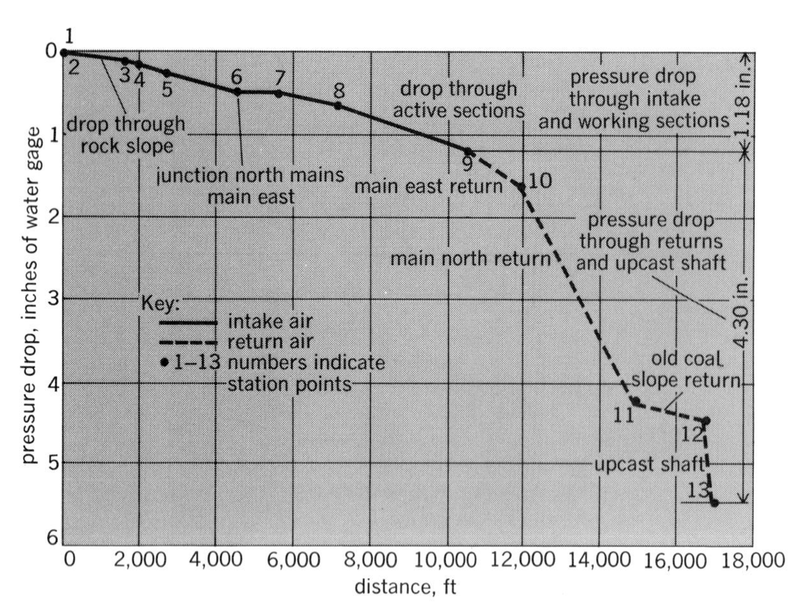

Fig. 6. Ventilation pressure gradient for a small slope mine. 1 ft = 0.305 m; 1 in. = 2.54 cm. (*U.S. Bureau of Mines*)

log computer designed especially for analyzing mine-ventilation distribution problems (Fig. 7). The heart of the analyzer is a nonlinear resistor known as a Fluistor which closely approximates the square-law resistance characteristics of mine airflow; that is, the voltage drop varies approximately as the square of the current.

In developing problem solution the ventilation circuit is divided into segments and the resistance factor for each segment is calculated. The system is reproduced electrically on the analog, using Fluistors of the proper coefficient to simulate airway resistance. Once the analog network is balanced to mine conditions represented by field data, any changes can be readily made and analyzed. The master meter can be connected directly to any segment in the circuit to measure voltage or current. The meter is a digital voltmeter which reads out pressure and airflow in units of inches of water and cubic feet per minute, respectively.

[E. J. HARRIS]

Fig. 7. The fluid network analyzer used to solve mine ventilation distribution problems. (*U.S. Bureau of Mines*)

MINE ILLUMINATION

The lighting of mines is accomplished by use of both movable and stationary lamps.

Mobile illumination. Cap lamps, flashlights, portable hand lamps, trip lamps for haulage cars, and lights in mobile machinery provide mobile lighting. The carbide lamp has been outmoded by the safer and more efficient electric cap lamp operated by a 4- or 5-volt battery. Such a lamp, using a polished reflector, can produce a beam candlepower of 1000 candles in a small spot. Less light with greater spread is obtained with a matte reflector.

Permissible continuous mining machines, shuttle cars, and other mining production equipment are fitted with explosion-proof 150-watt, 120-volt lights, front and rear. All permissible lights for use in gassy or dusty mines in the United States are approved by the U.S. Bureau of Mines.

Stationary or area illumination. This type of lighting is provided by incandescent lamps or dual 15-in. fluorescent lights at intervals of 20–90 ft. Incandescent types, 50–150 watts, dc-operated, may be connected singly or in groups in ventilated haulageways but not in gassy areas. Fluorescent lighting, first approved in 1957 for gassy or dusty mines, employs 14-watt lamps on ac power (Fig. 8).

Fluorescent lights have proved very effective in gassy areas, along beltways and shuttle-car roadways, and have been credited with reduction in accidents, better morale among the men, and better production records. Three-wire conventional grounding or two-wire isolated ungrounded systems provide shock protection. Special connectors facilitate the disconnection of lights without arcing. The U.S. Bureau of Mines approved a flashlamp for underground photographic use in 1957.

[ROBERT S. JAMES]

MINE DRAINAGE

It is necessary to prevent water infiltration into mines from the surface or from sources underground. Control, transportation, and ultimate disposal of such waters constitute mine drainage.

Mine drainage varies in facilities and importance, at different localities, according to conditions of the mine water. There must be a careful study of its occurrence, corrosive and erosive character, volume to be handled, and the handling and disposal facilities that can economically provide efficient drainage. Major facilities consist of flumes, dikes, and storage ponds to prevent or minimize seepage of surface waters, underground ditches and sumps to collect, store, and control mine water, and pumping plants and drainage tunnels for the disposal of the mine water.

Control of stream pollution by mine drainage is important for both esthetic and economic reasons.

Sources and types of mine waters. Physical characteristics of the individual mining areas create distinctive sources and types of mine waters. Water enters directly into open-pit or underground workings, seeps through pervious strata, drains from normal water channels both on the surface and underground, and infiltrates from impounded waters into mine workings through cracked, crushed, and broken formations.

Mine waters vary widely in character in different mines because of the different geological forma-

Fig. 8. An installation of permissible fluorescent lights in a coal mine. (*U.S. Bureau of Mines*)

tions. Waters from mines may be classed as alkaline (pH 7–14), neutral (pH 7), and acidic (pH 0–7). They may carry abrasive matter in suspension and be erosive to a degree dependent on the amount of abrasive substances. Although ground waters are normally alkaline, analyses from the majority of mines show considerable variation because the waters contain varying quantities of dissolved sulfate salts. Acidic water formed by such salts in the majority of mines has resulted in tremendous losses by corrosion, especially to pipes and pumping equipment.

Sumps, drains, and tunnels. The infiltration of water into mines, the volume of water to be handled, and pumping facilities vary with the seasons. The storage of mine water until it can be disposed of satisfactorily is of great importance. Ample sumpage where mine water can be stored is necessary in any drainage system. Sumps should be provided near pumping stations, have ample capacity, and be arranged so as to permit easy cleaning. If practicable, they should be designed to provide gravity feed to the pumps.

Drains, flumes, or ditches. These are used on the surface and underground to divert and convey water and thereby to prevent stream pollution and inflow through crop caves, strippings, cracks and fissures, and stream beds. Large flumes are used for those purposes in subsided areas, particularly where flash floods or quick runoff occur. Because ditches are usually made in mine haulage roads, they should be designed to carry the maximum drainage.

Drainage tunnels between different working levels are preferred means of handling and keeping mine water from reaching lower mine workings. Drainage tunnels from a mine or a group of mines to some disposal point situated favorably with respect to surface disposal areas are a means of draining mines economically and saving reserves of minerals that would be lost otherwise. They are advantageous over a long economic life of a mine for handling large quantities of water that sometimes could not be pumped to the surface. Although the initial cost is high, their upkeep is comparatively low.

Pumping. As long as mines are operated, mine drainage will most likely be done in whole or in part by pumps. Because mine water is usually acidic, suitable metals or alloys should be used for mine pumping equipment. The designing engineer should consult with qualified technicians before attempting to solve corrosion problems. Centrifugal pumps are both horizontal and vertical and known as standard, deepwell, and shaft. Displacement pumps, piston and plunger, are being discarded. Mine tailing pumping is sometimes done by means of centrifugal sand pumps. Deep-well pumps in a shaft or bore hole have proved successful for emergency use and for unwatering abandoned mines. Pumping systems and controls must be designed to handle the maximum inflow of water in mine workings. [SIMON H. ASH]

MINING SAFETY

The prevention of mine worker injury by precautionary practices in mining operations. This article discusses records of mine safety, methods of fire prevention, government regulations, training programs for mine workers, and mine environment problems causing disease.

Safety records. Vigilance is the price of safety in a mine, as it is in any other activity. Fortunately, the presentation of mining hazards in motion pictures and television bears little relation to fact. The Mine Safety Appliances Co. in the late 1930s originated the John T. Ryan trophies, to be awarded annually to the Canadian mines with the best safety records each year. These awards include two national trophies, one for metalliferous mining and one for coal mining, and six regional trophies, two for coal mining and four for metalliferous mining. The awards are based on the number of lost-time accidents during 1,000,000 worker-hours of exposure for metalliferous mines, and for 120,000 worker-hours of exposure in coal mines.

A lost-time accident is defined as one that results in a compensation payment for a total or partial disability that lasts more than 3 days. In Canada, before Aug. 1, 1968, a 3-day waiting period was required in some provinces before an injury became compensable. The records of the workmen's compensation boards are used to compile safety records, and all locations of employment are included, such as mine, mill, offices, and shops, but smelters and open-pit or strip-mine operations are not included.

The Canada John T. Ryan trophy for metalliferous mines is usually won by a mine with no lost-time accidents as defined. A mine with four or five accidents is likely to be disqualified from the countrywide competition. A fatal accident eliminates a mine. The coal mining record is not as good as that of the metalliferous mines; for one thing, there are fewer coal mines. The low incidence of accidents is not attained without effort. Supervisors, like traffic police, spend a lot of their time trying to keep people from acting foolishly. In North America governments have regulations regarding the safety of workers in mines and inspectors to see that they are observed.

Fire hazards. Government regulations for mine operation and the safety of workers in mines were established in North America during the 19th century, patterned on those of Europe and Great Britain, where there was a longer history of coal mining than in North America. The regulations were first directed to coal mining because it presents the hazard of combustible gas from bituminous coal and flammable dust. Coal is made up of volatile material and fixed carbon. If the ratio

$$\frac{\text{volatile}}{\text{volatile} + \text{fixed carbon}}$$

exceeds 0.12, the dust can burn explosively when mixed with air in dangerous proportions, and the range of dangerous proportions is quite wide. All bituminous coals have a ratio in excess of 0.12. Combustible gas is especially dangerous because it is mobile and if it is ignited the ensuing fire or explosion may affect men remote from the source of trouble. Other accidents, a fall of rock, for instance, do not affect individuals working outside the location of the accident.

Methane gas, the major component of the combustible gas, is dangerous in two ways. It dilutes the mine air and so reduces the amount of life-sustaining oxygen. Coal mines are classified as gassy, slightly gassy, and nongassy. A nongassy mine has less than 0.05% methane in the air. If methane is in a concentration of 5–13.9%, it may be ignited and will burn explosively. The most violent explosive mixture is 9.4% methane. At some point, as the location of the emission is approached, the explosive ratio will be reached; a single spark can set off an explosion. The resultant turbulence in the air will stir up the coal dust and the fire can spread with explosive violence.

A mine fire has side effects. Workers may be caught in the actual fire, but the burning produces carbon monoxide and carbon dioxide which will spread far beyond the fire area. If the roof is sulfurous shale, it may be brought down by the heat.

Prevention of ignition. The propagation of a fire in the dust stirred up by the air turbulence accompanying an explosion may be inhibited by dusting. An incombustible rock dust is spread throughout the workings. When stirred up, it dilutes the mine dust and absorbs heat, so combustion cannot be maintained. The initiation of combustion must be prevented.

Equipment. In the United States only equipment certified by the U.S.B.M. (United States Bureau of Mines) as permissible may be used where gas or combustible dust may be present. The restriction applies to electrical equipment, including cap lamps and machinery. The certification is based on the no-sparking qualities of switches and adequate current-carrying capacities for electrical equipment; and on ensurance that machinery has no operating parts that can heat to the ignition temperature for gas or dust. In Canada a British certification is also accepted.

Explosives. Only permitted explosives may be used. All explosives involve combustion and high temperatures. The U.S.B.M. has certified certain explosives as permissible for use in coal mines. They are compounded to produce a lower temperature and volume of flame than ordinary explosives, and certain salts are included in the formulas to quench the flame rapidly. Permitted explosives must be used as prescribed for prescribed conditions. Special attention is required for stemming to prevent blowouts.

Safety lamp. The biggest single advance in coal mine safety was probably the invention of the Davy safety lamp in 1815. A mantle of wire gauze around the flame dissipated the heat from the flame to below the ignition temperature of methane, which is about 650°C. It provided light and the miner no longer had to work with an open flame. The height and color of the flame gave a measure of the amount of methane in the air. Improved models are still in use, though they are being superseded by no-flame instruments that give a prompt reading or can monitor the methane content continually. The miner is no longer dependent on a flame for light since the electric cap lamp became available.

Ventilation. Coal mine operators have developed ventilation techniques, which until recently, surpassed those used in metal mines. Great care is taken to sweep working places with enough air to prevent dangerous accumulations of gas or of dust. The advent of mechanical coal-cutters and other mechanical equipment has made ventilation more difficult, but the problem has been overcome.

Other mine fires. Methane is not confined to coal mines. It is occasionally released from pockets in metalliferous mines. It is seldom troublesome unless it has accumulated in an unused unventilated working or in a sump at the bottom of a shaft.

Sulfide ores containing certain sulfide minerals may oxidize and generate sulfur dioxide. This may generate enough heat to ignite wood. Pyrrhotite should always be suspected as such an oxidizing sulfide.

Dry, timbered mines in which there is careless smoking are obvious fire hazards. Most metalliferous mines are damp and vigorous fires are not expected, but smoldering fires may be even more hazardous. They may burn for a long time without being detected and the incomplete combustion generates carbon monoxide which is heavier than air, odorless, and lethal. The fire is usually in rubbish accumulated in abandoned places. Good housekeeping is required. Many fires are started in old power cables when the insulation has broken down, and a surprising number start during locomotive battery charging. Sparks from acetylene torch burning are another common source of fire.

Mine rescue teams. The U.S.B.M. officials have accumulated a vast store of fire-fighting knowledge. Most mines have teams trained in mine rescue and in the use of U.S.B.M.-approved equipment in accordance with manuals prepared by the U.S.B.M. officials. They work under their own supervisors but U.S.B.M. officials are available for consultation.

Government regulations. Government regulations for safety in mine operations must be enforced. The higher the government authority and the less localized the enforcement agent, the more effective the enforcement will be. The basic enforcement unit in the United States is the state authority, and even obviously good regulations are often hard to enforce. In addition ot the human tendency to resist change, pressure may be brought to bear on the legislatures of the mining states. Some government inspectors may be appointed for political expediency; in the past the company safety inspector was too often an employee given a sinecure in lieu of a pension.

The U.S.B.M. has been assigned responsibility for the health and safety of United States miners. The period of 1907–1913 was formative. States are jealous of their prerogatives, and so it was 1941 before the U.S.B.M. officers had the right to enter a mine. They were chiefly involved by invitation after a disaster. They attained the right to order a coal mine closed for dangerous practices only in 1952. Legislation in 1969 resulted in a safety code to apply to all mines, coal and metalliferous. The initial responsibility continues to be with the state authority, and the U.S.B.M. will interfere only when the state regulations and enforcement do not meet the requirements of the Federal code.

Regulations in Canada are prepared and enforced by the provinces. The Northwest Territories, Yukon, and the Arctic Islands are not provinces and are under the authority of the federal Canadian government. The Ontario metal mining code is the most comprehensive and has been used as a model for the preparation of the codes in other provinces and countries. There is no coal mining in Ontario. The coal mining provinces based their regulations on British codes originally, and these regulations have since been modified in the light of experience in the United States.

Occurrence of accidents. The following discussion is based on data from the *Ontario Department of Mines Inspection Branch Annual Report for 1967*, which gives accident experience for an average of 17,461 individuals who put in 32,391,000 worker-hours in underground mines and open pits of all sizes and degrees of organization. There were 1887 lost-time accidents. The experience did not vary much from that of the preceding 4- or 5-year period.

The time distribution for lost-time accidents was 1 per 17,100 worker-hours worked or, using 1850 worker-hours per worker-year, 1 per 9.4 worker-years. Whether or not an individual has had an accident during any one year has no bearing on whether the person could have another in the same year or any other year, so this is a Poisson distribution. A person would have a 0.33 probability of having no lost-time accidents in 9.4 years.

Actually, the mathematical probability is misleading because an individual is more apt to have an accident in earlier years of employment before becoming mine-wise, or if employed in a developing mine before a safety program is well organized.

The 16 fatal accidents had an incidence of 1 per 2,020,000 worker-hours worked, or 1 per 1090 worker-years. Mathematically a group of 100 individuals working 10 years would have about 0.37 probability of having no fatal accident. That is subject to the same reservations as those set out above for the lost-time accidents.

Training. The necessity for training people to work safely cannot be overstated. Of the 1887 nonfatal accidents in Ontario in 1967 about 34% (641) were personal accidents—fall of persons, or strains while moving or lifting or handling material other than rock or ore. There is reason to suspect laxity in the use of safety equipment, such as gloves and safety belts and boots; ineffective instruction on how to lift; and insufficient emphasis on the importance of good housekeeping in working places. It must be emphasized that the data are from all sizes of mines and from mines that range from those operated under rather primitive conditions to those using modern equipment and management. Some of them would be close runners-up in the Canada John T. Ryan trophy competition, and one of them won the Regional trophy. Others were barely passing the government inspections. One runner-up for the Canada trophy had 2 accidents and the next had 3 accidents in 1967. The winner of the Ontario Regional trophy had 7 accidents per 1,000,000 worker-hours.

The most lethal class of accident is the fall of rock or ore. About 10% (185) of all accidents, including fatal accidents, in Ontario in 1967, were falls of rock or ore, but 31% (5) of the fatal accidents were from that cause. The real cause was probably the failure to recognize an unsafe condition due to neglect or lack of experience. Either cause requires better training of the supervisors and the workers. Nearly half of the accidents occurred during drilling or scaling of loose rock, clearly showing a lack of skill or the neglect of an unsafe condition.

The advent of mechanized mining in large deposits requires greater areas of exposure to provide room to maneuver equipment. Fortunately, it has been accompanied by the extended use of rock bolts as a means for rock reinforcement, and the incidence of falls of rock has been reduced. Figure 9 shows an experimental stope in a mine where U.S.B.M. engineers tested the effectiveness of rock bolting.

The ultimate in mine safety cannot be achieved without the complete support of top management, but the key person in the achievement is the workman, and in the chain of responsibility it is the supervisor at the lowest level of contact with the worker. Though having the best of intentions, the supervisor is handicapped because workers perform in scattered places, out of the supervisor's sight except for short intervals each day. That condition has improved in mines that work larger deposits with mechanical equipment. There is a concentration of working places and the crews are under less intermittent surveillance.

Most mines have a staff safety engineer responsible only to top management. The safety department usually has no line authority. The most important duty that the safety engineer has is the education of supervisors and workers by lectures and training sessions.

It must be recognized that any accident could be fatal, even a neglected scratch from a nail. Furthermore, an accident need not involve injury to a person. Any undesirable happening at any time or in any place that has not been foreseen is an accident. It may be only by chance that there is no personal injury.

Environmental problems. Instantly recognizable injuries to persons attract more attention, but for ages miners have been subject to pneumoconiosis. This includes all lung diseases, fibrotic or nonfibrotic, caused by breathing in a dusty atmosphere. It takes time to develop, and began to be recognized as something to be eliminated early in the 20th century. Most varieties are not fibrotic and will clear up when the person is out of the dust-laden atmosphere; silicosis and the effects of breathing radioactive dust or gas, however, will not.

Silicosis. South African mine doctors led in the diagnosis of silicosis. About 1920 mine operators in North America became aware of the magnitude of the problem and moved to do something about it. It is caused by inhaling silica dust, and possibly

Fig. 9. Rock bolting in a Michigan copper mine. *(Photograph by H. R. Rice)*

some other mineral dusts. It is not confined to mining. Any industry that produces a silica-dusty atmosphere is dangerous. Furthermore, a clear-looking atmosphere may still be dangerous because the particles that are harmful are less than 5 μm in size. At that size they settle slowly and do not show in a beam of light.

The small particles pass the filtering hairs and mucus in the nose and throat. They reach the lungs and by some action, mechanical or chemical, create fibrosis. The useful volume of the lung is reduced. The disease is seldom lethal in itself but the victim is susceptible to tuberculosis and pneumonia.

As soon as the cause was recognized, the operators and the government authorities increased ventilation requirements and insisted on water sprays, particularly after blasting, to knock down the dust. Wet drilling was well established at that time but there have been some improvements in the drills. The dust content of the mine air at working places and elsewhere was measured, especially in the fine sizes because the presence of coarse dust is easily detected. What are thought to be safe, or acceptable, working levels have been established.

All employees were examined by x-ray for fibrosis in the lungs, and for a proneness to develop silicosis. Some lung shapes are more likely to develop it than others are. Employees with developing fibrosis were given work in dust-free locations, or treatments and pensions. Employees are now examined at least yearly to detect the disease in the primary stage so that prompt action may be taken, and the incidence of silicosis is lessening. Aluminum dust sprayed into the air in mine change houses and underground has been found to have a prophylactic, and possibly therapeutic, effect. Ventilation is the only completely effective remedy.

Radioactivity. It was observed that in some mine areas silicosis seemed to be much more virulent than in others. This was thought to be due to some additive to the silica. It was found that coal dust speeded up the development of the disease.

When uranium mining got under way, it was found that the rocks emitted radon gas which broke down in the lungs and produced lung cancer. When the instruments developed for radioactivity research became available, it was learned that many rocks that had a low level of radioactivity emitted small amounts of radon gas, which in some cases were carried into the mine workings in the mine water and released under the reduced pressure. That provided one reason for the virulence of silicosis in some mining areas.

The government authorities moved quickly and established safe levels of radioactivity and safe exposure time in those levels. The regulations are enforced. The acceptable levels are attained by ventilation.

Other considerations. When diesel engines were introduced underground for motive power, they had to be certified for an acceptable level of carbon monoxide and oxides of nitrogen in the fumes. The required levels were attained by engine design and by scrubbers in the exhaust system. An engine must have a U.S.B.M. certificate of approval, and then it is only approved to travel certain routes for which the government inspectors consider the ventilation to be adequate. [A. V. CORLETT]

Bibliography: *Eng. Mining J.*, Centennial Issue, June, 1966; D. Frasche, *Mineral Resources*, NAS-NRC Publ. no. 1000, pt. C, 1962; I. A. Given et al., *SME Mining Engineering Handbook*, 1973; M. K.

Hubbert, *Energy Resources*, NAS-NRC Publ. no. 1000, pt. D, 1962; R. B. Lewis and G. B. Clark, *Elements of Mining*, 3d ed., 1964; T. A. Rickard, *Man and Metals*, 1932, reprint 1974; G. Robson, *Economics of Mineral Engineering*, 1977; Society of Mining Engineers, *Surface Mining*, 1968; U.S. Bureau of Mines, *Mineral Facts and Problems*, Bull. no. 667, 1975; W. A. Vogely et al., *Economics of the Mineral Industries*, 3d ed., 1976.

Mutation

Any alteration capable of being replicated in the genetic material of an organism. When the alteration is in the nucleotide sequence of a single gene, it is referred to as gene, or point, mutation; when it involves the structures or number of the chromosomes it is referred to as chromosome mutation, or rearrangement.

Mutations may be recognizable by their effects on the phenotype of the organism (mutant). In *Drosophila melanogaster*, for example, changes in wing size, body color, and eye color are a few of the phenotypic changes traceable to mutational changes within genes. The altered gene is inherited in the same way as the original unmutated gene.

GENE MUTATIONS

Two classes of gene mutations are recognized: point mutations and intragenic deletions.

Point mutations. Two different types of point mutation have been described. In the first of these, one nucleic acid base is substituted for another. The change may involve replacement of one purine by another purine, or a pyrimidine by another pyrimidine; it is known as a transition. Alternatively, a purine may be replaced by a pyrimidine, or vice versa. This is a transversion. The second type of change results from the insertion of a base into, or its deletion from, the polynucleotide sequence. These mutations are all called sign mutations or frame-shift mutations because of their effect on the translation of the information of the gene.

Intragenic deletions. In addition to the two types of point mutation, which involve changes in one or two bases at the most, more extensive deletions can occur within the gene. These are sometimes difficult to distinguish from mutants which involve only one or two bases. In the most extreme case, all the informational material of the gene is lost.

A single-base alteration, whether a transition or a transversion, affects only the codon or triplet in which it occurs. Because of code redundancy, the altered triplet may still insert the same amino acid as before into the polypeptide chain, which in many cases is the product specified by the gene. Such DNA changes pass undetected. However, many base substitutions do lead to the insertion of a different amino acid, and the effect of this on the function of the gene product depends upon the amino acid and its importance in controlling the folding and shape of the enzyme molecule. Some substitutions have little or no effect, while others destroy the function of the molecule completely.

Single-base substitutions may sometimes lead not to a triplet which codes for a different amino acid but to the creation of a chain termination signal. Premature termination of translation at this point will lead to an incomplete and generally inactive polypeptide. Although many mutations involve loss or diminution of enzyme activity, the gene products involved may also be structural proteins, such as components of ribosomes, and other functions may be affected, such as regulatory ones.

Sign mutations (adding or subtracting one or two bases to the nucleic acid base sequence of the gene) have a uniformly drastic effect on gene function. Because the bases of each triplet encode the information for each amino acid in the polypeptide product, and because they are read in sequence from one end of the gene to the other without any punctuation between triplets, insertion of an extra base or two bases will lead to translation out of register of the whole sequence distal to the insertion or deletion point. The polypeptide formed is at best drastically modified and usually fails to function at all. This sometimes is hard to distinguish from the effects of intragenic deletions. However, whereas extensive intragenic deletions cannot revert, the deletion of a single base can be compensated for by the insertion of another base at, or near, the site of the original change.

CHROMOSOMAL CHANGES

Some chromosomal changes involve alterations in the quantity of genetic material in the cell nuclei, while others simply lead to the rearrangement of chromosomal material without altering its total amount.

Polyploidy and aneuploidy. The presence of three or more full sets of chromosomes is referred to as polyploidy; that is, each chromosome homolog is represented at least three times.

The term "aneuploidy" describes all situations in which whole chromosomes are lost or gained. The loss of a single chromosome from a diploid complement gives rise to monosomy; that is, one chromosome is represented only once. The presence of an extra chromosome homologous with one of the chromosome pairs constitutes trisomy. Such changes are often deleterious, particularly in animals, although in plants they appear to be tolerated more readily. On occasion, aneuploidy has been exploited for genetic analysis.

Chromosome breakage. When a chromosome is broken, the broken ends may fail to rejoin. When a cell with a broken chromosome divides, acentric fragments of chromosomes are lost and the daughter cells are often nonviable. When the broken ends rejoin, they may either restitute, reforming the original chromosome; or, if breaks exist simultaneously elsewhere in the same nucleus, rearrangements may also be formed.

Deficiencies and duplications. The loss or repetition of a chromosomal segment may be represented diagrammatically as shown in Fig. 1.

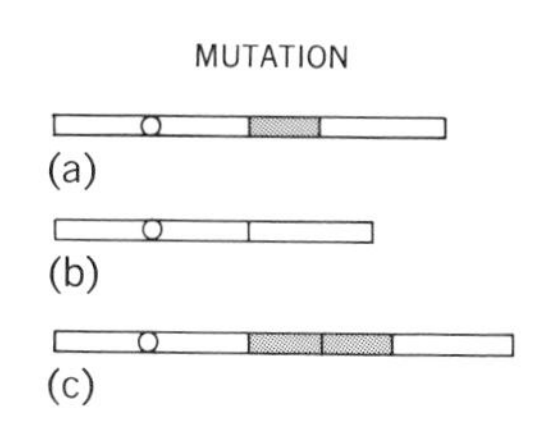

Fig. 1. Chromosome aberrations which involve segments. (*a*) Normal. (*b*) Deficiency. (*c*) Duplication.

Rearrangements. There are two main types of rearrangement: translocations and inversions. Translocations are sometimes reciprocal; nonhomologous chromosomes become broken simultaneously and switch broken ends in the process of rejoining (Fig. 2*a*). Inversions occur when two breaks take place in the same chromosome, and the chromosomal fragment between them is rotated through 180° before the ends rejoin (Fig. 2*b*).

In general, duplications are tolerated more readily than deletions, which are virtually always lethal when homozygous. Deletions may also have

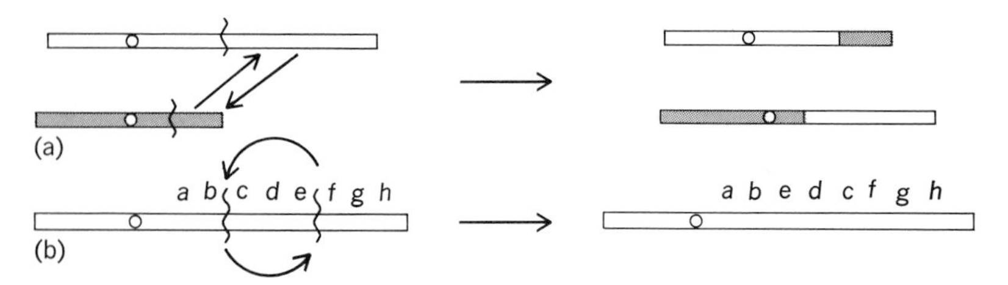

Fig. 2. Chromosome rearrangement mutations. (*a*) Translocation. (*b*) Inversion.

profound heterozygous effects since they unmask any recessive alleles in the corresponding undeleted part of the homologous chromosome.

With the exception of the "position effect" described below, the effect of translocation and inversion on gene function is, generally speaking, minimal. Linkage relationships between genes will be changed, and when meiosis takes place in a diploid cell which is heterozygous for a translocation, progeny are produced which have unbalanced genomes and are therefore nonviable. Inversions also have a profound effect on the viability of the meiotic products. When an exchange or crossing-over event takes place within a heterozygous inversion, dicentric and acentric products result as well as duplications and deficiencies which lead to loss in viability of the daughter cells carrying them.

Position effects. In some cases a change in the relative position of genes may affect the phenotype. When this occurs, there is said to be a "position effect." An example is found in the fruit fly *D. melanogaster.* As a result of an inversion, the nonmutant eye-color gene w^+ is moved from its normal position in an euchromatic region at the tip of the X chromosome to the heterochromatic region near the centromere. Although it can be shown that no change has taken place in the gene itself, its function is changed. The fly has mosaic eyes with patches of red and white tissues. It is hard to distinguish this effect from true mutation in some cases.

ORIGINS OF MUTATIONS

Mutations can be induced by various physical and chemical agents or can occur spontaneously without any artificial treatment with known mutagenic agents.

Spontaneous mutation. Until the discovery of x-rays as mutagens, all the mutants studied were spontaneous in origin; that is, they were obtained without the deliberate application of any mutagen. Spontaneous mutations occur unpredictably, and among the possible factors responsible for them are tautomeric changes occurring in the DNA bases which alter their pairing characteristics, ionizing radiation from various natural sources, naturally occurring chemical mutagens, and errors in the action of the DNA-polymerizing and correcting enzymes. In Table 1, a few examples are given of spontaneously occurring mutations together with their approximate frequencies. Reversions often have a lower incidence than forward mutations, presumably because fewer possibilities exist for correcting a damaged gene than for damaging a functional one.

Spontaneous chromosomal aberrations. These are also found infrequently. One way in which deficiencies and duplications may be generated is by way of the breakage-fusion-bridge cycle: During a cell division one divided chromosome suffers a break near its tip, and the sticky ends of the daughter chromatids fuse. When the centromere divides and the halves begin to move to opposite poles, a chromosome bridge is formed, and breakage may occur again along this strand (Fig. 3). Since new broken ends are produced, this sequence of events can be repeated. Unequal crossing over is sometimes cited as a source of duplications and deficiencies, but it is probably less important than often suggested.

Radiation-induced mutations. In the absence of mutagenic treatment, mutations are very rare. In 1927 H. J. Muller discovered that x-rays significantly increased the frequency of mutation in *Drosophila.* Subsequently, other forms of ionizing radiation, for example, gamma rays, beta particles, fast and thermal neutrons, and alpha particles, were also found to be effective. By employing the specific-locus mutation test to analyze the effects of x-rays, it has been estimated that the average mutation frequencies per locus in spermatogonia of *Drosophila* and in mice are 1.5×10^{-8} and 25×10^{-8} per roentgen, respectively. Since these observations were made, similar data have been collected for a range of organisms from bacteriophage to higher plants, and there is a positive correlation between the mutation frequency per locus per rad and the DNA content of the nucleus. This could have practical importance in assessing hazards for humans from ionizing radiation.

Induced gene mutation frequencies increase in proportion to dose: Extrapolation of the curve of mutation frequency versus dose back to zero dose gives zero mutations. There is no reason to believe that there is any dose of ionizing radiation which is too low to produce a mutagenic effect proportional to its magnitude. However, in mice spermatogonia and oocytes, there is evidence to suggest that doses administered at extremely low rates are less effective than those administered at higher rates. This may result from the operation of cellular repair mechanisms.

The relationship between radiation dose and chromosome aberration frequencies has been determined in the case of, say, *Tradescantia.* Individual chromosome breaks increase in direct proportion to dose, in the same way as point mutations. However, the increase in chromosomal aberrations

Table 1. Spontaneous mutation frequencies

Organism	Mutation	Incidence, 10^6 genes
Human	Huntington's chorea	1
	Retinoblastoma	4
	Muscular dystrophy	8
	Hemophilia (combined)	32
Maize	R (anthocyanin production)	490
	I (colorless aleurone)	110
	Pr (purple aleurone)	11
	Sh (shrunken endosperm)	1
Neurospora	Adenine reversion (allele no. 38701)	0.06
	Inositol reversion (allele no. 37401)	0.01
Bacteriophage T4	r101 (rapid lysis) reversion	0.3
	r51 (reversion)	11

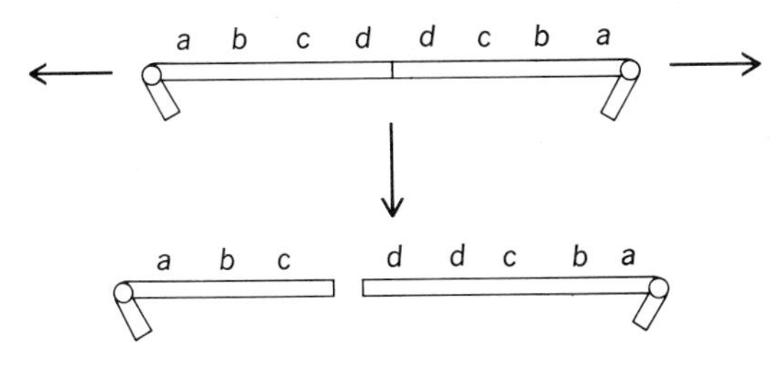

Fig. 3. Breakage-fusion-bridge cycle showing formation of mutant chromosomes following fusion of daughter chromatids prior to cell division.

is approximately proportional to the square of the dose. This is to be expected since each aberration requires two independent events to take place before it can be formed. Both intensity of treatment and fractionation have an effect on the yield of aberrations from a given dose of ionizing radiation, because breaks must be produced together temporally as well as spatially for rearrangements to occur. High intensities of treatment without fractionation achieve this more successfully than low intensities or fractionated exposures. The proportionality between the frequency of point mutations and chromosome breaks and the dose of ionizing radiation may be explained if each results from a single ionization event (target theory).

Ultraviolet light is also an effective mutagen. The wavelength most employed experimentally is 2537 A (253.7 nm), which corresponds to the peak of absorption of nucleic acids. Ultraviolet radiation produces excitations in the material which absorbs it. One of the first indications of the genetic role of DNA came from experiments which showed a close correspondence between the action spectrum for mutation induction and the ultraviolet absorption spectrum of nucleic acids. It is known that an important ultraviolet effect on DNA is the production of pyrimidine dimers, formed by covalent bonding of two adjacent pyrimidine bases on the DNA chain.

Modification of radiation damage. At least part of the mutagenic effect of radiation is direct. Evidence for this is obtained from experiments in which ancillary, nonmutagenic treatments are found to influence the yield of mutations from a given dose of radiation. For example, the concentration of oxygen is an important factor in determining the yield of x-ray-induced mutations, since within limits there is a direct relationship between oxygen tension and the amount of damage from a given dose. Infrared light and temperature also seem to be important, possibly because they alter the sensitivity of the chromosomes to breakage. Oxidative metabolism is necessary for chromosome rejoining to take place. Repair of ultraviolet damage is brought about by an enzyme system that works only in the visible light range, and restores pyrimidine monomers from dimers formed as ultraviolet photoproducts. In bacteria it has been shown that certain enzymes occur which cut out the damaged proportions of one of the DNA strands, while other enzymes then patch up the gap left using the undamaged DNA strand as a template. One other enzyme system operating partly after the replication of dimer-containing DNA is thought to mediate the production of ultraviolet-induced mutations in bacteria. Finally, the sensitivity to radiation is strongly dependent on both the type of cell used and the stage in the cell cycle at which the treatment is applied.

Chemical mutagens. Reports concerning the mutagenic effects of chemicals were first made in the 1930s, but it was not until a decade later that those claims were substantiated. Some of the chemicals which have been found to be effective are the alkylating agents (which appear to attack guanine principally, possibly producing base-pair substitutions by alkylation at the O^6 position, or by ionization of the base which follows N^7 alkylation), base analogs (which are incorporated into the DNA in place of normal bases and promote mispairing at DNA replication), nitrous acid (which deaminates certain of the DNA bases, altering their pairing properties), and peroxides (which form a link between radiation and chemical agents since they are formed in organic media upon irradiation). Some mutagens are shown in Table 2.

Some mutagens are known to have specific effects. Apart from base analogs, which cause base-pair substitutions of the transition type, acridines are believed to act by promoting the addition or deletion of bases in bacteriophage DNA. It is thus sometimes possible to classify a mutant as a base-pair substitution or a frame shift based on ability to revert with either base analogs or acridines.

There is strong correlation between the mutagenic activity of certain classes of mutagen, notably alkylating agents, and their carcinogenic potential. Recently the correlation between these two activities has been shown to be far wider than was first appreciated since it was not realized formerly that many carcinogens act only after conversion into an active form within the mammal. Many of these active carcinogenic metabolites have also been shown to be mutagenic. Mutagens are often carcinostatic agents also. The chromosome breaks which they produce are lethal to the rapidly dividing tumor cells.

TECHNIQUES FOR STUDYING MUTATION

While chromosome mutations may be studied directly by cytological means in suitable material, their effects, as well as the effects of gene mutation, are studied by genetic tests.

Sex-linked recessive lethal tests. Recessive lethal mutations may be induced in the X chromo-

Table 2. Some known chemical mutagens and type of effect

Mutagen	Effective in producing	
	Mutations	Chromosomal aberrations
Urethane	+	+
Diepoxide	?	+
Maleic hydrazide	?	+
Diepoxybutane	+	+
Triazine	+	+
Formaldehyde	+	+
Mustard gas	+	+
Nitrogen mustard	+	+
Hydrogen peroxide and organic peroxides	+	+
Caffeine	+	+

somes of *Drosophila*. Heterozygous females bearing such a lethal gene do not die, but all male offspring receiving the lethal will die since the gene is not masked. To obtain heterozygous females, males are treated with mutagenic agents and mated to untreated females. Each daughter from this cross bears one treated X chromosome and its untreated homolog. These females are mated individually to untreated males. The progeny of some of these matings will have only one-half the normal proportion of males. Cultures which display this unusual sex ratio originated from a female which bore a sex-linked recessive lethal. Many sex-linked recessive lethals are point mutations, but other changes, notably small deficiencies, also occur. Because so many genes are involved in this test, it is a particularly useful one for testing potential mutagens.

Specific locus test. In diploid organisms, induced mutations are often recessive and can be detected only by special techniques. One of these involves the use of a tester strain which carries several known visible recessive mutants in the homozygous condition. Germ cells from a nonmutant strain are treated to induce recessive mutations and are then used in crosses to the test strain. Any induced recessive mutations which are allelic with those of the test strain will be expressed in the progeny. This technique has been used effectively in the mouse, as well as in *Drosophila* and maize. Deletions as well as point mutations will be obtained using this means.

Dominant lethal mutations. When germ cells of an organism are exposed to a mutagen, fewer progeny may be produced. It has been shown that a proportion of dominant lethals which caused progeny inviability arise from chromosome breaks which have not healed and from nondisjunction. Death during the early developmental stages may result from other nonchromosomal anomalies.

Studies with microorganisms. Cytological evidence cannot be obtained from these organisms, and mutations must be recognized by an alteration in a function of the organism. The advantages of using microorganisms in mutation experiments lie in the large numbers of genetically uniform individuals which are readily available, and in the selective methods which can be used to screen mutants from nonmutants. Furthermore, because many microorganisms have an almost entirely haploid life cycle, the detection of mutants is greatly simplified.

The more frequently used selective techniques are the following.

Mutations to drug resistance. These can be selected readily by plating treated or untreated organisms on a medium containing levels of the drug which kill all but the resistant mutants. A variation of this technique for bacteria is to screen for resistance to a particular bacteriophage.

Reverse mutations. An auxotrophic mutant is one which requires the supply of a specific factor, say an amino acid, for its growth. It is sometimes possible to induce mutations in these mutants which restore the nonrequiring, prototrophic phenotype. Treated cells from the auxotroph are plated on a medium which lacks the specific requirement, and only mutants which have no need of the growth factor will grow. Although these mutants are referred to as revertants, it is necessary to test them genetically to be sure that they are not the result of further mutations elsewhere in the genome which compensate for the original mutation without reversing it; such mutants are called suppressor mutations.

Microorganisms which have been used extensively in mutational research are the fungi *Neurospora*, yeast, and *Aspergillus* and the bacteria *Escherichia* and *Salmonella*. Recently, methods have been developed which enable the products of the drug-metabolizing systems of mammals to be tested with microorganisms in liquid medium and on solid plates.

Mutation studies in humans. These include such indirect methods as sex ratio studies and pedigree analysis. Sex-linked recessive and dominant lethals affect the sex ratio. For example, if a man is exposed to radiation, any sex-linked dominant lethals induced in his X-bearing sperm will cause the death of his female offspring. Thus, on the average, more male than female offspring will be produced. Conversely, if a woman is exposed, dominant lethals will be distributed equally between her sons and daughters, but recessive lethals will be uncovered in her sons only, resulting on the average in more daughters than sons. Aside from sex ratio studies, mutation studies in humans are mainly limited to techniques of pedigree analysis, or to indirect techniques involving the analysis of gene frequencies and population structure. Simpler tests of mutation induction are becoming possible as suitable genetic techniques are developed in tissue culture systems and by screening for electrophoretic variants of blood proteins.

Cytological tests. It is possible to observe chromosomal aberrations following mutagenic treatment of suitable material. A few organisms which have proved useful are listed below.

The spiderwort (*Tradescantia*) is a higher plant which has been used extensively in studies of induction of chromosomal aberrations because the chromosomes are large and there are only six homologous pairs. Treatment is made by immersing the inflorescence in a solution of the mutagenic agent and then, after a predetermined period, fixing the material. The chromosome mutations are scored in the metaphase of the microspore division following meiosis.

The broad bean (*Vicia faba*) is also very favorable material for this type of study because the cells of the root tip have very large, easily visible chromosomes. The roots are treated by immersion in a mutagenic solution.

The fruit fly (*D. melanogaster*) is particularly useful since it allows the researcher to combine sophisticated genetic techniques with cytological study. The salivary glands of this organism contain massive polytene chromosomes which undergo precise pairing, making inversions and other aberrations easily observed. It has been possible in some cases to correlate genetical data with cytological observations.

Mammalian cytological techniques have been greatly facilitated by the discovery that certain fluorescent dyes bind to particular regions of chromosomes which can then be made to fluoresce. The method has provided additional morphological markers with which to identify and to trace specific chromosomes. This is particularly important in human cytology.

THE SIGNIFICANCE OF MUTATIONS

Mutations are the source of genetic variability, upon which natural selection has worked to produce organisms adapted to their present environments. It is likely, therefore, that most new mutations will now be disadvantageous, reducing the degree of adaptation. Harmful mutations will be eliminated after being made homozygous or because the heterozygous effects reduce the fitness of carriers. This may take some generations, depending on the severity of their effects. Chromosome alterations may also have great significance in evolutionary advance. Duplications are, for example, believed to permit the accumulation of new mutational changes, some of which may prove to be useful at a later stage in an altered environment.

Rarely, mutations may occur which are beneficial: Drug yields may be enhanced in microorganisms, the characteristics of cereals can be improved. However, for the few mutations which are beneficial, many deleterious mutations must be discarded. Because of the chemical nature of the DNA, it seems very unlikely that chemicals will be found which single out specific genes for action. However, accumulating evidence suggests that the metabolic conditions in the treated cell and the specific activities of repair enzymes may sometimes promote the expression of some types of mutation rather than others. This possibility has yet to be exploited in practical mutant production.

[BRIAN J. KILBEY]

Bibliography: C. Auerbach, *Mutation Research*, 1976; C. Auerbach and B. J. Kilbey, Mutation in eukaryotes, *Annu. Rev. Genet.*, vol. 5, 1971; J. W. Drake, *Molecular Basis of Mutation*, 1970; D. N. Walcher, *Mutations: Biology and Society*, 1978.

Nitrogen cycle

The collective term given to the natural biological and chemical processes through which inorganic and organic nitrogen are interconverted. It includes the processes of ammonification, ammonia assimilation, nitrification, nitrate assimilation, nitrogen fixation, and denitrification.

Nitrogen exists in nature in several inorganic compounds, namely N_2, N_2O, NH_3, NO_2^-, and NO_3^-, and in several organic compounds such as amino acids, nucleotides, amino sugars, and vitamins. In the biosphere, biological and chemical reactions continually occur in which these nitrogenous compounds are converted from one form to another. These interconversions are of great importance in maintaining soil fertility and in preventing pollution of soil and water.

Nitrogen reserves exist in five major sinks: the primary rocks, the sedimentary rocks, the deep-sea sediment, the atmosphere, and the soil-water pool. Although primary rocks contain as much as 97.8% of the Earth's total (1.9×10^{17} metric tons), their N_2 contributes little to the nitrogen cycle. Of the remaining nitrogen, 2% is in the atmosphere as N_2, and about 0.2% is in sedimentary rocks. The biosphere, consisting of the soil-water pool, contains only a small portion of the Earth's total nitrogen (2.4×10^{13} MT), and even here the predominant species (2.2×10^{13} MT) is N_2 dissolved in the sea. In spite of this, it is in this soil-water pool that the major reactions of the nitrogen cycle occur.

An outline showing the general interconversions of nitrogenous compounds in the soil-water pool is presented in Fig. 1. The reactions are much more complex than in the outline, and biological agents have evolved intricate ways to manipulate these nitrogenous compounds for their own use. There are three primary reasons why organisms metabolize nitrogen compounds: (1) to use them as a nitrogen source, which means first converting them to NH_3, (2) to use certain nitrogen compounds as an energy source such as in the oxidation of NH_3 to NO_2^- and of NO_2^- to NO_3^-, and (3) to use certain nitrogen compounds (NO_3^-) as terminal electron acceptors under conditions where oxygen is either absent or in limited supply. The reactions and products involved in these three metabolically different pathways collectively make up the nitrogen cycle and are discussed below.

Nitrogen compounds as nutrients. The syntheses of organic nitrogen compounds from inorganic nitrogen and carbon compounds begins with NH_3 incorporation. One major reaction, catalyzed by glutamic acid dehydrogenase, involves 2-ketoglutarate, NADH (reduced nicotinamide adenine dinucleotide) and ammonia, and the product is glutamic acid. There are two ways in which organisms obtain ammonia. One is to use nitrogen already in a form easily metabolized to ammonia. Thus, nonviable plant, animal, and microbial residues in soil are enzymatically decomposed by a series of hydrolytic and other reactions to yield biosynthetic monomers such as amino acids and other small-molecular-weight nitrogenous compounds. These amino acids, purines, and pyrimidines are decomposed further to produce NH_3 which is then used by plants and bacteria for biosynthesis, or these

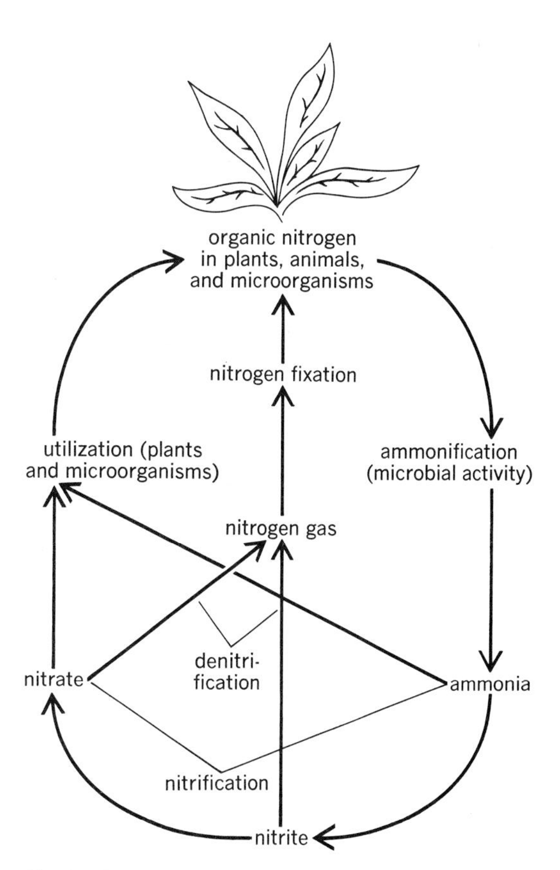

Fig. 1. Diagram of the nitrogen cycle.

biosynthetic monomers can be used directly by some microorganisms. The decomposition process is called ammonification. Not all organic nitrogen is ammonified easily, and resistant nitrogenous residues constitute humus, a complex component of great importance to soil structure and water-holding capacity.

The second way in which inorganic nitrogen is made available to biological agents is by nitrogen fixation (this term is maintained even though N_2 is now called dinitrogen), a process in which N_2 is reduced to NH_3. Since the vast majority of nitrogen is in the form of N_2, nitrogen fixation obviously is essential to life. The N_2-fixing process is confined to prokaryotes (certain photosynthetic and nonphotosynthetic bacteria). The major nitrogen fixers (called diazotrophs) are members of the genus *Rhizobium*, bacteria that are found in root nodules of leguminous plants (Fig. 2), and of the cyanobacteria (originally called blue-green algae). Even though rhizobia have recently been cultured so that they can fix nitrogen in the absence of the plant, the conditions needed for this fixation, such as a very low O_2 concentration supplied by oxyleghemoglobin and the necessary carbon and energy sources supplied by the plant, are ideal in the nodule and are not easily met in the laboratory or elsewhere in nature. There are many "free-living" diazotrophs, and even though they appear to contribute little to soil nitrogen, most knowledge of the biochemistry of nitrogen fixation comes from studies of three of them, *Clostridium pasteurianum*, *Azotobacter vinelandii*, and *Klebsiella pneumoniae*. The nitrogen-fixing system of all these organisms consists of two protein components. One, called the Fe protein, is a dimer of molecular weight about 60,000, and it contains an Fe_4S_4* center (similar to that of ferredoxin) that is involved in accepting and transferring the electrons needed for N_2 reduction. It is the Fe protein that initially binds the magnesium adenosinetriphosphate (MgATP) needed for N_2 reduction. The second component, called the MoFe protein, is a tetramer of about 220,000 daltons and contains four Fe_4S_4* centers and two $Fe_8Mo_1S_6$ centers, with the latter centers probably being the sites where N_2 is reduced to NH_3. For N_2 reduction the MoFe protein accepts electrons from the Fe protein, and to facilitate the transfer, ATP is hydrolyzed to adenosinediphosphate (ADP) and inorganic phosphate. The N_2 fixation process consumes as much as a third of the cell's energy supply, so for obvious reasons the N_2-fixing system is not synthesized if a usable nitrogen compound other than N_2 is available, that is, nitrogenase synthesis is repressed in the presence of NH_3.

Fig. 2. Effect of inoculation with two strains (cultures 175 and 378) of the nitrogen-fixing *Rhizobium leguminosarum* on the development of peas.

Nitrogen compounds as energy source. Many microorganisms can use organic nitrogen compounds as energy sources, but in most cases the nitrogen of the compound is first removed and excreted as NH_3, and then the reduced carbon compound remaining is catabolized to yield both energy and organic carbon intermediates. A relatively few microorganisms (including some fungi) are able to aerobically convert the excreted NH_3 to NO_2^-, and NO_2^- to NO_3^-, and couple these oxidations to the production of ATP and the creation of a membrane potential needed for biosynthetic reactions. Thus, NH_3 added to aerobic soils is rapidly converted by such organisms to NO_3^-. Two chemoautotrophs (organisms that grow using an inorganic compound as an energy source and CO_2 as a carbon source) are primarily responsible for NO_3^- production. First *Nitrosomonas* converts the NH_3 to NO_2^-, and then the relatively toxic NO_2^- is rapidly oxidized to NO_3^- by *Nitrobacter*. Plants and microorganisms readily use this NO_3^- as a nitrogen source by first reducing it to NH_3. The overall process of oxidation of NH_3 to NO_3^- is called nitrification. The process whereby NO_3^- is reduced to NH_3 is called nitrate assimilation.

Nitrogen compounds as electron acceptors. When NO_3^- accumulates in soils in which metabolizable carbon compounds are available and when such soils become anaerobic because growth of aerobic organisms exhausts the O_2, certain organisms such as *Pseudomonas*, *Micrococcus*, *Achromobacter*, and *Bacillus* use the NO_3^- either as a normal electron acceptor or as an electron acceptor in place of O_2. The electron acceptor is needed by such cells to allow electrons from cellular oxidations to flow through the array of electron carriers in the cell membrane. This electron flow is needed to facilitate proton (H^+) transfer across the membrane, which in turn creates a membrane potential and a pH gradient. The potential energy of the pH gradient is used in conjunction with the cells adenosinetriphosphatase (ATPase) to allow ATP to be synthesized. The NO_3^- is the terminal electron acceptor in the electron flow and becomes reduced to NO_2^-, and the NO_2^- in turn is further reduced to N_2 (some N_2O may also be produced). The N_2 (and some N_2O) is released into the atmosphere. This process, called denitrification or nitrate respiration, is responsible for ridding many bodies of water and soil of excess fixed nitrogen that could lead to pollution by overproduction and decay of algae and other bacteria.

N_2 cycle in the oceans. Most information on the N_2 cycle comes from studies of the soil-water system. Much less is known about the nitrogen cycle in seas, even though it is estimated that as much as 20% of the N_2 fixed on Earth occurs in the ocean. This N_2 seems to be primarily fixed by cyanobacteria, although in localized areas some contribution by other photosynthetic and nonphotosynthetic bacteria also occurs. One calculation estimates that a single bloom of the cyanobacterium *Trichodesmium* could fix as much as 1000 MT of N_2 per day. The other processes associated with the nitrogen cycle, ammonification, nitrification, and denitrification, although obviously present in oceans, have been studied even less. Heterotrophic organisms were not shown to be responsible for nitrification in seas, and therefore chemoautotrophic marine nitrifiers, such as *Nitrospira* and *Nitrococcus*, appear to be the major nitrifiers in oceans. The denitrification step of the ocean's nitrogen cycle may account for a third of the Earth's total denitrification, but there is relatively little information on this. One suspects that most of this denitrification takes place in the anaerobic conditions of marine mud.

Summary. The major microbial reactions discussed above are the three major contributors to the N_2 cycle. First N_2 is fixed (reduced) to ammonia, and the ammonia is used by N_2-fixing microorganisms and by plants harboring N_2-fixing microorganisms. Then these plants and bacteria, and animals and plants living off these plants and bacteria, die and lyse, and the nitrogen of their nitrogenous compounds is converted to ammonia by ammonification. The released ammonia is converted rapidly by nitrifying bacteria to NO_3^-, or it is used directly by microorganisms or plants. The NO_3^- produced by nitrification is used by plants and by bacteria as a nitrogen source; or if anaerobic conditions are created, the NO_3^- is denitrified (reduced) to N_2 and N_2O. This completes the nitrogen cycle. *See* SOIL MICROORGANISMS.

[LEONARD E. MORTENSON]

Bibliography: R. C. Burns, and R. W. F. Hardy, *Nitrogen Fixation in Higher Plants*, in A. Kleinzeller, G. F. Springer, and H. G. Wittmann (eds.), *Molecular Biology Biochemistry and Biophysics*, vol. 21, 1975.

Noise

Unwanted sound. This definition of acoustic noise, while purely general, implies that some criterion must exist before the sound can be termed unwanted. Whether a sound is noise, insofar as humans are concerned, is a subjective matter.

Criteria. Considerable effort has been expended to analyze unwanted sounds in an effort to specify objective criteria for the subjective human reactions to noise. These criteria have included annoyance, interference with speech, damage to hearing, and reduction in efficiency of work performance.

Noise is usually thought of in terms of its effect on humans; an equally important aspect, however, is its effect on the fatigue or malfunction of physical structures and equipment. In these instances criteria can in theory be established on a completely objective basis. The "unwanted" aspect in the definition of noise applies here to the fact that it is generally considered undesirable to have a structure such as an aircraft experience fatigue, or that it is undesirable to have electronic equipment guiding the aircraft fail because of malfunctions brought on by intense sound waves.

A third major criterion for describing sound as noise arises in conjunction with the perception or detection of a wanted sound in the presence of other sounds which tend to mask it. Thus in sonar the reflected sound from an object being detected is a signal which is wanted, whereas all other sound detected by the system is termed noise.

Physical specifications. The generality of the preceding definitions gives no clue to the physical specifications of sound waves called noise. The sound can be composed of definite pure-tone, or sine-wave components, or it can be a completely random phenomenon made up of an infinite number of components, each having purely random amplitude and phase characteristics. Automobile exhaust noise, for example, contains pure-tone components, related to the engine rotational speed, whereas an air jet hiss is a random noise.

The physical specification of such noises is given by their radiated intensity, frequency, and spatial distribution.

Random noises are usually described in terms of statistical values rather than in terms of the discrete variable which is used for single-frequency sounds.

The statistical description of random sound waves parallels that used for electrical noise. The magnitude of the noise is usually specified in terms of its radiated intensity in a 1-Hz frequency band, also known as the intensity spectrum level. If the random noise has a relatively uniform distribution of intensity as a function of frequency, it is often described as intensity in a frequency band more than 1 Hz wide. This may be done in terms of a constant bandwidth, such as 5, 50, or 500 Hz, or in terms of a constant percentage bandwidth, such as 1/10, 1/3, or 1 octave of frequency. The most common usage in industrial noise control is specification of intensity in octave frequency bands.

For most noise measurements the frequency range of practical interest can be covered by eight-octave frequency bands, the lowest band having a center frequency of 63 Hz, and the highest band a center frequency of 8000 Hz. The overall intensity of a random noise is the sum of the intensities in all the frequency bands by which it is specified.

It is often useful to convert the physical specification of acoustic noise to a psychophysical measure such as loudness or perceived noisiness. These measures are computed from the sound-pressure level values in 1/3- or 1-octave frequency bands.

White noise. Random noise having the same intensity, in a 1-Hz band, at every frequency in the range of interest is called white noise. Although white noise is a fairly common type of electrical noise, it is rarely encountered in acoustic noise. Most random acoustic noises tend to have a definite nonuniform distribution of intensity as a function of frequency.

Ambient noise. The residual noise present at any location of interest is called ambient noise. It is the sum of all noises present. Thus ambient noise in an office could be the result of ventilating systems, distant conversations, office machines, and so forth. Background noise is a term often used to describe the ambient noise when a particular source of sound being studied is not in operation.

[WILLIAM J. GALLOWAY]

Noise pollution

Unwanted sound at levels so high that permanent loss of hearing may occur after long-term exposure. Noise pollution can occur in industrial locations, in homes, in cities, and even in some recreational activities such as hunting. Wherever creative activity goes on, noise can interfere with both thought and communication. Regulation of noise at the workplace, in the home environment, and in urban areas has been initiated by all levels of government since 1969.

Noise can cause progressive loss of hearing in those employees exposed to the relatively high but common levels experienced in industry. In cities noise seldom reaches the high levels found in industry except for a few locations in which most city residents find themselves only for a short period each day, that is, subways. However, it is apparent that noise in the city is an unwelcome component of the environment, at least at the levels commonly experienced within residential buildings, in nonindustrial workplaces, and in recreational areas both indoors and out. One major contributor to noise in the city, as well as in the suburban-rural environments, is jet aircraft overflight. Late in 1969 the Federal Aviation Administration, authorized by Public Law 90-411, issued rules under an amendment to the Federal Aviation Act that regulated the noise levels of future subsonic aircraft which would be certificated for commercial flight. This was the beginning step in a long-range program to reduce aircraft noise exposure in the vicinity of existing and planned airports.

The effects of noise on persons not engaged in industrial tasks, whether at work or in leisure activities, are not clearly understood, although it is well known that distraction plays an important part in human response to noise when carrying out creative tasks. In any case, some specific ranges of noise levels can be defined. Figure 1 shows how to interpret noise levels in terms of voice level and distance between a speaker and a listener. For example, jet aircraft cabin noise is roughly 80 + 2 decibels A-weighted, or 82 dB(A). The term "A-weighted" means that the microphone and amplifying circuits of the sound-level meter are less sensitive to low-frequency noise compared to mid- and high-frequency noise. At 80 dB(A) at their expected (raised) voice level, seatmates can converse at 2 ft (60 cm) and by moving closer can lower their voices to normal level and talk at 1 ft (30 cm).

Interference with creative activity. Noise interferes with creative activity through distraction. Although it may not readily interfere with sleep, it can prevent onset of sleep, especially in sensitized individuals. At moderate noise levels, noise can interfere with speech communication, the audition of the sound portion of television programs, music listening, and audio-visual instruction. Thus, even

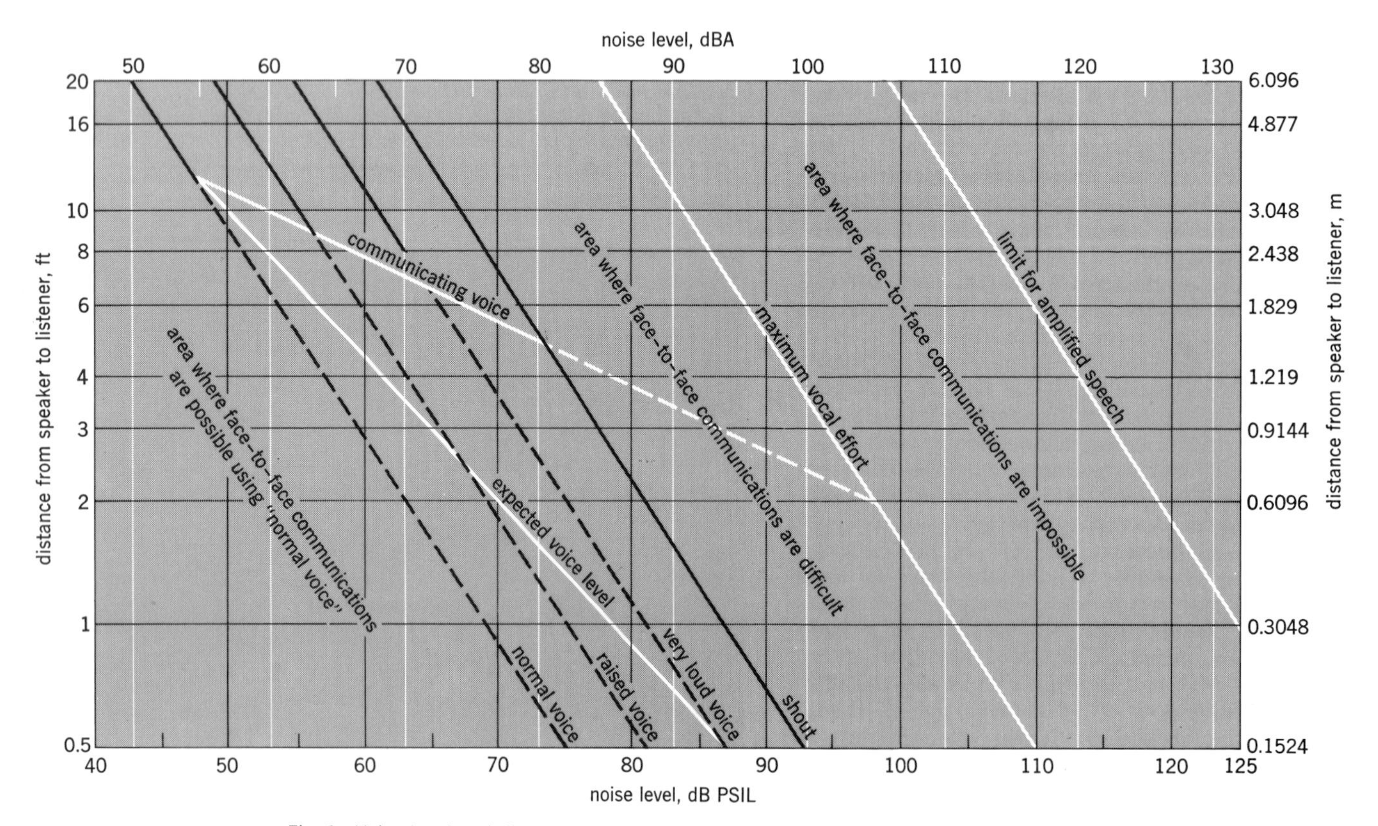

Fig. 1. Voice level and distance between talker and listener for satisfactory face-to-face speech communications as limited by ambient noise level. Along the abscissa are two generally equivalent objective measures of noise level: the average octave-band level in the octaves centered at 500, 1000, and 2000 Hz, called the three-band preferred octave speech-interference level, PSIL, and the A-weighted sound-level meter reading, in dB(A).

at relatively low sound levels, below 50 dB(A), noise may be undesirable. It was also indicated in several applied research efforts that noise may at low levels, in the neighborhood of 35–40 dB, screen people from less desirable sounds. Thus a quiet fountain or nearby ocean "roar" can shield the auditor from the sounds of traffic, children, and distant aircraft that otherwise might be distracting. The character of this "masking" sound must be bland and practically nonvarying or, if varying, must vary slowly and contain no threatening or unpleasant information.

Current activity in the area of defining the acceptable noise environment for humans away from the workplace is going on both in the United States and abroad. Aspects being considered include the intrusiveness of noise, or how much louder specific single occurrences are than the background or ambient noise; the spread in the statistical parameters known as the 10th and 90th percentile noise levels, and the standard deviation of the noise level as a function of time during the three socially important periods of the day: day, evening, and night; the effects of social and political factors; and the views on the part of the auditor as to whether the particular noise is necessary or not and whether the auditor believes that it may or may not be readily abated.

Noise levels in industry. Surveys in many industries show that noise levels at many work stations are in excess of the allowable exposure, although the provisions for higher-level exposure for shorter durations does cover probably one-third of the otherwise excessive level situations. Efforts by industries of all types are based on current technology in the noise control field using small and large mufflers on airflow system fans and blowers, engines, and pneumatic control valves. Some mufflers cause energy losses that are not readily made up by the system and thus require successive changes in order to provide the quieting without reducing machine efficiency, causing excessive temperatures to exist, or reducing the combustion air available to a process.

Protection for employed. Many work stations must be protected by the use of baffles between the noise source and the employee. These baffles reduce the noise by acting as acoustical barriers, and they also form visual barriers even when made of clear plastic. In some situations employees can be enclosed and thereby protected from a high-level-noise environment. Earmuff-type and ear-insert-type ear protectors are also used in high-noise-level conditions.

The relationship between exposure to high noise levels at the workplace and progressive loss of hearing has been well known for many years. Because of the known causal relationship, the Federal government has set strict limits on the levels to which industrial workers, in fact, any employee, may be exposed. Not so well known are the relationships between noise in the office, home, or community and any resultant annoyance, distraction, or interference with sleep. However, the fact that noise can cause annoyance, distraction, and sleep interference, even at fairly low levels, has led to considerable private and governmental effort to study the nature of noise, its sources, its effects on people, and methods for noise abatement. Some municipalities, responding to community pressures, have enacted statutes for the control of noise using modern instrumentation based on studies of the community and its needs, while others have adapted the codes for other cities in the hope that they can avoid the long and costly studies usually required to define the existing situation and evaluate proposed improvements. A major force in the program against noise pollution has been the Federal government. By the revision of existing laws, the enactment of others, and reports by the Environmental Protection Agency, the Department of Transportation, and the Department of Housing and Urban Development, the Federal government has established standards and guidelines in this field.

Occupational Health and Safety Act. The Williams-Steiger Act, known as the Occupational Safety and Health Act (OSHA) of 1970 (PL 91-596), limits an employee's daily exposure to 90 dB(A), measured with a sound-level meter. If the sound-level meter meets certain American National Standards Institute specifications, it will meet the requirements of the OSHA. The act specifies that the 8-hr daily exposure must not be greater than 90 dB(A), but that if the exposure is for shorter periods, the level may be higher. Thus the act ties the exposure to a total time history of each individual's exposure throughout each day. Table 1 shows the duration at each of a series of sound levels.

Enforcement of the act has begun with visits of the Department of Labor's inspectors to industries known to be noisy. Where exposure is in excess of the OSHA, the companies are given a citation and have a fixed time period in which to develop a compliance plan. The plan must include (1) a detailed survey of sound levels and sound spectra to determine the sources of excessive sound levels; (2) initiation of engineering studies to determine methods for reducing sound levels at their sources; (3) planning and initiation of feasible administrative controls, such as modifying production schedules to divide noisy jobs among enough people to bring each below the permissible limit or spreading part-time noisy operations; (4) initial audiograms for personnel excessively exposed; (5) installation of a personal protective equipment program; (6) follow-up audiograms at appropriate intervals to assess effectiveness of the personal protective equipment program and administrative controls; (7) installation of engineering controls or process changes to reduce noises at their source; and (8) repeated noise surveys to measure effectiveness of the engineering changes.

Proposed changes to the OSHA regulations

Table 1. Noise exposures permissible under the Occupational Health and Safety Act

Duration per day, hr	Sound level, dB(A) slow
8	90
6	92
4	95
3	97
2	100
1.5	102
1	105
0.5	110
0.25 or less	115

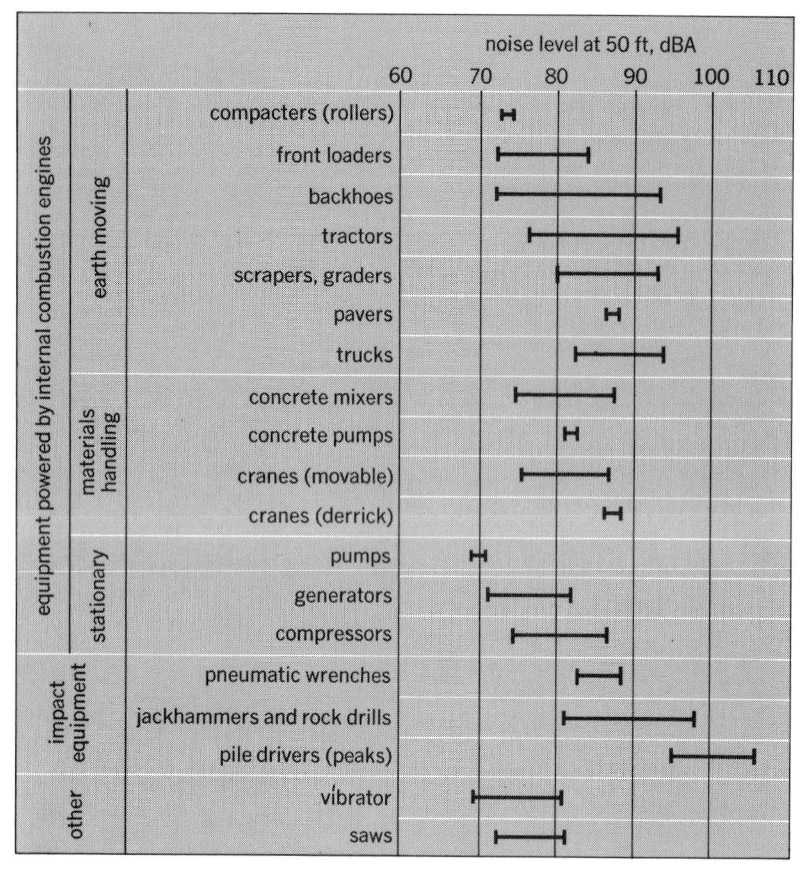

Fig. 2. Construction equipment noise ranges (based on limited available data samples). (*From U.S. Environmental Protection Agency, Noise from Construction Equipment and Operations, Building Equipment, and Home Appliances, Rep. no. NTID300 1, 1971*)

would lower the 8-hr exposure level to 85 dB(A) and would include limits for lower levels for exposure durations up to 16 hr. Hearings on the proposed changes were held in midsummer of 1975.

EPA report. In an effort not only to look at the noise levels themselves but to relate the noise exposure to the fabric of life within the United States, the Environmental Protection Agency (EPA) published in 1971 a major analysis of noise as it affects people at home, at work, and at play. The study included noise-generating sources in home appliances and in industry, construction, and transportation equipment, and assessed the effects of the noises on the individual and on the community. It examined existing levels in the community, the available technology for noise abatement, and the potential for noise abatement using future technologies. The levels of construction equipment noise in the study are shown in Fig. 2.

As part of the examination of the existing conditions at some representative locations in the United States and to establish the range of community noise levels, Fig. 3 shows the statistics of the noise levels for sites ranging from a noisy downtown apartment to the quiet noises of nature at the rim of the Grand Canyon. Going one major step further, a rating method was developed, based on a large number of case histories, relating community noise levels with the noise exposure and the relationship of the noise exposure to several factors that modify a community's response to noise at any given level. These modifying factors include the season of the year, the time of day, the ambient noise against which the intruding noises are heard,

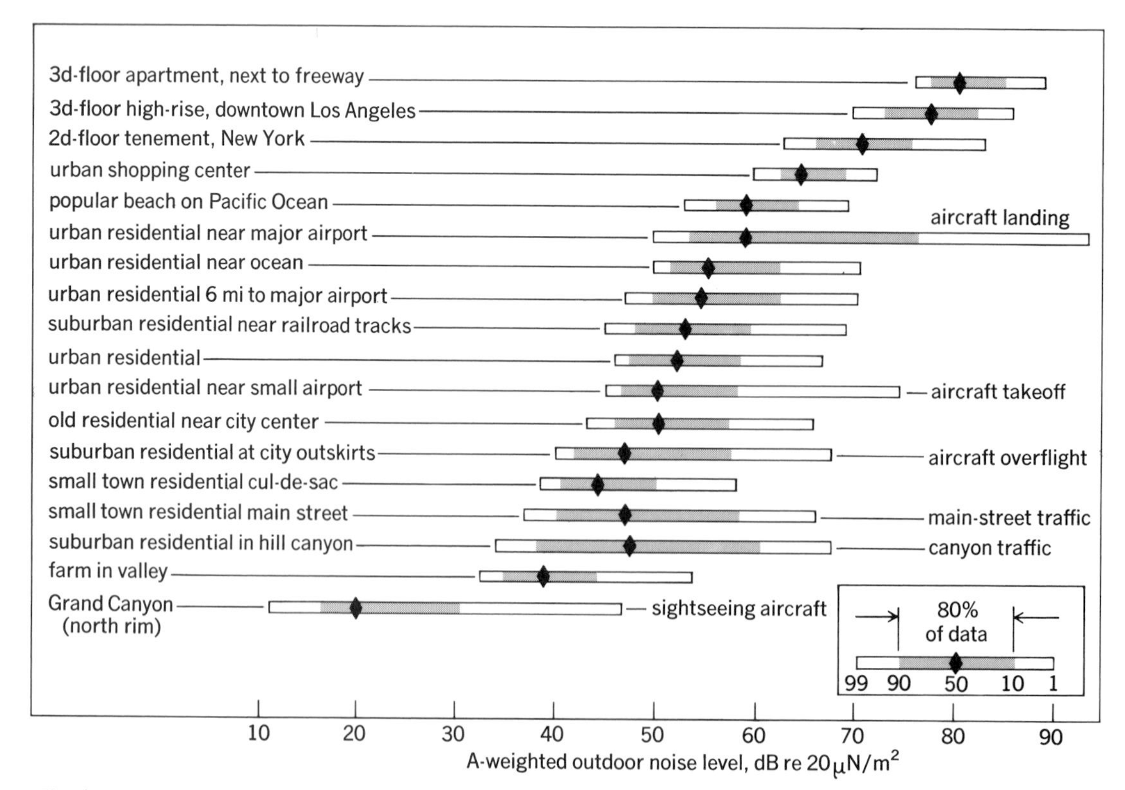

Fig. 3. Daytime outdoor noise levels found in 18 locations ranging from the wilderness to the downtown city, with significant intruding sources noted. Data are arithmetic averages of the 12 hourly values in the daytime period (7:00 A.M. to 7:00 P.M.) of the levels which are exceeded 99, 90, 50, 10, and 1% of the time. (*From U.S. Environmental Protection Agency, Community Noise, Rep. no. NTID300 3, 1971*)

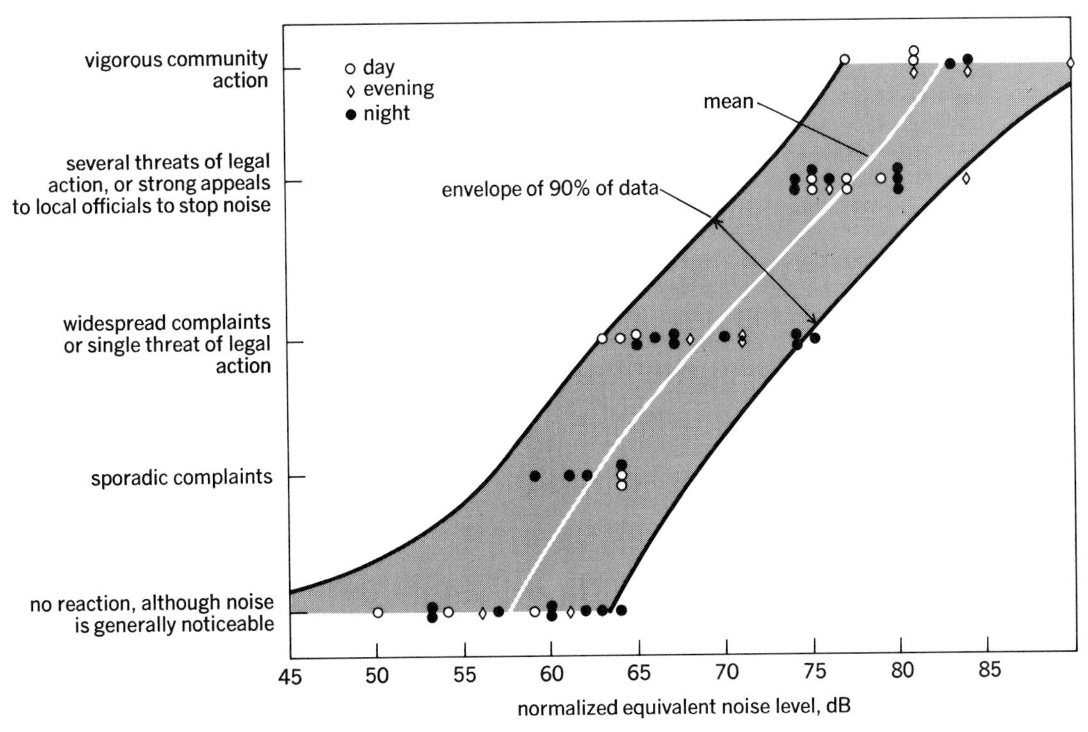

Fig. 4. Community reaction to many types of intrusive noises as a function of the normalized noise level, using original procedures of W. A. Rosenblith and S. S. Stevens. Data are normalized to residential urban residual noise, some prior exposure, windows partially open, and no pure tone or impulses. (*From U. S. Environmental Protection Agency, Community Noise, Rep. no. NTID 300.3, 1971*)

the attitude of the community toward the operator of the noise source or sources, and the nature of the noise. This last factor accounts for tonal and impulsive qualities which can be especially disturbing. The result of the analysis is presented in Fig. 4. The term "normalized" expresses the fact that the various modifying factors have been accounted for. If the noise levels are recorded on magnetic tape over a number of 24-hr periods, the normalized community noise equivalent level (NCNEL) may be calculated from the measured levels using computer-based data-analysis equipment. It is possible to estimate the NCNEL for many continuous noises. The NCNEL then permits prediction of the community response.

The study indicates that at present there are local situations where industrial noise is disturbing to the neighboring community, but that, in general, transportation noise usually masks the industrial noise. In most cases, industrial noise is controlled by the long-term interaction that takes place between the community and its industries. A process of accommodation occurs as a result of the desire of the industry not to abandon its investment or spend too much money to remain at a given location. This is balanced by the municipal government's wishing to maintain the industries as the local tax base, but knowing that the citizens whom the local government represents must be protected against "excessive" noise. How the term "excessive" is defined, and by whom, is often the factor that controls whether any given industry can build or remain in any specific community. The report also points out that many industries' neighbors will tolerate noise at high levels if (1) it is continuous; (2) it does not interfere with speech communication; (3) it does not include pure tones or impacts; (4) it does not vary rapidly; (5) it does not interfere with getting to sleep; and (6) it does not contain fear-producing elements.

HUD noise abatement and control circular. One major step toward improving the noise conditions within the housing for a large number of people came with the issuance of a circular by the Department of Housing and Urban Development (HUD) which set three categories for noise exposures at HUD project sites: (1) acceptable; (2) discretionary—normally acceptable, or normally unacceptable; (3) unacceptable.

Table 2 sets out the noise levels corresponding to the various categories. Also set by the circular is a minimum standard of sound transmission class (STC) 45 for isolation of sound between apartments in multiple dwellings. Detailed procedures are specified for compliance, including the requirement of an environmental impact statement where any request for an exception to the new policy is made. Also issued by HUD is a publication outlining the noise assessment guidelines. This permits builders and applicants for HUD support, as well as HUD field officials, to make preliminary assessments of proposed sites. Detailed assessments must be made using carefully prepared acoustical studies if the site appears to be in the discretionary or unacceptable categories in order to determine its exact status or to apply for an exception.

In 1974 the EPA, in a study of acceptable noise environments, proposed to use the day-night level, L_{dn}, the continuous A-weighted sound level having the same energy as the time-varying noise at the point under study, with a 10-dB penalty for nighttime exposures. The L_{dn} of 55 dB(A) was set as the maximum suitable for nonworkplace noise

Table 2. External noise exposure standards for new construction sites*

Acceptability level	General external exposures, dB(A)	Comments
Unacceptable	Exceeds 80 dB(A) 60 min per 24 hr Exceeds 75 dB(A) 8 hr per 24 hr	Exceptions are strongly discouraged and require a 102(2)C environmental statement and the Secretary's approval
Discretionary		
Normally unacceptable	Exceeds 65 dB(A) 8 hr per 24 hr Loud repetitive sounds on site	Approvals require noise attenuation measures, the Regional Administrator's concurrence, and a 102(2)C environmental statement
Normally acceptable	Does not exceed 65 dB(A) more than 8 hr per 24 hr	
Acceptable	Does not exceed 45 dB(A) more than 30 min per 24 hr	

*Measurements and projections of noise exposures are to be made at appropriate heights above site boundaries.
SOURCE: *Noise Abatement and Control: Department Policy, Implementation Responsibilities, and Standards*, Circ. 1390.2, U.S. Department of Housing and Urban Development, 1971.

exposures. The L_{dn} is quite controversial and has yet to achieve the wide acceptance of earlier environmental noise rating methods.

[LEWIS S. GOODFRIEND]

Urban noise. It is common knowledge that cities are noisier places to live in than rural areas, though it is not always fully appreciated that passenger cars and trucks are the predominant source of noise in the urban environment by virtue of their numbers and wide distribution. Ambient noise levels, rising and falling in a diurnal cycle in response to traffic, have also been rising year by year with the increase in the number of motor vehicles on the road. Attempts to control noise at the local municipal level by means of bylaws is nothing new, but these efforts have rarely been successful. Governments at all levels have shown interest in noise abatement as part of the wider concern over environmental pollution. This is particularly true in North America and Europe, and traffic noise is frequently the first target.

There is available a broad base of technical knowledge about noise, its measurement, effect on people, and control. The task of standardizing measurement procedures and establishing criteria is pursued by organizations such as the International Standard Organization (ISO), the Society of Automatic Engineers (SAE), and the American National Standards Institute (ANSI). The U.S. Environmental Protection Agency (EPA), the International Organization for Economic Cooperation and Development (OECD), national laboratories, and private acoustics consulting firms have studied the problem of noise and made recommendations for the guidance of governments and industries. Control of noise at the source, that is, motor vehicles, is clearly the most effective approach to the problem and benefits everyone. Noise-emission control on motor vehicles is preferably established at the Federal government level so that all manufacturers supplying the nationwide market will have the same uniform standards to meet. International agreement on noise-emission standards would be a further step forward. Provincial or state governments control directly, or through local governments, roads, highways, motor vehicle traffic, zoning and building regulations, and enforcement of noise-control legislation.

Measured motor vehicle noise. Measured noise levels of a given type of motor vehicle in a given range of speed exhibit a statistical variation. Some typical statistical distributions of motor vehicle noise levels versus numbers of vehicles are shown in Fig. 5. Noise levels have been measured at a distance of 50 ft (15 m) from passing vehicles along roadsides in open terrain. Vehicle categories are shown in the order of increasing weight, for a speed range of 30–39 mph (1 mi = 1.6 km). Similar distributions occur for other ranges of speed. It is found that the noise level of the average vehicle in each category, defined as the vehicle with the weighted mean dB(A) value (peak of the distribution), increases about 9 dB for a doubling of speed (30–39 mph to 60–69 mph) and about 3.5 dB for a doubling of weight from light to heavy vehicle categories. Insofar as the noise levels fall within one or another of these statistical distributions depending on vehicle type and speed, all measurements of noise level of individual motor vehicles are in essential agreement.

This predictable behavior of motor vehicles in everyday use on the roads has practical importance. Realistic maximum noise levels for each vehicle category and speed range can be derived from the statistical distributions, both for vehicles in use in the mid 1970s and for a specified future date when automobile manufacturers could be expected to produce quieter vehicles. A progressive reduction of noise-level limits based on anticipated advances in noise-control technology is an approach that is being adopted. As an example, from the statistical distributions for passenger cars in four speed ranges, namely, 30–39, 40–49, 50–59, and 60–69 mph, of which the first appears in Fig. 5, noise-level limits of 70, 73, 76, and 78 dB(A) respectively at 50 ft would exclude a few of the noisier vehicles now on the road. At some future date, the limits might be set at 65, 68, 71, and 73 dB(A) respectively, the present average vehicle noise levels. This would be reasonable in view of

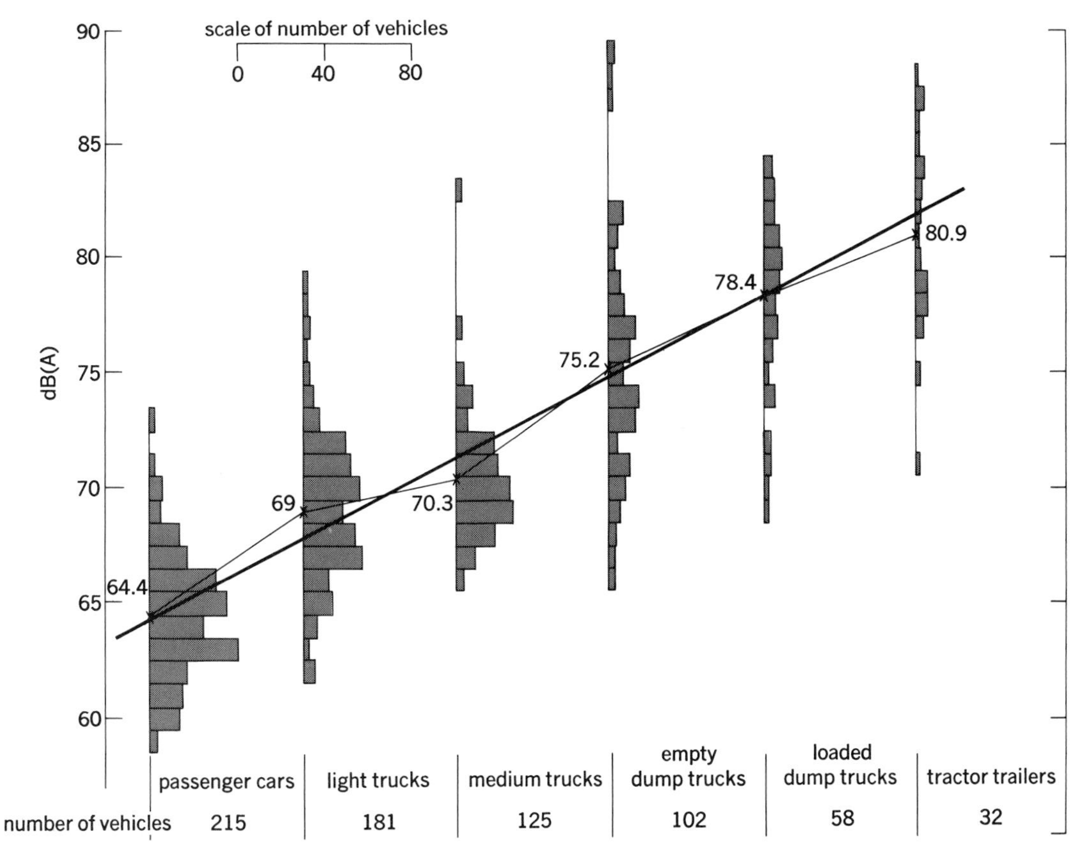

Fig. 5. Distributions, in dB(A), for passenger cars, light, medium, and heavy trucks (dump trucks, empty and loaded), and tractor trailers, in order of increasing weight. Slope of curve is 3.5 dB per doubling of weight. *(From N. Olson, Survey of Motor Vehicle Noise, JASA, vol. 52, no. 5, pt. 1, 1972)*

the fact that about one-half of the present passenger cars could meet this limit. It is well recognized that the North American passenger car is already a quiet vehicle at low road speed. At higher speeds, it is the tire-road interaction which becomes the predominant noise source. In the case of heavier vehicles, particularly heavy trucks and tractor trailers, with engines operating near maximum power, the noise problem is much greater. Direct radiation of noise from the engine due to high pressures in the combustion chamber, exhaust noise, and noise from supercharger fans are among the major noise problems associated with these vehicles. At higher speeds, tire-road interaction adds its quota of noise.

Legislative control of noise. The overall picture of noise-control legislation, as it applies to traffic noise in the mid 1970s, is one of variety—and some confusion. In North America a number of cities have enacted legislation to control motor vehicle noise. The states of New York and California have statewide legislation, and the province of Ontario is moving in this direction also. The state of Hawaii has legislation applying to the island of Oahu. In some cases, realistic limits have been adopted, but in others the limits are so high that there is virtually no control. The state of Hawaii had noise limits for light vehicles (up to 6000 lb, or 2700 kg, GVW, or gross vehicle weight) effective until Jan. 1, 1977, which were slightly higher than the hypothetical ones mentioned previously. A new set of limits effective after Jan. 1, 1977, corresponds closely to the present average vehicle noise levels. The city of Chicago and the state of California have limits of 76 dB(A) at 50 ft for speeds less than 35 mph, and after January 1978 the Chicago limit goes to 70 dB(A). The city of Calgary has limits of 70, 75, and 78 dB(A) for speeds of less than 30 mph, 30–45 mph, and greater than 45 mph respectively. Extreme examples are the limits of 87 dB(A) for the cities of Seattle and Cincinnati and 88 dB(A) for Anchorage, which clearly provide little or no control for light vehicles.

Heavy trucks and tractor trailers are a more formidable target for noise-control legislation. Present legislation is directed mainly at these two categories, generally ignoring the lighter trucks or lumping them all together as heavy vehicles. The state of Hawaii noise limits for heavy vehicles (greater than 6000 lb GVW) are shown in Table 3. The daytime limits affect a substantial number of heavy trucks and more than half the tractor trailers at higher speeds. At night these vehicles cannot operate at all except by special permit. Few of the present heavy trucks or tractor trailers could meet the limits effective after Jan. 1, 1977, but these limits are directed at vehicles first landed on Oahu on or after that date.

The statistical distributions for heavy trucks and tractor trailers at 30–39 mph and accelerating from a stop are shown side by side in Fig. 6, with some later legislation indicated by arrows. Given the additional information that the distributions move upward on the dB(A) scale about 7 dB for tractor trailers and 9 dB for heavy trucks as the speed increases from 30–39 mph through interme-

Table 3. Noise level limits for heavy vehicles*

Posted speed limit	Time periods when applicable	Measurement distance 20 ft (6.1 m)	25 ft (7.6 m)	50 ft (15.2 m)
Effective between Jan. 1, 1972, and Dec. 31, 1973				
35 mph or less†	Daytime	94 dB(A)	92 dB(A)	86 dB(A)
	Evening	92	90	84
	Night Holiday Sunday	81	79	73
More than 35 mph	All	94	92	86
Truck routes	All	96	94	88
Effective after Jan. 1, 1974				
35 mph or less	Daytime	92 dB(A)	90 dB(A)	84 dB(A)
	Evening	92	90	84
	Night Holiday Sunday	81	79	73
More than 35 mph	All	92	90	84
Truck routes	All	96	94	88
Vehicles first landed on Oahu on or after Jan. 1, 1977				
35 mph or less	Daytime	83 dB(A)	81 dB(A)	75 dB(A)
	Evening	75	73	67
	Night Holiday Sunday	73	71	65
More than 35 mph	All	83	81	75

*From Department of Health, State of Hawaii, *Public Health Regulations*, Chapter 44A: "Vehicular Noise Control for Oahu." †35 mph = 56 km/h = 15.6 m/s.

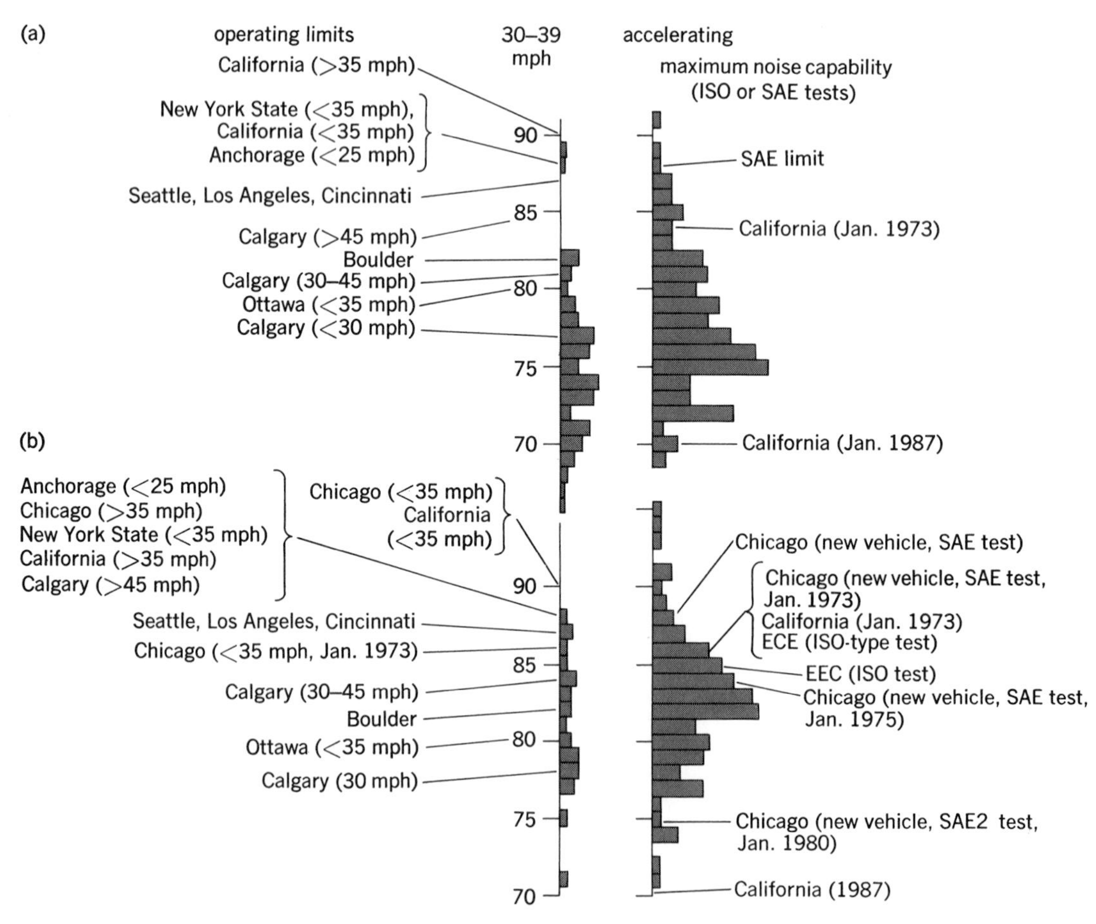

Fig. 6. Statistics of motor vehicle noises at 30–39 mph (48–63 km/h or 13.4–17.4 m/s) and accelerating, compared with bylaws specifying noise limits in dB(A). *(a)* Dump trucks. *(b)* Tractor trailers. All values are corrected to or specified at 50 ft (15.2 m). 1 mph = 1.61 km/h = 0.447 m/s. *(From National Research Council of Canada, Div. of Physics, Legislative Control of Noise: Feasibility and Recommendations, Rep. APS-500, 1972)*

diate speeds to 60–69 mph, the diagram in Fig. 6 is self-explanatory. The need for a uniform set of noise-level limits is apparent, from the point of view of both enforcement of limits and design of vehicles that will meet these limits.

Community noise levels. Community noise surveys have shown that motor vehicle traffic noise heads the list of noises that disturb people, whether they are at home or at work or indoors or outdoors, or whether the setting is urban or rural. It has also been shown that the noise level at which people are annoyed depends on the time of day, the location, whether the noise is uniform in level or punctuated by intermittent noises, and the intensity and duration of these intermittent noises. Generally speaking, home is where people expect quiet surroundings and are least tolerant of noise. Maximum acceptable noise levels are dictated largely by interference with communication by day and with sleep at night. It has been recommended that noise levels inside residential quarters should not exceed 40–45 dB(A) during the day and 30–35 dB(A) at night for more than about 10% of the time. Windows being the weakest acoustic link, these inside noise limits establish maximum outside limits depending on whether windows are open or closed. Expected noise-level reductions are 5–10 dB for open windows, 15–25 dB for closed single windows, and 20–35 dB for closed double windows. With open windows, a normal condition in the summer, the outside limits would be about 45–55 dB(A) during the day and 35–45 dB(A) at night. It is clear that these conditions are not always fulfilled in present situations in which heavy traffic and residential development have been mixed indiscriminately. Ambient noise levels in quiet residential areas are typically about 35 dB(A) during the early morning hours, when traffic is absent.

The current approach to noise control in the urban community recognizes three lines of attack: first, to hold the line and prevent further deterioration; second, to reduce existing noise levels where they are unacceptably high; and third, to accept noise as a major factor in urban planning. Enforcing maximum noise limits for motor vehicles will ensure that they do not become noisier, but if the number of vehicles continues to increase, so will ambient noise levels. Thus, to just hold the line may require reducing the noise emission of motor vehicles; and if the ambient noise levels are also to be reduced, then the noise emission of motor vehicles will have to be drastically reduced. In situations in which noise levels are unacceptably high, relief may be gained by rerouting heavy traffic away from sensitive areas, by restricting noisier vehicles as to the time of day when they may be operated, by reducing speed, or by using physical barriers. Many cities have already adopted one or more of these measures. Proper zoning of new developments to separate citizens from noise sources is an obvious measure which is being widely discussed but which is surprisingly difficult to implement. All the major studies on urban noise emphasize these measures. [N. OLSON]

Bibliography: American Speech and Hearing Association, *Noise as a Public Health Hazard*, 1970; C. R. Bragdon (ed.), *Noise Pollution: A Guide to Information Sources*, 1979; K. D. Kryter, *The Effects of Noise on Man*, 1970; A. Milne, *Noise Pollution: Impact and Countermeasures*, 1979; Organization for Economic Cooperation and Development, *Urban Traffic Noise*, 1971; U.S. Department of Transportation, Office of Noise Abatement, *A Community Noise Survey of Medford, Massachusetts*, Rep. no. DOT-TSC-OST-72-1, 1971; U.S. Department of Housing and Urban Development, *Noise Abatement and Control: Department Policy, Implementation Responsibilities and Standards*, Circ. 1390.2, 1971; U.S. Environmental Protection Agency, *An Assessment of Noise Concern in Other Nations*, Rep. no. NTID 300.6, vol. 1, 1971; U.S. Environmental Protection Agency, *Effects of Noise on People*, 1971; U.S. Environmental Protection Agency, *Information on Levels of Environmental Noise Requisite to Protect Public Health and Welfare with an Adequate Margin of Safety*, 1971; M. F. Young et al., *Highway Noise Reduction by Barrier Walls*, Rep. no. PB-208 398, Texas Transportation Institute, 1971.

Nuclear power

Power derived from fission or fusion nuclear reactions. More conventionally, nuclear power is interpreted as the utilization of the fission reactions in a nuclear power reactor to produce steam for electrical power production, for ship propulsion, or for process heat. Fission reactions involve the breakup of the nucleus of heavy-weight atoms and yield energy release which is more than a millionfold greater than that obtained from chemical reactions involving the burning of a fuel. Successful control of the nuclear fission reactions provides for the utilization of this intensive source of energy, and with the availability of ample resources of uranium deposits, significantly cheaper fuel costs for electrical power generation are attainable. Safe, clean, economic nuclear power has been the objective both of the Federal government and of industry's programs for research, development, and demonstration. Critics of nuclear power seek a complete ban or at least a moratorium on new commercial plants.

Considerations. Fission reactions provide intensive sources of energy. For example, the fissioning of an atom of uranium yields about 200 MeV, whereas the oxidation of an atom of carbon releases only 4 eV. On a weight basis, the 50,000,000 energy ratio becomes about 2,500,000. Only 0.7% of the uranium found in nature is uranium-235, which is the fissile fuel used. Even with these considerations, including the need to enrich the fuel to several percent uranium-235, the fission reactions are attractive energy sources when coupled with abundant and relatively cheap uranium ores. Although resources of low-cost uranium ores are extensive (see Tables 1 and 2), more explorations in the United States are required to better establish the reserves. Most of the uranium resources in the United States are in New Mexico, Wyoming, Colorado, and Utah. Major foreign sources of uranium are Australia, Canada, South Africa, and southwestern Africa; smaller contributions come from France, Niger, and Gabon; other sources include Sweden, Spain, Argentina, Brazil, Denmark, Finland, India, Italy, Japan, Mexico, Portugal, Turkey, Yugoslavia, and Zaire.

Government administration and regulation. The development and promotion of the peaceful uses of nuclear power in the United States was under the direction of the Atomic Energy Commis-

Table 1. United States uranium resources, in tons of U_3O_8 as of Jan. 1, 1979*

Production cost, $/lb U_3O_8†	Proved reserves	Potential reserves			Total reserves
		Probable	Possible	Speculative	
15	290,000	415,000	210,000	75,000	990,000
30	690,000	1,005,000	675,000	300,000	2,670,000
50	920,000	1,505,000	1,170,000	550,000	4,145,000

*1 short ton = 0.907 metric ton.
†Each cost category includes all lower-cost resources. $1/lb = $2.20/kg.
SOURCE: *Statistical Data of the Uranium Industry*, U.S. Department of Energy, 1979.

sion (AEC), which was created by the Atomic Energy Act of 1946 and functioned through 1974. The Atomic Energy Act of 1954, as amended, provided direction and support for the development of commercial nuclear power. Congressional hearings established the need to assure that the public would have the availability of funds to satisfy liability claims in the unlikely event of a serious nuclear accident, and that the emerging nuclear industry should be protected from the threat of unlimited liability claims. In 1957 the Price-Anderson Act was passed to provide a combination of private insurance and governmental indemnity to a maximum of $560,000,000 for public liability claims. The act was extended in 1965 and again in 1975, each time following congressional hearings which probed the need for such protection and the merits of having nuclear power. The Federal Energy Reorganization Act of 1974 separated the promotional and regulatory functions of the AEC, with the creation of a separate Nuclear Regulatory Commission (NRC) and the formation of the Energy Research and Development Administration (ERDA). In 1977 ERDA was absorbed into the newly formed Department of Energy.

Safety measures. The AEC, overseen by the Joint Committee on Atomic Energy (JCAE), a statutory committee of United States senators and representatives, had sought to encourage the development and use of nuclear power while still maintaining the strong regulatory powers to ensure that the public health and safety were protected. The inherent dangers associated with nuclear power which involves unprecedented quantities of radioactive materials, including possible wide-scale use of plutonium, were recognized, and extensive programs for safety, ecological, and biomedical studies, research, and testing have been integral with the advancement of the engineering of nuclear power. Safety policies and implementation reflect the premises that any radiation may be harmful and that exposures should be reduced to "as low as reasonably achievable" (ALARA); that neither humans nor their creations are perfect and that suitable allowances should be made for failures of components and systems and human error; and that human knowledge is incomplete and thus designs and operations should be conservatively carried out. The AEC regulations, inspections, and enforcements sought to develop criteria, guides, and improved codes and standards which would enhance safety, starting with design, specification, construction, operation, and maintenance of the nuclear power operations; would separate control and safety functions; would provide redundant and diverse systems for prevention of accidents; and would provide to a reasonable extent for engineered safety features to mitigate the consequences of postulated accidents.

Table 2. Foreign resources of uranium, in tons of U_3O_8 as of Jan. 1, 1979*

Production cost, $/lb U_3O_8†	Reasonably assured	Estimated additional	Total resources
30	1,460,000	870,000	2,330,000
50	2,010,000	1,350,000	3,360,000

*Excluding People's Republic of China, Soviet Union, and associated countries. 1 short ton = 0.907 metric ton.
†Each cost category includes all lower-cost resources. $1/lb = $2.20/kg.

Criteria for siting a nuclear power station involve thorough investigation of the region's geology, seismology, hydrology, meteorology, demography, and nearby industrial, transportation, and military facilities. Also included are emergency plans to cope with fires or explosions, and radiation accidents arising from operational malfunctions, natural disasters, and civil disturbances. The AEC Directorates of Licensing, Regulatory Operations, and Regulatory Standards thus functioned to achieve an extraordinary program of safety to be commensurate with the extraordinary risks involved with nuclear power.

Starting about 1970, regulatory safety measures have been significantly augmented in response to the introduction of nuclear power reactors with larger powers and higher specific power ratings; to improved technology, experiences, and more sophisticated analytical methods; to the National Environmental Policy Act of 1969 (NEPA) and the interpretations of its implementations; and to public participation and criticism. A variety of special assessments, studies, and hearings have been undertaken by such parties as congressional committees other than JCAE, the U.S. General Accounting Office (GAO), the American Physical Society, the National Research Council representing the National Academy of Sciences and the National Academy of Engineering, and by organizations which represent public interests in the environmental impacts of the continued use of nuclear power.

Nuclear Regulatory Commission. The independent NRC is charged solely with the regulation of nuclear activities to protect the public health and safety and the environmental quality, to safeguard nuclear materials and facilities, and to ensure conformity with antitrust laws. The scope of the activities include, in addition to the regulation of the nuclear power plant, most of the steps in the nuclear fuel cycle; milling of source materials; conversion, fabrication, use, reprocessing, and transportation of fuel; and transportation and management of wastes. Not included are uranium mining and operation of the government enrichment facilities.

Public issues. Public issues of nuclear power have covered many facets and have undergone some changes in response to changes being effected. Key issues include possible theft of plutonium, with threatening consequences; management of radioactive wastes; whether, under present escalating costs, nuclear power is economic and reliable; and protection of the nuclear industry from unlimited indemnity for catastrophic nuclear accidents.

Special nuclear materials. Guidance for improving industrial security and safeguarding special nuclear materials has been initiated. Scenarios studied include possible action by terrorist groups, and evaluations have been undertaken of effective methods for preventing or deterring thefts and for recovering stolen materials. Loss of plutonium by theft and diversionary tactics could pose serious dangers through threats to disperse toxic plutonium oxide particles in populated areas or to make and use nuclear bombs.

In the commercial nuclear fuel cycle which had been envisioned previous to 1977, the more critical segments in the safeguard program would involve the chemical reprocessing plants where the high-level radioactive wastes would be separated from the uranium and the plutonium, the shipment of the plutonium oxide to the fuel manufacturing plant, and the fuel manufacturing plant where plutonium oxide would be incorporated in the uranium oxide fuel. Plutonium is produced in the normal operation of a nuclear power reactor through the conversion of uranium-238. For each gram of uranium-235 fissioned, about 0.5–0.6 g of plutonium-239 is formed, and about half of this amount is fissioned to contribute to the operation of the power reactor. Reactor operations require refueling at yearly intervals, with about one-fourth to one-third of the irradiated fuel being replaced by new fuel. In the chemical reprocessing, most of the uranium-238 initially present in the fuel would be recovered, and about one-fourth of the uranium-235 initially charged would remain to be recovered along with an almost equal amount of plutonium-239.

The only commercial chemical reprocessing plant, the Nuclear Fuel Services, Inc., facility in West Valley, NY, opened in 1966, recovered uranium and plutonium as nitrates, and stored the high-level wastes in large, underground tanks. The facility was closed in 1972, and in 1976 Nuclear Fuel Services withdrew from nuclear fuels reprocessing because of changing regulatory requirements. The Midwest Fuel Recovery Plant at Morris, IL, was to have begun operations in 1974, but functional pretests revealed that major modifications would have to be undertaken before initiating commercial operations. Maximum use of the facility has been made to accommodate storage of irradiated fuel. A third plant, the Allied General Nuclear Services Barnwell Nuclear Fuel Plant at Barnwell, SC, whose construction was begun in 1971, has not been licensed. Thus, no commercial reprocessing plant is in operation in the United States, and there is only very limited use of test fuel assemblies containing mixed oxides of plutonium and uranium.

In April 1977 the Carter administration decided to defer indefinitely the reprocessing of spent nuclear fuels. This decision reflected a policy of seeking alternative approaches to plutonium recycling and the plutonium breeder reactor for the generation of nuclear power. This policy was motivated by a concern that the use of plutonium in other parts of the world might encourage the use of nuclear weapons.

Prior to the administration's decisions, the NRC had developed a system of reviews, including public participation through hearings and through comments received on draft regulations, to determine whether recycling of plutonium was to be licensed in a generic manner. Consideration has also been given to the possible collocation of chemical reprocessing and fuel manufacturing plants, and whether there are net gains achieved through the concentration of nuclear power reactors and fuel facilities in energy parks.

Some foreign countries have opposed the antiplutonium policies of the United States, and proceeded with the development of reprocessing plants and plutonium breeder reactors. A small commercial reprocessing plant for oxide fuel began operation in 1976 at La Hague, France, and construction of a much larger facility has been undertaken at this site. Design has begun for an oxide fuel reprocessing plant at Windscale in the United Kingdom to serve overseas markets as well as domestic markets.

Public participation in licensing. The procedures for licensing a nuclear facility for construction and for operation provide opportunities for meaningful public participation. Unique procedures have evolved from the Atomic Energy Act of 1954, as amended, which are responsive to public and congressional inquiry. The applicant is required to submit to the NRC a set of documents called the Preliminary Safety Analysis Report (PSAR), which must conform to a prescribed and detailed format. In addition, an environmental report is prepared. The docketed materials are available to the public, and with the Freedom of Information Act and the Federal Advisory Committee Act, even more public access to information is available. The NRC carries out an intensive review of the PSAR, extending over a period of about a year, involving meetings with the applicant and its contractors and consultants. Early in the review process, the NRC attempts to identify problems to be resolved, including concerns from citizens in the region involved with the siting of the plant. Formal questions are submitted to the applicant, and the replies are included as amendments to the PSAR.

Environmental Impact Statement. Major Federal actions that significantly affect the quality of human environment require the preparation of an Environmental Impact Statement (EIS) in accordance with the provisions of NEPA. The EIS presents (1) the environmental impact of the proposed action; (2) any adverse environmental effects which cannot be avoided should the proposal be implemented; (3) alternatives to the proposed action; (4) relationships between short-term uses of the environment and the maintenance and enhancement of long-term productivity; and (5) any irreversible and irretrievable commitments of resources which would be involved in the proposed action should it be implemented. To better achieve these objectives, the NRC staff supplements the applicant's submittal with its own investigations and analyses and issues a draft EIS so as to gain the benefit of comments from Federal, state, and local governmental agencies, and from all interested parties. A final EIS is prepared which reflects consideration of all comments.

Safety Evaluation Report. A second major report issued by the NRC staff is the Safety Evaluation Report (SER), which contains the staff's conclusions on the many detailed safety items, including discussions on site characteristics; design criteria for structures, systems, and components; design of the reactor, fuel, and coolant system; engineered safety features; instrumentation and control; both off-site and on-site power systems; auxiliary systems, including fuel storage and handling, water systems, and fire protection system; radioactive waste management; radiation protection for employees; qualifications of applicant and contractors; training programs; review and audit; industrial security; emergency planning; accident analyses; and quality assurance.

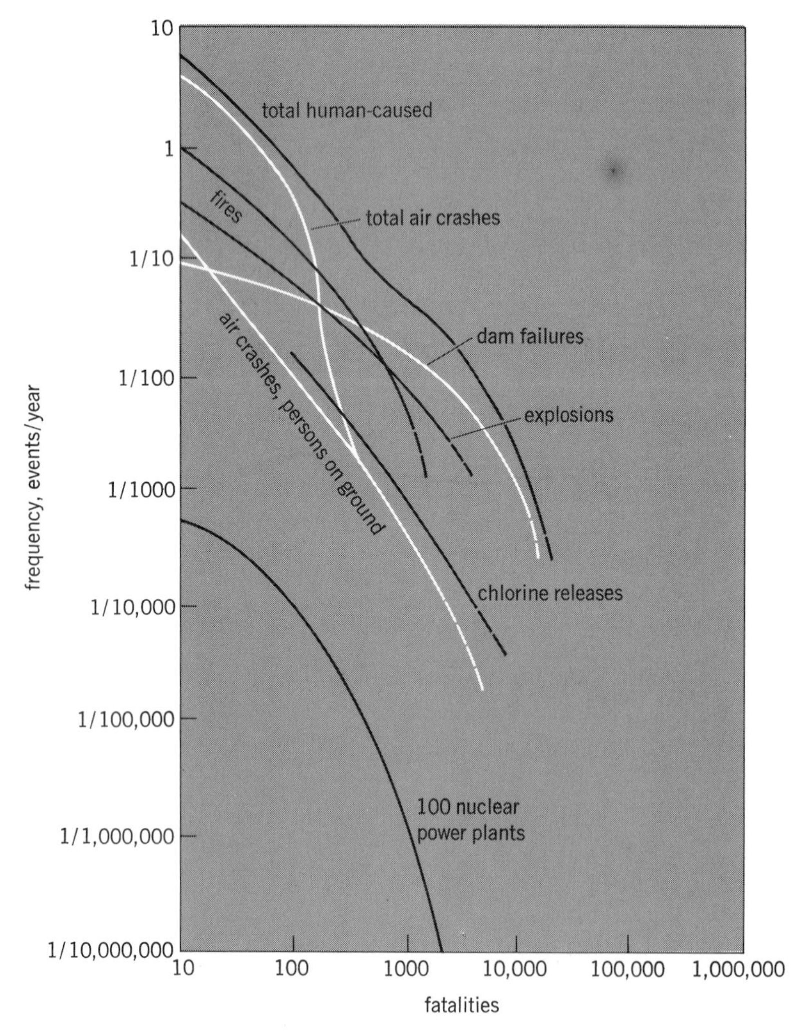

Fig. 1. Estimated frequency of fatalities due to human-caused events.

Independent review. Two additional steps are required before the decision is made regarding the construction license. An independent review on the radiological safety items is made and reported by the Advisory Committee on Reactor Safeguards (ACRS), and a public hearing is held by the Atomic Safety and Licensing Board (ASLB). The ACRS is a statutory committee consisting of a maximum of 15 members, covering a variety of disciplines and expertise. Appointments are made for this part-time activity by the NRC. Members are selected from universities, national laboratories and institutes, and industry, including experienced engineers and scientists who have retired, and, in each case, any possible conflicts-of-interest are carefully evaluated. The ACRS has a full-time staff and has access to more than 90 consultants. The ACRS conducts an independent review on nuclear safety issues and prepares a letter to the chairman of the NRC. Both subcommittee and full committee meetings are held to review the documents available and to discuss the applicant's and NRC staff's views on specific and generic issues.

Public hearing and appeal. Public participation is a major objective in the public hearing conducted by the ASLB. The ASLB is a three-member board, chaired by a lawyer, with usually two technical experts. For each application, a board is chosen from among the members of the Atomic Safety and Licensing Panel. Most members of the panel are part-time, and all members are appointed by the NRC. Prehearing conferences are held by the ASLB to identify parties who may wish to qualify and participate in the public hearing. Attempts are made to improve the understanding of the contentions, to see which contentions can be settled before the hearing, and to agree on the issues to be contested. The hearing may probe the need for additional electrical power, the suitability of the particular site chosen over possible alternative sites, the justification of the choice of nuclear power over alternate energy sources, and special issues regarding environmental impact and safety. The ASLB makes a decision on the construction application and may prescribe conditions to be followed. The decision is reviewed by and may be appealed to the Atomic Safety and Licensing Appeal Board. The appeal board is chosen from a panel completely separate from the ASLB panel. The NRC retains the authority to accept, reject, or modify the decisions rendered. Parties not satisfied by the review process and the decisions rendered can take their case to the courts. In several cases, resort to the U.S. Supreme Court has been utilized.

Authorization. A construction permit license is not issued until the NRC, ACRS, and ASLB reviews have been completed and the application has been approved, including conditions to be met during the construction review phase. Depending upon the justification of need, a Limited Work Authorization may be granted for limited construction activities following satisfactory review of the EIS, but prior to completion of the public hearings. Construction of a nuclear power plant may take 5 to 6 years or more, during which time the NRC Office of Inspection and Enforcement is involved in monitoring and inspection programs. Several years prior to the completion of construction, a

Final Safety Analysis Report (FSAR) is submitted by the applicant, and again an intensive review is undertaken by the NRC staff, and later by the ACRS. A detailed Safety Evaluation Report is prepared by the NRC, and a letter is prepared by the ACRS. If all items have been satisfactorily resolved and the construction and preoperational tests have been completed, a license for operation up to the full power is granted by the NRC. Technical specifications accompany the operating license and provide for detailed limits on how the plant may be operated. In some situations where additional information is sought, less than full power is authorized. A public hearing at the operation license stage is not mandatory and would be held only if an intervenor justifies sufficient cause.

Hearings on general matters. Public hearings have also been held on generic matters to establish rules for operation. The two rulemaking hearings conducted by the AEC that have attracted much attention were the Emergency Core Cooling Systems (ECCS) and the "As Low As Practicable" (ALAP) hearings. The hearing on the criteria and conditions for evaluating the effectiveness of the ECCS for a postulated loss of coolant accident lasted from January 1972 to July 1973, and provided more than 22,000 pages of transcript, with probably twice as much additional material in supporting exhibits. EISs on major activities, such as the liquid metal fast breeder reactor (LMFBR) program and the management of commercial high-level and transuranium-contaminated radioactive waste, have provided a process for public interactions and influence.

Reactor Safety Study. A detailed, quantitative assessment of accident risks and consequences in United States commercial nuclear power plants has been carried out (WASH-1400, October 1975). The final report has had the benefit of comments and criticisms on a previous draft from governmental agencies, environmental groups, industry, professional societies, and a broad spectrum of other interested parties. Although the study was initiated by the AEC and continued by the NRC, the ad hoc group directed by N. C. Rasmussen of the Massachusetts Institute of Technology carried out an independent assessment. Aside from the very significant technical advancements made in the risk assessment methodology, the "Reactor Safety Study" represents an approach to deal with one phase of the controversial impact of technology upon society. The study presents estimated risks from accidents with nuclear power reactors and compares them with risks that society faces from both natural events and nonnuclear accidents caused by people. The judgment as to what level of risk may be acceptable for the nuclear risks still remains to be made.

Figures 1 and 2 illustrate that the frequency of human-caused nonnuclear accidents and natural events is about 10,000 times more likely to produce large numbers of fatalities than accidents at nuclear plants. The study examined two representative types of nuclear power reactors from the 50 operating reactors, and has considered that extrapolation to a base for 100 reactors is a reasonable representation. Improvements in design, construction, operation, and maintenance would be expected to reduce the risks for later expansion in nuclear

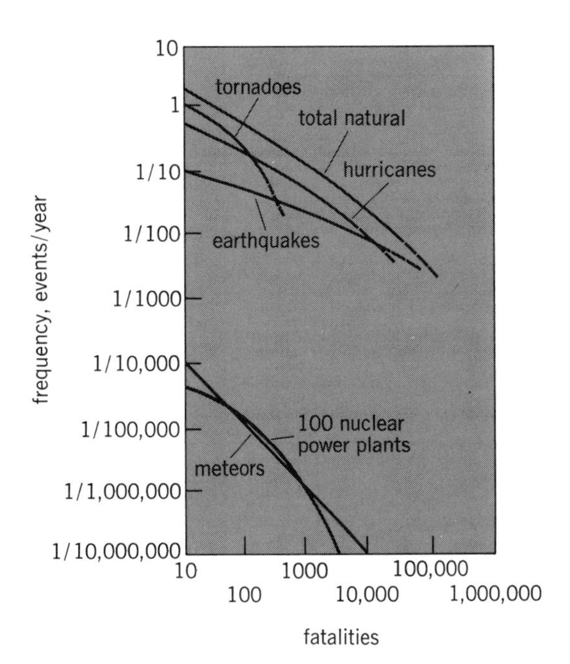

Fig. 2. Estimated frequency of fatalities due to natural events.

reactor operations. The fatalities shown in Figs. 1 and 2 do not include potential injuries and longer-term health effects from either the nonnuclear or nuclear accidents. For the nuclear accidents, early illness would be about 10 times the fatalities in comparison to about 8,000,000 injuries caused annually by other accidents. The long-term health effects such as cancer and genetic effects are predicted to be smaller than the normal incidence rates, with increases in incidence difficult to detect even for large accidents. Thyroid illnesses, which rarely lead to serious consequences, would begin to approach the normal incidence rates only for large accidents.

The likelihood and dollar value of property damage arising from nuclear and nonnuclear accidents are illustrated in Fig. 3. Both natural events (tornadoes, hurricanes, earthquakes) and human-

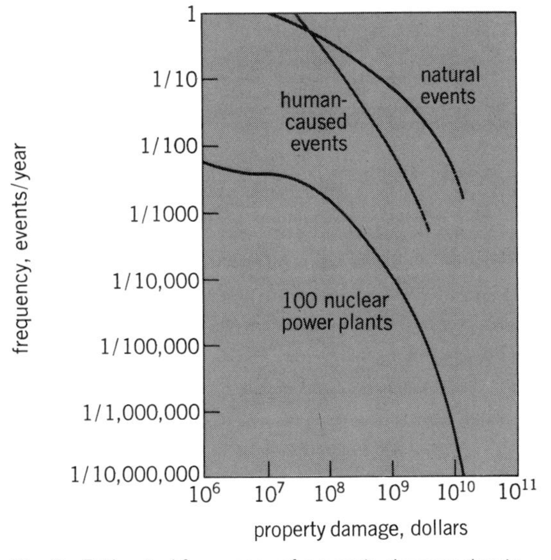

Fig. 3. Estimated frequency of property damage due to natural and human-caused events.

Table 3. Average risk of fatality by various causes

Accident type	Total number	Individual chance per year
Motor vehicle	55,791	1 in 4,000
Falls	17,827	1 in 10,000
Fires and hot substances	7,451	1 in 25,000
Drowning	6,181	1 in 30,000
Firearms	2,309	1 in 100,000
Air travel	1,778	1 in 100,000
Falling objects	1,271	1 in 160,000
Electrocution	1,148	1 in 160,000
Lightning	160	1 in 2,000,000
Tornadoes	91	1 in 2,500,000
Hurricanes	93	1 in 2,500,000
All accidents	111,992	1 in 1,600
Nuclear reactor accidents (100 plants)	–	1 in 5,000,000,000

caused events (air crashes, fires, dam failures, explosions, hazardous chemicals) might result in property damages in billions of dollars at frequencies up to 1000 times greater than that for accidents arising from the operation of 100 nuclear power plants.

Figures 1, 2, and 3 represent overall risk information. Risk to individuals being fatally injured through various causes is summarized in Table 3. The results of the study indicate that the predicted nuclear accidents are very small compared to other possible causes of fatal injuries.

The probability of an accident leading to the melting of the fuel core was estimated to be one chance per 20,000 reactor-years of operation, or for 100 operating reactors, one chance in 200 per year. The consequences of a core melt depend upon a number of subsequent factors, including additional failures leading to release of radioactivity, type of weather conditions, and population distribution at the particular site. The factors would have to occur in their worst conditions to produce severe consequences. Table 4 illustrates the progression of consequences and the likelihood of occurrence.

There has been considerable controversy concerning the risk estimates given in the Reactor Safety Study as a result of an NRC-sponsored review of them by a panel chaired by H. W. Lewis in 1977–1978. The Lewis panel argued that the Reactor Safety Study had, by its quantitative estimates, suggested a higher estimating precision than was justified by the data and procedures used. The Lewis group suggested that, rather than providing a single numerical estimate for a given risk, a numerical range should be quoted to more correctly reflect the uncertainties involved. However, Lewis also suggested that the Reactor Safety Study estimates were probably overconservative in that the most conservative choice had been made at each branch of the fault tree analyses whereas in an actual incident it would be anticipated that at least some of the branches would involve positive rather than negative choices.

Table 4. Approximate values of early illness and latent effects for 100 reactors

Chance per year	Consequences			
	Early illness	Latent cancer fatalities* per year	Thyroid illness* per year	Genetic effects† per year
1 in 200‡	< 1.0	< 1.0	4	< 1.0
1 in 10,000	300	170	1,400	25
1 in 100,000	3,000	460	3,500	60
1 in 1,000,000	14,000	860	6,000	110
1 in 10,000,000	45,000	1500	8,000	170
Normal incidence per year	4×10^5	17,000	8,000	8,000

SOURCE: Nuclear Regulatory Commission, *Reactor Safety Study*, WASH-1400, 1975.

*This rate would occur approximately in the 10–40-yr period after a potential accident.

†This rate would apply to the first generation born after the accident. Subsequent generations would experience effects at decreasing rates.

‡This is the predicted chance per year of core melt for 100 reactors.

Enrichment facilities. Only 0.7% of the uranium that is found in nature is the isotope uranium-235, which is used for the fuel in the nuclear power reactors. An enrichment of several percent is needed, and the Federal government (ERDA) owns the enrichment facilities. Expansion of the enrichment facilities is needed to meet expected demands.

Breeder reactor program. In a breeder reactor, more fissile fuel is generated than is consumed. For example, in the fissioning of uranium-235, the neutrons released by fission are used both to continue the neutron chain reaction which produces the fission and to react with uranium-238 to produce uranium-239. The uranium-239 in turn decays to neptunium-239 and then to plutonium-239. Uranium-238 is called a fertile fuel, and uranium-235, as well as plutonium-239, is a fissile fuel and can be used in nuclear power reactors. The reactions noted can be used to convert most of the uranium-238 to plutonium-239 and thus provide about a 60-fold extension in the available uranium energy source. Breeder power reactors can be used to generate electrical power and to produce more fissile fuel. Breeding is also possible using the fertile fuel thorium-233, which in turn is converted to the fissile fuel uranium-233. An almost inexhaustible energy source becomes possible with breeder reactors. The breeder reactors would decrease the long-range needs for enrichment and for mining more uranium.

Development. The development of the breeder nuclear power reactors had been initiated by the AEC, and the experimental breeder reactor I (EBR-I) was the first reactor to demonstrate production of electrical energy from nuclear energy (Dec. 20, 1951) and to prove the feasibility of breeding utilizing a fast reactor and a liquid metal coolant. In a special test in 1955, an operator error led to a substantial melting of the fuel, but with no off-site effects. The construction on EBR-II was begun in 1958, and it has been operating since 1963. The EBR-II installation is an integrated nuclear reactor power, breeding, and fuel cycle facility. The first commercial breeder licensed for operation, in 1965, was the Enrico Fermi Fast Breeder Reactor. In 1966, while the reactor was operating at a low power, partial coolant blockage occurred, leading to some melting of fuel and release of some radioactivity within the containment building. Although the operation of the reactor was resumed in 1970, the project was discontinued in 1972, primarily for economic considerations.

Environmental impact studies. The AEC had established that the priority breeder program would utilize the LMFBR concept. In response to a

1973 Court of Appeals ruling requiring an environment impact statement for the total LMFBR research and development program, the AEC issued a seven-volume *Proposed Final Environmental Statement* in December 1974 (WASH-1535), covering environmental, economic, social, and other impacts; alternative technology options; mitigation of adverse environmental impacts; unavoidable adverse environmental impacts; short- and long-term losses; irreversible and irretrievable commitments of resources; cost-benefit analysis; and responses to the many critical comments received during the development of the environmental statements in previous drafts. With due regard to the inherent hazards and the need to carefully manage plutonium, including a vigorous program to strengthen and improve safeguards, the conclusion reached was: "The LMFBR can be developed as a safe, clean, reliable and economic electric power generation system and the advantages of developing the LMFBR as an alternative energy option for the Nation's use far outweigh the attendant disadvantages."

A reexamination of the LMFBR program was undertaken by ERDA with its issuance of the 10-volume *Final Environmental Statement* (ERDA-1535) to cover additional and supporting information for ERDA's findings and responses to the comment letters on WASH-1535. The possible impact of a new technology upon society has never been so thoroughly questioned. The *Final Environmental Statement* addressed major uncertainties in nuclear reactor safety, fuel cycle performance, safeguards, waste management, health effects, and uranium resource availability. The environmental acceptability, technical feasibility, and economic advantages of the LMFBR cannot be ascertained until additional research, development, and demonstration programs are undertaken. The ERDA decision was to proceed with a plan to continue the research and supporting programs so as to provide sufficient data by 1986 to make the decision on commercialization of the LMFBR technology. The plan contemplated the licensing, construction, and operation of the Clinch River Breeder Reactor; the design, procurement, component fabrication and testing, and the licensing for construction of the prototype large breeder reactor; and the planning of a commercial breeder reactor (CBR-I). The heat transport and power generation system for the Clinch River Breeder Reactor is illustrated in Fig. 4.

United States policies. In April 1977 the Carter administration proposed to cancel construction of the Clinch River Breeder Reactor, reflecting the policies on plutonium recycling and breeding discussed above. However, Congress decided to continue funding for this project in fiscal years 1978 and 1979.

Foreign development. Prototype breeder reactors have been constructed in France, England, and the Soviet Union. Construction of a full-scale (1250 megawatts electric power, or MWe) commercial breeder, the Super Phénix, has been undertaken in France.

Radioactive waste management. In 1975 the responsibility for the development of a proposed Federal repository for high-level radioactive wastes was transferred from the AEC to ERDA (absorbed by the Department of Energy in 1977). The AEC had established regulations that the high-level radioactive wastes from the chemical reprocessing of irradiated fuel must be converted to a stable solid within 5 years of its generation at the reprocessing plant, and that the solids be sealed in high-integrity steel canisters and delivered to the AEC for subsequent management within 10 years of its generation at the reprocessing plant. Initially the AEC had sought to develop a salt mine near Lyons, KS, as the repository, but effective public intervention disclosed deficiencies for the site selected and the proposal was withdrawn. Subsequently, a more extensive study was undertaken to review permanent, safe, geologic formations and to locate possible sites. Stable, deep-lying formations could serve to isolate the wastes and dissipate their heat generation without need for maintenance or monitoring. During the period of time that might be used for the demonstration of a repository on a pilot plant scale, a retrievable surface storage facility (RSSF) would be placed in operation. The RSSF would require maintenance and monitoring, and could be engineered using known technology.

Critics of nuclear power consider the radioactive wastes generated by the nuclear industry to be too great a burden for society to bear. Since the high-level wastes will contain highly toxic materials with long half-lives, such as about a few tenths of one percent of plutonium that was in the irradiated fuel, the safekeeping of these materials must be assured for time periods longer than social orders have existed in the past. Nuclear proponents an-

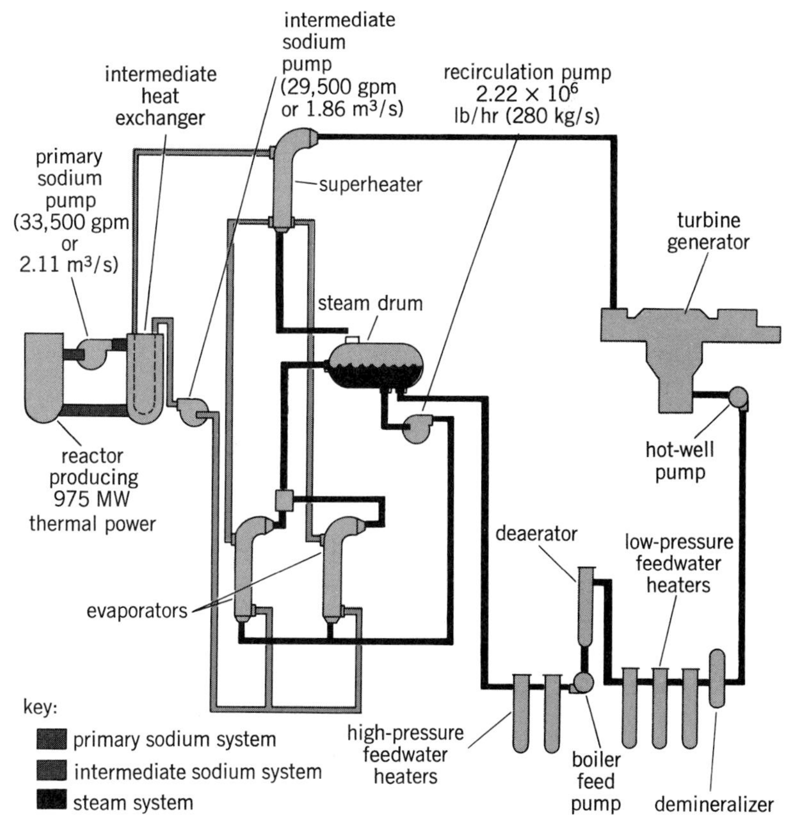

Fig. 4. Heat transport and power generation system for the Clinch River Breeder Reactor.

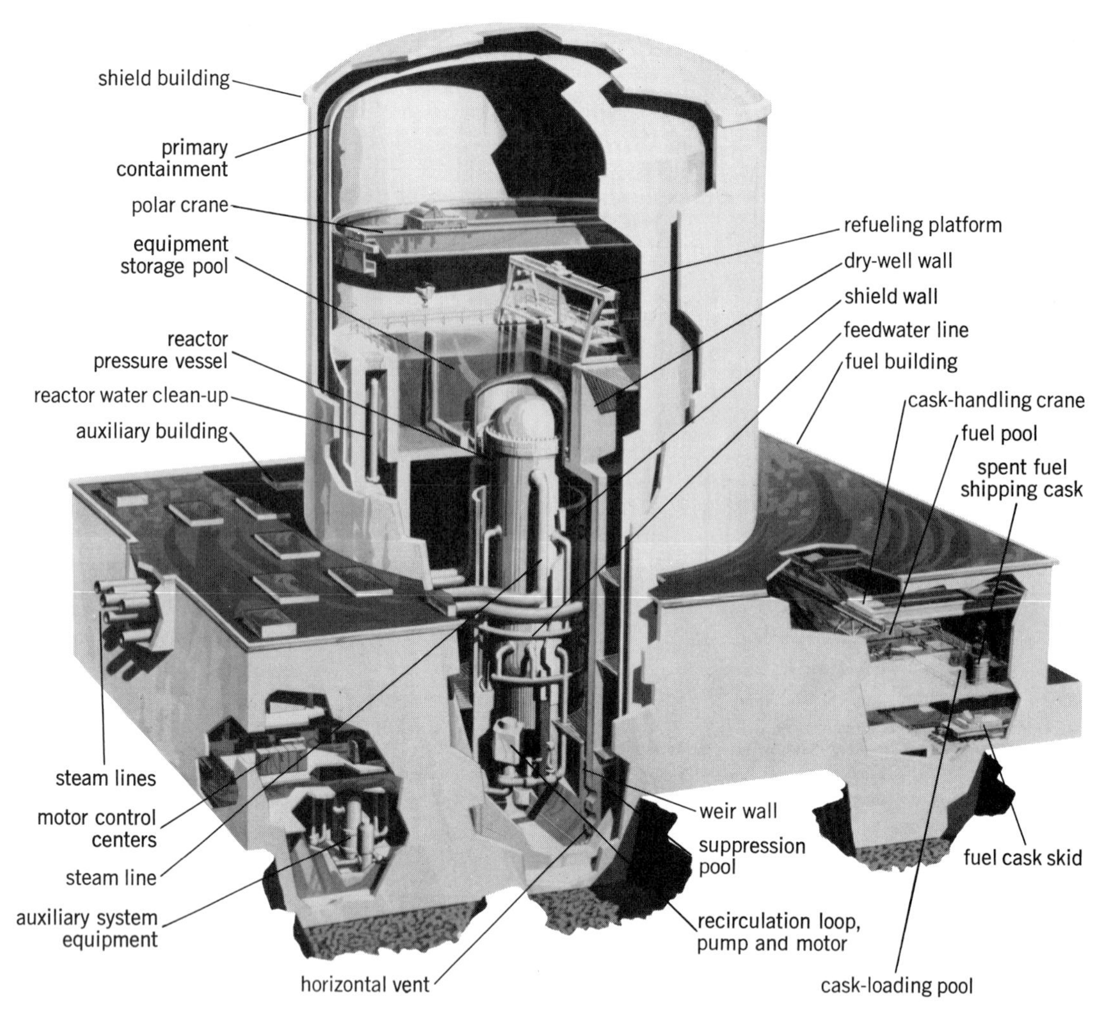

Fig. 5. Mark III containment of a boiling-water reactor, which illustrates safety features designed to minimize the consequences of reactor accidents. (*General Electric Co.*)

swer that the proposed use of bedded salts, for example, found in geologic formations that have prevented access of water and have been undisturbed for millions of years provide assurance that safe storage can be engineered. A relatively small area of several square miles would be needed for disposal of projected wastes.

Research and development has been underway since the 1950s on methods for solidifying the wastes; however, neither ERDA nor the NRC has provided the detailed criteria and guides needed by industry to carry forth the design of the waste solidification portions for the chemical reprocessing plants. Until such time as the chemical reprocessing plants begin operation, irradiated fuel will be retained at the reactor sites and at separate facilities away from reactors in appropriate water storage pools. Spent fuels can be stored in such pools for a period of decades if necessary. The water provides shielding and also carries away residual heat.

The decision by the Carter administration in 1977 to defer indefinitely the reprocessing of spent nuclear reactor fuels forced a reconsideration of the plans and schedules that had been under development for handling spent fuel from reactors in the United States. Utilities have expanded their on-site storage capabilities for spent fuel amid uncertainty about if and when reprocessing would be permitted. With indefinite deferral of reprocessing, some people are calling for geologic disposal of spent fuel assemblies. These assemblies are about the same size as the glass rods of processed waste and could also be placed in permanent geologic storage. However, this concept has drawbacks on a resource basis because it would mean throwing away the remaining fuel value of the uranium and plutonium.

Transportation. Transportation of nuclear wastes has received special attention. With increasing truck and train shipments and increased probabilities for accidents, the protection of the public from radioactive hazards is achieved through regulations and implementation which seek to provide transport packages with multiple barriers to withstand major accidents. For example, the cask used to transport irradiated fuel is designed to withstand severe drop, puncture, fire, and immersion tests.

Low-level wastes. Management of low-level wastes generated by the nuclear energy industry requires use of burial sites for isolation of the wastes and decay to innocuous levels. Operation of the commercial burial sites is subject to regulations by Federal and state agencies.

Routine operations of nuclear power stations

result in very small releases of radioactivity in the gaseous and water effluents. The NRC has adopted the principle that all releases should conform to the ALARA standard. Extension of ALARA guidance to other portions of the nuclear fuel cycle has been undertaken. *See* RADIOACTIVE WASTE MANAGEMENT.

Phase-out of governmental indemnity. In 1954 the Atomic Energy Act of 1946 was revised, making possible the possession and use of fissionable materials for industrial uses. As noted above, the Price-Anderson Act of 1957, extended in 1965 and again in 1975, protects the nuclear industry against unlimited liability in the unlikely event of a catastrophic nuclear accident. The total amount was set at $560,000,000, with private liability insurance starting at $60,000,000 in 1957 and increasing to $140,000,000 by 1977, and the government indemnity commensurately decreasing to $420,000,000. About 67% of the premiums paid to the private sector are placed in a reserve fund, and after a 10-year period, approximately 97–98% of this reserve fund has been returned. The smaller premiums paid to the Federal government are not returned. A 1966 amendment to the Price-Anderson Act provided features for no-fault liability and provisions for accelerated payment of claims. The 1975 extension of the act to 1987 provides for phasing out government indemnification, permits the $560,000,000 limit to float upward, and extends coverage to certain nuclear incidents that may occur outside the territorial limits of the United States. Each licensee is assessed a deferred payment, with a maximum level to be set per reactor. For example, in 1978 the licensees of the 66 operating reactors were each liable for a retrospective premium assessment of $5,000,000, making $330,000,000 available in addition to the base level of $140,000,000, and leaving $90,000,000 for the government indemnity. As the number of operating reactors increases, the government indemnity would phase out and would permit increases of total indemnity to exceed $560,000,000.

In addition to the indemnity insurance, private pools have provided property damage up to $175,000,000. New nuclear power stations would have capital costs in excess of $1,000,000,000.

Types of power reactors. There are five commercial nuclear power reactor suppliers in the United States: three for pressurized-water reactors, Combustion Engineering, Inc., Babcock and Wilcox, and Westinghouse Electric Corporation; one for boiling-water reactors, General Electric Company; and one for high-temperature gas-cooled reactors, General Atomic Company. Approximately two-thirds of the orders for nuclear power reactors are shared by General Electric and Westinghouse, and the remaining one-third by Combustion and Babcock and Wilcox. One high-temperature gas-cooled reactor (HTGR), Fort St. Vrain Nuclear, with 330 MW net electrical power, has been licensed for operation, but orders placed for the higher-power units (with up to 1160 MW net electrical power) have been canceled or deferred indefinitely.

Boiling-water reactor (BWR). In a General Electric BWR designed to produce about 3580 MW thermal and 1220 MW net electrical power, the reactor vessel is 238 in. (6.05 m) in inside diameter,

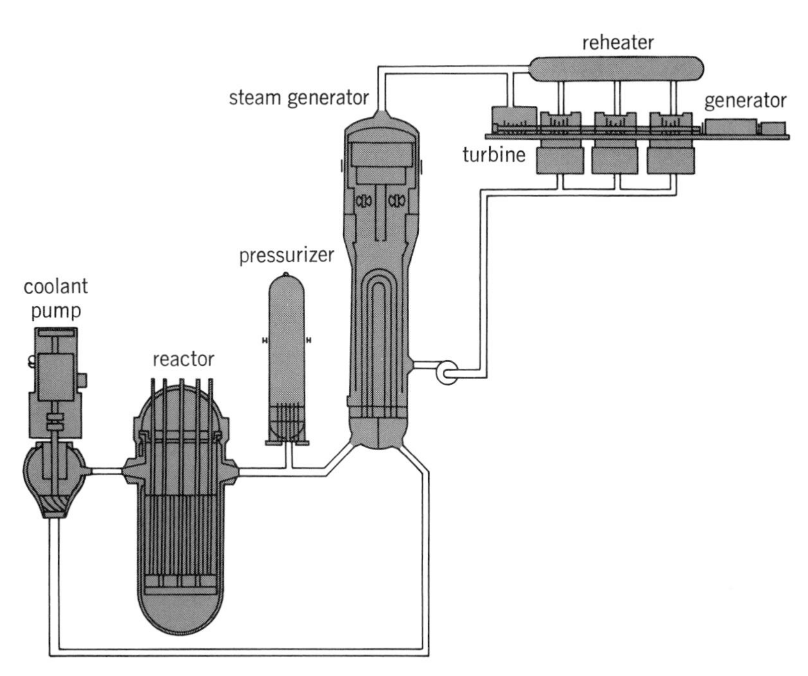

Fig. 6. Closed-cycle PWR. *(Westinghouse Electric Corp.)*

5.7 in. (14.5 cm) thick, and about 71 ft (21.6 m) in height. The active height of the core containing the fuel assemblies is 148 in. (3.76 m). Each fuel assembly contains 63 fuel rods, and 732 fuel assemblies are used. The diameter of the fuel rod is 0.493 in. (12.5 mm). The reactor is controlled by the cruciform-shape control rods moving up from the bottom of the reactor in spaces between the fuel assemblies (177 control rods are provided). The water coolant is circulated up through the fuel assemblies by 20 jet pumps at about 70 atm (7 MPa), and boiling occurs within the core. The steam is fed through four 26-in. diameter (66 cm) steam lines to the turbine. About one-third of the energy released by fission is converted into elec-

Fig. 7. Oconee Nuclear Power Station containment structures.

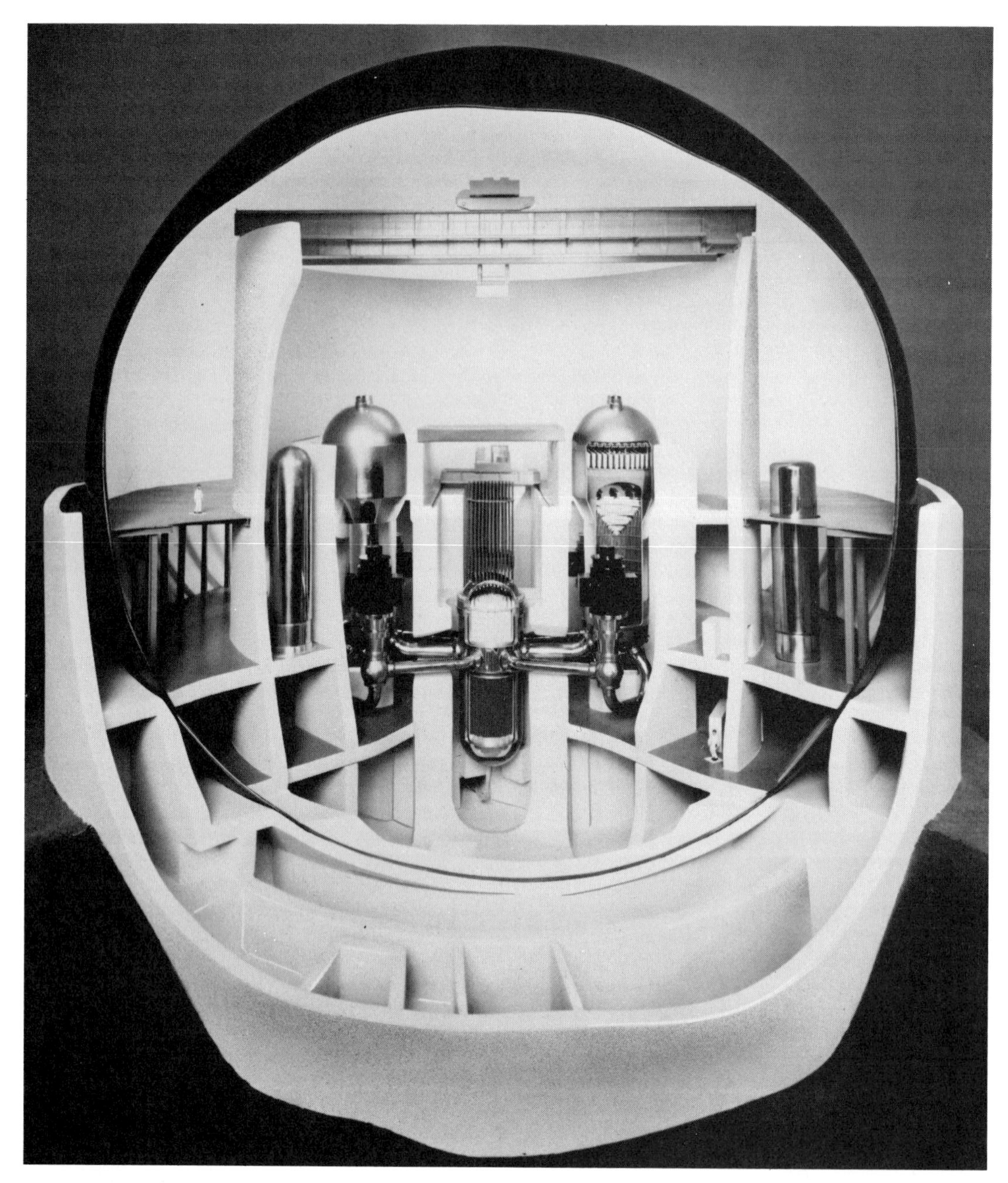

Fig. 8. Spherical containment design for a PWR. *(Combustion Engineering, Inc.)*

trical energy, and the remaining heat is removed in the condenser. The condenser operations are typical of both fossil and nuclear power plants, with heat being removed by the condenser having to be dissipated to the environment. Some limited use of the low-temperature heat source from the condenser is possible. The steam produced in the nuclear system will be at lower temperatures and pressures than that from fossil plants, and thus the efficiency of the nuclear plant in producing electrical power is less, leading to proportionately greater heat rejection to the environment.

Shielding is provided to reduce radiation levels, and pools of water are used for fuel storage and when access to the core is necessary for fuel transfers. Among the engineered safety features to minimize the consequences of reactor accidents is the containment. The function of the containment is to cope with the energy released by depressurization of the coolant should a failure occur in the primary piping, and to provide a secure enclosure to minimize leakage of radioactive material to the surroundings. The BWR utilizes a pool of water to condense the steam produced by the depressurization of the primary water coolant. Various arrangements have been used for the suppression pool. Other engineered safety features include the emergency core-cooling systems.

Pressurized-water reactor (PWR). Whereas in the BWR a direct cycle is used in which steam from the reactor is fed to the turbine, the PWR employs a closed system, as shown in Fig. 6. The water coolant in the primary system is pumped through the reactor vessel, transports the heat to a

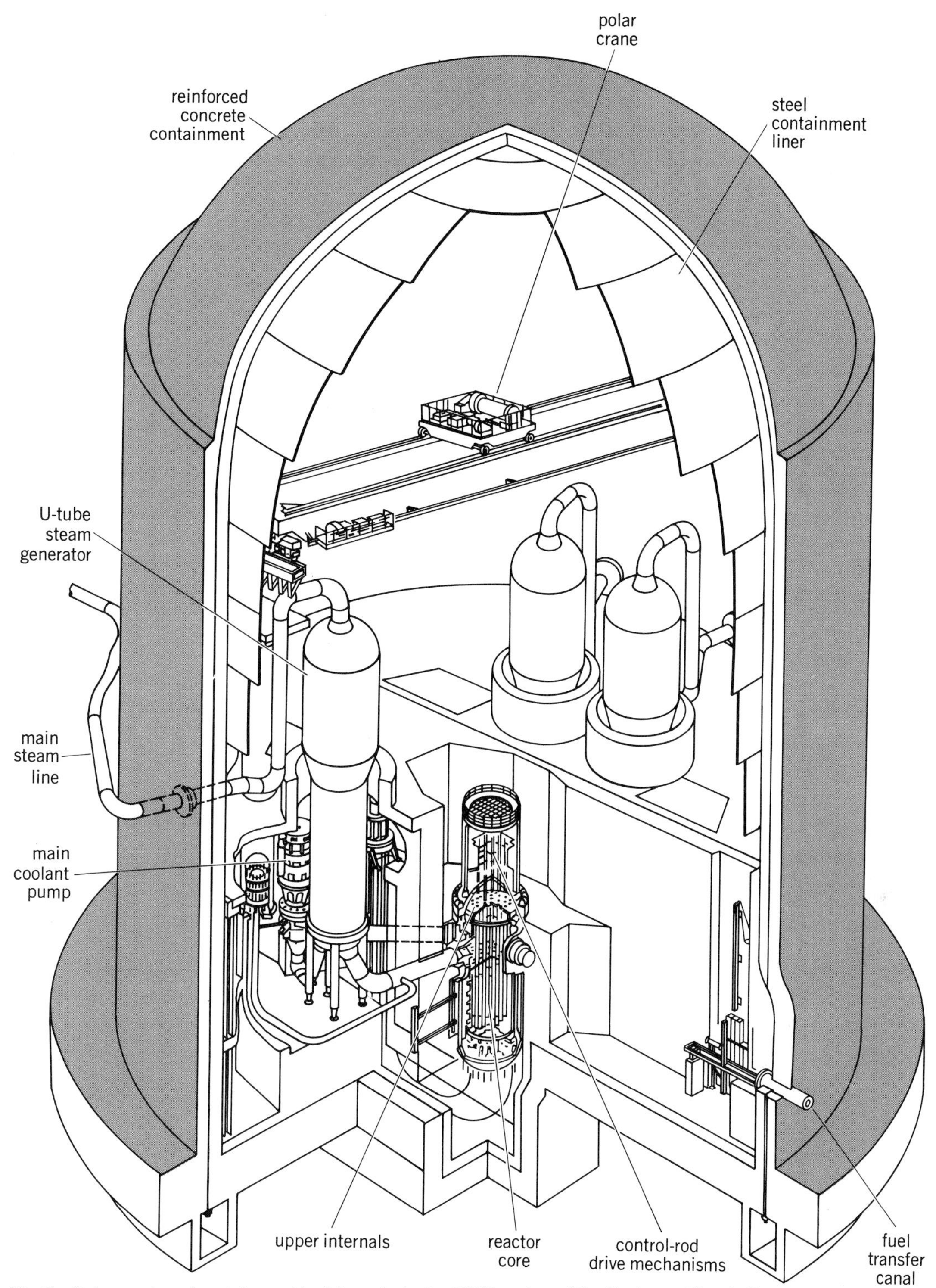

Fig. 9. Cutaway view of containment building of a typical PWR system. (*Westinghouse Electric Corp.*)

steam generator, and is recirculated in a closed primary system. A separate secondary water system is used on the shell side of the steam generator to generate steam, which is fed to the turbine, condensed, and recycled back to the steam generator. A pressurizer is used in the primary system to maintain about 150 atm (15 MPa) pressure to suppress boiling in the primary coolant. One loop is shown in Fig. 6, but up to four loops have been used.

The reactor pressure vessel is about 44 ft (13.4 m) in height, about 14.5 ft (4.4 m) in inside diameter, and has wall thickness in the core region at least 8.5 in. (22 cm). The active length of the fuel assemblies may range from 12 to 14 ft (3.7 to 4.3 m), and different configurations are used by the manufacturers. For example, one type of fuel assembly contains 264 fuel rods, and 193 fuel assemblies are used for the 3411-MW-thermal, four-loop plant. The outside diameter of the fuel rods is 0.374 in. (9.5 mm). For this arrangement, the control rods are grouped in clusters of 24 rods per cluster, with 61 clusters provided. In the PWR the control rods enter from the top of the core. Control of the reactor operations is carried out by using both the control rods and a system to increase or decrease the boric acid content of the primary coolant.

Figure 7 is an external view of the Oconee Nuclear Power Station with three reactors, each housed in a separate containment building. The prestressed concrete containment buildings are designed for a 4-atm (400 kPa) rise in pressure and have inside dimensions of about 116 ft (35.4 m) in diameter and 208.5 ft (63.6 m) in height. The walls are 45 in. (1.14 m) thick. The containments have cooling and radioactive absorption systems as part of the engineered safety features. Figure 8 is a view of a design for a 3800-MW-thermal nuclear power reactor utilizing a two-loop system placed in a spherical steel containment (about 200, or 60 m, in diameter), surrounded by a reinforced-concrete shield building. A cutaway view of the containment building is shown in Fig. 9.

The instrumentation and control for a nuclear power station involves separation of control and protection systems, redundant and diverse features to enhance the safety of the operations, and ex-core and in-core monitoring systems to ensure safe, reliable, and efficient operations. *See* ELECTRIC POWER GENERATION; ENERGY SOURCES.

[H. S. ISBIN]

Bibliography: Atomic Energy Commission, *Proposed Final Environmental Statement: Liquid Metal Fast Breeder Reactor Program*, WASH-1535, 1974; Energy Research and Development Administration, *Final Environmental Statement*, ERDA-1535 (including WASH-1535), 1975; International Atomic Energy Agency, *Nuclear Power and Its Fuel Cycle*, vols. 1–7, 1977–1978; H. S. Isbin (coordinator), *Public Issues of Nuclear Power*, 1975; Nuclear Regulatory Commission, *Operating Units Status Reports, such as Licensed Operating Reactors*, NUREG-75/020-12, December 1975; Nuclear Regulatory Commission, *Reactor Safety Study*, WASH-1400, 1975; Subcommittee on Energy and the Environment of the Committee on Interior and Insular Affairs, House of Representatives, 94th Congress, first session, pt. 1: *Overview of the Major Issues*, pt. 2: *Nuclear Breeder Development Program*, 1975; U.S. Congress, Office of Technology Assessment, *Nuclear Proliferation and Safeguards*, 1977.

Ocean-atmosphere relations

This field of investigation is concerned with the boundary zone between sea and air and the dynamic relationships between oceanographic and meteorologic studies. When it is considered that the largest fraction of the heat energy the atmosphere receives for maintaining its circulation is derived from the condensation of water vapor originating primarily from oceanic evaporation, it becomes evident that an understanding of processes occurring at the air-sea boundary is fundamental to an understanding of atmospheric behavior. The oceanic energy supply to the atmosphere is highly regionalized because of the character of ocean currents, which in turn implies that the atmospheric circulation itself (and resulting weather) is greatly influenced by the oceanic circulation. Conversely, the oceanic circulation represents a state of equilibrium in which the effects of the frictional stresses of the wind on the sea surface are balanced by changes in the distribution of density of oceanic waters. These compensating density changes are in turn related to time and space variations in radiation, heat conduction, evaporation, and precipitation. It is therefore equally manifest that an understanding of the oceanic circulation (and the resulting distribution of properties within the ocean) requires a thorough knowledge of atmospheric processes occurring at the air-sea boundary.

The conclusion is that neither ocean nor atmosphere should be treated independently, but that they should be considered together as a single dynamical-thermodynamical system. However, the interaction of ocean and atmosphere is so complicated that it is not yet possible to completely separate cause and effect. Therefore separate discussions are presented for each of the major classes of atmospheric influences upon the ocean, as well as the oceanic influences upon the atmosphere.

Effects of wind on ocean surface. The frictional stresses of the wind on the surface of the sea produce ocean waves (and storm surges) and ocean currents. The former are transitory phenomena and are not discussed in the present article because they are of little direct meteorological interest, even though waves at sea, coastal breakers, and storm tides are of considerable maritime, as well as oceanographic, importance. *See* SEA STATE.

Wind-induced ocean currents, on the other hand, are of large-scale significance from both the oceanographic and meteorological points of view. The wind exerts a twofold effect upon the surface layers of the ocean. In the first instance the stress of the wind leads to the formation of a shallow surface wind drift. The resulting transport of surface water by the wind drift leads in the second instance to pressure variations with depth and a changed distribution of mass (density) throughout the ocean. In the final analysis it is the resulting fields of density which account for the major current systems of the oceans. The total transport due to the wind drift is directed to the right of the wind (in the Northern Hemisphere), but the final density (slope) current which results from the sloping sea surface tends to flow in the direction of the prevailing wind, except where coastal configuration prevents the realization of such flow. Nevertheless, it should again be emphasized that the wind effect is not the sole meteorological factor that serves to determine the distribution of mass or the slope of isobaric surfaces in the oceans; heating and cooling, freezing and thawing, and evaporation and precipitation all exert their influences.

Major ocean current systems. For the reasons just outlined, the major ocean currents of the world conform closely with the prevailing anticyclonic wind circulatons of the oceans. With the exception of the northern Indian Ocean, warm currents flow poleward to about 40° lat in the western portion of all oceans, with easterly flow in the higher latitudes, equatorward drift (relatively cold) in the eastern portions of the oceans, and westerly flowing currents near the Equator. The low temperature of the waters in the eastern portions of the oceans is due partly to the high-latitude origin of the currents and partly to coastal upwelling of cold subsurface waters. Of all the currents the poleward-flowing warm currents of the western portions of the oceans are the best developed and the most important. Examples are the Gulf Stream of the North Atlantic and the Kuroshio of the North Pacific, each of which transports a tremendous volume of warm tropical water into higher latitudes.

Hydrologic and energy relations. For the Earth as a whole and for the entire year, the amounts of energy received and lost through radiation are in balance for all practical purposes. However, this is not true for any given portion of the Earth's surface, particularly during any given fraction of the annual solar cycle. The ratio of insolation to outgoing radiation decreases from the Equator toward the poles. Between the Equator and the 35th parallel, the Earth receives more energy through radiation than it loses; the reverse is true poleward from about the 35th parallel.

Because observations indicate that the lower latitudes are not becoming progressively warmer and the higher latitudes colder, it must be assumed that considerable heat is transported from lower to higher latitudes by both atmosphere and ocean. According to H. U. Sverdrup (see bibliography, G. P. Kuiper, 1954), the meridional transport of energy in the Northern Hemisphere reaches a maximum a little north of latitude 35°N. At latitude 30°N, Sverdrup computes the total energy transport across the latitude circle to be 6.5×10^{16} cal/min of which 1.9×10^{16} cal/min (or 29%) is accomplished by ocean currents, principally by the Gulf Stream and Kuroshio. The remaining energy transport is accomplished by the atmosphere.

The largest fraction of radiant energy absorbed by the oceans is utilized in evaporating sea water. A much smaller fraction, about 10%, is utilized in more direct heating of the atmosphere in contact with the sea surface. The latent energy of vaporization subsequently becomes available to the atmosphere as either sensible heat or gravitational potential energy when condensation takes place, often in a region far removed from the area where the evaporation occurred. The precipitation resulting from the condensation of atmospheric water vapor is then returned to the ocean, either directly as rainfall or snowfall, or indirectly as runoff and discharge from land areas. The hydrologic and energy cycle is thereby completed. *See* HYDROLOGY; HYDROSPHERIC GEOCHEMISTRY; TERRESTRIAL ATMOSPHERIC HEAT BALANCE.

The maximum evaporation and heat exchange between sea and atmosphere take place where cold air flows over warm-water surfaces. The ideal locations for maximum moisture or energy transfer are therefore those areas where cold continental air flows out over warm poleward-moving ocean currents. Such ideal conditions exist during winter off the eastern coasts of the continents and over warm currents such as the Gulf Stream and Kuroshio. Radiant energy that was absorbed and stored by the oceans at lower latitudes is given off to the atmosphere by this process at places and during seasons of marked deficiency in radiative energy. Any change in the oceanic transport by ocean currents must be reflected in corresponding changes in the rates of evaporation and must finally have significant effects upon the atmospheric circulation. *See* CLIMATIC CHANGE.

Relations with sea-water salinity. In the absence of horizontal flow, the surface salinity of any portion of the ocean is mainly determined by three processes: decrease of salinity by precipitation, increase of salinity by evaporation, and change of salinity by vertical mixing. Salinities thus tend to be high in regions where evaporation exceeds precipitation and low where precipitation exceeds evaporation. However, horizontal transport of surface waters by wind-induced ocean currents serves to displace the areas of maximum or minimum salinity in the direction of surface flow away from the areas of maximum differences (positive or negative) between evaporation and precipitation. The conclusion, of course, is that the distributions of surface salinities (as well as other properties) in the ocean are determined almost completely by atmospheric circumstances. *See* SEA WATER.

Other factors. In addition to the transfer of momentum, heat, and the latent energy involved in the evaporation-condensation-precipitation process, the oceans and the atmosphere interact in a multitude of other ways. These include radiation effects, flow of electrical charges, sea salts (a source of atmospheric condensation nuclei), and the flow of gases across the air-sea interface. Of the gases, the exchange of carbon dioxide between sea and atmosphere is perhaps the most important. Since the ocean absorption of carbon dioxide increases with salinity and decreases with temperature, it is suggested that in low latitudes the air is enriched with carbon dioxide given up by the ocean and that the general atmospheric circulation carries the carbon dioxide into higher latitudes, where it again dissolves in the sea water which in time brings it back toward the Equator. The combustion of fossil fuels has added considerable excess of carbon dioxide to the atmosphere during the past 100 years. In view of the great importance of carbon dioxide in determining the radiational balance of Earth and space, it is extremely important to determine whether this output of carbon dioxide has caused a significant increase in the total content of carbon dioxide in the atmosphere or whether most of it has been absorbed by the oceans. This is a controversial matter that is still under investigation by meteorologists and oceanographers.

[WOODROW C. JACOBS]

Bibliography: B. Bolin and E. Eriksson, Changes in the carbon dioxide content of the atmosphere and sea due to fossil fuel combustion, in B. Bolin (ed.), *The Atmosphere and Sea in Motion*, pp. 130–142, 1959; T. E. Graedel, The kinetic photochemistry of the marine atmosphere, *J. Geophys. Res.*, January 1979; W. C. Jacobs, The energy exchange between sea and atmosphere and some of its consequences, *Bull. Scripps Inst. Oceanogr. Univ. Calif.*, 6:27–122, 1951; G. P. Kuiper (ed.), *The Solar System*, vol. 2, 1954; P. S. Liss and P. G. Slater, Flux of gases across the air-sea interface, *Nature*, 247:181–184, 1974; J. S. Malkus, *Large-*

scale Interactions, in M. N. Hill (ed.), *The Sea*, vol. 1, 1962; G. Newman and W. J. Pierson, *Principles of Physical Oceanography*, 1966; H. U. Roll, *Physics of the Marine Atmosphere*, 1965; H. U. Sverdrup, *Oceanography for Meteorologists*, 1942.

Oceanography

The scientific study and exploration of the oceans and seas in all their aspects, including the sediments and rocks beneath the seas; the interaction of sea and atmosphere; the body of sea water in motion and subject to internal and external forces; the living content of the seas and sea floors and the behavior of these organisms; the chemical composition of the water; the physics of the sea and sea floor; the origin of ocean basins and ancient seas; and the formation and interaction of beaches, shores, and estuaries. Hence oceanography, sometimes called the science of the seas, consists of the marine aspects of several disciplines and branches of science: geology, meteorology, biology, chemistry, physics, geophysics, geochemistry, fluid mechanics, and in its more theoretical aspects, applied mathematics. Oceanography is also an environmental science which describes and attempts to explain all processes in the ocean, and the interrelation of the ocean with the solid and gaseous phases of the Earth and with the universe.

OCEAN RESEARCH

Because of the fluid nature of its contents, which permits vertical and horizontal motion and mixing, and because all the waters of the world oceans are in various degrees of communication, it is necessary to study the oceans as a unit. Further unification results from the technological necessity of studying the ocean from ships. Many phases of oceanic research can be carried out in a laboratory, but to study and understand the ocean as a whole, scientists must go out to sea with vessels adapted or built especially for that purpose. Furthermore, data must be obtained from the deepest part of the ocean and, if possible, scientists must go down to the greatest depths to observe and experiment. Another unifying influence is the fact that many oceanic problems are so complex that their geological, biological, and physical aspects must be studied by a team of scientists. Because of the unity of processes operating in the ocean, and because some writers have separated marine biology from oceanography (implying the term oceanography to embrace primarily physical oceanography, bottom relief, and sediments), the term "oceanology" is sometimes used as embracing all the science divisions of the marine hydrosphere. As used in this article, the term oceanography applies to the whole of sea science. *See* HYDROSPHERE.

Development. The early ocean voyages by Frobisher, Davis, Hudson, Baffin, Bering, Cook, Ross, Parry, Franklin, Amundsen, and Nordenskiold were undertaken primarily for geographical exploration and in search of new navigable routes. Information gathered about the ocean, its currents, sea ice, and other physical and biological phenomena was more or less incidental. Later, the polar expeditions of the Scoresbys, Parry, Markham, Greeley, Nansen, Peary, Scott, and Shackleton were also voyages of geographical discovery, although scientific observations about the sea and its inhabitants were made by some of them. William Scoresby took soundings and observed that discolored water containing living organisms (now known to be diatoms) was related to whale movements. Ross made dredge hauls of bottom-living animals. Nansen contributed to the improvement of plankton nets and suggested the existence of internal waves.

More closely related to the beginning of oceanography as comprehensive study of the seas are the 19th-century activities of naturalists Ehrenberg, Humboldt, Hooker, and Örstedt, all of whom contributed to the eventual recognition of plankton life in the sea and its role in the formation of bottom deposits. Charles Darwin's observations on coral reefs and Müller's invention of the plankton net belong to this phase of developing interest in marine science, in which men began to investigate ocean phenomena as biologists, chemists, and physicists rather than as oceanographers. In this group should also be included such physicists and mathematicians as Kepler, Vossius, Fournier, Varenius, and Laplace, who provided the background for the development of modern theories and investigations of ocean currents and air circulation.

Toward the middle of the 19th century a few scientists began to study the oceans as a whole, rather than as an incidental part of an established discipline. Forbes, as a result of his work at sea, first developed a scheme for vertical and horizontal distribution of life in the sea. On the physical side Matthew Fontaine Maury, developing and extending Franklin's earlier work, made comprehensive computations of wind and current data and set up the machinery for international cooperation. His book *Physical Geography of the Sea* has been regarded as the first text in oceanography.

Forbes and Maury were followed by a distinguished group of men whose interest in oceanography led them to make the first truly oceanographic expeditions. Most famous of these was the three-year around-the-world voyage of HMS *Challenger*, which followed earlier explorations of the *Lightning* and *Porcupine*. Instrumental in organizing these was Wyville Thompson, later joined by John Murray. Later in succession were the Norwegian Johan Hjort and the *Michael Sars* North Atlantic exploration; Louis Agassiz; and Albert Honoré Charles, Prince of Monaco, in a series of privately owned yachts named *Hirondelle I* and *II* and *Princess Alice I* and *II*. Other important contributions were made by Michael and G. O. Sars, Björn Helland-Hansen, Carl Chum, Victor Hansen, Otto Petterson, Gustav Ekman, and the vessels *Valdivia*, the Danish *Dana*, the British *Discovery*, the German *National* and *Meteor*, and the Dutch *Ingold*, *Snellius*, and *Siboga*, the French *Travailleur* and *Talisman*, the Austrian *Pola*, and the North American *Blake*, *Bache*, and *Albatross*. Among the North American pioneers were Alexander Agassiz, L. F. de Pourtales, and J. D. Dana. Pioneers in modern oceanographic work are M. Kunelsen, Sven Ekman, A. S. Sverdrup, A. Defant, Georg Wüst, Gerhard Schott, and Henry Bigelow.

Modern oceanography relies less upon single

explorations than upon the continuous operation of single vessels belonging to permanent institutions, such as *Atlantis II* of the Woods Hole Oceanographic Institution, *Argo* of Scripps Institution of Oceanography, *Vema* of the Lamont Geological Observatory, the French *Calypso*, and the large Soviet vessels *Vitiaz* and *Mikhail Lomonosov*. Single explorations continue to be made, as exemplified by the Swedish *Albatross* and Danish *Galatea*.

The reduction of data and study of collections from earlier expeditions were carried out generally in research institutions, museums, and universities not solely or primarily engaged in oceanography. The first marine laboratories were interested principally in fishery problems or were designed as biological stations to accommodate visiting investigators. Many of the former have extended their activities to cover chemical and physical oceanography during their growth and development. The latter, often active as extensions of university biological departments, are exemplified by the Naples Zoological Station and the Marine Biological Laboratory of Woods Hole. Visitors to such stations contribute greatly to the development of biology, generally in such fields as embryology and physiology.

The number of institutions devoted to organized oceanographic investigations with permanent scientific staffs has gradually grown. At first the requirements of fishery research provided the stimulus in countries adjacent to the North Sea, but in later years laboratories in other countries wholly or mainly devoted to oceanography have grown considerably in number. A few may be mentioned here. In England, among other important institutions, are the National Institute of Oceanography, the Marine Biological Laboratory at Plymouth, and the Fisheries Laboratory at Lowestoft. In the United States are the Woods Hole Oceanographic Institution in Massachusetts, the Scripps Institution of Oceanography in California, the Lamont-Doherty Geological Observatory in New York, the University of Miami Marine Laboratory in Florida, the Departments of Oceanography of Texas A & M College, the University of Washington at Seattle, and the University of Hawaii, and the School of Oceanography at Oregon State University. In Germany oceanographic laboratories are located at Kiel and at Hamburg. In Denmark the Danish Biological Station is at Copenhagen. Other European laboratories include those at Bergen, Norway; Göteborg and Stockholm, Sweden; Helsinki, Finland; and Trieste. Laboratories are located at Tokyo, Japan; Namaimio and Halifax, Canada; and Hawaii. This list is not inclusive and necessarily leaves out a considerable number of important institutions.

Surveys. Oceanographic surveys require careful planning because of high cost. Provision must be made for the proper type of vessel, equipment, and laboratory facilities, adapted to the nature and duration of the survey.

Research ships. Ships of all types and sizes have been gathering information about the oceans since earliest times. Vessels of less than 300 tons displacement seldom range farther than several hundred miles from land, whereas ships larger than 300 tons displacement may work in the open ocean for several months at a time. Research ships of all sizes must be seaworthy and must provide good platforms from which to work (Fig. 1). More specifically, a ship must have comfortable quarters, adequate laboratory and deck space for preliminary analyses, plus storage space for equipment, explosives, samples, and scientific data. Machinery, usually in the form of winches and booms, is necessary for handling the complex and often heavy scientific equipment needed to probe the ocean depths (Fig. 2).

A number of the larger oceanographic vessels

Fig. 1. Deep-sea drilling vessel *Glomar Challenger*, equipped with satellite navigation equipment and capable of holding position to within 100 ft (30 m) for several days, using bottom-mounted sonic beacons, tunnel thrusters, and computers.

Fig. 2. Trawl winch used on research vessel *Vema*. (*Lamont-Doherty Geological Observatory*)

are equipped with general-purpose digital computers, including tapes, disk files, and process-interrupt equipment. The result is that data can be reduced on board for experimental work, and all the operations taking place on the vessel can be centralized, including the satellite navigation equipment.

Standard oceanographic equipment includes collecting bottles (Nansen bottles) for obtaining water samples and thermometers (both reversing thermometers and bathythermographs) for measuring temperatures at all depths. In addition, there are various devices for obtaining samples of ocean bottom sediments and biological specimens. These include heavy coring tubes which punch cylindrical sediment sections out of the bottom, dredges which scrape rock samples from submerged mountains and platforms, plankton nets for collecting very small planktonic organisms, and trawls for collecting large free-swimming organisms at all oceanic depths. Echo sounders provide accurate profiles of the ocean floor.

Specialized equipment for oceanic exploration includes seismographs for measuring the Earth's crustal thickness, magnetometers for measuring terrestrial magnetism, gravimeters for measuring variations in the force of gravity, hydrophotometers for measuring the distribution of light in the sea, heat probes (earth thermometers) for measuring the flow of heat from the Earth's interior, deep-sea cameras to photograph the sea bottom, bioluminescence counters for measuring the amount of luminescent light emitted by organisms, salinometers for measuring directly the salinity of sea water, and current meters to clock the speed of ocean currents.

Positioning of a ship is very important for accurate plotting of data and detailed charting of the oceans. Celestial navigation is in wide use now as in the past. Navigational aides such as electronic positioning equipment (loran, shoran, and radar) are increasing the accuracy of positioning to within several tens of yards of the ship's true position. Navigational and radio communications equipment normally is situated near the captain's bridge, but often is duplicated in the scientific laboratories in order that complete communication between the ship's operators, scientists, and other participating ships can be carried on at all times.

Probably the single most important advance in deep-sea oceanography has been the introduction of satellite navigation. Combined with a computer, satellite navigation permits fixes to within several hundred yards while the ship is moving and to within several hundred feet when the ship is located with respect to bottom-mounted beacons, for example, in deep-sea drilling.

Ship's laboratories. Laboratories must be adaptable for a large number of operations. In general they are of two categories, namely, wet and special laboratories. Wet laboratories are provided with an open-drain deck so that surplus ample water can be drained out on deck. Such a laboratory is located near the winches used for running out and retrieving a long string of water-sample bottles (hydrocasts). Adjoining the wet laboratory are special laboratories equipped with benches for measuring chemical properties of the recovered water and for examination of biological and geological samples. Electronics laboratories are either part of, or adjacent to, the special laboratory, depending upon the size of the ship. Here, numerous recording devices, amplifiers, and computers are set up for a variety of purposes, such as measurements of underwater sound, measurement of the Earth's magnetic and gravity fields, and seismic measurements of the Earth's crustal thickness (Fig. 3).

Marine technology. The development of nuclear power plants permits extended voyages without the necessity of refueling. Nuclear power plants, used in conjunction with inertial guidance in the submarines *Nautilus* and *Skate*, made possible the first extended journey under the Arctic ice pack. Uncharted regions of the oceans are within the reach of exploration.

Direct visual observations of the ocean depths are fast becoming a reality, both by "manned" submersibles (bathyscaph) and by television cameras. Deep-sea cameras have been developed to the point whereby motion pictures of even the deepest parts of the sea bottom can be taken (Fig. 4). However, such observations yield no information about the subsurface material. Major crustal features are determined by seismic measurement. The shallower features of the subbottom structure and deposits were not reaily observed until the advent of the subbottom acoustic probe. This device is a very-high-energy echo sounder capable of penetrating below the sediment-water interface and yielding a continuous profile of the subbottom strata.

Information as to the physical and chemical makeup of the underlying material, however, is dependent upon penetration and actual recovery. Commonly used coring devices rarely penetrate more than 10–20 m (on occasion to about 33 m) below the surface. A new "incremental" coring device has been developed for taking successive 2-m sediment cores to depths of possibly 100 m or more.

The most impressive achievement in this area, however, has been in deep-sea drilling. The *Glomar Challenger*, operated by the Global Marine

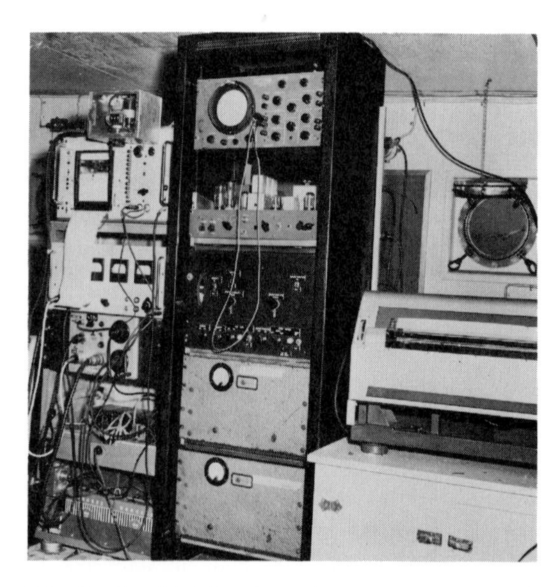

Fig. 3. Special laboratory aboard research vessel. (*Lamont-Doherty Geological Observatory*)

Fig. 4. Remote-controlled underwater television camera mounted in self-propelled vehicle, which can make visual surveys to depths of 1000 ft. Self-buoyant unit moves about or hovers at desired depth in currents or tides of several knots. (*Vare Industries, Roselle, N.J.*)

Corp., was designed primarily for the purpose of taking cores throughout the full column of sediment in mid-ocean. It is capable of handling 24,000 ft of drill string in mid-ocean and drilling through more than 2500 ft of sediment, taking 30-ft cores in the process. By mid-1969 the vessel had successfully operated at more than 40 sites in both the Atlantic and Pacific. This operation is sponsored by the National Science Foundation, with scientific guidance supplied by JOIDES (Joint Oceanographic Institutions Deep Earth Sampling), which include the Bundesanstalt für Geowissenschaften und Rohstoffe, Federal Republic of Germany; Scripps Institution of Oceanography, University of California at San Diego; Centre National pour l'Exploitation des Oceans; Lamont-Doherty Geological Observatory of Columbia University; Hawaii Institute of Geophysics, University of Hawaii; Rosentiel School of Marine and Atmospheric Sciences of the University of Miami; Natural Environment Research Council of London; School of Oceanography, Oregon State University; Graduate School of Oceanography, University of Rhode Island; Ocean Research Institute, University of Tokyo; P. P. Shirshov Institute of Oceanology, U.S.S.R. Academy of Sciences; Woods Hole Oceanographic Institution; and the Departments of Oceanography of the University of Washington and Texas A & M University. The Scripps Institution of Oceanography is the operating institution. Besides incorporating the necessary innovations in drilling and coring, the *Glomar Challenger* is an example of the many important advances in oceanography, such as satellite navigation and positioning.

Oceanographic stations. Work on station consists of sampling and measuring as many marine properties as possible within the limitations of an expedition. Water, sea-bottom, and biological samples are successively collected on the long cables extended to the ocean floor. In some surveys one cable lowering may include samplers for all these items, but this is not the usual procedure. Stations are systematically located at predetermined points along the ship's path. At hydrographic stations, observations of water temperature, salinity, oxygen, and phosphate content are determined upon sample recovery. Seismic stations generally are carried out by two ships, one running a fixed course and dropping explosives while the second remains stationary and records the returning subbottom reflected or refracted sound waves. Biological stations may consist of vertical net hauls or horizontal net tows depressed to sweep the ocean at a fixed depth. Geological stations are usually coring or bottom-dredging operations. Wherever possible the recovered data are given a preliminary reduction aboard ship so that interesting discoveries are not bypassed before sufficient information is obtained. Detailed analyses aboard ship are seldom possible because of the limitations of time, space, and laboratory equipment. Instead, the carefully processed, labeled, and stored material is preserved for intensive study ashore.

Home laboratory. This phase of the work may entail many months of careful examination and detailed analyses. Batteries of sophisticated scientific instruments are often necessary: data computers for reduction of physical oceanography information; spectrographic apparatus consisting of emission units; infrared, ultraviolet, x-ray, and mass spectrometers for chemistry; aquaria, pressure chambers, and chemostats for the biologist; electron microscopes and high-powered optical microscopes for examination of inorganic and organic constituents; radioisotope counters; and numerous standard physical, chemical, geological, and biological instruments. The great variety of measurable major and minor properties is reduced to statistical parameters which may then be integrated, correlated, and charted to increase the knowledge of sea properties and show their relationships. The essentially descriptive properties lead to an understanding of the principles which control the origin, form, and distribution of the observed phenomena. The present knowledge of the oceans is still fragmentary, but the increasing store of information already is being applied to a rapidly expanding number of man's everyday problems.

Research problems. Since the middle of the 19th century, man has learned more and more about that 71% of the Earth which is covered with sea water. The rate of increase of knowledge is being accelerated by improved tools and methods and increased interest of scientists and engineers. Some of the present and future research problems that are attracting scientists are mentioned below.

One of the oldest and still unsolved problems is the motion of ocean waters, involving surface currents, deep-sea currents, vertical and horizontal turbulent motion, and general circulation. New methods such as distribution of radioactive substances, deep-sea current meters and neutrally buoyant floats, high-precision determination of salt and gas content, and hot-wire anemometers for turbulence studies and current measurements have increased present knowledge considerably. The surface movement of water is wind-produced, and a general theory of the motion has been worked out. The deep-sea currents, which are known to be caused in part by variations in the thermohaline circulation, are still an open problem. Superimposed on these movements is turbulent motion, which ranges over the whole spectrum from large ocean eddies transporting millions of cubic meters of water per second to the tiniest vibrations of water particles. Very little is known about turbulence.

The mixing of water masses and the formation of new water masses cannot yet be completely and adequately described, as many of the thermodynamic parameters are not precisely known. Laboratory experiments and measurements of thermal expansion, saline contraction, and specific heat at constant pressure must be carried out. A further problem is the composition of sea water and the extent to which the ratio among the components is constant. In connection with these problems it has been urged that a library of water samples be established. Improved techniques of measuring sound velocity, electrical conductivity, refractive index, and density must be developed to enable scientists to follow many processes in the ocean. It is therefore necessary to study the small variations of these parameters in the sea. *See* SEA WATER.

The tides in the oceans are rather well known at the surface but are almost completely unknown in the deep sea; also, the influence of land boundaries on deep-sea tides is not yet understood. Further research also must be devoted to the interesting phenomenon of internal waves.

The study of ocean waves is one of the most advanced topics in oceanography, but the energy exchange between atmosphere and sea surface by friction must be studied further. Another problem is that of the heat exchange between ocean and atmosphere, an important link in the heat mechanism which determines the weather and the oceanic circulation. *See* OCEAN-ATMOSPHERE RELATIONS.

The climate of the past, in particular that of the last 1,000,000 years, is best studied in the ocean. Isotopic methods in paleoclimatologic research allow the determination of temperature variations in the ocean with a high degree of accuracy. The rapid growth of geochemistry and the increased sampling of deep-sea sediments through improved techniques have solved some of the problems of deep-sea sedimentation. At the same time a number of new ones have been created, such as: Why is the sediment carpet only about 300 m thick? What is the mechanism of sediment transport? What is the history of sea water? Of the ocean basin? What is the cause of the ice ages?

The results of the Deep Sea Drilling Project have confirmed the expectations of its most optimistic supporters. It has strikingly confirmed the sea-floor-spreading hypothesis of the development of the ocean basins and the newer concepts of plate tectonics. A core to the Moho (about 3 mi deep) may help answer many of the questions about the structure of the Earth's crust. The problem of the mechanism of formation of the ridges and island chains may be near solution.

The age determination of sediments by radioactivity methods, which was thought impossible 30 years ago, is now used on deep-sea sediments older than 10,000,000 years. Very little is known about the formation of minerals on the sea floor, the diffusion and adsorption of elements in and on sediments, and the reaction at slow rates in sediments. Certainly microbiological processes on the sea floor are an important factor, as they seem to produce chemical energy in sediments.

In marine biology the systematics and ecology remain the major aspects. It is still the science of the "naturalist." The interest of marine biology is many-sided and not grouped around a few central problems. Ocean life offers to the general biologist the best opportunities to study such complex problems as the structure of communities and the flux of energy through these communities. The zonation of animals on the shore and in the open ocean is not yet fully understood; the cause of patchiness in the distribution must be found. On the other hand, the distribution of species by currents and eddies must be studied, and large-scale experiments on behavior must be carried out. Observation at sea has been neglected to a large extent, and therefore the equilibrium between sea observation and laboratory experiment must be restored. The great advances in genetics, biochemistry, physiology, and microbiology also will advance the study of life in the sea. *See* MARINE ECOSYSTEM.

Applications of ocean research. Directly and indirectly the ocean is of great importance to humans. It is valuable as a reservoir of natural resources, an outlet for waste disposal, and a means of transport and communication. The ocean is also important as a harmful agent causing biological, chemical, and mechanical destruction of life and property. In addition to the peaceful exploration of the oceans, there are many military applications of surface and submarine phenomena. In all of these aspects oceanography provides basic information for engineers who seek to increase its benefits and to avoid its harmful effects. *See* MARINE RESOURCES.

Food resources. The food resources of the ocean are potentially greater than those of the land since its larger area receives a proportionately larger amount of solar radiation, the source of living energy. Nevertheless, this potential is only in part realized. Oceanographic studies provide information which can help to increase fishing yields through improved exploratory fishing, economical harvesting methods, fisheries forecasts, processing techniques at sea, and aquaculture.

Fishes are dependent in their distribution upon food organisms and plankton, vertical and horizontal currents which bring nutrients to the plankton, bottom conditions, and physical and chemical characteristics of the water. A knowledge of the relation of food fishes to these environmental conditions and of the distribution of these conditions in the oceans is vital to successful extension of fishing areas. Satisfactory measurements of the basic organic productivity of the sea may become essential in the selection of regions for extended fishery exploration. *See* SEA-WATER FERTILITY.

The catching of fishes may be facilitated and new and more efficient methods devised through a knowledge of the reaction of fishes to stimuli and of their habits in general. This knowledge may result in better design of nets and in the use of electrical, sonic, photic, and chemical traps or baits. The harvest also may be increased by using improved methods of locating schools of fish by sonic or other means.

The biology of fishes, their food preferences, their predators, their relation to oceanographic conditions, and the fluctuations in these conditions seasonally and from year to year are important factors in forecasting fluctuations in the fisheries. This information will aid in preventing the economic waste entailed in alternating glut and scarcity, and it is essential to good management of fisheries and to sound regulation by conservation agencies.

Other anticipated advances which require further oceanographic study include (1) the improvement of fishing by transplanting the young of existing stocks or by introducing new stocks; (2) farming or cultivation of sea fishes (although this does not seem feasible at present, scientific research has improved the cultivation of oysters and mussels in France and Japan); and (3) the use of planktonic vegetation as a source of food or animal nutrition.

Mineral resources. Although most of the valuable chemical elements in sea water are in very great dilution, the great volume of water of the oceans (about 300,000,000 mi^3 or 1.25×10^9 km^3) provides a limitless and readily accessible reservoir, if such dilute concentrations can be economically extracted. Magnesium is produced largely from sea water, and bromide also has been extracted commercially. High concentrations of manganese are found in manganese nodules, which are very common on certain areas of the sea floor.

Other elements occur in too great dilution to be extracted by present methods. Possibly a better understanding of the ability of certain marine plants and animals to accumulate and concentrate elements from sea water in their tissues may lead to new methods of recovering these elements. *See* MARINE MINING.

Energy and water source. Sea water contains deuterium and would be a limitless source of this element in the event of successful nuclear fusion developments. Further oceanographic knowledge and advances in engineering may lead to the increased utilization of tidal energy sources, or of the heat energy available from temperature differences in the ocean. The development of new methods for removing salt from sea water offers promise that the sea may become a practical source for potable water.

Disposal outlet. Because of its large volume, the sea is frequently used for the disposal of chemical wastes, sewage, and garbage. A knowledge of local currents and tides, as well as of the bottom fauna, is essential to avoid pollution of beaches or commercial fishing grounds. Radioactive waste disposal in offshore deeps poses problems of the rate of movement of deep waters and the transfer of radioactive materials through migration and food chains of marine organisms. A problem of rapidly growing concern is oil pollution, which is caused in part by the rapidly expanding exploitation of the continental-shelf-oil resources. The Santa Barbara incident in 1969 was one example. There must be considerable development in marine engineering and better ecological understanding if these oil resources are to be fully utilized. A second major cause of oil pollution is the breakup of giant oil tankers, such as the *Torrey Canyon*. Prevention of such accidents requires improved vessels and more stringent navigation controls.

Traffic and communication. The sea still remains an important highway; thus the knowledge and forecasting of waves, currents, tides, and weather in relation to navigation are of great practical importance. New developments include the continuous rerouting of ships at sea in order that they may follow the most economic paths in the face of changing weather conditions. A knowledge of submarine topography, geologic processes, and temperature conditions is important for the satisfactory location, operation, and repair of submarine cables. The use of Sofar in air-sea rescue operations is based upon submarine acoustics.

Defense requirements. Defense aspects of marine research involve not only the navigation of surface vessels but also undersea craft with special navigational problems related to submarine topography, echo sounding, and the distribution of temperature, density, and other properties. Research in submarine acoustics has improved communication between, and detection of, undersea craft. In spite of these advances natural conditions, such as warm water pockets and subsurface magnetic irregularities, can conceal submarines from conventional means of detection. Investigation of these conditions is essential to any defense against missile-carrying submarines.

Property and life. Damage to docks and ships by marine borers and fouling organisms is controlled by methods that utilize a knowledge of the biology, behavior, and physiology of the destructive organisms, and of the oceanographic conditions which control their distribution. Loss of life caused by the attacks of sharks and other fishes may be reduced through an understanding of their behavior and the development of repellants and other protective devices. The chemical characteristics of sea water pose special problems of corrosion of metals. Beach erosion, wave damage to harbor and offshore structures, the effects of tsunamis and internal waves, and storms cause loss of property and life. Much of this damage may be minimized by the application of oceanographic knowledge to forecasting methods and warning systems.

Indirect benefits of oceanography arise from the application of marine meteorology to weather prediction, not only over the sea areas but also over the land. The study of marine geology and marine ecology aid in the understanding of the character of oil-bearing sedimentary rocks found on land.

[WILLIAM A. NIERENBERG]

Bibliography: H. Barnes (ed.), *Oceanography and Marine Biology*, vol. 14, 1976; M. B. Deacon (ed.), *Oceanography: Concepts and History*, 1977; G. Dietrich, *General Oceanography*, 1963; W. A. Herdman, *Founders of Oceanography and Their Work*, 1923; G. L. Pickard, *Descriptive Physical Oceanography*, 1976; R. G. Pirie (ed.), *Oceanography: Contemporary Readings in the Ocean Sciences*, 1977; Scientific American Editors, *Ocean*, 1969; F. W. Smith and F. A. Kalber (eds.), *Handbook Series in Marine Science*, 2 vols., 1974.

THEORETICAL OCEANOGRAPHY

The basis for theoretical oceanographic studies is the known set of conservation equations for momentum, mass, and energy, supplemented by an equation of state for sea water and a conservation equation for dissolved salts. A general solution to the mathematical system is not possible. The aim of theoreticians is to develop simple mathematical models from the general set to explain observed oceanic features. A model may describe a process, such as the convective overturning of surface waters, or a phenomenon, such as the existence of the Gulf Stream on the western side of the North Atlantic. Whatever the purpose of the model, simplicity is important: the more directly that one can relate a feature to processes that are already understood, the better.

Oceanic flow. Oceanic flows with a horizontal scale of 100 km or more are strongly affected by the rotation of the Earth. Just as a tilted spinning top wobbles laterally instead of falling directly when acted upon by gravity, the rotation of the Earth causes a fluid flow to be deflected from the direction of the applied force. The deflecting (Coriolis) force is proportional to the angular rate of rotation of the Earth and to the sine of the latitude of the position of the fluid. Therefore, it is larger at high latitudes than at low.

The effect of rotation is easily incorporated into the conservation equation for momentum. When it is made to play a dominant role, the analysis is simplified because flows that are not affected by the rotation are effectively filtered out of the equations.

Wind effects. Simple theoretical analysis shows that the wind directly affects the surface waters of the ocean only in the top hundred meters or so and that rotation causes the net transport of water in this wind-driven layer to flow to the right of the direction of the wind stress. This indirect circulation in the surface layers induces a deeper flow in the direction of the wind, thereby setting up the wind-driven circulation. The Coriolis force exerts a dominating influence on the structure of the flow, and its variation with latitude gives rise to the Gulf Stream and the other observed western boundary currents.

Thermohaline circulation. Solar heating is most intense at low latitudes, and cooling of surface waters occurs in polar regions. The vertical circulation caused by the generated density differences has a global scale because the dense water that fills the abyssal ocean is formed in only a few polar locations. The rotation of the Earth affects the flow pattern of the deep circulation as well, giving rise to strong currents near coasts and underwater ridges that bound the basins on the west and to a relatively weak flow in the remaining areas. Theories of this thermohaline circulation verify the observation that the waters of the ocean are stably stratified. Light surface waters lie above the thermocline, a layer with a relatively sharp density change, and the abyss is filled with nearly homogeneous, dense water. *See* THERMOCLINE.

Waves. Given the stable stratification of the oceans, theory predicts the existence of a variety of large-scale waves. The periods of gravity waves are limited by rotation to be less than or equal to a day, or more precisely, the period of a Foucault pendulum. Longer-period, planetary waves owe their existence to the variation of the Coriolis force with latitude and to the variable depth of the ocean. These waves have periods ranging from several days to several years. Variable depth also gives rise to waves that are trapped in a layer near the bottom, the depth of the layer depending on stratification, rotation, and the scale of the waves. The periods here range from a few days to a few months. All of these waves can be identified in observational records, and some of the observed, nonwavelike features can be described by combinations of such wave motions.

Instability. Simpler wave studies are valid when the wave speeds are substantially larger than the fluid velocity, a condition that is often violated, especially for long-period waves. In such cases the current velocity must be included in the analysis. The energy associated with the currents then becomes available as a possible source of instability. The waves may grow in amplitude by drawing energy from the kinetic energy of the currents or from the potential energy of the stratification. Both types of instability occur in the ocean, though the more important source of energy seems to be the latter, in which case the basic flow is said to be baroclinically unstable. As the waves grow, they interact with each other and alter the current field from which they draw their energy. Eventually a state of statistically steady equilibrium may be achieved in which the basic flow is altered to the point where it provides enough energy to feed waves whose amplitudes are in equilibrium with the altered mean field.

Turbulence. In more extreme situations the instability may become so intense that it leads to turbulence. In that case traditional analysis breaks down, and a statistical-dynamical approach to the problem is required. The effect of turbulence is often parameterized in terms of properties of the mean field, somewhat in the manner in which the net effect of individual particle motions of a gas is parameterized in kinetic theory as a viscosity coefficient. However, the two situations are very different, and in the present setting that approach gives reasonable results only in some cases.

The latter 1970s saw the development of numerical oceanographic models that can resolve the unstable modes and follow their development as the turbulent regime is approached. The constraints of rotation and stratification help to restrict the structure of the motions essentially to two dimensions so that the problem can be handled with present computing facilities. Tentative results offer considerable hope that the turbulent effects can be parameterized. The role of bottom topography may be important in such a description.

The resolution of the large-scale turbulence problem is especially important in the oceanographic context because the fluctuating turbulent motions are normally much stronger than the velocity of the mean circulation. Hence, good understanding of the latter will require a knowledge of the effects of the former. A similar situation exists in the atmosphere, though the relative intensity of the fluctuations is somewhat smaller there.

Time-averaged circulation. A totally different theoretical approach can be used to determine the time-averaged circulation. Ocean water contains a number of dissolved chemicals, such as oxygen and silicates, which can be looked upon as tracers that indicate the flow from known source regions. Ideally one would like to be able to calculate the flow from the tracer distributions, but it is easy to show that this inverse problem generally does not have a unique solution. So far the most effective attacks on this problem have used a known velocity field, one that satisfies at least the major constraints, in the convective-diffusive equation for the tracer. The calculated distribution is then compared with the observed, and parameters are adjusted until an optimal fit is achieved. A more direct approach is to include the tracer in a numerical model of the circulation. [GEORGE VERONIS]

Bibliography: N. P. Fofonoff, Dynamics of Ocean Currents, in *The Sea*, vol. 1, pp. 323–395, 1962; National Academy of Sciences, *Numerical Models of Ocean Circulation*, 1975; P. B. Rhines. The Dynamics of Unsteady Ocean Currents, in *The Sea*, vol. 6 pp. 189–318, 1977; M. Stern, *Ocean Circulation Physics*, 1975; H. Stommel, *The Gulf Stream*, 1965; G. Veronis, Model of World Ocean Circulation, *J. Mar. Res.*, pt. I, 31:228–288, 1973, pt. III, 36:1–44, 1978.

Oil and gas, offshore

Oil and gas prospecting and exploitation on the continental shelves and slopes. Since Mobil Oil Co. drilled what is considered the first offshore well off the coast of Louisiana in 1945, exploration for petroleum and natural gas on the more than 8,000,000 mi² of the world's continental shelves and slopes, lying between the shore and 1000-ft water depth, has expanded rapidly to include exploration or drilling or both off the coasts of more than 75 nations. More than 119 national and private petroleum companies have joined in the worldwide offshore search for oil and gas.

Until early in 1975, most of the offshore rigs were employed in drilling on the United States Outer Continental Shelf. After that time, this changed dramatically, with more than 75% of the offshore drilling being conducted outside the United States waters. *See* CONTINENTAL SHELF AND SLOPE.

Paralleling the development of drilling vessels able to withstand the rigors of operation in the open sea have been remarkable technological developments in equipment and methods. More than 125 companies in the United States devote a large share of their efforts to the development and manufacture of material and devices in support of offshore oil and gas.

For many years petroleum companies stopped at the water's edge or sought and developed oil and gas accumulations only in the shallow seas bordering onshore producing areas. These activities were usually confined to water depths in which drilling and producing operations could be conducted from platforms, piers, or causeways built upon pilings driven into the sea floor. Major accumulations along the Gulf Coast of the United States, in Lake Maracaibo of Venezuela, in the Persian Gulf, and in the Baku fields of the Caspian Sea were developed from such fixed structures.

Exploration deeper under the sea did not begin in earnest until the world's burgeoning appetite for energy sources, coupled with a lessening return from land drilling, provided the incentives for the huge investments required for drilling in the open sea.

Geology and the sea. There is a sound geologic basis for the petroleum industry turning to the continental shelves and slopes in search of needed reserves. Favorable sediments and structures exist beneath the present seas of the world, in geologic settings that have proven highly productive onshore. In fact, the subsea geologic similarity—or in some cases superiority—to geologic conditions on land has been a vital factor in the rapid expansion of the free world's investment in offshore exploration and production.

Subsea geologic basins, having sediments considered favorable for petroleum deposits, total approximately 6,000,000 mi² (15,500,000 km²) out to a water depth of 1000 ft (300 m), or about 57% of the world's total continental shelf. This 6,000,000-mi² area is equivalent to one-third of the 18,000,000 mi² (46,600,000 km²) of geologic basins on land.

There has been a rapid growth of offshore production through the years. This production rate has been achieved despite the fact that only 166,000 mi² (430,000 km²), or about 2%, of the world's continental shelves out to 1000-ft water depths has been tested by the drill. Estimates indicate that only a small percentage of the seismic work necessary to evaluate the continental shelves has been accomplished, and that it would take another 127 years to complete the seismic survey of all those areas judged to be prospective. Surveying of the world's total favorable continental shelf area will take an estimated 68,400 seismic crew months, with 8300 months actually accomplished.

Virtually all of the world's continental shelf has received some geological study, with active seismic or drilling exploration of some sort either planned or in effect. Offshore exploration planned or underway includes action in the North Sea; in the English Channel off the coast of France; in the Red Sea bordering Egypt; in the seas to the north, west, and south of Australia; off Sumatra; along the Gulf Coast and east, west, and northwest coast of the continental United States; in the Cook Inlet and off the west coast of Alaska; in the Persian Gulf; off Mexico; off the east and west coasts of Central America and South America; and off Nigeria and North Africa. In addition, offshore exploration has been undertaken in the Caspian and Baltic seas off the coasts of East Germany, Poland, Latvia, and Lithuania; the China seas; Gulf of Thailand; Irish Sea; Arctic Ocean; and Arctic islands.

Successes at sea. The price of success in the offshore drilling is staggering in the amount of capital required, the risks involved, and the time required to achieve a break-even point.

Offshore Louisiana, long the center of much of the world's offshore drilling, has yielded a large number of oil and gas fields. Major oil and gas strikes have been made in the North Sea, and oil and gas production has been established in the Bass Straits off Australia. These have the potential for changing historical energy sources and the economies of the countries involved. Discoveries made since 1964 in the Cook Inlet of Alaska are making a major impact on the West Coast market of the United States; although they involve between 1.5×10^9 to 2×10^9 bbl (2.4 to 3.2×10^8 m³) of oil, the extremely high exploration and development costs will cause the break-even point to be

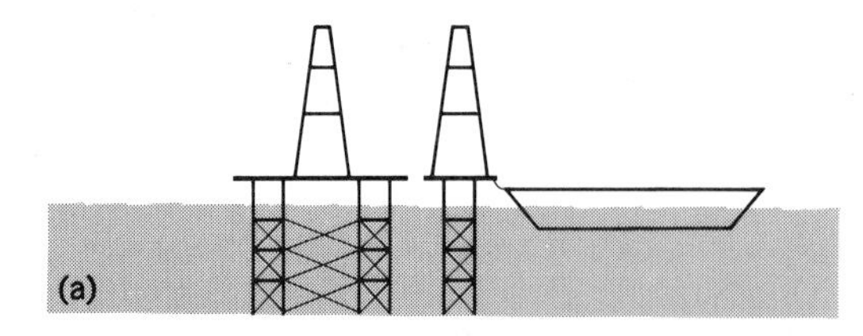

Fig. 1. Offshore fixed drilling platform. (*a*) Underwater design (*World Petroleum*). (*b*) Rig on a drilling site (*Marathon Oil Co., Findlay, Ohio*).

somewhere near 10 years. The most active offshore areas outside the United States are the North Sea, Southeast Asian seas, Nigeria, and the Persian Gulf.

The Persian Gulf area, where the world's largest fields on land have been found, has also produced the world's largest offshore field, Safaniya, in Saudia Arabia. This field is capable of producing more than 1,000,000 bpd (160,000 m^3/day).

Mobile drilling platforms. The underwater search has been made possible only by vast improvements in offshore technology. Drillers first took to the sea with land rigs mounted on barges towed to location and anchored or with fixed platforms accompanied by a tender ship (Fig. 1). A wide variety of rig platforms has since evolved, some designed to cope with specific hazards of the sea and others for more general work. All new types of platforms stress the characteristics of mobility and the capability for work in even deeper water.

The world's mobile platform fleet is of four main types: self-elevating platforms, submersibles, semisubmersibles, and floating drill ships.

The most widely used mobile platform is the self-elevating, or jack-up, unit (Fig. 2). It is towed to location, where the legs are lowered to the sea floor, and the platform is jacked up above wave height. These self-contained platforms are especially suited to wildcat and delineation drilling. They are best in firmer sea bottoms with a depth limit out to 300 ft (90 m) of water.

The submersible platforms have been developed from earlier submersible barges which were used in shallow inlet drilling along the United States Gulf Coast. The platforms are towed to location and then submerged to the sea bottom. They are very stable and can operate in areas with soft sea floors. Difficulty in towing is a disadvantage, but this is partially offset by the rapidity with which they can be raised or lowered, once on location.

Semisubmersibles (Fig. 3) are a version of submersibles. They can work as bottom-supported units or in deep water as floaters. Their key virtue is the wide range of water depths in which they can operate, plus the fact that, when working as floaters, their primary buoyancy lies below the action of the waves, thus providing great stability. The "semis" are the most recent of the rig-type platforms.

Floating drill ships (Fig. 4) are capable of drilling in 60-ft (18-m) to abyssal depths. They are built as self-propelled ships or with a ship configuration that requires towing. Several twin-hulled versions have been constructed to give a stable catamaran design. Floating drill ships use anchoring or ingenious dynamic positioning systems to stablize their position, the latter being necessary in deeper waters. Floaters cannot be used in waters much shallower than 70 ft (21 m) because of the special equipment required for drilling from the vessel subject to vertical movement from waves and tidal changes, as well as minor horizontal shifts due to stretch and play in anchor lines. Exploration in deeper waters necessitates building more semisubmersible and floating drill ships.

Production and well completion technology. The move of exploration into the open hostile sea has required not only the development of drilling vessels but a host of auxiliary equipment and techniques. A whole new industrial complex has developed to serve the offshore industry.

Of particular interest is the development of diving techniques and submersible equipment to aid in exploration and the completion of wells. Economics will soon force sea-bottom completions which require men or robots or both to make the necessary pipe and well connections. Such work is

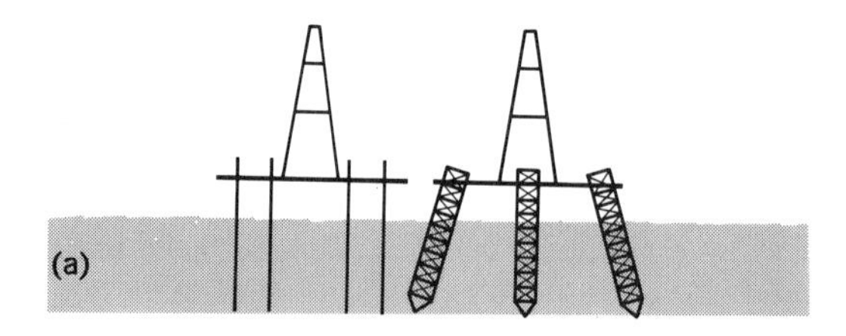

Fig. 2. Offshore self-elevating drilling platform. (*a*) Underwater design (*World Petroleum*). (*b*) Self-elevating drilling platform (*Marathon Oil Co., Findlay, Ohio*).

necessary even in the water depths now being developed.

A robot device has been developed which operates from the surface and uses sonar and television for viewing; it can excavate ditches for pipelines and make simple pipe connections and well hookups in water depths to 2500 ft (762 m). Limitations of robot devices are such that the more complex needs of well completions and service require the actions of men. To fill this need, diving specialists have been used to sandbag platform bases, recover conductor pipe, survey and remove wreckage, and make pipeline connections and well hookups. Diving depths have been increased to the point where useful work has been performed at depths in excess of 600 ft (180 m). Pressure chambers to take divers to the bottom and to return them to the surface to be decompressed are operational. One has been operating routinely in 425 ft (130 m) of water for Esso Exploration, Norway. This deep diving is made possible by the development of saturation diving, which uses a mixture of oxygen, helium, nitrogen, and argon. This technique has allowed divers to remain below 200 ft (60 m) for 6 days while doing salvage work on a platform. It has also allowed prolonged submergence at 600 ft in preparation for actual work on wells at this depth.

Miniature submarines have taken their place in exploration and completion work allowing the viewing of conditions, the gathering of samples, and some simple mechanical tasks. The depth range of these submarines is for all practical purposes unlimited.

Technical groups are experimenting with the design of drilling and production units that would be totally enclosed and be set on the sea bottom. Living in and working from these units, personnel would be able to carry out all the necessary oil field operations. In effect, such units would resemble a miniature city on the sea floor, from which a man would need to return to the surface only when his tour of work was completed. *See* OIL AND GAS FIELD EXPLOITATION.

Fig. 3. Offshore semisubmersible drilling platform. (*a*) Underwater design (*World Petroleum*). (*b*) Santa Fe Marine's Blue Water no. 3 drilling rig on a drilling site (*Marathon Oil Co., Findlay, Ohio*).

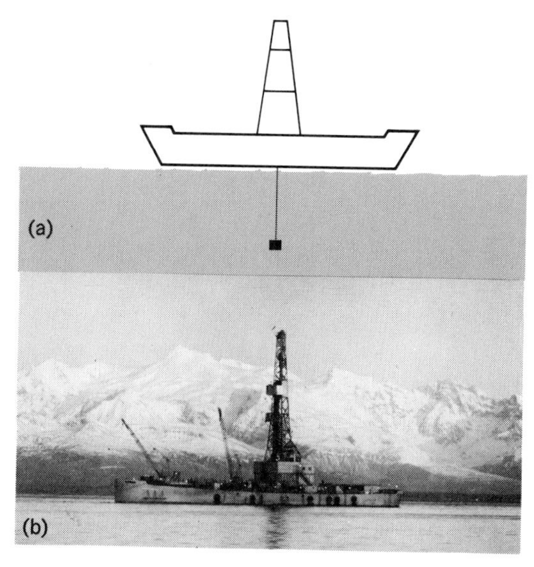

Fig. 4. Floating drill ship. Such ships can drill in depths from 60 to 1000 ft (18 to 300 m) or more. (*a*) Underwater design (*World Petroleum*). (*b*) Floating drill ship on a drilling site (*Marathon Oil Co., Findlay, Ohio*).

Concomitant with the progress of the petroleum industry in its venture into the open sea has been a vast increase in the knowledge of the sea and its contained wealth. Mining of the sea floor using some of the technology developed by the petroleum industry has started in several areas of the world, and actual farming or ranching of the life in the sea is being planned. *See* MARINE MINING; OCEANOGRAPHY.

Hazards at sea. As the petroleum industry pushed farther into the hostile environment of the sea, it sustained a series of disasters, reflected by the doubling of offshore insurance rates in April 1966. Between 1955 and 1968, for example, 23 offshore units were destroyed by blowout and 6 by hurricane and breakup and collapse at sea. The United States Gulf Coast, where a large percentage of the world's offshore drilling has taken place, was severely hit in 1964 and 1965 by hurricanes, which claimed over $7,500,000 in tow and service vessels, $21,000,000 in fixed platforms, and $28,000,000 in mobile platforms.

These figures do not include expenses sustained from loss of wells, removal of wrecked equipment, and loss of production. Such liabilities have raised insurance rates. Much current design work is aimed at engineering better safety features for the benefit of both the crews and structures.

Despite the hazards and monumental cost involved in extracting oil and gas from beneath the sea, the world's population explosion and its ever increasing demand for petroleum energy will force the search for new reserves into even deeper waters and more remote corners of the world. In truth, the search is only just beginning.

[G. R. SCHOONMAKER]

Bibliography: P. L. Baldwin and M. F. Baldwin,

Onshore Planning for Offshore Oil, 1975; *European Offshore Oil and Gas Yearbook 1976–77*, 1976; *OCS Statistics*, United States Geological Survey, 1973; Offshore news, *Offshore*, vol. 34, no. 10 and 13, 1974; Offshore 1973, *World Oil*, vol. 177, no. 1, 1973; L. G. Weeks, Petroleum resources potential of continental margins, in C. A. Burk and C. L. Drake (eds.), *The Geology of Continental Margins*, 1974.

Oil and gas field exploitation

In the petroleum industry, a field is an area underlain without substantial interruption by one or more reservoirs of commercially valuable oil or gas, or both. A single reservoir (or group of reservoirs which cannot be separately produced) is a pool. Several pools separated from one another by barren, impermeable rock may be superimposed one above another within the same field. Pools have variable areal extent. Any sufficiently deep well located within the field should produce from one or more pools. However, each well cannot produce from every pool, because different pools have different areal limits.

DEVELOPMENT

Development of a field includes the location, drilling, completion, and equipment of wells necessary to produce the commercially recoverable oil and gas in the field.

Related oil field conditions. Petroleum is a generic term which, in its broadest meaning, includes all naturally occurring hydrocarbons, whether gaseous, liquid, or solid. By variation of the temperature or pressure, or both, of any hydrocarbon, it becomes gaseous, liquid, or solid. Temperatures in producing horizons vary from approximately 60°F (16°C) to more than 300°F (149°C), depending chiefly upon the depth of the horizon. A rough approximation is that temperature in the reservoir sand, or pay, equals 60°F (16°C), plus 0.017°F/ft (0.031°C/m) of depth below surface. Pressure on the hydrocarbons varies from atmospheric to more than 11,000 psi (76 MPa). Normal pressure is considered as 0.465 psi/ft (10.5 kPa/m) of depth. Temperatures and pressure vary widely from these average figures. Hydrocarbons, because of wide variations in pressure and temperature and because of mutual solubility in one another, do not necessarily exist underground in the same phases in which they appear at the surface.

Petroleum occurs underground in porous rocks of wide variety. The pore spaces range from microscopic size to rare holes 1 in. or more in diameter. The containing rock is commonly called the sand or the pay, regardless of whether the pay is actually sandstone, limestone, dolomite, unconsolidated sand, or fracture openings in relatively impermeable rock.

Development of field. After discovery of a field containing oil or gas, or both, in commercial quantities, the field must be explored to determine its vertical and horizontal limits and the mechanisms under which the field will produce. Development and exploitation of the field proceed simultaneously. Usually the original development program is repeatedly modified by geologic knowledge acquired during the early stages of development and exploitation of the field.

Ideally, tests should be drilled to the lowest possible producing horizon in order to determine the number of pools existing in the field. Testing and geologic analysis of the first wells sometimes indicates the producing mechanisms, and thus the best development program. Very early in the history of the field, step-out wells will be drilled to determine the areal extent of the pool or pools. Step-out wells give further information regarding the volumes of oil and gas available, the producing mechanisms, and the desirable spacing of wells.

The operator of an oil and gas field endeavors to select a development program which will produce the largest volume of oil and gas at a profit. The program adopted is always a compromise between conflicting objectives. The operator desires (1) to drill the fewest wells which will efficiently produce the recoverable oil and gas; (2) to drill, complete, and equip the wells at the lowest possible cost; (3) to complete production in the shortest practical time to reduce both capital and operating charges; (4) to operate the wells at the lowest possible cost; and (5) to recover the largest possible volume of oil and gas.

Selecting the number of wells. Oil pools are produced by four mechanisms: dissolved gas expansion, gas-cap drive, water drive, and gravity drainage. Commonly, two or more mechanisms operate in a single pool. The type of producing mechanism in each pool influences the decision as to the number of wells to be drilled. Theoretically, a single, perfectly located well in a water-drive pool is capable of producing all of the commercially recoverable oil and gas from that pool. Practically, more than one well is necessary if a pool of more than 80 acres (32 hectares) is to be depleted in a reasonable time. If a pool produces under either gas expansion or gas-cap drive, oil production from the pool will be independent of the number of wells up to a spacing of at least 80 acres per well (1866 ft or 569 m between wells). Gas wells often are spaced a mile or more apart. The operator accordingly selects the widest spacing permitted by field conditions and legal requirements.

Major components of cost. Costs of drilling, completing, and equipping the wells influence development plans. Having determined the number and depths of producing horizons and the producing mechanisms in each horizon, the operator must decide whether he will drill a well at each location to each horizon or whether a single well can produce from two or more horizons at the same location. Clearly, the cost of drilling the field can be sharply reduced if a well can drain two, three, or more horizons. The cost of drilling a well will be higher if several horizons are simultaneously produced, because the dual or triple completion of a well usually requires larger casing. Further, completion and operating costs are higher. However, the increased cost of drilling a well of larger diameter and completing the well in two or more horizons is 20–40% less than the cost of drilling and completing two wells to produce separately from two horizons.

In some cases, the operator may reduce the number of wells by drilling a well to the lowest producible horizon and taking production from that level until the horizon there is commercially exhausted. The well is then plugged back to produce from a higher horizon. Selection of the plan for producing the various horizons obviously affects the cost of drilling and completing individual wells, as well as the number of wells which the operator will drill. If two wells are drilled at approximately the same location, they are referred to as twins, three wells at the same location are triplets, and so on.

Costs and duration of production. The operator wishes to produce as rapidly as possible because the net income from sale of hydrocarbons is obviously reduced as the life of the well is extended. The successful operator must recover from his productive wells the costs of drilling and operating those wells, and in addition he must recover all costs involved in geological and geophysical exploration, leasing, scouting, and drilling of dry holes, and occasionally other operations. If profits from production are not sufficient to recover all exploration and production costs and yield a profit in excess of the rate of interest which the operator could secure from a different type of investment, he is discouraged from further exploration.

Most wells cannot operate at full capacity because unlimited production results in physical waste and sharp reduction in ultimate recovery. In many areas, conservation restrictions are enforced to make certain that the operator does not produce in excess of the maximum efficient rate. For example, if an oil well produces at its highest possible rate, a zone promptly develops around the well where production is occurring under gas-expansion drive, the most inefficient producing mechanism. Slower production may permit the petroleum to be produced under gas-cap drive or water drive, in which case ultimate production of oil will be two to four times as great as it would be under gas-expansion drive. Accordingly, the most rapid rate of production generally is not the most efficient rate.

Similarly, the initial exploration of the field may indicate that one or more gas-condensate pools exist, and recycling of gas may be necessary to secure maximim recovery of both condensate and of gas. The decision to recycle will affect the number of wells, the locations of the wells, and the completion methods adopted in the development program.

Further, as soon as the operator determines that secondary oil-recovery methods are desired and expects to inject water, gas, steam, or, rarely, air to provide additional energy to flush or displace oil from the pay, the number and location of wells may be modified to permit the most effective secondary recovery procedures.

Legal and practical restrictions. The preceding discussion has assumed control of an entire field under single ownership by a single operator. In the United States, a single operator rarely controls a large field, and this field is almost never under a single lease. Usually, the field is covered by separate leases owned and operated by different producers. The development program must then be modified in consideration of the lease boundaries and the practices of the other operators who are in the field.

Oil and gas know no lease boundaries. They move freely underground from areas of high pressure toward lower-pressure situations. The operator of a lease is obligated to locate his wells in such a way as to prevent drainage of his lease by wells on adjoining leases, even though he may own the adjoining leases. In the absence of conservation restrictions, an operator must produce petroleum from his wells as rapidly as it is produced from wells on adjoining leases. Slow production on one lease results in migration of oil and gas to nearby leases which are more rapidly produced.

The operator's development program must provide for offset wells located as close to the boundary of his lease as are wells on adjoining leases. Further, the operator must equip his wells to produce as rapidly as the offset produces and must produce from the same horizons which are being produced in offset wells. The lessor who sold the lease to the operator is entitled to his share of the recoverable petroleum underlying his land. Negligence by the operator in permitting drainage of a lease makes the operator liable to suit for damages or cancellation of the lease.

A development program acceptable to all operators in the field permits simultaneous development of leases, prevents drainage, and results in maximum ultimate production from the field. Difficulties may arise in agreement upon the best development program for a field. Most states have enacted statutes and have appointed regulatory bodies under which judicial determination can be made of the permissible spacing of the wells, the rates of production, and the application of secondary recovery methods.

Drilling unit. Commonly, small leases or portions of two or more leases are combined to form a drilling unit in whose center a well will be drilled. Unitization may be voluntary, by agreement between the operator or operators and the interested royalty owners, with provision for sharing production from the well between the parties in proportion to their acreage interests. In many states the regulatory body has authority to require unitization of drilling units, which eliminates unnecessary offset wells and protects the interests of a landowner whose acreage holding may be too small to justify the drilling of a single well on his property alone.

Pool unitization. When recycling or some types of secondary recovery are planned, further unitization is adopted. Since oil and gas move freely across lease boundaries, it would be wasteful for an operator to repressure, recycle, or water-drive a lease if the adjoining leases were not similarly operated. Usually an entire pool must be unitized for efficient recycling, or secondary recovery operations. Pool unitization may be accomplished by agreement between operators and royalty owners. In many cases, difference of opinion or ignorance on the part of some parties prevents voluntary pool unitization. Many states authorize the regulatory body to unitize a pool compulsorily on application by a specified percentage of interests of operators and royalty owners. Such compulsory unitization is planned to provide each operator and each royalty owner his fair share of the petroleum products produced from the field regardless of the location of the well or wells through which these products actually reach the surface.

EXPLOITATION—GENERAL CONSIDERATIONS

Oil and gas production necessarily are intimately related, since approximately one-third of the gross gas production in the United States is produced from wells that are classified as oil wells.

However, the naturally occurring hydrocarbons of petroleum are not only liquid and gaseous but may even be found in a solid state, such as asphaltite and some asphalts.

Where gas is produced without oil, the production problems are simplified because the product flows naturally throughout the life of the well and does not have to be lifted to the surface. However, there are sometimes problems of water accumulations in gas wells, and it is necessary to pump the water from the wells to maintain maximum, or economical, gas production. The line of demarcation between oil wells and gas wells is not definitely established since oil wells may have gas-oil ratios ranging from a few cubic feet (1 cubic foot = 2.8×10^{-2} m^3) per barrel to many thousand cubic feet of gas per barrel of oil. Most gas wells produce quantities of condensable vapors, such as propane and butane, that may be liquefied and marketed for fuel, and the more stable liquids produced with gas can be utilized as natural gasoline.

Factors of method selection. The method selected for recovering oil from a producing formation depends on many factors, including well depth, well-casing size, oil viscosity, density, water production, gas-oil ratio, porosity and permeability of the producing formation, formation pressure, water content of producing formation, and whether the force driving the oil into the well from the formation is primarily gas pressure, water pressure, or a combination of the two. Other factors, such as paraffin content and difficulty expected from paraffin deposits, sand production, and corrosivity of the well fluids, also have a decided influence on the most economical method of production.

Special techniques utilized to increase productivity of oil and gas wells include acidizing, hydraulic fracturing of the formation, the setting of screens, and gravel packing or sand packing to increase permeability around the well bore.

Aspects of production rate. Productive rates per well may vary from a fraction of a barrel (1 barrel = 0.1590 m^3) per day to several thousand barrels per day, and it may be necessary to produce a large percentage of water along with the oil.

Field and reservoir conditions. In some cases reservoir conditions are such that some of the wells flow naturally throughout the entire economical life of the oil field. However, in the great majority of cases it is necessary to resort to artificial lifting methods at some time during the life of the field, and often it is necessary to apply artificial lifting means immediately after the well is drilled.

Market and regulatory factors. In some oil-producing states of the United States there are state bodies authorized to regulate oil production from the various oil fields. The allowable production per well is based on various factors, including the market for the particular type of oil available, but very often the allowable production is based on an engineering study of the reservoir to determine the optimum rate of production.

Crude oil production in the United States in 1978, as reported by the *Oil and Gas Journal*, averaged 8,680,000 barrels per day (bpd) or 1,380,000 m^3/day. This represents an increase of 6.1% over 1977 production. Imports amounted to 8,658,000 bpd (1,376,500 m^3/day) in 1977. World production in 1978 was estimated at 60,335,000 bpd (9,592,500 m^3/day). Net natural gas production during 1977 was estimated at $53{,}393.5 \times 10^6$ ft^3 per day (1,511.936 m^3/day).

Useful terminology. A few definitions of terms used in petroleum production technology are listed below to assist in an understanding of some of the problems involved.

Porosity. The percentage porosity is defined as the percentage volume of voids per unit total volume. This, of course, represents the total possible volume available for accumulation of fluids in a formation, but only a fraction of this volume may be effective for practical purposes because of possible discontinuities between the individual pores. The smallest pores generally contain water held by capillary forces.

Permeability. Permeability is a measure of the resistance to flow through a porous medium under the influence of a pressure gradient. The unit of permeability commonly employed in petroleum production technology is the darcy. A porous structure has a permeability of 1 darcy if, for a fluid of 1 centipoise (cp) [10^{-3} Pa·s] viscosity, the volume flow is 1 $cm^3/(sec)(cm^2)$ [10^{-2} $m^3/(sec)(m^2)$] under a pressure gradient of 1 atm/cm (1.01325×10^7 Pa/m).

Productivity index. The productivity index is a measure of the capacity of the reservoir to deliver oil to the well bore through the productive formation and any other obstacles that may exist around the well bore. In petroleum production technology, the productivity index is defined as production in barrels per day (1 barrel per day ≅ 0.1590 m^3/day) per pound per square inch (psi = 6.895 kPa) drop in bottom-hole pressure. For example, if a well is closed in at the casinghead, the bottom-hole pressure will equal the formation pressure when equilibrium conditions are established. However, if fluid is removed from the well, either by flowing or pumping, the bottom-hole pressure will drop as a result of the resistance to flow of fluid into the well from the formation to replace the fluid removed from the well. If the closed-in bottom-hole pressure should be 1000 psi, for example, and if this pressure should drop to 900 psi when producing at a rate of 100 bbl/day (a drop of 100 psi), the well in question would have a productivity index of one.

Barrel. The standard barrel used in the petroleum industry is 42 U.S. gal (approximately 0.1590 m^3).

API gravity. The American Petroleum Institute (API) scale that is in common use for indicating specific gravity, or a rough indication of quality of crude petroleum oils, differs slightly from the Baume scale commonly used for other liquids lighter than water. The table shows the relationship between degrees API and specific gravity referred to water at 60°F (15.6°C) for specific gravities ranging from 0.60 to 1.0.

Viscosity range. Viscosity of crude oils currently produced varies from approximately 1 cp (10^{-3} Pa·s) to values above 1000 cp (1 Pa·s) at temperatures existing at the bottom of the well. In some areas it is necessary to supply heat artificially down the wells or circulate lighter oils to mix with the produced fluid for maintenance of a relatively low viscosity throughout the temperature range to which the product is subjected.

In addition to wells that are classified as gas wells or oil wells, the term gas-condensate well has

Degrees API corresponding to specific gravities of crude oil at 60°/60°F*

Specific gravity, in tenths	Specific gravity, in hundredths									
	.00	.01	.02	.03	.04	.05	.06	.07	.08	.09
0.60	104.33	100.47	96.73	93.10	89.59	86.19	82.89	79.69	76.59	73.57
0.70	70.64	67.80	65.03	62.34	59.72	57.17	54.68	52.27	49.91	47.61
0.80	45.38	43.19	44.06	38.98	36.95	34.97	33.03	31.14	29.30	27.49
0.90	25.72	23.99	22.30	20.65	19.03	17.45	15.90	14.38	12.89	11.43
1.00	10.00									

*60°F = 15.6°C.

come into general use to designate a well that produces large volumes of gas with appreciable quantities of light, volatile hydrocarbon fluids. Some of these fluids are liquid at atmospheric pressure and temperature; others, such as propane and butane, are readily condensed under relatively low pressures in gas separators for use as liquid petroleum gas (LPG) fuels or for other uses. The liquid components of the production from gas-condensate wells generally arrive at the surface in the form of small droplets entrained in the high-velocity gas stream and are separated from the gas in a high-pressure gas separator.

PRODUCTION METHODS IN PRODUCING WELLS

The common methods of producing oil wells are (1) natural flow; (2) pumping with sucker rods; (3) gas lift; (4) hydraulic subsurface pumps; (5) electrically driven centrifugal well pumps; and (6) swabbing.

Numerous other methods, including jet pumps and sonic pumps, have been tried and are used to slight extent. The sonic pump is a development in which the tubing is vibrated longitudinally by a mechanism at the surface and acts as a high-speed pump with an extremely short stroke.

The total number of producing oil wells in the United States at the end of 1977 was reported to be 508,340, while the total number of producing gas wells was 145,453.

A total of 48,384 wells were drilled in the United States during 1978. Of this number, 17,747 were productive oil wells, 12,941 were classified as gas wells, 16,228 were nonproductive (dry holes), and 1468 were service wells. Service wells are utilized for various purposes, such as water injection for water flooding operations, salt-water disposal, and gas recycling.

A discussion of production methods, in approximate order of relative importance, follows.

Natural flow. Natural flow is the most economical method of production and generally is utilized as long as the desired production rate can be maintained by this method. It utilizes the formation energy, which may consist of gas in solution in the oil in the formation; free gas under pressure acting against the liquid and gas-liquid phase to force it toward the well bore; water pressure acting against the oil; or a combination of these three energy sources. In some areas the casinghead pressure may be of the order of 10,000 psi, so it is necessary to provide fittings adequate to withstand such pressures. Adjustable throttle values, or chokes, are utilized to regulate the flow rate to a desired and safe value. With such a high-pressure drop across a throttle valve the life of the valve is likely to be very short. Several such valves are arranged in parallel in the tubing head "Christmas tree" with positive shutoff valves between the chokes and the tubing head so that the wearing parts of the throttle valve, or the entire valve, can be replaced while flow continues through another similar valve.

An additional safeguard that is often used in connection with high-pressure flowing wells is a bottom-hole choke or a bottom-hole flow control valve that limits the rate of flow to a reasonable value, or stops it completely, in case of failure of surface controls. Figure 1 shows a schematic outline of a simple flowing well hookup. The packer is not essential but is often used to reduce the free gas volume in the casing.

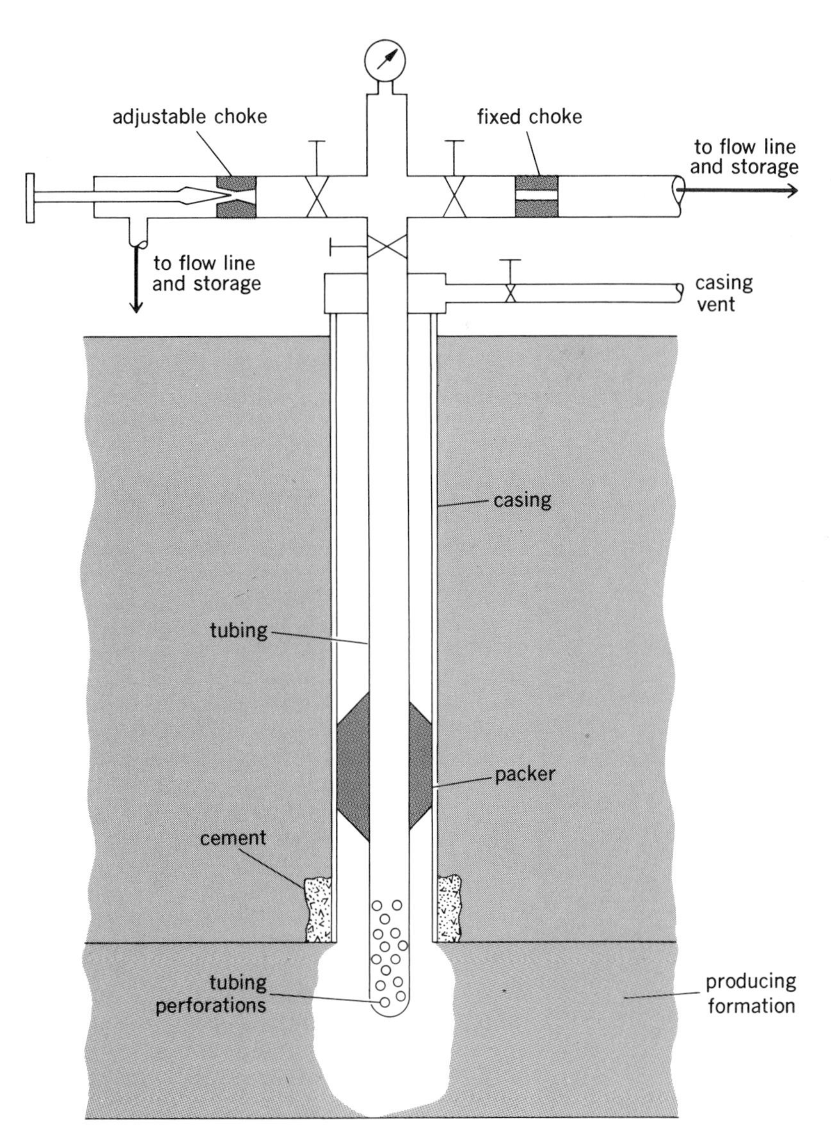

Fig. 1. Schematic view of well equipped for producing by natural flow.

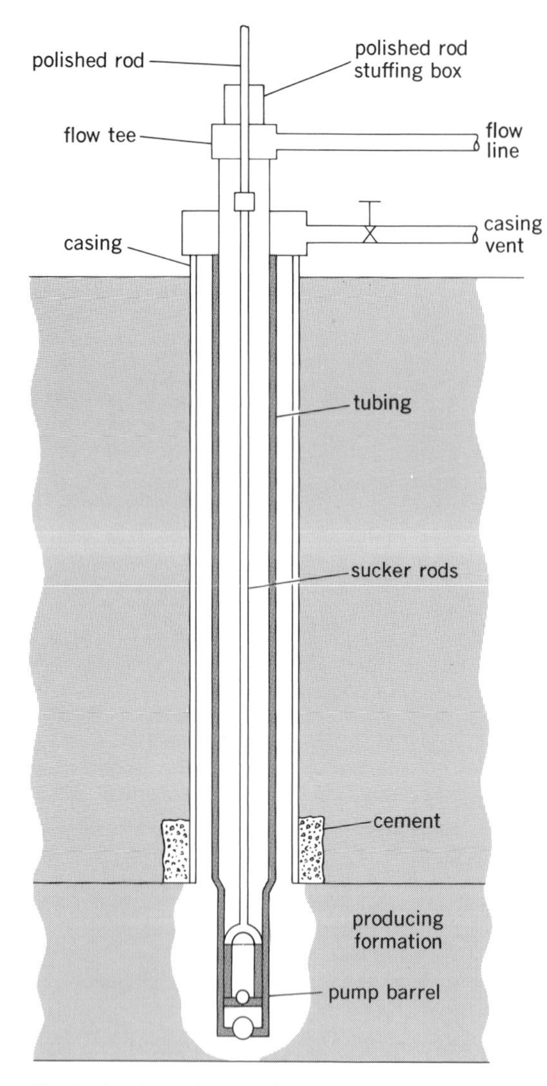

Fig. 2. A schematic view of a well which is equipped for pumping with sucker rods.

Flow rates for United States wells seldom exceed a few hundred barrels per day because of enforced or voluntary restrictions to regulate production rates and to obtain most efficient and economical ultimate recovery. However, in some countries, especially in the Middle East, it is not uncommon for natural flow rates to exceed 10,000 bpd/well [1590 m^3/(day)(well)].

Lifting. Most wells are not self-flowing. The common types of lifting are outlined here.

Pumping with sucker rods. Approximately 90% of the wells made to produce by some artificial lift method in the United States are equipped with sucker-rod–type pumps. In these the pump is installed at the lower end of the tubing string and is actuated by a string of sucker rods extending from the surface to the subsurface pump. The sucker rods are attached to a polished rod at the surface. The polished rod extends through a stuffing box and is attached to the pumping unit, which produces the necessary reciprocating motion to actuate the sucker rods and the subsurface pump. Figure 2 shows a simplified schematic section through a pumping well. The two common variations are mechanical and hydraulic long-stroke pumping.

1. Mechanical pumping. The great majority of pumping units are of the mechanical type, consisting of a suitable reduction gear, and crank and pitman arrangement to drive a walking beam to produce the necessary reciprocating motion. A counterbalance is provided to equalize the load on the upstroke and downstroke. Mechanical pumping units of this type vary in load-carrying capacity from about 2000 to about 43,000 lb (900 to 19,500 kg), and the torque rating of the low-speed gear which drives the crank ranges from 6400 in.-lb (720 N·m) in the smallest API standard unit to about 1,500,000 in.-lb (170,000 N·m) for the largest units now in use. Stroke length varies from about 18 to 192 in. (46 to 488 cm). Usual operating speeds are from about 6 to 20 strokes/min. However, both lower and higher rates of speed are sometimes used. Figure 3 shows a modern pumping unit in operation.

Production rates with sucker-rod–type pumps vary from a fraction of 1 bpd in some areas, with part-time pumping, to approximately 3000 bpd (480 m^3/day) for the largest installations in relatively shallow wells.

2. Hydraulic long-stroke pumping. For this the units consist of a hydraulic lifting cylinder mounted directly over the well head and are designed to produce stroke lengths of as much as 30 ft (9 m). Such long-stroke hydraulic units are usually equipped with a pneumatic counterbalance arrangement which equalizes the power requirement on the upstroke and downstroke.

Hydraulic pumping units also are made without any provision for counterbalance. However, these units are generally limited to relatively small wells, and they are relatively inefficient.

Gas lift. Gas lift in its simplest form consists of initiating or stimulating well flow by injecting gas at some point below the fluid level in the well. With large-volume gas-lift operations the well may be produced through either the casing or the tubing. In the former case, gas is conducted through the tubing to the point of injection; in the latter, gas may be conducted to the point of injection through the casing or through an auxiliary string of tubing. When gas is injected into the oil column, the weight of the column above the point of injection is

Fig. 3. Pumping unit with adjustable rotary counterbalance. (*Oil Well Supply Division, U.S. Steel Corp.*)

reduced as a result of the space occupied by the relatively low-density gas. This lightening of the fluid column is sufficient to permit the formation pressure to initiate flow up the tubing to the surface. Gas injection is often utilized to increase the flow from wells that will flow naturally but will not produce the desired amount by natural flow.

There are many factors determining the advisability of adopting gas lift as a means of production. One of the more important factors is the availability of an adequate supply of gas at suitable pressure and reasonable cost. In a majority of cases gas lift cannot be used economically to produce a reservoir to depletion because the well may be relatively productive with a low back pressure maintained on the formation but will produce very little, if anything, with the back pressure required for gas-lift operation. Therefore, it generally is necessary to resort to some mechanical means of pumping before the well is abandoned, and it may be more economical to adopt the mechanical means initially than to install the gas-lift system while conditions are favorable and later replace it.

This discussion of gas lift has dealt primarily with the simple injection of gas, which may be continuous or intermittent. There are numerous modifications of gas-lift installations, including various designs for flow valves which may be installed in the tubing string to open and admit gas to the tubing from the casing at a predetermined pressure differential between the tubing and casing. When the valve opens, gas is injected into the tubing to initiate and maintain flow until the tubing pressure drops to a predetermined value; and the valve closes before the input gas-oil ratio becomes excessive. This represents an intermittent-flow–type valve. Other types are designed to maintain continuous flow, proper pressure differential, and proper gas injection rate for efficient operation. In some cases several such flow valves are spaced up the tubing string to permit flow to be initiated from various levels as required.

Other modifications of gas lift involve the utilization of displacement chambers. These are installed on the lower end of the well tubing where oil may accumulate, and the oil is displaced up the tubing with gas injection controlled by automatic or mechanical valves.

Hydraulic subsurface pumps. The hydraulic subsurface pump has come into fairly prominent use. The subsurface pump is operated by means of a hydraulic reciprocating motor attached to the pump and installed in the well as a single unit. The hydraulic motor is driven by a supply of hydraulic fluid under pressure that is circulated down a string of tubing and through the motor. Generally the hydraulic fluid consists of crude oil which is discharged into the return line and returns to the surface along with the produced crude oil.

Hydraulically operated subsurface pumps are also arranged for separating the hydraulic power fluid from the produced well fluid. This arrangement is especially desirable where the fluid being produced is corrosive or is contaminated with considerable quantities of sand or other solids that are difficult to separate to condition the fluid for use as satisfactory power oil. This method permits use of water or other nonflammable liquids as hydraulic power fluid to minimize fire hazard in case of a failure of the hydraulic power line at the surface.

Centrifugal well pumps. Electrically driven centrifugal pumps have been used to some extent, especially in large-volume wells of shallow or moderate depths. Both the pump and the motor are restricted in diameter to run down the well casing, leaving sufficient clearance for the flow of fluid around the pump housing. With the restricted diameter of the impellers the discharge head necessary for pumping a relatively deep well can be obtained only by using a large number of stages and operating at a relatively high speed. The usual rotating speed for such units is 3600 rpm, and it is not uncommon for such units to have 50 or more pump stages. The direct-connected electric motor must be provided with a suitable seal to prevent well fluid from entering the motor housing, and electrical leads must be run down the well casing to supply power to the motor.

Swabs. Swabs have been used for lifting oil almost since the beginning of the petroleum industry. They usually consist of a steel tubular body equipped with a check valve which permits oil to flow through the tube as it is lowered down the well with a wire line. The exterior of the steel body is generally fitted with flexible cup-type soft packing that will fall freely but will expand and form a seal with the tubing when pulled upward with a head of fluid above the swab. Swabs are run into the well on a wire line to a point considerably below the fluid level and then lifted back to the surface to deliver the volume of oil above the swab. They are often used for determining the productivity of a well that will not flow naturally and for assisting in cleaning paraffin from well tubing. In some cases swabs are used to stimulate wells to flow by lifting, from the upper portion of the tubing, the relatively dead oil from which most of the gas has separated.

Bailers. Bailers are used to remove fluids from wells and for cleaning out solid material. They are run into the wells on wire lines as in swabbing, but differ from swabs in that they generally are run only in the casing when there is no tubing in the well. The capacity of the bailer itself represents the volume of fluid lifted each time since the bailer does not form a seal with the casing. The bailer is simply a tubular vessel with a check valve in the bottom. This check valve generally is arranged so that it is forced open when the bailer touches bottom in order to assist in picking up solid material for cleaning out a well.

Jet pumps. A jet pump for use in oil wells operates on exactly the same principle as a water-well jet pump. Advantage is taken of the Bernoulli effect to reduce pressure by means of a high-velocity fluid jet. Thus oil is entrained from the well with this high-velocity jet in a venturi tube to accelerate the fluid and assist in lifting it to the surface, along with any assistance from the formation pressure. The application of jet pumps to oil wells has been insignificant.

Sonic pumps. Sonic pumps essentially consist of a string of tubing equipped with a check valve at each joint and mechanical means on the surface to vibrate the tubing string longitudinally. This creates a harmonic condition that will result in several hundred strokes per minute, with the strokes being a small fraction of 1 in. in length. Some of these pumps are in use in relatively shallow wells.

Lease tanks and gas separators. Figure 4 shows a typical lease tank battery consisting of

Fig. 4. Lease tank battery with four tanks and two gas separators. *(Gulf Oil Corp.)*

four 1000-bbl tanks and two gas separators. Such equipment is used for handling production from wells produced by natural flow, gas lift, or pumping. In some pumping wells the gas content may be too low to justify the cost of separators for saving the gas.

Natural gasoline production. An important phase of oil and gas production in many areas is the production of natural gasoline from gas taken from the casinghead of oil wells or separated from the oil and conducted to the natural gasoline plant. The plant consists of facilities for compressing and extracting the liquid components from the gas. The natural gasoline generally is collected by cooling and condensing the vapors after compression or by absorbing in organic liquids having high boiling points from which the volatile liquids are distilled. Many natural gasoline plants utilize a combination of condensing and absorbing techniques. Figure 5 shows an overall view of a natural gasoline plant operating in western Texas.

PRODUCTION PROBLEMS AND INSTRUMENTS

To maintain production, various problems must be overcome. Numerous instruments have been developed to monitor production and to control production problems.

Corrosion. In many areas the corrosion of production equipment is a major factor in the cost of petroleum production. The following comments on the oil field corrosion problem are taken largely from *Corrosion of Oil- and Gas-Well Equipment* and reproduced by permission of NACE-API.

For practical consideration, corrosion in oil and gas-well production can be classified into four main types.

1. Sweet corrosion occurs as a result of the presence of carbon dioxide and fatty acids. Oxygen and hydrogen sulfide are not present. This type of corrosion occurs in both gas-condensate and oil wells. It is most frequently encountered in the United States in southern Louisiana and Texas, and other scattered areas. At least 20% of all sweet oil production and 45% of condensate production are considered corrosive.

2. Sour corrosion is designated as corrosion in oil and gas wells producing even trace quantities of hydrogen sulfide. These wells may also contain oxygen, carbon dioxide, or organic acids. Sour corrosion occurs in the United States primarily throughout Arbuckle production in Kansas and in the Permian basin of western Texas and New Mexico. About 12% of all sour production is considered corrosive.

3. Oxygen corrosion occurs wherever equipment is exposed to atmospheric oxygen. It occurs most frequently in offshore installations, brine-handling and injection systems, and in shallow producing wells where air is allowed to enter the casing.

4. Electrochemical corrosion is designated as that which occurs when corrosion currents can be readily measured or when corrosion can be mitigated by the application of current, as in soil corrosion.

Corrosion inhibitors are used extensively in both oil and gas wells to reduce corrosion damage to subsurface equipment. Most of the inhibitors used in the oil field are of the so-called polar organic type. All of the major inhibitor suppliers can furnish effective inhibitors for the prevention of sweet corrosion as encountered in most fields. These can be purchased in oil-soluble, water-dispersible, or water-soluble form.

Paraffin deposits. In many crude-oil–producing areas paraffin deposits in tubing and flow lines and on sucker rods are a source of considerable trouble and expense. Such deposits build up until the tubing or flow line is partially or completely plugged. It is necessary to remove these deposits to maintain production rates. A variety of methods are used to remove paraffin from the tubing, including the application of heated oil through tubular sucker rods to mix with and transfer heat to the oil being produced and raise the temperature to a point at which the deposited paraffin will be dissolved or melted. Paraffin solvents may also be applied in this manner without the necessity of applying heat.

Mechanical means often are used in which a scraping tool is run on a wire line and paraffin is scraped from the tubing wall as the tool is pulled back to the surface. Mechanical scrapers that attach to sucker rods also are in use. Various types of automatic scrapers have been used in connection with flowing wells. These consist of a form of piston that will drop freely to the bottom when flow is stopped but will rise back to the surface when flow is resumed. Electrical heating methods have

Fig. 5. Modern natural gasoline plant in western Texas. *(Gulf Oil Corp.)*

Fig. 6. Two pumping wells with tank battery. (*Oil Well Supply Division, U.S. Steel Corp.*)

been used rather extensively in some areas. The tubing is insulated from the casing and from the flow line, and electric current is transmitted through the tubing for the time necessary to heat the tubing sufficiently to cause the paraffin deposits to melt or go into solution in the oil in the tubing. Plastic coatings have been utilized inside tubing and flow lines to minimize or prevent paraffin deposits. Paraffin does not deposit readily on certain plastic coatings.

A common method for removing paraffin from flow lines is to disconnect the line at the well head and at the tank battery and force live steam through the line to melt the paraffin deposits and flow them out. Various designs of flow-line scrapers have also been used rather extensively and fairly successfully. Paraffin deposits in flow lines are minimized by insulating the lines or by burying the lines to maintain a higher average temperature.

Emulsions. A large percentage of oil wells produce various quantities of salt water along with the oil, and numerous wells are being pumped in which the salt-water production is 90% or more of the total fluid lifted. Turbulence resulting from production methods results in the formation of emulsions of water in oil or oil in water; the commoner type is oil in water. Emulsions are treated with a variety of demulsifying chemicals, with the application of heat, and with a combination of these two treatments. Another method for breaking emulsions is the electrostatic or electrical precipitator type of emulsion treatment. In this method the emulsion to be broken is circulated between electrodes subjected to a high potential difference. The resulting concentrated electric field tends to rupture the oil-water interface and thus breaks the emulsion and permits the water to settle out. Figure 6 shows two pumping wells with a tank battery in the background. This tank battery is equipped with a wash tank, or gun barrel, and a gas-fired heater for emulsion treating and water separation before the oil is admitted to the lease tanks.

Gas conservation. If the quantity of gas produced with crude oil is appreciably greater than that which can be efficiently utilized or marketed, it is necessary to provide facilities for returning the excess gas to the producing formation. Formerly, large quantities of excess gas were disposed of by burning or simply by venting to the atmosphere. This practice is now unlawful. Returning excess gas to the formation not only conserves the gas for future use but also results in greater ultimate recovery of oil from the formation.

Salt-water disposal. The large volumes of salt water produced with the oil in some areas present serious disposal problems. The salt water is generally pumped back to the formation through wells drilled for this purpose. Such salt-water disposal wells are located in areas where the formation already contains water. Thus this practice helps to maintain the formation pressure as well as the productivity of the producing wells.

Offshore production. Offshore wells present additional production problems since the wells must be serviced from barges or boats. Wells of reasonable depth on land locations are seldom equipped with derricks for servicing because it is more economical to set up a portable mast for pulling and installing rods, tubing, and other equipment. However, the use of portable masts is not practical on offshore locations, and a derrick is generally left standing over such wells throughout their productive life to facilitate servicing. There are a considerable number of offshore wells along the Gulf Coast and the Pacific Coast of the United States, but by far the greatest number of offshore wells in a particular region is in Lake Maracaibo in Venezuela. Figure 7 shows a considerable number of derricks in Lake Maracaibo with pumping wells in the foreground. These wells are pumped by electric power through cables laid on the lake bottom to conduct electricity from power-generating stations onshore. An overwater tank battery is visible at the extreme right. All offshore installations, such as tank batteries, pump stations, and the derricks and pumping equipment, are supported on pilings in water up to 100 ft (30 m) or more in depth. There are approximately 2300 oil derricks in Lake Maracaibo. A growing number of semipermanent platform rigs and even bottom storage facilities are being used in Gulf of Mexico waters at depths of more than 100 ft (30 m).

Instruments. The commoner and more important instruments required in petroleum production operations are included in the following discussion.

1. Gas meters, which are generally of the orifice type, are designed to record the differential pressure across the orifice, and the static pressure.

2. Recording subsurface pressure gages small enough to run down 2-in. ID (inside diameter) tubing are used extensively for measuring pressure gradients down the tubing of flowing wells, recording pressure buildup when the well is closed in,

Fig. 7. Numerous offshore wells located in Lake Maracaibo, Venezuela. (*Creole Petroleum Corp.*)

and measuring equilibrium bottom-hole pressures.

3. Subsurface samplers designed to sample well fluids at various levels in the tubing are used to determine physical properties, such as viscosity, gas content, free gas, and dissolved gas at various levels. These instruments may also include a recording thermometer or a maximum reading thermometer, depending upon the information required.

4. Oil meters of various types are utilized to meter crude oil flowing to or from storage.

5. Dynamometers are used to measure polished-rod loads. These instruments are sometimes known as well weighers since they are used to record the polished-rod load throughout a pumping cycle of a sucker-rod—type pump. They are used to determine maximum load on polished rods as well as load variations, to permit accurate counterbalancing of pumping wells, and to assure that pumping units or sucker-rod strings are not seriously overloaded.

6. Liquid-level gages and controllers are used. They are similar to those used in other industries, but with special designs for closed lease tanks.

A wide variety of scientific instruments find application in petroleum production problems. The above outline gives an indication of a few specialized instruments used in this branch of the industry, and there are many more. Special instruments developed by service companies are valued for a wide variety of purposes and include calipers to detect and measure corrosion pits inside tubing and casing and magnetic instruments to detect microscopic cracks in sucker rods.

[ROY L. CHENAULT]

Bibliography: American Petroleum Institute, *History of Petroleum Engineering*, 1961; K. E. Brown, *The Technology of Artificial Lift Methods*, 1977; E. L. DeGolyer (ed.), *Elements of the Petroleum Industry*, 1940; ETA Offshore Seminars, Inc., *The Technology of Offshore Drilling and Production*, 1976; L. L. Farkas, *Management of Technical Field Operations*, 1970; T. C. Frick (ed.), *Petroleum Production Handbook*, vol. 1: *Mathematics and Production Equipment*, 1962; L. M. Harris, *An Introduction Deep Water Floating Drilling Operations*, 1972; M. Muskat, *Physical Principles of Oil Production*, 1949; T. E. W. Nind, *Principles of Oil Well Production*, 1964; L. T. Stanley, *Practical Statistics for Petroleum Engineers*, 1973; L. C. Uren, *Petroleum Production Engineering: Oil Field Development*, 4th ed., 1956; L. C. Uren, *Petroleum Production Engineering: Oil Field Exploitation*, 3d ed., 1953; J. Zaba and W. T. Doherty, *Practical Petroleum Engineering*, 5th ed., 1970.

Oil mining

The surface or subsurface excavation of petroleum-bearing sediments for subsequent removal of the oil by washing, flotation, or retorting treatments. Oil mining also includes recovery of oil by drainage from reservoir beds to mine shafts or other openings driven into the oil rock, or by drainage from the reservoir rock into mine openings driven outside the oil sand but connected with it by bore holes or mine wells.

Surface mining consists of strip or open-pit mining. It has been used primarily for the removal of oil shale or bituminous sands lying at or near the surface. Strip mining of shale is practiced in Sweden, Manchuria, and South Africa. Strip mining of bituminous sand is conducted in Canada.

Subsurface mining is used for the removal of oil sediments, oil shale, and Gilsonite. It is practiced in several European countries and in the United States. Some authorities consider this the best method to recover oil when oil sediments are involved, because virtually all of the oil is recovered.

European experience. Subsurface oil mining was used in the Pechelbronn oil field in Alsace, France, as early as 1735. This early mining involved the sinking of shafts to the reservoir rock, only 100–200 ft (30–60 m) below the surface, and the excavation of the oil sand in short drifts driven from the shafts. These oil sands were hoisted to the surface and washed with boiling water to release the oil. The drifts were extended as far as natural ventilation permitted. When these limits were reached, the pillars were removed and the openings filled with waste.

This type of mining continued at Pechelbronn until 1866, when it was found that oil could be recovered from deeper, more prolific sands by letting it drain in place through mine openings, without removing the sands to the surface for treatment.

Subsurface mining of oil shale also goes back to the mid-19th century in Scotland and France. It is not so widely practiced now because of its high cost as compared with that of usual oil production, particularly in the prolific fields of the Middle East.

United States oil shale mining. The U.S. Bureau of Mines carried out an experimental mining and processing program at Rifle, Colo., between 1944 and 1955 in an effort to find economically feasible methods of producing oil shale.

One of the more important phases of this experimental program was a large-scale mine dug into what is known as the Mahagony Ledge, a rich oil shale stratum that is flat and strong, making it favorable for mining. This stratum lies under an average of about 1000 ft (300 m) of overburden and is 70–90 ft (21–27 m) thick.

The Bureau of Mines adopted the room-and-pillar system of mining, advancing into the 70-ft ledge face in two benches. The mine roof was supported by 60-ft (18-m) pillars staggered at 60-ft intervals and supplemented by iron roof bolts 6 ft (1.8 m) long.

Multiple rotary drills mounted on trucks made holes in which dynamite was placed to shatter the shale; the shale was then removed from the mine by electric locomotive and cars (Fig. 1).

The experimental mining program ended in February, 1955, when a roof fall occurred. Despite this occurrence, however, the Bureau is convinced that the room-and-pillar method used in coal, salt, and limestone mines is feasible for shale oil mining in Colorado.

Since 1955, several companies have conducted experimental efforts to produce shale oil. Those continuing to move toward commercialization include companies which plan to use traditional mining techniques combined with some form of surface retorting. One company is testing an onsite process in which both mining and retorting are done underground, and another firm is well along in development work using a method in-

Fig. 1. Mine locomotive and cars removing shale from U.S. Bureau of Mines Shale Mine at Rifle, Colo. At left, in middle ground, is the Colorado River, nearly 3000 ft (900 m) below. *(After R. Fleming, U.S. Bureau of Mines)*

volving a gas-combustion retort similar to that used by the Bureau of Mines in various pilot plants operated during the 1950s and 1960s.

Another possibility is a modified open-cast surface method, which proponents claim has a 95% recovery rate of the minable reserves, compared with a maximum 40% rate by room-and-pillar underground mining and 20% by on-site mining and retorting.

Oil shale does not contain oil, as such. Draining methods, therefore, are not applicable. It does, however, contain an organic substance known as kerogen. This substance decomposes and gives off a heavy, oily vapor when it is heated above 700°F (370°C) in retorts. When condensed, this vapor becomes a viscous black liquid called shale oil, which resembles ordinary crude but has several significant differences.

Colorado's Mahagony Ledge yields an average of about 30 gal of oil per ton (125 liters per metric ton). This means that large amounts of oil shale must be mined, transported, retorted, and discarded for production of commercial quantities of oil. Various types of retort also have been tested in Colorado, but none is in commercial use (Fig. 2).

Gilsonite. Gilsonite is a trade name, registered by the American Gilsonite Co., for a solid hydrocarbon found in the Uinta Basin of eastern Utah and western Colorado. The American Gilsonite Co. uses a subsurface wet-mining technique to extract about 700 tons (635 metric tons) of Gilsonite daily from its mine at Bonanza, Utah.

Conventional mining methods were found unsuitable for mass output of Gilsonite because it is friable and produces fine dust when so mined. This dust can be highly explosive. In the system now being used, tunnels are driven from the main shaft by means of water jetted through a 1/4-in. (6-mm) nozzle under pressure of 2000 psi (14 MPa). The stream of water penetrates tiny fissures and the ore falls to the bottom of the drift. The drifts are cut on a rising grade of about 2.5°. The ore is washed down to the main shaft where it is screened. Particles of sizes smaller than 3/4 in. (19 mm) are pumped to the surface in a water stream; larger pieces are hoisted in buckets. A long rotary drill with carbide-tipped teeth is used to remove ore that cannot be broken with water jets.

Gilsonite is moved through a pipeline in slurry form to a refinery, where it is dried and melted and then heated to about 450°F (232°C). The melted oil is fed to a coker and other processing units to make gasoline and other petroleum products.

[ADE L. PONIKVAR]

North American tar sands. The world's only tar sands mining operation is being conducted at the 50,000 barrels per day (bpd) [8000 m³/day] synthetic crude complex of Great Canadian Oil Sands (GCOS), 21 mi (34 km) north of Fort McMurray, Alberta. A second operation, the 125,000-bpd (20,000 m³/day) project of Syncrude Canada Ltd., located approximately 5 mi (8 km) away, was scheduled to begin production in 1978. Both projects are situated in the minable area of the gigantic Athabaska Tar Sands deposit, the world's largest

Fig. 2. Close-up view of Union Oil Co. of California shale-oil retort near Grand Valley, Colo. Right part of structure is portion of the system for removing oil vapors that would otherwise escape in the gas stream.

oil reservoir. The minable area, with an overburden thickness of less than 150 ft (45 m), embraces about 10% of the estimated 624×10^9 bbl (99×10^9 m^3) of total in-place bitumen reserves.

In both leases about half of the terrain is covered with muskeg, an organic soil resembling peat moss, which ranges anywhere from a few inches to 20 ft (6 m) in depth. The major part of the overburden, however, consists of Pleistocene glacial drift and Clearwater formation sands and shales. The total overburden varies from 20 to 120 ft (6 to 37 m) in thickness. Underlying this is the McMurray Formation (Lower Cretaceous) in which the oil-impregnated sands reside. The tar sands strata, in the region of the GCOS and Syncrude operations, are also variable in thickness; they average about 150 ft (45 m) although typically strata of 20–30-ft (6–9-m) thickness have a bitumen content below the economic cutoff grade of 6 wt %.

Composition. The bitumen content of tar sand can rage from 0 to 20 wt %, but feed-grade material normally runs between 10 and 12 wt %. The balance of the tar sand is composed, on the average, of 5 wt % water and 84 wt % sand and clay. The bitumen, which has an API gravity of 8°, is heavier than water and very viscous. Tar sand is a competent material, but it can be readily dug in the summer months; in the winter months, however, which see the temperature plunging to −50°F (−46°C), tar sand assumes the consistency of concrete. To maintain an acceptable digging rate in the winter, mining must proceed faster than the rate of frost penetration; if not, supplemental measures, such as blasting, are required.

Overburden removal. For muskeg removal, a series of ditches are dug two or three years in advance of stripping to permit as much of the water as possible to drain. Despite this, the spongy nature of the muskeg persists and removal is best accomplished after freeze-up.

Fig. 3. A Krupp bucket wheel excavator shown in the process of removing bituminous sand near Mildred Lake, Athabaska region, Alberta, Canada. (*Cities Service Co.*)

Great Canadian has tried several overburden removal methods, including shovel and trucks, scrapers, and front-end loaders and trucks. In 1974 the overburden removal equipment consisted of 5 Caterpillar D9G bulldozers for ripping and dozing, 7 Marathon LeToureau L700 15-yd^3 (11.5-m^3) front-end loaders, and a fleet of 21 WABCO 150-ton (136-metric ton) capacity trucks. Additional equipment is used for maintaining the haul roads and for spreading and compacting the spoiled material.

Bucket wheel excavators. Mining of tar sand is performed mainly by two large bucket wheel excavators, of German manufacture, each operating on a separate bench (Fig. 3). These units, weighing 1800 tons (1600 metric tons) have a 33-ft (10-m) diameter digging wheel on the end of a long boom. Each wheel has a theoretical capacity of 9500 tons/hr (8600 metric tons/hr), but the average output when digging is about 5000 tons/hr (4500 metric tons/hr). Because the availability of these machines is normally 55–60%, the extraction plant has been designed to accept a widely fluctuating feed rate. To facilitate digging of the highly abrasive tar sand and to achieve a reasonable bucket and tooth life, GCOS routinely preblasts the tar sand on a year-round basis. Tar sand at the rate of 140,000 tons/day (127,000 metric tons/day) is transferred from the mine to the plant by a system of 60-in. (1.52-m) face conveyor belts and 72-in. (1.83-m) trunk conveyors, operating at 1080 ft/min (5.5 m/s). Following extraction by a process using hot water, the bitumen is upgraded by coking and hydrogenation to a high-quality synthetic crude.

Dragline scheme. Syncrude has opted for an even more capital-intensive mining scheme. Four large draglines, each equipped with an 80-yd^3 (61-m^3) bucket at the end of a 360-ft (110-m) boom, will be employed to dig both the overburden, which will be free-cast into the mining pit, and the tar sand, which will be piled in windrows behind the machines. Four bucket wheel reclaimers, quite similar to the GCOS excavators except larger to handle the additional capacity, will load the tar sand from the windrows onto conveyor belts which will transfer it to the plant. With a peak tar sand mining rate of 336,000 tons/day (305,000 metric tons/day) the Syncrude project will be the largest mining operation in the world.

The advantages of the dragline scheme lie in its ability to handle overburden at a lower cost and its greater selectivity in rejecting lenses of low-grade tar sand and barren material. The disadvantages include the necessity of rehandling the tar sand and the probability of an increased percentage of lumps in the plant feed, which can damage conveyor belting.

Several years of comparative operation will be required to firmly establish whether one of the schemes will enjoy an economic advantage.

[G. RONALD GRAY]

Bibliography: A. R. Allen and E. R. Sanford, *The Great Canadian Oil Sands Operation*, Canadian Society of Petroleum Geology, Oil Sands Symposium, Calgary, Alberta, 1973; J. B. Jones, Jr. (ed.), *Hydrocarbons from Oil Shale, Oil Sands and Coal*, 1964; E. M. Perrini, *Oil from Shale and Tar Sands*, 1975; T. F. Yen (ed.), *Science and Technology of Oil Shale*, 1976.

Open-pit mining

The extraction of ores of metals and minerals by surface excavations. This method of mining is applicable for near-surface deposits which do not have a ratio of overburden (waste material that must be removed) to ore that would make the operation uneconomic. Where these criteria are met, large-scale earth-moving equipment can be used to give low unit mining costs. For a discussion of other methods of surface mining *see* COAL MINING; PLACER MINING; STRIP MINING.

Most open-pit mines are developed in the form of an inverted cone, with the base of the cone on the surface. Exceptions are open-pit mines developed in hills or mountains (Fig. 1). The walls of most open pits are terraced with benches to permit shovels, front-end loaders, or bucket-wheel excavators to excavate the rock and to provide access for trucks, trains, or belts to transport the rock out of the pit. The final depth of the pits ranges from less than 51 feet to nearly 3000 ft (15.5 to 914 m) and is dependent on the depth and value of the ore and the cost of mining. The cost of mining usually increases with the depth of the pit because of the increased distance that the ore must be hauled to the surface, the waste hauled to dumps, and the increased amount of overburden that must be removed. Numerous benches or terraces in the pit permit a number of areas to be worked at one time and are necessary for giving slope, typically less than a 2-to-1 slope, to the sides of the pit and for ore blending. Using multiple benches also permits a balanced operation in which much of the overburden is removed from upper benches at the same time that ore is being mined from lower benches. It is undesirable to remove all of the overburden from an ore deposit before mining the ore because the initial cost of developing the ore is then extremely high.

The principal operations in mining are drilling, blasting, loading, and hauling. These operations are usually required for both ore and overburden. Occasionally the ore or the overburden is soft enough that drilling and blasting are not required.

Fig. 1. Roads and rounds that have developed open cuts on Cerro Bolivar in Venezuela. (*U.S. Steel Corp.*)

Fig. 2. Aerial view of rotary drill. *(Kennecott Copper Corp.)*

Rock drilling. Drilling and blasting are interrelated operations. The primary purpose of drilling is to provide an opening in the rock for the placing of explosives. If the ore or overburden is weak enough to be easily excavated without blasting, it is not necessary for it to be drilled. The cuttings from the hole are often used to fill it after the explosive charge has been placed at the bottom. Blast holes drilled in ore serve a secondary function in that the cuttings can be sampled and assayed to determine the mineral content of the ore. Sampling of drill holes is frequently done for ores not readily identified by visual means.

The basic methods of drilling rock are rotary, percussion, and jet piercing. The drill holes are usually vertical and range in size from 11/2 to 15 in. (3.8 to 38 cm) in diameter and are drilled to varying depths and spacings as required for the particular type of rock being mined. The drill hole depth ranges from 20 to 50 ft (6 to 15 m), depending on the bench height used in the mine. The drill hole spacing (distance between holes) is governed by the depth of the holes and the hardness of the rock and generally ranges from 12 to 30 ft (3.7 to 9m). For short holes or hard rock, the hole spacing must be close; and for long holes and soft rock, the spacing may be increased.

Rotary drilling. Drilling is accomplished by rotating a bit under pressure. Compressed air is forced down the hollow drill shank or steel, and is allowed to escape through small holes in the bit for cooling the bit and blowing the rock cuttings out of the hole. Rotary drill holes range from a minimum diameter of 4 in. to a maximum diameter of 15 in. (10 to 38 cm) and are drilled by machines mounted on trucks or a crawler frame for mobility. The weight of these machines ranges from 30,000 to 300,000 lb (13,600 to 136,000 kg) and, in general, the heavier machines are required for the larger-diameter holes. The rotary method is the most common type of drilling used in open-pit mines because it is the cheapest method of drilling; however, it is restricted to soft and medium-hard rock.

Percussion drilling. Percussion drilling is accomplished with a star-shaped bit which is rotated while being struck with an air hammer operated by high-pressure air. Air is also forced down the drill steel to cool the bit and to blow the cuttings out of the hole. Small amounts of water are frequently added with the air to reduce the dust. The hole diameters range from 11/2 to 9 in. (3.8 to 23 cm), depending upon the type of drill. The drilling machine used for the smaller-diameter holes (1 1/2 to 41/2 in. or 3.8 to 11.4 cm) is small and usually mounted on a lightweight crawler or rubber-tired frame weighing a few thousand pounds (1 lb = 0.45 kg). The air hammer and rotating device are mounted on a boom on the machine, and the hammer blows are transferred by the drill steel to the drill bit. For the larger-diameter and deeper holes, the air hammer is attached directly to the bit, and the drill unit is lowered down the vertical hole because the impact of the hammer blows is dissipated in the larger and longer columns of drill steel. Percussion drilling is most applicable to brittle rock in the medium-hard to hard range. The depth of the smaller-diameter holes is limited to about 15 to 30 ft (4.6 to 9 m) but the larger holes can be drilled to 40 to 50 ft (12 to 15 m) without significant loss of efficiency.

Jet piercing. The jet piercing method was developed to drill the very hard iron ores (taconite). In this method, a hole is drilled by applying a high-temperature flame produced by fuel oil and oxygen to the rock. The holes range in size from 6 to 18 in. (15 to 46 cm), tend to be irregular in diameter, and require careful control of the flame to prevent overenlargement. The drilling machines are integrated units which control the fuel oil and oxygen mixture and lower the jet piercing bit down the hole. The jet piercing method is limited to very hard rock for which other types of drilling are more costly.

Blasting. The type and quantity of explosive are governed by the resistance of the rock to breaking. The primary blasting agents are dynamite and ammonium nitrate, which are detonated by either electric caps or a fuselike detonator called Primacord.

Dynamite is available in varying strengths for use with varying rock conditions. It is commonly used either in cartridge form or as a pulverzied, free-flowing material packed in bags. It can be used for almost any blasting application, including the detonation of ammonium nitrate, which cannot be detonated by the conventional blasting cap.

Commercial, or fertilizer-grade, ammonium nitrate has become a popular blasting medium because of its low cost and because it is safer to handle, store, and transport than most explosives. Granular or prill-size ammonium nitrate is commonly obtained in bulk, stored at the mine site in bins, and transported to the blasting site in hopper trucks that mix the agent and fuel oil as the explosive is discharged into the hole. It may also be obtained packed in paper, textile, or polyethylene bags. The carbon necessary for the proper detonation of ammonium nitrate is usually provided by the addition of the fuel oil. Granular or prilled ammonium nitrate is highly soluble and becomes insensitive to detonation if placed in water, and therefore its use is restricted to water-free holes. A portable blast hole dewatering pump is available for use with wet holes.

During the early 1960s ammonium nitrate slurries were developed by mixing calculated amounts of ammonium nitrate, water, and other ingredients such as TNT and aluminum. These slurries have several advantages when compared with dry ammonium nitrate: They are more powerful be-

Fig. 3. A 27-yd^3 (20.5-m^3) shovel loading a 170-ton (153-metric ton) haulage truck. *(Kennecott Copper Corp.)*

Fig. 4. An 8-yd³ (6.2-m³) shovel loading an ore train. *(Kennecott Copper Corp.)*

cause of higher densities and added ingredients, and they can be used in wet holes. Further, the slurries are generally safer and less costly than dynamite, and are used extensively in mines which are wet or have rock difficult to blast with the less powerful ammonium nitrate. Both ammonium nitrate and slurries are detonated by a small charge of dynamite.

Mechanical loading. Ore and waste loading equipment in common use includes power shovels for medium to large pits and tractor-type front-end loaders for the smaller pits, or a combination of these. The use of scrapers is becoming more common for overburden removal and in some cases for ore mining. The loading unit must be selected to fit the transportation system, but because the rock will usually be broken to the largest size that can be handled by the crushing plant, the size and weight of the broken material will also have a significant influence on the type of loading machine. Of equal importance in determining the type of loading equipment is the required production or loading rate, the available working room, and the required operational mobility of the loading equipment.

Power shovels. Shovels in open-pit mines range from small machines equipped with 2-yd³ (1.5-m³) buckets to large machines with 36-yd³ (27.5-m³) buckets. A 6-yd³ (4.6-m³) shovel will, under average conditions, load about 6000 tons (5400 metric tons) per shift, while a 12-yd³ (9.2-m³) shovel will load about 12,000 tons (10,900 metric tons) per shift. However, the nature of the material loaded will have a significant effect on productivity. If the material is soft or finely broken, the shovel productivity will be high; but if the material is hard or poorly broken with a high percentage of large boulders, the shovel production will be adversely affected. Power may be derived from diesel or gasoline engines or diesel electric or electric motors. The use of diesel or gasoline engines for power shovels is usually limited to shovels of up to 4-yd³ (3.1-m³) capacity, and diesel electric drives are not common in shovels of more than 6-yd³ (4.6-m³) capacity. Electric drives are used in shovels ranging up from 4-yd³ (3.1-m³) capacity and are the most widely used power sources in the open-pit mining operations (Figs. 3 and 4).

Draglines. A dragline is similar to a power shovel but uses a much longer boom. A bucket is suspended by a steel cable over a sheave at the end of the boom. The bucket is cast out toward the end of the boom and is pulled back by a hoist to gather a load of material which is deposited in an ore haulage unit or on a waste pile. Draglines range in size from machines with buckets of a few cubic yards' capacity to machines with buckets of 150-yd³ (115-m³) capacity. Draglines are extensively used in the phosphate fields of Florida and North Carolina. The overburden and the ore (called matrix in phosphate operations) are quite soft, and no blasting is required. The phosphate ore is deposited in slurrying pits, where it is mixed with water and transported hydraulically to the concentrating plant.

Front-end loaders. Front-end loaders are tractors, both rubber-tired and track-type, equipped with a bucket for excavating and loading material (Fig. 5). The buckets are usually operated hydraulically and range in capacity from 1 to 36 yd³ (0.76 to 27.5 m³). Front-end loaders are usually powered by diesel engines and are much more mobile and less costly than power shovels of equal capacity. On the other hand, they are not generally as durable as a power shovel, nor can they efficiently excavate hard or poorly broken material. A front-end loader has about one-half the productivity of a power shovel of equal bucket capacity; that is, a 10-yd³ (7.6-m³) front-end loader will load about the same tonnage per shift as a 5-yd³ (3.8-m³) shovel. However, the loader is gaining in popularity where mobility is desirable and where the digging is relatively easy.

Mechanical haulage. The common modes of transporting ore and waste from open-pit mines are trucks, railroads, and belt conveyors. The application of these methods or combination of them depends upon the size and depth of the pit, the production rate, the length of haul to the crusher or dumping place, the maximum size of the material, and the type of loading equipment used.

Truck haulage. Truck haulage is the most common means of transporting ore and waste from open-pit mines because trucks provide a more versatile haulage system at a lower capital cost than rail or conveyor systems in most pits (Fig. 6). Further, trucks are sometimes used in conjunction with rail and conveyor systems where the haulage distance is greater than 2 or 3 mi. (3 or 5 km). In these cases, trucks haul from the pit to a permanent loading point within the pit or on the surface, where the material is transferred to one or the other system. Trucks range in size from 20- to 350-ton (18- to 318-metric ton) capacity and are powered by diesel engines. Mechanical drives are used almost exclusively in trucks up to 75-ton (68-metric ton) capacity. Between 25- and 100-ton (23- and 91-metric ton) capacity, the drive can be either mechanical or electrical. In the larger trucks, the electric drive, in which a diesel engine drives a generator or alternator to provide power for electric motors mounted in the hubs of the wheels, is most common. Locomotive traction motors are used in some trucks of over 200-ton (181-metric ton) capacity. The diesel engines in the conventional drives range from 175 to 1200 hp (130 to 900 kW), while the engines in the diesel electric units range from 700 to 2400 hp (520 to 1800 kW). Most haulage trucks can ascend road grades of 8 to 12% fully loaded and are equipped with various braking devices, including dynamic electrical braking, to permit safe descent on steep roads. Tire cost is a major item in the operating cost of large trucks, and the roads must be well designed and maintained to enable the trucks to operate efficiently at high speed.

Fig. 5. A 3-yd³ (2.3-m³) front-end loader loading a haulage truck in a small open-pit mine. *(Eaton Corp.)*

The size of trucks used is primarily dependent upon the size of the loading equipment, but the required production rate and the length of haul are also factors of consideration. Usually 20- to 40-ton-capacity (18- to 36-metric ton) trucks are used with 2- to 4-yd³ (1.5- to 3.1-m³) shovels, while the 70- to 100-ton (64- to 91-metric ton) trucks are used with 8- to 10-yd³ (6.1- to 7.6-m³) shovels.

In the early 1960s major advances were made in truck design which put the truck haulage system in a favorable competitive position compared with the other methods of transportation. Trucks are used almost exclusively in small and medium-size pits and the majority of the large pits, but in some cases are used in conjunction with rail or conveyor haulage. Trucks have the advantage of versatility, mobility, and low cost when used on short-haul distances.

Rail haulage. When mine rock must be transported more than 3 or 4 mi (5 or 6 km), rail haulage is generally employed. Since rail haulage requires a larger capital outlay for equipment than other systems, only a large ore reserve justifies the investment. As a rough rule of thumb, the reserve should be large enough to support for 25 years a

Fig. 6. Copper mine waste is dumped from a 170-ton (153-metric ton) haulage truck. *(Kennecott Copper Corp.)*

Fig. 7. Aerial view of Kennecott's Bingham Mine. *(Kennecott Copper Corp.)*

production rate of 30,000 tons (27,000 metric tons) of ore per day and an equivalent or greater tonnage of stripping. Adverse grades should be limited to a maximum of 3% on the main lines and 4% for short distances on switchbacks. Good track maintenance requires the use of auxiliary equipment such as mechanical tie tampers and track shifters. The latter are required for relocating track on the pit benches as mining progresses and on the waste dumps as the disposed material builds up adjacent to the track. Ground movement in the pit resulting from disturbance of the Earth's crust or settling of the waste dumps makes track maintenance a large part of mining cost.

Locomotives in use range from 50 to 140 tons (45 to 127 metric tons) in weight, with the largest sizes coming into increased use for steeper grades and larger loads. Most mines operate either all electric or diesel electric models. The use of all-electric locomotives creates the problem of electrical distribution in the pit and on the waste dumps, and requires the installation of trolley lines adjacent to all tracks (Fig. 7).

Mine cars range in capacity from 50 to 100 tons (45 to 91 metric tons) of ore, and to 50 yd^3 (39 m^3) of waste. Ore is transported in various types of cars: solid-bottom, side-dump, or bottom-dump. The solid-bottom car is cheapest to maintain but requires emptying by a rotary dumper. Waste is mostly handled by the side-dump car. Truck haulage has replaced much of the rail haulage in recent years because of advances in truck design and reduction in truck haulage costs.

Conveyors. Rubber belt conveyors may be used to transport crushed material from the pit at slope angles up to 20°. Conveyors are especially useful for transporting large tonnages over rugged terrain and out of pits where ground conditions preclude building of good haulage roads, and where long haul distances are required. Improved belt design is permitting greater loading rates, higher speeds, and the substitution of single-flight for multiple-flight installations. The chief disadvantage of this transport system is that, to protect the belt from damage by large lumps, waste as well as ore must be crushed in the pit before loading on the belt.

Waste disposal problems. To keep costs at a minimum, the dump site must be located as near the pit as possible. However, care must be taken to prevent location of waste dumps above possible future ore reserves. In the case of copper mines, where the waste contains quantities of the metal which can be recovered by leaching, the ground on which such waste is deposited must be impervious to leach water. Where the creation of dumps is necessary, problems of possible stream pollution and the effect on farms and on real estate and land values must all be considered.

Slope stability and bench patterns. In open-cut and open-pit mining, the material ranges from unconsolidated surface debris to competent rock. The slope angle, that is, the angle at which the benches progress from bottom to top, is limited by the strength and characteristics of the material. Faults, joints, bedding planes, and especially groundwater behind the slopes are known to decrease the effective strength of the material and contribute to slides. In practice, slope angles vary from 22 to 60° and under normal conditions are about 45°. Steeper slopes have a greater tendency to fail, but may be economically desirable because they lower the quantity of waste material that must be removed to provide access to the ore.

Technology has been developed that permits an engineering approach to the design of slopes in keeping with measured rock and water conditions, making quantitative estimates of the factor of safety of a given design possible. Precise instrumentation can detect the boundaries of moving rock masses and the rate of their movement. Instrumentation to give warning of impending failure is being used in some pits.

Communication. Efficiency of mining operations, especially loading and hauling, is being improved by the use of communication equipment. Two-way, high-frequency radiophones are proving useful for communicating with haulage and repair crews and shovel operators.

[CARL D. BROADBENT]

Permafrost

Perennially frozen ground, occurring wherever the temperature remains below 0°C for several years, whether the ground is actually consolidated by ice or not and regardless of the nature of the rock and soil particles of which the earth is composed. Perhaps 25% of the total land area of the Earth contains permafrost; it is continuous in the polar regions and becomes discontinuous and sporadic toward the Equator. During glacial times permafrost extended hundreds of miles south of its present limits in the Northern Hemisphere.

Characteristics. Permafrost is thickest in that part of the continuous zone that has not been glaciated. The maximum reported thickness, about 1600 m, is in northern Yakutskaya, U.S.S.R. Average maximum thicknesses are 300–500 m in northern Alaska and Canada and 400–600 m in northern Siberia. In Alaska and Canada the general range of thickness in the discontinuous zone is 50–150 m and in the sporadic zone less than 30 m. Discontinuous permafrost in Siberia is generally 200–300 m.

Temperature of permafrost at the depth of no annual change, about 10–30 m, crudely approximates mean annual air temperature. It is below −5°C in the continuous zone, between −1 and −5°C in the discontinuous zone, and above −1°C in the sporadic zone. Temperature gradients vary horizontally and vertically from place to place and from time to time. Deep temperature profiles record past climatic changes and geologic events from several thousand years ago.

Ice is one of the most important components of permafrost, being especially important where it exceeds pore space. Physical properties of permafrost vary widely from those of ice to those of normal rock types and soil. The cold reserve, that is, the number of calories required to bring the material to the melting point and melt the contained ice, is determined largely by moisture content. Ice occurs as individual crystals ranging in size from less than 0.1 mm to at least 70 cm in diameter. Aggregates of ice crystals are common in dikes, layers, irregular masses, and ice wedges. These forms are derived in many ways in part when permafrost forms. Ice wedges grow later and characterize fine-grained sediments in continuous permafrost, joining to outline polygons. Microscopic study of thin sections of ice wedges reveals complex structures that change seasonally.

Permafrost develops today where the net heat balance of the surface of the Earth is negative for several years. Much permafrost was formed thousands of years ago but remains in equilibrium with present climates. Permafrost eliminates most groundwater movement, preserves organic remains, restricts or inhibits plant growth, and aids frost action. It is one of the primary factors in engineering and transportation in the polar regions.

Construction engineering problems. The construction of the Alaska Pipe Line to carry oil from Prudhoe Bay to Valdez has been the largest and most difficult project conducted in permafrost regions to date. The hot oil, at a temperature about 60°C, requires that special consideration be given to thawing of permafrost and consequent effects on the pipe line itself, as well as adjacent structures. For several short lengths the pipe line had to be buried, and artificial refrigeration is required to maintain the permafrost. The pipe was well insulated to reduce heat flow. About half the pipe line is elevated above the ground because of ice-rich permafrost. So much ice is present in fine-grained materials that liquefaction, flow, and slump would occur on melting. The elevated and insulated pipe dissipates some heat to the air, but with little effect on the ice below ground. However, the vertical supports for the horizontal crossbeams on which the pipe rests are prone to frost heaving in some places. To ensure that they remain securely frozen in place, cooling devices to minimize the thawing effects of the uprights are installed in them. One type is a metal tube filled with a refrigerant that evaporates whenever the ground temperature exceeds the air temperature. The evaporation keeps the ground frozen during warm summers. The devices contain no moving parts and are self-operating.

Buildings and other structures that could thaw ice-rich permafrost are also commonly put on piles, with air spaces separating the structures from the ground. Different kinds of insulation and the incorporation of heat-reflecting colors in road construction minimize thawing of permafrost. However, in places, important roads can be maintained only by eliminating ice-rich permafrost during construction and replacing it with nonfrostsusceptible granular aggregate. Because of the disturbance of surface and near-surface water flow by most structures, projects must be well planned to avoid icings, heaving, and erosion, and a variety of new engineering techniques have been developed. [ROBERT F. BLACK]

Bibliography: O. J. Ferrians, Jr., R. Kachadoorian, and G. W. Greene, *Permafrost and Related Engineering Problems in Alaska*, USGS Prof. Paper 678, 1969; *Proceedings of the 3d International Conference on Permafrost, Edmonton*, National Research Council of Canada, 1978; A. L. Washburn, *Periglacial Processes and Environments*, 1973.

Pesticide

A material useful for the mitigation, control, or elimination of plants or animals detrimental to human health or economy. Algaecides, defoliants, desiccants, herbicides, plant growth regulators, and fungicides are used to regulate populations of undesirable plants which compete with or parasitize crop or ornamental plants. Attractants, insecticides, miticides, acaricides, molluscicides, nematocides, repellants, and rodenticides are used principally to reduce parasitism and disease transmission in domestic animals, the loss of crop plants, the destruction of processed food, textile, and wood products, and parasitism and disease transmission in man. These ravages frequently stem from the feeding activities of the pests. Birds, mice, rabbits, rats, insects, mites, ticks, eel worms, slugs, and snails are recognized as pests.

MATERIALS AND USE

Materials used to control or alleviate disease conditions produced in humans and animals by plants or by animal pests are usually designated as drugs. For example, herbicides are used to control the ragweed plant, while drugs are used to alleviate the symptoms of hay fever produced in humans by ragweed pollen. Similarly, insecticides are used to control malaria mosquitoes, while drugs are used to control the malaria parasites—single-celled animals of the genus *Plasmodium*—transmitted to humans by the mosquito.

Sources. Some pesticides are obtained from plants and minerals. Examples include the insecticides cryolite, a mineral, and nicotine, rotenone, and the pyrethrins which are extracted from plants. A few pesticides are obtained by the mass culture of microorganisms. Examples are the toxin produced by *Bacillus thuringiensis*, which is active against moth and butterfly larvae, and the so-called milky disease of the Japanese beetle produced by the spores of *B. popilliae*. Most pesticides, however, are products of chemical manufacture. Two outstanding examples are the insecticide DDT and the herbicide 2,4-D.

Evaluation. The development of new pesticides is time-consuming. The period between initial discovery and introduction is frequently cited as being about 5 years. Numerous scientific skills and disciplines are required to obtain the data necessary to establish the utility of a new pesticide. Effectiveness under a wide variety of climatic and other environmental conditions must be determined, and minimum rates of application established. Insight must be gained as to the possible side effects on other animals and plants in the environment. Toxicity to laboratory animals must be measured and be related to the hazard which might possibly exist for users and to consumers. Persistence of residues in the environment must be determined. Legal tolerances in processed commodities must be set and directions for use clearly stated. Methods for analysis and detection must be devised. Economical methods of manufacture must be developed. Manufacturing facilities must be built. Sales and education programs must be prepared.

Regulation and restriction. By the mid-1960s, it became apparent that a number of the new pesticides, particularly the insecticide DDT, could be two-edged swords. The benefits stemming from the unmatched ability of DDT to control insect pests could be counterbalanced by adverse effects on other elements of the environment. Detailed reviews of the properties, stability, persistence, and impact upon all facets of the environment were carried out not only with DDT, but with other chlorinated, organic insecticides as well. Concern over the undesirable effects of pesticides culminated in the amendment of the Federal Inseticide, Fungicide, and Rodenticide Act (FIFRA) by Public Law 92-516, the Federal Environmental Pesticide Control Act (FEPCA) of 1972. The purpose behind this strengthening of earlier laws was to prevent exposure of either humans or the environment to unreasonable hazard from pesticides through rigorous registration procedures, to classify pesticides for general or restricted use as a function of acute toxicity, to certify the qualifications of users of restricted pesticides, to identify accurately and label pesticide products, and to ensure proper and safe use of pesticides through enforcement of FIFRA.

Selection and use. Pesticides must be selected and applied with care. Recommendations as to the product and method of choice for control of any pest problem—weed, insect, or varmint—are best obtained from county or state agricultural extension specialists. Recommendations for pest control and pesticide use can be obtained from each state agricultural experiment station office. In addition, it is necessary to follow explicitly the directions, restrictions, and cautions for use on the label of the product container. Insecticides are a boon to the production of food, feed, and fiber, and their use must not be abused in the home, garden, farm field, forest, or stream. *See* AGRICULTURAL CHEMISTRY; FUNGISTAT AND FUNGICIDE; HERBICIDE; INSECTICIDE. [GEORGE F. LUDVIK]

PERSISTENCE

An expanding use of pesticidal chemicals in agriculture, public health, forestry, warfare, and home gardens continues to be a source of controversy. For a wide variety of reasons many attempts have been made, particularly in North America, to restrict the use of such chemicals. Considerable scientific evidence has recently been accumulated that documents the deleterious effects on the environment of several of the chlorinated hydrocarbons. On this basis, a number of environmental scientists have attempted to eliminate the use of these compounds, especially DDT. Industry has responded with the considerable resources at its disposal, and the bureaucratic machinery of government has frequently been caught in the middle. Industry has accused the anti-DDT forces of being antitechnology, and environmental scientists have replied that the technology of 1945 is antiquated and that a new technology must be developed that will be compatible with the continued existence of life on Earth. The DDT controversy might therefore be considered a prologue to a much larger controversy of the future that will decide how increasing numbers of people, with decreasing resources, will achieve a technology that will be compatible with both human values and the long-term survival of man.

The term pesticide may be defined as any substance used to kill, or to inhibit the growth and reproduction of, the members of a species considered in a local context as pests. About 800 compounds are now used as pesticides and, with increasing technological sophistication, the list can be expected to grow. Until World War II pesticides consisted of inorganic materials containing sulfur, lead, arsenic, or copper, all of which have biocidal properties, or of organic materials extracted from plants such as pyrethrum, nicotine, and rotenone. During World War II a technological revolution began with the introduction of the synthetic organic biocides. DDT was widely used during World War II to control the vectors of diseases which until that time had been serious health hazards over much of the world. Since then, DDT has been extensively used with spectacular success to reduce the incidence of malaria in the tropics. In the public mind DDT has become almost synonymous with pesticide, yet it is important to point out that DDT is only one of an increasing number of organic biocides. They vary widely in toxicity, in specificity, in persistence, and in the production of undesirable derivatives. The factors that determine how, when, and where a particular pesticide should be used, or whether it should be used at all, are therefore necessarily different for each pesticide. During the recent controversy over pesticide usage, the industry spokesmen have consistently attempted to associate those who believe it is time to ban DDT with those who campaign against the use of all pesticides. The use of chemicals to achieve a measure of control over the environment, however, has become an integral part of technology, and there are no valid arguments to support the proposition that the use of all such chemicals should now cease.

Some take the view that pests include insects, fungi, weeds, and rodents. Many insects, however, particularly the predatory species that prey upon the insects that damage crops, are clearly not pests. Fungi are essential in the processes that convert dead plant and animal material to the primary substances that can be recycled through living systems. In many ecosystems rodents are also an important link in the recycling process. On the other hand, an increasing number of species traditionally considered desirable are becoming "pests" in local contexts. Thus chemicals are used to destroy otherwise useful vegetation along roadsides or in pine plantations, or vegetation that might provide food and shelter to undesirable human populations. The control of any component of the global environment through the use of biocidal chemicals is therefore a problem much more complex than is initially suggested by the connotation of the word pest. The highly toxic chemicals that kill many different species are clearly undesirable in situations in which control over only one or two is sought.

Like any other technological product, pesticides, once used, become waste products and some become pollutants. It can no longer be assumed that pollutants will disappear in the vastness of the sea, since the global environment has suddenly become small. The total amount of waste materials produced by man is now approximately as great as the amount of organic material cycled yearly through nature. Waste products that are readily degraded into elementary materials that can be recycled clearly do not threaten the stability of the environment. In traditional terms they might be compared with the biblical ashes that return to ashes and the dust that returns to dust.

Biodegradable compounds. Of the approximately 800 biocidal compounds used as pesticides, many obey this fundamental ecological law and are degraded to elementary materials that can be recycled. These compounds include the organophosphate insecticides and the carbamates, since no evidence now available suggests that any derivative of these compounds becomes a persistent pollutant. Malathion, diazinon, and parathion are examples of organophosphorous insecticides. Parathion is particularly toxic to many forms of life and has caused the deaths of many persons applying it to crops or who have been accidentally exposed. Other pesticides in this group, however, have been extensively used without causing human fatalities. They are readily metabolized in the mammalian body, and in the environment are degraded to elementary materials. Unwise use of these compounds may result in local fish kills or other loss of wildlife, but with careful use such losses can be minimized. Because of the chemical instability of these biocides they do not travel to areas for removed from the sites of application. Environmental damage, if any, is therefore local. The organophosphorous insecticides inhibit the enzyme cholinesterase, which normally breaks down acetylcholine following the transmission of the nerve impulse across the synapse.

The carbamates are another group of biocides that also inhibit the enzyme cholinesterase. Carbaryl, or sevin, is an example. These biocides are even less persistent than the organophosphates and pose fewer health hazards to man. Indiscriminate use, however, may cause local environmental problems. Carbaryl, for example, is particularly toxic to bees.

Other pest-control measures. Sophisticated methods of pest control are continually being developed. One technique involves the raising of a large number of insect pests in captivity, sterilizing them with radioactivity, and then releasing them into the environment, where they mate with the wild forms. Very few offspring are subsequently produced. Highly specific synthetic insect hormones are being developed. In an increasing number of pest situations, a natural predator of an insect has been introduced, or conditions are maintained that favor the propagation of the predator. The numbers of the potential pest species are thereby maintained below a critical threshold.

Chlorinated hydrocarbons. DDT belongs to a class of biocides that are chlorinated hydrocarbons. The chlorine-carbon bond is rare in nature and is comparatively stable. Few bacteria and fungi are equipped to break it. These biocides are therefore relatively persistent, much more so than are the carbamates and the organophosphates. The chlorinated hydrocarbons are also much less soluble in water but are very soluble in nonpolar media such as lipids. They are therefore readily concentrated by organisms from the environment, and accumulate in fatty tissues. Species higher in the food chains will subsequently accumulate the chlorinated hydrocarbons present in their food. Mobility is another property which makes some of the chlorinated hydrocarbons hazardous to the global environment. DDT and dieldrin become vapors and can be transported anywhere in the world to accumulate in distant ecosystems. Since the chlorinated hydrocarbon biocides are toxic to a wide spectrum of animal life, they have the capacity to cause harm to many nontarget species. *See* POLYCHLORINATED BIPHENYLS.

Several of the chlorinated hydrocarbons, however, do not appear to possess the combined properties of intensive use, persistence, and mobility that create the potential for environmental degradation. Lindane is more readily degraded than the others, and chlordane and toxaphene do not appear to possess the mobility of DDT and dieldrin, since they have not so far been detected in marine organisms away from contaminated estuaries.

Of all the pesticides now in use, only five have aroused the concern of environmental scientists who are attempting to consider the long-term effects upon the global environment. Aldrin, dieldrin, endrin, and heptachlor are extremely toxic, persistent, and mobile, and have the capacity to inflict harm upon nontarget species.

The fifth compound, DDT, has yet another property that renders it especially dangerous to the environment. The insecticidal compound, *p,p'*-DDT, is relatively persistent, but much of it is eventually degraded to another DDT compound, *p,p'*-DDE. This compound, DDE, is more persistent than the original DDT and appears to be the most abundant of the synthetic pollutants in the global environment. Most of the concern about DDT in the environment is actually about DDE.

Many marine biologists still find it difficult to accept the findings that DDE is now more abundant in marine birds and fish than in many terrestrial organisms. It is not the only pollutant to be so widely dispersed. Polychlorinated biphenyls, another group of chlorinated hydrocarbons that are industrial pollutants, have become similarly dispersed throughout the global ecosystem. The sea is no longer the vast entity that it was traditionally conceived to be. World production figures of many chemicals, including DDT, are now a significant fraction of the amount of organic material that is recycled each year through the world's biomass.

DDE was originally considered to be harmless because of its relative nontoxicity to organisms. Evidence is now rapidly accumulating that DDE has physiological effects that may not kill the organism but which are nevertheless deleterious and may impair the long-term survival of the species. Experiments have shown that DDE may cause several species of birds to lay eggs with abnormal amounts of calcium carbonate. Many species of raptorial and fish-eating birds have experienced low rates of reproductive success over the past 20 years, and in most cases one of the symptoms has been abnormally thin eggshells. In the most spectacular cases, such as the brown pelicans and other sea birds of the West Coast, the shells are so thin that they break during incubation. Increasing amounts of evidence link this phenomenon to DDE.

Physiological effects of DDE that may be associated with thin eggshells and other abnormalities include the inhibition of membrane enzymes that are adenosinetriphosphatases involved with ion transport, the inhibition of the enzyme carbonic anhydrase, and the induction of nonspecific liver enzymes that degrade a wide variety of lipid-soluble compounds, including steroids. Other effects of DDE are quite likely to be discovered in the future.

In 1969 fish caught in the Pacific Ocean were seized and condemned by the U.S. Food and Drug Administration because they had been found to contain high concentrations of the DDT compounds. A number of monitoring programs for pesticides in the environment had previously been initiated in North America, but none had been designed to find the accumulation site of a pesticide pollutant or to find the rate at which it accumulates there. It is now obvious that DDE has been accumulating in the sea, but since there are no data on how fast it is accumulating, it is difficult to know what to expect in the future.

At the present time the environmental science community is convinced that DDE is contributing to the extinction of a number of species of birds. Only 4 years ago the idea that DDE might cause thin eggshells would have been dismissed by most environmental scientists as somewhat preposterous. Moreover, there is no information on what additional effects DDE or another pollutant might have, once a critical threshold level is reached.

The spokesmen for the DDT industry have maintained that DDT has no effects upon man at dosages that might be encountered. No deleterious effects have so far been found in man. A preliminary screening test indicated, however, that *p,p'*-DDT or a metabolic derivative might be carcinogenic. Almost no work has been done with DDE, the DDT compound that is present in food and in the environment. For reasons that are not clear to an environmental scientist, almost all of the work with DDT has been done with the insecticidal compound, *p,p'*-DDT, which is applied to fields. Conclusions based upon work with this compound are of doubtful validity if they are meant to include DDE also. Conclusions about the effects of DDE upon man will require many analyses for DDE and other pollutants in human tissues that can be correlated with causes of death, similar to the work that has showed the correlation between cigarette smoking and lung cancer.

Future course of action. The confiscation of the marine fish is an instance of environmental food becoming unsuitable for man because of increasing pollution. The continued release of persistent pollutants into the environment will mean that more and more food will become unavailable.

In considering the expanding populations, the increasing amounts of waste materials produced per capita, and both the known and unknown effects of persistent pollutants, the environmental scientist is forced to assume a very conservative position. Any of the persistent pollutants that are released in amounts comparable to those of the DDT compounds has the capacity to produce an irreversible effect upon the global environment. The chances, therefore, simply are not worth taking if other compounds that will do the same job are available. To preserve the global environment as a fit habitat not only for pelicans and bald eagles but for man himself, a commitment must now be made to a technology consistent with the biblical and ecological law that ashes must return to ashes. Although laws have been passed to ban the use of DDT, the threat of a major insect infestation to forests or crops may bring about a relaxation of the law.

[ROBERT W. RISEBROUGH]

Bibliography: J. L. Apple and R. F. Smith, *Integrated Pest Management*, 1976; B. P. Beirne, *Pest Management*, 1966; G. L. Berg (ed.), *Farm Chemicals Handbook*, 1978; R. Carson, *Silent Spring*, 1962; R. L. Caswell, *Pesticide Handbook—Entoma 1977–78*, 1977; *Federal Insecticide, Fungicide and Rodenticide Act*, Public Law 92–516, 92d Congress, *H. R.* 10729, Oct. 21, 1972; *Federal Insecticide, Fungicide, and Rodenticide Act as Amended*, EPA, 1976; *Federal Insecticide, Fungicide, and Rodenticide Act as Amended*, OPA 17/9, EPA, 1978; *Federal Pesticide Act of 1978*, OPA 136/8, 1978; D. L. Fowler and J. N. Mahan, *The Pesticide Review—1977*, 1978; W. W. Kilgore and R. L. Doutt, *Pest Control: Biological, Physical and Selected Chemical Methods*, 1967; H. Martin and C. R. Worthing, *Pest Manual*, 4th ed., 1974; R. L. Metcalf and W. Luckman, *Introduction to Pest Management*, 1975; *Suspended and Cancelled Pesticides*, EPA, 1977; T. F. Watson, L. Moore, and G. W. Ware, *Practical Insect Pest Management*, 1975; W. J. Wiswesser, *Pesticide Index*, 1976.

Physiological ecology (plant)

The study of growth, biological processes, and reproduction within plant populations in natural or controlled environments. The five principal growth forms are unicellular, thallose, and herbaceous forms, and shrubs and trees. Life cycle processes include (1) germination, (2) growth, (3) absorption and loss of water, gases, and mineral nutrients, (4) photosynthesis and respiration, and (5) production of reproductive structures (flowers, cones, fruits, seeds, and spores). Each process is the result of a complex series of physical and chemical reactions within and between cells. All cellular processes are within the context of the whole plant as governed by the interaction of its genes and its environment through space and time.

Scope. Plant physiological ecology is part of the attempt to answer the question: Why does a plant grow where it does? An answer requires a knowledge of the genetic structure of the population, the population's physiological processes and their rates, its growth forms, the populational demography including seed sources, and the population's niche in the operation and structure of the ecosystem.

Actual and potential geographic ranges of populations or species seldom coincide. Environments change. Populations evolve, migrate, and become extinct in response to environmental changes. The potential range of a species or a local population depends upon the degree and nature of genetic variation among its individuals, the range of physiological and morphological adaptability within each individual (acclimatization and phenotypic plasticity), and the frequency, extent of occurrence, availability, and stability of suitable environments. These complex variables, operating together, determine the kinds and rates of physiological processes, and thus the degree of vegetative and reproductive success of a population within an ecosystem or in the biosphere.

All widespread species are genetically diverse. Within these species, and many others, ecotypes (ecological races) have evolved in response to environmental selection of local populations. Such ecotypes are particularly fitted genetically to the local or regional environment. Often, these ecotypes merge with each other to form a complex biological gradient (ecocline) coinciding with an environmental gradient. Some environmental gradients are simple and repeatable, with expected repeatability and simplicity in their ecoclines. But in broad space, and given enough time, environments exist as complex webs of gradients, and so also do the accompanying ecoclines which thus are displayed as ecoclinal networks.

Ecotypes and ecoclines consist of local populations usually with enough genetic diversity to allow some survival within certain limits of environmental change. Studies of process rates within local populations and ecotypes provide a measure of the environmental tolerance limits of a species. These data allow an understanding of the adaptability of plants and of the evolution of physiological processes. Such information may be obtained in natural environments under field conditions or by experiment in controlled and simulated environments.

Field measurements. The problems which physiological ecology attempts to solve exist in nature, and it is there that the solutions to such problems must begin. Field measurements are difficult because of logistics, and the vagaries of uncontrolled environments and their effects on process rates; but such measurements have the advantage of providing realistic values in an ecosystemic context. Such results should then be tested by experiment under controlled environments in the laboratory or phytotron. It is extremely difficult, if not impossible, to work with ecosystems or even most communities under controlled conditions. So, after going as far as possible by laboratory experiment, one must return to the field to measure and experiment again with more insight and precision than at the start. Such a research strategy not only is possible but is productive of answers.

The rates of most physiological processes can be measured in the field if portable instrument systems can be designed and made. In the past, field measurements were difficult logistically because of size and weight of such systems. Fortunately, as the result of other kinds of research, miniaturization of circuits and the use of solid-state components have helped to solve these problems. Mobile laboratories, with such systems in vans or trailers, make it easier now to get laboratory precision in the natural environment of the plant. These field laboratories also include data acquisition systems for the rapid recording and statistical handling of the data.

During the 1970s, as a result of the use of such new equipment by well-trained teams of ecologists and physiologists, there has been a tremendous growth in knowledge of how plants operate in field environments. The processes most often studied are germination, growth, leaf energy budgets, water movement, mineral nutrition, photosynthesis, respiration, flowering, fruiting, seed production, and dormancy of buds and seeds. Before approaching these processes in detail, it must be stated that the processes themselves are influenced not only by the physical and biological aspects of the environment but by the growth form of the plant. Processes and rates depend to a large extent on whether the plant is herbaceous or woody, a dwarf shrub or tall tree, an annual or a perennial, and upon the configurations of its leaves, stems, and roots. *See* PLANTS, LIFE FORMS OF.

Germination. Germination times and phenomena can easily be observed in natural populations if one knows young seedling characteristics. Percentage of germination can be measured by planting known numbers of seeds at known points and marking the seedlings by colored toothpicks or plastic markers as they emerge from the soil surface. Germination can be measured in phytotrons under controlled conditions. Alternating temperatures rather than constant temperatures result in higher germination percentages in many species. Seeds of some species require light for germination; others do not.

Growth. Growth of stems or leaves may be measured in terms of height or length, of fresh or dry weight, or of calorific values. Leaf production may be determined by marking each leaf as it emerges with a spot of nontoxic waterproof compound. These leaf counts should be tallied and retallied at daily or weekly intervals. Root growth in the field can be measured by planting or transplanting individual plants or sod blocks into wedge-shaped Plexiglas boxes which are embedded in the natural soil. These boxes can be lifted at frequent time intervals for marking on the Plexiglas the position of each root or rhizome tip.

Leaf energy budgets. The temperature of a leaf is the result of a balance between heat income and heat loss. The measurement of leaf energy budgets involves reflection, absorption, and transmission of visible and infrared radiation from sun, sky, and ground surface, evaporative heat losses, gains or losses by convection, heat storage, and metabolism. Leaf temperatures may be measured by thermocouples, thermistors, or noncontact infrared thermometers. Temperature of a leaf exposed to bright sun is largely controlled by convection and by transpiration rate (evaporative heat loss); the effect is greater in a large leaf than in a small one. In turn, leaf temperature affects transpiration rate itself and the important metabolic processes of photosynthesis and respiration. All of these energy budget processes are influenced by wind speed, boundary-layer effects, turbulent transfer, and thermal stability of the air.

Water loss and gain. A simple method of measuring transpiration rates and total water loss of whole herbaceous plants or young woody plants is to determine the loss of weight from systems consisting of a plant or sod block having its roots in a sealed container of moist soil. Also, a plant or leaf may be enclosed in a transparent cuvette or chamber and its water vapor losses measured in an air stream by an infrared gas analysis system or a dew-point hygrometer. Transpiration rates of whole trees are difficult to measure directly, but extrapolations may be made from single attached leaves, small branches, or whole tree "seedlings." Stomatal transpiration requires open stomates. Whether or not the stomates are open or closed, and for how long, is best measured by a diffusion porometer. Transpiration rates must be related to leaf surface areas, leaf fresh weight, or soil and plant water potentials.

The ability of a plant to absorb water from the soil is measured by obtaining quantitative data on a gradient in water potential from leaves to roots to soil. Water potential is the difference in free energy or chemical potential (per unit molal volume) between pure water and water in cells and solutions. The potential of pure water is set at zero; the potential of water in cells, in solutions, and in the soil therefore is less than zero and is expressed as a negative number. In rapidly transpiring plants there is a gradient toward decreasing water potential (more negative) from the soil through the roots, stems, and leaves to the air. This causes water to move from the soil through the plant to the atmosphere. All along this gradient, however, there are resistances to the diffusion and flow of water, and these must be considered in all calculations of the movement of water through the plant. A good technique for measuring leaf water potential in the field is to use one of the newer modifications of a Scholander pressure chamber. If this is not available, the Schardakov dye method will work on many leaves, but is relatively slow and somewhat prone to error. More precise measurements are possible in the laboratory with a sensitive thermocouple psychrometer; modifications of this technique are now available for field use with some success. Dew-point hygrometry is also being developed for such field measurement. It has the advantage of not destroying the leaves being measured; thus, their potentials can be measured again at a later date.

Mineral nutrition. Plants require some nutrients in fairly large quantities. These include N, P, S, K, Ca, Mg, and Fe. Other mineral elements, such as Mn, Zn, Cu, Mo, B, and Ce, are needed in only trace amounts. Among these, N is largely made available by microbial fixation in the familiar nitrogen cycle. Others are made available in various ways; soil pH, for example, affects the availability of certain nutrients across the pH scale. Plants and vegetation play a large role in cycling of nutrient elements (and some nonnutrient elements, including pollutants) in the biosphere. The utilization of nutrients in ecosystems can be estimated by analyzing soils, roots, leaves, flowers, and fruits. Also, fertilization experiments can be done relatively easily in both field and phytotron. *See* PLANT, MINERALS ESSENTIAL TO.

Photosynthesis and respiration. These two groups of processes control the gain and loss of carbon by green plants. Within the last 10 to 15 years, physiologists have found that several photosynthetic pathways or modes exist. Ecologists have been quick to relate these to different groups of species, some closely related and some not. Also, there appear to be some relationships of these pathways to certain kinds of environments. In brief, most plants appear to photosynthesize by the C_3 or pentose phosphate pathway (the Calvin-Benson cycle). Plants of other species, seemingly fewer in number, utilize a more complex C_4 or dicarboxylic acid pathway (the Hatch-Slack pathway). Leaves of C_4 plants are characterized by a green bundle sheath around each vein. The C_4 process involves malate synthesis in the mesophyll cells. Malate is transferred to the bundle sheath cells, where it is decomposed into CO_2 and utilized in a C_3 cycle. Thus, CO_2 is very efficiently used in C_4 plants, which results in high productivity.

Many succulent plants in families including the Cactaceae, Crassulaceae, Bromeliaceae, and several others display a third method of carbon capture. This photosynthetic mode is crassulacean acid metabolism (CAM). Plants exhibiting CAM have their stomates closed during the day and open at night. Carbon dioxide enters the plant at night through the open stomates and is fixed as malic acid. The stomates close as day comes; malic acid leaves the vacuoles and enters the chloroplasts, where it is decarboxylated in a pathway somewhat similar to that of C_4 plants. The carbon dioxide is converted to carbohydrate through ribulose bisphosphate so that during the daylight hours, with stomates closed, a succulent CAM plant essentially follows the C_3 pathway.

The ecological implications of these three photosynthetic modes have been studied extensively in the last few years; at this time, such research is one of the principal concerns of physiological plant ecology. Most plant species utilize the C_3 pathway; these include plants in all terrestrial climates. The C_4 pathway has evolved more recently, and has evolved independently in many relatively unrelated families, especially the grasses and sedges. In the saltbushes *(Atriplex)* some species are C_3 and some are C_4. The C_4 pathway appears to be restricted to regions having warm summers, and therefore is more common in the tropics and at low elevations. However, this pathway does occur in plants of the middle latitudes, particularly in grasslands and deserts that have hot or warm summers no matter how cold the winters. It is not known from plants in alpine or arctic conditions, where all plants so far examined have the C_3 pathway. Of course, C_3 plants are not restricted to cool summer environments; some of the principal dominants of hot deserts in North America use the C_3 mode. Creosote bush *(Larrea)*, for example, in the Mojave and Sonoran deserts is a C_3 species, as are other common plants there. On the other hand, in the hottest, driest parts of Death Valley, *Tidestromia*, a small, grayish perennial C_4 plant, reaches maximum photosynthesis in full sunlight at leaf temperatures ranging from 46 to 50°C in midsummer.

Plants with CAM are either true succulents or belong to families that are made up primarily of succulents. Cacti are good examples, as are members of the pineapple family. These usually occupy arid or semiarid habitats in warm regions with short daylengths: tropical or subtropical. The advantage of CAM is that the stomates are closed during the hot, dry days, and thus transpiration is reduced in an environment where water is available only occasionally.

Several attributes of C_3 and C_4 plants have ecological significance. Both kinds of plants have dark respiration, as do CAM plants. But C_3 plants also have photorespiration, which occurs only in the bundle sheath cells of C_4 plants and is thus negligible in these latter plants. Dark respiration rates are low compared to photosynthesis, but evidence now exists that photorespiration rates of C_3 plants are two to three times higher than dark respiration rates and that photorespiration can be half as great as the photosynthetic rate. The result is a lower net photosynthesis rate under high light intensities than the net photosynthesis rate in C_4 plants under similar conditions. Photosynthesis is also inhibited at atmospheric O_2 concentrations in C_3 plants but not in C_4 species. Optimum net photosynthesis rates in C_3 plants usually range from 10 to 25°C, but the optimum range in C_4 plants is higher, from 30 to 40°C.

Another difference which has been found between C_3 plants and C_4 plants seems to have great ecological importance. Carbon exists in two stable isotopic forms, namely, ^{12}C and ^{13}C. Carbon dioxide thus exists as $^{12}CO_2$ and $^{13}CO_2$, as well as the radioactive $^{14}CO_2$. Because of the nature of the photosynthetic pathways, C_3 plants discriminate to a greater extent against ^{13}C in assimilation of carbon than do C_4 plants. Therefore, C_4 plants have more ^{13}C in their compounds than do C_3 plants. These stable isotope concentrations are expressed as $\delta^{13}C$ values relative to ^{12}C and ^{13}C values in a standard substance. These δ^{13} values for C_3 plants are in the neighborhood of -28 parts per thousand (compared to the standard) in C_3 plants and about -12.5 ppt in C_4 plants. These values are determined by the use of a mass spectrometer. Not only is the $\delta^{13}C$ value of importance in identifying whether or not plants are C_3 or C_4, but the $\delta^{13}C$ value is transmitted into the ecosystem from plant to herbivore. Therefore, analyses of $\delta^{13}C$ in insect predators and also in rumen and fecal material from vertebrate herbivores can reveal which animals or invertebrate predators in an ecosystem prefer C_3 or C_4 plants. The importance of this new ecological tool cannot be overestimated in tropical and temperate zone ecosystems.

Photosynthesis and respiration rates in plants are usually measured in the field by one or both of two systems: an infrared gas analysis system or one using radioactive $^{14}CO_2$. In either system, a plant or leaf is enclosed in a transparent cuvette or chamber which is temperature-controlled. In the gas analyzer system, air with known content of CO_2 is passed through the chamber and then through the infrared analyzer. Decreased CO_2 content is a measure of the net CO_2 exchange or net photosynthesis. With the infrared analyzer system, this can be continuously recorded. It is important to measure flow rate of air through the system and also to measure leaf temperature, temperature of ambient air, humidity, and light intensity and quality. Covering the cuvette with a dark cloth allows measurement of dark respiration, which is indicated by an increase in CO_2 content of the controlled air stream.

In the $^{14}CO_2$ method, a plant or leaf is enclosed in a lighted cuvette and exposed to $^{14}CO_2$ of accurately known activity at a given flow rate for a precise period of time ranging from 30 s to 1 min. The ^{14}C-labeled leaf is then immediately cut, its area is measured, and the leaf is placed in a vial, frozen, and returned to the laboratory for processing and eventual counting in a liquid scintillation counter. Respiration cannot be measured with the $^{14}CO_2$ system. Therefore, it is measuring gross photosynthesis, not net.

By growing plants of the same population or clone under different temperature or light regimes in a phytotron and then measuring their metabolism across a span of temperature or light, it is possible to measure the ability to acclimate physiologically. This can be done with the use of an infrared analysis system. Using this approach to the measurement of photosynthesis and respiration, it has been found that acclimation is under genetic control and that acclimation ecotypes exist in some species.

Flowering and fruiting. These processes may be observed and measured in the field and correlated with environmental trends through time. Temperature, photoperiod, and soil moisture are the environmental factors most frequently measured in this regard. Biological pollinators and predators also should be counted. Flowering is easily studied also under controlled phytotron conditions.

The future. Even with present data from measurements and experiments in physiological ecology, it is possible to design mathematical models of plant performance in changing environments. Such models can be improved by increased knowledge of physiological processes as they occur under varying field environments in which individual plants and populations are subjected to stress. *See* ECOSYSTEM MODELS.

Beyond this, it is possible to carry out "giant" experiments by manipulating whole ecosystems. The results from experiments on an ecosystemic scale provide an integration of the physiological and ecological processes in entire ecosystems. Indeed, this next step has already been taken by the experiments and measurements at Hubbard Brook, New Hampshire, in the northern hardwood forest ecosystem, and also by the integrated results from several of the International Biological Programs. Such studies have led to a proposed national network of Experimental Ecosystem Reserves, where ecosystems can be manipulated and compared with nearby "control" ecosystems of the same type. In this kind of large-scale research, the techniques of physiological ecology are combined with those of other disciplines to measure ecosystem stress tolerance. From such results, it is possible to construct predictive mathematical models of ecosystemic performance in changing biospheric environments. *See* ECOLOGY.

[W. D. BILLINGS]

Bibliography: W. D. Billings et al., *Arctic Alpine Res.*, 3:277–289, 1971; O. Björkman et al., *Science*, 175:786–789, 1972; F. H. Bormann and G. E. Likens, *Pattern and Process in a Forested Ecosystem*, 1979; J. K. Detling et al., *Oecologia*, 38: 167–191, 1979; D. M. Gates and R. B. Schmerl (eds.), *Perspectives of Biophysical Ecology*, 1975; M. Kluge and I. P. Ting, *Crassulacean Acid Metabolism: Analysis of an Ecological Adaptation*, 1978; O. L. Lange, L. Kappen, and E.-D. Schulze, *Water and Plant Life*, 1976; W. Larcher, *Physiological Plant Ecology*, 1975; H. A. Mooney, *Annu. Rev. Ecol. System.*, 3:315–346, 1972; H. A. Mooney et al., *Carnegie Inst. Yearb.*, 73:793–805, 1974; L. L. Tieszen (ed.), *Vegetation and Production Ecology of an Alaskan Arctic Tundra*, 1979; L. L. Tieszen et al., *Oecologia*, 37:351–359, 1979.

Phytoplankton

Mostly autotrophic microscopic algae which inhabit the illuminated surface waters of the sea, estuaries, lakes, and ponds. Many are motile. Some perform diel (diurnal) vertical migrations, others do not. Some nonmotile forms regulate their buoyancy. However, their locomotor abilities are limited, and they are largely transported by horizontal and vertical water motions.

Energy-nutrient cycle. All but a few phytoplankton are autotrophic—they manufacture carbohydrates, proteins, fats, and lipids in the presence of adequate sunlight by using predominantly inorganic compounds. Carbon dioxide, water, inorganic nutrients such as inorganic phosphorus and nitrogen compounds, trace elements, and the Sun's energy are the basic ingredients. This organic matter is used by other trophic levels in the food web for nourishment. The energy in this organic matter ultimately is released as heat in the environment. The inorganic materials associated with this organic matter, on the other hand, are recycled. They are released back into the aquatic environment in inorganic form as a result of metabolic processes at all levels in the food web, and in large part again become available for reuse by the phytoplankton. Some structural materials, such as the silica in diatoms and silicaflagellates, and carbonates in the scales of certain chrysophytes, are returned to the aquatic environment by chemical dissolution rather than by biological processes. *See* FOOD CHAIN.

Varieties. A great variety of algae make up the phytoplankton. Diatoms (class Bacillariophyceae) are often conspicuous members of marine, estuarine, and fresh-water plankton. Reproducing mostly asexually by mitosis, they can divide rapidly under favorable conditions and produce blooms in a few days' time. Their external siliceous skeleton, termed a frustule, possesses slits, pores, and internal chambers that render them objects of great morphological complexity and beauty which have long attracted the attention of microscopists.

Dinoflagellates (class Dinophyceae) occur in both marine and fresh-water environments and are important primary producers in marine and estuarine environments. Dinoflagellates possess two flagella: one trails posteriorly and provides forward motion, while the other is positioned more or less transversely and often lies in a groove encircling the cell. This flagellum provides a rotary motion; hence the name dino (whirling) flagellate. In some dinoflagellates the cell wall is thin, while in others it consists of a complicated array of rather thick cellulosic plates, the number and arrangement of which are characters used in identifying genera and species.

The dinoflagellates are often conspicuous members of the marine plankton. Some taxa are bioluminescent. Others are one of the causes of red water, often called red tides, although their occurrence is not related to the tides. Depending on the causative species, red water may be an innocuous discoloration of the water or, if *Gymnodinium breve* is dominant, extensive fish mortalities will be experienced. Several species of *Gonyaulax* occur in bloom proportions in inshore marine waters. They contain a toxin which can accumulate in shellfish feeding upon *Gonyaulax* and which is the ultimate cause of paralytic shellfish poisoning in humans.

Coccolithophorids (class Haptophyceae) are also marine primary producers of some importance. They do not occur in fresh water. This class of algae possesses two anterior flagella. A third flagellumlike structure, the haptonema, is located between the two flagella. Calcium-carbonate-impregnated scales, called coccoliths, occur on the surface of these algae, and are sometimes found in great abundance in recent and ancient marine sediments.

Under certain conditions in subtropical and tropical seas, members of the nitrogen-fixing blue-green algal genus *Trichodesmium* (class Cyanophyceae) can occur in sufficient concentrations to strongly discolor the surface of the sea. Other nitrogen-fixing blue-greens commonly occur in great abundance in eutrophic and hypereutrophic fresh-water lakes and ponds.

Members of still other algal classes occur in marine and estuarine plankton. Their abundance will vary in different environments at different times, and they may even occasionally dominate the standing crop. Much remains to be learned about the identity, physiology, and ecology of the very small (<8 μm) flagellates which commonly occur in almost every marine phytoplankton sample.

Communities. Even though marine and fresh-water phytoplankton communities contain a number of algal classes in common (such as Bacillariophyceae, Chrysophyceae, and Dinophyceae), phytoplankton samples from these two environments will appear quite different. These habitats support different genera and species and groups of higher rank in these classes. Furthermore, fresh-water plankton contains algae belonging to additional algal classes either absent or rarely common in open ocean environments. These include the green algae (class Chlorophyceae), the euglenoid flagellates (class Euglenophyceae), and members of the Prasinophyceae.

In estuarine environments in which salinities vary from essentially zero to those of the local inshore ocean salinities, a mixture of organisms characteristic of both fresh-water and marine environments will be encountered. In addition, giving the estuarine plankton a somewhat distinctive appearance, some euryhaline planktonic taxa will be present.

Samples of phytoplankton in shallow areas of the sea, lakes, and estuaries often contain benthic and epiphytic microalgae. These algae become suspended in the water as a result of strong turbulent mixing so often found in shallow aquatic environments, and are called meroplankton to distinguish them from the holoplankton organisms which are truly planktonic.

While phytoplankton community composition reflects some complicated and as yet rather poorly understood series of biotic and abiotic interaction within ecosystems, the chemical composition of water is recognized as an important factor affecting phytoplankton communities. Many of the differences between the marine, estuarine, and fresh-water phytoplankton communities are associated with changes in salinity, major ion concentrations and ratios, pH, nutrients, and quite possibly trace-element concentrations. *See* COMMUNITY; ECOSYSTEM.

Eutrophication. Society's common use of lakes as receptacles for wastes, coupled with nutrient-rich runoff from cultivated fertilized land, has had pronounced biological effects not only on the phytoplankton but on other levels in the food web. Domestic sewage and agricultural runoff are quite rich in inorganic phosphorus, which is generally in short supply in most inland bodies of water. The addition of a nutrient which is limiting phytoplankton production, such as phosphorus, to fresh-water environments increases primary production, alters phytoplankton composition, and can lead to dense algal blooms which adversely affect water quality and esthetic and recreational values. The effects of nutrient enrichment, generally known as eutrophication, have been understood by limnologists since the 1920s – 1930s, but the American public became aware of the implication only in the 1960s with the widespread use of phosphate detergents and the associated and accelerated deterioration of the quality of lakes and ponds. Advanced techniques of sewage treatment are now available which remove phosphates sufficiently to greatly reduce the impact of sewage effluent upon receiving waters. Important in regulating these cycles are nutrient availability and the presence of a water density gradient, the pycnocline, at some relatively shallow depth (commonly 5 – 50 m) below the surface. Above the pycnocline, the location of which is usually well correlated with a zone of rapidly changing water temperature called the thermocline, is a well-mixed region of uniform density and temperature. This layer is called the mixed layer by oceanographers and the epilimnion by limnologists. Within this mixed layer, water motions caused by winds help keep the phytoplankton in suspension within the illuminated surface layer. The pycnocline also greatly reduces the rate of diffusion of nutrients upward into the mixed layer from nutrient-rich deeper waters and thus has a negative effect upon phytoplankton productivity. Any mechanism which introduces nutrients into the mixed layer that are limiting phytoplankton production will enhance primary production. *See* EUTROPHICATION.

Seasonal cycles. The phytoplankton in aquatic environments which have not been too drastically affected by human activity exhibit rather regular and predictable seasonal cycles. Coastal upwelling and divergences, zones where deeper water rises to the surface, are examples of naturally occurring phenomena which enrich the mixed layer with needed nutrients and greatly increase phytoplankton production. In the ocean these are the sites of the world's most productive fisheries.

In marine and fresh waters in temperate to arctic latitudes, the seasonal cycle of phytoplankton is to a large degree influenced by the pycnocline. During winter months a strong pycnocline is generally not present, and any planktonic plants in the water circulate to considerable depths. Even though nutrient concentrations are high during this period of deep vertical mixing, any phytoplankton present do not spend sufficient time in the illuminated surface layers for photosynthesis to exceed respiration. As spring approaches, incident solar radiation leads to the formation of a pycnocline. Phytoplankton then proliferate, giving rise to the annual spring phytoplankton bloom in which diatoms frequently dominate (thus it is often called the spring diatom bloom). The phytoplankton quickly exhaust the nutrients in the newly formed mixed layer, and throughout the summer phytoplankton concentrations generally remain low due to low mixed-layer nutrient concentrations resulting in part from the restricted upward movement of nutrients through the pycnocline. In the fall, incident solar radiation decreases and the pycnocline deepens. Nutrients or phytoplankton living in the pycnocline are incorporated into the deepening mixed layer, and a fall bloom occurs and then wanes as the pycnocline deepens, and finally disappears.

In the tropical oceans and lakes located near sea level, a permanent thermocline exists, and phytoplankton production remains low but may increase if nutrients are injected into the mixed layer as a result of strong wind mixing or other advective processes.

Fossilization. Under favorable conditions, some phytoplankton are incorporated into the sediments and become part of the fossil record. Diatoms, silicoflagellates, coccoliths, and the cysts of dinoflagellates are frequently sufficiently abundant and well preserved in lake and ocean sediments to permit paleontologists to reconstruct past environmental conditions and changes in environmental conditions through time. These fossilized remains are also used by stratigraphers to determine the age of sediments. [ROBERT W. HOLMES]

Bibliography: A. D. Boney, *Phytoplankton*, 1976; G. A. Cole, *Textbook of Limnology*, 1975; J. E. Raymont, *Plankton and Productivity in the Oceans*, 1963; W. D. Russell-Hunter, *Aquatic Productivity*, 1970; R. S. Wimpenny, *The Plankton of the Sea*, 1966.

Placer mining

The working of deposits of sand, gravel, and other alluvium and eluvium containing concentrations of metals or minerals of economic importance. For many years gold has been the most important product obtained, although considerable platinum, cassiterite (tin mineral), phosphate, monazite, columbite, ilmenite, zircon, diamond, sapphire, and other gems have been produced. Other valuables recovered include native bismuth, native copper, native silver, cinnabar, and other heavy weather-resistant metals or minerals.

In addition to onshore placer mining, offshore mining for gold, diamonds, iron ore, lime sand, and oyster shells is being done. Future possibilities exist for mining of manganese nodules containing manganese, iron, copper, nickel, and cobalt and phosphorite nodules which overlie large areas of deep ocean floors. Other apparent concentrations of metals and minerals exist in vast tonnages of clays, muds, and oozes which are found on the ocean floors.

Mining claims. In the United States placer mining claims are initially obtained by complying with mining and leasing laws of the Federal or state government or both. Concessions for mining ground in foreign countries can be obtained by following appropriate procedures of the country concerned. Specific information pertaining to acquiring mining claims in the United States can be obtained through offices of the U.S. Bureau of Land Management and the state land office having jurisdiction over the area in which the mining ground is located.

Prospecting, sampling, and valuation. A uniform grid pattern is best for sampling placers. This generally consists of equally spaced sample points in lines approximately perpendicular to the longest direction of the deposit. The distance between lines may be five times the distance between holes. If the apparent deposit seems equidistant in all directions, a grid of equally spaced samples is often used. Prior to actual laying out a prospecting grid, geophysical prospecting techniques may be used to determine approximate depth to bedrock and to outline probable old stream drainage systems. This information will allow more judicious planning of a sample point pattern. Samples are usually obtained by one or several of the following methods: shaft or caisson sinking, churn, and other drilling, and dozer or backhoe opencut trenches. Drilling in areas covered by water is accomplished by placing equipment on a barge or, in the case of the Northland, on ice. The information obtained allows the value of the deposit to be calculated and gives descriptions of the physical characteristics of the material in the deposit. The latter information is valuable for designing the mining method. Plotting unit values, obtained from sampling, on a map will indicate whether a concentration of mineral exists and the location of a pay streak. Calculations will indicate the total value and yardage in the pay streak. A decision can then be made as to whether or not the deposit can be mined on an economic basis.

Recovery methods. A form of sluice box is the type of unit most often used to separate and recover the valuable metals or minerals. Such a unit may vary in width from 12 to 60 in. (30 to 150 cm), 30 to 60 in. (75 to 150 cm) being common, and in

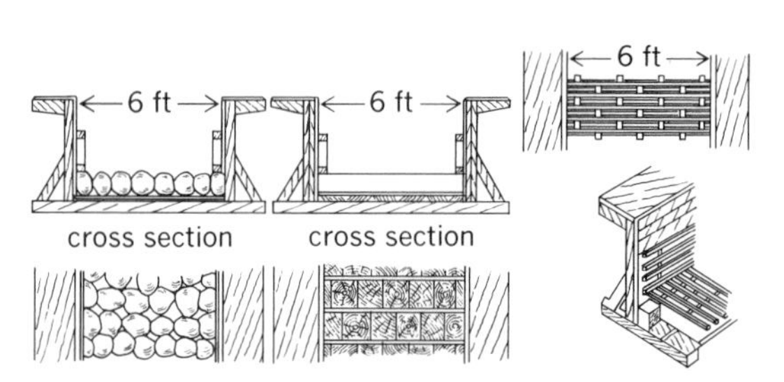

Fig. 1. Cross-sectional diagrams illustrating construction of sluices. 6 ft = 1.83 m. *(From G. J. Young, Elements of Mining, 4th ed., McGraw-Hill, 1946)*

lengths from 40 ft (12 m) to several hundred feet (1 ft = 0.3 m). The sluice box is placed on a grade of about 1½ in. (4 cm) vertical to 12 in. (30 cm) horizontal, but adjusted for the particular conditions. Recovery units used in the sluice include the following.

Riffles Rocks, 3–6-in. (8–15-cm) wood blocks, pole riffles, and Hungarian riffles (Figs. 1 and 2) form pockets in which valuable heavy particles will settle and be retained.

Amalgamation. Mercury is sprinkled at the head end of the sluice box to combine with and hold gold in various types of riffles and amalgam traps. Copper plates, mercury-coated and protected from coarse gravel by suitable screen, aid in collecting gold under some conditions.

Undercurrents. These are auxiliary sluices, parallel or at right angles to the main sluice. A variety of grizzly plates, screens, and mattings are used to separate various sizes of material and to catch small particles of concentrates that may be lost when using only a conventional sluice box.

Jigs. This is essentially a box with a screen top upon which rests a 3–6-in. (8–15-cm) bed, usually of steel shot, through which pulsating water currents act. Feed to the jig comes onto the bed in a water stream and the valuable heavy particles pass through the pulsating bed and are recovered as a concentrate. Lighter waterborne material passes over the bed and into other recovery units or emerges as a finished tailing.

Other. In some plants, units such as cyclones, spirals, and grease plates are used to effect recovery. More sophisticated equipment may be used to further concentrate and separate products. Details of design and application of various recovery units appear in technical literature.

Water supply. A large volume of water is essential to nearly all types of placer mining. Water may be brought into the mining area by a ditch, in which case the water is often under sufficient head to give ample gravity pressure for the mining operation. In other cases, water is pumped from its source. Where the water supply is limited, the water is collected after initial use in a settling pond and then pumped back to the mining operation for reuse. Water rights are obtained in accordance with Federal and state regulations in the United States.

Mining methods. A number of mining methods exist, but all have elements of preparing the ground, excavating the pay material, separating and recovering the valuable products, and stacking the tailings. A general classification of present placer mining methods is small-scale hand methods, hydraulicking, mechanical methods, dredging, and drift mining.

Small-scale hand methods. Perhaps most widely known because of gold rush notoriety are methods using pans, rockers, long toms, and other equipment that are responsible for a very small percent of all placer production. Other small-scale methods include shoveling into boxes, ground sluicing, booming, and the use of dip boxes, puddling boxes, dry washers, surf washers, and small-scale mechanical placer machines.

Pan and batea. Panning currently is mostly used for prospecting and recovering valuable material from concentrates. The pan is a circular metal dish that varies in diameter from 6 to 18 in. (15 to 46 cm), 16 in. (41 cm) being quite common. Many such pans are 2 or 3 in. (5 to 8 cm) deep and have 30–40° sloping sides. The pan with the mineral-bearing gravel, immersed in water, is shaken to cause the heavy material to settle toward the bottom of the pan, while the light surface material is washed away by swirling and over flowing water. These actions are repeated until only the heavy concentrates remain.

In some countries a conical-shaped wood unit called a batea (12–30-in. or 30–75-cm diameter with about 150° apex angle) is used to recover valuable metals from river channels and bars.

Rockers. Rockers are used to sample placer deposits or to mine high-grade areas (Nome, Alaska, beaches during the gold rush) when installation of larger equipment is not justified. At present, various types of engine-operated mechanical panning machines and mechanical sluices are used in concentrating samples.

Long tom. A long tom is essentially a small sluice box with various combinations of riffles, matting, and expanded metal screens and, in the case of gold, amalgamating plates.

Hydraulicking. Hydraulic mining utilizes water under pressure, forced through nozzles, to break and transport the placer gravel to the recovery plant (often sluice box), where it is washed. The valuable material is separated and retained in the sluice, and the tailings pass through and are stacked by water under pressure. Hydraulicking is a low-cost method of mining if a cheap, plentiful supply of water is available and streams are not objectionably polluted.

Mechanical equipment. In this category, equipment such as bulldozers, draglines, front-end loaders, pumps, pipelines, hydraulic elevators, and recovery units are used in a number of different combinations depending on the physical characteristics of placer deposits. A typical example is to use bulldozers to prepare the ground and push the placer material to the head end of a sluice box, from which the gravel is washed through the sluice box. The tailings are stacked with a dragline, and a pump is used to return the water for reuse.

Dredging. Large flat-lying areas are best suited for dredging because huge volumes of gravel can be handled fairly cheaply. Prior to dredging a deposit, the ground must be prepared. This preparation may consist of removing trees and other vegetation and overburden such as the frozen loess (muck) found in many parts of Alaska by use of equipment and water under pressure (stripping). Barren gravel may also be removed. If the placer gravel is frozen, the deposit may be thawed by the cold-water method, as has been done in northern Canada (Yukon Territory) and Alaska.

Bucket-ladder dredges (California type). For onshore placer mining, this mass-handling method now largely supersedes others, such as the chain-bucket dredge and one-bucket dredge for mining placer gold and other heavy placer material. Some large modern dredges can dig to depths of 150 ft (45 m) and handle 10,000 yd³ (7600 m³) or more/24 hr.

Modern dredges are for the most part steel hulls of the compartment or pontoon types. On the hull is mounted the necessary machinery to cause an endless bucketline to revolve and dig the placer gravels as the dredge swings from side to side in a

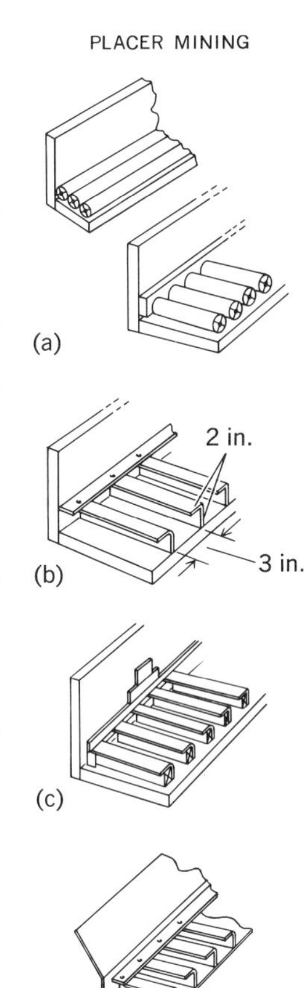

Fig. 2. Types of riffles in partial plan views. *(a)* Pole riffles. *(b)* Hungarian riffle. *(c)* Oroville Hungarian riffle. *(d)* All-steel sluice. 1 in. = 25.4 mm. *(From G. J. Young, Elements of Mining, 4th ed., McGraw-Hill, 1946)*

Fig. 3. Yuba bucket-ladder dredge in operation near Marysville, Calif. (*Yuba Manufacturing Co.*)

pond (Fig. 3). At the end of each swing, the buckets on the ladder are dropped and the cycle repeated. When the bucketline reaches the bottom of the pay streak, the ladder with buckets is raised, the dredge is moved ahead or into a parallel cut, and the cycle is repeated. The buckets discharge their load into a hopper, where it is washed with water into a long inclined revolving cylindrical screen (trommel) with holes that commonly vary in diameter from $\frac{1}{4}$ in. (0.6 cm) to $\frac{3}{4}$ in. (1.9 cm). In this revolving screen, a stationary manifold supplies water (commonly at 60 psi or 414 kPa) to nozzles spaced equally in a line the length of the screen. Water from these nozzles washes the sand and gravel and causes the fines and valuable minerals and metals to work through the holes in the revolving trommel and fall on the tables (series of parallel sluices), which may consist of various recovery units, such as mercury traps, jigs, riffles, matting, expanded metal screens, and undercurrents. The placer concentrates are held in the recovery units until cleanup time (the end of a day to 4 weeks, but commonly 2 weeks). The accompanying sand passes over the tables and into the dredge pond. The coarse material that will not pass through the holes in the screen is discharged at the lower end of the trommel screen onto an endless conveyor belt (stacker) which causes the coarse tailings to be deposited in back of the dredge. Overflow muddy pond water passes from the mining area through a drain (Fig. 4).

In the case of gold mining, during cleanup time, the gold amalgam is further concentrated by paddling in the sluices, then picked up, cleaned by panning, retorted to separate the gold from the mercury, and the remaining gold is cast into bricks for appropriate sale. The mercury is reused.

Dragline dredges. A washing and recovery plant, mounted on pontoons or crawler treads, is fed directly or by a conveyor from a shovel or dragline unit. Flow of material and recovery of products are essentially the same as for a bucketline dredge. Such dragline dredge installation is mostly used on deposits too small to justify the installation of a bucket dredge.

Offshore suction dredges. Exploration for offshore mineral deposits in the unconsolidated material of the ocean's floors is receiving a considerable amount of attention with the ultimate hope of increasing the efficiency of mining for gold, tin, diamonds, heavy mineral sands, iron sands, lime shells, and sand and gravel (also excavation for channel excavation). In addition to bucket-ladder dredges, grab dredges (clamshell buckets), hydraulic (suction) dredges, and airlift dredges are being used. In hydraulic dredges water under pressure is released near the intake and, helped by a suction created with pumps, returns to the surface, carrying with it sand and gravel from the ocean floor. In airlift dredges, air replaces water as described for the hydraulic dredge. Flow of material and recovery of products are essentially the same as for a bucketline dredge.

Small-scale venturi suction dredges have been used by skin divers to recover gold from river bottoms.

In the future, development of more sophisticated equipment for offshore mining of minerals can be anticipated.

Drift mining. Because of relatively high costs, drift mining of placer deposits is not used to a large extent. It consists of sinking a shaft to bedrock,

Fig. 4. Aerial oblique view up-valley over the site of a placer dredging, in which successive cuts, tailing patterns, and dredge pond are shown. The dredge appears in the left foreground. (*Pacific Aerial Survey, Inc.*)

driving drifts up- and downstream in the valley deposits for 250–300 ft (76–91 m), and then extending crosscuts the width of the pay streak. In thawed ground, timbering is necessary; in frozen ground, a minimum of timber is required and steam thawing plants are used to thaw the gravels before mining is done. Usually, the thawed gravel is hoisted to the surface and then washed to recover the valuable material. Research utilizing modern technique and equipment should make this method more economically competitive. *See* UNDERGROUND MINING.

Water pollution. Recent proposed regulations on both state and Federal levels tend to place more restrictions on discharge of water carrying sediments into stream drainages. Eventually, treatment of some type may be necessary to remove such sediments, and proposed new operations will no doubt consider this factor.

Tailing disposal. Increased restriction on tailings in placer mining is being proposed in various areas. Such regulations may require smoothing of tailing piles and in some cases resoiling. In Alaska, in many cases, dredge tailings have added to the value of the ground by removing permafrost and offering a solid foundation for road and building construction in place of the original swampland surface.

Power. Electric power for operation of equipment may be generated by several methods, such as utilizing coal, oil, gas, or water (hydro). Units may be permanently located or may be diesel power units aboard dredges or movable shore plants for dredges.

Developments in technology. In past years a number of improvements have been made to increase the efficiency of placer mining. These may be summarized as follows.

1. Development of the diesel engine allowed diesel-powered pumps, tractors, draglines, dredges, and the like to be manufactured and gave much more compactness, economy, and flexibility to placer mining methods. This in turn resulted in greater yardage at lower unit costs.

2. The adaptation and use of jigs in placer recovery plants have increased the recovery of valuable minerals and metals with less work and more efficiency of compact recovery units.

3. The use of the sluice plate in mechanical mining methods has been an important improvement in recent years. This method does away with the conventional nozzle setup in front of the sluice box and the accompanying labor expense. Gravel is pushed directly into the head of the sluice into a stream of free-flowing water. This results in a fast system of coordinated operation, especially for mining placer gravels that contain a minimum of clay and cemented gravels.

4. Development of more efficient transportation carriers has allowed relatively inaccessible areas to become accessible. These include large airplanes to haul heavy equipment and fuel and land machines for crossing swampy and difficult terrain.

5. Use of rippers on tractors to aid in the mining process is becoming successful.

6. Electronic devices and television cameras for control units further increase efficiency of operation.

7. Modern lightweight pipe with the "snap-on" type of couplings and continuously improving design of equipment such as the automatic moving nozzle (Intelligiant) tend to increase the efficiency of mining.

[EARL H. BEISTLINE]

Bibliography: American Institute of Mining, Metallurgical and Petroleum Engineers, *Surface Mining*, 1968; Institution of Mining and Metallurgy, London, *Opencast Mining, Quarrying and Alluvial Mining*, 1965; C. F. Jackson and J. B. Knaebel, *Small-Scale Placer-Mining Methods*, USBM Inform. Circ. no. 6611, 1932; R. S. Lewis and G. B. Clark, *Elements of Mining*, 3d ed., 1964.

Plant, minerals essential to

Beginning in the middle of the 19th century, botanists sought to determine the chemical elements required for the growth of plants. Elements such as nitrogen, phosphorus, and potassium, required in relatively large amounts, were among the first shown to be essential. The essentiality of certain elements, such as copper and molybdenum, was not established until chemical compounds of greater purity were produced. Earlier, many of the elements were present in sufficient amounts as impurities in a nutrient solution to preclude their detection as essential elements when they were "omitted" from the solution. Elements which are required in small amounts, such as copper, molybdenum, manganese, zinc, boron, iron, and chlorine, are sometimes called trace elements. They have also been called minor elements, but this term erroneously implies that these elements play only a minor role in plant nutrition. The term micronutrients is generally favored, since it implies that small amounts are required and that they perform a nutritive role.

Most plant scientists agree that the following elements are essential for higher plants: carbon, hydrogen, oxygen, phosphorus, potassium, nitrogen, sulfur, calcium, magnesium, iron, boron, manganese, zinc, copper, chlorine, and molybdenum. The last seven are micronutrients. Nitrogen, phosphorus, and potassium are the three most important and are the components of a 5-10-5 fertilizer (5% nitrogen, 10% phosphorus, calculated as phosphorus pentoxide, and 5% potassium, calculated as potassium oxide). Certain algae require vanadium, sodium, and cobalt. Additional elements may possibly be added to the present list of essential elements for plants.

Criteria of essentiality. In order to be considered essential, an element must meet the following criteria: (1) Absence of the element results in abnormal growth, injury, or death of the plant; (2) the plant is unable to complete its life cycle without the element; (3) the element is required for plants in general; and (4) no other element can serve as a complete substitute. Most scientists prefer to add a fifth criterion, namely, that the element have a specific and direct role in the nutrition of the plant. This last criterion is difficult to determine since known, direct roles for potassium and calcium, for example, are not as yet agreed upon; their essentiality, however, is unquestioned. Without exception, when the essential elements were experimentally removed from the external environment, drastically reduced growth ensued.

The direct, specific roles of the elements were then pursued, most of which have been elucidated.

Roles of essential elements. The following paragraphs discuss the roles in plants of carbon, hydrogen, oxygen, phosphorus, nitrogen, sulfur, potassium, calcium, magnesium, iron, manganese, copper, zinc, molybdenum, boron, and chlorine.

Carbon. Carbon may constitute as much as 44% of the dry weight of a typical plant such as corn. It is a component of all organic compounds, such as sugars, starch, proteins, and fats, in plants (and animals). One of its main roles, is as a constituent of a vast array of compounds which in turn are synthesized from sugar. Carbon may indeed be said to be involved in all the roles played by all the carbon-containing compounds. This would include such compounds as the plant hormones, which regulate plant growth, flowering, and reproduction.

Hydrogen. This element is also a constituent of all organic compounds, and what applies to carbon similarly applies to hydrogen. Hydrogen constitutes about 6% of the dry weight of a plant.

Oxygen. This element is a constituent of all carbohydrates, fats, and proteins and, in fact, of most organic compounds. Some organic compounds, such as carotene, which gives rise to vitamin A, are composed only of carbon and hydrogen. Since oxygen is a constituent of most organic compounds, it is involved in whatever functions those compounds perform. Oxygen constitutes about 43% of the dry weight of a plant.

Carbon, hydrogen, and oxygen are constituents of the bicarbonate ion (HCO_3^-), which is believed to be one of the chief anions (along with hydroxyl, OH^-) exchanged by the plant for anions absorbed from the soil. Because of this role, these three elements are involved in the process of salt absorption by roots.

In the process of oxidation (aerobic respiration) of foods by the cell, oxygen is the final acceptor of hydrogen. When a food, such as sugar, is respired by the removal of hydrogen and carbon dioxide, the latter and water appear as end products. Water is formed in the terminal step in these oxidation reactions when oxygen and hydrogen unite.

Phosphorus. Certain special proteins in all cells, the nucleoproteins, contain phosphorus. Nucleoproteins are in the nucleus of the cell and hence in the chromosomes which carry the hereditary units, the genes. Since phosphorus is a constituent of the nucleoproteins, cell division is dependent on phosphorus. The transmission of hereditary characteristics also depends on this element.

Cellular membranes are believed to consist, in part, of special phosphorus-containing fats or lipids called phospholipids. These, along with hydrated protein molecules, are very likely the chief components of cellular membranes. Differentially permeable membranes regulate the entry and exit of materials from the cells; phosphorus therefore plays a role in the permeability of cells to various substances and in the retention of substances by cells.

Phosphorus is a constituent of the special compounds diphosphopyridine nucleotide (DPN) and triphosphopyridine nucleotide (TPN). DPN and TPN are commonly called NAD (nicotinamide, adenine dinucleotide) and NADP (nicotinamide adenine dinucleotide phosphate), respectively. These compounds are involved in the transfer of hydrogen in aerobic respiration, and life itself depends on this important energy-liberating process.

The early stages in the combustion or utilization of sugar by cells involve the addition of phosphorus to the sugar molecule. Only after phosphorus is added to both ends of the sugar molecule is it cleaved and prepared for further transformations which release the chemical energy stored in the sugar molecule.

Also, phosphorus is a constituent of adenosinediphosphate (ADP) and adenosinetriphosphate (ATP). In ADP one of the two phosphate bonds is a high-energy bond, and in ATP two of the three phosphate bonds are high-energy bonds. This unique concentration of energy in these special phosphorus bonds is one of the ways in which potential energy is stored within the cell. The bond is important not only because it represents a form of stored energy, but also when it is broken the released energy can accomplish "work." Many synthetic reactions, such as the syntheses of sucrose and starch from glucose, require energy which the high-energy phosphate bond delivers to the reactions.

The reduced forms of diphosphorpyridine nucleotide ($DPNH_2$) and triphosphopyridine nucleotide ($TPNH_2$) constitute the other major form of energy storage in cells. This chemically stored energy is also available for work—driving certain chemical reactions that would otherwise proceed at imperceptibly low rates.

Nitrogen. Along with carbon, hydrogen, and oxygen, nitrogen is a constituent of all amino acids: the building blocks of proteins. Protoplasm usually has a high percentage of water, but the substance portion is primarily proteinaceous. Thus nitrogen and certain other elements such as carbon, hydrogen, and oxygen are a part of the living substance, protoplasm.

All enzymes thus far isolated are protein in nature. Therefore nitrogen is a constituent of these remarkable organic catalysts, which accomplish at room temperature, or below, chemical reactions that man can perform only with high temperature, pressure, or other special conditions.

As a constituent of chlorophyll (four nitrogen atoms in each molecule), nitrogen is required in photosynthesis, the food-manufacturing process which only plants can accomplish.

Sulfur. Certain amino acids, such as cystine, cysteine, and methionine, contain sulfur and are often components of plant proteins. Sulfur is also a constituent of the tripeptide, glutathione, which may function as a hydrogen carrier in the respiration of plants and animals. Biotin, thiamine, and coenzyme A are examples of still other sulfur-containing compounds in plants.

Potassium. Although potassium was one of the first elements shown to be essential and often accounts for 1% or more of the dry weight, there is no known potassium-containing compound in plants. Despite numerous researches, it still is not known why plants require potassium in such seemingly large amounts. Although it functions as a cofactor for certain enzyme systems, the need for such high concentrations is not known. Virtually all the potassium in plants appears to be water-soluble, emphasizing further that potassium is not a constituent of any compound, and certainly not of the larger, relatively insoluble and immobile compounds.

Calcium. Although calcium may typically be present in plants to the extent of 0.2% of the dry matter, it also is not clear why plants require so much calcium. Many workers consider the cementing substance between cells, the middle lamella, to be composed of calcium pectate. No other calcium-containing compounds of biological significance have been reported for plants, and yet they contain far more calcium than would be required for its postulated role in the middle lamella. Excesses of oxalic and other organic acids may appear in the cell as crystalline calcium salts of low solubilities. These salts, however, are considered waste products and serve no vital function. Their removal from solution by calcium may prevent a toxicity that would otherwise result from such acids.

Magnesium. Magnesium is a constituent of chlorophyll molecule, each molecule containing one atom of magnesium in the center. Although other metallic ions may be made to replace the magnesium, chlorophyll functions in photosynthesis only when it contains a magnesium atom. Despite much speculation, it has not yet been determined why only magnesium is effective in chlorophyll and in photosynthesis.

In addition to the unique role which magnesium plays in chlorophyll and photosynthesis, it is also required for the action of a host of enzymes. It may have two roles in protein synthesis: as an activator of some enzymes involved in the synthesis of nucleic acids, or as an important binding agent in microsomal particles where protein synthesis occurs. It is also apparently required for an enzyme concerned with oil formation in plants, since oil droplets are not formed in the alga *Vaucheria* in the absence of magnesium. The seeds of plants, which contain large amounts of oil, are consistently high in magnesium.

Iron. Iron is required for the formation of chlorophyll but is not a constituent of the molecule. In plant leaves about 80% of the iron is associated with chloroplasts, the chlorophyll-containing plastids. Iron is a constituent of cytochrome *f*, which may have a role in photosynthesis.

Iron is a constituent of the enzymes cytochrome oxidase, peroxidase, and catalase. Cytochrome oxidase is involved in respiration, and catalase catalyzes the breakdown of any hydrogen peroxide that forms in cells as a result of certain metabolic reactions. Why plants require so much iron is not known, since a smaller quantity would appear sufficient for its known roles. Iron is also a component of ferredoxin.

Manganese. There are no known compounds in plants of which manganese is a constituent. There is considerable evidence that the element may be a cofactor or an activator in certain enzyme systems. For example, manganese may be involved in nitrate reduction and hence in nitrogen metabolism. Manganese is known to be needed for photosynthesis, particularly by algae growing in an inorganic medium.

Copper. Copper is a constituent of laccase, ascorbic acid oxidase, plastocyanin, and tyrosinase (polyphenoloxidase). The last enzyme is believed to be involved in most plants with the terminal step in aerobic respiration, the transfer of hydrogen to oxygen to form water. This action thus links copper with energy release in plant cells.

Zinc. Although the enzyme has not been isolated, it has been shown that the enzyme which synthesizes the amino acid tryptophan requires zinc. Tryptophan in turn is the precursor from which the plant hormone indoleacetic acid is made. Zinc then is directly necessary for the formation of tryptophan and indirectly necessary for the production of indoleacetic acid.

Two zinc-containing enzymes have been isolated from plants, namely, carbonic anhydrase and alcohol dehydrogenase.

Molybdenum. Molybdenum is required in the external solution to the extent of 1 or 2 parts per billion. In the dry matter of the plant it may be present only to the extent of about 10 parts per billion. It has been calculated that the number of molybdenum atoms required per cell of *Scenedesmus obliquus* and *Azotobacter* is 3000 and 10,000, respectively.

Molybdenum is the metal component of xanthine oxidase and of the enzyme nitrate reductase, which effects reduction of nitrate nitrogen to the reduced form of nitrogen that is incorporated into amino acids and then into proteins. Molybdenum-deficient tomato plants may accumulate nitrate to the extent of 12% of the dry weight of the plant. If such plants are given a few parts per billion of molybdenum in the external medium, the nitrate content will drop to around 1% within 2 days. *Aspergillus niger*, *Scenedesmus obliquus*, and *Chlorella pyrenoidosa* also require molybdenum for the reduction of nitrate nitrogen. Fixation of atmospheric nitrogen by one of the free-living, nitrogen-fixing bacteria, *Azotobacter*, and by a blue-green alga, *Anabaena cylindrica*, requires molybdenum. The element is therefore intimately associated with nitrogen metabolism and synthesis of protein and hence, synthesis of protoplasm. *See* NITROGEN CYCLE.

Certain species of plants appear to require molybdenum for one or more unidentified roles other than nitrate reduction or nitrogen fixation. For example, cauliflower plants grown on urea and ammonium, as reduced nitrogen sources, nevertheless develop characteristic molybdenum-deficiency symptoms known as whip tail when molybdenum is withheld.

Boron. The essentiality of this element was established around 1910. Approximately 1/2 part per million (ppm) in the external solution suffices for growth of most plants. Garden and sugarbeets, as well as alfalfa, have a somewhat higher boron requirement, 5–10 ppm being optimal.

There are no known compounds in plants of which boron is a constituent, and no enzyme system has been shown to require boron. In most plants boron is immobile, suggesting that it is combined with large, immobile molecules, and plants therefore have to receive boron continually throughout the life cycle.

Numerous functions have been proposed for boron, including roles in carbohydrate and protein metabolism. One theory states that boron is required for the translocation of sugar from the

Fig. 1. Young tobacco seedling showing potassium-deficiency symptoms consisting of interveinal chlorosis and marginal and apical scorch of older leaves. *(From H. B. Sprague, ed., Hunger Signs in Crops: A Symposium, 3d ed., copyright © 1964 by Longman Inc., reprinted with permission of Longman)*

Fig. 2. Boron deficiency in branch of a grape plant, showing interveinal chlorosis of terminal leaves and necrotic terminal growing point. *(From J. A. Cook et al., Light fruit set and leaf injury from boron deficiency in vineyards readily corrected when identified, Calif. Agr., 15(3):3–4, 1961)*

leaves (where sugar is made) to the flowers, fruits, and the growing points of stems and roots. In the absence of boron, stem and root tips die, and flowering and fruiting are drastically reduced or altogether curtailed. A certain degree of deficiency of boron, for example, that which results in almost complete failure to set seeds in alfalfa, may not materially reduce the size of the plants. This well-established phenomenon signifies its unique role in flowering and fruiting. Successful germination of pollen grains and the production of the pollen tubes require boron.

Boron-deficient plants lose the normal response to gravity, indicating that boron is involved in the production, movement, or action of the natural hormones that cause the stem of a horizontally placed plant to turn up and the roots to turn down.

Chlorine. In the tomato plant, chlorine deficiency results in wilting of the leaf tips and chlorosis (yellowing), bronzing, and necrosis (death) of the leaves. If chlorine is added early enough, as little as 3 ppm banishes the symptoms and normal growth proceeds. Tomato plants show deficiency when they contain about 200 ppm of chlorine (dry-weight basis), whereas they show molybdenum deficiency when the concentration is about 0.1 ppm. Therefore, the tomato plant requires several thousand times more chlorine than molybdenum.

It should be made clear that plants cannot tolerate more than a few parts per million of chlorine in the molecular, gaseous state. Ordinarily plants absorb chlorine in the ionic form as chloride. Most plants tolerate 500 ppm or more of chloride without much effect upon growth, and certain halophytes (salt plants) can grow vigorously in high concentrations of chloride slats.

Other elements. Vanadium is required for the growth of *Scenedesmus obliquus*, and it plays a role in photosynthesis in *Chlorella*. There is no evidence of its essentiality for plants other than the green algae. Silicon is required for diatoms.

Sodium is an essential element for certain blue-green algae but is not required for green algae or higher plants.

Colbalt is required only for certain blue-green algae and the nitrogen-fixing bacteria in nodules of leguminous plants.

Deficiency symptoms. A deficiency of any one of the 16 essential elements results in stunted growth and reduced yield.

Deficiency symptoms are best identified by specially trained persons, since a deficiency of a particular element has different symptoms on different plants, for example, corn and beans. Furthermore, the application of nutrients to correct deficiencies, particularly of boron, copper, manganese, zinc, and molybdenum, requires specialists in plant nutrition.

The elements which are most likely to have a limiting effect on growth are nitrogen, phosphorus, and potassium; these are present in a typical, commercially available fertilizer, such as 5-10-5. Some generalizations can be made about the deficiency symptoms of these three main elements. When nitrogen moves out of the older, hence lower, leaves of a plant, the deficiency is generally characterized by yellowing of these leaves. Phosphorus deficiency is often characterized by a purpling of the stem, leaf, or veins on the underside of the leaves.

In corn, phosphorus deficiency causes purpling of the stem and sometimes of the leaf blades. Potash (potassium) deficiency results in burn or scorch of the margins of the leaves, particularly the older, lower leaves (Fig. 1). Recognition of the deficiency symptoms of these three elements can be corrected by the application of readily available commercial fertilizer.

Chemical tests (tissue tests) can often be made of key plant tissues to determine whether a particular element is lacking. These tissue tests can detect a near-deficiency state before the symptoms become manifest. In general, these tests should be made by persons trained to conduct them.

The best approach for the average homeowner or farmer, however, is to have the soil tested if there is any question as to its productive capacity. Commercial laboratories and state agricultural experiment stations provide this service. A soil test can predict in advance of planting what nutrients are lacking. By the time deficiencies appear, plant growth and yield are usually irretrievably retarded. If they are used early enough, tissue tests can detect an incipient deficiency in time for correction.

In addition to the widespread need for nitrogen, phosphorus, and potassium, it is often necessary to add other elements. The following elements have been found to be deficient in one or more areas of the United States: boron (Fig. 2), magnesium, copper, manganese, zinc, iron, calcium, sulfur, and molybdenum. A deficiency of chlorine has not been observed under field conditions. In one soil or another, a deficiency of every essential element except chlorine has been found. Considering the number of years that some soils have been under cultivation and the amounts of essential elements which have been removed by crops, it is not surprising that agricultural soils are becoming deficient in certain essential elements. When plants are unusually low in calcium or phosphorus, for example, people and particularly grazing animals may develop certain deficiency diseases. Thus there is a very intimate relationship between plant and animal nutrition that is receiving considerable attention. [HUGH G. GAUCH]

Bibliography: J. Bonner and J. E. Varner (eds.), *Plant Biochemistry*, 3d ed., 1976; R. M. Devlin, *Plant Physiology*, 3d ed., 1975; W. D. McElroy, *Cell Physiology and Biochemistry*, 3d ed., 1971; V. Sanchelli, *Trace Elements in Agriculture*, 1969; H. B. Sprague (ed.), *Hunger Signs in Crops: A Symposium*, American Society of Agronomy and National Fertilizer Association, 3d ed., 1964.

Plant community

An association of plants. Plants of various species are found growing together as vegetation, and certain combinations of species are found repeated in homogeneous areas of similar ecology, or biotopes, so often that generalizations can be made concerning these combinations. A plant community, then, has a certain species composition. A list of the plants occurring in a stand can be made by species names and by life forms. A list of all species is desirable; usually only vascular plants, bryophytes, and lichens can be recognized in the field. It is often necessary to take herbarium specimens, and such vouchers will document the study permanently. Ordinary taxonomic nomenclature is generally used, but a constant effort to improve this and to split the species into biotypes of more uniform relationships to environments must be made. The species list is limited, because within a given community the rate of increase of species number with increasing area is inversely proportional to the area investigated.

Characteristic species. Some kinds of plants are characteristic of a particular species combination; they are found only in one kind of combination wherever they occur, or regionally, or perhaps locally. Other plants are always found in a particular plant community. Still other plants occur in several kinds of communities; some are almost ubiquitous. Advantage is taken of such facts to classify the plants found into characteristic species which are exclusive to a given kind of vegetation or always found in it, differential species which occur in only one of two related communities, and accompanying species which show little or no preferences. The value of a given plant community as an indicator of habitat is determined largely by its characteristic and differential species, and it is by these species that plant communities are recognized.

Properties. It is possible to arrange the lists of species and associated ecological habitat data made for various stands of vegetation in other ways than into types of plant communities. They can be arranged to describe gradients, series, continua, or functions in correspondence with various habitat factors. Properties of the vegetation other than species composition can be used to help characterize plant communities. Total yields per unit area, such as tons of forage/hectare, life forms, dispersal or pollination spectra, and total contents of certain chemical elements, can be used as properties of plant communities.

Plants and animals which are associated also form a community, a biocoenose. Usually the plant community forms the fixed substratum for the animals, which may be mobile.

Structure. Plant communities have a structure varying in complexity from a many-layered forest to a unistratal polar or a hot desert cryptogam community. Moss, low and tall herb, low and tall shrub, and several tree layers may be present, and there may be epiphytic societies on the tree trunks and branches. In complex communities the aspect changes throughout the year as various groups of plants go through the stages of their life cycles at different times. Given the species list for a stand of vegetation, it is possible to predict much about the structure of the plant community from knowledge of the species concerned. However, one reason for studying plant communities in addition to individual plants is that in various communities individual species behave differently. Thus fireweed (*Epilobium angustifolium*) occurs in many forest communities of the Northern Hemisphere. It is usually sterile, but it flowers and proliferates abundantly when the forest is destroyed by fire or cutting, producing a new habitat and opportunity for a new plant community. *See* ECOLOGICAL SUCCESSION.

Dynamics. The functioning of plant communities is analogous to the physiological processes taking place in the individual plants of which the community is composed, but significant interactions between plants modify, for example, the water regime of forest floor plants and the carbon dioxide made available for photosynthesis by the green plants. The physiological tolerances of individual plants to features of their habitats are thus modified to ecological tolerances by competition with other plants in the community.

The relations of plant communities to environment are systematized under various factors of the environment. Thus, plants react to such features of the environment as regional climates, soil parent materials, topographical features as these condition local climates, ground water, wind, snow deposition, fire, and the biota available to the biotope concerned. Man is a most important part of this biota, both in uncivilized and in civilized states. Plant communities in different geographical regions differ perhaps first of all because the floras available in the different regions differ, even though ecologically the regions may be quite similar. The combined and interacting effects of all these groups of factors produce at a given time a particular ecosystem or combination of plant community in its environment in which the vegetation and environmental properties stand in functional relationship to each other. Thus, on a continental scale the change in regionally representative plant communities—from the shortgrass high plains of Colorado east through the tallgrass of Kansas to the deciduous forest of Ohio—can be interpreted as a reaction of decreasing moisture along a given annual isotherm, such as 11°C. In the mountain ranges of the western United States, transitions from shadscale (*Atriplex confertifolia*) desert to sagebrush (*Artemisia tridentata*) semidesert to oakbrush (*Quercus gambellii*) chaparral to spruce-fir (*Picea engelmanni–Abies lasiocarpa*) coniferous forest to alpine herbaceous vegetation are related to altitudinal changes in these continental climates. Precipitation increases from 100 to 1000 mm whereas temperatures drop from 10 to 0°C as annual means. *See* ECOSYSTEM.

Other factors. These regional changes in vegetation can be found when only climate changes; the other factors of the environment are fixed at some particular values. If they are fixed at another set of values, the sequence will be quite different. A modification of the relief factor in the case of the midwestern United States, a shift from the well-drained uplands to river floodplains, will result in riverine forest communities of various types all along the isotherm. If temperature is drastically lowered, as in the Arctic, even 100 mm of precipitation will result in bog vegetation. The sequence of plant communities which corresponds to a change in one factor of the ecosystem is a function of those other factors of the ecosystem which have been constant.

Climax. Static situations are described above. However, vegetation is dynamic; it evolves. Given a fixed set of the environmental factors operating on a bare area, this area will change in the types of plant communities it supports, at a constantly decreasing rate, until a steady state is attained. These equilibrium stages are climax plant communities. At such an equilibrium, which seems to be reached in a few hundred years depending on the ecosystem, the effects of climate in determining the kind of vegetation often become paramount. Although the effects of climate may become paramount in many ecosystems, in others with extremes of one of the other factors, the effect of this latter factor may persist indefinitely. Thus, very coarse-grained, sandy-soil parent material may continue to support a plant community quite different from that on the surrounding hard land, as in the Sand Hills of Nebraska with their tallgrass vegetation surrounded by climax mid- and shortgrass.

In addition to the short-term genesis mentioned above there are changes of the environmental factors themselves which result in historical changes in plant communities. Invasion of plants new to the flora, as the chestnut blight into the hardwood forest of eastern North America which almost totally killed one of the former leading dominants in this forest, or the postglacial climatic changes which have been so well documented by pollen analyses of bog sections, are examples. If one of the factors determining a plant community changes, it is an axiom of plant ecology that the community will change. *See* CLIMAX COMMUNITY.

Distribution. The distribution of plant communities over the face of the Earth has been studied more from a physiognomic than from a floristic viewpoint. Repetitions of physiognomically similar types of vegetation do occur in widely separated parts of the earth with similar climates. The evergreen sclerophyll chaparral, of winter-wet, summer-dry, mild climates in the Mediterranean region of Europe, and in Australia, South Africa, California, and Chile is an example of floristically completely diverse regions having at least superficial similarities in the appearance of plant communities because of their similar structure. *See* PLANT FORMATIONS, CLIMAX.

Classification. Finally, plant communities may be classified. The most widespread system is that developed by J. Braun-Blanquet and used extensively in Europe. Floristically similar stands of vegetation with some characteristic species in common are abstracted into associations denoted by the terminus -etum. Associations are combined into alliances (-ion), these into orders (-etalia), and these into classes (-etea). Classes in general coincide with broad, physiognomically defined kinds of vegetation or formations. The next higher unit is a floristic one recognizing such differences as those between the Mediterranean flora and that of central and northern Europe. Obviously, if two regions have different floras, they must also have different plant communities. *See* ANIMAL COMMUNITY; COMMUNITY.

[JACK MAJOR]

Bibliography: R. F. Daubemire, *Plant Communities: A Textbook of Syecology*, 1968; J. P. Grime, *Plant Strategies and Vegetation Processes*, 1979; H. P. Handson and E. D. Churchill, *The Plant Community*, 1961.

Plant disease

A great obstacle to the successful production of cultivated plants, plant disease is also sometimes destructive in natural forests and grasslands. Despite large expenditures for control measures, diseases annually destroy close to 10% of the crop plants in the United States, before and after harvest.

Diseases may destroy plant parts outright by rotting, or may cause stunting or other malformations. Most diseases are caused by parasitic microscopic organisms such as bacteria, fungi, algae, and nematodes or roundworms, although a few are caused by parasitic higher plants, such as dodder and mistletoe. Many are caused by viruses, and some are caused by poor soil conditions, unfavorable weather, or harmful gases in the air.

The living organisms and viruses which cause disease are called pathogens. Most pathogens can multiply extremely rapidly, the bacteria by simple division, the fungi by producing spores which behave as seeds but are much smaller and simpler in structure. Bacteria are about 0.0005 in. long, and fairly large fungus spores about 0.001 in. Virus particles are not visible with ordinary microscopes; they can multiply a millionfold in a short time. Roundworms reproduce by means of eggs.

Most pathogens can be disseminated quickly and widely by wind, water, insects, man, and other animals. They infect plants through wounds or pores (stomata) or by penetrating plant surfaces. Each kind of pathogen can attack only certain kinds of plants or plant parts. Once inside the plant, living pathogens obtain their nourishment from it in various ways, destroying plant tissues or weakening the plant by robbing it of its food substances. The rapidity of growth and reproduction of pathogens and of disease development varies with the kind of pathogen and host and with soil and weather conditions. Some pathogens thrive best in hot weather, others in cool weather. Extensive and destructive epidemics develop when all conditions favor the most rapid development of the pathogen.

Good cultural practices, chemical disinfestation of planting materials, spraying or dusting with appropriate chemicals to protect against airborne infection, and the use of resistant varieties are the principal control measures.

Discussed in the following sections are the economic importance, nature, and causes of plant diseases; the characteristics, growth, and reproduction of pathogens; the infection stage and development of diseases; the dissemination of pathogens; and the diseases to which plants are subject in storage. Discussion of other aspects can be found under separate titles or under the names of plants infected.

Economic importance. All plants and their parts are subject to diseases which may be caused at various stages of their life cycles not only by microorganisms, but also by higher plants, injurious salts in the soil, and harmful gases in the air. Diseases may rot the seed, kill plants, or make them poor and unsightly; they may cause root rots, stem cankers and rots, leaf spots and blights, blossom blights, and fruit scabs, molds, spots, and rots. In transit and storage they cause rots of fleshy fruits and vegetables; mold sickness of wheat, rice, corn, and other grains; and discoloration or rotting of wood and wood products.

When weather favors their development, some diseases become epidemic and ruin vast acreages of economically important plants. The historic potato famine in Ireland in the 1840s, resulting in the death of 1,000,000 people, was due to epidemics of potato late blight. Chestnut blight ruined the chestnut forests of the United States. Stem rust destroyed about 300,000,000 bu of wheat in the United States and Canada in 1916. In the United States it destroyed 60% of the spring wheat in 1935, and 75% of the macaroni wheat and 25% of the spring bread wheat in both 1953 and 1954. Stem rust, only one of more than 3000 kinds of plant rusts, has been similarly destructive in other wheat-growing areas of the world, and it continually menaces wheat, oats, barley, rye, and many grasses. The *Helminthosporium* disease of rice was the principal cause of a famine in which a million or more people died in India in 1943.

Plant diseases are a dangerous threat to man's future subsistence. Much of the world is now underfed, and acute food shortages often occur in many areas. The situation tends to become worse as population increases by many millions each year. Plant diseases, old and new, are a critical limiting factor in food production. The degree to which they can be controlled will help determine whether the world can feed its rapidly growing population.

[E. C. STAKMAN]

Nature of diseases. In the broad sense, disease in plants may be considered as any physiological abnormality which produces pathological symptoms, reduces the economic or esthetic value of plant products, or kills the plant or any of its parts. Damage, caused by wind or lightning or predation of insects or other animals, is not usually called disease, although such injury to living plants may result in a physiological disturbance which is truly disease. Decay of storage organs, such as tubers and roots, is disease because such plant parts are living; decay of lumber is disease only by extension of the definition, although the processes may be similar.

Disease in plants is usually evidenced by abnormalities in appearance, called symptoms, or by the presence of a pathogen in or on the plant. Some diseases, however, have no obvious symptoms; potato virus X, for example, reduces the yield of potatoes without apparent changes in the appearance of the plants.

The symptoms of plant diseases may be death (necrosis) of all or any part of the plant, loss of tur-

Fig. 1. Common bacterial blight of bean. (*From J. C. Walker, Plant Pathology, 3d ed., McGraw-Hill, 1969*)

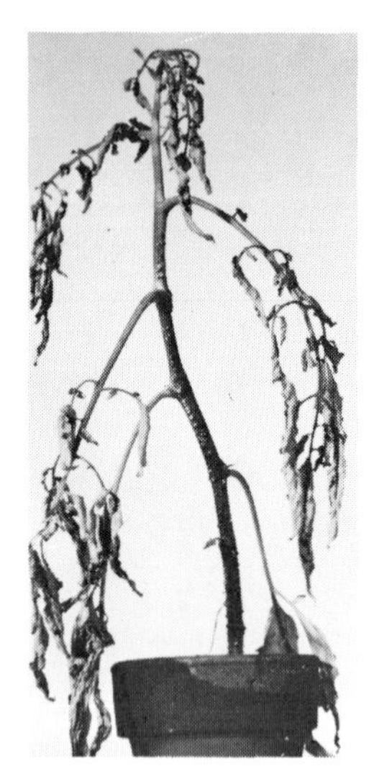

Fig. 2. Southern bacterial wilt of tomato. Plant shows leaf epinasty and wilt. *(After Kelman, from J. C. Walker, Plant Pathology, 3d ed., McGraw-Hill, 1969)*

gor (wilt), overgrowths (hypertrophy and hyperplasia), stunting (hypoplasia), or various other changes in the structure and composition of the plant. Necrosis may affect any part of the plant at any stage of growth. A rapid death of foliage is often called blight (Fig. 1), whereas localized necrosis results in leaf spots and fruit spots. Necrosis of stems or bark results in cankers. Wilting may be slow or rapid, and it is usually more pronounced in dry than in moist soil. Necrosis eventually follows persistent wilting (Fig. 2). Overgrowths composed primarily of undifferentiated cells are called galls (Fig. 3), the term tumor being less commonly used to designate these structures. A bunch of small, abnormal shoots is often referred to as a witches' broom. Underdevelopment or stunting may affect the entire plant or only certain of its parts.

Chlorosis (lack of chlorophyll in varying degree) is the most common nonstructural evidence of disease. For example, in leaves it may occur in stripes or in irregular spots (mosaic). Various degrees of curling and crinkling of the foliage often accompany chlorosis. Sometimes there is also other abnormal coloration, such as shades of red and brown.

A number of diseases may cause similar symptoms. These may be characteristic enough to permit diagnosis, but often it is necessary to identify the causal organism for exact diagnosis.

Causative agents. Usually two or more causes operate simultaneously to produce plant disease. For example, if a parasite is involved, the weather will influence the growth of the parasite as well as the plant's susceptibility to the parasite. The following subsections describe the influence on plant diseases of animals, plants, and viruses; soil conditions; weather; agricultural practices; industrial by-products; and plant metabolism.

Animals, plants, and viruses. Nematodes and insects are the animals that most commonly cause plant disease (Fig. 4). Although herbivorous animals, including many insects, bite off and swallow plant parts, the parts removed are not diseased and the animals are predators, not pathogens. However, the loss of the parts eaten may cause the rest of the plant to become diseased. Conversely, some insects are true pathogens because they remain on or in the plant and cause disease symptoms typically associated with the insects involved. Such symptoms may include yellowing, leaf curl, and overgrowths. Many nematodes are true parasites, hence pathogens, since they cause rots, overgrowths, and other plant abnormalities.

Certain algae, fungi, and bacteria are plant pathogens that cause disease. Most plant diseases are due to fungi; less than 200 are known to be caused by bacteria, and even fewer are caused by algae and parasitic seed plants, such as dodder and mistletoe (Fig. 5).

Many plant diseases are caused by viruses, which are neither plants nor animals but are similar to living things in many ways and may properly be called pathogens.

Soil conditions. Deficiencies of mineral nutrients in the soil are a frequent cause of plant disease (Fig. 6). Often the deficiency can be identified by characteristic plant symptoms. For example, yellowing of the leaf tip and midrib of corn indicates nitrogen deficiency; yellowing of the margins, potassium deficiency. However, the symptoms may vary somewhat in different plant species. In addition, deficiency diseases may be difficult to diagnose, since they sometimes resemble those caused by viruses.

Besides nitrogen, potash, and phosphorus, which plants need in relatively large amounts, smaller quantities of sulfur, calcium, and magnesium are required. Boron, iron, copper, manganese, molybdenum, zinc, and other minerals are used in such minute amounts that they are called trace elements. However, if one of the latter is missing, a typical disease may result, such as dry rot of rutabagas, which is due to boron deficiency. *See* PLANT, MINERALS ESSENTIAL TO.

Frequently deficiencies of minerals cannot be determined by soil analysis alone, because the minerals may be present in chemical combinations that plants cannot use. For example, iron is often unavailable on high-lime soils, even if it is present in the soil in appreciable quantities.

Besides lime, excess amounts of many other chemicals may be present in the soil and cause plant disease. Excess of soluble salts causes "alkali injury" and aggravates drought damage; too much nitrogen may stimulate abnormal growth; while an excess of boron may cause necrosis and stunting. Unfavorable chemical balance in the soil may also result in excess acidity (low pH) or alkalinity (high pH), either of which may inhibit normal plant growth.

The soil is the principal source of water, which all plants need in varying amounts, depending upon the species. Too little available water slows growth and, below certain limits, results in wilting of the plant. Plants can recover from limited (transient) wilting, but if it is prolonged the affected parts die.

Conversely, too much water in the soil results in oxygen deficiency, which causes suffocation of the tissues of roots and other underground parts of most plants. Water-inhabiting plants, such as rice, are exceptions. Excess water in the soil favors certain kinds of fungus and bacterial diseases, and these are often confused with the purely physiologic effects of too much water.

High soil temperature during the growing season may also cause disease, for instance, internal necrosis of potato. Lack of oxygen in storage may

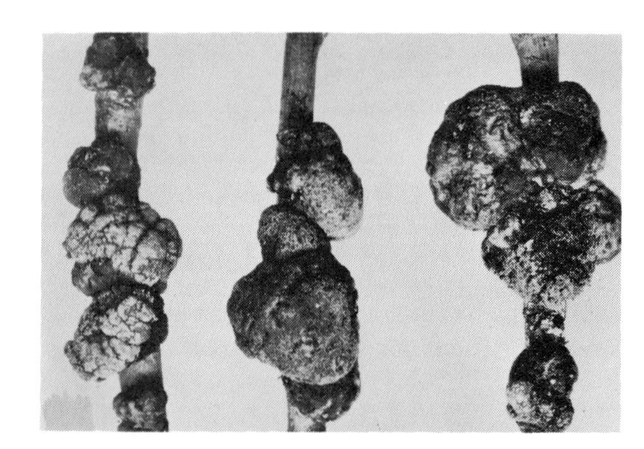

Fig. 3. Crown gall of apple. *(A. J. Riker, from J. C. Walker, Plant Pathology, 3d ed., McGraw-Hill, 1969)*

result in blackheart of potato tubers, especially at temperatures above 40°F (4°C) (Fig. 7).

The structure of soil (particle size and organic content) determines its water-holding capacity and hence affects both the conditions mentioned above and the ease with which plant roots penetrate the soil. *See* SOIL.

Weather conditions. Wind, lightning, and hail may injure plants and cause true diseases such as those resulting from unfavorable temperatures (Fig. 8). Temperature effects range from poor development of plants grown in climates too cold or too warm to actual frost or heat damage. For example, tomatoes grow poorly and drop their blossoms in cool weather; direct sunlight on the fruit may kill the tissue, causing sunscald; and the foliage is severely damaged by even light frosts that would not harm cabbage.

Although high temperatures may literally cook plant tissue, with such results as "heat canker" of young flax and sunscald of tomato fruit, the commonest effect is to increase water loss by transpiration, resulting in drought damage. Wind has the same effect, the degree depending upon its velocity and relative humidity.

In most green plants deficiency of light causes weak, spindly growth and chlorosis, although some species can endure much shade. House plants are frequently affected in this manner, but excess shading by buildings or other plants will produce the same effect outdoors.

Agricultural practices. Mismanagement of soil, including untimely applications of irrigation water and fertilizer, can cause plant disease, but other agricultural practices are frequently injurious. The more common of these injuries result from the improper use of chemicals such as fungicides, insecticides, and herbicides. *See* FUNGISTAT AND FUNGICIDE; HERBICIDE; INSECTICIDE.

Nearly all fungicides are injurious to plants as well as to fungi, although the damage to the plants is usually much less than the potential injury from the diseases controlled by the fungicides. Examples of effects are increased transpiration caused by Bordeaux mixture on tomatoes, russeting of fruits caused by lime sulfur on apples, and yield reductions without visible symptoms caused by other chemicals. Conversely, the fungicide may contain a nutrient such as zinc which, if deficient in the soil, may result in much better growth of the plant.

Chemicals used for seed treatment are frequently toxic, especially to some species of plants. For example, plants of the cabbage family are stunted by copper-containing seed treatment materials. Vegetative organs, such as potato tubers, are very susceptible to chemical injury, and strong poisons, such as mercuric chloride, often do more harm than good. Materials applied to the soil to control fungi, bacteria, and nematodes may injure plants grown in the soil too soon after treatment.

Some crop plants are very sensitive to herbicides, being affected by very minute amounts of such things as 2,4-D (2,4-dichlorophenoxyacetic acid). Tomatoes may be affected from sources far removed. Symptoms of 2,4-D are sometimes confused with those of virus diseases.

Industrial by-products. The fumes from ore smelters frequently cause widespread symptoms of plant disease, including stunting, yellowing, and necrosis. Where atmospheric inversion layers prevent their escape, even traffic and domestic fumes may be toxic. *See* AIR POLLUTION.

Plant metabolism products. Brown areas on

Fig. 4. Nematode galls incited by *Meloidogyne* sp. (*a*) On tomato. (*b, c*) On parsnip. (*After Cox and Jeffers, from J. C. Walker, Plant Pathology, 3d ed., McGraw-Hill, 1969*)

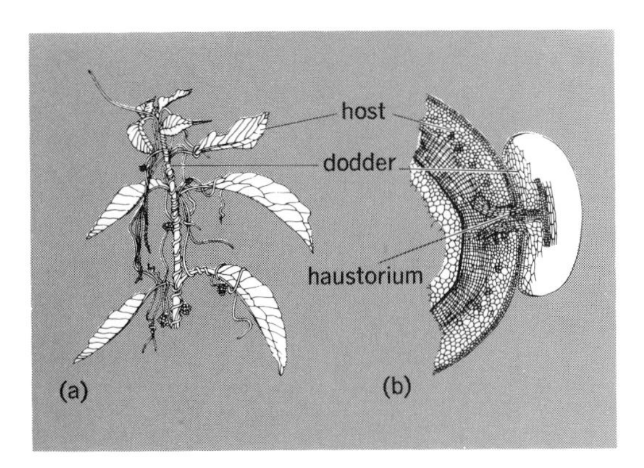

Fig. 5. Dodder (genus *Cuscuta*). (*a*) Plant attached to host. (*b*) Section through host showing haustorium of dodder extending into host. (*From F. W. Emerson, Basic Botany, 2d ed., Blakiston, 1954*)

Fig. 6. Mineral deficiencies. (*a*) Potassium-deficiency disease of cabbage. (*b*) Iron-deficiency disease of cabbage. (*c*) Magnesium-deficiency disease of bean. (*From J. C. Walker, Plant Pathology, 3d ed., McGraw-Hill, 1969*)

stored apples (scald) may be caused by ethylene gas produced by the apples (Fig. 9). This gas occurs in small quantities in many healthy plant tissues but is produced in greater amounts by diseased and aging cells. Ethylene gas may also cause yellowing in plants, and it accelerates ripening in certain fruits such as banana.

PLANT DISEASE PATHOGENS

Most pathogens are grouped primarily on the basis of their structure; but bacteria, being morphologically simple, are classified to a considerable extent by physiological characters. Viruses represent a special problem, and such considerations as means of transmission and host symptoms are used in classifying and naming them.

Fungi, bacteria, and a few seed plants are heterotrophs; that is, they lack chlorophyll and consequently are dependent, directly or indirectly, upon green plants (autotrophs) for carbohydrates. Animals and some fungi and bacteria are also dependent upon other organisms for nutrients such as amino acids and vitamins. Viruses multiply or are replicated by synthesizing virus nucleic acids and proteins from amino acids and other compounds present in the host cell.

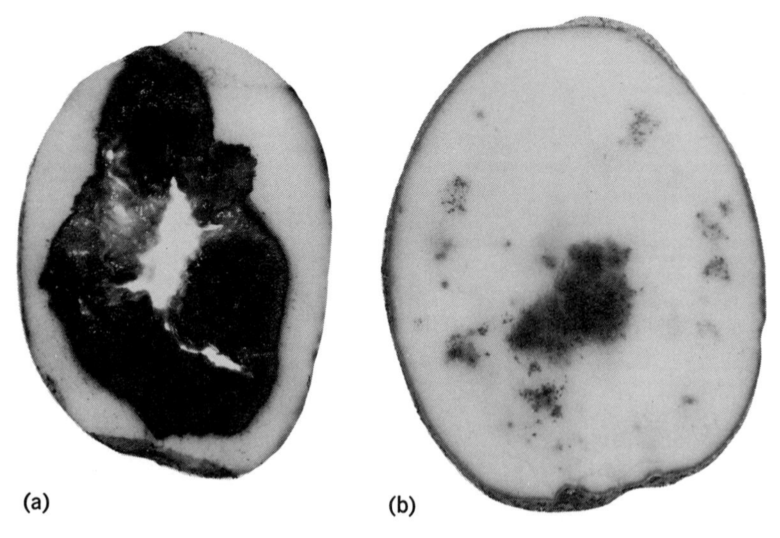

Fig. 7. Potato disease. (*a*) Blackheart. (*b*) Internal necrosis, due to high soil temperature. (*From J. C. Walker, Plant Pathology, 3d ed., McGraw-Hill, 1969*)

Plant pathogens usually penetrate into the host plant and grow within or between the cells. Viruses are usually intracellular, and some are confined to the phloem, whereas plant pathogenic bacteria are usually intercellular or occur in the xylem. Fungi are composed of microscopic tubes called hyphae, by means of which plant pathogenic species penetrate into or between the host cells. The powdery mildew fungi grow principally outside of the plant but send special absorptive organs (haustoria) into the host cells. Some intercellular species of fungi also produce haustoria. Pathogenic seed plants, such as mistletoe and dodder, usually penetrate the host by means of rootlike absorptive organs. Pathogenic insects and nematodes may be wholly within the plant, or they may remain superficial and penetrate the host with specialized mouthparts.

Most plant pathogens are parasites. Some, such as the rusts and powdery mildews, are obligate parasites, that is, can grow only on a living host plant. Viruses are also in this category, although they are not typical organisms. Fungi and bacteria that can use only nonliving food sources are called saprophytes.

Most fungi and all plant pathogenic bacteria can grow on nonliving organic matter as well as parasitically on living matter; these are called facultative saprophytes. Some organisms live primarily as saprophytes but also have the ability to parasitize weakened plants and are therefore called facultative parasites. Many plant pathogens have both a parasitic (or pathogenic) and a saprophytic phase of development.

Symbiotic relations of organisms. Parasitism is the one of a series of associations characterized by intimate physical union of taxonomically dissimilar organisms. Such relationships are known as symbiosis, and may be neutral, beneficial, or harmful to the symbionts. An association such as that of legumes and nodule bacteria, beneficial to both partners, is called mutualistic symbiosis. Parasitism is antagonistic symbiosis.

There are different degrees of parasitism. In the early stages, the association between rust fungi and their hosts may appear to be almost neutral, harming the plants little. Other fungi, such as those rotting fruit, can become established only in dead tissue, producing enzymes or toxins that kill adjacent living cells which they then inhabit. Some biologists say that such organisms are saprophytes, not parasites, because they never colonize living host tissue. But the term parasitism is generally used to refer to a relationship with the host plant as a whole, because the degree of intimate relationship is often difficult to determine.

Ecologic relations of organisms. Associations of organisms in the same environment without physical union are called ecologic and are often very important in plant disease. As in symbiosis, the effects may be beneficial, neutral, or harmful. Metabiosis occurs when one organism uses a substance for food and produces a by-product that enables another to grow. If the benefits are reciprocal, the relationship is called synergism, as when the fungus *Mucor ramannianus* produces pyrimidine and *Rhodotorula rubra*, a nonsporulating yeast, makes thiazole. These chemicals are components of thiamine, which both organisms need but which neither can produce alone. If deleterious substances (antibiotics) are produced, the relationship is called antibiosis. Usually the term antibiosis refers only to the deleterious effects of one microorganism on another, but similar relationships exist between plants of all kinds. *See* ECOLOGICAL INTERACTIONS.

All of these relationships may be important to the survival of certain plant pathogens, especially some of those which live in the soil part of the time. Metabiotic and synergistic relationships may help them to survive; antagonistic relationships hinder survival. One of the goals of the plant pathologist is to encourage antibiosis that eliminates certain soil-inhabiting pathogens.

Ecologic associations may exist between two or more pathogens inhabiting the same host plant as a common environment. When fire-blight bacteria parasitize apple twigs and permit the entrance of canker and wood-rotting fungi, the relationship between the bacteria and the fungi is metabiotic. The molds *Oospora citri aurantii* and *Penicillium digitatum* can rot fruit more rapidly together than either can alone; this is synergism. Antagonism seems to exist between races of the potato late blight fungus, and one will replace the other when they parasitize a potato plant together.

Even the relationship of host and pathogen may be ecologic at first. For example, *Rhizoctonia solani* in the soil causes visible injury to the roots of soybean before touching them. Accordingly, the fungus is at first toxic to soybean; later it becomes parasitic and pathogenic. [CARL J. EIDE]

Growth and reproduction. Many plant pathogens, especially among the bacteria, fungi, and viruses, multiply with amazing rapidity under favorable conditions. Viruses, although not generally considered living organisms, may increase a millionfold a few days after introduction into the right place in the right kind of living plant, when temperature and other environmental conditions are favorable to the virus.

Food requirements of bacteria and fungi. Although lack of chlorophyll prevents these organisms from using solar energy to synthesize basic carbohydrates from carbon dioxide and water as green plants do, their basic nutrient requirements are essentially the same as those of higher plants. They require carbon, hydrogen, oxygen, nitrogen, phosphorus, and sulfur as structural elements. In addition, they need the metallic elements potassium, magnesium, iron, zinc, copper, calcium, gallium, manganese, molybdenum, vanadium, and scandium. Potassium and magnesium, needed in relatively large amounts, are designated macroelements; the others, some of which are needed in minute amounts, are often designated microelements. Vitamins are also needed by some species for growth and reproduction.

For experimental purposes, pure cultures of facultative saprophytes are grown in the laboratory on sterilized synthetic media containing sugars or some other source of carbon, salts of the other necessary elements, and essential vitamins and amino acids for those organisms which cannot synthesize their own. Natural plant products, such as potato broth, steamed cornmeal, or oatmeal, often are used as nutrient bases. Liquid media are used for some purposes; for others, the nutrient solutions are solidified with gelatin or agar. Nutrient requirements for growth and reproduction are best determined by varying the composition of synthetic media. Studies on the effects of temperature, light, and other environmental factors are facilitated when organisms can be grown on culture media. Although all pathogenic organisms have some requirements in common, they differ greatly in special requirements, both on artificial media and on host plants. By growing pathogens artificially,

Fig. 8. Freezing injury of pea, several weeks after injury. (*a*) Enlargement of the injured growing point in *b*; in the youngest leaf the stipules and the first pair of leaflets have assumed abnormal shapes, and the second pair of leaflets did not form. (*b*) Following killing of the growing point at the left, a lower dormant bud grew out to form the main stem. (*c*) Necrotic bands in a pair of leaflets which were developing at the time of injury. (*From J. C. Walker, Plant Pathology, 3d ed., McGraw-Hill, 1969*)

Fig. 9. Apple scald. (*USDA*)

much is learned about them which enables the development of better control measures.

Host selectivity of bacteria and fungi. Among the approximately 150 species of pathogenic bacteria and the many thousands of fungi, there are wide differences with respect to the kinds of plants and plant parts on which they can grow. Flax rust (*Melampsora lini*) grows only on wild and cultivated flax, asparagus rust (*Puccinia asparagi*) principally on asparagus; *Xanthomonas campestris*, the bacterium which causes black rot of cabbage, cauliflower, and related plants, parasitizes members of the mustard family only. On the other hand, the fungus *Rhizoctonia solani* causes root rot of potatoes, alfalfa, clover, and hundreds of other species in many different plant families; the bacterium *Agrobacterium tumefaciens* causes crown gall on grape, raspberry, chrysanthemum, and numerous other plants; the bacterium *Erwinia carotovora* causes soft rot of almost all kinds of fleshy vegetables.

Some pathogens attack only a few plant parts or tissues, others attack many. Some attack roots only, others attack stems, others cause leaf spots, and still others attack fruits. Some attack young tissues, others attack old ones. There are diseases of youth and of age, of herbaceous plants, and of woody plants. Some pathogens parasitize all plant parts of susceptible hosts at all stages of development. To understand and control the numerous diseases of thousands of kinds of plants, it is necessary to learn the conditions under which each pathogen thrives.

Environmental factors. The rate and kind of growth and reproduction of pathogens are affected by nutrition, moisture, temperature, light, the acidity or alkalinity of the medium, the relative amounts of oxygen and carbon dioxide, and by other microorganisms with which they must compete. Most pathogens require free moisture for germination and infection, although some powdery mildews can germinate in dry air. Soil moisture sometimes is a determining factor in growth and reproduction. Some pathogens that live in the soil thrive best at high moisture content, some at low. Temperature determines the geographical and seasonal occurrence of many diseases, since the cardinal temperatures—the minimum, optimum, and maximum—differ for different pathogens. The peach leaf curl fungus, the potato late blight fungus, and yellow rust of wheat develop best at a relatively low temperature; the peach brown rot fungus, the potato wilt and brown rot bacterium, and stem rust of wheat develop best at a relatively high temperature. Light has less influence than temperature on the growth of pathogens in nature, but it strongly affects reproduction of some fungi. Some soil organisms, such as the potato scab bacterium, like an alkaline (high pH) soil; some, such as the cabbage clubroot fungus, like an acid soil (low pH).

Reproduction of bacteria and fungi. A single bacterium divides into 2, the 2 into 4, and so on. As division may occur every 20 to 30 min, a single bacterium could produce a progeny of 300,000,000,000 within 24 hr. The rate, however, varies with the kind of bacterium, its nutrition, temperature, and other environmental conditions.

Most fungi, however, reproduce both asexually and sexually. In many fungi asexual reproduction results in rapid multiplication, whereas sexual reproduction results in the production of spores that can survive unfavorable conditions. In general, fungi continue to grow and produce asexual spores while the environment is favorable and nutrients are easily available, but they tend to produce sexual spores when growth is checked. Thus an asexually produced urediospore (summer spore) of wheat stem rust (*Puccinia graminis* var. *tritici*) can cause infection, the resulting mycelium grows for a time and then forms a pustule containing 50 to 400,000 new urediospores. The time required is only about a week at 75°F (24°C), but it increases to a month at 50°F (10°C) and even longer as temperature decreases. Each new spore can cause a new infection, and this process continues at a rate that varies greatly with temperature, moisture, and light, until the wheat starts to ripen or growth is otherwise checked. Then the winter spores (teliospores) are produced; these differ from urediospores in appearance and cannot normally germinate until they have been exposed to winter weather. The apple scab fungus (*Venturia inaequalis*) may produce many successive crops of asexual spores (conidia) on the fruit and leaves during the growing season. But it does not produce sexual spores until the following spring, on infected leaves that have fallen to the ground the previous autumn. Some fungi, such as the ergot fungus (*Claviceps purpurea*), produce sclerotia, bodies made up of densely interwoven hyphae, which may survive winters or other unfavorable conditions and then produce fruiting structures under appropriate conditions in the spring.

Special stimuli are sometimes necessary to initiate the formation of fruit bodies; some fungi require the stimulus of light for fructification, although they grow well in darkness; some require special temperature; others require certain nutrients or vitamins.

How, where, when, and the rate at which fungi grow and reproduce depend on their inheritance and their environment. The inheritance determines the limits within which the behavior of each kind of fungus can vary, and the environment determines its behavior under particular combinations of conditions.

[CARL J. EIDE; E. C. STAKMAN]

INFECTION AND DEVELOPMENT OF DISEASE

Infection of plants by a pathogen terminates a series of events that begins with inoculation, which is the contact of a susceptible part of a plant by the inoculum. Inoculum is any infectious part of the pathogen, such as spores, bacterial cells, and virus particles. Typically, inoculation is followed by entrance into the host, and infection follows entrance. A plant is infected when the pathogen starts taking nourishment from it. However, some pathologists consider penetration part of the infection process.

The time between inoculation and infection is the incubation period. Because it is often difficult to tell when infection occurs, the incubation period is usually counted as the time between inoculation and the appearance of the first symptoms of infection.

The probability that infection will follow inoculation depends upon the susceptibility of the host, the virulence and amount of inoculum in contact with the host, and the duration of favorable environmental conditions. A single unit of inoculum (propagule) of many pathogens can infect a plant; this is probably true of all pathogens when conditions are ideal. Actually only part of the propagules on a plant cause infection, and hence the amount of inoculum on a plant helps to determine how many infections result. This fact has given rise to the concept of inoculum potential, which has been defined as the product of the number of propagules per unit area of the host plant and their physiological vigor. Because of the usually haphazard dissemination of inoculum of plant pathogens, only a small part of that produced actually reaches a susceptible plant. Consequently, most plant pathogens survive as species and are destructive partly because they produce fantastically large amounts of inoculum.

The inoculum. Inoculum of viruses and bacteria consists of the individual virus particles or bacterial cells, respectively; the inoculum of fungi may be spores, pieces of hyphae, or specialized structures, such as sclerotia. Pathogenic plants such as dodder produce true seed, and nematodes produce eggs, both of which function as inoculum.

Bacteria and viruses produce billions of cells or virus particles in infected plants, and each new unit theoretically can infect another plant. Fungi produce spores on the surface of hyphal growth or in a variety of specialized structures which may be large, as the giant puffball, or almost invisible to the unaided eye (Fig. 19). Some of the spore-producing structures function over a considerable period of time and, like bacteria, produce prodigious amounts of inoculum.

Bacteria and viruses are somewhat restricted as pathogens by having no special means of liberating themselves from the host, although the bacteria may ooze out in sticky droplets. For dissemination or transmission these pathogens depend chiefly upon plant contact, insects, or man, although bacteria may be spattered short distances by rain. Some fungi produce spores in sticky masses, like bacterial ooze, and are disseminated in much the same ways as bacteria. Other fungi have ways to liberate or forcibly eject spores into the air, where they can be carried by the wind. This gives fungus pathogens the potential of much farther and faster spread than the bacteria or viruses, although their arrival on a susceptible plant is much more a matter of chance than if insects carry the inoculum, because insects often seek similar plants for food. (Dissemination is considered in greater detail in a later section of this article.)

Dormant inoculum is one of the most important, but not the only, means by which plant pathogens survive during periods when parasitic life is impossible. If the pathogen is within a perennial host, it is usually quiescent during the rest period of the host. Sclerotia and even the vegetative hyphae of some fungi may survive periods of drought and cold independently of the host. Other pathogens require the protection of the dead host plant, not so much against cold and drought as against antagonistic organisms. This is especially true of plant pathogenic bacteria, few of which survive long if separated from host tissue. Some viruses can live only minutes apart from the living host; others, such as tobacco mosaic virus, remain infective for years in dried leaves.

At the beginning of the growing season, the first inoculum of a pathogen is called primary inoculum; that which is produced later on infected plants, secondary inoculum. The primary inoculum of fungi may be resting spores or the surviving hyphae or sclerotia; often the hyphae or sclerotia produce spores which function as primary inoculum.

Many fungi produce two or more kinds of spores. Those formed late in the growing season (resting spores) usually will not germinate until after a period of dormancy and will survive more cold and drought than spores produced during the growing season. Some, like the spores of the cabbage clubroot fungus and the chlamydospores of the onion smut fungus, stay dormant for several years, thus assuring the species of survival if susceptible hosts are not grown on the land for several seasons. Such diseases are difficult to control by crop rotation.

"Repeating" spores typically are morphologically distinct from the resting spores and are produced in great numbers on diseased plants during the growing season. They usually germinate rapidly whenever environmental conditions are favorable. Before germination, repeating spores can survive for periods ranging from several hours to several weeks, depending upon the species. This determines largely how far and under what conditions a pathogen will spread during the growing season.

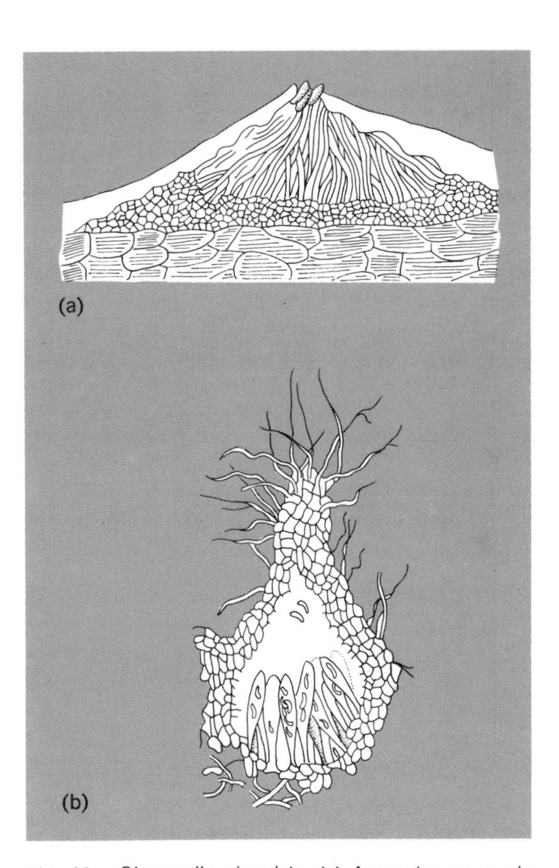

Fig. 10. *Glomerella cingulata.* (*a*) Acervulus on apple fruit. (*b*) Perithecium. (*From J. C. Walker, Plant Pathology, 3d ed., McGraw-Hill, 1969*)

Spore germination. Germination, as applied to spores or seeds, means the resumption of vegetative growth leading to the development of a new individual. In fungi this usually means the production of a hypha, called a germ tube (Fig. 11). Cell division of bacteria and the hatching of nematode and insect eggs are comparable processes, so far as their function as pathogens is concerned.

Germination occurs if the spore is not dormant and if environmental conditions are favorable. This usually requires a certain temperature range and liquid water, although a few species of fungi (powdery mildews) germinate in humid air. Certain species also require the presence of food substances, special stimulants associated with the host, absence of inhibitors that may be produced by the pathogen or associated organisms, or certain degrees of acidity. Such requirements limit germination but may be a benefit to the species. For example, the necessity for a host stimulant prevents wastage of spores in the absence of the host.

When a nondormant spore is placed under favorable conditions, germination may follow in 45 min or only after several days, depending upon the species, age of the spores, and variations in the environment. Since conditions change rapidly, germination is a critical time for a fungus, because if it does not penetrate the host quickly the germ tube may be killed, especially by dryness. It is at this stage that fungi are most easily killed by fungicides.

Establishment in the host. For bacteria and viruses, entering a host is a passive process. Bacteria accidentally get into injuries or are put there by insects or other agencies; they may also be drawn by water into stomata, hydathodes, or nectaries. Viruses often are placed in the host by insects, but many can be transmitted when the sap from infected plants comes in contact with minute wounds in healthy plants.

Spores of fungi may also be carried into plants by various agencies, but many species have active means of penetration, the method usually being characteristic of the species. In some, germ tubes enter stomata by producing a flat structure (appressorium) over the stoma from which a hypha grows through the opening. Others ignore the stomata; instead the appressorium adheres to the cuticle of the plant and forces a slender infection peg directly through the protective layer. This apparently is accomplished entirely by pressure, as no enzyme action has been demonstrated.

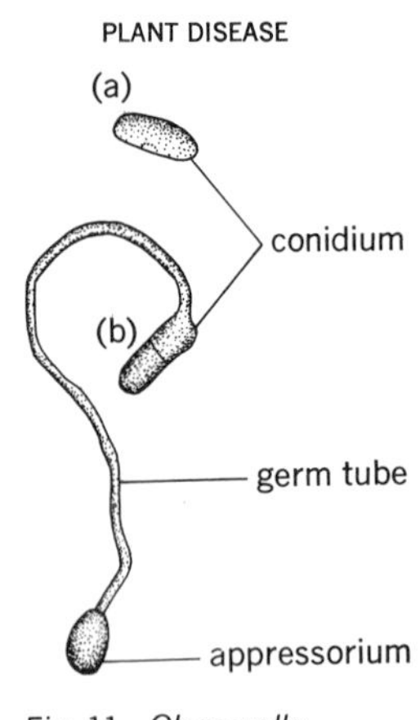

Fig. 11. *Glomerella cingulata.* *(a)* Conidium (spore). *(b)* Conidium that has become septate during germination; appressorium at tip of germ tube. *(From J. C. Walker, Plant Pathology, 3d ed., McGraw-Hill, 1969)*

Animal pathogens, such as nematodes, have special mouthparts that pierce the plant, and the nematode may remain external or it may actually enter the plant.

Even after penetration, pathogens may fail to invade the plant because of the presence of mechanical barriers, lack of proper nutrients, or the presence of inhibiting toxic substances. These factors depend not only on specific interactions between host and pathogen but also upon the environment. Successful establishment of the pathogen may mean killing the host cells and living upon the dead tissue, with or without the actual penetration of living cells.

[CARL J. EIDE]

Development within the host. After a pathogen has become established in a susceptible host, the rate of disease development under favorable conditions follows a sigmoid (S-shaped) curve with three major aspects: (1) the lag phase or incubation period, when infection is not evident externally; (2) the exponential phase, when the pathogen spreads rapidly in host tissues and symptoms and signs of disease appear; and (3) the senescence phase, when limiting mechanisms of either host or pathogen restrict further extension.

Disease development varies with genetic susceptibility of the host, genetic aggressiveness of the pathogen, and with many environmental factors that influence the host, the pathogen, and the interactions between the two. Environmental factors influence the growth rates and the metabolism of the host and the pathogen; and the interrelations between these activities determine the pattern of disease development. Furthermore, the effects of past environmental conditions on the host may affect disease development, a condition known as predisposition when host susceptibility is increased. The combined effects of these factors on growth and development of healthy crop plants in nature are poorly understood, and the problem becomes increasingly complex when the plants become diseased.

Climate often determines the adaptability of plant species to geographic areas and may also determine the geographic distribution of their diseases. For each disease there are minimum, optimum, and maximum values for each critical environmental factor. The mean measurements of weather, however, are often less important than the exact combinations of weather at critical times. Those environmental factors that deviate most from the optimum limit the development of a disease.

Effect of temperature. Temperature has a major effect on disease development and determines the seasonal and regional incidence of most diseases. For example, a succession of diseases attacks certain creeping bent grasses on golf greens in northern United States: snow mold occurs beneath melting snow in winter; red thread during the moderate temperatures of spring and fall; dollar spot during the warm temperatures of early and late summer; and brown patch during the hot weather of midsummer. The fungi causing these diseases may attack the leaves, grow into the crowns of the plants, and may kill the plants entirely in patches of characteristic size for each disease. These diseases produce similar effects, but temperature determines when each disease is most destructive.

The length of the incubation period of disease is governed by prevailing temperatures. Many diseases, such as rusts, mildews, and leaf spots, cause only small lesions on aboveground plant parts. The damage to the plant depends on the number of lesions, which in turn depends on the number of disease cycles. The elapsed time from infection to spore production—the length of the incubation period—determines the frequency of disease cycles. Thus temperature often determines whether pathogens can produce enough disease cycles for development of an epidemic. Temperature likewise may influence the symptom expression. Thus symptoms of many virus diseases disappear or are masked at high temperatures. Temperature also determines whether certain wheat varieties are susceptible or resistant to certain parasitic races of stem rust. The effect of temperature on disease development may be principally on the pathogen, or it may be on the host. When the cardinal temperatures are the same for growth of the pathogen in culture and for development of the disease, the effect is principally on the pathogen. However, when the optimum temperature for growth of the pathogen in culture differs from that for maximum disease development, the temperature probably predisposes the host plant by weakening it.

Effect of light. Light affects disease development principally by its effect on photosynthesis and the assimilative processes of the host. Obligate parasites, such as rusts and powdery mildews, generally develop best when assimilation is maximal, although the severity of the disease lesion caused by some pathogens on some hosts may be decreased by high light intensities. Low light often weakens plants and thus predisposes them to diseases caused by facultative saprophytes.

Effect of moisture. Moisture is a major factor in germination and entrance of pathogens into the host. The moisture requirements of the established pathogen are supplied by the host, since the osmotic value (water absorption capacity) of the hyphae of the pathogen is always greater than that of the parasitized host cells. Transpiration (water vapor loss) from diseased aboveground plant parts is greater than that of healthy parts. The water economy of the host is disrupted in wilt diseases by the effects of the pathogen on the translocation of water in the xylem and the osmotic permeability of foliage parenchyma, and in root diseases by the destruction of the tissues for water absorption and conduction. The rate of symptom development and death of the plant tissue in wilts and root rots is accelerated by excessively low atmospheric humidities and low soil moisture availability.

Relation of soil. Soil reaction, as regards hydrogen ion concentration, affects the development of many diseases in the soil. Potato scab is less severe in acid soils (below pH 5.2) while cabbage clubroot is not so severe in less acid soils (above pH 5.7). However, the extent to which the soil reaction affects the infectivity of these pathogens and the subsequent development of the diseases has not been determined. As the hydrogen ion concentration of the plant cell is relatively constant despite differences in the range of soil reaction, soil pH probably affects disease development indirectly by its effects on the availability to the host or pathogen of mineral nutritional elements in the soil.

Soil oxygen and carbon dioxide concentrations affect the development of root diseases. The effects on infectivity of the pathogen, predisposition of the host, and disease development have not been distinguished, although the development of the host is more adversely affected by high carbon dioxide and low oxygen tensions in the soil than is the growth of many fungal pathogens.

Effects of nutrients. The effects of nutrients are largely indirect since plants and their pathogens require the same essential mineral elements. However, the available amount of each mineral element and the balance between them affect the structure and physiology of the host and thus may be either favorable or unfavorable to the development of different pathogens. The principal mineral elements in fertilizers (nitrogen, phosphorus, potassium, and calcium) have the most pronounced effects. Diseases caused by obligate parasites such as rusts, powdery mildews, and many viruses develop best in "normal" plants having optimal mineral nutrition; while subnormal plant development due to inadequate or unbalanced mineral nutrition favors the development of many diseases, such as root rots, that are caused by facultative saprophytes. Some vascular pathogens are affected directly by the concentration of nitrogen compounds in the conductive tissues of the host.

[J. B. ROWELL]

Plant disease epidemics. When a disease spreads rapidly in a crop or other plant population, it constitutes an epidemic. Strictly speaking, the increase does not have to be spectacularly rapid to be an epidemic; the essential feature is that the pathogen, and hence the disease, is increasing in a population of host plants.

The population of plants may be small, as a single field of potatoes, or large as the total of all wheat fields from Texas to Canada. Thus epidemics may be local or regional in extent.

The time required for an epidemic to reach a destructive climax and subside may be a few days or weeks or it may go on for years. For example, an epidemic of stem rust (*Puccinia graminis* var. *tritici*) occupied roughly 3 weeks from its beginning until 100% of the plants were infected (Fig. 12). On the other hand, chestnut blight (*Endothia parasitica*) was introduced into the United States in 1904 and continued to spread through the chestnut forests for about 40 years, after which nearly all the chestnut trees were dead.

Typically the progress of an epidemic follows the same course as the growth of a population of any organism. The pathogen (for example, a fungus such as *Phytophthora infestans*, which causes late blight of potatoes) infects the host plants; in a few days the infected spots produce a crop of spores (propagules) which in turn infect new host tissue. As long as there is fresh host tissue to infect

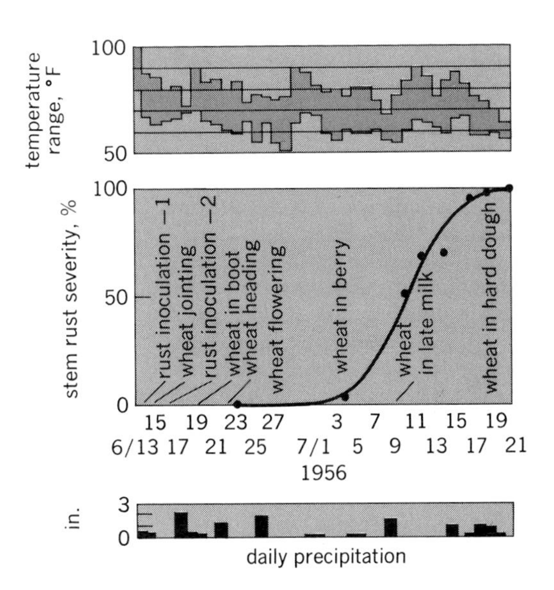

Fig. 12. Development of an epidemic of stem rust on Marquis wheat. *(From J. B. Rowell, Oil inoculation of wheat with spores of Puccinia graminis var. tritici, Phytopathology, 47(11):689–690, 1957)*

and the weather is favorable, this cycle will be repeated every 5 to 7 days. During the period of optimum development the population of the pathogen, and hence the severity of the disease, increases logarithmically, and it is possible to express the rate of increase mathematically. *See* POPULATION DYNAMICS.

Some pathogens, such as *Phytopthora infestans* and *Puccinia graminis*, may increase very rapidly, causing true population explosions. These two fungi may, on the average, double in numbers every 2 to 7 days during the logarithmic phase of development (Fig. 12). Disease of perennial plants may increase much more slowly, but still do so logarithmically.

The rate of logarithmic increase of a pathogen is dependent upon the existing population of individuals at any given time, that is, the numbers which are producing reproductive units or propagules. Other factors that affect the rate of increase can be put into three categories: (1) the number of host plants in a given area; (2) the susceptibility of the host plants; and (3) the inanimate environment.

Crops grown in fields of only one kind of plant provide ideal conditions for epidemics, because propagules of the pathogen (for example, fungus spores) can encounter more susceptible plants. Plants growing wild are usually (but not always) mixed with other species, and propagules of a pathogen, disseminated at random, have less chance to encounter susceptible plants. By growing pure stands of crop plants, disease problems are aggravated.

The pathogen itself affects the population of host plants. As an epidemic develops, plants or areas of tissue in plants become diseased and are no longer available for infection. Thus fewer of the propagules that are produced cause new infections, and the rate of increase of the pathogen and the disease slows down (Fig. 12).

Similarly, differences in host resistance influence reproduction of the pathogen. If a variety is immune, of course there is no disease, but such immunity is usually effective against only certain races of the pathogen. Varieties that are not immune may have different degrees of resistance that reduce penetration by the pathogen, increase its incubation time, and reduce the number of propagules produced. This slows down its rate of reproduction.

Environmental factors that affect individual organisms, such as temperature, light, moisture, and soil in which the plant is growing, naturally also affect the increase in the population of the pathogen and hence the rate at which the disease increases. Adverse environment tends to limit both the number of propagules that infect and the rate of development of the pathogen within the host, thus reducing the production of more propagules.

Fungicides applied to host plants reduce epidemics by killing some of the propagules of the pathogen before they can infect. Most fungicides are not completely effective, their essential effect is to reduce reproduction of the pathogen.

Epidemics, then, are characterized by the increase of successive generations of a pathogen in a population of host plants. Host and environmental factors affect the rate of increase, and anything that can be done to manipulate such factors may reduce the increase of the pathogen to relatively harmless proportions.

DISSEMINATION OF PLANT PATHOGENS

Insect and nematode plant pathogens can within limits move about and find plants on which to live; fungi, bacteria, viruses, and seed plants have no such means of locomotion and are dependent upon other agencies for dissemination. Most fungi produce spores that are adapted for transport by the wind, but bacteria and viruses are usually carried from place to place by insects, although bacteria are often spread by water. Occasionally other animals carry plant pathogens, and man, because of his ability to travel fast and far, often becomes the unwitting means of dissemination of destructive plant pathogens.

Dissemination, which means to disperse or spread abroad, is often closely associated with transmission, which refers to the inoculation of a new host plant, as well as to the transport of inoculum. This is often accomplished by insects. However, a pathogen such as *Xanthomonas phaseoli*, which causes common bean blight, usually penetrates the seed of the plant which it has parasitized. When such a seed is planted, the new plant becomes infected. Such a pathogen is said to be seed-transmitted. Therefore this bacterium can be disseminated long distances by transporting infected seed and can also be disseminated from plant to plant in the field by splashing rain or insects. [CARL J. EIDE]

Dissemination by wind. The plant spore is a form in which many pathogens of plants and animals spread between and within populations. Dispersal by air operates both locally and over great distances. Many viruses and bacteria that attack animals and man can be dispersed in exhaled droplets. Those that infect plants are mostly spread by vectors which range from fungi to nematodes, but the most common are insects, such as aphids or leafhoppers, which are also easily carried by wind. A few viruses can be spread within pollen and spores.

Local dissemination. Fungi are unique in the variety of physical processes they utilize to traverse the boundary layer of relatively still or smooth-flowing air adjacent to the substrates on which they grow. Powdery accumulations of detached spores, such as formed by many rust and smut fungi (or pollen), are removed by the shearing stress of moving air and, at least in the light winds usual among crops, in approximately logarithmically increasing numbers with linear increases in wind speed. Many ascomycetes and basidiomycetes have developed delicately adapted hydrostatic pressure mechanisms for releasing spores. Conidia were long assumed to be released without specialized discharge mechanisms, but some are now known to be released by the rupture of weak points in cell walls forced apart by internal pressures, by hygroscopic twisting, or (as D. S. Meredith showed) by drying mechanisms that produce great tensions in specially strengthened structures that are released when the cohesive force of the cell sap is abruptly overcome, with the formation of a gas phase within the cell. The kinetic energy of falling water drops is also used by some fungi with cup-shaped fructifications, but more often by the collision of raindrops with spore-bearing surfaces. Large transient increases of spore concentration can result from "rain tap and puff" the dual effects of the vibration of inpact and the fast radial displacement of air caused by the spreading of raindrops falling on dry surfaces. Once the surfaces are wetted, these processes are replaced by the dispersal of spores in splash droplets. Most splashed spores are in large droplets and are soon deposited, but those released in small droplets that quickly evaporate must become dry, airborne spores.

Weather and airflow factors. As so many fungi form spores on surfaces and need particular weather to liberate them, it is not surprising that the concentration of airborne spores changes dramatically with changing weather. Changes are abrupt, for example, when rain falls or cyclical in time with circadian rhythms. Spores of some pathogenic fungi are liberated only into dry air, others only into damp air, and so some are most abundant by day and others by night. Such properties have important effects on dispersal and establishment. Spores released during rain, although probably soon deposited, are more likely to find congenial conditions for germination than are spores liberated in hot dry weather; but if these can survive, they may be carried much greater distances.

Concentrations of airborne spores are determined both by the rate they are liberated and the rate they are diluted by diffusion when they reach turbulent air. The diffusion processes that meteorologists have described for aerosols show that velocities in eddies often greatly exceed the terminal velocity of spores (ranging from about 0.02 to 5.0 cm/sec for fungus spores and even up to 20 cm/sec for large pollen grains). Deposition by sedimentation is certainly important in sheltered places, but elsewhere it may be a less frequent method than impaction from moving air or washout by rain.

Range of dispersal. It is often necessary to measure the proportions of spores deposited at increasing distances from sources, propose safe isolation distances, explain the development of epidemics, and help forecast diseases. Typically, the intensity of deposition on the ground decreases fast within the first tens of meters from sources and even faster near sources close to the ground than around higher ones. P. H. Gregory drew attention to the paradox that, despite these steep deposition gradients, enough spores must escape local deposition to travel far and to establish disease hundreds or perhaps thousands of miles from their sources. Local deposition is much greater in the calmer air at night than on hot windy days with both thermal and frictional turbulence. As spores are carried higher, their distribution is increasingly determined by vertical temperature lapse rates and atmospheric pressure systems, although sedimentation and diffusion in small eddies continue. Aircraft sampling over the North Sea downwind of the British Isles has revealed the remnants of spore clouds produced over Britain on previous days and nights. Understanding these transport processes better helps to explain disease outbreaks and the dispersal tracks suggested by J. C. Zadoks.

Application of aerobiological information. Advances in aerobiology have given much information on the formation and transport of biological aerosols, but its relevance to disease attacks is often uncertain because too little attention has been paid to the viability of the spores being dispersed. This situation will not be remedied until there are better methods of growing the spores of obligate parasites and of fungi too featureless to be identified except in culture and better ways of relating contemporary spore concentration in the air to deposition on crops. [J. M. HIRST]

Dissemination by water. Nematodes and certain fungi and bacteria which inhabit the soil may be carried from place to place by surface water. This becomes important when a pathogen such as *Plasmodiophora brassicae*, which causes clubroot of cabbage, is introduced into a field. Spores from a small center of infection can be carried all over the field. Dissemination from field to field in drainage or irrigation ditches occurs, but such dissemination for long distances is much less frequent than by wind or insects.

Splashing rain, especially if accompanied by wind, is often responsible for local spread of pathogens, especially bacteria and those fungi which produce spores in sticky masses.

Dissemination by insects. Bacteria and fungi which produce sticky spores are often disseminated by insects that sometimes are attracted to them by sweet- or putrid-smelling substances. The spores or bacteria stick to the mouthparts or other parts of the insect body and are carried from plant to plant in an apparently incidental manner. Actually this is a more effective way of dissemination than by either wind or water, because through evolutionary adaptation many insects and pathogens have developed an affinity for the same kinds of plants. Consequently a greater proportion of insect-carried spores arrive on plants they can infect than if they were scattered by the wind.

In other instances the relationship between insect and fungus is much more highly specialized. Dutch elm disease is caused by *Ceratocystis ulmi*, the spores of which are introduced into the vascular system of the tree by bark beetles (*Scolytus multistriatus* or *Hylurgopinus rufipes*) when they feed on the small twigs of healthy elm. After feeding, the beetles lay their eggs on weak or dying trees. The larvae feed beneath the bark and then pupate; when the new adults emerge, they are contaminated with the spores of the fungus which are in the pupal chambers. If the beetles did not feed on the young twigs, the fungus probably would not cause such a destructive disease, although it might still be a relatively harmless bark fungus as are similar species found only in weak trees. This is a case of transmission being more important than dissemination, although the beetle is the agent of both.

Many plant viruses are completely dependent upon insects for transmission from one plant to another. In the process the virus is also disseminated. Although a few viruses are transmitted by chewing insects, by far the most important vectors are the sucking insects, including aphids, leafhoppers, thrips, whiteflies and mealy bugs.

The diversity of relationships between viruses, plants, and insects is very great. Some viruses are transmitted by only one species of insect, some by several, and a few by many. Some insects, for example, the peach aphid (*Myzus persicae*), transmit many viruses: some only one. In some instances the insect can transmit the virus immediately after feeding upon a virus-infected plant: in others hours or days elapse before it can do so. Some viruses persist in the insect only a few hours after it has fed on a diseased plant, others persist for the rest of the life of the insect, and in some the virus is transmitted from adult to offspring through the eggs for an indefinite number of generations. In such instances the virus increases in the insect as it does in the plant, and the insect can be considered an alternate host of the virus, not merely a carrier.

The epidemiology of insect-transmitted viruses depends upon the ease and frequency with which an insect transmits the virus, which varies a great deal, and upon the life habits of the insect. If it is active, like leafhoppers, the virus may be spread far and rapidly; if it is sedentary, like mealybugs, spread may be very slow.

Dissemination by other animals. Besides insects, mites and nematodes transmit and disseminate a number of plant viruses. Among higher animals few are important as agents of dissemination of plant pathogens, though occasionally they may be very important. *Endothia parasitica*, the fungus which causes chestnut canker, produces sticky spores which adhere to the feet of birds. These birds carry the spores over considerable distances; in fact, they made it impossible to eradicate the fungus after it had been accidentally introduced into the eastern United States from the Orient.

Dissemination by man. Man is an important disseminator of plant pathogens if they lack other means of transport. Some pathogens are not adapted for widespread dissemination by wind, water, insects, or other natural means. Others have natural means for widespread dissemination, but still may be unable to surmount natural barriers such as oceans, high mountain ranges, or deserts.

Plant pathogens, as other flora and fauna, were usually confined to certain areas of the Earth before man began to alter the terrestrial environment. Frequently in his commercial, agricultural, or recreational activities, man has carried pathogens to new localities where they have caused tremendous damage, even though they were relatively harmless in their native habitat. For example, the chestnut blight fungus (*Endothia parasitica*) is a relatively mild parasite on the chestnut (*Castanea mollissima*) in the Orient. When the fungus was brought to the United States about 1904 on imported chestnut trees, it spread to the native American chestnut (*Castanea dentata*), which is much more susceptible. The American chestnut forests were completely destroyed in about 40 years. Such lack of resistance in species of plants which have not evolved in the presence of the pathogen is very common, and it is the principal reason for the great danger involved in carrying pathogens to areas where they have not been before.

Dissemination by man may be hazardous even where natural barriers do not exist. Soil-borne pathogens such as *Plasmodiophora brassicae* and *Fusarium oxysporum* f. *conglutinans*, both of which attack cabbage, have been widely distributed on young cabbage seedlings which were grown in infested soil and sold for transplanting elsewhere. Neither fungus is disseminated by wind or insects, and water ordinarily carries them only to other parts of the field where they have already been introduced on infected transplants.

Plant pathogens disseminated by man are most frequently carried on infected plants. Some pathogens, such as certain cereal smuts (*Ustilago* sp.), bacterial blight (*Xanthomonas phaseoli*), and mosaic virus of bean, are transmitted and disseminated through seeds. However, this is less common and less hazardous than transmission in vegetative propagative parts such as potato seed tubers, flower bulbs, and nursery stock. Plant viruses in particular are more likely to be in the vegetative parts than in the seed.

Pathogens in either seed or planting stock have a good chance to become established in a new area because they have already infected the plant and need only to continue development when the plant starts to grow again. Similar plants and a favorable environment are likely to be present to permit further spread. Spores or bacteria disseminated independently are much less likely to survive or to find a host or favorable conditions for growth. The same may be said of pathogens in seeds, fruits, or vegetables to be used for food or in plant parts, such as lumber, to be used for manufacture.

Control of dissemination by humans is attempted by quarantine. [CARL J. EIDE]

PLANT DISEASES IN STORAGE

Tubers, fruits, and fresh vegetables are subject to spoilage by a variety of pathogenic and nonpathogenic agents during storage and transit, and often this hazard remains acute up to the time of consumption. Seeds such as those of wheat, corn, barley, soybeans, and flax, which often are stored in bulk for months or years, also are subject to deterioration. At times, the losses in transit and storage equal those that occur while the plants are growing. In general, storage diseases are divided into those caused by nonpathogenic factors and those caused by living organisms or pathogens.

Nonpathogenic storage diseases. Fruits and vegetables in storage suffer from a number of serious nonpathogenic or physiological diseases. Typically, these show up as discolored spots or areas on the surface of or within the affected parts, sometimes accompanied by collapse of the tissues, leaving pits on the surface or hollows within. These diseases are caused mainly by an excess of gases, such as certain esters or carbon dioxide, given off by the fruits or vegetables themselves or by chemicals introduced into the storage rooms. These diseases can be controlled by maintaining proper storage conditions, including temperature, humidity, and aeration. Fruits, vegetables, and seeds harbor abundant microflora, and damage beginning from nonpathogenic causes may be increased greatly by subsequent invasion of the tissues by bacteria and fungi able to cause rapid decay.

Pathogenic storage diseases. Common fungi, such as *Botrytis*, *Penicillium*, *Rhizopus*, and *Sclerotinia*, invade and rot many fruits and vegetables. Losses up to 25% of a shipment between harvest and consumption are common in fruits such as oranges, apples, peaches, pears, and plums and in vegetables such as potatoes, sweet potatoes, tomatoes, and peppers. Bacteria, or a combination of fungi and bacteria, often rot stored potatoes and root vegetables. These diseases may be controlled by harvesting only sound, disease-free products, careful handling to prevent bruising, the use of clean containers, maintenance of low (about 40°F) temperatures in transit and storage, and at times the use of fungicides.

Grains stored in bulk are subject to invasion by a number of fungi, principally those in the genus *Aspergillus*, which have the ability to grow at moisture contents in equilibrium with relative humidities above 70%. These reduce germinability of seeds, which is important in those to be used for malting or planting, and may reduce the quality of the grains or seeds for processing. Some storage fungi produce potent toxins; the resulting deterioration may not be detected until most of the damage has been done. Research is gradually making available to men in the grain trade the facts and principles that will enable them to reduce such losses. *See* PLANT DISEASE CONTROL.

[CLYDE M. CHRISTENSEN]

Bibliography: F. G. Bawden, *Plant Viruses and Virus Diseases*, 4th ed., 1964; C. H. Dickinson and A. J. Lucas, *Plant Pathology and Plant Pathogens*, 1977; W. J. Dowson, *Plant Diseases due to Bacteria*, 2d ed., 1957; W. R. Jenkins and D. P. Taylor, *Plant Nematology*, 1967; E. C. Stakman and J. G. Harrar, *Principles of Plant Pathology*, 1957; J. E. van der Plank, *Plant Diseases: Epidemics and Control*, 1963; J. C. Walker, *Plant Pathology*, 3d ed., 1969; R. K. S. Wood, *Physiological Plant Pathology*, 1976.

Plant disease control

Plant diseases may be controlled by a variety of methods which can be classified as cultural practices, chemicals, resistant plant varieties, eradication of pathogens, and quarantines. Often a combination of methods is most effective.

Cultural practices. Cultural practices help to control disease by eliminating or reducing the effectiveness of the pathogen or by altering the susceptibility of the host plant.

If the primary inoculum is seed-borne, its prevalence may be reduced by getting seed from an area where the disease is absent or is controlled by other means. For example, bean growers in the humid parts of the United States get seed from arid western states to avoid seed-borne *Xanthomonas phaseoli*, which causes bacterial blight. Potato growers buy certified seed which is almost free of viruses that are kept under control through isolation and rogueing, that is destroying infected plants.

Some pathogens, for example, the bacterium *X. phaseoli*, live in infected plant debris so long as the debris remains undecayed. Accordingly, bean growers not only buy western seed but also use a 2–3 year rotation plan, which allows the debris to decay, thereby eliminating the bacteria. Sometimes it is practical to remove or burn crop residue, thus reducing the amount of the pathogen that is returned to the soil.

Bacteria such as *X. phaseoli* that survive only so long as the crop residue is undecayed probably are killed by antagonistic microorganisms in the soil. Attempts have been made to reduce or eliminate some of the more persistent soil-borne plant pathogens by introducing antagonistic species into the soil or by increasing those already present. Most of these efforts have not been successful.

Certain plant pathogens, for example, viruses, infect biennial or perennial weeds which supply primary inoculum. Eradication of the weeds is often an effective means of controlling the disease in crop plants.

The elimination of primary inoculum usually is not complete by any of the above methods and is less effective as a means of control if the pathogen increases rapidly during the growing season from a few initial infections. Sometimes the rate of spread can be reduced. For example, the spread of viruses transmitted by insects can be delayed by controlling the insect. *See* INSECT CONTROL, BIOLOGICAL.

The inherent resistance of plants to disease is usually altered by environmental factors such as temperature, moisture, and available nutrients. Moisture sometimes can be controlled, as in irrigated areas, and soil nutrients can be regulated to some degree by fertilizer practice. For example, apple trees are more susceptible to fire blight, caused by *Erwinia amylovora*, if they are growing rapidly; growth can be reduced by withholding fertilizers, especially nitrogen. *See* PLANT, MINERALS ESSENTIAL TO.

Since some pathogens enter plants through wounds, avoiding injury often reduces infection. This applies particularly to fleshy vegetables, such as potato tubers, and to fruits which are subject to storage diseases caused by bacteria and fungi. Even seeds, for example, soybean, can be cracked if harvested when too dry and become infected by soil organisms when planted. [CARL J. EIDE]

Chemical control. Infectious plant diseases are caused by four major classes of agents: fungi (molds), nematodes (minute worms), bacteria, and viruses. These are arranged in the order of importance in chemical control, bacteria and viruses being the most difficult to control by chemicals. Hence the major tonnage of chemicals are fungicides and nematicides. *See* FUNGISTAT AND FUNGICIDE.

The last published list of antidisease chemicals showed 267 trade materials, many of which are duplicate trade names for the same chemicals.

The chemical control of plant disease differs sharply from the chemical control of human disease. Chemicals must be used to protect plants rather than to cure them, by killing the parasites while they are still in the environment and before they invade the plant.

External protection. Plant diseases attack all parts of the plant: seeds, roots, stems, leaves, and fruits. They attack plants in the field, packing shed, trains and trucks in transit, market, and home. This means that a variety of compounds and methods must be used to control parasitic forms.

Many fungi and a few nematodes attack seeds. If the organism is inside the seed, an organic mercury compound is sometimes effective. If the organism is merely on the surface of the seed, an organic sulfur compound, called a dithiocarbamate, may be used. Seeds are commonly treated by tumbling them with the dry compound in a drum or by mixing them in slurries and then drying.

Fig. 1. Chemical control of potato late blight, Toluca Valley, Mexico. (*Rockefeller Foundation*)

Nematodes and fungi which attack plant roots must be flushed out of the soil with fumigant gases that can seep into all the crevices and spaces in the soil. A common fumigant is a mixture of methylisothiocyanate and 1,3-dichloropropene. Soil fumigants are commonly inserted several inches deep in the soil through holes in the lower end of special thin "chisels" on a machine pulled with a tractor. Soil and seed may be treated together by spilling the chemical over the seed and adjacent soil as the seed is planted. Nematicides and fungicides for use in the soil are being developed rapidly.

The biggest tonnage of chemicals for plant disease control goes onto foliage and developing fruit for protection from fungi. Foliage fungicides are known as protectants; that is, they must be applied to the plant before an invading fungus arrives. Since the waiting period may be long, serious limitations are imposed on a compound. To work effectively, it must resist erosion by rain and dew, and it must withstand the heat of the summer sun.

Few chemicals have proved to be effective. The first was elemental sulfur, used since about 1803 to control powdery mildew diseases. Although it sublimes somewhat in the heat, it survives sufficiently for effective use. The next was Bordeaux mixture, a copper-containing substance that saved the French wine industry in the 1880s. It is both exceedingly resistant to rain erosion and does not sublime. It was over 60 years before another really good type of chemical was discovered, namely, the dithiocarbamates – organic sulfur compounds which are now used worldwide. Other fungicides were developed in rapid succession, including the imidazolines, dinitrocaprylphenyl crotonate, trichloromethylthioimides, and guanidines.

Fig. 2. Control of downy mildew of lima beans with a streptomycin sulfate compound, Agristrep. (*a*) Untreated, 100% infection. (*b*) Sprayed with 100 ppm of streptomycin, excellent control. (*W. J. Zaumeyer, USDA*)

Foliage and fruit fungicides are generally applied as sprays delivered by powerful machines that move through the crops. In areas where fields are large, farmers use airplanes as flying sprayers. Compounds are seldom effective if applied as dry dusts to foliage and fruit because the dusts do not stick well and consequently do not resist rain action (Fig. 1).

Fruits and vegetables are subject to disease and decay after picking. Much of this postharvest loss is not yet subject to chemical control, but oranges and other citrus fruits are widely treated in the packing shed with such compounds as diphenyl and sodium *o*-phenylphenate.

Chemotherapy. The newest frontier of the science of plant disease control is chemotherapy, the control of disease by internal treatment rather than by external protection (Fig. 2).

Although extensive researches have gone into chemotherapy of plant disease, the biological obstacles to success are great. Effective compounds have been discovered, but to move them inside the plant to the right place at the right time and to keep them there for a long enough time is extremely difficult. Sap in the sap stream passes a given point only once; therefore if the compound cannot cure the infection in one pass, it fails. An even more severe limitation is that plants have nothing even faintly similar to white blood cells. In animals the white cells clean up the few straggler germs left by the drug. Unless the chemical treatment kills every pathogen in a plant, however, the infection flares again and eventually kills the plant. *See* PESTICIDE. [JAMES G. HORSFALL]

Disease resistance. Plants are surrounded by a vast number of microorganisms that inhabit the air and soil. A few invade plants and injure them by disrupting their normal growth and development (disease). Such microorganisms are called pathogens. It is remarkable that plants are not injured by most microorganisms around them. Many microorganisms actually are essential for the growth and development of plants. The ability of plants to grow in an environment of microorganisms is evidence of how well they resist the harmful effects caused by pathogens (disease resistance). Most pathogens cause disease to only a certain plant, for example, pathogens of corn do not cause disease on wheat or geraniums. In addition, there are certain varieties of plants that are resistant and others that are susceptible, for example, a pathogen may attack a certain kind of wheat but not another.

What makes a plant resistant to disease? Each plant cell carries the genetic information that can be translated into chemical reactions and structures which determine whether a plant is resistant or susceptible. The mechanisms for disease resistance in plants are highly efficient since susceptibility is the rare exception in nature. Resistance may be due to the inability of the microorganism to penetrate the plant, inability to develop in the plant because the microorganism lacks nutrients or encounters compounds that inhibit its growth or reproduction, or the ability of the plant to resist microbial toxins.

Barriers to penetration. To cause disease, microorganisms must first establish physiological contact with the plant. Plants are covered by a coating of waxlike material called the cuticle, which serves as a nonspecific physical barrier. Plant storage organs also have outer protective tissues commonly referred to as peel. In addition to serving as a physical barrier, the outer coverings of many plants also contain chemical compounds that are toxic to many microorgansims. The peel of the Irish potato tuber contains chlorogenic acid, caffeic acid, α-solanine, and α-chaconine, whereas the outer scales of red onions contain protocatechuic acid and catechol. Natural openings (stomata and lenticels) exist in the barrier layer, and also this outer barrier is often broken by injury. Some pathogens invade plants through the natural openings and others only through wounds, but a few can grow through the barriers by mechanical pressure or by producing enzymes which dissolve parts of the protective barrier. The thickness of cuticle or peel is important in the resistance of certain plants to disease. In general, however, the contribution of cuticle or peel to resistance is thought to be small, except with microorganisms that invade plants only through wounds.

Nutrient factors. The absence of a factor necessary for microbial growth may be part of the disease-resistance mechanism of a plant. Certain bacteria can attack plants only if nutrients are available on the plant surface. Invasion of plum tree bark by *Rhodosticta quercina*, a fungus requiring myoinositol, is related to the myoinositol content of the bark. Susceptible varieties contain more than 10 times the amount of myoinostol found in resistant varieties. Except for microorganisms that can grow only in living plants, the presence or lack of nutrient factors is not considered a major mechanism for disease resistance. Most microorganisms grow on plant extracts or dead plant tissue from healthy susceptible or resistant plants.

Production of inhibitors. Once the microorganism has established physiological contact with the plant, it can influence the chemical reactions (metabolism) in the plant. Often this interaction of plant and microorganism results in the production of compounds (phytoalexins) around sites of penetration which inhibit the growth and reproduction of the microorganism. Phytoalexins include chlorogenic and caffeic acids, ipomeamarone, pisatin, phaseollin, 6-methoxy-mellein, orchinol, gossypol, oxidation products of phloretin, and an inhibitor not yet characterized from soybeans. Ian Cruickshank and coworkers established a relationship between the resistance of pea pods to fungi and the production of the phytoalexin, pisatin. Fungi unable to cause disease in pea pods were found to stimulate production of pisatin in amounts which markedly inhibited their growth, whereas pathogens of peas stimulated production of less pisatin. The amount of phytoalexin produced by the plant, therefore, is not the sole factor in disease resistance. The sensitivity of the microorganism to the amount of phytoalexin produced is also important. This suggests that at least two distinct biochemical mechanisms are involved in determining disease resistance—one controlling biosynthesis of phytoalexin and the other controlling the sensitivity of the microorganism to the inhibitor.

It is not surprising therefore to encounter situations where more of phytoalexin is produced by a plant after inoculation with a pathogen than with a nonpathogen. Some plants produce more than one phytoalexin after infection. Chlorogenic acid and 6-methoxy-mellein are produced by carrot and reach toxic levels around sites of microbial penetration within 24 hr after inoculation with microorganisms that do not cause disease on carrot. Microscopic examination of the inoculated tissue indicates that the inhibition of microbial growth in the carrot coincides with the production of 6-methoxy-mellein and chlorogenic acid at levels toxic to the microorganisms. It has also been demonstrated that the foliage of resistant but not susceptible varieties of soybeans produce a phytoalexin when inoculated with the pathogen *Phytophthora sojae*. Nonpathogens induce the production of the phytoalexin regardless of the susceptibility of the soybean variety of *P. sojae*. The appearance of phloretin in apple leaves following injury or inoculation with the pathogen *Venturia inaequalis* illustrates not the synthesis, but the liberation of a phenol from its nontoxic glycoside.

Phloridzin, the glucoside of phloretin, is found in leaves of apple varieties susceptible or resistant to attack by *V. inaequalis*. There is no correlation between the phloridzin content of leaves and resistance to the pathogen. When the pathogen penetrates the leaf of a highly resistant variety, the plant cells around the point of penetration immediately collapse, phloridzin is hydrolyzed by the enzyme β-glycosidase to yield phloretin and glucose, and phloretin is oxidized by phenol oxidases to yield highly fungitoxic compounds. In susceptible varieties the fungus penetrates the leaves and makes extensive growth beneath the cuticle for 10–14 days without causing collapse of plant cells. Thus phloridzin is not hydrolyzed and the pathogen is not inhibited. After 10–14 days the fungus sporulates on the leaves of susceptible varieties, and the affected tissue collapses. As with the resistance of soybeans to *P. sojae*, the potential for resistance appears to be present in all plants, and resistance may be determined by the ability of the microorganism to trigger synthesis or liberation of an inhibitor in the host. A similar series of reactions involving arbutin and hydroquinone may be important in determining the resistance of pear to the bacterium *Erwinia amylovora* that causes the disease fire blight.

Some compounds that accumulate around sites of infection or injury, for example, chlorogenic and caffeic acids, are widely distributed throughout the plant kingdom. The synthesis of others appears limited to a narrow host range, for example, 6-methoxy-mellein in carrot, phaseollin in the green bean, pisatin in the garden pea, and ipomeamarone in sweet potato. Apparently the plant has the potential for synthesis of the compound, and the microorganism used for inoculation determines the quantitative response of the plant. Where two or more compounds are produced in response to infection, the microorganism controls the relative concentration of each.

Resistance to toxins. The ability of plants to resist factors arising from plant-microbial interaction, which lead to tissue disintegration or impaired metabolic activity, may also be part of a disease-resistance mechanism. This resistance mechanism may include the presence of resistant structural components and metabolic pathways, mechanisms for detoxication, and the presence of alternate pathways for metabolism in the plant.

The resistance of immature apple fruit to many fungi that cause rots has been related to the resistance of pectic compounds in the cell walls of the green fruit to enzymes produced by the fungi. In mature fruit the microbial enzymes dissolve the cell walls, but the cell walls of immature fruit are not destroyed. During the growing season the amount of water-insoluble, resistant cell-wall material decreases as the apple matures. A major drop in the polyvalent cation content of cell-wall material occurs at about the time the fruit becomes susceptible. Conversely, the potassium content of the cell-wall material is higher in susceptibility than resistant fruit, with a major increase occurring with the onset of susceptibility. It appears that a pectin-protein polyvalent cation complex making up the cell-wall material of resistant fruits is responsible for the resistance of the fruits. The cell walls become susceptible as the polyvalent cations are lost.

Further evidence for the role of polyvalent cations is provided in the resistance of old bean seedlings to the fungus *Rhizoctonia*. Calcium ions accumulate in and immediately around developing lesions caused by the fungus. Barium, calcium, and magnesium ions inhibit tissue maceration by enzymes produced by the fungus, whereas potassium and sodium ions do not significantly influence the process. Bean tissue with lesions is more difficult to macerate with enzymes produced by the fungus than comparable healthy tissue. The cementing material (middle lamella) between plant cells around lesions is more difficult to dissolve than the middle lamella of cells more distant from lesions. The accumulation of polyvalent cations in advance of the fungus makes pectic substances of the tissue resistant to breakdown by the fungus.

In addition to producing enzymes which destroy plant cells, microorganisms have also been reported to produce toxic compounds that injure the plant without dissolving the plant tissue. The toxins that are produced are very specific for the plant in which the microorganism can cause disease. Varieties of the plant that are resistant to the microorganism are also resistant to the action of the toxin. The mechanism by which the plants are able to resist the toxin is not known; however, resistant plants may be able to change the toxin into nontoxic forms, or the toxin may be unable to penetrate into vital parts of the resistant cell. Microorganisms known to produce host-specific toxins include *Alternaria kikuchiana*, *Helminthosporium victoriae*, *Periconia circinata*, and *Helminthosporium carbonum*. [JOSEPH KUC]

Breeding and testing for resistance. The use of disease-resistant varieties that prevent or limit infection is one of the most widely used methods of disease control. Much of man's food and fiber supply depends upon the growth of disease-resistant crops.

In dealing with disease resistance, modern scientists must recognize not only the genetic system of the host plant but also the genetic system of the pathogen. Infectious diseases, in the final analysis, are the end result of gene-controlled chemical processes that are modified to some extent by the environment.

Use. The use of disease-resistant varieties with satisfactory yield and quality characteristics is the best and most economic means of disease control. It is the only feasible means of control for many virus and bacterial diseases, most soil-borne diseases, and many foliage diseases of extensively grown crops. Over 75% of the diseases of field crops and over 50% of the diseases of vegetable crops are controlled by means of host resistance.

The meat supply for human consumption depends on corn and sorghum hybrids resistant to blights, smuts, and other diseases, and the vegetable-processing industry depends on varieties of corn, peas, cucumber, beans, and tomatoes resistant to wilts, viruses, and other diseases. The list can be extended manyfold. Most of these crops are annual plants with which rapid breeding programs are possible. Resistance is less frequently used in crops with a long life-span or in high-value crops possessing special qualities or ornamental characteristics.

Disease tolerance. Tolerance is the ability of a variety to endure disease attack without suffering the same reduction in yield or quality as another variety. Although disease tolerance is a valuable attribute of a plant, tolerant varieties have the disadvantage of allowing pathogen reproduction to take place, thus exposing nearby nontolerant varieties to infection.

Disease escape. Some plant varieties, normally susceptible to disease, escape being inoculated and therefore are less damaged by disease than are other varieties grown in the same area at the same time. For example, early varieties of plants may mature before the pathogen reaches the area in which they are grown. Plants with an upright growth habit may be less frequently infected than plants with a prostrate type of growth. A plant may be resistant or unattractive to an insect that is a carrier for a virus and thereby may escape infection.

Genetics of resistance. The expression of resistance is an inherited character in the plant and can take several forms. The simplest mode of inheritance is a single gene. Gene action may be completely dominant, incompletely dominant, or recessive. When studied in detail, genes for resistance are frequently found to exist in a large number of separate functional forms or alleles. Resistance in a variety may also be due to two or more genes acting individually or through some form of gene interaction. When two or more genes must be present simultaneously in a variety for resistance to occur, gene action is spoken of as complementary. Modifier genes may individually show no effect on disease reaction, but together with another gene for resistance may either enhance or reduce the expression of resistance.

Disease reaction sometimes is under the control of a large number of gene loci. In these situations disease reaction in segregating populations usually grades continuously between, and sometimes beyond, the limits of the susceptible parent reactions. Gene action is largely additive, but sometimes dominant and epistatic effects are seen.

Stability of resistance. Resistance can fail to function because of certain environmental conditions or because of genetic changes in the pathogen which may enable it to overcome the resistance of the host.

Cabbage varieties with multiple-gene resistance

to yellows become susceptible at soil temperatures above 24°C, whereas varieties with single-gene resistance do not. Certain cereal varieties are resistant to rust in cool summers but may be susceptible when air temperatures are high. Resistance to root-invading fungi may break down when roots are injured by nematode feeding.

The most common failure of resistance, however, originates from genetic changes in the pathogen. Pathogens are living organisms with systems for the storage and release of genetic variation. Most rust, smut, and powdery mildew fungi occur in the form of races that differ from each other in ability to infect varieties of their host plants. New races that have the ability to attack a currently grown variety can appear in nature. Resistance to the new race must then be found and the breeding program repeated. Experiments indicate that this phenomenon is true for certain forms of resistance but not for others. Thus certain forms of generalized plant resistance give protection even against pathogens that are made up of many races. In other organisms, specialized races with reference to pathogenicity are not known. This is true for many organisms that cause root and culm rots, vascular wilts, and numerous foliage diseases. Resistance to these organisms seems to be lastingly effective.

Testing for disease resistance. In breeding for disease resistance or in selecting among established varieties, a reliable method for determining differences in disease reaction is necessary. Sometimes natural infection must be used; more commonly, however, carefully designed, artifical inoculation procedures are employed in field, greenhouse, or laboratory tests.

The inoculation method must be reliable so that escapes do not occur. Conditions for disease development should not be so severe, however, that different gradations of reaction to disease do not appear. It is desirable that a large number of plants be inoculated and that the method does not interfere with the breeding and selection scheme for the crop.

The inoculum should be a pure culture of the pathogen with an optimal degree of virulence. In many instances pathogenic races must be carefully selected so that plants with the desired type of resistance can be identified.

Environmental control, particularly of temperature and moisture, is needed during the inoculation and testing period. Comparable results from test to test can then be achieved, and different types of resistance can be distinguished.

Sources of resistance. Resistant germ plasm for use in breeding programs has been found in native and exotic varieties and even in wild species. When adapted native varieties are available, breeding objectives are more rapidly achieved. Resistant selections may need only to be identified and put into production. Exotic varieties have been valuable sources of resistance. For example, cottons from India and Africa have provided bacterial blight resistance; wheats from Australia and Kenya have furnished valuable genes for rust resistance; barleys from Ethiopia have provided yellow-dwarf-virus resistance; and sugarcanes from Java and India have provided the resistance to mosaic needed in American varieties of these crops. Noncultivated plant species frequently have more resistance to disease than cultivated species. Sterility, lack of chromosome pairing, and presence of undesirable characters are limitations in the use of wild plants in breeding programs. Nevertheless, wild relatives of tobacco, tomato, potato, sugarcane, wheat, oats, and other crops have contributed valuable genes for disease resistance to cultivated species.

Employment of resistance. Superior genes for disease resistance are usually exploited rapidly by plant breeders, and numerous varieties with similar resistances are put into production. With the recognition of different types of disease resistance, more careful attention is now given to the type of resistance used in the breeding program. Greater attention is directed to the less complete but more generalized types of resistance. Diversity of resistance and combinations of resistant types are also regarded as important. This has been necessary because of the repeated failure of varieties with only specific types of resistance to maintain this resistance in nature.

To maximize the effectiveness of genes for specific resistance, multiline varieties and hybrids are being developed. Multiline varieties are composed of a mechanical mixture of backcross-derived lines; each line contains different genes for resistance, but all are similar in maturity, appearance, and other respects. The components of the mixture can be varied from year to year, depending upon the pathogen races present. In addition, since some plants in the mixture are resistant each year, disease development on susceptible plants is delayed, enabling the plants to mature with light damage.

Breeding methods. Methods of breeding for disease resistance do not differ greatly from those employed for other characters. Selection, varietal hybridization and selection, and backcrossing are the methods most commonly employed. Simple selection procedures have resulted in the isolation of resistant varieties from heterogeneous crops. Where agricultural technology is more advanced and pure-line varieties of crops are grown, genetic variation must be achieved. Simple or complex crosses are made between sources of disease resistance and varieties possessing other desirable qualities. This is followed by careful selection and testing during the segregating generations so that superior disease-resistant varieties can be identified. As breeding programs have become more advanced, the only improvement needed in a variety may be additional disease resistance. In these situations the backcross breeding method is commonly employed. The method consists of repeatedly crossing each generation of resistant plants with the susceptible variety. After several generations, this is followed by selfing and selection. The breeding procedure allows for the recovery of nearly all the characteristics of the original variety but with resistance added.

Breeding for disease resistance has been a process of challenge and response. As each new disease has threatened the destruction of a crop, resistant varieties have been developed and put into production. Although breeding for resistance has been highly effective, it is expected that further gains in the efficiency and effectiveness of the method will be made. To achieve this, modern research is aimed at the identification of superior

Fig. 3. Hormone-type chemical sprays used for killing rust-susceptible barberry plants. This process is much faster and less costly than common salt. (*USDA*)

forms of disease resistance and a greater understanding of the genetical and biochemical nature of resistance. *See* BREEDING (PLANT).

[ARTHUR L. HOOKER]

Eradication campaigns. These are designed either to eliminate recently introduced pathogens completely or to protect economic plants by destroying alternate, or weed, hosts. Success in eliminating pathogens depends on early detection of the pathogen and on the efficiency of eradication measures.

Attempts made in the United States to eradicate chestnut blight and the Dutch elm disease were unsuccessful. The citrus canker disease, however, was eliminated from Florida by burning infected trees. Flag smut of wheat, which was introduced locally into Mexico, was also successfully burned out. Similarly, persistent eradication of infected plants has helped restrict many diseases.

Certain rusts can be controlled wholly or partly by eradicating alternate hosts. For example, the destruction of red cedars near apple orchards protects apples against the *Gymnosporangium* rust, because this rust cannot maintain itself on either host alone. To help control stem rust of wheat and other small grains, the growing of barberries, *Berberis* sp., has been prohibited by law in some countries. Denmark began a successful campaign

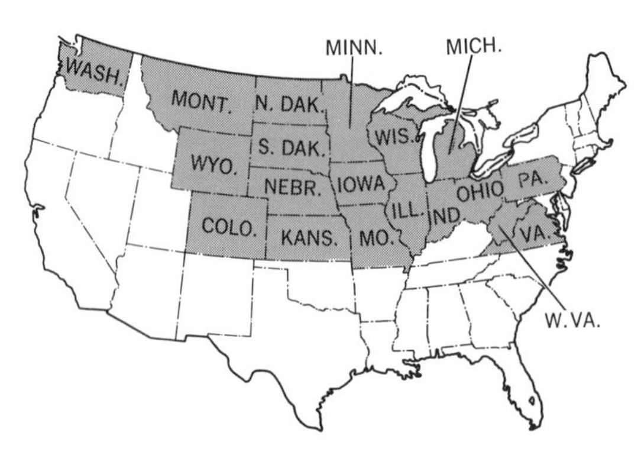

Fig. 4. More than 513,500,000 rust-susceptible barberry bushes were destroyed on 153,000 rural and urban properties in 19 barberry-eradication states. (*USDA*)

against barberries in 1904, and in the United States about 500,000,000 barberries have been destroyed since 1918, with substantial reduction of the stem-rust menace (Figs. 3 and 4). Likewise in the United States white pines and other susceptible species are partly protected from blister rust by eradicating nearby currants and gooseberries, *Ribes* sp.

Like legal public health measures for human beings and for domestic animals, those for plants are essential in keeping many diseases in check.

[E. C. STAKMAN]

Quarantines. Plant disease quarantines are legal measures taken by Federal or state governments to prevent the introduction of foreign plant diseases or pests into an area. Quarantines are based on the philosophy that government has the right and obligation to protect its agricultural resources and industry from the destructive effects of exotic plant diesases and pests.

The transportation of plants and plant parts was long a matter of private concern, with the result that many plant pathogens became widely distributed by international travelers or through unrestricted trade channels. The dangers of this situation became dramatically apparent following the accidental importation of the chestnut-blight fungus into the United States from Asia between 1900 and 1905 and the ultimate destruction of the American chestnut forests.

As a consequence of this and other bitter lessons, the United States government in 1912 passed the national Plant Quarantine Act. Today essentially all nations have enacted protective quarantine regulations. Quarantine laws authorize Federal or state officials to intercept and inspect shipments of plant materials and to release, fumigate, or confiscate the shipment in accordance with legal provisions. Quarantine inspectors are stationed at ports of entry, border stations, and at receiving and distributing points for freight and mail.

Value of quarantines. The value of quarantines has long been disputed. Antagonists claim that man is unable to prevent the movement of microscopic pathogens, that many quarantines are scientifically unsound, and that on occasions quarantines have been used as economic sanctions in restraint of free trade and have caused unnecessary economic losses. Supporters insist that even though not 100% effective, quarantines do prevent the introduction of many pests and diseases and retard the movement of others, giving scientists time to combat them before they become well established; that quarantines annually save the agricultural industry millions of dollars; and that these economic gains are many times greater than any possible business losses resulting from the application of quarantine measures.

Improvements in the practice of quarantining may provide assurance that all quarantines will be established on sound biological bases for maximum effectiveness, that they can be lifted with equal facility when it becomes clear that they are no longer necessary, and that, insofar as possible, new quarantine laws would be preceded by international consultation in an attempt to obtain mutual agreement and to ensure minimum disruption of the international exchange of commodities.

International disease protection. Joint efforts

are made by nations to protect their agricultural resources and industries without impairment of exchange of commodities. Ideally, available knowledge on plant pests and pathogens is utilized to devise methods to limit their geographic spread and to prevent the outbreak of epidemics. Changes in cropping patterns, trade agreements, and the distribution of pests and pathogens necessitate a continuing program consisting of (1) annual plant disease surveys by the several nations with free exchange of results, (2) rigorous practice of local sanitation and plant protection, (3) prompt distribution of resistant varieties of crop plants, (4) exchange of information in improved control measures, and (5) international consultation with respect to the establishment and enforcement of quarantines.

International plant protection can be successful only when regulatory activities are fortified by scientists investigating the etiology of plant diseases, life cycles of pathogens, host-parasite relationships, and chemical and other control measures. Exchange visits by scientific personnel further strengthen understanding and lead to logical and amicable agreements. Among international organizations active in plant protection are the Food and Agriculture Organization of the United Nations, and the International Commission on Plant Disease Losses.

[J. GEORGE HARRAR]

Bibliography: E. Evans, *Plant Diseases and their Chemical Control*, 1968; R. N. Goodman, Z. Kiraly, and M. Zaitlin, *The Biochemistry and Physiology of Infectious Plant Disease*, 1967; J. B. Harborne (ed.), *Biochemistry of Phenolic Compounds*, 1964; A. L. Hooker, The genetics and expression of resistance in plants to rusts of the genus *Puccinia*, *Annu. Rev. Phytopathol.*, 5:163–182, 1967; T. Johnson et al., The world situation of the cereal rusts, *Annu. Rev. Phytopathol.*, 5:183–200, 1967; J. Kuć, Resistance of plants to infectious agents, *Annu. Rev. Microbiol.*, 20:337–370, 1966; E. E. Leppik, Relation of centers of origin of cultivated plants to sources of disease resistance, *USDA Agr. Res. Serv. Plant Introd., Invest. Pap.*, no. 13, 1968; C. J. Mirocha and I. Uritani (eds.), *The Dynamic Role of Molecular Constituents in Plant-Parasite Interaction*, 1967; K. S. Quisenberry and L. P. Reitz (eds.), *Wheat and Wheat Improvement*, 1967; E. C. Stakman and J. G. Harrar, *Principles of Plant Pathology*, 1957; J. C. Walker, *Plant Pathology*, 3d ed., 1969; W. Williams, *Genetical Principles and Plant Breeding*, 1964; R. K. S. Wood, *Physiological Plant Pathology*, 1967.

Plant formations, climax

All the Earth's surface is occupied to some extent by vegetation except for polar and alpine regions of permanent snow and ice and local physiographically active areas such as sand dune complexes (Sahara desert, in part). Despite the hundreds of thousands of species of plants and the thousands of communities (associations) in which they occur, there is a limited number of major world formation types. Each formation occurs over wide areas, most of them on all continents, and each has a recognizable structure because of the characteristic life forms (tree form, grass form) of the dominant plants. Yet the floristic composition of a formation may be so different in its different areas of occurrence that no species is common to the major areas. Each formation occupies a distinctive type of general climate, although variable local climates occur within its region.

Plant geographers differ in their classifications of formations, particularly the subdivisions of the world formations as they are intensively analyzed in local regions. However, the several formations fall into three major groups of terrestrial vegetation: forests and woodlands, grasslands, and deserts. These formations are listed in the classification shown. *See* PLANT GEOGRAPHY.

Forest formations (including woodlands)
- Tropical rainforest (including subtropical types)
- Temperate rainforest (including mountain types)
- Tropical deciduous (monsoon) forest
- Summer-green deciduous forest
- Needle-leaved forest (circumboreal type especially)
- Evergreen hardwood (Mediterranean) forest
- Savanna (transitional to grassland)
- Thorn forest (transitional to desert)

Grassland formations
- Tall (prairie) grassland
- Short (steppe) grassland

Desert formations
- Dry desert (including semidesert)
- Cold desert (tundra, including alpine tundra)

Some minor types of stable vegetation are omitted from this general classification, such as the heather-dominated moors of oceanic northwest Europe, which are characterized by *Erica* and *Calluna*, and the mangrove thickets with species of *Avicennia* and *Rhizophora* of tropical ocean shores. Likewise the vegetation of inland waters and of the sea is not considered, and all animals are arbitrarily ignored, although characteristic ones occur with each formation. The major formations have typical climax, dynamically stable communities. However, within each formation area most of the land usually is occupied by vegetation of disturbed terrain, by communities of plants under edaphic control (conditions of the substratum, terrain, or associated microclimate that are extreme for the region), or by successional communities that have not had time to develop to climax status. The formation, when thought of in a regional sense, includes all of the great variety of such communities that are subordinate to the typical climax. Climax plant formations not only occupy vast areas, but they have been in existence during millions of years and have migrated as their controlling climates have shifted on the face of the Earth. *See* VEGETATION ZONES, WORLD.

Tropical rainforest. This formation occurs extensively in broad lowland areas in the Amazon basin of South America, in the Congo of West Africa, in Malaya, and in Indonesia, with lesser areas elsewhere that never have frost or a period of inadequate precipitation. The tropical rainforest (Fig. 1) is very rich in woody species (100 or more tree species per hectare) and is usually without dominance by one or a few species, which is in contrast with the monsoon forest and the temperate rainforest, in which dominance may occur. It is structured in several layers, the uppermost being a

Fig. 1. Tropical rainforest of upland type in the lower Amazon region, Belém, Pará, Brazil. Tall trees with crowns and plank-shaped buttresses are typical. Such forests may have 100 tree species on a hectare.

discontinuous layer of occasional emergent trees that extend far above the general canopy of the forest. Internally there usually are one or two layers of lower trees and shrubs. The ground cover often consists of reproduction of woody plants, although openings in the generally dark forest permit a luxuriant growth of ferns and other herbs. The tropical rainforest is famous for its climbing vines, such as *Bauhinia* and *Carludovica*, which vary from thin cords to massive stemmed lianas for the abundance of vascular epiphytes (many genera of Bromeliaceae, Orchidaceae, and Pteridophyta); and for cauliflory, or the bearing of flowers and fruits on old stems and even on the trunks of trees, as in *Theobroma* and *Diospyros*. Trees are commonly tall, with emergent ones very tall (250 ft or more in *Ceiba* and *Bertholletia*), and usually have columnar trunks which are unbranched up to the relatively small crowns. Jungle is a term more appropriate for the forest edge, such as areas along streams where the light has full effect, or where second growth has occurred. The interior of old, high, upland rainforest is comparatively open and easy to move through. Many species of trees have planklike buttresses rising along the trunk from the main roots, as seen in *Mora*, *Ceiba*, and *Terminalia*; and others, such as *Tovomita*, *Clusia*, *Symphonia*, and *Ficus*, stand on stilt roots. Many lianas and some trees (*Ficus*) start as epiphytes and later make contact with the ground. Tree leaves are medium to large, although many species, such as the Leguminosae, have compound leaves with medium to small leaflets. Leaves are typically evergreen, with entire margins and attenuated tips called drip points, as in *Catostemma*, and are somewhat coriaceous in upper layers and thinner in lower ones. Although the forest is luxurious in appearance, tree growth is slow because of the comparatively infertile, strongly leached soil and the intense competition for light and nutrients. In contrast, after clearings are abandoned, secondary growth of trees such as *Cecropia* is rapid, and many species have large leaves (Musaceae, Zingiberaceae, Marantaceae). The tropical rainforest is the home of aboriginal shifting cultivation, as cut and burned patches of the forest soon lose their fertility. It produces many valuable and beautiful hardwoods, such as ebony, mahogany, and rosewood; it is the region of industrial rubber and cacao plantations.

Temperate rainforest. The trees in this formation are medium or tall in height and have medium or small, hard, evergreen leaves, although some associated trees have scalelike or needle-shaped leaves. Tree ferns and bamboos are abundant in certain areas of the temperate rainforest, and moss and liverwort growth may cover the tree branches and the ground. The formation occurs in many moist warm regions, but does not occupy extensive areas because it is impinged upon from all sides and seems to be in delicate balance. With increasing heat and moisture, it grades into subtropical and tropical rainforest; with increased dryness or periodicity of precipitation, it grades into tropical deciduous forest; and with coolness, into summergreen and needle-leaved forest types. It occurs on all continents and is variously composed floristically, but has some widely occurring "binding" genera, such as *Podocarpus*, *Weinmannia*, *Drimys*, *Araucaria*, *Persea*, *Dacrydium*, *Laurelia*, *Cedrela*, and *Nothofagus* in the Southern Hemisphere. Some examples of the temperate rainforest are the Kauri pine (*Agathis*) and southern beech (*Nothofagus*) forests of New Zealand, the *Nothofagus* forest in Chile, the sweetgum (*Liquidambar*) forests of Mexico, the coastal redwood (*Sequoia*) forests of California, the Parana pine (*Araucaria*) forest of Brazil, and the laurel (*Laurus*) forest of the Canary Islands. Other types occur on tropical and subtropical mountains where the climate is temperate, and are called cloud forests when there is a high frequency of fogginess. *See* RAINFOREST.

Tropical deciduous (monsoon) forest. This is an intermediate type characteristically making transition to rainforest and savanna woodland. It is usually referred to as the monsoon forest because its leaves are borne during the rainy season and many of the trees are deciduous during the dry season. The leaves are small to medium and some evergreen elements are present. It usually is a tall stratified forest with many life forms present. Lianas and vascular epiphytes occur but are not as abundant as in the tropical rainforest, which lacks a dry season. With a more pronounced and usually drier season, this formation grades into savanna woods which may be closed or in more extreme forms may be open and grassy. Other transition may be with thorn woodland in Africa and South America. Typical monsoon forests occur in southern Asia, where teak (*Tectona grandis*) is important, and there are related forests in Africa and South America. When monsoon forests occur with a less pronounced dry period and with more evergreen elements, the semideciduous forest types are sometimes lumped with rainforest.

Summergreen deciduous forest. This formation occurs in cool to cold temperate regions of the Northern Hemisphere where there is a long winter and a summer of several months without a pronounced dry period (Fig. 2*a*). The rhythm of the forest is set by the contrast between the leafless, profusely branched trees ending in a high eventopped crown of fine twigs and the heavy summer canopy formed by broad, membranaceous, mostly simple leaves. This formation is comparatively rich in tree, shrub, and herbaceous species, and is typically structured in several layers by naturally tall and short trees, shrubs, and herbs (Fig. 2*b*). Further rhythm is produced by a pronounced spring

Fig. 2. Summergreen deciduous forest. (*a*) Dominated by broad-leaved hardwoods, including *Betula, Aesculus, Fagus, Acer, Quercus, Carya,* and *Ilex*. Southern state-line region, Great Smoky Mountains National Park, Tennessee–North Carolina. (*b*) Interior view with *Quercus alba*, numerous herbs, and flowering shrub *Azalea nudiflora*; Blue Ridge Mountains in Virginia.

flora that comes quickly into bloom from subterranean bulbs, rhizomes, and corms. Many of the plants of this spring flora are species of *Anemone, Erythronium, Viola, Geranium, Dicentra,* and *Claytonia* which complete their life cycles before the full shade of the forest canopy has developed in early summer. Taller summer and autumnal layers of herbs replace the spring flora and greatly change the interior appearance of the forest except where the shade is most dense. Climbing vines, species of *Vitis* and *Psedera*, are infrequent and epiphytes are usually mosses, liverworts, and lichens. This forest formation is well represented in the eastern United States, in Europe and contiguous Asia, except for high latitudes, in the Mediterranean region, and in southeastern Asia. Some minor extensions into the warm temperate and subtropical zones occur at higher altitudes in Mexico. It makes transition with grassland toward continental interiors, with conifer forests northward and at higher altitudes, and with various types of vegetation southward in warm temperate regions. It is absent from the Southern Hemisphere in regions with a pronounced dry season, and is poorly represented in western North America. In the United States the northern hardwood forest is mostly characterized by maple (*Acer*), beech (*Fagus*), birch (*Betula*), and basswood (*Tilia*), with lesser amounts of other genera such as ash (*Fraxinus*), oak (*Quercus*), and hickory (*Carya*). Toward the drier continental interior the forest is dominated by oaks and hickories, finally forming a transition with the tall-grass prairies. The southern aspect of the forest has yellow poplar (*Liriodendron*), gum (*Nyssa*), walnut (*Juglans*), holly (*Ilex*), buckeye (*Aesculus*), sweetgum (*Liquidambar*), and magnolia (*Magnolia*) as well as species of genera of the other regions of the formation. Chestnut (*Castanea*) was a prominent member of the eastern phase of the formation before it was eliminated by a blight. This forest produces many fine and commercially important hardwoods.

Needle-leaved forest. This is best represented by the circumboreal forest dominated by spruce (*Picea*) and fir (*Abies*) (Fig. 3*a*). It occurs south of the tundra and north of the summer-green deciduous forest or grassland (Fig. 3*b*). The taiga is the northern part of the formation and lacks a closed tree canopy. The tundra may be interpenetrated on higher and drier land. The southern part is penetrated by broad-leaved deciduous trees in the valleys and on warmer slopes or by grassland openings. Upper elevations of mountain masses to the south of the Canadian spruce-fire formation in the Northern Hemisphere have belts of needle-leaved forest with spruce and fir, or a variety of pines (*Pinus*) and other conifer genera, such as *Juniperus, Cupressus, Cedrus, Linocedrus, Larix, Tsuga,* and *Thuja*. Some representatives of this formation (*Pinus*) occur on mountains in the tropical zone and in the Southern Hemisphere. Some conifers belong to formations not dominated by needle-

Fig. 3. Needle-leaved (conifer) forest. (*a*) Dominated by *Picea glauca, P. mariana, P. rubens,* and *Abies balsamea*, as seen in Laurentides National Park, Quebec. (*b*) Dominated by *Picea engelmani* and *Abies lasiocarpa*, as seen in central Colorado, in the Silverton region. Alpine tundra region above timberline.

leaved species, such as the hemlock (*Tsuga canadensis*), or are dominant in successional stages earlier than the climax, such as the junipers (*Juniperus*) and pines (*Pinus*). The name of the forest comes from the needlelike or scalelike, hard, evergreen leaves. Mature trees, according to latitude or altitude and the nature of the species, may be comparatively low or tall, such as the Sitka spruce (*Picea sitchensis*) and redwood (*Sequoia sempervirens*). Typically the canopy is closed and the forest is dark, although the central trunk is usually unforked. Structure of the forest is simple because there are few layers in the vegetation and the dominant tree species are few in number, but the ground cover, although scanty in dry types, often is a dense mat of mosses and lichens with associated ferns, broad-leaved herbs, and low, often evergreen shrubs (*Vaccinium*, *Ledum*, *Kalmia*). The formation is typically moist, but conifer forests range into semiarid regions, which have species of *Pinus* and *Juniperus*, and into swamps, where *Taxodium* is common. In the tropical formation the climate is moist, without a dry season, and cool or cold. Winters are long and snow cover may be deep. Soils are acid and podzolized, comparatively shallow, and not fertile. The boreal forest occupies vast areas in Asia, Europe, and America (principally in Canada), which are also the zones of abundant acid bogs and peat formation. Other types of needle-leaved forests occur in different climates and on different soils, such as the juniper-piñion woodlands of the southern Rocky Mountains. This forest formation produces the world's most important commercial softwood lumber for building and pulp for paper manufacture.

Evergreen hardwood forest. Frequently, this is called the Mediterranean forest after the area where it was first studied, although it also occurs in the southwestern United States, Chile, southwestern Africa, and southwestern Australia. It occurs in regions where there is a Mediterranean-type climate with cool to cold moist winters and long dry summers. The result is that conditions are favorable for active growth in the spring, with many herbs rising from bulbs, and again in the autumn when the dry summer is over. In the Mediterranean region the forest dominants are mostly oaks (*Quercus ilex* and *W. suber*), but centuries of abuse have reduced most of the region to scrubby vegetation called maquis or garigue. In California the formation is represented by the oak-madrone type (*Quercus-Arbutus*), with fire and other disturbances producing a variety of types of chaparral (*Arctostaphylos, Rhamnus*); in Chile by Rosaceae (*Quillaja*) and Lauraceae (*Persea, Cryptocarya*); in Australia by gums (*Eucalyptus*) and many associated species; and in the Cape region of South Africa by genera of Ericaceae, Bruniaceae, Rutaceae, Rhamnaceae, and Thymelaceae. This forest formation is highly subject to fire and to degradation by grazing animals, such as sheep and goats, and by erosion, so that it is seldom seen in mature form. In the Mediterranean region it is the home of olive cultivation.

Savanna. This is a transitional formation between forest and grassland in which trees may occur generally but are widely spaced in islands of woodland or more or less as gallery woodland along drainage channels. It can occur in temperate regions as a rather narrow ecotone, with patches of grassland within the forest region and islands of trees within the grassland (Fig. 4*a*), as in Minnesota to Illinois. The formation may be extensive in tropical and subtropical regions, as in the big-game country of eastern Africa (*Acacia* savanna) and in the central plateau of Brazil, the Campo Cerrado. The savanna of Brazil (Fig. 4*b*) sharply abuts the rainforest in many places and varies from a closed or nearly closed woodland of short, contorted, thick-barked trees through more open structure with increasing density of grasses (Andropogoneae) and broadleaved herbs until there are no more trees present. The soil is generally infertile and lateritic, and sometimes there is a massive subterranean layer of stonelike laterite close to the surface. Because of the strong seasonal distribution of rain, the water table is fluctuating and sometimes very deep, and vegetational activity is limited to the rainy season. Common use of the savanna is for extensive animal husbandry, because the forage value and carrying capacity of the vegetation are low, and for subsistence agriculture along drainage channels. Fire is frequent, and some students of vegetation believe that all savanna has been produced from woodland by centuries or even millenniums of human interference, especially set fires. *See* SAVANNA.

Thorn forest. This formation is dominated by small trees and shrubs, many of which are armed with thorns and spines. Leaves are absent, succulent, or deciduous during the long dry periods, which may also be cool. Succulent plants of many

Fig. 4. Savanna. (*a*) *Juniperus-Pinus* savanna, as found in southwestern Colorado, with open-spaced trees and xerophytic grasses. (*b*) Tropical savanna near Pirapora, Minas Gerais, Brazil. The low, contorted, largely evergreen, broad-leaved trees and a good cover of semixerophytic grasses are typical.

forms usually accompany the thorny bushes and trees (Cactaceae in North America, Euphorbiaceae in Africa). There is an ephemeral layer of herbs that appears after rains, and sometimes a thin scattering of dry grasses. The type is mostly tropical and subtropical and is represented by the caatinga of northeastern Brazil and the South African thornbush. This type is intermediate between desert and steppe.

Tall grassland. In the United States this is called the prairie. It is a continental formation lying next to broad-leaved summer-green forest (oak-hickory type) on the more humid side and short-grass plains (steppe) on the drier side. Rainfall (20–40 in.) is strongly seasonal with a dry summer and a cool or cold winter, often with snow cover. True dominants of the prairie include such grasses as *Andropogon scoparius*, *Sporobolus asper* and *S. heterolepis*, *Stipa spartea* and species of *Koeleria*, *Agropyron*, *Muhlenbergia*, and *Panicum*, and the sedge genus *Carex*. Numerous associated broad-leaved herbs include many species of the families Leguminosae, Rosaceae, Scrophulariaceae, and Umbelliferae. Tall trees are sometimes present along the streams and some shrubs occur within the grassland, such as species of *Salix* and *Symphoriocarpus*. Most vegetative activity stops after the effects of the spring snowmelt and early summer rains are over. This is now a rich agricultural region concentrating on corn, small grains, and livestock. The Hungarian puszta and moist grasslands of Eurasia belong to this subformation.

Short grassland. This is generally called the steppe. In the United States it is the short-grass plains (Fig. 5). It makes contact with the tall-grass prairies on the moist side and with savanna, thorn-forest, and desert on the dry side. It is a continental formation with less precipitation (10–20 in.) than the prairie. The rainfall is less reliable and annual droughts are a hazard to agriculture. Summers may be long and dry, and winters long and cold at higher latitudes and cool to warm at lower ones. Winds are strong and wind erosion occurs where the vegetation is disturbed. The formation is typically dominated by short to medium-high grasses. Short-grass dominants in the United States include *Bouteloua hirsuta* and *B. gracilis*, *Buchloë dactyloides*, and species of *Muhlenbergia* and *Carex*. Mid-grasses include *Agropyron smithi*, *Hilaria jamesi*, *Sporobolus cryptandrus*, and species of *Stipa*, *Oryzopsis*, *Festuca*, and *Elymus*. Among the broad-leaved herbs, Compositae are especially abundant (*Aster*, *Solidago*, *Grindelia*, *Senecio*, *Artemisia*), as are Leguminosae (*Oxytropis*, *Psoralea*, *Petalostemon*, *Lupinus*). This formation was the main home of the American bison and now is largely devoted to the cattle industry and hazardous wheat raising in the less arid parts. It is the home of the man-induced dust bowls. In temperate regions the steppe plays a role corresponding to that of savanna in tropical regions. In addition to the North American grasslands, extensive areas occur in Argentina and Uruguay, in Africa on both sides of the Sahara desert, and in Asia mostly between latitudes 30 and 50°N from the Middle East to Mongolia and Manchuria. *See* GRASSLAND ECOSYSTEM.

Dry desert. The dry, or true, desert is both a climatic condition and the vegetation of dry regions. Precipitation is low, from essentially nothing to a variable upper limit of about 10 in., and is erratic. Atmospheric humidity is low in the day, but dew is common at night. Temperature varies from warm to hot for all months or may be cold in the winter at high latitudes and altitudes. There is a strong daily change in temperature from hot bright days to clear cool nights. In the southwestern United States the Chihuahua Desert extends from southern New Mexico and western Texas southward into Mexico, occupying the land between mountain ranges. Organ-pipe and columnar tree cacti, treelike yuccas, and species of *Agave*, *Dasylirion*, and *Hechtia* form a conspicuous part of the vegetation, along with ocotillo (*Fouquieria*), mesquite (*Prosopis*), and creosote bush (*Larrea*). The low rainfall occurs largely in the summer. The Sonoran Desert occupies the lowlands around the Gulf of California in Sonora and Baja California and the adjacent United States. It is characterized by the giant saguaro cactus (*Carnegiea gigantea*), many species of prickly-pear cactus (*Opuntia*), and the shrubby *Larrea* and *Franseria* (Fig. 6). Ephemeral annuals are abundant after late winter rains. Temperatures in the Mojave Desert of California may exceed 100°F for long periods of time, and precipitation around the Gulf is about 2 in. annually. Northward of these true severe desert areas, in the Great Basin region, the semideserts are less dry, but colder, and the northern part has severe winters. Species of sagebrush, especially *Artemisia tridentata* in the north and *A. spinescens* in the south, commonly are dominant, with shadscale (*Atriplex*) as an associate. Many parts of this

Fig. 5. Protected grassland in the Black Hills area east of Colorado Springs, Colo., showing tall grasses and mid-grasses in a generally short-grass (steppe) region.

Fig. 6. Sonoran Desert, in Arizona. Saguaro (*Carnegiea gigantea*) is mainly on south-facing slopes.

desert are salt encrusted, with halophytes such as *Kochia*, *Sarcobatus*, *Allenrolfea*, and *Salicornia* being important. Desert vegetation is typically open, with shrubs, cacti, and bushy herbs widely spaced, but often there is root competition for the scanty water supplies, and some species are deep-rooted. The long, severe dry periods are met variously by desert and semidesert plants. Some are annual, avoiding drought; others lose their leaves or are leafless all the time; and many store water in succulent organs. In addition to the American deserts described above, there is the coastal Atacama Desert in Pacific South America, the Kalahari Desert of southwestern Africa, the great Sahara desert of northern Africa, which crosses the continent between 15 and 30°N latitude, the great Arabian deserts of the Middle East, and the deserts of Asia which extend from Kara-Kum eastward to the Gobi. Much of central Australia is occupied by the Simpson, Victoria, Gibson, and other deserts. Desert mountain ranges are covered at appropriate elevations by other plant formations, often widely disjunct, culminating even in boreal-type needle-leaved forest and alpine tundra. Oases, where the water table is high or aquifers can be tapped by wells, may support luxuriant gardens of melons, vegetables, and date palms. With modern irrigation, desert soils often prove to be exceedingly fertile. Careful management is necessary to prevent the soil from becoming useless because of salt accumulation. Deserts sustain some browsing and semideserts can be useful country for cattle, horses, and camels, but overgrazing is easy and this can result in the spread of weedy shrubs, such as the spread of sagebrush in the Great Basin Desert and of mesquite around the Chihuahua Desert. *See* DESERT ECOSYSTEM.

Tundra. This formation comprises the vegetation of the treeless zone of high latitudes where winters are long and cold, with temperatures falling to −60°F or lower. Summers are short, with no month averaging above 50°F. Frost may occur in midsummer, although daytime maxima may reach 80°F. Precipitation usually is low, snow cover thin, and winds strong, although fogs may be frequent and cloudiness high. There are, according to the latitude, long winter nights when the Sun rises a few hours low on the horizon or not at all, and summer days when it never sets. The ground is perennially frozen to great depths (permafrost) below a variable surface layer subject to summer thaw. Although mountains, hills, and stream valleys occur, much of the Arctic tundra landscape is of low relief. However, the microrelief is extremely important in controlling the patterns of occurrence of the plants and the communities which they form in the tundra formation. Because all conditions of life are marginal, small changes in relief, gradient, and substratum produce striking changes in vegetational pattern. Part of this is the result of abrupt changes of dominant species with slight changes in temperature and moisture. The controlling forces in geomorphological processes, and consequently in vegetational development and patterning, are largely cryopedologic or caused by freezing-thawing disruptions that result in polygons, ice wedges, surface ridges, and various kinds of pools and ponds. Although the general cast of the vegetation is low and monotonous, there is intricate complexity and regional variation in community composition and patterning despite a relatively impoverished flora. Frequent and often dominant plants are sedges, grasses, and grasslike herbs, together with lichens, mosses, and liverworts. In places there is an abundance of low, creeping shrubs or densely compact ones, and many herbs have cushionlike forms. A few inches or feet of elevation produce dry stiuations with distinctive vegetation, but there are many moist or wet sites, and in places many ponds and lakes. Tundra typically makes transition with coniferous forest by an ecotonal zone sometimes called the taiga. Alpine tundra (Fig. 7) occurs above the altitudinal tree line on mountain masses all over the world, even near the Equator. Its lower limit is higher in lower latitudes. It is distinguished from Arctic tundra by greater precipitation and by the different light conditions which are the result of the latitude. Some Arctic species extend far southward, even in isolated alpine tundra. *See* TUNDRA.

Fig. 7. Alpine vegetation (tundra), Medicine Bow Peak (12,005 ft altitude), Snowy Range, Wyo. Dwarfed, matted trees at timberline are mostly the spruce *Picea engelmanni* and the fir *Abies lasiocarpa*.

[STANLEY A. CAIN]

Bibliography: S. A. Cain, *Foundations of Plant Geography*, 1944; P. Dansereau, *Biogeography*, 1957; H. A. Gleason and A. Cronquist, *The Natural Geography of Plants*, 1964; W. B. McDougall, *Plant Ecology*, 4th ed., 1949; H. J. Oosting, *The Study of Plant Communities*, 2d ed., 1956; P. W. Richards, *The Tropical Rain Forest*, 1952.

Plant geography

For botanists, the major subdivision concerned with all aspects of the spatial distribution of plants. This subdivision is also known as phytogeography, phytochorology, geographical botany, and geobotany, and is here restricted to terrestrial landscapes. By custom, it often involves some aspects of the distribution of plants in time, called historical plant geography, and is thus allied to paleoecology and paleobotany. In its historical development, the science has been intimately connected with the rise of evolution, ecology, and genetics. Furthermore, the field is not yet consistently segregated from natural history, ecology, and ecosystem analysis. For geographers, a second and equally logical viewpoint is that plant geography is a major subdivision of their field, and increasing interest is being shown by them. With growing concern for resource management, plant geography has opportunities for enlargement. Forestry, range management, and wildlife habitat management follow their own concepts and techniques.

The function of plant geography is to record the observed empirical facts of plant distribution, and also to understand and interpret these facts. Attention is directed at species, population, community, and ecosystem levels of integration. Where possible, the study includes the prediction and control of distributional phenomena, especially as these relate to plant pests, parasites, and diseases, and to the introduction and spread of desirable species and vegetation types. All such knowledge is pertinent to forestry, agriculture, range and pasture management, wildlife habitat management, horticulture, and soil and water conservation. Except mainly for genetical research on speciation and subsequent dispersal and survival into island and continental areas, plant geography is not an experimental or quantitative science, and in general does not involve laboratory procedures and technologic equipment.

Flora and vegetation. There are two major subdivisions of plant geography, focused upon the two most commonly employed levels of integration of plant life. Floristic plant geography embraces the spatial distribution of flora, while vegetational plant geography is the spatial distribution of vegetation.

Flora is a scientific term with no common usage. The flora of an area or period of time is the totality of all species within that geographical unit, independent of their relative abundances and their relationships to one another. The technical term "population," in this connection, refers collectively to all individuals of any one species within a locality.

Vegetation is a term of popular origin and refers to the mosaic of plant life that forms the natural or seminatural landscape. The vegetation of a region is the tapestry or carpet of plant life, developed by differential and varying combinations and growths of the numerous elements of the local flora. Technically, it is an organized and integrated whole, at a higher level of integration than the separate species, composed of those species and their populations. Sometimes vegetation is very weakly integrated, as pioneer plants of an abandoned field. Sometimes it is highly integrated, as in an undis-

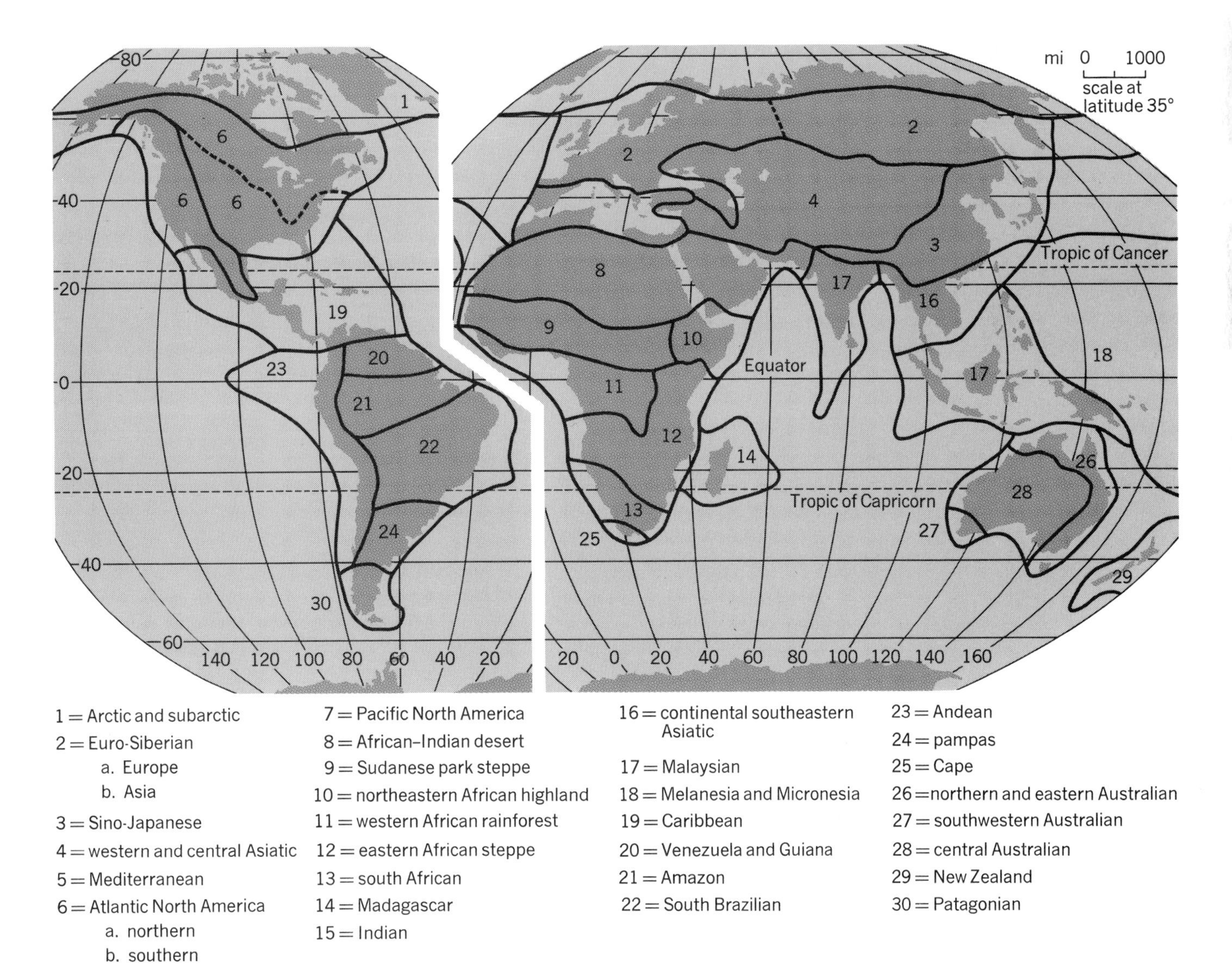

Fig. 1. Floristic regions of world. *(Modified from R. Good, Geography of Flowering Plants, 4th ed., Longmans, 1974)*

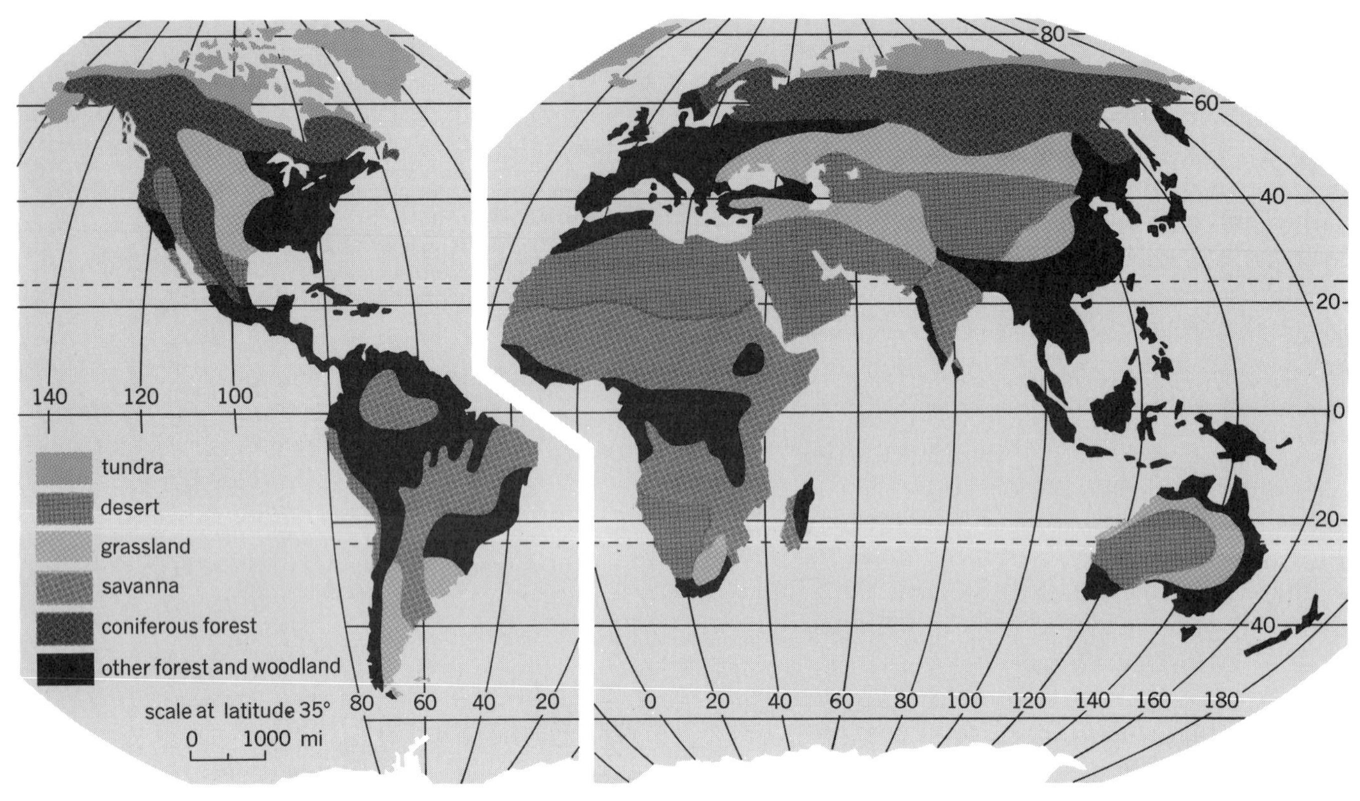

Fig. 2. Map of the world, showing the distribution of physiognomic vegetation types. (*After Brockmann-Jerosch, modified from R. Good, Geography of Flowering Plants, 4th ed., Longmans, 1974*)

turbed tropical rainforest. Vegetation possesses emergent properties not necessarily found in the species themselves, and is thus a holistic system in its own right. *See* VEGETATION.

In turn, vegetation is a component which together with all factors of the environment becomes the ecosystem, and the subject of study actively pursued since 1950. *See* ECOSYSTEM.

Floristic plant geography. The basic components of any flora are the kinds of plants composing it, commonly referred to as species. The species can be grouped and regrouped into various kinds of floral elements which are not mutually exclusive. For example, a genetic element has a common evolutionary origin; a migration element has a common route of entry into the territory; a historical element is distinct in terms of some past event; and an ecological element is related to an environmental preference. Aliens, escapes, and very widespread species are given special treatment. An endemic species is restricted to an area, usually small and of some special interest. *See* POPULATION DISPERSAL.

The idea of area is fundamental to the science and is itself the subject of a specialized section called areography. An area is the entire region of distribution or occurrence of any species, element, or even an entire flora. The local distribution within the area as a whole, as that of a swamp shrub, is the "topography" of that area. Areas are of interest in regard to their general size and shape, the nature of the margin, whether they are continuous or disjunct, and their relationships to other areas. Groups of areas are unicentric, or polycentric when they segregate into one or several geographically distinct territories. Areas of closely related plants that are mutually exclusive are said to be vicarious. A relict area is one surviving from an earlier and more extensive occurrence. On the basis of areas and their floristic relationships, the Earth's surface is divided into floristic regions (Fig. 1), each with a distinctive flora.

The understanding and interpretation of floras and their distribution have been predominantly in terms of their history and ecology. Historical factors, in addition to the evolution of the species themselves, include consideration of theories of shifting tectonic plates, moving continental masses, changing sea levels, and orographic and climatic variations in geologic time, all of which have affected migration and perpetuation of floras. Ecological factors, more amenable to observation and quantitative measurement (and thus unfortunately to causal and predictive reasoning), include the immediate and contemporary roles played by precipitation, humidity, water levels, temperature, wind, soil, animals, and humans.

Floristic plant geography as established by Ronald Good, Leon Croizat, and others is destined for significant reorganization of its facts and reconstruction of its concepts, when developments in geology with respect to mid-ocean rifts and shifting tectonic plates, are incorporated into the science.

Vegetational plant geography. The basic components of the vegetation of any landscape are the plant communities. In the United States the science is known as plant ecology, elsewhere as plant sociology, vegetation science, or phytocoenology, all of which terms have endless permutations. Many definitions of the plant community have been attempted, but none has gained universal acceptance. In part, this problem is inherent in the nature of the community itself, which is a natural phenomenon composed of individuals of the species, which themselves usually maintain a very high degree of independence, a concept completely grasped by H. A. Gleason before 1920. Thus, the community is often only a relative social continuity in nature, bounded by a relative discontinuity, as judged by competent botanists. Those workers who study only continuities are dubbed classificationists, while those who focus upon discontinuities are known as ordinationists. *See* PLANT COMMUNITY.

Vegetational plant geography has emphasized the mapping of so-called vegetation regions, and the interpretation of these overwhelmingly in terms of environment (ecological) influences. The literature and methods of vegetation mapping have been compendiously organized by A. W. Küchler.

There are many aspects of a mosaic of plant communities which could serve to identify a geographic unit of vegetation, but those which are still predominant in the literature had their origins in nonscientific folk knowledge. The physiognomic distinctions between grassland, forest, and desert, with such variants as woodland (open forest), savanna (scattered trees in grassland), and scrubland (dominantly shrubs), are most often emphasized. Within forest, the chief breakdown has been into coniferous evergreen forest, broad-leaved deciduous forest, and broad-leaved evergreen forests. Furthermore, the attempt is made to map original "virgin" vegetation, or "potential" vegetation, or "climax" vegetation (Fig. 2), as also actual "cover types" due to the influence of humans. *See* VEGETATION ZONES, WORLD.

There is increasing dissatisfaction with these inadequate approaches, but scientifically sound alternatives have not been accepted. Dissatisfaction arises from improved understanding of so-called virgin vegetation, frequently found to be influenced by ancient human populations, through people-caused fires, and Late Pleistocene large-herbivore extinctions. The segregation of coniferous from deciduous types is found to separate vegetations closely related in all other aspects, such as yellow birch and hemlock in eastern North America, and to unite types otherwise unrelated, like the pine stands which are found from the tropics to the tundra edge. In addition, disturbance of original grassland may allow the invasion of apparently self-perpetuating woody vegetation, or vice versa, in a manner that makes a physiognomic classification less fundamental. Unlike floristic botany, where evolution provides a single unifying principle for taxonomic classification, the nature of vegetation in its geographical distribution is such that many types of regions and many types of classifications may have equally valid significance in rationalizing the natural phenomena.

The interpretation of the distribution of vegetation has been overwhelmingly in terms of the existing average environment. Nevertheless, catastrophic factors such as fires, hurricanes, droughts, and other abnormal weather are receiving increasing attention. There has been relatively little emphasis on differences due to the genetic nature of the species, especially for isolated populations at the limits of their ranges.

Plant ages differ greatly. Bristlecone pine trees have a life span of over 4000 years. Australian eucalypts were absent from, but by nature amenable to, the environmentally similar but treeless California chaparral region. From one viewpoint, it is the varying genetic demands of different species upon their environments which permit and effect their segregation into communities. The ideas of chance and coincidence in the distribution of plant individuals, however, are not pertinent to quantitative methodologies. The fact that arboretums and botanical gardens are so successful in growing many species outside their normal ranges is being recognized as a refutation of the more extreme environmentalist views. Finally, the professional "gap" between the gardener who grows plants in bare soil and the ecologist who observes the same plants growing in communities has not been bridged.

This uniformitarian environmentalist interpretation of vegetation regions is the most completely documented. Climate is considered of primary importance. Numerous empirical formulas combining various features of temperature and moisture have been derived so as to correlate with the distribution of physiognomic vegetation types. Soil is accepted as of secondary importance. In addition, biotic factors, including both humans and other animals, have limiting effects. Although analysis of the normal environment is essential to the full understanding of the distribution of vegetation types, it is not likely that, except for trigger factors, direct and simple cause-and-effect relationships will be found between vegetation types and those elements of the total environment which scientists isolate and study. Much of the research since 1950 has not been directed toward flora or vegetation as such, but toward those higher levels of integration known as biomes and ecosystems. Albeit, the new emphasis has been on the "analysis" of those higher units, thus breaking them down to their components including, among other things, flora and vegetation. *See* HUMAN ECOLOGY; PHYSIOLOGICAL ECOLOGY (PLANT); TERRESTRIAL ECOSYSTEM. [FRANK E. EGLER]

Bibliography: P. Birot, *Les Formations Végétales du Globe*, 1968; S. A. Cain, *Foundations of Plant Geography*, 1944, repr. 1971; A. S. Collinson, *Introduction to World Vegetation*, 1978; L. Croizat, *Manual of Phytogeography*, 1952; S. R. Eyre, *Vegetation and Soils*, 2d ed., 1968; H. Gleason and A. Cronquist, *Natural Geography of Plants*, 1964; R. Good, *Geography of Flowering Plants*, 4th ed., 1974; A. Küchler, *International Bibliography of Vegetation Maps*, 1965, 1966, 1968, 1970; A. Küchler, *Vegetation Mapping*, 1967; F. J. Meyen, *Outlines of the Geography of Plants*, 1846, reprint 1978; N. Polunin, *Introduction to Plant Geography*, 1960.

Plant societies

Assemblages of plants which constitute structural parts of plant communities. They may be components in spatial arrangement, such as layers, life-form groups, or seasonally or locally prominent populations of plants. There is no agreement as to the precise usage of the term beyond the generally accepted notion that it should be used for structural vegetation elements of a rank below or within the plant community as a whole. A hierarchy of progressively larger units in vegetation structure with an example is as follows:

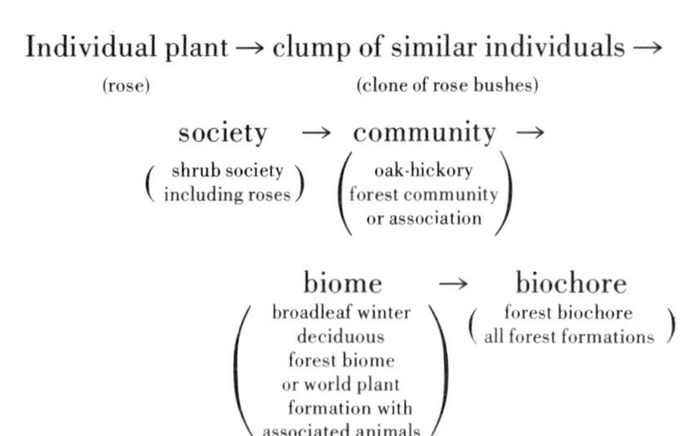

Societies can be defined on the basis of structure, dominance, season, and life form.

Structural societies. These societies are groups of plants within a community which attain approximately the same height and which bear their foliage at about the same level above ground. They are also known as layer societies or unions. Such societies may form a more or less continuous layer throughout the area occupied by the community. In a forest, for example, there is the canopy society of tall trees, low-tree society, shrub society, herb society, and ground-layer society. These may be refined further if necessary. Some authors have implied or stated that there is a certain cohesion among the component species of a layer society, giving it the status of a community within a community. However, since there is also often a strong interdependence between members of different layer societies, as in the influence of the canopy upon the density of lower vegetation in a forest, such societies should not be regarded as independent entities. A useful set of symbols for the recording of spatial arrangement of vegetation has been proposed by P. Dansereau.

Dominance societies. These societies were defined by J. Weaver and F. Clements as aspect societies or series of stands clearly belonging to a certain community. In addition to the characteristic dominants, these societies possess certain subdominants or codominants of another life form or aspect than the dominant elements of the community. The concept is useful especially in grasslands, marshes, heaths, and other vegetation types in which dominance is an important feature. For instance, prairie communities may be defined on

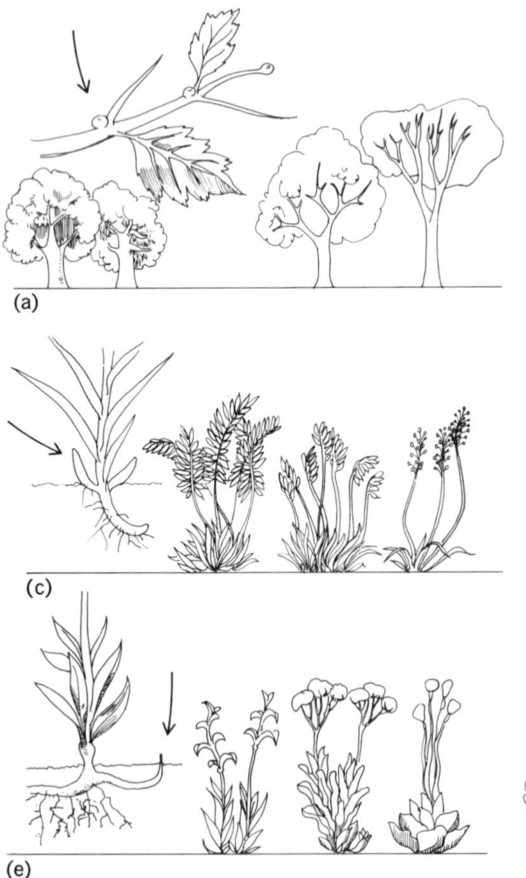

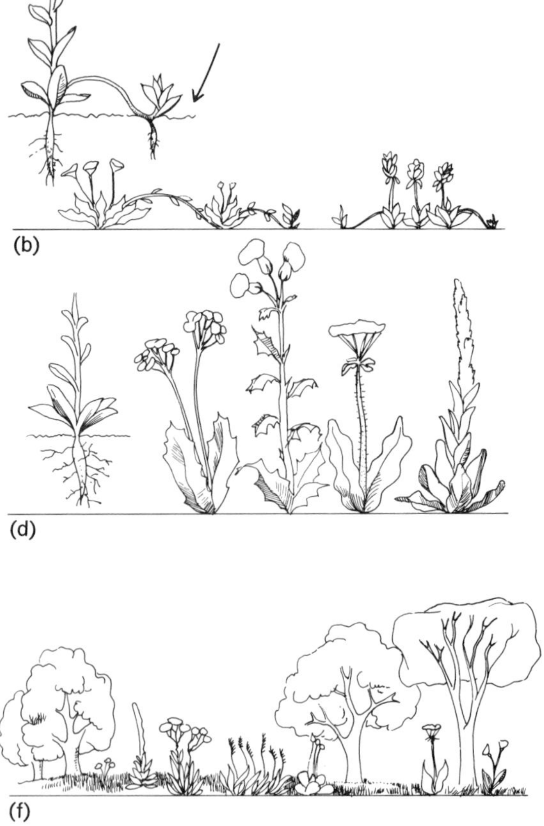

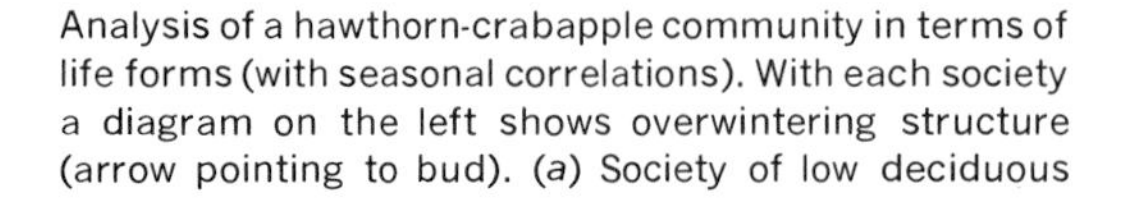
Analysis of a hawthorn-crabapple community in terms of life forms (with seasonal correlations). With each society a diagram on the left shows overwintering structure (arrow pointing to bud). (*a*) Society of low deciduous trees. (*b*) Society of sod-forming graminoids. (*c*) Society of rosette-forming hemicryptophytes. (*d*) Society of stoloniferous chamaephytes. (*e*) Society of rosette-biennials. (*f*) Structure of the entire community.

the basis of grass species. Within these communities, certain conspicuous, broadleaved herbs form local aspect societies. The extent and development of such societies have been used as indicators of the recent history of grassland stands, especially with regard to climatic fluctuations.

Seasonal societies. Weaver and Clements also applied the term society more precisely to groups of plants which determine the seasonal aspects of plant communities. Examples of such seasonal societies are the carpet of trillium, bloodroot, dogtooth violet, and spring beauty in deciduous forests in the spring; the asters and goldenrods of abandoned pastures in late summer; and the masses of short-lived annuals which suddenly develop after spring rains in the deserts of the western United States. Such societies are structural units by virtue of the uniform timing of phenologic response of the plant species involved.

Within a community there may be a progression of seasonal societies, such as prevernal → vernal → estival → serotinal → autumnal → hiemal. Each society makes its own demand upon the resources of the habitat. Therefore, a full understanding of a community requires observation of all its seasonal aspects.

Life-form societies. Plants in a community which have their permanent vegetative axes and buds at the same level in or above the soil constitute life-form societies, or synusiae. Members of such societies are therefore subject to similar growth conditions and frequently develop according to a similar pattern. The illustration shows how a plant community may be analyzed structurally, using life forms as a criterion to distinguish societies, such as the following: (1) society of clumped, low, deciduous trees (hawthorn); (2) society of sod-forming graminoids flowering in midsummer (bluegrass, redtop, and so forth); (3) society of short-rhizomatous hemicryptophytes with winter rosettes and autumnal flowering period (goldenrod and aster); (4) stoloniferous chamaephytes with winter rosettes, flowering in early summer (pussytoes, cinquefoil, and strawberry); and (5) society of biennial rosette plants (wild carrot, thistle, and mullen). *See* PLANTS, LIFE FORMS OF.

Such divisions into societies are useful because they demonstrate the arrangement in space of the aerial and underground parts of the species in the community as well as the timing of their vegetative and reproductive cycles.

From the enumeration of criteria it is evident that all types of societies, regardless of the criteria used to define them, have certain common features; and that there is a lack of agreement regarding the precise meaning and definition of the term "society." Pending a definitive, internationally acceptable vocabulary for ecology, the word society should be used only with a qualifying adjective, such as seasonal society or structural society. Where another term with a more specific meaning is available, such as union for structural society or synusia for life-form society, the word should be avoided.

The society concept remains useful to ecologists in the analysis of vegetation as a general term for structural elements. *See* PLANT COMMUNITY.

[KORNELIUS LEMS/ARTHUR W. COOPER]

Bibliography: S. A. Cain and G. M. de Oliveria Castro, *Manual of Vegetation Analysis*, 1959; J. R. Carpenter, *An Ecological Glossary*, 1956; R. Daubenmire, *Plant Communities*, 1968; H. C. Hanson and E. D. Churchill, *The Plant Community*, 1961; J. Miles, *Vegetation Dynamics*, 1979; J. E. Weaver and F. E. Clements, *Plant Ecology*, 2d ed., 1938.

Plants, life forms of

A term for the vegetative (morphological) form of the plant body. A related term is growth form but a theoretical distinction is often made: life form is thought by some to represent a basic genetic adaptation to environment, whereas growth form carries with it no connotation of adaptation and is a more general term applicable to structural differences.

Life-form systems are based on differences in gross morphological features, and the categories bear no necessary relationship to reproductive structures, which form the basis for taxonomic classification. Features used in establishing life-form classes include: deciduous versus evergreen leaves, broad versus needle leaves, size of leaves, degree of protection afforded the perennating tissue, succulence, and duration of life cycle (annual, biennial, or perennial). Thus the garden bean (family Leguminosae) and tomato (Solanaceae) belong to the same life form because each is an annual, finishing its entire life cycle in 1 year, while black locust (Leguminosae) and black walnut (Juglandaceae) are perennial trees with compound, deciduous leaves.

Climate and adaptation factors. There is a clear correlation between life forms and climates. For example, broad-leaved evergreen trees clearly dominate in the hot humid tropics, whereas broad-leaved deciduous trees prevail in temperate climates with cold winters and warm summers, and succulent cacti dominate American deserts. Although cacti are virtually absent from African deserts, members of the family Euphorbiaceae have evolved similar succulent life forms. Such adaptations are genetic, having arisen by natural selection. However, since there are no life forms confined only to a specific climate and since it is virtually impossible to prove that a given morphological feature represents an adaptation with survival value, some investigators are content to use life forms only as descriptive tools to portray the form of vegetation in different climates.

Raunkiaer system. Many life-form systems have been developed. Early systems which incorporated many different morphological features were difficult to use because of this inherent complexity. The most successful and widely used system is that of C. Raunkiaer, proposed in 1905; it succeeded where others failed because it was homogeneous and used only a few obvious morphological features representing important adaptations.

Reasoning that it was the perennating buds (the tips of shoots which renew growth after a dormant season, either of cold or drought) which permit a plant to survive in a specific climate, Raunkiaer's classes were based on the degree of protection afforded the bud and the position of the bud relative to the soil surface (see illustration). They ap-

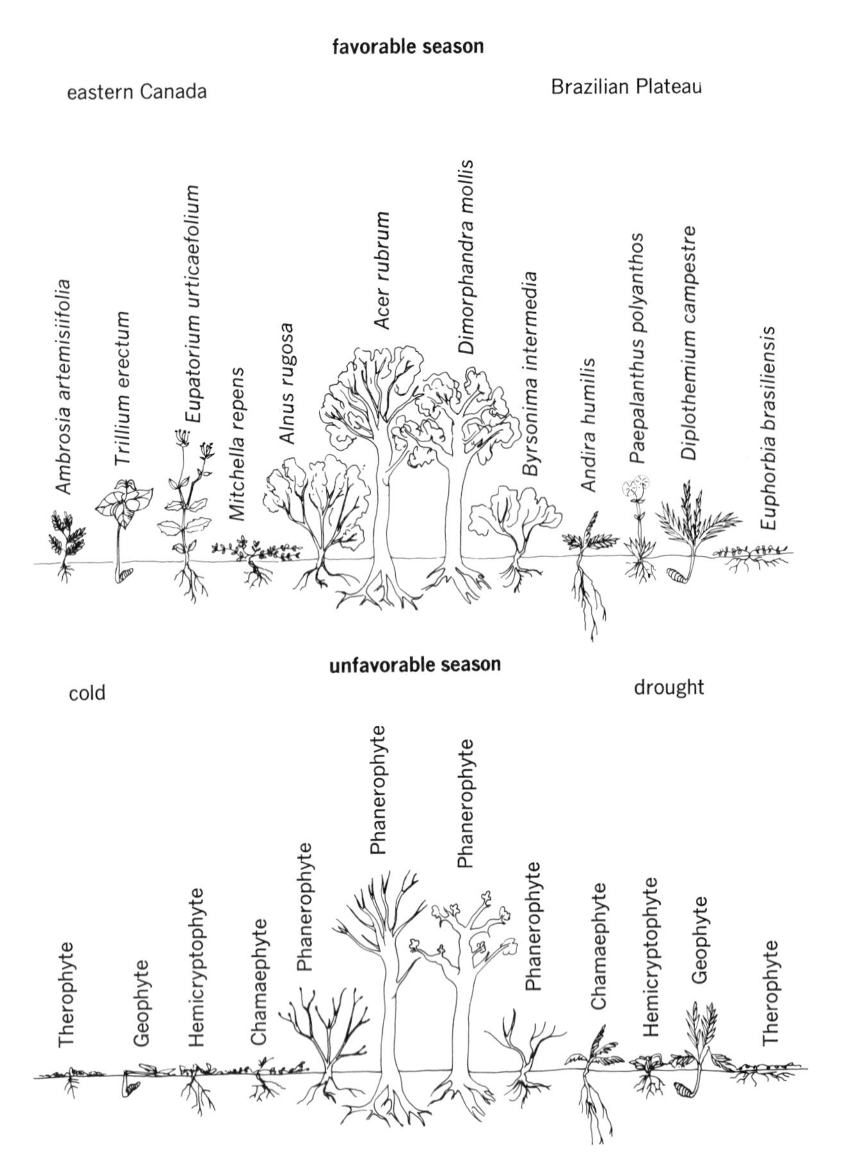

Life forms of plants according to C. Raunkiaer. (*From P. Dansereau, Biogeography: An Ecological Perspective, Ronald Press, 1957*)

plied to autotrophic, vascular, self-supporting plants. Raunkiaer's classificatory system is:

Phanerophytes: bud-bearing shoots in the air, predominantly woody trees and shrubs; subclasses based on height and on presence or absence of bud scales

Chamaephytes: buds within 25 cm of the surface, mostly prostrate or creeping shrubs

Hemicryptophytes: buds at the soil surface, protected by scales, snow, and litter

Cryptophytes: buds underneath the soil surface or under water

Therophytes: annuals, the seed representing the only perennating tissue

Subclasses were established in several categories, and Raunkiaer later incorporated leaf-size classes into the system.

By determining the life forms of a sample of 1000 species from the world's floras, Raunkiaer showed a correlation between the percentage of species in each life-form class present in an area and the climate of the area. The results (see table) were expressed as a normal spectrum, and floras of other areas were then compared to this. Raunkiaer concluded that there were four main phytoclimates: phanerophyte-dominated flora of the hot humid tropics, hemicryptophyte-dominated flora in moist to humid temperate areas, therophyte-dominated flora in arid areas, and a chamaephyte-dominated flora of high latitudes and altitudes.

Subsequent studies modified Raunkiaer's views. (1) Phanerophytes dominate, to the virtual exclusion of other life forms, in true tropical rainforest floras, whereas other life forms become proportionately more important in tropical climates with a dry season, as in parts of India. (2) Therophytes are most abundant in arid climates and are prominent in temperate areas with an extended dry season, such as regions with Mediterranean climate (for example, Crete). (3) Other temperate floras have a predominance of hemicryptophytes with the percentage of phanerophytes decreasing from summer-green deciduous forest to grassland. (4) Arctic and alpine tundra are characterized by a flora which is often more than three-quarters chamaephytes and hemicryptophytes, the percentage of chamaephytes increasing with latitude and altitude.

Most life forms are present in every climate, suggesting that life form makes a limited contribution to adaptability. Determination of the life-form composition of a flora is not as meaningful as determination of the quantitative importance of a life form in vegetation within a climatic area. However, differences in evolutionary and land-use history may give rise to floras with quite different spectra, even though there is climatic similarity. Despite these problems, the Raunkiaer system remains widely used for vegetation description and

Examples of life-form spectra for floras of different climates

Climate and vegetation	Area	Life form*				
		Ph	Ch	H	Cr	Th
Normal spectrum	World	46	9	26	6	13
Tropical rainforest	Queensland, Australia	96	2	0	2	0
	Brazil	95	1	3	1	0
Subtropical rainforest (monsoon)	India	63	17	2	5	10
Hot desert	Central Sahara	9	13	15	5	56
Mediterranean	Crete	9	13	27	12	38
Steppe (grassland)	Colorado, United States	0	19	58	8	15
Cool temperate (deciduous forest)	Connecticut, United States	15	2	49	22	12
Arctic tundra	Spitsbergen, Norway	1	22	60	15	2

*Ph = phanerophyte; Ch = chamaephyte; H = hemicryptophyte; Cr = geophyte (cryptophyte); and Th = therophyte.

for suggesting correlations between life forms, microclimate, and forest site index. *See* PLANT GEOGRAPHY.

Mapping systems. There has been interest in developing systems which describe important morphologic features of plants and which permit mapping and diagramming vegetation. Descriptive systems incorporate essential structural features of plants, such as stem architecture and height; deciduousness; leaf texture, shape, and size; and mechanisms for dispersal. These systems are important in mapping vegetation because structural features generally provide the best criteria for recognition of major vegetation units. *See* VEGETATION ZONES, ALTITUDINAL; VEGETATION ZONES, WORLD.

[ARTHUR W. COOPER]

Bibliography: S. A. Cain, Life forms and phytoclimate, *Bot. Rev.*, 16(1):1–32, 1950; S. A. Cain and G. M. de Oliveria Castro, *Manual of Vegetation Analysis*, 1959; P. Dansereau, *A Universal System for Recording Vegetation*, 1958; R. Daubenmire, *Plant Communities*, 1968; A. Kuchler, *Vegetation Mapping*, 1967; C. Raunkiaer, *The Life Forms of Plants and Statistical Geography*, 1934, reprint 1978.

Polychlorinated biphenyls

Polychlorinated biphenyls (PCBs) is a generic term covering a family of partially or wholly chlorinated isomers of biphenyl. The commerical mixtures generally contain 40–60% chlorine with as many as 50 different detectable isomers present (Fig. 1). The PCB mixture is a colorless, viscous fluid, relatively insoluble in water, that can withstand very high temperatures without degradation. However, they can be destroyed in a special industrial incinerator at 2700°F (1480°C). PCBs do not conduct electricity and the more highly chlorinated isomers are not readily degraded in the environment.

PCBs have become an important subject in national and international discussions concerned with man's pollution of his environment. Attention was drawn to PCBs by widespread reports of their presence in fish, poultry, humans, and packaging materials, even from remote areas of the world. PCBs have been in use since the 1930s in most industrial nations.

Production and uses. The major uses of PCBs are a result of the properties described above and can be grouped in three major categories: open uses, partially closed systems, and closed systems. Examples of uses in the first category are in paints, inks, plastics, and paper coatings. The PCBs in all these products are in direct contact with the environment and can be leached out by water or vaporize into the atmosphere. The so-called carbonless carbon paper contains PCBs in the encapsulated ink and has been claimed to be responsible for the PCBs found extensively in recycled paper. They have been used as plasticizers in polyvinyl chloride (PVC) and chlorinated rubbers. In an unusual industrial action the sole United States manufacturer of PCBs voluntarily released its production figures and stopped sales in 1971 to all users in this open system category, which amounted to about 30% of total production (Fig. 2). Swedish and Japanese producers followed this precedent, but manufacturers in Europe and the Soviet Union have not. It is generally believed that United States production is about one-half of world production.

Uses of PCBs regarded as partially closed systems include the working fluid in heat exchangers and hydraulic systems. These systems are subject to leakage of the PCB fluid either during use or after being discarded. PCB-contaminated poultry feed has been traced to a leaking heat exchanger used in the manufacturing process.

The electrical industry is the single major consumer of PCBs, mainly in transformers and capacitors. The fluid is generally sealed or welded into the unit so that loss is small. Of the 125,000 PCB-filled transformers put into service in the United States since 1932, over 99% are still in operation. Transformers and capacitors account for about 63% of all PCB uses.

Distribution in environment. It is not known exactly how PCBs are released into the environment or in what quantities. However, sewage outfalls, industrial and municipal disposal, leaking from dumps, and burning of refuse are certainly important sources. PCBs are found universally in the sewage of major cities, although the atmosphere is the major pathway of global transport.

In various instances, 50,000 turkeys in Minnesota, 88,000 chickens in North Carolina, and many thousands of eggs from various localities were destroyed after being found to contain very high concentrations of PCBs. The Swedish government has declared the cod liver products from the Baltic

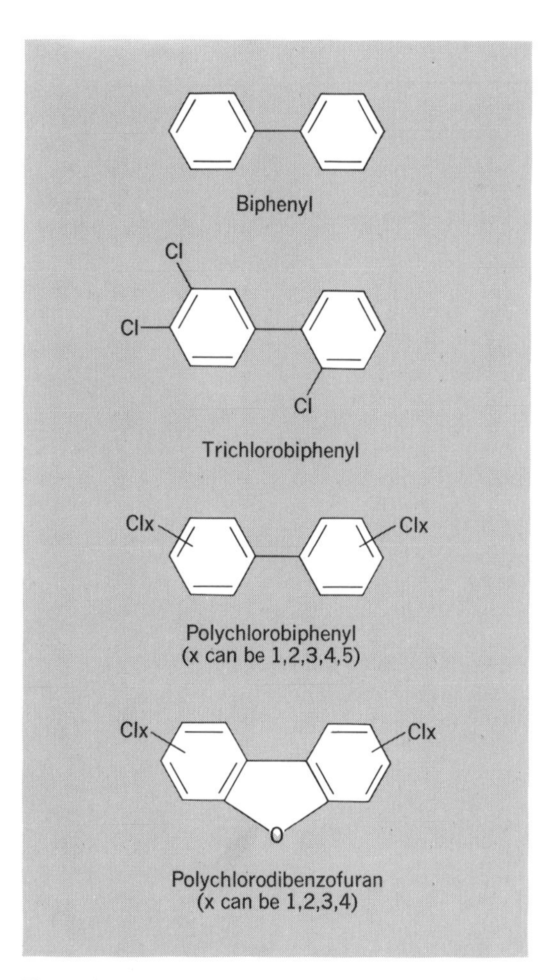

Fig. 1. Structural formulas of chlorinated biphenyls and related compounds.

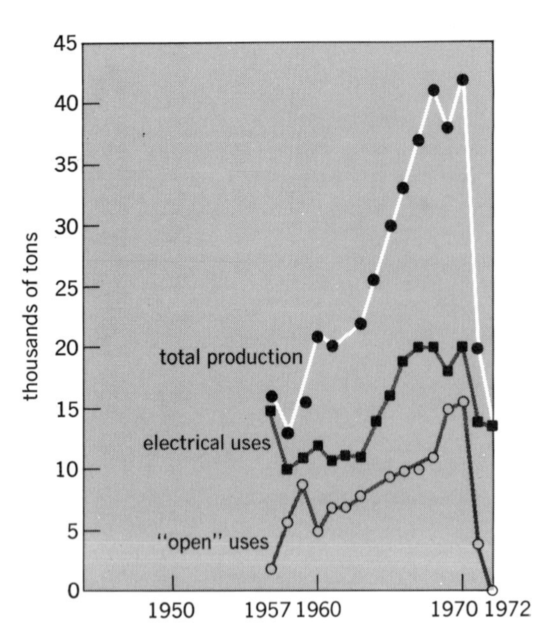

Fig. 2. United States production of polychlorinated biphenyls.

Sea contain several parts per million of PCB and are unfit for human consumption. Analyses of water samples from 30 major tributaries to the Great Lakes indicate widespread contamination, with 71% of all samples having detectable concentrations (greater than 10 parts per trillion). PCBs have been found in all organisms analyzed from the North and South Atlantic, even in animals living under 11,000 ft (3400 m) of water. The U.S. Environmental Protection Agency has reported that one-third of the human tissue sampled in the United States contains more than 1 part per million (ppm) of PCBs. According to Robert Risebrough of the University of California's Bodega Marine Laboratory, PCB concentrations in cormorants and ospreys may be higher than in any other wildlife in North America, reaching levels of 300–1000 ppm. Birds from the South Atlantic and Central Pacific generally have lower residues, while the wildlife of the Antarctic appear to have the lowest.

Once in the environment, PCBs appear to persist for a very long time. Evidence for this can be seen in the fact that in most areas of the continents and throughout the Atlantic Ocean more PCB than DDT is found in the animals, even though three times more DDT is produced each year and all of it is put directly into the environment. Based on available data, it seems safe to assume that PCBs are present in varying concentrations in every species of wildlife on Earth.

Toxicity. The U.S. Food and Drug Administration has placed a 5-ppm limit on PCBs in food products and has proposed even stricter limits. Studies of the toxic properties of PCBs have revealed some serious effects on humans. There is evidence that one class, the polybrominated biphenyls (PBBs), have caused abnormalities in the blood of exposed humans that indicate damage to the immune system.

The problem is complicated by the presence, in most PCB mixtures, of very toxic impurities believed to be polychlorodibenzofurans. Laboratory tests in the Netherlands have shown that fertile chicken eggs injected with chlorinated dibenzofurans produced seriously deformed chicks. Liver damage is a common effect of PCBs, while the occurrence of edema, skin lesions, and reproductive failure depends on the species. Hatchability of eggs is noticeably decreased by exposure.

In 1968 more than 1000 Japanese ate rice oil seriously contaminated with PCBs from a faulty heat exchanger during manufacture. The affected persons developed darkened skins, eye discharge, severe acne, and other symptoms. This came to be known as yusho, oil disease. The PCB was able to penetrate the placental barrier, since several infants born after the incident had yusho symptoms. In many cases of yusho, the symptoms were still present 3 years later. It is still a matter of controversy whether the subsequent deaths among the patients can be attributed to acute PCB poisoning, chlorinated dibenzofurans, or neither.

The history of the development and use of PCBs, followed by the discovery of its widespread occurrence in the environment, is very similar to the DDT story but with one exception: PCBs were seldom deliberately released into the environment. *See* ECOLOGY; OCEANOGRAPHY.

[GEORGE HARVEY]

Bibliography: H. Hays and R. W. Risebrough, *Nat. Hist.*, November 1971; A. V. Holden, *Nature*, 228:783, 1970; O. Hutzinger et al., *Chemistry of PCB's*, 1974; G. D. Veith and G. F. Lee, *Water Res.*, 4:265, 1970; V. Zitko and P. M. K. Choi, *PCB and Other Industrial Halogenated Hydrocarbons in the Environment*, Fisheries Research Board of Canada Rep. no. 272, 1971.

Population dispersal

The process by which groups of living organisms expand the space or range within which they live. Because of their reproductive capacity, all populations have a natural tendency to expand. As increased area supports more individuals, dispersal and reproduction are intimately correlated.

Distinction should be made between dispersal and seasonal migration. Birds, butterflies, salmon, and others migrate regularly without necessarily expanding their geographic range, since they usually return to their original areas or die out.

Dispersal phases. Dispersal consists of several phases: (1) the production of units, that is, of individuals or parts of individuals (disseminules) fit or adapted for dispersal; (2) the transportation of individuals or disseminules to the new habitat; and (3) ecesis, the process of becoming established through germination, rooting, or physiological and psychological adjustment.

Dispersal units. Certain disseminules (propagules or diaspores) may represent various stages of the life cycle of the individual. Many free-living animals do not produce special dispersal structures but rely upon the ability of the entire organism to move about (vagility). Organisms attached to a substratum, such as most plants and certain animals, produce disseminules adapted to certain agents of dispersal. In order to be effective, a disseminule must have the ability to develop into one or more complete individuals. The structures listed in Table 1 are examples of disseminules. Sperm cells, unfertilized eggs, and pollen grains, although capable of migration, are not true dissem-

Table 1. Examples of dispersal stages in life cycle of plants and animals

Environment	Organism	Disseminule	Dispersal by
Sea	Kelp	Zoospore	Currents
	Coral	Planula	Currents
	Sea worm	Trochophore	Currents
	Clam	Trochophore	Currents
	Barnacle	Adult	Driftwood, ships
	Crab	Zoea	Currents
	Sea urchin	Pluteus	Currents
	Fish	Adult	Autochory
	Lamprey	Adult	Fish
Terrestrial	Mushroom	Spore	Wind
	Fern	Spore	Wind
	Pine	Seed	Wind
	Blueberry	Fruit	Birds
	Tumbleweed	Entire plant	Wind
	Insect	Adult	Autochory, wind
	Spider	Young animal	Wind
	Reptiles, birds, mammals	Adult	Autochory
Parasitic	Bacteria	Entire cell	Water, food, air
	Intestinal ameba	Cyst	Water, food, man
	Malaria parasite	Gamete, sporozoite	Mosquito
	Tapeworm	Egg	Pig
	Blood fluke	Egg, cercaria	Water, snail

inules because they cannot give rise to new individuals.

It is possible to analyze plant communities on the basis of morphological features of the disseminules. By assigning species to dispersal types it is possible to construct dispersal spectra comparable to life form spectra in purpose and usefulness.

Transportation. Individuals or disseminules are transported in five general ways: self-dispersal (autochory), water dispersal (hydrochory), wind dispersal (anemochory), animal dispersal (zoochory), and dispersal by humans (anthropochory).

In active self-dispersal, or autochory, the organism spreads in the course of its normal activities. The flight of starlings resulting in their gradual spread through the United States and the motility of bacteria resulting in gradual spread through the nutrient media are examples. Certain plants possess mechanisms of self-dispersal, such as the auxochores and ballochores listed in Table 2. In passive dispersal, one or more agents carry the dispersal unit to a new location. These agents, or vectors, are water currents, wind, animals and any of man's vehicles, such as trains, ships, and airplanes.

Table 2. Plant dispersal types based upon morphological adaptations

Name	Definition	Example
Auxochores	Deposited by parent plant	Walking fern
Cyclochores	Spherical framework	Tumbleweed
Pterochores	Disseminules winged	Maple
Pogonochores	Disseminules plumed	Milkweed
Desmochores	Disseminules sticky or barbed	Cocklebur
Sarcochores	Disseminules fleshy	Cherry
Sporochores	Disseminules minute, light	Fern
Sclerochores	Disseminules without apparent adaptations	Violet
Barochores	Disseminules heavy	Oak
Ballochores	Shot away by parent plant	Touch-me-not

Water dispersal, or hydrochory, is prevalent in all marine and other aquatic populations. Plankton usually contains larval forms of bottom-dwellers (Fig. 1). Terrestrial forms associated with shore habitats are commonly dispersed by water. Buoyancy and resistance to salt water are a prerequisite for ocean dispersal. The first invaders of new islands such as Surtsey are often of this type. Transoceanic similarities in floras and faunas have been explained partly by ocean currents.

Wind dispersal, or anemochory, has various effects. It moves rolling disseminules in open deserts and grasslands (cyclochores, Fig. 2*a*); it deflects falling winged disseminules (pterochores, Fig. 2*b*–*d*); and it carries lightweight spores and disseminules with plumes for great distances (sporochores and pogonochores, Fig. 2*e* and *h*). Insects, spiders, and other light animals have been found many miles in the air, together with poplar seeds and other disseminules (Fig. 2*f* and *g*). Thus they may be carried hundreds of miles.

Animal dispersal, or zoochory, is divided into epizoochory (barbed or sticky disseminules, desmochores, Fig. 3*a*–*c*) and endozoochory (disseminules eaten and egested by animals). Disseminules adapted to endozoochory are those, such as arillate seeds (Fig. 3*d*), common in the tropics, and fruits with a fleshy mesocarp (Fig. 3*f* and *g*). Survival in the digestive tract of animals is a prerequisite. Bright fruit colors are frequent.

Dispersal by man, or anthropochory, involves purposely dispersed organisms such as domesticated animals and plants and those accidentally transported such as weeds along railroads, beetles in grain shipments, birds, rats, barnacles, and starfish on and in ships.

Ecesis. Success in population dispersal depends upon three factors: fitness of the new habitat, fitness of the migrating individuals, and the chance juxtaposition of these two which, in the long run, depends on the number of individuals invading the new habitat. The probability for a new habitat to be favorable is greatest close to the parent population. Spores blown over great distances have less

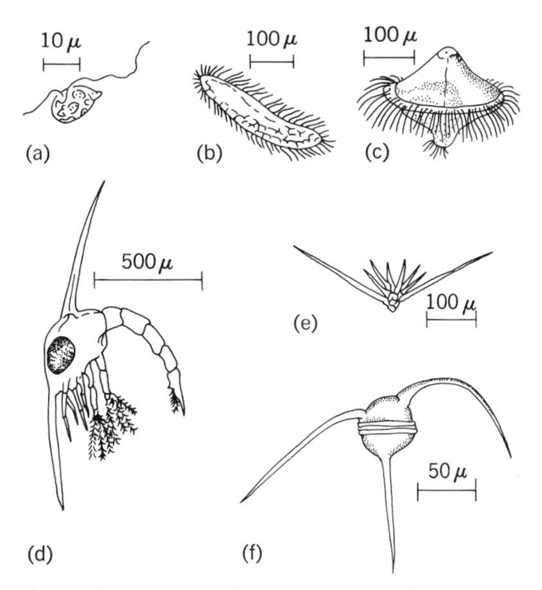

Fig. 1. Disseminules in plankton. (*a*) Kelp zoospore. (*b*) Coral planula. (*c*) Worm trochophore. (*d*) Crab zoea. (*e*) Brittle star pluteus. (*f*) Ceratium tripos.

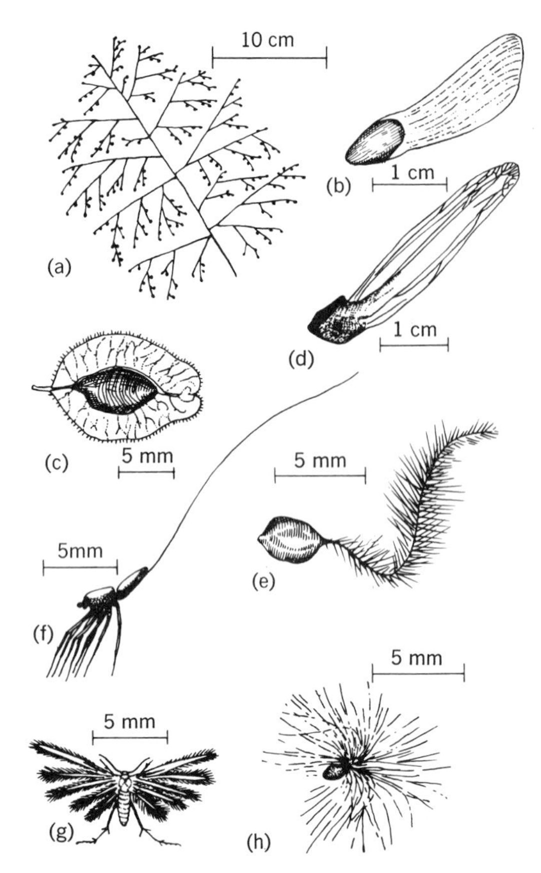

Fig. 2. Disseminules dispersed by wind. (*a*) Panic grass. (*b*) Pine seed. (*c*) Elm samara. (*d*) Tulip tree carpel. (*e*) Clematis carpel. (*f*) Spider. (*g*) Moth. (*h*) Cottonwood.

chance of landing in spots suited for germination than have seeds falling close to the parent plant. In wide-range dispersal larger numbers of disseminules are usually necessary than at close range, to insure ecesis.

The fitness of the individuals depends partly upon their genetic makeup. Offspring of organisms that reproduce without sexual union (apomictic), such as aphids, dandelions, and similar organisms, are likely to succeed only in identical habitats. Offspring from self- or cross-fertilizing parents may succeed in a variety of situations. However, some hybrids which are sexually sterile are known

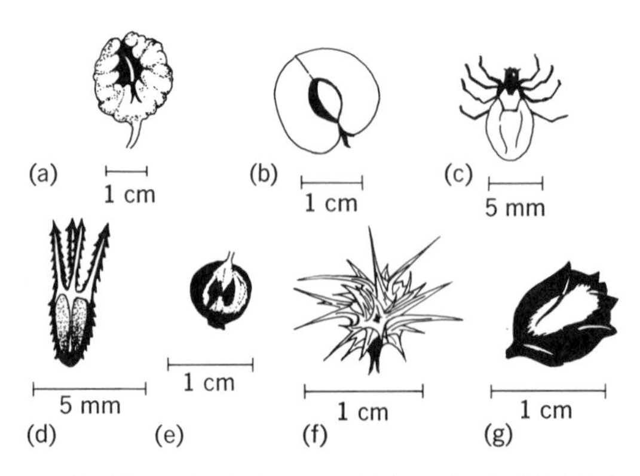

Fig. 3. Disseminules transported by animals. (*a*) Arillate legume seed. (*b*) Cherry. (c) Tick. (*d*) Beggar's tick fruit. (*e*) Currant berry. (*f*) Sandbur spikelet. (*g*) Juniper cone.

to perpetuate themselves through apomixis. These are usually very successful locally.

Barriers to dispersal. A barrier is any discontinuity in the habitat greater than the maximum distance traveled by organisms in their normal dispersal. Oceans separating terrestrial habitats, continents separating marine habitats, mountain ranges intercepting wind dispersal, and deserts interrupting the continuity of forested land are all effective major barriers. Through the intervention of man these barriers are broken down in many cases. Since the development of frequent world travel thousands of species have become established on new continents as a result of anthropochory. *See* POPULATION DISPERSION.

[KORNELIUS LEMS/HERBERT G. BAKER]

Bibliography: P. Dansereau and K. Lems, The grading of dispersal types in plant communities and their ecological significance, *Contrib. Inst. Bot. Univ. Montréal*, vol. 71, 1957; P. A. Fryxell, Mode of reproduction of higher plants, *Bot. Rev.*, 23:135–233, 1957; W. George, *Animal Geography*, 1962; P. A. Glick, *The Distribution of Insects, Spiders and Mites in the Air*, USDA Tech. Bull. no. 673, 1939; R. Hesse, W. C. Allee, and K. P. Schmidt, *Ecological Animal Geography*, 1937; H. N. Ridley, *The Dispersal of Plants Throughout the World*, 1931; E. J. Salisbury, *The Reproductive Capacity of Plants*, 1942.

Population dispersion

The spatial distribution at any particular moment of the individuals of a species of plant or animal. Under natural conditions organisms are distributed either by active movements, or migrations, or by passive transport by wind, water, or other organisms. The act or process of dissemination is usually termed dispersal, while the resulting pattern of distribution is best referred to as dispersion. Dispersion is a basic characteristic of populations, controlling various features of their structure and organization. It determines population density, that is, the number of individuals per unit of area, or volume, and its reciprocal relationship, mean area, or the average area per individual. It also determines the frequency, or chance of encountering one or more individuals of the population in a particular sample unit of area, or volume. The ecologist therefore studies not only the fluctuations in numbers of individuals in a population but also the changes in their distribution in space. *See* POPULATION DISPERSAL.

Principal types of dispersion. The dispersion pattern of individuals in a population may conform to any one of several broad types, such as random, uniform, or contagious (clumped). Any pattern is relative to the space being examined; a population may appear clumped when a large area is considered, but may prove to be distributed at random with respect to a much smaller area.

Random or haphazard. This implies that the individuals have been distributed by chance. In such a distribution, the probability of finding an individual at any point in the area is the same for all points (Fig. 1*a*). Hence a truly random pattern will develop only if each individual has had an equal and independent opportunity to establish itself at any given point. In a randomly dispersed population, the relationship between frequency and density can be expressed by Eq. (1), where F is percen-

$$F = 100(1 - e^{-D}) \quad (1)$$

tage frequency, D is density, and e is the base of natural or Napierian logarithms. Thus when a series of randomly selected samples is taken from a population whose individuals are dispersed at random, the numbers of samples containing 0, 1, 2, 3, . . . , n individuals conform to the well-known Poisson distribution described by notation (2).

$$e^{-D}, De^{-D}, \frac{D^2}{2!}e^{-D}, \frac{D^3}{3!}e^{-D}, \ldots, \frac{D^n}{n!}e^{-D} \quad (2)$$

Randomly dispersed populations have the further characteristic that their density, on a plane surface, is related to the distance between individuals within the population, as shown in Eq. (3),

$$D = \frac{1}{4\bar{r}^2} \quad (3)$$

where $\bar{r}$ is the mean distance between an individual and its nearest neighbor. These mathematical properties of random distributions provide the principal basis for a quantitative study of population dispersion. Examples of approximately random dispersions can be found in the patterns of settlement by free-floating marine larvae and of colonization of bare ground by airborne disseminules of plants. Nevertheless, true randomness appears to be relatively rare in nature, and the majority of populations depart from it either in the direction of uniform spacing of individuals or more often in the direction of aggregation.

Uniform. This type of distribution implies a regularity of distance between and among the individuals of a population (Fig. 1*b*). Perfect uniformity exists when the distance from one individual to its nearest neighbor is the same for all individuals. This is achieved, on a plane surface, only when the individuals are arranged in a hexagonal pattern. Patterns approaching uniformity are most obvious in the dispersion of orchard trees and in other artificial plantings, but the tendency to a regular distribution is also found in nature, as for example in the relatively even spacing of trees in forest canopies, the arrangement of shrubs in deserts, and the distribution of territorial animals.

Contagious or clumped. The most frequent type of distribution encountered is contagious or clumped (Fig. 1*c*), indicating the existence of aggregations or groups in the population. Clusters and clones of plants, and families, flocks, and herds of animals are common phenomena. The degree of aggregation may range from loosely connected groups of two or three individuals to a large compact swarm composed of all the members of the local population. Furthermore, the formation of groups introduces a higher order of complexity in the dispersion pattern, since the several aggregations may themselves be distributed at random, evenly, or in clumps. An adequate description of dispersion, therefore, must include not only the determination of the type of distribution, but also an assessment of the extent of aggregation if the latter is present.

Analysis of dispersion. If the type or degree of dispersion is not sufficiently evident upon inspection, it can frequently be ascertained by use of sampling techniques. These are often based on counts of individuals in sample plots or quadrats. Departure from randomness can usually be demonstrated by taking a series of quadrats and testing the numbers of individuals found therein for their conformity to the calculated Poisson distribution which has been described above. The observed values can be compared with the calculated ones by a chi-square test for goodness of fit, and lack of agreement is an indication of nonrandom distribution. If the numbers of quadrats containing zero or few individuals, and of those with many individuals, are greater than expected, the population is clumped; if these values are less than expected, a tendency towards uniformity is indicated. Another measure of departure from randomness is provided by the variance:mean ratio, which is 1.00 in the case of the Poisson (random) distribution. If the ratio of variance to mean is less than 1.00, a regular dispersion is indicated; if the ratio is greater than 1.00, the dispersion is clumped.

In the case of obviously aggregated populations, quadrat data have been tested for their conformity to a number of other dispersion models, such as Neyman's contagious, Thomas' double Poisson, and the negative binomial distributions. However, the results of all procedures based on counts of individuals in quadrats depend upon the size of the quadrat employed. Many nonrandom distributions will seem to be random if sampled with very small or very large quadrats, but will appear clumped if quadrats of medium size are used. Therefore the employment of more than one size of quadrat is recommended.

A measure of aggregation that does not depend on quadrat size or the mean density of individuals per quadrat and that can be applied to patterns consisting of a mosaic of patches with different densities has been developed by Morisita. His index of dispersion is a ratio of the observed probability of drawing two individuals randomly from the same quadrat to the expected probability of the same event for individuals randomly dispersed over the set of quadrats being studied. Index values greater than 1.0 indicate clumping, and values between 0 and 1.0 point to regularity of dispersion.

The fact that plot size may influence the results of quadrat analysis has led to the development of a number of techniques based on plotless sampling. These commonly involve measurement of the distance between a randomly selected individual and its nearest neighbor, or between a randomly selected point and the closest individual. At least four different procedures have been used (Fig. 2). The closest-individual method (Fig. 2*a*) measures the distance from each sampling point to the nearest individual. The nearest-neighbor method (Fig. 2*b*) measures the distance from each individual to its nearest neighbor. The random-pairs method (Fig. 2*c*) establishes a base line from each sampling point to the nearest individual, and erects a 90° exclusion angle to either side of this line. The distance from the nearest individual lying outside the exclusion angle to the individual used in the base line is then measured. The point-centered quarter method (Fig. 2*d*) measures the distance from each sampling point to the nearest individual in each quadrat.

In each of these four methods of plotless sampling, a series of measurements is taken which can

POPULATION DISPERSION

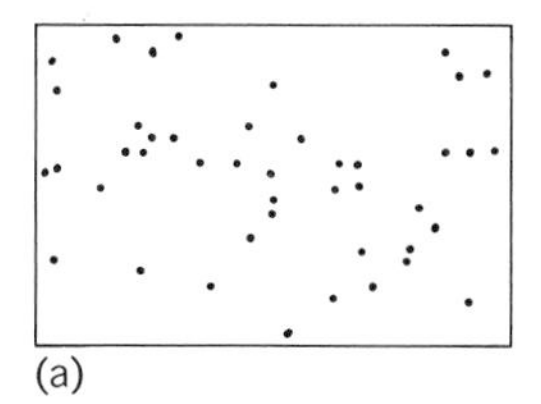

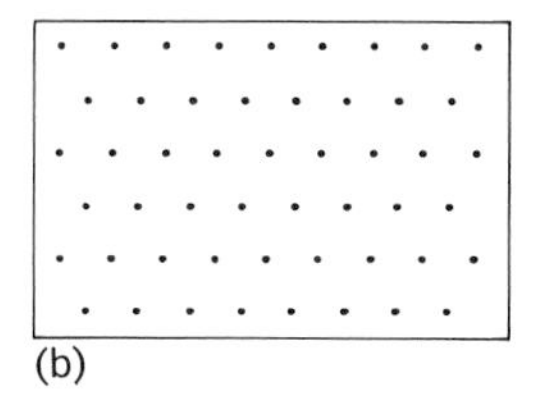

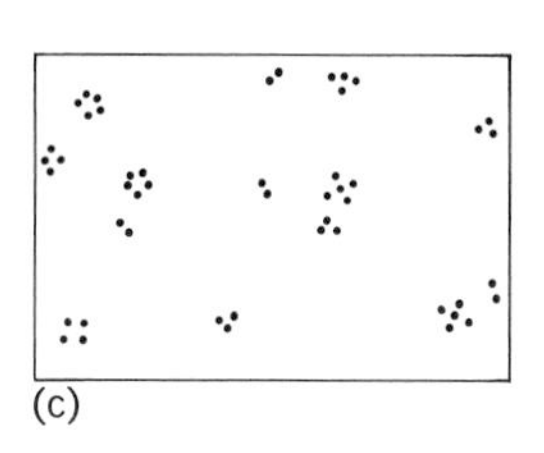

Fig. 1. Basic patterns of the dispersion of individuals in a population. (*a*) Random. (*b*) Uniform. (*c*) Clumped, but groups random. (*E. P. Odum, Fundamentals of Ecology, Saunders, 1953*)

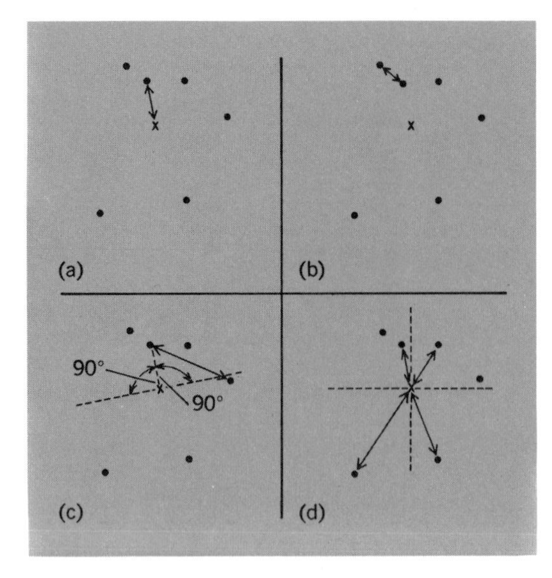

Fig. 2. Distances measured in four methods of plotless sampling. (*a*) Closest individual. (*b*) Nearest neighbor. (*c*) Random pairs, with 180° exclusion angle. (*d*) Point-centered quarter. The cross represents the sampling point in each case. (*P. Greig-Smith, Quantitative Plant Ecology, Butterworths, 1957*)

be used as a basis for evaluating the pattern of dispersion. In the case of the closest-individual and the nearest-neighbor methods, a population whose members are distributed at random will yield a mean distance value that can be calculated by use of the density-distance equation, Eq. (3). In an aggregated distribution, the mean observed distance will be less than the one calculated on the assumption of randomness; in a uniform distribution it will be greater. Thus the ratio $\bar{r}_A/\bar{r}_E$, where $\bar{r}_A$ is the actual mean distance obtained from the measured population and $\bar{r}_E$ is the mean distance expected under random conditions, affords a measure of the degree of deviation from randomness.

Students of human geography have used the nearest-neighbor measure as a basis for a highly sophisticated methodology to analyze the dispersion of towns, department stores, and other features of land-use patterns.

Additional information about the spatial relations in a population can be secured by extending these procedures to measurement of the distance to the second and successive nearest neighbors, or by increasing the number of sectors about any chosen sampling point. However, since all of these methods assume that the individuals are small enough to be treated mathematically as points, they become less accurate when the individuals cover considerable space.

Factors affecting dispersion. The principal factors that determine patterns of population dispersion include (1) the action of environmental agencies of transport, (2) the distribution of soil types and other physical features of the habitat, (3) the influence of temporal changes in weather and climate, (4) the behavior pattern of the population in regard to reproductive processes and dispersal of the young, (5) the intensity of intra- and interspecific competition, and (6) the various social and antisocial forces that may develop among the members of the population. Although in certain cases the dispersion pattern may be due to the overriding effects of one factor, in general populations are subject to the collective and simultaneous action of numerous distributional forces and the dispersion pattern reflects their combined influence. When many small factors act together on the population, a more or less random distribution is to be expected, whereas the domination of a few major factors tends to produce departure from randomness.

Environmental agencies of transport. The transporting action of air masses, currents of water, and many kinds of animals produces both random and nonrandom types of dispersion. Airborne seeds, spores, and minute animals are often scattered in apparently haphazard fashion, but aggregation may result if the wind holds steadily from one direction. Wave action is frequently the cause of large concentrations of seeds and organisms along the drift line of lake shores. The habits of fruit-eating birds give rise to the clusters of seedling junipers and cherries found beneath such perching sites as trees and fencerows, as well as to the occurrence of isolated individuals far from the original source. Among plants, it seems to be a general principle that aggregation is inversely related to the capacity of the species for seed dispersal.

Physical features of the habitat. Responses of the individuals of the population to variations in the habitat also tend to give rise to local concentrations. Environments are rarely uniform throughout, some portions generally being more suitable for life than others, with the result that population density tends to be correlated directly with the favorability of the habitat. Oriented reactions, either positive or negative, to light intensities, moisture gradients, or to sources of food or shelter, often bring numbers of individuals into a restricted area. In these cases, aggregation results from a species-characteristic response to the environment and need not involve any social reactions to other members of the population. *See* ENVIRONMENT.

Influence of temporal changes. In most species of animal, daily and seasonal changes in weather evoke movements which modify existing patterns of dispersion. Many of these are associated with the disbanding of groups as well as with their formation. Certain birds, bats, and even butterflies, for example, form roosting assemblages at one time of day and disperse at another. Some species tend to be uniformly dispersed during the summer, but flock together in winter. Hence temporal variation in the habitat may often be as effective in determining distribution patterns as spatial variation.

Behavior patterns in reproduction. Factors related to reproductive habits likewise influence the dispersion patterns of both plant and animal populations. Many plants reproduce vegetatively, new individuals arising from parent rootstocks and producing distinct clusters; others spread by means of rhizomes and runners and may thereby achieve a somewhat more random distribution. Among animals, congregations for mating purposes are common, as in frogs and toads and the breeding swarms of many insects. In contrast, the breeding territories of various fishes and birds exhibit a comparatively regular dispersion.

Intensity of competition. Competition for light, water, food, and other resources of the environment tends to produce uniform patterns of distribution. The rather regular spacing of trees in many forests is commonly attributed largely to competition for sunlight, and that of desert plants for soil moisture. Thus a uniform dispersion helps to reduce the intensity of competition, while aggregation increases it.

Social factors. Among many animals the most powerful forces determining the dispersion pattern are social ones. The social habit leads to the formation of groups or societies. Plant ecologists use the term society for various types of minor communities composed of several to many species, but when the word is applied to animals it is best confined to aggregations of individuals of the same species which cooperate in their life activities. Animal societies or social groups range in size from a pair to large bands, herds, or colonies. They can be classified functionally as mating societies (which in turn are monogamous or polygamous, depending on the habits of the species), family societies (one or both parents with their young), feeding societies (such as various flocks of birds or schools of fishes), and as migratory societies, defense societies, and other types. Sociality confers many advantages, including greater efficiency in securing food, conservation of body heat during cold weather, more thorough conditioning of the environment to increase its habitability, increased facilitation of mating, improved detection of, and defense against, predators, decreased mortality of the young and a greater life expectancy, and the possibility of division of labor and specialization of activities. Disadvantages include increased competition, more rapid depletion of resources, greater attraction of enemies, and more rapid spread of parasites and disease. Despite these disadvantages, the development and persistence of social groups in a wide variety of animal species is ample evidence of its overall survival value. Some of the advantages of the society are also shared by aggregations that have no social basis.

Optimal population density. The degree of aggregation which promotes optimum population growth and survival, however, varies according to the species and the circumstances. Groups or organisms often flourish best if neither too few nor too many individuals are present; they have an optimal population density at some intermediate level. The condition of too few individuals, known as undercrowding, may prevent sufficient breeding contacts for a normal rate of reproduction. On the other hand, overcrowding, or too high a density, may result in severe competition and excessive interaction that will reduce fecundity and lower the growth rate of individuals. The concept of an intermediate optimal population density is sometimes known as Allee's principle. *See* POPULATION DYNAMICS.

[FRANCIS C. EVANS]

Bibliography: P. Greig-Smith, *Quantitative Plant Ecology*, 1957; P. Haggett, *Locational Analysis in Human Geography*, 1966; K. A. Kershaw, *Quantitative and Dynamic Ecology*, 1965; L. J. King, *Statistical Analysis in Geography*, 1969; E. C. Pielou, *An Introduction to Mathematical Ecology*, 1969; E. C. Pielou, *Population and Community Ecology: Principles and Methods*, 1974; V. C. Wynne-Edwards, *Animal Dispersion in Relation to Social Behaviour*, 1962.

Population dynamics

The aggregate of processes that determine the size and composition of any population. In this context, a population is a group of organisms of a single species. The group is characterized by a definite schedule of birth and death rates, and often by a determined composition with respect to the numbers of individuals belonging to different age classes and by a determined ratio between the sexes. A fortuitous aggregation of individuals may or may not constitute a population in this sense, and the dynamics of a population will usually be influenced by local populations of other species.

Population size and density. Population size is normally measured in terms of numbers of individuals, while productivity is often expressed as the number of new individuals produced per unit time. There are exceptions such as stands of timber or populations of commercially valuable fish. In these instances, productivity and population size are appropriately measured in terms of mass or volume rather than numbers. The study of dynamics in such populations merely requires consideration of individual growth rates in addition to numbers and ages, and the discussion here will be limited to populations measured by enumeration.

In practice, it is difficult to define the limits of a population, and enumeration normally measures some type of population density. Crude density is the number of individuals per unit of selected space or volume. Examples of this are the number of deer in a county or other areal unit, or the mean number of fish per acre of water surface or per cubic meter of water. This measure suffers from the fact that the units of area and volume are heterogeneous and the population does not utilize all of the available space. Ecological or economic density refers to the mean number of individuals per unit of space actually utilized. Often, the only practicable measures of natural populations involve relative density and are designed to show whether a population is increasing or decreasing without determining its actual size. Thus, the number of birds seen per man-hour of walking and the number of squirrels treed per dog-hour have been used to compare population densities in different years and places. Formidable statistical and practical problems are involved in a mensuration of natural populations.

Population growth. A population can gain in numbers only by birth and immigration, and it can decrease only by death and emigration. In considerations of theoretical population dynamics, migratory movements are customarily ignored and attention is concentrated on the birth and death processes. Individuals of every species have the potentiality for producing more offspring than are required to replace the parents. Some organisms reproduce once per lifetime, others many times. Some produce tremendous numbers of gametes, others few. Also, the age at which reproductive maturity occurs varies tremendously—from a few minutes in bacteria to more than a century in the giant *Sequoia* tree. These life history features determine the potential growth rate of the population or the biotic potential of the species.

If these life history features remained the same in successive generations, the population would ultimately grow in accordance with the equation

$N_t = Ae^{rt}$, for which the growth rate at time t is $dN/dt = rN_t$. In this formula N_t is the population size at time t, A is a constant, e is the base of the natural logarithmic system, and r is a measure of population growth. The value of r, which is determined by natural history features, is commonly referred to as the intrinsic rate of natural increase, and it has been proposed to define biotic potential as the normal maximum value of r for a given population.

This exponential form of potential population growth implies exceedingly rapid expansion. A single individual of an annual plant in which each individual produces only 2 viable seeds would leave 1,000,000 descendants in 20 years if all seeds survived, and, in fact, every species is theoretically capable of overflowing the Earth. In actuality, shifts in natality (birth rates) and mortality inhibit unlimited exponential growth.

Population control or regulation. Since potential population growth is exponential, while real population size is limited, much interest and controversy has centered about the form of actual population growth. Commonly, when the size of a growing population is plotted against time, the result is a broad S-shaped (sigmoid) form of growth curve with a point of inflection, where the rate of growth shifts from increase to decrease (Fig. 1).

Attempts to give a generalized mathematical formulation of population growth have led to consideration of equations of the form $dN/dt = rNf(N)$. This equation differs from that describing exponential growth in that the factor $f(N)$ serves as a governor or damping factor which takes the value zero when N is very large, indicating that there is some finite upper limit to the size of a population that is capable of further growth.

Various proposals have been offered concerning the form of this governing factor. If $f(N)$ is assumed to be merely a decreasing function of time, the result is one form of the discredited doctrine of racial senescence, the supposition that populations age and die as do individual organisms. Numerous workers have supposed that $f(N)$ is a random variable, sometimes positive and sometimes negative, so that populations normally fluctuate about a steady-state, or equilibrium, size. Modern probability theory shows this position to be untenable in so simple a form. By this doctrine, random extingtion would be the eventual fate of all populations.

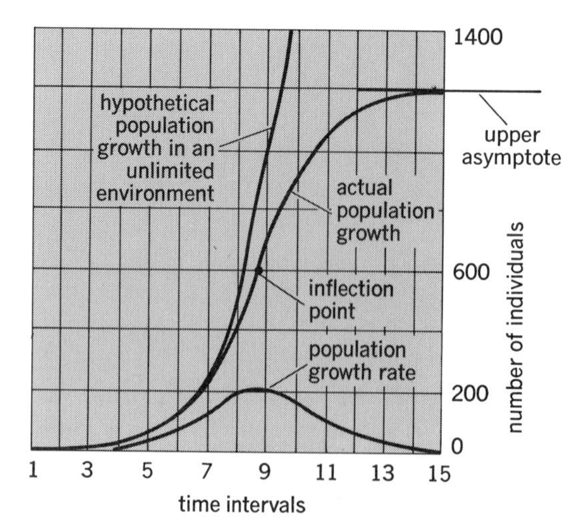

Fig. 1. Population dynamics. Three types of growth curves which use the same data.

In a somewhat more realistic approach, the governing factor is regarded as a decreasing function of population density so that the population inhibits its own growth beyond a certain size. It is obvious, however, that maximum population size must also be affected by the quality of the habitat. Populations of the same species are commonly more dense in some regions than in others owing to differences in the availability of essential resources such as food and nesting sites. Therefore, it is usual to speak of a carrying capacity for any given habitat and to define this as the maximum steady-state population of a given species that can be supported there.

Carrying capacity. No general agreement has been reached on a precise definition of carrying capacity, but it seems essential to recognize that this hypothetical upper limit may sometimes be exceeded temporarily. For example, populations as dense as 17 individuals per square yard have been observed in house mice, and the hordes of locusts, chinch bugs, lemmings, and other animals that occur in outbreak years far exceed the capacity of the occupied land to provide food and shelter. In practice, students of natural populations often think of population density not in absolute terms but as a density relative to some standard, such as the maximum that could be supported. Thus, the factor $f(N)$ governing population growth is most realistically regarded as a function of the unfulfilled possibilities for growth.

It is known from observations in the field and from laboratory experiments that crowding does promote increased death rates and inhibit reproduction. Pathogenic organisms can spread rapidly through populations where there is close contact between individuals, metabolic wastes may accumulate to toxic levels, malnutrition or other deficiency conditions may weaken the individuals, and in very crowded populations there may be interference with feeding and mating. Also, various symptoms of physiological stress apparently result from crowding. There are, therefore, sound reasons for considering increased population density to operate in limiting and finally inhibiting population growth; that is to say, the growing population exhibits negative feedback.

Density factors. Numerous students have tried to classify environmental influences into density-independent factors which are theoretically incapable of regulating population size and density-dependent factors which can exert a governing effect. Much controversy has surrounded these concepts and the varying definitions given to the various classes of factors. Since population density cannot ordinarily be considered to alter weather or climate, it has been contended that meteorological conditions are density-independent and cannot regulate population size. Others have recognized that weather does sometimes operate in a density-dependent manner, or they have maintained that the true density-dependent regulating factor in such cases is competition for shelter between the individuals which forces the losers to be exposed to unfavorable weather conditions. Another prominent school of thought is utterly opposed to this general approach and maintains that populations seldom attain a level where density effects are important but are normally held at lower levels by environmental inadequancies and mortality factors such as extremes of weather.

Underpopulation effects. At the other end of the scale of size there is also a great deal of evidence in many species for underpopulation effects on population growth. In sparse populations, females may have difficulty finding mates. A small population lacks the adaptability to changed conditions that is provided by a large stock of genetic variability, and this defect may be aggravated by inbreeding. Also, populations often condition their surroundings and modify the impact of environmental factors. The forest literally protects the trees from wind damage, excessive insolation, and evaporation. The advantages of maintaining the population above some minimum level are obvious in gregarious animals such as bird flocks and herds of ungulates, and even more so in the social insects where ants, bees, and termites, for example, exercise considerable control over the climate inside the colonial structure. In addition, many cases are known where, often for obscure reasons, populations seem unable to resist extinction if the numbers fall below some minimum level.

From these observations it follows that any generalized concept of the growth governor $f(N)$ must provide for this factor to be small in very small populations, to rise to a maximum at some optimum population size, and to decline to zero before the population overflows the Earth. In other words, the growing population may exhibit positive feedback up to a certain size range and negative feedback at higher levels.

Actual population growth. The logistic function has been the equation most employed for representing actual population growth. This equation in its differential form is

$$\frac{dN}{dt} = rN\left(1 - \frac{K}{N}\right)$$

where K represents the upper asymptote or maximum size attainable by the population in question. Here the intensity of the governing factor decreases linearly with population size so that no account is taken of underpopulation effects. The integrated logistic curve, however, is sigmoid-shaped and symmetrical about its central point of inflection. It often gives a very good representation of the course of population growth. It occupies a prominent, though controversial, position in modern theories of population dynamics.

Optimum yield. An important consequence of the sigmoid form of population growth is the fact that populations of intermediate size are capable of more rapid growth and greater productivity than either very large or very small populations. If growth were strictly logistic, the most rapid growth would occur at a population size of $K/2$ and the growth rate at this point would be $rK/4$.

When man begins to exploit a large population, as in commercial fishing, the effects of his catch will be to reduce population size. If exploitation is not too intense, the smaller population will lie on a steeper portion of the sigmoid growth curve and will therefore be more productive than the larger population. The population is said to compensate for the increased mortality. In theory, productivity will increase with rate of exploitation to the point where population size reaches the inflection in the growth curve. Hence the maximum possible sustained harvest would be obtained by reducing the population to the inflection point and harvesting at a rate just sufficient to maintain this size. There are many practical difficulties in all actual attempts to determine the optimum rate of harvest and the general problem has become widely known as the optimum-yield problem. It is noteworthy that if the population is "overfished" so that its size passes below the inflection point, productivity will decline with each further decrease in size. Then each increase in the effort to harvest a crop will have the effect of reducing the long-term yield. There is reason to believe that many commercial fisheries are reducing their total catch by fishing too intensively. This is shortsighted from a biological point of view, but the interest of a fishery is more economic than biological. From an economic point of view, the value of future catches must be discounted in comparison with the current catch, at a discount rate that is roughly comparable with the potential interest rate on alternative investments. The theory of optimal economic yields from exploited biological resources is in its infancy. It remains to be seen whether or not current practices of exploitation are optimal in any sense at all.

The same principles apply to attempts to control noxious species. Rodent populations, for example, compensate for mortality and it is possible to harvest a large annual crop of rats without actually reducing the population. Programs of killing are often discontinued before the population passes below the inflection point where control would become progressively easier. Consequently, the most effective way of dealing with noxious forms is often to reduce the carrying capacity of the environment; programs for improving garbage disposal and for ratproofing buildings will often be much more effective than programs of killing.

Fluctuations. Populations of many species fluctuate in size from year to year, and a very large literature exists on this subject. Plagues of rodents and locusts are recorded in the Old Testament and similarly ancient sources, and the migrations of the lemmings and the eruptions, outbreaks, or gradations of various insect populations have often attracted popular attention. There is a tendency for eruptions to be most conspicuous in high latitudes and other regions where the biota is composed of relatively few species of plants and animals. Outbreak years typically follow periods of buildup during which the population nearly realizes its potential of exponential growth. Eventually, population size exceeds the capacity of the environment to sustain it and the population "crashes," often dropping abruptly to a very low level.

Cycles. The most discussed of the fluctuating populations have been those of certain gallinaceous birds, rodents, rabbits, and fur-bearing mammals of northern regions. The records of the Hudson's Bay Company, for example, provide figures for a long series of annual catches, and many students of populations have considered that the rhythms, or cycles, in such records indicate a regular periodicity in the rise and fall of population size. Although cycles of various length have been postulated, most competent opinion has considered that there are two predominant cycle lengths: a short cycle of approximately 3 or 4 years and a longer one often referred to as the 10-year cycle.

Numerous explanations have been advanced. One of the most popular has been the belief that the populations follow some extraterrestrial rhythm, especially the "sunspot cycle." Such hypotheses suffer from numerous observations indicating that populations in different regions may be out of phase with each other. Others have based explanations on population dynamics, claiming, in effect, that the cyclic species are deficient in feedback mechanisms so that exponential growth is not inhibited until disastrously high densities are attained. In particular, if reproduction occurs in synchronous bursts each generation, and if the biotic potential is sufficiently high, a population could cycle or even behave chaotically in a continuously benign environment. Still other researchers have attributed the cycles to interactions between two species: herbivores and their food plants, or predators and their prey. The predator is visualized as growing until it exhausts its food supply and then undergoing violent decline until the prey population has time to recover. These hypotheses are not entirely satisfying because it is difficult to see why many species with diverse life histories should adhere to two basic cycle lengths. The Canadian lynx and the chinch bug, for example, are both considered to exhibit 10-year cycles.

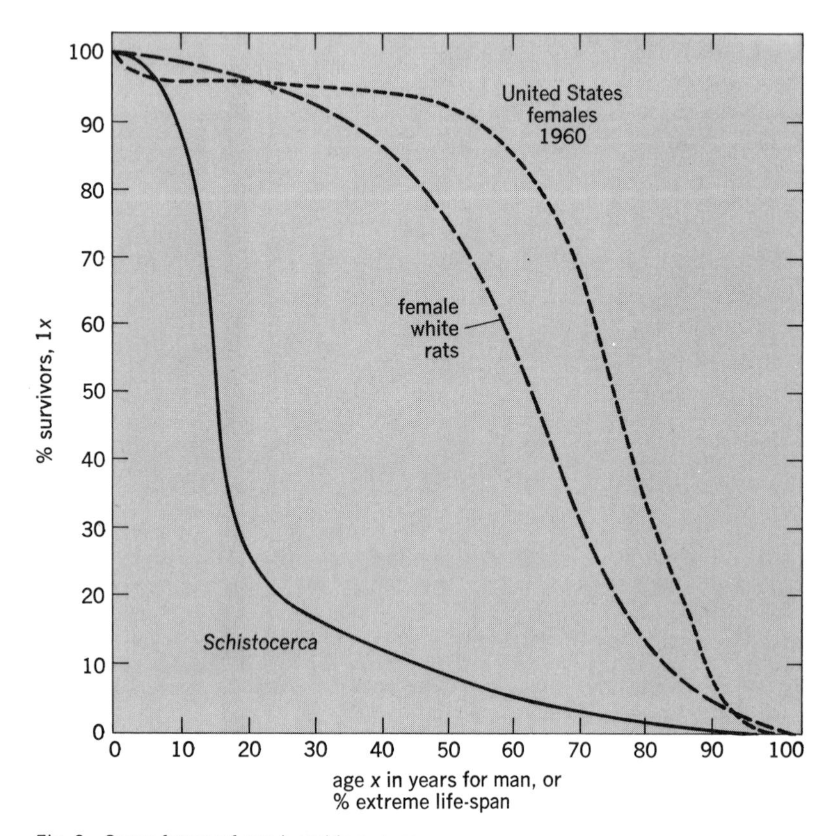

Fig. 2. Some forms of survivorship curves.

Whatever their causes, population fluctuations are often of great practical importance. Much remains to be learned about possibilities for predicting peak years in crop damage and in the harvest of food, game, and fur-bearing animals, or for minimizing the expectation of financial loss resulting from these fluctuations.

Natural selection. Some populations are continually buffeted by the effects of the physical environment on birth and death rates. Such populations remain well below the carrying capacity of their environment, and natural selection favors increases in the biotic potential for reproduction, even at the expense of increased sensitivity to the physical environment. Other populations persist for long periods very near the carrying capacity. Here selection favors survival and competitive ability even at the expense of lowered biotic potential. Thus different populations may be characterized by either of two self-reinforcing extremes of dynamic behavior: high biotic potential with high rates of turnover and population fluctuations versus steady and crowded population with very slow turnover. Human strategies for the control of noxious species may be profoundly different, depending on which of these extremes more realistically portrays the species in question.

Other factors. Of course, the logistic function is an overly simplistic caricature of population dynamics. It ignores migration and dispersal, as well as the interactions of a given population with the dynamics of its resources, its predators, and its competitors. In addition to influencing the biotic potential of a species, these factors may transform the simple "carrying capacity" into a multiplicity of equilibria, some stable and some not. This makes the study of population dynamics both challenging and very difficult. *See* ECOLOGICAL INTERACTIONS; POPULATION DISPERSAL.

Age structure. Not only the size but also the composition of a population is governed by the age schedules of natality and mortality. If the life history features and the death rates for individuals of each age remain constant, a population will eventually attain a stable age distribution such that the individuals of any particular age constitute a fixed proportion of the total. In human populations, about one-half of the individuals typically fall into the age range of 15–50 years, but the ratio of older individuals to very young differs greatly from one nation to another. Consequently, diseases of old age attain greatest importance in populations where life expectancy is greatest. Problems relating to age structure are also of great significance in nonhuman populations. It is apparent that a predator can benefit man by selectively killing superannuated game animals, thus increasing productivity, or that such a predator can work against man's efforts to control noxious species even while killing many of the undesirable forms.

In populations exploited by man, age structure is affected by the intensity of exploitation. It is obvious that imposing a new mortality factor on the population will reduce life expectancy and so reduce the average age of the population members. In some animals, including man, rats, and some insects, older individuals cease to be reproductive, and the lowering of average age by moderate exploitation leads to a direct increase in birth rate.

Also, by making additional resources available to promote the survival of young, a younger population results. With very intense exploitation an opposite result occurs: Reproduction is impaired by depletion of the reproductive age classes, and the remnant members become aged, producing an unproductive population of increased average age.

Demography. Demography is commonly and most concisely defined as the statistical study of populations. Demographers are usually students of human populations and are concerned not only with such fundamental and general phenomena as natality, mortality, migratory movements, and age and sex structure, but also with such factors as population composition by social status, ethnic groups, and income level; rates of marriage, divorce, and illegitimacy; the size of the labor force; and such esoteric questions as the motivation for a particular family size and survival rates in various occupations. However, many ecologists insist on recognizing a field of general demography embracing the study of the fundamental phenomena that are common to all populations of living things.

The basic technique for the statistical analysis of any population is the construction of a life table. The fundamental features of a life table is a column of numbers showing, for a "cohort" consisting of a large number of individuals born simultaneously, how many would be alive on each subsequent birthday until the cohort became extinct. This column is designated "survivorship" and is represented by the symbol l_x as the proportion surviving to the xth birthday.

For a long-lived species such as man it is impracticable actually to follow a cohort throughout its history and, in any case, mortality factors would probably change so drastically over the course of a century that a life table constructed by this straightforward technique would be worthless by the time the data for its construction became available. Consequently, actual tables are constructed indirectly from observed "age-specific death rates," which are the number of individuals dying at a particular age divided by the total number entering that age class.

Many life tables have been constructed for different species and for different environmental conditions. The survivorship curves vary from a convex to a very concave shape as viewed from above (Fig. 2). The convex shape indicates slight mortality until advanced age and the concave shape indicates heavy early mortaility followed by very gradual attrition of the few individuals that actually attain maturity. There is evidence that advances in medicine and sanitation have changed the survivorship curve for man from concave to convex.

By making certain assumptions about the temporal stability of mortality factors and the number of annual births it is possible to use the life table to derive "life expectancy" figures representing the average number of years of life remaining for an individual of a given age. This technique is fundamental for life insurance practice.

Given the form of the life table and a corresponding table of age-specific birth rates, it is possible to project the future of any population and to calculate the results of changing birth or death rates. *See* ECOLOGY. [LAMONT C. COLE]

Bibliography: A. J. Coale, *The Growth and Structure of Human Populations*, 1972; C. W. Clark, *Mathematical Bioeconomics*, 1976; P. R. Ehrlich and A. H. Ehrlich. *Human Ecology*, 1973; H. S. Horn, in J. R. Krebs and N. B. Davies (eds.). *Behavioural Ecology*, 1978; R. M. May, *Nature*, 272: 491–495, 1978; R. M. May (ed.), *Theoretical Ecology*, 1976; Population studies: Animal ecology and demography, *Cold Spring Harbor Symposium on Quantitative Biology*, vol. 22, 1957.

Predator-prey relationships

The natural history of predator-prey relationships (food habits and the like) in aquatic communities is moderately well known, except for environments such as the deep sea that are difficult to sample or observe. This article will discuss the quantitative implications of these relationships as follows: (1) the effect of predation on the population size of each species of prey, (2) the strategies that predatory populations have evolved through natural selection to ensure their own survival, and (3) the broader effects of predation on the species composition of the whole community. *See* FOOD CHAIN.

Species population size. In the simple case of two species, the predator (as distinct from a parasite or scavenger) affects the dynamics of the prey population by killing its members. The extent to which predation tends to stabilize the population size of a particular prey species depends on whether the intensity of predation—the fraction of the prey population removed by predation per unit of time—increases as the size of the prey population increases and on how much time lag is involved. The intensity of predation may increase through a disproportionate increase in abundance of the predators due to immigration or reproduction or both, and through a disproportionate increase in the number of prey eaten by each predator. Such an increase in the number of prey eaten occurs when the predator forms a "search image" and concentrates its feeding activities on the prey species in question (usually through learning on the part of the predator) or when increase in the body size of predators with indeterminate growth enables them to remove disproportionately more (or larger and hence more fecund) prey.

Carnivores that are not predators in the strictest sense may indirectly control the population dynamics of the organisms on which they feed. Young flatfish "prey" on buried bivalves by biting off the exposed siphons when bivalve density exceeds a threshold level. The bivalves are able to regenerate these respiratory-feeding tubes but do so at the expense of producing more gonadal tissue and hence produce fewer offspring.

Strategies for survival. The evolutionarily correct strategy for the predatory population is to feed so as to maintain the prey population at a size and age structure that will maximize the predator's sustainable intake of energy. Ideally, the pattern of predation should maximize the growth rate and food-chain efficiency (for example, for carnivorous predators, the ratio between the yield to the predator and the production of the prey species' food) of the prey population and minimize the metabolic cost to the predator of searching for, pursuing (if necessary), capturing, and digesting prey and the risk it experiences from its own predators. The intertidal predatory snail *Thais* attacks a limited number of barnacles per day but preferentially attacks large ones, in effect maximizing its ingestion per attack and removing slow-growing individuals from the prey population. However, the metabolic cost per attack probably also increases with increasing size of barnacle, and large barnacles are more fecund than small ones are. When prey species become infertile in later life, predation concentrated upon postreproductive individuals (which often have a low growth rate) also tends to maintain a high rate of replenishment of prey. For example, most of the mortality of pelagic salps (*Thalia democratica*) caused by the predacious copepod *Sapphirina* occurs with postreproductives, so that the very rapid rate of population growth of the salps is reduced rather slightly even by relatively high concentrations of this predator. Predation upon the surplus, territoryless individuals of a species of animal prey that requires possession of a territory for mating and successful rearing of young has the same effect. Removal of injured individuals is a more obvious example; carnivores preying on herds or schools often single out for attack individuals which are conspicuously different in appearance or behavior from their fellows.

Metabolic costs of predation are sometimes minimized, relative to the benefits, by varying the degree of selectivity for particular species or sizes of prey. As the abundance of potential prey increases relative to the searching range of the predator, so that a relatively large amount of energy is involved in pursuit (if any) and capture, some kinds of predators tend to be specialists, having a restricted search image and limited, but presumably efficient, feeding behavior. Predators whose prey is widely dispersed, so that most of the overall cost of predation is in searching, tend to be generalists. An individual predator may also change foraging behavior depending on the relative abundances of different prey; anchovies switch from a generalized filtration of small prey items, which is relatively ineffective at natural concentrations, to a directed, biting attack on individual, larger prey when such prey become sufficiently abundant.

Community species composition. Some predators markedly affect the composition of communities because those species on which they prey directly compete for resources with other species, or are predators themselves. Predation increases diversity (the number of species in a community and the distribution of individuals among species) when the species that would dominate the community is preyed upon preferentially because of its palatability, visibility, chemical attractiveness, or size. For example, moderate grazing by domestic sheep on rich pasture increases the diversity of the vegetation since the competitively superior plants are highly palatable; sea urchins have a similar effect on the seaweeds in tide pools; and grazing by limpets on young plants prevents the establishment of competitively superior macroscopic algae in the intertidal zone.

When planktivorous fish remove the larger and more visible zooplankters in a lake, smaller species which are competitively inferior may thrive, and the resulting change in composition of the zooplankton sometimes affects the composition of the phytoplankton as well, since larger zooplankters feed preferentially on larger phytoplankton cells. Mussels, which outcompete other sessile, intertidal species for space, are preyed upon preferentially by starfish, and space is thus made available to other species in the lower intertidal, where starfish are active. Predation by the coral-eating starfish *Acanthaster* apparently increases the diversity of coral in areas where the starfish is abundant but not so dense as to decimate all coral, although the total biomass of living coral is reduced by even moderate densities of *Acanthaster*.

There are also situations in which predation leads to a decrease in diversity. If the dominant species is relatively unpalatable, predators may increase its dominance by selectively removing more palatable species, as do sheep grazing on poor pasture, in which case diversity of vegetation may thus be reduced. Even in rich pasture, heavy overgrazing may lead to the dominance of a few toxic, spiny, or otherwise unpalatable species. *See* COMMUNITY; ECOLOGICAL INTERACTIONS.

[MICHAEL M. MULLIN]

Bibliography: E. Curio, *The Ethnology of Predation*, 1976; T. W. Schoener, *Annu. Rev. Ecol. System.*, 2:369–404, 1971.

Radiation biochemistry

The study of the response of the constituents of living matter to radiation, a specific injurious agent. Biochemistry, in the ordinary sense, deals with the chemistry of the building stones of living tissues and organisms, and with the balance and integrated metabolic reactions in which these take part. This article deals with the effect of ionizing radiations, radiations that ionize matter through which they pass.

The chance of exposure to radiation has increased in this atomic age. Not only is radiation more widely used in medicine for diagnostic and therapeutic purposes, but the applications of radiation in industry have increased. For example, the hazards connected with the development of atomic energy, and the possible use of atomic weapons in war make research in the effects of radiation important.

The capability of penetrating to every part of the interior of cells, without being obstructed by membranes or defensive barriers, puts ionizing radiations in a unique position as compared with other noxious agents, which are limited in penetration, or selectively active on special cell constituents.

The cells of living matter consist of a cell membrane surrounding protoplasmic protein, in which are embedded the nucleus and various small granular bodies, such as mitochondria and microsomes. The whole cell structure is permeated with water. The content of the cell is inhomogeneous, highly organized, and equipped with a series of enzymes which make complicated metabolic reactions possible.

The indirect and the direct modes of action of radiation have been proposed as mechanisms of radiation effects. These modes of action are discussed in the following paragraphs.

Indirect action. The water content of cells is about four times greater than their dry weight. The effect of radiation on the water has consequences for the solid matter contained in it.

When water is irradiated, it is split into the primary radiation products, hydroxyl (OH) radicals, hydrated electrons, and hydrogen (H) atoms, which are highly reactive, uncharged chemical entities. Oxidation and reduction reactions are brought about when the primary radiation products collide with solutes, the substances dissolved in water. In the absence of solutes, however, they quickly recombine to form water. The irradiation thus acts indirectly on solutes via the water. Consequences of this indirect action are the dilution effect, the protection effect, and the oxygen effect.

Dilution effect. This effect causes a dilute solution to appear more sensitive to radiation than a concentrated one, because a given dose of radiation results in the formation of a given number of radicals which will react with a corresponding number of solute molecules. Therefore, the proportion of solute molecules chemically changed will be large in dilute and small in concentrated solutions.

Protection effect. This comes into play when there are two or more solutes in solution which are capable of reacting with radicals. All these solutes will compete for the existing radicals and, therefore, fewer radicals will be available for each species of solute than would be the case if only these solutes were present in the solution. In other words, there is a mutual diminution of the radiation effect and each solute appears protected by the others against radiation. This protection effect will vary in accordance with the absolute amounts of solutes and with their specific capability of reacting with radicals, that is, with their capability of acting as acceptors for radicals.

The protection effect operates also in a one-solute solution if the irradiation products formed from this solute can still react with radicals. Before radiation starts, the solution of an enzyme (catalyst of living matter) contains the dissolved active enzyme molecules only. Each fraction of radiation dose delivered inactivates some enzyme molecules which, though no longer active, can still react with radicals. The inactive molecules then represent a second solute which competes with the still-active molecules for radicals. The result is that equal but subsequent increments of radiation become less and less effective, so that an activity-dose curve takes on an exponential shape (Fig. 1).

Oxygen effect. The primary OH radicals, hydrated electrons, and H atoms give rise to other radicals if oxygen is present in solution, namely to HO_2 or O_2^- radicals and to the stable product hydrogen peroxide (H_2O_2). These, as oxidative agents, can react with solutes and thereby enhance the radiation effect when compared with oxygen-free solutions, hence the term oxygen effect. Hydrogen peroxide may also lead to the formation of organic peroxides which can be responsible for an aftereffect, that is, a continuation of decomposition of solutes after irradiation has ceased. The interference of hydrogen peroxide complicates the clarification of reaction mechanisms. It has been found that nitric oxide increases sensitivity to radiation under conditions of anoxia to an extent similar to that observed with oxygen. The explanation which is put forward for this effect is that both nitric oxide and oxygen have an equal affinity for carbon radicals.

The target theory as originally proposed did not leave room for modification of radiation effects by chemical means. It was an all-or-none effect. Yet it has often been found that the presence of oxygen or of nitric oxide during irradiation increased the radiation effect not only in solutions but also in dry matter. It has now been proposed that the theory be modified by assuming that immediately after the passage of an ionizing particle the target molecule is left in a highly reactive state, facilitating chemical reaction. The reaction will depend on the chemical environment, the physical state, or both. Thus, one can visualize the possibility of modifying the effect of the primary dissipation of energy so as to restore the target molecule (a healing effect) or to cause its irreversible injury, depending on the reaction with the modifying agent.

The promising method of microwave spectroscopy for detecting and measuring concentrations of free radicals has established that radiation causes the formation of radicals of various lifetimes in biological matter. The subsequent interaction between these and the presence of gases like oxygen and nitric oxide, which are themselves radicals, offers an explanation of the oxygen effect and the nitric oxide effect that is alternative or supplementary to the explanation based on interaction with radical-reaction products.

Direct action. Since the dissipation of radiation energy is not confined to the solvent, direct ionization with subsequent chemical change will occur in solute molecules themselves. The frequency of such an event (single hit) increases in proportion to the concentration of the solution, and reaches its maximum when one hit is scored per molecule of a dry substance.

Some investigators believe that the nucleus of a cell contains a vital and sensitive structure in which the primary event, that is, an ionization, has to occur, or that an ionizing particle has to pass near or through it in order to cause a chemical alteration that results in the biological effect which is subsequently observed. This is the target-hit theory.

An important practical application of the direct-hit theory is the determination of the molecular weight—sometimes even the shape of large, biologically active molecules irradiated in the dry

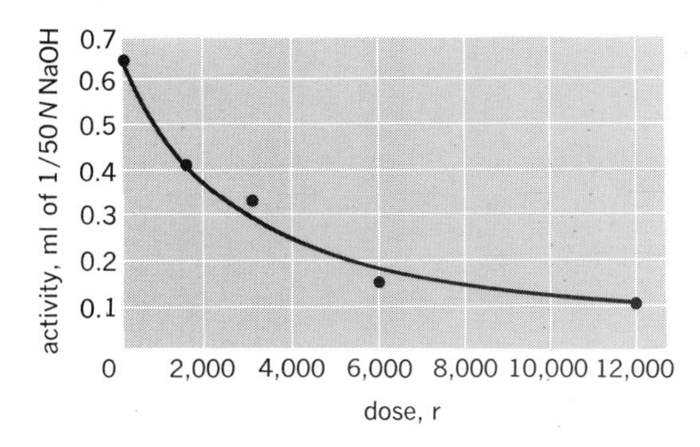

Fig. 1. Exponential relation between degree of inactivation and x-ray dose for carboxypeptidase solutions. (*W. M. Dale, 1940; reproduced by permission of the editors of Biochemical Journal*)

state. The underlying assumption is that the destruction of the biological function, for example, of enzymic activity, or of the ability of a virus to infect, is caused by a primary ionization produced by a fast charged particle passing through the molecule. It is possible to calculate the radiation dose which will produce, on the average, one ionization per molecule. From the number of ionizations occurring per unit volume the target size can be assessed. This method has been successful in a number of cases.

The protection effect which was previously the prerogative of the indirect action has been found to occur also in solid matter when relatively small amounts of additional substances are incorporated in the solid. Some form of intra- and intermolecular transfer of energy is assumed to occur whereby the energy is channeled preferentially to the added substances.

Direct plus indirect action. The share taken by the direct and indirect modes of action is illustrated in Fig. 2, which shows the inactivation of an aqueous solution of an enzyme (carboxypeptidase from pancreas) at various concentrations.

Since water is always in excess in living tissues, except in bone structure and fatty tissue, the opportunity of reactions with radicals is always preponderant. A distinction between the direct and indirect modes of action becomes increasingly trivial as the source of active radicals approaches the molecule or structure upon which they act. As a result, the two modes of action merge in the immediate vicinity of the target, constituting a direct hit.

Measurement of radiation sensitivity. The problem of a more detailed analysis of the radiation products from substances of biochemical importance has been solved in only a few instances. However, the protection effect has made it possible to assess overall radiation sensitivity over a wide range of substances, extending from small molecules to the large molecules of proteins. From the degree of inactivation of an enzyme solution of known concentration in the absence and in the presence of a protective substance, a value of the reactivity with radicals of the substance in question can be derived. In this way it has been found that specificity of radiation effect can be detected in small molecules. Thus, it has been shown that sulfur in organic molecules causes a high degree of

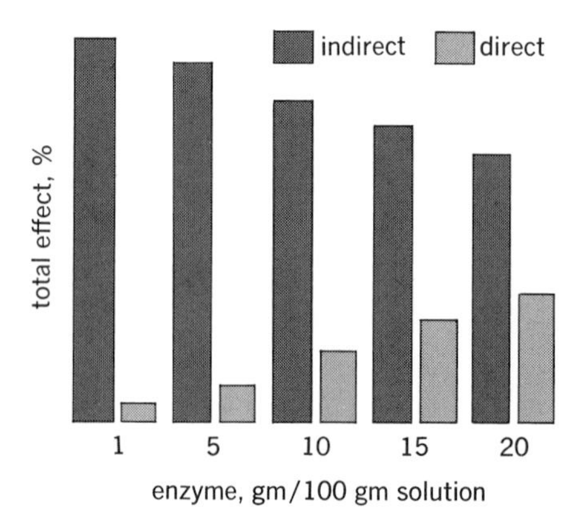

Fig. 2. Relative contributions of indirect and direct actions to total effect of x-rays on carboxypeptidase in solution. (*W. M. Dale, 1947; reproduced by permission of the editors of British Journal of Radiology*)

radiation sensitivity. In general, specific effects are not discernible in large molecules, because of the great number of their reactive groups, but it can be said that the protective power of large molecules is approximately proportional to their molecular weight.

Effect of radiation. The effect of radiation upon biological material is discussed in the following paragraphs.

Biological reduction-oxidation systems. Most reactions of solutes with radicals consist of oxidations and reductions, and it has been proposed that the OH radicals, hydrated electrons, and H atoms in irradiated water constitute a redox system, having an equivalent redox potential (ERP). This system reacts with a range of redox systems as solutes in such a way that for redox potentials greater than -0.52 volt, oxidation occurs, and for potentials less than -1.1 volts, only reduction takes place. This depends on such things as the type of radiation, pH, and presence or absence of oxygen. In the range between these values both oxidation and reduction are possible. The redox potentials of most redox systems in cells are in the oxidation range but will display different resistance to oxidative changes, according to their total concentration as well as to the respective ratio of their reduced state to their oxidized state. Biological redox systems are normal constituents in cells and form the link in many metabolic steps. There is opportunity, therefore, for interference with the normal metabolism. Some instances of biochemical redox systems are cysteine to cystine, sulfhydryl to disulfide (SH—SS), glutathione, prosthetic groups of enzymes, such as flavoprotein, coenzymes I and II, and ascorbic acid.

Redox systems may also occur in which the reversibility of the reaction is impaired. If the oxidized or the reduced state of the reacting compound suffers secondary changes, such as the formation of a polymer, or if the reduced form is capable of forming a molecular compound with the oxidized form, no proper equilibrium will be established.

Proteins. One of the most important constituents of living matter, proteins consist of long chains of amino acids occurring in characteristic proportions and in specific sequence, linked together by the peptide linkage CONH. Side chains protruding from the main chain can form cross linkages between neighboring chains. In solution, they form finely dispersed colloidal systems. The variety of existing proteins is very great, but all proteins have one reaction in common, namely, the property of denaturation. This is usually an irreversible change which occurs when proteins combine with certain chemicals or when heat or radiation is applied.

The denaturation manifests itself as coagulation with subsequent insolubility. It has been found that ionizing radiations lower the resistance of proteins to thermal denaturation. After irradiation, protein solutions contain different denatured protein derivatives and show a marked decrease in energy content, signifying deep-seated structural changes. It has been established that the oxidative attack by OH radicals is directed toward the peptide linkage. This results in the formation of high-molecular-weight carbonyl (C═O) compounds and of keto acids, and the release of ammonia. The radiation-sensitive SH group of cysteine, an amino acid occurring in egg albumin, can be oxidized to a disulfide (S-S), which can form a cross-linkage between neighboring chains. Oxidation can occur even further, beyond the S-S stage, without denaturation or marked instability of the protein. It has further been observed that after hydrolysis of irradiated serum albumin several amino acids are partly destroyed. Examination of irradiated protein solutions by spectrophotometry also reveals changes. Bovine serum albumin, serum globulin, and egg albumin show an increase in optical density which apparently is due to the action of radiation on their tyrosine component; if, however, the proteins contain more tryptophan than tyrosine, a decrease in density is observed.

Not only is protein, as such, changed by radiation but its building stones, the amino acids, are affected.

Amino acids. The principal effect is the loss of ammonia from, or deamination of, the amino acids. The extent of deamination varies with experimental conditions, but an important point is that it also varies with the chemical configuration of the amino acid itself. If the amino group is in the alpha (α) position, attached to the carbon atom next to the carboxyl group (COOH), as in α-alanine, $CH_3{\cdot}CH(NH_2){\cdot}COOH$, the loss of ammonia is nearly twice as great as for β-alanine, $CH_2(NH_2){\cdot}CH_2{\cdot}COOH$. In this compound, the amino group is attached to a second carbon atom, that is, in the β position. This is an example of the specificity of radiation effects. More ammonia is split from histidine, where the glyoxaline part may contribute to the yield.

In experiments in which radiation products other than ammonia were examined, it was found that alanine irradiated in a vacuum yielded acetaldehyde, pyruvic acid, propionic acid, ethylamine, and carbon dioxide. Products from glycine included glyoxylic acid, formaldehyde, acetic acid, formic acid, and carbon dioxide. It is, however, claimed that the appearance of the various products depends on whether radiation doses applied have been moderate or massive. This claim is made because some of the radiation products are not due to initial reaction, but rather such products are formed by further oxidation or decarboxylation of glyoxylic acid.

Enzymes. These are proteins which differ from other proteins in their ability to act as catalysts. Enzymes speed up specific chemical reactions. They either have special active groups in their makeup, enabling them to combine with their specific substrates on which they act, or they are more or less firmly linked to a nonprotein partner, or prosthetic group, which, in cooperation with the protein part, functions as a highly specific catalytic system.

As far as their protein nature is concerned, enzymes will undergo the same general changes as described for other proteins, with consequent loss of activity. There are some enzymes, the S-H enzymes, in which that group is essential for enzymic activity. This S-H group is particularly sensitive to radiation and may undergo changes before deeper-seated modification of the protein has taken place. If the inactivation of the S-H enzyme has not gone far, it can be restored to its original activity by the addition of the tripeptide glutathione, which contains S-H.

An example of an enzyme containing a prosthetic group is D-amino acid oxidase, which specifically oxidizes D-amino acids only. This enzyme can be split into flavin adenine dinucleotide, a nonprotein, and a specific protein. Neither part, on its own, has enzymic activity, but when combined they constitute the complete active enzyme. Each part can be chemically changed by radiation and, when rejoined, shows a lower activity than after irradiation of the complete enzyme.

In a comparative study of the effect of the densely ionizing alpha radiation versus x-radiation on the enzyme carboxypeptidase in solution, it was found that alpha radiation was only one-twentieth as effective as x-radiation. The effect appeared to be entirely due to the δ-rays, which branch off the α-ray track as spurs and which have an ion density similar to that of x-rays, not to the primary ionization column of the α-ray track. Quite generally, the lower efficiency of alpha radiation on substances in aqueous solution is in contrast to its higher efficiency on biological systems, such as the breakage of chromosomes in cells.

Nucleic acid and nucleoproteins. These are the important chemical building stones of the genetic material contained in the chromosomes of cell nuclei. The nucleoproteins are saltlike unions of a nucleic acid with basic proteins, such as protamine or a histone. Two types of nucleic acids exist, ribonucleic acid (RNA) and deoxyribonucleic acid (DNA), both of which are built up from nucleotides containing a nitrogenous base, a pentose sugar, and phosphoric acid. The base forms an ester with the phosphoric acid. Both nucleic acids contain the bases adenine, guanine, and cytosine but RNA has uracil and the sugar D-ribose, whereas DNA contains thymine and the sugar D-deoxyribose. RNA is formed predominately in the cytoplasm and DNA in the nucleus. Since the molecular weight is of the order of $6-8 \times 10^6$, each molecule must contain a great number of nucleotide units. Nucleic acids are structured as two complementary, helical, nucleotide threads, joined together by hydrogen bonds between the basic guanine and cytosine and between adenine and thymine. The nucleic acids are similar in organization to protein, with its amino acid units and its cross-linkages. Irradiation breaks down the hydrogen bonds between the DNA threads, and denaturation by heat is facilitated after irradiation. The instability of nucleic acids is evident from the short heating (15 min in boiling water) required for a marked decrease in viscosity of DNA accompanied by decrease in molecular weight. A similar decrease of viscosity in dilute DNA solution is effected by relatively small doses of radiation.

Chemical changes require large doses of radiation. Among the reactions observed are deamination, liberation of free purine, decrease in optical density, increase in amino nitrogen, breakage of the pyrimidine ring, oxidation of the sugar moiety (ribose portion), and liberation of some inorganic phosphate. The conditioning effect of radiation is shown by the fact that acid hydrolysis subsequent to irradiation liberates free phosphate more quickly than would have occurred without irradiation. In the irradiation of the breakdown products of nucleic acids, such as nucleotides, nucleosides, and the purine and pyrimidine bases, the radiation effects resemble those from nucleic acids.

[WALTER M. DALE/M. EBERT]

Bibliography: K. I. Altman (ed.), *Radiation Biochemistry*, 1970; M. Ebert and A. Howard (eds.), *Current Topics in Radiation Research*, vol. 1, 1964; M. Errera and A. Forssberg (eds.), *Mechanisms in Radiobiology*, vols. 1–2, 1961; A. Hollaender (ed.), *Radiation Biology*, vol. 1, 1954; A. M. Kuzin, *Radiation Biochemistry*, 1964; J. T. Lett and H. Edler (eds.), *Advances in Radiation Biology*, vol. 7, 1978.

Radiation biology

A study of the influence of light or ionizing radiation, such as x-rays or fast particles, on living systems. Radiation biology ranges from a consideration of the effects of visible light on metabolism to the effects of cosmic rays on whole organisms. Because of the breadth of the subject, it is studied to a large extent in separate areas. The first division separates the fields of ionizing radiation effects and photon effects. The term photon effects means the action of ultraviolet, visible, and infrared light in the region where ionization does not occur.

Ionizing radiation. The process of ionization is characterized not only by the release of an electron from an atom, with the formation of a positive residue (the whole being an ion pair), but also by the large amount of energy associated with the process. A typical chemical bond has an energy of 3 electron volts (ev); a typical primary ionization has an energy of 60 ev, or 20 times as large. In consequence, ionizing radiation exerts a powerful molecular action. Such action is not very specific; it occurs anywhere at random, and the ionizations are relatively far apart. Ionizing radiation therefore produces random energy releases of great size and generally of great disruptive effect. When such a disruptive effect is on functioning units of the organism, it is termed direct action. It is termed indirect action when it occurs on the water moiety through formation of active radicals, notably OH and H, which exert a gentler action, and which, because of radical diffusion, can act at some distance from the original ionization.

Dose. To measure any effect of ionizing radiation the amount of radiation must be measured. Such measurement is called dosimetry; the amount given is the dose. Three units are used: the roentgen (r), defined as that radiation which will release one electrostatic unit of separated charge in 0.001293 g of dry air; the rad, which is defined as that amount of radiation which will release 100 ergs in 1 g of representative biological tissue; and the roentgen equivalent physical (rep), which is meant to be an energy-defined unit equivalent to 1 r and which approximates 93 ergs per gram of tissue.

Some effects. A considerable variety of effects exists. A brief sampling follows.

1. Survival after whole-body irradiation. A whole animal, such as a mouse, which has been irradiated, shows no marked immediate effect. After a few days, however, a definite increase in mortality occurs if exposure exceeded 400 r, with a mean lethal dose at 500 r and a rapid increase in mortality rising to 100% at 1000 r. Such a dose-survival curve is called sigmoid. In such a process death is due to effects on a variety of organs, the blood-forming organs and any rapidly dividing tissue being the most sensitive.

2. Production of mutations. Ionizing radiation produces mutations in any living organism, with the possible exception of viruses. The majority of such mutations are of a lethal nature. *See* MUTATION.

3. Production of chromosome breaks. All kinds of chromosome abnormalities, including chromosome breaks, are produced by ionizing radiation. Such breaks seem to require of the order of 50 ion pairs to occur. Chromosome breaks can restitute and probably the majority do so. In the presence of dissolved oxygen the rate of chromosome break increases by a factor of more than 2. This is an example of the oxygen effect, which is one of the more important ways by which radiation action can be modified.

4. Delay of cell division. Ionizing radiation inhibits cell division to a marked degree. The degree of delay increases with the dose.

5. Formation of giant cells. Cells in which division is delayed may grow into giant cells many times the normal size in the case of mammalian cells, or into long filaments in the case of bacteria.

6. Reduction of survival. If means for studying the ability of a cell to divide are available, it is found that animal, plant, and bacterial cells and viruses are reduced in survival. For human cells the survival follows a nearly exponential course, having a shoulder at low doses which is related to repair processes described below. The dose to give 10% survival is approximately 250 r for such cells. For bacteria it approximates 8000 r, for viruses 60,000 r, and in this last case is single-hit, or simply exponential.

7. Action on biological macromolecules. All important biological macromolecules—deoxyribonucleic acid (DNA), ribonucleic acid (RNA), enzymes, and antigens—are destructively affected by ionizing radiation. Roughly speaking, if one ionization occurs within the molecule, or if one active radical in the case of nucleic acid, or 1–10 radicals in the case of protein reach the molecule, it loses its function. Such actions are affected by oxygen tension in general. The process in the case of nucleic acid appears to be breaking, cross linking, or damage to one of the bases, and in the case of protein, the opening of S-S bonds.

8. Action on metabolism and protein synthesis. Metabolism is reduced by ionizing radiation but much less than is division or the formation of lethal mutations. Also reduced are protein synthesis and formation of microsomal particles, or ribosomes, which are agents in protein synthesis.

9. There is evidence that ionizing radiation can precipitate the destructive action of enzymes already in the cell. One of these is a nuclease which very rapidly degrades the DNA, drastically damaging the cell.

10. The phenomenon of recovery from the effects of ionizing radiation has been established. Cellular DNA which has become fragmented by ionizing radiation is found to be restored to much larger size if the cells in which damage has occurred are held for 20 min (bacteria) or several

hours (mammalian cells). It is likely that there are multiple processes of repair and also that the error-prone repair system discussed below acts to restore DNA damaged by ionizing radiation.

11. Ionizing radiation, in rather high doses, acts to depress the immune system.

Action of ultraviolet light. Ultraviolet light differs from ionizing radiation in that it is less energetic and much more specific. Before any of its effects can occur, ultraviolet light must first be absorbed, a process which is wavelength-dependent. The most significant absorption of ultraviolet light is by nucleic acids, which have a broad absorption maximum at 260 nm. The absorption and action are predominantly in the pyrimidines of nucleic acid, and the chemical result may be the addition of H and OH to the pyrimidine, altering it chemically. One important alteration is the formation of a pyrimidine dimer in which two pyrimidines become bonded. In general, many photons must be absorbed for an effect to be observed, and the ratio of changes produced to photons absorbed is called the quantum yield. Quantum yields range from 0.01 for molecules to 10^{-5} for inactivation of viruses.

There is also absorption in protein, notably by aromatic amino acids and cystine. Although the cystine absorption is neither great nor specific, it seems to account for the action of ultraviolet light on proteins.

Light of high intensity (such as sunlight), though not readily absorbed, may yet produce marked biological action, the high intensity compensating for the lack of absorption. Light in the near-ultraviolet range, where absorption by nucleic acids is inefficient, has been found to have significant actions.

Because it must be absorbed, ultraviolet light rarely exerts lethal action on whole animals. The effects produced by ultraviolet radiation are discussed below.

Burns. Ultraviolet light produces drastic action on the skin, familiar as sunburn. Prolonged exposure to ultraviolet light can cause skin cancer, particularly in individuals with fair skin. Sunlight can cause skin cancer, the band of wavelength effective being in the near ultraviolet, from 300 to 330 nm. Between 100,000 and 300,000 new cases of skin cancer occur per year in the United States.

Production of chromosome breaks. Ultraviolet light produces chromosome breaks, though of a less drastic kind than ionizing radiation, being more commonly in one of two chromatids rather than the whole chromosome.

Delay of cell division and giant-cell formation. Cell division is inhibited by ultraviolet light, and such cells can grow to giant size.

Reduction of survival. Ultraviolet light reduces survival, though not so simply, in general, as ionizing radiation, there being a tendency for a fraction of the cells to show less sensitivity. The low-dose shoulder is very marked in some cases. Mutant forms of cells which lack the repair systems described below can be very sensitive.

Photoreactivation. If cells or virus-infected cells are subjected to illumination by light in the blue, or near-ultraviolet, range, the degree of survival is markedly increased, a phenomenon known as photoreactivation. Photoreactivation is explained as the action of a cellular enzyme which splits the pyrimidine dimers described above. This enzyme needs light to enable it to cause the splitting.

Excision repair. There is also a process whereby dimers and cross-links in the DNA can be enzymatically removed and the DNA then accurately patched. This occurs in the dark. Mutant cells lacking one or more of the enzymes needed have been isolated. They are characterized as *uvr*⁻.

Action on metabolism and protein synthesis. These are affected by ultraviolet light, though not in the same way as by ionizing radiation. More effect takes place on the DNA-synthetic process, which is less affected by ionizing radiation.

Recombination repair. A still only partly understood mechanism exists by which one intact genome can be assembled from two damaged genomes. This is related to genetic recombination, and the enzymes needed have not yet been characterized. Mutants lacking one or more of these enzymes are designated rec^-. Cell strains which are rec^-uvr^- are so sensitive that about one pyrimidine dimer is lethal.

Error-prone repair. It has been found that in some cells, which must be rec^+, the effect of radiation is to induce a new repair system which is less specific in its action than those described above. This can confer survival value but has consequences described below.

Production of mutations. Ultraviolet light, including near ultraviolet, produces mutation in all cells, including bacteria. It does so by a relatively small chemical alteration in the DNA. Where the error-prone repair system can be induced, the rate of mutation is higher at low doses. *See* RADIATION BIOCHEMISTRY.

[ERNEST C. POLLARD]

Bibliography: V. Arena, *Ionizing Radiation and Life*, 1971; H. Dertinger and H. Jung, *Molecular Radiation Biology*, 1970; W. Harm, *Biological Effects of Ultraviolet Radiation*, 1979; J. Jagger, *Introduction to Research in Ultraviolet Photobiology*, 1967; J. T. Lett and H. Edler (eds.), *Advances in Radiation Biology*, vol. 7, 1978; S. Okada, *Radiation Biochemistry*, 1970; D. W. Smith and P. C. Hanawalt, *Molecular Photobiology*, 1969.

Radiation damage to materials

Harmful changes in the properties of liquids, gases, and solids, caused by interaction with nuclear radiations. The interaction of radiation with materials often leads to changes in the properties of the irradiated material. These changes are usually considered harmful. For example, a ductile metal may become brittle. However, sometimes the interaction may result in beneficial effects. For example, cross-linking may be induced in polymers by electron irradiation leading to a higher temperature stability than could be obtained otherwise.

Radiation damage is usually associated with materials of construction that must function in an environment of intense high-energy radiation from a nuclear reactor. Materials that are an integral part of the fuel element or cladding and nearby structural components are subject to such intense nuclear radiation that a decrease in the useful lifetime of these components can result.

Radiation damage will also be a factor in thermonuclear reactors. The deuterium-tritium (D-T) fusion in thermonuclear reactors will lead to the production of intense fluxes of 14-MeV neutrons that will cause damage per neutron of magnitude two to four times greater than damage done by 1–2 MeV neutrons in operating reactors. Charged particles from the plasma will be prevented from reaching the containment vessel by magnetic fields, but uncharged particles and neutrons will bombard the containment wall, leading to damage as well as sputtering of the container material surface which not only will cause degradation of the wall but can contaminate the plasma with consequent quenching.

Superconductors are also sensitive to neutron irradiation, hence the magnetic confinement of the plasma may be affected adversely. Damage to electrical insulators will be serious. Electronic components are extremely sensitive to even moderate radiation fields. Transistors malfunction because of defect trapping of charge carriers. Ferroelectrics such as $BaTiO_3$ fail because of induced isotrophy; quartz oscillators change frequency and ultimately become amorphous. High-permeability magnetic materials deteriorate because of hardening; thermocouples lose calibration because of transmutation effects. In this latter case, innovations in Johnson noise thermometry promise freedom from radiation damage in the area of temperature measurement. Plastics used for electrical insulation rapidly deteriorate. Radiation damage is thus a challenge to reactor designers, materials engineers, and scientists to find the means to alleviate radiation damage or to develop more radiation-resistant materials.

Damage mechanisms. There are several mechanisms that function on an atomic and nuclear scale to produce radiation damage in a material if the radiation is sufficiently energetic, whether it be electrons, protons, neutrons, x-rays, fission fragments, or other charged particles.

Electronic excitation and ionization. This type of damage is most severe in liquids and organic compounds and appears in a variety of forms such as gassing, decomposition, viscosity changes, and polymerization in liquids. Rapid deterioration of the mechanical properties of plastics takes place either by softening or by embrittlement, while rubber suffers severe elasticity changes at low fluxes. Cross-linking, scission, free-radical formation, and polymerization are the most important reactions.

The alkali halides are also subject to this type of damage since ionization plays a role in causing displated atoms and darkening of transparent crystals due to the formation of color centers.

Transmutation. In an environment of neutrons, transmutation effects may be important. An extreme case is illustrated by reaction (1). The ^{6}Li

$$^6\mathrm{Li} + n \rightarrow {}^4\mathrm{He} + {}^3\mathrm{H} + 4.8\ \mathrm{MeV} \qquad (1)$$

isotope is approximately 7.5% abundant in natural lithium and has a thermal neutron cross section of 950 barns (1 barn $= 1 \times 10^{-24}$ cm^2). Hence, copious quantities of tritium and helium will be formed. (In addition, the kinetic energy of the reaction products creates many defects.). Lithium alloys or compounds are consequently subject to severe radiation damage. On the other hand, reaction (1) is crucial to success of thermonuclear reactors utilizing the D-T reaction since it regenerates the tritium consumed. The lithium or lithium-containing compounds might best be used in the liquid state.

Even materials that have a low cross section such as aluminum can show an appreciable accumulation of impurity atoms from transmutations. The capture cross section of ^{27}Al (100% abundant) is only 0.25×10^{-24} cm^2. Still the reaction

$$^{27}\mathrm{Al} + n \rightarrow {}^{28}\mathrm{Al}^{\beta-} \xrightarrow{2.3\ \mathrm{min}} {}^{28}\mathrm{Si}$$

will yield several percent of silicon after neutron exposures at fluences of 10^{23} n/cm^2.

The elements boron and europium have very large cross sections and are used in control rods. Damage to the rods is severe in boron-containing materials because of the ^{10}B(n,α) reaction. Europium decay products do not yield any gaseous elements. At high thermal fluences the reaction $^{58}\mathrm{Ni} + n \rightarrow {}^{59}\mathrm{Ni} + n \rightarrow {}^{56}\mathrm{Fe} + \alpha$ is most important in nickel-containing materials. The reaction $(n,n') \rightarrow \alpha$ at 14 MeV takes place in most materials under consideration for structural use. Thus, in many instances transmutation effects can be a problem of great importance.

Displaced atoms. This mechanism is the most important source of radiation damage in nuclear reactors outside the fuel element. It is a consequence of the ability of the energetic neutrons born in the fission process to knock atoms from their equilibrium position in their crystal lattice, displacing them many atomic distances away into interstitial positions and leaving behind vacant lattice sites. The interaction is between the neutron and the nucleus of the atom only, since the neutron carries no charge. The maximum kinetic energy ΔE that can be acquired by a displaced atom is given by Eq. (2), where M is mass of the primary

$$\Delta E = \frac{4Mm}{(M+m)^2} \cdot E_N \qquad (2)$$

knocked-on atom (PKA), m is the mass of the neutron, and E_N is the energy of the neutron.

The energy acquired by each PKA is often high enough to displace additional atoms from their equilibrium position; thus a cascade of vacancies and interstitial atoms is created in the wake of the PKA transit through the matrix material. Collision of the PKA and a neighbor atom takes place within a few atomic spacings or less because the charge on the PKA results in screened coulombic-type repulsive interactions. The original neutron, on the other hand, may travel centimeters between collisions. Thus regions of high disorder are dispersed along the path of the neutron. These regions are created in the order of 10^{-12} s. The energy deposition is so intense in these regions that it may be visualized as a temporary thermal spike.

Not all of the energy transferred is available for displacing atoms. Inelastic energy losses (electronic excitation in metals and alloys and excitation plus ionization in nonmetals) drain an appreciable fraction of the energy of the knocked-on atom even at low energies, particularly at the beginning of its flight through the matrix material. The greater the initial energy of the PKA, the greater is the inelastic energy loss; however, near the end of its range most of the interactions result in displacements. Figure 1 is a schematic representation of the various mechanisms of radiation damage that take place in a solid.

A minimum energy is required to displace an atom from its equilibrium position. This energy ranges from 25 to 40 eV for a typical metal such as iron; the mass of the atom and its orientation in the crystal influence this value. When appropriate calculations are made to compensate for the excitation energy loss of the PKA and factor in the minimum energy for displacement, it is found that approximately 500 stable vacancy-interstitial pairs are formed, on the average, for a PKA in iron resulting from a 1-MeV neutron collision. By multiplying this value by the flux of neutrons [10^{14-15} n/(cm²)(s)] times the exposure time [3×10^7 s/yr] one can easily calculate that in a few years each atom in the iron will have been displaced several times.

In the regions of high damage created by the PKA, most of the vacancies and interstitials will recombine. However, many of the interstitials, being more mobile than the vacancies, will escape and then may eventually be trapped at grain boundaries, impurity atom sites, or dislocations. Sometimes they will agglomerate to form platelets or interstitial dislocation loops. The vacancies left behind may also be trapped in a similar fashion, or they may agglomerate into clusters called voids.

Effect of fission fragments. The fission reaction in uranium or plutonium yielding the energetic neutrons that subsequently act as a source of radiation damage also creates two fission fragments that carry most of the energy released in the fission process. This energy, approximately 160 MeV, is shared by the two highly charged fragments. In the space of a few micrometers all of this energy is deposited, mostly in the form of heat, but a significant fraction goes into radiation damage of the surrounding fuel. The damage takes the form of swelling and distortion of the fuel. These effects may be so severe that the fuel element must be removed for reprocessing in advance of burn-up expectation, thus affecting the economy of reactor operation. However, fuel elements are meant to be ultimately replaced, so that in many respects the damage is not as serious a problem as damage to structural components of the permanent structure whose replacement would force an extended shut down or even reconstruction of the reactor.

Damage in cladding. Swelling of the fuel cladding is a potentially severe problem in breeder reactor design. The spacing between fuel elements is minimized to obtain maximum heat transfer and optimum neutron efficiency, so that diminishing the space for heat transfer by swelling would lead to overheating of the fuel element, while increasing the spacing to allow for the swelling would result in lower efficiencies. A possible solution appears to be in the development of low-swelling alloys.

Damage in engineering materials. Most of the engineering properties of materials of interest for reactor design and construction are sensitive to defects in their crystal lattice. The properties of structural materials that are of most significance are yield strength and tensile strength, ductility, creep, hardness, dimensional stability, impact resistance, and thermal conductivity. Metals and alloys are chosen for their fabricability, ductility, reasonable strength at high temperatures, and ability to tolerate static and dynamic stress loads. Refractory oxides are chosen for high-temperature stability and for use as insulators. Figure 2 shows the relative sensitivity of various types of materials to radiation damage. Several factors that enter into susceptibility to radiation damage will be discussed.

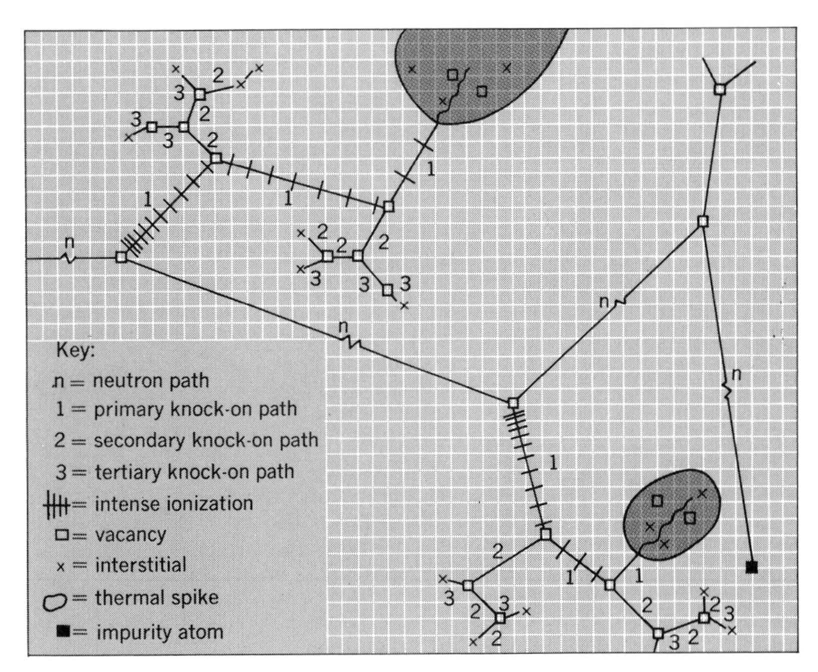

Fig. 1. The five principal mechanisms of radiation damage are ionization, vacancies, interstitials, impurity atoms, and thermal spikes. Diagram shows how a neutron might give rise to each in copper. Grid-line intersections are equilibrium positions for atoms. *(After D. S. Billington, Nucleonics, 14:54–57, 1956)*

Temperature of irradiation. Nuclear irradiations performed at low temperatures (4 K) result in the maximum retention of radiation-produced defects. As the temperature of irradiation is raised, many of the defects are mobile and some annihilation may take place at 0.3 to 0.55 of the absolute melting point T_m. The increased mobility, particularly of vacancies and vacancy agglomerates, may lead to acceleration of solid-state reactions, such as precipitation, short- and long-range ordering, and phase changes. These reactions may lead to undesirable property changes. In the absence of irradiation many alloys are metastable, but the diffusion rates are so low at this temperature that no significant reaction is observed. The excess vacancies above the equilibrium value of vacancies at a given temperature allow the reaction to proceed as though the temperature were higher. In a narrow temperature region vacancy-controlled diffusion reactions become temperature-independent. When the temperature of irradiation is above 0.55 T_m, most of the defects anneal quickly and the temperature-dependent vacancy concentration becomes overwhelmingly larger than the radiation-induced vacancy concentration. However, in this higher temperature region serious problems may arise from transmutation-produced helium. This gas tends to migrate to grain boundaries and leads to enhanced intergranular fracture, thereby limiting the use of many conventional alloys.

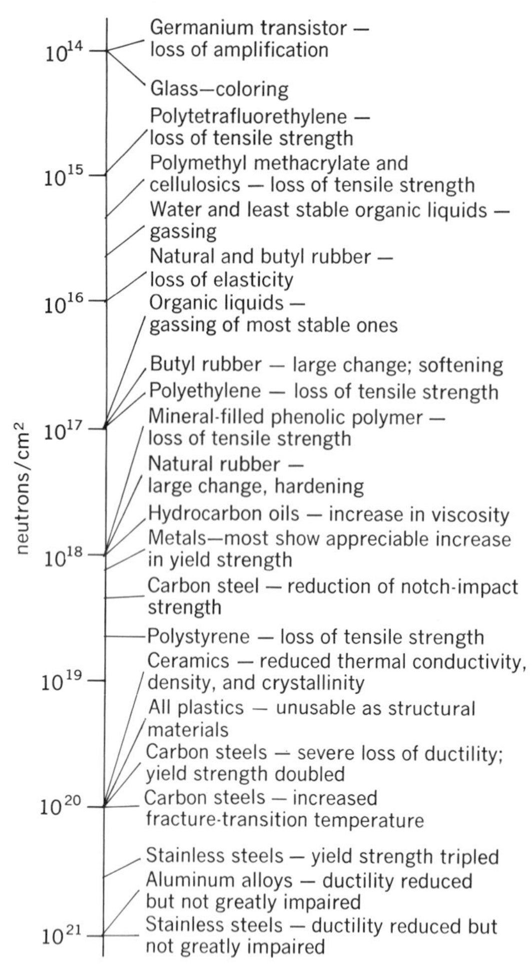

Fig. 2. Sensitivity of engineering materials to radiation. Levels are approximate and subject to variation. Changes are in most cases at least 10%. (*After O. Sisman and J. C. Wilson, Nucleonics, 14:58–65, 1956*)

Nuclear properties. Materials of construction with high nuclear-capture cross sections are to be avoided because each neutron that is captured in the structural components is lost for purposes of causing additional fissioning and breeding. The exception is in control rods as discussed earlier. Moderator materials, in particular, need to have low capture cross sections but high scattering cross sections. Low atomic weight is an important feature since moderation of fast neutrons to thermal energies is best done by those elements that maximize the slowing-down process. [See Eq. (2).] Beryllium and graphite are excellent moderators and have been used extensively in elemental form. Both elements suffer radiation damage, and their use under high-stress conditions is to be avoided.

Fluence. The total integrated exposure to radiation (flux × time) is called fluence. It is most important in determining radiation damage. Rate effects (flux) do not appear to be significant. The threshold fluence for a specific property change induced by radiation is a function of the composition and microstructure. One of the most important examples is the appearance of voids in metals and alloys. This defect does not show up in the microstructure of irradiated metals or alloys until a fluence of 10^{19} n/cm^2 or greater has been achieved. Consequently, there was no way to anticipate its appearance and the pronounced effect in causing swelling in structural components of a reactor. This and other examples point to the importance of lifetime studies in order to establish the appearance or absence of any unexpected phenomenon during this time.

Lifetime studies in reactors are time-consuming and are virtually impossible if anticipated fluences far exceed the anticipated lifetime of operating test reactors. A technique to overcome this impasse is to use charged-particle accelerators to simulate reactor irradiation conditions. For example, nickel ions can be used to bombard nickel samples. The bombarding ions at 5 to 10 MeV then simulate primary knocked-on atoms directly and create high-density damage in the thickness of a few micrometers. Accelerators are capable of produced beam currents of several $\mu\text{A/cm}^2$; hence in time periods of a few hours to a few days ion bombardment is equivalent to years of neutron bombardment. Correlation experiments have established that the type of damage is similar to neutron damage. Moreover, helium can be injected to approximate n,α damage when these reactions do not occur in accelerator bombardments. However, careful experimentation is required to obtain correlation between results obtained on thin samples and thicker, more massive samples used in neutron studies.

Pretreatment and microstructure. Dislocations play a key role in determining the plastic flow properties of metals and alloys such as ductility, elongation, and creep. The yield, ultimate and impact strength properties, and hardness are also expressions of dislocation behavior. If a radiation-produced defect impedes the motion of a dislocation, strengthening and reduced ductility may result. On the other hand, during irradiation point defects may enhance mobility by promoting dislocation climb over barriers by creating jogs in the dislocation so that it is free to move in a barrier-free area. Moreover, dislocations may act as trapping sites for interstitials and gas atoms, as well as nucleation sites for precipitate formation. Thus the number and disposition of dislocations in the metal alloy may strongly influence its behavior upon irradiation.

Heat treatment prior to irradiation determines the retention of both major alloying components and impurities in solid solution in metastable alloys. It also affects the number and disposition of dislocations. Thus it is an important variable in determining subsequent radiation behavior.

Impurities and minor alloying elements. The presence of small amounts of impurities may profoundly affect the behavior of engineering alloys in a radiation field. It has been observed that helium concentrations as low as 10^{-9} seriously reduce the high-temperature ductility of a stainless steel. Concentrations of helium greater than 10^{-3} may conceivably be introduced by the n,α reaction in the nickel component of the stainless steel or by boron contamination introduced inadvertently during alloy preparation. The boron also reacts with neutrons via the n,α reaction to produce helium. The addition of a small amount of Ti (0.2%) raises the temperature at which intergranular fracture takes place so that ductility is maintained at operating temperatures.

Small amounts of copper, phosphorus, and nitrogen have a strong influence on the increase in the ductile-brittle transition temperature of pressure vessel steels under irradiation. Normally these carbon steels exhibit brittle failure below room temperature. Under irradiation, with copper content above 0.08% the temperature at which the material fails in a brittle fashion increases. Therefore it is necessary to control the copper content as well as the phosphorus and nitrogen during the manufacture and heat treatment of these steels to keep the transition temperature at a suitably low level. A development of a similar nature has been observed in the swelling of type 316 stainless steel. It has been learned that carefully controlling the concentration of silicon and titanium in these alloys drastically reduces the void swelling. This is an important technical and economic contribution to the fast breeder reactor program.

Beneficial effects. Radiation, under carefully controlled conditions, can be used to alter the course of solid-state reactions that take place in a wide variety of solids. For example, it may be used to promote enhanced diffusion and nucleation, it can speed up both short- and long-range order-disorder reactions, initiate phase changes, stabilize high-temperature phases, induce magnetic property changes, retard diffusionless phase changes, cause re-solution of precipitate particles in some systems while speeding precipitation in other systems, cause lattice parameter changes, and speed up thermal decomposition of chemical compounds. The effect of radiation on these reactions and the other property changes caused by radiation are of great interest and value to research in solid-state physics and metallurgy.

Radiation damage is usually viewed as an unfortunate variable that adds a new dimension to the problem of reactor designers since it places severe restraints on the choice of materials that can be employed in design and construction. In addition, it places restraints on the ease of observation and manipulation because of the radioactivity involved. However, radiation damage is also a valuable research technique that permits materials scientists and engineers to introduce impurities and defects into a solid in a well-controlled fashion. [DOUGLAS S. BILLINGTON]

Bibliography: S. Amelinckx et al., *The Interaction of Radiation with Solids*, 1964; D. S. Billington and J. H. Crawford, Jr., *Radiation Damage in Solids*, 1961; E. E. Bloom et al., Austenitic stainless steels with improved resistance to radiation-induced swelling, *Scripta Met.*, 10:303, 1976; E. E. Bloom and J. R. Weir, *Tech. Publ. 457*, American Society for Testing and Materials, 1969; R. O. Bolt and J. G. Carroll (eds.), *Radiation Effects on Organic Materials*, 1963; C. J. Borokowski and T. V. Blalock, A new method of Johnson noise thermometry, *Rev. Sci. Inst.*, 45:151–162, 1974; J. W. Corbett and L. C. Ianniello (eds.), *Radiation Induced Voids in Metals*, CONF-71060 ERDA, 1972; G. J. Dienes (ed.), *Studies in Radiation Effects in Solids*, vol. 2, pp. 1–297, 1967; International Atomic Energy Agency, Vienna, *Interaction of Radiation with Condensed Matter*, 1978; International Atomic Energy Agency, Vienna, *Radiation Damage in Reactor Materials*, vol. 1, 1969; J. F. Kircher and R. E. Bowman (eds.), *Effects of Radiation on Materials and Components*, 1964; C. Lehmann, *Interaction of Radiation with Solids and Elementary Defect Production*, 1977; M. T. Robinson and F. W. Young (eds.), *Fundamental Aspects of Radiation Damage in Metals*, CONF-751006-P1 ERDA, 1976; F. Seitz and D. Turnball (eds.), *Solid State Physics*, vol. 2, 1956; L. E. Steele, *Neutron Embrittlement of Reactor Pressure Vessels*, Tech. Publ. 163, International Atomic Energy Agency, Vienna, 1975; Surface effects in controlled fusion, *J. Nucl. Mater.*, 53:1–357, 1974; F. W. Wiffen and J. S. Watson (eds.), *Radiation Effects and Tritium Technology for Fusion Reactors*, CONF-750989 ERDA, 1976.

Radioactive fallout

The radioactive material which results from a nuclear explosion in the atmosphere. In particular the term applies to the debris which is deposited on the ground, but common usage has extended its coverage to include airborne material as well.

Nuclear explosion. A nuclear explosion results when the fission of uranium-235 or plutonium-239 proceeds in a rapid and relatively uncontrolled way, as opposed to the controlled fission in a reactor. The energy release of an atomic (fission) bomb is usually expressed in terms of thousands of tons of TNT equivalent, and such explosions may have yields into the range of hundreds of kilotons. Still larger explosions can be produced by using the fission device as a trigger for a fusion reaction to produce a thermonuclear explosion. The yield of such thermonuclear devices can range up to hundreds of thousands of kilotons (hundreds of megatons).

The radioactivity from the explosion is produced by fission products and by activation products. The basic reaction in atomic fission is the splitting of an atom of the fissionable material (uranium or plutonium) into two lighter elements (fission products). These lighter elements are all unstable and emit beta and gamma radiation until they reach a stable state. During the fission reaction a number of neutrons are released, and they can interact with the surrounding materials of the device or of the environment to produce radioactive activation products. The fusion process does not produce fission products but does release neutrons which can add to the activation.

The fissioning of a single atom of uranium or plutonium yields almost 200 million electron volts (MeV) instantaneously, and just over 50 g of fissionable material are required to produce a yield of 1 kiloton. The tremendous energy release produces the explosive shock and temperatures ranging upward from 1,000,000°C. The device itself and the material immediately surrounding it are vaporized into a fireball which then rises in the atmosphere. The altitude at which the fireball cools sufficiently to stabilize depends very much on the yield of the explosion. In general, an atomic explosion fireball stabilizes as a cloud high in the troposphere, while a thermonuclear fireball tends to break through into the stratosphere and stabilize at altitudes above 10 mi.

As the fireball cools, the vaporized materials condense to fine particles. If the explosion has taken place high above the Earth, the only material present is that of the device itself, and the fine particulates are carried by the winds and distributed over a wide area before they descend to Earth.

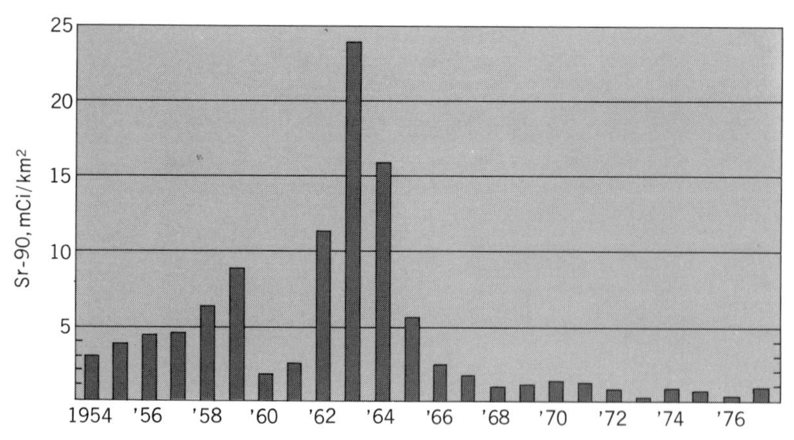

Fig. 1. Time distribution for strontium-90 fallout in New York City. 1 mCi = 3.7×10^7 disintegrations per second = 3.7×10^7 Bq. *(Data from Health and Safety Laboratory, U.S. Atomic Energy Commission)*

Bursts at or near the surface carry large amounts of inert material up with the fireball. A large fraction of the radioactive debris condenses out onto the large inert particles of soil and other material, and many small radioactive particulates attach themselves to the larger inert particles. The larger particles settle rapidly by gravity, and large percentages of the radioactivity may be deposited in a few hours near the site of the explosion. In contrast, the radioactivity from a thermonuclear explosion at high altitude takes months or years to reach the surface. It is also possible to test nuclear explosives deep in the earth. Such underground tests do not produce fallout as long as the explosion is completely contained, and they are not covered by the test ban treaty of 1962.

The first nuclear explosion took place in New Mexico in 1945 and was rapidly followed by the two bombings at Hiroshima and Nagasaki, Japan. Through 1978 six nations have tested a large number of nuclear and thermonuclear weapons. The United States has tested at its sites in Nevada and the Pacific; the United Kingdom has tested in Australia and Christmas Island. The Soviet Union has several areas, including Novaya Zemlya in the Arctic. Atmospheric testing was heaviest in 1954, 1958, and 1961 and was stopped following negotiation of the test ban treaty. France and China did not sign the treaty, however, and have carried out a number of tests since, France in the Sahara and the South Pacific and China at an inland test site at Lop Nor. India has also carried out one underground test.

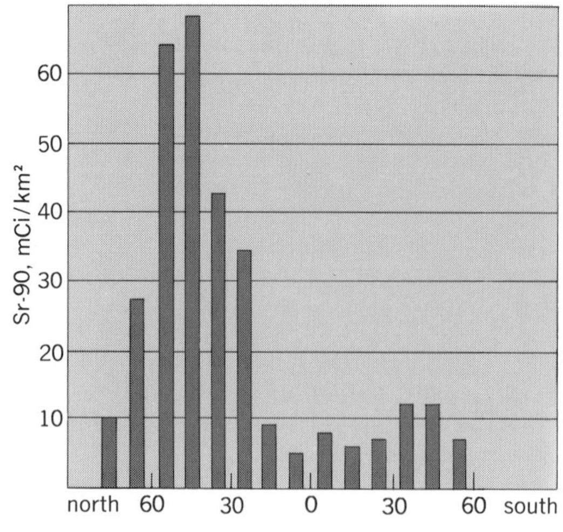

Fig. 2. Latitude distribution of cumulative strontium-90 deposit as of 1966. *(Data from Health and Safety Laboratory, U.S. Atomic Energy Commission)*

Radioactive products. The total fission yield for all tests has been about 200 megatons, the total explosive yield being over 500 megatons. The production of long-lived fission products has been about 20 megacuries of strontium-90 and 30 megacuries of cesium-137. Figure 1 shows the time distribution for strontium-90 fallout in New York City, which may be considered typical of the Northern Hemisphere mid-latitudes. Figure 2 shows the latitude distribution as of 1966. Testing through 1978 does not modify this picture because atmospheric explosions since the test ban have added less than 10% to the total.

Local fallout. Local fallout is the deposition of large radioactive particles near the site of the explosion. It may present a hazard near the test site or in the case of nuclear warfare. The smaller particulates, which are spread more widely, constitute the worldwide fallout. The radioactivity comes from the isotopes (radionuclides) of perhaps 60 different elements formed in fission, plus a few others formed by activation and any unfissioned uranium or plutonium. Each of the radioactive isotopes produced in the fission process goes through three or four successive radioactive decays before becoming a stable nuclide. The radioactive half-lives of these various fission products range from fractions of a second to about 100 years. The overall beta and gamma radioactivity decays according to the minus 1.2 power of the activity at any time. Thus decay is very rapid at first and then slows down as the short-lived isotopes disappear.

The hazard from local fallout is largely due to the external gamma radiation which a person would receive if this material were deposited near him; fallout shelters are designed to shield survivors of an explosion from this radiation. Actually, the effects of an explosion are so great that a shelter which would resist the shock would shield the occupants from fallout radiation.

In so-called clean nuclear weapons the ratio of explosive force to the amount of fission products produced is high. On this basis, small fission weapons are considered "dirty," although the absolute amount of fission products is much less than with a large clean bomb. It is also possible to increase the radioactivity produced, and thus the fallout, by surrounding the device with a material which is readily activated. Cobalt has been frequently mentioned in this connection.

Tropospheric and stratospheric fallout. As previously mentioned, the large particulates from a nuclear explosion are deposited fairly close to the site. The smaller particles are carried by the winds in the troposphere or stratosphere, generally in the direction of the prevailing westerlies. The debris in the troposphere circles the Earth in about 2 weeks, while material injected into the stratosphere may travel much faster. The horizontal distribution in the north-south direction is much slower, and it may take several months for radioactivity to cover the Northern Hemisphere, for example, following a test in the tropics. Transfer between the Northern and Southern hemispheres is slight in the troposphere but can take place in the stratosphere. Radioactive debris in the troposphere is brought down to the surface of the Earth mainly in precipitation, with perhaps 10–15% of the total amount brought down by dry deposition. Radioactive material remains about 30 days in the troposphere. The mechanism of transfer from the stratosphere to the troposphere is not completely understood, but it is obvious from measurements that the favored region for deposition is the midlatitudes (30–50°) of the Northern and Southern hemispheres. Since most of the testing took place in the Northern Hemisphere, the deposition there was about three times as great as in the Southern Hemisphere.

Hazards. Once the fallout has actually been deposited, its fate depends largely on the chemical nature of the radionuclides involved. Since the level of gamma radiation from worldwide fallout is negligible, attention has been paid to those nuclides which might possibly present some hazard when they enter the biosphere. Plants, for example, may be contaminated directly by foliar deposition or indirectly by deposition on the soil that is followed by root uptake. Animals may be contaminated by eating the plants, and man by eating plant or animal foods. The inhalation of radioactive material is not considered to be a significant hazard as compared to ingestion.

Based on the various considerations mentioned, the three radionuclides of greatest interest are strontium-90, cesium-137, and iodine-131. Strontium-90 is a beta emitter which follows calcium metabolically and tends to deposit in the bone. The radiation emitted could result in bone cancer or leukemia if the levels were to become sufficiently high. Cesium-137 is distributed throughout soft tissues and emits both beta and gamma radiation. It is considered to be a possible genetic hazard but not as dangerous to the individual as strontium-90. Iodine-131 is relatively short-lived and probably is a hazard only as it appears in tropospheric fallout. It is readily absorbed by cows on pasture and is transferred rapidly to their milk. The iodine in milk in turn is concentrated in the human thyroid and presents a possible high radiation dose, particularly to children, who consume larger amounts of milk and who have smaller thyroids than adults.

Sampling. Radioactive fallout is monitored by many countries. The usual systems involve networks that collect samples of airborne dust, deposition, and milk and other foods. The levels of iodine-131 or cesium-137 can be checked in living individuals by external gamma counting. Strontium-90 on the other hand can be measured only in autopsy specimens of human bone. These national data are also combined and evaluated by the Scientific Committee on the Effects of Atomic Radiation of the United Nations. Their reports are issued at intervals and offer broad reviews and summaries of the available information on levels of fallout and on possible hazards.

The studies of radioactive fallout have also included sampling in the stratosphere by balloons and high-flying aircraft, as well as taking aircraft samples in the troposphere. Very elaborate monitoring systems have been set up, and these have provided considerable scientific information in addition to their original monitoring purposes. The major benefits have been the understanding of stratospheric transfer processes as significant to meteorology and information on the uptake and metabolism of various elements by plants, animals, and humans. These studies are usually reported in the specialty journals for meteorology, agriculture, and biology.

[JOHN H. HARLEY]

Bibliography: *Reports of the United Nations Committee on the Effects of Atomic Radiation*, 1958, 1962, 1964, 1966, 1969, 1972, 1977; U.S. Public Health Service, *Radiological Health Data and Reports*, monthly (through 1974).

Radioactive materials, decontamination of

The removal of radioactive contamination which is deposited on surfaces or may have spread throughout a work area. Personnel decontamination is also included. The presence of radioactive contamination is a potential health hazard, and in addition, it may interfere with the normal functioning of plant processes, particularly in those plants using radiation detection instruments for control purposes. Thus, the detection and removal of radioactive contaminants from unwanted locations to locations where they do not create a health hazard or interfere with production are the basic purposes of decontamination.

There are four ways in which radioactive contaminants adhere to surfaces, and these limit the decontamination procedures which are applicable. The contaminant may be (1) held more or less loosely by such physical forces as electrostatic or surface tension, (2) absorbed in porous materials, (3) adsorbed on or by the surface in the form of ions, atoms, or molecules, or (4) mechanically bonded to surfaces through oil, grease, tars, paint, and so on.

Methods. Decontamination methods follow two broad avenues of attack, mechanical and chemical. Commonly used mechanical methods are vacuum cleaning, sand blasting, blasting with other abrasives, flame cleaning, scraping, and surface removal (for example, removal of concrete floors with an air hammer). The principal chemical methods of decontamination are water washing, steam cleaning, and scrubbing with detergents, acids, caustics, and solvents.

Another important method of handling contamination is to store the contaminated object, or temporarily abandon the contaminated space. This can be done when the use of the material or space is not necessary for a period of time and the half-life of the contaminant is relatively short. For example, tools contaminated with short-lived fission products may be stored, or a building contaminated with such material may be sealed off and barred from use, until the natural radioactive decay has reduced the contamination to an acceptable level.

Other methods involve covering the contamination by some means, such as painting, and disposing of part or all of the contaminated equipment or facility. Considerations which determine the methods used for decontamination or removal of contamination include (1) the hazards involved in the decontamination procedure, (2) the cost of removal of the contamination, and (3) the permanency of removal of the contamination (for example, painting over a surface contaminated with a long-lived radioactive material only postpones ultimate disposal considerations).

Personnel. Personnel decontamination methods differ from those used for materials primarily because of the possibilities of injury to the person being decontaminated. Procedures used for normal personal cleanliness usually will remove radioactive contaminants from the skin, and the method used will depend upon its form and associated dirt (grease, oil, soil, and so on). Soap and water (sequestrants and detergents) normally remove more than 99% of the contaminants. If it is necessary to remove the remainder, chemical methods which remove the outer layers of skin upon which the contamination has been deposited can be used. These chemicals—citric acid, potassium permanganate, and sodium bisulfite are examples—should be used with caution and preferably under medical supervision, because of the increased risk of injury to the skin surface. The use of coarse cleansing powders should be avoided for skin decontamination, because they may lead to scratches and abraded skin which can permit the radioactive material to enter the body. Similarly, the use of organic solvents should be avoided for skin decontamination because of the probability of penetration through the pores of the skin.

It is very difficult to remove radioactive material once it is fixed inside the body, and the ensuing hazard depends very little on the method of entry, that is, through wounds, through pores of the skin, by injection, or by inhalation. When certain of the more dangerous radioactive materials, such as radium or plutonium, have been taken into the body, various chemical treatments have been attempted to increase the body elimination, but the results of these treatments are not very encouraging. For plutonium and certain other heavy metals, the most effective treatment for removal from the body is the administration of chelating agents, such as calcium ethylenediaminetetraacetate (CaEDTA) or a sodium citrate solution of zirconyl chloride. In any case, the safest and most reliable procedure for preventing internal exposure from radioactive material is the application of health physics procedures to prevent entry of radioactive material into the body.

Air and water. Air contaminants frequently are eliminated by dispersion into the atmosphere. Certain meteorological conditions, such as prevailing wind velocities, wind direction, and inversion layers, seriously limit the total amount of radioactive material that may be released safely to the environment. Consequently, decontamination of the airstream by filters, cyclone separators, scrubbing with caustic solutions, and entrapment on charcoal beds is often resorted to. The choice of method used is guided by such things as the volume of airflow, the cost of heating and air conditioning, the hazards associated with the airborne radioactive material, and the isolation of the operation from other populated areas.

Water decontamination processes can use one or both of the two opposing philosophies of maximum dilution or maximum concentration (and subsequent removal) of the contaminant. Water concentration methods involve the use of water purification processes, that is, ion exchange, chemical precipitation, flocculation, filtration, and biological retention.

Certain phases of radioactive decontamination procedures are potentially hazardous to personnel. Health physics decontamination practices include the use of protective clothing, respiratory devices, localized shielding, isolation or restriction of an area, provisions for the proper disposal of the attendant wastes, and application of the recommended rules and procedures for limiting the internal and external doses of ionizing radiation.

[KARL Z. MORGAN]

Bibliography: J. A. Ayres (ed.), *Decontamination of Nuclear Reactors and Equipment*, 1970; *Control and Removal of Radioactive Contamination in Laboratories*, Natl. Bur. Std. Handb. no. 48, 1951; International Brotherhood of Electrical Workers Staff, *Radiation Hazards and Control*, 1965; S. Kinsman, *Radiological Health Handbook*, U.S. Department of Health, Education and Welfare, PB-121784, 1957.

Radioactive waste management

The treatment and containment of radioactive wastes. The requirement for radioactive waste management is present to some degree in all nuclear energy operations. Wastes in liquid, solid, or gaseous form are produced in the mining of ore, production of reactor fuel materials, reactor operation, processing of irradiated reactor fuels, and a great variety of related operations. Wastes also result from use of radioactive materials, for example, in research laboratories, industrial operations, and medical research and treatment. The magnitude of waste management operations undoubtedly will increase as the nuclear energy program is further extended and diversified and as a large and widespread nuclear power industry is developed.

In the safe handling and containment of radioactive wastes, the principal objective is the prevention of radiation damage to humans and the environment by controlling the dispersion of radioactive materials. Damage to humans may result from irradiation by external sources or by the intake (by ingestion, by inhalation, or through the skin) of radioactive materials, their passage through the respiratory and gastrointestinal tract, and their partial incorporation into the body. Radioactive waste contaminants in air, water, food, and other elements of the human environment must be kept below specified concentrations for the particular radionuclide or mixture of radionuclides present in the wastes. Liquid or solid waste products containing significant quantities of the more toxic radioactive materials require isolated and permanent containment media from which any potential reentry into human environment would be at tolerable levels. The radioactive materials of major concern are those that may be readily incorporated into the body and those that have relatively long half-lives, ranging from a few years to thousands of years.

Waste management operations are focused on those radioisotopes which originate in nuclear reactors. Here the fission products (other chemical elements formed by nuclear fragmentation of actinide elements such as uranium or plutonium, and others) accumulate in the nuclear fuel, along with plutonium and other transuranic nuclides. (Transuranic elements are those higher than uranium on the periodic table of chemical elements. They are also called actinide elements.) The concentrations of plutonium are substantially higher than those found in nature, ranging from 10 to 20 kg per metric ton (1 metric ton = 1000 kg) of uranium compared to a high of 17 g per metric ton of uranium in minerals from fumarole areas.

Reprocessing. Fuel discharged from the nuclear reactor is reprocessed to recover uranium and plutonium by chemical dissolution and treatment. During this step, favored treatment processes form high-level waste as an acidic aqueous stream. Other processes are being considered which would produce high-level waste in different forms. This high-level waste contains most of the reactor-produced fission products and actinides, with slight residues of uranium and plutonium (see Table 1). These waste products generate sufficient heat to require substantial cooling and emit large amounts of potentially hazardous ionizing radiation. Because the reprocessing step normally does not dissolve much of the nuclear fuel cladding, high-level waste normally contains only a small amount of the radionuclides formed as activation products within the cladding. This cladding hull waste is managed as a separate solid waste stream, as are several other auxiliary waste streams from the reprocessing plants.

Table 1. Typical materials in high-level liquid waste

Material[b]	Grams per metric ton from various reactor types[a]		
	Light water reactor[c]	High-temperature gas-cooled reactor[d]	Liquid metal fast breeder reactor[e]
Reprocessing chemicals			
Hydrogen	400	3800	1300
Iron	1100	1500	26,200
Nickel	100	400	3300
Chromium	200	300	6900
Silicon	–	200	–
Lithium	–	200	–
Boron	–	1000	–
Molybdenum	–	40	–
Aluminum	–	6400	–
Copper	–	40	–
Borate	–	–	98,000
Nitrate	65,800	435,000	244,000
Phosphate	900	–	–
Sulfate	–	1100	–
Fluoride	–	1900	–
SUBTOTAL	68,500	452,000	380,000
Fuel product losses[f,g]			
Uranium	4800	250	4300
Thorium	–	4200	–
Plutonium	40	1000	500
SUBTOTAL	4840	5450	4800
Transuranic elements[g]			
Neptunium	480	1400	260
Americium	140	30	1250
Curium	40	10	50
SUBTOTAL	660	1440	1560
Other actinides[g]	<0.001	20	<0.001
Total fission products[h]	28,800	79,400	33,000
TOTAL	103,000	538,000	419,000

SOURCE: From K. J. Schneider and A. M. Platt (eds.), *Advanced Waste Management Studies: High-Level Radioactive Waste Disposal Alternatives*, USAEC Rep. BNWL-1900, May 1974.

[a]Water content is not shown; all quantities are rounded.

[b]Most constituents are present in soluble, ionic form.

[c]U-235 enriched pressurized water reactor (PWR), using 378 liters of aqueous waste per metric ton, 33,000 MWd/MT exposure. (Integrated reactor power is expressed in megawatt-days [MWd] per unit of fuel in metric tons [MT].)

[d]Combined waste from separate reprocessing of "fresh" fuel and fertile particles, using 3785 liters of aqueous waste per metric ton, 94,200 MWd/MT exposure.

[e]Mixed core and blanket, with boron as soluble poison, 10% of cladding dissolved, 1249 liters per metric ton, 37,100 MWd/MT average exposure.

[f]0.5% product loss to waste.

[g]At time of reprocessing.

[h]Volatile fission products (tritium, noble gases, iodine, and bromine) excluded.

The recovered uranium and plutonium are reused by the nuclear industry by reconstitution into nuclear fuels using plutonium (instead of uranium-235) as the fissile material. The fabrication of such fuels, since they contain both uranium and plutonium mixed oxides (MOX), generates additional wastes that may contain plutonium.

Policy and treatment. The policy of the U.S. Energy Research and Development Administration (ERDA, which includes the former Atomic Energy Commission) is to assume custody of all commercial high-level radioactive wastes and to provide containment and isolation of them in perpetuity. Regulations require that the high-level wastes from nuclear fuels reprocessing plants be solidified within 5 years after reprocessing and then shipped to a Federal repository within 10 years after reprocessing.

Because of the anticipated increase in the quantities of waste-containing elements or those contaminated with transuranic (TRU) elements, and the long half-life and specific radiotoxicity of these elements, ERDA has also proposed that all transuranic wastes be solidified and transferred to ERDA as soon as practicable, but at most within 5 years after generation.

Both of these policies require that high-level, cladding and other transuranic wastes be converted to a solid. A variety of technologies exist for this conversion, including calcination, vitrification, oxidation, and metallurgical smelting, depending on the primary waste.

A typical solidification process, principally for high-level waste, is spray calcination-vitrification. In this process (see Fig. 1) atomized droplets of waste fall through a heated chamber, where flash evaporation results in solid oxide particles. Glassmaking solid frit or phosphoric acid can be added to provide for melting and glass formation in a continuous melter or directly in the vessel that will serve as the waste canister. The molten glass or ceramic is cooled and solidified.

Quantity of waste. The growth of nuclear power in the United States will result in increased quantities of high-level waste. Installed nuclear electrical generating capacity is projected, according to an ERDA Office of Planning and Analysis study, to increase to about 1,200,000 MW by the year 2000. The anticipated volume of solidified high-level waste accumulated by the year 2000 is about 13,000 m^3, the result of reprocessing almost 200,000 metric tons of fuel, about 80% of which is

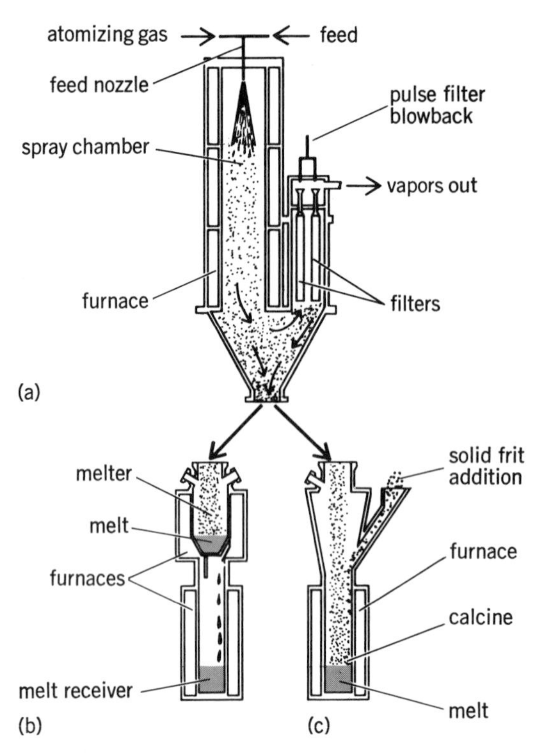

Fig. 1. Spray calcination-vitrification process. (*a*) Spray calciner, producing calcine that drops into either (*b*) continuous silicate glass melter, or (*c*) directly into waste canister vessel for for in-pot melting.

associated with light water reactor (LWR) plants. If this amount of solid waste were stacked as a solid cube, the cube would be about 25 m on a side. Approximately 150,000 megacuries (5.5×10^{21} disintegrations per second) of radioactivity and 700 MW of heat will be associated with this projected waste inventory in the year 2000. This heat content is equivalent to about one-third of the waste heat rejected from one LWR generating 1000 MW of electricity.

Table 2 shows the projected accumulation of solidified high-level waste, assuming that 0.5% of the fuel product (uranium and plutonium or thorium) is lost to waste during reprocessing, and that all other actinides are in the waste. The toxicity indices in Table 2 are base 10 logarithms of the quantity in cubic meters of air, for the inhalation index, or in cubic meters of water, for the ingestion hazard index, required to dilute radioactive material to limits stipulated in Federal regulations. Beyond the year 2000, fission products (primarily strontium) and transplutonium elements (primarily americium) are the chief potential hazards in drinking water up to about 350 years and 2×10^4 years, respectively. Radioactivity from plutonium losses during reprocessing then becomes the main factor until about 10^6 years. Finally, radioactivity remaining as the result of uranium losses during reprocessing becomes the predominant contribution to the ingestion toxicity index. For comparison with high-level waste projections, the estimated quantities of cladding waste and other TRU waste are shown in Table 3.

Alternative waste management concepts. Scientists are investigating many of the options for separating, treating, and otherwise managing radioactive waste from the time the material is formed in a fission reactor. Key considerations in the route to ultimate storage or disposal or elimination of the material are outlined in Fig. 2.

Constituents of the waste material are a mix of long- and short-lived radioisotopes. Some have radioactive decay half-lives of no more than tens of years, while others must be isolated from the biosphere for many thousands of years. Dividing the high-level waste into actinides and fission products (a process called partitioning), would permit managing the materials as separate classes.

Common to all waste management concepts is

Table 2. Projected accumulation of solidified high-level waste through end of year, 1977–2000

Fiscal year	Volume[a] of waste, m³	Actinide mass, metric tons	Radio-activity,[b,c] MCi[d]	Thermal power,[b,c] MW	Toxicity indices[b,c] Inhalation	Ingestion
1977	190	18	4,600	20	19.62	14.91
1980	550	50	10,200	50	20.26	15.38
1985	1720	190	26,300	140	21.30	15.85
1990	3900	410	50,300	250	21.59	16.18
1995	7650	760	90,500	420	21.71	16.48
2000	13,340	1270	149,000	660	21.86	16.70
Time elapsed after year 2000, years						
10^2			5700	20	21.29	15.55
10^3			30	<1	20.19	12.74
10^4			10	<1	19.70	12.14
10^5			4	<1	18.79	12.38
10^6			1	<1	18.60	11.86

SOURCE: From K. J. Schneider and A. M. Platt (eds.), *Advanced Waste Management Studies: High-Level Radioactive Waste Disposal Alternatives*, USAEC Rep. BNWL-1900, May 1974.

[a]Volume based on 0.057, 0.170, and 0.085 m³ of solidified waste per metric ton of heavy metal for LWR, high-temperature gas-cooled reactor (HTGR), and liquid-metal fast breeder reactor (LMFBR) fuels, respectively.

[b]Waste initially generated 150, 365, and 90 days after spent fuel discharged from LWR, HTGR, and LMFBR units, respectively.

[c]All tritium and noble-gas fission products and 99.9% of iodine and bromine fission products excluded.

[d]1 megacurie (MCi) = 3.7×10^{16} disintegrations per second.

Table 3. Projected accumulation of cladding and transuranic waste through end of year, 1985–2000

Fiscal year	Cladding: Volume, $m^3 \times 10^{-3}$	Cladding: Actinide mass, metric tons	Other TRU: Volume, $m^3 \times 10^{-3}$	Other TRU: Actinide mass, metric tons
1985	5	30	161	4
1990	9	55	275	10
1995	19	90	600	28
2000	36	135	1236	63

the possible need for interim storage in a retrievable surface storage facility. Three concepts have been evaluated by ERDA—a water basin concept, an air-cooled vault concept, and a concept for storage of wastes in sealed casks in the open air. The major differences among these proposals are in radiation shields, containment barriers, heat removal techniques, and relative dependence on utilities and maintenance. The canisters of solidified waste would be retrievable at all times for various waste management options or treatment by future techniques of disposal.

Many short-lived waste components will decay to unimportant radioactivity levels in relatively short periods of time. Their storage in artificial structures can be considered. For the longer-lived and highly toxic actinide fraction of radioactive waste, there appear to be only three basic management options: elimination of waste constituents by transmutation—the conversion to other, less undesirable isotopes by nuclear processes; transport off the Earth; and isolation from the human environment somewhere on Earth for periods of time sufficient to permit natural radioactive decay.

These potential alternative methods for long-term management of high-level radioactive waste provide the framework for a major study by the AEC's Division of Waste Management and Transportation. Included in the comprehensive review is a compilation of information relevant to the technical feasibility; the safety, cost, environmental, and policy considerations; the public response; and research and development needs for various waste management alternatives.

The basic requirement for the suitability of any environment for final storage or disposal of radioactive waste is its capability to safely contain and isolate it until decay has reduced the radioactivity to nonhazardous levels. Geologic formations exist which have been physically and chemically stable for millions of years. Ice sheets appear to offer some potential advantages as a disposal medium remote from the human environment. Both are

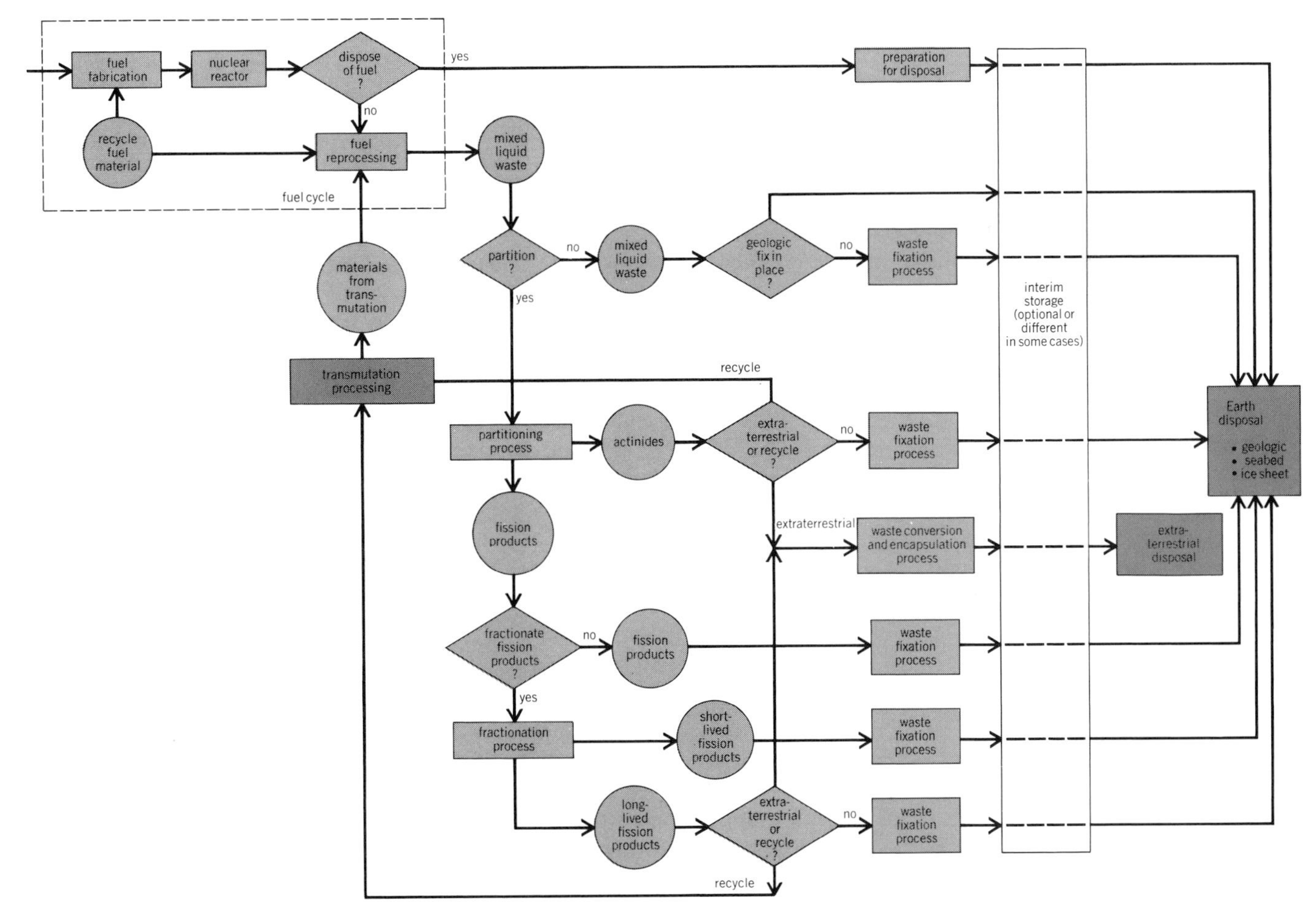

Fig. 2. High-level radioactive waste management options. (*K. J. Schneider and A. M. Platt, eds., Advanced Waste Management Studies: High-Level Radioactive Waste Disposal Alternatives, USAEC Rep. BNWL-1900, May 1974*)

under study as a potential future alternative. In the ice sheet disposal concept, a waste canister would either melt down through the ice sheet to bedrock, or be connected to the surface by cables or chains which would stop its descent through the ice, or be placed in a surface storage facility which would gradually become covered with snow and buried in the ice sheet.

The bedrock zones in the stable areas of the deep sea floor are the subject of another study concept, as is the process of tectonic plate movement which should carry waste material down into the Earth's mantle from areas known as subduction zones. Also under study is storage in bedrock beneath high-sedimentation-rate areas where major rivers are building deltas into the ocean. Although very high costs per unit of weight would be encountered, an in-depth analysis of extraterrestrial disposal is under way. Also being assessed for possible future applications are nuclear techniques for transmutation of the actinides to isotopes having lower toxicity or shorter half-lives or both. *See* HEALTH PHYSICS.

[ALLISON M. PLATT]

Bibliography: Milton H. Campbell (ed.), *High Level Radioactive Waste Management*, 1976; D. A. Deese, *Nuclear Power and Radioactive Waste*, 1978; *Draft Environmental Statement: Management of Commercial High-Level and Transuranium-Contaminated Radioactive Waste*, USAEC Rep. WASH-1539, September 1974; A. A. Moghissl et al. (eds.), *Nuclear Power Waste Technology*, 1978; K. J. Schneider and A. M. Platt (eds.), *Advanced Waste Management Studies: High-Level Radioactive Waste Disposal Alternatives*, USAEC Rep. BNWL-1900, May 1974; M. Willrich and R. K. Lester, *Radioactive Waste Management and Regulation*, 1977.

Radioecology

The study of the interaction of radioisotopes or ionizing radiation with populations, ecological communities, or ecosystems. Radioecology began with attempts to predict the impact of the developing nuclear industry on environmental systems: the fates of radioactive substances released into the environment, and the effects of ionizing radiation on ecological processes. The value of radioactive tracers for ecological studies emerged rapidly, as did the utility of ionizing radiation effects as a tool in ecological research. Radioecology now encompasses the use of radioactive tracers, as well as fates and effects of radioactive substances in the environment.

The terminology used in radioecology includes the type of radiation (alpha, beta, or gamma), the amount of radionuclide (usually measured in curies), the decay rate (half-life), and the dose of ionizing radiation (measured in rads).

Alpha radiation has little penetrating ability and is not often involved in radioecological studies. Beta radiation consists of low-mass, low-energy particles with medium penetrating ability. However, because beta radiation is rapidly attenuated by the air or protoplasm, its use in effects studies is limited to internal dosages or short-range work. Beta emitters such as ^{14}C and ^{32}P are useful radioecological tracers. Because of the penetrating ability of gamma radiation, radionuclides such as ^{137}Cs and ^{65}Zn have found great utility both as tracers and as sources of ionizing in effects studies.

Radioactive materials are measured in curies, a curie equaling 2.22×10^{12} radioactive disintegrations per minute (dpm). Various decimal fractions of the curie are commonly employed: A millicurie (mCi) is 2.22×10^9 dpm; and a microcurie (μCi) is 2.22×10^6 dpm. The dose of radiation received by an organism is a function of the size of the source (in curies), the distance to the source, the energy level of the radiation and its penetrability, and the physical characteristics of the organism. The standard of measure of radiation is the rad, the absorbed dose of 100 ergs of energy per gram of protoplasm. The rad is similar to an older unit, the roentgen, except that the roentgen is a measure of exposure dose.

RADIOECOLOGICAL TRACERS

Tracer studies provide information about basic structure and process. Animals can be located and monitored by the use of radioactive tags. Radionuclide tracers are also used to determine food uptake rates, metabolic rates, and cycling of mineral nutrients, and to identify food chains and ecosystems.

Marking studies. A major problem of ecological field studies is that of locating a given organism. However, small animals carrying radioactive tags can be easily located with a suitable radiation detector. Usually a radioactive pin or wire is introduced into the body, although sometimes a small capsule containing the radioisotope, or the radioisotope alone, may be ingested by or injected into the animal. The investigator carries a detector mounted on a long pole and sweeps the area traversed until a tagged individual is located. Such techniques used in conjunction with field enclosures have permitted the daily monitoring of the distribution of small rodent populations. Automatic monitoring of position and activity patterns has been accomplished by using a circular enclosure and mounting the detector on a rotating boom. A continuous graphic record of the count rate permits the investigator to pinpoint the position of the animal each time the boom rotates one full turn. Radionuclide tags have also proved useful in ascertaining the seasonal or annual movements of animals that are inaccessible—hibernating toads and 17-year cicadas, for example. Thus, radionuclide tracers provide information obtainable in no other way.

Uptake and elimination of tracers. Two important ecological processes difficult to measure in the field are ingestion rates of food and the metabolic rate of animals. Radionuclides used as tracers provide information about both of these processes. When an organism ingests food containing radionuclides, a certain portion, the "body burden," is absorbed into the body and then is excreted at a rate dependent on a number of variables, including the type of radionuclide and the species of animal.

The rate of excretion can be measured by periodic capture of the animals and measurement of decrease in the radioactive body burden. From this excretion rate, the amount of food ingested in nature can be estimated provided the following are also known: (1) the concentration of the radioiso-

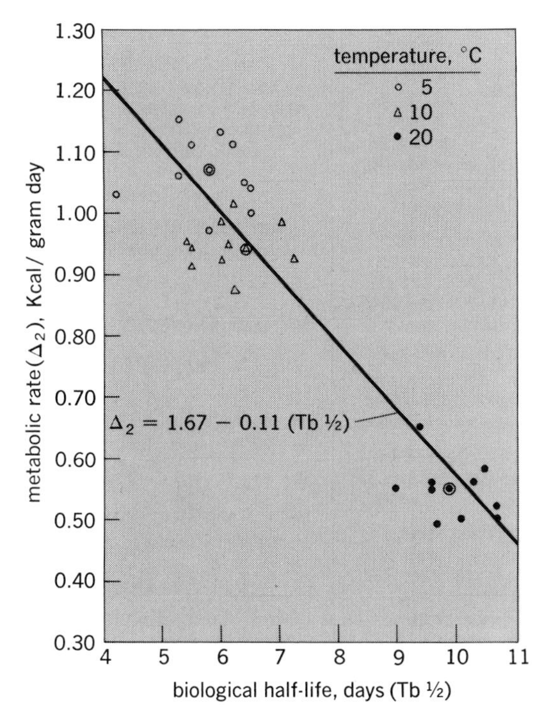

Fig. 1. Relationship of the loss rate of ^{65}Zn and the metabolic rate at three temperatures for feral house mice (*Mus musculus*). (*From R. Pulliam, G. Barrett, and E. P. Odum, Bioelimination of tracer Zn^{65} in relation to metabolic rates in mice, in Proceedings of the 2d National Symposium on Radioecology, U. S. Atomic Energy Comm., 1969*)

tope in the food plant, and (2) the percentage of absorbtion of the isotope in the gut of the organism. The former can be measured directly; the latter can be measured by suitable controlled experiments for it is less likely than is the excretion rate to vary between the laboratory and the field. In the procedure for measurement, an insect is introduced into an environment containing tagged food. As soon as a steady state has been reached (radionuclide elimination and absorption occurring at the same rate), the insect is removed and given untagged food, and the excretion rate is measured. The loss rate multiplied by the body burden must equal the intake under the steady-state conditions.

This method of estimating ingestion in the field is applicable only in areas in which plants have a measurable level of some radioactive material. Field environments tagged with radionuclides and available for ecological work are becoming more common.

The rate of excretion of radionuclides is also affected by temperature and activity factors. This led to the idea that the rate of excretion of certain radionuclides might provide an index to the general metabolic rate of the animal and that this could be estimated by periodic capture and counting of tagged organisms. Preliminary studies on the relationship between metabolic rate and loss rates of radionuclides have been encouraging (Fig. 1). Effects of temperature on excretion rates have been successfully used to extrapolate laboratory measurements to field situations.

Cycling and fallout. Direct monitoring of the uptake, storage, and recycling of mineral nutrients is an active field in radioecology. The injection of calcium-45 in various species of deciduous and evergreen trees, for example, has permitted a determination of the long-term retention time of this mineral nutrient, its turnover in each species, and the various pathways of loss, leaching, and other functions. Large differences have been found between species with respect to retention of essential nutrients. Nutrients also differ greatly in the rate of loss by leaching. Rain removes an appreciable amount of calcium from green leaves, but phosphorus is lost very slowly in this way.

By utilizing the radionuclides resulting from nuclear explosions as their own tracers, studies of their concentration through food chains have been made, such as the passage of iodine-131 through the simple food chain—grass, cow, milk, man. Such investigations have had immense practical value in predicting sites of concentrations of nuclear fallout. The high concentration of cesium-137 in the people of northern latitudes (for example, Alaska) who depend on the lichen-reindeer or the lichen-caribou food chains is a case in point. Recognition of the general principle of "biological concentration" of potential harmful materials (including pesticides) can be an important first step toward preventing serious contamination of the environment.

Food chain studies. Isotopes can be experimentally introduced, and the concentrations within each species measured, to identify functional food chains and natural ecosystems as well as the trophic or feeding positions of organisms composing these food chains. Such techniques, originally developed in the studies of stream ecosystems, have now been applied to simple terrestrial ecosystems. If an isotope (phosphorus-32 has been used with great success) is introduced into the plants or primary producers of a system, the radionuclide will be transferred most rapidly to the plant feeders (primary consumers), somewhat less rapidly to

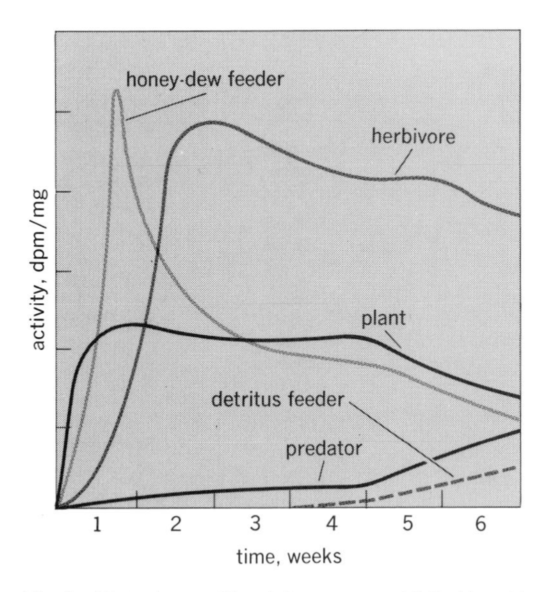

Fig. 2. Phosphorus-32 uptake curves exhibited by old-field organisms differing in trophic level, food source, or both. Phosphorus-32 levels in plants were relatively constant throughout 6-week period. (*From R. G. Wiegert, E. P. Odum, and J. H. Schnell, Ecology, vol. 48, copyright 1967 by the Ecological Society of America*)

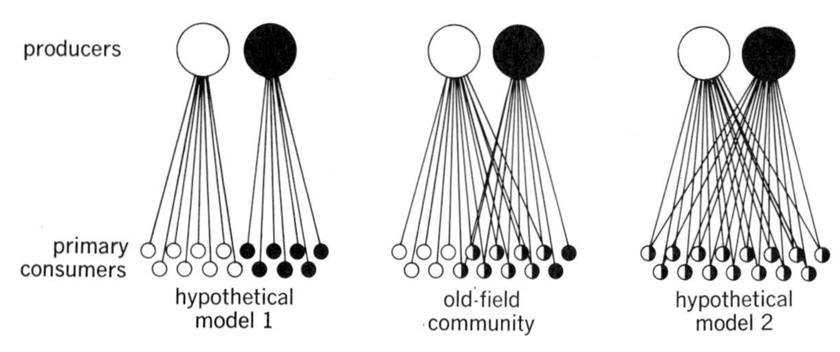

Fig. 3. Comparison of trophic relationships between two dominant species of producers and 15 species of primary consumers in a South Carolina old field with the theoretical maximum and minimum interaction models. *(From R. G. Wiegert and E. P. Odum, Radionuclide tracer measurement of food web diversity in nature, in Proceedings of the 2d National Symposium on Radioecology, U.S. Atomic Energy Commission, 1969)*

predators (secondary consumers), and so on. Thus the time that is necessary after tagging for the peak body burden or the steady-state situation to be reached will indicate the probable trophic position or feeding level of the group. In Fig. 2, uptake curves for plant, herbivore, predator, and detritus feeders are shown. Note the differences in both rate of uptake and relative time to the peak concentration. *See* FOOD CHAIN; FRESH-WATER ECOSYSTEM; TERRESTRIAL ECOSYSTEM.

A modification of this technique is to tag only a single species of plant or animal in the system and thus determine which feeding interactions or food chains are actually functional in the given natural community. The result may be surprising. Figure 3, for example, shows the actual number of plant-animal interactions which occurred between the two dominant species of plant and the most abundant species of animal in a first-year weed community. These food chain interactions are compared with the theoretical maximum and minimum numbers.

Fig. 4. The gamma source used in irradiating a forest ecosystem at Brookhaven National Laboratory. The source can be raised or lowered into a lead-shielded container through operation of a winch in a building a safe distance away. *(From G. M. Woodwell, Science, vol. 138, 1962)*

EFFECTS STUDIES

Investigations dealing with the reactions to the stress imposed by ionizing radiation are referred to as effects studies. Factors which must be considered are the duration of the radiation and the responses of specific populations and communities.

Duration of radiation. From an ecological standpoint, the effects of ionizing radiation on a population or an ecosystem are strongly influenced not only by the dose received, but by the time taken to deliver the dose. If the total dose of radiation is given in a few minutes or hours, the exposure is said to be acute. The same dose delivered over a few days may be short-term, over weeks or months long-term, and if a system is continuously subjected to radiation, exposure is chronic.

Population response. The effects of radiation delivered to a population of organisms can be assessed in many ways; often the simple question of survival is most important. Thus a criterion might be the average length of survival in nature, as compared to a control group or population that received no radiation. Animals that have been raised under stress (and this includes wild individuals that have been captured), can, under certain conditions, be more resistant to acute doses of radiation than are animals of the same species that have been raised in the laboratory. Naturally the response of populations that are exposed to different levels of chronic ionizing radiation may range from no effect to complete extinction. Within tolerable dose rates, that is, doses that do not produce extinction, responses to the radiation stress often take the form of changes in the population birth and death rates; individual metabolic or growth rates seldom change. In other words, the effect of chronic radiation often shows up first as an ecological effect at the population level, leaving the individual outwardly unaffected.

Community response. Drastic community effects such as death, desolation and destruction occur only after the most intense radiation. Interesting community effects studies are those which involve subtle changes in the structure or function or both of the ecological community or ecosystem. The response of even the simplest community to the stress of ionizing radiation is of course the net result of a series of direct effects on the component populations, plus some indirect changes which may be brought about because of the interaction of the radiation-affected populations in the ecosystem.

Natural communities have many homeostatic mechanisms; that is, they tend to persist unchanged if given a reasonably stable physical environment. This stable or climax state is the result of a process of development or replacement of species associations known as ecological succesion. Violent perturbations of the physical environment, fire for example, generally cause the community to return to some earlier preclimax state. If the change is temporary, succession proceeds to restore the steady-state system. If the change is permanent, annual fire for example, a successional process occurs until some steady state is established which is stable under the new environment. *See* ECOLOGICAL SUCCESSION; PLANT FORMATIONS, CLIMAX.

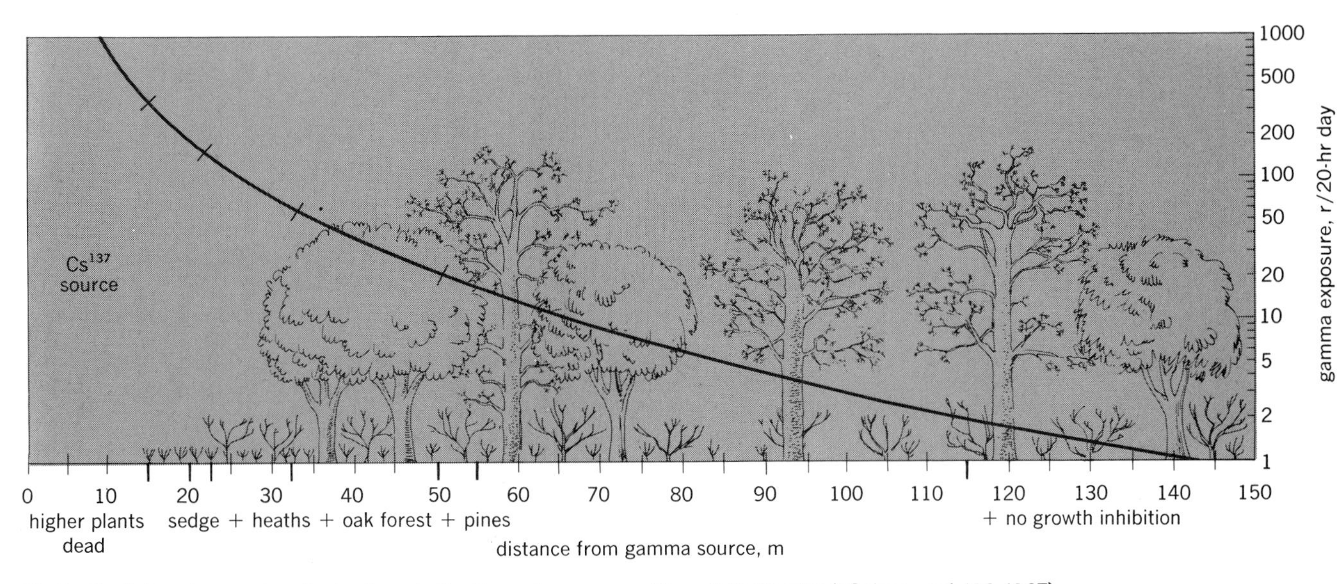

Fig. 5. Radiation damage to oak-pine forest after 6-month exposure. (*From G. M. Woodwell, Science, vol. 156, 1967*)

In the above sense ionizing radiation is an environmental stress. If the radiation is applied to a community over a period of days or weeks and then terminated, the response of the community is characterized by successional setback and subsequent recovery. Recovery rate depends upon the initial effect of the radiation. In the case of forests, for example, the recovery time depends on whether the trees are entirely killed or whether only the aboveground parts are killed so that the roots can sprout to quickly restore the forest community.

Radioecological effects studies at the community level usually employ some form of point radiation source which can be put out in the community, and can be opened or closed from a safe distance (Fig. 4).

If the radiation is of short duration (acute), subsequent study involves zones of radiation damage beginning at the center where the most drastic effects are observed, and continuing out to some peripheral limit at which only barely perceptible ecological effects are noted, and these only months or years after the event. Such experimental studies enable investigators to predict what would happen to a particular community type under given conditions of radiation. Because the radiation level is attenuated by distance, such studies are, in effect, several simultaneous experiments varying only in dose rate of radiation.

If the radiation stress is permanent (chronic), succession is permanently delayed and a new steady-state develops; the new community type is, as in the case of an acute dose, dependent on the radiation gradient with the most radiation-resistant species developing closest to the source (Fig. 5). Although much of the early work on the irradiation of natural communities was conducted in such permanent radiation fields under chronic conditions, portable irradiators have now been developed. Some of these are capable of delivering dose rates of several thousand rads per day within a few meters of the source. This has made possible the short-term (1–3 months) irradiation of natural ecosystems ranging from old fields to tropical rainforests. Studies of the response and recovery of these systems following the removal of the radiation stress not only are providing valuable data regarding basic ecological processes but are also helping to answer questions concerning the effects of irradiation on the natural environment. *See* ECOSYSTEM. [RICHARD G. WIEGERT]

Bibliography: C. E. Cushing, Jr. (ed.), *Radioecology and Energy Resources*, 1976. F. P. Hungate (ed.), *Radiation and Terrestrial Ecosystems*, 1966; D. J. Nelson and F. C. Evans (eds.), *Symposium on Radioecology*, USAEC TID-4500, 1969; G. A. Sacher (ed.), *Radiation Effects on Natural Populations*, 1966; V. Schultz and A. W. Klement (eds.), *Radioecology*, 1963.

Rainforest

A term used loosely in plant geography for forests of broad-leaved (dicotyledonous), mainly evergreen trees found in continually moist climates in the tropics, subtropics, and some parts of the temperate zones. Sometimes the term is unjustifiably extended to include other very wet forests such as the Olympic Rain Forest of the state of Washington in which the trees are mostly conifers.

The tropical rainforest includes the vast Amazon forest as well as large areas in western and central Africa, Malaysia, Indonesia, and New Guinea. Estimates of the total world area of rainforest vary from 2,000,000 to 3,600,000 mi^2 (5,500,000 to 9,400,000 km^2). It is the home of an enormous number of plant and animal species. The trees are of various heights, up to about 150 ft (1 ft = 0.3 m) or in some places to over 200 ft, and are arranged in several ill-defined strata. The total biomass, including roots, is larger than in most temperate forests, and is in the range 130–260 tons/acre (1 short ton = 0.9 metric ton), calculated as dry weight. As many as 200 different species of trees 1 ft or more in girth may be found in areas of 4 acres (1 acre = 4046 m^2). Tropical rainforests, unlike temperate forests of beech or oak, are usually mixed in composition, no one species forming a large proportion of the whole stand, but in some parts of the tropics there are rainforests dominated by a single species.

Certain structural features are characteristic of tropical rainforests, such as the thin, flangelike buttresses of the larger trees, and flowers and fruit that are produced, as in the cacao (*Theobroma cacao*), on the trunk (cauliflorous) instead of on the branches.

Orchids and other epiphytes and woody vines (lianes) are common. In a rainforest that has not been culled for timber or recently disturbed, the undergrowth is generally rather thin, and visibility is about 60 ft or more on the ground (see illustration).

In the rainforest, animals of most groups, like plants, show great species diversity. For example, there are several times as many species of birds as in an equal area of North American broad-leaved forest. Insects are far more numerous than other animals, in terms of species and sheer numbers. Interactions between plants and animals are extremely complex, and many very specialized relationships have evolved, such as those between certain types of ants and the trees in whose hollow twigs they live. In the tropical rainforest an unusually large variety of organisms live together in a state of balance, so that it is one of the most stable as well as most complex ecosystems on Earth.

Tropical rainforest of Brunei (Borneo). Buttressing and cauliflory are characteristic. (*P. S. Ashton, Ecological Studies on the Mixed Dipterocarp Forests of Brunei State, Oxford Forest. Mem. no. 25, Clarendon Press, Oxford, 1964*)

Climatic factors. In typical tropical rainforest climates there is no winter or severe dry season, and consequently plants can grow and reproduce all year. Nevertheless, many plants are not equally active at all times, leaf production and flowering often taking place at intervals that may be shorter or longer than a year or may be irregular. The behavior of different species, different individuals of the same species, or even different branches of the same tree is often not well synchronized. Rainforest trees seldom show annual growth rings, which makes their age difficult or impossible to determine.

There is little restriction of rainforest animal life by cold or drought, and species that feed on plants can always find food. As a result, groups of animals have evolved, such as hummingbirds, sunbirds, and certain kinds of bats, that depend on flowers being available all year. Rainforest animals are less seasonal in their breeding habits than those of temperate climates; birds raise several broods every year, but fewer eggs are laid in each clutch.

In tropical regions with a marked dry season the rainforest is replaced by deciduous and semideciduous forests and savannas. In these a considerable proportion of the trees becomes wholly or partly bare of leaves for some part of the year. The transition from the humid rainforest to these deciduoustype trees is usually gradual, and no clear dividing lines can be drawn.

Productivity. Rainforest timbers are mostly very hard and are sometimes too dense to float in water. Many, such as the mahoganies (Meliaceae) of Africa and tropical America, provide woods valued for furniture making and veneers. Exploitation of the timber is often difficult because the economically valuable trees are scattered among large numbers of less useful species. In some rainforest localities, wood chip factories have been established so that the entire timber crop can be converted into wood pulp for papermaking, though eucalypts and tropical pines, which are faster-growing than rainforest species, are more suitable for this purpose.

Besides timber, rainforests are an increasingly important source of rattans, fibers, drugs, pesticides, and other useful products.

Estimates of the net organic productivity of tropical rainforests vary from about 6 to 10 tons/acre/year, which is not much greater than for some temperate forests. The luxuriant appearance of the vegetation is deceptive and does not necessarily indicate a fertile soil. The limited supply of available plant nutrients circulates rapidly between the plants and the superficial layers of the soil. So long as the forest is intact, very little of these nutrients is lost in the drainage water, but when the forest is felled, especially if the trees are subsequently burned, there is a heavy loss of soil fertility.

Primary and secondary communities. Rainforests that have never been cleared (virgin forests), or which have been undisturbed long enough to be almost indistinguishable from such forest, are called primary. At the present time these are being replaced by secondary vegetation of various kinds at an ever-increasing rate. Young secondary forests are not so tall but more dense and tangled than primary rainforests. They are composed mainly of fast-growing, short-lived, soft-wooded trees with wind- or animal-dispersed fruits or seeds, enabling them to colonize natural or artificial clearings rapidly. *See* ECOLOGICAL SUCCESSION.

Exploitation. Large areas of tropical rainforest are being felled for lumber or for planting oil palm, rubber, pinetrees, and other industrial crops. More and more areas are also felled by native cultivators to grow manioc, rice, and other food crops under systems of shifting cultivation (slash-and-burn or swidden farming). After one or two harvests a new clearing is made, preferably in primary forest, and secondary forest is allowed to grow on the abandoned land. When the population pressure is heavy, this forest is often cleared again after too short an interval; the soil then deteriorates, especially if the vegetation is grazed and frequently burned. Under such conditions grasses such as *Imperata cylindrica* (alang-alang) invade the secondary forest, which eventually becomes replaced by grassland of little agricultural value, resembling the savannas of less humid climates. *See* SAVANNA.

Because of the rapidly increasing world demand for timber and pulpwood and the needs of fast-growing native populations for food, the tropical rainforest is disappearing so fast that, if present trends continue, little if any primary forest will remain by the end of the century. Efforts are being made by national and international organizations to conserve reserves for the future.

Distribution. The tropical rainforest occupies lowland areas where the annual rainfall is not less than about 80 in. (1 in. = 2.5 cm) and there are not more than three or four conseoutive months with less than about 4 in. At higher elevations it gives way to montane rainforest, and with increasing latitude it gradually merges into subtropical and temperate rainforests. These other types of rainforest are different in composition, less rich in species, and usually less tall than the lowland forest; features such as buttressing and cauliflory, which are so characteristic of the latter, are absent or less well developed. *See* VEGETATION ZONES, WORLD. [PAUL W. RICHARDS]

Bibliography: K. A. Longman and J. Jeník, *Tropical Forest and Its Environment*, 1974; H. T. Odum and R. F. Pigeon, *A Tropical Rain Forest: A Study of Irradiation and Ecology at El Verde, Puerto Rico*, 1970; P. W. Richards, *The Life of the Jungle*, Our Living World of Nature Series, 1970; P. W. Richards, *The Tropical Rain Forest*, 1975; J. P. Schulz, *Ecological Studies on Rain Forest in* Northern Surinam, *1960;* UNESCO/UNEP/FAO, *Tropical Forest Ecosystems: A State of Knowledge Report*, 1978; T. C. Whitmore, *Tropical Rain Forests of the Far East*, 1975.

Rangeland conservation

The major purpose of rangeland conservation in the United States is to secure maximum forage production for each site consistent with ecological stability of the vegetation and of the soil. Also, where applicable, rangeland conservation is concerned with the preservation of watersheds, with timber production, and with recreation. Range conservation is one of the youngest fields of natural resource management. *See* ECOLOGY; SOIL.

Grazing land economics. The public range is an integral part of many livestock operations in the West, where approximately 10,000,000 head of livestock receive about one-third of their annual forage requirements from public lands. Consequently, the public range is an important element in the national production of meat, wool, and leather. Great but unestimated value is added to the nation's economy by land management on the range which improves water yield and reduces erosion and downstream siltation.

Types of grazing lands. Artificial pastures and the tall-grass prairies (now largely in corn and wheat) are not considered rangeland. The mid-grass and short-grass plains west of the prairies and east of the Rocky Mountains, the arid and semiarid grasslands of the Southwest and the Great Basin, the grasslands of intermountain valleys, the brush and open woodlands of the mountains, and montane and subalpine meadows are all parts of the western range of the United States. Depending on the local climate, some of it provides only winter grazing whereas other parts, especially at higher elevations, provide from a few weeks to a few months of summer grazing. *See* GRASSLAND ECOSYSTEM.

Overgrazing, corrective legislation. Federally owned land suitable for grazing, under the administration of the Bureau of Land Management of the Department of Interior, forms nearly one-tenth of the United States. Previous to the Taylor Grazing Act (June 28, 1934), western grazing lands of about 170,000,000 acres (1 acre = 4047 m^2) were open to free access and use without public control and management. Now 156,000,000 acres are organized into grazing districts 3,000,000–9,000,000 acres in extent that are administered to conserve and regulate the public grazing land and to help stabilize the livestock industry. The public range is used by about 20,000 private stockmen under a system of permits and a code that seeks to guarantee proper use of the range and to return to the government fair compensation for use.

A long history of cutthroat competitive grazing and of conflict between cattle and sheep operators and between them and agricultural settlers had resulted in widespread deterioration of the range. Productivity, naturally low in more arid regions (averaging about 70 lb (1 lb = 0.454 kg) of air-dry forage per acre on the public range in contrast to about 2800 lb per acre for average hay land in the United States), was reduced by overgrazing to extremely low carrying capacity in many places. Erosion was accelerated and water loss by excessive runoff was increased on unprotected soils. Palatable and nutritious species, especially perennial grasses, were reduced in amount, and weeds and other undesirable species were increased. This is well illustrated in Texas, New Mexico, Arizona, and southwestern California by mesquite, which now occupies about 50,000,000 acres (twice what it did about 1900) and which cuts grass forage to one-third or less of full production when the mesquite bushes increase to 100 or more per acre. In the West, overgrazing also caused the spread of sagebrush, which now covers about 96,000,000 acres.

The Bureau of Land Management program since the passage of the Taylor Grazing Act has many accomplishments to its credit: increase of range forage, more livestock products, greater protection of public lands, more usable water, less erosion and downstream sedimentation, more flood control, more wildlife, and more recreational opportunities; yet the Bureau estimated that in the quarter century following the act not more than 10% of the needed range improvement work was done, largely because of inadequate Federal appropriations. The Bureau's program includes range inventories and management plans, improved fire protection, watershed treatment works (regrassing, water conservation and erosion control dams, and water spreading structures), pest and rodent control, and range management improvements (stock water developments, range fencing, corrals, and livestock and truck trails). In spite of this program, about 50% of Federal rangelands are still in a state of severe to critical erosion, 32% are suffering moderate erosion, and only 18% are in a condition of unaccelerated or no erosion.

Range inventory. The problem is to determine what the vegetation, soil, and climate will permit with regard to forage production as measured by the land's animal-carrying capacity. The range inventory includes a quantitative evaluation of the vegetation, its palatability, the nutritional value of each species, the status of the vegetation in the plant successional process, and the time and degree of permissible use.

The soil is evaluated to determine whether it is normal, eroded, or compacted and whether its productive potential is increasing or decreasing. It is helpful to know what the forage-producing capacity of the soil was originally, what it is at present, and what its possibilities are. The water infiltration rate and the moisture retention capacity of the soil are also measured, and hence its watershed management condition and its erosion potential are determined.

Existing water facilities and possibilities for water development are evaluated to determine whether artificial vegetation rehabilitation is feasible. Accessibility by roads and trails, fire history, and an analysis of fire potential are also determined. In addition, the presence or absence of poisonous plants, predators, and destructive rodents is noted.

Range management plan. The range management plan, developed from the inventory, sets forth the season of year that the range should be grazed, the duration (usually in days) of the grazing period, and the kinds and numbers of heavy livestock (cattle and horses) or lighter animals (sheep and goats) or both to be permitted. The range plan also states the need for and nature of fire protection, the need for additional trails and roads or the abandonment of old ones, the location and types of fences required to direct the control of trailing and herding, the water development needed, any special treatment of soil advisable for eroded or compacted areas, the steps required for control of poisonous plants, and the need for reduction of predatory animals and rodents. The range management plan also details the reseeding needed—the location, type of species, planting methods to be used, and use of the vegetation after planting—and considers whether rehabilitation may be expected without artificial seeding under a plan for temporary retirement from use and suitable protection.

Significant advances. Quantitative surveys of range vegetation, now a fairly well-developed technique, probably will be improved through use of aerial photos, greater field mobility of survey technicians, and better statistical techniques.

Since about 1935 much has been learned about the ecology and physiology of range plants. This has resulted in the classification of such plants with respect to their physiological condition and the ecological trends indicated—so-called condition and trend classification. *See* APPLIED ECOLOGY.

Through quantitative study of animal ingestion of various plant species, a significant advance has been made in understanding the relation of floristic composition to actual animal-carrying capacity of rangeland. One method contributing to this knowledge has been the use of the esophageal fistula and fecal collecting bags. Associated with this is an improved and intensive before-and-after range analysis. The actual nutritional value of the forage species and the usable portions of the plants under various methods of management are determined by laboratory study. This indicates to what extent supplemental feeding of livestock is necessary. *See* POPULATION DYNAMICS.

It is now accepted that the management of deer, elk, and antelope herds is also part of range management. Much has been learned concerning the problem of competition of domestic livestock with big-game animals, thereby contributing to improved management of rangelands for both kinds of animals.

There is growing knowledge of ways to improve range plant composition by using large machinery and selective sprays, often applied by airplane, for destruction of undesirable brush and weeds. Still to be studied are the adaptability of plants to various locations, methods of planting, physiological adaptiveness of species, time of planting, and the time and type of use after planting. The use of airplanes or helicopters for seeding has not yet solved all range planting problems.

Problems, present and future. There is still a question as to whether any significantly large, new range areas can be opened up by an improved or different type of management. The "discovery" for range use of ex-timber lands in the coastal plain of the Southeast is an example.

Many problems of multiple use of rangeland are yet to be resolved. In the western summer range there probably will always be the problem of correlating grazing and timber production. Demands for recreation and big-game hunting areas are becoming more important. With the increasing population of the West and the more intensive use of all its lands, the demand for water is growing rapidly, making efficient management of rangelands imperative for water conservation. *See* FISHERY CONSERVATION; FOREST MANAGEMENT AND ORGANIZATION; LAND-USE PLANNING; MINERAL RESOURCES CONSERVATION; SOIL CONSERVATION; WATER CONSERVATION; WILDLIFE CONSERVATION.

[LEWIS TURNER]

Bibliography: *See* CONSERVATION OF RESOURCES.

Remote terrain sensing

The gathering and recording of information about terrain surfaces without actual contact with the object or area being investigated. Remote terrain sensing is part of the large and rapidly expanding

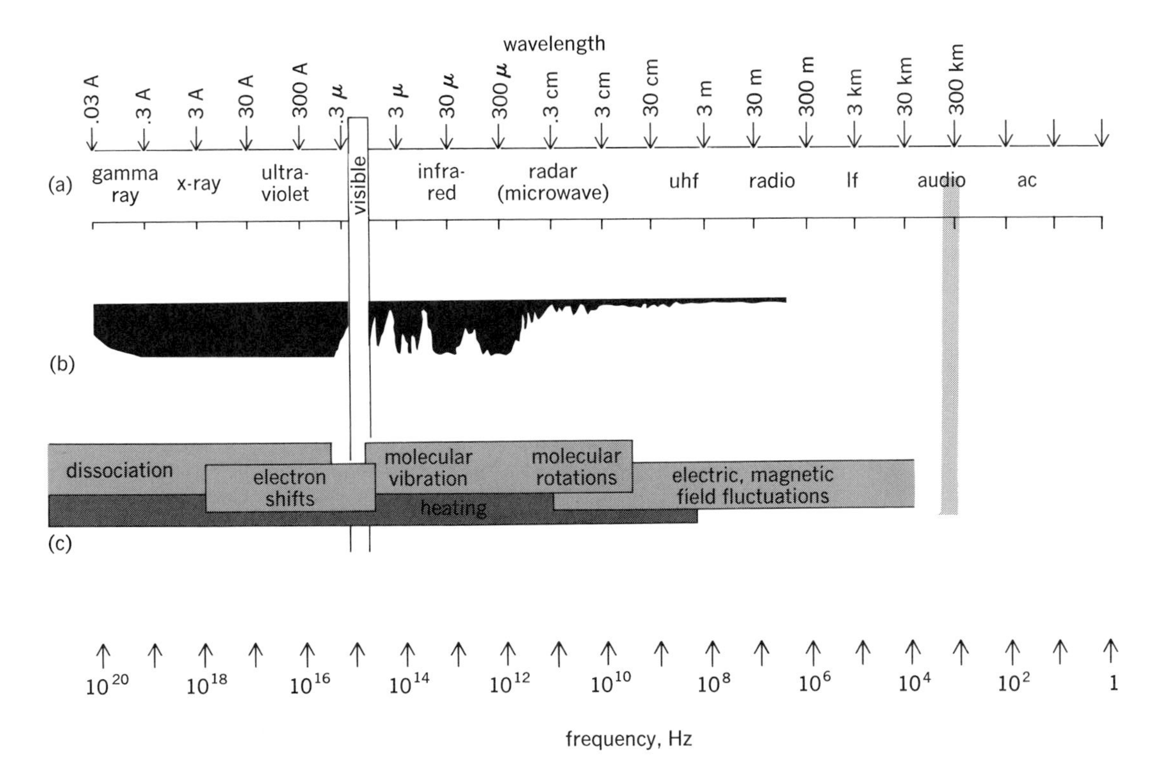

Fig. 1. Characteristics of the electromagnetic energy spectrum which are of significance in remote sensing. (*a*) Regions of electromagnetic spectrum. (*b*) Atmospheric transmission. The dark-colored areas are regions through which electromagnetic energy is not transmitted. (*c*) Phenomena detected.

subject of remote sensing, which deals with the gathering and recording of information on many types of natural phenomena from a distance (see table). This article will deal mainly with terrain sensing which uses the visual, infrared, and microwave portions of the electromagnetic spectrum (Fig. 1), with some discussion of oceanographic sensing also.

Humans have always used remote sensing in a primitive form. In ancient times, they climbed a tree or stood on a hill and looked and listened, or sniffed for odors borne on the wind. In fact, taste,

Some area of application for remote sensor instruments

Technique	Agriculture and forestry	Geology and planetology	Hydrology	Oceanography	Geography
Visual photography	Soil types, plant vigor, disease	Surface structure, surface features	Drainage patterns	Sea state, erosion, turbidity, hydrography	Cartography, land use, transportation, terrain and vegetation characteristics, thematic mapping
Multispectral imagery	Same as above	Lithological units, formation boundaries	Soil moisture	Sea color, biological productivity	Same as above
Infrared imagery and spectroscopy	Vegetation extent, soil moisture, conditions	Thermal anomalies, faults	Areas of cooling, soil moisture	Ocean currents, sea-ice type and extent, sea-surface temperature	Thermal activity in urban areas, land use
Radar: imagery, scatterometry, altimetry	Soil characteristics, plant conditions	Surface roughness, structural framework	Soil moisture, runoff slopes	Sea state, tsunami warning Sea-ice type and extent	Land-ice type and extent, cartography, geodesy
Passive microwave radiometry and imagery	Thermal state of terrain, soil moisture		Soil moisture, snow, ice extent	Sea state, sea-surface temperature, sea-ice extent	Snow and ice extent

touch, sight, and hearing are the only ways that humans can be aware of their environment without remote sensing. Although it is possible to sense the environment without the use of instruments, it is not possible to record the sensed information without the aid of instruments referred to as remote sensors. These instruments are a modern refinement of the art of reconnaissance, an early example being the first aerial photograph taken in 1858 from a balloon floating over Paris.

The eye is sensitive only to visible light, a very small portion of the electromagnetic spectrum (Fig. 1). Cameras and electrooptical sensors, operating like the eye, can sense and record in a slightly larger portion of the spectrum. For gathering invisible data, instruments operating in other regions of the spectrum are employed. Remote sensors include devices that are sensitive to force fields, such as gravity-gradient systems, and devices that record the reflection or emission of electromagnetic energy. Both passive electromagnetic sensors (those that rely on natural sources of illumination, such as the Sun) and active ones (those that utilize an artificial source of illumination such as radar) are considered to be remote sensors. Several remote-sensing instruments and their applications are listed in the table.

TERRAIN CHARACTERISTICS

Each type of surface material (for example, soils, rocks, and vegetation) absorbs and reflects solar energy in a characteristic manner depending upon its atomic and molecular structure (Fig. 1*c*). In addition, a certain amount of internal energy is emitted which is partially independent of the solar flux. The absorbed, reflected, and emitted energy can be detected by remote sensing instruments in terms of characteristic spectral signatures and images. These signatures can usually be correlated with known rock, soil, crop, and other terrain features. Chemical composition, surface irregularity, degree of consolidation, and moisture content are among the parameters known to affect the records obtained by electromagnetic remote sensing devices. Selection of the specific parts of the electromagnetic spectrum to be utilized in terrain sensing is governed largely by the photon energy, frequency, and atmospheric transmission characteristics of the spectrum (Fig. 1).

TYPES OF REMOTE SENSING

Because remote sensing is a composite term which includes many types of sensing, its meaning can be best understood by describing several of the types. Remote sensing is generally conducted by means of remote sensors installed in aircraft and satellites, and much of the following discussion refers to sensing from such platforms.

Photography. Photography is probably the most useful remote sensing system because it has the greatest number of known applications, it has been developed to a high degree, and a great number of people are experienced in analyzing terrain photographs. Much of the experience gained over the years from photographs of the terrain taken from aircraft is being drawn upon for use in space.

Results of conventional photography experiments carried out in the short-duration crewed spacecraft such as a Gemini, Apollo, and Skylab, and in the long-duration unmanned spacecraft such as ERTS, Landsat, Nimbus, NOAA, and DMSP, have vividly shown the applicability of these systems in space. These results indicate that space photography provides valuable data for delineating and identifying various terrain features such as those shown in Fig. 2.

The photomap in Fig. 2 was made by rectifying and fitting the photographs to the cultural plate of the existing line map of the area. The result combines all the original photographic detail with the cultural line data important for geographic orientation and location. This combination provides considerably more terrain data than conventional topographic or raised relief maps. A number of major copper deposits occur in this region, and their location is in part controlled by the presence of intense fracturing, as shown on the photomap. Also important is the presence of intrusive rocks, which sometimes show up as circular areas. Fault intersections, intrusive centers, and alteration halos are important clues in the search for new mineral deposits of this copper type. Detailed study of such high-quality spacecraft imagery is proving to be very useful for unraveling the structural framework of potential mineral districts.

Multispectral photography. Multispectral photographic techniques are also being used. Multispectral photography can be defined as the isolation of the reflected energy from a surface in a number of given wavelength bands and the recording of each spectral band separately on film. This technique allows the scientist to select the significant bandwidths in which a given area of terrain displays maximum tonal contrast and, hence, increases the effective spectral resolution of the system over conventional black-and-white or color systems.

Because of its spectral selectivity capabilities, the multispectral approach provides a means of collecting a great amount of specific information. In addition, it has less sensitivity to temperature, humidity, and reproduction variables than conventional color photography and retains the high resolution associated with broadband black-and-white mapping film.

Multispectral imagery. An advance in multispectral sensing is the use of multispectral scanning systems which record the spectral reflectance by photoelectric means (rather than by photochemical means as in multispectral photography) simultaneously in several individual wavelengths within the visual and near-infrared portions of the electromagnetic spectrum (0.47–1.1 μm). Such instruments have been used in aircraft since the early 1960s and are now being used in the earth resources technology satellite (*ERTS 1*) which was launched in 1972 and *Landsat B* which was launched January 1975.

In the satellite cases, optical energy is sensed by an array of detectors simultaneously in four spectral bands from 0.47 to 1.1 μm. As the optical sensors for the various frequency bands sweep across the underlying terrain in a plane perpendicular to the flight direction of the satellite, they record energy from individual areas on the ground. The size of these individual areas (instantaneous fields of view) is determined by the resolution capability (spot size) of the optical scanner. The smallest

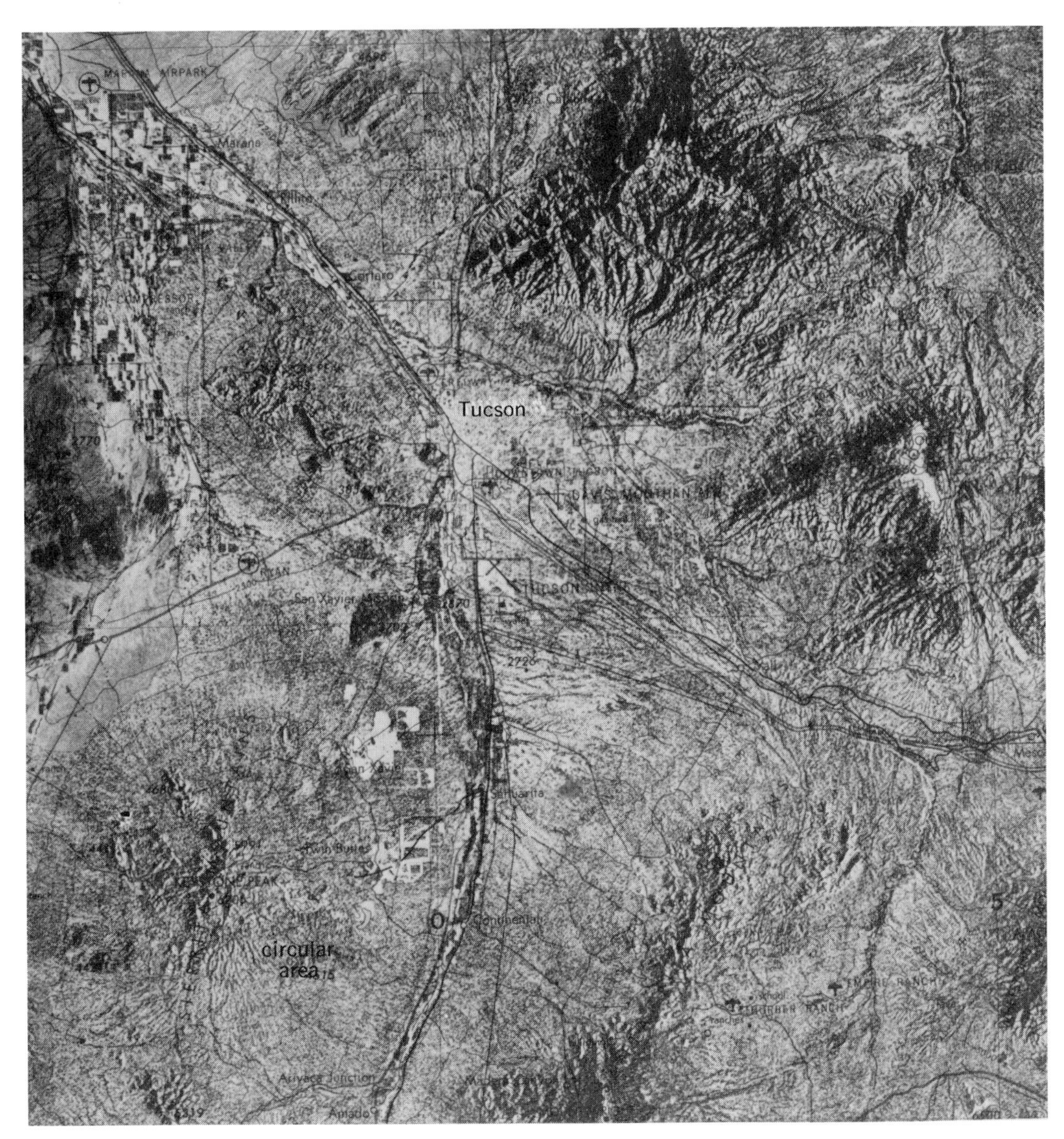

Fig. 2. Space photomap of the Tucson, AZ, area, using 80-mm-focal-length Hasselblad cameras on the *Apollo 9* flight. The strong northeast-trending fractures in the mountains east of Tucson are apparent. The white patches on the northeast flank of the circular area are open pits and mine dumps, associated with already discovered mineral deposits.

individual area distinguished by the scanner is called a picture element or pixel, and a separate spectral reflectance is recorded in analog or digital form for each pixel. A pixel covers about 1 acre (approximately 0.4 hectare, or 4047 m^2) of the Earth's surface in the case of the Landsat, with approximately 7,500,000 pixels composing each Landsat image (115 × 115 mi or 185 × 185 km in area).

Such multispectral scanning systems have a number of advantages over conventional photography: the spectral reflectance values for each pixel can be transmitted electronically to ground receiving stations in near-real time, or stored on magnetic tape in the satellite until it is over a receiving station. When the signal intensities are received on the ground, they can be reconstructed almost instantaneously into the virtual equivalent of conventional aerial photographs (provided the resolution size of the individual pixels is fine enough). In the case of conventional photography, the entire film cassette generally has to be returned physically to the ground, before the film can be developed and studied. Such physical return of film cassettes is not practical for long-duration satellites such as ERTS which are continuously recording vast amounts of data.

Computer enhancement of imagery. There are a number of other advantages of multispectral scanning systems. Since they record reflected radiation energy in a number of discrete wavelengths, these radiance values can be used singly or combined by digital computer processing to provide a response optimized for particular terrain features.

One type of such computer enhancement which is proving to be very useful involves the preparation of ratio images. The individual spectral responses are ratioed, picture element by picture element using a computer, and they can also be contrast-stretched to further enhance the spectral differences. Such ratio images are very useful in detecting such items as subtle soil changes and hydrothermally altered areas (important for mineral exploration), as well as in discriminating most major rock types.

Another advantage of the multispectral scanner is its ability to extend the spectral coverage (beyond the 0.9-μm cutoff available with conventional photographic films) into the thermal wavelengths of the electromagnetic spectrum.

Infrared. Infrared is electromagnetic radiation having wavelengths of 0.7 to about 800 μm. All materials continuously emit infrared radiation as long as they have a temperature above absolute zero in the Kelvin scale. This radiation involves molecular vibrations as modified by crystal lattice motions of the material. The total amount and the wavelength distribution of the infrared radiation are dependent on two factors: the temperature of the material and its radiating efficiency (called emissivity).

Thermal infrared radiation is mapped by means of infrared scanners similar to the multispectral scanners described previously, but in this case radiated energy is recorded generally in the 8–14-μm portion of the electromagnetic spectrum.

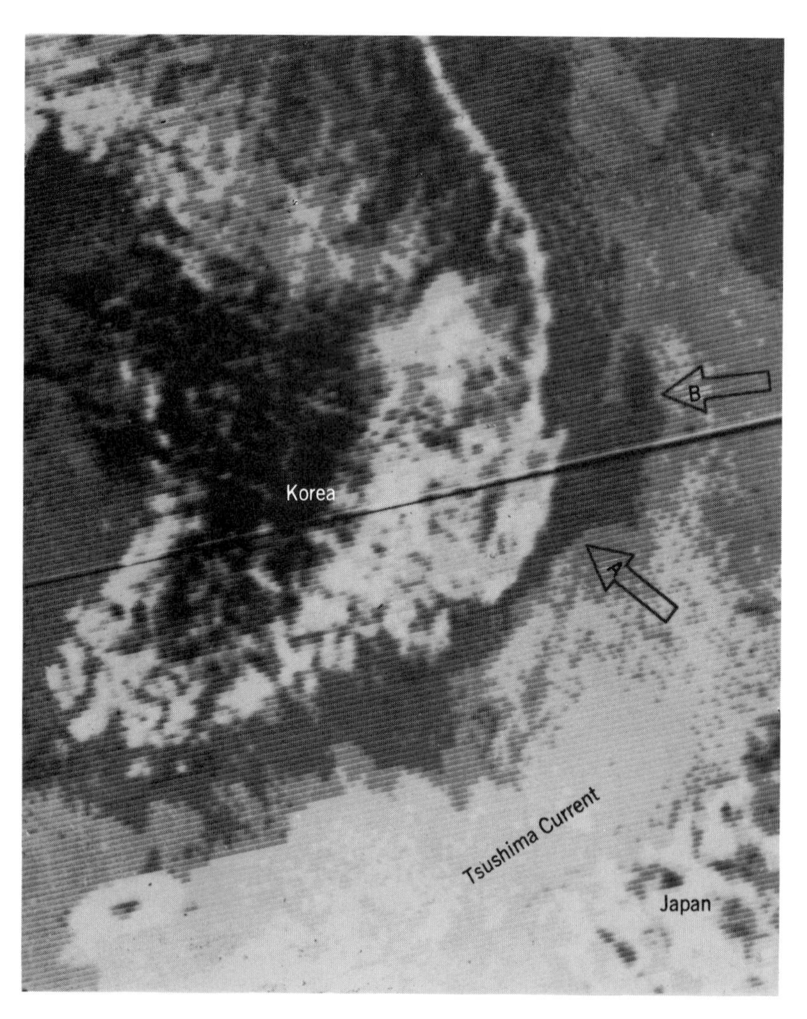

Fig. 3. Thermal infrared image of the Korea Straits, between South Korea and Japan. This image was acquired by the Defense Meteorological Satellite System (DMSS) on Oct. 13, 1972. DMSS infrared-sensor ground resolution is 2 nautical miles (3.7 km). The light-colored waters are part of the Tsushima Current, while the darker gray areas are the coastal shelf waters. An oceanic "front" forms the boundary between the continental shelf waters and the Tsushima Current. The front turns seaward at A, and a cyclonic eddy is located at B.

The imagery provided by an infrared scanning system gives information that is not available from ordinary photography or from multispectral scanners operating in the visual portion of the electromagnetic spectrum. The brightness with which an object appears on an infrared image depends on its radiant temperature. The hotter the object, the brighter it will appear on a positive image. Radiant temperature is largely dependent upon chemistry, grain size, surface roughness, and thermal properties of the material. The time of day or night and the season of the year at which the infrared energy is recorded are also important factors, particularly when water or moisture exists in the near-surface terrain materials, since the water has a different thermal inertia relative to rock and soil materials and can therefore appear as a thermal anomaly. Because moisture frequently collects in geologic faults and fractures, infrared images are useful in detecting such features.

Although it is possible to obtain relatively high resolutions (spot sizes of a few feet) from infrared imagers in low-flying aircraft, it has been more difficult to achieve such fine resolutions from satellite altitudes, partially because of the lower sensing-element sensitivity of infrared scanners (whose sensing elements must be cooled, in contrast to visible spectrum sensors), and partially because there is always a trade-off between field of view (generally large for satellite imagery as compared with aircraft imagery) and resolution. The larger the field of view, the more difficult it is to obtain fine resolution. Despite these restrictions, routine thermal mapping from long-duration satellites is now on the order of 700–900 m per picture element on the ground (*Nimbus 5* and *NOAA 2* satellites). The multispectral scanner on the short-duration Skylab space flight had an 80-m instantaneous field of view and recorded temperature differences of 0.40 K.

In the past, thermal infrared images were generally recorded on photographic film. Videotape records are now replacing film as the primary recording medium and permit better imagery to be produced and greater versatility in interpretation of data.

Thermal mapping from satellite altitudes is proving to be useful for a number of purposes, one of which is the mapping of thermal currents in the ocean (Fig. 3). Such currents change their position quite frequently; in many cases these changes occur daily or even more frequently. The positions of such thermal boundaries are important for a variety of uses (fishery, naval, and so forth).

Thermal infrared mapping (thermography) from aircraft and satellite altitudes has many other uses also, including the mapping of volcanic activity and geothermal sites, location of groundwater discharge into surface and marine waters, and regional pollution monitoring.

Microwave and radar. This type of remote sensing utilizes both active and passive sensors. The active sensors such as radar supply their own illumination and record the reflected energy. The passive microwave sensors record the natural radiation. A variety of sensor types are involved. These

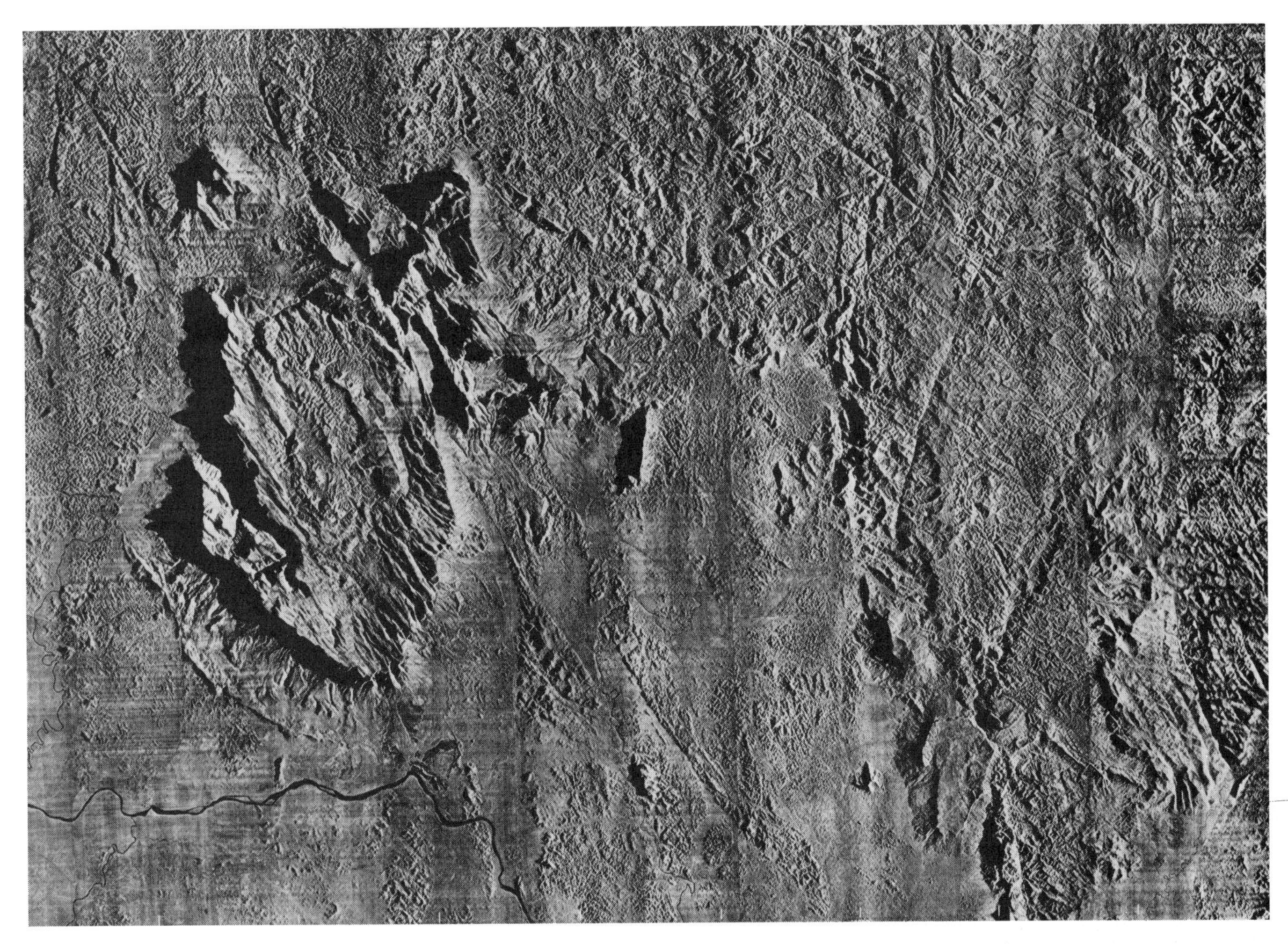

Fig. 4. Side-looking radar (SLAR) imagery of a portion of the jungle-covered Guayana Precambrian Shield of Venezuela. This imagery was acquired by a Goodyear synthetic-aperture X-band (3.12-cm) radar system mounted on board an Aero Service Caravelle aircraft. The flight altitude was 46,000 ft (14,030 m). The east-west dimension of the image mosaic is approximately 168 km.

include imaging radars, radar scatterometers and altimeters, and over-the-horizon radars using large ground-based antenna arrays, as well as passive microwave radiometers and imagers.

One of the most significant advantages of these instruments is their all-weather capability, both day and night. The active radars also possess a certain amount of foliage-penetration ability which is valuable in jungle terrains (Fig. 4). This imagery illustrates the unique capability of side-looking radar (SLAR) to penetrate both cloud cover and dense jungle vegetation and to map the structural fabric of the underlying terrain with considerable detail. It has been demonstrated that radar return amplitude is affected by the composition of the illuminated area, its moisture content, vegetation extent and type, surface roughness, and even temperature in certain circumstances.

Radar returns are recorded in various forms to aid in their analysis. The forms having primary geoscience interest can be placed in one of the following categories: (1) scattering coefficients, which are a powerful tool in studying the nature of radar return and which have been correlated with different terrain types and directly applied to oceanic surface studies; (2) altimetry data; (3) penetration measurements, which have been utilized for mineral exploration, for example, detection of faults through moisture associated with them; and (4) radar images.

Of these techniques, imagery generally presents the optimum geoscience information content of radar return, partly because well-developed photographic interpretation techniques are applicable to radar image analysis. These images are especially valuable for delineating various structural phenomena (Fig. 4).

The image record of the terrain return is affected by the frequency, angle of incidence, and polarization of the radar signal. For example, if the terrain being imaged is covered by vegetation, a *K*-band (35 GHz) signal records the vegetation, whereas a *P*-band (0.4 GHz) signal is likely to penetrate the vegetation and thus record a combination of vegetation and soil surface. In general, each frequency band represents a potential source of unique data. The angle of incidence of the incident wave affects the image because of radar shadowing on the backside of protruding objects. The angle of incidence can also show differences because of changes in orientation of the many facets on a surface which affect the return strength. Although information is lost in the shadow region, the extent of the shadow indicates the height of the object and hence has been useful in emphasizing linear features such as faults and lineations reflecting joint systems.

Although radar and microwave instruments have been largely confined to aircraft platforms, a radiometer/altimeter/scatterometer experiment was flown on the Skylab space flights in 1973, acquiring scatterometric, altitude, and passive microwave radiometric data at 13.8, 4, and 1.4135 GHz. The Seasat satellite system to be flown in the late 1970s is expected to include a radar altimeter (±10 cm precision), a microwave scatterometer, a microwave radiometer (±1°C precision, 14-km image resolution), and a synthetic aperture radar (25-m resolution for 100-km swath widths and 100-m resolution for 200-km swath widths).

Over-the-horizon radar. The over-the-horizon (OTH) radars, although well advanced for military tasks, are just beginning to be used for environmental science objectives. These radars utilize frequencies in the high-frequency (hf; 3–30 MHz) portion of the electromagnetic spectrum (median wavelength of about 20 m) and are thus not within the microwave part of the spectrum. The energy is transmitted by large ground-based antennas and is refracted by the various ionospheric layers back down to the Earth's surface some 800–3000 km (on a single-hop basis) away from the hf radar antenna site. This form of radar wave which is transmitted and reflected back via the ionosphere is referred to as a sky wave. The incident waves are reflected from such surface features as sea waves, and the reflections are enhanced when the wavelength of the sea waves is twice the wavelength of the incident radar waves (Bragg reflection effect). In addition, the Doppler principle is applied to determine the component of sea-wave velocities moving toward or away from the radar. The use of such OTH sky waves for sea-state determinations is in its infancy, and additional research needs to be done, particularly relative to interpretation of the data and their confirmation via ground-based sensors. Spot sizes observed on the ocean surface by this sky-wave method are now as fine as 15 km in azimuth by 3 km in range, and it is expected that these resolution cells can be reduced by further development work.

In addition to this experimental use of hf radars for remote sensing of sea state, they are also being used as telemetry (communication) and direction- and range-finding links in conjunction with disposable buoys or drogues. The buoys are used to detect the sea state, surface and subsurface temperature and salinity, and so forth. The information is relayed back to the interrogating site by either skywave or surface-wave modes. The surface waves are that portion of the hf wavefront that follows closely along the Earth's surface. Ranges of 100–300 km or so are possible with this mode, and the buoys can be interrogated and located by low-cost direction-finding loops and commercial battery-powered receivers. The big advantage of these ground-based hf radars, as compared to satellite-borne radars, is that they can maintain continuous monitoring of an ocean area, while satellite-borne radars can monitor only the areas that they happen to be flying over at the time. In the case of the Landsat system the satellite comes over (revisits) the same area about every 18 days, while in the case of the Defense Meteorological Satellite System (DMSS) the frequency of revisits can be up to several times daily, depending upon how many satellites are in orbit at one time. Neither the Landsat or DMSS carries an all-weather radar system, but the Seasat will carry imaging radars and scatterometers, and its revisit time is expected to be on the order of 6 days.

Future trends. Remote sensing of terrain has produced a number of significant benefits for humans, primarily as a result of United States goverment-sponsored satellite and aircraft remote sensing flights which were initiated in the mid-1960s. The developing countries will derive even greater benefits than the United States over the coming years since their natural resources (water, agriculture, forests, geology, and so on) are less well known and they have fewer alternate data-gathering systems available. Thus the international community will become more and more involved in the use of terrain data, remotely sensed by satellites, and will begin to share in the sponsorship of satellite launches. Geostationary satellites are being planned for launch by the European Space Research Organization, Japan, and the Soviet Union in the near future. These are part of a network of five geostationary operational environment satellites (GOES) which will observe the Earth's atmosphere continuously from geostationary altitudes. The first two GOES have already been launched by the United States. Although these particular satellites are primarily to be used for meteorological observations, they do have a sectorizing ("zoom") capability with ground resolutions of 2 × 4 km at nadir. This will provide some limited terrain observation capability. It is expected that further zoom capabilities will be introduced on both geostationary and orbiting satellites in coming years with accompanying expansion of the data transmission links. All-weather sensors (multispectral imaging radars in particular) will come into operational use on orbiting satellites. Reliable forecasts of global crop conditions will be one of the many user benefits to be expected from satellite sensing.

[PETER C. BADGLEY]

Bibliography: American Society of Photogrammetry, *Manual of Remote Sensing*, 1975; *Earth Resources Survey Benefit: Cost Study*, vols. 1–4, prepared under Contract 14-08-0001-13519 for the U.S. Department of Interior/Geological Survey, 1974; W. A. Fischer, *Status of Remote Sensing*, paper presented at the 12th Congress of International Society of Photogrammetry, Ottawa, July 1972; T. L. Lillesand and R. W. Kiefer, *Remote Sensing and Image Interpretation*, 1979; Remote sensing of earth resources, in *Proceeding before the Committee on Sciences and Astronautics*, U.S. House of Representatives, 92d Congress, 2d Session, Jan. 25, 26, 27, 1972; L. C. Rowan et al., *Discrimination of Rock Types and Detection of Hydrothermally Altered Areas in South Central Nevada by the Use of Computer-Enhanced ERTS Images*, USGS Prof. Pap. 883, 1974; R. Sisselman, Looking for minerals via satellite: A far-out approach to exploration, *Eng. Mining J.*, 176(5):87–94, 1975; W. L. Smith (ed.), *Remote-Sensing Applications for Mineral Exploration*, 1977; *Symposium of Significant Results Obtained from Earth Resources Technology Satellite 1*, NASA Spec. Publ. SP-327, X-650-127, X-650-73-155, 1973; *Third Earth Resources Technology Satellite Symposium*, NASA Spec. Publ. 351, 356, 357, 1973.

River

A natural, fresh-water surface stream that has considerable volume compared with its smaller tributaries. The tributaries are known as brooks, creeks, branches, or forks. Rivers are usually the main stems and larger tributaries of the drainage systems that convey surface runoff from the land. Rivers flow from headwater areas of small tributaries to their mouths, where they may discharge into the ocean, a major lake, or a desert basin.

Rivers flowing to the ocean drain about 68% of the Earth's land surface. The remainder of the land either is covered by ice or drains to closed basins (common in desert regions). Regions draining to the sea are termed exoreic, while those draining to interior closed basins are endoreic. Areic regions are those which lack surface streams because of low rainfall or lithologic conditions.

Sixteen of the largest rivers (see table) account for nearly half of the total world river flow of water. The Amazon River alone carries nearly 20% of all the water annually discharged by the world's rivers. Rivers also carry large loads of sediment. The total sediment load for all the world's rivers averages about 20×10^9 metric tons brought to the sea each year. Sediment loads for individual rivers vary considerably. The Yellow River of northern China is the most prolific transporter of sediment. Draining an agricultural region of easily eroded loess, this river averages about 2×10^9 tons of sediment per year, one-tenth of the world average.

River channels and patterns. The morphological features and the sizes of river channels depend on the supply of water and sediment from upstream. The dependency can be summarized in the proportions below, where Q_w is a measure of

$$Q_w \propto \frac{w, d, \lambda}{S}$$

$$Q_s \propto \frac{w, \lambda, S}{d, P}$$

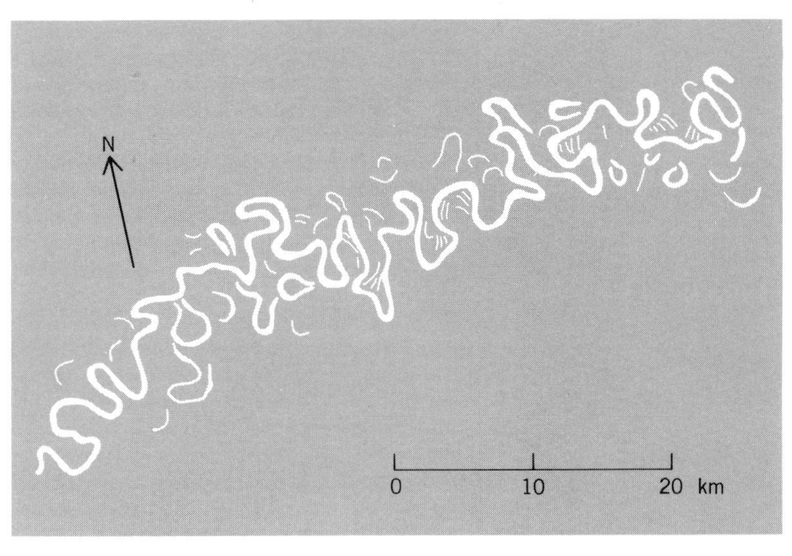

Fig. 1. Meandering pattern of the Juruá River in the western Amazon Basin of Brazil. Note the many abandoned meander loops created by this active river. *(From V. R. Baker, Adjustment of fluvial systems to climate and source terrain in tropical and subtropical environments, Can. Soc. Petrol. Geol. Mem. 5, pp. 211–230, 1978)*

mean annual water discharge (volume carried by the river per unit time) and Q_s is a measure of the type of sediment given by the proportion of coarse bedload (sand and gravel) to the total load (which includes considerable fine sediment). Q_w and Q_s are the controlling variables, while the other variables are dependent variables. The size of the river channel is indicated by its width w and depth d. Note that larger water discharges produce proportionally larger river channels. The slope or gradient of the river channel S is inversely proportional to the water discharge Q_w but directly proportional to the percentage of coarse load Q_s.

The sinuosity of a river P is the ratio of its channel length to its valley length. A perfectly straight river would have a sinuosity of 1. Such rivers are uncommon. Meandering is the most common river pattern (Fig. 1), and meandering rivers develop alternating bends that often have a regular spacing along the valley trend. The spacing of two successive bends is defined as the meander wavelength λ. Note that, like channel width, meander wave-

Characteristics of some of the world's major rivers

River	Average discharge, m³/s	Drainage area, 10³km²	Average annual sediment load, 10³ metric tons	Length, km
Amazon	181,000	7180	900,000	6275
Congo	39,620	3690	64,680	4670
Orinoco	22,640	1480	86,490	2570
Yangtze	21,790	1940	550,000	4990
Brahmaputra	19,980	935	800,000	2700
Mississippi-Missouri	17,830	3220	344,000	6260
Yenisei	17,380	2590	10,520	5710
Lena	15,480	3030	—	4600
Mekong	15,000	910	187,650	4180
Parana	14,890	3100	81,650	3940
St. Lawrence	14,150	1460	3,630	3460
Ganges	14,090	1170	1,600,000	2640
Irrawaddy	13,560	370	330,280	2300
Ob	12,480	2590	14,240	4500
Volga	9,900	1530	18,840	3740
Amur	9,570	2040	—	4670

length is directly proportional to both Q_w and Q_s. Sinuosity P is somewhat independent of Q_s, but it depends strongly on the percentage of coarse load Q_s. Highly sinuous streams often transport fine-grained loads and tend to be narrow and deep.

Braided rivers have channels that are divided into anastomosing branches by alluvial islands and bars. Braided rivers often have steeper gradients, more variable discharges, coarser sediment loads, and lower sinuosity than meandering rivers. The details of river pattern development are more complex than can be presented here. In addition to Q_w and Q_s, important controlling variables include the texture of floodplain and bank sediments, the absolute amount of sediment transported, the channel gradient, the channel stability, and the discharge variability.

River floods. River discharge varies over a broad range, depending on many climatic and geologic factors. The low flows of the river influence water supply and navigation. The high flows are a concern as threats to life and property. However, floods are also beneficial. Indeed, the ancient Egyptian civilization was dependent upon the Nile River floods to provide new soil and moisture for crops.

Floods are a natural consequence of the spectrum of discharges exhibited by a river (Fig. 2). Rivers in humid temperate regions often exhibit less variable flood behavior than rivers in semiarid regions of rugged terrain. Floods in the humid regions tend to occur on average once each year. Rarer, larger floods are often no more than one or two times the magnitude of more common annual floods. The semiarid floods, however, are usually very small for common events, such as the average annual flood. However, rare, high-magnitude floods may be catastrophic. In 1976 a flood wave over 6 m high swept down the Big Thompson River in Colorado, killing 139 people.

The greatest floods in the geologic record occurred during the Pleistocene about 13,000 years ago in the northwestern United States. Lake Missoula, an ice-dammed lake in western Montana, released several catastrophic floods across the Channeled Scabland region of eastern Washington. The largest of these floods discharged as much as 20×10^6 ft^3 (5.7×10^5 m^3) of water per second. Flow velocities in the Scabland channelways ranged from 10 to 30 m/s for water 30–100 m deep. These phenomenal floods created a bizarre landscape of anastomosing channels, abandoned cataracts, streamlined hills, and immense gravel bars. So much water entered the preflood river valleys that they filled to overflowing, and floodwater scoured the divide areas between the valleys.

The Missoula floods were certainly among the most spectacular fluvial phenomena of all time. However, it should be remembered that most rivers do their work very slowly. Rivers are mainly transport agents, removing debris produced by the prolonged action of rainsplash, frost action, and mass movement. The many smaller floods probably accomplish much more of this work of transport than the rare large flood.

Floods are but one attribute of rivers that affect human society. Means of counteracting the vagaries of river flow have concerned engineers for centuries. Today many of the world's rivers are managed to conserve the natural flow for release at times required by human activity, to confine flood flows to the channel and to planned areas of floodwater storage, and to maintain water quality at optimum levels. *See* RIVER ENGINEERING.

Geologic history. Some rivers possess a long heritage related to their location along relatively stable continental cratons or their correspondence to structural lows, such as rift valleys. More commonly, rivers have been disrupted through geologic time by the Earth's active tectonic and erosional processes. The most recent disruptions were caused by the glaciations of the Pleistocene. Many rivers, like the Mississippi, became heavily loaded with coarse glacial debris delivered to their headwaters by glaciers. Since the last glacial maximum about 18,000 years ago, most of the world's rivers have adjusted their channel sizes, patterns, and gradients to the new environmental conditions of postglacial time.

In tropical regions the effects of glaciation were indirect. During full-glacial episodes of the Pleistocene, tropical areas of the Amazon and Congo river basins experienced relative aridity. Forests were replaced by savanna and grassland. The greater erosion rates on the land contributed large amounts of coarse sediment to the rivers. Today these regions have returned to their high-rainfall condition. The rivers receive relatively little sediment from interfluves that are stabilized by a dense forest cover. However, it is alarming that human exploitation of the tropical forests is effectively returning the landscape to its glacial condition. This will undoubtedly induce profound changes in many tropical rivers.

Rivers on Mars? Orbital photographs of Mars obtained during the Viking space mission (Fig. 3) reveal that the planet possesses a remarkable variety of channeled terrains. Some channels form relatively small networks up to 100 km in length that resemble the dendritic valley systems of the Earth. The most spectacular channels, however, consti-

Fig. 2. Pedernales River near Fredericksburg, TX, during the flood of Aug. 3, 1978. Note the scour of the stream bank on the right side of this aerial photograph.

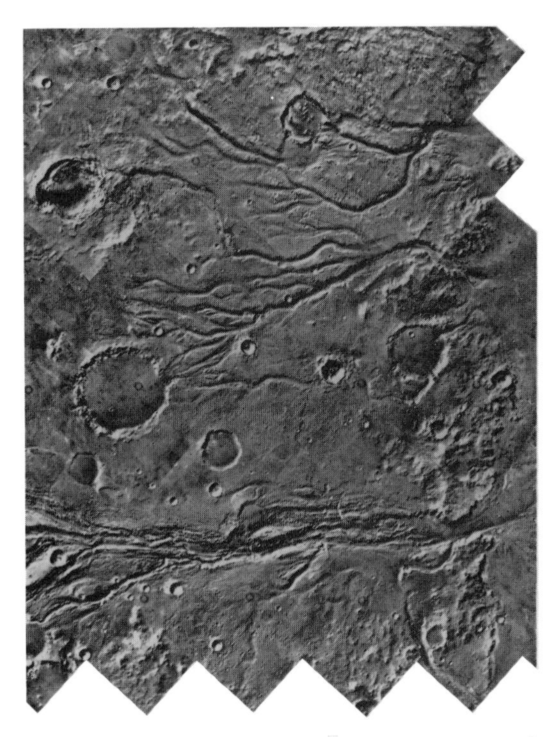

Fig. 3. Large channels in the Chryse Planitis region of Mars. This Viking photograph shows a scene about 150 × 200 km. *(NASA)*

tute great anastomosing complexes up to 100 km wide that can be traced 2000 km or more across the planet's surface. These zones are called outflow channels because they appear full-born at localized source regions, usually collapse zones. V. R. Baker and others suggested that immense floods of water were released from within the planet, emanating from troughs that receded headward to form the channels. The collapse zones at the heads of the channels mark the last points of fluid release.

The origin of the Martian channels is still somewhat controversial. Nevertheless, it is an exciting thought that rivers may be studied on other planets. [VICTOR R. BAKER]

Bibliography: V. R. Baker, The Spokane flood and the Martian outflow channels, *Science*, 202: 1249–1256, 1978; J. N. Holeman, The sediment yield of major rivers of the world, *Water Resources Res.* 4:737–747, 1968; L. B. Leopold, M. G. Wolman, and J. P. Miller, *Fluvial Processes in Geomorphology*, 1964; S. A. Schumm, *The Fluvial System*, 1977.

River engineering

A branch of transportation engineering consisting of the physical measures which are taken to improve the river and its banks.

Most centers of civilization developed in the valleys of the world's rivers. The people depended on alluvial plains for their agricultural economy and upon streams for domestic water and transportation. Subsequently, this reliance upon waterways has been expanded to include water for industrial as well as domestic consumption, to provide economical waterpower, and to utilize the river for waste disposal. With this expansion, use of the streams for transportation has continued, despite extensive developments of other transportation facilities. Today the improvement of rivers for inland navigation is actively prosecuted in all parts of the civilized world.

The measures that are taken to improve the river and its banks may include contraction of the river channel to improve navigation depths; bank stabilization to minimize erosion which would destroy farm land, cities, and bridges; creation of slackwater pools by means of locks and dams; or combinations of these means. It may also include improvement of the channels to assist them in carrying flood flows and regulation of the rivers' flows by upstream reservoir storage. In approaching the problems of river engineering, consideration must be given to the characteristics of the stream: the slope, meandering, sediment load, flow variations, and other factors (Figs. 1 and 2).

River characteristics. A stream is said to be in regimen if the major dimensions of the channel remain relatively constant and if it is neither aggrading (raising of the bed) nor degrading (lowering of the bed). The channel need not be fixed in position; however, many streams in regimen are constantly shifting their channels by eroding the banks at one location while building them at another. *See* RIVER.

Most natural streams are in regimen, with channel dimensions that are more or less characteristic of that stream. This implies a balance between the energy forces of the flow, the forces required to erode the bed and banks, the sediment load, and possibly other factors. There is no universally accepted theory relating these factors but, in general, a stream in erodible alluvium will be wide and

Fig. 1. Aerial view showing head of navigation project and uncontrolled and stabilized sections of Missouri River near Sioux City, Iowa. *(U.S. Army Corps of Engineers)*

Fig. 2. Uncontrolled river.

shallow if the banks are readily erodible or narrow and deep if the banks are erosion resistant.

Channels may be generally classified as straight, meandering (following an alignment consisting principally of pronounced bends), or braided (a number of interconnected channels presenting the appearance of a braid). These forms are influenced by many factors, including the stream discharge, the nature of the soils, and the sediment load; however, they may be best correlated with the slope of the valley in which they are located. Straight channels are found in valleys of either flat or steep slope; meandering channels occupy valleys of intermediate slope; and braided channels of streams in regimen occur on steep slopes. Braided channels of streams not in regimen may occur on either flat, intermediate, or steep slopes.

Technical knowledge is inadequate to explain fully the relationship between stream form and valley slope, but it is necessary in river engineering to recognize it. For example, many attempts to improve meandering channels by excavating straight channels have failed because the stream immediately began to erode its banks to reassert the meanders, meanwhile dumping excessive quantities of sediment into the channel downstream. In like manner, attempts to impose bends or curves on a channel in steep slopes must be considered with due caution.

River channel improvements. These may consist of revetting the banks to prevent erosion and shifting of the channel, realignment of the channel to provide smoother bends and a more regular alignment, contraction of an existent channel (particularly the contraction of a braided low-water reach to provide a single effective low-water channel), the provision of slack-water pools by the construction of low dams and ship locks, or the complete excavation of a new channel. Problems involving revetment, realignment, or contraction occur most frequently in meandering streams in erodible alluvium.

Contraction works. Used primarily to confine the low and moderate flows of a wide, shallow, or braided stream to a single effective channel, contraction works are required predominantly in meandering streams in erodible alluvium. It is important that the rectified channel be planned with due regard to maintenance of regimen. Alignment of the channel should be generally similar to that of the existent stream—in other words, following a series of smooth bends rather than straight lines and maintaining essentially the same channel slope.

Structures used in contraction consist of revetment in the concave portions of bends and of guide structures. The latter are normally of pile dike (Fig. 3) or other permeable fence-type construction designed to utilize the sedimentary and erosive characteristics of the stream in the initial shaping of the channel.

Where the position of the rectified channel deviates materially from that of the original channel, the concave banks of bends may be excavated and revetted in the dry, a pilot channel excavated, and the stream encouraged to scour the channel to the desired dimensions. In other cases the guide structures are constructed in stages, contracting the channel and causing the opposing bank to erode progressively to the desired location. The permeable guide structures serve to turn the current as desired yet permit deposition of sediments to build up the abandoned area behind them.

Locks and dams. In some streams, navigable depths are secured by relatively low-head dams, which create a series of slack-water pools. Locks, consisting of gated chambers, are provided to pass boats and barges around the dams (Fig. 4). A vessel is brought into the lock chamber from below the dam, the gates are closed, and the lock chamber is filled with water drawn from the upper reservoir. When the chamber has been filled to the level of the upper pool, the upper gates are opened and the vessel passes through into the upper pool. The reverse process is followed in going from the upper pool to the lower pool.

The lift of the lock may vary from a few feet to

Fig. 3. Pile dike contraction works.

Fig. 4. Lock and dam of Minneiska, Minn.

over 100 ft (30 m). The locks may be supplemented by gates through the dam. These gates may be lowered to the stream bed to permit free navigation during periods of adequate flow. *See* DAM.

Canals. Canals are constructed to provide connections between waterways or to bypass critical river reaches. They may range from essentially open waterways, such as the Suez Canal, to complex systems of excavated waterways, dams, and high lift locks such as the Panama Canal; or they may be included in a system with locks, dams, and open-river navigation as in the St. Lawrence River. They may also include channels excavated through low-slope braided streams or swamp areas, as in the Illinois River.

[WENDELL E. JOHNSON; DONALD C. BONDURANT]

Bibliography: T. Blench, *Regime Behavior of Canals and Rivers*, 1957; Committee on Channel Stabilization, Corps of Engineers, U.S. Army, *State of Knowledge of Channel Stabilization in Major Alluvial Rivers*, Tech. Rep. no. 7, October, 1969; M. M. Hufschmidt and M. B. Fiering, *Simulation Techniques for Design of Water Resource Systems*, 1966; J. V. Krutilla and O. Eckstein, *Multiple Purpose River Development*, 1958; A. Maass et al., *Design of Water Resources Systems*, 1962; H. M. Morris and J. M. Wiggert, *Applied Hydraulics in Engineering*, 1972; U.S. Office of Civil Engineers, *Seminars on River Basin Planning*, Fort Belvoir, Va., May 27–31, 1963.

Rural sanitation

Those procedures, employed in areas outside incorporated cities and not governed by city ordinances, that act on the human environment for the purpose of maintaining or improving public health. The purpose of these procedures is the furtherance of community cleanliness and orderliness for esthetic as well as health values.

Water. Purification of water supplies since 1900 has helped to prolong human life more than any other single public-health measure. Organisms

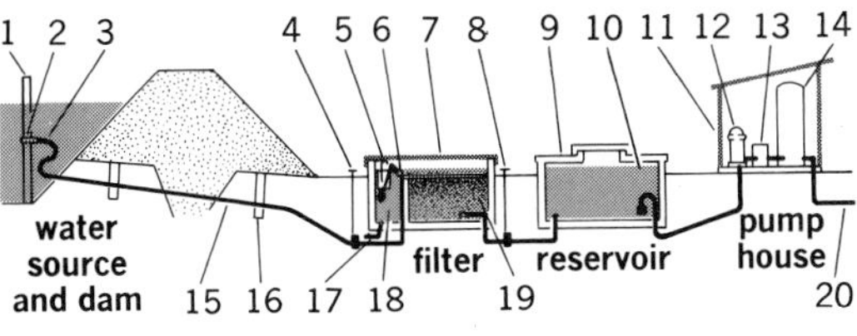

1 = hedge post or pipe
2 = screen suspended from post 3 ft under water
3 = flexible pipe
4 = hand valve
5 = aspirator (alum feeder)
6 = float valve
7 = hinged wood cover
8 = hand valve
9 = reinforced concrete top
10 = foot valve and strainer
11 = insulated pump house
12 = automatic pump
13 = automatic chlorinator
14 = pressure tank
15 = 2 in. iron pipe or plastic pipe muting
16 = concrete cutoff collar
17 = drain when needed
18 = coagulation sedimentation chamber
19 = washed river sand screened through 1/8 in. sieve
20 = purified water to house (below frost line)

Fig. 1. Farm-pond water-treatment system.

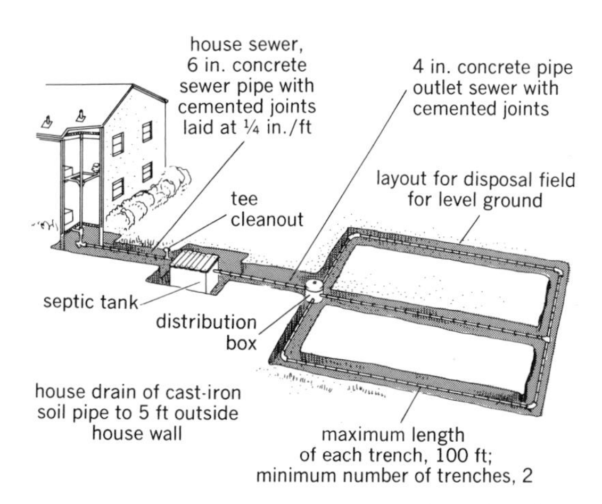

Fig. 2. Typical family-size sewage-disposal system. 1 in. = 2.54 cm; 1 ft = 0.3048 m.

which produce such diseases as typhoid, dysentery, and cholera may survive for a long time in polluted water, and prevention of contamination of water supplies is imperative to keep down the spread of such diseases. Watertight covers for wells are important means of preventing surface contamination to the water supply in rural areas.

Purification of a surface water supply, such as that from a lake or a farm pond, is accomplished by means of sedimentation, filtering, and chlorination. Sedimentation can be effected in a storage chamber by the addition of aluminum sulfate, which flocculates the finer particles of soil and other undesirable matter in suspension in the water. Filtering through fine sand removes the flocculated particles. Finally, the water is purified by means of a chlorine solution at the rate of $\frac{1}{2}$–1 part of chlorine to 1,000,000 parts of water. A system of treatment for farm-pond water is diagrammed in Fig. 1.

The colon bacillus is the usual indicator of pollution of water supplies by human waste. Chlorine kills such organisms, and therefore it is widely used for purification of water supplies.

Sewage disposal. The problem of safe disposal of sewage becomes more complex as population increases. The old practice of piping sewage to the nearest body of water has proved to be dangerous. Sanitary engineering techniques are now being used in rural areas, as well as in cities, for the disposal of household and human wastes. Where sewage-plant facilities are not available, the most satisfactory method of sewage disposal is by means of the septic tank system (Fig. 2).

The septic tank system makes use of a watertight tank for receiving all sewage. Bacterial action takes place in the septic tank and most of the sewage solids decompose, are given off as gases, or go out into the drainage lines as liquid. The gas and liquid are then released from the top 2 ft (0.6 m) of soil without odor or sanitary problems. The solids that do not decompose settle to the bottom of the tank, where they can be easily removed and disposed of safely. Such a sewage-disposal system has made it possible for all farm homes and rural communities to have modern bathroom equipment and sanitary methods of sewage disposal. *See* SEPTIC TANK; SEWAGE DISPOSAL; WATER TREATMENT.

[HAROLD E. STOVER]

Saline water reclamation

The partial demineralization of sea or brackish water sufficient to make the "fresh-water" product suitable for human or animal consumption, industrial uses, or irrigation. Brackish water is generally regarded as containing at least 1000 parts by weight of dissolved minerals in 1,000,000 parts of water, that is, 1000 parts per million, (ppm) or 1 gram per liter (g/liter), but less than sea water, which contains about 35,000 ppm (35 g/liter). Salinity requirements for the fresh-water product depend upon its use.

The U.S. Department of Health, Education, and Welfare Public Health Service Drinking Water Standards of 1962 require that the total dissolved solids of drinking water used on common carriers engaged in interstate commerce should not exceed 500 ppm where other more suitable supplies are or can be made available. The American Water Works Association endorses these standards for all public water supplies.

Water containing several thousand parts per million of dissolved minerals is consumed by man in many localities without noticeable ill effects, particularly where the rate of perspiration is high. Suitable salinities for irrigation waters depend upon the chemistry of the soil and the mineral requirements of the crop but generally should not exceed about 1200 ppm, particularly if the sodium content is high. Ruminant animals have developed tolerances for salinities up to 12,000 ppm. Industrial water requirements vary greatly, from 1 or 2 ppm for boiler waters up to 35,000 or more for some flushing and cooling.

Water can be separated from saline solutions by several means. Under a change of phase such as evaporation or freezing, part of the pure water will reach the second phase, but the salt remains in solution. Under the influence of a direct electrical current, salt ions in solution are forced toward the electrodes in electrodialysis. Chemicals may be added to cause exchange of ions in the solution or precipitation of the salts. Chemicals having greater affinity for water than for salt can be used to remove the water from a solution by solvent extraction. Under pressure exceeding the osmotic pressure, the fresh water in solution can be forced through an osmotic membrane by the process of reverse osmosis.

There is no one best process suitable for all applications. Selection of the best process for any specific application depends upon many factors, including plant capacity, relative costs of electric power and fuel, availability of waste heat, value placed on capital funds, analysis of the saline water to be used, and quality of the product desired.

The absolute minimum theoretical thermodynamic energy required for separating 1000 gal (3.785 m^3) of fresh water from sea water is 2.8 kilowatt-hours, but the energy requirements of all known processes is several times this figure. Current efforts in desalination are directed toward reduction of the total cost of the product water, including capital investment and operating and maintenance labor, as well as energy costs. Almost all manufacturers of desalting plants in the United States and abroad support research and development programs, but the greatest effort is maintained by the U.S. Department of the Interior, Office of Saline Water (OSW).

Multiple-effect distillation. In single-stage distillation (Fig. 1), water is evaporated and condensed once. In multiple-effect distillation, several evaporators or effects are utilized to recapture and reuse the latent heat of vaporization. At a given temperature and pressure, salt water is evaporated in the first effect. The vapor is led to a second effect, where additional evaporation occurs at a lower temperature and pressure, using the latent heat of the steam from the first effect, which condenses on the opposite side of a heat-transfer barrier such as a tube. Further evaporation of part of the water occurs in successive effects, each at lower temperature and pressure. The flow of water may be in either direction through the several effects, and either the water or the steam may be within the tubes.

The choice of the number of effects, and thus the reuse of the heat, depends upon the relative cost of fuel and equipment. Where equipment cost is high and fuel cost low, the number of effects will be smaller than where fuel is expensive, and a greater expenditure in equipment (number of effects) is justified for maximum reuse of the heat.

Salt deposits, known as scale, form on the evaporating surfaces where the phase change takes place at temperatures above about 160–180°F

vapor
condenser
evaporator
cooling water
salt water
steam
return to boiler
fresh-water distillate
brine
(a)

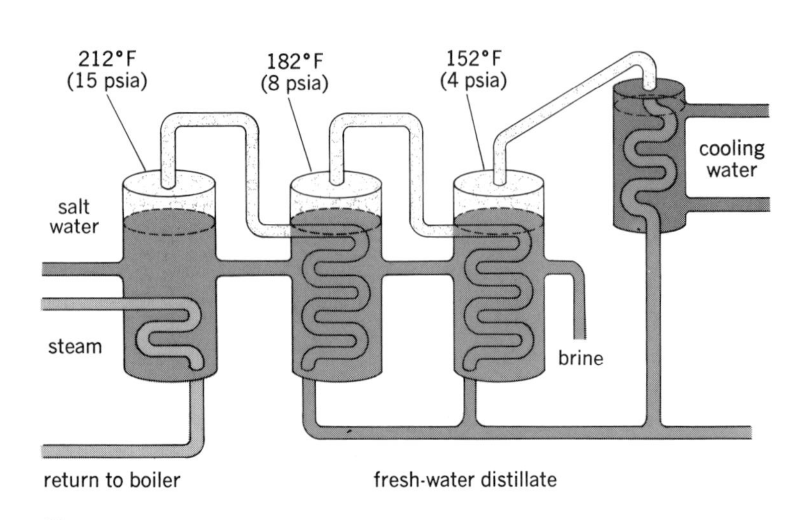

Fig. 1. Schematics of distillation process. (*a*) Single effect. (*b*) Multiple effect.

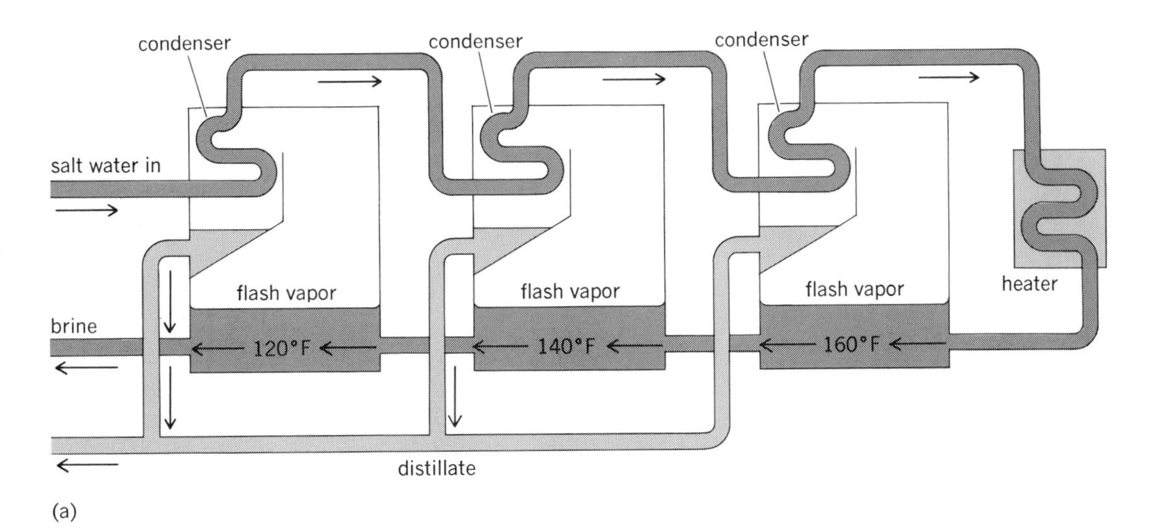

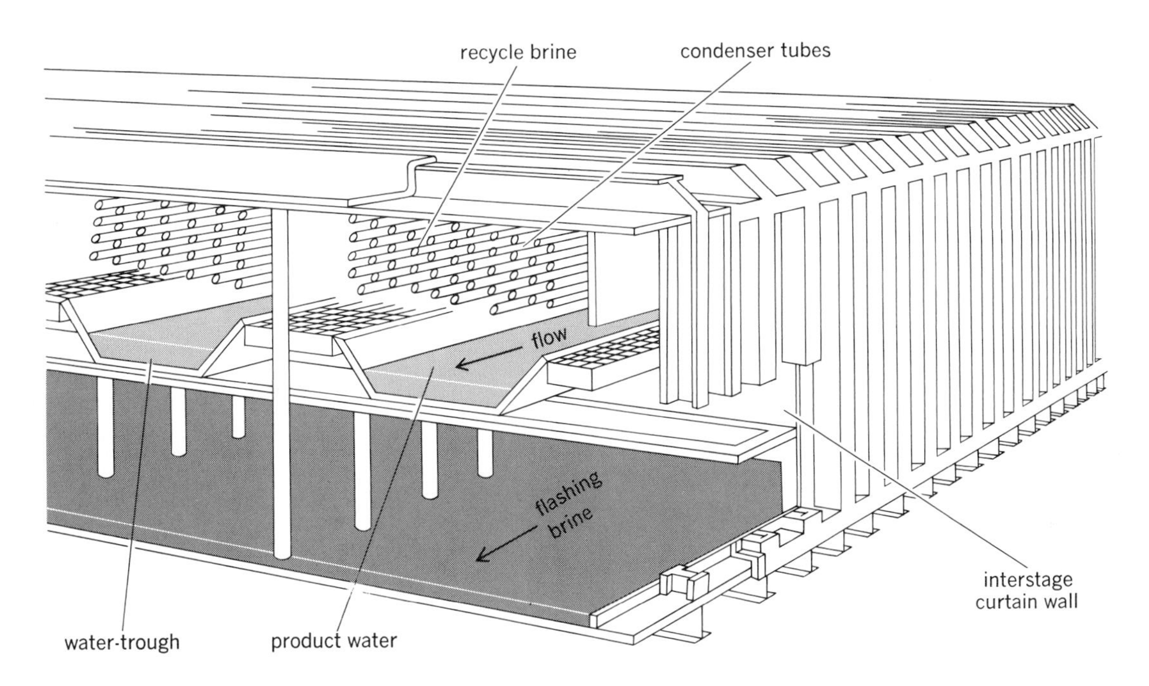

Fig. 2. Flash distillation. (*a*) Diagram of the process. (*b*) A large multistage flash plant (*from W. A. Homer, New concepts for desalting brackish water, Amer. Water Works J., August, 1968*).

(71–82°C) unless special measures are taken. Scale greatly impedes heat transfer and productivity.

Flash distillation. To circumvent the scaling problem occurring on the tubes of the submerged tube evaporator depicted in Fig. 1, the flash distillation process was developed. In this process water is progressively heated without phase change to a temperature of 180°F (82°C) or more. Temperatures up to 250°F (121°C) can be employed without scaling if acid is added to the brine to maintain a neutral pH. A portion is then evaporated quickly ("flashed") in successive chambers, each operating under slightly lower pressure. The flash vapor from each stage is condensed on tubes containing the incoming cooler sea water, and constitutes the fresh-water product. This is shown by Fig. 2. For desalination of sea water this process is now more widely used than any other in plants having a capacity over 10,000 gal per day (gpd) or 37.85 m^3 per day.

Vertical tube evaporators. In vertical tube evaporators (VTE) the incoming sea-water brine is distributed into the top of vertical tubes 3 in. (77 mm) in diameter, and allowed to fall as a film down the inside tube wall. Heat transfer through the falling film is excellent. A portion of the sea water evaporates, and this steam vapor is used to heat the outside of the tubes of the following effect, giving up its heat of vaporization as it condenses. The condensate constitutes the desired drinking water product.

The tubes originally had been 30 ft (9 m) or more in length, and the plants were identified as long-tube vertical evaporators (LTV). However, heat transfer through the tubes has been greatly increased by extrusion of flutes or fins on the inside and outside wall, and as a consequence a tube

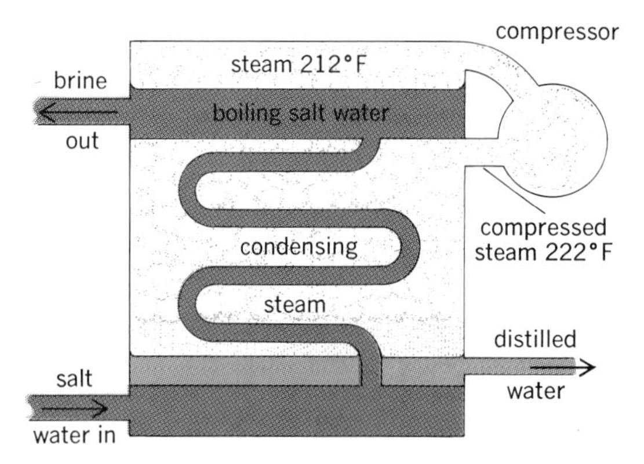

Fig. 3. Vapor-compression distillation.

length of 12 ft (3.7 m) or less has been found to be adequate.

VTE evaporators are employed as multiple-effect sea-water distillation plants having a capacity of over 1,000,000 gpd (3785 m^3/day), where they compete with multistage flash units.

Vapor compression. Salt water is boiled on one side of a heat-transfer barrier such as cylindrical tubes in a vertical tube evaporator (Fig. 3). The vapor is compressed, raising its pressure about 3 psi (21 kPa) and temperature about 9°F (5°C). The heated steam is returned to the outer side of the boiler tubes, where its latent heat of condensation evaporates additional water in the tubes. The condensed vapor is removed through a heat exchanger as distilled water. In this way, the energy is applied as power to drive the compressor, rather than as heat to boil the water directly.

Vapor compression stills have been widely used where the demand is small and a compact yet efficient desalination unit is required. Packaged skid-mounted units ranging in size from 1000 to 50,000 gpd (3.8 to 190 m^3/day) are available. The largest plant with a capacity of 1,000,000 gpd (3785 m^3/day) was built for OSW at Roswell, N. Mex.

Combination plants. Lowest sea-water desalination costs are achieved when the distillation plant is combined with a power generation plant. The exhaust steam from a turbine can be usefully absorbed in the evaporators, thereby avoiding wasteful rejection of large quantities of heat in a conventional power plant condenser. The advantages of merging power generators and distillation facilities hold both for fossil-fuel-fired plants and nuclear plants. There are many such combined distillation-power generation fossil-fuel-fired plants in the petroleum, chemical, paper, and other process industries.

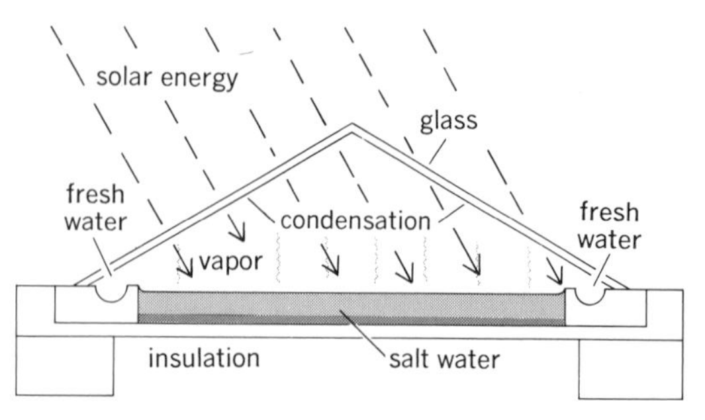

Fig. 4. Simple solar still.

Solar distillation. In this technique, the Sun's heat is used to evaporate salt water. Solar distillers are classified in three groups: those in which salt water is evaporated directly and condensed in one unit (Fig. 4); those employing focusing devices to acquire higher temperatures for use in various mechanisms; and those in which water is heated in one compartment with evaporation and condensation in another, including multiple-effect mechanisms. Since the magnitude of the incident solar energy is low and cannot be increased by focusing, practical application requires large areas for the collection of this heat energy. About 1 acre (4000 m^2) of collection area is necessary to produce 5000 gpd (19 m^3/day). Simplified construction methods, including the use of plastic films in place of the glass collectors, have been developed. Several new plastics, including the polyfluorocarbons, are relatively inert and resistant to the ultraviolet effects. However, plastics are hydrophobic, and the condensed vapor collects on them in drops and reduces their transparency. The use of wetting agents to prevent this is being developed.

A number of solar stills have been erected on arid islands in the Aegean Sea by the Greek government and by the Church World Service. The largest of these, installed on Patmos, covers 86,000 ft^2 (8000 m^2).

Freezing methods. Pure ice can be frozen from brine and melted to produce fresh water. Unfortunately, some of the brine clings tenaciously to the surface of the ice crystals or is entrapped in the interstices. Therefore, two operations are needed, one to form the pure ice and the second to separate the ice from the brine. One technique (Fig. 6) consists of admitting cold sea water to a chamber under high vacuum in which a portion of the water immediately vaporizes. The evaporation process absorbs heat from the remaining salt water, causing a portion of it to freeze to an ice-brine mixture. This slurry is passed through a separator, where it is washed with a portion of the product water. The water vapor which has been drawn off from the freezing chamber is compressed and then brought into contact with the brine-free ice crystals, where it condenses, melting the ice. This melted ice is the desired pure fresh-water product. The process has been successfully employed in pilot plants as large as 100,000 gpd (380 m^3/day).

The large water-vapor volume prevailing under vacuum conditions necessitates a compressor of large diameter. This limits the maximum feasible size of a single unit to about 200,000 gpd (760 m^3/day). Plants of larger capacity must consist of duplicate units. To circumvent this disadvantage, another process was developed called the secondary refrigerant freeze desalination process. In lieu of water vapor it employs a refrigerant which has a pressure above atmospheric at the freezing point of brine. Because of its low cost butane has been most often used. Liquid butane is mixed with the brine in the freezer and, after absorbing the heat of crystallization released by the growth of ice crystals, is removed as a gas. After compression the hot butane gas is used to melt the ice crystals. Development work on this refrigerant process was terminated in 1968 because projected costs for commercial plants were no lower than for distillation.

Reverse osmosis. When pure water and a salt solution are on opposite sides of a semipermeable

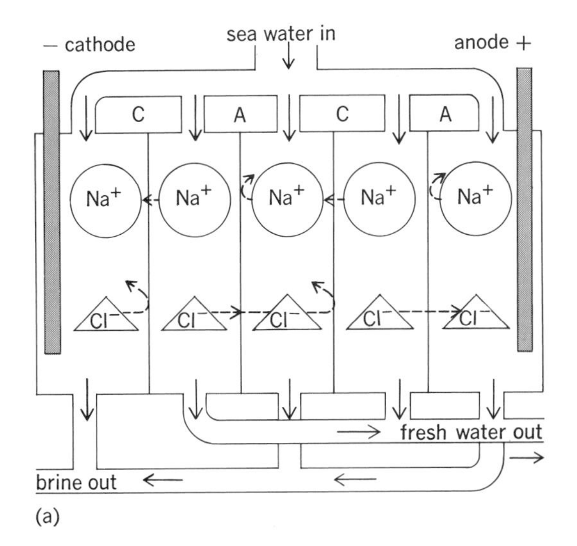

(a)

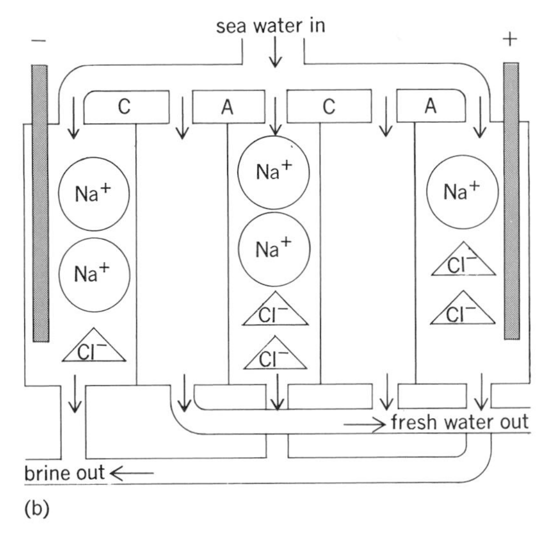

(b)

Key: C = cation permeable membrane
A = anion permeable membrane

Fig. 5. Schematic representation of electrodialysis. (a) Initial state. (b) After operating.

membrane in a vented container, the pure water diffuses through the membrane and dilutes the salt solution. At equilibrium, the liquid level on the saline-water side of the membrane will be higher than on the fresh-water side. This phenomenon is known as the process of osmosis. The effective driving force is called osmotic pressure. Its magnitude depends on the characteristics of the membrane, the temperature of the water, and the concentration of the salt solution. By exerting pressure on the salt solution, the osmosis process can be reversed. When the pressure of the salt solution is greater than the osmotic pressure, fresh water diffuses through the membrane in the opposite direction to normal osmotic flow.

The membranes are generally made of cellulose acetate and are usually fabricated in a cylindrical shape with the saline water pressurized to about 600 pounds per square inch (4.1 MPa) gage (psig) on the inside. Where the membrane tubes are relatively large, say 1/2 in. (13 mm) in diameter, it has been necessary to support the membranes on a rigid material such as fiber glass.

In 1967 a new design was announced utilizing hollow nylon fibers measuring only 25–250 micrometers in outside diameter. No supporting structure is necessary. Much more membrane surface can be packed into a given container volume. On the other hand, the flow through the membrane is only about 1/10 gpd/ft^2 (4×10^{-3} m^3/m^2-day) of membrane surface as opposed to 10–20 gpd/ft^2 (0.4–0.8 m^3/m^2-day) for the large-tube type.

Electrodialysis. When an electric current is transmitted through a saline solution, the cations in the solution migrate toward the cathode and the anions toward the anode. Membranes in sheets of cation- or anion-exchange material, a development since 1930, permit the passage of either cations or anions, respectively, in the solution. If a series of alternate cation- and anion-permeable membranes is placed between the electrodes (Fig. 5), the anions pass through the anion-permeable membrane toward the anode but are stopped at the next membrane, which is permeable only to cations. In the same way the cations moving in the opposite direction will pass the cation membrane but not the next. Thus, as anions and cations collect in alternate compartments, the water there is enriched with salt while that in the other compartments is depleted. The salt water flows parallel to the membranes, and the depleted and enriched streams are withdrawn separately as demineralized water and enriched brine. Electrodialysis units are in operation on brackish water containing up to 10,000 ppm, but are generally not economical on waters containing more than 5000 ppm.

In contrast to the various distillation processes the electrical energy required is a function of the amount of salt to be removed. For this reason the process does not appear to be economically attractive for sea-water desalination.

Solvent extraction. The solvent extraction process depends upon two characteristics of certain secondary and tertiary amines with water and salt water. First, the solubility of water in the solvent is very sensitive to temperature, and, second, the amine selectively dissolves water from saline solutions. The process is shown in Fig. 7. Saline water is mixed with solvent in the mixing vessel at 65°F (18°C). The fresh water dissolves in the solvent, and the resulting solution (30% water by weight) is then drawn off, heated to 122°F (50°C), and carried to a settling vessel where the solvent

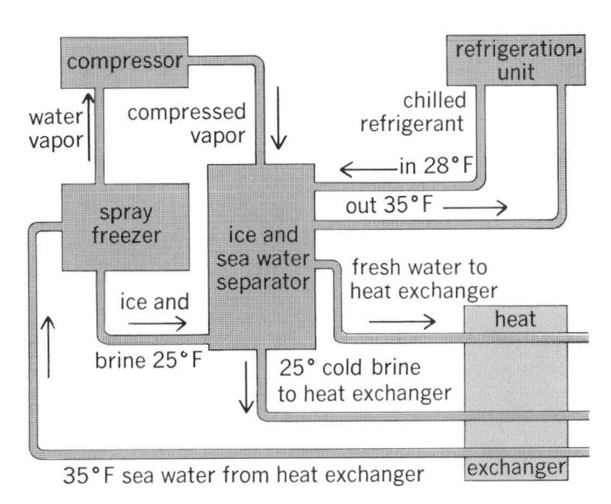

Fig. 6. Freeze-evaporation process.

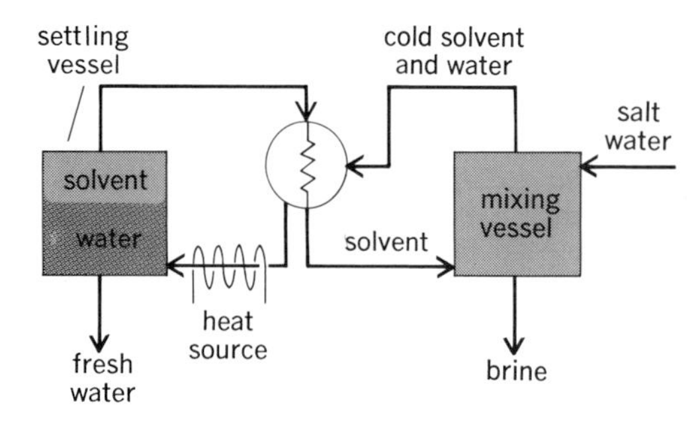

Fig. 7. Diagram of the solvent extraction process.

and water separate. The water remaining after removal of the last traces of solvent in a stripping vessel (not shown) is the desired fresh-water product. The solvent is returned to the mixing vessel after being cooled in the heat exchanger. The brine left behind in the mixing vessel is discarded to sewer.

Since in such single-stage apparatus the extractable substance approaches a concentration equilibrium in the effluents, nearly complete extraction requires a multiplicity of stages. The principal advantage of the process is that it can utilize low-level waste heat as an energy source.

Economics. There is more than sufficient fresh water on the Earth to fulfill all of mankind's needs for decades to come. Nature's distribution of this water, however, is not always equitable. Some localities have a plethora of water, some have a severe shortage. When an adequate supply of good fresh water is not available locally, the community may be forced to import fresh water of an acceptable quality from a distant point by the most economical means of transportation. If there is an abundant supply of saline water in the area, the community may instead elect to convert it to fresh water. For conversion to be selected, the total of all the cost components (see table) must be less than the total costs for the natural water supply.

Cost components of water supply

Component	Description
Source development and collection	Water sheds, intake facilities at rivers or lakes, reservoirs, wells
Transportation	Pipeline, conduit, or canal with pump stations or tank truck, ship, or barge with terminal facilities
Treatment and/or conversion	Treatment: coagulation, settling, filtration, softening, iron and manganese removal, chlorination, and so on; or desalination: distillation, reverse osmosis, electrodialysis, freeze desalination, and so on
Distribution	Distribution mains, laterals, hydrants, and service connections throughout the community as well as distribution systems, storage tanks, and pumping stations
Business functions	Meter reading, billing, collections, accounting, purchasing, engineering, financing, management, and administration

Desalination may be considered the ultimate water treatment. Its adoption may eliminate high transportation or source development expenditures. It will not, however, have any effect upon the capital and operating cost of the distribution system or on the business operating costs, which together account for well over half of the total delivered cost of water. *See* HYDROLOGY; WATER CONSERVATION; WATER TREATMENT.

[JOHN G. MULLER]

Bibliography: U.S. Department of the Interior, *Proceedings of the 1st International Symposium on Water Desalination*, Washington, D. C., Oct. 3–9, 1965; U.S. Department of the Interior, Office of Saline Water, *Saline Water Conversion Report for 1966*, 1967; R. G. Post and R. L. Seale (eds.), *Water Production Using Nuclear Energy*, 1966.

Saltmarsh

Generally, a maritime habitat characterized by special plant communities, which occur primarily in the temperate regions of the world; however, typical saltmarsh communities can be formed in association with mangrove swamps in the tropics and subtropics. In inland areas where saline springs emerge or where there are salt lakes, typical saltmarsh communities can be found, dominated sometimes by the same species that occur on maritime saltmarsh. The extensive areas of inland salt desert, while exhibiting some features of similarity with saltmarsh, are nevertheless best regarded as a separate entity. Excess sodium chloride is the predominant environmental feature of salt marsh (maritime or inland). In the case of salt deserts sodium chloride is only one of the alkali salts that may occur in excess.

Occurrence of different types. Maritime saltmarsh can be found on stable, emerging, or sinking coastlines. On emerging or sinking coasts the actual extent of saltmarsh depends upon an adequate degree of wave protection and also upon the rate of change of coast level in relation to the rate of silt deposition. Mud or sand flats are raised by silt deposition to a level at which the characteristic phanerogamic plants can colonize. Saltmarshes therefore are common features of estuaries and protected bays, provided the seabed is shallow and does not shelve steeply. They also occur behind spits, barrier beaches, and offshore sand, shell, and shingle islands.

On emerging coastlines the true saltmarsh zone tends to be narrow, though older saltmarsh areas, recognizable by remnant saltmarsh species, have subsequently been invaded and dominated by the local terrestrial species. On sinking coastlines the extent of saltmarsh depends essentially on the rate of accretion from silt deposition in relation to rate of sinking. The greater the former in relation to the latter, the more extensive the saltmarshes are likely to be. Because of the dependence of saltmarshes upon accretion, they are likely to be best developed in association with eroding soft rock coastlines and estuaries of rivers that bring down abundant silt from soft rock upland. Work on accretion rates has shown that previously calculated time scales for marsh development have been too great.

Physiography. Saltmarsh can form on mud, muddy sand, or sandy mud, but not on pure sand because mobile sand does not provide a sufficiently stable substrate. Typical physiographic features associated with saltmarsh are the creeks, which serve as drainage channels, and pools, which are known as pans. The type of creek system depends upon the initial substrate, whether muddy or sandy, local fresh-water drainage channels, and the type of primary colonist, for example, annual species of samphire (*Salicornia*) or clumps of ricegrass (*Spartina*) (Fig. 1). Various types of pan have been recognized, including primary, secondary, and creek (cutoff ends of creeks).

Plant zonation (succession). The colonizing plants are subject to considerable tidal inundation, but with the advent of plants the rate of accretion is hastened and the land level rises. The physical factors of the environment change, there are fewer submergences, and new species invade. A characteristic feature of saltmarsh vegetation, therefore, is the zonation of plant communities associated with changes in the environment. The zonation or succession for maritime saltmarsh is essentially dynamic since the saltmarsh is continuously, albeit slowly, building up toward the sea.

When maritime saltmarsh develops between two lateral ridges with only a narrow creek entry, the full succession may be passed through in a few hundred years, even on a subsiding coastline. The zonation of inland saltmarshes, a static zonation, is related to decreasing salinity in proportion to the distance from the source of salt.

The early stages on any saltmarsh are sufficiently rapid to be observed in 25–50 years. The final stages of the succession depend upon whether the saltmarsh borders sand dune, meadow, or fresh-water inflow. In the last case the succession continues into brackish communities and finally into fresh-water swamp.

Plant species and life forms. The plants that grow on salt marshes must tolerate the excess sodium chloride and are termed halophytes. These plants possess features associated with the halophytic environment, for example, development of succulence, waxy cuticle, and salt-excreting glands; overall there is a tendency for the vegetation to exhibit a drab grayness. The predominant life forms on saltmarshes are the hemicryptophytes and therophytes. Members of the Chenopodiaceae are common (*Salicornia*, *Suaeda*, and *Allenrolfia*) and also the Plumbaginaceae (*Limonium* or *Statice*). Among the grasses *Spartina* and *Puccinellia* are important genera in

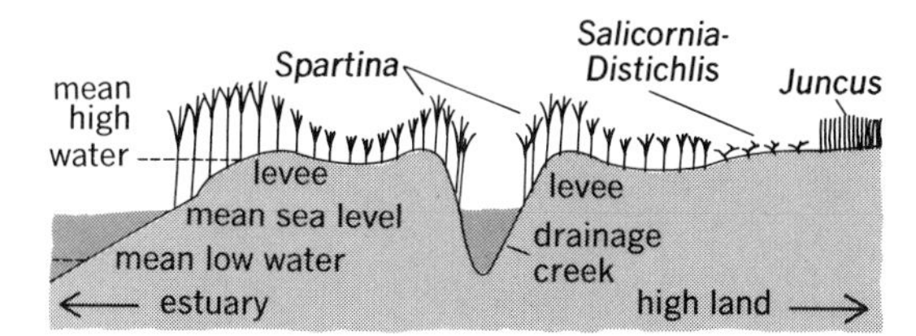

Fig. 1. Diagrammatic cross section of a marine marsh on the southern Atlantic coast of the United States; vertical exaggeration is about 10 to 1.

Fig. 2. View of *Spartina* marsh on the coast of Georgia.

different parts of the world (Fig. 2). Other genera that are widely represented in this habitat include *Plantago*, *Triglochin*, *Cotula*, *Scirpus*, and *Juncus*.

Inasmuch as there are characteristic phanerogams, so also there are characteristic algae associated with saltmarsh vegetation. Particular communities of green and blue-green algae are common, and in the Atlantic–North Sea area there may be extensive communities of free-living brown fucoids. Characteristic red algal communities are dominated by species of *Bostrychia* and *Catenella*. It has been shown that nitrogen fixation by blue-green algae may be significant. Work on the marine soil fungi and bacteria has been undertaken as well as on soil microfauna, but information on these groups is still far from adequate.

As the number of species capable of growing under the specialized conditions is limited, the type of succession tends to be similar for major areas. The following groups have been proposed for maritime saltmarshes: Arctic, European (with four subdivisions), Mediterranean, Atlantic North American (with three subdivisions), South American, Pacific North American, Japanese, and Australasian (with two subdivisions), each with a characteristic succession. Six groups have also been proposed for inland saltmarshes. These are Inland European (with two subdivisions), Inland Asia (with three subdivisions), Inland African (with three subdivisions), North American–Mexican (with five subdivisions), South American, and Australasian. The saltmarsh communities can be well distinguished by using the Montpellier system of classification; and on this basis 12 orders have been recognized.

Environment. The principal feature of the environment is the excess sodium chloride, with the resulting effect of the sodium ion upon the soil colloids and the chloride ion upon plant metabolism. Other specific ions may, in places, be important, and the ratio of chloride to sulfate can be significant. The phosphate ion appears to vary with season. So far as ion uptake by plants is concerned, two mechanisms appear to operate depending on whether there is a low or high salt concentration.

The frequency of tidal inundations is very important, with its effect upon salinity, as is also the water table and soil aeration. At lower marsh levels the existence of an aerated layer seems essential for the growth of the plants, and its absence may inhibit plant colonization. A lowering of the soil salinity at some season also appears necessary

Dollar value for various saltmarsh usages

	Annual return/acre*	Capital value/acre*
Fish and fish food	$ 100	$ 2,000
Oyster culture (maximum)	900	18,000
Sewage effluent treatment	2500	50,000
Life support	4150	83,000

*1 acre = 0.4046 hectare.

for successful seed germination, though temperature can also be involved. In respect to seed germination, the halophytes can be classed into three groups: the glycophyte group, where low salinity is essential; euryhalophyte group, with a wide salinity tolerance; and stenohalophytes, with a narrow salinity tolerance.

Productivity. From the scanty data available, it appears that wild saltmarsh is highly productive biologically and compares favorably with fresh-water reed swamps. Less than 10% of the net production appears to be consumed by grazing herbivores, most of it going into the detritus path of energy flow. For this reason, wild saltmarsh dominated by grasses forms excellent grazing for stock, except in spring tide periods. An evaluation of salt marshes has provided the estimates shown in the table.

Reclamation. In many parts of the world, saltmarshes have been enclosed by seawalls and converted to valuable agricultural land. About 25% of the saltmarsh on the eastern seaboard of Canada and the United States has been thus treated. In view of the importance of saltmarsh as the primary production agent for the coastal fisheries, concern has been rising over reclamations, and careful thought must be given before reclamation projects are approved. Thermal cooling water from power stations can inflict damage on the vegetation, as may industrial effluents. The use of saltmarsh for disposal of garbage should no longer be tolerated. Finally, before a reclamation is approved, very careful consideration needs to be given to the soil properties, the environmental conditions, and the economic return. [VALENTINE J. CHAPMAN]

Bibliography: V. J. Chapman, *Coastal Vegetation*, 2d ed., 1976; V. J. Chapman, *Salt Marshes and Salt Deserts of the World*, 2d suppl. reprint ed., 1974.

Sanitary engineering

A specialty field generally developed in civil engineering but not limited to that branch. The National Research Council defines the sanitary engineer as "a graduate of a full 4-year, or longer, course leading to a Bachelor's, or higher, degree at an educational institution of recognized standing with major study in engineering, who has fitted himself by suitable specialized training, study, and experience (1) to conceive, design, appraise, direct and manage engineering works and projects developed, as a whole or in part, for the protection and promotion of the public health, particularly as it relates to the improvement of man's environment, and (2) to investigate and correct engineering works and other projects that are capable of injury to the public health by being or becoming faulty in conception, design, direction, or management."

Sanitary engineering practice includes surveys, reports, designs, reviews, management, operation and investigation of works or programs for (1) water supply, treatment and distribution; (2) sewage collection, treatment and disposal; (3) control of pollution in surface and underground waters; (4) collection, treatment, and disposal of refuse; (5) sanitary handling of milk and food; (6) housing and institutional sanitation; (7) rodent and insect control; (8) recreational place sanitation; (9) control of atmospheric pollution and air quality in both the general air of communities and in industrial work spaces; (10) control of radiation hazards exposure; and (11) other environmental factors which have an effect on the health, comfort, safety, and well-being of people.

Sanitary engineers engage in research in engineering sciences and such related sciences as chemistry, physics, and microbiology and apply these in development of works for protection of man and control of his environment. *See* AIR-POLLUTION CONTROL; SEWAGE; SEWAGE DISPOSAL; SEWAGE TREATMENT; WATER POLLUTION.

[WILLIAM T. INGRAM]

Bibliography: H. E. Babbitt and E. R. Baumann, *Sewerage and Sewage Treatment*, 8th ed., 1958; H. Blatz (ed.), *Radiation Hygiene Handbook*, 1959; V. M. Ehlers and E. W. Steel, *Municipal and Rural Sanitation*, 6th ed., 1965; G. M. Fair, J. C. Geyer, and D. A. Okun, *Water and Wastewater Engineering*, vol. 2: *Water Purification and Wastewater Treatment and Disposal*, 1968; W. C. L. Hemeon, *Plant and Process Ventilation*, 1955; R. K. Linsley, Jr., and J. B. Franzini, *Elements of Hydraulic Engineering*, 1955; F. S. Merritt (ed.), *Standard Handbook for Civil Engineers*, sect. 22, Sanitary Engineering, 1968; *Municipal Refuse Disposal*, Commission on Refuse Disposal, APWA Research Foundation Project 104, 2d ed., 1966; F. A. Patty et al. (eds.), *Industrial Hygiene and Toxicology*, vol. 2, 2d ed., 1963; *Refuse Collection Practice*, APWA Commission on Solid Wastes, 3d ed., 1966; P. A. Sartwell, *Maxcy-Rosenau Preventive Medicine and Public Health*, 10th ed., 1973; E. W. Steel, *Water Supply and Sewerage*, 5th ed., 1979; H. H. Uhlig, *Corrosion and Corrosion Control*, 1971; M. Willrich and R. K. Lester, *Radioactive Waste: Management and Regulation*, 1977.

Savanna

The term savanna was originally used to describe a tropical grassland with more or less scattered dense tree areas. This vegetation type is very abundant in tropical and subtropical areas, primarily because of climatic factors. The modern definition of savanna includes a variety of physiognomically or environmentally similar vegetation types in tropical and extratropical regions. The physiognomically savannalike extratropical vegetation types (forest tundra, forest steppe, and everglades) differ greatly in environment and species composition.

In the widest sense savanna includes a range of vegetation zones from tropical savannas with vegetation types such as the savanna woodlands ("campo cerrado", Fig. 1) to tropical grassland and thornbush. In the extratropical regions it includes the "temperate" and "cold savanna" vegetation types known under such names as taiga, forest tundra, or glades. For further details on the synonyms and terminology *see* VEGETATION ZONES, WORLD.

During the growing season the typical tropical savanna displays a short-to-tall, green-to-silvery shiny cover of bunch grasses, with either single trees or groups of trees widely scattered. This is followed by a rest period of several months during which, because of severe drought, the vegetation appears quite different, with the brown-gray dead grasses bent over and the trees either without leaves or with stiff or wilted gray-green foliage. The heat and drought during this season of the year exert a high selective pressure upon the floral and faunal composition of the savanna.

Floral and faunal composition. The physiognomic similarity of the tropical savannas is underlined by the similarity among certain floristic components. All savannas contain members of the grass family (Gramineae) in the herbaceous layer. Most savannas of the world also have one or more members of the tree family (Leguminosae), particularly of the genus *Acacia*. Also included among the trees are the families Bombacaceae, Bignoniaceae, and Dilleniaceae, and the genera *Prosopis* and *Eucalyptus*; these are abundant when they occur. Palms are also frequently found. One of the most outstanding savanna trees is *Adansonia digitata* (Bombacaceae), which achieves one of the biggest trunk diameters known for all trees (Fig. 2). The grass species, although mostly from the genera *Panicum*, *Paspalum*, and *Andropogon*, include numerous other genera such as *Aristida*, *Eragrostis*, *Schmidtia*, *Trachypogon*, *Axonopus*, *Triodia*, and *Plectrachne*, all of regional importance.

The fauna of the savannas is among the most interesting in the world. Savannas shelter herds of mammals such as the genera *Antelopus*, *Gazellus*, and *Giraffus*, and the African savannas are especially famous for their enormous species diversity, including various members of the Felidae (for example, the lion).

Numerous species of birds are indigenous to the savannas. Among these is the biggest bird, the ostrich (*Struthio camelus*), found in Africa. Many birds from extratropical regions migrate into the tropical savannas when the unfavorable season occurs.

Among the lower animals, the ants and termites are most abundant. Termite colonies erect large, conical nests above the ground, which are so prominent in some savannas that they partly dominate the view of the landscape, especially during the dry season.

Environmental conditions. The climate of the tropical savannas is marked by high temperatures with more or less seasonal fluctuations. Temperatures rarely fall below 0°C. The most characteristic climatic feature, however, is the seasonal rainfall, which usually comes during the 3–5 months of the astronomic summertime. Nearly all savannas are in regions with average annual temperatures from 15 to 25°C and an annual rainfall of 32 in. (81 cm).

The soil under tropical savannas shows a diversity similar to that known from other semiarid regions. Black soils, mostly "chernozem," are common in the moister regions. Hardpans and occasional surface salinities are also found, and lateritic conglomerations occur along the rivers. The soils vary in mineral nutrient level, depending on geologic age, climate, and parent material. Deficiencies of minerals, specifically trace elements, are reported from many grassland areas in the continents of Africa, South America, and Australia. *See* SOIL.

Fig. 1. Savanna with gallery forest north of Guiaba, Campo Cerrado, Brazil.

Fig. 2. *Adansonia digitata*, a tree with large trunk, in the savanna in Senegal, Africa. (*Courtesy of H. Lieth*)

Certain savanna areas suffer from severe erosion and no soil can be accumulated. Plant growth in these areas is scarce, often depending on cracks and crevices in the ground material to support tree roots. The herbaceous cover opens up under these conditions, with many stones appearing on the land surface.

In the majority of the tropical savannas water is the main limiting environmental factor: total amount and seasonality of precipitation are unfavorable for tree growth. Additional stresses to forest vegetation are caused by frequent fires, normal activities of animals, excess of salt, or nutrient deficiencies. Wherever a river flows, most of these factors change in favor of tree growth. This explains the existence of extensive gallery forests along the rivers. The gallery forest is missing only where severe local floods after storms cause soil erosion and the formation of canyons.

Geographic distribution. Tropical savannas exist between the areas of the tropical forests and deserts; this is most apparent in Africa and Australia. In the New World tropics, different conditions exist because of the circumstances created by the continental relief, and the savannas are situated between tropical forests and mountain ranges. In Madagascar and India there is a combination of both conditions.

The transitional (ecotonal) position of the savannas between forest and grassland or semidesert is the basis for the differences in opinion among authors about the size and geographical distribution of savannas. Most authors include, however, the savanna in East Africa and the belt south of the Sahara, the bush veld in South Africa, the Llanos in northern South America, and some types of scrub vegetation in Australia. Some areas in southern Madagascar, on several tropical islands (in leeward position), in Central America, in southern North America, and in India are usually included. The two latter regions, however, are subtropical. Still other areas included in the savanna concept are the Campo Cerrado and parts of the Chaco in South America; portions of the Miombo in southern Africa; wide portions of northern Africa; south of the Sahara; and wide portions of Madagascar, the Indian peninsula, and Australia. Because of the variations in the savanna concept, it is difficult to give a correct estimate of the total surface area covered by savanna vegetation. The Food and Agriculture Organization (FAO) considers that about one-third of the total land surface is covered by predominantly grassland vegetation. Of this area, one-third can be assumed to be tropical grassland, most of which can be called savanna.

Agricultural practices. Most of the original savanna areas throughout the world are currently farmed. Ranch farming is the predominant type, with sheep being raised in the drier areas and cattle in the moister regions. In the hotter regions zebus are raised, along with several hybrids of zebus and European cattle, which are the preferred stock for this climate. The yield in meat per unit area is low, and even under extensive management it seems to be lower than the meat production of the natural animal herds of the savanna, including antelopes, giraffes, and zebras. Ranch farming does not change the character of the vegetation very much if it is well managed. The adjacent dry woodlands very often resume a physiognomy similar to the natural savannas if good farm management is applied.

Agricultural crops are of many varieties in the savanna areas, where with careful protection and management any crop can be cultivated, provided that enough water is available. Drought-tolerant crops are usually preferred among the perennials.

The majority of human settlements are small in the savanna regions of the world; the hot temperatures during part of or the entire year, together with problems of water supply, limit interest in larger settlements. Settlements are usually found along the rivers, close to the coast, or in the higher elevations. Most of the land is managed by small tribal villages or large plantation or ranch owners, with separate groups of tenants, sharecroppers or employees, or single families.

Extratropical types. The main structural character of tropical savannas is the scattered trees standing within a close cover of herbaceous vegetation. This structural character is also found in several extratropical vegetation types, but these differ greatly in the forces that limit a close tree cover, including drought (areas intermediate between steppe or prairie and forest); excess water or water combined with soil that has a shortage of oxygen and nutrients (peatbogs, marshes, or glades); short vegetation periods because of extended cold temperatures below the freezing point; excessive, long snow covers; and low light intensity (forest tundra, taiga, and cold savanna).

The forest tundra and the forest steppe are usually considered ecotonal units between their adjacent vegetation types which are, respectively, boreal coniferous forest and tundra, or deciduous forest and steppe. The ecology and species populations of these areas are so different from that of tropical savannas that it is hardly desirable to combine extratropical and tropical open woodlands under the heading savanna.

An intermediate condition is exhibited by the savannas and everglades of the southeastern United States. The intermittent soaked or dry conditions of a peaty soil, the tropically hot summers and mild winters (with frost periods, however), and the generally low nutrient level of the soil give these areas the characteristics of the cold savannalike vegetation types and the tropical and subtropical types.

The economic potential of the three extratropical savannalike areas also varies greatly. The majority of the forest tundra is beyond agricultural exploitation, while the potential for farming in the forest steppe is better than in the steppe. The conditions in the southeastern United States vary. Farming is potentially possible in most regions, but it must be determined by economic considerations whether a given area should be developed.

Some savanna regions in many parts of the world are the last survival territories for many plant and animal species. This implies the need for extratropical. *See* FOREST AND FORESTRY; GRASSLAND ECOSYSTEM; TAIGA; TUNDRA.

[HELMUT LIETH]

Bibliography: R. T. Coupland, ed., *Grassland Ecosystems of the World*, 1979; P. M. Dansereau, *Biogeography*, 1957; H. Walter, *Vegetation der Erde*, vol. 1, 1962; A. W. Kuechler, Natural vegetation of the world, *Goode's World Atlas*, 11th ed., 1960.

Sea ice

Ice formed by the freezing of sea water is referred to as sea ice. Ice in the sea includes sea ice, river ice, and land ice. Land ice is principally icebergs which are prominent in some areas, such as the Ross Sea and Baffin Bay. River ice is carried into the sea during spring breakup and is important only near river mouths. The greatest part, probably 99%, of ice in the sea is sea ice.

Properties. The freezing point temperature and the temperature of maximum density of sea water vary with salinity (Fig. 1). When freezing occurs, small flat plates of pure ice freeze out of solution to form a network which entraps brine in layers of cells. As the temperature decreases more water freezes out of the brine cells, further concentrating the remaining brine so that the freezing point of the brine equals the temperature of the surrounding pure ice structure. The brine is a complex solution of many ions. With lowering of temperature below −8°C, sodium sulfate decahydrate ($Na_2SO_4 \cdot 10H_2O$) and calcium sulfate dihydrate ($CaSO_4 \cdot 2H_2O$) are precipitated. Beginning at −24°C, sodium chloride dihydrate ($NaCl \cdot 2H_2O$) is precipitated, followed by precipitation of potassium chloride (KCl) and magnesium chloride dodecahydrate ($MgCl \cdot 12H_2O$) at −34°C, and the remaining ions with further lowering of temperature.

The brine cells migrate and change size with changes in temperature and pressure. The general downward migration of brine cells through the ice sheet leads to freshening of the top layers to near zero salinity by late summer. During winter the top surface temperature closely follows the air temperature, whereas the temperature of the underside remains at freezing point, corresponding to the salinity of water in contact. Heat flux up through the ice permits freezing at the underside. In summer freezing can also take place under sea ice in regions where complete melting does not occur. Surface melt water (temperature 0°C) runs down through cracks in the ice to spread out underneath and contact the still cold ice masses and underlying colder sea water. Soft slush ice forms with large cells of entrapped sea water which then solidifies the following winter.

The salinity of recently formed sea ice depends on rate of freezing; thus sea ice formed at −10°C (14°F) has a salinity from 4 to 6 parts per thousand (°/₀₀), whereas that formed at −40°C may have a salinity from 10 to 15°/₀₀. Sea ice is a poor conductor of heat and the rate of ice formation drops appreciably after 4–6 in. are formed. An undisturbed sheet grows in relation to accumulated degree-days of frost. Figure 2 shows an empirical relation between ice thickness and the sum of the mean diurnal negative air temperature (degrees Celsius). The thermal conductivity varies greatly with air bubble content, perhaps between 1.5 and 5.0×10^{-3} cal/(cm)(sec)(°C).

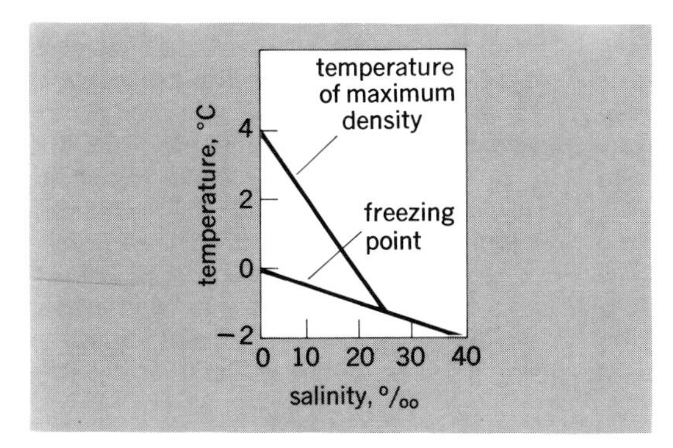

Fig. 1. Change of freezing point and temperature of maximum density with varying salinity of sea water.

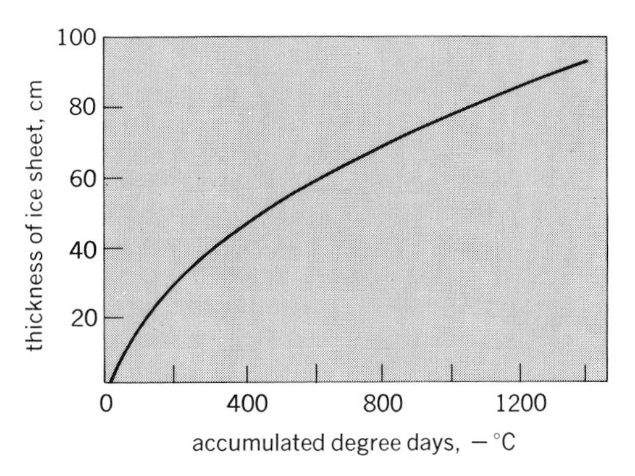

Fig. 2. Growth of undisturbed ice sheet.

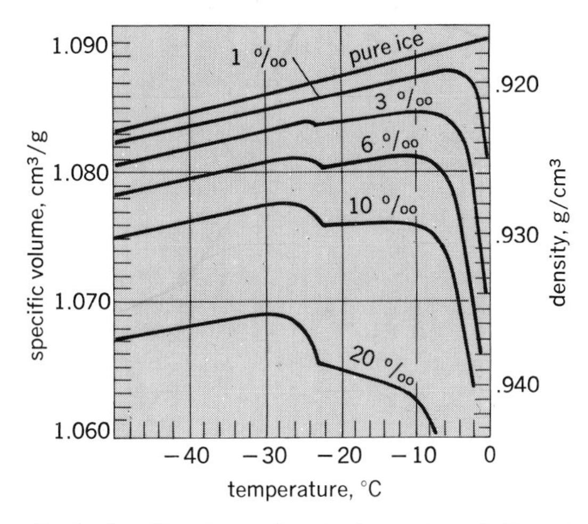

Fig. 3. Specific volume of sea ice for varying salinity and temperature, computed on basis of chemical model. *(By D. L. Anderson, based on data in Arctic Sea Ice, NAS-NRC Publ. no. 598, 1958)*

The specific gravity of sea ice varies between 0.85 and 0.95, depending on the amount of entrapped air bubbles. The specific heat varies greatly because changing temperature involves freezing or melting of ice. Near 0°C, amounts that freeze or melt at slight change of temperature are large and "specific heat" is anomalous. At low temperatures the value approaches that of pure ice; thus, specific heat for 4°/₀₀ saline ice is 4.6 cal/(g) (°C) at −2°C and 0.6 at −14°C; for 8°/₀₀ saline ice, 8.8 at −2°C and 0.6 at −14°C.

Sea ice of high salinity may expand when cooled because further freezing out occurs with attendant increase of specific volume, for example, ice of salinity 8°/₀₀ at −2°C expands at a rate of about 93×10^{-4} cm³/g per degree Celsius decrease in temperature, at −14°C expands 0.1×10^{-4}, but at −20°C contracts 0.4×10^{-4} per degree Celsius

Fig. 4. Pancake ice interspersed with blocks of young ice. *(U.S. Naval Oceanographic Office)*

decrease. Change of specific volume with temperature and salinity is illustrated in Fig. 3.

Sea ice is viscoelastic. Its brine content, which is very sensitive to temperature and to air bubble content, causes the elasticity to vary widely. Young's modulus measured by dynamic methods varies from 5.5×10^{10} dynes/cm^2 during autumn freezing to 7.3×10^{10} in winter to 3×10^{10} at spring breakup. Static tests give much smaller values, as low as 0.2×10^{10}. The flexural strength varies between 0.5 and 17.3 kg/cm^2 over salinity range of 7–16°/₀₀ and temperatures −2 to −19°C. Acoustic properties are highly variable, depending principally on the size and distribution of entrapped air bubbles.

Electrical properties vary greatly with frequency because of ionic migration within the brine cells. For example, for sea ice salinity of 10°/₀₀ at −22°C, the dielectric coefficient is very large, about 10^6 at 20 Hz, and decreases with increasing frequency to about 10^3 at 10 kHz and to 10, or less, at 50 MHz. The effective electrical conductivity decreases with lowering of temperature, for example, from less than 10^{-3} mho/cm at −5°C to 10^{-6} mho/cm at −50°C (frequency 1 to 10 kHz).

Types and characteristics. The sea ice in any locality is commonly a mixture of recently formed ice, old ice which has survived one or more summers, and possibly old ridges of ice that formed against a coast and contain beach material. The various descriptive forms are shown in Figs. 4–6. Except in sheltered bays, sea ice is continually in motion because of wind and current. The weaker parts of the sea ice canopy break when overstressed in tension, compression, or shear, pulling

Fig. 5. Hummocky floes that have weathered. *(U.S. Naval Oceanographic Office)*

Fig. 6. Unweathered pressure ridges, formed by rafting of floes. *(U.S. Naval Oceanographic Office)*

apart to form a lead (open water), or piling block on block to form a pressure ridge. Depending on the composition of the ice canopy, the ridges may form in ice of any thickness, from thin sheets (10 cm thick) to heavy blocks (3 m or more in thickness). The ridges may pile 13 m high above and extend 50 m, or more, below the sea surface. Massive ridges become grounded in coastal zones, further producing disruptive forces within the ice canopy.

[WALDO LYON]

Bibliography: T. Karlsson (ed.), *Sea Ice*, 1972; W. D. Kingery (ed.), *Ice and Snow: Properties, Processes, and Applications*, 1963; National Academy of Sciences, Division of Earth Sciences, *Beneficial Modifications of the Marine Environment*, 1972; *Proceedings of the Conference on Arctic Sea Ice*, NAS–NRC Publ. no. 598, 1958; J. C. Reed and J. E. Sater (eds.), *The Coast and Shelf of the Beaufort Sea*, 1974; World Meteorological Organization, *Sea-Ice Nomenclature*, 1971.

Sea state

The description of the ocean surface or state of the sea surface with regard to wave action. Wind waves in the sea are of two types: Those still growing under the force of the wind are called sea: those no longer under the influence of the wind that produced them are called swell. Differences between the two types are important in forecasting ocean wave conditions. Properties of sea and swell and their influence upon sea state are described in this article.

Sea. Those waves which are still growing under the force of the wind have irregular, chaotic, and unpredictable forms (Fig. 1*a*). The unconnected wave crests are only two to three times as long as the distance between crests and commonly appear to be traveling in different directions, varying as much as 20° from the dominant direction. As the waves grow, they form regular series of connected troughs and crests with wave lengths commonly ranging from 12 to 35 times the wave heights. Wave heights only rarely exceed 55 ft (17 m). The appearance of the sea surface is termed state of the sea (Table 1).

The height of a sea is dependent on the strength of the wind, the duration of time the wind has blown, and the fetch (distance of sea surface over which the wind has blown). *See* OCEAN WAVES.

Table 1. Sea height code*

Code	Height, ft†	Description of sea surface
0	0	Calm, with mirror-smooth surface
1	0–1	Smooth, with small wavelets or ripples with appearance of scales but without crests
2	1–3	Slight, with short pronounced waves or small rollers; crests have glassy appearance
3	3–5	Moderate, with waves or large rollers; scattered whitecaps on wave crests
4	5–8	Rough, with waves with frequent whitecaps; chance of some spray
5	8–12	Very rough, with waves tending to heap up; continuous whitecapping; foam from whitecaps occasionally blown along by wind
6	12–20	High, with waves showing visible increase in height, with extensive whitecaps from which foam is blown in dense streaks
7	20–40	Very high, with waves heaping up with long frothy crests that are breaking continuously; amount of foam being blown from the crests causes sea surface to appear white and may affect visibility
8	40+	Mountainous, with waves so high that ships close by are lost from view in the wave troughs for a time; wind carries off crests of all waves, and sea is entirely covered with dense streaks of foam; air so filled with foam and spray as to affect visibility seriously
9		Confused, with waves crossing each other from many and unpredictable directions, developing complicated interference pattern that is difficult to describe; applicable to conditions 5–8

*Modified from *Instruction Manual for Oceanographic Observations*, H.O. Publ. no. 607, 2d ed., U.S. Navy Hydrographic Office, 1955.

†1 ft = 0.3048 m.

Swell. As sea waves move out of the generating area into a region of weaker winds, a calm, or op-

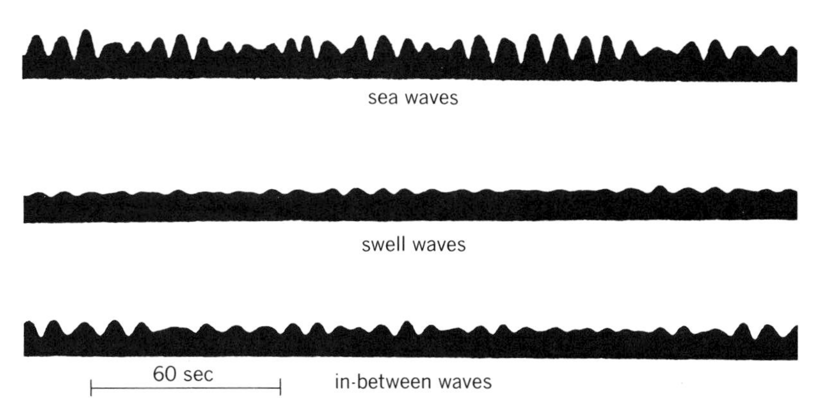

Fig. 1. Records of surface waves, for sea, swell, and in-between waves. *(Adapted from W. J. Pierson, Jr., et al., Observing and Forecasting Ocean Waves, H.O. Publ. no. 603, U.S. Navy Hydrographic Office, 1955)*

Table 2. Swell-condition code*

Code	Description	Height, ft†	Length, ft†
0	No swell	0	0
	Low swell	1–6	
1	Short or average		0–600
2	Long		600+
	Moderate swell	6–12	
3	Short		0–300
4	Average		300–600
5	Long		600+
	High swell	12+	
6	Short		0–300
7	Average		300–600
8	Long		600+
9	Confused		

**Instruction Manual for Oceanographic Observations*, H.O. Publ. no. 607, 2d ed., U.S. Navy Hydrographic Office, 1955.

†1 ft = 0.3048 m.

posing winds, their height decreases as they advance, their crests become rounded, and their surface is smoothed (Fig. 1*b*). These waves are more regular and more predictable than sea waves and, in a series, tend to show the same form or the same trend in characteristics. Wave lengths generally range from 35 to 200 times wave heights.

The presence of swell indicates that recently there may have been a strong wind, or even a severe storm, hundreds or thousands of miles away. Along the coast of southern California long-period waves are believed to have traveled distances greater than 5000 mi (8000 km) from generating areas in the South Pacific. Swell can usually be felt by the roll of a ship, and, under certain conditions, extremely long and high swells in a glassy sea may cause a ship to take solid water over its bow regularly.

A descriptive classification of swell waves is given in Table 2. When swell is obscured by sea waves, or when the components are so poorly defined that it is impossible to separate them, it is reported as confused.

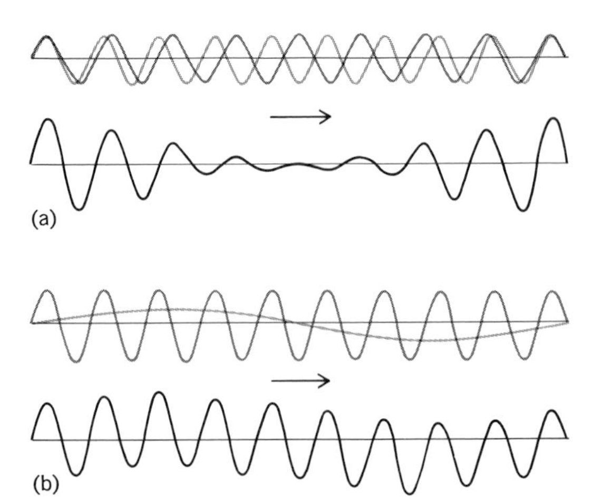

Fig. 2. Wave patterns resulting from interference. (*a*) Interference of waves of equal height and nearly equal length, forming wave groups. (*b*) Interference between short wind waves and long swell. (*From Techniques for Forecasting Wind Waves and Swell, H.O. Publ. no. 604, U.S. Navy Hydrographic Office, 1951*)

In-between state. Often both sea waves and swell waves, or two or more systems of swell, are present in the same area (Fig. 1*c*). When waves of one system are superimposed upon those of another, crests may coincide with crests and accentuate wave height, or troughs may coincide with crests and cancel each other to produce flat zones (Fig. 2). This phenomenon is known as wave interference, and the wave forms produced are extremely irregular. When wave systems cross each other at a considerable angle, the apparently unrelated peaks and hollows are known as a cross sea.

Breaking waves. The action of strong winds (greater than 12 knots) sometimes causes waves in deeper water to steepen too rapidly. As the height-length ratio becomes too large, the water at the crest moves faster than the crest itself and topples forward to form whitecaps.

Breakers. As waves travel over a gradually shoaling bottom, the motion of the water is restricted and the wave train is telescoped together. The wave length decreases, and the height first decreases slightly until the water depth is about one-sixth the deep-water wave length and then rapidly increases until the crest curves over and plunges to the water surface below. Swell coming into a beach usually increases in height before breaking, but wind waves are often so steep that there is little if any increase in height before breaking. Thus swell that is obscured by wind waves in deeper water often defines the period of the breakers.

Surf. The zone of breakers, or surf, includes the region of white water between the outermost breaker and the waterline on the beach. If the sea is rough, it may be impossible to differentiate between the surf inshore and the whitecaps in deep water just beyond. [NEIL A. BENFER]

Bibliography: W. Bascom, *Waves and Beaches: The Dynamics of the Ocean Surface*, 1964; H. J. McLellan, *Elements of Physical Oceanography*, 1966; G. Neumann and W. J. Pierson, Jr., *Principles of Physical Oceanography*, 1966.

Sea water

Water is most often found in nature as sea water (≅98%). The rest is found as ice, water vapor, and fresh water. Sea water is an aqueous solution of salts of a rather constant composition of elements. Its presence determines the climate and makes life possible on the Earth. The boundaries of sea water are the boundaries of the oceans, the mediterranean seas, and their embayments. The physical, chemical, biological, and geological events in the hydroplane within these boundaries are the studies which are grouped together and called oceanography. The basic properties of sea water, the distribution of these properties, the interchange of properties between sea and atmosphere or land, the transmission of energy within the sea, and the geochemical laws governing the composition of sea water and sediments are the fundamentals of oceanography.

The discussion of sea water which follows is divided into six sections: (1) physical properties of sea water; (2) interchange of properties between sea and atmosphere; (3) transmission of energy within the sea; (4) composition of sea water; (5) distribution of properties; and (6) sampling and measuring techniques. For further treatment of related aspects of physical character, composition, and

Table 1. Some physical properties of sea water (salinity, 35‰) at sea level*

Property	Temperature, °C			
	0	10	20	30
Specific volume, cm^3/g	0.9726	0.9738	0.9757	0.9784
Isothermal compressibility $\times 10^6$, $bars^{-1}$ ($\times 10^{11}$, Pa^{-1})	46.7	44.3	42.7	41.7
Thermal coefficient of volume expansion $\times 10^5$, $°C^{-1}$	5.4	16.6	25.8	33.4
Sound speed, m/sec	1449.4	1490.1	1521.7	1545.7
Electric conductivity $\times 10^3$, $(ohm\text{-}cm)^{-1}$	29.04	38.10	47.92	58.35
Molecular viscosity, centipoise	1.89	1.39	1.09	0.87
Specific heat, cal/g°C	0.953	0.954	0.955	0.956
Optical index of refraction $(n-1.333{,}338)\times 10^6$, $\lambda=0.5876\ \mu m$	6966	6657	6463	6337
Osmotic pressure, bars	23.4	24.3	25.1	26.0
Molecular thermal conductivity coefficient $\times 10^3$, cal/cm sec °C	1.27	1.31	1.35	1.38

*1 cal = 4.1868 J; 1 centipoise = 10^{-3} Pa·s; 1 bar = 10^5 Pa.

constituents *see* HYDROSPHERIC GEOCHEMISTRY; MARINE RESOURCES; SEA-WATER FERTILITY.

PHYSICAL PROPERTIES OF SEA WATER

Sea water is basically a concentrated electrolyte solution containing many dissolved salts. The ratio of water molecules to salt molecules is about 100 to 1. Since nearly all the salt exists as electrically conducting ions, this means that the ratio of water molecules to ions is about 50 to 1; consequently, the ions are on the average no farther than about 10^{-7}cm from each other, a distance equivalent to the diameter of about five water molecules. Because pure water has a relatively open structure with tetrahedral coordination, the water molecules by virtue of their electric dipole moment (arising from the separation of the positive and negative charges) can be readily oriented or polarized by an electric field. The polarizability manifests itself in the high dielectric constant of pure water. Since electrostatic attraction between ions is inversely proportional to the dielectric constant, a high dielectric constant facilitates ionization of electrolytes because of the reduced forces between ions of opposite sign.

Salinity effects. In the neighborhood of ions, extremely high electric fields exist (around 100,000 volts/cm) and water molecules near them become aligned; water molecules that remain in the vicinity of the ions for a long time constitute a hydration shell, and the ions are said to be solvated. The alignment of water molecules produces local dielectric saturation (that is, no further alignment is possible) around the ions, thereby lowering the dielectric constant of the solution below that of pure water.

Details of ion-ion and ion-solvent interactions and their effects on the physical property of solutions are treated in the theory of electrolyte solutions.

As a consequence of the salts in sea water, its physical properties differ from those of pure water, the difference being closely proportional to the concentration of the salts or the salinity. Salinity measurements (which can be conveniently made using electrical conductivity apparatus), along with pressure and temperature data, are used to differentiate water masses. In studying the movement of water masses in the oceans and their small- and large-scale circulation patterns, including geostrophic flow, properties such as density, compressibility, thermal expansion coefficients, and specific heats need to be known as functions of temperature, pressure, and salinity.

The large value of the osmotic pressure of sea water is of great significance to biology and desalination by reverse osmosis; for example, at a salinity of 35‰ (parts per thousand) the osmotic pressure relative to pure water is around 25 atm (2.5 MPa). Related to osmotic pressure and very important to the formation of ice is the reversal of the freezing point and temperature of maximum density of sea water compared to pure water: The freezing point temperature is lowered to −1.9°C, and the temperature of maximum density is decreased from just below 4°C for pure water to about −3.5°C for 35‰ salinity sea water. Some other properties which show significant changes between sea water (salinity 35‰) and pure water at atmospheric pressure are shown in Tables 1 and 2. Although the world oceans show a wide range in temperature and salinity, 75% by volume occurs within a range of 0 to 6°C and 34 to 35‰ salinity.

Pressure effects. Since the greatest ocean depths exceed 10,000 m and more than 54% of the oceans' area is at pressure above 400 bars (40 MPa), it is necessary to consider the effect of pressure, as well as temperature, on the physical properties of sea water. The pressure correspond-

Table 2. Sound attenuation coefficient α in sea water*

	Sound frequency, Hz			
	100	1000	10,000	100,000
Attenuation coefficient α, km^{-1}	0.00023	0.0069	0.15	6.3

*Depth ~ 1200 m, temperature ~ 4°C, and salinity = 35‰.

SOURCE: After W. H. Thorp, Deep ocean attenuation in the sub- and low-kilocycle-per-second region, *J. Acoust. Soc. Amer.*, 38:648–654, 1965.

Table 3. Percent change of sea water properties at elevated pressures*

Property	500 bars pressure		1000 bars pressure	
	0°C	20°C	0°C	20°C
Electrical conductivity	6.76	3.88	11.19	6.49
Specific volume	−2.16	−1.97	−4.00	−2.28
Compressibility	−13.0	−12.0	−24.0	−22.0
Thermal expansion coefficient	224.0	19.0	386.0	34.3
Sound speed	5.85	5.66	12.02	11.08
Sound absorption coefficient		−47.0		−65.0
Molecular viscosity	−3.8	−0.3	−4.7	0.5

*1 bar = 10^5 Pa.

ing to the maximum ocean depth is about 1100 bars. Table 3 shows the percent change in some properties at 500 and 1000 bars, corresponding to depths of about 5000 and 10,000 m at two temperatures, 0 and 20°C. The unusual pressure dependence of viscosity is a consequence of the open structure of water which is altered by pressure, temperature, and solutes.

Sound absorption. Because electromagnetic radiation can propagate in the ocean for only limited distances, sound waves are the principal means of communication in this medium. The equation for attenuation of the intensity of a plane wave (without geometrical spreading losses) is given by Eq. (1) where I_0 is the initial intensity, and I is the

$$I = I_0 e^{-2\alpha x} \tag{1}$$

intensity at a distance of x km. Sound absorption in sea water, shown in Table 2, is considerably greater than in fresh water, about 30 times greater between frequencies of 10 and 100 kHz. This arises from a pressure-dependent chemical reaction involving magnesium sulfate with a relaxation frequency around 100 kHz. Another relaxation frequency around 1 kHz has been found in the ocean; the origin of this phenomenon has not been determined. [F. H. FISHER]

Bibliography: A. Bradshaw and K. E. Schleicher, The effect of pressure on the electrical conductance of sea water, *Deep-Sea Res.*, 12:151–162, 1965; L. A. Bromley et al., Heat capacities of sea water solutions at salinities of 1 to 12% and temperatures of 2° to 80°, *J. Chem. Eng. Data*, 12:202–206, 1967; A. Defant, *Physical Oceanography*, vol. 1, 1961; F. H. Fisher, Ion pairing of magnesium sulfate in sea water: Determined by ultrasonic absorption, *Science*, 157:823, 1967; H. W. Menard and S. M. Smith, Hypsometry of ocean basin provinces, *J. Geophys. Res.*, 71:4305–4325, 1966; R. B. Montgomery, Water characteristics of Atlantic Ocean and of world ocean, *Deep-Sea Res.*, 5:134–148, 1958; G. Neumann and W. J. Pierson, Jr., *Principles of Physical Oceanography*, 1966; W. S. Reeburgh, Measurements of electrical conductivity of sea water, *J. Mar. Res.*, 23:187–199, 1965; M. Schulkin and H. W. Marsh, Sound absorption in sea water, *J. Acoust. Soc. Amer.*, 34:864–865, 1962; E. M. Stanley and R. C. Batten, *Viscosity of Sea Water at High Pressures and Moderate Temperatures*, Nav. Ship Res. Dev. Center Rep. no. 2827, 1968; W. D. Wilson, Speed of sound in sea water as a function of temperature, pressure and salinity, *J. Acoust. Soc. Amer.*, 32:641–644, 1960; W. D. Wilson and D. S. Bradley, Specific volume of sea water as a function of temperature, pressure and salinity, *Deep-Sea Res.*, 15:355–363, 1968.

INTERCHANGE BETWEEN SEA AND ATMOSPHERE

The sea and the atmosphere are fluids in contact with one another, but in different energy states—the liquid and the gaseous. The free surface boundary between them inhibits, but by no means totally prevents, exchange of mass and energy between the two. Almost all interchanges across this boundary occur most effectively when turbulent conditions prevail: a roughened sea surface, large differences in properties between the water and the air, or an unstable air column that facilitates the transport of air volumes from sea surface to high in the atmosphere.

Heat and water vapor. Both heat and water (vapor) tend to migrate across the boundary in the direction from sea to air. Heat is exchanged by three processes: radiation, conduction, and evaporation. The largest net exchange is through evaporation, the process of transferring water from sea to air by vaporization of the water.

Evaporation depends on the difference between the partial pressure of water vapor in the air and the vapor pressure of sea water. Vapor pressure increases with temperature, and partial pressure increases with both temperature and humidity; therefore, the difference will be greatest when the sea (always saturated) is warm and the air is cool and dry. In winter, off east coasts of continents, this condition is most ideally met, and very large quantities of water are absorbed by the air. On the average, 100 g water per square centimeter of ocean surface are evaporated per year.

Since it takes nearly 600 cal (2.5 kJ) to evaporate 1 g water, the heat lost to each square centimeter of the sea surface averages 150 cal/day (630 J/day). This heat is stored in the atmospheric volume but is not actually transferred to the air parcels until condensation takes place (releasing the latent heat of vaporization) perhaps 1000 mi (1600 km) away and 1 week later.

Radiation of heat from the water surface to the atmosphere and back again are both large—of the order of 800 cal/(cm²)(day) [33 MJ/(m²)(day)] according to E. R. Anderson (1953). However, the net flux is out from the sea; it averages about 100 cal/(cm²)(day) [4 MJ/(m²)(day)].

Conduction usually plays a much smaller role than either of the above; it may transfer heat in either direction, but usually it contributes a small net transfer from sea to air. *See* OCEAN-ATMOSPHERE RELATIONS.

Momentum. Momentum can be exchanged between these two fluids by a process related to evaporation, that is, migration of molecules of air or water across the boundary, carrying their momentum with them. However, in natural conditions the more effective mechanism is the collision of "parcels" of the fluids, as distinct from motions of individual molecules. Also, momentum is usually transferred from air to sea, not vice versa. Winds whip up waves; these irregular shapes are more easily attacked by wind action than is the flat sea surface, and both waves and currents are initiated and maintained by the push and stress of the wind on the water surface. [JUNE G. PATTULLO]

Isotopic relationships. The isotopic water molecules H_2O^{16}, HDO^{16}, and H_2O^{18} have different vapor pressures and molecular diffusion coefficients and therefore exchange at different rates between the atmosphere and sea. Variations in the relative proportions of the hydrogen isotope deuterium, D, and O^{18} can now be measured mass spectrometrically with high precision, and the isotopic fractionation effects can be studied both at sea and in the laboratory. The isotopic vapor pressures have been measured very accurately, and the ratios of the binary molecular diffusion coefficients for HDO-air and H_2O^{18}-air to that for H_2O^{16}-air have been calculated theoretically and confirmed experimentally. Since the relative transport properties of isotopic molecules are much better known than those of different chemical species, the isotopic variations observed in surface ocean water and atmospheric vapor and in experimental studies on evaporation in small wind tunnels provide a powerful method for the study of the air-sea interface and the molecular and turbulent transport processes controlling the moisture supply to the air.

Precipitation over sea and land varies in isotopic composition because of the effects of fractional condensation of liquid water and variation of the isotopic vapor pressure ratios with temperature. Local equilibrium however, is maintained, and kinetic effects do not occur; the deuterium and oxygen-18 variations in precipitation are linearly correlated, and the concentrations of both heavy isotopes decrease with increasing latitude because of their preferential concentration in the liquid precipitate, which strips them out in lower latitudes. These variations provide a continually varying liquid input into the sea which must be balanced against the direct molecular exchange.

Water vapor over the oceans is never in isotopic equilibrium with surface sea water. The deuterium and oxygen-18 concentrations are always lower than the two-phase equilibrium separation factors given by the vapor-pressure ratios. The deviations from equilibrium are correlated with latitude and go through a maximum in each hemisphere in the trade-winds regions of maximum evaporation to precipitation ratio. The vapor composition relative to surface sea water cannot be understood simply on the basis of multiple equilibrium stage processes in fractional condensation during precipitation, and the isotopic variations reflect the kinetic isotope effects in the molecular exchange of water at the interface.

Two types of kinetic processes affect the isotopic composition of the vapor. There is a fractionation at the interface between liquid and vapor, since the condensation coefficients for the isotopic species are not necessarily the same. (The condensation coefficient may be thought of as the fraction of molecules of a given type striking the liquid surface which actually condense into the liquid structure; conversely, it is also a measure of the probability of a molecule to surmount the energy barrier for evaporation and actually to escape from the liquid.) For such fractionation to occur, the vapor concentration at the liquid-vapor interface must be significantly lower than the equilibrium concentration.

The second kinetic process is molecular diffusion from the interface into the turbulent mixing zone in the atmosphere above the boundary layer. Two types of models have been postulated for the boundary layer. At low wind speeds it is generally postulated that a true laminar layer exists with a fixed thickness and vapor gradient for a given wind speed, the transport through this layer being by steady-state molecular diffusion. At high wind speeds, above a certain critical velocity, the water surface changes from a smooth surface to a hydrodynamically rough surface. H. U. Sverdrup proposed that small turbulent eddies extend down to the actual liquid surface when it is rough. In such a model the vapor flux into these eddies can be postulated to take place by unsteady-state diffusion. With isotopic measurements it is possible to distinguish between these two models, because in the first case the isotopic fractionation is governed by the ratio of molecular diffusion coefficients, whereas in the unsteady-state process the fractionation is governed by the square root of this ratio. The single-stage enrichments in the two processes thus differ by a factor of 2.

Experimentally, the following points are observed: The oxygen-18 fractionation in the exchange process can be predicted from the rough-surface model for conditions under which both smooth and rough surfaces should be present. These results thus indicate that interfacial fractionation (from differences in condensation coefficients) is not important for oxygen and that a "microeddy" transport process operates in all cases; no actual fixed laminar layer exists. Deuterium data from the same experiments show larger separation factors than can be obtained from diffusional effects, so that it appears that there is significant interfacial fractionation for this isotope. This is plausible because of the asymmetric character of the HDO molecule and the large effect of deuterium substitution on vibrational frequencies, both of which should produce marked fractionation effects. The deuterium data therefore indicate that in the experiments which have been made the vapor pressure at the interface is lower than the saturation vapor pressure and that there is a significant thermodynamic disequilibrium. The oceanic data on surface waters and atmospheric vapor and precipitation indicate that kinetic effects of the same type as observed in laboratory experiments are important over the oceans.

These investigations thus provide a way to determine directly the contributions of molecular and turbulent transport processes to the evaporation rate over lakes and oceans and to relate numerical measures of the transport coefficients to physical parameters, such as wind speed. In addition, if the interfacial fractionation factor for deuterium can be determined, a measure of the actual vapor pressure at the interface can be obtained. However, a great deal of work remains to be done on this subject, which is still in its infancy.

Tritium variations. The radioactive hydrogen isotope tritium occurred with an abundance of about 0.4×10^{-12} molecule of HTO per 10^6 molecules of H_2O in oceanic surface waters in the prenuclear era, representing steady-state production of tritium by cosmic rays in the atmosphere. In 1968 the concentration of this molecule was about 10 times higher in surface waters because of nuclear weapon testing. The concentration in deep

SEA WATER

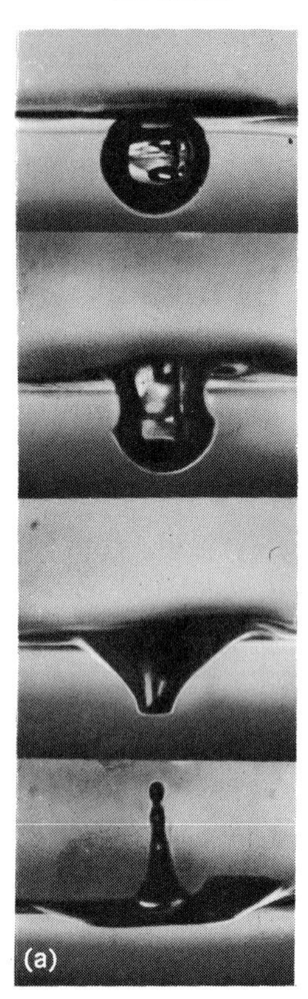

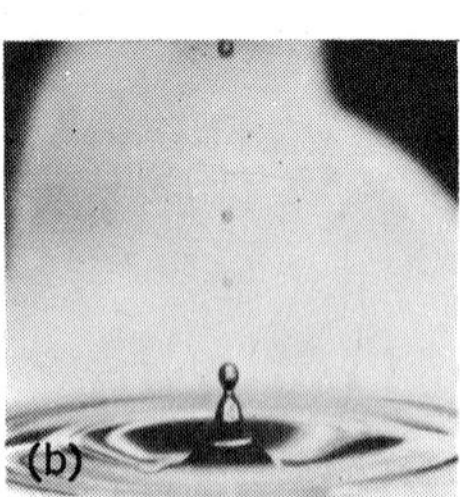

Fig. 1. Collapse of air bubble and formation of jet and droplets. (*a*) High-speed motion pictures of stages in the process. (*b*) Oblique view of jet and droplets from bubble 1.0 mm in diameter.

waters, below the thermocline, has always been too low to be measurable because of the short tritium half-life (12.5 years) compared to oceanic mixing times of the order of hundreds to thousands of years. Studies of tritium variations in present-day ocean waters have been made by Arnold Bainbridge, W. F. Libby, and others. Calculations of mixing rates of surface and deep waters and of atmospheric residence times for tritium can be made from these data. The models used in the various studies of this subject have neglected the molecular exchange of tritium between atmosphere and sea, which is important relative to the simple input by precipitation. Calculations, based on the stable isotope effects in molecular exchange, show that the molecular-exchange input of tritium into the ocean is about 2.5 times greater than the input by precipitation in both the prenuclear and postnuclear epochs. It is therefore necessary to reevaluate the tritium calculations, taking this finding into account. [HARMON CRAIG]

Projection of droplets. Water, salts, organic materials, and a net electric charge are transferred to the air through the ejection of droplets by bubbles bursting at the sea surface. The exchange of these properties between the sea and the atmosphere is of importance in meteorology and geochemistry. Upon evaporation of the water, the droplet residues are carried great distances by winds. These particles become nuclei for cloud-drop and raindrop formations and probably represent a large part of the cyclic salts of geochemistry.

Air bubbles are forced into surface waters of the sea by wave action, impinging raindrops, melting snowflakes, and other means. The larger bubbles rise to the surface, burst, and eject droplets. Many of the smaller bubbles dissolve before reaching the surface. The photomicrograph (Fig. 1) shows stages in the collapse of a bubble, and the jet and droplet formations which result. The graph (Fig. 2) shows the approximate relationships between the sizes of the bubbles, the sizes of the ejected droplets, and the weight of sea salt in these droplets.

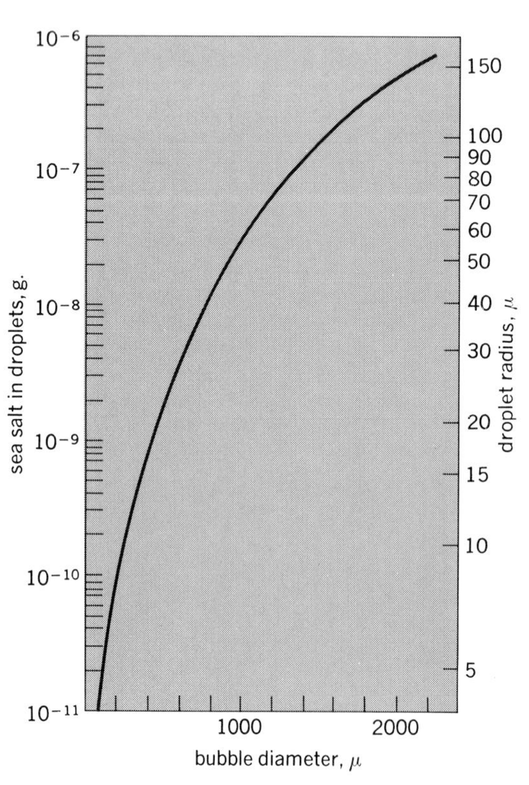

Fig. 2. Approximate relationships between the sizes of the bursting bubbles and the sizes and salt contents of the ejected droplets.

The amounts of water which become airborne as droplets near the sea surface are not known. The best estimates which can now be made (from the limited information about the number and size of bubbles in the sea) range from about 2 to 10 g/ (m^2) (day) during fresh winds.

The average amounts of sea salt which become airborne at considerable altitudes are shown in Table 4. The total range of observed amounts in individual samples is from about 4×10^{-13}g/ml in a wind of Beaufort force 1 to 10^{-9}g/ml in a wind of force 12. *See* WIND.

Parts of marine organisms are seen in droplets ejected from plankton-rich water. The droplets also become coated with organic monolayers when they arise through contaminated surfaces. During moderate winds in oceanic trade wind areas, organic materials can equal 20 to 30 percent of the airborne sea salt. [ALFRED H. WOODCOCK]

Electrification of the atmosphere. The traditional view states that the net positive space charge usually found in regions of fair weather is maintained by a charge separation process within thunderstorms. However, research has indicated that a charge separation mechanism at the surface of the ocean may contribute significantly to the atmospheric space charge over the oceans. It appears that this mechanism enables the oceans of the world to supply positive charge to the atmosphere at a rate of at least 10% of that supplied to the atmosphere over the oceans by thunderstorms. *See* ATMOSPHERIC ELECTRICITY.

The carriers of the charge separated at the ocean surface are the drops that emerge at the collapse of the jet that forms when a bubble breaks at the ocean surface (Fig. 1). Laboratory measurements have shown that the charge on these drops is a function of their size and of the age of the bubble from which they came. For drops in the size range commonly found in the atmosphere over the ocean, the charge per drop is positive and of the order of 10^{-6} electrostatic units (esu). From a consideration of the numbers and sizes of drops and the rate that they leave the sea surface, it has been computed that the net oceanic charge production is roughly proportional to the square of the wind speed, and is 5×10^{-8} esu/(cm^2) (sec) for winds of 10 knots. Measurements of space charge made on the windward shore of the island of Hawaii are in good agreement with these calculations. As the winds over the oceans attain a maximum mean

Table 4. Airborne sea salts in relation to wind force

Beaufort wind force	Concentrations,* μg/m³	Flux,* mg/(m²)(day)	Total,† mg/m²
2–3	2.7	0.42	6.0
4–5	9.9	3.9	11.6
6–7	21.3	24.0	21.8

*At about 500 m.
†Integrated through lowest 2000 m.

speed in each hemisphere at latitudes of 40–60°, a similarly located maximum should be found in the latitudinal distribution of the oceanic charge separation.

The normal oceanic fair-weather potential gradient does not have any significant influence on the magnitude of the charge of the drops from the bubble jet, but intense negative thunderstorm potential gradients of the order of 100 volts/cm at the sea surface can, by the process of induction charging, produce a positive charge on the drops that exceeds by many times the positive charge found on the drops in fair weather. Consequently, the normal positive space charge may be increased considerably in regions near oceanic thunderstorms.

[DUNCAN C. BLANCHARD]

Microlayer. The microlayer is the thin zone beneath the surface of the ocean or any free water surface within which physical processes are modified by proximity to the air-water boundary. It is characterized by suppression of vertical turbulence, a consequent decrease in diffusivity and increase in material, and an increase in kinetic and thermal gradients.

Because the microlayer at a free water surface is at least superficially similar to the boundary layer observed in the tangential flow of any viscous fluid near a rigid surface, it is tempting to identify one with the other. However, in view of the thermohydrodynamic complexity of the free ocean surface, such identification is unwarranted. It is not possible, in the present state of knowledge, to deal analytically with this problem. Consequently, all that can be described of the nature of the microlayer is gleaned from a few scattered observations. For the most part, these have been measurements of the effect of the microlayer on the flux of heat between water and air. The earliest determination appears to have been made by Alfred Merz in 1920. Subsequently A. H. Woodcock and H. Stommel measured thermal gradients at the surface of steaming ponds and estuaries, using a specially designed mercury thermometer of thin-stem diameter. G. C. Ewing and E. D. McAlister, using an infrared radiometer, observed the ocean surface to be as much as 0.6°C cooler than the underlying water. A somewhat related experiment, by W. S. Wise and others (1960), showed a measurable concentration gradient of solute at the surface of evaporating sugar solution. This suggests that a salinity gradient probably exists at the ocean surface, although it has yet to be described in the literature.

A different and striking manifestation of the microlayer can be observed when a gentle breeze blows over calm water. Specks of dust at the surface flow along noticeably faster than those 1 cm or so beneath the surface. The strongly developed vertical shear in the wind-driven flow which is thus revealed is possible only because the small eddy stresses in the microlayer permit the motion to remain nearly laminar at a relatively high Reynolds number. The reduction of the shear at higher wind speeds shows that the microlayer is thinner under these conditions.

The development and stability of the microlayer is enhanced by a contaminating film of surface-active agents, which is characteristically present on natural water surfaces. Such films quickly accumulate on any body of water exposed to the air. Even in the laboratory elaborate precautions are necessary to maintain a truly clean water surface.

The origin and composition of such films are complex. However, it can be asserted on thermodynamic grounds that substances will accumulate in a liquid surface if they reduce the surface tension and, hence, the free energy of the surface. Patches of such film can be observed on the sea under all normal conditions of wind and wave, although they are more strikingly visible at wind speeds under 3 m/sec, when they take the form of long, broad slicks. The most obvious effect of the film is to smooth the smallest ripples, giving the water a shiny appearance. The smoothing results from an altered boundary condition at the interface, substituting a sort of rubber-sheet elasticity for the relatively unrestricted freedom of a clean liquid surface. Such a stabilized surface more nearly approximates a rigid boundary, and therefore the associated microlayer becomes more nearly similar to the familiar boundary layer characterizing flow near a solid surface. Gradients, whether of substance, temperature, or momentum, are thus appreciably enhanced by surface films. In this purely mechanical manner, contaminating films can reduce the convective flux of heat across the air-water boundary independently of any throttling action they may have on the evaporation rate.

The flux of sensible heat across the air-water boundary of the ocean is usually in the upward direction, and hence the microlayer is cooler above than beneath. The sense of the flux results from the circumstance that most of the heating of the ocean is by solar radiation, which penetrates several meters into the sea before being absorbed, whereas heat balance is largely maintained by upward flux of sensible heat from a layer less than a few molecular diameters deep. This is a form of the well-known greenhouse effect. Thus, on the average, the microlayer is heated from below and cooled from above. The net upward flux is of the order of 150 g-cal/(cm^2) (day), varying with the latitude, the season, and the time of day. The flux is greatest in the tropics, in the autumn, and in the forenoon; least at the poles, in early summer, and in early afternoon.

The importance of the microlayer resides in the fact that most surface measurements are in reality volume measurements of a thin but finitely thick layer. Consequently, the value recorded depends on the method of measurement employed. For many purposes the differences are trivial and are ignored. However, where precision is required, the only way to arrive at a true value of any parameter at the exact surface is to calculate it from theoretical considerations or to estimate it by extrapolating some measured gradient to the boundary. As an example, one may assume intuitively that the surface temperature of water must approach the psychrometric, or so-called wet-bulb, temperature of the overlying air as a limit. However, the psychrometric temperature itself varies as the boundary is approached and therefore cannot be directly measured at the exact surface.

Hence, from a physical point of view, the concept of a surface is something of an abstraction which has precise meaning only when referred to a specific parameter. An estimate of the surface value of any physical quantity depends on arbitrary assumptions as to the pertinent gradient in the microlayer. The best that can be done at present is to ensure that the assumptions adopted make physical sense.

[GIFFORD C. EWING]

Exchange of gases. A sample of sea water taken from any location and depth in the oceans is found generally to contain in solution all the gaseous constituents of the atmosphere. The concentrations of the dissolved gases depend upon gas exchange processes at the air-sea interface and upon chemical, biological, and physical processes within the body of the oceans. Perhaps the most important fact about the dissolved gases is that their concentrations in near-surface waters are found to differ from saturated values by only a few percent. This makes possible the use of deviations from saturation as clues to the processes indicated above.

In solution, nitrogen, oxygen, and the noble gases do not react chemically with the water. Carbon dioxide, however, tends toward dissociation equilibrium with H_2CO_3, HCO_3^-, and CO_3^{--}. If air and sea water are equilibrated under normal conditions in the laboratory, the concentration of each dissolved gas is proportional to its partial pressure in the air above and independent of the presence of the other gases. The solubility coefficient of each gas (that is, the concentration of that dissolved gas when its partial pressure above the solution is 1 atm or 101.325 kPa) increases rapidly as the temperature is lowered and decreases with rising salinity. For the gases which do not react chemically with water, both the solubility coefficients and their variations with temperature generally increase with molecular weight.

Exchange of gases across the air-sea boundary may occur by molecular diffusion through a thin surface layer or by the motion of aggregates of molecules in the form of bubbles or gases dissolved in water droplets. The nearly saturated state of surface waters implies that molecular diffusion is the primary controlling process. The direction of net diffusion always is toward producing equilibrium. For example, upwelling warm water usually is undersaturated relative to the cold surface, and as the water cools net gas flow is into the water. Net transfer by molecular aggregates may be in either direction.

Surface water may sink and travel great distances beneath the surface. At all depths dissolved oxygen is consumed and carbon dioxide is produced by respiration and organic decomposition. The reverse of these effects is produced by photosynthesis when sufficient light is present. Consequently, these gases vary in concentration more than the others. A widespread characteristic of oxygen is the occurrence of a zone of minimum concentration at intermediate depths. This is accompanied by a maximum in the O^{18}/O^{16} ratio, which implies that O^{16} is consumed relatively more rapidly.

Nitrogen gas is produced in anoxic waters such as the Cariaco Trench, but no clear evidence has been obtained for significant biological influence on dissolved nitrogen in normal ocean water.

The concentration of carbon dioxide in the atmosphere probably plays a significant role in the heat budget of the Earth. Consequently, the partition between air and sea of the CO_2 produced by the combustion of fossil fuels has special importance. Although measurements have shown that a large proportion of the gas is absorbed by the oceans, precise calculations are difficult because of the complex solution chemistry of CO_2, mixing within the oceans, and influence of the biosphere.

Because helium is light and escapes from the Earth's gravitational field, its concentration in the atmosphere is very small (5 ppm). The amount in the sea is correspondingly small. Changes in dissolved He^4 concentrations from radioactive decay within the oceans are not measurable, but efflux through the sea floor from decay within the Earth should be detectable with sensitive techniques. There is controversial evidence for excess dissolved helium in deep waters. The observed world distribution of helium is the result of a steady state between production in the lithosphere and hydrosphere and escape from the atmosphere.

Similarly, radioactive decay of ^{40}K produces ^{40}Ar which does not escape but accumulates in the atmosphere. The relatively large abundance of argon in normal sea water precludes direct observation of concentration changes from radioactive decay.

The multiplicity of physical processes, such as atmospheric pressure changes, bubble trapping, upwelling, and internal mixing, complicates the problem of unraveling the history of a sample of water, but the differing solubility characteristics of the gases promise to be useful. For example, because the solubility coefficient of argon is approximately twice that of nitrogen, deviations from surface equilibrium which result from bubble entrainment will be accompanied by enhanced N_2/Ar ratios. Conversely, the physical processes influencing oxygen and argon are almost identical, and the O_2/Ar ratio depends primarily upon the biological history of the sample. For the same reason, the He/Ne ratio offers the most sensitive approach to the helium efflux problem. *See* ATMOSPHERIC CHEMISTRY. [BRUCE B. BENSON]

Bibliography: A. E. Bainbridge, Tritium in the North Pacific surface water, *J. Geophys. Res.*, 68: 3785, 1963; B. B. Benson, *Some Thoughts on Gases Dissolved in the Oceans*, Univ. Rhode Island Occas. Publ. no. 3, 1965; D. C. Blanchard, The electrification of the atmosphere by particles from bubbles in the sea, *Progr. Oceanogr.*, 1:71–202, 1963; H. Craig and L. I. Gordon, Deuterium and oxygen-18 variations in the ocean and marine atmosphere, in E. Tongiorgi (ed.), *Stable Isotopes in Oceanographic Studies and Paleotemperatures*, 1965; H. Craig, L. I. Gordon, and Y. Horibe, Isotopic exchange effects in the evaporation of water, *J. Geophys. Res.*, 68:5079, 1963; H. Craig, R. F. Weiss, and W. B. Clarke, Dissolved gases in the Equatorial and South Pacific Ocean, *J. Geophys. Res.*, 72:6165–6181, 1967; W. Dansgaard, Stable isotopes in precipitation, *Tellus*, 16:436, 1964; J. I. Drever (ed.), *Sea Water: Cycles of the Major Elements*, 1977; G. Ewing and E. D. McAlister, On the thermal boundary layer of the ocean, *Science*, 131(3410):1374–1376, 1960; I. Friedman et al., The variation of the deuterium content of natural waters in the hydrologic cycle, *Rev. Geophys.*, 2:177, 1964; C. E. Junge, *Air Chemistry and Radioactivity*, 1963; R. Revelle and H. E. Suess,

Gases, in M.N. Hill (ed.), *The Sea*, 1962; A. H. Woodcock, Salt nuclei in marine air as a function of altitude and wind force, *J. Meteorol.*, 10:362–371, 1953.

TRANSMISSION OF ENERGY

Electromagnetic and acoustic energy from various natural sources permeates the sea, supplying it with heat, supporting its ecology, and providing for sensory perception by its inhabitants; artificial sources afford man the means for underwater communication and detection.

Light. The primary source of energy which heats the ocean and supports its ecology is light from the Sun. On a clear day as much as 1 kW of radiant power from the Sun and sky may impinge on each square meter of sea surface. Of this power, 4–8% is reflected and the remainder is absorbed within the water as heat or as chemical potential energy due to photosynthesis. The peak of the irradiation is close to the wavelength of greatest transparency for clear sea water, 480 nm, but nearly half of the radiant power is infrared radiation which water absorbs so strongly that virtually none penetrates more than 1 m beneath the surface. As much as one-fifth of the incident power may be ultraviolet (below 400 nm), and this radiation may penetrate a few tens of meters if little or no "yellow substance" (humic acids and other materials associated with organic decomposition) is present. Only a narrow spectral band of blue-green light, representing less than 10% of the total irradiation, penetrates deeply into the sea. This radiation has been detected by multiplier-phototube photometers at depths of more than 600 m. Visibility, important to predators in the feeding grounds of the sea, is possible chiefly because of this blue-green light.

Irradiance. Irradiance on a flat surface oriented in any manner decreases exponentially with depth, as illustrated by Fig. 3, which depicts experimental values of irradiance on an upward-facing surface. Irradiance on any other surface could be represented by a curve parallel (within 5%) to the one shown; irradiance on downward-facing surfaces is approximately one-fiftieth of the irradiance on upward-facing surfaces at all depths.

Absorption. Light, to be useful for heating or for photosynthesis, must be absorbed. The quantity of radiant power absorbed per unit of volume depends upon the amount of power present and the magnitude of the absorption coefficient; to a useful (5%) approximation power absorbed per unit of volume at any depth can be calculated by multiplying the irradiance at that depth, as in Fig. 3, by the slope of the curve expressed in natural log-units per unit of depth, that is, the attenuation coefficient K. Thus in Fig. 3, at a depth of 64 m where the irradiance is 0.5 watt/m^2 and is decreasing with depth at the rate of 0.08 natural log-units/m, approximately $0.5 \times 0.08 = 0.04$ watt of radiant power is absorbed by every 1 m^3 of sea water.

Visibility. Visibility under water is accomplished by image-forming light (rays) which must pass from the object to the observer without being scattered. The transmission of water for image-forming light is less than for diffused light, since scattering in any direction constitutes a loss of image-forming light, whereas only scattering in rearward directions is a loss for diffused light. Image-forming light is exponentially attenuated with distance, but the attenuation coefficient α averages 2.7 times greater than the attenuation coefficient for irradiance K, defined above. Apparent contrast of an underwater object having deep water as a background is exponentially attenuated with distance, the effective attenuation coefficient being $\alpha + K \cos \theta$, where θ is the inclination angle of the path of sight, and $\cos \theta = 1$ when the observer looks straight down. See discussion of water color and transparency in section on sampling and measuring techniques. [SEIBERT Q. DUNTLEY]

Compensation intensity and depth. As daylight penetrating into the sea diminishes, the photosynthesis of plants is reduced but respiration remains approximately the same. The light value at which the rates of photosynthesis and respiration are equal is the compensation intensity. The depth at which the compensation intensity is found is the compensation depth. Both of the foregoing have also been termed compensation point, but since ambiguity may occur, it is best to avoid this term. The compensation intensity varies according to the species, the physiological condition of the plants, and other factors, particularly temperatures. Lowered temperature depresses respiration more than photosynthesis. The compensation depth depends upon the intensity of the incident radiation, the transparency of the water, and the period considered, since illumination varies with time. Compensation intensities of 10–200 footcandles (ftc) [110–2150 lux] have been measured for phytoplankton and of 17–45 ftc [180–480 lux] for filamentous algae. Compensation depths for 24-hr periods for phytoplankton range from less than 1 m in turbid water to more than 30 m in coastal areas and to 80 m or more in the clearest tropical waters, and for attached plants, to 50 m along the coast and to 160 m in especially clear water, as in the Mediterranean. Generally the

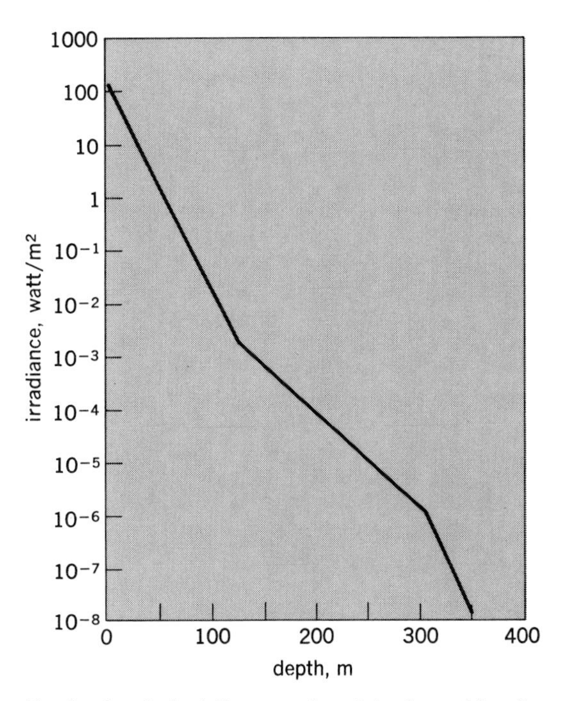

Fig. 3. Graph depicting experimental values of irradiance on an upward-facing surface.

compensation depth is found where daylight is reduced to about 1% of its value at the surface for phytoplankton or about 0.3% for bottom plants. The compensation depth is of particular significance since it marks the lower limit of the photic zone within which green plants can carry on primary production necessary as an energy source for the whole marine ecosystem.

[GEORGE L. CLARKE]

Electromagnetic fields. In sea water, as in any conductor, the electromagnetic behavior is determined by the magnetic permeability μ and the electrical conductivity σ. From these one may find the skin depth δ and the characteristic impedance η given by Eqs. (2) and (3).

$$\delta = (\pi f \mu \sigma)^{-1/2} \quad (2)$$

$$\eta = (2\pi f \mu / \sigma)^{1/2} \quad (3)$$

Both δ and η relate to electromagnetic waves of frequency f, for which the wavelength is $2\pi\delta$, the absorption over a path length x reduces the amplitude in the ratio $e^{-x/\delta}$, and the ratio of electric to magnetic field amplitude in a plane wave is η, the former leading in phase by 45°.

For sea water, magnetic permeability μ is nearly the same as for free space, and electrical conductivity σ is given to about 1% by Eq. (4), where t is

$$\sigma = [4.00 + a(t-12)][1 + .0269(S-35)] \quad (4)$$

temperature in °C, S is salinity in parts per thousand by weight, and a is .10 for $t > 12$ or $a = .092$ for $t < 12$ (Fig. 4). Using $\sigma = 4.0$ mho/m as a typical value, one obtains Eqs. (5) and (6). These formulas

$$\delta = 250 f^{-1/2} \text{ meter} \quad (5)$$

$$\eta = .0014 f^{1/2} \text{ ohm} \quad (6)$$

are expected to hold for all frequencies below about 900 MHz.

Absorption limits the penetration of a field, either inward from a boundary or outward from an electric or magnetic source, to a small multiple of the skin depth δ. A submerged horizontal dipole source near the surface will, however, have a more extensive field in the air and a shallow layer of water. At .01/Hz, δ is 2.5 km; this is a rough upper frequency limit for field fluctuations (such as those of the geomagnetic field) which can penetrate the entire thickness of the ocean layer. At 10 kHz, δ is 2.5 m. Fields of this and higher frequencies, existing, for example, in the natural atmospheric noise, can penetrate only a thin surface layer. For radio signals, the sea acts as an excellent ground plane, involving lower losses than transmission over land.

[PHILIP RUDNICK]

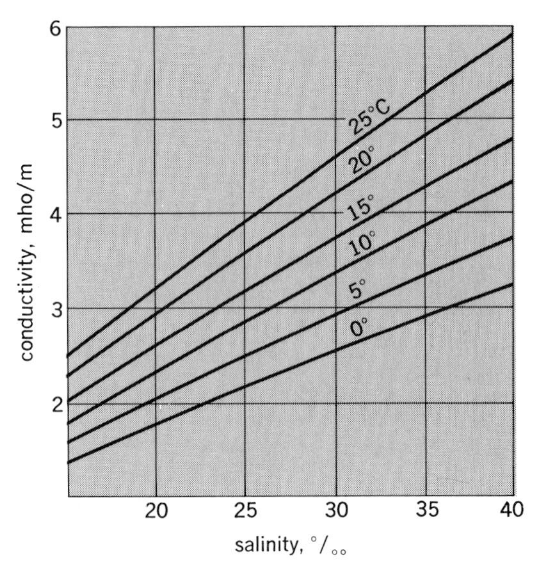

Fig. 4. Electrical conductivity of sea water. (*From B. D. Thomas, T. G. Thompson, and C. L. Utterback, The electrical conductivity of sea water, J. Cons. Perma. Int. Explor. Mer. 9:28–35, 1934*)

Sound. Sound in sea water travels four or five times its speed in air, or about 1500 m/sec. As sound propagates in the sea, its intensity diminishes inversely as the square of the distance from the source, in the absence of appreciable absorption, refraction, reflection, and scattering. Although losses by absorption are small compared with those that occur when sound of the same frequency travels through air, the increase in absorption with higher frequencies limits the effective range of ultrasonic waves, that is, waves having frequencies above those audible to the human ear. This is an important factor in submarine detection, where ultrasonic frequencies are used because of their desirable directional properties.

The velocity of sound in sea water varies with temperature, salinity, and pressure. Hence a beam of ultrasonic waves, when transmitted in a horizontal direction, may be refracted and then reflected one or more times from the surface, ocean bottom, or some layer within the vertical water structure. In this manner several different rays may be received at different intervals from a single source. The direct transmission is limited to specific distances, depending on depth of the bottom, and its theoretical velocity may be computed if the temperature and salinity are known. At greater distances the apparent horizontal velocity is less than the theoretical velocity because of such factors as distance between source and receiver, depth, bottom profile and character, and physical properties of the water.

The vertical velocity is a function of depth (pressure) and the distribution of temperature and salinity. Except in polar latitudes it generally decreases from the surface to some moderate depth (from 500 to 1500 m) because of decreasing temperature. Below this depth the velocity gradually increases again as the effect of increasing pressure becomes dominant. Because most sonic depth-finding instruments are calibrated for a constant velocity, when a very accurate depth reading is needed, it is necessary to correct readings to true depths.

Investigations of sound in the ocean offer many promising areas for further study, particularly studies of underwater noises produced by marine life, the relation of these noises to other sounds in the sea, the seasonal rhythm and geographic variation of the noise makers, and the ecological significance of sound-producing organisms. A variety of problems in underwater acoustics may result from the presence of sound-producing organisms.

[NEIL A. BENFER]

Bibliography: V. M. Albers, *Underwater Acoustics*, 1965; J. W. Caruthers, *Fundamentals of Marine Acoustics*, 1977; C. S. Clay and H.

Medwin, *Acoustical Oceanography: Principles and Applications*, 1977; R. W. Holmes, Solar radiation, submarine daylight, and photosynthesis, in *Treatise on Marine Ecology and Paleoecology*, Geol. Soc. Amer. Mem. no. 67, vol. 1, 1957; R. H. Lien, Radiation from a horizontal dipole in a semi-infinite dissipative medium, *J. Appl. Phys.* 24(1): 1–4, 1953; J. E. Tyler and R. C. Smith, *Measurements of Spectral Irradiation Underwater*, 1970; J. R. Wait, The radiation fields of a horizontal dipole in a semi-infinite dissipative medium, *J. Appl. Phys.*, 24(7):958–959, 1953.

COMPOSITION OF SEA WATER

The concentrations of the various components of sea water are regulated by numerous chemical, physical, and biochemical reactions.

Inorganic regulation of composition. The present-day compositions of sea waters (Table 5) are controlled both by the makeup of the ultimate source materials and by the large number of reactions, of chemical and physical natures, occurring in the oceans. This section considers the nonbiological regulatory mechanisms, most conveniently defined as those reactions occurring in a sterile ocean. For a discussion of the weathered and weathering substances that give rise to the waters of the world *see* HYDROSPHERIC GEOCHEMISTRY.

Interactions between the ions results in the formation of ion pairs, charged and uncharged species, which influence both the chemical and physical properties of sea water. For example, the combination of magnesium and sulfate to form the uncharged ion pair accounts for the marked absorption of sound in sea water. A model accounting for such interactions has been developed for the principal dissolved ions in sea water (Table 6).

pH and oxidation potential. The pH of surface sea waters varies between 7.8 and 8.3, with lower values occurring at depths. The pH normally goes through a minimum with increasing depth in the ocean, and this depth dependence shows a marked resemblance to the profiles of oxygen.

The oxidation potentials of sea-water systems are determined by oxygen concentration in aerobic waters and by hydrogen sulfide concentration in anoxic waters. For 25°C, 1 atm (101.325 kPa) pressure, and a pH of 8, the oxidation potential is about 0.75 volt for waters containing dissolved oxygen in a state of saturation. In anoxic waters the oxidation potentials can vary from about −0.2 to −0.3 volt.

Solubility. Only calcium, among the major cations of sea water, is present in a state of saturation, and such a situation generally occurs only in surface waters. Here its concentration is governed by the solubility of calcium carbonate. Barium concentrations in deep waters can be limited by the precipitation of barium sulfate. The noble gases and dissolved gaseous nitrogen have their marine concentrations determined by the temperature at which their water mass was in contact with the atmosphere and are in states of saturation or very nearly so.

Authigenic mineral formation. The formation and alteration of minerals on the sea floor apparently are responsible for controlling the concentrations of the major cations Na, K, Mg, and Ca. Such clay minerals as illite, chlorite, and montmorillonite are presumably synthesized from these dissolved species and the river-transported weathered solids (aluminosilicates, such as kaolinite) by reactions of the type shown in Eq. (7). The SiO_2

$$\text{Aluminosilicates} + SiO_2 + HCO_3^- + \text{Cations} = \text{Cation aluminosilicate} + CO_2 + H_2O \quad (7)$$

is introduced in part as diatom frustules. Such a process implies control of the carbon dioxide pressure of the atmosphere.

The formation of ferromanganese minerals, sea-floor precipitates of iron and manganese oxides, may govern the concentrations of a suite of trace

Table 5. Chemical abundances in the marine hydrosphere

Element	Concentration, mg/liter	Element	Concentration, mg/liter
H	108,000	Ag	0.0003
He	0.000007	Cd	0.00011
Li	0.17	In	0.000004
Be	0.0000006	Sn	0.0008
B	4.6	Sb	0.0003
C	28	Te	–
N	15	I	0.06
O	857,000	Xe	0.00005
F	1.2	Cs	0.0003
Ne	0.0001	Ba	0.03
Na	10,500	La	1.2×10^{-5}
Mg	1350	Ce	5.2×10^{-6}
Al	0.01	Pr	2.6×10^{-6}
Si	3.0	Nd	9.2×10^{-6}
P	0.07	Pm	–
S	885	Sm	1.7×10^{-6}
Cl	19,000	Eu	4.6×10^{-7}
A	0.45	Gd	2.4×10^{-6}
K	380	Tb	–
Ca	400	Dy	2.9×10^{-6}
Sc	<0.00004	Ho	8.8×10^{-7}
Ti	0.001	Er	2.4×10^{-6}
V	0.002	Tm	5.2×10^{-7}
Cr	0.00005	Yb	2.0×10^{-6}
Mn	0.002	Lu	4.8×10^{-7}
Fe	0.01	Hf	<0.000008
Co	0.0004	Ta	<0.000003
Ni	0.007	W	0.0001
Cu	0.003	Re	0.0000084
Zn	0.01	Os	–
Ga	0.00003	Ir	–
Ge	0.00006	Pt	–
As	0.003	Au	0.00001
Se	0.00009	Hg	0.0002
Br	65	Tl	<0.00001
Kr	0.0002	Pb	0.00003
Rb	0.12	Bi	0.00002
Sr	8.0	Po	–
Y	0.00001	At	–
Zr	0.00002	Rn	0.6×10^{-15}
Nb	0.00001	Fr	–
Mo	0.01	Ra	1.0×10^{-10}
Tc	–	Ac	–
Ru	0.0000007	Th	0.000001
Rh	–	Pa	2.0×10^{-9}
Pd	–	U	0.003

Table 6. Distribution of major cations as ion pairs with sulfate, carbonate, and bicarbonate ions in sea water of chlorinity 19°/₀₀ and pH 8.1*

Ion	Free ion, %	Sulfate ion pair, %	Bicarbonate ion pair, %	Carbonate ion pair, %
Ca^{++}	91	8	1	0.2
Mg^{++}	87	11	1	0.3
Na^{+}	99	1.2	0.01	–
K^{+}	99	1	–	–

Ion	Free ion, %	Ca ion pair, %	Mg ion pair, %	Na ion pair, %	K ion pair, %
SO_4^{--}	54	3	21.5	21	0.5
HCO_3^{-}	69	4	19	8	–
CO_3^{--}	9	7	67	17	–

*From R. M. Garrels and M. E. Thompson, A chemical model for sea water at 25°C and one atmosphere total pressure, *Amer. J. Sci.*, 260:57–66, 1962.

metals, including zinc, manganese, copper, nickel, and cobalt. These elements are in highly undersaturated states in sea water (Table 7) but are highly enriched in these marine ores, the so-called manganese nodules, which range in size from millimeters to about 1 m in the form of coatings and as components of the unconsolidated sediments.

Cation and anion exchange. Cation-exchange reactions between positively charged species in sea water and such minerals as the marine clays and zeolites appear to regulate, at least in part, the amounts of sodium, potassium, and magnesium, as well as other members of the alkali and alkaline-earth metals which are not major participants in mineral formations. High charge and large radius influence favorably the uptake on cation-exchange minerals. It appears, for example, that while 65% of the sodium weathered from the continental rocks resides in the oceans, only 2.5, 0.15, and 0.025% of the total amounts of potassium, rubidium, and cesium ions (ions increasingly larger than sodium) have remained there. Further, magnesium and potassium are depleted in the ocean relative to sodium on the basis of data obtained from igneous rock.

The curious fact that magnesium remains in solution to a much higher degree than potassium is not yet resolved but may be explained by the ability of such ubiquitous clay minerals as the illites to fix potassium into nonexchangeable or difficultly exchangeable sites.

Similarly, anion-exchange processes may regulate the composition of some of the negatively charged ions in the oceans. For example, the chlorine-bromine ratio in sea water of 300 is displaced to values around 50 in sediments. Such a result may well arise from the replacement of chlorine by bromine in clays; however, the meager amounts of work in this field preclude any unqualified statements.

Physical processes. Superimposed upon these chemical processes are changes in the chemical makeup of sea water by the melting of ice, evaporation, mixing with runoff waters from the continents, and upwelling of deeper waters. The net effects of the first three processes are changes in the absolute concentrations of all of the elements but with no major changes in the relative amounts of the dissolved species.

Changes with time. Changes in the composition of sea water through geologic time reflect not only differences in the extent and types of weathering processes on the Earth's surface but also the relative intensities of the biological and inorganic reactions. The most influential parameter controlling the inorganic processes appears to be the sea-water temperature.

Changes in the abyssal temperatures of the oceans from their present values of near 0 to 2.2, 7.0, and 10.4°C in the upper, middle, and lower Tertiary, respectively, have been postulated from studies on the oxygen isotopic composition of the tests of benthic foraminifera. Such temperature increases would of necessity be accompanied by similar ones in the surface and intermediate waters. One obvious effect from the recent cooling of the oceans is an increase in either or both of the calcium and carbonate ions, since the solubility product of calcium carbonate has a negative temperature coefficient. Similarly, the saturated amounts of gases that can dissolve in sea water in equilibrium with the atmosphere increase with decreasing water temperatures.

[EDWARD D. GOLDBERG]

Table 7. Observed concentrations of some trace metals in sea water*

Ion	Observed sea-water concentration, moles/liter	Limiting compound	Calculated limiting concentration, moles/liter
Mn^{++}	4×10^{-8}	$MnCO_3$	10^{-3}
Ni^{++}	4×10^{-8}	$Ni(OH)_2$	10^{-3}
Co^{++}	7×10^{-9}	$CoCO_3$	3×10^{-7}
Zn^{++}	2×10^{-7}	$ZnCO_3$	2×10^{-4}
Cu^{++}	5×10^{-8}	$Cu(OH)_2$	10^{-6}

*Calculated on the basis of their most insoluble compound.

Biological regulation of composition. In the open sea all the organic matter is produced by the photosynthesis and growth of unicellular planktonic forms. During this growth all the elements essential for living matter are obtained from the sea water. Some elements are present in great excess, such as the carbon of CO_2, the potassium, and the sulfur (as sulfate). Other elements—for example, phosphorus, nitrogen, and silicon—are present in small enough quantities so that plant growth removes virtually all of the supply from the water. During photosynthesis, as these elements are being removed from the water, oxygen is released.

The biochemical cycle. The organic matter formed by photosynthesis and growth of the unicellular plants may be largely eaten by the zooplankton, and these in turn form the food for larger organisms. At each step of the food chain a large proportion of the eaten material is digested and excreted, and this, along with dead organisms, is decomposed by bacterial action. The decomposition process removes oxygen from the water and returns to the water those elements previously absorbed by the phytoplankton.

The distribution of oxygen and essential nutrient elements in the sea is modified by the spatial separation of these biological processes. Photosynthesis is limited to the surface layers of the ocean, generally no more than 100 m or so of depth, but the decomposition of organic material may take place at any depth. Reflecting this separation of processes, the concentration of nutrient elements in the surface is low, rises to maximum values at intermediate depths (300–800 m), and decreases slightly to fairly constant values which extend nearly to the bottom. Frequently, a slight increase in the concentration of essential elements is observed near the bottom. The oxygen distribution is the opposite of the one just described, with high values at the surface, a minimum value at mid-depth, and intermediate values in the deep water. The oxygen-minimum–nutrient-maximum level in the ocean is the result of two processes working simultaneously. In part it is formed by the decomposition of organic matter sinking from the surface, and in part it results from the fact that this water was originally at the surface in high latitudes, where it contained organic matter and subsequently cooled, sank, and spread out over the oceans at the appropriate density levels.

Because of the nearly constant composition of marine organisms, the elements required in the formation of organic material vary in a correlated way. Analyses of marine organisms indicate that in their protoplasm the elements carbon, nitrogen, and phosphorus are present in the ratios of 100:15:1 by atoms. In the production of organic matter these elements are removed from the water in these ratios, and during the decomposition of organic matter they are returned to the water in the same ratios. However, since the decomposition of organic material is not an instantaneous process which releases all elements simultaneously, it is not unusual to find different ratios of concentration of these elements in the sea. Particularly in coastal waters and in confined seas the ratio of concentration of nitrogen to phosphorus, for example, may differ widely from the 15:1 ratio of composition within the organisms.

The biochemical circulation. Unlike the major elements in sea water, the concentrations of these nutrients are widely different in different oceans of the world. Pacific Ocean water contains nearly twice the concentration of nitrogen and phosphorus found at the same depth in the North Atlantic, and intermediate concentrations are found in the South Atlantic and Antarctic oceans. The lowest concentrations for any extensive body of water are found in the Mediterranean, where they are only about one-third of those in the North Atlantic.

These variations can be attributed to the ways in which the water circulates in these oceans and to the effect of the biological processes on the distribution of elements. The Mediterranean, for example, receives surface water, already low in nutrients, from the North Atlantic and loses water from a greater depth through the Straits of Gibraltar. While the water is in the Mediterranean, the surface layers are further impoverished by growth of phytoplankton, and the organic material formed sinks to the bottom water and is lost in the deeper outflow. A similar process explains the low nutrient concentrations in the North Atlantic, which receives surface water from the South Atlantic and loses an equivalent volume of water from greater depths. *See* SEA-WATER FERTILITY.

[BOSTWICK H. KETCHUM]

Buffer mechanism. The constituents of sea water include a number of cations, all of which are weak acids, and a smaller number of anions, some of which are strong bases. Thus sea water is always somewhat on the alkaline side of neutrality, ranging in pH roughly between the limits of 7.5 and 8.3.

In chemical oceanography the term alkalinity is used to denote not the concentration of hydroxyl ions, as might be expected, but the concentration of strong bases. Alkalinity can be defined as the number of equivalents of strong acid required to convert stoichiometrically all the strong bases to weak acids.

The addition or subtraction of weak acids therefore does not affect the alkalinity of sea water, although through the operation of the buffer mechanism it changes the pH.

The principal weak acid in sea water is carbonic, resulting from the hydrolysis of dissolved carbon dioxide. Boric acid is also present in significant amounts. Salts of these two weak acids are the strong bases which make up the alkalinity. Total combined boron, whether as acid or borate, is in virtually constant proportion to chlorinity, the ratio of boron to chlorinity being about 0.00024, which is equivalent to a specific ratio of boric acid to chlorinity of about 0.022 (concentrations in millimoles per liter).

The total alkalinity is more variable in the ocean. Near the surface it can be increased by addition of dissolved carbonates in river discharge or decreased by the precipitation of lime in the formation of coral and shells. In deeper layers it can be increased by the solution of calcareous debris sinking from the surface. F. Koczy (1956) gives a thorough discussion of the variation of specific alkalinity in the oceans. The average ratio of alkalinity to chlorinity is about 0.120 for surface sea water, increasing somewhat with depth (concentrations in milliequivalents per liter).

Even more variable than the alkalinity is the total dissolved CO_2. At the surface, CO_2 moves between the sea and the atmosphere, and in the euphotic zone CO_2 is removed by photosynthesis to be incorporated into organic matter. Below the euphotic zone, CO_2 is regenerated by biological oxidation of organic matter. Total CO_2 typically may vary from 2.0 or 2.1 mM/liter at the surface to 2.8 or more at the depth of the oxygen minimum.

Some of this CO_2 is in physical solution, and some is undissociated carbonic acid; but most of it is in ionic form, mainly bicarbonate ion with some carbonate. The proportions of the various forms are governed by the dissociation constants of carbonic and boric acids, which vary with temperature and pressure, and by the activity coefficients of the various ions concerned, which vary with temperature, pressure, and salinity. In practice, the activity coefficients and the dissociation constants are combined into apparent dissociation constants, which are tabulated by H. W. Harvey (1957) as functions of temperature and salinity at 1 atm pressure.

[JOHN LYMAN]

Bibliography: R. M. Garrels and M. E. Thompson, A chemical model for sea water at 25°C and one atmosphere total pressure, *Amer. J. Sci.*, 260: 47–66, 1962; H. W. Harvey, *Chemistry and Fertility of Sea Waters*, 1957; M. N. Hill (ed.), *The Sea*, vol. 2, 1963; F. Koczy, The specific alkalinity, *Deep-Sea Res.*, 3:279–288, 1956; J. Lyman and R. B. Abel, Chemical aspects of physical oceanography, *J. Chem. Educ.*, 35:113–115, 1958; F. T. MacKenzie and R. M. Garrels, Chemical mass balance between rivers and oceans, *Amer. J. Sci.*, 264:507–525, 1966; D. F. Martin, *Marine Chemistry*, vol. 1: *Analytical Methods*, 1972; A. C. Redfield, The biological control of chemical factors in the environment, *Amer. Sci.*, 46(3):205–221, 1958.

DISTRIBUTION OF PROPERTIES

The distribution of physical and chemical properties in the ocean is principally the result of the following: (1) radiation (of heat); (2) exchange with the land (of heat, water, and solids such as salts) and with the atmosphere (of water, salt, heat, and dissolved gases); (3) organic processes (photosynthesis, respiration, and decay); and (4) mixing and stirring processes. These processes are largely responsible for the formation of particular water types and ocean water masses.

Horizontal distributions. The general distribution of properties in the oceans shows a marked latitudinal effect which corresponds with radiation income and differences between evaporation and precipitation.

Temperature. Heat is received from the Sun at the sea surface, where parts of it are reflected and radiated back. Equatorward of 30° latitude the incoming radiation exceeds back radiation and reflection, and poleward it is less. The result is high sea-surface temperature (more than 28°C) in equatorial regions and low sea-surface temperatures (less than 1°) in polar regions.

Salinity. Various dissolved solids have entered the sea from the land and have been so mixed that their relative amounts are everywhere nearly constant, yet the total concentration (salinity) varies considerably. In middle latitudes the evaporation of water exceeds precipitation, and the surface salinity is high; in low and high latitudes precipitation exceeds evaporation, and dilution reduces the surface salinity.

Open ocean surface salinities range from lows of about 32.5‰ in the North Pacific, 34.0 in the Antarctic, 35.0 in the equatorial Atlantic, 34.0 in the equatorial Indian, and 33.5 in the equatorial Pacific, to highs in the great evaporation centers of the middle latitudes of 35.5 and 36.5 in the North and South Pacific, 37.0 in the North and South Atlantic, and 36.0 in the Indian Ocean.

Dissolved oxygen. Dissolved oxygen is both consumed (respiration and decay) and produced (photosynthesis) in the ocean, as well as being exchanged with the atmosphere at the sea surface. Above the thermocline the waters are nearly always near saturation in oxygen content (>7 ml/liter in the cold waters of high latitudes and <5 ml/liter in warm equatorial waters).

Density. The density of sea water depends upon its temperature, salinity, and depth (pressure) but can vary horizontally only in the presence of currents, and hence its distribution depends closely upon the current structure. In low and middle latitudes the effect of the high temperature exceeds that of the high salinity, and the surface waters are lighter than those in high latitudes, with values ranging from less than 1.022 g/ml to more than 1.027. The density at great depth (10,000 m) may exceed 1.065. The greatest vertical gradient is associated with the thermocline and the halocline and is therefore very near to the surface. The heavier surface waters from high latitudes move and mix underneath the lighter water at depths which depend upon their density. The difference in surface density is usually not great, since the high-latitude salinity is low, and the waters usually sink only a few hundred meters, forming intermediate water. But in cases where water of high salinity has been carried into high latitudes by the currents and cooled (as occurs in the North Atlantic) or where water of relatively high salinity freezes and gives up part of its water (as occurs on the continental shelf of Antarctica), deep and bottom waters with temperature less than 1°C are formed. *See* THERMOCLINE.

Vertical distributions. The subsurface distribution of properties is controlled largely by external factors, particularly those which influence surface density, and the type of deep-sea circulation.

Temperature. The thermohaline circulation results in a vertical distribution of temperature such that in low and middle latitudes the deeper waters are colder than the surface waters, and at very high latitudes where surface temperatures are low, the deeper waters are as warm as those at the surface, or warmer. Seasonal cooling in high latitudes may cause the temperature to be at a minimum at some intermediate depth, and the circulation may cause a temperature minimum or maximum at intermediate depths. Over a large part of the Pacific Ocean, where no bottom water is formed, the temperature increases downward from about 4000 m. Since the gradient is not greater than the adiabatic, the water is not unstable.

Density and salinity. Since much of the flow of the ocean is geostrophically balanced, surfaces of constant density slope in various ways with respect to the sea surface and other surfaces of constant pressure, and density varies in the east-west as well as the north-south direction. Mixing and movement of intermediate and deep water along these surfaces cause more complex distributions of other variables. The intermediate waters of the North and South Pacific and of the South Atlantic appear to have salinity minima at intermediate depths in middle and low latitudes, since they originate in the high-latitude regions of low salinity and pass between the high-salinity surface waters of middle latitudes and the bottom waters. In the North Pacific this minimum varies from 33.4 to 34.1‰ and in the South Pacific and South Atlantic, from 34.2 to about 34.6‰. In the North Atlantic Ocean there is no intermediate water of low salinity formed, but very saline water (36.5‰) flows in from the Mediterranean at depth and results in a more complicated distribution of salinity than is found in the other oceans. For the temperatures and salinities of the bottom waters, which are more homogeneous than the others, see the later discussion on ocean water masses.

Dissolved oxygen. Where the cold surface waters sink in high latitudes, quantities of oxygen are carried downward. Below the compensation depth consumption gradually reduces the concentration. The bottom waters of the Atlantic Ocean contain from 5 to 6 ml per liter of dissolved oxygen; the Indian and South Pacific Ocean, about 4; and the North Pacific, less than 4. Between these bottom waters and those waters at the surface, which are saturated, smaller values of oxygen are found, ranging from less than 0.10 in the eastern tropical Pacific and less than 1.0 in the eastern tropical Atlantic to greater than 4 ml per liter in the South Atlantic.

Nutrients. Other properties such as the nutrients, phosphate and nitrate, have low concentrations in the surface layer, where they are consumed by the plants, and high values at depths, where they are concentrated by the sinking and decay of organisms.

The maximum values of phosphate are found at intermediate depths, usually beneath the layer of minimum oxygen, and vary from less than 1 μg-atom/liter in the North Atlantic to more than 2 in

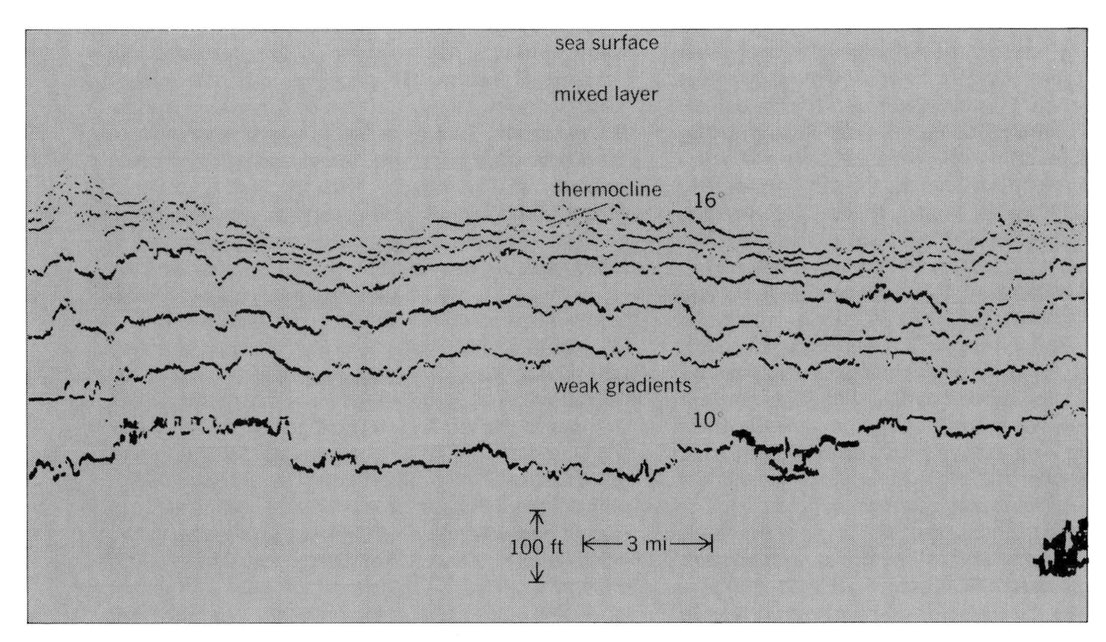

Fig. 5. Thermal structure most commonly found in the sea. 100 ft = 30.48 m; 3 mi = 4.83 km.

the Antarctic, Indian, and South Pacific, and more than 3.5 in the North Pacific. Surface values vary from less than 0.1 in the North Atlantic and 0.5 in the North Pacific to more than 1.5 in the Antarctic. Bottom values vary from about 1.0 in the North Atlantic and 2.0 in the Antarctic, South Pacific, and Indian oceans to 2.5 in the North Pacific.

Nitrate-nitrogen is present in a ratio of about 8:1 by weight to phosphate-phosphorus.

Silicate has no intermediate maximum but increases monotonically toward the bottom, where the values vary from more than 150 μg-atoms/liter in the North Pacific to less than 40 in the North Atlantic. Surface values are generally less than 10.

[JOSEPH L. REID]

Detailed thermal structure. A detailed knowledge of the sea's thermal structure and its relation to oceanic dynamic processes has been accumulated. The details of two-dimensional structure measurement of the sea have been provided by the temperature structure profiler. Although thermal changes occur during the towing of the profiler, the spatial plot is realistic since heat transfer at the surface is negligible, and advective changes proceed more slowly than the 6-knot (3 m/s) movement of the ship.

In Fig. 5 the vertical scale represents depth from the surface to 800 ft (240 m). The horizontal scale may be interpreted as either time or distance, because the towing ship has a constant speed of 6 knots. Since the vertical scale is about 100 times the horizontal, the isotherm lines appear to be steeper than they are. Each isotherm is a whole-degree Celsius from 16 to 10°. This example reflects the common, mixed layer, without isotherms, found from the surface to a depth, in this case, of 260 ft (79 m). Below these are the closely spaced isotherms that compose the sharp thermocline.

Isotherm oscillations. The profiler made possible a determination of detailed horizontal changes in isotherms. An important large-scale feature revealed is the vertical, wavelike undulation in the main thermocline, with wave heights of 50–100 ft (15–30 m) and crests about every 12 mi (19 km). Smaller oscillations are conspicuous on all isotherms. These are the ever-present internal waves, which occur on the average thermocline at the rate of 1–4 per mile (0.6–2.5 per km). Their height varies inversely with the strength of the thermocline and usually amounts to 3–20 ft (0.9–6 m).

The two-dimensional operation discloses the complexity of the thermal structure in the sea. From data acquired, it is possible to derive quantitative information from the slope of isothermal surfaces. By scaling the depth of isotherms every 304 ft (91 m), it is possible to obtain the angle the isotherm makes with the horizontal (Table 8).

The isotherms in the thermocline are unusually flat, with a median slope of 17 min. By use of a power spectrum, it was found that the wavelengths vary widely. There are frequently several peaks in the power spectrum corresponding to wavelengths between 0.25 and 1.0 mi (0.4 and 1.6 km). Repeated peaks occur at 0.3 and 0.7 mi (0.5 and 1.1 km). The greatest power is in the long-period waves. Other analyses from data collected in a single place reveal (Fig. 6) that vertical changes in the isotherms have definite cycles corresponding to the short-period internal waves (5–8 min), seiches (14 and 20 min), tidal phenomena (6, 8, 12, and 24 hr), meteorological (3–5 days), lunar cycles (14 and 28 days), and seasonal cycles (1 year).

Table 8. Slope of an isotherm*

Slope, minutes from horizontal	Observations, %
0–10	34
10–20	20
20–30	11
30–40	7
40–50	6
50–60	5
>60	17

*Based on 60,000 depth observations.

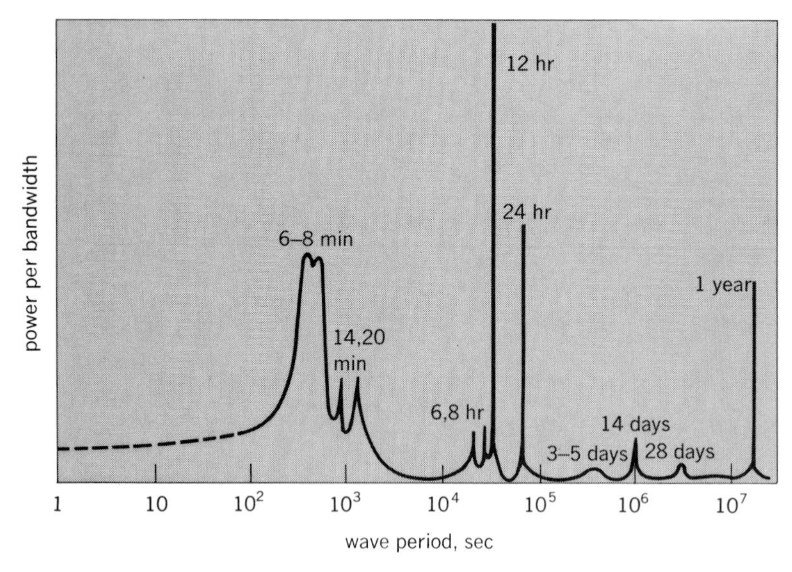

Fig. 6. Power spectrum showing the most common oscillation periods in the internal thermal structure of the ocean.

It is believed that the internal waves in the open sea move in different and changing directions. Analyses of tows in different directions have provided information on the direction of propagation of dominant waves. Up to 200 mi (320 km) off the coast of southern California the dominant direction is shoreward.

An attempt has been made to measure a three-dimensional structure. This is done by towing the profiler in circles and box patterns. Preliminary data indicate that the ocean-temperature structure possesses numerous small thermal domes or humps moving in different directions.

Circulation and boundary influences. The detailed vertical thermal sections reveal the dynamic processes going on in the sea. Domes or ridges in the thermal structure, caused by eddies or oppositely flowing adjacent currents, have been observed to be 150 ft (45 m) in height within a distance of 6 mi (10 km). A similar change in the depth and character of a thermocline has been found across major current systems. Areas of upwelling are apparent where the thermocline tilts upward toward the surface and generally becomes weaker.

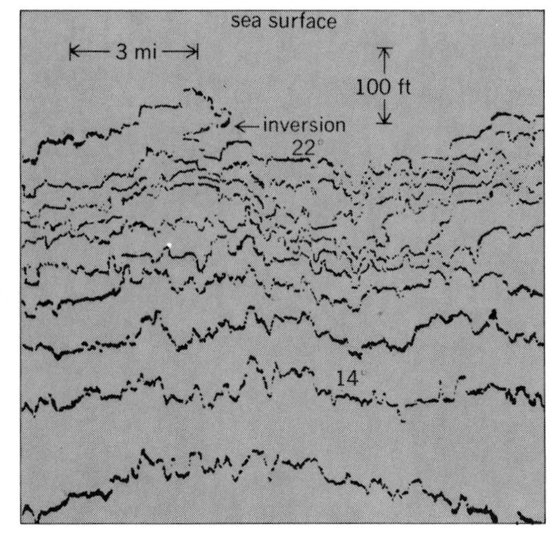

Fig. 7. Extensive turbulence with temperature inversion. 100 ft = 30.48 m. 3 mi = 4.83 km.

The structure is also influenced by the boundaries of water masses and land features. A strong horizontal thermocline may serve as a vertical boundary. If one water layer moves in a different direction from another on the opposite side of the thermocline, a large-scale turbulence, with temperature inversion, can occur (Fig. 7).

Vertical water mass boundaries, or fronts, are detected easily by taking detailed two-dimensional profiles. The isotherms extend to the surface to form a horizontal thermal gradient with turbulence along the thermal boundary.

Islands, points of land, and shoals also influence thermal structure and internal waves. The progressive nature of internal waves is slowed down over shoals, where they have a shorter wavelength. This happens when the waves approach shore, where they not only become more closely spaced but also refract and move shoreward with long crests. Another effect on the thermal structure is created by the tide, which causes a large mass of water to move in and out across the continental shelf off California. The colder, heavier water mass coming shoreward along the bottom is affected by the bottom friction to form a wall-like front (Fig. 8). This is essentially an internal tidal bore, which occurs at flooding spring tides. The front of this bore contains turbulence and sometimes weak thermal inversions. It is also evident that the strength of the thermocline changes with tide. When internal waves approach the surface boundary, their crests become flat and their troughs sharp. If they converge with the sea floor, the troughs grow flat, and the crests become steep and sharp.

[E. C. LA FOND]

Stirring and mixing processes. Stirring and mixing are processes of prime importance in determining the distribution of properties and in the formation of water masses in the ocean. Stirring refers here to motions which increase the average magnitude of the gradient of a property throughout a specified region. Mixing is employed in a narrow sense to denote the molecular diffusion or conduction processes by which gradients are decreased. Stirring and subsequent mixing can decrease the gradients in a region much more quickly than molecular processes acting alone. Combined stirring and mixing are often called turbulent diffusion.

Stirring motions involve shearing or more precisely, deformations of the water. Stirring scales range from those of the permanent currents down nearly to molecular scale. Smaller-scale motions, often highly complex, are the most effective in stirring.

With very important exceptions noted below, the preferred direction of stirring is along surfaces of constant potential density. (Potential density is the density of water when brought adiabatically to atmospheric pressure.) The preference is due to the fact that potential density almost always increases with depth. Under this condition, vertical displacement of a water parcel from a potential density surface is resisted by a stabilizing force proportional to the vertical gradient of potential density.

At the sea surface, powerful mechanisms exist

which induce stirring across potential density surfaces. These are the wind stresses and the thermal convections resulting from cooling and evaporation. Near the sea surface, they greatly outweigh the stabilizing forces to produce and maintain a shallow, homogeneous top layer over much of the ocean. Stirring induced at the surface penetrates across the potential density surfaces just below the homogeneous layer but is damped out as it extends deeper into the thermocline or halocline layer. Thus, at some depth in these layers, the more usual stirring along potential density surfaces becomes dominant. In great depths and in high latitudes, the dominance is less strong since the vertical gradient of potential density is small. Stirring across potential density surfaces may also be brought about by tidal currents, but these are strong only in the shallow, coastal parts of the ocean.

Because of the variety and complexity of stirring motions, no general method of treating them quantitatively has proved really satisfactory. It is often expedient to assume that stirring and mixing follow rules analogous to those for molecular diffusion. Sometimes analysis of the motions in detail is possible, although difficulties in both theory and observation are great. Statistical treatment of the detailed motion seems to provide the most realistic approach. [JOHN D. COCHRANE]

Ocean water masses. Ocean water masses are extensive bodies of subsurface ocean water characterized by a relatively constant relationship between temperature and salinity or some other conservative dissolved constituent. The concept was developed to permit identification and tracing of such water bodies. The assumption is made that the characteristic properties of the water mass were acquired in a region of origin, usually at the surface, and were subsequently modified by lateral and vertical mixing. The observed characteristics in place thus depend both on the original properties and on the degree of modification en route to the region where observed.

A water mass is usually defined by means of a characteristic diagram, on which temperature or some other thermodynamic variable is plotted against an expression for the amount of one component of the mixture. A point on such a diagram defines a water type (representing conditions in the region of origin), and the line between points observes, at least approximately, the property of mixtures; that is, on the line connecting the two points the proportion lying between the point representing one water type and the point for any mixture equals the proportion of the second water type in the mixture. The resulting curve for a vertical water column has been called a characteristic curve, because for a given water mass its shape is invariant, regardless of depth. The existence of such a curve implies continuous renewal of water types, since otherwise mixing would lead to homogeneous water, represented by a point on the diagram.

Temperature-salinity relationships. In oceanography the characteristic diagram of temperature against salinity is usually used in studies of water masses, and the resulting temperature-salinity curve (the *T-S* curve) is used to define a water mass (Fig. 9).

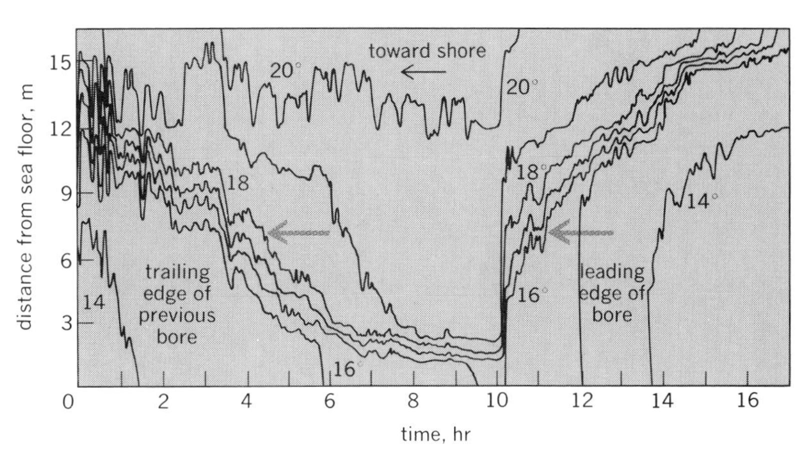

Fig. 8. Thermal structure changes caused by an internal tidal bore.

In drawing such a curve, data from the upper 100 m are usually omitted because of seasonal variation and local modification in the surface layer, so that strictly speaking, a water mass as defined extends only to within 100 m of the sea surface. Although ideally a water mass is defined by a single *T-S* curve, because of random errors in field measurements and perhaps fine structure in the water mass itself, in practice an envelope of values provides a more useful definition.

On the *T-S* diagram any property which is a function only of temperature and salinity can be represented by the appropriate family of isopleths (such as values at constant pressure of density expressed as σ_t, or thermosteric anomaly, sound speed, and saturation concentration of dissolved gases). Therefore, the *T-S* diagram with isopleths can be used to determine values of such temperature-salinity dependent functions. Since the ocean is inherently stable (that is, density increases monotonically with depth), examination of the slope of a *T-S* curve (on which depth is indicat-

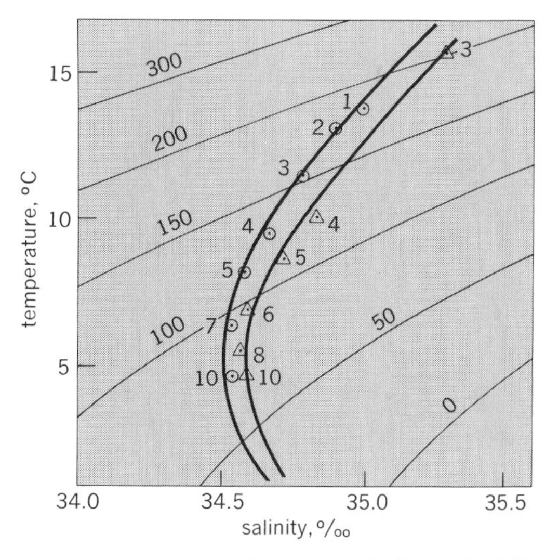

Fig. 9. Temperature-salinity values for Carnegie station 40 (circles) and Dana station 3756 (triangles); see Fig. 11 for locations. Depths of observations in hectometers (100 m). Light lines represent definition of Pacific equatorial water. Heavy lines represent specific volume as thermosteric anomaly in centiliters (.01 liter) per ton.

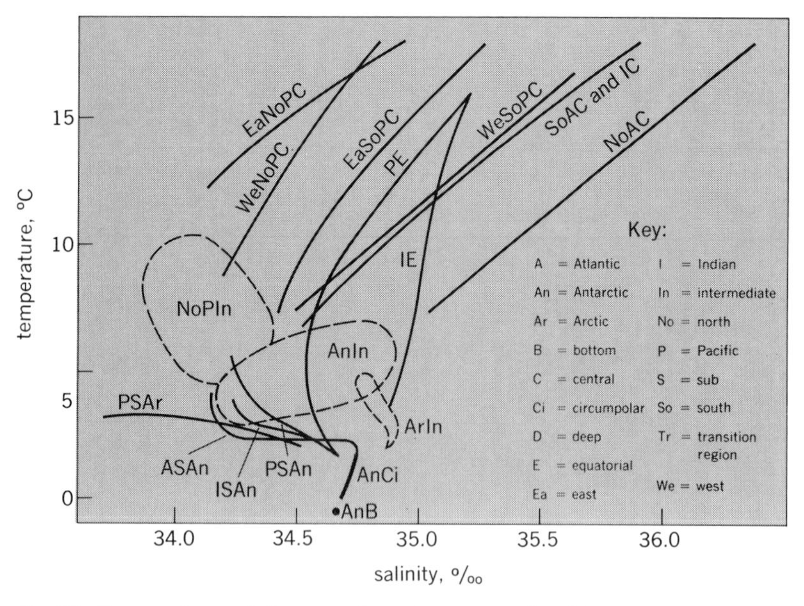

Fig. 10. Temperature-salinity curves for water masses of world ocean.

ed) relative to the isopleths of density permits an estimate of the vertical distribution of stability. The diagram is often useful for the detection of faulty observations and as a guide to interpolation on neighboring stations. When a uniform series of data is available, it can also be used for the quantitative representation of the frequency distribution of water characteristics.

Water-mass types. The most important and best-established water masses (characterized by *T-S* curves in Fig. 10; distribution shown in Fig. 11) occur in the upper 1000 m of the ocean. These are of three general types: (1) polar water, present south of 40°S in all oceans, and north of 40°N in the Pacific; (2) central water, occurring at mid-latitudes over most of the world ocean; and (3) equatorial water, present in the equatorial zones of the Pacific and Indian oceans.

Polar waters, including the Subarctic, Subantarctic, and Antarctic Circumpolar water masses, are formed at the surface in high latitudes and thus are cold and have relatively low salinity. Subantarctic water is bounded in the south by the well-defined Antarctic Convergence, south of which circumpolar water is found; Subarctic water has no clearcut northern boundary.

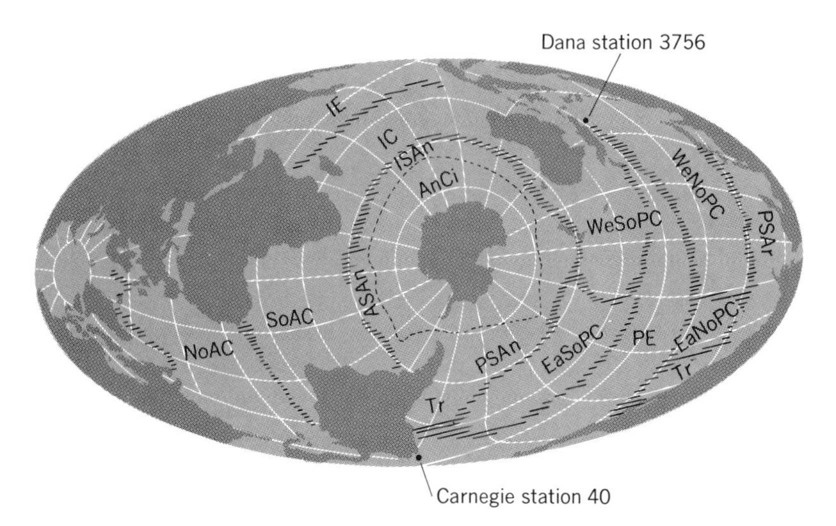

Fig. 11. Distribution of representative water masses of upper 1000 m (symbols as in Fig. 10). Dashed line around Antarctica represents Antarctic Convergence.

The central water masses appear to sink in the regions of the subtropical convergences (35–40°S and N), where during certain seasons of the year horizontal *T-S* relations at the surface are similar to the vertical distributions characteristic of the various water masses. The great differences in their properties are attributed to differences in the amounts of evaporation and precipitation, heating and cooling, atmospheric and oceanic circulation, and the distributions of land and sea in the source regions.

The widespread and well-defined equatorial waters (Fig. 9) separate the central water masses of the Indian and Pacific oceans. These equatorial water masses are apparently formed by subsurface mixing at low latitudes, although the place and manner of their formation is not well known.

Intermediate waters underlie the central water masses in all oceans. Antarctic Intermediate Water sinks as a water type along the Antarctic Convergence; the water mass then formed by subsequent mixing is characterized by a salinity minimum. Arctic Intermediate Water, of little importance in the Atlantic, is widespread in the Pacific and is apparently formed northeast of Japan. Other important intermediate water masses are formed in the Atlantic and Indian oceans by addition of Mediterranean and Red Sea water, respectively. Deep and bottom waters of the world ocean are formed in high latitudes of the North Atlantic, South Atlantic (Weddell Sea), and Indian oceans.

[WARREN S. WOOSTER]

Isotopic variations. Studies of the isotopic composition of ocean water have characterized the major deep-water masses and investigated their origin and mixing, have investigated isotopic exchange with dissolved ions, and have been concerned with the geological history of the sea. Other studies have been concerned with the exchange of moisture between the atmosphere and the sea, and the nature of the boundary layer at the air-sea interface.

Variations in the ratios of the stable isotopes are uniformly reported in terms of the ratio in an arbitrarily defined standard mean ocean water (SMOW), whose isotopic composition is numerically defined relative to a water standard distributed by the International Atomic Energy Agency in Vienna, Austria. Absolute concentrations of the isotopes are much more difficult to measure than variations in ratio, but the absolute concentrations of SMOW are believed to correspond to a D/H ratio of 1/6328, or 158 parts per million (ppm) of D, and an O^{18}/O^{16} ratio of 1993×10^{-6}, or 1989 ppm of O^{18}. The isotopic variations are reported as δ values relative to SMOW, defined from the relation $R/R_{SMOW} = 1 + \delta$, where R is the ratio D/H or O^{18}/O^{16}. The delta values are always given in parts per thousand (°/oo), the same units in which salinity is recorded by oceanographers.

The isotopic variations in surface ocean waters reflect the net effects resulting from the overall balance between precipitation, evaporation, molecular exchange, and mixing with other surface and deep waters. In equatorial latitudes and in high latitudes, precipitation exceeds evaporation, while the reverse is true in the latitudes of the trade winds. The isotopic relationships are best seen by plotting the δ values versus salinity, as shown in Fig. 12.

The δ-S relationships are seen to have different slopes in the North and South Pacific surface waters because of the varying intensities of the different processes operating. At higher latitudes the relationships are approximately linear in regions where precipitation exceeds evaporation, but in low latitudes the relationships become very complicated in passing through the trade-wind regions to the equatorial zone of high precipitation. These variations can be related quantitatively to the net evaporation, precipitation, and mixing rates in local regions if mixing models are used in which the number of equations does not exceed the variables, and preliminary models along these lines have been made.

Deuterium-O^{18} relationships. In equatorial and temperate latitudes, surface ocean waters show simple linear relationships between D and O^{18} of the form $\delta D = M\delta O^{18}$, with zero intercept and a slope M which decreases with increasing ratio of evaporation to precipitation in a region. Characteristic values of the slope M are 7.5 in the North Pacific, 6.5 in the North Atlantic, and 6.0 in the Red Sea.

Freezing effects. The freezing of sea water concentrates salt in the liquid so that water of high salinity tends to form and sink. This process is especially important in the formation of Antarctic bottom water on, for example, the Weddell Sea Shelf in the Atlantic portion of Antarctica. The isotopic separation factors in the freezing process are so small (1.0203 for D and 1.0027 for O^{18}) that very little change in isotopic composition is produced. Ocean waters generated by such a process will thus be expected to lie along an almost horizontal line, parallel to the abscissa in a plot, such as Fig. 12, of isotopic composition against salinity. This has been observed in measurements on both Arctic and Antarctic waters.

For example, in Fig. 12 the Antarctic Bottom Water plotted in the diagram is related to the points shown for samples of South Pacific surface waters in latitudes 59–64°S. Monsoon, Risepac, and Leapfrog refer to Scripps Institution of Oceanography (SIO) expeditions. Risepac samples were along an east-west track in the South Pacific between the Society Islands and South America. The Antarctic Bottom Water samples represented were actually taken from the Weddell Sea area. The surface waters in comparable latitudes in the Atlantic plot at the same points shown for the South Pacific surface waters at 59–64°S, because all these waters are restricted to the same isotopic composition and salinity by the rapid circumpolar circulation around Antarctica.

In the Weddell Sea region, however, the points for surface and intermediate waters connecting the high-latitude surface waters and the Antarctic Bottom Water point are parallel to the abscissa in Fig. 12, indicating that the bottom water is directly generated by the freezing process. This is completely in accord with the classical picture worked out by oceanographers. There is no evidence, either from classical or isotopic data, that this

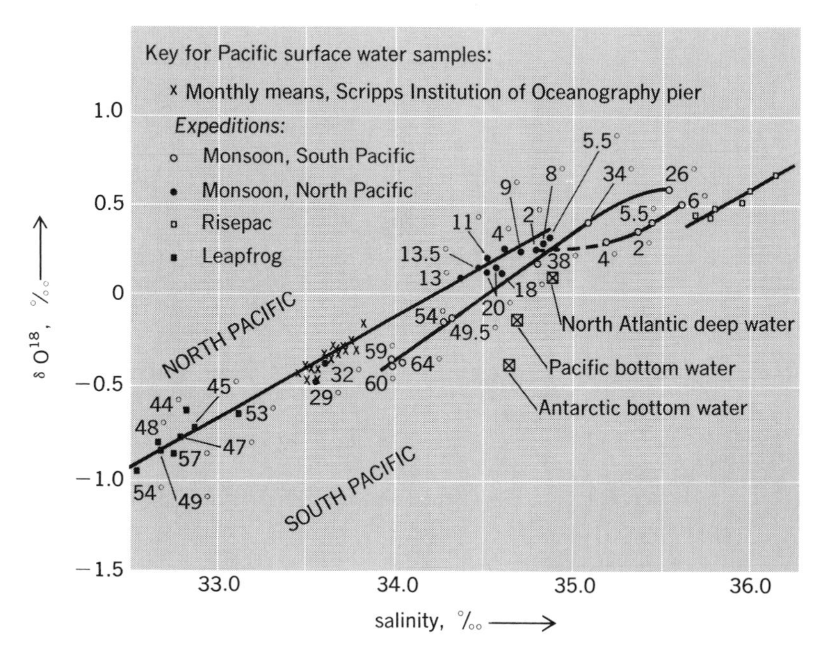

Fig. 12. Diagram illustrating the relationship between the O^{18} content and salinity in the surface and deep waters of the Pacific, Antarctic, and North Atlantic oceans.

process operates to a significant extent anywhere else in the oceans, although it may also operate on the Ross Sea Shelf, where no detailed winter studies have yet been made.

Deep-water relationships. The oxygen-18 values for the core waters in the major deep-ocean-water masses are shown in Fig. 12. Indian Ocean Deep Water is indistinguishable from Pacific Bottom Water, and the Pacific Deep Water has this composition over the entire Pacific Basin, except in very high southern latitudes where mixing with Antarctic Bottom Water is observed. The precision of the isotopic data shown is about ±0.02‰. The isotopic data are sufficiently precise that mixing between North Atlantic Deep Water and underlying, northerly flowing Antarctic Bottom Water can be seen directly on such a diagram.

The North Atlantic Deep Water point plots directly on the line relating the salinity and δ values of surface waters in the North Atlantic (not shown). Hence, the convective origin of this water from sinking of cooled surface water (discovered by F. Nansen) is directly verified by the isotopic data. However, Pacific and Indian Ocean deep and bottom waters nowhere plot on the lines relating the surface waters in the δ-S diagram. This is seen directly for the Pacific waters in Fig. 12. Thus, the deep water in these oceans cannot be significantly replenished by convective mixing with surface waters in the Pacific and Indian Ocean basins; they must originate by the mixing of other deep waters. This is in agreement with the classical oceanographic picture based on density and stability considerations.

The data in Fig. 12 further indicate that the Pacific and Indian deep waters do not lie on a line connecting the North Atlantic Deep Water and the Antarctic Bottom Water, the only major sources which can provide deep water for the Pacific and Indian oceans. Pacific and Indian deep waters are almost 50–50 mixtures of North Atlantic Deep Water and Antarctic Bottom Water, but there

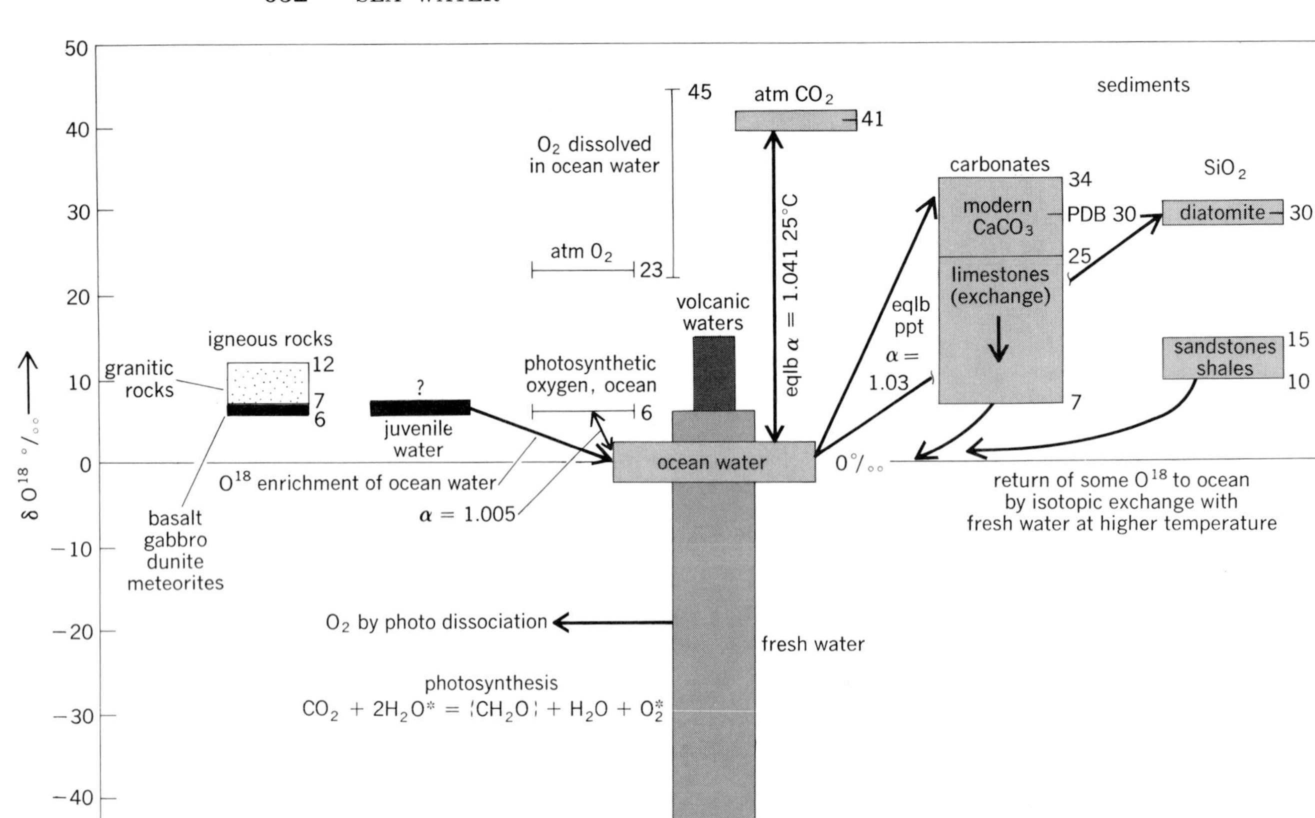

Fig. 13. The oxygen isotope geochemical cycle. The isotopic variations are the per mil deviations from standard mean ocean water (SMOW) with approximate isotopic abundances $O^{16}=1$, $O^{17}=1/2500$, and $O^{18}=1/500$. Single-stage isotopic fractionation factors (α) between oxygen in water and in the other substances are noted by connecting arrows.

must be a so-called "third component" in small amounts, which almost certainly represents intermediate water. It is not yet possible to decide whether the third component is intermediate water added in the Atlantic, where the major mixing of the two main sources takes place, or whether it represents Pacific and Indian Ocean intermediate water added locally in these latter oceanic basins; the former interpretation seems more probable and is supported by some measurements, but detailed studies are necessary. In any case, the major characteristics of the deep-ocean mixing processes seem to be understood.

Changes in isotopic composition. During the Pleistocene Epoch, the isotopic composition of the sea has oscillated about its present composition because of the periodic formation and disappearance of continental ice sheets greatly depleted in D and O^{18}. These variations are estimated to be of the order of 1°/∞ for O^{18} and 7°/∞ for D. On the geological time scale of hundreds of millions of years the changes may have been even greater. It is known that H and D atoms can escape from the Earth's gravitational field, and it is believed that H atoms will diffuse upward and escape at a greater rate than D. These atoms are formed by photolysis of water vapor in the atmosphere, and it thus seems likely that the D/H ratio of the ocean may have been increasing with time because of preferential loss of H atoms. Hydrogen found in igneous rocks generally averages about 80°/∞ lower in D than ocean water and is, therefore, at least consistent with the idea that this may be juvenile hydrogen which is of the same composition as the early oceans.

The geochemical cycle of O^{18} is much more complicated because of the variety of oxidation-reduction reactions in oxygen geochemistry. A schematic outline of the isotopic oxygen cycle is shown in Fig. 13. According to present knowledge of oxygen isotope fractionation effects, ocean water of 0°/∞ would be in isotopic exchange equilibrium with igneous rocks at about 250°C; at oceanic temperatures it is far removed from equilibrium. The ocean water observed today is therefore not a direct sample of juvenile water which has escaped from the interior of the Earth at high temperatures—such water would be about 9°/∞ on the SMOW scale. It is possible, however, that there is a continual input of juvenile water into the ocean, so that the ocean is changing to, or has reached, a steady state in which the net oxygen removed from the ocean has the composition of the incoming juvenile water.

Calcium carbonate and silica remove oxygen of about 30°/∞ from the ocean, but some of this O^{18} is cycled back into the sea by isotopic exchange with fresh water on the contentnts. The average oxygen removed in this way is probably about 20°/∞. It is difficult to see how other processes can account for significantly more back-cycling of O^{18}, and it seems likely that the ocean may have undergone a progressive depletion of O^{18} through time, especially since Cretaceous times when large amounts of limestone began to be deposited in the deep sea by organisms. Studies are being made of oxygen isotope ratios in ancient carbonates, phosphates, and sulfates to obtain definitive evidence on this problem. When more is known about the actual variations and secular changes in isotopic compositions of the ocean, this information can in turn be applied to questions of the origin and growth of the ocean itself. [HARMON CRAIG]

Bibliography: J. L. Cairns, Asymmetry of internal tidal waves in shallow coastal water, *J. Geophys. Res.*, 72:3563–3565, 1967; J. D. Cochrane, The frequency distribution of water characteristics in the Pacific Ocean, *Deep-Sea Res.*, 5:111–127, 1958; H. Craig, Abyssal carbon and radiocarbon in the Pacific, *J. Geophys. Res.*, 74:5491–5506, 1969; H. Craig, Isotopic composition and origin of the Red Sea and Salton Sea geothermal brines, *Science*, 154:1544, 1966; H. Craig and L. I. Gordon, Deuterium and oxygen 18 variations in the ocean and marine atmosphere, in E. Tongiorgi (ed.), *Stable Isotopes in Oceanographic Studies and Paleotemperatures*, 1965; H. Craig and B. Hom, Deuterium–oxygen 18–chlorinity relationships in the formation of sea ice (abstract), *Trans. Amer. Geophys. Union*, 49:216, 1968; C. Eckart, An analysis of the stirring and mixing processes in incompressible fluids, *J. Mar. Res.*, 7:265, 1948; S. Epstein and T. Mayeda, Variation of O^{18} content of waters from natural sources, *Geochim. Cosmochim. Acta*, 4:213, 1953; I. Friedman et al., Variation of deuterium content of natural waters in the hydrologic cycle, *Rev. Geophys.*, 2:177, 1964; Y. Horibe and N. Ogura, Deuterium content as a parameter of water mass in the ocean, *J. Geophys. Res.*, 73:1239, 1968; E. C. LaFond, in *Encyclopedia of Oceanography*, 1966; E. C. LaFond and K. G. LaFond, Internal thermal structure in the ocean, *J. Hydronautics*, 1:48–53, 1967; E. C. LaFond and K. G. LaFond, in *The New Thrust Seaward*; Transactions of the 3d Annual Marine Technology Society Conference, 1967; R. B. Montgomery, Water characteristics of Atlantic Ocean and of world ocean, *Deep-Sea Res.*, 5:134–148, 1958; D. Rochford, Total phosphorus as a means of identifying East Australian water masses, *Deep-Sea Res.*, 5:89–110, 1958.

SAMPLING AND MEASURING TECHNIQUES

Observations of conditions in the sea generally must be made in situ and the information transmitted back to the observer, or the result must be recorded in situ and retained for reading when withdrawn from the sea. Consequently, the marine scientist is faced with peculiar problems of technique and the use of special equipment to obtain much of his information about sea water. Some of the more commonly used methods and devices for obtaining observations relative to the physical properties of sea water are described here. For information pertaining to the collection of living organisms in the sea, the sampling of the ocean bottom, and the measurement of subsurface currents.

Temperature-measuring devices. Temperature measurements in the surface layer of the ocean, to depths of 900 ft (275 m), are usually made with the bathythermograph, or BT, a nonelectric device which gives a continuous record of water temperature and depth as it is lowered and raised. The instrument can be operated at frequent intervals while underway and therefore provides a rapid means of obtaining a detailed picture of temperature distribution within the surface layer.

Reversing thermometers. The most reliable and widely used temperature-measuring device is the deep-sea reversing thermometer. This mercury-in-glass thermometer records the temperature at the time it is inverted. It is reliable to about 0.01°C with proper corrections. These thermometers are usually in a pressure-proof glass tube. In this manner the thermometer is protected from the effect of pressure, and the true water temperature is given. These same thermometers are available with a mercury bulb which can be exposed to sea pressure. The compression of the bulb on these unprotected thermometers causes them to read about 1°C high for each 100 m depth. When protected and unprotected reversing thermometers are used in pairs, they give both temperature and depth (thermometric depth). Reversing thermometers usually are attached to a reversing water bottle which collects a water sample when it is inverted.

Electrical thermometers. Since 1950 a rapidly increasing number of electric temperature recorders have been developed around thermistor beads encased in glass. A typical thermistor will change resistance 4% or about 100 ohms/°C and will have a thermal time constant of a few tenths of 1 sec. Electric temperature recorders are usually made to plot temperature against time. They are often used to study microstructure and may have a sensitivity of 0.01–0.001°C. When measuring elements are to be lowered from a ship, the recorder is usually made to plot temperature against depth.

In 1958 a system was developed utilizing electrical thermometers attached to a special cable which could be towed behind a vessel at 500 ft (150 m) and at speeds of 10 knots (5 m/s). The thermometers are attached to the cable at 25-ft (7.6-m) intervals. A facsimile type of recorder draws continuous isotherms on a depth-distance plot. The depth of each degree or tenth-degree isotherm is plotted every 2 sec.

Radiation thermometers have been built and flown from low-flying aircraft. These measure changes in temperature of the water surface to about 0.1°C. Radiotelemetering buoys permit water and air temperatures to be observed on unattended buoys and transmitted to land.

Water samplers. For many chemical and gas analyses it is essential to obtain samples of 100–1000 ml of sea water. These usually are obtained by a series of Nansen bottles on $\frac{3}{16}$-in. (4.8-mm) hydrographic cable. The bottles are designed to flush continuously when lowered in an upright position. A weight messenger is then sent down the cable. As it strikes the tripping device of the Nansen bottle, the bottle is inverted and the lids are closed (Fig. 14). At the same time another messenger is released to trip the next lower bottle and so on. Usually two reversing thermometers are attached to each bottle. The thermometer and water bottle give accurate results, but the method is very time-consuming. To obtain a synoptic picture of temperature and salinity, or density, a number of

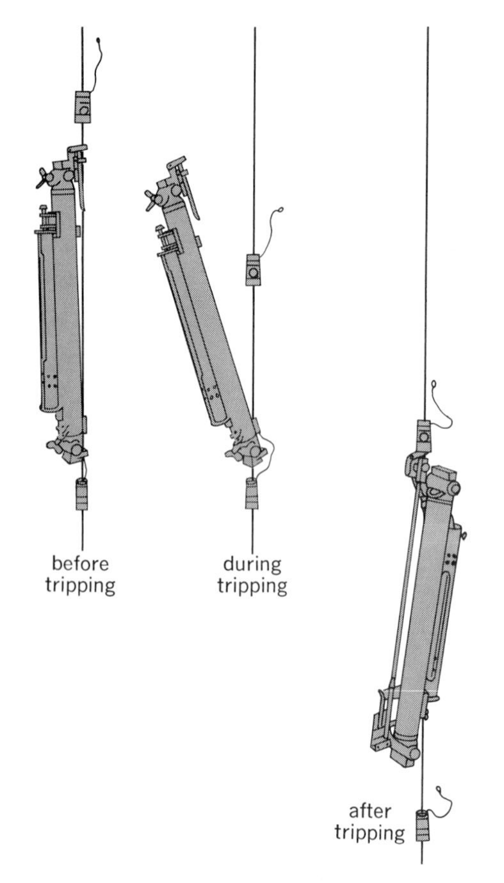

Fig. 14. Nansen bottle in three positions. (*From Instruction Manual for Oceanographic Observations, H.O. Publ, no. 607, 2d ed., U.S. Navy Hydrographic Office, 1955*)

closely spaced observations must be taken in a relatively short period of time.

In an effort to eliminate contamination from a metallic case, samplers such as the Van Dorn have been made from plastic tubing. Large rubber stoppers on each end are pulled shut by rubber bands. Simple open-tube-type samplers of $\frac{1}{10}$–10-liter capacity can easily be made.

Measurements such as carbon-14 require samples as large as 200–400 liters. Flushing of such large samplers requires either a large open-ended hose construction or a barrel with two flapper-type ports and water scoops to aid ventilation. After the sample is brought to the surface, some large samplers are retrieved on deck while full. Others are emptied while still in the water by means of a hose.

Continuous electrical temperature recorders that obtain temperatures with a single probe will almost certainly be used in the future. This will increase the need for a single collecting device that will take many water samples when used at the end of the cable. [ALLYN C. VINE]

Serial observations. These are measurements of temperature, salinity, and other properties at a series of depths at some location in the ocean (an oceanographic station), by which the distribution in space and time of these properties (and others computed from them such as density and geopotential) may be described.

Bottles in series. A number of water samplers with thermometers attached are usually lowered on the same cast. As many as 26 samplers have been lowered at once. The number depends on the number of levels to be sampled, the strength of the wire, and the extent of possible damage to the equipment which may result from the roll of the ship or from dragging the bottom.

After the first (deepest) bottle is attached, the wire is paid out and the next bottle and its messenger are attached. When all bottles have been lowered, it is necessary to wait for the thermometers to approach equilibrium (about 10 min) before releasing a messenger to trip the cast. If the wire is nearly vertical, the messenger falls about 200 m/min. At high wire angles (60° and more have occurred under high wind conditions or strong current shear) the messenger will fall more slowly and may stick. The angle can sometimes be reduced by maneuvering the ship.

After allowing time for the final messenger on the cable to trip the deepest bottle, the wire is pulled in and the bottles removed. Water samples are drawn into laboratory bottles and the thermometers are read.

Thermometric depths. When protected and unprotected thermometers are reversed at the same depth, the unprotected will give a higher reading because of the pressure on its bulb. The difference in the two readings depends upon the pressure at reversal, and since this is proportional to depth the "thermometric depth" can be computed. With information from both protected and unprotected thermometers at several of the levels, the shape of the wire can be estimated, and the depths of the other samplers computed. Unprotected thermometers are ordinarily used at depths greater than 200 m, since the depth of the upper bottles can be computed from the wire angle and length. Depth computations are estimated to be accurate to ±5 m in the upper 1000 m and to about 0.5% of wire length below that.

Standard depths. In 1936 the International Association of Physical Oceanography proposed certain standard levels at which observations should be made or values interpolated in reporting. They are (in meters) 0, 10, 20, 30, 50, 75, 100, 150, 200, (250), 300, 400, 500, 600, (700), 800, 1000, 1200, 1500, 2000, 2500, and 3000 and by 1000-m intervals at greater depths (depths in parentheses being optional). These values are recommended as a convenient standard of comparison, not as being sufficient for measuring the ocean everywhere. Where the precise level of maxima or minima in the various properties is to be determined, the standard depths may not be adequate and more depths must be sampled.

[JOSEPH L. REID]

Analysis of water samples. The development of analytical methods to measure the kind and quantities of dissolved and suspended substances in sea water has paralleled advances in analytical chemistry. In addition to the usual considerations of accuracy, precision, speed, and cost that control the choice of analytical methods in most applications, methods for sea-water analyses are further restricted by the necessities of performing some analyses on shipboard immediately after the samples are obtained and of storing other samples for analyses that can be performed only in a shore-based laboratory. For example, analyses for biologically active substances, especially those present in trace quantities, are performed on shipboard, whereas analyses for which the highest precision and accuracy are demanded, frequently those which require precision weighing, are performed in shore-based laboratories.

In-place techniques. During the 1960s considerable effort was expended in the development of in-place analytical techniques. All of the new techniques depend upon the generation of an electrical signal in which either a voltage or a frequency is made to be proportional to the concentration of the constituent or the intensity of the property being measured. All such methods produce a continuous record in which the property being measured is displayed either as a function of time or of depth of water.

Three kinds of in-place instruments have been used successfully. The first two utilize a transducer lowered on a cable beneath the ship. One version (the older) requires an electrical cable from transducer to ship with recording of the signal on shipboard; the second has the recorder and transducer on the cable end. With the recorder on shipboard there can be continuous monitoring of the signal, and the raising and lowering can be changed to rerecord or emphasize a feature. However, recording on shipboard requires the use of at least one electrical conductor insulated from the sea, a system which requires much maintenance. Placing the recording equipment on the end of the cable with the transducer eliminates the need for long lengths of insulated electrical conductors but introduces the possibility of damaging the entire instrument by flooding with sea water should a seal on the pressure case develop a leak under pressure. A third, the newest type of instrument, is a free-fall device in which all components are in a single container which is dropped into the water with no attachment to the ship. The device, which by itself has positive buoyancy, enters the water with jettisonable weights attached and is carried either to the bottom, where the weights are released, or to some predetermined depth, where a pressure-activated release drops the weights. Once back on the surface the device can be located with the aid of radar reflectors, flashing lights, and radio beacons. The primary advantage of free-fall devices, other than eliminating winches with long lengths of cable, is the possibility of a rapid survey of a region. Many devices can be set out in a predetermined pattern and later retrieved without the time-consuming "on station" intervals required by the older hydrographic casts with cable and sampling bottles.

In-place devices having transducers for electrical conductivity, temperature, and pressure (depth) are commercially available. Many of the so-called single-ion electrodes being developed and marketed have potential application with in-place devices. Included are electrodes for measuring dissolved oxygen, halides, fluoride, calcium, magnesium, and sulfide.

Examination of Table 5 shows that the dissolved substances in sea water fall into two main groups: the major constituents, which include sodium, potassium, calcium, magnesium, chloride, sulfate, bromide, and the sum of carbonate, bicarbonate, and carbonic acid; and the minor constituents, which include all of the other elements in the table. The major constituents show the unique property of being present in very nearly contant ratio in open sea water. For this reason, direct analyses for these substances are rarely performed today, except when estimates of small variations in concentrations that may be produced by biochemical and geochemical processes are sought. When analyses are made for these constituents (except for the CO_2 system constituents) samples are returned to a shore-based laboratory. Storage containers for such samples must be carefully chosen, because changes in concentration of some constituents may result from the interaction of sea water with the glass in many common, soft glass containers.

The most complete study on record of the concentrations of major sea-water constituents still is that conducted by W. Dittmar on 77 samples taken during the round-the-world cruise of the HMS *Challenger* in 1873–1876. Dittmar's analyses were made by what are now considered classical gravimetric and weight titration procedures: Chloride was analyzed by the Volhard method; calcium, by precipitation of the oxalate followed by ignition to, and weighing of, the oxide; magnesium, by precipitation of magnesium ammonium phosphate followed by ignition to, and weighing of, pyrophosphate; sulfate, by precipitation and weighing of barium sulfate; potassium, by precipitation of potassium chloroplatinate followed by weighing of metallic platinum after reduction with hydrogen; and sodium, indirectly, as the difference between the measured sum of all cations as sulfates and the sum of the magnesium, calcium, and potassium calculated as sulfates. Despite empirical corrections and analyses by difference in Dittmar's study, more recent analyses have produced only small changes in the mean values of the concentrations of the major constituents.

Activation analysis. Several modern analytical techniques are ideally suited for analyses of many of the minor constituents in sea water. Activation analysis is the production of radioactive isotopes of elements present in a sample by controlled exposure of the sample to the neutron flux in a nuclear reactor followed by measurement of the decay properties, such as half-life and rate of decay of the artificial activities produced. This method has been used to provide estimates of the concentrations of some 18 trace elements and minor constituents in both sea and fresh waters.

Two sample preparation procedures have been used in activation analysis studies. One uses scavenging or carrier techniques to concentrate minor elements by precipitating a naturally occurring major constituent. For example, trace quantities of iron can be carried by precipitation of magnesium hydroxide. Carrier techniques provide a means of concentrating minor elements from large volumes of sea water and in some cases allow selective removal of one or more elements.

The second method of sample preparation involves the freeze-drying of a sample and irradiation of all of the solids produced. This method has the advantage of low probability of sample contamination; however, the massive quantities of sodium and halide activities produced require "cooling" for many days to allow the short-lived activities of the major elements to die out before postirradiation chemistry can be conducted. Another distinct advantage of neutron-activation analysis of whole (freeze-dried) samples is that several elements can be determined in one sample and with one irradiation.

Gas chromatography. Gas-chromatographic methods have been adapted to the analyses of dissolved atmospheric gases and to a rapidly growing list of dissolved organic substances. As with the adaption of many other analytical systems to seawater analysis, sample preparation for gas-chromatographic analyses requires special attention. Not only does water interfere in many gas-chromatographic procedures but, in addition, the concentration of many gaseous and organic constituents is below the useful range of gas-chromatographic instruments. Sample preparation then always requires separation of water from the test substances and usually a concentrating step.

The development of gas-chromatographic methods for the analysis of dissolved oxygen, nitrogen, and argon gives a quick and convenient means of separating indicators of biological and physical processes from one another. This is possible because all three gases are atmospheric components, and the sea surface, under most circumstances, can be considered to be in equilibrium with the atmosphere. The concentration of argon in sea water, because it is not involved in any naturally occurring biological or chemical processes, provides a record of the physical conditions which existed at the sea surface when a sample of water now at some subsurface depth was last at the surface. Present-day argon concentration in subsurface samples then allows one to estimate what the oxygen and nitrogen concentrations would have been in the same samples had there been no changes of biological origin. The concentration of oxygen in subsurface waters, on the other hand, responds rapidly to increases by photosynthesis (in the upper 75–100 m) and to decreases by plant and animal respiration and bacteriological decomposition at all depths. Nitrogen concentration may be either increased or decreased during nitrification or denitrification processes, which are much slower than oxygen-controlling processes.

A single 5-mm sample is sufficient for gas-chromatographic analyses for oxygen, nitrogen, and argon. The three test gases are removed from sea water in an attachment to the chromatograph, which acts as a scrubber through which the carrier gas, usually helium, bubbles through the sample and scrubs the other gases from solution. A special sampling bottle, which can be interspersed among the usual Nansen-type samples on a cable and can be attached directly to the gas chromatograph, reduces sampling errors and eliminates the possibility of contamination by contact with the atmosphere.

Studies, using gas-chromatographic analyses, have shown some unique features concerning the distribution of carbon monoxide between the atmosphere and the sea.

Mass spectrometry. The high sensitivity of the mass spectrometer has been utilized in a method for the analyses of helium, neon, krypton, xenon, and argon in the sea. A single 5.5-ml sample is sufficient for these analyses. The method uses a unique sampling procedure, in which a piece of thin-walled metal tubing is crimped off at both ends at the sampling location. The spacing between the two crimping tools establishes the volume size.

The sample is released in a vacuum chamber within the mass spectrometer system by simply puncturing the thin-walled tubing.

Anodic stripping. Anodic-stripping techniques, which consist of modifications of the older polarographic method of analysis, provide measures of the extent to which some of the biochemically reactive elements—zinc, copper, cobalt, lead, and iron, for example—are complexed with both inorganic and organic ligands. Anodic-stripping methods have useful sensitivities down to 10^{-12} M for some elements and have shown complex formation to be a common occurrence and, therefore, a significant factor in biochemical and geochemical processes. [DAYTON E. CARRITT]

Water color and transparency. The physical relationships governing the penetration and absorption of light, the color of the water, and the transparency of the sea are of prime importance to physical and biological oceanography. Instruments for measuring the color (transmittance) and transparency of the water are discussed below.

Water color. The color of water in the visual sense is a phenomenon which has both objective and subjective aspects. From an objective point of view the color of water is primarily the result of selective scattering and absorption of visible light by the water itself or by the dissolved or suspended material in the water. The color which is brought about by these basic mechanisms can, however, be drastically altered by the color of the bottom when visible, by surface films, by the color of the sky or the reflected images of other objects, by the spectral quality of the source of light, and by the optical state of the water surface, as well as by subjective phenomena such as the chromatic adaptation of the observer.

From a physical point of view the color of water is the color of the hydrosol only and can be observed on an overcast day in deep water with the aid of a face mask.

Transmittance. The transmittance of water is closely related to its color. In measuring the transmittance of a path length R of water, consideration

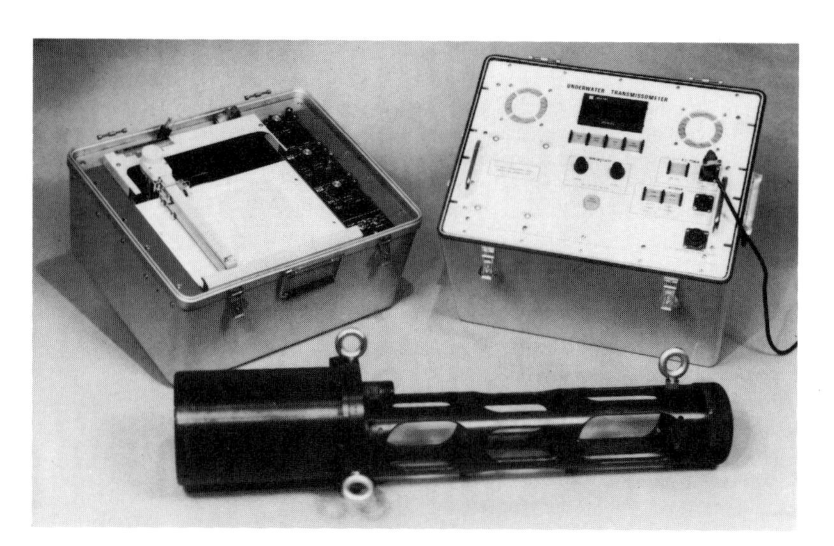

Fig. 15. Hydrophotometer for attenuation coefficient α. (*Visibility Laboratory, Scripps Institution of Oceanography*)

must be given to the directional distribution of the light, as well as to its spectral composition. For monochromatic light the general equation for transmittance is $T=e^{-CR}$. The attenuation coefficient C varies with the directional distribution of the light, being maximum for collimated light and minimum for completely diffuse light. Modern theory makes use of the attenuation coefficients for both collimated and diffuse light. Their independent measurement is therefore essential. The coefficient α, equal to C for collimated light, is commonly measured with an instrument having a collimated source of light (Fig. 15). The coefficient K, equal to C for diffuse light, is commonly measured under natural illumination by means of a photo detector and a diffusing plate (Fig. 16) having cosine collecting properties. Downwelling irradiance readings H_1 and H_2 are taken at two depths, d_1 and d_2, and the experimental coefficient K is computed from Eq. (8).

$$H_1/H_2 = e^{K(d_2-d_1)} \quad (8)$$

Data on the geographic and chronological variations of α and K are being collected. Some monochromatic data for α are available. Monochromatic data (1958) for K are given in Table 9.

Transparency. Measurements of water transparency or clarity are often made with a Secchi disk, an opaque white disk which is held horizontally and lowered into the water until it disappears. The greatest depth at which it can be visually detected is called the Secchi-disk reading and is related in a complex way to the optical properties of the water. Secchi-disk readings depend on the size and reflectance of the disk, the state of the sea surface, the state of the sky, the light adaptation level of the observer, and the technique of observing, as well as many other minor factors.

When the above factors are controlled, it can be shown that the apparent contrast of the disk is given by Eq. (9), where C_0 is the inherent contrast of

$$C_R = C_0 e^{-(\alpha+K)d} \quad (9)$$

the submerged disk. C_0 depends on the submerged reflectance r of the disk and the reflectance of the surrounding water r_b as follows: $C_0=(r-r_b)/r_b$. Definitions of α and K appear in previous sections; d is the depth of the disk below the surface.

Table 9. Spectral values of diffuse attenuation coefficient K per meter*

Wavelength, mμ	Near-surface values, 28–56 m	Deep-ocean values, 56–159 m
400	0.121	
420	0.139	0.0902
440	0.133	0.0820
460	0.118	0.0749
480	0.115	0.0611
500	0.106	0.0527
520	0.0995	0.0545
540	0.107	0.0806
560	0.105	0.0852
580	0.110	0.0908
600	0.107	0.0933

*Computed from measurements of relative downwelling irradiance obtained by Dr. J. C. Hubbard, Woods Hole Oceanographic Institution, with a submerged monochromater having a bandwidth of approximately 10 mμ. Location of measurements lat 39°38′N 68°42′W.

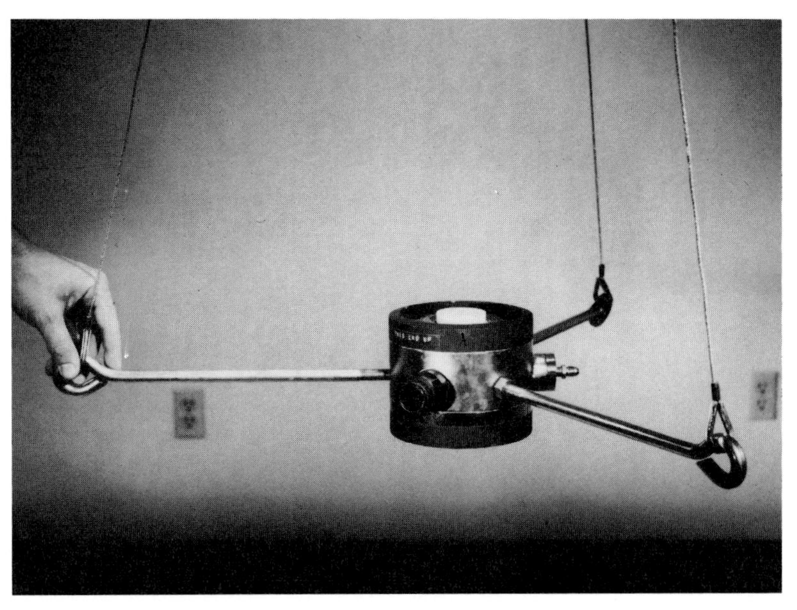

Fig. 16. Irradiance collector for obtaining experimental values of the diffuse attenuation coefficient *K*. (*Visibility Laboratory, Scripps Institution of Oceanography*)

If one uses the utmost precaution and careful technique and does not overlook the effect of human eye capabilities in the final computation, it is possible to obtain the sum $\alpha=K$ from a Secchi-disk reading. It is obvious that Secchi-disk measurements taken from the deck of a ship can be used only to describe the surface strata.

[JOHN E. TYLER]

Bibliography: H. Barnes, *Oceanography and Marine Biology: Annual Review*, vol. 14, 1976; D. E. Carritt, Analytical chemistry in oceanography, *J. Chem. Educ.*, 35:119–122, 1958; J. B. Hersey, Electronics in oceanography, *Advan. Electron. Electron Phys.*, 9:239–295, 1957; E. O. Hulburt, Optics of distilled and natural water, *J. Opt. Soc. Amer.*, 35(11):698–705, 1945; J. M. Kamenovich, *Fundamentals of Ocean Dynamics*, 1977; J. J. Myers, C. H. Holm, and R. F. McAllister (eds.), *Handbook of Ocean and Underwater Engineering*, 1969; H. Sverdrup et al., *Oceans: Their Physics, Chemistry and General Biology*, 1942.

Sea-water fertility

The fertility of a given area of the sea may be defined in terms of the production of living organic matter by the organisms which it contains. The primary productive process in the sea, as on land, is the photosynthetic reduction of carbon dioxide by plants, and the rate of this process for unit area or volume, expressed as mass of carbon fixed in unit time, is a measure of fertility. It may be measured directly in several ways, or it may be estimated from chemical changes in the water. The quantity of living organisms present at any time in a given place (standing stock or standing crop), although it may be related to fertility, is not a measure of it, being the result of a balance between production and removal. Coastal and oceanic waters differ considerably in the types of organisms which they support; the chief difference is in the bottom fauna, which is usually abundant in shallow water and sparse in the deep oceans. Along some shores there may be dense growths of kelp which may have exceptionally high rates of carbon fixation.

Fertility of the oceans. This depends on the production of phytoplankton in the upper layer, called the euphotic zone, which receives ample sunlight for the photosynthetic processes of plants. The plants are mostly diatoms and flagellates which reproduce by division, on average probably less than once per day. If not carried downward by water movement, they tend to remain for some time in the euphotic zone since they often have structures which offer resistance to motion through the water. They sink slowly but are usually eaten by the small zooplankton before reaching the bottom. It has been estimated that about 80% of these plants are so utilized; the remainder, comprising those uneaten, or eaten, partly digested, and excreted by zooplankton, descend to nourish the benthos.

For growth the phytoplankton need radiant energy, carbon dioxide and water, nitrogen, phosphorus, and a range of trace elements of which sea water usually contains enough, with possible occasional exceptions of iron and manganese. Diatoms also need silicon, which is present in the water as monomeric silicic acid, generally in sufficient quantity. There is also a requirement, varying in nature and amount for different species, for accessory organic nutrient factors, but this is not yet fully understood. There is always enough carbon dioxide and water, so that plant growth is normally limited by the supply of radiant energy, nitrogen, and phosphorus.

About one-half of the energy in the solar spectrum can be used for photosynthesis, since only the visible spectrum of 3800–7200 A (380–720 nm) is used. Carbon fixation (gross production) is proportional to light intensity up to about 20 g-cal/cm^2 per day (saturation intensity): above this amount photosynthetic efficiency decreases. It is inhibited at intensities about one-third that of full sunlight, so that carbon fixation at the surface may be less that at intermediate depths. Net production is positive when the radiation intensity is sufficient for carbon gained by photosynthesis to exceed that lost by respiration. This compensation intensity has been found to be somewhat less than 100 g-cal/cm^2 per day, and the maximum depth at which it is attained (compensation depth) varies from a few meters to more than 100 m, depending on the transparency of the water. Compensation depth corresponds very roughly to the depth at which the surface intensity in the middle of the day has been attenuated to about 1%. At low intensities it has been found that to convert 1 mole of carbon dioxide to carbohydrate, the algae require about 10 quanta of radiant energy: this corresponds, when various corrections are made, to a maximum energy-conversion efficiency of about 17%. Making further allowance for decreased efficiency at higher intensities and correcting for loss by respiration, it is possible to calculate the maximum possible yields of organic material at various levels of surface irradiation. These yields range from zero at the compensation intensity of 100 g-cal/cm^2 per day to some 25 g/m^2 (day weight) at 600 g-cal/cm^2 per day, which may be attained in spells of fine summer weather. Over longer periods intensities in most parts of the world average between 200 and 400 g-cal/cm^2 per day, and the calculated production comes out at some 10–20 g/m^2 organic matter daily. Such values are in fact approached by the highest values recorded for single days in shallow water. Open-ocean values in spring are usually around 2–5 g/m^2 per day, and annual averages are 0.4–1.0 g/m^2 per day.

When irradiation is adequate for any length of time, the limitation on growth is set by the availability of nitrogen and phosphorus. The amount of these elements depends on the locality. Deep water in the oceans contains relatively high concentrations of nitrate and phosphate ions, maintained by decomposition of descending dead organisms. However, these nutrients are hindered from reaching the surface by the stable vertical density gradient in the ocean. Surface concentrations of nitrogen and phosphorus normally range up to about 200 μg N/liter and 30 μg P/liter in temperate latitudes in winter. In spring, however, active plant growth can reduce these concentrations to undetectable levels, which may remain fairly constant throughout summer. New growth is then dependent upon nutrients regenerated from dead plants and from excretory matter from the animals. A similar state of affairs generally exists throughout the year in the tropics, and relatively low apparent rates of carbon fixation are normal. Nevertheless, the total yearly production is comparable with that of other regions since radiation intensities are adequate throughout the whole year and also because high transparency of the water allows light to penetrate deeply.

In regions such as the equatorial divergences and the west coasts of Africa or South America, where physical processes cause nutrient-rich deep water to be brought to the surface, high instantaneous productivities may occur, resulting in large standing stocks of all organisms in the food chain.

The chemical composition of marine phytoplankton, unlike that of land plants, shows high protein (40–50%) and high fat (20–27%) contents and is resembled by that of the animals which succeed it in the food chain. This food chain has more links than most terrestrial ones and is in fact a fairly complex web. Some estimates of production of each species in the food chain have been made in areas, such as the North Sea, where there are important fishing grounds, and it has been possible to make reasonably reliable estimates of fish populations. To account for these estimated rates of production of the various links in the chains leading to pelagic and to demersal fish, it is necessary to postulate transfer efficiencies well over 10%, a suggestion which is supported by experimental evidence. Indeed, feeding experiments with herbivorous zooplankton have shown efficiencies greater than 50%. Compared with those on land, marine food chains appear to be efficient. It is perhaps remark-

Annual production of living organisms, g/m^2

Organism	English Channel	North Sea
Phytoplankton	730–910	1000
Zooplankton	275	160
Pelagic fish	2.9	6
Demersal fish	1.9	1.7
Benthos	55	50

able that when the annual fishery yield of the North Sea is compared with the estimated primary production, it is found that nearly 0.5% of primary production is available for human use. The productivity of all the oceans has been put at 1.6–15.5 × 10^{10} tons (1.45–14.1 × 10^{10} metric tons) carbon/year, compared with 1.9 × 10^{10} tons (1.7 × 10^{10} metric tons) carbon/year for the land areas. The table gives estimates of the annual production of living material in two nearshore environments. The estimates are based on many assumptions. [FRANCIS A. J. ARMSTRONG]

Productivity and its measurement. The production of organic matter in the sea, as on land, is accomplished by the photosynthetic activity of autotrophic plants. In coastal waters, where sunlight penetrates to the bottom, both rooted plants and benthic algae contribute to this process. In the open sea, organic production is limited to unicellular algae, the phytoplankton, which live suspended in the upper layers of all ocean waters.

Productivity of benthic plant communities may be determined by periodic harvest and measurement of their growth over discrete time intervals. Such direct methods are impossible in the study of the short-lived plankton because of unmeasurable losses from natural death, predation, sedimentation, and advection.

A more satisfactory approach to both benthic and planktonic plant production is through measurement of chemical changes of the water accompanying photosynthesis and growth. These may be followed in the natural environment for periods ranging from 1 day to several weeks, or in the laboratory by exposing representative samples of the plant population to natural conditions for periods not exceeding 1–2 days.

Photosynthetic activity is indicated by the changes in the water of nitrogen and phosphorus salts, oxygen, and carbon dioxide, and by the degree of acidity (pH). Calculations based on natural-environment changes of these indicators must allow for gas exchanges across the water surface and the effects of vertical mixing between surface and deep waters. In such calculations the horizontal advection is generally neglected, and complete chemical recycling between sampling periods cannot be accounted for. For the last reason, the method tends to give conservative estimates of productivity.

Experimental laboratory studies include measurement of oxygen production, carbon dioxide assimilation, and pH change. Both natural-environment and laboratory studies of changes of these properties represent the net effect of photosynthesis and respiration (by both plants and animals), and hence measure net production. Respiration may be measured separately in the laboratory in dark-bottle experiments. This measurement, when added to the net change observed in transparent bottles, gives a measure of real photosynthesis or gross production. Oxygen-bottle experiments lack the necessary sensitivity for use in the open sea, where plankton are sparse; such experiments have been largely replaced by the extremely sensitive method of measuring CO_2 uptake using C^{14} as a tracer. $C^{14}O_2$ uptake appears to be equivalent to net production (photosynthesis minus respiration) by the plant community.

A third method for estimating productivity is based on the premise that photosynthesis is a function of two independent variables, the chlorophyll content of the plants and the light intensity which they receive. Production may be calculated from simultaneous measurements of these factors in the ocean and from their experimentally derived relationship to photosynthesis.

The few existing measurements of dense benthic plant communities indicate that they may produce as much as 20 g organic matter/(m^2) (day), an amount equivalent to the best agricultural yields on land. Plankton production seldom if ever attains this level, though values half as great are not uncommon. The productivity of shallow, inshore waters is generally higher than that of the open sea, but the seasonal range of most marine areas includes two orders of magnitude. The mean annual rate of production in the oceans as a whole is a matter of some controversy, but probably lies between 100 and 300 g organic matter/m^2 sea surface, which represents an efficiency of utilization of 0.1–0.2% of incident, visible solar energy.

[JOHN H. RYTHER]

Geographic variations in productivity. Strictly speaking, variations in productivity imply variations in the rate of entry of carbon into the organic cycle, or gross photosynthesis. The extent to which this takes place in the sea is determined by the amount of photosynthesizing plant tissue present, the temperature, and the available light energy. The net productivity is the rate of plant growth. This is of greater value as a measurement because it eliminates from the determination the respiratory and excretory losses of the plants and specifies the production of food for the planktonic animals. Variations in net productivity depend on the physiological and oceanographical factors affecting algal growth in the sea, to which the availability of nutrient salts is of prime importance.

The limited penetration of daylight into the sea restricts plant growth to the upper euphotic layer where the nutrients can be assimilated. The plants sink to deeper layers where the nutrients are released by decomposition of plant material by microorganisms and returned to the sea. Acting against this downward transport of nutrients is eddy diffusion (produced by turbulence), which brings nutrients up to the surface from the richer deeper waters. This process is facilitated where the water is homogeneous and it is suppressed by stratification. Currents perform the major transport of nutrients, and where surface divergences occur, upwelling currents bring deeper water to the surface (see illustration). In the illustration the northern Indian Ocean and the China Sea appear under southwestern monsoon conditions; these currents are reversed, with a shift in regions of divergence, in the northeastern monsoon.

In temperate and higher latitudes there is a pronounced seasonal cycle, with suppressed production in winter because of excessive turbulence and lack of light, a rapid burst of growth in spring, and limitation of this growth by lack of nutrients when the waters stabilize in summer. Frequently in autumn there is a subsidiary flowering before growth is again limited by lack of light.

In tropical and subtropical regions a more permanent stratification limits nutrient supply to the

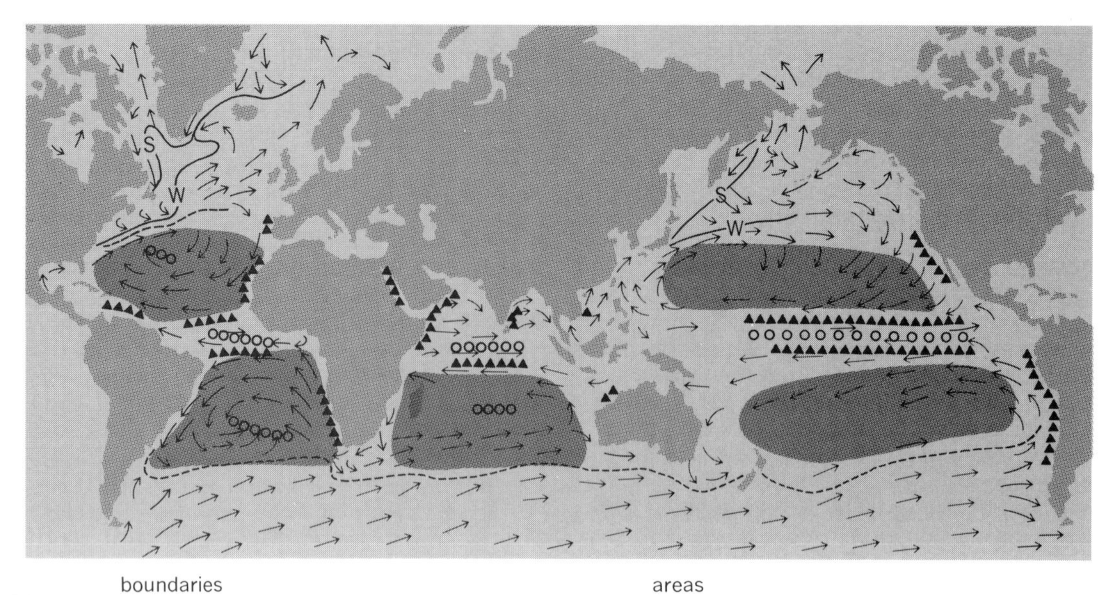

A world map (Mercator projection) which shows the ocean surface currents, the boundaries between areas of high and low productivity, and the principal regions of divergence and sinking.

euphotic zone, and there appears to be a low net production rate. Exceptions are found in regions of divergence, principally on the western coasts of the continents and to a lesser extent in the open ocean in the equatorial region, where upwelling of rich deeper waters permits a high productivity, often throughout the year.

The question of relative production in high and low latitudes is still undecided, for the high productivity of the polar seas is of short seasonal duration; the tropics may in fact equal it on an annual basis, although running at a lower instantaneous rate. As yet, measurements of production rates are inconclusive on this point.

Some estimate of the production of higher animals can be obtained from commercial fishing and whaling statistics. The correlation with net production rate appears to be fairly good; however, it is modified by feeding and breeding requirements of the animals in question and by the fact that many higher animals of no commercial interest may be produced in some areas. [RONALD I. CURRIE]

Size of populations and fluctuations. In temperate waters fish tend to spawn at the same place at a fixed season. The larvae drift in a current from spawning ground to nursery ground, and the adults migrate from the feeding ground to the spawning ground. Adolescents leave the nursery to join the adult stock on the feeding ground. The migration circuit is based on the track of the larval drift, and the population is isolated by its unique time of spawning. Because larvae suffer intense mortality, numbers in the population are perhaps regulated naturally during the period of larval drift.

Methods of measuring the size of marine populations are of three basic kinds. The first is by a census based on samples which together constitute a known fraction either of the whole population, as in the case of sessile species such as shellfish, or of a particular age range, as in the case of fish which have pelagic eggs whose total abundance can be measured by fine-meshed nets hauled vertically through the water column. The second is by marking or tagging, in which a known number of marked individuals are mixed into the population and the ratio of unmarked to marked individuals is subsequently measured from samples. The third is applicable to commercially exploited populations where the total annual catch is known; the mortality rate caused by exploitation is measured, based on the age composition of the populations, thus establishing what fraction of the population the catch is. The last two methods are most generally used.

The largest measured populations are of pelagic fish, particularly of the herring family and related species, Clupeidae. One of these is the Atlantic herring (*Clupea harengus*) which contains on the order of 1,000,000,000 mature individuals and ranges over hundreds of miles of the northeast Atlantic.

Populations of bottom-living fish tend to fluctuate slowly over long time periods of 50–75 years. However, those of pelagic fish, like the Atlantic herring, may fluctuate dramatically. If fish spawn at a fixed season, they are vulnerable to climatic change. During long periods there are shifts in wind strength and direction, with consequent delays or advancements in the timing of the production cycle. If the cycle becomes progressively delayed, the fish larvae, hatched at a fixed season, become progressively short of food, and the populations decline with time. Climatic change affects the population during the period of larval drift when isolation is maintained and when numbers are normally regulated. [D. H. CUSHING]

Biological species and water masses. Biogeographical regions in the ocean are related to the distribution of water masses. Their physical individuality and ecological individuality are derived from partly closed patterns of circulation and from amounts of incident solar radiation characteristic of latitudinal belts. Each region may be described in terms of its temperature-salinity property and of the biological species which are adapted to all or part of the relatively homogeneous physical-chemical environment.

Cosmopolitan species. The discrete distributions of many species are circumscribed by the regions of oceanic convergence bounding principal water masses. Other distributions are limited to current systems. Cosmopolitan species are distributed across several of the temperature-salinity water masses or oceans; their wider specific tolerances reflect adaptations to broadly defined water types. No pelagic distribution is fully understood in terms of the ecology of the species.

A habitat is integrated and maintained by a current system: oceanic gyral, eddy, or current, with associated countercurrents. This precludes species extinction that could occur if a stock were swept downstream into an alien environment. The positions of distribution boundaries may vary locally with seasonal or short-term changes in temperature, available food, transparency of the water, or direction and intensity of currents.

Phytoplankton species are distributed according to temperature tolerances in thermal water masses, but micronutrients (for example, vitamin B_{12}) are essential for growth in certain species. The cells of phytoplankton reproduce asexually and sometimes persist in unfavorable regions as resistant resting spores. New populations may develop in prompt response to local change in temperature or in nutrient content of the water. Such species are less useful in tracing source of water than are longer-lived, sexually reproducing zooplankton species.

Indicator organisms. The indicator organism concept recognizes a distinction between typical and atypical distributions of a species. The origin of atypical water is indicated by the presumed affinity of the transported organisms with their established centers of distribution.

Zooplankton groups best understood with respect to their oceanic geography are crustaceans such as copepods and euphausiids, chaetognaths (arrow worms), polychaetous annelids, pteropod mollusks, pelagic tunicates, foraminiferids, and radiolarians. Of these the euphausiids are the strongest diurnal vertical migrants (200–700 m). The vertical dimension of euphausiid habitat agrees with the thickness of temperature-salinity water masses, and many species distributions correspond with the positions of the masses. In the Pacific different species, some of which are endemic to their specific waters, occupy the subarctic mass (such as *Thysanoessa longipes*), the transition zone, a mixed mass lying between subarctic and central water in midocean and between subarctic and equatorial water in the California Current (for example, *Nematoscelis difficilis*), the barren North Pacific central (such as *Euphausia hemigibba*) and South Pacific central masses (for example, *E. gibba*), the Pacific equatorial mass (such as *E. diomediae*), a southern transition zone analogous to that of the Northern Hemisphere (represented by *Nematoscelis megalops*), and a circumglobal subantarctic belt south of the subantarctic convergence (such as *E. lucens*).

Epipelagic fishes and other strongly swimming vertebrates are believed to be distributed according to temperature tolerances of the species and availability of food. However, distributions of certain bathypelagic fishes (such as *Chauliodus*) have been related to water mass. See SEA WATER.

[EDWARD BRINTON]

Bibliography: R. J. H. Beverton and S. J. Holt, On the dynamics of exploited fish populations, *Fish Invest.* (London), vol. 19, 1957; R. J. Browning, *Fisheries of the North Pacific*, 1974; D. H. Cushing, Biological and hydrographic changes in British seas during the last thirty years, *Biol. Rev.*, 41:221–258, 1966; A. W. Ebeling, *Melamphaidae, I: Systematics and Zoogeography of the Species in the Bathypelagic Fish Genus Melamphaes*, Dana Rep. no. 58, 1962; F. E. Firth, *Encyclopedia of Marine Resources*, 1969; J. A. Gullond, *Population Dynamics of the World Fisheries*, 1972; H. W. Harvey, *The Chemistry and Fertility of Sea Waters*, 2d ed., 1957; M. N. Hill (ed.), *The Sea*, vol. 2: *The Composition of Sea Water, Biological Species, Water-Masses and Currents*, 1963; J. E. G. Raymont, *Plankton and Productivity in the Oceans*, 1963; J. P. Riley and G. Skirrow, *Chemical Oceanography*, vol. 1, 1965; W. D. Russel-Hunter, *Aquatic Productivity: An Introduction to Some Basic Aspects of Biological Oceanography and Limnology*, 1970.

Sensible temperature

The temperature at which air with some standard humidity, motion, and radiation would provide the same sensation of human comfort as existing atmospheric conditions. Of the many sensible temperature formulas thus far proposed, none is completely satisfactory or generally accepted. Most are intended for warm, moist conditions; a few, like "wind chill," are for cold weather. Some are purely empirical, modifying the actual temperature according to the humidity; others are theoretical, and express estimated heat loss rather than an equivalent temperature. See HUMIDITY.

Heat is produced constantly by the human body at a rate depending on muscular activity. For body heat balance to be maintained, this heat must be dissipated by conduction to cooler air, by evaporation of perspiration into unsaturated air, and by radiative exchange with surroundings. Air motion (wind) affects the rate of conductive and evaporative cooling of skin, but not of lungs; radiative losses occur only from bare skin or clothing, and depend on its temperature and that of surroundings, as well as sunshine intensity.

As air temperatures approach body temperature, conductive heat loss decreases and evaporative loss increases in importance. Hence, at warmer temperature, humidity is the second most important atmospheric property controlling heat loss and hence comfort, and the various sensible temperature formulas incorporate some humidity measure. Most used is C. P. Yaglou's effective temperature, represented by lines of equal comfort on a chart of dry-bulb versus wet-bulb temperature; it relates existing comfort to that in motionless, saturated air.

Effective temperature is approximated by E. Thom's temperature-humidity index (THI), called discomfort index until certain southern United States cities objected, given in °F by Eq. (1), where

$$\text{THI} = 15 + 0.4t \quad (1)$$

t is the sum of the dry-bulb and wet-bulb temperatures. The THI is routinely tabulated at major Weather Bureau stations, and accumulated into "cooling degree days." Similar is humiture, first applied by O. F. Heavener in 1937 to the average of temperature in °F and relative humidity, and redefined by V. E. Lally and B. F. Watson in 1960, as Eq. (2), where t is Fahrenheit temperature and e is

$$\text{Humiture} = t + e - 10 \quad (2)$$

vapor pressure in millibars. Many other "comfort temperatures" and "sultriness indices" have been proposed.

Under cold conditions, atmospheric moisture is negligible and wind becomes important in heat removal. P. A. Siple's wind chill, given by Eq. (3),

$$\text{Wind chill} = (10.45 + 10\sqrt{v} - v)(33 - t) \quad (3)$$

does not estimate sensible temperature as such, but heat loss in kcal/m^2 hr, for wind speed v in m/sec and air temperature t in °C. It is used extensively by the U.S. Army, government agencies, construction contractors, and others in cold areas. *See the feature article* ENVIRONMENTAL PROTECTION. [ARNOLD COURT]

Bibliography: S. Licht (ed.), *Medical Climatology*, 1978.

Septic tank

A single-story settling tank in which settled sludge is in immediate contact with sewage flowing through the tank while solids are being decomposed by anaerobic bacterial action. Such tanks have limited use in municipal treatment, but are the primary resource for the treatment of sewage from individual residences. There are probably well over 4,000,000 septic tanks in use in home disposal systems in the United States. Septic tanks are also used by isolated schools and institutions and for sanitary sewage treatment at small industrial plants.

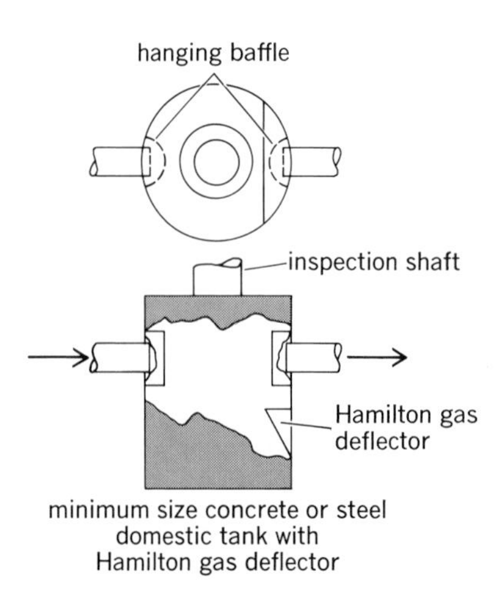

Fig. 1. Circular household septic tank. *(From H. E. Babbitt and E. R. Baumann, Sewerage and Sewage Treatment, 8th ed., copyright 1958 by John Wiley & Sons, Inc.; used with permission)*

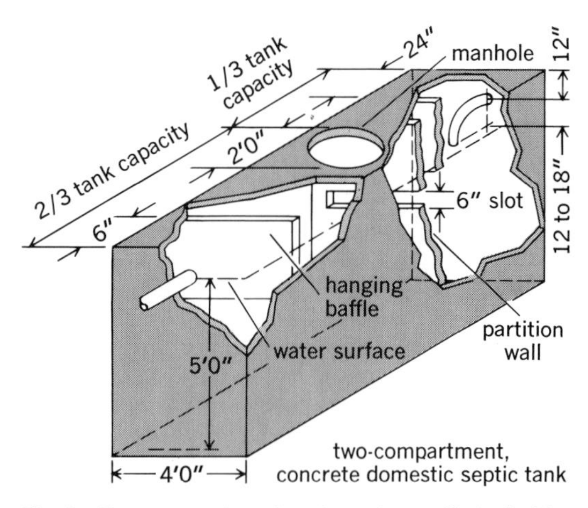

Fig. 2. Two-compartment rectangular septic tank. 1 in. = 25 mm. 1 ft. = 0.30 m. *(From H. E. Babbitt and E. R. Baumann, Sewerage and Sewage Treatment, 8th ed., copyright 1958 by John Wiley & Sons, Inc.; used with permission)*

Home disposal units. Septic tanks have a capacity of approximately 1 day's flow. Since sludge is collected in the same unit, additional capacity is provided for sludge. One formula for sludge storage that has been used is $Q = 17 + 7.5y$, where Q is the volume of sludge and scum in gallons per capita per year, and y is the number of years of service without cleaning. About one-half of a 500-gal (1.89 m^3) tank is occupied by sludge in 5 years in an ordinary household installation. The majority of states require a minimum capacity of 500 gal in a single tank. Some states require a second compartment of 300-gal (1.14-m^3) capacity. Single- and double-compartment tanks are shown in Figs. 1 and 2. Such units are buried in the ground and are not serviced until the system gives trouble because of clogging or overflow. Commercial scavenger companies are available in most areas. A tank truck equipped with pumps is brought to the premises, and the tank content is pumped out and taken to a sewer manhole or a treatment plant for disposal. In rural areas the sludge may be buried in an isolated place.

Municipal and institutional units. These are designed to hold 12–24 hours' flow, with additional sludge capacity provided. Provision is made for sludge withdrawal about once a year. Desirable features of design are (1) watertight and corrosion-resistant material (concrete and well-protected metal have been used); (2) a vented tank; (3) manhole openings in the roof of the tank to permit inspection; (4) baffles at the inlet and the outlet to a depth below the probable scum line, usually 18–24 in. below the water surface; (5) sludge draw-off lines—although seldom used, they should be designed so that they can be rodded or unplugged by some positive mechanism; (6) hoppers or sloped bottoms so that digested sludge can be withdrawn as required; and (7) provision for safe handling of septic tank effluent by disposal underground or by chlorination before discharge to a stream, or both.

Tank efficiency. Septic tank effluent is dangerous and odorous. It will contain pathogenic bacteria and sewage solids. Particles of sludge and scum are trapped in the flow and will cause nuisance at the point of discharge unless properly handled. Efficiency in removal of solids is less than that for plain sedimentation. While 60% suspended solids removal is used theoretically, it is seldom obtained in practice. Improvement is noted when tanks are built with two compartments. Shallow tanks give somewhat better results than very deep tanks. *See* SEWAGE TREATMENT.

[WILLIAM T. INGRAM]

Bibliography: U.S. Public Health Service, *Manual of Septic Tank Practice*, 1960; P. Warshall, *Septic Tank Practices*, 1979.

Sewage

A combination of (1) liquid wastes conducted away from residences, institutions, and business buildings and (2) the liquid wastes from industrial establishments with (3) such surface, ground, and storm water as may find its way or be admitted into the sewers. Category 1 is known as sanitary or domestic sewage; 2 is usually referred to as industrial waste; 3 is known as storm sewage.

Relation to water consumption. Sewage is the waste water reaching the sewer after use; hence it is related in quantity and in flow fluctuation to water use. The quantity of sewage is generally less than the water consumption since some portion of water used for fire fighting, lawn irrigation, street washing, industrial processing, and leakage does not reach the sewer. These losses are compensated for partly by the addition of water from private wells, groundwater infiltration, and illegal connections from roof drains. Water consumption increases with size of community served and many other community characteristics. Characteristics of each city must be studied and analyzed for specific information. As a general average estimate, communities with population under 1000 use about 60 gal per capita per day (gcd) (230 liters per capita per day), while communities of 100,000 use about 140 gcd (530 liters per capita per day). In a study of large cities of the United States, the median consumption was 154 gcd (583 liters per capita per day) and the median population was 658,000 (Fig. 1). An accepted unit flow for domestic sewage as shown in the table is 100 gcd (380 liters per capita per day).

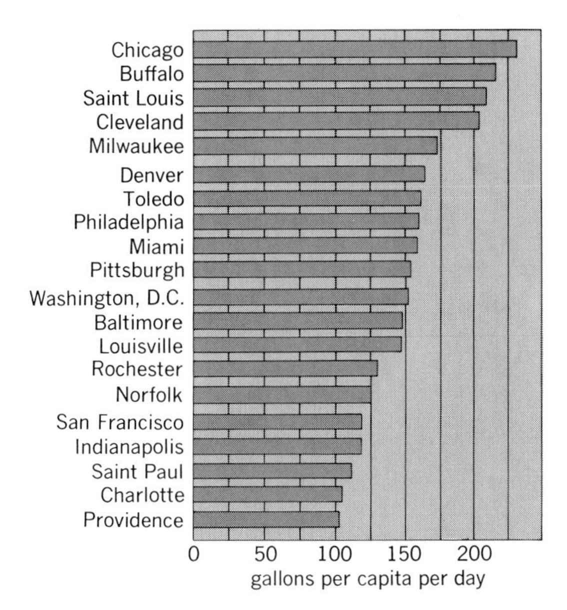

Fig. 1. Estimated water use in 20 major cities of the United States. 1 gal = 3.785 liters. (*Research Division, New York University College of Engineering*)

Rates of sewage flow from various sources*

Character of district	Gal per capita per day	Gal per acre per day	Source of sewage	Gal per capita per day
Domestic			Trailer courts	50†
Average	100		Motels	53†
High-cost dwellings	150	7,500	State prisons	
Medium-cost dwellings	100	8,000	Maximum	280‡
Low-cost dwellings	80	16,000	Average	176
Commercial			Minimum	104
Hotels, stores, and			Mental hospitals	
office buildings		60,000	Maximum	216‡
Markets, warehouses,			Average	123
wholesale districts		15,000	Minimum	38
Industrial			Grade school	4.4§
Light industry		14,000	High school	3.9§

SOURCE: From H. E. Babbitt and E. R. Baumann, *Sewerage and Sewage Treatment*, 8th ed., copyright 1958 by John Wiley & Sons, Inc.

*1 gal = 3.785 liters, 1 gal per acre = 9.35 liters per hecture.

‡From J. C. Frederick, *Public Works*, p. 112, April, 1957.

§Average of 4.4 gal per day per pupil between 7:30 A.M. and 5:30 P.M. The average for the high school is spread over more hours per day. From C. H. Coberly, *Public Works*, p. 143, May, 1957.

Infiltration of groundwater should be held to a minimum. It may be expected to be equal to or less than 30,000 gal/(day)(mile) [70 m³/(day)(km)] of sewer including house connections. Much depends on the quality of sewer construction. Water may enter through poorly made joints and, in quantity, through poorly constructed, leaky manholes and illegal and abandoned sewers. Sewers in wet ground with a high water table will have more infiltration. Sewers under pressure may have infiltration or leakage to the surrounding ground. The danger of groundwater pollution from leaky sewers should be avoided.

Fluctuations in sewage flow are related to water use characteristics but tend to dampen out since there is a time lag from the time of use to appearance in the sewer mains and trunks (Fig. 2). Hourly, daily, and seasonal fluctuations affect design of sewers, pumping stations, and treatment plants.

Daily and seasonal variations depend largely on community characteristics. Weekend flows may be lower than weekday. Industrial operations of seasonal nature influence the seasonal average. The seasonal average and annual average are about equal in May and June. The seasonal average is about 124% in late summer and may drop to about 87% at the end of winter. Peak flows may reach 200% of average at the treatment plant and may be more than 300% of average in the laterals. Laterals are designed for 400 gcd (1500 liters per capita per day) and mains and trunks for 250 gcd (950 liters per capita per day).

Design periods. These are dependent on the proposed sewer construction. Lateral sewers may be designed for ultimate flow of the area to be sewered. Mains may be designed for periods of 10–40 years. Trunk sewers may be planned for long periods with provision made in design for parallel or separate routings of trunks of smaller size to be constructed as the need arises. Economics, available funds, and engineering judgment affect selection of the design periods. Appurtenances may have a different life, since replacement of mechanical equipment will be necessary. A span of 20–25 years is often selected and a timetable of additions during that period is then scheduled in the overall improvement plan.

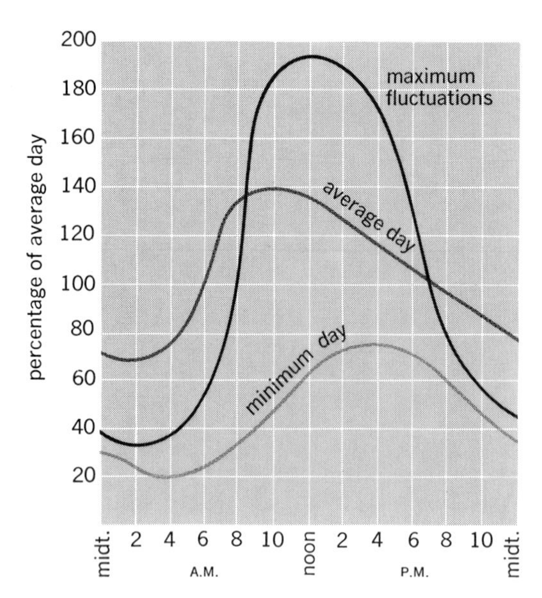

Fig. 2. Typical hourly variations in sewage flow. *(From H. E. Babbitt and E. R. Baumann, Sewerage and Sewage Treatment, 8th ed. copyright 1958 by John Wiley & Sons, Inc.; used with permission)*

Storm sewage. Storm sewage is liquid flowing into sewers during or following a period of rainfall and resulting therefrom. An estimate of the quantity of storm sewage is necessary in sewage design.

Estimating quantity of storm sewage requires a knowledge of intensity and duration of storms, distance water travels before reaching sewer, permeability and slope characteristics of the surface over which water flows to sewer inlet, and shape and amount of area to be drained to inlet. These general considerations are included in the equation $Q=CAIR$ expressing the runoff from a watershed having no retention or storage of water. Q is expressed in cubic feet per second (cfs), A is area, I is the relative imperviousness of the surface expressed as a decimal, and R is the rate of rainfall in inches per hour. C is a coefficient permitting the expression of the factors in convenient units; in the above units it may be taken as 1 so that the equation becomes $Q=AIR$.

Time of concentration is a combination of the theoretical time required for a drop to run from the most distant point to the inlet and time from sewer inlet to point of concentration. The inlet time may range from 3 min for a steep slope on an impervious area to 20 min on a city block. The time of flow is assumed to be the velocity in the full flowing sewer divided by the length of sewer from inlet to point of concentration. Flood crest and storage time while the sewer is filling are usually neglected, the effect being that of assuming a larger rate of flow and so providing a safety factor in design.

Values of I, the runoff coefficient, range from 0.01 in wooded areas to 0.95 on roofed surfaces. A common value used in residential areas with considerable land in lawn, garden, and shrubbery is 0.30–0.40. In built-up areas, values of 0.70–0.90 may be used.

Rainfall intensity values are selected on the basis of frequency and duration of storms. In some sewer design it is necessary to select a value for the expected occurrence of maximum runoff. This is done by using one of the several formulas which will allow a prediction of R for 5, 10, or 15 years. The element of calculated risk is combined with engineering judgment in deciding which R to choose. For lesser structures in residential areas a 5-year frequency may be used with reasonable safety. Where failure would endanger property, the 10-, 15-, or 25-year frequency of occurrence provides a more conservative design basis; 50-year frequency may be selected where flooding could cause lasting damage and disrupt facilities. In such instances cost-benefit studies may be made to guide the selection of a suitable frequency. *See* HYDROLOGY.

Pumping sewage. Not all sewage will flow by gravity without unnecessary expense in circuitous routing or deep excavation; therefore, pumping stations may be advantageous. Pumping stations may be required in the basements of large buildings. Pumping stations are provided with two or more pumps of sufficient capacity so that with one unit out of service the remaining unit or units will pump the maximum flow. Motive power is required from at least two sources, usually electric motors and auxiliary fuel-fed motor drive. Care must be taken to have motive power above flood level and protected from the elements. Screening is usually required ahead of pumping stations, unless the pumps themselves are self-cleaning. Many states require that pumping units be installed in a dry well and that sewage be confined to a separate wet well. Buildings above ground should fit the surroundings. Small pumping stations are often one unit and made fully automatic so that minimum attendance is required. Safety measures must be considered. Centrifugal pumps are used almost exclusively in larger stations. Air ejector units may be installed in smaller stations.

Examination of sewage. Sewage is actually water with a small amount of impurity in it. Examination of sewage is required to know the effects of these impurities. Various tests are used to aid in determining the characteristics, composition, and condition of sewage. These include physical examination, solids determination, tests for determining the oxygen requirement of organic matter, chemical and bacteriological tests, and examination under the microscope.

Physical tests for turbidity, odor, color, and temperature are made. Normal fresh sewage is gray and somewhat opaque, has little odor, and has a temperature slightly higher than the water supply. Decomposition of organic matter darkens the sewage, and odors are characteristic of stale or septic sewage.

Tests for residue or solid matter provide an indication of the types of solids, the strength of the sewage, and the physical state of the solids. Total solids determinations measure both suspended and dissolved solids. A sample of the sewage is filtered. The suspended solids can be determined by drying the material recovered on the filter. The dissolved solids can be determined by evaporation of the filtered portion. Heating the solids residue until organic matter gasifies separates volatile solids from fixed solids or inorganic ash. Loss on ignition represents the volatile or organic fraction and is a good measure of sewage strength.

Measurement of the part of the suspended solids heavy enough to settle is made in an Imhoff cone. The settleable-solids test is useful in determining the sludge-producing characteristics of sewage.

Tests for organic matter are made principally to determine the oxygen requirement of sewage. These tests include the biochemical oxygen demand test (BOD), the chemical oxygen demand test, the oxygen consumed test, and the relative stability test. Organisms in sewage require oxygen for growth, and the BOD measures the amount of dissolved oxygen required for decomposition of organic solids for a measured time at a constant temperature. The standard measurement is made for 5 days at 20°C and is a good measure of sewage strength. Since the BOD measurement includes both biological and chemical oxygen requirement, another test, the chemical oxygen demand, is sometimes used to measure the chemical oxygen requirement. Sewage is heated in the presence of an oxidizing agent such as potassium dichromate. The oxygen requirement is that of chemical digestion since all organisms have been killed. This test is increasing in use. The oxygen-consumed test uses potassium dichromate as the oxidizing agent. The result offers some index of the readily oxidizable carbonaceous material. The relative stability test indicates when the oxygen present in plant effluent or polluted water is exhausted. The data express as a percentage the approximate amount of oxygen available in water in relation to the amount required for complete stability. The test is a color test using methylene blue. Reducing agents, precipitation of color, concentration of dye, amount of dissolved oxygen in the sample, and other factors affect the reliability of this test, and it is considered generally as a rough or screening test of the condition of plant effluent. Tests for nitrogen include those for free ammonia, albuminoid ammonia, organic nitrogen, nitrites, and nitrates. The latter are indications of oxidation change and stabilization and are used in checking condition of plant effluent.

Bacteriological tests are made primarily to determine the presence of organisms of the coliform group. The organisms exist in the intestines of warm-blooded animals and are used as an index of the presence of fecal material. The coliform test on chlorinated effluents determines efficiency of chlorination. Occasionally other bacteriological determinations are made to determine the presence of organisms of the *Salmonella* group or dysentery group in polluted water and sewage.

Microscopic tests are not normally made on raw sewage. They are used as part of plant operator control in treatment processes. Examinations for the presence of algae, protozoa, bacteria, fungi, rotifers, and worms are made when necessary.

[WILLIAM T. INGRAM]

Bibliography: APHA-AWWA-WPCA Joint Committee, *Standard Methods for the Examination of Water, Sewage, and Industrial Wastes*, 12th ed., 1965; R. E. Bartlett, *Public Health Engineering: Sewage*, 1979; G. M. Fair, J. C. Geyer, and D. A. Okun, *Water Supply and Waste-water Removal*, vol. 1, 1958; W. T. Ingram, *Water Fluoridation Practices in Major Cities of the United States*, pt. 1, Research Division, New York University College of Engineering, 1958; F. A. Kristal and F. A. Annett, *Pumps*, 2d ed., 1953; H. F. Seidel and J. L. Cleasby, Statistical analysis of water works data for 1960, *J. AWWA*, 58 (12):1507, 1966; E. W. Steel, *Water Supply and Sewerage*, 1979.

Sewage disposal

The discharge of waste waters into surface-water or groundwater courses, which constitute the natural drainage of an area. Most waste waters contain offensive and potentially dangerous substances, which can cause pollution and contamination of the receiving water bodies. Contamination is defined as the impairment of water quality to a degree that creates a hazard to public health. Pollution refers to the adverse effects on water quality that interfere with its proper and beneficial use.

In the past, the dilution afforded by the receiving water body was usually great enough to render waste substances innocuous. Since the turn of the century, however, the dilution of many rivers has been inadequate to absorb the greater waste discharges caused by the increase in population and expansion of industry.

The principal sources of pollution are domestic sewage and industrial wastes. The former includes the used water from dwellings, commercial establishments, and street washings. Industrial wastes constitute acids, chemicals, oils, and animal and vegetable matter carried by cleaning or used process waters from factories and plants. For a discussion of sources of wastes *see* SEWAGE.

Regulation of water pollution. This is primarily a responsibility of the state, in cooperation with the Federal and local governments. The health departments of many states are given statutory power and responsibility for the control of water pollution, and they have established specific water quality standards. There are two basic types of standards—stream standards, dealing with the quality of the receiving water, and effluent standards, referring to strength of wastes discharged. Both types are based on the capacity of the receiving waters to absorb waste substances and on the beneficial uses made of the water.

The self-purification capacity is determined by the available dilution, the biophysical environment of the stream, and the strength and characteristics of the wastes. Beneficial uses include drinking, bathing, recreation, fish culture, irrigation, industrial uses, and disposal of wastes without creation of pollution.

Adjustment of these conflicting interests and equitable distribution of water resources is complex from the technical, economic, and political viewpoints. These considerations have led to the establishment of interstate commissions, which provide a means of coordinated control of the larger rivers.

Water-quality criteria deal with the physical, chemical, and biological parameters of pollution. The most common standards are concerned with physical appearance, odor production, dissolved-oxygen concentration, pathogenic contamination, and potentially toxic or harmful chemicals. The allowable quantity and concentration of these characteristics and substances vary with the water usage.

Absence of odor and unsightliness, and the presence of some dissolved oxygen are common minimum standards. Preliminary or primary treatment of waste waters is usually required for the maintenance of these standards. Highest-quality waters require clarity, oxygen saturation, low bacteriological counts, and absence of harmful substances. In these cases, intermediate or complete treatment may be required. *See* SEWAGE TREATMENT; WATER ANALYSIS; WATER POLLUTION.

Stream pollution. Biological, or bacteriological, pollution is indicated by the presence of the coliform group of organisms. While nonpathogenic itself, this group is a measure of the potential presence of contaminating organisms. Because of temperature, food supply, and predators, the environment provided by natural bodies of water is not favorable to the growth of pathogenic and coliform organisms. Physical factors, such as flocculation and sedimentation, also help remove bacteria. Any combination of these factors provides the basis for the biological self-purification capacity of natural water bodies.

When subjected to a disinfectant such as chlorine, bacterial die-away is usually defined by Chick's law, which states that the number of organisms destroyed per unit of time is proportional to the number of organisms remaining. This law cannot be directly applied in natural streams because of the variety of factors affecting the removal and death rates in this environment. The die-away is rapid in shallow, turbulent streams of low dilution, and slow in deep, sluggish streams with a high dilution factor. In both cases, higher temperatures increase the rate of removal.

The concentration of many physical characteristics and chemical substances may be calculated directly if the relative volumes of the waste stream and river flow are known. Chlorides and mineral solids fall into this category. Some substances in waste discharges are chemically or biologically unstable, and their rates of decrease can be predicted or measured directly. Sulfites, nitrites, some phenolic compounds, and organic matter are examples of this type of waste.

These simple relationships, however, do not apply to the concentration of dissolved oxygen. This factor depends not only on the relative dilutions, but upon rate of oxidation of the organic material and rate of reaeration of the stream.

Nonpolluted natural waters are usually saturated with dissolved oxygen. They may even be supersaturated because of the oxygen released by green water plants under the influence of sunlight. When an organic waste is discharged into a stream, the dissolved oxygen is utilized by the bacteria in their metabolic processes to oxidize the organic matter. The oxygen is replaced by reaeration through the water surface exposed to the atmosphere. This replenishment permits the bacteria to continue the oxidative process in an aerobic environment. In this state, reasonably clean appearance, freedom from odors, and normal animal and plant life are maintained.

An increase in the concentration of organic matter stimulates the growth of bacteria and increases the rates of oxidation and oxygen utilization. If the concentration of the organic pollutant is so great that the bacteria use oxygen more rapidly than it can be replaced, only anaerobic bacteria can survive and the stabilization of organic matter is accomplished in the absence of oxygen. Under these conditions, the water becomes unsightly and malodorous, and the normal flora and fauna are destroyed. Furthermore, anaerobic decomposition proceeds at a slower rate than aerobic. For maintenance of satisfactory conditions, minimal dissolved oxygen concentrations in receiving streams are of primary importance.

Figure 1 shows the effect of municipal sewage and industrial wastes on the oxygen content of a stream. Cooling water, used in some industrial processes, is characterized by high temperatures, which reduce the capacity of water to hold oxygen in solution. Thermal pollution, however, is significant only when large quantities are concentrated in relatively small flows. Municipal sewage requires oxygen for its stabilization by bacteria. Oxygen is utilized more rapidly than it is replaced by reaeration, resulting in the death of the normal aquatic life. Further downstream, as the oxygen demands are satisfied, reaeration replenishes the oxygen supply.

Any organic industrial waste produces a similar pattern in the concentration of dissolved oxygen. Certain chemical wastes have high oxygen demands which may be exerted quickly, producing a sudden drop in the dissolved oxygen content. Other chemical wastes may be toxic and destroy the biological activity in the stream. Strong acids and alkalies make the water corrosive, and dyes, oils, and floating solids render the stream unsightly. Suspended solids, such as mineral tailings, may settle to the bed of the stream, smother purifying microorganisms, and destroy breeding places. Although these latter factors may not deplete the oxygen, the polutional effects may still be serious.

Deoxygenation. Polluted waters are deprived of oxygen by the exertion of the biochemical oxygen demand (BOD), which is defined as the quantity of oxygen required by the bacteria to oxidize the organic matter. The rate of this reaction is assumed to be proportional to the concentration of the re-

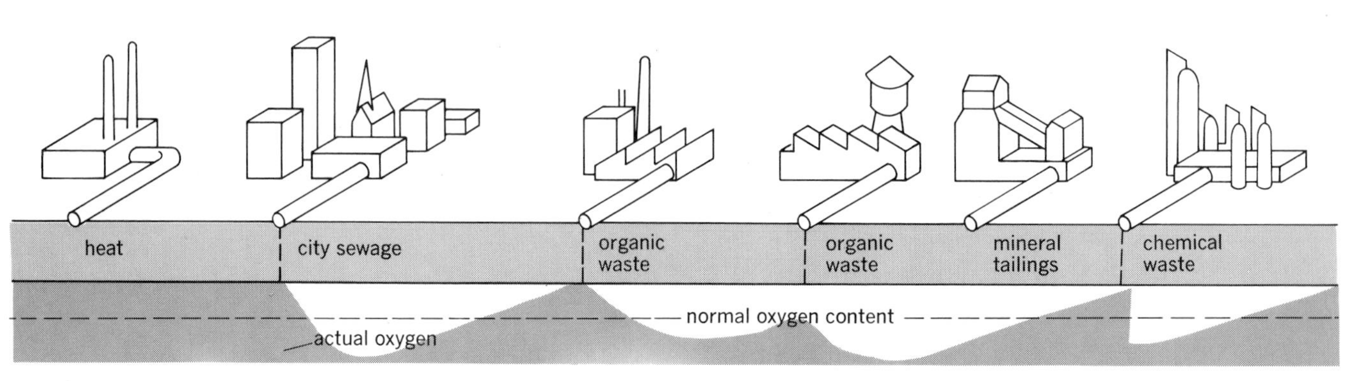

Fig. 1. Variation of oxygen content of polluted stream.

maining organic matter, measured in terms of oxygen. This reaction may be expressed as Eq. (1),

$$\frac{dL}{dt}=-K_1L \tag{1}$$

which integrates to give Eq. (2) or Eq. (3), in which

$$L_t=L_0e^{-K_1t} \tag{2}$$

$$y=L_0(1-e^{-K_1t}) \tag{3}$$

L_t is BOD remaining at any time t, L_0 is ultimate BOD, y is BOD exerted at end of t, and K_1 is coefficient defining the reaction velocity. The coefficient is a function of temperature given by Eq. (4), in which T is temperature in degrees Cel-

$$K_T=K_{20}\cdot 1.047^{T-20} \tag{4}$$

sius, K_T is value of the coefficient at T, and K_{20} is value of the coefficient at 20°C.

The BOD of a waste is determined by a standard laboratory procedure and is reported in terms of the 5-day value at 20°C. From a set of BOD values determined for any time sequence, the reaction velocity constant K_1 may be calculated. Knowledge of this coefficient permits determination of the ultimate BOD from the 5-day value in accordance with the above equations. For municipal sewages and many industrial wastes the value of K_1 at 20°C is between 0.15 and 0.75 per day. A common value for sewage is 0.4 per day.

The coefficient determined from laboratory BOD data may be significantly different from that calculated for stream BOD data. The determination of the stream rate may be made from a reexpression of Eqs. (1)–(4) in the form of Eq. (5),

$$K_r=\frac{1}{t}\log\frac{L_A}{L_B} \tag{5}$$

where L_A is the BOD measured at an upstream station, L_B is the BOD at a station downstream from A, and t is the time of flow between the two stations. Values of K_r range from 0.10 to 3.0 per day. The difference between the laboratory rate K_1 and the stream rate K_r is due to the turbulence of the stream flow, biological growths on the stream bed, insufficient nutrients, and inadequate bacteria in the river water. These factors influence the rate of oxidation in the stream as well as the removal of organic matter. Such processes as flocculation, sedimentation, and scour of the organic material in the river affect the removal rate but do not necessarily influence the rate of oxidation and the associated dissolved oxygen concentration. Field surveys are usually required to determine the pollution assimilation capacity of a stream.

When a significant portion of the waste is in the suspended state, settling of the solids in a slow-moving stream is probable. The organic fraction of the sludge deposits decomposes anaerobically, except for the thin surface layer which is subjected to aerobic decomposition due to the dissolved oxygen in the overlying waters. In warm weather, when the anaerobic decomposition proceeds at a more rapid rate, gaseous end products, usually carbon dioxide and methane, rise through the supernatant waters. The evolution of the gas bubbles may raise sludge particles to the water surface. Although this phenomenon may occur while the water contains some dissolved oxygen, the more intense action during the summer usually results in depletion of dissolved oxygen.

Reoxygenation. Water may absorb oxygen from the atmosphere when the oxygen in solution falls below saturation. Dissolved oxygen for receiving waters is also derived from two other sources: that in the receiving water and the waste flow at the point of discharge, and that given off by green plants. The latter source is restricted to daylight hours and the warmer seasons of the year and, therefore, is not usually used in any engineering analysis of stream capacity.

Unpolluted water maintains in solution the maximum quantity of dissolved oxygen. The saturation value is a function of temperature and the concentration of dissolved substances, such as chlorides. When oxygen is removed from solution, the deficiency is made up by the atmospheric oxygen, which is absorbed at the water surface and passes into solution. The rate at which oxygen is absorbed, or the rate of reaeration, is proportional to the degree of undersaturation and may be expressed as in Eq. (6), in which D is dissolved oxy-

$$\frac{dD}{dt}=-K_2D \tag{6}$$

gen deficit, t is time, and K_2 is reaeration coefficient.

The reaeration coefficient depends upon the ratio of the volume to the surface area and the intensity of fluid turbulence. An approximate value of the coefficient may be obtained from Eq. (7), in

$$K_2=\frac{D_LU^{1/2}}{H^{3/2}} \tag{7}$$

which D_L is coefficient of molecular diffusion of oxygen in water, U is average velocity of the river flow, and H is average depth of the river section.

The effect of temperature on this coefficient is identical with its effect on the deoxygenation coefficient. A common range of K_2 is from 0.20 to 5.0 per day. Many waste constituents, such as surface-active substances, interfere with the molecular diffusion of oxygen and reduce the value of the reaeration rate from that of pure water. Winds, waves, rapids, and tidal mixing are factors which create circulation and surface renewal and enhance reaeration.

Oxygen balance. The oxygen balance in a stream is determined by the concentration of organic matter and its rate of oxidation, and by the dissolved oxygen concentration and the rate of reaeration. The simultaneous action of deoxygenation and reaeration produces a pattern in the dissolved oxygen concentration known as the dissolved oxygen sag. The differential equation describing the combined action of deoxygenation and reaeration is given in Eq. (8), which states that the rate of

$$\frac{dD}{dt}=K_1L-K_2D \tag{8}$$

change in the dissolved oxygen deficit D is the result of two independent rates. The first is that of oxygen utilization in the oxidation of organic matter. This reaction increases the dissolved oxygen deficit at a rate that is proportional to the concentration or organic matter L. The second rate is that of reaeration, which replenishes the oxygen utilized by the first reaction and decreases the deficit. Integration of this equation yields Eq. (9), where L_0

$$D_t=\frac{K_1L_0}{K_2-K_r}(e^{-K_rt}-e^{-K_2t})+D_0e^{-K_2t} \qquad (9)$$

and D_0 are the initial biochemical oxygen demand and the initial dissolved oxygen deficit, respectively, and D_t is the deficit at time t. The proportionality constants K_1 and K_2 represent the coefficients of deaeration and reaeration, respectively, and K_r the coefficient of BOD removal in the stream.

Figure 2 shows a typical dissolved oxygen sag curve resulting from a pollution of amount L_0 at $t=0$. The sag curve is shown to result from the deoxygenation curve and the reaeration curve. A point of particular significance on the sag curve is that of minimum dissolved oxygen concentration, or maximum deficit. At this location, the rate of change of the deficit is zero, which results in the numerical equality of the opposing rates of deoxygenation and reoxygenation. The balance at this critical point may be written as Eq. (10), where the BOD at

$$K_2D_c=K_1L=K_1L_0e^{-K_rt_c} \qquad (10)$$

the critical point has been replaced by its equivalent at zero time (the location of the waste discharge). The value of the time t_c may be calculated from Eq. (11).

$$t_c=\frac{1}{K_2-K_r}\log\frac{K_2}{K_r}\left[1-\frac{D_0(K_2-K_r)}{K_1L_0}\right] \qquad (11)$$

Allowable pollutional load. The pollutional load L_0 that a stream may absorb is a function of the dissolved oxygen deficit D_c, the coefficients K_1, K_r, and K_2, and the initial deficit D_0. The dissolved oxygen deficit is usually established by water pollution standards of the health agency, and the initial deficit is determined by upstream pollution. The engineering problem is usually associated with the assignment of representative values of the coefficients K_1, K_r, and K_2 for a given flow and temperature condition.

Seasonal temperatures influence the saturation of oxygen and the rates of deaeration and reaeration. Variation in stream flow with the seasons affects the dilution factor. The most critical conditions occur during the summer when the stream runoff is low and the temperatures are high.

Pollution in lakes and estuaries. In lakes self-purification is slower than in streams because of the low rates of dispersion of the waste waters.

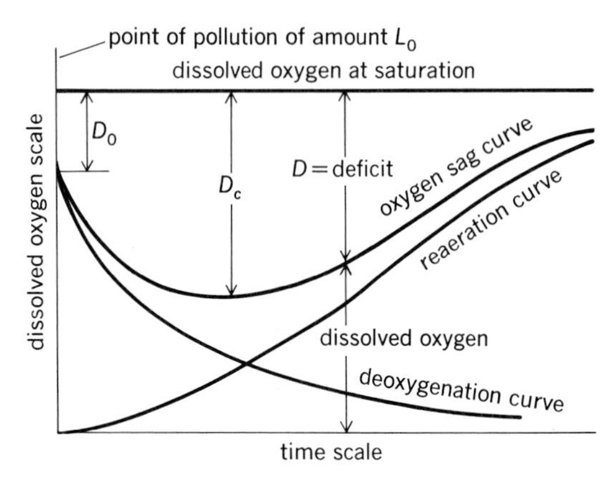

Fig. 2. Dissolved-oxygen sag curve and its components.

There is no turbulence characteristic as in flowing rivers, and mixing depends primarily on winds, waves, and currents. Waste-water outfalls are designed to take advantage of the dispersion induced by these factors and to prevent the development of concentrated sewage fields.

In estuaries, the dispersion of waste waters is complicated by the tides, which carry various portions of the pollutant back and forth over many cycles, and by the difference of density in fresh water, waste water, and salt water. The equation defining the oxygen balance must be modified to allow for the greater time that an average particle of pollution is detained within the estuary; the flushing mechanism of such bodies is therefore of primary concern. *See* ESTUARINE OCEANOGRAPHY.

Each estuary presents problems of density currents, configuration, and exchange that distinguish it from others. Field measurements of salinities, currents, and cross sections, in addition to the measurement of physical, chemical, and biological characteristics, are necessary to evaluate the pollution capacity of these watercourses. Dilution and dispersion in ocean waters is complicated by many of the same factors as in estuaries. The death rates of the coliform bacteria are greater in sea water. The outfalls must be designed and located to promote effective dispersion and to prevent the accumulation of sewage fields.

Oxidation ponds and land disposal. The forces of natural purification are utilized in shallow ponds, called oxidation ponds. Successful operation of these basins usually requires relatively high temperatures and sunshine. Carbon dioxide is released by means of the bacterial decomposition of the organic matter. Algae growth develops, consuming the carbon dioxide, ammonia, and other waste products and releasing oxygen under proper climatic conditions.

Oxidation ponds are efficient and relatively economical.

Instead of relying on the algae as a primary source of oxygen, mechanical aeration of the pond contents may be employed. Lagoons aerated in this manner are not as susceptible to climatic conditions as the oxidation ponds.

Land disposal of sewage is occasionally practiced by surface or flood irrigation. The former is the discharge of sewage upon the ground, from which it evaporates and through which it percolates. However, a significant portion remains which must be collected in surface drainage channels. Although this method is not particularly efficient for domestic sewage, a modification of it, spray irrigation, has been successfully employed in the treatment of a few industrial wastes. In flood irrigation, all the sewage is permitted to seep through the ground and is usually collected in underdrains. This method takes advantage of the mechanical filtration and biological purification afforded by the soil. Unless the sewage is treated before irrigation, odors and clogging usually occur and possible contamination of ground or surface water can result. *See* SANITARY ENGINEERING.

[DONALD J. O'CONNOR]

Bibliography: E. P. Anderson, *Audel's Domestic Water Supply and Sewage Disposal Guide*, 1960; D. A. Okun and G. Ponghis, *Community Wastewater Collection and Disposal*, 1975; H. W. Streeter and E. B. Phelps, *A Study of the Pollution and Natural Purification of the Ohio River*, Public Health Bull. no. 146, 1925; Texas Water and Sewage Works Association, *Manual for Sewage Plant Operators*, 3d ed., 1964; U.S. Public Health Service, *Oxygen Relationships in Stream*, Tech. Rep. no. W58–2, 1958.

Sewage treatment

Any process to which sewage is subjected in order to remove or alter its objectionable constituents and thus render it less offensive or dangerous. These processes may be classified as preliminary, primary, secondary, or complete, depending on the degree of treatment accomplished. Preliminary treatment may be the conditioning of industrial waste prior to discharge to remove or to neutralize substances injurious to sewers and treatment processes, or it may be unit operations which prepare the water for major treatment. Primary treatment is the first and sometimes the only treatment of sewage. It is the removal of floating solids and coarse and fine suspended solids. Secondary treatment utilizes biological methods of treatment, that is, oxidation processes following primary treatment by sedimentation. Complete treatment removes a high percentage of suspended, colloidal, and organic matter.

Septic tanks and Imhoff tanks are considered secondary treatment methods because sedimentation is combined with biological digestion of the sludge. *See* IMHOFF TANK; SEPTIC TANK.

Coarse solids removal. This is accomplished by means of racks, screens, grit chambers, and skimming tanks. Racks are fixed screens composed of parallel bars placed in the waterway to catch debris. The bars are usually spaced 1 in. or more apart. Screens are devices with openings usually of uniform size 1 in. or less placed in the line of flow. Screens may be fixed or movable and vary in construction as bar screens, band screens, or cage screens. Such screens are hand cleaned or mechanically cleaned. Grit chambers remove inorganic solids but may also trap heavier particles of organic nature such as seeds (Fig. 1). Grit chambers are designed so that the flow in the chamber is at 1 ft/s (0.3 m/s) or more. At less than that velocity, organic material also settles. Removal of grit is done either by hand or mechanically. Devices are added to mechanically cleaned units which wash most of the organic material out of the grit. Skimming chambers are devices for removing floating solids and grease. Air has been used to coagulate greases which then float and are skimmed off mechanically or by hand.

Fine solids removal. This is accomplished by screens with openings 1/16 or 1/32 in. (1.5875 or 0.79375 mm) wide, by sedimentation, or by both.

Fine screens are set in the line of flow and are operated mechanically. Band screens, drum screens, plate screens, and vibratory screens are in use and the finer particles of floating solids are removed as well as coarse solids passing a rack. In some treatment plants screenings are passed through a grinder and returned to the flow so that they will settle out in the sedimentation tank. Another device, the comminutor, barminutor, or griductor, has high-speed rotating edges working in the flow of sewage (Fig. 2). These blades cut, chop, and shred the solids, which then pass on to the sedimentation unit.

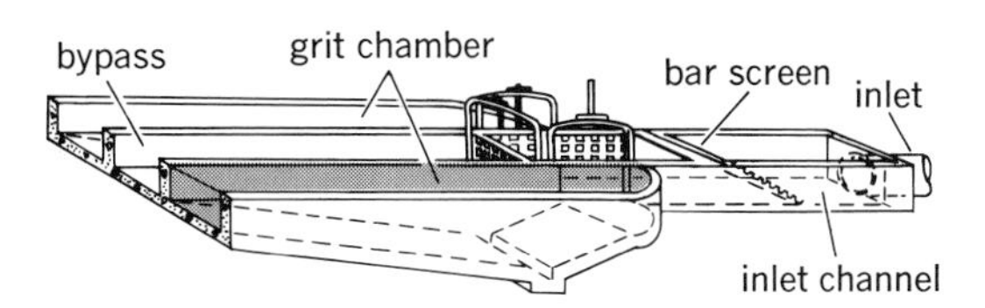

Fig. 1. Grit chamber. (*From Engineering Extension Department of Iowa State College, Bull. no. 58, 1953*)

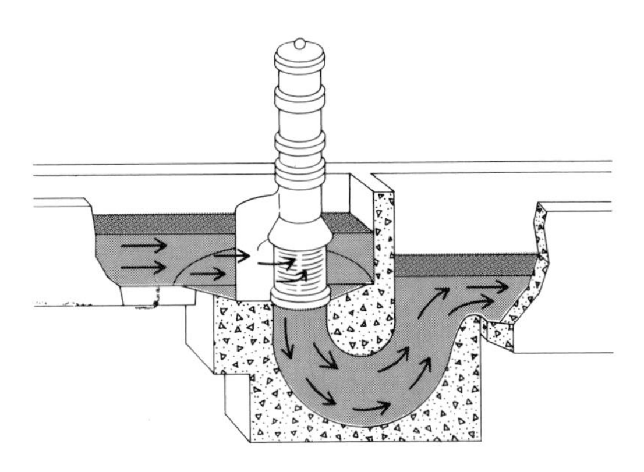

Fig. 2. A comminutor in place.

Sedimentation. Sedimentation has one objective, the removal of settleable solids. Some floating materials are also removed by skimming devices, called clarifiers, built into sedimentation units. The basins are either circular or rectangular. In the circular unit sewage flows in at the center and out over weirs along the circumference (Fig. 3). In

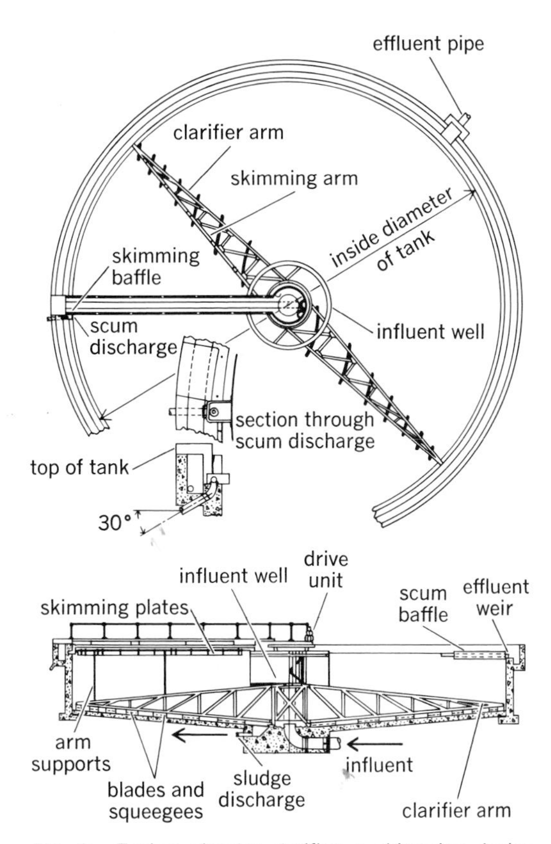

Fig. 3. Typical circular clarifier, a skimming device. (*From H. E. Babbitt and E. R. Baumann, Sewerage and Sewage Treatment, 8th ed., copyright 1958 by John Wiley & Sons, Inc.; used with permission*)

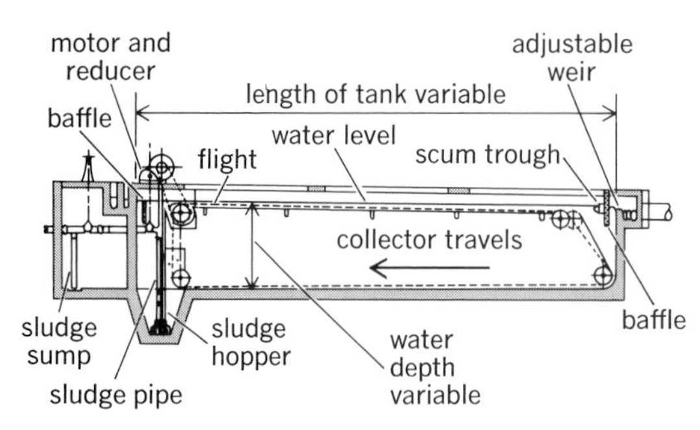

Fig. 4. Longitudinal section of typical rectangular clarifier. *(From E. W. Steel, Water Supply and Sewerage. 4th ed., McGraw-Hill, 1960)*

the rectangular tanks sewage flows into one end and out the other (Fig. 4).

The efficiency of a settling basin is dependent on a number of factors other than particle size, specific gravity, and settling velocity. Concentration of suspended matter, temperature, retention period, depth and shape of basin, baffling, total length of flow, wind, and biological effects all have an effect on solids removal. Density currents and short-circuiting may negate theoretical detention computations. Improper baffling may have the effect of reducing the effective surface area and creating dead or nonflow areas in the tank. A settling tank of good design with surface settling rates of 600 gal/(ft²)(day) [24.5 m³/(m²)(day)] and a 2-hr detention period will remove 50–60% of the suspended solids and at the same time remove 30–35% of the biochemical oxygen demand.

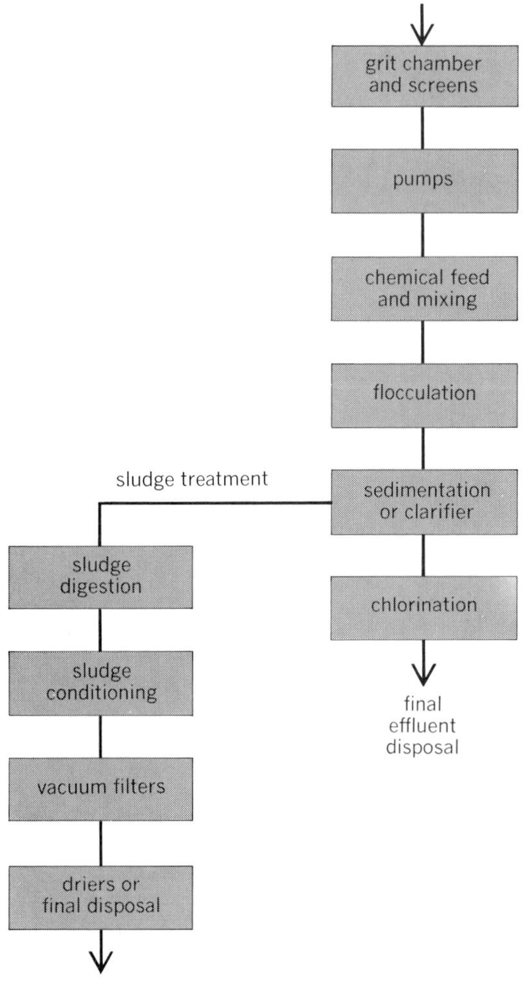

Fig. 5. Flow-through diagram of a chemical treatment plant. *(From H. E. Babbitt and E. R. Baumann, Sewerage and Sewage Treatment, 8th ed., copyright 1958 by John Wiley & Sons, Inc.; used with permission)*

The settling velocity of a particle is a function of specific gravity of the particle, specific gravity of water, viscosity of liquid, and particle diameter. Settling rates of particles larger than 0.1 mm are determined empirically. Sizes less than 0.1 mm settle in accordance with Stokes' law. Theoretically, if the forward motion of the water is less than the vertical settling rate of the particle, a particle at the surface will settle some distance below the surface in a given time interval. After that time interval the surface layer of water could be removed and it would contain no solids. The term surface settling rate is introduced as a practical measure of the rate of flow through the basin, if the rate of flow is equal to the surface area times the settling velocity of the smallest particle to be removed. Hence the selection of an overflow or surface settling rate expressed as gallons per day per square foot of surface area (or cubic meters per day per square meter of surface area) establishes a relationship between flow and area.

Flocculent suspensions have little or no settling velocity. These may occur in raw sewage but occur more frequently in secondary settling of effluents from activated sludge units. Such suspensions may be removed by passing the inflowing water upward through a blanket of the material. Theoretically there is a mechanical sweeping action in which smaller particles are attached to larger particles which then have sufficient weight to settle. Another type of treatment for such material is provided by an inner chamber equipped with baffles which rotate and stir the liquor and aid the formation of larger and heavier floc. The same purpose is also achieved by agitation with air. Some of the settled sludge is raised by airlift and mixed in with the material, thus forming a mixture with improved settling characteristics.

Sedimentation basin design. Practical considerations and engineering judgment must be applied in designing sedimentation basins. Depth is usually held at 10-ft (3-m) sidewall depth or less. The surface area requirement is usually 600 gal/(ft²)(day) [24.5 m³/(m²)(day)] for primary treatment alone and 800–1000 [32.6–40.7] for all other tanks. The detention period is normally 2 hr. These three parameters of design must be adjusted since each is dependent on the other for a given design flow (average daily flow at a plant). When mechanical sludge-removal equipment is used, the tank dimensions are usually sized to a conventional equipment specification. Rectangular tanks are built-in units with common walls between units and unit width up to 25 ft (7.6 m). The length-width ratio, frequently determined by economical design dimension, should not be greater than 5:1. The minimum length should be 10 ft (3.0 m). Final sizing may be fitted to convenient equipment dimensions.

Sludge removal on a regular schedule is mandatory in separate sedimentation tanks. If sludge is not removed, gasification occurs and large blocks of sludge begin to appear on the surface. These must then be removed by scum-removal mechanism or broken up so that they will settle. In circular tanks radial blades move the sludge to a center sludge hopper. In rectangular tanks the hopper is located at the inlet end and blades on a traveling chain move sludge in reverse of sewage flow. The heavier solids settle at the inlet and have a short travel path. These same blades may rise to the surface and move scum with the sewage flow to the outlet end where it is held by a baffle and removed by some form of scum-removal device. Sludge-removal mechanisms are often operated intermittently by time-clock relay mechanisms.

Appurtenances in the form of skimmers, scrapers, and other mechanical devices are many. Manufacturers have variants to offer, and competition is keen. Manufacturers' literature should be studied carefully and specifications should be carefully written to procure equipment meeting the requirements of engineering design.

Detention periods are theoretical. The actual flowthrough time is influenced by the inlet and outlet construction. On circular tanks inlets are submerged. Water rises inside a baffle extending downward to still the currents. Rectangular tank inlets may be submerged or, more commonly, sewage is brought to a trough which has a weir extending the width of the tank. The flow then moves forward with less short-circuiting. The outlet device on circular tanks is nearly always a circumferential weir adjusted to level after installation. The weir may be sharp-edged and level or provided with a sawtoothlike series of V-notches. On rectangular tanks, in order to provide enough weir length, a device known as a launder is used. A launder is a series of fingerlike shallow conduits set to water level and receiving flow from both sides of the conduit. Each of the fingers is connected to a common exit trough. The normal weir loading should not exceed 10,000 gal/linear ft (124 m^3/linear m) of weir per day in small plants, or 15,000 (186) in units handling more than 1,000,000 gal per day (1.0 mgd) [3785 m^3/day].

Chemical precipitation. Many attempts have been made to utilize chemical coagulants in the flocculation of sewage. The process, if used, is similar to that used in water treatment. The cost of chemicals and the somewhat intermediate treatment obtained with chemicals have kept this process out of general use. Its principal use today is in the preparation of sludge for filtration. Various steps in chemical precipitation are shown in Fig. 5. Alum, ferric sulfate, ferric chloride, and lime are used to form an insoluble precipitate which adsorbs colloidal and suspended solids. The entire floc settles and is removed as sludge. The Guggenheim process employs ferric chloride and aeration. The Scott-Darcy process employs ferric chloride made by treating scrap iron with chlorine solution.

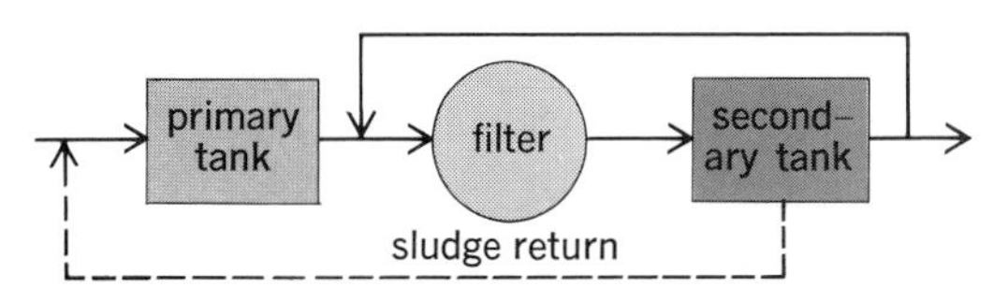

Fig. 6. Flow diagram of a single-stage high-rate trickling filter plant. *(From ASCE-FSIWA Joint Committee, Sewage Treatment Plant Design, 1959)*

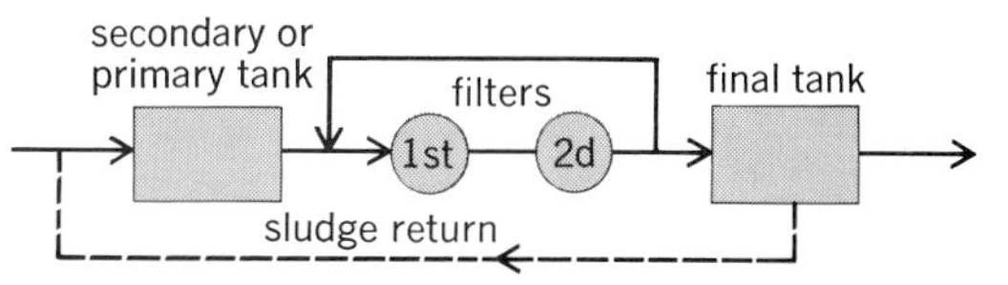

Fig. 7. Flow diagram of a two-stage high-rate trickling filter plant. *(From ASCE-FSIWA Joint Committee, Sewage Treatment Plant Design, 1959)*

Oxidation processes. These are secondary treatment processes, although a few activated-sludge plants have been built without primary sedimentation. Oxidation process methods are (1) filtration by intermittent sand filters, contact filters, and trickling filters; (2) aeration by the activated-sludge process or by contact aerators; and (3) oxidation ponds. There are three basic oxidation methods, all depending on biological growth. Each provides a method of bringing organic matter in suspension or solution in sewage into immediate contact with a population of microorganisms living under aerobic conditions. The processes are called filtration, activated sludge, and contact aeration.

Filtration. Intermittent sand filters are sand beds provided with underdrains. Sewage is dosed intermittently by siphon or by pump, at rates from 20,000 gal per acre per day (gad) [187 m^3 per hectare per day] to a maximum of 125,000 gad (1.17×10^3 m^3 per hectare per day) when operated as a secondary treatment process. Rates may go to 500,000 gad (or 0.5 mgad) [4.7×10^3 m^3 per hectare per day] when operated as a tertiary process. Beds are usually 2 1/2–3 ft (0.76–0.91 m) deep and are constructed with 6–12 in. (15–30 cm) of gravel at the bottom. The sand is sized to a uniformity coefficient of 5.0 or less (3.5 preferred), with effective size of 0.2–0.5 mm. The uniformity coefficient is the ratio between the sieve size that will pass 60% and the effective size.The effective size is the sieve size in millimeters that permits 10% of the sand by weight to pass. A mat of solids is formed in the surface layer of sand and must be removed periodically. The dry surface mat can be scraped clean, but periodically the top 6 in. (15 cm) or so of mat must be removed and replaced. Plants with sand filters operate at better than 95% removal of biochemical oxygen demand (BOD).

Trickling filters are beds of media, usually rock, over which settled sewage is sprayed. Microorganisms form a slime layer on the media surface and the water passes down over the surface in a thin

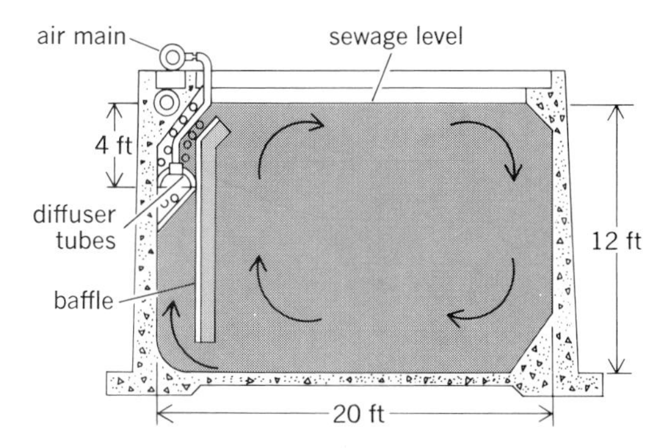

Fig. 8. Cross section of a spiral-flow activated-sludge tank with cylindrical diffusers. 1 ft = 0.30 m. *(From E. W. Steel, Water Supply and Sewerage, 4th ed., McGraw-Hill, 1960)*

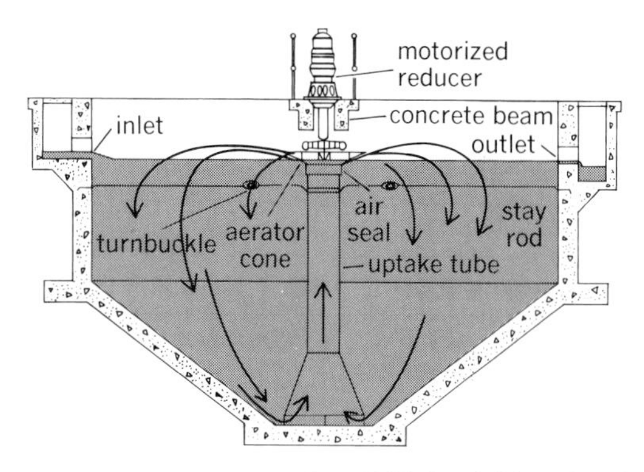

Fig. 9. Simplex aerator. *(From E. W. Steel, Water Supply and Sewerage, 4th ed., McGraw-Hill, 1960)*

film. Nutrients from the sewage are adsorbed in the slime layer and absorbed as food by organisms. Filters are ventilated through the underdrainage system or by other means, and thus oxygen, sewage, and organisms are brought together. Plants with trickling filtration have been operated at 90–95% efficiency of BOD removal.

Filter media include various materials such as stone, crushed rock, ceramic shapes, slag, and plastics. Preferred media are stone and crushed rock which do not fragment, flour, or soften on exposure to sewage. Rock sizes range from 1 to 6 in. (2.5 to 15 cm); however, current practice employs sizes between 2- and 4-in. (5- and 10-cm) nominal diameter. Plastic corrugated sheets have been employed on very deep filters. Pretreatment of sewage is normally required. When the waste contains a concentration of dissolved solids, as with milk waste, without any great concentration of settleable solids, the waste may be applied directly to the filter. Some advantage is gained by preaeration so that the waste applied to the filter has some dissolved oxygen.

Filters are classified as standard or low-rate filters, high-rate filters, and controlled filters. The filter introduced in the United States early in the 20th century was a bed of stone 6–8 ft (1.8–2.4 m) deep with a distribution system of fixed nozzles. This type of filter is called a standard or low-rate filter. The allowable organic loading is about one-third that of a high-rate filter having 3- to 6-ft (0.9- to 1.8-m) depth introduced during 1930–1940 and developed with many variations of recirculation and application of sewage since that time. In 1956 controlled filtration on sectionalized units composing a deep filter was introduced. The loading rate with no recirculation on such filters is 10–12 times that of low-rate filters.

Low-rate filters are dosed at a rate of 1–4 mgad ($9-37 \times 10^3$ m^3 per hectare per day) by siphon through nozzles so spaced that water reaches every part of the filter surface during a dosing cycle. The application of water by this method is intermittent. The rotary distributor may also be operated by siphon. This type of distributor has two or four radial arms supported on a center pedestal. Hydraulic force of water passing through the nozzles fixed to the arm causes the arm to rotate. The distributor may be operated in continuous rotation by feeding from a weir box. In either case the filter is sprayed as the arm passes over a given section and the dosing is intermittent with a short time interval between doses. With the fixed-nozzle method the interval may be 5 min, but with the rotary distributor the dosing interval may be no more than 15 sec.

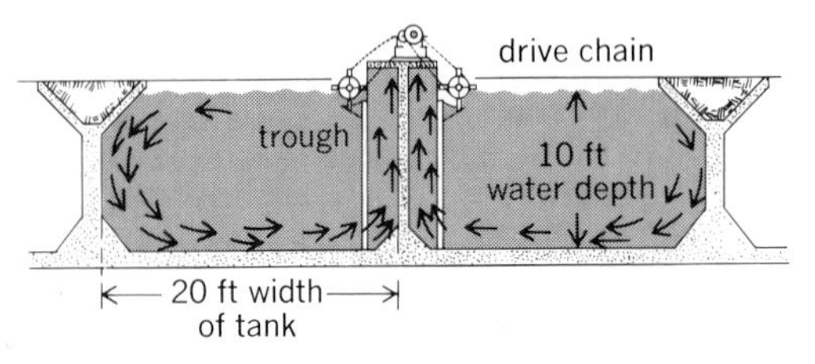

Fig. 10. Cross section of a Link-Belt mechanical aerator. 1 ft = 0.30 m. *(From E. W. Steel, Water Supply and Sewarage, 4th ed., McGraw-Hill, 1960)*

High-rate filters depend on recirculation. The hydraulic loading rate is about 20 mgad (187×10^3 m^3 per hectare per day) with a range of 9–44 mgad ($84-412 \times 10^3$ m^3 per hectare per day). Rotary distributors are used. Pumps pick up settled effluent and return it. Filters are often set up as primary and secondary filters with recirculation of water to each. Several alternative flow arrangements are demonstrated in Figs. 6 and 7. Recirculation ratios range from 1:1 to about 5:1. Final sedimentation is required for both low- and high-rate filters as filter slime and organic debris are washed free.

Aeration. Aeration is accomplished in tanks in which compressed air is diffused in liquid by various devices: filter plates, filter tubes, ejectors, and jets; or in which air is mixed with liquid by mechanical agitation. The high degree of treatment possible with conventional activated sludge, 95–98% BOD removal, has made it a popular method of treatment. Sewage organisms seeded in sludge which has passed through treatment are returned to incoming sewage and mixed thoroughly with the liquor. In this way the biota, oxygen supply, and sewage are brought together. Contact aeration utilizes air diffusion to keep a biota suspension thoroughly mixed; however, the biota are also maintained in active growth on plates of impervious material such as cement-asbestos suspended in the mixed liquor of the aeration tank. Slime growth forms on the plates, and liquid passing by them furnishes the plate biota with nutrients.

Activated-sludge process, the conventional process, requires an aeration period of 4–8 hr. Much of the oxidation takes place in the first 3 hr of detention. Aeration tanks are usually long, narrow, rectangular tanks with porous plates or diffusers along the length to keep the liquor well agitated throughout (Fig. 8). Widths are 15–30 ft (4.6–9.1 m) and depths about 15 ft (4.6 m). Length-width ratio is about 5:1.

Air requirements are 0.2–1.5 ft^3 air/gal (1.5–11.2 m^3 air/m^3) of sewage treated. It is necessary to maintain dissolved oxygen (DO) levels at 2 ppm or higher.

Mechanical aeration is done in square or rectangular aeration tanks, depending on the mechanism. In the Simplex method liquor is drawn by impeller up a draft tube and expelled over the tank surface (Fig. 9). In the Link-Belt unit, brushes introduce a spiral motion with considerable agitation. The period of aeration may be up to 8 hr with this method (Fig. 10). Modifications of the aeration process include modified aeration, step aeration, tapered aeration, stage aeration, biosorption, bioactivation, dual aeration, and others.

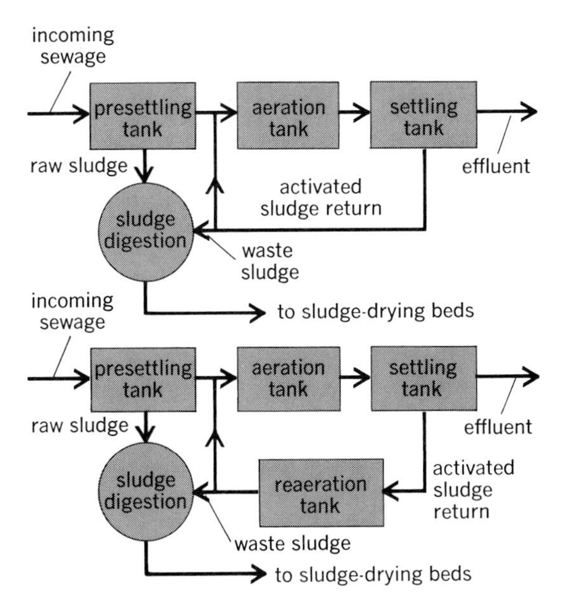

Fig. 11. Flow diagrams of typical activated-sludge plants. *(From E. W. Steel, Water Supply and Sewerage, 4th ed., McGraw-Hill, 1960)*

Recirculation of sludge is one of the essentials of the process. About 25–35% of the sludge settled in the final sedimentation tank is returned to the aeration tank (Fig. 11). Concentration of solids in mixed liquor may be about 3000 mg/liter in diffused air units and a little less in mechanical aeration units. The ratio of sludge volume settled to suspended solids is the Mohlmann index:

$$\text{Mohlmann index} = \frac{\text{volume of sludge settled in 30 min, \%}}{\text{suspended solids, \%}}$$

A good settling sludge has an index below 100. Sludge age, another important factor, is the average time that a particle of suspended solids remains under aeration and is the ratio of the dry weight of sludge in the tank in pounds to the suspended solids load in pounds per day.

Contact aerators provide an aeration period of 5 hr or more. Aeration is usually preceded by preaeration of the raw sewage before primary settling. The preaeration lasts 1 hr. Loadings are based on two factors: pounds per day per 1000 ft^2 of contact surface (6.0 or less, or 29.3 kg or less per day per 1000 m^2 of contact surface), and pounds per day per 1000 ft^2 per hour of aeration (1.2 or less, or 5.9 kg per day per 1000 m^2 per hour of aeration). Air supply of 1.5 ft^3/gal (11.2 m^3/m^3) of flow is required. The process has an overall plant efficiency of about 90% BOD removal.

Chlorination. Chlorination of treated sewage has one major purpose: to reduce the coliform group of organisms. Sufficient chlorine to satisfy demand and provide a residual of 2.0 mg/liter should be added. The following magnitude of dosage is possible: primary effluent, 20 mg/liter; trickling filter plant effluent, 15 mg/liter; activated sludge plant effluent, 8 mg/liter; sand filter effluent, 6 mg/liter. The contact period should be at least 15 min at peak hourly flow.

Oxidation ponds. These are ponds 2–4 ft (0.6–1.2 m) in depth designed to allow the growth of algae under suitable conditions in sewage media. Oxygen is absorbed from the air, but the conversion of CO_2 to O_2 by *Chlorella pyrenoidosa* and other algae provides an additional source of oxygen of great value. Oxidation ponds should be preceded by primary treatment. A loading figure of 50 lb BOD/acre (56 kg BOD/hectare) is recommended. BOD removal efficiency may range from 40 to 70%. [WILLIAM T. INGRAM]

Bibliography: R. W. James, *Sewage Sludge Treatment and Disposal*, 1977; D. Sundstrom and H. E. Klei, *Wastewater Treatment*, 1979; G. Tchobanoglous, *Wastewater Engineering: Collection, Treatment, and Disposal*, 1979.

Smog

The word smog can be defined in several ways. It was coined near the beginning of the century by H. A. Des Voeux to refer to a combination of coal smoke and fog, the particles of the smoke often serving as nuclei on which the water vapor condensed. Later, the term was used to refer to any dirty urban atmosphere. A unique type of dirty atmosphere was recognized in the late 1940s and early 1950s and came to be known as photochemical smog. It can be defined as that type of air pollution which owes many of its properties to the products of photochemical reactions involving the vapor of various organic substances, especially hydrocarbons, oxides of nitrogen, and atmospheric oxygen. Occasionally fog may also be involved in the formation of such smog.

Photochemical smog is particularly apparent in cities where little coal is burned, there is little industrial pollution, and there are large concentrations of automobiles. Undoubtedly, photochemical smog contributed to the pollution in many cities long before it was recognized as a special type of pollution. Smog in any given locality is generally a highly complex mixture of various pollutants, photochemically formed and otherwise. However, photochemical smog has received much attention in recent years.

History. Laws to attempt to control smog have existed at least since 1661, and many techniques for controlling smog at the source, such as electrostatic precipitation, were employed before the discovery of photochemical smog. Photochemical smog was first recognized in Los Angeles, and most of the early research was undertaken there. Ozone in high concentrations (0.2 part per million by volume, ppmv, is typical) was first detected in Los Angeles and other western United States cities by A. W. Bartel and J. W. Temple. A. J. Haagen-Smit, C. E. Bradley, and M. M. Fox demonstrated that the ozone is formed photochemically from a mixture of oxides of nitrogen, hydrocarbon vapor, and air. At about this time a group at Stanford Research Institute demonstrated that automobiles were the principal contributors to the reactive organic vapor such as hydrocarbons and much of the oxides of nitrogen in the Los Angeles atmosphere. R. D. Cadle and H. S. Johnston pointed out that the reaction that initiated the ozone formation was almost certainly the photochemical decomposition of nitrogen dioxide to form nitric oxide and atomic oxygen followed by reaction of the latter with hydrocarbons and with molecular oxygen of the air. While a large amount of very competent, valuable research into the causes, effects, and cures of photochemical smog has been undertaken since the early 1950s, it has all been based on that early research.

Meteorology. The intensity of photochemical smog at any time and place, like that of any type of air pollution, is markedly influenced by meteorological conditions and the geographical features of the region. If a city is in a bowl defined by mountains or hills, and an atmospheric inversion clamps a lid over the area, smog of any kind is apt to be intense. Even so, the smog intensity in a particular part of a city subject to such conditions may be very different from that in another part of the same city, due to a number of factors such as local air motions. *See* TEMPERATURE INVERSION.

Effects. Usually the most obvious and immediately annoying aspect of urban pollution is the murkiness of the air which it produces. This visibility decrease is mainly caused by the suspended particles. The median particle size in photochemical smog is so small that the particles preferentially scatter blue light, and depending upon the nature and angle of illumination, the air may appear blue or brown. The brown appearance is accentuated by the presence of nitrogen dioxide (NO_2), which is itself brown in color and results from various combustion operations, including driving automobiles.

Another especially annoying feature of photochemical smog is the eye irritation and lacrimation it produces. This irritation seems to result from the combined effects of several gases, especially formaldehyde, acrolein, peroxyacetyl nitrate (PAN), and possibly peroxybenzoyl nitrate.

As mentioned above, photochemical smog produces an unusual type of plant damage (phytotoxic effect). It has been variously described as silvering, bronzing, or glazing, and usually first appears on the underside of the leaves. This type of damage has resulted in serious economical loss, the extent of damage varying markedly with the type of plant. For example, spinach is especially susceptible. The damage is, of course, not limited to agricultural crops. Forests in the vicinity of Los Angeles have been severely damaged.

The phytotoxicants almost certainly are formed by the photochemical reactions mentioned above. Ozone, PAN, and various organic peroxides that may be produced by smog reactions all damage plants.

Although the main acute physiological effect of photochemical smog on humans—eye irritation—has been studied extensively, little is known about possible chronic effects, and the little that is known is mainly circumstantial evidence obtained by statistical comparisons of people living in rural and urban environments. Such studies appear to implicate smog of both types in causing heart disease as well as various types of cancer. Furthermore, various carcinogens, such as 3,4-benzpyrene, have been isolated from Los Angeles smog. Skin cancers have been produced by applying extracts of gasoline engine and diesel engine exhaust gases to the skin of mice. Although certainly not proved, photochemical smog may be considerably less carcinogenic than that of the coal-burning variety; the concentrations of 3,4-benzpyrene in western United States cities where the smog is primarily photochemical are lower than in the highly industrialized, coal-burning, eastern areas.

Typical concentrations of some pollutants in photochemical smog parts per hundred million by volume (pphmv)

Pollutant	Concentration
CO	3000
O_3	25
Hydrocarbons (paraffins)	100
Hydrocarbons (olefins)	25
Acetylene	25
Aldehydes	60
SO_2	20
Oxides of nitrogens	20
NH_3	2

Chemical reactions. The primary and secondary photochemical reactions leading to unpleasant products in photochemical smog were mentioned earlier. Although these reactions and their products render the smog much more unpleasant, the atmosphere would be very unpleasant even in the absence of these reactions because of the accumulations of automobile exhaust and other anthropogenic emissions to the atmosphere. In simplified form, these are reactions (1) through (11).

$$NO_2 + h\nu \rightarrow NO + O \quad (1)$$
$$O + O_2 + M \rightarrow O_3 + M \quad (2)$$
$$O_3 + NO \rightarrow NO_2 + O_2 \quad (3)$$
$$O + \text{Olefins} \rightarrow R + R^1O \text{ or } R{-}R^1 \text{ (bridged by O)} \quad (4)$$
$$O_3 + \text{Olefins} \rightarrow \text{Products} \quad (5)$$
$$R + O_2 \rightarrow RO_2 \quad (6)$$
$$RO_2 + O_2 \rightarrow RO + O_3 \quad (7)$$
$$RO_2 + NO \rightarrow RO + NO_2 \quad (8)$$
$$O_2(^1\Delta) + RH \rightarrow ROOH \quad (9)$$
$$RO_2 + SO_2 \rightarrow RO + SO_3 \quad (10)$$
$$SO_3 + H_2O \rightarrow H_2SO_4 \quad (11)$$

Several of these reactions are highly speculative. Reaction (9) has often been suggested but is probably too slow to be important. Reactions (1), (2), and (3) lead to a pseudosteady-state concentration of ozone that is much smaller than that occurring in photochemical smog. Two general processes have been proposed to explain the large accumulation of ozone. One (the more likely) is that some reaction, such as (8), removes NO so rapidly that reaction (3) is less effective in destroying ozone. The other is that some process in addition to (2), such as (7), produces ozone.

The chemical reactions produce large quantities of particles. A large percentage of these is produced by reactions involving free radicals such as R, RO, and RO_2. These radicals undergo complicated sequences of reactions producing high-molecular-weight substances of low vapor pressure that condense as particles from the air. Sulfuric acid droplets are also present in smog, and in the photochemical variety they are largely produced by rapid oxidation of sulfur dioxide (SO_2) emitted by the burning of sulfur-containing fuels followed by reaction (11). This oxidation seems to be intimately associated with the other smog reactions. The SO_2 may be oxidized to SO_3 by reaction (10) or a similar reaction involving the hydroperoxy radical (HO_2).

The free-radical reactions produce a large variety of organic compounds, such as PAN, organic acids, and aldehydes, in addition to the high-molecular-weight products. There is mounting evidence that chain reactions involving hydroxyl radicals (OH) may be important, the chains being initiated by reactions such as (12).

$$O + RCHO \rightarrow OH + RCO \qquad (12)$$

Obviously the chemistry is very complex, and details will remain unresolved for many years. The chemical reactions occurring in smog are not limited to cities, but are taking place in the atmosphere all over the globe. The trace atmospheric constituents are both natural and man-made—hydrocarbons, for example, being emitted in vast amounts by many species of plants.

Control. The alleviation of the unpleasant effects of smog remains primarily control at the source. Many suggestions, some highly ingenious, have been made for removing smog, once it has formed. The usual difficulty is the tremendous weight of air that must be removed or treated. Since automobile exhaust is the major precursor of photochemical smog, methods for greatly decreasing the rates of emission of harmful constituents in such exhaust have received much attention. Three such constituents are especially undesirable: organic vapors, particularly hydrocarbons; oxides of nitrogen; and carbon monoxide (CO). Control is achieved by engine modification, by afterburners, and by combinations of these. Automobiles would contribute to smog even in the absence of photochemical reactions, by emitting CO and particles, for example.

Of course, automobiles are not the only contributors to photochemical smog. Almost all combustion produces oxides of nitrogen. Various industrial operations, especially refineries, emit organic vapors; evaporation of gasoline occurs at service stations; and backyard incinerators can contribute appreciably.

In principle, all pollution of the air could be stopped by eliminating all pollutant emissions, but this is not economically or technologically feasible. Much research on smog is designed to provide the basis for sound decisions concerning where to place the emphasis for control. The results of such research have already been very effective. Attempts are being made to make the results of the chemical, meteorological, and economic studies even more useful by combining them into mathematical models which can be used to predict the amount of pollution at any given place depending upon the input into the models. Such models can be used to predict the effectiveness of various control measures. *See* AIR POLLUTION; AIR-POLLUTION CONTROL. [RICHARD D. CADLE]

Bibliography: R. D. Cadle, *J. Coll. Interface Chem.*, 39:25–31, 1972; R. D. Cadle and E. R. Allen, *Science*, 167:243–249, 1970; A. C. Stern (ed.), *Air Pollution*, 3d ed., 1976. C. S Thesday (ed.), *Chemical Reactions in Urban Atmospheres*, 1971.

Soil

Freely divided rock-derived material containing an admixture of organic matter and capable of supporting vegetation. Soils are independent natural bodies, each with a unique morphology resulting from a particular combination of climate, living plants and animals, parent rock materials, relief, the groundwaters, and age. Soils support plants, occupy large portions of the Earth's surface, and have shape, area, breadth, width, and depth. Soil, as used here, differs in meaning from the term as used by engineers, where the meaning is unconsolidated rock material. *See* SOIL MECHANICS.

This article is divided into four parts: origin and classification of soils, physical properties of soil, soil management, and soil erosion.

ORIGIN AND CLASSIFICATION OF SOILS

Soil covers most of the land surface as a continuum. Each soil grades into the rock material below and into other soils at its margins, where changes occur in relief, groundwater, vegetation, kinds of rock, or other factors which influence the development of soils. Soils have horizons, or layers, more or less parallel to the surface and differing from those above and below in one or more properties, such as color, texture, structure, consistency, porosity, and reaction (Fig. 1). The horizons may be thick or thin. They may be prominent, or so weak that they can be detected only in the laboratory. The succession of horizons is called the soil profile. In general, the boundary of soils with the underlying rock or rock material occurs at depths ranging from 1 to 6 ft, though the extremes lie outside of this range.

Origin of soils. Soil formation proceeds in stages, but these stages may grade indistinctly from one into another. The first stage is the accumulation of unconsolidated rock fragments, the parent material. Parent material may be accumulated by deposition of rock fragments moved by glaciers, wind, gravity, or water, or it may accumu-

Fig. 1. Soil profile showing horizons. The dark crescent-shaped spots at the soil surface are the result of plowing. The dark horizon lying 9–18 in. (23–46 cm) below the surface is the principal horizon of accumulation of organic matter that has been washed down from the surface. The thin wavy lines were formed in the same manner.

late more or less in place from physical and chemical weathering of hard rocks.

The second stage is the formation of horizons. This stage may follow or go on simultaneously with the accumulation of parent material. Soil horizons are a result of dominance of one or more processes over others, producing a layer which differs from the layers above and below.

Major processes. The major processes in soils which promote horizon differentiation are gains, losses, transfers, and transformations of organic matter, soluble salts, carbonates, silicate clay minerals, sesquioxides, and silica. Gains consist normally of additions of organic matter, and of oxygen and water through oxidation and hydration, but in some sites slow continuous additions of new mineral materials take place at the surface or soluble materials are deposited from groundwater. Losses are chiefly of materials dissolved or suspended in water percolating through the profile or running off the surface. Transfers of both mineral and organic materials are common in soils. Water moving through the soil picks up materials in solution or suspension. These materials may be deposited in another horizon if the water is withdrawn by plant roots or evaporation, or if the materials are precipitated as a result of differences in pH (degree of acidity), salt concentration, or other conditions in deeper horizons.

Other processes tend to offset those that promote horizon differentiation. Mixing of the soil occurs as the result of burrowing by rodents and earthworms, overturning of trees, churning of the soil by frost, or shrinking and swelling. On steep slopes the soil may creep or slide downhill with attendant mixing. Plants may withdraw calcium or other ions from deep horizons and return them to the surface in the leaf litter.

Saturation of a horizon with water for long periods makes the iron oxides soluble by reduction from ferric to ferrous forms. The soluble iron can move by diffusion to form hard concretions or splotches of red or brown in a gray matrix. Or if the iron remains, the soil will have shades of blue or green. This process is called gleying, and can be superimposed on any of the others.

The kinds of horizons present and the degree of their differentiation, both in composition and structure, depend on the relative strengths of the processes. In turn, these relative strengths are determined by the way man uses the soil as well as by the natural factors of climate, plants and animals, relief and groundwater, and the period of time during which the processes have been operating.

Composition. In the drier climates where precipitation is appreciably less than the potential for evaporation and transpiration, horizons of soluble salts, including calcium carbonate and gypsum, are normally found at the average depth of water penetration.

In humid climates some materials normally considered insoluble may be gradually removed from the soil or at least from the surface horizons. A part of the removal may be in suspension. The movement of silicate clay minerals would be an example. The movement of iron oxides is accelerated by the formation of chelates with the soil organic matter. Silica is removed in appreciable amounts in solution or suspension, though quartz sand is relatively unaffected. In warm humid climates free iron and aluminum oxides and silicate clays accumulate in soils, apparently because of low solubility relative to other minerals.

In cool humid climates solution losses are evident in such minerals as feldspars. Free sesquioxides tend to be removed from the surface horizons and to accumulate in a lower horizon, but mixing by animals and falling trees may counterbalance the downward movement.

Structure. Concurrently with the other processes, distinctive structures are formed in the different horizons. In the surface horizons, where there is a maximum of biotic activity, small animals, roots, and frost action keep mixing the soil material. Aggregates of varying sizes are formed and bound by organic matter, microorganisms, and colloidal material. The aggregates in the immediate surface tend to be loosely packed with many large pores among them. Below this horizon of high biotic activity, the structure is formed chiefly by volume changes due to wetting, drying, freezing, thawing, or shaking of the soil by roots of trees swaying with the wind. Consequently, the sides of any one aggregate, or ped, conform in shape to the sides of adjacent peds.

Water moving through the soil usually follows root channels, wormholes, and ped surfaces. Accordingly, materials that are deposited in a horizon commonly coat the peds. In the horizons that have received clay from an overlying horizon, the peds usually have a coating or varnish of clay making the exterior unlike the interior in appearance. Peds formed by moisture or temperature changes normally have the shapes of plates, prisms, or blocks.

Horizons. Pedologists have developed sets of symbols to identify the various kinds of horizons commonly found in soils. The nomenclature originated in Russia, where the letters A, B, and C were applied to the main horizons of the black soils of the steppes. A designated the dark surface horizon of maximum organic matter accumulation, C the unaltered parent material, and B the intermediate horizon. The usage of the letters A, B, and C spread to western Europe, where the intermediate or B horizon was a horizon of accumulation of free sesquioxides or silicate clays or both. Thus the idea developed that a B horizon is a horizon of accumulation. Some, however, define a B horizon by position between A and C. Subdivisions of the major horizons have been shown by either numbers or letters, for example, Bt or B2. No internationally accepted set of horizon symbols has been developed. In the United States the designations shown in Fig. 2 have been widely used since about 1935, with minor modifications made in 1962. Lowercase letters were added to numbers in B horizons to indicate the nature of the material that had accumulated. Generally, "h" is used to indicate translocated humus, "t" for translocated clay, and "ir" for translocated iron oxides. Thus, B2t indicates the main horizon of clay accumulation.

Classification. Systems of soil classification are influenced by concepts prevalent at the time a system is developed. Since ancient times, soil has

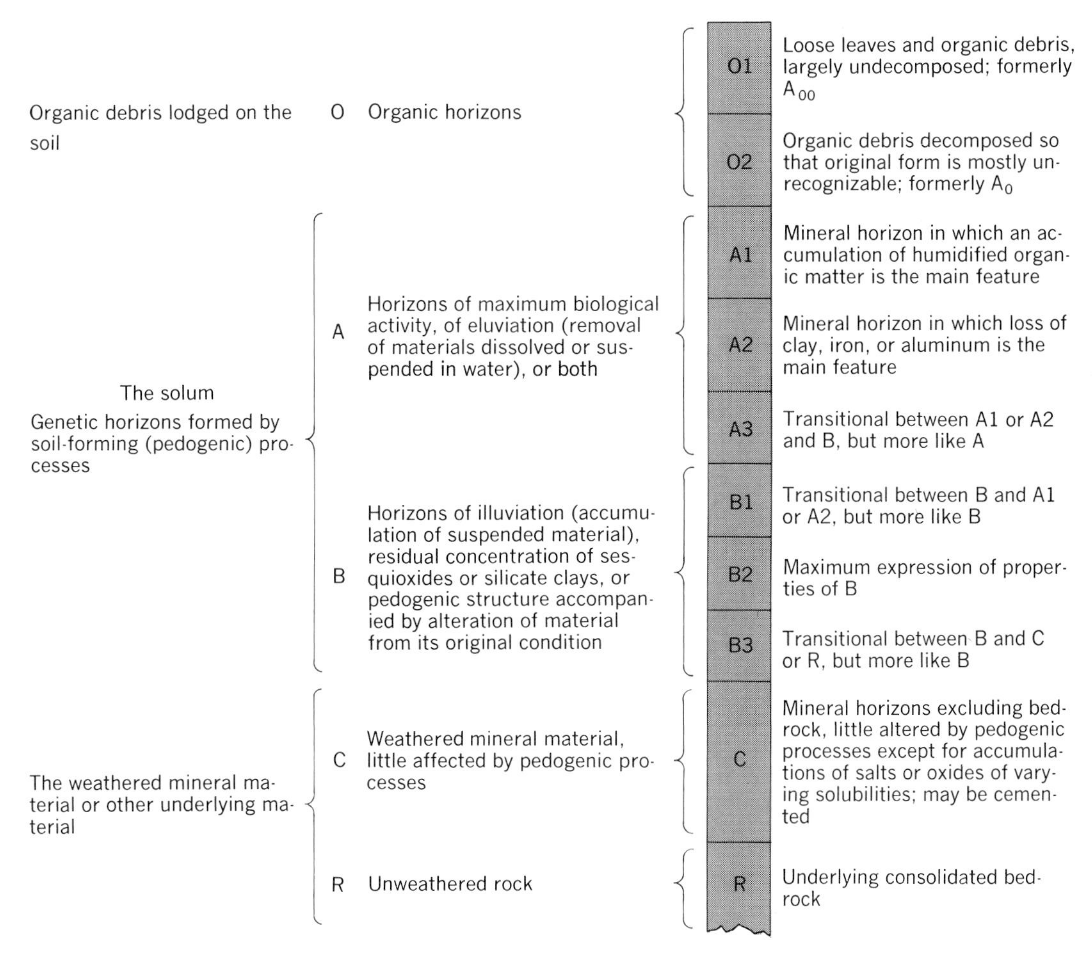

Fig. 2. A hypothetical soil profile having all principal horizons. Other symbols are used to indicate features subordinate to those indicated by capital letters and numbers. The more important of these are as follows: ca, as in Cca, accumulations of carbonates; cs, accumulations of calcium sulfate; cn, concretions; g, strong gleying (reduction of iron in presence of groundwater); h, illuvial humus; ir, illuvial iron; m, strong cementation; p, plowing; sa, accumulations of very soluble salts; si, cementation by silica; t, illuvial clay; x, fragipan (a compact zone which is impenetrable by roots).

been considered as the natural medium for plant growth. Under this concept, the earliest classifications were based on relative suitability for different crops, such as rice soils and vineyard soils.

Early American agriculturists thought of soil chiefly as disintegrated rock, and the first comprehensive American classification was based primarily on the nature of the underlying rock.

In the latter part of the 19th century, some Russian students noted relations between the steppe and black soils and the forest and gray soils. They developed the concept of soils as independent natural bodies formed by the influence of environmental factors operating on parent materials over time. The early Russian classifications grouped soils at the highest level, according to the degree to which they reflected the climate and vegetation. They had classes of Normal, Abnormal, and Transitional soils, which later became known as Zonal, Intrazonal, and Azonal. Within the Normal or Zonal soils, the Russians distinguished climatic and vegetative zones in which the soils had distinctive colors and other properties in common. These formed classes that were called soil types. Because some soils with similar colors had very different properties that were associated with differences in the vegetation, the nature of the vegetation was sometimes considered in addition to the color to form the soil type name, for example, Gray Forest soil and Gray Desert soil. The Russian concepts of soil types were accepted in other countries as quickly as they became known. In the United States, however, the name soil type had been used for some decades to indicate differences in soil texture, chiefly texture of the surface horizons; so the Russian soil type was called a Great Soil Group. *See* SOIL ZONALITY.

Many systems of classification have been attempted but none has been found markedly superior; most systems have been modifications of those used in Russia. Two bases for classification have been tried. One basis has been the presumed genesis of the soil; climate and native vegetation were given major emphasis. The other basis has been the observable or measurable properties of the soil. To a considerable extent, of course, these are used in the genetic system to define the great soil groups. The morphologic systems have not used soil genesis as such, but have attempted to use properties acquired through soil development.

The principal problem in the morphologic systems has been the selection of the properties to be used. Grouping by color, tried in the earliest systems, produces soil groups of unlike genesis.

The Soil Survey staff of the U.S. Department of Agriculture and the land-grant colleges adopted a new classification scheme in 1965. Although the new system has been widely tested, only time can tell how much more useful it will be than earlier systems. As knowledge of soil genesis increases, modifications of classification systems will continue to be necessary.

The system differs from earlier systems in that it may be applied to either cultivated or virgin soils. Previous systems have been based on virgin profiles, and cultivated soils were classified on the presumed characteristics or genesis of the virgin soils. The new system has six categories, based on both physical and chemical properties. These categories are the order, suborder, great group, subgroup, family, and series, in decreasing rank.

The nomenclature. The names of the taxa or classes in each category are derived from the classic languages in such a manner that the name itself indicates the place of the taxa in the system and usually indicates something of the differentiating properties. The names of the highest category, the order, end in the suffix "sol," preceded by formative elements that suggest the nature of the order. Thus, Aridisol is the name of an order of soils that is characterized by being dry (Latin *aridus*, dry, plus *sol*, soil). A formative element is taken from each order name as the final syllable in the names of all taxa of suborders, great groups, and subgroups in the order. This is the syllable beginning with the vowel that precedes the connecting vowel with "sol." Thus, for Aridisols, the names of the taxa of lower classes end with the syllable "id," as in Argid and Orthid.

Suborder names have two syllables, the first suggesting something of the nature of the suborder and the last identifying the order. The formative element "arg" in Argid (Latin argillus, clay) suggests the horizon of accumulation of clay that defines the suborder.

Great group names have one or more syllables to suggest the nature of the horizons and have the suborder name as an ending. Thus great group names have three or more syllables but can be distinguished from order names because they do not end in "sol." Among the Argids, great groups are Natrargids (Latin *natrium*, sodium) for soils that have high contents of sodium, and Durargids (Latin *durus*, hard) for Argids with a hardpan cemented by silica and called a duripan.

Subgroup names are binomial. The great group name is preceded by an adjective such as "typic," which suggests the type or central concept of the great group, or the name of another great group, suborder, or order converted to an adjective to suggest that the soils are transitional between the two taxa.

Family names consist of several adjectives that describe the texture (sandy, silty, clayey, and so on), the mineralogy (siliceous, carbonatic, and so on), the temperature regime of the soil (thermic, mesic, frigid, and so on), and occasional other properties that are relevant to the use of the soil.

Series names are abstract names, taken from towns or places near where the soil was first identified. Cecil, Tama, and Walla Walla are names of soil series.

Order. In the highest category 10 orders are recognized. These are distinguished chiefly by differences in kinds and amounts of organic matter in the surface horizons, kinds of B horizons resulting from the dominance of various specific processes, evidences of churning through shrinking and swelling, base saturation, and lengths of periods during which the soil is without available moisture. The properties selected to distinguish the orders are reflections of the degree of horizon development and the kinds of horizons present.

The orders, the formative elements in the names, and the general nature of the included soils are given in Table 1.

Suborder. This category narrows the ranges in soil moisture and temperature regimes, kinds of horizons, and composition, according to which of these is most important. Moisture or temperature or soil properties associated with them are used to define suborders of Alfisols, Mollisols, Oxisols, Ultisols, and Vertisols. Kinds of horizons are used for Aridisols, compositions for Histosols and Spodosols, and combinations for Entisols and Inceptisols.

Great group. The taxa (classes) in this category group soils that have the same kinds of horizons in

Table 1. Soil orders

Order	Formative element in name	General nature
Alfisols	alf	Soils with gray to brown surface horizons, medium to high base supply, with horizons of clay accumulation; usually moist, but may be dry during summer
Aridisols	id	Soils with pedogenic horizons, low in organic matter, and usually dry
Entisols	ent	Soils without pedogenic horizons
Histosols	ist	Organic soils (peats and mucks)
Inceptisols	ept	Soils that are usually moist, with pedogenic horizons of alteration of parent materials but not of illuviation
Mollisols	oll	Soils with nearly black, organic-rich surface horizons and high base supply
Oxisols	ox	Soils with residual accumulations of inactive clays, free oxides, kaolin, and quartz; mostly tropical
Spodosols	od	Soils with accumulations of amorphous materials in subsurface horizons
Ultisols	ult	Soils that are usually moist, with horizons of clay accumulation and a low supply of bases
Vertisols	ert	Soils with high content of swelling clays and wide deep cracks during some season

the same sequence and have similar moisture and temperature regimes. Exceptions to horizon sequences are made for horizons so near the surface that they are apt to be mixed by plowing or lost rapidly by erosion if plowed.

Subgroup. The great groups are subdivided into subgroups that show the central properties of the great group, intergrade subgroups that show properties of more than one great group, and other subgroups for soils with atypical properties that are not characteristic of any great group.

Family. The families are defined largely on the basis of physical and mineralogic properties of importance to plant growth.

Series. The soil series is a group of soils having horizons similar in differentiating characteristics and arrangement in the soil profile, except for texture of the surface portion, and developed in a particular type of parent material.

Type. This category of earlier systems of classification has been dropped but is mentioned here because it was used for almost 70 years and many references to it are found in the literature about soils. The soil types within a series differed primarily in the texture of the plow layer or equivalent horizons in unplowed soils. Cecil clay and Cecil fine sandy loam were types within the Cecil series. The texture of the plow layer is still indicated in published soil surveys if it is relevant to the use of the soil, but it is now considered as one kind of soil phase. Soil surveys are discussed below.

Classifications of soils have been developed in several countries based on other differentia. The principal classifications have been those of the Soviet Union, Germany, France, Canada, Australia, and New Zealand, and the United States. Other countries have modified one or the other of these to fit their own conditions. Soil classifications have usually been developed to fit the needs of a government that is concerned with the use of its soils. In this respect soil classification has differed from classifications of other natural objects, such as plants and animals, and there is no international agreement on the subject.

Many practical classifications have been developed on the basis of interpretations of the usefulness of soils for specific purposes. An example is the capability classification, which groups soils according to the number of safe alternative uses, risks of damage, and kinds of problems that are encountered under use.

Surveys. Soil surveys include those researches necessary (1) to determine the important characteristics of soils, (2) to classify them into defined series and other units, (3) to establish and map the boundaries between kinds of soil, and (4) to correlate and predict adaptability of soils to various crops, grasses, and trees; behavior and productivity of soils under different management systems; and yields of adapted crops on soils under defined sets of management practices. Although the primary purpose of soil surveys has been to aid in agricultural interpretations, many other purposes have become important, ranging from suburban planning, rural zoning, and highway location, to tax assessment and location of pipelines and radio transmitters. This has happened because the soil properties important to the growth of plants are also important to various engineering applications.

Soil surveys were first used in the United States in 1898. Over the years the scale of soil maps has been increased from 1/2 or 1 in. to the mile (1:126,720 or 1:63,360) to 3 or 4 in. to the mile (1:21,120 to 1:15,840) for mapping humid farming regions, and up to 8 in. to the mile (1:7920) for maps in irrigated areas. After the advent of aerial photography, planimetric maps were largely discontinued in favor of aerial photographic mosaics. The United States system has been used, with modifications, in many other countries.

Two kinds of soil maps are made. The common map is a detailed soil map, on which soil boundaries are plotted from direct observations throughout the surveyed area. Reconnaissance soil maps are made by plotting soil boundaries from observations made at intervals. The maps show soil and other differences that are of significance for present or foreseeable uses.

The units shown on soil maps usually are phases of soil series. The phase is not a category of the classification system. It may be a subdivision of any class of the system according to some feature that is of significance for use and management of the soil, but not in relation to the natural landscape. The presence of loose boulders on the surface of the soil makes little difference in the growth of a forest, but is highly significant if the soil is to be plowed. Phases are most commonly based on slope, erosion, presence of stone or rock, or differences in the rock material below the soil itself. If a legend identifies a phase of a soil series, the soils so designated on a soil map are presumed to lie within the defined range of that phase in the major part of the area involved. Thus, the inclusion of lesser areas of soils having other characteristics is tolerated in the mapping if their presence does not appreciably affect the use of the soil. If there are other soils that do affect the use, inclusions up to 15% of the area are tolerated without being indicated in the name of the soil.

If the pattern of occurrence of two or more series is so intricate that it is impossible to show them separately, a soil complex is mapped, and the legend includes the word "complex," or the names of the series are connected by a hyphen and followed by a textural class name. Thus the phrase Fayette-Dubuque silt loam indicates that the two series occur in one area and that each represents more than 15% of the total area.

In places the significance of the difference between series is so slight that the expense of separating them is unwarranted. In such a case the names of the series are connected by a conjunction, for example, Fayette and Downs silt loam. In this kind of mapping unit, the soils may or may not be associated geographically.

It is possible to make accurate soil maps only because the nature of the soil changes with alterations in climatic and biotic factors, in relief, and in groundwaters, all acting on parent materials over long periods of time. Boundaries between kinds of soil are made where such changes become apparent. On a given farm the kinds of soil usually form a repeating pattern related to the relief (Fig. 3).

Because concepts of soil have changed over the years, maps made 30–50 years ago may use the

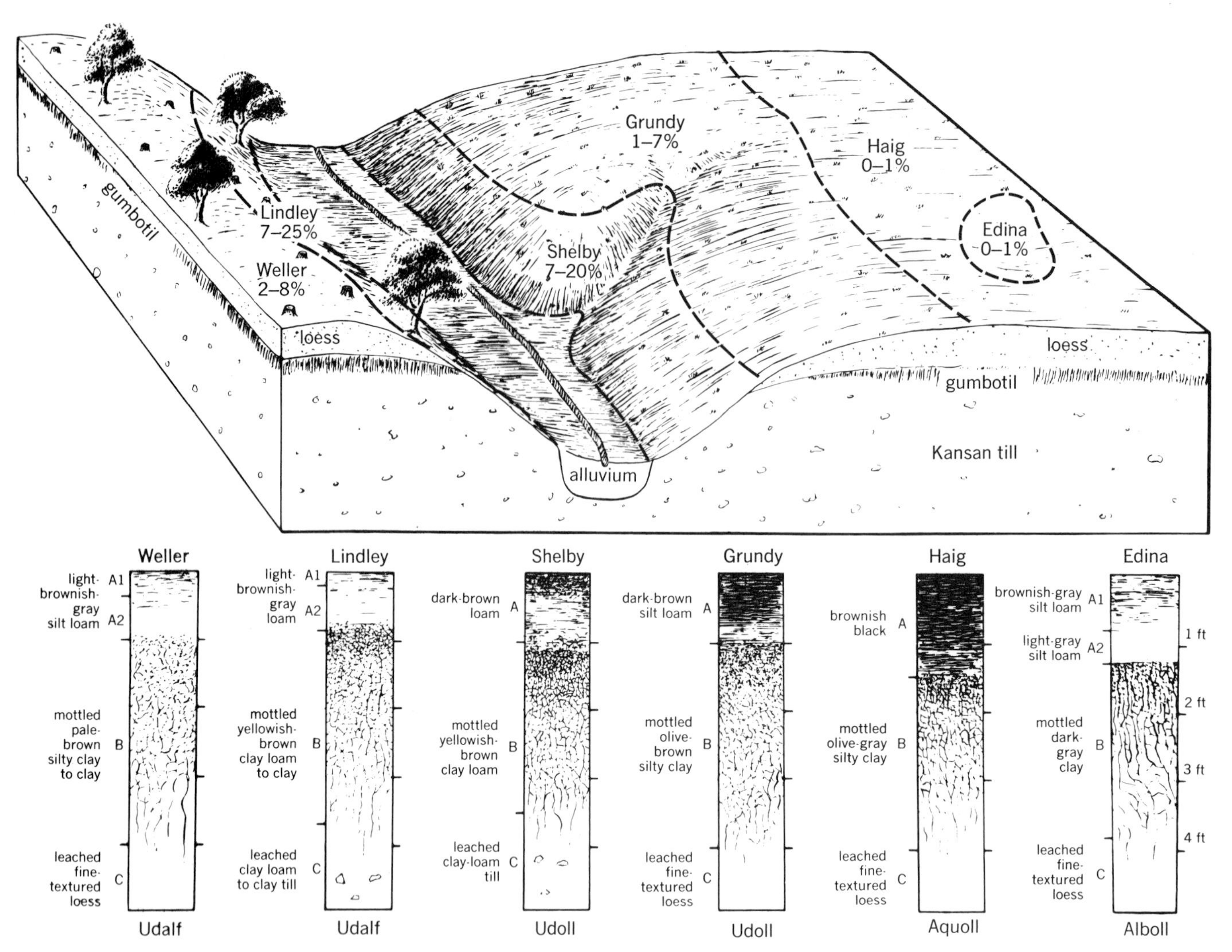

Fig. 3. Sketch showing the relation of the soil pattern to relief, parent material, and native vegetation on a farm in south-central Iowa. The soil slope gradient is expressed as a percentage. (*Modified from R. W. Simonson, F. F. Riecken, and G. D. Smith, Understanding Iowa Soils, Brown, 1952*)

same soil type names as maps made in recent years, but with different meanings. The older maps must therefore be interpreted with caution.

[GUY D. SMITH]

Nutrient element losses. Losses of most elements are normal to soil formation. Losses for a pair of soil orders are described below and the magnitudes indicated for others.

Ultisols are formed in strongly weathered regoliths from a variety of rocks, chiefly in warm, humid regions. The soils occupy old land surfaces. Major areas are in southeastern Asia and the United States.

Chemical and mineralogical composition of specimen Ultisols and their source rocks suggests that as much as 90% of the calcium, magnesium, and potassium disappears at the weathering front, where the rock decomposes. Losses continue as the soils form. Quantities are eventually reduced to very low levels. Because of the low levels, people have thought that Ultisols were worn out by long cropping, whereas they were really "worn out" while being formed.

The approximate amounts of four nutrient elements carried by a pair of streams draining Ultisols in North Carolina are given in Table 2. Amounts of Ca and Mg are very low, whereas those of K, Na, and N are moderate to low.

Mollisols are formed in slightly weathered regoliths, chiefly in cool-temperate grasslands. The soils occupy young land surfaces. Major areas are in the north-central United States and adjacent Canada, the Ukraine and adjacent parts of the Soviet Union, and the pampas of Argentina.

Roughly a third of the Ca and K, a larger share of the Mg, and a very small part of the P in the source rocks disappear during formation of Molli-

Table 2. Amounts of four nutrient elements carried by four rivers in one year*

River (state)	Ca	Mg	K	K + Na	N
Neuse (NC)	20	7	–	32	1.0
Hiwassee (NC)	33	11	9	–	–
Cedar (IA)	86	32	–	22	1.4
Iowa (IA)	78	29	–	22	1.0

*Expressed as pounds per acre of watershed. 1 lb per acre = 1.12 kg per hectare.

sols. Hence, the soils have relatively high levels of these elements. Moreover, they have high levels of exchangeable Ca, Mg, and K, which are readily available to plants. Expressed as milliequivalents per 100 g of soil, average figures are 15 of Ca, 6 of Mg, and 0.8 of K. Amounts of phosphorus are also high, with much in plant-available form.

High levels of nutrient elements in Mollisols are reflected in data for a pair of rivers in Iowa, given in Table 2. Quantities of Ca and Mg are about three times as large as in the streams in North Carolina. Quantities of K and Na are slightly lower, whereas that of N is about the same. A share of the amounts in the streams comes from the underlying rock, especially for Ultisols.

Ultisols and Mollisols are opposite extremes in losses of nutrient elements during soil formation. If the average chemical composition of the soils to a depth of 5 ft is compared, the ratios between Mollisols and Ultisols are 10:1 for Ca and Mg, 3:1 for K, 2:1 for P, and 5:1 for N. The ratios for elements in exchangeable form are even larger. As the ratios suggest, Mollisols are much more naturally fertile than Ultisols. Mollisols are at the top of the list, Ultisols near the bottom.

Losses of nutrient elements during formation of soils is well bracketed by the Mollisols and Ultisols. Similar to Mollisols in losses during their formation are Aridisols, Inceptisols of cold or dry regions, and Vertisols. If anything, losses are smaller for these soils than for Mollisols. Collectively, these groups of soils occupy about 40% of the Earth's land surface. Also similar to Ultisols in losses are the Oxisols and Inceptisols of the tropics and subtropics. Collectively, these soils plus Ultisols occupy about 25% of the land surface.

The remaining groups of soils, that is, Alfisols, Entisols, Spodosols, and some Inceptisols, fall between the two extremes. Mountain regions with their great variety of soils belong to this middle class. These soils are all intermediate in losses and also in present fertility levels. Collectively, these soils occupy about 35% of the land surface.

Losses of nutrient elements generally occur during formation of soils. The losses are small enough to be negligible for a few kinds of soils, very large for others, and intermediate for still others. The magnitude of past losses is reflected in the present fertility levels of all soils. The magnitude also directly affects soil usefulness for food and fiber production as well as the contributions of dissolved substances to lakes and streams.

[ROY W. SIMONSON]

PHYSICAL PROPERTIES OF SOIL

The physical properties of soil are important in agriculture because of their influence on plant growth and on the management requirements of the land. They influence plant growth from seeding to maturity by regulating the supply of air, water, and heat. The absorption of essential nutrients by plant roots is dependent upon an available supply of oxygen, water, and heat. Thus, physical properties indirectly regulate the nutrition of plants and their response to liming and fertilization. The more favorable the supply of air, water, and heat in each soil layer or horizon, along with the absence of mechanical impedance to root growth, the greater is the potential rooting system zone for plants.

Physical properties of the soil also determine the kind, amount, and ease of tillage, the runoff and erosion potential, and the type of plants which can or should be grown on a given soil.

Many people use the word tilth in referring to the physical condition of the soil. Tilth has been defined as the physical condition of the soil in its relation to plant growth. The physical condition of the soil is controlled by, or is the result of, whatever set of physical properties the soil has at any given time.

Soil physics is that branch of soil science which is concerned with the study of the physical properties of the soil. These physical properties include texture, particle density, structure, bulk density, porosity, water, air, temperature, consistency, compactibility, and color. Just as important as the amount of water, air, and heat in the soil at any one time is the soil's conductivity for these constituents. All of these properties are interrelated.

The four major components of the soil are inorganic particles, organic matter, water, and air. The proportions of these components vary greatly from place to place in a field, from one layer or horizon to another, and in different parts of the world. The amount of air, water, and heat in the soil changes from day to day and from season to season.

Soil texture. About one-half of the total volume of mineral soils consists of solid matter, of which 80–99% is inorganic and 1–20% is organic material. The inorganic fraction consists of rock and mineral particles of many sizes and shapes. They are classified into five major size groups called separates. The two largest separates are stone and gravel. Stone particles are greater than 76 mm (3 in.) and gravel particles are 2–76 mm along their greatest diameter. Sand particles are 0.05–2.00 mm in diameter. Sand particles may be graded by size as very fine, fine, medium, coarse, and very coarse. Silt has particles 0.002–0.005 mm in diameter. Clay, the smallest of the soil particles, has a diameter of less than 0.002 mm.

After separating the coarser separates by sieving, the amount of silt and clay is determined by methods that depend upon the rate of settling or sedimentation (based on Stoke's law) of these two separates from a water suspension in which they have been well dispersed with the aid of a dispersing agent. The stone, gravel, and sand separates of a soil can be seen with the naked eye. Clay can be examined only with an electron microscope.

Determination of the particle-size distribution in a soil is called a particle-size or mechanical analysis. The texture of a soil is determined by its content of sand, silt, and clay. The percentages of sand, silt, and clay in the 12 textural classes are shown in Fig. 4. With this texture triangle one can determine the textural class of a soil from its percentages of sand, silt, and clay. The textural class is combined with the series name of a soil to give the soil type, such as Sassafras sandy loam, Miami silt loam, or Houston clay loam.

The stone, gravel, and larger sand particles usually act as separate particles. They may be rounded, angular, or platelike in shape. They are composed of rock fragments and of primary minerals such as quartz. Soils with large amounts of stone, gravel, and coarse sand have low plant-nutrient and water-holding capacities, and permit rapid air,

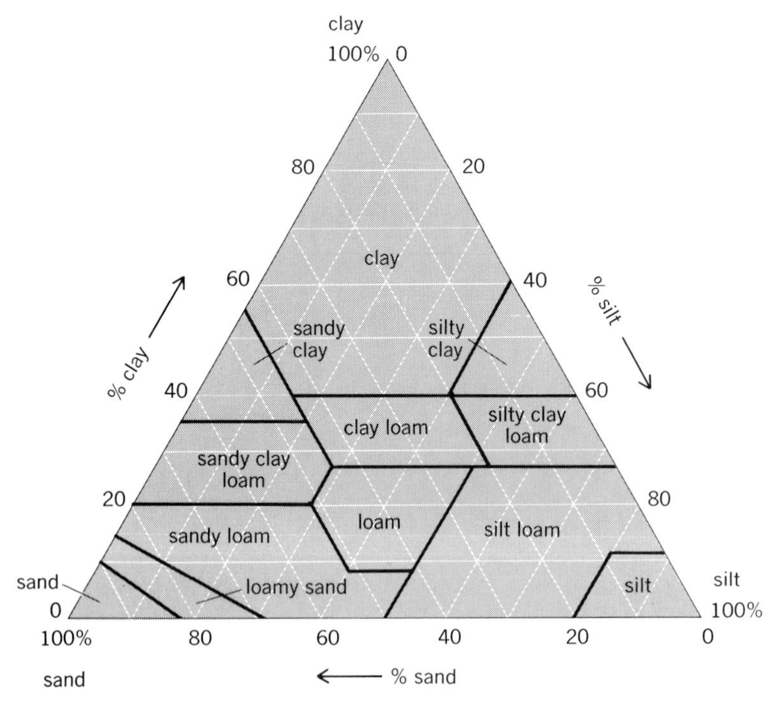

Fig. 4. Triangle showing percentage of sand, silt, and clay in each textural class.

water, and heat movement through them. A high content of fine sand may increase water-holding capacity, but it also often increases a soil's susceptibility to wind and water erosion. Sand imparts a grittiness to the feel of a soil.

The clay fraction controls most of the important properties of a soil. In soils of the cold and temperate regions, clay is composed chiefly of secondary crystalline alumina silicates. These consist of the kaolinite, illite, and montmorillonite groups of clay minerals. Hydrated oxides of iron and aluminum are the main components of the clay in the more highly weathered soils typical of many parts of the tropics.

Because of their extremely small size, clay particles have a very large specific surface which is responsible for the great adsorptive capacity of clay soils for water, gases, ions, and organic molecules. Clays are well known for their plasticity and stickiness when wet. They also expand or swell with wetting and contract or shrink upon drying. Movement of air and water through clay soils is often very slow because of the small size of the pores between the clay particles. Clay particles are platelike in shape.

Silt particles exhibit some of the properties of sand and clay. They are usually angular in shape with quartz being the dominant mineral. The available water-holding capacity of soils often is proportional to their silt content. Many silt particles have a coating of clay particles. Without this clay coating silt has a floury or a talcum-powder feel when dry and loose. Soils with large amounts of silt and clay have very poor air and moisture relations and are very difficult to manage. They are often very erodible. The loam soils generally have the most desirable texture for crop growth and ease of management.

It is seldom feasible to try to change the texture of a soil in the field. However, sand often is mixed with clay soils to change their texture to a sandy loam for special uses, as in greenhouses. The texture of surface soils may change as a result of removal of the smaller particles by wind and water erosion or by eluviation (movement within the soil).

Organic matter. The organic matter in the soil is made up of the partially decomposed remains of plant and animal tissues as well as the bodies of living soil microorganisms and plant roots. Humus is the more or less stable fraction of the organic matter or its decomposition products remaining in the soil. Many good and some bad effects accompany the decomposition of organic matter. During decomposition of organic matter by the soil microorganisms, gluelike soil-aggregate bonding substances are produced. With knowledge of the great importance of these natural soil-conditioning materials, the chemical industry has produced a number of synthetic soil conditioners.

Much of the soil organic matter has colloidal properties. It has two to three times the absorptive capacity for water, gases, ions, and other colloids as the same amount of clay. Its superior water- and nutrient-holding capacity makes it an ideal substitute for clay in improving droughty, infertile sandy soils, and its good tilth-promoting qualities make it the universally recognized ameliorator of tight, sticky, or hard and lumpy clay soils.

Density of particles. The inorganic soil particles may consist of many kinds of minerals with a wide range in particle density. The average particle density for most mineral soils varies between 2.60 and 2.75 g/cm^3. The average density of humus particles ranges between 1.2 and 1.4 g/cm^3. For general calculations the average particle density of soil is taken to be 2.65 g/cm^3. The pycnometer is used to determine soil particle density. The plowed layer weighs about 2,000,000 lb/acre (2,240,000 kg/hectare).

Soil structure. Soil structure refers to the arrangement of soil particles into aggregates or peds of different sizes and shapes. Pure sands have single grain structure. Because of the adhesive and cohesive properties of clay and organic matter, the inorganic and organic particles have been combined to form the following types of structure as found in the A and B horizons of most soils: platy, prismatic, columnar, blocky, subangular blocky, granular, and crumb (Fig. 5).

These types of structure have been developed from the bonding together of the individual particles (accretion) or the breakdown of large massive mixtures of gravel, sand, silt, clay, and organic matter (disintegration). The formation or genesis of a given type of structure and the stability of the aggregate produced seem to be associated with (1) the contraction and expansion resulting from hydration and desiccation of the clay-organic matter upon wetting and drying, as well as freezing and thawing; (2) the physical activity of roots and soil animals; (3) the influence of humus and decomposing organic matter and of the slimes and mycelia of the microorganisms that provide bonding substances with which aggregates are held together; (4) the effect of absorbed cations which cause flocculation or dispersion of the colloidal matter.

The prism, columnar, block, and sometimes the platelike types of structure are found mostly in subsoils. Granules and crumbs are found in largest numbers in surface soils (Fig. 6). Compacted layers in the soil often have a platy structure.

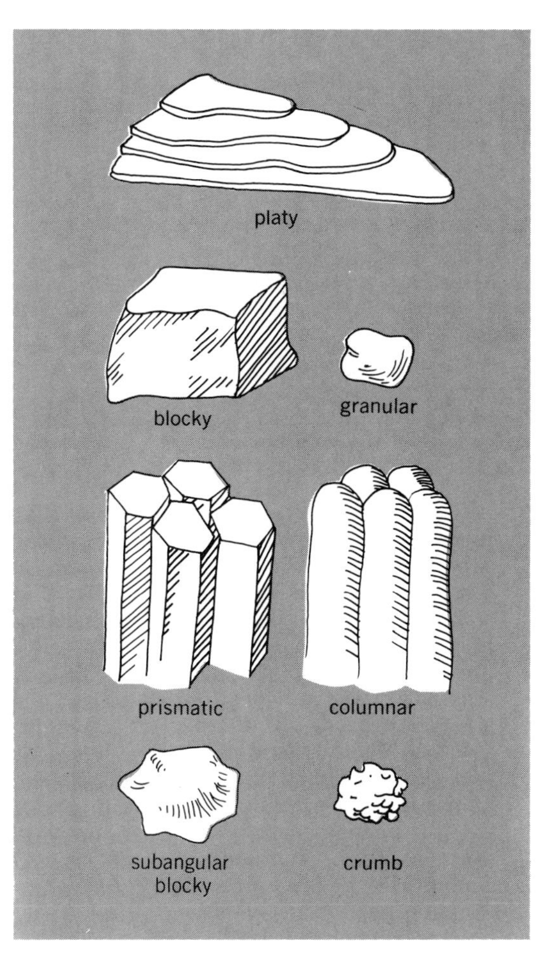

Fig. 5. Types of soil structure.

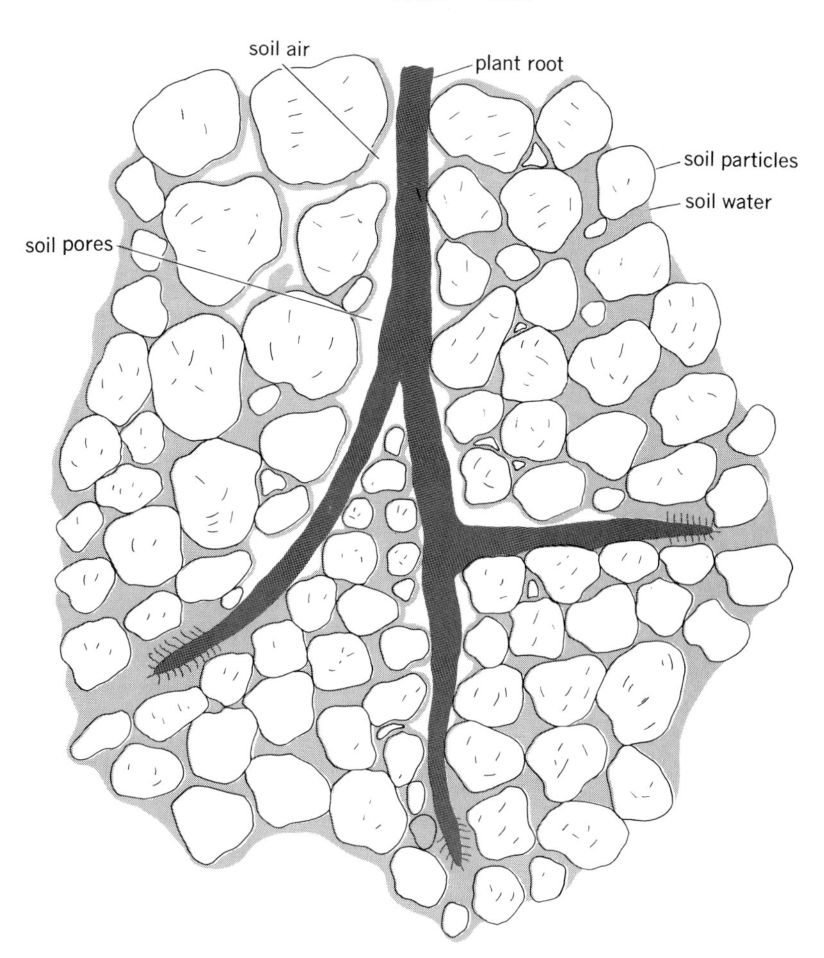

Fig. 6. Portion of surface soil with granular structure.

The size, shape, arrangement, and particularly the amount of overlap of soil aggregates and individual sand, gravel, and stone particles are extremely important because they largely determine the size, shape, arrangement, and continuity of pores in the soil.

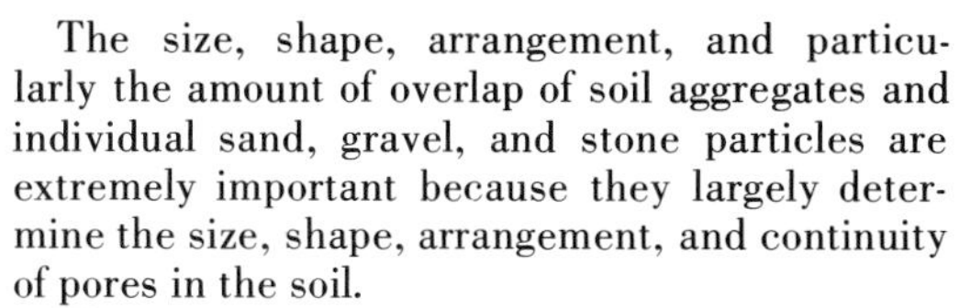

There are a number of ways of attempting to characterize the structure of a soil. The first and most direct is by visual examination of an undisturbed section of soil. Much can be learned about the size, shape, and arrangement of the soil particles, and the pore space, by close inspection of each horizon with the naked eye or with a magnifying glass. Micromorphology is the microscopic observation and photography of soil structure.

A second method is to measure how much of the soil has been aggregated into granules or crumbs with diameters above a given dimension, 0.25 mm being the most common. In well-granulated soils 70–80% of the total mass may be aggregated into granules or crumbs greater than 0.25 mm, as determined by wet-sieving a sample of soil through a 60-mesh screen. Aggregation values of 40–50% are more commonly found in soils under ordinary management. Sandy soils or clay soils having poor structure may have only 10–20% aggregation. Except in very sandy soils, such a low amount of aggregation usually forecasts a physical condition very unsatisfactory for plant growth.

Measurement of the permeability of the soil to water and air provides another means of evaluating its structure.

Bulk density. Bulk density is the mass (weight) of a unit volume of dry soil usually expressed in grams per cubic centimeter. It is determined by the density of the particles and by their arrangement.

The soil structure is the major factor in accounting for changes in the bulk density of a soil from time to time or from layer to layer in the profile. Soils with many particles closely packed together have high bulk densities and correspondingly low total pore space. Bulk density is a measure of the amount of compaction in soils. Traffic by farm machinery in intensively cultivated soils, trampling of cattle in heavily grazed pastures, and foot traffic on lawns and recreational areas result in severe compaction as reflected in bulk densities of 1.7–2.0 g/cm^3. Bulk densities of uncompacted, porous soils are about 1.2–1.3 g/cm^3. In undisturbed forest or grassland soils densities may be 0.9–1.0 g/cm^3. High amounts of organic matter will lower the bulk density.

Pore space. The voids or openings between the particles of the soil are spoken of collectively as the pore space. It makes up roughly one-half the volume of the soil. In very loose, fluffy soils with low bulk density it may occupy 60–65% of the total volume. In very compact soil layers it may be reduced to 35–40%. Pore space is calculated by Eq. (1).

$$\text{\% total pore space} = 100 - \left(\frac{\text{bulk density}}{\text{particle density}} \times 100\right) \quad (1)$$

Pore space in a soil is occupied by air and water in reciprocally varying amounts. Very dry soils have most of their pore spaces filled with air. The opposite is true for very wet soils.

There is considerable variation in the size, shape, and arrangement of pores in the soil. The effective size of a pore can be estimated by the amount of force required to withdraw water from the pore. These suction values, expressed in centimeters of water, can be translated into equivalent pore diameters using the capillary rise equation, Eq. (2), where r= radius of pore in centimeters,

$$r = \frac{2T}{hdg} \tag{2}$$

T= surface tension of water, d= density of water, g= acceleration of gravity, and h= suction force in centimeters of water.

The ideal soil should have the proper assortment of large, medium, and small pores. A sufficient number of large or macro pores (with diameters greater than 0.06 mm), connected with each other, are needed for the rapid intake and distribution of water in the soil and for the disposal of excess water by drainage into the substratum or into artificial drains. When without water they serve as air ducts. Cracks, old root channels, and animal burrows may serve as large pores. Soils with insufficient functional macro porosity lose a great deal of rainfall and irrigation water as runoff. They drain slowly and often remain poorly aerated after wetting. One of the first effects of compaction is the reduction of the size and number of the larger pore spaces in the soil.

The primary purpose of the small pores (less than 0.01 mm in diameter) is to hold water. It is through medium-sized pores (0.06–0.01 mm in diameter) that much of the capillary movement of water takes place. Loose, droughty, coarse, sandy soils have too few small pores. Many tight clay soils could well afford to have a greater number of larger pores.

Soil water. The movement and retention of water in the soil is related to the size, shape, continuity, and arrangement of the pores, their moisture content, and the amount of surface area of the soil particles. Movement and retention of water may be characterized by the energy relationships or forces which control these two phenomena.

Water retention. Some water is held in the soil pores by the force of adhesion (the attraction of solid surfaces for water molecules) and by the force of cohesion (the attraction of water molecules for each other). Water held by these two forces keeps the smaller pores full of water and also maintains relatively thick films on the walls of many larger pores. Not until the pores in one layer of soil are filled with all the water they can hold does water move into the layer below.

Water is found in the soil in both the liquid and vapor state. The soil air in all the pores, except those in the surface inch or so of very hot, dry soils is saturated with water vapor.

The liquid water may be characterized by the suction force or tension with which it is held in the soil by adhesion or cohesion. These suction or tension values may be expressed as (1) height in centimeters of a unit water column whose weight just equals the force under consideration; (2) pF, the logarithm of the centimeter height of this column; (3) atmospheres or bars; or (4) pounds per square inch (psi). For example, 1000 cm water tension= pF_3= 1 atm= 14.7 psi. The moisture content of the soil is determined by drying the soil at 105°C until it reaches constant weight and then dividing the weight of water lost by the weight of oven-dry soil. This value times 100 equals the percentage of water in the soil on a dry-weight basis. The percentage of moisture on the volume basis for a given depth of soil is calculated by Eq. (3). Tensiometers,

$$\begin{aligned}&\%\ \text{soil moisture (dry weight basis)} \div 100\\ &\quad \times \text{bulk density} \times \text{depth soil (in.)}\\ &\quad\quad = \text{in. of water per in. soil depth}\end{aligned} \tag{3}$$

electrical resistance blocks, and neutron gages are used to measure the moisture content of the soil in place. Thus changes in moisture content in the soil can be followed within the effective range of each instrument.

There are several soil-moisture "equilibrium" points. Water remaining at oven dryness is held at tensions above 10,000 atm (1.0 GPa). The hygroscopic coefficient is a rough measurement of the water held by air-dry soil at a tension of about 31 atm (3.1 MPa). The wilting point, or wilting percentage, represents that moisture content or moisture tension (15 atm or 1.5 MPa) at which plant roots cannot absorb water rapidly enough to offset losses by transpiration, causing the plant to wilt, first temporarily and then permanently. Certain plants of desert and dry farming regions are able to stay alive and even grow on water held at tensions up to 25–30 atm (2.5–3.0 MPa) by the soil. The field capacity represents the water remaining in a soil layer 2 or 3 days after having been saturated by rain or irrigation when the rate of downward movement of water held at low suction forces (0–1 atm or 0–100 kPa) has decreased to a progressively slower rate of water removal. A definite tension value cannot be assigned to this equilibrium point, although the water held at a tension of 0.33 atm (33 kPa) often is used to estimate the upper limit of a soil's available water-holding capacity. The maximum retentive capacity is the moisture content of a soil when all of its pores are filled or saturated with water, and under zero tension.

The moisture in a soil which is available for plant use is usually assumed to be that held between field capacity and the permanent wilting percentage. This is called the available soil moisture. Sandy loams hold 1–1½, loams 1½–2, and clay loams 1¾–2½ in. of available water per foot of soil. The retentivity of soils of different textures for moisture at different tensions is shown in Fig. 7. These are called soil moisture tension curves. Much of the water in sandy soils is held at low tensions. The opposite is true for clay soils.

Water held at tensions less than 1–2 atm (100–200 kPa) is the most easily available for root absorption and plant growth. An adequate supply of this water should be maintained in the root zone by rainfall or irrigation, especially during periods of critical water need by plants.

Water movement. Water moves in the soil as a gas and as a liquid. Vapor transfer takes place by diffusion in response to a vapor pressure gradient. Vapor movement is through air-filled pores from a moist to a dry layer and from a warm to a cool layer. Drying of a wet soil surface on a hot, dry, windy day and the condensation of water droplets on the undersurface of a plastic mulch on a cool, summer morning are the result of vapor transfer.

Liquid movement may be expressed by the equation $V = Ki$. V is the volume of water crossing the unit area perpendicular to the flow in the unit time. The proportionality factor K is the hydraulic conductivity or the permeability of the soil to water. It is controlled by the size, shape, arrangement, and moisture content of the soil pores. The value i is the water-moving force. It has two force components, the force of gravity and a suction or tension gradient force. The force or pull of gravity is of constant magnitude and always acts in a downward direction. The drainage or removal of most of the water from large pores is by this force, and the drainage is sometimes referred to as gravitational water. The suction or tension gradient force may vary both in magnitude and direction. The flow or capillary conduction of water in unsaturated soil is due to the gradient or difference in suction between two points in the soil. The flow is in the direction of the increase in suction. This accounts for the movement of water toward roots which have depleted the supply of water held at low tension in the soil at the soil-root interface, for the upward capillary conduction of water from an underlying saturated layer (water table), and for the slow downward movement of water after a rain or irrigation. Rate and amount of water movement by capillary conduction to the root system is not usually sufficient to meet the demands of the plant for water, except when a sufficient amount of water held at suctions less than 1 atm (100 kPa) is present around the root. Since capillary conductivity decreases rapidly as the soil becomes drier, the water needs of plants are satisfied also by an extension of their root systems into fresh supplies of water held at low tension in hitherto untapped or recently refilled soil pores. It is very important, therefore, that soil structure be such as to permit the rapid extension of the root system through the whole soil mass.

Soil air and aeration. Soil air differs from the atmosphere above the soil in that it usually contains 5–100 times as much carbon dioxide (0.15–0.65%) and slightly less oxygen, and is saturated with water vapor. In deep, poorly drained soil layers or in heavily manured soils the CO_2 content may reach 10% and the O_2 content decrease to 1%. In water-logged soils, anaerobic conditions may result in methane and hydrogen sulfide production.

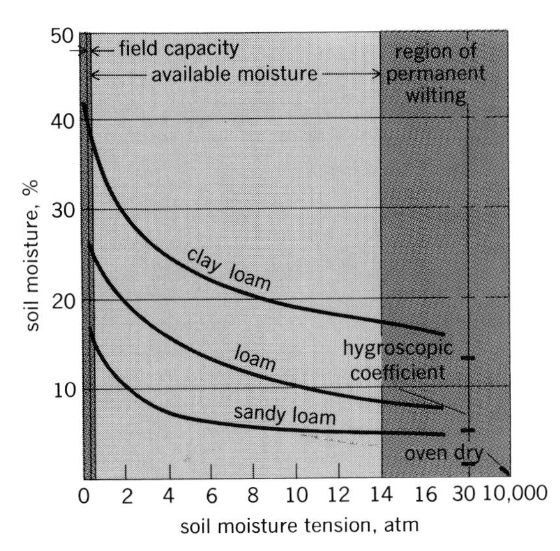

Fig. 7. Soil moisture tension curves. 1 atm= 101 kPa.

Aeration refers to the movement of gases in and out of each soil layer or horizon. The movement of gas within the soil as well as to and from the atmosphere is by diffusion in a direction determined by its own partial pressure. The rate of diffusion of each gas in and out of the soil depends on differences in concentration of each gas in the soil and in the atmosphere, and on the ability of the soil to transmit the gases. Diffusion or aeration is proportional to the volume of air-filled pores in any soil layer.

Temperature. The temperature of field soils shows rather definite changes at different depths, at different times during the day and night, and at different seasons of the year. These changes are determined by the amount of radiant energy that reaches the soil surface and by the thermal properties of the soil. Only that part of the heat energy which is absorbed causes changes in soil temperature. Heat produced by intense microbial decomposition of fresh organic matter mixed with the soil will also increase soil temperature. Dark-colored soils capture a much higher proportion of the radiant energy than do light-colored soils. The insulating effects of vegetative cover and surface mulches keeps the soil cooler than bare, fallow soil. Wet soils warm more slowly than dry soils because the heat capacity of water is five times that of the mineral soil particles. The energy absorbed by the soil surface is disposed of in one or more of the following ways: radiation to the atmosphere, heating of the air above the soil by convection, increasing the temperature of the surface soil, or conduction to the deeper soil layers.

Consistency and compactibility. As the moisture content of a soil changes from air dryness to saturation, its consistency varies from a state of hardness or brittleness, to loose, soft, or friable, to tough or plastic, to sticky or viscous. The reaction of the soil to physical manipulation, as in tillage, is primarily an expression of the properties of cohesion, adhesion, and plasticity. These properties are largely determined by the structure, organic matter content, kind and amount of clay, nature of adsorbed bases, and moisture content, which regulates the thickness of the water films around the soil particles. Tillage should be done only after the soil attains a soft, friable condition—when it breaks apart or can be worked into granules 1–5 mm in diameter. This is a very desirable range of particle size to provide good seed and root bed conditions.

Each soil has a critical moisture range, often near field capacity, at which pressure by foot or machinery traffic results in maximum compaction. Bulk density, permeability, porosity, and penetrometer measurements are used to indicate the degree of compaction as found in traffic pans in the surface soil, the plow sole, or natural hardpans. Soil compaction is a very serious problem because it reduces the permeability of the soil to air and water, and increases the resistance of the soil to root penetration. Hard, dry surface crusts may also prevent seedling emergence.

Color. Soil color may be influenced by, and indicates the kind of, parent material, chemical composition, organic matter content, drainage, aeration, or oxidation. A blotched or mottled yellow, gray, and blue subsoil indicates poor drainage, aeration, or oxidation. A clear red, yellow, or brown color indicates good drainage. The color of some soils is inherited from the parent material. Organic matter gives a brown to black color to that horizon where it is concentrated. Color of soils is determined by comparison with standard colors of known hue, value, and chroma in the Munsell soil color charts.

[RUSSELL B. ALDERFER]

SOIL MANAGEMENT

Soil management may be defined as the preparation, manipulation, and treatment of soils for the production of crops, grasses, and trees. Good soil management involves practices which will maintain a high level of production on a sustained basis. Ideally, these practices should provide the crop with an adequate supply of air, water, and nutrients; maintain or improve the fertility of the soil for subsequent crops; and prevent the development of conditions which might be injurious to plants.

Several systems of land-use classification have been developed which help a farmer to know the kinds of soils he has on his farm and their suitability for various types of farming. One of these systems, developed by the U.S. Soil Conservation Service, involves land-use capability ratings.

Land capability survey maps. Land capability survey maps are worked out in conjunction with the soil survey and serve as a guide to the suitability of land for cultivation, grazing, forestry, wildlife, watersheds, or recreation, with primary consideration given to erosion control. There are eight capability classes which describe the characteristics of the land and the difficulty or risk involved in using it for one kind of crop production or another. These eight classes are sometimes distinguished on land capability survey maps by roman numerals as well as by standard colors. Four of the eight classes include land that is suited for regular cultivation with varying degrees of erosion control measures and management practices required; three classes of land are not suited for cultivation but require permanent vegetation and impose severe limitations on land use; and one class includes lands suited only for wildlife, recreation, or watershed purposes. For a description of the land capability classes and the management practices recommended for each class *see* SOIL CONSERVATION.

Cropping system. A cropping system refers to the kind and sequence of crops grown on a given area of soil over a period of time. It may be a regular rotation of different crops, in which the sequence of crops follows a definite order, or it may consist of a single crop grown year after year in the same location. Other cropping systems include different crops but have no definite or planned sequence.

Cropping systems that involve the systematic rotation of different crops generally include hay and pasture crops, small grains, and cultivated row crops. Legumes, such as alfalfa, clover, and vetch, are usually grown alone or mixed with grasses in the hay and pasture sequence in the rotation because they supply nitrogen and contribute to good soil tilth. The beneficial effect of legumes and grasses on tilth may be attributed to the fact that (1) the soil is not tilled while these crops are being grown, and (2) the organic matter returned to the soil by the extensive root systems and in the plowed-under top growth is particularly suited to the development of a stable, porous soil structure.

Small grains function somewhat like legumes and grasses in giving protection against soil erosion, but they add no nitrogen and remove moderate quantities of plant food from the soil. Since small grains do not provide maximum economic return from the high nitrogen residues left in the soil by legumes, and are likely to lodge owing to the stimulation of growth from these residues, they are not planted in the rotation following legumes. Small grains are generally planted either at the end of a rotation following row crops or as a companion crop for legumes.

Row crops, such as corn, potatoes, cotton, and sugarbeets, are an excellent choice to follow legumes because they utilize the nitrogen supplied by the legumes and bring good cash returns. Since row crops in early stages of growth provide little protection against erosion and require considerable cultivation which breaks down soil structure, it is not considered desirable to plant them continuously.

A cropping system that involves growing the same crop year after year generally depletes the soil and results in lower crop yields. This is particularly true if the crop is cultivated frequently and returns little crop residue to the soil. Weeds, diseases, and insects also become more of a management problem when the cropping system does not involve rotation. Thus the farmer who intends to grow one crop year after year becomes completely dependent on disease-resistant varieties of plants, chemical insecticides and fungicides, soil fumigation, and other methods of controlling diseases, insects, and pests. Through the appropriate use of improved varieties, pesticides, and adequate amounts of fertilizers, farmers have succeeded in maintaining a high level of production on land repeatedly planted with the same crop. While such results are causing farmers to take another look at cropping systems that do not involve rotation, they are well aware that more intensive practices and costly supplements are required to maintain production.

Organic matter and tilth. The value of adding organic matter to the soil in the form of animal manures, green manures, and crop residues for producing favorable soil tilth has been known since ancient times. Research has provided information that helps to explain the mechanisms for this effect.

Experiments reveal that during the decomposition of organic matter in the soil, microorganisms synthesize a variety of gumlike substances, at least partly polysaccharide in nature which, when dried with the soil, bind the soil particles together into a porous, water-stable structure. While these binding substances are produced in relatively large quantities during stages of rapid organic-matter decomposition, they may in turn be decomposed by other organisms. Thus, to maintain a continuous supply of binding substances, organic matter must be added to the soil frequently.

In addition to the beneficial action of microbially synthesized binding substances, roots and fungal mycelia also contribute to the development of favorable soil tilth by molding the smaller gum-cemented granules into still larger aggregates. The aggregates adhering together form large pores that permit the rapid movement of air and water and form small pores that store water. Both conditions are essential features of good tilth.

Unfortunately not all the effects of organic matter on tilth are desirable. The growth of organisms in a fine-textured soil may interfere with the downward movement of water whenever the soil pores become clogged with microbial bodies. This condition is of particular significance where water is ponded to recharge underground water supplies or to leach excessive amounts of soluble salts from the soil.

Even the characteristic property of organic matter of promoting aggregate formation is not always desirable. Some surface mulches during decomposition induce the formation of a layer of small surface aggregates which are more susceptible to wind erosion than the fine soil particles initially present. In such cases, aggregation intensifies the hazard of severe wind erosion.

In spite of these negative effects of organic matter on the physical properties of soil, the incorporation of organic matter with soil is the most suitable and practical way of developing and maintaining good tilth.

Conditioners and stabilizers. Soil conditioners and stabilizers include a wide variety of natural and synthetic compounds that, upon incorporation with the soil, improve its physical properties. The term soil amendment also is applied to these compounds but is a more general term since it includes any material, exclusive of fertilizers, that is worked into the soil to make it more productive, regardless of whether it benefits the physical, chemical, or microbiological properties of the soil.

Soluble salts of calcium, such as calcium chloride and gypsum, or acid and acid formers, including sulfur, sulfuric acid, iron sulfate, and aluminum sulfate, have been used as conditioners to improve the physical properties of soils that were made unfavorable by excessive quantities of sodium ions adsorbed on the soil colloids.

The tilth of dense clay soils, which are slow to take water and have a marked tendency to become cloddy, may be improved by the addition of gypsum and by-product lime from sugarbeet processing factories. Limestone has improved the physical condition of acid soils, apparently by stimulating the activity of microorganisms to synthesize substances that bind soil particles into aggregates.

The discovery that soil microorganisms synthesize substances that improve soil structure stimulated the search for synthetic compounds that would be more effective than the natural products. While a wide variety of compounds have improved soil structure temporarily, three water-soluble, polymeric electrolytes of high molecular weight which are very resistant to microbial decomposition have been developed commercially for use in ameliorating poor soil structure. These are modified hydrolyzed polyacrilonitrile (HPAN), modified vinyl acetate maleic acid (VAMA), and a copolymer of isobutylene and maleic acid (IBMA). High cost relative to yield increase has limited these materials to experimental use.

Mixed with the soil in amounts ranging from 0.02 to 0.2% of soil by weight, these compounds are readily adsorbed by moist soils and tend to stabilize or fix the existing structure. They are therefore synthetic binding agents and should be added only to soils that have previously been worked into a desirable physical condition. These materials are not equally effective on all soils, and if improperly used can stabilize a poor physical condition.

Fertility. Soil fertility may be defined as that quality of a soil which enables it to provide nutrient elements and compounds in adequate amounts and in proper balance for the growth of specified plants, when other growth factors such as light, moisture, temperature, and the physical condition of the soil are favorable.

Testing. Even though relatively fertile and of good physical condition, a soil may be lacking in one or more of 16 elements presently known to be essential to plant growth, or it may be strongly acidic, alkaline, or salty, and thus unsuitable for plant growth. Fortunately, soil tests are available that indicate the existence of possible deficiencies or excesses in the soil. In most instances, these tests involve the use of various reagents for extracting from the soil the total or proportionate amount of the nutrient or compound in question. The amount of material extracted is then compared with values that have been correlated previously with crop response on the same or similar soil. No single test is reliable for all crops on all soils. *See* PLANT, MINERALS ESSENTIAL TO.

Control of pH. The availability of soil nutrients for plants is influenced greatly by the reaction of the soil. Soils may be classified as acid, neutral, or alkaline in reaction. The method commonly used in measuring and expressing degrees of acidity or alkalinity is in terms of pH. The pH value of soil may range from less than 4 to more than 8; the lower the value, the more acid the soil. Under most conditions lime is applied to acid soils to maintain their pH between 6.5 and 7.0. Under special conditions it may be desirable to maintain pH values either higher or lower than these. In any case the desirability of applying lime should be determined by the pH of the soil and the requirements of the plants to be grown.

It is occasionally necessary to make soils more acid. Materials commonly used to decrease pH are sulfur, sulfuric acid, iron sulfate, and aluminum sulfate.

Control of salinity. Restricted drainage caused by either slow permeability or a high water table is the principal factor in the formation of saline soils. Such soils may be improved by establishing artificial drainage, if a high water table exists, and by subsequent leaching with irrigation water to remove excess soluble salts.

Soils can be leached by applying water to the surface and allowing it to pass downward through the root zone. Leaching is most efficient when it is possible to pond water over the entire surface.

The amount of water required to leach saline soils depends on the initial salinity level of the soil and the final salinity level desired. When water is ponded over the soil about 50% of the salt in the root zone can be removed by leaching with 6 in. (15 cm) of water for each foot of root zone; about 80% can be removed with 1 ft (30 cm) of water per foot of soil to be leached; and 90% can be removed with 2 ft (60 cm) of water per foot of soil to be leached.

Because all irrigation waters contain dissolved salts, nonsaline soils may become saline unless water is applied in addition to that required to replenish losses by plant transpiration and evaporation, to leach out the salt that has accumulated during previous irrigations and through the addition of fertilizer.

Regulating nutrient supplies. The nutrients supplied to crops can be regulated by modifying the availability of nutrients already present in the soil. This can be accomplished by changing soil reaction, turning under green manure crops, including legumes which add nitrogen, and adding fertilizers. *See* FERTILIZER.

By changing soil reaction through the addition of lime, acidulating agents such as sulfur, or residually acid fertilizers such as ammonium sulfate, solubility and availability of compounds of phosphorus, iron, manganese, copper, zinc, boron, and molybdenum can be increased or decreased. Phosphorus compounds are generally more available in the slightly acid to neutral pH range, whereas compounds of iron, manganese, zinc, and copper become more available as the acidity of the soil increases. The activity of microorganisms responsible for the transformation of nitrogen, sulfur, and phosphorus compounds into forms available for plants also is influenced by soil reaction. A reaction which is too acid or too alkaline retards the activities of these organisms.

The decay of turned-under green manures and plant residues produces carbon dioxide, rendering soluble the nutrients from soil particles, and the nutrients which were absorbed from the soil during the growth of these crops are also made available. Although the turning under of a green manure affects the availability of nutrients, it does not add to the total nutrient supply unless the green manure is a legume which fixes atmospheric nitrogen.

The system of farming determines to a considerable extent the manner in which fertilizers are used to regulate nutrient supplies. Each system of farming depends upon the crop, soil, climate, kinds and rates of fertilizers applied, and available equipment; and for each system of farming there are many ways of applying fertilizers.

Common methods of applying fertilizers include broadcasting, banding, deep placement, and foliar applications. Broadcasting fertilizer on the soil is usually less desirable than localized placement of the fertilizer in relation to the seed or plant. Banding fertilizers to the side of the rows in furrow bottoms or beds and drilling fertilizer with the seed give the best response from limited quantities of fertilizer. Deep placement of fertilizers is effective in arid regions where soils dry out to a considerable depth or where deep-rooted crops are grown. Foliar applications of fertilizers, particularly those containing micronutrients, circumvent soil interactions. Such interactions within the soil may render the applied fertilizer unavailable to the crop. Foliar applications also make it possible for the farmer to supply his crops directly with a number of essential plant nutrients at critical stages of growth.

[DANIEL G. ALDRICH, JR.]

Soils of the tropics. The tropics may be defined as that part of the Earth's surface between the Tropic of Cancer and the Tropic of Capricorn, about 23-1/2° north and south of the Equator. In this region three broad ecological zones may be recognized: evergreen and deciduous forests which cover about one-fourth of the land area; savanna and grasslands, about one-half; and semidesert and desert areas, about one-fourth. Because of the wide range of climate, vegetation, parent materials, and other factors affecting soil formation, there are as many if not more different kinds of soils in the tropics as in temperate regions. They range from fertile soils of alluvial valleys, through deep, highly weathered infertile acid soils of uplands through shallow stony soils of steep mountain slopes, to high lime soils of deserts. Though many soils of the tropics have formed from recently deposited alluvium and volcanic ash, most have developed from older weathered parent materials, and consequently are generally much less fertile than soils in temperate regions which are mostly formed from relatively recent glacial and loessial deposits.

Resources. Of the 12×10^9 acres (5×10^9 hectares) of land area in the tropics, about 2×10^9 acres (8×10^8 hectares) are estimated to be potentially arable but as yet uncultivated. This is more than is now cultivated and represents about half of the world's uncultivated land which is potentially arable. Most of the potential for bringing new land under cultivation is in Africa and South America, where only about 22 and 11%, respectively, of the potentially arable soils are cultivated. Probably most potentially arable soils in the tropics not under cultivation are relatively infertile and require addition of nutrients and good management for economic crop production. The UN Food and Agriculture Organization and UNESCO have published soil maps of South America and Africa at a scale of 1:5,000,000 with accompanying texts. These are two volumes in a ten-volume series, *Soil Map of the World.* More detailed soil resource appraisals and expanded research efforts are required to provide necessary information for development of new lands and for improvement of soil management practices for intensification of production on presently cultivated soils. Substantial progress has been made in many tropical countries in making inventories of their soil resources.

Misconceptions. There are many misconceptions about soils of the tropics. One relates to the localized occurrence of laterite (irreversibly hardened ironstone) associated with the extensive red soils common in the tropics. These red soils have been called lateritic soils, and the mistaken notion arose that these soils when cleared of vegetation and cultivated would turn to hardened laterite and become worthless in a short time. Actually, extensive areas of red (so-called lateritic) soils in the tropics have been farmed for centuries without laterite forming. Probably only about 5% of soils in the tropics have laterite, and most are relics of previous geologic eras and not related to present use. Laterite forms under a particular set of conditions in localized areas and is of minor importance in the tropics.

A second prevalent misconception is that soils of the tropics have much less organic matter than those of temperate regions. This mistaken idea arose because of (1) warmer temperature in most of the tropics which accelerates organic matter decomposition and (2) widespread occurrence of reddish-colored soils in contrast to the generally dark-colored soils of the temperate regions. Analyses of soils from many different areas in the tropics have shown that, generally, the organic matter content is as high if not higher than that of soils in temperate regions. The long dry season in much of the tropics inhibits biological activity, just as winter temperatures do in temperate climates. In tropical areas with abundant rainfall there is much greater vegetative growth, which balances the increased biological activity in organic matter decomposition because of the warmer temperatures. Hence the soil organic matter content is maintained.

A third common misconception is that luxuriant forest growth in the low humid tropics is an indicator of fertile soils. More often than not, soils in tropical forests are relatively infertile. The nutrients are contained mostly in the vegetation and recycled as the decomposing organic matter releases them. The extensive roots of the forest vegetation prevent leaching of nutrients, and with abundant rainfall throughout the year luxuriant vegetation can be maintained on relatively infertile soils.

Agricultural systems. Though large areas of soils in the tropics are intensively cultivated for commercial export crops, most of the food for much of the population is produced under traditional, largely subsistence, shifting cultivation (slash-and-burn) systems. These are extensive systems supporting low-density populations, as the soil is usually cultivated with mixed crops for only a year or two followed by natural fallow for 5 to 15 years or more. With increased population pressure in many areas, the introduction of modern technology for more intensive use of the soils becomes necessary. Though substantial progress in this direction has been made, the intensification of traditional agriculture in the tropics within the limited resources of the farmers poses a major challenge to agricultural scientists. [MATTHEW DROSDOFF]

Determining irrigation needs. Ideally, farmers should irrigate periodically before drought cuts crop growth and yield and then apply only the water required to refill the soil occupied by roots to its capacity to retain water against drainage. When necessary, more water (about 10%) is applied to leach excess salts to maintain low salinity in the root zone.

Three general approaches have been followed to assist farmers in the irrigation decision-making process: (1) measure soil water content or suction, (2) use plants as indicators, and (3) maintain a soil water budget.

Soil water content or suction. Various methods have been described for measuring soil water content or suction: the tensiometer and sorption block, the electrical resistance unit, the electrothermal unit, and the neutron moderation technique. The neutron meter measures soil water on a volume basis (cm^3/cm^3), which means that readings can be expressed in surface centimeters or inches of water per acre, a distinct advantage when considering water budgets for scheduling irrigations. However, the meter embodies a radiation hazard from fast neutrons and requires a competent technician to use it effectively.

Tensiometers equipped with vacuum gages measure soil water suction up to 0.8 bar (where 1 bar is equivalent to 10^{-5} Pa) and can be read quickly. Irrigation water is applied when the vacuum gage registers a prescribed limit of soil water suction at a specified depth for a given crop. For example, tensiometers used to irrigate avocados are placed in the active root zone and near the bottom of the root zone. Irrigation applied when the soil water suction at the 30-cm depth approaches 50 centibars does not penetrate to the 60-cm depth until the application time is nearly doubled (July 5, August 26, and September 16). Between irrigations, soil water suction at the lower depth gradually increases, indicating that loss of water below root zone is being controlled.

Plants as indicators. Farmers are using change of leaf color (from light to bluish green) as a practical guide for scheduling irrigations on field beans, cotton, and peanuts.

Measurement of the internal plant water condition by sophisticated techniques as a criterion for scheduling irrigation is impractical for two reasons: plant water status is difficult to measure, and variation in plant water stress with time of day is difficult to interpret.

Soil water budget. Use of the soil water budget approach to determine need for irrigation requires a knowledge of: (1) short-term evapotranspiration (ET) rates at various stages of plant development. (2) soil water retention characteristics. (3) permissible soil water deficits in relation to evaporative demand, and (4) the effective rooting depth of the crop grown.

Evaporation pans are being used to develop soil water budget schemes for scheduling irrigations for sugarcane in Hawaii and for cotton and orchard and vegetable crops in Israel. Evaporation from pans to schedule irrigations must be calibrated for a specific crop and geographic area.

Another advance in techniques to schedule irrigations has been made by using the modified Penman equation to estimate potential ET. Four basic meteorological parameters are required: solar radiation, mean temperature, wind speed, and vapor pressure (dew point). Crop characteristic curves are required.

Sophisticated solid-set and traveling-type sprinkler systems, dead-level automated surface irrigation systems, and graded furrow systems with tail-water reuse facilities provide a high degree of water control and are readily adapted to the irrigation-scheduling techniques described. Farmer acceptance of the computerized meteorological approach to irrigation scheduling has been far greater than any of the other methods discussed. Professional scheduling services were provided for a fee on about 250,000 acres of irrigated lands in Arizona, California, Idaho, Nevada, Washington, Nebraska, Kansas, and Colorado in 1974.

[HOWARD R. HAISE]

Soil conservation. Numerous studies continue to indicate that most herbicides and insecticides, when applied to soils at recommended dosage rates, exert little effect on most soil microorganisms or on soil properties. However, certain fumigants and fungicides may temporarily kill or reduce the numbers of nonparasitic soil organisms and may temporarily influence soil chemical properties. Generally the side effects of most pesticides are not detrimental or they may actually be beneficial, but occasionally they may retard or inhibit plant growth for a few weeks or a few months. The magnitude of these effects depends on dosage, soil type, temperature, moisture, and other factors.

Influence on soil organisms. When applied to soils or crops at normal field application rates, most herbicides or insecticides have little effect on the soil microbes, or they may slightly stimulate growth of some species. However, herbicides may kill soil algae, and insecticides may kill nonparasitic soil insects. Volatile soil fumigants, such as D-D (a mixture of dichloropropane and dichloropropene) and methyl bromide and many soil fungicides, on the other hand, kill numerous nonparasitic soil organisms, including bacteria, fungi, actinomycetes, yeasts, algae, protozoa, earthworms, and insects. After the initial kill, numbers of soil bacteria soon reach much higher numbers than were present in the original soil. In acid soil, fungi may quickly return in greater numbers than were originally present, whereas in alkaline soils they may return quickly or their numbers may be reduced for a year or longer. Although the total number of species may be greatly reduced, with time additional species recolonize the soil and total numbers slowly decline to normal. Fumigants and insecticides exert a similar effect on soil insects.

Factors responsible for the increased numbers of certain microbes following a soil treatment which initially kills large numbers of soil organisms are: (1) The pesticide chemical may be utilized as a food source by one or more organisms; (2) The bodies of the organisms killed by the treatment are utilized as a food source by surviving forms or by species that first recolonize the soil. (3) The organisms which survive or first become reestablished can reach very high numbers because the environment is less competitive.

Influence on soil chemistry. Pesticides which reach the soil or which are applied directly to the soil may influence soil chemical properties in the following ways: (1) The pesticide chemical or partial decomposition products represent a change in the chemical composition of the soil. (2) Upon decomposition of the pesticide chemical, the constituent elements are released as simple inorganic compounds such as ammonia, phosphate, hydrogen sulfide, sulfate, chloride, and bromide. Carbon is released as carbon dioxide. Some of the elements, especially carbon and nitrogen, are utilized for synthesis of cells and organic products of the soil population. Eventually about 10–20% or more of the added carbon is incorporated into the relatively resistant soil humus. The new humus, however, decomposes faster than older, stabilized humus. (3) Pesticidal chemicals which kill an appreciable percentage of the soil organisms often increase the water-soluble salt content of soils. Soluble calcium is especially increased; usually more soluble magnesium, potassium, and phosphorus are noted. (4) Fumigants and fungicides may increase soluble or extractable micronutrient elements, including manganese, copper, and zinc. (5) Most fumigants and fungicides and certain other pesticides kill or reduce numbers of the relatively sensitive nitrifying bacteria in the soil. These organisms oxidize ammonia to nitrites and nitrates. Until these bacteria become reestablished, relatively more of the available soil nitrogen will be in the form of ammonia. Reestablishment of nitrifying bacteria occurs generally within a few weeks to a few months. (6) Pesticides containing the benzene ring may be detoxified and altered with respect to side chains or groups, and may then undergo polymerization reactions with plant and microbial phenolic substances. In this way, parts of pesticide molecules may serve as constituent units in the formation of the beneficial soil humus.

Increased growth response. The microbiological and chemical effects (side effects) of pesticides in soils generally exert little influence on plant growth. Sometimes, however, growth may be temporarily enhanced or retarded. Increased growth may be related to fertilizer value of nitrogen or phosphorus released during decomposition of specific pesticides and cells of organisms killed by the pesticide treatments; increased availability of soil manganese, phosphorus, and other plant nutrient elements; a plant auxin-type action of some pesticide chemicals or of their partial decomposition products; and recolonization of the soil by microbial species that exert a strong antagonism against plant root parasites.

Reduced plant growth. Occasionally something seems to go wrong following application of a pesticide to soils, and plants may be injured or growth may be retarded for short periods of time. These unexpected results may be caused by various factors:

1. Many pesticide chemicals are toxic to some plant species, and a few, such as the fumigants, are toxic to most plants. If time is not allowed for these chemicals to decompose in the soil or volatilize from the soil, the residual chemical may injure or even kill crop plants. Continuous or frequent use of pesticides which decompose very slowly in the soil may increase soil levels of the pesticide to a point that growth of sensitive crops will be retarded.

2. Simple inorganic substances released during decomposition of certain pesticides may injure sensitive plants. Avocado plants, for example, are highly sensitive to soil chloride. Treatment of the soil with D-D, chloropicrin, or other chloride-containing chemicals may cause or increase chloride injury to this plant. Similarly, several plant species, including onions, carnations, and citrus, are sensitive to bromide. Bromide residues from certain pesticides may temporarily retard growth of these species. Other possibly toxic inorganic decomposition residues of pesticides include arsenic, iodine, copper, and mercury.

3. In greenhouse or ornamental operations, in which fertilization rates are high, the killing of bacteria which oxidize ammonia to nitrites and nitrates may result in the accumulation of toxic concentrations of ammonia from decomposing organic nitrogenous fertilizers.

4. In soils high in manganese, fumigation may

increase the soluble manganese to toxic levels for a short period of time. Although extractable soil manganese is generally increased, soil fumigation may sometimes cause manganese deficiencies of cauliflower, brussels sprouts, and broccoli which may be related to increases in available potassium or other elements which reduce manganese uptake.

5. Sometimes treatment of the soil with any pesticide which kills large numbers of microbes will cause a temporary plant growth inhibition manifested by reduced absorption of phosphorus, zinc, and sometimes copper. The growth inhibition is quite spotty, and healthy plants may grow next to severely injured ones. Young citrus, peach, and certain other tree seedlings are especially sensitive to this phenomenon.

Studies have shown that a primary factor in this type of plant growth inhibition is the killing of endotrophic mycorrhizal fungi which aid the plant roots in absorbing certain plant nutrient elements, especially phosphorus. The condition can be corrected, or partially corrected, by proper fertilization with phosphorus, zinc, and sometimes copper, by inoculation of seed or seedlings with an effective mycorrhiza strain, or by delaying the planting of a sensitive crop until the mycorrhizal fungi have become reestablished in the soil.

[JAMES P. MARTIN]

Fertilizer in semiarid grasslands. In the semiarid grasslands of temperate regions, lack of available nitrogen often limits production as much as does lack of available water. Consequently, use of nitrogen fertilizer is increasing on millions of acres of grasslands, particularly in the northern Great Plains of the United States and Canada. This practice immediately raises concern about the ecological impact of extensive nitrogen fertilization on pollution of surface and ground waters with nitrate. Results of recent research on the fate of fertilizer nitrogen applied to semiarid grasslands have greatly reduced the uncertainty that has surrounded this subject.

In addition to the nitrogen absorbed and translocated into plant tops, a semiarid grassland ecosystem can immobilize a fairly definite quantity of fertilizer nitrogen in the roots, mulch, residues, and soil organic matter. The quantity of nitrogen immobilized in these pools varies with soil type and texture, water availability, and possibly temperature, but is not influenced greatly by either grass species or most management schemes. Fertilizer nitrogen immobilized in these organic forms may later be mineralized by soil microorganisms and recirculated through the ecosystem. A relatively small quantity (10 to 40 lb per acre; 1 lb per acre equals 1.12 kg per hectare) of fertilizer nitrogen seems to be absorbed directly into the cells of the soil microbes and in a few weeks or months is mineralized and recirculated as successive generations of microbes are produced and die. Much more fertilizer nitrogen (up to about 200 pounds per acre) is immobilized in grass roots and mulches and seems to be recycled in 3 to 5 years. Typically, up to 350 pounds of fertilizer nitrogen per acre can be immobilized in various organic pools in the soil-plant system.

Nitrogen cycle. The nitrogen cycle in semiarid grassland ecosystems is essentially a closed system; that is, losses of nitrogen from the soil-plant system are relatively low. Ordinarily, no fertilizer nitrogen is leached below the root zone in semiarid grasslands, so leaching losses are generally inconsequential. Losses of fertilizer nitrogen in gaseous form (by ammonia volatilization or denitrification) also seem usually to be relatively small, except perhaps where urea-containing fertilizers are applied to semiarid grasslands at rates exceeding 80 to 100 pounds of nitrogen per acre. In such instances, available data suggest that volatilization losses may be relatively high. *See* NITROGEN CYCLE.

Typical data on the fate of fertilizer nitrogen applied to semiarid grasslands emerged in an experiment in which 80 pounds of nitrogen per acre (as ammonium nitrate) was applied annually for 11 years to mixed prairie (primarily *Agropyron smithii, Stipa viridula,* and *Bouteloua gracilis*) grazed by yearling steers. After 11 years, approximately 35% of the fertilizer nitrogen applied was found in the roots (19%) and vegetative mulch (16%) on the soil surface. A slightly larger quantity remained in the soil as inorganic (ammonium 2%, nitrate 39%) nitrogen, indicating that the fertilizer applied exceeded the nitrogen required by the ecosystem. Less than 3% of the fertilizer nitrogen was physically removed from the pasture in the form of beef. In total, about 82% (including standing tops 2%, crown 1%) of the fertilizer nitrogen applied was accounted for. The 18% not accounted for was immobilized in soil organic matter or lost to the atmosphere as gaseous nitrogen. Other research suggests that the gaseous loss was about 5% of that applied. Therefore, losses from the nitrogen cycle are relatively low, and a major part of the nitrogen applied to grasslands remains in forms that can be recycled and used for plant growth in later years. All of the nitrogen in the inorganic pool plus a major part of that in roots, residues, and mulches, and some in the soil organic nitrogen pool (included in the unaccounted-for fraction), may be recycled.

Plant debris. Research using the ^{15}N isotope shows that within hours after application, the isotope is found primarily in the senescent or dead plant material—mulches and decaying root materials. This suggests that fertilizer nitrogen is absorbed into the cells of the microorganisms as they rapidly multiply after addition of nitrogen fertilizer. The increased microbial population then decomposes the senescent and dead plant materials, mobilizing the nitrogen they contain. Thus, nitrogen immobilized in plant material is recirculated through the nitrogen cycle, illustrating the importance of plant debris both above and below the soil surface as a pool of potential plant-available nitrogen in semiarid grasslands.

[J. F. POWER]

SOIL EROSION

Soil erosion is that physical process by which soil material is weathered away and carried downgrade by water or moved about by wind. Two categories of erosion are recognized. The first, called geologic erosion, is a natural process that takes place independent of man's activities. This kind of erosion is always active, wearing away the surface features of the Earth. The second kind, referred to as accelerated erosion, occurs when man disturbs the surface of the Earth or quickens the pace of erosion in any way. It produces conditions that are

Fig. 8. Sheet erosion showing how soil has been brought from entire cultivated hillside. A large soil deposit has collected on flat area at bottom of slope. (*USDA*)

Fig. 9. Rill erosion showing how water has followed the old corn rows. (*USDA*)

Fig. 10. Large gully could be repaired by plowing in and seeding to grass. (*USDA*)

abnormal and poses a problem for the future food supply of the world. To combat erosion successfully, it is important that man recognize the erosion processes and have a knowledge of the factors which affect erosion.

Types. Erosion by running water is usually recognized in one of three forms: sheet erosion, rill erosion, and gully erosion.

Sheet erosion. The removal of a thin layer of soil, more or less uniformly, from the entire surface of an area is known as sheet erosion. It usually occurs on plowed fields that have been recently prepared for seeding, but may also take place after the crop is seeded. Generally only the finer soil particles are removed. Although the depth of soil lost is not great, the loss of relatively rich topsoil from an entire field may be serious (Fig. 8). If continued for a period of years, the entire surface layer of soil may be removed. In many parts of the world only the surface layer is suited for cultivation.

Rill erosion. During heavy rains runoff water is concentrated in small streamlets or rivulets. As the volume or velocity of the water increases, it cuts narrow trenches called rills. Erosion of this type can remove large quantities of soil and reduce the soil fertility rapidly (Fig. 9). This type of erosion is particularly detrimental because all traces of the rills are removed after the land is tilled. The losses which occurred are often forgotten and adequate conservation measures are not taken to prevent further loss of soil.

Gully erosion. This type of erosion occurs where the concentrated runoff is sufficiently large to cut deep trenches, or where continued cutting in the same groove deepens the incision. Gullying often develops where there is a water overfall. The stream bed is cut back at the overfall and the gully lengthens headward or upslope. Once started, gullying may proceed rapidly, particularly in soils that do not possess much binding material. Gully erosion requires intensive control measures (Fig. 10), such as terracing or the use of diversion ditches, check dams, sod-strip checks, and shrub checks.

Affective factors. The rate and extent of soil erosion depend upon such interrelated factors as type of soil, steepness of slope, climatic characteristics, and land use.

Type of soil. Soil types vary greatly in physical and chemical composition. The amounts of sand, silt, and clay constituents, colloidal material, and organic matter all have a bearing upon the ease with which particles or aggregates can be detached from the body of the soil. Such detachment is caused chiefly by the beating action of raindrops. The particles are then transported downgrade by moving water. Sandy or gravelly soils often have little colloidal material to bind particles together, and hence these materials are easily detached. However, because of their size, sand particles are more difficult to move than fine particles. For this reason sand particles are moved chiefly by rapidly flowing water on steep slopes or by streams at flood stage. Finer particles, such as silt, clay, and organic matter, can be carried by water moving at a slower rate. On gently sloping fields there is a tendency for more of the fine particles to be carried away, leaving the heavier sand particles behind. However, if rainfall is intense and the volume of runoff great, sand may be moved even on gentle slopes.

Slope. The relation of slope to the amount of erosion on different classes of soil is illustrated in Fig. 11. The amount of total runoff from rainfall increases only slightly with increase in the slope of the land above 1–2%, but the speed of the flowing water, or rate of runoff, may increase greatly. Since the capacity of moving water to transport soil particles increases in geometric ratio to the rate of flow, the amount of erosion increases greatly with increase in the slope of the land (Fig. 12).

Climate. In cold climates the frozen soil is not subject to erosion for several months of the year. However, if such areas receive heavy snow, serious erosion may take place when the snow melts. This is particularly true if the snow melts as the ground gradually thaws. As the water moves over the thin unfrozen layer of soil, it transports much of this soil material downgrade.

In warm climates soils are susceptible to erosion any time there are heavy rains. Such soils are particularly vulnerable to erosion if rains fall in winter and there is little vegetative cover.

The amount of rainfall is an important factor in determining the erosion that occurs in a given region. However, the character of the rainfall is usually a much more important factor than the total amount in determining the seriousness of erosion. A rain falling at the rate of 2 in./hr (50 mm/hr) may cause three to five times as much erosion as a rainfall of 1 in./hr (25 mm/hr). Regions where most of the precipitation comes in the form of mist or gentle rain may undergo little erosion, even though the total rainfall may be high and other conditions conducive to erosion.

In some areas of dry climates strong winds cause soil movement and serious loss of soil. Wind erosion is more common on sandy soils, but it is by no means confined to them. Heavier soils, which have a fluffy physical condition produced by freezing and thawing or drying, may be moved in great quantities by the wind.

Land use. The type of crops and the system of management influence the amount and type of erosion. Bare soils, clean uncultivated soils, or land in intertilled row crops permit the greatest amount of erosion. Crops that give complete ground cover throughout the year, such as grass or forests, are most effective in controlling erosion. Small-grain crops, or those that provide a fairly dense cover for only part of the year, are intermediate in their effect on erosion. Table 3 gives results of some of the earliest experiments in the United States on differences in land use and the effect on runoff and erosion. These results show that cultivated land, especially without a crop or protective cover of vegetation, is particularly vulnerable to erosion. In addition, excessive erosion usually occurs where cultivated crops like corn, cotton, and tobacco are grown on hilly or sloping land that is subjected to increased rates of runoff. In some areas where row crops have been grown continually the soil has been removed to the depth of the plow layer within a lifetime.

Pastures in humid areas usually have a tough continuous sod that prevents or greatly reduces sheet or surface erosion. Natural range cover, if in good condition, is usually effective in controlling erosion, but in areas of limited rainfall, where bunch grasses form most of the cover on range land, occasional heavy rain may cause severe erosion of the bare soil exposed between the bunches of grass. Forest lands, with their overhead canopies of trees and surface layers of decaying organic matter, have much greater water intake and much less surface runoff and erosion.

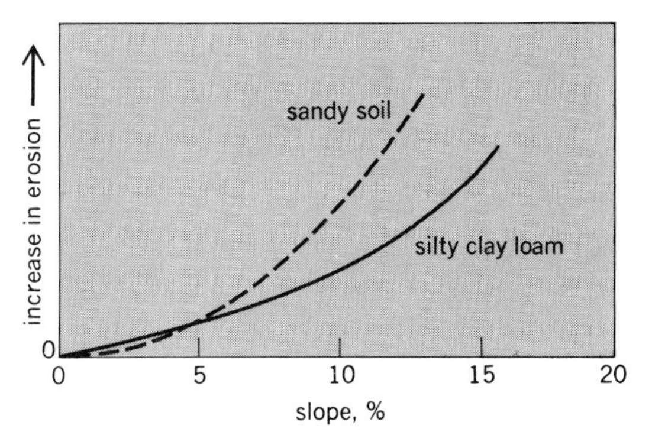

Fig. 11. Generalized diagram illustrating greater loss of fine-grained soil (silty clay loam) on gentle slopes (0–5%) and greater loss of sandy soil on steeper slopes.

Wind erosion. In the western half of the United States and in many other parts of the world, great quantities of soil are moved by the wind. This is particularly so in arid and semiarid areas. Sandy soils are more subject to wind erosion than silt loam or clay loam soils. The latter, however, are easily eroded when climatic conditions cause the soil to break into small aggregates, ranging from 0.4 to 0.8 mm in diameter. The coarse particles usually are moved relatively short distances, but the fine dust particles may be carried by strong winds for hundreds or even thousands of miles.

In some areas the coarse, or sand, particles are moved by the wind and deposited over extensive areas as dunes. The dunes move forward in the same direction as the prevailing winds, the particles being moved from the windward side of the dune to the lee side. If dunes become covered with grass or other vegetation, they cease to move. The sandhill region of Nebraska is a good example.

Control. The following are a few fundamental principles which will help control erosion and greatly reduce the damage done by soil erosion.

1. Keep land covered with a growing crop or grass as much time as possible. Cover increases

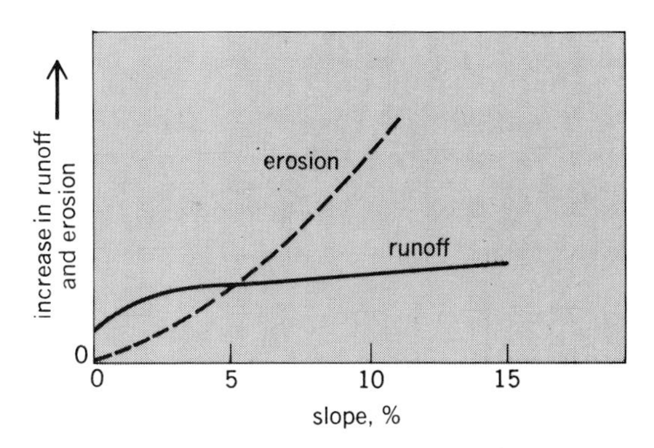

Fig. 12. Effect of slope on total amount of runoff and on rate of runoff and soil erosion.

Table 3. Relative runoff and erosion from soil under different land uses, with mean rainfall of 35.87 in. (910.1 mm)*

Land use and treatment	Runoff, %	Tons soil per acre eroded annually	Years required to lose surface 7 in. of soil
Plowed 8 in. (203 mm) deep, no crop; fallowed to keep weeds down	28.4	35.7	28
Plowed 8 in., corn annually	27.4	17.8	56
Plowed 8 in., wheat annually	25.2	6.7	150
Rotation; corn, wheat, red clover	14.1	2.3	437
Bluegrass sod	11.6	0.3	3547

**Missouri Res. Bull*, no. 63, 1923.

intake of water and reduces runoff. The extent of erosion control will be roughly in proportion to the effective cover.

2. When there is no growing crop, retain a cover of stubble or crop residue on the land between crops and until the next crop is well started. This can be done by using a system known as stubble-mulch farming. It ultilizes the idea of preparing a seedbed for a new crop without burying the residue from the previous crop. Tillage tools that work beneath the surface and pulverize the soil without necessarily inverting it or burying the residue are used instead of moldboard plows. This system is best adapted to regions of low rainfall or warm climates.

3. Avoid letting water concentrate and run directly downhill. By doing this the soil is protected against water at its maximum cutting power. Construct terraces with gentle grades to carry the runoff water around the hill at slow speeds. These diversions should empty onto grassed waterways or on meadowland to prevent creation of gullies.

4. Plant crops and till the soil along the contours.

5. Control wind erosion by keeping land covered with sod or planted crops as much of the time as possible. Maintain crop residue on the land between crops and while the next crop is getting started.

6. If wind erosion begins on a bare field or where a crop is just getting started, the soil drifting may be stopped temporarily by cultivation. An implement with shovels that will throw up clods or chunks of soil to give a rough surface is usually effective. Often, only strips through the field need be so treated to stop erosion on the whole area.

7. Moving dunes may require artificial cover or mechanical obstructions on the windward side, followed by vegetative plantings, depending on climatic conditions. Along shorelines, beach grasses followed by woody plants and forests may be required.

For a discussion of the physical, economic, and social effects of soil erosion *see* SOIL CONSERVATION.

[FRANK L. DULEY]

Bibliography: L. D. Baver and W. H. Gardner, *Soil Physics*, 4th ed., 1972; P. W. Birkeland, *Pedology, Weathering, and Geomorphological Research*, 1974; E. T. Cleaves, A. E. Godfrey, and J. K. Coulter, Soil management systems, in *Soils of the Humid Tropics*, 1972; Food and Agriculture Organization–UNESCO, *Soil Map of the World*, 1971–1976; H. D. Foth and L. M. Turk, *Fundamentals of Soil Science*, 5th ed., 1972; C. E. Kellogg and A. C. Orvedal, Potentially arable soils of the world and critical measures for their use, *Advan. Agron.*, 21:109–170, 1969; H. Kohnke, *Soil Physics*, 1968; J. P. Martin, in C. A. I. Goring (ed.), *Organic Chemicals in the Soil Environment*, vol. 2, pp. 733–792, 1972; National Academy of Sciences, *Soils of the Humid Tropics*, 1972; F. N. Ponnamperuma, in *Advances in Agronomy*, 1972; R. W. Simonson, Loss of nutrient elements in soil formation, in O. P. Englestad (ed.), *Nutrient Mobility in Soils: Accumulation and Losses*, Soil Sci. Soc. Amer. Spec. Publ. no. 4, 1970; Soil Survey Staff, *Soil Taxonomy: A Basic System of Soil Classification for Making and Interpreting Soil Surveys*, USDA Handb. 436, 1975; L. M. Thompson and F. R. Troeh, *Soils and Soil Fertility*, 4th ed., 1978.

Soil conservation

The practice of arresting and minimizing artificially accelerated soil deterioration. Its importance has grown because cultivation of soils for agricultural production, deforestation and forest cutting, grazing of natural range, and other disturbances of the natural cover and position of the soil have increased greatly in the last 100 years in response to the growth in world population and man's technical capacity. Accelerated soil deterioration has been the consequence.

Erosion extent and intensity. Accelerated erosion has been known throughout history wherever men have tilled or grazed slopes or semiarid soils. There are many evidences of the physical effects of accelerated erosion in the eastern and central parts of the Mediterranean basin, in Mesopotamia, in China, and elsewhere. Wherever the balance of nature is a delicate one, as on steep slopes in regions of intense rainstorms, or in semiarid regions of high rainfall variability, grazing and cultivation eventually have had to contend with serious or disabling erosion. Irrigation works of the Tigris and Euphrates valleys are thought to have suffered from the sedimentation caused by quickened erosion on the rangelands of upstream areas in ancient times. The hill sections of Palestine, Syria, southern Italy, and Greece experienced serious soil losses from grazing and other land use mismanagement many centuries ago. Accelerated water erosion on the hills of southern China and wind erosion in northwestern China also date far back into history. Exactly what effects these soil movements may have had on history has been a debated question, but their impact may have been serious on some cultures, such as those of the Syrian and Palestinian areas, and debilitating on others, as in the case of classical Rome and the China of several centuries ago.

The exact extent of accelerated soil erosion in the world today is not known, particularly as far as the rate of soil movement is concerned. However, it may be safely said that nearly every semiarid area with cultivation or long-continued grazing, every hill land with moderate to dense settlement in humid temperate and subtropical climates, and all cultivated or grazed hill lands in the Mediterranean climate areas suffer to some degree from such erosion. Thus recognized problems of erosion are found in such culturally diverse areas as southern China, the Indian plateau, south Australia, the South African native reserves, the Soviet Union, Spain, the southeastern and midwestern United States, and Central America.

Within the United States the most critical areas have been the hill lands of the Piedmont and the interior Southeast, the Great Plains, the Palouse area hills of the Pacific Northwest, southern California hills, and slope lands of the Midwest. The high-intensity rainstorms of the Southeast, and the cyclical droughts of the Plains have predisposed the two larger areas to erosion. The light-textured A horizon formed under the Plains grass cover was particularly susceptible to wind removal, while the high clay content of many southeastern soils predisposed them to water movement. These natural susceptibilities were repeatedly brought into play by agricultural systems which stressed corn and cotton in the Southeast, corn in the Midwest, and intensive grazing and small grains on the Plains, Palouse, and California. The open soil surface left in the traditional cotton, corn, and tobacco cultivation of the Southeast furnished almost ideal conditions for water erosion, and at the same time caused heavy nutrient depletion of soils thus cropped. The open fields of seasons between crops have also been susceptible to soil depletion. Open fields have been especially disastrous to maintenance of soil cover during the droughts of the Plains. Soil mismanagement thus has been a common practice in parts of the United States where the stability of soil cover hangs in delicate balance.

Types of soil deterioration. Soil may deteriorate either by physical movement of soil particles from a given site or by depletion of the water-soluble elements in the soil which contribute to the nourishment of crop plants, grasses, trees, and other economically usable vegetation. The physical movement generally is referred to as erosion. Wind, water, glacial ice, animals, and man's tools in use may be agents of erosion. For purposes of soil conservation, the two most important agents of erosion are wind and water, especially as their effects are intensified by the disturbance of natural cover or soil position. Water erosion always implies the movement of soil downgrade from its original site. Eroded sediments may be deposited relatively close to their original location, or they may be moved all the way to a final resting place on the ocean floor. Wind erosion, on the other hand, may move sediments in any direction, depositing them quite without regard to surface configuration. Both processes, along with erosion by glacial ice, are part of the normal physiographic (or geologic) processes which are continuously acting upon the surface of the Earth. The action of both wind and water is vividly illustrated in the scenery of arid regions (Fig. 1). Soil conservation is not so much concerned with these normal processes as with the new force given to them by man's land use practices. *See* LAND-USE PLANNING.

Depletion of soil nutrients obviously is a part of soil erosion. However, such depletion may take place in the absence of any noticeable amount of erosion. The disappearance of naturally stored ni-

Fig. 1. Erosion of sandstone caused by strong wind and occasional hard rain in an arid region. (*YSDA*)

trogen, potash, phosphate, and some trace elements from the soil also affects the usability of the soil for man's purposes. The natural fertility of virgin soils always is depleted over time as cultivation continues, but the rate of depletion is highly dependent on management practices. *See* SOIL.

Accelerated erosion may be induced by any land use practice which denudes the soil surfaces of vegetative cover. If the soil is to be moved by water, it must be on a slope. The cultivation of a corn or a cotton field is a clear example of such a practice. Corn and cotton are row crops; cultivation of any row crop on a slope without soil-conserving practices is an invitation to accelerated erosion. Cultivation of other crops, like the small grains, also may induce accelerated erosion, especially where fields are kept bare between crops to store moisture. Forest cutting, overgrazing, grading for highway use, urban land use, or preparation for other large-scale engineering works also may speed the natural erosion of soil (Fig. 2).

Where and when the soil surface is denuded, the movement of soil particles may proceed through splash erosion, sheet erosion, rill erosion, gullying, and wind movement (Fig. 3). Splash erosion is the minute displacement of surface particles caused by the impact of falling rain. Sheet erosion is the gradual downslope migration of surface particles, partly with the aid of splash, but not in any defined rill or channel. Rills are tiny channels formed where small amounts of water concentrate in flow. Gullies are V- or U-shaped channels of varying depths and sizes. A gully is formed where water concentrates in a rivulet or larger stream during periods of storm. It may be linear or dendritic (branched) in pattern, and with the right slope and soil conditions may reach depths of 50 ft (15 m) or more. Gullying is the most serious form of water erosion because of the sharp physical change it causes in the contour of the land, and because of its nearly complete removal of the soil cover in all horizons. On the edges of the more permanent stream channels, bank erosion is another form of soil movement.

Fig. 2. Rill erosion on highway fill. The slopes have been seeded (horizontal lines) with annual lespedeza to bind and stabilize soil. *(USDA)*

Causes of soil mismanagement. One of the chief causes of erosion-inducing agricultural practices in the United States has been ignorance of their consequences. The cultivation methods of the settlers of western European stock who set the pattern of land use in this country came from a physical environment which was far less susceptible to erosion than North America, because of the mild nature of rainstorms and the prevailing soil textures in Europe. Corn, cotton, and tobacco, moreover, were crops unfamiliar to European agriculture. In eastern North America the combination of European cultivation methods and American intertilled crops resulted in generations of soil mismanagement. In later years the plains environment, with its alternation of drought and plentiful moisture, was also an unfamiliar one to settlers from western Europe.

Conservational methods of land use were slow to develop, and mismanagement was tolerated because of the abundance of land in the 18th and 19th centuries. One of the cheapest methods of obtaining soil nutrients for crops was to move on to another farm or to another region. Until the 20th century, land in the United States was cheap, and for a period it could be obtained by merely giving assurance of settlement and cultivation. With low capital investments, many farmers had little stimulus to look upon their land as a vehicle for permanent production. Following the Civil War, tenant cultivators and sharecroppers presented another type of situation in the Southeast where stimulus toward conservational soil management was lacking. Management of millions of acres of Southeastern farmland was left in the hands of men who had no security in their occupancy, who often were illiterate, and whose terms of tenancy and meager training forced them to concentrate on corn, cotton, and tobacco as crops.

On the Plains and in other susceptible western areas, small grain monoculture, particularly of wheat, encouraged the exposure of the uncovered soil surface so much of the time that water and wind inevitably took their toll (Fig. 4). On rangelands, the high percentage of public range (for whose management little individual responsibility could be felt), lack of knowledge as to the precipitation cycle and range capacity, and the urge to maximize profits every year contributed to a slower, but equally sure denudation of cover.

Finally, the United States has experienced extensive erosion in mountain areas because of forest mismanagement. Clearcutting of steep slopes, forest burning for grazing purposes, inadequate fire protection, and shifting cultivation of forest lands have allowed vast quantities of soil to wash out of the slope sites where they could have produced timber and other forest values indefinitely. In the United States the central and southern Appalachian area and the southern part of California have suffered severely in this respect, but all hill or mountain forest areas, except the Pacific

Northwest, have had such losses. *See* FOREST MANAGEMENT AND ORGANIZATION; FOREST RESOURCES.

Economic and social consequences. Where the geographical incidence of soil erosion has been extensive, the damages have been of the deepest social consequence. Advanced stages of erosion may remove all soil and therefore all capacity for production. More frequently it removes the most productive layers of the soil—those having the highest capacity for retention of moisture, the highest soil nutrient content, and the most ready response to artificial fertilization. Where gullying or dune formation takes place, erosion may make cultivation physically difficult or impossible. Thus, depending on extent, accelerated erosion may affect productivity over a wide area. At its worst, it may cause the total disappearance of productivity, as on the now bare limestone slopes of many Mediterranean mountains. At the other extreme may be the slight depression of crop yields which may follow the progress of sheet erosion over short periods. In the case of forest soil losses, except where the entire soil cover disappears, the effects may not be felt for decades, corresponding to the growth cycle of given tree species. Agriculturally, however, losses are apt to be felt within a matter of a few years.

Moderate to slight erosion cannot be regarded as having serious social consequences, except over many decades. As an income drepressant, however, it does prevent a community from reaching full productive potentiality. More severe erosion has led to very damaging social dislocation. For those who choose to remain in an eroding area or who do not have the capacity to move, or for whom migration may be politically impossible, the course of events is fateful. Declining income leads to less means to cope with farming problems, to poor nutrition and poor health, and finally to family existence at the subsistence level. Communities made up of a high proportion of such families do not have the capacity to support public services, even elementary education. Unless the cycle is broken by outside financial and technical assistance or by the discovery of other resources, the end is a subsistence community whose numbers decline as the capacity of the land is further reduced under the impact of subsistence cultivation. This has been illustrated in the hill and mountain lands of southeastern United States, in Italy, Greece, Palestine, China, and elsewhere for many millions of peasant people. Illiteracy, short life-spans, nutritional and other disease prevalence, poor communications, and isolation from the rest of the world have been the marks of such communities. Where they are politically related to weak national governments, idefinite stagnation and decline may be forecast. Where they are part of a vigorous political system, their rehabilitation can be accomplished only through extensive investment contributed by the nation at large. In the absence of rehabilitation, these communities may constitute a continued financial drain on the nation for social services such as education, public health, roads, and other public needs.

Effects on other resources. Accelerated erosion may have consequences which reach far beyond the lands on which the erosion takes place

Fig. 3. Two most serious types of erosion. *(a)* Sheet erosion as a result of downhill straight-row cultivation. Note onions washed completely out of ground. *(b)* Gully erosion destroying rich farmland and threatening highway. *(USDA)*

and the community associated with them. During periods of heavy wind erosion, for example, the dust fall may be of economic importance over a wide area beyond that from which the soil cover has been removed. The most pervasive and widespread effects, however, are those associated with water erosion. Removal of upstream cover changes the regimen of streams below the eroding area. Low flows are likely to be lower and their period longer where upper watersheds are denuded than where normal vegetative cover exists. Whereas flood crests are not necessarily higher in eroding areas, damages may be heightened in the valleys below eroding watersheds because of the increased deposition of sediment of different sizes, the rapid lifting of channels above floodplains, and the choking of irrigation canals.

A long chain of other effects also ensues. Because of the extremes of low water in denuded areas during dry seasons, water transportation is made difficult or impossible without regulation,

Fig. 4. Wind erosion. Accumulation of topsoil blown from bare field on right. *(USDA)*

fish and wildlife support is endangered or disappears, the capacity of streams to carry sewage and other wastes safely may be seriously reduced, recreational values are destroyed, and run-of-the-river hydroelectric generation reaches a very low level. Artificial storage becomes necessary to derive the services from water which are economically possible and needed. But even the possibilities of storage eventually may disappear when erosion of upper watersheds continues. Reservoirs may be filled with the moving sediment and lose their capacity to reduce flood crests, store flood waters, and augment low flows. For this reason plans for permanent water regulation in a given river basin must always include watershed treatment where eroding lands are in evidence. *See* WATER CONSERVATION.

Conservation measures and technology. Measures of soil management designed to reduce the effects of accelerated erosion have been known in both the Western world and in the Far East since long before the time of Christ. The value of forests for watershed protection was known in China at least 10 centuries ago. The most important of the ancient measures on agricultural lands was terrace construction, although actual physical restoration of soil to original sites also has been practiced. Terrace construction in the Mediterranean countries, in China, Japan, and the Philippines represents the most impressive remaking of the face of the Earth before the days of modern earth-moving equipment (Fig. 5). Certain land management practices which were soil conserving have been a part of western European agriculture for centuries, principally those centering on livestock husbandry and crop rotation. Conservational management of the soil was known in colonial Virginia and by Thomas Jefferson and others during the early years of the United States. However, it is principally since 1920 that the technique of soil conservation has been developed for many types of environment in terms of an integrated approach. The measures include farm, range, and forest management practices, and the building of engineered structures on land and in stream channels.

Fig. 5. The Ifugao rice terraces, Philippines. *(Philippine Embassy, Washington, D.C.)*

Farm, range, and forest. A first and most important step in conservational management is the determination of land capability—the type of land use and economic production to which a plot is suited by slope, soil type, drainage, precipitation, wind exposure, and other natural attributes. The objective of such determination is to achieve permanent productive use as nearly as possible. The United States Soil Conservation Service has developed one of the more easily understood and widely employed classifications for such determination (Fig. 6). In it eight classes of land are recognized within United States territory. Four classes represent land suited to cultivation, from the class 1 flat or nearly flat land suited to unrestricted cultivation, to the steeper or eroded class 4 lands which can be cultivated only infrequently. Three additional classes are grazing or forestry land, with varying degrees of restriction on use. The eighth

Fig. 6. Land capability classes. Suitable for cultivation: 1, requires good soil management practices only; 2, moderate conservation practices necessary; 3, intensive conservation practices necessary; 4, perennial vegetation—infrequent cultivation. Unsuitable for cultivation (pasture, hay, woodland, and wildlife); 5, no restrictions in use; 6, moderate restrictions in use; 7, severe restrictions in use; 8, best suited for wildlife and recreation. (*USDA*)

class is suited only to watershed, recreation, or wildlife support. The aim in the United States has been to map all lands from field study of their capabilities, and to adjust land use to the indicated capability as it becomes economically possible for the farm, range, or forest operator to put conservational use into force.

Once the capability of land has been determined, specific measures of management come into play. For class 1 land few special practices are necessary. After the natural soil nutrient minerals begin to decline under cultivation, the addition of organic or inorganic fertilizers becomes necessary. The return of organic wastes, such as manure, to the soil is also required to maintain favorable texture and optimum moisture-holding capacity. Beyond these measures little need be added to the normal operation of cultivation.

On class 2, 3, and 4 lands, artificial fertilization will be required, but special measures of conservational management must be added. The physical conservation ideal is the maintenance of such land under cover for as much of the time as possible. This can be done where pasture and forage crops are suited to the farm economy. However, continuous cover often is neither economically desirable nor possible. Consequently, a variety of devices has been invented to minimize the erosional results from tillage and small grain or row crop growth. Tillage itself has become an increasingly important conservational measure since it can affect the relative degree of moisture infiltration and soil grain aggregation, and therefore erosion. Where wind erosion is the danger, straw mulches or row or basin listing may be employed and alternating strips of grass and open-field crops planted. Fields in danger of water erosion are plowed on the contour (not up- and downslope), and if lister cultivation is also employed, water-storage capacity of the furrows will be increased. Strip-cropping, in which alternate strips of different crops are planted on the contour, may also be employed. Crop rotations that provide for strips of closely planted legumes and perennial grasses alternating with grains such as wheat and barley and with intertilled strips are particularly effective in reducing soil erosion. Fields may also be terraced, and the terraces strip-cropped. The bench terrace, which interrupts the slope of the land by a series of essentially horizontal slices cut into the slope, is now comparatively little used in the United States in contrast to the broad-base terrace, which imposes comparatively little impediment to cultivation. A broad-base terrace consists of a broad, shallow surface channel, flanked on the downward slope by a low, sloping embankment. If the terrace is constructed with a slight gradient (channel-type or graded terrace), it serves to reduce erosion by conducting excess water off the slope in a controlled manner. If it is constructed on the contour (level or ridge-type terrace), its primary purpose is to conserve moisture. The embankment on the downslope side of a broad-base terrace may be constructed from soil taken from the upper side only (sometimes referred to as a Nichols terrace) or from both sides (Mangum terrace).

Design of conservational cultivation also includes provision for grass-covered waterways to collect drainage from terraces and carry it into stream courses without erosion. Where suitable conditions of slope and soil permeability are found, shallow retention structures may also be constructed to promote water infiltration. These are of special value where insufficient soil moisture is a problem at times. Additional moisture always encourages more vigorous cover growth.

The measures just described may be considered preventive. There are also measures of rehabilitation where fields already have suffered from erosion and offer possibilities of restoration. Grading with mechanical equipment and the construction of small check dams across former gullies are examples.

For the remaining four classes of land, whose principal uses depend on the continuous maintenance of cover, management is more important than physical conditioning. In some cases, however, water retention structures, check dams, and other physical devices for retarding erosion may be applied on forest lands and rangelands. In the United States such structures are not often found in forest lands, although they have been commonly employed in Japanese forests. In forestry the conservational management objective is one of maximum production of wood and other services while maintaining continuous soil cover. The same is true for grass and other forage plants on managed grazing lands. For rangelands, adjustment of use is particularly difficult because grazing must be tolerated only to the extent that the range plants still retain sufficient vitality to withstand a period of drought which may arrive at any time.

A last set of erosion-control measures is directed toward minimizing stream bank erosion which may be large over the length of a long stream. This may be done through revetments, retaining walls, and jetties, which slow down current undercutting banks and hold sand and silt in which soil-binding willows, kudzu, and other vegetation may become established. Sediment detention reservoirs also reduce the erosive power of the current, and catchment basins or flood-control storage helps reduce high flows (Fig. 8).

Conservation agencies and programs. Whereas excellent soil-conserving soil management was maintained for generations by some farmers and farm groups, as in Lancaster County, Pa., a major amount of the soil-conservation activities in the United States is derived from Federal government assistance. The Soil Conservation Service of the USDA has been a focal agency in spreading knowledge of soil conservation in farmland and rangeland management and aiding in its application. In practice, the local administration of a soil-conserving program is within a Soil Conservation District, which usually is coincident with a county, and is organized under state law. The district is the liaison unit between the farmer and public assistance agencies at the state and Federal levels. It is managed by a board or committee, generally composed of five members, and usually elected by farmers within the district. Other local public bodies which may have soil-conservation objectives include conservancy districts, wind-erosion districts, drainage or irrigation districts, Agricultural Stabilization and Conservation Service County Committees, and Farmers' Home Administration County Committees. In addition there are private groups with conservational interests, such as the farmers' cooperatives and national farm organizations like the Farm Bureau Federation.

The local districts may be aided technically and financially in their program. Much of the financial aid stems from Federal sources, and theoretically it is on a matching fund basis. In actual practice, however, a major part of the expenditures for special soil-conserving programs is from Federal funds. Technical aid is provided throughout the nation by the Soil Conservation Service, and also by the U.S. Forest Service for its special fields of forestry and grazing-land management. Technical aid also has been provided by the Agricultural Extension Services and the Land Grant Colleges of the several states. The Tennessee Valley Authority has maintained a program of its own design, with the cooperation of the colleges and the Extension Services. The Soil and Moisture Conservation Operations Office of the Indian Service, U.S. Department of the Interior, likewise has conducted a program limited to specific Indian lands.

Financial assistance for soil-conservation measures has been provided by the Federal government through the Soil Conservation Service, the Agricultural Stabilization and Conservation Service, the TVA, and the Farmers' Home Administration. Assistance has been particularly in the form of loans from the FHA, in low-cost fertilizer from the TVA, and as direct cash outlay from other agencies. Over the years, the program of the Agricultural Stabilization and Conservation Service has been the largest single source of financial aid for these purposes.

In addition to technical and financial aid, the farmers or other land operators of the United States are given valuable indirect assistance through the many research programs, basic and applied, which treat the fields related to soil conservation. The work of the Agricultural Research Service, of the Soil Conservation Service, of the Tennessee Valley Authority, and of the Land Grant Colleges has been especially helpful. Through these works new soil-conserving plants, new fertilizers, improved means of physical control, and new methods of management have been developed. Through them soil conservation has not only become important but also an increasingly efficient public activity in the United States. *See* FOREST AND FORESTRY; RANGELAND CONSERVATION.

[EDWARD A. ACKERMAN; DONALD J. PATTON]

Bibliography: I. Burton and R. W. Kates, *Readings in Resource Management and Conservation*, 1965; R. B. Held, *Soil Conservation in Perspective*, 1965; R. M. Highsmith, J. G. Jensen, and R. D. Rudd, *Conservation in the United States*, 1978; G. O. Schaub et al., *Soil and Water Conservation Engineering*, 2d ed., 1966.

Soil mechanics

The application of the laws of solid and fluid mechanics to soils and similar granular materials as a basis for design, construction, and maintenance of stable foundations and earth structures. Soil mechanics differs from other applications of mechanics to engineering, such as the design of concrete and steel structures, in that soil and similar materials have a much wider range of mechanical and other physical properties than concrete and steel. In addition, soil materials are usually present in layers and strata that vary widely in composition and physical character.

Soil mechanics as an applied science has a relatively small system of legitimate theory and a large and important methodology concerned with soil exploring, sampling, and testing. It also makes use of information from other fields, such as geology and soil science.

Soil mechanics will always be an important part of soil engineering. However, the tools of soil engineering are no longer predominantly the principles of solid and fluid mechanics but to an ever-increasing extent those of other subdivisions of physics, such as thermodynamics, electricity and magnetism, acoustics, optics, and chemical, atomic, and nuclear physics. This is true also in the present development of soil exploration and testing and in endeavors to improve soil properties.

Soil, in the engineering sense, comprises all accumulations of solid particles in the earth mantle that are loose enough to be moved with spade or shovel. Soils range from deep-lying geologic deposits to agricultural surface soils. Soil mechanics is also applicable to similar granular assemblies of artificial origin, such as fills of mineral waste materials. *See* SOIL.

SOIL TYPES AND COMPOSITION

Soils vary widely in composition and physical properties. At one extreme are the inert, granular, cohesionless sands and gravels; at the other are clay soils of great water affinity and well-developed cohesion in the moist and dry state. According to the ease or difficulty of working them, rather than their density, soils are called light when predominantly granular and noncohesive and heavy when predominantly clayey and cohesive.

Mineral origins of soils. The mineral composition of gravels, stones, and boulders is essentially that of the parent rock from which they have been derived, predominantly by mechanical weathering action.

Sands are largely quartzitic and siliceous in humid climates but may be any kind of mineral in dry climates or under special circumstances. The white sands of New Mexico consist of gypsum particles; coral and shell beach sands may have more than 90% of calcium carbonate particles; the black sand of Yellowstone Park, Wyo., and some of the blue and purple beaches of the Pacific consist of obsidian particles.

The silt particles resemble quite closely the minerals in the parent rock, with feldspars, micas, and quartz usually well represented. In certain tropical soils, however, the silt and even larger particles may have been formed by stable agglomeration of smaller-sized chemical rock decomposition products.

Clay particles are submicroscopic, plate-shaped crystalline minerals that have been divided into three main groups: the kaolinite, the hydrous mica or illite, and the montmorillonite group. The clays possess great affinity for water as a result of the large amount of surface per volume of particle and as a function of the number and kind of adsorbed cations that neutralize unbalanced electrical charges in the clay mineral structures.

The boulder, gravel, and sand fractions are called coarse-grained or granular; they may be considered as the bones, and the combined silt, clay, and water fractions as the meat of a soil. Depending on their size, composition, or granulometry, soils may have a continuous granular skeleton, with the pores between the sand and gravel particles either empty or filled to various degrees with the silt-clay-water phase; they may possess a matrix of the latter in which the granular materials are discontinuously dispersed, or they may consist entirely of silt components, clay components, or both. The physical and mechanical properties of soils with sand and gravel skeletons are markedly different from those without. In practice, the limiting size between granular and nongranular (silt-clay) fractions is the opening of a 200-mesh sieve (74 μ).

Types by deposition. Engineering soils include unconsolidated sediments transported to their present place by glaciers, water, and air and soils formed at their present site by climatic and biologic forces from solid igneous, sedimentary, or metamorphous rock or from loose sediments transported by the above-named agents. The different agents have different carrying capacities and also affect the properties of their loads in different ways. It is, therefore, important to recognize and name such soils in accordance with the means of their transportation.

Glacial soils have been transported by glaciers whose action may be likened to giant bulldozers that push all sorts of materials ahead, with droppings on the side and grinding underneath. Spring melting of the glaciers stops their forward movement and permits the settling out of finely ground material. Thus, glacial soils may vary in size composition from boulders to varved clays. Typical for these soils is a disordered landscape, often with inhibited drainage and development of bogs and peat.

Aeolian soils range from sand dunes to loess deposits whose particles are predominantly of silt size. The valley loesses (Missouri, Mississippi, Rhine, and others) are formed from glacier-ground particles stirred up and dissipated by wind from the dry bottoms of glacier-draining rivers during the winter when the glacier supplies no water.

Fluvial soils are river deposits of relatively uniform particle size within the range from gravel to clay. The size itself depends upon the speed of water flow at the specific location.

Lacustrine soils are sediments formed on the bottom of lakes, and marine soils are sediments in ocean and other salt-water bodies. The salt content of the latter often flocculates the fine sediments and gives them a special type of structure. Because the carrying capacity of the respective agents varies with their speed, thus with weather and season, sedimentary deposits are usually stratified, that is, built up in layers of similarly sized particles.

Soils developed in place from either solid- or loose-rock parent material by the action of climate, plant growth, and animal life are usually shallow in temperate and cold regions. They are of importance mainly in the case of shallow foundations, as in highway and airport engineering. In tropical regions, however, they may extend to considerable depth and acquire importance for deep foundations. The science of pedology is concerned with their formation and characteristics.

Soil structure. Natural soil systems are characterized not only by the sizes and types of their component particles but also by the arrangement of these particles in relation to each other. For granular, noncohesive soils the porosity, or its supplement, the portion of the total volume filled by the solid particles, often suffices as an indicator of structure or packing. In natural clay and silt soils, typical secondary structures are formed whose disturbance can greatly alter their mechanical properties. The structure may be caused by and be typical of flocculation, as in marine clays, or may have been developed by wetting and drying and freezing and thawing cycles. Since soils may be employed in engineering in the undisturbed, partly disturbed, or greatly disturbed condition, the existence of soil structure must be taken into account.

For each specific purpose, the soil must be tested in a condition as close as possible to the one in which it is to be used.

Particle size and weight. The grain-size distribution of soils lacking or having negligible amounts of particles smaller than 74 μ is determined by dry sieving, and that of materials with all particles smaller than 74 μ by sedimentation methods. The latter are based on Stokes' law for the rate of fall in a viscous medium of lesser density, Eq. (1). Here ΔG = difference in specific gravity

$$v = \Delta G g d^2/1800n \quad \text{cm/sec} \qquad (1a)$$

$$d = \sqrt{1800\, nv/\Delta G g} \qquad (1b)$$

between particle and viscous medium, g = acceleration of gravity, d = diameter of particle in millimeters, n = viscosity of medium in poises, and v = rate of fall in centimeters per second. Practical methods utilize the decrease of suspended particles with time at a definite distance from the surface of a soil-in-water suspension. This decrease is usually calculated from hydrometer measurements of the density of the suspension. For soils having both coarse and fine constituents, sieving (both dry and wet) is combined with sedimentation analysis.

At least four different sieve sizes are used, with openings of 4760, 2000, 420, and 74 μ, respectively. For specific purposes, other sieves are used to advantage. The results of mechanical analysis are best presented in the form of a grain-size accumulation curve as in Fig. 1. The naming of soils in accordance with their texture is shown in Fig. 2.

Determination of the specific gravity of soil particles is necessary for the sedimentation analysis and also for calculating (from the weight per unit volume and the moisture content) the actual volume relationships in the soil, which are important for its mechanical-strength properties. The ratio of the volume of water- and air-filled voids to the volume of solids in the sample is called the voids ratio e. Because of the variability of water content of

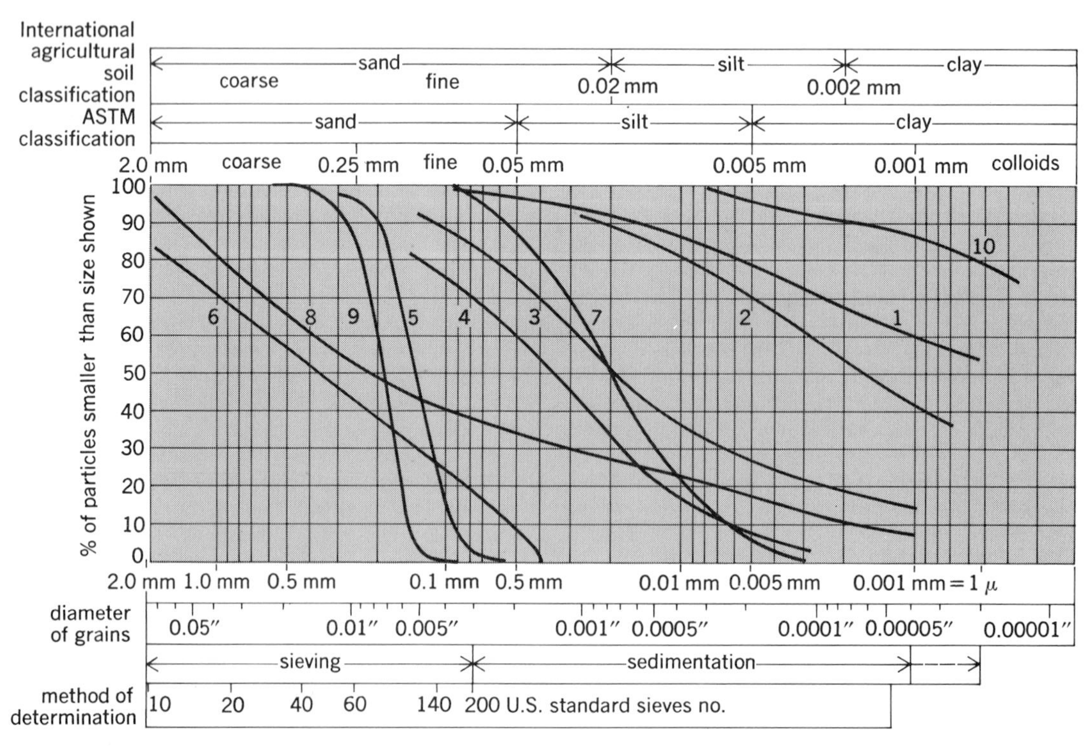

Fig. 1. Grain-size accumulation curves, plotted on semilogarithmic paper. Coarser fractions (left) measured by sieving; finer fractions (right) determined by sedimentation. Two of several systems of classification by grain size are shown (top). Curves 1 and 2, clay soils of the Nile Delta; 3 and 4, silts from the Nile Delta; 5, Port Said beach sand; 6, sand artificially graded for maximum density; 7, Vicksburg loess; 8, New Mexico adobe brick; 9, Daytona Beach sand; 10, Wyoming bentonite. *(From G. P. Tschebotarioff, Soil Mechanics, Foundations and Earth Structures, McGraw-Hill, 1951)*

soils, their unit weight is usually given in pounds per cubic foot of dry soil.

Engineering soil classifications. Soils have been classified for engineering purposes by several public agencies and individuals. The most widely used system at the present time is the so-called unified classification of the U.S. Bureau of Reclamation. While the different systems vary in details and nomenclature, the physical classification principles are essentially the same. The first division is made between soils that contain a coarse granular skeleton of gravel, sand, or both and those without such a skeleton, the silt-clay soils. The coarse granular materials are then subdivided into gravel and sand soils, which are further differentiated by the degree of water affinity of silt-clay material found in the intergranular spaces. The silt-clay soils which may or may not contain separate occluded coarse particles are subdivided into groups that differ in water affinity, plasticity, and swelling and shrinkage characteristics. A separate class of boulders and cobbles may be added to the coarse granular category and one of fibrous organic soil (peat) of great compressibility to the fine-grained category. The physical distinctions underlying engineering soil classifications entrain definite differences in other important physical properties, such as mechanical strength, elasticity, compressibility, shear and flow behavior, and permeability. The classification of a soil thus may serve to indicate its suitability for various engineering uses.

Soil-water relationships. Relationships between soil and water are of primary importance in soil mechanics. Their understanding and utilization presupposes a thorough knowledge of the properties of the water substance and of the physical and physicochemical characteristics of the surfaces of the soil minerals. Water is a peculiar substance. According to its molecular weight, it should be a gas at room temperature, but it is a liquid. Even as a liquid, it possesses structural properties commonly associated with solids and provable by means of x-rays. Its peculiarities, which are the consequences of the geometric-electric structure of the H_2O molecule, assert themselves also in its interactions with any type of surface.

Hygroscopic water. Dry soil minerals adsorb water from the atmosphere; the actual amounts depend on the physicochemical character of the surfaces and increase with rising relative humidity. The amount adsorbed is called hygroscopic water; it is usually measured at room temperature and expressed in percent of the dry soil weight. Very small amounts of water may be in solid solution within the surface of the minerals. Additional increments build up water films whose consistency may range from solid near the particle, through plastic, to that of normal water at a certain distance from the particle surface. A plastic water film 10^{-6} cm thick may not even equalize the roughness of a sand or gravel particle, but if it is around a plate-shaped clay mineral 10^{-5} cm thick, the water film represents more than 20% of the total clay-water volume. Therefore, the smaller the soil minerals, the more important are their water affinity and its consequences, such as swelling and shrinkage.

Gravitational and capillary water. In addition to this physicochemically restrained water, there may be free water in the soil pores, which is called gravitational water if it moves freely under the force of gravity, or capillary water if it is controlled by the forces of capillarity. The height at which capillary water can exist above the groundwater level depends on the effective pore radius. The water affinity and the capillarity of clay and silt soils cause the entrance of water in either the liquid or vapor phase and also affect the ease with which water moves through a soil, or the difficulty of its removal.

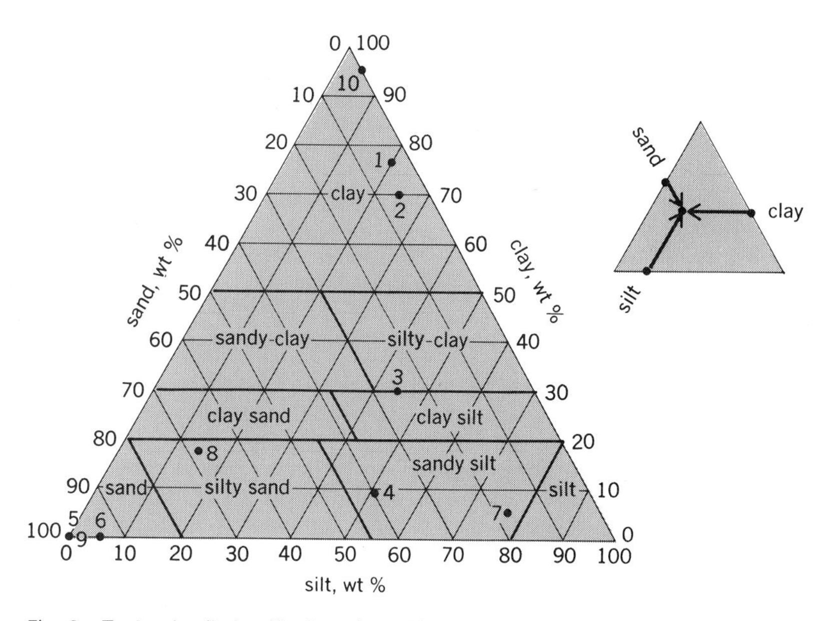

Fig. 2. Textural soil classification chart. The nomenclature of soil is determined by the percentages of sand, clay, and silt that are contained. The 10 numbered soils refer to those in Fig. 1. (*From G. P. Tschebotarioff, Soil Mechanics, Foundations and Earth Structures, McGraw-Hill, 1951*)

SHEAR AND PLASTICITY

Soil under stress will deform by rupture, plastic flow, or creep. The mechanism will depend on factors such as stress conditions, water content, and soil cohesiveness.

Shear in soils. Soils are composed of many separate particles of great range in size and shape. The particles may or may not be held together by water films or by a clay-water cement. Analogous systems are encountered in the molecular world. Systems composed of a single kind of molecule or atom have, at constant pressure, definite temperatures of transition from the solid to the liquid state. This transition generally involves an expansion, that is, an increase in the interparticle spacing, which represents the only real difference between the solid and the liquid phases at the melting point.

Introduction of molecules of different size and character into a pure substance lowers the melting point; also, the mixture will soften over a range of temperature instead of melting sharply at a single temperature. In multicomponent materials, such as asphalts and pitches, a wide softening and liquefaction range replaces a definite melting point. The same phenomenon occurs in macroparticle systems, such as soils. Densely packed gravels and sands are macromeritic (large-particle) solids. If submitted to shear stresses, they must expand in the shear zone to voids ratios characteristic of the molten state, in order that shear may take place without breaking of the individual particles. At voids ratios above the critical (melting range) ratio, gravel and sand soils can be "liquefied" by vibration that reduces interparticle friction or, if sheared, they may collapse to the critical voids ratio.

In granular noncohesive soils, the shear resistance obeys the equation $S = N \tan \varphi$, in which S= shear resistance in psi, N= effective normal pressure on the shear plane in psi, and $\tan \varphi$= coefficient of friction, or tangent of angle of friction related to angle of repose of a pile of the granular material. If the soil possesses a granular skeleton bonded together by moist or dry clay, the shear resistance can be approximated by $S = C + N \tan \varphi$, in which C denotes the cohesion of the system in psi. The numerical value of C depends on the amount, type, and water content of the clay binder, its physicochemical interaction with the coarse particles, and the history of the system. In soils without granular skeleton, S approaches C.

Plasticity. The property of plasticity is possessed by many crystalline solids within a certain temperature range below and adjoining the melting point.

Conditions for plasticity. Plasticity denotes the ability of a body to deform permanently without rupture under applied stresses. This presupposes that during deformation (1) no marked expansion takes place in the shearing zones which would alter the cohesive forces; (2) the particulate components (molecules, atoms, micro- and macroparticles) are in similar geometric arrangement after as before deformation; and (3) the rate of deformation and of breaking existing bonds does not exceed that of forming new bonds between atoms or molecules. Even in plastic bodies, very rapid deformation produces brittle fracture. In solid crystals, these requirements are best fulfilled if their building units are arranged with the highest degree of symmetry, as is the case with many metals. In macroparticle systems, such as soils, the requirements for plasticity are essentially the same. Large masses of relatively uniform sands and gravels may deform plastically at a slow rate and without change in voids ratio.

Small soil masses, and especially laboratory samples, show plasticity only if they are moist and cohesive. If so, the reforming of bonds broken during deformation of the mass and the reestablishment of the original symmetry of the bonding elements pertain essentially to the molecules in the water films around the particles. These water films play the same role in soils (increasing the distances between gliding planes and decreasing cohesion) as elevated temperature does in crystalline solids. Because plasticity increases with increasing total area of gliding planes per unit volume, the greatest plasticity is possessed by moist systems of pure clays of smallest particle size and greatest force of interparticle attraction. Admixture of coarser-grained material to clay interferes with the normal development of gliding planes and "shortens" the soil to an extent that depends on the amount, type, and size distribution of the coarser components.

Consistency limits. The term consistency is preferentially employed for mechanical resistance properties in the twilight zone between true elastic solid and simple liquid behavior. Since several physical phenomena are usually involved in consistency, this property is normally defined by the use of specific apparatus and standardized procedure. Typical are the slump test for fresh concrete, the penetration test for asphalt, and the consistency limits tests for soils. The last indicate what water content (percent of dry weight of soil) will bring soils to states of analogous consistency. The liquid limit is the water content at the transition between the plastic and the liquid state, the plastic limit is the water content at the solid-plastic transition, and the shrinkage limit is the moisture content that will just fill the soil pores in the solid state if that state is reached by the drying out of a soil paste. The difference between the liquid and plastic limits is the plasticity index. It indicates the moisture range in which a soil shows plastic behavior. These consistency indices are valuable tools if properly understood, but they do not as such make it possible to predict under what circumstances large soil systems will show rupture, plastic flow, or creep. This depends on complex factors of granulometry; amount, type, and water content of soil fines; and stress conditions. Analogous factors govern the behavior of all building materials.

SEEPAGE AND FROST

Water movement and frost action in soil must be considered in projects such as dam building and construction on seasonally and permanently frozen soils.

Seepage. Water movement in soils is normally caused by hydraulic pressure or tension gradients. This flow is ordinarily viscous or laminar rather than turbulent. Darcy's law is fundamental: $V = Akit$, where V= volume of water flowing in time t through soil with a cross section A and where pressure gradient $i = \Delta p / \Delta l$ (pressure drop per unit length of flow) and k= coefficient of permeability. The permeability coefficient varies from 100 cm/

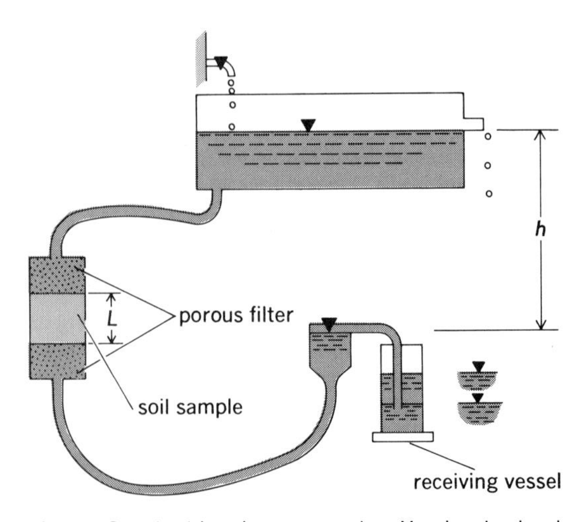

Fig. 3. Constant-head permeameter. Head water level is kept constant. Water percolates through soil sample of thickness L. Porous filters hold soil in place. The tail water level is kept constant by overflow. Volume of discharge is measured in receiving vessel. (*From G. P. Tschebotarioff, Soil Mechanics, Foundations and Earth Structures, McGraw-Hill, 1951*)

sec for clean gravel to 10^{-9} cm/sec for heavy clay.

Since water possesses an electric structure that interacts with the electrically charged soil minerals, it also moves in soil capillaries upon application of electric potentials. In this case, k of the Darcy equation is replaced by k_e, the electroosmotic transmission coefficient, and i by the electric potential gradient. Since the electric soil-mineral surface interaction structure is temperature-susceptible, water also moves in dense clay soils upon application of a thermal gradient with coefficient of thermoosmosis k_{th}. In unsaturated soils of high porosity, thermal gradients also cause water movement in the vapor phase.

Measurement of permeability. The coefficient of permeability may be determined either in the laboratory on disturbed or undisturbed samples by means of the constant- or falling-head permeameters, or in the field by pumping or injection tests. In the constant-head permeameter, used for materials of high permeability, water maintained at a constant level flows through a soil sample. Permeability is computed from the rate of flow. In the falling-head permeameter, used for materials of low permeability, the permeating water is supplied by means of a standpipe with cross-sectional area a in which the hydraulic head falls from h_1 to h_2 during the time t while the water volume $a(h_1 - h_2)$ flows through a soil specimen of thickness L (Figs. 3 and 4). Then k is given by Eq. (2).

$$k = 2.3 \frac{La}{At} \log \frac{h_1}{h_2} \qquad (2)$$

In calculating k from pumping and injection tests in the field, spherical symmetry is assumed in the pressure distribution around the well point or the injection nozzle. Moving water transmits momentum to the contacting soil particles. If in noncohesive soils the flow is upward and the hydrostatic uplift is sufficient to sustain the weight of the individual soil particles, a quick condition results. The minimum hydraulic gradient i_{cr} to cause quick condition can be estimated from Eq. (3), where G= specific gravity and e= the voids ratio.

$$i_{cr} = \frac{G-1}{1-e} \qquad (3)$$

Flow nets. The study of seepage through natural and man-made soil structures is greatly facilitated by the use of flow nets. These are two nests of curves, one representing the flow lines and the other the equipotential lines. In accordance with the laws of laminar flow, flow lines, which follow the path of the water, may not intersect each other; and the equipotential lines, each connecting points of equal hydraulic head, must cross the flow lines at right angles. While an infinite number of flow lines exists, only a few are drawn and in such a manner that the quantities of flow between adjacent lines are equal. Then the equipotential lines are drawn so that they intersect the flow lines at right angles and form areas that resemble squares as closely as possible (Fig. 5).

A conventional flow net has the following properties: (1) All paths, each of which lies between two adjacent flow lines, carry the same seepage quantities; (2) potential differences are the same between all adjacent equipotential lines; (3) flow lines intersect equipotential lines at right angles;

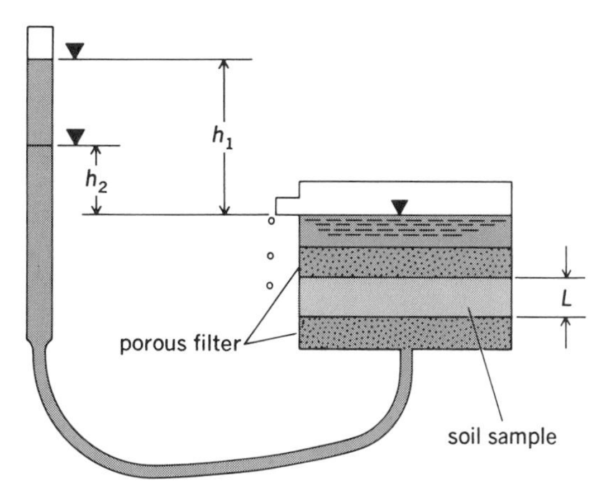

Fig. 4. Falling-head permeameter. Water level h_1 in thin glass tube decreases to h_2 as water percolates through soil sample of thickness L, restrained between porous filters. *(From G. P. Tschebotarioff, Soil Mechanics, Foundations and Earth Structures, McGraw-Hill, 1951)*

(4) all figures formed by adjacent pairs of lines resemble squares as closely as possible if the soil is homogeneous; and (5) at any point of the net, the spacing of equipotential lines is inversely proportional to the hydraulic gradient and the spacing of flow lines inversely proportional to the seepage velocity. For the drawing of flow nets, the boundary conditions must be known. Two-dimensional presentation is of course applicable only if the profile remains the same in the third dimension. From a satisfactory flow net, the seepage can be calculated using Eq. (4), where Q= quantity of

$$Q = \frac{n_f}{n_d} k h_i \qquad (4)$$

the seepage in cubic feet per second per linear foot of the length of the structure, n_f= number of flow paths, n_d= number of spaces between equipotential lines in the net, k= coefficient of permeability in feet per second, and h_i= head loss in feet be-

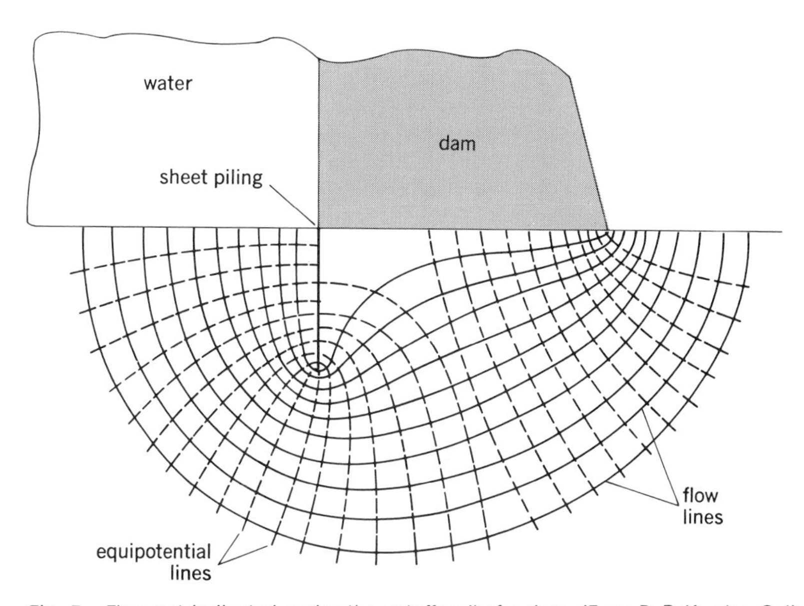

Fig. 5. Flow net indicated under the cutoff wall of a dam. *(From D. P. Krynine, Soil Mechanics, 2d ed., McGraw-Hill, 1947)*

tween surfaces of the head and tail waters. Methods are available to correct for differences in permeability in different directions. Seepage in complicated structures is studied on models employing, instead of water, electric or other energy forms whose transmission obeys the same mathematical laws expressed in the Laplace equations for conjugate functions.

Frost action in soils. Frost action is of great engineering importance. The forceful expansion of confined water when it freezes can destroy porous building materials and loosen soil. Of greater importance is the accumulation of water in the form of ice lenses, with resulting frost heave in winter and morass formation during thawing.

A freezing front penetrating into moist soil starts only a limited number of crystallization centers. Water moves to these from the surrounding soil, especially from lower depths. As it freezes, it gives off about 80 cal/g (335 J/g) of water. If the heat released by the freezing water just balances the heat lost by conduction to the earth surface, the freezing front becomes stationary and forms thick ice lenses, especially if the groundwater level or other water reservoir is close by. Without such close supply of free water, the frost may advance until a new moist layer is encountered. Here nuclei and ice formation begin anew.

Daily and other short-period temperature variations make the larger lenses grow at the expense of the smaller. In large areas of Siberia and also in Canada and Alaska, permanently frozen soil (permafrost) is encountered at depths that depend on the climate and the thermal conductivity of the surface soil. This frozen ground prevents the drainage of water from spring thawing. Structures that are founded on permanently frozen soil must be separated from it by insulating materials. Some of the permafrost is not in equilibrium with the present climate and if once melted would not be reformed.

EXPLORATION, SAMPLING, AND TESTING

The extent to which field testing is necessary depends on the type, magnitude, and importance of the job. Before a sampling and testing program is started, all readily available information should be utilized. Such information may be in the form of construction experience records in the same general location; air photos on which recognizable erosion and vegetation features indicate the types of surface and subsurface soil to be encountered, as on a proposed highway route; geological maps that show the profiles of solid and loose rock and serve as reminders of troubles usually associated with certain rock types, as sinkholes are with limestone; and pedologic maps, which are especially useful in highway soil work. Valuable information is often obtained from old maps which show soil and drainage patterns that have been covered up in urban or industrial developments.

Site investigations. These are made either to find out whether the soil in its natural-site condition will support the planned structure or to establish the qualities of soils as construction materials for dams, embankments, subgrades, and other uses. For the first case, the natural-site condition and the samples taken for laboratory testing must be disturbed as little as possible. For the second case, disturbed samples are useful, but they should not be permitted to dry out before being tested.

For exploration of relatively large sites or extended strips, such as airports and highways, electrical and mechanical energy-transmission phenomena may be employed. The electric method furnishes information on the electric resistivity of the soil at various depths. Previous determination of relationships between soil type, water content, salinity, and electric resistivity permits the plotting of a soil profile and the detection of strata of specific materials.

Mechanical energy is employed in seismic and vibration tests. These tests utilize the reflection and refraction of earth waves at interfaces of different strata and the variation of their speeds of propagation with the character of the conducting medium. The seismic velocity of compression waves varies for different soils within a range of 500–8000 ft/sec and for solid rock from 6000 to more than 25,000 ft/sec. The velocity in water is about 4700 ft/sec. In refraction shooting, one usually explodes a blasting cap or a small charge of dynamite and records the first signal picked up by the seismographs located at three different distances. The methods employing excited vibration utilize either the difference in rate of propagation in different soils or the characteristic frequency of the soil-vibrator system.

Sampling. Soil sampling may be divided into shallow and deep and disturbed and undisturbed. Shallow disturbed samples of cohesive soils are usually obtained with a soil auger. Slightly disturbed samples of both cohesive and noncohesive soils may be obtained by carefully pushing thin-walled metal cylinders into the ground. The cylinder is subsequently capped on both ends to prevent moisture loss.

Deep samples can be taken by digging trenches and pits so that the actual profile is exposed. Also, machines are available for making open holes into the ground 10–36 in. wide and well over 10 ft deep. The most common method for taking deep samples is by driving a casing into the ground. The casing is at least $2\frac{1}{2}$ in. wide. A record is kept of the hammer weight, height of drop, and the number of blows required to drive through each linear foot. While the boring proceeds, the casing is cleaned out with water conducted to the bottom of the casing by a pipe of about 1-in. diameter. Inspection of washings can give an idea of the character of the soil layer reached. If an undisturbed sample is to be taken at a certain depth, then the wash water is shut off above that depth and the wash pipe replaced by or changed into a pushing rod, by which a thin-shelled sampler (Shelby tube) is pushed into the stratum to be sampled. The tubes with the samples are then extracted, sealed on both ends, and sent to the laboratory for testing. A number of modifications of procedure and of type of sampler exists, but the essence of the method remains the same.

For important construction, the casing is driven to bedrock and through it for 5–10 ft in order to make sure that a boulder is not mistaken for a rock stratum. Because of the possibility of sinkholes, core drilling in limestone and dolomite should be as close to the actual construction site as possible.

Field tests. Static or dynamic tests at the site may be made for mechanical properties or other physical characteristics, such as density, moisture content, permeability, and thermal resistivity. The in-place density and moisture content may be measured by digging a clean hole, measuring its volume, and weighing and determining the water content of the earth removed. Densities and moisture contents of surface soils and of deeper layers reached by boreholes may also be determined by nondestructive methods, employing γ-rays for the former and neutrons for the latter. Thermal resistivity is usually determined with the thermal needle.

Common mechanical-resistance tests include sounding, that is, pushing a steel rod or $\frac{3}{4}-1$ in.-wide pipe into the soil and noticing the resistance to driving as a function of depth; penetration tests, where the resistance to penetration by a cone of standardized form is measured at various levels reached by a borehole; and vane shear tests, where a vane is pushed into a soil layer with subsequent application of horizontal shear forces. These tests possess only indicative value.

Load or bearing-power tests are performed in open pits at foundation level. Round or square plates of practically undeformable material are employed for transmission of the applied load. The pressure bulb, or zone, created under the loaded plate bears a definite geometrical relationship to the shape and dimensions of the plate. Thus, the loading of small plates will not detect layers of low bearing power and large compressibility that may be reached by the pressure bulb of the actual foundations. Within limitations, load tests are valuable tools, especially for shallow foundations. To determine the effect of surface loading by a foundation on deeper soil layers, the stress acting on them is calculated from theory, and undisturbed samples taken from the location and depth concerned are submitted in the laboratory to the calculated stresses.

Stresses in soil masses may be due to their own weight or to outside forces. In either case, the geometry of the system is an important factor in the resulting stress distribution. The other decisive factor is the physical state of the soil mass. This physical state may range from an elastic solid to a macromeritic liquid like quicksand. Over the entire range, plasticity may be observable to various degrees. Every type of internal structure and response to stresses of different intensity, known from other construction materials, may be encountered in undisturbed and disturbed soils. This emphasizes the importance of obtaining representative soil samples and testing them carefully in the laboratory. Even then the results must be used with engineering judgment, since large masses of a material behave differently from small ones.

THEORETICAL STRESSES

Calculation of stresses would become too cumbersome if all pertinent soil factors were carried along. Theories are therefore based on idealized earth masses, which for calculation of stress distribution are homogeneous, isotropic, elastic bodies and for nonelastic deformation or stability problems are idealized masses endowed with friction, cohesion, or both. The methods of stress analysis are the same as those in statics and strength of materials.

Pressure bulb. Stress distribution in a perfectly elastic continuum under a vertical point load at the horizontal boundary was first derived by J. Boussinesq. If the point of application of a force P serves as the origin of a coordinate system with vertical coordinates z and horizontal coordinates x and y, and $x^2+y^2=r^2$, then vertical pressure is given by Eq. (5). By connecting points of equal vertical

$$\sigma_z = \frac{3P}{2\pi} z^3 (x^2+y^2+z^2)^{-5/2} \tag{5}$$

stress, a pressure bulb is obtained. Stress distribution in the elastic medium under a loaded area, rather than a point, is obtained by resolving the distributed force into a large number of point forces and then employing the principle of superposition where stresses from different points overlap. Such solutions are available only for simple forms, such as line, strip rectangle, square, and circle. However, the form of most foundations can be approximated by summation, subtraction, or both, of shapes for which solutions are available, and the stresses calculated, using the principle of superposition. A pressure bulb under a loaded area of simple shape is shown in Fig. 6.

Accuracy of analysis. Despite the drastic simplifications in theoretical treatment, loading tests have shown that, except in the immediate vicinity of the loaded areas, the calculated data agree quite well with the experimental results.

Direct loading tests have shown differences in load acceptance by different soils. Contrary to theoretical assumptions, movement of soil material occurs close to the plate where stresses are intensive. Coarse-grained soils in which shear resistance is directly proportional to normal stress resist movement most in the zone under the center of the plate and have lower resistance and more pronounced outward movement as the plate edges are

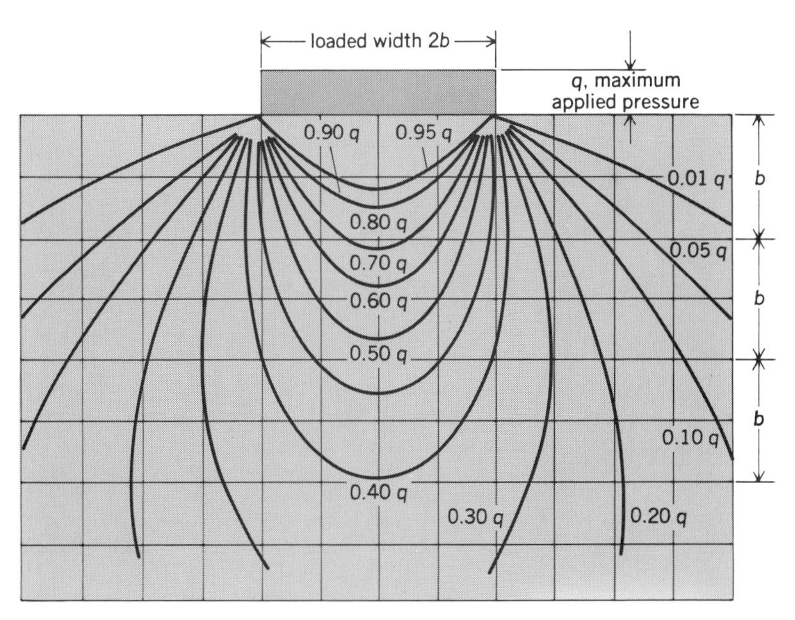

Fig. 6. Stress distribution for loaded areas of simple shape. Distribution pattern of vertical pressures through soil from vertical load on surface is called pressure bulb, from shape of equal pressure curves. (*From Road Research Laboratory of Department of Scientific and Industrial Research, London, Soil Mechanics for Road Engineers, 1954*)

approached. As a result, the center cone accepts most of the applied load, and the stress intensity under the center of the plate has been found in the case of sand to be two or three times the average stress under the plate. With plastic clay, where the shear resistance is pressure-independent, material moves from the central position outward until it encounters the resistance of the earth outside the plate edge. This equalizes the pressure under the plate and produces a maximum under the rim (Fig. 7).

CONSOLIDATION AND SETTLEMENT

The consequences of soil loading are affected by the presence of pore water. Normal pore water may be under simple hydraulic pressure γz if located at a distance z below the groundwater level and having a density γ. It may be stressless, if located at the groundwater level, or under tension, if in capillaries above the latter. Sudden decrease in pore volume, occasioned by sudden loading, may result in a pressure increment which is called excess hydrostatic or neutral pressure. Presence of hydrostatic water decreases the weight of the solid soil particles by the weight of an equal volume of water. This must be considered in calculating the total effective stress due to the weight of overburden; it affects the friction part of shear resistance, which is proportional to the normal pressure on the shear plane. Excess hydrostatic pressure as well as hydraulic uplift must be subtracted from the total interparticle pressure to obtain the effective pressure which governs frictional resistance. Excess hydrostatic pressure is dissipated by expulsion of water, at a rate determined by the hydraulic gradient and the permeability, with consolidation of the system until the total stress is carried by the skeleton of solid particles.

Consolidation theory. Consolidation has been studied theoretically only for systems of simple geometry, such as a soft-clay stratum bounded by one or two pervious sand or gravel strata, under the assumptions that the soil is saturated, soil particles and water are incompressible, Darcy's law is valid and the coefficient of permeability remains constant during consolidation, the time lag of consolidation is due entirely to low permeability, the soil is laterally confined, the total and effective normal stresses are the same for any point on a horizontal plane and water flows out of the voids only in a vertical direction, and the change in effective pressure results in a corresponding change in voids ratio.

Loading of sand within its elastic range produces an immediate elastic deformation supplemented by some consolidation due to grain adjustment. Loading of a saturated clay produces an immediate, though small, elastic deformation and a slow deformation that can be called elastoid since, though water is expelled, no shear takes place in the system.

Settlement analysis. For actual structures, settlement analysis involves the following: (1) detection of a soil stratum of high compressibility by borings; (2) determination of pressure-voids ratio relationships and time factor in the laboratory on undisturbed soil samples from borings; (3) calculation of existing pressures due to natural overburden and of pressure increase by added foundation loads; (4) initial condition of consolidation; (5) theoretical curve of time factor versus consolidation; (6) predicted time-settlement curve assuming instantaneous loading and curve corrected for construction period; and (7) comparison with settlement experience of other buildings in location and follow-up by actual settlement observations on structure.

If plastic under the applied stresses, soil will flow sideways, resulting in secondary consolidation that may, like creep in metals, continue at a constant rate or may decrease with time. The rigidity of the foundation influences the stress acceptance of the contracting soil and thereby the consolidation behavior. Proper design of a foundation minimizes total settlement and eliminates as much objectionable differential settlement as possible.

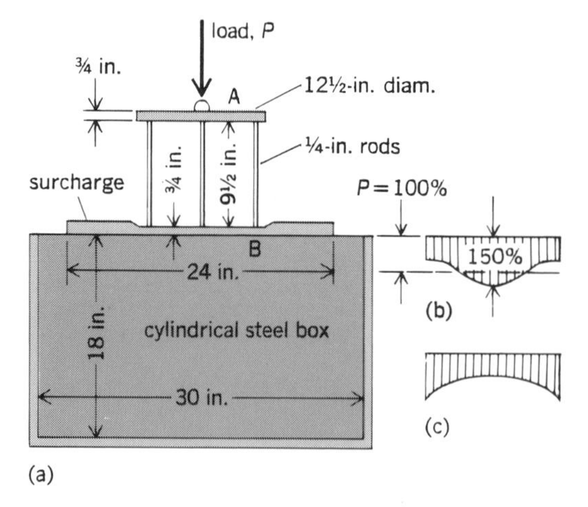

Fig. 7. Stress acceptance by sand and clay. O. Faber's experiments in 1933, in diagram *a*, showed that under a load applied through a circular plate B, stresses in sand would follow the pattern shown in *b*, reaching a maximum at the center of the plate and that stresses in clay would follow the pattern shown in *c*, reaching a maximum around the edges. Soil stresses were determined from contraction of the rods between the two disks (A and B) when load *P* was applied. 1 in. = 25.4 mm. *(From D. P. Krynine, Soil Mechanics, 2d ed., McGraw-Hill, 1947)*

STABILITY

Stability of an earth structure is primarily resistance to shear failure. Such resistance involves geometrical and soil-physical factors. Wide and deep foundations are more stable than narrow and shallow ones.

Bearing power. Several building codes contain so-called safe bearing values, which may have ranges of 25–40 tons/ft^2 (24–38 MPa) for rock to 1 ton/ft^2 (0.10 MPa) for soft clay. The minimum load causing failure is the ultimate bearing power which, divided by an arbitrary factor of safety, is called safe bearing power. More meaningful values are obtained from loading tests, shear tests, and unconfined and triaxial compression tests. Safe construction must be designed for the weakest possible condition of the soil on site. There exist a number of semiempirical laws with respect to foundation stability against shear failure.

Pile foundations are used for carrying building loads through soft compressible layers to layers strong enough to carry them. They may be end-bearing if resting on solid rock or floating if driven into sand or clay; if floating, the load is transmitted by skin and point resistance. Groups of floating piles develop, by superposition, pressure bulbs characteristic of the geometry of the group rather than of the individual pile. *See* PILE FOUNDATION.

Landslides. These may be due to slow soil flow, such as creep, to fast liquid flow, to sinking into caves or mines, and to sliding. Often they occur along water-lubricated boundaries of inclined rock strata. Soil mechanics is particularly concerned with ensuring cuts, embankments, earth dams, and other earth structures against failure by sliding.

Stability analysis employs the concept of a sliding wedge, one surface of which is the surface of maximum shear stress in the soil. This surface is found from the geometry of the system. Next, the shear resistance on this surface is determined as a function of soil character and normal stresses. Specific calculation methods for plane and curved sliding surfaces have been developed. Since soil cohesion remains constant, while friction increases with pressure, slopes of cohesive soils possess critical heights, while slopes of purely frictional materials may be stable up to any height. This demonstrates the importance of proper choice of materials and densification to greatest possible shear resistance of soil used for embankment and earth dam construction (Fig. 8).

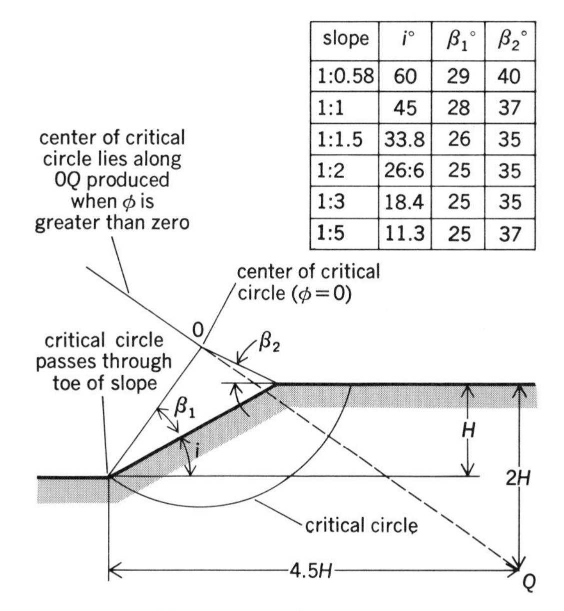

slope	$i°$	$\beta_1°$	$\beta_2°$
1:0.58	60	29	40
1:1	45	28	37
1:1.5	33.8	26	35
1:2	26:6	25	35
1:3	18.4	25	35
1:5	11.3	25	37

Fig. 8. Stability analysis of embankments. Fellenius' construction for location of critical slip circle. Tan ϕ is coefficient of internal friction. *(From Road Research Laboratory of the Department of Scientific and Industrial Research, London, Soil Mechanics for Road Engineers, 1954)*

Retaining walls. Stability analysis of retaining walls, that is, determination of the magnitude, distribution, and direction of soil pressure on such walls, is also based on the concept of a sliding wedge. One differentiates between neutral pressure (if no movement at all has taken place), active pressure (in the case of sufficient movement to mobilize the shear resistance of the sliding surface), and passive pressure (if the wall has pushed the soil back sufficiently to mobilize shear in a wedge which for geometric reasons involves a larger sliding surface than the active pressure). Special methodology for calculating earth pressures with varying simplifying assumptions is available. Further progress appears to depend more on a better understanding of soil-physical properties than on employing more complicated mathematics.

Conduits. Conduits are rigid or flexible pipes for the conduction of liquids, especially water, and are placed in ditches with backfilling or under embankments. Present knowledge and design formulas for conduits are based mainly on experimentation, and the closer the actual conditions resemble those of the experiment, the more reliable are the formulas.

LUNAR AND PLANETARY SOILS

Lunar and planetary soils, the same as their terrestrial counterparts, are products of environmental forces acting on the original rock substrate plus acquired meteoritic material and the substance of incident corpuscular radiation. The latter is especially important in the case of the Moon, which because of its low gravity ($\frac{1}{6}$ of the terrestrial) has an extremely thin atmosphere (10^{-14} torr) with very little shielding power against electromagnetic and corpuscular radiation. The protons (H^+) of the solar wind impinging on the lunar surface can react chemically with the oxides of the lunar rock and form hydroxides or even H_2O, which subsequently evaporates into space. This may account for the vesicular character of lunar rock and soil particles and for their dark color. Because of the practical absence of atmosphere on the Moon, dry soil particles have no adsorbed air films, which on Earth satisfy their attractive force fields. As a result, the "raw" lunar soil particles have a tendency to cohere and behave somewhat like moist soils, even in the absence of water. Modern developments in remote sensing permit the collection of compositional and environmental data by surveying spacecraft to a degree that is sufficient for prediction of the most significant features of planetary soils on the basis of present knowledge of the physics and chemistry of matter. [H. F. WINTERKORN]

Bibliography: American Society for Testing and Materials, *Procedures for Testing Soils*, 1958; D. P. Krynine and W. R. Judd, *Principles of Engineering Geology and Geotechnics*, 1957; National Research Council, *Water and Its Movement in Soils*, Highway Res. Board Spec. Rep. no. 40, 1958; F. F. Scott, *Principles of Soil Mechanics*, 1963; K. Terzaghi and R. B. Peck, *Soil Mechanics in Engineering Practice*, 1967; G. P. Tschebotarioff, *Foundations, Retaining, and Earth Structures*, 1973; H. F. Winterkorn, Principles and practice of soil stabilization, *Colloid Chem.*, 6:459–492, 1946.

Soil microbiology

A study of the microorganisms in soil, their functions, and the effect of their activities on the character of soil and the growth and health of plant life, particularly cultivated crops. It embraces the biology of microorganisms—their morphology, physiology, and taxonomy—as well as their biochemistry. It is related to soil chemistry and physics and, in its application, to agronomy, plant physiology, and plant pathology.

Microorganism characteristics. The soil microorganisms are viruses, Myxomycetes (slime fungi), protozoa, algae, yeasts, fungi, actinomycetes, and bacteria (see illustration). Soil is distinguished from subsoil chiefly by the presence of organic matter in the soil. The organic matter is composed of dead plant and animal tissues and the products of their decomposition by microorganisms. The microorganisms derive their energy by oxidizing plant and animal residues. Plants depend upon nutrients made available by microorganisms which form an essential link in the cycle of food in nature.

The soil is the greatest natural reservoir of microorganisms. These take part in many reactions in the soil. Some microorganisms are specific, taking part only in certain reactions (*Nitrosomonas*, which oxidize ammonia to nitrite). Some are more general in their metabolism (heterotrophic types, which depend on organic material for energy source) and take part in many reactions. One group of organisms may use the products of another, for example, during the course of decomposition of complex nitrogenous material to nitrate. In addition to forming simpler degradation products, microorganisms synthesize many complex substances. These are bound up in microbial protoplasm, while excess amounts may be liberated to provide essential substances for other groups of microorganisms. In contrast to this associative action are antagonisms caused by competition for the same food or by the ability of many organisms to produce antibiotics.

The net result of such associations and antagonisms is the establishment of a microbial equilibrium, which varies with the geological origin and evolutionary history of the soil. Although the equilibrium is ever shifting under the influences of season, temperature, moisture, and state of cultivation, the micropopulation of a soil of definite type has a characteristic composition, consisting of organisms that have become adapted to the particular environment.

Microorganisms are not uniformly distributed throughout the soil, because soil is not homogeneous but comprises a variety of microenvironments. The organisms congregate in colloidal films about the surface of soil particles and are particularly abundant around fragments of decaying plant and animal debris.

The majority of soil microorganisms are most active at pH 6.0–6.8. Some sulfur-oxidizing bacteria tolerate a pH as low as 2.0 and other organisms a pH as high as 10. A supply of free oxygen is important to most organisms. Almost all fungi, actinomycetes and protozoa as well as most bacteria, are aerobic. With abundant oxygen, decomposition proceeds rapidly to completion. However, in close-textured and wet soils where anaerobic conditions (molecular oxygen not available) prevail, organic matter is decomposed slowly and incompletely. The number of microorganisms is highest in spring and fall. Proximity to growing plants markedly increases the number of soil organisms.

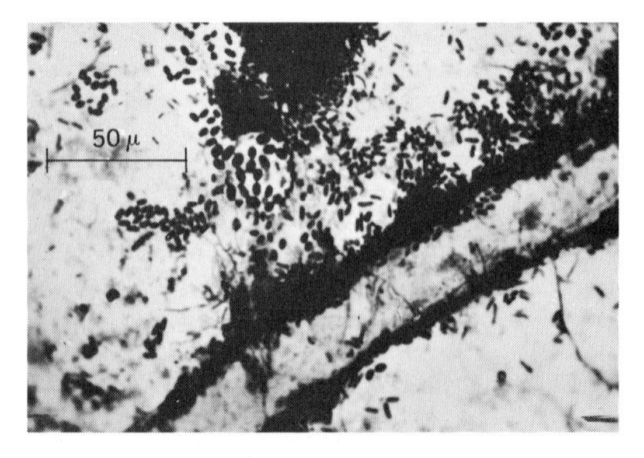

Bacteria attacking fragments of dead plant material in soil. (*T. Gibson, World Crops, vol. 3, no. 4, 1951*)

The soil micropopulation may be divided into two groups, autochthonous and zymogenous microorganisms. The autochthonous organisms, making up the great majority, are the indigenous forms responsible for the processes occurring under normal conditions. Their character is relatively uniform in soil of definite type, and they are little affected by amendments (changes in soil organic matter). In contrast, the zymogenic microorganisms, much less numerous and normally quiescent, increase temporarily to participate in primary decomposition processes upon addition of readily decomposable organic material.

Microbiological soil analysis. The enumeration of soil microorganisms is carried out by culture and microscopic methods. Culture methods depend upon the growth of organisms and permit their isolation for detailed study. In microscopic methods counts are made from stained films, but since the organisms are killed no cultures can be made.

Culture methods. These include plate and elective culture procedures. Plate cultures are used for counts of bacteria, actinomycetes, and fungi, and are prepared by suspending a definite weight of soil in sterile water. Dilutions are made of the soil suspension with sterile water. Aliquots (representative portions) of these dilutions are placed in petri plates with an agar medium which will support the growth of the microorganisms in the soil sample. For counts of protozoa the soil aliquots are placed on the surface of agar plates, and after incubation wet mounts are examined microscopically. In order to group soil bacteria on the basis of taxonomy, nutritional requirements, and ability to synthesize substances such as vitamins or antibiotics, isolates (individual bacterial colonies) are taken nonselectively from plate cultures. The relative incidence in soil of a given category of bacteria can be estimated from a study of the individual isolates.

Elective culture methods are used to count microorganisms of special physiological groups, such as algae and the nitrifying, denitrifying, sulfur-oxidizing, and cellulose- or protein-decomposing bacteria. Portions of successive soil dilutions are added to liquid media selected to promote growth of the group in question. The presence of organisms is determined by visual or microscopic observation or by chemical tests for by-products. The use of replicates (that is, more than one sample of the particular soil) permits the calculation of the most probable numbers.

Microscopic methods. Such methods are based on the examination of a definite amount of soil spread over a definite area of a slide and stained. From the known area of the microscope field numbers of organisms may be estimated. As this procedure involves mixing the soil, the "contact-slide" method is used to observe the localized distribution of the organisms. A glass slide is inserted in the soil and later removed, stained, and examined for organisms adhering to the surface.

Plating methods, which are more commonly used for counting soil microorganisms, give too low values because there is no medium on which all will grow. Microscopic methods give far higher numbers than plate counts but suffer from the disadvantage that they do not distinguish readily between living and dead cells. The two methods are complementary. *See* NITROGEN CYCLE; SOIL MICROORGANISMS. [ALLAN G. LOCHHEAD]

Bibliography: M. Alexander, *Introduction to Soil Microbiology*, 1977; A. Burges and F. Raw (eds.), *Soil Biology*, 1967; S. D. Garrett, *Soil Fungi and Soil Fertility*, 1963; T. R. Gray and S. T. Williams, *Soil Microorganisms*, 1975; N. A. Krasilnikov, *Soil Microorganisms and Higher Plants*, transl. by Y. Halperin, 1958; W. Kuhwelt, *Soil Biology*, 1976; N. Walker (ed.), *Soil Microbiology*, 1975.

Soil microorganisms

Microorganisms in the soil include protozoa, fungi, slime molds, green algae, diatoms, blue-green algae, and bacteria. The bacteria are a heterogeneous group and include the procaryotic mycelial forms called actinomycetes as well as the simple unicellular forms called eubacteria.

Bacteria, actinomycetes, and fungi are the groups most active in decomposing organic residues and in rendering inorganic nutrients soluble. The final result of this activity is the liberation of such elements as carbon, nitrogen, phosphorus, potassium, and sulfur in forms available to plants.

Eubacteria. The eubacteria exceed all other soil microorganisms in numbers and in the variety of their activities. Numbers may surpass 100,000,000/g of soil by plate count or 1,000,000,000/g by microscopic count. The bacteria vary in size, shape, growth requirements, energy utilization, and function. Morphologically they are divided into straight or irregular rods, of both spore-forming and non-spore-forming types, thin flexible rods, cocci, vibrios, and spirilla. Short rods and cocci are most frequent, but many of the cocci are the coccoid stage of pleomorphic (varying in shape and size) rods or spores of actinomycetes.

Members of most taxonomic groups, with the exception of certain animal and human parasites, occur in soil. Some taxonomic groups are characteristic of soil alone. Although the identity of many bacteria engaged in specific processes, such as nitrification and nitrogen fixation, is known, a large proportion of the indigenous (autochthonous) organisms have not been classified. Of these, many have not been grown in any culture medium, a requisite for systematic study. Consequently, taxonomic knowledge of the autochthonous microflora is imperfect.

On the basis of their nutrition, soil bacteria are divided into autotrophs and heterotrophs. Autotrophs are able to use carbon dioxide as the sole source of carbon for their body tissues; heterotrophs must obtain carbon from organic foods.

Autotrophic bacteria. These comprise two groups (photosynthetic and chemosynthetic) according to their source of energy. The purple and the green sulfur bacteria are photosynthetic because of the presence of bacteriochlorophyll or chlorobium chlorophyll pigments. Like chlorophyll-containing algae and higher plants, they obtain energy from sunlight.

Other autotrophs are chemosynthetic, deriving energy from various oxidation reactions. Their requirements for food and energy are met by inorganic sources. These autotrophs carry out the process of nitrification, in which two stages are distinguished, the oxidation of ammonia to nitrite and that of nitrite to nitrate. Autotrophic sulfur bacteria derive energy from the oxidation of elemental sulfur, sulfides, sulfites, thiosulfates, and thiocyanates to sulfuric acid, which reacts with soil bases to form sulfates.

Heterotrophic bacteria. Heterotrophic bacteria, which constitute the great majority of soil bacteria, derive both food and energy from the decomposition of organic substances. They embrace a wide variety of morphological and taxonomic types including spore-formers or zymogenous forms, which are bacteria that develop in soil in response to the addition of certain substances like organic matter, or certain processes like aeration. Also included are the far more numerous non-spore-formers which make up the great majority of the autochthonous microflora. The most abundant forms are short rods and pleomorphic rods.

The majority of the heterotrophs require combined nitrogen to build cell substance. The nitrogen-fixing bacteria utilize elemental nitrogen of the air. These bacteria include symbiotic organisms, such as species of the genus *Rhizobium* that live in symbiosis with leguminous plants, and nonsymbiotic organisms, such as species of the aerobic genera *Azotobacter*, *Beijerinckia*, and *Azotomonas*, and the anaerobic genus *Clostridium*. The indigenous heterotrophic bacteria may be classified according to their nutritional requirements. Though all require a source of energy, such as a simple sugar, the additional needs of some are satisfied by inorganic salts. Other soil bacteria require amino acids or more complex food sources, and some more exacting bacteria require factors present in soil extract. The proportion of each of the nutritional groups is fairly constant in soil of definite type. However, increased fertility, such as that resulting from fertilizer treatment, is reflected in an increase in the proportion of bacteria with complex requirements at the expense of those with simple nutritional needs. Although there is no precise correlation between nutritional requirements and morphological type or taxonomic grouping, *Pseudomonas* species are more abundant among the nutritional group with simpler needs; the pleomorphic types, particularly *Arthrobacter* species, are relatively more numerous in the group with the most complex requirements.

As much as 25% of the indigenous soil bacteria capable of being isolated may require one or more vitamins for growth. The vitamins most essential are, in order of frequency, thiamine, biotin, and vitamin B_{12}. A smaller percentage require other B vitamins as well as the terregens factor, a substance found only in soil that promotes bacterial growth. *See* NITROGEN CYCLE.

Actinomycetes. Next to the bacteria in numbers, the actinomycetes range from hundreds of thousands to several millions per gram of soil. They are more abundant in dry and warm soils than in wet and cold soils. With increasing depth of soil, their numbers are reduced proportionately less than those of bacteria. Actinomycetes are particularly abundant in grassland soil. The three genera of Actinomycetales occurring most commonly in soil are *Nocardia, Streptomyces*, and *Micromonospora*.

Thermophilic forms, represented by the genus *Thermoactinomyces*, are active in rotting manure and also may be present, though inactive, in normal soil. *Streptomyces* species are the dominant types. Although they are largely saprophytes, a few species, such as those associated with potato scab, are parasitic.

Less is known of the function of the actinomycetes in soil than of the bacteria. They are heterotrophic and are nutritionally an adaptable group, less demanding in growth requirements than many bacteria. They take part in the decomposition of a wide range of carbon and nitrogen compounds, including the more resistant celluloses and lignins, and are important in humus formation. Actinomycetes are responsible for the earthy or musty odor characteristic of soil rich in humus.

As much as 60% of the actinomycetes isolated from soil by plating methods may show antagonism toward bacteria or fungi in artificial culture. The importance of this antibiotic-producing capacity under normal soil conditions is not known. However, although antibiotics can rarely be detected in soil and then only under abnormal conditions, they may be important in microenvironments where intense microbial activity is taking place.

Fungi. Fungi are present in numbers ranging from several thousand to several hundred thousand per gram of soil. They occur extensively in the mycelial state, as well as in the form of spores. Since plate colonies may develop from fragments of mycelium or from spores, plate counts give only an approximation of the abundance of fungi in soil. As with bacteria, some fungi do not grow on plates; consequently plating methods give minimum counts. Most fungi require humid, aerobic conditions for growth and spore formation. They are most common near the surface of soil and are more abundant in lighter, well-aerated soils than in heavier soils. Because the optimum pH range for fungi is 4.5–5.5, they are more prevalent in acid soils which are less favorable to bacteria and actinomycetes.

Ecologically, two broad groups of soil fungi may be recognized, the soil-inhabiting and the root-inhabiting types. The soil-inhabiting fungi are able to survive indefinitely as saprophytes and have a general distribution in soil. They include not only obligate saprophytes but also some unspecialized parasites which are able to infect plant roots, but whose parasitism is only incidental to their saprophytic existence. Root-inhabiting fungi are specialized parasites that invade living root tissues. Their distribution in soil is localized and depends upon the presence of the host plant. Their activity diminishes following death of the plant and they persist in soil only as resting spores or sclerotia. Mycorrhizal fungi are included among the root-inhabiting fungi.

Soil fungi are heterotrophic and have a wide variety of food requirements. All obtain their carbon entirely from simple carbohydrates, alcohols, or organic acids. But although some fungi can utilize inorganic nitrogen, others require more complex forms or vitamins, chiefly thiamine and biotin.

Soil fungi do not comprise as many physiological groups as do the bacteria, but as a group they are more versatile in their ability to decompose a great variety of organic compounds. The saprophytes, the true soil inhabitants, may be divided into groups depending upon the nature of the substrate favoring their development. Two such groups are the sugar fungi and the cellulose-decomposing fungi. Other groups attack some of the most resistant substances, such as lignins, vegetable gums, and waxes. When plants die, or when fresh plant material is added to soil, the growth of fungi is greatly stimulated. Those able to attack the more soluble constituents, such as sugars and other simpler carbon compounds, develop rapidly. Chief among such forms are the Phycomycetes. As the special substrate is exhausted, other types flare up and attack progressively more resistant components of organic residues. Cellulose and hemicelluloses are decomposed by a variety of fungi including species of *Penicillium, Aspergillus, Sporotrichum*, and *Fusarium*. Fungi are the predominant lignin-decomposing organisms. Various simple fungi can attack the lignin of straw and leafy plant material, although the higher Basidiomycetes are most active in decomposing lignin-rich residues.

Although polysaccharide-forming bacteria play a part, fungi are chiefly responsible for improving the physical structure of soil by exerting a binding effect on loose particles, thus forming water-stable aggregates. This binding effect is caused by the growth of mycelia which form fine networks that entangle the smaller particles. The soil-binding effect is favored by addition of fresh organic material whose decomposition products provide cementing substances.

Yeasts. A group of simple fungi, yeasts occur in soil only to a limited extent in the surface layers. In field soils their numbers are small, some samples being devoid of yeasts. They are found most frequently in the soils of orchards, vineyards, and apiaries where special conditions, particularly the presence of sugars, favor growth of yeasts which invade the soil. Soil is not a favorable medium for the growth of yeasts, and they do not play a significant part in soil processes.

Algae. These are widely distributed in soils, developing most abundantly in moist, fertile soils well supplied with nitrates and available phosphates. They contain chlorophyll and in the soil surface layers, where they are chiefly confined, function as green plants converting carbon dioxide and inorganic nitrogen into cell substance by means of energy derived from sunlight. Smaller numbers occur at lower depths where, in the absence of sunlight, they exist heterotrophically. The soil algae comprise the green algae (Chlorophyceae), the blue-green algae (Myxophyceae), and the diatoms (Bacillariaceae). In acid soils green algae predominate, whereas in neutral or alkaline soils the other groups are more prominent. Numbers of algae in soil vary widely, ranging from a few hundred to several hundred thousand per gram of soil.

As autotrophs, algae are of importance in adding to the organic matter of soils. They play a fundamental ecological role on barren and eroded lands by colonizing such areas and synthesizing protoplasm from inorganic substances. Several blue-green algae are able to fix atmospheric nitrogen and are of agricultural significance, particularly in rice culture. Under the water-logged conditions needed for this crop, blue-green algae develop abundantly and may increase the nitrogen supply by as much as 20 lb/acre (22 kg/hectare).

Protozoa. Occurring in all arable soils, protozoa are largely confined to the surface layers although in drier, sandy soils they may penetrate more deeply. Numbers usually range from a few hundred in dry soil to several hundred thousand per gram in moist soils rich in organic matter. Most soil protozoa are flagellates and amebas; ciliates are less frequent although they are often found in wet soils and swamps. Protozoa are active in soil only when living in a water film. The majority are able to form cysts, and in this inactive state they can withstand desiccation.

Although a few flagellates, such as *Euglena*, have chlorophyll and are autotrophic and others can live saprophytically by absorbing nutrients from solution, the majority of soil protozoa feed by ingesting solid particles, mainly bacteria. Not all bacteria are suitable as food. Amebas have decided preferences for certain bacterial species and will not ingest others, particularly pigmented bacteria. The formation of cysts by protozoa is favored by some bacterial types, not by others. Excystment of some amebas requires the presence of bacteria; others are independent of bacteria. Though protozoa are a factor in maintaining the microbial equilibrium in soil through their selective action on bacteria, their effect is limited and is not considered detrimental to the activities of the micropopulation as a whole.

Myxomycetes. Myxomycetes, or slime fungi, form a minor group of soil microorganisms intermediate in character between the flagellated protozoa and the fungi. They possess a motile, flagellated stage in their life cycle, and later form large aggregates of cells or coalesce into jellylike masses of naked protoplasm. These eventually form spores which give rise to flagellated forms. Like the protozoa, the myxomycetes feed on bacteria.

Viruses and phages. Although these ultramicroscopic organisms exist in soil, little is known of the part they play in soil processes. Viruses that attack plants and animals can in some cases be transmitted from the soil. Phages, which are active against bacteria and actinomycetes, limit susceptible microorganisms and thus affect the microbial balance. Those that attack the various species of symbiotic nitrogen-fixing bacteria may prevent effective inoculation of legumes, particularly in soils in which the same crop has been repeatedly grown. The deleterious effect of phage on the nitrogen-fixing bacteria was formerly ascribed to direct lytic action (dissolution of the bacterial cell). However, this is now considered to result from the development of phage-resistant mutants which are less effective in fixing nitrogen than are the parent strains.

[ALLAN G. LOCHHEAD]

Soil zonality

Many soils that are geographically associated on plains have common properties that are the result of formation in similar climates with similar vegetation. Because climate determines the natural vegetation to a large extent and because climate changes gradually with distance on plains, there are vast zones of uplands on which most soils have many common properties. This was first observed in Russia toward the end of the 19th century by V. V. Dokuchaev, the father of modern soil science. He also observed that on floodplains and steep slopes and in wet places the soils commonly lacked some or most of the properties of the upland soils. In mountainous areas, climate and vegetation tend to vary with altitude, and here the Russian students observed that many soils at the same altitude had many common properties. This they called vertical zonality in contrast with the lateral zonality of the soils of plains.

Zonal classification of soils. These observations led N. M. Sivirtsev to propose about 1900 that major kinds of soil could be classified as Zonal if their properties reflected the influence of climate and vegetation, as Azonal if they lacked well-defined horizons, and as Intrazonal if their properties resulted from some local factor such as a shallow groundwater or unusual parent material.

This concept was not accepted for long in Russia. It was adopted in the United States in 1938 as a basis for classifying soil but was dropped in 1965. This was because the Zonal soils as a class could not be defined in terms of their properties and because they had no common properties that were not shared by some Intrazonal and Azonal soils. It was also learned that many of the properties that had been thought to reflect climate were actually the result of differences in age of the soils and of past climates that differed greatly from those of the present.

Zonality of soil distribution is important to students of geography in understanding differences in farming, grazing, and forestry practices in different parts of the world. To a very large extent, zonality is reflected but is not used directly in the soil classification currently being used in the United States. The Entisols include most soils formerly called Azonal. Most of the soils formerly called Intrazonal are included in the orders of Vertisols, Inceptisols, and Histosols and in the aquic suborders such as Aquolls and Aqualfs. Zonal soils are mainly included in the other suborders in this classification. For a discussion of this classification *see* SOIL.

The soil orders and suborders have been defined largely in terms of the common properties that result from soil formation in similar climates with similar vegetation. Because these properties are important to the native vegetation, they have continuing importance to farming, ranching, and forestry. Also, because the properties are common to most of the soils of a given area, it is possible to make small-scale maps that show the distribution of soil orders and suborders with high accuracy.

Zonal properties of soils. A few examples of zonal properties of soils and their relation to soil use follow. The Mollisols, formerly called Chernozemic soils, are rich in plant nutrients. Their natural ability to supply plant nutrients is the highest of any group of soils, but lack of moisture often limits plant growth. Among the Mollisols, the Udolls are associated with a humid climate and are used largely for corn (maize) and soybean production. Borolls have a cool climate and are used for spring wheat, flax, and other early maturing crops. Ustolls have a dry, warm climate and are used largely for winter wheat and sorghum without irrigation. Yields are erratic on these soils. They are moderately high in moist years, but crop failures are common in dry years. The drier Ustolls are used largely for grazing. Xerolls have a rainless summer, and crops must mature on moisture stored in the cool seasons. Xerolls are used largely for wheat and produce consistent yields.

The Alfisols, formerly a part of the Podzolic soil group, are lower in plant nutrients than Mollisols, particularly nitrogen and calcium, but supported a permanent agriculture before the development of fertilizers. With the use of modern fertilizers, yields of crops are comparable to those obtained on Mollisols. The Udalfs are largely in intensive cultivation and produce high yields of a wide variety of crops. Boralfs, like Borolls, have short growing seasons but have humid climates. They are used largely for small grains or forestry. Ustalfs are warm and dry for long periods. In the United States they are used for grazing, small grains, and irrigated crops. On other continents they are mostly intensively cultivated during the rainy season. Population density on Ustalfs in Africa is very high except in the areas of the tsetse fly. Xeralfs are used largely for wheat production or grazing because of their dry summers.

Ultisols, formerly called Latosolic soils, are warm, intensely leached, and very low in supplies of plant nutrients. Before the use of fertilizers, Ultisols could be farmed for only a few years after clearing and then had to revert to forest for a much longer period to permit the trees to concentrate plant nutrients at the surface in the leaf litter. With the use of fertilizers, Udults produce high yields of cotton, tobacco, maize, and forage. Ustults are dry for long periods but have good moisture supplies during a rainy season, typically during monsoon rains. Forests are deciduous, and cultivation is mostly shifting unless fertilizers are available.

Aridisols, formerly called Desertic soils, are high in some plant nutrients, particularly calcium and potassium, but are too dry to cultivate without irrigation. They are used for grazing to some extent, but large areas are idle. Under irrigation some Aridisols are highly productive, but large areas are unsuited to irrigation or lack sources of water. [GUY D. SMITH]

Bibliography: M. Baldwin, C. E. Kellogg, and J. Thorp, Soil classification, in USDA, *Soils and Men: The Yearbook of Agriculture*, 1938; R. M. Basile, *Geography of Soils*, 1971; B. T. Bunting, *The Geography of Soil*, 1965; D. Steila, *Geography of Soils: Formation, Distribution and Management*, 1976.

Solar energy

The energy transmitted from the Sun. This energy is in the form of electromagnetic radiation. The Earth receives about one-half of one-billionth of the total solar energy output. In 1971, based on radiation measurements in space, the National Aeronautics and Space Administration proposed a new space solar constant of 1353 watts per square meter (W/m^2), and a standard spectral irradiance in W/m^2 over a small range of wavelengths (bandwidth) centered at the wavelength (in millionths of meters, or μm) shown in Fig. 1. Accordingly, the solar radiation energy in the ultraviolet is 105.8 W/m^2 (7.82% of the solar constant), in the visible 640.4 W/m^2 (47.33%), and in the infrared 606.8 W/m^2 (44.85%). The solar radiation energy output is essentially constant. However, because of the ellipticity of the Earth's orbit, the solar constant varies between 1398 W/m^2 at the winter solstice and 1308 W/m^2 at the summer solstice, or 3% about the mean value. Based on its cross-sectional area, the rotating Earth receives therefore 751.10^{15} kWhr annually.

The following are the most frequently used metric units for the radiative input area: langley (1 l= 1 cal/cm^2=0.001163 Whr/cm^2=4.186 J/cm^2); calorie (1 cal/cm^2 min=0.0697 W/cm^2=1 l/min= 0.00418 J/cm^2 min); kilowatt-hour (1 kWhr/m^2= 860,000 cal/m^2=86.2 l=3.6 · 10^6 J/m^2); and joule (1 J=0.239 cal=2.78 · 10^{-4} Whr=0.239 l-cm^2).

Actually, passage through the atmosphere splits the radiation reaching the surface into a direct and a diffuse component, and reduces the total energy through selective absorption by dry air molecules, dust, water molecules, and thin cloud layers, while heavy cloud coverage eliminates all but the diffuse radiation. Figure 2 specifies the conditions for a surface perpendicular under a clear sky at mid- and low latitudes within 4 hr on either side of high noon. If these conditions were to prevail for 12 hr each day of the year (4383 hr), the energy received would lie between some 4200 and 5200 kWhr/m^2 yr. Actually, the number of sunshine hours even in high-insolation areas ranges from 78 to 89% of the possible, resulting in a reduction in radiative energy ("solar crude") received to the values shown in Fig. 2. Since atmospheric absorption and scattering increase strongly at low solar elevation, the average solar crude received in the most favorable areas by a horizontal surface is about 2550 kWhr/m^2 yr or 2.55 terawatt-hours per square kilometer per year (TWhr/km^2 yr). Figure 3 shows the

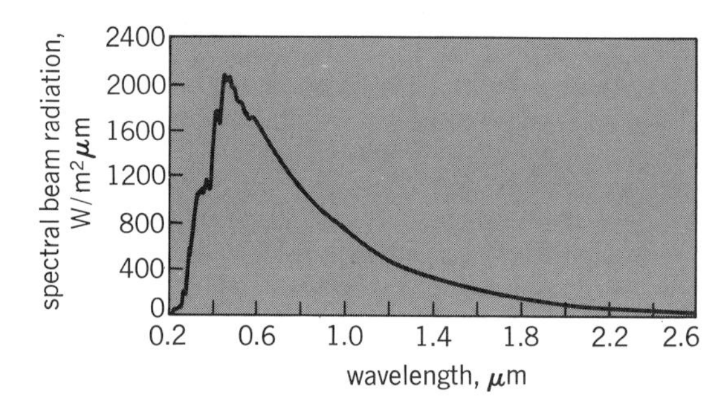

Fig. 1. NASA standard spectral irradiance at 1 astronomical unit (AU) and a solar constant of 1353 W/m^2. *(From Solar Electromagnetic Radiation, NASA, SP-8005, May 1971)*

global distribution of the average solar radiation energy incident on a horizontal surface. By following the Sun's diurnal and seasonal motion, thus facing it from near-sunrise to sunset, an instrument such as a heliostat can attain values between 3 and 3.5 TWhr/km² yr.

On a global basis, about 50% of the total incident radiation of 751.10[15] kWhr/yr is reflected back into space by clouds, 15% by the surface, and about 5.3% is absorbed by bare soil. Of the remaining 29.7%, only about 1.7% ($3.79 \cdot 10^{15}$ kWhr/yr) is absorbed by marine vegetation and 0.2% ($4.46 \cdot 10^{14}$ kWhr/yr) by land vegetation. By far the largest portion is used to evaporate water and lift it into the atmosphere. The evaporation energy is radiated into space by vapor condensation to clouds. The solar energy spent to lift the water can be partly recovered in the form of waterpower (hydraulic energy). Solar energy can be utilized in the form of heat, organic chemical energy through photosynthesis, and wind power, and also in the form of photovoltaic power (generating electricity by means of solar cells). The two greatest problems in utilizing solar energy are its low concentration and its irregular availability due to the diurnal cycle and to seasonal and climatic variations. Improved technology and investment capital are also needed.

In 1974, the U.S. Congress established the Energy Research and Development Agency (ERDA) and charged it with the development of energy conservation techniques and new technologies for extracting less readily accessible fossil energy reserves (such as shale oil) and for broadening the use of coal (coal gasification and liquefaction); with the advancement of nuclear (fission) technology; and with spearheading new energy technologies (geothermal, fusion, and all forms of

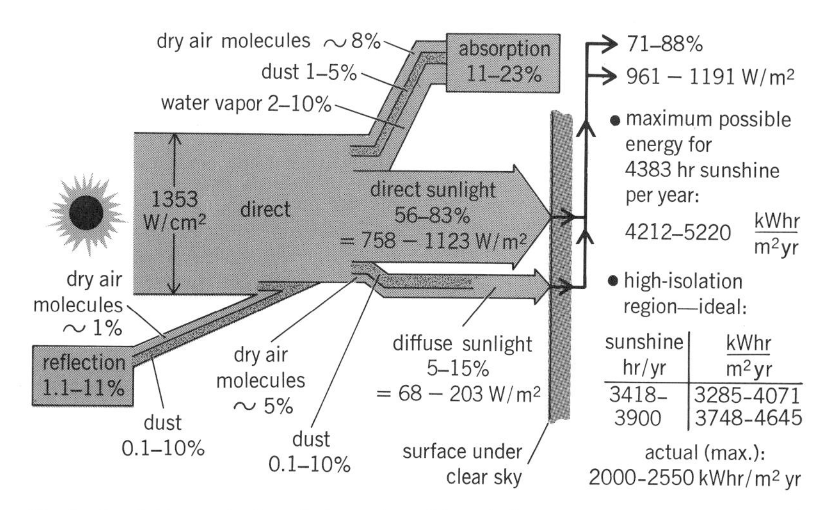

Fig. 2. Sunlight penetration of atmosphere (clear sky).

solar energy). The basic energy systems associated with solar power generation are surveyed in Fig. 4.

Waterpower. Of the world power consumption, only about 2% is derived from waterpower, while contributions by wind power are negligible. In Europe, about 23% of available waterpower is utilized, in North America about 22%. This contrasts sharply with the lower utilization level of waterpower in other parts of the globe.

Waterpower is continuously resupplied by the Sun. Its use does not diminish a given reservoir as in the case of fossil fuels (oil, coal, gas). Waterpower is concentrated solar power and is more regularly available. It can be readily regulated and stored in the form of reservoirs. Today hydroelectric conversion efficiency from waterpower to electric power approaches 80%, compared to

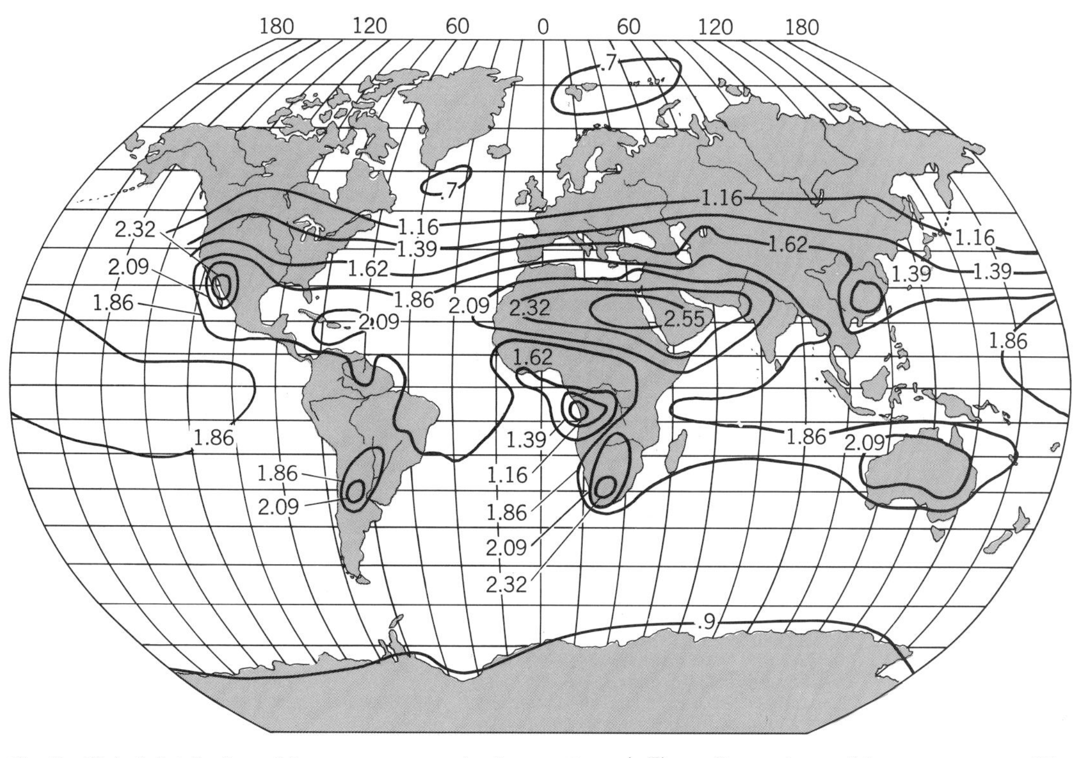

Fig. 3. Global distribution of the average annual solar radiation energy incident on a horizontal surface at the ground. The units are terawatt-hours per square kilometer year (TWhr/km² yr).

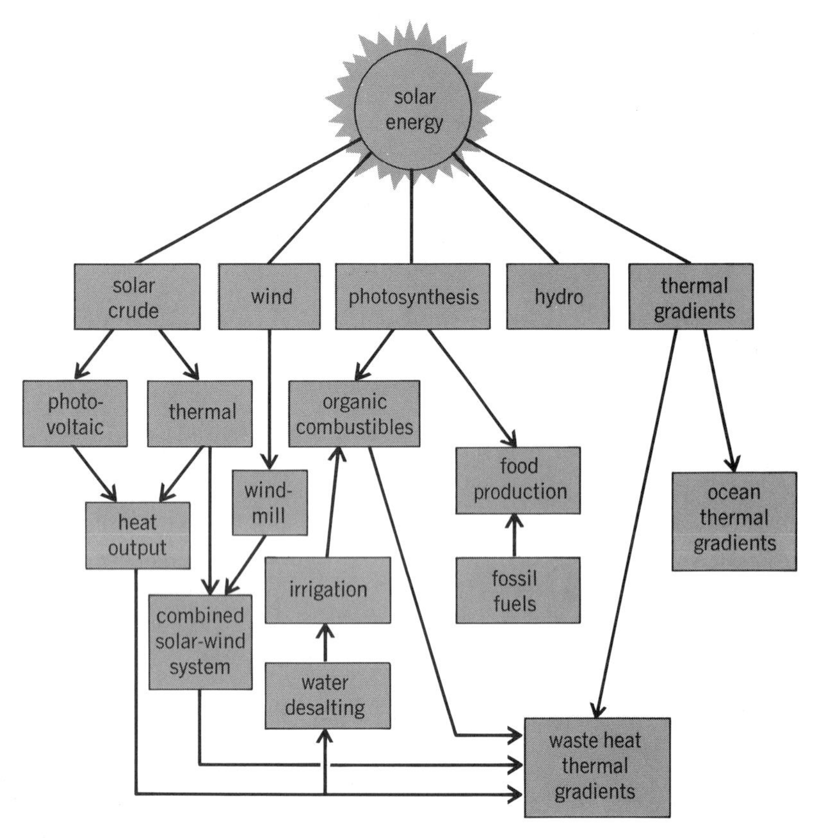

Fig. 4. Basic energy systems in solar power generation.

about 33% conversion efficiency from coal or oil. Utilization of waterpower avoids air pollution.

These are compelling reasons for increasing utilization of the world's waterpower. It is entirely within man's grasp to raise the hydroelectric power supply by a factor of 10, to some 6,400,000,000 kWhr, by the year 2000, benefiting primarily the developing areas. The main obstacle is the availability of investment capital to build the large installations required and, in some cases, the transmission facilities to distant load centers.

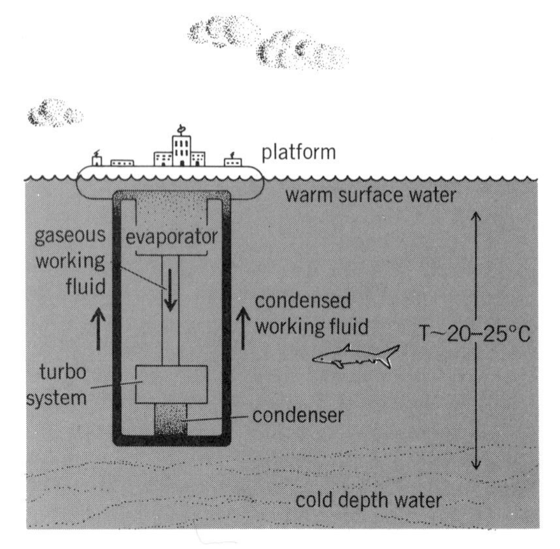

Fig. 5. Closed Rankine cycle, ocean thermal energy conversion system.

The development of waterpower is being advanced particularly in the Soviet Union and the People's Republic of China. Large unused waterpower resources exist still in New Guinea, Africa, South America, and Greenland.

Another form of waterpower generation is through utilization of the thermal gradient in oceans, which serve as the storage system for a vast amount of solar energy. The temperature difference between surface and bottom waters is a potentially very large source of electric power. Tropical regions, particularly between 10°N and 10°S latitude, are especially suitable, because relatively high surface temperatures provide a larger temperature gradient. Temperature differences of at least 20–23°C are available between surface and depth throughout the year, wind speeds are moderate (25 knots or less; no hurricanes), and currents are below 1 knot at all depths. One method (Fig. 5) to extract energy is to heat a suitable working fluid (for example, propane or ammonia) which is evaporated in the warm surface water and ducted to an underwater turbine, where it is allowed to expand, driving a turbogenerator system, and to condense in cool depth-waters. From there it is returned to the surface, and the process is repeated (closed Rankine cycle).

Wind power. Wind is the next largest solar-derivative power, after solar crude itself and the energy contained in the oceans. Its utilization is environmentally even more benign than fresh waterpower, because no dams and land floodings are involved.

The bulk of wind power, which lies in the upper troposphere and the lower stratosphere, is not accessible to present-day technological potential. However, there are large areas with moderate to strong surface winds in the United States, particularly along the Aleutian chain, through the Great Plains, and along portions of the East and West Coasts. This is shown in Fig. 6 for the contiguous 48 states. A study conducted at Oklahoma State University showed that the average wind energy in the Oklahoma City area is about 0.2 kW/m^2 (18.5 W/ft^2) of area perpendicular to the wind direction. This is roughly equivalent to the solar energy received by the same area, averaging the sunlight over 24 hr per day, all seasons, and all weather conditions. However, in contrast to solar energy, the wind energy could be converted at an efficiency of some 40%.

Agricultural utilization. The basis of the biological utilization of solar energy is the process called photosynthesis, in which solar energy provides the power within plants to convert carbon dioxide (CO_2) and water (H_2O) into sugars (carbohydrates) and oxygen. The prime conversion mechanism is the chlorophyll molecule. Organic-chemical solar energy conversion operates at very low efficiency of 0.1–0.2%; that is, for every light quantum used, 1000–500 quanta are reflected by the vegetation. However, research on algae, especially the alga *Chlorella*, has shown that higher efficiencies can be achieved. On the basis of extensive experimentation, the practically achievable yield of "chlorella farms," using sunlight as the energy source, has been estimated to be at least of the order of 35 tons of dry algae per acre. This corresponds to about 0.6% efficiency and compares very favorably even with the highest agricultural yields (10–15 tons per acre), let alone the much lower yields in less developed countries (2–2.5 tons per acre). By building large algae farms on nonarable ground, a growing portion of the solar energy presently absorbed by bare soil, that is, about 5.3% or 400×10^{14} kWhr per annum, could be utilized for the production of organic matter for food and for conversion into synthetic liquid fuels as a complement to the world's oil supply. Again, capital and local or regional requirements determine the feasibility and worthwhileness of such endeavors.

Industrial utilization. In Japan, the United States, the southern Soviet Union, and other countries, solar energy is utilized for drying of fruits and vegetables by means of solar-heated air. Water evaporation in solar stills is an attractive method only where fresh water is extremely expensive, as in isolated arid regions with ready access to brackish or sea water, or under special conditions such as emergency provisions for downed flyers or astronauts, or for shipwrecked sailors. Chile, Greece, Australia, and Israel are operating and developing solar stills. Outputs may run as high as 0.15 gal of fresh water per square foot per day (6 liters per square meter per day). Generally, it is of the order of 0.1 to 0.12 gal per square foot per day (4 to 5 liters per square meter per day).

The most important direct use of solar radiation can be divided into two basic categories: solar heating and cooling for residential and commercial buildings, and electric power generation. As of 1975–1976, the four principal systems for buildings—water heating, space heating, space cooling, and combined system—were at widely different stages of development. Only water-heating systems had reached a stage of commercial readiness.

For solar-electric power plants, the annual sunshine hours should be as large as possible, and the

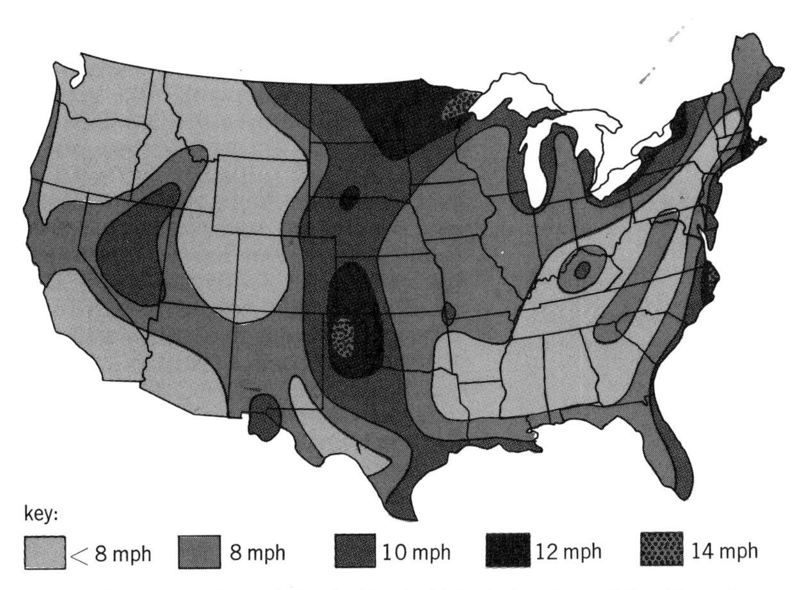

Fig. 6. Average surface wind velocities in high-wind regions of the 48 contiguous states. 1 mph = 0.447 m/sec.

humidity, which causes absorption and scattering, should be low. With its "sun bowl" (Fig. 7), the United States is the only major industrial country with high-insolation (2–2.5 TWhr/km² yr) territory within its borders; the highest values lie around 2.2 TWhr/km² yr (Fig. 3). The table shows that if in such high-insolation areas the incident solar radiation over 1000 km² is converted to electric energy at only 20% efficiency, the output is equivalent to the annual consumption of 734×10^6 bbl of crude oil or 161×10^6 metric tons of coal.

At a received solar energy level of 2.2×10^9 kWhr/km² yr and 20% conversion efficiency to electricity, a 100-GWe-yr plant requires 20,000

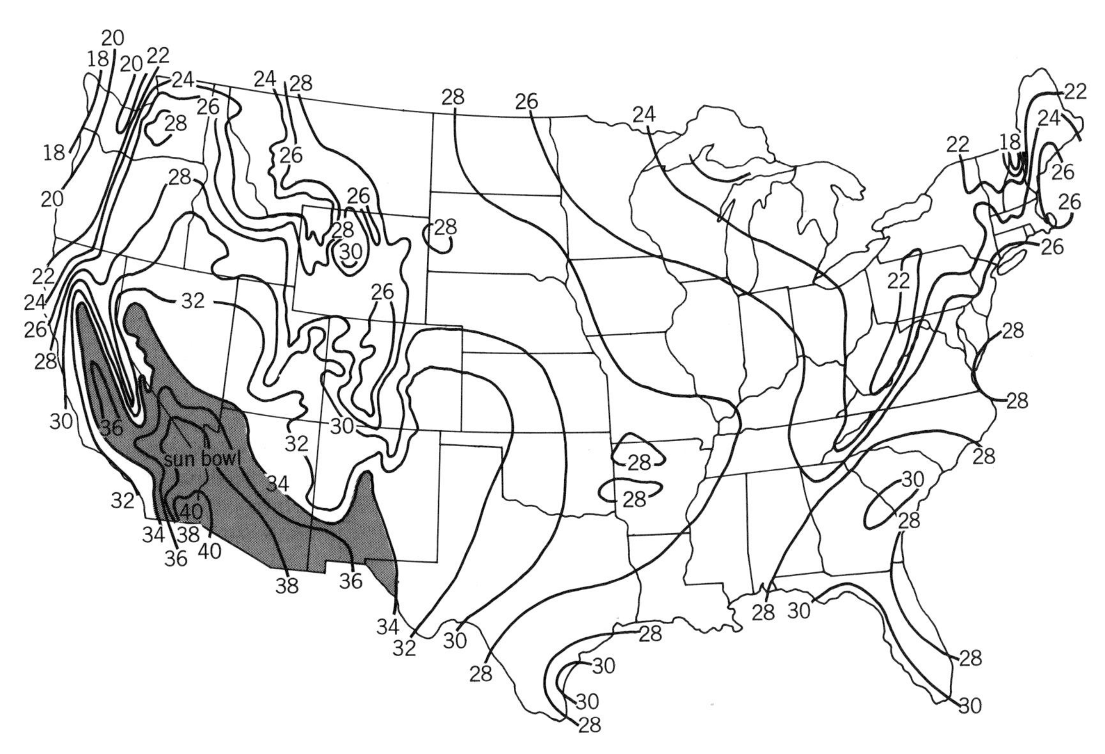

Fig. 7. National climatic center annual sunshine hours with "sun bowl" accented; values in hundreds.

Comparison of energy sources in terms of electric energy output at the bus-bar

Energy source	at electric conversion efficiency of	yields the following electric energy in kilowatt-years*
1000 metric tons of oil	35%	490 = 1.0
1000 tons of coal†	35%	312 = 0.637
1000 tons of enriched uranium in light-water reactor	31.7%	28,500,000 = 58,160
1000 tons of plutonium in liquid-metal fast-breeder reactor	40.4%	36,000,000 = 73,470
1000 km² solar absorber area‡ (solar crude: 2.2×10^9 kWhr/km² yr)	20%	50,300,000§ = 102,700

*The numbers at the right are indices, based on the yield of 1000 tons of oil as unity (1.0). For example, 1000 km² solar absorber area under specified conditions yields 102,700 times as much electric energy as 1000 tons of oil, and 1000 tons of coal yields 63.7% of the energy from 1000 tons of oil.

†Based on mean heating value of 26×10^6 Btu/ton.

‡Since 1000 km² solar crude yields the equivalent of 102.7×10^6 tons of oil and since a ton of oil corresponds to 7.14 British barrels, the yield of the 1000 km² corresponds to $102.7 \times 10^6 \times 7.14 = 734 \times 10^6$ bbl. Coal yield equivalent per 1000 km² solar follows from 102,700/0.637 = 161,000 per 1000 tons of coal, or the yield of 1000 km² corresponds to the yield of 161×10^6 tons of coal.

§United States production in 1969 = 177.1×10^6 kW-yr.

km² net collector area. With an arrangement of the linear parabolic reflectors such that they can follow the Sun from 30° elevation when rising to 30° elevation when setting, the total occupied area is larger. Adding 20% reserve, the collector area is 24,000 km², or about 28% of the overall occupied area of 86,000 km². Such a system would be highly modularized and would require an area that is small (Fig. 8). At 90% thermal efficiency and 40% electric efficiency, that is, a total efficiency of 36%, the required collector area is reduced from 24,000 to 13,300 km² and the overall occupied area is about 48,000 km². The desert area west of Phoenix, AZ, alone covers some 70,000 km², only a fraction of the overall available territory in the Southwest.

The energy budget of a solar station is compared with that of average desert ground in Fig. 9. Of the solar irradiation, the desert surface reflects about 40% and retains 60%, whereas the absorber area retains about 90%. Thus, 30% more of the incident solar energy is trapped. Of this, about two-thirds appears as electric energy and one-third as true, that is, extrinsic, heat production at 20% overall conversion. At 30% overall conversion from solar crude to electricity, no extrinsic heat would be generated. At still higher conversion efficiency, the area occupied by the power station would be cooled, compared to the natural environment, rather than slightly heated as with the 20% system.

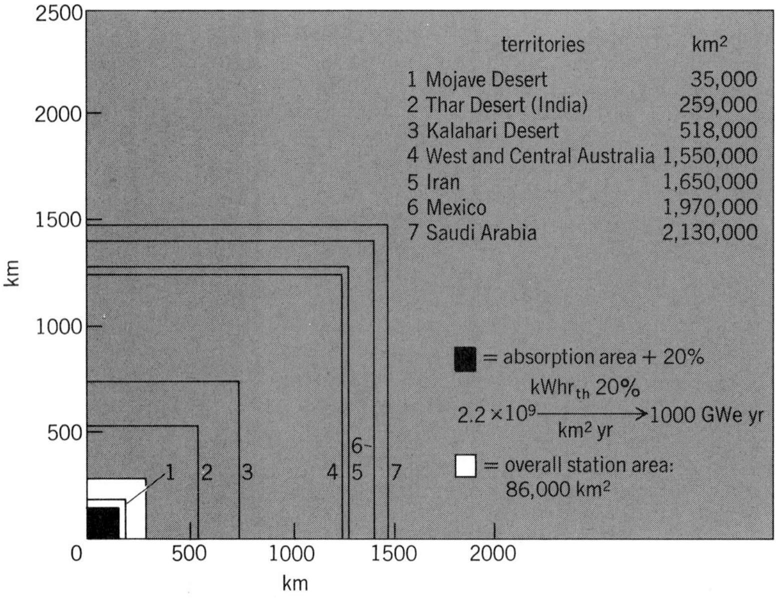

The overall unused thermal energy is concentrated in the power station's thermal storage system and ultimately in its electric conversion system. If the electric conversion system uses cooling water rather than air cooling, advantage can be taken of the large amount of thermal energy concentrated in the water. For this, two alternatives are available: desalination, and power generation by temperature-gradient utilization.

Fig. 8. Comparison of a 1000 GW-yr medium-temperature solar power complex with high-insolation territories (1 GW-yr = 8766×10^6 kWhr).

Three methods can be used to generate solar-electric power: (1) the solar-thermal distributed receiver system (STDS); (2) the solar-thermal central receiver system (STCRS); and (3) the photovoltaic system (PVS). In each case, the overall system may serve as backup, that is, operating only when the sun shines, equipped only with a minor energy storage capacity (for example, for 1 hr full output) to bridge temporary cloud coverage; alternatively, the overall system may include a conventional fuel system to replace solar energy at night or during cloudy days; and finally, the independent solar-electric system includes a storage system to ensure continuous power-generating capacity based on solar energy only.

In the STDS, solar radiation is absorbed over a large area covered with flat-plate (nonconcentrating) collectors or parabolic-trough concentrators, which focus the sunlight on a heat pipe carrying the working fluid (Fig. 10). The heat pipes are covered with selective coating characterized by high absorptivity to solar radiation and low emissivity at the temperature of the heat pipe. The flat-plate collector operates at turbine inlet temperatures of 250–500°F (121–260°C). With the para-

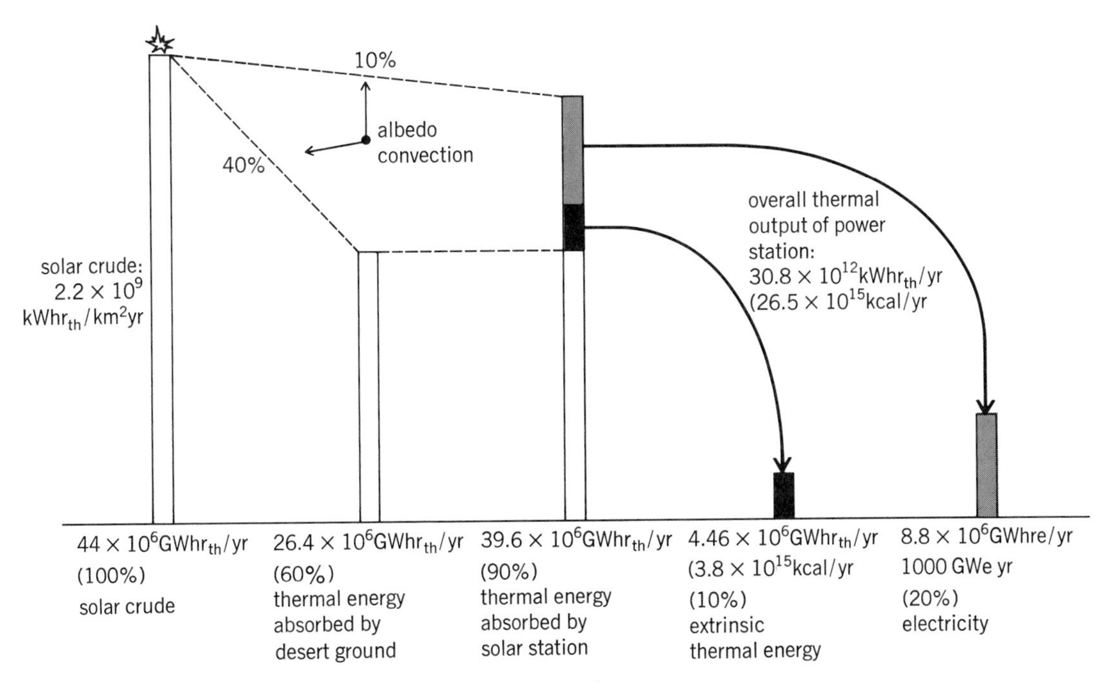

Fig. 9. Typical energy budget of a 1000 GWe-yr solar power station.

bolic trough, temperatures between 550–1000°F (288–538°C) at turbine inlet are attainable. The higher the temperature, the higher the efficiency and the smaller the land area needed for a given power level, but the more expensive is the system, especially the piping and the coating.

In the STCRS, sunlight is concentrated on a receiver by a large number of mirrors designed to follow the Sun (heliostats). The receiver is a heater located atop a tower (Fig. 11) which is served by a certain collector area. The power output of such a module depends on the collector array area which, in turn, determines the tower height. The horizontal distance of the farthest mirror from the foot of the tower is about twice the tower height, which may range from 250 m to 500 m. A given solar power plant may consist of an arbitrary number of these modules.

Both the STDS and STCRS require large areas, due to the nature of the energy source. It is possible, of course, to subdivide the STDS into small modules, each with a small standardized electric power station, and to collect the electric current in an overall power-conditioning station. In the STCRS, the working fluid can be heated to higher temperatures, since the collector area acts as a giant parabolic mirror consisting of individual facets (heliostats). The higher temperatures attainable with the STCRS yield higher overall efficiencies (25–35%), or 150,000–200,000 kWe per square kilometer of *collector* area (not total area covered), compared to about 1.5 km² for the same power output in an STDS. The working fluid can be water (steam), sodium, or helium (in the order of increasing system temperature). In the simplest case, superheated steam is generated in a heater atop the tower. The steam is ducted to the ground and used in a high-pressure and low-pressure turbogenerator system.

To provide power during off-radiation hours, part of the energy generated during sunshine hours must be stored. This means that part of a solar power module's (SPM) power output, or the output of entire modules, is not available during sunshine hours. Several options are available for storage: the energy can be stored as heat, as mechanical energy (pumping water to elevated storage basins), or as chemical energy (for example, electrolytic decomposition of water to hydrogen and oxygen). It may be possible to design the central receiver towers so that wind generators can be attached, causing the towers to generate power in the absence of sunshine as well. Figure 12 compares the cost effects of several storage modes in terms of magnitude of storage capacity (hours of full power output) and type of storage.

In the photovoltaic system (PVS), solar cells are used to produce the necessary electric energy. They are spread over a large area, as in a distributed system, to collect sunlight for the desired electric output which is in direct current. This is advantageous if long-distance transmission is involved, since losses are lower than in transmission of alternating current. For most uses, dc must be converted into ac, and if output is fed into existing distribution grids, dc-to-ac conversion equipment must be available at the plant.

The principal advantage of the PVS over the

Fig. 10. Solar-thermal distributed power station system.

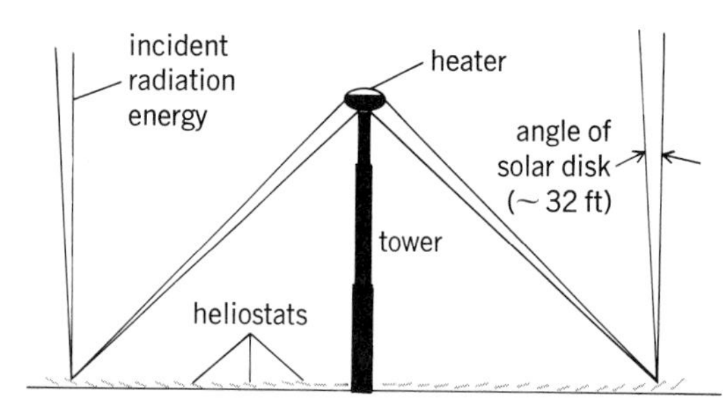

Fig. 11. Solar-thermal central-receiver system. 32 ft = 9.75 m.

thermal systems is the absence of moving parts and fluids at high temperatures. No cooling is needed and probably no Sun-tracking. Its major disadvantages are its comparatively low efficiency (probably below 20% even after extensive development) and the fact that thermal storage cannot be used. Considerable development is required before the system can be economically competitive with the thermal systems.

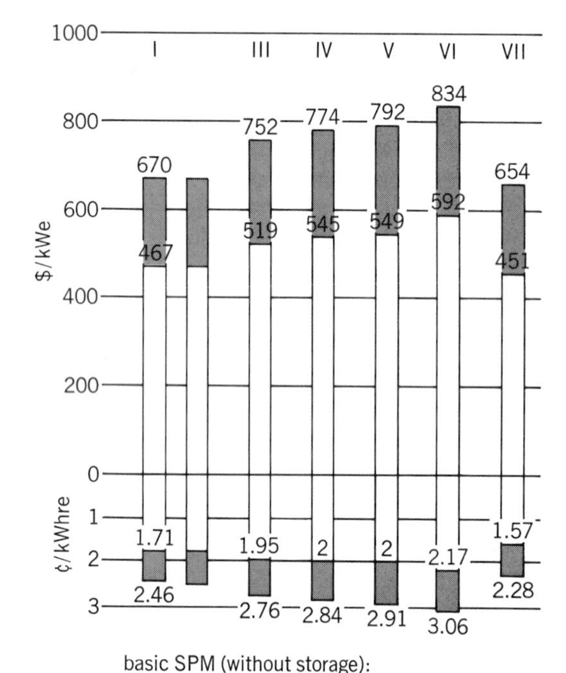

basic SPM (without storage):
area 2.6 km²
collector area 1.3 km²
number of towers 2
sunshine hours per year/day = 3800/10.4
concentration and cycle efficiency = 0.3
oil equivalent (0.35 cycle efficiency) annually:
182,000 tons
1.3×10^6 bbl

I basic SPM
II SPM with thermal storage capacity – 1 hr
III – 5.2 hr
IV – 10.4 hr
V SPM with flywheel storage capacity – 5.2 hr
VI – 10.4 hr
VII SPM with wind power storage capacity – 1 hr
uncertainty range

Fig. 12. Solar-power module (SPM), performance data, and cost summary for different energy storage types and capacities. 182,000 tons = 165,000 metric tons; 1.3×10^6 bbl = 2.1×10^5 m³. *(K. A. Ehricke)*

Solar energy and space. Extensive use of solar energy can be anticipated in spacecraft and by extraterrestrial communities in orbit or on the Moon. Solar cells are the primary power supply for unmanned spacecraft. They are the most attractive source of electric power for manned space stations. In the 1970s and 1980s, solar cells will power tiny electric thrust units, propelling unmanned probes into the asteroid belt and on other far-flung missions in the solar system. On the Moon, undiminished solar energy can be soaked up by solar cells and solar concentrators during 14 days for immediate use, for storage in large banks of rechargeable batteries, and in fuel cells to be used during the long lunar night.

Another important relation between space and solar energy is provided by weather satellites. With improving long-term weather prediction, the practical aspects of utilizing solar energy on Earth will improve also. Advancements in the industrial utilization of space beyond the application satellites will make it possible to intercept large amounts of solar radiation in space for use on Earth or in extraterrestrial production facilities. For terrestrial applications, sunlight can be transmitted optically by reflectors (Space Light), or it can be converted to electric energy in space and transmitted to Earth via microwave beam (power generation satellites, PGS). Moreover, electric energy generated at one point on Earth can be transmitted by microwave beam to a distant load center via power relay satellite (PRS).

The general objective of Space Light is to transmit sunlight to selected areas of the Earth's night side. The key objectives can be divided into three categories: low light level for night illumination (Lunetta satellite); strong light (up to half the Sun's brightness) for stimulating food growth by enhancing photosynthetic production (photosynthetic production enhancement, or PSPE, Soletta); and light up to about 70% of the Sun's brightness as a "night sun" to provide around-the-clock solar energy for industrial purposes in a selected area of about 90,000 km².

Lunetta is designed to illuminate with the brightness of 10 to 100 full moons on a clear night (about 1/6 of that on a cloudy night), where one full moon nominally corresponds to 1/400,000 of the Sun's brightness on a clear day. In agriculture Lunetta provides the necessary brightness for sowing and harvesting at night. Lunetta's light can also accelerate large construction projects in remote areas or during long polar nights. The Lunetta can be economically placed in a 4-hr inclined orbit (about 1 earth-radius altitude) and still illuminate a fixed area (minimum of 2800 km²) if the reflector is rotated appropriately. To ensure an 8-hr illuminative period per night for a given area, six Lunettas, totaling 4.2 km² in area, are required.

The PSPE Soletta provides daylight extension (dusk and dawn illumination), especially in northern regions, and brief nighttime illumination. In a 4-hr orbit, a PSPE Soletta of a nominal 40% of full solar brightness would consist of a "swarm" of reflectors of a total area of 1040 km², all trained on the same focal area.

The night sun Soletta is in a fixed position with respect to the focal area, which preferably is located within 30° latitude on either side of the Equator. This means that the Soletta stays near the zenith of the irradiated area and does not rise or set like the Sun. Consequently, even at about 60% solar brightness the overall energy delivered per night by the Soletta is about equal to the energy delivered by the Sun per day (at equal sky conditions). The overall reflecting area of the Soletta swarm, therefore, is about 54,000 km². A solar power station in this area could operate around the clock, as if in space.

PRS and PGS are part of an overall system in which the primary electric-power station is located either on the ground or in space. The electricity is converted to microwave energy which is shaped to a controlled beam by a transmitter array antenna, which in turn is trained on a large receiver array where the microwave energy is reconverted to dc electricity by rectifiers. In the absence of optical line-of-sight connection between transmitter and receiver, a relay is required to redirect the beam.

Many costly technological problems must be solved before either optical- or microwave-beam systems can be realized. In addition, research on possible environmental effects of large numbers of microwave beams, each carrying millions of kilowatts, must be completed and evaluated for added technological requirements. [KRAFFT A. EHRICKE]

Bibliography: AMETEK, *Solar Energy Handbook: Theory and Application*, 1979; Battelle Columbus Laboratories, *Survey of the Applications for Solar Thermal Energy Systems to Industrial Process Heat*, 1979; F. Daniels, *Direct Use of the Sun's Energy*, 1965; J. A. Duffie and W. A. Beckman, *Solar Energy Thermal Processes*, 1974; K. A. Ehricke, *Space Industrial Productivity: New Options for the Future*, Future Space Programs, Hearings of House Subcommittee on Space Science and Applications, 1975; K. A. Ehricke, *The Power Relay Satellite*, Rockwell International Rep. E74-3-1, 1974; P. E. Glaser, The satellite solar power station, *Proc. IEEE: International Microwave Symposium*, 1973; D. K. McDaniels, *The Sun: Our Future Energy Source*, 1979; National Aeronautics and Space Administration, *Feasibility Study of a Satellite Solar Power Station System*, NASA CR-2357, 1974; National Aeronautics and Space Administration, *Solar Electromagnetic Radiation*, NASA SP-8005, May 1971; N. Robinson, *Solar Radiation*, 1966; C. Zener, Solar sea power, *Phys. Today*, 1973.

Solid waste management

The systematic administration of activities that provide for collection, processing, and disposal of solid wastes, including the recovery of materials and energy. Solid wastes are produced by municipal, industrial, mining, and agricultural sources.

Wastes have always been present, they have been casually discarded, and in the past they have caused little concern. Solid waste management today is made difficult and costly by the increasing volumes of waste produced; by the need to control what are now recognized as serious environmental and health effects of disposal; and by the lack of land in urban areas for disposal purposes, partly due to public opposition to proposed sites. The drain of potential energy and materials that is represented by disposal of large waste flows is also of increasing concern. Thus waste management, once strictly a local or private-sector matter, now involves regional, state, and Federal authorities. The Resource Conservation and Recovery Act (RCRA) of 1976 is the main Federal law directed toward safe management of solid wastes and conservation of the resources they represent.

WASTE VOLUMES

The volumes of waste produced in the United States each year tend to grow with increasing population, production, and consumption, and the greater amounts of pollutants being held back from discharge into rivers, lakes, oceans, and the air.

Municipal solid waste (mainly trash and garbage from residential, commercial, and institutional sources) amounted to about 140,000,000 metric tons in 1978. (This is enough to fill the New Orleans Superdome from floor to ceiling, twice a day, weekends and holidays included.) Per capita generation is about 650 kg per year. In 1960 the tonnage was 79,000,000, and per capita generation was 460 kg. Growth in municipal solid waste generation seems to have slowed in the past few years; nevertheless, it appears that by 1990 the yearly total will be up to 178,000,000 metric tons.

Industrial wastes generated in 1977 totaled about 344,000,000 metric tons. Roughly 30–40% was in solid form, the rest liquids and sludges. Industrial waste generation is growing at a rate of about 3% per year. An increasing percentage of the waste is resulting from pollution control processes (Fig. 1). About 10–15% of industrial wastes may be considered potentially hazardous, requiring special safeguards in handling and disposal because of the substantial danger they pose to health and the environment. Discarded toxic chemicals, explosives, acids, caustics, radioactive material, and infectious material are types of hazardous waste. Preliminary estimates indicate that such waste, from industrial and other sources, will total about 57,000,000 tons in 1980. Major hazardous waste generators, among 17 industries the U.S.

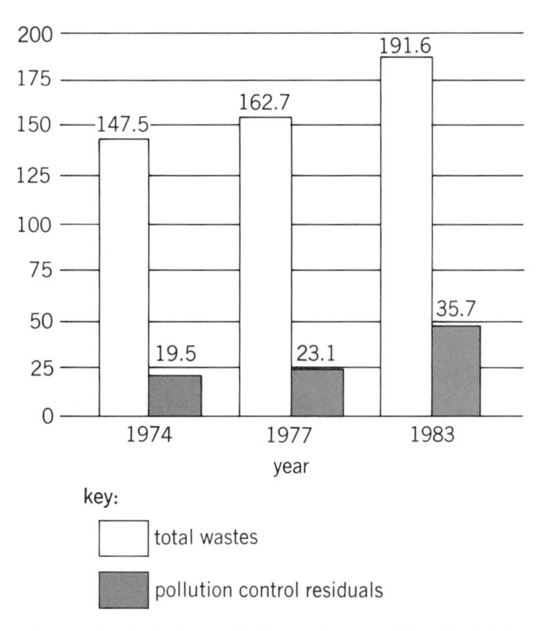

Fig. 1. Projected growth in waste quantities for 14 industries, in 10^6 metric tons (dry weight).

Major hazardous waste generators (1977 estimates)

Source	10^6 metric tons (wet basis)
Organic chemicals	11.7
Primary metals	9.0
Electroplating	4.1
Inorganic chemicals	4.0
Textiles	1.9
Petroleum refining	1.8
Rubber and plastics	1.0
Other	1.0
Total	34.5

Environmental Protection Agency (EPA) has studied in detail, are shown in the table. *See* INDUSTRIAL WASTES.

The nation's 18,000 municipal wastewater treatment plants produce 5,000,000 metric tons (dry weight) of sludge per year; the amount is expected to increase greatly in the coming years due to higher levels of treatment.

Agricultural wastes, including feedlot wastes, poultry manure, and crop residues, total about 430,000,000 metric tons (dry weight) per year. Mining waste generation is roughly estimated at over 3,000,000,000 tons a year. As the richer deposits of minerals are depleted, less rich deposits are worked, and greater amounts of waste material are generated. *See* AGRICULTURAL WASTES.

WASTE MANAGEMENT PRACTICES

In the main, solid wastes have been disposed of with inadequate care to prevention of effects on the environment, public health, and safety. Methods for sound management are available but have not been widely used, largely because of their higher costs and the lack of requirements that would compel their use.

Disposal practices and their effects. Most solid waste is disposed of on the land through landfilling (burial), open dumping, spreading (as with sewage sludge or other organic wastes), impounding, or related methods.

Rough estimates of land disposal sites in the United States include the following: 18,500 landfills accepting municipal solid wastes, 23,000 sites disposing of sewage sludge, 76,000 industrial landfills, and 272,000 impoundments for industrial, agricultural, and municipal wastes (mainly semisolids and liquids). Hazardous wastes are stored or disposed of at 32,000 to 50,000 sites.

Disposal of waste on land can affect public health and environmental quality in many ways. Improper disposal practices have led to deaths and injuries to workers, direct exposure of nearby residents to toxic wastes, contamination of groundwaters and surface streams, air pollution, damage to wetlands and other environmentally sensitive areas, explosions of landfill gas, contamination of croplands with heavy metals, and other effects. Potentially, the most widely significant environmental effect is contamination of groundwater. Probably well over half of all disposal facilities leak contaminants into groundwater. *See* AIR POLLUTION; WATER POLLUTION.

About 90% of hazardous wastes are being disposed of without adequate precautions. Numerous severe cases of damage from inadequate management came to national attention in 1978 and 1979. Perhaps the most serious of these was at Love Canal in Niagara Falls, N.Y. The area was declared eligible for Federal disaster relief in August 1978 because of the danger from toxic chemicals seeping into basements and backyards. The former canal site had long been used for burial of industrial chemical wastes prior to 1953. The state identified 82 chemical compounds at the site, including 1 known carcinogen (benzene) and 11 suspected carcinogens. The incidence rates for miscarriages and birth defects are elevated among families adjacent to the canal. Over 200 families were relocated, extensive remedial construction is planned, and additional monitoring and testing are taking place.

Other recent incidents include: numerous cases of illegal dumping of hazardous wastes in marshes, streets, vacant lots, and sewers in New Jersey, Rhode Island, and Connecticut; dumping of polychlorinated biphenyl (PCB) wastes along 200 mi (322 km) of North Carolina's roads; severe contamination of wells near a chemical waste disposal site in Tennessee; and discovery of many thousands of barrels of unidentified industrial wastes in the "Valley of the Drums" in Kentucky. A preliminary EPA survey in 1979 indicated that 1200 to 2000 hazardous waste storage and disposal sites may pose significant dangers.

In addition to land disposal methods, incineration is employed in disposing of approximately 4% of municipal solid wastes, 10% of industrial wastes, and 35% of sewage sludge. The ash and other residues are landfilled. At one time a greater portion of municipal solid waste was incinerated, but the majority of municipal incinerators were closed down because the older designs could not meet air-pollution standards. Many industrial and other incinerators in current use still contribute unduly to air pollution.

Ocean dumping accounts for disposal of about 15% of sewage sludge and small fractions of other types of waste. Under the Marine Protection, Research, and Sanctuaries Act, ocean dumping of sewage sludge must cease by the end of 1981. Ocean disposal is controlled by a permit system of the EPA and the U.S. Corps of Engineers.

Environmentally sound disposal. Land disposal sites can be located, engineered, and operated so as to avoid or control adverse effects on public health, safety, or the environment. The type or combination of wastes to be disposed of is a major determinant in the siting, design, operating, and monitoring needs of a disposal site. Thus, to reach the same level of environmental protection, the requirements are likely to be more demanding in the disposal of certain industrial wastes, for example, than in the disposal of municipal refuse alone because of the greater potential for environmental damage. Avoidance of groundwater pollution, control of surface runoff, and prevention of air pollution and migration of explosive gases are major concerns. To protect groundwater in some areas, landfills and impoundments must be lined with impermeable material. The leachate (fluid formed by water percolating through the wastes) from landfills may have to be collected and treated.

In the daily operation of a municipal solid waste

landfill, the wastes are spread in thin layers and compacted; at the end of each operating day, an earth cover is applied and compacted. In a few cities the waste is shredded or baled to reduce waste volume; while adding significantly to disposal costs, such processing facilitates handling and transport of the waste, extends the life of the landfill, and usually reduces requirements for soil cover.

Sewage sludge is utilized in many areas as a soil conditioner following stabilization through anaerobic digestion, composting, or other methods. Clearly, such utilization is to be generally encouraged; however, avoidance of environmental and health problems requires consideration of the composition of the sludge (which varies significantly from place to place), the analysis of the soil to which the sludge will be applied, and the use to which the land will be put. The cadmium content of sludges from industrial areas is a particularly important consideration, since plants can take up this toxic element from the soil. *See* SEWAGE TREATMENT.

Wherever feasible, it is clearly preferable to utilize waste, process it to reduce its potential for environmental damage, or reduce its volume, rather than dispose of it directly on land. Land available for disposal purposes is decreasing, waste generation is increasing, and disposal in landfills and impoundments incurs long-term responsibilities for preventing the release of contaminants into groundwaters and other environmental media.

Adequately controlled incineration is a proved, safe method of destroying organic wastes. It is an important tool in management of hazardous wastes, since approximately 60% by weight of hazardous wastes generated in the United States are organics. An EPA demonstration project completed in 1977 matched seven commercial incinerator types with 13 chemical wastes and obtained very high destruction rates (99.9+%) in rotary kiln, fluidized-bed, and liquid injection incinerators, as well as cement-manufacturing kilns.

Numerous techniques are available for detoxifying, neutralizing, and concentrating hazardous wastes. Such methods include: physical processes such as carbon or resin adsorption, distillation, centrifugation, flocculation, sedimentation, reverse osmosis, and ultrafiltration; chemical processes such as fixation into solids, neutralization (pH control), ion exchange for removal of heavy metals, oxidation, and precipitation; and biological processes such as activated sludge treatment to destroy organics, composting of high-organic wastes, trickling filters to promote decomposition, and controlled application on land for degradation of organics.

COSTS OF SOLID WASTE MANAGEMENT

The available estimates of solid waste management costs are rough but indicative of the magnitudes involved.

Estimated costs of municipal solid waste management, including collection, transfer stations, incineration, and landfilling, but excluding resource recovery, average $43 a ton and total more than $6,000,000,000 a year nationally, according to 1978 EPA estimates.

Municipal solid waste management systems have three basic cost centers associated with collection, transfer station usage, and disposal. Collection costs include the cost of pickup from households and commercial establishments and of hauling to a transfer station, processing plant, or disposal site. Transfer station costs include waste handling and hauling to landfills. Disposal costs are largely landfill costs. On a nationwide average, collection costs are by far the largest component, representing 75–85% of total costs, with transfer stations and disposal accounting for the remaining 15–25%. For certain communities, however, transfer and disposal costs may exceed cost of collection. The relatively high cost of collection service is largely due to its labor-intensiveness—about three-quarters of the estimated 227,000 workers in municipal solid waste management (1973) are needed for collection. Equipment and fuel costs are also important factors.

The manufacturing industries spent approximately $962,000,000 in 1976 for solid waste management, according to Bureau of the Census data. EPA has estimated that the 17 industries believed to include the major producers of hazardous wastes now spend about $155,000,000 annually for hazardous waste management.

Municipal sludge management costs about $635,000,000 a year (1977).

All these costs are climbing with increased waste generation, inflation, shortage of disposal space in many areas, rising land values, longer haul distances to disposal sites, and more stringent environmental and safety requirements.

The costs of environmentally adequate disposal are substantially in excess of open dumping or other inadequate practices. For example, some cost estimates of hazardous waste disposal techniques are: chemical waste landfill, $30–55 per metric ton; incineration, $75–265 per metric ton ($110 typical); landspreading, $2–25 per metric ton ($6 typical); chemical fixation, $10–30 per metric ton.

The final cost of inadequate practices can also be high, however. Containment costs alone at hazardous waste dump sites that are in a dangerous condition have been estimated at $3,600,000 on the average. Full cleanup (not including compensation for damages) may average $26,000,000 per site.

RESOURCE RECOVERY AND CONSERVATION

Reducing waste generation and recovering usable materials and energy from waste can help lessen solid waste problems while conserving resources. Industries are increasingly looking to ways to reduce wastes and utilize waste material and energy, not only for the savings in materials and energy costs, but also to keep down costs of waste management and pollution control. Reduction of waste may be achieved through reformulating or redesigning products or through modification of processes and equipment. For example, consumer products that are more durable, made with less material, or designed for repeated use rather than as "throwaways" reduce waste and save material and energy.

Recycling can take place at any of several points in the flow of materials through the economic system (Fig. 2). At the material-refining stage, efforts have been directed to producing by-product raw

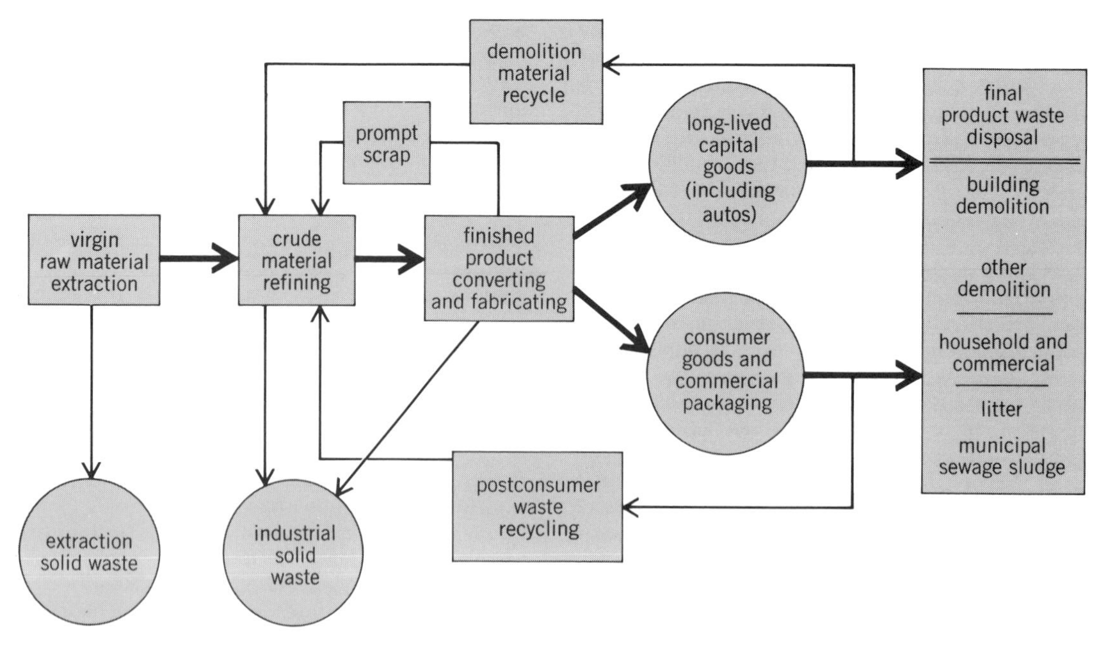

Fig. 2. Material flow diagram.

materials (from such wastes as slags, sawdust, and pulping liquors) and to reclaim and recycle processing chemicals and other materials within the plant.

In certain product-manufacturing sectors, especially the metalworking and paper product–converting industries, a very large percentage (possibly over 90%) of the scrap waste generated is recycled as "prompt" or "new industrial" scrap. Estimates place scrap metal recycling from this sector at around 20,000,000 metric tons, and paper and paperboard converting scrap recycling at 5,000,000 metric tons per year.

A mechanism to encourage industrial recycling is the "waste exchange"—a clearinghouse, usually run by a chamber of commerce or trade association, which facilitates transfers and sales of waste materials by companies that generate the wastes to others that want them as raw materials. Approximately 20 waste exchanges have been established since 1976.

By far the greatest volumes of final products are in the form of long-lived capital goods. Current estimates indicate that about 24,000,000 metric tons of metals is recovered from salvaging capital goods, including junked autos and other transportation equipment, railroad rails, and other structures and equipment.

Out of the 140,000,000 metric tons of municipal solid waste (household, commercial, and government office sources), approximately 10% was recovered in 1978—7% for recycling and 3% for energy production.

Resource recovery from municipal solid waste. Activity in municipal resource recovery has been increasing, primarily because of the cities' need for an alternative to land disposal. However, the United States rate of waste utilization is still low compared with potential recovery and levels achieved by some other industrialized nations. Several West European countries (West Germany, Sweden, the Netherlands, Switzerland, and Denmark), for example, process 20–60% of their municipal solid waste for energy recovery.

A number of interrelated factors have held back more rapid expansion of resource recovery: The traditional forms of waste disposal, dumping or landfilling, have generally been cheap in the United States, at least in terms of direct costs—environmental damage has been ignored and, compared with many European countries, land has been plentiful. Most communities lack the experience, expertise, and organization required to plan effectively for resource recovery. There are technological uncertainties. Markets for recovered materials have been limited and highly unstable. National policies have tended to encourage use of virgin resources. Until recently, fossil fuels have been plentiful and cheap, dampening interest in wastes as fuels.

Source separation. Source separation accounts for 95% of materials recovery from municipal solid waste. Source separation is the setting aside of recyclable material, such as paper, glass, and cans, at their points of generation (home, office). Mechanical processes for recovering materials from mixed wastes are in the developmental stage except for ferrous metal extraction, and only a small amount of ferrous metal is being mechanically recovered today.

The most common form of source separation is the local recycling center, where segregated waste materials are accepted for recycling. Several thousand of these are sponsored by local civic organizations of all types as well as by municipal governments. Also, aluminum companies purchase used aluminum at numerous collection centers.

Curbside collection of recyclables is being practiced in about 220 American cities, as compared with only 2 cities in 1970. Although the majority of these programs collect only newspaper, approximately 40 cities are also collecting glass and cans.

More than 125 office paper recovery programs have been established in Federal agencies. More than 4000 tons of white ledger paper were recovered through these programs in 1978. Many state and local government agencies and over 600 private companies also recover their office paper.

Energy recovery. There is high interest and activity in establishing refuse-to-energy plants. According to a 1979 EPA survey, there are now 29 operating plants with a total design capacity of over 13,800 metric tons per day (compared with 19 plants in 1977 with a total design capacity of 7400 metric tons per day).

Among the energy technologies currently being used or tested, the most promising are: mass burning of unprocessed solid waste in heat recovery furnaces (waterwall combustion units); mass burning of unprocessed waste in small modular incinerators; and mechanical conversion of the organic portion of municipal solid waste into a refuse-derived fuel that is combusted as a primary fuel in a dedicated boiler or as a supplementary fuel along with coal in an existing boiler. In addition, conversion of solid waste to other fuel forms through pyrolysis (destructive distillation) is being tested in several systems.

Waterwalls. The waterwall combustion unit is a furnace enclosed by closely spaced water-filled tubes. Water circulating through the tubes recovers heat radiated from the burning waste. The first sophisticated American system, patterned after the European designs, was constructed in 1971 at Braintree, MA. Shortly thereafter, systems were built at Chicago and Harrisburg, PA. Although all were designed with the possibility in mind of selling the steam generated, none of these plants did so until after the 1973 oil embargo. More recently, however, waterwall combustion units have been constructed at Nashville, TN, and Saugus, MA. Markets for the recovered steam (and chilled water at Nashville) were secured prior to construction of these units, assuring sale of their products.

Modular incinerators. Small modular incinerators have been in use for years in the United States, where some 20 companies manufacture them. Their greatest use in the past has been for volume reduction of the waste, mostly in large retail or commercial complexes, apartment houses, institutions, and industrial sites. Heat recovery units to recover energy either as hot water, steam, or heated air have been incorporated only during the past few years. There are currently about 20 such heat recovery units in operation, mostly at industrial sites. The several municipal operations using one or more of these incinerators range from 15 to 100 metric tons per day in total capacity. At least one industrial site is burning both municipal and industrial waste. The relatively low capital and operating costs of the modular incinerators make them feasible in some instances for communities as small as 10,000 to 15,000 people.

Mechanical conversion of waste to fuel. Refuse-derived fuel can be produced in many different ways, but in general the process entails primary shredding of the waste to reduce particle size to about 10 cm. At this point, ferrous metal is usually removed, and the remaining material is separated into a lighter, mostly combustible fraction and a heavier, mostly noncombustible fraction by injecting the waste into a strong, vertically rising airstream. The light fraction is then further processed to improve its fuel characteristics (additional shredding and screening). The fuel that is produced can be burned as a supplement to coal or can be burned in a dedicated spreader-stoker boiler as a primary fuel. Plants are in operation in Ames, IA, Milwaukee, WI, and Baltimore County, MD. Others are under construction in Bridgeport, CT, Chicago, IL, and Monroe County, NY.

Pyrolysis systems. Pyrolysis occurs when organic material is exposed to heat in the absence or near absence of oxygen. Processes under development use part of the waste to provide the heat absorbed during pyrolysis, and recover the remaining heat in the form of steam (from immediate combustion of pyrolytic gas) or a gaseous or liquid fuel. Ideally, pyrolysis would convert solid waste into a more readily marketable fuel which could be readily transported, stored, and used compatibly with conventional fuels. The concept is still in the developmental stage, however, and not ready for commercial use.

Codisposal with sewage sludge. Energy recovered from municipal refuse, by means of waterwall combustion or other means, can be used to dewater sewage sludge to the point where it will burn by itself. Thus two disposal problems can be alleviated in one system, and energy can be produced for use in the wastewater treatment plant or other purposes. Such thermal codisposal projects are under way in Contra Costa County, CA, Duluth, MN, Glen Cove, NY, Harrisburg, PA, and Memphis, TN.

Beverage container deposits. A beverage container deposit is a fee added to the price of a beverage which is refunded when the container is returned. The containers may then be recycled or reused. Mandatory container deposits for beer and soft drinks have been supported as a means of reducing solid waste and litter, materials use, energy use, and pollution.

The first state law concerning deposits, passed in Vermont in 1953, simply banned the sale of beer and ale in nonreturnable bottles. That law was allowed to expire in 1957, but in 1972 Vermont passed a new law requiring deposits on all beer and soft-drink containers. Oregon had passed a deposit law the year before. Five other states have followed with similar laws (Maine, Michigan, Connecticut, Iowa, and Delaware); however, the law in Delaware is contingent on the passage of similar laws in neighboring Pennsylvania and Maryland.

Because of congressional and public interest in legislation that would create a nationwide retail trade requirement that all beer and soft-drink bottles and cans carry a minimum 5-cent refundable deposit, the interagency Resource Conservation Committee, established by the RCRA, made such legislation a major subject of its studies. The committee found that mandatory deposits would effectively reduce beverage container litter; eliminate up to 2% of municipal solid waste; result in significant conservation of material and energy if at least 85–90% of the containers are returned; cause some inconvenience to consumers; and result in a net increase in jobs, although some industries would experience significant dislocations.

FEDERAL-STATE PROGRAMS

All levels of government are involved in dealing with the problems related to solid waste. The actual operation of municipal solid waste management systems, including collection, processing, and disposal, is the responsibility of local governments. The state governments have responsibility for setting and enforcing standards and for planning the development of solid waste management in the state. All states have some authority over disposal of municipal solid waste, and many states have enacted legislation for regulation of hazardous waste in recent years. The states are also increasingly active in the planning and funding of resource recovery systems.

At the Federal level, the EPA has the major responsibility for dealing with solid waste management under authority of the Solid Waste Disposal Act. The act was passed in 1965, with major amendments in 1970 and 1976.

Until the 1976 amendments, titled the Resource Conservation and Recovery Act, the program consisted mainly of developing ways to improve management of municipal solid waste (including resource recovery), technical assistance, and financial aid to states for statewide planning in solid waste management. RCRA reflected a greatly expanded awareness of environmental and health problems associated with improper disposal of solid wastes. The chief objectives of the act are: (1) regulation of the management of hazardous wastes from point of generation through disposal, by EPA or by state programs authorized by EPA; (2) regulation of the disposal on land of all other solid wastes by the states in accordance with minimum Federal criteria; and (3) establishment of resource recovery and conservation as the preferred solid waste management approach.

[It should be noted that the act defines solid waste broadly as follows: "Any garbage, refuse, sludge from a waste treatment plant, water supply treatment plant, or air pollution control facility and other discarded material, including solid, liquid, semisolid, or contained gaseous material resulting from industrial, commercial, mining, and agricultural operations, and from community activities, but does not include solid or dissolved material in domestic sewage, or solid or dissolved materials in irrigation return flows or industrial discharges which are point sources subject to permits under section 402 of the Federal Water Pollution Control Act, as amended (86 Stat. 880), or source, special nuclear, or byproduct material as defined by the Atomic Energy Act of 1954, as amended (68 Stat. 923)."]

The act requires or authorizes a number of activities directed toward achieving these objectives: Federal regulations and guidelines; financial and technical assistance to state and local governments; research, demonstrations, and studies; and public participation and education.

Hazardous waste management. RCRA requires creation of a regulatory system for hazardous wastes to prevent serious risk of injury to health or the environment from the mismanagement of such wastes. The key provisions are directed to identification of hazardous wastes, institution of a manifest system to track wastes through their life cycle (Fig. 3), and establishment of a permit system for hazardous waste treatment, storage, and disposal facilities. Seven sets of standards and regulations that will form the basis of the system were due to go into effect in 1980.

Federal aid ($15,000,000 in 1979) is provided to states for development of programs that will qualify for EPA authorization. EPA must administer the regulatory program in states that do not seek or qualify for authorization.

To deal with the problem of existing hazardous waste storage and disposal sites that pose substantial danger, EPA and the Justice Department, in cooperation with the states, began in 1978 a program of investigations and legal actions. This pro-

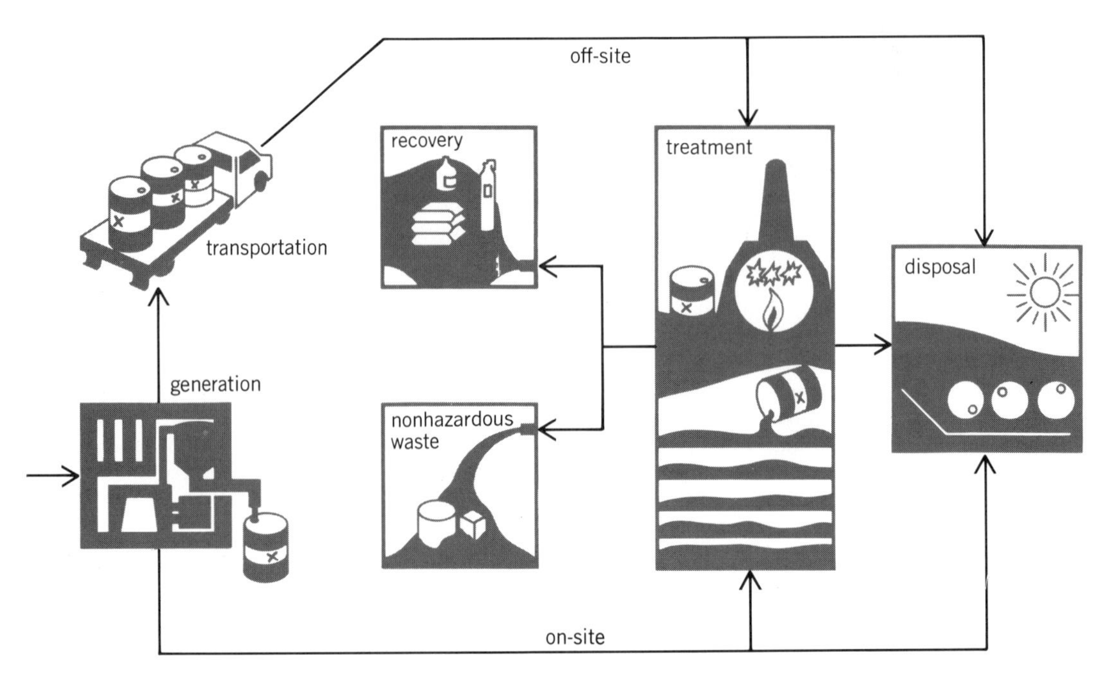

Fig. 3. Hazardous waste control, cradle to grave.

gram is viewed as an initial phase of a comprehensive plan to deal with problem sites and hazardous spills. The Administration proposed legislation in June 1979 that would establish a system of notification, emergency response, enforcement, liability, and limited economic compensation (for spills only), with the financial responsibility shared by Federal, state, and local governments as well as industry.

Land disposal of other solid wastes. Under RCRA, EPA is required to establish criteria that will define environmentally acceptable disposal of solid wastes not covered by the hazardous waste regulations. These criteria were issued in September 1979. All land disposal sites will be evaluated against the criteria by the states. Sites that fail to meet the criteria are to be closed or upgraded in accordance with compliance schedules in the state solid waste management plan. States must implement this program to continue eligibility for Federal financial aid ($15,200,000 in 1979) for their overall solid waste management programs. Federally approved state plans must provide not only for application and enforcement of environmentally sound disposal practices, but also for the identification of state, local, and regional responsibilities for solid waste management, and the encouragement of resource recovery and conservation.

Resource recovery and conservation. Federal activity in resource recovery and conservation includes financial assistance to state and local programs, technical assistance, technology development and evaluation, development of markets for recovered resources, policy studies, and guidelines. RCRA is also expected to further resource recovery in an indirect but highly important way by eliminating environmentally unacceptable land disposal as a standard option in waste management. The availability of this unrealistically low-cost alternative has heretofore limited the attractiveness of resource recovery systems. Similarly, the regulation of hazardous waste management should encourage reduction and recovery of hazardous wastes.

A financial assistance program for resource recovery projects in urban areas was organized in 1978 as part of President Jimmy Carter's Urban Policy and under authority of RCRA provisions for aid to states and localities. Aid totaling $15,000,000 was to be awarded to over 60 communities by EPA in 1979 to support planning, feasibility studies, and other preparatory steps.

The many other Federal agencies involved in resource recovery and conservation include the Department of Energy, which has primary responsibility for energy recovery technology development, and the Department of Commerce, whose role is to develop specifications for recovered materials and encourage commercialization of new uses for recovered materials.

The interagency Resource Conservation Committee was established by RCRA to study and make recommendations on Federal policies that affect material consumption and conservation. In its final report, the committee presented findings and recommendations regarding 10 existing and proposed policies: beverage container deposits, solid waste disposal charges (tax on consumer products and packaging), existing Federal tax subsidies for virgin materials, extraction taxes on virgin materials, subsidies for resource recovery, railroad freight rate discrimination against secondary materials, product regulations, deposits and bounties on durable and hazardous goods, litter taxes on consumer products, and local user fees for solid waste management services.

Supporting programs. In all the current major areas of the Federal program—hazardous waste control, improvement of land disposal practices, and resource conservation—there are supporting research and development projects, technical assistance to state and local governments provided through 10 EPA regional offices, and public information and participation activities. Particularly because public rejection of proposed disposal sites is an important obstacle to rapid improvement in disposal practices, widespread understanding of the issues and opportunities to participate in constructive solutions are considered vital.

[STEFFEN W. PLEHN]

Bibliography: T. Fields and A. W. Lindsey Office of Solid Waste, U.S. Environmental Protection Agency, *Landfill Disposal of Hazardous Wastes: A Review of Literature and Known Approaches*, 1975; OSW EPA, *Decision-Makers Guide in Solid Waste Management*, 1976; OSW, EPA, *Resource Recovery and Waste Reduction: Fourth Report to Congress*, 1977; OSW, EPA, *Solid Waste Facts*, 1978; OSW, EPA, *State Decision-Makers Guide for Hazardous Waste Management*, 1977; Resource Conservation Committee, *Choices for Conservation: Final Report to the President and the Congress*, July 1979; D. B. Sussman and S. J. Levy, OSW, EPA, *Recovering Energy from Municipal Solid Waste: A Review of Activity in the United States*, March 1979.

Sonic boom

Strong pressure waves (shock waves) generated by aircraft in supersonic flight and heard along the ground as explosivelike sounds called booms or bangs. Contrary to popular belief, a sonic boom does not occur at only one location at the instant the aircraft exceeds the local speed of sound. Instead, aircraft flying faster than the speed of sound (approximately 1100 ft/s or 330 m/s) generate shock waves that radiate away from and behind the aircraft and are dragged with it as long as it is flying supersonic. Where this trailing shock wave intercepts the ground (Fig. 1), it is heard as an impulsive type of sound. In some situations, depending on the shock-wave pattern, an observer may hear two distinct booms or a double boom rather than hear the single sound.

Some factors affecting the sonic boom signal heard on the ground, such as speed, altitude, and route of the aircraft, can be reasonably well controlled. Others, such as meteorological conditions, topography, and ground-level air turbulence, cannot be modified. Consequently, the extent to which sonic booms can be predicted and controlled for known flight profiles and flying conditions is limited by those environmental factors beyond human control. During certain flight conditions, sonic booms much greater in magnitude (superbooms) than those created in straight, level flight may be generated. Sonic booms may be magnified two to three times as the vehicle accelerates from sub-

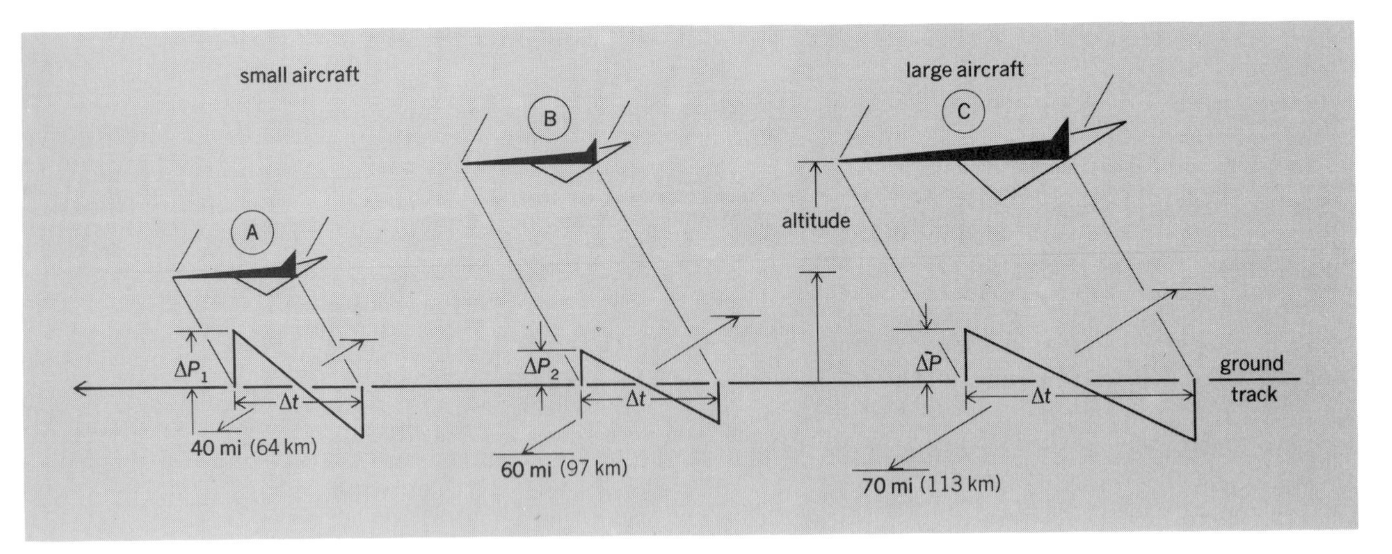

Fig. 1. Duration, intensity, and lateral spread of sonic booms as a function of aircraft size and altitude. 1 mi = 1.6 km. (*From H. E. von Gierke and C. W. Nixon, Sonic boom, McGraw-Hill Yearbook of Science and Technology of 1967*)

sonic to supersonic speeds, and as much as two to five times by various flight maneuvers at supersonic speeds.

Typical pressure-time functions or signatures for sonic booms generated by large and small aircraft in supersonic flight are shown in Fig. 1. Because of the N-like shape of the pressure signatures, they are frequently referred to as "N-waves." Aircraft A and B are identical, on the same course flying at the same speed, but differing in altitude. Aircraft B and C differ only in size as represented. The durations of the sonic boom pressure waves Δt are directly related to the length of the aircraft. The range of durations of sonic booms for current military operations and for commerical supersonic transport (SST) operations varies from about 0.05 s for a fighter aircraft to about 0.5 s for a large commercial supersonic transport. Duration varies only slightly with altitude (compare aircraft A and B in Fig. 1), being shorter and more directly related to aircraft length for lower altitudes where shock waves have less opportunity to disperse. The significant difference between Δt's of the two different-sized aircraft is easily seen.

The loudness of the sonic boom perceived by an observer is a function of, among other things, the initial rise time of the primary shock-wave signature. Signals with a steep or fast rise time have been expected and shown in actual test to sound louder than signals with the same peak overpressure and duration and slower rise times. Sonic booms with fast rise times have much more energy in the frequency bands where the ear is sensitive and, therefore, seem to give louder acoustic signals than the others. A sonic boom with a slower rise time is a more effective stimulus inside buildings. The shaking and rattling of the building and its contents due to the shock wave add to the overall acoustic stimulus caused by the boom and it may be perceived as more intense or more objectionable then a sonic boom of equal intensity heard outdoors.

The intensity of the sonic boom, that is, the magnitude of its pressure peak ΔP, and the lateral distance from the ground track at which it will be heard are dependent upon the size (lift and drag) of the aircraft and its altitude. Increasing the altitude of the aircraft reduces the magnitude of the overpressure on the ground; however, at the same time it increases the lateral spread or width of the area which is being exposed to the sonic boom. This altitude effect is shown by the two identical aircraft A and B, with the boom at the lower altitude being higher in intensity (ΔP_1 compared to ΔP_2) and with a narrower sonic boom path (40 mi or 64 km as compared to 60 mi or 97 km).

Sonic booms are measured in pounds per square foot (1 psf = 47.88 Pa) or dynes/cm^2 (1 dyne/cm^2 = 0.1 Pa) of overpressure or pressure above the normal atmospheric pressure, or in decibels (dB) of sound pressure level referenced to 20 μPa (0.00002 Pa). Peak pressure level of the impulsive sonic boom in decibels should not be confused with continuous noise levels in decibels. The intensity of sonic booms typically experienced on the ground from aircraft above 30,000 to 40,000 ft (9 to 12 km) is seldom above 2.0 psf (96 Pa) or about 134 dB, and rarely as high as 5.0 psf (239 Pa) or about 142 dB. The maximum sonic booms ever experienced by humans of 120–144 psf (5.75–6.9 kPa) or about 170–171 dB produced by aircraft flying at 50–100 ft (15–30 m) above the ground are about 100 times greater than the usual community sonic boom exposures.

Aircraft altitude is the primary contributor to the magnitude of the sonic boom and is one of the factors most accessible to control measures. Current U.S. Air Force aircraft are restricted from supersonic flight over inhabited areas below an altitude of 30,000 ft (9 km) except for special missions or in the event of emergency.

Maximum nominal overpressures during transoceanic flight of commercial SST-type aircraft are about 2.5 psf (120 Pa) or 136 dB during acceleration, and 1.7 psf (81 Pa) or 133 dB during cruise and descent phases. Minimum altitudes of 50,000–

60,000 ft (15–18 km) during cruise produce slightly lower booms of about 1.5 psf (72 Pa) or 132 dB as fuel is consumed and the weight of the aircraft is decreased. It has been determined that these levels of sonic boom are unacceptable to the general population for frequent operations, so that flights over land are not currently permitted.

It appears that there is no possibility to significantly reduce or eliminate the sonic boom from current and next-generation aircraft which fly at supersonic speeds. Longer booms from the larger aircraft may be psychoacoustically no less acceptable assuming a constant magnitude ΔP. No major breakthrough in the reduction of sonic booms appears on the horizon. The most promising approach to minimizing boom exposure is that of regulating flights through operational control. Commercial supersonic transportation requires regulation in a manner that will neither annoy or disturb the general population nor financially penalize the SST to an unbearable extent. Judicious scheduling of supersonic flights, care in acceleration and maneuver, cruise at high altitudes, and avoidance of population centers are major factors of practical significance. The operational controls and flight regulations necessary to ensure acceptable levels of sonic booms on the ground have not been achieved for present-day aircraft flying over land. Commercial transports are prohibited by law in the United States from supersonic operations over land or over water where the sonic booms may be propagated to impact on land.

Human responses. The complex problem of describing man's reactions to sonic booms is represented in Fig. 2. The occurrence of a sonic boom can affect the environment in many ways. The two major situations experienced by residents in a community are the outdoor and the indoor sonic boom exposures. The outdoor exposure is nominally a clean signal or N-wave with fast initial rise time. The indoor sonic boom exposure is quite different because it is affected by the transmission properties of the building, that is, the cavity resonance and absorption properties of the room and mechanical excitation of the main building modes. Indoor booms are lower in magnitude, longer in duration, have slower rise times, and visibly vibrate objects within the building which in turn generate noises. Indoor sonic boom exposures, with the added visual, vibratory, and acoustic cues, are generally less acceptable than those out of doors.

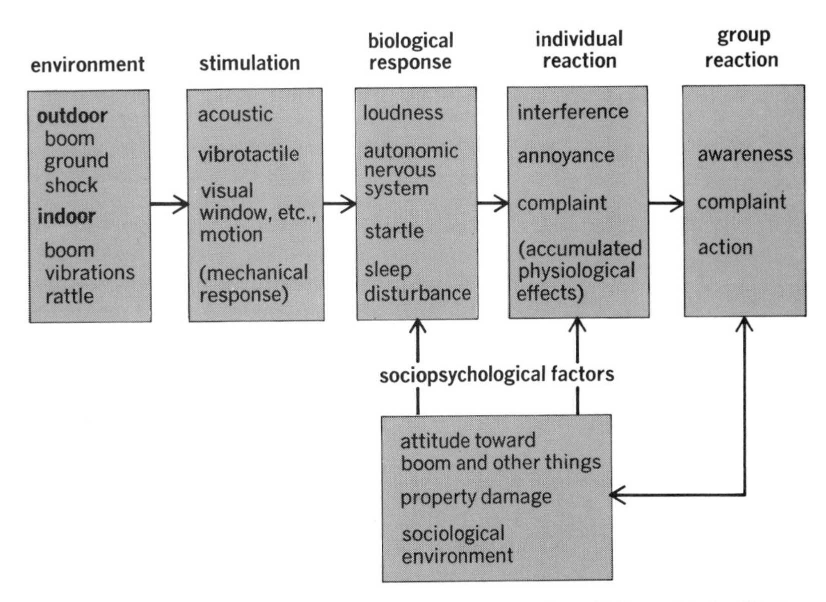

Fig. 2. Individual and group reactions to sonic booms. *(From H. E. von Gierke, Effects of sonic boom on people: Review and outlook, J. Acoust. Soc. Amer., vol. 39, no. 5, pt. 2, 1966)*

The auditory component of the booms which results in loudness sensations is the primary stimulus affecting the individual. When people are not used to sonic boom exposure, they express concern for this sudden loud sound which can visibly shake and rattle buildings and their contents, and they may worry about its possible harmful effects. Close analysis of considerable experience from human and animal exposures to blast waves of varying duration (which are similar to sonic boom

Table 1. Estimates and observations of effects of sonic boom exposures of various peak overpressures*

Peak overpressure			Predicted or measured effects	
lb/ft²	dynes/cm²	Pa		
0–1	0–478	0–47.8	Sonic booms from normal operational altitudes: typical community exposures (seldom above 2 lb/ft²)†	No damage to ground structures; no significant public reaction day or night
1.0–1.5	478–717	47.8–71.7		Very rare minor damage to ground structures; significant public reaction
1.5–2.0	717–957	71.7–95.7		Rare minor damage to ground structures; significant public reaction particularly at night
2.0–5.0	957–2393	95.7–239.3		Incipient damage to structures
20–144	9.57×10^3 to 6.8×10^4	9.57×10^2 to 6.8×10^3	Measured sonic booms from aircraft flying at supersonic speeds at minimum altitude; experienced by humans without injury	
720	3.44×10^5	3.44×10^4	Estimated threshold for eardrum rupture (maximum overpressure)	
2160	1.033×10^6	1.033×10^5	Estimated threshold for lung damage (maximum overpressure)	

*From H. E. von Gierke, Effects of sonic boom on people: Review and outlook, *J. Acoust. Soc. Amer.*, vol. 39, no. 5, pt. 2, 1966. †2 lb/ft² = 95.78 Pa.

waves but much more intense) as well as to rather intense sonic booms leaves little doubt that present-day sonic booms and those expected in the future do no direct physiological harm to humans. A widespread margin of safety exists between what appears to be a loud boom in a community and corresponding pressures required to do any possible damage.

This fact is demonstrated by the data in Table 1. Data on mechanical response phenomena leading to pathological damage such as eardrum rupture and lung damage are taken from blast literature. These pressures are 5 times higher for eardrum rupture and 15 times higher for lung damage than the highest booms ever observed in special test flights from aircraft at minimum altitudes and many hundred times higher than the booms observed in communities from typical supersonic flying.

If the auditory system, which is extremely sensitive to changes in pressure, is not injured or harmed by sonic booms, the rest of the human organism may be expected and has been proven safe for such exposures. Specific auditory observations made during overflight programs, essentially all negative findings, are given in Table 2. As indicated, no temporary or permanent hearing loss has been observed following boom exposures as intense as 120 psf (5.75 kPa) or about 170 dB, and as frequent as 30 booms per day for 30 consecutive days at levels above 5 psf (239 Pa) or 142 dB. The probability of a measurable temporary hearing loss from exposures to booms at 2-5 psf (96-239 Pa) or 134-142 dB is nearly zero. It has been estimated that it would take 75 booms within 3 min repeated daily for many years to produce a modest permanent hearing loss. Consequently, both theoretical and empirical data demonstrate that pressure variations thousands of times greater than those experienced in communities are necessary to even approach the threshold of damage to the human ear. The margin of safety is obvious.

Table 2. Observed and predicted auditory responses to sonic booms*

<table>
<tr><th>Nature of auditory response</th><th colspan="2">Sonic boom experience or prediction†</th></tr>
<tr><td>Rupture of the tympanic membrane</td><td colspan="2">None expected below 720 psf; none observed up to 144 psf</td></tr>
<tr><td>Aural pain</td><td colspan="2">None observed up to 144 psf</td></tr>
<tr><td>Short temporary fullness, tinnitus</td><td colspan="2">Reported above 95 psf</td></tr>
<tr><td>Hearing loss: permanent</td><td colspan="2">None expected from frequency and intensity of boom occurrence</td></tr>
<tr><td>Hearing loss: temporary</td><td colspan="2">None measured:
1. 3–4 hr after exposure up to 120 psf
2. Immediately after boom up to 30 psf</td></tr>
<tr><td>Stapedectomy</td><td>No ill effects reported</td><td rowspan="2">After booms up to 3.5 psf</td></tr>
<tr><td>Hearing aids</td><td>No ill effects reported</td></tr>
</table>

*From C. W. Nixon, *Proceedings of Noise as a Public Health Hazard*, ASHA Rep. no. 4, February 1969.

†1 psf = 47.88 Pa.

Startle is another psychophysiological response resulting from sudden, loud sonic boom exposures. The possibility of completely eliminating the startle response to occasional sonic booms is not promising. The consistency of this reaction from one person to another suggests that it is an inborn reaction that is modified very little by learning and experience. Although it is hoped that such reactions may be greatly minimized in situations where sonic booms occur with scheduled regularity, the extent to which such adaptive behavior would occur has not been determined. When an adequate startle stimulus occurs a startle reaction may be expected. It is noted that these same startle reactions also occur in response to sharp, loud noises resulting from slamming doors, thunder, dropping of heavy objects, or even automotive backfires.

It is now clear that humans need not be concerned about direct physiological harm from sonic booms. Nevertheless, it must also be acknowledged that the possibility of indirect harm does exist; for example, a person may be struck by a falling object shaken off a shelf by a strong sonic boom. These indirect effects, although rarely observed, cannot be totally eliminated as factors. In addition, the fact that repeated exposure to sonic booms, particularly at night, would constitute another increment to the overall environmental stress of modern living, such as neighborhood and traffic noise and the five o'clock rush hour, must be recognized.

Individual reaction to sonic booms, which in turn is the basis for the reaction of larger groups, has been studied in depth during community overflight studies in St. Louis, MO, in 1961–1962; in Oklahoma City, OK, in 1964; and at Edwards Air Force Base, CA, in 1966–1967. During the program in Oklahoma City a total of 1253 sonic booms were generated over the area at a rate of eight per day during a 6-month period. The intensity of the booms was scheduled for 1.5 psf (72 Pa) or 132 dB during most of the study and 2.0 psf (96 Pa) or 134 dB during the latter stage. Almost 3000 adults were personally interviewed three times during the 6-month period to determine their reactions to the sonic booms.

Substantial numbers of residents reported interferences with ordinary living activities, such as house rattles and vibrations, startle, and interruption of sleep and rest, and annoyance with these interruptions. Although the majority felt they could learn to live with the numbers and kinds of booms experienced, at the end of the study 27% felt they could not accept the booms.

Annoyance is a key factor in human acceptance of sonic booms. Individuals with feelings of annoyance are more prone to complain and to take some action against the sonic booms than are others. The belief that sonic booms have damaged personal property and homes (see response of structures below) contributes strongly to feelings of annoyance. Individuals who ordinarily would not complain about booms do so freely when they believe their personal property is damaged. Clearly, any commercial supersonic transport, to be acceptable to the public, must generate sonic booms that are at levels below the thresholds of damage to personal property and homes as well as the threshold of annoyance, that is, levels which do not interfere with normal living activities.

Answers to many questions in the physiological, psychoacoustic, behavioral, and structures-response areas require a description of the physical stimulus that correlates with and allows predictions of these various responses. A scale was developed in the laboratory which expresses the noisiness or acceptability of sound exposures, and this scale serves as the basis of various criteria and guidelines currently in use for estimating community reaction to aircraft. The level of the noisiness or the perceived noise level (PNL) may be calculated from physical measures of the sound and expressed in units of PN dB. Appropriate psychoacoustical studies have demonstrated, and experience has confirmed, the ranges of acceptability which correspond to the various values along the PN dB scale. Sonic boom exposures have been empirically equated with this well-established noise acceptability scale (PN dB), as demonstrated in Fig. 3. With this procedure a sonic boom of 2.0 psf (96 Pa) is judged equivalent in noisiness or acceptability to an aircraft noise of about 120 PN dB. The PNL concept appeared to be a more appropriate means of representing acceptability of sonic booms than the physical unit of pounds per square foot, which was customarily employed. However, a number of advances have occurred in this technical area to cause this acceptability concept (PN dB) to become less viable.

Recent developments in instrumentation have produced small, portable measurement units with overall response characteristics and meter ballistics of sufficient speed to accurately capture the impulsive sonic boom stimulus. These instruments employ the decibel as the physical descriptor instead of pounds per square foot or PNL, and are fully adequate for describing the required peak sound pressure level. This is a highly significant advancement in sonic boom measurement and monitoring because it eliminates the need for complex oscillographic equipment required to define the pressure-time signature of the sonic boom for some purposes.

In an effort to provide a universal descriptor for all high-energy impulse noises, which include the sonic boom, a new concept has been implemented. The new descriptor appropriate for use with sonic booms is the C-weighted sound exposure level (CSEL) which is defined for use with impulsive noises that exceed 85 dB in the daytime and 75 dB at night. In general, the CSEL will be numerically about 20-26 dB less than the peak sound pressure level of the impluse.

Many preliminary answers to practical questions are now available, and more will be obtained in the years ahead by ongoing research. However, an understanding of the overall long-term human response to stimuli of this type and potential health hazards will require long and continuing efforts, and may be realized only if SSTs ever fly regularly over populated land, which they do not do at the present time.

Response of structures. Supersonic flights over populated areas must generate sonic booms at levels that will not damage structures and personal property. Sonic booms above this level will not be

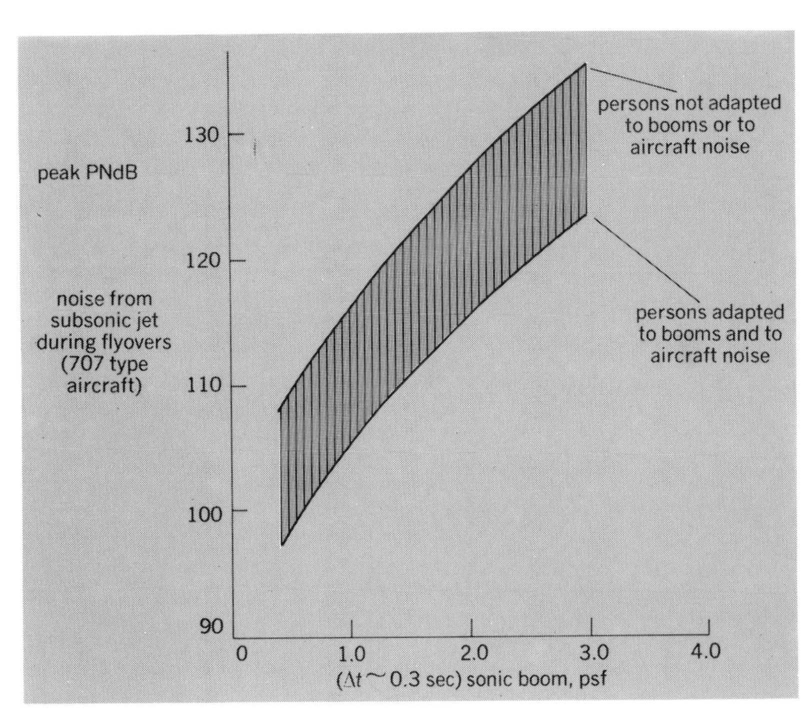

Fig. 3. Sonic boom ($\Delta t \cong 0.3$ s) peak overpressures (psf; 1 psf = 47.88 Pa) judged equal in noisiness or acceptability (PNdB) to noise from subsonic jet aircraft (Boeing 707) during flyovers at altitudes of approximately 400 ft (120 m). (*Adapted from K. Kryter, P. J. Johnson, and J. R. Young, Psychological Experimentation on Noise from Subsonic Aircraft and Sonic Booms at Edwards Air Force Base, Contract no. AF 49(638)-1758, Final Report, NSBEO-4-67, SRI International, 1967*)

accepted by the general public. Window glass and other minor damage to isolated structures may appear as a result of sonic boom exposures and such damage can be an important factor in community response. The types of damage to property most often reported (Fig. 4) relate to secondary or decorative elements and consist of cracks in brittle surface treatments, such as plaster, tile, window glass, and masonry. Such damage is noted to be superficial in nature, is restricted to non-load-bearing members, and thus does not affect the strength of the primary structure. Further, it is judged that the superficial damage usually reported is, in large measure, associated with stress concentrations in the structure.

Stress concentrations in buildings may be due to such factors as curing of green lumber, dehydration of cementious materials, settling of foundations, and poor workmanship. Such factors exist in varying degrees in all structures and can contribute to failures when a triggering load is applied. The overpressure of a sonic boom has this triggering action capability as do vehicle traffic, thunder, windstorms, heavy falling objects, and even many routine household operations. Well-constructed buildings in good repair would not be expected to experience serious damage. Superficial damage would not be expected either, except in situations where critical load concentrations existed.

Concorde. The only commercial supersonic transport is the Concorde, an aircraft developed and constructed jointly by France and the United Kingdom. Concorde operations began into Dulles International Airport, Washington, DC, in May 1976 and into Kennedy International Airport, New York, in November 1977. Each of these times rep-

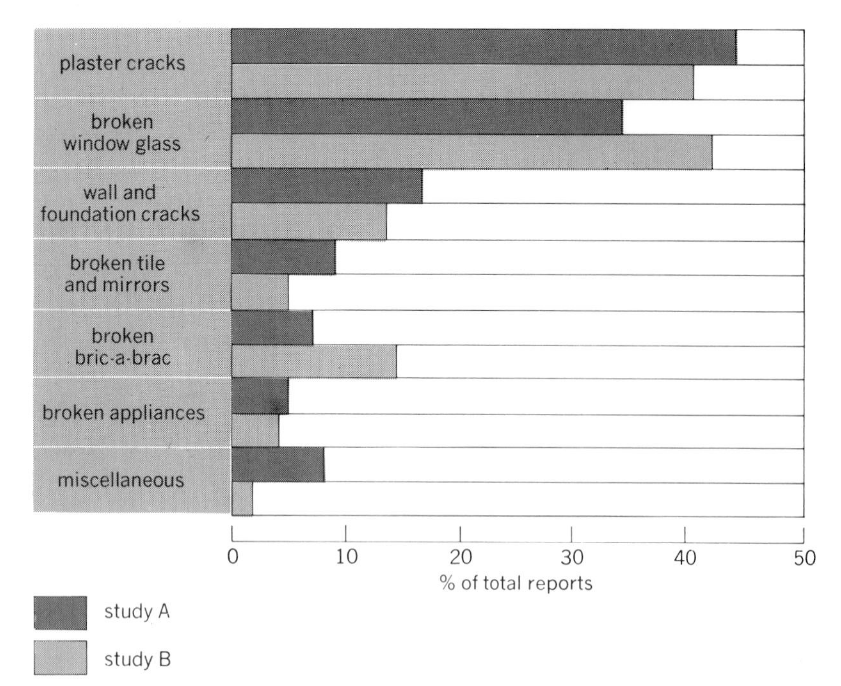

Fig. 4. Types of damage to property reported during two different community overflight programs. *(From C. W. Nixon and P. Borsky, Effects of sonic boom on people: St. Louis, Missouri, 1961–1962, J. Acoust. Soc. Amer., vol. 39, no. 5, pt. 2, 1966)*

resents the beginning of a 12-month demonstration period at the respective airports, during which flight operations were closely monitored to assess all aspects of the environmental impact of the Concorde. The U.S. Secretary of Transportation authorized the demonstration periods for the Concorde, whose takeoff and landing noise levels exceed those permitted at United States airports, and directed that noise and other emissions be monitored to provide sufficient data to allow a realistic assessment of the environmental impact.

During the Dulles International Airport demonstration period, one sonic boom was recorded with no accompanying complaints from residents in the affected area. Subsequent flight operations required the aircraft to slow to subsonic speeds at a sufficient distance from the coast to ensure that a sonic boom would not reach United States territory. During the Kennedy International Airport demonstration period, no direct sonic booms were recorded. However, numerous reports of barely audible booms were received by Federal authorities from residents of eastern New England. Low-amplitude secondary pressure disturbances were measured at times along the south shore of Long Island at a level of about 0.8 psf (38 Pa) or 126 dB. Concorde operations into Dulles and Kennedy airports continued beyond the demonstration periods. Flight controls imposed by the Federal government on the operation of the Concorde into these airports essentially eliminated its threat of sonic boom exposure to the United States mainland.

Conclusions. On the basis of an overview of the pool of available knowledge and experience with sonic boom phenomena, current-generation supersonic transport aircraft clearly create sonic booms that are not acceptable to the general public. Contemporary factors such as mobility in the community, technological advances, political climate, national and international affairs, as well as many other things, motivate beliefs, attitudes, and opinions that influence community response at any one time period. Whether technology will develop a supersonic vehicle that does not produce a sonic boom on the ground or whether the level of acceptability of sonic booms by the population will change in future generations cannot be predicted.

[HENNING E. VON GIERKE; CHARLES W. NIXON]

Bibliography: H. H. Hubbard, Sonic booms, *Phys. Today*, February 1968; C. W. Nixon, Human response to sonic boom, *J. Aerosp. Med.*, 36:399, 1965; National Academy of Sciences, *Guidelines for Preparing Environmental Impact Statements on Noise*, 1977; NSBEO, *Sonic Boom Experiments at Edwards Air Force Base*, AD no. 655310, July 27, 1967; L. J. Runyan and E. J. Kane, *Sonic Boom Literature Survey*, DOT Rep. FAA-RD-73-129-1, September 1973; Sonic Boom Symposium, *J. Acoust. Soc. Amer.*, vol. 51, no. 2, pt. 3, 1972, and vol. 39, no. 5, pt. 2, 1966; H. A. Wilson, Sonic *Sci. Amer.* vol. 206, January 1962.

Species diversity

The taxonomic variability of an ecological community. A community with many species is more diverse than one with few; also, for a given number of species, a community in which the species are equally (evenly) represented has higher diversity than one in which the species abundances are very unequal (uneven).

Measurement of diversity. In the same way that dispersion denotes the variability of a set of observations that vary quantitatively (such as weights, lengths, ages, or numbers of eggs), diversity denotes the variability of a set of objects that vary qualitatively. In the case of species diversity, the variability is taxonomic.

The crudest measure of diversity is s, the number of species. Several measures that take account of the evenness of the species abundances, as well as of s, have been devised. The best known are the information indexes and the Simpson indexes. There are two forms of each; they are applicable, respectively, to fully censused collections and to "large" (conceptually infinite) communities whose diversities must be estimated from samples, with inevitable sampling errors. The formulas are:

1. Information measure for a censused collection [the Brillouin index; Eq. (1)]:

$$H = (1/N)\,(\log N! - \sum_j \log N_j!) \qquad (1)$$

2. Information measure for a large community [the Shannon index; Eq. (2)]:

$$H' = -\sum_j p_j \log p_j \qquad (2)$$

3. Simpson index for a fully censused collection [Eq. (3)]:

$$D = -\log \sum_j N_j(N_j - 1)/N(N-1) \qquad (3)$$

4. Simpson index for a large community [Eq. (4)]:

$$D' = -\log \sum_j p_j^2 \qquad (4)$$

N_j is the number of members of species j in a collection of $\Sigma_j N_j = N$ individuals with $j = 1, \ldots, s$; p_j is the proportion of a large community that belongs to species j; $\Sigma_j p_j = 1$.

These indexes do not keep separate the two components of diversity, namely the number of species, s, and the evenness of their abundances. Evenness can be measured by itself in various ways. Three much-used indexes, all based on information measures, are J [for fully censused collections; Eq. (5)], and J' and s' [for large communities; Eqs. (6)].

$$J = \frac{H(\text{observed}) - H(\text{min})}{H(\text{max}) - H(\text{min})} \quad (5)$$

$$J' = H'(\text{estimated})/\log s \quad (6a)$$

$$s' = \exp\,[H'(\text{estimated})] \quad (6b)$$

H(max) and H(min) are the maximum and minimum possible values of the Brillouin index given the observed N and s; s' is the number of species that would be found in a perfectly even community with the estimated H'.

The units of diversity depend on the bases of the logs used to compute H, H', D, and D'. With logs to base e, the units are nats; with logs to base 10 or 2, they are decits or bits. J and J' are dimensionless ratios.

There are several unsolved statistical difficulties, often ignored by fieldworkers, in estimating H' and J' and their standard errors for large, incompletely known communities.

For a diversity measure to be informative, it is essential that the community it relates to be precisely defined. Its spatial and taxonomic boundaries must be exactly specified. Diversity indexes are not comparable with one another unless they relate to communities having similar spatial extent and composed of taxocenes of comparable inclusiveness. A taxocene is the set of organisms living together and forming the community under study—for example, all the trees in a tract of forest; all the invertebrates on a stretch of shore; all the breeding birds in a marsh; all the insects in a light trap; or all the diatoms in a sediment sample.

Alpha and gamma diversity. It is important to distinguish between the diversities of communities in the narrow sense (alpha diversities) and of regions (gamma diversities). An alpha diversity is a characteristic of a single, integrated community, that is, a taxocene of interacting individuals occupying a continuous tract in a single habitat type. A gamma diversity is a characteristic of a taxocene (or the whole biota) in an extensive region, in which many habitat types may be represented.

Factors controlling alpha diversities. Different communities differ greatly in their alpha diversities. A great many possible causes for these differences have been suggested. A community of high alpha diversity contains many coexisting species populations. These share, and presumably compete for, the same or similar resources. High alpha diversity therefore implies a high degree of "species packing"; equivalently, it implies that species with narrow ecological amplitudes have not been excluded because of the competitive success of a few wide-amplitude species. Explanatory theories can aim at two different levels: they may explain why numerous species have evolved despite competition; or, taking for granted that the species have evolved, they may explain how they can successfully coexist. Factors that are believed to lead to high alpha diversity include: (1) Marked environmental constancy, especially weak or absent seasonality, which permits species to be narrow specialists in their environmental requirements. (2) Great biotic productivity, which ensures that species populations will grow large and have large gene pools, the raw material for speciation. (3) The subjection of a taxocene of low trophic level to intense predation; this fosters high diversity by ensuring that population sizes are kept low and hence that interspecific competition remains mild. (4) Heterogeneity of the habitat on a small, intracommunity scale; patchiness of the inorganic substrate promotes plant diversity; in turn, a diverse plant community favors high diversity in the animal communities that depend on it for food and shelter.

Factors controlling gamma diversities. The gamma diversity of a whole region depends on the alpha diversities of its component communities. Thus, factors such as low seasonality and high productivity cause both alpha and gamma diversities to be high. Also, the more variable a region's topography, the larger the number of distinct communities, each with its own species complement, that will be included in it, and hence the larger the region's gamma diversity. Further, the biota of a given area of uniform habitat tends to be more diverse if the area is part of a large continent than if it is part of an isolated oceanic island (this accords with the theory of island biogeography of R. H. MacArthur and E. O. Wilson).

The diversity of a regional biota is also the outcome of historical causes. For example, the diversities of the littoral, shelf, and deep-sea biotas of the North Atlantic are much less than those of the North Pacific. The probable reason is that during the Pleistocene glaciations, Atlantic waters (which are continuous with those of the Arctic Ocean) underwent great temperature fluctuations, which caused many extinctions. In contrast, the Pacific was protected from much of the cooling caused by the glaciations; when continental ice sheets formed and sea level fell, the Bering land bridge emerged and formed a barrier which prevented the flow of cold waters from the Arctic Ocean into the Pacific.

Latitudinal gradient in gamma diversities. On a worldwide scale, gamma diversities exhibit a clear trend. Notwithstanding small-scale exceptions, they tend to increase with decreasing latitude, and attain their greatest values in the wet tropics. This is probably due to a combination of modern and historical causes.

Modern causes: Among terrestrial communities, productivity is greatest and seasonality least in tropical rainforest. Moreover, the long growing seasons of low latitudes permit more trophic levels to become established.

Historical causes: The rate of evolution of new species is believed to be greater in warm than in cool climates. If extinction rates are independent of latitude, therefore, and if the number of species in a region is the resultant of evolutions and extinctions, then more species are to be expected at low latitudes than at high. Given this mechanism, the average geological age of taxa should be less at low than at high latitudes; there is some evidence (for example, in the benthic foraminifera and the bivalve mollusks of western North Atlantic coastal shelf communities) that this is in fact so.

The reasons why evolutionary rates should be higher in the tropics are unclear. Two necessary conditions for speciation to occur are: (1) An ancestral population must become fragmented into two or more separate demes (intrabreeding populations) among which gene flow is absent or at least reduced; this is a necessary preliminary to allopatric or parapatric speciation. (2) The demes must undergo evolutionary divergence; that is, they must become mutually intersterile and (if their status as new species is to be perceived) they must become morphologically differentiated. For differentiation to occur, a variety of separate, locally optimum phenotypes must be selected in separate localities. It is not known why these conditions are more fully met in warm climates than in cold.

Diversity changes through geologic time. It is entirely unknown whether worldwide biotic diversity has yet reached a saturation level. Perhaps saturation was achieved in the distant past, and since then species numbers have remained, approximately, at an equilibrium that has varied through time only because of changing world climate and geography. Or perhaps the species richness of the biosphere is still increasing. The problem is unsolved. It seems reasonable to suppose that the equilibrium diversity of the biosphere depends on the configuration of the Earth's land masses and on the world climate. Thus, fragmentation of the continents and a steep temperature gradient from Equator to poles would both be expected to be accompanied by great diversity in the biosphere as a whole. Conversely, in periods when the continents were fewer and larger, or when the world climate was more uniform, biotic diversity may have been much less than it is now. *See* BIOGEOGRAPHY; COMMUNITY. [E. C. PIELOU]

Bibliography: R. H. MacArthur and E. O. Wilson, *The Theory of Island Biogeography*, 1967; R. M. May, *Stability and Complexity in Model Ecosystems*, 1973; E. C. Pielou, *Biogeography*, 1979; E. C. Pielou, *Ecological Diversity*, 1975; E. C. Pielou, *Mathematical Ecology*, 1977; E. C. Pielou, *Population and Community Ecology*, 1974.

Spring

A place where a concentrated flow of groundwater reaches the land surface or discharges into a body of surface water. Springs issue where the water table intersects the land surface. Likewise, where the piezometric surface of an artesian aquifer stands above the land surface, water may issue as a spring if a suitable conduit such as a fault or a solution channel is available through which the groundwater can reach the land surface. *See* GROUNDWATER.

Springs, including some less concentrated flows called seeps, are the chief source of the dry-weather flow of most streams. They range from mere zones of seepage along river and pond banks to single or multiple orifices that discharge hundreds of millions of gallons of water per day. Springs may be classified also by such characteristics as mineral content of the water, temperature, geologic structure, and periodicity of flow.

The largest single spring in the United States is probably Silver Springs in Florida, which has an average flow of more than 700,000,000 gal/day (2,650,000 m^3/day). The water issues from limestone into a pool and then flows to the sea in a sizable river. Giant Springs in Montana issues from sandstone and has a discharge of nearly 400,000,000 gal/day (1,500,000 m^3/day). Big Spring in Missouri discharges nearly 300,000,000 gal/day (1,100,000 m^3/day) from a limestone aquifer. Comal Springs in Texas discharges a similar amount from a limestone aquifer, the water being brought to the surface by a large fault which provides a conduit. Groups of springs such as Malad Springs in Idaho discharge more than 700,000,000 gal/day (2,600,000 m^3/day), and the Thousand Springs in Idaho discharge about 550,000,000 gal/day (2,100,000 m^3/day) from lavas of the Snake River Plain.

[ALBERT N. SAYRE/RAY K. LINSLEY]

Bibliography: H. Bower, *Groundwater Hydrology*, 1978; S. N. Davis and R. J. M. DeWiest, *Hydrogeology*, 1966.

Squall line

A line of thunderstorms, near whose advancing edge squalls occur along an extensive front. The thundery region, 20–50 km wide and a few hundred to 2000 km long, moves at a typical speed of 15 m/sec (30 knots) for 6–12 hr or more and sweeps a broad area. In the United States, severe squall lines are most common in spring and early summer when northward incursions of maritime tropical air east of the Rockies interact with polar front cyclones. Ranking next to hurricanes in casualties and damage caused, squall lines also supply most of the beneficial rainfall in some regions.

A squall line may appear as a continuous wall of cloud, with forerunning sheets of dense cirrus, but severe weather is concentrated in swaths traversed by the numerous active thunderstorms. Their passage is marked by strong gusty winds, usually veering at onset, rapid temperature drop, heavy rain, thunder and lightning, and often hail and tornadoes. Turbulent convective clouds, 10–15 km high, present a severe hazard to aircraft, but may be circumnavigated with use of radar.

Formation requires an unstable air mass rich in water vapor in the lowest 1–3 km, such that air

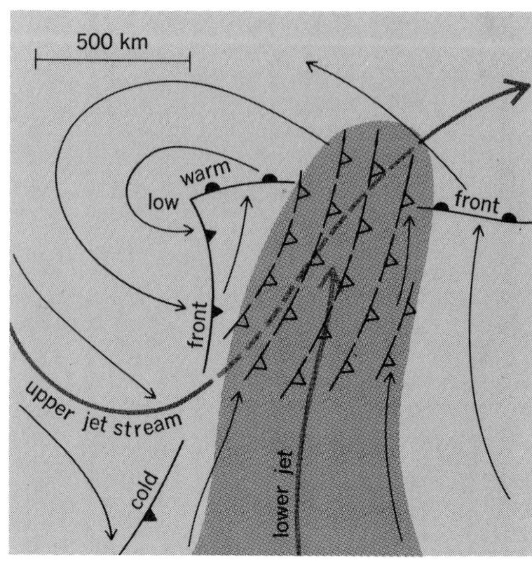

Fig. 1. Successive locations of squall line moving eastward through unstable northern portion of tongue of moist air in warm sector of a cyclone. Thin arrows show general flow in low levels; thick arrows, axes of strongest wind at about 1 km aboveground and at 10–12 km.

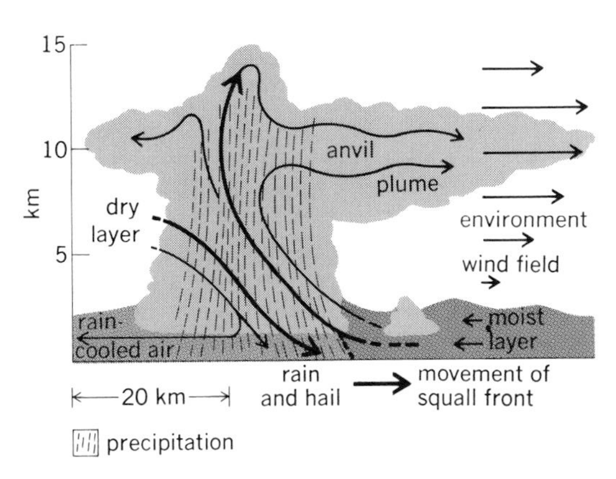

Fig. 2. Section through squall-line-type thunderstorm.

rising from this layer, with release of heat of condensation, will become appreciably warmer than the surroundings at upper levels. Broad-scale flow patterns vary; Fig. 1 typifies the most intense outbreaks. In low levels, warm moist air is carried northward from a source such as the Gulf of Mexico. This process, often combined with high-level cooling on approach of a cold upper trough, can rapidly generate an unstable air mass. *See* THUNDERSTORM.

The instability of this air mass can be released by a variety of mechanisms. In the region downstream from an upper-level trough, especially near the jet stream, there is broad-scale gentle ascent which, acting over a period of hours, may suffice; in other cases frontal lifting may set off the convection. Surface heating by insolation is an important contributory mechanism; there is a marked preference for formation in midafternoon although some squall lines form at night. By combined thermodynamical and mechanical processes, they often persist while sweeping through a tongue of unstable air (Fig. 1). Squall lines forming in midafternoon over the Plains States often arrive over the midwestern United States at night. *See* STORM.

Figure 2 shows, in a vertical section, the simplified circulation normal to the squall line. Slanting of the drafts is a result of vertical wind shear in the storm environment. Partially conserving its horizontal momentum, rising air lags the foot of the updraft on the advancing side. In the downdraft, air entering from middle levels has high forward momentum and undercuts the low-level moist layer, continuously regenerating the updraft. Buoyancy due to release of condensation heat drives the updraft, in whose core vertical speeds of 30–60 m/sec are common near tropopause level. Rain falling from the updraft partially evaporates into the downdraft branch, which enters the storm from middle levels where the air is dry, and the evaporatively chilled air sinks, to nourish an expanding layer of dense air in the lower 1–2 km that accounts for the region of higher pressure found beneath and behind squall lines. In a single squall-line thunderstorm about 20 km in diameter, 5–10 kilotons/sec of water vapor may be condensed, half being reevaporated within the storm and the remainder reaching the ground as rain or hail. [CHESTER W. NEWTON]

Bibliography: L. J. Battan, *The Thunderstorm*, 1964; C. W. Newton, Severe convective storms, in *Advances in Geophysics*, vol. 12, 1967.

Storm

An atmospheric disturbance involving perturbations of the prevailing pressure and wind fields on scales ranging from tornadoes (1 km across) to extratropical cyclones (2–3000 km across); also, the associated weather (rain storm, blizzard, and the like). Storms influence human activities in such matters as agriculture, transportation, building construction, water impoundment and flood control, and the generation, transmission, and consumption of electric energy.

The form assumed by a storm depends on the nature of its environment, especially the large-scale flow patterns and the horizontal and vertical variation of temperature; thus the storms most characteristic of a given region vary according to latitude, physiographic features, and season. This article is mainly concerned with extratropical cyclones and anticyclones, the chief disturbances over roughly half the Earth's surface. Their circulations control the embedded smaller-scale storms. Large-scale disturbances of the tropics differ fundamentally from those of extratropical latitudes. *See* HURRICANE; SQUALL LINE; THUNDERSTORM; TORNADO.

Extratropical cyclones mainly occur poleward of 30° latitude, with peak frequencies in latitudes 55–65°. Those of appreciable intensity form on or near fronts between warm and cold air masses. They tend to evolve in a regular manner, from small, wavelike perturbations (as seen on a sea-level weather map) to deep waves to occluded cyclones. *See* FRONT; METEOROLOGY.

Dynamical processes. The atmosphere is characterized by regions of horizontal convergence and divergence in which there is a net horizontal inflow or outflow of air in a given layer. Regions of appreciable convergence in the lower troposphere are always overlaid by regions of divergence in the upper troposphere. As a requirement of mass conservation and the relative incompressibility of the air, low-level convergence is associated with rising motions in the middle troposphere, and low-level divergence is associated with descending motions (subsidence).

Fields of marked divergence are associated with the wave patterns seen on an upper-level chart (for

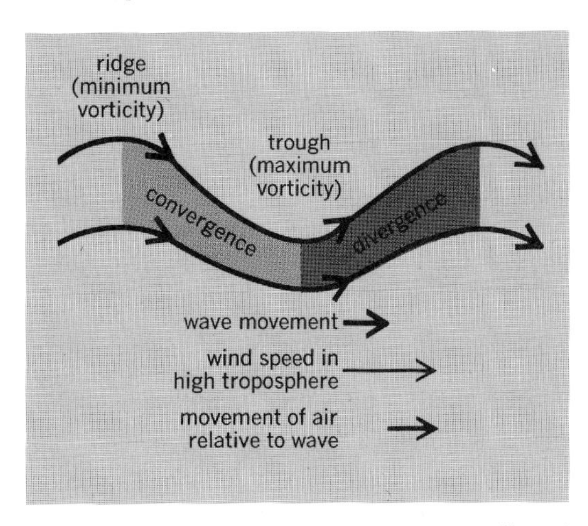

Fig. 1. Convergence and divergence in a wave pattern at a level in the upper troposphere or lower stratosphere. The lengths of arrows indicate the speed and relative motion of wind and wave.

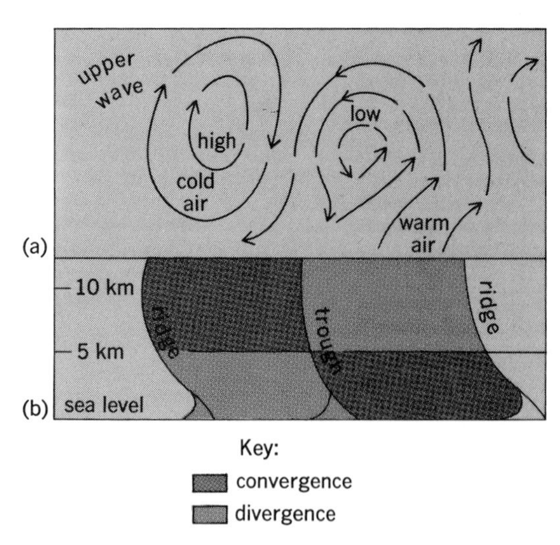

Fig. 2. Air motion associated with a wave pattern. (*a*) Circulation pattern in low levels in relation to upper wave. (*b*) West-east vertical section showing simplified regions of convergence and simplified regions of divergence. (*Adapted from J. Bjerknes*)

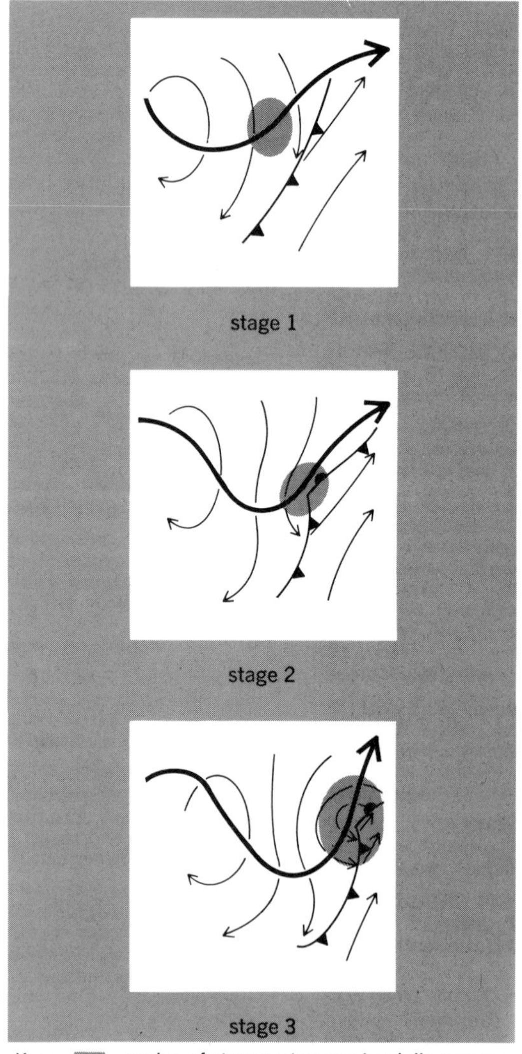

Fig. 3. Upper-level trough advancing relative to surface front and initiating cyclone. (*After S. Petterssen*)

example, at the 300-millibar or 30-kPa level), in the manner shown in Fig. 1. The flow curvature indicates a maximum of vorticity (or cyclonic rotation about a vertical axis) in the troughs and minimum vorticity in the ridges. An air parcel moving from trough to ridge would undergo a decrease of vorticity, which by the principle of conservation of angular momentum implies horizontal divergence. In the upper troposphere, where the wind exceeds the speed of movement of the wave pattern, the divergence field is as shown in Fig. 1. To sustain a cyclone, a relative arrangement of upper and lower flow patterns as shown in Fig. 2 is ideal. This places upper divergence over the region of low-level convergence that occupies the central part and forward side of the cyclone, and upper convergence over the central and forward parts of the anticyclone where there is lower-level divergence.

The upper- and lower-level systems, although broadly linked, move relative to one another. Cyclogenesis commonly occurs when an upper-level trough advances relative to a slow-moving surface front (stages 1 and 2 in Fig. 3). The region of divergence in advance of the trough becomes superposed over the front, inducing a cyclone that develops (stage 3 in Fig. 3) in proportion to the strength of the upper divergence. In a pattern such as Fig. 1, the divergence is strongest if the waves have short lengths and large amplitudes and if the upper-tropospheric winds are strong. For the latter reason, cyclones form mainly in close proximity to the jet stream, that is, in strongly baroclinic regions where there is a large increase of wind with height. *See* JET STREAM.

Frontal storms and weather. Weather patterns in cyclones are highly variable, depending on moisture content and thermodynamic stability of air masses drawn into their circulations. Warm and occluded fronts, east of and extending into the cyclone center, are regions of gradual upgliding motions, with widespread cloud and precipitation but usually no pronounced concentration of stormy conditions. Extensive cloudiness also is often present in the warm sector.

Passage of the cold front is marked by a sudden wind shift, often with the onset of gusty conditions, with a pronounced tendency for clearing because of general subsidence behind the front. Showers may be present in the cold air if it is moist and unstable because of heating from the surface. Thunderstorms, with accompanying squalls and heavy rain, are often set off by sudden lifting of warm, moist air at or near the cold front, and these frequently move eastward into the warm sector.

Middle-latitude highs or anticyclones. Extratropical cyclones alternate with high-pressure systems or anticyclones, whose circulation is generally opposite to that of the cyclone. The circulations of highs are not so intense as in well-developed cyclones, and winds are weak near their centers. In low levels the air spirals outward from a high; descent in upper levels results in warming and drying aloft.

Anticyclones fall into two main categories, the warm "subtropical" and the cold "polar" highs. The large and deep subtropical highs, centered over the oceans in latitudes 25–40° and separating the easterly trade winds from the westerlies of middle latitudes, are highly persistent.

Cold anticyclones, forming in the source regions of polar and arctic air masses, decrease in intensity with height. Such highs may remain over the region of formation for long periods, with spurts of cold air and minor highs splitting off the main mass, behind each cyclone passing by to the south. Following passage of an intense cyclone in middle latitudes, the main body of the polar high may move southward in a major cold outbreak.

Blizzards are characterized by cold temperatures and blowing snow picked up from the ground by high winds. Blizzards are normally found in the region of a strong pressure gradient between a well-developed arctic high and an intense cyclone. True blizzards are common only in the central plains of North America and Siberia and in Antarctica.

Principal cyclone tracks. Principal tracks for all cyclones of the Northern Hemisphere are shown in Fig. 4. In middle latitudes cyclones form most frequently off the continental east coasts and east of the Rocky Mountains.

Movements of cyclones, both extratropical and tropical, are governed by the large-scale hemispheric wave patterns in the upper troposphere. The character of these waves is reflected in part by large circulation systems such as the subtropical highs, and greatest anomalies from the principal cyclone tracks occur when these highs are displaced from the mean positions shown in Fig. 4. Warm highs occasionally extend into high latitudes, blocking the eastward progression suggested by the average tracks and causing cyclones to move from north or south around the warm highs.

Over the Mediterranean, cyclones form frequently in winter but rarely in summer, when this area is occupied by an extension of the Atlantic subtropical high. Both the subtropical highs and the cyclone tracks in middle latitudes shift northward during the warmer months; on west coasts cyclones are infrequent or absent in summer south of latitudes 40–45°.

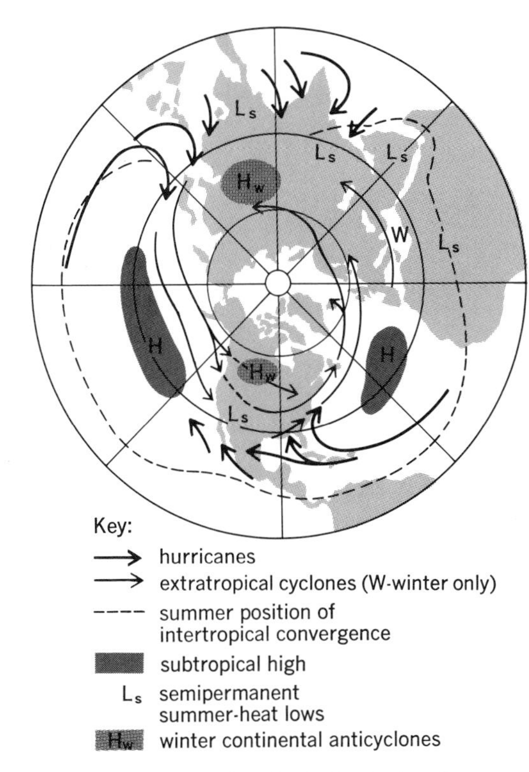

Fig. 4. Principal tracks of extratropical cyclones and hurricanes with significantly associated features in the Northern Hemisphere.

Role in terrestrial energy balance. Air moving poleward on the east side of a cyclone is warmer than air moving equatorward on the west side. Also, the poleward-moving air is usually richest in water vapor. Thus both sensible- and latent-heat transfer by disturbances contribute to balancing the net radiative loss in higher latitudes and the net radiative gain in the tropics. Air that rises in a disturbance is generally warmer than the air that sinks (Fig. 2), and latent heat is released in the ascending branches by condensation and precipitation. Hence the disturbance also transfers heat upward, as is required to balance the net radiative loss in the upper part of the atmosphere. *See* ATMOSPHERIC GENERAL CIRCULATION.

[CHESTER W. NEWTON]

Bibliography: E. Palmén and C. W. Newton, *Atmospheric Circulation Systems*, 1969; S. Petterssen, *Introduction to Meteorology*, 3d ed., 1969; H. Riehl, *Introduction to the Atmosphere*, 3d ed., 1978.

Storm detection

Microbarographs, radars, satellite-borne instruments, and sferics detectors (radio receivers) are used to detect storms and to assess their potential for destruction.

Microbarographs. Certain severe thunderstorms emit an identifiable kind of infrasound, or ultralow-frequency acoustic wave. Traveling at the speed of sound and ducted between the ground and high-temperature layers in the upper atmosphere, these waves are often so powerful that they can be detected by sensitive pressure detectors, called microbarographs, more than 1500 km from the emitting storm. At such distances, the pressure fluctuations of the waves are only about one-millionth of average atmospheric pressure. It takes 10–60 s for one wave cycle to pass, but the oscillations can last for hours (Fig. 1).

Similar storm-related waves travel into the ionospheric F region, 200 km above the earth. These waves were discovered by looking with a high-frequency radar for a certain kind of oscillation in the radar reflection height. Though there appears to be a causal connection between tornadic storms and ionospheric infrasound, there is no statistical evidence of the warning value of the ionospheric waves. The mechanism causing the emissions is not known, but observations of both ground-level and ionospheric waves have established that only a small fraction of all storms emit detectable infrasound, and that most of the emissions appear to come from tornadic storm systems.

In the United States, storm detection exercises have been carried out using direction-finding arrays of microbarographs for ground-level infrasound. Within a 14-state test area, 65% of the tornadic storms that occurred during the 1973 storm season were considered "detected" by three infrasound observatories, and the storm emissions had enough distinctive characteristics that false-alarm rates were considered acceptable.

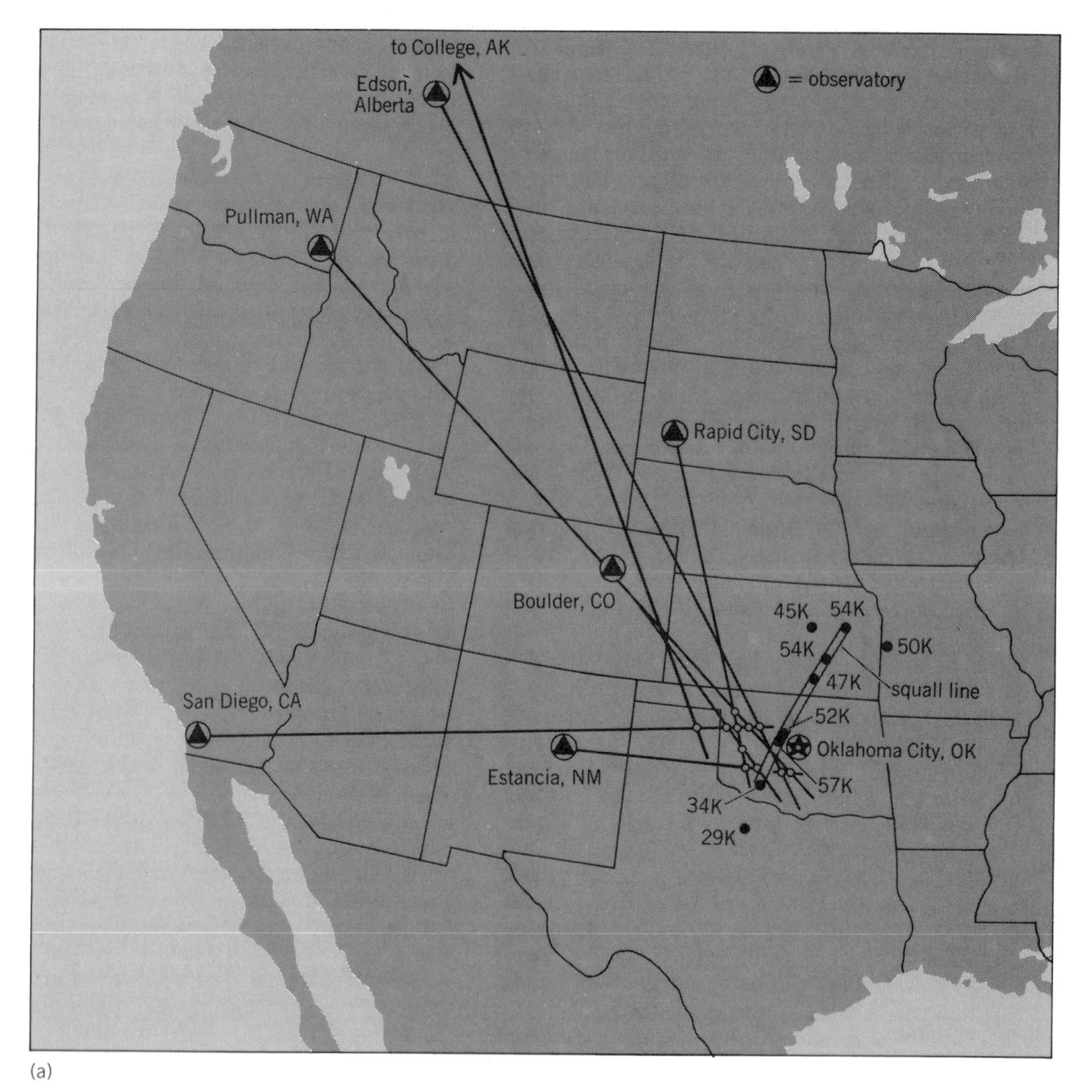

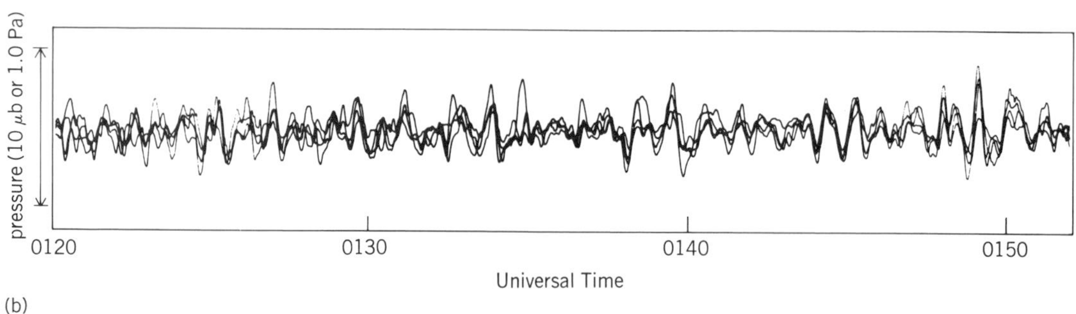

Fig. 1. Infrasound and microbarograph data during a severe-storm outbreak in Oklahoma. (*a*) Map of the western United States showing the intersecting infrasound bearings measured at seven observatories (the seventh being in College, AK). The numbered dots show radar-indicated storm cells whose tops reach the heights shown (K = 10^3 ft or 305 m). The most violent storms often occur at the southern end of a squall line. (*b*) Superimposed pressure records from four microbarographs at Boulder, CO, during this event. (*Wave Propagation Laboratory, NOAA*)

However, triangulation, when possible, is inaccurate, and even though the waves are often emitted prior to the observed tornadoes, their relatively slow sonic travel speed (1000 km/hr) diminishes their warning value.

Acoustical detection is thus not presently used for storm warning, and its main value lies in storm research.

[T. M. GEORGES]

Radar. Radars emit pulses of electromagnetic radiation in a wavelength region that penetrates storm clouds to provide a three-dimensional, inside view of the storm with angular resolution of about 1°. Advanced weather radars (Fig. 2) provide accurate images of both precipitation intensity inside a storm's shield of clouds and precipitation velocity. Velocity is provided by Doppler radar, using the same principle that produces the familiar change in pitch of a horn on a passing car or train. Precipitation reflects the radar's transmitted pulse and produces change in the microwave pitch proportional to the radial (Doppler) velocity toward or away from the radar. Precipitation reflectivity is proportional to the intensity of rain, snow, or hail.

Contoured reflectivity maps are routinely displayed by radars of the U.S. National Weather Service on widely used plan position indicator (PPI) scopes giving range and azimuth to precipitation targets whose reflectivity (intensity) is indicated by a stepped brightness scale (Fig. 3*a*). While a storm's reflectivity image is valuable for rainfall assessment and severe weather warnings, it is not a highly reliable tornado indicator. Highest reflectivity areas often signify hail. Radar warnings are primarily based on reflectivity values, on storm top heights, and sometimes on circulatory features of hook echoes seen in the patterns of reflectivity (Fig. 3*b* and *d*).

One of the earliest images of a storm's Doppler velocity field was obtained when a pulse-Doppler radar was mated to the PPI. Significant dynamic meteorological events such as tornado cyclones, whose visual sightings are often blocked by rain showers and nightfall, produce telltale signatures in the Doppler velocity field. Such a swirling vortex signature is composed of constant Doppler velocity contours (isodops) forming a symmetric couplet of closed contours of opposite sign, that is, velocity either toward or away from the radar (Fig. 3*c* and *d*). This pattern portends tornadoes, damaging winds, and hail.

The tornadic storm depicted in Fig. 3 produced a particularly large cyclone, and its reflectivity and isodop signatures are clearly seen. A signature pattern for circularly symmetric convergence of air is similar to the cyclone pattern but is rotated clockwise by 90°. The reflectivity spiral suggests some convergence, as well as rotation, a conclusion supported by the clockwise angular displacement of isodop maxima about the cyclone center with positive maximum somewhat closer to the radar (Fig. 3*c*).

Color displays of reflectivity and Doppler velocity allow easier quantitative evaluation and better resolution of signatures. Color-coded velocities in shades of blue (toward) and red (away) of increasing brightness to signify Doppler speed provide quick and reliable detection and assessment of cyclones. Doppler radar can sort from many seemingly severe storm cells the ones that have circulation and hence potential for tornado development.

Programs have been conducted to assess the improvement in severe storm advisories when Doppler velocity is given to forecasters. The probability of detection (POD) of tornadoes with Doppler radar was found to be 0.69 and the false-alarm rate (FAR) 0.25. A weather forecast office, covering the same area with then standard techniques, showed a POD of 0.47 and an FAR of 0.4. More significant is Doppler detection of tornadoes 20 min before their touch-down on the ground, whereas the warning system dependent on visual sightings generally shows a negative lead time. These findings support plans for a national network of Doppler weather radars, to replace present non-Doppler facilities by the late 1980s.

Hurricanes have a spiraling inflow of air like that shown in Fig. 3 for the severe thunderstorm. However, the scale size of a hurricane (Fig. 4) is an order of magnitude larger than a thunderstorm. Thunderstorms are most common in hurricane outskirts and again near the storm center or "eye,"

Fig. 2. Radar facilities at the National Severe Storms Laboratory in Norman, OK. The large hemispherical radome houses a Doppler weather radar. A conventional non-Doppler radar housed in the radome on the tall tower is the type (WSR-57) used by the National Weather Service. (*National Severe Storms Laboratory, NOAA*)

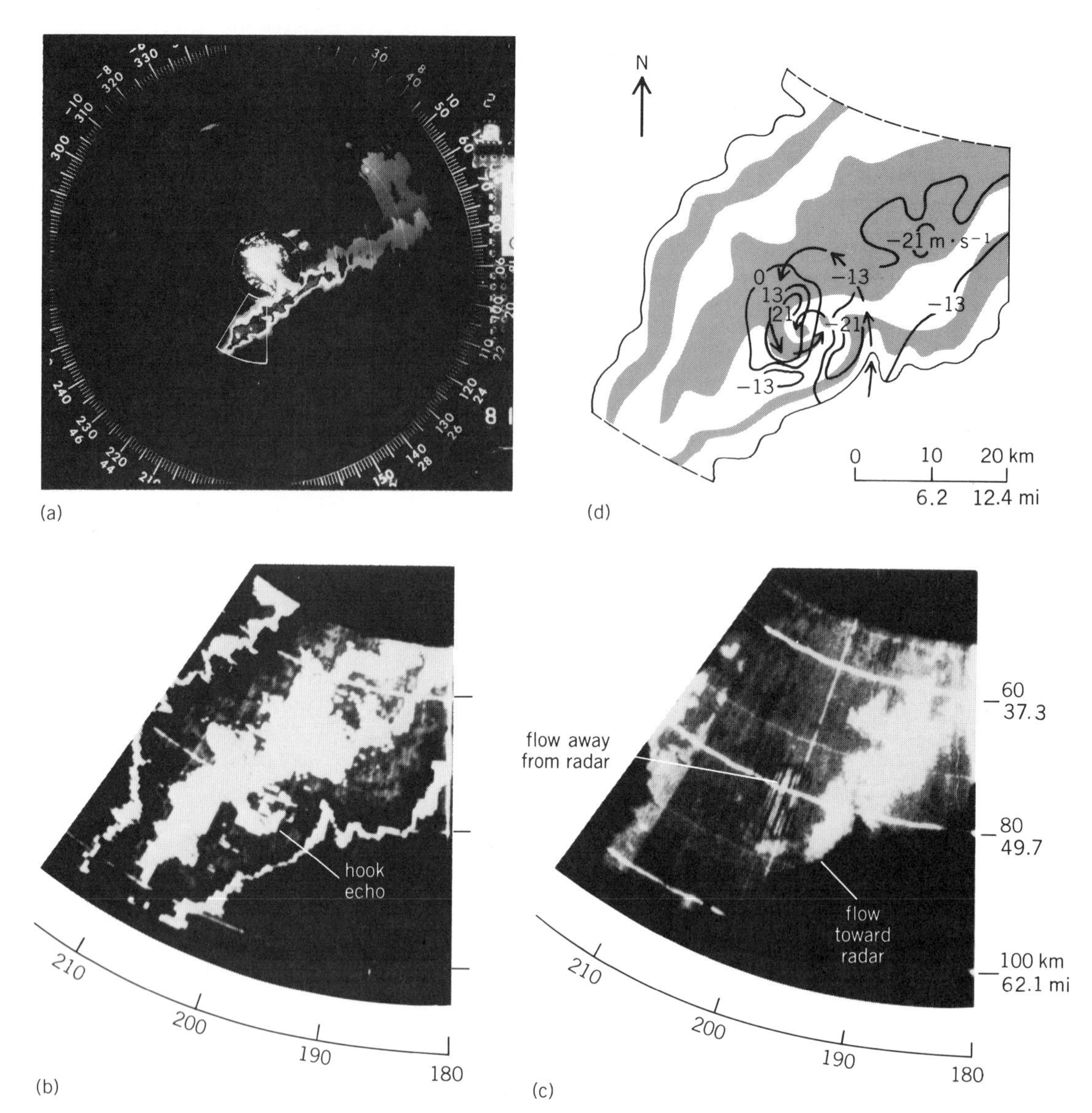

Fig. 3. Radar images of a storm. (*a*) WSR-57 PPI display of squall line showing its reflectivity (rainfall rate) categories as constant brightness. Dim brightness area surrounding the squall line corresponds to rainfall rate between .05 and .30 mm/hr. Then in sequence to the storm interior: bright, dark, dim, and bright areas are rainfall rates 0.3–1.7, 1.7–9.1, 9.1–48.8, and 48.8–262 mm/hr. Circles are range marks spaced 40 km apart. (*b, c*) Magnifications of area outlined in *a* to show Doppler radar reflectivity and Doppler velocity signatures of a storm cell. In *b*, reflectivity pattern from the Doppler radar shows a hook or spiral convergence (see arrows in *d*) that signifies the presence of a cyclone. In *c*, isodop pattern shows cyclone signature. Brightness levels are velocity categories: dim ($<13\ m \cdot s^{-1}$), bright (13–21), and brightest (>21). Positive (away) radial velocities are angularly strobed. The Doppler signature of a tornado cyclone is 193–203° and 75–90 km. (*d*) Schematic overlay of reflectivity and isodop patterns of *b* and *c*, showing the coincidence of the reflectivity spiral and cyclone signature. (*National Severe Storms Laboratory, NOAA*)

which is clearly revealed and tracked by radar when the storm is within about 200 mi (320 km) of stations which line the Gulf and Atlantic coasts of the United States. [R. J. DOVIAK]

Satellites. On Apr. 1, 1960, the first United States weather satellite *(TIROS 1)* was placed in orbit and began to send televised photographs of cloud systems back to Earth. Since then, satellite data have become more comprehensive and timely. Photographs of nearly an entire hemisphere are now disseminated to users throughout the United States within 30 min of scan time. Scans are made routinely every 30 min. Intensity and development of weather systems are easily monitored by combining sequential pictures into motion picture loops. In addition, infrared data allow display of cloud systems at night, and indications of both temperature and water vapor by both night and day.

Satellite photos reveal components of storm systems heretofore inaccessible. For example, frontal boundaries can be located precisely in most cases, since sharp cloud lines usually accompany fronts. The location and configuration of jet streams can often be identified through analysis of cirrus cloud streaks. (Much of large-scale air motion which initiates convection and precipitation is associated with the jet stream.) Low-level moisture can be tracked via low clouds moving from the Gulf of Mexico into the central plains of the United States in advance of severe thunderstorm and tornado outbreaks. Developing low-pressure systems are identifiable as soon as they generate even the smallest of cloud bands. Small but intense weather disturbances of a scale less than the resolution of the widely spaced (about 150 km) surface observing stations often produce telltale cloud patterns which are amenable to analysis by satellite meteorologists.

Two of the most destructive weather phenomena in terms of life and property are the hurricane and the severe thunderstorm. Both events are much better monitored as a consequence of satellite technology.

Hurricanes usually form from westward-moving tropical disturbances called easterly waves. It is not known exactly why this occurs, and before satellite technology a tropical disturbance, or even a hurricane, could exist for days before discovery. With weather satellites, the hurricane is spotted and tracked routinely when it is still far at sea. The configuration of the deadly storm is unmistakable (Fig. 5). What is perhaps more significant is that the storm is becoming more recognizable in its embryonic stage (the easterly wave). Meteorologists watch easterly waves carefully as the waves evolve from mild disturbances, to tropical storms, and finally into hurricanes.

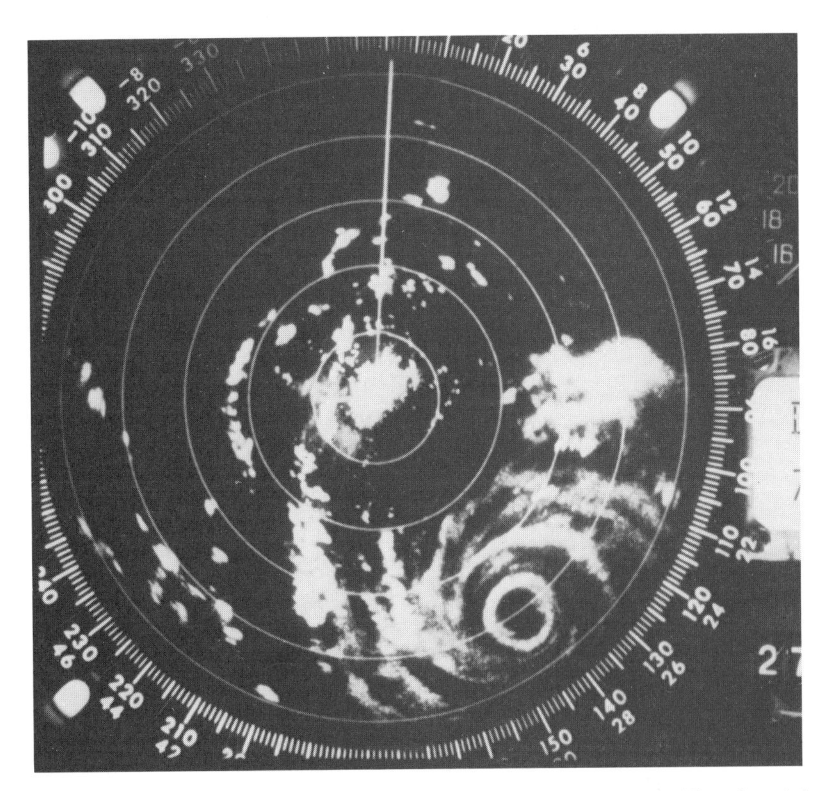

Fig. 4. Radar reflectivity image of a hurricane. (*National Hurricane and Experimental Meteorology Laboratory, NOAA*)

Fig. 5. Hurricane Belle, Aug. 8, 1976, off Florida. (*National Environmental Satellite Service, NOAA*)

Fig. 6. Satellite photo of a line of thunderstorms (arrow locates a line of individual thunderstorms extending northward into a mass of cloud cover with embedded storms) along a cold front. The much larger, "comma"-shaped cloud band which encompasses most of the photo is associated with the large low-pressure system which helped trigger these storms. (*National Environmental Satellite Service, NOAA*)

Since thunderstorms occur on a smaller time and space scale, the thunderstorm forecast problem has always involved obtaining data detailed enough to specify small zones of high storm potential. It has been known for some time, for example, that storms form along boundaries (such as cold fronts) where the convergence of two different air masses may force air to rise rapidly. But where precisely is the front at any given instant? Exactly how strong is its lifting capacity? These and other such questions are receiving at least partial answers through use of satellite photography.

Figure 6 illustrates a developing tornadic situation. Notice the distinct north-south-oriented line of thunderstorms in northern Texas. Conventional surface data indicated the existence of a nearby cold front, but it can be located only with accuracy of about 50 mi (80 km) without the supplementary satellite data. Notice also the large "comma"-shaped cloud encompassing most of the print. This identifies an intense upper-air disturbance, having the capacity to trigger thunderstorms and intensify normal storms into the severe or tornadic category. By observing these patterns, invaluable lead time is gained in gearing up public warning systems. Three hours after this photo was made, a large tornado, accompanied by 3-in.-diameter hail, struck Dallas, TX. [JOHN WEAVER]

Sferic detectors. Lightning discharges produce a wide spectrum of electromagnetic signals that are detected by radio receivers and provide effective means for locating and tracking thunderstorms. The lower frequency portions (lower than 1 MHz) of the radio spectrum are generally selected to detect cloud-to-ground and major channels of intracloud discharges, while the higher frequencies (higher than 10 MHz) are used to sense small branches and fine structure of the discharge paths.

Both the electric and the magnetic components of lightning-produced radio signals have been used to locate and track thunderstorms through triangulation methods from the directions of arrival at two or more spaced stations. Directions in both azimuth and elevation are generally obtained through the directional responses of particular antennas or by the use of spaced antennas to measure either the simultaneous phase differences of waves or time differences separating impulse arrivals. Other methods have been devised to couple the direction of arrival with measurements of radio signal waveform or spectrum to give an estimate of range to the lightning so that it can be located from a single station.

While radar and satellite technologies provide the principal bases nowadays for storm warnings around the world, simple directional radio receivers are an inexpensive and useful alternate for thunderstorm monitoring where the more advanced facilities are unavailable. Advanced facilities for sferics detection and analysis are essential, of course, for research on storm electrical processes. *See* HURRICANE; STORM; THUNDERSTORM.

[WILLIAM L. TAYLOR]

Bibliography: American Meteorological Society, *Proceedings of the 5th Conference on Severe Storms*, 1967; D. Atlas (ed.), *Severe Local Storms*, Amer. Meteorol. Soc. Monogr., vol. 5, no. 27, 1964; L. J. Battan, *Radar Observation of the Atmosphere*, 1973; W. J. Kotsch and E. T. Harding, *Heavy Weather Guide*, 1965; J. S. Marshall (ed.), *Proceedings of the 13th Conference on Radar Meteorology*, American Meteorological Society, 1968.

Strip mining

A surface method of mining by removing the material overlying the bed and loading the uncovered mineral, usually coal. It is safer than underground mining because neither the workers nor the equipment is subjected to such hazards as roof falls and explosions caused by gas or dust ignitions. Coal near the outcrop or at shallow depth can be stripped not only more cheaply but more completely than by deep mining, and the need for leaving pillars of coal to support the mine roof is eliminated. The roof over coal at shallow depth is weak and difficult to support in underground workings by conventional methods, yet this same weakness, of cover and of coal seam, makes stripping less difficult.

Stripping techniques. Power shovels, draglines, bulldozers, and other types of earth-moving equipment slice a cut through the overburden down to the coal. The cut ranges from 40 to 150 ft (12 to 46 m) wide, depending on the type and size of equipment used. The stripped overburden (spoil) is stacked in a long ridge (spoil bank) parallel with the cut and as far as possible from undisturbed overburden (high wall). The slope of a spoil bank is approximately 1.4:1 and that of a high wall under average conditions is 0.3:1. The uncovered coal (berm) is then fragmented, loaded, and transported from the pit. Spoil from each succeeding cut is stacked overlapping and parallel with the previous ridge and also fills the space left by the coal removed (Fig. 1).

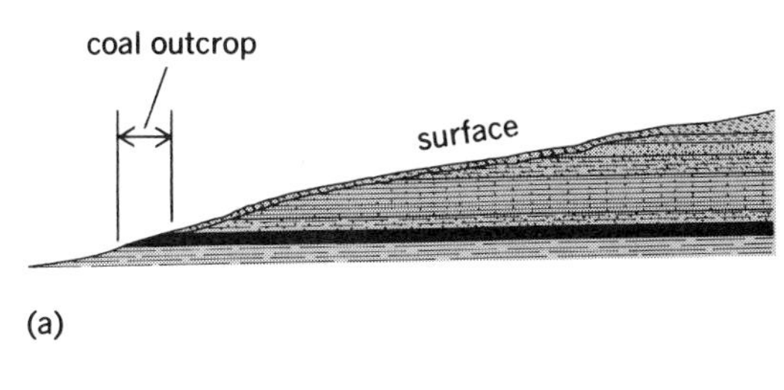

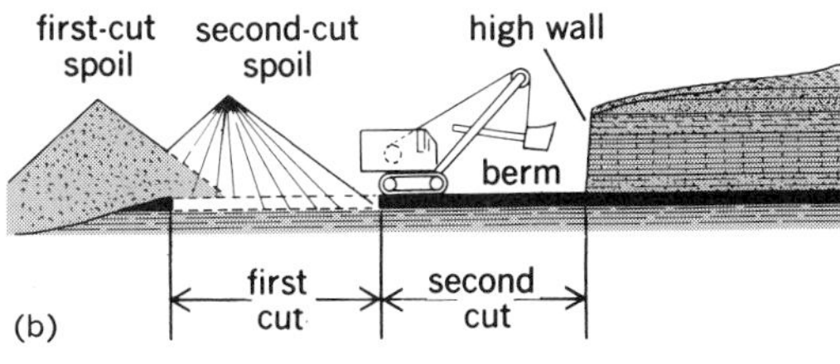

Fig. 1. Representative cross-section profile diagrams of contour strip mining of coal. (*a*) Section before stripping. (*b*) Section after second cut.

Techniques of stripping methods are similar, but the size of equipment used depends on whether the mine is in prairie or hill country. In prairie areas the thickness of overburden is nearly uniform, the coal bed is extensive, and equipment can be used for years at one mine without dismantling and moving to another location. Large-capacity shovels costing $1,000,000 or more and requiring many months to erect on the site are used at prairie mines. A unit of this type is the 60-yd^3 (46-m^3) rig shown in Fig. 2.

In 1958 the world's largest power shovel was the 70-yd^3 (54-m^3) unit at Peabody Coal Co.'s River King mine near Freeburg, Ill. By 1968 the 180-yd^3 (138-m^3) shovel at Southwestern Illinois Coal Corp.'s Captain mine near Percy, Ill., was the largest.

Most coal underlying hills is mined by underground methods, but where the working approaches the outcrop and the overburden is thin, the roof becomes difficult and expensive to support. The coal between the actual or potential underground workings and the outcrop is then more suitable for stripping. Usually, only two or three cuts 40–50 ft (12–15 m) wide can be made on the contour of the coal bed, after which the shovel has to be moved to another site. Thus, in contour stripping, mobility of shovels up to 5-yd^3 (3.8-m^3) capacity is more important than those of larger capacity.

Large draglines are used instead of shovels to strip pitching beds of anthracite to depths surpassing 400 ft; however, this use of large draglines could more properly be classed as open pit. *See* OPEN-PIT MINING.

Removal of unconsolidated overburden by hydraulic monitoring is a technique used especially in Alaska. Water under a high-pressure head is directed through a nozzle against the overburden to wash it into deep valleys where swift streams carry it away.

Although the character of the overburden determines the thickness of overburden that can be stripped, the maximum for shovels up to 5-yd^3 (3.8-m^3) capacity is about 50 ft (15 m) and for the largest equipment about 110 ft (33.5 m). To reach these goals it frequently is necessary to use a dragline, carryall, or bulldozer on the high wall or a dragline in tandem with the shovel on the berm to strip the upper few feet of the overburden.

Digging equipment known as the wheel can be used ahead of a large power shovel to remove the upper 20–40 ft (6–12 m) of unconsolidated soil, clay, or weathered strata. This spoil is discharged onto a belt conveyor, then onto a stacker, and finally deposited several hundred feet from the high wall. Overburden thus removed improves the shovel productivity rate materially. Also, coal reserves can be mined that would have been too deep for the power shovel alone to handle.

Rocks overlying coal beds present some diversity of conditions for removal. Materials generally comprise shale, sandstone, and limestone with shale predominating. Proper fragmentation before stripping may be necessary to produce sizes that are smaller than the shovel dipper. Probably more research has been done on overburden drilling and blasting than on any other phase of stripping. The diameter, depth, and spacing of drill holes, the type of drill (whether vertical or horizontal), and the amount and type of explosive for each blast hole are the variables that must be determined for optimum production. Truck-mounted rotary drills have replaced churn drills for drilling vertical holes, and for horizontal drilling, auger drills are used. Package explosives, Airdox, Cardox, and commercial ammonium nitrate mixed with diesel fuel are used for blasting. The ammonium nitrate–diesel fuel mixture is one of the cheaper explosives and has gained favor rapidly. Equipment for mixing this explosive and automatically injecting it through a plastic tube into horizontal drill holes is being tested. If perfected, it will mechanize the only manual operation remaining in the stripping cycle.

Fig. 2. Large electric shovel of the type used in prairie regions, high wall (right), and spoil bank (left) at Hanna Coal Co.'s Georgetown mine, eastern Ohio. (*U.S. Bureau of Mines and Marion Power Shovel Co.*)

Before coal is loaded, spoil remaining on the berm is removed by bulldozer, grader and rotary brooms, or at small mines by hand brooms. Small-diameter vertical holes are augered into the bed and blasted with small charges of explosive to crack the coal. A ripper pulled by a bulldozer is effective in replacing coal drilling and blasting. Broken coal is loaded by $\frac{1}{2}$–5-yd^3 (0.38–3.8-m^3) capacity shovels into trucks of 5–80-ton (4.5–73-metric ton) capacity and transported from the stripping.

Land reclamation. Surface mining for coal can create many problems. The drastic disturbance of the overburden severely changes the chemical and physical properties of the resulting spoils. These altered properties often create a hostile environment for seed germination and subsequent plant growth. Unless vegetative cover is established almost immediately, the denuded areas are subject to both wind and water erosion that pollute surrounding streams with sediment.

Topsoil. Stripmining removes the developed soils and vegetative cover and creates a heterogeneous mass of topsoil, subsoil, and substrata rock fragments. The 1977 United States stripmine law requires that topsoil be removed and reapplied on the spoil surface during regrading and reclamation. Even when topsoil is reapplied over the spoil surface, an organic mulch is generally required for good seed germination and development. These mulching materials alter the surface microclimate and help to conserve soil moisture during the critical seedling establishment period.

The removal of topsoil before mining and its replacement on the spoil surface after final grading have aided materially in the reclamation process. Many of the chemical and physical limitations previously associated with mine spoils have been alleviated or eliminated. Surface grading techniques and seedbed preparation are very important in obtaining good vegetative cover for erosion and sedimentation control. Research has indicated that the surface should be left rough—preferably with small contour furrrows perpendicular to the slope. These furrows catch seed and fertilizer carried in runoff water, thus increasing germination and seedling development and decreasing soil and fertilizer loss from erosion.

Plant species adaptation. Many plant species of economic importance can be used to produce hay, pasture, various horticultural crops, and major row crops. Commercial varieties of grasses that have shown promise on United States eastern stripmine spoils with moderate pH (5.0–6.0) include orchard grass, tall fescue, bromegrass, ryegrass, and timothy. Other grasses that have shown promise on low-pH (4.5 or less) stripmine spoils are weeping lovegrass, bermuda grass, switch grass, bent grass, deer's-tongue, and redtop. Legumes that have shown promise on eastern acid stripmine spoil areas include alfalfa, white clover, crimson clover, bird's-foot trefoil, lespedeza, red clover, crown vetch, hairy vetch, flat pea, kura clover, zigzag clover, and white and yellow sweet clovers.

In the western United States a number of species have been tested. In general, several of the wheatgrass varieties, green needle grasses, side oats grama grass, smooth brome, wheat, and wild rye species have been used with varying degrees of success. Legumes species tested include bird's-foot trefoil, sweet clover, alfalfa, flat pea, and crown vetch. Most spoils in the western United States are returned to perennial grasses for eventual use by grazing livestock and wildlife.

Several woody species can be used on stripmine areas where rainfall is adequate. Tree species should be planted only with or after herbaceous species, such as grasses and legumes, that have stabilized the soil. Almost any species of trees can be grown on stripmine areas as long as their nutrient and environmental needs are met.

Many of the high-organic waste materials such as sewage sludge and composted sewage and garbage materials, are highly beneficial for establishing plant material on stripmine areas, especially with a low pH. These materials can contribute significantly to the plant nutrition on stripmine spoils for the production of vegetative cover. *See* LAND RECLAMATION. [ORUS L. BENNETT]

Bibliography: W. S. Doyle, *Strip Mining of Coal: Environmental Solutions*, 1976; R. F. Munn, *Strip Mining: An Annotated Bibliography*, 1973; R. Peele, *Mining Engineers' Handbook*, 3d ed., 1948; U.S. Department of the Interior, *Surface Mining and Our Environment*, 1967; R. A. Wright (ed.), *The Reclamation of Disturbed Arid Lands*, 1978.

Swamp, marsh, and bog

Wet flatlands, where mesophytic vegetation is areally more important than open water, are commonly developed in filled lakes, glacial pits and potholes (see illustration), or poorly drained coastal plains or floodplains. Swamp is a term usually applied to a wet land where trees and shrubs are an important part of the vegetative association, and bog implies lack of solid foundation. Some bogs consist of a thick zone of vegetation floating on water.

Unique plant associations characterize wet lands in various climates and exhibit marked zona-

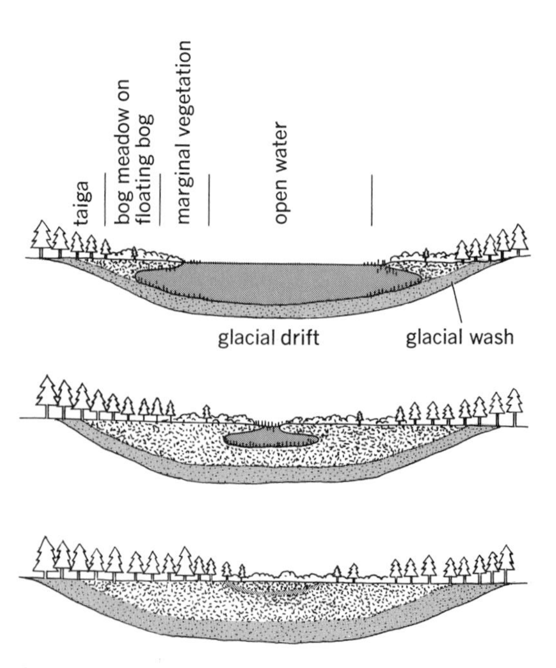

Progressive filling by vegetation of a pit lake in recently glaciated terrain. (*After C. A. Davis*)

tion characteristics around the edge in response to different thicknesses of the saturated zone above the firm base of soil material. Coastal marshes covered with vegetation adapted to saline water are common on all continents. Presumably many of these had their origin in recent inundation due to post-Pleistocene rise in sea level.

The total area covered by these physiographic features is not accurately known, but particularly in glaciated regions many hundreds of square miles are covered by marsh. *See* MANGROVE SWAMP. [LUNA B. LEOPOLD]

Taiga

A zone of forest vegetation encircling the Northern Hemisphere between the arctic-subarctic tundras in the north and the steppes, hardwood forests, and prairies in the south. The chief characteristic of the taiga is the prevalence of forests dominated by conifers. The taiga varies considerably in tree species from one major geographical region to another, and within regions there are distinct latitudinal subzones. The dominant trees are particular species of spruce, pine, fir, and larch. Other conifers, such as hemlock, white cedar, and juniper, occur locally, and the broad-leaved deciduous trees, birch and poplar, are common associates in the southern taiga regions. Taiga is a Siberian word, equivalent to "boreal forest." *See* FOREST AND FORESTRY; TUNDRA.

Climate. The northern and southern boundaries of the taiga are determined by climatic factors, of which temperature is most important. However, aridity controls the forest-steppe boundary in central Canada and western Siberia. In North America there is a broad coincidence between the northern and southern limits of the taiga and the mean summer and winter positions of the arctic air mass. In the taiga the average temperature in the warmest month, July, is greater than 50°F (10°C), distinguishing it from the forest-tundra and tundra to the north; however, less than four of the summer months have averages above 50°F, in contrast to the summers of the deciduous forest further south, which are longer and warmer. Taiga winters are long, snowy, and cold—the coldest month has an average temperature below 32°F (0°C). Permafrost occurs in the northern taiga. It is important to note that climate is as significant as vegetation in defining taiga. Thus, many of the world's conifer forests, such as those of the American Pacific Northwest, are excluded from the taiga by their high precipitation and mild winters.

Subzones. The taiga can be divided into three subzones (see illustration) in almost all of the regions which it occupies; these divisions are recognized mainly by the particular structure of the forests rather than by changes in tree composition. These subdivisions are the northern taiga, the middle taiga, and the southern taiga.

Northern taiga. This subzone is characterized on moderately drained uplands by open-canopy forests, dominated in Alaska, Canada, and Europe by spruce and in Siberia by spruce, larch, and pine. The well-spaced trees and low ground vegetation, usually rich lichen carpets and low heathy shrubs, yield a beautiful parkland landscape; this is exemplified best in North America by the taiga of Labrador and Northern Quebec. This subzone is seldom reached by roads and railways, in part because the trees seldom exceed 30 ft (9 m) in height and have limited commercial value. These forests are important as winter range of Barren Ground caribou, but in many parts of the drier interior of North America their area has been decreased by fire, started both by lightning and humans.

Middle taiga. This subzone is a broad belt of closed-canopy evergreen forests on uplands. The dark, somber continuity is broken only where fires, common in the drier interiors of the continents, have given temporary advantage to the rapid colonizers, pine, paper birch, and aspen poplar. The deeply shaded interior of mature white and black spruce forests in the middle taiga of Alaska and Canada permits the growth of few herbs and shrubs; the ground is mantled by a dense carpet of mosses. Here, as elsewhere in the taiga, depressions are filled by peat bogs, dominated in North America by black spruce and in Eurasia by pine. Everywhere there is a thick carpet of sphagnum moss associated with such heath shrubs as bog cranberry, Labrador tea, and leatherleaf. Alluvial

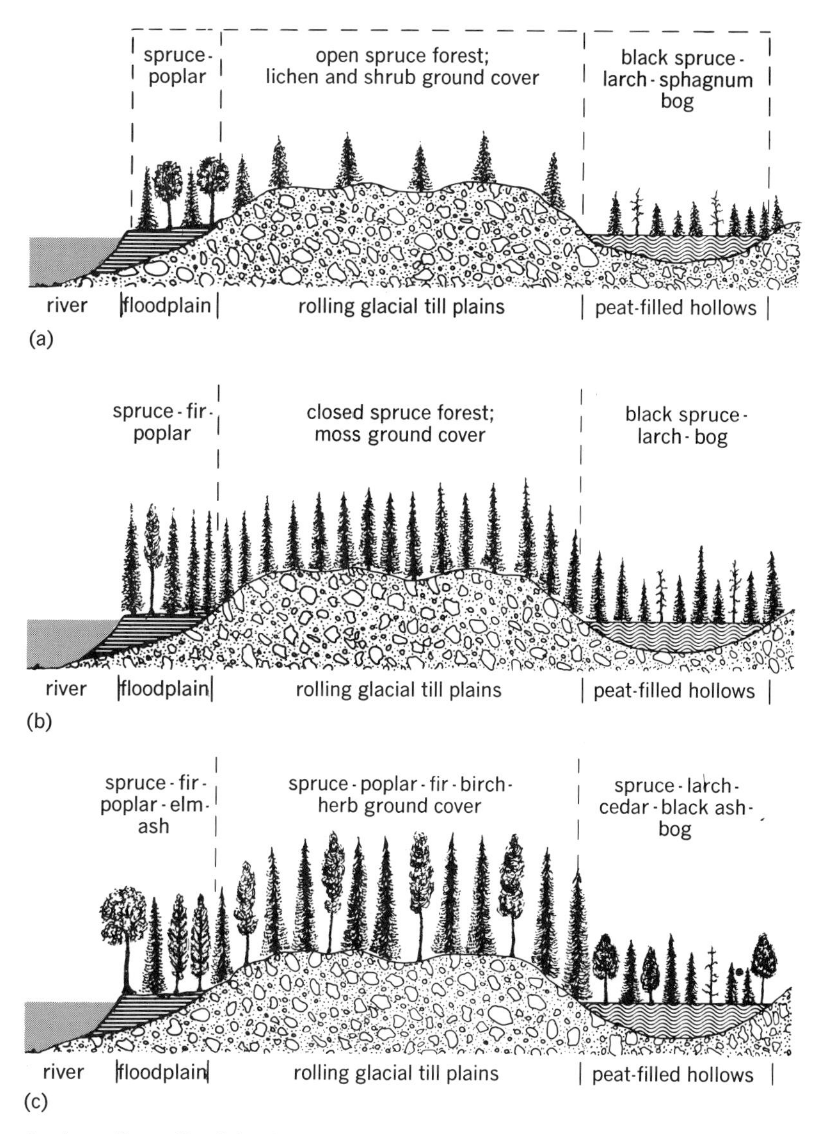

A schematic profile of the three main subzones of the North American taiga, showing the main forest assemblages on three of the more important landform types. (*a*) Northern taiga. (*b*) Middle taiga. (*c*) Southern taiga.

sites bear a well-grown forest yielding merchantable timber, with fir, white spruce, and black poplar as the chief trees.

Southern taiga. This subzone is characterized on moderately drained soils throughout the Northern Hemisphere by well-grown trees (mature specimens up to 95 ft or 29 m) of spruce, fir, pine, birch, and poplar. The trees are represented by different species in North America, Europe, and Siberia. Of the three taiga zones, this has been exploited and disturbed to the greatest extent by humans and relatively few extensive, mature, and virgin stands remain. In northwest Europe this subzone has been subject to intensive silviculture for several decades and yields forests rich in timber and pulpwood. In Alaska and Canada the forests have a much shorter history of forest management, but they yield rich resources for the forest industries. *See* FOREST MANAGEMENT AND ORGANIZATION.

Fauna. In addition to caribou, the taiga forms the core area for the natural ranges of black bear, moose, wolverine, marten, timber wolf, fox, mink, otter, muskrat, and beaver. The southern fringes of the taiga are used for recreation. *See* VEGETATION ZONES, WORLD. [J. C. RITCHIE]

Bibliography: A. Bryson, *Geographical Bulletin*, vol. 8, 1966; *Good's School Atlas*, 1950; *New Oxford Atlas*, 1975; W. Scotter, *Effects of Forest Fires on the Winter Range of Barren-ground Caribou in Northern Saskatchewan*, Wildlife Manage. Bull. no. 18, Canadian Wildlife Service, 1965.

Temperature inversion

The increase of air temperature with height; an atmospheric layer in which the upper portion is warmer than the lower. Such an increase is opposite, or inverse, to the usual decrease of temperature with height, or lapse rate, in the troposphere of about 6.5°C/km or 3.3°F/1000 ft, and somewhat less on mountain slopes. However, above the tropopause, temperature increases with height throughout the stratosphere, decreases in the mesosphere, and increases again in the thermosphere. Thus inversion conditions prevail throughout much of the atmosphere much or all of the time, and are not unusual or abnormal. *See* ATMOSPHERE.

Inversions are created by radiative cooling of a lower layer, by subsidence heating of an upper layer, or by advection of warm air over cooler air or of cool air under warmer air. Outgoing radiation, especially at night, cools the Earth's surface, which in turn cools the lowermost air layers, creating a nocturnal surface inversion a few centimeters to several hundred meters thick. Over polar snowfields, inversions may be a kilometer or more thick, with differences of 30°C or more. Solar warming of a dust layer can create an inversion below it, and radiative cooling of a dust layer or cloud top can create an inversion above it. Sinking air warms at the dry adiabatic lapse of 10°C/km, and can create a layer warmer than that below the subsiding air. Air blown from over cool water onto warmer land or from snow-covered land onto warmer water can cause a pronounced inversion that persists as long as the flow continues. Warm air advected above a colder layer, especially one trapped in a valley, may create an intense and persistent inversion.

Inversions effectively suppress vertical air movement, so that smokes and other atmospheric contaminants connot rise out of the lower layer of air. California smog is trapped under an extensive subsidence inversion; surface radiation inversions, intensified by warm air advection aloft, can create serious pollution problems in valleys throughout the world; radiation and subsidence inversions, when horizontal air motion is sluggish, create widespread pollution potential, especially in autumn over North America and Europe. *See* AIR POLLUTION; SMOG. [ARNOLD COURT]

Terrain areas, worldwide

Subdivisions of the continental surfaces distinguished from one another on the basis of the form, roughness, and surface composition of the land. These areas of distinctive landforms are the product of various combinations and sequences of events involving both deformation of the Earth's crust and surficial erosion and deposition by water, ice, gravity, and wind. The pattern of landform differences is strongly reflected in the arrangement of such other features of the natural environment as climate, soils, and vegetation. These regional associations must be carefully reckoned with by man in his planning of activities as diverse as agriculture, transportation, city development, and military operations.

The illustration distinguishes among eight classes of terrain, on the basis of steepness of slopes, local relief (the maximum local difference in elevation), cross-sectional form of valleys and divides, and nature of the surface material. Approximate definitions of terms used and percentage figures indicating the fraction of the world's land area occupied by each class are as follows: (1) flat plains: nearly level land, slight relief, 4%; (2) rolling and irregular plains: mostly gently sloping, low relief, 30%; (3) tablelands: upland plains broken at intervals by deep valleys or escarpments, moderate to high relief, 5%; (4) plains with hills or mountains: plains surmounted at intervals by hills or mountains of limited extent, 15%; (5) hills: mostly moderate to steeply sloping land of low to moderate relief, 8%; (6) low mountains: mostly steeply sloping, high relief, 14%; (7) high mountains: mostly steeply sloping, very high relief, 13%; and (8) ice caps: surface material, glacier ice, 11%.

The continents differ considerably. Australia, the smoothest continent, has only one-fifth of its area occupied by hill and mountain terrain as against one-third of North America and more than one-half of Eurasia. Antarctica is largely ice covered; the only other great ice cap is on Greenland.

North America, South America, and Eurasia are alike in that most of their major mountain systems are linked together in extensive cordilleran belts. These form a broken ring about the Pacific basin, with an additional arm extending westward across southern Asia and Europe. The principal plains of Eurasia and the Americas lie on the Atlantic and Arctic sides of the cordilleras, but are in part separated from the Atlantic by lesser areas of rough terrain.

Most of Africa and Australia, together with the eastern uplands of South America and the peninsulas of Arabia and India, show great similarity to

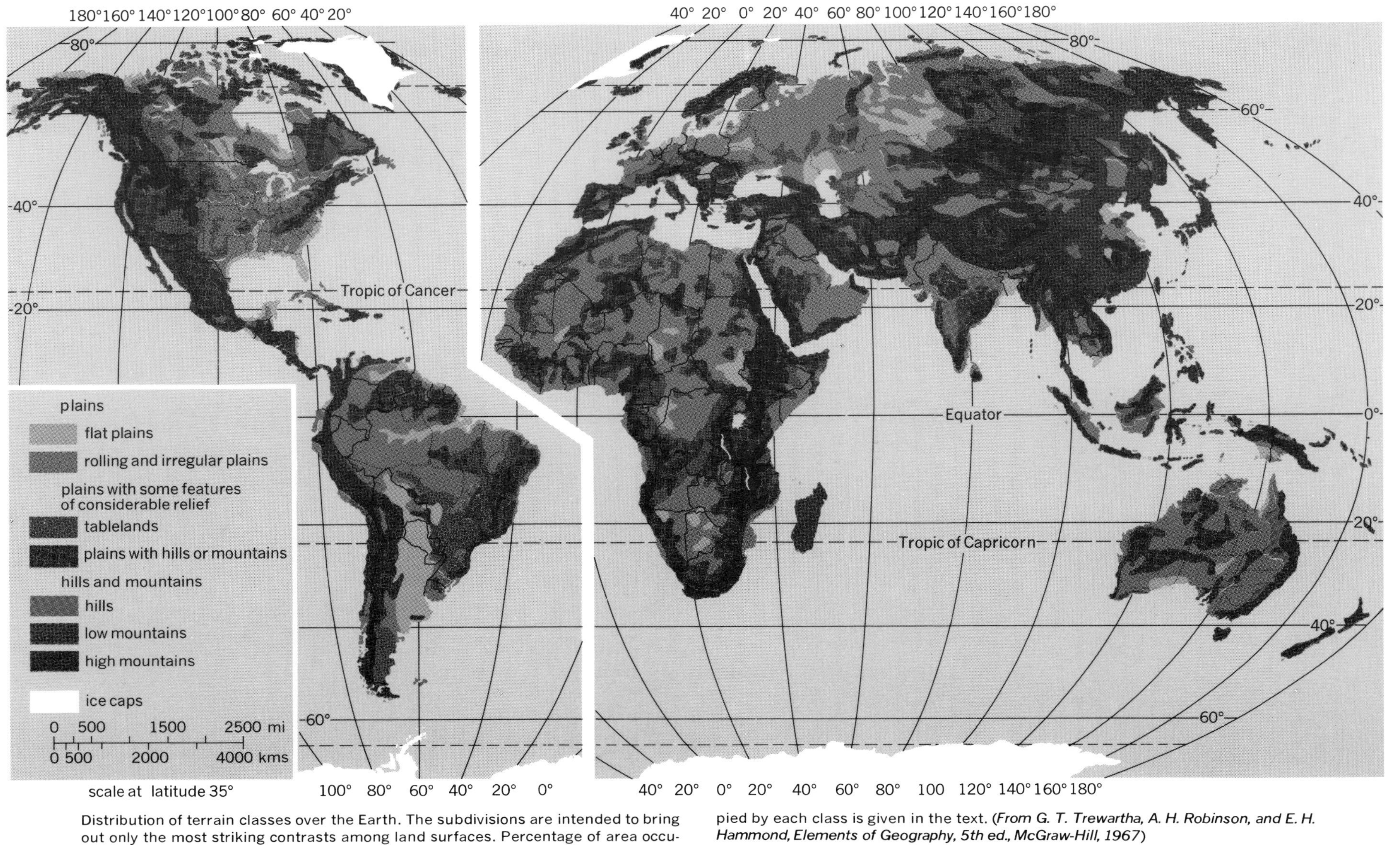

Distribution of terrain classes over the Earth. The subdivisions are intended to bring out only the most striking contrasts among land surfaces. Percentage of area occupied by each class is given in the text. (*From G. T. Trewartha, A. H. Robinson, and E. H. Hammond, Elements of Geography, 5th ed., McGraw-Hill, 1967*)

one another. They lack true cordilleran belts, and are composed largely of upland plains and tablelands, locally surmounted by groups of hills and mountains, and in many places descending to the sea in rough, dissected escarpments.

[EDWIN H. HAMMOND]

Terrestrial ecosystem

A term that distinguishes the complex of ecosystems of the land surfaces of the Earth from fresh-water and marine ecosystems. It encompasses ecosystems that exist on the continents and islands of the world and comprehends a series of dynamic open interaction systems that include living forms (animals, plants, and microorganisms) and their nonliving environment (soils, geological formations, and atmospheric constituents) and the activities, interrelations, chemical reactions, physical changes, and all other phenomena of each. Energy that enters these systems, chiefly in the form of sunlight, circulates through the systems and powers the life processes of the organisms, influences the rate and nature of chemical reactions and physical changes, and is partially accumulated in the bodies of organisms and in other chemical and physical states. *See* ECOSYSTEM; FRESH-WATER ECOSYSTEM; MARINE ECOSYSTEM.

Comparison of ecosystems. Terrestrial ecosystems differ from aquatic ecosystems in several important respects. The most obvious difference in the abiotic components of the systems is the basic physical contrast between the media. Terrestrial organisms are surrounded by air, a mixture comprised of gaseous elements and compounds. Water vapor is present in the atmosphere but forms only a small portion of the total volume of the air.

Liquid water, the medium of the aquatic environment, is much more dense, less fluid, and less transparent, and has a much greater thermal stability than air. The aquatic environment is marked by a superabundance of moisture, a lack of temperature extremes, slow changes in temperature, the dispersion of most of the essential mineral elements in the engulfing medium (usually in low concentration), a relatively short supply of oxygen and a high content of carbon dioxide, and a sharp decrease in illumination with slight increase in depth. On the other hand, in the terrestrial environment water is often in critically short supply, extremes and rapid changes of temperature are common, mineral elements are limited in occurrence to the substrate (generally in relatively high concentrations), and light is intense, at least at upper levels of the vegetation.

Adaptations to the habitat. The contrast in density, composition, and physical properties of the media in terrestrial and aquatic habitats is accompanied by and is responsible for, contributory to, or correlated with, contrasts in the geomorphologic development of the habitat, climatological features of the environment, pedological development, and morphologic and physiologic characteristics of the biota. Many organisms that live in the water have not developed, or have lost, devices that afford protection against water loss; indeed, many are able to absorb moisture and nutrients through the entire surface of their bodies. Many terrestrial organisms have evolved impervious cuticular coatings, surficial cuticular coverings, behaviorial adaptations, and physiologic adaptations that reduce water loss or make exceptionally efficient use of available water. Land plants, particularly the ferns and seed plants, have developed various water- and nutrient-collecting structures like rhizoids and roots. Correlated with these is the development of a complex translocation system through which water and dissolved minerals move from root to leaves and food circulates through the plant. Water plants generally have poorly developed roots and translocation systems.

Structural adaptations. Particularly in animals, the body shape of terrestrial species generally presents a small surface area per unit of volume, whereas marine animals, especially the mobile forms, are often elongate or flattened and hence present a large amount of surface per unit of volume. To some extent, body shape in motile forms is correlated with the density of the media and represents a streamlining adaptation that facilitates the movement of aquatic animals. Another structural adaptation to the contrasting density of the media is the development of rigid or semirigid supporting tissues in terrestrial plants and the virtual lack of such development in aquatic plants, which are buoyed up by the water. Virtually all woody plants are terrestrial or semiaquatic. Indeed, in both plants and animals, there is a tremendous taxonomic (evolutionary) division between aquatic and terrestrial forms. Seed plants, ferns, mosses, liverworts, fungi, and lichens are most characteristic and abundantly represented on land, while water is the chief domain of the algae. Thus the diversity in gross structure and life form, as well as in species, is much greater on land. A tremendous variety of forests, grasslands, savannas, scrubs, succulent deserts, tundras, and other forms of vegetation characterize the terrestrial environment. The range of physiognomy of aquatic vegetation is much less and stratification or layering is absent or weakly developed in aquatic communities. *See* PLANT FORMATIONS, CLIMAX.

Habitat distribution of organisms. Warm-blooded animals and snakes and lizards typically are land dwellers. Insects are abundant and ecologically important on land, less so in fresh water, and virtually are absent from salt-water habitats. Fish, bivalves, and various lower animals are limited to or at least most abundant in aquatic environments. The bulk of aquatic vegetation, particularly where the water is deep, is floating rather than rooted, while that of land is almost entirely rooted and stationary. In contrast, many aquatic animals are immobile and depend on water currents to deliver food to their vicinity; most terrestrial animals are mobile. Land animals are generally capable of much more rapid movement in their dispersed medium than are comparable aquatic animals, which occupy a very dense medium.

Extent of terrestrial ecosystem. Although the aerial environment is well lighted from the upper limits of the atmosphere to the ground, the thickness of the terrestrial ecosystem is limited by two sets of circumstances. First the vertical development of land vegetation is restricted by the characteristics and limitations of supporting tissues, by the periodic stresses and breakage produced by strong winds, by the limitations of water-raising systems, and various other factors. Secondly, the soil-air interface is a tremendously important physical boundary. Below this interface there is no light, oxygen becomes a limiting factor, and water may saturate the soil at a depth of a few inches or feet or it may be in critically short supply. Thus the tallest living tree is a redwood 364 ft (111 m) high, and the roots of only a few species of plants extend to a depth of 100 ft (30 m). Discounting airborne spores, insects, and similar material accidentally carried into the upper atmosphere, the maximum thickness of the terrestrial ecosystem in a given locality is less than 400 ft (122 m). Usually, it does not exceed 150 ft (46 m) in thickness in forested areas and is much thinner in areas that support only grassland or tundra. On the other hand, the inhabited zone of a marine environment may be greater than 30,000 ft (9000 m) thick, although food-producing plants are limited to the surface layers due to limits of light penetration.

Temperature. The great seasonal temperature variations and rapid temperature changes in most terrestrial regions force periods of dormancy in plants and poikilothermic animals whose body temperatures approximate the environment. Homeothermic or warm-blooded animals, on the other hand, can remain active even during very cold weather if sufficient food is available. Aquatic plants and animals are not exposed to rapid thermal changes and the amplitude of the seasonal thermal cycle is much less in the water than in most terrestrial habitats. Not only is temperature regulation a greater problem in terrestrial habitats, but the functioning of the entire terrestrial ecosystem is controlled to a much greater extent by temperature changes. The fluidity of the atmosphere, the cyclonic circulation of huge air masses, and the variations in solar energy result in considerable diurnal thermal variations as well as seasonal and secular variations. During the growing season, or nondormant periods, these thermal variations impose corresponding variations in the rate of productivity of the green plants (through their effect on the rate of photosynthesis), on the rate of herbivore and carnivore harvesting activity (through their effect on the physiology of the organisms, especially as they impose dormancy due to low temperatures or estivation due to high temperatures), and on the energy expenditures of homeothermic animals (through their effect on the temperature gradient between the air and the bodies of the animals and the respiratory activity required to maintain body temperature). *See* HOMEOSTASIS.

Seasonal effects. During the cold or dry dormant seasons, the primary production of the terrestrial ecosystem is drastically reduced or even halted. Most herbivorous animals are inactive. The activities of herbivores that remain operative and the activities of saprophytes and scavengers that continue to feed during the unfavorable seasons result in a net decrease in the standing crop, for they feed on materials elaborated during the preceding favorable season or seasons. Most poikilothermic carnivores are forced to remain dormant or die during cold seasons, but homeothermic carnivores may remain active. In the latter instance, the carnivores' food preferences may change during the cold season, or their methods of hunting may be changed to compensate for the cold-season habits of their prey. Migrations of populations are related closely to seasonal changes. Migration of some birds, for example, has been found to be triggered by changes in day-length. But avian populations that are dependent on insects migrate to and from an area at times that correspond closely with the vernal increase and autumnal decrease in insect populations. Other species that are dependent on plant foods migrate at times when the daylight period becomes too short to allow the consumption of sufficient plant food to carry the animals through the longer dark period. Many grazing and browsing mammals migrate vertically in mountainous areas or horizontally in level areas to obtain food and to benefit from more favorable climatic conditions.

Unfavorable seasons also affect aquatic communities, especially fresh-water communities, but the seasonal changes generally are less intense, the activity of primary producers generally does not ebb as low as in terrestrial communities, and the limiting factors are most often a lack of oxygen and nutrients rather than extremely low temperatures or the unavailability of water. Animal activity in aquatic habitats is also reduced during unfavorable seasons, but the level of minimum activity in the aquatic environment of deeper fresh-water bodies and in the oceans is considerably higher than that in the terrestrial environment.

Catastrophic agents. Fire, windthrow, and many other catastrophic agents are peculiar to the terrestrial environment and play important roles in the functioning of the ecosystem. Furthermore, the terrestrial environment is the habitat of humans and is more subject to their interference than the aquatic environment, particularly the marine segment of it. Human influence is so great that man has substituted artificially maintained ecosystems for natural ecosystems. The artificial ecosystems are represented on land by farmed areas, managed forests, urbanized areas, and similar developments. *See* APPLIED ECOLOGY.

Energy. Terrestrial ecosystems differ from aquatic ecosystems in another significant aspect. They are dependent on stored nutrients that may have been residual for hundreds of thousands of years. They are marked by a continual net loss of nutrients and other physical components through erosion and leaching, while there is a concomitant net gain to the aquatic ecosystems. Except in certain restricted areas, the use of aquatic food by terrestrial animals, the phenomenon of salt spray, the filling of lakes and bogs, and the harvest of marine and fresh-water organisms by human activities represent a small return of materials to the terrestrial environment. Much larger returns are made by crustal upheavals that expose sections of the ocean floor and once again make available to terrestrial organisms the materials accumulated on the ocean bottom. The reverse also takes place, and former land areas may be depressed beneath the surface of the oceans. These crustal movements result in a long-term recirculation of mineral elements as well as a periodic renewal of erosion cycles. The terrestrial and aquatic ecosystems are also connected in a great many other direct ways, so that it is most logical to consider the entire Earth and its envelope of gases to be the ultimate ecosystem. However, this world ecosystem is an open system and is dependent upon an external supply of energy, chiefly from the Sun, and is influenced by a variety of forces of external origin.

Trophic levels. If the weight (biomass) of the living components of an ecosystem at a given moment is determined, separated into weights for the various trophic levels (green plants, herbivores, predators, scavengers, and saprophytes), and figured graphically, a pyramidal form of graph results. In the terrestrial ecosystem, the green plants are relatively long-lived; some live for several weeks, others for years, a few for centuries. Because the green vegetation is the primary producer level and thus limits the bulk of organisms that can feed upon it and because the green vegetation is long-lived, its biomass is greater than that of any other trophic level or of all other trophic levels combined. In aquatic ecosystems, however, this is not necessarily the case. Where short-lived plankton forms are the predominant green vegetation and longer-lived fish and bottom organisms are the herbivores and carnivores, the biomasses of the higher trophic levels often exceed those of the primary producers. Of course, when the annual production of the various trophic levels in either ecosystem is calculated, the production of the green vegetation is greatest.

Productivity. From the meager quantitative data available on the primary productivity of various ecosystems, it appears that no generalization can be made concerning the comparative productivity of terrestrial and aquatic communities. Further, it is apparent that productivity in either environment is not a regional characteristic, but is dependent wholly upon the ecosystem involved. According to summaries made by E. Odum (1953), an Ohio cornfield was found to produce 862 metric tons of organic carbon per square kilometer per year, the trees in a New York apple orchard produce 526, European forests produce an average of 225, cultivated land on the average produces 160, a dry grassland produces 48, and a desert produces only 6 metric tons. H. Odum and coworkers (1958) present data that indicate that a montane rainforest in Puerto Rico produces approximately 1100 metric tons, and data presented by A. Krogh (1934) indicate that a Danish beech forest may produce 1500–2000 metric tons.

For comparison, data cited by E. Odum (1953) show that several lakes in the north-central United States average 111–480 metric tons per square kilometer per year and that several marine communities average 60–1000 metric tons. Significantly higher annual primary production has been reported by H. Odum (1957) for a fresh-water spring in Florida, 6390 metric tons, and by E. Odum and H. Odum (1955) for a coral atoll in the Pacific Ocean, 8328 metric tons.

Trophic structure. The trophic structure of terrestrial and aquatic ecosystems is strikingly similar (see table). There is more variability within the series of analyzed terrestrial communities than there is between those communities and the aquatic types. In addition, the relative efficiency of the vegetation, in terms of the proportion of incident solar radiation converted to stored form in organic carbon compounds, is virtually identical for the terrestrial and fresh-water communities listed, although the efficiency of the tropical atoll community was considerably greater.

In all the ecosystems, a relatively low proportion of available solar energy is utilized directly by the biota.

Energy transfer. The energy transfers that occur in a beech-maple forest community in the central United States have been estimated by J. S. McCormick (1959) and are an example of terrestrial systems in general. The energy utilized each year by the organisms in an acre of this forest is approximately equivalent to the electricity required to supply an average New York City household with power for nearly half a century. Virtually all the energy that enters the system directly is in the form of sunlight, but only about 1% of the available solar energy is actually transformed by green plants into the chemical energy of food. Approximately 10% of the solar energy is reflected from the plant surfaces, 15% is passed through the leaves, and 74% is dissipated as heat.

A portion of the energy stored in basic foods or in photosynthetic products manufactured by green plants is utilized by the plants in respiration and the remainder is stored in the form of plant tissue. Part of the energy of plant tissues harvested and utilized by herbivores such as insects, rodents, and deer is dissipated as heat by respiration, and part is stored in the body tissues of the herbivores. The energy contained in the tissues of herbivores is utilized by predators and the tissues of predators are, in turn, utilized by secondary predators. At each step, some energy is lost to the community through respiration and part is unharvested or unassimilated. The unharvested and unassimilated energy accumulates chiefly on the forest floor in the form

Trophic structure of terrestrial and aquatic communities on basis of biomass or energy content of standing crop

	Terrestrial communities			Fresh-water communities			Marine communities	
Category	Blue-grass meadow, Michigan	Beech forest, Denmark	Montane rainforest, Puerto Rico	Cedar Bog Lake, Minnesota	Webster's Lake, Wisconsin	Silver Springs, Florida	Coral reef, Eniwetok Atoll	Eel-grass, North Sea
Primary producers, %	78	93	98.4	89.3	86.8	94.3	83.1	79.7
Herbivores, %	16	7	1.6	9.0	10.2	4.3	15.6	19.9
Carnivores, %	6			1.6	3.4	1.5	1.3	0.3
Efficiency of primary producers*	1.2	1.0				1.2	5.8†	
Basis of figures	Energy	Biomass	Biomass	Energy	Biomass	Biomass	Biomass	Biomass

*Ratio of energy transformation to incident solar energy.
†Computed on basis of light reaching community rather than light reaching water surface.

of dead leaves, twigs, flowers, fruits, fallen trunks, dead bodies, feces, and liquid wastes and is utilized by scavengers and saprophytes. Ultimately, all the energy that enters the forest community is dissipated as heat by respiratory processes, is lost in the bodies of plants and animals that leave or are taken from the forest, or is lost in the form of heat evolved by forest fires. In natural communities, these losses are balanced, or at least offset, by materials that enter the community from other sources. *See* ECOLOGY. [JACK S. MC CORMICK]

Bibliography: J. McCormick, *The Living Forest*, 1959; E. P. Odum, *Fundamentals of Ecology*, 3d ed., 1971; H. T. Odum, Trophic structure and productivity of Silver Springs, *Ecol. Monogr.*, 27:55–112, 1957; H. T. Odum, W. Abbott, and R. Selander, Studies on the productivity of the lower montane rainforest of Puerto Rico (Abstract), *Bull. Ecol. Soc. Amer.*, 39:85, 1958; H. T. Odum and E. P. Odum, Trophic structure and productivity of a windward coral reef community on Eniwetok Atoll, *Ecol. Monogr.*, 25(3):291–320, 1955.

Thunderstorm

A convective storm accompanied by lightning and thunder and a variety of weather such as locally heavy rainshowers, hail, high winds, sudden temperature changes, and occasionally tornadoes. The characteristic cloud is the cumulonimbus or "thunderhead," a towering cloud, often with an anvil-shaped top. A host of accessory clouds, some attached and some detached from the main cloud, are often observed in conjunction with cumulonimbus. The height of a cumulonimbus base above the ground ranges from 1000 to over 10,000 ft (300–3000 m), depending on the relative humidity of air near the Earth's surface. Tops usually reach 30,000–60,000 ft (9000–18,000 m), with the taller storms occurring in the tropics or during summer in mid-latitudes. Thunderstorms travel at speeds from near zero to 60 miles per hour (mph; 27 m/s). In many tropical and temperate regions, thunderstorms furnish much of the annual rainfall.

Development. Thunderstorms are manifestations of convective overturning of deep layers in the atmosphere and occur in environments in which the decrease of temperature with height (lapse rate) is sufficiently large to be conditionally unstable and the air at low levels is moist. In such an atmosphere, a rising air parcel, given sufficient lift, becomes saturated and cools less rapidly than it would if it remained unsaturated because the released latent heat of condensation partly counteracts the expansional cooling. The rising parcel reaches levels where it is warmer (by perhaps as much as 10°C) and less dense than its surroundings, and buoyancy forces accelerate the parcel upward. The convection may be initiated by surface heating or by air flowing over rising terrain or by converging airflow in atmospheric disturbances such as fronts. The rising parcel is decelerated and its vertical ascent arrested at altitudes where the lapse rate is stable, and the parcel becomes denser than its environment. The forecasting of thunderstorms thus hinges on the identification of regions where the lapse rate is unstable, low-level air parcels contain adequate moisture, and surface heating or uplift of the air is expected to be sufficient to initiate convection.

Occurrence. Thunderstorms are most frequent in the tropics, and rare poleward of 60° latitude. In the United States, the Florida peninsula has the maximum activity with 60 thunderstorm days (days on which thunder is heard at a given observation station) per year. Thunderstorms occur at all hours of day and night, but are most common during late afternoon because of the influence of diurnal surface heating. The weak nighttime maximum of thunderstorms in the broad Mississippi Valley of the central United States is still a topic of debate.

Structure. Radar is used to detect thunderstorms at ranges up to 250 mi (400 km) from the observing site. Much of present-day knowledge of thunderstorm structure has been deduced from radar studies, supplemented by visual observations from the ground and satellites, and in-place measurements from aircraft, surface observing stations, and weather balloons. Thunderstorms occur in isolation, in chaotic patterns over wide areas, in the walls and spiral bands of hurricanes, in clusters within large-scale weather systems, and in squall lines perhaps several hundred miles long. An individual thunderstorm typically covers a surface area of 200–1000 mi² (500–2500 km²) and consists of one or more distinct cells, each of which is several miles across, lasts about an hour, and undergoes a characteristic life cycle. In the cumulus or growing stage, a cell consists primarily of updrafts (vertical speeds of 10–40 m/s or 20–90 mph) with precipitation suspended aloft; in the mature stage, updrafts and downdrafts coexist and heavy rain falls to the ground; in the dissipating stage, a cell contains weakly subsiding air and only light precipitation. During the mature stage, downdrafts may reach 35 mph or 15 m/s. The downdraft air is denser than its surroundings due to evaporational cooling, which occurs as clear air is entrained into the cloud from outside and forced downward by gravitational pull and by the drag of falling precipitation. The downflowing air spreads outward in all directions as it nears the surface, and forms a cold, gusty wind which is directed away from the precipitation area. This advancing cold air may provide the necessary lift in neighboring warm moist air for the formation of new updraft cells.

In an environment where the winds increase and veer with height, and mid-level air is dry enough to provide the potential for strong downdrafts, a thunderstorm may become organized so as to maintain a steady state for hours. In such strong vertical shear of the horizontal wind, the updraft is tilted so that precipitation falls out of the updraft instead of through it, and updraft and downdraft can coexist for several hours in the configuration shown in Fig. 1. A long-lived storm in a sheared environment may consist of a single intense cell (supercell) or of many cells with an organized growth of new cells on one side of the storm (generally, the southwest in the Northern Hemisphere) and decay of old cells on the opposite flank.

Severe storms. Thunderstorms are considered severe when they produce winds greater than 58 mph (26 m/s or 50 knots), hail larger than 3/4 in. (19mm) in diameter, or tornadoes. While thunderstorms are generally beneficial because of their needed rains (except for occasional flash floods),

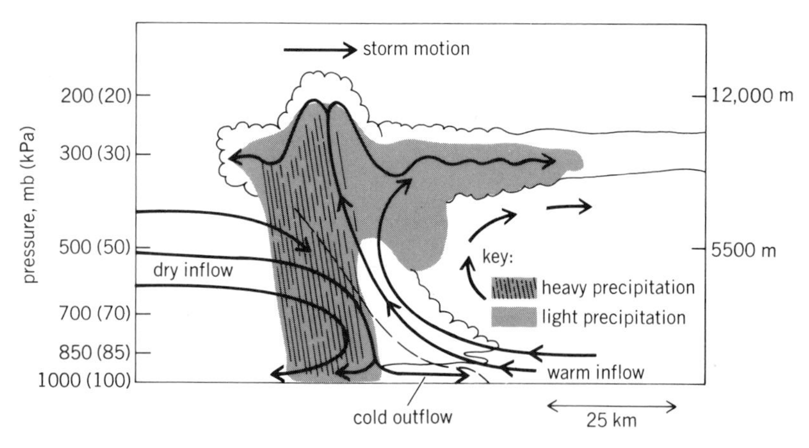

Fig. 1. Cloud boundaries and simplified circulation (arrows denote flow) of a typical mature thunderstorm in winds which blow from left to right and increase with height. Vertical scale has been exaggerated fivefold compared with the horizontal scale.

severe storms are the atmospheric equivalent of a rogue elephant with the capacity of inflicting utter devastation over narrow swaths of the countryside. Severe storms are most frequent in the Great Plains region of the United States, but even there only about 1% of the thunderstorms are severe. Severe storms are most frequently supercell storms which form in environments with high convective instability and moderate to large vertical wind shears.

Since severe storms constitute a hazard to aircraft, their internal dynamics has been deduced largely from radar measurements. Doppler radar is specialized to measure the velocity of radar targets parallel to the radar beam, in addition to the intensity of precipitation. Doppler radar studies and analysis of surface pressure falls have shown that large hail, high winds, and tornadoes often develop from a rotating thunderstorm cell known as a mesocyclone (or tornado cyclone, if it spawns at least one tornado). Large hail, high winds, and weak tornadoes may form from nonrotating (on broad scale) multicellular storms, but are less likely.

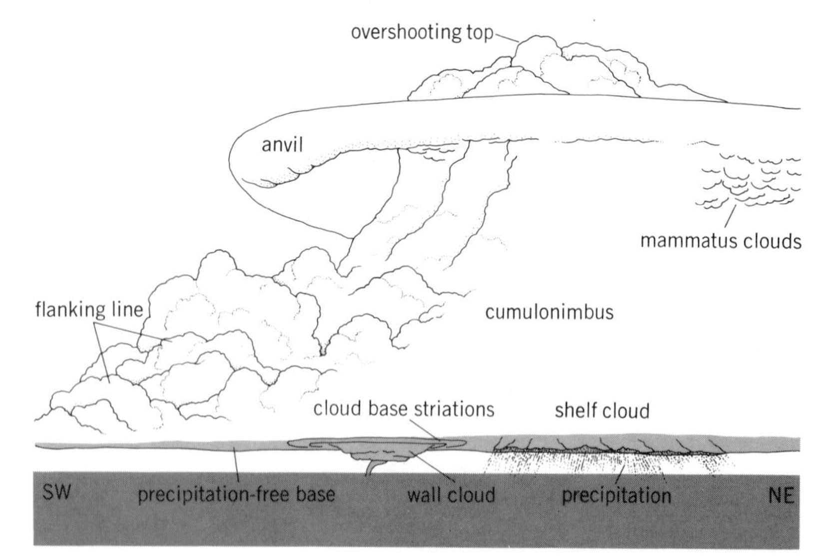

Fig. 2. Composite view of a typical tornado producing cumulonimbus as seen from a southeasterly direction. Horizontal scale is compressed, and all the features shown could not be seen from a single location. (*NOAA picture by C. Doswell and B. Dirham*)

Maximum tangential winds around the typical mesocyclone are roughly 50 mph (20 m/s) and are located in a circular band which is 1–6 mi (2–10 km) in radius. A surface pressure deficit of several millibars exists at the mesocyclone center. Identification of a mesocyclone "signature" on radar has been used experimentally to issue severe weather warnings. On conventional radar displays, hook-shaped appendages to echoes are also good indications of mesocyclones, but unfortunately a large percentage of tornadic storms never exhibit such a hook. A mesocyclone sometimes is recognizable visually by rotation of a wall cloud, a discrete and distinct lowering of the cumulonimbus base (Fig. 2). The wall cloud is often seen visually to be rotating as an entity. The wall cloud is frequently the seat of intense vertical motions at low levels. The mesocyclone's rotation is apparently a combination of two effects. First, since the initial air-mass condition is one of rotation (at least the Earth's), inward motion of air parcels at low levels toward the base of an updraft is associated with increasing spin in rough accord with the conservation of angular momentum. Second, differential vertical air motion in an environment where the wind varies with height causes a tilting of vortex tubes from the horizontal into the vertical and thus also amplifies vertical vorticity (that is, vertical spin).

Attempts have been made to modify thunderstorms to increase areal rainfall and suppress hail. The results of such experiments have been inconclusive. *See* STORM DETECTION; TORNADO.

[ROBERT DAVIES-JONES]

Storm electricity. A thunderstorm produces lightning and is thus highly electrified. A thunderstorm can be considered as a current generator that separates charge, resulting in strong electric fields and electricity for lightning and various conduction currents within, below, and above the storm. Electric currents from thunderstorms to the earth and ionosphere maintain the fair weather state of the earth and atmosphere. Typical maximum electric fields are 6 to 15 kV/m at the earth below the storm and 100 to 400 kV/m or more inside thunderstorms. Electric fields at the ground are less intense because of corona discharge from pointed objects. Lightning is the most obvious and deadly manifestation of storm electrification. Flashing rates are extremely variable. A small isolated storm produces about 3 flashes per minute, but rates as high as 100 have been observed. Lightning in squall lines and severe storms can appear nearly continuous. Lightning can propagate long distances, with at least one flash 170 km long having been observed with radar.

Initial and subsequent cloud electrification mechanisms are unknown, but charged hydrometeors and air motion are key elements. The gross charge distribution within a thunderstorm is an upper positive charge, which spreads into the anvil, and a lower negative charge. Magnitudes of these charges are unknown; generally accepted estimates range from a few tens of coulombs to a few hundred coulombs. Most of the charge is thought to be on cloud and not on precipitation particles. Research with balloons and aircraft penetrating thunderstorms shows complex charge distributions on a smaller scale, and that the dominant charge polarity can change over distances as small as a kilometer. Knowledge of thunderstorm electrification processes awaits complete measurements. Modern research is characterized by simultaneous measurements of electrical parameters, microphysical processes, and storm dynamics in efforts to understand the complex interrelationships that result in the thunderstorm. *See* ATMOSPHERIC ELECTRICITY. [W. DAVID RUST]

Tornado

An intense rotary storm of small diameter, the most violent of weather phenomena. Tornadoes always extend downward from the base of a convective-type cloud, generally in the vicinity of a severe thunderstorm.

Appearance ranges from a broad funnel with smallest diameter at the ground, to a narrow rope-like vortex which may not reach the ground or may intermittently lift and dip. An ill-defined cloud of dust or debris often surrounds the true tornado cloud near the ground (Fig. 1). In surface layers, air spirals inward toward the vortex, generally rotating in a counterclockwise sense, rising rapidly in the funnel.

The visible funnel consists of cloud droplets condensed because of expansional cooling resulting from markedly lower (probably by 100–200 millibars or 10–20 kPa) pressure in the vortex than in the surroundings. Height of the visible funnel depends upon the cloud base and may be 1000–10,000 ft (300–3000 m); however, the tornado vortex probably extends a considerable distance upward within the accompanying cloud.

The path of destruction varies from a few yards to over a mile in width, and from very short to about 300 mi (500 km) in length; in about 80% of cases the path is less than 50 yd (46 m) wide and 3 mi (5 km) long. Movement may be from any direction, most commonly from the southwest, but many tornadoes move from the northwest. Speed of movement averages 35 mph (16 m/s) but is variable. Tornadoes may occur at any time of day, but are most frequent from midafternoon to early evening.

In the United States an annual average of over 700 tornadoes have been reported in recent years, with monthly percentages ranging from 20 in May to 3 in January. Maximum frequency is observed in the "tornado alley" belt, from the Texas Panhandle across Oklahoma and Kansas, thence across the Midwestern states (Fig. 2). Greatest activity is over the Southern states in winter and early spring, migrating to the tornado alley region in May, and to the northern Plains and the Midwest states in late summer. Other regions with fairly high tornado frequency include a belt from southern England across northern Europe; also Japan, the Ganges valley, southernmost Africa, southwest and southeast Australia, New Zealand, and the la Plata basin of South America.

According to T. Fujita, tornadoes in the United States east of the Rocky Mountains typically have broader and longer tracks than those in other parts of the world, and are also more intense. Based on estimates from damage to structures, 1 in 1000 may have winds exceeding 260 mph (116 m/s), although more than half have winds less than 100 mph (45 m/s). Deaths and property damage mainly result from the 1–2% of the total number of tornadoes that are the largest and most intense and touch down in built-up areas. Damage results both from the force of the wind and from outward collapse of building walls when the atmospheric pressure is suddenly reduced.

Fig. 1. Tornado, at Fargo, ND. (*Fargo Forum photograph by C. Gerbert, Grand Prize Winner, 11th Annual Graflex Contest*)

Requisite conditions for tornado formation are pronounced thermodynamic instability combined with sufficient amounts of water vapor to produce thunderstorms, along with the presence of strong winds in the upper troposphere. These are the conditions that favor development of a squall line; at times 10–30 tornadoes occur as parasites to such a disturbance. Tornado probability increases

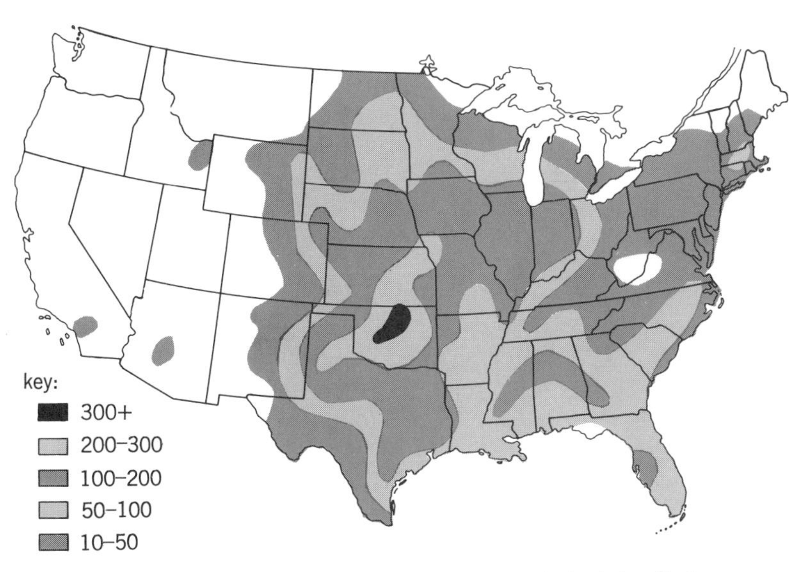

Fig. 2. Distribution of tornadoes in the United States by 2° latitude-longitude squares (totals for 1955–1967).

with the thermal instability, and it is greatest with thunderstorms that have an intense radar reflectivity (large precipitation content) and high tops penetrating into the stratosphere. However, some tornadoes are not connected to the thunderstorm itself, but to vigorously developing cumulus clouds adjacent to the thunderstorm. *See* SQUALL LINE.

Both laboratory simulations and observations of natural vortexes suggest that rising motions (up to 150 mph or 67 m/s observed) take place mainly in the sheath of the funnel, and that there is likely to be descending motion in or near its core. The intense winds are accounted for by inward-moving rings of air increasing in rotary motion under conservation of angular momentum, and speeds are consistent with movement toward the inner region of very low pressure, allowing for some frictional loss. The cause of the low pressure is inadequately understood.

Tornadoes in the United States are mostly found on the south sides of the parent thunderstorms. Heavy rain and hail often follow (but sometimes precede) passage of a tornado. The heaviest rain is likely to fall a few miles north of the tornado track; sometimes no rain falls along the track itself. Widespread thundersqualls are often observed outside the actual tornado path.

Over the past 20 years, property damage from tornadoes has averaged $75,000,000, with 113 deaths, half the previous 40-year average. Decreased fatalities (despite increased population density) are partly due to improved prediction and communications. Detection of severe thunderstorms by radar has been important. Doppler radar, which measures cloud circulations that may later develop into tornadoes, may come into widespread use.

[CHESTER W. NEWTON]

Bibliography: R. Davies-Jones and E. Kessler, Tornadoes, in W. N. Hess, (ed.), *Weather and Climate Modification*, 1974; S. D. Flora, *Tornadoes of the United States*, 1973; T. T. Fujita, Tornadoes around the world, *Weatherwise*, 26: 56–62, 78–83, 1973; G. Grigsby, *Tornado Watch*, 1977.

Tree diseases

Forest and shade tree diseases are discussed separately here, although the same causal agents can be involved with both groups of trees. In forests, generally the concern is with the effect of diseases on stands of trees rather than with individuals, and although the value of individual trees in forests can be high, usually this is much less than the value of shade trees. Diseases that result in disfiguration of trees, such as leaf spots, are important on shade trees but not on forest trees and, in contrast, shade trees can serve very well with substantial amounts of heart rot, which in forest trees is of major consequence. Thus the emphases on control are different, and the corrective measures themselves are often quite distinct.

Diseases of forest trees. From seed to maturity forest trees are subject to a succession of diseases. Losses due to pests and fire amount to 92% of the net saw timber growth, and diseases account for 45% of these losses. Young succulent seedlings, especially conifers, are killed by soil-inhabiting fungi (damping off); the root systems of older seedlings may be attacked by complexes of fungi and nematodes, including in particular fungi in the genera *Cylindrocladium*, *Sclerotium*, and *Fusarium*. Chemical treatment of soil with biocides, such as formulations containing methyl bromide, and fungicides, as well as cultural practices unfavorable to the development of pathogens, will help control seedling diseases. *See* FOREST ECOLOGY; FOREST PLANTING AND SEEDING; FUNGISTAT AND FUNGICIDE.

Leaf diseases. In forests, leaf diseases usually cause negligible losses, but brown spot needle blight, caused by *Scirrhia acicola*, can cause excessive defoliation in nurseries and plantations and prevent longleaf pine from starting height growth. Fungicides in the nurseries and prescribed burning in plantations have been used successfully for control. There are other important leaf diseases, such as Dothistroma needle blight, which have caused losses in plantations of Austrian pine (*Pinus nigra*) and ponderosa pine (*P. ponderosa*) in the central and northwestern United States. Cop-

Fig. 1. Oak wilt. (*a*) Oaks killed by the oak wilt fungus. (*b*) Mycelial mat that the oak wilt fungus produces beneath the bark of a wilted tree.

Fig. 2. Chestnut blight in American chestnut tree (*USDA*). Inset is mycelial fan of chestnut blight caused by *Endothia parasitica*, advancing through bark of American chestnut. The tip of the fan on the left is surrounded by cortical tissue. The contents of the cortical cells back from the tip of the fan are discolored to a yellowish brown, as indicated by the darkened cells (*after W. C. Bramble, from J. S. Boyce, Forest Pathology, 2d ed., McGraw-Hill, 1948*).

per fungicides applied at the right time control some of these diseases.

Wilt diseases. Diseases such as oak wilt (Fig. 1) are systemic (infecting entire plant), killing susceptible species in a matter of weeks or months by

Fig. 4. Shelf fungus (*Fomes applanatus*) on a dead aspen tree. This wood-rotting fungus enters through roots and wounds and decays the heartwood and sapwood of both hardwoods and softwoods.

plugging the tree's vascular system. The fungus *Ceratocystis fagacearum* is disseminated to nearby healthy trees through root grafts and to longer distances by unrelated insects, including the Nitidulidae. The mycelial mats of the fungus are covered with spores that can be disseminated by insects. Control is possible by eradicating infected trees, especially the recently wilted red oaks on which spores are produced, and by disrupting root grafts with the chemical called Vapam, or VPM.

Canker diseases. These result from localized killing of the cambium and range from chestnut blight, which since its introduction in 1904 has practically annihilated the American chestnut (Fig. 2), to less important cankers caused by species of *Nec-*

Fig. 3. White pine blister rust (*Cronartium ribicola*). (*Photograph by Robert Campbell*)

Fig. 5. Pistillate shoots of ponderosa pine mistletoe (*Arceuthobium vaginatum forma cryptopodum*) on branch of sapling. Note unripe fruits. (*U.S. Forest Service*)

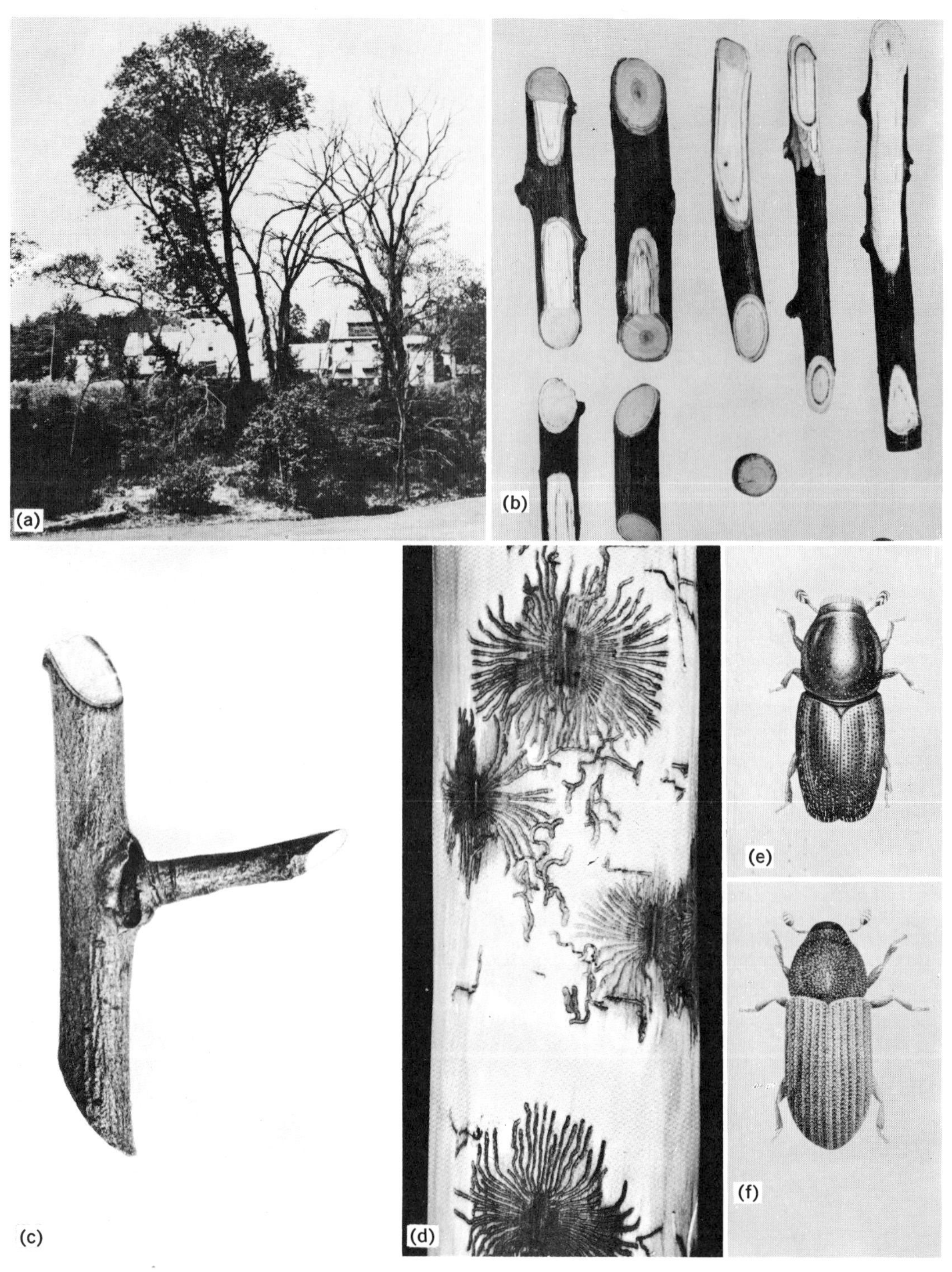

Fig. 6. Dutch elm disease. (*a*) Group of trees affected by the disease. (*b*) Discoloration in the sapwood of infected trees. (*c*) Feeding scar in small elm crotch made by an adult of the smaller European elm bark beetle (*Scolytus multistriatus*). (*d*) Brood galleries made by female beetles and larvae. (*e*) European elm bark beetle, the most important carrier of the Dutch elm disease. (*f*) Native elm bark beetle. (*a,c,d, USDA; b,e,f, Michigan State University*)

tria and *Eutypella* which seldom result in death of the tree. Large living chestnut trees are rarely seen in northern states, and they are vanishing rapidly in the Smoky Mountains of the southern Appalachians. Satisfactory resistant varieties of chestnut are being developed, but this fine tree species may never regain its prominent place in eastern forests. Control measures for other canker diseases usually consist of removing infected trees while thinning stands.

Rust diseases. Rusts, such as white pine blister rust and southern fusiform rust, which result in death of trees are very important diseases (Fig. 3), while other rusts which involve only the leaves are not usually serious problems. Many of the important rust fungi occur on two unrelated hosts, such as the currant for white pine blister rust and the oak for fusiform rust. Eradication of the alternate host, resistant varieties, pruning, and selection of planting sites less favorable for the fungus are the major control measures.

Heart rots. All tree species, even woods resistant to decay such as redwood, are subject to one or more fungi which can decay the center or heartwood portion of the main stem (Fig. 4). Some of these fungi enter through wounds and branch stubs and decay the heartwood; others enter roots and decay these roots and then the lower portion of the main stem. Most heart-rotting fungi invade trees approaching maturity, and losses by these fungi can be avoided by shortening rotation and by avoiding wounds. A few decay fungi such as *Fomes annosus*, *Polyporus tomentosus*, and *Poria weirii* can invade young, vigorous trees and kill them. Thining plantations increases the incidence of *F. annosus*, which can quickly invade fresh stumps and move from there through the roots to surrounding healthy trees. Losses caused by this fungus can be reduced by less thinning and by treating fresh stumps with chemicals (borax, urea).

Miscellaneous diseases. Dwarf mistletoe is a small, partially parasitic plant that causes its host to produce witches'-brooms (dense clusters of branches) and eventually kills ponderosa pine, other western conifers, and black spruce (Fig. 5). Control consists of eradicating infected trees.

In the 1930s birch dieback, little-leaf disease of shortleaf pine, and pole blight of western white pine appeared, and as yet the causes are not completely known. Birch dieback may be due in part to increased soil temperature which causes excessive root mortality. Little-leaf results from a combination of poor soil drainage and the root parasite *Phytophthora cinnamomi.* Pole blight is due in part to soil moisture deficiencies. These are examples of other similar diseases, which are all due in part to unfavorable weather and which may or may not involve parasites.

Diseases of shade trees. Shade trees are often grown in abnormal habitats, and because appearance is more important than the wood produced, shade tree problems are different from those in the forest. The desire for new and different trees has resulted in the introduction of exotic species, many of which are unsuited to the new climate. Even native species are often moved from their normal habitats and placed where they are predisposed to secondary pathogens.

Fungus diseases. A tree species such as Colorado blue spruce in its natural environment is seldom attacked by the spruce canker fungus (*Cytospora kunzei*), yet this fungus can kill this tree species when it is planted in other parts of the world. The effects of this secondary parasite can be minimized by selecting protected sites with better soils and avoiding dry, southwest-facing slopes. As a general rule, many diseases which can be attributed to an unfavorable environment as well as those caused by weakly parasitic fungi can be avoided by selecting native tree species from a local seed source.

Dutch elm disease, introduced to the United States in 1930 or earlier, is an important shade tree disease because of the value of the elm for lining streets (Fig. 6). The Dutch elm disease fungus (*Ceratocystis ulmi*) is introduced to the vascular system of healthy trees by the smaller European elm bark beetle (*Scolytus multistriatus*) and the American species (*Hylurgopinus rufipes*). Resistant varieties of elm and possibly systemic chemicals to prevent bark beetle feeding may be developed, but until that time sanitation (destruction of dead and dying elm wood) and a protective insecticide are the best means of control.

Virus diseases. The elms are also attacked by a virus, and the disease is known as phloem necrosis. This is the best known of the virus diseases of trees. Infected trees wilt and die, the inner bark turning brown and emitting a wintergreen odor. The virus is transmitted by the elm leafhopper (*Scaphoides luteolus*).

Bacterial diseases. One of the few known bacterial diseases of trees is wetwood of elm, caused by *Erwinia nimipressuralis* (Fig. 7). The wetwood condition occurs in many other tree species and presumably is caused by bacteria. Satisfactory control measures have not been developed.

Fig. 7. Elm infected with bacterial wetwood. The bacterial ooze is coming from a small wound.

Leaf spot and heart rot. Diseases of minor importance in the forest, such as leaf spots, may be objectionable on shade trees. Leaf spots occur in years when the weather is favorable for the fungi involved, usually coinciding with a wet spring season. Such diseases can be controlled by application of the right fungicides, usually in early spring. Heart rot, on the other hand, is a major problem in the forest but is of minor importance in shade trees.

Many trees may remain alive for decades with extensive heart rot. Expensive cavity work is of questionable value since it does not stop the decay process and may not add much strength to the tree. *See* PLANT DISEASE. [DAVID W. FRENCH]

Tundra

An area supporting some vegetation beyond the northern limit of trees, between the upper limit of trees and the lower limit of perennial snow on mountains, and on the fringes of the Antarctic continent and its neighboring islands. The term is of Lapp or Russian origin, signifying treeless plains of northern regions. Biologists, and particularly plant ecologists, sometimes use the term tundra in the sense of the vegetation of the tundra landscape. Tundra has distinctive characteristics as a kind of landscape and as a biotic community, but these are expressed with great differences according to the geographic region.

Patterns. Characteristically tundra has gentle topographic relief, and the cover consists of perennial plants a few centimeters to a meter or a little more in height. The general appearance during the growing season is that of a grassy sward in the wetter areas, a matted spongy turf on mesic sites, and a thin or sparsely tufted lawn or lichen heath on dry sites. In winter, snow mantles most of the surface with drifts shaped by topography and surface objects including plants; vegetation patterns are largely determined by protecting drifts and local areas exposed to drying and scouring effects of winter winds. By far, most tundra occurs where the mean annual temperature is below the freezing point of water, and perennial frost (permafrost) accumulates in the ground below the depth of annual thaw and to depths at least as great as 500 m. A substratum of permafrost, preventing downward percolation of water, and the slow decay of

Fig. 1. Fjell-field tundra of the high Arctic. Sedges, mosses, and lichens form a thin and discontinuous sod. Late-persisting snowbanks are withdrawing from surfaces that are lighter in color because they lack many of the common plants, including dark-colored species of lichens. (*Photograph by W. S. Benninghoff, U.S. Geological Survey*)

water-retaining humus at the soil surface serve to make the tundra surface moister during the thaw season than the precipitation on the area would suggest. Retention of water in the surface soils causes them to be subject to various disturbances during freezing and thawing, as occurs at the beginning and end of, and even during, the growing season. Where the annual thaw reaches depths of less than about 50 cm, the soils undergo "swelling," frost heaving, frost cracking, and other processes that result in hummocks, polygonal ridges or cracks, or "soil flows" that slowly creep down slopes. As the soils are under this perennial disturbance, plant communities are unremittingly disrupted and kept actively recolonizing the same area. Thus topography, snow cover, soils, and vegetation interact to produce patterns of intricate complexity when viewed at close range.

Fig. 2. Alpine tundra in French Alps. Altitudinal limit of trees occurs in valley behind building in middle distance. Although similar in vegetation structure to tundra of polar regions, Alpine tundras of lower latitudes are usually richer in vascular plant species than tundras of polar regions, and structure and composition of the vegetation have been modified by pasturing. (*Photograph by W. S. Benninghoff*)

Plant species, life-forms, and adaptations. The plants of tundra vegetation are almost exclusively perennial. A large proportion have their perennating buds less than 20 cm above the soil (chamaephytes in the Raunkiaer life-form system), especially among the abundant mosses and lichens. Another large group has the perennating organs at the surface of the soil (hemicryptophytes in the Raunkiaer system). Vegetative reproduction is common – by rhizomes (many of the sedges), stolons (certain grasses and the cloudberry, *Rubus chamaemorus*), or bulbils near the inflorescence (*Polygonum viviparum*, *Poa vivipara*, *Saxifraga hirculis*); thus clone formation is common in plant populations. Apomixis, the short-circuiting of the sexual reproduction process, is found frequently among flowering plants of tundra. Seed is set regularly by agamospermy, for example, in many dandelions (*Taraxacum* sp.), hawkweeds (*Hieracium* sp.), and grasses (*Calamogrostis* sp., *Poa* sp., *Festuca* sp.). The high incidence of apomixis in tundra flowering plants is coincident with high frequency of polyploidy, or multiple sets of chromosomes, in some circumstances a mechanical cause of failure of the union of gametes by the regular sexual process. Asexual reproduction and polyploidy tend to cause minor variations in plant species populations to become fixed to a greater extent than in populations at lower latitudes, and evolution tends to operate more at infraspecific levels without achieving major divergences. Adaptations are more commonly in response to physical factors of the stressful cold environment rather than to biotic factors, such as pollinators or dispersal agents, of the kinds that exert such control in the congenial warm, moist climates.

Soil conditions. Tundra soils are azonal, without distinct horizons, or weakly zonal. Soils on all but very dry and windswept sites tend to accumulate vegetable humus because low temperatures and waterlogging of soils inhibit processes of decay normally carried out by bacteria, fungi, and minute animals. Where permafrost or other impervious layers are several meters or more beneath the surface in soils with some fine-grained materials, leaching produces an Arctic Brown Soil in which there is moderately good drainage and cycling of mineral nutrients. In the greater part of tundra regions not mantled by coarse, rocky "fjell-field" materials (Fig. 1), the soils are more of the nature of half-bog or bog soils. These are characterized by heavy accumulations of raw or weakly decayed humus at the surface overlying a waterlogged or perennially frozen mineral horizon that is in a strongly reduced state from lack of aeration. Such boggy tundra soils are notoriously unproductive from the standpoint of cultivated plants, but they are moderately productive from the standpoint of shallowly rooted native plants. In Finland, forest plantations are being made increasingly productive on such soils by means of nutrient feeding to aerial parts. *See* SOIL ZONALITY.

Productivity. By reason of its occurrence where the growing season is short and where cloudiness and periods of freezing temperatures can reduce growth during the most favorable season, tundra vegetation has low annual production. Net radiation received at the Earth's surface is less than 20 kg-cal/cm^2/year for all Arctic and Antarctic tundra regions. Assuming a 2-month growing season and 2% efficiency for accumulation of green plant biomass, 1 cm^2 could accumulate biomass equivalent to 66.6 g-cal/year. This best value for tundra is not quite one-half the world average for wheat production and about one-eighth of high-yield wheat production. The tundra ecosystem as a whole runs on a lower energy budget than ecosystems in lower latitudes; in addition, with decomposer and reducer organisms working at lower efficiency in cold, wet soils, litter and humus accumulate, further modifying the site in unfavorable ways. Grazing is one of the promising management techniques (Fig. 2) because of its assistance in speeding up the recycling of nutrients and reducing accumulation of raw humus. *See* BIOMASS; ECOSYSTEM; VEGETATION MANAGEMENT.

Fauna. The Arctic tundras support a considerable variety of animal life. The vertebrate herbivores consist primarily of microtine mammals (notably lemmings), hares, the grouselike ptarmigan, and caribou (or the smaller but similar reindeer of Eurasia). Microtine and hare populations undergo cyclic and wide fluctuations of numbers; these fluctuations affect the dependent populations of predators, the foxes, weasels, hawks, jaegers, and eagles. Alpine tundras generally have fewer kinds of vertebrate animals in a given area because of greater discontinuity of the habitats. Arctic and Alpine tundras have distinctive migrant bird faunas during the nesting season. Tundras of the Aleutian Islands and other oceanic islands are similar to Alpine tundras with respect to individuality of their vertebrate faunas, but the islands support more moorlike matted vegetation over peaty soils under the wetter oceanic climate. Tundras of the Antarctic continent have no vertebrate fauna strictly associated with it. Penguins and other sea birds establish breeding grounds locally on ice-free as well as fringing ice-covered areas. The only connection those birds have with the tundra ecosystem is the contribution of nutrients from the sea through their droppings. All tundras, including even those of the Antarctic, support a considerable variety of invertebrate animals, notably nematode worms, mites, and collembola on and in the soils, but some other insects as well. Soil surfaces and mosses of moist or wet tundras in the Arctic often teem with nematodes and collembola. Collembola, mites, and spiders have been found above 20,000 ft in the Himalayas along with certain molds, all dependent upon organic debris imported by winds from richer communities at lower altitudes. *See* TAIGA; VEGETATION ZONES, WORLD. [WILLIAM S. BENNINGHOFF]

Bibliography: Arctic Institute of North America, *Arctic Bibliography*, 16 vols., 1953–1975; G. A. Doumani (ed.), *Antarctic Bibliography*, 2 vols., 1965, 1966; M. J. Dunbar, *Ecological Development in Polar Regions*, 1968; H. P. Hansen (ed.), *Arctic Biology*, 18th Annual Biology Colloquium, Oregon State College, 1957; J. D. Ives and R. G. Barry, *Arctic and Alpine Environments*, 1974; N. Polunin, *Introduction to Plant Geography and Some Related Sciences*, 1960; J. C. F. Tedrow (ed.), *Antarctic Soils and Soil Forming Processes*, American Geophysical Union, Antarctic Research Series, vol. 8, 1966; H. E. Wright, Jr., and W. H. Osburn, *Arctic and Alpine Environments*, 1968.

Vegetation

The total mass of plant life that occupies a given area. The plant cover in any landscape consists of a matrix of individuals usually belonging to many different species. The different species (irrespective of their abundance) are collectively referred to as the flora. Thus the flora of Walden Pond (Massachusetts), of Cheboygan (Michigan), of Manitoba (Canada), or of Switzerland is merely a list of all the different kinds of plants that have been found in a certain pond, township, province, or country. The flora of wheat fields, city streets, woodlots, and the like may also be inventoried. About 350,000 species of plants have been described. The tropics contain by far the largest numbers; the Lower Amazon Basin, for instance, may have as many as 42,000 higher plants, many arctic areas have less than 100, some high mountains have 10 or so, a large tract of desert in Mauritania is reported to have 4, and vast Antarctica has only 2. In this respect, some areas are considered to be floristically rich and others poor.

Flora provides the raw materials or building blocks of vegetation. The pieces of this mosaic are the communities, or plant societies, which can be recognized and described in terms of their composition (flora), their appearance or physiognomy (structure), and their ecology or site requirements. To this list may be added some consideration of origin and migration and of relation to man's influence.

Origin of flora. In any one region, plant species (and whole masses of vegetation, for that matter) are likely to have come from the four points of the compass, and quite often the stamp of past migrations is quite visible on the land. Thus, at high altitudes in the southern Appalachians, spruces and firs bear witness to a colder period in the past, as do the hemlocks that have taken refuge on the cool slopes at lower altitudes. Similarly, in Spain and Portugal, evergreen cherries and rhododendrons in moist ravines are relicts of a wetter climatic period. In all parts of the world, local floras can be analyzed to estimate, for instance, the importance of species of prairie origin in a now forested area, of species of desert origin in an alpine area, of species of montane origin in the plain, or of species of tropical origin in a temperate area.

Structure. The organization in space of the mass (or structure) of vegetation confers a particular appearance or physiognomy upon the countryside: The height, branching, and spread or coverage of the plants, the size, texture, and density of their foliage, the ratio of woody to herbaceous to mossy tissues, and the emergence of definite layers contrast rather strongly from place to place. Also, the times of development and withering impose an alternation of physical conditions that is sometimes extreme. For instance, the deciduous forests lose their whole canopy of leaves for a long period, allowing the light and warmth of the Sun to reach the soil. At that time a thick layer of herbs frequently develops. Conversely, the evergreenness of the boreal forest of Canada retards the wasting of snow on the forest floor in the spring. *See* PLANT COMMUNITY; PLANTS, LIFE FORMS OF.

In any one landscape, vegetation masses vary in structure: A hillside may be forested at the top and have a savannalike shrubby field lower down and a grassy swamp at the base; emerging sandbanks or sharp rock outcrops may be almost completely barren. Flooded or very wet places harbor a vegetation very different from that of the driest areas. In fact, it is often possible to assign a particular community (or group of communities) to each site: Dunes, rocky outcrops, riverbanks, and poorly drained flats all support characteristic assemblages of plants. Some species indicate the presence of lime (the Canada violet), poor drainage (the cardinal flower), the passage of fire (the fireweed), the prevalence of acidity (the pitcher plant), or the periodic recurrence of floods (the skunk cabbage). These are called indicator species. It is in their physiology that these plants differ and thereby draw attention to contrasting ecosystems, since the condition of each habitat imposes its particular stresses. *See* ECOSYSTEM.

Not only are some kinds exclusive to wet or dry soils, or to acid or salt soils, but also the overall development of the plant mass (structure) is different: Only herbs and low shrubs appear in a marsh; grasses and other herbs occur in a field; and a multilayered distribution of trees, shrubs, herbs, and mosses characterizes a forest.

Ecology. Even a summary investigation of site conditions reveals that slope, exposure, drainage, chemical composition, texture and structure of soil, and relative acidity are major factors that most visibly influence the local distribution of vegetation and induce the emergence of different ecosystems. Thus, if sampled in many stands, dune communities, floodplain forests, bog heaths, and ravine scrubs are likely to have the same composition and the same physiognomy or at least to fluctuate within a given pattern in any one region.

These landforms are not static, and as the dunes become stabilized, as the marshes fill in, as the slopes erode, and as the riverbanks are silted, the vegetation itself changes. This process, known as succession, makes for a more or less gradual replacement of one plant community by another, until a climax is reached. Such relative stability (or dynamic equilibrium) features, as a rule, the best-developed vegetation on the most completely stratified soil of the well-drained uplands. *See* CLIMAX COMMUNITY; ECOLOGICAL SUCCESSION; PLANT FORMATIONS, CLIMAX.

Energy cycles. Especially where green plants are involved, vegetation is a transformer of environmental resources: Water, oxygen, and carbon dioxide in the air and water, and various organic and inorganic compounds in the soil are taken up by plants able to utilize solar energy and are transformed into living tissue, and substances are often accumulated as reserves (starch, sugar, and cellulose), eventually to be returned to the air or soil. In the fulfillment of this process the potential of different kinds of vegetation contrasts greatly: Algal or lichen crusts on rocks have a very slow uptake, whereas a tall forest or a deeply rooted grassland effects a tremendous turnover.

This energy relationship, which expresses the productivity of the plant cover, is of course related to environmental controls, mostly to climate and soil, but also to historical factors, such as dissemination, migration, and physiographic change, as well as to human and animal action.

Management. Human interference (lumbering, plowing, damming, burning, and other activities) creates new habitats and new resources for plant and animal life. A landscape modified by people and later abandoned tends to revert to its primeval conditions, but frequently cannot because of the permanent damage or change to the soil or to the other factors of the site. *See* VEGETATION MANAGEMENT; VEGETATION ZONES, ALTITUDINAL; VEGETATION ZONES, WORLD. [PIERRE DANSEREAU]

Bibliography: W. D. Billings, *Plants and the Ecosystem*, 1964; S. A. Cain, *Foundations of Plant Geography*, 1944; P. Dansereau, *Biogeography: An Ecological Perspective*, 1957; R. E. Daubenmire, *Plants and Environment*, 1974; F. E. Egler, Vegetation as an object of study, *Phil. Sci.*, 9(3):245–260, 1942; R. Good, *The Geography of the Flowering Plants*, 4th ed., 1974; H. J. Oosting, *The Study of Plant Communities*, 2d ed., 1956.

Vegetation management

The art and practice of manipulating vegetation, such as forest, grasslands, and deserts, so as to produce and harvest a desired part or aspect of that material in higher quantity or quality, such as timber forage or wildlife. The term "vegetation" refers to the complex of plant communities in the botanical landscape. It is a technical term not to be confused with the layman's use of the word in reference to vegetable life and to plants in general. *See* VEGETATION.

The integrated field of vegetation management lies between the broader and more inclusive fields of natural resource management (concerned both with renewable biological resources and with nonrenewable mineral resources) and of land-use planning on the one hand, and on the other hand such specialties as forestry, range management, and wildlife management.

There continues to be relatively little communication between foresters, range managers, wildlife specialists, fundamental scientists, and others concerning their common interests in the behavior of the same vegetation types, presumably since they are interested in very different components and products of those plant communities. Consequently the subject is not treated as the integrated and unified sphere of knowledge which it is.

In the total landscape, aside from urban and cropland areas, vegetation is the most easily manipulatable element. Although fauna, flora, climate, and soil are part of the natural environment, and although these parameters are being subjected to increasingly effective technologies, the process is generally uncertain, hazardous, and costly to do so. The vegetation, however, can often be changed relatively cheaply and effectively.

HUMANS AND MANAGEMENT

Vegetation management involves an interrelationship between vegetation and humans, in which humans become an external factor of the environment, acting upon, interrelating with, and harvesting the vegetation in ways which are presumably purposeful and beneficial.

Mismanagement. The history of the human race is commingled with the use and abuse of forests and grasslands. Prehistoric humans, dependent upon edible roots, fruits, and nuts, "gathered" the produce. When too many people were too hungry, there was no thought for the next year. They ate at once, before someone else took the food. Only 13,000 years ago, the invention of improved spears and spearpoints resulted in the complete extinction of many large herbivores, with effects on the vegetation that have not yet been assessed. In later periods, the Biblical cedars of Lebanon and the ginkgo trees that have survived only in Chinese monastery gardens are symbolic of the all but total destruction of forest lands by early civilizations. The areas around the Mediterranean, with their distinctive semiarid winter-rain climate, have witnessed an almost complete destruction of forest cover, which resulted in replacement by various kinds of scrubs and grasslands, the loss of soil blankets, and alterations of the water and climatic regimes. In turn, these changes have had an impact on the survival of civilizations. In such cases, mismanagement reached the point of essential irreversibility.

Overexploitation of vegetation resources was common in North America, but did not receive ameliorative attention until the start of the 20th century. Ranges had been overgrazed, so as to carry the highest numbers of stock. These practices permanently destroyed desirable species of plants, which were replaced by others of less desirable quality and of reduced quantity, not adequate for feeding the animals. Changes in rodent populations, in avian and mammalian predators, soil deterioration and erosion, and other alterations make recovery extremely slow. Comparable situations occur in forest management, in which highly profitable short-term methods of lumber exploitation lead to increasingly lower productivity of the forest lands.

As the last quarter of the 20th century moves ahead, the nations are faced with new dilemmas. The scientific knowledge exists to rectify and reverse the deterioration of renewable biological resources. Nevertheless, the uncontrolled increasing human populations, the increased demands for decreasing resources, the temporary economic factors of jobs and profits, and international factors of balances of payments all combine to force short-term economic solutions to long-term ecologic problems, solutions that lead toward collapses as disastrous as those which destroyed ancient civilizations, though this time on a worldwide scale.

Effective practices. Although increased scientific knowledge has brought, and can bring, many improvements in vegetation management, contemporary practices have not been in operation long enough to prove themselves immune to short-term economic pressures or capable of maintaining continued high levels of resource productivity without unexpected side effects and ultimate deterioration of the total ecosystems.

Furthermore, technologies are being developed that, while they are shortcuts to the objectives, may in turn produce destructive side effects. For example, the widespread use of herbicides and insecticides has led to spectacular immediate gains, but it took more than a decade for extremely undesirable alterations to the ecosystem to become evident.

In turn, the development on an unprecedented scale of clear-cutting equipment, coupled with the building of logging roads, has tended to turn the forestry profession into an exploitive industry. The problem is not so much a matter of silvicultural principles as of incidental destruction of the land surface. Costs, benefits, values, judgments, and hazards of vegetation management in the last analysis must be evaluated not in the light of short-term profits to the individual or to a minor social unit, but in relation to long-term gains to society itself, qualitatively as well as quantitatively, leading to a balanced continuing ecosystem.

MANAGEMENT PRINCIPLES

All management is "change." Such artificial and anthropogenic changes in vegetation are basically not different, in their scientific nature, from "natural change," which occurs essentially independently of human influence. An understanding of one allows for an understanding and control of the other. Natural changes may be autogenic, induced directly by the vegetation itself, or heterogenic, induced by some other factor such as drought, flood, disease, or fire. Autogenic change is exemplified by the chemical reactions in the soil which result from leaf fall and decay, and by allelopathy.

Natural changes may be very rapid, as on old abandoned agricultural lands where a forest may develop from bare soil within a century; or changes may be very slow, involving thousands of years, as when new volcanoes arise from the seas, when granite islands are left by a receding ocean, or when rocky talus slopes remain for 10,000 years after the recession of a glacier. Furthermore, any one plant community may vary greatly in duration, from 1 year for a particular annual weed type to 4000 years for a redwood forest.

Those autogenic changes which may occur on completely abandoned ungrazed croplands typify important principles of vegetation management. On such old fields the human influence had been dominant until abandonment, but the influence may continue after this. For example, high nutrient levels in the soil from heavy fertilization, or accumulations of pesticidal arsenic, may long persist, recycling themselves in the new ecosystem. The actual vegetational changes are often inadequately described merely in terms of external appearances, such as the physiognomy of the various stages, with respect to floristic composition and density. That is, the original crop plant is succeeded by annual weeds, then in turn by grasses, heavy forbs, and shrubs; finally trees are predominant. Such a superficial description, however, gives no indication of the time when certain species first appear on the land, and when they finally disappear. Since the entire goal of vegetation management is the production of particular plant species, it is of great importance to know when each species does, or can, enter or leave the community.

Theory of floristic relays. According to this theory, physiognomic change is brought about by a succession of plants invading in relays, each relay being a stage of "plant succession." As traditionally described, this process begins with hardy pioneer species which invade an exposed or plowed area; these plants are adapted to survive in a harsh environment, and their survival causes the environment to be altered. They bind and change the soil and thus create new conditions to which they are less well adapted but which are suitable for new species with other adaptations. The new organisms slowly replace the pioneers and then dominate the community until their activities alter the environment, allowing replacement by better-adapted forms. The stages continue until theoretically the most highly adapted forms succeed and dominate, and reproduce, resulting in the so-called climax. *See* CLIMAX COMMUNITY; ECOLOGICAL SUCCESSION.

These simplistic classical ideas of "plant succession to climax," as the sole interpretation of vegetation change, were introduced at the start of the 20th century, but have never been accepted by a certain group of ecologists.

Theory of initial floristic composition. According to this theory, physiognomic change can be interpreted in terms of a floristic composition determined largely or entirely at the time the land was abandoned. Physiognomic stages are thus related not to newly invading plants, but to plants present from the start that develop into prominence only later. The early-predominant plants eventually terminate their short-lived existence, or are overtopped, or fail in competition with still other plants which also had been there from the start. It is important to remember that initial floristic composition was never stated as being the sole determining factor in vegetation change, but only as being a component in the total complex of change.

Actual change. Not only in the well-studied temperate summer-rain regions of the world, but from the tropics to the tundra, and from the deserts to rainy lands, there is usually a predominance of species that entered initially or early into once-bare land, including most of the shrubs and also some of the trees. In general, it may be said that few shrubs invade dense grassland as seedlings (though underground root-invasion may be significant from established plants). Solid shrub land is not often invaded by trees (if trees occur with the shrubs, they are likely to be as old as, or older than, the shrubs). Many forests are composed of trees that invaded original bare soil, or spots of later bare soil. Some trees, however, are able to invade grassland as seedlings, such as certain elms, ashes, and pines. In other cases, one invading tree species can totally alter the floristic composition. *See* GRASSLAND ECOSYSTEM.

On the basis of natural vegetation change, vegetation management involves altering the natural time of invasion, the time of disappearance, and the degree of predominance of the different species in the normal patterns of vegetation change, so as to increase the harvestable or desirable plant products, through several generations. So-called weed species are to be minimized or eliminated, while desirable species, those that produce the resource, are to be encouraged.

MANAGEMENT PRACTICES

Desirable alterations in the processes of natural vegetation change are accomplished by a wide variety of techniques and methods which directly and indirectly affect the behavior of vegetation. The direct methods are commonly segregated into fire, physical methods, and chemical methods.

Fire. Fire has probably been used to manage, and mismanage, vegetation as long as it has been used by humans to cook food. The effects of old anthropic fires are all but inseparable from those of natural fires, such as are caused by lightning and by spontaneous combustion. When humans entered North America, they found a continent of vegetation already conditioned by fire. Nevertheless, following a later period of total and unnatural fire protection, the present development of the wise use of fire in forest and range practice has been a very slow and painful process with respect to the U.S. Forest Service. Fire has now become a recognized tool in the management of an ever-increasing number of forest types. *See* FOREST FIRE CONTROL.

Physical methods. Physical methods can be symbolically expressed as the ax, the sickle, and the plow. When primitive humans first used the ax, they were selective, choosing only those kinds of trees which bore the fruits they wished to eat, the best hearth wood, the best timber for building, and the best foliage for their beasts to eat. This selectivity quickly changed the species composition of the forest, sometimes encouraging the advance of species that bore little if any resemblance to those of the original condition. For example, parts of the northern forest fringe in the Canadian Arctic were thought to be composed naturally of deciduous hardwoods. Eventually it was realized that the native peoples had completely removed the more desirable conifers.

The ax was soon supplemented by the saw, and as technology developed, massive machines were produced that moved over the land, snipping trees, stripping the branches, and neatly stacking the poles in huge carriers.

The sickle is to grassland what the ax is to woodland. Though originally used to harvest the heads from grains, this selective act turned to overexploitation which tended to destroy the very species that were desirable. As technology led to the development of machines, cutting tended to be indiscriminate. Larger and larger woody species were cut, ground, chipped, or otherwise harvested or "controlled." A pattern emerged in which there was a sharp segregation between mowed grasslands and unmowed brushy lands; this is found in many parts of the world today. Mowing machines, steadily increasing in size and efficiency, now cut swaths 20 m wide, and can handle woody stems 3 cm or more in diameter.

Although the plow is essentially an agricultural tool, its use is important in the history of much land that has been abandoned and now bears seminatural vegetation. The plow not only turns over and homogenizes the upper layers of the soil which previously might have been distinctly stratified, but also chops up the rhizomes and roots of certain species of plants, tending to make them far more abundant than they otherwise would be. The temporarily bare soil, relatively free from competing plants, can be ideal for the establishment of many species, especially trees which would otherwise not become established. The effects of plowing may thus be discernible in managed vegetation for decades, even for centuries.

Chemical methods. Since World War II a direct method of affecting vegetation has been the use of chemical herbicides. Chemicals have been used, however, since the cave dwellers first noticed that salt water killed vegetation, a technique improved by the Romans as they plowed the Carthage croplands with salt. Military research during World War II developed the present potent, but still very imperfect, herbicides. The effectiveness and safety of these chemicals depend largely on their mode of application, that is, whether this practice is indiscriminate and nonselective as by airplane, or whether it is a discriminate, selective local application. A classic case involving the indiscriminate aerial application of herbicides was the spraying in South Vietnam that started in 1965. This was done by the United States, with the idea of depriving the guerrillas of forest cover and food resources. As of 1975, there had been no serious scientific studies of the effects of these herbicides, including their poisonous contaminant dioxin, on the vegetation or people. *See* HERBICIDE.

Insecticides also play their role. The indiscriminate aerial application of some broad-spectrum insecticides such as DDT, more aptly termed ecocides, in order to control a single species of insect has vast and ramifying effects throughout that total ecosystem, and far beyond its borders. Since all pollinating insects may be affected, as well as animals which feed on these insects, the indirect effects on vegetation are felt for years beyond the time of applications. *See* INSECTICIDE.

Indirect methods. The major indirect methods of managing vegetation may be classified as those affecting water, soil, air, and animals.

Water. Among the more obvious methods of managing water is the technology of dam building. Beavers have long dominated the events of many upstream tributaries throughout the world, even to the extent of being recognized as a geological force shaping the local physiography. Yet the water effects of beavers become insignificant when compared with those of dam engineers, who are turning many rivers into a series of lakes, with enormous consequences to the entire landscape, including depletion of the deposits of sands and silts at ocean mouths, and thus to coastal phenomena related to long-shore currents. *See* DAM.

Controlled changes in lake levels can be undesirable, as in the wide bare shores during seasonal drawdowns, but can also be desirable, as in certain wildlife-management practices in which temporary drawdowns control excessive growth of undesirable aquatic plants. However, the lowering of water tables by excessive "mining" of water for human populations, such as for irrigation and urban use, is having subtle effects on natural and seminatural vegetation types, and there are further problems of polluted water. Overfertilization by phosphates and nitrates from sewage disposal systems and septic tanks leads to population explosions of undesirable aquatic plants. Balances are further upset by the addition of herbicides to control these plants, while the basic nutrient problem which causes the imbalance is unaltered. Pollution by persistent insecticides, washed in from the land, carried in the atmosphere, or transported by organisms which accumulate them, is becoming ever more hazardous. If, for example, such biocides affected the diatom populations of the ocean, the oxygen balance would be upset, in turn affecting the oxygen content of the atmosphere, with effects on all animal life, including humans. *See* WATER CONSERVATION; WATER POLLUTION.

Soil. Fertilizing is one of the most common methods of managing soil, and thus of indirectly managing the vegetation. Fertilizing is generally thought of as only an agricultural practice, but it has growing potentialities in the management of forest lands and rangelands. Another approach in soil management, the analog of agricultural contour plowing, is the terracing of rangelands and watershed lands. The terraces greatly reduce superficial runoff of precipitation and increase infiltration into the soil, thus leading to marked changes in the nature of the local vegetation. One of the problems in soil management is polluted soil, that is, soil that because of previous agricultural practices has accumulated significant amounts of arsenic, cadmium, copper, lead, and zinc, and thus affects the composition and structure of vegetation. For example, not only have smelters destroyed considerable vegetation in their vicinities, but the land has continued bare, or with stunted plants, for several decades. In addition, the accumulation of persistent pesticides in the soil has definite but still largely unknown effects on soil fauna and flora, and thus on the associated grassland and woodland vegetation.

Air. Temperature and precipitation control is a primary consideration among the approaches to managing the air. Temperature control was long accomplished locally in certain agricultural practices, such as the use of smudge pots in orange groves to prevent radiation at night during otherwise freezing temperatures. The technology of precipitation control is rapidly advancing, accomplished largely by seeding silver iodide at cloud-forming levels, or by electric discharges. Such questions as whether such procedures cause subsequent floods, or deprive other regions of rain which they need, are part of menacing social problems. Pollution and smog in urban air are known not only to limit the growth of plants that can satisfactorily be grown in cities, but also to eradicate such plants as lichens from the vegetation. However, the delayed chronic effects of minor pollution in the landscape are not easily open to scientific study. In aerial spraying with pesticides, for example, large percentages of the chemical do not fall on the target areas but are wafted away to points unknown, or can volatize and can be later precipitated with rain. Smog effects on pine forests many miles east of Los Angeles have attracted considerable attention. Even more significant is the increasing acidification of normal rainfall, first studied in New Hampshire, and now being found worldwide. Such rainfall acidifies the soil solution, with distinct and undesirable effects on the related vegetation. *See* AIR-POLLUTION CONTROL; CLIMATE MODIFICATION.

Animals. Scientists have long observed that concentrations of animals, such as birds on oceanic islands, have totally altered what would otherwise be the natural vegetation, in the process of producing guano deposits that have been measured up to 100 m thick. It is only one step removed to consider flocks of roosting exotic birds, like the starling in America. Such roosting birds have been known to kill coniferous stands. Rabbits, pigs, goats, and deer introduced to oceanic islands have decimated the natural vegetation, before they themselves have starved. England never fully realized the importance of its "rabbit pressure," until the population collapsed by an introduced disease. Relatedly, grazing of the slowly abandoned New England farmlands produced the economically valuable pine forests. Silviculturists, ignoring the animals but wanting pine regeneration, spent several decades in research before deciding that "pine does not succeed pine." Elephants, proliferating under the protection in East African national parks, can thoroughly destroy large acreages of forest. Animals, used wisely and knowledgeably, can be extremely important in managing vegetation for products other than themselves.

Management products. Various methods applied to managing vegetation can accompany or be linked with natural changes and alterations, to lead to desirable or undesirable "products." The undesirable products are called weeds, and their "control" promptly triggers still further management practices. It is convenient to distinguish two groups of desirable products: specific products such as wood derived from vegetation, and complex products inherent in the landscape itself, such as watershed vegetation.

Specific products. Among the specific products of economic value arising from vegetation are wood, forage, and wildlife habitat. Emphasis on these three products has given rise to the three major fields of forest management, range management, and wildlife-habitat management.

Wood. In forestry, timber production maintains precedence, while pulp and paper manufacture is secondary. The art and practice of forest management is often known as silviculture, while the science behind silviculture is known as silvics. Water pollution from the pulp and paper industry has become critical in many states. Timber harvesting methods have become more destructive of land surfaces, causing soil erosion. Silviculturists are tending to adopt industry-promoted aerial-spraying with herbicides, although there have been no studies fully documenting the effects of these procedures on the total vegetation or on the ecosystem. Silvics has remained essentially static, despite the vast amount of research done by government agencies. *See* FOREST MANAGEMENT AND ORGANIZATION; FOREST RESOURCES.

Forage. Range management is concerned with the natural and seminatural grasslands, scrub lands, and woodlands with a ground layer of herbaceous vegetation. The desirable forage is viewed primarily in terms of cattle and sheep. Goats, camels, asses, and other domesticated grazing mammals are locally important. Conflicts arise between the food preferences of these different animals and those of native wildlife such as antelope, deer, and bison. Long-term continued high productivity of rangeland is the goal of range mangement. Excessive destructive grazing increases the rodent populations, which take their toll of the remaining forage, but in turn attract eagles and other predators which control the numbers of rodents and thus help to maintain the balance of this ecosystem. Attempts to control real or supposed predators such as coyotes and eagles with poisons that directly or indirectly affect many other animals create greater problems. Attempts to control invading unpalatable brush by aerial spraying with herbicides also causes additional problems. On overgrazed unburned rangelands, unpalatable woody species quickly become abundant, and aggressively removing these only opens the way to other undesirable species. *See* RANGELAND CONSERVATION.

Wildlife habitat. In the early decades of the 20th century, it was found that wildlife is essentially a product of its habitat, and the habitat is to be considered in terms of food plants and cover plants. The herbivores increase directly in suitable habitat; and the predators increase as the herbivores increase. The cliché is well established that to preserve a rare or endangered animal species the habitat must be preserved. Unfortunately the practice has rarely extended beyond the cliché. Wildlife professionals have yet to attain an adequate botanical knowledge of the habitat needs of critical species, and of how to attain those vegetational habitats. In the meantime, it is the explosion of the human population, with demands for cities, tree plantations, and farms, which is the most critical single factor. *See* WILDLIFE CONSERVATION.

Complex products. Four important plant-controlled landscapes that demand purposive vegetation management practices are the watersheds, rights-of-way and roadsides, scenic and landscape areas, and wilderness areas.

Watershed vegetation. The purpose of watershed management is to produce the most water for human consumption by controlling the vegetation to obtain the greatest infiltration into the soil and the least transpiration back into air, with the greatest amounts at the sources of water supply. Depending on local circumstances, these goals are effected by raising water tables, by impounding water into reservoirs, or by equalizing river flow throughout the years. At all times the engineers seek to control floods (sometimes by quickly channeling off the water, so as to flood downstream areas) and to minimize soil erosion. Riparian vegetation, encouraged several decades ago for its soil-protecting qualities, is now often destroyed by aerial application of herbicides on the rationale that it transpires needed water. Such herbicides, however, often have undesirable effects on the grazing and human populations of those riparian areas. Basic research is still needed. In forest regions, for example, it has always been assumed, but without comparative data, that infiltration, runoff, and other factors favor watershed forests rather than substituted watershed herblands. *See* WATER SUPPLY ENGINEERING.

Right-of-way and roadside vegetation. In this domain are included the seminatural unmowed sides of highways and railroad systems, the land under electric power and telephone lines, the land above pipelines, and various other rights-of-way for the transportation of people and materials. In 1953 the total area of these lands was estimated at over 50,000,000 acres (1 acre = 4047 m^2) in the United States alone, more acreage than all six New England states combined. It has been growing ever since, especially with the development of the Interstate Highway System, said to have been taking 1,000,000 acres a year. Vegetation on these lands is largely a problem of disposing of unwanted trees by root-killing and by developing a stable low cover of herbs or shrubs that is also highest in scenic, recreational, wildlife, and conservation values.

Unfortunately the field has been dominated by the unwise use of herbicides, and much unwanted brush remains after 25 years of repeated applications. *See* LAND-USE PLANNING.

Scenic vegetation. Wild-landscape management is a field closely akin to landscape gardening, but differs from it in the minimal importance of initial clearing and subsequent planting with nursery-grown horticultural materials. To the contrary, the process involves an intaglio effect of removing—by physical or chemical methods—those particular native individual plants that are not wanted. The goal is to produce a terrain of varied plant communities that forms a scenic mosaic of vegetation, one of high esthetic values. The ideas are applicable to the margins of parkways, to the median strips of divided highways, to commercial resort areas, and to small country estates, all situations where timber and forage are not of primary importance, but where conservation, wildlife, low costs, and esthetic values are paramount.

Wilderness vegetation. A final concern is the management of complex vegetation-dominated areas commonly known as wilderness or natural areas, although both terms are inappropriate. The scientific and educational values of these lightly exploited lands are unquestioned, both for themselves and as ecological buffers and reservoirs of diversity for a finite and reasonably balanced world ecosystem. Nevertheless, private organizations owning acreages of such lands have no established programs for basic scientific studies. These programs in turn would also serve as applied and practical research, for the natural areas would be "controls" or "standards" for heavily exploited surrounding areas. It was originally thought that a natural area would stay natural if it were "preserved," as by legal ownership and by the essential elimination of people. Increasing sophistication forced the realization that not only was the human influence all-pervasive in virgin areas, but nature itself was in constant change.

Current vegetation management policies are little more than decisions by advisory scientists concerning the allowance or use of fire, control or attempted eradication of exotic "unnatural" plants or animals, and the control of overpopulation of herbivorous species such as deer in the absence of natural controls such as predators or famine. Finally, certain plant communities are recognized as ephemeral or temporary. As such, they may have been born in a catastrophe, and to "preserve" them may be impossible.

The paradox exists that by total and complete protection, natural areas become "super-natural," of tremendous scientific interest, but not representative of any area that ever existed before on the face of the Earth. The absence of reasonable vegetation management programs for these lands is merely a reflection of the immaturity of an adequate vegetation science, and in turn of the absence of reasonable research programs on these wild areas that would eventually fill that deficit.

Summary. Vegetation management as a scientific discipline continues to look forward to an integration that will involve the fields of forestry, range management, wildlife management, and other vegetation-related practices. In turn, it waits upon a mature, integrated vegetation science. Since terrestrial vegetation is part of a larger whole, involving soil, climate, oceans, and humans, scientists must watch for side effects, long-term alterations, hazards, and polluting technologies that are detrimental to the Earth ecosystem, upon which humans are dependent. *See* CONSERVATION OF RESOURCES; ECOSYSTEM. [FRANK E. EGLER]

Bibliography: F. E. Egler, *The Nature of Vegetation. Its Management and Mismanagement*, 1976; F. E. Egler et al., *The Right-of-Way Domain*, 1975; H. F. Heady, *Rangeland Management*, 1975; W. R. Humphrey, *Range Ecology*, 1962; W. G. Kenfield, *Wild Gardener in the Wild Landscape*, 1972; D. M. Smith, *Practice of Silviculture*, 7th ed., 1962; S. H. Spurr and B. V. Barnes, *Forest Ecology*, 2d ed., 1973; L. A. Stoddard and A. D. Smith, *Range Management*, 1975.

Vegetation zones, altitudinal

The extensive, even transcontinental, bands of physiognomically similar vegetation that range from tropical rainforest to arctic tundra; the more or less distinct belts of vegetation that occur, one above another, on high mountains; and the concentric or linear bands of distinctive vegetation around ponds, along rivers and coasts, and in similar sites. Such zones are correlated with vertical and horizontal gradients of environmental conditions. In most situations, these graded conditions are related directly or indirectly to solar radiation (heat and light) and moisture. *See* ENVIRONMENT.

Latitudinal variations in solar radiation. The two sensible components of solar radiation are heat and light; the Sun is the principal source of the heat and light energy impinging upon the Earth. If the Earth were a plane oriented perpendicularly to the axis of the Sun's rays, its entire surface would be exposed equally to radiation. Because the Earth is spherical, however, only a small section, known as the tropics, is exposed regularly to more or less vertical radiation. As one approaches the Earth's poles, the angle of incidence of the Sun's rays decreases and their energy is diluted by being spread over a greater area. In addition, the Earth is enveloped by layers of gases collectively known as the atmosphere. These gases and the materials suspended in them reflect and absorb solar radiation. The path of solar rays through the atmosphere is shortest in the tropics and lengthens toward the poles as the angle of incidence decreases. Thus, from the tropics poleward the intensity of solar radiation generally decreases. Because moist air absorbs and reflects more of radiation than dry air, however, the pattern of decrease is not uniform. Through their effects on atmospheric humidity and their action as heat exchangers, large bodies of water, including the oceans and major lakes, modify the climatic patterns of the Earth. The pattern is further modified because the Earth's rotational axis is not perpendicular to the plane of the Earth's orbit around the Sun. In effect, this inclination causes the directly vertical noontime rays of the Sun to migrate annually from the Tropic of Cancer to the Tropic of Capricorn and back again and produces the familiar seasonal changes in day length and temperatures in extratropical regions. These factors and others combine to form an intricate pattern of climatic regions over the Earth's surface. These regions generally are oriented in parallel bands, from the tropics around the Equator to the arctic and antarctic regions around the poles, but there are many irregularities occurring in the pattern.

Altitudinal variations in solar radiation. Within any major region there are variations in solar radiation associated with elevation and, outside the tropics, with aspect. With an increase in elevation above sea level, air becomes less dense and is not able to hold as much moisture per unit volume. The lessening of density results in a decrease of barometric pressure of about 1 in./1000 ft. With a greater dispersion of gas molecules and a lower concentration of water vapor, the heat-absorbing power of the air is diminished. In addition, as altitude increases, the distance through the atmosphere that the Sun's rays must travel to reach the Earth diminishes. Therefore, more light and more heat energy reach a surface at a high elevation than reach a similar surface at a low elevation. It has been estimated that about 20% of solar radiation is reflected or absorbed by the atmosphere before it reaches sea level, whereas only 11% is lost before it reaches a 14,000-ft-high peak. The difference in light intensity is not of sufficient importance to photosynthesis to be a critical ecological factor, but the heating of the soil may produce very high surface temperatures during the day. The heat is lost rapidly by reradiation at night, and because the air is not able to absorb appreciable quantities of heat, nocturnal temperatures are relatively low. Measurements on Pike's Peak, Colo., for example, occasionally indicate surface soil temperatures near 140°F (60°C) at a time when the air temperature 5 ft (1.5 m) above the soil is about 70°F (21°C). Nocturnal reradiation also prevents the heat from being conducted into the deeper soil layers. On Pike's Peak the soil at a depth of 10 in. (25 cm) may be 85°F (47°C) cooler than the surface soil.

Altitudinal variations in temperature. Air masses that encounter mountains are forced upward and over the barrier. The rising air expands and is cooled adiabatically at a rate that averages about 1°F decrease in temperature per 330 ft (or 1°C per 181 m) rise in altitude in summer and 1°F per 400 ft (or 1°C per 219 m) rise in winter. The thermic rate varies considerably because of moisture, wind, slope exposure, and other factors.

The reduction of temperature as altitude increases also brings about later frosts in the spring and earlier frosts in the autumn and thus shortens the frost-free or growing season. In Arizona, for example, a station at an elevation slightly above sea level has an average frost-free season of 322 days, a station at an elevation of 2660 ft (811 m), has an average frost-free season of 246 days, and another station at an elevation of 9000 ft (2743 m), Mount Lemmon, has a frost-free season that averages only 122 days. However, there is a considerable variation in the length of the frost-free season from place to place in a given vegetation type. Also, temperature alone may not suffice to determine the growing season, because many low plants may be covered by snow for several days or even weeks after the air temperature becomes favorable. Apparently, the absolute length of the season of favorable temperatures is not closely correlated with altitudinal changes in vegetation.

Mountainous precipitation and evaporation. A portion of the water vapor contained in rising air masses condenses as the air mass is cooled and may form clouds or fall as rain or snow. Thus, mountains are often referred to as islands of greater precipitation. In the mountains, the annual precipitation generally increases in amount with an increase in altitude, at least up to a certain elevation. Mountains subject to moist winds from a prevailing direction may have heavy precipitation on windward slopes and a rain shadow or dry belt on the lee side. If the mountains are high enough, there is generally a maximum of precipitation on the intermediate slopes and a gradual decrease in average precipitation from that point to the mountain crests. Furthermore, the lower temperatures at high altitudes result in a lessening of evaporation and transpiration, so that precipitation may be more effective at higher elevations. The rate of diminution of the evaporation rate caused by cooling is partly offset by the fact that the evaporating power of the air increases with altitude as atmospheric pressure decreases. Decreases of evaporation losses that result from cooling have been demonstrated. However, these losses are also offset by lower soil temperatures, which reduce water uptake, and by increases in wind velocity.

The decreased atmospheric pressure at higher altitudes also is accompanied by reductions in the partial pressures of oxygen and carbon dioxide. Thus, the oxygen concentration in the soil atmosphere decreases with altitude and may become a limiting factor for some species. Likewise, the lower partial pressure of carbon dioxide may be limiting to photosynthesis in many species.

Mountain environments and vegetation. From the preceding summary of altitudinal gradation of climatic phenomena, it can be seen that mountain environments are complex and highly variable. The conditions are never the same on two mountain masses, because many factors, such as slope gradient, exposure, direction of prevailing winds, nearness to oceans, geologic structure, and local topography, lead to individuality. In addition, plant life may respond to different limiting factors at various elevations. For example, studies in eastern Washington and northern Idaho have indicated that ecotones between vegetation types are related to critical deficiencies of heat at upper elevations, and to critical water shortages at lower elevations.

Generally, mountains in tropical, subtropical, and temperate areas are covered by forests, regardless of the nature of the surrounding vegetation. The forests, often of several types, such as tropical rainforest, summergreen deciduous forest, and needle-leaved evergreen forest, are found at different elevations, and will extend upslope to a point at which they yield to meadows composed of grasses, perennial herbs, and low shrubs. If the mountain is high enough, there may be an alpine tundra and an area of perennial snow without vegetation (Fig. 1). Even in widely separated areas with different floras, similar environmental conditions tend to support similar vegetation types. *See* LIFE ZONES.

The variety of altitudinal zones thus increases with the height of the mountain. However, it also increases with the nearness of the mountain to the Equator (Fig. 1), because as one travels farther

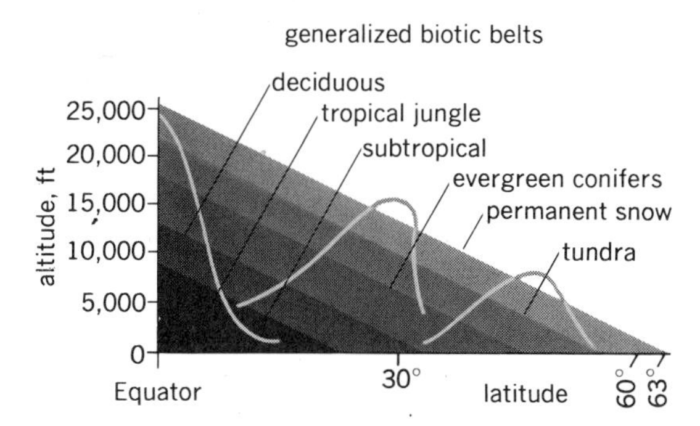

Fig. 1. Diagram which illustrates the variations of altitudinal positions of some vegetational zones with latitudinal changes. 1 ft = 0.3 m. (*From A. M. Woodbury, Principles of General Ecology, McGraw-Hill, 1954*)

from the Equator, he experiences climatic and vegetational changes parallel to those experienced when one climbs a mountain. Thus, fewer low-elevation forms of vegetation are available in higher latitudes. Ultimately, in arctic regions, only tundra vegetation may occur on the mountains, and in the antarctic there are many snow-covered ranges with no vegetation.

Altitudinal zonation. The altitudinal zonation of vegetation in the northern intermountain region of Idaho and eastern Washington can be taken as an example of the arrangement of zones in a temperate mountainous area (Fig. 2). In this area, the lower elevations are dominated by sagebrush (*Artemisia tridentata*). At higher elevations, several grassland types may occur, and above these may be a poorly developed woodland of juniper (*Juniperus scopulorum*), with or without limber pine (*Pinus flexilis*). The ponderosa pine zone is the lowest of the well-formed conifer forests. It is often composed entirely of ponderosa pine (*P. ponderosa*), with an undergrowth of prairie grasses and forbs or shrubs. Douglas fir (*Pseudotsuga taxifolia*) occupies the zone above the ponderosa pine, but is often found in mixed stands with the pine or with species from the arborvitae-hemlock zone that occurs above. The highest coniferous forest zone is the spruce-fir zone in which the Engelmann spruce (*Picea engelmanni*), alpine fir (*Abies lasiocarpa*), and mountain hemlock (*Tsuga heterophylla*) occur in various mixtures. The sedge-grass zone, or tundra, occupies all of the areas above the limits of forest growth.

Topographic effects. Most mountainous areas are composed of a variety of geological formations, are interrupted by deep canyons, ravines, or passes, and have slopes with various exposures. In such areas, vegetation zones, therefore, generally are very irregular and may have projections or outliers that extend hundreds of feet in elevation above or below the main body. Usually, a given zone is higher on a warm, Equator-facing slope than on a cooler, pole-facing slope. Measurements in several areas have shown that temperatures on an Equator-facing slope are usually a few degrees warmer than those on a pole-facing slope as a result of the greater intensity of insolation on the former.

In narrow valleys, ravines, or canyons, inversions of temperature may occur frequently as masses of cold, more dense airflow down the valley walls and along the valley bottom. These cold air masses often accumulate to some depth in places where their paths are obstructed by constrictions in the valley or by other barriers. Thus, the minimum temperatures increase upslope from the valley bottom to a point at which the inversion ends, sometimes as high as 300 m above the valley floor; then they decrease in accord with the normal thermic gradient. As a response to the lower temperatures and greater moisture in such valleys, they are often occupied by pennantlike projections of vegetation zones that occur at much higher elevations on the mountain slopes. Thus, inversions of vegetation zones can be seen in many narrow valleys. In the northern intermountain region, for

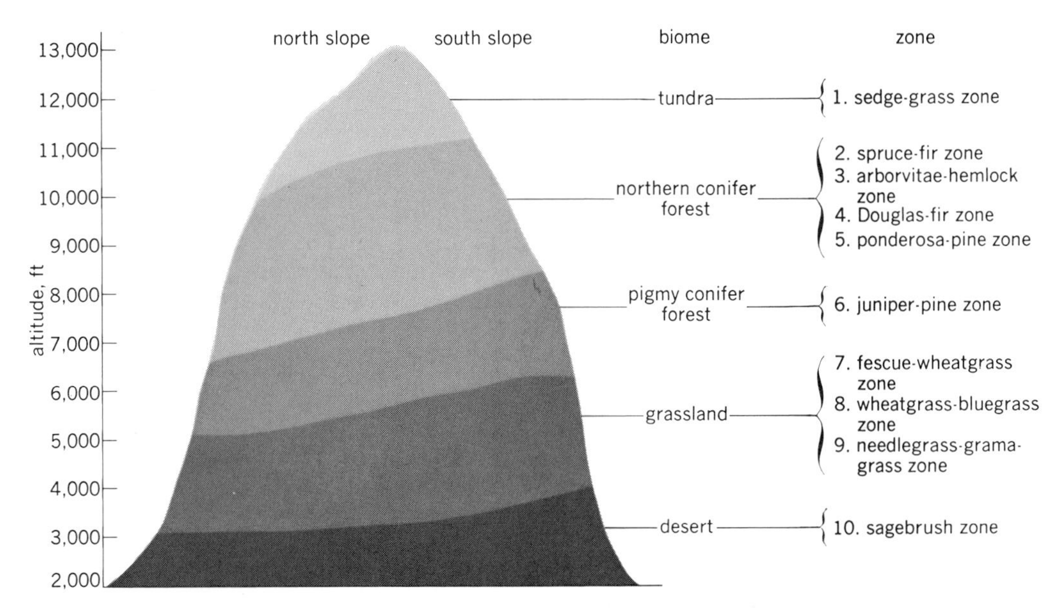

Fig. 2. The altitudinal zonation of vegetation in the northern intermountain region of Idaho and eastern Washington can be used as an example of the arrangement of zones in a temperate mountainous area. 1 ft = 0.3 m. (*Based on data by R. F. Daubenmire, from E. Odum, Fundamentals of Ecology, Saunders, 2d ed., 1959*)

example, the bottom of a cool canyon may be occupied by a spruce-fir forest, the lower slope may be covered with a Douglas fir forest that intergrades with ponderosa pine; and the upper slopes and adjacent mountain sides may be covered with ponderosa pine. Thus, the ponderosa pine zone is higher in altitude in such localities than is the spruce-fir zone. *See* CLIMAX COMMUNITY; PLANT COMMUNITY; TERRESTRIAL ECOSYSTEM; VEGETATION ZONES, WORLD.

[JACK MC CORMICK]

Bibliography: R. F. Daubenmire, The life-zone problem in the northern intermountain region, *Northwest Sci.*, 20(2):28–38, 1946; R. F. Daubenmire, Vegetational zonation in the Rocky Mountains, *Bot. Rev.*, 9(6):325–393, 1943; F. Shreve, *The Vegetation of a Desert Mountain Range as Conditioned by Climatic Factors*, Carnegie Inst. Wash. Publ. no. 217, 1915.

Vegetation zones, world

Plant communities assembled into regional patterns by the region's physiography, geological parent material, and history. The physiography determines the drainage of the region; the geological parent material provides a definite physicochemical composition for the plant community; and the history of the region reflects the major climatic shifts and the more or less disturbing interference of humans.

The vegetation of the world forms a mosaic that can be viewed at different orders of magnitude. A contemplation of the concentric belts that circle a pond reveals that, from the deep to the shallow water and on to permanently flooded ground, different species of plants, and therefore different communities, dominate: The floating pads of water lilies and pondweeds are strewn across the open water and rooted in the ooze below; the rushes form an emergent belt of stiff culms rising to an even height; and, nearer the shore, arrowheads expand their broad halberd-shaped leaves. Similar zonations can be seen in many places where the water level varies or where some other feature of the substratum undergoes a change induced by soil or light. These individual and closely knit plant communities respond in a characteristic way to site factors, to the quality and quantity of resources and conditions necessary to their life: light, heat, water, oxygen, carbon dioxide, nutrients, penetrability, and relative acidity of the substratum. *See* PLANT COMMUNITY; VEGETATION.

On the broader geographical scale, where vegetation can actually be mapped, it is impossible to show such features as the rings of vegetation that surround a pond or even to account for the many variations of height and density of forest on a mountain or for the interwoven scheme of marsh and upland in a river basin. Some generalization must therefore be made.

As far as land is concerned, it is generally agreed that the prevailing upland vegetation is the most characteristic of the region as a whole; it very often covers the greatest percentage of the total surface. It is also argued that the plant cover of well-drained land holds the key to the whole regional dynamics, in that the gradual improvement of drainage in both the wet and dry sites is conducive to a replacement of marshy (hygrophytic) or desertlike (xerophytic) types by intermediate (mesophytic) types, which are better balanced and more stable. This concept is no longer adhered to in such a simplified form: The theoretical convergence of all vegetation changes toward a single climax type does not operate with uniform efficiency throughout. For instance, in regions of low or irregular rainfall, the arrests in succession are more pronounced, and consequently the effect of soil becomes as great as that of climate in determining the plant cover. *See* ECOLOGICAL SUCCESSION; PLANT FORMATIONS, CLIMAX.

There remains some validity, however, in retaining the prevalent upland vegetation as an indicator of total regional conditions, primarily climate. It is in this connection that one of the following 20 vegetation classes or formation classes can be recognized as a characteristic of each region. It may be added that they differ very much in their features of productivity: The huge mass of plant tissue which the tropical rainforest elaborates daily from soil and air materials is in strong contrast to the long dormancy of the desert and the tundra and their explosive but rather limited growth. *See* CLIMAX COMMUNITY; PHYSIOLOGICAL ECOLOGY (PLANT).

Important as each of these climax types may be to its region, their assignment to an area hardly provides a true geographical description. Even when the prevailing climatic control, as well as the predominant soil-forming mechanism, has been detected, and the principal historical influences have been traced, other forms of vegetation, especially in the xerophytic and hygrophytic situations, remain to be considered. The prevalence of dunes, cliffs, and permanent floods may also be such as to allow overwhelmingly more space to the xerophytic or hygrophytic than to the mesophytic. Finally, so much of the surface of the Earth has been modified by man that the vegetation of the man-made and man-maintained landscape, as well as the main crops that serve the regional economy, must be brought into focus.

Therefore, the following 20 vegetation classes are to be considered as areal formations. In the summary outlines hereafter, the following information will be given for each class: a description of the vegetation that is typical of the class on a regional basis; the kind of climate that prevails and the obvious soil-forming process; the principal kinds of vegetation to be expected in drier or wetter habitats; and some actual geographical examples of each one. It is certainly of major interest to point also to the peculiar play of climate-soil-vegetation as it operates in each situation. This permits an evaluation of quantity and diversity of environmental resources, of their availability, of the relative adequacy of the tapping method of the plants, and of the limitations of plants in resource utilization. Thus, formation of peat, susceptibility to fire, succulence, and other responses resulting from the reaction of vegetation as a whole to environmental forces are very unevenly, although predictably, distributed.

Much of the prevailing vocabulary used to describe vegetation is misleading, for two reasons. First, it all too frequently evokes a characteristic of the terrain that is assumed to induce a particular plant formation (such as plains, tundra, barrens, or shallows) rather than a feature of the plants themselves (such as forest, meadow, or palm brake). Thus rainforest is a type prevalent in areas of constantly high rainfall. It would be better to refer to it as "a tall, dense, broad-leaved evergreen forest," a descriptive expression which evokes qualities of the vegetation. The second reason is that many expressions are given a different meaning in different areas. Such is the case for bush, woodland, jungle, savanna, and steppe. Nothing short of a new approach (involving a new vocabulary) would do away with such equivocations. Synonyms are usually partial, since they have often been applied to a regional phase and not to the formation class as a whole.

Tropical rainforest. A class also known as selva, hylaẹa, forêt dense, forêt ombrophile, forêt tropicale humide, Regengehölze, and pluviilignosa. It consists of tall, close-growing trees, often buttressed at the base, their columnar trunks more or less unbranched in the lower two-thirds, and forming a spreading and frequently flat crown. Some of the outstanding ones protrude above the canopy. The number of species in any one area is so great that often no more than one individual of each can be seen from any given observation point. Several families are represented, among them the laurel, fig, brazil nut, locust, and myrtle families. However, they resemble one another in that their leaves are large and evergreen and that they often support a mass of epiphytic plants that rest upon their boughs but are not parasitic. These are ferns and orchids and, in America, bromeliads. Lianas are also common. They are twining rather than climbing, their wood often weak and needing the support of trees and shrubs. They can therefore attain great heights in the forest only if they are carried by a growing tree. The dark floor of the forest supports rather sparse plant life, and thus it is easy to travel through. The tropical rainforest prevails over large tracts of the wet and warm lowlands near the Equator in South America (the Lower Amazon Valley), Asia (parts of India and Indonesia), and Africa (Congo Basin). *See* RAINFOREST.

Such vegetation occurs only in areas of high temperature and high rainfall where both are also evenly distributed and constant; that is, no shocks are experienced by the plants and the seasons are rather poorly defined. The soils are lateritic (latosolic); they accumulate no humus at the surface, the silicates are washed down, and the iron compounds remain near the surface, lending a red color.

Coastal areas harbor mangrove, a forest of stilted trees rooted in the fine silt of tidal swamps. The flooded and riparian forests, for example, the Amazonian igapó, contrast with the uplands by the thickness of their undergrowth and the abundance of palms. Coconut groves are common on strands, where they alternate with screw pines, casuarinas, and other drought- and wind-resisting plants. But where the rainforest itself has been disturbed, the low growth becomes very dense, with tangles of lianas, bristling bamboo scrub, thorny palms, and thickly branching shrubs. This secondary condition is known as jungle. The soils of these regions are quite vulnerable once they have been stripped of their native cover. Therefore, clearing agriculture, with shifting plots of small dimensions, is not a fundamentally bad practice. Subsistence on upland rice, squashes, and corn is possible. However, large tracts of land are now cultivated to coffee, bananas, sugarcane, pineapple, rubber trees, and other crops, or they are grazed. It is a great problem to maintain the fertility of these lands at a productive level. Cassava thrives on very poor sites. Rice must be grown under periodic artificial flooding, although the less productive upland varieties are also in use.

Temperate rainforest. Terms used to describe this class are laurisilva, laurel forest, cloud forest, notohylaea, moss forest, and sometimes subtropical forest. It differs from the tropical rainforest in that it has comparatively few species and therefore large populations of one kind. The trees are somewhat shorter, with smaller, evergreen leaves (especially the conifers, podocarps, and araucarioids, which are frequently present). Large tree ferns are usually abundant, in fact more so than in the tropical rainforest. Palms are sometimes present, as are lianas and epiphytes (with a prevalence of bryophytes over vascular plants). The herb layer of the forest floor is frequently well developed. Sometimes a sheath of mosses invests the trunk of trees at a certain level (moss line).

Areas of high and evenly distributed rainfall also having relatively little change in temperature and no frost (much of New Zealand, Madeira, southern Japan, and the Paraná Plateau of Brazil) are favorable to this kind of vegetation. It is probably more accurate to class the western hemlock – red cedar forest of Pacific Northwest America here than in the needle-leaved evergreen forest. Tropical mountains commonly grow stands of temperate rainforest (then called cloud forest) at the cloud level (Mexico, West Indies, Congo, and East Indies). This vegetation prevailed over very large portions of the Earth during the Tertiary, and some uninterruptedly equable areas (such as Madeira and the Canaries or the coastal strip of central California) retain vestiges of ancient communities (laurel or redwood forests). Some podocarps or araucarioids, several laurels, Myrsinaceae, the southern beeches, and trees of the *Weinmannia*, *Clethra*, and *Drimys* genera are typical of temperate rainforests.

Pedogenesis (soil formation) oscillates between laterization and podzolization. A blocking of drainage often results not only in marsh formation but in bog formation. Dune, scree, cliff vegetation, and practically all the very abundant second growth are evergreen. Forest yield is excellent in quantity and quality, although exotic species often do better than native ones. Much of the land is grazed or given to mixed farming. The permanent moisture is extremely favorable to tropical and subtropical garden crops as well as to temperate ones. Thus in Madeira oranges and apples, bananas and grapes, and sugarcane and wheat are grown side by side.

Tropical deciduous forest. Also known as monsoon forest, subtropical forest, semideciduous forest, forêt tropophile, forêt mésophile, and Monsundwald, this type consists of trees, sometimes almost as tall as those of the rainforest, which shed their leaves periodically. This condition makes for rather a conspicuous development of one or more lower layers. Thus the teak forest of Burma shows good stands of bamboo underbrush that retain their foliage all year. Lianas and epiphytes are locally abundant. *See* FOREST AND FORESTRY.

The shedding of leaves is a response to the long dry season which usually alternates with a very wet one, as in monsoon Asia. It may result, in very warm areas, from a shorter dry season (West Indies). However, the light period is often involved also, and many trees lose their leaves even in exceptionally wet years. The soil processes tend toward laterization. West Africa and Malaysia probably have the greatest areas of this vegetation, which is also present more sparsely in South and Central America, often in contact with the boundary of tropical rainforest.

The galleria forests that grow along the streams are quite different, inasmuch as moisture is available to them constantly. Jungle, a thicker, more obstructed vegetation, is common as second growth. However, where lateritic processes have been extreme or where soil degradation is otherwise very advanced, savanna is likely to predominate. These are areas of tropical agriculture (with the exception of the higher-moisture-demanding crops such as pineapple and cacao) where the yam, sweet potato, and sugarcane are cultivated and where pasturing is a common practice.

Summergreen deciduous forest. Other terms used to describe this class are temperate deciduous forest, forêt décidue, forêt à feuilles caduques, and Sommergrün Laubholzwald. This forest is tall and close-growing, the trees unbranched in their lower half and shading a rather scattered underbrush of small trees and shrubs. The herbs are exceedingly abundant in the spring, before the leaves fully develop in the canopy, but many of them disappear completely from above the soil surface to be replaced in the summer by well-distanced tufts of taller ones, many of which are short-day plants.

This kind of forest (dominated by oaks, maples, chestnuts, beeches, ashes, cherries, hickories, walnuts, and lindens) is practically restricted to the Northern Hemisphere (eastern North America, western Europe, and eastern Asia) under a continental climate of cold winters and very warm summers. The warm summers partly inhibit podzolization, so that the zonal soil is brown podzolic or gray-brown podzolic. With the possible exception of a small belt of alder forest in Argentina, this formation is absent from the Southern Hemisphere.

After lumbering, in this type of forest pines very often take over in a spectacular way, spreading from a former position on ridges, sandy, or other dry sites. The periodically flooded lowlands that accumulate a rich muck are also normally forested, although the tree species (poplars, willows, ashes, and elms) are not the same as on the upland. Poorly drained areas, on the other hand, harbor extensive swamps and marshes, both fresh- and salt-water. Completely blocked drainage retains peat, which has often accumulated to a great depth in previous (often colder) times. *See* FOREST MANAGEMENT AND ORGANIZATION.

Since some of the most densely populated parts of North America, Europe, and Asia occur in this area, few vestiges of primeval forest remain; besides the domesticated groves, there are many second-growth stands of timber (mostly pine, oak, and birch). The nonurbanized or unindustrialized land is usually devoted to cereals, mixed farming, and horticulture.

Needle-leaved evergreen forest. This class, also termed coniferous forest, Nadelwald, softwood forest, and resinous forest, consists of stiff, upright trees (conifers), usually conical at the top, with rather short lateral branches, growing quite close together and casting deep shade on very sparse shrubs, patches of herbs, and a sometimes continuous moss layer on the ground. The leaves are small, hard, and evergreen. Specific composition is most homogeneous in spite of the large circumboreal range. The tree flora is richest around the Pacific (*Pseudotsuga*, *Keteleeria*, and *Libocedrus*), poorer in the continental masses (*Picea* and *Abies*), and poorest on some of the European mountain ranges. The subordinate plants are often ericads, orchids, and sedges.

The podzolized soil is acid as a combined result of a cold, humid climate and the slow decomposition of the rather stable resins. Where blocked drainage causes anaerobic decomposition, peat accumulates indefinitely, so that bogs are an outstanding feature of rough, glaciated terrain such as the Canadian and the Scandinavian Shields. The Canadian muskeg shows all stages of development, from reed or sedge meadows to ericaceous scrub on *Sphagnum*-moss mats to closed larch and spruce bog forest. Rocky outcrops are commonly covered at first by lichen crusts and later by small grasses and trailing shrubs.

The needle-leaved evergreen forest is extremely flammable, and natural causes, as well as human carelessness, induce large-scale destruction in the western United States, northern Canada, Siberia, and northwestern Europe. This is often followed by a rapid invasion of deciduous-leaved trees (poplars and birches) that last hardly more than a generation.

The production of lumber for construction or pulpwood, hydraulic power, fishing, and hunting dominate. However, subsistence agriculture is scattered throughout, and some of the broader alluvial lands have higher fertility and harbor a prosperous dairy industry. *See* FOREST RESOURCES.

Evergreen hardwoods. This class, also termed sclerophyll forest, Hartlaubgehölze, and durilignosa, occurs as low forest or woodland characterized by hard- and rather small-leaved trees with a thick bark and sometimes an open crown. The undergrowth consists largely of evergreen broad-leaved shrubs, with some lianas (often spiny) and a spring herb layer. In the Northern Hemisphere, live oaks predominate, whereas proteaceous types do in the Southern Hemisphere.

The yearly growth cycle has the two peaks (spring and autumn) that distinguish the Mediterranean regime (whether in Spain, coastal California and Chile, South Africa, or southwestern Australia): cool, rainy winters, and warm (but not tropical) dry summers. The tendency in soil formation oscillates between laterization and podzolization.

Although bogs are relictual, marshes are extremely well developed (Pontine, Camargue, and Doñana) and are in contact with extensive salt-marshes.

A great deal of the vegetation in these areas has been modified by combinations of lumbering, fire, grazing, and agriculture, and the secondary types are mostly maquis or chaparral, a rather dense scrub, often spiny and interspersed with aromatic herbs. Garigue or matorral are even lower and more open and show ultimate degradation. Some areas are also occupied by ericaceous scrub (heath) or by pine or oak barrens (landes).

Mediterranean lands, where they are not too badly eroded by grazing goats and sheep, grow olives, citrus fruits, vines, and also cereals, especially wheat and corn.

Tropical woodland. This class has also been referred to as savanna-woodland, parkland, forêt claire, parc, and Savannenwald. It differs from the forest not so much in the size of the trees as in the wider spacing that allows more growth in the subordinate layers. Thus, the *Eucalyptus* woodlands in Australia, the broad-leaved-evergreen woodlands of western Africa, and a wavering belt in Brazil between the Amazonian rainforest and the campo cerrado follow the same pattern of tree-shrub-herb distribution. The lower strata are usually sparse but almost always characterized by evergreen shrubs and seasonal graminoids. *See* SAVANNA.

The climate is warm and moist enough for vigorous tree growth and yet too dry to allow closure of the canopy. These conditions prevail between tropical-wet and tropical-dry zones. The climatic cycle involved, however, may be one of short-term shift between these extremes or of relatively stabilized alternation. Pedogenesis reflects these opposing tendencies, with the drier sites tending to calcification (or even salinization), and the moister to laterization.

In ravines and other places where a more constant water supply is available, true forest is likely to develop. On the contrary, on ridges, dunes, and cliffs, the plant cover is reduced to low scrub. These areas are vulnerable to fire, which, combined with grazing, creates extensive erosion hazards.

Temperate woodland. Woodland, parkland, forêt-parc, and Savannenwald are other names used to describe this vegetation class. Similar to the tropical woodland in spacing, height, and stratification, it can be either deciduous or evergreen, broad-leaved or needle-leaved. The California yellow pine zone is fairly typical: It stands between the sagebrush scrub and the needle-leaved evergreen forest. The parkland of Saskatchewan offers a very different physiognomy with its medallions of aspen trees in a matrix of grassland, a sort of "emulsion" of forest and prairie.

The climate shows a moist and a dry season and, as with the tropical woodland, there may be short-term fluctuations. Soil formation here is also unstable and highly susceptible to conflicting chernozem-podzol, glei-podzol, or saline-podzol influences.

In washes and flats, grassy or shrubby formations tend to prevail, whereas in ravines and on river edges a curtain of trees appears, generally poplars and willows.

Tropical savanna. This is defined as a vegetation of widely scattered or bunched small trees growing out of lower vegetation, and it undergoes many variations. It is also known as sabana, Savannen, campo cerrado, savane, and savane arboree. The trees may be deciduous or evergreen, and the low growth, shrubby or grassy. The seasonal rhythm is very pronounced and induces a pulsation of vegetative activity which strongly marks the aspect of the landscape. Thus the grasses in the Brazilian campo cerrado and the Kenyan acacia savanna rise, green and dense, and eventually fade to a straw mat. Savanna plants are frequently very deeply rooted: Some shrubs are able to tap the enormous reserve at the phreatic (underground water) level itself.

The tropical savanna regime rests upon a pronounced dry season, often two. The fluctuation of water availability is due to decreased and even interrupted precipitation, usually accompanied by the highest temperatures. An extreme swing of laterization is likely to develop hardpan.

Savanna regions generally exhibit a forest on stream edges (galleria) and harbor rather sparse scrub on dry topography. In Brazil, as in many parts of Africa, relict forests are still included within savanna territory. At all events, the exact boundary of climatic savanna is not always reliably traced, as this formation is extremely subject to recurrent fires. Some of the woody elements are especially resistant, whereas the extensive mats of seasonally dry graminoids are extremely flammable.

Pasturing by large herds (elephants, giraffes, and antelopes in Africa) also has the effect of reducing local coverage of plants over the years. The tropical savanna lends itself well to many extensive crops (millet, sorghum, cassava, coffee, and several types of fruit trees).

Temperate and cold savanna. Forest-tundra, lichen woodland, boreal woodland, and taiga are terms used to describe the temperate and cold savanna. This regional formation, very extensively represented in North America and in Eurasia at high latitudes, consists of scattered or clumped trees (very often conifers and mostly needle-leaved evergreens) and a shrub layer of varying coverage. Mosses and, even more abundantly, lichens often form an almost continuous carpet.

The very peculiar high-altitude vegetation known as páramo in South America (and duplicated in Africa) is also a cold savanna, which is characterized by rosette trees and cushion plants.

In most areas where cold savanna prevails, especially at high latitudes, the extreme continentality (possibly the world record of temperature ranges) effectively reduces total growth and precludes agglomeration above the shrub level. The balance of gleization and podzolization shows many oscillations.

Congested depressions harbor muskeg; there is relatively little marsh formation; and the drier ridges, especially the rocky ones, harbor scrub or tundra.

Large herds of reindeer and caribou which feed upon the lichens and the buds of woody plants exert a strong modifying influence. Fire is less important than in the needle-leaved forest. Although some urban centers have developed around the mines of northern Canada and Siberia, agricultural exploitation is virtually impossible. *See* TAIGA.

Thornbush. Thorn scrub, savane armée, savane épineuse, Dorngeholz, Dorngestrauch, and dornveld are other names for this class. The dominants are principally of two kinds: tall succulents and profusely branching smooth-barked deciduous hardwoods which vary a great deal in density, from an inextricable mesquite bush in the Caribbean to an open spurge thicket in Central Africa. The northeastern Brazilian caatinga consists of thorny tall shrubs and cacti in its upper layer and mostly of annuals among the herbs. The annuals tend to be quite ephemeral. The South African dornveld is rather similar, except that spurges replace the cacti.

The climate is that of warm desert, except for a rather short but intense rainy season that causes much lateral transport of soil and strongly interferes with the pedogenic processes.

In certain favored positions, palm savannas or grasslands appear, whereas the soils that can hold no moisture have desert vegetation.

The hard soils are not favorable to cultivation, although certain fiber plants (sisal in Mexico and caroá in Brazil) do very well. They are extensively if unproductively pastured.

Tropical scrub. This class is also known as bush, brush, thicket, mallee, and fourré. It is composed of low woody plants (shrubs), sometimes growing quite close together, but more often separated by large patches of bare ground. Clumps of herbs are scattered throughout. This formation class is not very frequent, for it tends to be replaced by tropical savanna or tropical woodland. The Ghanaian evergreen coastal thicket and the Australian mallee are good examples of a rather narrowly confined zone, a wedge between better-developed formations. The pedological units are thus formed as a result of conflicting tendencies.

Grassy marshes and meadows are common in these areas; islands of forest and savanna also occur. The natural thickness is often reduced, especially for pasturing, but the woody plants always tend to reinvade.

These areas, like their temperate analogs, are most often rangelands.

Temperate and cold scrub. Synonymous terms for this class are heath, bosque, and fourré. The density of this formation varies a good deal, as does its periodicity. The shrubs may be evergreen (Tierra del Fuego beeches and Irish heathers) or deciduous (subarctic willows and alders). An undergrowth of ferns and other large-leaved herbs is quite frequent, especially at the subalpine level.

A considerable amount of moisture is necessary, whether in the soil (snowmelt), in the atmosphere (mist), or as a result of great seasonal downpour. Wind shearing or very cold winters prevent tree growth, although occasionally widespread soil conditions or historic factors are responsible (Azorean laurel–juniper scrub).

Closed depressions allow bog formation which, in the superhumid regions, tends to creep upland.

The warmer scrub areas (sagebrush zone of Wyoming and Azorean laurel heath) are most often converted to pasture and are excellent grass producers. This is also true of moist subalpine regions (mugo pine and *Rhododendron* scrub in the Pyrenees).

Tundra. This formation grows very low, and consists of trailing or matted shrubs (willows, heaths, and sometimes conifers), some of them evergreen, compact, and small-leaved; of cushion plants (with foreshortened stems and deep taproots as in the alpine campion); of tufted herbs; and of a great abundance of mosses and lichens. The geographical extent of many arctic-alpine species is almost coextensive with the greatest development of tundra at the high latitudes of the Northern Hemisphere, descending to the 60th parallel and lower in some places, and reaching all the major mountain systems. In many mountain systems, however, grassy formations (meadows) are more prevalent. Alpine areas of the Southern Hemisphere (New Zealand, Chile, and Falklands), having few floristic analogies with the boreal, nevertheless harbor a similarly structured tundra.

A short growing season and extremely severe winters are the determining climatic factors, whether at high latitudes or altitudes. The higher altitudes, however, usually have higher summer temperatures and do not always exhibit a soil with a permanently frozen layer (permafrost) which inhibits deep rooting; but the seasonal freezing is often drastic enough to cause a churning of soil particles which results in polygonal structures at the surface. The main soil-forming process is gleization, the development of a bluish-gray layer of soil at the bottom, with organic accumulation at the surface.

The apparent uniformity of tundra regions, which is due to the very restricted flora, is broken by the aridity of certain fell-fields (or felsenmeer) where crusts of lichens clothe the evenly spaced stones or boulders, by small mossy bogs and meadows, and by dunes and cliffs. An occasional ravine harbors willow scrub. The fauna show adaptive features of several kinds: The migrant birds form a very rich and abundant group, whereas migrants of intra-arctic amplitude or local residents (barren-ground caribou, ptarmigan, arctic hare, and fox) modify their fur, plumage, and feeding habits with the season.

There is virtually no direct utilization of vegetation by humans, and no cultivation even at radio, aviation, military, or mining settlements.

Prairie. Tall-grass prairie, Wiesen, and high tussockland are terms that describe the vegetation of this class. This formation is dominated by tall grasses and contains almost no woody elements, at least none that overtop the grasses. Frequently, the seasonal turnover can be very striking: Grasses leaf and flower in the spring and early summer, whereas the forbs (broad-leaved herbs, most of them composites) remain in leaf until the late summer, when the days become shorter, and then flower. This is a phenomenon of middle and high latitudes, as in Iowa and the Hungarian puszta. The Argentine pampa also belongs to this formation, although it contains some shrubs, as does part of the New Zealand tussockland.

Typically continental temperature extremes, seasonal rains, and cold winters combine to induce a development of deep black soil (chernozem) overlying an accumulation of calcium in depth. River edges and bluffs are often wooded, since they have more available water during the drought and because wind action is somewhat reduced. The immediate river edge has a floodplain regime with willows, poplars, and elms (in the Northern Hemisphere), but the humid bluffs may harbor extensions of upland forest from the adjacent climatic zone. On the other hand, marshy ground is occupied by "wet prairie," that is, by graminoid formations. Stripped areas are easily windblown and quickly turn to steppe and even desert or dune. Fire is so frequent that it appears to be a normal agent in the natural cycle.

Large bands of grazing animals (bison in North America and Europe) were part of the original balance. In some areas (northeastern Oklahoma) they have been replaced by domesticated cattle. But for the most part prairie country is considered the best cereal land in the world and has been used for growing wheat, barley, and maize.

Steppe. The steppe, or short-grass prairie, differs from the prairie in that the grasses are of lower stature and bunched or sparsely distributed and also in that low shrubs are somewhat conspicuous. Much bare ground remains exposed which may be used by annuals after the rains. Low-growing succulents are not unusual. *See* GRASSLAND ECOSYSTEM.

The mid-latitude continental regime which prevails, for instance, over the extensive flat plains of the Soviet Union and of central North America, is favorable to the development of steppe. Drought, which usually follows winter rains, is greater than in the prairie and does not allow the accumulation of a deep black horizon at the surface of the soil, but rather a characteristic chestnut layer. Excessively drained sites lend themselves to alkaline accumulations.

Salt lakes and saltmarshes are fairly common with their concentric belts of saltwort, shadscale, and other halophytes. In protected areas, small groves of deciduous woodland (aspen groves in Saskatchewan) or patches of savanna prevail, and in badly eroded areas an open scrub virtually without grasses is prevalent.

The steppe is pastured and cultivated (cereals and cotton) and with proper exploitation is quite fertile. However, where abusive practices have prevailed, eolian erosion has advanced very far, and sandstorms have resulted in "badlands" on which rehabilitation is very slow.

Meadow. This class, also known as pelouse and Wiesen, is a low grassland, dense and continuous, variously interspersed with forbs (and even with mosses, which locally prevail) but very few if any shrubs.

It is characteristic of the formation class only in areas of high moisture, its optimum being in some alpine regions, where relatively high day temperatures in summer offset the rigor of winter. Some parts of the Arctic (for example, the Mackenzie Lowlands eastward to Hudson Bay) also have meadow rather than tundra. The rather dense "short-grass prairie" of Saskatchewan probably should be classed here also. The growing season is necessarily rather short (or else quite cold, as in the subantarctic islands and parts of Patagonia), and the yearly increment is not very abundant. The soil consists of a well-developed humus layer resting on gravel or on some other parent material which is usually not strongly modified.

On the rockier sites (fell-fields) low woody vegetation becomes more important and tundra is rather likely to take over. In some of the sheltered ravines either a megaphorbia (a prairie dominated by forbs) or a scrub will develop.

Natural alpine meadows are used for pasturing, although they are much less productive than secondary meadows at the subalpine level.

Warm desert. This is a vegetation zone, also named désert, Wüst, and Trockenwüst, in which little vegetation of any kind grows, although succulent and small broad-leaved shrubs are characteristic. In a sense, tiny annuals (sometimes called "belly plants") are even more typical in that they crop up in large numbers shortly after one of the infrequent and usually inconstant rains. Less than 10% of the space is covered by plants, and they economize water in a variety of ways: succulence (cacti), reduction of leaf surface (creosote bush), and great extension of root system (smoke tree), among others. Accordingly, deserts are of many kinds: the dunes and rocky wastes of the Sahara that harbor hard grasses (*Stipa*) and gnarled evergreen shrubs, the hard caliche that bears creosote bush in the Sonoran region of the United States and Mexico, the coarse pebbly surfaces of the Karroo with its small succulent plants (*Lithops*, *Faucaria*, and *Fenestraria*), and the saltbush areas found in Australia, Algeria, and the southwestern part of the United States. *See* DESERT ECOSYSTEM.

The climate is one of extremely low rainfall (possibly no rain at all for years) and very high temperatures (in fact, much higher than in the equatorial zone). Yearly amplitudes of temperature are considerable. Pedogenesis leads to salinization, and even to the formation of alkali crusts on the soil surface.

The rockier sites are usually taken over by succulent "gardens." The stream beds (known as draws, arroyos, and oueds) usually have a characteristic vegetation, closer or taller than those of the rocky sites, consisting of shrubs or even trees very different in kind from those of other habitats. Whereas the salt lakes are bordered with a belt of halophytic vegetation (such as *Salicornia* and *Atriplex*), the fresh-water bodies have palm copses (*Washingtonia* in California and *Phoenix* in Africa) and grass.

When irrigated, desert soils are frequently quite rich and will grow big crops of dates, grapefruit, oranges, and even fodder such as alfalfa.

Cold desert. The cold desert is also known as rock desert and frigorideserta. Extreme cold combined with adverse edaphic conditions such as a very coarse rocky substratum allow neither tundra nor meadow to develop. Instead, tufts of herbs (often graminoids) occupy shallow crevices, open sandy areas, or loose schistose screes. Mosses and lichens are not very conspicuous.

This condition is frequently found on nunataks, on emergent mountain peaks, and at very high latitudes (northern Ellesmere Island).

Crust vegetation. Although it is to be seen literally everywhere, crust vegetation does not prevail over large enough sections of the Earth's surface sufficiently often to be considered characteristic of a regional formation-class. It consists of algae, lichens, mosses, or liverworts and has a thickness of only a few centimeters and a rather variable coverage, ranging from the full carpeting by mosses of some Antarctic rocky islands to the sparse dotting by lichens of some Saharan rocky ranges.

The climate must be one of very sparse precipitation and of extremely high or extremely low temperature. The ice margin and nunataks of Greenland, part of northern Ellesmere Island, all of ice-free Antarctica, and some of the snow-free high altitudes of various mountain ranges have practically no other vegetation, with the exception of a few herbs and shrubs in protected sites. The contrastingly richer fauna (for example, migrant birds) occasionally provides good soil fertilizer and permits small patches of lush growth.

Such regions are uninhabited, and it cannot be said that their soil or vegetation are of any utility to man. See VEGETATION ZONES, ALTITUDINAL.

[PIERRE DANSEREAU]

Bibliography: P. Dansereau, *A Universal System For Recording Vegetation*, 1958; P. Dansereau, *Biogeography: An Ecological Perspective*, 1957; P. Dansereau, *Vegetation of the Continents*, in preparation; S. R. Eyre, *Vegetation and Soils: A World Picture*, 1963; S. R. Eyre (ed.), *World Vegetation Types*, 1971; R. W. J. Keay et al., *Vegetation Map of Africa South of the Tropic of Cancer*, 1959; A. W. Küchler, National vegetation, in E. B. Espenshade, Jr. (ed.), *Goode's World Atlas*, 11th ed., 1960; D. L. Linton, Vegetation, in C. G. Lewis (ed.), *Oxford Atlas*, 1951; A. Strahler, *Introduction to Physical Geography*, 1973.

Water analysis

A broad field that deals with the specific applications of chemical, physical, and biological methods to the analysis of aqueous samples ranging from the purest distilled water to the most polluted wastewater. One major area of concern is the analysis of drinking water to ensure that the water does not transmit organisms causing human disease, and does not contain chemicals hazardous to human health. Another concern is that the water is esthetically acceptable to consumers in terms of taste, odor, and appearance. Raw or treated wastewaters are analyzed to determine proper treatment and to monitor their discharge to assure public safety and environmental protection. Since the early 1970s, Federal, then state, laws and regulations have led to a more formalized approach to the development and utilization of analytical methods, although the first standardized analytical methods applicable to water appeared in 1905.

CHEMICAL ANALYSIS

The analytical chemistry of water deals with major mineral constituents, both anions and cations, trace organic and inorganic substances, and radionuclides. Insofar as possible, contemporary procedures are instrumental and do not depend on conventional colorimetric or titrimetric methods. Foremost among analytical instruments are the atomic absorption spectrophotometer (for metals) and the gas-liquid chromatograph (for organic constituents). Inorganic nonmetals typically are measured by noninstrumental methods, and radionuclides are determined by a combination of chemical separation followed by instrumental measurement of radioactivity.

Metals. In atomic absorption spectrophotometry, a sample is aspirated into a flame and atomized. Monochromatic light, usually produced by a hollow cathode tube specific for each metal, is passed through the flame. The unabsorbed light goes through a monochromator for wavelength selection, it is measured with a photoelectric detector, and the results are recorded. In the absence of interferences, light absorption is related directly to metal concentration. Proper selection of light source, wavelength, sample pretreatment, and combustion temperature (a function of gases used, such as air-acetylene, nitrous oxide-acetylene, or nitrogen-hydrogen) makes it possible to measure most metals in the microgram-per-liter range. To further increase sensitivity and permit direct sample analysis, the electrothermal (heated graphite furnace) atomic absorption technique may be used. For mercury analysis, a cold vapor (flameless) variation is used.

Less preferable than atomic absorption spectrophotometry are spectrophotometric (colorimetric) methods that are available for many metals. These usually are based on the use of organic chromogenes such as eriochrome cyanine (for aluminum), aurintricarboxylic acid (for beryllium), and dithizone, or diphenylthiocarbazone (for cadmium, lead, mercury, silver, and zinc). Flame photometric methods that are based on the emission of light of specific wavelength in a flame (light emission rather than absorption) may be used for lithium, potassium, sodium, and strontium. A titrimetric method using ethylenediaminetetraacetic acid as an indicator may be used for calcium and magnesium.

Inorganic nonmetals. Included in this group are such constituents as acidity and alkalinity (measured by titration with standard base or acid); boron (spectrophotometry); the halogens (measured by iodometric titration; amperometrically, that is, by a modification of the polarographic principle which involves voltage changes in the solution; or colorimetrically); cyanide (by titration, colorimetrically, or by means of an ion-selective electrode coupled with an expanded-scale pH meter); ammonia and organic nitrogen (by distillation and acidimetry or colorimetrically); oxygen (by iodometric titration or with an ion-selective membrane electrode); pH (electrometrically); phosphate (colorimetrically); silica (colorimetrically or gravimetrically); sulfate (turbidimetrically or gravimetrically); and sulfide (colorimetrically). This extensive list is given to show both the diversity of analytical methods in use and the variety of constituents that may be of importance.

Organics. Gas chromatography is the method of choice for specific organic constituents such as various pesticides, among which are several chlorinated hydrocarbons and phenoxy acid herbicides, phenols, digestor gases including carbon dioxide and methane, and trihalomethanes such as chloroform and bromoform. In gas chromatography, a sample is carried by a gas (nitrogen, argon, helium, or hydrogen) through a column packed with an inert, granular solid that holds the liquid stationary phase. The column is heated, and at its outlet is placed a detector, often a flame ionization detector, connected to a recorder. By modifying column size and stationary liquid, temperature, carrier gas, and detector, it is possible to separate chemical compounds on the basis of their mobility through the system and to identify them by the time required for each to reach the detector and cause a peak (the height of which is concentration-related) to be recorded.

Less specifically identified organic substances are measured gravimetrically (oil and grease), by acidimetry (organic, volatile acids), or by colorimetry (phenols, surfactants, tannins, and lignins). In wastewater analysis, one of the most important tests is the measurement of oxygen consumption. This may be done by using a biological system (biochemical oxygen demand) or by using a chemical oxidant such as dichromate and measuring its residual after a standardized oxidation under reflux conditions (chemical oxygen demand).

Radionuclides. Instrumental methods alone are suitable for measuring gross alpha and beta particles and gamma rays. For alpha and beta measurements, the instrument of choice is an internal proportional counter in which the sample is enclosed within the counting cell. Alternatively, a Geiger counter may be used. For gamma counting, a scintillation detector is suitable. The detector emits photons of light when hit by gamma rays, and the light emission is measured with a photomultiplier tube and recording device. Specific radionuclides, strontium-89 and -90 and total radium (radium-226), are chemically separated and counted. The beta particles emitted by tritium (hydrogen 3) are counted in a liquid scintillation spectrometer.

PHYSICAL ANALYSIS

Physical analyses are made to assess important characteristics of both water and wastewater. Organoleptic tests are made for taste and odor which are significant in the consumer's perception of drinking-water quality. Color is measured by direct visual comparison with standards or by a spectrophotometric method. Conductivity, a numerical estimate of the ability of water to carry an electric current, is a function of the mobility of dissolved ions. It is measured electrometrically and serves as an indirect indicator of the dissolved residue. Residue also may be measured gravimetrically by determining the weight of a dried sample. The suspended residue (suspended-solids test) is important in evaluating the strength of wastewater and the efficacy of its treatment. Turbidity relates to water clarity and is important in the analysis of drinking water and, for safety reasons, in swimming pools. It is measured visually by determining the depth of water required to obscure the image of a standard candle flame or, better, by nephelometry in which light scatter under standard conditions is measured.

BIOLOGICAL ANALYSIS

Biological analyses include a broad spectrum of tests for organisms found in water and for the effects of substances in water on aquatic life. The special case of microbiological analyses, because of their importance to human health, is dealt with separately below.

In making biological analyses, the classical methods of visual field observations are coupled with laboratory examinations, often microscopic, for identification and enumeration of plants and animals in water. The organisms of interest are grouped as plankton, microscopic free-floating or suspended organisms; periphyton, organisms living attached to or on submerged surfaces; macrophyton, aquatic flowering plants and other macroscopic plant forms; and benthic macroinvertebrates, animals living in bottom muds. Analytical results, especially those that are historical or comparative, are important in assessing the effect of pollution on natural waters. Algae, which are included among the plankton, may affect the quality of drinking water by modifying taste, odor, and color, and usually require control in water supply reservoirs.

Standardized bioassay tests are available for assessing the effect of pollutants. Historically, the test organisms that were exposed under specified conditions were fish. Because fish have been recognized as only a single component of the biological environment of concern, comparably standardized procedures using other organisms have been developed. Tests are available using algae, ciliated protozoa, *Daphnia magna*, *Acartia tonsa*, scleractinian coral, marine polychaete worms, amphipods, crayfish, crabs, the American lobster, shrimp, aquatic insects, and mollusks.

MICROBIOLOGICAL ANALYSIS

Microbiological tests are used to determine sanitary quality and suitability of water for general use. In the routine analysis of water, including drinking water, enteric pathogens are not enumerated; rather, indicators of fecal pollution are determined. The most commonly used indicator is the coliform group, which includes normal inhabitants of the gastrointestinal tract of warm-blooded animals. Microbiological standards of water quality have been based on coliform counts for more than 50 years and have served well to protect the public health.

Coliform testing. The coliform group is defined as all aerobic and anaerobic, gram-negative, nonspore-forming rod-shaped bacteria that ferment lactose with gas formation within 48 hr at 35°C. Two quantitative procedures are used to enumerate the coliform group; both are officially recognized.

Multiple-tube fermentation or MPN technique. This test has three parts. In the first part ("presumptive test"), measured volumes of serial dilutions of sample are inoculated into tubes of lactose or lauryl tryptose broth and incubated at 35°C for up to 48 hr. Each tube contains an inverted vial that serves as a gas trap. Production of gas in any quantity is recorded only as a positive presumptive test because some organisms other than coliform bacteria may produce gas under these conditions. To confirm that positive tubes result from the growth of coliform bacteria, a transfer is made from each positive tube to a tube of brilliant green bile lactose broth. The tubes, which also contain an inverted vial, are incubated at 35°C for up to 48 hr. Production of gas is a positive "confirmed test." The confirmed test eliminates most false positives, particularly those due to gram-positive organisms. To make the "completed test," a streak is made from each positive confirmed tube to Endo or eosin methylene blue agar plates. From typical isolated colonies, a picking is made to a tube of the presumptive medium and a nutrient agar slant. Production of gas from lactose within 48 hr at 35°C, and, on a gram-stained preparation, the presence of nonsporing gram-negative rods, is a positive completed test. The completed test usually is not used for more than about 10% of samples analyzed to verify that the confirmed test is adequate. Using statistical tables, the most probable number (MPN) of coliform bacteria per 100 ml is determined from the pattern of positive sample dilutions.

Membrane filter technique. This procedure gives relatively rapid results (within 24 hr) of greater accuracy because it provides a direct colony count rather than a statistically derived MPN. The sample is filtered through a cellulosic filter that collects bacteria on its surface. The filter is placed on a selective-nutrient-saturated pad or an agar plate. Upward diffusion of nutrients permits growth of coliform bacteria at 35°C. Colonies are counted, and the number of coliforms per 100 ml are reported.

Fecal coliform bacteria. Fecal coliforms, otherwise known as thermotolerant coliforms, are that subgroup of coliform bacteria able to produce gas from lactose at 44.5°C within 24 hr. There is a high correlation between fecal origin and thermotolerance positive results. Either an MPN procedure or a membrane filter technique may be used. These tests are seldom used in the analysis of drinking water because no coliform bacteria should be present in water delivered to consumers. The tests are useful, however, in analyzing wastewater or polluted waters.

Testing for other microorganisms. Organisms other than coliform bacteria may be tested for under special circumstances, but routine analysis for pathogens in water is not recommended.

Standard plate count. The standard plate count, of aerobic heterotrophic mesophilic bacteria, is a colony count of those microorganisms able to grow under defined conditions of medium, temperature, and time. It is a useful test in evaluating water treatment plant efficiency. High plate counts have been associated with suppression of gas production by coliform bacteria, which may lead to an erroneous conclusion of drinking water safety, and to the possibility of opportunistic infections in users of dialysis machines or in other hospital-type uses (nosocomial infections).

Fecal streptococci. With use of selective media, either an MPN or membrane filter technique is available. As these streptococci are relatively host-specific, speciation is useful in identifying wastes of nonhuman origin. In assessing water quality, fecal streptococcus data seldom are used alone, but usually in conjunction with coliform or fecal coliform results.

Pathogenic microorganisms. Qualitative and quantitative procedures for various pathogens are available. Sample concentration on gauze swabs or by filtration through diatomaceous earth or membrane filters usually is required. Included are an immunofluorescence technique for *Salmonella*, and biochemical procedures for *Salmonella* (specifically *S. typhi*), *Shigella*, and pathogenic leptospires.

Enteric viruses. Concern for possible waterborne transmission of viruses such as polioviruses, coxsackieviruses, echoviruses, reoviruses, and the viruses of infectious hepatitis has led to development of special sample concentration techniques. These may be based on the use of microporous filters through which samples of up to 4000 liters may be drawn. Sample concentrates are then analyzed by classical virological techniques. No water quality standards for viruses exist, but considerable research is under way, particularly in the context of wastewater reclamation.

Miscellaneous organisms. In evaluating water used for recreational purposes, tests for *Pseudomonas aeruginosa* or staphylococci may be made; in drinking water or wastewater, examinations may be made for fungi, yeasts, actinomycetes, and nematodes. These procedures include both direct microscopy and biochemical testing. Similarly, special procedures are available for iron and sulfur bacteria that produce objectionable slimes in water or are involved with pipe corrosion.

Maximum contaminant levels (MCL) allowable for drinking water

Contaminant	MCL, mg/liter
Arsenic	0.05
Barium	1.
Cadmium	0.010
Chromium	0.05
Fluoride	1.4–2.4*
Lead	0.05
Mercury	0.002
Nitrate (as N)	10.
Selenium	0.01
Silver	0.05
Endrin	0.0002
Lindane	0.004
Methoxychlor	0.1
Toxaphene	0.005
2,4-D	0.1
2,4,5-TP Silvex	0.01
Turbidity	1 turbidity unit
Coliform bacteria	1 per 100 ml

*Level varies inversely with annual average of the maximum daily air temperature.

WATER QUALITY STANDARDS

Standards of water quality have been promulgated by Federal and state authorities. These are available for drinking water as well as wastewater. Because wastewater requirements are highly variable depending in part on the nature and volume of the waste and the conditions of the receiving water, they will not be discussed in detail. The most commonly specified constituents are biochemical oxygen demand (BOD), suspended solids, and oil and grease.

For drinking water, regulations prescribe the maximum contaminant level (MCL) for a number of chemical species, turbidity, and coliform bacteria (see table). Experience has demonstrated that water that meets these specifications will, with high probability, not produce human disease. The standards are subject to revision as new information becomes available. Under consideration at the present time is the addition of an MCL for trihalomethanes, and possibly other chloroorganic compounds.

[ARNOLD E. GREENBERG]

Bibliography: American Public Health Association, *Standard Methods for the Examination of Water and Wastewater*, 14th ed., 1976; U.S. Environmental Protection Agency, *Fed. Reg.*, 41:232: 52780–52786, Dec. 1, 1976; U.S. Environmental Protection Agency, *National Interim Primary Drinking Water Regulations*, EPA-570/9-76-003, 1976.

Water conservation

The protection, development, and efficient management of water resources for beneficial purposes. Water occupies more than 71% of the Earth's surface. Its physical and chemical properties make it essential to life and modern civilization. Water is combined with carbon dioxide by green plants in the synthesis of carbohydrates, from which all other foods are formed. It is a highly efficient medium for dissolving and transporting nutrients through the soil and throughout the bodies of plants and animals. It can also carry deadly organisms and toxic wastes. Water is an indispensable raw material for a multitude of domestic and industrial purposes.

UNDERGROUND WATER

Water occurs both underground and on the surface. Usable groundwater in the United States is estimated to be 47,500,000,000 acre-feet (1 acre-foot = 1.233×10^3 m^3). Annual runoff from the land averages 1,299,000,000 acre-feet. The volume of groundwater greatly exceeds that of all fresh-water lakes and reservoirs combined. It occurs in several geologic formations (aquifers) and at various depths. Groundwater under pressure is known as artesian water, and it may become available either by natural or artificial flowing wells. Groundwater, if abundant, may maintain streams and springs during extended dry periods. It originates from precipitation of various ages as determined by measurements of the decay of tritium, a radioisotope of hydrogen found in groundwater. The water table is the upper level of saturated groundwater accumulation. It may appear at the surface of the Earth in marshes, swamps, lakes, or streams, or hundreds of feet down. Seeps or springs occur where the contour of the ground intercepts the water table. In seeps the water oozes out, whereas springs have distinct flow. Water tables fluctuate according to the source and extent of recharge areas, the amount and distribution of rainfall, and the rate of extraction. The yield of aquifers depends on the porosity of their materials. The yield represents that portion of water which drains out by gravity and becomes available by pumping. Shallow groundwater (down to 50 ft or 15 m) is trapped by dug or driven wells, but deep sources require drilled wells. The volume of shallow wells may vary greatly in accordance with fluctuations in rainfall and degree of withdrawal. *See* GROUNDWATER.

SURFACE WATER

Streams supply most of the water needs of the United States. Lakes, ponds, swamps, and marshes, like reservoirs, represent stored streamflow. The natural lakes in the United States are calculated to contain 13,000,000,000 acre-feet. Swamps and other wet lands along river deltas, around the borders of interior lakes, and in coastal regions add millions more to the surface supplies. The oceans and salty or brackish sounds, bays, bayous, or estuaries represent almost unlimited potential fresh-water sources. Brackish waters are being used increasingly by industry for cooling and flushing. Reservoirs, dammed lakes, farm ponds, and other small impoundments have a combined usable storage of 300,000,000 acre-feet. The smaller ones furnish water for livestock, irrigation, fire protection, flash-flood protection, fish and waterfowl, and recreation. However, most artificial storage is in reservoirs of over 5000 acre-feet. Lake Mead, located in Arizona and Nevada and formed by Hoover Dam, is the largest (227 mi^2 or 588 km^2) of the 1300 reservoirs, and it contains 10% of the total stored-water capacity, or over 31,000,000 acre-feet. These structures regulate streamflow to provide more dependable supplies during dry periods when natural runoff is low and demands are high. They store excess waters in wet periods, thus mitigating damaging floods.

Water use. Water withdrawals for all purposes totaled 1740 billion gallons per day (bgd = 3.785×10^6 m^3 per day) in 1955 in the United States, but actual consumption was only 94 bgd. Exclusive of reservoir evaporation, daily use was 71 bgd, 444 gal (1.681 m^3) per capita or 6% of the nation's runoff. Not counted is an additional 20 bgd transpired by western phreatophytes, plants growing along stream banks, canals, and reservoirs. Because reservoirs evaporate 23 bgd, interest has developed in underground storage and in coating water surfaces with monomolecular films of oil-derived detergents, such as hexadecanol, to reduce evaporation. *See* ECOLOGY.

The 115,000,000 people in the United States served by public water supplies in 1955 withdrew 11.3 bgd, and rural dwellers and their livestock used 2.5 bgd. Crop irrigation practiced in all 48 states withdrew 110 bgd, the heaviest use being in the 17 drier Western states. Only 40% of irrigation water is returned to streams or aquifers, whereas 90% of other public water supplies is returned. However, pollution prevents reuse of part of the return flow of public water supplies.

Withdrawals for waterpower account for 1500 bgd out of the total in the United States, but this use consumes practically none of the water. Industry takes nearly 116 bgd, mostly for condenser cooling and for boilers, sanitary services, and cooling machinery in power plants, but industry consumes only 2% of the water utilized. Air conditioning averages only 1.1 bgd but may constitute a severe drain on available supplies during hot dry spells in local areas.

Fish, wildlife, and recreation water requirements are nonconsumptive. Clean natural waters and the aquatic environment they create constitute major attractions in undeveloped wilderness areas, national parks, and even in more highly developed agricultural, forest, or suburban localities. However, artificial impoundments, unless properly located, designed, and operated, can destroy or depreciate priceless natural environment.

Water pollution. Streams have traditionally served for waste disposal. Towns and cities, industries, and mines provide thousands of pollution sources. Pollution dilution requires large amounts of water. Treatment at the source is safer and less wasteful than flushing untreated or poorly treated wastes downstream. However, sufficient flows must be released to permit the streams to dilute, assimilate, and carry away the treated effluents. *See* WATER POLLUTION.

Hydrologic cycle. This term refers to the continuous circulation of the Earth's moisture. Because of this characteristic, water is considered a renewable natural resource, but underground water is "mined" when it is pumped out faster than its natural renewal and thus may be like a fund resource. The oceans furnish most of the moisture for evaporation and precipitation. Part of the precipitation evaporates, part is returned directly to the oceans, part runs off quickly into watercourses and lakes, part enters the soil or other porous material where it is retained, and the balance enters deep aquifers where it is stored or flows along impermeable underground layers into streams or springs. Solar radiation provides the energy for the hydrologic cycle. Unequal heating of the Earth's surface creates air currents of varying temperatures and pressures. Warm air masses carry moisture from the oceans. When these air masses are cooled, their capacity to retain moisture is reduced and precipitation results. The source of the air masses and the pressure and temperature gradients aloft determine the form, intensity, and duration of precipitation. Precipitation may occur as cyclonic or low-pressure storms (mostly a winter phenomenon responsible for widespread rains), as thunderstorms of high intensity and limited area, or as mountain storms wherein warm air dumps its moisture when lifted and cooled in crossing high land barriers (see illustration).

Precipitation is characteristically irregular. Generally, the more humid an area and the nearer the ocean, the more evenly distributed is the rainfall. Whatever the annual average, rainfall in arid regions tends to vary widely from year to year and to fall in a few heavy downpours. Large floods and active erosion result from heavy and prolonged rainfall or rapid melt of large volumes of snow.

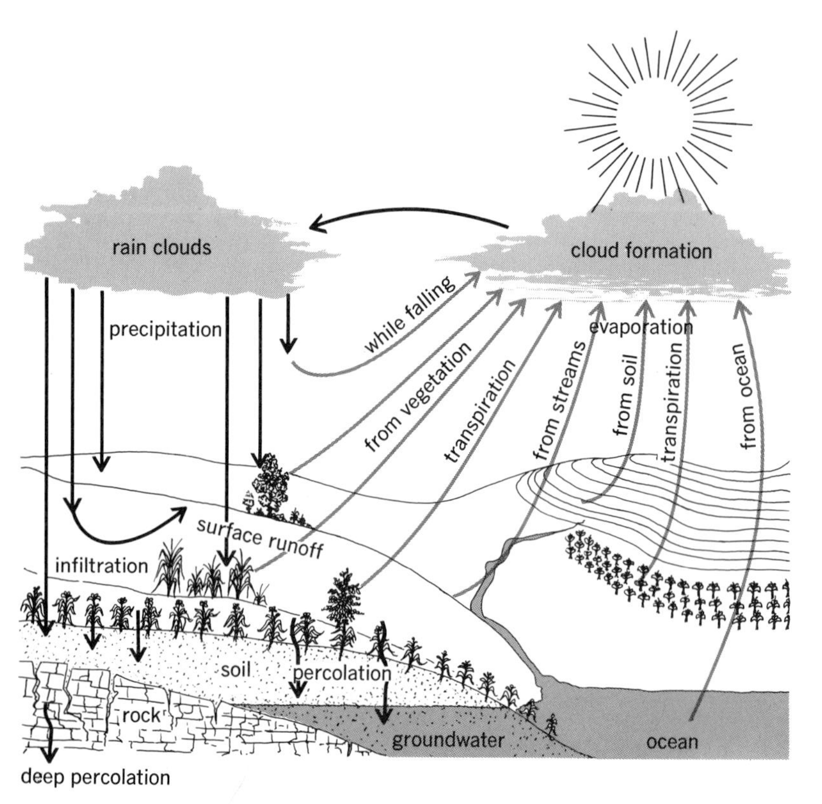

The hydrologic cycle. *(Water, USDA Yearbook, 1955)*

Flash floods often follow local intense thunderstorms.

Runoff depends on depth, porosity, and compactness of the soil and the underlying material, steepness and configuration of the surface, and character and density of the vegetation. Plant crowns, ground cover, litter, and humus dissipate the force of rainfall, thus reducing its power to compact and dislodge mineral soil particles and seal the surface pores. The quantity of water entering the soil during a given time depends on the rate of rainfall in relation to the size and distribution of the pores in each soil horizon and the thickness of each horizon. Surface runoff follows when rainfall exceeds the rate at which water is absorbed and transmitted downward. Soil water is retained by adhesion against the pull of gravity in the capillary pores of the soil until their capacity is filled. Such water may evaporate or is available for plant use including photosynthesis and transpiration. Excess moisture in the large pores drains slowly into watercourses or wet-weather springs, or enters rock crevices, limestone sinks, and shales, sands, or other permeable materials. *See* HYDROLOGY; SOIL.

Land management vitally influences the distribution and character of runoff. Inadequate vegetation or surface organic matter; compaction of farm, ranch, or forest soils by heavy vehicles; frequent crop-harvesting operations; repeated burning; or excessive trampling by livestock, deer, or elk all expose the soil to the destructive energy of rainfall or rapid snowmelt. On such lands little water enters the soil, soil particles are dislodged and quickly washed into watercourses, and gullies may form. *See* LAND-USE PLANNING; SOIL CONSERVATION.

Water management problems. These involve economic, social, and intangible values. Efforts to plan and develop river systems for multiple purposes often generate conflicts among different water uses, for example, irrigation versus navigation on the Missouri River, or hydropower versus salmon or trout fisheries, wildlife, national park, wilderness, or historic resources on such rivers as the Columbia; Colorado, and Potomac. Other conflicts stem from actual or threatened dumping of municipal, agricultural (silt), industrial, or acid mine wastes into streams, as occurs upstream from Philadelphia, Pa.; Washington, D.C.; Cumberland, Md.; and other cities, or from operations of power dams, irrigation projects, or other uses which restrain or divert flows to the extent of destroying fish habitats or impairing recreational or wildlife resources. Another source of conflict is the mining of groundwaters in areas of critically low groundwater supplies.

Water management technology. This involves the application of biological and engineering principles to attain desired goals. Biological methods for growing upland vegetation having low moisture requirements are being studied by the U.S. Forest Service and other agencies. Mechanical methods include the practice of water spreading, which is utilized to desilt floodwaters and to promote the percolation of water into the soil for crop use and groundwater recharge. Water that would be a strong pollutant of streams and lakes may be spread on the land. Seabrook Farms in southern New Jersey releases 50 ft/acre (38 m/ha) of food-processing wash water annually onto an 80-acre (32 ha) oak woodland on sandy soil. About 103,000,000 acres (41,700,000 ha) of wet lands have been drained in 40 states. Drainage has failed, however, where the suitability of soils for such practice was not adequately determined, or where erosion from adjacent slopes of improperly farmed land silted up the drainage structures. In some instances drainage has drastically reduced waterfowl habitat or aggravated downstream flood damages. *See* FOREST AND FORESTRY.

The avoidance of water waste takes several forms. Recycling has permitted huge savings of water, especially in petroleum plants, chemical factories, and steel mills. In some cases reductions have amounted to 96%. Artificial groundwater recharge is successfully practiced in Long Island, N.Y., where over 311 injection wells conduct water used in air conditioning to underground storage areas for cooling and reuse. A National Association of Manufacturers survey reported that 45% of 3343 industrial establishments applied some kind of water purification treatment. Flows in pipes can be reduced, warmer water can be used, or several grades can be applied by means of separate pipelines. Metering stimulates more economical use and encourages repair of leaky connections.

Water rights. In the United States early rights to water followed the riparian doctrine, which grants the property owner reasonable use of surface waters flowing past his land unimpaired by upstream landowners. The drier West, however, has favored the appropriation doctrine, which advocates the prior right of the person who first applied the water for beneficial purposes, whether or not his land adjoins the stream. Rights to groundwater are generally governed by the same doctrines. Both doctrines are undergoing intensive study.

State laws generally are designed to protect riparian owners against pollution. States administer the regulatory provisions of their pollution-control laws, develop water quality standards and waste-treatment requirements, and supervise construction and maintenance standards of public service water systems. Some states can also regulate groundwater use to prevent serious overdrafts. Artesian wells may have to be capped, permits may be required for drilling new wells, or reasonable use may have to be demonstrated.

Federal responsibilities consist largely of financial support or other stimulation of state and local water management. Federal legislation permits court action on suits involving interstate streams where states fail to take corrective action following persistent failure of a community or industry to comply with minimum waste-treatment requirements. Federal legislation generally requires that benefits of water development projects equal or exceed the costs. It specifies that certain costs be allocated among local beneficiaries but that most of the expense be assumed by the Federal government. In 1955, however, the Presidential Advisory Committee on Water Resources Policy recommended that cost sharing be based on benefits received, and that power, industrial, and municipal water-supply beneficiaries pay full cost. These phases of water resource development present difficult and complex questions, because many imponderables enter into the estimates of probable monetary and social benefits from given projects as well as into the cost allocation aspect.

Watershed control. This approach to planning, development, and management rests on the established interdependence of water, land, and people. Conditions on the land are often directly reflected in the behavior of streamflow and in the accumulation of groundwater. The integrated approach on smaller watersheds is illustrated by projects under the Watershed Protection and Flood Prevention Act of 1954 (Public Law 566) as amended by P. L. 1018 in 1956. This act originally applied to floods on the smaller tributaries whose watersheds largely are agricultural, but more recently the application of the act has been broadened to include mixed farm and residential areas. Damages from such frequent floods equal half the national total from all floods. Coordination of structures and land-use practices is sought to prevent erosion, promote infiltration, and retard high flows. The Soil Conservation and Forest Services of the U.S. Department of Agriculture administer the program. The Soil Conservation Service also cooperates with other Federal and state agencies and operates primarily through the more than 2000 soil conservation districts.

River basins may be large and complex watersheds. For example, the Tennessee River Basin comprises 40,000 mi^2 (103,600 km^2) in contrast to the 390-mi^2 (1010-km^2) upper limit specified in Public Law 1018. Basin projects may involve systems of multipurpose storage reservoirs, intensive programs of watershed protection, and improvement and management of farm, forest, range, and urban lands. They may call for scientific research, industrial development, health and educational programs, and financial arrangements to stimulate local initiative. The most complete development to date is the Tennessee River Basin, where well-planned cooperative activities have encompassed a wide variety of integrated land and water developments, services, and research.

Water conservation organizations. Organizations for meeting water problems take various forms. Local or intrastate drainage, irrigation, water-supply, or flood-control activities may be handled by special districts, soil conservation districts, or multipurpose state conservancy districts with powers to levy assessments. Interstate compacts have served limited functions on a regional level. To date Congress has not given serious consideration to proposals for establishing a special Federal agency with powers to review and coordinate the recommendations and activities of development services such as the Corps of Engineers, Bureau of Reclamation, Fish and Wildlife Service, Soil Conservation Service, and Forest Service and to resolve conflicts among agencies and citizen groups. Some national and regional civic groups, such as the League of Women Voters, are studying alternative approaches to the administration of river basin programs.

International agreements. Cooperation in the control, allocation, and utilization of international waters is authorized by treaties with Canada and

Mexico. Permanent commissions have been established to deal with specific streams such as the Rio Grande and Colorado, or with boundary waters generally, as provided in the treaty with Canada. The United Nations, through its Technical Assistance Program and through regional commissions, is promoting cooperative studies and developments among underdeveloped nations having common boundaries. [BERNARD FRANK]

COASTAL WATERS

Most coastal waters less than 100 m deep were dry land 15,000 years ago. The North Sea was a peat bog, for example, and one could walk from England to France or from Siberia to Alaska. As the glaciers retreated, these exposed continental shelves began to fill with water until they now constitute 10% of the world's ocean area. The average depth of the present continental shelf is 60 m, with a width extending 75 km from shore. The salinity of coastal water ranges from 35‰ (100% ocean water) at the seaward edge of the continental shelf (200 m depth contour) to 0‰ (100% fresh water) within coastal estuaries and bays at the shoreward edge of the shelf. The annual temperature range of mid-latitude shelf water is 20°C off New York and 10°C off Oregon, with less temperature change in tropical waters. *See* CONTINENTAL SHELF AND SLOPE.

Resources. Because of the shallow bottom, compared with the deep ocean of 5000 m depth, organic matter is transformed to nutrients and recycled (returned to the water column) faster on the continental shelves. The growth of plants in the sunlit regions of these relatively shallow areas is thus 10 times that of the open ocean, and the rest of the shelf food web is similarly more productive. At present, 99% of the world's fish catch is taken from these rich shelves. As a result of their accessibility and commercial value, coastal waters have been the object of extensive scientific studies for the last 100 years.

Withdrawal of the Wisconsin glacier and buildup of the native Amerindic populations about 10,000 years ago led to simple harvesting of the living resources of the United States continental shelves. Since colonial days, however, this coastal region has been the focus of increased exploitation with little thought given to the impact of these activities. After the discovery of codfish in the New World by Cabot in 1497, a "foreign" fishing fleet was inaugurated by the French in 1502, the Portuguese in 1506, the Spanish in 1540, and the English in 1578; the first "domestic" fishery of the United States was initiated by the ill-fated Roanoke colony in 1586. The adjacent human population then grew from a few Indian settlements scattered along the coast to the present east coast megalopolis of 45,000,000 people, housed in an almost continuous urban development from Norfolk, VA, to Portland, ME. By 1970, continued fishing pressure of the foreign and domestic fleets had reduced the fish stocks of the northeast continental shelf to about 25% of their virgin biomass.

Pollutant impacts. At the same time, attempts at waste control in colonial days began as early as 1675 with a proclamation by the governor of New York against dispersal of refuse within the harbor; yet the New York urban effluent expanded until the percent saturation of dissolved oxygen of the harbor halved between 1910 and 1930. The amount of trace metals in the New York Bight Apex sediments now exceeds that of the outer shelf by as much as a hundredfold, and the cumulative solid waste disposal from 1890 to 1971 of the New York area was larger than the suspended sediment discharge of all the Atlantic coast rivers. Questions about the impact of extended offshore United States jurisdiction of fisheries, construction of ocean sewage outfalls, dredging, beach erosion, and emplacement of pipelines are hotly debated issues in coastal communities that depend on revenue from commercial fishing, tourism, and other forms of recreational activities. As a result of these possibly conflicting uses of the coastal zone, multidisciplinary research on this ecosystem has been intensified over the last 10 years by the U.S. Department of Energy (DOE), the National Oceanic and Atmospheric Administration (NOAA), the U.S. Geological Survey (USGS), the Environmental Protection Agency (EPA), the Bureau of Land Management (BLM), and the National Science Foundation (NSF).

People have come to realize that dilution of wastes by marine waters can no longer be considered a simple or permanent removal process within either the open ocean or nearshore waters. The increasing utilization of the continental shelf for oil drilling and transport, siting of nuclear power plants, and various types of planned and inadvertent waste disposal, as well as for food and recreation, requires careful management of human activities in this ecosystem. Nearshore waters are presently subject, of course, to both atmospheric and coastal input of pollutants in the form of heavy metals, synthetic chemicals, petroleum hydrocarbons, radionuclides, and other urban wastes.

However, overfishing is an additional human-induced stress. For example, the sardine fishery collapsed off the California coast, herring stocks are down off the east coast, and the world's largest fishery, for anchovy off Peru, has now been reduced to less than 10% of its peak harvest in the late 1960s. Determination of what is the cause and which is the direct effect within a perturbation response of the food web of this highly variable continental shelf ecosystem is a difficult matter. One must be able to specify the consequences of human impact within the context of natural variability; for example, populations of sardines exhibited large fluctuations within the geological record off California before a fishery was initiated.

Furthermore, physical transport of pollutants, their modification by the coastal food web, and demonstration of transfer to humans are sequential problems of increasing complexity on the continental shelf. For example, after 30 years of discharge of mercury into the sea, the origin of the Minimata neurological disease of Japan was finally traced to human consumption of fish and shellfish containing methyl mercuric chloride. The Itai itai disease is now attributed to ingestion of food with high cadmium levels. Discharges of chlorinated hydrocarbons, such as DDT off California, polychlorinated biphenyl (PCB) in both the Hudson River, New York, and within Escambia Bay, Florida, mirex in the Gulf of Mexico, and vinyl chloride in the North Sea, have also led to inhibition of algal photosynthesis, large mortality of shrimp, and reproductive failure of birds and fish.

Oil spills, such as in the Santa Barbara oil field off California and the Ekofisk fields within the North Sea and from the tankers *Torrey Canyon*, *Argo Merchant*, *Amoco Cadiz*, *Metula*, *Florida*, and *Arrow*, constitute an estimated annual input of 2,000,000 metric tons of petroleum to the continental shelves with an unresolved ecological impact; another 2,000,000 tons of petrochemicals is added each year from river and sewer runoff. Fission and neutron-activation products of coastal reactors, such as San Onofre (California), Hanford (Washington), and Windscale (United Kingdom), are concentrated in marine food chains with, for example, cesium-137 found in muscle tissue of fish, ruthenium-106 in seaweed, zinc-65 in oysters, and cerium-144 in phytoplankton; their somatic and genetic effects on humans are presumably minimal. Finally, disposal of dissolved and floatable waste material from New York City has recently been implicated as a possible factor in both a $60,000,000 shellfish loss off New Jersey and a $20,000,000 closure of Long Island beaches during the summer of 1976.

Simulation models. One approach to quantitatively assess the above pollutant impacts is to construct simulation models of the coastal food web in a systems analysis of the continental shelf. Models of physical transport of pollutants have been the most successful, for example, as in studies of beach fouling by oil. Incorporation of additional biological and chemical terms in a model, however, requires dosage response functions of the natural organisms to each class of pollutants, and a quantitative description of the "normal" food web interactions of the continental shelf. *See* ECOSYSTEM MODELS.

Toxicity levels in terms of median lethal concentrations (LC50) of metals, pesticides, biofouling agents (such as chlorine), PCB, and petroleum fractions have been determined only for organisms that can be cultured in the laboratory. The actual form of the pollutant, such as methyl mercuric chloride or chloramine, and its concentration in the marine environment, however, are not always known. Furthermore, the actual contribution of a pollutant to mortality of organisms within a coastal food web is additionally confounded by the lack of understanding of natural mortality. Natural death on the continental shelf is a poorly known process. Nevertheless, there are some clear-cut examples of pollutant impacts on the coastal zone and, in these cases, management decisions are continually being made to correct these situations.

Sewage. For example, raw sewage contains pathogenic bacteria that cause human diseases such as typhoid, typhus, and hepatitis. These and various gastroenteric diseases may be contracted from eating raw shellfish that live in sewage-polluted waters. Health authorities have closed more than a million acres (4000 km^2) of the best shellfish beds and put segments of the shellfish industry abruptly out of business, with economic losses running to tens of millions of dollars per year. Purification of sewage is thus absolutely necessary for healthy coastal waters. The cost increases as the degree of treatment is intensified, however. Most treatments remove only a fraction of dissolved fertilizing minerals such as nitrates and phosphates. These nutrients from sewage plants overfertilize coastal waters where the effluent is discharged, and can at times lead to oxygen depletion of bottom waters. The cost of this additional removal of nutrients must now be weighed against their potential damage to the coastal ecosystem. *See* SEWAGE TREATMENT.

Toxic materials. Insecticides also reach coastal waters via runoff from the land, often causing fish kills. Any amount above one-tenth part of insecticide to 1,000,000 parts of water can be lethal to some fish for most of the following: DDT, parathion, malathion, endrin, dieldrin, toxaphene, lindane, and heptachlor. Contamination of fish eggs by DDT is fatal to a high proportion of young. Insecticides function mainly as paralytic nerve poisons with resulting lack of coordination, erratic behavior, loss of equilibrium, muscle spasms, convulsions, and finally suffocation. Federal and state legislation has all but eliminated DDT, the worst of the pesticides, from future use in the United States. *See* AGRICULTURAL WASTES.

Other chemical pollutants such as metals, acids, and gases result from industrial activities. Paper and pulp mills discharge wastes that are dangerous to aquatic life because the wastes have a high oxygen demand and deplete oxygen. Other factories discharge lead, copper, zinc, nickel, mercury, cadmium, and cobalt, which are toxic to coastal life in concentrations as low as 0.5 part per million (ppm). Cyanide, sulfide, ammonia, chlorine, fluorine, and their combined compounds are also poisonous. To prevent chemical pollution, factories are now being required to remove contaminants from their wastes before discharging them into coastal waters or into local sewage systems. *See* INDUSTRIAL WASTES.

Oil pollution arises from various sources. Most cases of fish poisoning are from accidental spillage from tankers, storage depots, or wells. However, slow but constant leakage from refineries ruins waterways and is difficult to remedy. Oysters seem unable to breed in the vicinity of refineries. Enclosed ocean regions take longer to recover from oil spills than open coastal areas. Careless handling at plants also results in water pollution by poisonous by-products, such as cresols and phenols that are toxic in amounts of 5–10 ppm. In past years tankers used to pump oil into the water while cleaning their tanks, but this and cleanup procedures after oil spills are being corrected by stronger Federal laws.

Thermal pollution. Thermal pollution is caused by the discharge of hot water from power plants or factories and, recently, from desalination plants. Power plants are the main source of heated discharges. They are placed at the coast or on bays to secure a ready source of sea water coolant. A large power installation may pump in 1,000,000 gal/min (63 m^3/s) and discharge it at a temperature about 10°C above that of the ambient water. Although temperatures of coastal waters range from summer highs of 35°C in southern lagoons to winter lows of −1°C in northern estuaries, each has a typical pattern of seasonal temperature to which life there has adapted. In a shallow bay with restricted tidal flow, the rise in temperature can cause gross alterations to the natural ecology. Federal standards prohibit heating of coastal waters by more than 0.5°C.

Dredging. Finally, dredging waters to fill wetlands for house lots, parking lots, or industrial sites destroys the marshes that provide sanctuary for waterfowl and for the young of estuarine fishes. As the bay bottom is torn up, the loosened sediments shift about with the current and settle in thick masses on the bottom, suffocating animals and plants. In this way, the marshes are eliminated and the adjoining bays are degraded as aquatic life zones. The northeast Atlantic states have lost 45,000 acres (182 km^2) of coastal wetlands in only 10 years, and San Francisco Bay has been nearly half obliterated by filling. Dredging to remove sand and gravel has the same disruptive effects as dredging for landfill or other purposes, whether the sand and gravel are sold for profit or used to replenish beach sand eroded away by storms. The dredging of boat channels adds to the siltation problem, and disposal of dredge spoils is being regulated in coastal areas.

Management for the future. Human populations have grown to a level where they now can have serious impacts on coastal waters. Past experience suggests that human-induced stress is most likely to lead to species replacement by undesirable forms rather than a decrease in the organic production of the ecosystem. Any societal action must now be considered in the context of what is known about the shelf ecosystem, what management decision is required, what perturbation events are likely to ensue, and what the societal costs are in using renewable and nonrenewable coastal water resources. Prediction of such perturbation events has both immediate and future value to humans in terms of management and conservation options, such as removal of shellfish before depletion of bottom oxygen, the best mode of sewage treatment, preservation of coastal species, and a decrease of toxicant levels within the coastal food web. As one moves from prediction of meteorological events to biological changes of the coastal food web, however, increasing sources of error emerge in the predictions. In the last 100 years, humans have introduced more sources of environmental variability to the continental shelf than this coastal ecosystem has encountered during the last 10,000 years. Nevertheless, sufficient information on continental shelf processes is emerging to suggest that specification of management options by delineation of cause and effect within a perturbation response of the coastal zone is a feasible goal over the next decade.

[JOHN J. WALSH]

Bibliography: J. Clark, *Fish and Man: Conflict in the Atlantic Estuaries*, American Littoral Society, Highlands, N.J., 1967; F. M. D'Itri, *Wastewater Renovation and Reuse*, 1977; N. T. Kottegoda, *Stochastic Water Resources Technology*, 1978; National Technical Advisory Committee, *Water Quality Criteria*, Federal Water Pollution Control Administration, 1969.

Water pollution

Any change in natural waters which may impair their further use, caused by the introduction of organic or inorganic substances, or a change in temperature of the water. The growth of population and the concomitant expansion in industrial and agricultural activities have rapidly increased the importance of the field of water-pollution control. In the attack on environmental pollution, higher standards for water cleanliness are being adopted by state and Federal governments, as well as by interstate organizations.

Historical developments. Ancient humans joined into groups for protection. Later, they formed communities on watercourses or the seashore. The waterway provided a convenient means of transportation, and fresh waters provided a water supply. The watercourses then became receivers of wastewater along with contaminants. As industries developed, they added their discharges to those of the community. When the concentration of added substances became dangerous to humans or so degraded the water that it was unfit for further use, water-pollution control began. With development of wide areas, pollution of surface water supplies became more critical because wastewater of an upstream community became part of the water supply of the downstream community.

Serious epidemics of waterborne diseases such as cholera, dysentery, and typhoid fever were caused by underground seepage from privy vaults into town wells. Such direct bacterial infections through water systems can be traced back to the late 18th century, even though the germ or bacterium as the cause of disease was not proved for nearly another century. The well-documented epidemic of the Broad Street Pump in London during 1854 resulted from direct leakage from privies into the hand-pumped well which provided the neighborhood water supply. There were 616 deaths from cholera among the users of the well within 40 days.

Eventually, abandoning wells in such populated locations and providing piped water to buildings improved public health. Further, sewers for drainage of wastewater were constructed, but then infections between communities rather than between the residents of a single community became apparent. Modern public health protection is provided by highly refined and well-controlled plants both for the purification of the community water supply and treatment of the wastewater.

Relation to water supply. Water-pollution control is closely allied with the water supplies of communities and industries because both generally share the same water resources. There is great similarity in the pipe systems that bring water to each home or business property, and the systems of sewers or drains that subsequently collect the wastewater and conduct it to a treatment facility. Treatment should prepare the flow for return to the environment so that the receiving watercourse will be suitable for beneficial uses such as general recreation, and safe for subsequent use by downstream communities or industries.

The volume of wastewater, the used water that must be disposed of or treated, is a factor to be considered. Depending on the amount of water used for irrigation, the amount lost in pipe leakage, and the extent of water metering, the volume of wastewater may be 70–130% of the water drawn from the supply. In United States cities, wastewater quantities are usually 75–200 gal (284–757 liters) per capita daily. The higher figure applies to large cities with old systems, limited metering, and comparatively cheap water; the lower figure to smaller communities with little leakage and good metering. Probably the average in the United States for areas served by sewers is 125–150 gal (473–568 liters) of wastewater per person per day. Of course, industrial consumption in larger cities increases per capita quantities.

Related scientific disciplines. The field of water-pollution control encompasses a part of the broader field of sanitary or environmental engineering. It includes some aspects of chemistry, hydrology, biology, and bacteriology, in addition to public administration and management. These scientific disciplines evaluate problems and give the civil and sanitary engineer basic data for the designing of structures to solve the problems. The solutions usually require the collection of domestic and industrial wastewaters and treatment before discharge into receiving waters. *See* HYDROLOGY; SANITARY ENGINEERING.

Self-purification of natural waters. Any natural watercourse contains dissolved gases normally found in air in equilibrium with the atmosphere. In this way fish and other aquatic life obtain oxygen for their respiration. The amount of oxygen which the water holds at saturation depends on temperature and follows the law of decreased solubility of gases with a temperature increase. Because water temperature is high in the summer, oxygen dissolved in the water is then at a low point for the year.

Degradable or oxidizable substances in wastewaters deplete oxygen through the action of bacteria and related organisms which feed on organic waste materials, using available dissolved oxygen for their respiration. If this activity proceeds at a rate fast enough to depress seriously the oxygen level, the natural fauna of a stream is affected; if the oxygen is entirely used up, a condition of oxygen exhaustion occurs which suffocates aerobic organisms in the stream. Under such conditions the stream is said to be septic and is likely to become offensive to the sight and smell.

Domestic wastewaters. Domestic wastewaters result from the use of water in dwellings of all types, and include both water after use and the various waste materials added: body wastes, kitchen wastes, household cleaning agents, and laundry soaps and detergents. The solid content of such wastewater is numerically low and amounts to less than 1 lb per 1000 lb of domestic wastewater. Still, the character of these waste materials is such that they cause significant degradation of receiving waters, and they may be a major factor in spreading waterborne diseases, notably typhoid and dysentery.

Characteristics of domestic wastewater vary from one community to another and in the same community at different times. Physically, community wastewater usually has the grayish colloidal appearance of dishwater, with floating trash apparent. Chemically, it contains the numerous and complex nitrogen compounds in body wastes, as well as soaps and detergents and the chemicals normally present in the water supply. Biologically, bacteria and other microscopic life abound. Wastewaters from industrial activities may affect all of these characteristics materially.

Industrial wastewaters. In contrast to the general uniformity of substances found in domestic wastewaters, industrial wastewaters show increasing variation as the complexity of industrial processes rises. Table 1 lists major industrial categories along with the undesirable characteristics of their wastewaters.

Table 1. General nature of industrial wastewaters

Industry	Processes or waste	Effect
Brewery and distillery	Malt and fermented liquors	Organic load
Chemical	General	Stable organics, phenols, inks
Dairy	Milk processing, bottling, butter and cheese making	Acid
Dyeing	Spent dye, sizings, bleach	Color, acid or alkaline
Food processing	Canning and freezing	Organic load
Laundry	Washing	Alkaline
Leather tanning	Leather cleaning and tanning	Organic load, acid and alkaline
Meat packing	Slaughter, preparation	Organic load
Paper	Pulp and paper manufacturing	Organic load, waste wood fibers
Steel	Pickling, plating, and so on	Acid
Textile manufacture	Wool scouring, dyeing	Organic load, alkaline

Because biological treatment processes are ordinarily employed in water-pollution control plants, large quantities of industrial wastewaters can interfere with the processes as well as the total load of a treatment plant. The organic matter present in many industrial effluents often equals or exceeds the amount from a community. Accommodations for such an increase in the load of a plant should be provided for in its design.

Discharge directly to watercourses. The industrial revolution in England and Germany and the subsequent similar development in the United States increased problems of water-pollution control enormously. The establishment of industries caused great migrations to the cities, the immediate result being a great increase in wastes from both population and industrial activity. For some years discharges were made directly to watercourses, the natural assimilative power of the receiving water being used to a level consistent with the required cleanliness of the watercourse. Early dilution ratios required for this method are shown in Table 2. Because of the more rapid absorption of oxygen from the air by a turbulent stream, it has a high rate of reaeration and a low dilution ratio; the converse is true of slow-flowing streams.

Development of treatment methods. With the passage of time, the waste loads imposed on streams exceeded the ability of the receiving water to assimilate them. The first attempts at wastewater treatment were made by artificially provid-

Table 2. Dilution ratios for waterways

Type	Stream flow, ft^3/(sec) (1000 population)*
Sluggish streams	7–10
Average streams	4–7
Swift turbulent streams	2–4

*1 $ft^3/sec = 28.3 \times 10^{-3}$ m^3.

ing means for the purification of wastewaters as observed in nature. These forces included sedimentation and exposure to sunlight and atmospheric oxygen, either by agitated contact or by filling the interstices of large stone beds intermittently as a means of oxidation. However, practice soon outstripped theory because bacteriology was only then being born and there were many unknowns about the processes.

In later years testing stations were set up by municipalities and states for experimental work. Notable among these were the Chicago testing station and one established at Lawrence by the state of Massachusetts, a pioneer in the public health movement. From the results of these direct investigations, practices evolved which were gradually explained through the mechanisms of chemistry and biology in the 20th century.

Thermal pollution. An increasing amount of attention has been given to thermal pollution, the raising of the temperature of a waterway by heat discharged from the cooling system or effluent wastes of an industrial installation. This rise in temperature may sufficiently upset the ecological balance of the waterway to pose a threat to the native life-forms. This problem has been especially noted in the vicinity of nuclear power plants. Thermal pollution may be combated by allowing the wastewater to cool before it empties into the waterway. This is often accomplished in large cooling towers.

Current status. Modern water-pollution engineers or chemists have a wealth of published information, both theoretical and practical, to assist them. While research necessarily will continue, they can draw on established practices for the solution to almost any problem. A challenging problem has been the handling of radioactive wastes. Reduction in volume, containment, and storage constitute the principal attack on this problem. Because of the fundamental characteristics of radioactive wastes, the development of other methods seems unlikely.

Public desire for complete water pollution control continues, but there is an increasing realization that solution to the problem is costly. While cities have had little concern about the initial construction cost because of the large Federal share, the expenditures for operation fall entirely on the local community. During the life of a project, operating costs may exceed the initial construction outlay, and with present rates of increase, they may become a major financial burden.

The control of 100% of the organic pollution reaching the watercourses is the goal of many people, but such an ideal cannot even be approached, since about one-third of the total is from nonpoint sources. Essentially, this third is from vegetable matter carried by surface drains and direct runoff. To achieve the public health protection which is the primary purpose of collection and treatment of wastewater, proper measures are essential, and should be the principal focus of a program of water pollution control. Interestingly, such was the original purpose of sewers and drains employed by ancient civilizations, as manifested by the Romans, who gave Venus the title of Goddess of the Sewers, in addition to her other titles associated with health and beauty. The pediment of her "lost" statue in the Roman Forum identifies her as *Venus Cloacinae.*

Table 3. Federal funds for wastewater treatment plant construction, 1973–1979, in $\$10^9$

Fiscal year	Authority	Appropriated	Obligated	Outlays
1973	5	2	1.531	0
1974	6	3	1.444	.159
1975	7	4	3.616	.874
1976*	0	9	4.814	2.563
1977†	1.48	1.48	6.664	2.710
1978	4.5	4.5	2.301	2.960
1979‡	5	4.2	.953	1.984

*Includes transition quarter, July-September 1976.
†Includes $480,000,000 under Public Works Employment Act.
‡Obligated and Outlays as of Apr. 30, 1979.

However, there are strong manifestations of improved quality in the waters throughout the United States. This is apparent not only in chemical and biological measurements, but in more readily observed effects such as better appearance, eliminated smells, and the return of fish life to watercourses which had become "biological deserts" because of the effects of pollution from municipal and industrial wastewaters. The overall results are a living tribute to the cooperative efforts of local citizens and local, state, and Federal agencies working together to improve the quality of the nation's waters. A measure of the activity in the field is indicated by the employment of nearly 90,000 in the local wastewater collection and treatment works of the United States.

Federal aid. Because of public demands and the actions of state legislatures and the Congress of the United States, there has been a surge of interest in, and a demand for, firm solutions to water-pollution problems. Although the Federal government granted aid for construction of municipal treatment plants as an employment relief measure in the 1930s, no comprehensive Federal legislation was enacted until 1948. This was supplemented by a major change in 1956, when the United States government again offered grants to municipalities to assist in the construction of water-pollution control facilities. These grants were further extended to small communities for the construction of both water and sewer systems.

Since 1965, Federal activity in water-pollution control has advanced from a minor activity in the Public Health Service, through the Water Pollution Control Administration in the Department of the Interior, to a major activity in the Environmental Protection Agency. In the 1972 act (P. L. 92-500) Congress authorized a massive attack on municipal pollution problems by a grant-in-aid program eclipsing any previous effort. Federal funds for 1973–1979 are given in Table 3.

State and Federal regulations are increasing constantly in severity. This tendency is expected to continue until the problem of water pollution is brought under complete control. Even then, water quality will be monitored to make certain that actual control is achieved on a day-to-day or even an hour-to-hour basis. *See* SEPTIC TANK; SEWAGE; SEWAGE DISPOSAL; SEWAGE TREATMENT.

[RALPH E. FUHRMAN]

OIL SPILL

The problem of oil spillage came to the public's attention following the grounding of the tanker *Torrey Canyon* in March 1967 at the southwest coast of England near the entrance to the English Channel. Subsequent major oil spills such as the Santa Barbara channel California oil spill in January 1969 have further raised the level of concern until today the terminology "oil spill" has become a household word. Few problems have had greater impact on the petroleum industry than those associated with oil spills. This industry in the United States is faced with the problem of supplying the ever-increasing demand for oil and petroleum products to customers who are demanding that the oil be supplied without a risk of oil spills. This large disparity between the production and use of oil throughout the world results in the requirement that enormous quantities of oil be transported large distances, primarily by tanker.

To counter the threat of environmental damage as a result of oil spills, extensive research is being performed in the United States by private industry as well as by the Environmental Protection Agency and the Coast Guard. This research is primarily directed at developing methods to combat oil spills which minimize the damage to the environment. Treating the spilled oil with dispersants was the primary method used to fight oil spills at the time of the *Torrey Canyon*. Dispersants cause oil to spread farther and disperse in a manner similar to the way soap removes oil from one's hands, allowing the oil to be emulsified and washed away with the water. The dispersants used during the *Torrey Canyon* cleanup effort were not developed specifically for use in waters containing marine life and contained aromatic solvents which are toxic. Since that time specific dispersants less toxic to marine life and biota have been developed. Today it is generally accepted that the most extensive damage to marine life resulting from the *Torrey Canyon* incident was caused by the excessive use of dispersants in the coastal zone. In fact, the areas of the shore where dispersants were not used, but which were heavily polluted with oil alone, showed very minimum damage according to J. E. Smith, director of the Plymouth Laboratory, who has studied the biological effects of the *Torrey Canyon* oil spill. At present in the United States regulations severely limit the use of dispersants, and research efforts place emphasis on containment and recovery of oil by mechanical means.

Effects of oil pollution. When oil is spilled on water, it spreads rapidly over the surface. The forces which cause the oil to spread include the force of gravity, which results in the lighter oil seeking constant level by spreading horizontally on the heavier water. A second force is the surface-tension force, which acts at the edge of the oil slick as shown in Fig. 1. It is the surface-tension force which can result in the oil spreading to a thickness approaching a monomolecular layer. This limiting thickness is almost never achieved in large oil spills, however, because the oil interfacial surface tensions change, and the net surface tension becomes negative. The interfacial surface-tension forces change as a result of the natural processes that affect oil. One of the most important natural processes is the evaporation of the oil. Evaporation occurs rapidly, the rate depending upon the nature of the oil, the rate of thinning of the slick, wave intensity, strength of the wind, temperature, and so forth. Crude oil is a mixture of a very large number of components, each with its own properties. The most volatile components evaporate first, but with all crude oils there will undoubtedly be a residue left which is virtually involatile. In addition to evaporation, some of the oil goes into solution with the water, some is oxidized, and some is utilized by microorganisms. The most important of these processes, and the one receiving the most extensive research, is the process of microbial degradation. Many microorganisms present in seas, fresh-water lakes, and rivers have a great capacity to utilize hydrocarbons. The hydrocarbons are used as an energy source and are incorporated into new cell mass. Seeding oil slicks with special bacterial cultures has been suggested to accelerate the rate of microbial decay. However, the rate of microbial degradation of oil is limited not only by the quantity of organisms but by the availability of the oxygen and nutrients needed to support the metabolic process. Acceleration of the natural process by adding nutrients such as phosphorus or nitrogen compounds, particularly in open seas where the nutrients are not naturally available, is presently being considered. Unfortunately the rate of bacterial degradation (accelerated or natural) of floating oil is slow and is therefore not effective if the oil is threatening a coastline.

As with most types of problems, the short-term deleterious effects associated with an oil spill are better understood than the long-term effects. Marine birds, especially diving birds, appear to be the most vulnerable of the living resources to the effects of oil spillage. Harm to birds from contact with oil is reported to be a result of breakdown of the natural insulating oils and waxes which shield the birds from water, as well as due to plumage damage and ingestion of oil. Efforts to cleanse or rehabilitate birds have been generally unsuccessful because of the excessive stress that the bird experiences. If treatment is prolonged for any reason, most if not all of the birds will die. Shellfish are another segment of marine life directly affected by oil spillage in the coastal zone. Many shellfish have a relatively high tolerance to oil, but their flesh can become tainted for a period subsequent to heavy pollution. Shellfish are particularly vul-

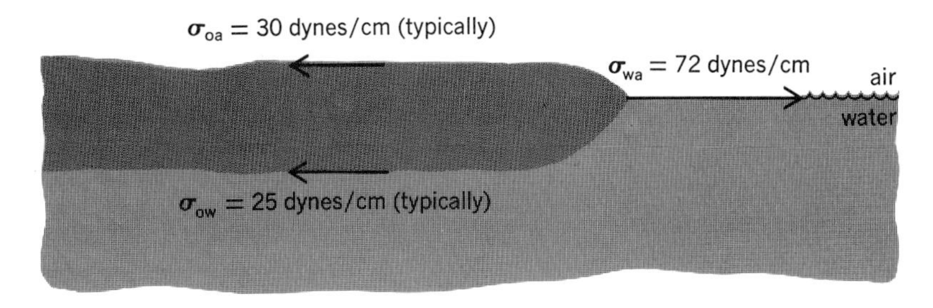

Fig. 1. Cross section of oil-water-air interface of a spreading oil slick showing the relevant surface-tension forces; 1 dyne is equivalent to 10^{-5} N.

nerable to most chemical dispersants. Fish are not generally affected by an oil spill because of their mobility which allows them to avoid heavily contaminated areas. The effects of oil on the marine food chain which consists of plants, bacteria, and small organisms is not well understood because of its complexity and because of the wide fluctuations that occur naturally and are independent of the effects of oil. In contrast to the ecological damage, the damage caused by an oil spill which is associated with recreational beach areas, coastline areas used for water sports, and areas where personal property such as docks or boats are located is well understood. A large expense is associated with loss of use of these areas for even a few days.

Cleanup procedures. Experience in attempting to clean up an oil spill has shown that no perfect method exists for all situations. Cleanup methods must be evaluated and chosen on a case-by-case basis. Spills can be more easily dealt with if they are confined to a small area on the water surface. At present, however, confinement devices have not been developed which are successful in all situations. The methods being studied and which are being used to dispose of oil floating on the surface of the sea include mechanical removal of the floating oil, the use of absorbents to facilitate the removal of the oil, sinking the oil, dispersion of the oil, and burning the oil.

Mechanical removal of oil. The primary ingredient of many mechanical oil spill cleanup systems is the use of mechanical booms or barriers. Containment of the oil spill at its source is the most important single action which can be taken when the oil spill is first detected. The use of booms has only had limited success to date because at moderate currents as low as 1 knot many booms have failed by allowing oil to pass below the boom. Figure 2 shows the types of failure which booms can experience in the presence of currents. At low currents, approximately 1.5 knots or less, oil will pile down against the barrier until the boom reaches its capacity. Additional oil will cause the boom to fail as shown in the figure. At higher currents the situation is less manageable since the boom will fail before a large quantity of oil has been collected within the barrier. For the low-current situation booms will perform satisfactorily if the oil is continually skimmed from the region in front of the barrier. Booms also are satisfactory for directing or sweeping oil, provided the angle between the boom and the current or drift direction of the oil is small so that oil does not accumulate along the boom. In addition, the relative velocity of the water at right angles to the boom must be less than the critical velocity for boom failure.

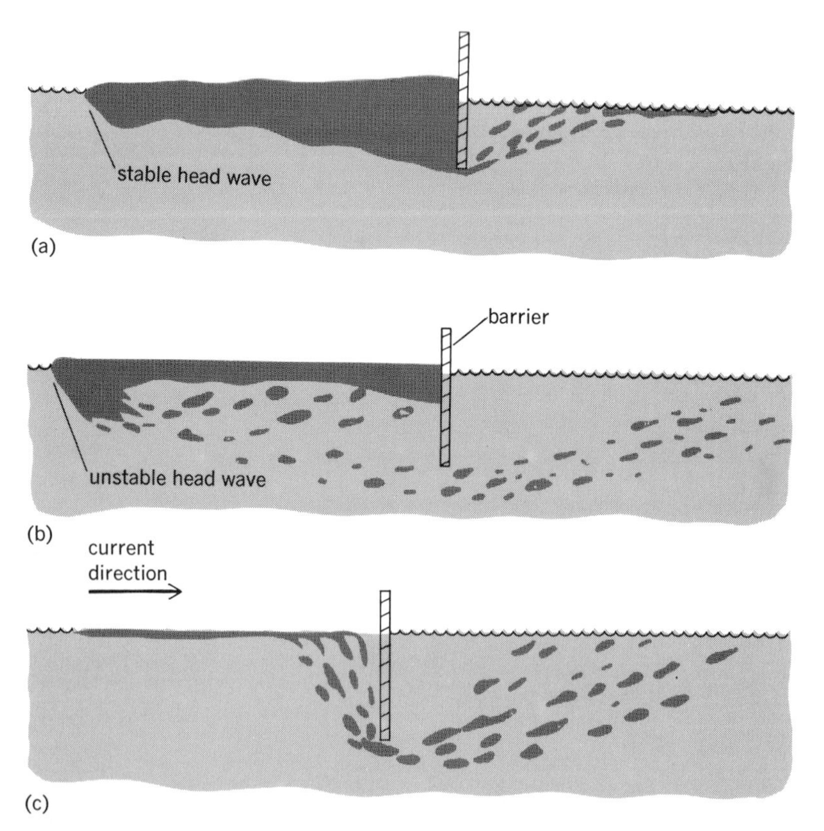

Fig. 2. Types of mechanical boom failure. (*a*) Low current. (*b*) Moderate current. (*c*) High current.

The performance of skimmers depends upon the thickness of the oil. When the film thickness is below 1/4 in. (6 mm), many techniques require pumping large amounts of water and very small amounts of oil. However, once the oil-and-water mixture is removed from the water surface, the separation of the oil from the water is easily accomplished by gravity when the oil and water are allowed to settle in a tank. Skimmers now available generally fall into one of two categories. The first, mechanical surface skimming, removes the top layers of the water and oil from the surface. These devices suffer particularly in wave action, where they gulp large amounts of water unless some provision is provided to allow the weir or suction port to follow the water surface. A second type of skimmer operates on the principle of selective wetting of a surface by oil rather than by water. Rotating metal disks or conveyor belts dip into the water surface through the oil slick. When the moving surface is drawn from the water, a surface layer of oil is removed.

Absorbents. Absorbents are used to facilitate the cleaning up of oil spills. When they are applied to the slick, they absorb the oil and prevent it from spreading, and when the absorbent material is removed from the water, the oil is removed. A class of absorbents which is commonly used consists of natural materials such as peat moss, straw, sawdust, pine bark, talc, and perlite. A second class of absorbents is derived from synthetics or plastics such as high-molecular-weight polyethylene and polystyrene, polypropylene, and polyurethane. Of all synthetic absorbents, polyurethane is generally accepted to be the most promising. The natural absorbents are generally less expensive and are attractive whenever there is a chance of losing the absorbent material. The natural absorbents are either inert or biodegrade more quickly than the synthetic materials. Alternatively the synthetic material has a greater buoyancy and a higher affinity for oil. One of the problems with absorbents is distributing them in large enough quantities on the slick. Unless the absorbents can be applied and recovered from the shore, special equipment must be available. One of the most recent concepts including absorbents involves recycling the absorbent material by wringing the oil from the absorbent and returning the absorbent material to the slick. Synthetic absorbents are most suitable for this application, and a feasibility study of such a system for operation in offshore conditions is presently being conducted by the Environmental Protection Agency.

Sinking the oil. A sinking agent which consisted of 3000 tons (2700 metric tons) of calcium carbonate with about 1% of sodium stearate was applied to an oil slick which originated from the *Torrey Canyon* and reportedly resulted in the sinking of about 20,000 tons (18,000 metric tons) of oil. The oil was sunk in the Bay of Biscay off the coast of France in 60 to 70 fathoms (110–128 m) of water. The sinking of the oil prevented the French coast from being contaminated, and after a period of 14 months no sign of the oil was found. Other materials such as specially treated sand, fly ash, and similar synthetic material have also been used to sink oil. Opinion is still divided as to the possible environmental effects of treating the oil with a dense material and sinking it. Opposition to sinking centers around the fact that sinking the oil reduces the contact surfaces between the oil and air and between the oil and water by preventing natural diffusion of the oil. Hardening of the oil subsequent to sinking would lead to a more persistent and concentrated pollution of the sea bed compared with a lower level of more dispersed pollution on the surface. In any case, utilization of this technique would be most advantageous in deeper waters outside the heavy fishing zones and where there will be a minimum of adverse effects to productive biological life in the coastal zones.

Dispersants. A dispersant is a substance which, when applied to an oil slick, causes the oil to spread farther and disperse. A dispersant contains a surfactant, a solvent, and a stabilizer. The solvent usually comprises the bulk of the dispersant and enables the surface-active agent or surfactant to mix with, and penetrate into, the oil slick and thus form an emulsion. The stabilizer fixes the emulsion and prevents it from coalescing once it is formed. The process of dispersion of the slick is shown in Fig. 3. Dispersion is similar to sinking in that it simply displaces the oil from the water surface rather than removing it from the water altogether. Dispersion has one advantage over sinking in that it increases the slick surface area and allows a rapid increase in the rate of microbial decomposition. Dispersants are not useful in coastal regions because the process of dispersion leads to an increased extent of contamination. In addition, the most effective dispersants use solvents which are toxic to marine life. To reduce this toxic effect, reduced effectiveness must be accepted. Primarily due to the question of toxicity, the use of dispersants in the open sea appears doubtful, although their use there has potential pending additional study and field data.

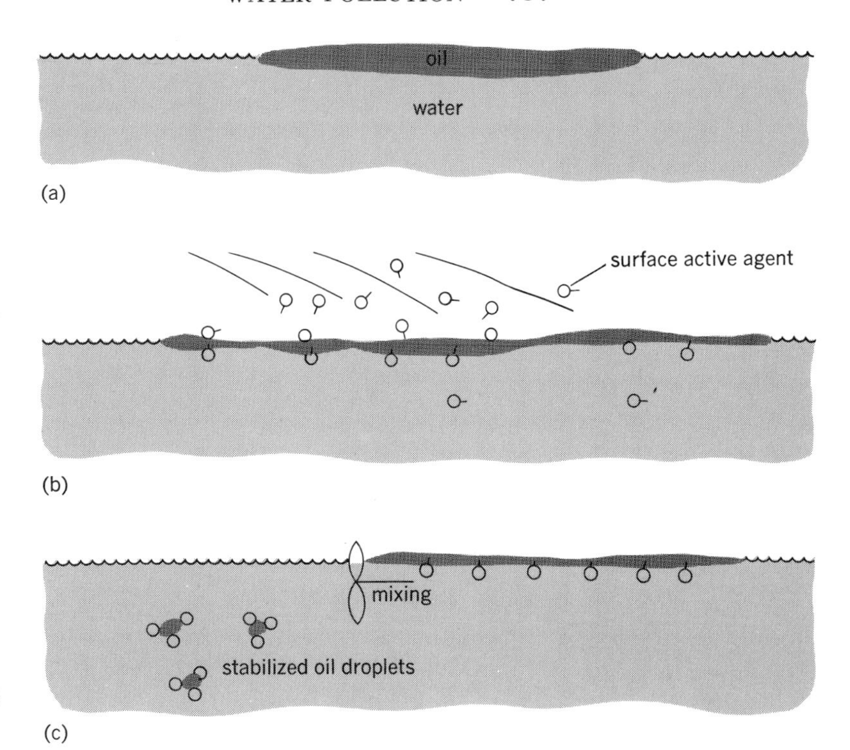

Fig. 3. Mechanism of oil slick dispersion. (*a*) Initial slick. (*b*) Application of chemical dispersant. (c) Mixing forms droplets which become stable emusion.

Burning. Burning oil slicks on the open sea has generally met with little success because the more volatile light ends evaporate quickly from the oil slick. Also, the water generally can remove heat faster than it can be created to support the combustion. Various attempts have been made to treat oil slicks to facilitate burning. The most promising of these techniques involves spreading a wicking material over the oil slick which acts to physically separate the flame from the water. Wicking agents also aid in confining the fire to a particular location, but air pollution must be expected when oil is burned in this fashion.

Prevention of oil spills. Although oil spills can probably not be eliminated entirely, steps are being taken to reduce the probability that they will occur. The United States Environmental Protection Agency and others have sponsored studies that apply reliability engineering principles to the problem and that will result in recommending procedures to be adopted to reduce the oil-spill threat. Steps are also being taken to quickly discover oil spills and to develop methods to continually monitor the water where spills are likely so that, in the event of a spill, it can be discovered quickly and proper action can be taken. The United States Coast Guard is sponsoring considerable research toward the development of remote sensing techniques. These studies have shown that, using radar and passive microwave techniques, large areas can be surveyed on a 24-hr basis even in adverse weather conditions. For the purpose of cleaning up oil spills, 67 private cooperatives are now in operation throughout the United States. These cooperatives will become more numerous in the future. Cooperatives operate in coordination with the Coast Guard, the Army Corps of Engineers, and the Environmental Protection Agency. Assessing the ability of these cooperatives, or of any group for that matter, to clean up a large spill is difficult, but it appears that near the shore mechanical techniques can be used effectively to remove oil from the water. In the offshore situation, oil recovery is more complicated and will depend upon the given situation and other environmental conditions encountered. Most current research efforts are directed toward developing systems which will operate effectively in offshore conditions. *See* MARINE ECOSYSTEM.

[ROBERT A. COCHRAN]

Bibliography: G. M. Fair et al., *Elements of Water Supply and Wastewater Disposal*, 1971; Federal Water Quality Administration, *Santa Barbara Oil Pollution, 1969: A Study of the Biological Effects of the Oil Spill Which Occurred at Santa Barbara, California, in 1969*, University of California, Santa Barbara, October 1970; R. E. McKinney, *Microbiology for Sanitary Engineers*, 1962; N. L. Nemerow, *Liquid Waste of Industry*, 1971; *Proceedings of the 1975 Conference on Prevention and Control of Oil Pollution*, San Francisco, Mar. 25–27, 1975, American Petroleum Institute, 1975; C. N. Sawyer and P. L. McCarty, *Chemistry for Sanitary Engineers*, 1962; J. E. Smith (ed.) *"Torrey Canyon" Pollution and Marine Life: A Report by the Plymouth Laboratory of the Marine Biological Association of the United Kingdom*, 1968; J. Snow, *Mode of Communication of Cholera*, 1855; Water Pollution Control Federation, *Careers in Water Pollution Control*.

Water supply engineering

A branch of civil engineering concerned with the development of sources of supply, transmission, distribution, and treatment of water. The term is used most frequently in regard to municipal waterworks, but applies also to water systems for industry, irrigation, and other purposes.

SOURCES OF WATER SUPPLY

Underground waters, rivers, lakes, and reservoirs, the primary sources of fresh water, are replenished by rainfall. Some of this water flows to the sea through surface and underground channels, some is taken up by vegetation, and some is lost by evaporation.

Groundwater. Water obtained from subsurface sources, such as sands and gravels and porous or fractured rocks, is called groundwater. Groundwater flows toward points of discharge in river valleys and, in some areas, along the seacoast. The flow takes place in water-bearing strata known as aquifers. The velocity may be a few feet to several miles per year, depending upon the permeability of the aquifer and the hydraulic gradient or slope. A steep gradient or slope indicates relatively high pressure, or head, forcing the water through the aquifer. When the gradient is flat, the pressure forcing the water is small. When the velocity is extremely low, the water is likely to be highly mineralized; if there is no movement, the water is rarely fit for use. *See* GROUNDWATER.

Permeability is a measure of the ease with which water flows through an aquifer. Coarse sands and gravels, and limestone with large solution passages, have high permeability. Fine sand, clay, silt, and dense rocks (unless badly fractured) have low permeability.

Water table. In an unconfined stratum the water table is the top or surface of the groundwater. It may be within a few inches of the ground surface or hundreds of feet below. Normally it follows the topography. Aquifers confined between impervious strata may carry water under pressure. If a well is sunk into such an aquifer and the pressure is sufficient, water may be forced to the surface, resulting in an artesian well. The water table elevation and artesian pressure may vary substantially with the seasons, depending upon the amount of rainfall recharging the aquifer and the amount of water taken from the aquifer. If pumpage exceeds recharge for an extended period, the aquifer is depleted and the water supply lost.

Salt-water intrusion. Normally the groundwater flow is toward the sea. This normal flow may be reversed, however, by overpumping and lowering of the water table or artesian pressure in an aquifer. Salt water flowing into the fresh-water aquifer being pumped is called salt-water intrusion.

Springs. Springs occur at the base of sloping ground or in depressions where the surface elevation is below the water table, or below the hydraulic gradient in an artesian aquifer from which the water can escape. Artesian springs are fed through cracks in the overburden or through other natural channels extending from the confined aquifer under pressure to the surface. *See* SPRING.

Wells. Wells are vertical openings, excavated or drilled, from the ground surface to a water-bearing stratum or aquifer. Pumping a well lowers the water level in it, which in turn forces water to flow from the aquifer. Thick, permeable aquifers may yield several million gallons daily with a drawdown (lowering) of only a few feet. Thin aquifers, or impermeable aquifers, may require several times as much drawdown for the same yield, and frequently yield only small supplies.

Dug wells, several feet in diameter, are frequently used to reach shallow aquifers, particularly for small domestic and farm supplies. They furnish small quantities of water, even if the soils penetrated are relatively impervious. Large-capacity dug wells or caisson wells, in coarse sand and gravel, are used frequently for municipal supplies. Drilled wells are sometimes several thousand feet deep.

The portion of a well above the aquifer is lined with concrete, stone, or steel casing, except where the well is through rock that stands without support. The portion of the well in the aquifer is built with open-joint masonry or screens to admit the water into the well. Metal screens, made of perforated sheets or of wire wound around supporting ribs, are used most frequently. The screens are galvanized iron, bronze, or stainless steel, depending upon the corrosiveness of the water and the expected life of the well. Plastic screens are sometimes used.

The distance between wells must be sufficient to avoid harmful interference when the wells are pumped. In general, economical well spacing varies directly with the quantity of water to be pumped, and inversely with the permeability and thickness of the aquifer. It may range from a few feet to a mile or more.

Infiltration galleries are shafts or passages extending horizontally through an aquifer to intercept the groundwater. They are equivalent to a row of closely spaced wells and are most successful in thin aquifers along the shore of rivers, at depths of less than 75 ft (23 m). The galleries are built in open cuts or by tunneling, usually with perforated or porous liners to screen out the aquifer material and to support the overburden.

Ranney wells consist of a center caisson with horizontal, perforated pipes extending radially into the aquifer. They are particularly applicable to the development of thin aquifers at shallow depths.

Specially designed pumps, of small diameter to

fit inside well casings, are used in all well installations, except in flowing artesian wells or where the water level in the well is high enough for direct suction lift by a pump on the surface (about 15 ft or 5 m maximum). Well pumps are set some distance below the water level, so that they are submerged even after the drawdown is established. Well-pump settings of 100 ft (30 m) are common, and they may exceed 300 ft (90 m) where the groundwater level is low. Multiple-stage centrifugal pumps are used most generally. They are driven by motors at the surface through vertical shafts, or by waterproof motors attached directly below the pumps. Wells are sometimes pumped by air lift, that is, by injecting compressed air through a pipe to the bottom of the well.

Surface water. Natural sources, such as rivers and lakes, and impounding reservoirs are sources of surface water. *See* DAM.

Water is withdrawn from rivers, lakes, and reservoirs through intakes. The simplest intakes are pipes extending from the shore into deep water, with or without a simple crib and screen over the outer end. Intakes for large municipal supplies may consist of large conduits or tunnels extending to elaborate cribs of wood or masonry containing screens, gates, and operating mechanisms. Intakes in reservoirs are frequently built as integral parts of the dam and may have multiple ports at several levels to permit selection of the best water. The location of intakes in rivers and lakes must take into consideration water quality, depth of water, likelihood of freezing, and possible interference with navigation. Reservoir intakes are usually designed for gravity flow through the dam or its abutments.

TRANSMISSION AND DISTRIBUTION

The water from the source must be transmitted to the community or area to be served and distributed to the individual customers.

Transmission mains. The major supply conduits, or feeders, from the source to the distribution system are called mains or aqueducts.

Canals. The oldest and simplest type of aqueducts, especially for transmitting large quantities of water, are canals. Canals are used where they can be built economically to follow the hydraulic gradient or slope of the flowing water. If the soil is suitable, the canals are excavated with sloping sides and are not lined. Otherwise, concrete or asphalt linings are used. Gravity canals are carried across streams or other low places by wooden or steel flumes, or under the streams by pressure pipes known as inverted siphons.

Tunnels. Used to transmit water through ridges or hills, tunnels may follow the hydraulic grade line and flow by gravity or may be built below the grade line to operate under considerable pressure. Rock tunnels may be lined to prevent the overburden from collapsing, to prevent leakage, or to reduce friction losses by providing a smooth interior.

Pipelines. Pipelines are a common type of transmission main, especially for moderate supplies not requiring large aqueducts or canals. Pipes are of cast iron, ductile iron, steel, reinforced concrete, cement-asbestos, or wood. Pipeline material is determined by cost, durability, ease of installation and maintenance, and resistance to corrosion. The pipeline must be large enough to deliver the required amount of water and strong enough to withstand the maximum gravity or pumping pressure. Pipelines are usually buried in the ground for protection and coolness.

Distribution system. Included in the distribution system are the network of smaller mains branching off from the transmission mains, the house services and meters, the fire hydrants, and the distribution storage reservoirs. The network is composed of transmission or feeder mains, usually 12 in. or more in diameter, and lateral mains along each street, or in some cities along alleys between the streets. The mains are installed in grids so that lateral mains can be fed from both ends where possible. Mains fed from one direction only are called dead ends; they are less reliable and do not furnish as much water for fire protection as do mains within the grid. Valves at intersections of mains permit a leaking or damaged section of pipe to be shut off with minimum interruption of water service to adjacent areas.

House services. The small pipes, usually of iron, copper, or plastic material, extending from the water main in the street to the customer's meter at the curb line or in the cellar are called house services. In many cities each service is metered, and the customer's bill is based on the water actually used.

Fire hydrants. Fire hydrants have a vertical barrel extending to the depth of the water main, a quick-opening valve with operating nut at the top, and connections threaded to receive fire hose. Hydrants must be reliable, and they must drain upon closing to prevent freezing.

Distribution reservoirs. These are used to supplement the source of supply and transmission system during peak demands, and to provide water during a temporary failure of the supply system. In small waterworks the reservoirs usually equal at least one day's water consumption; in larger systems the reservoirs are relatively smaller but adequate to meet fire-fighting demands. Ground storage reservoirs, elavated tanks, and standpipes are used for distribution reservoirs.

Ground storage reservoirs, if on high ground, can feed the distribution system by gravity, but otherwise it is necessary to pump water from the reservoir into the distribution system. Circular steel tanks and basins built of earth embankments, concrete, or rock masonry are used. Earth reservoirs are usually lined to prevent leakage and entrance of dirty water. The reservoirs should be covered to protect the water from dust, rubbish, bird droppings, and the growth of algae, but many older reservoirs without covers are in use.

Elevated storage reservoirs are tanks on towers, or high cylindrical standpipes resting on the ground. Storage reservoirs are built high enough so that the reservoir will maintain adequate pressure in the distribution system at all times.

Elevated tanks are usually of steel plate, mounted on steel towers. Wood is sometimes used for industrial and temporary installations. Standpipes are made of steel plate, strong enough to withstand the pressure of the column of water. The required capacity of a standpipe is greater than that of an elevated tank because only the upper portion of a standpipe is sufficiently elevated for normal use.

Distribution-system design. To assure the proper location and size of feeder mains and laterals to meet normal and peak water demands, a distribution system must be expertly designed. As the water flows from the source of supply or distribution reservoir across a city, the water pressure is lowered by the friction in the pipes. The pressures required for adequate service depend upon the height of buildings, need for fire protection, and other factors, but 40 psi (275 kPa) is the minimum for good service. Higher pressures for fire fighting are obtained by booster pumps on fire engines which take water from fire hydrants. In small towns adequate hydrant flows are the controlling factor in determining water-main size; in larger communities the peak demands for air conditioning and lawn sprinkling during the summer months control the size of main needed. The capacity of a distribution system is usually determined by opening fire hydrants and measuring simultaneously the discharge and the pressure drop in the system. The performance of the system when delivering more or less water than during the test can be computed from the pressure drops recorded.

An important factor in the economical operation of municipal water supplies is the quantity of water lost from distribution because of leaky joints, cracked water mains, and services abandoned but not properly shut off. Unaccounted-for water, including unavoidable slippage of customers' meters, may range from 10% in extremely well-managed systems to 30–40% in poor systems. The quantities flowing in feeder mains, the friction losses, and the amount of leakage are frequently measured by means of pitometer surveys. A pitometer is a portable meter that can be inserted in a water main under pressure to measure the velocity of flow, and thus the quantity of flow.

Pumping stations. Pumps are required wherever the source of supply is not high enough to provide gravity flow and adequate pressure in the distribution system. The pumps may be high or low head depending upon the topography and pressures required. Booster pumps are installed on pipelines to increase the pressure and discharge, and adjacent to ground storage tanks for pumping water into distribution systems. Pumping stations usually include two or more pumps, each of sufficient capacity to meet demands when one unit is down for repairs or maintenance. The station must also include piping and valves arranged so that a break can be isolated quickly without cutting the whole station out of service.

Centrifugal pumps have displaced steam-driven reciprocating pumps in modern practice, although many of the old units continue to give good service. The centrifugal pumps are driven by electric motors, steam turbines, or diesel engines, with gasoline engines frequently used for standby service. The centrifugal pumps used most commonly are designed so that the quantity of water delivered decreases as the pumping head or lift increases. Both horizontal and vertical centrifugal pumps are available in a wide capacity range. In the horizontal type, the pump shaft is horizontal with the driving motor or engine at one end of the pump. Vertical pumps are driven by a vertical-shaft motor directly above the pump or are driven by a horizontal engine through a right-angle gear head.

Automatic control of pumping stations is provided to adjust pump operations to variations in water demand. The controls start and stop pumps of different capacity as required. In the event of mishap or failure of a unit, alarms are sounded. The controls are activated by the water level in a reservoir or tank, by the pressure in a water main, or by the rate of flow through a meter. Remote control of pumps is often used, with the signals transmitted over telephone wires. *See* WATER POLLUTION; WATER TREATMENT.

[RICHARD HAZEN]

Bibliography: R. W. Abbett, *American Civil Engineering Practice*, 3 vols., 1956; H. E. Babbitt, J. J. Doland, and J. H. Cleasby, *Water Supply Engineering*, 6th ed., 1962; C. V. Davis, *Handbook of Applied Hydraulics*, 2d ed., 1952; G. M. Fair and H. L. Geyer, *Water Supply and Waste-Water Disposal*, 1954; W. A. Hardenbergh and E. B. Rodie, *Water Supply and Waste Disposal*, 1961; R. K. Lensley and J. B. Franzini, *Water Resources Engineering*, 1964; L. S. Nielsen, *Standard Plumbing Engineering Design*, 1963; E. W. Steel, *Water Supply and Sewerage*, 4th ed., 1960.

Water table

The upper surface of the zone of saturation in permeable rocks not confined by impermeable rocks. It may also be defined as the surface underground at which the water is at atmospheric pressure. Saturated rock may extend a little above this level, but the water in it is held up above the water table by capillarity and is under less than atmospheric pressure; therefore, it is the lower part of the capillary fringe and is not free to flow into a well by gravity. Below the water table, water is free to move under the influence of gravity. The position of the water table is shown by the level at which water stands in wells penetrating an unconfined water-bearing formation.

Where a well penetrates only impermeable material, there is no water table and the well is dry. But if the well passes through impermeable rock into water-bearing material whose hydrostatic head is higher than the level of the bottom of the impermeable rock, water will rise approximately to the level it would have assumed if the whole column of rock penetrated had been permeable. This is called artesian water, and the surface to which it rises is called the piezometric surface.

The water table is not a level surface but has irregularities that are commonly related to, though less pronounced than, those of the land surface. Also, it is not stationary but fluctuates with the seasons and from year to year. It generally declines during the summer months, when vegetation uses most of the water that falls as precipitation, and rises during the late winter and spring, when the demands of vegetation are low. The water table usually reaches its lowest point after the end of the growing season and its highest point just before the beginning of the growing season. Superimposed on the annual fluctuations are fluctuations of longer period which are controlled by climatic variations. The water table is also affected by withdrawals, as by pumping from wells. *See* GROUNDWATER.

[ALBERT N. SAYRE/RAY K. LINSLEY]

Bibliography: S. N. Davis and R. J. M. DeWiest, *Hydrogeology*, 1966.

Water treatment

Physical and chemical processes for making water suitable for human consumption and other purposes. Drinking water must be bacteriologically safe, free from toxic or harmful chemical or substances, and comparatively free of turbidity, color, and taste-producing substances. Excessive hardness and high concentration of dissolved solids are also undesirable, particularly for boiler feed and industrial purposes. The more important treatment processes are sedimentation, coagulation, filtration, disinfection, softening, and aeration.

Plain sedimentation. Silt, clay, and other fine material settle to the bottom if the water is allowed to stand or flow quietly at low velocities. Sedimentation occurs naturally in reservoirs and is accomplished in treatment plants by basins or settling tanks. The detention time in a settling basin may range from an hour to several days. The water may flow horizontally through the basin, with solids settling to the bottom, or may flow vertically upward at a low velocity so that the particles will settle through the rising water. Settling basins are most effective if shallow, and rarely exceed 10–20 ft (3–6 m) in depth. Plain sedimentation will not remove extremely fine or colloidal material within a reasonable time, and the process is used principally as a preliminary to other treatment methods.

Coagulation. Fine particles and colloidal material are combined into masses by coagulation. These masses, called floc, are large enough to settle in basins and to be caught on the surface of filters. Waters high in organic material and iron may coagulate naturally with gentle mixing. The term is usually applied to chemical coagulation, in which iron or aluminum salts are added to the water to form insoluble hydroxide floc. The floc is a feathery, highly absorbent substance to which color-producing colloids, bacteria, fine particles, and other substances become attached and are thus removed from the water.

The coagulant dose is a function of the physical and chemical character of the raw water, the adequacy of settling basins and filters, and the degree of purification required. Moderately turbid water coagulates more easily than perfectly clear water, but extremely turbid water requires more coagulant. Coagulation is more effective at higher temperatures. Lime, soda ash, or caustic soda may be required in addition to the coagulant to provide sufficient alkalinity for the formation of floc, and regulation of the pH (hydrogen-ion concentration) is usually desirable for best results. Powdered limestone, clay, bentonite, or silica are sometimes added as coagulant aids to strengthen and weight the floc, and a wide variety of polymers developed in recent years are used for the same purpose.

Filtration. Suspended solids, colloidal material, bacteria, and other organisms are filtered out by passing the water through a bed of sand or pulverized coal, or through a matrix of fibrous material supported on a perforated core. Filtration of turbid or highly colored water usually follows sedimentation or coagulation and sedimentation. Soluble materials such as salts and metals in ionic form are not removed by filtration.

Slow sand filters. Used first in England about 1850, slow sand filters consist of beds of sand 20–48 in. (51–122 cm) deep, through which the water is passed at fairly low rates—2,500,000 to 10,000,000 gal per acre (24,000–94,000 m^3/ha). The size of beds ranges from a fraction of an acre in small plants to several acres in large plants. An underdrain system of graded gravel and perforated pipes transmits the filtered water from the filters to the point of discharge. The sand is usually fine, ranging from 0.2 to 0.5 mm in diameter. The top of the filter clogs with use, and a thin layer of dirty sand is scraped from the filter periodically to maintain capacity.

Slow sand filters operate satisfactorily with reasonably clear waters but clog rapidly with turbid waters. The filters are covered in cold climates to prevent the formation of ice and to facilitate operation in the winter. In milder climates they are often open. Slow sand filters have a high bacteriological efficiency, but few have been built since the development of water disinfection, because of the large area required, the high construction cost, and the labor needed to clean the filters and to handle the filter sand. Slow sand filters are still used in many English and European cities, but have not been built in the United States since 1950, and few remain in operation.

Rapid sand filters. These operate at rates of 125,000,000 to 250,000,000 gal per acre (1,170,000–2,340,000 m^3/ha) per day; or 25 to 50 times the slow-sand-filter rates. The high rate of operation is made possible by the coagulation and sedimentation ahead of filtration to remove the heaviest part of the load, the use of fairly coarse sand, and facilities for backwashing the filter to keep the bed clean. The filter beds are small, generally ranging from 150 ft^2 (14 m^2) in small plants to 1500 ft^2 (140 m^2) in the largest filter plants. The filters consist of a layer of sand or, occasionally, crushed anthracite coal 18–24 in. (46–62 cm) deep, resting on graded layers of gravel above an underdrain system. The sand is coarse, 0.4–1.0 mm in diameter, depending upon the raw water quality and pretreatment, but the grain size must be fairly uniform to assure proper backwashing. The underdrain system serves both to collect the filtered water and to distribute the wash water under the filters when they are being washed. Several types of underdrains are used, including perforated pipes, perforated false bottoms of concrete, and tile and porous plates.

Filters are backwashed at rates 5–10 times the filtering rate. The wash water passes upward through the sand and out of the filters by way of wash-water gutters and drains. Washing agitates the sand bed and releases the dirt to flow out of the filter with the wash water. The quantity of water used for washing ranges from 1 to 10% of the total output, depending upon the turbidity of the water applied to the filters and the efficiency of the filter design. Combination air and water filter washes are popular in Europe, but are not often used in the United States.

Municipal and large-capacity filters for industry usually are built in concrete boxes or in open tanks of wood and steel. The flow through the sand may be caused by gravity, or the water may be forced through the sand under pressure by pumping. Pressure filters can be operated at higher rates than gravity filters, because of the greater head available to force the water through the sand. However, excessive pressure may increase the effluent turbidity, and bacteria may appear in the discharge water. For this reason, and because pressure filters are difficult to inspect and keep in good order, open gravity filters are favored for public water supplies.

Diatomaceous earth filters. Swimming-pool installations and small water supplies frequently use this type of filter. The filters consist of a medium or septum supporting a layer of diatomaceous earth through which the water is passed. A filter layer is built up by the addition of diatomaceous earth to the water. When the pressure loss becomes excessive, filters must be backwashed and a fresh layer of diatomaceous earth applied. Filter rates of $2\frac{1}{2}$–6 gal per minute per square foot ($1.7-4.1 \times 10^{-3}$ $m^3/s \cdot m^2$) are attained.

Disinfection. There are several methods of treatment of water to kill living organisms, particularly pathogenic bacteria; the application of chlorine or chlorine compounds is the most common. Less frequently used methods include the use of ultraviolet light, ozone, or silver ions. Boiling is the favorite household emergency measure.

Chlorination is simple and inexpensive and is practiced almost universally in public water supplies. It is often the sole treatment of clear, uncontaminated waters. In most plants it supplements coagulation and filtration. Chlorination is used also to protect against contamination of water in distribution mains and reservoirs after purification.

Chlorine gas is most economical and easiest to apply in large systems. For small works, calcium hypochlorite or sodium hypochlorite is frequently used. Regardless of which form is used, the dose varies with the water quality and degree of contamination. Clear, uncontaminated water can be disinfected with small doses, usually less than one part per million; contaminated water may require several times as much. The amount of chlorine taken up by organic matter and minerals in water is known as the chlorine demand. For proper disinfection the dose must exceed the demand so that free chlorine remains in the water.

Chlorination alone is not reliable for the treatment of contaminated or turbid water. A sudden increase in the chlorine demand may absorb the full dose and provide no residual chlorine for disinfection, and it cannot be assumed that the chlorine will penetrate particles of organic matter. Chlorine is applied before filtration, after filtration, and sometimes at both times.

Chlorine sometimes causes objectionable tastes or odors in water. This may be due to an excessive chlorine dose, but more frequently it is caused by a combination of chlorine and organic matter, such as algae, in the water. Some algae, relatively unobjectionable in the natural state, produce unbearable tastes after chlorination. In other cases, strong chlorine doses oxidize the organic matter completely and produce odor-free water. Excessive chlorine may be removed by dechlorination with sulfur dioxide. Also, ammonia is often added for taste control to reduce the concentration of free chlorine. Activated carbon is also effective in the reduction of natural and chlorine tastes, harmful organic compounds, heavy metals, and such.

Water softening. The "hardness" of water is due to the presence of calcium and magnesium salts. These salts make washing difficult, waste soap, and cause unpleasant scums and stains in households and laundries. They are especially harmful in boiler feedwater because of their tendency to form scales.

Municipal water softening is common where the natural water has a hardness in excess of 150 parts per million. Two methods are used: (1) The water is treated with lime and soda ash to precipitate the calcium and magnesium as carbonate and hydroxide, after which the water is filtered; (2) the water is passed through a porous cation exchanger which has the ability of substituting sodium ions in the exchange medium for calcium and magnesium in the water. The exchange medium may be a natural sand known as zeolite, or may be manufactured from organic resins. It must be recharged periodically by backwashing with brine.

For high-pressure steam boilers or some other industrial processes, almost complete deionization of water is needed, and treatment includes both cation and anion exchangers. Lime-soda plants are similar to water purification plants, with coagulation, settling, and filtration. Zeolite or cation-exchange plants are usually built of steel tanks with appurtenances for backwashing the media with salt brine. If the water is turbid, filtration ahead of zeolite softening may be required.

Aeration. Aeration is a process of exposing water to air by dividing the water into small drops, by forcing air through the water, or by a combination of both. The first method uses jets, fountains, waterfalls, and riffles; in the second, compressed air is admitted to the bottom of a tank through perforated pipes or porous plates; in the third, drops of water are met by a stream of air produced by a fan.

Aeration is used to add oxygen to water and to remove carbon dioxide, hydrogen sulfide, and taste-producing gases or vapors. Aeration is also used in iron-removal plants to oxidize the iron ahead of the sedimentation or filtration processes. *See* WATER POLLUTION; WATER SUPPLY ENGINEERING. [RICHARD HAZEN]

Bibliography: R. W. Abbett, *American Civil Engineering Practice*, 3 vols., 1956; C. V. Davis (ed.), *Handbook of Applied Hydraulics*, 3d ed., 1969; G. M. Fair and J. C. Geyer, *Water Supply and Waste Water Disposal*, 1954; G. V. James, *Water Treatment*, 1965; L. G. Rich, *Unit Processes of Sanitary Engineering*, 1963.

Weather forecasting and prediction

Procedures for extrapolation of the future character of weather on the basis of present and past conditions. Accurate weather prediction requires a knowledge of the past state of the atmosphere, an understanding of the physical laws governing atmospheric behavior, and the availability of necessary technological aids for the rapid dissemination of meteorological information and the preparation of the forecast. The historical development of methods for forecasting the weather can be traced to innovations in these three areas. For introductory discussions of atmospheric science *see* ATMOSPHERE; METEOROLOGY.

This article is divided into five sections. The first part emphasizes the current status of the whole field of weather forecasting and prediction, and concentrates on short-range prediction, up to 48 hr in advance. The second portion deals with long-range prediction. The third section considers statistical forecasting procedures. A fourth section summarizes the bases for the developing techniques of numerical prediction. The last portion examines the weather offices and centers through which are funneled the data of observation for analysis and processing into forecast and prediction.

DEVELOPMENT

Information on the state of the atmosphere has been greatly expanded since 1939, when the demands of aircraft in World War II led to the installation of radio-sounding stations and the development of weather reconnaissance systems. These sources have since been supplemented by radar aids, rockets, and satellites. When these data are recorded on charts, the meteorologist has a three-dimensional picture of atmospheric structure. A series of such charts at 6-, 12-, or 24-hr intervals shows the development of weather systems in terms of the changes in wind, pressure, temperature, humidity, cloudiness, precipitation, and visibility. These maps permit an analysis of the dominant long-wave patterns (discussed under long-range forecasting) with linear dimensions of 10^3-10^4 mi (1 mi = 1.6 km), the migratory cyclones and anticyclones with dimensions of 10^2-10^3 mi, and the weather conditions averaged over an area on the order of 10^4 mi^2 (1 mi^2 = 2.6 km^2). The resolution of weather detail is less exact over oceanic regions, in the polar areas, and in most tropical areas. Small-scale weather systems such as land-sea breezes, mountain-valley winds, and convective showers cannot be depicted on the conventional weather map and can be studied only by means of a dense network of observing stations. Radar information gives precipitation detail within the 10^4-mi^2 area.

During the 1960s satellites provided the most significant developments in data acquisition. Besides cloud pictures, satellite technology will yield much needed information on upper-air winds and temperatures on a global scale.

Data to forecasting. The step from data to forecast is achieved by a variety of methods which may be classified as follows.

Semiempirical techniques. The forecaster first synthesizes the raw information into dynamically meaningful models of the atmosphere (as best exemplified by the polar-front model of J. Bjerknes and H. Solberg). Then, by a combination of methods including the extrapolation of past trends, expected changes based upon qualitative physical reasoning, and recollection of the behavior of similar situations in the past, he arrives at an estimate of the position of the prominent features on tomorrow's weather charts. These features include the location and intensity of cyclones, anticyclones, fronts, upper-level pressure ridges and troughs, and the like. Certain formulas can be applied to this phase of the forecast but, because of simplifying assumptions involved in their use, they yield questionable results. The details of weather such as types of clouds, rainfall, and temperature are deduced from this prediction by again referring to models and by considering such effects as advection, radiation, topographical influences, and expected stability changes.

The success of these techniques is limited by their semiempirical character and by the ability of an individual to handle the tremendous mass of significant information. Data are now so voluminous that this method of prediction utilizes only a fraction of the available information, and it becomes somewhat a matter of personal choice as to the selection process.

Numerical methods. The advent of an expanded network of weather observations and of high-speed computers has greatly stimulated meteorological research so that forecasting is passing from the pre-1945 qualitative phase to a quantitative era in which predictions are based upon computation, guided by physical principles. These methods are discussed in the sections on the numerical prediction of weather and the statistical prediction of weather.

The large-scale aspects of the atmospheric flow pattern are most accurately predicted by numerical methods. In addition, numerical methods are superseding the qualitative techniques of the past in the prediction of cloudiness, precipitation, wind, and temperature.

Evaluating forecasting. To analyze the accuracy of weather forecasts, it is necessary to recognize the statistical nature of the element being predicted. Precipitation, even when it covers wide areas, is rarely uniform in intensity. Small convective showers, which develop and dissipate rapidly, are often embedded in the general rain or snow and cause significant variations over distances of the order of 10 mi and over time periods of less than 1 hr. The shower or thunderstorm cells which are associated with air-mass showers or with cold fronts exhibit a maximum variability in time and space. This turbulent behavior presents a major forecasting problem. Consequently, the prediction must give wide range to the estimates of precipitation intensity in order to cover the probable variability over an area on the order of 10^4 mi^2. Only in those areas where the local variations can be attributed to orographical influences is it possible to present a more definitive estimate.

In contrast to the variability of precipitation, the turbulent fluctuations of wind and temperature are so rapid that they are not of general interest to the forecaster or the public. The specialized problem of predicting atmospheric pollution is an exception. Local variations in temperature and wind within the 1-hr, 10-mi scale can be attributed to such well-understood influences as the nature of the underlying surface, proximity to a water area, and a valley or mountain effect. Hence, detailed forecasts of these elements can be made with considerable reliability.

The final consideration with all types of forecast is the length of the time step. The larger the scale of the atmospheric system, the more persistent the phenomenon. For example, an individual summer shower has a life-span of approximately 1 hr; a cyclone is ordinarily identifiable for at least 3 days, and a particular long-wave pattern may persist for weeks. The major problem in weather prediction is the forecasting of a new development; it may be as difficult to pinpoint a summer shower a few hours in advance as to predict, a week in advance, the broad-scale features of the weather associated with a long-wave pattern. For similar reasons, the accuracy with which the weather may be predicted in detail decreases rapidly with the time elapsed since the observations were made.

Probable future developments. A major handicap in the past has been the paucity of dynamically significant information on the atmosphere over the ocean areas and over the less-populated regions of the globe. The developments in satellite technology offer real promise for filling these data gaps. With the globe's atmosphere well charted and with developments in numerical prediction it is anticipated that quantitative forecasts, via the computer, will replace completely the older qualitative techniques. The major improvement can be expected in the longer-range aspects of weather prediction. The isolated summer afternoon rain shower, which plagues many a picnic, will probably remain an unsolved forecast problem for many years to come. [JAMES M. AUSTIN]

LONG-RANGE FORECASTING

Long-range weather forecasts are of two types. Medium- or extended-range forecasts cover periods of from 48 hr to a week in advance. Forecasts for longer periods generally extend over periods of a month, a season, or possibly two or more seasons in advance. Although meteorologists have been working on long-range prediction problems for more than 100 years, the degree of accuracy is small for all predictions exceeding a week in advance. This seems particularly true when the predictions are examined under rigid statistical controls, such as climatological probability. There is little or no evidence to indicate any sustained success for forecasts embracing periods of more than a season in advance. The reason for such limited ability lies in the utter complexity of the atmosphere's behavior—the vicissitudes of a compressible fluid responding to changing external stimuli, such as the Sun, and changing characteristics of the Earth's surface, both in space and time.

Medium or extended ranges. Scientific methods of extended-range prediction take for granted that the further out in time the forecast is projected, the more general must be the nature of the prediction. Short-range forecasts for 48 hr or less in advance specify the detail of weather in space and in time. Medium-range forecasts for a week in advance cover average conditions and trends within the week in intervals of a couple of days. Forecasts for a month or more, however, can indicate only the broad-scale (for example, areas of several hundred thousand square miles) features of average or prevailing weather. Such broad-scale aspects are usually expressed in terms of departures from seasonal norms for elements such as temperature and precipitation.

Medium-range forecast methods are apt to use one or a combination of dynamic, statistical, and synoptic techniques. Dynamic methods capitalize upon the best physical knowledge of meteorological phenomena. Statistical methods employ empirically derived equations as substitutes for physical knowledge. In the synoptic technique, various hemispheric wind and weather charts are surveyed and interpreted by an experienced meteorologist. An important part of modern dynamic methods is the principle that the vertical component of absolute vorticity remains fairly constant as air columns of the middle troposphere move from one area to another (discussed below under numerical weather prediction). When instantaneous wind and pressure charts for midtropospheric levels are averaged in time or space, certain small-scale perturbations, including short waves or vortexes in the horizontal, are suppressed. What remains are smooth, long, or planetary waves which in effect constitute a special class of motions; they are not only of larger scale (often being composed of a family of cyclones or anticyclones), but they also evolve more slowly than the individual wind charts from which they are constructed. The planetary waves which these time-averaged charts reveal are responsible for variations in the position and intensity of the well-known sea-level centers of action (like the Bermuda high, one of the subtropical oceanic highs) which largely determine prevailing weather abnormalities. For purposes of extended forecasting, the averaging process is performed on past (observed) data as well as on numerically predicted charts to 4 days in advance. Various methods of comparison of such time-averaged charts enable the synoptician to assay the continuity and trend of large-scale systems and to extrapolate them into the future for some reasonable period. Dynamic methods of prediction may also be used with some success on time-averaged charts, but the physical reasons for this are not clear.

Procedures for extended forecasting vary around the world largely in accordance with facilities and availability of scientific manpower. Many countries do not have available the high-speed computing equipment necessary to prepare the dynamic component of the forecast, nor have they even the statistical components. In these cases extended forecasts are either not prepared at all or are made by educated synoptic guesswork.

After predictions of average planetary wind flows at upper levels have been made, it is possible to infer the accompanying types of weather in different areas as well as to estimate the general regions for breeding and movement of storms and air masses. Here again the statistics of the motions and weather are more predictable than is the day-to-day detail. In fact, the translation of average wind circulation into average weather is amenable to statistical stratification procedures and is fairly objective. These methods employ as input numerically predicted charts for the 700-millibar (mb) level and also surface temperature predictions to 48 hr made by field forecasters throughout the United States. From these data multiple-regression formulas make possible high-speed computation of temperature forecasts for the weather forecast centers in the United States. A similar procedure gives precipitation estimates. Other weather elements are inferred from the predicted charts for the period.

The boundary between the domain of short- and extended-range forecasting techniques has shifted: Dynamic predictions performed by computer now form the basis for 72-hr prognoses. It is possible that before 1980 dynamic predictions will form a new base for daily forecasts up to a week in advance and that time-averaging methods will no longer be necessary in that range.

Longer-period forecasting. For periods more than a week in advance, methods of forecasting rely more upon statistics and less upon physical reasoning. Concentrated efforts to explore the physics of long-range weather phenomena began about 1955 as a result of increased availability of computing facilities and hemisphere-wide data coverage, particularly from the upper air. However, attempts are being made to prepare forecasts a month ahead by employing dynamic principles in conjunction with statistical and synoptic techniques. For these purposes, another class of mean motions is defined by construction of mean maps for 30-day periods. Although real understanding of these methods is remote, experiments indicate that such objective, machine-produced prognoses are helpful and contain a large part of the accuracy of 30-day forecasts. These prognoses take rough account of the net effect of changes in insolation associated with the change of season, the tendency of certain branches of the general circulation to persist, and the compatibility of the positions of certain large-scale features like the Bermuda high and Icelandic low. Once the circulation pattern for the Northern Hemisphere is prognosticated, the average temperature and precipitation anomalies are computed with the help of elaborate statistical specification equations. The numerical results are then adjusted subjectively by attempting to consider factors not in the equations, like snow cover and wet or dry soil.

Another less expensive and less time-consuming method of longer-range prediction involves the use of statistical analogs. The historical files of weather maps for past periods (mean maps may also be used) are searched, and a wind and weather pattern is sought which is as similar as possible to the one which has been operative, on the assumption that what transpired in the earlier case will repeat. For best results the analogy should be good for large areas of the hemisphere, should hold for upper levels as well as for sea level, and should stand up for a sequence of periods preceding the forecast. The logic of this method is appealing; similar patterns under the same stimuli (such as season of the year) tend to repeat. However, the relatively short span of time for which meteorological records have been kept makes it difficult to find good analogs.

For periods beyond a month, statistical techniques seem to be the only ones sufficiently accurate; even here, there is some question as to whether the samples of data (length of record) are adequate to assure the stability of discovered relationships. However, experiments in seasonal forecasting indicate some reliability in predicting departures from normal of temperature for the contiguous United States.

Another avenue of approach to the long-range forecast problem which shows promise involves large-scale interactions between the ocean and atmosphere. J. Bjerknes has produced work which indicates that seasonally variable ocean temperatures near the Equator affect the rainfall there, and that the resulting variable release of heat of condensation controls the Hadley circulation, which in turn determines the strength of the subtropical anticyclones and the prevailing westerlies. Complementary work by J. Namias suggests that the longitudinal positioning and intensities of the long waves of the planetary circulation are also determined by air-sea interactions. Although these ideas are not developed to the point of utilizing them objectively in long-range forecasting, attempts are being made to consider them subjectively. *See* ATMOSPHERIC GENERAL CIRCULATION.

Finally, there is hope that numerical forecasts iterated day by day for weeks in advance may yield economically valuable statistics on the weather of the forthcoming month or season. A truly adequate global network of observations is necessary before this will be possible.

[JEROME NAMIAS]

STATISTICAL WEATHER FORECASTING

Statistical weather forecasting is the prediction of weather by rules based upon the statistics of weather behavior. A prediction may state the expected value of a specific weather element, such as a wind speed, or the probability of occurrence of a specific weather event, such as a thunderstorm. In the former case the prediction is understood to contain an error whose probable value may or may not be stated. The choice of the form of prediction may depend upon the intended audience; for example, the statement that there are 2 chances in 10 that tonight's temperature will fall below 32°F (0°C) might aid a fruit grower more than the statement that tonight's expected minimum temperature will be 36°F (2°C).

Basic premises. Statistical forecasting is based upon the premise that the future worldwide state of the atmosphere is determined, at least approximately, by the present state, together with the intervening influences of the Sun and the underlying ocean and land, according to immutable physical laws. In theory, forecasting is equivalent to solving the equations representing these laws, but the equations are rather intractable, and because there are vast gaps between observing stations, the present weather is only partially known. It is sometimes more feasible to ascertain how future weather must evolve from present weather by studying how the observed portions of the atmosphere have previously behaved.

Prediction rules established from such study often relate future weather to present and past weather, instead of present weather alone, since past knowledge may partially compensate for incomplete present knowledge. A rule is commonly expressed as a mathematical formula. Sometimes the same information is more conveniently presented as a graph or a table.

A rule established for one location does not generally apply at another location. For example, a table established for San Francisco, showing the probability of occurrence of nighttime fog following various combinations of midafternoon temperature and relative humidity, would not be valid for predicting fog in New York. A new rule is usually needed for each new weather prediction in any particular area.

Statistical procedures. A general kind of procedure is used to establish a formula. The meteorologist chooses a set of weather elements for "predictors" and selects, commonly from past records, a set of data consisting of corresponding observed values of the predictors and the predictand. As the next step, he chooses a mathematical form with a limited number of degrees of freedom, ordinarily appearing as undetermined constants, and restricts the formula to this form. He specifies a process, ordinarily the minimization of the sum of squares of the prediction errors, by which the chosen data shall determine the constants. Evaluating the constants is then an objective and usually routine mathematical task.

The meteorologist may modify this procedure

and classify combinations of values of the chosen predictors into categories. He may then construct a table by choosing, for each category, the average observed value of the predictand as the expected value or, alternatively, by choosing the observed frequency of occurrence as the probability.

The preparation of a forecast once the rule is established is objective and usually simple. The forecaster evaluates the formula after introducing the appropriate numerical values of the predictors, or reads the forecast from the appropriate location in the graph or table.

The data selected for establishing a formula constitute a finite sample of the total history of the weather. The formula is likely to succeed, when applied to future weather, only if the number of degrees of freedom is small compared to the number of values of each predictor in the sample, since virtually any finite set of numbers will fit a sufficiently complicated formula.

Ideally the sample should be made very large. When this is not feasible, because of the excessive labor involved or the absence of extensive past records, the degrees of freedom must be restricted. This is accomplished by limiting the number of predictors or restricting the formula to a more highly specialized form. Meteorological experience or physical reasoning should be used as a guide, since a blind choice of predictors, or of a mathematical form, is unlikely to yield a successful formula.

Statistical linear regression. The simplest mathematical form, and the one whose theory is most highly developed, is the linear formula. When many predictors have been chosen, the number may be reduced either by factor analysis, which selects a few linear combinations of the predictors in order of their ability to represent all the predictors, or by a procedure which selects a few predictors in order of their independent contribution to the prediction. Widespread investigation of linear formulas has followed the advent of high-speed electronic computing machines.

Appropriate nonlinear formulas are theoretically superior to linear formulas, but since they usually involve many degrees of freedom, they are more difficult to discover.

Statistical methods are highly suitable for predicting special local phenomena such as the occurrence of fog. For preparing prognostic weather maps one or two days in advance, statistical formulas are useful, but are frequently inferior to conventional subjective forecasts. For forecasting several days in advance, linear statistical formulas show a slight positive utility and compare favorably with other methods. [EDWARD N. LORENZ]

NUMERICAL WEATHER PREDICTION

Numerical weather prediction is the prediction of weather phenomena by the numerical solution of the equations governing the motion and changes of condition of the atmosphere. More generally, the term applies to any numerical solution or analysis of the atmospheric equations of motion.

The laws of motion of the atmosphere may be expressed as a set of partial differential equations relating the instantaneous rates of change of the meteorological variables to their instantaneous distribution in space. These are developed in dynamic meteorology. A prediction for a finite time interval is obtained by summing the succession of infinitesimal time changes of the meteorological variables, each of which is determined by their distribution at the preceding instant of time. Although this process of integration may be carried out in principle, the nonlinearity of the equations and the complexity and multiplicity of the data make it impossible in practice. Instead, one must resort to finite-difference approximation techniques in which successive changes in the variables are calculated for small, but finite, time intervals at a finite grid of points spanning part or all of the atmosphere. Even so, the amount of computation is vast, and numerical weather prediction remained only a dream until the advent of the modern high-speed electronic computing machine. These machines are capable of performing the millions of arithmetic operations involved with a minimum of human labor and in an economically feasible time span. Numerical methods are gradually replacing the earlier, more subjective methods of weather prediction in many United States government weather services. This is particularly true in the preparation of prognoses for large areas. The detailed prediction of local weather phenomena has not yet benefited greatly from the use of numerico-dynamic methods, as indicated above in the general section on weather forecasting.

Short-range numerical prediction. By the nature of numerical weather prediction, its accuracy depends on (1) an understanding of the laws of atmospheric behavior, (2) the ability to measure the instantaneous state of the atmosphere, and (3) the accuracy with which the solutions of the continuous equations of motion are approximated by finite-difference means. The greatest success has been achieved in predicting the motion of the large-scale (>1000 mi or 1600 km) pressure systems in the atmosphere for relatively short periods of time (1–3 days). For such space and time scales, the poorly understood energy sources and frictional dissipative forces may be largely ignored, and rather coarse space grids may be used.

The large-scale motions are characterized by their properties of being quasi-static, quasi-geostrophic, and horizontally quasi-nondivergent, as dicussed in another article. *See* METEOROLOGY.

These properties may be used to simplify the equations of motion by filtering out the motions which have little meteorological importance, such as sound and gravity waves. The resulting equations then become, in some cases, more amenable to numerical treatment.

A simple illustration of the methods employed for numerical weather prediction is given by the following example. Consider a homogeneous, incompressible, frictionless fluid moving over a rotating, gravitating plane in such a manner that the horizontal velocity does not vary with height. For quasi-static flow the equations of motion are Eqs. (1), and the equation of mass conservation is Eq.

$$\begin{aligned} \frac{\partial u}{\partial t}+u\frac{\partial u}{\partial x}+v\frac{\partial u}{\partial y}&=-g\frac{\partial h}{\partial x}+2\omega v \\ \frac{\partial v}{\partial t}+u\frac{\partial v}{\partial x}+v\frac{\partial v}{\partial y}&=-g\frac{\partial h}{\partial y}-2\omega u \end{aligned} \tag{1}$$

(2), where u and v are the velocity components in

$$\frac{\partial h}{\partial t}+u\frac{\partial h}{\partial x}+v\frac{\partial h}{\partial y}=-h\left(\frac{\partial u}{\partial x}+\frac{\partial v}{\partial y}\right) \qquad (2)$$

the directions of the horizontal rectangular coordinates x and y, t is the time, g is the acceleration of gravity, ω is the angular speed of rotation, and h is the height of the free surface of the fluid. Let the variables u, v, and h be defined at the points $x=i\Delta x$, $y=j\Delta x$ $(i=0,1,2,\ldots,I;\ j=0,1,2,\ldots,J)$ and at the times $t=k\,\Delta t$ $(k=0,1,2,\ldots,K)$, and denote quantities at these points and times by the subscripts i, j, and k. Derivatives such as $\partial u/\partial t$ and $\partial u/\partial x$ may be approximated by the central difference quotients given by Eqs. (3). In this way

$$\frac{\Delta_k u_{i,j}}{2\,\Delta t}\equiv\frac{u_{i,j,k+1}-u_{i,j,k-1}}{2\,\Delta t}$$
$$\frac{\Delta_i u_{j,k}}{2\,\Delta x}\equiv\frac{u_{i+1,j,k}-u_{i-1,j,k}}{2\,\Delta x} \qquad (3)$$

Eqs. (4), the finite-difference analogs of the continuous equations, are obtained.

$$u_{i,j,k+1}=u_{i,j,k-1}-\frac{\Delta t}{\Delta x}\left(u_{i,j,k}\,\Delta_i u_{j,k}+v_{i,j,k}\,\Delta_j u_{i,k}+g\,\Delta_i h_{j,k}\right)+4\omega v_{i,j,k}\,\Delta t$$
$$v_{i,j,k+1}=v_{i,j,k-1}-\frac{\Delta t}{\Delta x}\left(u_{i,j,k}\,\Delta_i v_{j,k}+v_{i,j,k}\,\Delta_j v_{i,k}+g\,\Delta_j h_{i,k}\right)-4\omega u_{i,j,k}\,\Delta t \qquad (4)$$
$$h_{i,j,k+1}=h_{i,j,k-1}-\frac{\Delta t}{\Delta x}\left[u_{i,j,k}\,\Delta_i h_{j,k}+v_{i,j,k}\,\Delta_j h_{i,k}+h_{i,j,k}\left(\Delta_i u_{j,k}+\Delta_j v_{i,k}\right)\right]$$

Equations (4) give u, v, and h at the time $(k+1)\,\Delta t$ in terms of u, v, and h at the times $k\,\Delta t$ and $(k-1)\,\Delta t$. It is then possible to calculate u, v, and h at any time by iterative application of the above equations.

It may be shown, however, that the solution of the finite-difference equations will not converge to the solution of the continuous equations unless the criterion $\Delta s/\Delta t > c\sqrt{2}$ is satisfied, where c is the maximum value of the speed of long gravity waves $\sqrt{gh}$. Under circumstances comparable to those in the atmosphere, Δt is found to be so small that a 24-hr prediction requires some 200 time steps and approximately 10,000,000 multiplications for an area the size of the Earth's surface. The computing time on a machine with a multiplication speed of 100 μsec, an addition speed of 10 μsec, and a memory access time of 10 μsec would be about 30 min. The magnitude of the computational task may be comprehended from the fact that the more accurate atmospheric models now envisaged will require some 100–1000 times this amount of computation.

A saving of time is accomplished by utilizing the quasi-nondivergent property of the large-scale atmospheric motions. If, in the above example, the horizontal divergence $\partial u/\partial x+\partial v/\partial y$ is set equal to zero, the motion is found to be completely described by the equation for the conservation of the vertical component of absolute vorticity, as developed in another article.

The solution of this equation may be obtained in far fewer time steps since gravity wave motions are filtered out by this constraint and the velocity c in the Courant-Friedrichs-Lewy criterion becomes merely the maximum particle velocity instead of the much greater gravity wave speed.

Cloud and precipitation prediction. If, to the standard dynamic variables u, v, w, p, and ρ, a sixth variable, the density of water vapor, is added, it becomes possible to predict clouds and precipitation as well as the air motion. When a parcel of air containing a fixed quantity of water vapor ascends, it expands adiabatically and cools until it becomes saturated. Continued ascent produces clouds and precipitation.

To incorporate these effects into a numerical prediction schema one adds Eq. (5), which governs

$$\frac{Dr}{Dt}\equiv\frac{\partial r}{\partial t}+u\frac{\partial r}{\partial x}+v\frac{\partial r}{\partial y}+w\frac{\partial r}{\partial z}=S \qquad (5)$$

the rate of change of specific humidity r. Here S represents a source or sink of moisture. Then it is necessary also to include as a heat source in the thermodynamic energy equation a term which represents the time rate of release of the latent heat of condensation of water vapor. The most successful predictions made by this method are obtained in regions of strong rising motion, whether induced by forced orographic ascent or by horizontal convergence in well-developed depressions. The physics and mechanics of the convective cloud-formation process make the prediction of convective cloud and showery precipitation more difficult.

Large-scale numerical weather prediction. In 1955 the first operational numerical weather prediction model was introduced at the National Meteorological Center (NMC). This simplified barotropic model consisted of only one layer and therefore could model only the temporal variation of the mean vertical structure of the atmosphere. By the late 1960s, the speed of computers had increased sufficiently to permit the development of multilevel (usually about 6–10) models which could resolve, at least in part, the vertical variation of the wind, temperature, and moisture. These multilevel models predicted the fundamental meteorological variables mentioned above for large scales of motion. The characteristic grid size was about 400 km on a side, and the model's domain covered most of the Northern Hemisphere. Because the boundary of the domain was located in the tropics where the horizontal gradients of the atmospheric properties were weak compared with those in middle latitudes, it was possible to treat the model variables on the boundary in rather simple ways (such as holding the variables temporally constant during the forecast).

Numerical calculation of climate. While hemispheric models were being implemented for operational weather prediction 1–3 days in advance, similar research models were being developed which covered the entire Earth. These global models (also called general circulation models, or GCMs) could, in principle, be used to simulate the long-term variation of weather, that is, the climate.

The extension of numerical predictions to long time intervals requires a more accurate knowledge than now exists of the energy transfer and turbulent dissipative processes within the atmosphere and at the air-earth boundary, as well as greatly augmented computing-machine speeds and capacities. However, predictions of mean conditions over large areas may well become possible before such developments have taken place, for it is now possible to incorporate into the prediction equations estimates of the energy sources and sinks—estimates which may be inaccurate in detail but correct in the mean. In fact, several mathematical experiments which have involved such simplified energy sources have yielded predictions of mean circulations that strongly resemble those of the atmosphere.

The above-mentioned experiments lead to a hope that it will be possible to explain the principal features of the Earth's climate, that is, the average state of the weather, well before it becomes possible to predict the daily fluctuations of weather for extended periods. Should these hopes be realized it would then become possible to undertake a rational analysis of paleoclimatic variation and changes induced by artificial means. If the existing climate could be understood from a knowledge of the existing energy sources, atmospheric constituents, and Earth surface characteristics, it might also be possible to predict the effects on the climate of natural or artificial modifications in one or more of these elements.

Specialized prediction models. Although the coarse grids in the hemispheric and global models are necessary for economical reasons, they are sources of two major types of forecast error. First, the truncation errors introduced when the continuous differential equations are replaced with finite difference approximations cause erroneous behavior of the scales of motion that are resolved by the models. Second, the neglect of scales of motion too small to be resolved by the mesh (for example, thunderstorms) may cause errors in the larger scales of motion. In an effort to simultaneously reduce both of these errors, models with considerably finer meshes have been tested. However, the price of reducing the mesh has been the necessity of covering smaller domains in order to keep the total computational effort within current computer capability. Thus the limited-area fine-mesh model (LFM) run at NMC has a mesh length of approximately 120 km on a side, but covers a region only slightly larger than North America. Because the side boundaries of this model lie in meteorologically active regions, the variables on the boundaries must be updated during the forecast. A typical procedure is to interpolate these required future values on the boundary from a coarse-mesh model which is run first. Although simple in concept, there are mathematical problems associated with this method, including overspecification of some variables on the fine mesh. Nevertheless, limited-area models have made significant improvements in the accuracy of short-range numerical forecasts over the United States.

Even the small mesh sizes of the LFM are far too coarse to resolve the detailed structure of many important atmospheric phenomena, including hurricanes, thunderstorms, sea- and land-breeze circulations, mountain waves, and a variety of air-pollution phenomena. Considerable effort has gone into developing specialized research models with appropriate mesh sizes to study these and other small-scale systems. Thus, fully three-dimensional hurricane models with mesh sizes of 20 km simulate many of the features of real hurricanes. On even smaller scales, models with horizontal resolutions of several kilometers reproduce many of the observed features in the life cycle of thunderstorms and squall lines. It would be entirely misleading, however, to imply that models of these phenomena differ from the large-scale models only in their resolution. In fact, physical processes which are negligible on large scales become important for some of the phenomena on smaller scales. For example, the drag of precipitation on the surrounding air is important in simulating thunderstorms, but not for modeling large scales of motion. Thus the details of precipitation processes, condensation, evaporation, freezing, and melting are incorporated into sophisticated cloud models.

In another class of special models, chemical reactions between trace gases are considered. For example, in models of urban photochemical smog, predictive equations for the concentration of oxides of nitrogen, oxygen, ozone, and reactive hydrocarbons are written. These equations contain transport and diffusion effects by the wind as well as reactions with solar radiation and other gases. Such air-chemistry models become far more complex than atmospheric models as the number of constituent gases and permitted reactions increases.

[RICHARD A. ANTHES]

CENTERS AND OFFICES OF FORECASTING

Weather forecasts for all parts of the United States are prepared by the National Weather Service (NWS), a part of the National Oceanic and Atmospheric Administration (NOAA). Various phases of the forecast work are performed at five working levels: the National Meteorological Center (NMC); National Severe Storms Forecast Center (NSSFC); National Hurricane Center (NHC); River Forecast Centers (RFCs); and Weather Service Forecast Offices (WSFOs).

National Meteorological Center. The NMC, located in Camp Springs, MD, collects weather observations, prepares meteorological analyses, and provides forecast guidance for use by NWS field offices. The aerial coverage of NMC products includes the entire globe, with most products covering the Northern Hemisphere and the tropical regions of the Southern Hemisphere. The World Meteorological Organization (WMO) has designated the NMC as the analysis and forecast arm of the Washington World Meteorological Center, which requires global responsibilities as part of the international efforts and cooperation known as the World Weather Watch. To carry out these responsibilities, all of the NMC divisions are extending coverage of their products to the entire globe.

Divisions. The NMC has four divisions as follows:

The Forecast Division applies a combination of numerical and manual techniques to produce analyses and prognoses up to 240 hr into the future, emphasizing the advance time period of 2 to 72 hr. The meteorologists manually adjust computer output to reduce errors, and interpret areas of weather and cloudiness for the guidance of NWS field offices (Fig. 1).

The Automation Division operates the NMC's computers and their interface to various NWS (Fig. 2) and Federal Aviation Administration (FAA) communication links.

The Development Division adapts research and

Fig. 1. Meteorologists in NMC Forecast Division manually adjust computer surface prognostic charts to reduce errors, and interpret areas of weather and cloudiness for guidance of NWS field offices. (*NOAA*)

development results in numerical weather prediction to the NMC output and conducts research into improved numerical weather prediction, stratospheric research, and the use of satellite-derived data.

The Climate Analysis Center analyzes current and short-range climate fluctuations and prepares short-term forecasts and long-term outlooks. The forecasts and projections are used by a wide variety of government and private users.

Operations. The centralized preparation of data, analyses, and forecasts is designed to eliminate most requirements for hand-charting and independent meteorological analysis in the NWS offices as well as user groups such as airline and private meteorologists. The NMC, through the use of a large computer facility and together with numerical forecast methods, provides the NWS and other government agencies with daily forecasts and outlooks out to 10 days in advance. In the course of a day, the NMC receives 25,000 hourly aviation reports, 14,000 synoptic land station reports, 2500 ship reports, 3500 aircraft reports, and 2500 atmospheric soundings. All available cloud and temperature data derived from weather satellites are integrated into the analyses. The NMC makes 785 facsimile and 819 teletypewriter transmissions daily to the field offices.

National Severe Storms Forecast Center. The NSSFC, located in Kansas City, MO, is responsi-

Fig. 2. Cathode-ray tube display system used throughout NWS for forecast preparation and weather watch. (*NOAA*)

Fig. 3. Master "storm room" at NHC. (*NOAA*)

ble for preparing and releasing forecasts of areas of expected severe local storms, including tornadoes. The long-term outlook, prepared shortly after midnight, gives the geographic areas most likely to have severe thunderstorms or tornadoes during the next 24 hr. The convective forecast is used for planning purposes by preparedness groups and as input to the local forecast program by NWS offices throughout the contiguous United States. As the possibility of severe thunderstorms increases, the NSSFC severe storms meteorologist prepares tornado or severe thunderstorm watches, typically for an area of 25,000 mi^2 (65,000 km^2) and for time periods 1–7 hr in advance. A second group of NSSFC meteorologists and technicians monitor weather surveillance radar reports and satellite photos to locate areas of potential hazard for aircraft inflight. The resulting warnings, called convective SIGMETs, are relayed immediately to various FAA facilities and to the airline operations offices. In a typical year, the NSSFC logs 700–900 tornadoes and 1500–2000 severe thunderstorms. Four hundred tornado or severe thunderstorm watches are issued along with 10,000–15,000 convective SIGMETs. *See* STORM DETECTION.

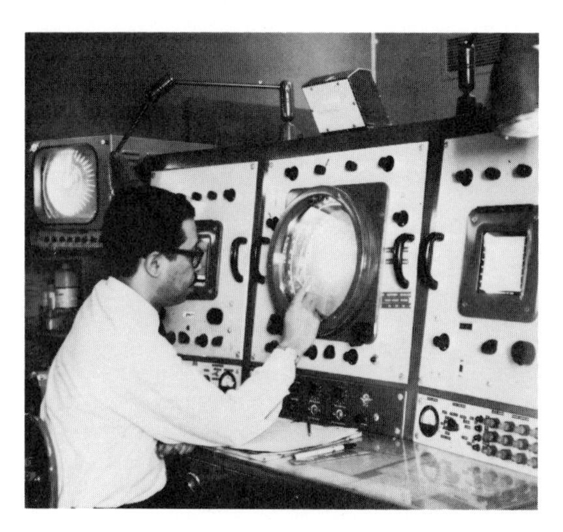

Fig. 4. Meteorological technician monitoring weather on NWS long-range radar. (*NOAA*)

National Hurricane Center. The NHC, located in Miami, FL, prepares hurricane and tropical storm watches and advisories for most of the tropical Atlantic Ocean, the Caribbean, and the Gulf of Mexico. The NHC delegates parts of its public warning responsibilities to three Hurricane Watch Offices (HWO) when the United States coastal areas are threatened (Fig. 3). The HWOs are WSFOs at Boston, Washington, and San Juan. WSFOs at San Francisco and Honolulu prepare hurricane and tropical storm advisories for the eastern Pacific and central Pacific, respectively.

River Forecast Centers. There are 13 RFCs, which prepare river and flood forecasts and warnings for approximately 2500 communities. This includes forecasts for height of the flood crest as well as times when the river is expected to overflow its banks and when it will recede into its banks. At many points along larger streams or rivers, daily forecasts of river stage or discharge are routinely prepared for activities such as navigation and water management. Forecasts of seasonal snowmelt or water-year runoff are prepared for major rivers such as the Columbia, Missouri, and Mississippi and their tributaries.

Weather Service Offices. There are an additional 238 WSOs, which represent the third echelon of the system. They issue local forecasts which are adaptations of the zone forecasts. They are important in the warning and observation program and are generally located in smaller cities.

Weather Service Forecast Offices. There are 52 WSFOs within the NWS. There are no WSFOs in Rhode Island, Connecticut, Vermont, and New

Hampshire, while California and Pennsylvania have two and Alaska, Texas, and New York three. Each WSFO provides whatever weather service is required for the geographical areas it serves, utilizing output from the NMC, NSSFC, NHC, and a RFC where appropriate. Each WSFO provides routine issuances of state and zone forecasts for the public plus terminal and route forecasts for aviation. Other specialized services are handled where there is a need and where resources and workload permit. These include a Disaster Preparedness Program designed to save lives and mitigate the social and economic impacts of natural disasters; Severe Local Storms Warning Program, where the weather radar is the principal method for monitoring the storms (Fig. 4); Winter Weather Warning Service; Coastal Flood Warning Program; Agricultural Weather Service; warnings of low temperatures for winter and spring crops; forecasts for flights from the United States and its possessions to other countries; meteorological support to control and combat air pollution; specialized forecasts and warnings to fire control agencies; flash flood watches and warnings; tsunami watches and warnings; warnings for high seas, coastal waters, and inland waterways and offshore marine forecasts for recreational boating and fishing. Forecasts and warnings are transmitted via the NOAA Weather Wire, a dedicated teletype circuit, and over the NOAA Weather Radio, a continuous weather and river information broadcast on one of three high-band frequencies—162.40, 162.475, or 162.55 MHz.

[ALLEN PEARSON]

Bibliography: J. Bjerknes and H. Solberg, Life cycle of cyclones and polar front theory of atmospheric circulation, *Geofys. Publ.*, no. 3, 1922; G. I. Marchuk, *Numerical Methods in Weather Prediction*, 1973; A. S. Monin, *Weather Forecasting as a Problem in Physics*, 1972; J. Namias, Long range weather forecasting: History, current status, and outlook, *Bull. Amer. Meteorol. Soc.*, 49(5), 1968; S. Petterssen, *Weather Analysis and Forecasting*, 2 vols., 2d ed., 1956.

Weather modification

The changing of natural weather phenomena by humans. Weather is the product of the interaction of atmospheric processes on many scales, reaching from the planetary circulation to the microphysical processes in the evolution of cloud droplets and ice crystals. So far, only on the microscale of condensation and freezing nuclei have humans begun to exert modifying influences on weather. These influences may expand to a scale of several hundreds or thousands of square kilometers, that is, to what is called in meteorology the meso scale of meteorological phenomena.

There are actually four techniques by which humans may affect the natural course of weather: (1) by injection of large amounts of heat into the atmosphere in order to burn off warm fog; (2) by utilizing metastable states in the atmosphere as they happen—for instance, if clouds are being cooled below the freezing point, seeding can be effective; (3) by altering the surface condition, as by deforestation or urbanization; (4) by influencing the weather inadvertently by pollution—for example, the exhaust gases from metal smelters contain many freezing nuclei, or smog affects the atmospheric radiation balance and, consequently, also cloud formation, which is also affected by the "heat island" over cities. Most frequently used nowadays is the second technique, the seeding of clouds with certain types of nuclei. Figure 1 shows the locations of research and operational weather modification projects in the United States during 1973–77. In 1977 about 7% of the area of the United States was covered by cloud seeding. There are 131 operational projects and 11 research projects.

Experimental principles. The numerous physical and meteorological processes which are involved in the experimental approach to weather modification make it complex and difficult. In many cases it is virtually impossible to design a classical physical experiment for determining a cause-effect relationship, and it is necessary to adopt a statistically designed experiment. Here one need not know all physical processes, feedback mechanisms, and interactions in order to derive the influence of the one artificially modified parameter, but one is required to conduct a great number of identical experiments. This calls for experimental periods which have to be counted in decades. In the meantime the environmental conditions of the experiment may change, so the applicability of the statistical approach is limited. In view of these difficulties, Fig. 2 reflects the approach which should be taken during research projects by combining the classical and the statistical experimental principles.

This approach can be further sharpened by predicting the experimental result by numerical simulation of the experiment. Strangely enough, this is not the approach science has taken: it has relied more and more on statistics and given less and less weight to the physics involved in the experiment. Design, execution, and analysis were determined by statistical principles with disregard of the grave uncertainties to which the statistical approach is liable. The statistical approach postulates that one has a population of identical experimental situations which one can subdivide into control events and target events. The target events are treated with the real seeding agent, the control events with a placebo. The scientist in charge has to make the decision whether a certain cloud situation meets the experimental criteria for seeding, but in order to avoid bias in the scientist's observations the seeding decision is made from a set of random numbers and the decision is not communicated to the scientist. In rainmaking experiments, one finds the ratio of target rainfall to control rainfall and determines the significance of the results. Usually, a 5% significance is accepted as proof that the effect is not caused accidentally.

Developments in experimental meteorology have caused an air of optimism in the approach to weather modification. In addition, radar, pulse Doppler radar, aircraft with sophisticated instrumentation serving as observing platforms, and satellites are providing unprecedented observation and analysis facilities.

Cloud seeding. Clouds being cooled below the freezing point remain liquid down to very low temperatures and only those droplets will freeze which contain a so-called freezing nucleus. It was the

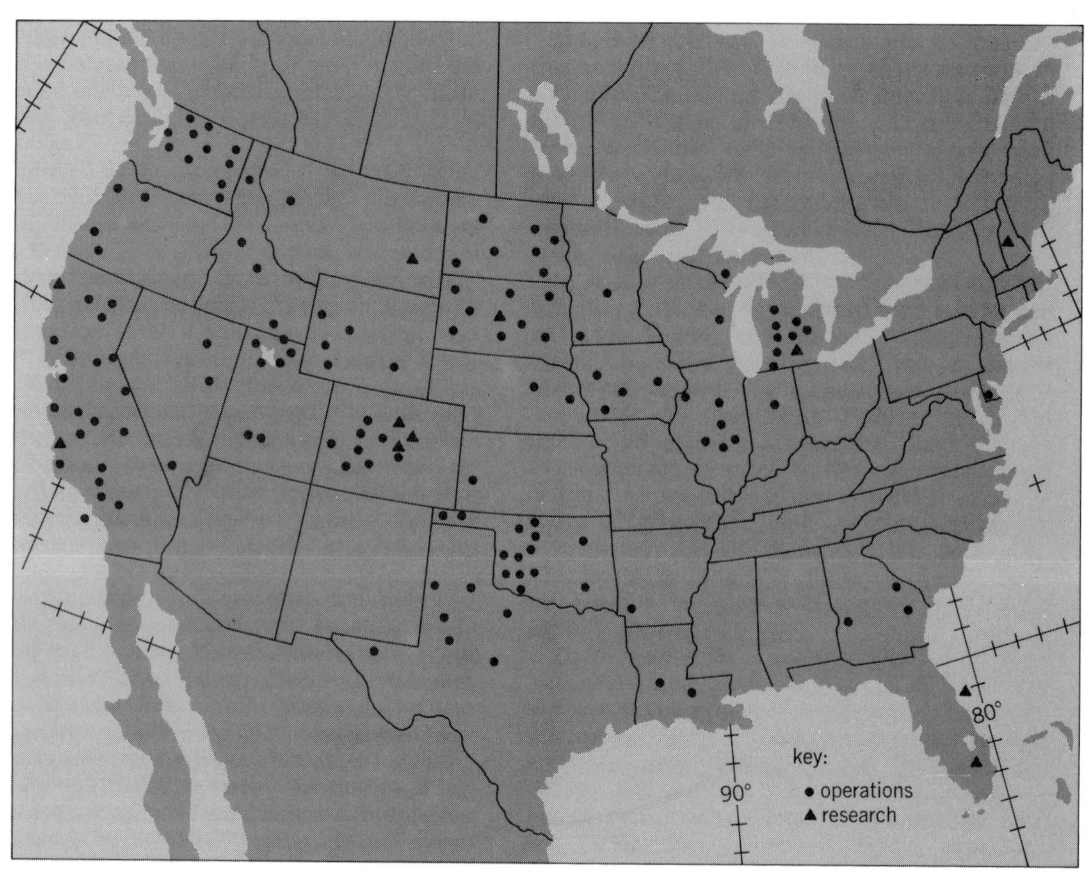

Fig. 1. Locations of research and operational weather projects in the United States, 1973–77. Two operational projects were in south-central Alaska, and no projects in Hawaii. (*NOAA Weather Modification Reporting Program, Rockville, MD*)

great discovery in the 1940s of Nobel prize winner I. Langmuir and his two assistants, V. Schaefer and B. Vonnegut, that clouds could be dissipated by seeding either with dry ice or with certain chemicals (such as silver iodide) and that, artificially, the cloud could be changed from a water cloud into an ice cloud. Judicious seeding would permit generation of only a certain number of artificial freezing nuclei and thus would influence only certain cloud conditions, such as the precipitation process or cloud dynamics.

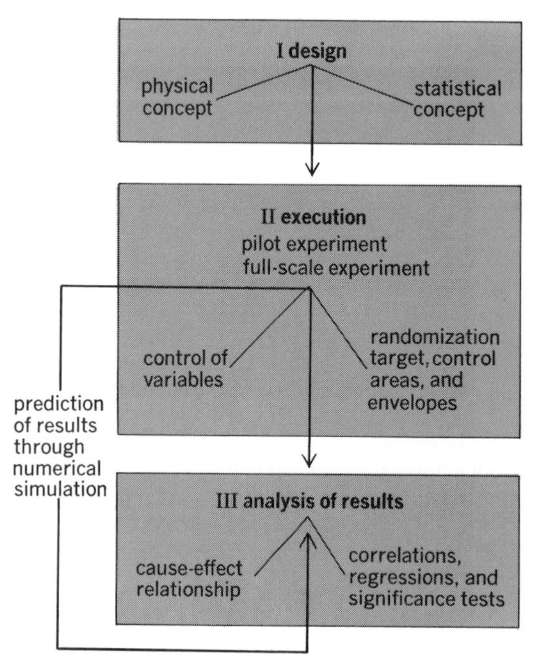

Fig. 2. Scheme for approach to weather modification experiments. Experiments must be in agreement with physical and statistical principles of experimentation.

Fog dispersal. The method of seeding is now applied operationally for the dispersal of supercooled fogs from airports all over the world. Systems have been developed for seeding with liquid propane in France (Orly Airport) and for seeding with dry ice, silver iodide, and lead iodide in the United States and the Soviet Union.

Dispersal of warm fog (temperature warmer than frost point) is accomplished by injection of heat into the atmosphere, as it is done alongside the runways of Orly Airport.

Cloud modification. Some striking results of artificial weather modification have been obtained in cloud modification. Figure 3 shows an area of about 100 km² in which supercooled stratocumulus clouds were dissipated by seeding each of three parallel 16-km-long lines 5 km apart with 100 lb (45 kg) of dry ice. In experiments over populated regions the weather has been modified conspicuously on the mesoscale of meteorological events. Sunshine has been made for several thousand people on a dull overcast winter day.

A contrasting case is illustrated in Fig. 4. Here, after seeding for dissipation, a miniature squall line developed over Lake Michigan consisting of heavy snow showers which lasted for about 1 hr. Similar results have been repeatedly obtained during seeding experiments over Lake Erie.

Figure 5 shows another spectacular result of cloud seeding. A cumulus cloud reaching to the −8°C level was overseeded with silver iodide. Overseeding caused glaciation and hence heat of fusion was released by the glaciating cloud water. This heat increased the buoyancy of the cloud, and the cloud "exploded," growing in height and also in width. This experiment has a clear physical concept which tested a cause-effect relationship; however, randomization with seeded and control clouds contributed materially to sharpening the experimental result. Of 22 test clouds 14 were seeded and 8 were not; only 1 of the clouds not seeded grew comparatively. Of the 14 seeded clouds, 4 exploded and 10 increased in height with no increase in width.

The great complexity of cloud response to feedback mechanisms between microphysics and cloud dynamics was observed when an intensely growing cumulus was seeded in the updraft with 1500 g of AgI. Fifteen pyrotechnic devices were released at −20°C into the updraft at about 6000 m (mean sea level), and seeding occurred over a depth of 3000 m. At this temperature, snow crystals formed which very efficiently converted the cloud water to precipitation, so that the cloud snowed out and did not grow.

These experiments indicate that it is possible to affect materially the life history of clouds through seeding. *See* CLOUD PHYSICS.

Rainmaking. Water is one of the most abundant but also one of the most wanted substances on Earth. The interest in artificial rainmaking is

Fig. 3. Cloud dissipation. *(a)* Three lines in stratocumulus cloud layer 15 min after seeding. *(b)* Opening in stratocumulus layer 70 min after seeding. *(U.S. Army ECOM, Fort Monmouth, NJ)*

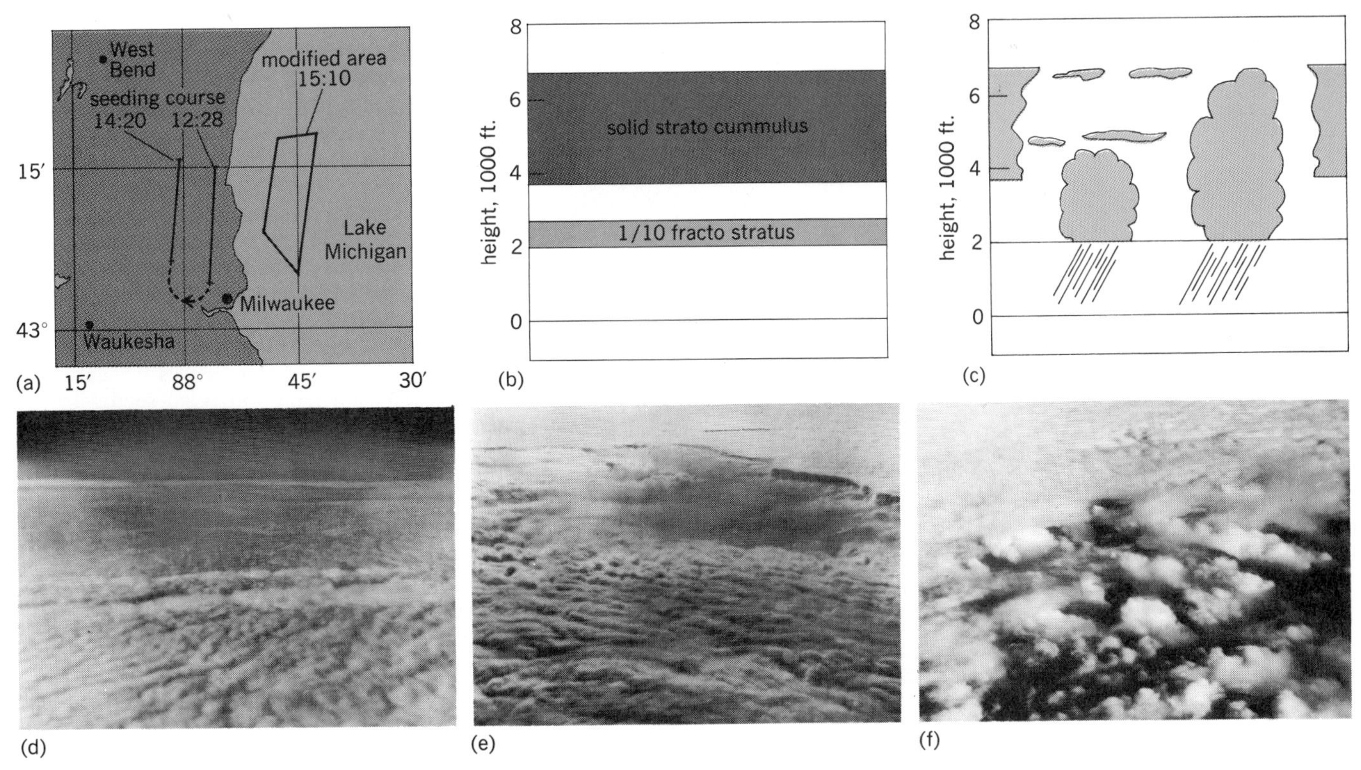

Fig. 4. Development of heavy snow showers after a seeding experiment for cloud dissipation over Milwaukee, WI, on Nov. 24, 1953. *(a)* Location of experiment. *(b)* Vertical section before seeding and *(c)* 63 min after seeding. *(d)* Cloud appearance 15 min after seeding, *(e)* 31 min after seeding, and *(f)* 63 min after seeding. 1 ft = 0.3 m. *(U.S. Army ECOM, Fort Monmouth, NJ)*

Fig. 5. "Explosion" of cumulus cloud following release of heat of fusion caused by seeding with silver iodide. *(a)* Time of seeding. *(b)* Views at 9 min, *(c)* at 19 min, and *(d)* at 38 min after the seeding. *(Courtesy of J. Simpson, ESSA)*

therefore understandable. A panel of the U.S. National Academy of Sciences expressed restrained optimism by stating: "There is increasing but still somewhat ambiguous statistical evidence that precipitation from some types of cloud and storm systems can be moderately increased or redistributed by seeding techniques." However, the panel also recommended the early establishment of several carefully designed, randomized, seeding experiments, planned in such a way as to permit assessment of the seedability of a variety of storm types. These recommendations reflect that much corroborative evidence is still missing.

Indeed, the evidence is often controversial, particularly where the experiments can only be based on statistical design. In all such experiments it is tacitly assumed that the seeding agent gets to the right location in the cloud system and acts as desired. Seeding with ground-based generators has uncertainties because of the unknown diffusion properties of the seeding agent in the atmospheric boundary layer. Seeding from aircraft has yielded positive results in Israel, inconclusive results in Missouri, and negative results in Arizona.

The analysis of seeding data often presents great difficulties because it must be based on point measurements. Take, for instance, rainfall data. Rainfall can be measured only on discrete points whose number is limited because a balance has to be struck between the desire to have a network as dense as possible and to have one that can economically be analyzed and maintained. It has been attempted to measure rainfall by means of radar, but the relationship between the radar return and the rate of rainfall is very unreliable and changes from cloud to cloud. The variabilities introduced by these analysis problems may be greater than the seeding effect which should be measured.

The most noteworthy progress in this area comes from increased theoretical understanding of the precipitation process, particularly for convective clouds. Well-designed seeding experiments on 33 pairs of cumulus clouds in Australia indicated in the statistical analysis that seeding at cloud base with 20 g of silver iodide produced significant rain increases, while seeding with 0.2 g was ineffective. A seedability index can be derived from these data: lifetime of cloud in hours times depth of cloud in kilometers. Figure 6 illustrates that, for an index below 3, clouds do not rain naturally or after seeding. However, the cases which statistically determined the success through seeding were all connected with high seedability indexes.

Downwind effects. Sporadic statistical analyses indicate that downwind effects occur as far away as 300 mi (500 km). Observations have been reported from the United States and Australia, with precipitation increases in most cases. In view of the rate of decay of silver iodide in daylight, the mechanism of such effects is dubious, and a strong case can be made that, downwind from any location, positive anomalies can be found simply because of the natural variability of the rate of rainfall.

Hail suppression. Scientists of the Soviet Union reported the development of an operational hail suppression project which reduces hail damage by

80–90%. It involved the detection with radar of the hail-spawning cloud regions and the delivery into them of seeding material by means of grenades. However, Soviet scientists have admitted to the difficulty of suppressing hail in very severe storms. In the United States a physical concept was developed which differed somewhat from the Soviet concept, and delivery of the seeding agent from aircraft was used rather than from the ground. The aircraft traversed the storm at the base level and discharged the seeding material, by means of rockets, vertically into the updraft. The project had to be terminated prematurely. The analysis of 3 years of field experiments gave the result that an average of 60% more hail mass was measured on days when seeding was carried out (16 seeded, 16 unseeded days) than on days when seeding was not attempted. However, the 90% confidence limits for the ratio of hail mass on seeded to unseeded days range from −48 to 531% for a log normal fit of the data distribution. These results, then, permit exclusion of a suppression effect in excess of 50% at the 5% confidence limit.

There are mainly two physical properties of hailstorms which make the failure to suppress hail fall (or even to increase it) plausible. First, hail does not fall in uninterrupted long swathes, but in long swathes which consist of individual hail streaks. These streaks are roughly between 10 and 20 km long and a few kilometers wide. The fine structure of a hail swath, which may be 50 km wide, is made up of numerous hail streaks whereby each streak may originate in a separate storm cell. Unless all potential hail cells are seeded at the correct time of hail formation, seeding may be ineffective. Second, as discovered by French scientists, a proportionality exists between the number of hailstones falling per square meter of surface area and their size: the more hailstones fall, the bigger they are. The reason for this relationship is unknown, but one will not go much wrong with the assumption that it is due to a feedback mechanism between the cloud dynamics and the cloud microphysics. Since the tolerance limits of the cloud are not known, it is quite possible that the artificial generation of hail embryos will allow them to grow to hailstones as large as the natural ones, and, consequently, increase the hail mass.

A project in France conducted surface seeding on a large scale, discharging tons of silver iodide into the air during the hail season. The analysis of the nonrandomized experiments indicated success. Other projects have been conducted in Germany (Bavaria), Italy, and Argentina. While the theory of hail formation suggests that hail may be suppressed by a comparatively small seeding effort, there is great difficulty in designing a field experiment which is in agreement with the scheme presented in Fig. 2. The sporadic nature of hailstorms and their large natural variability practically exclude effective randomization of the experiments. For the analysis phase it is therefore necessary to have a store of excellent historical data on the experimental area, such as in the Canadian Hail Research Project in Alberta.

Lightning suppression. Two concepts have been tested in the United States. The first, developed by the Department of Agriculture, makes use of overseeding of thunderstorms with silver iodide. While physical relationships in the suppression mechanism are not fully developed, it appears that the method decreases ground strokes and increases intracloud strokes. Another approach was developed by the U.S. Army jointly with the Environmental Science Services Administration (ESSA). Discharge of the charge centers of a storm by corona discharges initiated through the introduction of metallic needles, so-called chaff. It can be shown that for thunderstorm fields, 10^6-10^7 chaff particles (weighing 5 to 50 lb or 2.3 −23 kg) can discharge several amperes, a result in agreement with the magnitude of the thunderstorm-charging current.

The result of field experiments is shown in Fig. 7. The figure is based on aircraft measurements of the number of lightning discharges in 5-min intervals of seeded and nonseeded storms. The chaff fibers were made of aluminized nylon 10 cm long and 25 μm in diameter. They initiated corona in an electric field of about 35,000 V/m (lightning requires about 300,000 V/m), thus draining away the charge that had been built up by the storm. The figure shows that the number of lightning discharges in seeded storms goes quickly to zero, while in unseeded storms lightning proceeds uninterrupted.

Modification of severe storms. Modification of severe storms such as tornadoes or hurricanes is in its infancy, essentially because of incomplete understanding of the dynamic structure of these storms. ESSA and the U.S. Navy developed under Project Stormfury a concept of hurricane modifi-

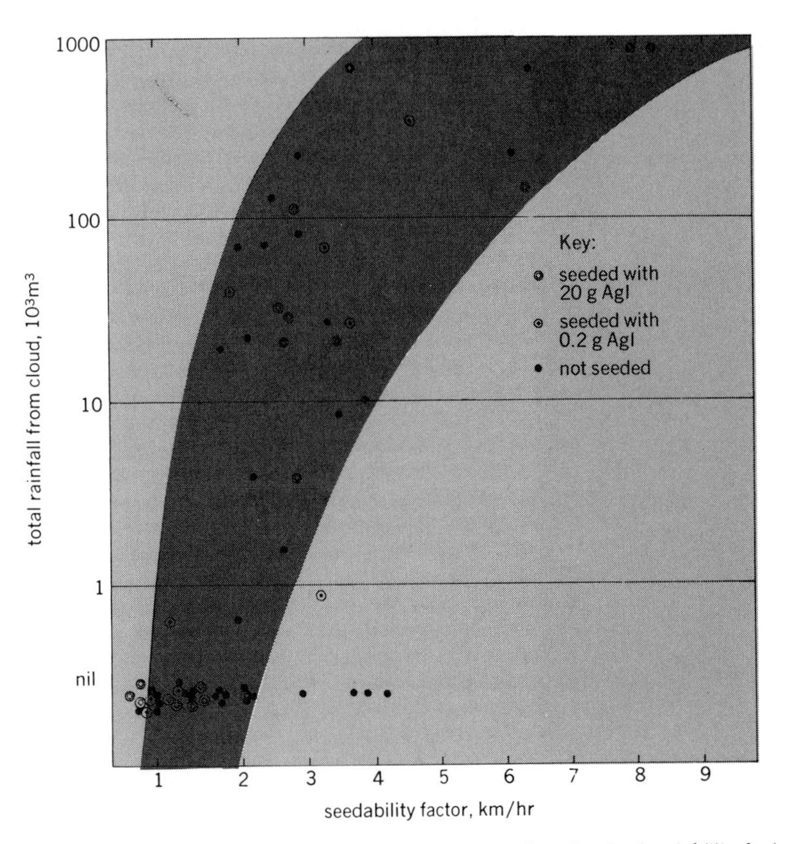

Fig. 6. Seedability factor for rainmaking from convective clouds. Seedability factor is product of cloud depth in kilometers and cloud lifetime in hours.

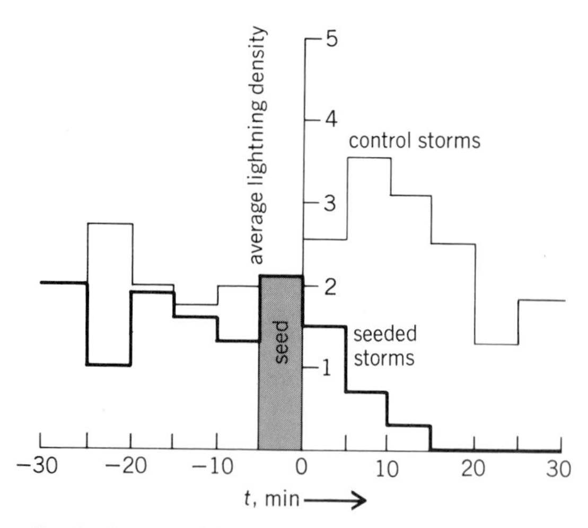

Fig. 7. Average lightning density for each time interval is for 10 seeded and 18 control storms during 1972 and 1973. *(From H. W. Kasemir et al., Lightning suppression by chaff seeding at base of thunderstorms, J. Geophys. Res., 81:1965–1970, 1976)*

cation based upon overseeding of the wall clouds. Release of the heat of fusion is believed to alter the pressure gradient in such a way as to diminish the destructive winds near the storm center. As wind damage is proportional to the square of the wind velocity, a relatively small reduction of the velocity may mean a large reduction of damage.

Seeding experiments were conducted on hurricanes Esther in 1961 and Beulah in 1963. A 10% reduction of wind speed was observed; however, the reduction is well within the natural wind variability of the storm.

Inadvertent weather modification. Modern civilization affects weather in many ways. The great industrial centers of the world are plagued with haze and smog due to effluents from factories and combustion of hydrocarbons. Haze and smog affect the heat radiation budget of the atmosphere as well as precipitation processes. Aerosols discharged by metal foundries act as freezing nuclei, and strong evidence now appears for such inadvertent weather modification. H. K. Weickmann has repeatedly observed in the Great Lakes region the formation of snow showers downwind of steelmills after measuring large numbers of freezing nuclei in the exhaust plume. It is likely that this effect is more widespread than is presently known, but systematic investigations are lacking.

The most famous case, the La Porte anomaly, is somewhat controversial. La Porte, IN, is located near the southeastern shore of Lake Michigan, 30 mi (48 km) downwind of the Gary industrial complex. The rainfall measured at this site and its environment indicates that the rainfall at this station has increased with the increasing industrialization (smoke-haze days in Chicago), whereas neighboring stations remained unaffected. The difference developed particularly after about 1930, or after the last observer change for this station had taken place. The observers voluntarily measure the rate of rainfall. It appears that a systematic error in these observations caused the apparent singularity of rainfall at this location, since after this observer left in 1963, the La Porte record agrees again with the neighboring stations. For this reason, the National Oceanic and Atmospheric Administration claims observer error for the anomaly.

While the combustion of hydrocarbons causes formation of smog in large urban areas, a more subtle climate modification takes place on a global scale. The combustion process not only liberates aerosols but also carbon dioxide gas, an important constituent of the air that contributes to the greenhouse effect of the atmosphere. The gas is transparent for the Sun's visible radiation but traps the Earth's thermal radiation; thus the more carbon dioxide released, the warmer the climate becomes. Computer calculations project an average temperature increase of 1–2°C within two to three generations from this effect. Though this change may appear small, it would have profound influences on weather and climate. Glaciers and ice caps would melt, and coastal cities might become inundated. It is possible that other changes in the atmosphere will counteract the effect of carbon dioxide. *See* AIR POLLUTION; CLIMATE MODIFICATION.

Effect on ecology and society. While influences on ecological systems appear to be often exaggerated in view of the great natural variability of weather, socioeconomic considerations are important. Fog dispersal methods at airports enable many passengers to travel, effective hail suppression would save millions of dollars, and rainmaking would be the least expensive answer to the water dilemma of modern civilization. But as interests of individuals differ, rainmaking may be a blessing to one and blight for another; socioeconomic considerations are therefore important at a time when weather modification is still in its infancy.

[HELMUT K. WEICKMANN]

Wildlife conservation

The science and art of making decisions and taking actions to manipulate the structure, dynamics, and relations of wild animal populations, habitats, and people to achieve specific human benefits from the wildlife resource. Although still called wildlife conservation by some people, wildlife management, now the preferred term, has more than its earlier connotations of preservation. Wildlife management is concerned not only with preserving and increasing populations of threatened or endangered species, such as the whooping crane (Fig. 1), but also with stabilizing certain populations, such as the American bison, increasing or stabilizing desirable species taken for sport recreation (game), and decreasing some populations of birds and harmful mammals.

The problems of management are complex and involve conflicts between groups with opposing interests. For example, orchard owners want fewer bark-eating rabbits, while hunters want more rabbits of all kinds; fox hunters want increased populations, while other people wish to reduce foxes (Fig. 2) as vectors of rabies; hunters who want only native game animals differ with hunters who would stock or import foreign species into any area; hunters and farmers are at odds on the proper number of raccoons to have in forests near cornfields; foresters managing their forests to supply deer to hunters do not want deer in their nurseries; city dwellers want birds in parks but not on buildings.

Efforts to resolve or to balance these oppos-

Fig. 1. The whooping crane, a rare North American migratory species, is in danger of extinction. There were only 55 individuals in existence in 1969. (*Bureau of Sport Fisheries and Wildlife*)

ing interests require the best possible individual and group decisions. These decisions, to be satisfying for many people, require a great deal of information, including information about wildlife values. *See* CONSERVATION OF RESOURCES.

WILDLIFE VALUES

Wildlife has many values. It contributes to the natural "character" of land; that is, land of whatever type does not seem really wild if one knows animals are absent. Animals contribute to a sense of completeness or wholeness of the landscape. Wildlife has great esthetic values—geese flying in spring, bighorn sheep silhouetted on the horizon, even the chipmunk on a mossy log—all produce real, though hard-to-measure, values. Wild creatures have a great capacity for inspiring creative reactions to their beauty as evidenced in works of art, music, and poetry. Their past values in motivating explorers and pioneers to open the way westward and hasten the development of America are well known.

A measurable value of wildlife is its contribution to personal, community, and national economy. Over 95,800,000 Americans take part in wildlife-related activities. Furs are an obvious monetary resource, but the market fluctuates with women's and men's clothing styles. In 1975, some 20,600,000 hunters spent $5,800,000,000 on hunting activities, and 53,900,000 fishers spent 15,200,000,000. The money is a rough measure of the recreational values attached to the wildlife resource, since such values go beyond the actual taking of game animals, birds, and fish, but also include the time and thought spent anticipating and reliving the outdoor experience. Other ways of estimating wildlife values are (1) willingness of people to travel to participate in wildlife-related recreation; (2) money spent in birdseed, books, and binoculars; (3) money spent on active wildlife management works, including refuges and producing food;

Fig. 2. The fox is a valuable fur animal. It eats animals that damage crops and fruit trees, but it also eats rabbits and is a vector of rabies in some areas.

and (4) money foregone from forests or cropland when a harvest is left specifically for benefitting wildlife. Wildlife also has the potential of providing insights for investigators into behavior and other scientific phenomena not similarly available from studying other species. Finally, there is the contribution of animals to the proper functioning of natural systems. Their function has been considered a major part of the balance of nature, an idea that suggests the interdependence of plants on animals, animals on plants, and both to each other and their environment. *See* ECOSYSTEM.

Management task. To achieve a desired abundance of a particular species of wildlife, modern wildlife managers work intensively with the environment in which the animals live. Wildlife populations naturally respond very sensitively to their habitats. When water levels are down, ducks produce fewer young; when dams of beaver (Fig. 3) are torn out and the rich land within beaver ponds is planted to crops, beaver disappear; where the woody food of deer and elk is not abundant or after the supply of forage has been eaten, populations of these animals are reduced or they have fewer young.

Land has a carrying capacity. Fish ponds can support only a given number of fish. A particular pond may carry 500 fish, each weighing 1/2 lb (0.23 kg), or 250 fish weighing 1 lb (0.46 kg), based on the size of the pond, food in the water, temperature, and other factors. Only by fertilization, planting, or other cultural practices can carrying capacity be increased. Mammal and bird populations are similar; each area has a limited ability to support animals.

Most wildlife managers now believe that energy is the primary factor limiting that ability. Animals require large amounts of energy to survive. It is provided to them by food, sunlight, and reradiation of sun energy from plants and the ground. It is being removed from them at a rapid rate by natural forces, primarily wind, contact with the cold ground, and evaporation of moisture; of course, movement and producing young require large expenditure of energy. Animals have developed means for conserving this energy so that outgo does not exceed income. Wildlife managers may

Fig. 3. Beaver fur was the "gold" that once lured trappers to first explore America. The beaver is still a valuable animal, not only for fur but for its beneficial influence on water supplies. However, it is a forest and agricultural pest in some areas.

attempt to supply food energy or reduce energy losses. By providing protection from the wind, preserving areas where animals may sun, protecting them from harassment by dogs, predators, or people, and providing high-energy foods, the modern wildlife manager attempts to improve the energy balance and budgets of wild animal populations.

Where soils are poor, wildlife will be sparse. Richer soils produce more vigorous plants that store sun energy better and make it available to wildlife. The types of plants and animals must be balanced by the wildlife manager, for on very rich soil, plants can grow to heights out of reach of animals. Rich soils can encourage large populations with high food consumption that can temporarily destroy the vegetation on areas and even result in soil erosion. These changes can result in population fluctuations. *See* POPULATION DYNAMICS.

Wildlife managers are concerned with the natural ability of populations to reproduce. The reproduction potential is set by genetics but strongly influenced by the quality of food. When food supplies are low or the quality of food is reduced (as a result of soil erosion or leaching of nutrients from soil), reproduction is reduced. When food quantity or quality is low, animals have fewer young per year. Another factor called stress influences populations in several complex ways. Research on stress is now in progress, but the evidence is that, as populations become more crowded, reproduction is reduced, more food is required, and diseases have greater effects. Thus there are several interacting natural limits on wildlife population growth imposed both by the environment and by the population itself. Overcoming these limits is the job of the wildlife manager interested in increasing populations. Working with these factors, learning more about them, and trying to influence the role of each factor in the context of other values are the job of the wildlife manager responsible for decreasing damage to wildlife.

History. Past efforts at species preservation were too few and too late to save animals such as the passenger pigeon, heath hen, Labrador duck, Carolina parakeet, several species of grizzly bears, the eastern bison, great sea mink, plains wolf, Badlands bighorn sheep, and several species of fish. The problem of preventing extinction is quite real today, and threatened species include the whooping crane, black-footed ferret, Florida Everglade kite, California condor, Atwater's prairie chicken, and Hawaiian goose.

Threats to populations of wildlife usually result from a very complex set of factors. The factor usually at the top of the list is destruction of wildlife homes and habitat. Interest in habitat has appeared late in the history of efforts to stabilize or increase wildlife or game populations. As species decreased, protective laws were passed first, then stocking was done from elaborate game farms, and then predators were controlled. Following these efforts, protected areas or refuges were set aside, and then finally habitat was managed as an ecosystem. Establishing areas of complete wildlife protection, stocking, and broad-scale predator control have proved over and over to be insufficient. Habitat management has been a more efficient technique.

Now the trend is away from an approach of any

single technique for influencing a population. The newer approach, a systems approach, is that of selecting a group of techniques specifically designed for benefiting one target species for each area and for achieving the specific objectives of the landowner. The system involves (1) intensive study and research by experts to isolate problems, (2) definition of goals, (3) education of the public (and specialists) by many media, (4) programs to retain and improve the habitat, (5) goal-oriented game harvest regulations or population control programs, (6) manipulation of sex and age ratios, as well as density of animals, to achieve desired population change, (7) actions by agents to deter or apprehend poachers or game-law violators, (8) control of disease and parasites, and of select individual predators and the causes of accidents, and (9) well-researched transplanting of wildlife from areas where it has bred successfully to areas with high potential for population buildups.

Wildlife management or conservation began in the late 1800s when sensitive men saw that many wildlife populations could not take care of themselves. The tedious, costly, and often dangerous experiences of early conservationists is well told by James B. Trefethen in *Crusade for Wildlife*. The picture of success is confusing, for what worked in one place failed in others; what was an acceptable practice for one species was harmful to another. Only after a long period of trial and error with game species did the need for scientific wildlife research become obvious.

The ring-necked pheasant is an example of an introduced bird that succeeded well over much of the northern part of the United States. Hundreds of other species have been tried under almost every conceivable program of introduction. Where the habitat is good for pheasants they will reproduce abundantly and fill every available space. Where the conditions of the environment are not right, where there are one or more limiting factors (such as inadequate natural calcium in the soil or critical moisture-temperature relations in the nest), there will be no natural populations of birds. Continued stocking is wasteful. There are yet no naturalized ring-necked pheasant populations in the southeastern states although millions of dollars have been spent in work to achieve such populations. Extra birds stocked onto areas where birds occur naturally are also wasted unless immediately harvested by hunters; stocking "before the gun" is practiced in many states. The cost to the public of each pheasant taken by a hunter in some areas exceeds $35. The cost of producing some bobwhite quail has exceeded $150 per bird taken by the hunter (Fig. 4). Nature quickly cuts populations to its carrying capacity.

Sanctuaries, or areas of complete protection for upland game, are generally considered useless since animals do not "spill out" of such areas as once thought. On the other hand, protected breeding areas and special areas are considered essential for migratory waterfowl (Fig. 5). Some refuges concentrate populations, stop their normal migration, and increase hunter harvests along the route. Canada geese in the Atlantic and Mississippi flyways are an example.

Predator control may be needed on select individual animals, but is generally undesirable and even very hazardous when practiced indiscriminately by individual or public impulse. Often there are no net benefits from control. Extra costs may be incurred as rodents or rabbits and their damage increase without the presence of the animals that once exercised some restriction on their populations.

Fig. 4. Game farm with pens for quail. *(Virginia Commission of Game and Inland Fisheries)*

Winter feeding attracts wildlife to some areas for increased viewing and appreciation by man, but generally does little good for most game populations in the long run. Artificial winter feeding of elk or deer is frequently harmful because it may allow more animals to survive the winter than the environment can support. The range is damaged by the extra animals; plants lose their vigor; erosion occurs; poaching, predation, and disease are increased; and mass die-offs occur in subsequent years. Also, winter feeding of high-quality food may even kill starving animals.

Today's needs. The public and the professional wildlife staffs of state and Federal conservation agencies can influence wildlife populations by the following major practices: (1) laws to protect endangered species, (2) regulations to allow a recreational harvest of game animals and manipulation of other populations that put each population in balance with its food supply, (3) protection of wildlife areas, both large land holdings of the government and small areas, such as woodlots and fencerows, needed by certain species of birds and animals, (4) research into the needs of and responses of animals to each other and their environment, (5) investment in select areas for maximum wildlife production, maintaining moderate production cheaply under existing conditions, and (6) education of the public: increasing their appreciation of wildlife (and thus the benefits which they derive from it) and also their awareness of how actions (hunting, voting, destroying food and cover, changing the environment, and contributing to

Fig. 5. Migratory wildlife need protection as well as year-around habitat if they are to retain their present abundance or to increase.

research and law enforcement) influence wildlife and the larger complex of man's own environment. *See* LAND-USE PLANNING.

Government programs. Every state now has employees with university degrees in wildlife management. The public can gain assistance, advice, or referral to appropriate experts or agents from such people. In addition, the Federal government has professional wildlife managers in the Soil Conservation Service, the U.S. Forest Service, the U.S. Park Service, the U.S. Bureau of Land Management, and the U.S. Fish and Wildlife Service. The last, an agency of the Department of the Interior, has extensive activities in the areas of management and enforcement, research, education, wildlife damage and depredation control, allocation of sportsmen's taxes for fish and wildlife restoration, and operation of the U.S. National Wildlife Refuge System. The Fish and Wildlife Service works with the states and other countries to develop practices for the wisest use of the wildlife resource. The Migratory Bird Treaty Act brings Canada, Mexico, the United States, and other countries together in work to secure the protection of all migratory birds, particularly the protection of waterfowl.

Many private groups work alone or with these agencies to aid in the best use of the wildlife resource. The *Conservation Directory* (National Wildlife Federation) lists such groups as the Wildlife Society, the Wildlife Management Institute, Sport Fishing Institute, Ducks Unlimited, Trout Unlimited, and the International Association of Fish and Game Commissioners. The organizations, as well as individuals throughout the United States, attach different values to wildlife and desire different benefits. Some have short range goals; others, long-range. Some are well educated and have abundant information about wildlife populations; others have little information but have strong emotional (or commercial) attachments to certain animals or inflexible ideas about them. *See* ENVIRONMENT; FISHERY CONSERVATION; FOREST AND FORESTRY.

The problem of wildlife management in a democracy is that of resolving conflicts between groups, such as those arising from the above differences. Recently the difficulty has been formulated as an optimizing problem and thus is subject to the methods of wildland operations research.

WILDLAND OPERATIONS RESEARCH

There has emerged within the past 30 years an art called operations research (OR) based on the science of analyzing large and complex systems of all types and designing means to achieve specified goals in some optimal fashion. Wildland operations research (WOR) is the application of the concepts and methods of OR to wildland problems. The term wildlands usually refers to all of the resources (space, water, soil, air, minerals, forests, and wildlife) of nonagricultural lands, such as forests, rangelands, marshes, deserts, and undeveloped parks. The differences within and between these categories may be great or indistinct (for example, differences between highly managed tree farms, farm woodlots, and wilderness forests). Every system can be considered a subsystem of a larger system. *See* TERRESTRIAL ECOSYSTEM.

At the root of OR is the concept of an optimum system. A system is almost anything that can be logically analyzed in the categories of inputs, processes, outputs, feedback and feedforward (Fig. 6). It is possible to conceive of machines, decisions, organs of the body, or even the entire body or populations of animals in the five components of systems. OR is concerned with analyzing, designing, or manipulating systems in order to understand them, to be able to predict their function, or to be able to manipulate them to achieve goals and objectives according to certain criteria.

Wildland operations research deals with natural resource systems such as the ecosystem, with populations of fish or other animals, with forests and wood-dependent industries, with rangelands and marshlands, with migratory-bird flyways, and with pest species and their effects. *See* ECOLOGY.

WOR is a natural, evolutionary response of science to the recognition that certain problems have no simple solutions, perhaps no solutions at all, and that choices between alternatives, even wrong ones, may at times be necessary. The problems of natural resources are often of this type. A solution to a timber harvest problem may have profound effects that cause "vibrations" throughout the forest system. A typical large problem arises when roads must be built to get the logs from a remote timber tract to market; the local economics may be such that profit cannot be made if a high-quality road is built; if the timber is not harvested, it will be "wasted" and local mills will suffer; low-grade roads cause silt; silt fills dams, reducing their net long-term effectiveness; silt reduces stream suitability for fisheries; sport fishing influences the local economy and thus impinges on the cost-benefit decisions of the road; logging roads increase hunter access to areas as well as providing some recreational sites; roads decrease the wild character of land; logging usually increases deer food and deer populations; logging reduces gray squirrel and wild turkey populations. In the long run logging may increase air and water pollution from mills; increased numbers of people have increased needs for all types of wood and wood products.

The problem is large and complex with many decisions throughout, all requiring inputs of both information and value judgments. The problems become increasingly more complex and difficult, and each decision more risk-laden. The latter is

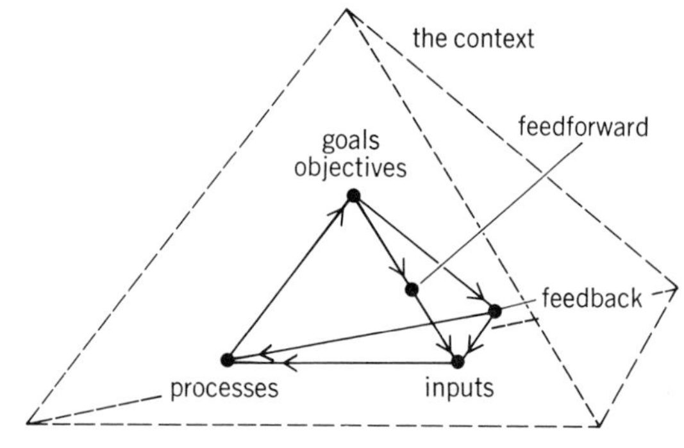

Fig. 6. Wildland systems, as well as others, can be described with the components shown here. Arrows represent the flow of thought and action.

true since each decision that is made today by wildland managers influences resources that are becoming increasingly limited; each decision influences more people; and more money or values now ride on each choice. WOR is thus an interdisciplinary scientific way of arriving more objectively at optimal solutions to large problems concerning the use of natural resources.

The ability of the computer to handle thousands of calculations rapidly, to make repetitive calculations, to modify its own instructions, and to help make decisions have made it the major tool for coping with the strategy and tactics of analysis and decision making. Among the useful methods are modeling, linear programming, heuristic programming, game theory, simulation, network analysis, and decision theory.

Modeling. To work with complex realities one's ideas must be reduced to a manageable form. Some ideas, once written, can be satisfactorily dealt with as a "word model". Some ideas cannot be communicated adequately by words, and thus mathematical formulations are needed. These mathematical models, often equations or a series of inequalities, are ways of communicating the size and relations of certain variables. For example, it can be said that a population of game animals is the result of the interaction of the proportion of females in the previous population, the life expectancy of that population, and the average number of young produced by females in the population. Mathematically this statement can be expressed as

$$P = S \cdot e_1 \cdot m$$

where P is the population, S is the proportion of females, e_1 is the life expectancy of the first age class, and m is the birth rate. A graphical model (Fig. 7) can be developed from such expressions to provide game managers with a tool to help them understand and make decisions for modifying the factors that influence the size and changes in populations.

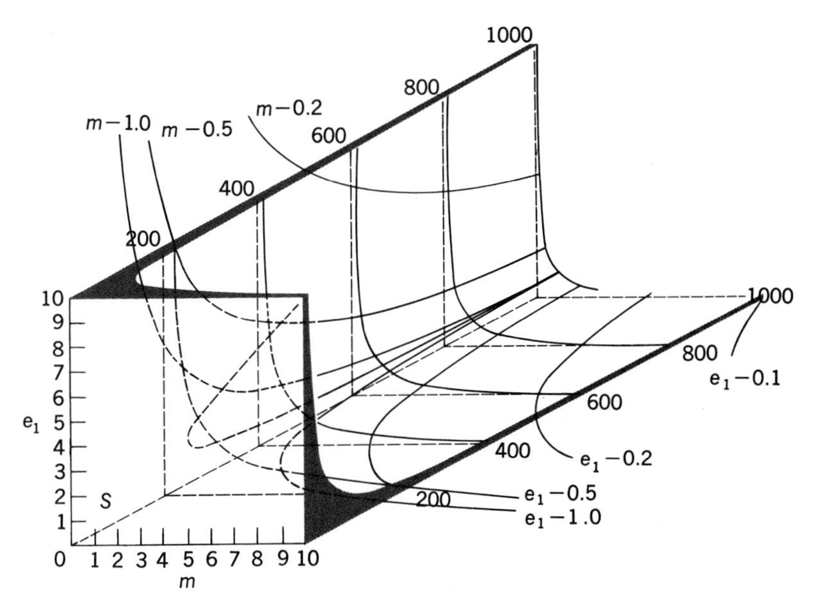

Fig. 7. A working model of population stability. The surface described by the curves represents population stability. When values of P fall on the surface, the population is stable; when they fall below the surface, a decreasing population is indicated; and when values are above the surface and not within the volume, an increasing population is indicated.

Usually mathematical models are developed to represent all known or major features of the reality of a situation. These models are needed by computers for making calculations or generating alternatives for a vast array of decisions made in wildland systems. Models can be based on theoretical relations or can be generalizations of quantities of data. Much research in the wildland sciences is devoted to collecting better data in order to build models or to explain the differences between predictions and real-world events. One exciting aspect of developing mathematical abstractions of reality is that during the process unexpected new results or insights are frequently gained. Model building still requires compromises; natural resource systems have so many variables that only a part of them can be dealt with at any time. Some models are now accurate enough to allow predictions in which users can be very confident. Certain risks are involved in decisions based on models, but the risks are usually not as great as when no models are made clear in words, symbols, and pictures.

Linear programming. Linear programming (LP) is a technique for arriving at an optimum solution to certain problems—maximizing a desired outcome subject to constraints which limit that outcome. If these constraints can be described as linear equations or inequalities, simple problems can be handled graphically but most problems require a computer to search within the "space" described by all of the equations to find an optimal solution.

For example, given a forest with existing distributions of ages and types of trees by acreage, how can a specific timber harvest schedule be satisfied and how can the resulting forest types be distributed in an acceptable fashion within a predefined time and at the least possible cost? The implications of such a decision and its impact on the habitat of forest game animals, such as elk, deer, and wild turkey, are obvious.

The following simple example presents a graphical LP solution (Fig. 8) to a problem of how to achieve an optimum mix of the number of elk and cattle on a range to maximize the animal units produced. Both elk and cattle can be produced on a given unit of land. Elk yield 0.5 animal unit; cows yield 1 animal unit. There is only range enough (because of different feeding habits) to support 400 elk or 300 cows. The capacity of the land is such that not more than a total of 600 animals can be supported.

Linear programming has been used to relate yield of forest or rangeland vegetation to soil and topographic variables, and these are used to determine optimum yields.

Efforts are now underway to perform LP and related techniques on many Federal, state, and private lands. The current limitations are (1) that objective functions (goals) of complex wildland systems are difficult to specify, and (2) most wildland equations are not linear. For an industry, the objective functions may be to maximize profit or minimize costs. The objective function for most public wildlands is to maximize the long-term total benefits from the wildlands. The benefits are so

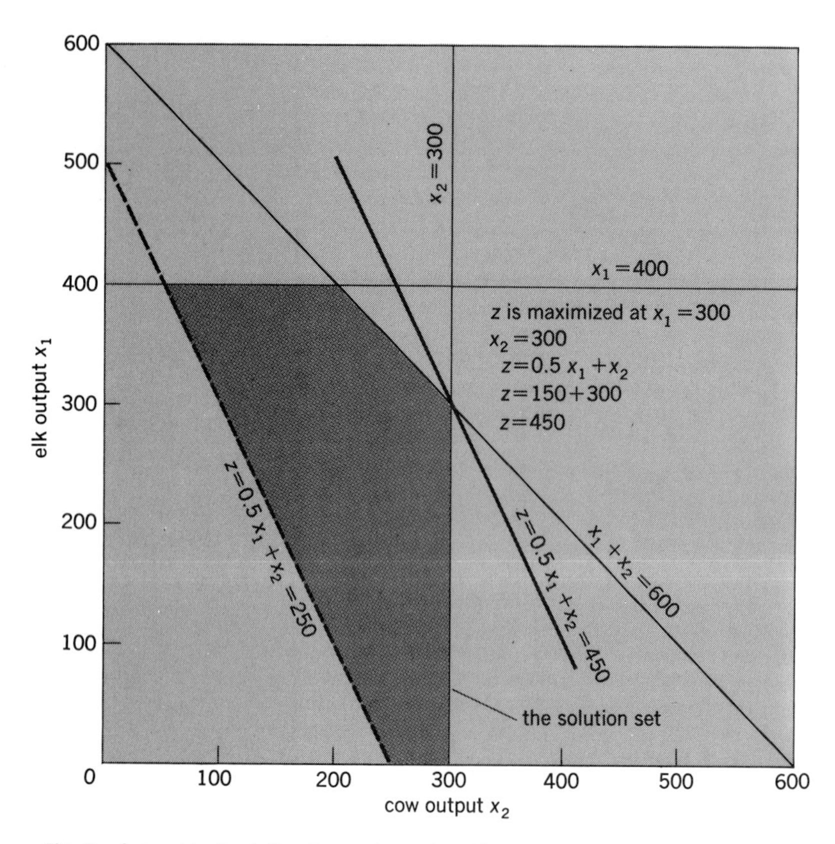

Fig. 8. A graphical solution to an elementary linear-programming problem. Within the colored area lie all possible solutions; the point $x_1 = 300$; $x_2 = 300$ is optimal and thus the optimum mix is 150 elk and 300 cows.

interrelated, abundant, and variable among people that the task is very difficult. Wildlife management problems might be formulated: Given a region with different deer populations, each with different sex and age ratios, and each consuming food at different rates, what combination of regulations of hunting-season opening dates, season length, legal weapons, sex limits, and number of deer taken will maximize the kill while stabilizing the residual population at a level compatible with available forage? Many of these relations are not linear. Present results encourage continuance of the task.

Heuristic programming. LP requires that the relations with which it deals have no exponents. It also typically solves a problem for some point in time. Often these requirements cannot be met-over a period of time, changes are rarely constant but often curvilinear, fluctuating, or cyclic. Heuristic programming provides several techniques for handling such changes. Dynamic programming for example carries equations to a point in time; new calculations are made based on shifting rules or criteria that are part of the program; it moves to the next point in time and makes continuing calculations until the optimal solutions are found. The technique even allows for random developments to occur and a sequence of decisions to be made.

As with LP, large dynamic-programming problems require a computer, and many programs are now available.

Game theory. The possibilities for employing game theory to game populations are intriguing and have been scarcely touched. Game theory is the mathematical formulation of conflicts of interest. The types of wildland conflicts already outlined may depend on questions like: who are the players, what are the probabilities of various moves or plays of the game, what are the risks and uncertainties involved with each decision or play, what are appropriate strategies or plans of play, and what are the game payoffs or outcomes? In so-called war games, the quest is the formulation and selection of strategies to win most often in the long run with complete or incomplete information. Industries have similarly used game theory to analyze markets, to make investment decisions, and to win against their competitors. In games against nature, in decisions of what types of crops and how much to plant in the face of drought or insect attacks, and in flood and forest-fire loss reduction game theory deals with situations of various sorts of conflict. The players have limited control over the variables of the situation; in conflicting human interests each player may wish to maximize his selfish payoff, but all may realize that a better outcome is to allow the game to continue and not press for termination.

Games have been constructed for university wildlife management students in which the students make many decisions about the setting of hunting seasons or the manipulation of populations. They play against a computer simulation of nature, and in so doing, learn whether they win or lose without ever harming the resource. Game theory has been used to decide on the proper allocation of marshes to certain types of waterfowl-food planting, to determine suggested moves for law-enforcement agents against poachers, and for allocating equipment and personnel for standby for forest-fire fighting.

Decision theory is grossly defined and includes decision making based on sampling, statistics, and probabilities; the construction and analysis of decision trees or branching networks of related alternatives; and network analysis employing the critical path method (CPM) or project evaluation and review technique (PERT). Decision theory overlaps game theory. A decision tree is presented (Fig. 9) showing the method. A tree is a model, and when employed experimentally, it is a simulator. CPM and PERT are techniques used by management to plan and schedule events. They are used to schedule the expected progress and possible delays in programs that must meet deadlines. Wildland planting practices on wildlife refuges would be an example.

Simulation. Simulation is the activity of building a model and then performing experiments on it by asking questions such as "what if I change this factor?" Computers are often used and elaborate systems can be studied without disrupting ongoing physical operations and without hazards or great costs that may be associated with experimentation or change of the real world. Simulation allows explanation and prediction. It is the dynamic use of models but it is not an optimizing technique. It may display thousands of alternatives, the selection among which will be dependent upon some outside criteria or strategy of choice.

Ecosystems, forests, bogs, rangelands, grasslands, industries, physiological systems, populations, and socioeconomic systems have now been satisfactorily simulated. Systems exist for simulating the elk and big-game forage production in national forests of the Pacific Northwest. Physical and computer-based models of major waterways now exist. Management practices for forests are now simulated as are locations of power line corridors and their impact on wildlife.

When objectives are poorly defined, when variables are still being selected and studied, or when decisions are needed on problems for which data are scarce, simulation may provide the technique for evaluation alternatives through computer assistance.

Simulation is one of the more powerful tools of WOR. Associated with this power is the surprise discovery (serendipity) that is associated with both the building of models and examining of results.

Only a few universities now offer strong programs of study linking operations research and systems analysis with ecology and wildlife management. Both the analytical abilities of the wildlands scientist and manager and his manipulative and decision-making activities in nature need to be combined in predicting the consequences of various actions taken with man's environment. As an aid to making this combination, WOR will undoubtedly be crucial for designing and maintaining an environment fit for people's material and intangible needs.

[ROBERT H. GILES, JR.]

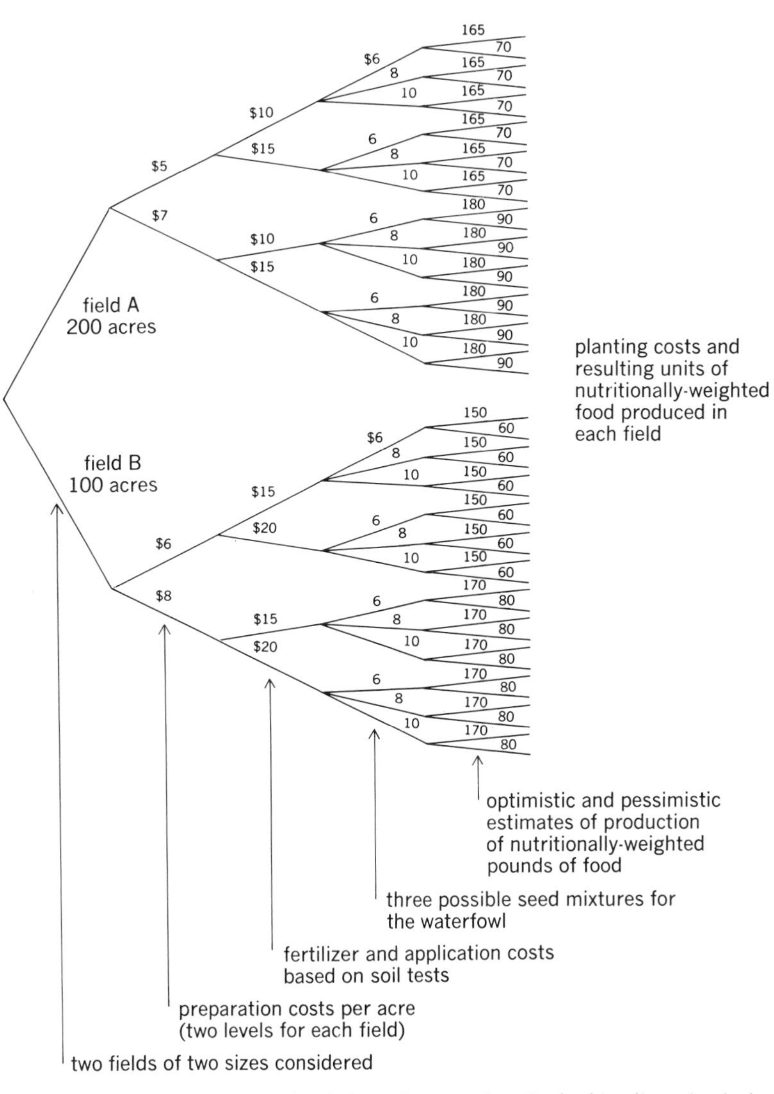

Fig. 9. A decision tree by which relative values can be attached to alternate strategies or methods for producing waterfowl foods. 1 acre= 0.4047 hectare.

PRESERVATION OF WILDERNESS

Advocates of the preservation of wilderness have been active in the United States since the early 1900s, with Theodore Roosevelt's administration bestowing an official blessing on the concept of natural area conservation. The acknowledged leader of the contemporary wilderness movement was the forester Robert Marshall, who wrote in 1930: "A thorough study should forthwith be undertaken to determine the probable wilderness needs of the country. Of course no precise reckoning could be attempted, but a radical calculation would be feasible. It ought to be radical for three reasons: because it is easy to convert a natural area to industrial or motor usage, impossible to do the reverse; because the population which covets wilderness recreation is rapidly enlarging; and because the higher standard of living which may be anticipated should give millions the economic power to gratify what is today merely pathetic yearning. Once the estimate is formulated, immediate steps should be taken to establish enough tracts to insure everyone who hungers for it a generous opportunity of enjoying wilderness isolation."

A wilderness system formed on the nucleus of administratively designated national forest lands, most of them high mountain country, was the dream of Marshall and his friends who created the Wilderness Society in 1935, a national organization with the goal of wilderness preservation. Since then, it has also become deeply involved with the broad complex of environmental problems brought on by the demands of technological progress and a pyramiding population.

Shaping the programs. To carry on its purpose, the Wilderness Society conducts an educational program concerning the value of wilderness in its relationship to the problems of society, the overall objective being to increase knowledge and appreciation of wilderness in the context of man's ecological perspective, as well as his spiritual and cultural needs. Wilderness, as the Society sees it, is a valuable natural resource that belongs to all, and its preservation for educational, scientific, and recreational use is part of a balanced conservation program essential to the survival of man's culture.

The founders of the Wilderness Society, knowing that the impact of civilization had already changed the face of the United States, hoped that many of the remaining Federally owned wild lands would be protected through administrative practices such as those already set up by the U.S. Forest Service to designate wild, scenic, and primitive areas and the National Park Service with its mandate from Congress to preserve and protect natural phenomena such as the Grand Canyon, Yellowstone, and Yosemite.

The work of Olaus Murie in wildlife biology helped to round out the concept of a national wilderness preservation system through consideration of refuge areas as ecosystems in which wild-

A list of the locations of wilderness areas, their acreage, and date of inclusion in the Wilderness System*

State areas	Acreage	Year of origin
Alaska		
Bering Sea	41,113	1970
Bogoslof	390	1970
Forrester Island	2,630	1970
Hazy Island	42	1970
St. Lazaria	62	1970
Tuxedni	6,402	1970
Arizona		
Chiricahua	18,000	1964
Galiuro	52,717	1964
Mazatzal	205,346	1964
Mount Baldy	6,975	1970
Petrified Forest	50,260	1970
Pine Mountain	19,500	1970
Sierra Ancha	20,850	1964
Superstition	124,140	1964
Sycamore Canyon	48,500	1970
California		
Caribou	19,080	1964
Cucamonga	9,022	1964
Desolation	41,343	1969
Dome Land	62,561	1964
Lassen Volcanic National Park	78,982	1972
Lava Beds National Monument	8,460	1972
Hoover	42,800	1964
John Muir	504,263	1964
Marble Mountain	214,543	1964
Minarets	109,559	1964
Mokclumne	50,400	1964
San Gabriel	36,137	1968
San Gorgonio	34,718	1964
San Jacinto	21,955	1964
San Rafael	142,722	1968
South Warner	69,547	1964
Thousand Lakes	16,335	1964
Ventana	54,857	1969
Yolla-Bolly Middle Eel	111,091	1964
Colorado		
La Garita	48,486	1964
Maroon Bells – Snowmass	71,329	1964
Mount Zirkel	72,472	1964
Rawah	26,674	1964
West Elk	61,412	1964
Florida		
Island Bay Wilderness	20	1970
Passage Key	20	1970
Pelican Island	75	1970
Idaho and Montana		
Selway-Bitterroot	1,243,669	1964
Idaho		
Craters of the Moon	43,243	1970
Sawtooth	216,400	1970
Maine		
Moosehorn	2,782	1970
Massachusetts		
Monomoy	2,340	1970
Michigan		
Huron Islands	147	1970
Michigan Islands	12	1970
Seney	25,150	1970
Minnesota		
Boundary Waters Canoe Area	1,029,257	1964
Montana		
Anaconda-Pintlar	159,086	1964
Bob Marshall	950,000	1964
Cabinet Mountains	94,272	1964
Gates of the Mountains	28,562	1964
Scapegoat	240,000	1972
Nevada		
Jarbidge	64,827	1964
New Hampshire		
Great Gulf	5,552	1964
New Jersey		
Great Swamp National Wildlife Refuge	3,750	1968
New Mexico		
Gila	433,916	1964
Pecos	167,416	1964
Salt Creek	8,500	1970
San Pedro Parks	41,132	1964
Wheeler Peak	6,029	1964
White Mountain	31,283	1964
Oklahoma		
Witchita Mountains	8,900	1970
Oregon		
Diamond Peak	35,440	1964
Eagle Gap	221,355	1964
Eagle Gap (Minam River Canyon)	72,420	1972
Gearhart Mountain	18,709	1964
Kalmiopsis	76,900	1964
Mount Hood	14,160	1964
Mount Jefferson	100,000	1968
Mount Washington	46,655	1964
Mountain Lakes	23,071	1964
Oregon Islands	21	1970
Strawberry Mountain	33,653	1964
Three Arch Rocks	17	1970
Three Sisters	196,708	1964
Washington		
Glacier Peak	464,741	1964
Glacier Peak	10,000	1968
Goat Rocks	82,680	1964
Mount Adams	42,411	1964
Pasayten	520,000	1968
Washington Islands	5,174	1970
Wisconsin		
Wisconsin Islands	29	1970
Wyoming		
Bridger	383,300	1964
North Absoraka	351,104	1964
Stratified Primitive Area	208,000	1972
South Absoraka	483,678	1964
Teton	563,500	1964
Total acreage in National Wilderness Preservation System		11,284,372

*1 acre = 0.4047 ha.

life conservation purposes operated for the benefit of a whole community of species, rather than for the stocking or farming of one "target" game or fish species.

Through books, articles, and lectures, Sigurd F. Olson's interpretation of wilderness increased public awareness and appreciation of the natural scene. When he said "Wilderness to the people of America is a spiritual necessity, an antidote to the pressures of modern life and a means of regaining serenity and equilibrium," he spoke for millions, and helped lay the groundwork for congressional action.

The three decades following the creation of the Wilderness Society were punctuated by fierce struggles to withstand the encroachment of developments on Echo Park and Dinosaur National Monument (Utah), the wilderness lake country of the Quetico-Superior, the New York State Forest Preserve, and many others, demonstrating the growing concern of the public in protecting and setting aside wilderness regions.

Government regulation. Wilderness preservation was adopted as a national policy with the signing on Sept. 3, 1964, of the Wilderness Law after 8 years of public discussion and debate in Congress and across the land. In the process of its passage the attitudes of many Americans became firm on the subject of conservation in general, and groundwork was laid for the present nationwide interest in environmental quality and the close scrutiny of development activities and practices which would damage the country's diminishing resource of wild land, free-flowing water, and clean air.

Such tracts of roadless land as are already in public ownership are to be protected as wilderness upon their inclusion in the National Wilderness Preservation System, established by the act of 1964. The law defines wilderness as follows:

"A wilderness in contrast with those areas where man and his own works dominate the landscape, is hereby recognized as an area where the earth and its community of life are untrammeled by man, where man himself is a visitor who does not remain. An area of wilderness is further defined to mean . . . an area of undeveloped Federal land retaining its primeval character and influence, without permanent improvements or human habitation, which is protected and managed so as to preserve its natural conditions and which (1) generally appears to have been affected primarily by the forces of nature with the imprint of man's work substantially unnoticeable; (2) has outstanding opportunities for solitude or a primitive and unconfined type of recreation; (3) has at least five thousand acres of land or is of sufficient size as to make practicable its preservation and use in an unimpaired condition; and (4) may also contain ecological, geological, or other features of scientific, educational, scenic, or historical value."

The Wilderness Law established an initial 54 units of national forest land as wilderness areas, provided for review of roadless areas and islands within the remaining national forest primitive areas, the national parks and monuments, and the national wildlife refuges and ranges. These reviews were to be undertaken on a 10-year schedule by the agency in charge of administering the area (Forest Service of the Department of Agriculture for forests; National Park Service of the Department of the Interior for parks and monuments; Bureau of Sport Fisheries and Wildlife of the Interior Department's Fish and Wildlife Service for refuges and ranges). Other Federal lands, including wilderness within the extensive public domain under the Bureau of Land Management, are not mentioned in the law, although they could be included in the System through direct enactment by Congress at any time.

Public hearings are provided for in the Wilderness Law. These are to be held in the states where the candidate wilderness areas are located and are to be conducted by the administering agency. At these hearings local citizens and groups, as well as government officials, may comment on or suggest improvements in the agency's proposal for wilderness boundaries. A 10-year deadline, to expire Sept. 3, 1974, was established, during which time the agencies' review of roadless areas was to be completed and their proposals transmitted to the President and thence to Congress.

During the first 3-year period after passage of the Wilderness Law, the "wilderness agencies" framed their regulations for the management of the wilderness areas which would be under their jurisdictions. These were published in the Federal Register of Feb. 17 and May 31, 1966. In 1968 the first additions to the Wilderness System were made through enactments relating to five new wilderness units and an addition to one existing area (Glacier Peak); two more additions had been made by Oct. 15, 1969. Toward the end of 1969, the schedule of agency reviews and hearings had been completed on 45 wilderness areas in 18 states, while 130 more, totaling more than 40,000,000 acres, remained to be dealt with. In 1970 Congress designated 23 new wilderness areas in 12 states, a total of 201 212 acres (81,428 ha) (see table). Proposals to include about 15,000,000 acres (6,000,000 ha) of forests, mainly in Alaska, were being actively considered by Congress in 1979. These generated a great deal of controversy, particularly in Alaska.

An important feature of the Wilderness Law is its avoidance of any contradiction of the basic purposes for which the Federal lands had originally been set aside. The national forest primitive areas had been administratively protected from development and mechanized equipment since 1929. The national parks, with their many scenic treasures, had been dedicated to be preserved for the benefit and enjoyment of future generations. The wildlife conservation practices in refuges and ranges were seen as requiring in many cases an unspoiled natural environment, large enough to meet the need for a complete community of associated species. Thus the Wilderness System was set up to strengthen the authority of the wilderness administrator in resisting pressure for declassifying or downgrading the protection afforded such areas.

The all-important result of the long effort to give wilderness official recognition was public acceptance of the idea, so well expressed in the oft-quoted statement of Wallace Stegner: "Something will have gone out of us as a people if we ever let the remaining wilderness to be destroyed; if we permit the last virgin forests to be turned into comic books or plastic cigarette cases; if we drive the few remaining species into zoos or extinction; if we pollute the last clear air and dirty the last clean stream and push our paved roads through the last of the silences." *See* CONSERVATION OF RESOURCES.

[SIGURD F. OLSON]

Bibliography: D. L. Allen, *Our Wildlife Legacy*, 1962; C. W. Churchman, R. L. Ackoff, and E. L. Arnoff, *Introduction to Operations Research*, 1957; R. F. Dasmann, *Wildlife Biology*, 1964; R. H. Giles (ed.), *Wildlife Management Techniques*, Wildlife Society, 1969; J. B. Trefethen, *Crusade for Wildlife*, 1961; G. M. Van Dyne, Application and integration of multiple linear regression and linear programming in renewable resource analysis, *J. Range Manage.* 19(6):356–362, 1966; K. E. F. Watt, *Systems Analysis in Ecology*, 1966; K. E. F. Watt, *Ecology and Resource Management*.

Wind

The motion of air relative to the Earth's surface. The term usually refers to horizontal air motion, as distinguished from vertical motion, and to air motion averaged over a chosen period of 1–3 min. Micrometeorological circulations (air motion over periods of the order of a few seconds) and others small enough in extent to be obscured by this averaging are thereby eliminated. The choice of the 1- to 3-min interval has proven suitable for the study of (1) the hour-to-hour and day-to-day changes in the atmospheric circulation pattern; and (2) the larger-scale aspects of the atmospheric general circulation.

The direct effects of wind near the surface of the Earth are manifested by soil erosion, the character of vegetation, damage to structures, and the production of waves on water surfaces. At higher levels wind directly affects aircraft, missile and rocket operations, and dispersion of industrial pollutants, radioactive products of nuclear explosions, dust, volcanic debris, and other material. Directly or indirectly, wind is responsible for the production and transport of clouds and precipitation and for the transport of cold and warm air masses from one region to another. *See* ATMOSPHERIC GENERAL CIRCULATION.

Cyclonic and anticyclonic circulation. Each is a portion of the pattern of airflow within which the streamlines (which indicate the pattern of wind direction at any instant) are curved so as to indicate rotation of air about some central point of the cyclone or anticyclone. The rotation is considered cyclonic if it is in the same sense as the rotation of the surface of the Earth about the local vertical, and is considered anticyclonic if in the opposite sense. Thus, in a cyclonic circulation, the streamlines indicate counterclockwise (clockwise for anticyclonic) rotation of air about a central point on the Northern Hemisphere or clockwise (counterclockwise for anticyclonic) rotation about a point on the Southern Hemisphere. When the streamlines close completely about the central point, the pattern is denoted respectively a cyclone or an anticyclone. Since the gradient wind represents a good approximation to the actual wind, the center of a cyclone tends strongly to be a point of minimum atmospheric pressure on a horizontal surface. Thus the terms cyclone, low-pressure area, or "low" are often used to denote essentially the same phenomenon. In accord with the requirements of the gradient wind relationship, the center of an anticyclone tends to coincide with a point of maximum pressure on a horizontal surface, and the terms anticyclone, high-pressure area, or "high" are often used interchangeably.

Cyclones and anticyclones are numerous in the lower troposphere at all latitudes. At higher levels the occurrence of cyclones and anticyclones tends to be restricted to subpolar and subtropical latitudes, respectively. In middle latitudes the flow aloft is mainly westerly, but the streamlines exhibit wavelike oscillations connecting adjacent regions of anticyclonic circulation (ridges) and of cyclonic circulation (troughs).

Although the atmosphere is never in a completely undisturbed state, it is customary to refer to cyclonic and anticyclonic circulations specifically as atmospheric disturbances. Cyclones, anticyclones, ridges, and troughs are intimately associated with the production and transport of clouds and precipitation, and hence convey a connotation of disturbed meteorological conditions.

A more rigorous definition of circulation is often employed, in which the circulation over an arbitrary area bounded by the closed curve S is given by Eq. (1), where the integration is taken complete-

$$C = \oint v_t \, dS \qquad (1)$$

ly around the boundary of the area. Here v refers to the wind at a point on the boundary, the subscript t denotes the component of this wind parallel to the boundary, and dS is a line element of the boundary. The component v_t is considered positive or negative according to whether it represents cyclonic or anticyclonic circulation along the boundary S. In this context, the circulation may be positive (cyclonic) or negative (anticyclonic) even when the streamlines within the area are straight, since the distribution of wind speed affects the value of C. *See* ATMOSPHERE; CLOUD; STORM.

Convergent or divergent patterns. These are said to occur in areas in which the (horizontal) wind flow and distribution of air density is such as to produce a net accumulation or depletion, respectively, of mass of air. Rigorously, the mean horizontal mass divergence over an arbitrary area A bounded by the closed curve S is given by Eq. (2),

$$D = \frac{1}{A} \oint \rho v_n \, dS \qquad (2)$$

where the integration is taken completely around the boundary of the area. Here ρ is the density of air, v refers to the wind at a point on the boundary, the subscript n denotes the component of this wind perpendicular to the boundary, and dS is an element of the boundary. The component v_n is taken positive when it is directed outward across the boundary and negative when it is directed inward. Convergence is thus synonymous with negative divergence. If spatial variations of density are neglected, the analogous concept of velocity divergence and convergence applies.

The horizontal mass divergence or convergence is intimately related to the vertical component of motion. For example, since local temporal rates of change of air density are relatively small, there must be a net vertical export of mass from a volume in which horizontal mass convergence is taking place. Only thus can the total mass of air within the volume remain approximately constant. In particular, if the lower surface of this volume coincides with a level ground surface, upward motion must occur across the upper surface of this volume. Similarly, there must be downward motion immediately above such a region of horizontal mass divergence.

The horizontal mass divergence or convergence is closely related to the circulation. In a convergent wind pattern the circulation of the air tends to become more cyclonic; in a divergent wind pattern the circulation of the air tends to become more anticyclonic.

Regions which lie in the path of an approaching cyclone are characterized by a convergent wind pattern in the lower troposphere and by upward vertical motion throughout most of the troposphere. Since the upward motion tends to produce condensation of water vapor in the rising air current, abundant cloudiness and precipitation typically occur in this region. Conversely, the area in advance of an anticyclone is characterized by a divergent wind pattern in the lower troposphere and by downward vertical motion throughout most of the troposphere. In such a region, clouds and precipitation tend to be scarce or entirely lacking.

A convergent surface wind field is typical of fronts. As the warm and cold currents impinge at the front, the warm air tends to rise over the cold air, producing the typical frontal band of cloudiness and precipitation. *See* FRONT.

Zonal surface winds. Such patterns result from a longitudinal averaging of the surface circulation. This averaging typically reveals a zone of weak variable winds near the Equator (the doldrums) flanked by northeasterly trade winds in the Northern Hemisphere and southeasterly trade winds in the Southern Hemisphere, extending poleward in each instance to about latitude 30°. The doldrum belt, particularly at places and times at which it is so narrow that the trade winds from the two hemispheres impinge upon it quite sharply, is designated the intertropical convergence zone, or ITC. The resulting convergent wind field is associated with abundant cloudiness and locally heavy rainfall. A westerly average of zonal surface winds prevails poleward of the trade wind belts and dominates the middle latitudes of both hemispheres. The westerlies are separated from the trade winds by the subtropical high-pressure belt, which occurs between latitudes 30 and 35° (the horse latitudes), and are bounded on the poleward side in each hemisphere between latitudes 55 and 60° by the subpolar trough of low pressure. Numerous cyclones and anticyclones progress eastward in the zone of prevailing westerlies, producing the abrupt day-to-day changes of wind, temperature, and weather which typify these regions. Poleward of the subpolar low-pressure troughs, polar easterlies are observed.

The position and intensity of the zonal surface wind systems vary systematically from season to season and irregularly from week to week. In general the systems are most intense and are displaced toward the Equator in a given hemisphere during winter. In this season the subtropical easterlies and prevailing westerlies attain mean speeds of about 15 knots (8 m/s), while the polar easterlies are somewhat weaker. In summer the systems are displaced toward the pole by 5 to 10° of latitude and weaken to about one-half their winter strength.

When the pattern of wind circulation is averaged with respect to time instead of longitude, striking differences between the Northern and Southern Hemispheres are found. On the Southern Hemisphere, variations from longitude to longitude are relatively small and the averaged pattern is described quite well in terms of the zonal surface wind belts. On the Northern Hemisphere there are large differences from longitude to longitude. In winter, for example, the subpolar trough is mainly manifested in two prominent low centers, the Icelandic low and the Aleutian low. The subtropical ridge line is drawn northward in effect over the continents and is seen as a powerful and extensive high-pressure area over Asia and as a relatively weak area of high pressure over North America. In summer the Aleutian and Icelandic lows are weak or entirely absent, while extensive areas of low pressure over the southern portions of Asia and western North America interrupt the subtropical high-pressure belt. *See* CLIMATOLOGY.

Upper air circulation. Longitudinal averaging indicates a predominance of westerly winds. These westerlies typically increase with elevation and culminate in the average jet stream, which is found in lower middle latitudes near the tropopause at elevations between 35,000 and 40,000 ft (10,700 and 12,200 m). The subtropical ridge line aloft is found equatorward of its surface counterpart and easterlies occur at upper levels over the equatorward portions of the trade wind belts. In high latitudes, weak westerlies aloft are found over the surface polar easterlies. Seasonal and irregular fluctuations of the circulation aloft are similar to those which characterize the surface winds. *See* JET STREAM.

Minor terrestrial winds. In this category are circulations of relatively small scale, attributable indirectly to the character of the Earth's surface. One example, the land and sea breeze, is a circulation driven by pronounced heating or cooling of a given area in comparison with little heating or cooling in a horizontally adjacent area. During the day, air rises over the strongly heated land and is replaced by a horizontal breeze from the relatively cool sea. At night, air sinks over the cool land and spreads out over the now relatively warm sea.

Another example is formed by the mountain and valley winds. These result from cooling and heating, respectively, of the mountain slopes relative to the horizontally adjacent free air above the valley floor. During the day, air flows up from the valley along the strongly heated mountain slopes, but at night, air flows down the relatively cold mountain slopes toward the valley bottom. A similar type of descending current of cooled air is often observed along the sloping surface of a glacier. This nighttime air drainage, under proper topographical circumstances, can lead to the accumulation of a pool of extremely cold air in nearby valley bottoms.

Local winds. These commonly represent modifications by local topography of a circulation of large scale. They are often capricious and violent in nature and are sometimes characterized by extremely low relative humidity. Examples are the mistral which blows down the Rhone Valley in the south of France, the bora which blows down the gorges leading to the coast of the Adriatic Sea, the foehn winds which blow down the Alpine valleys, the williwaws which are characteristic of the fiords of the Alaskan coast and the Aleutian Islands, and the chinook which is observed on the eastern slopes of the Rockies. Local names are also given sometimes to currents of somewhat larger scale which are less directly related to topography.

Some examples of this type of wind are the norther, which represents the rapid flow of cold air from Canada down the plains east of the Rockies and along the east coast of Mexico into Central America; the nor'easter of New England, which is part of the wind circulation about intense cyclones centered offshore along the Middle Atlantic coastal states; and the sirocco, a southerly wind current from the Sahara which is common on the coast of North Africa and sometimes crosses the Mediterranean Sea. [FREDERICK SANDERS]

Zoogeographic region

A major unit of the Earth's surface characterized by faunal homogeneity. This definition is more of a concept than a reality. Because different classes of animals arose and dispersed at different times during Earth history and because animals vary greatly in their vagilities and tolerances to environmental conditions, no two major groups of animals display complete coincidence in their geographic limits. As a result, delimitation of the several regions is difficult and such patterns as have been defined are not universally applicable to the entire animal kingdom.

Fortunately, insofar as existing animals are concerned, there appears to be a rough average geographic pattern. This pattern is best mirrored in the distribution of birds and mammals. Considerable coincidence in the geographic patterns of these two groups has possibly resulted from the fact that they arose and dispersed relatively recently in geologic time, and, as a result, their extant distributions have been controlled by major geographic barriers of the not too distant past. These patterns reflect to a lesser degree the distributions of other major groups. It appears, therefore, that several parts of the Earth's surface have, through isolation, served as major centers of evolution, and in the final analysis, it is these centers that are recognized as zoogeographic regions. Broad zones of transition have developed between the major zoogeographic regions owing either to the ability of some animals to transcend major barriers or to the disappearance of the barriers. The occurrence of such transitional areas has led to most of the difficulties in plotting zoogeographic boundaries.

Classification. Although earlier attempts had been made to partition the Earth into zoogeographic regions, it was in 1858 that P. Sclater, on the basis of bird distribution, presented the first practical regional classification. This was extended by A. Wallace in 1876 and has since undergone some further modification, but the pattern described by Sclater has remained basically sound. The table correlates the classical Sclater-Wallace system with K. Schmidt's 1954 review of the problem.

Although the Palearctic and Nearctic regions share many faunal elements, particularly insofar as eastern Asia and eastern North America are concerned, they do display important faunal differences. The tropical elements of the two regions, especially, show considerable divergence. That of Palearctica has been derived from the Ethiopian and Oriental regions, whereas the Nearctic tropical element has stemmed, for the most part, from Neotropica. Thus, although the two northern regions are frequently combined as Holarctica in order to express their many resemblances, they are more generally treated individually.

Zoogeographic regions

Schmidt (1954)			Classical system regions
Realms	Regions	Subregions	
Arctogaean	Holarctic	Arctic	Nearctic-Palearctic
		Nearctic	Nearctic
		Caribbean	
		Palearctic	Palearctic
	Paleotropical	Oriental	Oriental
		Ethiopian	Ethiopian
		Malagasy	
Neogaean	Neotropical	Neotropical	Neotropical
Notogaean	Australian	Australian	Australian
		Papuan	
	Oceanian	New Zealandian	
		Oceanic	
		Antarctic	Unclassified

Palearctic region. This region includes all Eurasia north of the Tropic of Cancer, excepting parts of the Arabian Peninsula and northwestern India. Both the East (China) and the West (Europe) originally supported deciduous forest. To the North, the boreal forest and the tundra are circumpolar. The vast interior of Asia is largely grasslands with considerable desert on the high plateaus.

The Palearctic fauna is difficult to characterize. As a generalization, it is composed of three elements. One is worldwide or almost so, the second is more or less circumpolar, and the third has been derived from the Old World tropics. At the family and subfamily levels there are few exclusives. Furthermore, the extensive longitudinal spread of the region that leaves the temperate forests of East and West disjunct, because of the subhumid lands of the interior, produces a number of east-west faunal differences.

Fishes. The fish fauna is dominated by the widely distributed cyprinids or minnows. With North America, the Palearctic shares the percids or perches, paddlefish, and suckers. The last two are restricted to the Asian area of the Paleartic. The cobitids (loaches) and a few catfishes are shared with the Old World tropics.

Amphibians. Among the amphibians the hynobiid salamanders are endemic to eastern Asia. Newts are Holarctic in their distribution. The plethodontid salamander of Europe is a single representative of a large, New World group, whereas, in contrast, the hellbender of China is related to that of eastern North America. The tailless amphibians are largely the widely distributed bufonids or toads, and the hylid and ranid frogs.

Reptiles. The reptilian fauna is poor. The emydid turtles and the veranid and agamid lizards have all been derived from the South. Among the snakes, the majority belong to the harmless and almost worldwide colubrids. True vipers in the West have been derived from the South, and the pit vipers of the East are shared with the New World.

Birds. The Palearctic avifauna reflects the general faunal picture outlined above. A more or less worldwide element is represented by the hawks, woodpeckers, swallows, and finches. Circumpolar groups include the grouse, waxwings, and creepers. Tropical elements, largely migratory, include the Old World flycatchers, larks, and starlings. Such Oriental derivatives as pheasants, cuckoos, shrikes, and white-eyes occur only in eastern Asia. Only a single family, the hedge sparrows, is endemic to the region.

Mammals. The mammalian fauna resembles the avifauna in its geographic relationships. Worldwide (except for Australia) groups include rabbits, cricetid mice and rats, squirrels, and cats. Circumpolar types include the beaver, jumping mice, and the pikas. The murids (Old World mice and rats), the dormice (also African), and the panda (restricted to eastern Asia) are essentially Old World groups.

Nearctica. Nearctica includes all of North America north of the edge of the Mexican Plateau. The vegetation pattern is similar to that of

Palearctica. Across the North, the boreal forest and the tundra are both transcontinental. The entire East and the coastal fringe of the West support temperate forest cover. The central region, although bisected by the forested Rocky Mountains, is largely grasslands and deserts.

In general, the faunal picture is also similar to that of Palearctica. There is considerable east-west diversity and a mixture of worldwide, tropical, mostly from South America, and holarctic groups. Endemism is, however, better marked.

Fishes. The fish fauna is richest east of the Mississippi River. Cyprinids are abundant. Holarctic or at least Nearctic and Asian groups include the suckers, paddlefishes, and perches. A few South American elements, notably characins and cichlids, extend up to the Rio Grande River. Exclusive groups are represented by the bowfin, mooneyes, trout, and basses.

Amphibians. Among the amphibians, the salamanders, especially the plethodontids, are particularly characteristic. Most tailless amphibians, like those of Palearctica, are divided among the widely distributed bufonids, the hylid tree frogs, and the ranids. A rather interesting, primitive frog, *Ascaphus*, is known only from the high mountains of the western coast. It belongs to a group that is otherwise restricted to a small island off the coast of New Zealand.

Reptiles. The reptile fauna includes such almost worldwide groups as the gekkonid and scincid lizards, and the colubrid snakes. It shares with Eurasia the emydid turtles, the anguid lizards, and the pit vipers. More restricted in their distributions (mainly in South America) are the coral snakes, the teiid lizards, the iguanid lizards (also on Fiji and in Madagascar), and the peculiar, poisonous gila monster.

Birds. For the most part, the avifauna comprises worldwide and Holarctic groups and a number of migratory neotropical elements. Ducks, pigeons, hawks, kingfishers, swallows, and finches are all widely distributed. Holarctic groups include grouse, creepers, and waxwings. Strictly New World elements are represented by the New World flycatchers, vireos, orioles, and hummingbirds. The turkey is almost endemic, extending as far south as northern Central America.

Mammals. Of the mammals, the cats, bovids, rabbits, cricetid mice, and squirrels are wide-ranging. Mammals more typical of Holarctica include the jumping mice, microtine mice, and the beaver. Two exclusives occur in the region, the pronghorn of the western deserts and the mountain beaver of the Northwest.

Ethiopian region. In the modern parlance, this region refers to Africa south of about the mid-Sahara. Some zoogeographers also include the Arabian Peninsula, which with Mediterranean Africa is considered by others to be a transition province. The region is entirely tropical, except for the subtropics of the extreme south, although temperatures are moderated by the continent's upland nature. Tropical forests fringe the Gulf of Guinea and extend inland through the Congo Basin and occur also in the eastern mountains. Deserts include the Sahara, the Kalahari, and the Somaliland coast. Most of the remainder of the continent supports scrub forest and grasslands.

Faunally, the Ethiopian region is most closely related to the Oriental region. Nevertheless, it shares a number of groups with Holarctica, whereas South American relationships are evident in its fish and turtle faunas.

Fishes. Among the fishes, the widely distributed cyprinids are abundant. The lungfish has South American affinities, as do a wealth of characins and the cichlids which have undergone extensive differentiation in the East African lakes. The primitive birchirs as well as several groups of primitive bony fishes are exclusives.

Amphibians. The amphibian fauna is not particularly remarkable. It includes the widely distributed caecilians, bufonids, and ranids. It shares the primitive pipids with South America. The true tree frogs are lacking, but their niche is filled by the rhacaphorids, which are also Oriental; some authorities consider the African representatives of this group to be a separate family restricted to the continent. A family of frogs related to the worldwide narrow-mouth frogs is exclusively African.

Reptiles. Although it includes such widely distributed groups as the skinks among the lizards and many harmless colubrid snakes, the reptile fauna is essentially pantropical. Notable examples of these are the side-necked turtles, crocodiles, gekkos, and worm snakes. Oriental affinities are evident in the chameleons, an egg-eating snake, the cobras, and the true vipers.

Birds. The avifauna, although large, is represented by few endemics above the generic level. The bulk of the avifauna is composed of widely distributed groups such as the owls and hawks, kingfishers, swallows, and true finches. The region has much in common with both Europe (Old World flycatchers, starlings, and orioles) and the Orient (broadbills, hornbills, and honey guides). Exclusives include the ostrich, the secretary bird, certain genera of guinea fowl, widow birds, and tick birds.

Mammals. The spectacular big game, such as the elephant, rhinoceros, hippopotamus, a variety of antelopes, horses (zebras), cats, and giraffes, hardly serve to bring out the true nature of the mammalian fauna. The region supports such widely distributed groups as mustelids, rabbits, and many rodents. Oriental affinities are indicated by rhinoceroses, Old World monkeys, great apes, scaly anteaters, and bamboo rats. Africa also possesses a number of endemic or near-endemic mammals of which the golden mole, elephant shrews, hyraxes, and the giraffe are best known.

The fauna of Madagascar and the islands of the Indian Ocean presents a number of interesting problems that are not likely to be settled in the immediate future. Madagascar supports a fauna that has much in common with Africa—side-necked turtles, rhacophorid frogs, chameleons, and many birds and bats. In contrast, many mainland groups are absent, such as primary fresh-water fishes, pythons and cobras, the ostrich and tickbirds, hystricomorph rodents, Old World monkeys, and great apes. The island possesses a number of endemics including a group of narrow-mouthed frogs, some rollers and vangas among the birds, and a variety of lemurs. A few Oriental relationships are also apparent.

The Seychelles and Mascarene Island groups of the Indian Ocean support small but interesting faunas. The flightless birds of the Mascarenes, especially the now extinct dodo, are particularly noteworthy.

Oriental region. This region encompasses tropical Asia from the Iranian Peninsula eastward through the East Indies to and including Borneo and the Philippines. Its exact boundaries, however, are difficult to define because of broad areas of transition between it and adjacent regions. Aside from the Thar Desert and isolated areas of semidesert, especially along the eastern coast of India, the region originally supported forest growth.

The region appears to have served both as a center of dispersal for cold-blooded vertebrates and as a crossroads through which various other groups have passed. Although its fauna bears much in common with the Ethiopian region, it shows both Palearctic and Australian affinities.

Fishes. Although ancient groups of fishes that are so characteristic of tropical regions are absent, its fish fauna is, nevertheless, large and diversified. Cypriniform fishes, though widely distributed, are especially characteristic of the region, and one group, the loaches, is almost exclusively Oriental. Several families of catfishes are endemic to it, and still others of this same group are shared with Africa. The climbing perch make up another distinctive group and further emphasize African affinities.

Amphibians. The amphibian fauna is large. Caecilians are present, although the salamanders barely enter the region. Among the tailless amphibians are the widely distributed bufonids, ranids, and narrow-mouthed frogs. It has many rhacophorid frogs which further attest to African faunal affinities.

Reptiles. The reptilian fauna is largely shared with other regions. Widely distributed groups include the skinks and harmless colubrid snakes, whereas such pantropical groups as the gekkos, worm snakes, and pythons are all well represented. It shares chameleons, egg-eating snakes, and the cobras with Africa, while the agamid lizards and true vipers are widely distributed through the Old World. The pit vipers are shared with the New World. The Oriental region also supports several small endemic families of snakes, and the surrounding seas are well populated with sea snakes.

Birds. The avifauna, like that of Africa with which it shares many groups, is poor in endemics above the generic level. Widely distributed groups such as woodpeckers, pigeons, and jays make up the bulk of the fauna, and pheasants are particularly characteristic of the region. A few Australian groups such as the frogmouths and wood swallows enter the Orient from the southeast. The region boasts only a single exclusive family, the fairy bluebirds.

Mammals. The mammal fauna, although represented by such wide-ranging groups as weasels, rabbits, and squirrels, includes a number of animals shared with Africa as well as a good representation of endemics. African affinities are evident in the Old World monkeys, the scaly anteaters, the rhinoceros, the elephant, and the fruit bats. Among the endemics may be mentioned the flying lemurs, tree shrews, and tarsiers.

The island archipelago to the east and southeast, which includes Sumatra, Java, and the Philippines, supports a vertebrate fauna that is very definitely Oriental in character. As is generally true of archipelagoes, the geographic patterns of various animal groups form a complex mosaic that is accompanied by depauperization, which is particularly evident in the northernmost islands of the Philippines.

Neotropical region. This area includes Mexico south of the Mexican Plateau, the West Indies, Central America, and South America. The first three are frequently treated as a transition zone between Nearctica and Neotropica.

Although about two-thirds of the region lies within the tropics, its plateau and montane character is responsible for considerable areas with nontropical temperatures. Generally speaking, the Amazon Basin and the eastern coasts support high rainforests. Scrub forests and grasslands clothe much of the interior of South America, whereas desert occupies much of the continent's western coast and Patagonia. The Andes and the Central American mountain systems provide a variety of vegetation belts.

Owing to a history of isolation through much of the Cenozoic Era, the South American fauna is composed of two very distinct elements. One is of considerable age with some pantropical groups. The other is a younger element that has invaded the region relatively recently from North America.

Fishes. The fish fauna, although represented by only a few major groups, is, nevertheless, very rich and with many endemics at the family level. Representatives of more ancient types include the pantropical osteoglossids and a lungfish which is shared with Africa. Other African affinities are evidenced by the characins, cichlids, and nandids. The widely distributed cyprinids are absent, and a number of other northern types barely enter Central America. Among the more notable endemics are the gymnotid (electric) eels and many families of catfishes.

Amphibians. The amphibian fauna includes a wealth of the widely distributed hylids and leptodactylids (these are especially abundant in Australia) and such pantropical or near-pantropical groups as the caecilians and brachycephalid frogs. With Africa it shares the pipid frogs and with the Orient the narrow-mouthed frogs. The wide-ranging ranids and salamanders are poorly represented. The region supports only a single exclusive frog family, the primitive *Rhinophrynus*, which is restricted to southern Mexico.

Reptiles. The reptile fauna includes skinks and harmless colubrid snakes among the near-cosmopolitan groups. Pantropical representatives include the side-necked turtles, some related to those of Africa and others to those of Australia, gekkos, worm snakes, and the coral snake family. It shares the pit vipers with Nearctica and the Orient. Other groups confined to the New World include the many iguanid lizards, also on Madagascar and one on Fiji, and the teiids. Endemic reptiles include several families of turtles that are restricted to northern Central America, the caimans, and the boas.

Birds. The rich avifauna has led to the designation of South America as the bird continent. Although many widely distributed groups such as herons, ducks, hawks, parrots, trogons, and thrushes are well represented, about 50% of the Neotropical families are endemic. About one-third of the bird species belong to the exclusive furnaroid group, antbirds, ovenbirds, and woodhewers, whereas the wood warblers, hummingbirds, and flycatchers, although shared to a greater or lesser extent with North America, make up most of the remaining endemic species. Other exclusives include the flightless rheas, tinamous, toucans, and cotingas.

Mammals. The mammal fauna further emphasizes South America's history of isolation. The list of endemics is long and includes several groups of marsupials, sloths, New World monkeys, and most of the hystricomorph families. With the Old World it shares the camels (llama) and tapir, and with North America the armadillos, peccaries, and pocket gophers. More wide-ranging groups include weasels, cats, squirrels, cricetid mice and rats, and some bats.

Central America and lowland Mexico is an area of faunal overlap between the Nearctic and Neotropical elements. The region is a pathway of considerable environmental diversity, and it has been utilized by northern groups dispersing southward and by southern groups dispersing northward, both to varying degrees. As a result a complex zoogeographic mosaic, still largely unstudied, obtains throughout the region.

The West Indies support a depauperate fauna of both Nearctic and Neotropical affinities. The representation of the various groups through the archipelago appears to be in direct proportion to their ability to cross water barriers. Although many schemes have been presented to explain the populating of the islands, most evidence indicates that most groups were transported across water barriers from Central America.

Australian region. Included in this region are continental Australia, New Guinea, Tasmania, and lesser islands through the Solomon group. The boundary between it and the Oriental region has long been debated. It was originally placed by Wallace (Wallace's Line) well to the West between Bali and Lombok, between the Celebes and Borneo, and south of the Philippines. The easternmost boundary now generally accepted lies just west of Aru and New Guinea. Modern zoogeographers recognize that the fauna of the archipelago between the two extremes is transitional. Schmidt refers to it as the Celebesian transition subregion, the Wallacea of many authors.

Continental Australia possesses both tropical and subtropical climates. Probably no less than 75% of its area supports desert or grassland cover. Forests and forested grasslands are restricted to the northern, eastern, and southwestern fringes of the continent. Eucalyptus forest, almost endemic to Australia, predominates in the southeast. New Guinea has a cover of considerable rainforest and wet mountain forest.

As a result of its long history of isolation, the region has a fauna that includes some very ancient endemic groups. These have survived here and in South America and have undergone considerable adaptive radiation. In addition, Australia also possesses a more recent element of Oriental affinities.

Fishes. The fish fauna is poor. Aside from an osteoglossid, an ancient group shared with South America, Africa, and southeastern Asia, and an exclusive family of lungfish, only distantly related to that of South America and Africa, the fishes of Australia are salt-tolerant and of wide distributions.

Amphibians. Of the amphibians, only the frogs are represented and these by only four families. In Australia proper the leptodactylids and hylids, both of which are almost cosmopolitan, make up about 90% of the amphibian fauna. New Guinea, in contrast, supports a wealth of the narrow-mouthed frogs.

Reptiles. Reptiles are well represented throughout the region. Pantropical groups include crocodiles, side-necked turtles, and gekkos, whereas skinks and colubrid snakes are even more wide-ranging. General Old World affinities are expressed by the agamid and veranid lizards and the pythons. Among the poisonous snakes the region possesses only the elapids. A single reptilian family, the pygopod lizards, is endemic.

Birds. The avifauna, although lacking many widely distributed families such as pheasants, woodpeckers, and finches, is represented by such near-cosmopolitan groups as the pigeons, kingfishers, and parrots. Endemics include the spectacular emu and cassowary (both flightless), the bird of paradise (especially characteristic of New Guinea), and the strictly Australian scrubbirds and lyrebirds.

Mammals. The mammalian fauna of the region is spectacular in the adaptive radiation displayed by the marsupials. Although not exclusive, they constitute the bulk of the mammal fauna. These and the monotremes testify to the long-continued isolation of Australia proper. Bats and the murid mice and rats are the only other mammalian groups native to the region. The dingo (wild dog) and the rabbit are introductions.

To the east of Australia and New Guinea, the faunas of the islands of the Pacific (Oceanic region of some zoogeographers) suffer gradual impoverishment. New Zealand to the southeast lacks strictly fresh-water fishes, turtles, snakes, and mammals, aside from bats and a few introduced species such as the deer. The islands have served as a refuge for such ancient groups as the lizard-like tuatara *Sphenodon*, a tailed frog which is shared with northwestern North America, and the moas and kiwis.

Toward the northeast strictly fresh-water fishes do not extend beyond New Guinea, terrestrial mammals and frogs reach only to the Solomons, snakes extend to Fiji, and lizards fall just short of Samoa. The Hawaiian Islands, aside from introduced forms, include among their vertebrate fauna only a single bat and a variety of endemic birds of New World affinities. *See* BIOGEOGRAPHY; BIOTIC ISOLATION; ISLAND BIOGEOGRAPHY; PLANT GEOGRAPHY.

[L. C. STUART]

Bibliography: P. J. Darlington, *Zoogeography*, 1957; K. P. Schmidt, Faunal realms, regions, and provinces, *Quart. Rev. Biol.*, 29(4):322–331, 1954; P. L. Sclater, On the general geographical distribution of the members of the class Aves, *J. Proc. Linnaean Soc. (Zool.)*, 2:130–145, 1858; A. R. Wallace, *The Geographical Distribution of Animals*, 2 vols., 1876.

Zoogeography

The subdivision of the science of biogeography that is concerned with the detailed description of the distribution of animals and how their past distribution has produced present-day patterns. This field attempts to formulate theories that explain the present distributions as elucidated by geography, physiography, climate, ecological correlates (especially vegetation), geological history, the canons of evolutionary theory, and an understanding of the evolutionary relationships of the particular animals under study. Zoogeographical theories are then tested by new data from all germane fields to amplify, verify, or falsify the constructs. In this sense, zoogeography is an integrative science that synthesizes data from other disciplines to apply to the realities of animal distribution.

Realities. The field of zoogeography is based upon five observations and two conclusions. The observations are: (1) Each species and higher group of animals has a discrete nonrandom distribution in space and time (for example, the gorilla occurs only in two forest areas in Africa). (2) Different geographical regions have an assemblage of distinctive animals that coexist (for example, the fauna of Africa south of the Sahara with its monkeys, pigs, and antelopes is totally different from the fauna of Australia with its platypuses, kangaroos, and wombats). (3) These differences (and similarities) cannot be explained by the amount of distance between the regions or by the area of the region alone (for example, the fauna of Europe and eastern Asia is strikingly similar although separated by 11,500 km of land, while the faunas of Borneo and New Guinea are extremely different although separated by a tenth of that distance across land and water. (4) Faunas strikingly different from those found today previously occurred in all geographical regions (for example, dinosaurs existed over much of the world in the Cretaceous). (5) Faunas resembling those found today or their antecedents previously occurred, sometimes at sites far distant from their current range (for example, the subtropical–warm temperate fauna of Eocene Wyoming, including many fresh-water fish, salamander, and turtle groups, is now restricted to the southeastern United States).

The conclusions are: (1) There are recognizible recurrent patterns of animal distribution that are not solely the product of areal extent, distance of separation, or climatic distinctiveness of a geographical region. (2) These patterns represent faunas composed of species and higher groups that have evolved through time in association with one another.

Approaches. Two rather different approaches have dominated the study of zoogeography since the beginning of the 19th century: ecological and historical. Ecological zoogeography attempts to explain current distribution patterns principally in terms of the ecological requirements of animals, with particular emphasis on environmental parameters, physiological tolerances, ecological roles, and adaptations. The space and time scales in this approach are narrow, and emphasis is upon the statics and dynamics of current or very recent events.

Historical zoogeography recognizes that each major geographical area has a different assemblage of species, that certain systematic groups of organisms tend to cluster geographically, and that the interaction of geography, climate, and evolutionary processes over a long time span is responsible for the patterns or general tracks. Emphasis in this approach is upon the statics and dynamics of major geographical and geological events ranging across vast areas and substantial time intervals of up to millions of years. The approach is based on concordant evolutionary association of diverse groups through time.

Ecological zoogeography is the study of animal distributions in terms of their environments; historical zoogeography is the study of animal distributions in terms of evolutionary history.

Ecological zoogeography. Ecological zoogeography is rooted within the discoveries of 19th-century plant ecologists and physiologists. They found that under similar conditions of temperature and moisture, terrestrial plants develop similar growth forms, regardless of their evolutionary relationships, to produce one of the following vegetation forms: forest, woodland, savanna, grassland, or scrub. Subsequently it was demonstrated that these units showed differentiation associated with major soil types and broad latitudinal climatic regions. Each of these main kinds of vegetation is called a formation type (such as tropical lowland evergreen forest), and its geographical subdivisions are formations (such as South American tropical lowland evergreen forest; African tropical lowland evergreen forest).

Later investigators realized that within each formation a series of animals had evolved to undertake homologous ecological functions in the dynamics of the community, so that the concept of biome-type (vegetation and associated animals) was developed. The biome-type is a series of major geographical climatic regions characterized by similar ecological adaptations in plants and animals. For example, the grassland biome-type may be typified by the comparisons in the table. The biome-types are distinctive from the zoological point of view in that ecological equivalents (such as top predator) in the different biomes are usually from phylogenetically nonrelated stocks. M. D. F. Udvardy provides the best recent summary of the general distribution of world terrestrial biomes.

The concept of biomes is a method of generalizing the distribution of animals by major environments on a latitudinal basis. The ecological zoogeographer is also interested in vertical (altitudinal and

Comparison of three grassland biomes

Ecological roles	North American	African	Australian
Top predator	Wolf (dog)	Lion (cat)	Tasmanian wolf (marsupial)
Large herbivore	Bison (cattle)	Zebra (horse)	Red kangaroo (marsupial)
Small predator-scavenger	Coyote (dog)	Hyaena (hyaena)	Tasmanian devil (marsupial)

bathymetric) zonation as a feature of animal distribution. The idea of zonation is based upon the recognition that there are ecoclinal gradients in the parameters of the environment (especially temperature) with an increase in altitude or in ocean depth. The composition of species distributions along these ecoclines, because of differences in ecological requirements, produces recognizable and characteristic life zones. As an example, Fig. 1 shows the principal life zones in the ocean. These zones are divided into two groups: benthos or substrate zones and pelagic or free-swimming zones. Altitudinal life zones on land have been similarly described, most recently and completely by L. R. Holdridge.

Historical zoogeography. This approach has its origins in systematic biology. Workers in this field recognized very early in the 19th century that different geographical regions support different faunas and that representatives of these faunas occur in a wide variety of environments (for example, apes and monkeys in Africa in rainforest to desert environments). In addition, they noted that different systematic groups tend to cluster geographically (kangaroos in Australia and anteaters in South America). With the development of the canons of evolutionary theory later in the century, a framework for understanding the evolution of faunas through time became the backbone of historical zoogeography.

The raw data of historical zoogeography are the distributions or tracks of individual species of animals in space (geographical ecology) and time. Because each species has its own set of peculiar ecological requirements and its own unique evolutionary history, each has a discrete nonrandom ecogeographical distribution. As a consequence, no species is universally present, and many species have small or unique tracks. The first level of generalization in zoogeography is based on the recognition that, in spite of the unique nature of individual species distributions, many tracks are concordant or show a common pattern. Determination of patterns involving the coincident distribution of species or monophyletic groups (genera, families, and so on) of species (generalized tracks) is the fundamental step in zoogeographical analysis.

The second level of generalization in this process is to cluster the strongly recurrent generalized tracks involving extensive geographical areas, whose components are then regarded as the major modern faunas. A third level of generalization attempts to tentatively identify the historical source units (ancestral faunas) that have contributed to the modern patterns. The terrestrial biogeographic realms in Fig. 2 are generally accepted to show clustering of major modern land faunas. J. C. Briggs has provided an up-to-date summary of marine biogeographical units.

Zoogeographical dynamics. The early workers in zoogeography concentrated their effort on the description of patterns of animal distribution and the clustering of patterns into large units of distribution by major environment or major fauna. These static approaches, while useful in phrasing and structuring data, have proved inadequate models for understanding scientific zoogeography since they address only the first of four key elements: (1) recognizing common patterns of distribution; (2) analyzing these patterns to determine common ecological or evolutionary processes that produced the patterns; (3) using the patterns and processes as a prediction of patterns for as yet unstudied groups and (4) for as yet undiscovered geographical and evolutionary events.

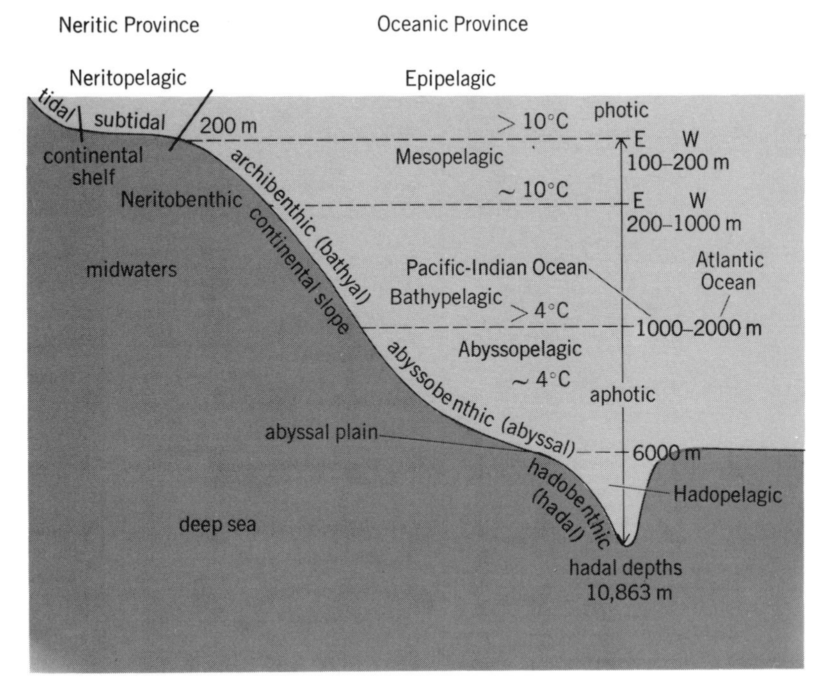

Fig. 1. Life zones in the sea.

The static approach ultimately had a stultifying effect on the development of zoogeography since by the middle of the 20th century descriptive and narrative zoogeography had run out of new ideas. Fortunately during the last decade the science has been revitalized by a resurgence of interest in the dynamics of animal distribution. This interest has led to the development of two major new theories of zoogeographical explanation, one ecological (the dynamic equilibrium theory) and one historical (the vicariance theory). The development of the latter theory has forced a reexamination of the previously dominant historical zoogeographical explanation (the dispersion theory), most effectively expounded earlier in the century by W. D. Matthew and G. G. Simpson.

Dynamic equilibrium theory. This theory of zoogeography was developed during the 1960s primarily by R. H. MacArthur and E. O. Wilson. It originally was aimed at developing predictive mathematical models that would explain the differences in numbers of species on islands of differing sizes and differing distances from the closest mainland source areas. Since almost all habitats on the mainland are patchy in distribution as well (meadows, lakes, mountaintops, and so forth), the theory can also be applied to any discontinuous (insular) segments of the same environment type.

The central axiom of the theory is that the fauna of any disjunct ecological area is a dynamic equilibrium between immigration of new species into the area and extinction of species already present. Species number is thus constant over ecological time, while evolution will act gradually over geo-

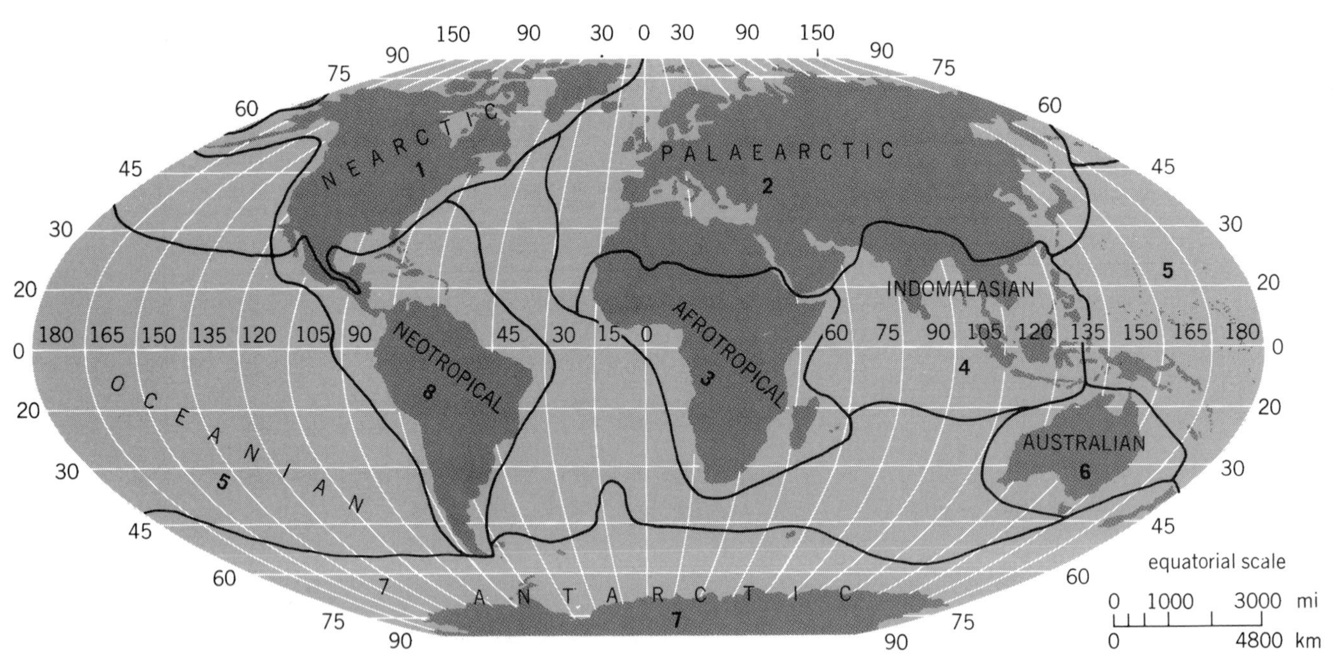

Fig. 2. Terrestrial biogeographic realms of the world. *(From M. D. F. Udvardy, A classification of the biogeographical provinces of the world, Int. Union Conserv. Nat. Occas. Pap., 18:1–49, 1975)*

logical time to increase the equilibrium number of species. From this base it is possible to construct equilibrium models that predict the interactions of distance from the source area, areal extent of the disjunct area, and immigration versus extinction, as seen in Fig. 3. Data from well-known insular faunas supported the value of these models. Subsequently, controlled defaunization experiments on small islands and study of the immigration process through time confirmed the predictive power of the theory.

Several conclusions may be derived from these models and have been confirmed in the field: (1) Distant disjunct areas will have fewer species than those close to a source area. (2) Small disjunct areas will have fewer species than large ones. (3) Distant disjunct areas take longer to reach equilibria when originally sterile or defaunated than do those close to the source area. (4) The smaller the area and the closer the area source to the area, the higher the turnover rate (change in species composition). The essential insight of equilibrium theory is that local extinctions and immigrations are relatively frequent events. Development of more complex equilibrium models and extension of the approach to other areas of zoogeography are actively under way and promise to produce exciting new views of distribution events.

Dynamic historical biogeography. This field is currently undergoing a major revolution of thought stimulated by new knowledge of global tectonics based on the now generally accepted theory of continental drift as outlined by R. S. Dietz and J. C. Holden. Simply stated, theorists now believe that the major continents were formerly welded together as a supercontinent (Pangaea) that began to rift apart in the Triassic, about 190,000,000 years before present. By the Early Jurassic, northern (Laurasia) and southern (Gondwanaland) land masses had drifted apart. During the Cretaceous these masses fragmented further, and in the Cenozoic several southern segments became attached to the northern continents (Africa to Eurasia, India to Eurasia, and South America to North America).

Previously, historical zoogeography had been dominated by the ideas of Matthew and Simpson, who believed in the permanency of the ocean basins and continents. These authors developed the idea that major groups originated on the northern continents and dispersed southward. P. J. Darlington developed a slightly different point of view and suggested that major groups arose in the Asian tropics and dispersed elsewhere on the continental masses across land bridges or marine barriers by island hopping. According to these kinds of ideas, the present-day distribution of lungfishes (tropical Africa, South America, and Australia) involved origin in the old world (Asia?) and immigration across land bridges or by island hopping to the southern land masses.

Recently a group of zoogeographers, L. Croziat, G. Nelson, and D. E. Rosen, much influenced by the role of continental drift in geography, developed and extended some ideas originally put forward by Croziat into a new theoretical construct. This theory focuses not on the dispersion of organisms from centers of origin, but on the idea that current zoogeographical patterns are the result of the fragmentation of previously continuous tracks by major physiographical change. Thus the present distribution of lungfishes is the result of the fragmentation of a previously continuous track by the breakup of Gondwanaland in the Cretaceous.

Essentially, this last development creates a controversial dichotomy in zoogeographical thought. One view emphasizes the active movement (dispersion) of animals as the principal agent re-

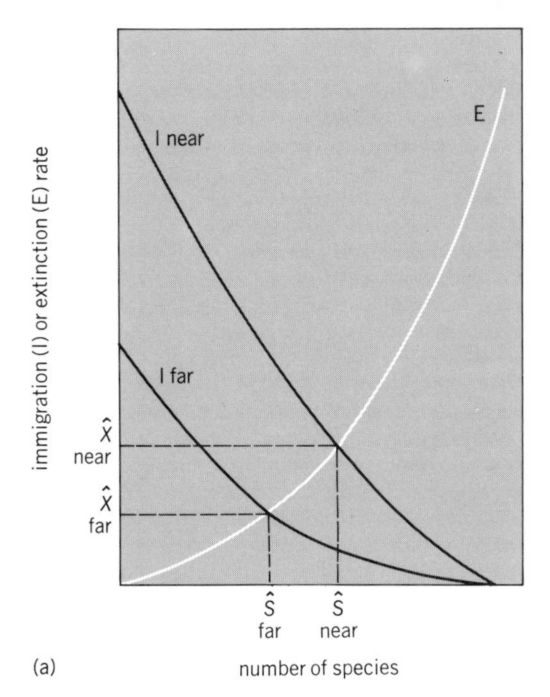

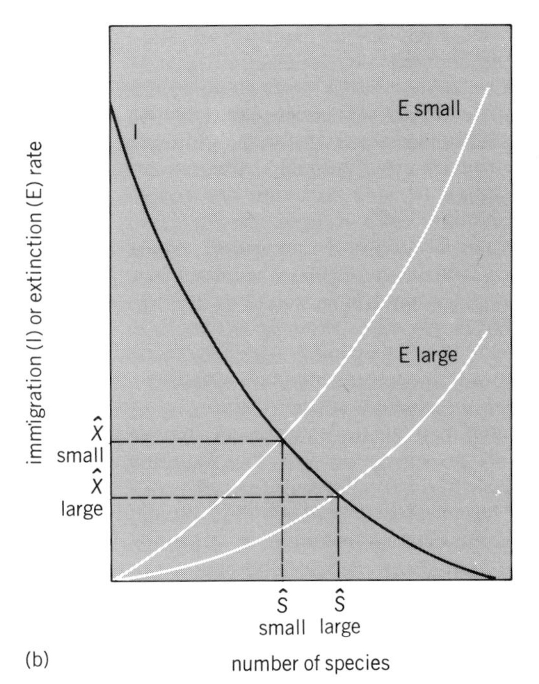

Fig. 3. A disjunct fauna is an equilibrium in ecological time between immigration of new species and extinction of those already present. *(a)* Distance effect; a near island has large equilibrium number of species ($\hat{S}$) and turnover rate ($\hat{X}$). *(b)* Area effect; a large island has larger $\hat{S}$ and smaller $\hat{X}$. *(After D. S. Simberloff, Equilibrium theory of island biogeography and ecology, Ann. Rev. Ecol. Syst. 5:161–182, 1974)*

sponsible for patterns; the other regards dispersion as unimportant and emphasizes the movement and fragmentation of the land masses and the relative immobility of animals as responsible for patterns. The latter position is called the vicariance theory, in distinction to the dispersion theory. An active, ongoing, vigorous interchange of ideas, with serious reexamination and critique of both schools, is the present theme of zoogeography.

The essential features of the dispersion and vicariance theories are as follows.

Dispersion theory: (1) A monophyletic group arises at a center of origin. (2) Each group disperses from this center. (3) Substantial numbers of monophyletic groups followed the same dispersal route at about the same time to contribute to the composition of a modern fauna. (4) A generalized track corresponds to a dispersion route. (5) Each modern fauna represents an assemblage derived from one to several historical source units. (6) Direction of dispersal may be deduced from tracks, evolutionary relations, and past geodynamic and climatic history. (7) Climate or physiographical change provides the major impetus or opportunity for dispersal. (8) Faunas were shaped by dispersion across barriers and subsequent evolution in isolation. (9) Dispersion is the key to explaining modern patterns: related groups separated by barriers have dispersed across them.

Vicariance theory: (1) Vicariants (allopatric species) arise after barriers separate parts of a formerly continuous population. (2) Substantial numbers of monophyletic groups are simultaneously affected by the same vicariating events (geographical barrier formation). (3) A generalized track estimates the faunal composition and geographical distribution of an ancestral biota before it subdivided (vicariated) into descendant faunas. (4) Each generalized track represents a historical source unit. (5) Sympatry of generalized tracks reflects geographical overlap of different faunas due to dispersal. (6) The primary vicariating events are changes in world geography (geodynamics) that subdivided ancestral faunas. (7) Faunas evolve in isolation after barriers arise. (8) Vicariance is of primary significance in understanding modern patterns: related groups separated by barriers were fragmented by the appearance of the barriers. *See* ECOLOGY, ZOOGEOGRAPHIC REGION.

[JAY M. SAVAGE]

Bibliography: J. C. Briggs, *Marine Zoogeography*, 1974; L. Croizat, *Space, Time, Form: The Biological Synthesis*, 1962; L. Croizat, G. Nelson, and D. E. Rosen, Centers of origin and related concepts, *Syst. Zool.*, 23: 265–287, 1974; P. J. Darlington, *Zoogeography: The Geographical Distribution of Animals*, 1957; R. S. Dietz and J. C. Holden, The breakup of Pangaea, *Sci. Amer.*, 223(4):30–41, 1970; J. W. Hedgepeth, Classification of marine environments, *Mem. Geol. Soc. Amer.*, 67(1): 17–27, 1957; L. R. Holdridge, *Life Zone Ecology*, 1967; R. H. MacArthur and E. O. Wilson, *The Theory of Island Biogeography*, 1967; W. D. Matthew, Climate and evolution, *Ann. N.Y. Acad. Sci.*, 24: 171–318, 1915; P. Müller, *The Dispersal Centers of Terrestrial Vertebrates in the Neotropical Realm*, 1973; J. M. Savage, The Geographic Distribution of Frogs: Patterns and Predictions, *Evolutionary Biology of the Anurans: Contemporary Research on Major Problems*, chap. 13, 1973; D. S. Simberloff, Equilibrium theory of island biogeography and ecology, *Ann. Rev. Ecol. Syst.*, 5:161–182, 1974; G. G. Simpson, *The Geography of Evolution*, 1965; M. D. F. Udvardy, A classification of the biogeographical provinces of the world, *Int. Union Conserv. Nat. Occas. Pap.*, 18:1–49, 1975.

Zooplankton

The zooplankton is the assemblage of animals in an aquatic environment that is composed of the forms which occur suspended in the water and have such limited powers of locomotion that they drift passively with the water movements. Except in certain weak swimming larval stages, the fish and the few other animals that actively control their own movements in the water are excluded from the zooplankton and are referred to collectively as nekton. Many animal species occur only as part of the zooplankton and spend their entire life cycle in this assemblage. Many other species occur in the zooplankton during part of their life cycle but spend the rest of their time as part of a different assemblage, usually the bottom community.

The zooplankton is an important component in most aquatic environments and is present in a variety of freshwater and marine habitats, from small ponds to large lakes, from the upper reaches of brackish estuaries to the deep oceans. In flowing-water environments, such as streams and rivers, the zooplankton forms are much less important because they can not maintain themselves and are continuously swept downstream. Often, however, a river will contain appreciable numbers of zooplankton organisms because there are bodies of standing water along its course that contribute a steady supply of such forms in the outflow waters.

Composition. The zooplankton of the oceans is composed of a great variety of forms. Almost every animal phylum is represented by at least a few species. Single-celled animals of many types (Protozoa), jellyfish and their relatives (Coelenterata), comb jellies (Ctenophora), polychaete worms (Annelida), snails with reduced or absent shells (Mollusca), transparent arrow worms (Chaetognatha), primitive members of man's own phylum (Chordata), and especially crustaceans of a great variety of species (Arthropoda) are the most commonly encountered forms in the open-ocean zooplankton. In fact, two of these groups, the arrow worms and the comb jellies, are found only as part of this assemblage.

Especially in nearshore areas the eggs and immature stages of bottom-dwelling animals form an important element in the marine zooplankton. Many bottom animals produce eggs or young which join the zooplankton for part of their existence and then settle to the bottom, where they complete their development. The early stages of starfish, oysters, snails, clams, barnacles, crabs, lobsters, worms of various kinds, and many other forms are found in the zooplankton. The existence of a planktonic stage for these animals aids their dispersion and explains how these forms can quickly develop on newly available habitats such as dock pilings and buoys that may be many miles from the nearest adults.

In contrast to the oceanic assemblage, the freshwater zooplankton is much more limited in variety. This is because many of the groups that are prominent in the seas are unable to survive in fresh water. Further, freshwater bottom-dwelling forms seldom possess planktonic immature stages. The limited variety of the freshwater assemblage is shown by the fact that three groups provide the great majority of the species. Two of these groups, the protozoans and the crustaceans, are also major components of the marine assemblage. The third group, the wheel animals or rotifers (Rotifera), are also present in the seas but usually only as a minor component.

One noteworthy observation concerning the zooplankton is the almost complete absence of representatives from that most abundant and diversified group, the insects. Insects are, with very minor exceptions, entirely lacking from the marine zooplankton. In fresh water, although the larvae of the phantom midge *(Chaoborus)* are of some importance in ponds and small lakes, the insects as a group are also of very minor significance.

Occurrence. Although the zooplankton is well represented in most aquatic habitats, the exact species present vary greatly in relation to the specific environmental conditions. Physical-chemical factors, such as temperature, light intensity, amount of dissolved solids (salinity), concentrations of heavy metals, and pH, play important roles in determining what species will occur in an environment. Biotic factors, such as availability of food, prevalence of predators, and abundance of competitors, also play vital roles in controlling species composition.

A physical factor of special importance in determining species composition is depth. Many species are present only near the surface, while other forms, often rather bizarre in appearance, are limited to a particular stratum of the subsurface waters. In both lakes and the oceans the depth range within which most of the individuals of a population occur may vary somewhat during the course of a day due to vertical migrations. These movements are controlled by fluctuations in light intensity as well as other factors and often result in the individuals' ascending at night and descending during the daylight hours.

Collection and identification. With the exception of certain of the jellyfish and related forms, the members of the zooplankton are almost all microscopic. Their small size in conjunction with their dispersion in great volumes of water concealed the existence of this assemblage until the early 1800s. The development that revealed the vast community of planktonic animals and plants was the introduction of small tow nets for catching and concentrating the individuals. Such nets were made of fine silk gauze and were towed from boats in such a way that the organisms were strained from the water and retained in a small jar attached to the end of the net.

Such tow nets are still the most common equipment for collecting zooplankton, although newer materials such as nylon have largely replaced silk in the gauze. The organisms collected in a net haul are usually identified and counted. The meaning of these counts is often difficult to interpret, however, for it is usually somewhat unclear how much water was strained during the haul and thus what concentration of the animals in their environment. Also, many of the smaller individuals may have escaped through the meshes of the net, while larger forms may have seen or felt its approach and avoided capture altogether. A further problem with the use of nets is that the process of capture often severely damages some of the delicate forms such as the jellyfish, comb jellies, and primitive chordates. Newer techniques of collection are being developed to improve scientists' ability to study the zooplankton populations. Pumping large volumes of water aboard a ship and then concentrating the animals, or determining the amount of zooplankton at different depths by measuring the strength of the acoustical signals reflected from the different layers of the water column, are examples of more modern methods.

After capture the zooplankton animals are usually preserved for later examination under a microscope. Formalin and ethyl alcohol are the most commonly used preservatives. The identification to species of the preserved animals in a zooplankton sample is usually an arduous undertaking. Many tens of thousands of zooplankton species occur in the sea, a sizable number of them as yet undescribed. Even when the adults are known, the young stages are often undescribed. For general ecological studies it is usually just not possible to carry the identifications to the species level, and the animals are segregated into more general groupings. For special studies or where specific identifications are absolutely necessary, experts with many years of experience are usually called upon.

Feeding relationships. The zooplankton organisms play a vital role in the ecology of aquatic environments, especially large bodies of water such as lakes and the oceans. Members of the zooplankton are the major herbivores in these ecosystems. They filter out or otherwise capture the algae and other plants in the plankton and consume these cells as their food. The herbivorous individuals are preyed upon, in turn, by larger zooplankters (often themselves taken by larger predators), small fish, certain of the bottom-dwelling forms, and even some of the largest aquatic organisms such as basking sharks and whalebone or baleen whales. Thus, the zooplankton functions as an intermediate link in the food chain, linking the plants that produce organic matter with the larger forms such as fish. *See* FOOD CHAIN.

The herbivorous zooplankters tend to be concentrated close to the surface in large bodies of water. In the deeper waters there is not enough light for plants to grow. Thus, the herbivorous forms cannot find food there. This does not mean, however, that the deep waters of large lakes and the oceans are devoid of zooplankton. On the contrary, there are many marine forms and a few freshwater species that occur only at these depths. Dead and dying individuals from the plankton of the surface layers tend to sink into the depths. These organic particles furnish a food source for many of the deepwater forms. Other species are carnivorous and feed on the scavengers or each other.

Reproduction. As might be expected with an assemblage that contains a great variety of forms, the modes of reproduction vary greatly among the zooplankton groups. One general characteristic, however, is the production of large numbers of young. In general, the young zooplankters receive very little parental care. Further, the characteristics of the zooplankter's environment leave the young with no place to hide from predators. Thus, a large number must be produced to ensure that a few survive to maturity. Many zooplankton forms of the surface layers are also aided in survival by being nearly transparent and thereby difficult for the predators to detect.

The zooplankton forms, at least in fresh water and the upper layers of the temperate and polar seas, tend to have restricted periods of reproduction. During these periods one or more generations are produced, whereas reproduction stops or is at a very low level for the rest of the year. The periods of reproduction seem to be at least partially related to seasonal fluctuations in temperature and availability of food. Such seasonal fluctuations of environmental factors are much less prominent in the deep oceans and tropical seas, and there reproduction often seems to be more evenly spaced throughout the year.

[ANDREW ROBERTSON]

Bibliography: A. C. Hardy, *The Open Sea: The World of Plankton*, 1956; G. E. Hutchinson, *A Treatise on Limnology*, vol. 2: *Introduction to Lake Biology and the Limnoplankton*, 1967; T. R. Parsons et al., *Biological Oceanographic Processes*, 1978; W. D. Russell-Hunter, *Aquatic Productivity*, 1970.

LIST OF CONTRIBUTORS

List of Contributors

A

Ackerman, Dr. Edward A. *Deceased; formerly, Carnegie Institution of Washington.* LAKE; SOIL CONSERVATION — coauthor.

Alderfer, Dr. Russell B. *Professor of Soils and Crops Department, College of Agriculture and Environmental Science, Rutgers University.* SOIL — in part.

Aldrich, Dr. Daniel G., Jr. *Chancellor, University of California, Irvine.* SOIL — in part.

Allen, Dr. L. H., Jr. *University of Florida.* CROP MICROMETEOROLOGY.

Anthes, Richard A. *Department of Meteorology, College of Earth and Mineral Sciences, Pennsylvania State University.* WEATHER FORECASTING AND PREDICTION — in part.

Armstrong, Francis A. J. *Freshwater Institute, Winnipeg.* SEA-WATER FERTILITY — in part.

Armstrong, Dr. Richard L. *Institute of Arctic and Alpine Research, University of Colorado.* AVALANCHE.

Ash, Col. Simon H. *Consulting Civil and Mining Engineer, Santa Rosa, CA.* MINING — in part.

Austin, Prof. James M. *Professor of Meteorology, Massachusetts Institute of Technology.* WEATHER FORECASTING AND PREDICTION — in part.

Avallone, Prof. Eugene A. *Department of Mechanical Engineering, City College of the City University of New York.* ENGINEERING, SOCIAL IMPLICATIONS OF.

B

Badgley, Dr. Peter C. *Director, Earth Sciences Division, Office of Naval Research, U.S. Department of the Navy.* REMOTE TERRAIN SENSING.

Baker, Dr. Herbert G. *Department of Botany, University of California, Berkeley.* Validator of POPULATION DISPERSAL.

Baker, Dr. Victor R. *Department of Geological Sciences, University of Texas at Austin.* GEOMORPHOLOGY; RIVER.

Barker, Dr. Joseph W. *Chairman of the Board (retired), Research Corporation, New York.* ENGINEERING.

Barnes, Dr. Burton V. *School of Natural Resources, University of Michigan.* FOREST ECOLOGY.

Beistline, Dr. Earl H. *Dean, College of Earth Sciences and Mineral Industry, University of Alaska.* PLACER MINING.

Benfer, Neil A. *BOMAP Scientific Editor, Barbados Oceanographic and Meteorological Analysis Project.* SEA STATE; SEA WATER — in part.

Bennett, Dr. Orus L. *Supervisory Soil Scientist, Science and Education Administration, U. S. Department of Agriculture, and Department of Plant Science, West Virginia University.* LAND RECLAMATION; STRIP MINING.

Benninghoff, Dr. William S. *Department of Botany, University of Michigan.* AEROBIOLOGY; TUNDRA.

Benson, Dr. Bruce B. *Department of Physics, Amherst College.* SEA WATER — in part.

Bergeron, Prof. Tor. *Deceased; formerly, Institute of Meteorology, Uppsala.* METEOROLOGY — in part.

Berkofsky, Prof. Louis. *Institute for Desert Research, Ben-Gurion University of the Negev, Beersheva.* DESERTIFICATION.

Billings, Dr. William D. *Department of Botany, Duke University.* PHYSIOLOGICAL ECOLOGY (PLANT).

Billington, Dr. Douglas S. *Senior Staff Advisor for Materials Science, Metals and Ceramics Division, Oak Ridge National Laboratory.* RADIATION DAMAGE TO MATERIALS.

Bjerknes, Dr. Jacob. *Deceased; formerly, Professor Emeritus, Department of Meteorology, University of California, Los Angeles.* CLIMATIC CHANGE — in part.

Black, Dr. Robert F. *Department of Geology, University of Connecticut.* PERMAFROST.

Blad, Prof. Blaine L. *Institute of Agriculture and Natural Resources, University of Nebraska.* EVAPOTRANSPIRATION.

Blair, Lewis N. *Anaconda Company, Butte.* ENVIRONMENTAL ENGINEERING — coauthor.

Blanchard, Dr. Duncan C. *State University of New York.* SEA WATER — in part.

Bondurant, Donald C. *U.S. Army Engineer Division, Missouri River, Omaha.* RIVER ENGINEERING — coauthor.

Borgstrom, Dr. Georg. *Department of Food Science and Human Nutrition, Michigan State University.* AGRICULTURAL GEOGRAPHY.

Brinton, Dr. Edward. *Scripps Institution of Oceanography, La Jolla.* SEA-WATER FERTILITY — in part.

Broadbent, Carl D. *Manager, Mining Engineering, Kennecott Copper Corporation, Salt Lake City.* OPEN-PIT MINING.

Buettner, Dr. Konrad J. K. *Deceased; formerly, Pierce Foundation, Yale University.* BIOCLIMATOLOGY.

C

Cadle, Dr. Richard D. *Department of Atmospheric Chemistry, National Center for Atmospheric Research, Boulder.* SMOG.

Cain, Dr. Stanley A. *Director, Institute for Environmental Quality, and Charles Lathrop Pack Professor, Department of Resource Planning and Conservation, University of Michigan.* CONSERVATION OF RESOURCES; PLANT FORMATIONS, CLIMAX.

Carritt, Dr. Dayton E. *American Dynamics International, Inc., Fort Lauderdale.* SEA WATER — in part.

Carter, Archie N. *President, Carter, Krueger and Associates, Inc., Minneapolis.* HIGHWAY ENGINEERING.

Chanlett, Emil T. *Professor of Sanitary Engineering, Department of Environmental Sciences & Energy, University of North Carolina, Chapel Hill.* ENVIRONMENTAL PROTECTION — feature.

Chapman, Dr. Valentine J. *University of Auckland.* SALTMARSH.

Chenault, Roy L. *Chief Research Engineer (retired), Oilwell Division, U. S. Steel Corporation.* OIL AND GAS FIELD EXPLOITATION.

Christensen, Dr. Clyde M. *Professor of Plant Pathology, University of Minnesota.* PLANT DISEASE — in part.

Christian, C. S. *Member of the Executive, CSIRO, Canberra City, Australia.* LAND-USE CLASSES.

Clarke, Dr. George L. *Professor of Biology, Harvard University, and Marine Biologist, Woods Hole Oceanographic Institution.* SEA WATER — in part.

Cochran, Robert A. *Shell Development Company, Bellaire Research Center, Houston.* WATER POLLUTION — in part.

Cochrane, Dr. John D. *Department of Oceanography, Texas A&M University.* SEA WATER — in part.

Cole, Dr. Lamont C. *Professor of Ecology, Division of Biological Sciences, Cornell University.* POPULATION DYNAMICS.

Cooper, Dr. Arthur W. *Department of Botany, North Carolina State University, Raleigh.* ECOLOGICAL SUCCESSION; LIFE ZONES; PLANT SOCIETIES—coauthor; PLANTS, LIFE-FORMS OF.

Corlett, A. V. *Deceased; formerly, Mining Engineer, Kingston, Ontario.* MINING—in part.

Coroniti, S. C. *Climatic Impact Assessment Program, Office of the Secretary of the Treasury.* AIR POLLUTION—coauthor.

Coupland, Dr. Robert T. *Department of Plant Ecology, University of Saskatchewan.* GRASSLAND ECOSYSTEM—coauthor.

Court, Dr. Arnold. *Professor of Climatology, Department of Geography, San Fernando Valley State College.* FROST; SENSIBLE TEMPERATURES; TEMPERATURE INVERSION.

Crafts, Dr. Alden S. *Department of Botany, University of California, Davis.* DEFOLIANT AND DESICCANT.

Craig, Dr. Harmon. *Professor of Geochemistry, Scripps Institution of Oceanography, La Jolla.* SEA WATER—in part.

Crow, Loren W. *Certified Consulting Meteorologist, Denver.* INDUSTRIAL METEOROLOGY.

Cruickshank, Michael J. *Geological Survey, U. S. Department of the Interior, Reston, VA.* MARINE MINING.

Currie, Ronald I. *Secretary, Scottish Marine Biological Association, and Director, Dunstaffnage Marine Research Laboratory, Oban.* SEA-WATER FERTILITY—in part.

Cushing, Dr. D. H. *Fisheries Laboratory, Ministry of Agriculture, Fisheries and Food, Lowestoft, England.* SEA-WATER FERTILITY—in part.

D

Dale, Dr. Walter M. *Deceased; formerly, Professor and Head, Department of Biochemistry, Christie Hospital, Manchester, England.* RADIATION BIOCHEMISTRY.

Dana, Dr. Samuel T. *Dean Emeritus, School of Natural Resources, University of Michigan.* FOREST AND FORESTRY.

Dansereau, Dr. Pierre. *Professor of Ecology, Institut d'Urbanisme, Université de Montréal.* VEGETATION; VEGETATION ZONES, WORLD.

Davies-Jones, Robert. *National Severe Storms Laboratory, National Oceanic and Atmospheric Administration, Norman, OK.* THUNDERSTORM—in part.

Davis, C. H. *Assistant Director of Chemical Development, Tennessee Valley Authority.* FERTILIZER.

Davis, Dr. John H. *Professor Emeritus of Biology, University of Florida.* DUNE VEGETATION; MANGROVE SWAMP—in part.

Davis, Prof. Kenneth P. *School of Forestry, Yale University.* FOREST MANAGEMENT AND ORGANIZATION.

DeBach, Dr. Paul. *Professor of Biological Control, University of California, Riverside.* INSECT CONTROL, BIOLOGICAL.

Deland, Dr. Raymond J. *Department of Meteorology and Oceanography, New York University.* AIR PRESSURE.

Dice, Dr. Lee R. *Professor Emeritus of Zoology, University of Michigan.* HUMAN ECOLOGY.

Dietz, Dr. Robert S. *Atlantic Oceanographic and Meteorological Laboratories, Environmental Science Services Administration, U. S. Department of Commerce, Miami.* CONTINENTAL SHELF AND SLOPE.

Doty, Dr. Robert. *Intermountain Forest and Range Experiment Station, Ogden, UT.* HYDROLOGY—in part.

Doviak, R. J. *National Severe Storms Laboratory, National Oceanic and Atmospheric Administration, Norman, OK.* STORM DETECTION—in part.

Drosdoff, Dr. Matthew. *Professor of Soil Science, Department of Agronomy, Cornell University.* SOIL—in part.

Duley, Dr. Frank L. *Principal, Agricultural College, University of Peshawar.* SOIL—in part.

Dunn, Dr. James R. *Dunn Geoscience Corporation, Latham, NY.* ENVIRONMENTAL GEOLOGY.

Duntley, Dr. Seibert Q. *Director, Visibility Laboratory, Scripps Institution of Oceanography, La Jolla.* SEA WATER—in part.

Dyer, Dr. K. R. *Institute of Oceanographic Sciences, Somerset, England.* ESTUARINE OCEANOGRAPHY.

E

Easterbrook, Don J. *Department of Geology, Western Washington University, Bellingham.* AEOLIAN LANDFORMS.

Ebert, Dr. M. *Holt Radium Institute and Christie Hospital, Manchester, England.* Validator of RADIATION BIOCHEMISTRY.

Egler, Dr. Frank. *Director (retired), Aton Forest, Norfolk, CT.* PLANT GEOGRAPHY; VEGETATION MANAGEMENT

Ehricke, Dr. Krafft A. *Autonetics Division, North American Rockwell Corporation, Anaheim, CA.* SOLAR ENERGY.

Eide, Dr. Carl J. *Department of Plant Pathology, Institute of Agriculture, University of Minnesota.* PLANT DISEASE; PLANT DISEASE CONTROL—both in part.

Eliassen, Prof. Arnt. *Institute of Geophysics, University of Oslo.* METEOROLOGY—in part.

Ellis, Dr. A. J. *Director, Chemistry Division, Department of Scientific and Industrial Research, Petone, New Zealand.* GEOTHERMAL POWER—in part.

Evans, Prof. Francis C. *Professor of Zoology, University of Michigan.* ANIMAL COMMUNITY; POPULATION DISPERSION.

Ewing, Dr. Gifford. *Woods Hole Oceanographic Institution.* SEA WATER—in part.

F

Fisher, Dr. F. H. *Scripps Institution of Oceanography, La Jolla.* SEA WATER—in part.

Fosberg, Dr. F. R. *Smithsonian Institution.* ISLAND BIOGEOGRAPHY.

Frank, Prof. Bernard. *Deceased; formerly, Professor of Watershed Management, Colorado State University.* WATER CONSERVATION—in part.

French, Prof. David W. *Department of Plant Pathology, University of Minnesota.* TREE DISEASES.

Frey, Prof. David G. *Department of Zoology, Indiana University.* FRESH-WATER ECOSYSTEM.

Fuhrman, Dr. Ralph E. *Black and Veatch, Consulting Engineers, Washington, DC.* WATER POLLUTION—in part.

Fulks, J. R. *National Weather Service, Chicago.* DEW POINT; HUMIDITY.

G

Galloway, Dr. William J. *Bolt, Beranek and Newman, Inc., Canoga Park, CA.* NOISE.

Gauch, Dr. Hugh G. *Department of Botany, College of Agriculture, University of Maryland.* PLANT, MINERALS ESSENTIAL TO.

Georges, T. M. *Wave Propagation Laboratory, National Oceanic and Atmospheric Administration, Boulder.* STORM DETECTION—in part.

Giles, Prof. Robert H., Jr. *Division of Forestry and Wildlife Science, College of Agriculture, Virginia Polytechnic Institute.* WILDLIFE CONSERVATION—in part.

Goldberg, Dr. Edward D. *Scripps Institution of Oceanography, La Jolla.* HYDROSPHERE; HYDROSPHERIC GEOCHEMISTRY; SEA WATER—in part.

Golley, Dr. Frank B. *Institute of Ecology, Athens, GA.* APPLIED ECOLOGY.

Goodfriend, Lewis S. *Lewis S. Goodfriend and Associates, Cedar Knolls, NJ.* NOISE POLLUTION—in part.

Grava, Dr. Sigurd. *Technical Director for Planning, Parsons, Brinckerhoff, Quade & Douglas, Inc., New York.* URBAN PLANNING—feature.

Gray, G. Ronald. *Director, Syncrude Canada Ltd., Edmonton, Alberta.* OIL MINING—in part.

Greenberg, Arnold E. *Assistant Chief, Laboratory Services Branch, State of California Department of Health Services, Berkeley.* WATER ANALYSIS.

Gregory, Dr. G. R. *Department of Forestry, School of Natural Resources, University of Michigan.* FOREST RESOURCES.

Grobecker, Dr. Alan J. *Project Manager, Climatic Impact Assessment Program, Office of the Secretary of the Treasury.* AIR POLLUTION—coauthor.

H

Hader, Rodney N. *Secretary, American Chemical Society, Washington, DC.* AGRICULTURAL CHEMISTRY.

Hagood, Mel A. *Irrigated Agriculture Research and Extension Center, Washington State University.* IRRIGATION OF CROPS.

Haise, Dr. Howard R. *Northern Plains Branch Headquarters, Soil and Water Conservation Research Division, Agriculture Research Service, U. S. Department of Agriculture, Fort Collins, CO.* SOIL—in part.

Haley, Dr. Leonor D. *Chief, Mycology Training Unit, Department of Health, Education, and Welfare, National Communicable Disease Center, Atlanta.* FUNGISTAT AND FUNGICIDE—in part.

Hammond, Dr. Edwin H. *Professor and Head, Department of Geography, University of Tennessee.* HILL AND MOUNTAIN TERRAIN; TERRAIN AREAS, WORLDWIDE.

Hanson, Prof. Allen L. *Professor of Chemistry, Saint Olaf College, Northfield, MN.* DETERGENT.

Harley, Dr. John H. *Director, Health and Safety Laboratory, U. S. Atomic Energy Commission.* RADIOACTIVE FALLOUT.

Harrar, J. George. *Rockefeller Foundation, New York.* PLANT DISEASE CONTROL—in part.

Harris, E. J. *Assistant Chief, Health and Safety Technical Support Center, U. S. Bureau of Mines, Pittsburgh.* MINING—in part.

Harvey, Dr. George R. *Woods Hole Oceanographic Institution.* POLYCHLORINATED BIPHENYLS.

Hasler, Dr. Arthur D. *Laboratory of Limnology, University of Wisconsin.* EUTROPHICATION.

Hayes, William C. *Editor in Chief, "Electrical World," McGraw-Hill Publications Company, New York.* ELECTRIC POWER SYSTEMS.

Hazen, Richard. *Hazen and Sawyer, Consulting Engineers, New York.* WATER SUPPLY ENGINEERING; WATER TREATMENT.

Heald, Dr. Walter R. *Northeast Watershed Research Center, U. S. Department of Agriculture, University Park, PA.* AGRICULTURAL WASTES.

Hewson, Dr. E. W. *Chairman, Department of Atmospheric Sciences, Oregon State University.* AIR POLLUTION—coauthor.

Hirst, Dr. J. M. *Rothamsted Experimental Station, Harpenden, England.* PLANT DISEASE—in part.

Holmes, Dr. Robert W. *Department of Biological Sciences, University of California, Santa Barbara.* PHYTOPLANKTON.

Hooker, Dr. Arthur L. *Department of Plant Pathology, University of Illinois.* PLANT DISEASE CONTROL—in part.

Horsfall, Dr. James G. *Director, Connecticut Agriculture Experiment Station, New Haven.* FUNGISTAT AND FUNGICIDE; PLANT DISEASE CONTROL—both in part.

I

Ingram, William T. *Consulting Engineer, Whitestone, NY.* AIR-POLLUTION CONTROL; IMHOFF TANK; SANITARY ENGINEERING; SEPTIC TANK; SEWAGE; SEWAGE TREATMENT.

Irving, Dr. Laurence. *Institute of Arctic Biology, University of Alaska.* ARCTIC BIOLOGY.

Isbin, Prof. Herbert S. *Department of Chemical Engineering and Materials Science, University of Minnesota.* NUCLEAR POWER.

J

Jacobs, Dr. Woodrow C. *Senior Scientist, Ocean Data Systems, Inc., Rockville, MD.* OCEAN-ATMOSPHERE RELATIONS.

James, Robert S. *Pennsylvania State University.* MINING—in part.

Johnson, Prof. Ronald R. *Department of Animal Science, Institute of Agriculture, University of Tennessee.* AGRICULTURAL SCIENCE (ANIMAL).

Johnson, Wendell E. *Consulting Engineer, McLean, VA.* RIVER ENGINEERING—coauthor.

Just, Prof. Evan. *Department of Mining and Geology, Stanford University.* MINING—in part.

K

Kaplan, Prof. Lew D. *Professor of Meteorology, University of Chicago.* GREENHOUSE EFFECT

Keller, Prof. W. D. *Department of Geology, University of Missouri.* EROSION.

Kendeigh, Dr. S. Charles. *Department of Zoology, University of Illinois.* BIOME; COMMUNITY.

Ketchum, Dr. Bostwick H. *Woods Hole Oceanographic Institution.* SEA WATER—in part.

Kilbey, Dr. Brian J. *Department of Genetics, University of Edinburgh.* MUTATION.

Kingery, Donald S. *Director, Health and Safety, Research and Testing Center, U. S. Bureau of Mines.* MINING—in part.

Kuc, Dr. Joseph. *Department of Biochemistry, Purdue University.* PLANT DISEASE CONTROL—in part.

Kutzbach, Prof. John E. *Department of Meteorology, University of Wisconsin.* CLIMATIC CHANGE—in part.

L

LaFond, Dr. E. C. *Senior Scientist and Consultant for Oceanography, Naval Underseas Center, San Diego.* SEA WATER—in part.

Lagler, Prof. Karl F. *School of Natural Resources, University of Michigan.* FISHERY CONSERVATION.

Landsberg, Dr. H. E. *Professor and Director, Institute for Fluid Dynamics and Applied Mathematics, University of Maryland.* Validator of BIOCLIMATOLOGY

Lapple, Charles E. *Senior Scientist, Chemical Engineering Department, Stanford Research Institute, Menlo Park, CA.* DUST AND MIST COLLECTION.

Law, Dr. Stephen L. *Research chemist and Coordinator for Elemental Analysis for the Particulate Mineralogy Unit, Bureau of Mines, U. S. Department of the Interior, Avondale, MD.* INDUSTRIAL WASTES.

Lems, Dr. Kornelius. *Deceased; formerly, Associate Professor and Chairman, Department of Biological Sciences, Goucher College.* PLANT SOCIETIES; POPULATION DISPERSAL—coauthor for both.

Leopold, A. Starker. *School of Forestry and Conservation, University of California, Berkeley.* BIOGEOGRAPHY.

Leopold, Dr. Luna B. *U. S. Geological Survey.* SWAMP, MARSH, AND BOG.

Lettau, Prof. Heinz H. *Department of Meteorology, University of Wisconsin.* MICROMETEOROLOGY.

Lieth, Dr. Helmut. *Department of Botany, University of North Carolina, Chapel Hill.* SAVANNA.

Likens, Prof. Gene E. *Section of Ecology and Systematics, Division of Biological Sciences, Cornell University.* MEROMICTIC LAKE.

Linsley, Prof. Ray K. *Professor Emeritus of Civil Engineering, Stanford University.* AQUIFER; ATMOSPHERIC WATER VAPOR; GROUNDWATER. Validator of ARTESIAN SYSTEMS; HYDROLOGY; SPRING; WATER TABLE.

Lochhead, Dr. Allan G. *Canada Department of Agriculture, Microbiology Research Institute, Ottawa.* SOIL MICROBIOLOGY; SOIL MICROORGANISMS.

Lorenz, Dr. Edward N. *Department of Meteorology, Massachusetts Institute of Technology.* WEATHER FORECASTING AND PREDICTION—in part.

Ludlam, Prof. Frank. *Deceased; formerly, Department of Meteorology, Imperial College, London.* CLOUD.

Ludvík, Dr. George F. *Environmental Protection Agency, Washington, DC.* INSECTICIDE; PESTICIDE—in part.

Luthin, Prof. J. N. *Department of Civil Engineering, University of California, Davis.* LAND DRAINAGE (AGRICULTURE).

Lyman, Dr. John. *Professor of Oceanography, University of North Carolina, Chapel Hill.* SEA WATER—in part.

Lyon, Dr. Waldo K. *Arctic Submarine Research Laboratory, Naval Undersea Warfare Center, San Diego.* SEA ICE.

M

McClain, Dr. E. Paul. *Director, Environmental Sciences Group, National Environmental Satellite Service, National Oceanic and Atmospheric Administration, Washington, DC.* ENVIRONMENTAL SATELLITES—feature.

McClelland, Richard H. *Manager, SNG Project Planning, CNG Energy Company, Pittsburgh.* COAL GASIFICATION.

McCormick, Dr. Jack S. *Chairman, Department of Ecology and Land Management, Academy of Natural Sciences, Philadelphia.* TERRESTRIAL ECOSYSTEM; VEGETATION ZONES, ALTITUDINAL.

Macfadyen, Dr. Amyan. *Professor of Biology, New University of Ulster.* BIOLOGICAL PRODUCTIVITY; BIOMASS; ECOLOGICAL SYSTEMS, ENERGY IN; FOOD CHAIN.

Major, Dr. Jack. *Department of Botany, University of California, Davis.* PLANT COMMUNITY.

Martin, Prof. James P. *Department of Soils and Plant Nutrition, College of Biological and Agricultural Sciences, University of California, Riverside.* SOIL—in part.

Mason, Dr. Basil J. *Director General, Meteorological Office, Bracknell, England.* CLOUD PHYSICS.

Mitchell, Dr. J. Murray, Jr. *Senior Research Climatologist, National Oceanic and Atmospheric Administration, McLean, VA.* CLIMATOLOGY.

Mitterer, Dr. Richard M. *Department of Geosciences, University of Texas at Dallas.* BIOSPHERE.

Morgan, Dr. Karl Z. *Neely Professor, School of Nuclear Engineering, Georgia Institute of Technology.* HEALTH PHYSICS; RADIOACTIVE WASTES, DECONTAMINATION OF.

Mortenson, Prof. Leonard E. *Department of Biological Sciences, Purdue University.* NITROGEN CYCLE.

Mottet, Dr. N. Karle. *Professor of Pathology and Director of Hospital Pathology, University Hospital, University of Washington, Seattle.* DISEASE; HEAVY METALS, PATHOLOGY OF.

Muffler, Dr. L. J. Patrick. *Geologist, Branch of Field Geochemistry and Petrology, U. S. Geological Survey, Department of the Interior, Menlo Park, CA.* GEOTHERMAL POWER—in part.

Muller, John G. *Vice President, Struthers Research and Development Corporation, Washington, DC.* SALINE WATER RECLAMATION.

Mullin, Dr. Michael M. *Department of Oceanography, University of California, La Jolla.* PREDATOR-PREY RELATIONSHIPS.

Murgatroyd, Dr. R. J. *Meteorological Office, Bracknell, England.* ATMOSPHERE.

N

Namias, Dr. Jerome. *Scripps Institution of Oceanography, La Jolla.* DROUGHT; WEATHER FORECASTING AND PREDICTION—in part.

Natusch, Dr. David F. S. *Chairman, Department of Chemistry, Colorado State University.* ENVIRONMENTAL ANALYSIS—feature.

Nelson, Dr. Curtis J. *Department of Agronomy, University of Missouri.* AGRICULTURAL SCIENCE (PLANT).

Newcombe, Dr. Curtis L. *Director, San Francisco Bay Marine Research Center, San Francisco State College.* MARINE ECOSYSTEM.

Newton, Dr. Chester Whittier. *National Center for Atmospheric Research, Boulder.* ATMOSPHERIC GENERAL CIRCULATION; SQUALL LINE; STORM; TORNADO.

Nierenberg, Prof. William A. *Director, Scripps Institution of Oceanography, La Jolla.* OCEANOGRAPHY—in part.

Nixon, Dr. Charles W. *Consultant, Kettering, OH.* SONIC BOOM—coauthor.

O

O'Connor, Dr. Donald J. *Department of Civil Engineering, Manhattan College.* SEWAGE DISPOSAL.

Odum, Dr. Eugene P. *Director, Institute of Ecology, University of Georgia.* ECOLOGY.

Olson, Dr. Jerry S. *Oak Ridge National Laboratory.* CLIMAX COMMUNITY; LAND-USE PLANNING.

Olson, Nils. *National Research Council of Canada.* NOISE POLLUTION—in part.

Olson, Sigurd F. *President, Wilderness Society, Ely, MN.* WILDLIFE CONSERVATION—in part.

P

Park, Prof. Charles F., Jr. *Department of Earth and Planetary Sciences, Massachusetts Institute of Technology.* MINERAL RESOURCES CONSERVATION.

Parker, David E. *Meteorological Office, Bracknell, England.* PRECEDENTS FOR WEATHER EXTREMES—feature.

Patton, Dr. Donald J. *Florida State University.* EARTH RESOURCE PATTERNS; SOIL CONSERVATION—coauthor.

Pattullo, Dr. June G. *Department of Oceanography, Oregon State University.* SEA WATER—in part.

Pearson, Allen. *Director, National Severe Storms Forecast Center, National Oceanic and Atmospheric Administration, Kansas City, MO.* WEATHER FORECASTING AND PREDICTION—in part.

Pielou, Prof. E. C. *Department of Biology, Dalhousie University, Halifax.* SPECIES DIVERSITY.

Pimental, Dr. David. *Professor of Insect Ecology, Department of Entomology and Section of Ecology and Systematics, New York State College of Agriculture and Life Sciences, Cornell University.* ECOLOGICAL ENTOMOLOGY.

Platt, Allison M. *Manager, Nuclear Waste Technology Department, Battelle–Pacific Northwest Laboratories, Richland, WA.* RADIOACTIVE WASTE MANAGEMENT.

Platt, Dr. Robert B. *Chairman, Department of Biology, Emory University.* ENVIRONMENT.

Plehn, Steffen W. *Deputy Assistant Administrator for Solid Wastes, U. S. Environmental Protection Agency.* SOLID WASTE MANAGEMENT.

Pollard, Dr. Ernest C. *Department of Biophysics, Pennsylvania State University.* RADIATION BIOLOGY.

Pomeroy, Lawrence R. *Department of Zoology, University of Georgia.* BIOGEOCHEMICAL BALANCE.

Ponikvar, Ade L. *Formerly, "Modern Plastics," McGraw-Hill Publications Company, New York.* OIL MINING—in part.

Power, Dr. J. F. *U. S. Department of Agriculture, Agricultural Research Service, Northern Great Plains Research Center, Mandan, ND.* SOIL—in part.

Priester, Gayle B. *Consulting Engineer, Baltimore.* COMFORT HEATING.

Prinn, Prof. Ronald. *Department of Meteorology, Massachusetts Institute of Technology.* ATMOSPHERIC CHEMISTRY.

Prosser, Dr. C. Ladd. *Professor Emeritus of Physiology, University of Illinois.* HOMEOSTASIS.

R

Radke, Dr. Rodney O. *Agricultural Product Research Laboratory, Monsanto Company.* HERBICIDE.

Rasmusson, Dr. Eugene M. *Geophysical Fluid Dynamics Laboratory, Environmental Science Services Administration, Princeton, NJ.* HYDROMETEOROLOGY.

Reed, Prof. Richard J. *Department of Atmospheric Sciences, University of Washington, Seattle.* FRONT—in part.

Reid, Joseph L. *Scripps Institution of Oceanography, La Jolla.* SEA WATER—in part.

Reifsnyder, Dr. William E. *Professor of Forest Meteorology, School of Forestry, Yale University.* FOREST FIRE CONTROL.

Rich, Dr. Saul. *Senior Plant Pathologist, Connecticut Agricultural Experiment Station, New Haven.* FUNGISTAT AND FUNGICIDE—in part.

Richards, Dr. Paul W. *School of Plant Biology, University College of North Wales.* RAINFOREST.

Riley, Dr. Ralph. *Plant Breeding Institute, Cambridge, England.* BREEDING (PLANT).

Risebrough, Dr. Robert W. *Research Ecologist, Bodega Marine Laboratory, University of California, Bodega Bay.* PESTICIDE—in part.

Ritchie, Dr. J. C. *Department of Biology, Trent University, Peterborough, Ontario.* TAIGA.

Robertson, Dr. Andrew. *National Oceanic and Atmospheric Administration, Rockville, MD.* ZOOPLANKTON.

Robinson, Dr. G. D. *Center for the Environment and Man, Hartford, CT.* CLIMATE MODIFICATION.

Rosenberg, Prof. Norman J. *Department of Agricultural Meteorology, University of Nebraska.* AGRICULTURAL METEOROLOGY.

Rowell, Dr. J. B. *U.S. Department of Agriculture, and Department of Plant Pathology and Botany, University of Minnesota.* PLANT DISEASE—in part.

Rudnick, Dr. Philip. *Scripps Institution of Oceanography, La Jolla.* SEA WATER—in part.

Rudolf, Paul O. *Principal Silviculturist (retired), North Central Forest Experimental Station, St. Paul, MN.* FOREST PLANTING AND SEEDING.

Rust, W. David. *National Severe Storms Laboratory, National Oceanic and Atmospheric Administration, Norman, OK.* THUNDERSTORM—in part.

Ryther, Dr. John H. *Woods Hole Oceanographic Institution.* SEA-WATER FERTILITY—in part.

S

Sanders, Dr. Frederick. *Department of Meteorology, Massachusetts Institute of Technology.* JET STREAM; WIND.

Savage, Dr. Jay M. *Department of Biological Sciences, University of Southern California, Los Angeles.* ZOOGEOGRAPHY.

Sayre, Dr. Albert N. *Deceased; formerly, Consulting Groundwater Geologist, Behre Dolbear and Company.* ARTESIAN SYSTEMS; HYDROLOGY; SPRING; WATER TABLE.

Schindler, Dr. James. *Department of Zoology, University of Georgia.* LIMNOLOGY.

Schmitt, Walter R. *Scripps Institution of Oceanography, La Jolla.* MARINE RESOURCES.

Schoener, Dr. Thomas W. *Department of Zoology, University of Washington, Seattle.* ECOLOGICAL INTERACTIONS.

Scholz, Christopher H. *Lamont-Doherty Geological Observatory, Palisades, NY.* EARTHQUAKE.

Schoonmaker, G. R. *Vice President, Production-Exploration, Marathon Oil Company, Findlay, OH.* OIL AND GAS, OFFSHORE.

Schrock, Prof. Robert R. *Department of Earth and Planetary Sciences, Massachusetts Institute of Technology.* EARTH SCIENCES.

Sheppard. Prof. P. A. *Deceased; formerly, Department of Meteorology, Imperial College, London.* METEOROLOGY—in part.

Silberman, Dr. Edward. *Civil Engineering Department, University of Minnesota.* HYDROLOGY—in part.

Simonson, Dr. Roy W. *Director (retired), Soil Classification and Correlation, U. S. Department of Agriculture, Hyattsville, MD.* SOIL—in part.

Simpson, Dr. Joanne. *Department of Environmental Sciences, University of Virginia.* HURRICANE—coauthor.

Simpson, Prof. Robert H. *Research Professor, Department of Environmental Sciences, University of Virginia.* HURRICANE—coauthor.

Smith, Dr. Guy D. *Soil Conservation Service, U. S. Department of Agriculture.* SOIL—in part; SOIL ZONALITY.

Sorensen, Dr. Robert M. *Coastal Engineering Research Center, Department of the Army, Fort Belvoir, VA.* COASTAL ENGINEERING.

Spindler, John C. *Anaconda Company, Butte.* ENVIRONMENTAL ENGINEERING—coauthor.

Stakman, Dr. E. C. *Professor Emeritus, Institute of Agriculture, University of Minnesota.* PLANT DISEASE; PLANT DISEASE CONTROL—both in part.

Starr, Dr. Eugene C. *U. S. Department of the Interior, Bonneville Power Administration, Portland, OR.* ELECTRIC POWER GENERATION.

Stover, Prof. Harold E. *Professor and Extension Agricultural Engineer (retired), Kansas State University.* RURAL SANITATION.

Stuart, Prof. L. C. *Professor of Zoology, Panajachel, Guatemala.* ZOOGEOGRAPHIC REGION.

T

Taylor, William L. *National Severe Storms Laboratory, National Oceanic and Atmospheric Administration, Norman, OK.* STORM DETECTION—in part.

Thompson, Jack. *U. S. Army Corps of Engineers, Office of the Secretary of the Army.* DAM.

Turner, Dean Lewis. *Formerly, Utah State University.* RANGELAND CONSERVATION.

Tyler, John E. *Scripps Institution of Oceanography, La Jolla.* SEA WATER—in part.

U

Unger, Walter H. *Anaconda Company, Butte.* ENVIRONMENTAL ENGINEERING—coauthor.

V

Van Dyne, Dr. George. *Natural Resource Ecology Laboratory, Colorado State University.* GRASSLAND ECOSYSTEM—coauthor.

Van Vleck, Dr. Dale. *Department of Animal Science, New York State College of Agriculture & Life Science, Cornell University.* BREEDING (ANIMAL).

Veronis, Dr. George. *Chairman, Department of Geology, Yale University.* OCEANOGRAPHY—in part.

Vine, Allyn C. *Oceanographer, Woods Hole Oceanographic Institution.* SEA WATER—in part.

Vohra, Dr. F. C. *School of Biological Sciences, University of Malaya.* MANGROVE SWAMP—in part.

Voigt, Dr. Garth K. *Acting Dean, School of Forestry and Environmental Studies, Yale University.* FOREST SOIL.

Von Gierke, Dr. Henning E. *Consultant, Yellow Springs, OH.* SONIC BOOM—coauthor.

Vonnegut, Dr. Bernard. *Department of Atmospheric Science, State University of New York.* ATMOSPHERIC ELECTRICITY.

W

Wagner, Prof. Frederic H. *College of Natural Resources, Utah State University.* DESERT ECOSYSTEM.

Walsh, John J. *Brookhaven National Laboratories, Upton, NY.* WATER CONSERVATION—in part.

Weaver, John. *National Severe Storms Laboratory, National Oceanic and Atmospheric Administration, Norman, OK.* STORM DETECTION—in part.

Weickmann, Dr. Helmut K. *Director, Atmospheric Physics and Chemistry Laboratory, Environmental Science Services Administration, Boulder.* WEATHER MODIFICATION.

White, Prof. Colin. *Department of Epidemiology and Public Health, School of Medicine, Yale University.* EPIDEMIOLOGY.

Whittaker, Dr. Robert H. *Ecology and Systematics, Cornell University.* BIOTIC ISOLATION; ECOSYSTEM.

Wiegert, Dr. Richard G. *Department of Zoology, University of Georgia.* ECOSYSTEM MODELS; RADIOECOLOGY.

Willett, Prof. Hurd C. *Department of Meteorology, Massachusetts Institute of Technology.* AIR MASS.

Williams, Dr. Roger T. *Department of Meteorology, Naval Postgraduate School, Monterey.* FRONT—in part.

Winchester, Prof. John W. *Chairman, Department of Oceanography, Florida State University.* HALOGENS, ATMOSPHERIC CHEMISTRY OF.

Winterkorn, Dr. H. F. *Department of Civil and Geological Engineering, Princeton University.* SOIL MECHANICS.

Woodcock, Dr. Alfred H. *Institute of Geophysics, University of Hawaii.* SEA WATER—in part.

Wooster, Dr. Warren S. *Scripps Institution of Oceanography, La Jolla.* SEA WATER—in part.

INDEX

Index

A

B

C

D

E

F

G

H

P

X, Y

Z